Animal Physiology
From Genes to Organisms

Lauralee Sherwood
Department of Physiology and Pharmacology
School of Medicine
West Virginia University

Hillar Klandorf
Division of Animal and Veterinary Sciences
West Virginia University

Paul H. Yancey
Department of Biology
Whitman College

THOMSON
™
BROOKS/COLE

Australia • Canada • Mexico • Singapore • Spain
United Kingdom • United States

THOMSON

★

TM

BROOKS/COLE

Publisher: Peter Marshall
Editor-in-Chief: Michelle Julet
Development Editor: Elizabeth Howe
Assistant Editor: Elesha Feldman
Editorial Assistant: Lisa Michel
Technology Project Manager: Travis Metz
Marketing Manager: Ann Caven
Marketing Assistant: Leyla Jowza
Advertising Project Manager: Kelley McAllister
Project Manager, Editorial Production: Teri Hyde
Print/Media Buyer: Barbara Britton

Permissions Editor: Sarah Harkrader
Production Service: Thomas E. Dorsaneo, Publishing Consultant
Text Designer: Tani Hasegawa
Photo Researcher: Myrna Engler
Copy Editor: Linda Purrington
Illustrator: Precision Graphics, Thompson Type
Cover Designer: Larry Didona
Cover Image: Konrad Wothe/Minden Pictures
Cover Printer: CTPS
Compositor: Thompson Type
Printer: CTPS

Printed in China
4 5 6 7 08 07

For more information about our products, contact us at:
Thomson Learning Academic Resource Center
1-800-423-0563

For permission to use material from this text or product, submit a request online at **http://www.thomsonrights.com.**
Any additional questions about permissions can be submitted by email to **thomsonrights@thomson.com.**

Library of Congress Control Number: 2004100794

Student Edition: ISBN-13: 978-0-534-55404-0
ISBN-10: 0-534-55404-0

Instructor's Edition: ISBN-13: 978-0-534-40994-4
ISBN-10: 0-534-40994-6

Thomson Brooks/Cole
10 Davis Drive
Belmont, CA 94002
USA

Asia
Thomson Learning
5 Shenton Way #01-01
UIC Building
Singapore 068808

Australia/New Zealand
Thomson Learning
102 Dodds Street
Southbank, Victoria 3006
Australia

Canada
Nelson
1120 Birchmount Road
Toronto, Ontario M1K 5G4
Canada

Europe/Middle East/Africa
Thomson Learning
High Holborn House
50/51 Bedford Row
London WC1R 4LR
United Kingdom

Latin America
Thomson Learning
Seneca, 53
Colonia Polanco
11560 Mexico D.F.
Mexico

Spain/Portugal
Paraninfo
Calle Magallanes, 25
28015 Madrid, Spain

■ *Dedication*

*My family, for all they have done for me in the past, all they mean to me
in the present, and all I hope will yet be in the future.   Lauralee Sherwood*

*Britt, Alex and Emma
And thanks to Laurie for asking and Paul for seeing it through.
Here's to a Grinch and Rodent!   Hillar Klandorf*

My wife, son, and parents for all their support.   Paul Yancey

Brief Contents

Contents

SECTION II: WHOLE-BODY REGULATION AND INTEGRATION

SECTION III: SUPPORT AND MOVEMENT

CHAPTER 8

Preface

This book focuses on the evolution and mechanisms of animal body functions from genes to organ systems to the whole organism interacting with its environment. It is designed for mid- to upper-level undergraduate courses in general biology and zoology programs, and for pre-professional programs in animal husbandry, veterinary science, marine biology, and wildlife and fisheries biology.

The text follows an "integrative systems" approach, which involves both vertical and horizontal integration. **Vertical integration** concerns the hierarchical nature of physiology, in which organ systems and their functions arise from cells and their molecular components, and from the evolutionary forces that gave rise to those components. The chapter sequence and content reflect this: after an introduction to the major themes of physiology, the chapters proceed from molecular and cellular physiology to classical organ systems. Organs systems are grouped according to universal functions of whole-body regulation and integration, support and movement, maintenance, and reproduction. For each organ system, both the evolutionary forces and the cellular and molecular underpinnings are discussed, including new research in genomics, which is having major impacts on all areas of biology. Where necessary, sufficient relevant anatomy is included to make the inseparable relation between structure and function meaningful.

Horizontal integration concerns the interactions of all organ systems to yield a whole functioning organism. Too often, students view the subject of physiology as separate entities (systems) unclearly linked to each other and to the functioning of the intact organism. The text works to overcome this in two ways. First, the text's central themes are homeostasis and other integrated forms of regulation—how processes from the genetic to the behavioral level allow an animal's body to meet changing demands while maintaining the internal consistency necessary for all cells and organs to function. At the end of each chapter, a perspective on homeostasis and integration is provided to ensure that the student understands how the components in that chapter interact with other components and contribute to the whole body. Second, special integrative chapters—Fluid and Acid-Base Balance, and Energy Balance and Thermal Physiology—focus on crucial whole-body phenomena that are dependent on more than one organ system.

In addition to an integrated systems organization, the text also incorporates a **comparative approach** throughout. Mechanisms employed by major and exemplary groups of invertebrates and vertebrates are included, for three purposes. First, comparisons among types of animals are selected to illuminate important universal functions and principles, emphasizing the unity of life. Second, unique or striking adaptations are featured that reveal the diversity that can result from evolutionary adaptation. Finally, the text incorporates detailed coverage of animal species of relevance to those preparing for animal-related careers, especially vertebrates.

The scope of this text has been limited by judicious selection of pertinent content that a student can reasonably be expected to assimilate in a one-semester course. To keep pace with today's rapid advances in the biological sciences, students interested in careers related to animal physiology must be able to draw on their overall understanding of concepts instead of merely recalling isolated facts. For this reason the text is designed to promote understanding of the basic principles and concepts of physiology rather than the memorization of details. Yet depth, where needed, is not sacrificed. All aspects of physiology receive coverage, including separate chapters on immune and reproductive systems, and detailed coverage of pre- and postgastric digestion. Furthermore, new information based on recent discoveries has been included in all chapters.

Even the most tantalizing of subject matters can be drudgery to study and difficult to comprehend if not effectively presented. The text maintains a logical, straightforward, understandable style that emphasizes how each individual concept is an integral part of the whole subject matter. Every effort has been made to assure smooth reading through good transitions, logical reasoning, and integration of ideas throughout the text.

Features and Learning Aids

Homeostasis and integration

Following the integrated systems approach, each chapter concludes with a narrative, **Chapter in Perspective: Homeostasis and Integration**, which helps the student put into perspective how the system just discussed contributes to the whole animal.

Feedforward statements as subsection titles

Instead of traditional short, topic titles for each major subsection (for example, "Evolution of Vertebrate Circulation"), the text begins each chapter section with a summary statement of a major concept (e.g., "The vascular system evolved from one circuit to two separate circuits in vertebrates.").

■ Cross-referencing

Many chapters build on material presented in immediately preceding chapters, yet each chapter is designed to stand on its own to allow the instructor flexibility in curriculum design. With extensive cross-references provided, the sequence of presentation can be varied at the instructor's discretion.

■ Unanswered Questions

The physiological sciences are changing constantly due to new discoveries and challenges of old ideas. In most chapters, Unanswered Question boxes present short overviews and questions on controversial ideas and new hypotheses to illustrate that physiology is a dynamic, changing discipline, and to stimulate students to propose their own ideas.

■ Box features

Four other, more detailed box features are inserted within chapters where appropriate. *Beyond the Basics* exposes students to high-interest, tangentially relevant information on diverse topics including historical perspectives, environmental issues, and common diseases. *Challenges and Controversies* boxes introduce the frontiers of physiological research with new hypotheses and unsolved questions, in more depth than the Unanswered Questions boxes. *A Closer Look at Adaptation* illustrates the scope and limits of evolutionary adaptation with selected illustrative examples. *Molecular Biology and Genomics* provides examples of cutting-edge research resulting from the animal genome projects and related research to uncover the physiological functions of newly discovered genes.

■ Cellular, molecular and genomic foundations

Even though the text is primarily organized according to body systems, it provides coverage of cellular, molecular, and genomic topics in early chapters and incorporates these throughout as a basis for understanding organ function. The current tools used to explore molecular physiology are also outlined and serve as a backbone for each chapter. This is the cutting edge of physiology research.

■ Evolutionary foundations

Each chapter has an overview of major evolutionary events and selective forces. All topics are placed in a true biological context that melds both the proximate (mechanistic) approach of traditional physiology with the evolutionary explanations behind them, to explain not only the power of natural selection but also the often illogical features of life.

■ Pathophysiology

The text is also designed to help students realize that they are learning worthwhile and applicable material. Because many students using this text will have a clinical component to their careers, reference to important pathophysiological issues demonstrates the content's relevance to their professional goals. For example, the text includes the effects of environmental toxins on some body systems, an issue of increasing concern.

■ Full-color illustrations

The anatomic illustrations, schematic representations, photographs, tables, and graphs are designed to complement and reinforce the written material. Flow diagrams are used extensively to help students integrate the written information presented.

■ Key terms, word derivations, and glossary with phonetic pronunciations

Key terms are defined as they appear in the text. A glossary at the end of the book enables students to quickly review key termswhen they occur later in the book, includes phonetic pronunciations of the entries.

■ End of chapter questions, suggested readings, Infotrac articles, and web sites

Each chapter concludes with a set of review questions, suggested readings, many of which are on Infotrac College edition, and helpful physiological web site addresses.

■ Appendixes

The appendixes on the book's web site are designed for the most part to help students who need to brush up on some foundation materials that they are assumed to already have had in prerequisite courses.

- *Appendix A,* **The Metric System,** is a conversion table between metric measures and their English equivalents.
- *Appendix B,* **A Review of Chemical Principles.** Most undergraduate physiology texts have a chapter on chemistry, yet physiology instructors rarely teach basic chemistry concepts. The decision was made, therefore, to reserve valuable text space for physiological concepts and to provide instead this Appendix as a handy reference for students who need a review of basic chemistry concepts that are essential to understanding physiology.
- *Appendix C,* entitled **Storage, Replication, and Expression of Genetic Information,** includes a discussion of DNA and chromosomes, protein synthesis, cell division, and mutations.

Instructor and Student Ancillaries

A number of helpful resources are available to instructors and students. The Multimedia Manager includes over 400 illustrations and figures from the textbook to create custom Power-Point presentations. Also included are many animations that illuminate key concepts in the textbook such as carrier-mediated transport, neuronal structure and function and cross-bridge interaction during muscle contraction are included to facilitate classroom discussion. The Multimedia Manager also includes a thorough test bank to aid in class preparation.

Students will appreciate the book's robust web site that contains many animations to reinforce chapter concepts as well as quizzes, flash cards, glossary terms, hypercontents, and thorough chapter summaries.

Acknowledgments

It takes the effort of many individuals to create a textbook. The authors would like to thank Dana Garcia for preparing the testbank to the textbook, and the following reviewers who provided excellent feedback during the various stages of the writing process.

Joe Bastian – The University of Oklahoma

Frank Blecha – Kansas State University

Ken Blemings – West Virginia University

Charles E. Booth – Eastern Connecticut State University

Gerrit de Boer – University of Kansas

Randal K. Buddington- Mississippi State University

David Bunick – University of Illinois at Urbana

Mark L. Burleson – University of Texas at Arlington

Craig Cady – Bradley University

Randy Cohen – California State University, Northridge

Victoria Connaughton – American University

Robin L. Cooper – University of Kentucky

Bob Dailey – West Virginia University

William L. Diehl-Jones – University of Waterloo

Michael E. Dorcas – Davidson College

David P. Froman – Oregon State University

David S. Gonzales – Arizona State University

Bill Hoover – West Virginia University

Keith Inskeep – West Virginia University

Merideth Kamradt Krevosky – Bridgewater State College

Gabriel Kass-Simon – University of Rhode Island

Phil Keeting – West Virginia University

Jack R. Layne, Jr – Slippery Rock University

James Larimer – University of Texas at Austin

John C. Lee – Virginia-Maryland Regional College of Veterinary Medicine (Virginia Tech)

James A. Long – Boise State University

Catherine Loudon – University of Kansas

David A. Lovejoy – University of Toronto

Margaret MacNeil – York College, City University of New York

Bill Martin, West Virginia University

Sarah L. Milton – Florida Atlantic University

Robert J. Omeljaniuk, PhD – Lakehead University

Sanford E. Ostroy – Purdue University

Valerie M. Pasztor – McGill University

John W. Parrish, Jr. – Georgia Southern University

Donald R. Powers – George Fox University

Bill Radke – University of Central Oklahoma

Chris R. Ross – Kansas State University

Dean D. Schwartz – Auburn University

Peter Sharp – Roslin Research Institute, Scotland

Donald E. Spiers – University of Missouri

Andrea R. Tilden – Macalester College

Alexa Tullis – University of Puget Sound

Norma Venable, West Virginia University

Carol Vleck, PhD – Iowa State University

Linda Vona-Davis – West Virginia University

Melvin Weisbart – University of Regina

Matt Wilson – West Virginia University

Jianbo Yao – West Virginia University

Yong Zhu – East Carolina University

■ *About the Authors*

L. Sherwood is a Professor of Physiology at West Virginia University. Having earned a D.V.M. degree from Michigan State University, she has continued to have an interest in animals even though her faculty appointment for the past thirty-seven years has been in the School of Medicine. Throughout this time, she has participated as a guest lecturer in Animal Science classes. Dr. Sherwood has authored two human physiology textbooks: *Human Physiology: From Cells to Systems, Fifth Edition,* Brookes/Cole Publishing, 2004 and *Fundamentals of Physiology: A Human Perspective,* Third Edition, Brookes/Cole Publishing, 2005. She brings to this animal physiology textbook her knowledge of veterinary medicine, her years of classroom teaching experience, and her text-writing expertise. She has received numerous teaching awards, including an Amoco foundation Outstanding Teacher Award, a Golden Key National Honor Society Outstanding Faculty Award, two listings in WHO'S WHO AMONG AMERICA'S TEACHERS, and the Dean's Award for Excellence in Education.

H. Klandorf is a Professor of Animal and Veterinary Science at West Virginia University. He completed his Ph.D. degree at the Poultry Research Center in Edinburgh, Scotland in Avian Physiology. After completing two post-doctoral research programs he spent $3\frac{1}{2}$ years at UCLA studying agents that affect the onset of diabetes. Currently he is investigating factors that limit the accelerated tissue aging associated with the elevated plasma glucose concentrations in birds. He instructs animal physiology and animal behavior classes at West Virginia University. The text incorporates current physiological issues and ideas generated from students enrolled in the Animal Physiology class.

P. H. Yancey is a Professor of Biology at Whitman College. After earning a Ph.D. in Marine Biology (Animal Physiology and Biochemistry emphasis) from U.C.S.D., he has conducted research on invertebrates, fishes and mammals at the University of St. Andrews (Scotland), the National Institutes of Health, the Mt. Desert Island Biological Laboratory, the Monterey Bay Aquarium Research Institute, and the University of Otago (New Zealand), as well as at Whitman. His research is on biochemical and physiological adaptations to water stress, from marine animals to the mammalian kidney in health and disease (including diabetes). He teaches animal physiology, human anatomy and physiology, marine biology, and bioethics, and has won several teaching awards. He brings to this text a broad evolutionary perspective, and 21 years of teaching animal physiology using texts supplemented with a wide variety of external materials.

Homeostasis and Integration: The Foundations of Physiology

Size Matters. Tiny bandicoots (left) have a much greater surface-area:volume ratio than the huge elephants. This ratio profoundly affects their physiologies.

Introduction

Life. The word *life* is the essence of the biological sciences, and yet has no clear definition. We seem to recognize instinctively what is alive and what is not alive on Earth, but we founder when contemplating how alien life on another planet could be recognized. How might we define life? Most attempts to define it focus on the **functions** of living systems, that is, on the dynamic processes that life forms have and that nonliving things do not. Living things organize themselves using energy and raw materials from their surroundings (a set of processes called *metabolism*), maintain integrity in the face of disturbances (a process called *homeostasis*), and reproduce. Such abilities are what physiology is all about—the study of the functions of organisms, or how life works.

◼ Physiological processes arise through evolution.

Physiologists often view organisms as organic machines whose mechanisms of action can be explained in terms of cause-and-effect sequences of physical and chemical processes—the same types of processes that occur in other components of the universe. However, biological features differ from the rest of the universe in one crucial way: They are the result of millions of years of evolution through random variation and **natural selection**: the process in which members of a species having genetic characteristics that enhance survival are able to produce more surviving offspring than other members not having those characteristics. (In contrast, although astronomers may speak of "stellar evolution," historical change in the nonliving universe does not appear to involve selection.) Therefore, scientists must recognize that biological phenomena always have two different levels of explanation or description:

- *The **mechanistic**—also called the **proximate**—explanation:* This is the answer to "How does it work?" This focus emphasizes the mechanism or composition of a function or structure, and is the traditional core of the physiological sciences.

- *The **evolutionary explanation:*** This is the answer to "How did it evolve to be this way?" This emphasis recognizes that biological features are the result of variation and natural selection. A species must cope with a variety of *selective pressures* (environmental stresses, mating and other societal interactions, and so on). Many genes within a species differ among individuals, and under a particular selective pressure some gene variants may help an individual survive and reproduce better than do other variants. Those gene variants thus get passed on to more offspring than do other variants, and thus the species changes genetically over time. This process never ends, because selective pressures (such as climate and predator–prey interactions) change over time.

Because of this process of natural selection, physiologists assume that most organismal features are **adaptations,** that is, features that evolved to *enhance survival* of the species. Natural selection is indeed a powerful force, which often results in exquisite adaptations, but it is important to recognize that it is also constrained by the past. That is, evolution works primarily through modifications of previously evolved features and variations of these features, and so does not always result in the most logical or optimal design.

In addition, it is important to distinguish between evolutionary and purely teleological approaches to explaining the various features of organisms. In a **teleological approach,** phenomena that occur in organisms are explained in terms of their particular **purpose** (a function that is useful or beneficial) in fulfilling an organismal need, without necessarily considering how this outcome is accomplished (mechanism) or how it evolved. Teleology is commonly used, but it is dangerous thinking, as you will see. It is often used because most biological mechanisms do serve a useful purpose (having been selected through evolutionary time to do so). If you try to find the thread of "purpose" in what you are studying, you can avoid a good deal of pure memorization.

However, be aware of a pitfall: Teleology can be misleading because it assumes that features are always logical, having evolved for idealized purposes. But in fact adaptations are often a record of historical compromises; that is, they are not necessarily the most logical solution to a problem. This inevitable constraint on natural selection has been often overlooked in the physiological sciences, especially since the modern physiological sciences started in the 1600s, long before Charles Darwin presented his theory of evolution (in 1859, the year he published *On the Origin of Species*). However, a recent movement called *evolutionary physiology,* and its applied counterpart *Darwinian medicine,* is re-evaluating many teleological explanations in traditional physiology. As we shall see in later chapters, this has often profound implications for the understanding of physiology.

Simple examples can help you to distinguish among these approaches (mechanistic/proximate, evolutionary, teleological). First, let us take a reasonably logical, useful adaptation: shivering in mammals. The *mechanistic* or *proximate* explanation of why a mammal shivers is that, when temperature-sensitive nerve cells detect a fall in body temperature, they signal the hypothalamus, the part of the brain responsible for temperature regulation. In turn, the hypothalamus activates nerve pathways that trigger rapid oscillating muscle contractions (that is, shivering) to produce heat. A *teleological* explanation of why a mammal shivers when it is cold is "to keep warm" to maintain a relatively constant, warm interior (in a process called *negative feedback homeostasis,* a topic we explain in detail shortly). But this does not answer the question of why it is useful for most mammals to keep constantly warm. It is not necessary, for example, for fish in a variable temperate habitat or for a hibernating mammal in winter. To explain this, we must analyze the course of *evolution* among mammals to try to determine why regulation of body temperature enhanced survival in this group, and why 37° to 39°C (about 99° to 102°F) was the ideal temperature selected for most placental (but not other) mammals.

Now let us take a less logical example, the human spinal column and its series of bony vertebrae joined by cushioning pads. The *proximate* description is that the spine is a movable support device for the upper body and a conduit for the major nerve cords. *Teleologically,* we might want to say that vertebrae are a logical way to provide both flexibility for movement and rigidity for support and protection of the nerves. But this ignores the fact that the spine in humans is a suboptimal design! Back problems—pinched nerves, damaged cushioning pads, and so on—are among the most common ailments of humankind (check your telephone directory to see the long lists of chiropractors!). To explain this, we must look at the course of evolution. The spine originally evolved as a *horizontal* flexible swimming device in fishes, which are not concerned with gravity, because they float. Later the spine became adapted into a (still horizontal) support system against gravity in four-legged land vertebrates. But it only recently began evolving into a *vertical* support in hominid ancestors, and thus has not been optimized. Indeed, it may never be, because the spine is fundamentally a horizontal system in its inception.

■ Physiology is an integrative discipline.

As the example of the spine shows, physiology is closely interrelated with *anatomy,* the study of the structure of organisms. Just as the functioning of an automobile depends on the shapes, organization, and interactions of its various parts, the structure and function of an animal's body are inseparable. But physiology also integrates many other scientific fields. *Physics* is necessary to understand processes in organisms such as electrical conduction, fluid dynamics, and leverage exerted by musculoskeletal systems. Of course, all life processes depend on chemical reactions, so a solid understanding of *chemistry* is crucial. Also, biological functions depend on large biological molecules such as DNA and proteins (see Chapter 2), and therefore *molecular biology* provides a major foundation for physiological understanding.

Molecular biology is the most rapidly growing biological science, and its integration with physiology is now the dominant force in many physiological studies. Driving much of this growth are organismal **genome projects** (the complete analysis of genetic codes of key species, including humans). The genetic code ultimately gives rise to crucial parts of the functioning organism, and those functions in turn affect how that code is used (see Chapter 2 for details). Many physiologists are dedicating their research to understanding how genes are regulated and how they are used in physiological processes.

Genome studies are also having a major impact on human understanding of evolution. In particular, these studies are revealing the stunningly complex and often illogical nature of evolution at the most fundamental level. The human genome, for example, appears riddled with formerly functional genes that have mutated into unused forms, and a large percentage of code that seems to have no function at all now or in the past (although certain recent studies hint at some functions for these codes). Furthermore, at least some genes seem to have arisen from bacteria and viruses!

■ Physiology includes comparative as well as integrative approaches.

As we have just seen, complete physiological understanding must integrate in a "vertical" sense: from the atomic level to the whole organism to the evolutionary forces that have acted on that organism's ancestors. But another major approach in physiology—*comparative physiology*—works in a "horizontal" sense, by comparing physiological features in different types of organisms. Such research is important in two ways. First, we learn about the uniqueness and diversity of life on Earth, by discovering the amazing functions that one or a few organisms have evolved and others have not. This basic knowledge is fascinating in its own right, but can sometimes have practical applications. For example, very few animals can climb up smooth vertical surfaces and walk upside down on glass ceilings—but geckos can! Only recently has the mechanism for this lizard's phenomenal abilities been uncovered (see the box How Do Geckos Cling to Glass Ceilings?), and human-made adhesive tape based on its mechanism is now being developed.

(a) A tokay gecko. (b) The foot of the gecko. Each pad consists of many lamellae (folds) which in turn have millions of tiny bristles that adhere to surfaces through van der Waals interactions.

?

How Do Geckos Cling to Glass Ceilings? The feet of geckos have pads with millions of tiny bristles. Flies and spiders have similar pads. Researchers have recently found that these provide a huge surface area for weak molecular attractions called *van der Waals* forces (about which you probably learned in a chemistry class). Here is a fascinating topic for integrating basic chemistry with a whole-animal function. A suggestion: Try searching on the Internet for "gecko feet mechanism."

Second, comparative studies help us find out what physiological functions are universal rather than unique. This broadens our understanding of the basic nature of life. For example, from studying the adaptations of other animals to environmental temperature we learn that the human need to maintain a nearly constant body temperature is not universal, but we also learn (as we will see in Chapter 13) that there are universal effects of temperature on biological molecules to which all life must adapt.

Much of comparative physiology is motivated by the **August Krogh principle**, formulated by German physiologist Hans Kreb in 1975: "For a large number of problems there will be some animal of choice, or a few such animals, on which it can be most conveniently studied." This principle was based on remarks made by the great Danish physiologist August Krogh (1874–1949), who championed the integrative and compara-

tive approaches to physiology and who made major discoveries about (among other things) the use of radioactive isotopes in metabolic research, metabolism in insect flight, and the regulation of capillary blood flow (for which he was awarded the Nobel Prize in Physiology or Medicine, in 1920). He believed—as many comparative physiologists have ever since—that the diversity of adaptations on Earth is such that, for any particular physiological process and challenge to that process, some species will have adapted in such a way that it provides an ideal model system for studying that process. In turn, this knowledge can provide insight into that process in a much wider variety of organisms. For example, to understand the regeneration of damaged body parts, and why mammals are so poor at this, some researchers use a salamander called the *axolotl*, which like other salamanders can regenerate whole body parts, including entire limbs. The axolotl is ideal in many ways, because it lives and breeds well in the laboratory (unlike other salamanders), and is more closely related to mammals (such as humans) than are other animals (such as seastars) that can regenerate readily.

The inherently broad nature of physiology presents a major challenge to learning. To help handle this challenge, as we tell the story of how animals work, we integrate the physical, chemical, molecular/genetic, anatomic, and evolutionary features necessary for fully understanding function. Also, to provide breadth in the comparative sense we discuss a variety of animal types. To fully understand these comparisons, you should develop a grasp of the basic evolutionary relationships of animal phyla. If you have not learned about these, or need a review, look at ● Figure 1–1. In particular, be aware that the dichotomy of "vertebrates" and "invertebrates," widely used for the animal kingdom, is quite misleading. In fact, there are at least 30 animal phyla (groups distinguished by major body plans), and vertebrates are only a subphylum within one of those, the phylum Chordata (unified by all members having a dorsal *notochord* at some stage in their life cycles). Also, the phylum Arthropoda ("jointed legs"), which includes insects, is by far the most successful in terms of numbers of species and sheer mass of animal life. We follow tradition and use the vertebrate–invertebrate dichotomy in this text, but we also name the pertinent phyla.

Methods in Physiology

Physiology is of course a science, and thus employs the same philosophical logic that other scientists use to study the world. But because of physiology's broad, integrative nature, we can fairly say it employs a wider range of specific techniques (from genetic to behavioral) than most other biological disciplines. Here we provide a brief overview of these essential aspects of "doing science."

■ **The hypotheticodeductive method is the most widely accepted version of "the scientific method."**

Science is above all else a *way of thinking* about the world. First, it assumes that consistent processes in nature explain structures and phenomena, and that we can uncover and understand these processes. Second, science provides a logical approach (the "scientific method") for this investigation of

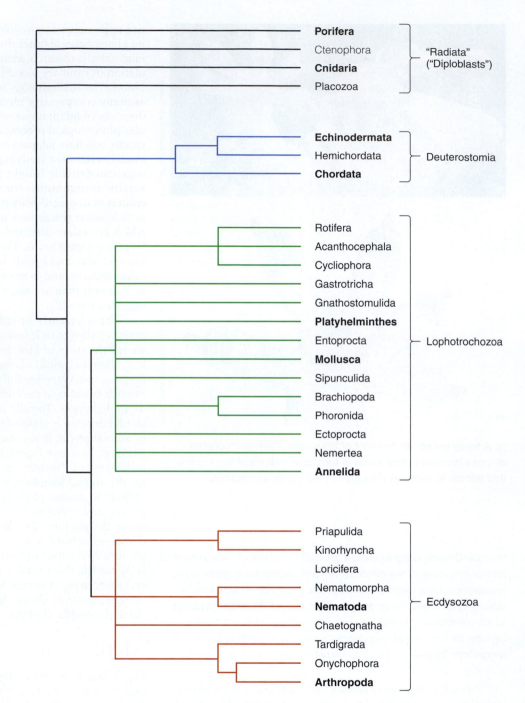

Figure 1–1 ● Animal evolutionary tree, showing the relationship among phyla as deduced from morphology coupled with molecular analyses. The nine most successful phyla, to which we refer most often in this text, are in **bold** letters; these are the Porifera (sponges), Cnidaria (jellies, hydras, corals, anemones, etc.), Echinodermata (seastars, sea urchins, sea cucumbers, etc.), Chordata (tunicates, vertebrates, etc.), Platyhelminthes (flatworms), Mollusca (snails, octopods, bivalves, etc.), Annelida (segmented worms), Nematoda (round worms), and Arthropoda (crustacea, insects, arachnids, etc.).

Source: http://www.mhhe.com/biosci/pae/zoology/animalphylogenetics/section02.mhtml, from *Teaching Animal Molecular Phylogenetics* by C. Leon Harris. Used by permission of The McGraw-Hill Companies.

the universe. Several versions of scientific method have been put forth, but the most widely used is termed the **hypotheticodeductive method.** It can be stated in these steps:

1. *Ask a question about nature.* This is sometimes called the **discovery phase** of science. The scientist finds an aspect of interest that has not been studied or fully explained. This step requires gathering all accessible information that has been discovered previously, any existing hypotheses, and (if there is little knowledge yet) *exploration* of the unknown. Exploration involves the gathering of data (by both observation and experimentation) about the world without necessarily having an hypothesis. For example,

biologists exploring the deep sea today are often asking, "What lives in the deep?" and are finding new species on a regular basis.

2. *Propose alternative hypotheses to explain the phenomenon.* Hypotheses are, in essence, tentative explanations about some aspect of nature. Unless all reasonable hypotheses already exist, the scientist must create them. This process is called **induction:** taking specific information and creating a general explanation. There are two important corollaries to making hypotheses. First, alternative explanations or predictions should be made, or else the testing steps (which follow) may focus on an erroneous explanation and miss a correct one. For example,

a deep-sea physiologist might ask how life survives the crushing pressures of the deep, which are known to inhibit the functions of biomolecules such as proteins. He or she might propose that deep sea organisms (1) have evolved proteins that are resistant to pressure, or (2) suffer from pressure inhibition and thus have sluggish metabolisms. (Can you think of other hypotheses?)

Second, scientific hypotheses must be **testable** (amenable to experiments or observation), and, in principle, **falsifiable** (capable in principle of being disproved by experiment or observation), or they are of little use in science (whether they are true or not!). For example, *vitalism*, a view of life prominent before the 20th century, hypothesized that living entities get their dynamic natures from an unseen life force or spirit. No experiment could be devised that detected such a force, and vitalists proposed that the force is beyond ordinary nature and thus unmeasurable. Thus, vitalism is today considered nonscientific, not because it is false (it may actually be true, although living processes have been found to be consistent with and explainable by physicochemical laws), but because it cannot be tested or (even in principle) falsified.

3. *Design experiments or observations that test the hypothesis or hypotheses by making testable predictions.* This process is called **deduction**: taking a general explanation, making specific testable predictions based on it, and testing those predictions. In physiological experiments, an organism or natural process is often perturbed in a clearly defined way, and the outcome measured. It is vital to have *controls* in such testing. Controls come in different forms, but typically involve organisms or processes that are not altered or perturbed or that are not adapted to an aspect of nature being tested. In our deep-sea example, proteins from both deep-sea and shallow-living animals are studied in the laboratory with and without high pressure to determine whether pressure resistance has in fact evolved in deep-living (but not shallow-living) animals.

4. *Conduct the observations or experiments.* This can require sophisticated equipment and techniques (for example, pressure chambers in which scientists can analyze protein functions), though not always. We briefly discuss this aspect in the next section.

5. *Using the outcome of these tests, refine the earlier questions and hypotheses and design new tests.* Often this process can take months or even decades, often because results are not clear or test only one small aspect of a larger question. For example, deep-sea physiologists have found that many proteins from deep-sea animals are indeed more resistant to pressure than are proteins from shallow-living animals. The researchers are now trying to find out how those proteins can resist pressure. However, some deep-sea proteins are *not* pressure resistant, and now researchers are asking whether other protective mechanisms exist.

Once a single hypothesis has been consistently supported by test after test, and all alternative (and testable) hypotheses have been falsified, an hypothesis may be elevated to a scientific *theory*.

Physiological observations and experiments use techniques from the molecular to the behavioral level, for both discovery and hypothesis testing. As we noted earlier, the integrative nature of physiology presents a challenge to the researcher. The range of physiology requires that scientists conduct observations and experiments from the functions of specific molecules to the behaviors that animals use with internal processes. Understanding the methods can be important to understanding the knowledge they yield. Although this text does not focus heavily on such methods, we cover selected ones in the appropriate chapters. For example, Chapter 2 discusses a number of important techniques scientists use in genomic and other molecular studies.

It is worth repeating that scientific experiments and observations are used for both discovery (before there are detailed hypotheses) and hypothesis testing. The analysis of organismal genomes, for example, is largely a discovery process, at least in initial phases.

Levels of Organization in Organisms

One triumph of scientific study in the 19th century was the development of the cell theory of life. Regardless of the difficulties in defining "life," the foundation of all functions of life, and thus of all living organisms, is the **cell**—the smallest unit capable of carrying out the processes associated with life on Earth. A cell consists of an outer barrier called a **membrane** and an internal (semigelatinous) fluid called the **cytosol**, which contains water, salts, small organic molecules, and **macromolecules**. Macromolecules include **deoxyribonucleic** and **ribonucleic acid** (**DNA** and **RNA**), which carry the genetic instructions for making most cell components, and **proteins**, molecules with complex shapes, which carry out those instructions by assembling into many cell structures, regulating most cell functions, and catalyzing most cell reactions (see Chapter 2).

Cells are found in two distinct forms: **prokaryotic cells** (in two groups, the Eubacteria and Archaea), which lack complex internal membranous structures such as nuclei; and **eukaryotic cells,** which do have these structures, found in Protista (such as protozoa and algae), Fungi, Plantae (higher plants), and Animalia. The simplest life forms are **unicellular** (single-cell) or colonial (loose collections of similar cells), typically prokaryotes and many protists such as amoebae. More complex organisms such as animals are **multicellular** (many-cell)—an aggregate of hundreds (in certain worms) to trillions (in mammals) of cooperating eukaryotic cells with distinct specializations in structure and function. As you will see later (p. 8), physical constraints such as *diffusion* limit the size of a single cell, so that large body size depended on the evolution of multicellularity.

All cells, whether they exist as solitary cells or as part of a multicellular organism, perform certain basic functions essential for survival of the cell and, in turn, survival of the organism. These basic cell functions include the following:

1. *Self-organization:* Using resources from the environment to create the cell. This function includes many steps, including:
   - Obtaining energy and raw materials from the environment surrounding the cell (for plants, these are light and inorganic nutrients; for most animals, these come

from food and oxygen). Initially, this occurs through the cell's membrane.

- Performing various chemical reactions that use energy and raw materials to provide energy stores and building blocks for the cell's other needs. Most cell reactions are catalyzed by large molecules called **enzymes,** which in most cases are complex proteins that increase the rate of specific chemical reactions. In most animal cells, the overall process can be roughly summarized thus:

$$\text{Food} + O_2 \rightarrow CO_2 + H_2O + \text{energy stores} + \text{building-block molecules}$$

- Eliminating to the cell's surrounding environment (through the membrane) any waste products of metabolism, such as carbon dioxide and ammonia or other nitrogenous waste.
- Synthesizing proteins and other components needed for cell structure, for growth, and for carrying out particular cell functions (using energy stores and building blocks).

2. *Self-regulation:* Maintaining self-integrity in the face of disturbances, including
   - Being sensitive and responsive to changes in the environment surrounding the cell.
   - Controlling the exchange of materials between the cell and its surrounding environment.
   - Repairing damage to the cell.
   - Correcting deviations in internal conditions, which threaten other functions (*homeostasis*).

3. *Self-support and movement:* Having structures that give specific form to the cell, and (in some cases) the ability to move materials within the cell, and (in some cases) the ability to move the whole cell through the surrounding environment.

4. *Self-replication:* Reproducing to carry on the species, and to repair damage. Some animal cells, such as nerve cells and muscle cells in mammals, are unable or have greatly restricted abilities to replicate. When trauma or disease processes destroy these cells, they are generally not replaced.

Cells are remarkable in the similarity with which they carry out these functions. Thus, all cells share many common characteristics. In multicellular organisms, each cell also performs a specialized function, which is usually a modification or elaboration of a basic cell function. Here are a few examples in animals:

- As a specialization of basic cell protein-synthesizing ability, *gland cells* of *digestive systems* secrete digestive enzymes.
- By enhancing the basic ability of cells to respond to changes in their environments, *neurons* generate and transmit to other regions electrical impulses that relay information about stimuli such as light, temperature, nutrient concentrations, and so on.
- Through specializations in the basic ability of cells to transport molecules through membranes, *kidney cells and tubules* selectively retain substances needed by animals while eliminating unwanted substances in the urine.
- Using a highly specialized elaboration of the inherent capability of eukaryotic cells to produce intracellular move-

ment, *muscles* use movement of internal protein filaments to bring about shortening of the whole organ.

It is important to recognize that cells perform these specialized activities in addition to carrying on the unceasing, fundamental activities required of all cells. The fundamental cell activities are essential for the survival of each individual cell, whereas the specialized contributions and interactions among the cells of a multicellular organism are essential for the survival of the whole organism.

◼ Cells are progressively organized into tissues, organs, systems, and finally the whole body.

Just as a machine does not function unless all its various parts are properly assembled, the cells of organisms must be specifically organized to carry out the life-sustaining processes of the whole, such as (in animals) digestion, respiration, and circulation. Multicellular organisms have four levels of organization: cells, tissues, organs, and systems.

In animals, cells of similar structure and function are organized into **tissues** (groups of cells with similar structures and functions), of which there are four primary types: muscular, nervous, epithelial, and connective. Each tissue consists of specialized cells along with varying amounts of extracellular ("outside the cell") material.

- **Muscular tissue** consists of cells specialized for contraction and force generation. Vertebrates have three types of muscle tissue: **skeletal muscle,** which causes movement of the skeleton; **cardiac muscle,** which is responsible for pumping blood out of the heart; and **smooth muscle,** which encloses and controls movement of contents through hollow tubes and organs, such as the digestive tract.
- **Nervous tissue** consists of cells specialized for initiation and transmission of electrical impulses, sometimes over long distances. These electrical impulses act as signals that relay information from one part of organisms to another. Nervous tissue in vertebrates is found in (1) the brain; (2) the spinal cord; (3) epithelial linings that measure information about the external environment and about the status of various internal factors that are subject to regulation, such as temperature, blood pressure, and gut contents; and (4) muscles, glands, and other *effector* organs (see p. 148).
- **Epithelial tissue** is made up of cells specialized in the exchange of materials. This tissue is organized into two general types of structures: sheets and secretory glands. Epithelial cells are joined together very tightly to form **epithelial sheets** that cover and line various organs. For example, the outer layer of the skin is epithelial tissue, as is the inner lining of the digestive tract. In general, these epithelial sheets serve as boundaries that separate animals from the external environment and from the contents of cavities that communicate with the external environment, such as the digestive tract lumen. (A **lumen** is the cavity within a hollow organ or tube, and in the case of digestive tracts, is technically outside of the body!). Only selected transfer of materials is permitted between the regions separated by an epithelial barrier. The type and extent of controlled exchange varies, depending on the location and

function of the epithelial tissue. For example, very little can be exchanged across the skin and the external environment in most animals, whereas the epithelial cells lining the digestive tract are specialized for absorbing water and nutrients.

- **Glands** are epithelial tissue derivatives that are specialized for secretion. Secretion is the release from a cell, in response to appropriate stimulation, of specific products that have in large part been synthesized by the cell. Glands are formed during embryonic development by pockets of epithelial tissue that dip inward from the surface. There are two categories of glands: *exocrine* and *endocrine* (● Figure 1–2). If during development the connecting cells between the epithelial surface cells and the secretory gland cells within the depths of the invagination remain intact as a duct between the gland and the surface, an exocrine gland is formed. **Exocrine glands** (*exo* means "external"; *crine* means "secretion") secrete through ducts to the outside of organisms (or into a cavity that communicates with the outside). Examples are sweat and digestive glands. If, in contrast, the connecting cells disappear during development and the secretory gland cells are isolated from the surface, an endocrine gland is formed. **Endocrine glands** (*endo,* "internal") lack ducts and release their secretory products, known as **hormones,** internally into the blood within the animal (see Chapter 7). For example, the vertebrate parathyroid gland secretes parathyroid hormone into the blood, which transports this hormone to its sites of action at the bones and kidneys.

■ **Connective tissue** is distinguished by having relatively few cells dispersed within an abundance of extracellular material that they secrete. As its name implies, connective tissue connects, supports, and anchors various body parts. It includes such diverse structures as the **loose connective tissue** that attaches epithelial tissue to underlying structures; **tendons,** which attach skeletal muscles to bones; **bone,** which gives vertebrates shape, support, and protection; and **blood** (or **hemolymph**), which transports materials from one part of a body to another, and in a sense, connects all cells of the body. Except for blood, the cells within connective tissue produce specific molecules that they release into the extracellular spaces between the cells. One such molecule is the rubber-band–like protein fiber **elastin,** whose elastic properties facilitate the stretching and recoiling of structures such as lungs, which alternately inflate and deflate during breathing (see p. 484).

Muscle, nervous, epithelial, and connective tissue are the primary tissues in a classical sense; that is, each is an integrated collection of cells of the same specialized structure and function. The term *tissue* is also frequently used to refer to the aggregate of various cellular and extracellular components that make up a particular organ (for example, lung tissue or liver tissue).

Organs consist of two or more types of primary tissue organized to perform a particular function or functions. The stomach is an example of an organ made up of all four primary tissue types (● Figure 1–3), which function collectively to store ingested food and move it forward into the rest of the digestive tract as well as to begin digestion. Vertebrate stomachs are lined with epithelial tissue that restricts the transfer of harsh

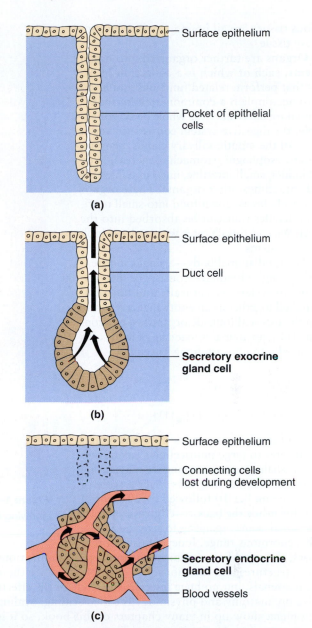

Figure 1–2 ● **Exocrine and endocrine gland formation.**
(a) Glands arise during development from the formation of pocket-like invaginations of surface epithelial cells. (b) If the cells at the deepest part of the invagination become secretory and release their product through the connecting duct to the surface, an exocrine gland is formed. (c) If the connecting cells are lost and the deepest secretory cells release their product into the blood, an endocrine gland is formed.

digestive chemicals and undigested food from the stomach lumen into the blood. Gland cells, derived from epithelia in the stomach, include exocrine cells, which secrete digesting juices into the lumen, and endocrine cells, which secrete hormones that help regulate the stomach's exocrine secretion and muscle contraction. The walls of the stomach contain smooth muscle tissue whose contraction mixes ingested food with the digestive juices and propels the mixture forward into the intestine. Also within the walls is nervous tissue, which, along with hormones, controls muscle contraction and gland secretion. These various tissues are all bound together by connective tissue.

Organs are further organized into **organ systems,** each of which is a collection of organs that perform related functions and interact to accomplish a common activity that is essential for survival of the whole body. For example, the digestive system consists (in vertebrates) of the mouth, salivary glands, pharynx (throat), esophagus, stomach, pancreas, liver, gallbladder, small intestine, and large intestine (and sometimes other organs). These organs collectively break down food into small nutrient molecules that can be absorbed into the blood. We end this chapter with a closer look at these systems.

The total animal body—a single, independently living individual—consists of the various organ systems structurally and functionally linked together as an entity that is separate from the external (outside organisms) environment. Thus, an animal is made up of living cells organized into life-sustaining systems.

Size and Scale Among Organisms

Because life on Earth ranges from unicellular prokaryotes to large multicellular eukaryotes such as whales, organisms range in mass (using the metric weights in grams) over a scale of 10^{20} (the number 10 followed by 20 zeroes)! The table inside the back cover gives examples of the size of some organisms to give you a sense of this enormous range. Independent of other factors, the size of an organism has important implications for its structures and functions. For example, why are there huge mammals, but no huge insects? The study of the effects of size on anatomy and physiology is called **scaling.** Scaling phenomena show up in many chapters of this book, so it is important to examine this issue now. The most important size effect is a simple mathematical concept known as the **surface-area-to-volume ratio.**

■ The larger the organism, the smaller the surface-area-to-volume ratio.

Consider a small spherical cell of minute size. It has a certain volume and mass, and a certain surface area of its outer boundary (● Figure 1–4a). From simple mathematics, we know the volume is related to the *cube of the radius* of the object. The surface area is related to the *square of the radius*. Now consider a cell (or organism) 10 times bigger (Figure 1–4b). Because of the cubed factor, the volume of it is 1000 (10^3) times bigger than the first object, but because of the squared factor, the surface area is only 100 times (10^2) bigger. Its surface-area-to-volume ratio is one tenth that of the smaller sphere. This has tremendous implications for a variety of living functions. For instance, the entire mass of both cells needs to obtain nutrients and get rid of wastes through the surface bound-

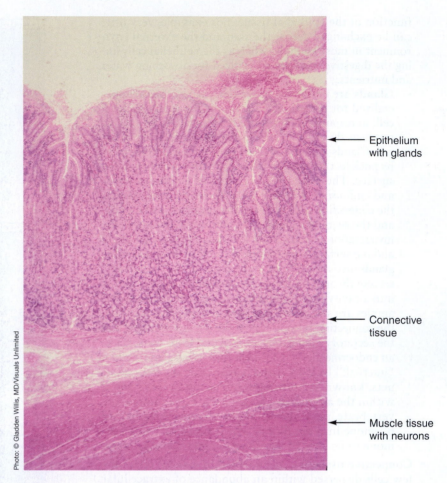

Photo: © Gladden Willis, MD/Visuals Unlimited

Epithelium with glands

Connective tissue

Muscle tissue with neurons

Figure 1–3 ● Low-power microscope view of the mammalian stomach, showing the primary tissue layers.

ary, but the larger object is at a 10-fold disadvantage because of its relatively smaller ratio (that is, relatively smaller surface area to support the volume). This is one reason why larger animals have evolved highly branched (that is, high surface area) circulatory and respiratory systems. We examine this development in Chapters 9 and 11.

However, if there is something that each cell or organism needs to retain, such as heat, the larger object is at a 10-fold advantage because heat is typically produced by the volume of the cell and lost through its surface area. In Chapter 15 we examine this aspect of physiology in detail.

Scaling also affects skeletal mechanics. The ability of a skeleton (or a tree trunk, for that matter) to support mass against gravity increases with the cross-sectional area of the skeletal element (such as a leg bone or insect limb exoskeleton). But the weight being supported increases as the cube of the animal size. Thus a large animal such as an elephant must have massive limbs in comparison to its body, whereas a tiny spider can support itself with limbs that are much thinner in comparison to its body.

It is important to remember this basic phenomenon. Note that larger objects have larger surface areas on an absolute scale, but it is the ratio that matters. Put simply, all other things being equal, *larger organisms have smaller surface-area-to-volume ratios.*

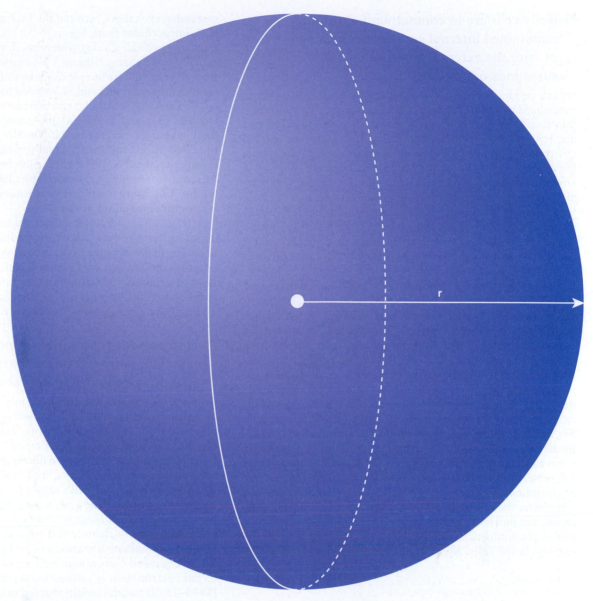

If r = 1, then
Surface area = $4\pi r^2$ = 12.6
Volume = $^4/_3\pi r^3$ = 4.2
Ratio = 3

(a) Small sphere

If r = 10, then
Surface area = $4\pi r^2$ = 1257
Volume = $^4/_3\pi r^3$ = 4189
Ratio = 0.3

(b) Large sphere

Figure 1–4 ● The effects of size on surface area and volume.

Homeostasis: Basic Mechanisms and Enhancements

Earlier we discussed several functions of cells (and thus life): self-organization, self-regulation, support and movement, and self-replication. To many biologists, the ability to self-regulate is the most crucial of these functions for distinguishing living from nonliving things. For example, a fire can arise spontaneously (given an initial energy source such as lightning) and can maintain itself by using combustible materials in the environment, can move by spreading to nearby fuel sources, and can replicate through sparks. But it takes no corrective action when water is dumped on it. Living systems, in contrast, can correct for at least some disturbances. Indeed, one recent definition of life that has been proposed is "a system that tries to regulate itself to preserve its identity."

Maintenance of a desired state in the face of disturbances has been given a term that is central to physiology: **homeostasis** (*homeo,* "similar"; *stasis,* "to stand or stay," although used here to mean "state"). Unicellular organisms have some homeostatic abilities, such as to maintain energy levels, acid–base balance, and volume. In animals, although individual cells also have some homeostatic abilities, many regulatory processes occur at the level of the whole organism. Let's now examine this crucial concept in detail.

■ **Body cells are in contact with a privately maintained internal environment instead of with the external environment that surrounds organisms.**

If each cell has basic survival skills, why must cells perform specialized tasks and be organized according to specialization into systems dedicated to the whole body's survival? The vast majority of cells are not in direct contact with the external environment and cannot function without contributions from the other cells. In contrast, a unicellular organism such as an amoeba can directly obtain nutrients and O_2 from its immediate external surroundings and eliminate wastes back into those surroundings. In a multicellular animal, a muscle cell has the same need for life-supporting nutrients and O_2 uptake and waste elimination, yet the muscle cannot directly make these exchanges with the external environment, because the cells are physically isolated from it.

How can a muscle cell make vital exchanges with the external environment with which it has no contact? The key is the presence of an aqueous internal environment with which body cells are in direct contact. This internal environment is outside the cells but inside bodies. It consists of the **extracellular** (*extra* means "outside of") **fluid** or **ECF**, which in vertebrates is made up of *plasma*, the fluid portion of the blood, and *interstitial fluid*, which surrounds and bathes the cells (● Figure 1–5). Various epithelial layers accomplish exchanges between the external environment and the internal environment. For example, a digestive system transfers the nutrients required by all body cells from the external environment into the ECF (such as blood, which is part of the internal body environment). Likewise, a respiratory system transfers O_2 from the external environment into the ECF. A circulatory system, if present, distributes these nutrients and O_2 throughout the body. As a result, the nutrients and O_2 originally obtained from the external environment are delivered to the interstitial fluid that surrounds the cells. Membranes of cells, in turn, make life-sustaining exchanges between the ECF and their interiors, the **ICF** (**intracellular fluid**, Figure 1–5). (Note that although the term *intracellular fluid* is widely used, the fluid inside cells is actually somewhat gelatinous.) No matter how remote a cell is from the external surface, it can take in from the internal environment the nutrients and O_2 needed to support its own existence. Similarly, metabolic end products produced by the cells (such as CO_2 and ammonia) are extruded into the interstitial fluid, where (in more complex animals) they are picked up by the circulation, and transported to organs that specialize in eliminating these wastes from the internal environment to the external environment. For example, lungs (in air breathers) remove CO_2 from the blood, and kidneys remove many other wastes for elimination in urine.

Thus, a body cell takes in essential nutrients from and eliminates wastes into its watery surroundings, just as an amoeba does. The major difference is that body systems must help maintain the composition of the internal environment so that this fluid continuously remains suitable to support the existence of all body cells. In contrast, an amoeba does nothing to regulate its surroundings, which are essentially infinite.

■ **Homeostasis is essential for proper cell function, and each cell, as part of an organized system, contributes to homeostasis.**

Animal cells can live and function only when they are bathed by an ECF that is compatible with their survival; thus many aspects of the chemical composition and physical state of the internal environment can be allowed to deviate only within narrow limits. As cells remove nutrients and O_2 from the internal environment, these essential materials must constantly be replenished for each cell's ongoing maintenance of life processes to continue. Likewise, wastes must be removed from the internal environment so they do not reach toxic levels. Other elements in the internal environment that are important for maintaining life also must be kept relatively constant. As we discussed earlier, maintenance of a relatively stable internal environment is termed *homeostasis*. Claude Bernard (1813–1878) had the insight that led to the concept of the constancy of the internal environment (*le milieu interieur*). His ambition was to be a playwright, but he turned to medicine and then to physiological research, which he termed "experimental medicine." Subsequently, Walter B. Canon (1871–1945) took an integrative approach to physiological function, which further developed Bernard's ideas and which led Canon to coin the term *homeostasis*.

The functions performed by the various body systems contribute to homeostasis, thereby maintaining within animals the environment required for the survival and function of all the cells of which organisms consist. This is one of the most important themes of physiology and of this book: *Homeostasis is essential for the survival of each cell, and each cell, through its specialized activities, contributes as part of a body system to the maintenance of the internal environment shared by all cells* (● Figure 1–6).

The factors of the internal environment that are often homeostatically regulated include the following:

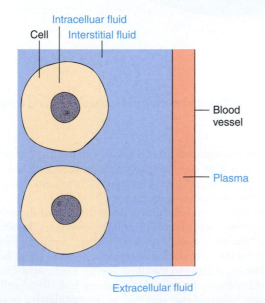

Cell   Intracelluar fluid
        Interstitial fluid

Blood vessel

Plasma

Extracellular fluid

Figure 1–5 ● Components of the extracellular fluid (internal environment).

1. *Concentration of energy-rich molecules.* Nondormant cells need a consistent supply of such molecules (both ex-

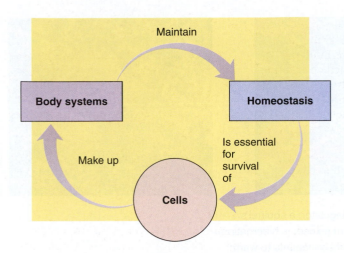

Figure 1-6 ● Interdependent relationship of cells, body systems, and homeostasis. The depicted interdependent relationship serves as the foundation for modern-day physiology: *Homeostasis is essential for the survival of cells, body systems maintain homeostasis, and cells make up body systems.*

ternally and internally) to serve as a metabolic fuel for life-sustaining and specialized cell activities.

2. *Concentration of O_2 and CO_2.* Cells need O_2 to perform chemical reactions that extract from nutrient molecules the most energy possible for use by the cell. The CO_2 produced during these chemical reactions must be balanced by removal of CO_2 from the body (for example, by lungs or gills) so that CO_2 does not increase the acidity (H^+ concentration) of the internal environment. This balance is accomplished through these reactions:

$$CO_2 + H_2O \leftrightarrow H_2CO_3 \leftrightarrow H^+ + HCO_3^-$$

3. *Concentration of waste products.* Various chemical reactions produce end products that can be toxic to cells if these wastes are allowed to accumulate beyond a certain limit.

4. *pH.* Among the most pronounced effects of changes in the pH (acidity) of the internal fluid environment are alterations in the electrical signaling mechanism of nerve cells and in the enzyme activities of all cells.

5. *Concentration of water, salt, and other electrolytes.* Because the relative concentrations of salt ions (mainly Na^+ and Cl^-), organic solutes, and water in the ECF influence how much water enters or leaves the cells, these concentrations may be regulated to maintain the proper volume of the cells. Cells do not function normally when they are swollen or shrunken. Other ions perform a variety of vital functions. For example, the rhythmic beating of a vertebrate heart depends on a relatively constant concentration of potassium (K^+) in the extracellular fluid.

In addition, some animals regulate the following:

6. *Volume and pressure.* The circulating component of the internal environment, the plasma in animals with circulatory systems, must be maintained at an adequate volume and pressure to ensure bodywide distribution of this important link between the external environment and the cells.

7. *Temperature.* Cells function optimally within a narrow temperature range. Chemical reactions slow down if they are too cold, and, worse yet, proteins are impaired if they get too hot.

8. *Social parameters.* Sometimes homeostasis can extend beyond the individual to a social level. This has been documented in social insects—in termite mounds, for example. In what has been termed *social homeostasis*, the colony of termites acts as a single "superorganism," regulating its population density and number of different termite types (such as workers, soldiers) to nearly constant levels.

The fact that the internal environment has some stable features does not mean that its composition, volume, and other characteristics are absolutely unchanging. First, external and internal factors continuously threaten to disrupt homeostasis. When this occurs, appropriate opposing reactions (or behaviors) are initiated to restore these conditions. For example, exposure to a cold environmental temperature tends to reduce a mammal's internal temperature. To counter this, compensatory shivering may be initiated, which internally generates heat that restores body temperature to normal. (Alternatively or in conjunction with shivering, the animal may seek a warmer location.) Likewise, addition of CO_2 into the internal environment as a result of energy-generating chemical reactions tends to raise the concentration of this gas (and the acid level) within organisms. This triggers an increase in breathing rate in many animals. As a result, the extra CO_2 (or HCO_3^-) is released to the external environment, restoring the CO_2 concentration in the ECF to normal. Thus homeostasis should be viewed not as a fixed state but as a dynamic steady state in which the changes that do occur are minimized by compensatory physiological responses. This is why the prefix *homeo* ("similar") is used instead of *homo* ("same"). For each factor in the internal environment, the small fluctuations around the optimal level are normally kept within the narrow limits compatible with life by carefully regulated mechanisms.

Regulators, Conformers, and Enantiostasis

There is a second reason why the concept of homeostasis does not mean that the characteristics of the internal environment are absolutely unchanging. The term *homeostasis* was developed primarily from research on mammals, which do exhibit remarkably consistent internal conditions. However, many kinds of multicellular organisms exhibit far more variability in their internal fluids than do mammals. Indeed, there are two categories of organisms from the viewpoint of internal environments: *conformers*, whose internal state matches that of the environment, and *regulators*, which defend a relatively constant state. Some physiologists recognize a third category, *avoiders*, which may not be capable of internal regulation, but which nevertheless can minimize internal variations by avoiding environmental disturbances (● Figure 1–7). For example, most organisms (including nonanimals) are thermoconformers (body temperatures equal to that of the environment), whereas birds and mammals are generally thermoregulators, maintaining a stable internal temperature regardless of the environmental temperature. But some thermoconformers (such

Figure 1–7 ● **Examples of regulators, conformers and avoiders.** (a) Regulator: a squirrel's internal body temperature is relatively constant regardless of environment (except in hibernation). (b) Avoider: monarch butterflies migrate in the autumn from freezing northern habitats to warm southern habitats to avoid extreme body temperatures. (c) Conformer: a marine snail's body temperature matches that of the external aquatic environment

as some fish and insects) can avoid large changes in body temperature by changing their locations in the environment (see Chapter 15).

Some animals also often achieve a kind of homeostasis termed **enantiostasis,** also called **allostasis.** This term refers to consistency of *function* achieved by changing one physiological variable to counteract a change in another. The term was first used for blue crabs, which suffer a decrease in internal salt composition when they leave the ocean and enter a lower-salinity environment (an estuary). Researchers have found that reduced salt-ion levels inhibit oxygen binding by *hemocyanin,* a large biomolecule (a *protein* similar to hemoglobin in your red blood cells), which transports oxygen in the crabs' circulatory fluid. Yet the crabs do not suffer from this, because they

make their internal fluids more alkaline (less acidic) by increasing production of ammonia, a strong base. Higher alkalinity in turn increases oxygen binding by hemocyanin and thus offsets the ion effects. Thus functional homeostasis of hemocyanin is achieved. In this text, we use the term *homeostasis* in a broader sense to include enantiostasis.

■ Negative feedback is the main regulatory mechanism for homeostasis.

To maintain homeostasis, organisms must be able to detect deviations in internal environmental factors that need to be held within narrow limits, and must be able to control the various body systems responsible for adjusting these factors. For

---

CHALLENGES *AND* CONTROVERSIES

Can a Planet Have Physiology?

As we have discussed, one of the main features that seems to define life itself is the ability to react to, and adjust for, environmental disturbances. In recent years, a number of scientists headed by James Lovelock have proposed that planet Earth itself exhibits this key property of life. Called the Gaia hypothesis, this concept is based on the observation that the conditions of weather and nutrient cycling that favor life have been relatively constant for hundreds of millions, perhaps billions of years. Although periods of prolonged cold (ice ages) have been interspersed with eons of global warmth, at no time have conditions deviated from the range necessary for life. Lovelock and his supporters propose that life itself, in conjunction with larger geologic and atmospheric processes, maintains this favorable state in homeostatic feedback processes some collectively call "geophysiology." As an example of a such a feedback process, researchers have found that cloud formation can be triggered by an atmospheric gas called *dimethylsulfide (DMS).* DMS in turn

is a breakdown product of an *osmolyte* (a molecule used to regulate cell water content; see Chapter 13) in marine phytoplankton (single-cell algae such as diatoms that are Earth's most common photosynthesizers). The hypothesis is that, as Earth's climate enters any warming trend, plankton growth increases and so, therefore, does DMS production. In turn, more clouds form that, by blocking sunlight, oppose the heating trend. Thus the temperature of the planet's atmosphere is prevented from extreme changes:

Increasing atmospheric temperature → more DMS release from phytoplankton → more clouds form → less sunlight reaches Earth → atmospheric cooling

If this is correct, would it be fair to think of Earth itself as a giant living organism (called Gaia)? The other properties of life such as reproduction and evolution may influence your answer to this intriguing question.

example, to maintain body temperature at an optimal value mammals must be able to detect a change in internal temperature and then appropriately alter heat production or loss so that internal temperature returns to the desirable level.

Homeostatic control mechanisms primarily operate on the principle of negative feedback. **Negative feedback** occurs when a change in a controlled variable triggers a response that opposes the change, driving the variable in the opposite direction of the initial change. Feedback systems can be very simple *unreferenced* ones, with imprecise control, or *referenced* with a built-in **set point** that gives the ideal state. An example of an unreferenced system has been proposed for Earth's climate

and global warming (see box, "Challenges and Controversies: Can a Planet Have Physiology?"). As Earth's atmosphere begins to heat up from excess carbon dioxide, evaporation may increase and thus produce more clouds. Clouds may reduce input from the sun and so oppose the warming trend. But there is no precise regulation, because there is no set point.

In contrast, a referenced negative-feedback system typically has these components: a *sensor* to measure the variable being regulated, an *integrator* that compares the sensed information with a set point, and an *effector*, the device or process that actually makes the corrective response (● Figure 1–8a). A common example of a negative feedback is control of room

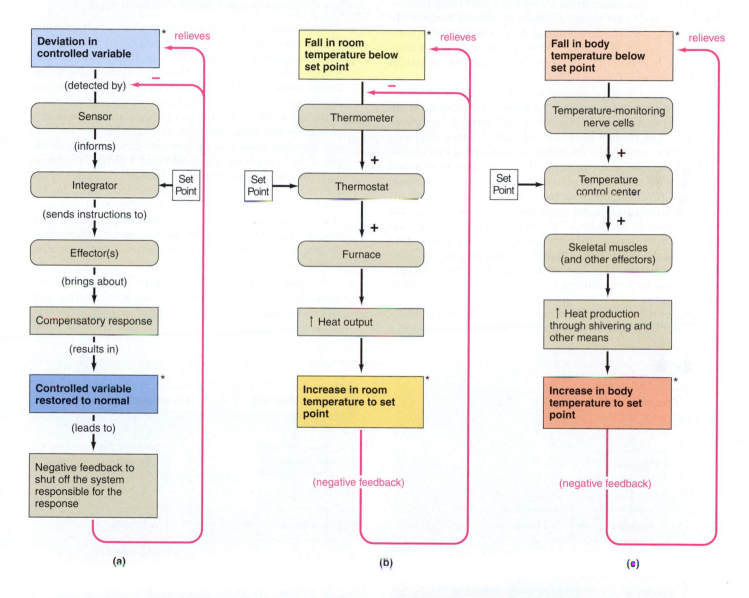

(a)   **(b)**   **(c)**

For flow diagrams throughout the text:
  + = Stimulates or activates
  – = Inhibits or shuts off
  ⬭ = Physical entity, such as body structure or a chemical
  ▭ = Actions
  ❘ = Compensatory pathway
  ❘ = Turning off of compensatory pathway (negative feedback)
  * Note that lighter and darker shades of the same color are used to denote respectively a decrease or an increase in a controlled variable.

Figure 1–8 ● **Negative feedback.** (a) Components of a negative-feedback control system. (b) Negative-feedback control of room temperature. (c) Negative-feedback control of mammalian body temperature.

temperature (the controlled variable) by a system that includes a sensing and integrating thermostatic device, a furnace and an air conditioner, and all their electrical connections. The room temperature is altered by the activity of the effectors—the furnace (a heat source) and air conditioner (a heat remover). Suppose someone opens the room door, and too much cold air enters. To switch the effectors on or off appropriately, the control system must "know" what the actual room temperature is, "compare" it with the desired room temperature (set point), and adjust the output of the effectors to bring the actual temperature to the desired level. A thermosensing device in the thermostat, which monitors the magnitude of the controlled variable, provides information about the actual room temperature. The thermostat setting provides the desired temperature level, or *set point*. The thermostat acts as an integrator: It compares the sensor's input with the set point and, in this case, adjusts the heat output of the furnace to oppose the deviation from the set point (Figure 1–8b).

When cold is being compensated for, the heat produced by the furnace counteracts or is "negative" to the original fall in the temperature. Note that this is still termed "negative" feedback even though heat is being added. Once the room temperature reaches the set point, the activating mechanism in the thermostat and consequently the effector is switched off. If heat production were to continue unabated, the room temperature would be increased above the set point. Overshooting far beyond the set point does not occur because the air temperature "feeds back" to shut off the thermostat that triggered heat output, thereby limiting its own production by altering the signal that initiated the heat production. Thus the control system takes corrective actions to prevent the controlled variable from drifting too far below or too far above the set point.

Homeostatic systems in organisms operate very similarly with negative feedback to maintain a controlled factor in a relatively steady state. For an example, you can see the similarities between the thermostat system of Figure 1–8b and one aspect of the mammalian thermoregulatory system, shown in Figure 1–8c. This is the system we described earlier (p. 2), in which a region of the brain, the hypothalamus, acts as an integrator for controlling body temperature.

■ Feedback effectors can be antagonistic and can include behaviors as well as internal organs.

The basic negative feedback mechanism we have just discussed can be expanded in several ways. For effectors, there are two alternative design features to keep in mind:

- *Antagonistic control.* If corrections can occur in either direction of change, by having two effectors with opposite effects, or by regulation that can increase or decrease a single effector's output, the system is said to have **antagonistic control** (● Figure 1–9a). Note the importance of the antagonistic design in the home system just described: If the controlled variable can be regulated to oppose a change in one direction only, the variable can move in uncontrolled fashion in the opposite direction. For example, if the house is equipped only with a furnace to oppose a fall in room temperature, no mechanism is available to prevent the house from getting too hot in summer. In contrast, the room temperature can be kept relatively constant through two opposing mechanisms, one that heats and one that cools the room (that is, an air conditioner), despite wide variations in the temperature of the external environment.

- *Behaviors as effectors.* An effector for a disturbance is typically an organ that directly adjusts some internal state (such as a muscle that shivers to produce heat). However, **behaviors** (specific animal movement patterns used in spe-

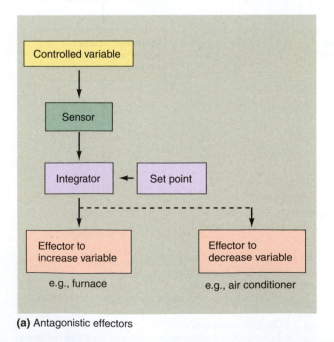

(a) Antagonistic effectors

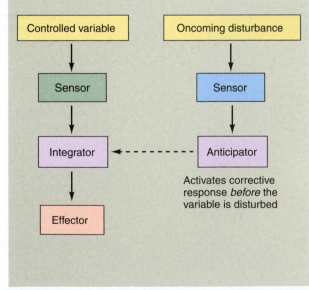

(b) Anticipation or feedforward control

Figure 1–9 ● Expansion and enhancement of basic negative feedback system.

cific situations; typically requiring coordinated actions of many neural and muscular components) can also serve as corrective processes, and are thus also effectors. Recall that some animals are *avoiders;* these typically use behavioral effectors. For example, monarch butterflies (Figure 1–7) migrate from the northern United States and Canada to California and Mexico for the winter, thus avoiding extreme cold. (This navigational behavior is genetically programmed to an amazing degree: The monarchs that migrate south are hatched in the north and are several-generation descendants of the monarchs that migrate north.)

Let's look at another example. Killifish are remarkable intertidal bony fishes that can survive in salinities ranging from freshwater to well above seawater concentrations, while maintaining a relatively consistent internal salinity at about 35 to 45% that of normal seawater. When the fish face a change in external salinity (such as when rainfall dilutes an exposed tide pool), cells in their gills (physiological effectors) adjust the rate of salt intake into the blood to a level necessary for homeostasis. At the same time, if the tide pool has a salinity gradient the fish actively seek a patch of water that has about 35 to 45% salinity (a behavioral effector process).

■ Inadequacies in negative feedback systems can be improved by anticipation and acclimatization.

It is important to note that, even with the added sophistication of antagonistic control, a basic referenced negative-feedback system (sensor, integrator with set points, and effectors) has two inherent flaws. First, it cannot be perfectly *homo*static ("same state") because it must first suffer a disturbing change before it can react and make corrections, resulting in a **delayed response.** Similarly, there is another delay in shutting off an effector, and so basic systems tend to overshoot the set point somewhat. Thus, many regulated parameters—body temperature, hormone levels, water balance, for example—fluctuate continuously. Second, some components may not work well in a new environment or situation, such as after an animal migrates from low to high altitude. Evolution (and human engineers) have come up with two ways to enhance regulation to overcome these problems (which exist to varying degrees among systems and animals).

■ *Anticipation or feedforward system.* An anticipation or feedforward system reduces the delay phenomenon by detecting or predicting an oncoming disturbance *before* a regulated state is changed (for example, perturbed beyond the set point) (Figure 1–9b). It may also predict the results of an effector's output before reaching the set point. The anticipator then activates the appropriate response in advance of the change or suppresses the effector before it overshoots the set point. For example, modern electronic thermostats have anticipator circuits that shut off the furnace just before the set point is actually reached, so that the hot air already on its way from the furnace to the thermostat brings the room to the set point without overshooting. As an organismal example, temperature sensors in a mammal's skin detect external heat or cold,

and can trigger corrective responses *before* the internal "core" temperature is disturbed. Similarly, when a meal is still in the digestive tract, a feedforward mechanism increases secretion of a hormone (insulin) that promotes cellular uptake and storage of ingested nutrients after they have been absorbed from the digestive tract. This anticipatory response helps limit the rise in blood nutrient concentration that follows nutrient absorption.

■ *Acclimatization systems.* Acclimatization systems are mechanisms that alter existing feedback and other components, usually over many days, to a new situation. Examples of new situations include migrating to high altitude and entering a different season. To continue the temperature theme, mammals may increase their hair or fat layers as winter approaches and shed them in the spring, thus augmenting the basic internal negative-feedback system.

Some processes that are technically forms of acclimatization are not designated as such, because they do not involve climate. For example, the muscle mass of an animal increases over time if locomotory activity undergoes a prolonged increase, as during migration. This form of acclimatization is often called **upregulation.** The opposite effect, **downregulation,** can also occur, as when muscle mass declines during more sedentary periods.

A note on terminology: The term *acclimation* refers to acclimatization processes that take place in a controlled situation, such as a laboratory. Acclimation and acclimatization processes, as well as up- and downregulation, are often called *adaptations* by many people. For example, a mountain climber may say he or she has "adapted" to high altitude. However, this term can be misleading, because evolutionary biologists use "adaptation" to refer to genetic changes that occur over many generations through variation and natural selection (p. 1). You will encounter both terms—*acclimatization, adaptation*—frequently, and for the latter term, you should always note

Newly hatched brine shrimp. Before hatching, they were in an embryonic cyst for several years in a dormant state.

Photo: Kevin Johnson, U.S. Geological Survey

whether it is being used in the between-generations (evolutionary) or the within-organism (physiological) sense.

Regulated Change

Not everything in organisms is homeostatic by any means, not even some of those traditionally considered to be classic examples. Mammalian body temperature is again a good example: Although normally quite narrowly regulated in most mammals most of the time, body temperature varies considerably with fever as well as between wake and sleep cycles.

■ Some internal processes are not always homeostatic but may be changed by reset and positive feedback systems.

There are many nonhomeostatic outcomes of acclimatization to changing environmental stresses. For many animals, negative feedback regulation cannot be modified to keep working at an optimum, and acclimatization involves switching to a state of **dormancy** (a state of greatly reduced metabolism), such as hibernation by a small mammal in which body temperature falls dramatically. Or consider the embryos of a brine shrimp (sold commercially as "sea monkeys") in the Great Salt Lake in Utah. When the embryos are washed up on a dry shoreline, they can lose over 98% of their internal water, yet remain viable by entering a resistant cyst state (which can hatch out years later if put in water). These animals maintain homeostasis only in the broadest sense of "survival." (Compare this to mammals, who die if they lose more than 10 to 15% body water!)

Furthermore, many functions such as locomotion, growth and development, and neural signals are not homeostatic at all (they may contribute to homeostasis in other parameters, but not always). Some regulatory systems are simply switched on when stimulated for a particular need, and are then switched off. Examples of such "on-demand" regulation are muscles and glands in the digestive tract, which are locally activated to churn and digest the food when present, and shut off when not. This can contribute to other homeostatic needs such as nutrient concentrations in the blood, but this regulation is not homeostatic in itself. Some physiologists have termed nonhomeostatic regulation **rheostasis** ("variable state").

Evolution has also come up with two other mechanisms for regulating useful nonhomeostatic change, which we mention frequently in later chapters:

- *Reset system:* one that changes the set point of a negative feedback system, in a temporary, permanent, or cyclic fashion (● Figure 1–10a). Fever is an example of a temporary reset of body temperature. At sexual maturity in a mammal, sex hormone concentrations are permanently reset to higher levels. Annual hibernation (such as by the squirrel in Figure 1–7) and many reproductive cycles are additional examples of cyclical resets regulated by internal biological clocks (see Chapter 7). You can probably think of both temporary and cyclical mechanisms for the room thermostat system of Figure 1–8b.

- *Positive feedback system.* A mechanism to create rapid change when conditions demand a rapid change from a set point. In negative feedback, a control system's output is regulated to resist change, so that the regulated variable is maintained at a relatively steady set point. With positive feedback, however, the output is continually enhanced so that the controlled variable continues to move toward the initial change (Figure 1–10b). Instead of bringing about a response that counteracts the initial change, positive feedback reinforces the change in the same direction. Such action is comparable to the heat generated by a furnace triggering the thermostat to call for even more heat output from the furnace so that the room temperature continuously rises. Because positive feedback moves the controlled variable even farther from a steady state, it does not occur purposefully very often in organisms, where the major goal is maintenance of stable, homeostatic conditions. Indeed, unintended positive feedback can lead to rapid death. For example, in congestive heart failure a weak mammalian heart results in a drop in blood pressure, triggering negative feedback systems that retain fluid in the body in an attempt to increase blood pressure (more fluid in blood vessels increases pressure). However, more fluid increases stress on the heart, weakening it further, dropping pressure again, and so the cycle continues until the heart fails completely.

Useful positive feedback does occur in certain instances, for example, neuron action potentials, lactation, blood clotting, bile release, many mating behaviors and orgasms, ovulation, and some reactions of immune systems are all positive feedback. These all have one thing in common: A relatively rapid change is needed because of a new situation in which the homeostatic set point is no longer appropriate. As a detailed example, consider the birth of a mammalian infant. The hormone *oxytocin* causes powerful contractions of the uterus. As uterine contractions push the baby against the cervix (the exit from the uterus), the resultant stretching of the cervix triggers a sequence of events that releases even more oxytocin, which causes even stronger uterine contractions, triggering the release of more oxytocin, and so on (Figure 1–10c). Positive feedback here is useful to help the infant be born relatively quickly once it reaches a key stage of developmental maturity.

Except for some cyclical reset systems, all these nonhomeostatic "change" mechanisms (on-demand, temporary reset, and positive feedback) need to be shut off once no longer needed. The mechanisms for this vary but typically involve the loss of the initiating signal. For example, the oxytocin positive-feedback loop stops once the fetus is no longer pushing on the cervix. Like negative feedback systems, some of these change systems can be activated in feedforward fashion and improved or adjusted up or down by acclimatization. For example, the activation of digestive muscles and glands (described earlier) can be feedforward-activated by the taste and smell of food. And the internal clock that cyclically resets a human's body temperature each night and day can be acclimatized to a new time zone by local sunlight exposure.

■ Disruptions in regulation can lead to illness and death.

When one or more of an animal's systems fail to function properly, homeostasis (or the ability to regulate change) is disrupted, and all the cells may suffer, because they no longer have a proper environment in which to live and function. Var-

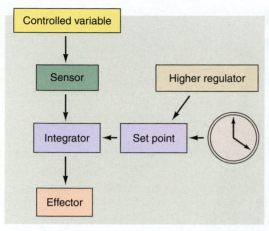

(a) Reset control of negative feedback by a higher system or clock

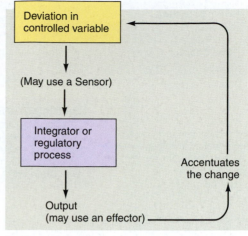

(b) Positive feedback

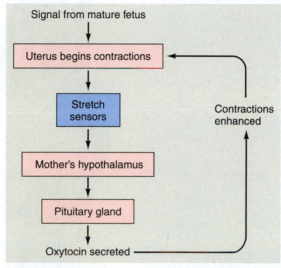

(c) Example of positive feedback: birth of a mammal

Figure 1–10 • Mechanisms for regulated change.

ious pathophysiological states ensue, depending on the type and extent of homeostatic disruption. The term *pathophysiology* refers to the abnormal functioning of organisms (altered physiology) associated with disease. When a homeostatic disruption becomes so severe that it is no longer compatible with survival, death results. An example is that of congestive heart failure, mentioned earlier.

Organization of Regulatory and Organ Systems

Control of an animal's functions is often—erroneously—portrayed as a centralized command system with the brain and endocrine systems regulating everything. In fact, such a design would be needlessly cumbersome and slow, because every tiny response needed by each cell or tissue for homeostasis would require these events: a signal to a central command organ; a decision; a return signal; and finally a local response. For many simple responses, the time delay could be harmful; for example, you would not want to wait for some central endocrine gland to signal the need for a blood clot when your foot is in-

jured and bleeding. Therefore, control in an animal is actually **hierarchical,** much like governmental systems in many countries!

■ Homeostasis (and other regulation) is hierarchically distributed.

The United States and Canada, for example, have at least four levels in this hierarchy in order of increasing priority: city, county, state/province, and federal (or national) governments. Basic local needs are delegated to city governments, because local knowledge of the situation is better, and it would be unnecessarily slow to wait for interactions with the national level (considerable delays occur if a city resident requires new federal permission for construction work on his or her home). But needs that involve the whole country (such as distributing food supplies and defending the country) often require the highest level of control. Similarly, many simple, common homeostatic responses in animals are controlled locally, at the level of individual cells, tissues, or organs. Only when more complex responses are needed is a whole-body ("federal" or "national") system needed.

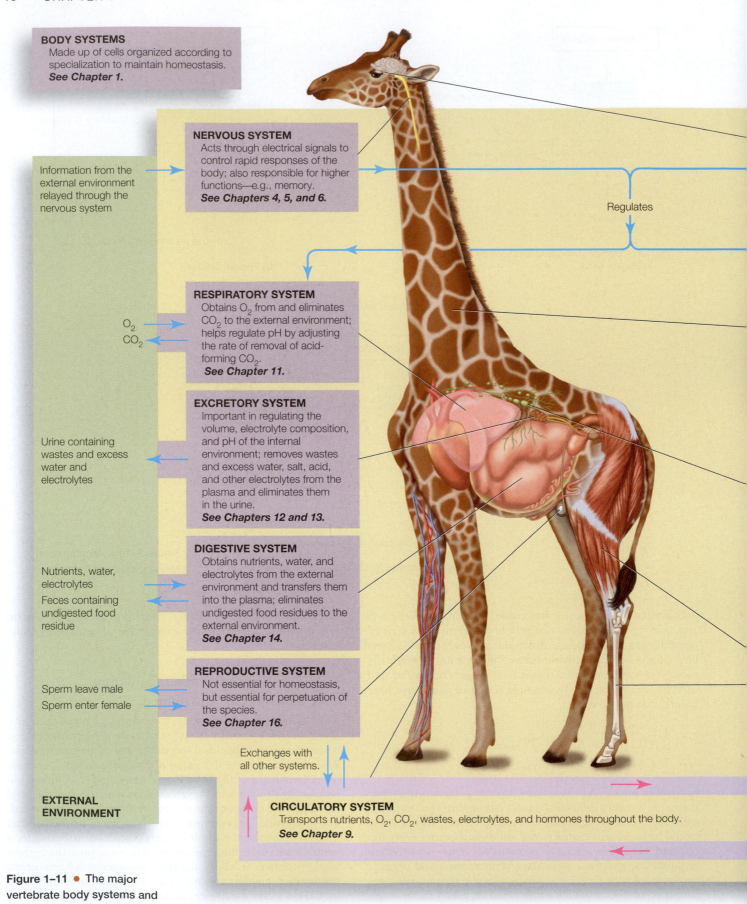

BODY SYSTEMS
Made up of cells organized according to specialization to maintain homeostasis. *See Chapter 1.*

NERVOUS SYSTEM
Acts through electrical signals to control rapid responses of the body; also responsible for higher functions—e.g., memory. *See Chapters 4, 5, and 6.*

Information from the external environment relayed through the nervous system

Regulates

RESPIRATORY SYSTEM
Obtains O_2 from and eliminates CO_2 to the external environment; helps regulate pH by adjusting the rate of removal of acid-forming CO_2. *See Chapter 11.*

O_2
CO_2

EXCRETORY SYSTEM
Important in regulating the volume, electrolyte composition, and pH of the internal environment; removes wastes and excess water, salt, acid, and other electrolytes from the plasma and eliminates them in the urine. *See Chapters 12 and 13.*

Urine containing wastes and excess water and electrolytes

DIGESTIVE SYSTEM
Obtains nutrients, water, and electrolytes from the external environment and transfers them into the plasma; eliminates undigested food residues to the external environment. *See Chapter 14.*

Nutrients, water, electrolytes

Feces containing undigested food residue

REPRODUCTIVE SYSTEM
Not essential for homeostasis, but essential for perpetuation of the species. *See Chapter 16.*

Sperm leave male

Sperm enter female

Exchanges with all other systems.

EXTERNAL ENVIRONMENT

CIRCULATORY SYSTEM
Transports nutrients, O_2, CO_2, wastes, electrolytes, and hormones throughout the body. *See Chapter 9.*

Figure 1–11 ● The major vertebrate body systems and their roles.

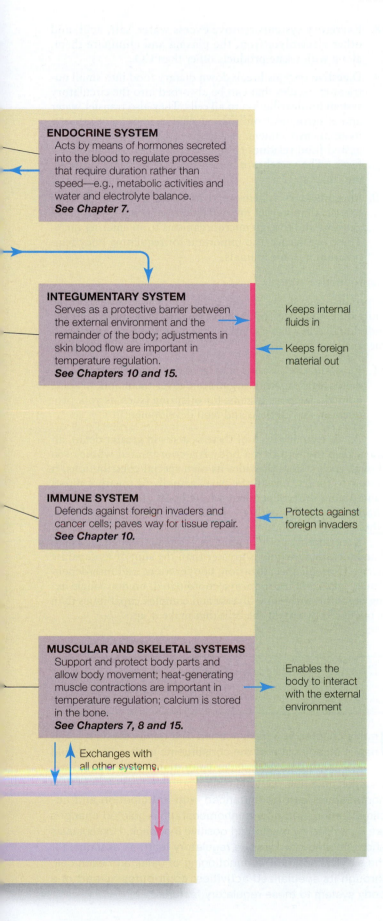

ENDOCRINE SYSTEM
Acts by means of hormones secreted into the blood to regulate processes that require duration rather than speed—e.g., metabolic activities and water and electrolyte balance.
See Chapter 7.

INTEGUMENTARY SYSTEM
Serves as a protective barrier between the external environment and the remainder of the body; adjustments in skin blood flow are important in temperature regulation.
See Chapters 10 and 15.

Keeps internal fluids in

Keeps foreign material out

IMMUNE SYSTEM
Defends against foreign invaders and cancer cells; paves way for tissue repair.
See Chapter 10.

Protects against foreign invaders

MUSCULAR AND SKELETAL SYSTEMS
Support and protect body parts and allow body movement; heat-generating muscle contractions are important in temperature regulation; calcium is stored in the bone.
See Chapters 7, 8 and 15.

Enables the body to interact with the external environment

Exchanges with all other systems.

Let us examine this hierarchy in more detail. At the most basic level, an individual cell can regulate its internal energy supplies, ion concentrations, and cell volume, at least to some extent. Beyond the cellular level, control systems that operate to maintain homeostasis are often grouped into two broad classes in this hierarchical design—intrinsic and extrinsic controls. **Intrinsic controls** are those regulated by a single tissue or organ on its own (akin to a county or state/province government). For example, as an exercising muscle rapidly uses up O_2 and produces CO_2 to generate energy to support its contractile activity, the O_2 concentration falls and the CO_2 concentration increases within the muscle. By acting directly on the smooth muscle in the walls of the blood vessels that supply the exercising muscle in a vertebrate, these local chemical changes cause the smooth muscle to relax and the vessels to dilate (open widely) to increase blood flow into the exercising muscle. This local mechanism contributes to the maintenance of an optimal level of O_2 and CO_2 in the internal fluid environment surrounding the exercising muscle's cells. As we discussed earlier, this response is generated at the local level, which avoids unnecessary delays and possibly avoids the reduced precision of generalized whole-body responses.

However, many factors in the internal environment are maintained by **extrinsic controls,** which are regulatory mechanisms initiated outside an organ to alter its activity (the "national government" level). Often this involves coordination of two or more organs. Extrinsic control of the various organs and systems is accomplished by the nervous, endocrine, and (to some extent) immune systems. Extrinsic control permits coordinated regulation of several organs toward a common goal; in contrast, intrinsic controls are self-serving for the organ in which they occur. Coordinated, overall regulatory mechanisms are critical for maintaining the needs of the internal environment as a whole. For example, the nervous system can take priority over an individual organ's regulation of blood flow during exercise (diverting blood from digestive to muscular systems; just as the national government can sometimes override state/province and city governments).

■ Organ systems can be grouped according to whole-body contributions.

Now that you have considered the basic features of living systems and the importance of homeostasis, let's briefly examine the actual components of animals that carry out these functions. These components govern the organization of most of this book. As we said earlier, physiologists typically categorize organs in animals into functional **systems** based on their contributions to the whole organism. In turn, these systems can be grouped into four categories based on the broad view of life itself: whole-body control systems (for self-regulation of the whole); maintenance systems (for self-organization); support and movement systems; and reproductive systems. There are about 11 major body systems (see ● Figure 1–11 for vertebrates), as follows.

Whole-body control systems

Whole-body control systems regulate and coordinate homeostatic and other functions of the other systems for the good of the whole body. Keep in mind that these do not control all activities in the hierarchical control design.

1. **Nervous systems** control and coordinate bodily activities that require swift responses. They are especially important in detecting and initiating reactions to changes in the external environment. Furthermore, they are responsible for more complex "higher" functions such as consciousness, learning and memory, and creativity. In animals with central brains, this system is also hierarchical, for example, with short **reflex** responses (automatic, unlearned reactions to a stimulus) and longer more complex responses.

2. The hormone-secreting glands of **endocrine systems** regulate activities that require duration rather than speed. These systems are especially important in controlling reproductive cycles, the concentration of nutrients, and the internal environment's volume and electrolyte composition.

Support and movement systems

From a purely homeostatic view, support and movement are not necessarily directed toward maintaining homeostasis, although holding position ("posture") is a type of homeostasis. And the ability to protect from harm and obtain food contributes to the broad homeostasis of body integrity and survival.

3. **Skeletal systems** provide support and protection for the soft tissues and organs. In vertebrates, the skeleton also functions in homeostatic regulation of calcium (Ca^{++}), an electrolyte whose plasma concentration must be maintained within very narrow limits. Together with the muscular system, the skeletal system also enables movement of animals and their parts.

4. **Muscular systems** move the skeletal components to which the skeletal muscles are attached. Furthermore, the heat generated by muscle contraction is important in temperature regulation in some animals, especially birds, mammals, and some insects and fishes.

Maintenance systems

These are organ systems that do most of the actual work of maintaining the internal environment. The systems and their most important contributions to homeostasis are as follows:

5. **Circulatory systems** are the transport systems that carry materials such as nutrients, O_2, CO_2, nitrogenous wastes, electrolytes, heat, and hormones from one part of an animal to another.

6. **Defense** or **immune systems** defend against small foreign invaders (viruses, bacteria, parasites) and body cells that have become cancerous. They also pave the way for repair or replacement of injured or worn-out cells. Because some defensive responses can coordinate activities of organs throughout the body (at least in mammals), in recent years many researchers have come to regard the mammalian immune system as the third whole-body control system (along with nervous and endocrine systems). For now, we will leave it under maintenance systems.

7. **Respiratory systems** obtain O_2 from and eliminate CO_2 to the external environment. By adjusting the rate of removal of acid-forming CO_2, a respiratory system is also important in maintaining the proper pH (acid–base balance) of the internal environment.

8. **Excretory systems** remove excess water, salt, acid, and other electrolyte from the plasma and eliminate them, along with waste products other than CO_2.

9. **Digestive systems** break down dietary food into small nutrient molecules that can be absorbed into the circulatory system for distribution to all cells. They also transfer water and electrolytes from the external environment into the internal environment. A digestive system eliminates undigested food residues to the external environment in the feces. (The vertebrate liver also excretes some wastes—see Chapter 12.)

10. **Integumentary systems** serve as outer, protective physical barriers that prevent internal fluid from being lost from organisms and foreign microorganisms from entering them. These organs are also important for regulating body temperature in some animals.

Reproductive systems

Reproductive systems are not essential for homeostasis of the individual and therefore are not essential for survival of the individual. They are essential, however, for perpetuation of the species.

11. **Reproductive systems** consist of organs that produce gametes (eggs, sperm), organs to deliver them and, in some animals, support systems for offspring (such as yolk production, and embryo and fetal development).

As we examine each of these systems in greater detail, always keep in mind that a body is a coordinated whole even though each system provides its own special contributions. It is easy to forget that all the parts actually fit together into a functioning, interdependent whole body. Accordingly, each chapter concludes with a brief overview of the integrative, whole-body contributions of the system.

Keep another point in mind as you read through the book: The functioning whole is greater than the sum of its separate parts. Through specialization, cooperation, and interdependence, cells combine to form a coordinated, unique, single living organism with more diverse and complex capabilities than is possessed by any of the cells that make it up.

Chapter in Perspective:
HOMEOSTASIS AND INTEGRATION

In this chapter, you have learned that homeostasis is a dynamic steady state of the constituents in the internal fluid environment (the extracellular fluid) that surrounds and exchanges materials with the cells. You have also learned how homeostasis can be enhanced with anticipation and acclimatization, and how nonhomeostatic regulated changes are controlled by resets and positive feedback. Homeostasis and regulated changes are regulatory features essential for the survival and normal functioning of cells, and each cell, through its specialized activities, contributes as part of a body system to these regulatory features.

This integrated relationship among cells, organ systems, and the whole organism is the foundation of physiology and the central theme of this book. We have already described how cells are organized according to specialization into body systems. How the key functions of life and body systems maintain this internal consistency and necessary changes are the topics covered in the remainder of this book. Each chapter concludes with this capstone feature to facilitate your understanding of (1) how the systems and functions under discussion contribute to whole-body homeostasis and regulated change, and (2) the integrated interactions and interdependency of body systems. ■

REVIEW QUESTIONS *(Answers are on p. A-1.)*

Additional study tools for this chapter, including chapter summaries and practice tests, are available online at *www.biology .brookscole.com*

1. Physiology is about
   a. the functions of life forms
   b. the purpose of evolution
   c. teleology
   d. adaptation and homeostasis
   e. a and d
2. The hypothetic-deductive scientific method begins by
   a. designing experiments
   b. conducting experiments
   c. asking a question about nature
   d. testing a hypothesis
3. A scientific hypothesis must be
   a. physiologically based
   b. mechanistic
   c. teleological
   d. testable
   e. proximate
4. Which of the following is not a type of tissue?
   a. epithelial
   b. scalar
   c. nervous
   d. muscular
   e. connective
5. Organs are composed of two or more types of
   a. exocrine glands
   b. epithelial tissue
   c. primary tissue
   d. hemolymph
   e. systems
6. Which of the following is not a term that is associated with epithelial tissue?
   a. glands
   b. sheets
   c. endocrine
   d. exocrine
   e. elastin
7. If one cell has a larger radius than another cell, it has
   a. a larger surface-area-to-volume ratio
   b. a more difficult time retaining heat
   c. a smaller surface-area-to-volume ratio
   d. an advantage over the smaller cell in obtaining nutrients through its surface boundary
   e. none of the above

8. The most general term for maintenance of a desired state by a living organism in the face of disturbance is called
   a. deregulation
   b. upregulation
   c. acclimation
   d. homeostasis
   e. delayed response
9. Which of the following internal factors are homeostatically regulated in at least some animals?
   a. concentration of water
   b. concentration of waste products
   c. concentration of electrolytes
   d. social parameters
   e. all of the above
10. Homeostatic mechanisms operate primarily on the fundamental principle of
    a. enantiostasis
    b. negative feedback
    c. avoidance
    d. feedforward systems
    e. rheostasis
11. Which of the following is not a component of a referenced negative feedback system?
    a. a sensor
    b. a set point
    c. an integrator
    d. a replicator
    e. an effector
12. Antagonistic control mechanisms have
    a. no set points
    b. two effectors with opposite effects
    c. only delayed responses
    d. an effector that can increase or decrease output
    e. b and d
13. An effector for a disturbance is typically an organ that directly adjusts an internal state. But an effector can also be
    a. set point
    b. an integrator
    c. a whole organism's behavior
    d. a sensor
    e. a reset system
14. A basic referenced feedback system cannot be perfectly homeostatic, because
    a. there is a delay time in its response
    b. it must reset its set point each time a change takes place
    c. basic systems tend to undershoot the set point

d. its integrated feedforward systems increase delays in response

e. upregulation systems are antagonistic to the referenced feedback

15. Non-homeostatic regulation has been termed
    a. downregulation
    b. acclimatization
    c. rheostasis
    d. upregulation
    e. antagonistic control

16. In a positive feedback system,
    a. change is resisted
    b. change is enhanced
    c. an initial change is counteracted
    d. change is created slowly
    e. a sensor is deregulated

17. In the regulation hierarchy, intrinsic controls
    a. are regulated by entire organ systems

b. are regulated only by the endocrine system

c. are regulated only by the muscular and skeletal systems

d. are regulated by a single tissue or organ on its own

e. none of the above

18. In the regulatory hierarchy, extrinsic controls
    a. are regulatory mechanisms initiated only by an organism's external environment
    b. are regulatory mechanisms initiated outside of an organ
    c. are regulatory mechanisms that involve only positive feedback
    d. are always antagonistic
    e. almost never involve the endocrine system

19. Which of the following systems is not essential for homeostasis?
    a. immune system
    b. reproductive system
    c. digestive system
    d. integumentary system
    e. excretory system

SUGGESTED READINGS AND INTERNET SITES

Boyd, C. A. R., & D. Noble, eds. 1993. *The Logic of Life: The Challenge of Integrative Physiology.* Oxford, UK: Oxford University Press. Discusses recent ideas in the evolution and regulation of physiological processes.

Burggren, W. W. 2000. Developmental physiology, animal models, and the August Krogh principle. *Zoology* 102:148–156. On using the Krogh principle.

Calder, W. A. (1996). *Size, Function and Life History.* Cambridge, MA: Harvard University Press. A classic book on the scaling of organisms.

Kelly, K. 1995. *Out of Control: The New Biology of Machines, Social Systems and the Economic World.* Reading, MA: Addison-Wesley. Discusses how biological concepts of feedback and hierarchical regulation are being applied to human society and industry.

McGowan, C. 1994. *Diatoms to Dinosaurs: The Size and Scale of Living Things.* Washington, DC: Island Press.

Mangum, C., & D. Towle. 1977. Physiological adaptation to unstable environments. *American Scientist* 65:67–75. Introduces the concept of enantiostasis.

Moore, J. A. 1999. *Science as a Way of Knowing.* Cambridge, MA: Harvard University Press. A detailed, highly readable look at the scientific method and scientific thinking.

Mrosovsky, N. (1997). *Rheostasis: The Physiology of Change.* Oxford, UK: Oxford University Press. Discusses how the concept of homeostasis is overemphasized at the expense of regulated change or rheostasis.

Nesse, R. M., & G. C. Williams. 1996. *Why We Get Sick: The New Science of Darwinian Medicine.* New York: Vintage Books. A classic book introducing this new field arising out of evolutionary physiology.

Schulkin, J. 2003. *Rethinking Homeostasis: Allostatic Regulation in Physiology and Pathophysiology.* Cambridge, MA: MIT Press. Discusses important nonhomeostatic and enantiostatic processes of mammalian bodies.

Volk, T. (1997). *Gaia's Body: Toward a Physiology of Earth.* New York: Copernicus Books. Discusses feedback and other regulatory processes in the planetary geo- and biosphere.

INTERNET SITES

Patrick, J. 1999. *Homeostasis.* **physioweb.med.uvm.edu/ homeostasis/.** Includes links to online interactive simulations of simple and complex negative feedback.

Full, R. 2002. *Animals of the Poly-PEDAL Lab.* **polypedal .berkeley.edu/Animals/Animals.html.** Provides information on a laboratory that applies the Krogh principle to investigating how animals move with limbs.

Cellular and Molecular Physiology

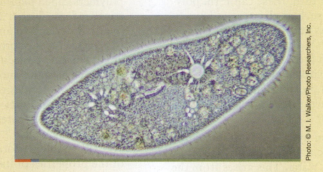

Photo: © M. I. Walker/Photo Researchers, Inc.

A single-celled protozoan, a Paramecium, *carries out all the necessary functions for survival and reproduction.*

Introduction

As we noted in Chapter 1, cells are distinct from their inanimate constituents because they have the ability to grow, replicate, perform complex metabolic reactions, and respond to environmental stimuli. Modern cellular physiologists are unraveling many of the broader mysteries of how cells work, both as living systems individually and in their specialized contributions to multicellular organisms. Comparative physiologists study how an individual species is adapted to a specific environment, in part by analyzing the molecular structure and organization of its cells. Analyses of the structures of living cells, and of the molecular components of which they are made, show the deep interconnectedness of all life through evolution.

It is important to realize that, although researchers have analyzed structures and functions of most of the molecular constituents of cells, it has not been possible to organize these chemicals artificially into a living system in a laboratory environment. It is not merely the presence of various molecules that confers the unique characteristics of life; it is the complex organization and interaction of these molecules within the cell that is critical. Much remains to be learned about the molecular organization necessary for life, and tools that enable scientists to do so are continuously increasing in sophistication and power.

■ **Water, other inorganic chemicals, and four types of organic molecules are the universal components of cells.**

Living systems are constructed from two broad classes of chemicals called *inorganic* and *organic molecules*. Scientists originally designated organic chemicals as those from living or once-living sources, but today the term *organic* refers to almost all molecules made from the element carbon (and invariably hydrogen, with the smallest organic molecule being methane or natural gas, CH_4). Based on this definition, organic material has now been confirmed on a distant planet, Osiris, by the Hubble Space Telescope. Located 150 light-years from Earth (one light-year is approximately 6 billion miles), Osiris also has oxygen in its atmosphere. However, the development of life on Osiris is unlikely, because of its close proximity (4.3 million miles) to its sun. All other chemicals are termed *inorganic*, even small carbon-containing molecules such as carbon dioxide (CO_2). Although cells contain a large variety of inorganic and organic components, there are some fundamental, universal types of molecules. We assume you are familiar with the basic inorganic and organic molecules of life; for readers who are not, details are summarized here.

Water (H_2O) is the universal inorganic molecule of life on Earth. Its properties such as its polarity (unequal sharing of electrons between hydrogen and oxygen) and hydrogen-bond formation make it unique among molecules in the universe for serving as a medium for the reactions of life. In particular, polarity allows water molecules to bind to any molecule that is charged or polar. Indeed, water is sometimes called "the universal solvent." In addition, other small inorganic chemicals such as potassium (K^+) and phosphate (PO_4^{3-}) appear to be universal solutes (dissolved molecules) in cells; we explore their roles and those of other inorganic chemicals later.

Because of the unique bonding ability of carbon, millions of different organic molecules are possible. However, it is possible to simplify this tremendous diversity into a few categories. The building blocks of organic molecules are called *monomers*, and chains of monomers linked together are called *polymers*. The largest polymers, called *macromolecules*, are the most important organic components of life (examples being proteins and nucleic acids such as deoxyribonucleic acid, or DNA). The basic categories of organic molecules are as follows:

1. **Carbohydrates** are made of carbon, hydrogen, and oxygen. The smallest building-block units are simple sugars or *monosaccharides* such as glucose, the dominant energy molecule in most animals. In turn, these can form polymers called *polysaccharides*, such as glycogen (a branching chain of glucose molecules and an important energy-storage molecule in animals). In addition to energy, carbohydrates can be used to form important structures such as *chitin* in arthropod exoskeletons. Sugars are also often found polymerized with proteins into *glycoproteins*, which are important components of cell membranes and signaling systems (see Chapter 3).

2. **Lipids** are made of carbon and hydrogen (and sometimes oxygen) and are hydrophobic ("water fearing"); that is, they do not easily dissolve in water. This arises from the fact that C–H bonds are nonpolar because they share electrons equally and cannot form hydrogen bonds with water. You probably know that oil, a hydrophobic hydrocarbon, floats on water. This is because water molecules make hydrogen bonds with each other, but not with oil molecules, which are in effect squeezed out of the network of water. (Lipids then rise to the top because they are less dense than water.) This lipid–water interaction is one of the most important chemical forces in living systems, as we will show. Key examples of lipids are *triglycerides*, which store energy as fat in animals; *phospholipids*, used to make membranes; and *cholesterol*, which make up membranes, steroid hormones, and bile.

3. **Amino acids** and **proteins** are made of carbon, hydrogen, oxygen, and nitrogen in a specific arrangement. Amino acids have an amino group ($-NH_2$), an acid group (carboxyl or $-COOH$), and a side group that gives each amino acid unique properties. Amino acids may be used for energy, as osmolytes (Chapter 13), or as signal molecules (Chapter 4), but their most crucial role is to form polymers called *peptides* and *proteins*. Peptides and proteins are formed from 20 carboxylic amino acids, arranged in any order and to any length, providing an endless list of possible types of proteins. These polymers are the principal dynamic molecular agents of life, forming regulators, such as signal molecules and receptors; transporters that move molecules through membranes; and enzymes, the primary catalysts of life's chemical reactions (● Figure 2–1a). Some proteins also form large structures such as muscle filaments.

   Ultimately a protein obtains its structure as a result of the properties of its amino acid types, especially their side groups, and how they interact with each other and the surrounding medium. The highly complex three-dimensional structures that result from these molecular interactions determine a protein's function; that is, shape determines function. There are up to four levels of structure involving peptide chains and proteins.

4. **Nucleotides** and **nucleic acids** are made of carbon, nitrogen, oxygen, hydrogen, and phosphate in a particular arrangement. The building blocks are called *nucleotides*, which have three components; a sugar (ribose or deoxyribose), phosphate, and a *base* which is one of five major types designated A, G, C, T, and U (adenine, guanine, cytosine, thymine, and uridine). These can form two types

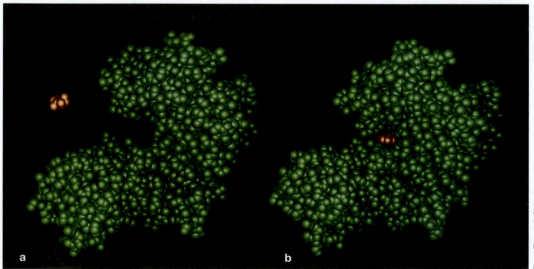

Photo: Thomas A. Steitz

Figure 2–1 ● The complex structures of proteins are mainly determined by weak bonds between amino-acid side groups, allowing proteins to be dynamic and flexible. Here, a model of an enzyme, hexokinase, is at work. Hexokinase catalyzes phosphorylation of glucose. (a) A glucose molecule (color-coded *red*) is heading toward the active site, a cleft in the enzyme (*green*). (b) When the glucose molecule makes contact with the site, parts of the enzyme briefly close in around it and prod the molecule to enter into the reaction.

of polymers differing in the sugar component and one of the bases:

- **Ribonucleic acid** or **RNA,** made of A, G, C, and U.
- **Deoxyribonucleic acid** or **DNA,** made of A, G, C, and T.

DNA is the basis of genetic information storage and inheritance, whereas RNA is an intermediate that helps convert that information into cell structures. These small molecules and macromolecules are organized in highly specific ways to construct a living cell.

Recall that cells come in two distinct forms, prokaryotic and eukaryotic (Chapter 1). Even though there is no such thing as a "typical" cell (because of diverse structural and functional specializations), these two types share many common features. All cells have two major subdivisions: the plasma membrane and the cytoplasm. Eukaryotic cells have a third subdivision, the nucleus; the cytoplasm of eukaryotic cells also has other membrane-bound structures called *organelles.*

The plasma or cell membrane, is a very thin structure that encloses each cell, separating cell contents from the surrounding environment. It is made out of phospholipids (which form the membrane because of their hydrophobic property), plus proteins and sugars (Chapter 3). The fluid contained within all the cells of the body is known collectively as *intracellular fluid (ICF),* and the fluid outside the cells is referred to as *extracellular fluid (ECF).* The plasma membrane does not merely serve as a mechanical barrier to hold in the contents of the cells; it also can selectively control movement of many molecules between the ICF and ECF. The plasma membrane can be likened to the gated walls that enclosed ancient cities. Through this structure, the cell can regulate the entry of food and other needed supplies and the export of products manufactured within, while at the same time guarding against unwanted traffic into or out of the cell. However, not all molecules are selectively regulated; for example, water molecules (because they are so small) and hydrophobic molecules (except very large ones) can typically diffuse through most membranes.

The cytoplasm of cells contains water, many small inorganic and organic molecules, and macromolecules such as enzymes and nucleic acids (although DNA in eukaryotes is primarily contained within the nucleus). Within this complex mixture (and within organelles in eukaryotic cells), the storage and replication of genetic information occurs, and the chemical reactions that convert raw materials into cell structures and specialized functions take place. The chemical reactions of cells are generally *catalyzed* by enzymes. (A *catalyst* is an agent that increases the rate of a chemical reaction without itself being permanently changed.)

■ Many macromolecular structures need to be flexible to function and to be regulated, and protecting those structures is the basis for many forms of homeostasis.

Large molecular complexes, the framework of living cells, can be very vulnerable, because their higher-level structures are often held together with weak bonds such as hydrogen bonds, as opposed to the strong or covalent bonds that join atoms within single molecules. Reliance on weak bonds is important, because they give macromolecular structures dynamic features that could not occur with stronger, more rigid covalent bonds. Membranes, for example, are held together primarily by the result of hydrophobic interactions (Chapter 3) and thus can be flexible. Flexibility is necessary for cell shape changes and for many transport proteins to move molecules through membranes. The DNA double helix relies on hydrogen bonding so it can be readily disassembled for RNA formation and DNA replication. In addition, the complex three-dimensional structures of proteins are mainly determined by weak bonds between amino-acid side groups, allowing many proteins to be dynamic and flexible. Dynamic changes in proteins are crucial for everything from transport to enzyme catalysis. An example of the latter is shown in ● Figure 2–1b.

Another reason why many protein molecules are inherently flexible is that this characteristic allows their shapes to alter. This in turn regulates their activities, that is, activates or inhibits them. Useful conformational changes are induced in at least four ways: (1) from the binding of a regulatory molecule to a special binding site, a process called *allosteric* ("other site") *modulation;* (2) from a change in the electric field; (3) from physical deformation; and (4) by addition of a phosphate group (phosphorylation) by an enzyme called a *kinase* (or by removal of phosphate by a *phosphatase*). We present examples of all these methods in later chapters.

But there is a price to be paid for having flexible macromolecules. The cost is that the shapes of such molecules can be disturbed by changes in common environmental factors. Temperature, for example, can disrupt weak bonds so that proteins lose their shapes necessary to their functions. Similarly, inorganic ions can bind to charged side chains of amino acids, disrupting their weak bond interactions. *Thus, one of the major forces in the evolution of homeostasis*—maintaining a relatively constant interior environment (for example, in pH level or salt content)—*has been the provision of an optimal environment for the functioning of macromolecules.*

■ Prokaryotic cells have a simpler organization than eukaryotes.

Prokaryotic cells have an outer plasma membrane but do not contain a separate nucleus or (for the most part) other organelles per se. (However, in 2003 researchers discovered a true organelle in some bacteria. The organelles—called *acidocalcisomes*—seem to store energy and regulate acidity.) The prokaryotic cytoplasm is the site where the cell converts energy and synthesizes the molecules necessary for cell growth and function. Even though there is no distinct nucleus, prokaryotes have a distinct nuclear area with a single, circular chromosome.

The two prokaryotic groups are called *archaea* and *eubacteria* both often loosely called "bacteria," which are significantly different in many ways, such as in their genetic sequences and cell-wall components. (Indeed, some researchers place cells into three rather than two fundamental categories—*archaeal, eubacterial,* and *eukaryotic.*) However, we use the term "*bacteria*" for both prokaryotic groups in this text, in part because this usage is still quite widespread, and because not all prokaryotes have been studied to determine which category they fall into. Eubacteria include many of the familiar pathogens that cause animal diseases, but there are even more species that are not pathogenic. Archaea are noted for their ability to colonize extreme environments, such as hot springs and highly saline lakes. Some archaea have unusual metabolisms as well. One group of particular importance to animals is the methanogens, methane-producing bacteria that populate digestive tracts of many animals (see Chapter 14) as well as low-oxygen sediments of aquatic habitats.

■ Eukaryotic cells are subdivided into the plasma membrane, nucleus, and cytoplasm.

Eukaryotes are all based on a complex cell type bounded by the plasma membrane, with a distinct nucleus and other organelles in the cytoplasm. The cytoplasm is the cell interior, containing the organelles dispersed within the *cytosol*, a com-

plex, semigelatinous watery mass filling the cell interior. ▌ Table 2–1 summarizes the main division of a cell and the functions of the major components. Nearly all eukaryotic cells contain six main types of organelles—the *endoplasmic reticulum, Golgi complex, lysosomes, proteasomes, peroxisomes,* and *mitochondria* (● Figure 2–2). These organelles are similar in all eukaryotic cells, although there are some variations depending on the specialized capabilities of each cell type. Organelles increase the efficiency of metabolism (especially in cells larger than bacteria) by acting as scaffolds for macromolecules (such as enzymes) participating together in a series of reactions, and by serving as intracellular "specialty shops." They contain a specific set of chemicals for carrying out a par-

Table 2–1 ▌ Summary of Cell Structures and Functions

Cell Part	Number Per Cell	Structure	Function
Plasma membrane	1	Lipid bilayer studded with proteins and small amounts of carbohydrate	Acts as selective barrier between cellular contents and extracellular fluid; controls traffic in and out of the cell
Nucleus	1	DNA and specialized proteins enclosed by a double-layered membrane	Acts as control center of the cell, providing storage of genetic information
			Nuclear DNA provides codes for the synthesis of structural and enzymatic proteins that determine the cell's specific nature
			Nuclear DNA serves as blueprint for cell replication
Cytoplasm			
Organelles			
Endoplasmic reticulum	1	Extensive, continuous membranous network of fluid-filled tubules and flattened sacs, partially studded with ribosomes	Forms new cell membrane and other cell components and manufactures products for secretion
Golgi complex	1 to several hundred	Sets of stacked, flattened membranous sacs	Modifies, packages, and distributes newly synthesized proteins
Lysosomes	300	Membranous sacs containing hydrolytic enzymes	Serve as digestive system of the cell, destroying unwanted material, such as foreign substances and cellular debris
Peroxisomes	200	Membranous sacs containing oxidative enzymes	Perform detoxification activities
Mitochondria	100–2,000	Rod- or oval-shaped bodies enclosed by two membranes, with the inner membrane folded into cristae that project into the interior matrix	Act as energy organelles; major site of ATP production; contain enzymes for citric acid cycle and electron transport chain
Vaults	Thousands	Shaped like octagonal barrels	Unclear; may transport messenger RNA from nucleus to cytoplasm; may be important in cellular contractile systems
Proteasome		Shaped like hollow dumbbells	Breakdown of unneeded proteins
Cytosol			
Intermediary metabolism enzymes	Many	Sequential arrangement within the cytoskeleton	Facilitate intracellular reactions involving the degradation, synthesis, and transformation of small organic molecules
Ribosomes	Many	Granules of RNA and proteins—some attached to rough endoplalsmic reticulum, some free in the cytoplasm	Serve as workbenches for protein synthesis
Transport vesicles	Varies	Membrane-enclosed packages of products for construction of new cellular membrane or organelle components	Transport selected proteins and lipids from one organelle to another or to the plasma membrane

ticular cell function. This compartmentalization is evolutionarily advantageous because it permits chemical activities that would not be compatible with each other to occur simultaneously within the cell. For example, the enzymes that destroy unwanted proteins in the cell do so within the protective confines of the lysosomes and proteasomes without the risk of destroying essential cell proteins. About half of the total cell volume is occupied by organelles.

The rest of the cytoplasm (the part not occupied by organelles) consists of cytosol, which contains many small molecules, proteins such as enzymes, and critical large nonmembrane molecular complexes: ribosomes, proteasomes, vaults, and several types of cytoskeleton—filamentous proteins that control cell shape and movements (Figure 2–2). We discuss these later. Many fundamental metabolic chemical reactions take place in the cytosol. The cytoskeletal network, which elaborately laces the cytosol, gives the eukaryotic cell its shape, provides for its internal organization (including location and movement of organelles), and regulates its movements.

The nucleus, which is typically the largest single organized cell component, can be seen under magnification as a distinct spherical or oval structure, usually located near the center of the cell. Surrounding it is a double-layered membrane, which separates the nucleus from the remainder of the cell. Sequestered within the nucleus is the cell's genetic material, which we discuss in the next section.

Table 2–1 ▌ Summary of Cell Structures and Functions *(continued)*

Cell Part	Number Per Cell	Structure	Function
Secretory vesicles	Varies	Membrane-enclosed packages of secretory products	Store secretory products until signaled to empty to the outside
Endocytotic vesicles	Varies	Membrane-enclosed packages of material engulfed by the cell	Transfer selected, large, extracellular materials into the cell
Inclusions	Varies	Glycogen granules, fat droplets	Store excess nutrients
Cytoskeleton			
Microtubules	Many	Long, slender, hollow tubes composed of tubulin molecules	Maintain asymmetric cell shapes Coordinate complex cell movements 　Facilitate transport of secretory vesicles within cell, such as axonal transport 　Serve as dominant structural and functional component of cilia and flagella 　Form mitotic spindle during cell division
Micro-filaments	Many	Helically intertwined chains of actin molecules; microfilaments composed of myosin molecules also present in muscle cells	Play a vital role in various cellular contractile systems 　Play a dominant role in muscle contraction 　Form nonmuscle contractile assemblies, such as in white blood cells during amoeboid movement Serve as a mechanical stiffener for microvilli 　Increase the surface area available for absorption in intestine and kidneys 　Are specialized to detect sound and positional changes in ear
Intermediate filaments	Many	Irregular, threadlike proteins	Play a structural role in parts of the cell subject to mechanical stress
Microtrabecular lattice	1	Meshwork of exceedingly fine interlinked filaments	Suspends and functionally links larger cytoskeletal elements and various organelles Organizes cytosolic enzymes *Integrated Functions of Entire Cytoskeleton:* Is responsible for the shape, rigidity, and spatial geometry of each type of cell; the "bone" of the cell Is responsible for directing intracellular transport and for regulating cellular movements; the "muscle" of the cell Appears to play a role in regulating growth and division of cells

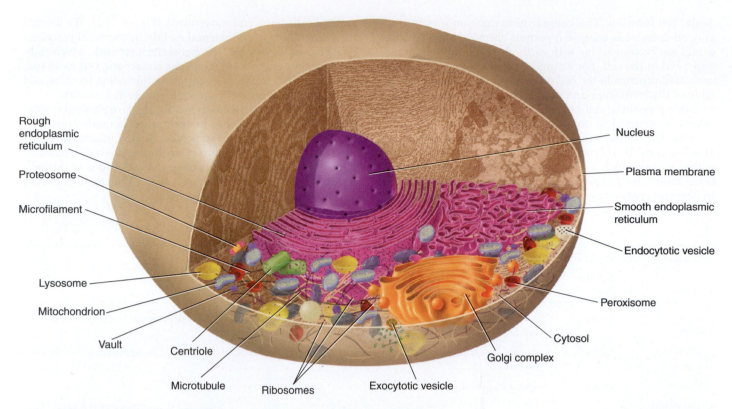

Figure 2–2 ● Schematic three-dimensional diagram of cell structures visible under an electron microscope.

Scientists have discovered that the eukaryotic cell evolved from a *symbiosis* (two different species living intimately together) between two or more prokaryotic cell types. Mitochondria and chloroplasts (organelles for photosynthesis in plants and algae) are descended from free-living eubacterial ancestors that took up residence in a larger host cell. Many lines of evidence support this conclusion, among the most telling of which are the occurrence inside chloroplasts and mitochondria of bacterial-type ribosomes (complexes that synthesize proteins that we examine later) and DNA in genes (the units of inheritance to which we turn our attention next).

Let's now examine each of the eukaryotic components in more detail.

Nucleus, Chromosomes, and Genes

The **nucleus** of eukaryotic cells (as well as the cytosol of prokaryotic cells) contains the materials for genetic instructions and inheritance. The key molecule is the double helix formed by DNA, which is packaged with proteins called *histones* in the nucleus to form chromosomes. DNA has two important functions: providing a code of information for RNA and protein synthesis and serving as a genetic blueprint during cell replication (for genetic inheritance). By coding for the kinds and amounts of various enzymes and other proteins that are produced, the nucleus indirectly governs most cell activities and serves as the cell's control center. Information critical to animal physiology is summarized here (● Figure 2–3).

■ **DNA contains codes in the form of genes for making proteins through the processes of transcription and translation.**

Recall that DNA is constructed from the four nucleotide bases A, G, T, and C. Because these bases exhibit complementary binding—that is, they bind to each other in pairs of A to T, G to C—the double helix of DNA serves as a genetic blueprint during cell replication to ensure that the cell produces additional cells just like itself, thus continuing the identical type of cell line within the body. Furthermore, in the reproductive cells, the DNA blueprint passes on genetic characteristics to future generations.

The sequence of A, G, T, and C is not random or fixed, but varies to form numerous codes, or "instructions." An individual coding sequence, along with regulatory codes that we discuss shortly, is called a *gene*. Within any single species, there may be tens of thousands of genes; and the variations that occur among all species on Earth are limitless. This is analogous to using the relatively few letters of an alphabet to make unlimited numbers of sentences. Gene codes ultimately direct the synthesis of specific RNA and proteins within the cell. Three types of RNA play a role in protein synthesis. First, DNA's genetic code (gene) for a particular protein is *transcribed* (copied) by an enzyme called *RNA polymerase* into a premessenger RNA molecule, a complementary copy of the gene that contains coding sequences (called *exons*) and noncoding sequences (*introns*). The premessenger RNA in turn is processed into a final messenger RNA (mRNA) by removal of introns and splicing together of the exons, and by special noncoding signal sequences being added to the leading and

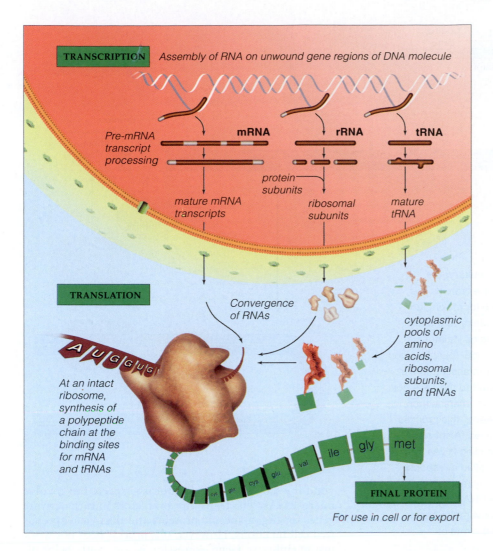

TRANSCRIPTION *Assembly of RNA on unwound gene regions of DNA molecule*

Pre-mRNA transcript processing

mRNA   **rRNA**   **tRNA**

protein subunits

mature mRNA transcripts   *ribosomal subunits*   *mature tRNA*

TRANSLATION

Convergence of RNAs

cytoplasmic pools of amino acids, ribosomal subunits, and tRNAs

At an intact ribosome, synthesis of a polypeptide chain at the binding sites for mRNA and tRNAs

A U G G U G

cys gln cys glu val ile gly met

FINAL PROTEIN

For use in cell or for export

Figure 2–3 • Summary of the flow of genetic information from DNA to proteins in eukaryotic cells. The DNA is transcribed into RNA in the nucleus; RNA is translated in the cytoplasm. Prokaryotic cells do not have a nucleus; transcription and translation proceed in their cytoplasm.

trailing ends of the molecule. The mRNA exits the nucleus through the nuclear pores in the nuclear membrane. Within the cytoplasm, mRNA delivers the coded message to a ribosome (p. 35), which "reads" the code and translates it into the appropriate amino acid sequence for the designated protein being synthesized. Finally, transfer RNA (tRNA) transfers the appropriate amino acids within the cytoplasm to their designated site in the protein under construction (Figure 2–3). (Note that rRNA and tRNA are themselves the products of gene transcription, thus not all genes code for mRNAs.)

These steps constitute the basic processes of **transcription** (copying DNA into RNA) and **translation** (converting RNA code into protein sequence). However, there are a number of alternative modifications of these steps. One is *alternative splicing*: By using different combinations of exons, splicing can lead to more than one mRNA code and hence, to more than one type of protein. Another involves the transcription and, in some cases, translation of introns. Once thought largely useless because they do not contribute to the protein specified by the main gene, some introns are now being found to code for separate proteins. In addition, some untranslated introns and other noncoding RNAs become *micro-RNAs,* a newly discovered class of RNAs that may play major roles in regulating how and when genes and mRNAs are used.

■ **Different genes are expressed in different tissues and organs.**

Because each body cell in a multicellular eukaryote usually has identical DNA sequences, you might assume that all cells would produce the same proteins. This is not the case, however, because—in addition to common or "housekeeping" functions that all cells require—different cell types transcribe different sets of genes and thus synthesize different proteins for their specialized roles. For example, all mammalian cells contain the genes for hemoglobin proteins—which serve to bind and carry oxygen—but these genes are only expressed (that is, transcribed and translated to produce proteins) in erythrocytes (red blood cells). In addition, many genes are active only under certain conditions or life stages. For example, many genes involved in animal sexual maturation are inactive in preadults.

How is this **differential gene expression** accomplished? There appear to be two levels of control:

1. *Regulation of individual genes with promoters and transcription factors.* A complete gene does not just consist of a code for an RNA strand, but also has regulatory DNA sequences called *promoters* that help regulate the gene's expression (● Figure 2–4a). Special regulatory proteins—

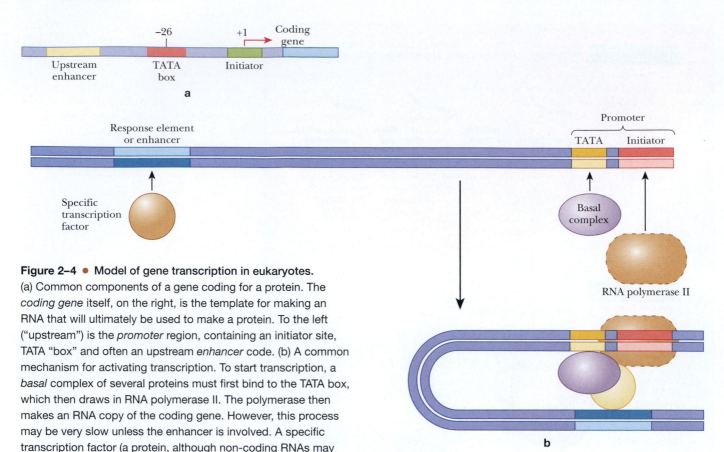

Figure 2–4 ● Model of gene transcription in eukaryotes. (a) Common components of a gene coding for a protein. The *coding gene* itself, on the right, is the template for making an RNA that will ultimately be used to make a protein. To the left ("upstream") is the *promoter* region, containing an initiator site, TATA "box" and often an upstream *enhancer* code. (b) A common mechanism for activating transcription. To start transcription, a *basal* complex of several proteins must first bind to the TATA box, which then draws in RNA polymerase II. The polymerase then makes an RNA copy of the coding gene. However, this process may be very slow unless the enhancer is involved. A specific transcription factor (a protein, although non-coding RNAs may be involved) must bind to the enhancer (also called a *response element* if the transcription factor is activated by metabolic or hormonal signals). This causes looping of the DNA to bring the specific factor-enhancer complex in alignment with the basal promoter complex. This can greatly enhance the rate of transcription.

themselves the products of other genes—control the process by regulating the packaging of the DNA and the action of RNA polymerase. One DNA sequence, the TATA box (so called because it contains the nucleotide sequence TATAA), must be bound by proteins forming a *basal transcription* complex (● Figure 2–4b), which recognizes the promoter code and then activates RNA polymerase to make the pre-mRNA copy of the nearby protein-coding sequences. However, this does not provide specificity for different cell types, because these general factors are often the same for numerous genes in numerous cell types. Other promoter sequences, often called *enhancers,* typically must be activated first. In one common process, specific transcription factors that are different in different cell types must bind to a particular DNA code in an enhancer. This bound complex of proteins then activates the basal complex and ultimately RNA polymerase (Figure 2–4b).

2. *Regulation of transcription factors in different tissues and at different stages.* Through the process of cell differentiation in embryonic development, each specialized cell type expresses its own set of genes for specific transcription factors. This process is poorly understood and be-

yond the scope of this text, but is suspected to involve micro-RNAs. The transcription factors in turn activate genes for organ-specific proteins and thus yield specialized cell functions. However, this process does not work just in embryos. Gene expression is constantly being altered in adult animals by various metabolic factors and signaling molecules. Enhancers are also called *response elements* if they respond (via specific transcription factors) to metabolic factors and signal molecules such as hormones. As we discuss later, many hormones exert their effects by controlling specific transcription factors that control gene expression through response elements (Figure 2–4b).

■ Telomeres protect chromosome ends, and their loss is associated with aging.

Eukaryotic chromosomes are linear, and researchers have found that the ends lose pieces of DNA every time a chromosome is replicated during cell division. To prevent loss of coding genes, the chromosomes have stretches of repeated, noncoding DNA sequences called *telomeres.* Each time a cell divides, a piece of telomere is lost, but presumably no harm is done because the lost DNA nucleotide sequences contain no genes. However, that changes once the telomeres are gone. At that point, useful genes begin to be lost during cell replication, causing cell malfunctions, and in fact, most cells cease dividing. Scientists have found that this feature of chromosome structure, which sets an upper limit on the number of times a cell can divide, correlates with the aging of individuals. Indeed, in 2003 researchers found a linear correlation between the length of telomere lost per year and the life span of five species of birds and eight species of mammals. For example, zebra finches lose about 500 nucleotides of telomere each year and live for about 5 years. Common terns, in contrast, lose only about 50 nucleotides per year and live for about 26 years.

The rate of telomere loss is not random or uncontrollable: Telomeres can be repaired by an enzyme called *telomerase*. For reasons not fully understood, the rate of telomere repair, and thus presumably the longevity of species, varies widely among organisms and is subject to natural selection. One possibility that limits the evolution of higher repair rates in short-lived species is cancer. Researchers have found that high telomerase activity does increase the number of times a cell can divide but also correlates with a greater chance of that cell becoming cancerous, that is, repeatedly dividing without stopping and thus forming a tumor.

Methodology

Many of the processes of molecular biology we have just discussed are used throughout this text. Because this is the most rapidly growing area of biology, it is useful to understand how molecular knowledge is acquired. Here we introduce some of the major molecular techniques mentioned later in the text.

Driving much of the growth in knowledge about molecular biology are the *genome projects*. In the last years of the 20th century, scientists undertook an international effort to decipher the complete genetic sequence—the arrangement of the A, G, T, and C "letters"—of humans. Dubbed the Human Genome Project, its goal is to yield insight into the codes responsible for human genetics, development, physiological functions, evolution, and diseases. At about the same time, researchers started genome projects for various microbes (some of which were the first to be completed), some plants, and several other animal species; and other species have been added to the list. By 2004, animal genomes were largely or fully described for human, mouse, rat, zebrafish, pufferfish, tunicate, fruit fly, mosquito, nematode worm, and one researcher's pet poodle. The deciphering of genomes is an example of the exploration or discovery phase of science (p. 4). See the box Molecular Biology and Genomics: Genomics and Evolution.

Although the wealth of genome data available is enormous and growing daily, that does not mean we are close to achieving a genetic understanding of physiology. The role of any individual gene in an organism's physiology is sometimes readily apparent—think of hemoglobin genes, for example, which code for the proteins that carry oxygen in blood—but often it is not. Many genes being discovered by genome analysis have functions that are either speculative or fully unknown, and many proteins have been found that are similarly mysteri-

---

MOLECULAR BIOLOGY *AND* GENOMICS

Genomics and Evolution

The animal genome data are producing many exciting findings important for understanding animal physiology and its evolution. Evolutionary change is primarily based on changes—mutations—in the DNA code that increase the fitness of the organism. Mutations have been the subject of genetic analysis for more than 100 years, yielding many important discoveries about the evolution of new features via changes in genetic codes. The new technologies have enormously increased the pace of discovery. For example, the traditional view holds that DNA mutates randomly. But genome analysis is revealing that some organisms have stretches of DNA that mutate at extraordinarily high rates, resulting in apparently useful evolutionary diversity. One example has been found in an interesting group of tropical mollusks, the predatory cone snails. Not only do these snails have colorful shells prized by collectors around the world, they also have venom proteins used to paralyze prey fish, shrimp, and other mol-

The cone shell, Conus textile.

lusks (the venom blocks ion channels, described in Chapter 4). These venoms are being used medically, for example, to deaden nerves in people who feel chronic pain. Each of the 500 species of cone snails has a unique set of genes coding for unique venom proteins. Analysis reveals that sections of the venom genes have mutated at rates far faster than the rest of the snail genome. The causes of this rapid mutation are uncertain but have played a role in the evolutionary diversity of cone snail species and their predatory physiologies.

This type of "deliberate" rapid mutation in a selective section of DNA is called *focused variation.* It may be a widespread phenomenon, thought to occur in the evolution of snake and scorpion venom, for example. It probably occurs in the evolution of parasite antigens (proteins on the surface that an animal's immune system recognizes as foreign), allowing some parasites to evade host immune defenses previously induced against them.

Photo: © Roger Steene/Imagequestmarine.com

ous. Because of the enormous complexity of animal genomes, it clearly will be many decades before scientists know what each and every gene does, how each evolved, how each is regulated throughout an organism's life cycle, and how each and its products interact with other genes. Uncovering gene functions is the goal of the new field of *genomics,* which uses a variety of new technologies to reveal the connections between genes and cell (or organismal) function. A companion field, *proteomics* (also called *proteonomics*), is focused on discovering the functions of proteins.

In later chapters, we describe some of the recent findings of genomics and proteomics. Here we briefly describe some of the key methods being used.

Immunofluorescence   An important technique that predates modern genomics and proteomics, but that is used widely in conjunction with them, is immunofluorescence. This method relies on the ability of mammals and birds to generate specific antibodies (immune proteins) that bind to foreign proteins entering their bodies. A target protein of interest in species A is purified and injected into species B, typically a laboratory rabbit. The rabbit's immune system makes antibodies to the protein. The antibodies are purified, and a fluorescent molecule is covalent bound to each. The new fluorescent antibody is then injected into species A, or added to a solution bathing a tissue slice from species A. The antibody binds only to places where the target protein occurs, and the fluorescent molecule allows the researchers to visualize its locations through a microscope (● Figure 2–5). In this way, proteins of known functions, and "new" proteins discovered by proteomics, can be located in specific tissues and at specific developmental stages of a particular species.

Gene Knockout Methodology

One major technique being used to elucidate gene functions is called the **knockout method,** in which a laboratory animal is raised from a fertilized egg that has had a specific gene disrupted. Defects in the animal's physiology are then studied to gain clues about gene (and protein) function. This method is not always useful, however, particularly if a missing protein causes aborted development or widespread aberrations. An analogy is studying an automobile's internal workings by selectively removing instructions in the factory for assembling key components—components you know little about as you begin. If you removed instructions for attaching a tube between the fuel tank and the engine, the engine would turn over but not start, and you would find fuel leaking. This gives some insight into that tube's function. In contrast, if you destroy instructions for welding steel together, the resulting loss of the automobile's structure might not tell you what the instructions were for.

DNA Microarrays

Another powerful new technology is called the **DNA microarray,** or colloquially, the "gene chip." In its basic form, a gene chip is a small plate (membrane or glass slide) onto which is bound thousands of different genes from an organism. This technique takes advantage of the fact that DNA and RNA sequences can bind in complementary fashion to one another. RNA from a particular organ is isolated and converted into corresponding complementary DNA (called *cDNAs*). The cDNA is labeled with a fluorescent molecule or radioactive

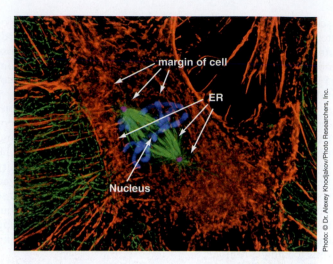

Figure 2–5 ● Immunofluorescence. Immunofluorescent light micrograph of a section through a kangaroo rat kidney epithelial cell during mitosis (cell division). Antibodies have been used to attach fluorescent dyes to specific cell structures. The nuclear membrane (not seen) that encloses the chromosomes (blue) is breaking down. The actin microfilaments (red) and tubulin microtubules (green) of the cytoskeleton maintain the structure of the cell. The two centrioles (pink dots, upper left and lower right) are microtubule-organizing centers (see p. 59).

atoms and used as a "probe": The cDNAs are exposed to the chip, and any genes active in that organ are revealed by the binding of the cDNA to the genes from which the original RNAs were transcribed. Different levels of fluorescent intensities indicate the relative activities of these genes. In this way, researchers can determine what genes are active at any given time in a life cycle, and during stress conditions and disease (● Figure 2–6). Much of microarray research is "discovery science" (p. 4) in that gene expression patterns are often conducted without a detailed hypothesis, but data for devising new hypotheses are arising from this exploration.

Databases

Genomics also relies heavily on the fact that all life is related via evolution. Thus, the nucleotide sequences of related genes in different organisms are similar. Analysis based on this principle makes extensive use of computers and the Internet, which provide access to databases containing most sequenced genes to date. There are at least two important ways to use these data. First, if a gene is discovered that has an unknown function, its sequence can be compared to all related genes in all analyzed species. This should give a clue to the "new" gene's function, because some of the related genes in the database are likely to have known functions. Second, the sequence of a gene with a known function can be used to look for similar (undiscovered) genes in the genome databases. In this way, previously unsuspected genes (and possibly proteins) can be found in other tissues in the same organisms, or in other species.

Gene Therapy and Transgenics

The ultimate goal of genomics is to understand genes so that they can be used in practical ways. An increasingly important

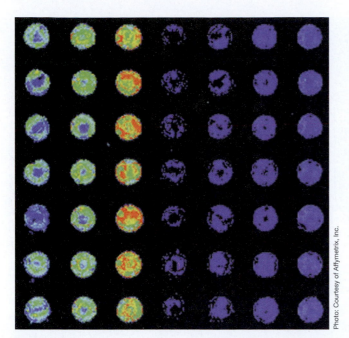

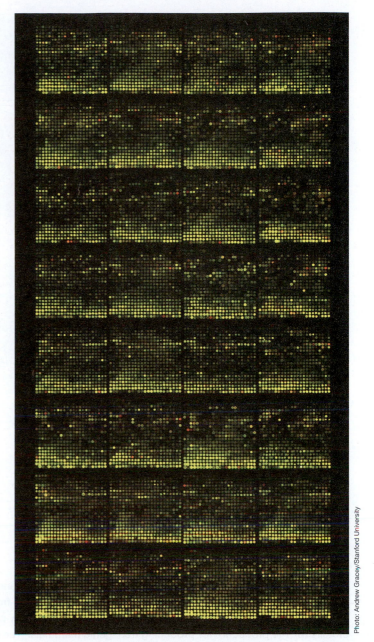

Figure 2–6 ● (a) A microarray chip. Thousands of DNA samples are placed on a 1-cm square slide by robotics. These slides are then bound by complementary DNA or RNA and visualized using fluorescent markers. The location and intensity of the fluorescent signal is analyzed by computer to establish which genes are currently being transcribed and to what extent. (b) A DNA microarray from DNA of a carp, hybridized with liver cDNAs from control animals (labeled green) and from animals subjected to cold temperatures (red). Yellow spots indicate genes with expressions equal in the two groups; green spots are genes expressed more strongly in control animals and red spots are those expressed more strongly in cooled animals.

application is that of *gene therapy,* in which a defect arising from a mutated gene is corrected by inserting a normal gene (● Figure 2–7a). This can be done in a fertilized egg, or in an adult tissue. For example, researchers recently found that one form of canine blindness—Leber congenital amaurosis—can arise from a single defective gene that codes for a protein called RPE65. This protein helps synthesize a pigment used by the retina to receive light. Researchers created a modified nonpathological virus carrying a normal RPE65 gene, and injected it into the right retina of blind dogs (the left retina received a control injection). Amazingly, the dogs regained some visual function and could navigate around obstacles that they had formerly bumped into repeatedly in a dimly lit room.

The technology used in gene therapy can also be used to create **transgenic** animals (Figure 2–7b). Here, a gene from one species is inserted into a fertilized egg or adult tissue of another. This may confer commercially useful properties into an agricultural animal; for example, sheep have been created that carry genes for human proteins with medical uses. The sheep produce milk containing these human proteins. Also, cows have been engineered to produce human antibodies against pathogens. The process may also be used to create markers or tracers of gene activity for basic research. In 2000,

the first transgenic primate—a rhesus monkey named ANDi— was created, with a jellyfish gene that produces a protein for bioluminescence. This gene, when attached to a host species' gene, provides a visual marker of that gene's activity in development and in various organs.

Cloning

Finally, another new technology based on the nucleus is that of whole-animal cloning. Because most adult cells contain the same DNA sequences on chromosomes throughout an organism's life cycle, the adult DNA, in theory, can code for a genetically identical offspring—a clone. In one common method of cloning called *nuclear transplantation,* the nucleus of a fertilized egg is removed or destroyed, and a carefully selected nucleus from an adult cell is injected in its place. If the adult DNA receives the proper signals (such as specific transcription factors) to start the genetic program for development, a

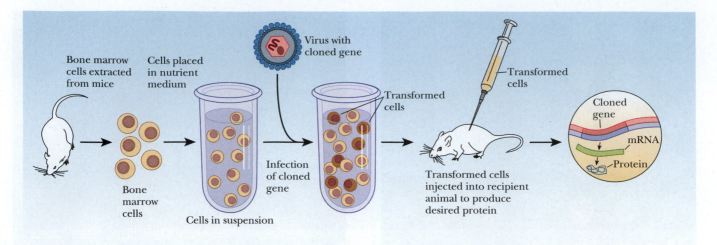

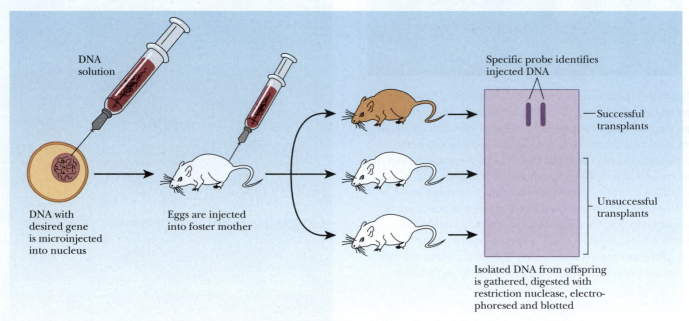

Figure 2–7 ● (a) Gene therapy in bone marrow cells. A cloned gene that directs the synthesis of a missing protein is introduced into bone marrow cells from the mouse. The transformed cells are replaced in the mouse's body, where they produce the desired protein. **(b) The procedure for producing transgenic mammals.** The two mice on the bottom have not received new genes. The mouse on the top has received a new gene, which it can pass on to its offspring. The mouse on the top is a transgenic organism and the altered gene is a transgene.

(*Source:* (a) Adapted from *Dealing with Genes: The Language of Heredity,* by Paul Berg and Maxine Singer, © 1992 by University Science Books. (b) Campbell, *Biochemistry,* p. 381.)

new animal grows (● Figure 2–8). Frogs were cloned in this manner in the 1960s, but only in the late 1990s was a mammal—a sheep named Dolly, in Scotland—produced by cloning. Since then, many other mammals have been cloned, including rodents, cows, goats, cats, a mule, and a horse. Strains of medically and agriculturally useful animals might be maintained by cloning. In addition, this technology might be used to save endangered species. However, it is not clear how successful the process will be in the long run. Many of the cloned mammals have shown signs of premature aging, probably because they developed from adult DNA that had suffered damage during the parent's life. In addition to the physiological problems, the use of this and other genetic technologies remains fraught with ethical dilemmas. See box, Are Cloning and Genetic Engineering Ethical?

Note that cloning is not simply an artificial technique. Basic cell division is a form of cloning. Many multicellular organisms, including many "primitive" animals, reproduce by cloning (although not by nuclear transplantation). Adult sea anemones and corals (Cnidarians), for example, can split in two to produce two genetically identical individuals (see Chapter 16, Reproduction).

Now that we have examined some key techniques of molecular investigation, we will turn our attention to organelles, other molecular complexes, and metabolism. Recall that in addition to the nucleus, eukaryotic cells contain several other large structures, including other membrane-bound structures or organelles.

Ribosomes

Ribosomes are ribosomal RNA-protein complexes that synthesize proteins, indirectly under the direction of nuclear DNA. A messenger RNA carries the genetic message from the nucleus to the ribosome "workbench," where protein synthesis takes place. Note that prokaryotes also have ribosomes, with a slightly different structure.

Ribosomes are found in two locations. Unattached or "free" ribosomes are dispersed throughout the cytosol. Bound ribosomes are found on membranes of a major organelle, the endoplasmic reticulum.

Endoplasmic Reticulum

The endoplasmic reticulum (ER) is an elaborate, fluid-filled membranous system distributed extensively throughout the cytosol, where it is primarily a protein-manufacturing factory. Two distinct types of endoplasmic reticulum—the smooth ER and the rough ER—can be distinguished. The smooth ER is a meshwork of tiny interconnected tubules, whereas the rough ER projects outward from the smooth ER as stacks of relatively flattened sacs (● Figure 2–9). Even though these two regions differ considerably in appearance and function, scientists think they are continuous with one another. The outer surface of the rough ER membrane is studded with small, dark-staining particles that give it a "rough" or granular appearance. These particles are the ribosomes. The relative amount of smooth and rough ER varies between cells, depending on the activity of the cell.

■ The rough endoplasmic reticulum synthesizes proteins for secretion and membrane construction.

The rough ER, in association with its ribosomes, synthesizes and releases a variety of new proteins into the ER lumen, the fluid-filled space enclosed by the ER membrane. These proteins serve one of two purposes: (1) Some proteins are destined for export to the cell's exterior as secretory products, such as proteinaceous hormones or enzymes. (2) Other proteins are transported to sites within the cell for use in constructing new cell membrane (either new plasma membrane or new organelle membrane) or other protein components of organelles. Cell membranes consist mostly of lipids (fats) and proteins. The membranous wall of the ER also contains enzymes essential for the synthesis of nearly all the lipids needed for producing new membranes. These newly synthesized lipids enter the ER lumen along with the proteins. Predictably, the rough ER is most abundant in cells specialized for protein secretion (for example, cells that secrete digestive enzymes) or in cells that require extensive membrane synthesis (for example, rapidly growing cells such as immature egg cells).

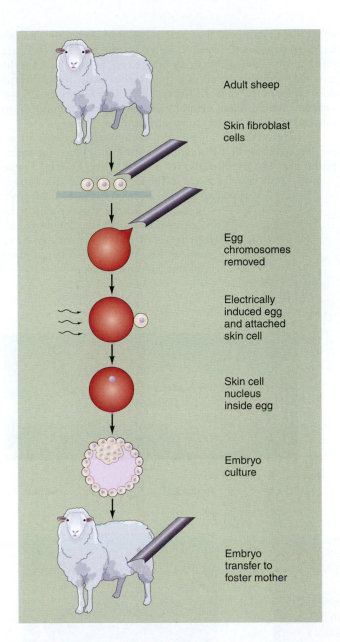

Figure 2–8 ● Nuclear-transplantation techniques in mammals. The genetic material is removed from the recipient cell (an egg), then replaced by a nucleus from a donor cell. The resulting embryo is then transferred to a surrogate mother. The clones are genetically identical to the donor.

Source: From *Nature,* December 16, 1999, p. 743. Used by permission of Nature Publishing Group. www.nature.com

?

Are Cloning and Genetic Engineering Ethical? As we have noted, the fields of genomics and proteonomics have far more questions than answers. In addition, the techniques involved raise serious bioethical questions for human society. What are the ethical pros and cons in cloning animals or creating transgenic animals?

Within the ER lumen, a newly synthesized protein is folded into its final conformation and may also be modified in other ways, such as being pruned or having sugar molecules attached to it. After this processing within the ER lumen, a new protein cannot pass out through the ER membrane and therefore becomes permanently separated from the cytosol as soon as it has been synthesized. In contrast to the rough ER ribosomes, the free ribosomes synthesize proteins that are used intracellularly within the cytosol. In this way, newly produced molecules that are destined for export out of the cell or for synthesis of new cell components (those synthesized by the ER) are physically separated from those that belong in the cytosol (those produced by the free ribosomes).

How do the newly synthesized molecules within the ER lumen get to their destinations at other sites inside the cell or to the outside of the cell, if they cannot pass out through the ER membrane? They do so through the action of the smooth endoplasmic reticulum.

■ The smooth endoplasmic reticulum packages new proteins in transport vesicles.

The smooth ER does not contain ribosomes; hence it is "smooth." Lacking ribosomes, it is not involved in protein synthesis. Instead, it serves a variety of other purposes that vary in different cell types. In most cells, the smooth ER is rather sparse and serves primarily as a central packaging and discharge site for molecules that are to be transported from the ER. Newly synthesized proteins and lipids pass from the rough ER to gather in the smooth ER. Portions of the smooth ER then "bud off" (that is, bulge on the surface, then are pinched off), giving rise to *transport vesicles* that contain the new molecules enclosed in a spherical capsule of membrane derived from the smooth ER membrane (● Figure 2–10). (A **vesicle** is a fluid-filled, membrane-enclosed intracellular cargo that has been segregated from surrounding cytosol.)

Newly synthesized membrane components are rapidly incorporated into the ER membrane itself to replace the membrane that was used to "wrap" the transport vesicle. Transport vesicles move to the Golgi complex for further processing of their cargo (discussed shortly).

In contrast to the sparseness of the smooth ER in most cells, some specialized types of cells have an extensive smooth ER, which confers additional cell functions as follows:

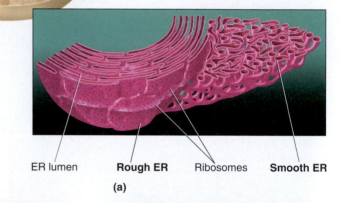

ER lumen   **Rough ER**   Ribosomes   **Smooth ER**

(a)

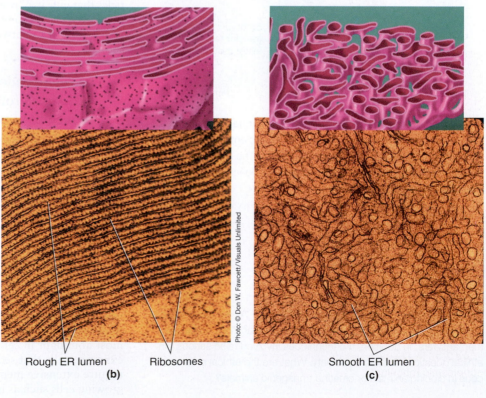

Figure 2–9 ● Endoplasmic reticulum (ER). (a) Schematic three-dimensional representation of the relationship between the smooth ER, which is a meshwork of tiny interconnected tubules, and the rough ER, which is studded with ribosomes and projects outward from the smooth ER as stacks of relatively flattened sacs. (b) Electron micrograph of the rough ER. Note the layers of flattened sacs studded with small dark-staining ribosomes. (c) Electron micrograph of the smooth ER.

Rough ER lumen   Ribosomes

(b)

Smooth ER lumen

(c)

Photo: © Don W. Fawcett/Visuals Unlimited

■ The smooth ER is abundant in cells that specialize in lipid metabolism—for example, cells that secrete steroid hormones in arthropods and vertebrates. (A steroid is a special type of lipid derived from cholesterol.) The membranous wall of the smooth ER, like that of the rough ER, contains enzymes for synthesis of lipids. These cells have an expanded smooth ER compartment to house the additional enzymes necessary to keep pace with demands for hormone secretion.

■ In vertebrate liver cells, the smooth ER has a special capability. It contains enzymes that detoxify harmful substances produced within the body by metabolism of substances that enter the body from the outside (such as drugs and plant toxins). These detoxification enzymes alter toxic substances so that the latter can be eliminated more readily via excretory organs. The amount of smooth ER available in liver cells for the task of detoxification can vary dramatically, depending on the need. For example, if phenobarbital (a barbiturate drug used as a sedative) is administered in large quantities to a mammal, the amount of smooth ER with its associated detoxification enzymes doubles within a few days (a form of *acclimatization* or *upregulation*, p. 15), only to return to normal within five days after drug administration ceases. The mechanisms involved in regulating these changes are not well understood.

■ Muscle cells have developed another specialized use for the smooth ER. They have an elaborate, modified smooth ER known as the *sarcoplasmic reticulum*, which stores calcium and plays an important role in muscle contraction (see Chapter 8).

Golgi Complex

Closely associated with the endoplasmic reticulum is the **Golgi complex,** which consists of sets of flattened, slightly curved, membrane-enclosed sacs, or *cisternae,* stacked in layers (● Figure 2–11 on the next page). Note that the flattened sacs are thin in the middle but have dilated edges. The number of Golgi stacks varies, depending on the cell type. Some cells have only one stack, whereas cells highly specialized for protein secretion may have hundreds of stacks.

■ Transport vesicles carry their cargo to the Golgi complex for further processing.

The majority of the newly synthesized molecules that have just budded off from the smooth ER enter a Golgi stack. When a transport vesicle carrying its newly synthesized cargo reaches a Golgi stack, the vesicle membrane fuses with the membrane of the sac closest to the center of the cell. The vesicle membrane opens up and becomes a new part of the Golgi membrane, and the contents of the vesicle are released to the interior of the sac (Figure 2–10). (Note that vesicles are typically associated with cytoskeletal filaments, which are discussed later in this chapter.)

These newly synthesized raw materials from the ER travel through the layers of the Golgi stack, from the innermost sac closest to the ER to the outermost sac near the plasma membrane. During this transit, two important interrelated functions take place:

1. *Processing the raw materials into finished products.* Within the Golgi complex, the "raw" proteins from the ER are

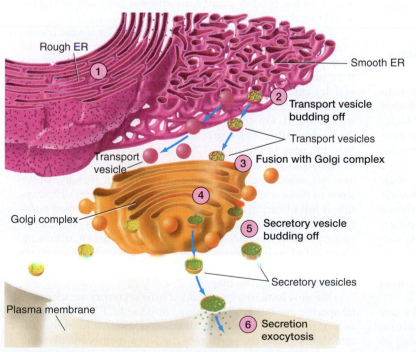

1 The rough ER synthesizes proteins to be secreted to the exterior or to be incorporated into the cellular membrane.

2 The smooth ER packages the secretory product into transport vesicles, which bud off and move to the Golgi complex.

3 The transport vesicle fuses with the Golgi complex, opens up, and empties its contents into the closest Golgi sac.

4 As the newly synthesized proteins from the ER travel by vesicular transport through the layers of the Golgi complex, this complex modifies the raw proteins into final form and sorts and directs the finished products to their final destination by varying their wrappers.

5 Secretory vesicles containing the finished protein product bud off the Golgi complex and remain in the cytosol, storing the product until signaled to empty.

6 On appropriate stimulation, the secretory vesicles fuse with the plasma membrane, open, and empty their contents to the cell's exterior. Secretion has occurred by exocytosis, with the secretory product never having come into contact with the cytosol.

Labels in figure: Rough ER; Smooth ER; Transport vesicle budding off; Transport vesicles; Transport vesicle; Fusion with Golgi complex; Golgi complex; Secretory vesicle budding off; Secretory vesicles; Plasma membrane; Secretion exocytosis

Figure 2–10 ● **Overview of the secretion process for proteins synthesized by the endoplasmic reticulum.**

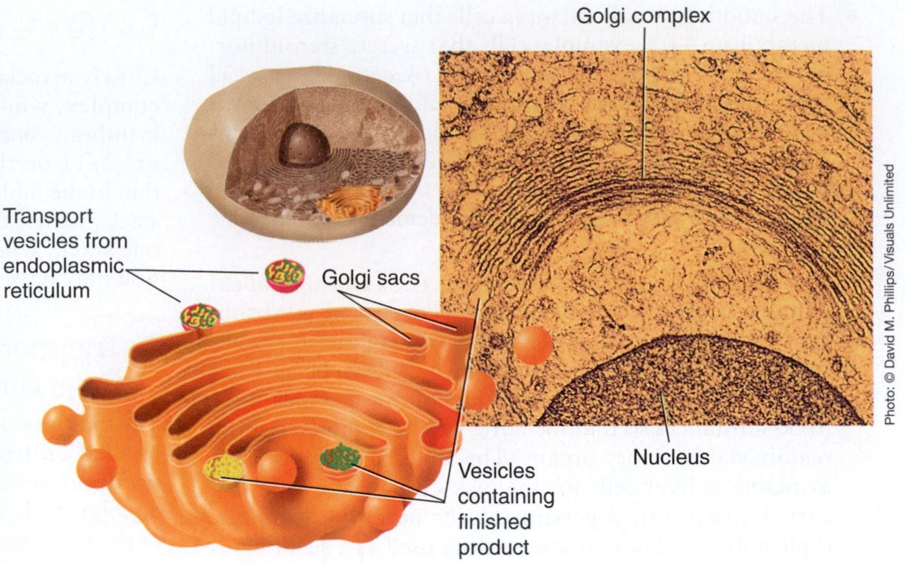

Figure 2–11 ● Golgi complex. (a) Schematic three-dimensional diagram of a Golgi complex. (b) Electron micrograph of a Golgi complex. The vesicles at the dilated edges of the sacs contain finished protein products packaged for distribution to their final destination.

modified into their final form, largely by adjustments made in the sugars attached to the protein. The biochemical pathways that the proteins undergo during their passage through the Golgi complex are elaborate, precisely programmed, and specific for each final product.

2. *Sorting and directing the finished products to their final destinations.* The Golgi complex sorts and segregates different types of products toward one of three destinations according to their function and destination, namely, (a) molecules that are destined for secretion to the exterior, (b) molecules that will become part of the plasma membrane, and (c) molecules that will become incorporated into other organelles, especially lysosomes.

■ **The Golgi complex packages secretory vesicles for release by exocytosis.**

How does the Golgi complex sort and direct finished proteins to the proper destinations inside and outside the cell? Finished products are collected within the dilated edges of the Golgi complex's sacs. The dilated edge of the outermost sac then pinches off to form a membrane-enclosed vesicle containing the finished product. For each type of product to read its appropriate site of function, each distinct type of vesicle takes up a specific product before budding off. This vesicle with its selected cargo then fuses only with the membrane of a particular destination site. Vesicles destined for different sites are wrapped in membranes containing different surface protein molecules that serve as specific "docking markers," which ensure that the vesicles "dock" and "unload" their cargo only at the appropriate "address," or destination within the cell.

Let's examine how the Golgi complex of specialized secretory cells sorts and packages proteins destined for secretion from the cell, as an example of how this organelle handles its molecular traffic. Specialized secretory cells include endocrine cells that secrete protein hormones and digestive gland cells that secrete digestive enzymes. Secretory cells package their products destined for export out of the cell in **secretory vesi-**

cles. Secretory vesicles are about 200 times larger than other transport vesicles. The product remains stored within the secretory vesicles until the cell is stimulated by a specific signal that indicates a need for release of that particular product. On appropriate stimulation, the vesicles move to the cell's periphery. Their contents are quickly released to the cell's exterior as the vesicles fuse with the plasma membrane, open, and empty their contents to the outside (Figure 2–10). This mechanism—extrusion to the exterior of substances originating within the cell—is called *exocytosis* (*exo*, "out of "; *cyto*, "cell"). This constitutes the process of secretion. Secretory vesicles fuse only with the plasma membrane and not with any of the internal membranes that bound organelles.

Exocytosis can be an explosive event, because vesicle contents are often ejected. The most extreme example is in Cnidaria (anemones, coral, hydras, jellies, and so on). All Cnidaria sting prey by using epithelial cells with special Golgi-derived vesicles called *nematocysts,* capsules each having a double-layered wall and an inverted tube studded with barbs (often with a toxin). On contact with prey, the capsule exocytoses the barbed tube, which inverts, exposing the spines and jabbing the prey (● Figure 2–12). This exocytosis is one of the fastest biological actions known. It appears driven in part by *osmosis,* which we cover in the next chapter. (Briefly, water tends to move from dilute to concentrated solutions.) The capsule is full of cations and long amino-acid chains called *polyglutamate,* serving as *osmolytes* (Chapter 13) that create a solution much more concentrated than the external environment. The rigid capsule prevents premature discharge. Prey contact triggers a lid on the capsule to open, and water rushes in by osmosis, inverting the tube.

We now look in more detail at how secretory vesicles take up specific products to be released into the ECF and then dock only at the plasma membrane. Before budding off from the outermost Golgi sac, the portion of the Golgi membrane that will be used to enclose the secretory vesicle becomes "coated" with a layer of specific proteins from the cytosol (● Figure 2–13 on page 40). These and associated membrane proteins serve three important functions:

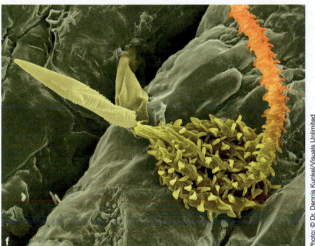

(a)

(b)   (c)   (d)   (e)

Figure 2–12 ● Nematocyst formation in Cnidarian. "C" indicates the capsule of the forming nematocyst, "G" is the Golgi complex, "rER" is the rough ER, "N" is the nucleus, "t" is tubule, "s" are stylets. (a): a newly formed nematocyst grows from Golgi vesicles. (b): proteins are added to the nematocyst to strengthen the capsule wall. (c): an external tubule grows and elongates. (d): the tubule inverts, and spines (stylets) are added. (e): the final mature nematocyst. (f) Micrograph of a nematocyst (ejected tubule) from a tentacle of the box jellyfish (*Carybdea alata*), 600× magnification.

(*Source:* (a–e) Dr. Jürgen Engel, Olivier Pertzl, Charlotte Fauser, Jürgen Engel, Charles N. David and Thomas W. Holstein (2001), A switch in disulfide linkage during minicollagen assembly in Hydra nematocysts, *EMBO Journal,* 20:3063–3073.)

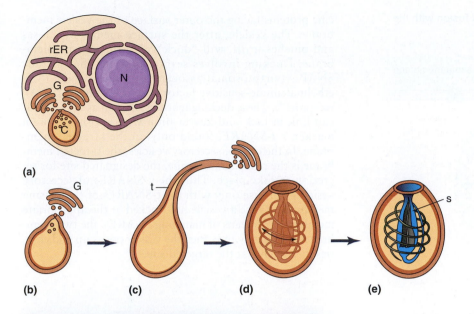

Photo: © Dr. Dennis Kunkel/Visuals Unlimited

- First, specific proteins on the interior surface of the membrane facing the Golgi lumen act as *recognition markers* for recognizing and attracting specific molecules that have been processed in the Golgi lumen. The newly finished proteins destined for secretion contain a unique sequence of amino acids known as a *sorting signal.* Recognition of the right protein's sorting signal by the complementary membrane marker ensures that the proper cargo is captured and packaged as a secretory vesicle is forming and budding off of the outermost Golgi sac.

- Second, *coat proteins* from the cytosol bind with another specific protein facing the outer surface of the membrane. The linking together of these coat proteins causes the surface membrane of the Golgi sac to curve and form a dome-shaped bud around the captured cargo. Eventually the surface membrane closes and pinches off the vesicle.

- After budding off, the vesicle sheds its coat proteins. This uncoating exposes *docking markers*, which are other spe-

Photo: Florida Keys National Marine Sanctuary, Courtesy of NOAA

Coral polyps (on a purple sea whip) with tentacles extended for feeding. These and all of their cnidarian relatives (hydras, anemones, jellies, etc.) sting their prey using cellular capsules called *nematocysts.* These are specialized vesicles made by the Golgi complex in epithelial cells.

Figure 2–13 ● Secretory vesicle formation and fusion with the plasma membrane.

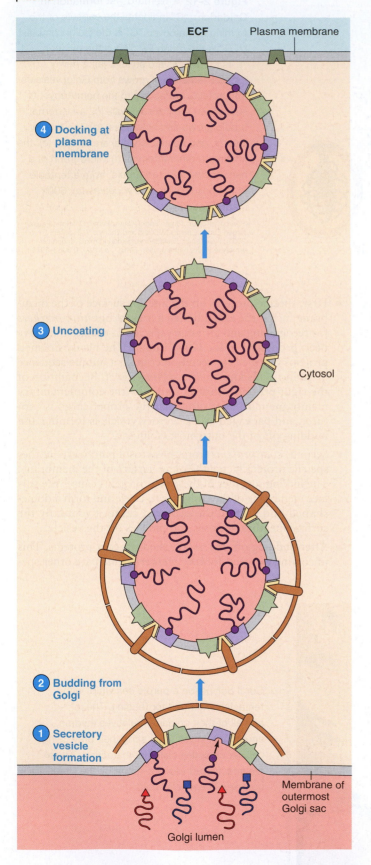

cific proteins facing the outer surface of the vesicle membrane. The vesicle, after the surface membrane closes and pinches it off, will "dock" with the plasma membrane. Docking involves several proteins that include SNAPs (synaptosomal-associated proteins), NSF ("N-ethylmaleimide-sensitive factor"), and SNAREs ("SNAP receptors"). These docking markers, known as *v-SNAREs,* can link in lock-and-key fashion with another protein marker, a *t-SNARE,* found only on the targeted membrane. In the case of secretory vesicles, the targeted membrane is the plasma membrane, the designated site for secretion to take place. Thus the v-SNAREs of secretory vesicles can fuse only with the t-SNAREs of the plasma membrane. Once a vesicle has docked at the appropriate membrane by means of matching SNAREs, the two membranes completely fuse, then the vesicle opens up and empties its contents at the targeted site.

① During secretory vesicle formation, recognition markers in the membrane of the outermost Golgi sac capture the appropriate cargo from the Golgi lumen by binding lock-and-key fashion with the sorting signals of the designated protein molecules to be secreted. The membrane that will wrap the vesicle is coated with a molecule that causes the membrane to curve, forming a dome-shaped bud.

② The membrane closes beneath the bud, pinching off the secretory vesicle.

③ The vesicle loses its coating, exposing v-SNARE docking markers on the vesicle surface of the membrane.

④ The v-SNAREs bind lock-and-key fashion only with the t-SNARE docking-marker acceptors of the targeted plasma membrane. This specificity ensures that secretory vesicles fuse only with the surface membrane of the cell and empty their contents to the cell's exterior.

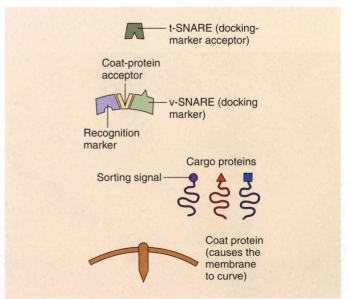

By manufacturing its particular secretory protein ahead of time and storing this product in secretory vesicles, a secretory cell has a readily available reserve from which to secrete large amounts of this product on demand.

In a similar way, the Golgi complex of these and other cell types sorts and packages newly synthesized products for different destinations within the cell. In each case, a particular vesicle captures a specific kind of cargo from among the many proteins in the Golgi lumen, then addresses each shipping container for a distinct destination.

Lysosomes and Proteasomes

Lysosomes are membrane-enclosed sacs containing powerful **hydrolytic** enzymes capable of digesting and thereby removing various unwanted cellular debris and foreign material, such as bacteria that have been internalized within the cell. Thus lysosomes serve as the intracellular "digestive system." Lysosomes can also be viewed as recycling centers that break down old, damaged, or dead cell parts or extracellular materials that have been engulfed by the cell, into reusable raw ingredients that can be refashioned into new and useful cell parts.

On the average, a mammalian cell contains about 300 lysosomes. Instead of having a uniform structure, as is characteristic of all other organelles, lysosomes vary in size and shape, depending on the contents they are digesting. Most commonly, lysosomes are small (0.2 to 0.5 μm in diameter) oval or spherical bodies that appear granular during inactivity. These granules are protein aggregates of the powerful digestive enzymes contained within. The surrounding membrane that confines these enzymes normally prevents them from destroying the cell that houses them. Both the membrane and the enzymes are derived from the Golgi complex. A specific collection of newly synthesized hydrolytic enzymes is captured in a coated vesicle that is pinched off from the Golgi complex to become a new lysosome.

■ Extracellular material is brought into the cell by endocytosis for attack by lysosomal enzymes.

Extracellular material to be attacked by lysosomal enzymes is brought inside the cell through the process of endocytosis (*endo*, "within"). Endocytosis can be accomplished in three ways—*pinocytosis, receptor-mediated endocytosis,* and *phagocytosis*—depending on the contents of the internalized material.

Pinocytosis

With **pinocytosis** ("cell drinking"), a small droplet of extracellular fluid is internalized. First, the plasma membrane dips inward, forming a pouch that contains a small bit of ECF (● Figure 2–14a on the next page). The endocytotic pouch is formed as a result of membrane-deforming coat proteins attaching to the inner surface of the plasma membrane. These coat proteins are similar to those involved in the formation of secretory vesicles. The linking together of the coat proteins causes the plasma membrane to dip inward. The plasma membrane then seals at the surface of the pouch, trapping the contents in a small, intracellular *endocytotic vesicle.* Recently, the molecule responsible for pinching off an endocytotic vesicle has been identified as *dynamin,* which forms rings that wrap around and "wring the neck" of the pouch, severing the vesicle from the surface membrane. Besides bringing ECF into a cell, pinocytosis provides a means to retrieve extra plasma membrane that has been added to the cell surface during exocytosis.

Receptor-Mediated Endocytosis

Unlike pinocytosis, which involves the nonselective uptake of the surrounding fluid, **receptor-mediated endocytosis** is a highly selective process that enables cells to import specific large molecules that the cell needs from its environment. Receptor-mediated endocytosis is triggered by the binding of a specific molecule such as a protein to a surface membrane receptor site specific for that protein. This binding causes the plasma membrane at that site to sink in, then seal at the surface, trapping the protein inside the cell (Figure 2–14b on the next page). Cholesterol complexes, vitamin B_{12}, the hormone insulin, and iron are examples of substances selectively taken into cells by receptor-mediated endocytosis.

Unfortunately, some viruses can sneak into cells by exploiting this mechanism. For instance, flu viruses and HIV, the virus that causes human and feline AIDS (see p. 451), gain entry to cells via receptor-mediated endocytosis. They do so by binding with membrane receptor sites normally designed to trigger internalization of a needed molecule.

Phagocytosis

During **phagocytosis** ("cell eating"), large multimolecular particles are internalized. Most body cells perform pinocytosis, many carry out receptor-mediated endocytosis, but only a few specialized cells are capable of phagocytosis. The latter are the "professional" phagocytes, the most notable being certain types of immune cells that play an important role in defense mechanisms in all animals (see Chapter 10). When an immune cell encounters a large multimolecular particle, such as a bacterium or tissue debris, it extends surface projections known as *pseudopodia* ("false feet"), or *pseudopods,* that completely surround or engulf the particle and trap it within an internalized vesicle (● Figure 2–14c and d). A lysosome fuses with the membrane of the internalized vesicle and releases its hydrolytic enzymes into the vesicle, where they attack the trapped material without damaging the rest of the cell. The enzymes largely break down the material into reusable raw ingredients, such as amino acids, glucose, and fatty acids. These small products pass through the lysosomal membrane to enter the cytosol.

■ Lysosomes remove useless but not useful parts of the cell.

Lysosomes can also fuse with aged or damaged organelles to remove parts of the cell. This selective self-digestion makes way for new replacement parts. All organelles are renewable. If the whole cell is severely damaged or dies, the lysosomes rupture and release their destructive enzymes into the cytosol so that the cell digests itself entirely. In most tissues, elimination of a nonfunctional cell clears the way for its replacement with a healthy new one through cell division. However, in tissues in which cell reproduction rarely occurs, such as mammalian heart and brain tissue, scar tissue replaces the self-destructed dead cells.

In specific instances, lysosomes cause intentional self-destruction of healthy cells. This happens as a normal part of embryonic development when certain unwanted tissues that

form are programmed for destruction. For example, embryonic ducts in mammals that are capable of forming a male reproductive tract are deliberately destroyed during the development of a female fetus. Similarly, many larval (caterpillar) tissues must be destroyed in a butterfly's pupal stage. Lysosomes also play an important role in tissue regression, such as

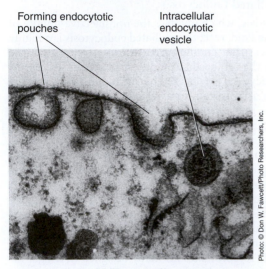

Forming endocytotic pouches

Intracellular endocytotic vesicle

Photo: © Don W. Fawcett/Photo Researchers, Inc.

(a) Pinocytosis

Figure 2–14 ● Forms of endocytosis. (a) Electron micrograph of pinocytosis. The surface membrane dips inward to form a pouch, then seals the surface, forming an intracellular vesicle that nonselectively internalizes a bit of ECF. (b) Receptor-mediated endocytosis. When a large molecule such as a protein attaches to a specific surface receptor site, the membrane dips inward and pinches off to selectively internalize the molecule in an intracellular vesicle. (c) Phagocytosis. White blood cells internalize multimolecular particles such as bacteria by extending surface projections known as *pseudopods* that wrap around and seal in the targeted material. A lysosome fuses with the internalized vesicle, releasing enzymes that attack the engulfed material within the confines of the vesicle. (d) Scanning–electron-micrograph series of a white blood cell phagocytizing an old, worn-out red blood cell.

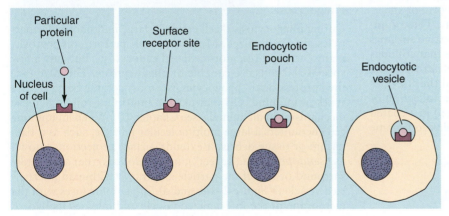

Particular protein

Surface receptor site

Endocytotic pouch

Endocytotic vesicle

Nucleus of cell

(b) Receptor-mediated endocytosis

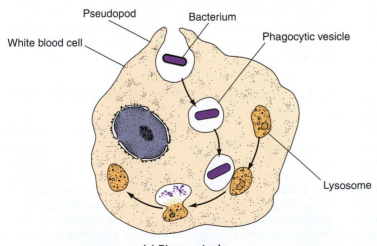

Pseudopod

Bacterium

Phagocytic vesicle

White blood cell

Lysosome

(c) Phagocytosis

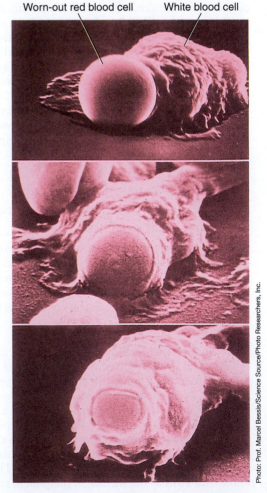

Worn-out red blood cell

White blood cell

Photo: Prof. Marcel Bessis/Science Source/Photo Researchers, Inc.

(d)

during the normal reduction in the uterine lining following pregnancy. Such *programmed* cell death is termed *apoptosis*. (Other cellular enzymes and organelles, especially mitochondria, are also involved in apoptosis; see Chapter 3.)

An inherent danger even in a healthy, intact cell is that lysosomal membranes may inadvertently rupture. Because these organelles have the potential ability to self-destruct the cell, their discoverer named them "suicide bags." However, two factors make this threat less severe than imagined. First, the lysosomal hydrolytic enzymes function best in an acidic environment. The lysosomal membrane transports hydrogen ions (acid formers) into the lysosome, making it considerably more acidic than the rest of the cell. Should the enzymes accidentally leak into the cytosol, they would be less potent than they are in their acidic home. Second, in most instances cells could tolerate the limited damage that would occur if only one or two lysosomes inadvertently ruptured because most parts of the cell are renewable (the major exception being the DNA).

■ Proteasomes destroy internal proteins.

The other important intracellular digestive apparatus is the **proteasome**, a large tunnel-like structure made of numerous proteins (● Figure 2–15). The proteasome, of which there are many thousands per cell, has the job of taking in internal cell proteins (rather than external ones) and chopping them up into reusable amino acids. Why would a cell want to destroy its own proteins? Although cell proteins are made for specific purposes, they outlive their usefulness if that purpose is only needed temporarily. For example, during cell division, many regulatory proteins have to be made, then destroyed, for the cell to move through the various stages of mitosis. Also, in extreme starvation proteins must be broken down to provide emergency energy supplies. Some cell proteins are not useful at all, because of errors in their production or even genetic mutations. Without proteasomes, unneeded proteins could accumulate in cells to harmful levels.

Proteins are not randomly destroyed by proteasomes. Instead, special enzymes recognize unwanted proteins (in ways

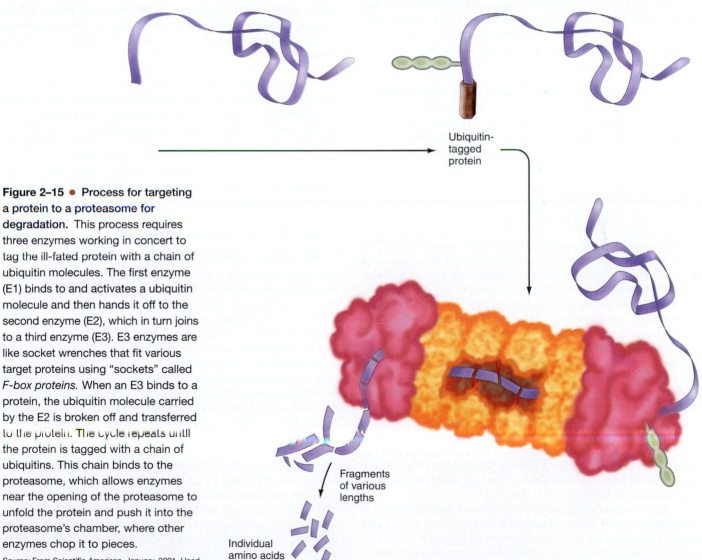

Figure 2–15 ● **Process for targeting a protein to a proteasome for degradation.** This process requires three enzymes working in concert to tag the ill-fated protein with a chain of ubiquitin molecules. The first enzyme (E1) binds to and activates a ubiquitin molecule and then hands it off to the second enzyme (E2), which in turn joins to a third enzyme (E3). E3 enzymes are like socket wrenches that fit various target proteins using "sockets" called *F-box proteins.* When an E3 binds to a protein, the ubiquitin molecule carried by the E2 is broken off and transferred to the protein. The cycle repeats until the protein is tagged with a chain of ubiquitins. This chain binds to the proteasome, which allows enzymes near the opening of the proteasome to unfold the protein and push it into the proteasome's chamber, where other enzymes chop it to pieces.

Source: From *Scientific American,* January, 2001. Used by permission.

Ubiquitin-tagged protein

Fragments of various lengths

Individual amino acids

not completely understood) and "tag" them with a tiny protein called *ubiquitin*. The ubiquitin-tagged protein is then recognized by the proteasome and drawn in to its disassembly tunnel. However, as in the case of lysosomes, this destruction process can go awry and cause harm. For example, some viruses have evolved their own proteins that can tag useful cellular defense proteins for proteasome destruction.

Peroxisomes

Typically, several hundred small **peroxisomes** about one third to one half the average size of lysosomes are present in a cell. Peroxisomes are similar to lysosomes in that they are membrane-enclosed sacs containing enzymes, but unlike the lysosomes, which contain hydrolytic enzymes, peroxisomes house several powerful oxidative enzymes and contain most of the cell's catalase. *Oxidative enzymes,* as the name implies, use oxygen (O_2) to strip hydrogen from specific molecules. Such a reaction is important in detoxifying various wastes produced within the cell or foreign toxic compounds that have entered the cell, such as the ethanol consumed in alcoholic beverages. The major product generated in the peroxisome is hydrogen peroxide (H_2O_2), which is formed by molecular oxygen and the hydrogen atoms stripped from the waste. Hydrogen peroxide, itself a powerful oxidant, is potentially destructive if it is allowed to accumulate or escape from the confines of the peroxisome. However, peroxisomes also contain an abundance of catalase, an antioxidant enzyme that decomposes potent H_2O_2 into harmless H_2O and O_2. This latter reaction is an important safety mechanism that destroys the potentially deadly peroxide at the site of its production, thereby preventing its possible devastating escape into the cytosol.

Mitochondria and Energy Metabolism

Mitochondria are the energy organelles or "power plants" of the cell; they extract energy from the nutrients in food and transform it into a usable form to energize cell activities. About 90% of the energy that cells—and, accordingly, that tissues, organs and the whole body—need to survive and function is generated by mitochondria. The actual number of mitochondria per cell varies greatly, depending on the energy needs of each particular cell type. A single cell may contain as few as a hundred or as many as several thousand mitochondria. In some cell types, the mitochondria are densely compacted in cell regions that use most of the cell's energy. For example, mitochondria are packed between the contractile units in the muscle cells of the vertebrate heart. The mammalian organ most dependent on mitochondrial energy production is the brain, followed by the heart and some forms of skeletal muscle, and the kidneys and hormone synthesizing tissues. (As you will see later, muscles within and among various animals can vary considerably in mitochondrial density.)

Mitochondria are generally rod- or oval-shaped structures about the size of bacteria. However, they can vary in size and shape depending on the tissue type, even within a given animal. In some cells such as arthropod motor neurons, mitochondria are rodlike, with branches extending 1 μm or more.

The similarity in basic size of mitochondria and bacteria is not a coincidence; as we noted, considerable evidence indicates that mitochondria are descendants of bacteria that invaded or were engulfed by primitive cells early in evolutionary history. Paleobiologists generally believe that a symbiotic ("living together") relationship subsequently evolved, with the bacteria eventually becoming permanently associated with their host (an early eukaryotic or pre-eukaryotic cell). As part of their separate heritage, mitochondria have their own bacterial-type ribosomes, and their own DNA, distinct from the DNA housed in the cell's nucleus. Mitochondrial DNA, exclusively inherited from the maternal side, contains genes for producing many of the molecules mitochondria need to transduce energy. In surprising new research, flaws that gradually accumulate in mitochondrial DNA over time, and the resultant progressive impairment in mitochondrial energy transduction, have been implicated in a wide variety of disorders, including diabetes, heart failure, and degenerative illnesses associated with aging.

Each mitochondrion is enclosed by a double membrane—a smooth outer membrane that surrounds the mitochondrion itself and an inner membrane that forms a series of infoldings or shelves called *cristae*, which project into an inner cavity filled with a gel-like solution known as the *matrix* (● Figure 2–16). Cristae contain crucial proteins that ultimately are responsible for converting much of the energy in food into a usable form (the electron transport proteins, described shortly). The generous folds of the inner membrane greatly increase the surface area available for housing these important proteins. (This is a common theme in physiology: Increasing the surface area of various body structures by means of surface projections or infoldings makes more surface area available to participate in the structure's functions.) The matrix consists of a concentrated mixture of hundreds of different dissolved enzymes (the citric acid cycle enzymes, soon to be described) that are important in preparing nutrient molecules for the final extraction of usable energy by the cristae proteins.

■ **Aerobic metabolism relies on oxygen to convert energy in food into ATP.**

Every cell of an animal relies on specific biochemical pathways to generate energy from the food it consumes. One unifying principle of biological chemistry is the similarity in the major pathways of metabolism that organisms use. Some of these pathways are termed *aerobic* ("with air" or "with O_2") and require the consumption of oxygen, whereas others are referred to as *anaerobic* ("without air," specifically "without O_2"), and can proceed in the absence of oxygen. Most cell types can switch from one pathway to the other depending on the physiological state of the organism and availability of O_2. Aerobic metabolism requires mitochondria; anaerobic may or may not.

The source of energy for the cell is the chemical energy stored in the carbon–hydrogen bonds of ingested food. However, for most organisms the ultimate source of energy is the sun (exceptions are noted in Chapter 14). Solar energy is trapped by photosynthetic organisms and used to convert CO_2 into carbohydrates, which pass through food chains to herbivores (plant-eating animals) and ultimately to carnivores (meat-eating animals). Animal cells are not equipped to use solar energy directly. Instead, they must extract energy from food nutrients and convert it into energy forms that they can use—primarily

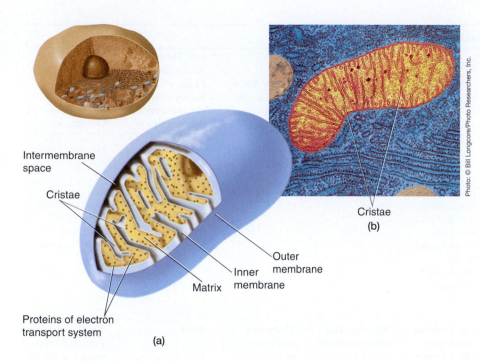

Intermembrane space

Cristae

Cristae
(b)

Outer membrane

Inner membrane

Matrix

Proteins of electron transport system

(a)

Photo: © Bill Longcore/Photo Researchers, Inc.

Figure 2–16 ● Mitochondrion.
(a) Schematic representation of a mitochondrion. The electron transport proteins embedded in the cristae folds of the mitochondrial inner membrane are ultimately responsible for converting much of the energy of food into a usable form. (b) Electron micrograph of a mitochondrion.

the high-energy phosphate bonds of adenosine triphosphate (ATP), which consists of adenosine with three phosphate groups attached, and nicotinamide adenine dinucleotide (NAD$^+$, which can bind two electrons and a hydrogen to become NADH). ATP and NADH are the two most prominent examples because these energized biomolecules contain chemically useful forms of stored energy. When these molecules are used in chemical reactions, the energy released can drive the energy-requiring activities within the cell. For example, when a high-energy bond such as that binding the terminal phosphate in ATP is split, a substantial amount of energy is released. This energy can power the contraction of a muscle, transport ions across a membrane or be used to synthesize a new protein. Similarly, high-energy hydrogen atoms are transferred from fuel molecules to NAD$^+$. NADH carries the energy-rich electrons just as ATP carries energy-rich phosphate groups. Each NADH molecule is actually worth almost three ATPs!

Adenosine triphosphate is the universal energy carrier—the common energy "currency" of life on Earth. Cells can use the energy in ATP to pay the energy "price" for running the cellular machinery. To obtain immediate usable energy, cells split the terminal phosphate bond of ATP, which yields adenosine diphosphate (ADP)—adenosine with two phosphate groups attached—plus inorganic phosphate (P_i) plus energy:

$$ATP \rightarrow ADP + P_i + \underset{\text{use by cell proteins}}{\text{splitting energy for}} + \text{heat}$$

In this energy scheme, food might be thought of as the "crude fuel," whereas ATP is the "refined fuel" for operating the cell's machinery. ATP is constantly being used by cell processes and must constantly be replenished. Let us elaborate on this fuel conversion process (■ Table 2–2). In the vast majority of animals, the cell has several options available to it when its supply of ATP molecules becomes depleted.

■ Nutrients stored within the cell can be broken down into forms capable of generating ATP.

■ Nutrients can be mobilized from specialized storage cells in the animal and transferred, via a circulatory system or diffusion, to energy-depleted cells.

■ Dietary food can be digested, or broken down, by the digestive system into smaller absorbable units that can be transferred (again, by circulation or diffusion) from the lumen of the digestive tract to cells whose energy supplies are depleted (Chapter 14).

The main endogenous fuels stored by most animal cells are **glycogen** (a storage form of glucose molecules, made from covalently linked glucose units) and **triglycerides** (a storage form of fatty acids, made from three fatty acids covalently linked to a glycerol molecule). Vertebrates store glycogen in the liver, muscles, and glia, whereas triglycerides are stored in adipose (fat) tissue. Insects store huge reserves of carbohydrate and triglycerides in their fat body, the insect equivalent of a liver. Generally, periods of short-term strenuous activity rely on the combustion of glycogen whereas during prolonged activity there is a transition to the combustion of fat. For example, during foraging flight locusts (flying insects that migrate long distances) rely on carbohydrate sources, but during migratory flight they rely on the combustion of fat. In ruminants such as cows, short-chain fatty acids represent the primary energy source, because these molecules are produced by rumen microbes (Chapter 14). However, in salmon (fish that migrate between rivers and the ocean), so-called "expendable proteins" can also be converted into amino acids and used as fuel during migration. Similarly, amino acids can be used to fuel energy needs in some species of marine invertebrates as well as in insects. In many vertebrates, the use of protein stores as a fuel to generate ATP represents a "last ditch" effort for survival. For example, Emperor penguins, who fast for prolonged periods in Antarctica during the winter months when nearshore seas are frozen over, rely on stored triglycerides for fuel generation. When these supplies are exhausted and the body is forced to burn its protein stores, the penguin returns to the sea to replenish its fuel.

Table 2–2 ■ Overview of Cellular Energy Production from Glucose

Reaction	Substance Processed	Location	Energy Yield (per glucose molecule processed)	End Products Available for Further Energy Extraction (per glucose molecule processed)	Need for Oxygen
Glycolysis	Glucose	Cytosol	Two molecules of ATP	Two pyruvic acid molecules	No; anaerobic
Citric acid cycle	Acetyl CoA, which is derived from pyruvic acid, the end product of glycolysis; two acetyl CoA molecules result from the proccessing of one glucose molecule	Mitochondrial matrix	Two molecules of ATP	Eight NADH and two $FADH_2$ hydrogen carrier molecules	Yes, derived from molecules involved in citric acid cycle reactions
Electron transport chain	High-energy electrons stored in hydrogen atoms in the hydrogen carrier molecules NADH and $FADH_2$ derived from citric acid cycle reactions	Mitochondrial inner-membrane cristae	Thirty-two molecules of ATP	None	Yes, derived from molecular oxygen acquired from breathing

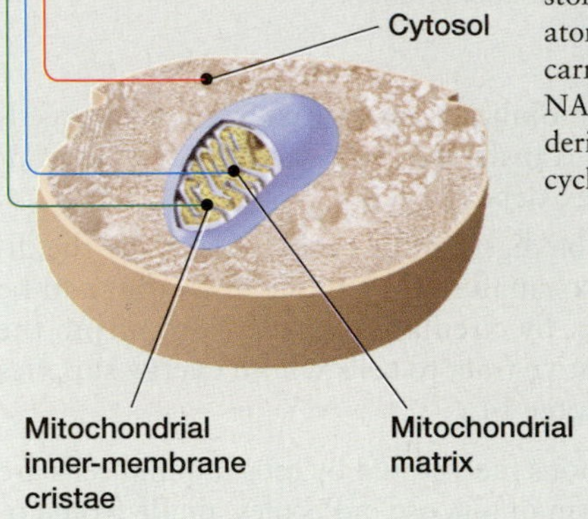

Cytosol

Mitochondrial inner-membrane cristae

Mitochondrial matrix

The mitochondrion is exaggerated and other intracellular components are omitted for better visualization.

Glycolysis

The most common energy pathway begins with glucose, which can be derived from most dietary carbohydrates (such as starch, a plant storage polymer composed of glucose units). Stored carbohydrates and some amino acids can also be converted into glucose. When delivered to the cells by a circulatory system, the nutrient molecules are transported across the plasma membrane into the cytosol. Among the thousands of enzymes within the cytosol are those responsible for glycolysis, a chemical process that breaks down the simple 6-carbon sugar molecule, glucose, into two pyruvic acid molecules, each of which contains 3 carbons. There are actually nine separate sequential reactions in glycolysis, each catalyzed by a separate enzyme. Glycolysis is common to almost every cell (prokaryotic and eukaryotic) and is believed to be the most ancient of the metabolic pathways, having arisen before oxygen accumulated in the atmosphere. During glycolysis, some of the energy stored in the chemical bonds of glucose is used to convert ADP into ATP

(● Figure 2–17). Importantly, although glycolysis is widespread, it is not very efficient in terms of energy extraction: *One molecule of glucose yields only two molecules of ATP in glycolysis.* Much of the energy originally contained in the glucose molecule is still locked in the chemical bonds of the pyruvic acid molecules. The low-energy yield of glycolysis is sufficient to sustain the needs of many single-celled anaerobic organisms and temporarily supports some aerobic cells through periods of reduced oxygen supply, but is grossly insufficient to sustain the long-term energy requirements of large, active, multicellular animals. This is where the mitochondria come into play.

Citric Acid Cycle

The pyruvic acid produced by glycolysis in the cytosol can be selectively transported into the mitochondrial matrix. Here it is further broken down into a 2-carbon molecule, acetic acid, by enzymatic removal of one of the carbons in the form of carbon dioxide (CO_2), which eventually is eliminated from the

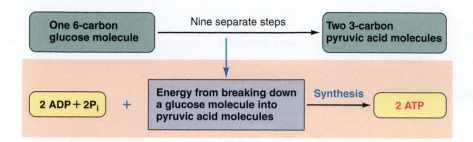

Figure 2–17 ● A simplified summary of glycolysis. Glycolysis involves the breakdown of glucose into two pyruvic acid molecules, with a net yield of two molecules of ATP for every glucose molecule processed.

body as an end product, or waste (● Figure 2–18). During this breakdown process, a carbon–hydrogen bond is disrupted, so a hydrogen atom is also released. A hydrogen carrier molecule, the function of which we discuss shortly, holds this hydrogen atom. The acetic acid thus formed combines with coenzyme A, a derivative of pantothenic acid (a B vitamin), producing the compound acetyl coenzyme A (acetyl CoA).

Acetyl CoA then enters the **citric acid cycle,** which consists of a series of eight separate biochemical reactions that are directed by the enzymes of the mitochondrial matrix. This cycle is also known as the *Krebs cycle* in honor of its principal discoverer (p. 3), or the *tricarboxylic acid (TCA) cycle,* because citric acid contains three carboxylic acid groups. Notably, the pyruvic acid formed in glycolysis can be further oxidized to generate more ATP *only* when oxygen is available to the animal. This cycle of reactions can be compared to one revolution around a Ferris wheel. (Keep in mind that Figure 2–18 is highly schematic. It depicts a cyclical series of biochemical reactions. The molecules themselves are not physically moved around in a cycle.) On the top of the Ferris wheel, acetyl CoA, a 2-carbon molecule, enters a seat already occupied by oxaloacetic acid, a 4-carbon molecule. These two molecules link together to form a 6-carbon citric acid molecule, and the trip around the citric acid cycle begins. As the seat moves around the cycle, at each new position, matrix enzymes modify the passenger molecule to form a slightly different molecule. These molecular alterations have the following important consequences:

1. Two carbons are sequentially removed from the 6-carbon citric acid molecule, converting it back into the 4-carbon oxaloacetic acid, which is now available at the top of the cycle to pick up another acetyl CoA for another revolution through the cycle.

2. The released carbon atoms, which were originally present in the acetyl CoA that entered the cycle, are converted into two molecules of CO_2. This CO_2, as well as the CO_2 produced during the formation of acetic acid from pyruvic acid, passes out of the mitochondrial matrix and subsequently out of the cell to enter the ECF. In animals with circulatory systems such as blood, the gas (and its derivative, bicarbonate; p. 11) is carried to the lungs or gills, where it is eliminated. (Note that the oxygen in the CO_2 molecule is derived from the molecules that were involved in the reactions, not from free molecular oxygen supplied by breathing.)

3. Hydrogen atoms and their associated electrons are also removed during the cycle at four of the chemical conversion steps. These hydrogens and electrons are "cap-

tured" by two *cofactors* (nonprotein components that act as hydrogen-electron carrier molecules)—nicotinamide adenine dinucleotide (NAD), a derivative of the B vitamin niacin, and flavine adenine dinucleotide (FAD), a derivative of the B vitamin riboflavin. These compounds are converted by the transfer of hydrogen and electrons to NADH and $FADH_2$, respectively.

4. One more molecule of ATP is produced for each molecule of acetyl CoA processed. Actually, ATP is not directly produced by the citric acid cycle. The released energy is used to directly link inorganic phosphate to guanosine diphosphate (GDP) to form guanosine triphosphate (GTP), a high-energy molecule similar to ATP. The energy from GTP can then be transferred to ATP as follows:

$$ADP + GTP \rightarrow ATP + GDP$$

Because each glucose molecule is converted into two acetic acid molecules, thus permitting two turns of the citric acid cycle, two more ATP molecules are produced from each glucose molecule. These two additional ATPs are still not much of an energy profit. However, the citric acid cycle is important in preparing the hydrogen carrier molecules for their entry into the electron transport chain, which produces far more energy than the sparse amount of ATP produced by the cycle itself.

Electron Transport Chain

Considerable untapped energy is still stored in the released hydrogen atoms, which contain electrons at high-energy levels. So far we have not mentioned any role for free oxygen in these reactions. That is because free (or molecular) oxygen—O_2—is not a reactant in the pathways of either glycolysis or the citric acid cycle. However, oxygen is essential, as you will see, in the disposition of the reduced cofactors, NADH and $FADH_2$. The "big payoff" comes when NADH and $FADH_2$ enter the electron transport chain, which consists of electron carrier molecules located in the inner mitochondrial membrane lining the cristae (● Figure 2–19a on page 50). The high-energy electrons are extracted from the hydrogens held in NADH and $FADH_2$ and transferred sequentially to the electron carrier molecules, freeing NAD and FAD to pick up more hydrogen atoms. The electron transport molecules are arranged in a specific order on the inner membrane so that the high-energy electrons are progressively transferred through a chain of reactions, with the electrons falling to successively lower energy levels at each step.

This electron transport chain is also called the **respiratory chain,** because it relates to use of O_2: Ultimately, the electrons are passed to O_2 derived from the environment (air or

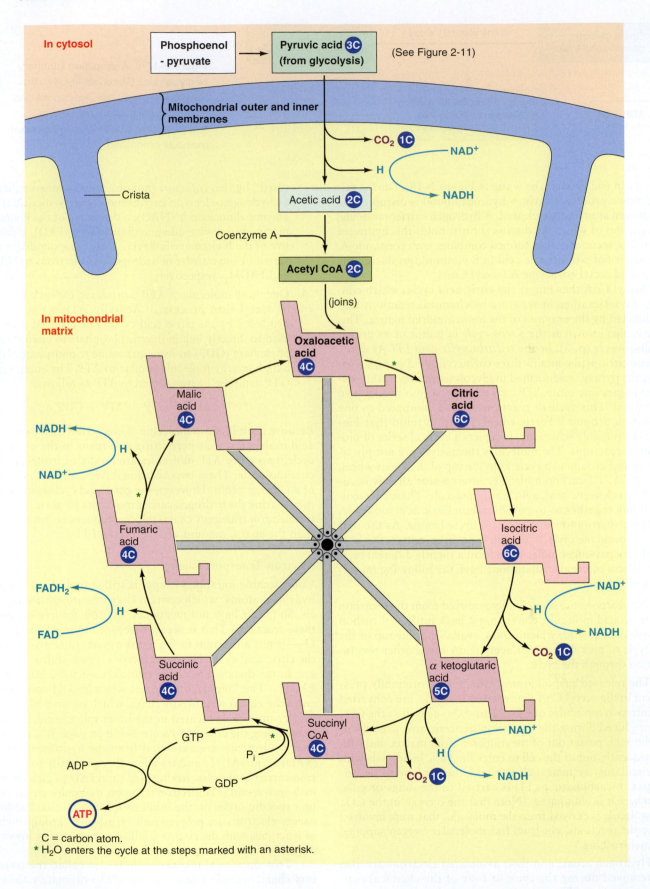

In cytosol

Phosphoenol - pyruvate → Pyruvic acid **3C** (from glycolysis)   (See Figure 2-11)

Mitochondrial outer and inner membranes

Crista

CO_2 **1C**

H

NAD⁺ → NADH

Acetic acid **2C**

Coenzyme A

Acetyl CoA 2C

In mitochondrial matrix

(joins)

Oxaloacetic acid 4C

Malic acid **4C**

*

Citric acid 6C

NADH ← H ← NAD⁺

Fumaric acid **4C**

*

Isocitric acid **6C**

NAD⁺ → H → NADH

CO_2 **1C**

FADH₂ ← H ← FAD

Succinic acid **4C**

α ketoglutaric acid **5C**

NAD⁺ → H → NADH

CO_2 **1C**

GTP

Succinyl CoA **4C**

*

P_i

ADP

GDP

ATP

C = carbon atom.
* H_2O enters the cycle at the steps marked with an asterisk.

Figure 2–18 ◄ ● **Citric acid cycle.** A simplified version of the citric acid cycle, showing how the two carbons entering the cycle by means of acetyl CoA are eventually converted to CO_2, with oxaloacetic acid, which accepts acetyl CoA, being regenerated at the end of the cyclical pathway. Also denoted is the release of hydrogen atoms at specific points along the pathway. These hydrogens bind to the hydrogen carrier molecules NAD and FAD for further processing. One molecule of ATP is generated for each molecule of acetyl CoA that enters the citric acid cycle, for a total of two molecules of ATP for each molecule of processed glucose.

water). Electrons bound to O_2 are in their lowest energy state. Oxygen enters the mitochondria to serve as the final electron acceptor of the electron transport chain. This negatively charged oxygen (negative because it has acquired additional electrons) then combines with the positively charged hydrogen ions (positive because they have donated the electrons at the beginning of the electron transport chain) to form two waters (H_2O).

As the electrons move through this chain of reactions to ever-lower energy levels, they release energy. Part of the released energy is lost as heat, but some is harnessed by the mitochondrion to synthesize ATP through the following steps, which are collectively known as the *chemiosmotic mechanism:*

1. At three sites in the electron transport chain, the energy released during the transfer of electrons is used to transport hydrogen ions across the inner mitochondrial membrane from the matrix to the space between the inner and outer mitochondrial membranes (the intermembrane space) (Figure 2–19b on page 51).

2. As a result of this transport process, hydrogen ions are more concentrated in the mitochondrial intermembrane space than in the matrix representing a source of potential energy. Because of this difference in concentration, or chemical gradient, the transported hydrogen ions have a strong tendency to flow back into the matrix through channels or passageways formed by special proteins within the inner mitochondrial membrane.

3. The channels through which the transported hydrogen ions return to the matrix bear the enzyme ATP synthase, which is activated by the flow of hydrogen ions from the intermembrane space to the matrix.

4. On activation, ATP synthase converts ADP + P_i to ATP, providing a rich yield of 32 more ATP molecules for each glucose molecule thus processed. ATP is subsequently transported out of the mitochondrion into the cytosol for use as the cell's energy source.

The harnessing of energy into a useful form as the electrons flow from a high-energy state to a low-energy state can be likened to a power plant converting the energy of water tumbling down a waterfall (while turning a turbine) into electricity. Because O_2 is used in these final steps of energy conversion when a phosphate is added to form ATP, this process is known as *oxidative phosphorylation.*

The series of steps that lead to oxidative phosphorylation may at first seem unnecessarily complicated. Why not just directly oxidize, or "burn," food molecules to release their en-

ergy? When this process is carried out outside an organism, all the energy stored in the food molecule is released explosively as heat (● Figure 2–20 on page 52). In the animal, oxidation of food molecules occurs in many small, controlled steps so that the food molecule's chemical energy is gradually made available for convenient packaging in a storage form useful to the cell. (Similarly, the fuel in an automobile's tank is not normally burned all at once—that would produce an explosion. Rather, fuel is fed a little at a time into the engine combustion chambers.) Approximately 70% of the chemical energy released in the oxidation of glucose to CO_2 and H_2O is recovered as energy in the form of ATP. The cell, by means of its mitochondria, efficiently captures the energy from the food molecules within ATP bonds when it is released in small quantities. In this way, much less of the energy is converted to heat. The heat produced is not completely wasted energy in endotherms such as birds and mammals; it is used to help maintain body temperature, with any excess heat being lost to the environment. (Endotherms are organisms that generate internal heat for thermoregulation; see Chapter 15.)

When enough O_2 is present—an aerobic condition—mitochondrial processing (that is, the citric acid cycle in the matrix and the electron transport chain on the cristae) harnesses sufficient energy to generate 34 more molecules of ATP, for a total net yield of 36 ATPs per molecule of glucose processed. The overall reaction for oxidation of food molecules to yield energy is as follows:

$$\text{Food} + O_2 \rightarrow CO_2 + H_2O + \text{ATP}$$

(necessary for oxidative phosphorylation)   (produced primarily by the citric acid cycle)   (produced primarily by the electron transport chain)

Glucose, the principal nutrient derived from dietary carbohydrates, is the fuel preference of most cells. The overall equation for glucose is

$$C_6H_{12}O_6 + 6\,O_2 \rightarrow 6\,CO_2 + 6\,H_2O + 36\,\text{ATP}$$

However, nutrient molecules derived from fats (fatty acids) and, if necessary, from protein (amino acids) can also participate at specific points in this overall chemical reaction to eventually produce energy. Amino acids are usually used for protein synthesis instead of energy production, but they can be used as fuel if insufficient glucose and fat are available.

■ **Mitochondrial metabolism can create oxidative stress.**

Unfortunately, leakage of some of the electrons associated with mitochondrial metabolism can potentially damage the tissues of the host animal. This requires cellular strategies to detoxify and limit the production of certain metabolites of molecular oxygen known as *reactive oxygen species (ROS).* ROS encompass a variety of diverse chemicals, which includes superoxide anions, hydroxyl radicals, peroxynitrite, and hydrogen peroxide (● Figure 2–21 on page 53). Some ROS are free radicals (molecules containing one or more unpaired electrons and thus react readily with other molecules, acquiring or giving up an electron to achieve stability). Radicals, such as hydroxyl radicals and superoxide, are extremely unstable.

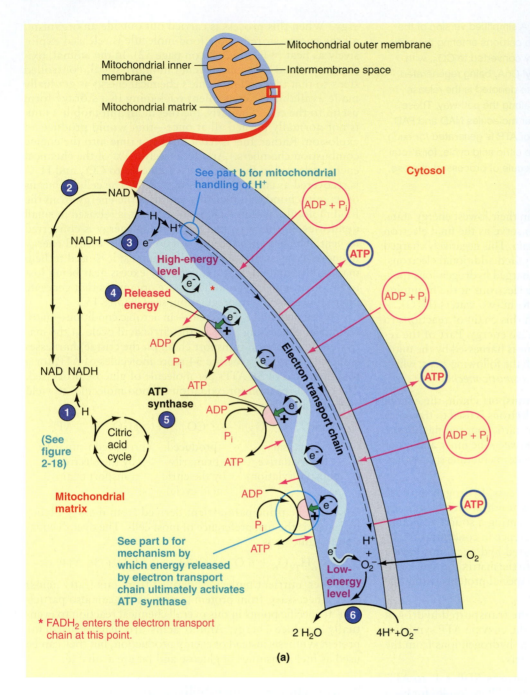

1. Hydrogen (H) that is released during the degradation of carbon-containing nutrient molecules by the citric acid cycle in the mitochondrial matrix is carried to the mitochondrial inner membrane by hydrogen carriers such as NADH.

2. After releasing hydrogen at the inner membrane, NAD shuttles back to pick up more hydrogen generated by the citric acid cycle in the matrix.

3. Meanwhile, high-energy electrons extracted from the hydrogen are passed through the electron transport chain located on the mitochondrial inner membrane.

4. Energy is gradually released as the electrons fall to successively lower energy levels by moving through the electron transport chain of reactions.

5. The released energy triggers a sequence of steps (shown in part b) that ultimately results in the activation of the enzyme ATP synthase within the mitochondrial inner membrane.

6. Molecular oxygen, after serving as the final electron acceptor, combines with the hydrogen ions (H⁺) generated from hydrogen on extraction of high-energy electrons to produce water.

(a)

In contrast, other forms such as hydrogen peroxide are comparatively long-lived. ROS production is countered by an elaborate antioxidant system that includes the enzymatic scavengers *SOD* (superoxide dismutase), *catalase*, and *glutathione peroxidase*. SOD increases the conversion of superoxide to hydrogen peroxide, while catalase and glutathione peroxidase convert hydrogen peroxide to water. A variety of other nonenzymatic, low-molecular-weight molecules are also important in scavenging ROS and include *ascorbate* (vitamin C), *flavenoids, carotenoids,* and perhaps most important in birds, *uric acid,* which is present in high concentrations within avian cells. Oxidative stress occurs when reactive oxygen species production overwhelms the antioxidant defenses, as with inflammation. Problems arise when these reactive species

not scavenged by antioxidants react instead with other components of the cell.

ROS are not always useless by-products. ROS generation by phagocytic cells is a critical host defense mechanism used to combat infection (Chapter 10). Further, cytosolic ROS production under conditions of metabolic stress may trigger specific signaling pathways within the cell, activating mechanisms to cope with stress.

■ Mitochondrial densities vary among tissue and organ types such as muscles.

Concentrations of ATP normally remain fairly constant in any particular tissue, even during high metabolic activity. How,

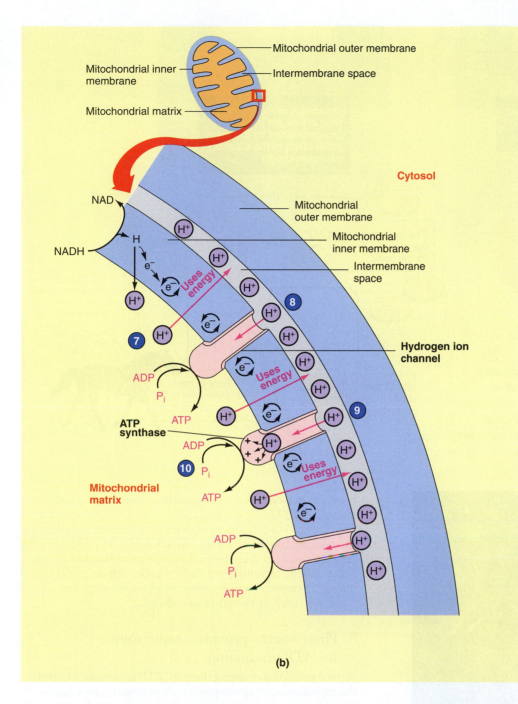

7 Energy released during the transfer of electrons by the electron transport chain is used to transport hydrogen ions from the matrix to the intermembrane space.

8 The resultant buildup of hydrogen ions in the intermembrane space brings about the flow of hydrogen ions from the intermembrane space to the matrix through special channels in the mitochondrial inner membrane.

9 The flow of hydrogen ions through the channel activates ATP synthase, which is located at the matrix end of the channel.

10 Activation of ATP synthase brings about the formation of ATP from ADP and P_i.

Figure 2–19 ● **ATP synthesis by the mitochondrial inner membrane.** (a) ATP synthesis resulting from the passage of high-energy electrons through the mitochondrial electron transport chain. (b) Activation of ATP synthase by movement of H^+.

then, can cells maintain ATP supplies during the transition from rest to work, particularly in skeletal muscles? Consider the different rates of ATP use in animal muscles at maximum energy output. For example, the flight muscle of the locust uses ATP at about 5400 mole/g/min. In comparison, ATP use is about 600 mole/g/min in hummingbird flight muscle and only 30 mole/g/min in limb muscles of an exercising human. How are ATP demands met in these different muscles? One solution is to alter the total number of mitochondria per unit volume—that is, the *density* of mitochondria—within each cell. Indeed, muscle cells generally have mitochondrial densities corresponding to their energy loads (Chapter 8). Skeletal muscles of terrestrial vertebrates have 1 to 10% of their volume as mitochondria. (In comparison, vertebrate hearts have

mitochondria at 30 to 50% of their volumes.) As befits its enormous energy demand, the locust flight muscle has up to 40 to 50% of its volume as mitochondria. As an extreme example, specialized modified muscles called *heater cells*, located at the base of the brain in billfish (such as swordfish), keep temperatures in the brain and eye above ambient temperature (see Chapter 15). These cells contain one of the highest mitochondrial densities of any animal cell, ranging from 55 to 70% of total cell volume. Mitochondrial density is not necessarily fixed; it can be upregulated in many skeletal muscles after a prolonged increase in muscular activity (see Chapter 8). Acclimation to cold temperatures (Chapter 15) also involves change in mitochondrial numbers. For example, the mitochondrial cell volume increases from 2.9 to 4.5% in the aero-

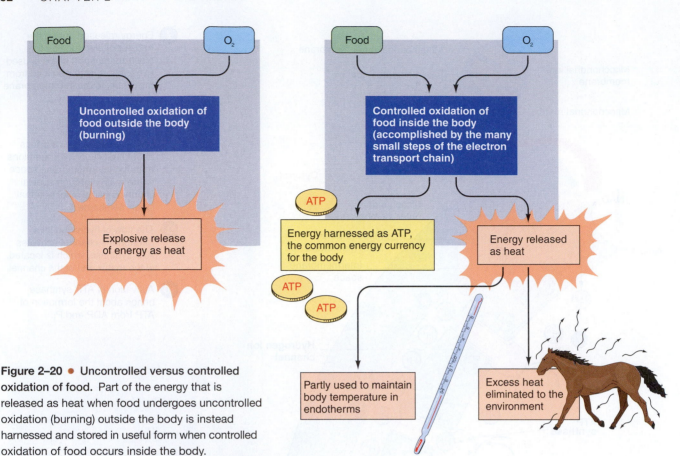

Figure 2–20 ● **Uncontrolled versus controlled oxidation of food.** Part of the energy that is released as heat when food undergoes uncontrolled oxidation (burning) outside the body is instead harnessed and stored in useful form when controlled oxidation of food occurs inside the body.

A Carolina locust. The flight muscles of this animal are densely packed with mitochondria, which in turn have an unusually high packing density of cristae for ATP production.

bic (red) muscle fibers in striped bass following acclimation from 25° to 5°C.

The second solution to high energy demands is to increase the *packing of cristae* within mitochondria. In particular, the packing is two to three times denser in mitochondria of insect flight muscle, hummingbird flight muscle, skipjack tuna swimming (red) muscle, and leg muscles of mammalian endurance "athletes" such as pronghorn antelopes.

■ Phosphogens provide a rapid source for ATP production.

A third solution to meeting the cells' ATP requirements involves the mobilization of tissue-specific phosphogens (see Chapter 8). Phosphogens are organic phosphate compounds present in high concentrations in the muscle cell that can transfer a high-energy phosphate group to ADP in order to regenerate ATP. They are made from ATP in times of plenty (when the cell is resting and has plenty of energy intake). In mammals, phosphogens are in sufficient supply to produce ATP for only a short time, such as up to about 10 seconds in humans. Thereafter the concentration of phosphogens must also be replenished by oxidation of sugars and fatty acids. The highest concentrations of phosphogens are in white skeletal muscle fibers, cells that rely heavily on anaerobic energy processes (Chapter 8). There are somewhat lower concentrations in red skeletal muscle fibers (those that rely on aerobic metabolism), heart and brain. The phosphogen of vertebrates is *creatine phosphate,* whereas invertebrates use a variety of organic phosphate com-

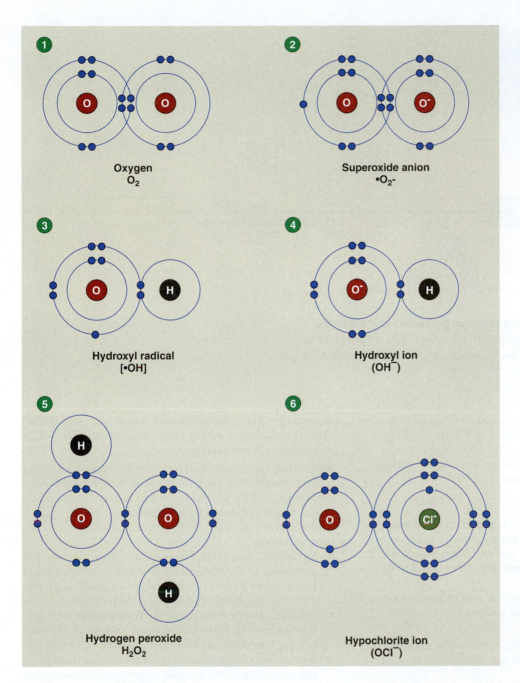

Figure 2–21 • Examples of reactive oxygen species (ROS).
Shown: ② superoxide anion (shown next to "normal" oxygen) ① for comparison, ③ hydroxyl radical (shown next to hydroxyl ion) ④ for comparison, ⑤ hydrogen peroxide, and ⑥ hypochlorite ion.

(*Source:* http://users.rcn.com/jkimball.ma .ultranet/BiologyPages/R/ROS.html.)

pounds, the best investigated being *arginine phosphate*. Insect flight muscle contains a surprisingly low amount of arginine phosphate, suggesting that the efficiency of oxygen delivery to the muscle tissue is adequate to meet the requirements of aerobic metabolism.

■ **Oxygen deficiency forces cells to rely on glycolysis and other anaerobic reactions, producing lactic acid, propionic acid, octopine, or other end products.**

Phosphogens can produce ATP in the absence of oxygen, but for a very limited time. Another metabolic adaptation for ATP

production in the absence of molecular oxygen works for longer periods. In the absence of O_2, the electron transport chain cannot accept electrons from NADH and $FADH_2$ and so it "shuts down." This occurs during bursts of high activity (such as a cheetah running), in low-oxygen habitats (such as stagnant water or deep within mud), and when animals must cease breathing (as in an oyster closing its shell tightly during low tide). In such cases, glycolysis is usually the primary pathway for ATP replenishment. Recall that glycolysis takes place in the cytosol and involves the breakdown of glucose into pyruvic acid, producing a low yield of 2 molecules of ATP per molecule of glucose. Oxidative phosphorylation, the source of 34 out of every 38 ATP molecules, ceases, and the untapped energy of the glucose molecule may remain locked in the bonds of the py-

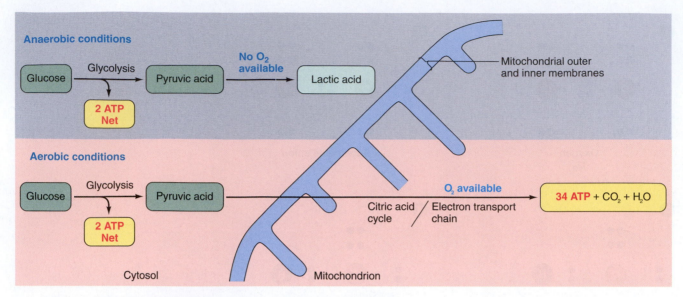

Figure 2–22 ● **Comparison of energy yield and products under anaerobic and aerobic conditions.** In anaerobic conditions, only 2 ATPs are produced for every glucose molecule processed, but in aerobic conditions a total of 36 ATPs are produced per glucose molecule.

ruvic acid molecules. In some groups such as most vertebrates, pyruvic acid is converted into lactic acid (● Figure 2–22) by an enzyme called *lactic dehydrogenase (LDH)*, in a reaction that regenerates NAD^+, the hydrogen-electron carrier needed for earlier steps in glycolysis to proceed (p. 47):

2 Pyruvic acid + 2 NADH → 2 lactic acid + 2 NAD^+

Lactic acid thus steadily accumulates in the tissues and gradually reduces the pH. Lactic acid production is necessary because the electron transport chain can no longer regenerate the cofactors NAD^+. We will encounter this reaction and LDH in later chapters.

Significant modifications to anaerobic end products have evolved at two points in the glycolytic pathway.

1. *Higher ATP yield.* The first modification is at the *phosphoenolpyruvate (PEP)* step, the one just before pyruvic acid formation in glycolysis:

2 PEP + 2 ADP → 2 pyruvic acid + 2 ATP

Many invertebrate groups (bivalve mollusks such as oysters, nematodes, parasitic helminths, and the swamp-dwelling annelid *Alma emeni*) gain more than 2 ATPs per glucose by channeling PEP into a partial set of reactions in the citric acid cycle, in reverse direction. The end product is usually either succinic acid, yielding 4 ATPs per glucose, or propionic acid, yielding 6 ATPs per glucose (compare the following with Figure 2–18):

2 PEP → 2 oxaloacetic acid (+ 2 ATP) → 2 malic acid → 2 fumaric acid → 2 succinic acid (+ 2 ATP) → 2 succinyl CoA → 2 methylmalonyl CoA → 2 propionyl CoA (+ 2 ATP) → 2 propionic acid

Some invertebrates (such as bivalve mollusks) can use **aspartate** (an amino acid) as an anaerobic fuel. Aspartate

is particularly high in oyster and mussel hearts. It is converted to oxaloacetate (by removal of ammonia), which enters the pathway just described.

2. *Alternative end products.* The second modification point is the conversion of pyruvic acid for the purposes of regenerating NAD^+. The acid problem created by lactic acid production is avoided in some organisms, in which pyruvic acid is converted to a variety of nonacidic organic products, including

■ **Strombine, tauropine, lysopine, octopine, alanine,** and **alanopine** in mollusks. The first four compounds are produced by joining pyruvic acid to an amino acid: glycine, taurine, lysine and arginine, respectively.

■ **Ethanol** (alcohol) and **acetate.** Yeasts metabolize glucose anaerobically to two ethanol and two CO_2 molecules, a process used in beer making, for example. Interestingly, researchers have reported that wild goldfish wintering at the bottom of ice-covered ponds and lakes produce ethanol as the oxygen levels drop. This is the only animal known to use this pathway.

Ultimately, end products such as lactic acid that accumulate in body fluids are converted back to pyruvic acid using NAD^+ as the electron acceptor. Pyruvic acid can then either enter into the citric acid cycle or reconverted back into glucose or glycogen (polymer of glucose molecules). Regardless of the disposition of pyruvic acid, oxygen is required. A functional electron transport system is essential to capture the energy from the NADH and $FADH_2$ during these reactions. The conversion of lactic acid to glucose is an example of *gluconeogenesis* ("new formation of glucose"). Basically, it is a reversal of glycolysis, because the principal intermediates of the pathway are the same. However, the gluconeogenic pathway involves several different enzymes and requires the expenditure of six molecules of ATP for each molecule of glucose generated. The ATP is necessary because glucose synthesis is energetically "uphill," that is, an energy-requiring process (in contrast to glycolysis, which is "downhill," in terms of energy).

■ Tolerance of oxygen deficiency varies widely among organisms.

Many animals require oxygen continuously as the final electron acceptor for survival, and are classified as obligate aerobes. In mammals, most cells cannot survive more than a few minutes without oxygen. Anaerobic pathways are, however, important during intense bursts of activity. Skeletal muscle cells in particular take advantage of this ability during short bursts of strenuous exercise, when energy demands for contractile activity outstrip the animal's ability to bring adequate O_2 to the exercising muscles to support oxidative phosphorylation. For example, cheetahs are noted for their remarkable sprint speeds. However, the metabolic costs incurred to the animal include an enormous *oxygen debt* and a reduction in blood and tissue pH (acidosis; see Chapter 13) caused by the accumulation of lactic acid. Cheetahs can sustain this intense activity for only brief periods of time and must rest for lengthy periods before resuming any activity. In some species of fish, anaerobic pathways can fuel intermediate swimming speeds (between cruising and sprint), for sustained periods of time (tens of minutes). Lactate accumulates in large quantities in their white muscle fibers, and, unlike in mammals, removing these end products requires many hours of recovery. Because of the depletion of glucose stores and the poor return in energy yield on glucose, animals can rely on anaerobic metabolism for only relatively brief periods. When energy stores become depleted, vertebrate muscles fatigue and lose their ability to contract.

Some species of organisms—the **facultative anaerobes**—can adapt to anaerobic conditions for periods ranging from days to months by substituting other electron acceptors for oxygen. These include the bivalve mollusks that live in the intertidal zone. For extended periods of time each day, these animals close their shells when exposed to air. Limiting desiccation by the sun in this way also renders them unable to exchange respiratory gases. Thus they rely on the alternative pathways described earlier. Snapping turtles, the longest hibernating reptile in North America, bury themselves in the mud for up to eight months each year. Anaerobic glycolysis supplies the energy needs of these animals, at the expense of a buildup of lactic acid (particularly in the shell; see Chapter 13). However, in these and most vertebrates the brain and the heart normally continue their reliance on oxygen, albeit at a reduced rate.

A third group of organisms, called **obligate anaerobes**, are inhibited or killed in the presence of oxygen and thrive in anaerobic environments. These include some invertebrate parasites and dormant stages of animals such as brine shrimp (p. 15), which can survive without oxygen indefinitely. Some bacteria (such as *Clostridium*, the source of botulism in improperly canned foods) and protozoa are also obligate anaerobes.

Animal cells that rely on anaerobic metabolism must have available to them an immediately available reservoir of stored energy supplies. For example, in vertebrate white skeletal muscle fibers, fuel is stored within the cell in the form of glycogen. Tissue glycogen can then be readily mobilized to glucose in times of energy expenditure. In many inverte-

brates, glycogen stores can form a significant proportion of the actual muscle tissue weight (see Chapter 8), and some have alternative fuels such aspartate (described in the preceding section). Phosphogens (p. 52) can also be used anaerobically, though generally for only a short time before they are exhausted.

■ The energy stored within ATP is used for synthesis, transport, mechanical work, and light and heat production.

Once formed, ATP is available as an energy source as needed within the cell. If produced by oxidative phosphorylation, ATP is transported out of the mitochondria. Cellular activities that require energy expenditure fall into three common categories, and two rarer ones:

1. *Synthesis of new chemical compounds,* such as protein synthesis by the endoplasmic reticulum. All cells use ATP for this general purpose. Some cells, especially cells with a high rate of secretion and cells in the growth phase, use up to 75% of the ATP they generate just to synthesize new chemical compounds.

2. *Membrane transport,* such as the selective transport of molecules across the digestive tract wall. Researchers think all cells use ATP for at least some forms of transport. This can reach extremes in specialized transporters such as kidney tubule cells, which can expend as much as 80% of their ATP currency to operate their selective membrane-transport mechanisms.

3. *Mechanical work,* such as contraction of the heart muscle to pump blood or contraction of skeletal muscles to lift an object. These activities require tremendous quantities of ATP.

4. *Bioluminescence,* in which ATP reacts with large carbon–nitrogen molecules (a luciferin) to produce photons in the visible range of wavelength. This use of ATP is relatively rare on land, but is common in marine organisms, especially in the deep.

5. *Heat production,* generally for the purposes of thermoregulation and other uses of high temperature. This occurs for useful purposes only in a few groups of organisms. Skeletal muscles and special heater tissues, described in Chapter 15, convert chemical-bond energy into heat. (Note that most biological reactions produce some heat, but these heater processes produce more.)

As a result of cellular energy expenditure to support these various activities, large quantities of ATP are produced. These energy-depleted ADP molecules enter the mitochondria for "recharging" and then cycle back into the cytosol as energy-rich ATP molecules after participating in oxidative phosphorylation. A single ADP/ATP molecule may shuttle back and forth between the mitochondria and cytosol for this recharging/expenditure cycle thousands of times per day.

The high demands for ATP render glycolysis alone an insufficient as well as inefficient supplier of power for most cells. If it were not for the mitochondria, which house the metabolic machinery for oxidative phosphorylation, the energy capability of a cell would be very limited. However, glycolysis does provide cells with a sustenance mechanism that can produce at least some ATP under anaerobic conditions.

Vaults

Vaults, which are three times as large as ribosomes, are shaped like octagonal barrels (● Figure 2–23). Their name comes from their multiple arches, which reminded their discoverers of vaulted or cathedral ceilings. Like barrels, vaults have a hollow interior. Sometimes vaults are seen in an open state, appearing like pairs of unfolded flowers with each half of the vault bearing eight "petals" attached to a central ring. A cell may contain thousands of vaults. These numerous, relatively large organelles have been elusive until recently because they do not show up with ordinary staining techniques. Two clues to the function of vaults may be their octagonal shape and their hollow interior. Intriguingly, the nuclear pores are also octagonal and the same size as vaults, leading to speculation that vaults may be cellular "trucks." According to this proposal, vaults would dock at nuclear pores, pick up molecules synthesized in the nucleus, and deliver their cargo elsewhere in the cell. Ongoing research supports vaults' role in nucleus-to-cytoplasm transport, but what cargo they are carrying is uncertain. Possibilities include messenger RNA, carried from the nucleus to the ribosomal sites of protein synthesis, or the two subunits that make up ribosomes. These two subunits are produced in the nucleus, then exit through the nuclear pores by unknown means to reach their sites of action—either attached to the rough ER or in the cytosol.

Cytosol

Occupying about 55% of the total cell volume, the cytosol is the semiliquid portion of the cytoplasm that surrounds the organelles. Its amorphous appearance under an electron microscope belies the fact that the cytosol is not a uniform liquid mixture, but is actually more like a highly organized, gelatinous mass with differences in composition and gelatinous consistency between various regions and states of the cell.

▇ The cytosol is important in intermediary metabolism, ribosomal protein synthesis, and storage of fat and glycogen.

Four general categories of activities are associated with the cytosol: (1) enzymatic regulation of intermediary metabolism; (2) ribosomal protein synthesis; and (3) storage of fat and carbohydrate; and (4) temporary storage of vesicles. Dispersed throughout the cytosol is a cytoskeleton that gives shape to the cell, provides an intracellular organizational framework, and is responsible for various cell movements.

Enzymatic Regulation of Intermediary Metabolism

The term *intermediary metabolism* refers collectively to the large set of intracellular chemical reactions that involve the degradation, synthesis, and transformation of small organic molecules such as simple sugars, amino acids, and fatty acids. These reactions are critical for ultimately capturing energy to be used for cellular activities and for providing the raw materials needed for maintaining cell structure and function and for cell growth. All intermediary metabolism occurs in the cytoplasm, with most being accomplished in the cytosol. Thousands of enzymes involved in glycolysis and other intermediary biochemical reactions are found in the cytosol.

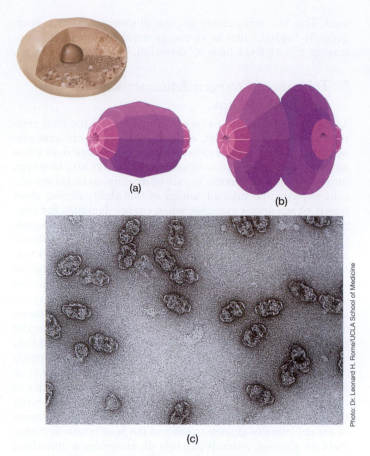

Figure 2–23 ● **Vaults.** (a) Schematic three-dimensional representation of a vault, an octagonal barrel–shaped organelle believed to transport either messenger RNA or the ribosomal subunits from the nucleus to the cytoplasmic ribosomes. (b) Schematic representation of an opened vault, showing its octagonal structure. (c) Electron micrograph of vaults.

Ribosome Protein Synthesis

Also dispersed throughout the cytosol are the free ribosomes, which synthesize proteins for use in the cytosol itself. In contrast, rough ER ribosomes synthesize proteins for secretion and for construction of new cell components. Often, cytosolic ribosomes that are synthesizing identical proteins are clustered together in "assembly lines" known as polyribosomes.

Storage of Fat and Glycogen

Excess nutrients not immediately used for ATP production are converted in the cytosol into storage forms that are readily visible even under a light microscope. Such nonpermanent masses of stored material are known as *inclusions*. The main forms, as we noted earlier, are glycogen and triglycerides. The latter form the largest and most important storage inclusions. Storage of energy in the form of triglyceride lipids represents the preferred form of storing extra energy because lipids have almost twice the energy density of carbohydrates and are channeled directly into the tricarboxylic cycle of mitochondria, yielding proportionately greater amounts of ATP. (Curiously, mitochondria of cartilaginous fishes cannot use fatty acids, but those of all other vertebrates can.) Through a microscope, small fat droplets can be seen within the cytosol in various cells. In insect fat bodies and vertebrate adipose tissue, the stored triglycerides can occupy almost the entire cytosol, coalescing to form one large fat droplet (● Figure 2–24a). The other visible storage product is glycogen, which appears as aggregates or clusters dispersed throughout the cell (Figure 2–24b). Cells vary in their ability to store glycogen, with liver and muscle cells having the greatest stores. When food is not available to provide fuel for the citric acid cycle and electron transport chain, stored glycogen and fat are broken down to release glucose and fatty acids, respectively, which can feed the mitochondrial energy-producing machinery.

Storage of Vesicles

Secretory vesicles that have been processed and packaged by the endoplasmic reticulum and Golgi complex also remain in the cytosol, where they are stored until signaled to empty their contents extracellularly. In addition, transport vesicles are transiently present in the cytosol as they carry their selected cargo from one membrane-enclosed organelle compartment to another or to the plasma membrane. Likewise, endocytotic vesicles can be found transiently in the cytosol until the cell disposes of these packages it has internalized.

Cytoskeleton

Permeating the cytosol is the **cytoskeleton,** a complex protein network that acts as the "bone and muscle" of the cell. The distinct shape, size, complexity, and intracellular specialization of the various body cells necessitate intracellular scaffolding to support and organize the cell components into an appropriate arrangement and to control their movements. These functions are performed by the cytoskeleton. This elaborate network has at least three distinct elements: (1) microtubules, (2) microfilaments, and (3) intermediate filaments (Table 2–1). The different parts of the cytoskeleton are structurally linked and functionally coordinated to provide certain integrated functions for the cell. Because of the complexity of this network and the variety of functions it serves, we discuss its elements separately.

■ Microtubules are essential for maintaining asymmetric cell shapes and are important in complex cell movements.

The **microtubules** are the largest of the cytoskeletal elements. They are very slender (22 nanometer, or nm, in diameter; 1 nm = 1 billionth of a meter), long, hollow, unbranched tubes composed primarily of tubulin, a small, globular protein molecule (6 nm in diameter) (● Figure 2–25a on the next page). Microtubules are essential for maintaining an asymmetric cell shape, such as that of a nerve cell, whose elongated axon may extend a meter in length or more in a large vertebrate, from the origin of the cell body in the spinal cord to the termination of the axon at a muscle (● Figure 2–26 on page 59). Microtubules, along with specialized intermediate filaments, stabilize this asymmetric axonal extension.

Microtubules also play an important role in coordinating numerous complex cell movements, including (1) transport of secretory vesicles from one region of the cell to another, (2) movement of specialized cell projections such as cilia and

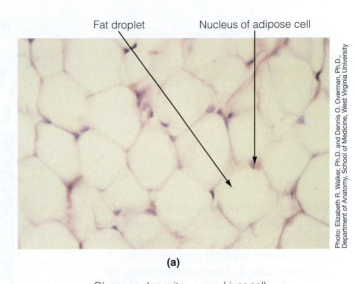
Fat droplet   Nucleus of adipose cell

(a)

Photo: Elizabeth R. Walker, Ph.D. and Dennis O. Overman, Ph.D., Department of Anatomy, School of Medicine, West Virginia University

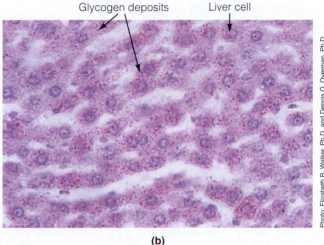
Glycogen deposits   Liver cell

(b)

Photo: Elizabeth R. Walker, Ph.D. and Dennis O. Overman, Ph.D., Department of Anatomy, School of Medicine, West Virginia University

Figure 2–24 ● **Inclusions.** (a) Light micrograph depicting fat storage in a mammalian adipose cell. Note that the fat droplet occupies almost the entire cytosol. (b) Light micrograph depicting glycogen storage in a mammalian liver cell. The red-staining granules throughout the liver cell's cytosol are glycogen deposits.

flagella, and (3) distribution of chromosomes during cell division through formation of a mitotic spindle.

Transport of Secretory Vesicles

Axonal transport provides a good example of the importance of an organized system for moving secretory vesicles. In a nerve cell, specific chemicals are released from the terminal end of the elongated axon to influence a muscle or another structure that the nerve cell controls. These chemicals are largely produced within the cell body, where the nuclear DNA blueprint, endoplasmic reticular factory, and Golgi packaging and distribution outlet are located. Yet these chemicals ultimately function at the end of the axon, which may be a meter away. If these chemicals had to diffuse on their own from the cell body to a distant axon terminal, it would take them years to get there in a large animal—obviously an impractical solution. The microtubules provide a "highway" for vesicular traffic along the axon, with the driving force dependent on ATP (Figure 2–26). Specific protein molecules attach to the particle

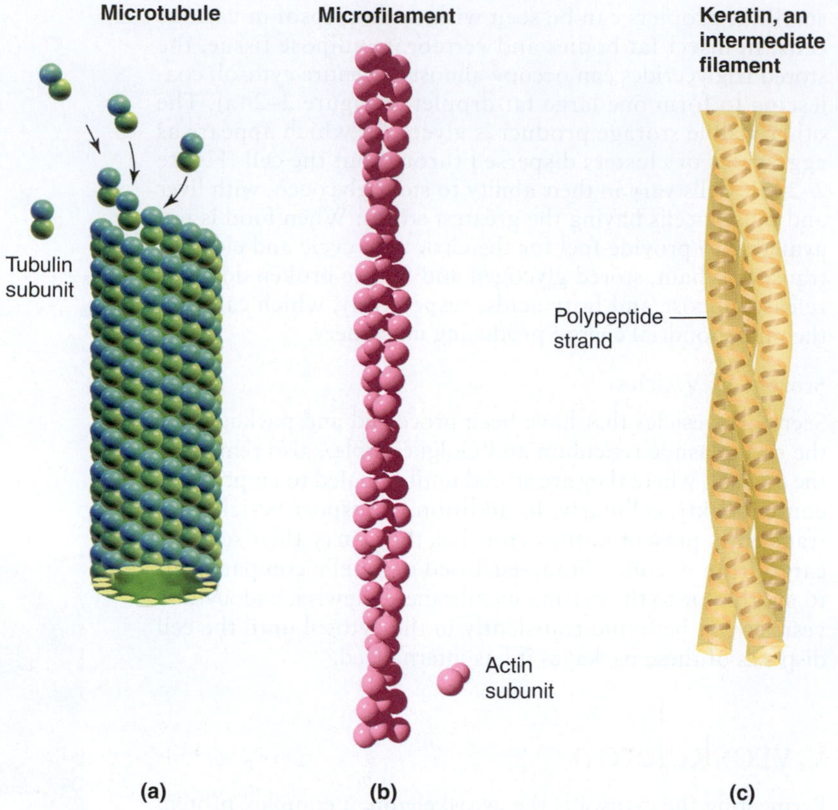

Microtubule   **Microfilament**   **Keratin, an
                                                       intermediate
                                                       filament**

Tubulin
subunit

Polypeptide
strand

Figure 2–25 • Components of the cytoskeleton.
(a) Microtubules, the largest of the cytoskeletal
elements, are long, hollow tubes formed by two
slightly different variants of globular-shaped
tubulin molecules. (b) Most microfilaments, the
smallest of the cytoskeletal elements, consist of
two chains of actin molecules wrapped around
each other. (c) The intermediate filament keratin
found in skin is made up of three polypeptide
strands wound around each other. The
composition of intermediate filaments, which
are intermediate in size between the microtubules
and microfilaments, varies among different
cell types.

Actin
subunit

(a)   (b)   (c)

to be *transported* and "walk" along the microtubule, powered
by ATP. **Kinesin,** one such motor protein, consists of two globular heads, a stalk, and a fanlike tail. Kinesin's tail binds to the
secretory vesicle to be moved, and its globular heads act like
little feet that move one at a time, like the way you walk. The
feet alternately attach to one tubulin molecule on the microtubule, bend and push forward, then let go. During this process, the back foot is yanked forward so that it swings ahead of
what was the front foot and then attaches to the next tubulin
molecule farther down the microtubule. The process is repeated
over and over as kinesin moves its cargo to the end of the axon
by using each of the tubulin molecules as a stepping-stone.

Reverse vesicular traffic also occurs along these microtubular highways. Vesicles that contain debris are transported
by different motor proteins, most commonly members of the
dynein superfamily, from the axon terminal to the cell body
for degradation by lysosomes, which are confined within the
cell body.

Movement of Cilia and Flagella

Microtubules are also the dominant structural and functional
components of cilia and flagella. These specialized protrusions from the cell surface allow a cell to move materials across
its surface (in the case of a stationary cell) or to propel itself
through its environment (in the case of a motile cell). Cilia
are numerous tiny, hairlike protrusions, whereas a flagellum
is a single, long, whiplike appendage. Even though they project
from the surface of the cell, cilia and flagella are both intracellular structures covered by the plasma membrane.

Cilia beat or stroke in unison, much like the coordinated
efforts of a rowing team. Each cilium exerts a rapid active
stroke, which moves material on the cell surface forward. This

stroke is followed by a recovery phase in which the cilium
more slowly returns to its original position with a kind of un-
rolling, backward movement that does not exert much force.
In this way, the material just pushed forward is not futilely
pushed backward but instead is moved only in the direction
of the forward stroke.

Mammals have ciliated cells that line the respiratory tract
and the oviduct of the female reproductive tract. Respiratory
cilia help move foreign particles out of the lungs. The thousands of cilia lining the respiratory airways project into a layer
of sticky mucus that traps dust and other inspired (breathed-
in) particles. The coordinated stroking action of these cilia
sweeps this dust-laden mucus up to the throat, where it can
be expectorated (spit out) or swallowed and eventually eliminated in the feces. In the female reproductive tract, the sweeping action of the cilia that line the oviduct draws the egg
(ovum) released from the ovary during ovulation into the
oviduct and then guides it toward the uterus (womb).

The only vertebrate cells that bear flagella are sperm. The
whiplike motion of the flagellum or "tail" enables a sperm to
move through its environment. This ability is particularly useful when the sperm maneuvers for final penetration of the
ovum during fertilization. Flagella are found in a few other
animal cell types, such as the choanocytes (cells that create
feeding current) of sponges that we discuss in Chapter 9.

Cilia and flagella have the same basic internal structure.
Both consist of nine fused pairs of microtubules (doublets)
arranged in an outer ring around two single unfused microtubules in the center (• Figure 2–27). This characteristic "9 +
2" array of microtubules extends throughout the length of
the motile appendage. A cilium or flagellum originates from a
specialized cytoplasmic structure, the basal body, which is lo-

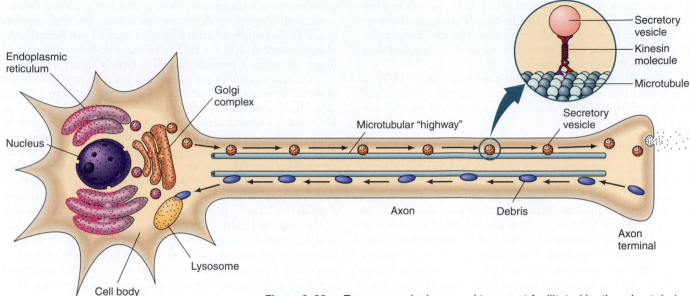

Figure 2–26 ● **Two-way vesicular axonal transport facilitated by the microtubular "highway" in a nerve cell.** Schematic illustration of a neuron depicting secretory vesicles being transported from the site of production in the cell body along a microtubular "highway" to the terminal end for secretion. Vesicles containing debris are being transported in the opposite direction for degradation in the cell body. The enlargement depicts kinesin, a motor protein, carrying a secretory vesicle down the microtubule by using its "feet" to "step" on one tubulin molecule after another.

cated inside the main part of the cell and provides a base from which the microtubules grow to form the appendage. Each basal body is a short cylinder composed of a parallel microtubular symmetry similar to that of the cilium or flagellum.

Associated with these microtubules are accessory proteins that maintain the microtubules' organization and play an essential part in the microtubular movement that causes the entire structure to bend. The most important of these accessory proteins is dynein, which forms a set of armlike projections from each doublet of microtubules (Figure 2–27a). The sliding of adjacent microtubule doublets past each other produces the bending movements of cilia and flagella. Sliding is accomplished by the dynein arms, which are motor proteins such as kinesin. The dynein arms have the capability of splitting ATP and then using the released energy to "crawl" along the neighboring microtubule doublet (similar to kinesin movement) to cause relative displacement of the doublets. Groups of cilia working together are oriented to beat in the same direction and contract in a synchronized manner through controlling mechanisms that are poorly understood. These mechanisms appear to involve the single microtubules at the cilium's center and their surrounding accessory proteins, the inner sheath (Figure 2–27).

Formation of the Mitotic Spindle

Cell division involves two discrete, but related activities: **mitosis** (nuclear division) and **cytokinesis** (cytoplasmic division). Mitosis involves replication of the DNA-containing chromosomes, which are then evenly distributed in the two halves of the cell. In cytokinesis, the plasma membrane is constricted in the middle of the cell, and the two halves separate into two new daughter cells, each with a full complement of chromosomes.

Microtubules are transiently assembled to form a **mitotic spindle,** which organizes and directs the movement of the replicated chromosomes away from each other toward opposite ends of the cell so that the genetic material is evenly distributed when the cell divides. The mitotic spindle is formed by the **centrioles,** a pair of short, cylindrical structures that lie at right angles to each other near the nucleus (Figure 2–2). The centrioles also duplicate during cell division. After self-replication, the centriole pairs move toward opposite ends of the cell and form the spindle apparatus between them through a precisely organized assemblage of microtubules.

Besides their role in mitotic-spindle formation, the centrioles and surrounding complex of densely staining proteinaceous material together assemble the many microtubules that normally radiate throughout the cytoskeleton. The centrioles are identical in structure to basal bodies. In fact, under some circumstances, the centrioles and basal bodies are interconvertible. During development of ciliated mammalian cells, the centriole pair migrates to the region of the cell where the cilia will be formed and duplicates itself to produce the many basal bodies that will form the cilia.

■ Microfilaments are important to cellular contractile systems and as mechanical stiffeners.

The **microfilaments** are the smallest (6 nm diameter) elements of the cytoskeleton visible with a conventional electron microscope. The most obvious microfilaments in most cells are those composed of actin, a protein molecule that has a globular shape similar to tubulin. Unlike tubulin, which forms a hollow tube, actin is assembled into two twisted strands, much like two strings of pearls twisted into a helix (spiral) to form a microfilament (Figure 2–25b). In muscle cells, another protein called *myosin* forms a different kind of microfilament (see Chapter 8). In most cells, myosin is not as abundant and does not form such distinct filaments.

Microfilaments serve at least two functions: (1) They play a vital role in various cellular contractile systems, and (2) they act as mechanical stiffeners for several specific cellular projections.

Microfilaments in Cellular Contractile Systems

Actin-based assemblies are involved in muscle contraction, cell division, and cell locomotion. The most obvious, best organized, and most clearly understood cell contractile system is that found in muscle. Muscle contains an abundance of actin and myosin filaments, which are organized to slide past each other at the expense of splitting ATP, generating a contractile force in the process. This ATP-powered microfilament sliding and force development is triggered by a complex sequence of electrical, biochemical, and mechanical events initiated when the muscle cell is stimulated to contract (see Chapter 8 for details).

Nonmuscle cells can also contain "musclelike" assemblies. Some of these microfilament contractile systems are transiently assembled to perform a specific function when needed. A good example is the contractile ring that forms during cytokinesis to split apart the duplicate cell halves. The ring consists of a beltlike bundle of actin filaments located just beneath the plasma membrane in the middle of the cell. When this ring of fibers contracts, it pinches the cell in two (● Figure 2–28a).

Complex actin-based assemblies are also responsible for most cell locomotion. Four types of adult mammalian cells are capable of moving on their own—sperm, white blood cells, fibroblasts, and skin cells. Sperm move by the flagellar mechanism already described. Motility for the other cells is accomplished by amoeboid movement, a cell-crawling process that depends on the activity of their actin filaments, in a mechanism similar to that used by amoebae to maneuver through their environment. When crawling, the motile cell forms *pseudopods* at the "front" or leading edge of the cell in the direction of the target (● Figure 2–29). For example, the target that triggers amoeboid movement might be the proximity of food in the case of an amoeba, or a bacterium in the case of a white blood cell. Pseudopods are formed as a result of the organized assembly and disassembly of branching actin networks. During amoeboid movement, actin filaments continuously grow at the cell's leading edge through the addition of actin molecules at the front of the actin chain. This filament growth pushes that portion of the cell forward as a pseudopod extension. Simultaneously, actin molecules at the rear of the filament are

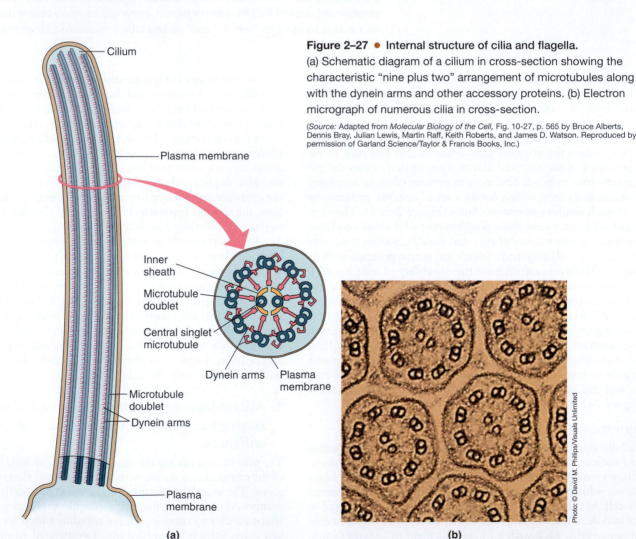

Figure 2–27 ● Internal structure of cilia and flagella. (a) Schematic diagram of a cilium in cross-section showing the characteristic "nine plus two" arrangement of microtubules along with the dynein arms and other accessory proteins. (b) Electron micrograph of numerous cilia in cross-section.

(*Source:* Adapted from *Molecular Biology of the Cell,* Fig. 10-27, p. 565 by Bruce Alberts, Dennis Bray, Julian Lewis, Martin Raff, Keith Roberts, and James D. Watson. Reproduced by permission of Garland Science/Taylor & Francis Books, Inc.)

Cilium

Plasma membrane

Inner sheath

Microtubule doublet

Central singlet microtubule

Dynein arms

Plasma membrane

Microtubule doublet

Dynein arms

Plasma membrane

(a)

Photo: © David M. Phillips/Visuals Unlimited

(b)

being disassembled and transferred to the front of the line. Thus the filament does not get any longer; it stays the same length but moves forward through the continuous transfer of actin molecules from the rear to the front of the filament in what is termed *treadmilling* fashion. The cell progressively moves forward through repeated cycles of pseudopod formation at the leading edge.

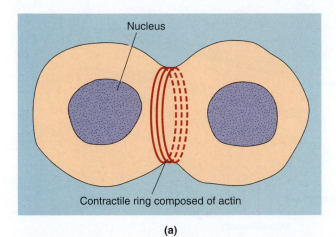

Nucleus

Contractile ring composed of actin

(a)

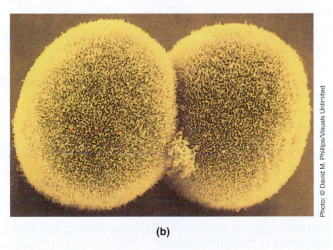

(b)

Figure 2–28 ● **Cytokinesis.** (a) Schematic illustration of the actin contractile ring squeezing apart the two duplicate cell halves during cytokinesis. (b) Photograph of a cell undergoing cytokinesis.

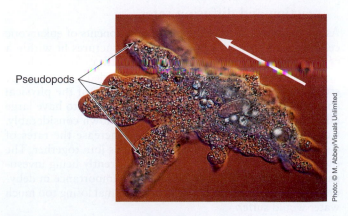

Pseudopods

Figure 2–29 ● **An amoeba moving.**

White blood cells and amoebocytes are the most active crawlers in vertebrates and most invertebrates, respectively. These cells travel by amoeboid movement to areas of infection or inflammation, where they engulf and destroy microorganisms and cellular debris. Fibroblasts ("fiber formers"), another type of motile cell, move amoeboid fashion into a wound from adjacent connective tissue to help repair the damage and are responsible for scar formation.

Microfilaments as Mechanical Stiffeners

Besides their role in cell contractile systems, the actin filaments' second major function is to serve as mechanical supports or stiffeners for several cellular extensions, of which the most common are microvilli. Microvilli are microscopic, nonmotile, hairlike projections from the surface of many epithelial cells linings, such as the vertebrate small intestine and kidney tubules. A single small-intestinal cell may have several thousand microvilli, which are packed together like the bristles of a brush, projecting from the cell's free surface (● Figure 2–30). This bristly appearance has given these microvilli the alternative name of *brush border*. Their presence greatly increases the surface area available for transferring material across the plasma membrane, such as for absorbing digested nutrients in an intestine. Within each microvillus, a core consisting of parallel actin filaments linked together forms a rigid mechanical stiffener that keeps these valuable surface projections intact.

The hair cells of the mammalian inner ear show a remarkable specialization of microvilli. In the part of the inner ear that governs hearing, the actin-stiffened projections on the surface of the hair cells are exquisitely sensitive to vibrations produced by incoming sound. In the part of the inner ear that plays a key role in equilibrium and balance, the specialized microvilli respond to changes in head movement and position (see Chapter 6).

Both microtubules and microfilaments form both stable structures, such as microvilli, and transient structures, such as mitotic spindles and contractile rings, as the need arises. Pools of unassembled tubulin and actin subunits in the cytosol can be rapidly assembled into organized structures to

Microvilli

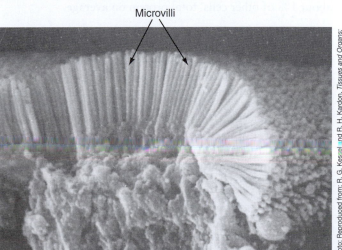

Figure 2–30 ● Scanning electron micrograph of intestinal microvilli.

perform specific activities and then can be disassembled when they are no longer needed.

■ Intermediate filaments are important in regions of the cell subject to mechanical stress.

The **intermediate filaments** are intermediate in size between the microtubules and the microfilaments (7 to 11 nm in diameter)—hence their name. The proteins that compose the intermediate filaments vary between cell types, but in general they appear as irregular, threadlike molecules. These proteins form tough, durable fibers that play a central role in maintaining the structural integrity of a cell and in resisting mechanical stresses externally applied to a cell. In contrast to the other cytoskeletal elements, the intermediate filaments are highly stable structures. No evidence exists for a reversible pool between unassembled and assembled intermediate-filament proteins. The different types of intermediate filaments are tailored to suit their structural or tension-bearing role in specific cell types. In general, only one class of intermediate filament is found in a particular cell type. Several important examples follow:

- Neurofilaments are intermediate filaments in nerve cell axons. Together with microtubules, neurofilaments strengthen and stabilize these elongated cellular extensions.

- Intermediate filaments in skeletal muscle cells hold the actin–myosin contractile units in proper alignment.

- Skin cells contain irregular networks of intermediate filaments made of the protein keratin (Figure 2–25c). These intracellular filaments interconnect with extracellular filaments that tie adjacent cells together, creating a continuous filamentous network that extends throughout the skin and gives it strength. When the surface skin cells die, their tough keratin skeletons persist to form a protective waterproof outer layer. Hair and nails are also keratin-based structures.

Emphasizing the importance of intermediate filaments in some specialized cell types, intermediate filaments account for up to 85% of the total protein in nerve cells and keratin-producing skin cells, whereas these filaments constitute only about 1% of other cells' total protein on average.

■ The cytoskeleton functions as an integrated whole and links other parts of the cell together.

With high-voltage electron microscopy, which provides a three-dimensional view of the internal organization of the cell, a meshwork of exceedingly fine, interlinked filaments can be seen extending throughout the cytoplasm and connecting to the inner layer of the plasma membrane. Some cell biologists believe this lattice is an artifact that arises during preparation of the specimen, but others think it constitutes intricate interconnections between the cytoskeletal structures as well as various organelles (● Figure 2–31). Collectively, the cytoskeletal elements and their interconnections support the plasma membrane and are responsible for the particular shape, rigidity, and spatial geometry of each different cell type. This internal framework thus acts as the cell's "skeleton." New studies hint that the cytoskeleton as a whole is not merely a supporting structure that maintains the tensional integrity of the cell but

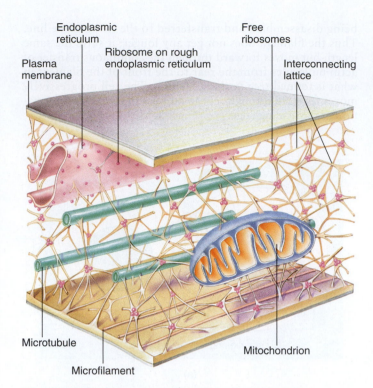

Figure 2–31 ● Interconnections between cytoskeletal structures and organelles.

may serve as a mechanical communications system as well. Various components of the cytoskeleton behave as if they were structurally connected or "hard-wired" to each other as well as to the surface plasma membrane and to the nucleus. Researchers speculate that this force-carrying network serves as a mechanism for mechanical forces acting on the cell surface to reach all the way from the plasma membrane through the cytoskeleton to ultimately influence gene regulation in the nucleus. Furthermore, as you have learned, the coordinated action of the cytoskeletal elements directs intracellular transport and regulates numerous cellular movements and thereby also serves as the cell's "muscle."

■ The cytoskeleton, along with other cellular macromolecules, creates a very crowded environment that alters reaction and diffusion rates.

As we finish this look at the major components of eukaryotic cells, you may wonder how all these structures fit within a cell. In fact, the total number of organelles along with the cytoskeleton do fit, but they create a crowded cytosol. These structures are estimated to take up to 40% of the physical space inside a cell. Such crowding is predicted to have large effects on diffusion rates, slowing them down considerably. Conversely, crowding is predicted to increase the rates of chemical reactions, where two molecules join together. The overall importance of these effects is currently being investigated and is likely to be of even greater importance in dehydration stresses, such as the cells of an animal losing too much water would suffer.

Cell-to-Cell Adhesions

In multicellular organisms, plasma membranes not only serve as the outer boundaries of all cells but also participate in cell-to-cell adhesions, allowing groups of cells to bind together into tissues and to be packaged further into organs. Organization of cells into appropriate groupings may be at least partially attributable to the carbohydrate chains on the membrane surface. Once arranged, cells are held together by three different means: (1) cell adhesion molecules (CAMs) in the cells' plasma membranes, (2) the extracellular matrix, and (3) specialized cell junctions.

◼ The extracellular matrix serves as the biological "glue."

Many cells within a tissue are not in direct physical contact with neighboring cells. Instead, the extracellular matrix holds them together, an intricate meshwork of fibrous proteins embedded in a watery, gel-like substance composed of complex carbohydrates. The watery gel provides a pathway for diffusion of nutrients, wastes, and other water-soluble traffic between the blood and tissue cells. Interwoven within this gel are three major types of protein fibers: collagen, elastin, and fibronectin.

1. **Collagen** forms cablelike fibers or sheets that provide tensile strength (resistance to longitudinal stress). In scurvy, a condition caused by vitamin C deficiency, these fibers are not properly formed. As a result, the tissues, especially those of the skin and blood vessels, become very fragile. This leads to bleeding in the skin and mucous membranes, which is especially noticeable in the gums.

2. **Elastin** is a rubberlike protein fiber most abundant in tissues that must be capable of easily stretching and then recoiling after the stretching force is removed. It is found, for example, in the lungs, which stretch and recoil as air moves in and out.

3. **Fibronectin** promotes cell adhesion and holds cells in position. Researchers have found reduced amounts of this protein within certain types of cancerous tissue, possibly explaining why cancer cells do not adhere well to each other but tend to break loose and metastasize (spread elsewhere in the body).

Local cells, most commonly **fibroblasts** present in the matrix, secrete the extracellular matrix. Often the matrix and the cells within it are known collectively as *connective tissue* because they connect cells together into tissues and tissues into organs. The exact composition of extracellular matrix components varies for different tissues, providing distinct local environments for the various cell types in the body. In some tissues, the matrix becomes highly specialized to form such structures as cartilage or tendons or, on appropriate calcification, the hardened structures of bones and teeth.

New information suggests that the extracellular matrix is not just a passive scaffolding for cellular attachment but also helps regulate the behavior and functions of the cells with which it interacts. Investigators have shown that cells can function normally only in association with their normal matrix components. The matrix is especially influential in cell growth and differentiation.

◼ Some cells are directly linked together by specialized cell junctions.

In addition to the tissue cohesion provided by CAMs and by the extracellular matrix, some cells are directly linked together by one of three types of specialized cell junctions: (1) desmosomes (adhering junctions), (2) tight junctions (impermeable junctions), or (3) gap junctions (communicating junctions).

Desmosomes

Desmosomes act like "spot welds" that anchor together two closely adjacent but nontouching cells. A desmosome consists of two components: (1) a pair of dense, buttonlike cytoplasmic thickenings known as *plaque* located on the inner surface of each of the two adjacent cells; and (2) strong glycoprotein filaments that extend across the space between the two cells and attach to the plaque on both sides (● Figure 2–32). These intercellular filaments bind adjacent plasma membranes together so they resist being pulled apart. Thus desmosomes are adhering junctions. Desmosomes are distributed widely throughout animal bodies. They are most abundant in tissues that are subject to considerable stretching, such as the epithelial tissue of skin and cardiac muscle tissue of the heart.

Tight Junctions

Tight junctions join sheets of epithelial tissue. Epithelial tissue covers the surface of animal bodies and lines all their internal cavities. Usually these epithelial sheets serve as highly selective barriers between two compartments that have con-

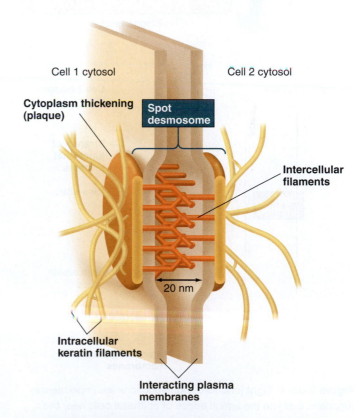

Cell 1 cytosol

Cell 2 cytosol

Cytoplasm thickening (plaque)

Spot desmosome

Intercellular filaments

20 nm

Intracellular keratin filaments

Interacting plasma membranes

Figure 2–32 ● Spot desmosome. Spot desmosomes are adhering junctions that anchor cells together in tissues subject to considerable stretching.

siderably different chemical compositions. For example, the epithelial sheet lining a digestive tract separates the food and potent digestive juices within the inner cavity (lumen) from the ECF that lies on the other side, keeping undigested food particles and digestive juices from moving into the ECF. To accomplish this, the lateral (side) edges of the adjacent cells in the epithelial sheet are joined together in a tight seal near their

luminal border by "kiss sites," sites of direct fusion of proteins on the outer surfaces of the two interacting plasma membranes (● Figure 2–33). These tight junctions are impermeable and thus prevent materials from passing between the cells. Passage across the epithelial barrier, therefore, must take place through the cells, not between them. This transcellular ("through the cell") traffic is regulated by means of the channels and carriers present in the plasma membrane. If the cells were not joined by tight junctions, uncontrolled exchange of molecules could take place between the compartments by unregulated traffic through the spaces between adjacent cells (paracellular pathway). Thus, tight junctions prevent undesirable leaks within epithelial sheets. (See Chapters 12 and 14 for specific details on epithelial transport in kidneys and digestive tracts.)

Gap Junctions

At a **gap junction,** a gap exists between adjacent cells that are linked by small connecting tunnels known as *connexons.* Connexons are formed by the joining of proteins that extend outward from each of the adjacent plasma membranes (● Figure 2–34). The small diameter of the tunnels permits small water-soluble particles such as ions (electrically charged particles) to pass between the connected cells but precludes pas-

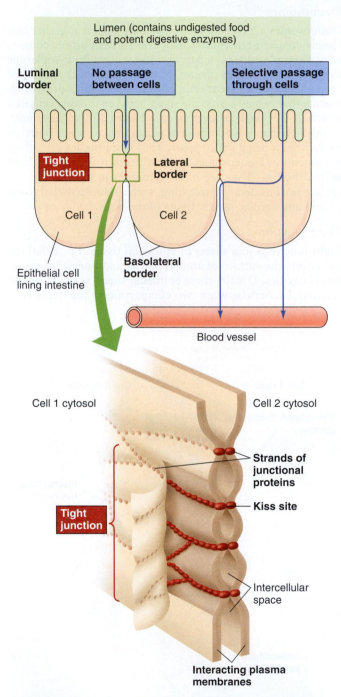

Figure 2–33 ● **Tight junction.** Tight junctions are impermeable junctions that join the lateral edges of epithelial cells near their luminal borders, thus preventing materials from passing *between* the cells. Only regulated passage of materials can occur *through* these cells, which form highly selective barriers that separate two compartments of highly different chemical composition.

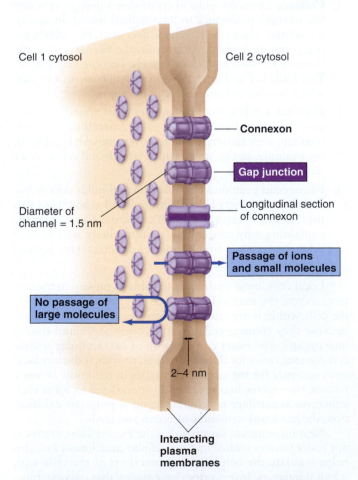

Figure 2–34 ● **Gap junction.** Gap junctions are communicating junctions consisting of connexons, small connecting tunnels that permit movement of charge-carrying ions and small molecules between two adjacent cells.

sage of large molecules such as proteins. Gap junctions are found in heart muscle and smooth muscle. Movement of ions between cells through gap junctions plays an important role in transmitting electrical activity throughout an entire muscle mass. Because this electrical activity brings about contraction, the presence of gap junctions enables synchronized contraction of a whole muscle mass, such as the heart (see Chapter 9). Some nerve cells are also connected in this way (see Chapter 4). For example, crayfish escape attack by rapidly flipping their tails. Gap junctions allow for rapid conduction of the electrical signal (p. 122).

Many tissues that are not electrically active also are connected by gap junctions. These junctions in these nonmuscle tissues link the involved cells together metabolically by permitting unrestricted passage of small, metabolically active molecules between the cells. Furthermore, these gap junctions also serve as avenues for the direct transfer of small signaling molecules from one cell to the next. Such transfer permits the cells connected by gap junctions to directly communicate with each other. This communication provides one possible mechanism by which cooperative cell activity may be coordinated.

As we noted early in this chapter, the cytosol is contained by a surrounding barrier, the plasma membrane. The properties of this major structure are explored in the next chapter, including other ways by which cells "talk to each other."

Chapter in Perspective

HOMEOSTASIS AND INTEGRATION

The ability of cells to perform functions essential for their own survival as well as specialized tasks that contribute to homeostasis and regulated changes within an animal ultimately depends on the successful, cooperative operation of the intracellular components. For example, genes must be regulated at the right time and place to construct cell structures. Also, to support life-sustaining activities, all cells must convert energy into a usable form from nutrient molecules. Energy is converted intracellularly by chemical reactions that take place within the cytosol and mitochondria.

In addition to being essential for basic cell survival, the organelles and cytoskeleton also participate in many cells' specialized tasks that contribute to homeostasis. Here are several examples:

- Nerve and endocrine cells both release chemical messengers that are important in regulatory activities aimed at maintaining homeostasis—for example, chemical messengers released from nerve cells stimulate the respiratory muscles, which accomplish life-sustaining exchanges of O_2 and CO_2 between the body and environment through breathing. These chemical messengers (neurotransmitters in nerve cells and hormones in endocrine cells) are all produced by the endoplasmic reticulum and Golgi complex and released by exocytosis from the cell when needed.

- The ability of muscle cells to contract depends on their highly developed cytoskeletal microfilaments sliding past each other. Muscle contraction is responsible for many homeostatic activities, including (1) contraction of heart muscle, which pumps life-supporting fluids throughout most animal bodies; (2) contraction of the muscles attached to skeletal elements, which enables behaviors, such as procuring food, fleeing predators, and mating; and (3) contraction of the muscle in the walls of digestive tracts, which moves the food along the tracts so that ingested nutrients can be progressively broken down into a form that can be absorbed for delivery to the cells. Muscles are also critical to behaviors, which often serve as corrective effectors.

- White blood cells and amoebocytes help animal bodies resist infection by making extensive use of lysosomal destruction of engulfed particles as they police the body for microbial invaders. these cells are able to roam the body fluids by means of amoeboid movement, a cell-crawling process accomplished by alternate assembly and disassembly of actin, one of their cytoskeletal components.

As we begin to examine the various organs and systems, keep in mind that proper cell functioning is the foundation of all organ activities. ∎

REVIEW QUESTIONS *(Answers are on p. A–1.)*

Additional study tools for this chapter, including chapter summaries and practice tests, are available online at *www.biology.brookscole.com*

1. A kinase is an enzyme that
   a. removes triglycerides from cells
   b. adds phosphate groups to proteins
   c. removes phosphate groups from DNA
   d. is an allosteric modulator
   e. is also called a phosphatase

2. Which of the following are found in both prokaryotic and eukaryotic cells?
   a. plasma membranes
   b. organelles
   c. chromosomes
   d. nuclei
   e. a and c

3. The flow of genetic information in cells goes from
   a. histones to DNA to RNA
   b. protein to RNA to DNA

c. DNA to RNA to protein

d. RNA to DNA to protein

e. chromosomes to protein to DNA

4. The process of transcription is
   a. copying tRNA to ribosomes
   b. copying mRNA to form tRNA
   c. converting the mRNA code into protein
   d. copying DNA codes into RNA messages
   e. splicing exons

5. For multicellular organisms which of the following statements is false?
   a. Each cell in a multicellular organism usually has identical DNA sequences.
   b. All genes must be constantly expressed, even in different types of cells.
   c. Genes are differentially expressed in different types of cells.
   d. A liver cell makes exactly the same proteins as a muscle cell.
   e. b and d

6. Which of the following techniques uses "gene chips" and cDNA to locate active genes?
   a. immunofluorescence
   b. knock-out genes
   c. DNA microarrays
   d. databases
   e. cloning

7. Ribosomes are "factories" for synthesizing
   a. triglycerides
   b. proteins
   c. DNA
   d. carbohydrates
   e. genes

8. Rough endoplasmic reticulum differs from smooth endoplasmic reticulum in that it is studded with
   a. vesicles
   b. histones
   c. peroxisomes
   d. ribosomes
   e. mitochondria

9. The majority of newly synthesized molecules contained in vesicles that bud off from the ER move to the
   a. nucleus
   b. plasma membrane
   c. mitochondria
   d. Golgi complex
   e. ribosomes

10. The Golgi complex
    a. processes cellular raw materials into finished products
    b. moves completed proteins to the nucleus
    c. transports proteins to the ER
    d. sorts and directs finished products to their final destinations
    e. a and d above

11. The Golgi complex packages secretory vesicles for release by
    a. lysosomes
    b. exocytosis
    c. proteasomes
    d. endocytosis
    e. peroxisomes

12. Lysosomes and proteasomes serve as
    a. vesicles to export cellular contents
    b. vesicles to transport molecules from cell to cell
    c. intracellular digestive systems
    d. transporters of DNA from the cytosol to the nucleus
    e. the sites of translation of the genetic code into proteins

13. Which of the following describes a means of bringing material into cells?
    a. endocytosis
    b. pinocytosis
    c. receptor-mediated endocytosis
    d. phagocytosis
    e. all of the above

14. Which of the following organelles is called the "energy organelle"?
    a. endoplasmic reticulum
    b. Golgi complex
    c. nucleus
    d. peroxisome
    e. mitochondrion

15. The Krebs cycle takes place in
    a. the Golgi complex
    b. the nucleus
    c. the endoplasmic reticulum
    d. the lysosomes
    e. the mitochondria

16. The final electron acceptor in the respiratory chain is
    a. ATP
    b. glucose
    c. oxygen
    d. phosphate
    e. ATP synthase

17. Gluconeogenesis is essentially the reverse of
    a. the Krebs cycle
    b. glycolysis
    c. electron transport
    d. the tricarboxylic acid cycle
    e. chemiosmosis

18. The cytosol is important in which of the following?
    a. intermediary metabolism
    b. ribosomal protein synthesis
    c. storage of fat
    d. storage of glycogen
    e. all of the above

19. Mitosis involves
    a. cytoplasmic division only
    b. cytoplasmic and mitotic division
    c. amoeboid movement
    d. cytokinesis
    e. nuclear division only

20. Which of the following is not a type of cell junction?
    a. desmosome
    b. tight
    c. fibronectic
    d. gap
    e. a and b

SUGGESTED READINGS AND INTERNET SITES

Alberts, B., A. Johnson, J. Lewis, M. Raff, K. Roberts, & P. Walter. 2002. *Molecular Biology of the Cell.* New York: Garland. A best-selling textbook on this topic.

Dhand, R., ed. 2000. Nature insight: Functional genomics. *Nature* 405:819–865. Several articles on the state of genomics as of 2000.

Ellis, R. J., & A. P. Minton. 2003. Join the crowd. *Nature* 425: 27–28. A science news story on crowding in the cell cytosol.

Mattick, J. S. 2003. Challenging the dogma: The hidden layer of non–protein-coding RNAs in complex organisms. *BioEssays* 25:930–939. An update on non-coding RNAs.

Moyes, C. D., & D. A. Hood 2003. Origins and consequences of mitochondrial variation in vertebrate muscle. *Annual Review of Physiology* 65:177–201.

Stryer, L. 2001. *Biochemistry.* New York: W. H. Freeman. A best-selling textbook on this topic.

Wells, W. A. 2001. Containing the bullet. *Journal of Cell Biology* 154:13. Available online at www.jcb.org/cgi/content/full/154/1/13. Accessed on February 2, 2004. A review of nematocyst physiology.

Wilmut, I. 1998. Cloning for medicine. *Scientific American* 279:59–63.

INFOTRAC READINGS

Goldberg, A. L. 1995 . Functions of the proteasome: The lysis at the end of the tunnel. *Science* 268:522–523.

INTERNET SITES

sciencegenomics.org. Information on current genomics research.

nature.com/genomics. Information on current genomics research.

Rome, L. **vaults.arc.ucla.edu/sci/sci_home.htm.** Discusses history and current research on vaults.

animalbiotechnology.org/. Current news in animal genetics.

Membrane Physiology

An arctic fish, Thymallus arcticus. *The membranes of this animal have a high proportion of unsaturated lipids, allowing the membrane to remain flexible in the cold.*

Membrane Structure and Composition

The survival of every animal cell depends on the maintenance of an intracellular content unique for that cell type despite the remarkably different composition of the extracellular fluid surrounding it. This difference in fluid composition inside and outside a cell is maintained by the **plasma membrane,** an extremely thin layer of lipids and proteins that forms the outer boundary of every cell and encloses the intracellular contents. In addition to serving as a mechanical barrier that traps needed molecules within the cell, the plasma membrane plays an active role in determining the composition of the cell by selectively permitting specific substances to pass between the cell and its environment. Besides controlling the entry of nutrient molecules and the exit of waste products, the plasma membrane maintains differences in ion concentrations between the cell's interior and exterior. These ionic differences, as you will learn, are important in the electrical activity of cells. Also, the plasma membrane plays a key role in the ability of a cell to respond to changes, or signals, in the cell's environment. No matter what the cell type, these common functions accomplished by the plasma membrane are crucial to the cell's survival and to the cell's ability to perform specialized homeostatic activities.

In this chapter, we examine the common structural and functional patterns shared by plasma membranes of all cells. We also explore how many of the functional differences between cell types are due to subtle variations in the composition of their plasma membranes. For example, slight modifications in specific protein components of the plasma membranes of various cell types enable different cells to interact in different ways with essentially the same extracellular fluid environment. To illustrate, thyroid gland cells are the only cells in a verte-

brate to use iodine. Appropriately, a membrane protein unique to the plasma membranes of thyroid gland cells permits these cells (and no others) to take up iodine from the blood.

■ The plasma membrane is a fluid lipid bilayer embedded with proteins.

The plasma membrane is too thin to be seen under an ordinary light microscope, but with an electron microscope it appears as a **trilaminar** (three-layered) **structure** consisting of two dark layers separated by a light middle layer (● Figure 3–1). Scientists believe the specific arrangement of the molecules that make up the plasma membrane is responsible for this three-layered "sandwich" appearance.

All plasma membranes consist mostly of lipids (fats) and proteins plus small amounts of carbohydrates. The most abundant membrane lipids are phospholipids, with lesser amounts of cholesterol. **Phospholipids** have a *polar* (see Appendix Figure B–10) head containing a negatively charged phosphate group and two *nonpolar* (electrically neutral) fatty acid tails (● Figure 3–2a). The polar end is *hydrophilic* ("water loving") because it can interact with water molecules, which are also polar; the nonpolar end is *hydrophobic* ("water fearing") and will not mix with water. Such two-sided molecules self-assemble into a **lipid bilayer,** a double layer of lipid molecules, when in contact with water (Figure 3–2b). The hydrophobic tails bury themselves in the center away from the water, whereas the hydrophilic heads line up on both sides in contact with the water. The outer surface of the bilayer is exposed to extracellular fluid (ECF), whereas the inner surface is in contact with the intracellular fluid (ICF).

This lipid bilayer is not a rigid structure at normal body temperatures but instead is fluid, with a consistency more like liquid cooking oil than like solid shortening. The phospholipids,

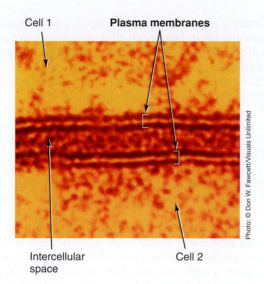

Cell 1 — Plasma membranes

Intercellular space — Cell 2

Photo: © Don W. Fawcett/Visuals Unlimited

Figure 3–1 ● **Trilaminar appearance of a plasma membrane in an electron micrograph.** Depicted are the plasma membranes of two adjacent cells. Note the trilaminar structure (that is, two dark layers separated by a light middle layer) of each membrane.

which are not held together by chemical bonds, can twirl around rapidly as well as move about within their own half of the layer, much like skaters on a crowded skating rink. A certain degree of fluidity is required, for example, for the proper action of transport and channel proteins (p. 71), and for cell shape changes. Conversely, an overly fluid membrane is too leaky and can harm or kill cells.

Also contributing to the fluidity as well as the stability of the membrane is *cholesterol*. By being tucked in between the phospholipid molecules, cholesterol molecules prevent the fatty acid chains from packing together and crystallizing, a process that would drastically reduce membrane fluidity. Cholesterol affects the functional attributes of cell membranes such as the activities of various membrane proteins. In animals exposed to varied environmental temperatures, cholesterol provides some rigidity to the membrane and can counter some of the temperature-induced perturbations in membrane fluidity (which increases to detrimental levels at high temperatures). For example, studies have shown that the concentrations of cholesterol in the plasma membranes of rainbow trout (*Oncorhynchus mykiss*) increase with increasing environmental temperature. Cholesterol also reduces the permeability of the gill membrane to water and the flow of ions. For this reason the concentration of cholesterol is approximately twice as great in the gill plasma membranes as in other tissues.

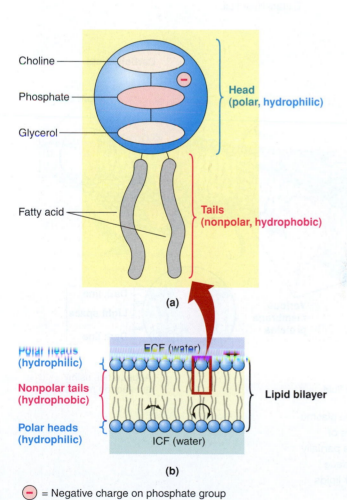

Choline

Phosphate

Glycerol

Fatty acid

Head (polar, hydrophilic)

Tails (nonpolar, hydrophobic)

(a)

Polar heads (hydrophilic)

Nonpolar tails (hydrophobic)

Polar heads (hydrophilic)

ECF (water)

ICF (water)

Lipid bilayer

(b)

⊖ = Negative charge on phosphate group

Lipid bilayer

Intracellular fluid

Extracellular fluid

(c)

Figure 3–2 ● **Structure and organization of phospholipid molecules in a lipid bilayer.** (a) Phospholipid molecule. (b) When in contact with water, phospholipid molecules organize themselves into a lipid bilayer with the polar heads interacting with the polar water molecules at each surface and the nonpolar tails all facing the interior of the bilayer. (c) An exaggerated view of the plasma membrane enclosing a cell, separating the ICF from the ECF.

(*Source:* Adapted from Cecie Starr and Ralph Taggart, *Biology: The Unity and Diversity of Life,* Eighth Edition, Fig. 4-26, p. 56. © 1998 Wadsworth Publishing Company.)

Also contributing to membrane flexibility are differences in membrane composition. For example, fish, as well as other poikilotherms (organisms whose body temperatures vary with the environment; Chapter 15), can alter the composition of their lipids to synthesize membranes with physical properties that let them adapt and function at low ambient temperatures. Cold temperatures cause the membrane lipids to become more rigid (like butter placed in a refrigerator). Membrane lipids of animals adapted to cold environments are enriched in **polyunsaturated fatty acids (PUFAs)** at the expense of a reduction in **saturated fatty acids (SFAs).** Addition of double bonds (points of unsaturation) into the fatty acid molecule offsets the effects of low temperatures on membrane fluidity; that is, the chains do not stack together well and so do not become too rigid in the cold (see Chapter 15 for more details, p. 684). There are two distinctly different groups of PUFAs: the omega-6 series, in which the first double bond is found between the sixth and seventh carbon atoms in the fatty acid chain, and the omega-3 series, in which the first double bond is located between the third and fourth carbon atoms. For every 20-degree C decrease in ambient temperature there is a 1.3- to 1.4-fold increase in the ratio of UFAs to SFAs. Characteristics of lipids containing PUFAs include a reduction in membrane order because of their more expanded conforma-tion. Under frigid conditions these lipids pack less efficiently. For example, brain fatty acids of Antarctic fish are enriched in 24-carbon PUFAs. The proportion of PUFAs in membranes of cold-adapted marine fish may be as high as 42%, compared to 33% in species inhabiting more temperate waters. (See Chapter 15 for more on temperature adaptation.)

The most prominent long-chain omega-3 PUFAs in fish oils are **eicosapentaenoic acid (EPA),** which is 22 carbons long, with five double bonds (20:5, omega 3) and **docosahexaenoic acid (DHA),** which is 22 carbons long, with six double bonds (22:6, omega-3). The highest amounts of EPA and DHA are in fatty fish such as mackerel, herring, and salmon; white fish such as cod and flounder contain only small amounts of omega-3 PUFAs.

Attached to or inserted within the lipid bilayer are the **membrane proteins** (● Figure 3–3). Some of these proteins extend through the entire thickness of the membrane; they have polar regions at both ends joined by a nonpolar central portion. Other proteins stud only the outer or inner surface; they are anchored by interactions with a protein that spans the membrane or by attachment to the lipid bilayer. The fluidity of the lipid bilayer enables many membrane proteins to float freely like "icebergs" in a moving "sea" of lipid, although the mobility of proteins that perform a specialized function in a specific

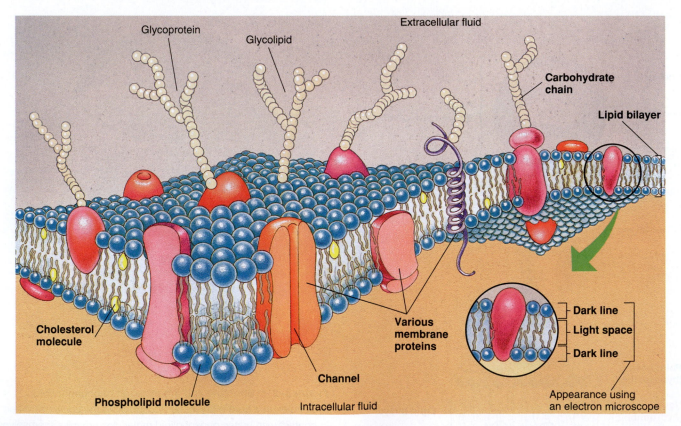

Figure 3–3 ● Fluid mosaic model of plasma membrane structure. The plasma membrane is composed of a lipid bilayer embedded with proteins. Some of these proteins extend through the thickness of the membrane, some are partially submerged in the membrane, and others are loosely attached to the surface of the membrane. Short carbohydrate chains are attached to proteins or lipids on the outer surface only.

area of the cell is restricted. This view of membrane structure is known as the **fluid mosaic model,** in reference to the membrane fluidity and the ever-changing mosaic pattern of the proteins embedded within the lipid bilayer.

The small amount of **membrane carbohydrate** is located only at the outer surface. Short-chain carbohydrates, which protrude from the outer surface, are bound primarily to membrane proteins and to a lesser extent to lipids, forming **glycoproteins** and **glycolipids,** respectively (Figure 3–3).

The different components of the plasma membrane carry out a variety of functions. As a quick preview, the lipid bilayer forms the primary barrier to diffusion, the proteins perform the most of the specific membrane functions, and the carbohydrates play an important role in "self"-recognition processes.

■ The mosaic model and membrane-skeleton fence model describes membrane structure and function.

The fluid mosaic model can account for the trilaminar appearance of the plasma membrane. When researchers use stains to help them visualize the plasma membrane under an electron microscope, the two dark lines represent the hydrophilic polar regions of the lipid and protein molecules that have taken up the stain. The light space between corresponds to the poorly stained hydrophobic core formed by the nonpolar regions of these molecules.

Even though the outer and inner layers have the same appearance when viewed with an electron microscope, our description makes clear that the plasma membrane actually is asymmetric; that is, the surface of the membrane facing the extracellular fluid is strikingly different from the surface facing the cytoplasm. Carbohydrate is located only on the outer surface; the outer and inner surfaces bear different amounts and types of protein; and even the lipid composition of the outer half of the bilayer varies somewhat from the inner half. This distinct sidedness of the membrane is related to the different functions carried out at the outer and inner surfaces.

Furthermore, considerable heterogeneity exists within each half of the membrane. For example, recent information suggests that not all membrane proteins are as freely mobile and uniformly distributed as proposed by the fluid mosaic model. To explain this phenomenon, researchers proposed the **membrane-skeleton fence model.** According to this model, the mobility of proteins that perform a specialized function in a specific area of the membrane is restricted by several means. Whereas some proteins can drift randomly within the membrane (similar to a dog who freely roams the neighborhood), scientists think other proteins are tethered to the cytoskeleton (see p. 59) (much like a dog restricted by a leash). These so-called *non-diffusing* or *immobile* proteins constitute a sizable proportion of the total membrane fraction. Still other proteins appear restricted to **confinement zones** by fine cytoskeletal *meshwork fences* within the membrane (similar to a dog confined by a fence to its own backyard). Sometimes these barriers to mobility temporarily open so that proteins can move between confinement zones. Finally, a few proteins undergo rapid, highly directed transport toward a specific region of the membrane, perhaps being carried by motor proteins similar to kinesin (see p. 58). (This is analogous to a dog being carried by its owner

to a particular destination.) Thus the emerging view is that the plasma membrane is a highly complex, dynamic, regionally differentiated structure.

■ The lipid bilayer forms the basic structural barrier that encloses the cell.

The lipid bilayer serves at least three important functions:

1. It forms the basic structure of the membrane. (The phospholipids can be visualized as the "pickets" that form the "fence" around the cell.)
2. Its hydrophobic interior serves as a barrier to passage of water-soluble substances between the ICF and ECF. Most water-soluble substances cannot dissolve in and pass through the lipid bilayer. (However, water molecules themselves and some other small uncharged molecules such as CO_2 are small enough to pass between the molecules that form this barrier.)
3. It is largely responsible for the fluidity of the membrane.

■ The membrane proteins perform a variety of specific membrane functions.

A variety of different proteins within the plasma membrane serve the following specialized functions:

1. *Channels.* Some proteins that span the membrane form water-filled pathways, or channels, across the lipid bilayer. Their presence enables water-soluble ions that are small enough to enter a channel, to pass through the membrane without coming into direct contact with the hydrophobic lipid interior (Figure 3–3). The channels are highly selective. Not only does their small diameter preclude passage of particles greater than 0.8 nanometer (nm) in diameter (1 nm = 1 billionth of a meter or 40 billionths of an inch), but a given channel can also selectively attract or repel particular ions. For example, sodium (Na^+) channels can accommodate the passage of only Na^+, whereas only K^+ can pass through potassium (K^+) channels. This selectivity is due to size and specific arrangements of various chemical groups on the interior surfaces of the proteins that form the channel walls. Cells vary in the number, kind, and activity of channels they have. It is even possible for a given channel to be *open* or *closed* to its specific ion as a result of changes in channel shape in response to a controlling mechanism. Channels with this property are called **gated channels.**

2. *Carriers.* Another group of proteins that span the membrane serve as **carrier proteins,** which transfer specific substances across the membrane that cannot cross on their own. Thus channels and carrier molecules are both important in the transport of substances between the ECF and ICF. Each carrier can transport only a specific molecule or closely related molecules. Variation in the kinds of carriers different cells have lets them selectively transport different substances across their membranes. For example, the plasma membranes of vertebrate thyroid gland cells uniquely have carriers for iodide, enabling this essential element to be transported from the blood into thyroid gland cells, a capability other body cells lack.

3. *Receptors.* Many proteins on the outer surface serve as **receptor sites** or **receptors** that "recognize" and bind with specific molecules in the environment of the cell. Receptors are proteins with a binding site for the specific molecule. The binding initiates a series of membrane and intracellular events that alter the activity of the particular cell. (The postreceptor pathways involved in altering cell function are discussed in a later section.) In this way, chemical messengers in the blood, such as hormones, can influence only the specific cells that have receptors for the messenger while having no effect on other cells, even though every cell is exposed to the same messenger via its widespread distribution by the blood. To illustrate, the anterior pituitary gland secretes into the blood thyroid-stimulating hormone (TSH), which can attach to the surface of thyroid gland cells to stimulate secretion of thyroid hormone. Receptor sites for TSH are also found in a number of locations, including the heart and gonads, although the function of these sites remains unclear.

4. *Docking-marker acceptors.* Still other proteins on the inner surface are t-SNAREs (p. 40), which serve as **docking-marker acceptors** that bind lock-and-key fashion with the v-SNAREs of secretory vesicles (see p. 41). Secretion is initiated as stimulatory signals trigger the fusion of the secretory vesicle membrane with the inner surface of the plasma membrane through interactions between their SNARE proteins. The secretory vesicle subsequently opens up and empties its contents to the outside by exocytosis.

5. *Enzymes.* Another group of proteins function as **membrane-bound enzymes** that control specific chemical reactions at either the inner or the outer cell surface. Cells display specialization in the types of enzymes embedded within their plasma membranes. For example, the outer layer of the plasma membrane of skeletal muscle cells contains an enzyme that destroys the neurotransmitter that triggers muscle contraction, enabling the muscle to relax.

6. *CAMs.* Still other proteins serve as **cell adhesion molecules (CAMs)**. Many CAMs protrude from the outer membrane surface and form loops or other appendages that the cells use to grip each other and to grasp the connective tissue fibers that interlace between cells. For example, on the surface of adjacent cells *cadherins*, one type of CAM, interdigitate in zipper fashion to help hold the cells within tissues and organs together. Some CAMs such as the *integrins* span the plasma membrane. Integrins not only serve as a mechanical link between the outer membrane and its extracellular surroundings but also connect the inner membrane surface to the intracellular cytoskeletal scaffolding. These CAMs mechanically link the cell's external environment and intracellular components. Furthermore, integrins can relay regulatory signals through the plasma membrane in either direction. Investigators have now learned that some CAMs also act as "signaling molecules" in cell growth and in inflammatory responses and wound healing, among other things.

7. *Self-identity markers.* Finally, still other proteins on the outer membrane surface, especially in conjunction with carbohydrates, are important in cells' ability to recognize "self" (that is, cells of the same type) and in cell-to-cell interactions, as described next.

■ The membrane carbohydrates serve as self-identity markers.

The short sugar chains on the outer-membrane surface serve as self-identity markers that enable cells to identify and interact with each other in the following ways.

1. Different cell types have different markers. The unique combination of sugar chains projecting from the surface membrane serves as the "trademark" of a particular cell type, enabling a cell to recognize others of its own kind. Thus these carbohydrate chains play an important role in recognition of "self" and in cell-to-cell interactions. Cells can recognize other cells of the same type and join together to form tissues. This is especially important during embryonic development, where, for example, developing muscle cells must join together to form muscle organs. "Self"-recognition is also crucial in all animal immune responses (Chapter 10).

2. Carbohydrate-containing surface markers also appear involved in tissue growth, which is normally held within certain limits of cell density. Cells do not "trespass" across the boundaries of neighboring tissues; that is, they do not overgrow their own territory. Abnormal surface carbohydrate markers have been identified in certain tumor cells, suggesting that this abnormality may underlie the uncontrolled growth of tumor cells.

Membrane Transport

Anything that passes between a cell and the surrounding extracellular fluid must be able to penetrate the plasma membrane. If a substance can cross the membrane, the membrane is said to be **permeable** to that substance; if a substance cannot pass, the membrane is **impermeable** to it. The plasma membrane is **selectively permeable** in that it lets some particles pass through while excluding others. In some instances permeability can be controlled, as exemplified by the opening or closing of specific channels.

■ Lipid-soluble substances and small polar molecules can passively diffuse through the plasma membrane down their electrochemical gradient.

Two properties of particles influence whether they can permeate the plasma membrane without any assistance: (1) the relative solubility of the particle in lipid and (2) the size of the particle. Highly lipid-soluble particles can dissolve in the lipid bilayer and pass through the membrane. Uncharged or nonpolar molecules (such as O_2, CO_2, alcohol, steroids, and fatty acids) are very lipid-soluble and readily permeate the membrane. Charged particles (ions such as Na^+ and K^+) and polar molecules (such as glucose and proteins) have low lipid solubility but are very soluble in water. The lipid bilayer serves as an impermeable barrier to particles poorly soluble in lipid. For water-soluble ions less than 0.8 nm in diameter, the protein channels serve as an alternate route for passage across the membrane. Only ions for which specific channels are available can permeate the membrane. Particles that have low lipid solubility and are too large for channels cannot permeate the membrane on their own.

Even though a particle might be capable of permeating the membrane by virtue of its lipid solubility or its ability to fit through a channel, some force is needed to produce its movement across the membrane (a particle does not just decide it wants to be on the other side). Two general types of forces are involved: (1) forces that do not require the cell to expend energy to produce movement (**passive forces**) and (2) forces requiring cellular energy (ATP) to be expended to transport a substance across the membrane (**active forces**). Let's now examine the various methods of membrane transport, indicating whether each uses a passive or active mechanism.

■ Diffusion follows a concentration gradient.

All molecules are in continuous random motion at temperatures above absolute zero as a result of heat (thermal) energy. This motion is most evident in liquids and gases, where the individual molecules have more room to move before colliding with another molecule. Each molecule moves separately and randomly in any direction. Because of this Brownian motion, the molecules frequently collide, bouncing off each other in different directions like billiard balls striking each other. **Diffusion** (from *diffusus*, spread out) is ultimately directional, not random. For example, the greater the molecular concentration of a substance in a solution, the greater the likelihood of collisions. For example, in ● Figure 3–4a, the concentration differs between area A and area B in a solution. Such a difference in concentration between two adjacent areas is called a **concentration gradient** (or **chemical gradient**). Random molecular collisions occur more frequently in area A because of its greater concentration of molecules. As a result, more molecules bounce from area A into area B than in the opposite direction. In both areas, the individual molecules move randomly and in all directions, but the net movement of molecules by diffusion is from the area of higher concentration to the area of lower concentration.

Net diffusion is the difference between two opposing movements. If 10 molecules move from A to B while 2 molecules simultaneously move from B to A, the net diffusion is 8 mol-

ecules moving from A to B. Molecules spread in this way until the substance is uniformly distributed between the two areas and a concentration gradient no longer exists (Figure 3–4b). At this point, even though movement is still taking place, no *net* diffusion is occurring, because the opposing movements exactly counterbalance each other, that is, are in equilibrium. Movement of molecules from A to B is exactly matched by movement of molecules from B to A. This situation is known as a **steady state** or **equilibrium.**

What happens if a plasma membrane separates different concentrations of a substance? If the substance can permeate the membrane, net diffusion of the substance occurs through the membrane down its concentration gradient from the area of high concentration to the area of low concentration (● Figure 3–5a). No energy is required for this movement, so it is a **passive** mechanism of membrane transport. For example, O_2 is transferred across the lung membrane by this means. The blood carried to the lungs is low in O_2, having given up O_2 to the body tissues for cellular metabolism. The air in the lungs, in contrast, is high in O_2 because it is continuously exchanged with fresh air by the process of breathing. Because of this concentration gradient, net diffusion of O_2 occurs from the lungs into the blood as blood flows through the lungs. Thus, as blood leaves the lungs for delivery to the tissues, it is high in O_2.

Several factors in addition to the concentration gradient influence the rate of net diffusion across a membrane (**Fick's law of diffusion**) (■ Table 3–1) and can be quantified as follows.

1. *The magnitude (or steepness) of the concentration gradient.* The greater the differences in concentration, the greater the rate of diffusion. For example, let's follow CO_2 in a running lizard. The working muscles produce CO_2 more rapidly as a result of burning additional fuel to produce the extra ATP they need to power the stepped-up, energy-demanding contractile activity. The resultant increase in CO_2 level in the muscles creates a greater-than-normal CO_2 difference between the muscles and the blood. Because of this larger gradient, more CO_2 than usual enters the blood. As this blood reaches the lizard's lungs, a greater-

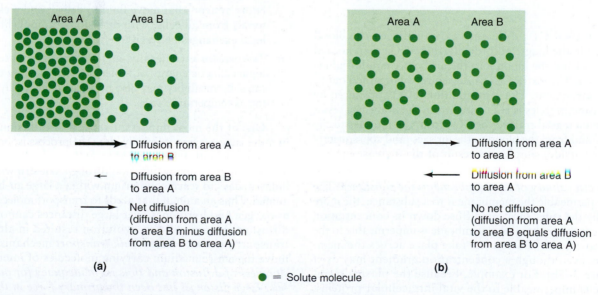

(a)

→ Diffusion from area A to area B

← Diffusion from area B to area A

→ Net diffusion (diffusion from area A to area B minus diffusion from area B to area A)

(b)

→ Diffusion from area A to area B

← Diffusion from area B to area A

No net diffusion (diffusion from area A to area B equals diffusion from area B to area A)

● = Solute molecule

Figure 3–4 ● **Diffusion.** (a) Diffusion down a concentration gradient. (b) Steady state, with no net diffusion occurring.

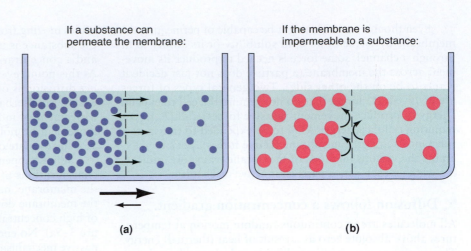

Figure 3–5 ● Diffusion through a membrane. (a) Net diffusion across the membrane down a concentration gradient. (b) No diffusion through the membrane despite the presence of a concentration gradient.

If a substance can permeate the membrane:

If the membrane is impermeable to a substance:

(a)

(b)

Table 3–1 ▮ Factors Influencing the Rate of Net Diffusion of a Substance across a Membrane (Fick's Law of Diffusion)

Factor	Effect on Rate of Net Diffusion
↑ Concentration gradient of substance (ΔC)	↑
↑ Permeability of membrane to substance (P)	↑
↑ Surface area of membrane (A)	↑
↑ Molecular weight of substance (MW)	↓
↑ Distance (thickness) (ΔX)	↓

Modified Fick's equation:

$$\text{Net rate of diffusion } (Q) = \frac{\Delta C \cdot P \cdot A}{MW \cdot \Delta X}$$

$$\left[\frac{P}{\sqrt{MW}} = \text{diffusion coefficient (D)} \right]$$

$$\text{Restated: } Q \propto \frac{\Delta C \cdot A \cdot D}{\Delta X}$$

than-normal CO_2 difference exists between the blood and air in the lungs. Accordingly, more CO_2 than normal diffuses from the blood into the air before equilibrium is achieved. This extra CO_2 is subsequently breathed out to the environment. Thus any additional CO_2 produced by exercising muscles is eliminated from the body as a result of the increased transfer of CO_2 from the muscles to the blood and from the blood to the air sacs (and subsequently to the outside) simply as a result of the increase in CO_2 concentration gradient.

2. *The permeability of the membrane to the substance.* The more permeable the membrane is to a substance, the more rapidly the substance can diffuse down its concentration gradient. Note that if the membrane is impermeable to the substance, no diffusion can take place across the membrane, even though a concentration gradient may exist (Figure 3–5b). For example, because the plasma membrane is impermeable to the vital intracellular proteins, they cannot escape from the cell, even though they are in much greater concentration in the ICF than in the ECF.

3. *The surface area of the membrane across which diffusion is taking place.* Obviously, the larger the surface area available, the greater the rate of diffusion it can accommodate. For example, absorption of nutrients across the membranes in the small intestine is enhanced by surface foldings and projections (microvilli) that greatly increase the available absorptive surface area.

4. *The molecular weight of the substance.* Lighter molecules such as O_2 and CO_2 bounce farther on collision than do heavier molecules. Consequently, O_2 and CO_2 diffuse rapidly, permitting rapid exchanges of these gases across the respiratory membranes.

5. *The distance through which diffusion must take place.* The greater the diffusion barrier thickness, the slower the rate of diffusion. Accordingly, membranes across which diffusing particles must travel are normally relatively thin, such as the membranes separating the air and blood in vertebrate lungs. Furthermore, diffusion is efficient only for short distances between cells and their surroundings. It becomes an inappropriately slow process for distances of more than a few millimeters. This makes it imperative that all but the smallest and most sluggish animals have circulatory systems that move fluid to deliver and pick up materials from all cells, with diffusion accomplishing short local exchanges between the ECF and cells.

6. *Temperature.* Generally, as temperatures increase, the higher kinetic energy of molecules will increase diffusion rates. Permeability may also be affected, generally increasing as temperature rises.

One of the most important aspects of diffusion to keep in mind is how extraordinarily *slow* the process is, because of its inherent randomness and molecular scale. For example, oxygen moving only by diffusion through a nostril would take hours to days to reach equilibrium within a large air-breathing animal. Thus an animal that needs to transport molecules such as oxygen and hormones over large distances cannot rely on diffusion. As you will see, evolution resulted in alternative transport systems, termed *bulk transport* mechanisms, that move the entire medium carrying molecules of interest. *The slowness of diffusion and thus its inadequacy for movement over large distances has been the primary force in the evolution of circulatory systems* and other bulk transport processes in multicellular animals.

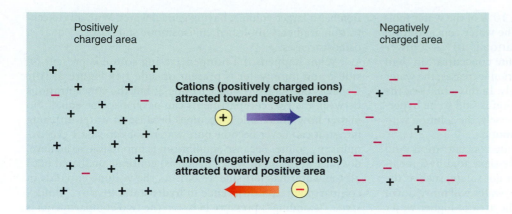

Figure 3–6 • Movement along an electrical gradient.

Conduction along an Electrical Gradient

Movement of **ions** (electrically charged particles that have either lost or gained an electron) is also affected by their electrical charge. Like charges (those with the same kind of charge) repel each other whereas opposite charges attract each other. If a relative difference in charge exists between two adjacent areas (● Figure 3–6), the positively charged ions (**cations**) tend to move toward the more negatively charged area, whereas the negatively charged ions (**anions**) tend to move toward the more positively charged area. A difference in charge between two adjacent areas thus produces an **electrical gradient** that passively induces ion movement—a process called *conduction*. When an electrical gradient exists between the ICF and ECF, only ions that can permeate the plasma membrane can conduct down this gradient. The simultaneous existence of an electrical gradient and concentration (chemical) gradient for a particular ion is called an **electrochemical gradient.** Later in this chapter you will learn how electrochemical gradients contribute to the electrical properties of the plasma membrane.

A common and major mistake is to call ion movement "diffusion," but this is not diffusion when charge differences are involved. Recall that diffusion is based on random molecular motions, and is very slow. Ion conduction, in contrast, is much faster, because of charge interaction, which is a strong directional force (think about how fast two magnets can move toward each other, an analogous situation). If the electrical systems in your house relied on diffusion (rather than conductance) of electrons, it would take many *years* for the electrons to reach your house from a power plant!

■ Osmosis is the net movement of water through a membrane.

Water can readily permeate most plasma membranes. It can pass between the phospholipid molecules and through **aquaporins,** which are channel proteins evolved specifically for the passage of water molecules, found in some cell types such as kidney tubules (p. 556). This simple fact has enormous implications for cell physiology and evolution. At some early point in the evolution of the first cells, there would have been an unequal distribution of molecules: a high concentration of organic molecules on the inside and a low concentration on the outside. Why might such differences be a potential problem for cells? The answer lies with the difference in total **solute** (dissolved substance) concentrations on either side of the mem-

brane, coupled with water's ability to penetrate most membranes. As a membrane-enclosed system, a cell must contend with a powerful force, **osmosis**—the result of movement of water down its own concentration gradient through a membrane (note that diffusion of water in an open system without a membrane is not osmosis, but simply diffusion). Osmosis occurs if the total concentration of solutes is not equal inside and outside the cell. Osmotic imbalances can be inward (leading to cell swelling) or outward (leading to cell shrinkage). Left unchecked, osmotic movement of water can impair cellular function and, if extreme, result in cell destruction.

Osmosis with a Nonpenetrating Solute

In simplest terms, the driving force for diffusion of water across the membrane is the same as for any other diffusing molecule, namely, its concentration gradient. Usually, the term **concentration** refers to the density of the solute in a given volume of water. It is important to recognize, however, that the addition of a solute to pure water in essence decreases the water concentration, because solute molecules displace water molecules.

Compare the water and solute concentrations in the two containers in ● Figure 3–7. The container in part (a) is full of pure water, so the water concentration is 100% whereas the

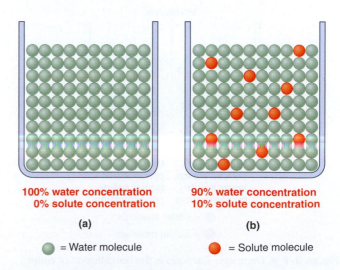

100% water concentration
0% solute concentration

(a)

90% water concentration
10% solute concentration

(b)

● = Water molecule   ● = Solute molecule

Figure 3–7 • Relationship between solute and water concentration in a solution. (a) Pure water. (b) Solution.

solute concentration is 0%. In part (b), 10% of the water molecules have been replaced by solute. The water concentration is now 90%, and the solute concentration is 10%—a lower water concentration and a higher solute concentration than in (a). Note that as the solute concentration increases, the water concentration decreases correspondingly. In this situation, the solution with higher solute concentration is said to be **hyperosmotic** with respect to the other. Conversely, the solution with higher water concentration is **hypo-osmotic**.

Let's say solutions of unequal solute concentration (and hence unequal water concentration) are separated by a membrane that permits passage of water but not solute, such as the plasma membrane (● Figure 3–8). Here, water will move down its own concentration gradient from the area of higher water concentration (lower solute concentration) to the area of lower water concentration (higher solute concentration). Because solutions are always referred to in terms of concentration of solute, *water moves by osmosis to the area of higher solute concentration.* The solute can be thought of as "drawing," or attracting, water, but osmosis is almost always portrayed as a special case of diffusion, in which water molecules move by this inherently random process. But when we actually measure the rate of osmosis across membranes, we find that for many solutes the rate is much faster than that predicted by Fick's law. In fact, the physical chemistry of osmosis is not really fully understood (see box, "Is Osmosis Really the Simple Diffusion of Water Down Its Concentration Gradient?" p. 78). However, we will treat it as a form of diffusion.

Let's follow what happens in this situation (● Figure 3–9). At the onset, the concentration gradients are identical to those in the previous example. However, even though net diffusion of water takes place from side 1 to side 2, the solute cannot move. As a result of water movement alone, side 2's volume increases while the volume of side 1 correspondingly decreases. Loss of water from side 1 increases the solute concentration on side 1, whereas addition of water to side 2 reduces the solute concentration on that side. Eventually, the concentrations of water and solute on the two sides of the membrane become equal, and net diffusion of water ceases. The side originally containing the greater solute concentra-

tion has a larger volume, having gained water. With all concentration gradients abolished, an **isosmotic** state (*isos,* "equal"), osmosis ceases.

What happens if a nonpenetrating solute is present on side 2 and pure water is present on side 1 (● Figure 3–10)? Osmosis occurs from side 1 to side 2, but the concentrations between the two compartments can never become equal. No matter how dilute side 2 becomes because of water diffusing into it, it can never become pure water, nor can side 1 ever acquire any solute. Because equilibrium is impossible to achieve, does net diffusion of water (osmosis) continue unabated until all the water has left side 1? No. As the volume expands in compartment 2, a difference in **hydrostatic** (*hydro,* "fluid;" *static,* "standing") **pressure** between the two compartments is created, and it opposes osmosis. (Hydrostatic pressure is the pressure exerted by a standing, or stationary, fluid on an object—in this case the plasma membrane.) The hydrostatic pressure exerted by the larger volume of fluid on side 2 is greater than the hydrostatic pressure exerted on side 1. This

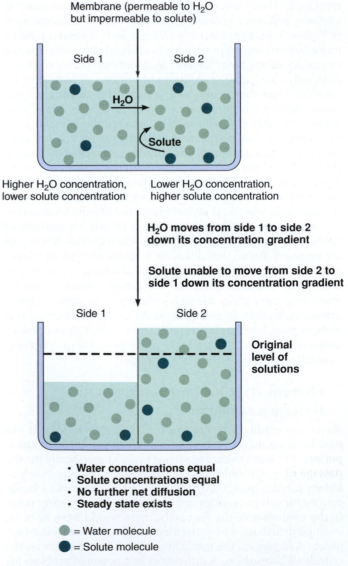

Figure 3–9 ● Osmosis in the presence of an unequally distributed nonpenetrating solute.

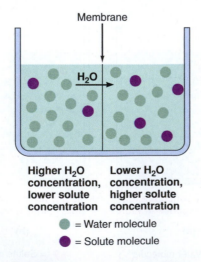

Figure 3–8 ● **Osmosis.** Osmosis is the net diffusion of water down its own concentration gradient (to the area of higher solute concentration).

hydrostatic pressure difference tends to push fluid from side 2 to side 1. The magnitude of opposing pressure necessary to completely stop osmosis is equal to the osmotic pressure of the solution on side 2. The **osmotic pressure** can be related directly to the concentration of nonpenetrating solute. The greater the concentration of nonpenetrating solute → the lower the concentration of water → the greater the drive for water to move by osmosis from pure water into the solution → the greater the opposing pressure required to stop the osmotic flow → the greater the osmotic pressure of the solution. Therefore, a solution with a high solute concentration exerts greater osmotic pressure than does a solution with a lower solute concentration.

Colligative Properties and Tonicity

Osmotic pressure is one of the so-called **colligative properties** of solutes. These properties depend solely on the number of

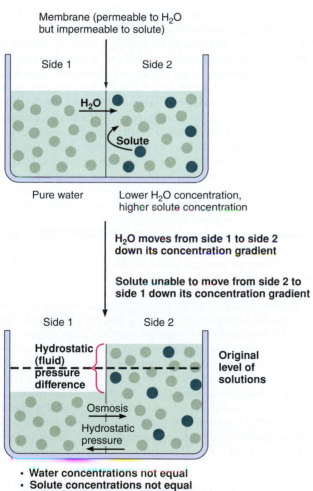

Figure 3–10 ● Osmosis when pure water is separated from a solution containing a nonpenetrating solute.

dissolved particles in a given volume of solution, not on the chemical properties of the solute. An idealized solute (one that fits this description) at 1 mole in 1 kg of water (defined as 1 molal) has an osmotic pressure of 22.4 atmospheres (atm), raises the boiling point by 0.54 degrees C, depresses freezing point by 1.86 degree C, and also reduces vapor pressure. The effect of solutes on freezing point is important in the context of animals adapting to freezing temperatures (Chapter 15).

Osmotic pressure can be measured in atmospheres, but is more conveniently expressed as osmoles/liter (Osm) or milliosmoles/liter (**mOsm**), because for an ideal solute this value is equal to its molar concentration. That is, an ideal solute dissolved at a concentration of 1 molar (M) is also 1 Osm or 1000 mOsm. Note that a 1-molar solution (1 mole/liter) is not quite equal to a 1-molal solution (1 mole/1000g water, the solution value on which colligative properties are based), but in the range of concentrations for most biological fluids, the two are nearly equal. Two examples of important osmotic pressures in the biological world are about 1000 mOsm for average seawater, and 300 mOsm (approximately) for most mammalian internal fluids. In fact, the basic dissolved constituents of cells—inorganic ions plus metabolites, proteins, and so forth—typically yield an osmotic pressure of roughly 300 mOsm in most organisms.

Whereas the *osmotic pressure* of a solution refers to a colligative property, the **tonicity** of a solution refers to the effect on cell volume of the concentration of nonpenetrating solutes in the solution surrounding the cell. Solutes that can penetrate the plasma membrane quickly become equally distributed between the ECF and ICF (intracellular fluid) and thus do not contribute to long-term osmotic or tonicity differences. An **isotonic solution** has the same concentration of nonpenetrating solutes as do normal body cells. When a cell is bathed in an isotonic solution, no water enters or leaves the cell by osmosis, so cell volume remains constant. Thus the ECF in animals is normally isotonic, so that no net diffusion of water occurs across the plasma membranes of body cells. This is important because cells, especially neural cells, do not function properly if they are swollen or shrunken. Any change in the concentration of nonpenetrating solutes in the ECF produces a corresponding change in the water concentration difference across the plasma membrane. The resultant osmotic movement of water brings about changes in cell volume.

The easiest way to demonstrate this phenomenon is to place mammalian red blood cells in solutions with varying concentrations of nonpenetrating solutes. Normally the plasma in which red blood cells are suspended has the same osmotic activity as the fluid inside these cells, enabling the cells to maintain a constant volume. If red blood cells are placed in a hypoosmotic solution, water enters the cells by osmosis. Net gain of water by the cells causes them to swell, perhaps to the point of rupturing, or lysing. In this situation causing swelling, the solution is called **hypotonic**. If, in contrast, red blood cells are placed in a concentrated or hyperosmotic solution, the cells shrink as they lose water by osmosis. In this situation, the solution is called **hypertonic**. Thus it is critical that the concentration of nonpenetrating solutes in the ECF quickly be restored to normal should the ECF become hypotonic (as with ingesting too much water) or hypertonic (as with losing too much water through severe diarrhea). (See Chapters 12 and 13 for further details about the important homeostatic mechanisms that regulate ECF osmolarity.)

Given that water penetrates membranes readily, whereas many solutes do not, a top priority of every animal cell is **osmotic homeostasis**, or **cell volume regulation**. This is the ability of a cell to adapt to an inequality of total solutes between its internal composition and the surrounding fluid. Inequalities can readily arise because a cell has a minimal requirement for internal organic solutes necessary for life, but the total concentration of these solutes may not match that of the external environment. Imbalances can occur from selective transport of solutes in and out of the cell as well (see next section). And imbalances may change in magnitude and direction within an animal's life cycle. For example, in the ocean the epithelial cells covering the gills of a salmon are responsible for limiting water loss from their body fluids, whereas during migration into the spawning grounds in fresh water, the cells are involved with the net movement of water into the body fluids. How cells and whole animals cope with these demands is covered in Chapter 13.

?

Is Osmosis Really the Simple Diffusion of Water Down Its Concentration Gradient? Although osmosis is generally characterized as water diffusing down its concentration gradient, created by differences in dissolved solutes that reduce water concentration, considerable evidence shows this is an oversimplification. In particular, some dissolved salts actually *increase* the effective concentration of water, and yet water in a lower-salt solution always osmoses through a membrane into a higher-salt solution. G. Hulett in 1902 first proposed an alternative explanation for osmosis. He hypothesized that the *pressure of the dissolved molecules colliding with the membrane* causes water molecules to move. The solution with the higher concentration has more collisions and thus exerts more pressure. This appears to alter the bonding forces between water molecules such that these molecules are pulled through the membrane into the side with higher salt. In recent decades, H. T. Hammel has provided considerable evidence that Hulett was correct. The phenomenon of osmosis still remains incompletely understood.

Special mechanisms are used to transport selected molecules unable to cross the plasma membrane on their own.

All kinds of transport we have discussed thus far—diffusion down concentration gradients, conduction along electrical gradients, and osmosis—produce net movement of molecules capable of permeating the plasma membrane by virtue of their lipid solubility or small size. Large, poorly lipid-soluble molecules such as proteins, glucose, and amino acids cannot cross the plasma membrane on their own no matter what forces are acting on them. This impermeability ensures that the large polar intracellular proteins cannot escape from the cell. It is important that these proteins stay in the cell where they belong and can carry out their functions, such as serving as metabolic enzymes.

However, because large, poorly lipid-soluble molecules cannot cross the plasma membrane on their own, the cell must

provide mechanisms for deliberately transporting these types of molecules into or out of the cell as needed. For example, the cell must usher into the cell essential nutrients, such as glucose for energy and amino acids for the synthesis of proteins, and transport out of the cell metabolic wastes and secretory products, such as water-soluble protein hormones and digestive enzymes. Furthermore, passive diffusion alone does not suffice for the movement of small, charged molecules (ions). Cells use two different mechanisms to accomplish selective transport processes for poorly permeable molecules: *carrier-mediated transport* for transfer of small water-soluble substances across the membrane and *vesicular transport* for movement of large molecules and even multimolecular particles. Let's now examine these processes.

Carrier-mediated transport is generally accomplished by a membrane protein changing its shape.

Carrier proteins span the thickness of the plasma membrane and can undergo reversible changes in shape so that specific binding sites can alternately be exposed at either side of the membrane. That is, the carrier "flip-flops" so that binding sites located in the interior of the carrier are alternately exposed to the ECF and the ICF. ● Figure 3–11 is a schematic representation of how this **carrier-mediated transport** process takes place. As the molecule to be transported attaches to a binding site on the carrier on one side of the membrane (step 1), it triggers a change in the carrier's shape that causes the same site to be exposed to the other side of the membrane (step 2). Then, having been moved in this way from one side of the membrane to the other, the bound molecule detaches from the carrier (step 3). After the passenger detaches, the carrier reverts back to its original shape (step 4).

Carrier-mediated transport systems display three important characteristics that determine the kind and amount of material that can be transferred across the membrane: *specificity*, *saturation*, and *competition*.

1. *Specificity.* Each carrier protein is specialized to transport a specific substance, or at most a few closely related chemical compounds. For example, amino acids cannot bind to glucose carriers, although several structurally similar amino acids may be able to use the same carrier. Cells vary in the types of carriers they have, thus permitting transport selectivity among cells. A number of inherited diseases involve defects in transport systems for a particular substance. *Cysteinuria* (cysteine in the urine) is such a disease, involving defective cysteine carriers in mammalian kidney membranes. This transport system normally removes cysteine from the fluid destined to become urine and returns this essential amino acid to the blood. When this carrier is malfunctional, large quantities of cysteine remain in the urine, where it is relatively insoluble and tends to precipitate. This is one cause of urinary or kidney stones.

2. *Saturation.* A limited number of carrier binding sites are available within a particular plasma membrane for a specific substance. Therefore, there is a limit to the amount of a substance that a carrier can transport across the membrane in a given time. This limit is known as the **transport maximum** (T_m). Until the T_m is reached, the number of

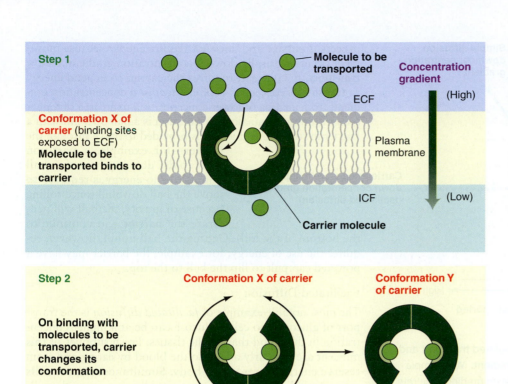

Step 1

Molecule to be transported

Concentration gradient

ECF

(High)

Conformation X of carrier (binding sites exposed to ECF)
Molecule to be transported binds to carrier

Plasma membrane

ICF

(Low)

Carrier molecule

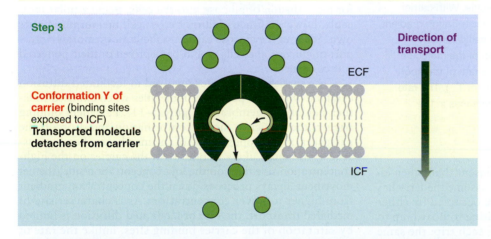

Step 2

Conformation X of carrier

Conformation Y of carrier

On binding with molecules to be transported, carrier changes its conformation

Step 3

Direction of transport

ECF

Conformation Y of carrier (binding sites exposed to ICF)
Transported molecule detaches from carrier

ICF

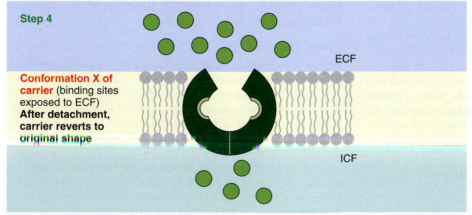

Step 4

ECF

Conformation X of carrier (binding sites exposed to ECF)
After detachment, carrier reverts to original shape

ICF

Figure 3–11 ● Schematic representation of carrier-mediated transport: Facilitated diffusion.

carrier binding sites occupied by a substance and, accordingly, the substance's rate of transport across the membrane are directly related to its concentration. The more of a substance available to be transported, the more will be transported. When the T_m is reached, the carrier is saturated (all binding sites are occupied), and the rate of the substance's transport across the membrane is maximal. Further increases in the substance's concentration are not accompanied by corresponding increases in the rate of transport (● Figure 3–12).

As an analogy, consider a ferry boat that can maximally carry 100 people across a river during one trip in an hour. If 25 people are on hand to board the ferry, 25 will be transported that hour. Doubling the number of people on hand to 50 doubles the rate of transport to 50 people that hour. Such a direct relationship exists between the number of people waiting to board (the concentration) and the rate of transport until the ferry is fully occupied (its T_m is reached). The ferry can maximally transport 100 people per hour. Even if 150 people are waiting to board, still only 100 will be transported per hour.

Saturation of carriers is a critical rate-limiting factor in the transport of selected substances across vertebrate kidney membranes during urine formation and across the intestinal membranes during absorption of digested foods. Furthermore, it is sometimes possible to regulate (for example, by hormones) the rate of carrier-mediated transport by varying the affinity (attraction) of the binding site for its passenger or by varying the number of carriers ("ferry boats"). For example, the hormone insulin greatly increases the carrier-mediated transport of glucose into most cells by promoting an increase in the number of glucose carriers in the cell's plasma membrane. Deficiency of insulin (diabetes mellitus) drastically impairs a mammal's ability to use glucose as the primary energy source.

3. *Competition.* Several closely related compounds may compete for a ride across the membrane on the same carrier. If a given binding site can be occupied by more than one type of molecule, the rate of transport of each substance

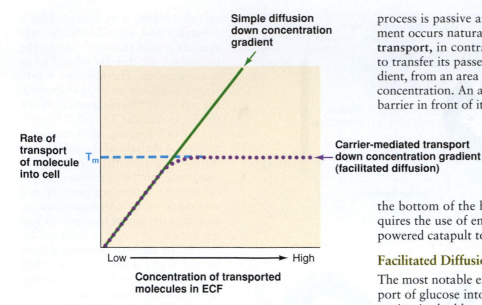

Figure 3–12 ● Comparison of carrier-mediated transport and simple diffusion down a concentration gradient. With simple diffusion of a molecule down its concentration gradient, the rate of transport of the molecule into the cell is directly proportional to the extracellular concentration of the molecule. With carrier-mediated transport of a molecule down its concentration gradient, the rate of transport of the molecule into the cell is directly proportional to the extracellular concentration of the molecule until the carrier is saturated, at which time the rate of transport reaches a maximal value (transport maximum, or T_m). The rate of transport does not increase with further increases in the ECF concentration of the molecule.

is lower when both molecules are present than when either is present by itself. To illustrate, assume the ferry has 100 seats (binding sites) that can be occupied by either men or women. If only men are waiting to board, up to 100 men can be transported during each trip; the same holds true if only women are waiting to board. If, however, both men and women are waiting to board, they will compete for the available seats so that fewer men and fewer women will be transported than when either group is present alone. Fifty of each might make the trip, although the total number of people transported will still be the same, 100 people. In other words, when a carrier can transport two closely related substances, such as the amino acids glycine and alanine, the presence of both diminishes the rate of transfer of either.

■ **Carrier-mediated transport may be passive or active.**

Carrier-mediated transport takes two forms, depending on whether energy must be supplied to complete the process: **facilitated diffusion** (not requiring energy) and **active transport** (requiring energy). Facilitated diffusion uses a carrier to facilitate (assist) the transfer of a particular substance across the membrane "downhill" from high to low concentration. This

process is passive and does not require energy because movement occurs naturally down a concentration gradient. **Active transport**, in contrast, requires the carrier to expend energy to transfer its passenger "uphill" *against* a concentration gradient, from an area of lower concentration to an area of higher concentration. An analogous situation is a ball on a hill with a barrier in front of it. If a way is provided to move through the barrier—for example, a partition that folds down when the ball bumps it—no energy is required to move the ball downhill (once rolling because of gravity, the ball will penetrate the barrier and continue to the bottom of the hill). Getting the ball uphill, however, requires the use of energy; for example, the barrier may have a powered catapult to flip the ball to the top.

Facilitated Diffusion

The most notable example of *facilitated diffusion* is the transport of glucose into cells. Glucose can be at a higher concentration in the blood than in the tissues. Fresh supplies of this nutrient are regularly added to the blood by eating and from reserve energy stores in the body. Simultaneously, the cells may metabolize glucose almost as rapidly as it enters the cells from the blood. As a result, there can be a continuous gradient for net diffusion of glucose into the cells. Being a polar molecule, however, glucose cannot cross cell membranes on its own. Without the glucose carrier to facilitate membrane transport of glucose, the cells would be deprived of their preferred source of fuel.

The carrier binding sites involved in facilitated diffusion can bind with their passenger molecules when exposed to either side of the membrane (Figure 3–11). Passenger binding triggers the carrier to flip its conformation and drop off the passenger on the opposite side of the membrane. Because passengers are more likely to bind with the carrier on the high-concentration side than on the low-concentration side, the net movement always proceeds down the concentration gradient from higher to lower concentration. As is characteristic of mediated transport, the rate of facilitated diffusion is limited by saturation of the carrier binding sites, unlike the rate of simple diffusion, which is always directly proportional to the concentration gradient (Figure 3–12).

Active Transport

Active transport also involves the use of a protein carrier to transfer a specific substance across the membrane, but in this case the carrier transports the substance against its concentration gradient. For example, the uptake of iodide by the vertebrate thyroid gland cells necessitates active transport because most of the iodine (99% in a mammal) is concentrated in the thyroid. To move iodide from the blood, where its concentration is low, into the thyroid, where its concentration is high, requires expenditure of energy to drive the carrier. Specifically, energy in the form of ATP is required in active transport to vary the affinity of the binding site when exposed on opposite sides of the plasma membrane. In contrast, the affinity of the binding site in facilitated diffusion is the same when exposed to either the outside or the inside of the cell.

With active transport, the binding site has a greater affinity for its passenger on the low-concentration side as a result

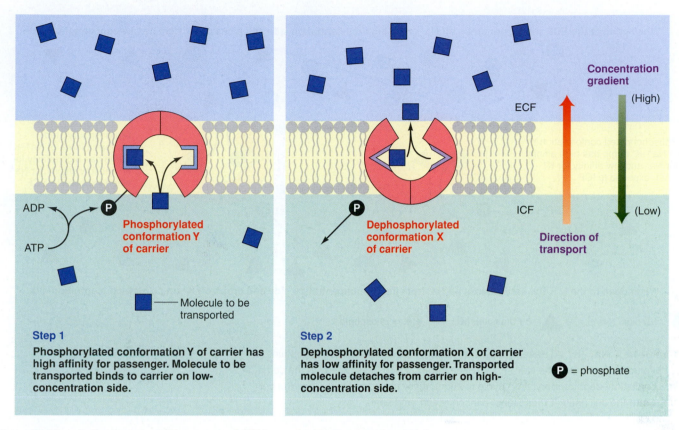

Figure 3–13 ● **Active transport.** The energy of ATP is required in the phosphorylation–dephosphorylation cycle of the carrier to transport the molecule uphill from a region of low concentration to a region of high concentration.

of *phosphorylation* of the carrier on this side (● Figure 3–13, step 1). The carrier exhibits ATPase activity in that it splits the terminal phosphate from an ATP molecule to yield ADP plus a free inorganic phosphate (see Chapter 2). The phosphate group is then attached to the carrier. This phosphorylation and the binding of the passenger on the low-concentration side cause the carrier protein to flip its conformation so that the passenger is now exposed to the high-concentration side of the membrane (Figure 3–13). The change in carrier shape is accompanied by *dephosphorylation;* that is, the phosphate group detaches from the carrier. Removal of phosphate reduces the affinity of the binding site for the passenger, so the passenger is released on the high-concentration side. The carrier then returns to its original conformation. Thus ATP energy is used in the phosphorylation–dephosphorylation cycle of the carrier. It alters the affinity of the carrier's binding sites on opposite sides of the membrane so transported particles are moved uphill from an area of low concentration to an area of higher concentration. These active transport mechanisms are frequently called **pumps,** analogous to water pumps that require energy to lift water against the downward pull of gravity. There are numerous pump proteins, especially those that move protons and other ions essential to life.

The simplest active-transport systems pump a single type of passenger. An example is the **hydrogen-ion** (proton or H^+) **pump.** One type of this pump—the **F-ATPase**—may have been the first ATPase pump to evolve; as we saw in Chapter 2, they

are used to generate energy by creating a hydrogen gradient. Another widespread type, an **H^+ V-ATPase** pump, is used to remove acid from body fluids, such as by the mammalian kidney, insect Malpighian tubules, and crustacean gills for excretion (see Chapter 12).

Other more complicated active transport mechanisms involve the transfer of two different passengers, either simultaneously in the same direction (called **symporting**) or sequentially in opposite directions (**antiporting**). For example, specialized vertebrate stomach cells use another type of proton pump to transport H^+ into the stomach lumen in exchange for a K^+ ion. This pump moves H^+ against a tremendous gradient: The concentration of H^+ in the stomach lumen is three to four million times greater than in the blood.

The Na^+–K^+ ATPase Pump

The Na^+–K^+ pump (Figure 3–14) plays three important roles:

1. It maintains Na^+ and K^+ concentration gradients across the plasma membrane of all cells; these gradients are critically important in the ability of nerve and muscle cells to generate electrical signals essential to their functioning (a topic discussed in the next chapter).

2. It helps regulate cell volume by controlling the concentrations of solutes inside the cell and thus minimizing osmotic effects that would induce swelling or shrinking of the cell.

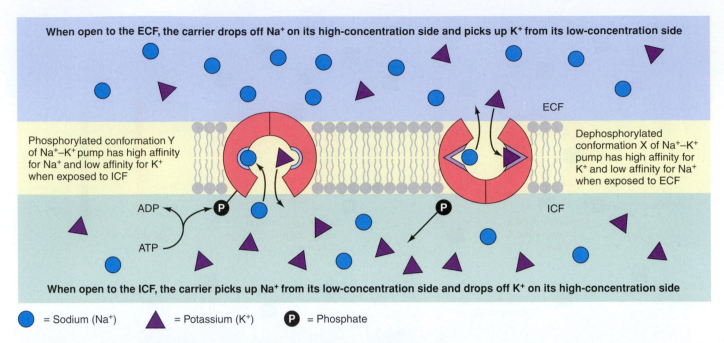

When open to the ECF, the carrier drops off Na⁺ on its high-concentration side and picks up K⁺ from its low-concentration side

Phosphorylated conformation Y of Na⁺–K⁺ pump has high affinity for Na⁺ and low affinity for K⁺ when exposed to ICF

ECF

Dephosphorylated conformation X of Na⁺–K⁺ pump has high affinity for K⁺ and low affinity for Na⁺ when exposed to ECF

ADP

ATP

P

P

ICF

When open to the ICF, the carrier picks up Na⁺ from its low-concentration side and drops off K⁺ on its high-concentration side

● = Sodium (Na⁺)   ▲ = Potassium (K⁺)   Ⓟ = Phosphate

Figure 3–14 ● Na⁺–K⁺ ATPase pump. The plasma membrane of all cells contains an active transport carrier, the Na⁺–K⁺ ATPase pump, which uses energy in the carrier's phosphorylation–dephosphorylation cycle to sequentially transport Na⁺ out of the cell and K⁺ into the cell against these ions' concentration gradients.

3. The energy used to run the Na⁺–K⁺ pump also indirectly serves as the energy source for the cotransport of glucose and amino acids across intestinal and kidney cells. This process is known as *secondary active transport.*

The activity of the Na⁺–K⁺ pump is inhibited by the highly toxic metabolic inhibitor **ouabain.** Ouabain is isolated from the bark and root of the ouabaio tree and has been used by tribes in Africa since antiquity as an arrow poison. Oxygen deprivation can also reduce activity of the Na⁺–K⁺ pump because of the decline in ATP production. In vertebrates the number of Na⁺–K⁺ pump units in the cell membrane is regulated, in part, by thyroid hormone (see p. 280). For example, an increase in plasma thyroid hormone in the salmonids (such as chum and pink salmon) is a part of the adaptive osmoregulatory response as they migrate to seawater. Concentrations of gill Na⁺–K⁺ pump units markedly increase in order to achieve tolerance to seawater.

Secondary Active Transport

Many cells can actively transport certain solutes by moving them uphill from low to high concentration using **cotransport carriers (symporters)** that have two binding sites, one for Na⁺ and one for the nutrient molecule. Cell energy (in the form of ATP) is not directly used by the carrier protein, but rather the energy stored in the sodium gradient (maintained with ATP) is used. Let us examine this more closely in mammals, in which (unlike most cells of the body) the intestinal and kidney cells transport glucose and amino acids by cotransport. Intestinal cells transport these nutrients from inside the intestinal lumen into the blood, concentrating them in the blood until none of these molecules are left in the lumen to be lost in the feces. The kidney cells save these nutrient molecules for the body by trans-

porting them out of the fluid that is to become urine, moving them against a concentration gradient into the blood.

The carriers that transport glucose against its concentration gradient from the lumen in the intestine and kidneys are distinct from the glucose facilitated-diffusion carriers. In these cells the Na⁺–K⁺ pumps lie in the basolateral membrane (the membrane at the base of the cell opposite the lumen and along the lateral edge of the cell below the tight junction; see Figure 2–33). More Na⁺ is present in the lumen than inside the cells because of the energy-requiring Na⁺–K⁺ pump that transports Na⁺ out of the cell at the basolateral membrane, keeping the intracellular Na⁺ concentration low. Because of this Na⁺ concentration difference, more Na⁺ binds to the luminal cotransport carrier when it is exposed to the outside (● Figure 3–15). Binding of Na⁺ to the cotransport carrier increases the carrier's affinity for its other passenger (for example, glucose), so the carrier has a high affinity for glucose when exposed to the outside. When both Na⁺ and glucose are bound to the carrier, it undergoes a conformational change and opens to the inside of the cell. Both Na⁺ and glucose are released to the interior, Na⁺ because of the lower intracellular Na⁺ concentration, and glucose because of the reduced affinity of the binding site on release of Na⁺.

The movement of Na⁺ into the cell by this cotransport carrier is downhill, because the intracellular Na⁺ concentration is low, but the movement of glucose is uphill, because glucose becomes concentrated in the cell. The released Na⁺ is quickly pumped out by the active Na⁺–K⁺ transport mechanism, keeping the level of intracellular Na⁺ low. The energy expended in this process is not directly used to run the cotransport carrier, because phosphorylation is not required to alter the affinity of the binding site to glucose. Instead, the establishment of a Na⁺ concentration gradient by a primary

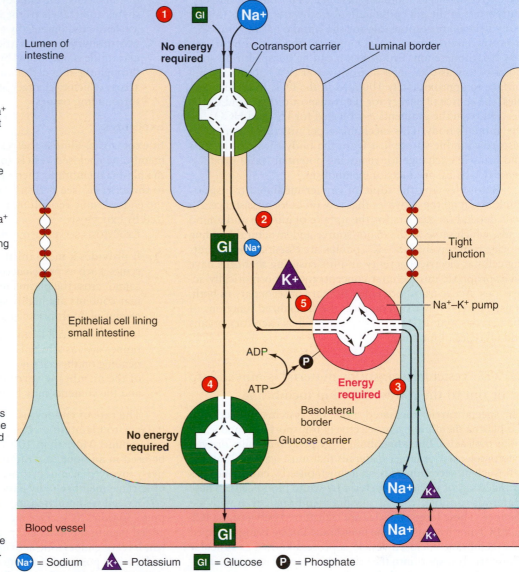

1. A cotransport carrier at the luminal border simultaneously transfers glucose against a concentration gradient and Na$^+$ down a concentration gradient from the lumen into the cell.

2. No energy is directly used by the cotransport carrier to move glucose uphill. Instead, operation of the cotransport carrier is driven by the Na$^+$ concentration gradient (low Na$^+$ in ICF compared to lumen) established by the energy-using Na$^+$–K$^+$ pump.

3. The Na$^+$–K$^+$ pump actively transports Na$^+$ out of the cell at the basolateral border, keeping the ICF Na$^+$ concentration lower than the luminal concentration.

4. After entering the cell by secondary active transport, glucose is transported down its concentration gradient from the cell into the blood by facilitated diffusion, mediated by a passive glucose carrier at the basal border.

5. The Na$^+$–K$^+$ pump also actively transports K$^+$ into the cell, maintaining a high intracellular K$^+$ concentration, but this action has no influence on secondary active transport.

Figure 3–15 ● Secondary active transport. Glucose (as well as amino acids) is transported across intestinal and kidney cells against its concentration gradient by means of secondary active transport.

active transport mechanism (the Na$^+$–K$^+$ pump) drives this secondary active-transport mechanism (Na$^+$–glucose cotransport carrier) to move glucose against its concentration gradient. With primary active transport, energy is *directly* required to move a substance uphill. With **secondary active transport,** energy is required in the entire process, but it is *not directly* required to run the pump. Rather it uses "secondhand" energy stored in the form of an **ion electrochemical gradient** (for example, a Na$^+$ gradient) to move the cotransported molecule uphill. This is a very efficient interaction. The cotransported molecule is essentially getting a free ride, because Na$^+$ must be pumped out anyway to maintain the electrical and osmotic integrity of the cell.

The glucose that is carried into the cell across the luminal border by secondary active transport then passively moves out of the cell across the basolateral border down its concentration gradient and enters the blood. This movement is accomplished by facilitated diffusion, mediated by another carrier in the plasma membrane. This passive carrier is identical to the one that transports glucose into other cells, but in intestinal and mammalian kidney cells it transports glucose out of the cell. The difference depends on the direction of the glucose concentration gradient. In the case of intestinal and kidney cells, glucose is in higher concentration inside the cells.

In the previous example, the symporter is responsible for the inward movement of both Na$^+$ and the substrate molecule. Molecules can also be moved in *opposing* directions by transport carriers (**antiporters**) as long as there is an established concentration or electrical (or electrochemical) gradient for one of the molecules. For example, the uptake of Na$^+$ through the skin of frogs is associated with the extrusion of H$^+$. Adding *amiloride* to the water inhibits both the uptake of Na$^+$ and the excretion of H^+. Similarly, the Na^+/H^+ and Cl^-/HCO_3^- antiporters in the gill epithelium of freshwater teleosts (higher bony fishes) are responsible for the uptake of Na^+ and Cl^- from the environment in exchange for the loss of hydrogen and bicarbonate ion. Chloride ion can be pumped from solutions as low as 0.02 mM in fresh water (FW) across the gill surface to approximately 100 mM in the ECF. In these examples, the source of energy for the transport of Na^+ and Cl^- is derived from the gradients established for H^+ and HCO_3^- in the animal's body. This mechanism permits these animals to restore valuable solute lost from the body through the gills and urine in exchange for removing potential waste-products generated as by-products of metabolism.

Before leaving the topic of carrier-mediated transport, think about all the activities that rely on carrier assistance. All cells depend on carriers for the uptake of glucose and amino acids, which serve as the major source and structural building blocks, respectively. Na^+–K^+ pumps are essential for generating cellular electrical activity and for ensuring that cells have an appropriate intracellular concentration of osmotically active ions. Active transport, either primary or secondary or both, is used extensively to accomplish the specialized functions of the nervous and digestive systems as well as excretory organs and all types of muscle.

■ With vesicular transport, material is moved into or out to the cell wrapped in membrane.

The special carrier-mediated transport systems embedded in the plasma membrane can selectively transport ions and small polar molecules. But what about large polar molecules or even macromolecular or larger materials that must leave or enter the cell, such as secretion of protein hormones by endocrine cells or ingestion of invading bacteria by white blood cells? These materials cannot cross the plasma membrane, even with assistance. These large particles are transferred between the ICF and ECF not by crossing the membrane but by being wrapped in a membrane-enclosed vesicle, a process known as **vesicular transport**. Transport into the cell in this manner is termed *endocytosis;* transport out of the cell is *exocytosis*. An important feature of vesicular transport is that the materials sequestered within the vesicles never mix with the cytosol. The vesicles are designed to recognize and fuse only with a targeted membrane, ensuring a directed transfer of designated large polar molecules or multimolecular materials between the cell's interior and exterior.

Endocytosis

In endocytosis, the plasma membrane surrounds the substance to be ingested, then fuses over the surface, pinching off a membrane-enclosed vesicle so that the engulfed material is trapped within the cell (see p. 42). The transported material has not actually passed through the surface membrane but has gained entrance to the interior of the cell by being wrapped in a piece of the membrane. Endocytosis of fluid is called **pinocytosis** (cell drinking), whereas endocytosis of a large multimolecular particle is **phagocytosis** (cell eating).

Once inside the cell, an engulfed vesicle has two possible destinies:

1. In most instances, lysosomes fuse with the vesicle to degrade and release its contents into the intracellular fluid.

2. In some cells, the endocytotic vesicle bypasses the lysosomes and travels to the opposite side of the cell, where it releases its contents by exocytosis. This provides a pathway to shuttle intact particles through the cell. Such vesicular traffic is one means by which materials are transferred through the thin cells lining the capillaries of vertebrates, across which exchanges are made between the blood and surrounding tissues.

Exocytosis

In **exocytosis,** almost the reverse of endocytosis occurs. A membrane-enclosed vesicle formed within the cell fuses with the plasma membrane, then opens up and releases its contents to the exterior (see p. 37). Materials packaged for export by the endoplasmic reticulum and Golgi complex are externalized by exocytosis.

Exocytosis serves two different purposes:

1. It provides a mechanism for secreting large polar molecules, such as proteinaceous hormones and enzymes, that cannot cross the plasma membrane. In this case, the vesicular contents are highly specific and are released only on receipt of appropriate signals.

2. It enables the cell to add specific components to the membrane, such as selected carriers, channels, or receptors, depending on the cell's needs. In such cases, the composition of the membrane surrounding the vesicle is important, and the contents may be merely a sampling of ICF.

The rate of endocytosis and exocytosis must be kept in balance to maintain a constant membrane surface area and cell volume. More than 100% of the plasma membrane may be used in an hour to wrap internalized vesicles in a cell actively involved in endocytosis, necessitating rapid replacement of surface membrane by exocytosis. In contrast, when a secretory cell is stimulated to secrete, it may temporarily insert up to 30 times its surface membrane through exocytosis. This added membrane must be specifically retrieved by an equivalent level of endocytotic activity. Thus, through exocytosis and endocytosis portions of the membrane are constantly being restored, retrieved, and generally recycled.

There may be more than one way in which cells can exchange materials with their environment. Caveolae, recently investigated membranous structures, are believed to accomplish another method of membrane transport. Let's see how.

■ Caveolae may play roles in membrane transport and signal transduction.

The outer surface of the plasma membrane is not smooth but instead is dimpled with tiny, cavelike indentations known as **caveolae** ("tiny caves"). Researchers have observed these small, flask-shaped pits for more than 40 years through electron microscopy (● Figure 3–16). The scientists did not consider caveolae to have functional significance, however, until further investigation in the mid-1990s suggested that they (1) provide another route for transport into the cell and (2) serve as a "switchboard" for relaying signals from many extracellular chemical messengers into the cell's interior.

An abundance of proteins, including a variety of receptors, cluster in these tiny chambers. Some of these receptors appear to play a role in cellular uptake of small molecules and ions. The best-studied example is the transport of the B vita-

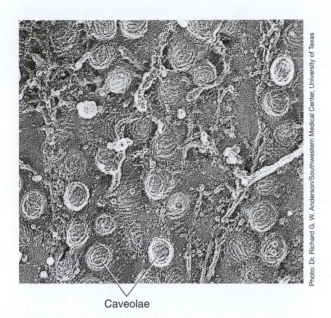

Caveolae

Photo: Dr. Richard G. W. Anderson/Southwestern Medical Center, University of Texas

Figure 3–16 ● **Caveolae.** Using a new technology involving making microscopic replicas of molds of rapidly frozen membranes, the caveolae, which are small cavelike indentations in the surface membrane, appear as small bumps on the surface of the mold. This is analogous to the "valleys" in a Jell-O mold becoming the "ridges" of the molded gelatin.

min *folic acid* into the cell. When folic acid binds with its receptors, which are concentrated in the portions of the plasma membrane that form the caveolae, the extracellular openings of these tiny caves close off. The high concentration of folic acid within a closed caveolar compartment encourages the movement of this vitamin across the caveolar membrane into the cytoplasm. Cellular uptake through the cyclic opening and closing of caveolae has been termed **potocytosis**. Scientists believe potocytosis is an uptake mechanism for selected small molecules and ions, in contrast to receptor-mediated endocytosis, which transports selected large molecules into the cell (see p. 41). Furthermore, caveolar vesicles do not fuse with lysosomes, thus their contents remain intact within the cell.

Besides serving as uptake vehicles, caveolae also appear to be important sites for signal transduction, a topic we cover in detail in the next section. Briefly, many membrane receptors important in signal transduction are concentrated in the caveolae, leading one investigator to call these tiny caves "signaling organelles." The functions of these organelles are now being analyzed by genomics (see box, "Can Genomics Solve the Mystery of Caveolae?").

?

Can Genomics Solve the Mystery of Caveolae? In 2001, genetic "knockout" mice (see Chapter 2) lacking a key protein for caveolar formation were created. In addition to a complete lack of caveolae, these mice showed cardiovascular defects related to impaired signal transduction pathways that use nitric oxide and calcium (see following section). Their lungs were also damaged by overgrowth of epi-

thelial cells, possibly from lack of proper signals that would suppress such growth. But the mice were fertile and not noticeably unhealthy. Many mysteries remain about the roles of these structures.

Our discussion of membrane transport is now complete; ▮ Table 3–2 summarizes the established pathways by which materials can pass between the ECF and ICF. Note the common patterns of transport available to all cells. All exchanges of substances between cells and the extracellular fluid environment is accomplished by one of these several means of membrane transport. Different cell types vary in the amounts and kinds of substances permitted to enter or leave because of their differences in the numbers and kinds of channels, carriers, and vesicular transport mechanisms.

Intercellular Communication and Signal Transduction

Communication is critical for the survival of the society of cells that collectively compose the animal. The ability of cells to communicate with each other is essential for coordination of their diverse activities to maintain homeostasis as well as to control growth and development of the animal as a whole.

▮ Communication between cells is largely orchestrated by extracellular chemical messengers.

There are three types of intercellular ("between cell") communication (● Figure 3–17), as follows.

1. The most intimate and rapid means of intercellular communication is through **gap junctions**, minute tunnels that bridge the cytoplasm of neighboring cells in some types of tissue. Through these specialized anatomic arrangements (Figure 2–34), small molecules and ions are directly exchanged between interacting cells without ever entering the ECF. Gap junctions are especially important in permitting electrical signals to spread from one cell to the next in cardiac and smooth muscle. However, in response to an increase in intracellular Ca^{++} or H^+ ion concentrations, gap junctions rapidly close.

2. The presence of signaling molecules on the surface membrane of some cells lets them directly link up transiently and interact with certain other cells in a specialized way. This is the means by which the phagocytes of the body's immune system specifically recognize and selectively destroy only undesirable cells, such as microbial invaders, while leaving the body's own cells alone.

3. The most common means by which cells communicate with each other is through intercellular chemical messengers, of which there are six types: **paracrines, neurotransmitters, hormones, neurohormones, pheromones,** and **cytokines.** In each case, a specific chemical messenger is synthesized by specialized cells to serve a designated purpose. On being released into the ECF by appropriate stimulation, these signaling agents act on other particular cells, the messenger's **target cells,** in a prescribed manner. To exert its effect, an intercellular messenger, also known as a **ligand,** must bind with target-cell **receptors** specific for it.

Table 3–2 ▮ Characteristics of the Methods of Membrane Transport

Methods of Transport	Substances Involved	Energy Requirements and Force-Producing Movement	Limit to Transport
Diffusion			
Through lipid bilayer	Nonpolar molecules of any size (e.g., O_2, CO_2, fatty acids)	Passive; molecules move down concentration gradient (from high to low concentration)	Continues until the gradient is abolished (steady state with no net diffusion)
Through protein channel	Specific small ions (e.g., Na^+, K^+, Ca^{2+}, Cl^-	Passive; ions move down electro-chemical gradient (from high to low concentration and attraction of ion to area of opposite charge)	Continues until there is no net movement and a steady state is established
Special case of osmosis	Water only	Passive; water moves down its own concentration gradient (water moves to area of lower water concentration, i.e., higher solute concentration)	Continues until concentration difference is abolished or until stopped by an opposing hydrostatic pressure or until cell is destroyed
Carrier-Mediated Transport			
Facilitated diffusion	Specific polar molecules for which a carrier is available (e.g., glucose)	Passive; molecules move down concentration gradient (from high to low concentration)	Displays a transport maximum (T_m); carrier can become saturated
Primary active transport	Specific ions or polar molecules for which carriers are available (e.g., Na^+, K^+, amino acids)	Active; ions move against concentration gradient (from low to high concentration); requires ATP	Displays a transport maximum; carrier can become saturated
Secondary active transport	Specific polar molecules and ions for which cotransport carriers are available (e.g., glucose, amino acids, some ions, including Ca^{2+})	Active; molecules move against concentration gradient (from low to high concentration); driven directly by ion gradient (usually Na^+) established by ATP-requiring primary pump	Displays a transport maximum; cotransport carrier can become saturated
Vesicular Transport			
Endocytosis			
Pinocytosis	Small volume of ECF fluid, perhaps with specific bound molecules (e.g., protein); also important in membrane recycling	ATP required	Control poorly understood
Phagocytosis	Multimolecular particles (e.g., bacteria and cellular debris)	ATP required	Necessitates binding to specific receptor site on membrane surface
Exocytosis	Secretory products (e.g., hormones and enzymes) as well as large molecules passing through cell intact; also important in membrane recycling	ATP required; increase in cytosolic Ca^{2+} induces fusion of vesicle with plasma membrane	Secretion triggered by specific neural or hormonal stimuli; other controls involved in transcellular traffic and membrane recycling not known

These six types of chemical messengers differ in their source and the distance and means by which they get to their site of action, as follows (Figure 3–17):

■ **Paracrines** are local chemical messengers whose effect is exerted only on neighboring cells in the immediate environment of their site of secretion. (If their action affects the same cell that secreted them, they are called **autocrines.** Paracrines released by neurons are **neuromodulators.**) Because paracrines are distributed by simple diffusion, their action is restricted to short distances. They do not gain entry to the blood in any significant quantity, because they are rapidly inactivated by locally existing

enzymes. One example of a paracrine is *histamine,* which is released from a specific type of mammalian connective tissue cell during an inflammatory response within an invaded or injured tissue (see p. 434). Among other things, histamine dilates (opens more widely) the blood vessels in the vicinity to increase blood flow to the tissue. This action brings additional blood-borne immune-response supplies into the affected area.

Paracrines must be distinguished from chemicals that influence neighboring cells after being nonspecifically released during cellular activity. For example, an increased local concentration of CO_2 in an exercising muscle promotes local dilation of the blood vessels supplying

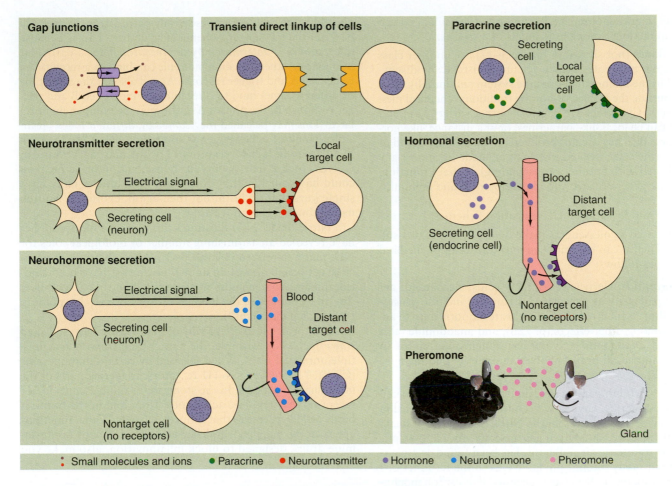

Figure 3–17 ● Types of intercellular communication. Gap junctions and transient \direct linkup of cells are both means of direct communication between cells. Paracrines, neurotransmitters, hormones, neurohormones, and pheromones are all extracellular chemical messengers that accomplish indirect communication between cells. These chemical messengers differ in their source and the distance they travel to reach their target cells.

the muscle. The resultant increased blood flow helps meet the more active tissue's increased metabolic demands. However, CO_2 is produced by all cells and is not specifically released to accomplish this particular response, so it and similar nonspecifically released chemicals are not considered paracrines.

- **Neurotransmitters** are used by neurons (nerve cells), which communicate directly with the cells they innervate (their target cells) by releasing these signal molecules in response to electrical signals (see p. 123). Like paracrines, neurotransmitters are very short-range chemical messengers, which diffuse from their site of release across a narrow extracellular space to act locally on only an adjoining target cell, which is usually another neuron, a muscle, or a gland. (Note that neurons may also release true paracrines, called *neuromodulators,* as noted in Chapter 4).

- **Hormones** are long-range chemical messengers that are specifically secreted into the circulation by endocrine (ductless) glands (see p. 250) in response to an appropriate signal. The blood or hemolymph carries the messengers to other sites in the body, where they exert their ef-

fects on their target cells some distance away from their site of release. Only the target cells of a particular hormone bear receptors to bind with this hormone. Nontarget cells are not influenced by any blood-borne hormones that reach them. As you'll see later, some neurotransmitters can also be used as hormones and vice versa.

- **Neurohormones** are hormones released into the blood specifically by neurosecretory neurons. Like ordinary neurons, neurosecretory neurons can respond to and conduct electrical signals. Instead of directly innervating target cells, however, a neurosecretory neuron releases its chemical messenger, a neurohormone, into the circulatory fluid on appropriate stimulation. The neurohormone is then distributed to the target cells. Thus, like endocrine cells, neurosecretory neurons release blood- (or hemolymph-) borne chemical messengers, whereas ordinary neurons secrete short-range neurotransmitters into a confined space.

- **Pheromones** are chemical signals released into the environment, usually by glands, and travel through the air or water to sensory cells in another animal. Such signals are used for sexual activity (such as signaling of readiness to mate), marking of territories, and other behaviors related to interactions among individuals of a species.

- **Cytokines** (not shown in Figure 3–17) are another class of signal molecules, which scientists historically discovered in the mammalian immune system and initially studied outside the framework of the traditional chemical messengers. Cytokines are peptides with a variety of effects that can be local, like paracrines, or at a distance, like hormones. But they are not produced by specialized gland or paracrine cells. Instead, they can be made by almost any body cell with another (nonsignaling) primary function, and are generally involved in development and immunity.

Structural and Functional Classification Schemes

Molecules used as messengers make up a long list that is still growing. The list can be simplified into categories of chemical structures, as follows:

- *Eicosanoids:* Derivatives of fatty acids, **eicosanoids** are used primarily as paracrines. Examples include **prostaglandins,** which regulate vertebrate smooth-muscle action among other effects (p. 726), and **anandimide** (also spelled *anandamide*), a paracrine neuromodulator in the mammalian brain involved in appetite and memory (p. 188).

- *Gases:* Two inorganic gases, though normally considered toxic pollutants, have now been found to be produced in animals as natural messengers. **Nitric oxide (NO),** a small, highly reactive, short-lived gas molecule once known primarily as a toxic air pollutant, has been found as a paracrine in many animal phyla. We examine its functions later. Also, researchers have suggested that **carbon monoxide (CO)** is a paracrine and neuromodulator.

- *Purines:* The primary messengers in this chemical group are **adenosine** and **ATP.** They are used as paracrines and neurotransmitters (p. 123) in addition to their roles in nucleic acids and cellular energy. Adenosine is one major target of **caffeine** (see later discussion).

- *Amines:* This large category consists of a few amino acids and numerous derivatives. It includes many paracrines, such as histamine and serotonin, and many hormones, such as epinephrine (Chapter 7).

- *Peptides and proteins:* Numerous amino-acid sequences are produced as messengers, as paracrines, cytokines, neurohormones, and hormones (p. 252, Chapter 7).

- *Steroids:* These hydrophobic molecules are derived from cholesterol. They are primarily hormones and sometimes pheromones (Chapter 7). Their lipid nature permits them to diffuse directly through cell membranes.

- *Retinoids:* These messengers are derived from vitamin A and are mainly paracrines involved in development and differentiation. These will not be considered further.

■ Chemical messengers can have multiple, unrelated effects because of the "same key, different locks" principle.

As we have noted, chemical messengers generally exert their effects by binding to receptor proteins. (We discuss the details of this shortly.) It is important to realize that one cannot de-

duce what the cellular response to any particular messenger will be from the messenger's chemical structure. This is because the *response is determined by the receptor.* Thus a particular hormone or other signaling molecule can bind to a receptor on target cell A and trigger response A, and also (at the same time or different time) bind to a receptor on target cell B and trigger response B. Responses A and B may be completely unrelated. This may be called the *"same key, different locks"* principle. Think of a messenger as something like a key that can fit a variety of locks (receptors). Lock A may start your car, whereas lock B may open your house, and in principle, you could have one key that does both. (We don't usually have single keys that do both, but then again, evolution is often more economical than we are.)

Nitric oxide (NO) illustrates this principle nicely. Studies have revealed an astonishing number of biological roles for NO, many unrelated (■ Table 3–3). It has a variety of paracrine actions in almost all animal phyla that have been studied, ranging from memory in neurons to control of blood flow (for an example, see the box "A Closer Look at Adaptation: Lighting Up the Night").

This principle explains why so many drugs have so-called *side effects:* For example, the paracrine *histamine* has multiple unrelated roles. When released by certain immune cells (Chapter 10), it binds to H1 receptors, triggering parts of the inflammation process. In the stomach, histamine binds to H2 receptors that regulate acid production; and in the brain, it binds to H3 receptors that induce alertness. These functions appear to be independent. They explain why many antihistamine drugs, designed to block inflammation, also make you

---

Table 3–3 ■ Functions of Nitric Oxide (NO)

- Causes relaxation of arteriolar smooth muscle. By means of this action, NO plays an important role in controlling blood flow through the tissues and in maintaining mean arterial blood pressure.

- Dilates the arterioles of the mammalian penis and clitoris, thus serving as the direct mediator of erection of these reproductive organs. Erection is accomplished by rapid engorgement of these organs with blood.

- Used as chemical warfare against bacteria and cancer cells by macrophages, large phagocytic cells of the immune system.

- Interferes with platelet function and blood clotting at sites of vessel damage.

- Serves as a novel type of neurotransmitter in the brain and elsewhere. Unlike classical neurotransmitters, NO is not stored in vesicles and released by exocytosis, and it does not bind with receptors on its target cell. NO is synthesized on demand in the neuron terminal, then diffuses out of the terminal and into the adjacent target cell, where it brings about its effect.

- Plays a role in the changes underlying memory.

- By promoting relaxation of digestive-tract smooth muscle, helps regulate peristalsis, a type of contraction that pushes digestive tract contents forward.

- Relaxes the smooth muscle cells in the airways of the lungs, helping keep these passages open to facilitate movement of air in and out of the lungs.

- May play a role in relaxation of skeletal muscle.

drowsy: Those drugs bind to and block histamine from binding both H1 and H3 receptors.

■ Extracellular chemical messengers bring about cell responses primarily by signal transduction.

Most extracellular chemical messengers, such as protein hormones delivered by the blood or neurotransmitters released from nerve endings, cannot actually enter their target cells to bring about the desired intracellular response. Instead, these first messengers issue their orders by binding with receptors (with specific binding sites for one type of messenger) on the surface membrane, triggering a biochemical chain of events. This process is called **signal transduction,** in which incoming signals (instructions from extracellular chemical messengers) are conveyed to the target cell's interior for execution. (A *transducer* is a device that receives energy or information from one system and transmits it in a different form to another system. For example, your radio receives radio waves sent out from the broadcast station and transmits these signals in the form of sound waves that can be detected by your ears.) The combination of extracellular messenger with membrane receptor triggers a sequence of intracellular events that ultimately controls a particular cellular activity important in maintaining homeostasis or activating a regulated change. Despite the wide range of possible responses, binding of the membrane receptor with the extracellular chemical messenger (the **first messenger**) brings about the desired intracellular response by only three general means: (1) by activating an enzyme that phosphorylates a cell protein, (2) by opening or closing specific channels in the membrane to regulate the movement of particular ions into or out of the cell, or (3) by transferring the signal to an intracellular chemical messenger (the **second messenger**), which in turn triggers a preprogrammed series of biochemical events within the cell. Because of the universal nature of these events, let us examine each more closely.

■ Some first messengers activate a phosphorylating enzyme in the membrane.

Some receptors have an associated enzyme, called a **kinase,** that phosphorylates a target cell protein. This activity is activated when a first messenger binds to the receptor. Recall that *phosphorylation* refers to the transfer of a phosphate group from ATP to the protein at the expense of degrading ATP to ADP. Attachment of a phosphate group to a protein induces the protein to change its shape and function (either activating or inhibiting it) to bring about the desired response. Insulin and growth factors generally work by this mechanism (see Chapter 7).

■ Some first messengers open chemically gated channels.

Some first messengers bind to receptors that open membrane channels. Membrane channels may be either *leak channels* or *gated channels.* **Leak channels** are open all the time, thus permitting unregulated leakage of their chosen ion across the mem-

A CLOSER LOOK AT ADAPTATION

Lighting Up the Night

One of the most memorable natural displays that humans encounter are the summertime light shows of lightning bugs or fireflies. Male fireflies (actually a type of beetle) produce yellowish light in their abdomens as a way to attract mates; the longer and brighter a male produces light, the more likely he will attract females. Light is produced when an enzyme, called *luciferase,* catalyses a reaction between oxygen and a molecule called *luciferin.* This takes place in organelles called *peroxisomes* (p. xxx) in light-organ cells. Scientists have long known that the beetle's brain triggers light by neurons that extend to the light organ in the abdomen. But the nerves attach to the air tubes (trachea) leading into the organ rather than on the light-producing cells. How does the signal actually get to those cells? Light production requires oxygen, which is normally too low for light production in the light organs because mitochondria consume oxygen in the course of normal aerobic metabolism

(p. xxx). The secret to light regulation is to let the oxygen get to the light cells. New studies show that the nerve signal triggers an enzyme to produce **nitric oxide (NO),** a paracrine signal researchers discovered in mammals as a local regulator of blood flow. In fireflies, NO appears to inhibit mitochondrial processes, allowing oxygen from the trachea to diffuse to the light cells. A crucial experiment consisted of exposing a dissected light organ to NO in the lab. It lit up!

The control of light production illustrates two major mechanisms of intercellular communication. First, a neuron carrying a signal from a central regulator (the brain) releases a neurotransmitter molecule; second, a local or paracrine signal (NO) is triggered to affect the target system (in this case, mitochondria in the light cells). We'll mention many examples of such long-distance and local signaling mechanisms throughout our study of animal physiology.

Photo: © Michael Jeffords

brane through the channels down electrochemical gradients. **Gated channels,** in contrast, behave as if they have "gates" that can be opened or closed in response to various controlling stimuli. Thus ionic passage through a gated channel is regulated. There are three different types of gated channels, depending on which of the following mechanisms bring about channel opening and closing: (1) **chemically gated** or **ligand gated,** by binding of an extracellular chemical messenger to a specific membrane receptor that is in close association, or an integral part, of the channel; (2) **voltage gated,** by changes in electrical current in the plasma membrane; and (3) **mechanically gated,** by stretching or other mechanical deformation of the channel. For now we will concentrate on chemical gating and defer the other two methods of controlling channels until later (for example, see p. 125).

An extracellular (first) messenger can alter a channel through one of two mechanisms. For some channels, the receptor binding site on the plasma membrane is part of the channel itself. When first messenger binds with its receptor site, the channel opens (a mechanism we discuss in Chapter 4). For other channels, the receptor is a separate protein located near the channel. In this case, binding of the extracellular messenger with its receptor activates membrane-bound intermediaries known as **G proteins,** which in turn open (or in some instances close) the appropriate adjacent channel. (The mechanism of action of G proteins is discussed shortly.)

By opening or closing specific channels, extracellular messengers can regulate the flow of particular ions across the membrane. This ionic movement can be responsible for two different cellular events:

1. A small, short-lived movement of Na^+, K^+, or both across the membrane alters the electrical activity of cells that are capable of generating electrical signals (or impulses), such as nerve and muscle cells.

2. A transient flow of calcium (Ca^{++}) into the cell through opened Ca^{++} channels triggers an alteration in shape and function of specific intracellular proteins, which leads to the cell's response. Illustrative is the increase in cytosolic Ca^{++} responsible for triggering the release of secretory product from many gland cells.

On completion of the response, the extracellular messenger is removed from the receptor site, and the chemically gated channels close once again. The ions that moved across the membrane through opened channels to trigger the response are returned to their original location by special membrane carriers. Cytosolic Ca^{++} may be increased in another way besides entry from the ECF through opened membrane channels: Intracellular Ca^{++} stores can be released in a second-messenger pathway, as described in the next section.

■ Many first messengers activate second-messenger pathways.

Finally, many first messengers serve as signals for activation of intracellular *second messengers* that ultimately relay the orders to particular intracellular proteins that carry out the dictated response. The intracellular pathways activated by a second messenger in response to binding of the first messenger to a receptor are remarkably similar among different cells despite the diversity of ultimate responses to that signal. The variability in response depends on the specialization of the cell, not on the mechanism used.

There are two major second-messenger pathways: One uses **cyclic adenosine monophosphate (cyclic AMP or cAMP)** as a second messenger, and the other employs **diacylglycerol, inositol triphosphate,** and Ca^{++} in this role. The two pathways have much in common. The initial stages are similar in that binding of a first messenger to a membrane receptor leads to activation of an enzyme on the cytoplasmic side of the membrane. This enzyme, in turn, produces intracellular second messengers that diffuse throughout the cell. In both pathways, the cellular response is accomplished by altering the structure and subsequent function of particular cell proteins. For example, a particular enzymatic protein regulating a specific metabolic event may be modified so that its activity is increased or decreased. Let's now look at the mechanisms by which second messengers are created and by which they act.

Cyclic AMP Second-Messenger Pathway

In the **cyclic AMP** (or **cAMP**) pathway, binding of an appropriate first messenger to its surface receptor (step ① in ● Figure 3–18) eventually activates (or in some instances inhibits) the enzyme **adenylyl cyclase** (step ②) on the inner surface of the membrane. A membrane-bound "middleman," a **G protein,** acts as an intermediary between the receptor and adenylyl cyclase. G proteins, often called *transducer proteins,* are so named because they are bound to guanine nucleotides—**guanosine triphosphate (GTP)** or **guanosine diphosphate (GDP).** G proteins are found on the inner surface of the plasma membrane. An unactivated G protein consists of a complex of alpha (α), beta (β), and gamma (γ) subunits, with a GDP molecule bound to the alpha subunit. A number of different G proteins with varying α subunits have been identified. The different G proteins are activated in response to binding of various first messengers to surface receptors.

When a first messenger binds with its receptor, the receptor attaches to the appropriate G protein, resulting in the release of GDP from the G protein complex. GTP then attaches to the site on the α subunit (step ① in Figure 3–18) vacated by the released GDP, activating the α subunit. Once activated, the α subunit breaks away from the G protein complex and moves along the inner surface of the plasma membrane until it binds to an **effector protein.** An effector protein is either an ion channel or an enzyme, and the α subunit alters its activity. There are both stimulatory and inhibitory G proteins. Depending on the outcome signaled by the first messenger, the G protein either opens or closes a particular channel or activates or inhibits a particular enzyme. Researchers have identified more than 300 different receptors that convey instructions of extracellular messengers through the membrane to effector proteins by means of G proteins.

In the cyclic AMP pathway, the effector protein is adenylyl cyclase. Adenylyl cyclase induces the conversion of intracellular ATP to cAMP by cleaving off two of the phosphates (step ②). (This is the same ATP used as the common energy currency in the body.) In turn, cAMP activates a specific intracellular enzyme, **cAMP-dependent protein kinase,** also known as **protein kinase A (PKA)** (step ③). PKA in turn phosphorylates a specific intracellular protein (step ④), such as an enzyme important in a particular metabolic pathway. For example, a particular enzymatic protein regulating a specific metabolic

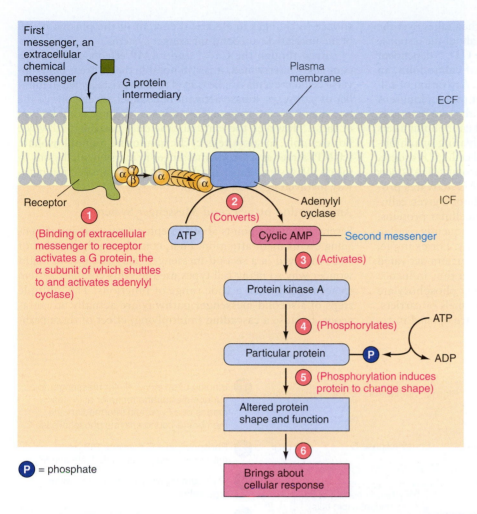

First messenger, an extracellular chemical messenger

G protein intermediary

Plasma membrane

ECF

Receptor

(Binding of extracellular messenger to receptor activates a G protein, the α subunit of which shuttles to and activates adenylyl cyclase)

② (Converts)

ATP

Adenylyl cyclase

ICF

Cyclic AMP — Second messenger

③ (Activates)

Protein kinase A

④ (Phosphorylates)

ATP

Particular protein

P

ADP

⑤ (Phosphorylation induces protein to change shape)

Altered protein shape and function

⑥

P = phosphate

Brings about cellular response

① Binding of an extracellular messenger, the *first messenger*, to a surface membrane receptor activates by means of a G protein intermediary the membrane-bound enzyme adenylyl cyclase.

② Adenylyl cyclase converts intracellular ATP into cyclic AMP.

③ Cyclic AMP acts as an intracellular *second messenger*, triggering the desired cellular response by activating protein kinase A.

④ Protein kinase A in turn phosphorylates a particular intracellular protein.

⑤ Phosphorylation induces a change in the shape and function of the protein.

⑥ The altered protein then accomplishes the cellular response dictated by the extracellular messenger.

Figure 3–18 • Activation of the cyclic AMP second-messenger system by an extracellular messenger.

event may be modified so that its activity is increased or decreased (step ⑤). After the response is accomplished, the α subunit converts GTP to GDP by cleaving off a phosphate, in essence shutting itself off, then rejoins the β and γ subunits to restore the inactive G protein complex. Intracellular enzymes inactivate the other participating chemicals so that the response can be terminated, so that the response does not go on indefinitely. To terminate the actions of cAMP the enzyme **phosphodiesterase (PDE)** converts cAMP into **adenosine 5′-phosphate (AMP)**. The concentration of cAMP in the cell is thus a balance between the rate of synthesis of cAMP and the rate at which it is hydrolyzed.

Note that in this signal transduction pathway, the steps involving the extracellular first messenger, the receptor, the G protein complex, and the effector protein *occur in the plasma membrane* and lead to activation of the second messenger. The extracellular messenger cannot gain entry into the cell to "personally" deliver its message to the proteins that carry out the desired response. Instead, it initiates events that create an intracellular messenger. This second messenger then triggers a chain reaction of biochemical events *inside the cell* that leads to the cellular response.

It is important to recognize that different types of cells have different proteins available for phosphorylation and modification by PKA. Therefore, a *common second messenger, cAMP, can induce widely differing responses in different cells,* depending on what proteins are modified. You can think of cyclic AMP as a commonly used molecular "key" that can "turn on" (or "turn off") different cellular events depending on the unique specialization of a particular cell type. This is the "same key, different locks" principle we mentioned earlier (p. 88). For example, activation of the cAMP system modifies heart rate (p. 386), stimulates the formation of female sex hormones in the ovaries, breaks down stored glucose in the liver (p. 296), controls water conservation during urine formation in the kidneys (p. 556), creates some simple memory traces in the brain (p. 190), and mediates signal transduction for a "sweet" molecule by a taste bud (p. 238).

Calcium Second-Messenger Pathway

Instead of cAMP, some cells use a three-part system including Ca^{++} as a second messenger. In such cases, binding of the first messenger to the surface receptor eventually leads by means of G proteins to activation of the enzyme **phospholipase C**, a protein effector that is bound to the inner side of the membrane (step ① in • Figure 3–19). This enzyme breaks down **phosphatidylinositol bisphosphate** (abbreviated PIP_2), a component of the tails of the phospholipid molecules within the membrane itself. The products of PIP_2 breakdown include the membrane-bound **diacylglycerol (DAG)** and the water-soluble **inositol trisphosphate** (IP_3) (step ②). IP_3 is the fragment responsible for mobilizing Ca^{++} from intracellular stores to increase cytosolic Ca^{++} (step ③). Calcium then takes over the role of second messenger, ultimately bringing about the response dictated by the first messenger. Many of the Ca^{++}-dependent cellular events are triggered by activation of **calmodulin,** an intracellular Ca^{++}-binding protein (step ④). Activation of calmodulin by Ca^{++} is similar to activation of protein kinase A by cAMP. From here onward the patterns of the two pathways are similar. Once bound to Ca^{++}, activated calmodulin alters other cellular proteins (step ⑤), either activating or inhibiting them, to bring about the ultimate desired cellular response. Simultaneously, the other PIP$_2$ breakdown product, DAG, sets off another second-messenger pathway. DAG activates **protein kinase C (PKC),** which in turn brings about a given cellular response by phosphorylating particular cellular proteins.

The cAMP and DAG- Ca^{++} pathways frequently overlap in bringing about a particular cellular activity. For example, cAMP and Ca^{++} can influence each other. Calcium-activated calmodulin can regulate adenylyl cyclase and thus influence cAMP, whereas cAMP-dependent kinase may phosphorylate and thereby change the activity of Ca^{++} channels or carriers. Also, remember that some Ca^{++} channels are opened in re-

sponse to changes in electrical current in the plasma membrane unrelated to second-messenger systems.

Even though the Ca^{++} and cAMP effects can become complexly intertwined, it is still notable that a great many diverse cellular events can be traced to a surprisingly small number of pathways: kinase activation, channel effects, one of the two major second-messenger pathways, or some combination of these. However, these are not the only possible pathways. For example, in some cells **cyclic guanosine monophosphate (cGMP)** serves as a second messenger in a system analogous to the cAMP system. An example is the signal transduction pathway involved in vision (p. 212).

Amplification by a Second-Messenger System

Several remaining points about receptor activation and the ensuing events merit attention. First, considering the number of steps involved in a second-messenger relay chain, you might wonder why so many cell types use the same complex system to accomplish such a wide range of functions. The multiple steps of a second-messenger pathway are actually advantageous, because a **cascading** (multiplying) effect of these path-

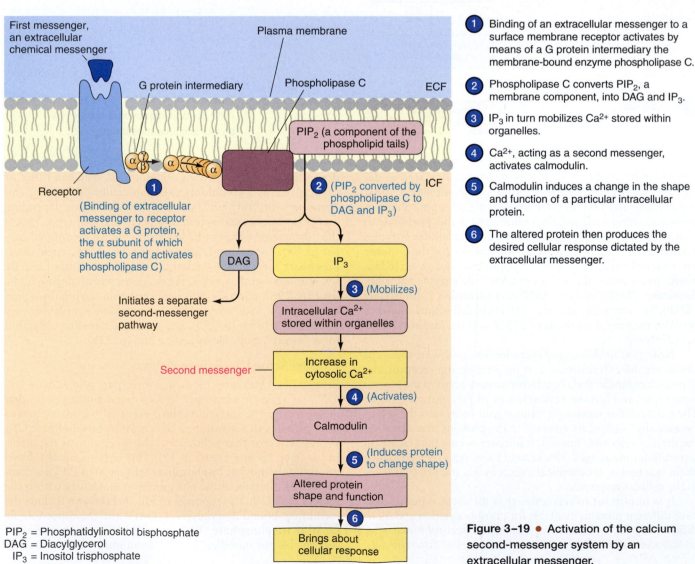

1. Binding of an extracellular messenger to a surface membrane receptor activates by means of a G protein intermediary the membrane-bound enzyme phospholipase C.

2. Phospholipase C converts PIP$_2$, a membrane component, into DAG and IP$_3$.

3. IP$_3$ in turn mobilizes Ca^{2+} stored within organelles.

4. Ca^{2+}, acting as a second messenger, activates calmodulin.

5. Calmodulin induces a change in the shape and function of a particular intracellular protein.

6. The altered protein then produces the desired cellular response dictated by the extracellular messenger.

Figure 3–19 ● Activation of the calcium second-messenger system by an extracellular messenger.

PIP$_2$ = Phosphatidylinositol bisphosphate
DAG = Diacylglycerol
IP$_3$ = Inositol trisphosphate

ways greatly amplifies the initial signal (● Figure 3–20). Amplification means the magnitude of the output of a system is much greater than the input. Binding of one first-messenger molecule to a receptor activates a number of adenylyl cyclase molecules (let us arbitrarily say 10), each of which produces many (say, 100) cAMP molecules. Each cAMP molecule then acts on a single PKA, which phosphorylates and thereby influences many (again, less us say 100) specific proteins, such as enzymes. Each enzyme, in turn, is responsible for producing many (perhaps 100) molecules of a particular product, such as a secretory product. The result of this cascade of events, with one event triggering the next event in sequence, is a tremendous amplification of the initial signal. In the hypothetical example in Figure 3–20, one chemical messenger molecule has been responsible for inducing a yield of 10 million molecules of a secretory product. In this way, very low concentrations of hormones and other chemical messengers can trigger pronounced cellular responses. Therefore, multistep second-messenger systems are very efficient. Such amplification through a cascading effect is another common functional pattern seen throughout the animal kingdom.

Modifications of the Second-Messenger Pathways

Although membrane receptors can serve as links between extracellular first messengers and intracellular second messengers in the regulation of specific cellular activities, the receptors themselves are also frequently subject to regulation. In many instances, the number and affinity (attraction of a receptor for its chemical messenger) can be altered, depending on the circumstances. For example, the number of receptors for the hormone insulin can be decreased in response to a chronic ele-

vation of insulin in mammalian blood. This can be thought of as a form of acclimatization (p. 15) but is usually called *down-regulation* if receptors are decreased in number, or *upregulation* if increased.

■ Signal transduction may trigger a normal cell function or death.

In the vast majority of cases, the signal transduction pathways triggered by the binding of an extracellular first messenger to a membrane receptor are aimed at promoting useful functions, growth, survival, or reproduction of the cell. By contrast, every cell has an unusual built-in pathway that, if triggered, causes the cell to commit suicide by activating intracellular protein-snipping enzymes, which slice the cell into small, disposable pieces. Such deliberate programmed cell death is termed **apoptosis** (pronounced with a silent "p" as "ap-oh-TOE-sis"). (This term means "dropping off," in reference to the dropping off of cells that are no longer useful, much as autumn leaves drop off trees.) Apoptosis differs from the other form of cell death, **necrosis** (meaning "make dead"). Necrosis is uncontrolled, accidental, messy murder of useful cells that have been severely injured by an agent external to the cell, as by a physical blow, O_2 deprivation, or disease.

Apoptosis is a normal part of life: Individual cells that have become superfluous or disordered are triggered to self-destruct for the greater good of maintaining the whole body's health. Here are some examples:

1. *Development.* Certain unwanted cells produced during development are programmed to kill themselves as a body is sculpted into its final form. For example, apoptosis de-

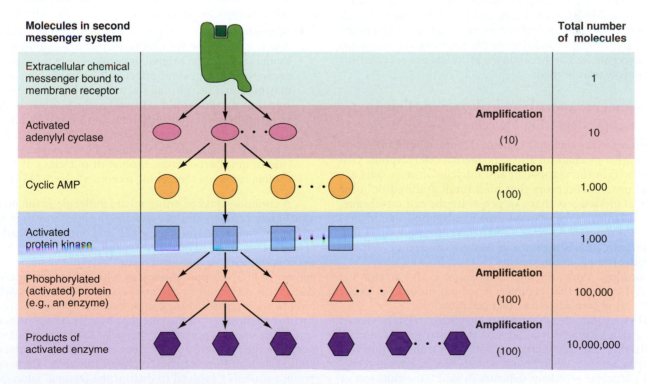

Figure 3–20 ● Amplification of the initial signal by a second-messenger system.
Through amplification, very low concentrations of extracellular chemical messengers, such as hormones, can trigger pronounced cellular responses.

liberately prunes the embryonic ducts capable of forming a mammalian male reproductive tract during the development of a female. Likewise, apoptosis destroys many larval tissues when a caterpillar pupates and becomes a butterfly.

2. *Turnover.* Optimal functioning of most tissues depends on a balance between controlled production of new cells and regulated cellular self-destruction. This balance maintains the proper number of cells in a given tissue while ensuring a controlled supply of fresh cells that are at their peak of performance.

3. *Defense.* Apoptosis provides a means to remove cells infected with harmful viruses. Furthermore, infection-fighting white blood cells (in vertebrates) that have finished their prescribed function and are no longer needed, execute themselves. Also, apoptosis often removes cells that have suffered irreparable damage by exposure to radiation or other poisons, and cells that are becoming cancerous.

A cell signaled to commit suicide detaches itself from its neighbors, then shrinks instead of swelling and bursting. As its lethal weapon, the suicidal cell activates a cascade of normally inactive intracellular protein-cutting enzymes, the **caspases,** which kill the cell from within. When a cell has been signaled to undergo apoptosis, the mitochondria become leaky, permitting *cytochrome c* to leak out into the cytosol. Cytochrome c, a component of the electron transport chain, usually participates in oxidative phosphorylation to produce ATP. Outside its typical mitochondrial environment, cytochrome c activates the caspase cascade. Once unleashed, the caspases act like molecular scissors to systematically dismantle the cell. Snipping protein after protein, they chop up the nucleus, disassembling its life-essential DNA, then break down the internal shape-holding cytoskeleton, and finally fragment the cell itself into disposable membrane-enclosed packets. Cells in the vicinity swiftly engulf and destroy the apoptotic cell fragments by phagocytosis. The breakdown products are recycled for other purposes as needed.

Internal receptors bind hydrophobic first messengers entering the cell.

In addition to the signaling systems that use membrane receptors, another mechanism also uses internal cell receptors. Dispersed within the cytoplasm and nucleus of most cells are special protein receptors that bind small and hydrophobic chemical messengers that can penetrate the cell membrane, primarily steroids, eicosanoids, the hydrophobic amine messengers of the thyroid gland, and gases. Following the binding of messenger and receptor, at least two different events can ensue, depending on the system:

For nitric oxide, the internal receptor is also an enzyme, **guanylyl cyclase,** that produces a second messenger, **cyclic GMP.** This in turn activates other cell proteins.

Thyroid and steroid hormones have what might be viewed as a more profound effect, because they alter gene transcription. In this case the internal receptor is a **transcription factor,** a protein that, when bound with its messenger, can recognize a particular promoter sequence of genes (the **enhancer or response element** sequence). The adjoining coding gene is then transcribed, an mRNA is translated, and new proteins are produced in the cell. Details of this mode of action are cov-

ered at the gene level in Chapter 2, and at the hormonal level in Chapter 7.

Pharmacological agents and toxins are often agonists or antagonists affecting communication mechanisms.

As we have seen, cell communication mechanisms involve a wide variety of small organic signal molecules and large proteins such as receptors and G-protein transducers. The steps involving these molecules are vulnerable to external chemicals entering the animal body. In particular, signaling steps are major targets of chemicals that have evolved for defending against predators and for capturing prey. Plants in particular, because they cannot actively fight or run away, have evolved an enormous variety of defensive chemicals that interfere with animal communication systems. Many animals (such as the curare frog of South America) also have evolved such defensive toxins, and also "offensive" toxins such as snake venom that are used to capture prey. Chemicals that affect a communication mechanism fall into two broad categories:

- **Antagonists** block a step in a communication pathway. Often these are chemicals that are similar in molecular structure to a signal molecule, but different enough to cause interference. Frequently, an antagonist binds to a receptor without activating that receptor, while at the same time blocking the binding of the normal signal molecule. For example, the active agent of the curare frog, *D-tubocurarine,* binds to and blocks the acetylcholine receptor at nerve-muscle synapses. As we shall see later (p. 127), acetylcholine normally triggers receptors that initiate muscle contraction; the toxin causes paralysis. Another example are *antihistamines,* discussed earlier (p. 88).

- **Agonists** activate a step in a communication pathway. Again, often these are chemicals that are similar in molecular structure to a signal molecule, but in this case they cause the same general effect; for example, an agonist may bind to a receptor and activate it, just as the normal signal molecule would. However, the agonist is often structurally somewhat different, so that it may bind for an abnormally long time, or it may not be broken down effectively to shut off the stimulus. For example, *nicotine* (found in tobacco) is an agonist of one class of acetylcholine receptors ("nicotinic receptors"; see p. 154).

A common class of antagonistic molecule is the **methyl xanthines,** *caffeine* and *theophylline,* molecules that have structures similar to those of cAMP and adenosine. Caffeine is found in coffee and cacao beans (chocolate). In tea leaves the concentration of theophylline can be as high as 3.5%. The activity of PDE (phosphodiesterase, which shuts off the cAMP cascade) is decreased by these antagonistic chemicals, which bind to the enzyme. These compounds also block the receptors for **adenosine,** a paracrine that usually dampens neural activity by opening potassium channels. Methyl xanthines are noted for their powerful stimulatory effects on the central nervous system and (in vertebrates) heart function. These compounds probably evolved to disrupt the feeding behavior of herbivorous insects. When administered artificially to animals, methyl xanthines exert unpredictable effects on their behavior. For example, caffeine affects spiders so that they cannot spin a functional web. In many cases scientists do not know whether inhibition of PDE or adenosine receptors or both are the basis of these effects.

Much medicine and physiological research is based on using natural agonists and antagonists (or laboratory-designed ones) as pharmacologic agents (or drugs). Medicine, of course, uses these for treating various diseases, whereas physiologists use them as probes of signal functions.

Membrane Potential

Earlier we noted that solutes are unequally distributed between the interior and exterior of cells. We have also discussed how some cells can exploit these differences in solute concentration on either side of the membrane and can selectively change the permeability of the membrane to particular solutes. In most cells, inorganic solutes are unevenly distributed as well as organic ones, with **sodium (Na^+)** and **chloride (Cl^-)** dominating outside and **potassium (K^+)** (along with organic molecules) dominating inside. Put another way, Na^+ and Cl^- are usually the major **osmolytes** (solutes involved in osmotic pressure regulation) in the **ECF (extracellular fluid)** in animals; whereas K^+ and organic molecules are the dominant osmolytes in the **ICF (intracellular fluid)**. Indeed, the occurrence of K^+ as the dominant intracellular cation appears universal from prokaryotes on up and suggests that this feature originated with the earliest life forms. **Excitable cells** such as neurons regulate permeability of these ions (especially Na^+ and K^+) as a signaling mechanism, as we discuss later in this chapter and the next.

■ Membrane potential is a separation of opposite charges across the plasma membrane.

The unequal distribution of a few key ions between the ICF and ECF and their selective movement through the plasma membrane govern the electrical properties of the membrane. All plasma membranes have a membrane potential, or are polarized electrically. The term **membrane potential** refers to a separation of charges across the membrane, or to a difference in the relative number of cations and anions in the ICF and ECF. Recall that opposite charges tend to attract each other and like charges tend to repel each other. Work must be performed (energy expended) to separate opposite charges after they have come together. Conversely, when oppositely charged particles have been separated, the electrical force of attraction between them can be harnessed to perform work when the charges are permitted to come together again. This is the basic principle underlying electrically powered devices. Potential is measured in units of volts (the same unit used for the voltage in electrical devices), but because the membrane potential is relatively low the unit used is millivolts (mV) (1 mV = 1/1,000 volt).

Because the concept of potential is fundamental to understanding nerve and muscle physiology, it is important to understand clearly what this term means. The membrane in ● Figure 3–21a is electrically neutral. An equal number of positive (+) and negative (−) charges are on each side of the membrane, so no membrane potential exists. In Figure 3–21b, some of the + charges from the right side have been moved to the left. Now the left has an excess of + charges, leaving an excess of − charges on the right. In other words, there is a separation of opposite charges across the membrane, or a difference in the relative number of + and − charges between the two sides (that is, a membrane potential exists). The attractive force between these separated charges causes them to accumulate in a thin layer along the outer and inner surfaces of the plasma membrane (Figure 3–21c). These separated charges represent only a small fraction of the total number of charged particles (ions) present in the ICF and ECF. The vast majority of the fluid inside and outside the cells is electrically neutral (Figure 3–21d). The electrically balanced ions can be ignored, because they do not contribute to membrane potential. Thus an almost insignificant fraction of the total number of charged particles present in the body fluids is responsible for the membrane potential.

The magnitude of the potential depends on the degree of separation of the opposite charges: The greater the number of charges separated, the larger the potential. Therefore, in Figure 3–21e membrane B has more potential than A and less potential than C.

■ Membrane potential is primarily due to differences in the distribution and permeability of key ions.

All living cells have a membrane potential characterized by a slight excess of positive charges outside and a corresponding slight excess of negative charges on the inside. The cells of *excitable tissues*—namely nerve cells and muscle cells—have the ability to produce rapid, transient changes in their membrane potential when excited. These brief fluctuations in potential serve as electrical signals. The constant membrane potential present in nonexcitable tissues and in excitable tissues when they are at rest—that is, when they are not producing electrical signals—is known as the **resting membrane potential.** We concentrate now on the generation and maintenance of the resting potential, and in later chapters we examine the changes that take place in excitable tissues during signaling.

The unequal distribution of a few key ions between the ICF and ECF and their selective movement through the plasma membrane govern the electrical properties of the membrane. In an animal body, electrical charges are carried by ions. The ions primarily responsible for generating the resting membrane potential are Na^+, K^+, and A^-. The last refers to the large, negatively charged (anionic) intracellular proteins. Other ions (calcium, magnesium, chloride, bicarbonate, and phosphate, to name a few) do not make a direct contribution to the electrical properties of the plasma membrane in most cells, even though they play other important roles in the body. The concentrations of these ions differ from one animal to another, but the factors that regulate their distribution are the same.

The concentrations and relative permeabilities of the ions critical to membrane electrical activity are compared in ■ Table 3–4. Note that Na^+ *is in greater concentration in the extracellular fluid and* K^+ *is in much higher concentration in the intracellular fluid.* These concentration differences are maintained in two very distinct ways:

1. *The Na^+–K^+ pump* (p. 81) *at the expense of energy (ATP).*
2. *Different solubilities in cell water and affinity for cell proteins.* Due to attraction of water molecules by proteins, water inside cells appears to form a hydrogen-bonded net-

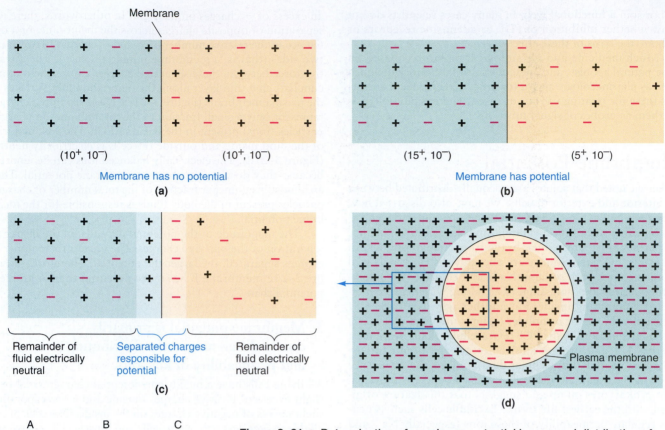

Figure 3–21 ● **Determination of membrane potential by unequal distribution of positive and negative charges across the membrane.** (a) When the positive and negative charges are equally balanced on each side of the membrane, no membrane potential exists. (b) When opposite charges are separated across the membrane, membrane potential exists. (c) The unbalanced charges responsible for the potential accumulate in a thin layer along opposite surfaces of the membrane. (d) The vast majority of the fluid in the ECF and ICF is electrically neutral. The unbalanced charges accumulate in a thin layer along the plasma membrane. (e) Membrane B has more potential than membrane A and less potential than membrane C.

Table 3–4 ▍ Concentration and Permeability of Ions Responsible for Membrane Potential in a Resting Nerve Cell

Ion	Concentration (millimoles/liter)		Relative Permeability
	Extracellular	Intracellular	
Na^+	150	15	1
K^+	5	150	50–75
A^-	0	65	0

work that is different from the network of water by itself. Experimentally, researchers have found that K^+ is more soluble in this internal water than is Na^+ and that this leads to K^+ preferentially entering a cell. Also, the negative charges of proteins attract K^+ more strongly than Na^+, because K^+ has a smaller **hydration shell** (a sphere of water molecules attracted by the ion's positive charge). The ion must lose its hydration shell before it can bind

to a protein, and it is easier for K^+ to do this than for Na^+. This effect also reinforces the accumulation of K^+ over Na^+.

Because the plasma membrane is virtually impermeable to A^-, these large, negatively charged proteins are found *only inside* the cell. After they have been synthesized from amino acids transported into the cell, they remain trapped within the cell.

In addition to the active carrier mechanism, Na^+ and K^+ can passively cross the membrane through protein channels specific for them. It is usually much easier for K^+ than for Na^+ to get through the membrane because typically the membrane has many more K^+ leak channels than it has Na^+ leak channels. Thus, more channels are open for passive K^+ traffic across the membrane. In a nerve cell at rest (that is, when it is not conducting a nerve impulse), the membrane is about 50 to 75 times more permeable to K^+ than to Na^+.

Armed with a knowledge of the relative concentrations and permeabilities of these ions, we can now analyze the forces acting across the plasma membrane. This analysis is broken down as follows: First we consider the direct contributions of the Na^+–K^+ pump to membrane potential; second, the effect the movement of K^+ alone has on membrane potential; third, the effect of Na^+ alone; and finally, the situation that exists in the cells when both K^+ and Na^+ effects are taking place concurrently. ▮ Table 3–5 summarizes the concentration and electrical gradients that exist for K^+ and for Na^+ under various conditions. Remember that the concentration gradient for K^+ is always outward and the concentration gradient for Na^+ is always inward, because the Na^+–K^+ pump maintains a higher concentration of K^+ inside the cell and a higher concentration of Na^+ outside the cell. Also, note that because K^+ and Na^+ are both cations (positively charged), the *electrical gradient for both is always toward the negatively charged side of the membrane.*

Effect of sodium–potassium pump on membrane potential

About 20% of the membrane potential is directly generated by the Na^+–K^+ pump. This active transport mechanism pumps three Na^+ out for every two K^+ it transports in. Because Na^+ and K^+ are both positive ions, this unequal transport generates a membrane potential, with the outside becoming relatively more positive than the inside as more positive ions are transported out than in. However, most of the membrane potential—the remaining 80%—is caused by the passive diffusion of K^+ and Na^+ down concentration gradients. Thus most of the Na^+–K^+ pump's role in producing membrane potential is indirect, through its critical contribution to maintaining the concentration gradients directly responsible for the ion movements that generate most of the potential.

Effect of the movement of potassium alone on membrane potential

Let's consider a hypothetical situation characterized by (1) the concentrations that exist for K^+ and A^- across the plasma membrane, (2) free permeability of the membrane to K^+ but not to A^-, and (3) no potential as yet present. The concentration gradient for K^+ would tend to move this ion out of the cell (● Figure 3–22). Because the membrane is permeable to

Table 3–5 ▮ Concentration and Electrical Gradients for K^+ and Na^+ at Equilibrium Potential (E) and Resting Potential

Ion	Condition	Gradient	Direction of Gradient
K^+	E_{K^+} (−90 mV)	Concentration gradient	Outward
		Electrical gradient	Inward
Na^+	E_{Na^+} (+60 mV)	Concentration gradient	Inward
		Electrical gradient	Outward
K^+	Resting potential (−70 mV)	Concentration gradient	Outward
		Electrical gradient	Inward
Na^+	Resting potential (−70 mV)	Concentration gradient	Inward
		Electrical gradient	Inward

K^+, this ion would readily pass through. As potassium ions moved to the outside, they would carry their positive charge with them, so more positive charges would be on the outside whereas negative charges in the form of A^- would be left behind on the inside, similar to the situation shown in Figure 3–21b. (Remember that the large protein anions cannot diffuse out, despite a tremendous concentration gradient.) A membrane potential would now exist. Because an electrical gradient would also be present, K^+, being a positively charged ion, would be attracted toward the negatively charged interior and repelled by the positively charged exterior. Thus, two opposing forces would now be acting on K^+: the concentration gradient tending to move K^+ out of the cell and the electrical gradient tending to move these same ions into the cell

Initially, the concentration gradient would be stronger than the electrical gradient, so net diffusion of K^+ out of the cell would continue and the membrane potential would increase. As more and more K^+ moved down its concentration gradient and out of the cell, however, the opposing electrical gradient would also become greater as the outside became increasingly more positive and the inside more negative. You might think that the outward concentration gradient for K^+

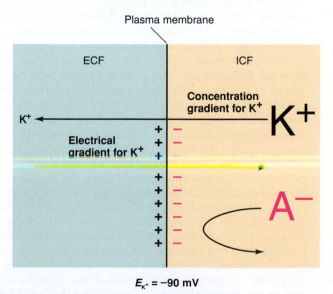

Plasma membrane

ECF | ICF

Concentration gradient for K^+ K^+

K^+ ←

Electrical gradient for K^+

A^-

E_{K^+} = −90 mV

① The concentration gradient for K^+ tends to push this ion out of the cell.

② The outside of the cell becomes more + as the positively charged K^+ ions move to the outside down their concentration gradient.

③ The membrane is impermeable to the large intracellular protein anion (A^-). The inside of the cell becomes more − as the positively charged K^+ ions move out, leaving behind the negatively charged A^-.

④ The resulting electrical gradient tends to move K^+ into the cell.

⑤ No further net movement of K^+ occurs when the inward electrical gradient exactly counterbalances the outward concentration gradient. The membrane potential at this equilibrium point is the equilibrium potential for K^+ (E_{K^+}) at −90mV.

Figure 3–22 ● Equilibrium potential for K^+.

would gradually decrease as K^+ leaves the cell down this gradient. Surprisingly, however, the K^+ concentration gradient would remain essentially constant despite the outward movement of K^+. The reason is that only infinitesimal movement of K^+ out of the cell would bring about rather large changes in membrane potential. Accordingly, such an extremely few K^+ ions present in the cell would have to leave to establish an opposing electrical gradient that the K^+ concentration inside and outside the cell would remain essentially unaltered. As K^+ would continue to move out down its unchanging concentration gradient, the inward electrical gradient would continue to increase in strength. Net outward diffusion would gradually be reduced as the strength of the electrical gradient approached that of the concentration gradient. Finally, when these two forces exactly balanced each other, no further net movement of K^+ would occur. The potential that would exist at this equilibrium is known as the **equilibrium potential** for K^+ (E_{K^+}). At this point, a large concentration gradient for K^+ would still exist, but no more net movement of K^+ would occur out of the cell down this concentration gradient because of the exactly equal opposing electrical gradient (Figure 3–22).

The membrane potential at E_{K^+} is -90 mV. It is not really a negative potential. By convention, *the sign always designates the polarity of the excess charge on the inside of the membrane.* A membrane potential of -90 mV means that the potential is of a magnitude of 90 mV, with the inside being negative relative to the outside. A potential of $+90$ mV would have the same strength, but in this case the inside would be more positive than the outside.

The equilibrium potential for a given ion of differing concentrations across a membrane can be calculated by means of the **Nernst equation** as follows:

$$E = 61/z \log C_0/C_i$$

where

E = equilibrium potential for ion in mV

61 = a constant that incorporates the universal gas constant (R), absolute temperature (T), the ion's valence (z), and an electrical constant known as Faraday (F); 61 = RT/zF

C_0 = concentration of the ion outside the cell (ECF) in millimoles/liter (millimolar, mM)

C_i = concentration of the ion inside the cell (ICF) in mM

z = the charge on the ion (for divalent ions such as Ca^{2+}, z = 2)

Given the mammalian ECF concentration of K^+ at 5 mM and the ICF concentration at 150 mM:

$$E_{K^+} = 61 \log 5 \text{ mM}/150 \text{ mM}$$
$$= 61 \log 1/30$$

Because the log of $1/30 = -1.477$,

$$E_{K^+} = 61(-1.477) = -90 \text{ mV}$$

Because 61 is a constant (for a mammal with a 37°C body temperature), the equilibrium potential is essentially a measure of the membrane potential (that is, the magnitude of the electrical gradient) that exactly counterbalances the concentration gradient that exists for the ion (that is, the ratio between the ion's concentration outside and inside the cell). For a poi-

kilotherm with a lower body temperature of 18°C, the constant is reduced to 58. Note that the larger the concentration gradient for an ion, the greater the ion's equilibrium potential. A comparably greater opposing electrical gradient would be required to counterbalance the larger concentration gradient.

Effect of movement of sodium alone on membrane potential

A similar hypothetical situation could be developed for Na^+ alone (● Figure 3–23). The concentration gradient for Na^+ would move this ion into the cell, building up positive charges on the interior of the membrane and leaving negative charges unbalanced outside (primarily in the form of chloride, Cl^-; Na^+ and Cl^- are the predominant ECF ions). Net diffusion inward would continue until equilibrium was established by the development of an opposing electrical gradient that exactly counterbalanced the concentration gradient. At this point, given the concentrations for Na^+, the **Na^+ equilibrium potential** (E_{Na^+}) would be $+60$ mV. In this case the inside of the cell would be positive, in contrast to the equilibrium potential for K^+. The magnitude of $E Na^+$ is somewhat less than for E_{K^+} (60 mV compared to 90 mV) because the concentration gradient for Na^+ is not as large (Table 3–4); thus the opposing electrical gradient (membrane potential) is not as great at equilibrium.

Concurrent potassium and sodium effects on membrane potential

Neither K^+ nor Na^+ exists alone in the body fluids, so equilibrium potentials are not present in the body cells. They exist only in hypothetical or experimental conditions. In a living cell, the effects of both K^+ and Na^+ must be taken into account. *The greater the permeability of the plasma membrane for a given ion, the greater the tendency for that ion to drive the membrane potential toward the ion's own equilibrium potential.* Because the membrane at rest is 50 to 75 times more permeable to K^+ than to Na^+, K^+ passes through more readily than Na^+; thus, K^+ influences the resting membrane potential to a much greater extent than does Na^+. Recall that K^+ acting alone would establish an equilibrium potential of -90 mV. The membrane is somewhat permeable to Na^+, however, so some Na^+ enters the cell in a limited attempt to reach its equilibrium potential. This Na^+ influx neutralizes, or cancels, some of the potential produced by K^+ alone.

To facilitate an understanding of this concept, assume that each separated pair of charges in ● Figure 3–24 represents 10 mV of potential. (This is not technically correct, because in reality, many separated charges must be present to account for a potential of 10 mV.) In this simplified example, nine separated pluses and minuses, with the minuses on the inside, would represent the E_{K^+} of -90 mV. Superimposing the slight influence of Na^+ on this K^+-dominated membrane, assume that two sodium ions enter the cell down the Na^+ concentration and electrical gradients. (Note that the electrical gradient for Na^+ is now inward in contrast to the outward electrical gradient for Na^+ at E_{Na^+}. At E_{Na^+}, the inside of the cell is positive as a result of the inward movement of Na^+ down its concentration and electrical gradients. In a resting nerve cell, however, the inside is negative because of the dominant influence of K^+ on membrane potential. Thus, both the concentration and electrical gradients now favor the inward movement of Na^+). The inward movement of these two positively

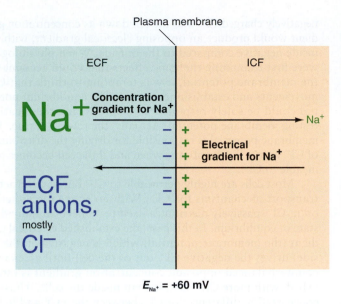

Plasma membrane	

ECF | **ICF**

Na^+

Concentration gradient for Na^+

→ Na^+

− + Electrical gradient for Na^+

ECF anions, mostly **Cl⁻**

E_{Na^+} = +60 mV

① The concentration gradient for Na^+ tends to push this ion into the cell.

② The inside of the cell becomes more positive as the positively charged Na^+ ions move to the inside down their concentration gradient.

③ The outside becomes more negative as the positively charged Na^+ ions move in, leaving behind in the ECF unbalanced negatively charged ions, mostly Cl⁻.

④ The resulting electrical gradient tends to move Na^+ out of the cell.

⑤ No further net movement of Na^+ occurs when the outward electrical gradient exactly counterbalances the inward concentration gradient. The membrane potential at this equilibrium point is the equilibrium potential for Na^+ (E_{Na^+}) at +60 mV.

Figure 3–23 ● Equilibrium potential for Na^+.

charged sodium ions neutralizes some of the potential established by K^+, so now only seven pairs of charges are separated, and the potential is −70 mV. This is the **resting membrane potential** of a typical nerve cell. All cells have a negative resting membrane potential with values that range from −30 to −100 mV. The resting potential of a cell is much closer to E_{K^+} than to E_{Na^+} because of the greater permeability of the membrane to K^+, but it is slightly less than E_{K^+} (−70 mV is a lower potential than −90 mV) because of the weak influence of Na^+.

At resting potential, neither K^+ nor Na^+ is at equilibrium. A potential of −70 mV does not exactly counterbalance the concentration gradient for K^+; it takes a potential of −90 mV to do that. Thus there is a continual tendency for K^+ to passively *leak* out through its channels. In the case of Na^+, the concentration and electrical gradients do not even oppose each other; they both favor the inward movement of Na^+. Therefore, Na^+ continually leaks inward down its electrochemical gradient, but only slowly because of its low permeability; that is, because of the sparseness of Na^+ leak channels.

① The Na^+–K^+ pump actively transports Na^+ out of and K^+ into the cell, keeping the concentration of Na^+ high in the ECF and the concentration of K^+ high in the ICF.

② Given the concentration gradients that exist across the plasma membrane, K^+ tends to drive the membrane potential to K^+'s equilibrium potential (−90 mV), whereas Na^+ tends to drive the membrane potential to Na^+'s equilibrium potential (+60 mV).

③ However, K^+ exerts the dominant effect on the resting membrane potential because the membrane is more permeable to K^+. As a result, the resting potential (−70 mV) is much closer to E_{K^+} than to E_{Na^+}.

④ During the establishment of resting potential, the relatively large net diffusion of K^+ outward does not produce a potential of −90 mV because the resting membrane is slightly permeable to Na^+ and the relatively small net diffusion of Na^+ inward neutralizes (in gray shading) some of the potential that would be created by K^+ alone, bringing the resting potential to −70 mV, slightly less than E_{K^+}.

⑤ The negatively charged intracellular proteins (A⁻) that cannot permeate the membrane remain unbalanced inside the cell during the net outward movement of the positively charged ions, so the inside of the cell is more negative than the outside.

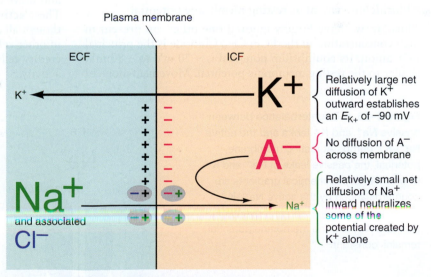

Plasma membrane

ECF | **ICF**

K^+ ←

Relatively large net diffusion of K^+ outward establishes an E_{K^+} of −90 mV

A^-

No diffusion of A⁻ across membrane

Na^+ and associated Cl^-

→ Na^+

Relatively small net diffusion of Na^+ inward neutralizes some of the potential created by K^+ alone

Resting membrane potential = −70 mV

(A⁻ = Large intracellular anionic proteins)

Figure 3–24 ● Effect of concurrent K^+ and Na^+ movement on establishing the resting membrane potential.

Because such leaking goes on all the time, why doesn't the intracellular concentration of K^+ continue to fall and the concentration of Na^+ inside the cell progressively increase? This does not happen, because of the Na^+–K^+ pump. This active transport mechanism counterbalances the rate of leakage (● Figure 3–25). At resting potential, the pump transports back into the cell essentially the same number of potassium ions that have leaked out and simultaneously transports to the outside the sodium ions that have leaked in. Because the pump offsets the leaks, the concentration gradients for K^+ and Na^+ remain constant across the membrane. Thus not only is the Na^+–K^+ pump initially responsible for the Na^+ and K^+ concentration differences across the membrane, but it also maintains these differences.

As just discussed, it is the presence of these concentration gradients, together with the difference in permeability of the membrane to these ions, that accounts for the resting membrane potential. In this resting state, the potential remains constant. There is no net movement of any ions. All passive forces are exactly balanced by active forces. A steady state exists, even though there is still a strong concentration gradient for both K^+ and Na^+ in opposite directions, as well as a slight excess of positive charges in the ECF accompanied by a corresponding slight excess of negative charges in the ICF (enough to account for a potential of the magnitude of 70 mV). At this point, although movement across the membrane is taking place by means of passive leaks and active pumping, the exchange of charges between the ICF and ECF is exactly balanced, with the potential that has been established by these forces remaining constant.

Consider the effect of a twofold increase in the ECF concentration of Na^+ or K^+ ions. Which increase in ion concentration is the more deadly and why? Does the calculated decrease in E_{K^+} or the increase in E_{Na^+} bring the resting membrane potential of the nerve or muscle cell closer to threshold potential? Why would this change in ion concentration stop the heart from beating?

Chloride movement at resting membrane potential

Thus far, we have largely ignored one other ion present in high concentration in the ECF: Cl^-. Chloride is the principal ECF anion. Its equilibrium potential is −70 mV to −80 mV, similar to the resting membrane potential. Movement alone of negatively charged Cl^- into the cell down its concentration gradient would produce an opposing electrical gradient, with the inside negative compared to the outside. When physiologists were first examining the ionic effects that could account for the membrane potential, it was tempting to think that Cl^- movements and establishment of the Cl^- equilibrium potential could be solely responsible for producing the identical resting membrane potential. Actually, the reverse is true. The membrane potential is responsible for driving the distribution of Cl^- across the membrane when and if the cell becomes permeable to Cl^-.

Most cells are highly permeable to Cl^- but have no active transport mechanisms for Cl^-. With no active forces acting on it, Cl^- passively distributes itself to achieve an individual state of equilibrium. In this case, the established electrical gradient (the membrane potential, which is negative on the inside) drives the negative Cl^- out of the cell until an exactly counterbalanced opposing concentration gradient is established, with more Cl^- outside than inside the cells. Thus the concentration difference for Cl^- between the ECF and ICF is brought about passively by the presence of the membrane potential, rather than being maintained by an active pump, as is the case for K^+ and Na^+. Therefore, in most cells Cl^- does not influence membrane potential; instead, membrane potential passively influences the Cl^- distribution. (Some specialized cells have an active Cl^- pump, with subsequent movement of Cl^- accounting for part of the potential.)

Specialized use of membrane potential in nerve and muscle cells

Nerve and muscle and cells have developed a specialized use for membrane potential. They can rapidly and transiently alter their membrane permeabilities to the involved ions in response to appropriate stimulation, thereby bringing about fluctuations in membrane potential. The rapid fluctuations in potential are responsible for producing nerve impulses in sensory and nerve cells and for triggering contraction in muscle cells. These activities are the focus of the next five chapters. Even though all cells display a membrane potential, its significance in other cells is uncertain, although the potential of some secretory cells may somehow be linked to their level of secretory activity.

Figure 3–25 ● **Counterbalance between passive Na^+ and K^+ leaks and the active Na^+–K^+ pump.** At resting membrane potential, the passive leaks of Na^+ and K^+ down their electrochemical gradients are counterbalanced by the active Na^+–K^+ pump, so that there is no net movement of Na^+ and K^+, and the membrane potential remains constant.

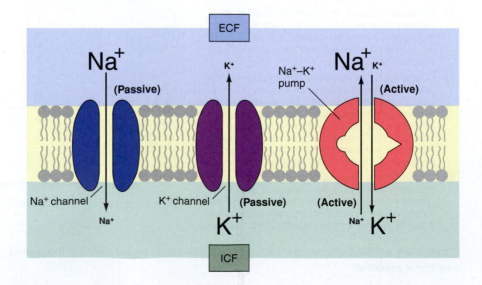

Chapter in Perspective
HOMEOSTASIS AND INTEGRATION

All cells of the body must obtain vital materials such as nutrients and O_2 from the surrounding ECF and must transfer to the ECF wastes to be eliminated as well as secretory products such as chemical messengers and digestive enzymes. Thus, transport of materials across the plasma membrane between the ECF and ICF is essential for cell survival, and to support these life-sustaining exchanges the constituents in the ECF must be homeostatically maintained.

Many cell types use membrane transport to carry out their specialized activities geared toward maintaining homeostasis. Following are several examples:

1. Absorption of nutrients from the digestive tract lumen involves the transport of these energy-giving molecules across the membranes of the cells lining the tract.

2. Exchange of O_2 and CO_2 between the air and blood in the lungs involves transport of these gases across the membranes of cells lining the lungs' air sacs and blood vessels.

3. Urine formation is accomplished by selective transfer of materials between blood and fluid within the kidney tubules across the membranes of cells lining the tubules.

4. Secretion of chemical messengers such as neurotransmitters from nerve cells and hormones from endocrine cells involves transport of these regulatory products to the ECF on appropriate stimulation.

5. The plasma membrane also contains receptor sites for binding with specific chemical messengers that regulate various cell activities, many of which are specialized activities aimed toward maintaining homeostasis. For example, the hormone vasopressin, which is secreted in response to a water deficit in the body, binds with receptor sites in the plasma membrane of a specific type of kidney cell. This binding triggers these cells to conserve water during urine formation, thus helping alleviate the water deficit that initiated the response.

All living cells have a membrane potential, with the cell's interior being slightly more negative than the fluid surrounding the cell when the cell is electrically at rest. The specialized activities of nerve and muscle cells depend on these cells' ability to change their membrane potential rapidly on appropriate stimulation. These transient, rapid changes in potential in nerve cells serve as electrical signals or nerve impulses, which provide a means to transmit information along nerve pathways. This information is used to accomplish homeostatic adjustments, such as restoring blood pressure to normal when signaled that it has fallen too low. ■

REVIEW QUESTIONS *(Answers are on p. A–1.)*

Additional study tools for this chapter, including chapter summaries and practice tests, are available online at *www.biology.brookscole.com*

1. The plasma membrane
   a. is a lipid bilayer with embedded proteins
   b. contains cholesterol
   c. selectively permits movement of specific substances between the cell and its environment
   d. can be described as a fluid mosaic of proteins in a lipid bilayer
   e. all of the above

2. Proteins in the plasma membrane serve as channels, carriers, receptors, docking-marker acceptors, cells adhesion markers (CAMs), self-identity markers, and
   a. hydrophobic bilayers
   b. hydrophilic trilaminates
   c. membrane-bound enzymes
   d. confinement zones
   e. cholesterol storage vesicles

3. One of the more important aspects of diffusion is
   a. how rapidly it occurs
   b. that it is selectively permeable
   c. that it requires ATP
   d. how slowly it occurs
   e. that it requires a large surface area

4. An electrochemical gradient
   a. requires only cations
   b. provides the forces needed to move ions by diffusion
   c. allows more rapid movements of materials than diffusion
   d. requires bulk transport
   e. occurs only through aquaporins

5. Osmosis involves the movement of
   a. solute molecules across a membrane
   b. water molecules down a concentration gradient
   c. water molecules across a membrane into a solution of lower concentration
   d. water molecules and solute molecules across a membrane
   e. membrane-bound proteins

6. Colligative properties
   a. depend solely on the concentration of dissolved particles
   b. reflect the chemical properties of dissolved particles
   c. depend on a combination of the concentration and the chemical properties of dissolved particles
   d. are determined by the properties of the membrane separating two solutions
   e. determine the tonicity of a solution

7. Carrier-mediated transport
   a. is dependent on the amount of lipids in a membrane
   b. is generally carried out by membrane proteins
   c. occurs only down concentration gradients

d. is always passive

e. always requires energy

8. Active transport
   a. always requires energy
   b. occurs against concentration gradients
   c. involves carriers with ATPase activity
   d. involves phosphorylation and dephosphorylation of the carrier
   e. all of the above

9. The Na^+–K^+ ATPase pump
   a. is found only in mitochondria
   b. becomes more active under the influence of ouabain
   c. maximizes the effects of osmotic forces
   d. moves three Na^+ ions out of the cell for every two K^+ it pumps in
   e. is found only in nerve cells

10. Endocytosis
    a. moves material out of the cell
    b. moves materials from place to place only within the cell
    c. is the opposite of phagocytosis
    d. moves material into the cell
    e. provides a mechanism for secreting large polar molecules

11. Exocytosis
    a. moves material into the cell
    b. moves material out of the cell
    c. requires lysosomes
    d. is almost always more rapid than endocytosis

12. Which of the following is not a type of chemical messenger?
    a. cytokines
    b. hormones
    c. caveolines
    d. pheromones
    e. paracrines

13. Which of the following is not involved directly in first messenger dynamics?
    a. ligand gated channels
    b. mechanically gated channels

c. kinase

d. cAMP

e. voltage gated channels

14. Which of the following is not a component of second messenger pathways?
    a. protein kinase A
    b. cAMP
    c. neurohormones
    d. calcium
    e. adenylyl cyclase

15. Intentional programmed cell death is
    a. caused by cascading
    b. called apoptosis
    c. promoted by gap junctions
    d. initiated by calmodulin
    e. dependent on G proteins

16. A cell's resting membrane potential is due
    a. to the equal distribution of key ions between the ICF and the ECF
    b. to the unequal distribution of key ions between the ICF and the ECF
    c. solely to the distribution of the K^+ ions
    d. mainly to the distribution of Na^+ ions
    e. only to the presence of negatively charged intracellular proteins

17. At the equilibrium potential for K^+
    a. the concentration of K^+ inside of the cell and outside of the cell are equal
    b. the concentration of K^+ outside of the cell is equal to the concentration of Na^+ inside the cell
    c. no net movement of K^+ out of the cell would occur
    d. negatively charged intracellular proteins would begin to leak out of the cell to balance the charges on each side of the membrane
    e. the flow of Na^+ out of the cell would equal the flow of K^+ into the cell

---

SUGGESTED READINGS AND INTERNET SITES

Hammel, H. T. (1999). Evolving ideas about osmosis and capillary fluid exchange. *FASEB Journal* 13:213–231. Describes an alternative explanation for osmosis.

Ling, G. N., & G. Bohr. 1970. Studies on ion distribution in living cells. Part 2. Cooperative interaction between intracellular K^+ and Na^+. *Biophysical Journal* 10:519. Discusses why K^+ universally accumulates more than Na^+ in cells.

Singer, S. J., & G. Nicholson. 1972. The fluid mosaic model of the structure of cell membranes. *Science* 175:720–731. The classic paper establishing the modern membrane model.

INFOTRAC READINGS

Drab, M., P. Verkade, M. Elger, et al. 2001. Loss of caveolae, vascular dysfunction, and pulmonary defects in caveolin-1 gene-disrupted mice. *Science* 293:2449–2455.

Parton, R. G. 2001. Life without caveolae. *Science* 293:2404. Discusses the effect of knockout genes for caveolae.

INTERNET SITES

Bowen, R. 2000, July 9. *Osmosis and Hydrostatic Pressure Simulator.* **arbl.cvmbs.colostate.edu/hbooks/cmb/cells/pmemb/hydrosim.html.** An interactive simulation of osmosis.

Maciver, S. 2002, December 10. *Caveolae.* **bms.ed.ac.uk/research/others/smaciver/cyto-topics/caveolae.htm.** History and current research on caveolae.

Neuronal Physiology

Photo: © Keith Clements, Photographer/
www.H₂Ophotography.com

The squid Loligo *has giant nerve axons for rapid locomotion. Because of their large size (≈ 1 mm), these neurons were initially mistaken for blood vessels. It was the British investigator J. Z. Young who, in 1936, identified them as nervous tissue based on their ability to conduct a signal called an action potential. The comparatively large size of these neurons permitted the intracellular placement of recording electrodes for measurement of voltage changes during the course of an action potential. These studies, by A. Hodgkin and A. Huxley (in Britain) and by K. Cole and H. Curtis (in America) in the 1930s, yielded the first detailed analysis of ion movements and voltage changes during nerve signaling. The squid axon studies are a classic example of the August Krogh Principle (p. 3).*

Introduction

Every cell in an animal displays a membrane potential, which refers to a separation of positive and negative charges across the membrane, as discussed in the preceding chapter. This potential is related to the uneven distribution of Na^+, K^+, and large intracellular protein anions between the intracellular fluid (ICF) and extracellular fluid (ECF), and to the differential permeability of the plasma membrane to these ions (see p. 95).

■ Neurons and muscles are excitable tissues.

Two types of cells, *neurons* (or *nerve cells*) and *muscle cells,* have developed a specialized use for this membrane potential. They can undergo transient, rapid changes in their membrane potentials. These fluctuations in potential serve as electrical signals. The constant membrane potential that exists when a neuron or muscle cell is not displaying rapid changes in potential is called the *resting potential* (although the membrane is far from "resting," because of the balanced leak–pump activity constantly going on). In Chapter 3 you learned that the resting membrane potential of a typical neuron is −70mV.

Neurons and muscle cells are **excitable tissues;** that is, they can produce electrical signals when stimulated. Evolution of these signaling devices enabled both long-distance communication and coordination of activities between cells. Neurons use these electrical signals to receive, process, initiate, and transmit messages. In muscle cells, these electrical signals initiate contraction. Thus electrical signals are critical to the function of the nervous system as well as all muscles. In this chapter, we consider how neurons undergo changes in potential to accomplish their function. Muscle cells are discussed in later chapters (especially Chapter 8).

Terminology and Methodology

Before understanding what these electrical signals are and how they are created, you must become familiar with the following terms used to describe changes in potential, as pictured in ● Figure 4–1:

1. **Polarization:** Charges are separated across the plasma membrane, so that the membrane has potential to do work. Any time the value of the membrane potential is not 0 mV, in either the positive or negative direction, the membrane is in a state of polarization. Recall that the magnitude of the potential is directly proportional to the number of positive and negative charges separated by the membrane and that the sign of the potential (+ or −) always designates whether excess positive or excess negative charges are present, respectively, on the inside of the membrane.

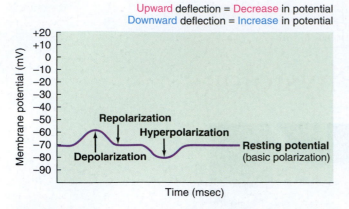

Figure 4–1 ● Types of changes in membrane potential.

2. **Depolarization:** A change in potential that makes the membrane less polarized (less negative) than at resting potential. Depolarization decreases membrane potential, moving it closer to 0 mV (for example, a change from −70 mV to −60 mV); fewer charges are separated than at resting potential.

3. **Repolarization:** The membrane returns to resting potential after having been depolarized.

4. **Hyperpolarization:** A change in potential that makes the membrane more polarized (more negative) than at resting potential. Hyperpolarization increases membrane potential, moving it even farther from 0 mV (for instance a change from −70 mV to −80 mV); more charges are separated than at resting potential.

It is also important to understand how researchers obtained much of the information in this chapter (and the end of Chapter 3). There are two major types of techniques. The first involves the use of various **microelectrodes** that can be inserted into a neuron with little damage. With these electrodes (in conjunction with electrodes outside the cell), local potentials across membranes can be measured accurately (see the caption to the chapter opening photograph). In addition, such electrodes can be used to produce a **voltage clamp,** in which the potential across a membrane is held at a constant value by an electronic circuit. This allows the researcher to investigate, for example, what kinds of ions move across the membrane at any given voltage. The second group of methodology involves **patch clamping,** in which a tiny pipette is attached to a plasma membrane with gentle suction. This forms a tight seal around one patch of a membrane. With a fine-enough pipette, a researcher can isolate a *single* ion channel or receptor protein and measure its properties. Through the pipette, various substances can be added (such as signal molecules to bind to receptors); and electrodes within and external to the pipette can be used to measure or control (for example, clamp) potentials.

In addition, much research on neuron function, including all the pioneering work, has been done on invertebrate preparations (see the opening photograph of the chapter). Over the years, researchers have found that neurons work by nearly identical membrane and molecular mechanisms in all animal phyla that have been studied. Thus there has been

a high degree of conservation since the first animal neurons evolved.

■ Electrical signals are produced by changes in ion movement through ion channels across the plasma membrane.

Changes in membrane potential are brought about by changes in ion movement across the membrane. For example, if the net inward flow of positively charged ions increases compared to the resting state, the membrane becomes depolarized (less negative inside). By contrast, if the net outward flow of positively charged ions increases compared to the resting state, the membrane becomes hyperpolarized (more negative inside).

Changes in ion movement in turn are brought about by changes in membrane permeability in response to *triggering events*. Depending on the type of electrical signal, a triggering event might be (1) a stimulus, such as sound waves stimulating specialized neural endings in an animal's ear; (2) a change in the electrical field in the vicinity of an ion channel within the membrane of an excitable cell; (3) an interaction of a chemical messenger with a surface receptor on a neuron or muscle cell membrane: or (4) a spontaneous change of potential caused by inherent imbalances in the leak–pump cycle. (You will learn more about the nature of these various triggering events as our discussion of electrical signals continues.)

Because the water-soluble ions responsible for carrying charge cannot penetrate the plasma membrane's lipid bilayer, these charges can only cross the membrane through channels specific for them. There are two types of channels: **leak channels** or nongated channels, which are open all the time, and **gated channels,** which can be opened or closed in response to specific triggering events. Thus, triggering events alter membrane permeability and consequently ionic flow across the membrane by opening or closing the gates guarding particular ionic channels. These ionic movements redistribute charge across the membrane, causing membrane potential to fluctuate.

Gated channels have movable folds in the protein that can alternately be open, permitting ion passage through the channel, or closed, preventing ion passage through the channel. Like many proteins, these channels are inherently flexible molecules whose conformations can be altered in response to external factors (Chapter 3, p 71). There are at least three kinds of gated channels, depending on the factor that induces the change in channel conformation: (1) **voltage-gated ion channels,** which open or close in response to changes in membrane potential; (2) **chemically gated channels** (also called **ligand gated**), which change conformation *allosterically* (p. 25) in response to the binding of a specific chemical messenger with a membrane receptor that is in close association with the channel, as in channels involved in signal transduction (see p. 89); and (3) **mechanically gated channels,** which respond to stretching or other mechanical deformation. There is also evidence of a fourth kind of channel, which opens or closes when a phosphate group is covalently added by a kinase (p. 25).

There are two basic forms of electrical signals: (1) the more primitive **graded potentials,** which serve as short-distance signals, and (2) **action potentials,** which signal over longer distances. Let's examine these types of signals in more detail and then explore how neurons use these signals to convey messages.

Graded Potentials

Graded potentials are local changes in membrane potential that occur in varying grades or degrees of magnitude or strength. For example, membrane potential could change from -70 mV to -60 mV (a 10-mV graded potential change) or from -70 mV to -50 mV (a 20-mV graded potential change).

■ The stronger a triggering event, the larger the resultant graded potential.

Graded potentials are usually produced by a specific triggering event that causes gated ion channels to open in a specialized region of the excitable cell membrane. Most commonly, gated Na^+ channels open in response to the triggering event, leading to the inward movement of Na^+ down its concentration and electrical gradients. The resultant depolarization—the graded potential—is confined to this relatively small specialized region of the total plasma membrane.

The magnitude of this initial graded potential (that is, the difference between the new potential and the resting potential) is related to the magnitude of the triggering event: *The stronger the triggering event, the more gated channels that open, the greater the positive charge entering the cell, and the larger the depolarizing graded potential at the point of origin.* Also, the longer the duration of the triggering event, the longer the duration of the graded potential (● Figure 4–2). Graded potentials can either be depolarizing or hyperpolarizing.

When a graded potential occurs locally in a neuron or muscle cell membrane, the potential in this area differs from that in the remainder of the membrane, which is still at resting potential. The temporarily depolarized region is called an *active area*. Note from ● Figure 4–3 that inside the cell, the active area is relatively more positive than the neighboring *inactive areas* that are still at resting potential. Outside the cell, the active area is relatively less positive than these adjacent areas. Because of this difference in potential, electrical charges, in this case carried by ions, passively flow between the active and adjacent resting regions on both the inside and outside of the membrane. Any flow of electrical charges is called a *current*. By convention, the direction of current flow is always designated by the direction in which the positive charges are moving (Figure 4–3c). On the inside, positive charges flow through the ICF away from the relatively more positive depolarized active region toward the more negative adjacent resting regions. Similarly, outside the cell, positive charges flow through the ECF from the more positive adjacent inactive regions toward the relatively more negative active region. Ion movement (that is, current) is occurring *along* the membrane between regions next to each other on the same side of the membrane. This flow is in contrast to ionic movement *across* the membrane through ionic channels with which you are more familiar.

As a result of local current flow between an active depolarized area and an adjacent inactive area, potential in the previously inactive area alters. Positive charges have flowed into this adjacent area on the inside, while simultaneously positive charges have flowed out of this area on the outside. Thus at this adjacent site the inside is more positive (or less negative) than before (Figure 4–3c). Stated differently, the previously inactive adjacent region has been depolarized; thus the graded potential has spread. This area's potential now differs from that of the inactive region immediately next to it on the other side, inducing further current flow at this new site, and so on. In this manner, current spreads in both directions away from the initial site of the potential change.

The amount of current that flows between two areas depends on the potential difference between the areas and on the resistance of the material through which the charges are moving. **Resistance** is the hindrance to electrical charge movement. The greater the difference in potential, the greater the current flow. Similarly, the lower the resistance, the greater the current flow. *Conductors* have low resistance and provide little hindrance to current flow. Electrical wires and the intracellular and extracellular fluid are all good conductors, so current readily flows through them. *Insulators* have high resistance and greatly hinder movement of charge. The plastic surrounding electrical wires have high resistance, as do body lipids. Thus current does not flow across the plasma membrane's lipid bilayer. Current, carried by ions, can move across the membrane only through ion channels.

■ Graded potentials die out over short distances.

The passive current flow between active and adjacent inactive areas is similar to the means by which current is carried through electrical wires. We know from experience that current leaks out of an electrical wire unless the wire is covered with an insulating material such as plastic. Likewise, current is lost across the cell membrane as charge-carrying ions leak out through the uninsulated parts of the membrane, that is, through open channels. Because of this current loss, the magnitude of the local current progressively diminishes with increasing distance from the initial site of origin (● Figure 4–4). Thus the magnitude of the graded potential continues to decrease the farther it moves away from the initial active area.

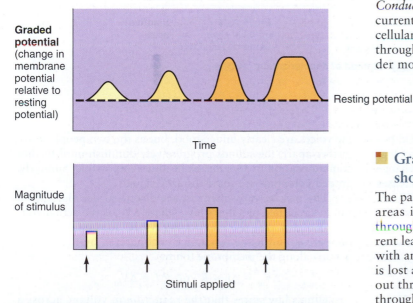

Figure 4–2 ● **The magnitude and duration of a graded potential.** The magnitude and duration of a graded potential depend directly on the strength and duration of the triggering event, such as a stimulus.

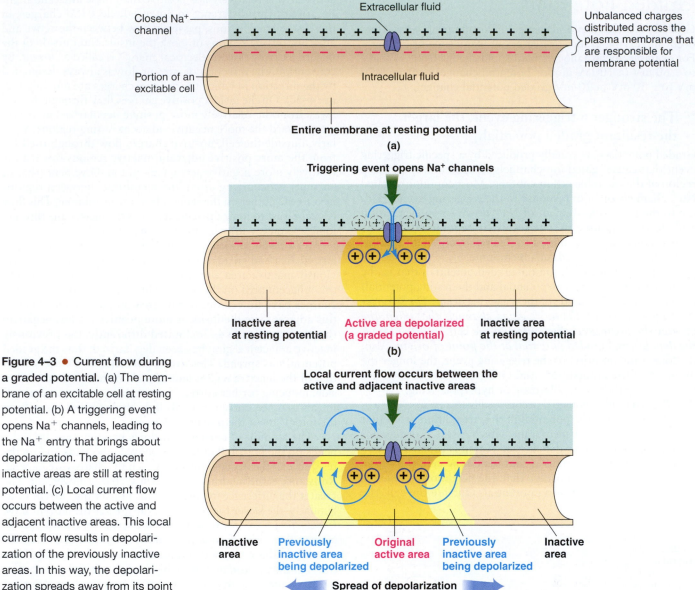

Figure 4–3 ● Current flow during a graded potential. (a) The membrane of an excitable cell at resting potential. (b) A triggering event opens Na$^+$ channels, leading to the Na$^+$ entry that brings about depolarization. The adjacent inactive areas are still at resting potential. (c) Local current flow occurs between the active and adjacent inactive areas. This local current flow results in depolarization of the previously inactive areas. In this way, the depolarization spreads away from its point of origin.

Labels in figure:
(a) Closed Na$^+$ channel; Extracellular fluid; Unbalanced charges distributed across the plasma membrane that are responsible for membrane potential; Portion of an excitable cell; Intracellular fluid; **Entire membrane at resting potential**

(b) **Triggering event opens Na$^+$ channels**; Inactive area at resting potential; Active area depolarized (a graded potential); Inactive area at resting potential

(c) **Local current flow occurs between the active and adjacent inactive areas**; Inactive area; Previously inactive area being depolarized; Original active area; Previously inactive area being depolarized; Inactive area; **Spread of depolarization**

Another way of saying this is that the spread of a graded potential is **decremental** (gradually decreases). Note that in ● Figures 4–4 and 4–5, the magnitude of the initial potential change is 15 mV (a change from −70 mV to −55 mV), then decreases as it moves along the membrane to a potential change of 10 mV (from −70 mV to −60 mV), and continues to diminish the farther it moves away from the initial active area until there is no longer a potential change. In this way, these local currents die out within a few millimeters from the initial site of potential change and consequently can function as signals only for very short distances.

For this reason, this form of electrical depolarization is termed **passive conduction.** An analogy is sound waves. The magnitude of the sound waves is greatest (loudest) at the point of origin, then progressively decreases and eventually dies off as the sound waves spread out from this initial site. Consider two people speaking directly to each other. Over short distances,

the voices are clearly understood, but as the two people move farther apart, the sounds progressively diminish until further communication by this means becomes impossible. Similarly, graded potentials are short-distance signals.

The passive movement of the electrical depolarization can be described by **Ohm's law,** where ΔV_m is the actual voltage change across the membrane, ΔI the amount of current (amperes), and R the resistance encountered by the current passing along the membrane (ohms).

$$\Delta V_m = \Delta I \times R \quad \text{or} \quad \Delta I = \Delta V_m \times g$$

Ohm's law states that the *reduction* in voltage across a membrane in response to a current flowing through the cell is directly related to the amount of current multiplied by the resistance of the membrane. As current flows inside the cell, a constant proportion leaks through the membrane at each point. Membrane resistance is also the reciprocal of membrane con-

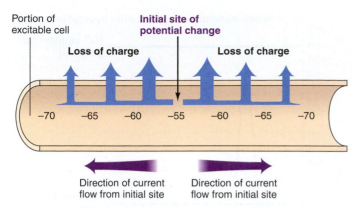

Portion of
excitable cell

**Initial site of
potential change**

Loss of charge   Loss of charge

−70   −65   −60   −55   −60   −65   −70

Direction of current   Direction of current
flow from initial site   flow from initial site

* Numbers refer to the local potential in mV
  at various points along the membrane.

Figure 4–4 ● Current loss across the plasma membrane.
Leakage of charge-carrying ions across the plasma membrane
results in progressive loss of current with increasing distance
from the initial site of potential change.

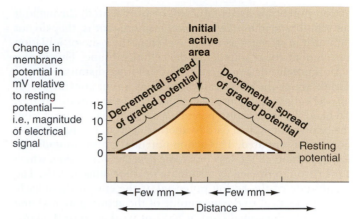

Change in
membrane
potential in
mV relative
to resting
potential—
i.e., magnitude
of electrical
signal

Initial
active
area

Decremental spread
of graded potential

Decremental spread
of graded potential

15
10
5
0   Resting
                                         potential

←Few mm→   ←Few mm→
          Distance

Figure 4–5 ● Decremental spread of graded potentials.
Because of leaks in current, the magnitude of a graded potential
continues to decrease as it passively spreads from the initial
active area. The potential dies out altogether within a few
millimeters of its site of initiation.

ductance (*g*), where conductance depends on both the permeability and driving force for the movement of ions. The passive movement of current across the cell membrane is termed *electrotonic conduction*. Comparable to the flow of electricity through wires, the speed of electrotonic spread is extremely rapid. The extent of electrotonic spread is determined in part by membrane resistance and membrane capacitance. **Capacitance** is a measure of the amount of charge that can be maintained across an insulating gap. The lipid bilayer is an effective capacitor because it can maintain the separation of charged ions across a relatively narrow space. Because of these two properties of a cell membrane, there is, for every excitable cell, only a finite distance that current can passively flow from the activated region. The fraction that leaks through the membrane flows back in a path to the source, so less current is available to influence the membrane potential at a more distant site. The distance from the active area that a signal can be passively transmitted is best described by the following relationship:

$$V_x = V_0 e^{-x/\lambda}$$

In this equation, V_x is the potential measured at a distance x from the active area whereas V_0 is a measurement of the original amount of potential generated at the active area. The length constant (λ) is related to the resistance of the membrane, cytoplasm, and the ECF by the following equation:

$$\lambda = \sqrt{\frac{R_m}{R_i}}$$

In this equation R_m is the resistance of the membrane and R_i is a measurement of the internal and external resistances. Under normal circumstances the contribution of the external resistance to R_i is comparatively small. The length constant can also be defined as the distance from V_0 that the potential is reduced 63% from its initial value. Thus, when comparing the rate of signal propagation between neurons, the one with the higher value of λ indicates the higher velocity. Neurons that rely on the electrotonic flow of current for communication purposes are rarely more than a few millimeters in length and are noted for their high membrane resistance. These neu-

rons are termed **nonspiking** or *local-circuit* neurons. Determination of the length constant also provides some insight into structural adaptations that species have evolved to increase λ.

Note that λ increases as the square root of (1) an increase in the membrane resistance or (2) a decrease in the internal resistance. How can an animal accomplish either of these strategies in order to increase λ? As we note later (p. 119), vertebrates as well as some invertebrates can increase R_m by wrapping an insulating material around the cell, which minimizes leakage of ions through the membrane. An insulating sheath also confers fast conduction by lowering the capacitance. Alternatively, R_i can be decreased by forcing more of the current to travel inside the cell by increasing the diameter of the cell. This solution is used by many invertebrates, which rely on a large fiber diameter to conduct important messages such as escape responses.

Although graded potentials have limited signaling distance, they are critically important to neural function, as we explain in later chapters. The following are all graded potentials: *postsynaptic potentials, receptor potentials, end-plate potentials, pacemaker potentials,* and *slow-wave potentials.* These terms may be unfamiliar to you now, but you will become well acquainted with them as we continue discussing nerve and muscle physiology. We are including this list here because it is the only place in this book that we group all these graded potentials together. For now it is enough to say that for the most part, excitable cells produce one of these types of graded potentials in response to a triggering event. In turn, graded potentials can initiate *action potentials,* the long-distance signals, in an excitable cell.

Action Potentials

Action potentials are brief, rapid, large (100 mV) changes in membrane potential during which the potential actually reverses so that the inside of the excitable cell transiently becomes more positive than the outside. As with a graded potential, a single action potential involves only a small portion of the total excitable cell membrane. Unlike graded potentials, however, action potentials are conducted, or propagated, throughout the membrane in *nondecremental* fashion; that is, they do not diminish in strength as they travel from their site of initiation throughout the remainder of the cell membrane. This is why action potentials can serve as faithful long-distance signals. Think about the nerve cell that brings about contraction of muscle cells in a mammal's leg for flexing the toes (see p. 59). To bend the toes, action potentials are sent via a neuron from the brain to the spinal cord to another neuron, where an action potential is initiated. This action potential travels in undiminishing fashion all the way down the neuron's long axon, which runs through the leg to terminate on foot muscle cells. The magnitude of the action potential at the end of the muscle axon is identical to the magnitude of the action potential first initiated in the spinal cord; it has not weakened or died off.

Let's now consider the changes in permeability and ionic movements responsible for generating an action potential, before turning our attention to the means by which action potentials spread throughout the cell membrane in undiminishing fashion.

■ During an action potential, the membrane potential rapidly and transiently reverses.

If of sufficient magnitude, a graded potential can initiate an action potential before the graded potential dies off. You will discover the means by which this initiation is accomplished for the various types of graded potentials as we examine each graded potential in more detail later. Typically the portion of the excitable membrane where graded potentials are produced in response to a triggering event does not undergo an action potential. Instead, the graded potential, by electrical or chemical means, depolarizes adjacent portions of the membrane where action potentials can take place. For convenience in this discussion, we will now jump from the triggering event to depolarization of the portion of the membrane that is to undergo an action potential, without considering the involvement of the intervening graded potential.

To initiate an action potential, a triggering event causes the membrane to depolarize from the resting potential of -70 mV (● Figure 4–6). Depolarization proceeds slowly at first until it reaches a critical level known as **threshold potential**, typically between -50 and -55 mV or $+10$ to $+15$ mV above resting potential. At threshold potential, an explosive depolarization takes place. A recording of the potential at this time shows a sharp upward deflection to $+30$ mV as the potential rapidly moves toward 0 mV, then reverses itself so that the inside of the cell becomes positive compared to the outside. Just as rapidly, the potential drops back to resting potential as the membrane repolarizes. Often the forces responsible for driving the membrane back to resting potential push it too far, causing a transient **hyperpolarization**, during which the inside of the membrane briefly becomes even more negative than normal (for example, -80 mV) before the resting potential is restored.

The *action potential* is an electrical waveform composed of depolarization, repolarization, and hyperpolarization. Unlike the variable duration of a graded potential, the duration of an action potential is always the same in a given excitable cell. In a neuron, an action potential lasts for only 1 msec (0.001 sec). It lasts longer in muscle, with the duration varying depending on the muscle type. The portion of the action potential during which the potential is reversed (between 0 mV and

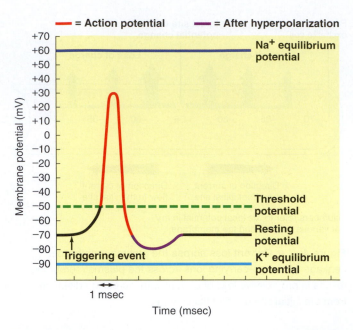

Figure 4–6 ● Changes in membrane potential during an action potential.

$+30$ mV) is called the **overshoot**. Often an action potential is called a **spike** because of its spiky appearance on an oscilloscope. Alternatively, when an excitable membrane is triggered to undergo an action potential, it is said to **fire**. Thus the terms *action potential, spike,* and *firing* all refer to the same phenomenon of rapid potential reversal.

If threshold potential is not reached by the initial triggering depolarization, no action potential takes place. Thus threshold is a critical all-or-none event. Either the membrane is depolarized to threshold and an action potential takes place, or threshold is not reached in response to the depolarizing event and no action potential occurs.

■ Marked changes in membrane permeability and ion movement lead to an action potential.

How is the membrane potential, which is usually maintained at a constant resting level, thrown out of balance to such an extent as to produce an action potential? Recall that K^+ contributes the most to establishing the resting potential because the membrane at rest is much more permeable to K^+ than to Na^+. During an action potential, marked changes in membrane permeability to Na^+ and K^+ take place because of voltage-gated channels, permitting rapid fluxes of these ions down their electrochemical gradients. These ion movements carry the current responsible for the potential changes that occur during an action potential.

Voltage-Gated Na^+ and K^+ Channels

Two specific types of channels are of major importance in developing an action potential: **voltage-gated Na^+ channels** and **voltage-gated K^+ channels.** Voltage-gated channels consist of proteins that have a number of charged groups. The electric field (potential) surrounding the channels can distort the channel structure as charged portions of the channel proteins are

Voltage-Gated Sodium Channel

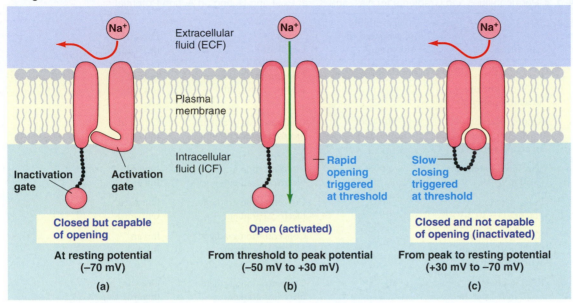

Inactivation gate | Activation gate

Closed but capable of opening

At resting potential (−70 mV)

(a)

Na⁺

Rapid opening triggered at threshold

Open (activated)

From threshold to peak potential (−50 mV to +30 mV)

(b)

Na⁺

Slow closing triggered at threshold

Closed and not capable of opening (inactivated)

From peak to resting potential (+30 mV to −70 mV)

(c)

Extracellular fluid (ECF)

Plasma membrane

Intracellular fluid (ICF)

Figure 4–7 ● Conformations of voltage-gated sodium and potassium channels.

Voltage-Gated Potassium Channel

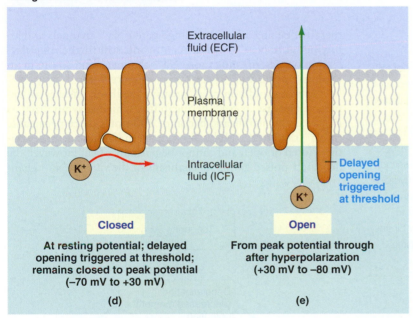

Extracellular fluid (ECF)

Plasma membrane

Intracellular fluid (ICF)

K⁺

Closed

At resting potential; delayed opening triggered at threshold; remains closed to peak potential (−70 mV to +30 mV)

(d)

Delayed opening triggered at threshold

K⁺

Open

From peak potential through after hyperpolarization (+30 mV to −80 mV)

(e)

catching prey or avoiding predators (note that sluggish animals have other defenses).

The voltage-gated Na⁺ channel has two gates: an *activation gate* and an *inactivation gate* (● Figure 4–7). The activation gate guards the channel by opening and closing like a hinged door. The inactivation gate consists of a ball-and-chain–like sequence of amino acids. This gate is open when the ball is dangling free on its chain and closed when the ball binds to its receptor located at the channel opening, thus blocking the opening. Both gates must be open to permit passage of Na⁺ through the channel, and closure of either gate prevents passage. This voltage-gated Na⁺ channel can exist in three different conformations: (1) *closed but capable of opening* (activation gate closed, inactivation gate open, Figure 4–7a; (2) *open, or activated* (both gates open, Figure 4–7b); and (3) *closed and not capable of opening* (activation gate open, inactivation gate closed, Figure 4–7c).

The basic voltage-gated K⁺ channel of excitable tissues is simpler. It has only one gate, which can be either open or closed (Figure 4–7d and e). Actually, there are two broad groups of voltage-gated K⁺ channels, called (1) **ether a go-go (EAG) K-channels** (because mutant forms in the fruit fly *Drosophila* produce dancelike movements in ether-exposed flies), and (2) **shaker K-channels** (because mutant forms in fruit flies cause uncontrollable leg shaking). The shaker channels are responsible for the basic neuronal function we will delineate here. They represent an ancient family found in all animal neurons. For example, the simple nervous systems of jellies (Cnidaria) express at least two shaker-type genes; and the genome of the electric fish *Apteronotus* has at least 10 such genes (electric fish are discussed in Chapter 6). The EAG type occurs in other excitable tissues such as heart (Chapter 9).

electrically attracted or repelled by charges in the fluids surrounding the membrane. Unlike the majority of membrane proteins, which remain stable despite fluctuations in membrane potential, the voltage-gated channel proteins are especially sensitive to voltage changes. Small distortions in channel shape induced by potential changes can cause them to alternate to another conformation. Here is another example of how subtle changes in structure can have a profound influence on function. For example, all vertebrates and those invertebrates evolved for speed have comparatively fast-opening Na⁺ channels; in contrast, sluggish invertebrates such as sea slugs have Na⁺ channels that open at rates approximately 10 times slower. Fast-opening channels increase survival rates, because electrical impulses are more rapidly transmitted to muscles used in

Table 4–1 ▮ Permeabilities and Gradients for Na$^+$ and K$^+$ at Rest and During an Action Potential

Ion	Condition	Permeability Compared to Resting P_{Na^+}	Gradient	Direction of Gradient
Na$^+$	Resting potential (−70 mV) before an action potential	1	Concentration gradient Electrical gradient	Inward Inward
K$^+$	Resting potential (−70 mV) before an action potential	50–75×	Concentration gradient Electrical gradient	Outward Inward
Na$^+$	Threshold potential (−50 mV)	600×	Concentration gradient Electrical gradient	Inward Inward
K$^+$	Peak of action potential (+30 mV)	300×	Concentration gradient Electrical gradient	Outward Outward
Na$^+$	Resting potential (−70 mV) after an action potential	1	Concentration gradient Electrical gradient	Inward Inward
K$^+$	Resting potential (−70 mV) after an action potential	50–75×	Concentration gradient Electrical gradient	Outward Inward

?

How Do Voltage Gates Actually Open and Close? It is still not fully understood how the so-called gates on neural voltage-gated channels actually work. But recent studies on a bacterial voltage-gated K$^+$ channel are revealing.

Researchers used x-ray crystallography to get a detailed picture of this protein's structure. The channel contains a central pore for K$^+$ ions, surrounded by what the researchers term "voltage-sensor paddles." These are folds on the outside of the protein with hydrophobic and cationic (positively charged) amino acids, and with flexible hinge regions. These features suggest that paddles react to voltage changes in the nearby membrane and move on their hinges in response. Whether this bacterial mechanism applies to animal neurons is not yet known.

Changes in Permeability and Ion Movement During an Action Potential

At resting potential (−70 mV), all the voltage-gated Na$^+$ and K$^+$ channels are closed, with the Na$^+$ channels' activation gates being closed and their inactivation gates being open; that is, the voltage-gated Na$^+$ channels are in their "closed but capable of opening conformation." Therefore, passage of Na$^+$ and K$^+$ is prevented through these channels at resting potential. However, because of the presence of many K$^+$ leak channels, the resting membrane is 50 to 75 times more permeable to K$^+$ than to Na$^+$ (▮ Table 4–1).

When a membrane starts to depolarize toward threshold as a result of a triggering event, the activation gates of some of its voltage-gated Na$^+$ channels open. Now both gates of these activated channels are open. Because both the concentration and electrical gradients for Na$^+$ favor its movement into the cell, Na$^+$ starts to move in. The inward movement of positively charged Na$^+$ depolarizes the membrane further,

thereby opening adjacent voltage-gated Na$^+$ channels and allowing more Na$^+$ to enter. As a result, still further depolarization occurs, opening more Na$^+$ channels, and so on, in an explosive *positive-feedback* cycle (● Figure 4–8).

At threshold potential, positive feedback ensures an explosive increase in Na$^+$ permeability, which is symbolized as P_{Na^+}, as the membrane becomes 600 times more permeable to Na$^+$ than to K$^+$ (Table 4–1). Each individual channel is either closed or open and cannot be partially open. However, the delicately poised gating mechanisms of the various Na$^+$ chan-

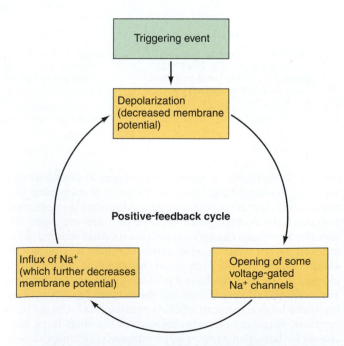

Figure 4–8 ● Positive-feedback cycle responsible for opening Na$^+$ channels at threshold.

nels are jolted open by slightly different voltage changes. During the early depolarizing phase, Na^+ channels are gradually opened as the potential progressively decreases. At threshold, enough Na^+ gates have opened to set off the positive feedback cycle that rapidly causes the remaining Na^+ gates to swing open. Now Na^+ permeability dominates the membrane potential, in contrast to the K^+ domination at resting potential. Thus, at threshold Na^+ rushes into the cell, rapidly eliminating the internal negativity and even making the inside of the cell more positive than the outside in an attempt to drive the membrane potential to the Na^+ equilibrium potential (see p. 114) (● Figure 4–9). The potential reaches +30 mV, close to the Na^+ equilibrium potential. The potential does not become any more positive, because at the peak of the action potential the Na^+ channels close to the inactivated state, and P_{Na^+} falls to its low resting value.

What causes the Na^+ channels to close? When the membrane potential reaches threshold, two simultaneous events take place in the gates of each Na^+ channel. First, the activation gates are triggered to *open rapidly* in response to the depolarization, converting the channel to its open (activated) conformation (Figure 4–7b). The conformational change on channel opening enables the inactivation gate's ball to bind to its receptor at the channel opening, thereby physically blocking the opening of the channel. However, this closure process takes time, so the inactivation gate *closes slowly* compared to the rapidity of channel opening. Meanwhile, during the 0.5-msec delay after the activation gate opens and before the inactivation gate closes, both gates are open and Na^+ rushes into the cell through these open channels, bringing the action potential to its peak of +30 mV. Then the inactivation gate closes, membrane permeability to Na^+ plummets to its low resting value, and further Na^+ entry is prohibited. The channel remains in this inactivated conformation with the inactivation gate closed until the membrane potential has been restored to its resting value.

Simultaneous with inactivation of the Na^+ channels at the peak of the action potential, the voltage-gated K^+ channels (shaker-type) are opened. This opening of the K^+ channel gate is a delayed voltage-gated response triggered by the initial depolarization to threshold. Thus, three action potential–related events occur at threshold: (1) the rapid opening of the Na^+ activation gates, which permits Na^+ to enter, moving the potential from threshold to its positive peak; (2) the slow closing of the Na^+ inactivation gates, which halts further Na^+ entry after a brief time delay, thus keeping the potential from rising any further; and (3) the slow opening of the K^+ gates, which, as you will see, is responsible for returning the potential from its peak back to resting.

The membrane potential would gradually return to resting after closure of the Na^+ channels as K^+ continued to leak out. However, the return to resting is hastened by the opening of K^+ gates at the peak of the action potential. Opening of the voltage-gated K^+ channels greatly increases the K^+ permeability (P_{K^+}) to about 300 times the resting P_{Na^+} (Table 4–1). This marked increase in P_{K^+} causes K^+ to rush out of the cell down its concentration and electrical gradients, carrying positive charges back to the outside. Note that at the peak of the action potential, the internal positivity of the cell tends to repel the positive K^+ ions, so the electrical gradient for K^+ is outward, unlike at resting potential. The outward movement of K^+ rapidly restores the negative resting potential (Figure 4–9g).

To review (● Figure 4–10), *the rising phase of the action potential* (from threshold to +30 mV) *is due to Na^+ influx* (Na^+ entering the cell) induced by an explosive increase in P_{Na^+} at threshold. *The falling phase* (from +30 mV to resting potential) *is brought about by K^+ efflux* (K^+ leaving the cell) caused by the marked increase in P_{K^+} occurring simultaneously with the inactivation of the Na^+ channels at the peak of the action potential.

As the potential returns to resting, the changing voltage shifts the Na^+ channels to their "closed but capable of opening" conformation, with the activation channel closed and the inactivation channel open. Now the channel is reset, ready to respond to another triggering event. The newly opened voltage-gated K^+ channels also close, so the membrane returns to the resting number of open K^+ leak channels. Typically, the voltage-gated K^+ channels are slow to close. As a result of this persistent increased permeability to K^+, more K^+ leaves than is necessary to bring the potential to resting. This slight excessive K^+ efflux makes the interior of the cell transiently even more negative than resting potential, causing the hyperpolarization afterward.

Several metabolic poisons have provided evidence for the specificity of the voltage-dependent ion channels and their role in the propagation of an action potential. One of the best characterized of these poisons is **tetrodotoxin (TTX)**, which occurs in the tropical marine Japanese pufferfish (*Fugu rubripes*) and in several other species (see box, "Molecular Biology and Genomics: Neurotoxins in War and Peace" on page 114). Because of the toxicity of this compound, chefs in Japan require years of training before they are permitted to serve pufferfish to their guests. TTX is specific for voltage-gated Na^+ channels and binds to the extracellular region of the channel. However, if injected inside the neuron, TTX has no effect on action potential propagation, demonstrating that TTX is specific in blocking the entry of Na^+. Researchers have studied K^+ channels using the synthetic compound **tetraethylammonium (TEA)**. TEA selectively blocks most types of gated K^+ channels when placed at either the ICF or ECF. Subtypes of Ca^{++} ion channels can be blocked by **2-conotoxins**, which can be isolated from the cone snail (*Conus geographus*; see box on p. 31, Chapter 2).

■ The Na^+–K^+ ATPase pump gradually restores the concentration gradients disrupted by action potentials.

At the completion of an action potential, the membrane potential has been restored to its resting condition, but the ion distribution has been altered slightly. Sodium has entered the cell during the rising phase, and a comparable amount of K^+ has left during the falling phase. Recall that the bulk of the cell is electrically neutral; the potential difference between the ECF and ICF is measurable only nm (nanometers) deep along the surface of the membrane. It is the task of the Na^+–K^+ pump to restore these ions to their original locations in the long run, but not after each action potential.

The active pumping process takes much longer to restore Na^+ and K^+ to their original locations than it takes for the passive fluxes of these ions during an action potential. However, the membrane does not need to wait until the Na^+–K^+ pump slowly restores the concentration gradients before it can undergo another action potential. Actually, the movement of

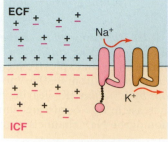

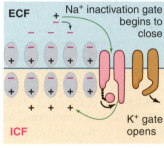

At resting potential, all voltage-gated Na⁺ and K⁺ channels are closed. Only the unbalanced charges separated across the membrane contribute to potential.

(a)

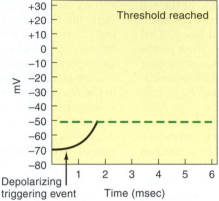

After a depolarizing triggering event brings the membrane to the threshold potential of −50 mV, the Na⁺ activation gates open.

(b)

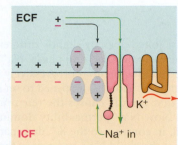

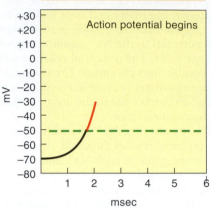

The resultant movement of Na⁺ inward neutralizes negative charges inside the cell. Inward movement of Na⁺ leaves behind on the outside the negative charges (primarily Cl⁻) with which Na⁺ had been paired. These negative charges neutralize positive charges that had been contributing to membrane potential, making the outside progressively less positive.

(c)

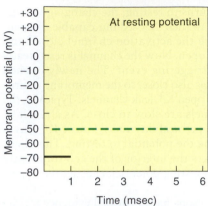

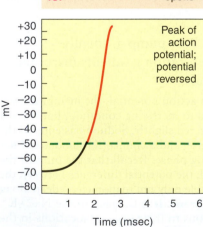

Further inward movement of Na⁺ reverses the potential, with the inside becoming positive and the outside becoming negative as the action potential peaks. At the peak of the action potential, the Na⁺ inactivation gates begin to close and the K⁺ gates open. Entry of Na⁺ ceases, and K⁺ starts to leave the cell.

(e)

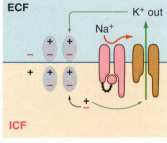

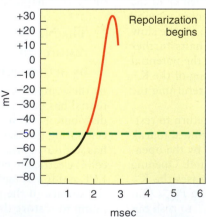

Outward movement of K⁺ leaves behind negative charges (A⁻) inside the cell and neutralizes negative charges outside; as a consequence, the inside becomes progressively less positive and the outside less negative until 0 mV is reached.

(f)

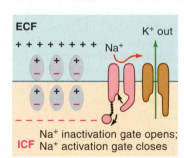

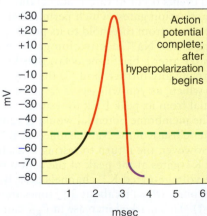

Continued outward movement of K⁺ restores the resting membrane potential, with the potential reversing back, so that the inside is once again negative and the outside positive. At resting potential, the Na⁺ inactivation gates open, and the activation gates close, reset to respond to another triggering event. Further outward movement of K⁺ through the still-open K⁺ gates briefly hyperpolarizes the membrane.

(g)

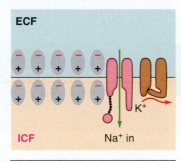

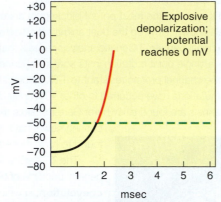

Continued inward movement of Na⁺ progressively reduces the potential (the inside becoming less negative and the outside less positive) until 0 mV is reached.

(d)

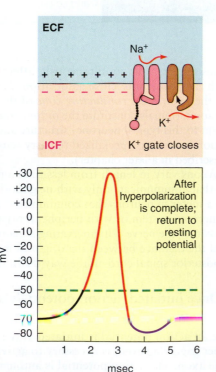

Then the K⁺ gates close, and the membrane returns to resting potential.

(h)

Figure 4–9 ● Ion movements responsible for changes in membrane potential during an action potential.

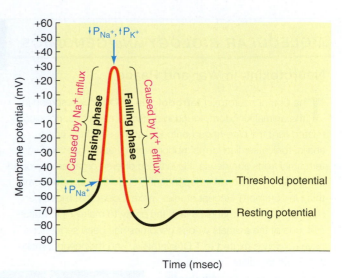

Figure 4–10 ● Permeability changes and ion fluxes during an action potential.

only relatively few of the total number of Na⁺ and K⁺ ions present is responsible for the dramatic swings in potential that occur during an action potential. Only about 1 out of every 100,000 K⁺ ions present in the cell leaves during an action potential, while a comparable number of Na⁺ ions enter from the ECF. The shifting locations of these extremely small percentages of the total Na⁺ and K⁺ during a single action potential produce only infinitesimal changes in their ICF and ECF concentrations. There is still much more K⁺ inside the cell than outside, and Na⁺ is still predominantly an extracellular cation. Consequently, the Na⁺ and K⁺ concentration gradients still exist, so repeated action potentials can occur without the pump having to keep pace to restore the gradients.

Were it not for the pump, of course, after a finite number of fluxes equilibrium would be reached and generation of action potentials would be impossible. If the concentrations of Na⁺ and K⁺ were equal between the ECF and ICF, changes in permeability to these ions would not bring about ionic fluxes, so no change in potential would occur. Thus the Na⁺–K⁺ pump is critical to maintaining the concentration gradients in the long run. However, it does not have to perform its role between action potentials, nor is it directly involved in the ion fluxes or potential changes that occur during an action potential.

■ Action potentials are propagated from the axon hillock to the axon terminals.

A single action potential involves only a small patch of the total surface membrane of an excitable cell. If action potentials are to serve as long-distance signals, obviously they cannot be merely isolated events occurring in a limited area of a neuron or muscle cell membrane. Mechanisms must exist to conduct or spread the action potential throughout the entire cell membrane. Furthermore, the signal must somehow be transmitted from one cell to the next cell (for example, along specific nerve pathways). Let's first examine how an action potential (nerve impulse) is conducted throughout a neuron, before turning our attention to how the impulse is passed to another cell.

A single nerve cell, or **neuron**, consists of three basic parts: the cell body, the dendrites, and the axon, although there are

MOLECULAR BIOLOGY AND GENOMICS

Neurotoxins in War and Peace

The potent neurotoxin **tetrodotoxin (TTX)** is widespread in the animal kingdom, found in pufferfish and other, not closely related fish; the blue-ringed octopus, some seastars, crabs, snails, tunicates, flatworms, ribbon worms, and algae; and some terrestrial frogs and newts. Have these diverse organisms evolved the genes for this toxin independently, or do they all have something in common? Recent studies make it clear that most, perhaps all of these organisms lack the ability to make TTX. Pufferfish, one of the animals whose genomes has been sequenced, have no genes related to TTX synthesis, and they cannot make it if grown in isolated aquaria. It now seems clear that **symbiotic bacteria** manufacture the toxin for their hosts. The blue-ringed octopus of Australia, for example, has been found to harbor these bacteria. Thus it now seems that a peaceful, beneficial coexistence has independently evolved between these bacteria and a number of animal hosts. Presumably, the bacteria are given nourishment and protection in the symbiosis. In turn, the TTX made by the bacteria is used by some of these organisms as defense (such as algae and pufferfish) and by others to kill prey (such as carnivorous snails).

Photo: © Masa Ushioda/Imagequestmarine.com

TTX is one of the most potent blockers of animal sodium channels known. How do the host animals themselves survive their bacterial partners? Genetic analysis reveals that the pufferfish has a single point mutation in its sodium-channel gene, rendering the channel protein resistant to TTX binding. Presumably similar mutations have occurred in other TTX hosts. But resistant channels are found elsewhere: Garter snakes that eat TTX-bearing newts have recently been found to have these as well.

The TTX story illustrates a general principle of biology called **coevolution,** in which adaptations in one species select for new adaptations in other species—those that are prey, predator, competitor, or symbiote. In turn, these changes in the second species may lead to yet newer adaptations in the first species . . . and so on.

---

variations in structure, depending on the location and function of the neuron. (The distinctions of other specialized neurons are described later.) The nucleus and organelles are housed in the **cell body** (● Figure 4–11), from which numerous extensions known as **dendrites** typically project like antennae to increase the surface area available for receiving signals from other nerve cells. Dendrites carry signals *toward* the cell body. In most neurons the plasma membrane of the cell body and dendrites contains protein receptors for binding chemical messengers (neurotransmitters) from other neurons. Therefore the dendrites and cell body are the neuron's *input zone,* because these components receive and integrate incoming signals. This is the region where graded potentials are produced in response to triggering events, in this case incoming chemical messengers.

The **axon,** or **nerve fiber,** is a single, elongated, tubular extension that conducts action potentials *away from* the cell body and eventually terminates at other cells. A **nerve** (as opposed to a neuron or nerve cell) is defined as a bundle of axons outside the CNS, whereas a **fiber tract** is a bundle of axons inside the CNS. The axon frequently gives off side branches, or **collaterals,** along its course. The first portion of the axon plus the region of the cell body from which the axon leaves is known as the **axon hillock.** The axon hillock is the neuron's *trigger zone,* because it is the site where action potentials are triggered, or initiated by the graded potential if it is large enough. The action potentials are then conducted along the axon from the axon hillock to the typically highly branching endings at the **axon terminals.** These terminals release chemical messengers that simultaneously affect target cells with which they come into close association. Functionally, therefore, the axon is the *conducting zone* of the neuron, whereas the axon terminals constitute the *output zone.* (The major exception to this typical neuronal structure and functional organization is neurons specialized to carry sensory information, as described in a later chapter).

Axons vary in length from less than a millimeter in neurons that communicate only with neighboring cells to longer than a meter in neurons that communicate with distant parts of a nervous system or with peripheral organs. For example, the axon of the nerve cell innervating a mammalian foot must travel the distance between the origin of its cell body within the posterior spinal cord all the way down the leg to the foot.

■ Once initiated, action potentials are conducted over the surface of an axon.

Once an action potential is initiated at the axon hillock, no further triggering event is necessary to activate the remainder of the axon. The action potential is automatically conducted throughout the neuron without further stimulation, by one of two methods of propagation: *contiguous conduction* or *saltatory conduction.*

Contiguous conduction involves the spread of the action potential along every patch of membrane down the length of the axon (*contiguous* means "touching" or "next to in sequence") This process is illustrated in ● Figure 4–12. You are

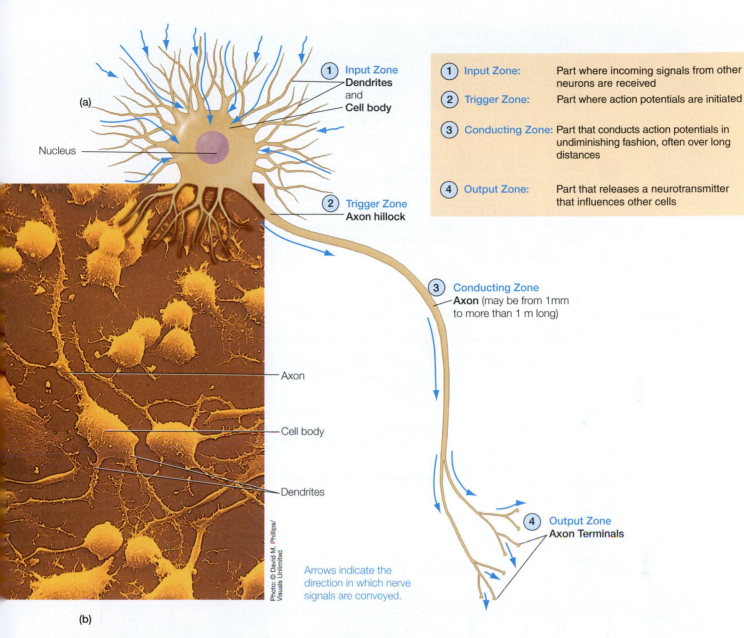

①	Input Zone:	Part where incoming signals from other neurons are received
②	Trigger Zone:	Part where action potentials are initiated
③	Conducting Zone:	Part that conducts action potentials in undiminishing fashion, often over long distances
④	Output Zone:	Part that releases a neurotransmitter that influences other cells

(a)

Nucleus

① Input Zone
Dendrites
and
Cell body

② Trigger Zone
Axon hillock

③ Conducting Zone
Axon (may be from 1mm to more than 1 m long)

④ Output Zone
Axon Terminals

Axon

Cell body

Dendrites

Photo: © David M. Phillips/
Visuals Unlimited

Arrows indicate the
direction in which nerve
signals are conveyed.

(b)

Figure 4–11 ● Anatomy of a neuron (nerve cell). (a) Most but not all neurons consist of the basic parts schematically represented in the figure. (b) An electron micrograph highlighting the cell body, dendrites, and part of the axon of a neuron within the central nervous system.

viewing a schematic representation of a longitudinal section of the axon hillock and the portion of the axon immediately beyond it. The membrane at the axon hillock is at the peak of an action potential. The inside of the cell is positive in this active area because Na^+ has already rushed into the neuron at this point. The remainder of the axon, still at resting potential and negative inside, is considered inactive. For the action potential to spread from the active to the inactive areas, the inactive areas must somehow be depolarized to threshold before they can undergo an action potential. This depolarization is accomplished by local current flow between the area already undergoing an action potential and the adjacent inactive area, similar to the current flow responsible for the spread of graded potentials. Recall that the rate at which the region ahead of the active area is depolarized depends on λ. Because opposite charges attract, current can flow locally between the active area and the neighboring inactive area on both the inside and the outside of the membrane. This local current flow in effect neutralizes or eliminates some of the unbalanced charges in the inactive area; that is, it reduces the number of opposite charges separated across the membrane or reduces the potential in this area. This depolarizing effect quickly brings the involved inactive area to threshold, at which time the voltage-gated Na^+ channels in this region of the membrane are all thrown open, leading to an action potential in this previously inactive area. Meanwhile, the original active area returns

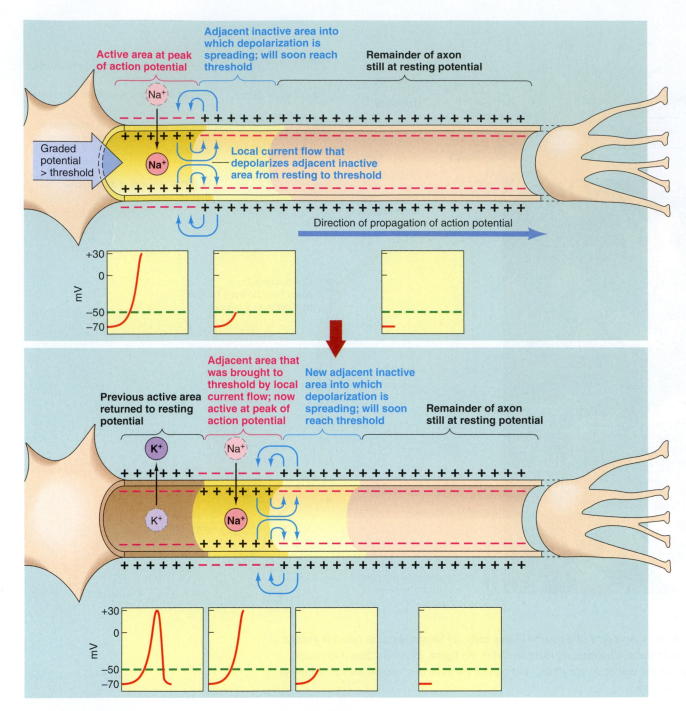

Figure 4–12 ● Contiguous conduction. Local current flow between the active area at the peak of an action potential and the adjacent inactive area still at resting potential reduces the potential in this contiguous inactive area to threshold, which triggers an action potential in the previously inactive area. The original active area returns to resting potential, and the new active area induces an action potential in the next adjacent inactive area by local current flow as the cycle repeats itself down the length of the axon.

to resting potential as a result of K^+ efflux. (Local current flow does *not* make a significant contribution to returning the original active area to resting potential).

In turn, beyond the new active area is another inactive area, so the same thing happens again. Local current flow brings this next inactive area to threshold, causing it to fire and become

a new active area. This cycle repeats itself in a chain reaction until the action potential has spread to the end of the axon. *Once an action potential is initiated in one part of a neuron membrane, a self-perpetuating cycle is initiated so that the action potential is propagated throughout the rest of the fiber automatically.* In this way, the axon is like a firecracker fuse that needs to be lit at only one end. Once ignited, the fire spreads down the fuse; it is not necessary to hold a match to every separate section of the fuse.

Note that the original action potential waveform does not travel along the membrane. Instead, it triggers an identical new action potential in the adjacent area of the membrane, with this process being repeated along the axon's length. An analogy is the spectator "wave" often seen at a sports stadium. Each section of spectators stands up (the rising phase of an action potential), then sits down (the falling phase) in sequence one after another as the wave moves around the stadium. The wave, not individual spectators (individual action potentials), travels around the stadium (conduction of the action potential along the axon). Because each new action potential in the conduction process is a fresh event dependent on the induced permeability changes and electrochemical gradients, which are virtually identical down the length of the axon, the last action potential at the end of the axon is identical to the original one, no matter how long the axon. Thus an action potential is spread throughout the axon in undiminished fashion. In this way, action potentials can serve as faithful long-distance signals without attenuation or distortion.

The nondecremental propagation of an action potential is in contrast to the decremental spread of a graded potential, which dies out over a very short distance because it is not able to regenerate itself. Action potentials can be initiated only in portions of the membrane that have an abundance of voltage-gated Na$^+$ channels, which can be triggered to open by a depolarizing event. Typically, regions of excitable cells where graded potentials take place do not have an achievable threshold for undergoing action potentials because of a scarcity of voltage-gated Na$^+$ channels. (There is no threshold phenomenon for the occurrence of graded potentials themselves.) Therefore, sites specialized for graded potentials do not undergo action potentials, even though they might be depolarized considerably. However, graded potentials can, before dying out, trigger action potentials in adjacent portions of the membrane by bringing these more sensitive regions to threshold through local current flow spreading from the site of the graded potential. In a typical neuron, for example, graded potentials are generated in the cell body and dendrites in response to incoming signals. If of sufficient magnitude by the time they have

spread to the axon hillock, these graded potentials initiate an action potential at this triggering zone. Once initiated, the action potential is propagated to the axon terminals. ▮ Table 4–2 summarizes the differences between graded potentials and action potentials, some of which are yet to be discussed.

▮ The refractory period ensures unidirectional propagation of the action potential and limits the frequency of action potentials.

What ensures the one-way propagation of an action potential away from the initial site of activation? Note from ● Figure 4–13 that once the action potential has been regenerated at a new neighboring site (now positive inside) and the original

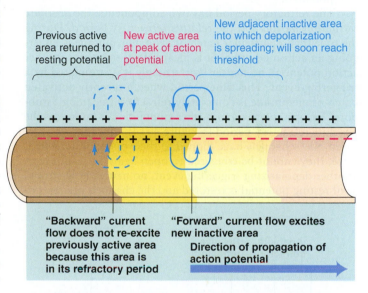

Figure 4–13 ● Value of the refractory period. "Backward" current flow is prevented by the refractory period. During an action potential and slightly beyond, an area cannot be restimulated by normal events to undergo another action potential. Thus the refractory period ensures that an action potential can be propagated only in the forward direction along the axon.

Table 4–2 ▮ Comparison of Graded Potentials and Action Potentials

Graded Potentials	Action Potentials
Graded potential change; magnitude varies with magnitude of triggering event	All-or-none membrane response; magnitude of triggering event coded in frequency rather than amplitude of action potentials
Decremental conduction; magnitude diminishes with distance from initial site	Propagated throughout membrane in undiminishing fashion
Passive spread to neighboring inactive areas of membrane	Self-regenerating in neighboring inactive areas of membrane
No refractory period	Refractory period
Can be summed	Summation impossible
Can be a depolarization or hyperpolarization	Always depolarization and reversal of charges
Triggered by a stimulus, by combination of neurotransmitter with receptor, or by spontaneous shifts in leak–pump cycle	Triggered by depolarization to threshold, usually through the spread of a graded potential
Occurs in specialized regions of membrane designed to respond to the triggering event	Occurs in regions of membrane with an abundance of voltage-gated Na$^+$ channels

active area has returned to resting (once again negative inside), the close proximity of opposite charges between these two areas is conducive to local current flow taking place in the backward direction as well as in the forward direction (into as yet unexcited portions of the membrane). If such backward current flow were able to bring the just inactivated area to threshold, another action potential would be initiated here, which would spread both forward and backward, initiating another action potential, and so forth. The situation would be chaotic, with numerous action potentials bouncing back and forth along the axon until the nerve cell eventually fatigued. Fortunately, neurons are saved from this fate of oscillating action potentials by the existence of the **refractory period,** during which a new action potential cannot be initiated by normal events in a region that has just undergone an action potential.

The refractory period has two components: the *absolute refractory period* and the *relative refractory period*. During the time that a particular patch of axonal membrane is undergoing an action potential, it is incapable of initiating another action potential, no matter how strongly it is stimulated by a triggering event. This time period when a recently activated patch of membrane is completely refractory (meaning "stubborn" or unresponsive) to further stimulation is known as the **absolute refractory period** (● Figure 4–14). Once the voltage-gated Na^+ channels have flipped to their open, or activated, state, they cannot be triggered to open again in response to another depolarizing triggering event, no matter how strong, until resting potential is restored and the channels are reset to

their original positions. Accordingly, the absolute refractory period lasts the entire time from opening of the voltage-gated Na^+ channels' activation gates at threshold, through closure of their inactivation gates at the peak of the action potential, until the return to resting potential when the channels' activation gates close and inactivation gates open once again; that is, until the channels are in their "closed but capable of opening" conformation. Only then can they respond to another depolarization with an explosive increase in P_{Na^+} to initiate another action potential. Because of this absolute refractory period, one action potential must be over before another can be initiated at the same site. Action potentials cannot overlap or be added one on top of another "piggyback fashion."

Following the absolute refractory period is a **relative refractory period** during which a second action potential can be produced only by a triggering event considerably stronger than is usually necessary. The relative refractory period occurs after the action potential is complete because of a twofold effect: lingering inactivation of the voltage-gated Na^+ channels and slowness of the voltage-gated K^+ channels that opened at the peak of the action potential to close. During this time, fewer than normal voltage-gated Na^+ channels are in a position to be jolted open by a depolarizing triggering event. Simultaneously, K^+ is still leaving through its slow-to-close channels during the after hyperpolarization. The less-than-normal Na^+ entry in response to another triggering event is opposed by a persistent hyperpolarizing outward leak of K^+ through its still-open channels, and thus a greater-than-normal depolarizing triggering event is needed to bring the membrane to threshold during the relative refractory period.

By the time the original site has recovered from its refractory period and is capable of being restimulated by normal current flow, the action potential is so far away that it can no longer influence the original site. Thus *the refractory period ensures the unidirectional propagation of the action potential down the axon away from the initial site of activation.*

■ The refractory period also limits the frequency of action potentials.

The refractory period also sets an upper limit on the frequency of action potentials; that is, it determines the maximum number of new action potentials that can be initiated and propagated along the fiber in a given period of time. The original site must recover from its refractory period before a new impulse can be triggered to follow the first impulse. The length of the refractory period varies for different types of neurons. The longer the refractory period, the greater the delay before the new action potential can be initiated and the lower the frequency with which a neuron can respond to repeated or ongoing stimulation.

■ Action potentials occur in all-or-none fashion.

If any portion of the neuronal membrane is depolarized to threshold, an action potential is initiated and relayed throughout the membrane in undiminished fashion. Furthermore, once threshold has been reached, the resultant action potential always goes to maximal height. The reason for this effect is that the changes in voltage during an action potential result from ion movements down concentration and electrical gradients, and these gradients are not affected by the strength of

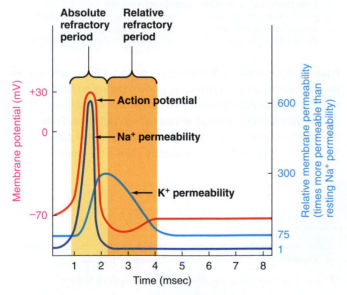

Figure 4–14 ● Absolute and relative refractory periods. During the absolute refractory period, the portion of the membrane that has just undergone an action potential cannot be restimulated. This period corresponds to the time during which the Na^+ gates are not in their resting conformation. During the relative refractory period, the membrane can be restimulated only by a stronger stimulus than is usually necessary. This period corresponds to the time during which the K^+ gates opened during the action potential have not yet closed.

the depolarizing triggering event. A triggering event stronger than one necessary to bring the membrane to threshold does not produce a larger action potential. However, a triggering event that fails to depolarize the membrane to threshold does not trigger an action potential at all. Thus *an excitable membrane either responds to a triggering event with a maximal action potential that spreads nondecrementally throughout the membrane, or it does not respond with an action potential at all.* This is called the **all-or-none law.**

This all-or-none concept is like firing a gun. Either the trigger is not pulled enough to fire the bullet at all (threshold is not reached), or it is pulled hard enough to elicit the full firing response of the gun (threshold is reached). Squeezing the trigger harder does not produce a greater explosion. Just as it is not possible to fire a gun halfway, it is normally not possible to have a halfway action potential.

The threshold phenomenon allows some discrimination between important and unimportant stimuli or other triggering events. Stimuli too weak to bring the membrane to threshold do not initiate an action potential and therefore do not clutter up the nervous system by transmitting insignificant signals.

■ The strength of a stimulus is coded by the frequency of action potentials.

How is it possible to differentiate between two stimuli of varying intensities if both bring the membrane to threshold and generate action potentials of the same magnitude? For example, how can an animal distinguish between a warm and a hot surface if both trigger identical action potentials in a nerve fiber relaying information about skin temperature to the central nervous system? The answer lies in the *frequency* with which the action potentials are generated. A stronger stimulus does not produce a larger action potential, but it does trigger a greater *number* of action potentials per second to be propagated along the fiber. In addition, a stronger stimulus in a region will result in more neurons reaching threshold, thus increasing the total information sent to the central nervous system.

Once initiated, the velocity, or speed, with which an action potential travels down the axon depends on two factors: (1) whether the fiber is myelinated and (2) the diameter of the fiber. Contiguous conduction occurs in unmyelinated fibers. In this case, as we just said, each individual action potential initiates an identical new action potential in the next contiguous (bordering) segment of the axon membrane so that every portion of the membrane undergoes an action potential as this electrical signal is conducted from the beginning to the end of the axon. A faster method of propagation, *saltatory conduction,* takes place in myelinated fibers. Next let's see how a myelinated fiber compares with an unmyelinated fiber, then see how saltatory conduction compares with contiguous conduction.

■ Myelination increases the speed of conduction of action potentials and conserves energy in the process.

Myelinated fibers, found primarily in vertebrates, are covered with myelin at regular intervals along the length of the axon (● Figure 4–15a). **Myelin** consists primarily of lipids. Because the water-soluble ions responsible for carrying current across

the membrane cannot permeate this thick lipid barrier, the myelin coating acts as an insulator, just like rubber around an electrical wire, to prevent current leakage across the myelinated portion of the membrane. Myelin is not actually a part of the nerve cell but consists of separate myelin-forming cells that wrap themselves around the axon in jelly-roll fashion (Figure 4–15b and c). These myelin-forming cells are **oligodendrocytes** in the central nervous system (the brain and spinal cord) and **Schwann cells** in the peripheral nervous system (the nerves running between the central nervous system and the various regions of the body). The lipid composition of myelin is due to the presence of layer on layer of the lipid bilayer that composes the plasma membrane of these myelin-forming cells. Between the myelinated regions, the axonal membrane is bare and exposed to the ECF. Only at these bare spaces, called **nodes of Ranvier,** can membrane potential exist and current flow across the membrane (Figure 4–15d). Voltage-gated Na^+ channels are concentrated at the nodal areas; the myelin-covered regions are almost devoid of these special passageways. By contrast, an unmyelinated fiber has a uniform density of voltage-gated Na^+ channels throughout its entire length. As you know, action potentials can be generated only at portions of the membrane furnished with an abundance of these channels. Spread of action potentials in myelinated fibers is called *saltatory conduction,* which we will now examine.

The nodes are usually about 1 mm apart, close enough that local current from an active node can reach an adjacent node before dying off. When an action potential occurs at one of the nodes, opposite charges attract from the adjacent inactive node, reducing its potential to threshold so that it undergoes an action potential, and so on. Consequently, in a myelinated fiber the impulse "jumps" from node to node, skipping over the myelinated sections of the axon (● Figure 4–16); this process is called **saltatory conduction** (the Latin word *saltere* means "to jump or leap"). Saltatory conduction propagates action potentials more rapidly than does conduction by local current flow, because the action potential leaps over myelinated sections but must be regenerated within every section of an unmyelinated axonal membrane from beginning to end. Myelinated fibers conduct impulses about 50 times faster than unmyelinated fibers of comparable size. As a general rule for vertebrates, the most urgent types of information are transmitted via myelinated fibers, whereas the nervous pathways carrying less urgent information are unmyelinated.

In addition to permitting action potentials to travel faster, a second advantage of myelination is that it conserves energy. Because the ion fluxes associated with action potentials are confined to the nodal regions, the energy-consuming Na^+–K^+ pump must restore fewer ions to their respective sides of the membrane after propagation of an action potential.

Insects do not have a myelin sheath per se but, rather, loosely coat the axon with a cellular *nerve sheath.* There are no nodes for enhancing conduction speed. The purpose of the sheath becomes evident when you examine the concentration of ions in the hemolymph of herbivorous (plant-eating) insects. Plants and fruits are high in K^+ but low in Na^+ content, so animals that feed exclusively on plant material potentially may consume insufficient Na^+. In herbivorous insects, concentrations of Na^+ in the hemolymph range from 2 to 15 mM lower than that measured in the ICF—too low to sustain propagation of an action potential. Compared to vertebrates, concen-

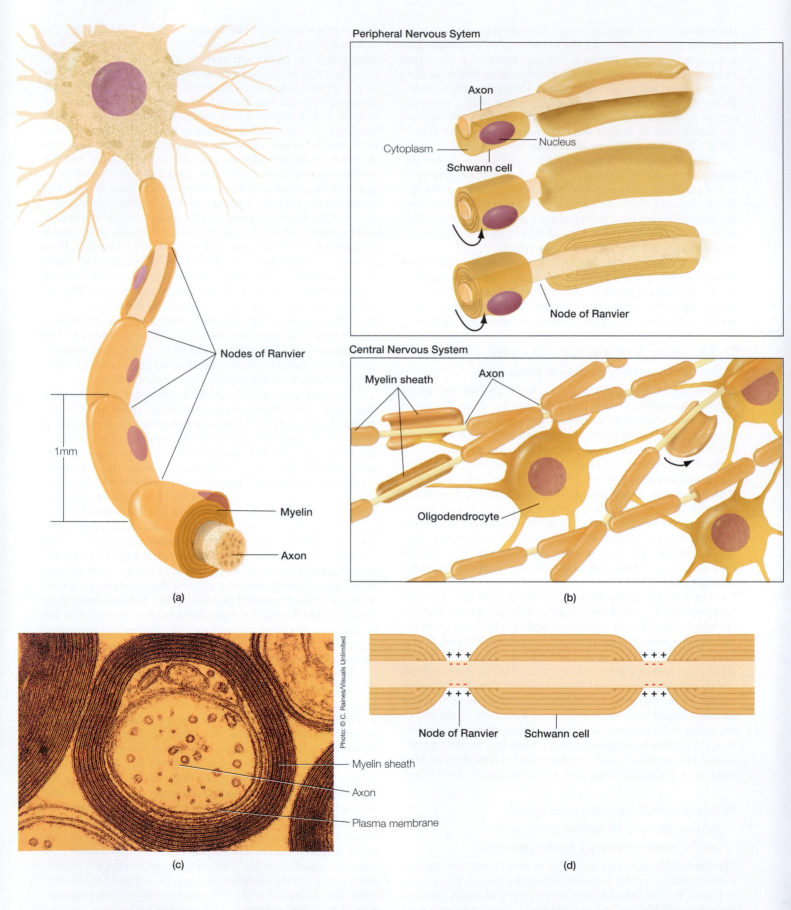

Peripheral Nervous Sytem

Axon

Nucleus

Cytoplasm

Schwann cell

Node of Ranvier

Nodes of Ranvier

1mm

Myelin

Axon

(a)

Central Nervous System

Myelin sheath

Axon

Oligodendrocyte

(b)

Photo: © C. Raines/Visuals Unlimited

Myelin sheath

Axon

Plasma membrane

(c)

+ + +

+ + +

+ + +

+ + +

Node of Ranvier   Schwann cell

(d)

Figure 4–15 ◀ ● **Myelinated fibers.** (a) A myelinated fiber is surrounded by myelin at regular intervals. The intervening unmyelinated regions are known as *nodes of Ranvier.* (b) In the peripheral nervous system, each patch of myelin is formed by a separate Schwann cell that wraps itself jelly-roll fashion around the nerve fiber. In the central nervous system, each of the several processes of a myelin-forming oligodendrocyte forms a patch of myelin around a separate nerve fiber. (c) An electron micrograph of a myelinated fiber in cross section. (d) Membrane potential exists only at the nodes of Ranvier, where the bare axon is exposed to the ECF. No charges exist across the insulated myelinated regions.

Figure 4–16 ●
Saltatory conduction.
The impulse "jumps"
from node to node
in a myelinated fiber.

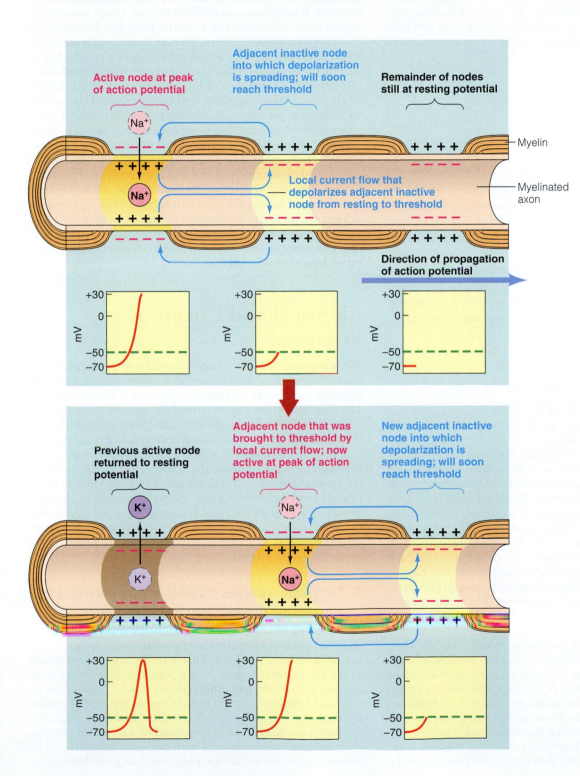

trations of K^+ are elevated in the hemolymph (20 to 60 mM) although still lower than in the ICF. These insects have solved the dietary ion problem by evolving the nerve sheath, which aids in propagation of an action potential. Because the nerve sheath is only loosely wrapped around the axon, it can regulate the local fluid bathing the axon. Measurement of Na^+ concentrations in this fluid reveals concentrations about 10-fold higher than in the hemolymph, concentrations now high enough to sustain propagation of an action potential. In contrast, herbivorous mammals such as deer meet this need by locating natural salty minerals ("licks"), whereas farmers give dietary mineral supplements to domesticated species or provide access to artificial salt licks.

?

Does Myelin Occur Only in Vertebrates? Traditionally, scientists thought only vertebrates accelerate action potentials with insulation. In recent years, however, researchers have found such insulation in earthworms and in shrimp, and, most recently and unexpectedly, in copepods. Copepods are tiny (almost microscopic) crustaceans that inhabit all oceans and, as the primary ocean herbivores, are thought to be the most numerous animals on Earth. But scientists always assumed their tiny size meant unmyelinated axons would work fast enough. In 1998, a chance microscopic observation by researchers at the University of Hawaii led to the discovery that copepods have myelin-like insulation. What purpose might this serve in the survival of this most important but tiny animal? See Suggested Reading by Lenz, Hartline, and Davis (2000).

◼ Fiber diameter also influences the velocity of action potential propagation.

Besides the effect of myelination, the diameter of the fiber also influences the speed with which an axon can conduct action potentials. The magnitude of current flow (that is, the amount of charge that moves) depends not only on the difference in potential between two adjacent electrically charged regions but also on the resistance to electrical charge movement between the two regions. When fiber diameter increases, the resistance to local current decreases. Thus the larger the diameter of the nerve fiber, the faster it can propagate action potentials. Fast conduction of an impulse increases the possibility of an animal escaping from a predator. Other uses include prey capture or a response to stress. Compared to slow conduction pathways, fast conduction requires considerably greater energy expenditure.

In some species of animal, fiber diameters have become so large that they are called **giant nerve axons**. They are characterized as axons with diameters that are markedly larger than the diameters of other axons in the same animal and by their role in the coordination of abrupt withdrawal or escape movements. Unicellular giant axons are found in cockroaches, crayfish, and some annelids. Examples of giant axons in vertebrates include **Mauthner neurons** that in fish and larval amphibians arise from the brain stem and run to tail muscles. They are important in the rapid tail-flips needed for escaping preda-

tors or catching prey. Similarly, in a crayfish the perception of danger is relayed by a giant axon to muscles for the tail, which rapidly flips, propelling the animal away from the danger.

Giant axons may also arise from the fusion of smaller fibers (up to 1500 cell bodies). Multicellular forms occur in numerous other invertebrate groups, including annelids, decapod crustaceans, and cephalopods and in many vertebrates. Giant axons were first identified in the squid *Loligo;* indeed, researchers first used these axons (see opening picture of this chapter). Investigators established that an action potential in a giant axon is conducted at velocities from 5 to 25 meters/sec.

Large *myelinated* fibers, such as those supplying skeletal muscles, can conduct action potentials at a speed of up to 120 meters/sec (360 miles/hr), compared with a conduction velocity of 0.7 meters/sec (2 miles/hr) in small unmyelinated fibers such as those supplying the digestive tract. This variability in speed of propagation of action potentials is related to the urgency of the information being conveyed. A signal to skeletal muscles to execute a particular movement (for example, to prevent you from falling as you trip on something) must be transmitted more rapidly than a signal to modify a slow-acting digestive process. Were it not for myelination, axon diameters within urgent nerve pathways would have to be very large to achieve the necessary conduction velocities, as in many invertebrates. Vertebrates have escaped the necessity of very large fibers by wrapping the axons in myelin to permit economical, rapid, long-distance signaling.

Electrical and Chemical Synapses

What happens once an action potential reaches the end of an axon? Where the action potential waveform must somehow transmit its message to another neuron or effector cell (such as a muscle) is at the region termed the **synapse** (meaning "the clasper"). Charles Sherrington in 1897 first recognized that a different form of communication occurred at this juncture.

To preview, when the action potential waveform reaches the axon terminal, it alters the activity of the cells on which the neuron terminates. There are only two basic kinds of transmission of signals at this point: electrical (direct) or chemical (indirect). Although electrical transmission is not as common as chemical, it is much simpler. Initially let's consider electrical transmission, to set the stage for the more complex chemical transmission process.

◼ Electrical synapses transfer action potential waveforms through gap junctions.

Action potential waveforms are essentially transmitted across **electrical synapses** unperturbed, moving along the axon as if the synapse wasn't present. This is so because the cytoplasm of the presynaptic neuron is in direct contact with the postsynaptic neuron via **gap junctions** (see p. 65). Because the resistance of the gap junction membranes to the flow of current is low, this permits a nearly full-strength signal to be induced in the postsynaptic cell. Also, compared to a chemical synapse, electrical transmission occurs with negligible time delay. This feature makes electrical synapses useful as a mechanism for escape responses in invertebrates. Electrical synapses were first discovered in the abdominal nerve cord of crayfish in 1959 by E. Furshpan and D. Potter and have since been identified in numerous other invertebrates as well as in certain vertebrate excitable cells including the retinal neurons and smooth and cardiac muscle fibers. The rapidity of electrical communication is a feature exploited by these vertebrate tissues to coordinate the activity of a group of cells (for cardiac cells, see Chapter 9). In addition, in some neural pathways, signal transmission is both electrical and chemical. This is found in the escape response of some fishes, who flip their tails quickly when startled.

■ Chemical synapses convert action potentials into organic chemical messengers that are exocytosed into the synaptic gap.

Nerves and their target cells do not actually make direct contact at the second type of synapse, the **chemical synapse.** The gap, or **synaptic cleft,** between these two structures is too large (approximately 20 to 40 nm wide) to permit electrical transmission of an impulse between them (that is, the action potential cannot "jump" that far). Instead, an organic *chemical messenger*—generically called a **neurotransmitter**—is used to carry the signal between the neuron terminal and the target cell. Briefly, the messenger is exocytosed from the neuron and diffuses to the target, where it binds to **receptor** proteins. The receptors in turn alter the activity of the target cell. (Again, as with ions in action potentials, the giant neurons of squid provided the first evidence that transmission of a neural signal across the synapse required the release of a transmitter substance.) Because chemical synapses rely on diffusion of the neurotransmitter, they are much slower (by several orders of magnitude) than electrical synapses; but they have two advantages over electrical synapses. (1) They typically operate in one direction only, so that false messages do not inadvertently travel from effectors to integrators; that is, the *presynaptic* neuron brings about changes in membrane potential of the *postsynaptic* (target) membrane, but the postsynaptic membrane does not influence the potential of the presynaptic neuron (however, we later discuss *retrograde* messengers in which a signal is sent "backward."). (2) They allow for various kinds of signaling events other than simply triggering action potentials in the target. How these features arise and how they are used will shortly become apparent when we examine the events that occur at a synapse.

Neurotransmitters come in several chemical classes of signal molecules (Chapter 3, p. 88), but the most common ones are *amines* (■ Table 4–3, listing classical neurotransmitters). Examples you will encounter later in this chapter include glutamate, norepinephrine, dopamine, and serotonin. Some of these are highly conserved in evolution, serving as neurotransmitters in many different animal phyla. For example, **serotonin** is found in vertebrate nervous systems; it is used in synapses involved with sleep, pain, aggression, sexual behavior, and food intake. The transmitter **dopamine** is found in mammalian neurons involved with control of locomotion and reward. In snails (a mollusk), serotonin has been found to stimulate foot locomotion, whereas dopamine inhibits it. Serotonin has even been found (using immunofluorescence, p. 32) in synaptic vessels of a sea anemone (a cnidarian).

Table 4–3 ■ Some Known or Suspected Neurotransmitters and Neuropeptides

Classical Neurotransmitters (small, rapid-acting molecules)	
Acetylcholine	Histamine
Dopamine	Glycine
Norepinephrine	Glutamate
Epinephrine	Aspartate
Serotonin	Gamma-aminobutyric acid (GABA)

Neuropeptides (large, slow-acting molecules)	
β-endorphin	Motilin
Adrenocorticotropic hormone (ACTH)	Insulin
α-melanocyte-stimulating hormone (MSH)	Glucagon
Thyrotropin-releasing hormone (TRH)	Angiotensin II
Gonadotropin-releasing hormone (GnRH)	Bradykinin
Somatostatin	Vasopressin
Vasoactive intestinal polypeptide (VIP)	Oxytocin
Cholecystokinin (CCK)	Carnosine
Gastrin	Bombesin
Substance P	
Neurotensin	
Leucine enkephalin	
Methionine enkephalin	

A neuron may terminate with a chemical synapse at one of two broad categories of target structures: another neuron, or an effector organ—muscle, gland, and a few other types. Therefore, depending on where a neuron terminates, it can cause another neuron to convey an electrical message along a nerve pathway, a muscle cell to contract, a gland cell to secrete, or some other function. When a neuron terminates on an effector, the neuron is said to **innervate** the structure. The junctions between nerves and glands that they innervate are described later. For now we will concentrate on the junction between two neurons, and between a neuron and a skeletal muscle. Let's begin with two neurons.

Neuron-to-Neuron Synapses

Typically, a neuron-to-neuron synapse involves a junction between an axon terminal of one neuron, known as the **presynaptic neuron,** and the dendrites or cell body of a second neuron. Less frequently, axon-to-axon and dendrite-to-dendrite connections occur. Most neuronal cell bodies and associated dendrites receive thousands of synaptic inputs, which are axon terminals from many other neurons. Neurologists have estimated that some neurons within the central nervous system receive as many as 100,000 synaptic inputs (● Figure 4–17).

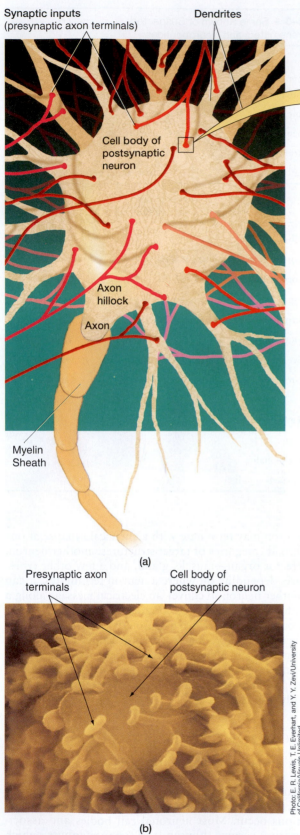

Synaptic inputs (presynaptic axon terminals)

Dendrites

See Figure 4-18

Cell body of postsynaptic neuron

Axon hillock

Axon

Myelin Sheath

(a)

Presynaptic axon terminals

Cell body of postsynaptic neuron

(b)

Photo: E. R. Lewis, T. E. Everhart, and Y. Y. Zevi/University of California/Visuals Unlimited

Figure 4–17 ● Synaptic inputs to a postsynaptic neuron.
(a) Schematic representation of synaptic inputs (presynaptic axon terminals) to the dendrites and cell body of a single postsynaptic neuron. (b) Electron micrograph showing multiple presynaptic axon terminals to a single postsynaptic cell body.

The anatomy of one of these thousands of synapses is shown in ● Figure 4–18. The axon terminal of the presynaptic neuron, which conducts its action potentials *toward* the synapse, ends in a slight swelling, the **synaptic knob.** The synaptic knob contains synaptic vesicles, which store neurotransmitters synthesized and packaged by the Golgi apparatus (p. 37) of the presynaptic neuron. The synaptic knob comes into close proximity to, but (because of the cleft or gap) does not actually directly contact, the **postsynaptic neuron,** the neuron whose action potentials are propagated *away* from the synapse. The part of the postsynaptic membrane immediately underlying the synaptic knob is the **subsynaptic membrane** (*sub,* "under").

■ A neurotransmitter carries the signal across a fast synapse and opens a chemically gated channel.

Most, but not all, neurotransmitters function by changing the conformation of *chemically gated channels,* thereby altering membrane permeability and ionic fluxes across the postsynaptic membrane. Synapses involving these rapid responses are considered **"fast"** synapses. The events that occur at these common type of synapse are summarized next and illustrated in Figure 4–18 (we investigate **"slow"** synapses later):

1. When an action potential in a presynaptic neuron has been propagated to the axon terminal (step ① in Figure 4–18), this change in potential triggers the opening of voltage-gated Ca^{++} channels in the synaptic knob.

2. Because Ca^{++} is in much higher concentration in the ECF, and because the cell has a negative charge, this ion flows into the synaptic knob through the opened channels (step ②).

3. Ca^{++} induces the release of a neurotransmitter from some of the synaptic vesicles into the synaptic cleft (step ③). The release is accomplished by exocytosis (see p. 38).

4. The released neurotransmitter diffuses across the cleft and combines with specific protein receptor sites on the subsynaptic membrane (step ④).

5. This binding triggers the opening of specific ion channels in the subsynaptic membrane, changing the permeability of the postsynaptic neuron (step ⑤). These are chemically gated channels, in contrast to the voltage-gated channels responsible for the action potential and for the Ca^{++} influx into the synaptic knob.

Because only the presynaptic terminal can release a neurotransmitter and only the membrane of the postsynaptic neuron has receptor sites for the neurotransmitter, the synapse can operate only in the direction from presynaptic to postsynaptic neuron. This advantage over electrical synapses prevents signals from moving backward, which would interfere with the ability of neurons to control their targets.

■ Some neurotransmitters excite the postsynaptic neuron, whereas others inhibit the postsynaptic neuron.

Each presynaptic neuron usually releases only one neurotransmitter (exceptions are noted later); however, different neurons vary in the neurotransmitter they release. On binding with

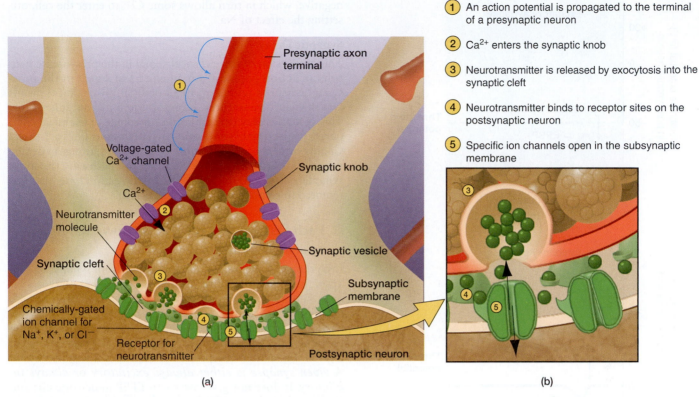

1. An action potential is propagated to the terminal of a presynaptic neuron

2. Ca²⁺ enters the synaptic knob

3. Neurotransmitter is released by exocytosis into the synaptic cleft

4. Neurotransmitter binds to receptor sites on the postsynaptic neuron

5. Specific ion channels open in the subsynaptic membrane

(a)

(b)

Figure 4–18 ● **Synaptic structure and function.** (a) Schematic representation of the structure of a single synapse. The circled numbers designate the sequence of events that take place at a synapse. (b) A blowup depicting the release by exocytosis of neurotransmitter from the presynaptic axon terminal and its subsequent binding with receptor sites specific for it on the subsynaptic membrane of the postsynaptic neuron.

their subsynaptic receptor sites, different neurotransmitters cause different permeability changes. There are two broad types of chemical synapses, depending on the permeability changes induced in the postsynaptic neuron by the combination of neurotransmitter with receptor sites: *excitatory synapses* and *inhibitory synapses*. This situation is unlike that found at the vertebrate neuromuscular junction (as you will see later) where the communication is always excitatory.

Excitatory Synapses

At an **excitatory synapse,** the response to the neurotransmitter–receptor combination is the opening of nonspecific cation channels within the subsynaptic membrane that permit simultaneous passage of Na⁺ and K⁺ through them. Thus permeability to both these ions is increased at the same time. How much of each ion diffuses through an open channel depends on the electrochemical gradients. At resting potential, both the concentration and electrical gradients for Na⁺ favor its movement into the postsynaptic neuron, whereas only the concentration gradient for K⁺ favors its movement outward (Table 4–1). Therefore, the permeability change induced at an excitatory synapse results in the simultaneous movement of a few K⁺ ions out of the postsynaptic neuron while a relatively larger number of Na⁺ ions enter this neuron. The result is a net movement of positive ions into the cell. This makes the inside of the membrane slightly less negative than at resting potential, thus producing a *small depolarization* of the postsynaptic neuron.

Activation of one excitatory synapse can rarely depolarize the postsynaptic neuron sufficiently to bring it to threshold. Too few channels are involved at a single subsynaptic membrane to permit adequate depolarizing fluxes to reduce the potential to threshold. This small depolarization, however, does bring the membrane of the postsynaptic neuron closer to threshold, increasing the likelihood that threshold will be reached (in response to further excitatory input) and an action potential will occur. That is, the membrane is more excitable (easier to bring to threshold) than when at rest. Accordingly, such a postsynaptic potential change occurring at an excitatory synapse is called an **excitatory postsynaptic potential,** or **EPSP** (● Figure 4–19a).

Inhibitory Synapses

The second advantage of a chemical over an electrical synapse is the ability to trigger responses other than simple stimulation. One of these other responses is inhibition. At an **inhibitory synapse,** the combination of a chemical messenger with its receptor site increases the permeability of the subsynaptic membrane to either K⁺ or Cl⁻ by altering these ions' respective channel conformations. In either case, the resulting ion movements bring about a *small hyperpolarization* of the postsynaptic neuron (greater internal negativity). In the case of increased P_{K^+}, more positive charges leave the cell via K⁺ efflux, leaving more negative charges behind on the inside; in the case of increased P_{Cl^-}, negative charges enter the cell in the

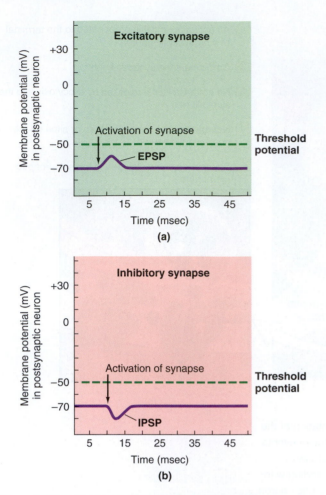

Figure 4–19 • Postsynaptic potentials. (a) Excitatory synapse. An excitatory postsynaptic potential (EPSP) brought about by activation of an excitatory presynaptic input brings the postsynaptic neuron closer to threshold potential. (b) Inhibitory synapse. An inhibitory postsynaptic potential (IPSP) brought about by activation of an inhibitory presynaptic input moves the postsynaptic neuron farther from threshold potential.

form of Cl^- ions, because Cl^- concentration is higher outside the cell. This small hyperpolarization moves the membrane potential even farther away from threshold (Figure 4–19b), lessening the likelihood that the postsynaptic neuron will reach threshold and undergo an action potential. That is, the membrane would be harder to bring to threshold by excitatory input than when it is at resting potential. Under these circumstances the membrane is said to be inhibited, and the small hyperpolarization of the postsynaptic cell is called an **inhibitory postsynaptic potential,** or **IPSP.**

In cells where the equilibrium potential for Cl^- exactly equals the resting potential, an increased P_{Cl^-} does not result in a hyperpolarization because there is no driving force to produce Cl^- movement. Nevertheless, opening of Cl^- channels in these cells is still inhibitory, because it tends to hold the membrane at resting potential, thus reducing the likelihood that threshold will be reached. That is, if a small amount of Na^+ enters from another channel, the cell becomes slightly less

negative, which in turn allows some Cl^- to enter the cell, offsetting the effect of Na^+.

Synaptic Delay

This conversion of the electrical signal in the presynaptic neuron (an action potential) to an electrical signal in the postsynaptic neuron (either an EPSP or IPSP) by chemical means (via the neurotransmitter–receptor combination) takes time. This **synaptic delay** (the major disadvantage of chemical compared to electrical synapses) is usually about 0.5 to 1 msec. Chains of neurons often must be traversed along a specific neural pathway. The more complex the pathway, the more synaptic delays and the longer the *total reaction time* (the time required to respond to a particular event).

■ Each fast synapse is either always excitatory or always inhibitory.

Even though transmitter substances vary from synapse to synapse, the same transmitter is always released at a given synapse. Furthermore, binding of a specific neurotransmitter with the appropriate subsynaptic receptors always leads to the same change in permeability and resultant change in potential of a particular postsynaptic membrane. That is, the response to a given transmitter–receptor combination is always constant. *A given synapse is either always excitatory or always inhibitory.* It does not give rise to an EPSP under one circumstance and produce an IPSP at another time. Some neurotransmitters (for example, **glutamate**) always induce EPSPs, whereas others (for example, **glycine**) always induce IPSPs. Still other neurotransmitters (for example, **norepinephrine**) may produce an EPSP at one synapse and an IPSP at another synapse; that is, different permeability changes in the postsynaptic neuron can occur in response to the binding of the same neurotransmitter to the subsynaptic receptor sites of different postsynaptic neurons. This is a result of having different types of receptors (an example of the "same key, different locks" principle, p. 88).

Most of the time each axon terminal releases only one neurotransmitter. Recent evidence suggests, however, that in some instances two different neurotransmitters can be released simultaneously from a single axon terminal. For example, glycine and γ-**aminobutyric acid (GABA),** both of which produce inhibitory responses, can be packaged and released from the same synaptic vesicles. Scientists speculate that the fast-acting glycine and the more slowly acting GABA may complement each other in the control of activities that depend on precise timing, for example, coordination of complex movements.

■ Neurotransmitters are quickly removed from the synaptic cleft.

As long as the neurotransmitter remains bound to the receptor sites, the alteration in membrane permeability responsible for the EPSP or IPSP continues. It is important that the neurotransmitter be inactivated or removed after producing the appropriate response in the postsynaptic neuron so that the postsynaptic "slate" is "wiped clean," ready to receive additional messages from the same or other presynaptic inputs. Thus, after combining with the postsynaptic receptor chemical transmitters are removed and the response is terminated. This occurs in three general ways: (1) The transmitter may diffuse away from the synaptic cleft, (2) it is inactivated by specific enzymes within the subsynaptic membrane, or (3) it is taken back into the axon terminal by transport mechanisms in the presynaptic membrane. Once within the synaptic knob, the transmitter can be stored and released another time (recycled) in response to a subsequent action potential or destroyed by enzymes within the synaptic knob. The method employed depends on the particular synapse. We describe a detailed example at neuromuscular synapses shortly.

■ Neurotransmitters in slow synapses function through intracellular second-messenger systems.

As we have seen, neurotransmitters usually function by changing the conformation of chemically gated channels, in so-called **"fast" synapses.** Another mode of synaptic transmission used by some neurotransmitters, such as *serotonin,* involves the activation of intracellular second messengers, such as cAMP (see p. 90), within the postsynaptic neuron. Synapses that lead to responses mediated by second messengers are known as **"slow" synapses,** because these responses take longer and often last longer (because of second-messenger cascades, p. 90) than those accomplished by fast synapses. Activation of the intracellular messenger, cyclic AMP, can induce both short- and long-term effects. In the short term, cAMP can lead to a prolonged opening of specific ionic gates. The gating effects can be either excitatory or inhibitory. In addition, cAMP may trigger more long-term changes in the postsynaptic cell, even to the extent of altering the cell's genetic expression by (indirectly) activating transcription factors. Such long-term cellular changes are linked to neuronal growth and development and play a role in learning and memory (see p. 186).

Neuromuscular Synapses

Now let's turn to a second example of chemical synapse. Action potentials are conducted to skeletal muscle through large, myelinated **motor neurons.** As the axon approaches a muscle, it divides into many terminal branches and loses its myelin sheath. Each of these axon terminals forms a special synaptic junction, a **neuromuscular junction,** with one of the many muscle cells that compose the whole muscle (● Figure 4–20). A single muscle cell, called a **muscle fiber,** is long and cylindrical in shape. The axon terminal is enlarged into a knoblike structure, the **terminal button,** which fits into a shallow depression, or groove, in the underlying muscle fiber (● Figure 4–21). Some scientists alternatively call the neuromuscular junction a *motor end plate.* However, we reserve the term **motor end plate** for the specialized portion of the muscle cell membrane immediately under the terminal button.

Normally, each vertebrate muscle fiber is innervated at one or two end plates, although in amphibians and lizards specialized "slow" muscle fibers receive multiterminal innervation; that is, they are innervated by numerous chemical synapses along the length of each fiber. Interestingly, these fibers lack all-or-none action potentials and rather depend on the copious distribution of neuromuscular junctions to generate graded contractions. Tension produced in these muscles thus

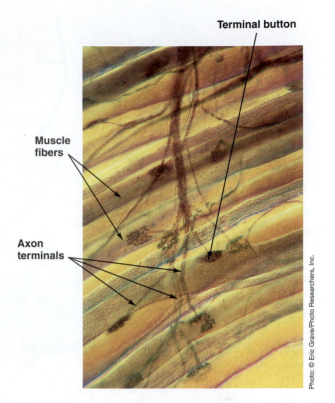

Figure 4–20 ● Motor neuron innervating skeletal muscle cells. When a motor neuron reaches a skeletal muscle, it divides into many terminal branches, each of which forms a neuromuscular junction with a single muscle cell (muscle fiber).

depends on the frequency of motor neuron output and are associated with slow, sustained contractions. (See Chapter 8 for details on typical neuromuscular control.)

■ Acetylcholine, a fast excitatory neurotransmitter, links electrical signals in motor neurons with electrical signals in skeletal muscle cells.

Here let's take a look at the events at a muscle synapse in a vertebrate, following Figure 4–21. Note that these are similar and often identical to steps at the neuron-to-neuron synapse, and so our description for most steps is brief.

Each terminal button contains thousands of vesicles that store many molecules of the neurotransmitter **acetylcholine (ACh).** Arrival of an action potential at the axon terminal triggers the following events:

1. Opening of voltage-gated Ca^{++} channels in the terminal button (step ①, Figure 4–21)
2. Ca^{++} conduction into the terminal button (step ②)
3. Release of ACh by exocytosis from several hundred of the vesicles into the cleft, and diffusion of ACh across the cleft (step ③)
4. Binding of two ACh molecules to **receptors** in the postsynaptic membrane (step ④) (these **cholinergic receptors** are of the *nicotinic* type; see p. 154)

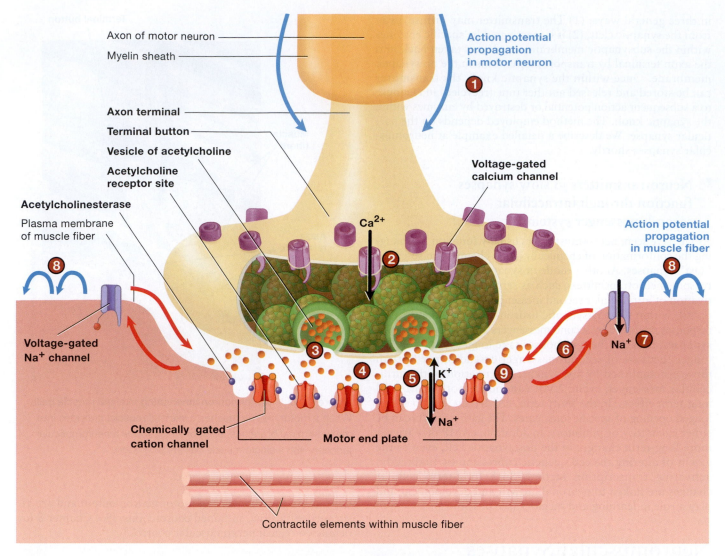

Axon of motor neuron

Myelin sheath

Action potential propagation in motor neuron ①

Axon terminal

Terminal button

Vesicle of acetylcholine

Acetylcholine receptor site

Acetylcholinesterase

Plasma membrane of muscle fiber

Voltage-gated calcium channel

Ca²⁺

Action potential propagation in muscle fiber ⑧

②

Voltage-gated Na⁺ channel

③

⑧

④ ⑤ K⁺ ⑨

⑥

Na⁺ ⑦

Na⁺

Chemically gated cation channel

Motor end plate

Contractile elements within muscle fiber

① An action potential in a motor neuron is propagated to the terminal button.

② The presence of an action potential in the terminal button triggers the opening of voltage-gated Ca²⁺ channels and the subsequent entry of Ca²⁺ into the terminal button.

③ Ca²⁺ triggers the release of acetylcholine by exocytosis from a portion of the vesicles.

④ Acetylcholine diffuses across the space separating the nerve and muscle cells and binds with receptor sites specific for it on the motor end plate of the muscle cell membrane.

⑤ This binding brings about the opening of cation channels, leading to a relatively large movement of Na⁺ into the muscle cell compared to a smaller movement of K⁺ outward.

⑥ The result is an end-plate potential. Local current flow occurs between the depolarized end plate and adjacent membrane.

⑦ This local current flow opens voltage-gated Na⁺ channels in the adjacent membrane.

⑧ The resultant Na⁺ entry reduces the potential to threshold, initiating an action potential, which is propagated throughout the muscle fiber.

⑨ Acetylcholine is subsequently destroyed by acetylcholinesterase, an enzyme located on the motor end-plate membrane, terminating the muscle cell's response.

Figure 4–21 ● Events at a neuromuscular junction.

5. Opening of chemical-messenger–gated channels in the motor end plate, which depolarizes the motor end plate (step ⑤)

This potential change is known as the **end-plate potential (EPP).** It is similar to an EPSP (p. 125), except that the magnitude of an EPP is much larger, because (1) more transmitter is released from a terminal button than from a *presynaptic knob* in re-

sponse to an action potential; (2) the motor end plate has a larger surface area bearing a higher density of transmitter receptor sites, and so, accordingly, has more sites for binding with transmitter than does a subsynaptic membrane ; and (3) many more ion channels are opened in response to the transmitter–receptor complex in the motor end plate. This permits a greater net influx of Na⁺ ions and a larger depolarization. An EPP is a graded potential, whose magnitude depends on the amount and duration of ACh at the end plate.

The development of patch clamp recording technology (p. 104) enabled researchers to measure current flow through a single ACh receptor channel. Generally, the ACh channel remains open for 1 to 3 msec and permits approximately 1 to 5×10^4 ions to flow through it.

The motor end-plate region itself does not have a threshold potential, so an action potential cannot be initiated at this site. However, an EPP brings about an action potential in the rest of the muscle fiber, as follows (continuing with Figure 4–21):

6. The neuromuscular junction is usually located in the middle of the long cylindrical muscle fiber. When an EPP takes place at the motor end plate, local current flow occurs between the depolarized end plate and the adjacent, resting cell membrane in both directions, reducing the potential to threshold in the adjacent areas (step ⑥).

7. The subsequent action potential initiated at these sites is propagated throughout the muscle fiber membrane by opening voltage-gated Na^+ channels (step ⑦) and thus conduction (see p. 115).

8. The spread occurs in both directions, away from the motor end plate toward both ends of the fiber. This electrical activity initiates contraction of the muscle fiber (step ⑧).

Thus by means of ACh, an action potential in a motor neuron brings about an action potential and subsequent contraction in the muscle fiber. Unlike synaptic transmission among neurons, the magnitude of a single EPP is normally sufficient to cause an action potential in the muscle cell. Typically, therefore, one-to-one transmission of an action potential occurs at a neuromuscular junction; one action potential in a neuron triggers one action potential in a muscle cell that it innervates.

Unlike most vertebrate neuromuscular systems (as you will see later), neurons that innervate arthropod muscles are both inhibitory as well as excitatory. For example, the neuromuscular systems regulating claw opening and closing in decapod crustaceans (such as crabs) contains both excitatory and inhibitory motor input. Inhibitory motor neurons also have endings on the presynaptic terminals of the excitatory motor neurons. This feature specifically protects the excitatory synapse from neurotransmitter depletion by limiting the amount released.

■ Acetylcholinesterase terminates acetylcholine activity at the neuromuscular junction.

To ensure purposeful movement, electrical activity and the resultant contraction of the muscle cell must be turned on by motor neuron action potentials and must be switched off promptly when there is no longer a signal from the motor neuron. The muscle cell's electrical response is turned off by an enzyme present in the motor end-plate membrane, **acetylcholinesterase (AChE)**, which inactivates ACh. This occurs because ACh binding to the receptor is reversible (Figure 4–21):

9. Two molecules of acetylcholine bind very briefly (for about 1 millionth of a second) with the receptor, then detach. Some of the ACh molecules quickly rebind with receptor sites, keeping the end-plate channels open, but some diffuse deeper into the folds of the motor end plate, where AChE is located. AChE cleaves ACh into fragments, de-

activating it (step ⑨). As this process is repeated, more and more ACh is inactivated until it has been virtually removed from the cleft within a few milliseconds after its release.

Removal of ACh terminates the EPP so the remainder of the muscle cell membrane returns to resting potential. Now the muscle can relax. Or, if sustained contraction is essential for the desired movement, another motor neuron action potential leads to the release of more ACh, which keeps the contractile process going (see p. 321). By removing contraction-inducing ACh from the motor end plate, AChE permits the choice of allowing relaxation to take place (no more ACh released) or keeping the contraction going (more ACh released), depending on the body's momentary needs.

ACh receptors were first isolated from the electric organs of marine electric fish (see p. 245). Because the electric organs of electric rays, for example, consist of high concentrations of modified neuromuscular junctions with a high density of ACh receptors, this feature aided investigators in characterizing these proteins.

Synapses and Integration

Let's end this exploration of synapses with one of the most important functions found in animals: that of **summation** at chemical synapses. Summation is the ultimate basis of decision making (that is, whether to act or not, and if so, how strongly) in all animal activities involving neural control. Summation is most evident at neuron-to-neuron synapses, to which we now return.

■ The grand postsynaptic potential depends on the sum of the activities of all presynaptic inputs.

As we showed earlier, the events that occur at a single neuron-to-neuron synapse (and some other kinds) result in either an EPSP or an IPSP at the postsynaptic neuron. If a single EPSP is inadequate to bring the postsynaptic neuron to threshold and an IPSP moves it even farther from threshold, how is it possible to initiate an action potential in the postsynaptic neuron? Recall that a typical neuronal cell body receives thousands of presynaptic inputs from many other neurons. Some of these presynaptic inputs may be carrying sensory information brought from the environment; some may be signaling internal changes in homeostatic balance; others may be transmitting signals from control centers in the brain; and still others may arrive carrying other bits of information. At any given time, any number of these presynaptic neurons (probably hundreds) may be firing and thus influencing the postsynaptic neuron's level of activity. The total potential in the postsynaptic neuron, the **grand postsynaptic potential (GPSP)**, is a composite of all EPSPs and IPSPs occurring at approximately the same time.

The postsynaptic neuron can be brought to threshold in two ways: (1) *temporal summation* and (2) *spatial summation*. To illustrate these methods of summation, we will examine the possible interactions of three presynaptic inputs—two excitatory (Ex1 and Ex2) and one inhibitory (In1)—on a hypothetical postsynaptic neuron (● Figure 4–22). The recording represents the potential in the postsynaptic cell. Bear in mind during our discussion of this simplified version that many

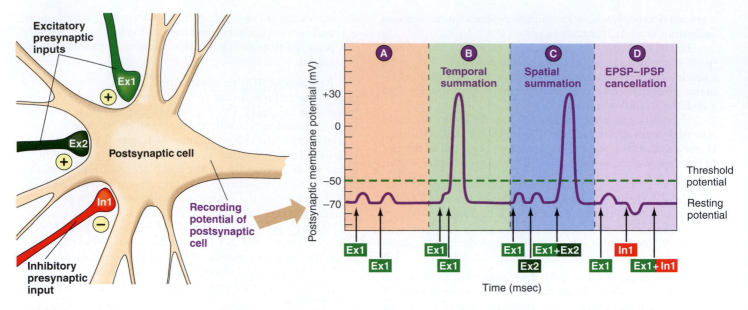

Panel Ⓐ If an excitatory presynaptic input (Ex1) is stimulated a second time after the first EPSP in the postsynaptic cell has died off, a second EPSP of the same magnitude will occur.

Panel Ⓑ If, however, Ex1 is stimulated a second time before the first EPSP has died off, the second EPSP will add onto, or sum with, the first EPSP, resulting in *temporal summation*, which may bring the postsynaptic cell to threshold.

Panel Ⓒ The postsynaptic cell may also be brought to threshold by *spatial summation* of EPSPs that are initiated by simultaneous activation of two (Ex1 and Ex2) or more excitatory presynaptic inputs.

Panel Ⓓ Simultaneous activation of an excitatory (Ex1) and inhibitory (In1) presynaptic input does not change the postsynaptic potential, because the resultant EPSP and IPSP cancel each other out.

Figure 4–22 ● **Determination of the grand postsynaptic potential by the sum of activity in the presynaptic inputs.** Two excitatory (Ex1 and Ex2) and one inhibitory (In1) presynaptic inputs terminate on this hypothetical postsynaptic neuron. The potential of the postsynaptic neuron is being recorded.

thousands of synapses are actually interacting in the same way on a single cell body and its dendrites.

Temporal Summation

Suppose Ex1 has an action potential that causes an EPSP in the postsynaptic neuron. Because EPSPs (as well as IPSPs) are graded potentials, the EPSP spreads only a short distance before dying off. If another action potential subsequently occurs in Ex1, an EPSP of the same magnitude takes place (panel A in Figure 4–22). Next suppose that Ex1 has two action potentials in close succession to each other (panel B). The first action potential in Ex1 produces an EPSP in the postsynaptic neuron. Although the postsynaptic membrane is still partially depolarized from this first EPSP (before it has returned to resting), the second presynaptic action potential produces a second EPSP in the postsynaptic neuron. The second EPSP adds on to the first EPSP, bringing the membrane to threshold, so that an action potential can occur in the postsynaptic neuron. Graded potentials do not have a refractory period, so this additive effect is possible.

The summing of several EPSPs occurring very close together in time because of successive firing of a single presynaptic neuron is known as **temporal summation** (the Latin word *tempus* means "time"). In reality, the situation is much more complex than just described. The sum of up to 50 EPSPs might be needed to bring the postsynaptic membrane to threshold. Each action potential in a presynaptic neuron triggers the emptying of a certain number of synaptic vesicles. The amount of neurotransmitter released and the resultant magnitude of the change in postsynaptic potential are thus directly related to the frequency of presynaptic action potentials. One way, then, in which the postsynaptic membrane can be brought to threshold is through rapid, repetitive excitation from a single, persistent input.

Spatial Summation

Let's now see what will happen in the postsynaptic neuron if both excitatory inputs are stimulated simultaneously (panel C). An action potential in either Ex1 or Ex2 will produce an EPSP in the postsynaptic neuron; however, neither of these alone will bring the membrane to threshold to elicit a postsynaptic action potential. Simultaneous action potentials in Ex1 and Ex2, however, will produce EPSPs that add to each other, bringing the postsynaptic membrane to threshold, so that an action potential occurs. Such summation of EPSPs originating simultaneously from several different presynaptic inputs (that is, from different points in "space") is known as **spatial summation**. A second way, therefore, to elicit an action potential in a postsynaptic cell is through concurrent activation of several excitatory inputs. Again, in reality up to 50 EPSPs arriving simultaneously on the postsynaptic membrane are required to bring it to threshold.

Similarly, IPSPs can undergo temporal and spatial summation. As IPSPs add together, however, they progressively move the potential farther from threshold.

Cancelation of Concurrent EPSPs and IPSPs

If an excitatory and an inhibitory input are simultaneously activated, the concurrent EPSP and IPSP more or less cancel each other out. The extent of cancelation depends on their respective magnitudes. In most cases, the postsynaptic membrane potential remains close to resting (panel D).

Importance of Postsynaptic Neuronal Integration

The magnitude of the GPSP depends on the sum of activity in all the presynaptic inputs and in turn determines whether or not the neuron will undergo an action potential to pass information on to the cells at which the neuron terminates. The following oversimplified real-life example demonstrates the benefits of this neuronal integration. The explanation is not completely accurate technically, but the principles of summation are accurate.

Assume for simplicity's sake that urination is controlled by a postsynaptic neuron supplying the urinary bladder in a mammal. When this neuron fires, neurotransmitter release induces the bladder to contract or relax. (Actually, voluntary control of urination is accomplished by postsynaptic integration at the neuron controlling the external urethral sphincter rather than the bladder itself.) As the bladder starts to fill with urine and becomes stretched, a reflex is initiated that ultimately produces EPSPs in the postsynaptic neuron responsible for causing bladder contraction. Partial filling of the bladder does not cause sufficient excitation to bring the neuron to threshold, so urination does not take place (panel A of Figure 4–22). As the bladder becomes progressively filled, the frequency of action potentials is progressively increased in the presynaptic neuron that signals the postsynaptic neuron about the extent of bladder filling (Ex1 in panel B of Figure 4–22). When the frequency becomes great enough that the EPSPs are temporally summed to threshold, the postsynaptic neuron has an action potential that stimulates bladder contraction.

What if the time is inopportune for urination? Consider a pet dog with its eye set on a favorite fire hydrant or a bobcat moving to the boundary of its territory, to mark it. IPSPs can be produced at the bladder postsynaptic neuron by presynaptic inputs originating in higher levels of the brain responsible for voluntary control (In1 in panel D of Figure 4–22). These "voluntary" IPSPs in effect cancel out the "reflex" EPSPs triggered by stretching of the bladder. Thus the postsynaptic neuron remains at resting potential and does not have an action potential, so the bladder is prevented from contracting and emptying even though it is full.

What if the bladder is only partially filled, so that the presynaptic input originating from this source is insufficient to bring the postsynaptic neuron to threshold to cause bladder contraction, yet the dog has found another inviting fire hydrant? The animal can voluntarily activate an excitatory presynaptic neuron (Ex2 in panel C of Figure 4–22), which spatially summates with the reflex-activated presynaptic neuron (Ex1) to bring the postsynaptic neuron to threshold. This achieves the

action potential necessary to stimulate bladder contraction, even though the bladder is not full.

This example illustrates the importance of postsynaptic neuronal integration. Each postsynaptic neuron in a sense "computes" all the input it receives and makes a "decision" about whether to pass the information on (that is, whether threshold is reached and an action potential is transmitted down the axon). In this way, neurons serve as complex computational devices, or integrators. The dendrites function as the primary processors of incoming information. They receive and tally the signals coming in from all the presynaptic neurons. Each neuron's output in the form of frequency of action potentials to other cells (muscle cells, gland cells, or other neurons) reflects the balance of activity in the inputs it receives via EPSPs or IPSPs from the thousands of other neurons that terminate on it. Each postsynaptic neuron filters out and does not pass on information it receives that is not significant enough to bring it to threshold. If every action potential in every presynaptic neuron that impinges on a particular postsynaptic neuron caused an action potential in the postsynaptic neuron, the neuronal pathways would be overwhelmed with trivia. Only if an excitatory presynaptic signal is reinforced by other supporting signals through summation is the information passed on. Furthermore, interaction of postsynaptic potentials provide a way for one set of signals to offset another set (IPSPs negating EPSPs). This allows a fine degree of discrimination and control in determining what information is acted on.

■ Action potentials are initiated at the axon hillock because it has the lowest threshold.

Threshold potential is not uniform throughout the postsynaptic neuron. The lowest threshold is present at the axon hillock because this region has a much greater density of voltage-gated Na^+ channels than anywhere else in the neuron. This greater density of these voltage-sensitive channels makes the axon hillock considerably more sensitive to changes in potential than the dendrites or the remainder of the cell body. The latter regions have a significantly higher threshold than the axon hillock. Because of local current flow, changes in membrane potential (EPSPs or IPSPs) occurring anywhere on the cell body or dendrites spread throughout the cell body, dendrites, and axon hillock. When summation of EPSPs takes place, the lower threshold of the axon hillock is reached first, whereas the cell body and dendrites at the same potential are still considerably below their own much higher thresholds. Therefore, an action potential originates in the axon hillock and is propagated from there to the end of the axon.

■ Neuropeptides act primarily as neuromodulators.

Researchers recently discovered that in addition to the classical neurotransmitters just described, some neurons also release *neuropeptides* (Table 4–3). Neuropeptides differ from classical neurotransmitters in several important ways (■ Table 4–4). **Classical neurotransmitters** are small, rapid-acting molecules that typically bring about a response in the postsynaptic neuron within a few milliseconds or less, usually a change in postsynaptic potential (an EPSP or IPSP) as a result of triggering the opening of specific ion channels. Most classical neurotransmitters are synthesized and packaged locally in synaptic vesicles in the cytosol of the axon terminal. These chemical messengers are primarily amino acids or closely related compounds.

Neuropeptides are larger molecules made up of anywhere from 2 to about 40 amino acids. They are synthesized in the neuronal cell body in the endoplasmic reticulum and Golgi complex (see p. 37) and are subsequently moved by axonal transport along the microtubular highways to the axon terminal (see p. 59). Neuropeptides are not stored within the small synaptic vesicles with the classical neurotransmitter but instead are packaged in large **dense-core vesicles,** which are also present in the axon terminal. The dense-core vesicles undergo Ca^{++}-induced exocytosis and release neuropeptides at the same time that the neurotransmitter is released from the synaptic vesicles. An axon terminal typically releases only a single classical neurotransmitter, but the same terminal may also contain one or more neuropeptides that are cosecreted along with the neurotransmitter.

Even though neuropeptides are currently the subject of intense investigation, knowledge about their functions and control is still sketchy. They are known to diffuse locally and act on other adjacent neurons at much lower concentrations than classical neurotransmitters, and they bring about slower, more prolonged responses. Some neuropeptides released at synapses may function as true neurotransmitters, but most are believed to function as neuromodulators.

Neuromodulators are chemical messengers that do not cause the formation of EPSPs and IPSPs but rather bring about long-term changes that subtly *modulate*—depress or enhance—the action of the synapse. They bind to neuronal receptors at nonsynaptic sites—that is, not at the subsynaptic membrane—and they often activate second-messenger systems. Neuromodulators may act on either presynaptic or postsynaptic sites. For example, a neuromodulator may influence the level of an enzyme critical in the synthesis of a neurotransmitter by a presynaptic neuron, or it may alter the sensitivity of the postsynaptic neuron to a particular neurotransmitter by causing long-term changes in the number of subsynaptic receptor sites for the neurotransmitter. Thus, neuromodulators delicately fine-tune the synaptic response. The effect may last for days or even months or years. Like some slow neurotransmitters, neuromodulators are involved with more long-lasting events such as learning and motivation.

Interestingly, vertebrate neuromodulators include many substances that also have distinctly different roles in other body regions as hormones released into the blood from neural and nonneural tissues. For example, *cholecystokinin (CCK)* is a well-known hormone released from the small intestine that causes the gallbladder to contract and release bile into the intestine, among other digestive activities (p. 639). Researchers have also found CCK in axon terminal vesicles in various areas of the rat brain, where they believe it mediates satiety (the feeling of no longer being hungry) and perhaps decreases the perception of painful stimuli, among other unknown activities. In fact, in many instances neuropeptides are named for their first-discovered role as hormones, as is the case with cholecystokinin (*chole*, "bile"; *cysto*, "bladder"; *kinin*, "contraction"). A number of chemical messengers are apparently quite versatile and can assume different roles, depending on their source, distribution, and interaction with different cell types.

■ Presynaptic inhibition or facilitation can selectively alter the effectiveness of a given presynaptic input.

Besides neuromodulation, presynaptic inhibition or facilitation is another means of depressing or enhancing synaptic effectiveness. Sometimes a third neuron influences activity between a presynaptic ending and a postsynaptic neuron. A presynaptic axon terminal (labeled A in ● Figure 4–23) may itself be innervated by another axon terminal (labeled B). The neurotransmitter released from modulatory terminal B binds with receptor sites on terminal A. This binding alters the amount of transmitter released from terminal A in response to action potentials. If the amount of transmitter released from A is reduced, the phenomenon is known as **presynaptic inhibition.** If the release of transmitter is enhanced, the effect is **presynaptic facilitation.**

Let's look more closely at how this process works. You know that Ca^{++} entry into an axon terminal induces the release of neurotransmitter by exocytosis of synaptic vesicles. The amount of neurotransmitter released from terminal A depends on how much cytosolic Ca^{++} enters this terminal in response to an action potential. Ca^{++} entry into terminal A, in turn, can be influenced by activity in modulatory terminal B. Let's use presynaptic inhibition to illustrate (Figure 4–23).

Table 4–4 ■ Comparison of Classical Neurotransmitters and Neuropeptides

Characteristic	Classical Neurotransmitters	Neuropeptides
Size	Small; one amino acid or similar chemical	Large: 2 to 40 amino acids in length
Site of synthesis	Cytosol of synaptic knob	Endoplasmic reticulum and Golgi complex in cell body; travel to synaptic knob by axonal transport
Site of storage	In small synaptic vesicles in axon terminal	In large dense-core vesicles in axon terminal
Site of release	Axon terminal	Axon terminal; may be cosecreted with neurotransmitter
Speed and duration of action	Rapid, brief response	Slow, prolonged response
Site of action	Subsynaptic membrane of postsynaptic cell	Nonsynaptic sites on either presynaptic or postsynaptic cells at much lower concentrations than classical neurotransmitters
Effect	Usually alter potential of postsynaptic cell by opening specific ion channels	Usually enhance or suppress synaptic effectiveness by long-term changes in neurotransmitter synthesis or postsynaptic receptor sites

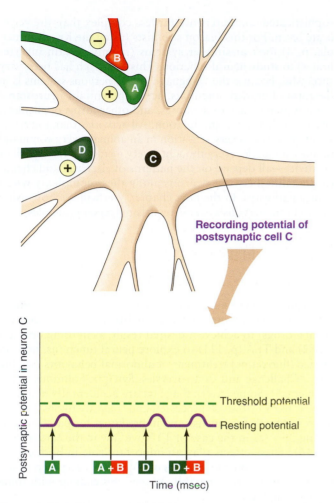

Recording potential of postsynaptic cell C

Figure 4–23 • **Presynaptic inhibition.** A, an excitatory terminal ending on postsynaptic cell C, is itself innervated by inhibitory terminal B. Stimulation of terminal A alone produces an EPSP in cell C, but simultaneous stimulation of terminal B prevents the release of excitatory neurotransmitter from terminal A. Consequently, no EPSP is produced in cell C despite the fact that terminal A has been stimulated. Such presynaptic inhibition selectively depresses activity from terminal A without suppressing any other excitatory input to cell C. Stimulation of excitatory terminal D produces an EPSP in cell C even though inhibitory terminal B is simultaneously stimulated because terminal B only inhibits terminal A.

The amount of transmitter released from presynaptic terminal A, an excitatory input in our example, influences the potential in the postsynaptic neuron on which it terminates (labeled C in the figure). Firing of A alone will bring about an EPSP in postsynaptic neuron C. Now consider that B is stimulated simultaneously with A. Binding of the neurotransmitter from modulatory terminal B on excitatory terminal A causes opening of chemical messenger-gated Cl$^-$ channels in terminal A. The subsequent entry of Cl$^-$ into terminal A blunts the size of the action potentials taking place in A. That is, the inside of terminal A doesn't reach the positivity of +30 mV, because of the opposing entry of negatively charged Cl$^-$. Because of the smaller action potentials in A, the voltage-dependent Ca^{++} entry into this terminal is reduced. Less Ca^{++} entry

means less neurotransmitter release from A. Obviously, modulatory neuron B can suppress transmitter release from A only when A is firing. If A is prevented from releasing its transmitter by presynaptic inhibition via B, the formation of EPSPs on postsynaptic membrane C will specifically be prevented from input A. As a result, no change in the potential of the postsynaptic neuron will occur in spite of action potentials in A.

Could the same thing be accomplished by a simultaneous production of an IPSP through activation of an inhibitory input to negate an EPSP produced by activation of A? Not quite. Activation of an inhibitory input to cell C in Figure 4–23 would produce an IPSP in cell C, but this IPSP could cancel out not only an EPSP from excitatory presynaptic input A, but also any EPSPs produced by other excitatory terminals, such as terminal D in the figure. The entire postsynaptic membrane is hyperpolarized by IPSPs, negating (canceling) excitatory information fed into any part of the cell from any presynaptic input. By contrast, presynaptic inhibition (or presynaptic facilitation) works in a much more specific way than does the action of inhibitory inputs to the postsynaptic cell.

Presynaptic inhibition provides a means by which certain inputs to the postsynaptic neuron can be *selectively* inhibited without affecting contributions of any other inputs. For example, firing of B specifically prevents formation of an EPSP in the postsynaptic neuron from excitatory presynaptic neuron A but does not influence other excitatory presynaptic inputs. Excitatory input D can still produce an EPSP in the postsynaptic neuron even when B is firing. This type of neuronal integration is another means by which electrical signaling between neurons can be carefully fine-tuned.

■ Retrograde messengers travel "backward" and also modulate synaptic function.

Although the primary message being relayed through a synapse travels in one direction in most animals, researchers are finding an increasing number of examples of **retrograde** signals. In such cases, a stimulated postsynaptic cell may produce and release a signal molecule that diffuses backward, that is, to the presynaptic terminal. Once there, it does not trigger a message to travel up the axon; rather, it modifies the sensitivity of the terminal in a prolonged or long-term way, similar to the actions of neuromodulators. One newly discovered example is the neuromuscular synapse of the laboratory fruit fly (*Drosophila*). The primary neurotransmitter from the neuron to the muscle is glutamate (not ACh as in vertebrates). The (postsynaptic) muscle cell also produces a peptide that travels to the (presynaptic) neuron. There it enhances release of glutamate. The system appears to work in negative feedback fashion for "synaptic homeostasis." The muscle produces *more* peptide when it is unstimulated and *less* when stimulated. This tends to keep the amount of glutamate released at a consistent level. In the next chapter, we describe how *nitric oxide* in a vertebrate can be used in a positive feedback retrograde system.

■ Neurons are linked to each other through convergence and divergence to form complex nerve pathways.

Two important relationships exist between neurons: convergence and divergence. A given neuron may have many other neurons synapsing on it. Such a relationship is known as **con-**

vergence (● Figure 4–24). Through this converging input, a single cell is influenced by thousands of other cells. This single cell, in turn, influences the level of activity in many other cells by divergence of output. The term **divergence** refers to the branching of axon terminals so that a single cell synapses with and influences many other cells.

Note that a particular neuron is postsynaptic to the neurons converging on it but presynaptic to the other cells on which it terminates. Thus the terms **presynaptic** and **postsynaptic** refer only to a single synapse. Most neurons are presynaptic to one group of neurons and postsynaptic to another group.

There are an estimated 100 billion neurons and 10^{14} (100 quadrillion) synapses in the human brain alone. When you consider the vast and intricate interconnections possible between these neurons through converging and diverging pathways, you can begin to imagine how complex the wiring mechanism of an animal's nervous system can be. Even the most

sophisticated computers are far less complex than the vertebrate brain. For this reason scientists (using the Krogh principle, p. 3) look at simpler systems (mainly certain invertebrates) to study neural function. This approach has been very successful, because the "language" of *all* nervous systems is in the form of graded potentials, action potentials, neurotransmitter signaling across synapses, and other forms of chemical chatter such as neuromodulation. All activities that a nervous system governs—every sensation in an animal, every command to move a muscle, every thought, every memory, every spark of creativity—all depend on the patterns of electrical and chemical signaling between neurons along these complexly wired neural pathways. In the next chapter, you will see how using invertebrate nervous systems reveals universal principles.

Neural Transmission and External Agents

The signaling systems of animal neurons can be affected by many external agents. Some are deliberate pharmacologic agents (drugs) to achieve a desired result, as in using ouabain (p. 82) and TEA (p. 111) to explore neural functions, or as in Prozac (fluoxetine) treatment of abnormal behaviors (see the box, "Challenge and Controversies: Synaptic Solutions to Pet Problems"). Some external agents are accidentally acquired as poisons or by disease processes, and some are toxins evolved in prey animals for defense and in predators for paralyzing or killing prey (as in the case of TTX we saw in the earlier box, p. 114). These effects play an important role in both evolutionary adaptation and in medicine. Virtually every step of signal transmission is vulnerable. We end this chapter with some more examples.

■ Drugs, disease toxins, and pollutants can modify synaptic transmission between neurons.

Most drugs that influence the nervous system perform their function by altering synaptic mechanisms. Synaptic drugs may block an undesirable effect (antagonist) or enhance a desirable effect (agonist). Possible drug actions include (1) altering the synthesis, axonal transport, storage, or release of a neurotransmitter; (2) modifying neurotransmitter interaction with the postsynaptic receptor; (3) influencing neurotransmitter reuptake or destruction; and (4) replacing a deficient neurotransmitter with a substitute transmitter.

For example, the drug **cocaine** blocks reuptake of the neurotransmitter **dopamine** at presynaptic terminals by competitively binding with the dopamine reuptake transporter. It does so by binding competitively with the dopamine reuptake transporter, a protein molecule that picks up released dopamine from the synaptic cleft and shuttles it back to the axon terminal. With cocaine occupying the dopamine transporter, dopamine remains in the synaptic cleft longer than usual and continues to interact with its postsynaptic receptor sites. The result is prolonged activation of neural pathways that use this chemical as a neurotransmitter. Among these pathways are those that play a role in emotional responses, especially feelings of reward. In essence, when cocaine is present the neural switches in the reward pathway are locked in the "on" position.

Cocaine is addictive to humans and laboratory rodents (who will choose cocaine over food and water even to the point

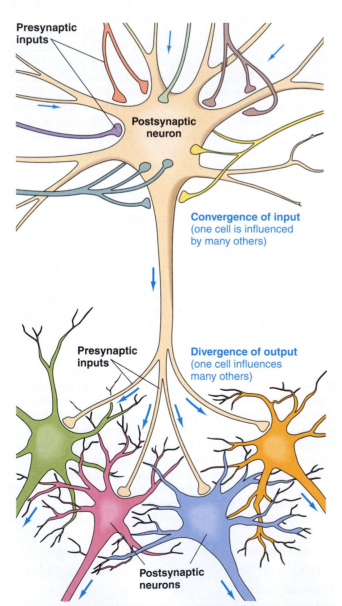

Presynaptic inputs

Postsynaptic neuron

Convergence of input (one cell is influenced by many others)

Presynaptic inputs

Divergence of output (one cell influences many others)

Postsynaptic neurons

Arrows indicate direction in which information is being conveyed.

Figure 4–24 ● Convergence and divergence.

CHALLENGES AND CONTROVERSIES

Synaptic Solutions to Pet Problems

Ever since the identification of the synapse as the functional junction between two neurons, the synapse has been regarded as the most vulnerable physiological site for manipulation and artificial regulation. Behavioral pharmacotherapy, the use of drugs to treat behavioral problems, is an example of a science that has evolved from this original precept. Interestingly, scientists observed that during the course of treatment with antihistamines for congestion that some human patients noted a significant improvement in emotional state. In fact, many psychoactive drugs are chemically similar to antihistamines. Psychoactive drugs exert their behavioral effects by acting on neurotransmitters in the brain (see p. 175). For example, treatment with drugs that selectively inhibit reuptake of **serotonin** into the neuron (such as **Prozac**) results in a higher level of activity in any part of the nervous system that uses serotonin as a chemical signal.

Serotonin is the common name for **5-hydroxytryptamine (5-HT)**. It is synthesized in the brain from the amino acid tryptophan. Because there are at least nine different serotonin receptor subtypes, each localized separately in the mammalian brain and responsible for different behavioral responses, regulation of the serotonin receptor system is particularly complex. In mammals, serotonin plays an important role in the control of sleep, pain, aggression, sexual behavior, and food intake. This diverse range of responses is not surprising, considering that serotonin neurons project fibers to virtually all parts of the central nervous system. Documented effects of serotonin-reuptake inhibitors in dogs include a reduction in the intensity of obsessive-compulsive disorders such as excessive licking behavior, tail mutilation, separation anxieties, dominance-related aggression, and so forth. In cats, serotonin-reuptake inhibitors are used for the treatment of psychogenic alopecia (unnatural loss of hair), offensive aggression as well as urine spraying. Prozac has been used to treat cribbing

(force-swallowing gulps of air) in horses and bulimia (episodes of binge eating that continue until terminated by abdominal pain or vomiting) in pigs.

Tricyclic antidepressants (so called because of their three-ringed chemical structure) also block the reuptake of serotonin and catecholamines from synapses, which promotes a state of calmness. Tricyclics can be used to treat urine spraying in cats, urine marking in dogs, and separation anxiety in both cats and dogs. These drugs are also useful in the treatment of stereotypies. **Stereotypies** are repetitive behaviors that perform no function (such as pacing in zoo animals). They arise from chronic confinement, chronic conflict, or sensory deprivation, and researchers suggest they provide some relief. Chemical intervention is normally warranted to eliminate stereotypies because, once incorporated into an animal's behavioral repertoire, the stereotypy continues to be performed even in the absence of the initiating stressor.

Neuroleptics represent a family of drugs whose mode of action is the antagonism of **dopamine** (p. 175) and are effective in treating certain phobias, fears, and anxieties. Phobias are extreme and apparently irrational fears whose origins can normally be identified. For example, animals may be fearful of people or of other animals, sensory stimuli such as sounds (thunderstorms), odors, or touch, or exhibit phobias under specific circumstances such as visiting the veterinarian's office. These compounds are also effective in preventing anxiety-related inappropriate urine marking and spraying in cats. Neuroleptics are used in the treatment of stereotypic disorders such as flank biting in horses, fur sucking in cats, and feather picking in birds. Although many of these and other drugs show promise in the treatment of fear-based conditions in pets, veterinarians recommend that chemical therapy be combined with behavioral modification in order to achieve long-lasting results.

of death) because it causes permanent molecular changes of the involved neurons such that they cannot transmit normally across synapses without increasingly higher doses of the drug. After the neurons have been incessantly stimulated for an extended time by cocaine, a transcription factor (p. 30) called δ-*FosB* is activated by unknown means. This factor, in turn, activates the gene for an enzyme called *cyclin-dependent kinase 5* or *Cdk5*. Cdk5 greatly reduces sensitivity for cocaine (again, by unknown mechanisms). Whatever the mechanism, the synapses require more and more cocaine to achieve a stimulus and thus a reward feeling. When cocaine is no longer available, the sense of reward cannot be achieved, because the normal level of dopamine release is no longer sufficient to stimulate the postsynaptic neuron. Note that this is a form of *acclimatization* or *downregulation* (p. 15).

Synaptic transmission is also vulnerable to numerous toxins, which may cause nervous system disorders by acting at either presynaptic and postsynaptic sites. For example, two

different neural poisons, strychnine and tetanus toxin, act at different synaptic sites to block inhibitory impulses while leaving the excitatory inputs unchecked. **Strychnine,** a poison used to kill rodents that infest buildings, competes with an inhibitory neurotransmitter, glycine, at the postsynaptic receptor site. This antagonist combines with the receptor but does not directly alter the potential of the postsynaptic cell. Instead, strychnine blocks the receptor so that glycine cannot bind when it is released from the inhibitory presynaptic ending. Thus postsynaptic inhibition (formation of IPSPs) is abolished in nerve pathways that use glycine as an inhibitory transmitter. Unchecked excitatory pathways lead to convulsions, muscle spasticity, and death.

Tetanus toxin, in contrast, prevents release of another inhibitory transmitter, γ-aminobutyric acid (GABA), from inhibitory presynaptic inputs terminating on neurons that supply skeletal muscles. Unchecked excitatory inputs to these neurons result in uncontrolled muscle spasms. These spasms occur especially in the jaw muscles early in the disease, giving rise to the common name of **lockjaw** for this condition. Later they progress to the muscles responsible for breathing, at which point the disease proves fatal.

Strychnine and tetanus toxin poisoning have similar outcomes, but one poison (strychnine) blocks specific postsynaptic inhibitory receptors, whereas the other (tetanus toxin) prevents the presynaptic release of a specific inhibitory neurotransmitter.

Some pollutants can inhibit neural signaling. Lead (Pb^{++}), for example, is an important environmental toxicant because it competes with Ca^{++} for entry into the terminal button. Once ingested by an animal, Pb^{++} at high doses inhibits neural transmission, which can kill the animal.

Other drugs and diseases that influence synaptic transmission are too numerous to mention, but as these examples illustrate, any site along the synaptic pathway is vulnerable to interference, either pharmacologic (drug induced) or pathologic (disease induced).

The neuromuscular junction is vulnerable to several chemical agents and diseases.

Several chemical agents and diseases affect the neuromuscular synapse by acting at different sites in the transmission process. Two well-known toxins—**black widow spider venom** and **curare**—alter the function of ACh, but in different ways.

The venom of black widow spiders exerts its deadly effect on small prey animals (not humans) by causing an explosive release of ACh from the storage vesicles, not only at neuromuscular junctions but at all cholinergic sites. Thus the toxin is an *agonist* (p. 94). All cholinergic sites undergo prolonged depolarization, the most detrimental consequence of which is respiratory failure in a bitten mammal. Breathing is accomplished by alternate contraction and relaxation of skeletal muscles, particularly the diaphragm. Respiratory paralysis occurs as a result of prolonged depolarization of the mammalian diaphragm. During this so-called *depolarization block*, the voltage-gated Na^+ channels are trapped in their inactivated state (see p. 109), prohibiting the initiation of new action potentials and resultant contraction of the diaphragm. As a consequence, the spider's victim cannot breathe.

Other chemicals are *antagonists* (p. 94) that interfere with neuromuscular junction activity by blocking the effect of released ACh. The best-known example is **curare**, obtained from the skin of certain frogs in South America. Curare reversibly binds to the ACh receptor sites on the motor end plate. Unlike ACh, however, curare does not alter membrane permeability, nor is it inactivated by AChE. When ACh receptor sites are occupied by curare, ACh cannot combine with these sites to open the channels that would permit the ionic movement responsible for an EPP. Consequently, because muscle action potentials cannot occur in response to nerve impulses to these muscles, paralysis ensues. The toxin presumably evolved to protect these frogs from predators: when sufficient curare is present to effectively block a significant number of ACh receptor sites, the predator dies from respiratory paralysis caused by an inability to contract the diaphragm. Some people in South America use curare as a deadly arrowhead poison. Curare and related drugs have also been used medically (at very low doses!) during surgery to help achieve more complete skeletal muscle relaxation with less anesthetic.

An example of a disease that affects the junction is *myasthenia gravis,* characterized by extreme muscle weakness. It is caused by antibodies of an animal's own immune system that attack ACh receptors of the junction. This *autoimmune* (p. 453) disease has been documented in cats, dogs, and humans. The drug **neostigmine**, which inhibits AChE, is used to treat this, because its action prolongs the activity of whatever amount of ACh is released at the synapse.

Temperature and pressure also influence the propagation of action potentials.

Finally, evolutionary adaptation of animals has been greatly affected by physical factors that alter neural functions. In particular, temperature as well as pressure exert pronounced effects on membrane processes. Recall from Chapter 3 (p. 70) that cold temperatures can make membranes too rigid, and high temperatures can make them too fluid. Temperature also directly affects the rates of all reactions (p. 683), including channel opening and closing. The effects of temperature are discussed in detail in Chapter 15.

Effects of an increase in pressure on cells is more complex, and most are not discussed here. Briefly, high pressure can inhibit the proper folding of proteins, reduce movement of proteins during functional shape changes, and reduce membrane fluidity. Such effects presumably explain the effect of pressure on propagation of action potentials and synaptic transmission in both invertebrates and vertebrates. In animals that are *unadapted* to pressure, these effects include the following:

- A reduction in the activity of the Na^+–K^+ pump
- A decrease in the rate at which an action potential is propagated
- A decrease in the rate of axonal repolarization and an increase in axon potential duration
- Inhibition of excitatory synaptic transmission, because of reduced neurotransmitter release from the presynaptic terminal
- Reduced binding of ACh to its receptor on the muscle membrane

All these effects are potentially detrimental to animals living permanently in the deep sea, However, adaptation has occurred: Some of these effects are minimized or not apparent in animals adapted to high pressures (such as fish and shrimp in the deep sea). Adaptations to elevated pressures include membranes with higher fluidity, achieved by a decrease in the ratio of saturated to unsaturated fatty acids (as you saw in Chapter 3 for temperature, p. 70). We discuss another form of adaptation in Chapter 13.

Such adaptations have not occurred in shallow-living animals that occasionally experience high pressure—in particular, deep-diving marine mammals such as Weddell seals and sperm whales. However, these effects may be useful to these divers, contributing to an energy-saving reduction in heart rate when the animals descend to depth (see Chapter 11). For example, at high pressures (150 atm) the conduction velocity of an impulse in a seal's heart is reduced by 40%.

Interestingly, high temperature can reverse the effects of high pressure; that is, a moderate increase in temperature restores fluidity to the membrane and increases the activity of the Na^+–K^+ pump. However, at elevated temperatures there is an associated increase in membrane permeability, and K^+ leaks into the ECF and Na^+ enters the cell. Generally, there is a reduction in synaptic transmission at higher temperatures, although there is little effect on the rate of axon propagation.

Chapter in Perspective
HOMEOSTASIS AND INTEGRATION

Neurons are specialized to receive, process, encode, and rapidly transmit information from one part of the body to another. The information is transmitted over intricate nerve pathways by propagation of action potentials along the neuron's length as well as by chemical transmission of the signal from neuron to neuron and from neuron to muscles and glands through neurotransmitter–receptor interactions at synapses and from neuron to muscles and glands through other neurotransmitter–receptor interactions at these junctions.

Collectively, neurons make up the nervous system, one of the two major control systems of the body. Many of the activities controlled by the nervous system are geared toward maintaining homeostasis. Some neuronal electrical signals convey information about changes to which the body must respond in order to maintain homeostasis—for example, these signals convey information about a fall in blood pressure. Other neuronal electrical signals convey messages to muscles and glands to stimulate appropriate responses to counteract these changes—for example, the nervous system through its electrical signals initiates adjustments in heart and blood vessel activity to restore blood pressure to normal when it starts to fall, or to initiate behaviors that aid survival such as feeding when low internal energy supplies are sensed.

The specialization of muscle cells, contraction, also depends on these cells' ability to undergo action potentials. Action potentials trigger muscle contractions, many of which are important in maintaining homeostasis. For example, beating of the heart, mixing of ingested food with digestive enzymes, and shivering to generate heat when a mammal is cold are all accomplished by muscle contractions. ■

REVIEW QUESTIONS   *(Answers are on p. A–1.)*

Additional study tools for this chapter, including chapter summaries and practice tests, are available online at *www.biology.brookscole.com*

1. Which of the following is true of graded potentials?
   a. They are local changes in membrane potential.
   b. They occur in varying degrees of strength.
   c. The graded potential is related to the magnitude of the triggering event.
   d. Gated Na^+ channels open in the area of a triggered graded potential.
   e. all of the above

2. Any flow of electrical charges is called
   a. a resistance
   b. a capacitance
   c. a current
   d. an action potential
   e. hyperpolarization

3. Which of the following is not true of action potentials?
   a. They diminish in strength as they travel along the cell membrane.
   b. They are brief and rapid.
   c. At threshold potential, depolarization occurs with explosive rapidity.
   d. They are often graded responses.
   e. a and d above

4. Ether-a-go-go and Shaker are terms used to describe
   a. types of voltage-gated Na^+ channels
   b. types of voltage-gated K^+ channels
   c. activation gates
   d. inactivation gates
   e. action potentials

5. At the peak of an action potential,
   a. Na^+ channels open and K^+ channels close
   b. Na^+ channels close and K^+ channels close
   c. Na^+ channels open and K^+ channels open
   d. Na^+ channels close and K^+ channels open
   e. The membrane is 600 times more permeable to Na^+ than to K^+

6. In order for another action potential to be generated,
   a. the concentrations of Na^+ and K^+ must once again become equal on each side of the membrane
   b. the Na^+–K^+ pump must immediately restore the original resting concentration gradients of these ions
   c. there must be less K^+ inside the cell than outside the cell
   d. Na^+ must once again become the predominant intracellular ion
   e. none of the above

7. Dendrites carry signals
   a. away from the cell body
   b. toward the cell body
   c. away from the axon hillock
   d. from the axon to the cell body
   e. from the axon to the axon terminals

8. Once an action potential is initiated in one part of a neuron's membrane,
   a. it is propagated only along the axon
   b. it is propagated only along the dendrite
   c. it is propagated from the axon to the dendrite

d. action potentials become weaker as they travel along a nerve fiber

e. action potentials remain equally strong as they propagate along the rest of the fiber

9. The "all or none" law in regard to action potentials means
   a. a stronger triggering event results in a larger action potential
   b. all refractory periods are absolute
   c. once triggered, an action potential always goes to a maximal height
   d. relative refractory periods either trigger an action potential or do not trigger an action potential
   e. refractory periods determine the direction of the propagation of an action potential

10. Perception of the strength of a stimulus is determined by
    a. saltatory conduction
    b. myelination of the fiber
    c. the diameter of the fiber
    d. the frequency of action potentials
    e. Schwann cells

11. In myelinated fibers, the membrane potential can exist and current can flow only
    a. in Schwann cells
    b. in oligodendrocytes
    c. at nodes of Ranvier
    d. within the myelin lipid layers
    e. in Mauthner cells

12. Electrical synapses send signals
    a. through synaptic clefts
    b. through gap junctions
    c. using neurotransmitters
    d. using dopamine
    e. using serotonin

13. In "fast" synapses, the arrival of an action potential triggers a chain of events including
    a. the opening of voltage-gated Ca^{++} channels in the synaptic knob
    b. release of neurotransmitters from synaptic vesicles by exocytosis
    c. diffusion of neurotransmitters across the synaptic cleft
    d. neurotransmitters combining with receptors in the postsynaptic neuron
    e. all of the above

14. Neurotransmitters
    a. always excite the postsynaptic neuron
    b. always inhibit the postsynaptic neuron
    c. can excite or inhibit the postsynaptic neuron
    d. can cause an EPSP or an IPSP
    e. c and d

15. Slow synapses involve
    a. the activation of intracellular second messengers
    b. the activation of cAMP
    c. the neurotransmitter, serotonin
    d. the neurotransmitter, norepinephrine
    e. a, b, and c

16. Which of the following events occurs first at a muscle synapse when an action potential arrives?
    a. ACh binds to receptors.
    b. Chemical–messenger-gated channels in the motor end plate open.
    c. Acetylcholinesterase inactivates Ach.
    d. voltage-gated Ca^{++} channels open in the terminal button.
    e. Ca^{++} ions are conducted into the terminal button.

17. The grand synaptic potential acting on a neuron can be described as
    a. the composite of all EPSPs
    b. the composite of all IPSPs
    c. the composite of all EPSPs and IPSPs
    d. end plate potential
    e. none of the above

18. In which of the following ways do neuropeptides differ from neurotransmitters?
    a. They are synthesized in the cell body rather than the cytosol of the axon terminal.
    b. They are packaged in dense core vesicles.
    c. As neuromodulators, they do not cause EPSPs or IPSPs.
    d. In general, they are larger molecules than neurotransmitters.
    e. All of the above

19. Presynaptic inhibition and facilitation act
    a. only as retrograde signals
    b. only after an action potential has occurred in the postsynaptic membrane
    c. through release of neurotransmitters
    d. through release of neuromodulators
    e. by canceling concurrent EPSPs and IPSPs

20. The drug cocaine blocks the reuptake of the neurotransmitter
    a. norepinephrine
    b. serotonin
    c. GABA
    d. dopamine
    e. acetylcholine

SUGGESTED READINGS AND INTERNET SITES

Cowan, W. M., T. C. Südhoff, & C. F. Stevens, eds. 2001. *Synapses*. Baltimore: Johns Hopkins University Press. A compendium of reviews on synaptic physiology.

Doyle, D. A., J. Morais Cabral, R. A. Pfuetzner, et al. 1998. The structure of the potassium channel: Molecular basis of K^+ conduction and selectivity. *Science* 280:69–77. The first detailed structural analysis of K-channels.

Hodgkin, A. L. 1994. *Chance and Design, Reminiscences of Science in Peace and War*. New York: Cambridge University Press. A personal view of the pioneering work on neurons.

Kandel, E. R., J. H. Schwartz, T. M. Jessell, eds. 2000. *Principles of Neural Science,* 4th ed. New York: McGraw-Hill/Appleton & Lange. An excellent textbook on neuroscience.

Katz, P. S. 1999. *Beyond Neurotransmission: Neuromodulation and Its Importance for Information Processing.* New York: Oxford University Press. A review of neuromodulation and how it differs from classic neurotransmission.

Lenz, P. H., D. K. Hartline, & A. D. Davis. 2000. The need for speed. Part 1. Fast reactions and myelinated axons in copepods. *Journal of Comparative Physiology* 186:337–345. The discovery of myelin in copepods.

Levitan, I. B., & L. K. Kaczmarek. 2001. *The Neuron: Cell and Molecular Biology,* 3rd ed. Oxford, UK: Oxford University Press. A textbook on neuroscience at the molecular level.

Matthews, G. G. 2001. *Neurobiology: Molecules, Cells and Systems.* Palo Alto, CA: Blackwell Science. An excellent textbook on neurobiology.

Randall, D., W. Burggren, & K. French. 2002. *Eckert Animal Physiology. Mechanisms and Adaptation,* 5th ed. New York: W. H. Freeman. A classic animal physiology text with comprehensive coverage of many of the topics discussed in this book.

Salinas, P. C. 2003. Backchat at the synapse. *Nature* 425:464–466. On retrograde transmission and a new discovery in *Drosophila.*

Unwin, N. 1995. Acetylcholine receptor channel imaged in the open state. *Nature* 373: 37–43. A detailed structural analysis of receptor channel function.

INTERNET SITES

Matthews, G. G. *Neurobiology.* 2001. **blackwellscience.com/ matthews/neurotrans.html.** Animations of excitatory and inhibitory synapses, neuromuscular junction, and cardiac muscle synapse, associated with the Matthews 2001 reference.

Songer, G. *Botulinum Toxin.* **microvet.arizona.edu/Courses/ MIC420/lecture_notes/clostridia/clostridia_neurotox/movie/ botulinum_movie.html.** An animation of the effect of botulism on a synapse.

Nervous Systems

Photo: © Stephen Frink/CORBIS

Photo: © Thomas D. Mangelsen/Images of Nature

An octopus and a gorilla. The former has the largest brain:body ratios among invertebrates, and the primate has one of the highest ratios in the animal kingdom.

Evolutionary Considerations and Invertebrate Nervous Systems

With this chapter, we begin examining the organ systems of animals, starting with those involved in whole-body regulation. The nervous system is one of two such control systems of an animal body, the other being the endocrine system (although some researchers include the immune system as a third). In general, the nervous system coordinates rapid responses, whereas the endocrine (hormonal) system (Chapter 7) regulates activities that require duration rather than speed.

■ Nervous systems evolved from simple reflex arcs to centralized brains with distributed, hierarchic regulation.

In all animals the ultimate function of the nervous system is to translate sensory information into action potentials (APs), which can be processed (integrated) into an adaptive response via activation of effector organs. The adaptive response is often one that, in negative-feedback fashion, corrects in some way for a disturbance that initiated the sensory information. Although feedback regulation can occur locally within cells and tissues, nervous systems have the advantage of providing rapid regulation over greater distances. If the regulatory response involves movement via one or more effector skeletal muscles, the response is called a **behavior.** The most basic pathway for accomplishing this is known as a **reflex arc,** with the simplest of these having a sensory neuron that also controls an effector cell directly, as shown in ● Figure 5–1a.

However, most reflex arcs have two or more neurons involved, with information being transmitted through one or more intermediate synapses (Figure 5–1b, c). These intermediate synapses constitute the simplest form of a *central nervous system (CNS),* where signals from more than one neuron can be integrated. The ability to have multiple synaptic connections allowed for the evolution of complex control of responses.

For example, an integrator neuron can have inputs from a "higher" integrator such as a brain (Figure 5–1d). The higher regulatory center modifies the basic reflex response: It may suppress the reflex, alter its sensitivity, or activate it in advance of a disturbance in *anticipation* fashion. The brain may modify the reflex based on past experience, that is, on *memory.* Integrating neurons in a CNS also allows for sophisticated behaviors, which involve multiple sensory inputs and multiple controls of effector muscles.

Because of these advantages, the evolution of a CNS is a major trend in animal evolution. This has been generally associated with two other significant trends: the evolution of longitudinal *centralization,* the association of neuronal cell bodies with a distinct longitudinal *nerve cord,* as well as *cephalization,* the concentration of neurons in the head, the leading part of an animal's body that typically deals with more environmental information than do other parts. Neurons in more complex animals became localized within nerve cords, and neural control pathways became routed through cords. Associated with this evolution was the acquisition of more specialized receptors for sensing the external environment and the selective routing of this information through the CNS for processing.

Let's examine the relationship between simple reflex circuits and a CNS with cephalization in more detail. Although more advanced animals do indeed have centralized brains, it is important to note that no animal has fully centralized cephalic control of all its functions. As you saw in Chapter 1, many functions in advanced animals remain regulated at cell/organ levels and at "lower" neural reflex levels, that is, regulation is *distributed* in a hierarchy from simple functions at the local level to complex functions at the cephalic level. There is a very important reason for this: the time-delay phenomenon for feedback regulation, as discussed in Chapter 1 (p. 15). A centralized brain is not always necessary and may even be detrimental for some kinds of neural regulation. A simple example—the *withdrawal reflex* in a mammal (Figure 5–1e)—illustrates this. Here, a potentially harmful stimulus to the skin

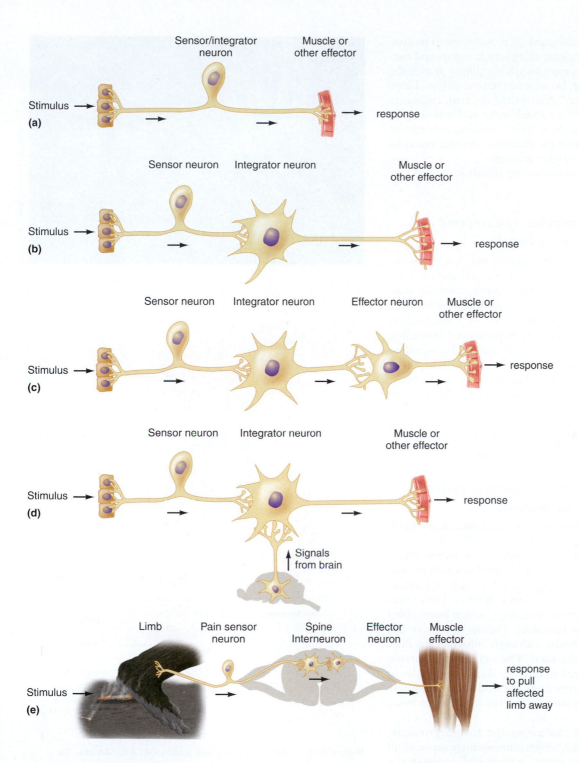

(a) Stimulus → response
Sensor/integrator neuron   Muscle or other effector

(b) Stimulus → response
Sensor neuron   Integrator neuron   Muscle or other effector

(c) Stimulus → response
Sensor neuron   Integrator neuron   Effector neuron   Muscle or other effector

(d) Stimulus → response
Sensor neuron   Integrator neuron   Muscle or other effector
Signals from brain

(e) Stimulus → response to pull affected limb away
Limb   Pain sensor neuron   Spine Interneuron   Effector neuron   Muscle effector

such as intense heat triggers heat and pain sensors, which send action potentials to a spinal (CNS) integrating synapse and neuron) that then sends signals back to a skeletal muscle, the effector, which then rapidly pulls the stimulated area away from the heat. Consider what would happen were this reflex regulated by fully centralized cephalic control. The signal would need to go all the way to the brain, presumably passing through many more synapses, before the muscle could be activated. By that time, the skin would be seriously burned. Thus, advanced neural regulation involves distributed con-

trol, with simple, *common homeostatic functions typically remaining at the reflex level* through a central nerve cord (or physically close brain regions) with a minimal number of synapses. However, the brain's usefulness comes into play in other situations, such as anticipation. For example, based on past experience of a burn, the brain may detect a potentially hot object from a distance and trigger the limb (or body) to move away before the skin sensors register painful heat.

The logic of this distributed control hierarchy is being used by human engineers building robots that mimic animal locomotion. Initially, engineers designed all movements to be controlled by a centralized computer using sensor inputs and outputs to motors. These robots were invariably clumsy at even the simplest walking movement, because of the time delays. Later, using "biologically inspired" distributed control, engineers began to build robots that were much more stable and agile. To do so, they incorporated the simplest walking movements into the legs themselves, with the central computer reserved for dealing with unusual, complex motions.

Now let's examine the evolutionary trends in animal nervous systems in more detail.

■ Sponges have no nerves but can respond to stimuli using electrical signals.

Sponges are interesting because they are the only animals without neurons. Yet some sponges react to external stimuli in reflex fashion. For example, the deep-sea sponge *Rhabdocalyptus dawsoni* stops its feeding currents if sediment-laden water enters its pores (threatening to clog them) or if the body is touched. Action potentials have been measured spreading from the outer cells to the rest of the sponge body, including the flagellate cells that create feeding currents (see Chapter 9, p. 360). The electrical pathway is thought to be the *trabecular reticulum,* a syncytium (a group of joined cells acting as one functional unit) that connects all parts of the sponge. The reticulum is not made of true neurons. However, it appears that long-distance electrical signaling evolved before neurons arose and is fundamental to the animal kingdom.

■ Nerve nets are the simplest nervous systems and are found in most animals.

The simplest existing nervous system (with true neurons) is the *nerve net* (● Figure 5–2) in cnidarians (such as hydra, sea anemones, and jellies). Of course, cnidarian nervous-system evolution has occurred over a period of perhaps 700 million years, so nerve nets, although comparatively simple and diffuse in structure, should not be considered primitive. Nerve nets represent a successful evolutionary strategy, which has enabled survival of these widespread animals. The nerve net extends throughout the animal's body, although information flow through it is relatively unfocused. The net controls simple movements of the body wall and tentacles; for example, in jellies the net primarily commands the body wall to slowly contract and expand for swimming and the tentacles to move through the water.

Of particular interest (following the Krogh principle, p. 3) is the freshwater hydra, which continuously replaces all its tissues, and thus, unless eaten, is immortal. It also has a nerve net but no brain. Therefore it is an excellent model animal for studying neural growth and development and the most basic neural functions and behaviors. Hydra neurons are continually sloughed at the extremities, so that individual neurons continually change their relative position to each other. New neurons are formed by differentiation from *stem cells* (a reserve of undifferentiated cells that can develop into specialized cells if given proper signals), and synapses are formed where neurons cross each other. For this reason, excitations originating in one neuron spread in all directions along multiple paths. Sensory cells are situated in the epidermis such that the animal can respond to external stimuli from any di-

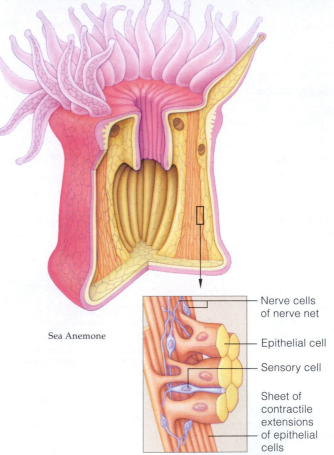

Sea Anemone

Nerve cells of nerve net
Epithelial cell
Sensory cell
Sheet of contractile extensions of epithelial cells

Figure 5–2 ● **Nerve net of a sea anemone, a cnidarian.** Its nerve cells interact with sensory and contractile cells. Both cell types are embedded in epithelium between the outer epidermis and a jellylike midlayer (mesoglea) of the body wall.

(*Source:* C. Starr & R. Taggart, 2004, *Biology: The Unity and Diversity of Life,* 10th ed. (Belmont, CA: Brooks/Cole), Figure 34.13)

rection. Branches of the nerve net are connected to these sensors and to epithelial cells that have contractile properties.

Neuron density in the nerve net is higher in the hydra's head and foot region compared to its body column, but it is insufficient to label either of these regions as an integrating center in the sense of a CNS. However, not all hydra neurons are in diffuse nets. Some neurons are concentrated in well-defined bundles, which are believed important in simple reflex behavioral responses. Feeding behavior, for example, is associated with interconnections between sensory neurons in the tentacles and contractile epithelial cells lining the tentacles and surrounding the mouth. When the sensory receptors come into contact with a potential food item, the ensuing signals generate a reflex response that ultimately draws the food item into the mouth. These bundles may represent the evolutionary precursors of central nervous systems.

Nerve nets are found not just in cnidarians. Every phylum, including the chordates, has diffuse nerve nets, which innervate various organs. In vertebrates, nets take the form of a **nerve plexus** (network of interwoven nerves), located in the intestinal wall, that is responsible for propelling digesta (food being processed) along the intestine (Chapter 14, p. 619). In some animals the nerve net may have a sensory function, whereas in others nerve nets are used to modulate the activity of an organ by releasing neurotransmitters or neuromodulators (see p. 123). And in some instances nerve nets may be used to coordinate behavior such as locomotion. In mollusks, for example, motor nerve nets are associated with local reflex pathways.

■ Simple ganglia and nerve rings evolved for more complex behavior.

More complicated neural circuitry can be found in swimming cnidarians or *medusae* (such as true jellies, box jellies, and hydromedusae, the free-swimming, jellylike larvae of hydroids). In these radially symmetric animals, because there is little polarity in the direction from which sensory information is received, a distribution of sensory structures over a full 360 degrees is advantageous. Unlike sessile hydra, however, swimming cnidarians must coordinate more complex behaviors, especially swimming and orienting in the open water. In these animals, radial symmetry is often associated with a **nerve ring** or "ring brain." Also, in addition to having diffuse nerve nets, some of these animals have certain neurons condensed into simple **ganglia.** A **ganglion** (singular of *ganglia*) is a group or cluster of nerve cell bodies, often with related functions. Ganglia and rings represent steps of centralization between nerve nets and true brains.

Medusae have primitive sensory organs and integrating ganglia in a complex called the **rhopalium,** which has **ocelli** (simple eye structures, p. 203) and **statocysts** (fluid-filled sacs that help indicate position as the animal moves, p. 222). The rhopalium ganglia in scyphomedusae (true jellies) also function as **pacemakers** (excitable cells that fire spontaneously), which generate the swimming rhythm: Activity in the nerve net results in a wave of contraction. Transmission of action potentials in either direction down an axon is achieved by the presence of gap junctions as well as **bidirectional chemical synapses,** which allow neurons to transmit in either direction. Each synaptic ending is simultaneously pre- and postsynaptic, because neurotransmitter vesicles are present on both sides of the synaptic junction. Hydromedusae and cubomedusae (box jellies) have nerve rings as well as rhopalia. The hydromedusa *Aglantha* uses two nerve rings: an inner pacemaker ring that controls normal, slow swimming, and an outer ring in the bell margin that controls escape swimming behavior in response to mechanical stimuli to the body. The outer ring has a single giant axon (35 μm) that is unique: It forms a perfect *annulus* (ringlike body part) with neither a beginning nor an end. Action potentials are conducted at rates up to 4 m/sec with synaptic delays as low as 0.7 msec at neuromuscular junctions. Firing of the ring giant axons generates action potentials (approximately 95 mV), which lead to forceful contractions of the swimming muscle and pulsation of the bell. In contrast, slow swimming is generated by smaller-amplitude action potentials (approximately 27 mV), which originate from the pacemaker system.

Nerve rings are also found in the radially symmetric echinoderms (seastars, sea urchins, and so on). These animals are considered more advanced than cnidarians, and their ancestors had bilateral symmetry with a CNS. But presumably because they "re-evolved" radial symmetry, they have also independently evolved a nerve ring system. In seastars, the parallel with hydromedusae is even more striking because seastars have primitive eyespots on the tips of their arms that send information about light into the central ring.

■ A true CNS first evolved with bilateral symmetry.

As you have seen, radially symmetric animals do not have a true CNS, although they may have some centralization in the form of ganglia and rings. Only when we progress from Cnidaria to bilaterally symmetric animals such as platyhelminths (flatworms, tapeworms, and so on) do we find the beginnings of a true CNS, longitudinal nerve cords that coordinate nervous activity, in conjunction with a **peripheral nervous system (PNS),** a communication network that extends between the CNS and all parts of the animal's body. Flatworms represent an early stage in the evolution of a CNS (● Figure 5–3a). They have two anterior ganglia from which branch two principal nerve cords with lateral neurons connecting the two cords and projecting to the various regions of the body. Collectively, the ganglia can be thought of as a rudimentary brain, because they serve as an integrating center for particular nerve pathways. Ganglia within the head of a flatworm coordinate signals arising from sensory organs such as the eyespots. The longitudinal nerve cord evolved for transmitting motor signals to effectors located bilaterally along the animal's elongated body.

Complex ganglionic nervous systems are characteristic of higher invertebrates. The next evolutionary advance is found in a more complex ganglionic CNS, a characteristic of higher invertebrates, which include annelids, arthropods, and some mollusks (Figure 5–3b, c, d, e). In segmented animals, the CNS consists of a chain of *segmental ganglia*. For example, the CNS of annelids (Figure 5–3b) consists of a bilobed brain (cerebral ganglion) with a double nerve cord with ganglia (one in most body segments), connecting with separate sensory and motor neurons (in the PNS). These networks rely on one-way conduction along the nerve axons. Ganglia are larger in arthropods (Figure 5–3c, d) and are associated with more highly developed sense organs. Generally there is one ganglion for each thoracic and abdominal body segment, although some arthropods show secondary fusion of some of these segmental ganglia. Ganglia in each segment are specialized in coordinating the regional function of each body segment, that is, legs, wings, tail or abdomen, and structures on the head. *Decentralized* brain function best describes the neural circuitry of these animals. For example, the ability to copulate is not in the least

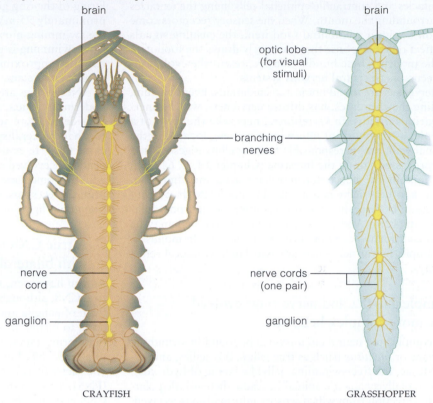

Figure 5–3 ● **Bilateral nervous systems of a few invertebrates.** (a) Flatworm, (b) earthworm, (c) crayfish, (d) grasshoppers, (e) a cuttle (or cuttlefish), a cephalopod mollusk.

(*Source for parts a, b, c, d:* C. Starr & R. Taggart, 2004, *Biology: The Unity and Diversity of Life,* 10th ed. Belmont, CA: Brooks/Cole, Figure 34.14. *Source for part e:* © 1999 Richard E. Young, Michael Vecchione, Katharina M. Mangold. http://www.fortunecity.com/emachines/ e11/86/graphics/cephpod/CEPHPOD6.gif)

brainlike structure (ganglion)

pair of nerve cords cross-connected by lateral nerves

branching nerves

FLATWORM
(a)

rudimentary brain

branching nerves

nerve cord

ganglion (one in most body segments)

EARTHWORM
(b)

brain

nerve cord

ganglion

CRAYFISH
(c)

brain

optic lobe (for visual stimuli)

branching nerves

nerve cords (one pair)

ganglion

GRASSHOPPER
(d)

impaired in a decapitated male praying mantis. The female can be steadily consuming her mate while being inseminated at the same time! (Nutrients provided by the male in this manner ultimately help ensure the development of the ova within the female.) You will soon see that elements of segmental organization are also found in vertebrates and govern the regulation of many reflex actions.

Arthropod ganglia consist of an outer rind and an inner core. In some ganglia, virtually all the neuron cell bodies are located in the rind. The central region of each ganglion contains the synaptic contacts and bundles of nerve processes (called the **neuropil**). The packing of pre- and postsynaptic neuronal branches into a dense neuropil reduces transmission delays that are characteristic of animals with diffuse nerve nets. The degree to which the nervous system is organized into a cellular rind varies between animals, with some species exhibiting neuronal cell bodies intermixed with the neuropil.

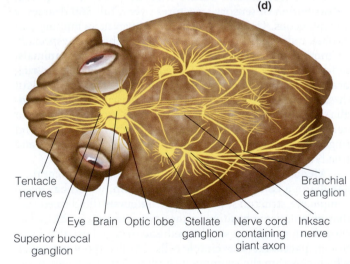

Tentacle nerves

Eye Brain Optic lobe Stellate ganglion

Nerve cord containing giant axon

Inksac nerve

Branchial ganglion

Superior buccal ganglion

(e) CUTTLEFISH

■ True brains evolved at the anterior end of advanced animals.

There has been an evolutionary trend for the most anterior region of the CNS to be larger and contain sizable clusters of neurons, which contributed to the evolution of a so-called *superganglion* or **brain.** These are found in the higher invertebrates that we just discussed (Figure 5–3c, d, e). For example, the brain of insects consists of three pairs of lobes, the **proto-**

cerebrum, which innervates the compound eyes and ocelli (see p. 219), the **deutocerebrum,** which innervates the antennae, and the **tritocerebrum,** which innervates the foregut and labrum (upper lip). As we noted earlier, the relative position of the brain is believed to be an evolutionary adaptation resulting from the tendency of the head of bilaterally symmetric animals to move forward, so that newly acquired sensory information about the environment is initially detected in the anterior region. Some sensory information is then integrated in the brain and command signals are sent to modulate the functioning of particular effectors, especially skeletal muscles. Again, in the logic of distributed hierarchic control, the simplest basic responses remain regulated at "lower" levels of the CNS in most animals, but cephalization allows for more complex functions such as processing complex sensory information (such as images), learning and memory, and even self-awareness.

The most advanced brains are found in the cephalopod mollusks (squids, cuttles, nautiloids, and octopods) and in vertebrates (which we examine in detail later). In general, evolution has favored the addition of new structures to brains, rather than modification of existing ones. More primitive structures have retained their basic functions and, in vertebrates, are layered beneath the newly evolved ones. This preserves the important distributed hierarchy of control we discussed earlier. The relative proportion of a particular region of the brain indicates the relative importance of that function to the animal. For example, in vertebrates relying primarily on tactile stimuli for perception, the regions of the brain that process this information are significantly larger. The large *optic lobe* of birds and *visual cortex* of primates are classically viewed as examples of the importance of visual information for survival. Weakly electrical fish use electrical currents for intraspecies communications and exhibit relative hypertrophy of the *cerebellum* and *rhombencephalon,* those areas of the brain involved in processing electrosensory information.

■ Cephalopod brains have complex structures supporting complex behaviors.

A similar trend occurred in cephalopods. Mollusks in general have two nerve cords running down each side of the body (in contrast to the single spinal cord in vertebrates, which, however, has left and right halves). Along the cords are several pairs of ganglia. The most anterior ganglia are joined and enlarged in some species, forming a brain that in most mollusks is small. In cephalopods, however, these joined ganglia are greatly enlarged and organized into a series of specialized, tightly packed lobes (Figure 5–3e). The brain has many similarities to those of vertebrates: Both have lobes with complex folds, and both generate similar electrical wave patterns. Distinct visual and tactile memory centers have also been identified in cephalopod brains; and one area, the vertical lobe, is involved in learning tasks. Another vertebrate attribute, specialization in the brain's hemispheres (p. 166), has also been demonstrated in octopods. For example, octopuses, which rely on monocular vision, favor one eye over the other.

The cephalopod CNS shows the hierarchic distributed arrangement that we have discussed earlier. Indeed, one recent study revealed that the octopus's arms contain simple motor programs that are completely independent of the brain. That is, certain stereotypical movements used daily, such as reaching out and then curling up around prey animals, occur even in a stimulated arm removed from the animal. Thus the brain probably sees the prey, sends a "go" command to the arms, and the local "reach out and grab" program in the arm itself does the rest. Again, this avoids feedback delays to the brain that might allow the prey to escape. (As you will see, many motor programs in vertebrates are similarly arranged.)

The large brain of cephalopods is probably involved more directly in other, more complex processes. Controversial evidence suggests that octopods can learn as effectively as some mammals and may even be able to learn by observing other octopods (individual octopods and cuttles seem to "invent" unique behaviors not found in others of their species, suggesting that such behaviors are probably not innate). Squid, cuttle, and octopod brains also control an elaborate array of **chromatophores** in their skins. These pigment cells of varying colors can be neuronally (and therefore rapidly) expanded or contracted to display different color patterns. In some cuttles, up to three dozen distinct color patterns have been observed being used in various behaviors such as mating, territorial displays, and possibly other social interactions such as alarm warnings. Some researchers have even proposed that these patterns form a kind of language. Color patterns are also used for camouflage, often in complex ways. In some octopods, camouflage patterns are coupled with sophisticated mimicry behavior, for example, a Malaysian octopod species has been seen using its arms and colors to simulate sea snakes, and molding its entire body to mimic a flounder or a seastar.

Despite similarities between cephalopod and vertebrate brains, they clearly evolved independently, because detailed structures are quite different. For example, the cephalopod brain is wrapped around the esophagus instead of lying in a cranium. This is not a very good design, because spines of prey animals can actually pierce the cephalopod brain if they get caught in the esophagus!

■ Vertebrate brain size varies up to 30-fold for a given body size.

The total number of neurons in nervous systems ranges from only several hundred in simple invertebrates such as nematodes (approximately 300 neurons), to approximately 10^8 in cephalopods and 10^{11} in large mammals. Because of these numbers as well as the complexity of their neural circuitry, brain function in cephalopods rivals that of some vertebrates, as evidenced by the intricacy of their behavior compared to other invertebrates. In the human brain, the number of neurons is approximately 8 to 9×10^{10}, whereas larger mammals such as elephants and whales show a further twofold increase in number. This example illustrates that the total number of neurons does not correspond to intellect! The exhibition of specific behavioral patterns is far more dependent on the internal wiring of the brain rather than on its overall size, and depends on the ratio of the brain size to body size.

The logarithm of brain weight as a function of body weight is typically linear for a variety of vertebrates, with a positive slope indicating that the overall size of an animal is related to its relative brain size (● Figure 5–4). (The slope is less than 1.0, meaning that brain size does not increase as much as body weight does. This *scaling* phenomenon is explored for metabolism in Chapter 15.) However, as you have just seen, the various regions of the brain do not necessarily vary directly with body weight. For example, the forebrain shows a disproportionately large size relative to body weight in primates, whereas there is a relatively large cerebellum and rhombencephalon in bony fishes.

There are also significant exceptions to the basic scaling of total brain size with some species exhibiting five times the predicted size, whereas others have approximately one fifth the predicted size. Brain size varies approximately 30-fold for a given body size when considered across all vertebrate radia-

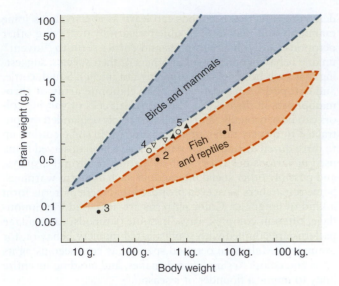

Figure 5–4 ● Weights of adult cephalopod brains compared those of various vertebrates. Upward triangle *Sepia* squid; downward triangle *Loligo* squid; 1, *Octopus vulgaris;* 2, *O. salutii;* 3, *O. defillipi;* 4 and 5, squids *Illex* and *Todarodes.*

(*Sources:* Modified from H. J. Jerison (1976), Paleoneurology and the evolution of mind, *Scientific American* 234: 90–100; and K. Mangold-Wirz (1966), Cerebralisation und Ontogenesemodus bei Eutherien, *Acta Anatomica* 63: 449–508)

tions. For example, agnathans (hagfish, lampreys) generally have the smallest brains for their body size, but hagfishes have brains two to three times larger than lampreys of the same body size. Many cartilaginous fishes (sharks and rays) have brains as large for their body size as those of many birds and mammals. Among amphibians, frogs generally have larger brains than salamanders, whereas the brains of reptiles are two to three times larger than the brains of most amphibians of the same body size. The most advanced vertebrate classes, birds and mammals, have brains that are up to 10-fold larger than the brains of reptiles of the same body size. Interestingly, the brains of octopods and squids are larger on average (relative to body size) than in most animals except some cartilaginous fishes, birds, and mammals (Figure 5–4).

The largest brains relative to body size among birds are found in the perching birds (woodpeckers and parrots), whereas the smallest brains are in the granivores (quail and pigeons). Brain size of primates, in general, is much greater than the average for mammals of the same body size, although the second largest relative brain size (after humans) is found in dolphins (cetaceans). In contrast, the insectivores, rodents, and nonplacental mammals (marsupials, monotremes) have the smallest brains.

For primates, numerous theories have been put forward to explain differences in relative brain size. For example, primates that consume leaves generally have smaller brains than those that eat fruits. This observation leads to the idea that a larger brain is required of animals that must use their intellect to locate ripened fruits, compared to those that simply find leaves. Carnivores (who must hunt wary prey) also tend to have larger brains than herbivores. Also, selection for an increase in the size of a single brain center leads to an overall increase in brain size. For example, dolphins with their rela-

tively large brains are carnivores that use echolocation, a process requiring considerable neural power, corresponding to a relative large auditory cortex. In contrast, baleen whales, which do not echolocate and do not hunt wary prey, have brains that are relatively small for their body size. A relationship between relative size of the brain and the complexity of social organization has also been suggested.

■ Relative brain size may be explained by the expensive-tissue hypothesis.

More recently, scientists have proposed the **expensive-tissue hypothesis** to explain relative brain size. The metabolic cost of neural tissue is comparatively very high because of the energy required to maintain the membrane potential across axonal membranes. Thus neurons contain numerous energy-consuming ion pumps. Energy is also expended by the continuous synthesis and reuptake of neurotransmitters. The larger the brain, the greater the expenditure of energy. In this model, the cost of maintaining increasingly larger brains is offset by a corresponding reduction in the size of the intestinal tract. The splanchnic organs (liver, gastrointestinal tract) are equally as expensive, in terms of energy expenditure, as neural tissue. Because gut size is highly correlated with the quality of diet (see Chapter 14), and relatively small intestinal tracts are associated with high-quality foodstuffs, a reduction in the size of one of these metabolically expensive organs is "traded off" against a greater degree of cephalization. Thus the increase in brain size could not have been achieved without a concomitant increase in quality of the diet, and high-quality diets are associated with reduced size (and energy cost) of the intestinal tract. This pattern is seen in a general comparison between carnivores and herbivores: The former are large-brained, with high-quality diets, whereas the latter are small-brained and often eat low-quality fibrous plants that require large digestive tracts to process.

This largely *proximate* (p. 1) hypothesis does not explain which evolved first: a higher-quality diet (which then favored greater intelligence), or nondietary selective forces favoring a larger brain (which then gave the ability to find higher-quality food). Also, an exception to this relationship has been found in bats: Insect-eating bats have smaller brains than their fruit-eating relatives. In this example researchers found no correlations either with social complexity or with the relative size of the digestive tract. Finally, an additional explanation for the relative brain size of mammals takes into account the relative brain size of the mother and the brain size of the developing offspring. The **maternal-energy hypothesis** states that ultimate brain size depends on the allocation of maternal resources to fetal and postnatal development. Because development of the brain begins at an early stage of development, resources provided by the mother can be directly linked to brain development in her offspring. Note, though, that this is also a proximate hypothesis that does not explain what selective forces would be at work in maternal allocation.

■ Nervous systems can exhibit plasticity.

Ultimately, the way animals act and react depends on organized, discrete neuronal processing. In addition to brain–body size ratios, another key factor in complexity of this processing is change during development and aging. Many inverte-

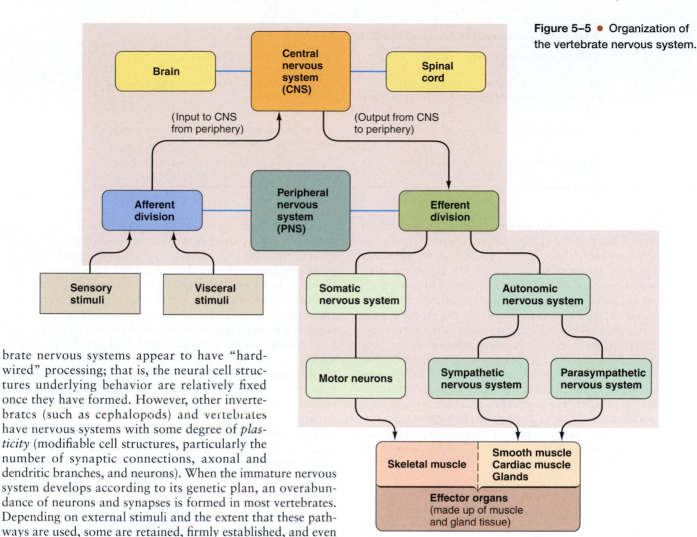

Figure 5–5 ● Organization of the vertebrate nervous system.

brate nervous systems appear to have "hard-wired" processing; that is, the neural cell structures underlying behavior are relatively fixed once they have formed. However, other invertebrates (such as cephalopods) and vertebrates have nervous systems with some degree of *plasticity* (modifiable cell structures, particularly the number of synaptic connections, axonal and dendritic branches, and neurons). When the immature nervous system develops according to its genetic plan, an overabundance of neurons and synapses is formed in most vertebrates. Depending on external stimuli and the extent that these pathways are used, some are retained, firmly established, and even enhanced with new branches and connections, whereas others are eliminated. Once the nervous system has matured, ongoing modifications still occur as an animal continues to learn from its unique set of experiences. The ability of neurons to respond structurally to usage and incoming stimuli is the basis for certain forms of learning as well as for nervous system regeneration. The nervous system in advanced animals thus represents a network of living cells that are modifiable throughout a lifetime.

The Vertebrate Nervous System

As with bilateral animals in general, the nervous system of vertebrates is functionally organized into a **central nervous system** (**CNS**) and a **peripheral nervous system** (**PNS**). In vertebrates, the CNS consists of the brain and spinal cord, and the PNS consists of nerve fibers that carry information between the CNS and other parts of the body (the periphery) (● Figure 5–5). In contrast to the organization of the arthropod CNS, which consists of a brain and discontinuous ganglia linked by connectives, the vertebrate CNS arises from a single neural tube and is organized into a continuous column of neuron cell bodies and bundles of diverse axons. There is some similarity, however, in that branches of the vertebrate PNS have a segmented pattern (❚ Table 5–1). Most of the neuronal cell bodies are

contained within the CNS. However, the PNS also includes ganglia (also in a segmented pattern) that contain the cell bodies of most sensory neurons and some nonmotor effector neurons. The PNS is further subdivided into *afferent* and *efferent* divisions. The afferent division (*a* is from *ad*, "toward," as in *advance*; and *ferent*, "carrying") carries information from **sensors** to the CNS, appraising it of the external environment and providing status reports on internal activities being regulated by the nervous system. In the next chapter, on sensory physiology, we focus on the afferent division of the peripheral nervous system. The **efferent division** (*e* is from *ex*, "from," as in *exit*) transmits instructions from the CNS to **effector organs**—the muscles, glands, and other organs that carry out the orders to bring about the desired action. The efferent nervous system in turn has two components (Figure 5–5):

1. The **somatic nervous system,** which consists of the fibers of the **motor neurons** that supply the skeletal muscles

2. The **autonomic nervous system** fibers, which innervate smooth muscle, cardiac muscle, glands, and other nonmotor organs (such as brown adipose tissue, p. 697, and immune organs, p. 456). In most vertebrates (mammals, birds, amphibians, teleost fish) the latter system is further subdivided into (1) the **sympathetic nervous system** and

Table 5–1 ▪ Comparison of Arthropod and Vertebrate Nervous Systems

Arthropod Nervous System	Vertebrate Nervous System
Anterior brain or large ganglion, with ganglia in body segments linked by paired bundles of connective fibers	Anterior brain with spinal cord having segmental branching pattern
Nerve cords are ventral, solid, and double in origin	Central nerve cord is dorsal, hollow, and develops from a single neural tube
Ganglia are discrete entities within the CNS	Ganglia are a collection of neuronal cell bodies outside the CNS

(2) the **parasympathetic nervous system,** both of which innervate most of the visceral organs supplied by the autonomic system. The somatic subdivision is sometimes called the *voluntary* system, whereas the autonomic division is often called the *involuntary* system. However, these terms are based on human biology. For other vertebrates, it is more accurate to say simply that the motor system controls skeletal muscle actions, whereas the autonomic system operates all other innervated effector organs, generally independently of higher brain regions involved in locomotion, learning, and memory. Not all vertebrates have clearly subdivided autonomic systems, as evidenced in lampreys where the sympathetic and parasympathetic divisions are difficult to distinguish. In hagfish the autonomic nervous system is poorly developed. Reptiles, as well, do not have clearly partitioned subdivisions of the autonomic nervous system.

It is important to recognize that all these "nervous systems" are really subdivisions of a single, integrated nervous system. They are arbitrary divisions based on differences in the structure, location, and functions of the various diverse parts of the whole nervous system.

▪ The three classes of neurons are afferent neurons, efferent neurons, and interneurons.

Three classes of neurons make up the nervous system: **afferent neurons, efferent neurons,** and **interneurons.** The afferent division of the peripheral nervous system consists of **afferent neurons,** which are shaped differently from efferent neurons and interneurons (● Figure 5–6). At its peripheral ending, an afferent neuron has a **sensory receptor** that generates action potentials in response to a particular type of stimulus. (This stimulus-sensitive afferent neuronal receptor should not be confused with the special protein receptors that bind chemical messengers and that are found in the plasma membrane of all cells.) The afferent-neuron cell body, which lacks dendrites and presynaptic inputs, is located adjacent to the spinal cord in a ganglion (in the dorsal root of a spinal nerve). A long *peripheral axon,* commonly called the *afferent fiber,* extends from the receptor to the cell body, and a short *central*

axon passes from the cell body into the spinal cord. Action potentials are initiated at the receptor end of the peripheral axon in response to a stimulus and are propagated along the peripheral axon and central axon toward the spinal cord. The terminals of the central axon diverge and synapse with other neurons within the spinal cord, disseminating information about the stimulus. Afferent neurons lie primarily within the peripheral nervous system: only a small portion of their central axon endings project into the spinal cord to relay peripheral signals.

Efferent neurons also lie primarily in the peripheral nervous system (Figure 5–6). The cell bodies of efferent neurons originate in the CNS, where many centrally located presynaptic inputs converge on them to influence their outputs to the effector organs. Efferent axons (*efferent fibers*) leave the CNS to course their way to the muscles or glands they innervate, conveying their integrated output for the effector organs to

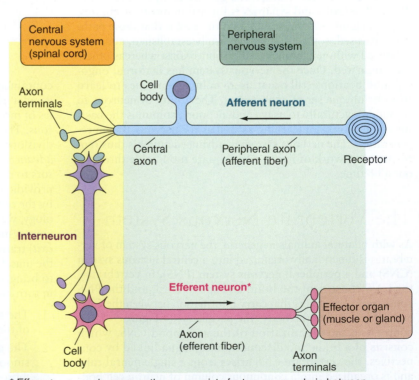

* Efferent autonomic nerve pathways consist of a two-neuron chain between the CNS and the effector organ.

Figure 5–6 ● Structure and location of the three classes of neurons.

put into effect. (An autonomic pathway actually consists of a two-neuron chain between the CNS and the effector organ.)

Interneurons lie entirely within the CNS. About 99% of all neurons belong to this category. The human CNS is estimated to have 100 billion interneurons! These neurons, which are the integrator neurons we saw at the beginning of this chapter, serve two main roles. First, as their name implies, they lie between (*inter,* "between") the afferent and efferent neurons and are important in the integration of peripheral responses to peripheral information, as you saw with the withdrawal reflex earlier (Figure 5–1e). The more complex the required action, the greater the number of interneurons interposed between the afferent message and efferent response. Second, in the most advanced vertebrates, interconnections between interneurons themselves are responsible for the abstract phenomena associated with the "mind," such as planning, memory, creativity, intellect, and motivation. These activities are the least understood functions of the nervous system. With this brief introduction to the types of neurons and their location in the various divisions of the nervous system, let's now turn our attention to the autonomic nervous system and CNS, followed in the next chapter by a discussion of the sensory function of the peripheral nervous system—converting an environmental stimulus into action potentials.

The Efferent Division of the Vertebrate Peripheral Nervous System

The efferent division of the peripheral nervous system is the communication link by which the central nervous system controls the activities of muscles, glands, and other effector organs that carry out actions, such as making feedback corrections. Efferent output typically influences either movement or secretion, as illustrated by the examples of the effects of neural control on various types of effector organs in ▮ Table 5–2. Much of this efferent output is directed toward maintaining homeostasis, but some is not. For example, the efferent output to skeletal muscles is also directed toward nonhomeostatic behaviors, such as play. (It is important to realize that many effector organs are also subject to hormonal control and/or intrinsic control mechanisms.)

How many different neurotransmitters would you guess are released from the various efferent neuronal terminals to elicit essentially all the neurally controlled effector organ responses? Only two—**acetylcholine (ACh)** and **norepinephrine (NE)**! Acting independently, these neurotransmitters bring about such diverse effects as salivary secretion, bladder contraction, and voluntary motor movements. These effects are a prime example of how the same chemical messenger may elicit a multiplicity of responses from various tissues, depending on specialization of the effector organs (the "same key, different locks" principle, p. 88).

Vertebrate Autonomic Nervous System

Functionally, the autonomic nervous system is involved in many basic, physiological functions that are critical to the survival of a vertebrate. Specifically, it regulates visceral activities normally outside the realm of higher brain control, such as circulation, digestion, thermoregulation, and pupil size (to name a few).

▮ The sympathetic and parasympathetic nervous systems both innervate most visceral organs.

Most visceral organs are innervated by both sympathetic and parasympathetic nerve fibers (● Figure 5–7). ▮ Table 5–3 summarizes the major effects of these autonomic branches. Although the details of this wide array of autonomic responses are described more fully in later chapters that discuss the individual organs involved, several general concepts can be derived now. As can be seen from the table, the sympathetic and parasympathetic nervous systems generally exert opposite effects in a particular organ. Sympathetic stimulation increases the heart rate, whereas parasympathetic stimulation decreases it; sympathetic stimulation slows down movement within the digestive tract, whereas parasympathetic stimulation enhances digestive motility. Note that both systems increase the activity of some organs and reduce the activity of others.

Rather than memorizing a list such as that presented in Table 5–3, you will do better to logically deduce the actions of the two systems based on an understanding of the circumstances under which each system dominates. Usually, both systems are partially active; that is, normally some level of ac-

Table 5–2 ▮ Examples of the Influence of Efferent Output on Movement and Secretion by Effector Organs in a Mammal		
Category of Influence	Examples of Effector Organs with Different Types of Tissues	Sample Outcome in Response to Efferent Output
Influence on Movement	Heart (cardiac muscle)	Increased pumping of blood when the blood pressure falls too low
	Stomach (smooth muscle)	Delayed emptying of the stomach until the intestine is ready to process the food
	Diaphragm—a respiratory muscle (skeletal muscle)	Augmented breathing in response to exercise
Influence on Secretion	Sweat glands (exocrine glands)	Intitiation of sweating on exposure to a hot environment
	Endocrine pancrease (endocrine gland)	Increased secretion of insulin, a hormone that puts excess nutrients in storage following a meal

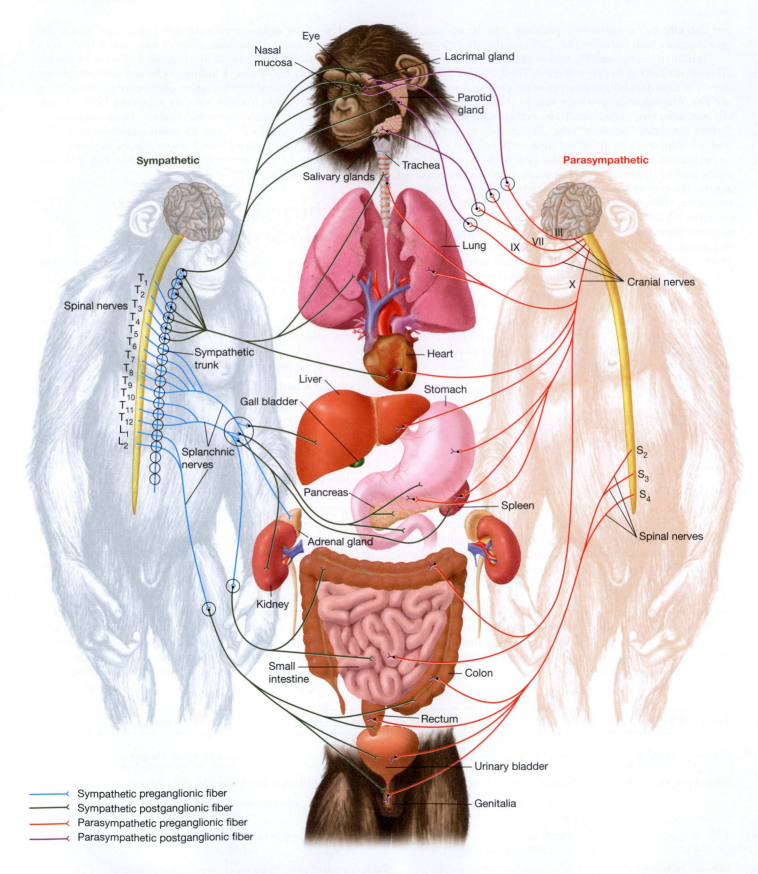

Sympathetic

Eye
Nasal mucosa
Lacrimal gland
Parotid gland
Salivary glands
Trachea

Parasympathetic

Lung
III
IX
VII
X
Cranial nerves

Spinal nerves
T₁
T₂
T₃
T₄
T₅
T₆
T₇
T₈
T₉
T₁₀
T₁₁
T₁₂
L₁
L₂

Sympathetic trunk

Heart
Liver
Gall bladder
Stomach

Splanchnic nerves

Pancreas
Spleen

Adrenal gland

Kidney

Small intestine
Colon

Rectum

Urinary bladder

Genitalia

S₂
S₃
S₄

Spinal nerves

——< Sympathetic preganglionic fiber
——< Sympathetic postganglionic fiber
——< Parasympathetic preganglionic fiber
——< Parasympathetic postganglionic fiber

Figure 5–7 ● Schematic representation of the structures innervated by the sympathetic and parasympathetic nervous systems.

Table 5–3 ▌ Effects of the Autonomic Nervous System on Various Organs

Organ	Effect of Sympathetic Stimulation	Effect of Parasympathetic Stimulation
Heart	Increased rate, increased force of contraction (of whole heart)	Decreased rate, decreased force of contraction (of atria only)
Blood Vessels	Constriction	Dilation of vessels supplying the penis and clitoris only
Lungs	Dilation of bronchioles (airways)	Constriction of bronchioles
	Inhibition (?) of mucus secretion	Stimulation of mucus secretion
Digestive Tract	Decreased motility (movement)	Increased motility
	Contraction of sphincters (to prevent forward movement of contents)	Relaxation of sphincters (to permit forward movement of contents)
	Inhibition (?) of digestive secretions	Stimulation of digestive secretions
Gallbladder	Relaxation	Contraction (emptying)
Urinary Bladder	Relaxation	Contraction (emptying)
Eye	Dilation of pupil	Constriction of pupil
	Adjustment of eye for far vision	Adjustment of eye for near vision
Liver (Glycogen Stores)	Glycogenolysis (glucose released)	None
Adipose Cells (Fat Stores)	Lipolysis (fatty acids released)	None
Exocrine Glands		
Exocrine pancreas	Inhibition of pancreatic exocrine secretion	Stimulation of pancreatic exocrine secretion (important for digestion)
Sweat glands	Stimulation of secretion by most sweat glands	Stimulation of secretion by some sweat glands
Salivary glands	Stimulation of small volume of thick saliva rich in mucus	Stimulation of large volume of watery saliva rich in enzymes
Endocrine Glands		
Adrenal medulla	Stimulation of epinephrine and norepinephrine secretion	None
Endocrine pancreas	Inhibition of insulin secretion; stimulation of glucagon secretion	Stimulation of insulin and glucagon secretion
Genitals	Ejaculation and orgasmic contractions (males); orgasmic contractions (females)	Erection (caused by dilation of blood vessels in penis [male] and clitoris [female])
Brain Activity	Increased alertness	None

tion potential activity exists in both the sympathetic and the parasympathetic fibers supplying a particular organ. This ongoing activity is called **sympathetic** or **parasympathetic tone** or **tonic activity.** Under given circumstances, activity of one division can dominate the other. *Sympathetic dominance* to a particular organ exists when the sympathetic fibers' rate of firing to that organ increases above tonic level, coupled with a simultaneous decrease below tonic level in the parasympathetic fibers' frequency of action potentials to the same organ. The reverse situation is true for *parasympathetic dominance.* Shifts in balance between sympathetic and parasympathetic activity can be accomplished discretely for individual organs to meet specific demands (for example, sympathetically induced dilation of the pupil in dim light; see p. 205), or a more generalized, widespread discharge of one autonomic system in favor of the other can be elicited to control body-wide functions. Massive widespread discharges take place more fre-

quently in the sympathetic system. The value of this potential for massive sympathetic discharge is evident, considering the circumstances during which this system usually dominates.

▌ The sympathetic system dominates in times of "fight or flight."

The sympathetic system promotes responses that prepare a vertebrate animal for strenuous physical activity in the face of emergency or other stressful situations, such as a physical threat from the environment. This response is typically referred to as a **fight-or-flight response,** because the sympathetic system readies the body to fight against or flee from the threat. The changes are not homeostatic in every case; rather, many physiological systems are *reset* (p. 15) to different levels. In other cases, the changes are anticipatory, to maintain homeostasis with little or no delays. Think about the body resources needed in such circumstances. The heart beats more rapidly and more forcefully; the respiratory airways open wide to permit maximal air flow; glycogen (stored sugar) and fat stores are broken down to release extra fuel into the blood; and blood vessels supplying skeletal muscles dilate (open more widely). All these responses are aimed at providing increased flow of oxygenated, nutrient-rich blood to the skeletal muscles in anticipation of strenuous physical activity. Furthermore, the pupils dilate and the eyes adjust for far vision, enabling the animal to make a quick visual assessment of the entire threatening scene. In some species sweating or panting is promoted, in anticipation of excess heat production by the physical exertion. Because digestive and urinary activities are inessential in meeting the threat, the sympathetic system inhibits these activities.

◼ The parasympathetic system dominates in times of "rest and digest."

The parasympathetic system, in contrast, dominates in quiet, relaxed situations. Under such nonthreatening circumstances, the body can be concerned with its own "general housekeeping" activities, such as digestion and emptying of the urinary bladder. The parasympathetic system promotes these types of bodily functions while slowing down those activities that are enhanced by the sympathetic system. In a tranquil setting, for example, the animal does not need to have its heart beating rapidly and forcefully

◼ Dual innervation gives precise, antagonistic control.

What is the advantage of dual innervation of organs with nerve fibers whose actions oppose each other? It enables precise control over an organ's activity, like having both an accelerator and a brake to control the speed of a car. If an animal suddenly darts across the road as you are driving, you could eventually stop if you simply took your foot off the accelerator, but you might stop too slowly to avoid hitting the animal. If you simultaneously apply the brake as you lift up on the accelerator, however, you can come to a more rapid, controlled stop. In a similar manner, a sympathetically accelerated heart rate could gradually be slowed to normal after a stressful situation by decreasing the rate of firing in the cardiac sympathetic nerve (letting up on the accelerator), but the heart rate

can be reduced more rapidly by simultaneously increasing activity in the parasympathetic supply to the heart (applying the brake). Indeed, the two divisions of the autonomic nervous system are usually reciprocally controlled; increased activity in one division is accompanied by a corresponding decrease in the other. Thus dual innervation of an organ by the two branches of the autonomic system permits more precise control over the organ. This is the *antagonistic negative feedback* design we saw in Chapter 1 (p. 12).

There are several exceptions to the general rule of dual reciprocal innervation by the two branches of the autonomic nervous system (Table 5–3); the most notable are the following:

- *Innervated blood vessels* (most arterioles and veins are innervated, arteries and capillaries are not) for the most part receive only sympathetic nerve fibers. Regulation is accomplished by increasing or decreasing the firing rate above or below the tonic level in these sympathetic fibers. The only blood vessels to receive both sympathetic and parasympathetic fibers are those supplying the penis and clitoris. The precise vascular control this dual innervation affords these organs is important in accomplishing erection.

- *Most sweat glands* are innervated only by sympathetic nerves. The postganglionic fibers of these nerves are unusual because they secrete ACh rather than NE.

- *Salivary glands* are innervated by both autonomic divisions, but unlike elsewhere, sympathetic and parasympathetic activity is not antagonistic. Both stimulate salivary secretion, but the saliva's volume and composition differ, depending on which autonomic branch is dominant.

◼ An autonomic nerve pathway consists of a two-neuron chain.

Each autonomic-nerve pathway extending from the CNS to an innervated organ consists of a two-neuron chain (● Figure 5–8). The cell body of the first neuron in the series is located in the CNS. Its axon, the **preganglionic fiber,** synapses with the cell body of the second neuron, which lies within a ganglion outside of the CNS. The axon of the second neuron, the **postganglionic fiber,** innervates the effector organ.

Sympathetic nerve fibers originate in the thoracic and lumbar regions of the spinal cord (see p. 150). Most sympathetic preganglionic fibers in birds and mammals are very short, syn-

Figure 5–8 ●
Autonomic nerve pathway.

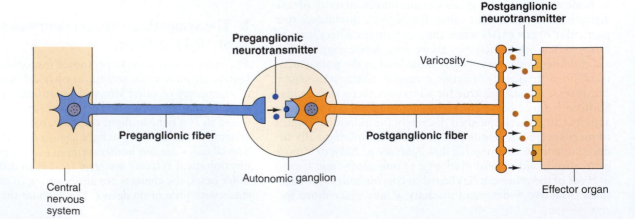

apsing with cell bodies of postganglionic neurons within paravertebral (*para*, "near") ganglia that lie in a **sympathetic ganglion chain** (the **sympathetic trunk**) located along either side of the spinal cord (● Figure 5–9). Long, postganglionic fibers originating in the ganglion chain terminate on the effector organs. Some preganglionic fibers pass through the ganglion chain without synapsing and terminate later in sympathetic *collateral ganglia* located about halfway between the CNS and the innervated organs, with postganglionic fibers traveling the remainder of the distance. An unusual group of ganglia found in birds, referred to as the *intestinal* or *Remak's* nerve, consists of both sympathetic and parasympathetic components. This nerve is located along the entire length of the alimentary canal from the duodenum to the cloaca.

Parasympathetic preganglionic fibers arise from the cranial (brain) and sacral areas (lower spinal cord) of the CNS.

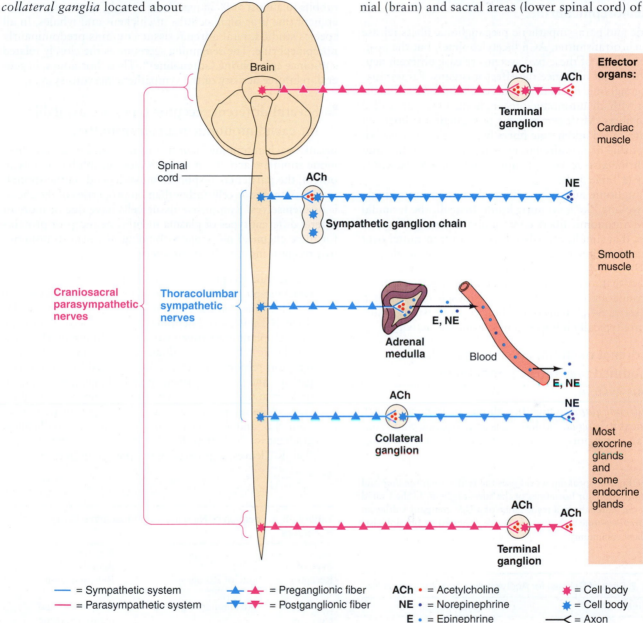

Figure 5–9 ● **Autonomic nervous system.** The sympathetic nervous system, which originates in the thoracolumbar regions of the spinal cord, has short cholinergic (acetylcholine-releasing) preganglionic fibers and long adrenergic (norepinephrine-releasing) postganglionic fibers. The parasympathetic nervous system, which originates in the brain and sacral region of the spinal chord, has long cholinergic preganglionic fibers and short cholinergic postganglionic fibers. In most instances, sympathetic and parasympathetic postganglionic fibers both innervate the same effector organs. The adrenal medulla is a modified sympathetic ganglion, which releases epinephrine and norepinephrine into the blood.

(Some cranial nerves contain parasympathetic fibers.) These fibers are long in comparison to sympathetic preganglionic fibers because they do not end until they reach **terminal ganglia** that lie in or near the effector organs. Very short postganglionic fibers terminate on the cells of the organ itself.

Parasympathetic postganglionic fibers release acetylcholine; sympathetic ones release norepinephrine.

Sympathetic and parasympathetic preganglionic fibers release the same neurotransmitter, **ACh** (acetylcholine), but the postganglionic endings of these two systems release different neurotransmitters that influence the effector organs. Parasympathetic postganglionic fibers release acetylcholine. Accordingly, they, along with all autonomic preganglionic fibers, are called **cholinergic fibers**. Most sympathetic postganglionic fibers, in contrast, are called **adrenergic fibers** because they release NE (norepinephrine or noradrenaline).* Both acetylcholine and norepinephrine also serve as chemical messengers elsewhere in the body (▌ Table 5–4).

Postganglionic autonomic fibers do not end in a single terminal swelling like a synaptic knob. Instead, the terminal branches of autonomic fibers contain numerous swellings, or **varicosities,** that simultaneously release neurotransmitter over a large area of the innervated organ rather than on single cells (Figure 5–8). This diffuse release of neurotransmitter, coupled with the fact that any resulting change in electrical activity is spread throughout a smooth or cardiac muscle mass via gap junctions (see p. 64), means that whole organs instead of discrete cells are typically influenced by autonomic activity.

The adrenal medulla, an endocrine gland, is a modified part of the sympathetic nervous system.

The *adrenal gland,* which lies above the kidney on each side (*ad,* "next to"; *renal,* "kidney"), is an endocrine gland consisting of an outer portion, the *adrenal cortex,* and an inner

portion, the *adrenal medulla.* The **adrenal medulla** is considered a modified sympathetic ganglion that does not give rise to postganglionic fibers. Instead, it secretes hormones directly into the blood on stimulation by the preganglionic fiber that originates in the CNS (Figures 5–9 and 5–10). Not surprisingly, the hormones are identical or similar to postganglionic sympathetic neurotransmitters. In mammals, norepinephrine ranges from about 2% of the adrenal medullary hormone output in rabbits to almost 50% in pigs. Interestingly, the output of norepinephrine is as high as 80% in chickens and whales. In all species studied, fetal adrenal tissue contains predominantly norepinephrine. The remaining secretion is the closely related substance **epinephrine (adrenaline)**. These hormones, in general, reinforce activity of the sympathetic nervous system.

Several different receptor types are available for each autonomic neurotransmitter.

Because each autonomic neurotransmitter and medullary hormone stimulate activity in some tissues but inhibit activity in others, the particular responses must depend on the specialization of the tissue cells rather than on properties of the chemicals themselves. Responsive tissue cells have one or more of several different types of plasma membrane receptor proteins for these chemical messengers. Binding of a neurotransmitter to a receptor induces the tissue-specific response.

1. **Cholinergic receptors.** Two types of acetylcholine (cholinergic) receptors—*nicotinic* and *muscarinic* receptors—have been identified on the basis of their response to particular drugs (▌ Table 5–5). **Nicotinic receptors** (activated by the tobacco plant derivative nicotine) are found on the postganglionic cell bodies in all autonomic ganglia. Nicotinic receptors respond to acetylcholine released from both sympathetic and parasympathetic preganglionic fibers. Binding of ACh to these receptors brings about opening of cation channels in the postganglionic cell that permit passage of both Na^+ and K^+. Because of the greater electrochemical gradient for Na^+ than for K^+, more Na^+ enters the cell than K^+ leaves, depolarizing the postganglionic cell.

---

*Most of the English-speaking world uses the terms *noradrenaline* and *adrenaline* for two major hormones of the adrenal gland. In the United States, however, Adrenalin is a trade name of a U.S. company's drug, so the terms *norepinephrine* and *epinephrine* are preferred by physiologists and the medical community.

Table 5–4 ▌ Sites of Release for Acetylcholine and Norepinephrine

Acetylcholine	Norepinephrine
All preganglionic terminals of the autonomic nervous system	Most sympathetic postganglionic terminals
All parasympathetic postganglionic terminals	Adrenal medulla
Sympathetic postganglionic terminals at sweat glands and some blood vessels in skeletal muscle	Central nervous system
Terminals of efferent neurons supplying skeletal muscle (motor neurons)	
Central nervous system	

Table 5–5 ▌ Location of Nicotinic and Muscarinic Cholinergic Receptors

Type of Receptor	Site of Receptor	Respond to Acetylcholine Released from:
Nicotinic receptors	All autonomic ganglia	Sympathetic and parasympathetic preganglionic fibers
	Motor end plates of skeletal muscle fibers	Motor neurons
	Some CNS cell bodies and dendrites	Some CNS presynaptic terminals
Muscarinic receptors	Effector cells (cardiac muscle, smooth muscle, glands)	Parasympathetic postganglionic fibers
	Some CNS cell bodies and dendrites	Some CNS presynaptic terminals

2. **Muscarinic receptors** (activated by the mushroom poison muscarine) are found on effector cell membranes (smooth muscle, cardiac muscle, and glands). They bind with ACh released from parasympathetic postganglionic fibers. There are five subtypes of muscarinic receptors, all linked to G proteins that activate second-messenger systems that lead to the target cell response.

3. **Adrenergic receptors.** There are two major classes of adrenergic receptors for norepinephrine and epinephrine based on the ability of various drugs to either initiate or prevent responses in the effector organ. These receptors are designated as **alpha (α)** and **beta (β) receptors,** with further subclassifications of α_1 and α_2 and β_1 and β_2 **receptors.** These various receptor types are distinctly distributed among the effector organs. Receptors of the β_2 type bind primarily with epinephrine, whereas β_1 receptors have about equal affinities for norepinephrine and epinephrine, and α receptors of both subtypes have a greater sensitivity to norepinephrine than to epinephrine ($\bullet$ Figure 5–10).

All adrenergic receptors are coupled to G proteins, but the ensuing pathway differs for the various receptor types. Activation of both β_1 and β_2 receptors brings about the target cell response by means of the cyclic AMP second-messenger system (see p. 90). Stimulation of α_1 receptors elicits the desired response via the Ca^{++} second-messenger system (see p. 91). By contrast, binding of a neurotransmitter to an α_2 receptor blocks cyclic AMP production in the target cell. Accordingly, activation of α_2 receptors brings about an inhibitory response in the effector organ, such as decreased smooth muscle contraction in the digestive tract. Activation of α_1 receptors, in contrast, usually brings about an excitatory response in the effector organ—for example, arteriolar constriction caused by increased contraction of the smooth muscle in the walls of these blood vessels. Alpha$_1$ receptors are present in most sympathetic target tissues. Stimulation of β_1 receptors, which are found primarily in the heart, also causes an excitatory response, namely, increased rate and force of cardiac contraction. The response to β_2 receptor activation is generally inhibitory, such as arteriolar or bronchiolar (respiratory airway) dilation caused by relaxation of the smooth muscle in the walls of these tubular structures.

Autonomic Agonists and Antagonists in Vertebrates

Because activation of various receptor types brings about different responses to the same autonomic messenger, these receptors can be manipulated fairly selectively by drugs. Drugs are available that selectively enhance or mimic (**agonists**) or block (**antagonists**) (p. 94) autonomic responses at each of the receptor types. Some are only of experimental interest, but others are very important therapeutically. For example, **atropine** blocks the effect of ACh at muscarinic receptors but does not affect nicotinic receptors. Because the ACh released at both parasympathetic and sympathetic preganglionic fibers combines with nicotinic receptors, blockage at nicotinic synapses would knock out both of these autonomic branches. By acting selectively to interfere with ACh action only at muscarinic junctions, which are the sites of parasympathetic postganglionic action, atropine effectively blocks parasympathetic effects but does not influence sympathetic activity at all. This principle is used to suppress salivary and bronchial secretions before surgery to reduce the risk of a patient (human or veterinary) inhaling these secretions into the lungs.

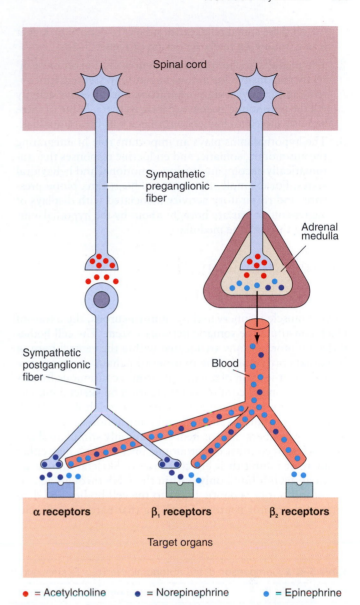

Figure 5–10 $\bullet$ **Comparison of the release and binding to receptors of epinephrine and norepinephrine.** Norepinephrine is released both as a neurotransmitter from sympathetic postganglionic fibers and as a hormone from the adrenal medulla. Beta$_1$ (β_1) receptors bind equally with both norepinephrine and epinephrine, whereas beta$_2$ (β_2) receptors bind primarily with epinephrine and alpha (α) receptors of both subtypes have a greater affinity for norepinephrine than for epinephrine.

Many regions of the central nervous system are involved in the control of autonomic activities.

Messages from the CNS are delivered to cardiac muscle, smooth muscle, glands, and other non–skeletal-muscle effectors via the autonomic nerves, but what regions of the CNS regulate autonomic output?

1. Some autonomic reflexes, such as urination, defecation, and erection, are integrated at the spinal cord (reflex) level,

but all these spinal reflexes are subject to control by higher levels (following the basic design of Figure 5–1e).

2. The medulla within the brain stem is the region most directly responsible for autonomic output. Centers for controlling cardiovascular, respiratory, and digestive activity via the autonomic system are located there. Some of these are also subject to higher control (such as breath holding).

3. The hypothalamus plays an important role in integrating the autonomic, somatic, and endocrine responses that automatically accompany various emotional and behavioral states. For example, the increased heart rate, blood pressure, and respiratory activity associated with displays of aggression or fear are brought about by the hypothalamus acting through the medulla.

Vertebrate Somatic Nervous System

Skeletal muscle is innervated by **motor neurons,** the axons of which constitute the **somatic nervous system.** The cell bodies of these motor neurons are located within the ventral horn of the spinal cord. Unlike the two-neuron chain of autonomic nerve fibers, the axon of a motor neuron is continuous fromits origin in the spinal cord to its termination on skeletal muscle. Motor-neuron axon terminals release acetylcholine (ACh), which brings about excitation and contraction of the innervated muscle fibers. Motor neurons can only stimulate skeletal muscles, in contrast to autonomic fibers, which may either stimulate or inhibit their effector organs. Skeletal muscle activity can be inhibited only within the CNS through activation of inhibitory synaptic input to the cell bodies and dendrites of the motor neurons supplying that particular muscle.

■ Motor neurons are the final common pathway.

Motor neuron dendrites and cell bodies are influenced by many converging presynaptic inputs, both excitatory and inhibitory. Some of these inputs are part of spinal-reflex pathways originating with peripheral sensory receptors. Others are part of descending pathways originating within the brain. Areas of the brain that exert control over skeletal muscle movements include the motor regions of the cortex, the basal nuclei, the cerebellum, and the brain stem (see p. 170).

Motor neurons are considered the **final common pathway,** because the only way any other parts of the nervous system can influence skeletal muscle activity is by acting on these motor neurons. The level of activity in a motor neuron and its subsequent output to the skeletal muscle fibers it innervates depend on the relative balance of EPSPs and IPSPs (see p. 125) brought about by its presynaptic inputs originating from these diverse sites in the brain.

The somatic system is considered under cerebral control, but much of the skeletal muscle activity involving posture, balance, and stereotypic movement is controlled at lower levels, following the logic of a hierarchical system. A vertebrate animal's cerebrum may make the decision to start walking, but the cerebrum itself does not usually control the alternate contraction and relaxation of the involved muscles, because these movements are coordinated by lower brain centers such as the cerebellum, brain stem, or spine (hence the infamous tales of a chicken running around even after its head has been severed. Similarly, a shark's body will continue to swim even if its brain is destroyed).

■ Table 5–6 summarizes the features of the two divisions of the efferent nervous system discussed in this chapter.

Table 5–6 ■ Comparison of the Autonomic Nervous System and the Somatic Nervous System

Feature	Autonomic Nervous System	Somatic Nervous System
Site of origin	Brain or lateral horn of spinal cord	Ventral horn of spinal cord for most; those supplying muscles in head originate in brain
Number of neurons from origin in CNS to effector organ	Two-neuron chain (preganglionic and postganglionic)	Single neuron (motor neuron)
Organs innervated	Cardiac muscle, smooth muscle, exocrine and some endocrine glands	Skeletal muscle
Type of innervation	Most effector organs dually innervated by the two antagonistic branches of this system (sympathetic and parasympathetic)	Effector organs innervated only by motor neurons
Neurotransmitter at effector organs	May be acetylcholine (parasympathetic terminals) or norepinephrine (sympathetic terminals)	Only acetylcholine
Effects on effector organs	Either stimulation or inhibition (antagonistic actions of two branches)	Stimulation only (inhibition possible only centrally through IPSPs on cell body of motor neuron)
Types of control	Under involuntary control; may be voluntarily controlled with biofeedback techniques and training	Subject to voluntary control; much activity subconsciously coordinated
Higher centers involved in control	Spinal cord, medulla, hypothalamus, prefrontal association cortex	Spinal cord, motor cortex, basal nuclei, cerebellum, brain stem

Vertebrate Central Nervous System

About 90% of the cells within the CNS are not neurons but **glial cells** or **neuroglia,** which serve as the connective tissue of the CNS and as such help support the neurons both physically and metabolically. Despite their large numbers, the glial cells occupy only about half the volume of the brain, because they do not branch as extensively as the neurons do.

■ Glial cells support the interneurons physically, metabolically, and functionally.

Unlike neurons, glial cells do not initiate or conduct nerve impulses. They are important in the viability of the CNS, however. For much of the time since discovering them in the 19th century, scientists thought the glial cells were passive "mortar" that physically supported the functionally important neurons. In the last decade, however, the varied and important roles of these dynamic cells have become apparent. The glial cells serve as the connective tissue of the CNS and as such help support the neurons both physically and metabolically. They homeostatically maintain the composition of the specialized extracellular environment surrounding the neurons within the narrow limits optimal for normal neuronal function. Certain glial cells can also take up and destroy neurotransmitters released from neighboring neurons. The four major types of glial cells in the CNS are **astrocytes, oligodendrocytes, ependymal cells,** and **microglia** (● Figure 5–11 and ▌Table 5–7).

Astrocytes

Named for their starlike shape (*astro,* "star"; *cyte,* "cell") (Figure 5–11), **astrocytes** are the most abundant glial cells. They fill a number of critical functions:

Table 5–7 ▌ Functions of Glial Cells

Type of Glial Cell	Functions
Astrocytes	Physically support neurons in proper spatial relationships
	Serve as scaffold during fetal brain development
	Induce formation of blood–brain barrier
	Form neural scar tissue
	Take up and degrade released neurotransmitters into raw materials for synthesis of more neurotransmitters by neurons
	Take up excess K^+ to help maintain proper brain ECF ion concentration and normal neural excitability
	Possess receptors for neurotransmitters, which may be important in a chemical signaling system
Oligodendrocytes	Form myelin sheaths in CNS
Microglia	Play a role in defense of brain as phagocytic scavengers
Ependymal cells	Line internal cavities of brain and spinal cord
	Contribute to formation of cerebrospinal fluid
	Serve as neural stem cells with the potential to form new neurons and glial cells

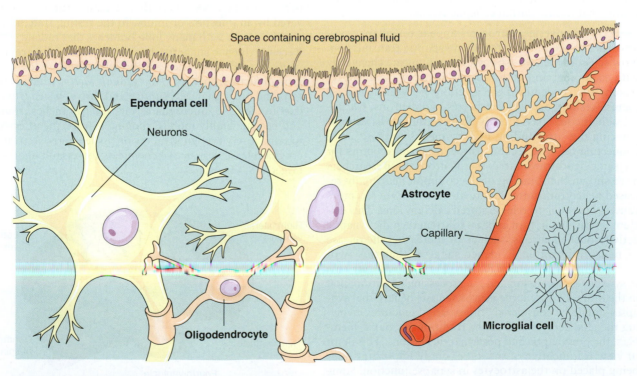

Figure 5–11 ● **Glial cells of the mammalian central nervous system.** The glial cells include the astrocytes, oligodendrocytes, microglia, and ependymal cells.

1. First, as the main "glue" (*glia,* "glue") of the CNS, they hold the neurons together in proper spatial relationships.

2. Astrocytes serve as a scaffold to guide neurons to their proper final destination during fetal brain development.

3. These glial cells induce the small blood vessels of the brain to undergo the anatomic and functional changes that are responsible for the establishment of the blood–brain barrier, a highly selective barricade between the blood and brain (soon to be described in greater detail).

4. Astrocytes are important in the repair of brain injuries and in neural scar formation.

5. Astrocytes play a role in neurotransmitter activity. They take up glutamate and gamma-aminobutyric acid (GABA), excitatory and inhibitory neurotransmitters, respectively, halting the actions of these chemical messengers. Furthermore, they degrade the absorbed chemical messengers into raw materials for the neurons to use in making more of these neurotransmitters.

6. Astrocytes take up excess K^+ from the brain ECF when high action-potential activity outpaces the ability of the Na^+–K^+ pump to return the effluxed K^+ to the neurons. (Recall that K^+ leaves a neuron during the falling phase of an action potential; see p. 113.) By taking up excess K^+, the astrocytes help maintain the proper brain ECF ion concentration to sustain normal neural excitability. If brain ECF K^+ levels were allowed to rise, the resultant lower K^+ concentration gradient between the neuronal ICF and surrounding ECF would reduce the neuronal membrane closer to threshold, even at rest. This would increase the excitability of the brain. In fact, an elevation in brain ECF K^+ concentration may be one of the factors responsible for the brain cells' explosive convulsive discharge that occurs during epileptic seizures.

7. In recent discoveries, astrocytes appear to enhance synapse formation and to strengthen synaptic transmission. Now scientists believe astrocytes communicate chemically with each other and with neurons in two ways. First, researchers have identified gap junctions between astrocytes themselves and between astrocytes and neurons. Chemical signals could pass directly between these cells without entering the surrounding ECF by means of these small connecting tunnels. Second, astrocytes have receptors for the same neurotransmitters that neurons do. Researchers have shown that binding of the common neuronally released neurotransmitter glutamate to astrocyte receptors causes these cells to release stored calcium ions. This calcium unleashing in turn appears to strengthen the synaptic activity of the neurons, such as by increasing the release of neurotransmitter. Evidence suggests that this two-directional extracellular signaling plays an important role in synaptic transmission and the processing of information in the brain. In fact, some neuroscientists believe that synapses should be considered "three-party" junctures involving the glial cells as well as the traditional synapse members, the presynaptic and postsynaptic neurons. This point of view is indicative of the increasingly important roles being placed on the astrocytes in synapse function. Some researchers even speculate that glial modulation of synaptic activity may be important in memory and learning.

Other Glial Cells

Oligodendrocytes form the insulating myelin sheaths around axons in the CNS. An oligodendrocyte has several elongated projections, each of which is wrapped jelly-roll fashion around a section of an interneuronal axon to form a patch of myelin (see Figure 4–15, p. 120, and Figure 5–11).

Ependymal cells line the internal cavities of the vertebrate CNS. As the nervous system develops embryonically from a hollow neural tube, the original central cavity of this tube is maintained and modified to form the ventricles of the brain (● Figure 5–12) and the central canal of the spinal cord. The ependymal cells lining the ventricles contribute to the formation of cerebrospinal fluid, to be discussed shortly. Ependymal cells are one of the few cell types to bear cilia, whose beating action contributes to the flow of cerebrospinal fluid throughout the ventricles.

Ependymal cells also serve as neural stem cells with the potential of forming not only other glial cells but new neurons as well. Traditional view has long held that new neurons are not produced anywhere in the mature mammalian brain. Then, in the late 1990s, scientists discovered that new neurons are produced in one restricted site, namely in a specific part of the hippocampus, a structure important for learning and memory. Neurons in the rest of the brain are considered irreplaceable. But the discovery that ependymal cells are precursors for new neurons suggests the adult brain has more potential for repairing damaged regions than previously assumed.

Microglia are the immune defense cells of the CNS. These scavengers are "cousins" of monocytes, a type of white blood cell that exits the blood and sets up residence as front-line defense agents in various tissues throughout the body (see Chapter 10). Microglia are derived from the same tissue that gives rise to monocytes, and during embryonic development they migrate to the CNS. There they remain stationary until activated by an infection or injury. In the resting state, microglia are wispy cells with many long branches that radiate outward. Recent evidence suggests that resting microglia are not just waiting watchfully. They are thought to release low levels of growth factors, such as *nerve growth factor,* which help neurons and other glial cells survive and thrive. When trouble occurs in the CNS, microglia retract their branches, round up, and become highly mobile, moving toward the affected area to remove any foreign invaders or tissue debris. Activated microglia release destructive chemicals for assault against their target.

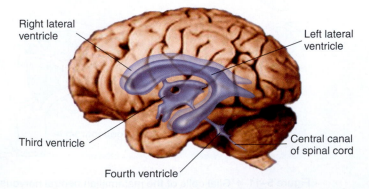

Right lateral ventricle

Left lateral ventricle

Third ventricle

Central canal of spinal cord

Fourth ventricle

Figure 5–12 ● **The ventricles of the mammalian brain.**

■ The delicate central nervous tissue is well protected.

Central nervous tissue is very delicate. This characteristic, coupled with the fact that damaged nerve cells cannot be replaced because neurons cannot divide, makes it imperative that this fragile, irreplaceable tissue be well protected. Four major features help protect the vertebrate CNS from injury:

1. It is enclosed by hard, bony structures. The **cranium (skull)** encases the brain of vertebrates, and the vertebral column surrounds the spinal cord.

2. Three protective and nourishing membranes, the **meninges,** lie between the bony covering and the nervous tissue. From the outermost to the innermost layer, these are the dura mater, the arachnoid mater, and the pia mater (*mater,* "mother").

3. The brain "floats" in a special cushioning fluid, the **cerebrospinal fluid (CSF).**

4. A highly selective **blood–brain barrier** limits access of blood-borne materials into the vulnerable brain tissue.

The role of the bony covering is self-evident. The latter three protective mechanisms warrant further discussion (● Figure 5–13).

The meninges wrap the central nervous system. The **dura mater** (*dura,* "tough") is a tough, inelastic covering made of two layers. Usually these layers adhere closely, but in some regions they are separated to form blood-filled cavities or sinuses. Venous blood draining from the brain empties into these sinuses to be returned to the heart. Cerebrospinal fluid also re-enters the blood at these sinus sites.

The **arachnoid mater** (arachnoid, "spiderlike") is a delicate, richly vascularized layer with a "cobwebby" appearance. The space between the arachnoid layer and the underlying pia mater, the **subarachnoid space,** is filled with CSF. Protrusions of arachnoid tissue, the arachnoid villi, penetrate through gaps in the overlying dura and project into the dural sinuses. It is across the surfaces of these villi that CSF is reabsorbed into the blood passing through the sinuses.

The innermost meningeal layer, the **pia mater** (*pia* means "gentle"), is the most fragile. It is highly vascular and closely adheres to the surfaces of the brain and spinal cord, following every ridge and valley. In certain areas it dips deeply into the brain to bring a rich blood supply into close contact with the ependymal cells lining the ventricles. This relationship is important in the formation of CSF, a topic to which we now turn.

■ The brain floats in its own special cerebrospinal fluid.

Cerebrospinal fluid (CSF) surrounds and cushions the brain and spinal cord. The CSF has about the same density as the brain itself, so the brain essentially floats or is suspended in its special fluid environment. A major function of CSF is to serve as a shock-absorbing fluid to minimize damage of the brain when subjected to sudden, jarring movements.

Cerebrospinal fluid is formed primarily by the **choroid plexi** found in particular regions of the ventricle cavities of the brain. Choroid plexuses consist of richly vascularized pia mater tissue that dips into pockets formed by ependymal cells. Once CSF is formed, it flows through the four interconnected ventricles within the interior of the brain and through the spinal cord's narrow central canal, which is continuous with the last ventricle. Cerebrospinal fluid escapes from the fourth ventricle to enter the subarachnoid space and subsequently flows between the meningeal layers over the entire surface of the brain and spinal cord (Figure 5–13). When the CSF reaches the upper regions of the brain, it is reabsorbed into the venous blood through the arachnoid villi.

Flow of CSF through this system is facilitated by circulatory and postural factors that result in a CSF pressure of about 10 mm Hg in mammals. CSF is formed as a result of selective transport mechanisms across the membranes of the choroid plexuses, and its composition differs from that of plasma. For example, CSF is lower in K^+ and higher in Na^+, making it an ideal environment for the movement of these ions down concentration gradients, a process essential for conduction of nerve impulses (see p. 116).

■ A highly selective blood–brain barrier carefully regulates exchanges between the blood and brain.

The vertebrate brain is carefully shielded from harmful changes in the blood by the blood–brain barrier. Throughout the body, exchange of materials between the blood and the surrounding interstitial fluid can take place only across the walls of capillaries, the smallest of blood vessels (Chapter 9, p xxx). Unlike the rather free exchange across most capillaries in the body, permissible exchanges across brain capillaries are strictly limited. Changes in most plasma constituents do not easily influence the composition of brain interstitial fluid because only selected exchanges can be made. For example, even if the K^+ level in the blood is doubled, little change occurs in the K^+ concentration of the fluid bathing the central neurons. This is beneficial, because alterations in interstitial fluid K^+ would be detrimental to neuronal function.

The blood–brain barrier consists of both anatomic and physiologic factors. A single layer of endothelial cells forms capillary walls throughout the body. Usually, holes or pores between the cells making up the capillary wall (Figure 9–52) permit free exchange of all plasma components, except the large plasma proteins, with the surrounding interstitial fluid. In the brain capillaries, however, the cells are joined by **tight junctions,** which completely seal the capillary wall so that nothing can be exchanged across the wall by passing *between* the cells (● Figure 5–14). The only possible exchanges are through the capillary cells themselves. Lipid-soluble substances such as O_2, CO_2, alcohol, and steroid hormones penetrate these cells easily by dissolving in the lipid plasma membrane. Small water molecules also diffuse through readily, apparently by passing between the phospholipid molecules that compose the plasma membrane. All other substances exchanged between the blood and brain interstitial fluid, including such essential materials as glucose, amino acids, and ions, are transported by highly selective membrane-bound carriers. For example, the uptake of glucose is accomplished by primary active transport by a specific glucose transporter. Accordingly, transport across the capillary walls between the cells is anatomically prevented, and transport through the cells is physiologically restricted.

The brain capillaries are surrounded by astrocyte processes, which at one time were thought to be physically responsible

Figure 5–13 ● **Relationship of the meninges and cerebrospinal fluid to the mammalian brain and spinal cord.** (a) Brain, spinal cord, and meninges in sagittal section. The arrows and circled numbers with accompanying explanations indicate the direction of flow of cerebrospinal fluid (in yellow). (b) Frontal section in the region between the two cerebral hemispheres of the brain, depicting the meninges in greater detail.

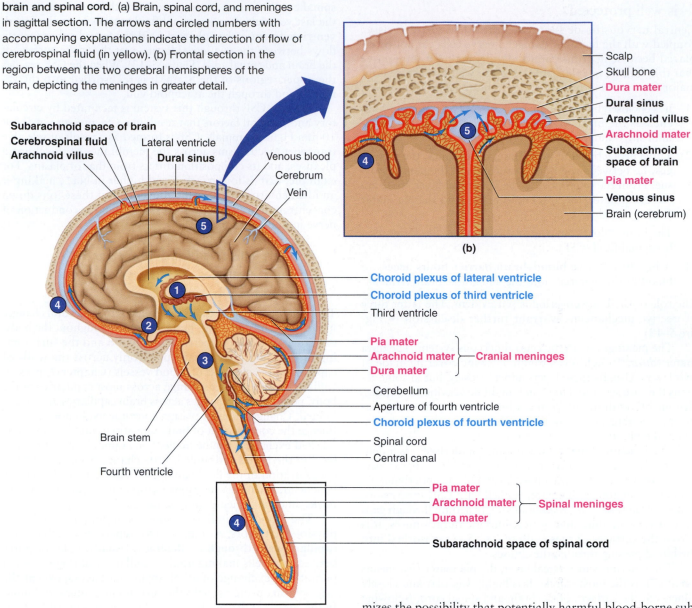

Cerebrospinal fluid

① is produced by the choroid plexuses,

② circulates throughout the ventricles,

③ exits the fourth ventricle at the base of the brain,

④ flows in the subarachnoid space between the meningeal layers, and

⑤ is finally reabsorbed from the subarachnoid space into the venous blood across the arachnoid villi.

for the blood–brain barrier. As we noted earlier (p. 158), astrocytes are believed to participate in regulating transport of some solutes such as K^+, and to stimulate the formation of the blood–brain barrier.

The blood–brain barrier protects the delicate brain and spinal cord from chemical fluctuations in the blood and minimizes the possibility that potentially harmful blood-borne substances might reach the central neural tissue. It further prevents certain circulating hormones that could also act as neurotransmitters from reaching the brain, where they could produce uncontrolled nervous activity or suppression. The blood–brain barrier limits the use of drugs for treating brain and spinal cord disorders because many drugs cannot penetrate this barrier. To "fool" the brain, some drugs can be attached to specific delivery vehicles, such as proteins or fatty acids, that are normally transported across the barrier. Once in the brain the protein or fatty acid is metabolized, leaving the drug to pursue its therapeutic task.

Certain areas of the brain are not subject to the blood–brain barrier, most notably a portion of the hypothalamus. Functioning of the hypothalamus depends on its "sampling" the blood and adjusting its controlling output accordingly to maintain homeostasis. Part of this output is in the form of hormones that must enter hypothalamic capillaries to be transported to their sites of action. Appropriately, these hypothalamic capillaries are not sealed by tight junctions.

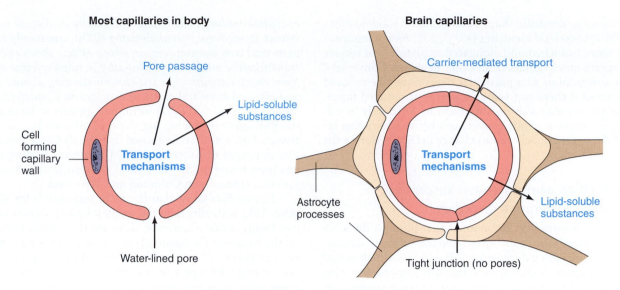

Most capillaries in body

Brain capillaries

Pore passage

Carrier-mediated transport

Lipid-soluble substances

Cell forming capillary wall

Transport mechanisms

Transport mechanisms

Astrocyte processes

Lipid-soluble substances

Water-lined pore

Tight junction (no pores)

Capillaries in cross section

Figure 5–14 • Blood–brain barrier. Unlike most capillaries in the body, the cells forming the walls of brain capillaries are joined by tight junctions that prevent materials from passing between the cells. The only passage across brain capillaries is through the cells that form the capillary walls. With the exception of lipid-soluble substances and water, passage of all other materials through these cells is physiologically regulated by carrier-mediated systems, which are not present in capillaries elsewhere.

■ The brain depends on delivery of oxygen and glucose by the blood.

Even though many substances in the blood never actually come in contact with the brain tissue, the brain, more than any other tissue, is highly dependent on a constant blood supply. Unlike most tissues, which can resort to anaerobic metabolism to produce ATP in the absence of O_2 for at least short periods (see p. 53), most vertebrate brains cannot produce ATP in the absence of O_2. Furthermore, in contrast to most tissues, which can use other sources of fuel for energy production in lieu of glucose, the mammalian brain uses only glucose under normal physiological conditions, and glucose plus ketone bodies during starvation, but does not store any of these nutrients. Therefore, the brain is absolutely dependent on a continuous, adequate blood supply of O_2 and glucose. Accordingly, brain damage results if this organ (particularly in birds and mammals) is deprived of its critical O_2 supply for more than a few minutes or if its glucose supply is cut off for more than 10 to 15 minutes. This is in contrast to insect neurons, where a significant amount of carbohydrate is stored in the nervous system as glycogen, particularly in glial cells, where it can be transported to nerve cells in the form of alanine (after breakdown of glycogen to pyruvate and transamination to alanine). Although neurons of honey bees lack enzymes to metabolize fatty acids, the brains of moths have a high capacity to oxidize fatty acids as food. Further, in many insects, as lipid fuel is mobilized on flight, much of it reaches the brain via the hemolymph and is used by neurons during this extraordinary energy-requiring activity.

However, some vertebrates can cope without O_2 for extended periods; for example, some fish and turtles spend the winter living under the ice without O_2 (see Chapter 11). At low temperatures their cells consume ATP at a substantially lower rate.

■ Brains of virtually all vertebrates display a degree of plasticity.

For the most part, in adult vertebrates mature neurons cannot divide. Destroyed CNS neurons, with the exception of those lining the nasal cavities and those in taste buds, cannot be replaced through cell division. Scientists think replacement of neurons in the taste buds and nasal passages is an adaptation to the wear and tear on these particular body surfaces. The known exceptions include fish, which continue to grow throughout their lifetime (for example, olfactory bulb in salmon increases 70% in size before migrating to the spawning grounds). A remarkable regeneration of neurons involved in regulating the *song control system* of songbirds has also been established (see box, "Neural Plasticity: A Song for All Seasons"). Normally, damaged axons cannot regenerate within the CNS as they can in the peripheral nervous system, because the myelin-forming oligodendrocytes of the CNS release nerve growth–inhibiting proteins. However, the brains of virtually all vertebrates, and many invertebrates, display a degree of plasticity (as we discussed briefly on p. 146). This ability is more pronounced in early development, but even adult animals retain some plasticity. When an area of the brain associated with a particular activity is destroyed, other areas of the brain may gradually assume some or all of the responsibilities of the damaged region. The underlying molecular mechanisms responsible for the brain's plasticity are only beginning to be unraveled. Current evidence suggests that the formation of new neural pathways (new connections between existing neurons) in response to changes in experience are mediated in part by alterations in dendritic shape resulting from modifications in certain cytoskeletal elements (see p. 57). As its dendrites become more branched and elongated, a neuron can receive and integrate more signals from other neurons. Conversely, synaptic connections to a particular neuron may be reduced or eliminated if there is a sustained level of reduced input. This mechanism may also be important in seasonally breeding animals, where synaptic connections between differing regions of the brain are believed to change cyclically during the breeding cycle.

■ Mammalian neural tissue is susceptible to neurodegenerative disorders.

Although the mammalian brain and other parts of the CNS are well protected, they can suffer from a number of damaging disorders, including degenerative diseases (in which neural tissue dies) such as Alzheimer's in humans. Among the most perplexing of these diseases are the *transmissible spongiform*

encephalopathies (TSE). The TSE family of disease includes *bovine spongiform encephalopathy (BSE),* commonly referred to as mad cow disease; *scrapie,* which affects sheep and goats, *transmissible mink encephalopathy* in minks; *feline spongiform encephalopathy* in cats; *chronic wasting disease* in deer and elk; and *kuru* and *classical* and *variant Creutzfeldt-Jacob disease* in humans (*vCJD*) form that appears to be caused by the BSE agent following ingestion of infected bovine neural tissue). BSE was first diagnosed in 1986 in Great Britain and first appeared in America in 2004. Chronic wasting disease has been found mainly in central western United States and Canada, especially in Colorado and Wyoming.

Researchers know that the agent responsible for BSE and other TSEs is smaller than the smallest known viruses and that it is highly stable. Currently, there are three main hypotheses on the nature of the agent: (1) It is a small virus with unusual characteristics. (2) It is a **virino,** a partial virus made of nucleic acid coated with a protein from the infected animal. (3) It is a **prion**—a normal protein of the infected animal (host) that is

---

CHALLENGES AND CONTROVERSIES

Neural Plasticity: A Song for All Seasons

Traditional physiology tells us that vertebrate neurons in the brain cannot regenerate. Songbirds are unusual in that there is a constant turnover of neurons in certain regions of their brains. The canary's forebrain (which contains about 7 million neurons), adds approximately 20,000 new neurons each day to replace the ones that die. Within the forebrain the pathway for song production is anatomically distinct from the rest of the brain. Discrete song-responsive nuclei in the high vocal center (HVc) connect to the robust nucleus of the archistriatum (RA), which then connects to the motor neurons controlling the musculature of the syrinx and vocal tract muscles, in conjunction with the respiratory and postural systems. Male canaries sing only during the spring breeding season and then again in the fall as they begin learning a new song repertoire for the subsequent breeding season. During singing, the pattern of firing of neurons in the HVc can be associated with the production of syllables, that is, a group of notes. Each pattern of notes is associated with a stable and unique pattern of neuronal activity. In contrast, the RA exhibits precisely timed bursts of activity that are uniquely associated with note identity.

Canaries are sexually dimorphic, that is, the HVc and the RA are anatomically distinct between males and females. However, females can be induced to sing if they are administered the hormone testosterone. Associated with song production, both the HVc

and RA in the treated females increase in size. Male songbirds with the largest HVc and RA are also the best singers and so are the most likely to attract a mate successfully during the breeding season. In another species, the marsh wren, there is a 40% greater size of the HVc and 30% greater size of the RA nucleus of males living along the west coast of North America compared to males living on the east coast. Associated with this, west coast wrens have song repertoires three times greater than their east coast cousins.

The neural circuitry associated with song learning is initially formed in response to both listening and practicing of the adult song. The HVc receives auditory input from the primary forebrain auditory nucleus. During this developmental period, neurogenesis is restricted to the song nuclei. Song development normally begins one month after hatch. At this time the size of the HVc is approximately 10 to 15% of the adult. In the fall, the number of RA-projecting neurons increase as the birds modify their song in preparation for the next breeding season. Even in old birds, new neurons have been shown to project an axon exactly the distance between the HVc and the RA, successfully linking these two brain regions. At the same time some of the RA-projecting neurons die. This continual turnover of neurons indicates an enormous plasticity in the brain of birds, a finding that suggests the possibility of regeneration and repair of neural circuitry in humans.

Photo: Ron Austing; Frank Lane Picture Agency/CORBIS

modified to an abnormal form after exposure to an abnormal form from an external source. The term *prion* is short for *proteinaceous infectious particle.*

The best current explanation for BSE is that it is caused by a prion. It has been suggested that the use of protein feed supplements made from meat and bonemeal of carcasses of scrapie-infected animals first caused the disease in cattle. BSE appears to start when molecules of a normal neuronal protein called *PrP* become abnormally folded. The normal form of PrP has an uncertain function, but new evidence suggests that it plays a role in memory formation. When an abnormal PrP comes into contact with normal PrP's the latter refold to match the abnormal ones, forming new prions. Eventually neurons become clogged with prions, which severely impairs their function. Ultimately the assault leads to **spongiform damage:** a stipple of microscopic holes that accumulate as the brain reacts to the infection.

Unfortunately, the BSE agent is extremely resistant to heat and is partially resistant to proteases (enzymes that degrade other proteins), and does not evoke any detectable immune responses or inflammatory reactions in the infected animal. The incubation period (the time from when an animal becomes infected until it first shows disease signs) is from two to eight years. Following the onset of clinical signs, the animal's condition steadily declines until it either dies or is destroyed. This period normally takes from two to six months. At present there is no treatment for the disease or vaccine to prevent it.

Cattle affected by BSE experience progressive degeneration of the nervous system, leading to nervousness or aggression, abnormal posture, lack of coordination, difficulty in rising, and low milk production. The BSE agent has been identified only in brain tissue, spinal cord, and retina. In experimentally infected cattle, BSE has been localized in the dorsal root ganglion, trigeminal ganglion, distal ileum, and the bone marrow. To date no evidence of infection has been detected in milk or in muscle tissue.

Brain Evolution in Vertebrates

At the beginning of this chapter, we discussed the evolution of brains among animals. Here we examine the specific details of vertebrate brains.

◾ Newer, more sophisticated regions of the vertebrate brain are piled on top of older, more primitive regions.

Even though the vertebrate brain is a functional whole, it is organized into several different regions. The parts of the brain can be arbitrarily grouped in various ways based on anatomic distinctions, functional specialization, and evolutionary origin. A vertebrate brain is classically organized into three distinct regions—the hindbrain, midbrain, and forebrain—which form from three successive portions of the embryonic neural tube (● Figure 5–15). But these do not readily follow functional aspects of the brain. Therefore, we use the following regions (❚ Table 5–8):

1. Brain stem
2. Cerebellum

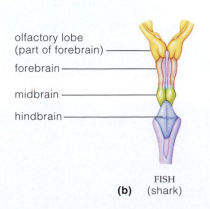

FOREBRAIN. Receives, integrates sensory information from nose, eyes, and ears; in land-dwelling vertebrates, contains the highest integrating centers

MIDBRAIN. Coordinates reflex responses to sight, sounds

HINDBRAIN. Reflex control of respiration, blood circulation, other basic tasks; in complex vertebrates, coordination of sensory input, motor dexterity, and possibly mental dexterity

(start of spinal cord)

(a) Expansion of the dorsal, hollow nerve cord into more complex, functionally distinct regions in certain lineages.

Figure 5–15 ● Evolutionary trend toward an expanded, more complex brain. The trend became apparent after morphological comparisons were made of the brains of some vertebrates. These dorsal views are not to the same scale.

(*Source:* C. Starr & R. Taggart, 2004, *Biology: The Unity and Diversity of Life,* 10th ed. (Belmont, CA: Brooks/Cole), Figure 34.15)

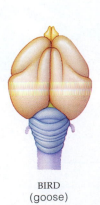
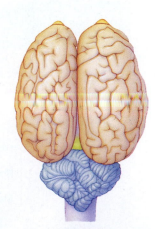

olfactory lobe (part of forebrain)

forebrain

midbrain

hindbrain

(b)

FISH (shark)

AMPHIBIAN (frog)

REPTILE (alligator)

BIRD (goose)

MAMMAL (horse)

3. Forebrain
    a. Diencephalon
        i. Hypothalamus
        ii. Thalamus
    b. Cerebrum
        i. Basal nuclei
        ii. Cerebral cortex

The order in which these components are listed generally represents both their anatomic location (from bottom to top in upright animals, or front to back in horizontal animals) and their complexity and sophistication of function (from the least specialized to the most specialized level). As you saw at the beginning of this chapter, the more sophisticated layers of the brain were added on to the more primitive layers, so the order in the list also represents an evolutionary sequence of oldest to newest. The fundamental regional structure of the brain remains similar between all modern vertebrates, with the exception of jawless vertebrates, which lack a defined cerebellum. (However, as we noted earlier, the relative sizes of each region vary considerably between species.) The relatively conservative nature of these major brain divisions suggests that much of the organization of these structures must have arisen

with the origin of vertebrates or shortly thereafter. Let's examine the brain regions and their evolution in more detail.

The brain stem, the smallest and least changed region of the vertebrate brain, is continuous with the spinal cord (Figure 5–15). It is the most ancient part of the brain, having evolved approximately 500 million years ago. It encompasses the **medulla oblongata** (also called just the *medulla*), and **pons**: brain centers that control many of the life-sustaining processes, such as breathing and circulation, as well as certain baseline activities of skeletal muscle. In teleosts (which are the majority of bony fishes), one notable specialization is a region devoted to taste, whereas in electric fishes there is an area that governs activation of the electric organs.

Attached at the top rear portion of the brain stem is the **cerebellum**, another structure that has changed comparatively little with evolution. Only the jawless vertebrates, hagfishes and lampreys, lack a cerebellum. A cerebellum-like structure comprised of a central body and paired lobes arose with the origin of the jawed fishes. The cerebellum is concerned with maintaining proper position of the body in space and subconscious coordination of motor activity (movement). For example, signals from the balance organs and the auditory and visual systems are integrated in the cerebellum and the information

Table 5–8 ■ Overview of Structures and Functions of the Major Components of the Brain

used to orient an animal relative to its surroundings. Flying mammals (such as bats) and birds have a comparatively large cerebellum because of the complexity associated with flight.

Anterior to the brain stem and cerebellum is the **forebrain.** The forebrain has changed more than other regions of the brain during vertebrate evolution. It has two major subdivisions—an inner **diencephalon** and outer **cerebrum.** The diencephalon houses two regions of specific nuclei: the **hypothalamus,** which controls many homeostatic functions important in maintaining stability of the internal environment, and the **thalamus,** which performs some primitive sensory processing and is a relay station for almost all sensory input and for some outgoing commands. In fishes and amphibians, the forebrain is primarily devoted to olfactory input.

On top of this "cone" of lower brain regions is the **cerebrum,** which became progressively larger and more highly convoluted (that is, has tortuous ridges delineated by deep grooves or folds) the more advanced the vertebrate species was. The number of interconnections also increased; for example, in the cortex of the primate brain, each neuron can receive inputs from as many as 10,000 other neurons. The cerebrum is most highly developed in humans, where it constitutes about 80% of the total brain weight. The outer layer of the cerebrum

Major Functions

1. Sensory perception
2. Voluntary control of muscle
3. Language
4. Personality traits
5. Sophisticated mental events, such as thinking, memory decision making, creativity, and self-consciousness

1. Inhibition of muscle tone
2. Coordination of slow, sustained movement
3. Supression of useless patterns of movement

1. Relay station for all synaptic input
2. Crude awareness of sensation
3. Some degree of consciousness
4. Role in motor control

1. Regulation of many homeostatic functions, such as temperature control, thirst, urine output, and food intake
2. Important link between nervous and endocrine systems
3. Extensive involvement with emotion and basic behavioral patterns

1. Maintenance of balance
2. Enhancement of muscle tone
3. Coordination and planning of skilled voluntary muscle activity

1. Origin of majority of peripheral cranial nerves
2. Cardiovascular, respiratory, and digestive control centers
3. Regulation of muscle reflexes involved with equilibrium and posture
4. Reception and integration of all synaptic input from spinal cord; arousal and activation of cerebral cortex
5. Role in sleep-wake cycle

is the highly convoluted **cerebral cortex,** which caps an inner core that houses the **basal nuclei.** The myriad convolutions of the cerebral cortex in higher mammals give it the appearance of a much-folded walnut. The cortex is perfectly smooth in many lower mammals. Without these surface wrinkles, the primate cortex would take up to three times the area it does, and accordingly, would not fit like a cover over the underlying structures. The cortex plays a key role in the most sophisticated neural functions, such as voluntary initiation of movement, final sensory perception, self-awareness, language, personality traits, and other factors we associate with the sophisticated behavior and intellect. It is the highest, most complex integrating area of the brain.

Each of these regions of the vertebrate brain is discussed in turn, starting with the highest level, the cerebral cortex, and moving down to the lowest level, the spinal cord.

Mammalian Cerebral Cortex

The cerebrum, by far the largest region of mammalian brains, is divided into two halves, the right and left **cerebral hemispheres.** They are connected to each other by the **corpus callosum,** a thick band consisting of an estimated 300 million neuronal axons traversing between the two hemispheres (● Figure 5–16a). The corpus callosum is the body's "information highway." The two hemispheres communicate and cooperate with each other by means of constant information exchange through this neural connection.

■ The cerebral cortex is an outer shell of gray matter covering an inner core of white matter.

Each hemisphere consists of a thin outer shell of *gray matter,* the **cerebral cortex,** covering a thick central core of *white matter,* the cerebral medulla (Figure 5–16b). Located deep within the white matter is another region of gray matter, the basal nuclei. Throughout the entire CNS, **gray matter** consists predominantly of densely packaged cell bodies and their dendrites as well as glial cells. Bundles or tracts of myelinated nerve fibers (axons) constitute the **white matter;** the white appearance is due to the lipid (fat) composition of the myelin. The gray matter can be viewed as the "computers" of the CNS and the white matter as the "wires" that connect the computers to each other. The fiber tracts in the white matter transmit signals from one part of the cerebral cortex to another or between the cortex and other regions of the CNS. Such communication between different areas of the cortex and elsewhere facilitates integration of their activity. This integration of neural input is essential for even a relatively simple task such as selecting a food item to eat. Vision and smell of the food is received by one area of the cortex, reception of its palatability takes place in another area, and movement is initiated by still another area. More complex functions may also come into play; for example, a particular food item may have made an animal sick, and its memory therefore may make it reject this particular item even though it might be nutritious. How such complex processing actually works is still poorly understood.

■ The cerebral cortex is organized into layers and functional columns.

The cerebral cortex is organized into six well-defined layers based on varying distributions of the cell bodies and locally

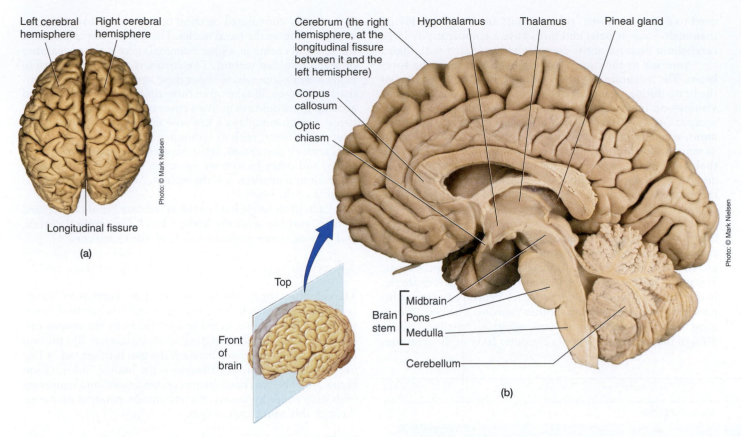

Figure 5-16 ● Brain of human cadaver. (a) Dorsal view looking down on the top of the brain. Note that the deep longitudinal fissure divides the cerebrum into the right and left cerebral hemispheres. (b) Sagittal view of the right half of the brain. All major brain regions are visible from this midline interior view. The corpus callosum serves as a neural bridge between the two cerebral hemispheres.

associated fibers of several distinctive cell types. These layers are organized into functional vertical columns that extend perpendicularly about 2 mm from the cortical surface down through the thickness of the cortex to the underlying white matter. The neurons within a given column are believed to function as a "team," with each cell being involved in different aspects of the same specific activity—for example, perceptual processing of the same stimulus from the same location.

The functional differences between various areas of the cortex result from different layering patterns within the columns and from different input–output connections, not from the presence of unique cell types or different neuronal mechanisms. For example, those regions of the cortex responsible for perception of senses have an expanded layer 4, a layer rich in **stellate cells,** which are responsible for initial processing of sensory input to the cortex. In contrast, the cortical areas that control output to skeletal muscles have a thickened layer 5, which contains an abundance of large **pyramidal cells.** These cells send fibers down the spinal cord from the cortex to terminate on the efferent motor neurons that innervate the skeletal muscles.

■ **The four pairs of lobes in the cerebral cortex are specialized for different activities.**

We are now going to consider the locations of the major functional areas of the cerebral cortex. It is important to recognize

that even though a discrete activity is ultimately attributed to a particular region of the brain, no part of the brain functions in isolation. Each part depends on complex interplay among numerous other regions for both incoming and outgoing messages. With this in mind, let's now consider the locations of the major functional areas of the brain.

The anatomic landmarks used in cortical mapping are certain deep folds that divide each half of the cortex into four major lobes: the *occipital, temporal, parietal,* and *frontal* lobes (● Figure 5-17). During the following discussion of the major activities attributed to various regions of these lobes, refer to the basic functional map of the cortex in ● Figure 5-18a.

The **occipital lobes,** which are located posteriorly (at the back of the head), are responsible for initial processing of visual input in the cortex. Sound sensation is initially received by the **temporal lobes,** located laterally (on the sides of the head) (Figure 5-18a and b). You will learn more about the functions of these regions in Chapter 6 when we discuss vision and hearing.

The parietal lobes and frontal lobes, located on the top of the head, are separated by the **central sulcus,** a deep infolding that runs roughly down the middle of the lateral surface of each hemisphere. The **parietal lobes** lie to the rear of the central sulcus on each side, and the **frontal lobes** lie in front of it. The **parietal lobes** are primarily responsible for receiving and processing sensory input. The frontal lobes are responsible for three main functions: (1) voluntary motor activ-

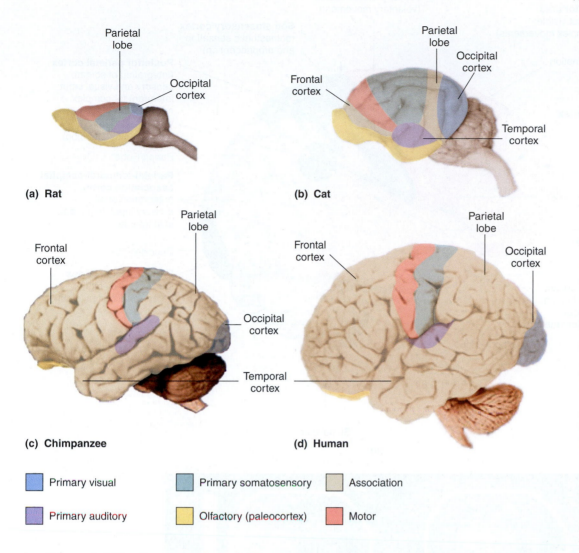

(a) **Rat**

(b) **Cat**

(c) **Chimpanzee**

(d) **Human**

▪	Primary visual	▪	Primary somatosensory	▪	Association
▪	Primary auditory	▪	Olfactory (paleocortex)	▪	Motor

Figure 5–17 ● Major functional divisions of the mammalian cerebrum, shown for a rat, cat, chimpanzee, and human. Regions devoted to motor control are colored red. Regions devoted to processing sensory information are colored as indicated in the legend boxes. Regions dedicated to "higher" associative functions are shaded in cream. Associative functions (those involved in complex memory, planning, and so on) have expanded in certain mammalian groups such as primates.

ity, (2) vocal ability (in mammals with this capability), and (3) higher mental functions such as planning.

We are next going to examine the role of the parietal lobes in sensory perception, then turn our attention to the frontal lobes. Note first how the relative sizes of the various lobes have changed in mammalian evolution (Figure 5–17).

■ **The parietal lobes are responsible for somatosensory processing.**

Sensations from the surface of the body, such as touch, pressure, heat, cold, and pain, are collectively known as **somesthetic sensations** (*somesthetic,* "body feelings"). Within the CNS of mammals, this information is "projected" (transmitted along specific neural pathways to higher brain levels) to the **somatosensory cortex.** The somatosensory cortex is located at the front of each parietal lobe immediately behind the central sulcus (● Figures 5–18 and 5–19a). It is the site for initial cortical processing and perception of somesthetic input as well as proprioceptive input. **Proprioception** is the awareness of body position. In nonmammalian vertebrates, sensory input generated from proprioceptive receptors are sent to several regions of the brain. For example, in fishes the cerebellum and tectum respond to stimulation initiated from the skin and fins, whereas in turtles and frogs, thalamic nuclei generate somatotopic responses.

Each region within the somatosensory cortex receives sensory input from a specific area of the body. The greater the sensory input density from a specific area of the body, the greater the area of the cortex devoted to processing this information. This orderly distribution of cortical sensory processing is depicted in Figure 5–19b. Note that on this so-called **sensory homunculus** in humans (*homunculus,* "little man"), the body is represented upside down on the somatosensory cortex and, more importantly, different parts of the body are not equally represented. For each species, the extent of each body part is indicative of the relative proportion and location of the somatosensory cortex devoted to that area (● Figure 5–20). For example, hand and face areas are large in monkeys, mouthparts in rabbits, claws and forelimbs in cats, face in dogs. In cats this cortical representation is referred to a *felunculus* and in dogs it is a *canunculus*. In humans the exaggerated size of the face, tongue, hands, and genitalia is indicative of the high degree of sensory perception and density of nerves associated with these body parts.

These maps of the sensory world of an animal were put together by experiments on several species of animals. Electrodes were placed in the cerebral cortex of anesthetized animals, and recordings of clusters of neurons were made in response to visual, auditory or somatosensory stimulation. For example, tactile stimulation of the animal's body activated neurons in the rostral regions of the cortex. These studies es-

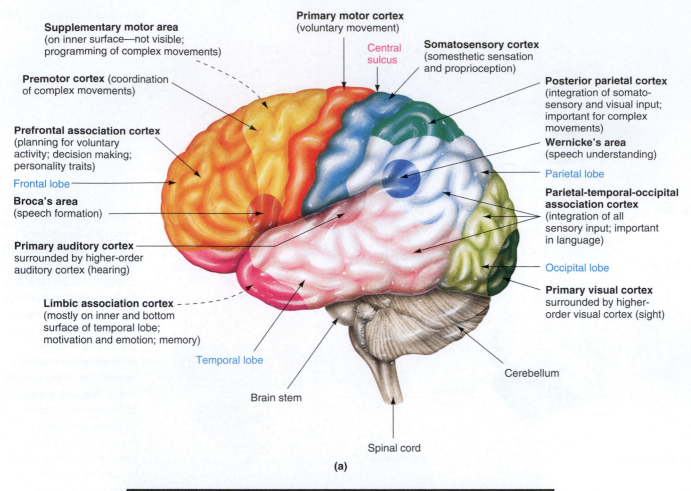

Supplementary motor area (on inner surface—not visible; programming of complex movements)

Primary motor cortex (voluntary movement)

Central sulcus

Somatosensory cortex (somesthetic sensation and proprioception)

Premotor cortex (coordination of complex movements)

Posterior parietal cortex (integration of somatosensory and visual input; important for complex movements)

Prefrontal association cortex (planning for voluntary activity; decision making; personality traits)

Wernicke's area (speech understanding)

Frontal lobe

Parietal lobe

Broca's area (speech formation)

Parietal-temporal-occipital association cortex (integration of all sensory input; important in language)

Primary auditory cortex surrounded by higher-order auditory cortex (hearing)

Occipital lobe

Primary visual cortex surrounded by higher-order visual cortex (sight)

Limbic association cortex (mostly on inner and bottom surface of temporal lobe; motivation and emotion; memory)

Temporal lobe

Cerebellum

Brain stem

Spinal cord

(a)

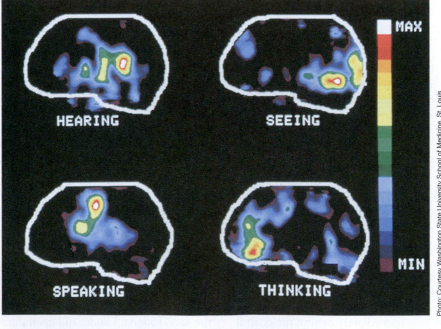

HEARING

SEEING

SPEAKING

THINKING

MAX

MIN

(b)

Figure 5–18 ● **Functional areas of the human cerebral cortex.** (a) Various regions of the cerebral cortex are primarily responsible for various aspects of neural processing, as indicated in this schematic lateral view of the brain. (b) Different areas of the brain "light up" on positron emission tomography (PET) scans as a person performs different tasks. PET scans detect the magnitude of blood flow in various regions of the brain. Because more blood flows into a particular region of the brain when it is more active, neuroscientists can use PET scans to "take pictures" of the brain at work on various tasks.

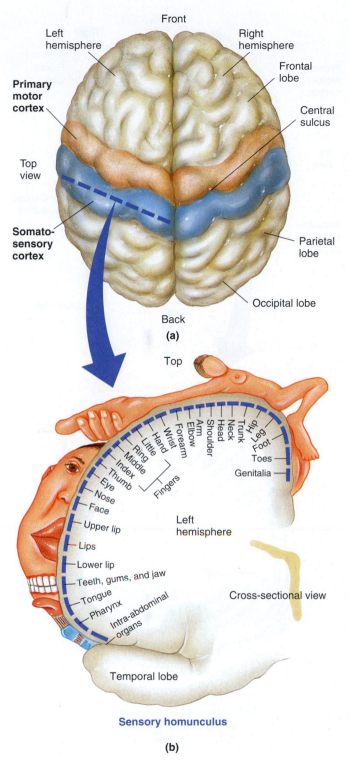

Front

Left hemisphere

Right hemisphere

Frontal lobe

Primary motor cortex

Central sulcus

Top view

Somato-sensory cortex

Parietal lobe

Occipital lobe

Back

(a)

Top

Hip
Leg
Foot
Toes
Genitalia

Trunk
Neck
Head
Shoulder
Arm
Elbow
Forearm
Wrist
Hand
Little
Ring
Middle
Index
Thumb
Eye
Nose
Face
Fingers

Upper lip

Lips

Lower lip

Teeth, gums, and jaw

Tongue

Pharynx

Intra-abdominal organs

Left hemisphere

Cross-sectional view

Temporal lobe

Sensory homunculus

(b)

Figure 5–19 ◄ ● Somatotopic map of the somatosensory cortex. (a) Top view of cerebral hemispheres. (b) Sensory homunculus showing the distribution of sensory input to the somatosensory cortex from different parts of the body. The distorted graphic representation of the body parts indicate the relative proportion of the somatosensory cortex devoted to reception of sensory input from each area.

tablished that there are maps of the sensory surfaces in the cerebral cortices of all vertebrates. The rhombencephalon of the catfish, for example, contains extensive sensory maps of the taste receptors found on the body surface as well as those from the mouth and pharynx.

The somatosensory cortex on each side of the brain for the most part receives sensory input from the opposite side of the body, because most of the ascending pathways carrying sensory information up the spinal cord cross over to the opposite side before eventually terminating in the cortex. Thus damage to the somatosensory cortex in the left hemisphere produces sensory deficits on the right side of the body, whereas sensory losses on the left side are associated with damage to the right half of the cortex.

Simple awareness of touch, pressure, or temperature is detected by the thalamus, a lower level of the brain, but the somatosensory cortex goes beyond pure recognition of sensations to fuller sensory perception. The thalamus detects changes in temperature but does not provide information on the intensity. The somatosensory cortex localizes the source of sensory input and perceives the level of intensity of the stimulus. It also is capable of spatial discrimination, so it can discern shapes of objects being held and can distinguish subtle differences in similar objects that come into contact with the skin or outer body surfaces.

The somatosensory cortex, in turn, projects this sensory input via white matter fibers to adjacent higher sensory areas for even further elaboration, analysis, and integration of sensory information. These higher areas are important in the perception of complex patterns of somatosensory stimulation—for example, simultaneous perception of the texture, firmness, temperature, shape, position, and location of an object the ani-

Figure 5–20 ▼ ● Somatosensory projections of four mammals. Each is distorted to represent the extent of innervation to different body parts. Disproportionately large projection areas in the somatosensory cortex are associated with densely innervated tissue.

Rabbit

Cat

Monkey

Human

mal is touching. (Note that we can only study perception in humans, who can report to the investigator.)

■ The primary motor cortex is located in the frontal lobes.

The posterior part of the frontal lobe immediately in front of the central sulcus and adjacent to the somatosensory cortex is the **primary motor cortex** (● Figures 5–18a and 5–21a). It confers voluntary control over movement produced by skeletal muscles. As in sensory processing, the motor cortex on each side of the brain primarily controls muscles on the opposite side of the body. Neuronal tracts originating in the motor cortex of the left hemisphere cross over in the medulla oblongata before passing down the spinal cord to terminate on efferent motor neurons that trigger skeletal muscle contraction on the right side of the body. Accordingly, damage to the motor cortex on the left side of the brain produces paralysis on the right side of the body, and the converse is also true.

Stimulation of different areas of the primary motor cortex brings about movement in different regions of the body. Like the sensory homunculus for the somatosensory cortex, the **motor homunculus**, which depicts the location and relative amount of motor cortex devoted to output to the muscles of each body part, is upside down and distorted (Figure 5–21b). For humans, the fingers, thumbs, and muscles important in vocalization, especially those of the lips and tongue, are grossly exaggerated in size, indicative of the fine degree of motor control with which these body parts are endowed. Compare this to how little brain tissue is devoted to the trunk, arms, and lower extremities, which are not capable of such complex movements. Thus the extent of representation in the motor cortex, as seen in the somatosensory cortex, is proportional to the precision and complexity of motor skills required of the respective part.

Recent evidence suggests that the "map" of the primary motor cortex (the depiction of which cortical areas control the muscles in which parts of the body) is not as orderly as the motor homunculus suggests; that is, there is not a neat one-to-one correspondence between a cortical area and a specific body part. Instead, it appears that individual muscles receive controlling input from multiple spots on the motor cortex and that individual neurons in the motor cortex control more than one muscle. Even though neurons that control movement of a particular body part, such as the arm, are concentrated in one large zone of the primary motor cortex, other arm-specific neurons are located diffusely in other areas of the motor cortex. Furthermore, the arm zone overlaps considerably with the hand zone, and arm and hand neurons are intermingled. Neurons that activate wrist and shoulder muscles are even sprinkled among the intermingled arm and hand neurons. Not using a particular appendage over time reduces the amount of motor cortex dedicated to its control. The emerging view of the primary motor cortex is that the neurons are functionally linked, so they can work in concert to activate multiple muscles in a coordinated fashion to produce a given movement, such as raising your arm. Thus physiologists will probably eventually replace the spatially organized motor homunculus map of the primary motor cortex with a functionally organized map of cortical regions governing combinations of muscle activity coordinated to choreograph particular movements.

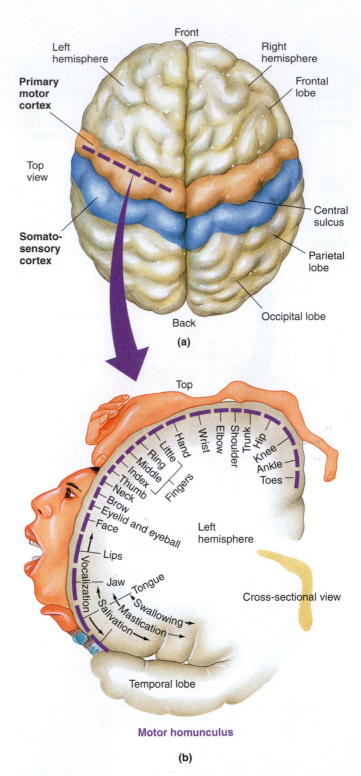

Figure 5–21 ● **Somatotopic map of the human primary motor cortex.** (a) Top view of cerebral hemispheres. (b) Motor homunculus showing the distribution of motor output from the primary motor cortex to different parts of the body. The distorted graphic representation of the body parts is indicative of the relative proportion of the primary motor cortex devoted to controlling skeletal muscles in each area.

■ Other regions of the nervous system besides the primary motor cortex are important in motor control.

Even though signals from the primary motor cortex terminate on the efferent neurons that trigger voluntary skeletal muscle contraction, the motor cortex is not the only region of the brain involved with motor control. First, lower brain regions and the spinal cord control involuntary skeletal muscle activity, such as the maintenance of posture. Some of these same regions also play an important role in monitoring and coordinating voluntary motor activity that the primary motor cortex has set in motion. Second, although fibers originating from the motor cortex can activate motor neurons to bring about muscle contraction, the motor cortex itself does not *initiate* voluntary movement. The motor cortex is activated by a widespread pattern of neuronal discharge, the **readiness potential,** which occurs about 750 msec before specific electrical activity is detectable in the motor cortex. The higher motor areas of the brain believed to be involved in this voluntary decision-making period include the *supplementary motor area,* the *premotor cortex,* and *the posterior parietal cortex* (Figure 5–18a). These higher areas all command the primary motor cortex. Furthermore, the cerebellum appears to play an important role in the anticipatory planning and timing of certain kinds of movement by sending input to the motor areas of the cortex. This is particularly true for learned skilled movements. We investigate the cerebellum later in the chapter.

These four regions of the brain are important in programming and coordinating complex movements involving simultaneous contraction of many muscles. Even though electrical stimulation of the primary motor cortex brings about contraction of particular muscles, no purposeful coordinated movement can be elicited, just as pulling on isolated strings of a puppet does not produce any meaningful movement. A puppet displays purposeful movements only when a skilled puppeteer manipulates the strings in a coordinated manner. In the same way, these four regions (and perhaps other areas as yet undetermined) develop a **motor program** for the specific voluntary task and then "pull" the appropriate pattern of "strings" in the primary motor cortex to bring about the sequenced contraction of appropriate muscles to accomplish the desired complex movement.

■ The cerebral hemispheres have some degree of specialization.

The cortical areas described thus far appear equally distributed in both the right and left hemispheres (except for the language areas, which are found only on one side, usually the left). The left side is also most commonly the dominant hemisphere for fine motor control, at least in primates. Thus most humans are right-handed, because the left side of the brain controls the right side of the body. In contrast, most parrots are left-footed, with the right side of the brain controlling the left side of the body. Each hemisphere is somewhat specialized in the types of mental activities it carries out best. In humans, the **left cerebral hemisphere** excels in the performance of logical, analytical, sequential, and verbal tasks, such as math, language forms, and philosophy. In contrast, the **right cerebral hemisphere** excels in nonlanguage skills, especially spatial perception and artistic and musical endeavors. Whereas the left hemisphere tends to process information in a fragmentary way, the right hemisphere views the world holistically. Normally, much sharing of information occurs between the two hemispheres so that they complement each other, but in many individuals the skills associated with one hemisphere appear to be more strongly developed. Left cerebral-hemisphere dominance tends to be associated with "thinkers," whereas the right-hemispheric skills dominate in "creators."

Subcortical Structures and Their Relationship with the Cortex in Higher Brain Functions

The **subcortical** ("under the cortex") **regions** of the brain interact extensively with the cortex in the performance of their functions. These regions include the *basal nuclei,* located in the cerebrum, and the *thalamus* and *hypothalamus,* located in the diencephalon.

■ The basal nuclei play an important inhibitory role in motor control.

The **basal nuclei** (also known as **basal ganglia**) consist of several masses of gray matter located deep within the cerebral white matter (Table 5–8 and ● Figure 5–22). In the nervous system, a **nucleus** (plural, **nuclei**) refers to a functional aggregation of neuronal cell bodies (thus the term *ganglion* is probably more appropriate, but *nucleus* is unfortunately more widely used). The basal nuclei play a complex role in the control of movement in addition to having nonmotor functions that are less well understood. In particular, the basal nuclei are important in (1) inhibiting muscle tone throughout the body (proper muscle tone is normally maintained by a balance of excitatory and inhibitory inputs to the neurons that innervate skeletal muscles); (2) selecting and maintaining purposeful motor activity while suppressing useless or unwanted patterns of movement; and (3) helping monitor and coordinate slow, sustained contractions, especially those related to posture and support. The basal nuclei do not directly influence the efferent motor neurons that bring about muscle contraction but act instead by modifying ongoing activity in motor pathways.

To accomplish these complex integrative roles, the basal nuclei receive and send out much information, as is indicated by the tremendous number of fibers linking them to other regions of the brain. One important pathway consists of strategic interconnections that form a complex feedback loop linking the cerebral cortex (especially its motor regions), the basal nuclei, and the thalamus (● Figure 5–23). It is speculated that the thalamus positively reinforces voluntary motor behavior initiated by the cortex, whereas the basal nuclei modulate this activity by exerting an inhibitory effect on the thalamus to eliminate antagonistic or unnecessary movements. The basal nuclei also exert an inhibitory effect on motor activity by acting through neurons in the brain stem.

■ The thalamus is a sensory relay station and is important in motor control.

Deep within the brain near the basal nuclei is the diencephalon, a midline structure that forms the walls of the third ven-

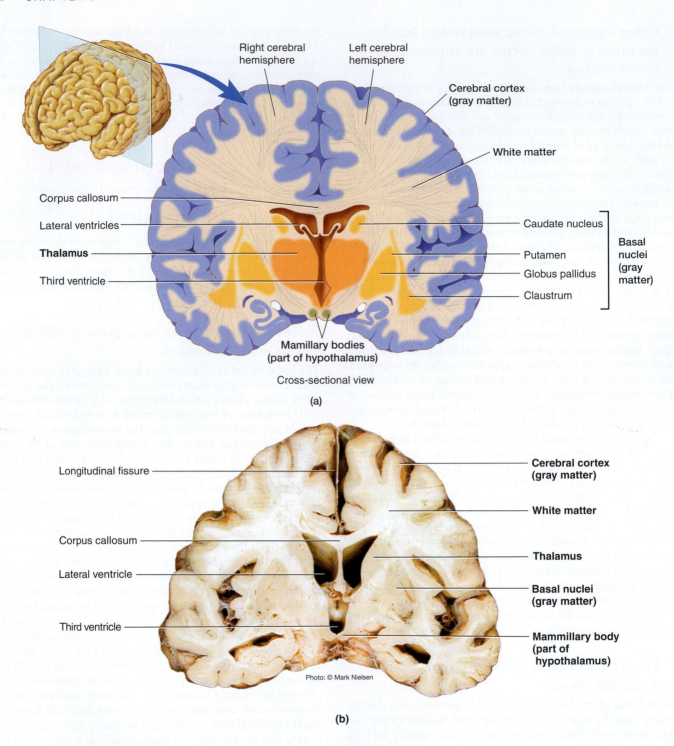

(a)

Right cerebral hemisphere

Left cerebral hemisphere

Cerebral cortex (gray matter)

White matter

Corpus callosum

Lateral ventricles

Thalamus

Third ventricle

Caudate nucleus

Putamen

Globus pallidus

Claustrum

Basal nuclei (gray matter)

Mamillary bodies (part of hypothalamus)

Cross-sectional view

Longitudinal fissure

Corpus callosum

Lateral ventricle

Third ventricle

Cerebral cortex (gray matter)

White matter

Thalamus

Basal nuclei (gray matter)

Mammillary body (part of hypothalamus)

Photo: © Mark Nielsen

(b)

Figure 5–22 ● **Frontal section of the human brain.** (a) Schematic frontal section of the brain. The cerebral cortex, an outer shell of gray matter, surrounds an inner core of white matter. Deep within the cerebral white matter are several masses of gray matter, the basal nuclei. The ventricles are cavities in the brain through which the cerebrospinal fluid flows. The thalamus forms the walls of the third ventricle. (b) Photograph of a frontal section of the brain of a cadaver.

(Photo: Mark Nielsen, Department of Biology, University of Utah)

tricular cavity, one of the spaces through which cerebrospinal fluid flows. Its two major parts, the thalamus and the hypothalamus, have distinct roles (Table 5–8 and ● Figures 5–16b, 5–22, and 5–24).

The **thalamus** serves as a "relay station" and synaptic integrating center for preliminary processing of all sensory input on its way to the cortex. It screens out insignificant signals and routes the important sensory impulses to appropriate areas of the somatosensory cortex, as well as to other regions of the brain. Along with the brain stem and cortical association areas,

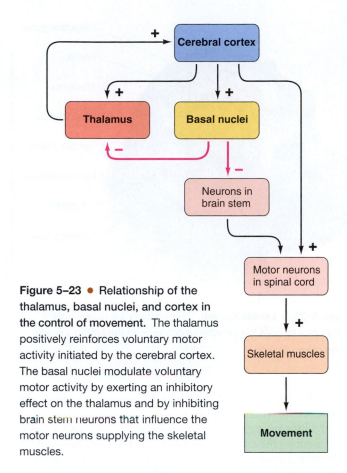

Figure 5–23 ● Relationship of the thalamus, basal nuclei, and cortex in the control of movement. The thalamus positively reinforces voluntary motor activity initiated by the cerebral cortex. The basal nuclei modulate voluntary motor activity by exerting an inhibitory effect on the thalamus and by inhibiting brain stem neurons that influence the motor neurons supplying the skeletal muscles.

the thalamus is important in an animal's ability to direct attention to stimuli of interest. For example, mammalian parents can sleep soundly through the noise of wind or rain but be instantly aware of their offspring's slightest whimper. The thalamus is also capable of crude awareness of various types of sensation but cannot distinguish their location or intensity. As described in the preceding section, the thalamus also plays an important role in motor control by positively reinforcing voluntary motor behavior initiated by the cortex.

■ The hypothalamus regulates many homeostatic functions.

The **hypothalamus** is a collection of specific nuclei and associated fibers that lie beneath the thalamus. It is an integrating center for many important homeostatic functions and serves as an important link between the autonomic nervous system and the endocrine system. Specifically, the hypothalamus (1) controls body temperature (although this ability varies considerably among vertebrates, as you will see in Chapter 15); (2) controls thirst and urine output; (3) controls food intake; (4) controls uterine contractions and milk ejection in mammals; (5) serves as a major autonomic nervous system coordinating center, which in turn affects all smooth muscle, cardiac muscle, and exocrine glands; and (6) plays a role in emotional and behavioral patterns. Some of these functions are accomplished by controlling anterior pituitary hormone secretion and by producing posterior pituitary hormones.

The hypothalamus (along with the brain stem) is the area of the brain most notably involved in regulating the direct homeostasis of the internal environment. For example, when

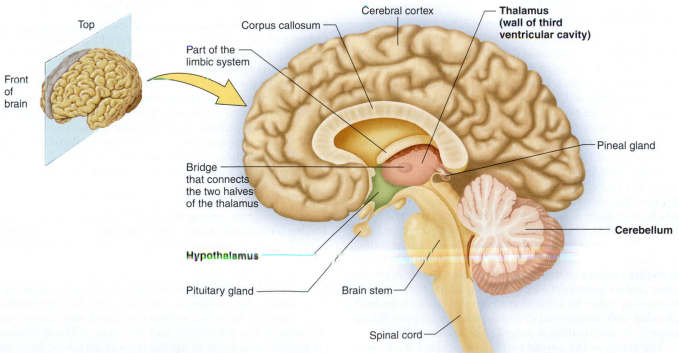

Figure 5–24 ● Location of the thalamus, hypothalamus, and cerebellum in sagittal section.

a mammalian body is cold, its hypothalamus initiates internal responses to increase heat production (such as shivering) and to decrease heat loss (such as constricting the skin blood vessels to reduce the flow of warm blood to the body surface, where heat could be lost to the external environment). Other areas of the brain, such as the cerebral cortex, act more indirectly to regulate the internal environment. For example, a cat that feels cold is motivated to seek a warm, sunny place to nap. Even these voluntary behavioral activities are strongly influenced by the hypothalamus, which, as a part of the *limbic system* (see the next section), functions in conjunction with the cortex in controlling emotions and motivated behavior.

The hypothalamus represents an intermediate level in the distributed hierarchy of neural control and is considered a reflex integrator. That is, it receives input from various internal sensors and sends out effector commands with fewer synapses than would be required if the cerebrum were involved. The hypothalamus is also physically closer to body effectors than is the cerebrum.

■ The limbic system plays a key role in animal motivation.

The **limbic system** is not a separate structure but refers to a ring of forebrain structures that surround the brain stem and are interconnected by intricate neuronal pathways (● Figure 5–25). The limbic system includes portions of each of the following: the lobes of the cerebral cortex, the basal nuclei, the thalamus, and the hypothalamus (although in birds the limbic system remains poorly defined). This complex interacting network is associated with emotions, basic survival and sociosexual behavioral patterns, motivation, and learning.

■ The amygdala processes inputs that give rise to the sensation of fear.

The **amygdala** of mammalian forebrain is located on the interior underside of the temporal lobe. It is believed to be the homologue of the reptilian and avian **paleostriatum**. It is an especially important region for processing inputs that give rise to the sensation of fear, a sense that normally helps animals avoid danger. Research studies have shown that the amygdala is the region where the link between an unconditioned stimulus and the conditioned stimulus is formed (see p. 189). For example, when a rat hears a sound and receives a shock, there is a pronounced increase in the firing rate of neurons in the amygdala. This leads to a strengthening of synaptic connections in this region, which, once laid down, are retained for the lifetime of the animal. The amygdala, in turn, activates the animal's fight-or-flight stress system that is controlled by the hypothalamus (p. 291), in anticipation of strenuous escape (or fight) activity. Thus the next time the rat's ear receives that sound, the amygdala gears up the body for fight-or-flight even before the higher sensory centers "perceive" the sound. After the higher centers process the information, actual fight-or-flight behavior can either be initiated at a high rate (because the body is flooded with oxygen and glucose), or it can suppress the response if the information is judged to be nondangerous.

Normally, in the amygdala the circuits that learn these hard-wired *emotional memories* are held in check by the release of the inhibitory neurotransmitter GABA (p. 125), so that fight-or-flight responses are activated only by the strongest danger

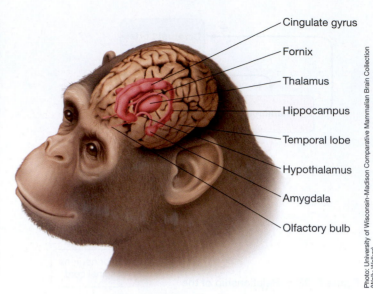

Figure 5–25 ● **Limbic system.** This partially transparent view of the brain reveals the structures composing the limbic system.

(*Source:* brainmuseum.org)

Photo: University of Wisconsin-Madison Comparative Mammalian Brain Collection (Wally Welker)

Labels: Cingulate gyrus, Fornix, Thalamus, Hippocampus, Temporal lobe, Hypothalamus, Amygdala, Olfactory bulb

signals. GABA is believed to permit the amygdala to filter out unthreatening stimuli. If, however, the amygdala fails to adequately filter enough incoming information, and, like a leaky valve, permits the entry of too much information, even mildly frightening stimuli can be perceived as terrifying. This syndrome is seen in overly anxious humans and domestic mammals.

■ The limbic system and higher cortex participate in the control of basic behavioral patterns.

Behavioral patterns controlled at least in part by the limbic system include those aimed at survival of the animal (attack, searching for food) and those directed toward perpetuation of the species (sociosexual behaviors conducive to mating). In experimental mammals, stimulation of the limbic system brings about complex and even bizarre behaviors. For example, stimulation in one area can elicit responses of anger and rage in a normally docile animal, whereas stimulation in another area results in placidity and tameness, even in an otherwise vicious animal. Stimulation in yet another limbic area can induce sexual behaviors such as copulatory movements in rodents.

Role of the Hypothalamus in Basic Behavioral Patterns

The relationships among the hypothalamus, limbic system, and higher cortical regions regarding emotions and behavior are still not well understood. It appears that the extensive involvement of the hypothalamus in the limbic system is responsible for the involuntary internal responses of various body systems in preparation for appropriate action to accompany a particular emotional state. For example, the hypothalamus controls the increased heart rate and respiratory rate, elevation of blood pressure, and diversion of blood to skeletal muscles that occur in anticipation of attack or when angered. These preparatory changes in internal state require no conscious control, and as you have seen, emotional memories in the amygdala can trigger them via the hypothalamus.

Role of the Higher Cortex in Basic Behavioral Patterns

In executing complex behavioral activities such as attack, flight, or mating, an animal must interact with the external environment. Higher cortical mechanisms are called into play to connect the limbic system and hypothalamus with the outer world so that appropriate overt behaviors are manifested. At the simplest level, the cortex provides the neural mechanisms necessary for implementing the appropriate skeletal muscle activity required to approach or avoid an adversary, participate in sexual activity (mating display), or display emotional expression. For example, the stereotypic sequence of movement for the universal human emotional expression of smiling is apparently preprogrammed in the cortex and can be called forth by the limbic system. People can also voluntarily call forth the smile program, as when posing for a picture. Even individuals blind from birth have normal facial expressions; that is, they do not need to learn to smile by observation. Smiling means the same thing in every culture, despite widely differing environmental experiences. Thus smiling is an *innate* (genetically determined) behavior. In contrast, so-called smiling in chimpanzees is actually an innate, limbic-controlled show of fear; but it too can be voluntarily initiated, as chimpanzees in human entertainment shows are trained to do, to make human viewers believe (falsely) that the chimpanzee is happy or amused. Such behavior patterns shared by all members of a species are believed to be more abundant in lower animals.

Reward and Punishment Centers

An animal tends to reinforce behaviors that have proved gratifying and to suppress behaviors that have been associated with unpleasant experiences. Certain regions of the limbic system have been designated as "reward" and "punishment" centers, because stimulation in these respective areas gives rise to pleasant or unpleasant sensations in humans, and responses that appear to evoke similar sensations in other mammals. When a self-stimulating device is implanted in a reward center, an experimental rodent will self-deliver up to 5000 stimulations per hour and will even shun food when starving, in preference for the pleasure derived from self-stimulation. In contrast, when the device is implanted in a punishment center animals will avoid stimulation at all costs. Reward centers are found most abundantly in regions mediating the highly motivated activities of eating, drinking, and sexual activity.

■ Motivated behaviors are goal directed.

Motivation is the ability to direct behavior toward specific goals. The concept of animal motivation encompasses many factors, which include subjective emotional feelings and moods (such as rage and fear) that only human subjects can report, plus the overt physical responses that occur in association with these feelings. Animals must continually evaluate the value or worth of a particular stimulus and make decisions weighing the beneficial versus the painful consequences of their choices. These responses include specific behavioral patterns, for example, preparation for attack or defense when approached by an adversary. Evidence (from stimulating or inactivating various brain regions) points to a central role for the brain stem and the limbic system in all aspects of emotion. Some goal-directed behaviors are aimed at satisfying specific identifiable physical needs related to homeostasis.

Homeostatic drives represent the subjective urges associated with specific bodily needs that motivate appropriate behavior to satisfy those needs. As an example, the sensation of thirst accompanying a water deficit in the body drives an animal to drink to satisfy the homeostatic need for water. Animal behavior is influenced by experience, learning and habit, shaped in a complex framework of unique personal gratifications.

■ Norepinephrine, dopamine, and serotonin are neurotransmitters in pathways for emotion and behavior.

The underlying neurophysiological mechanisms responsible for the psychological observations of motivated behavior and emotions largely remain a mystery, although the neurotransmitters **norepinephrine, dopamine,** and **serotonin** all have been implicated. Norepinephrine and dopamine, both chemically classified as *catecholamines,* are known to be transmitters in the regions that elicit the highest rates of self-stimulation in animals equipped with self-administering devices. Imbalances in these are associated with various mental disorders in humans, and possibly other mammals. For example, a functional deficiency of serotonin or norepinephrine or both is implicated in **depression,** a disorder characterized by a pervasive unpleasant mood accompanied by a generalized loss of interests and inability to experience pleasure. All effective antidepressant drugs increase the available concentration of these neurotransmitters in the CNS. **Prozac,** the most widely prescribed drug in American psychotherapy, is illustrative. It blocks the reuptake transporters (p. 127) for released serotonin, thus prolonging serotonin activity at synapses. It is often prescribed for humans and companion animals suffering from certain forms of depression and other behavior disorders (see the box "Concepts and Controversies: Synaptic Solutions to Pet Problems," in Chapter 4, p. 135). Researchers are optimistic that as our understanding of the molecular mechanisms of mental disorders is expanded in the future, many psychological problems can be corrected or managed through drug intervention.

Cerebellum, Brain, and Spinal Cord

The **cerebellum,** which is attached to the back of the upper portion of the brain stem, lies behind the occipital lobe of the cortex (Table 5–8 and Figures 5–16b and 5–24). The cerebellum functions as an integrative portion of the brain that consists of circuitry that is functionally similar in all classes of vertebrates that have this region. The surface of the cerebellum is smooth in lower vertebrates but becomes increasingly convoluted in higher vertebrates, increasing available surface area for additional neurons (as in the cerebrum).

■ The cerebellum is important in balance as well as in planning and execution of voluntary movement.

In birds and mammals the cerebellum consists of three functionally distinct parts, which are believed to have developed successively during evolution (● Figure 5–26). These parts have different sets of inputs and outputs, and, accordingly, each

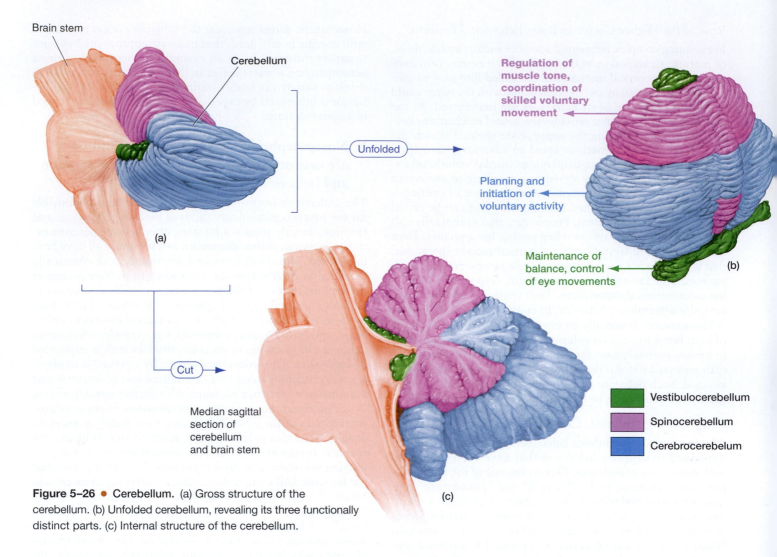

Figure 5–26 ● **Cerebellum.** (a) Gross structure of the cerebellum. (b) Unfolded cerebellum, revealing its three functionally distinct parts. (c) Internal structure of the cerebellum.

has different functions, although collectively they are concerned primarily with subconscious control of motor activity.

■ The **vestibulocerebellum** is important for the maintenance of balance and controls eye movement.

■ The **cerebrocerebellum** plays a role in planning and initiating voluntary activity by providing input to the cortical motor areas. This is also the cerebellar region involved in *procedural memories* (which we discuss later, p. 188).

■ The **spinocerebellum** enhances muscle tone and coordinates skilled, voluntary movements. This brain region is especially important in ensuring the accurate timing of various muscle contractions to coordinate movements involving multiple joints. This region also receives input from peripheral receptors that appraise it of the body movements and positions that are actually taking place. The spinocerebellum essentially acts as "middle management," comparing the "intentions" or "orders" of the higher centers (especially the **motor cortex**) with the "performance" of the muscles and then correcting any "errors" or deviations from the intended movement (● Figure 5–27). The spinocerebellum even appears able to predict the position of a body part in the next fraction of a second and to make adjustments accordingly.

?

Is the Cerebellum a "Smith Predictor"? The cerebellum has long been known to be important for smooth, coordinated muscle movements, but scientists still do not entirely understand how it does this. One hypothesis that explains its actions well is that this brain region is a biological version of a Smith Predictor. Smith Predictors originated in industry for situations with long feedback delays, such as a grasping robotic arm in a factory that crushes a soft object before feedback sensors can inform the controller that the object is soft. The Smith Predictor has an "internal model" of the action it is about to take that allows it to initiate certain motions *before* feedback information can inform whether it is correct or not. The cerebellum seems to act this way, particularly once it has learned a skilled motor task. If you are reaching for a pencil, for example, the spinocerebellum "puts on the brakes" soon enough to stop the forward movement of your hand at the intended location rather than allowing you to overshoot your target. Neural recordings suggest that it makes such smoothing, coordinating commands even in the absence of feedback from sensors. These ongoing *anticipatory* adjustments are especially important for rapidly changing activities such as climbing, swinging through trees, and running, in which delays in sending

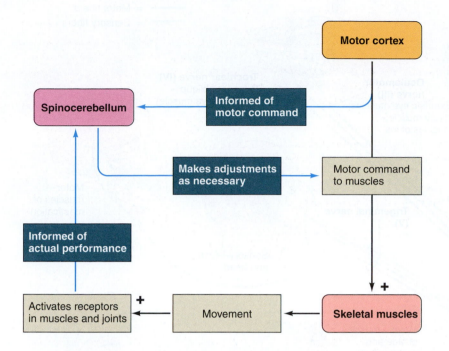

Figure 5–27 ● **Role of the spinocerebellum in subconscious control of voluntary motor activity.** In coordinating rapid, phasic motor activity, the spinocerebellum compares the "intentions" of higher motor centers with the "performance" of the muscles and corrects any "errors" by making the necessary adjustments to accomplish the intended movement.

feedback information would make the action clumsy. See the Suggested Reading by Miall, Weir, Wolpert, and Stein (1993), at the end of this chapter.

The cerebellum and basal nuclei both monitor and adjust motor activity commanded from the motor cortex, and like the basal nuclei, the cerebellum does not have any direct influence on the efferent motor neurons. Both function indirectly by modifying the output of major motor systems of the brain. Even though the cerebellum and basal nuclei both help to subconsciously coordinate skeletal motor activity, they play different roles. The cerebellum helps maintain balance; helps coordinate fast, phasic motor activity; and enhances muscle tone. The basal nuclei are important in coordinating slow, sustained movement related to posture and support, and they also function in inhibiting muscle tone.

The motor command for a particular voluntary activity arises from the motor cortex, but coordination of the actual execution of that activity is accomplished subconsciously by these subcortical regions. You will learn more about motor control when we cover muscle physiology in Chapter 8. For now we are going to move on to the remaining part of the brain, the brain stem.

■ The brain stem is a critical connecting link between the remainder of the brain and the spinal cord

The brain stem is a vital link between the spinal cord and the higher brain regions. The **brain stem,** which consists of the **medulla, pons,** and **midbrain** (see Table 5–8), is a critical connecting link between the remainder of the brain and the spinal cord. All incoming and outgoing fibers traversing between the periphery and higher brain centers must pass through the brain stem (with exception of the olfactory and optic nerves), with incoming fibers relaying sensory information to the brain and outgoing fibers carrying command signals from the brain for

efferent output. A few fibers merely pass through, but most synapse within the brain stem for important processing. The functions of the brain stem include the following:

1. *Sensation input and motor output in the head and neck via cranial nerves.* Twelve pairs of **cranial nerves** arise from the brain stem (● Figure 5–28); most supply structures in the head and neck with both sensory and motor fibers. They are important in sight, hearing, taste, smell, sensation of the face and scalp, eye movement, chewing, swallowing, facial expressions, and salivation. An exception is cranial nerve X, the **vagus nerve,** which innervates organs in the thoracic and abdominal cavities. The vagus is the major nerve of the parasympathetic nervous system.

2. *Reflex control of heart, blood vessels, respiration, and digestion.* Collected within the brain stem are integrating neuronal clusters, or "centers," that control these functions primarily for homeostasis.

3. *Modulating the sense of pain* (see p. 242).

4. *Regulation of muscle reflexes involved in equilibrium and posture.*

5. *Receives and integrates all synaptic input via the reticular formation.* A widespread network of interconnected neurons called the **reticular formation** runs throughout the entire brain stem and into the thalamus. Ascending fibers originating in the reticular formation carry signals upward to arouse and activate the cerebral cortex (● Figure 5–29). These fibers compose the **reticular activating system (RAS),** which controls the overall degree of cortical alertness and is important in the ability to direct attention. In turn, fibers descending from the cortex, especially its motor areas, can activate the RAS.

6. The centers responsible for sleep traditionally have been considered to be housed within the brain stem, although recent evidence suggests that the sleep-promoting centers may be located in the hypothalamus.

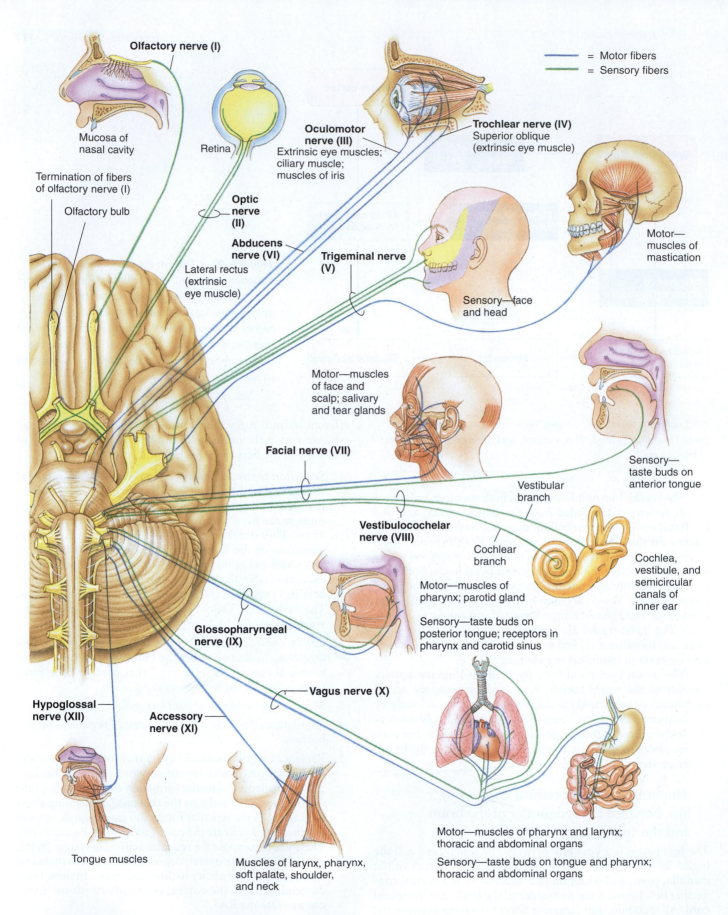

= Motor fibers
= Sensory fibers

Olfactory nerve (I)

Mucosa of nasal cavity

Termination of fibers of olfactory nerve (I)

Olfactory bulb

Retina

Optic nerve (II)

Abducens nerve (VI)

Lateral rectus (extrinsic eye muscle)

Oculomotor nerve (III)
Extrinsic eye muscles; ciliary muscle; muscles of iris

Trochlear nerve (IV)
Superior oblique (extrinsic eye muscle)

Trigeminal nerve (V)

Sensory—face and head

Motor—muscles of mastication

Motor—muscles of face and scalp; salivary and tear glands

Facial nerve (VII)

Sensory—taste buds on anterior tongue

Vestibular branch

Cochlear branch

Vestibulocochelar nerve (VIII)

Cochlea, vestibule, and semicircular canals of inner ear

Motor—muscles of pharynx; parotid gland

Sensory—taste buds on posterior tongue; receptors in pharynx and carotid sinus

Glossopharyngeal nerve (IX)

Vagus nerve (X)

Hypoglossal nerve (XII)

Accessory nerve (XI)

Tongue muscles

Muscles of larynx, pharynx, soft palate, shoulder, and neck

Motor—muscles of pharynx and larynx; thoracic and abdominal organs

Sensory—taste buds on tongue and pharynx; thoracic and abdominal organs

Figure 5–28 ● Cranial nerves. Inferior (underside) view of the human brain, showing the attachments of the 12 pairs of cranial nerves to the brain and many of the structures innervated by those nerves.

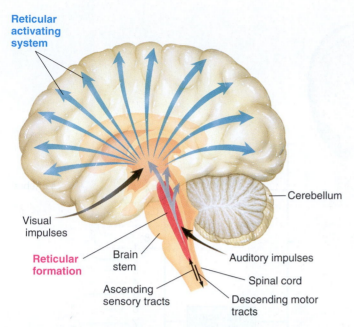

Reticular
activating
system

Cerebellum

Visual
impulses

**Reticular
formation**

Brain
stem

Auditory impulses

Spinal cord

Ascending
sensory tracts

Descending motor
tracts

Figure 5–29 • **The reticular activating system (RAS).** The reticular formation, a widespread network of neurons within the brain stem (in red), receives and integrates all synaptic input. The reticular activating system, which promotes cortical alertness and helps direct attention toward specific events, consists of ascending fibers (in blue) that originate in the reticular formation and carry signals upward to arouse and activate the cerebral cortex.

◼ The spinal cord retains an inherent segmental organization characteristic of invertebrates.

The **spinal cord** is a long, slender cylinder of nerve tissue that extends from the brain stem. Comparable to the organization of segmented ganglia in some invertebrates, the spinal cord and its associated nerves have retained a fundamental segmental organization. In humans, it is about 45 cm long (18 inches) and 2 cm in diameter (about the size of your thumb). Exiting through a large hole in the base of the skull, the spinal cord is enclosed by the protective vertebral column as it extends posteriorly through the vertebral canal. Paired *spinal nerves* emerge from the spinal cord through spaces formed between the bony, winglike arches of adjacent vertebrae (• Figure 5–30). The actual number of spinal cord segments differs between species, with elongated animals having a greater number of spinal cord segments, whereas short animals have fewer.

The spinal nerves are named according to the region of the vertebral column from which they emerge (• Figure 5–31): Humans have eight pairs of **cervical** (neck) nerves (that is, C1–C8), 12 **thoracic** (chest) nerves, 5 **lumbar** (abdominal) nerves, 5 **sacral** (pelvic) nerves, and 1 **coccygeal** (tailbone) nerve. The spinal cord extends only to the level of the first or second lumbar vertebra (about waist level in humans), so the nerve roots of the remaining nerves are greatly elongated in order to exit the vertebral column at their appropriate space. The thick bundle of elongated nerve roots within the lower vertebral canal is known as the **cauda equina** ("horse's tail") because of its appearance (Figure 5–31). Thoracic and lumbar regions of the avian spinal cord are highly fused, because of their adaptation for flight. For this reason individual vertebrae and their associated ganglia are not numbered.

Figure 5–30 • Location of the spinal cord relative to the vertebral column.

(*Source:* Adapted from Cecie Starr and Ralph Taggart, *Biology: The Unity and Diversity of Life,* Eighth Edition, Fig. 35.9a, p. 577. Copyright 1998 Wadsworth Publishing Company.)

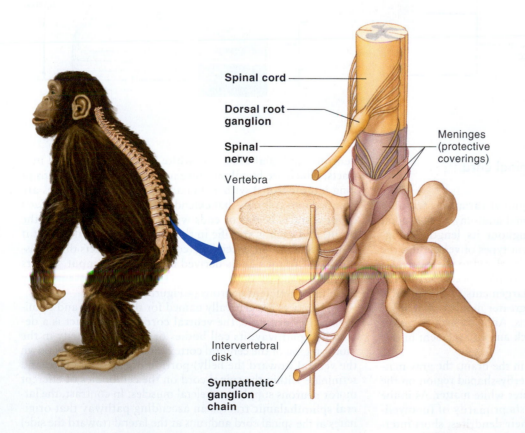

Spinal cord

Dorsal root ganglion

Spinal nerve

Vertebra

Meninges (protective coverings)

Intervertebral disk

Sympathetic ganglion chain

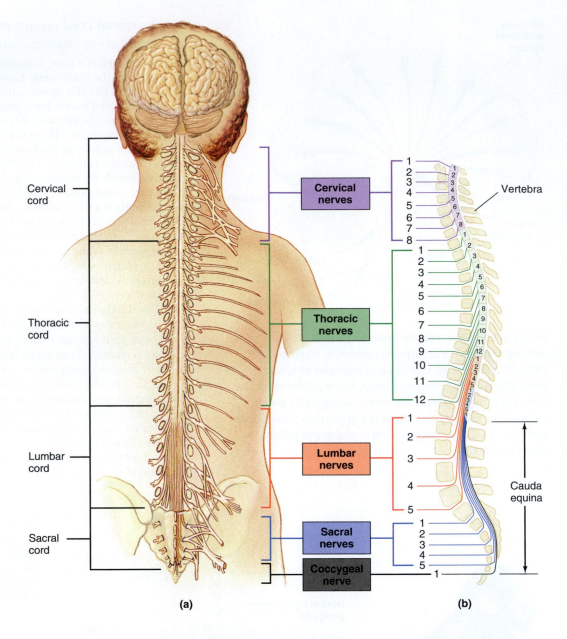

Figure 5–31 • Spinal nerves. There are 31 pairs of spinal nerves named according to the region of the vertebral column from which they emerge. Because the spinal cord is shorter than the vertebral column, spinal nerve roots must descend along the cord before emerging from the vertebral column at the corresponding intervertebral space, especially those beyond the level of the first lumbar vertebra (L1). Collectively these rootlets are called the cauda equina, literally "horse's tail." (a) Posterior view of the human brain, spinal cord, and spinal nerves (on the right side only). (b) lateral view of the spinal cord and spinal nerves emerging from the vertebral column.

Cervical cord

Thoracic cord

Lumbar cord

Sacral cord

Cervical nerves

Thoracic nerves

Lumbar nerves

Sacral nerves

Coccygeal nerve

Vertebra

Cauda equina

(a)   (b)

The white matter of the spinal cord is organized into tracts.

Although there are some slight regional variations within a particular species, the cross-sectional anatomy of the spinal cord is generally the same throughout its length (• Figure 5–32). However, among different types of vertebrates the size and number of cord segments is related to the particular locomotory strategies used. For example, species with well-developed limbs show distinct enlargements of the thoracic and lumbar regions, reflecting the greater number of neurons within the spinal cord at these levels. Animals such as snakes that completely lack limbs also lack any enlargement in this region of the spinal cord.

In contrast to the gray matter in the brain, the gray matter in the spinal cord forms a butterfly-shaped region on the inside and is surrounded by the outer white matter. As in the brain, the cord gray matter consists primarily of (unmyelinated) neuronal cell bodies and their dendrites, short inter-

neurons, and glial cells. The white matter is organized into **tracts,** which are bundles of myelinated nerve fibers (axons of long interneurons) with a similar function. The bundles are grouped into columns that extend the length of the cord. Each of these tracts begins or ends within a particular area of the brain, and each is specific in the type of information that it transmits. Some are **ascending** (cord to brain) **tracts** that transmit to the brain signals derived from afferent input. Others are **descending** (brain to cord) **tracts** that relay messages from the brain to efferent neurons (• Figure 5–33).

The tracts are generally named for their origin and termination. For example, the **ventral corticospinal** tract is a descending pathway; its cell bodies originate primarily in the motor region of the cerebral cortex, and its axons travel down the ventral (toward the belly) portion of the spinal cord and terminate within the spinal cord on the cell bodies of efferent motor neurons supplying skeletal muscles. In contrast, the **lateral spinothalamic** tract is an ascending pathway that originates in the spinal cord and runs at the lateral (toward the side)

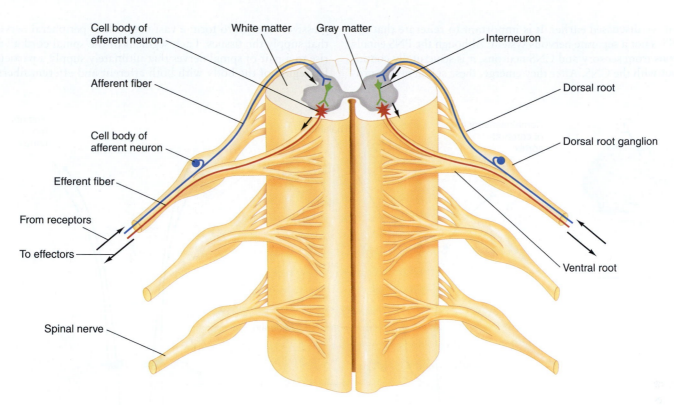

Cell body of efferent neuron
White matter
Gray matter
Interneuron
Afferent fiber
Dorsal root
Cell body of afferent neuron
Dorsal root ganglion
Efferent fiber
From receptors
To effectors
Ventral root
Spinal nerve

Figure 5–32 • **Spinal cord in cross section.** Schematic representation of the spinal cord in cross section showing the relationship between the spinal cord and spinal nerves. The afferent fibers through the dorsal root, and the efferent fibers exit through the ventral root. Afferent and efferent fibers are enclosed together within a spinal nerve.

margin of the cord until it synapses in the thalamus. Sensory information about pain and temperature that has been delivered to the spinal cord by afferent input from various regions of the body is carried within the spinothalamic tract up the spinal cord to the thalamus, which in turn sorts and relays the information to the somatosensory cortex. Because various types of signals are carried in different tracts within the spinal cord, damage to particular areas of the cord can interfere with some functions while other functions remain intact.

■ Each horn of the spinal cord gray matter houses a different type of neuronal cell body.

The centrally located gray matter is also functionally organized (● Figure 5–34). The central canal, which is filled with CSF, lies in the center of the gray matter. Each half of the gray matter is divided into a **dorsal (posterior)** (toward the back) horn, a **ventral (anterior) horn,** and a **lateral horn.** The dorsal horn contains cell bodies of interneurons on which afferent neurons terminate. The ventral horn contains cell bodies of the efferent motor neurons supplying skeletal muscles. Autonomic nerve fibers supplying cardiac and smooth muscle and exocrine glands originate at cell bodies found in the lateral horn.

■ Spinal nerves contain both afferent and efferent fibers.

Spinal nerves (bundles of axons) connect with each side of the spinal cord by a **dorsal root** and a **ventral root** (Figure 5–32).

Afferent fibers carrying incoming signals enter the spinal cord through the dorsal root; efferent fibers carrying outgoing signals leave through the ventral root. The cell bodies for the afferent neurons at each level are clustered together in a **dorsal root ganglion.** (A collection of neuronal cell bodies or a group of interneurons located outside of the CNS is called a **ganglion,** whereas a functional collection of cell bodies within the CNS is referred to as a **center** or a **nucleus.**) The cell bodies for the efferent neurons originate in the gray matter and send axons out through the ventral root.

The dorsal and ventral roots at each level join to form a **spinal nerve** that emerges from the vertebral column (Figure 5–32). A spinal nerve contains both afferent and efferent fibers traversing between a particular region of the body and the spinal cord. Note the difference between a **nerve** and a **neuron.** A nerve is a bundle of peripheral neuronal axons, some afferent and some efferent, enclosed by a connective tissue covering and following the same pathway (● Figure 5–31). A nerve does not contain a complete nerve cell, only the axonal portions of many neurons. (By this definition, there are no nerves in the CNS! Bundles of axons in the CNS are called **tracts.**) The individual fibers within a nerve generally do not have any direct influence on each other. They travel together for convenience, just as many individual telephone lines are carried within a telephone cable, yet any particular connection can be private without interference or influence from other lines in the cable.

The spinal nerves, along with the cranial nerves that arise from the brain, constitute the peripheral nervous system (PNS) that we discussed earlier. It is important to reiterate that the PNS is not a separate nervous system: Although the PNS entails axons from sensory and CNS neurons, it is structurally continuous with the CNS. After they emerge, these spinal nerves progressively branch to form a vast network of peripheral nerves that supply the tissues. Each segment of the spinal cord gives rise to a pair of spinal nerves that ultimately supply a particular region of the body with both afferent and efferent fibers.

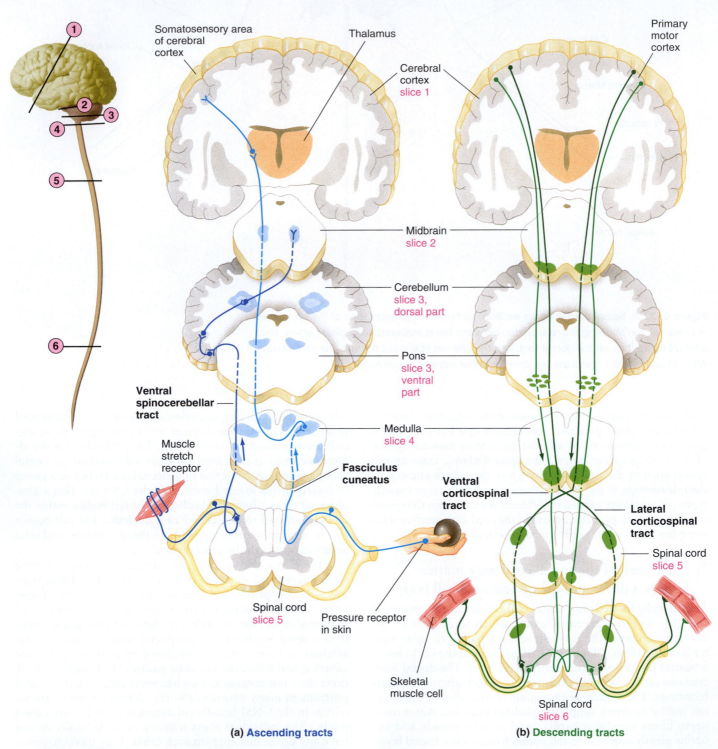

(a) Ascending tracts

(b) Descending tracts

Figure 5–33 • **Ascending and descending tracts in the white matter of the spinal cord.** (a) Cord-to-brain pathways of several ascending tracts (fasciculus cuneatus and ventral spinocerebellar tract). (b) Brain-to-cord pathways of several descending tracts (lateral corticospinal and ventral corticospinal tracts).

(*Source:* Adapted from Cecie Starr and Ralph Taggart, *Biology: The Unity and Diversity of Life*, Eighth Edition, Fig. 34.10, p. 566. Copyright 1998 Wadsworth Publishing Company.)

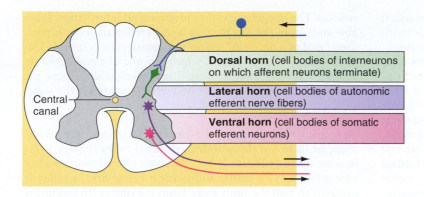

Figure 5–34 ● Regions of the gray matter.

With reference to sensory input, each specific region of the body surface supplied by a particular spinal nerve is called a **dermatome**. These same spinal nerves also carry fibers that branch off to supply internal organs, and sometimes pain originating from one of these organs is "referred" (that is, perceived as arising from the wrong location) to the corresponding dermatome supplied by the same spinal nerve. **Referred pain** originating in the heart, for example, may appear to come from the left shoulder and arm in humans or from the withers region (on the back at the base of the neck) in a ruminant. Pain appears to arise from the withers in response to the penetration of a nail, for example, through the ruminant gut into the pericardium (p. 376), resulting in pericardial inflammation called **traumatic pericarditis**. This is relatively common in cattle allowed to graze in old pastures. The mechanism responsible for referred pain is not completely understood. Presumably, inputs arising from the heart share a pathway to the brain in common with inputs from the upper extremities. The higher perception levels, being more accustomed to receiving sensory input from the withers than from the heart,

may interpret the pain input from the heart as having arisen from the withers.

◼ Many reflex responses and patterned movements in vertebrates are integrated in the spinal cord.

The spinal cord is strategically located between the brain and afferent and efferent fibers of the peripheral nervous system; this location enables the spinal cord to fulfill its two primary functions: (1) serving as a link for transmission of information between the brain and the remainder of the body and (2) integrating reflex activity between afferent input and efferent output without involving the brain. This type of reflex activity is known as a **spinal reflex**. Often neuronal connections that produce patterned or stereotyped movements find their origins in the spinal cord. For example, a rudimentary walking rhythm is retained in chickens in the absence of any input from the brain. In fishes, the spinal cord can mediate near-normal swimming and visceral functions. Although the brain can over-

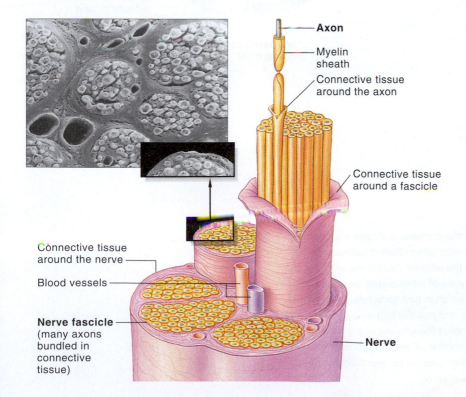

Figure 5–35 ● Structure of a nerve. Diagrammatic view of a nerve, showing neuronal axons (both afferent and efferent fibers) bundled together into connective-tissue–wrapped fascicles. A nerve consists of a group of fascicles enclosed by a connective-tissue covering and following the same pathway. The photograph is a scanning electron micrograph of several nerve fascicles in cross section.

ride specific behavioral patterns that originate from the spinal cord, the complexity of the spinal neuron circuitry is sufficient to generate these complex and coordinated behaviors.

As you saw at the beginning of this chapter, a **reflex** is an innate response that occurs without higher neural functions (anticipation and learning), such as pulling a limb away from a hot object (Figure 5–1e) or the scratch reflex so elegantly performed by a pet dog. Some physiologists suggest that some responses that are acquired, such as hunting or flying skills, should be called conditioned reflexes, as long as they become "automatic." However, this is misleading, because these learned reactions usually involve higher brain regions such as the motor cortex and cerebellum, and this suggestion ignores the importance of hierarchic control in neural evolution. In some cases, though, some true reflexes through the spine, brain stem, or hypothalamus may improve through *potentiated* synapses, in which case these can be considered true acquired reflexes. We explore potentiation later in the chapter.

Recall that the neural pathway involved in accomplishing reflex activity is the **reflex arc** (Figure 5–1). The spinal cord and brain stem are responsible for integrating basic reflexes, whereas higher brain levels usually process so-called acquired

reflexes. The integrating center processes all information available to it from the receptor as well as from all other inputs, then "makes a decision" about the appropriate response. Unlike higher-level behaviors, in which any one of a number of responses is possible, a reflex response is predictable because the pathway between the receptor and effector is always the same.

Withdrawal Reflex

Now that you have learned the details of vertebrate nervous systems, let us revisit the withdrawal spinal reflex (Figure 5–1e) in more detail (● Figure 5–36). When an animal touches a hot object (such as an ember following a forest fire), a reflex is initiated to pull the limb away from the object (to withdraw from the painful stimulus). The skin has different receptors for warmth, cold, light touch, pressure, and pain. Even though all information is sent to the CNS by way of action potentials, the CNS can distinguish between various stimuli because specific receptors and consequently different afferent pathways are activated by different stimuli. When a receptor is stimulated sufficiently to reach threshold, an action potential is generated in the afferent neuron. The stronger the stimulus, the greater the frequency of action potentials generated and

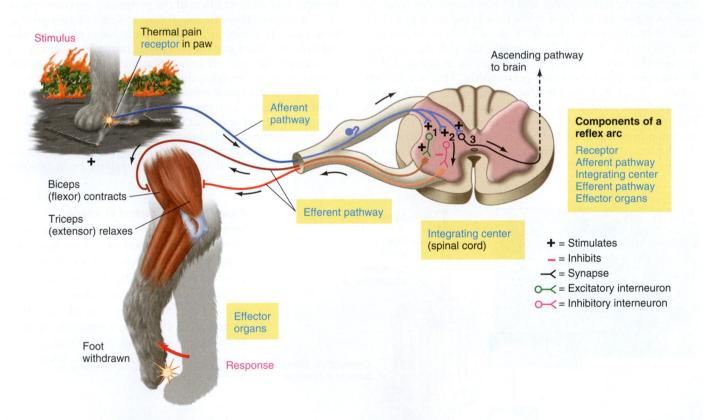

Figure 5–36 ● **The withdrawal reflex.** When a painful stimulus activates a receptor in the paw, action potentials are generated in the corresponding afferent pathway, which propagates the electrical signals to the CNS. Once the afferent neuron enters the spinal cord, it diverges and terminates on three different types of interneurons (only one of each type is depicted): (1) excitatory interneurons, which in turn stimulate the efferent motor neurons to the biceps, causing the leg to flex and pull the foot away from the painful stimulus; (2) inhibitory interneurons, which inhibit the efferent motor neurons to the triceps, thus preventing counterproductive contraction of this antagonistic muscle; and (3) interneurons that carry the signal up the spinal cord via an ascending pathway to the brain for awareness of pain, memory storage, and so on.

propagated to the CNS. Once the afferent neuron enters the spinal cord, it diverges to synapse with the following different interneurons (the numbers correspond to those on Figure 5–36).

1. An excited afferent neuron stimulates excitatory interneurons that in turn stimulate the efferent motor neurons supplying the flexor, the muscle in the limb that bends the knee joint. The resultant contraction of the flexor pulls the foot away from the hot ember.

2. The afferent neuron also stimulates inhibitory interneurons that in turn inhibit the efferent neurons supplying the triceps to prevent it from contracting. The extensor is the muscle in the leg that straightens out the knee joint. When the flexor is contracting, it would be counterproductive for the extensor to be contracting. Therefore, built into the withdrawal reflex is inhibition of the muscle that antagonizes (opposes) the desired response. This type of neuronal connection involving stimulation of the nerve supply to one muscle and simultaneous inhibition of the nerves to its antagonistic muscle is known as **reciprocal innervation.**

3. The afferent neuron stimulates still other interneurons that carry the signal up the spinal cord to the brain via an ascending pathway. Only when the impulse reaches the sensory area of the cortex is the mammal aware of the pain, its location, and the type of stimulus. (Or at least we assume it is aware, based on their responses being similar to humans'.) Also, when the impulse reaches the brain, the information can be stored as memory, and other actions can be initiated. All this activity at the cortical level is above and beyond the basic reflex.

As is characteristic of all spinal reflexes, the brain can modify the withdrawal reflex. Impulses may be sent down descending pathways to the efferent neurons supplying the involved muscles to override the input from the receptors, actually preventing the biceps from contracting despite the painful stimulus. For example, when a young animal is suckling its mother's teat, pain receptors may be stimulated, initiating the withdrawal reflex. However, because the animal must be fed, in the mother IPSPs are sent via descending pathways to the motor neurons supplying the musculature of the leg. The activity in these efferent neurons depends on the sum of activity of all their synaptic inputs. Because the neurons supplying the muscles are now receiving more IPSPs from the brain (voluntary) than EPSPs from the afferent pain pathway (reflex), these neurons are inhibited and do not reach threshold. Therefore, the muscles are not stimulated to contract and the mother remains stationary. In this way, the withdrawal reflex has been overridden.

Stretch Reflex

Only one reflex is simpler than the withdrawal reflex: the **stretch reflex,** in which an afferent neuron originating at a stretch-detecting receptor in a skeletal muscle terminates directly on the efferent neuron supplying the same skeletal muscle to cause it to contract and counteract the stretch. This is the monosynaptic ("one synapse") reflex of Figure 5–1b, so named because the only synapse in the reflex arc is the one between the afferent neuron and the efferent neuron. The withdrawal reflex and all other reflexes are polysynaptic ("many synapses"), be-

cause interneurons are interposed in the reflex pathway and, therefore, a number of synapses are involved.

Other Reflex Activity

Spinal reflex action is not necessarily limited to motor responses on the side of the body to which the stimulus is applied. Assume an animal steps on a sharp rock fragment. The ensuing reflex response includes the same neurons as those used for the withdrawal reflex but also recruits neurons that control extension of the opposite limb. A reflex arc is initiated to withdraw the injured foot from the painful stimulus, while the opposite leg simultaneously prepares to suddenly bear the extra weight so that the animal does not lose balance (● Figure 5–37). Unimpeded bending of the injured extremity's knee is accomplished by concurrent reflex stimulation of the muscles that flex the knee and inhibition of the muscles that extend the knee. At the same time, unimpeded extension of the opposite limb's knee is accomplished by activation of pathways that cross over to the opposite side of the spinal cord to reflexly stimulate this knee's extensors and inhibit its flexors. This **crossed extensor reflex** ensures that the opposite limb will be in a position to bear the weight of the body as the injured limb is withdrawn from the stimulus.

Besides protective reflexes such as the withdrawal reflex and simple postural reflexes such as the crossed extensor reflex, basic spinal reflexes also mediate emptying of pelvic organs (for example, urination, defecation, and expulsion of semen).

One final point, concerning the term *reflex:* Not all reflex activity is based on neurons. Simple hormonal and paracrine responses can also be called *reflexes,* as you will see in later chapters.

Reflexes and Higher Regulation

Most spinal reflexes can be overridden at least temporarily by higher brain centers. As we discussed at the beginning of the chapter, this evolved for higher-level functions such as anticipation (activating a reflex before it is activated by its own sensors) and suppression in learned contexts (such as suppressing the urination reflex when urination is not desired, or, in humans, conscious breath holding while swimming). The mechanisms involved were illustrated in Chapter 4 (Figure 4–22).

■ Activity of command fibers elicits a fixed action pattern.

Activation of certain **command fibers** located within the CNS of both invertebrates and vertebrates can generate *coordinated behaviors of varying degrees of complexity, that is, fixed action patterns.* Fixed action patterns are stereotyped behaviors that are generated in response to specific stimuli and are often characteristic for a particular species or group of animals; and they involve more motions than the simple reflexes we have been considering. Because they are automatic and innate, these patterns are considered reflexes. For example, the giant **Mauthner neurons** (p. 122) in the brains of teleost fish generate the startle response, characterized by a vigorous muscle contraction on one side of the body, which work to move the body quickly out of harm's way. There are two "giant" Mauthner cells located on each side of the brain. Tapping the glass of an aquarium creates a mechanical disturbance in the water, which generates sensory input into the Mauthner cells. The neuron that first receives the signal sends inhibitory input to its counter-

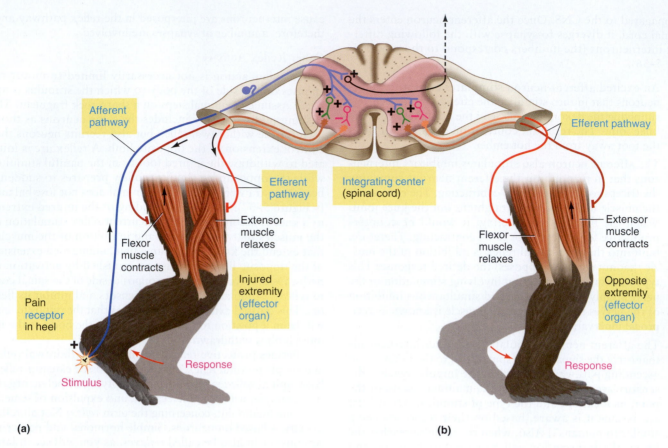

Figure 5–37 ● The crossed extensor reflex coupled with the withdrawal reflex.
(a) The withdrawal reflex, which causes flexion of the injured extremity to withdraw from a painful stimulus. (b) The crossed extensor reflex, which extends the opposite limb to support the full weight of the body.

part Mauthner cell while at the same time activating muscles on one side of the body, permitting the escape response to occur.

Fixed action patterns are widespread among animals, not just in vertebrates. Command fibers regulate a variety of activities in arthropods and mollusks including feeding, flight, walking and gill ventilation. For example, the crayfish tail flip (in which the tail is repeatedly flipped for rapid swimming) is illustrative of a rapid escape behavior. Because the tail flip is generated in specific giant neurons (medial giant and lateral giant neurons), these neurons are considered to function as command fibers. More complex behavior such as the patterned movements insects use in molting their old exoskeletons are also driven by command fibers. *Ecdysis behavior* of grasshoppers and crickets involves a series of synchronous muscular contractions in each of the body segments, which loosens the old exoskeleton. Subsequently, the large muscles of the body wall contract in a peristaltic fashion from posterior to anterior, which then forces the animal out of the exoskeleton.

Memory and Learning

As we have repeatedly noted, one reason for the evolution of a complex CNS is the ability to *learn*. **Learning** is the acquisition of abilities or knowledge as a consequence of experience, instruction, or both. It is widely believed that rewards and

punishments are integral parts of many types of learning, at least in higher vertebrates (and perhaps cephalopods). If an animal feels rewarded on responding in a particular way to a stimulus, the likelihood increases that it will respond in the same way again to the same stimulus as a consequence of this experience. Conversely, if a particular response is accompanied by punishment, the animal is less likely to repeat the same response to the same stimulus. When behavioral responses that give rise to pleasure are reinforced or those accompanied by punishment are avoided, learning has taken place. Housebreaking a puppy is an example. If the puppy is praised when it urinates outdoors but scolded when it wets the carpet, it soon learns the acceptable place to empty its bladder. Wild mammals, except those specifically marking territorial boundaries, have no such training experience, but they can learn about other aspects of their environments through reward (such as finding and remembering of a source of tasty food) and punishment (such as encountering and remembering the den of a predator). (Note again that the perception of "reward" and "punishment" are derived solely from human subjects; we infer similar feelings in some animals because of similar reactions.) Thus learning is a change in behavior that occurs as a result of experiences. It is highly dependent on the animal's interaction with its environment. The only limits to the effects that environmental influences can have on learning are the biological constraints imposed by species-specific and individual genetic endowments.

■ Memory comes in two forms—declarative and procedural—and is laid down in stages.

Memory is the storage of acquired knowledge or abilities for later recall. It is widely recognized that there are at least two categories of memory: (1) *declarative* or *explicit* memory—the learning of events, places, and so on; and (2) *procedural* or *implicit* memory—the learning of skilled motor movements. Learning and both types of memory form the basis by which animals adapt their behaviors to their particular external circumstances. Without these mechanisms, planning for successful interactions and intentional avoidance of predictably disagreeable circumstances would be impossible. Not all forms of learning provide well for this flexibility, however, as evidenced by the phenomenon of **imprinting**, in which newborns are programmed to learn that situations and objects encountered early in life are both normal and important for life. For example, many newly hatched birds are programmed to learn that the first moving object they encounter is a parent. Konrad Lorenz first demonstrated this programmed learning by exposing newly hatched goslings to a substitute mother. Rather than follow their biological parent, the goslings could be induced to follow a variety of inanimate objects or another animal, including a human. In birds and some species of mammals there is a narrowly defined period in which the nervous system is responsive to this form of learning. For example, foals exposed to a variety of handling and grooming procedures within 24 hours after birth are less fearful of these procedures later in life (such as the vibration and sound of clippers, and rhythmic palpation of both the inside and outside of the ears, nostrils, mouth, anus, genitals). If properly desensitized, these animals are more easily examined and treated by veterinarians and other professionals dealing with horses. Conversely, infant chimpanzees raised in highly stressful situations have hair-trigger fight-or-flight responses for the rest of their lives.

The neural change responsible for retention or storage of knowledge is known as the **memory trace**. Generally concepts, not *verbatim* information, seem to be stored. As you read this page, you are storing the concept discussed, not the specific words. Later, when you retrieve the concept from memory, you will convert it into your own words. It is possible, however, to memorize bits of information word by word. Storage of acquired information is accomplished in at least two stages: short-term memory and long-term memory (● Figure 5–38). **Short-term memory** lasts for seconds to hours, whereas **long-term memory** is retained for days to years. The process of transferring and fixing short-term memory traces into long-term memory stores is known as **consolidation**. Stored knowledge is of no use unless it can be retrieved and used to influence current or future behavior. A recently developed concept is that of a **working memory**, or what has been called "the blackboard of the mind." Working memory involves comparing current sensory data with relevant stored knowledge and manipulating that information, as in being able to plan a future action based on a current situation coupled with past experience.

Newly acquired information is initially deposited in short-term memory, which has a limited capacity for storage. Information in short-term memory has one of two eventual fates. Either it is soon forgotten (for example, forgetting a telephone number after you have looked it up and finished dialing), or it is transferred into the more permanent long-term memory mode through *active practice*. The recycling of newly acquired information through short-term memory increases the likelihood of long-term memory consolidation. (Therefore, when you cram for an exam, your long-term retention of the information is poor!) This relationship can be likened to developing photographic film. The originally developed image (short-term memory) will rapidly fade unless it is chemically fixed (consolidation) to provide a more enduring image (long-term memory). Sometimes only parts of memories are fixed, whereas others fade away. Information of interest or importance to the individual is more likely to be recycled and fixed in long-term stores, whereas less important information is quickly erased.

Short-term memory has very limited capacity; indeed, there is evidence that all vertebrates have only about 7 to 10 distinct short-term "registers." That is, a vertebrate brain cannot deal with more than 7 to 10 distinct items on a short-term basis (think about how many numbers you can remember briefly, as in a telephone number). Prey fish that school are

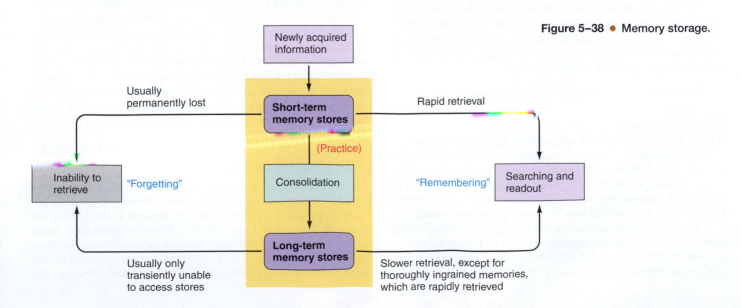

Figure 5–38 ● Memory storage.

thought to take advantage of this. A predator such as a barracuda approaching a school of prey herring may initially focus on one individual to pursue. But by initiating certain stereotypical maneuvers (such as the flash expansion, in which the fish move outward radially with each individual moving off at a unique angle from the predator), the barracuda's brain seems to become confused by having too many moving targets for its short-term memory to cope with.

The storage capacity of the long-term declarative memory bank is much larger than the capacity of short-term memory. Different informational aspects of long-term declarative memory traces seem to be processed and codified, then stored in conjunction with other memories of the same type; for example, visual memories are stored separately from auditory memories. This organization facilitates future searching of memory stores to retrieve desired information.

Because long-term memory stores are larger, it often takes longer to retrieve information from long-term declarative memory than from short-term memory. *Remembering* is the process of retrieving specific information from memory stores; forgetting is the inability to retrieve stored information. Information lost from short-term memory is permanently forgotten, but information in long-term storage is frequently forgotten only transiently. Forgetting short-term memory is important, to make room for new information. There is evidence that short-term memories are in fact actively erased in the sense that there are neuromodulators associated with forgetting. One type of neuromodulator, **anandimide** (also spelled *anandamide*), appears to reset short-term memory neurons to their unpotentiated state, particularly after some stressful events. (The active ingredient in marijuana, tetrahydrocannabinol, is an anandimide agonist, explaining why this drug interferes with short-term memory).

■ **Memory traces are present in multiple regions of the brain.**

What parts of the CNS govern memory? We here focus primarily on vertebrates but consider invertebrates for some aspects of memory later. There is no single "memory center" in the vertebrate brain. Instead, the neurons involved in memory traces are widely distributed throughout the subcortical and cortical regions of the brain. The regions of the brain implicated in memory include the hippocampus and other parts of the limbic system, and associated structures of the medial (inner) temporal lobes, the prefrontal cortex, the cerebellum and other regions of the cerebral cortex.

The Hippocampus and Declarative Memories

The **hippocampus,** the elongated, sea-horse–shaped medial portion of the temporal lobe that is part of the limbic system (Figure 5–25), plays a vital role in short-term memory involving the integration of various related stimuli and is also crucial for consolidation into long-term memory (*hippocampus,* "horse-headed sea monster"). The hippocampus is believed to store new long-term memories only temporarily and then transfer them to other cortical sites for more permanent storage.

Accessing and manipulating these long-term stores through operation of the working memory appear to be carried out by the prefrontal region of the cerebral cortex. The hippocampus and surrounding regions play an important role in **declar**ative memories—memories of specific facts and events that often result after only one experience.

The Cerebellum and Procedural Memories

In contrast to the role of the hippocampus and nearby forebrain regions in declarative memories, the cerebellum and relevant cortical regions such as the primary motor cortex, somatosensory cortex, and visual processing areas play an essential role in the "how to" **procedural memories** involving motor skills gained through repetitive training. The cortical areas important for a given procedural memory are the specific motor of sensory systems engaged in performing the routine. As you saw earlier (p. 170), learned skilled motor behaviors may be initiated by higher brain centers (that is, the frontal lobes or premotor cortex), but after that they are taken over by the lower centers that have learned the detailed movements.

■ **Short-term and long-term memory involve different molecular mechanisms.**

The major orchestrator of working memory is the prefrontal association cortex. The prefrontal cortex not only serves as a temporary storage site for holding relevant data "online" but also is largely responsible for the so-called executive functions involving manipulation and integration of the information for planning, problem-solving, and organizing activities. The prefrontal cortex carries out these complex reasoning functions in cooperation with all the brain's sensory regions, which are linked to the prefrontal cortex through neural connections. Researchers have identified different storage bins in the primate prefrontal cortex, depending on the nature of the current relevant data. For example, working memory involving spatial cues is in a prefrontal location distinct from working memory involving auditory cues or cues about an object's appearance.

Another question besides the "where" of memory is the "how" of memory. Despite a vast amount of psychological data, only a few tantalizing scraps of physiological evidence concerning the cellular basis of memory traces are available. Obviously, some change must take place within the neural circuitry of the brain to account for the altered behavior that follows learning. A single memory does not typically reside in a single neuron but rather in changes in the pattern of signals transmitted across synapses within a neuronal network. Different mechanisms are responsible for short-term and long-term memory. Short-term memory involves transient modifications in the function of pre-existing synapses, such as a temporary alteration in the amount of neurotransmitter released in response to stimulation within affected nerve pathways. Long-term memory, in contrast, involves relatively permanent functional or structural changes between existing neurons in the brain. Let's look at each of these types of memory in more detail.

■ **Short-term memory involves transient changes in synaptic activity.**

To investigate short-term memory, we depart from vertebrates for the moment. Following the Krogh Principle (p. 3), ingenious experiments were undertaken by Eric Kandel using the sea slug *Aplysia* (● Figure 5–39). These animals have large neurons easily identified and recorded, relatively small nervous systems (only about 20,000 neurons), and simple behaviors with rudimentary learning, making them ideal subjects. The

Photo: © Herve Chaumeton
Agence Nature

Figure 5–39 ● Gill of a sea hare (*Aplysia*), one of the gastropods. The gill is withdrawn into the mantle cavity in response to a gentle tap on the siphon or mantle.

(*Source:* C. Starr & R. Taggart, 2004, *Biology: The Unity and Diversity of Life*, 10th ed., Belmont, CA: Brooks/Cole, Figure 40.4)

experiments have shown that two forms of short-term memory—habituation and sensitization—are due to modification of different membrane channel proteins in presynaptic terminals of specific afferent neurons. This modification in turn brings about changes in transmitter release.

These studies have exploited the defensive withdrawal reflexes of the external organs of the mantle cavity. This reflex is analogous to vertebrate defensive escape and withdrawal responses. In mollusks, the mantle cavity is a respiratory chamber housing the gill. Surrounding the cavity is a protective covering, the *mantle shelf,* which terminates in a fleshy spout, the *siphon.* When the siphon or mantle shelf is stimulated by touch, the siphon and gill are vigorously withdrawn into the mantle cavity under the shelf. Afferent neurons (presynaptic neurons) responding to touch of the siphon or mantle shelf directly synapse on efferent motor neurons (postsynaptic neurons) controlling gill withdrawal. This reflex can be modified by experience in two ways:

- **Habituation** represents a decreased responsiveness to repetitive presentations of an indifferent stimulus—that is, one that is neither rewarding nor punishing. The snail becomes habituated when its siphon is repeatedly touched; that is, it learns to ignore an indifferent stimulus and no longer withdraws its gill in response. (A natural example would be an ocean current.)

- The term **sensitization** refers to increased responsiveness to mild stimuli following a strong or noxious stimulus. Sensitization, a more complex form of learning, takes place in *Aplysia* when it is given an electric shock to the tail followed immediately by a touch on the siphon, or repeated shocks to the tail (a natural analogue might be a fish nibbling on the tail). Subsequently, the snail withdraws its gill more vigorously in response to even mild stimuli it once failed to react to. The animal in effect has learned that something truly dangerous is in the vicinity.

Interestingly, these different forms of learning affect the same site—the synapse between a siphon afferent and a gill efferent—in opposite ways. Habituation depresses this synaptic activity, whereas sensitization enhances it.

Mechanism of Habituation

In habituation, scientists have found that repeated stimulation leads to closing of the Ca^{++} channels, reducing Ca^{++} entry into the presynaptic terminal, which leads to a decrease in quantity of neurotransmitter released. As a consequence, the postsynaptic potential is reduced compared to normal, resulting in a decrease or absence of the behavioral response controlled by the postsynaptic efferent neuron (gill withdrawal). The animal has learned to ignore the stimulus and after 10 to 15 such stimuli, exhibits markedly reduced reflex responses. Thus the memory for habituation in *Aplysia* is stored in the form of modification of specific Ca^{++} channels. With no further training, this reduced responsiveness lasts for several hours. The synaptic events responsible for habituation are also observed in amphibians and crayfish. This suggests that Ca^{++} channel modification may be a general mechanism of habituation, although in advanced species the involvement of intervening interneurons makes the process somewhat more complicated. Habituation is probably the most common form of learning; by learning to ignore indifferent stimuli, the animal is free to focus on other more important stimuli.

Mechanism of Sensitization

Sensitization in *Aplysia* has also been shown to involve channel modification, although a different channel and mechanism are involved. In contrast to habituation, Ca^{++} entry into the presynaptic terminal is enhanced in sensitization. The *nociceptors* (pain receptors) activated by the electric shocks to the tail connect with the presynaptic terminal of the sensory neuron in the gill-withdrawal reflex pathway. The subsequent increase in neurotransmitter release produces a larger postsynaptic potential (EPSP) in the motor neuron, resulting in a more vigorous gill-withdrawal response. Sensitization does not have a direct effect on the presynaptic Ca^{++} channels. Instead, it indirectly enhances Ca^{++} entry via presynaptic facilitation (see p. 132), that is, enhanced neurotransmitter release from the terminals of the sensory neurons. Sensitization thus involves a more complex response at the molecular level. For this form of learning to occur, noxious stimulation of the tail (unconditioned stimulus) must be either repeated, or paired with stimulation of the sensory neurons in the siphon or mantle shelf (conditioned stimulus). The conditioned stimulus normally produces only a small response, whereas the unconditioned stimulus generates a powerful response. After pairing of these two stimuli, the gill withdrawal response to stimulation of the siphon is enhanced. For this form of learning to occur, the conditioned stimulus *must* precede the unconditioned stimulus by a narrow critical period of 0.5 sec. Biochemical events initiated in the sensory neuron are then exaggerated by neurotransmitter (serotonin) released from the modulatory neuron. Here is a summary of the events associated with sensitization (● Figure 5–40):

1. Arrival of an action potential in the sensory neuron within 0.5 sec of activation of the nociceptor pathway. Release of *serotonin* by the modulatory neuron.

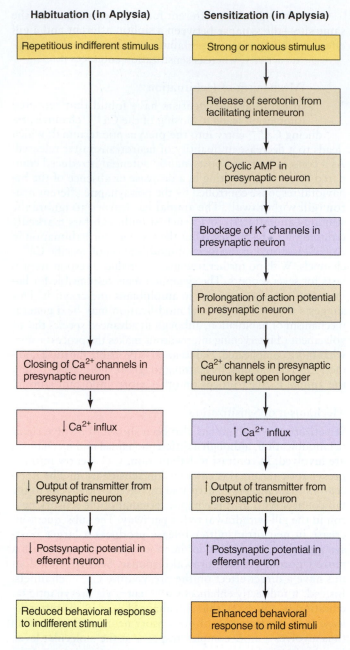

Habituation (in Aplysia)

Repetitious indifferent stimulus

↓

Closing of Ca²⁺ channels in presynaptic neuron

↓

↓ Ca²⁺ influx

↓

↓ Output of transmitter from presynaptic neuron

↓

↓ Postsynaptic potential in efferent neuron

↓

Reduced behavioral response to indifferent stimuli

Sensitization (in Aplysia)

Strong or noxious stimulus

↓

Release of serotonin from facilitating interneuron

↓

↑ Cyclic AMP in presynaptic neuron

↓

Blockage of K⁺ channels in presynaptic neuron

↓

Prolongation of action potential in presynaptic neuron

↓

Ca²⁺ channels in presynaptic neuron kept open longer

↓

↑ Ca²⁺ influx

↓

↑ Output of transmitter from presynaptic neuron

↓

↑ Postsynaptic potential in efferent neuron

↓

Enhanced behavioral response to mild stimuli

Figure 5–40 ● **Habituation and sensitization in *Aplysia*.** Researchers have shown that in the sea snail *Aplysia* two forms of short-term memory—habituation and sensitization—result from opposite changes in neurotransmitter release from the same presynaptic neuron, caused by different transient channel modifications.

2. Binding of serotonin to a specific receptor in the sensory neuron and activation of adenylyl cyclase and generation of *cAMP* from ATP (see p. 90).

3. Activation of a *cAMP-dependent protein kinase* (see p. 91).

4. Protein kinase phosphorylation of a specific class of K⁺ channels, which reduces the efflux of K⁺ from the cell, prolonging the action potential and permitting

the Ca⁺⁺ channels to remain open for an extended period of time.

5. Increased influx of Ca⁺⁺, causing a greater quantity of neurotransmitter release from the sensory neuron.

6. Ca⁺⁺ activation of the protein *calmodulin*, priming adenylyl cyclase to generate additional cAMP.

7. Duration of the action potential extended by 20%, which leads to a more vigorous withdrawal of the gill.

Thus existing synaptic pathways may be functionally interrupted (habituated) or enhanced (sensitized) during simple learning. Habituation persists for about an hour in *Aplysia*, as does the memory for sensitization, the time it takes for cAMP concentrations to return to normal. However, if noxious stimuli (4 to 5 tail shocks) are repeatedly administered to *Aplysia* behavioral sensitization can last from one to several days. This phenomenon is known as **long-term potentiation (LTP)**—a prolonged increase in the strength of existing synaptic connections in activated pathways following brief periods of repetitive stimulation. Sensitization alerts animals to predators and other potentially harmful stimuli and so is critical for their survival. Researchers speculate that much of short-term memory in vertebrates is similarly a temporary modification of already existing processes. Several lines of research suggest that the cyclic AMP cascade, especially activation of protein kinase, plays an important role in elementary forms of learning and memory in mammals, to which we now turn our attention.

Mechanism of Long-Term Potentiation in Mammals

LTP has been shown to last for days or even weeks in vertebrates—long enough for this short-term memory to be consolidated into more permanent long-term memory. LTP is especially prevalent in the mammalian hippocampus, a site critical for converting short-term memories into long-term memories. When LTP occurs, simultaneous activation of both the presynaptic and the postsynaptic neurons at a given excitatory synapse results in long-lasting modifications that enhance the ability of the presynaptic neuron to excite the postsynaptic neuron (● Figure 5–41). Because both the presynaptic neuron and the postsynaptic neurons must be activated at the same time, the development of LTP is restricted to the pathway that is stimulated. Pathways between other inactive presynaptic inputs and the same postsynaptic cell are not affected. With LTP, signals from a given presynaptic cell to a postsynaptic neuron become stronger with repeated use. Keep in mind that strengthening of synaptic activity results in the formation of more EPSPs in the postsynaptic neuron in response to chemical signals from this particular excitatory presynaptic input. This increased excitatory responsiveness is ultimately translated into more action potentials being sent along this postsynaptic cell to other neurons.

Enhanced synaptic transmission could theoretically result from either changes in the postsynaptic neuron (such as increased responsiveness to the neurotransmitter) or in the presynaptic neuron (such as increased release of neurotransmitter). The underlying mechanisms for LTP are still the subject of much research and debate. Most likely, multiple mechanisms are involved in this complex phenomenon. Based on current scientific evidence, here are two plausible proposals, one involving a postsynaptic change and the other a presynaptic modification.

- The first proposed mechanism involves increased responsiveness of the postsynaptic cell. In this proposal, the presynaptic neuron releases glutamate, an excitatory neurotransmitter that binds with **NMDA receptors** (named after the chemical N-methyl D-aspartate, one of several types of glutamate receptors in the postsynaptic neuron's plasma membrane). Because these receptors act as Ca^{++} channels, their activation and opening leads to Ca^{++} entry into the postsynaptic cell. Activation of NMDA receptors does not lead to EPSP formation. Calcium activates a Ca^{++}-dependent second-messenger pathway in the postsynaptic neuron resulting in phosphorylation and thus increased availability of yet another type of glutamate receptor, the **AMPA receptor** (α-amino-3-hydroxy-5-methylisoaxazole-4-propionic acid). The AMPA receptor is the one that results in the formation of EPSPs in response to glutamate activation. Because of this array of newly active AMPA receptors, the postsynaptic cell exhibits a greater response to subsequent release of glutamate from the presynaptic cell. This heightened sensitivity of the postsynaptic neuron to glutamate from the presynaptic cell helps maintain LTP.

- Another proposal involves increased neurotransmitter release from the presynaptic cell as a major contributor to LTP. In this proposed mechanism, activation of the Ca^{++}-dependent second-messenger pathways in the postsynaptic neuron cause this postsynaptic cell to release nitric oxide (NO, p. 88), a so-called **retrograde messenger** that diffuses to the presynaptic neuron and activates a second messenger system in the presynaptic neuron, ultimately enhancing transmitter release from this neuron. This *positive feedback* strengthens the signaling process at this synapse, helping sustain LTP (contrast this with the retrograde negative feedback process we saw in Chapter 4, p. 133). In this way, signals from the presynaptic neuron to the postsynaptic neuron become stronger with repeated use. Note that in the development of LTP, a chemical factor from the postsynaptic neuron influences the presynaptic neuron, just the opposite direction of neurotransmitter activity at a synapse. Synapses in most animals have been traditionally considered to operate unidirectionally, because the cell body and dendrites of the postsynaptic neuron do not contain transmitter vesicles. Thus the retrograde factor is distinct from classical neurotransmitters or neuromodulators.

Long-term memory involves formation of new, permanent synaptic connections.

Whereas short-term memory involves transient strengthening of pre-existing synapses, long-term memory storage requires the activation of specific genes that control protein synthesis. These proteins, in turn, are needed for the formation of new synaptic connections. Thus long-term memory storage involves permanent physical changes in the brain.

Studies comparing the brains of experimental rodents reared in a sensory-deprived environment with those raised in a sensory-rich environment demonstrate readily observable microscopic differences. The animals afforded more environ-

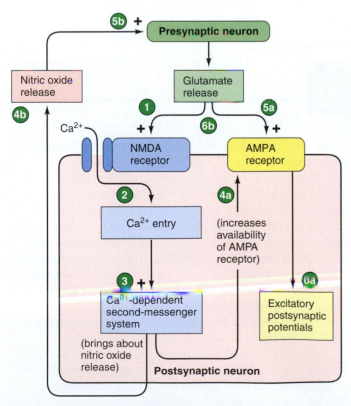

1. Glutamate released from an activated presynaptic neuron binds with NMDA receptors on the postsynaptic neuron, resulting in opening of receptor-related Ca^{2+} channels.

2. Calcium enters through these opened channels.

3. Calcium entry activates a second-messenger system that can initiate two separate intracellular pathways that enhance the effectiveness of this synapse.

4a. In the first pathway, the Ca^{2+}-dependent second-messenger system increases the availability of AMPA receptors.

4b. In an alternate pathway, the Ca^{2+}-dependent second-messenger system causes the postsynaptic neuron to release nitric oxide.

5a. These receptors also bind with glutamate.

5b. This retrograde messenger enhances the release of more glutamate from the presynaptic neuron.

6a. This binding results in the formation of EPSPs. Thus this pathway strengthens the synapse by increasing postsynaptic excitatory responsiveness.

6b. This positive feedback strengthens the signaling process at this synapse.

Figure 5–41 ● Possible pathways for long-term potentiation.

mental interactions—and supposedly, therefore, more opportunity to learn—have greater branching and elongation of dendrites in nerve cells in regions of the brain suspected to be involved with memory storage. Greater dendritic surface area presumably provides more sites for synapses. Thus long-term memory may be stored at least in part by a particular pattern of dendritic branching and synaptic contacts.

A transcription factor (p. 30), **cAMP response-element binding protein (CREB),** is apparently the molecular switch that activates (turns on) the genes important in long-term memory storage (recall that response elements are the switches of genes, p. 30). Another related molecule, **CREB2,** has recently been identified. CREB2 is a repressor of CREB-facilitated protein synthesis. Formation of enduring memories involves not only the activation of positive regulatory factors (CREB) that favor memory storage but also the turning off of inhibitory, constraining factors (CREB2) that prevent memory storage. The shifting balance between positive and repressive factors is believed to ensure that only information relevant to the individual, not everything encountered, is put into long-term storage.

How is CREB activated (and CREB2 inhibited)? No one knows for sure, but some researchers suggest the interesting possibility that the conversion from short-term to long-term memory may involve the turning on of CREB by cAMP. This second-messenger system plays a regulatory role in LTP as well as in simpler forms of short-term memory such as sensitization. Furthermore, cAMP can activate CREB, which ultimately leads to new protein synthesis and the consolidation of long-term memory. Consolidation of both declarative and procedural memories depends on CREB (and presumably CREB2). CREB serves as a common molecular switch for converting both these categories of memory from the transient short-term to the permanent long-term mode.

The CREB transcription factors regulate a group of genes, the **immediate early genes (IEGs),** which play a critical role in memory consolidation. These genes govern the synthesis of the proteins that encode long-term memory. The exact role that these critical new long-term memory proteins might play remains speculative. They may be needed for structural changes in dendrites or used for synthesis of more neurotransmitters or additional receptor sites. Alternatively, they may accomplish long-term modification of neurotransmitter release by sustaining biochemical events initially activated by short-term memory processes.

To complicate the issue even further, numerous hormones and neuropeptides are known to affect learning and memory processes. We discussed one of these, anandimide, earlier (p. 188).

Can Intelligence and Memory Be Changed by Single Genes?
We have just discussed a number of proteins and genes associated with learning and memory. To complicate the issue even further, numerous hormones and neuropeptides are known to affect learning and memory processes. Thus many researchers have believed that complex higher brain functions such as intelligence are the result of many interacting genes (as well as environmental influences) and thus cannot be readily manipulated. But others believe that single genes may have profound influences on memory and perhaps intelligence. See the box, "Molecular Biology and Genomics: A Smarter Rodent," for some of these studies.

MOLECULAR BIOLOGY *AND* GENOMICS

A Smarter Rodent

Genetic engineering is being used to test hypotheses about cellular molecules involved in higher brain functions. Some studies focus on changes in gene expression during learning. For example, one research team put rats to a standard learning task, and used a gene chip (p. 32) to analyze gene expression in the hippocampus at various times before, during and after the task. They found that learning was associated with significant changes in expression of about 30 genes. The roles of these genes are mostly still uncertain, but one of them, Fibroblast Growth Factor (FGF) 18, was tested further because it increased at every stage of learning. Significantly, injections of FGF-18 into rats improved the ability of rats to learn a new task.

Other studies approach the problem by genetically engineering rodents with excess genes suspected of being involved in learning, or by knocking out such genes. One study engineered mice to overexpress growth-associated protein GAP-43, a target of the second-messenger transducer protein kinase C, and which is suspected of being crucial to the growth of new synapses

during learning. The mice exhibited significantly greater long-term learning abilities.

Another study knocked out a gene called NR1, which codes for part of the NMDA receptor associated with memory formation (see main text). The gene was knocked out in the hippocampus of mice, and the mice indeed could not learn very well. The next study engineered overexpression with another NMDA gene, called NR2B, that is active only in the forebrain. The mice's forebrains then had an NMDA receptor with a major difference: the channel of the receptor stayed open (during learning) about 2.5 times longer than normal (250 msec instead of 100 msec). These mice were smarter—they solved problems such as mazes about 40% faster than normal mice—and their memories were better.

Other such studies have revealed a variety of genes associated with learning. These are leading to the development of drugs that might manipulate such genes and proteins, and thus may help with certain human mental disorders.

■ Complex memories and learning may reside in neuronal networks.

Although we have seen how the beginnings of learning and memory probably begin with changes in synaptic functions and connections, we have not addressed the larger question of how such changes actually encode a retrievable complex memory such as an image of an object or the location of a home den or lair. This question about neuronal function remains largely unsolved. Working hypotheses and models that have made some headway in explaining complex learning are based on **neuronal networks** (not to be confused with nerve nets, p. 142). These are largely computer models that simulate a number of neurons with synaptic connections that excite, inhibit, modulate, and change with usage. Crude forms of image recognition and other simple learning have been demonstrated in these models. Whether these represent the way a real nervous system works is still under investigation.

■ The term consciousness refers to a subjective awareness of the world and self.

This ends our coverage of nervous system evolution and mechanisms. There is one final topic of complex brain function we have not analyzed: *consciousness*. This term refers to subjective awareness of the external world and self. Even though it is clear that the final level of awareness resides in the cerebral cortex, and although there is evidence that a crude sense of awareness depends on the thalamus, conscious experience depends on the integrated functioning of many parts of the nervous system. But how it does so remains unknown (though there are a number of hypotheses). Although consciousness is a thorny topic, some researchers believe that many primates (in addition to humans), African parrots, and dolphins have a sense of self and personal experience over time. Perhaps other animals do as well.

Chapter in Perspective:
HOMEOSTASIS AND INTEGRATION

To interact in appropriate ways with the external environment to sustain an animal's viability, such as in the acquisition of food, and to make the internal adjustments necessary to maintain homeostasis, the animal's body must be informed about any changes taking place in the external and internal environment and must be able to process this information and send messages to various muscles and glands to accomplish the desired results. The nervous system, one of the body's two major control systems, plays a central role in this life-sustaining communication. The central nervous system (CNS), which in advanced animals consists of a brain and longitudinal cord(s), receives information about the external and internal environment by means of the afferent peripheral nerves. After sorting, processing, and integrating this input, the CNS (acting as an integrator) sends directions by means of efferent peripheral nerves to bring about appropriate muscular contractions, glandular secretions, and other effector functions.

With its swift electrical signaling system, the nervous system is especially important in controlling the rapid responses of the body. Nervous systems have become progressively more complex during evolution. Newer, more complicated, and more sophisticated layers of the brain have been piled on top of older, more primitive regions. Many of the basic activities necessary for survival are built into the older parts of the brain. The newer, higher levels progressively modify, enhance, or nullify actions coordinated by lower levels in a hierarchy of command, and they also add new capabilities. In vertebrates, the spinal cord integrates many basic protective and evacuative reflexes that do not require conscious participation, such as withdrawal from a painful stimulus and emptying of the urinary bladder. In addition to serving as a more complex integrating link between afferent input and efferent output, the brain is also responsible for initiating nonreflex skeletal muscle movements, for complex perceptual awareness of the external environment and for abstract neural phenomena such as learning, remembering, and consciousness.

Many neurally controlled effector activities are aimed toward maintaining homeostasis, whereas others are for regulated changes. For example, when informed by the afferent nervous system that blood pressure has fallen in a mammal, the CNS sends appropriate commands to the heart and blood vessels to increase the blood pressure to normal. When a predator is detected through the eyes, and recognized in the brain, commands may be sent to initiate rapid running. Were it not for this processing and integrative ability of the CNS, coordinated regulation of multiple organs in complex animals would be impossible.

The autonomic nervous system in vertebrates, which is the efferent branch that innervates smooth muscle, cardiac muscle, and glands, plays a major role in the following range of homeostatic activities:

- Regulation of blood pressure.
- Control of digestive juice secretion and of digestive tract contractions that mix ingested food with the digestive juices.
- Control of sweating to help maintain body temperature.

The somatic nervous system of vertebrates, the efferent branch that innervates skeletal muscle, contributes to homeostasis by stimulating the following activities:

- Skeletal muscle contractions that enable the body to move in relation to the external environment contribute to homeostasis by, for example, moving the body toward food or away from harm.
- Skeletal muscle contraction also accomplishes breathing to maintain appropriate levels of O_2 and CO_2 in the body.
- Shivering is a skeletal muscle activity important in the maintenance of body temperature. ■

REVIEW QUESTIONS (Answers are on p. A–1.)

Additional study tools for this chapter, including chapter summaries and practice tests, are available online at *www.biology.brookscole.com*

1. A reflex arc
   a. consists of a nerve net
   b. consists of a nerve ring
   c. consists of a sensory and a motor neuron
   d. must include longitudinal centralization
   e. must include an integrator neuron

2. True central nervous systems evolved in the context of
   a. radial symmetry
   b. nerve ring ganglia
   c. bi-directionl chemical synapses
   d. rhopallia
   e. bilateral symmetry

3. Relative brain size may be explained by
   a. neuron density in nerve nets
   b. the expensive tissue hypothesis
   c. the evolution of radial symmetry and nerve rings
   d. the maternal energy hypothesis
   e. b and d

4. In the peripheral nervous system (PNS) of vertebrates
   a. the afferent division carries information to the CNS
   b. the afferent division carries information from the CNS
   c. the afferent division carries information to effector organs
   d. the afferent division carries information from the efferent division to the CNS
   e. the efferent division carries information to sensory organs

5. The autonomic nervous system
   a. is subdivided into the sympathetic and parasympathetic nervous systems
   b. is an efferent system
   c. uses only acetylcholine and norepinephrine as neurotransmitters
   d. regulates visceral activities
   e. all of the above

6. The sympathetic system dominates
   a. during "rest and digest" situations
   b. in the "fight-or-flight" response
   c. when the autonomic system is turned off
   d. antagonistic control mechanisms
   e. the parasympathetic control system

7. Which of the following are types of cholinergic receptors?
   a. alpha receptors
   b. nicotinic receptors
   c. beta receptors
   d. muscarinic receptors
   e. b and d

8. Which of the following is not a type of glial cell?
   a. astrocyte
   b. oligodendrocyte
   c. ependymal
   d. interneuronal
   e. microglial

9. Which of the following does not play a role in protecting vertebrate central nervous tissue?
   a. meninges
   b. cerebrospinal fluid
   c. the cerebellum
   d. the cranium
   e. the blood-brain barrier

10. Which of the following vertebrate brain structures has changed least over evolutionary time?
    a. cerebrum
    b. cerebral cortex
    c. brain stem
    d. forebrain
    e. diencephalon

11. Somatosensory processing is the responsibility of the
    a. primary motor cortex
    b. cerebral hemispheres
    c. thalamus
    d. parietal lobes
    e. limbic system

12. Which of the following plays a key role in sociosexual behaviors related to mating?
    a. the hypothalamus
    b. the limbic system
    c. the thalamus
    d. the amygdala
    e. the motor homunculus

13. The widely prescribed drug Prozac acts as an antidepressant, because it
    a. binds dopamine
    b. negates the action of norepinephrine
    c. prolongs the activity of serotonin
    d. supplements low levels of norepinephrine
    e. blocks the reuptake of transporters for released dopamine

14. Which of the following plays a role in maintaining balance and controlling eye movement?
    a. the motor cortex
    b. the cerebrocerebellum
    c. the spinocerebellum
    d. the cerebral cortex
    e. the vestibulocerebellum

15. Which pair of spinal nerves is closest to the brain?
    a. sacral
    b. coccygeal
    c. lumbar
    d. cervical
    e. thoracic

16. Which of the following is true of a nerve?
    a. It is composed of bundled nerve cells.
    b. It is composed solely of afferent neuronal axons.
    c. It is composed of afferent and efferent neurons.
    d. It is composed of bundled afferent and efferent neuronal axons.
    e. It is composed of bundles of axons called tracts.

17. Which of the following is true of spinal reflexes?
    a. They cannot be overridden by higher brain centers.
    b. They result in fixed action patterns.
    c. They are limited to responses only on the side of the body that is stimulated.
    d. The neural pathway involved is called the reflex arc.
    e. The neural pathway involved is called a corticospinal tract.

18. Which of the following terms does not apply to memory?
    a. short term
    b. imprinting
    c. fixed action patterns
    d. procedural
    e. declarative

19. In the sea slug *Aplysia*,
    a. habituation and sensitization enhance synaptic activity
    b. habituation and sensitization depress synaptic activity
    c. habituation depresses synaptic activity and sensitization enhances synaptic activity
    d. habituation enhances synaptic activity and sensitization depresses synaptic activity
    e. none of the above

20. Long-term memory storage involves
    a. CREB
    b. activation of specific genes
    c. CREB2
    d. formation of new synaptic connections
    e. all of the above

SUGGESTED READINGS AND INTERNET SITES

Beer, R. D., H. J. Chiel, R. D. Quinn, K. S. Espenschied, & P. Larsson. 1992. A distributed neural network architecture for hexapod robot locomotion. *Neural Computation* 4: 356–365. Shows how "biological inspiration" is aiding robot engineering.

Delcomyn, F., R. Gillette, C. L. Prosser, W. T. Greenough, & A. Spencer. 1991. In C. L. Prosser, ed., *Neural and Integrative Animal Physiology*, 4th ed. New York: Wiley-Liss. A classic encyclopedic textbook of comparative physiology.

Matthews, G. G. 2001. *Neurobiology: Molecules, Cells and Systems*. Palo Alto, CA: Blackwell Science. An excellent textbook on neurobiology at all levels.

Miall, R. C., D. J. Weir, D. M. Wolpert, & J. F. Stein. 1993. Is the cerebellum a Smith Predictor? *Journal of Motor Behavior* 25:203–216. The first article to make this detailed hypothesis.

Northcutt, R. G. 2002. Understanding vertebrate brain evolution. *Integrative and Comparative Biology* 42:743–756. A recent review of this topic.

Randall, D., W. Burggren, & K. French. 2002. *Eckert Animal Physiology. Mechanisms and Adaptation*, 5th ed. New York: W. H. Freeman. A classic animal physiology text with comprehensive coverage of many of the topics discussed in this book.

Satterlie, R. A. 2002. Neuronal control of swimming in jellyfish: A comparative story. *Canadian Journal of Zoology* 80:1654–1669. A review of evolution and physiology of jelly nervous systems and swimming.

INFOTRAC READINGS

Matynia, A., S. A. Kushner, & A. J. Silva. 2002. Genetic approaches to molecular and cellular cognition: A focus on LTP and learning and memory. *Annual Review of Genetics* 2002: 687–724. A recent review of molecular research on learning.

INTERNET SITES

Miall, R. C., D. J. Weir, D. M. Wolpert, & J. F. Stein. 1993. **physiol.ox.ac.uk/~rcm/pdf_files/SmithPred_93.PDF.** Discusses the Smith Predictor model of the cerebellum.

Roach, J. 2001. **news.nationalgeographic.com/news/2001/09/0920_octopusmimic.html.** Octopus mimicry, including photographs.

Kimball, J. 2003. **users.rnc.com/jkimball.ma.ultranet/BiologyPages/M/Memory.html.** Molecular events in *aplysia* memory.

Sensory Physiology

The great white shark, Carcharinus carcharius. *Like other sharks, it has electrical sensors in its snout that allow it to find prey even in murky waters and when its eye is covered by a protective flap during an attack.*

Introduction

Sensory cells, those specialized cells designed to obtain information about the environment, are inextricably linked to **receptor proteins** in their membranes and to the opening and closing of ion channels.

■ Sensory cells have ion channels and receptors that respond to cues in the environment.

As you have seen (Chapter 3), ion channels represent a basic component of living systems and are found even in single-celled organisms such as bacteria, yeasts, and protozoa. The first ion channels to evolve were most likely stretch-sensitive channels (called *mechanically gated* channels) that served as osmotic stress sensors, activated when a cell was deformed by changes in cell volume. But other channel types such as *voltage-gated* ones (p. 108) are also ancient. For example, in the common gut and laboratory bacterium *Escherichia coli,* two types of membrane channels have been identified, one of which responds to deformation, and another that is voltage-gated. Comparable channels are also found in yeast cells, which also have voltage-gated K^+ channels. Voltage-gated channels permeable to cations have also been identified in the plasma membranes of various plant cells. For example, action potentials are generated in response to touch in the insectivorous Venus flytrap.

Although Ca^{++} and K^+ channels are ubiquitous, Na^+ channels first appeared in the Metazoa ("true animals"). Most if not all animal voltage-gated channels probably evolved from a common ancestral protein. Cnidaria (such as sea anemones) are a possible exception: They have Na^+ channels that are in-

sensitive to tetrodotoxin (p. 111), suggesting that their structure differs from that of other animals' channels.

In addition to mechanical and voltage gating of channels, other stimuli can also be "sensed" by membrane proteins. Higher unicellular eukaryotes, such as the ciliate *Paramecium* (protozoa with cilia that cover certain cell surfaces and facilitate movement; p. 23) can actually be considered free-swimming sensory cells, because they can respond to a variety of environmental stimuli (mechanical, thermal, chemical, photic, and ionic) through a change in their membrane potentials. They respond behaviorally by changing either their cell shape or the pattern of ciliary beating. They have multiple types of ion-selective membrane channels that are voltage-, calcium-, and mechanically gated, and they have *receptor proteins* that react to certain stimuli and indirectly affect membrane potential. For example, ciliates, other protozoa, bacteria, and other unicellular organisms (including slime molds, some algae, and some fungi), can respond to chemical changes in their environments, using *chemically gated* sensing. Bacteria and protozoa residing in the stomach of ruminants are attracted toward "desirable" molecules, such as sugars, amino acids, and small peptides, which are associated with the arrival of a fresh bolus of food. They also can move away from repellents. Both responses follow the binding of the molecules to surface-membrane receptor proteins, which transduce (lead) the signal to the cell interior. In some species of bacteria, these *chemoreceptors* are strategically localized at the forward-moving poles of the cell.

Chemoreception is at the heart of *intercellular communication,* classically studied in the slime mold (*Dictyostelium discoideum*). Starvation induces the single-celled amoeba to secrete **cAMP** (p. 90) into the environment, which serves as a chemoattractant for neighboring amoeba. Through the process of **chemotaxis** (movement of an organism toward or away from a chemical substance), the amoebas aggregate into large-bodied slugs, which can move to a new food source or turn into a reproductive stage to spread spores to new habitats. Activation of the membrane-bound cAMP receptor triggers a G protein–linked biochemical cascade (p. 91), which leads to the developmental changes. Similar mechanisms occur in higher animals in sensing chemicals, both external (such as odors) and internal (such as many hormones), as you will see in this chapter.

Interestingly, some of the mechanisms for sensing chemical signals in microbes are similar to the mechanisms used by neurons in higher metazoan animals. Thus much of the machinery associated with the function of sensory neurons arose long before the evolution of nervous systems.

■ Sensing in animals can be classified into three roles: sensing the external environment, the internal environment, and body motion and position.

In higher animals, the number and types of sensory cells are quite extensive, far more than the classic "five senses" students are frequently taught. Many physiologists recognize three primary functions or roles of sensors:

- **Exteroreceptors** are the "classic" senses that detect external stimuli such as light, chemicals, touch, temperature, and sound.

- **Interoreceptors** detect information about the internal body fluids usually crucial to homeostasis, such as blood pressure and the concentration of O_2.

- **Proprioceptors** send information about movement and position of an animal's body or certain parts such as limbs.

Vertebrate interoreceptors are located at numerous sites, mainly in association with blood vessels and gut fluids. The incoming pathway for information derived from visceral interoreceptors is called a **visceral afferent**. Exteroreception information is sometimes categorized as (1) *somesthetic sensation* arising mainly from the body surface (touch, pressure, temperature, pain, and electrical fields), and (2) *special senses*, including *vision, hearing, taste,* and *smell*. Somesthetic sensation is detected by widely distributed receptors for a variety of stimuli. These receptors in an organ (usually the integument) that has many other functions that provide information about an animal's interaction with the environment across its entire surface. In contrast, each of the special senses has highly localized, extensively specialized receptors (usually in a distinct, dedicated sensory organ) that respond to unique environmental stimuli. Each animal group has evolved specific senses that are essential for survival in its particular environment. Lastly, vertebrate proprioception is located mainly in the muscles, tendons, and joints, and also in the inner ear.

The peripheral nervous system (PNS) carries information from the sensory receptors to the central nervous system CNS (spine and brain in vertebrates). Processing of sensory input by the CNS, followed by efferent activation of appropriate effectors, is essential for interaction with the environment for basic survival (for example, food procurement and defense from predators). The afferent division of the PNS sends information about the internal and external environment, body position and motion to the CNS, where it may be "perceived" if it reaches the brain.

Perception is an animal's interpretation of the external world as created by the brain from a pattern of nerve impulses delivered to it from sensory receptors. Perception is primarily a human concept, because we cannot know what another animal truly perceives.

?

Do Fish Perceive Pain? Recent studies with fish illustrate the difficulties in demonstrating perception in nonhuman animals. Higher fish (teleosts) have skin sensors that are similar to pain sensors in mammals, whereas more primitive fish (cartilaginous fishes, lampreys) do not. Researchers recently injected acetic acid (an irritant) or saline (a control) into the lips of trout. The trout with acetic acid exhibited complex nonreflex behaviors, such as rubbing their lips on the aquarium walls, that indicate a perception of pain. Similar experiments in more primitive fish did not induce such behaviors. However, other researchers note that fish lack the neocortical centers associated with pain perception in humans. Thus it is still not certain whether trout actually perceive pain. See the Suggested Readings by Rose (2002) and by Sneddon, Braithwaite, and Gentle (2003).

Is the environment, as we humans perceive it, reality? The answer is a resounding no. Our perception is different from what is really "out there" for several reasons. Humans, for example, have receptors that detect only a limited number of existing energy forms. We perceive sounds, colors, shapes, textures, smells, tastes, and temperature but are not informed of magnetic and electric forces, polarized and ultraviolet light waves (stimuli that some animals can sense), because none of our receptors respond to the latter energy forms. Our response range is limited even for the energy forms for which we do have receptors. For example, dogs can hear a dog whistle pitched above our level of detection. Second, the information channels to our brains are not high-fidelity recorders. During precortical processing of sensory input, some features of stimuli are *filtered:* Some features are accentuated, and others are suppressed or ignored. For example, the retina of mammals, horseshoe crabs, and probably many other animals actively enhances contrast, making fuzzy boundaries appear as sharp edges. Furthermore, new research has found that the mammalian retina also enhances light signals from any object that is moving relative to the background light.

Third, the cerebral cortex further manipulates the data, comparing the sensory input with other incoming information as well as with memories of past experiences to extract the significant features—for example, sifting out a friend's words from the hubbub of sound in a school cafeteria. In the process, the cortex often fills in or distorts the information to abstract a logical perception; that is, it "completes the picture." As a simple example, you "see" a white square in ● Figure 6–1 even though there is no white square but merely right-angle wedges taken out of four purple circles. Optical illusions illustrate how the brain interprets reality according to its own

Figure 6–1 ● Do you "see" a white square that is not really there?

rules. Thus our human perceptions do not replicate reality. Other species equipped with different types of receptors and sensitivities and numbers of receptors and with different neural processing presumably perceive a markedly different (and filtered) world from ours.

Receptor Physiology

A **stimulus** is a change detectable by the body. Stimuli exist in a variety of energy forms, or **modalities,** such as heat, electromagnetic, sound, pressure, and chemical changes. Afferent neurons have **receptors** at their peripheral endings that respond to stimuli in both the external world and internal environment. Because the only way afferent neurons can transmit information to the CNS about these stimuli is via action potential propagation, receptors must convert these other forms of energy into electrical energy (action potentials). This energy conversion process is known as **transduction.** Receptors specialized for reacting to a particular form of stimulus are generally grouped together into specialized **sense organs.**

■ Receptors have differential sensitivities to various stimuli.

Each type of receptor is specialized to respond more readily to one type of stimulus, its **adequate stimulus,** than to other stimuli. For example, receptors in the eye are most sensitive to light, certain receptors in the ear to sound waves, and warmth receptors in the skin to heat energy. An animal cannot "see" with its ears or "hear" with its eyes because of this differential sensitivity of receptors, a principle Johannes Muller formulated over 150 years ago and entitled the **doctrine of specific nerve energies.** Some receptors can respond weakly to stimuli other than their adequate stimulus, but even when activated by a different stimulus a receptor still gives rise to the sensation usually detected by that receptor type. As an example, the adequate stimulus for eye receptors (photoreceptors) is light, to which they are exquisitely sensitive, but these receptors can also be activated to a lesser degree by mechanical stimulation. When hit in the eye, a person often "sees stars" because the mechanical pressure stimulates the photoreceptors. Thus the sensation perceived depends on the type of receptor stimulated rather than on the type of stimulus. However, be-

cause receptors typically are activated by their adequate stimulus, the sensation usually corresponds to the stimulus modality.

In addition to categorizing receptors by their roles—extero, intero, and proprio—we can also categorize them according to the type of energy to which they ordinarily respond, as follows:

■ **Photoreceptors** are sensitive to light, including ultraviolet and infrared wavelengths. They are primarily exteroreceptors.

■ **Mechanoreceptors** are sensitive to mechanical energy. Examples include touch, skeletal muscle proprioceptors sensitive to stretch, the exteroreceptors in the mammalian ear containing fine hair cells that are bent as a result of sound waves, the fish lateral-line system, and blood pressure–monitoring baroreceptors (extero- and interoreceptors, respectively).

■ **Chemoreceptors** are sensitive to specific chemicals. Chemoreceptors include the exteroreceptors for smell and taste, as well as interoreceptors that detect O_2 and CO_2 concentrations in the blood, the chemical content of the digestive tract, and the osmotic concentration of key fluid molecules.

■ **Thermoreceptors** are sensitive to heat and cold.

■ **Nociceptors,** or pain receptors, are sensitive to tissue damage such as pinching or burning and to distortion of tissue. Intense stimulation of any receptor is also perceived as painful.

■ **Electroreceptors** are exteroreceptors responsive to electric fields.

■ **Magnetoreceptors** are responsive to magnetic fields.

Some sensations in humans are compound sensations in that their perception arises from central integration of several, simultaneously activated, primary sensory inputs. For example, the perception of wetness comes from touch, pressure, and thermal receptor input; there appears to be no such thing as a "wet receptor."

Uses for Information Detected by Receptors
The information detected by receptors is conveyed via afferent neurons to the CNS, where it is used for a variety of purposes:

■ First, afferent input is essential for control of efferent output, both for regulating motor behavior in accordance with external circumstances and for coordinating internal activities to maintain homeostasis. At the most basic level, afferent input provides information (of which an animal may or may not be consciously aware) for the CNS to use in directing activities necessary for survival.

■ Second, processing of sensory input by the reticular activating system in the brain stem is critical for cortical arousal and consciousness (see p. 177).

■ Third, central processing of sensory information gives rise to an animal's perceptions of the environment.

■ Finally, selected information delivered to the CNS may be stored for future reference.

Next let's examine how adequate stimuli initiate action potentials that ultimately are used for these purposes.

A stimulus alters the receptor's permeability, leading to a graded receptor potential.

A receptor may be either (1) a specialized ending of the afferent neuron or (2) a separate cell closely associated with the peripheral ending of the neuron. Stimulation of a receptor alters its membrane permeability, usually by causing a nonselective opening of all small ion channels. The means by which this permeability change takes place is individualized for each receptor type. Because the electrochemical driving force is greater for Na$^+$ than for other small ions at resting potential, the predominant effect is an inward flux of Na$^+$, which depolarizes the receptor membrane. (There are exceptions; for example, photoreceptors are hyperpolarized on stimulation.)

This local depolarizing change in potential is known as a **receptor potential** in the case of a separate receptor or as a **generator potential** if the receptor is a specialized ending of an afferent neuron. The receptor (or generator) potential is a graded potential (see p. 104) whose amplitude and duration can vary, depending on the strength and the rate of application or removal of the stimulus. The stronger the stimulus, the greater the permeability change and the larger the receptor potential. As is true of all graded potentials, receptor potentials have no refractory period, so summation in response to rapidly successive stimuli is possible. Because the receptor region has a very high threshold, action potentials do not take place at the receptor itself. For long-distance transmission, the receptor potential must be converted into action potentials that can be propagated along the afferent fiber.

Receptor potentials may initiate action potentials in the afferent neuron.

If of sufficient magnitude, a receptor (or generator) potential initiates an action potential in the afferent neuron membrane adjacent to the receptor by triggering the opening of Na$^+$ channels in this region. The means by which the Na$^+$ channels are opened differ depending on whether the receptor is a separate cell or a specialized afferent ending.

- In the case of a separate receptor, such as in the retina, a receptor potential triggers the release of a chemical messenger—a neurotransmitter—that diffuses across the small space separating the receptor from the ending of the afferent neuron, essentially a synapse (● Figure 6–2a). Binding of the chemical messenger with specific protein receptor sites on the afferent neuron opens chemical messenger–gated Na$^+$ channels (see p. 123).

- In the case of a specialized afferent ending (Figure 6–2b), local current flow between the activated receptor ending undergoing a generator potential and the cell membrane adjacent to the receptor brings about opening of voltage-gated Na$^+$ channels in this adjacent region.

In either case, if the magnitude of the resulting ionic flux is sufficient to bring the adjacent membrane to threshold, an action potential is initiated that is self-propagated along the afferent fiber to the CNS. (For convenience, from here on we refer to both receptor potentials and generator potentials as *receptor potentials*.)

Note that in an afferent neuron the site of initiation of action potentials differs from that in an efferent neuron or interneuron. In the latter two types of neurons, action potentials are initiated at the axon hillock located at the beginning of the axon adjacent to the cell body (see p. 113). By contrast, action potentials are initiated at the peripheral end of an afferent nerve fiber adjacent to the receptor, a long distance from the cell body (● Figure 6–3).

The intensity of the stimulus is reflected by the magnitude of the receptor potential. In turn, the larger the receptor potential, the greater the frequency of action potentials generated in the afferent neuron. A larger receptor potential cannot bring about a larger action potential (this is the all-or-none law), but it can induce more rapid firing of action potentials.

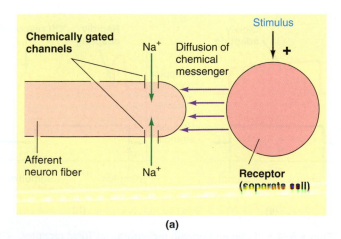 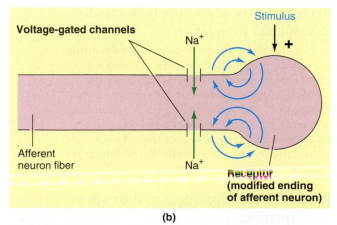

Figure 6–2 ● Conversion of receptor and generator potentials into action potentials. (a) Receptor potential. The chemical messenger released from a separate receptor initiates an action potential in the fiber by opening chemically gated Na$^+$ channels. (b) Generator potential. The local current flow between the depolarized receptor ending and the afferent fiber initiates an action potential in the fiber by opening voltage-gated Na$^+$ channels.

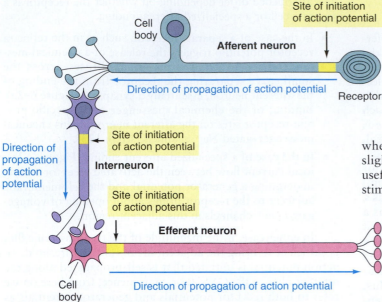

Figure 6–3 • Comparison of the site of initiation of an action potential in the three types of neurons.

This is one way stimulus strength is coded: The stronger the stimulus, the greater the frequency of action potentials. Stimulus strength is also reflected by the size of the area stimulated. Stronger stimuli usually affect larger areas, so a correspondingly greater population of receptors responds. For example, a light touch does not activate as many pressure receptors in the skin as a more forceful touch applied to the same area. Stimulus intensity is therefore distinguished both by the frequency of action potentials generated in the afferent neuron (**frequency code**) and by the number of receptors activated within the area (**population code**).

■ Receptors may adapt slowly or rapidly to sustained stimulation.

Stimuli of the same intensity do not always bring about receptor potentials of the same magnitude from the same receptor. Some receptors can diminish the extent of their depolarization despite sustained stimulus strength, a phenomenon known as **adaptation**. Subsequently, the frequency of action potentials generated in the afferent neuron decreases. That is, the receptor "adapts" to the stimulus by no longer responding to it to the same degree.

■ Receptors vary according to their speed of adaptation.

The two types of receptors—*tonic receptors* and *phasic receptors*—differ in speed of adaptation. **Tonic receptors** do not adapt at all, or adapt slowly (● Figure 6–4a). These receptors are important in situations where maintained information about a stimulus is valuable. Examples are muscle stretch proprioceptors, which monitor muscle length, and joint proprioceptors, which measure the degree of joint flexion. To maintain

posture and balance, the CNS must continually be informed about the degree of muscle length and joint position. It is important, therefore, that these receptors do not adapt to a stimulus but continue to generate action potentials to relay this information to the CNS.

Phasic receptors, in contrast, adapt rapidly. These receptors often exhibit an "off" response (Figure 6–4b). The receptor rapidly adapts by no longer responding to a maintained stimulus, but when the stimulus is removed the receptor responds with a slight depolarization, the **off response.** Phasic receptors are useful in situations where it is important to signal a *change* in stimulus intensity rather than to relay status quo information.

Rapidly adapting receptors include *tactile (touch)* receptors in the skin that signal changes in pressure on the skin surface. Because these receptors adapt rapidly, a horse does not remain conscious of a saddle on its back or a bit in its mouth. You, for example, are not continually conscious of wearing your watch, rings, and clothing. When you put something on, you soon become accustomed to it, because of these receptors' rapid adaptation. When you take the item off, you are aware of its removal because of the off response.

Mechanism of Adaptation in the Pacinian Corpuscle

The mechanism by which adaptation is accomplished varies for different receptors and has not been fully elucidated for all receptor types. One receptor type that has been extensively studied is the vertebrate **Pacinian corpuscle,** a rapidly adapting skin receptor that detects pressure and vibration. A comparable sensory receptor, the *Herbst corpuscle,* is the most widely distributed receptor in the skin of birds and can be found in the bills of some aquatic species. Adaptation in a Pacinian corpuscle is believed to involve both mechanical and electrochemical components. The mechanical component depends on the physical properties of this receptor. A Pacinian corpuscle

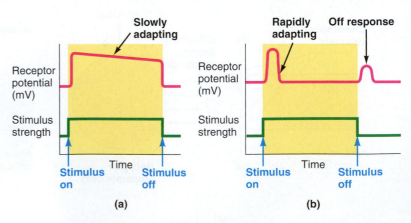

Figure 6–4 • Tonic and phasic receptors. (a) Tonic receptor. This receptor type does not adapt at all or adapts slowly to a sustained stimulus and thus provides continuous information about the stimulus. (b) Phasic receptor. This receptor type adapts rapidly to a sustained stimulus and frequently exhibits an off response when the stimulus is removed. Thus the receptor signals changes in stimulus intensity rather than relaying status quo information.

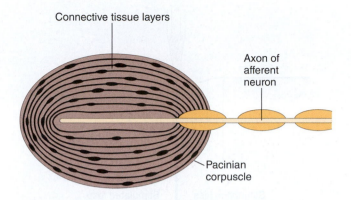

Connective tissue layers

Axon of afferent neuron

Pacinian corpuscle

Figure 6–5 ● **Pacinian corpuscle.** The Pacinian corpuscle consists of concentric layers of connective tissue wrapped around the peripheral terminal of an afferent neuron.

is a specialized receptor ending that consists of concentric layers of connective tissue resembling layers of an onion wrapped around the peripheral terminal of an afferent neuron (● Figure 6–5). When pressure is first applied to the Pacinian corpuscle, the underlying terminal responds with a receptor potential of a magnitude that reflects the intensity of the stimulus. As the stimulus continues, the pressure energy is dissipated, because it causes the receptor layers to slip (just as steady pressure on a peeled onion causes its layers to slip). Because this physical effect filters out the steady component of the applied pressure, the underlying neuronal ending no longer responds with a receptor potential; that is, **accommodation** has occurred. Also contributing to adaptation is the electrochemical component, which involves changes in ionic movement across the receptor membrane. In a Pacinian corpuscle, for reasons unknown, the Na^+ channels that opened in response to the stimulus are slowly inactivated, thus reducing the inward flow of Na^+ ions that was largely responsible for the depolarizing receptor potential. With prolonged depolarization, the voltage-gated Na^+ channels cannot open, regardless of stimulus intensity.

Olfactory (smell) receptors can also adapt, as you might recall from the last time you entered a musty room—at first you notice the musty odor, but after a few minutes you don't. As you will see later, binding of an odorant molecule to a receptor leads to the opening of cation channels. Researchers now think that calcium ions entering these open channels slowly activate a protein that in turn closes the ion channels. Interestingly, some olfactory cells—particularly those that detect odors of dangerous substances—do not adapt. Rotting food, for example, produces sulfurous odors that do not lead to adaptation. The skunk takes advantage of this by producing a defensive cocktail of seven compounds, six of them based on a sulfur atom combined with a hydrogen atom, which slowly react with water to release more of the strong-smelling substances. Potential predators sprayed with this cocktail are continually bothered by a noxious smell until the odor molecules decay.

Accommodation or adaptation should not be confused with habituation (see p. 189). Although both these phenomena involve decreased neural responsiveness to repetitive stimuli, they operate at different points in the neural pathway. Accommodation is a receptor adjustment in the PNS, whereas habituation involves a modification in synaptic effectiveness in the CNS.

■ Each somatosensory pathway is "labeled" according to modality and location.

On reaching the spinal cord of vertebrates, afferent information has two possible destinies: (1) It may become part of a reflex arc, bringing about an appropriate effector response, or (2) it may be relayed upward to the brain via ascending pathways for further processing (and possible conscious awareness). Pathways conveying conscious somatic sensation, the **somatosensory pathways,** consist of discrete chains of neurons synaptically interconnected in a particular sequence to accomplish progressively more sophisticated processing of the sensory information. The afferent neuron with its peripheral receptor that first detects the stimulus is known as a **first-order sensory neuron.** It synapses on a **second-order sensory neuron,** either in the spinal cord or the medulla, depending on which sensory pathway is involved. This neuron then synapses on a **third-order sensory neuron** in the thalamus, and so on. With each step, the input is processed further. A particular sensory modality detected by a specialized receptor type is sent over a specific afferent and ascending pathway (a neural pathway committed to that modality) to excite a defined area in the somatosensory cortex. That is, a particular sensory input is "projected" to a specific region of the cortex. Thus information is kept separated within specific labeled lines between the periphery and the cortex. In this way, even though all information is propagated to the CNS via the same type of signal (action potentials), the brain can decode the type and location of the stimulus. ■ Table 6–1 summarizes how the CNS is informed of the type (what?), location (where?), and intensity (how much?) of a stimulus.

■ Acuity is influenced by receptive field size and lateral inhibition.

Each somatosensory neuron responds to stimulus information only within a circumscribed region of the skin surface surrounding it; this region is known as its **receptive field.** The size of a receptive field varies inversely with the density of receptors in the region; the more closely receptors of a particu-

Table 6–1 ■ Coding of Sensory Information	
Stimulus Property	Mechanism of Coding
Type of Stimulus (stimulus modality)	Distinguished by the type of receptor activated and the specific pathway over which this information is transmitted to a particular area of the cerebral cortex
Location of Stimulus	Distinguished by the location of the activated receptor field and the pathway that is subsequently activated to transmit this information to the area of the somatosensory cortex representing that particular location
Intensity of Stimulus (stimulus strength)	Distinguished by the frequency of action potentials initiated in an activated afferent neuron and the number of receptors (and afferent neurons) activated

lar type are spaced, the smaller the area of skin each monitors. The smaller the receptive field in a region, the greater its **acuity** or **discriminative ability**. Compare the tactile (touch) discrimination in your fingertips with that in your elbow by "feeling" the same object with both. For example, you can discern more precise information about the object with your richly innervated fingertips because the receptive fields there are small; as a result, each neuron signals information about small, discrete portions of the object's surface. An estimated 17,000 tactile mechanoreceptors are present in the fingertips and palm of each hand. In contrast, the skin over the elbow is served by relatively few sensory endings with larger receptive fields. Subtle differences in pressure within each large receptive field cannot be detected. The distorted cortical representation of various body parts in the sensory homunculus (see p. 167) corresponds precisely with the innervation density; more cortical space is allotted for sensory reception from areas with smaller receptive fields and, accordingly, greater tactile discriminative ability.

Besides receptor density, a second factor influencing acuity is **lateral inhibition.** You can appreciate the importance of this phenomenon by slightly indenting the surface of your skin with the point of a pencil (● Figure 6–6a). The receptive field is excited immediately under the center of the pencil point where the stimulus is most intense, but the surrounding receptive fields are also stimulated, only to a lesser extent because they are less distorted. If information from these marginally excited afferent fibers in the fringe of the stimulus area were to reach the cortex, localization of the pencil point would be blurred. To facilitate localization and sharpen contrast, lateral inhibition occurs within the CNS (Figure 6–6b). The most strongly activated signal pathway originating from the center of the stimulus area inhibits the less excited pathways from the fringe areas. This occurs via inhibitory interneurons that pass laterally between ascending fibers serving neighboring receptive fields. Blocking further transmission in the weaker inputs increases the contrast between wanted and unwanted information so that the pencil point can be precisely localized.

The extent of lateral inhibitory connections within sensory pathways varies for different modalities. Those with the most lateral inhibition—touch and vision—bring about the most accurate localization. As we mentioned earlier, retinas enhance the contrast of edges, making blurry boundaries appear sharp, by using lateral inhibition.

Now we turn to the major organs and modalities of senses, primarily the extero- and proprioceptors. The roles of most interoceptors are discussed in other chapters.

Photoreception: Eyes and Vision

Light is one of the most useful signals in the environment, and the ability to detect it is certainly one of the most useful senses. In some animals (such as birds) it may be the dominant sense in day-to-day life (and in some species vision is considered responsible for an evolutionary decline in acuity in the olfactory system, which is often the dominant sense).

■ Light detection in dedicated organs uses photopigments in eyespots or eyes.

Almost all organisms are sensitive to light, but their sensitivities and acuities vary tremendously. At the most primitive

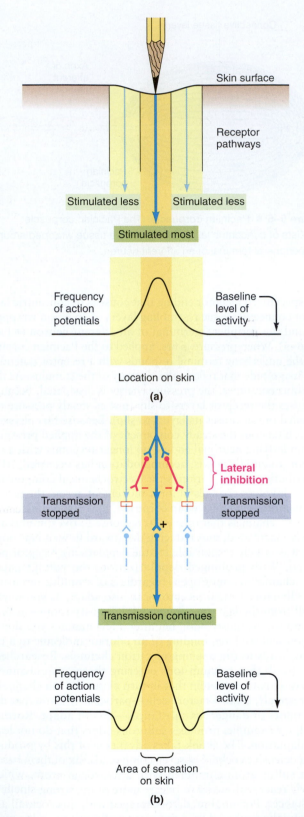

Figure 6–6 ● **Lateral inhibition.** (a) The receptor at the site of most intense stimulation is activated to the greatest extent. Surrounding receptors are also stimulated but to a lesser degree. (b) The most intensely activated receptor pathway halts transmission of impulses in the less intensely stimulated pathways through lateral inhibition. This process facilitates localization of the site of stimulation.

level, light detection is based on structures not dedicated solely to this purpose. For example, **chromatophores** (pigment-containing skin cells) in many phyla of invertebrates and classes of lower vertebrates are directly sensitive to light, responding with color and shade changes in the skin. Even a single-celled *Amoeba* responds to a flash of light by ceasing to move. However, special organs dedicated to light detection have evolved numerous times in a variety of structures. As varied as these are (as you will see shortly), the transduction of light energy into an action potential is based on a highly conserved group of **photopigment molecules** (see p. 211). Thus even though the light-capturing organ of an animal may be structurally very different, the chemical pathways that transduce the light energy are similar. Further, most biological organisms that have evolved use light energies in a narrow spectral band between 400 and 700 nm (nanometers, or billionths of a meter between peaks), matching the major output frequencies of the sun as well as the biochemical limits of the pigments, because too little energy would not excite, whereas energy frequencies that are too high would potentially break down the chemical constituents necessary for phototransduction. However, as you will see, a number of invertebrate and vertebrate species have expanded the range into the ultraviolet region of the spectrum below 400 nm.

As noted, considerable evolutionary inventiveness exists in the design of light detecting structures (• Figure 6–7). Indeed, scientists have discovered at least 10 separate designs for eyes. **Eyespots,** the simplest structure in the evolution of eyes, consist of a small number of photoreceptor cells (less than 100) lining an open cup or pit (Figure 6–7a). Each pit samples a relatively large area of the visual world. Representatives with these simple photoreceptor plates that have received study include planarians (flatworms), jellies (cnidarians), and seastars (echinoderms); but similar structures are found in many animal phyla. These photoreceptors are lined by pigment on one side so that they receive light from only one direction. The pigment cells enclose the dendritic endings of the sensory neurons; the neuronal axons are bundled into nerves that travel from each cup, usually to integrating nerves such as a ganglion or brain. This particular arrangement permits the animal to locate a light source and, in some cases, to orient movement.

However, recognition of predator/prey movement requires a more complex optical system that permits formation of an image—a true **eye.** By reducing the size of the cup aperture to produce a **pinhole eye,** an eye can actually form an image, even though it may not be very light sensitive. The evolution of a **lens** in some animals enhances the light-gathering power of the eye, forming what is called a **camera eye** (Figure 6–7b, c, d). This familiar arrangement appears in many phyla; the most ancient version is probably that evolved by the Cubozoa (box jellies), a surprisingly advanced structure for a cnidarian (Figure 6–7e). More complex camera eyes are the optical solution evolved by vertebrates and cephalopods (the most intelligent invertebrates). The major alternative evolutionary solution is the **compound** eye (see p. 213), with densely packed units or **ommatidia,** each having its own lens and pigment-shielded photoreceptors. Fossil evidence of ommatidia is very ancient, found in arthropods such as trilobites living during the mid-Cambrian period. Although compound eyes provide comparatively poor image formation, they are superior for detecting movement and normally have a wide field of view.

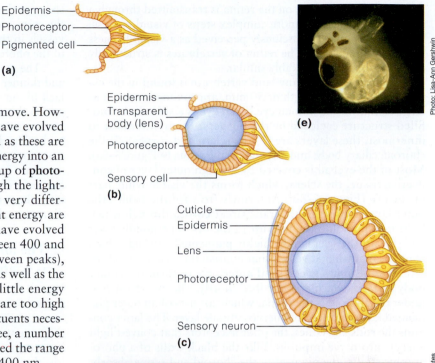

Figure 6–7 • **Organization of invertebrate eyes, longitudinal section.** There are far more photoreceptors than can be shown in these diagrams. (a) Limpet ocellus, a shallow depression in the epidermis that incorporates light-sensitive receptors. (b) Abalone eye, with its spherical, transparent lens. (c) Eye of a land snail. (d) Well-developed eyes of a conch, which is peering into the waters of the Great Barrier Reef along the east coast of Australia. (e) The camera eye of a Cubozoan.

(*Source:* a–d C. Starr & R. Taggart, 2004, *Biology: The Unity and Diversity of Life,* 10th ed., Belmont, CA: Brooks/Cole, (e) gladstone.uoregon.edu/~ghale/pdf/jelly2.pdf)

A final improvement in eye evolution was the origin of two or more different photoreceptors sensitive to different wavelengths, a fundamental condition for color vision. Many vertebrates, for example, have this adaptation to at least some degree. Let's now examine the vertebrate eye in detail, and then take a look at some invertebrate eyes.

■ The vertebrate eye is a fluid-filled sphere enclosed by three specialized tissue layers.

The eyes capture the patterns of illumination in the environment as an "optical picture" on a layer of light-sensitive cells, the **retina,** much as a camera captures an image on film. Just as film can be developed into a visual likeness of the original image, the coded image on the retina is transmitted through a series of progressively more complex steps of visual processing until it is finally consciously perceived as a visual likeness of the original image. The retina of vertebrates is structurally and functionally remarkably similar.

The largest eye of any land vertebrate is found in the ostrich with an axial length of 50 mm, approximately twice as large as that of the human eye. Each **eye** is a spherical, fluid-filled structure enclosed by three layers. From outermost to innermost, these layers are (1) the sclera and cornea; (2) the choroid, ciliary body, and iris; and (3) the retina (● Figure 6–8a). Most of the eyeball is covered by a tough outer layer of connective tissue, the **sclera,** which forms the visible white part of the eye (Figure 6–8b). Anteriorly (toward the front), the outer layer consists of the transparent **cornea** through which light rays pass into the interior of the eye. The middle layer underneath the sclera is the highly pigmented choroid, which contains many blood vessels that nourish the retina. The choroid layer becomes specialized anteriorly to form the **ciliary body** and **iris,** which are described shortly. The innermost coat under the choroid is the retina, which consists of an outer pigmented layer and an inner nervous tissue layer. The latter contains the **rods** and **cones,** the photoreceptors that convert light energy into nerve impulses. Like the black walls of a photographic studio, the pigment in the choroid and retina absorbs light after it strikes the retina to prevent reflection or scattering of light within the eye.

The interior of the eye consists of two fluid-filled cavities, separated by a lens, all of which are transparent to permit light to pass through the eye from the cornea to the retina. The anterior (front) cavity between the cornea and lens contains a clear, watery fluid, the **aqueous humor,** and the larger posterior (rear) cavity between the lens and retina contains a semifluid, jellylike substance, the **vitreous humor** (Figure 6–8a). The vitreous humor is important in maintaining the spherical shape of the eyeball. The aqueous humor carries nutrients for the cornea and lens, both of which lack a blood supply. Blood vessels in these structures would impede the passage of light to the photoreceptors.

The mammalian retina is comparatively well vascularized and thinner compared to birds and reptiles, which completely lack blood vessels within the retina. These species rely on a **pecten** (birds) or **conus papillaris** (reptiles) to provide oxygen and nutrients to the retina by diffusion through the vitreous body. The pecten is a thin, black structure, which projects from the retina toward the lens. The endothelial cells, whose structure is folded to form a series of minute folds, enhance the surface area available for nutrient exchange. In nocturnal birds the pecten tends to be simplified in structure, whereas in diurnal birds it is elongated and more elaborate in structure.

■ The amount of light entering the eye is controlled by the iris.

Not all the light passing through the cornea reaches the light-sensitive photoreceptors, because of the **iris,** a thin, pigmented, smooth muscle that forms a visible ringlike structure within the aqueous humor (Figure 6–8a and b). The pigment in the iris governs eye color. The round opening in the center of the iris, through which light enters the interior portions of the eye, is the **pupil.** The size of this opening can be adjusted by variable contraction of the iris muscles to admit more or less light as needed, much as the shutter controls the amount of light entering a camera. The iris contains two sets of smooth muscle networks, one *circular* (the muscle fibers run in a ringlike fashion within the iris) and the other *radial* (the fibers project outward from the pupillary margin like bicycle spokes) (● Figure 6–9). Because muscle fibers shorten when they contract, the pupil gets smaller when the **circular** (or **constrictor**) **muscle** contracts and forms a smaller ring while the radial muscles

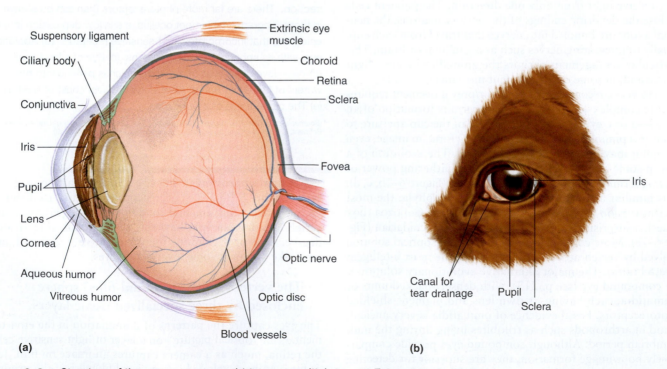

(a)

Suspensory ligament
Ciliary body
Conjunctiva
Iris
Pupil
Lens
Cornea
Aqueous humor
Vitreous humor
Extrinsic eye muscle
Choroid
Retina
Sclera
Fovea
Optic nerve
Optic disc
Blood vessels

(b)

Iris
Canal for tear drainage
Pupil
Sclera

Figure 6–8 ● **Structure of the mammalian eye.** (a) Internal sagittal view. (b) External front view.

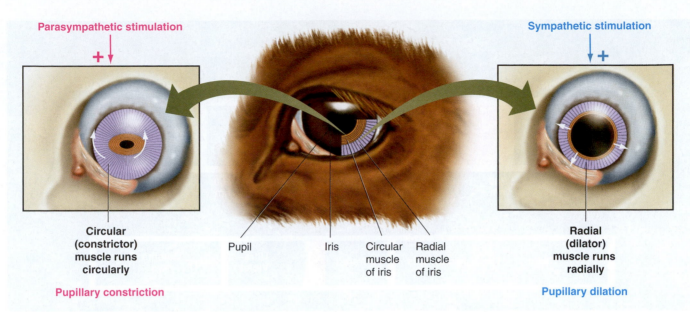

Parasympathetic stimulation

Sympathetic stimulation

Circular (constrictor) muscle runs circularly

Pupil   Iris   Circular muscle of iris   Radial muscle of iris

Radial (dilator) muscle runs radially

Pupillary constriction

Pupillary dilation

Figure 6–9 ● Control of pupillary size.

relax. This reflex pupillary constriction occurs in bright light to decrease the amount of light entering the eye. When the **radial** (or **dilator**) **muscle** shortens, the size of the pupil increases and the circular muscles now relax. Such pupillary dilation occurs in dim light to allow more light to enter.

Iris muscles are controlled by the autonomic nervous system. Parasympathetic nerve fibers innervate the circular muscle, and sympathetic fibers supply the radial muscle. Acting via the autonomic nervous system, conditions other than light can induce changes in pupillary size (Figure 6–9). For example, dilation of the pupils accompanies generalized discharge of the sympathetic nervous system in response to actual or perceived danger.

■ The eye refracts the entering light to focus the image on the retina.

Light is a form of electromagnetic radiation consisting of particle-like individual packets of energy, called **photons**, that travel in wavelike fashion. The distance between two wave peaks is known as the **wavelength** (● Figure 6–10). Depending on the species of animal, the photoreceptors in the vertebrate eye are sensitive only to wavelengths between 360 and 700 nm. The visual range of an insect's perception is shifted to shorter wavelengths than those perceived by vertebrates. However, this violet and visible light is only a small portion of the total electromagnetic spectrum (● Figure 6–11), but it does cover the major output of the sun reaching Earth's surface. Light of different wavelengths in this frequency band is perceived as different color (hue) sensations, at least by humans. Short wavelengths are sensed as violet and blue; long wavelengths are interpreted as orange and red.

In addition to having variable wavelengths, light energy also varies in **intensity**; that is, in the amplitude, or height, of the wave (Figure 6–10). Dimming a bright red light does not change its color; it just becomes less intense or less bright.

Light waves *diverge* (radiate outward) in all directions from every point of a light source. The forward movement of a light wave in a particular direction is known as a *light ray*.

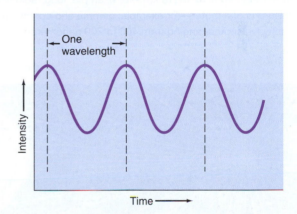

Figure 6–10 ● Properties of an electromagnetic wave. A wavelength is the distance between two wave peaks. The term *intensity* refers to the amplitude of the wave.

Divergent light rays reaching the eye must be bent inward to be focused back into a point on the light-sensitive retina to provide an accurate image of the light source (● Figure 6–12). The bending of a light ray (**refraction**) occurs when the ray passes from a medium of one density into a medium of a different density (● Figure 6–13). Light travels faster through air than through other transparent media such as water and glass. When a light ray enters a medium of greater density, it slows down (the converse is also true). The ray's course of direction changes if it strikes the surface of the new medium at any angle other than perpendicular.

Two factors contribute to the degree of refraction: the comparative densities of the two media (the greater the difference in density, the greater the degree of bending) and the angle at which the light strikes the second medium (the greater the angle, the greater the refraction).

With a curved surface such as a lens, the greater the curvature, the greater the degree of bending and the stronger the lens. When a light ray strikes the curved surface of any object

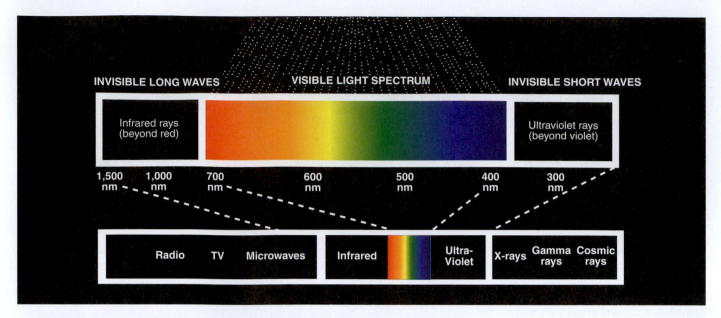

Figure 6–11 ● **Electromagnetic spectrum.** The wavelengths in the electromagnetic spectrum range from 10^4 m (10 km—for example, long radio waves) to less than 10^{-14} m (quadrillionths of a meter—for example, gamma and cosmic rays). The visible spectrum includes wavelengths ranging from 400 to 700 nanometers (nm; billionths of a meter).

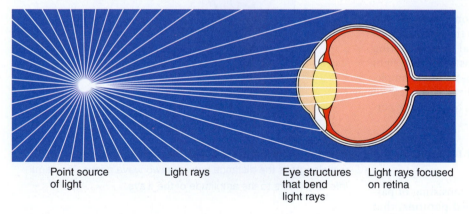

Figure 6–12 ● **Focusing of diverging light rays.** Diverging light rays must be bent inward to be focused.

of greater density, the direction of refraction depends on the angle of the curvature (● Figure 6–14a). A lens with **convex** surfaces converges light rays, bringing them closer together, a requirement for bringing an image to a focal point. Refractive surfaces of the eye are therefore convex. (A lens with **concave** surfaces, not shown in the figure, diverges light rays.)

The two structures most important in the eye's refractive ability are the *cornea* and the *lens.* The curved corneal surface, the first structure light passes through as it enters the eye, contributes most extensively to the refractive ability of a terrestrial animal's eye because the difference in density at the air/corneal interface is much greater than the differences in density between the lens and the fluids surrounding it. In contrast, the refractive power of the fish cornea is negligible, because the surrounding medium of the fish cornea is water and not air.

Thus in fish, most of the refractive power is present in the lens. Because the fish lens is extremely dense (high refractive index), focusing requires physical movement of the lens. Variable compression of the lens fibers creates a gradient of refractive indices throughout the lens such that the more peripheral region has a lower refractive index (a decreased ability to bend light) than does the more central region. Dogs, for example, also have a negative spherical aberration of the lens, which results in the peripheral rays being brought to focus behind the more central rays. Biologists believe this feature compensates for the positive spherical aberration of the peripheral portion of the cornea. The lens of birds is softer than that of mammals, which facilitates the rapid accommodation associated with the avian eye.

In primates, yellow pigment in the lens acts as a filter, which absorbs wavelengths below 400 nm, eliminating ultraviolet light. In contrast, the lens of animals that perceive ultraviolet wavelengths is clear and permits the transmission of wavelengths as low as 310 nm. The lens thus absorbs only the more destructive region of the spectrum (nucleic acids and proteins begin to absorb light strongly at wavelengths shorter than 310 nm). In some species of fish the cornea is pigmented, which can also limit the penetration of short-wavelength light. In some species of fishes, birds, reptiles, and amphibians, colored oil droplets effectively limit the passage of short wavelengths to the photoreceptor containing the oil droplet. These oil droplets lie in the inner segments of the cone receptors.

For clear vision, the refractive structures of the eye must bring light images into focus on the retina. If an image is fo-

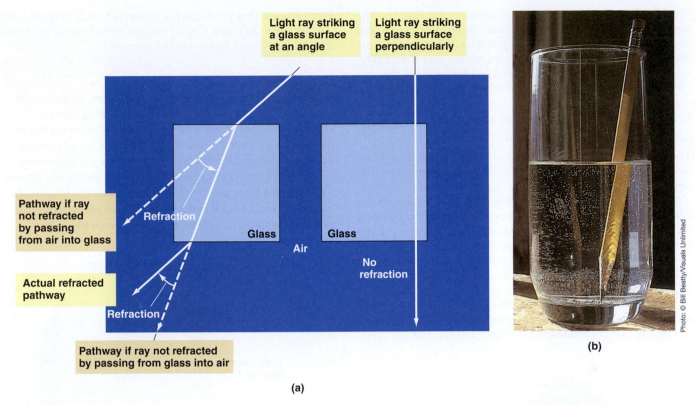

Light ray striking a glass surface at an angle

Light ray striking a glass surface perpendicularly

Pathway if ray not refracted by passing from air into glass

Refraction

Glass

Glass

Air

No refraction

Actual refracted pathway

Refraction

Pathway if ray not refracted by passing from glass into air

(a)

(b)

Photo: © Bill Beatty/Visuals Unlimited

Figure 6–13 ● **Refraction.** A light ray is bent (refracted) when it strikes the surface of a medium of different density from the one in which it had been traveling (for example, moving from air into glass) at any angle other than perpendicular to the new medium's surface. (b) The pencil in the glass of water *appears* to bend. What is happening, though, is that the light rays coming to the camera (or your eyes) are bent as they pass through the water, then the glass, and then the air. Consequently the pencil appears distorted.

cused before it reaches the retina, or is not yet focused when it reaches the retina, it will be blurred. Light rays originating from near objects are more divergent when they reach the eye than are rays from distant sources. By the time they reach the eye, rays from light sources more than 6 meters away are considered parallel. For a given refractive ability of the eye, a near source of light requires a greater distance behind the lens for focusing than does a distant source, because the near-source rays are still diverging when they reach the eye (Figure 6–14a and b).

Figure 6–14 ➤ ● Focusing of distant and near sources of light with a concave lens. The rays from a distant (far) light source (more than 6 meters from the eye) are parallel by the time the rays reach the eye. (b) The rays from a near light source (less than 6 meters from the eye) are still diverging when they reach the eye. A longer distance is required for a lens of a given strength to bend the diverging rays from a near light source into focus compared to the parallel rays from a distant light source. (c) To focus both a distant and a near light source in the same distance (the distance between the lens and retina), a stronger lens must be used for the near source.

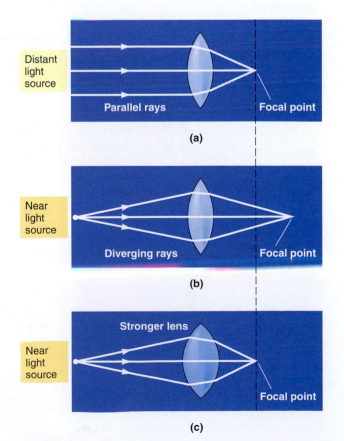

Distant light source

Parallel rays   **Focal point**

(a)

Near light source

Diverging rays   **Focal point**

(b)

Stronger lens

Near light source

Focal point

(c)

In species where the distance between the lens and the retina always remains the same in order to bring both near and distant light sources into focus on the retina (that is, in the same distance), a stronger lens must be used for the near source (Figure 6–14c). The strength of the lens can be adjusted through the process of accommodation, to which we now turn.

■ Accommodation increases the strength of the lens for near vision.

The ability to adjust lens strength to focus both near and distant sources of light on the retina is called **accommodation.** The strength of the lens depends on its shape, which in turn is regulated in mammals, birds, and some species of reptiles by the ciliary muscle. However, if the lens has a fixed focal length, then accommodation can be achieved either by moving the lens (in fish) or by moving the photoreceptor layer. For example, some species of annelids can increase or decrease the fluid volume within the optic chamber to alter distance between the lens and the photoreceptors.

In mammals, the **ciliary muscle** is part of the ciliary body, an anterior specialization of the choroid layer. The ciliary body has two major components: the ciliary muscle and the capillary network that produces the aqueous humor (Figure 6–8). The ciliary muscle is a ring of smooth muscle attached to the lens by suspensory ligaments (● Figure 6–15).

Figure 6–15 ● Mechanism of accommodation. (a) Schematic representation of suspensory ligaments extending from the ciliary muscle to the outer edge of the lens. (b) Scanning electron micrograph showing the suspensory ligaments are attached to the lens. (c) When the ciliary muscle is relaxed, the suspensory ligaments are taut, putting tension on the lens so that it is flat and weak. (d) When the ciliary muscle is contracted, the suspensory ligaments become slack, reducing the tension on the lens. The lens can then assume a stronger, rounder shape because of its elasticity.

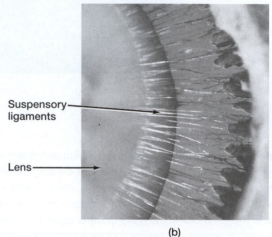

Suspensory ligaments

Lens

(b)

Photo: Patricia N. Farnsworth, Ph.D., Professor of Physiology and Ophthalmology, University of Medicine and Dentistry of New Jersey, New Jersey Medical School

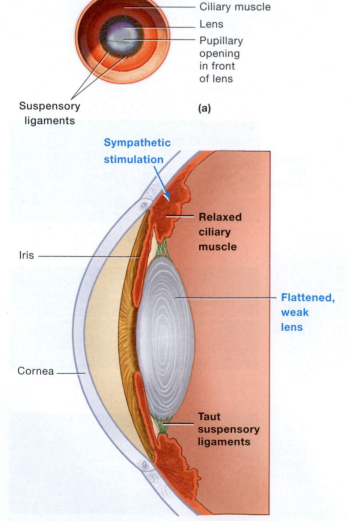

Ciliary muscle

Lens

Pupillary opening in front of lens

Suspensory ligaments

(a)

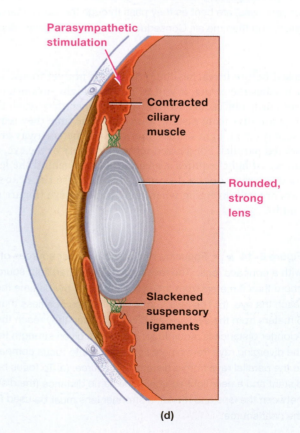

Sympathetic stimulation

Iris

Cornea

Relaxed ciliary muscle

Flattened, weak lens

Taut suspensory ligaments

(c)

Parasympathetic stimulation

Contracted ciliary muscle

Rounded, strong lens

Slackened suspensory ligaments

(d)

When the ciliary muscle is relaxed, the suspensory ligaments are taut, and they pull the lens into a flattened, weakly refractive shape (Figure 6–15). As the muscle contracts, its circumference decreases, slackening the tension in the suspensory ligaments (Figure 6–15). When the lens is subjected to less tension by the suspensory ligaments, it assumes a more spherical shape because of its inherent elasticity. The greater curvature of the more rounded lens increases its strength, bending light rays more strongly.

In the normal mammalian eye, the ciliary muscle is relaxed and the lens is flat for distant vision, but the muscle contracts to let the lens become more convex and stronger for near vision. The ciliary muscle is controlled by the autonomic nervous system. Sympathetic nerve fibers induce relaxation of the ciliary muscle for distant vision, whereas the parasympathetic nervous system causes the muscle's contraction for near vision. Occasionally the fibers of the lens in older animals progressively become so opaque that light rays cannot pass through, a condition known as a **cataract.**

■ **Light must pass through several retinal layers in vertebrates before reaching the photoreceptors.**

The major function of the eye is to focus light rays from the environment on the *rods* and *cones*, the photoreceptor cells of the retina. The photoreceptors then transform the light energy into electrical signals for transmission to the CNS.

The receptor-containing portion of the vertebrate retina is actually an extension of the CNS and not a separate peripheral organ. During embryonic development, the retinal cells "back out" of the nervous system, so the retinal layers, surprisingly, are facing backward! The neural portion of the retina consists of three layers of excitable cells (● Figure 6–16): (1) the outermost layer (closest to the choroid) containing the **rods** and/or **cones,** whose light-sensitive ends face the choroid (away from the incoming light); (2) a middle layer of **bipolar cells;** and (3) an inner layer of **ganglion cells.** Rods and cones were named for their appearance under a light microscope. In lower vertebrates, rods and cones converge via chemical synapses onto the same bipolar neuron, in contrast to the separate rod bipolar and cone bipolar pathways of higher vertebrates. Axons of the ganglion cells join together to form the optic nerve, which leaves the retina slightly off center. The point on the retina at which the optic nerve leaves and through which blood vessels pass is the **optic disc** (● Figures 6–8a and 6–17). This region is often called the **blind spot;** no image can be detected in this area because it lacks rods and cones. We are normally not aware of the blind spot, because central processing somehow "fills in" the missing spot. You can discover your own blind spot by a simple demonstration (● Figure 6–18).

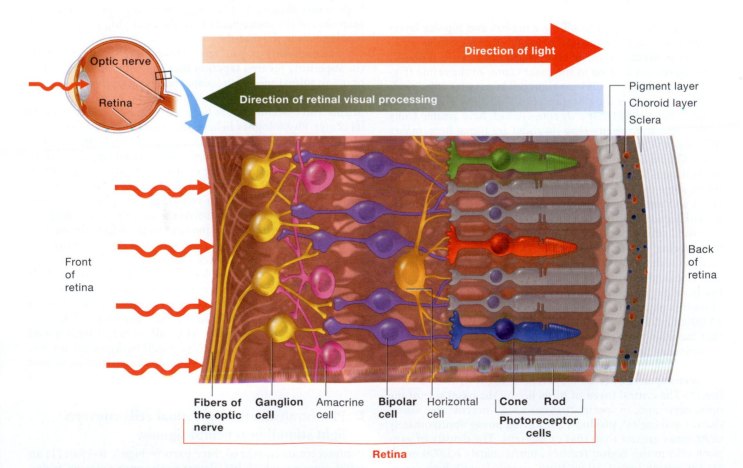

Figure 6–16 ● Retinal layers. The retinal visual pathway extends from the photoreceptor cells (rods and cones, whose light-sensitive ends face the choroid *away from* the incoming light) to the bipolar cells to the ganglion cells. The horizontal and amacrine cells act locally for retinal processing of visual input.

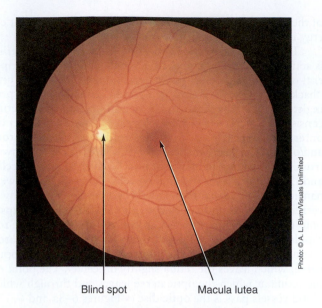

Blind spot   Macula lutea

Figure 6–17 • View of the retina seen through an
ophthalmoscope. With an ophthalmoscope, a lighted viewing
instrument, it is possible to view the optic disc (blind spot)
and macula lutea within the retina of the rear of the eye.

Figure 6–18 • Demonstration of the blind spot. Discover the
blind spot in your left eye by closing your right eye and holding the
book about 4 inches from your face. While focusing on the cross,
gradually move the book away from you until the circle vanishes
from view. At this time, the image of the circle is striking the blind
spot of your left eye. You can similarly discover the blind spot in
your right eye by closing your left eye and focusing on the circle.
The cross will disappear when its image strikes the blind spot
of your right eye.

Light must pass through the ganglion and bipolar layers
before reaching the photoreceptors in all areas of the retina
except the **fovea.** In the primate fovea, which is a pinhead-
sized depression located in the exact center of the retina (Figure 6–8a), the bipolar and ganglion cell layers are pulled aside
so that light strikes the photoreceptors directly. This feature,
coupled with the fact that *only* cones (which have greater acuity
or discriminative ability than do rods) are found here, makes
the fovea the point of greatest visual acuity (most distinct vision). Thus you turn your eyes so that the object at which you
are looking is focused on the fovea. Fovea are present in teleost
fishes, some snakes, and birds. Dogs (and wolves) lack a fovea
per se and instead have a **visual streak,** which serves as the region of highest visual acuity. In addition to a dense central area,
the visual streak extends into the temporal and nasal portions
of the retina. Physiologists believe these extensions enable animals to scan the horizon with enhanced visual acuity. The visual streak also contains a higher density of ganglion cells.
For example, the maximum density of ganglion cells is greater
in wolves as compared to domesticated canines (12,000 to
14,000/mm^2 versus 6,400 to 14,000/mm^2). Survival pressures
that maintain the elevated density of ganglion cells are likely
evolutionary factors in wolves, whereas breeding programs in
dogs place little selective pressure on maximizing visual acuity.

Some species of birds may have as many as two or three
foveas. The **central fovea** of birds lies on the nasal side of the
optic nerve and, in species that need high stereoscopic vision
(hawks and eagles), produces a resolving power approximately
eight times greater than that of humans. The density of ganglion cells in this region reaches approximately 65,000 mm^2,
far surpassing foveal values from mammals with high acuity
(human, 38,000 mm^2; macaque, 33,000 mm^2). A second **lateral fovea** is located above the optic nerve and a third area,
the so-called **linear area,** extends horizontally across the central portion of the retina. Nocturnal predators, such as owls,

must summate light from both eyes under dim conditions and
are characterized by a single temporal fovea. In mammals, the
area immediately surrounding the fovea, the **macula lutea** (yellow spot), also has a high concentration of cones and fairly
high acuity (Figure 6–17). Macular acuity, however, is less
than that of the fovea because of the overlying ganglion and
bipolar cells in the macula.

In conditions of dim light, a layer of reflecting material,
the superiorly located **tapetum lucidum,** enhances the ability
of many animal species to locate and detect distant objects. In
dogs the tapetum lucidum is a highly cellular structure enriched
in zinc and cysteine that is between 9 and 20 layers thick at
its center. Physiologists believe the tapetum reflects light that
has previously passed through the retina back through it a
second time, giving photoreceptors a second opportunity to
capture the light stimuli. For example, the feline eye reflects
approximately 130 times more light than does the human eye.
However, there is a price for reflecting this light: The ability
of the eye to accurately resolve the details of an image is
somewhat compromised by the scattering of light during this
process. This reflecting layer accounts for the eye shine you
can see in animals staring back into a source of light.

The tapetum of cats and lemurs is rich in riboflavin, which
can absorb light in the shorter wavelengths (blue, 450 nm) and
then emit fluorescent light at a longer wavelength (520 nm)
more closely approximating the maximal sensitivity of rhodopsin in the rod photoreceptors. This shift in wavelength would
brighten the appearance of a blue-black background and enhance the contrast between objects in the environment and
the background.

Phototransduction by retinal cells converts light stimuli into neural signals.

Photoreceptors consist of three parts (● Figure 6–19a): (1) an
outer segment, which lies closest to the eye's exterior, facing
the choroid, and detects the light stimulus; (2) an *inner segment,* which lies in the middle of the photoreceptor's length
and contains the metabolic machinery of the cell; and (3) a
synaptic terminal, which lies closest to the eye's interior, fac-

ing the bipolar cells, and transmits the signal generated in the photoreceptor on light stimulation to these next cells in the visual pathway. The outer segment, which is rod-shaped in rods and cone-shaped in cones (Figure 6–19b), consists of stacked, flattened, membranous discs containing an abundance of photopigment molecules. Each retina has about 150 million photoreceptors, and over a billion photopigment molecules may be packed into the outer segment of each photoreceptor.

Photopigments within the photoreceptor undergo chemical alterations when activated by light. A photopigment consists of an enzymatic protein called **opsin** combined with **retinene**, a derivative of vitamin A. In birds, there are five different photopigments, one in the rods and one in each of four types of cones. Retinene is identical in all five photopigments, but the photoreceptors' opsins vary slightly, enabling the photopigments to differentially absorb various wavelengths of light. **Rhodopsin**, the vertebrate rod photopigment, cannot discriminate between various wavelengths in the visible spectrum; it absorbs all visi-

ble wavelengths. Therefore, rods provide vision only in shades of gray by detecting different intensities, not different colors. The rod pigment found in the microvilli of insect rhabdomeric photoreceptors is called **porphyropsin** (3-dehydroretinal), in part because it is more purple in hue than the vertebrate rhodopsin. Porphyropsin and rhodopsin can be found together in the retina of many vertebrate animals, including amphibians and newts. The photopigments in the four types of vertebrate cones—**red, green, blue,** and **ultraviolet** cones—respond selectively to various wavelengths of light, making color vision possible. Some species of insects also can distinguish color, including ultraviolet and polarized light. For example, the honey bee can distinguish blue and yellow but not red wavelengths. The mechanism by which the bee perceives color is less well understood than in vertebrates but is believed to result from retinal cells with differing sensitivities to light of different wavelengths.

Phototransduction, the mechanism of excitation, is basically the same for all vertebrate photoreceptors (Figure 6–19b).

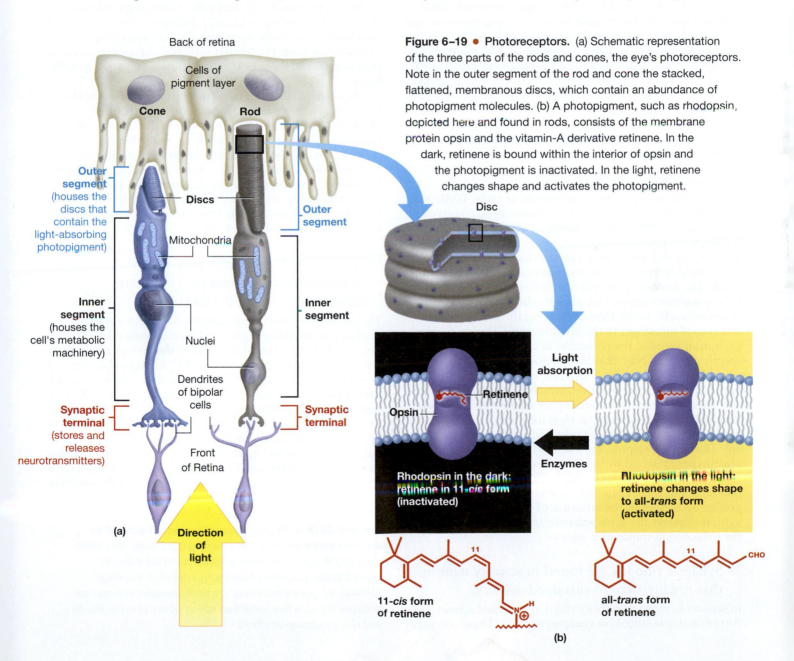

Figure 6–19 ● Photoreceptors. (a) Schematic representation of the three parts of the rods and cones, the eye's photoreceptors. Note in the outer segment of the rod and cone the stacked, flattened, membranous discs, which contain an abundance of photopigment molecules. (b) A photopigment, such as rhodopsin, depicted here and found in rods, consists of the membrane protein opsin and the vitamin-A derivative retinene. In the dark, retinene is bound within the interior of opsin and the photopigment is inactivated. In the light, retinene changes shape and activates the photopigment.

When a photopigment molecule absorbs light it dissociates into its retinene and opsin components, and its retinene portion changes shape, triggering the enzymatic activity of opsin. Through a series of steps, this light-induced breakdown and subsequent activation of the photopigment bring about a hyperpolarizing receptor potential that influences transmitter release from the synaptic terminal of the photoreceptor. Ultimately, as you will see, phototransduction is a graded response, as well as a hyperpolarization, and as such is distinctly different from the "all-or-none" response of typical nerves.

Photoreceptor Activity in the Dark

The plasma membrane of a photoreceptor's outer segment contains chemical messenger–gated Na^+ channels. Unlike other chemical-gated channels that respond to external chemical messengers, these channels respond to an internal second messenger, **cyclic GMP** or **cGMP** (cyclic guanosine monophosphate). Binding of cGMP to these Na^+ channels keeps them open. In the absence of light, the concentration of cGMP is high (● Figure 6–20a). Unlike most receptors, therefore, the Na^+ channels of a vertebrate photoreceptor are open in the absence of stimulation, that is, in the dark. The resultant passive inward Na^+ leak depolarizes the photoreceptor. The passive spread of this depolarization from the outer segment (where the Na^+ channels are located) to the synaptic terminal (where the photoreceptor's neurotransmitter is stored) keeps the synaptic terminal's voltage-gated Ca^{2+} channels open. Calcium entry triggers the release of transmitter from the synaptic terminal while in the dark.

Photoreceptor Activity in the Light

On exposure to light, the concentration of cyclic GMP is decreased through a series of biochemical steps triggered by photopigment activation (Figure 6–20b). When retinene absorbs light, it changes shape but still remains bound to opsin. This change in conformation activates the photopigment. Rod and cone cells contain a G protein (see p. 90) called **transducin**. The activated photopigment activates transducin, which in turn activates the enzyme **phosphodiesterase**. This enzyme degrades cyclic GMP, thus decreasing the concentration of this second messenger in the photoreceptor. During the light-excitation process, the reduction in cyclic GMP permits the chemically gated Na^+ channels to close. This channel closure stops the depolarizing Na^+ leak and hyperpolarizes the membrane. This hyperpolarization, which is the receptor potential, passively spreads from the outer segment to the synaptic terminal of the photoreceptor. Here the potential change leads to a reduction in transmitter release from the synaptic terminal. Thus photoreceptors are inhibited by their adequate stimulus (hyperpolarized by light) and excited in the absence of stimulation (depolarized by darkness). The hyperpolarizing potential and subsequent decrease in transmitter release are graded according to the intensity of light. The brighter the light, the greater the hyperpolarizing response and the greater the reduction in transmitter release.

■ Synaptic ribbons are found in sensory neurons that are involved in sustained activity.

In sensory neurons that are involved in sustained activity, neurotransmitter is stored on **synaptic ribbons**. These compara-

tively large "docking sites" for neurotransmitter provide a source of approximately 100 readily releasable vesicles per ribbon. Each terminal contains 50 to 60 ribbons, which together tether between 5000 to 6000 vesicles. Each vesicle is primed by ATP and moves down the synaptic ribbon as if carried along on a conveyer belt, toward the presynaptic membrane, where they are exocytosed at the presynaptic membrane.

Further Retinal Processing of Light Input

How does the retina signal the brain about light stimulation through such an inhibitory response? The photoreceptors synapse with bipolar cells. These cells in turn terminate on the ganglion cells, whose axons form the optic nerve for transmitting signals to the brain. The answer to our seeming paradox lies in the fact that the transmitter released from the photoreceptors' synaptic terminal has an *inhibitory* action on the bipolar cells. The reduction in transmitter release that accompanies light-induced receptor hyperpolarization decreases this inhibitory action on the bipolar cells. Removing inhibition has the same effect as directly exciting the bipolar cells. The greater the illumination on the receptor cells, the greater the removal of inhibition from the bipolar cells and the greater in effect the excitation of these next cells in the visual pathway to the brain.

Bipolar cells display graded potentials similar to the photoreceptors. Action potentials do not originate until the ganglion cells, the first neurons in the chain that must propagate the visual message over long distances to the brain.

The altered photopigments are restored to their original conformation in the dark by enzyme-mediated mechanisms (Figure 6–19b). Subsequently, the membrane potential and rate of transmitter release of the photoreceptor are returned to their unexcited state, and no action potentials are transmitted to the visual cortex (Figure 6–20a).

■ Rods provide indistinct gray vision at night, whereas cones provide sharp color vision during the day.

The earliest mammals are believed to have been nocturnal and thus characterized by a pure rod retina, comparable to that found in living nocturnal animals. In the human retina there are 30 times more rods than cones (100 million rods compared to 3 million cones per eye). Cones, however, are most abundant in the center of the retina in the macula, although the proportion of cones in this region varies between species. For example, in this region of the canine retina there are fewer than 10% cones, whereas diurnal squirrels have an all-cone retina. Other rodents, including the American flying squirrel (*Glaucomys volans*), have an all-rod retina. Diurnal birds have a greater number of cones, as well as a greater cone density, than

Figure 6–20 ▶ ● **Phototransduction and initiation of an action potential in the vertebrate visual pathway.** (a) Events occurring in a photoreceptor in response to the dark that prevent action potentials from being initiated in the visual pathway. (b) Events occurring in a photoreceptor in response to a light stimulus that initiate an action potential in the visual pathway (phototransduction).

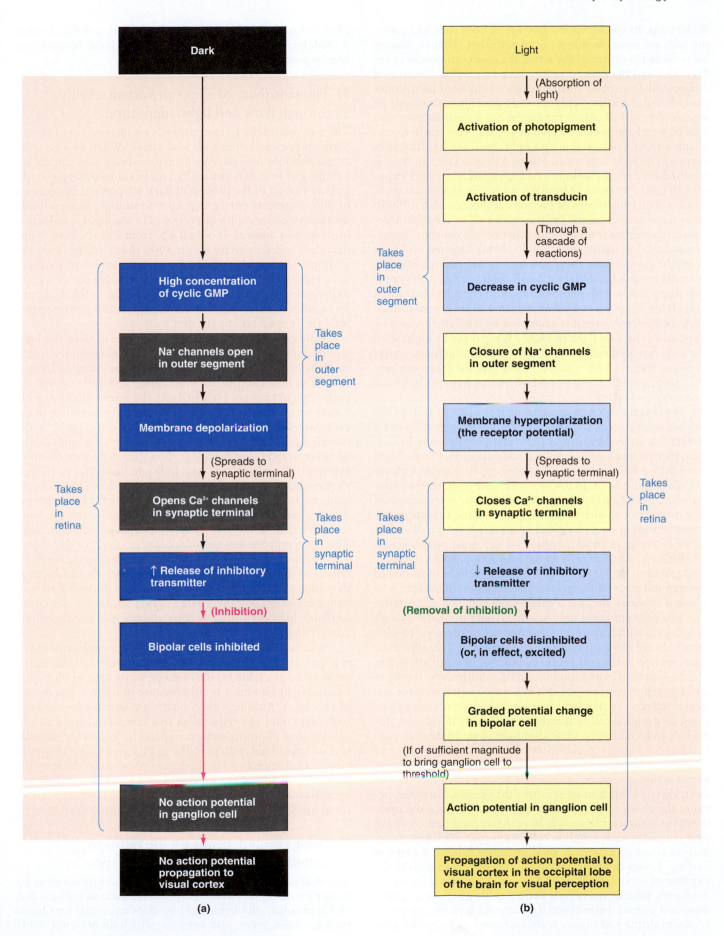

Dark

High concentration
of cyclic GMP

Na⁺ channels open
in outer segment

Membrane depolarization

(Spreads to
synaptic terminal)

Opens Ca²⁺ channels
in synaptic terminal

↑ Release of inhibitory
transmitter

↓ (Inhibition)

Bipolar cells inhibited

No action potential
in ganglion cell

No action potential
propagation to
visual cortex

Takes
place
in
outer
segment

Takes
place
in
retina

Takes
place
in
synaptic
terminal

(a)

Light

(Absorption of
light)

Activation of photopigment

Activation of transducin

(Through a
cascade of
reactions)

Decrease in cyclic GMP

Closure of Na⁺ channels
in outer segment

Membrane hyperpolarization
(the receptor potential)

(Spreads to
synaptic terminal)

Closes Ca²⁺ channels
in synaptic terminal

↓ Release of inhibitory
transmitter

(Removal of inhibition)

Bipolar cells disinhibited
(or, in effect, excited)

Graded potential change
in bipolar cell

(If of sufficient magnitude
to bring ganglion cell to
threshold)

Action potential in ganglion cell

Propagation of action potential to
visual cortex in the occipital lobe
of the brain for visual perception

Takes
place
in
outer
segment

Takes
place
in
synaptic
terminal

Takes
place
in
retina

(b)

do humans. In species with a mixture of both rods and cones, the rods are most abundant in the periphery. That is, moving away from the center of the retina the concentration of cones decreases and the concentration of rods increase. Because of differential absorption of specific wavelengths of light, cones provide color vision, whereas rods provide vision only in shades of gray. In humans rhodopsin has a peak sensitivity to light of wavelengths in the region of 496 nm, whereas in species adapted to function well in dim light, such as dogs, the peak sensitivity to light is between 506 to 510 nm. Over an hour is required for canine rhodopsin to regenerate itself after exposure to light, whereas in humans it regenerates much more quickly. The capabilities of the rods and cones also differ in other respects because the "wiring patterns" in these photoreceptor types differ from those in other retinal neuronal layers. Cones have low sensitivity to light, being "turned on" only by bright daylight, but they have high acuity (sharpness; ability to distinguish between two nearby points). Thus cones provide sharp vision with high resolution for fine detail. Diurnal animals use cones for day vision, which is in color and distinct. Rods, in contrast, have low acuity but high sensitivity, so they respond to the dim light of night. We can see at night with our rods but at the expense of color and distinctness, unless of course there is such a bright moon that the cones can perceive some color. Let's now see how wiring patterns influence sensitivity and acuity.

There is little convergence of neurons (see p. 133) in the retinal pathways for cone output. Each cone generally has a private line connecting it to a particular ganglion cell. In contrast, there is much convergence in rod pathways. Output from more than a hundred rods may converge via bipolar cells on a single ganglion cell. In lower vertebrates (fish, turtles, salamanders, toads), many thousands of rods are interconnected by gap junctions, whereas in mammals (cats, rabbits, monkeys) the rods are not electrically connected to each other but rather converge in small numbers via gap junctions on cones.

Before a ganglion cell can have an action potential that is propagated to the CNS, the cell must be brought to threshold through influence of the graded potentials in the receptors to which it is wired. Because a single-cone ganglion cell is influenced by only one cone, only bright daylight is intense enough to induce a sufficient receptor potential in the cone to ultimately bring the ganglion cell to threshold. The abundant convergence in the rod visual pathways, in contrast, offers good opportunities for summation of subthreshold events in a rod ganglion cell (see p. 129). Whereas a small receptor potential induced by dim light in a single cone is not enough to bring its ganglion cell to threshold, similar small receptor potentials induced by the same dim light in multiple rods converging on a single ganglion cell have an additive effect to bring the rod ganglion cell to threshold. Because rods can bring about action potentials in response to very small amounts of light, they are much more sensitive than cones. However, because the cones of primates have private lines into the optic nerve (in contrast, researchers have determined that cats have four cones for each ganglion cell), each cone transmits information about an extremely small receptive field on the retinal surface. Cones can thus provide highly detailed vision at the expense of sensitivity. With rod vision, acuity is sacrificed for sensitivity. Because many rods share a single ganglion cell, once an action potential is initiated it is impossible to discern which of the multiple rod inputs were activated to bring the gan-

glion cell to threshold. Objects appear fuzzy when rod vision is used, because of this poor ability to distinguish between two nearby points.

■ The sensitivity of eyes can vary markedly through dark and light adaptation.

The eyes' sensitivity to light depends on the amount of photopigment present in the rods and cones. When you go from bright sunlight into darkened surroundings, you cannot see anything at first, but gradually you begin to distinguish objects as a result of the process of **dark adaptation.** Breakdown of photopigments during exposure to sunlight tremendously decreases photoreceptor sensitivity. For example, a reduction in rhodopsin content of only 0.6% from its maximum value decreases rod sensitivity about 3000 times. In the dark, the photopigments broken down during light exposure are gradually regenerated. As a result, the sensitivity of your eyes gradually increases so that you can begin to see in the darkened surroundings. However, only the highly sensitive, rejuvenated rods are "turned on" by the dim light.

Conversely, when you move from the dark to the light (for example, leaving a movie theater and entering the bright sunlight), your eyes are very sensitive to the dazzling light at first. With little contrast between lighter and darker parts, the entire image appears bleached. As the intense light rapidly breaks some of the photopigments down, the sensitivity of the eyes decreases and normal contrasts can once again be detected, a process known as **light adaptation.** The rods are so sensitive to light that sufficient rhodopsin is broken down to essentially "burn out" the rods in bright light; that is, the rod photopigments, having already been broken down by the bright light, are no longer able to respond to the light. Furthermore, a central neural adaptive mechanism switches the eye from the rod system to the cone system on exposure to bright light. Therefore, only the less-sensitive cones are used for day vision.

Researchers estimate that human eyes' sensitivity can change as much as 1 million times with each change event as they adjust to various levels of illumination through dark and light adaptation. Pupillary reflexes that adjust the amount of available light permitted to enter the eye also enhance these adaptive measures.

Because retinene, one of the photopigment components, is a derivative of vitamin A, adequate amounts of this nutrient must be available for the ongoing resynthesis of photopigments. **Night blindness** occurs because of dietary deficiencies of vitamin A. Although photopigment concentrations in both rods and cones are reduced in this condition, there is still enough cone photopigment to respond to the intense stimulation of bright light, except in the most severe cases. However, even modest reductions in rhodopsin content can decrease the sensitivity of rods so much that they cannot respond to dim light. Such an animal can see in the day using cones but cannot see at night because the rods are no longer functional.

■ Color vision depends on the ratios of stimulation of the various cone types.

Vision depends on stimulation of retinal photoreceptors by light. Certain objects in the environment such as the sun, fire, and light bulbs, emit light. But how does an animal see objects such as chairs, trees, and animals, which do not emit light?

The pigments in various objects selectively absorb particular wavelengths of light transmitted to them from light-emitting sources, and the unabsorbed wavelengths are reflected from the objects' surfaces. These reflected light rays enable animals to see the objects. With an object perceived as blue (in humans), such as the open ocean, water molecules absorb the longer red and green wavelengths of light and scatter the shorter blue wavelengths back to the observer, which can be absorbed by the photopigment in the eyes' blue cones, thereby activating them.

Color vision occurs in some nonprimate mammals, birds, reptiles, amphibians, fish, and insects. Most orders of mammals have been shown to have color vision in that they can discriminate based on color, although the number and specificity of each cone type depend on the species. For example, whales completely lack blue cones, dogs and cats have two types of cones (and are called *dichromatic*), and primates have three types. Baboons have primarily green (63%) and red (33%) cones, whereas only 4% are blue. Most mammals are in fact thought to be dichromatic. Many invertebrates and some species of birds, fish, marsupial mammals, and rodents have a fourth type of cone, which detects **ultraviolet** light (although there are not necessarily four types of cones in any one species). Honey bees, for example, can detect the pattern of ultraviolet, **polarized** light (light in which the waves are all oriented at the same angle) in the sky and use this information to communicate the location of a food source to the worker bees in the hive. Polarized light is that which is vibrating predominantly in one plane, the angle of which changes with respect to the movement of the sun. (This form of communication can be eliminated experimentally by using filters that specifically block the passage of ultraviolet wavelengths, to determine the role of polarized light in animal behavior.) The detection of ultraviolet wavelengths may also give birds and fish spatial cues that they can use in orientation. For example, ultraviolet light may play a role in navigation in birds. Short-wavelength gradients vary depending on the sun's angle in the sky. These gradients increase in saturation as the sun moves from directly overhead to an angle of 90 degrees (that is, ultraviolet light is a greater proportion of the incoming sunlight at twilight). Some species of salmon migrate over 3000 kilometers and rely on color gradients in the sky to locate the position of the sun on days when clouds obscure it.

Food detection is another use of ultraviolet sensing. Because scent markings (urine and feces) of small mammals are visible in ultraviolet light, kestrels can use this information to locate areas of food abundance. Recently, researchers have shown that a flower bat (*Glossophaga soricina*) senses ultraviolet; although it cannot distinguish colors, it may be able to see flowers better at twilight, when ultraviolet radiation is relatively more prevalent. Polarized light can also be used to find potential mates, prey, and predators by the pattern of polarized light reflected from the body surfaces. For example, researchers have recently shown that one species of butterfly can detect patterns of polarized light reflecting off the wings of others of its kind, a signal used in mating.

Each cone type is most effectively activated by a particular wavelength of light in the range of color indicated by its name—ultraviolet, blue, green, or red. In the human retina, cones that respond to short wavelengths of light in the region of 420 nm are termed **short cones** and they contribute to the perception of blue. Cones that respond best to more intermediate wavelengths of 530 nm are termed **middle cones**, and

contribute to the perception of green, whereas cones that are most sensitive to wavelengths in the region of 560 nm are termed **long cones** and contribute to the perception of red. Insects cannot perceive the color red and confuse it with black or gray. Animals that can perceive ultraviolet waves have a fourth type of cone, the **ultraviolet cone**, with a peak response in the region of 380 nm (fishes living in shallow waters contain UV-sensitive cones with a peak at about 360 nm). However, cones also respond in varying degrees to other wavelengths (● Figure 6–21). An animal's perception of the many colors of the world depends on the cone types' various *ratios of stimulation* in response to different wavelengths. A wavelength perceived as blue does not stimulate red or green cones at all but excites blue cones maximally (the percentages of maximal stimulation for red, green, and blue cones, respectively, are 0:0:100). The sensation of yellow, in comparison, arises from a stimulation ratio of 83:83:0, red and green cones each being stimulated 83% of maximum, whereas blue cones are not excited at all. The ratio for green is 31:67:36, and so on, with various combinations giving rise to the sensation of all the different colors. Interestingly, the absorption spectra of all the cones overlap in the ultraviolet wavelengths. For this reason, ultraviolet light stimulates not only the ultraviolet cones, but also the long cones, the middle cones, and the short cones. White is a mixture of all wavelengths of light, whereas black is the absence of light.

The extent that each of the cone types is excited is coded and transmitted in separate parallel pathways to the brain. A distinct color vision center in the primary visual cortex has recently been identified. This center combines and processes these inputs to generate the perception of color, taking into consideration the object in comparison with its background.

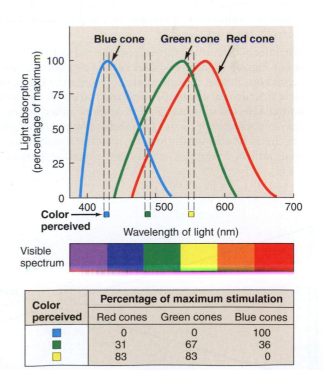

Figure 6–21 ● **Sensitivity of the three types of cones to different wavelengths.** The ratios of stimulation of the three cone types are shown for three sample colors.

?

How Many Types of Cones Do Animals Have? The number of distinct types of cones in vertebrates has long been thought quite limited. For example, for more than two centuries scientists have accepted the three-cone system in humans as the standard model of color vision. But new evidence suggests that reception of color may be more complex. Genomic analysis has shown that men with normal color vision have a variable number of genes coding for cone pigments. For example, many had multiple genes (from two up to four) for red-light detection and could distinguish more subtle differences in colors in this long-wavelength range than those with single copies of red-cone genes. This finding will undoubtedly lead to a re-evaluation of how the various photopigments contribute to color vision in humans, and suggests that geneticists should examine genes for vision in other species, to see if similar genetic diversity occurs.

■ Visual information is separated and modified within the visual pathway before it is integrated into a perceptual image of the visual field by the cortex.

The field of view that can be seen without moving the head is known as the **visual field.** The information that reaches the visual cortex in the occipital lobe is not a replica of the visual field for several reasons.

1. The image detected on the retina at the onset of visual processing is upside down and backward, because the light rays bend (● Figure 6–22). Once projected to the brain, the animal interprets the inverted image as being correctly oriented.

2. The information transmitted from the retina to the brain is not merely a point-to-point record of photoreceptor activation. Before the information reaches the brain, the retinal neuronal layers beyond the rods and cones reinforce selected information and suppress other information to enhance contrast. One mechanism of retinal processing is lateral inhibition, by which strongly excited cone pathways suppress activity in surrounding pathways of weakly stimulated cones. This increases the dark–bright contrast to enhance the sharpness of boundaries.

Another mechanism of retinal processing involves differential activation of two types of ganglion cells, **on-center** and **off-center ganglion cells.** The receptive field of a cone ganglion cell is determined by the field of light detection by the cone with which it is linked. On-center and off-center ganglion cells respond in opposite ways, depending on the relative comparison of illumination between the center and periphery of their receptive fields. Think of the receptive field as a doughnut. An on-center ganglion cell increases its rate of firing when light is most intense at the center of its receptive field (that is, when the doughnut hole is lit up). In contrast, an off-center cell increases its firing rate when the periphery of its receptive field is most intensely illuminated (that is, when the doughnut itself is lit up). This is useful for enhancing the difference in light level between one small area at the center of a receptive field and the illumination immediately around it. By emphasizing differences in relative brightness, this mechanism helps define contours of images, but in so doing, information about absolute brightness is sacrificed (● Figure 6–23).

3. Various aspects of visual information such as form, color, depth, and movement are separated and projected in parallel pathways to different regions of the cortex. Only when these separate bits of processed information are integrated by higher visual regions is a reassembled picture of the visual scene perceived. This is similar to the blobs of paint on an artist's palette versus the finished portrait; the separate pigments do not represent a portrait of a face until they are appropriately integrated on a canvas.

4. Because of the pattern of wiring between the eyes and the visual cortex, the left half of the cortex receives information only from the right half of the visual field as detected by both eyes, and the right half receives input only from the left half of the visual field of both eyes.

As a result of refraction, light rays from the left half of the visual field fall on the right half of the retina of both eyes (the medial or inner half of the left retina and the lateral or outer half of the right retina) (● Figure 6–24). Similarly, rays from the right half of the visual field reach the left half of each retina (the lateral half of the left retina and the medial half of the right retina). Each optic nerve exiting the retina carries information from both halves of the retina it serves. The human optic nerve contains 1.2 million nerve fibers; compare this to 167,000 in canines and 116,000 to 165,000 in the optic nerve of cats. This information is separated as the optic nerves meet at the **optic chiasm** (*chiasm* means "cross") located underneath the hypothalamus. Within the optic chiasm, the fibers from the medial half of each retina cross to the opposite side, but those from the lateral half remain on the original side. The reorganized bundles of fibers leaving the optic chiasm are known as *optic tracts.* Each optic tract carries information from the lateral half of one retina and the medial half of the

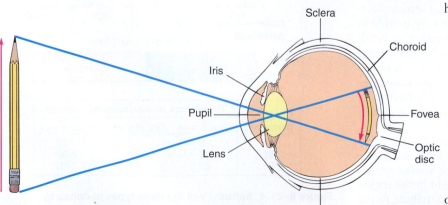

Figure 6–22 ● Inversion of the image on the retina.

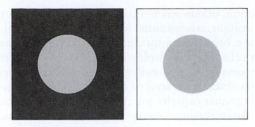

Figure 6–23 ● Example of the outcome of retinal processing by on-center and off-center ganglion cells. Note that the gray circle surrounded by black appears brighter than the one surrounded by white, even though the two circles are identical (same shade and size). Retinal processing by on-center and off-center ganglion cells is largely responsible for enhancing differences in relative (rather than absolute) brightness, which helps define contours.

other retina. Therefore, this partial crossover brings together from the two eyes fibers that carry information from the same half of the visual field. Each optic tract, in turn, delivers to the half of the brain on its same side information about the opposite half of the visual field.

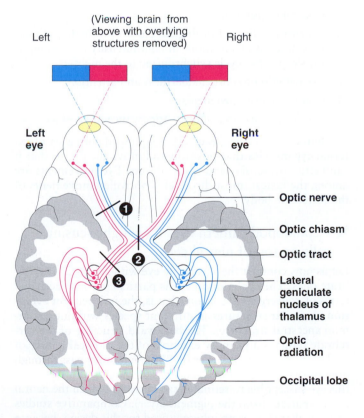

Figure 6–24 ● The mammalian visual pathway and visual deficits associated with lesions in the pathway. Note that the left half of the visual cortex in the occipital lobe receives information from the right half of the visual field of both eyes (in blue), and the right half of the cortex receives information from the left half of the visual field of both eyes (in red).

■ The thalamus and visual cortices elaborate the visual message.

The first stop in the mammalian brain for information in the visual pathway is the **lateral geniculate nucleus** in the thalamus (Figure 6–24) (**optic tectum** in fish and amphibians). It separates information received from the eyes and relays it via fiber bundles known as **optic radiations** to different zones in the cortex, each of which processes different aspects of the visual stimulus (for example, color, form, depth, movement). This sorting process is no small task, because each optic nerve contains more than 1 million fibers carrying information from the photoreceptors in one human retina. This is more than all the afferent fibers carrying somatosensory input from all the regions of the body! Researchers estimate that hundreds of millions of neurons, occupying about 30% of the human cortex, participate in visual processing, compared to 8% devoted to touch perception and 3% to hearing. Yet the connections in the visual pathways are precise. The lateral geniculate nucleus and each of the zones in the cortex that processes visual information have a topographic map representing the retina point for point. As with the somatosensory cortex, the neural maps of the retina are distorted. The fovea, the retinal region capable of greatest acuity, has much greater representation in the neural map than do the more peripheral regions of the retina. Interestingly, in animals born with immature visual systems such as the ferret and cats, during the first three to four weeks after birth while their eyes remain closed, the retinal ganglion cells are still developing connections among themselves as well as to other retinal cells. The long axons of the ganglion cells grow along the optic tract to each lateral geniculate nucleus of the thalamus. These ganglion cells eventually grow into layers in the lateral geniculate nucleus, although axons from each eye remain separate from each other.

Depth Perception

Although each half of the visual cortex receives information simultaneously from the same part of the *visual field* as received by both eyes, the messages from the two eyes are not identical. Depending on the species of animal, there is a potential area of overlap (● Figure 6–25). The overlapping area seen by both eyes at the same time is known as the **binocular** ("two-eyed") **field** of vision, which is important for **depth perception**. Binocular vision is achieved in part by routing axons from one eye together with axons from the other eye to synapse in the same layers of the lateral geniculate nucleus. Depending on placement of the eyes in the skull, both the extent of the visual field and the amount of binocular overlap can vary considerably. For example, there is no evidence for binocular vision in teleost fish, where eyes are located on opposite sides of the head. Dog eyes are laterally directed approximately 20 degrees from the midline, and human eyes look straight ahead. Thus with each eye the average dog has a monocular field of view from 135 to 150 degrees, approximately 60 to 70 degrees greater than that of humans. Although this gives dogs a greater ability to scan the horizon, the degree of binocular overlap is much less (about 30 to 60 degrees compared to 130 in cats and 140 degrees in humans). Interestingly, in some species of birds the effect of two or more foveas enables each eye to have completely separate visual fields. Such an arrangement increases the total visual field and may also permit some binocular vision.

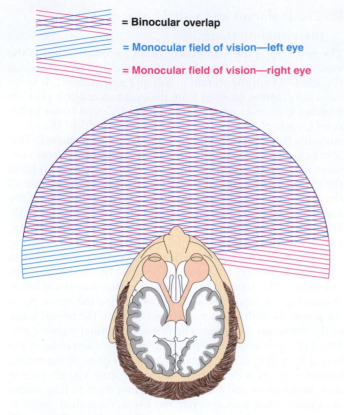

= Binocular overlap

= Monocular field of vision—left eye

= Monocular field of vision—right eye

Figure 6–25 • Visual fields.

Because most adult insects have a pair of compound eyes (as we describe shortly) that bulge out of each side of the head, the field of vision is markedly enhanced. For example, in the water bug *Notonecta* the field of view for the two eyes provides binocular vision in front (almost 250 degrees in the horizontal plane), above and below the head.

Like other areas of the vertebrate cortex, the primary visual cortex is organized into functional columns, each processing information from a small region of the retina. Independent alternating columns are devoted to information about the same point in the visual field from the right and left eyes. The brain uses the slight disparity in the information received from the two eyes to estimate distance, allowing an animal to perceive three-dimensional objects in spatial depth.

Within the cortex, visual information is first processed in the primary visual cortex, then is projected to higher-order visual areas for even more complex processing and abstraction. The cortex contains a hierarchy of visual cells that respond to increasingly complex stimuli. Three types of visual cortical neurons have been identified based on the complexity of stimulus requirements needed for the cell to respond; these are called **simple, complex,** and **hypercomplex cells.** Simple and complex cells are stacked on top of each other within the cortical columns of the primary visual cortex, whereas hypercomplex cells are found in the higher visual processing areas. Unlike a retinal cell, which responds to the amount of light, a cortical cell fires only when it receives a particular pattern of illumination for which it is programmed. These patterns are built up by converging connections that originate from closely aligned photoreceptor cells in the retina. For example, some simple cells fire only when a bar is viewed vertically in a spe-

cific location, others when a bar is horizontal, and others at various oblique orientations. Movement of a critical axis of orientation becomes important for response by some of the complex cells. Hypercomplex cells add a new dimension to visual processing by responding only to particular edges, corners, and curves. Each level of cortical visual neurons has increasingly greater capacity for abstraction of information built up from the increasing convergence of input from lower-order neurons. In this way, the dotlike pattern of photoreceptors stimulated to varying degrees by varying light intensities in the retinal image is transformed in the cortex into information about depth, position, orientation, movement, contour, and length. Other aspects of this information, such as color perception, are processed simultaneously through a similar hierarchical organization. How and where the entire image is finally put together is still unresolved.

■ Visual input goes to other areas of the vertebrate brain not involved in vision perception.

Not all fibers in the visual pathway terminate in the visual cortices. Some are projected to other regions of the brain for purposes other than direct vision perception. The following are examples of nonsight activities dependent on input from the retina:

1. Control of pupil size
2. Synchronization of biological clocks (in the hypothalamus) to cyclical variations in light intensity (for example, the sleep–wake cycle synchronized to the night–day cycle)
3. Contribution to cortical alertness and attention
4. Control of eye movements
5. Startle response to a sudden appearance of an object

In the latter regard, each eye is equipped with a set of **external eye muscles** that position and move the eye so that it can better locate, see, and track objects. Eye movements are among the fastest, most discretely controlled movements of the vertebrate body.

■ Cephalopod camera eyes have light-sensing cells on top of neural cells.

Let us now contrast the vertebrate eye with advanced invertebrate ones. The cephalopod eye is particularly instructive in terms of *convergent evolution* (that is, the independent evolution of similar structures in different species without a common ancestral structure). The backward structure of the vertebrate retina is a primary example of the illogical nature of evolution. As you saw earlier (p. 209), light must pass through nonsensory cell layers first; and this arrangement also makes this eye susceptible to retinal detachment (in which the neural layer separates from the pigment layer). Comparative studies show us that there is no inherent need for this design, because the cephalopod eye has evolved the reversed, more "logical" arrangement. The eye of the squid, octopus, cuttlefish, and nautiloids is remarkably similar to the vertebrate eye in having a cornea, lens, and retina, but there the resemblance ends. In these eyes, the pigment (sensory) cells are on top of the retina and receive light directly from the lens. The difference between cephalopod and vertebrate retinas seems to be an accident of evolution, arising from the different ways in which this sensory system evolved out of the nervous system.

Compound eyes found in some invertebrates consist of multiple image-forming units.

The functional unit of the arthropod faceted, or compound, eye is termed the **ommatidium** (● Figure 6–26). In addition to arthropods, compound eyes occur in some annelids and mollusks. Each ommatidium consists of an optical, light-gathering part as well as a sensory portion, which transduces light into an action potential. The number of ommatidia can vary from one, in the worker ant *Ponera punctatissima*, to over 10,000 in the eye of a dragonfly. Generally, in animals with only a few ommatidia the **facets** (external surface of an individual compound eye unit or ommatidium) are circular in appearance, whereas with increased numbers the facets are more densely packed together and assume a hexagonal shape (● Figure 6–27a). Each of these optical units focuses on a separate portion of the visual field, with the conglomerate image appearing as a mosaic rather than the sharp visual image perceived by vertebrates (Figure 6–27b). This is because vertebrate photoreceptors sample approximately 0.02 degrees of the visual field whereas each ommatidium views as much as 2 to 3 degrees. Consequently, the visual fields of adjacent ommatidia overlap. Compound eyes also have lower visual acuity than do vertebrate eyes, because of the comparatively larger receptive field of each individual unit. In addition, compound eyes form an upright retinal image, in contrast to the inverted image displayed on the vertebrate retina.

The cuticular lens of arthropod compound eyes forms a biconvex corneal lens located at the outer end of each ommatidium. Normally the cuticle forming the lens is transparent and colorless.

In addition to having its own lens system, each ommatidia contains a cluster of at least eight **retinular** or **photoreceptor cells** (Figures 6–26b and 27a). The retinular cells are arranged in a circle with the periphery lined with pigment to shield against leakage of light from neighboring ommatidia. At the center of the circle lies a dendrite from a single, modified retinular cell, the so-called **eccentric cell.** The photopigment rhodopsin is localized in a specialized region of the retinular cell called the **rhabdomere** (Figure 6–26c). Normally the rhabdomeres of each retinular cell are tightly packed along this central region of the ommatidium. Structurally, rhabdomeres appear as microvilli that increase the surface area available for light absorption. This elongated region of photosensitive microvilli, with contributions from each of the eight or more rhabdomeres, is termed the **rhabdom** of the ommatidium (Figure 6–26c). The body of the eccentric cell lies on the periphery of the ommatidium, with the dendrite making direct connections with the rhabdom.

To enhance visual contrast at a light–dark edge, the eccentric cells interact with each other by the process of lateral inhibition that we briefly noted earlier. Lateral inhibition also diminishes the apparent difference in brightness between areas that are uniformly illuminated, whereas in darkness the ommatidia are less stimulated by light but are comparatively less inhibited by their neighbors. Each eccentric cell sends collaterals, which make inhibitory synaptic contacts with its neighbor cells lying approximately three to five ommatidia away. In this manner, strongly activated cells directly receiving a light

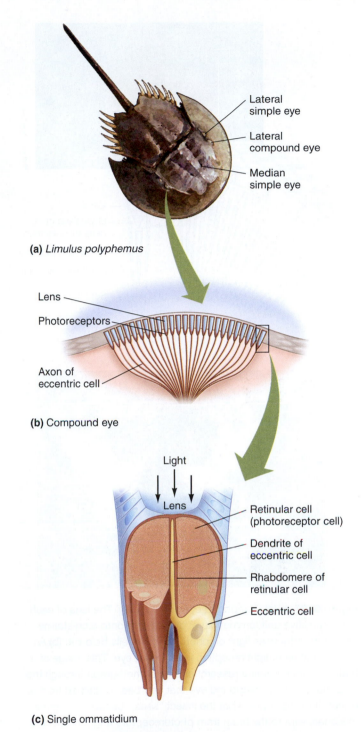

(a) *Limulus polyphemus*

(b) Compound eye

(c) Single ommatidium

Figure 6–26 ● Anatomy of the compound eyes of the horseshoe crab, *Limulus polyphemus*.

stimulus generate impulses, which *decrease* impulse generation in its comparatively weakly stimulated neighbors. At the boundary between light and dark illumination, the net effect is to enhance discrimination between the two zones because the eccentric cells lying in this region receive different degrees of inhibition from the two sides. Thus ommatidia lining the border of a light–dark edge and not receiving a light stimulus to activate them, are strongly inhibited by their neighboring cells, which are better illuminated. The net effect is that the

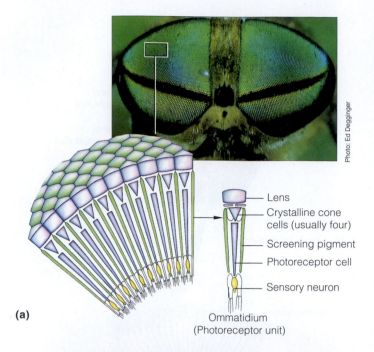

(a)

Lens
Crystalline cone cells (usually four)
Screening pigment
Photoreceptor cell
Sensory neuron

Ommatidium
(Photoreceptor unit)

Photo: Ed Degginger

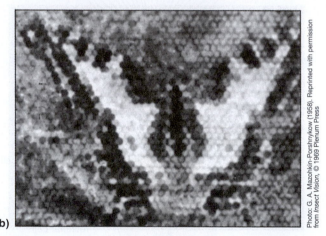

(b)

Photo: G. A. Mazohkin-Porshnykow (1969). Reprinted with permission from *Insect Vision*, © 1969 Plenum Press

Figure 6–27 ● (a) Compound eyes of a deerfly. The lens of each photosensitive unit (ommatidium) directs light onto a crystalline cone, which focuses light on photoreceptor cells below it. (b) An approximation of light reception in an insect eye. This image of a butterfly formed after a researcher took a photograph through the outer surface of a compound eye that had been detached from an insect. It might not be what the insect "sees," because integration of signals sent to the brain from photoreceptors might produce a sharper image. But this representation is useful insofar as it suggests how the separate ommatidia sample the overall visual field.

(*Sources*: C. Starr & R. Taggart, 2004, *Biology: The Unity and Diversity of Life*, 10th ed., Belmont, CA: Brooks/Cole, (a) Figure 35.14, (b) Figure 35.15)

activity in these cells is comparatively low as compared to ommatidia lying in either the light or dark zones.

Two basic principles govern the design of compound eyes. Most diurnal insects have **apposition eyes**, characterized by ommatidium that are optically isolated from each other. Thus each ommatidium receives light from a comparatively

narrow region of the visual field. However, a wide, optically clear zone between the rhabdoms and the lens characterizes **superposition eyes**, which nocturnal arthropods have. Light from different ommatidia can mix together so that when the light finally reaches the rhabdom it represents a composite from many ommatidia. The adaptive value of this arrangement lies with its ability to increase light energy under dim lighting conditions. Light striking the eye surface can be brought to focus on a single rhabdom.

■ **Phototransduction in compound eyes differs from that in vertebrate eyes.**

The process of phototransduction in compound eyes differs remarkably from that discussed for vertebrates. Recall that vertebrate Na^+ channels *close* in response to an increase in light. Consequently, the reduction in the entry of Na^+ contributes to the hyperpolarization in rods and cones. In contrast, most invertebrate photoreceptors *depolarize* in response to light. The receptor potential in turn activates rhodopsin. Rhodopsin, via a G protein, activates phospholipase C, which in turn leads to an increased production of inositol triphosphate (IP_3) and diacylglycerol (Chapter 3, p. 91). The increase in IP_3 triggers release of Ca^{2+} ions from intracellular stores. Together these two second messengers, intracellular Ca^{2+} and diacylglycerol, govern the opening of cation channels and thus generate a depolarizing potential. Graded potentials generated in the retinular cells spread through gap junctions into the dendrite of the eccentric cell. The eccentric cell thus depolarizes when light falls on the rhabdomeres and an action potential is conducted through the optic nerve to the optic lobe of the brain.

Insects, such as the blowfly, use the same set of photoreceptors for vision at all light intensities. At absolute threshold, these photoreceptors are similar in function to vertebrate rods in that they generate discrete responses to single photons. However, in full daylight these same cells reduce their sensitivity approximately 1000-fold, easily outperforming vertebrate cones in terms of signal-to-noise ratio and frequency response. For example, light-adapted blowfly photoreceptors have the fastest known photoreceptor response. To achieve these remarkable dynamics, two subtypes of K^+ channels, one activated more rapidly than the other, are available. The *rapidly activating* type operates close to resting potential, leading to a K^+ current that is activated virtually instantaneously on depolarization. The membrane pumps that counteract the large ion fluxes generated by fast photoreceptors account for as much as 10% of the oxygen consumption of a resting blowfly. A second, distinctive, *slow* component appears with larger depolarizations. In the dark, the membrane potential of the blowfly photoreceptor is approximately −65 mV and most of the K^+ channels are closed. This results in an increased sensitivity; that is, absorption of only a few photons results in the generation of a 1- to 2-mV response. The inward current generated by the light depolarizes the cell and rapidly activates an outward K^+ current. However, under daylight conditions the K^+ conductance approximately equals the light-gated conductance, which reduces sensitivity almost 10-fold. This is possible only because the powerful K^+ conductance enables the light-gated conductance to increase nearly 100-fold. In bright light, the slow component dominates, producing a transient response to depolarization.

Comparative studies of photoreceptor K$^+$ conductance have shown that rapidly moving diurnal species have fast photoreceptor responses, whereas species of crane-flies that are active at night (nocturnal) and fly weakly, have slow photoreceptors. Interestingly, locusts (*Schistocerca*) are diurnal but migrate at night. In this species changes in K$^+$ conductance are associated with the functioning of the photoreceptors as they switch between a day state, with high acuity and low sensitivity, to a night state, with low acuity and high sensitivity.

Mechanoreception: Touch and Pressure

Detection of physical forces in the environment is one of the most ancient senses, as we noted in the chapter's beginning. Virtually all single-celled motile organisms react to surface contact with an external object. In animals, this mode of ex-

ternal sensing is most prominent in the skin, giving animals the somesthetic senses of *touch* and *pressure*. Mechanoreception is also used in proprioception and hearing.

■ Mechanically gated channels transduce touch and pressure into electrical signals.

Touch and pressure rely on *mechanically gated* channel proteins, usually in sensory dendrites (● Figure 6–28a). Researchers have recently investigated these channels in the exoskeleton of insects, specifically the sensors located at the hinge of leg bristles. The channel proteins have external protein fibers attached to them, and when these fibers are stretched or distorted (as when vertebrate skin is touched, or when the insect bristle moves), they essentially pull open the gate of the ion channel. Cations entering the sensory dendrite then create a receptor potential and, if the signal is strong enough, action potentials.

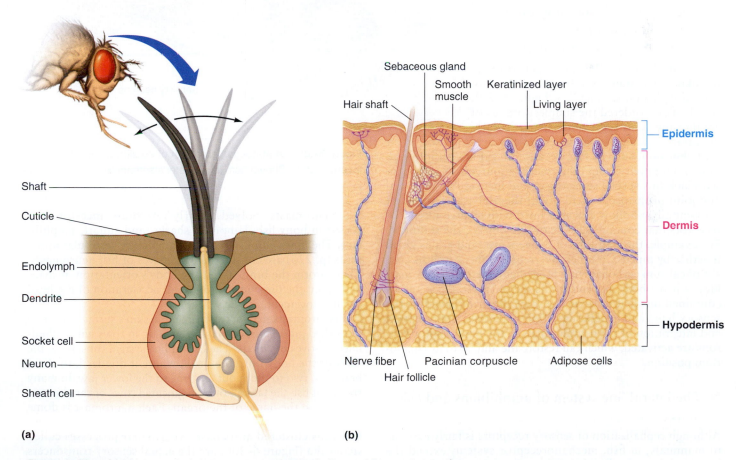

(a) **(b)**

Figure 6–28 ● (a) Diagram of a bristle mechanoreceptor in *Drosophila,* consisting of a bristle and three cells: a socket cell, a sheath cell, and a neuron. When the bristle moves, it distorts the membrane of the neuron's dendrites, opening up mechanogated channels. (b) Vertebrate skin mechanoreceptors include free nerve endings that wrap around the base of the hair shaft.

(*Source:* (a) R. C. Walker, A. T. Willingham, & C. S. Zucker, 2000, "A *Drosophila* mechanosensory transduction channel," *Science* 187: 2229–2234, Figure 1A. Copyright © 2000 AAAS. Reprinted by permission.)

The skin of vertebrates has several different types of mechanoreceptors. Earlier we saw the Pacinian corpuscle, which detects deep pressure. Closer to the skin surface are highly sensitive touch sensors consisting of numerous dendrites with a large surface area. There are also mechanosensors at the base of hairs, much like those associated with insect bristles (Figure 6–28b).

Proprioception: the Mechanoreception of Motion and Position

Another major use of mechanoreception is the detection of motion and position. This sense is critical to most animals' ability to navigate and orient properly in the environment, and to move in a coordinated manner.

■ The statocyst is the simplest organ that can monitor an animal's position in space.

Numerous classes of animals have gravity receptors, known as **statocysts** (from Greek *statos*, "standing"; *kystis*, "bag") (● Figure 6–29), which are considered the simplest organs of equilibrium. This sense organ is particularly important in animals that are essentially neutrally buoyant, such as fish, because they don't get information about gravity from other sensory sources. (Notably, insects lack this special sense organ and instead rely on visual and joint proprioceptors to orient themselves.) A statocyst is essentially a hollow chamber lined with ciliated mechanoreceptors that contain dense, movable objects, the **statoliths**. For example, the statoliths of lobsters are sand grains glued together by mucus. In other invertebrates the statoliths consist of calcium carbonate fused together to resemble tiny marbles. When a body movement tilts the statocyst, the statoliths contained inside it move in the same direction, bending the sensory hairs and generating action potentials. On reception and integration of the neural signals in the brain, motor neurons are activated, which can restore the animal to its equilibrium position.

■ The lateral line system of amphibians and fish detects motion in the surrounding water.

Although cephalization of sensory receptors is fairly extreme in mammals, in fish, mechanoreceptor systems extend the length of the animal's body. Information about the fish's orientation with respect to gravity, its swimming velocity and details about water currents and vibrations are collected by a string of **neuromast cells** (● Figure 6–30a and b), the basic sensory unit of the **lateral line system**. Fish behavior linked to lateral line function includes object and predator avoidance and the ability to school. Cavefishes as well as species active at night rely on the lateral line system for detecting prey.

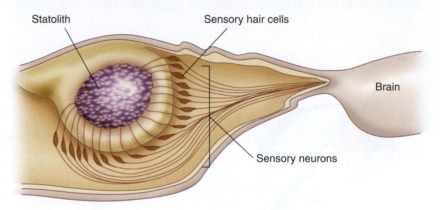

Figure 6–29 ● **A statocyst, an organ of equilibrium in invertebrates.** Shown here in a lobster's antennule.

Neuromasts evolved in early vertebrate ancestors and persist in bony fish, lampreys, sharks, and aquatic amphibians. The inner *hair cells* of the mammalian vestibular apparatus (p. 223) are considered derivatives of fish neuromasts. They inform the brain about the body's orientation in space and about the directions and rate of acceleration of the head during movement, as you will see.

Neuromasts may occur free on the skin's surface, in small pits called *pit organs,* or may be lined up in rows within fluid-filled grooves or canals. The lateral line system runs along the sides of the body onto the head, where it divides into three branches, two to the snout and one to the lower jaw. In many species of fish the lateral, often-pigmented rows of canal pores, give rise to the name of the organ. Each neuromast is dome shaped, with up to several hundred mechanoreceptor sensory **hair cells** clustered at its base. Microvillar processes called **stereocilia** (Figure 6–30b) are the actual sensory transducers that protrude from the sensory hair cells into a jellylike substance (at the base of the dome). The cilia of the hair cells may actually extend up to 50 microns into the surrounding water, usually embedded in an overlying caplike, gelatinous layer, the **cupula**, stereocilia are arranged together into ciliary bundles and are oriented according to size. The tallest row of stereocilia lie closest to an elongated cilium, the **kinocilium** (Figure 6–30b). Signals are sent to the brain in response to

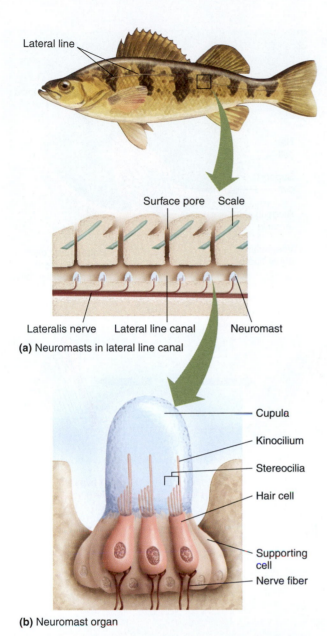

Lateral line

Surface pore   Scale

Lateralis nerve   Lateral line canal   Neuromast

(a) Neuromasts in lateral line canal

Cupula

Kinocilium

Stereocilia

Hair cell

Supporting cell

Nerve fiber

(b) Neuromast organ

Figure 6–30 ● **The lateral line system and inner ears constitute the acousticolateralis system in fishes.** (a) The lateral line system consists of a row of neuromasts set in pits and canals along each side of the body. (b) Vertical section through the skin and lateral line canal.

(*Source:* Modified from K. Liem, W. Bemis, W. Walker, & L. Grande, 2001, *Functional Anatomy of Vertebrates: An Evolutionary Perspective*, 3rd ed., Belmont, CA: Brooks/Cole, p. 406, Figure 12-8.)

system. As the fish approaches a solid object, such as a rock or the glass wall of an aquarium, the pressure waves around its body are distorted enough that the lateral line system registers these changes and results in the fish swerving away. Because sound waves are waves of pressure, the hair cells can also detect very-low-frequency sounds of 100 Hz or less.

■ The vestibular apparatus of vertebrates detects position and motion of the head and is important for equilibrium and coordination of head, eye, and body movements.

The advanced vertebrate inner ear has two specialized sensory components: the **cochlea,** a coiled structure in reptiles and mammals, which is used in hearing (which we discuss later); and the **vestibular apparatus,** which provides information essential for the sense of equilibrium and for coordinating head movements with eye and postural movements. The vestibular apparatus consists of two sets of structures lying within a tunneled-out region of the temporal bone near the cochlea—the *semicircular canals* and the *otolith organs* (these organs are the *utricle* and *saccule*) (Figure 6–31a). Vertebrates without a cochlea (fish and amphibians) also use the vestibular apparatus for hearing.

As in fish neuromasts, all the inner-ear senses begin with hair cells each having 20 to 50 stereocilia and one kinocilium (Figure 6–31b). Hair cells (in neuromasts or inner ear) normally send out continuous bursts of nerve impulses when they are at rest (that is, the cilia are unbent). When a force causes the cilia ("hairs") to bend (by pressure waves, sound waves, or gravity), the signal to the brain changes (Figure 6–31c). If the ciliary bundle bends in the direction of the kinocilium, this results in an excitatory depolarization of the hair cell, whereas bending in the opposite direction generates an inhibitory hyperpolarization. The actual magnitude of the response depends on the degree of bending. If the bundle is bent in directions other than along the major axis, then the response is concomitantly lower. Depolarization increases the rate of firing in the afferent fibers; conversely, when the hair cells are hyperpolarized the frequency of action potentials in the afferent fibers is reduced. The hair cells form a chemically mediated synapse with terminal endings of afferent neurons whose axons join with those of the other vestibular structures to form the **vestibular nerve.** This nerve unites with the auditory nerve from the cochlea to form the **vestibulocochlear nerve.**

Role of the Semicircular Canals

Let us now examine the vestibular senses. The **semicircular canals** (Figure 6–31a) detect *rotational or angular acceleration or deceleration of the head,* such as when starting or stopping spinning, diving, or turning the head. Each ear contains three semicircular canals arranged three-dimensionally in planes at right angles to each other. The receptive hair cells of each semicircular canal lie on top of a ridge located in the **ampulla,** a swelling at the base of the canal (● Figures 6–32a and b). The hairs are embedded in a cupula (as in a neuromast), which protrudes into the endolymph within the ampulla. The cupula sways in the direction of fluid movement, much like seaweed leaning in the direction of the prevailing tide.

Acceleration or deceleration during rotation of the head in any direction causes endolymph movement in at least one

the bending of the kinocilium and stereocilia, using a mechanism identical to that of the mammalian inner ear, which we describe shortly (● Figure 6–31). The cells synapse with sensory neuron dendrites associated with either the **facial** or **vagus nerves.**

A swimming fish sets up a pressure wave in the water that can be detected by the lateral line systems of other fishes. It also sets up a bow wave in front of itself, the pressure of which is higher than that of the wave flow along its sides. These minute differences in pressure are detected by its own lateral line

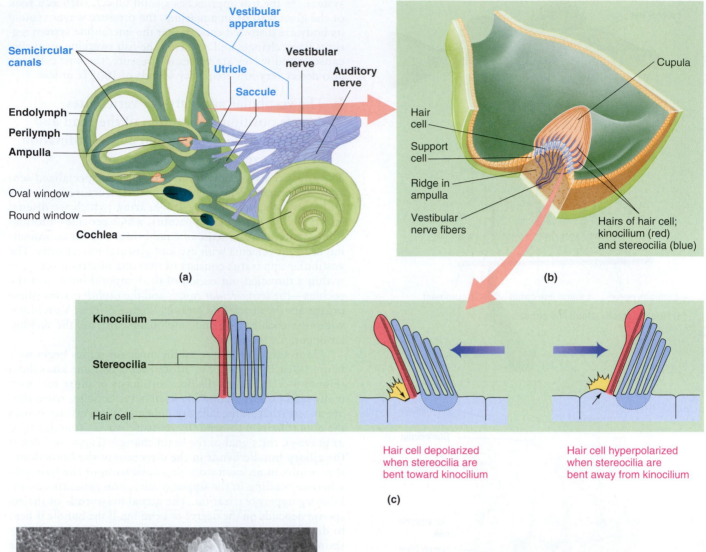

(a)

Vestibular apparatus

Semicircular canals

Utricle

Saccule

Vestibular nerve

Auditory nerve

Endolymph

Perilymph

Ampulla

Oval window

Round window

Cochlea

(b)

Cupula

Hair cell

Support cell

Ridge in ampulla

Vestibular nerve fibers

Hairs of hair cell; kinocilium (red) and stereocilia (blue)

(c)

Kinocilium

Stereocilia

Hair cell

Hair cell depolarized when stereocilia are bent toward kinocilium

Hair cell hyperpolarized when stereocilia are bent away from kinocilium

Kinocilium

Stereocilia

(d)

Figure 6–31 ● **Vestibular apparatus.** (a) Gross anatomy of the vestibular apparatus. (b) Receptor cell unit in the ampulla of the semicircular canals. (c) Schematic representation of the "hairs" on the sensory hair cells of the semicircular canals. (d) Scanning electron micrograph of the kinocilium and stereocilia on the hair cells within the vestibular apparatus.

(*Source:* Figure 6.41b adapted from Cecie Starr and Ralph Taggart, *Biology: The Unity and Diversity of Life,* Eighth Edition, Fig. 36.10b, p. 595. Copyright 1998 Wadsworth Publishing Company.)

of the semicircular canals, because of their three-dimensional arrangement (Figure 6–32a and b). As the head starts to move, the bony canal and the ridge of hair cells embedded in the cupula move with the head. Initially, however, the fluid within the canal, not being attached to the skull, does not move in the direction of the rotation but lags behind because of its inertia. (Because of inertia, a resting object remains at rest, and

a moving object continues to move in the same direction unless acted on by some external force.) Thus the endolymph that is in the same plane as the head movement is shifted in the opposite direction from the movement (like your body tilting rightward as the car in which you are riding suddenly turns left). This relative fluid movement causes the cupula to lean in the opposite direction from the head movement, bend-

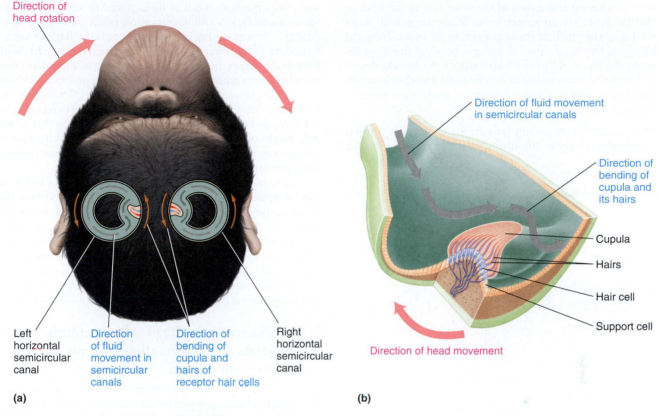

Figure 6-32 ● Activation of the hair cells in the semicircular canals.

ing its embedded sensory hair. If the head movement continues at the same rate in the same direction, the endolymph catches up and moves in unison with the head so that the hairs return to their unbent position. When the head slows down and stops, the reverse occurs. The endolymph briefly continues to move in the direction of the rotation while the head decelerates to a stop. When the endolymph gradually comes to a halt, the hairs straighten again. Thus the semicircular canals detect changes in the rate of rotational movement of the head. They do not respond when the head is motionless or during circular motion at a constant speed.

The size of the semicircular canals has changed greatly with evolutionary adaptation. In primates that swing through trees and flying vertebrates, the canals are very large, evolved for processing complex motion in three dimensions. They are of moderate size in terrestrial animals, including extinct terrestrial ancestors of whales (whose canals can be reconstructed from fossil skulls). Modern whales, in contrast, have very small canals, perhaps to reduce their sensitivity, to prevent seasickness, or because they do not make rapid head motions.

Role of the Otolith Organs

Whereas the semicircular canals provide the CNS with information about rotational changes in head movement, the **otolith organs** provide information about position of the head relative to gravity and also detect changes in rate of linear motion (moving in a straight line regardless of direction). The otolith organs, the **utricle** and **saccule**, are saclike structures housed within a bony chamber situated between the semicircular canals and the cochlea (Figure 6–31a). The hairs of the receptive hair cells in these sense organs also protrude into an

overlying gelatinous sheet, whose movement displaces the hairs and results in changes in hair cell potential. Many tiny crystals of calcium carbonate—the **otoliths** ("ear stones")—similar to the statoliths associated with statocysts in invertebrates, are suspended within the gelatinous layer, making it heavier and giving it more inertia than the surrounding fluid (● Figure 6–33). When an animal is in an upright position, the hairs within the utricles are oriented vertically and the saccule hairs are lined up horizontally.

Let's look at the utricle as an example. Its otolith-embedded, gelatinous mass shifts positions and bends the hairs in two ways, described as follows.

1. When the head is tilted in any direction other than vertical (that is, other than straight up and down), the hairs are bent in the direction of the tilt, because of the gravitational force exerted on the top-heavy gelatinous layer (Figure 6–33b). Within the utricles on each side of the head, some of the hair cell bundles are oriented to depolarize and others to hyperpolarize when the head is in any position other than upright. The CNS thus receives different patterns of neural activity, depending on head position with respect to gravity.

2. The utricle hairs are also displaced by any change in horizontal linear motion (such as moving straight forward, backward, or to the side). As an animal starts to walk forward (Figure 6–33c), the top-heavy otolith membrane at first lags behind the endolymph and hair cells because of its greater inertia. The hairs are thus bent to the rear, in the opposite direction of the forward head movement. If the walking pace is maintained, the gelatinous layer

soon catches up and moves at the same rate as the head so that the hairs are no longer bent. When the animal stops walking, the otolith sheet continues to move forward briefly as the head slows and stops, bending the hairs toward the front. Thus the hair cells of the utricle detect horizontally directed linear acceleration and deceleration, but they do not provide information about movement in a straight line at constant speed.

The saccule functions similarly to the utricle, except that it responds selectively to the tilting of the head away from a horizontal position (such as diving) and to vertically directed linear acceleration and deceleration (such as leaping up and down). An interesting variation occurs in flatfish such as flounders (*Pleuronectiformes*): Fry (immature fish), which have body forms and orientation similar to most fish, metamorphose into adults that lie and swim on one side. Although one eye migrates across the top of the head, leaving the bottom side eyeless, the vestibular system does not change in orientation, and so functions at 90 degrees relative to that of other fish. Ichthyologists believe the saccule replaces the utricle in posture regulation, suggesting that central nervous pathways have reorganized.

Signals arising from the various components of the vestibular apparatus are carried through the vestibulocochlear nerve to the cerebellum and the **vestibular nuclei,** a cluster of neuronal cell bodies in the brain stem. Here the vestibular information is integrated with input from the skin surface, eyes, joints, and muscles for (1) maintaining balance and desired posture; (2) controlling the external eye muscles so that the eyes remain fixed on the same point, despite movement of the head; and (3) perceiving motion and orientation (● Figure 6–34).

Proprioceptors in the muscles, tendons, and joints give information on limb position and motion.

A final group of mechanoreceptors used in proprioception are the stretch and related sensors associated with muscle movement. These **muscle spindles, Golgi tendon organs,** and other such sensors are discussed in Chapter 8. In humans, these sensors provide feedback to the brain that is sometimes called the **kinesthetic sense,** a conscious "sixth sense" that lets you know where your limbs are even when you have your eyes closed. These sensors also provide feedback to the cerebellum while it is learning motor skills to produce smooth coordination (p. 175).

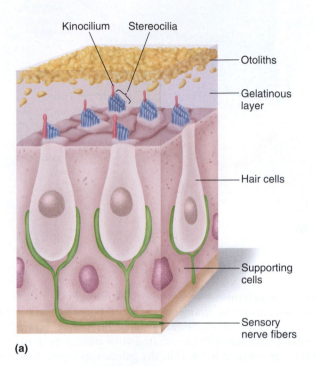

(a)

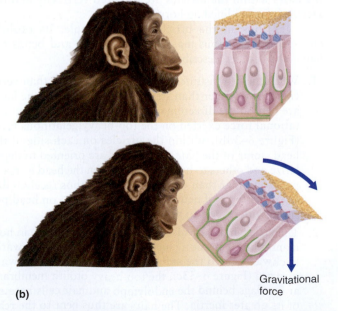

(b)   (c)

Figure 6–33 ● **Utricle.** (a) Receptor unit in utricle. (b) Activation of the utricle by a change in head position. (c) Activation of the utricle by horizontal linear acceleration.

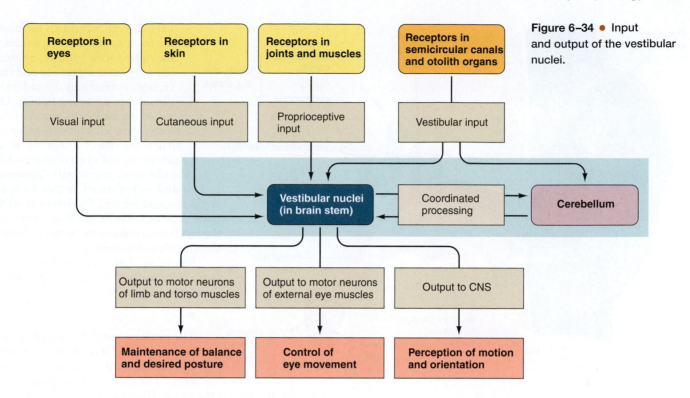

Figure 6–34 ● Input and output of the vestibular nuclei.

Ears, Hearing, and the Mechanoreception of Sound Waves

Detection of sound waves is less common than photoreception and touch; for example, there is no evidence for sound sensing in single-celled organisms. But many animals do detect sound; like light but unlike touch, sound is a long-range "early warning" signal used to detect predators, competitors, physical threats, potential prey, and mating calls. Fishes, for example, can hear noises for some of these purposes, such as the intense "whistle" of the male toadfish used as a mating call during the breeding season. Researchers think fish and amphibians detect sounds that penetrate the body and cause vibrations in the otoliths in their inner ears. Some fish (such as minnows and catfish) have a **Weberian apparatus** (● Figure 6–35), consisting of connective tissue, muscle, and bone that transfer sound from the gas bladder to the inner ear. This is advantageous, because the swim bladder is much more sensitive to sound vibrations than are the otoliths. Whether directly stimulated by sound waves or indirectly by the swim bladder, otoliths in motion result in a shearing action on the cilia of the sensory hair cells and thus in generation of action potentials.

Compared to animals living in aquatic habitats, animals living in the air evolved different structures capable of receiving and processing sounds traveling through the air. In mammals, each **ear** has three parts: *external, middle,* and *inner* ear (● Figure 6–36). In brief, the external and middle portions of the ear transmit airborne sound waves to the fluid-filled inner ear, amplifying the sound energy. The inner ear of terrestrial vertebrates evolved a separate sound-sensing organ, the *cochlea,* which contains the receptors for converting sound waves into nerve impulses. We explore these structures in more detail shortly.

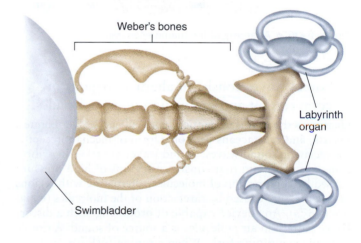

Figure 6–35 ● Weberian apparatus for fish hearing.

Among nonmammalian vertebrates, birds have the most highly evolved auditory system; comparable to some mammalian species (such as elephants), some species of birds can hear infrasound as well. The avian external ear is inconspicuous, although ear-flaps or specialized feathers may reflect or amplify sound, as in the barn owl (*Tyto alba*). In some species of owls, the external openings in their left and right ears differ in height and direction on the head in order to more accurately locate a particular sound source. Remarkably, barn owls can successfully hunt prey (such as a field mouse making a faint noise) in total darkness. In the owl's midbrain, certain cells fire only when sounds are received from a particular spot in the external environment. In principle, this could give the owl the ability to "see" (that is, visually perceive) its prey's location, using sound!

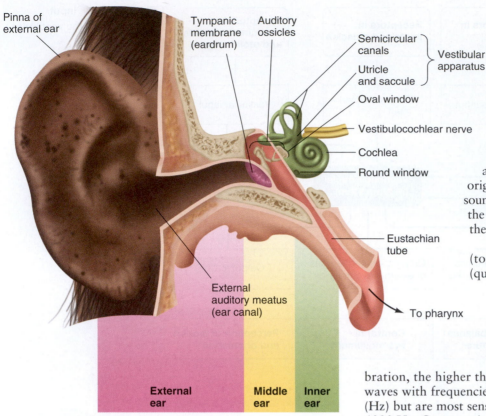

Pinna of
external ear

Tympanic
membrane
(eardrum)

Auditory
ossicles

Semicircular
canals

Utricle
and saccule

Vestibular
apparatus

Oval window

Vestibulocochlear nerve

Cochlea

Round window

Eustachian
tube

External
auditory meatus
(ear canal)

To pharynx

**External
ear**

**Middle
ear**

**Inner
ear**

Figure 6–36 ● Anatomy of the mammalian ear.

Even though individual molecules are moved only short distances as the tuning fork vibrates, alternating waves of compression and rarefaction spread out considerable distances in a rippling fashion. Disturbed air molecules disturb other molecules in adjacent regions, setting up new regions of compression and rarefaction, and so on (Figure 6–37c). Sound energy is gradually dissipated as sound waves travel farther from the original sound source. The intensity of the sound decreases, until it finally dies out when the last sound wave is too weak to disturb the air molecules around it.

Sound is characterized by its pitch (tone), intensity (loudness), and timbre (quality) (● Figure 6–38).

Pitch

The **pitch** or **tone** of a sound (for example, whether it is a C or a G note) is determined by the *frequency* of vibrations. The greater the frequency of vibration, the higher the pitch. Human ears can detect sound waves with frequencies from 20 to 20,000 cycles per second (Hz) but are most sensitive to frequencies between 1000 and 4000 Hz. Comparatively, goldfish (*Carassius auratus*) detect

▪ Sound waves consist of alternate regions of compression and rarefaction of air molecules.

Hearing is the neural perception of sound energy. Vertebrates, insects, and crustaceans use some form of mechanoreceptor for receiving **sound waves.** Sound waves are traveling vibrations of air or fluid that consist of regions of high pressure caused by compression of molecules, alternating with regions of low pressure caused by rarefaction of the molecules (● Figure 6–37a). Any device capable of producing such a disturbance pattern in air molecules is a source of sound. A simple example is a tuning fork. When a tuning fork is struck, its prongs vibrate. As a prong of the fork moves in one direction (Figure 6–37b), air molecules ahead of it are pushed closer together, or compressed, increasing pressure in this area. Simultaneously, as the prong moves forward, the air molecules behind the prong spread out or are rarefied, lowering the pressure in that region. As the prong moves in the opposite direction, an opposite wave of compression and rarefaction is created.

▶ **Collared scops owl (*Otus lettia*).** This bird hunts at night and has enhanced two major sensory systems to aid in this behavior. First, it has large eyes with a reflective layer, the tapetum lucidum, behind the retina. This layer reflects photons back into the retina that were not detected initially. Second, it has a specialized auditory system that amplifies sounds and localizes prey in three dimensions (see text).

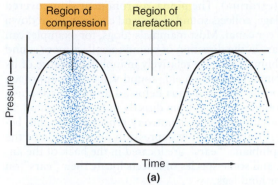

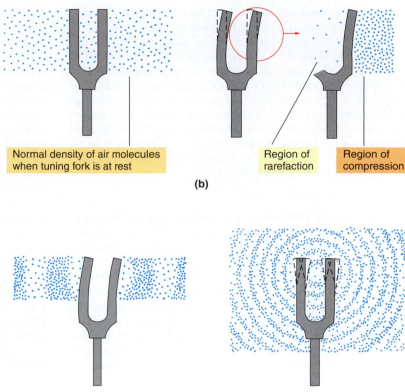

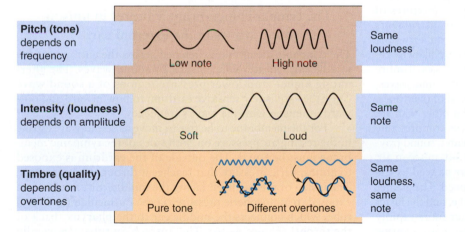

Figure 6–37 ● **Formation of sound waves.** (a) Sound waves are alternating regions of compression and rarefaction of air molecules. (b) A vibrating tuning fork sets up sound waves as the air molecules ahead of the advancing arm of the tuning fork are compressed while the molecules behind the arm are rarefied. (c) Disturbed air molecules bump into molecules beyond them, setting up new regions of air disturbance more distant from the original source of sound. In this way, sound waves travel progressively farther from the source, even though each individual air molecule travels only a short distance when it is disturbed. The sound wave dies out when the last region of air disturbance is too weak to disturb the region beyond it.

Figure 6–38 ● **Properties of sound waves.**

frequencies between 50 to 3000 Hz with the best sensitivity between 200 to 1000 Hz. Dogs can hear sound frequencies up to 40 kHz. These higher frequencies or *ultrasonic* sounds consist of frequencies greater than that audible by the human ear, whereas *infrasound* consists of sounds below the range of frequencies audible by humans. For example, insectivorous bats locate prey by emitting successive pulses of sound; the duration of each varies from about 10 to 15 ms (5 pulses per second) and consists of a descending frequency spectrum that varies from approximately 100 kHz to 20 kHz. Faint reflections, or echoes from the insect are then detected by enlarged **pinnae** on the bat, and the bat's brain uses the information to locate the insects. This is the process of **echolocation**: the

use of sound echoes to detect objects in the environment.

Other vertebrates that can obtain information from echoes of the sounds that they produce include dolphins, shrews, and several species of birds. Toothed whales (Odontoceti) also generate patterned arrangements of clicks with sound frequencies between 200 Hz to 32 kHz for communicating between whales or for echolocating prey items (such as squid) and other objects in the environment. In dolphin echolocation, for example, audiologists think click sound waves (1) are produced by lips near the blowhole; (2) are focused by the melon, a fat-filled organ in the forehead, much as a lens focuses light; (3) travel until they hit a sound-reflecting object (such as another animal); (4) reflect back ("echo") toward the dolphin; and (5) are received by a fat body in the lower jaw, and (6) focused into the inner ear. The dolphin's brain then computes the distance of the object by the time delay of the echo. By changes in timbre of the echoed sound, the dolphin can also learn about the object's composition, and by making repeated clicks, can judge the size and speed of the object. Bats employ a similar system, allowing them to chase and catch tiny insects at high flight speeds without vision.

Whales may also use infrasound for long-distance communication, broadcasting, for example, the location of feeding grounds or spacing mechanisms by relying on frequencies less than 60 Hz. This form of acoustic communication is particularly useful for animals active at night or living in either deep or murky waters.

Low-frequency sound is subject to very little attenuation in the environment and so can function in long-distance communication, and it travels well in the ground as well as air. For example, elephants use frequencies in the range of 14 to 25 Hz, which can travel distances as great as several kilometers. This is valuable because once every four years and for only four days, a female elephant makes a special call that signals mature elephant bulls in the area that she is in estrus. Movement of separate elephant herds can also be coordinated by this form of communication. (Interestingly, recent research suggests that elephants can pick up these sound wave vibrations in the ground through their feet!)

Intensity of Sound

The **intensity**, or loudness, of a sound depends on the *amplitude* of the sound waves, or the pressure differences between a high-pressure region of compression and a low-pressure region of rarefaction. Within the hearing range, the greater the amplitude, the louder the sound. An animals' ear can detect a wide range of sound intensities, from the slightest rustle of a leaf to the comparatively loud roar of a waterfall. Loudness is expressed in **decibels (dB)**, which is a logarithmic measure of intensity compared with the faintest sound that can be heard—the **hearing threshold**. Because of the logarithmic relationship, every 10 decibels indicates a 10-fold increase in loudness.

Sound Timbre

The **timbre**, or **quality**, of a sound depends on its *overtones*, which are additional frequencies superimposed on top of the fundamental pitch or tone. A tuning fork has a pure tone, but most sounds lack purity. For example, complex mixtures of overtones impart different sounds to different instruments playing the same note (a C note sounds different on a trumpet from the way it sounds on a piano). Overtones are likewise responsible for the characteristic differences in voices. Timbre enables the listener to distinguish the source of sound waves, because each source produces a different pattern of overtones.

Sound waves can also travel through media other than air, such as solids and liquids. Because water is less compressible than air, sound attenuation is not as great. In water, sound travels at an average of 1500 m/sec, compared to 340 m/sec in air. However, both the temperature of the water as well as the hydrostatic pressure can affect the measured speed. The greater the depth, the heavier the column of water above, and the faster sound waves are propagated. Salts in seawater more rapidly absorb high-frequency sounds, whereas low-frequency acoustic waves are not affected by the molecular structure of the salts and may travel for thousands of kilometers.

■ The external ear and middle ear convert airborne sound waves into fluid vibrations in the inner ear.

The specialized receptors for sound are located in the fluid-filled inner ear. Airborne sound waves must therefore be channeled toward and transferred into the inner ear, compensating in the process for the loss in sound energy that naturally occurs as sound waves pass from air into water. This function is performed by the external ear and the middle ear.

In mammals the **external ear** (Figure 6–36) consists of the *pinna* (ear), *external auditory meatus* (ear canal), and *tympanic*

membrane (eardrum). The **pinna**, a prominent skin-covered flap of cartilage, collects sound waves and channels them down the external ear canal. Most mammals (dogs, for example) can cock their ears in the direction of sound to collect more sound waves, but human ears are relatively immobile. Because of its shape, the pinna partially shields sound waves that approach the ear from the rear and thus helps an animal distinguish whether a sound is coming from directly in front or behind. In reptiles the auditory system is not very sensitive to sound, because there is only a shallow depression on each side of the head. In amphibians, "ears" are located in the back of the animal's head, and some species of insects locate their "ears" on their front or hind legs.

Localization for sounds approaching from the right or left is determined by two cues. First, the sound wave reaches the ear closer to the sound source slightly before it arrives at the farther ear. Second, the sound is less intense as it reaches the farther ear, because the head acts as a sound barrier that partially disrupts the propagation of sound waves. Fishes (such as cod, *Gadus morhua*) also can localize sound, although the precise mechanism is not known. The auditory cortex integrates all these cues to determine the location of the sound source. It is difficult to localize sound with only one ear. Recent evidence suggests that the auditory cortex pinpoints the location of a sound by distinguishing differences in the timing of neuronal firing patterns, not by using any spatially organized map such as the one projected on the visual cortex point-for-point from the retina that enables the animal to locate a visual object.

■ The tympanic membrane vibrates in unison with sound waves in the external ear.

The **tympanic membrane**, stretched across the entrance to the middle ear, vibrates when struck by sound waves. The alternating higher- and lower-pressure regions of a sound wave cause the exquisitely sensitive eardrum to bow inward and outward in unison with the wave's frequency.

For the membrane to be free to move as sound waves strike it, the resting air pressure on both sides of the tympanic membrane must be equal. The outside of the eardrum is exposed to atmospheric pressure that reaches it through the ear canal. The inside of the eardrum facing the middle-ear cavity is also exposed to atmospheric pressure via the **Eustachian (auditory) tube**, which connects the middle ear to the **pharynx** (back of the throat) (Figure 6–36). The Eustachian tube is normally closed, but it can be pulled open by yawning, chewing, and swallowing. Such opening permits air pressure within the middle ear to equilibrate with atmospheric pressure so that pressures on both sides of the tympanic membrane become equal. During rapid external pressure changes (for example, during air flight), the eardrum can bulge painfully as the pressure outside the ear changes while the pressure in the middle ear remains unchanged. Opening the Eustachian tube by yawning allows the pressure on both sides of the tympanic membrane to equalize, relieving the pressure distortion as the eardrum "pops" back into place. Infections originating in the throat sometimes spread through the Eustachian tube to the middle ear. The resulting fluid accumulation in the middle ear not only is painful but also interferes with conduction of sound across the middle ear.

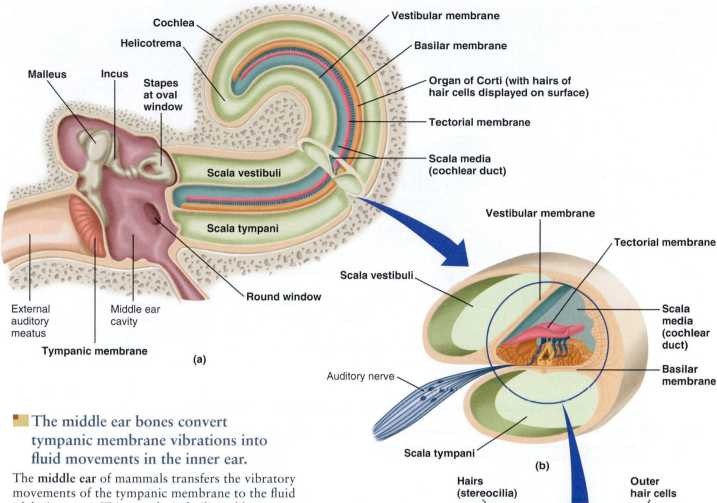

Figure 6–39 ● Middle ear and cochlea. (a) Gross anatomy of the middle ear and cochlea, with the cochlea "unrolled." (b) Cross section of the cochlea. (c) Enlargement of the organ of Corti.

The middle ear bones convert tympanic membrane vibrations into fluid movements in the inner ear.

The **middle ear** of mammals transfers the vibratory movements of the tympanic membrane to the fluid of the inner ear. This transfer is facilitated by a movable chain of three small bones or **ossicles** (the **malleus, incus,** and **stapes**) that extend across the middle ear (● Figure 6–39a). The first bone, the malleus, is attached to the tympanic membrane, and the last bone, the stapes, is attached to the **oval window,** the entrance into the fluid-filled cochlea. As the tympanic membrane vibrates in response to sound waves, the chain of bones is set into motion at the same frequency, transmitting this frequency of movement from the tympanic membrane to the oval window. The resulting pressure on the oval window with each vibration produces wavelike movements in the inner ear fluid at the same frequency as the original sound waves. However, as noted earlier, greater pressure is required to set fluid in motion. Two mechanisms related to the ossicular system amplify the pressure of the airborne sound waves to set up fluid vibrations in the cochlea. First, because the surface area of the tympanic membrane is much larger than that of the oval window, pressure is increased as force exerted on the tympanic membrane is conveyed to the oval window (pressure = force/unit area). Second, the lever action of the ossicles provides an additional mechanical advantage. Together, these mechanisms increase the force exerted on the oval window by 20 times what it would be if the sound wave struck the oval window directly. This additional pressure is sufficient to set the cochlear fluid in motion. The tympanic membrane of reptiles, anuran amphibians, and birds is connected to the inner ear by a single, bony

ossicle, the **columella,** whose function is comparable to that in mammals.

Several tiny muscles in the middle ear contract reflexively in response to loud sounds (over 70 dB), tightening the tympanic membrane and limiting movement of the ossicular chain. This reduced movement of middle ear structures di-

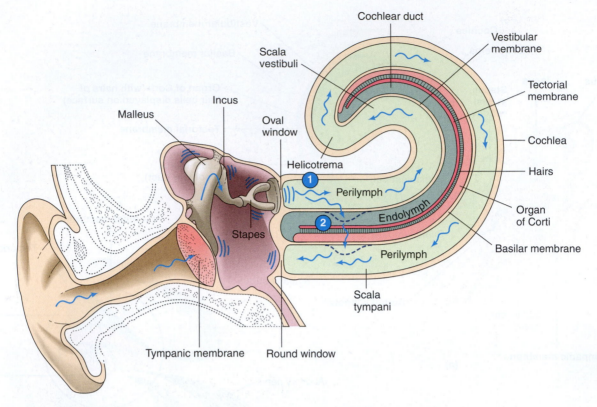

Fluid movement within the perilymph set up by vibration of the oval window follows two pathways:

Pathway ① : Through the scala vestibuli, around the heliocotrema, and through the scala tympani, causing the round window to vibrate. This pathway just dissipates sound energy.

Pathway ② : A "shortcut" from the scala vestibuli through the basilar membrane to the scala tympani. This pathway triggers activation of the receptors for sound by bending the hairs of hair cells as the organ of Corti on top of the vibrating basilar membrane is displaced in relation to the overlying tectorial membrane.

(a)

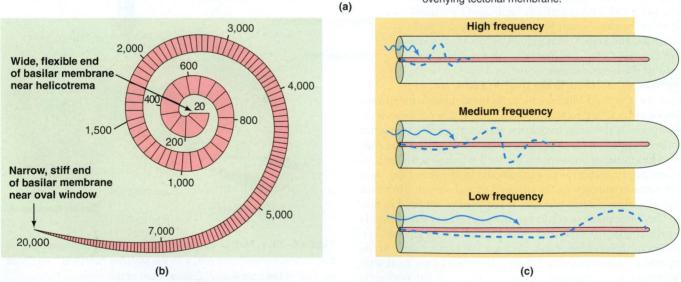

The numbers indicate the frequencies in cycles per second with which different regions of the basilar membrane maximally vibrate.

(b)   **(c)**

Figure 6–40 ● Transmission of sound waves. (a) Fluid movement within the perilymph set up by vibration of the oval window follows two pathways, one dissipating sound energy and the other initiating the receptor potential. (b) Different regions of the basilar membrane vibrate maximally at different frequencies. (c) The narrow, stiff end of the basilar membrane nearest the oval window vibrates best with high-frequency pitches. The wide, flexible end of the basilar membrane near the helicotrema vibrates best with low-frequency pitches.

minishes the transmission of loud sound waves to the inner ear, to protect the delicate sensory apparatus from damage. This reflex response is relatively slow, however, happening at least 40 msec after exposure to a loud sound. It thus provides protection only from prolonged loud sounds, not from sudden sounds like an explosion. Taking advantage of this reflex, World War II antiaircraft guns were designed to make a loud pre-firing sound, to protect the gunner's ears from the much louder boom of the actual firing.

■ The cochlea contains the organ of Corti, the sense organ for hearing.

The snail-shaped cochlear portion of the inner ear (in terrestrial vertebrates) is a coiled, tubular system lying deep within the temporal bone (Figures 6–36 and 6–39). The cochlea is divided throughout most of its length into three fluid-filled longitudinal compartments. A blind-ended **cochlear duct,** which is also known as the **scala media,** constitutes the middle compartment. It tunnels lengthwise through the center of the cochlea, almost but not quite reaching its end. The upper compartment, the **scala vestibuli** (Figures 6–39a and b), follows the inner contours of the spiral, and the **scala tympani,** the lower compartment, follows the outer contours. The fluid within the cochlear duct is called **endolymph** (● Figure 6–40a). The scala vestibuli and scala tympani both contain a slightly different fluid, the **perilymph.** The region beyond the tip of the cochlear duct where the fluid in the upper and lower compartments is continuous is the **helicotrema.** The scala vestibuli is sealed from the middle ear cavity by the oval window, to which the stapes is attached. Another small, membrane-covered opening, the **round window,** seals the scala tympani from the middle ear. The thin **vestibular membrane** forms the ceiling of the cochlear duct and separates it from the scala vestibuli. The **basilar membrane** forms the floor of the cochlear duct, separating it from the scala tympani. The basilar membrane is especially important because it bears the **organ of Corti,** the sense organ for hearing.

■ Hair cells in the organ of Corti transduce fluid movements into neural signals.

The organ of Corti, which rests on top of the basilar membrane throughout its full length, contains **hair cells** that are the receptors for sound. The 16,000 hair cells within each human cochlea are arranged in four parallel rows along the length of the basilar membrane: one row of *inner hair cells* and three rows of *outer hair cells* (Figure 6–39c). Protruding from the surface of each hair cell are about 100 hairs known as **stereocilia,** which are actin-stiffened microvilli (see p. 61). Hair cells generate neural signals when their surface hairs are mechanically deformed in association with fluid movements in the inner ear. These stereocilia are mechanically embedded in the **tectorial membrane, an awninglike projection overhanging the** organ of Corti throughout its length (Figure 6–39b and c). Birds, however, have no organ of Corti to separate the inner and outer hair cells. In contrast, the hair cells of reptiles project into the endolymph without an overlying tectorial membrane, and amphibians lack both tectorial and basilar membranes.

The pistonlike action of the stapes against the oval window sets up pressure waves in the upper compartment. Because fluid is incompressible, pressure is dissipated in two ways as the stapes causes the oval window to bulge inward: (1) displacement of the round window and (2) deflection of the basilar membrane (Figure 6–40a). In the first of these pathways, the pressure wave pushes the perilymph forward in the upper compartment, then around the helicotrema, and into the lower compartment, where it causes the round window to bulge outward into the middle-ear cavity to compensate for the pressure increase. As the stapes rocks backward and pulls the oval window outward toward the middle ear, the perilymph shifts in the opposite direction, displacing the round window inward. This pathway does not result in sound reception; it just dissipates pressure.

Pressure waves of frequencies associated with sound reception take a "shortcut." Pressure waves in the upper compartment are transferred through the thin vestibular membrane, into the cochlear duct, and then through the basilar membrane into the lower compartment, where they cause the round window to alternately bulge outward and inward. The main difference in this pathway is that transmission of pressure waves through the basilar membrane causes this membrane to move up and down, or vibrate, in synchrony with the pressure wave. Because the organ of Corti rides on the basilar membrane, the hair cells also move up and down as the basilar membrane oscillates.

■ The inner hair cells convert sound waves into electrical impulses.

The inner and outer hair cells differ in function. The inner hair cells transform the mechanical forces of sound (cochlear fluid vibration) into the electrical impulses of hearing (action potentials propagating auditory messages to the cerebral cortex) Because the stereocilia of the receptor cells are embedded in the stiff, stationary tectorial membrane, they bend back and forth when the oscillating basilar membrane shifts their position in relationship to the tectorial membrane (● Figure 6–41).

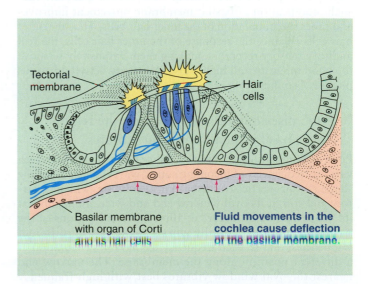

Figure 6–41 ● Bending of hairs on deflection of the basilar membrane. The stereocilia (hairs) from the hair cells of the basilar membrane are embedded in the overlying tectorial membrane. These hairs are bent when the basilar membrane is deflected in relation to the stationary tectorial membrane. This bending opens channels, leading to ion movements that result in a receptor potential.

This back-and-forth mechanical deformation of the hairs alternately opens and closes mechanically gated ion channels (see p. 104) in the hair cell, resulting in alternating depolarizing and hyperpolarizing potential changes—the receptor potential—at the same frequency as the original sound stimulus.

The inner hair cells communicate via a chemical synapse, with the terminals of afferent nerve fibers making up the **auditory (cochlear) nerve.** Depolarization of these hair cells (when the basilar membrane is deflected upward) increases their rate of transmitter release, which steps up the firing rate in the afferent fibers. Conversely, the firing rate decreases as these hair cells release less transmitter when they are hyperpolarized on displacement in the opposite direction.

Thus the ear converts sound waves in the air into oscillating movements of the basilar membrane that bends the hairs of the receptor cells back and forth. This shifting mechanical deformation of the hairs alternately opens and closes the receptor cells' channels, bringing about graded potential changes in the receptor that lead to changes in the rate of action potentials propagated to the brain. In this way, sound waves are translated into neural signals that the brain can perceive as sound sensations (● Figure 6–42).

■ The outer hair cells are thought to enhance sound discrimination.

Whereas the inner hair cells send auditory signals to the brain over afferent fibers, the outer hair cells do not signal the brain about incoming sounds. Instead, the outer hair cells actively and rapidly elongate in response to changes in membrane potential, a function known as **electromotility.** These changes in length are believed to amplify or accentuate the motion of the basilar membrane. An analogy is a person deliberately pushing the pendulum of a grandfather clock in time with its swing, to accentuate its motion. Researchers speculate that such modification of basilar membrane movement improves and tunes the stimulation of the inner hair cells. Thus the outer hair cells enhance the response of the inner hair cells, the real auditory sensory receptors, making them exquisitely sensitive to sound intensity and highly discriminatory between various pitches of sound.

■ Pitch discrimination depends on the region of the basilar membrane that vibrates.

Pitch discrimination (that is, the ability to distinguish between various frequencies of incoming sound waves) depends on the shape and properties of the basilar membrane, which is narrow and stiff at its oval-window end and wide and flexible at its helicotrema end (Figure 6–40b). Different regions of the basilar membrane naturally vibrate maximally at different frequencies; that is, each frequency displays peak vibration at a different position along the membrane. The narrow end nearest the oval window vibrates best with high-frequency pitches, whereas the wide end nearest the helicotrema vibrates maximally with low-frequency tones (Figure 6–40c; note the cochlea has been diagrammatically "unrolled" in this figure to aid understanding, but the real cochlea does not do this!). The pitches in between are sorted out along the length of the membrane from higher to lower frequency. As a sound wave of a particular frequency is set up in the cochlea by oscillation of

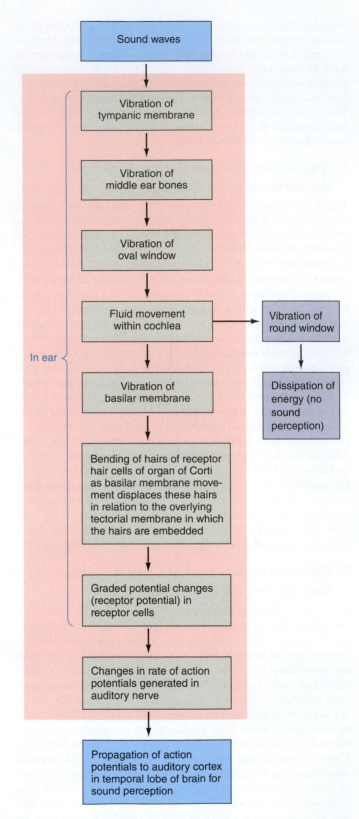

Figure 6–42 ● Sound transduction.

the stapes, the wave travels to the region of the basilar membrane that naturally responds maximally to that frequency. The energy of the pressure wave is dissipated with this vigorous membrane oscillation, so the wave dies out at the region of maximal displacement.

The hair cells in the region of peak vibration of the basilar membrane undergo the most mechanical deformation and, accordingly, are the most excited. This information is propagated to the CNS, which interprets the pattern of hair cell stimulation as a sound of a particular frequency. Modern techniques have determined that the basilar membrane is so fine-tuned that the peak membrane response to a single pitch probably extends no more than the width of a few hair cells.

Overtones of varying frequencies cause many points along the basilar membrane to vibrate simultaneously but less intensely than the fundamental tone, enabling the CNS to distinguish the timbre of the sound (**timbre discrimination**).

■ Loudness discrimination depends on the amplitude of vibration.

Intensity (loudness) discrimination depends on the amplitude of vibration. As sound waves originating from louder sound sources strike the eardrum, they cause it to vibrate more vigorously (that is, bulge in and out to a greater extent) but at the same frequency as a softer sound of the same pitch. The greater tympanic membrane deflection is converted into a greater amplitude of basilar membrane movement in the region of peak responsiveness. The CNS interprets greater basilar membrane oscillation as a louder sound.

■ The auditory cortex is mapped according to tone.

Just as various regions of the basilar membrane are associated with particular tones, the auditory cortex is also **tonotopically** organized. Each region of the basilar membrane is linked to a specific region of the auditory cortex in the temporal lobe. Accordingly, specific cortical neurons are activated only by particular tones; each region of the auditory cortex becomes excited only in response to a specific tone detected by a selected portion of the basilar membrane.

The afferent neurons that pick up the auditory signals from the inner hair cells exit the cochlea via the auditory nerve. The neural pathway between the organ of Corti and the auditory cortex involves several synapses en route, the most notable of which are in the brain stem and medial geniculate nucleus of the thalamus. The brain stem uses the auditory input for alertness and arousal. The thalamus sorts and relays the signals upward. Unlike the visual pathways, auditory signals from each ear are transmitted to both temporal lobes because the fibers partially cross over in the brain stem. For this reason, a disruption of the auditory pathways on one side beyond the brain stem does not affect hearing in either ear to any extent.

Like other regions of the cortex, the **primary auditory cortex is organized into columns.** In contrast to the visual system, however, no hierarchy of cells within the columns that respond to increasingly complex auditory signals has been identified. The primary auditory cortex appears to perceive discrete sounds, whereas the surrounding higher-order auditory cortex integrates the separate sounds into a coherent, meaningful pattern. Consider the complexity of the task accomplished by the auditory system. An animal's ability to discriminate the sound of an approaching predator from that of background noises can be a matter of survival.

■ Insect ears are derived from the respiratory tracheal system.

Let's conclude our examination of hearing with the most successful group of all animals. Insects use auditory information for a variety of behaviors, including social dominance as well as communication with the opposite sex. In crickets the prothoracic walking leg contains a specialized auditory organ, which is derived from a modified trachea. In contrast, locusts' ears are located on the first abdominal segment and consist of modified spiracles (see p. 471). Sensory cells are aggregated into clusters directly coupled to the **tympanum**, a structure similar to the mammalian tympanic membrane. The tympanum is exposed to air on both sides of the membrane. Thus, as in mammals, any changes in air pressure induced by sound waves causes the tympanum to vibrate. Vibration of the tympanum activates specific receptors, generating action potentials, which are sent to the central nervous system. Depending on the source of the sound, differences in the time it takes the sound waves to reach the right or left tympani can be used to establish the source. In addition, because the tracheal system is interconnected, sounds arriving at one tympanum are conducted throughout the animal and cause the opposite tympanum to move from the inside. Interestingly, rather than the ability to discriminate pitch, as in the case of vertebrates, insects' hearing organs are specialized for determining patterns, duration, and intensity of sound waves.

Chemoreception: Taste and Smell

The sensing of chemicals in the environment is another very ancient sense, found in unicellular organisms as well as almost all animals. Animal receptors for taste and smell, plus numerous interoreceptors, are chemoreceptors, which generate neural signals on binding with particular chemicals in their environment. The functions of interoreceptors such as those monitoring blood chemistry are discussed in other chapters. Here we focus on the exteroreceptors for **gustation** (**taste,** or detection of molecules in objects in contact with the body) and **olfaction** (**smell,** or detection of molecules released from a distant object) exteroreceptors. Chemoreception can take place on the skin, or in special organs such as antennae of insects and the olfactory bulbs in nasal cavities of air-breathing vertebrates.

The sensations of taste and smell in association with food intake in some species can influence the flow of digestive juices and affect appetite. In many species of animals they permit social recognition of kin. Furthermore, stimulation of taste or smell receptors induces pleasurable or objectionable sensations (at least in humans). Thus the chemical senses provide a "quality control" checkpoint for substances available for ingestion.

The sense of smell is highly variable in sensitivities among species. In mammals, for example, smell is a relatively poor sense in humans (see box, "Molecular Biology and Genomics: The Smell and Taste of Evolution") and is much less important in influencing our behavior, compared to most mammals. Bird species are also variable. Some species of pelagic birds, which until recently ornithologists thought had no sense of smell at all, can detect the odor of dimethylsulfide (DMS) from up to

MOLECULAR BIOLOGY AND GENOMICS

The Smell and Taste of Evolution

Until the advent of modern genetic techniques, the receptor proteins involved in many senses were unknown. But with these new methods, genes can be found once a single receptor's structure has been found, by searching the genome for similar codes, and then seeing where (if anywhere) those genes are expressed (see Chapter 2). In one interesting study, by searching the mouse genome for codes related to a known taste receptor, researchers recently found that there are many (not one) genes for variants of bitter receptors. Multiple sweet receptors have also been found. Furthermore, the researchers found these genes were being expressed in the mouse tongue. They predict that 50 to 100 taste-receptor genes will be found using genome analysis. The sense of taste may be more sophisticated than originally thought, particularly for bitter substances, which can be dangerous and which have evolved into an enormous array of defensive molecules in plants.

The most recent studies on taste involve "knocking out" genes (p. 32) for taste receptors. In one study, mice were engineered by knocking out a sweet-receptor gene (called *T1R2*) and replacing it with the human equivalent. The mice were highly attracted to the artificial sweetener *aspartame,* which humans taste as sweet but which mice normally do not seem to sense.

Genomic database searches have revealed odorant receptors of humans, dogs, mice, catfish, salamanders, and insects. A surprisingly large family of vertebrate odor receptors (OR)

has been found, with approximately 20 OR genes in birds, 50 in fish, and about 1000 in mammals. In mice, animals highly reliant on their sense of smell, 80% of these receptor genes are functional; the other 20% have mutated into unusable **pseudogenes.** In comparison, the proportion of pseudogenes in chimpanzees is about 30%, and in humans it is 60%! This is an excellent example of **vestigial structures,** which, like remnant hindlimbs in snakes and whales, provide some of the strongest evidence for the theory of evolution. The loss of 60% of functional OR genes in humans accords with what people have long known about human sense of smell: It has clearly degenerated in comparison to other mammals.

Finally, the first insect taste-receptor genes have been recently isolated in *Drosophila* (fruit flies). Here again, the fly genome was scanned for codes of known taste receptors, and the results were about 75 potential taste genes. So far, only 19 have been analyzed further, and 18 of these were found expressed in the fly's taste organ, the proboscis **labellum.** The researchers in this study hope the information will allow the design of bad-tasting (to an insect) chemicals for use in repelling pests that eat crops and damage trees.

Clearly, genomic analysis continues to produce many surprising and exciting results for physiologists. Many long-standing mysteries such as sensory mechanisms are now becoming amenable to further physiological analysis.

---

4 kilometers away. White-chinned petrels, bloodhounds of the Antarctic skies, rely on the odor of DMS to locate zooplankton (such as krill, shrimp-like crustaceans) on which to feed. DMS is produced when zooplankton feed on phytoplankton (single-celled plants), which contain an **osmolyte** (p. 77) called dimethylsulfonioproprionate that breaks down into DMS. Vultures and kiwis also locate food sources by smell and have unusually large proportions of their brains devoted to the sense of smell.

Smell can play a major role in finding direction, finding shelter, seeking prey or avoiding predators, and sexual attraction to a mate. Examples include the location of living prey by carnivorous gastropods such as the cone snail (p. 31), shell availability in hermit crabs, and the induction of larval settlement in barnacles. Crabs can detect virtually all classes of biologically available molecules, including petroleum hydrocarbons and the odors of decaying flesh.

In the social insects, including ants, honey bees, and termites, olfactory communication maintains the very fabric of their complex societies. Exocrine glands located on the bodies of these insects release volatile pheromones into the environment. These olfactory signals, **pheromones,** are lightweight chemical compounds made of between 5 to 20 carbon atoms that are passed from one animal to another and trigger a characteristic behavioral or physiological response. Pheromones

function as powerful regulators of social and reproductive behaviors in both vertebrates and invertebrates; Among the specific behaviors attributed to pheromones in insects include alarm reactions, orientation, swarming, food source tracking, fighting, and recognizing members of a colony. The first pheromone to be identified was **bombykol,** a pheromone released from sexually receptive female moths to attract the male. Chemoreceptors of insects can be exquisitely sensitive to their ligands: The binding of only one pheromone molecule to a receptor on the antennae of the male gypsy moth (*Lymantria dispar*) is sufficient to trigger an action potential. Relying on this remarkable degree of sensitivity, the male flies "upwind" in pursuit of the flightless, sexually receptive female.

We will continue to examine these senses in detail (again, primarily in vertebrates, with some invertebrate mechanisms for comparison). First we examine the mechanism of taste (**gustation**) and then turn our attention to smell (**olfaction**).

■ Taste sensation is coded by patterns of activity in various taste receptors.

In higher vertebrates the chemoreceptors for taste (**gustatory**) sensation are packaged in **taste buds** (● Figure 6–43), about 9000 of which are present in the human oral cavity and throat, with the greatest percentage on the upper surface of the tongue.

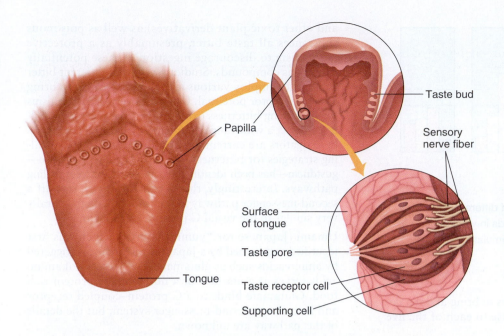

Figure 6–43 ● **Location and structure of human taste buds.** Taste buds are located primarily along the edges of moundlike papillae on the upper surface of the tongue. The receptor cells and supporting cells of a taste bud are arranged like slices of an orange.

This compares to over 15,000 in pigs and rabbits, 550 in lizards, and only 24 in chickens. In other species, such as catfish, taste buds may be located over the entire surface of the animal's body and number over 100,000. In most species of bony fish, taste buds are located in the mouth, pharyngeal region, and gill arches. The extraoral taste receptors help in detecting and capturing the foodstuff, whereas the function of the intraoral taste system is similar to that of air-breathing animals. A typical taste bud consists of about 50 long, spindle-shaped receptor cells packaged with supporting cells in an arrangement like slices of an orange. Each taste bud has a small opening, the **taste pore,** through which fluids in the mouth come into contact with the surface of its receptor cells.

In insects, **sensilla,** specialized projections from the cuticle, permit signal molecules to reach the sensory endings through minute pores. In flies and moths, for example, these chemoreceptors are located in sensory hairlike projections on the terminal segment of the leg as well as in the tip of the proboscis, which is used for feeding and drinking. Appropriate stimuli results in extension of the proboscis and elicitation of feeding or drinking behavior. Dendrites, containing the sensory receptors, project to the apex of each sensillum. In some species of insects (such as ants, bees, and wasps), taste organs are also located on the antennae. Insects rely on their chemical senses for feeding and mating behavior, habitat selection, and parasite–host relationships.

Vertebrate taste receptor cells are modified epithelial cells with many surface folds, or microvilli, that protrude slightly through the taste pore, greatly increasing the surface area exposed to the oral contents. The plasma membrane of the microvilli contains receptor sites that bind selectively with chemical molecules in the environment. Only chemicals in solution—either ingested liquids or solids that have been dissolved in saliva—can attach to receptor cells and evoke the sensation of taste. Binding of a taste-provoking chemical, a **tastant,** with a receptor cell alters the cell's ionic channels to produce a depolarizing receptor potential. This receptor potential, in turn, initiates action potentials within terminal endings of afferent nerve fibers with which the receptor cell synapses.

Most receptors are carefully sheltered from direct exposure to the environment, but the taste receptor cells, by virtue of their task, frequently come into contact with potent chemicals. Unlike the eye or ear receptors, which are irreplaceable, taste receptors have a life span of about 10 days in humans. Epithelial cells surrounding the taste bud differentiate first into supporting cells and then into receptor cells to constantly renew the taste bud components.

Terminal afferent endings of several cranial nerves synapse with taste buds in various regions of the mouth. Signals in these sensory inputs are conveyed via synaptic stops in the brain stem and thalamus to the **cortical gustatory area,** a region in the parietal lobe adjacent to the "tongue" area of the somatosensory cortex. Unlike most sensory input, the gustatory pathways are primarily uncrossed. The brain stem also projects fibers to the hypothalamus and limbic system, presumably to add affective dimensions, such as whether the taste is pleasant or unpleasant, and to process behavioral aspects associated with taste and smell.

Humans can discriminate among thousands of different taste sensations, yet all tastes are thought to be varying combinations of the five tastant categories, the **primary tastes: salty, sour** (or acidic), **sweet, bitter,** and **umami,** a meaty or savory taste. Each receptor cell responds in varying degrees to all the primary tastes but is generally preferentially responsive to one of the taste modalities (● Figure 6–44). (You may have learned that specific areas of your tongue respond to specific tastes; for example, sweet taste is sometimes shown as located at the tip of your tongue. Recent studies show that these common taste maps are erroneous. Human taste types are scattered over most of the tongue, and the map varies from individual to individual.)

The richness of fine taste discrimination beyond the five primary tastes depends on subtle differences in the stimulation patterns of all the taste buds in response to various substances, similar to the variable stimulation of the three cone types that gives rise to the range of color sensations.

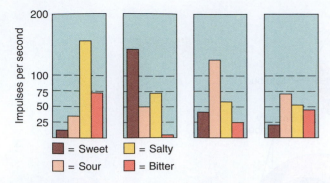

Figure 6–44 ● Relative responsiveness of different taste buds to different stimuli. Each taste bud responds in varying degrees to the four primary tastes. The pattern of stimulation for four different taste buds is depicted.

Receptor cells use different pathways to bring about a depolarizing receptor potential in response to each of the five primary tastant categories:

■ **Salt taste** is stimulated by chemical salts, especially NaCl (table salt). Direct entry of positively charged Na^+ ions through specialized Na^+ channels in the receptor cell membrane of frogs, rats, hamsters, and primates, a movement that reduces the cell's internal negativity, is believed to be responsible for receptor depolarization in response to salt. These ionic channels are regulated by ADH and aldosterone (see p. 544). Amiloride, a specific blocker of these receptor channels, when added to a NaCl stimulus solution placed on the tongue, diminishes the nerve response to the NaCl solution. Taste transduction is different across species of animals, as evidenced by the observation that frog taste cells have nonspecific cation channels whereas rat taste cells are highly specific for Na^+.

■ **Sour taste** is caused by acids, which contain a free hydrogen ion, H^+. The citric acid content of lemons, for example, accounts for their distinctly sour taste. Depolarization of the receptor cell by sour tastants is thought to occur because H^+ blocks K^+ channels in the receptor cell membrane. The resultant decrease in the passive movement of positively charged K^+ ions out of the cell reduces the internal negativity, producing a depolarizing receptor potential.

■ **Sweet taste** is evoked by the particular configuration of glucose. Depending on the species of animal, other organic molecules, such as saccharin, aspartame, and other artificial sweeteners, can also interact with sweet receptor binding sites. In some species of animals, binding of glucose or another chemical with the taste cell receptor activates a G protein, which turns on the cAMP second-messenger pathway in the taste cell (see p. 91), whereas in others a given sweetener may activate the IP_3 pathway. The second-messenger pathway ultimately results in phosphorylation and blockage of K^+ channels in the receptor cell membrane, leading to a depolarizing receptor potential.

■ **Bitter taste** is elicited by a more chemically diverse group of tastants than the other taste sensations. For example, alkaloids (such as caffeine, nicotine, strychnine, morphine,

and other toxic plant derivatives) as well as poisonous substances all taste bitter, presumably as a protective mechanism to discourage ingestion of these potentially dangerous compounds. Studies reveal that different bitter substances employ various signaling pathways to bring about a receptor potential. Using a variety of mechanisms for perceiving bitterness expands the taste receptor's ability to detect a wide range of potentially harmful chemicals. Investigators are currently trying to sort out these signaling strategies for bitterness. The first G protein in taste—**gustducin**—has been identified in one of the bitter-signaling pathways. Interestingly, this G protein, which sets off a second-messenger pathway in the taste cell, is structurally very similar to the visual G protein, transducin.

■ **Umami** (Japanese for "yummy") **taste**, which was first identified and named by a Japanese researcher, is triggered by amino acids such as glutamate. The presence of amino acids serves as a marker for a nutritionally protein-rich food. Glutamate binds to a G protein–coupled receptor and activates a second-messenger system, but the details of this pathway are unknown.

?

How Many Taste Receptors Are There? Traditionally, it has been thought that mammalian taste was a crude discriminator, with only four mammalian taste receptors (sweet, sour, bitter, salt), with a fifth one (umami) recently added to the list. But genome studies are causing us to rethink this dogma. See the box, "Molecular Biology and Genomics: The Smell and Taste of Evolution."

Taste perception is also influenced by information derived from other receptors, especially odor. Any impairment in the sense of smell—for example, swollen nasal passageways as the result of a cold—also markedly reduces the human sense of taste, even though the taste receptors are unaffected by the cold. Other factors affecting taste include temperature and texture of the food as well as psychological factors associated with past experiences with the food. How the cortex accomplishes the complex perceptual processing of taste sensation is currently not known.

Now that we have examined the reception of taste in animals, we turn to the sensation of smell.

■ The olfactory receptors in the nose are specialized endings of renewable afferent neurons.

The **olfactory (smell) mucosa** of vertebrates, located in the **nasal fossae** (upper tract of the respiratory airways), contains three cell types: *olfactory receptors, supporting cells,* and *basal cells* (● Figure 6–45). These supporting cells (also called **Bowman's glands**) secrete mucus, which coats the nasal passages. The basal cells are precursors for new olfactory receptor cells, which are replaced about every two months. This is remarkable because, unlike the other special sense receptors, **olfactory receptors** are specialized endings of afferent neurons, not separate cells. The entire neuron, including its afferent axon projecting into the brain, is replaced. These are the

Figure 6–45 ● Location and structure of the olfactory receptors.

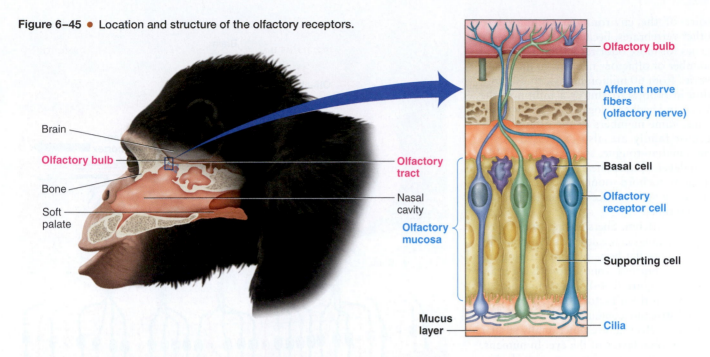

only neurons that undergo cell division (although recent evidence suggests that new neurons can be produced in the hippocampus, a part of the brain important in forming long-term memory; see p. 188).

The axons of the receptor cells collectively form the olfactory nerve.

The receptor portion of the olfactory receptor cells consists of an enlarged knob bearing several long cilia that extend like a tassel to the surface of the mucosa. These cilia contain the binding sites for attaching **odorants,** molecules that can be smelled. During quiet breathing in air breathers, odorants typically reach the sensitive receptors only by diffusion because the olfactory mucosa is above the normal path of airflow. The act of sniffing enhances this process by drawing the air currents upward within the nasal cavity so that a greater percentage of the odoriferous molecules in the air contact the olfactory mucosa. Odorants also reach the olfactory mucosa during eating by wafting up to the nose from the mouth through the pharynx (back of the throat).

The nasal cavity in both fish and aquatic amphibians is generally a pit or saclike structure on either the dorsal or ventral surface of the snout. In aquatic vertebrates water is drawn across the olfactory epithelium and odor signals become absorbed in the mucous layer, where they bind to specific receptor sites on the surface membranes of olfactory cilia in the outer mucous layer.

To be smelled, a substance must be (1) sufficiently volatile (easily vaporized or dissolved) that some of its molecules can enter the nose in the inspired air or water; and (2) sufficiently soluble that it can dissolve in the mucus layer coating the olfactory mucosa. As with taste receptors, molecules must be dissolved to be detected by olfactory receptors. Several thousand odors can be distinguished by the sense of smell, although it has been difficult to categorize particular odors based on their chemical structure. Some odors quickly dissipate in the atmo-

sphere, such as alarm substances released by ants, whereas others are intended to persist for long periods of times (territorial markings). For example, lipid added to the scent markings of lions and tigers extends the release time of the pheromonal message released by the anal sacs. Interestingly, bacterial action on the urine marks of these carnivores contributes to the generation of additional small, volatile pungent-smelling compounds.

Various parts of an odor are detected by different olfactory receptors and sorted into "smell files."

The human nose contains 5 million olfactory receptors (about 10-fold less than in rodents and dogs). Physiologists think mammals have about 1000 different types of odor receptors, far more than the two to five receptors that code for color vision and the five that code for taste (see box, "Molecular Biology and Genomics: The Smell and Taste of Evolution"). Smell detection "dissects" an odor into various components. Each receptor responds to only one discrete component of an odor rather than to the whole odorant molecule. Accordingly, different receptors detect the various parts of an odor, and a given receptor can respond to a particular odor component shared in common by different scents, ultimately permitting spatial summation. Basically the gustatory sense determines whether the food is good or not (that is, is it safe?), whereas the olfactory sense is more discriminatory and picks up the selected differences in food items.

Binding of an appropriate scent signal to an olfactory receptor activates a G protein, triggering a cascade of intracellular reactions that leads to opening of Na^+ channels. The resultant ion movement brings about a depolarizing receptor potential that generates action potentials in the afferent fiber. The frequency of the action potentials depends on the concentration of the stimulating chemical molecules. Although chemoreception is important in all animals, mammals can sample a larger repertoire of the environment compared to other vertebrates, because they have more genes for more odor receptors. The actual number of olfactory neurons in mammals varies from 60 million in rats, to 100 million in rabbits to almost a billion in the German Shepherd dog. For unknown reasons, some members of the olfactory receptor family are also expressed in the mammalian tongue, sperm cells (p. 744), and developing rat heart, as well as in the avian notochord (nonneuronal embryonic tissue, which induces formation of the neural tube).

The afferent fibers arising from the receptor endings in the nose pass through tiny holes in the flat bone plate separating the olfactory mucosa from the overlying brain tissue (● Figure 6–46). They immediately synapse in the **olfactory bulb,** a complex neural structure containing several different layers of cells that are functionally similar to the retinal layers of the eye. In humans, the twin olfactory bulbs, one on each side, are about the size of small grapes. Each olfactory bulb is lined by small, ball-like neural junctions known as **glomeruli** ("little balls") (Figure 6–46). Comparatively, fish are noted for having few glomeruli. Within each glomerulus, the terminals of receptor cells carrying information about a particular scent component synapse with the next cells in the olfactory pathway, the **mitral cells.** Because each glomerulus receives signals only from receptors that detect a particular odor component, the glomeruli serve as "smell files." The separate components of an odor are sorted into different glomeruli, one component per file. Thus the glomeruli, which serve as the first relay station in the brain for processing olfactory information, play a key role in organizing scent perception.

The mitral cells on which the olfactory receptors terminate in the glomeruli refine the smell signals and relay them to the brain for further processing. Fibers leaving the olfactory bulb travel in two different routes (● Figure 6–47): (1) a *subcortical route* going primarily to regions of the limbic system, especially the lower medial sides of the temporal lobes (considered to be the **primary olfactory cortex**), and (2) a *thalamic-cortical route.* Until recently, physiologists thought the subcortical route was the only olfactory pathway. This route, which includes hypothalamic involvement, permits close coordination between smell and behavioral reactions associated with feeding, mating, and direction orienting. As with the other senses, the thalamic-cortical route is important for conscious perception and fine discrimination of smell.

Odor discrimination is coded by patterns of activity in the olfactory bulb glomeruli.

Because each given odorant activates multiple receptors and glomeruli in response to its various odor components, odor

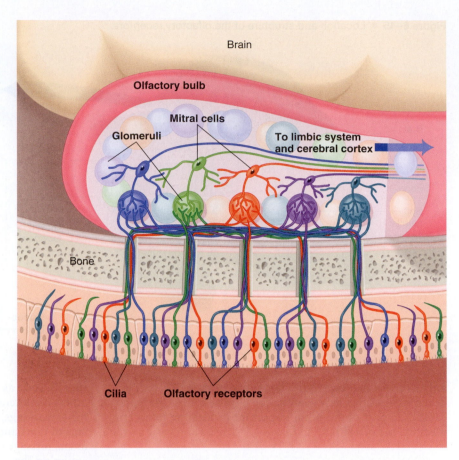

Figure 6–46 ● Processing of scents in the olfactory bulb. Each of the glomeruli lining the olfactory bulb receives synaptic input from only one type of olfactory receptor, which, in turn, responds to only one discrete component of an odorant. Thus the glomeruli sort and file the various components of an odiferous molecule before relaying the smell signal to the mitral cells and higher brain levels for further processing.

discrimination is based on different patterns of glomeruli activated by various scents. In this way, the cortex can distinguish about 20,000 different scents. This mechanism for sorting out and distinguishing different odors is very effective, even in humans, who have a poor sense of smell compared to other species. (By comparison, dogs' sense of smell is hundreds of times more sensitive than that of humans). A noteworthy example is our ability to detect methyl mercaptan (garlic odor) at a concentration of 1 molecule per 50 billion molecules in the air! This substance is added to odorless natural gas to enable us to detect potentially lethal gas leaks.

The olfactory system adapts quickly, and odorants are rapidly cleared.

Although the olfactory system is sensitive and highly discriminating, it is also quickly adaptive. Our sensitivity to most new odors diminishes rapidly after a short period of exposure to it, even though the odor source continues to be present. This reduced sensitivity or habituation does not involve receptor adaptation, as researchers thought for years; actually, the olfactory receptors themselves adapt slowly. It apparently involves some sort of adaptation process at the receptor as well

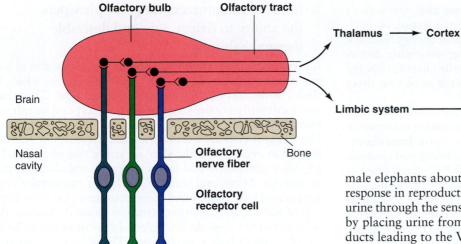

Figure 6–47 ● Olfactory pathways. See Figures 6–45 and 6–46 for a location reference point.

as in the CNS. The initial desensitization of the odor receptors to the odor takes approximately a minute. This is followed by a slower adaptation of neuronal pathways in the central nervous system. Adaptation is specific for a particular odor, and responsiveness to other odors remains unchanged.

What clears the odorants away from their binding sites on the olfactory receptors so that the smell doesn't "linger" after the source is removed? Several "odor-eating" enzymes have recently been discovered in the olfactory mucosa that may serve as molecular janitors, clearing away odoriferous molecules so they don't continue to stimulate olfactory receptors. Interestingly, these odorant-clearing enzymes are very similar chemically to detoxification enzymes in the liver. (These liver enzymes inactivate potential toxins absorbed from the digestive tract; see p. 640.) This resemblance may not be coincidental. Researchers speculate that these enzymes may serve the dual purpose of clearing the olfactory mucosa of old odorants and transforming potentially harmful chemicals into harmless molecules. Such detoxification would be very useful, considering the open passageway between the olfactory mucosa and the brain.

■ The vomeronasal organ detects pheromones.

In addition to the olfactory mucosa, the vertebrate nose may contain another sense organ: the **vomeronasal organ (VNO)**, primarily in reptiles and mammals, including humans (most aquatic vertebrates and birds lack a VNO). The VNO is located just inside the nose next to the vomer bone, hence its name. The VNO is called the "sexual nose" for its role in governing reproductive and social behaviors, such as identifying and attracting a mate. For example, male ungulates, such as stallions, put their lips into the urine of an estrous mare and curl their upper lip in the **Flehman position** ("horse's laugh"), which partially blocks the nostril opening. Breathing deeply carries the urine into the VNO, where the stallion determines sexual receptivity of the mare by the concentration of pheromone. Sex pheromones are also produced in the urine of fe-

male elephants about to ovulate, which initiates the Flehman response in reproductively active males. Males first identify the urine through the sense of smell and then detect the pheromone by placing urine from the tip of their trunk to the opening of ducts leading to the VNO. Interestingly, this pheromone, (Z)-7-dodecenyl acetate, is identical to a substance used by more than a hundred species of moths as one of their major sex pheromones. One of the few other mammalian pheromones that has been characterized is **androstenone**, a steroid in the saliva of male pigs and human male sweat. Release of this potent pheromone in pigs causes sows to assume the **lordosis (mating) position**. Androstenone elicits this behavior only in estrous sows, which suggests that in some species, the response to a *releaser* pheromone is context dependent and not necessarily automatic.

Although the role of the VNO in human behavior has not been validated, researchers suspect it governs spontaneous "feelings" between people, either "good chemistry," such as "love at first sight," or "bad chemistry," such as "getting bad vibes" from someone you just met. Evidence from individuals whose VNO has been surgically removed supports the existence of a subtle role for the VNO in group behavior, sexual activity, and compatibility with others, similar to the role they play in animals. Because messages conveyed by the VNO are believed to bypass cortical consciousness, the response to the largely odorless pheromones is not a distinct, discrete perception, such as smelling a favorite fragrance, but more like an inexplicable impression.

Thermoreception

The ability to detect environmental temperature has clear survival value, given that biological processes are often highly temperature sensitive (Chapter 15). Vertebrates have at least three kinds of thermoreceptors, best studied in mammals and the rattlesnake: cold sensors, warmth sensors, and infrared sensors.

■ Warmth and cold thermoreceptors give information about the temperature of the environment near the body.

Mammalian skin and tongue (technically an extension of the skin) have two distinct kinds of sensors for temperature. One responds to increased temperature (warmth sensors), primarily above the skin's temperature, whereas the other responds to decreasing temperature (cold sensors) below skin temperature. (Skin temperatures tend to be lower than body temperatures, around 30 to 33°C in humans.) These are used primarily for thermoregulation, informing the brain about possible disturbances to core temperature (there are also pain sensors that respond to extreme heat or cold; see the section on nociception that follows). These sensors are *anticipatory* feedback components: They inform the brain of an oncoming disturbance *before* the core temperature is actually altered, allowing corrective measures to be activated with little or no time delay (Chapter 1, p. 15, and Chapter 15, p. 699).

The molecular receptors for these senses have proven remarkably elusive until recently. However, distinct mammalian heat-gated and cold-gated ion channels have now been discovered, and their genes cloned. At least three heat-gated channels are known, called **TRPV1, -2, and -3.** TRPV3 starts firing at about 33°C, perhaps providing the signal for thermoregulation. But TRPV1 only fires above 42°C, a point at which heat starts to become painful. As you will see later, this receptor also reacts to spicy food! And TRPV2 fires at 52°C and higher, clearly for pain response rather than thermoregulation.

To date, two cold-gated channels are known, called **CMR1** (which is also activated by menthol, the "cooling" chemical of mint) and **ANKTM1.** CMR1 opens between 8 and 28°C and may be the receptor used in thermoregulation. ANKTM1, in contrast, is only opened at colder temperatures and is associated more with pain.

Recently, researchers have discovered thermal sensors in sharks. The sensory cells are those that have long been known as electroreceptors, the topic of the next section, where we return to this discovery.

Baja California rattlesnake, *Crotalus enyo enyo.* This and other pit vipers have infrared sensors in their snouts, allowing them to detect mammalian prey by their body heat, and perhaps to locate warm microhabitats for thermoregulation.

Photo: Courtesy of American International Rattlesnakes Museum

■ Infrared thermoreceptors give pit vipers the ability to detect prey and desirable thermal habitats.

A very different use of thermal stimuli has evolved in at least two groups of animals, the pit vipers (Viperidae, which includes rattlesnakes), and some pythons and boas. These have extraordinarily sensitive receptors that respond to **infrared radiation,** the low-energy radiation that carries heat energy. The receptor cells are simply branched dendrites of neurons, located in small pits in the skin, on either side of the head in pit vipers (anterior to and below the eyes), and along the jaws of pythons. Behavioral studies (in which a rattlesnake's eyes are covered with black tape) show that these pits allow detection of warm mammalian prey (such as rodents). A mouse that is 10°C warmer than the background environment can be detected at about half a meter away. The sensors are particularly important for use in the first rapid strike to capture the prey. Because there are two pits on vipers, a type of binocular infrared vision is possible, allowing a rattlesnake to determine prey distance.

Physiologists do not know how these receptor cells detect infrared radiation. Remarkably, neurophysiological recordings suggest that different sensors respond to different infrared wavelengths. Thus the snakes may be able to see infrared light in color!

Recent behavioral studies suggest that pit vipers can also use thermal information to find desirable thermal habitats and to avoid undesirable ones. These animals, like all ectotherms, cannot regulate body temperature by internal mechanisms, and instead must rely on the environment (see Chapter 15).

Nociception: Pain

■ Stimulation of nociceptors elicits the perception of pain plus motivational responses.

Pain is primarily a protective mechanism meant to bring to awareness (or the higher brain centers) the fact that tissue damage is occurring or is about to occur. Unlike other somatosensory modalities, pain is accompanied by motivated behavioral responses (such as withdrawal or defense) as well as emotional reactions (such as anger or fear). Also, unlike other sensations, the subjective perception of pain can be influenced by other past or present experiences (for example, heightened pain perception in a dog that has been abused, or lowered pain signaling induced by stress responses in an injured animal. Recall, though, that perception is only reportable by humans; we infer perception in other animals by their reactions, but cannot know what, if anything, they actually perceive).

There are three categories of pain receptors:

- **Mechanical nociceptors** respond to mechanical damage such as cutting, crushing, or pinching.
- **Thermal nociceptors** respond to temperature extremes, especially heat.
- **Polymodal nociceptors** respond equally to all kinds of damaging stimuli, including irritating chemicals released from injured tissues.

None of the nociceptors have specialized receptor organs; they are all naked nerve endings. Because of their value to survival, most nociceptors do not adapt to sustained or repetitive stimulation. However, at least in mammals, to some extent pain occurs in two phases: a quick response to mechanical damage (which may subside), followed by a slower, prolonged chemical response to inflammation and cell contents released by damage. Researchers have recently isolated mammalian nociceptor proteins responsible for the persistent burning pain associated with inflammation. One ion channel is called TRPV1 (see p. 242) but is also known as "the chili pepper receptor" because it is also activated by **capsaicin,** the chemical of chili peppers that causes burning sensation in the mouth. Inflammation triggers a massive increase in the number of these receptors, leading to persistent pain. And as you saw earlier in the section on thermoreception, the channel TRPV2 fires at 52°C and higher.

Table 6–2 ▮ Characteristics of Pain	
Fast Pain	Slow Pain
Occurs on stimulation of mechanical and thermal nociceptors	Occurs on stimulation of polymodal nociceptors
Carried by small, myelinated A-delta fibers	Carried by small, unmyelinated C fibers
Produces sharp, prickling sensation	Produces dull, aching, burning sensation
Easily localized	Poorly localized
Occurs first	Occurs second; persists for longer time; more unpleasant

?

Why Do Mammals, But Not Birds, Taste Capsaicin? The chemical capsaicin is found in the seed-bearing fruits of pepper plants. Most mammals find the chemical highly irritating, and thus avoid eating the fruit. But many birds eat them with impunity. Why? Recall that all biological phenomena have both proximate and evolutionary explanations (Chapter 1). The proximate explanation has been recently revealed: Whereas one mammalian pain receptor is triggered by capsaicin, the equivalent receptors in birds are completely insensitive. Thus birds do not sense the painful burning that mammals do on eating peppers. In evolutionary terms, researchers speculate that the plants evolved this chemical to prevent mammals from eating their fruits, while allowing birds to do so and thus disperse their seeds. The seeds pass unharmed through a bird's digestive tract, and their flight capabilities ensure a more widespread dispersal than mammalian dispersal would.

All nociceptors in mammals can be sensitized by **prostaglandins,** which greatly enhance the receptor response to noxious stimuli (that is, it hurts more when prostaglandins are present). Prostaglandins are a special group of fatty acid derivatives that are cleaved from the lipid bilayer of the plasma membrane and act locally on being released (that is, in paracrine fashion; see p. 86). Aspirin-like drugs inhibit the synthesis of these paracrines, accounting at least in part for the analgesic (pain-relieving) properties of these drugs.

Pain impulses originating at nociceptors are transmitted to the mammalian CNS via one of two types of afferent fibers (▮ Table 6–2). Signals arising from mechanical and thermal nociceptors are transmitted over large, myelinated **A-delta fibers** at rates of up to 30 meters/sec (the fast pain pathway). Impulses from polymodal nociceptors are carried by small, unmyelinated **C fibers** at a much slower rate of 12 meters/sec (the **slow pain pathway**). Think about the last time you cut or burned your finger. You undoubtedly felt a sharp twinge of pain at first, with a more diffuse, disagreeable pain shortly thereafter. Pain typically is perceived initially as a brief, sharp, prickling sensation that is easily localized; this is the fast pain pathway from specific mechanical or heat nociceptors. This feeling is followed by a dull, aching, poorly localized sensation that persists for a longer time and is more unpleasant; this is the slow pain

pathway, activated by chemicals, especially **bradykinin,** a normally inactive substance that is activated by enzymes released into the ECF from damaged tissue. Bradykinin and related compounds not only provoke pain, presumably by stimulating the polymodal nociceptors, but they also contribute to the inflammatory response to tissue injury (Chapter 10). The persistence of these chemicals might explain the long-lasting, aching pain that continues after removal of the mechanical or thermal stimulus that caused the tissue damage.

The primary afferent pain fibers synapse with specific second-order interneurons in the dorsal horn of the spinal cord (● Figure 6–48a). In response to stimulus-induced action potentials, afferent pain fibers release neurotransmitters that influence these next neurons in line. The two best known of the pain neurotransmitters are **substance P** and **glutamate.** Substance P activates ascending pathways that transmit nociceptive signals to higher levels for further processing. Ascending pain pathways have poorly understood destinations in the *somatosensory cortex,* the *thalamus,* and the *reticular formation.* The role of the cortex in pain perception is not clear, although it is probably important at least in localizing the pain. Pain can still be perceived in the absence of the cortex, presumably at the level of the thalamus. The reticular formation increases the level of alertness associated with the noxious encounter. Interconnections from the thalamus and reticular formation to the hypothalamus and limbic system elicit the behavioral responses accompanying the painful experience. The limbic system appears particularly important in perceiving the unpleasant aspects of pain.

Glutamate, the other neurotransmitter released from primary afferent pain terminals, is a major excitatory neurotransmitter. Glutamate acts on two different plasma membrane receptors on the dorsal horn neurons, with two different outcomes. First, binding of glutamate with its AMPA receptors leads to permeability changes that ultimately result in the generation of action potentials in the dorsal horn cell. These action potentials transmit the pain message to higher centers. Second, binding of glutamate with its NMDA receptors leads to Ca^{++} entry into the dorsal horn cell. This pathway is not involved in transmitting pain messages. Instead, Ca^{++} initiates second-messenger systems that make the dorsal horn neuron more excitable than usual. This hyperexcitability contributes in part to the exaggerated sensitivity of an injured area to subsequent exposure to painful or even normally nonpainful stimuli, such as light pressure. The exaggerated sensitivity presum-

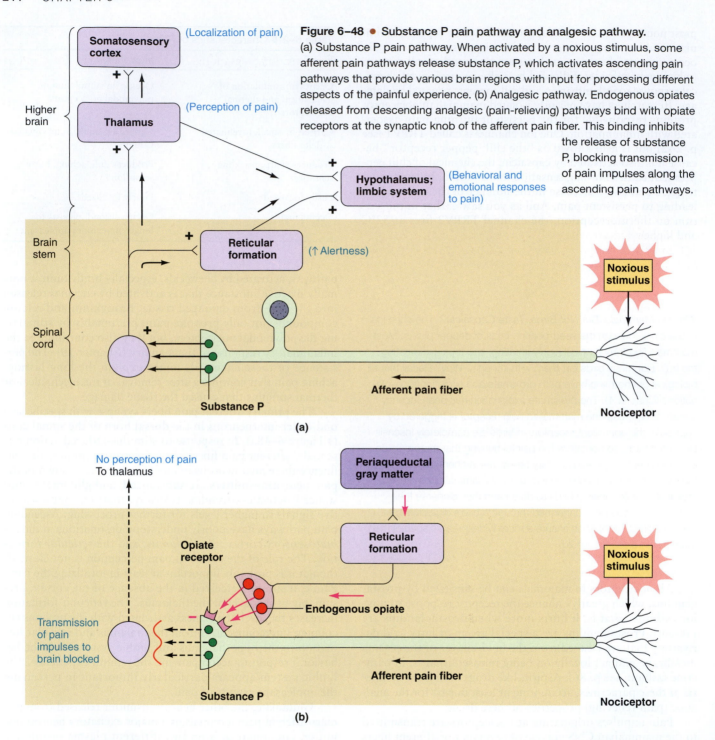

Figure 6–48 ● Substance P pain pathway and analgesic pathway. (a) Substance P pain pathway. When activated by a noxious stimulus, some afferent pain pathways release substance P, which activates ascending pain pathways that provide various brain regions with input for processing different aspects of the painful experience. (b) Analgesic pathway. Endogenous opiates released from descending analgesic (pain-relieving) pathways bind with opiate receptors at the synaptic knob of the afferent pain fiber. This binding inhibits the release of substance P, blocking transmission of pain impulses along the ascending pain pathways.

ably discourages the animal from engaging in activities that could cause further damage or interfere with healing of the injured area.

■ The vertebrate brain has a built-in analgesic system.

In addition to the chain of neurons connecting peripheral nociceptors with higher CNS structures for pain perception, the CNS also contains a neuronal system that suppresses pain. Our knowledge about this built-in **analgesic system** is still fragmentary. Neural mechanisms apparently suppress transmission in

the pain pathways as they enter the spinal cord. Electrical stimulation of the **periaqueductal gray matter** (gray matter surrounding the cerebral aqueduct, a narrow canal that connects the third and fourth ventricular cavities) results in profound analgesia, as does stimulation of the reticular formation within the brain stem. Physiologists think these two regions are part of a descending analgesic pathway (Figure 6–48b) that block, by presynaptic inhibition (see p. 132), the release of substance P from afferent pain fiber terminals.

The built-in analgesic system depends on the presence of **opiate receptors.** People have long known that **morphine,** a derivative of the opium poppy, is a powerful analgesic. It seemed highly unlikely that the mammalian nervous system would evolve opiate receptors only to interact with chemicals derived from a flower! Researchers therefore undertook a search to discover the substances that normally bind with these opiate receptors. The result was the discovery of *endogenous opiates* (morphinelike substances)—the **endorphins, enkephalins,** and **dynorphin**—that are important in the body's natural analgesic system. These endogenous opiates serve as analgesic neurotransmitters; they are released from a descending analgesic pathway and bind with opiate receptors on the afferent pain fiber terminal. This binding suppresses the release of substance P, blocking further transmission of the pain signal. Morphine binds to these same opiate receptors, which accounts for its analgesic properties.

The endogenous opiate system is normally dormant, and it is unclear how this natural pain-suppressing mechanism is activated. Factors known to modulate pain include heavy exercise, acupuncture, hypnosis, and stress (explaining why an injured animal can ignore pain when escaping or fighting). Eating capsaicin in chili peppers is also thought to lead to endorphin release, explaining why some humans come to like eating peppers. There is evidence that some types of stress induce analgesia via the opiate pathway and other, less-understood, nonopiate mechanisms. Acupuncture is another phenomenon that may trigger this system.

Electroreception and Magnetoreception

Echolocation represents a highly evolved sensory system for environmental navigation in bats, some cave-dwelling birds, and dolphins (p. 229), useful in situations where light is not adequate. An alternative solution for navigation in light-limited areas evolved in a number of bony and cartilaginous fishes and at least two mammals: **electroreception.**

■ Electroreception can be passive or active, and can be used for navigation, prey detection, and communication.

Passive electroreception serves to detect extraneous electric fields, not an animal's own electrical currents, and is used primarily for *electrolocation*. **Ampullary electroreceptors** (found in almost all nonteleost fishes and in some teleosts as well as in several amphibian species) respond to low-frequency electric signals that are typical of electrical output from animal nerves and hearts. (Most animals emit a direct current or DC field in seawater because of ion currents in their excitable tissues, and a wound or even a scratch can markedly alter these electrical fields). In elasmobranchs (sharks, skates, and rays) and certain teleosts, the ampullae of Lorenzini, a modified group of neuromasts (p. 222), act as electroreceptors and are used to locate prey even if it is located below the sediment surface (● Figure 6–49a). For example, a shark (*Scyliorhinus*) or skate (*Raja*) can easily detect the faint potentials associated with the gill muscles (from minute leakage of ions) of a buried flounder. The ampullae are filled with a clear, salty gel (a glycoprotein complex) that readily transmits minute electrical currents. New studies show that very small temperature changes induce voltage changes in the gel, suggesting that it serves in both electrosensing and thermosensing.

Passive electroreception has been also demonstrated in two mammals: the platypus and the star-nosed mole. See the Suggested Readings on the Internet by Zimmer (1993) and by Bethge (1997).

In contrast, *active electroreception* resembles echolocation in that the animal assesses its environment by actively emitting signals and receiving the feedback signal. Found in some freshwater fishes in the murky Amazon and some (also murky) African rivers, this system uses an **electric organ** in the tail section of the fish's body that discharges a current field, which emanates from its anterior body and then converges on the tip of the tail (Figure 6–49b), a transepidermal current flow. Momentarily, the tip of the tail becomes negatively charged so that an electric current is produced in the surrounding water. This momentary change is induced repeatedly, at a high frequency. **Tuberous electroreceptors** on the anterior body surface in the lateral line system monitor changes in the local transepidermal current flow. These receptors respond to the high-frequency signals from the fish's electric organ discharge.

The reception is used in both *electrolocation* and *electrocommunication* with conspecifics (members of the same species). Active electroreceptive fish test the conductivity of the water around them and locate perturbations by detecting distortions produced in their own electric field. Any object that differs in impedance from the surrounding water distorts the electric field and alters the pattern of transepidermal current intensities in the area of body surface closest to the object. The South American electric fish *Eigenmannia* generates electric organ discharges at frequencies from 250 to 600 Hz. When two fish with similar discharges meet, they shift their frequencies away from each other in order to create a larger frequency difference and to avoid "jamming" or interfering with its conspecific neighbor's own discharge pattern.

In marine fish, the electroreceptors (mainly ampullary) are connected to the body surface via long, low-resistance channels, whereas the ducts of freshwater fish receptors (mainly tuberous) are comparatively short. Because skin resistance of freshwater fish is much higher than skin resistance of marine fish, transepidermal voltage differences are greater, which means that current flow can be detected with the much shorter ducts. When current enters an electroreceptor cell of a fish, the resulting depolarization opens voltage-gated Ca^{++} channels in the membrane, which triggers the efflux of neurotransmitter from the receptor cell. Consequently, the frequency of action potentials in the sensory fiber that innervates the receptor is increased. Although one important disadvantage of having an electric sense is its very limited range (only several meters) electrolocation permits an animal to actively function in darkness or in murky waters.

■ Electric organs use specialized electrocytes to generate signals.

Electric organs, located within active electroreceptive animals, are mostly derived from muscle or nerve tissue and consist of thin, waferlike cells called **electrocytes** stacked in columns of several thousand surrounded by a gelatinous insulating material (one side of the electrocyte is highly innervated, whereas the other consists of deep folds). When the organ is at rest, the *opposite* faces of the generating cells are at the same positive potential (that is, the extracellular fluid is positive on opposite surfaces of the electrocyte at approximately +80 mV).

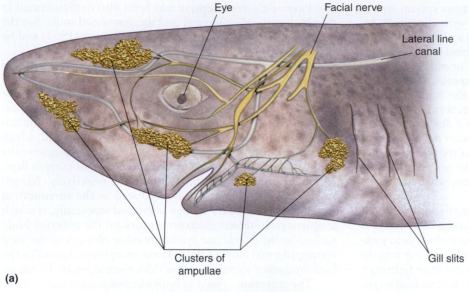

(a)

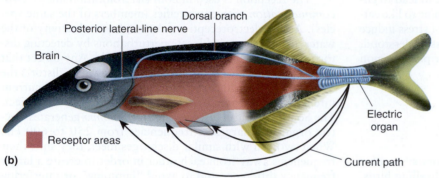

(b)

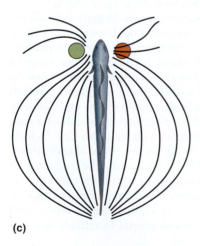

(c)

Figure 6–49 ● (a) Ampullary electroreceptors in a shark. Sensory nerve endings in the ampullae detect electrical currents produced by other animals; they also appear to be sensitive to changes in temperature and salinity. (b) Active electroreception in a weakly electric fish (*Gymnotus*). An electric field is set up by an organ in the tail and received by sensors in the head. (c) An external object (round circle) in the electric field distorts the field (in this case, a nonconductive object diverts the field around it), allowing the fish to detect it (a conductive object such as another animal would concentrate the field lines).

command center is in the medulla drives the electric organ. Gap junctions assure precision in the firing of the electrocytes, maximizing signal intensity to achieve close electrotonic coupling of the medullary and spinal neurons. An additional factor that determines signal intensity is the number of columns of electrocytes capable of generating current.

A few species, such as the electric eel (*Electrophorus*) can produce discharges up to several hundred volts *outside* their bodies, whereas in most species the discharge is limited in range to millivolts to volts. Strong voltages are emitted to stun or kill prey, whereas weak discharges are emitted continually for electrolocation and social communication.

■ Some animals can detect magnetic fields for long-range navigation.

The highly sensitive electroreceptors of elasmobranchs also can detect Earth's magnetic field. Movement of the fish across magnetic field lines produces distortions in the electric currents, which are monitored by the electroreceptors in the ampullae of Lorenzini. Further processing and integration of the sensory information occur in the electrosensory maps located in the midbrain and cerebellar structures.

A variety of other organisms, including some bacteria, insects, amphibians, fish, reptiles, birds, and possibly some mammals, can also detect magnetic fields and are capable of using this information for orientation. The lines of Earth's field are vertical at the poles and horizontal at the equator. Migratory

However, when the potential on one side of the electrocyte is reversed, current flows in a circuit that involves the two membranes, the cell cytoplasm and the external medium. During a discharge, the potential on the innervated side is briefly reversed. The electrocytes are excited in synchrony by spinal nerves that generate small individual voltage gradients around each electrocyte (approximately 150 mV). Because the electrocytes are stacked in series and surrounded by insulating material, the voltage adds arithmetically, in a similar manner to batteries connected in series. A pacemaker potential whose

birds are known to be able to use this magnetic field to help locate established nesting sites thousands of miles away from the wintering grounds. Similarly, sea turtles have been shown in the laboratory to respond to artificial magnetic fields.

How do these animals detect the field? Magnetic materials (magnetite crystals, Fe_3O_4) have been identified in some bacterial cells, honey bee thorax, and the brains of the homing pigeons, rainbow trout, and salmon. For the magnetic sense to function, magnetic crystals are arranged in a chain called a **magnetosome** within a bacterial or an animal magnetoreceptor cell. Single crystals of magnetite (50 nm in size) do not interact strongly enough with Earth's magnetic field to overcome the randomizing effects of thermal buffeting. However, once arranged in chains their individual movements add together such that they can align with a magnetic field and interact with the receptor. How this leads to cellular signals is unknown.

Other strategies for detecting magnetic fields come from studies investigating photosensitive molecules, such as rhodopsin, which are subject to magnetic influences and thus may be linked to signal transduction. When the rhodopsin molecules are aligned in a receptor, the cell only fires off when it picks up light that is polarized in the same direction as the rhodopsin molecules are aligned. Thus if the light is polarized in a particular direction, only a few cells will be aligned to detect the light. The brain then must unscramble those signals to extract additional information about the light, such as its color and brightness, to add to the polarization information. Although every animal that is sensitive to polarized light uses the same basic receptor mechanisms to detect the information, they've all developed different decoding strategies.

?

How Do Magnetoreceptors Work? Scientists don't yet know how magnetoreceptors work. A complication is the observation that specific receptors need not be aggregated into a complex sense organ, although crystals have been localized near the terminal end of the magnetically responsive branch of the trigeminal nerve of trout. Current studies are investigating how the chains of magnetite crystals transduce a magnetic field into an electrical signal in the nervous system. One hypothesis is that the sensor uses a mechanically gated channel. Can you propose how this might work? After that, read the Suggested Reading by Kirschvink, Walker, and Diebel (2001).

Chapter in Perspective:
HOMEOSTASIS AND INTEGRATION

As you saw in Chapter 1, biological regulation begins with the sensing of a regulated state or of a potential threat to that state. In other words, to maintain a life-sustaining stable internal environment, animal nervous systems must be able to sense the myriad external and internal factors that continuously threaten to disrupt homeostasis, such as external exposure to cold, the impending attack of a predator, or internal acidity (pH) changes. Regulated (nonhomeostatic) changes such as reproduction also involves senses, such as in finding a mate.

Sensory information is sent to the central nervous system (CNS), the integrating and decision-making component of the nervous system in "advanced" animals. The CNS in turn commands appropriate effector responses in organ systems to maintain the body's viability. Almost every system in an animal body can be involved in regulatory responses, including physiological (such as altering respiration in response to sensing an abnormal blood CO_2 concentration) and behavioral effector processes (such as running rapidly on sensing of a predator).

In some cases, sensory information from interoreceptors does not use the nervous system. For example, low blood-oxygen levels in a mammal are sensed by the kidney, which in turn secretes a hormone that increases erythrocyte production. Similarly, immune responses typically begin with the sensing of foreign objects without involving neurons. Whether it be neural, endocrine, or immune system regulation, sensory input is the starting point. ■

REVIEW QUESTIONS *(Answers are on p. A–1.)*

Additional study tools for this chapter, including chapter summaries and practice tests, are available online at *www.biology .brookscole.com*

1. The aroma of fresh baked bead would be detected by
   a. nociceptors
   b. proprioceptors
   c. mechanoreceptors
   d. exteroceptors
   e. electroreceptors

2. A more intense stimulus can
   a. bring about larger action potentials in the afferent neuron
   b. cause a higher frequency of action potentials in the afferent neuron
   c. bring about an increase in the number of receptors activated
   d. directly stimulate the interneuron
   e. b and c

3. Tonic receptors
   a. become phasic once they have adapted
   b. adapt more slowly than phasic receptors
   c. adapt more rapidly than phasic receptors
   d. exhibit an "off response"
   e. are important to the sense of touch

4. The type of eye found in arthropods is
   a. an eyespot
   b. a camera eye
   c. a chromatophore eye
   d. a compound eye
   e. a retinal eye

5. Which of the following is not part of the vertebrate eye?
   a. retina
   b. ommatidium
   c. sclera
   d. rod
   e. cone

6. The ability to adjust strength of the lens so that both near and far light sources can be focused on the retina is known as
   a. refraction
   b. phototransduction
   c. filtration
   d. accommodation
   e. light adaptation

7. Which of the following are not found in a layer of the retina?
   a. rods
   b. lens
   c. bipolar cells
   d. cones
   e. ganglion cells

8. When a photopigment in a vertebrate eye absorbs light, it
   a. dissociates into its retinene and opsin components
   b. activates the tapetum lucidum
   c. moves to a bipolar cell
   d. produces an action potential in a bipolar cell
   e. none of the above

9. You would expect the retina of a nocturnal vertebrate to have
   a. more cones than rods
   b. approximately equal numbers of rods and cones
   c. more rods than cones
   d. more ganglion cells than rods
   e. sharp color vision at night

10. The perception of the color yellow arises from
    a. stimulation of middle cones
    b. stimulation of short cones
    c. stimulation of middle and long cones
    d. stimulation of middle and short cones
    e. stimulation of long and ultraviolet cones

11. Which of the following is affected by visual input?
    a. pupil size
    b. synchronization of biological clocks
    c. control of eye movements
    d. startle response
    e. all of the above

12. Which of the following are not part of an ommatidium?
    a. retinular cell
    b. eccentric cell
    c. rhabdomere

d. fovea
e. biconvex lens

13. Which of the following is not involved in the ability of an animal to perceive its position in space?
    a. statocysts
    b. neuromast cells
    c. hair cells
    d. vestibular apparatus
    e. ossicles

14. If someone were scientifically correct in telling you that you had rocks in your head, they would be referring to your
    a. statocysts
    b. otoliths
    c. cochlea
    d. pinna
    e. malleus

15. In our ear, which structure encounters sound waves first?
    a. the oval window
    b. the organ of Corti
    c. the incus
    d. the tympanum
    e. the Eustachian tube

16. A pheromone would be detected by
    a. nociceptors
    b. gustatory receptors
    c. olfactory receptors
    d. vomeronasal organs
    e. c and d

17. Which of the following would lead to the perception of a salty taste?
    a. citric acid
    b. Na^+
    c. aspartame
    d. caffeine
    e. glutamate

18. Which of the following would make an injury "hurt more"?
    a. A-delta fibers
    b. C fibers
    c. enkephalins
    d. dynorphin
    e. prostaglandin

19. Ampullae of Lorenzini serve as
    a. magnetoreceptors
    b. nociceptors
    c. electroreceptors
    d. opiate receptors
    e. a and c

20. Electric organs in electroreceptive animals consist of
    a. periaqueductal gray matter
    b. electrocytes
    c. polymodal nociceptors
    d. glomeruli
    e. none of the above

SUGGESTED READINGS AND INTERNET SITES

Axel, R. 1995. The molecular logic of smell. *Scientific American* 273:154–159.

Brown, B. R. 2003. Sensing temperature without ion channels. *Nature* 421:495.

Hill, R. W., & G. A. Wyse 1989. *Animal Physiology*, 2nd ed. New York: Harper & Row.

Kirschvink, J. L., M. W. Walker, & C. E. Diebel, 2001. Magnetite-based magnetoreception. *Current Opinion in Neurobiology* 11:462–467. Available online at www.gps.caltech.edu/users/jkirschvink/pdfs/COINS.pdf. Accessed on February 9, 2004.

Rose, J. D. 2002. The neurobehavioral nature of fishes and the question of awareness and pain. *Reviews in Fisheries Science* 10:1–38.

Smith, D. V., & R. F. Margolskee. 2001. Making sense of taste. *Scientific American*. Available online at www.sciam.com/article.cfm?articleID=000641D5-F855-1C70-84A9809EC588EF21. Accessed on February 9, 2004.

Sneddon, L. U, V. A. Braithwaite, & M. J. Gentle. 2003. Do fish have nociceptors: Evidence for the evolution of a vertebrate sensory system. *Proceedings of the Royal Society—Biological Sciences* 270:1115–1121.

Stokstad, E. 2003. Peering into ancient ears. *Science* 302:770–771.

Todd, J. L., & T. C. Baker. 1997. The cutting edge of insect olfaction. *American Entomologist* 43:174–82.

Walker, R. G., A. T. Willingham, & C. S. Zucker. 2000. A *Drosophila* mechanosensory transduction channel. *Science* 287:2229–2234.

Young, J. M., & B. J. Trask. 2002. The sense of smell: Genomics of vertebrate odorant receptors. *Human Molecular Genetics* 11:1153–1160.

INTERNET SITES

Bethge, P. 1997. *Platypus physiology.* **healthsci.utas.edu.au/medicine/research/mono/Platpage1.html.**

Ritchison G. 2003. *Bird Brain II.* **biology.eku.edu/RITCHISO/birdbrain2.html.** A detailed site on special senses in birds, with animations.

INFOTRAC READINGS

Ascribe Higher Education News Service. 2003. Taste receptor cells share common pathway.

Zimmer, C. 1993. The electric mole. *Discover* 14:16.

7

Endocrine Systems

Photo: Paul Yancey

*An Australian ibis (*Threskiornis spinicollis*), from the northern rainforest. In the past, scientists thought this and other tropical birds did not have seasonal hormonal and reproductive cycles; rather, hormonal changes for reproduction seemed to be triggered by rainfall. But new research on some tropical birds show that even an increase by 1 hour of daylight time in the 24-hour cycle can trigger a surge in reproductive hormones. This slight increase in day length corresponds to the approach of the annual rainy season, and this response in the birds prepares them in advance for its onset.*

Introduction:
Principles of Endocrinology

The ability of cells to "communicate" with each other is ancient, preceding the evolution of multicellularity. Protozoan cells, for example, communicate in order to mate through pheromones, labile chemicals released by an individual that elicit specific behavioral responses in other individuals of the same species (see p. 87). In animals, pheromones and internal signal molecules used locally within a tissue (that is, paracrine action, p. 86) appear in all phyla, whereas signal molecules used by neurons at synapses appear in all phyla except the sponges. With the evolution of circulatory systems, which provided an effective route for delivering signal molecules to distant tissues, there arose **endocrine systems,** which consist of ductless endocrine glands (see p. 7), often scattered throughout an animal's body (such as ● Figure 7–1 for a mammal). These glands secrete **hormones** (signal molecules delivered by circulatory fluids), providing a second mechanism (in addition to nervous systems) for regulating and coordinating distant organs. The endocrine and nervous systems are specialized for controlling different types of activities. In general, the nervous system coordinates rapid, precise responses and is especially important in mediating interactions with the external environment. The endocrine system, by contrast, primarily controls activities that require duration rather than speed. In the course of animal evolution, there has been an increase in the number and complexity of processes that have come under hormonal regulation.

Even though the endocrine glands for the most part are not connected anatomically, they constitute a system in a functional sense, because they all accomplish their functions by secreting hormones into the blood or hemolymph, and many functional interactions can take place among the various endocrine glands. Hormonal synthesis is both widely distributed and expressed in many tissue types of invertebrates and vertebrates.

Endocrinology is the study of the homeostatic chemical adjustments and other activities that hormones accomplish. We begin this chapter by examining general principles of endocrinology, and then we examine specific hormones in invertebrates (primarily insects) and vertebrates.

■ Hormones exert a variety of regulatory effects throughout the body.

Endocrine systems have regulatory designs similar to that of nervous systems. These are cells that sense (usually with receptor proteins) some state in the body that is being regulated, integrator cells (which may be neurons or glands), and effector organs and behaviors that carry out responses. Hormones are the messengers that transmit commands, similar to the roles of action potentials and neurotransmitters in nervous systems.

Even though hormones are distributed throughout an animal body by blood or hemolymph, only specific target cells can

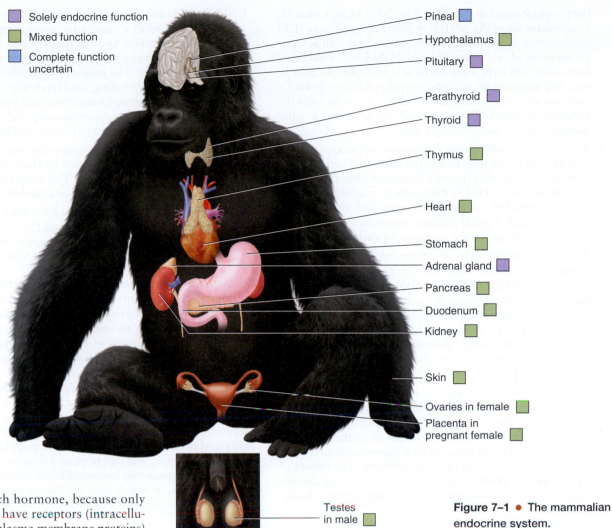

Solely endocrine function
Mixed function
Complete function uncertain

Pineal
Hypothalamus
Pituitary
Parathyroid
Thyroid
Thymus
Heart
Stomach
Adrenal gland
Pancreas
Duodenum
Kidney
Skin
Ovaries in female
Placenta in pregnant female
Testes in male

Figure 7–1 ● The mammalian endocrine system.

respond to each hormone, because only the target cells have receptors (intracellular or integral plasma membrane proteins) for binding with the particular hormone. Binding of a hormone with its specific target cell receptors initiates a chain of events within the target cells to bring about the hormone's final effect. Some hormones have a single target cell type; others have many.

Tropic Hormones

The sole function of some hormones is regulating the production and secretion of another hormone. A hormone that has as its primary function the regulation of hormone secretion by another endocrine gland is classified functionally as a **tropic hormone** (*tropic* means "causing to turn" or "attracting," here meaning "stimulating"). (Do not confuse these with *trophic hormones,* because "trophic" means "nourishing." These hormones are involved in triggering cell growth and development.) Tropic hormones stimulate and maintain their endocrine target tissues. For example, the tropic hormone thyroid-stimulating hormone (TSH) from the anterior pituitary stimulates thyroid hormone secretion by the thyroid gland and also maintains the structural integrity of this gland. In the absence of TSH, the thyroid gland atrophies (shrinks) and produces very low levels of its hormone. A **nontropic hormone,** in contrast, primarily exerts its effects on nonendocrine target tissues. Thyroid hormone, which increases the rate of O_2 consumption and the metabolic activity of almost every cell of mammals and birds,

is an example of a nontropic hormone. **Trophic** hormones are usually nontropic!

Complexity of Endocrine Function

The following points add to the complexity of an endocrine system:

■ A single endocrine gland may produce multiple hormones. The mammalian anterior pituitary, for example, secretes six different hormones, usually made by different specialized cells in that gland; each hormone is under different control mechanisms and has different functions, some being tropic and others nontropic.

■ A single hormone may be secreted by more than one endocrine gland. For example, the vertebrate hypothalamus and pancreas both secrete the hormone *somatostatin.* In amphibians *thyrotropin-releasing hormone* is synthesized in the hypothalamus and the skin as well as other organs. The functional significance of this arrangement has not been established.

■ Frequently, a single hormone has more than one type of target cell and therefore can induce more than one type of effect. This is the "same key, different locks" concept that we discussed in Chapter 3 (p. 88). As an example, *vasopressin* from the posterior pituitary promotes H_2O reabsorption by mammalian kidney tubules as well as vasoconstriction of arterioles throughout the body. Sometimes hormones that have multiple target-cell types can coordinate and integrate the activities of various tissues toward a common end. For example, the effects of *insulin* from the pancreas on muscle, liver, and fat all act in concert to store nutrients after absorption of a meal.

- The secretion rate of some hormones varies considerably over the course of time in a cyclic pattern. Therefore, endocrine systems also provide temporal (time) coordination of function. This is particularly evident in the endocrine control of reproductive cycles, such as the estrous cycle, in which normal function requires highly specific patterns of change in the secretion of various hormones.

- A single target cell may be influenced by more than one hormone. Some cells contain an array of receptors for responding in different ways to different hormones. To illustrate, *insulin* promotes the conversion of glucose into glycogen within vertebrate liver cells by stimulating one particular hepatic enzyme, whereas another hormone, *glucagon,* by activating a different set of hepatic enzymes, enhances the degradation of glycogen into glucose within liver cells.

- The same chemical messenger may be either a hormone or a neurotransmitter, depending on its source and mode of delivery to the target cell. A prime example is norepinephrine, which is secreted as a hormone by the vertebrate adrenal medulla and released as a neurotransmitter from sympathetic postganglionic nerve fibers. The effects of the identical hormone and neurotransmitter are generally different, again illustrating the "same key, different locks" principle.

- Some organs are exclusively endocrine in function (they specialize in hormonal secretion alone, the anterior pituitary and thyroid glands being examples), whereas other organs of the endocrine system perform nonendocrine functions in addition to secreting hormones. For example, vertebrate testes produce sperm and also secrete the male sex-hormone *testosterone.*

This has been a brief overview of the general functions of the endocrine system. Certain hormones are introduced elsewhere and are not discussed in this chapter; these (in vertebrates) are the *gastrointestinal hormones* (Chapter 14), the renal hormones (*erythropoietin* in Chapter 9, and *renin* in Chapter 13), *atrial natriuretic peptide* from the heart (Chapter 12), and *thymosin* (Chapter 10). Most of the remainder of the major hormones are described in greater detail in this chapter.

Hormones are chemically classified into three categories: peptides and proteins, amines, steroids.

Hormones are not all similar chemically, but instead fall into three distinct classes according to their biochemical structure (Table 7–1; compare this table to the list of all chemical messengers on p. 123):

1. The **peptide** and **protein hormones** consist of specific amino acids arranged in a chain of varying length; the shorter chains are peptides and the longer ones are categorized as proteins. For convenience, we refer to this entire category as *peptides*. The majority of animal hormones fall into this class, including (in vertebrates) those secreted by the hypothalamus, anterior pituitary, posterior pituitary, pineal gland, pancreas, parathyroid gland, gastrointestinal tract, kidneys, liver, thyroid C cells, and heart.

2. The **amines** are derived from the amino acid tyrosine and include the hormones secreted by the vertebrate thyroid gland and adrenal medulla. The adrenomedullary hormones are specifically known as **catecholamines.**

3. The **steroids,** which include the hormones secreted by the molting glands of arthropods and the vertebrate adrenal cortex and gonads, as well as some mammalian placental hormones, are neutral lipids derived from cholesterol.

Minor differences in chemical structure between hormones within each category often result in profound differences in biological response. For example, note the subtle difference between testosterone, the male sex hormone responsible for inducing the development of masculine characteristics, and estradiol, a form of estrogen, which is the feminizing female sex hormone (● Figure 7–2).

There is also a fourth category of nonnative hormonelike substances termed **endocrine-disrupting chemicals (EDCs),** whose structures are derived from the by-products of manufactured organic compounds. EDCs range across all continents and oceans, are found in animal populations from the poles to the tropics, and can be passed from generation to generation. Also, nonanimal hormonelike substances are produced by some plants, in which they may serve as plant regulators or as endocrine-disrupting defenses against herbivores, especially insects.

Solubility Characteristics of Hormone Classes

The structural classification of hormones is of more than biochemical interest. The chemical properties of a hormone, most notably its solubility, determine the means by which the hormone is synthesized, stored, and secreted; the way it is transported in the blood; and the mechanism by which it exerts its effects at the target cell, and typically, the type of processes regulated (rapid versus long-term) and their half-lives (turnover) in the circulation. The following differences in the solubility of the various types of hormones are critical to their function (Table 7–1):

All peptides and catecholamines are *hydrophilic* (water loving) and *lipophobic* (lipid fearing); that is, they are highly H_2O soluble and have low lipid solubility.

All steroid and thyroid hormones are *lipophilic* (lipid loving) and *hydrophobic* (water fearing); that is, they have high lipid solubility and are poorly soluble in H_2O.

We are first going to consider the different ways in which these hormone types are processed at their site of origin, the endocrine cell, before comparing their means of transport and mechanisms of action.

Table 7–1 ▍ Chemical Classification of Vertebrate Hormones

Properties	Peptides	Amines		Steroids
		Catecholamines	*Thyroid Hormone*	
Structure	Chains of specfic amino acids, for example:	Tyrosine derivative, for example:	Iodinated tyrosine derivative, for example:	Cholesterol derivative, for example:
	Cys¹—s—s—Cys⁶—Pro⁷—Arg⁸—Gly⁹NH₂ Tyr² Asn⁵ Phe³ — Gln⁴ (vasopressin)	(epinephrine)	(thyroxine, T₄)	(cortisol)
Solubility	Hydrophilic (lipophobic)	Hydrophilic (lipophobic)	Lipophilic (hydrophobic)	Lipophilic (hydrophobic)
Synthesis	In rough endoplasmic reticulum; packaged in Golgi complex	In cytosol	In colloid, an inland extra-cellular site	Stepwise modification of cholestrol molecule in various intra-cellular compartments
Storage	Large amounts in secretory granules	In chromaffin granules	In colloid	Not stored; cholesterol precursor stored in lipid droplets
Secretion	Exocytosis of granules	Exocytosis of granules	Endocytosis of colloid	Simple diffusion
Transport in blood	As free hormone	Half bound to plasma proteins	Mostly bound to plasma proteins	Mostly bound to plasma proteins
Receptor site	Surface of target cell	Surface of target cell	Inside target cell	Inside target cell
Mechanism of action	Channel changes or activation of second-messenger system to alter activity of pre-existing proteins that produce the effect	Activation of second messenger system to alter activity of pre-existing proteins that produce the effect	Activation of specific genes to produce new proteins that produce the effect	Activation of specific genes to produce new proteins that produce the effect
Hormones of this type	All hormones from the hypothalamus, anterior pituitary, posterior pituitary, pineal gland, pancreas, parathyroid gland, gastrointestinal tract, kidneys, liver, thyroid C cells, heart	Only hormones from the adrenal medulla	Only hormones from the thyroid follicular cells	Hormones from the adrenal cortex and gonads plus most placental hormones (vitamin D is steroidlike)

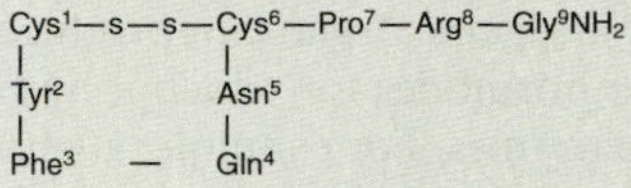

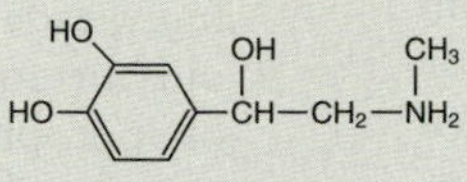

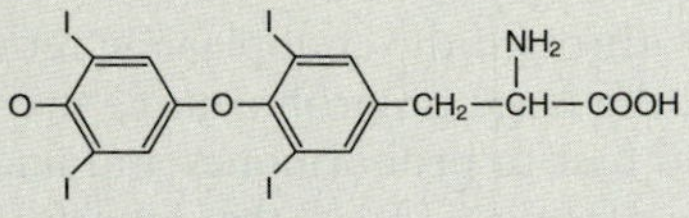

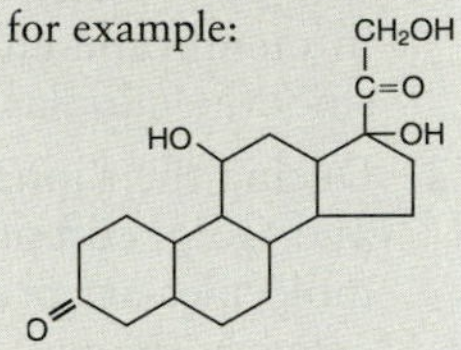

▍ The mechanisms of hormone synthesis, storage, and secretion vary according to the class of hormone.

Because of their chemical differences, the means by which the various classes of hormones are synthesized, stored, and secreted differ as follows:

Peptide Hormones

Peptide hormones are synthesized by the same method used for the manufacture of any protein that is to be exported (see p. 35). Because they are destined to be released from the endocrine cell, the synthesized hormones must be segregated from

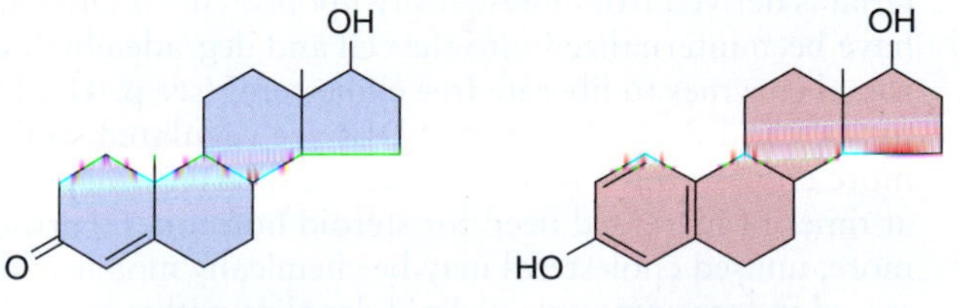

Testosterone, a masculinizing hormone

Estradiol, a feminizing hormone

Figure 7–2 ● Comparison of two steroid hormones, testosterone and estradiol.

intracellular proteins by being sequestered in a membrane-enclosed compartment until they are secreted. Briefly, synthesis of peptide hormones requires the following steps:

1. Large precursor proteins, or **preprohormones,** are synthesized by ribosomes on the rough endoplasmic reticulum. The preprohormones then migrate to the Golgi complex in membrane-enclosed vesicles that pinch off from the smooth endoplasmic reticulum.

2. During their journey through the endoplasmic reticulum and Golgi complex, the large preprohormone precursor molecules are pruned first to **prohormones** and finally to **active hormones.** The peptide "scraps" that are left over as a large preprohormone molecule is cleaved to form the classic hormone are often stored and co-secreted along with the hormone. This raises the possibility that these other peptides may also exert biological effects that differ from the traditional hormonal product; that is, the cell may actually be secreting multiple hormones, but the functions of the other peptide products are for the most part unknown. A known example is the cleavage of the large precursor molecule pro-opiomelanocortin in vertebrates into three active products: adrenocorticotropic hormone (ACTH), melanocyte-stimulating hormone (MSH), and β-endorphin.

3. The Golgi complex concentrates the finished hormones, then packages them into secretory vesicles that are pinched off and stored in the cytoplasm until an appropriate signal triggers their secretion. By storing peptide hormones in a readily releasable form, the gland can respond rapidly to any demands for increased secretion without first needing to increase hormone synthesis.

4. On appropriate stimulation, the secretory vesicles fuse with the plasma membrane and release their contents to the outside by the process of exocytosis (see p. 40). Such secretion usually does not go on continuously; it is triggered only by specific stimuli. The blood subsequently picks up the secreted hormone for distribution.

Steroid Hormones

All steroidogenic (steroid-producing) cells perform the following steps to produce and release their hormonal product:

1. Cholesterol is the common precursor for all steroid hormones; however, numerous species including insects cannot synthesize cholesterol and must obtain it from other sources. Vertebrate steroidogenic cells can synthesize some cholesterol on their own, although most of this raw material is derived from low-density lipoproteins (LDLs) that have been internalized into the cell and degraded by lysosomal enzymes to liberate free cholesterol (see p. 41). The uptake and degradation of LDLs are regulated so that more cholesterol is made available to steroidogenic cells at times of increased need for steroid hormones. Furthermore, unused cholesterol may be chemically modified and stored in large amounts as lipid droplets within steroidogenic cells. Conversion of this major storage form of cholesterol into free cholesterol for use in steroid hormone production is also subject to control. Thus the provision of free cholesterol for use by steroidogenic cells can be closely coordinated with the body's overall need for the hormone product.

2. Synthesis of the various steroid hormones from cholesterol requires a series of enzymatic reactions that modify the basic cholesterol molecule—for example, by varying the type and position of side groups attached to the cholesterol framework or the degree of saturation within the rings (● Figure 7–3). Each conversion from cholesterol to a specific steroid hormone requires the help of a number of enzymes that are limited to the mitochondria or endoplasmic reticulum of certain steroidogenic organs. Accordingly, each steroidogenic organ can produce only the steroid hormone or hormones for which it has a complete set of appropriate enzymes. For example, a key enzyme necessary for the production of *cortisol* is found only in the adrenal cortex. In crustaceans and insects, molting glands synthesize the steroid hormone *ecdysone*. The steroid molecule is therefore shuttled back and forth by unknown means between different compartments within the steroidogenic cell for step-by-step modification until the final secretory product is formed.

3. Unlike peptide hormones, steroid hormones are not stored after their formation. Once formed, the lipid-soluble steroid hormones immediately diffuse through the steroidogenic cell's lipid plasma membrane to enter the circulatory system. Only the hormone precursor cholesterol is stored in significant quantities within steroidogenic cells. Accordingly, the rate of steroid hormone secretion is controlled entirely by the rate of hormone synthesis. In contrast, peptide hormone secretion is controlled primarily by regulating the release of presynthesized, stored hormone.

4. After their secretion into the circulation, some steroid hormones undergo further interconversions within the blood or other organs, where they are converted into more potent or different hormones.

Amines

The amine hormones of vertebrates—thyroid hormone and adrenomedullary catecholamines—have unique synthetic and secretory pathways that are thoroughly described when each of these hormones is specifically addressed. However, in brief, the amines share the following features:

- They are derived from the naturally occurring amino-acid tyrosine.
- None of the enzymes directly involved in the synthesis of either of these hormone types are located in organelle compartments within the secretory cells.
- Both types of amines are stored until they are secreted.

■ Water-soluble hormones are transported dissolved in the plasma, whereas lipid-soluble hormones are almost always transported bound to plasma proteins.

All hormones are carried by the blood, but they are not all transported in the same manner:

1. The hydrophilic peptide hormones are transported simply dissolved in the plasma or, in some cases, bound to a specific carrier protein.

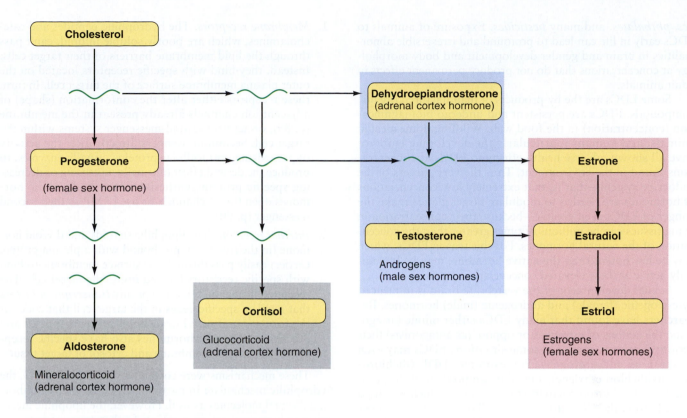

Figure 7–3 ● Steroidogenic pathways for the major steroid hormones. All steroid hormones are produced through a series of enzymatic reactions that modify cholesterol molecules, such as by varying the side groups attached to them. Each steroidogenic organ can produce only those steroid hormones for which it has a complete set of the enzymes needed to appropriately modify cholesterol. For example, the testes have the enzymes necessary to convert cholesterol into testosterone (male sex hormone), whereas the ovaries have the enzymes needed to yield progesterone and the various estrogens (female sex hormones).

2. Lipophilic steroids and thyroid hormone (in vertebrates), which are poorly soluble in water, cannot dissolve in the aqueous plasma in sufficient quantities to account for their known plasma concentrations. Instead, most of these hormones circulate to their target cells reversibly bound to plasma proteins. Some plasma proteins are designed to carry only one type of hormone, whereas other plasma proteins, such as albumin, indiscriminately pick up any "hitchhiking" hormone.

Only the small, unbound, freely dissolved fraction of a lipophilic hormone is biologically active (that is, *free* to diffuse and bind with target cell receptors to exert an effect). Once a hormone has interacted with a target cell, it is rapidly inactivated or removed so that it is no longer available to interact with another target cell. Because the *carrier-bound* hormone is in dynamic equilibrium with the *free hormone* pool, the bound form of lipophilic hormone provides a large reserve that can be called on to replenish the active free pool. To maintain normal endocrine function, the magnitude of the small, free, effective pool, rather than the total plasma concentration of a particular lipophilic hormone, is monitored and adjusted.

Catecholamines in vertebrates are unusual in that only about 50% of these hydrophilic hormones circulate as free hor-

mone, whereas the other 50% are loosely bound to the plasma protein albumin. Because catecholamines are water soluble, the importance of this protein binding is unclear.

■ Endocrine-disrupting chemicals can mimic the effects of native hormones.

Endocrine-disrupting chemicals (EDCs) are human-made substances that are released into the environment and interfere with the endocrine modulation of neural and behavioral maturation of animals. The discovery of hormonally mediated toxic effects in fish downstream of sewage discharge points led to the realization that wastewater contains EDCs. Some EDCs or their breakdown products are nearly as potent as natural hormones and may have substantially longer biological half-lives than their native counterparts. For this reason they are more likely to be stored in an animal's body and accumulated in tissues to concentrations that become toxic to the animal. Occasionally they may be converted in the liver to more toxic compounds, or, if an animal is exposed simultaneously to many chemicals at nontoxic levels, potentiating and additive interactions can lead to toxic effects. The full range of substances interfering with endocrine function has not been entirely established but includes such compounds as *dioxins, PCBs, pheno-*

lics, phthalates, and many *pesticides.* Exposure of animals to EDCs early in life can lead to profound and irreversible abnormalities in brain and gender development and body morphology at concentrations that do not produce permanent effects in adult animals.

Some EDCs are the by-products of manufactured organic compounds. EDCs are persistent and undergo biomagnification (concentration) in the food web. Wildlife, domesticated animals, and humans (particularly the developing embryo) have all shown adverse health effects to EDCs at concentrations found in the environment. Thus the very success of the endocrine system—that is, that extremely low concentrations of hormones are needed to modulate target glands make the danger of EDCs even greater—because their concentration in living tissues can be millions of times greater than the concentration of the natural hormones. For this reason EDCs, which may be less potent than the native hormone, may be biologically active at these elevated concentrations.

Sexual development of the vertebrate brain is influenced by estrogenic (female) and androgenic (male) hormones. Researchers have found that many EDCs either mimic (as agonists) the actions of estrogens or oppose (as antagonists) their actions (antiestrogenic); in some situations EDCs may even function as antiandrogens. For example, DDE (dichlorodiphenyldichloroethylene), a breakdown product of the pesticide DDT (dichlorodiphenyltrichloroethane), which was used globally from the 1940s to 1960s to kill mosquitoes and other insect pests, can be found in almost all living tissue on Earth and acts as an antiandrogen in mammals. More recent discoveries have demonstrated that common household products (for example, heavy-duty laundry powders and liquid detergents, personal care products, and household cleaners) contain nonionic surfactants that break down in the environment to form estrogen agonists (which interact with estrogen receptors). EDCs that interfere with sex hormone activity and production have the potential to disturb normal brain sexual development, as has been demonstrated in wildlife studies of birds, fishes, whales, porpoises, alligators, and turtles. These effects have been linked with direct exposure to sewage and industrial effluents and pesticides, and indirectly to accumulation through aquatic food webs. Similarly, EDCs that impair thyroid function are believed to contribute to learning difficulties in animals, in addition to other neurological abnormalities.

■ Hormones produce their effects by altering intracellular proteins through ion fluxes, second messengers, and transcription factors.

To induce their effects, hormones must bind with target cell receptors specific for them. Each interaction between a particular hormone and a target cell receptor produces a highly characteristic response that differs among hormones and among different target cells influenced by the same hormone.

General Mechanisms of Hydrophilic and Lipophilic Hormone Action

The location of the receptors within the target cell, and the mechanism by which binding of the hormone with the receptors induces a response, both vary, depending on the hormone's solubility characteristics. Receptor–hormone interactions can be grouped into two broad categories based on the location of their receptors (Table 7–1):

1. *Membrane receptors.* The hydrophilic peptides and catecholamines, which are poorly soluble in lipid, can't pass through the lipid membrane barriers of their target cells. Instead, they bind with specific receptors located on the outer plasma membrane surface of the target cell. In turn, these receptors either alter the conformation (shape) of adjacent ion channels already present in the membrane (p. 89), or activate second-messenger systems within the target cell. Second messengers directly alter the activity of pre-existing intracellular proteins, usually enzymes, to produce the desired effect (normally increasing or decreasing specific protein synthesis). Many hydrophilic hormones open Ca^{++} channels or use cAMP as their second messenger (p. 90).

2. *Internal receptors.* The lipophilic steroids and thyroid hormone (in the free form, not bound with a plasma protein carrier) easily pass through the surface membrane to bind with specific receptors located *inside* the target cell. The receptors inside the cell are typically *transcription factors* that regulate specific genes in the target cell that code for the formation of new intracellular proteins (pp. 29 and 30). Some lipophilic hormones also have specific receptors in the plasma membrane and cytosol of target tissues.

These mechanisms were covered in Chapters 2 and 3, the hydrophilic mechanism in particular used by many nonhormonal signal molecules as well. However, the lipophilic mechanism of hormonal action warrants further examination.

■ By stimulating genes, lipophilic hormones promote synthesis of new hormones.

Lipophilic hormones (steroids and thyroid hormone) typically produce their effects on the target cell's genes as follows (● Figure 7–4):

1. Most lipophilic hormones, after diffusing into the cell, bind to receptors located in the nucleus, although researchers have identified some receptors in the plasma membrane and cytosol.

2. Each receptor has a specific region for binding with its hormone and another region for binding with DNA. Most receptors cannot bind with DNA unless it first binds with the hormone.

3. Once the hormone is bound to the receptor, the hormone–receptor complex binds with DNA at a specific attachment site on DNA known as the **hormone response element (HRE)** (Chapter 2, p. 30). Different steroid hormones and thyroid hormone, once bound with their respective receptors, attach at different HREs on DNA. For example, the estrogen–receptor complex binds at DNA's estrogen response element (ERE).

4. Binding of the hormone–receptor complex with DNA ultimately "turns on" or accelerates specific genes within the target cell, leading to the synthesis of new cell proteins, or more of existing cell proteins (see Chapter 2, p. 36).

5. The newly synthesized proteins produce the target cell's ultimate physiological response to the hormone.

By means of this mechanism of nuclear receptors, different lipophilic hormones activate different genes, resulting in

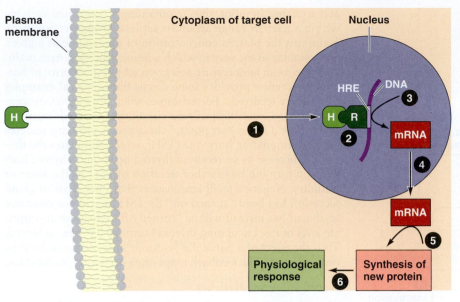

Plasma membrane
Cytoplasm of target cell
Nucleus

HRE   DNA

H R

mRNA

mRNA

Physiological response

Synthesis of new protein

1. A lipophilic hormone diffuses through the plasma and nuclear membranes of its target cells and binds with a nuclear receptor specific for it.

2. The hormone receptor complex in turn binds with the hormone response element, a segment of DNA specific for the hormone receptor complex.

3. DNA binding activates specific genes, which produce complementary messenger RNA.

4. Messenger RNA leaves the nucleus.

5. In the cytoplasm, messenger RNA directs the synthesis of new proteins.

6. These new proteins, either enzymatic or structural, accomplish the target cell's ultimate physiologic response to the hormone.

H = Free lipophilic hormone
R = Lipophilic hormone receptor
HRE = Hormone response element
mRNA = Messenger RNA

Figure 7–4 ● Mechanism of action of lipophilic hormones.

(*Source*: Adapted from George A. Hedge, Howard D. Colby, and Robert L. Goodman, *Clinical Endocrine Physiology,* Philadelphia: W. D. Saunders Company, 1987, Figure 1-9, p. 20)

different biological effects. Genomic and evolutionary analysis of nuclear receptors has revealed that they share an extensive homology and so are grouped into a large superfamily divided into six subfamilies. One subfamily places the vertebrate thyroid-hormone receptor in the same class as the vertebrate vitamin-D receptor and the arthropod ecdysone receptor. Another subfamily contains the vertebrate steroid receptors; vertebrate genomes have six nuclear steroid receptors: two for estrogens, and one each for testosterone/androgens, glucocorticoids, mineralocorticoids (such as aldosterone), and progesterone (hormones we discuss later). Recently, researchers have found that the genome of a mollusk, the seaslug *Aplysia* (p. 189), has an estrogen-type receptor. Although this receptor is not activated by steroids, its existence suggests that steroid receptors evolved before chordates arose.

Onset and Duration of Hormonal Responses

Compared to neural responses that are brought about within milliseconds, hormone action is relatively slow and prolonged, taking minutes to hours for the response to take place after the hormone binds to its receptor. The variability in time of onset for hormonal responses depends on the mechanism employed. Hormones that act through a second-messenger system to alter a pre-existing enzyme's activity elicit full action within a few minutes. In contrast, hormonal responses that require the synthesis of new protein may take up to several hours before any action is initiated.

Also in contrast to neural responses that are quickly terminated once the triggering signal ceases, hormonal responses persist for a period of time after the hormone is no longer bound to its receptor. Once an enzyme is activated in response to hydrophilic hormonal input, it no longer depends on the presence of the hormone. Thus the response lasts until the enzyme is inactivated. As a result, a hormone's effect usually lasts for some time after its withdrawal. Predictably, the responses that depend on protein synthesis last longer than do those stemming from enzyme activation.

Furthermore, as we explain shortly, lipophilic hormones themselves usually persist longer after secretion before being inactivated than hydrophilic hormones do.

■ Hormone actions are greatly amplified at the target cell.

The actions of both hormones are greatly amplified at the target cell. As we noted earlier, hormones are greatly diluted and exert their effect at incredibly low concentrations—as low as 1 picogram (10^{-12} gram; 1 millionth of a millionth of a gram) per mL—as opposed to the much higher localized concentration of neurotransmitter at the target cell during neural communication. Interaction of one hormonal molecule with its receptor can result in the formation of many active protein products that ultimately carry out the physiological effect. For example, one peptide hormone results in the production of many cAMP messengers, each in turn activating many latent enzymes (see p. 91). Similarly, one steroid hormone–activated gene induces formation of many messenger RNA molecules, each of which is used to make many new proteins.

■ The effective plasma concentration of a hormone is normally regulated by changes in its rate of secretion.

The primary function of hormones is the regulation of various homeostatic activities. Because hormones' effects are proportional to their concentrations in the blood, it follows that these concentrations must be subject to control according to homeostatic need (● Figure 7–5). The plasma concentration

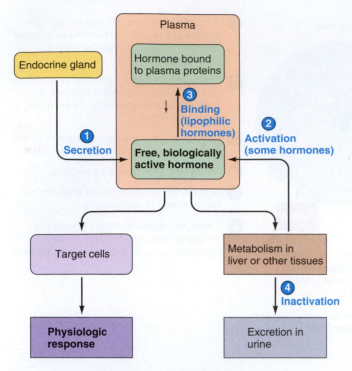

Figure 7–5 ● Factors affecting the plasma concentration of free, biologically active hormone. The plasma concentration of free, biologically active hormone, which can interact with its target cells to produce a physiological response, depends on ① the hormone's rate of secretion by the endocrine gland, ② its rate of metabolic activation (for a few hormones), ③ its extent of binding to plasma proteins (for lipophilic hormones), and ④ its rate of metabolic inactivation and excretion.

of free, biologically active hormone—and thus the hormone's availability to its receptors—depends on several factors: (1) the hormone's rate of secretion into the blood by the endocrine gland; (2) for a few hormones, rate of metabolic activation; (3) for lipophilic hormones, extent of binding to plasma proteins; and (4) rate of removal from the circulation by metabolic inactivation and excretion in urine. Furthermore, the magnitude of the hormonal response depends on the availability and sensitivity of the target cell's receptors for the hormone. Let's first examine the factors that influence the plasma concentration of the hormone before turning our attention to the target cells' responsiveness to the hormone.

Normally, the effective plasma concentration of a hormone is regulated by appropriate adjustments in the rate of its secretion. Secretion rates of all hormones are subject to control, often by a combination of several complex mechanisms. The regulatory system for each hormone is considered in detail in later sections. For now, we address these general mechanisms, which are common to many different hormones: negative-feedback control, neuroendocrine reflexes, and diurnal (circadian) rhythms.

Negative-Feedback Control

Negative feedback is a prominent feature of most biological control systems (see p. 12), including hormonal. Recall that negative feedback exists when the output of a system counter-

acts a change in input, thus maintaining a controlled variable within a narrow range around a set level. Negative feedback maintains the plasma concentration of a hormone at a given level, similar to the way in which a home heating system maintains the room temperature at a given set point. Control of hormonal secretion provides some classic physiological examples of negative feedback. For example, when the plasma concentration (in a mammal) of free circulating thyroid hormone falls below a given "set point," the anterior pituitary secretes thyroid-stimulating hormone (TSH), which stimulates the thyroid to increase its secretion of thyroid hormone. Thyroid hormone in turn inhibits further secretion of TSH by the anterior pituitary. Negative feedback ensures that once thyroid gland secretion has been "turned on" by TSH, it will not continue unabated but instead will be "turned off" when the appropriate level of free circulating thyroid hormone has been achieved. Thus the effect of a particular hormone's actions can inhibit its own secretion. The feedback loops often become quite complex.

Neuroendocrine Reflexes

Many endocrine control systems involve **neuroendocrine reflexes,** which include neural as well as hormonal components. The purpose of such reflexes is to produce a sudden increase in hormone secretion (that is, a *reset* mechanism, p. 16, that "turns up the thermostat setting") in response to a specific stimulus, frequently a stimulus external to the body. In some instances, neural input to the endocrine gland is the only factor regulating secretion of the hormone. For example, secretion of epinephrine by the adrenal medulla is solely controlled by the sympathetic nervous system. Some endocrine control systems, in contrast, include both feedback control, which maintains a constant basal level of the hormone, and neuroendocrine resetting reflexes (which cause sudden bursts in secretion in response to a sudden increased need for the hormone). An example is the increased secretion of cortisol by the adrenal cortex during a stress response (see p. 285).

Diurnal (Circadian) and Other Biological Rhythms

Although some form of negative feedback usually regulates hormone secretion rates, this does not imply that hormones are always maintained at a constant level. Instead, the secretion rates of most hormones rhythmically fluctuate up and down as a function of time. The most common endocrine rhythm is the **diurnal** ("day–night"), or **circadian** ("around a day"), **rhythm,** which is characterized by repetitive oscillations in hormone levels that are very regular and have a frequency of one cycle every 24 hours. This rhythmicity is caused by endogenous oscillators, called **biological clocks,** similar to the self-paced respiratory neurons in the brain stem that govern the rhythmic motions of breathing, except the time-keeping oscillators cycle on a much longer time scale. Furthermore, unlike the rhythmicity of breathing, endocrine rhythms are locked on, or *entrained,* to external cues, called **zeitgebers** ("time givers"), such as the light–dark cycle; that is, the inherent 24-hour cycles of peak and ebb of hormone secretion are set to "march in step" with cycles of light and dark. For example, as you will see later, *cortisol* secretion (p. 286) in a diurnal mammal rises during the late night, reaching its peak secretion in the early morning, and then falls throughout the day to its lowest level at dusk (● Figure 7–6). This prepares the animal for the stresses associated with waking up and initiating activity such as finding food.

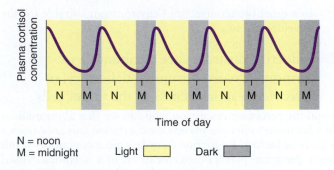

Figure 7–6 ● Diurnal rhythm of cortisol secretion.

N = noon
M = midnight   Light [ ]   Dark [ ]

(*Source:* Adapted from George A. Hedge, Howard D. Colby, and Robert L. Goodman, *Clinical Endocrine Physiology,* Philadelphia: W. D. Saunders Company, 1987, Figure 1-13, p. 28)

Inherent hormonal rhythmicity and the entrainment of zeitgebers are not accomplished by the endocrine organs themselves, but instead result from the central nervous system changing the set point of these organs. (We examine these neural biological clocks in detail later; p. 263.) Negative-feedback control mechanisms maintain whatever set point is established for that time of day.

Some endocrine cycles operate on time scales other than a circadian rhythm—some much shorter than a day and some much longer. A well-known example of the latter is the monthly menstrual cycle in humans, and a tidal rhythm called **circa-lunadian** ("about a lunar day") in many shoreline animals. Another common cycle is the **circannual** (yearly) rhythm in seasonally reproducing organisms and hibernating animals. Beard growth in human males follows a less obvious circannual rhythm.

■ The effective plasma concentration of a hormone can be influenced by the hormone's transport, metabolism, and excretion.

Even though the effective plasma concentration of a hormone is normally regulated by appropriate adjustments in the rate of secretion, alterations in the transport, metabolism, or excretion of a hormone can also influence the size of its effective pool, sometimes inappropriately. For example, the vertebrate liver synthesizes plasma proteins, so liver disease may result in abnormal endocrine activity because of a change in the balance between free and bound pools of certain hormones.

Eventually, all hormones are metabolized by enzyme-mediated reactions that modify the hormonal structure in some way. In most cases this inactivates the hormone. Hormone metabolism is not always a mechanism for removal of used hormones, however. In some cases, a hormone is activated by metabolism; that is, the hormone's product has greater activity than the original hormone. For example, after the thyroid hormone thyroxine is secreted it is converted to a more powerful hormone by enzymatic removal of one of the iodine atoms it contains. Usually the rate of such hormone activation is itself under hormonal or metabolic control.

Metabolic Inactivation and Urinary Excretion of Hormones

The liver is the most common site for metabolic hormonal inactivation (or activation, if that be the case) in vertebrates, but some hormones are also metabolized in the kidneys, blood, or target cells. In contrast to hormone activation, which is typically regulated, hormonal inactivation and excretion are not subject to control. The primary means of eliminating hormones and their metabolites from vertebrate blood is by urinary excretion. When liver and kidney function are normal, measuring urinary concentrations of hormones and their metabolites provides a useful, noninvasive way to assess endocrine function, because the rate of excretion of these products in urine directly reflects their rate of secretion by the endocrine glands. This type of information may provide the only clue that an animal has come into breeding condition and can participate in a captive breeding program. Because the liver and kidney are important in removing hormones from the blood by means of metabolic inactivation and urinary excretion, animals with liver or kidney disease may suffer from excess activity of certain hormones solely as a result of reduced hormone elimination.

The amount of time after a hormone is secreted before it is inactivated and the means by which this is accomplished differ for different classes of hormones. In general, the hydrophilic peptides and catecholamines are easy targets for blood and tissue enzymes, so they remain in the blood only briefly (a few minutes to a few hours) before being enzymatically inactivated. In the case of some peptide hormones, the target cell actually engulfs the bound hormone by endocytosis and degrades it intracellularly. In contrast, binding of lipophilic hormones to plasma proteins renders them less vulnerable to metabolic inactivation and prevents them from escaping into the urine. Therefore, lipophilic hormones are removed from the plasma much more slowly. They may persist in the blood for hours (steroids) or up to a week in humans (thyroid hormone). In general, lipophilic hormones undergo a series of reactions that reduces their biological activity and enhances their H_2O solubility so that they can be freed of their plasma protein carriers and be eliminated in the urine.

■ The responsiveness of a target cell to its hormone can be varied by regulating the number of its hormone-specific receptors.

In contrast to endocrine dysfunction caused by unintentional receptor abnormalities, the target cell receptors for a particular hormone can be deliberately altered as a result of physiological control mechanisms. A target cell's response to a hormone is correlated with the number of the cell's receptors occupied by molecules of that hormone, which in turn depends on the number of receptors present in the target cell for that hormone as well as on the plasma concentration of the hormone. Thus the response of a target cell to a given plasma concentration can be fine-tuned up or down by varying the number of receptors available for hormone binding.

Downregulation

As an illustration, when the plasma concentration of insulin is chronically elevated, the total number of target cell receptors for insulin is reduced as a direct result of the effect that an elevated level of insulin has on the insulin receptors. This phenomenon, known as **downregulation,** constitutes an important locally acting negative-feedback mechanism that prevents the target cells from overreacting to the high concentration of insulin; that is, the target cells are desensitized to insulin, helping blunt the effect of insulin hypersecretion. This is a form of *acclimatization,* one of the key regulatory mechanisms we discussed in Chapter 1 (p. 15). Downregulation of insulin is accomplished by the following mechanism. The binding of insulin to its surface receptors induces endocytosis of the hormone–receptor complex, which is subsequently attacked by intracellular lysosomal enzymes. This internalization serves a twofold purpose: It provides a pathway for degradation of the hormone, and it also plays a role in regulating the number of receptors available for binding on the target cell's surface. At high plasma-insulin concentrations, the number of surface receptors for insulin is gradually reduced as a result of the accelerated rate of receptor internalization and degradation brought about by increased hormonal binding. The rate of synthesis of new receptors within the endoplasmic reticulum and their insertion in the plasma membrane do not keep pace with their rate of destruction. Over time, this self-induced loss of target cell receptors for insulin reduces the target cell's sensitivity to the elevated hormone concentration.

Permissiveness, Synergism, and Antagonism

A given hormone's effects are influenced not only by the concentration of the hormone itself but also by the concentrations of other hormones that interact with it. Because hormones are widely distributed through the blood, target cells may be exposed simultaneously to many different hormones, giving rise to numerous complex hormonal interactions on target cells. Hormones frequently alter the receptors for other kinds of hormones as part of their normal physiological activity. A hormone can influence the activity of another hormone at a given target cell in one of three ways: permissiveness, synergism, and antagonism.

- With **permissiveness,** one hormone must be present in adequate amounts for the full exertion of another hormone's effect. In essence, the first hormone, by enhancing a target cell's responsiveness to another hormone, "permits" this other hormone to exert its full effect. For example, thyroid hormone increases the number of receptors for epinephrine in epinephrine's target cells, increasing the effectiveness of epinephrine. Epinephrine is only marginally effective in the absence of thyroid hormone.

- **Synergism** occurs when the actions of several hormones are complementary and their combined effect is greater than the sum of their separate effects. An example is the synergistic action of follicle-stimulating hormone and testosterone, both of which are required to maintain the normal rate of sperm production. Synergism probably results from each hormone's influence on the number or affinity of receptors for the other hormone.

- **Antagonism** occurs when one hormone causes the loss of another hormone's receptors, reducing the effectiveness of the second hormone. To illustrate, progesterone (a mammalian hormone secreted during pregnancy that decreases contractions of the uterus) inhibits uterine responsiveness to estrogen (another hormone secreted during pregnancy that increases uterine contractions). By causing loss of estrogen receptors on uterine smooth muscle, progesterone prevents estrogen from exerting its excitatory effects during pregnancy and thus keeps the uterus

in a quiet (noncontracting) environment suitable for the developing fetus.

■ Endocrine disorders are attributable to hormonal excess, hormonal deficiency, or decreased responsiveness of the target cells.

From the preceding discussion, you can see that abnormalities in a hormone's effective plasma concentration can arise from a variety of factors. Endocrine disorders most commonly result from abnormal plasma concentrations of a hormone caused by inappropriate rates of secretion—that is, too little hormone secreted (**hyposecretion**) or too much hormone secreted (**hypersecretion**). Occasionally endocrine dysfunction arises because target cell responsiveness to the hormone is abnormally low, even though the plasma concentration of the hormone is normal.

Hyposecretion

If an endocrine gland is secreting too little of its hormone because of an abnormality within that gland, the condition is referred to as *primary hyposecretion.* If, in contrast, the endocrine gland is normal but is secreting too little hormone because of a deficiency of its tropic hormone, the condition is known as *secondary hyposecretion.* The following are among the many different factors (each listed with an example) that may be responsible for hormone deficiency: (1) *genetic* (inborn absence of an enzyme that catalyzes synthesis of the hormone); (2) *dietary* (lack of iodine, which is necessary for synthesis of thyroid hormone); (3) *chemical or toxic* (certain insecticide residues may destroy the adrenal cortex); (4) *immunologic* (autoimmune antibodies may cause self-destruction of the animal's own thyroid tissue); (5) *other disease processes* (cancer or tuberculosis may coincidentally destroy endocrine glands); (6) *iatrogenic* (physician or veterinarian-induced, such as surgical removal of a cancerous thyroid gland); and (7) *idiopathic* (meaning the cause is not known).

The most common method of treating hormone hyposecretion (in humans and domestic mammals) is to administer a hormone that is the same as (or similar to, such as from another species) the one that is deficient or missing. The sources of hormone preparation for clinical use include (1) endocrine tissues from domestic livestock; (2) placental tissue and urine of pregnant women; (3) laboratory synthesis of hormones; and (4) genetically engineered "hormone factories": bacteria into which genes coding for the production of mammalian hormones have been introduced. The method of choice for a given hormone is determined largely by its structural complexity and degree of species specificity.

Hypersecretion

Like hyposecretion, hypersecretion by a particular endocrine gland is designated as primary or secondary depending on whether the defect lies in that gland or is due to excessive stimulation from the outside, respectively. Hypersecretion may be caused by (1) tumors that ignore the normal regulatory input and continuously secrete excess hormone and (2) immunologic factors, such as excessive stimulation of the thyroid gland by an abnormal antibody (long-acting thyroid stimulator, or LATS) that mimics the action of TSH, the thyroid tropic hormone. Excessive levels of a particular hormone may also arise from substance abuse, such as the outlawed practice among athletes of using certain steroids that increase muscle mass by promoting protein synthesis in muscle cells, and the injections of these and other hormones into livestock to enhance production (such as GH, p. 276).

There are several ways of treating hormonal hypersecretion. If a tumor is the culprit, it may be surgically removed or destroyed with radiation treatment. In some instances, drugs that block hormone synthesis or inhibit hormone secretion can limit hypersecretion. Sometimes giving drugs that inhibit the action of the hormone without actually reducing the excess hormone secretion may treat the condition.

Having completed our discussion of the general principles of endocrinology, we now look at the function of invertebrate and vertebrate endocrine glands.

Invertebrate Endocrinology

Considerable diversity is evident in the function of endocrine systems in invertebrates, and currently researchers are seeking to determine how homologous these systems are among different animal groups. The nonvertebrate hormonal systems that have been documented include the following.

Mollusks

Gastropods produce hormones that regulate such functions as egg laying and growth. For example, the **juxtaganglionar organ (JO)** consists of scattered cells in the connective tissue around the cerebral ganglion; the cells release peptides of uncertain function into the hemolymph during egg laying. Also, **bag cells** (two clusters in connective tissue above the abdominal ganglion) produce **egg-laying hormone (ELH)**. Injected into the seaslug *Aplysia* (p. 189), ELH elicits ovulation and egg-

laying behaviors such as head tamping on the sand to cover eggs. This illustrates the important point that both endocrine systems and single effector organs can trigger neural-based behaviors.

Annelids

Earthworms and leeches have ganglionic tissues that have been found to contain molecules closely related to vertebrate hormones. These molecules are also in the circulatory fluid. One of these, *annetocin* (related to vertebrate oxytocin, p. 266), elicits egg-laying behavior when injected into a leech or earthworm.

Crustaceans

Decapods (crabs, shrimp, and so on) have a number of endocrine glands. The best studied are the Y- and X-organs:

- **Y-organs** in the head make **crustecydsone** to promote molting. This hormone is identical to ecdysone in insects, which we discuss shortly

- *Neurosecretory* (p. 266) **X-organs** (in the eyestalks) send their axons into **sinus glands** (also in eyestalks) to release several neurohormones, including (an incomplete list): (1) **chromatophorotropins** that activate *chromatophores* (p. 145) for changing color of the cuticle, (2) **crustacean hyperglycemic hormone** that triggers release of glucose from glycogen in the hepatopancreas (similar to vertebrate *glucagon*, p. 299); and (3) **vitellogenesis-inhibiting hormone,** which inhibits egg development.

Insects

The most intensely studied of invertebrate endocrine systems is that of insects, which we discuss in detail. Several tissues in insects produce hormones, including (● Figure 7–7a):

- The paired **corpora allata,** located behind the brain, with one lying on either side of the esophagus or fused into a

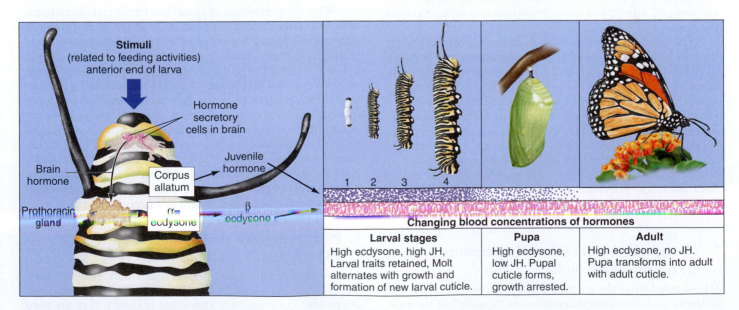

Larval stages	Pupa	Adult
High ecdysone, high JH, Larval traits retained, Molt alternates with growth and formation of new larval cuticle.	High ecdysone, low JH. Pupal cuticle forms, growth arrested.	High ecdysone, no JH. Pupa transforms into adult with adult cuticle.

Figure 7–7 ● Developmental hormones in insects. (a) Hormone-producing tissues of an insect. (b) Ecdysone and juvenile hormone (JH) are two of the hormones involved in larval development and metamorphosis.

single organ (as in higher Diptera), which secrete hormones, including **juvenile hormone (JH)**

- The prothoracic glands, a pair of diffuse glands that are usually located at the back of the head or in the thorax, and that secrete **ecdysone** into the circulatory system

- Neurosecretory cells in the brain that produce one or more neurohormones, including **prothoracicotropic hormone (PTTH)**

- The paired **corpora cardiaca**, which form a part of the wall of the dorsal longitudinal circulatory vessel (aorta)

The function of insect hormones has been studied with regard to their role in molting, osmoregulation, and reproduction, with molting processes receiving the most study. It is that process that we now examine.

Molting is the process of replacing one exoskeleton with another.

To develop from larvae to adult, arthropods must periodically shed their chitin exoskeleton and replace it with a larger one. This cyclical pattern of replacement of the exoskeleton is essential for increasing body size as well as for any changes in body form (such as the acquisition of wings in adults). **Molting** is the process of replacing the exoskeleton with another, whereas the shedding of the old exoskeleton is termed **ecdysis**. In most insects, a final molt occurs when the adult characteristics are expressed and the juvenile characteristics are lost. **Ecdysone** and **juvenile hormone (JH)** are two of the regulatory hormones involved in larval development and **metamorphosis**, the change from the juvenile form to the adult stage (Figure 7–7b). Ecdysone, a member of the steroid family of **ecdysteroids**, is secreted by the prothoracic glands or their equivalent homologs, and governs the initiation of the molting process. Ecdysteroid hormones (as with other steroid hormones) bind to a receptor in the nucleus, in this case the **ecdysone receptor (EcR)**, which activates gene transcription. In addition, all arthropods use ecdysteroids as a type of growth hormone.

The secretion of ecdysteroids in insects is principally regulated by a tropic neurohormone, **prothoracicotropic hormone (PTTH)**. PTTH is synthesized by neurosecretory cells in the brain and travels along axons to peripheral neurohemal organs, the corpora cardiaca and corpora allata, where it is released into the circulation. At the time of molt, neural signals trigger the release of PTTH, which then is transported by the hemolymph to the prothoracic glands, where it stimulates the release of ecdysone, the molt-inducing hormone. In most species the prothoracic glands break down after the final molt to the adult stage. Ecdysone is later peripherally converted into another ecdysteroid, **20-OH-ecdysone**, in various tissues, and this form is recognized as the molt-inducer.

The quantity of JH released determines the quality of the molt.

JH, secreted from the corpora allata, partners the release of ecdysone and ultimately determines the "quality" of the molt. JH has two broad actions in insects: It has a "priming" or activating action whereby it prepares a tissue for hormonal response by directing the synthesis and assembly of the cell machinery necessary to carry out the responses elicited by the binding of JH to its receptor. It also performs a regulatory role by controlling the rate of functioning of the primed tissue. For example, the fat body of virtually all insects cannot synthesize the protein **vitellogenin** (protein taken up by the developing oocyte as **vitellin**, the principal yolk protein) until it has been primed by exposure to JH. Thus, in addition to controlling metamorphosis, JH regulates many aspects of reproduction as well.

The quantity of JH released at each molt determines whether molting will ultimately produce a larva or an adult, because when released, JH ensures that larval characteristics are retained. For example, in insects such as moths and beetles JH is released at progressively decreasing concentrations during each larval stage (or instar) until the final stage, when its release is inhibited and negligible concentrations of JH are detectable. JH release is regulated by the cerebral neuropeptide **allostatin**, which suppresses the release of JH, as well as by **allotropin**, which stimulates the release of JH. During the time of molt the tissues are exposed to a pulse of ecdysteroids in the absence of JH, which enables metamorphosis of the tissues into the adult form. However, if the corpora allata are prematurely removed from an insect, metamorphosis occurs, producing a small adult. Conversely, if JH is administered at each molt the juvenile fails to develop into the adult form (Figure 7–7b).

There is also evidence that JH induces the production and release of pheromones in several species of insects (such as cockroaches, some coleopterans, and lepidopterans) and in the boll weevil. JH enhances the production of sex pheromone in the male fat body.

Concentrations of JH are regulated by the proteins **juvenile hormone–binding protein (JHBP)** and **juvenile hormone esterase (JHE)**. JHBP is produced by the fat body and released into the hemolymph, where it binds JH for transport and protection from degrading enzymes, including JHE. Fat bodies make JHE at increasing levels before the final larval molt to degrade JH; at high concentrations, JHE can degrade the JH bound to the transporter JHBP.

With the prevalence of insects in terrestrial ecosystems, their endocrine system is a prominent target of evolutionary adaptation in plants that are eaten by insects. Substances closely related to ecdysone occur in certain varieties of plants and are believed to protect the plants from feeding by insects. Plant substances called **precocenes** block JH production by the corpora allata, triggering a premature metamorphosis in species in which the larval stage (such as caterpillars) and not the adult is the herbivore. Similarly, to protect crops humans have targeted insect endocrine and pheromone systems. EcR, for example, is now an important target in the design of novel, environmentally safe insecticides. Also, the genes for JHBP and JHE have recently been cloned and are being studied as a mechanism for controlling development of insect pests.

Pheromones are used in mating and colonial interactions.

Let's end our examination of insect endocrinology with a brief look at pheromones. Although, strictly speaking, pheromones are neither hormones nor endocrine in origin, they are used in very similar ways, traveling from one body to another and binding to receptors to stimulate specific behaviors. The most widespread use of pheromones in all animals including insects is in mating. This was first discovered in the 1870s by the

French naturalist Jean-Henri Fabré. A female Great Peacock moth that hatched from a cocoon in his laboratory was soon surrounded by dozens of male Great Peacocks coming in through the windows. He postulated the existence of an attractive odorant, which we now know is the case. Males, using receptors on their antennae, are incredibly sensitive to these female **sex pheromones,** and some species can detect just a single molecule that has traveled a great distance from the female.

Pheromone use beyond reproduction has probably diversified more in insects than in any other animal group. Social insects, such as ants, termites. and bees, rely heavily on pheromones to control and coordinate colony functions and development of castes such as workers and soldiers. For example, **alarm pheromones** are released when a colony is attacked; these signal molecules trigger rapid and violent defense behaviors. Many biologists consider insect colonies "superorganisms," with individual insects acting like cells and organs and with pheromones acting like true hormones.

Vertebrate Endocrinology: An Overview

The central endocrine glands of vertebrates include the **hypothalamus,** the **pituitary gland,** and **pineal gland.** The hypothalamus, a part of the brain, and the posterior pituitary gland act as a unit to release hormones essential for maintaining water balance and for giving birth and lactation (milk production). The hypothalamus also secretes regulatory hormones that control the hormonal output of the **anterior pituitary gland,** which secretes six hormones that in turn largely control the hormonal output of several peripheral endocrine glands. For example, one anterior pituitary hormone, growth hormone, promotes growth and influences nutrient homeostasis. The pineal gland is a part of the brain that secretes a hormone important in establishing biological rhythms, a topic we now discuss in detail.

Biological Clocks: The Role of the Suprachiasmatic Nuclei and the Pineal Gland

Earlier, we discussed the concept of biological clocks, zeitgebers, and rhythmicity in hormone production. Clocks appear in virtually all eukaryotes, including unicellular ones. In animals, clocks are based on neural circuits. The master biological clock that serves as the pacemaker for circadian rhythms in mammals is the **suprachiasmatic nucleus (SCN),** *mediobasal hypothalamus* in birds, a cluster of nerve cell bodies in the hypothalamus above the optic chiasm (the point at which part of the nerve fibers from each eye cross to the opposite half of the brain; see p. 217) (*supra,* "above"; *chiasm,* "cross"). The self-induced rhythmic firing of the SCN neurons plays a major role in establishing many of the mammalian body's inherent 24-hour rhythms.

■ **The biological clock must be synchronized with environmental cues.**

On its own, the circadian biological clock generally cycles a bit slower or faster than the 24-hour environmental cycle. With-

out any external cue (zeitgeber), the SCN sets up cycles that average about 25 hrs in some mammals, including humans. If this master clock were not continuously adjusted to keep pace with the world outside, the body's circadian rhythm would lag progressively out of synchrony with the daily cycles of light and dark. The existence of internal clocks was revealed by the discovery that circadian rhythms persist even if the light–dark cycle is artificially reverted to constant light (LL) or constant dark (DD). Under these conditions the rhythm is allowed to **free-run,** and animals may entrain on other zeitgebers, including environmental temperature, or the noises associated with individuals providing food at particular time in the day. Thus the SCN must be reset daily by external cues so that the body's biological rhythms synchronize with the activity levels driven by the surrounding environment. The SCN works in conjunction with the pineal gland and its hormonal product melatonin to synchronize the various circadian rhythms with the 24-hour day–night cycle.

Biological clocks are ultimately regulated by genes and proteins that must somehow respond to sunlight, but until recently these were almost completely unknown. Scientists have recently unraveled *clock genes* and other underlying molecular mechanisms responsible for the SCN's circadian oscillations and synchronization to the environment. See the box, "Molecular Biology and Genomics: Clocks and Genes," for these findings. However, note that many mysteries remain. For example, researchers still do not know how the synchronizing systems send signals to the clock genes, nor how activity of clock genes regulates other physiological processes.

■ **The pineal gland produces melatonin for circadian regulation, keeping the body's circadian rhythms in time with the light–dark cycle.**

The **pineal gland** is located in the center of the vertebrate brain. The pineal secretes the hormone **melatonin.** (Do not confuse melatonin with the skin-darkening pigment, melanin—see p. 265). Melatonin was named after its lightening action on amphibian skin: Adding a pineal extract to water that contained tadpoles blanched their skin.

One of melatonin's most widely accepted roles is helping to keep the vertebrate body's inherent circadian rhythms in synchrony with the light–dark cycle (although studies in the quail and chicken demonstrated that the pineal and its melatonin rhythm are not essential for circadian function). Melatonin is the hormone of darkness: Its secretion increases up to 10-fold during the darkness of night and then falls to low levels during the light of day. Furthermore, cycles of melatonin are used to synchronize breeding with the seasons, usually spring or fall, in seasonally breeding animals.

In mammals, pineal and melatonin output are under the control of the SCN master clock. Daily change in light intensity is the major environmental cue used to adjust this master clock. Special photoreceptors in the retina pick up light signals and transmit them directly to the SCN. These photoreceptors are distinct from the rods and cones used to perceive, or see, light (see p. 209). Basing their conclusions on recent evidence, scientists suspect that **light-sensitive proteins** found in a special retinal ganglion cell are the likely receptors for light that keeps the body in tune with external time (see box, "Molecu-

MOLECULAR BIOLOGY AND GENOMICS

Clocks and Genes

Circadian biological rhythms have been detected in virtually all organisms, from bacteria to mammals. These clocks regulate reproductive and feeding and other behaviors, and are a classic example of *anticipatory regulation* (Chapter 1): Once synchronized with a regular environmental cycle such as day–night or summer–winter, clocks allow organisms to prepare for these cyclical changes in advance of their actual occurrence.

Such clocks have been known for decades. But the molecular mechanisms have only recently been uncovered, with the fullest details uncovered in the fungus *Neurospora,* certain plants, fruit flies, and mice. Genes that have been named *PERIOD, CLOCK, TIMELESS,* and *Bmal1* (among others) have been discovered and their expression activities tracked. In mice, specific self-starting **clock genes** within the nuclei of SCN neurons set in motion a series of events that brings about the synthesis of **clock proteins** in the cytosol surrounding the nucleus. As the day wears on, these clock proteins continue to accumulate, finally reaching a critical mass, at which time they are transported back into the nucleus. Here they block the genetic process responsible for their own production. The level of clock proteins gradually dwindles as they degrade within the nucleus, thus removing their inhibitory influence from the clock-protein genetic machinery. No longer being blocked, these genes once again rev up the production of more clock proteins, as the cycle repeats itself. Each cycle takes about a day. The fluctuating levels of clock proteins bring about cyclical changes in neural output from the SCN that in turn lead to cyclic changes in effector organs throughout the day. An example is the diurnal variation in cortisol secretion (see Fig-

ure 7–6). Circadian rhythms are thus linked to fluctuations in clock proteins, which use a feedback loop to control their own production at the transcriptional (gene) level. In this way, internal timekeeping is a self-sustaining mechanism built into the genetic makeup of the SCN neurons.

These clock protein cycles allow the clock mechanism to run without any input from the environment. But at some point all clocks need to be synchronized with external signals (see text). This appears to be the role of one or more light-sensitive proteins. One group of these are called **cryptochromes,** found first in plants and later in the retinas and SCN of mice and humans. Cryptochromes in the retina react to blue light and are probably activated at dawn. Knockout mice lacking the genes for two cryptochromes become completely arrhythmic, as if their clocks were shut down. The other possible synchronizing protein is **melanopsin,** found in retinas of fishes, frogs, and (most recently) mammals. It is a type of *opsin,* a family of genes used to make proteins for light absorption in vision (p. 211). Melanopsin's gene, though related to visual opsins, is distinctly different enough to suggest a function other than vision. Knockout mice unable to make melanopsin were found to have 50 to 80% less sensitivity to light in setting their SCN clocks.

Both melanopsin and the cryptochromes are made in the inner layer of the retina, and not in the sensory cells (rods and cones) used for vision. These light-sensitive proteins are thought to trigger an as-yet-unclear chain of events that activate those self-starting genes in the SCN noted earlier. This would ensure that the self-running cycle stays synchronized to the light–dark cycle.

lar Biology and Genomics: Clocks and Genes"). This is the major way the internal clock is coordinated to a 24-hour day. In mammals the eyes cue the pineal gland about the absence or presence of light by means of a neural pathway that passes through the SCN first. This pathway is distinct from the neural systems that result in vision perception.

Fluctuations in melatonin secretion in turn help entrain the body's biological rhythms with the external light–dark cues. Short-day breeders (such as sheep and deer) rely on steadily increasing concentrations of melatonin, whereas long-day breeders (such as many birds) rely on the progressively decreasing concentrations of melatonin to trigger reproduction. In fish, reptiles, birds, and amphibians, the pineal is directly photosensitive and is not directly controlled by nervous input, but rather by sunlight penetrating the skull. Pinealocytes in these vertebrates are modified rod cells with reduced outer segments, containing photosensitive pigments structurally related to rhodopsin. As in mammals, there is a diurnal rhythm of melatonin, which is independent of cues from the eye. Some studies also show that melatonin controls color changes in certain ectothermic vertebrates independently of the control pathway of MSH regulation (p. 265).

The Vertebrate Hypothalamus and Pituitary

The vertebrate **pituitary gland,** or **hypophysis,** is a small endocrine gland in a bony cavity at the base of the brain just below the hypothalamus (● Figure 7–8). The pituitary is connected to the hypothalamus by a thin stalk, the **infundibulum,** which contains nerve fibers and small blood vessels.

■ The pituitary gland consists of anterior and posterior lobes.

The pituitary has two anatomically and functionally distinct lobes, the **posterior pituitary** and the **anterior pituitary.** The posterior pituitary, being derived embryonically from an outgrowth of the brain, consists of nervous tissue and thus is also termed the **neurohypophysis.** The anterior pituitary, in contrast, consists of glandular epithelial tissue derived embryonically from an outpouching of ectoderm that buds off from the roof of the mouth. Accordingly, the anterior pituitary is also known as the **adenohypophysis** (*adeno,* "glandular"). The an-

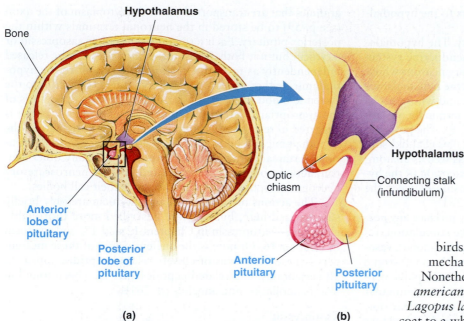

Figure 7–8 ● **Anatomy of the pituitary gland.** (a) Relation of the pituitary gland to the hypothalamus and to the rest of the brain. (b) Schematic enlargement of the pituitary gland and its connection to the hypothalamus.

terior and posterior pituitaries have nothing more in common than their location. The posterior pituitary is connected to the hypothalamus by a neural pathway, whereas the anterior pituitary is connected to the hypothalamus by a vascular link.

The Intermediate Lobe and Melanocyte-Stimulating Hormone

In agnathans (jawless fish—lampreys and hagfish), amphibians, reptiles, and most mammals, the adenohypophysis includes a third, well-defined **intermediate lobe.** Although absent in birds and cetaceans (whales and dolphins), an intermediate lobe exists in the human fetus but becomes rudimentary after birth. Compared to the other regions of the pituitary gland and endocrine tissues in general, the intermediate lobe is poorly vascularized.

The intermediate lobe secretes several **melanocyte-stimulating hormones** or **MSHs.** MSH is under primarily inhibitory control, because ectopic transplantation of the pituitary gland leads to hypertrophy of the intermediate lobe and an increase in MSH secretion. Inhibition of MSH secretion by the hypothalamus is by the catecholamine dopamine. α-MSH and β-MSH are derived from a larger precursor protein, **proopiomelanocortin (POMC).** β-MSH has no known physiological role and may only be a structural component of POMC, whereas α-MSH controls skin coloration via the dispersion of granules containing the pigment **melanin.** These granules are in **melanocytes** (melanin-containing cells; p. 432), found in the skin of most vertebrate species.

Melanocytes in the epidermis synthesize the colored substance melanin using the precursor amino acid tyrosine. Melanins that appear brown or black are referred to as **eumelanins,** whereas red or lighter colored melanins are termed **phaeomelanins.** Melanin is stored in melanin granules or **melanosomes,** where it is released into the skin cells in response to environmental photic cues. By causing variable skin darkening in certain amphibians, reptiles, and fishes, MSHs play a vital role in the camouflage of these species. Some species of mammals and birds also rely on seasonal changes in melanin deposition in the pelage or feathers to minimize detection by predators or prey. MSH and steroid hormones can also differentially affect melanin deposition in the hairs and feathers of mammals and birds, although scientists do not understand the mechanism of melanin pigmentation of feathers. Nonetheless some mammals (such as the hare *Lepus americanus*) and some birds (such as the ptarmigan *Lagopus lagopus*) can change from a brown summer coat to a white winter coat in response to the dramatic changes in photoperiod.

In humans, the anterior pituitary secretes small amounts of MSH. It is not involved in differences in the amount of melanin deposited in the skin, nor with the process of skin tanning although excessive MSH secretion does darken skin. Instead, MSH helps regulate food intake and influences the excitability of the nervous system. MSH has also been shown to suppress the immune system, possibly serving in a check-and-balance way to prevent excessive immune responses.

■ The hypothalamus and posterior pituitary form a neurosecretory system that secretes vasopressin and oxytocin.

The release of hormones from both the posterior and the anterior pituitary is directly controlled by the hypothalamus, but the nature of the relationship is entirely different. The posterior pituitary connects to the hypothalamus by a neural path-

Dispersion of pigment granules in the melanocytes by α-MSH permits the skin coloration of the California lizard, *Sceloporus,* to approximate that of the background

way, whereas the anterior pituitary connects to the hypothalamus by a unique vascular link.

Let's look first at the posterior pituitary. The hypothalamus and posterior pituitary form a neuroendocrine system that consists of a population of neurosecretory neurons whose cell bodies lie in two well-defined clusters in the hypothalamus (the **supraoptic** and **paraventricular nuclei**) and whose axons pass down through the connecting stalk to terminate on capillaries in the posterior pituitary (● Figure 7–9). The posterior pituitary consists of these neuronal terminals plus glial-like supporting cells called **pituicytes.** Functionally as well as anatomically, the posterior pituitary is simply an extension of the hypothalamus. This is similar to the relationship between the X-organ and sinus gland in crustaceans (p. 261).

The posterior pituitary does not actually produce any hormones. It simply stores and, on appropriate stimulation, releases into the blood two small peptide neurohormones, **vasopressin** and **oxytocin,** which are synthesized by the neuronal cell bodies in the hypothalamus. Both these hydrophilic peptides are made in both the supraoptic and the paraventricular nuclei, but a single neuron can produce only one of these hormones. The synthesized hormones are packaged in secretory

granules that are transported down the cytoplasm of the axon (see p. 59) to be stored in the neuronal terminals within the posterior pituitary. Each terminal stores either vasopressin or oxytocin but not both. Thus, these hormones can be released independently as needed. On stimulatory input to the hypothalamus, either vasopressin or oxytocin is released into the systemic blood from the posterior pituitary by exocytosis of the appropriate secretory granules. This hormonal release is triggered in response to action potentials that originate in the hypothalamic cell body and sweep down the axon to the neuronal terminal in the posterior pituitary. As in any other neuron, action potentials are generated in these neurosecretory neurons in response to synaptic input to their cell bodies.

The actions of vasopressin and oxytocin are only briefly summarized here, but they are described more thoroughly elsewhere—vasopressin in Chapters 12 and 13, and oxytocin in Chapter 16. Of note is the ancient origin of these messengers—these hormones or closely related peptides appear in all vertebrates, and related peptides have also been found in insects, mollusks, and annelids (p. 261).

Vasopressin

In most mammals vasopressin (also called **antidiuretic hormone, ADH**) has two major effects that correspond to its two names: (1) it enhances retention of H_2O by kidneys (an antidiuretic effect), and (2) it causes contraction of arteriolar smooth muscle (a vessel pressor effect). The first effect is of greater physiological importance. Under normal conditions, vasopressin is the primary endocrine factor that regulates urinary H_2O loss and overall H_2O balance. In contrast, typical levels of vasopressin play only a minor role in regulating blood pressure by means of the hormone's pressor effect.

The major control for hypothalamic-induced release of vasopressin from the posterior pituitary is input from hypothalamic osmoreceptors, which increase vasopressin secretion in response to a rise in plasma osmolarity. In Chapters 12 and 13, you will see how this serves as negative-feedback regulation.

In lower vertebrates, including agnathans, the antidiuretic hormone is called **arginine vasotocin (AVT).** In addition to its role in osmoregulation, AVT produces vasoconstriction and an increase in blood pressure in most poikilotherms. In flounder, during the initial phase of adaptation from fresh water to seawater there is a transitory rise in plasma osmolarity, which increases plasma AVT. During metamorphosis of tadpoles into frogs, the increase in AVT stimulates ACTH-induced interrenal steroidogenesis, which raises concentrations of corticosterone (see p. 270). In birds it also governs *oviposition,* the physical process of laying an egg; administering AVT into a laying hen causes contraction of the uterus (shell gland) and expulsion of the egg. AVT also causes uterine contractions that accompany birth in viviparous ("live birth") snakes (*Nerodia*).

Oxytocin

Oxytocin in mammals stimulates contraction of the uterine smooth muscle to help expel the fetus during childbirth, and it promotes ejection of the milk from the mammary glands. Appropriately, oxytocin secretion is increased by reflexes that originate within the birth canal of mammals during parturition and by reflexes that are triggered by suckling. Milk production is regulated by prolactin and involves the synthesis of milk constituents and their transport from the alveolar cells into the lumen. Milk removal is regulated by oxytocin and in-

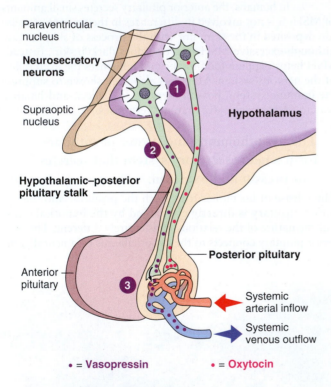

• = **Vasopressin**   • = **Oxytocin**

1 The paraventricular and supraoptic nuclei both contain neurons that produce vasopressin and oxytocin. The hormone, either vasopressin or oxytocin depending on the neuron, is synthesized in the neuronal cell body in the hypothalamus.

2 The hormone travels down the axon to be stored in the neuronal terminals within the posterior pituitary.

3 On excitation of the neuron, the stored hormone is released from these terminals into the systemic blood for distribution throughout the body.

Figure 7–9 ● **Relationship of the hypothalamus and posterior pituitary.**

volves contraction of myoepithelial cells, increased intramammary pressure, and milk ejection (see p. 754).

The nonmammalian homologue, **mesotocin (MT),** does not affect the uterus but, rather, influences the blood flow to some organs (for example, in poultry an increased concentration of MT reduces the temperature of the shank and comb) in addition to reducing the circulating concentrations of aldosterone.

■ The anterior pituitary secretes six established hormones, many of which are tropic.

Unlike the posterior pituitary, which releases hormones that are synthesized by the hypothalamus, the anterior pituitary itself synthesizes the hormones that it releases into the blood. Different cell populations within the anterior pituitary produce and secrete six established peptide hormones. The action of each of these hormones is discussed in detail in subsequent sections. For now, we briefly state their primary effects, to explain their names (● Figure 7–10):

1. **Growth hormone (GH,** also called both **somatotropin** and **somatotrophin**), the primary hormone responsible for regulating overall body growth, is also important in intermediary metabolism.

2. **Thyroid-stimulating hormone (TSH, thyrotropin)** stimulates secretion of thyroid hormone and growth of the thyroid gland.

3. **Adrenocorticotropic hormone (ACTH, adrenocorticotropin)** stimulates cortisol secretion by the adrenal cortex and promotes growth of the adrenal cortex.

4. **Follicle-stimulating hormone (FSH)** has different functions in females and males. In females it stimulates growth and development of ovarian follicles, within which the ova, or eggs, develop. Furthermore, FSH promotes secretion of the hormone estrogen by the ovaries. In males, FSH is required for sperm production.

5. **Luteinizing hormone (LH)** also functions differently in females and males. In females LH governs ovulation, luteinization (that is, formation of a postovulatory hormone-secreting corpus luteum in the ovary), and regulation of ovarian secretion of the female sex hormones, estrogen and progesterone. In males the same hormone stimulates the interstitial cells of Leydig in the testes to secrete the male sex hormone testosterone, giving rise to its alternate name of **interstitial cell–stimulating hormone (ICSH).**

6. **Prolactin (PRL)** is structurally similar to growth hormone and exhibits some overlapping functions. In female mammals prolactin stimulates milk production whereas in males, evidence indicates it may induce production of testicular LH receptors. Furthermore, recent studies suggest that prolactin may enhance the immune system and support the development of new blood vessels at the tissue level in both sexes—both actions totally unrelated to its known roles in reproductive physiology. In other vertebrates, prolactin has remarkably diverse roles, some bearing no relationship to reproduction (see Chapter 16, p. 757). No other polypeptide hormone has such a wide repertoire of biological actions.

TSH, ACTH, FSH, and LH are all tropic hormones, because they each regulate the secretion of another specific endocrine gland. FSH and LH are collectively referred to as **gonadotropins** because they control secretion of the sex hormones by the gonads (ovaries and testes; see Chapter 16). Because growth hormone has been shown to exert its growth-promoting effects indirectly by stimulating release of the liver hormones, the somatomedins, it too is sometimes categorized as a tropic hormone. Among the anterior pituitary hormones, prolactin is the only one that does not stimulate secretion of another hormone. Of the tropic hormones, FSH, LH, and growth hormone differ from TSH and ACTH in that the former exert nontropic functions in addition to stimulating secretion of other hormones.

■ Hypothalamic releasing and inhibiting hormones are delivered to the anterior pituitary by the hypothalamic-hypophyseal portal system to control anterior pituitary hormone secretion.

None of the anterior pituitary hormones are secreted at a constant rate. Even though each of these hormones has a unique control system, there are some common regulatory patterns. The two most important factors that regulate secretion of anterior pituitary hormone are (1) hypothalamic hormones and (2) feedback by target gland hormones.

Because the anterior pituitary secretes hormones that control the secretion of various other hormones, it has long had the undeserved title of "master gland." It is now known that the release of each of the anterior pituitary hormones is largely controlled by still other hormones produced by the hypothalamus, and that secretion of these regulatory neurohormones in turn is controlled by a variety of neural and hormonal inputs to the hypothalamic neurosecretory cells.

Role of Hypothalamic Releasing and Inhibiting Hormones

The secretion of each anterior pituitary hormone is stimulated or inhibited by one or more of the seven generally accepted hypothalamic **hypophysiotropic** (*hypophysis* means "pituitary") **hormones.** These small peptide hormones are listed in ▌ Table 7–2. Depending on their actions, hypophysiotropic hormones are called **releasing hormones** or **inhibiting hormones.** In each case, the primary action of the hormone is apparent from its name. For example, **thyrotropin-releasing hormone (TRH)** stimulates release of TSH (alias thyrotropin) from the anterior pituitary, whereas **prolactin-inhibiting hormone (PIH)** inhibits release of prolactin from the anterior pituitary. Several substances, including serotonin, TRH, vasoactive intestinal peptide (VIP), oxytocin, and angiotensin II, elicit release of prolactin in certain species. In several vertebrates, including quail (*Coturnix japonica*), amphibians, and fishes, there is evidence of a gonadotropin-inhibitory hormone (GnIH). Note that hypophysiotropic hormones in most cases are involved in a three-hormone hierarchical chain of command (● Figure 7–11): The hypothalamic hypophysiotropic hormone (hormone 1) controls the output of an anterior pituitary tropic hormone (hormone 2). This tropic hormone in turn regulates secretion of the target endocrine gland's hormone (hormone 3), which exerts the final physiological effect.

Although scientists originally proposed a neat one-to-one correspondence—one hypophysiotropic hormone for each anterior pituitary hormone—it is now clear that many of the hypothalamic hormones have more than one effect, so their names indicate only the function that was initially attributed to them. Moreover, a single anterior pituitary hormone may be regulated by two or more hypophysiotropic hormones, which may even exert opposing effects. For example, **growth hormone–releasing hormone (GHRH)** stimulates growth hormone secretion, whereas **growth hormone–inhibiting hormone (GHIH)**,

Figure 7–10 ● **Functions of the anterior pituitary hormones.** Five different endocrine cell types produce the six anterior-pituitary hormones—TSH, ACTH, growth hormone, LH and FSH (produced by the same cell type), and prolactin—which exert a wide range of effects throughout the body.

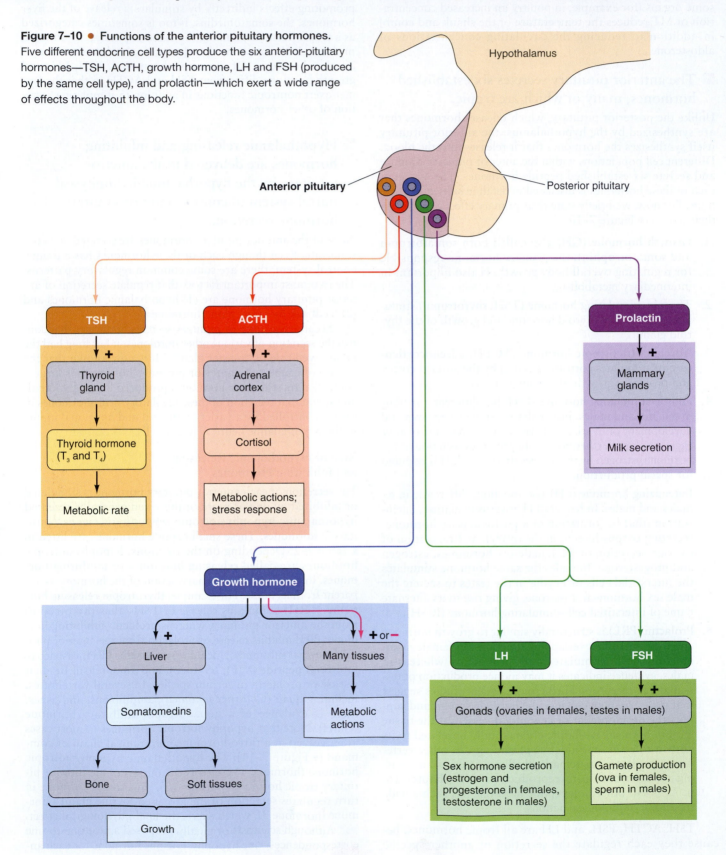

also known as **somatostatin,** inhibits it. The output of the anterior pituitary growth-hormone–secreting cells (that is, the rate of growth hormone secretion) in response to two such opposing inputs depends on the relative concentrations of these hypothalamic hormones as well as on the intensity of other regulatory inputs. This is analogous to the output of a nerve cell (that is, the rate of action potential propagation) depending on the relative magnitude of excitatory and inhibitory synaptic inputs (EPSPs and IPSPs) to it (see p. 129).

Role of the Hypothalamic-Hypophyseal Portal System

The hypothalamic regulatory hormones reach the anterior pituitary by means of a unique vascular link. In contrast to the direct neural connection between the hypothalamus and posterior pituitary, the link between the hypothalamus and anterior pituitary is an unusual capillary-to-capillary connection, the **hypothalamic-hypophyseal portal system.** A portal system is a vascular arrangement in which venous blood flows directly from one capillary bed through a connecting vessel to another capillary bed without passing through the usual route. The largest and best-known portal system is the hepatic portal system (see p. 641). Although much smaller, the hypothalamic-hypophyseal portal system is no less important, because it provides a critical link between the brain and much of the endocrine system. It begins in the base of the hypothalamus with a group of capillaries that recombine into small portal vessels, which pass down through the connecting stalk into the anterior pituitary. Here they branch to form most of the anterior pituitary capillaries, which in turn drain into the systemic venous system (● Figure 7–12).

As a result, almost all the blood supply to the anterior pituitary must first pass through the hypothalamus. Because materials can be exchanged between the blood and surrounding tissue only at the capillary level, the hypothalamic-hypophyseal portal system provides a route where releasing and inhibiting hormones can be picked up at the hypothalamus and delivered immediately and directly to the anterior pituitary at relatively high concentrations, completely bypassing the general circulation.

The axons of the neurosecretory neurons that produce the hypothalamic regulatory hormones terminate on the cap-

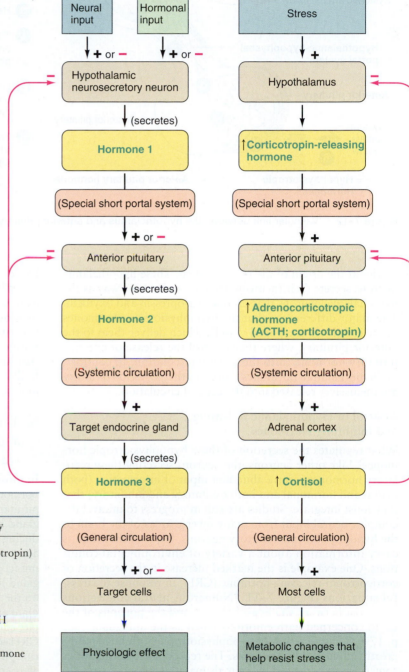

Figure 7–11 ● **Hierarchic chain of command in endocrine control.** The general pathway involved in the hierarchic chain of command among the hypothalamus, anterior pituitary, and peripheral target endocrine gland is depicted on the left. The pathway on the right leading to cortisol secretion provides a specific example of this endocrine chain of command.

Table 7–2 ▮ Major Hypophysiotropic Hormones

Hormone	Effect on the Anterior Pituitary
Thyrotropin-releasing hormone (TRH)	Stimulates release of TSH (thyrotropin) and prolactin
Corticotropin-releasing hormone (CRH)	Stimulates release of ACTH (corticotropin)
Gonadotropin-releasing hormone (GnRH)	Stimulates release of FSH and LH (gonadotropins)
Growth hormone-releasing hormone (GHRH)	Stimulates release of growth hormone
Growth hormone-inhibiting hormone (GHIH)	Inhibits release of growth hormone and TSH
Prolactin-releasing hormone (PRH)	Stimulates release of prolactin
Prolactin-inhibiting hormone (PIH)	Inhibits release of prolactin

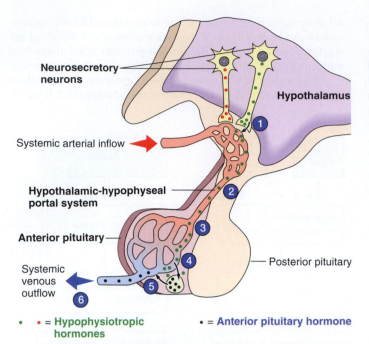

1. Hypophysiotropic hormones (releasing hormones and inhibiting hormones) produced by neurons in the hypothalamus enter the hypothalamic capillaries.

2. These hypothalamic capillaries rejoin to form the hypothalamic-hypophyseal portal system. This vascular link passes to the anterior pituitary.

3. Here it branches into the anterior pituitary capillaries.

4. The hypophysiotropic hormones leave the blood across the anterior pituitary capillaries and control the release of anterior pituitary hormones.

5. On stimulation by the appropriate hypothalamic releasing hormone, a given anterior pituitary hormone is secreted into these capillaries.

6. The anterior pituitary capillaries rejoin to form a vein, through which the anterior pituitary hormones leave for ultimate distribution throughout the body by the systemic circulation.

Figure 7–12 ● Vascular link between the hypothalamus and anterior pituitary.

illaries at the origin of the portal system. These hypothalamic neurons secrete their (neuro)hormones in the same way as the hypothalamic neurons that produce vasopressin and oxytocin. The major difference is that the hypophysiotropic hormones are released into the portal vessels, which deliver them to the anterior pituitary, where they control the release of anterior pituitary hormones into the general circulation. In contrast, the hypothalamic hormones stored in the posterior pituitary are themselves released into the general circulation.

Control of Hypothalamic Releasing and Inhibiting Hormones

What regulates the secretion of these hypophysiotropic hormones? Like other neurons, the neurons secreting these regulatory hormones receive abundant input of information (both neural and hormonal and both excitatory and inhibitory) that they must integrate. Studies are still in progress to unravel the complex neural input from many diverse areas of the brain to the hypophysiotropic secretory neurons. Some of these inputs carry information about a variety of environmental conditions. One example is the marked increase in the secretion of corticotropin-releasing hormone (CRH) in response to stressful situations (Figure 7–11). Numerous neural connections also exist between the hypothalamus and the portions of the brain concerned with emotions (such as the amygdala—see p. 174). Thus secretion of hypophysiotropic hormones can be greatly influenced by emotions. The reproductive irregularities sometimes experienced by animals maintained under stressful conditions are a common manifestation of this relationship.

In addition to being regulated by different regions of the brain, the hypophysiotropic neurons are also controlled by various chemical inputs that reach the hypothalamus through the blood. Unlike other regions of the brain, portions of the hypothalamus are not guarded by the blood–brain barrier, so the hypothalamus can easily monitor chemical changes in the

blood. In some instances, hypophysiotropic secretion is influenced by the immediate metabolic state of the individual. For example, one or more of the metabolic responses produced by growth hormone (such as an elevated blood-glucose level) negatively feed back to the hypothalamus to control growth hormone secretion. The more common blood-borne factors that influence hypothalamic neurosecretion are the negative-feedback effects of either anterior pituitary or target gland hormones, to which we now turn our attention.

■ Target gland hormones inhibit hypothalamic and anterior pituitary hormone secretion via negative feedback.

In most cases, hypophysiotropic hormones initiate a three-hormone sequence: (1) hypophysiotropic hormone, (2) anterior pituitary tropic hormone, and (3) peripheral target–endocrine gland hormone (Figure 7–11). Typically, in addition to producing its physiologic effects, the target gland hormone also acts to suppress secretion of the tropic hormone that is driving it. This negative feedback is accomplished by the target gland hormone acting either directly on the pituitary itself or on the release of hypothalamic hormones, which in turn regulate anterior pituitary function (Figure 7–11). As an example, consider the **CRH-ACTH-cortisol system.** Hypothalamic CRH (corticotropin-releasing hormone) stimulates the anterior pituitary to secrete ACTH (adrenocorticotropic hormone, alias corticotropin), which in turn stimulates the adrenal cortex to secrete cortisol. The final hormone in the system, cortisol, inhibits the hypothalamus to reduce CRH secretion and also reduces the sensitivity of the ACTH-secreting cells to CRH by acting directly on the anterior pituitary. Through this double-barreled approach, cortisol exerts negative-feedback control to stabilize its own plasma concentration. If plasma cortisol levels start to rise above a prescribed set level, cortisol suppresses its own further secretion by its inhibitory actions at the hypothalamus and anterior pituitary. If plasma cortisol levels fall below the desired set point, cortisol's inhibitory actions at the hypothalamus and anterior pituitary are reduced, so the driving forces for cortisol secretion (CRH-ACTH) increase accordingly. The other target-gland hormones act by similar negative-feedback loops to maintain their plasma levels relatively constant at a set point.

Diurnal rhythms are superimposed on this type of stabilizing negative-feedback regulation; that is the set point changes as a function of the time of day. Furthermore, other controlling inputs may break through the negative-feedback control to alter hormone secretion (that is, change the set point level) at times of special need. For example, stress raises the set point for cortisol secretion.

The one exception to the negative-feedback relationship just described is the preovulatory long-loop positive-feedback effect of estrogen on LH secretion in mammals. This causes an extremely dramatic rise in LH secretion that triggers ovulation. This relationship is discussed further in Chapter 16.

In addition, other hormones outside a particular sequence may also exert important influences, either stimulatory or inhibitory, on the secretion of hypothalamic or anterior pituitary hormones within a given sequence. For example, even though estrogen is not in the direct chain of command for prolactin secretion, this sex steroid notably enhances prolactin secretion by the anterior pituitary. This is but one example of the common phenomenon in the endocrine system that one seemingly unrelated hormone can have pronounced effects on the secretion or actions of another hormone.

■ Hypothalamic hormones are produced in other body regions, where they have unrelated functions including learning and pair-bonding.

In accordance with the "same key, different locks" principle (p. 88), chemical messengers that are identical in structure to the hypothalamic releasing and inhibiting hormones, and to oxytocin and vasopressin as well, are produced in many areas of the brain outside the hypothalamus. Instead of being released into the blood, these messengers act locally as neurotransmitters and neuromodulators in these extrahypothalamic sites. For example, PIH is now known to be identical to dopamine, a major neurotransmitter in the basal nuclei and elsewhere (see p. 171), whereas the gut peptide vasoactive intestinal peptide (VIP) is the major hypothalamo-hypophysiotropic factor stimulating prolactin release in birds. The prolactin response to VIP is enhanced during incubation and in the presence of estrogen. Other chemical messengers are thought to modulate a variety of functions that range from motor activity (such as TRH) to libido (such as GnRH) to learning (vasopressin) to various social behaviors (oxytocin and vasopressin).

Perhaps the most dramatic discovery in this regard was the discovery of the roles of oxytocin and vasopressin in pair-bonding ("falling in love," in human terms). In one study, researchers compared prairie and montane voles, which are 99% genetically identical but which have very different social behaviors: The montane vole nests alone and mates promiscuously, whereas the prairie vole forms monogamous male–female couples (pair bonds) for mating and infant care. Oxytocin and vasopressin appear to be important signals in this process: When oxytocin is injected into a female prairie vole's brain, she will immediately pair-bond with a nearby male, even if she is not ready to mate. In contrast, oxytocin antagonists prevent her from pair bonding. Interestingly, vasopressin, not oxytocin, has the corresponding effects in male prairie voles. Montane voles are not affected in these ways by the peptides. Prairie voles but not montane voles have receptors for the peptides in their brain's reward centers, possibly explaining why the peptides have such different effects in the two closely related species.

Endocrine Control of Growth in Vertebrates

Growth requires net synthesis of proteins and includes lengthening of the long bones (the bones of the extremities) as well as increases in the size and number of cells in the soft tissues throughout the body. Growth is an example of a regulated change (p. 16) as opposed to a homeostatic process. Weight gain alone is not synonymous with growth, because weight gain may occur as a result of retention of excess H_2O or fat without true structural growth of tissues. If the pituitary of a young vertebrate animal is surgically removed, the animal ceases growth; only by the administration of extracts of the pituitary gland can this defect can be partially remedied.

■ Growth depends on growth hormone but is influenced by other factors as well.

Although, as the name implies, **growth hormone (GH)** is absolutely essential for growth, it alone is not wholly responsible for determining the rate and final magnitude of growth in a given animal. The following factors also affect growth:

- *Genetic determination* of an animal's maximum growth capacity. Attainment of this full growth potential further depends on the other factors listed here.

- *An adequate diet,* including sufficient total protein and ample essential amino acids to accomplish the protein synthesis necessary for growth. Malnourished animals never achieve their full growth potential. The growth-stunting effects of inadequate nutrition are most profound when they occur in infancy. In severe cases, the animal may be locked into irreversible stunting of body growth and neural development. For example, about 70% of the total growth of the human brain occurs in the first two years of life. In contrast, mammals cannot exceed the genetically determined maximum by eating more than an adequate diet. The excess food intake produces obesity instead of growth in stature (this is not true of fish, as you will see).

- *Freedom from chronic disease and stressful environmental conditions.* Stunting of growth under adverse circumstances is due in large part to the prolonged stress-induced secretion of glucocorticoid (e.g., cortisol) secretion from the adrenal cortex. Glucocorticoids exert several potent antigrowth effects, such as promoting protein breakdown, inhibiting growth in the long bones in land vertebrates, and blocking the secretion of GH. This applies to aquatic vertebrates also; fish grown in modern aquaculture facilities do not maximize their growth potential, because of the metabolic costs associated with crowding. Even though sickly or stressed animals do not grow well, if the condition is rectified before adult size is achieved they can rapidly catch up to their normal growth curve through a remarkable spurt in growth.

- *Normal concentrations of growth-influencing hormones.* In growing animals, numerous genes code for particular proteins and protein regulators of growth. In addition to the absolutely essential GH, other hormones, including thyroid hormone, insulin, and the sex hormones, play secondary roles in promoting growth.

Growth hormone is essential for growth, but it also exerts metabolic effects not related to growth.

In addition to promoting growth, GH has important metabolic effects and enhances the immune system. In fishes GH has partially retained ancestral osmoregulatory features independent of its growth-promoting functions. We briefly describe GH's metabolic actions before turning our attention to its growth-promoting actions.

Metabolic Actions Unrelated to Growth

GH increases fatty acid levels in the blood by enhancing the breakdown of triglyceride fat stored in adipose tissue, and it increases blood glucose levels by decreasing glucose uptake by muscles. Muscles use the mobilized fatty acids instead of glucose as a metabolic fuel. Thus the overall metabolic effect of GH is to mobilize fat stores as a major energy source while conserving glucose for glucose-dependent tissues such as the brain. The vertebrate brain can use only glucose as its metabolic fuel, yet mammalian nervous tissue cannot store glycogen (stored glucose) to any extent. This metabolic pattern is suitable for maintaining the body during prolonged fasting or other situations when the body's energy needs exceed available glucose stores.

Growth-Promoting Actions on Soft Tissues

When tissues are responsive to its growth-promoting effects, GH stimulates growth of both soft tissues and the skeleton. GH promotes growth of soft tissues by (1) increasing the number of cells (**hyperplasia**) and (2) increasing the size of cells (**hypertrophy**). GH increases the number of cells by stimulating cell division and by preventing apoptosis (programmed cell death). GH increases the size of cells by favoring synthesis of proteins, the main structural component of cells. GH stimulates almost all aspects of protein synthesis while it simultaneously inhibits protein degradation. It promotes the uptake of amino acids (the raw materials for protein synthesis) by cells, decreasing blood amino-acid levels in the process. Furthermore, it stimulates the cell machinery responsible for accomplishing protein synthesis according to the cell's genetic code.

Growth-Promoting Actions on Bones

Growth of the long bones, resulting in increased height, is the most dramatic effect of GH. **Bone** is a living tissue. Being a form of connective tissue, it consists of cells and an extracellular organic matrix that is produced by the cells. The bone cells that produce the organic matrix are known as **osteoblasts** ("bone-formers"). The organic matrix is composed of colla-

gen fibers (see p. 63) in a mucopolysaccharide-rich semisolid gel called *ground substance*. This matrix has a rubbery consistency and is responsible for the tensile strength of bone (the resilience of bone to breakage when tension is applied). Bone is hardened by impregnation with **hydroxyapatite crystals**, which consist primarily of precipitated $Ca_3(PO_4)_2$ (calcium phosphate) salts. These inorganic crystals provide the bone with compressional strength (the ability of bone to hold its shape when squeezed or compressed). Bones have structural strength approaching that of reinforced concrete, yet they are not brittle and are much lighter in weight, as a result of the structural blending of an organic scaffolding hardened by inorganic crystals.

A long bone basically consists of a fairly uniform cylindrical shaft, the **diaphysis,** with a flared articulating knob at either end, an **epiphysis.** In a growing bone, the diaphysis is separated at each end from the epiphysis by a layer of cartilage known as the **epiphyseal plate** (● Figure 7–13a). The central cavity of the bone is filled with bone marrow, which is the site of blood cell production (see p. 365).

Bone Growth

Growth in *thickness* of bone is achieved by the addition of new bone on top of the already existing bone on the outer surface. This growth occurs through activity of osteoblasts within the **periosteum,** a connective tissue sheath that covers the outer bone surface. As new bone is being deposited by osteoblast activity on the external surface, other cells within the bone, the osteoclasts ("bone breakers"), dissolve the bony tissue on the inner surface adjacent to the marrow cavity. In this way, the marrow cavity is enlarged to keep pace with the increase in the circumference of the bone shaft.

Growth in *length* of long bones is accomplished by a different mechanism from growth in thickness. Bones grow in length as a result of proliferation of the cartilage cells in the epiphyseal plates (Figure 7–13b). During growth, new cartilage cells (**chondrocytes**) are produced through cell division on the outer edge of the plate adjacent to the epiphysis. As new chondrocytes are formed on the epiphyseal border, the older cartilage cells toward the diaphyseal border are enlarging. This combination of proliferation of new cartilage cells and hypertrophy of maturing chondrocytes causes the epiphyseal plate to temporarily widen. This thickening of the intervening cartilaginous plate pushes the bony epiphysis farther away from the diaphysis. Soon the matrix surrounding the oldest hypertrophied cartilage calcifies. Because cartilage lacks its own capillary network, the survival of cartilage cells depends on diffusion of nutrients and O_2 through the ground substance, a process prevented by the deposition of calcium salts. As a result, the old nutrient-deprived cartilage cells on the diaphyseal border die. As osteoclasts clear away the dead chondrocytes and the calcified matrix that imprisoned them, the area is invaded by osteoblasts, which swarm upward from the diaphysis, trailing their capillary supply with them. These new tenants lay down bone around the persisting remnants of disintegrating cartilage until the inner region of cartilage on the diaphyseal side of the plate is entirely replaced by bone. When this **ossification** ("bone-forming") process is completed, the bone on the diaphyseal side has lengthened, and the epiphyseal plate has returned to its original thickness. The cartilage that has been replaced by bone on the diaphyseal end of the plate is

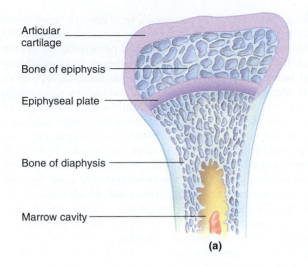

Articular cartilage

Bone of epiphysis

Epiphyseal plate

Bone of diaphysis

Marrow cavity

(a)

Figure 7–13 ● **Anatomy and growth of long bones.** (a) Anatomy of long bones. (b) Two sections of the same epiphyseal plate at different times, depicting the lengthening of long bones.

Cartilage

Calcified cartilage

Bone

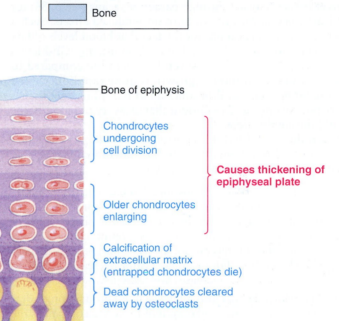

Bone of epiphysis

Bone of epiphysis

Resting chondrocytes

Chondrocytes undergoing cell division

Causes thickening of epiphyseal plate

Older chondrocytes enlarging

Epiphyseal plate

Calcification of extracellular matrix (entrapped chondrocytes die)

Dead chondrocytes cleared away by osteoclasts

Diaphysis

Osteoblasts swarming up from diaphysis and depositing bone over persisting remnants of disintegrating cartilage

(b)

equivalent in thickness to the new cartilaginous growth on the epiphyseal end of the plate. Thus bone growth is made possible by the growth and death of cartilage, which acts like a "spacer" to push the epiphysis farther out while it provides a framework for future formation of bone on the end of the diaphysis.

GH promotes growth of bone in both thickness and length. It stimulates the proliferation of epiphyseal cartilage, thereby making space for more bone formation, and also stimulates osteoblast activity. GH can promote lengthening of long bones as long as the epiphyseal plate remains cartilaginous, or is "open." At the end of adolescence, under the influence of the sex hormones, these plates completely ossify, or "close," so that the bones can grow no further in length despite the presence of

GH. Thus, after the plates are closed, the individual animal does not grow any taller.

■ Growth hormone exerts its growth-promoting effects indirectly by stimulating somatomedins.

GH does not act directly on its target cells to bring about its growth-producing actions (increased cell division, enhanced protein synthesis, and bone growth). These effects are directly brought about by peptide mediators known as **somatomedins.** These peptides are also referred to as **insulin-like growth factors (IGF)** because they are structurally and functionally similar to insulin. Two somatomedins—IGF-I and IGF-II—have been identified, each having evolved from a common ancestral gene through the process of gene duplication, point mutations and divergence. Fish are the first vertebrate group that has a complete system of hormones and receptors for the insulin, IGF-I and IGF-II molecules.

IGF-I

IGF-I synthesis in mammals is stimulated by GH and mediates most of this hormone's growth-promoting actions. The major source of circulating IGF-I is the liver, which releases this peptide product into the blood in response to GH stimulation. However, IGF-I is produced in many if not most tissues, although it is not released into the blood from these other sites.

Scientists have proposed that IGF-I produced locally in target tissues may act through paracrine means (see p. 86) for at least some of the growth hormone–induced effects. Such a mechanism could account for the fact that blood levels of GH in humans are no higher, and indeed circulating IGF-I levels are lower, during the first several years of life compared to adult values, even though growth is quite rapid during the postnatal period. Local production of IGF-I in target tissues may possibly be more important than delivery of blood-borne IGF-I during this time.

Production of IGF-I is controlled by a number of factors other than GH in humans, including nutritional status, age, and tissue-specific factors as follows:

- IGF-I production depends on adequate nutrition. Inadequate food intake results in reduced IGF-I production, apparently brought about by decreased sensitivity to GH of the tissues that produce IGF-I. As a result, changes in circulating IGF-I levels do not always coincide with changes in GH secretion. For example, fasting decreases IGF-I levels even though it increases GH secretion.

- Age-related factors also influence IGF-I production. A dramatic increase in circulating IGF-I levels accompanies the moderate increase in GH at puberty, which may, of course, be an important factor in the pubertal growth spurt.

- Finally, various tissue-specific stimulatory factors can increase IGF-I production in particular tissues. To illustrate, the gonadotropins and sex hormones stimulate IGF-I production within reproductive organs such as the testes in males and the ovaries and uterus in females.

Thus, control of IGF-I production is complex and subject to a variety of systemic and local factors.

IGF-II

In contrast to IGF-I, **IGF-II** production does not depend on GH. IGF-II is known to be important during fetal development. Unlike IGF-I, IGF-II does not increase during the pubertal growth spurt. IGF-II continues to be produced during adulthood and may be involved in muscle growth. Recently, a study found that a simple mutation that increased IGF-II gene activity in pig muscle resulted in a 3 to 4% growth in meat (muscle) mass.

From the preceding discussion of the factors involved in controlling growth, you can see that many gaps still exist in the state of knowledge in this important area.

Rapid-Growth Periods

The rate of growth is not continuous in animals, nor are the factors responsible for promoting growth the same through-

out the growth period. In mammals, fetal growth is promoted largely by certain hormones from the placenta (see p. 748), with the size at birth being determined principally by genetic and environmental factors. GH and other nonplacental hormonal factors begin to play an important role in regulating growth after birth. Genetic and nutritional factors also strongly affect growth during this period. In addition, day length affects growth in juvenile animals because day length influences energy intake and energy expenditure. For example, exposure of an animal to short day lengths, as occurs seasonally in northern and southern latitudes, results in long nocturnal periods of food deprivation. During the comparatively short light period, large amounts of food are consumed, which makes available to the tissues high concentrations of amino acids. In a growing animal, these dietary amino acids are particularly important for protein synthesis (repair, growth, and maintenance) as well as storage for use in subsequent periods of food deprivation. Normally, when the supply of amino acids exceeds the protein synthesis capacity of the animal, the excess amino acids are oxidized and thus lost to the animal. However, during periods of food unavailability such as overnight fasting, protein synthesis rates drop significantly, which may severely limit growth rates.

Unlike birds and mammals, which reach a maximum body size and then cease growing, most cultured fish species continue to grow if food supplies remain available. However, species such as trout show periodic growth spurts, which depend on season and stage of development. For example, growth is reduced during the winter months as well as during the reproductive period. This is due in part to the lowered energy requirements of fishes because they are poikilotherms (Chapter 15). Some fishes also need a higher dietary protein intake than do other vertebrates. For example, carnivorous species such as bass and salmon require diets containing 40 to 55% dietary protein compared to the diets of birds and mammals, which typically obtain maximal growth rates when fed diets of only 12 to 25% protein. Superimposed on the seasonal variations in growth rate in many fishes is a 14-day semilunar cycle of growth rate, which correlates with a rhythm in food consumption. Slow growth rates are associated with either the full or new moon, which are associated with reduced food requirements.

In contrast, most mammals display two periods of rapid growth—a postnatal growth spurt during their first years of life and a pubertal growth spurt during adolescence. Much of the growth observed in young animals is due to the increase in body protein, with skeletal muscle showing the greatest percentage change in body composition. Growth is achieved when protein synthesis rates are higher than protein breakdown rates. The productive efficiency of growing farm animals is determined by the proportion of nutrients partitioned to fat relative to muscle, and by the rate at which growth occurs.

Before puberty there is little sexual difference in height or weight. During puberty, a marked acceleration in linear growth takes place because of the lengthening of the long bones. The mechanisms responsible for the pubertal growth spurt are not clearly understood. Apparently both genetic and hormonal factors are involved. During rapid mammalian growth in puberty, there is a marked increase in the magnitude and frequency of GH release, which induces IGF-I expression in the liver and other tissues, including the skeleton. In turn, systemic and local IGF-I contributes to the acceleration of growth during this time. Furthermore, **androgens** ("male" sex hormones), whose secretion increases dramatically at puberty, also contribute to the pubertal growth spurt by promoting protein synthesis and bone growth. The potent androgen from the male testes, testosterone, is of greatest importance in promoting a sharp increase in height in adolescent males, whereas the less potent adrenal androgens from the adrenal gland, which also show a sizable increase in secretion during adolescence, are most likely important in the female pubertal growth spurt. Although estrogen secretion by the ovaries also begins during puberty, it is unclear what role this "female" sex hormone may play in the pubertal growth spurt in females. There is no doubt that testosterone and estrogen both ultimately act on bone to halt its further growth so that full adult height is attained by the end of adolescence.

■ **Growth hormone secretion is regulated by two hypophysiotropic hormones and influenced by a variety of other factors.**

The control of GH secretion is complex, with two hypothalamic hypophysiotropic hormones playing a key role.

Growth Hormone–Releasing Hormone and Growth Hormone–Inhibiting Hormone

Two antagonistic regulatory hormones from the hypothalamus are involved in controlling GH secretion: growth hormone–releasing hormone (GHRH), which is stimulatory, and growth hormone–inhibiting hormone (GHIH or somatostatin), which is repressive (● Figure 7–14). (Note the distinctions among somatotropin/trophin, alias GH; somatomedin, the liver hormone that directly mediates the effects of GH; and somatostatin, which inhibits GH secretion.) Any factor that increases GH secretion could theoretically do so either by stimulating GHRH release or by inhibiting GHIH release. Researchers do not know which of these pathways is used in each specific case.

As with the other hypothalamus–anterior pituitary axes, negative-feedback loops participate in regulating GH secretion. Both GH and the somatomedins inhibit pituitary secretion of GH, presumably by stimulating GHIH release from the hypothalamus. The somatomedins may also exert direct effects on the anterior pituitary to inhibit the effects of GHRH on GH release.

Other Factors That Influence GH Secretion

A number of factors influence GH secretion by acting on the hypothalamus. GH secretion displays a well-characterized diurnal rhythm. In a diurnal mammal, through most of the day GH levels tend to be low and fairly constant. About one hour

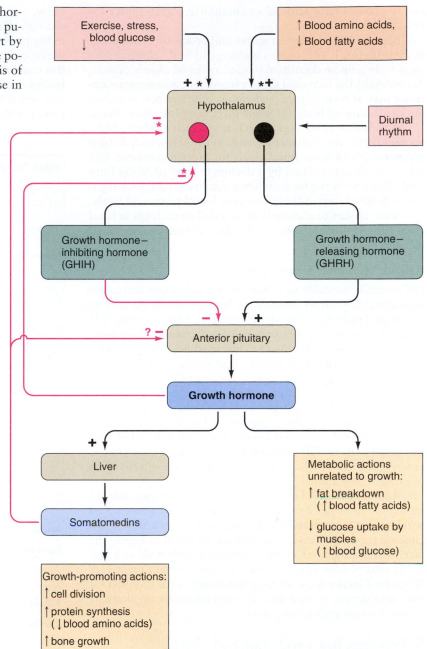

*These factors all increase growth hormone secretion, but it is unclear whether they do so by stimulating GHRH or inhibiting GHIH, or both.

*These factors inhibit growth hormone secretion in negative-feedback fashion, but it is unclear whether they do so by stimulating GHIH or inhibiting GHRH or inhibiting the anterior pituitary itself.

Figure 7–14 ● Control of growth hormone secretion.

after the onset of deep sleep, however, GH secretion markedly increases, then rapidly drops over the next several hours.

Superimposed on this diurnal fluctuation in GH secretion are further bursts in secretion, in response to exercise, stress, and hypoglycemia (low blood glucose), the major stimuli for increased secretion. The benefit of increased GH secretion during these situations when energy demands outstrip the body's glucose reserves is presumably to conserve glucose for the brain and to provide fatty acids as an alternative energy source for muscle.

Because GH uses up fat stores and promotes synthesis of body proteins, it encourages a change in body composition away from adipose deposition toward increased muscle protein. Accordingly, the increased GH secretion that accompanies exercise may at least in part mediate exercise's effects in reducing percentage of body fat while increasing lean body mass.

An abundance of blood amino acids, such as after a high-protein meal, also enhances secretion of GH, which in turn promotes use of these amino acids for protein synthesis. GH release is also stimulated by a decline in level of blood fatty acids. Because of the fat-mobilizing actions of GH, such regulation helps maintain fairly constant blood fatty-acid levels.

Some species of mammals show a decline with age in basal levels for GH as well as the amplitude of the peak in nocturnal increase of GH.

Growth Hormone Administration

GH is readily available for administering to humans and to domestic and laboratory animals. In humans, it is used to treat dwarfism (and abused by some athletes). Administering GH to swine dramatically increases protein **accretion** (deposition) and muscle mass in addition to reducing adipose tissue growth. However, deposition of greater lean tissue requires an increased amount of amino acids in the diet of treated animals. Uptake of glucose into target tissues is also reduced, which results in increased concentrations of both plasma glucose and insulin. In swine, concentrations of GH peak between 4 and 7 hours after subcutaneous injection and are associated with increased plasma IGF-I concentrations as well. However, in poultry, unless GH is administered in pulses, there is no effect of exogenously administered GH on growth.

Administering GH to dairy cows results in an exquisite coordination of metabolism to meet the nutrient needs associated with increased synthesis of milk components (see ▪ Table 7–3). Glucose production and oxidation are modified to supply the increased glucose needs of the mammary glands with such precision that plasma glucose concentrations remain unaltered. When feed intake does not meet the energy needs required for milk production, treated animals must capitalize on alternate energy sources such as fatty acids.

▪ Prolactin has a wide range of effects, including lactogenesis, reproductive behaviors, and water regulation.

The first suggestion that a hormone of the anterior pituitary might affect lactation (milk production) in mammals arose from the experiments of Stricker and Grueter in 1928, who found that lactation could be elicited in the pseudopregnant rabbit, before and after spaying, by injections of an anterior pituitary extract. Prolactin was eventually isolated and found to be essential for initiating and maintaining lactation (only in mammals, whose very name is based on this). Lactogenesis, the initiation of lactation, involves mammary cell differentiation and increased activity of the enzymes responsible for the production of milk components. We explore this process in Chapter 16.

Lactogenic ("milk generating") effects have been found (by administering prolactin to laboratory mammals) in the pituitary extracts of mammals, fish, amphibians, reptiles, and birds, showing that all vertebrates have this hormone. Obviously, it is not involved in lactogenesis in nonmammals! For example, Riddle and coworkers in 1931 found that administering pituitary extract to pigeons resulted in development of the crop sac (see p. 626). Subsequent work demonstrated that this response in birds is due to the same hormone that induces lactation in mammals. In fact, prolactin is involved in a remarkable spectrum of functions, including nurturing of young, reproduction, osmoregulation, promotion of growth, support

Table 7–3 ▪ Effect of Bovine Somatotropin Supplement of Lactating Cows on Specific Tissues and Physiological Processes*

Tissue		Process Affected during First Few Days and Weeks of Supplement
Mammary	↑	Synthesis of milk with normal composition
	↑	Uptake of all nutrients used for milk synthesis
	↑	Activity per secretory cell
	↓	Loss of secretory cells (i.e., enhanced persistence)
	↑	Blood flow consistent with increase in milk yield
Liver	↑	Basal rates of gluconeogenesis
	↓	Ability of insulin to inhibit gluconeogenesis
	φ	Glucagon effects on gluconeogenesis and/or glycogenolysis
Adipose	↓	Basal lipogenesis if in positive energy balance
	↑	Basal lipolysis if in negative energy balance
	↓	Ability of insulin to stimulate lipogenesis
	↑	Ability of insulin to inhibit lipolysis
	↑	Ability of catecholamines to stimulate lipolysis
Muscle	↓	Uptake of glucose
Pancreas	φ	Basal or glucose-stimulated secretion of insulin
	φ	Basal or insulin/glucose-stimulated secretion of glucagon
Kidney	↑	Production of 1,25 vitamin D_3
Intestine	↑	Absorption of Ca, P, and other minerals required for milk
	↑	Ability of 1,25 vitamin D_3 to stimulate calcium-binding protein
	↑	Calcium-binding protein
Whole body	↓	Oxidation of glucose
	↑	Fatty-acid oxidation if in negative energy balance
	φ	Insulin and glucagon clearance rates
	φ	Energy expenditure for maintenance
	↑	Energy expenditure consistent with increase in milk yield (i.e., heat per unit of milk not changed)
	↑	Cardiac output consistent with increases in milk yield
	↑	Productive efficiency (milk per unit of energy intake)

Note: Changes (↑ = increased, ↓ = decreased, φ = no change).

of metabolism, water drive, metamorphosis, brood patch formation (see the accompanying box, "A Closer Look at Adaptation: Brood Patch Development: Some Have It, Others Don't!"), and migratory and parental behavior in birds and mammals. Prolactin's effects on behavior are explored in more detail in Chapter 16. Prolactin's osmoregulatory roles have been primarily demonstrated in freshwater fishes, a topic we explore in Chapter 13.

Further evidence of the complex role of prolactin is suggested by the finding that the gene encoding prolactin receptor is widely expressed in avian peripheral tissues, including kidneys, skin, brood patch, gut, gonads, adrenal glands, liver, adipose tissue and spleen, whereas in the CNS the gene is highly expressed in the anterior pituitary gland and in the hypothalamus. Thus, unlike other pituitary hormones, prolactin was not committed early in evolution to the control of one or few re-

lated processes, but remained diversified and adaptive in nature. This multiplicity of functions is likely why lactotrophs constitute the largest category of the anterior pituitary cell types in addition to being regulated by multiple factors (see also p. 267).

In the previous sections, we discussed the major central endocrine glands of vertebrates, those associated directly with the central nervous system. We now turn our attention to the so-called peripheral glands.

Vertebrate Thyroid Gland

The **thyroid gland** is a major regulator of metabolism and other metabolic and developmental processes. In mammals the thyroid gland consists of two lobes of endocrine tissue joined in

---

A CLOSER LOOK AT ADAPTATION

Brood Patch Development: Some Have It, Others Don't!

The vast majority of birds incubate their eggs by transfer of heat between parts of their body, usually but not always their ventral surface, and the clutch of eggs. That region of the incubating bird in contact with the egg is commonly termed the incubation or **brood patch,** and analysis of the hormonal regulation of this patch of skin in different sexes and species of birds is instructive.

Although the evolution of this structure is open to conjecture, one possibility is that the appearance of a brood patch was a definitive step in the evolution of birds from reptiles, and it has been thus suggested to share a common ancestry with the mammalian mammary gland. The nature, formation and structure of this patch varies enormously between species, reflecting the diverse mechanisms and methods adopted by different species in the incubation of eggs. In some birds such as the domestic pigeon and ring dove, there is a form of pocket or apterium on the ventral surface of both sexes, which is devoid of feathers throughout adult life. In other species, patch formation occurs periodically and closely related to the period of egg laying and incubation. Furthermore, although both sexes may have the potential to develop a brood patch, it is a general observation that the patch only develops in the sex actually involved in incubation. There is also a relationship between the gender that incubates, and the form of steroid that is most effective, in combination with prolactin, to produce a full brood patch. In those species in which the female alone incubates, it is estrogen, whereas in those species in which the male incubates it is androgen. Consistent with this, species

that might be considered as intermediate, such as the California quail and laughing gull, are sensitive, in terms of brood patch development, to both estrogen plus prolactin and androgen plus prolactin.

In most species, brood patch development involves a number of morphological changes to the ventral body area: defeathering (mainly down feathers); a significant increase in the folding of the skin, together with an infiltration of leukocytes (edema formation); a thickening of the cornified layer of the ventral skin surface (epidermal hyperplasia); and an increase in both size and number of local blood vessels (vascularization). The structure is surprisingly sophisticated in that the musculature of arterioles supplying blood to the patch also increases and thus can shut down blood flow to this region when the parent is off the nest. In some species, such as the house sparrow (*Passer domesticus*), there is also an increase in the amount of underlying defatted tissue.

Taken together, these dramatic changes are superbly designed to facilitate a closer contact and more effective heat transfer between the parenting bird and the surface of the eggs, while minimizing any possible damage caused to the skin by the sustained period of contact time. The function of the brood patch during the incubatory period of the breeding cycle is complex and multifactorial. Much still remains to be understood, and one early suggestion for a possible purpose of the patch, that the bird finds relief from the peripheral irritation of the developing brood patch by sitting on eggs, remains plausible.

Photo: Hillar Kandorf

The Northern Gannett (Sula Bassanus) does not develop a brood patch and instead incubates its egg by increasing blood flow (and heat exchange) through the webbing in its feet. Because these birds dive into the cold waters of the Atlantic Ocean, this adaptation reduces potential heat loss to the environment.

the middle by a narrow portion of the gland, giving it a bow-tie shape (● Figure 7–15a). The gland is even located in the appropriate place for a bow tie in a human, lying over the trachea just below the larynx. In contrast, the thyroid gland of most nonmammalian vertebrates consists of discrete structures lying at varying distances lateral to the esophagus; in agnathans and most teleosts the thyroid consists of patchy clusters of cells.

■ The major thyroid hormone secretory cells are organized into colloid-filled spheres.

The major thyroid secretory cells are arranged into hollow spheres, each of which forms a functional unit called a **follicle.** Consequently, these secretory cells are often referred to as **fol-**

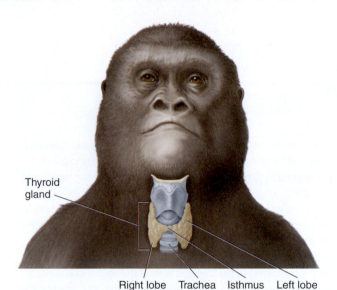

Thyroid gland

Right lobe   Trachea   Isthmus   Left lobe

(a)

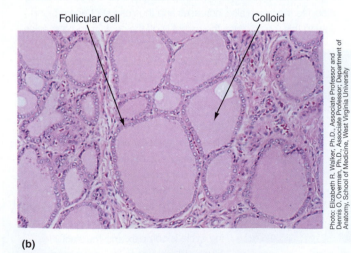

Follicular cell   Colloid

Photo: Elizabeth R. Walker, Ph.D., Associate Professor and Dennis O. Overman, Ph.D., Associate Professor; Department of Anatomy, School of Medicine, West Virginia University

(b)

Figure 7–15 ● **Anatomy of the thyroid gland.** (a) Gross anatomy of the thyroid gland, anterior view. The thyroid gland lies over the trachea just below the larynx and consists of two lobes connected by a thin strip called the *isthmus.* (b) Light-microscope appearance of the thyroid gland. The thyroid gland consists primarily of colloid-filled spheres enclosed by a single layer of follicular cells.

licular cells. On a microscopic section (Figure 7–15b), the follicles appear as rings of follicular cells enclosing an inner lumen filled with colloid, a substance that serves as an "inland" extracellular storage site for thyroid hormones. Extracellular hormone storage is unique to the thyroid gland and is considered both an adaptation related to the scarcity of the trace element iodine, which is an essential component of the thyroid gland, as well as a means of preventing diffusion of the steroidlike thyroid hormone out of the colloid.

The chief constituent of the colloid is a large, species-specific complex molecule known as **thyroglobulin (Tg),** within which are incorporated the thyroid hormones in their various stages of synthesis. The follicular cells of vertebrates produce two iodine-containing hormones derived from the amino acid tyrosine: **tetraiodothyronine (T_4 or thyroxine)** and **tri-iodothyronine (T_3).** The prefixes *tetra* and *tri* and the subscripts 4 and 3 denote the number of iodine atoms incorporated into each of these hormones. These two hormones, together referred to as **thyroid hormone,** are important regulators of development and overall basal metabolic rate.

In mammals, interspersed in the interstitial spaces between the follicles is another secretory cell type, the **C cells,** so called because they secrete the peptide hormone **calcitonin,** which plays a role in calcium metabolism. Calcitonin is not related in any way to the two other major thyroid hormones. We restrict our discussion of thyroid hormones to the secretions of the follicular cells, deferring coverage of calcitonin until a later section dealing with endocrine control of calcium balance.

■ Thyroid hormone synthesis and storage occur on the thyroglobulin molecule.

The basic ingredients for thyroid hormone synthesis are tyrosine and iodine, both of which must be taken up from the blood by the follicular cells. Tyrosine, an amino acid, is synthesized in sufficient amounts by the body, so it is not an essential dietary requirement. The iodine needed for thyroid hormone synthesis, in contrast, must be obtained from dietary intake. The synthesis, storage, and secretion of thyroid hormone involve the following steps:

1. All steps of thyroid hormone synthesis take place on the thyroglobulin molecules within the colloid. Thyroglobulin itself is produced by the endoplasmic reticulum/Golgi complex of the thyroid follicular cells. Tyrosine becomes incorporated in the much larger thyroglobulin molecules as the latter are being produced. Once produced, tyrosine-containing thyroglobulin is exported from the follicular cells into the colloid by exocytosis (step ① in ● Figure 7–16).

2. The thyroid captures iodine from the blood and transfers it into the colloid by means of a very active "iodine pump" or "iodine-trapping mechanism"—powerful, energy-requiring carrier proteins located in the outer membranes of the follicular cells (step ②). Almost all the iodine in the body is moved against its concentration gradient to become trapped in the thyroid for the purpose of thyroid hormone synthesis. Iodine serves no other known purpose in the vertebrate body.

3. Within the membrane–colloid interface, iodine is quickly attached to a tyrosine within the thyroglobulin molecule by the enzyme *peroxidase.* Attachment of one iodine to

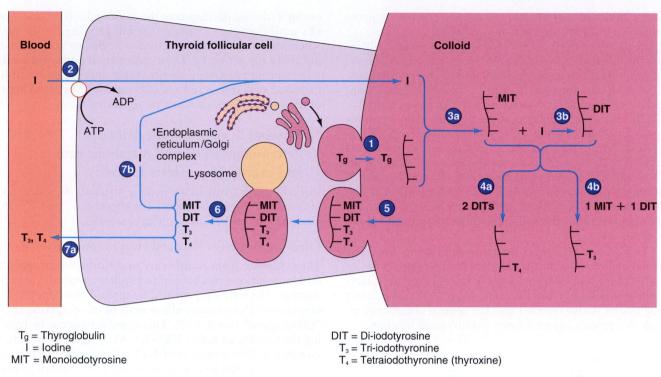

Tg = Thyroglobulin
I = Iodine
MIT = Monoiodotyrosine

DIT = Di-iodotyrosine
T_3 = Tri-iodothyronine
T_4 = Tetraiodothyronine (thyroxine)

* Organelles not drawn to scale. Endoplasmic reticulum/Golgi complex are proportionally too small.

1 Tyrosine-containing T_g produced within the thyroid follicular cells is transported into the colloid by exocytosis.

2 Iodine is actively transported from the blood into the colloid by the follicular cells.

3a Attachment of one iodine to tyrosine within the T_g molecule yields MIT.

3b Attachment of two iodines to tyrosine yields DIT.

4a Coupling of two DITs yields T_4.

4b Coupling of one MIT and one DIT yields T_3.

5 On appropriate stimulation, the thyroid follicular cells engulf a portion of T_g-containing colloid by phagocytosis.

6 Lysosomes attack the engulfed vesicle and split the iodinated products from T_g.

7a T_3 and T_4 diffuse into the blood.

7b MIT and DIT are deiodinated, and the freed iodine is recycled for synthesizing more hormone.

Figure 7–16 • Synthesis, storage, and secretion of thyroid hormone.

tyrosine yields monoiodotyrosine (MIT) (step ③a). Attachment of two iodines to tyrosine yields di-iodotyrosine (DIT) (step ③b).

4. Next, a coupling process occurs between the iodinated tyrosine molecules to form the thyroid hormones. Coupling of two DITs (each bearing two iodine atoms) yields tetraiodothyronine (T_4 or thyroxine), the four-iodine form of thyroid hormone (step ④a). Coupling of one MIT (with one iodine) and one DIT (with two iodines) yields triiodothyronine or T_3 (with three iodines) (step ④b). Coupling does not occur between two MIT molecules.

Because these reactions occur within the thyroglobulin molecule, all the products remain attached to this large protein. Immobilized in this fashion they, contrary to their steroid

hormone–like behavior, cannot freely diffuse into the circulation after synthesis. Thyroid hormones remain stored in this form in the colloid until they are split off and secreted. Researchers estimate that sufficient thyroid hormone to supply a mammal's needs for several months is normally stored in the colloid.

◼ To secrete thyroid hormone, the follicular cells phagocytize thyroglobulin-laden colloid.

The release of the thyroid hormones into the systemic circulation requires a rather complex process, for two reasons. First, before their release T_4 and T_3 are still bound within the thyroglobulin molecule. Second, these hormones are stored in the follicular lumen, so they must be transported completely across the follicular cells to reach nearby capillaries.

The process of thyroid hormone secretion essentially involves the follicular cells "biting off" a piece of colloid, breaking the thyroglobulin molecule down into its component parts, and "spitting out" the freed T_4 and T_3 into the blood. On appropriate stimulation for thyroid hormone secretion, the follicular cells internalize a portion of the thyroglobulin–hormone complex by phagocytizing a piece of colloid (step ⑤ of Figure 7–16). Within the cells, the membrane-enclosed droplets of colloid coalesce with lysosomes, whose enzymes split off the biologically active thyroid hormones, T_4 and T_3, as well as the inactive iodotyrosines, MIT and DIT (step ⑥). The thyroid hormones, being very lipophilic, pass freely through the outer membranes of the follicular cells and into the blood (step ⑦a).

The MIT and DIT are of no known endocrine value. The follicular cells contain an enzyme that swiftly removes the iodine from MIT and DIT, allowing the freed iodine to be recycled for synthesis of more hormone (step ⑦b). Some intact thyroglobulin is also released into the general circulation, although this protein has no known physiological function.

■ For the most part, both T_4 and T_3 are transported bound to specific plasma proteins.

Once released into the blood, the highly lipophilic thyroid hormone molecules very quickly bind with several plasma proteins. Less than 1% of the T_3 and less than 0.1% of the T_4 remain in the **unbound (free)** form. This is remarkable, considering that only the free portion of the total thyroid hormone pool has access to the target cell receptors and thus can exert a biological effect.

Three different plasma proteins synthesized within the liver are important in thyroid hormone binding: In mammals (excluding cats) **thyroxine-binding globulin (TBG)** selectively binds all thyroid hormones even though its name specifies only "thyroxine" (T_4); binding proteins in other species include **albumin,** which nonselectively binds many lipophilic hormones, including T_4 and T_3; and **thyroxine-binding prealbumin (or transthyretin),** which binds the remaining T_4. Transthyretin evolved independently in birds and mammals.

■ Most of the secreted T_4 is converted into T_3 outside the thyroid.

In general, the proportion of T_4 and T_3 present in the thyroid gland of vertebrates is variable, although avian thyroglobulin has an essentially undetectable amount of T_3. About 90% of the secretory product released from the thyroid gland of mammals is in the form of T_4, yet T_3 is about four times more potent in biological activity. Regardless, most of the secreted T_4 is converted into T_3, or *activated,* by being stripped of one of its iodines outside of the thyroid gland, primarily in the liver and kidneys. About 80% of the circulating T_3 is derived from secreted T_4 that has been peripherally stripped. Therefore, T_3 *is the major biologically active form of thyroid hormone at the cellular level,* even though the thyroid gland secretes mostly T_4. T_4 can also be *inactivated* by being converted into the metabolically inactive **reverse tri-iodothyronine or rT_3.** For example, during development the chick embryo produces only rT_3 until the time of piping (the bill piercing the air sack), whereon the enzymatic machinery begins to synthesize the metabolically active T_3. These inactivation pathways are also important in preserving energy stores during periods of limited food availability and are associated with reduced activity of the enzymatic machinery that generates T_3.

■ Thyroid hormone is the primary determinant of overall metabolic rate and exerts other effects as well.

Virtually every tissue in the body is affected either directly or indirectly by thyroid hormone. The effects of T_3 and T_4 can be grouped into several overlapping categories.

Effect on Metabolic Rate and Heat Production

The evolution from ectothermy to endothermy necessitated development of a mechanism to regulate metabolic heat production. The solution provided by thyroid hormones is to increase a bird's or mammal's overall basal metabolic rate or "idling speed" (see p. 697). This appears to occur by regulating the number of active $Na^+–K^+$ ATPase pump units in the membrane. The transport of Na^+ relies on the hydrolysis of ATP, which yields heat as a by-product. Considering that as much as 20 to 40% of the total cell energy supply is required to maintain the pump activity, a considerable amount of heat is liberated in the process. Thyroid hormone is thus the most important regulator of the rate of O_2 consumption and energy expenditure under resting conditions. In addition, during selected periods of development, elevated thyroid hormone concentrations are associated with increased metabolic activity such as during the perihatch period in birds. Inhibition of the $Na^+–K^+$ ATPase pump activity by ouabain (p. 82) markedly reduces the effect of thyroid hormone on heat production and oxygen consumption.

During periods of prolonged fasting, generation of rT_3 is important to conserve body resources. That is, similar to a car at a stoplight, the body "idles" at a slower rate and therefore consumes less fuel.

Compared to other hormones, the action of thyroid hormone is "sluggish." Only after a delay of several hours is the metabolic response to an increase in thyroid hormone detectable, and the maximal response is not evident for several days. The response also lasts quite long, partially because thyroid hormone is not rapidly degraded in mammals but also because the response to an increase in secretion continues to be expressed for days or even weeks after the plasma thyroid-hormone concentrations have returned to normal.

Effect on Intermediary Metabolism

In addition to increasing the general metabolic rate, thyroid hormone modulates the rates of many specific reactions involved in fuel metabolism. The effects of thyroid hormone on the metabolic fuels are multifaceted; not only can it influence both the synthesis and degradation of carbohydrate, fat, and protein, but small or large amounts of the hormone may induce opposite effects. For example, conversion of glucose to glycogen, the storage form of glucose, is facilitated by small amounts of thyroid hormone, but the reverse—breakdown of glycogen into glucose—occurs with large amounts of the hormone. Similarly, adequate amounts of thyroid hormone are essential for the protein synthesis needed for normal bodily growth, yet protein degradation effects predominate at high doses. In general, at abnormally high plasma levels of thyroid hormone, as in thyroid hypersecretion, the overall effect is to favor consumption rather than storage of fuel, as shown by depleting liver glycogen stores, depleting fat stores, and muscle wasting from protein degradation. (The main structural component of cells is protein. Muscle cells are particularly rich in structural protein because they are packed full of contractile elements made of protein—that is, the actin and myosin filaments.)

The effects of concentration are important, because thyroid hormone levels vary on daily and/or seasonal bases in many birds and mammals. For example, serum levels of T_3 and T_4 decrease during winter sleep (often loosely called *hibernation*; see p. 701) in the American black bear, in association with reduced metabolism. In birds, daily changes in circulating levels of thyroid hormone are driven in part by food intake. For example, during the day concentrations of T_3 increase in association with a decline in concentrations of T_4, whereas at night concentrations of T_3 decline while those of T_4 increase. These daily changes are not observed in fasted animals and are attenuated at high temperatures by reduced food intake.

Sympathomimetic Effect

Any action similar to one produced by the sympathetic nervous system is known as a **sympathomimetic** ("sympathetic-mimicking") **effect**. Thyroid hormone increases target cell responsiveness to catecholamines (epinephrine and norepinephrine), the chemical messengers used by the sympathetic nervous system and its hormonal reinforcements from the adrenal medulla. Thyroid hormone presumably accomplishes this permissive action by causing a proliferation of specific catecholamine target-cell receptors (see p. 288). Because of this action, many of the effects observed when thyroid hormone secretion is elevated are similar to those that accompany activation of the sympathetic nervous system (a sympathomimetic effect).

Effect on the Cardiovascular System

By increasing the heart's responsiveness to circulating catecholamines, thyroid hormone increases heart rate and force of contraction, thus increasing cardiac output. In addition, in response to the heat load generated by the calorigenic effect of thyroid hormone, peripheral vasodilation occurs to carry the extra heat to the body surface for elimination to the environment.

Effect on Growth and the Nervous System

Thyroid hormone is essential for normal growth. Thyroid hormones act *permissively* (indirectly) in concert with other hormones in stimulating the growth process. Thyroid hormone is required for GH secretion in mammals and also promotes the effects of GH (or somatomedins) on the synthesis of new structural proteins and on skeletal growth. Thyroid-deficient animals have stunted growth that is reversible with thyroid replacement therapy. Unlike excess GH, however, excess thyroid hormone does not result in excessive growth.

Thyroid hormone plays a crucial role in normal development of the nervous system, especially the CNS, an effect impeded in animals with thyroid deficiency from birth (or hatch). Thyroid hormone is also essential for normal CNS activity in adult animals. Furthermore, the conduction velocity of peripheral nerves varies directly with the availability of thyroid hormone.

Thyroid hormone levels are associated with behavioral regulation as well. In salmonids an increase in thyroid hormone is linked to migratory behavior. The mammalian brain has numerous, widespread receptors for T_3; although the functions of T_3 in the brain are uncertain, abnormally low levels are associated with depression in humans.

Thyroid hormones are also important for development of the reproductive system and reproductive function, although high concentrations of thyroid hormone exert an antigonadal effect. An increase in thyroid hormone is also associated with the molt process in mammals and birds as well as the formation of new feathers and growth of horns or hair.

Developmental Effects of Thyroid Hormone

Thyroid hormone also controls **metamorphosis** in amphibians (● Figure 7–17). Metamorphosis is associated with radical developmental changes in body structure that permit transition of an amphibian from an aquatic to a terrestrial habitat. The enzyme *hyaluronidase*, is normally induced when tissues mature and thus serves as a signal for differentiation. Concentrations of hyaluronic acid are elevated in proliferating tissue and decline as the cells become increasingly differentiated (specialized). Researchers have recognized that at the time of tadpole differentiation generation of T_3 increases, paralleled by a decrease in hyaluronic acid. Administering thyroid hormone to a tadpole also accelerates metamorphosis of the juvenile stage into the adult stage. Conversely, if the thyroid gland is prematurely removed from a tadpole, metamorphosis does not occur and the animals eventually develop into giant juveniles.

In at least one species of fish, an increase in thyroid hormone is associated with metamorphosis. In flounder, migration of the eye from one side of the head to the other is triggered by increased thyroid hormone secretion.

■ Thyroid hormone is regulated by the hypothalamus–pituitary–thyroid axis.

Thyroid-stimulating hormone (TSH), the thyroid tropic hormone from the anterior pituitary, is the most important physiological regulator of thyroid hormone secretion (● Figure 7–18). TSH stimulates almost every step of thyroid hormone synthesis and release.

In addition to enhancing thyroid hormone secretion, TSH maintains structural integrity of the thyroid gland. In the absence of TSH, the thyroid atrophies (decreases in size) and secretes its hormones at a very low rate. Conversely, it undergoes hypertrophy (increase in the size of each follicular cell) and hyperplasia (increase in number of follicular cells) in response to excess TSH stimulation, which results in thyroid enlargement or **goiter**.

The hypothalamic **thyrotropin-releasing hormone (TRH)**, in tropic fashion, "turns on" TSH secretion by the anterior pituitary, whereas thyroid hormone, in negative-feedback fashion, "turns off" TSH secretion. In the hypothalamus–pituitary–thyroid axis, inhibition is exerted primarily at the level of the anterior pituitary. Like the other negative-feedback loops, the

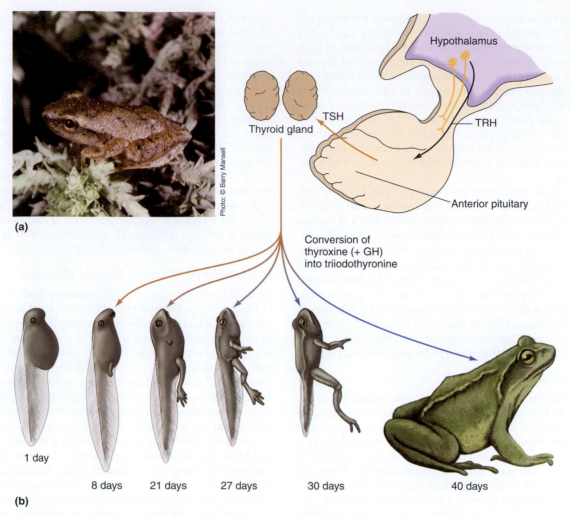

Photo: © Barry Mansell

(a)

Hypothalamus

TSH

Thyroid gland

TRH

Anterior pituitary

Conversion of
thyroxine (+ GH)
into triiodothyronine

1 day

8 days   21 days   27 days   30 days   40 days

(b)

Figure 7–17 • Thyroid hormones and amphibian metamorphosis. (a) A grass frog. The frog spends the first weeks of its life as an aquatic herbivore—a tadpole—in which the thyroid gland and other organs gradually mature. During this period its skeletal muscles and other tissues are relatively insensitive to thyroxine, which is being synthesized in the thyroid although little is released. At about 21 days (b) a spurt in thyroxine secretion begins to trigger major remodeling of the amphibian body. These developmental changes only occur if growth hormone also is present. The entire process may be triggered by changing environmental conditions. As metamorphosis gets underway, shifts in gene expression in cells of various tissues leads to major changes in body structures and physiological functions. Among other alterations, the tail regresses, digested away by newly synthesized lysosomal enzymes, and limbs grow. Lungs develop and gills degenerate, and the digestive tract becomes more suited to processing animal foods such as insects. The kidneys shift from excreting nitrogenous wastes as ammonia—an adaptation typical of aquatic animals—to eliminating urea.

one between thyroid hormone and TSH tends to maintain a stable thyroid-hormone output.

Day-to-day regulation of free thyroid-hormone levels is apparently accomplished by negative feedback between the thyroid and anterior pituitary, whereas long-range adjustments are mediated by the hypothalamus. Unlike most other hormonal systems, hormones in the thyroid axis in an adult mammal normally do not undergo sudden, wide swings in secre-

tion. The relatively steady rate of thyroid hormone secretion is in keeping with the sluggish, long-lasting responses that this hormone induces; there would be no adaptive value in suddenly increasing or decreasing plasma thyroid-hormone levels. Seasonal changes in TRH are known to occur in some mammals, however the only known consistent factor that increases TRH secretion in mammals (and, accordingly, TSH and thyroid hormone secretion) is exposure to cold. This mechanism

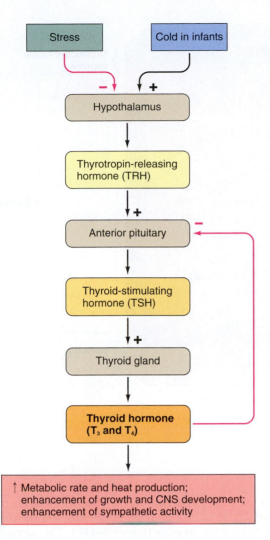

Figure 7–18 • Regulation of thyroid hormone secretion.

is highly adaptive in newborn mammals. The dramatic increase in heat-producing thyroid hormone secretion is thought to help maintain body temperature in the abrupt drop in surrounding temperature at birth, as the infant passes from the mother's warm body to the cooler environmental air. A similar TSH response to cold exposure does not occur in adult mammals.

Various types of stress inhibit TSH and thyroid hormone secretion, presumably through neural influences on the hypothalamus, although the adaptive importance of this inhibition is unclear.

■ Abnormalities of thyroid function include both hypothyroidism and hyperthyroidism.

Normal thyroid function is called **euthyroidism.** Abnormalities of thyroid function are among the most common of all endocrine disorders in humans and domestic (and some wild) birds and mammals. Disorders fall into two major categories—*hypothyroidism* and *hyperthyroidism*—reflecting deficient and excess thyroid-hormone secretion, respectively. A number of specific causes can give rise to each of these conditions. Whatever the cause, the consequences of too little or too much thy-

roid hormone secretion are largely predictable, given a knowledge of the functions of thyroid hormone.

Hypothyroidism

Hypothyroidism can result (1) from primary failure of the thyroid gland itself; (2) secondary to a deficiency of TRH, TSH, or both; or (3) from an inadequate dietary supply of iodine. The symptoms of hypothyroidism largely stem from reduced overall metabolic activity. Among other things, a mammal with hypothyroidism has a reduced basal metabolic rate; displays poor tolerance of cold (lack of the calorigenic effect); has a tendency to gain excessive weight (not burning fuels at a normal rate); is easily fatigued (lower energy production); has a slow, weak pulse (caused by reduced rate and strength of cardiac contraction and lowered cardiac output); and exhibits slow reflexes and slow mentation (because of the effect on the nervous system), characterized by diminished alertness, and poor memory.

In most endotherms, hypothyroidism diminishes quality of fur or feathers. For example, without adequate thyroid-hormone concentrations the seasonal growth in their pelage fails to occur, whereas in birds plumage fails to develop. In horses, hypothyroidism is associated with increased plasma concentrations of VLDL, LDL, and plasma triglycerides (see p. 362).

Hyperthyroidism

As you would expect, the hyperthyroid animal has an elevated basal metabolic rate. The resulting increase in heat production leads to excessive perspiration or panting and poor tolerance of heat. Despite the increased appetite and food intake in response to the increased metabolic demands, body weight typically falls, because the body is burning fuel abnormally fast. Net degradation of endogenous carbohydrate, fat, and protein stores occurs. The loss of skeletal muscle protein results in weakness. Various cardiovascular abnormalities are associated with hyperthyroidism, caused both by the direct effects of thyroid hormone and by its interactions with catecholamines. Heart rate and strength of contraction may rise to dangerous levels. In severe cases, the heart may fail to meet the body's metabolic demands in spite of increased cardiac output. Nervous system involvement is manifested by excessive mental alertness to the point where the animal is irritable, tense, and anxious.

Three general methods of treatment are available for suppressing excess thyroid-hormone secretion: surgical removal of a portion of the oversecreting thyroid gland; administration of radioactive iodine, which, after being concentrated in the thyroid gland by the iodine pump, selectively destroys thyroid glandular tissue; and antithyroid drugs that specifically interfere with thyroid hormone synthesis and generation of T_3.

Vertebrate Adrenal Glands

The **adrenal gland** (*adrenal,* "next to the kidney") (● Figure 7–19a) of higher vertebrates consists of two distinct cell types: **chromaffin** (*chroma,* "color"; *affinis,* "affinity") cells, which are derived from the neural crest, and **steroidogenic cells,** which are of mesodermal origin.

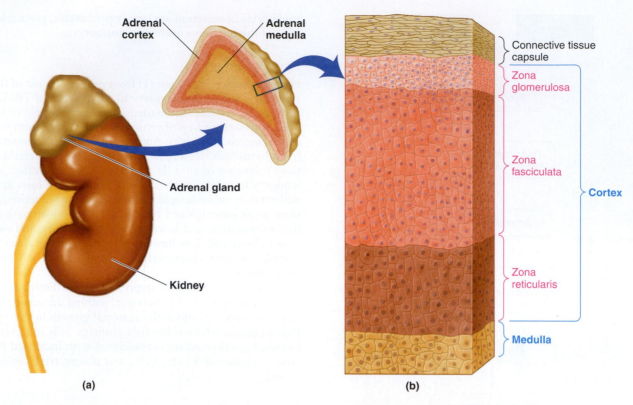

Figure 7–19 ● **Anatomy of the adrenal glands.** (a) Location and structure of the adrenal glands. (b) Layers of the adrenal cortex.

■ In most vertebrates the adrenal gland consists of a steroid-secreting cortex intermingled with chromaffin tissue.

The term *chromaffin* arises from the observation that the tissue stains the color brown when reacted with oxidizing agents such as chromate. In mammals there are two adrenal glands, one embedded above each kidney in a capsule of fat. The shape of the adrenal varies considerably and in some species actually fuses with the kidney. In most mammals the adrenal gland consists of an outer, steroid-secreting **adrenal cortex** and an inner, catecholamine-secreting **adrenal medulla.** For this reason the chromaffin tissue is referred to as the adrenal medulla. However, in most nonmammalian species the chromaffin tissue is not associated with any surrounding cortex and in many instances, rather than forming distinct zones, the two tissue types are intermingled in the adrenal gland. In elasmobranchs, **interrenal tissue,** which is homologous with the adrenal cortex of higher vertebrates, is organized into glands situated between the posterior regions of the kidneys.

The steroid-secreting adrenal cortex and catecholamine-secreting medulla produce hormones belonging to different chemical categories, whose functions, mechanisms of action, and regulation are entirely different.

■ The adrenal cortex secretes mineralocorticoids, glucocorticoids, and sex hormones.

About 80% of the adrenal gland of most mammals is composed of the cortex, which consists of three different layers or zones: the **zona glomerulosa,** the outermost layer; the **zona fasciculata,** the middle and largest portion; and the **zona reticularis,** the innermost zone (Figure 7–19b). The adrenal cortex produces a number of different adrenocortical steroid hormones, all steroids derived from the common precursor molecule, cholesterol, each dependent on the particular enzymatic machinery located in each layer.. Slight variations in structure confer different functional capabilities on the various adrenocortical hormones. On the basis of their primary actions, the adrenal steroids can be divided into three categories:

1. **Mineralocorticoids,** mainly *aldosterone,* which influence mineral (electrolyte) balance (produced exclusively in the zona glomerulosa).

2. **Glucocorticoids,** primarily *cortisol* and *corticosterone,* which play a major role in glucose metabolism as well as in protein and lipid metabolism (synthesized in the two inner layers, with the zona fasciculata being the major source of this glucocorticoid).

3. **Sex hormones** identical or similar to those produced by the gonads (testes in males, ovaries in females). The most abundant and physiologically important of the adrenocortical sex hormones is *dehydroepiandrosterone (DHEA),* a "male" sex hormone (produced by the two inner zones).

The production of adrenocortical and other steroid hormones requires a series of steps in which the cholesterol molecule undergoes various enzymatic modifications (see p. 255). The enzymes that carry out the synthesis of the major steroids are highly conserved among vertebrates and homologous to those characterized in mammals. The different functional types of adrenal steroids are produced in anatomically distinct portions of the adrenal cortex, as noted in the preceding list.

Being lipophilic, the adrenocortical hormones are all carried in the blood extensively bound to plasma proteins. About 60% of circulating aldosterone is protein bound, primarily to nonspecific albumin, whereas approximately 90% of the glucocorticoids are bound, mostly to a plasma protein specific for it called **corticosteroid-binding globulin (transcortin).** Likewise, 98% of dehydroepiandrosterone is bound, in this case exclusively to albumin.

The actions and regulation of the primary adrenocortical mineralocorticoid, **aldosterone,** are described thoroughly elsewhere (Chapter 12). To highlight aldosterone activity in vertebrates, its principal site of action is on the kidney tubules, where it promotes Na^+ retention and enhances K^+ elimination during urine formation. The promotion of Na^+ retention by aldosterone secondarily induces osmotic retention of H_2O, thereby causing expansion of the ECF volume, which is important in the long-term regulation of blood pressure.

Mineralocorticoids are essential for life. Without aldosterone, an animal rapidly dies from circulatory shock because of the marked fall in plasma volume caused by excessive losses of H_2O-holding Na^+. With most other hormonal deficiencies, death is not imminent, even though a chronic hormonal deficiency may eventually lead to a premature death.

■ Glucocorticoids exert metabolic effects and have an important role in adaptation to stress.

Cortisol and **corticosterone,** the primary species-specific glucocorticoids, play an important role in carbohydrate, protein, and fat metabolism; execute significant permissive actions for other hormonal activities; and help animals cope with stress.

Metabolic Effects

The overall effect of glucocorticoids' metabolic actions is to *increase the concentration of blood glucose* at the expense of protein and fat stores. Specifically, cortisol performs the following functions:

- It stimulates hepatic **gluconeogenesis** (p. 54), the conversion of noncarbohydrate sources (namely, amino acids) into carbohydrate within the liver. Between meals or during periods of fasting, when no new nutrients are being absorbed into the blood for use and storage, the glycogen (stored glucose) in the liver tends to become depleted as it is broken down to release glucose into the blood. Gluconeogenesis is an important factor in replenishing hepatic glycogen stores and thus in maintaining normal blood-glucose levels between meals. This is essential because the vertebrate brain can use only glucose as its metabolic fuel, yet nervous tissue cannot store glycogen to any extent. The concentration of glucose in the blood must therefore be maintained at an appropriate level to adequately supply the glucose-dependent brain with nutrients.

- It *inhibits glucose uptake* and use by many tissues, but not the brain, thus sparing glucose for use by the brain, which absolutely requires it as a metabolic fuel. This action contributes to the increase in blood glucose concentration brought about by gluconeogenesis.

- It *stimulates protein degradation* in many tissues, especially muscle. By breaking down a portion of muscle proteins into their constituent amino acids, cortisol increases the blood amino acid concentration. These mobilized amino acids are available for use in gluconeogenesis or wherever else they are needed, such as for repair of damaged tissue or synthesis of new cellular structures. For example, an anuran amphibian's cortisol is markedly elevated during metamorphosis of tadpole into frog (metamorphic climax). This is associated with increased mRNA levels of *pro-opiomelanocortin* (see p. 265) during this developmental stage.

- It *facilitates lipolysis (lysis,* "breakdown"), the breakdown of lipid (fat) stores in adipose tissue, releasing free fatty acids into the blood. The mobilized fatty acids are available as an alternative metabolic fuel for tissues that can use this energy source in lieu of glucose, conserving glucose for the brain.

- It stimulates *acclimatization to seawater* in euryhaline fishes (see p. 587).

Role in Adaptation to Long-Term Stress

Glucocorticoids play a key role in adaptation to stress, particularly *long-term* stress (you will learn in the section on adrenal medulla hormones about *short-term* stress responses). **Stress** is the generalized, nonspecific response to any factor that overwhelms, or threatens to overwhelm, the body's compensatory abilities to maintain homeostasis. Contrary to popular usage, the agent inducing the response is correctly called a *stressor,* whereas *stress* refers to the state induced by the stressor. Stressors include those that are *physical* (trauma, intense heat or cold); *chemical* (reduced O_2 supply, acid–base imbalance, nutritional deficit); *physiological* (hemorrhagic shock, pain); *psychological* or *emotional* (anxiety, fear); and *social* (conflict, isolation from a group). The following types of noxious stimuli illustrate the range of factors that can induce a stress response in fish: poor water quality, handling, confinement, pollutants, and water acidification.

Dramatic increases in glucocorticoid secretion, mediated by the central nervous system, occur in response to all kinds of stressful situations. For this reason, measurement of cortisol/corticosterone in an animal's blood is one of the best indicators of stress. For example, cortisol levels skyrocket in infant chimps separated from their mothers, and conversely, cortisol levels decline below average when their mothers are grooming these chimps. Similarly, a horse that is separated from its herd exhibits increased cortisol and other stress symptoms, showing the importance of social contact in this herd animal. The magnitude of the increase in plasma glucocorticoid concentration is generally proportional to the intensity of the stressful stimulation; a greater increase in glucocorticoid levels is evoked in response to severe stress than to mild stress.

Although the precise role of glucocorticoids in adapting to stress is not known, a speculative but plausible explanation might be as follows. An animal wounded or faced with a life-threatening situation must forgo eating. A glucocorticoid-induced enhancement of breakdown of carbohydrate stores (which increases the availability of blood glucose) would help protect the brain from malnutrition during the imposed fasting period. Also, the amino acids liberated by protein degradation would provide a readily available supply of building blocks for tissue repair should physical injury occur. Thus an increased pool of glucose, amino acids, and fatty acids is available for use as needed.

Anti-inflammatory and Immunosuppressive Effects

When cortisol or synthetic cortisol-like compounds are administered to yield higher-than-physiological concentrations of glucocorticoids (that is, *pharmacological levels*), not only are all the metabolic effects increased, but several important new actions not evidenced at normal physiological levels are seen. The most noteworthy of glucocorticoids' pharmacological effects are anti-inflammatory and immunosuppressive effects in addition to their negative effects on growth and reproduction. Researchers have found cortisol activates a gene in white blood cells that codes for a protein that suppresses interleukin 2, an activator of immune functions in these cells (see Chapter 10).

Scientists have developed synthetic glucocorticoids that maximize the anti-inflammatory and immunosuppressive effects of these steroids while minimizing the metabolic effects. Although their use is often beneficial, glucocorticoids can be dangerous because they suppress the normal inflammatory and immune responses: A glucocorticoid-treated animal has limited ability to resist infections. Furthermore, high levels of exogenous glucocorticoids act in negative-feedback fashion to suppress the hypothalamus–pituitary axis that drives normal glucocorticoid secretion and maintains the integrity of the adrenal cortex. Prolonged suppression of this axis can lead to irreversible atrophy of the glucocorticoid-secreting cells of the adrenal gland and thus to permanent inability of the body to produce its own glucocorticoids. (Note that male humans who take testosterone-like steroids can experience an atrophy in testicular sperm production.)

Although these actions are traditionally considered to occur only at pharmacological levels, recent studies suggest that cortisol may exert anti-inflammatory effects even at normal physiological levels. Indeed, cortisol may serve as a natural antagonist to keep the immune system in check after it has been activated by trauma, disease, and so on. Unfortunately, because cortisol is also increased by prolonged social stresses, susceptibility to disease increases maladaptively in these situations (which we discuss on p. 290).

Permissive Actions

Glucocorticoids are extremely important for their permissiveness (see p. 260). For example, in mammals glucocorticoids must be present in adequate amounts to permit the catecholamines to induce vasoconstriction. An animal lacking cortisol, if untreated, may go into circulatory shock in a stressful situation that demands immediate widespread vasoconstriction.

Effects on the Brain

Glucocorticoids also affect neural functions such as memory. For example, elevated concentrations of plasma cortisol in Pacific salmon during home-stream migration aids in recalling the imprinted memory of the home-stream chemical aroma. Mammalian memory centers (such as the hippocampus) have cortisol receptors of uncertain function. In contrast to salmon, prolonged elevation in cortisol from stress is associated with memory loss in humans and laboratory mammals, but this is considered an abnormal situation.

■ Glucocorticoid secretion is directly regulated by the hypothalamic–pituitary–adrenal axis.

Glucocorticoid secretion by the adrenal cortex is regulated by a long-loop negative-feedback system involving the hypothalamus and anterior pituitary (● Figure 7–20). **Adrenocorticotropic hormone (ACTH)** from the anterior pituitary stimulates the adrenal cortex to secrete cortisol. In higher vertebrates ACTH is derived from a large precursor molecule, **proopiomelanocortin (POMC)**, produced within the endoplasmic reticulum of the anterior pituitary's ACTH-secreting cells (see p. 265). Before secretion, this large precursor is pruned into ACTH and several other biologically active peptides, namely, *melanocyte-stimulating hormone (MSH)* (see p. 265) and a morphinelike substance, *β-endorphin* (see p. 290). The possible significance of these multiple secretory products from a single precursor molecule is addressed later.

Being tropic to the zona fasciculata and zona reticularis, ACTH stimulates both the growth and the secretory output of these two inner layers of the cortex. In the absence of ade-

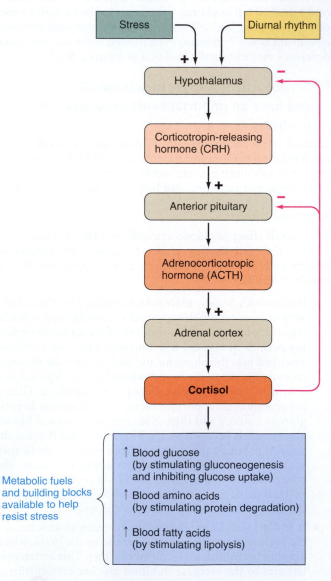

Figure 7–20 ● Control of cortisol secretion.

quate amounts of ACTH, these layers shrink considerably, and glucocorticoid secretion is drastically reduced. Like the actions of TSH on the thyroid gland, ACTH enhances many steps in the synthesis of glucocorticoids.

The ACTH-producing cells in turn secrete only at the command of corticotropin-releasing hormone (CRH) from the hypothalamus. The feedback control loop is completed by glucocorticoid's inhibitory actions on CRH and ACTH secretion by the hypothalamus and anterior pituitary, respectively.

The negative-feedback system for glucocorticoids maintains a relatively constant average level of glucocorticoid secretion that is punctuated by alternating bursts of modest secretion separated by "silent" periods of little or no secretion. The amount of glucocorticoid secreted with each burst does not vary much from one secretory episode to the next. However, the total amount of glucocorticoid secreted during a given period of time can be changed by varying the *frequency* of secretory bursts. Reminiscent of summation in skeletal muscle contraction (p. 331), one burst "adds on" to the previous burst, increasing the plasma concentration of glucocorticoid beyond that achieved by a single burst.

Superimposed on the basic negative-feedback control system are two additional factors that influence plasma glucocorticoid concentrations by changing the set points: These are *stress* and *diurnal rhythms,* both of which act on the hypothalamus to vary the secretion rate of CRH. We have already discussed the elevation of cortisol during prolonged stress.

We also noted earlier that there is a characteristic diurnal rhythm in plasma cortisol concentration, with the highest level occurring in the morning and the lowest level at night (see Figure 7–6, p. 259). This diurnal rhythm, which is intrinsic to the hypothalamus–pituitary control system, is related primarily to the sleep–wake cycle. The peak and low levels are thus reversed in animals that are nocturnal.

■ The adrenal cortex secretes both male and female sex hormones in both sexes.

In both sexes, the adrenal cortex produces both *androgens,* or "male" sex hormones, and *estrogens,* or "female" sex hormones. The main site of production for the sex hormones is the gonads: the testes for androgens and the ovaries for estrogens. Accordingly, males have a preponderance of circulating androgens, whereas in females estrogens are predominant. However, no hormones are unique to either males or females (except those from the placenta during pregnancy in a mammal), because small amounts of the sex hormone of the opposite sex are produced by the adrenal cortex in both sexes.

Under normal circumstances in mammals, the adrenal androgens and estrogens are not sufficiently abundant or powerful to induce masculinizing or feminizing effects, respectively. The only adrenal sex hormone that has any biological importance is the androgen **dehydroepiandrosterone (DHEA)**. The testes' primary androgen product is the potent testosterone, but the most abundant adrenal androgen is the much weaker DHEA. Adrenal DHEA is overpowered by testicular testosterone in males but is of physiological significance in females, who otherwise lack androgens. This adrenal androgen is responsible for androgen-dependent processes in the female such as growth of pubic and axillary (armpit) hair, enhancement of the pubertal growth spurt, and development and maintenance of the female sex drive. Because the enzymes required

for producing estrogens are found in very low concentrations in the adrenocortical cells, estrogens are normally produced in very small quantities from this source.

In addition to controlling glucocorticoid secretion, ACTH (not the pituitary gonadotropic hormones, p. 267) controls adrenal androgen secretion. In general, glucocorticoid and DHEA output by the adrenal cortex parallel each other. However, adrenal androgens feed back outside the hypothalamus–pituitary–adrenal cortex loop. Instead of inhibiting CRH, DHEA inhibits gonadotropin-releasing hormone, just as testicular androgens do. Furthermore, sometimes adrenal androgen and cortisol output diverge from each other—for example, at the time of puberty adrenal androgen secretion undergoes a marked surge, but cortisol secretion does not change. This enhanced secretion initiates the development of androgen-dependent processes in females. In males the same thing is accomplished primarily by testicular androgen secretion, which is also aroused at puberty. The nature of the pubertal inputs to the adrenals and gonads is still unresolved.

■ The catecholamine-secreting adrenal medulla is a modified sympathetic postganglionic neuron.

The adrenal medulla is actually a modified part of the sympathetic nervous system. A sympathetic pathway consists of two neurons in sequence—a *preganglionic neuron* originating in the CNS, whose axonal fiber terminates on a second peripherally located *postganglionic neuron,* which in turn terminates on the effector organ (see p. 153). The neurotransmitter released by sympathetic postganglionic fibers is norepinephrine, which interacts locally with the innervated organ by binding with specific target receptors known as *adrenergic receptors.*

The adrenal medulla consists of modified postganglionic sympathetic neurons. Unlike ordinary postganglionic sympathetic neurons, those in the adrenal medulla do not have axonal fibers that terminate on effector organs. Instead, the ganglionic cell bodies within the adrenal medulla release their chemical transmitter directly into the circulation on stimulation by the preganglionic fiber (see Figure 5–8, p. 152). In this case, the transmitter qualifies as a hormone instead of a neurotransmitter. Like sympathetic fibers, the adrenal medulla does release norepinephrine, but in mammals its most abundant secretory output is a similar chemical messenger known as **epinephrine**. Both epinephrine and norepinephrine belong to the chemical class of *catecholamines,* which are derived from the amino acid tyrosine (see p. 175). Epinephrine is the same as norepinephrine except that it has a methyl group added to it. In species where the chromaffin tissue is separated from the steroidogenic tissue, such as the dogfish shark, norepinephrine is the only catecholamine produced. However, in frogs, where chromaffin tissue is loosely intermingled with steroidogenic tissue, norepinephrine is reduced to about 55 to 70% of the total catecholamine content. Researchers have suggested that the degree of intermingling of chromaffin tissue with steroidogenic tissue determines the extent to which norepinephrine is converted to epinephrine.

Storage of Catecholamines in Chromaffin Granules

Catecholamine synthesis in vertebrates is accomplished almost entirely within the cytosol of the adrenomedullary secretory cells, with only one step taking place within hormonal storage granules. Once produced, epinephrine and norepinephrine are stored in **chromaffin granules,** which are similar to the transmitter storage vesicles in sympathetic nerve endings. The chromaffin granules have a very efficient, active transport system for catecholamine uptake; as a result, the concentration of epinephrine in the chromaffin granules is at least 25,000 times greater than in the cytosol. Segregation of catecholamines in chromaffin granules protects them from being destroyed by cytosolic enzymes during storage. Furthermore, packaging of these hormones into granules is a prerequisite for their export from the adrenomedullary cells. Catecholamines are secreted into the circulation by exocytosis of chromaffin granules; their release is analogous to the release mechanism for secretory vesicles that contain stored peptide hormones or the release of norepinephrine at sympathetic postganglionic terminals.

■ **Epinephrine and norepinephrine vary in their affinities for the different adrenergic receptor types.**

Epinephrine and norepinephrine have differing affinities for four distinctive receptor types: α_1, α_2, β_1, and β_2 adrenergic receptors (see p. 155). Most sympathetic target cells have α_1 receptors, some have only α_2, some only β_2, some have both α_1 and β_2, and β_1 receptors are found almost exclusively in the heart. In general, the responses elicited by activation of α_1 and β_1 receptors are excitatory in nature, whereas the responses to stimulation of α_2 and β_2 receptors are typically inhibitory.

Neither catecholamine is totally specific for particular receptor types, but each has markedly different affinities. Norepinephrine binds predominantly with α and β_1 receptors located in the vicinity of postganglionic sympathetic-fiber terminals. Hormonal epinephrine can reach all α and β_1 receptors via its circulatory distribution, and interacts with these receptors with approximately the same potency as neurotransmitter norepinephrine. Thus epinephrine and norepinephrine exert similar effects in many tissues, with epinephrine generally reinforcing sympathetic nervous activity. In addition, epinephrine activates β_2 receptors, over which the sympathetic nervous system exerts little influence. Many of the essentially epinephrine-exclusive β_2 receptors are located at tissues not even supplied by the sympathetic nervous system but reached by epinephrine through the blood. An example is skeletal muscle, where epinephrine exerts metabolic effects such as promoting the breakdown of stored glycogen.

Sometimes epinephrine, through its exclusive β_2-receptor activation, brings about a different action from that elicited by norepinephrine and epinephrine action through their mutual activation of other adrenergic receptors. As an example, epinephrine and norepinephrine bring about a generalized vasoconstrictor effect mediated by α_1-receptor stimulation. By contrast, epinephrine promotes vasodilation of the blood vessels that supply skeletal muscles and the heart through β_2-receptor activation.

Realize, however, that epinephrine functions only at the bidding of the sympathetic nervous system, which is solely responsible for stimulating its secretion from the adrenal medulla. Epinephrine secretion always accompanies a generalized discharge of the sympathetic nervous system, so sympathetic activity indirectly controls actions of epinephrine. By having the more versatile circulating epinephrine at its call, the sympa-

thetic nervous system has a means of reinforcing its own neurotransmitter effects plus a way of exerting additional actions on tissues that it does not directly innervate.

■ **Epinephrine reinforces the sympathetic nervous system in "fight-or-flight" short-term stress, and exerts additional metabolic effects.**

Adrenomedullary hormones are not essential for life, but virtually all organs in the body are affected by these catecholamines. They play important roles in stress responses, regulation of arterial blood pressure, and control of fuel metabolism. In teleost fish, catecholamines increase oxygen delivery to the tissues, change gill diffusion capacity, increase erythrocyte release from the spleen, increase the blood oxygen capacity by elevating intracellular pH, and increase blood flow and ventilation rate. Together, the sympathetic nervous system and adrenomedullary epinephrine mobilize the body's resources to support peak physical exertion in the face of impending danger ("short-term" stress). The sympathetic and epinephrine actions constitute a "fight-or-flight" response that prepares the animal to deal with a range of physical and environmental disturbances. The insect equivalent of norepinephrine is **octopamine,** a biogenic amine derived from the amino acid tyrosine. In many invertebrates octopamine is a neurotransmitter, neuromodulator, and neurohormone. In response to stressful situations, such as handling or during the early stage of flight, octopamine mobilizes stores from the fat body for release into the hemopymph.

The following sections discuss epinephrine's major effects, which it accomplishes in collaboration with the sympathetic transmitter norepinephrine or performs alone to complement direct sympathetic response.

Effects on Organ Systems

The sympathetic system and epinephrine both exert widespread effects on organ systems that are ideally suited for fight-or-flight responses (see p. 151). Under the influence of epinephrine and the sympathetic system, the rate and strength of cardiac contraction increase, resulting in increased cardiac output, and their generalized vasoconstrictor effects increase total peripheral resistance. Together, these effects cause an increase in arterial blood pressure, thus ensuring an appropriate driving pressure to force blood to the organs that are most vital for meeting the emergency. Meanwhile, vasodilation of coronary and skeletal-muscle blood vessels induced by epinephrine and local metabolic factors shifts blood to the heart and skeletal muscles from other vasoconstricted regions of the body. In fishes, the physiological effects of catecholamines are also used to maintain levels of energy and oxygen supply under conditions such as hypoxia. In teleosts, adrenergic control of the heart is mainly neuronal, and it is only under severe stress that circulating levels of epinephrine increase sufficiently to affect cardiac function.

Epinephrine (but not norepinephrine) dilates the respiratory airways to reduce the resistance encountered in moving air in and out of the lungs. Epinephrine also reduces digestive activity and inhibits bladder emptying, both activities that can be "put on hold" during a fight-or-flight situation.

Metabolic Effects

Epinephrine exerts some important metabolic effects, even at blood hormone concentrations lower than required for eliciting the cardiovascular responses. In general, epinephrine prompts mobilization of stored carbohydrate and fat to provide immediately available energy to fuel muscular work. Specifically, epinephrine increases the blood glucose level by several different mechanisms. First, it stimulates both hepatic (liver) gluconeogenesis and **glycogenolysis,** the latter being the breakdown of stored glycogen into glucose, which is released into the blood. Epinephrine also stimulates glycogenolysis in skeletal muscles. Because of the difference in enzyme content between liver and muscle, however, muscle glycogen is not converted to glucose for the blood. Instead, muscle glycogen is broken down via glycolysis into lactic acid, providing a burst of ATP for muscles in the emergency situation. Lactic acid is then released into the blood. The liver removes lactic acid from the blood and converts it into glucose, so epinephrine's actions on skeletal muscle indirectly contribute to an increase in blood glucose concentrations. Epinephrine and the sympathetic system may further add to this hyperglycemic effect by inhibiting insulin secretion, the pancreatic hormone primarily responsible for removing glucose from the blood, and by stimulating glucagon, another pancreatic hormone that promotes hepatic glycogenolysis and gluconeogenesis (see p. 299). In addition to raising blood glucose levels, epinephrine also raises the blood fatty-acids level by promoting lipolysis.

Epinephrine's metabolic effects are appropriate for transient fight-or-flight situations. The elevated levels of glucose and fatty acids provide additional fuel to power the muscular movement required by the situation and also ensure adequate nourishment for the brain during the crisis when no new nutrients are being consumed. Muscles can use fatty acids for energy production, but the brain cannot.

Because of its other widespread actions, epinephrine also increases overall metabolic rate. Under the influence of epinephrine, many tissues metabolize faster. For example, the work of the heart and respiratory muscles is increased, and the pace of liver metabolism is stepped up. Thus epinephrine as well as thyroid hormone can increase metabolic rate.

Other Effects

Epinephrine affects the central nervous system to promote a state of arousal and increased CNS alertness, also useful to fight-or-flight situations.

Both epinephrine and norepinephrine cause sweating in mammals that have sweat glands, which helps the body rid itself of extra heat generated by increased muscular activity. Also, epinephrine acts on smooth muscles within the eyes to dilate the pupil and flatten the lens. These actions adjust the eyes for more encompassing vision so that the whole threatening scene can be quickly viewed.

Similar to the effects observed in farm animals after the administration of GH, synthetic analogues of epinephrine and norepinephrine have proven effective in causing shifts in protein and lipid accretion of meat animals, including poultry. These derivatives are termed **β-adrenergic agonists** because they bind to the β-receptors on target cell membranes consistent with β-adrenergic activity. However, unlike GH they can be administered orally. This differs from a protein, which is digested and inactivated if orally consumed. However, some β-agonists exhibit side effects and can cause transitory increases in a variety of physiological processes including heart and respiration rate.

■ Sympathetic stimulation of the adrenal medulla is solely responsible for epinephrine release.

Catecholamine secretion by the adrenal medulla is controlled entirely by sympathetic input to the gland. When the sympathetic system is activated under conditions of fear or stress, it simultaneously triggers a surge of adrenomedullary catecholamine release, flooding the circulation with up to 300 times the normal concentration of epinephrine. Although a number of different factors influence adrenal catecholamine secretion, they all act by increasing preganglionic sympathetic impulses to the adrenal medulla. Among the major factors that stimulate increased adrenomedullary output are a variety of stressful conditions such as physical or environmental disturbances, hemorrhage, illness, exercise, hypoxia (low arterial O_2), cold exposure, and hypoglycemia (low blood glucose).

■ The stress response is a generalized, nonspecific pattern of neural and hormonal reactions to any situation that threatens homeostasis.

Because both components of the adrenal gland play an extensive role in responding to stress, this is an appropriate place to pull together the various major factors involved in the stress response. Recall that a variety of noxious physical, chemical, physiological, and psychosocial stimuli that threaten to overwhelm the body's compensatory ability to maintain homeostasis all can elicit a stress response. For example, most research on stress physiology in fish during the past few decades has focused on aquaculture, because of the interest in managing for stress while maximizing production in artificial environments. Different stressors may produce some specific responses characteristic of that stressor; for example, in some mammals the specific response to cold exposure is shivering and skin vasoconstriction, whereas the specific response to bacterial invasion includes increased phagocytic activity and antibody production. In addition to their specific response, however, all stressors also produce a similar nonspecific, generalized response regardless of the type of stressor.

In the 1930s, Hans Selye was the first to recognize this commonality of responses to noxious stimuli in what he called the **general adaptation syndrome.** When a stressor is recognized, both nervous and hormonal responses are called into play to bring about defensive measures to cope with the emergency. Across the entire vertebrate lineage, the result is a state of intense readiness and mobilization of biochemical resources.

To appreciate the value of the multifaceted stress response, imagine a gazelle that has just seen a lion lurking in the grass. The major neural response to such a stressful stimulus is generalized activation of the sympathetic nervous system. The resultant increase in cardiac output and ventilation as well as diversion of blood from vasoconstricted regions of suppressed activity, such as digestive tract and kidneys, to the more active vasodilated skeletal muscles and heart prepare the body for a fight-or-flight response. Simultaneously, the sympathetic system calls forth hormonal reinforcements in the form of a massive outpouring of epinephrine from the adrenal medulla. Epinephrine strengthens sympathetic responses and reaches places not innervated by the sympathetic system to perform additional functions, such as mobilizing carbohydrate and fat stores.

Besides epinephrine, a number of other hormones are involved in the overall stress response (▌Table 7–4). As you have seen, the other predominant hormonal response is activation of the CRH-ACTH-glucocorticoid system, usually for longer-term stress. In our gazelle, this system would dominate if it escapes the lion but receives a serious wound (such as a bite in its leg). Note that a major difference between epinephrine and cortisol is that the former does not promote muscle protein breakdown, whereas the latter does. It would be maladaptive to cannibalize muscles during short-term fight-or-flight, but during long-term trauma it is useful to mobilize amino acids for tissue repair in the event of an injury.

In addition to the effects of cortisol in the hypothalamus–pituitary–adrenal cortex axis, there is much evidence that ACTH may play a role in resisting stress. ACTH suppresses release of GH, TSH, and gonadotropins, suppressing growth, metabolism, and reproduction, respectively. This helps divert energy toward stress needs. ACTH is one of several peptides that facilitate learning and behavior. Thus, it is possible that an increase in ACTH during psychosocial stress might help the body cope more readily with similar stressors in the future by facilitating the learning of appropriate behavioral responses. Furthermore, ACTH is not released alone from its anterior pituitary storage vesicles. Pruning of the large pro-opiomelanocortin precursor molecule yields not only ACTH but also morphinelike β-endorphin and similar compounds. These compounds are co-secreted with ACTH on stimulation by CRH during stress. As a potent endogenous opiate, β-endorphin may exert a role in mediating **analgesia** (reduced pain perception) should physical injury be inflicted during stress (see p. 244), allowing an animal to ignore pain.

Role of the Renin-Angiotensin-Aldosterone System and Vasopressin in Stress

In addition to the hormonal changes that mobilize energy stores during stress, other hormones are simultaneously called into play to sustain blood volume and blood pressure during the emergency. The renin-angiotensin-aldosterone system is activated as a consequence of a sympathetically induced reduction of blood supply to the kidneys (see p. 543). Vasopressin secretion is also increased during stressful situations. Collectively, these hormones expand the plasma volume by promoting retention of salt and H_2O. Presumably, the enlarged plasma volume serves as a protective measure to help sustain blood pressure should acute loss of plasma fluid occur through hemorrhage or heavy sweating during the impending period of danger. Vasopressin and angiotensin also have direct vasopressor effects, which would be of benefit in maintaining an adequate arterial pressure in the face of acute blood loss. As we noted earlier, vasopressin is further believed to facilitate learning, which has implications for future adaptation to stress.

■ The multifaceted stress response is coordinated by the hypothalamus.

All the individual responses to stress just described are either directly or indirectly influenced by the hypothalamus (● Figure 7–21). The hypothalamus receives input concerning physical and emotional stressors from many areas of the brain and from many receptors throughout the body. In response, the hypothalamus directly activates the sympathetic nervous system, secretes CRH to stimulate ACTH and cortisol release, and triggers the release of vasopressin. Sympathetic stimulation in turn brings about the secretion of epinephrine, with which it has a conjoined effect on the pancreatic secretion of insulin and glucagon. Furthermore, vasoconstriction of the renal afferent arterioles by the catecholamines indirectly triggers the secretion of renin by reducing the flow of oxygenated blood through the kidneys. Renin in turn sets in motion the renin-angiotensin-aldosterone mechanism. In this way, the hypothalamus integrates the responses of both the sympathetic nervous system and the endocrine system during stress.

■ Activation of the stress response by chronic psychosocial stressors may be harmful.

In every animal in which it has been investigated, severe psychological or physical stress reduces appetite and food intake. Chronic stress, such as social subordination, may completely shut down reproductive behavior in many vertebrates, where diminished sex-hormone concentrations correlate with elevated levels of ACTH, POMC, and corticosterone. Even in the acute phase of the stress response, behaviors necessary for dealing with an imminent threat (avoidance, escape) emerge,

Table 7–4 ▌ Major Hormonal Changes during the Stress Response		
Hormone	Change	Purpose Served
Epinephrine	↑	Reinforces the sympathetic nervous system to prepare the body for "fight or flight"
		Mobilizes carbohydrate and fat energy stores; increases blood glucose and blood fatty acids
CRH-ACTH-cortisol	↑	Mobilizes energy stores and metabolic building blocks for use as needed; increases blood glucose, blood amino acids, and blood fatty acids
		ACTH facilitates learning and behavior
		β-Endorphin cosecreted with ACTH may mediate analgesia
Glucagon	↑	Act in concert to increase blood glucose and blood fatty acids
Insulin	↓	
Renin-angiotensin-aldosterone	↑	Conserve salt and H_2O to expand the plasma volume; help sustain blood pressure when acute loss of plasma volume occurs
Vasopressin	↑	
		Angiotensin II and vasopressin cause arteriolar vasoconstriction to increase blood pressure
		Vasopressin facilitates learning

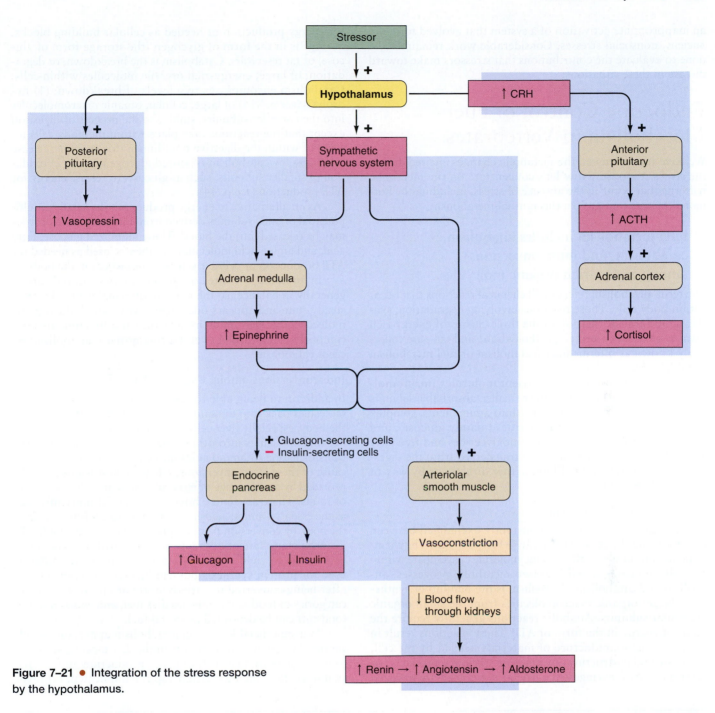

Figure 7–21 • Integration of the stress response by the hypothalamus.

while feeding and reproductive behaviors are temporarily suspended. Acceleration of cardiovascular and respiratory activity, retention of salt and H₂O, and mobilization of metabolic fuels and building blocks can be of benefit in response to a physical stressor. Many stressors in an animal's life are psychosocial in nature, however, yet they induce these same magnified responses. Although the mobilization of body resources is appropriate in the face of real or threatened physical injury, it is generally inappropriate in response to nonphysical stress. If no extra energy is demanded, no tissue damaged, and no blood lost, body stores are being broken down and fluid retained needlessly. Perhaps most importantly, prolonged glucocorticoid elevation also suppresses the immune system

(p. 456). Overall, chronic stress responses (distress) are often to the detriment of the stressed animal, which will have a much greater susceptibility to diseases. You may be aware that overly stressed humans are more prone to diseases; so are zoo animals in poorly designed enclosures, as are social animals in the wild in certain situations. For example, a male baboon entering a troop for the first time must become highly aggressive to be accepted in the social hierarchy. During prolonged aggressive states, cortisol levels rise in this baboon, but his actions also elevate cortisol levels in the rest of the troop. As a result, the number of white blood cells in all the males declines.

The suppression of immunity in these social stresses suggests that this phenomenon is not an adaptation but is rather an inappropriate activation of a system that evolved for more ancient, nonsocial stresses. Considerable work remains to be done to evaluate the contributions that stressors make toward disease in these situations.

Endocrine Control of Fuel Metabolism in Vertebrates

We have just discussed the metabolic changes elicited during the stress response. Now let's concentrate on the metabolic patterns that occur in the absence of stress, including the hormonal factors that govern this normal metabolism.

■ Fuel metabolism includes anabolism, catabolism, and interconversions among energy-rich organic molecules.

The term **metabolism** refers to all chemical reactions that occur within body cells. Those reactions involving degradation, synthesis, and transformation of the three classes of energy-rich organic molecules—protein, carbohydrate, and fat—are collectively known as **intermediary metabolism** or **fuel metabolism** (▌ Table 7–5).

During digestion, large nutrient molecules (**macromolecules**) are broken down into their smaller, absorbable subunits as follows: Proteins are converted into amino acids, complex carbohydrates into monosaccharides (mainly glucose), and triglycerides (dietary fats) into monoglycerides and free fatty acids. These absorbable units are transferred from the digestive tract lumen into the blood, either directly or by way of lymph (Chapter 14).

Anabolism and Catabolism

These organic molecules are constantly exchanged between the blood and the cells of the body (● Figure 7–22). The chemical reactions in which the organic molecules participate within the cells are categorized into two metabolic processes: anabolism and catabolism. **Anabolism** is the buildup or synthesis of larger organic macromolecules from the small organic molecular subunits. Anabolic reactions generally require the input of energy in the form of ATP. These reactions result in either (1) the manufacture of materials needed by the cell, such as cellular structural proteins or secretory products, or (2) storage of excess ingested nutrients not immediately needed

for energy production or needed as cellular building blocks. Storage is in the form of glycogen (the storage form of glucose) or fat reservoirs. **Catabolism** is the breakdown, or degradation, of large, energy-rich organic molecules within cells. Catabolism encompasses two levels of breakdown: (1) hydrolysis (see p. 614) of large, cellular, organic macromolecules into their smaller subunits, similar to the process of digestion except that the reactions take place within the body cells instead of within the digestive tract lumen (for example, release of glucose by catabolism of stored glycogen), and (2) oxidation of smaller subunits, such as glucose, to release energy for ATP production (see p. 44).

As an alternative to energy production, the smaller, multi-potential organic subunits derived from intracellular hydrolysis may be released into the blood. These mobilized glucose, fatty acid, and amino acid molecules can then be used as needed for ATP production or cellular synthesis elsewhere in the body.

In an adult, the rates of anabolism and catabolism are generally in balance, so the adult body remains in a dynamic steady state and appears unchanged even though the organic molecules that determine its structure and function are continuously being turned over. During growth, anabolism exceeds catabolism.

Interconversions among Organic Molecules

In addition to being able to resynthesize catabolized organic molecules back into the same type of molecules, many cells of the body, especially liver cells, can convert most types of small organic molecules into other types—as in, for example, the transformation of amino acids into glucose or fatty acids. Because of these interconversions, adequate nourishment can be provided by a wide range of molecules present in different types of foods. There are limits, however. **Essential nutrients,** such as the essential amino acids and vitamins, cannot be formed in the body by conversion from another type of organic molecule.

The major fate of both ingested carbohydrates and fats is catabolism to yield energy. Amino acids are predominantly used for protein synthesis but can be used to supply energy after being converted to carbohydrate or fat. Thus all three categories of foodstuff can be used as fuel, and excesses of any foodstuff can be deposited as stored fuel.

At a superficial level, fuel metabolism appears relatively simple: The amount of nutrients in the diet must be sufficient to meet the body's needs for energy production and cellular synthesis. This simple relationship is complicated, however,

Table 7–5 ▌ Summary of Reactions in Fuel Metabolism		
Metabolic Process	Reaction	Consequence
Glycogenesis	Glucose → glycogen	↓Blood glucose
Glycogenolysis	Glycogen → glucose	↑Blood glucose
Gluconeogenesis	Amino acids → glucose	↑Blood glucose
Protein synthesis	Amino acids → protein	↓Blood amino acids
Protein degradation	Protein → amino acids	↑Blood amino acids
Fat synthesis (lipogenesis or triglyceride synthesis)	Fatty acids and glycerol → triglycerides	↓Blood fatty acids
Fat breakdown (lypolysis or triglyceride degradation)	Triglycerides → fatty acids and glycerol	↑Blood fatty acids

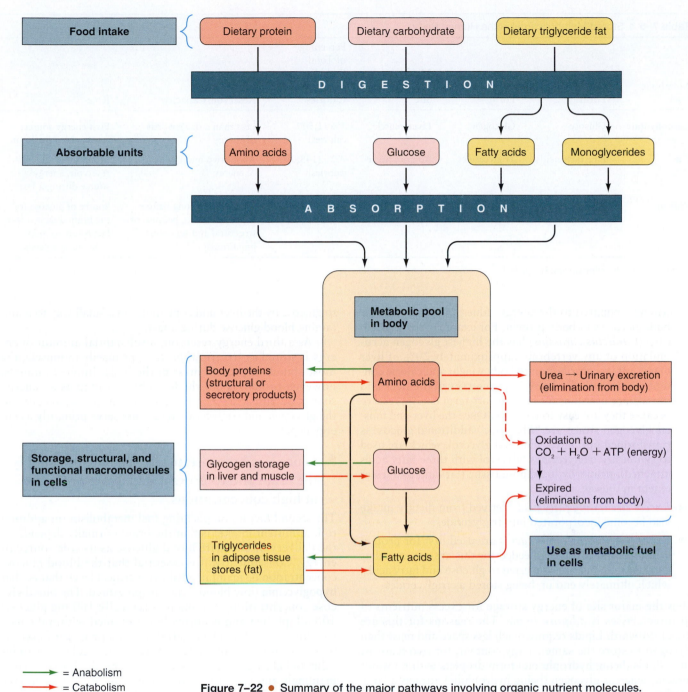

= Anabolism
= Catabolism

Figure 7–22 ● Summary of the major pathways involving organic nutrient molecules.

by two important considerations: (1) Nutrients taken in at meals must be stored and then released between meals, and (2) the brain must be continually supplied with glucose. Let us examine the implications of each of these considerations.

■ **Because food intake is intermittent, nutrients must be stored for use between meals, primarily as adipose tissue.**

Dietary fuel intake is generally intermittent, not continuous; even an animal that grazes on grass for many hours straight

must walk, rest and sleep. As a result, excess energy must be absorbed during meals and stored for use during fasting periods between meals, when dietary sources of metabolic fuel are not available (■ Table 7–6). The goal of these storage processes is *energy homeostasis:* maintaining an adequate supply of energy for both short- and long-term needs.

■ *Excess circulating glucose* is stored in the liver, glial cells, insect fat body and muscle as *glycogen,* a large polymer of glucose insect fat body molecules (p. 45). For many species, researchers ignore the importance of glycogen storage, because its contribution to energy metabolism is

Table 7–6 ▮ Stored Metabolic Fuel in the Human Body

Metabolic Fuel	Circulating Form	Storage Form	Major Storage Site	Percentage of Total Body Energy Content (and Calories*)	Reservoir Capacity	Role
Carbohydrate	Glucose	Glycogen	Liver, muscle	1% (1,500 calories)	Less than a day's worth of energy	First energy source; essential for the brain
Fat	Free fatty acids	Triglycerides	Adipose tissue	77% (143,000 calories)	About two months' worth of energy	Primary energy reservoir; energy source during a fast
Protein	Amino acids	Body proteins	Muscle	22% (41,000 calories)	Death results long before capacity is used, because of structural and functional impairment	Source of glucose for the brain during a fast; last resort to meet other energy needs

*Actually refers to kilocalories; see p. 671.

trifling compared to the energy value of lipid deposits or bulk increases in body protein. For example, the Crucian carp (*Carassius carassius*) has the highest glycogen accumulation of any vertebrate (approximately 20% of liver mass), yet its contribution to body weight and caloric content is minimal. Carbohydrate reserves in the liver and muscle are useful as a limited short-term energy reserve because they are easy to mobilize. Once the liver and muscle glycogen stores are "filled up," additional glucose is transformed into fatty acids and glycerol, which are used to synthesize *triglycerides* (glycerol with three fatty acids attached), primarily in adipose tissue (fat) and to a lesser extent in muscle.

- *Excess circulating fatty acids* derived from dietary intake also become incorporated into triglycerides.
- *Excess circulating amino acids* are used first for protein synthesis, but those not needed for this are not stored as extra protein but are converted to glucose and fatty acids, which ultimately end up being stored as triglycerides.

Thus the major site of energy storage for excess nutrients of all three classes is *adipose tissue*. The reasons for this are straightforward. Lipids require much less space and mass than glycogen to store the same energy content, for two reasons. First, lipids (being hydrophobic) form droplets with no water content, whereas glycogen (being hydrophilic) attracts a large number of water molecules, taking up space and weight with no usable energy. Second, lipids are dominated by high-energy C–H bonds, which can be used to make NADH (see p. 47), whereas glycogen has a large number of C–O bonds that are already oxidized.

Sufficient triglyceride is stored to provide energy for long periods (such as about two months in an average human); during any prolonged period of fasting, the fatty acids released from triglyceride catabolism serve as the primary source of energy for most tissues. The catabolism of stored triglycerides frees glycerol as well as fatty acids, but quantitatively speaking, the fatty acids are far more important. (Recall that a triglyceride is made from one glycerol bonded three fatty acids; p. 24.) Catabolism of stored fat yields 90% fatty acids and 10% glycerol by weight. Glycerol can be converted

to glucose by the liver and contributes in a small way to maintaining blood glucose during a fast.

As a third energy reservoir, a substantial amount of energy is stored as *structural protein,* primarily in muscle, the most abundant protein mass in the body. In most animals, however, protein is not the first choice to tap as an energy source, because it serves other essential functions; in contrast, the glycogen and triglyceride reservoirs serve primarily as energy depots.

▮ **Glucose is homeostatically regulated to supply the brain and to prevent damaging processes at high concentrations.**

The second factor complicating fuel metabolism (in addition to intermittent intake) is that the brain normally depends on the delivery of adequate blood glucose as its sole source of energy. Consequently, it is essential that the blood glucose concentration be maintained above a critical level, that is, that **hypoglycemia** (low blood sugar) be prevented. The blood glucose concentration of humans is typically 100 mg glucose/100 mL plasma and is normally maintained within the narrow limits of 70 to 110 mg/100 mL in a process of *glucose homeostasis* (you will see later how this is regulated with insulin and glucagon). Blood glucose is typically lower in large mammals and two- to sixfold higher in birds, but all have some general set-point regulation. Liver glycogen is an important reservoir for maintaining blood glucose levels during a short fast, and is the primary reservoir in daily life. However, liver glycogen is depleted relatively rapidly, so during a longer fast than that between daily meals, other mechanisms must ensure that the energy requirements of the glucose-dependent brain are met. First, when new dietary glucose is not entering the blood, tissues not obligated to use glucose shift their metabolic gears to burn fatty acids instead, thus sparing glucose for the brain. Fatty acids are made available by catabolism of triglyceride stores as an alternative energy source for non–glucose-dependent tissues. Second, amino acids can be converted to glucose by gluconeogenesis, whereas fatty acids cannot. Thus, once glycogen stores are depleted despite glucose sparing, new glucose supplies for the brain are provided by catabolism of body proteins and conversion of the freed amino acids into glucose.

An alternative fuel for the brain that is derived from fats are ketones, generally during starvation. We investigate ketones shortly (p. 296).

A third complication of fuel metabolism is the potential damage of excessively high blood glucose (**hyperglycemia**). At levels significantly higher than an animal's set point, the osmotic effect of the glucose leads to cell dehydration. Also, glucose can spontaneously react—in a process called **glycation**—with proteins such as hemoglobin and collagen, forming a covalent bond that alters the structure and function of the proteins (see p. 526). These effects are relatively slow, so that temporary rises in blood glucose are not necessarily dangerous (if not too extreme). Thus glucose homeostasis is not as precise as other regulation such as mammalian body-temperature homeostasis. In summary, glucose homeostasis works to prevent short-term hypoglycemia and long-term hyperglycemia.

■ Metabolic fuels are stored during the absorptive state and are mobilized during the postabsorptive state.

From the preceding discussion, you can see that the disposition of organic molecules depends on the body's metabolic state. Two functional metabolic states are related to eating and fasting cycles—the absorptive state and the postabsorptive state, respectively (■ Table 7–7).

Absorptive State

Following a meal, ingested nutrients are being absorbed and entering the blood during the **absorptive**, or **fed, state**. During this time, glucose is plentiful (assuming the meal includes carbohydrates) and serves as the major energy source. Very little of the absorbed fat and amino acids is used for energy during the absorptive state because most cells are programmed to use glucose first, when available. (This situation does not apply if the meal has low or no carbohydrate content.) Extra nutrients not immediately used for energy or structural repairs are channeled into storage as glycogen or triglycerides.

Postabsorptive State

In humans the average meal is completely absorbed in about four hours. Therefore, on a typical three-meals-a-day diet, no nutrients are being absorbed from the digestive tract during late morning and late afternoon and throughout the night. These times constitute the **postabsorptive**, or **fasting, state**. There are few daily fasting periods in nonruminant herbivores, which may graze most of the day, and more in carnivores (such as a cheetah), which may gorge on only a few large prey animals per month.

During the fasting state, endogenous energy stores are mobilized to provide energy, whereas gluconeogenesis and glucose sparing are used to maintain the blood glucose at an adequate level to nourish the brain. The synthesis of protein and fat is curtailed. Instead, stores of these organic molecules are catabolized for glucose formation and energy production, respectively. Carbohydrate synthesis does occur through gluconeogenesis, but the use of glucose for energy is greatly reduced.

Note that blood concentration of nutrients does not fluctuate markedly between absorptive and postabsorptive states. During the absorptive state, the glut of absorbed nutrients is swiftly removed from the blood and placed into storage; during the postabsorptive state, these stores are catabolized to maintain blood concentrations at levels necessary to sustain tissue energy demands.

Roles of Key Tissues in Metabolic States

During these alternating metabolic states, various tissues play different roles as summarized here:

- The *liver* plays the primary role in maintaining normal blood-glucose levels. It stores glycogen when excess glucose is available, releases glucose into the blood when needed, and is the principal site for metabolic interconversions such as gluconeogenesis.
- *Adipose* tissue serves as the primary energy-storage site and is important in regulating fatty acid levels in the blood.
- *Muscle* is the primary site of amino acid storage and is the major energy user.
- The *brain* normally needs glucose as its main energy source, yet it has only minor glycogen stores in glial cells, making it critical that blood glucose levels be maintained.

■ Lesser energy sources are tapped as needed.

Several other organic intermediates play a lesser role as energy sources—namely, glycerol, lactic acid, and ketones (or ketone bodies):

- As mentioned earlier, *glycerol* derived from triglyceride hydrolysis (it is the backbone to which the fatty acid chains are attached) can be converted to glucose by the liver.
- Similarly, *lactic acid*, which is produced by the incomplete catabolism of glucose via glycolysis in muscle (see p. 54), can also be converted to glucose in the liver.

Table 7–7 ■ Comparison of Absorptive and Postabsorptive States		
Metabolic Fuel	Absorptive State	Postabsorptive State
Carbohydrate	Glucose providing major energy source	Glycogen degradation and depletion
	Glycogen synthesis and storage	Glucose sparing to conserve glucose for the brain
	Excess converted to and stored as triglyceride fat	Production of new glucose through gluconeogenesis
Fat	Triglyceride synthesis and storage	Triglyceride catabolism
		Fatty acids providing the major energy source for non–glucose-dependent tissues
Protein	Protein synthesis	Protein catabolism
	Excess converted to and stored as triglyceride fat	Amino acids used for gluconeogenesis

■ **Ketone bodies** are a group of compounds produced by the liver during glucose sparing. Unlike other tissues, when the liver uses fatty acids as an energy source, it oxidizes them only to acetyl coenzyme A (acetyl CoA, p. 47), which it cannot process through the citric acid cycle for further energy extraction. Thus the liver does not degrade fatty acids all the way to CO_2 and H_2O for maximum energy release. Instead, it partially extracts the available energy and converts the remaining energy-bearing acetyl CoA molecules into ketone bodies, which it releases into the blood. Ketone bodies serve as an alternative energy source for tissues capable of oxidizing them further by means of the citric acid cycle.

During long-term starvation, the vertebrate brain starts using ketones instead of glucose as a major energy source, because glucose homeostasis cannot be maintained. Because death from starvation is usually due to protein wasting rather than to hypoglycemia, prolonged survival without any caloric intake requires that gluconeogenesis be minimized, although the energy needs of the brain are not compromised. A sizable portion of cell protein can be catabolized without serious cell malfunction, but a point is finally reached at which a cannibalized cell can no longer function adequately. To ward off the fatal point of failure so long as possible during prolonged starvation, the brain starts using ketones as a major energy source, correspondingly decreasing its use of glucose. The brain's use of ketones from the liver limits the necessity of mobilizing body proteins for glucose production to nourish the brain.

Locusts and cockroaches also rely on ketone bodies during periods of starvation. The activities of ketone body–oxidizing enzymes are also elevated in brain tissue of insects that make extensive use of fatty acids during flight and of insects that do not feed as adults.

Both the major metabolic adaptations to prolonged starvation—a decrease in protein catabolism and use of ketones by the brain—are attributable to the high levels of ketones in the blood at the time. The brain only uses ketones when the blood ketone level is high. The high blood levels of ketones also directly inhibit protein degradation in muscle. Thus ketones are responsible for sparing body proteins while satisfying the brain's energy needs.

Failure of the Glucose Homeostatic Mechanisms

As you have seen, glucose homeostasis fails during extreme starvation. Another failure can occur in dairy cows, which synthesize lactose, a milk sugar, from glucose. In high-producing dairy cows nearly all the glucose produced is committed to lactose generation, requiring other tissues to rely on alternative fuels to support metabolic activities. Sheep can experience a similar imbalance in glucose homeostasis in late gestation because the developing fetuses (usually twins or triplets) and placenta rely on glucose and amino acids to support their energy needs. Sheep are particularly sensitive to this imbalance in glucose homeostasis because they have a higher-than-average ratio of fetal mass to ewe's body size, and in late gestation the energy demands of the fetus are maximal. What happens, then, in dairy cows at peak lactation and sheep during late gestation, when the demands for glucose are maximal? In both these situations, **lactational ketosis** in dairy cows and **pregnancy toxemia** in sheep can arise from the failure of the glu-

cose homeostatic mechanisms to generate sufficient glucose supplies.

A disease-induced failure occurs with diabetes mellitus, which we discuss later.

■ The pancreatic hormones, insulin and glucagon, are most important in regulating fuel metabolism.

How does the body "know" when to shift its metabolic gears from one of net anabolism and nutrient storage to one of net catabolism and glucose sparing? The flow of organic nutrients along metabolic pathways is influenced by a variety of hormones, including *insulin, glucagon, epinephrine, glucocorticoids* (cortisol or corticosterone), and *GH*. Under most circumstances, the pancreatic hormones, **insulin** and **glucagon**, are the dominant hormonal regulators responsible for glucose homeostasis; that is, they shift the metabolic pathways back and forth from net anabolism to net catabolism and glucose sparing, depending on whether the body is feasting or fasting, respectively.

The vertebrate **pancreas** is an organ composed of both exocrine and endocrine tissues. The exocrine portion of the pancreas secretes a watery alkaline solution and digestive enzymes through the pancreatic duct into the digestive tract lumen (Chapter 14). Scattered throughout the pancreas between the exocrine cells are clusters, or "islands," of endocrine cells known as the **islets of Langerhans** (see Figure 14–12, p. 638). These cells are the integrators of endocrine regulatory responses controlled by the pancreas. The most abundant pancreatic-endocrine cell type in mammals is the β (**beta**) **cell**, the site of *insulin* synthesis and secretion. Insulin has been isolated from representative species in all classes of vertebrates including hagfish and lampreys, the living representatives of the jawless vertebrates (Agnatha) and the most evolutionarily ancient of extant vertebrates. Next most important are the α (**alpha**) **cells**, which produce *glucagon*. The δ (**delta**) **cells** are the pancreatic site of *somatostatin* synthesis, whereas the least common islet cells, the **PP cells**, secrete *pancreatic polypeptide* (which has an uncertain function, and is not discussed further).

Somatostatins

Somatostatins have been found in the central and peripheral nervous systems, urogenital tract, endocrine glands, and placenta, as well as in cerebral spinal fluid, blood, and saliva. Somatostatin is produced by the hypothalamus, where it inhibits GH secretion (the effect on GH release is the basis for its name) and TSH. Pancreatic somatostatin inhibits the digestive system in a variety of ways, the overall effect of which is to inhibit digestion of nutrients and to decrease nutrient absorption. Somatostatin is released from the pancreatic δ cells in direct response to an increase in blood glucose and blood amino acids during absorption of a meal. By exerting its inhibitory effects, pancreatic somatostatin acts in negative-feedback fashion to put the brakes on the rate at which the meal is being digested and absorbed, thereby preventing excessive plasma levels of nutrients.

In addition, pancreatic somatostatin may play an important role in local regulation of pancreatic hormone secretion. Secretion of insulin, glucagon, pancreatic polypeptide, and somatostatin itself is decreased by the local presence of somatostatin, but the physiological importance of such paracrine function has not been determined.

Insulin lowers blood glucose, amino acid, and fatty acid levels and promotes their storage.

In vertebrates **insulin** has important effects on carbohydrate, fat, and protein metabolism. It lowers blood levels of glucose, fatty acids, and amino acids and promotes their storage. As these nutrient molecules enter the blood during the absorptive state, insulin promotes their cellular uptake and conversion into glycogen, triglycerides, and protein, respectively. Insulin exerts its many effects either by altering transport of specific blood-borne nutrients into cells or by altering the activity of the enzymes involved in specific metabolic pathways.

Insulin consists of two peptide chains (A and B) cross-linked by disulfide bonds. Insulin-like peptides and genes have also been found in many invertebrates. Invertebrate versions also consist of two peptide chains cross-linked by highly conserved insulin-type disulfide bonds.

Vertebrate insulin belongs to a superfamily consisting of a number of peptide hormones including **insulin-like peptides (ILPs)** in some invertebrates (mollusks, nematodes, insects). Some insect ILPs are neuroendocrine hormones as exemplified by locust insulin-related peptide (LIRP), which was isolated from the neurohemal lobes in the corpora cardiaca (p. 261). Functionally, these invertebrate insulin-like peptides act as growth factors and developmental regulators, much like vertebrate IGF. For example, in fruit flies and *C. elegans* (nematode) mutations in the insulin-like receptor genes result in defects in growth and development.

The insulin *receptor* consists of two α subunits located externally to the membrane and two β subunits that transverse the membrane. Disulfide bonds connect the two α subunits to the two β subunits. The portion of the β subunit that protrudes into the cytoplasm contains a kinase domain that undergoes autophosphorylation as a result of insulin binding to the external α subunits. Autophosphorylation of tyrosine residues on the insulin receptor results in a widely diverse response, and depending on the tissue involved, the actions are as follows.

Actions on Carbohydrates

The maintenance of blood glucose homeostasis is a particularly important function of the pancreas. Circulating glucose concentrations are determined by the balance among the following processes (● Figure 7–23): glucose absorption from the digestive tract; transport of glucose into cells; cellular (primarily hepatic) glucose production; and (abnormally) urinary excretion of glucose. Insulin exerts four effects to lower blood-glucose levels and promote carbohydrate storage:

1. *Facilitation of glucose transport* into most cells. However, one notable exception is the observation that insulin does not affect glucose uptake in fish adipocytes. (The mechanism of this increased glucose uptake is explained in the next section).

2. *Stimulation of glycogenesis,* the production of glycogen from glucose, in both skeletal muscle and the liver.

3. *Inhibition of glycogenolysis,* by inhibiting breakdown of glycogen into glucose. By inhibiting degradation of glycogen, insulin likewise favors carbohydrate storage and decreases glucose output by the liver.

4. *Inhibition of gluconeogenesis* in the liver. By inhibiting gluconeogenesis, the conversion of amino acids into glucose in the liver, hepatic glucose output is reduced. Insulin accomplishes this in two ways: by decreasing the amount of amino acids in the blood that are available to the liver for gluconeogenesis, and by inhibiting the hepatic enzymes required for converting amino acids into glucose.

Thus insulin decreases concentration of blood glucose by promoting the cells' uptake of glucose from the blood for use and storage, while simultaneously blocking the two mechanisms by which the liver releases glucose into the blood (glycogenolysis and gluconeogenesis). Note that the cells stimulated

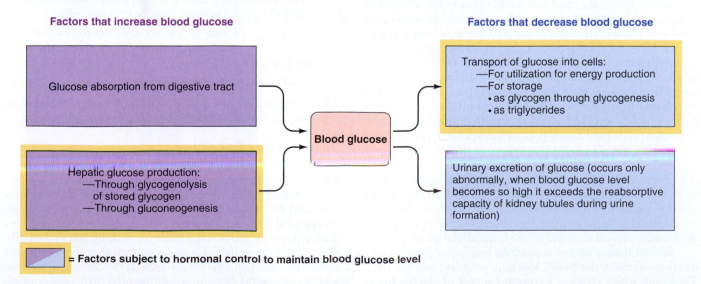

Factors that increase blood glucose

Glucose absorption from digestive tract

Hepatic glucose production:
—Through glycogenolysis of stored glycogen
—Through gluconeogenesis

Blood glucose

Factors that decrease blood glucose

Transport of glucose into cells:
—For utilization for energy production
—For storage
• as glycogen through glycogenesis
• as triglycerides

Urinary excretion of glucose (occurs only abnormally, when blood glucose level becomes so high it exceeds the reabsorptive capacity of kidney tubules during urine formation)

= Factors subject to hormonal control to maintain blood glucose level

Figure 7–23 ● Factors affecting blood glucose concentration.

by glucose are the effectors in a negative-feedback sense, because they make the actual corrective changes, whereas insulin is a signal molecule (akin to the electrical signals running from a thermostat to an air conditioner). Because lowering blood glucose concentrations below a certain threshold value can result in death, insulin is the *only* hormone capable of lowering glucose concentrations in the blood.

These functions have been primarily explored in depth in mammals, but in nonmammalian vertebrates experiments suggest that insulin has similar roles. For example, removing islet cells from either the teleost *Gillichthys mirabilis* (mudsucker) or the Southern lamprey *Geotria australis* dramatically increases blood glucose concentrations. In general, the hypoglycemic action of insulin in ectotherms develops more slowly compared to endotherms and can require days instead of hours to achieve a full effect.

Glucose Transporter Recruitment. Glucose transport between the blood and cells is accomplished by means of plasma membrane carriers known as **glucose transporters (GLUT)**. Researchers have identified six forms of glucose transporters, named in the order discovered—GLUT-1, GLUT-2, and so on. The amino acid sequence of each GLUT isoform is nearly identical when the structure is compared with different species, which suggests that all glucose transporters originated from a common ancestor. For example, GLUT-1 is over 95% identical between different mammalian species. These glucose transporters all accomplish passive facilitated diffusion of glucose across the plasma membrane (see p. 80) and are distinct from the Na$^+$–glucose cotransport carriers responsible for secondary active transport of glucose across the kidney and intestinal epithelia (see p. 83). Each member of the GLUT family performs slightly different functions. For example, *GLUT-1* transports glucose across the blood–brain barrier, *GLUT-2* transfers into the adjacent bloodstream the glucose that has entered the kidney and intestinal cells by means of the cotransport carriers, and *GLUT-3* is the main transporter of glucose into neurons. The glucose transporter responsible for most glucose uptake by most body cells is *GLUT-4,* which operates only at the bidding of insulin. Glucose molecules cannot readily penetrate most cell membranes in the absence of insulin, making most tissues highly dependent on insulin for uptake of glucose from the blood and for its subsequent use.

GLUT-4 is the only type of glucose transporter that is responsive to insulin. Unlike the other types of GLUT molecules, which are always present in the plasma membranes at the sites where they perform their functions, GLUT-4 is excluded from the plasma membrane in the absence of insulin. Insulin promotes glucose uptake by **transporter recruitment.** Insulin-dependent cells maintain a pool of intracellular vesicles containing GLUT-4. Insulin induces these vesicles to move to the plasma membrane and fuse with it, inserting GLUT-4 molecules into the plasma membrane. In this way, increased insulin secretion promotes a rapid, 10- to 30-fold increase in glucose uptake by insulin-dependent cells. When insulin secretion decreases, these glucose transporters are retrieved from the membrane and returned to the intracellular pool.

Several tissues do not depend on insulin for their glucose uptake—namely, the brain, working muscles, and the liver. The brain, which requires a constant supply of glucose for its minute-to-minute energy needs, is freely permeable to glucose at all times via GLUT-1 and GLUT-3 molecules. Skeletal muscle cells do not depend on insulin for glucose uptake during exercise, even though they are dependent at rest. Muscle contraction triggers insertion of GLUT-4 into the plasma membranes of exercising muscle cells in the absence of insulin. The liver is also not dependent on insulin for glucose uptake, because it does not use GLUT-4; however, insulin does enhance the metabolism of glucose by the liver by stimulating the first step in glucose metabolism, the phosphorylation of glucose to form glucose-6-phosphate. The phosphorylation of glucose as it enters the cell keeps the intracellular concentration of "plain" glucose low so that a gradient favoring facilitated diffusion of glucose into the cell is maintained.

Insulin also exerts important actions on fat and protein.

Actions on Fat

Insulin exerts multiple effects to lower blood fatty acids and promote triglyceride storage:

1. *Increase in the transport of glucose into adipose tissue cells* by means of GLUT-4 recruitment. Glucose serves as a precursor for the formation of fatty acids and glycerol, which are the raw materials for triglyceride synthesis.

2. *Activation of enzymes* that catalyze the production of *fatty acids* from glucose derivatives.

3. *Promotion of the uptake of fatty acids* from the blood into adipose tissue cells.

4. *Inhibition of lipolysis* (fat breakdown), thus reducing release of fatty acids from adipose tissue into the blood, although in birds insulin is not antilipolytic.

Collectively, these actions favor removal of glucose and fatty acids from the blood and promote their storage as triglycerides. Adipose tissue here serves as an effector.

Actions on Protein

Insulin lowers blood amino-acid levels and enhances protein synthesis through several effects:

1. *Promotion of the active transport of amino acids* from the blood into muscles and other tissues. This effect lowers the circulating amino-acid level and provides the building blocks for protein synthesis within the cells.

2. *Enhancing the rate of amino acid incorporation into protein* by stimulating the cells' protein-synthesizing machinery.

3. *Inhibition of protein degradation.*

The collective result of these actions is a protein anabolic effect. For this reason, insulin is essential for normal growth.

Summary of Insulin's Actions

In short, insulin stimulates biosynthetic pathways that increase glucose use, carbohydrate and fat storage, and protein synthesis. In so doing, this hormone lowers the blood glucose, fatty acid, and amino acid levels. This metabolic pattern is characteristic of the absorptive state. Indeed, insulin secretion rises during this state and is responsible for shifting metabolic pathways to net anabolism. Consider as well that the number of insulin receptors on a target cell reflects the carbohydrate content of the natural diet and the differential involvement of insulin in the regulation of protein or carbohydrate metabolism in different species. For example, in salmonids a carnivorous group of fish whose diet consists mainly of protein, the number of insulin receptors is lower compared to either herbivorous (such as the African fish tilapia) or omnivorous (such as carp) species. In general, the number of insulin receptors in fish, amphibians, and reptiles is higher than that of IGF-I receptors, whereas in birds and mammals the opposite holds true. The increase in ratio between insulin and IGF-I receptors during vertebrate evolution is probably related to the higher physiological response of skeletal muscles to insulin in endothermic vertebrates.

When insulin secretion is low, the opposite effects occur. The rate of glucose entry into cells is reduced, and net catabolism rather than net synthesis of glycogen, triglycerides, and protein occurs. This pattern is reminiscent of the postabsorptive state; indeed, insulin secretion falls during the postabsorptive state. However, the other major pancreatic hormone, glucagon, also plays an important role in shifting from absorptive to postabsorptive metabolic patterns, as we describe shortly.

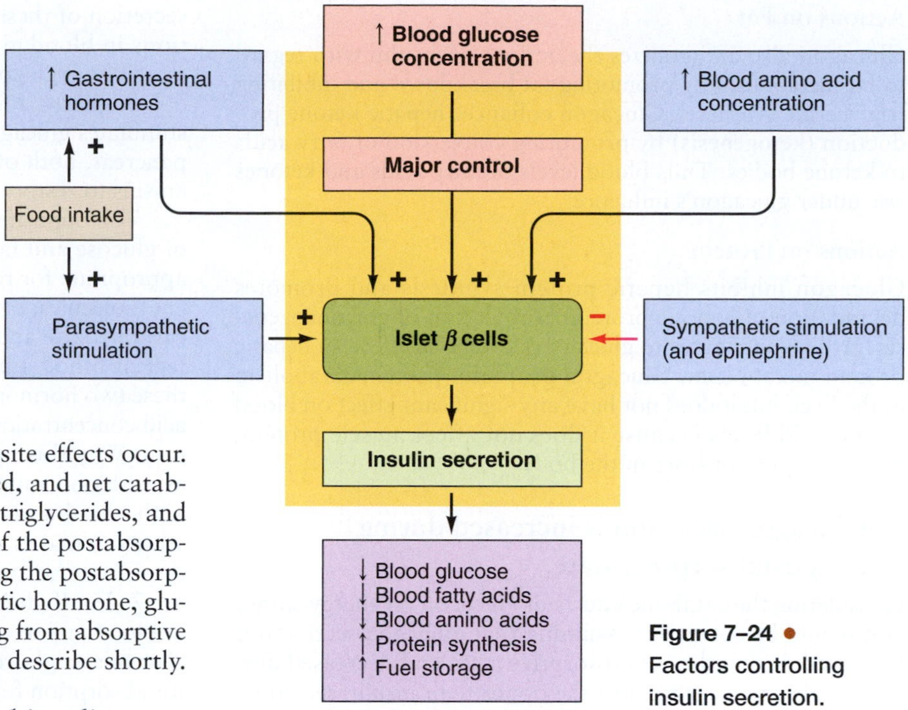

■ The primary stimulus for increased insulin secretion is an increase in blood glucose concentration.

The primary control of insulin secretion is a direct negative-feedback system between the pancreatic β cells and the concentration of glucose in the blood flowing to them. An elevated blood-glucose level, such as during absorption of a meal, directly stimulates synthesis and release of insulin by the β cells. The increased insulin in turn reduces blood glucose to normal while it promotes use and storage of this nutrient. Conversely, a fall in blood glucose below normal, such as during fasting, directly inhibits insulin secretion. Lowering the rate of insulin secretion shifts metabolism from the absorptive to the postabsorptive pattern. Thus this simple negative-feedback system can maintain a relatively constant supply of glucose to the tissues without requiring the participation of nerves or other hormones.

In addition to plasma glucose concentration, other inputs are involved in regulating insulin secretion, as follows (● Figure 7–24):

- An elevated blood amino-acid level, such as after eating a high-protein meal, directly stimulates β cells to increase insulin secretion. In negative-feedback fashion, the increased insulin enhances entry of these amino acids into the cells, lowering blood amino-acid level while promoting protein synthesis.

- The major gastrointestinal hormones secreted by the digestive tract in response to food, especially gastric inhibitory peptide (see p. 667), stimulate pancreatic insulin secretion in addition to having direct regulatory effects on the digestive system. Through this control, insulin secretion is increased in "feed-forward," or anticipatory, fashion even before nutrient absorption increases the concentration of glucose and amino acids in the blood.

- The autonomic nervous system also directly influences insulin secretion. The islets are richly innervated by both

Figure 7–24 ●
Factors controlling insulin secretion.

parasympathetic (vagal) and sympathetic nerve fibers. The increase in parasympathetic activity that occurs in response to food in the digestive tract stimulates insulin release. This, too, is a feed-forward response in anticipation of nutrient absorption. In contrast, sympathetic stimulation and the concurrent increase in epinephrine both inhibit insulin secretion. The reduction in insulin allows the blood glucose concentration to increase, an appropriate response to the circumstances under which generalized sympathetic activation occurs—namely, stress (fight or flight) and exercise. In both these situations, extra fuel is needed for increased muscle activity.

■ Glucagon in general opposes the actions of insulin.

Even though insulin plays a central role in controlling the metabolic adjustments between the absorptive and postabsorptive states, the secretory product of vertebrate pancreatic-islet α cells, **glucagon**, is also very important. Many physiologists view the insulin-secreting β cells and the glucagon-secreting α cells as a coupled antagonistic endocrine system whose combined secretory output is a major factor in regulating fuel metabolism.

Glucagon affects many of the same metabolic processes that are influenced by insulin, but in most cases glucagon's actions are opposite to those of insulin. The major site of action of glucagon is the liver (serving as an effector), where it exerts a variety of effects on carbohydrate, fat, and protein metabolism.

Actions on Carbohydrate

The overall effects of glucagon on carbohydrate metabolism increase hepatic glucose production and release and thus raise blood glucose levels. Glucagon exerts its hyperglycemic effects by decreasing glycogen synthesis, promoting glycogenolysis, and stimulating gluconeogenesis.

Actions on Fat

Glucagon also antagonizes the actions of insulin with regard to fat metabolism by promoting fat breakdown and inhibiting triglyceride synthesis. Glucagon enhances hepatic ketone production (**ketogenesis**) by promoting conversion of fatty acids to ketone bodies. Thus blood levels of fatty acids and ketones rise under glucagon's influence.

Actions on Protein

Glucagon inhibits hepatic protein synthesis and promotes degradation of hepatic protein. Stimulation of gluconeogenesis further contributes to glucagon's catabolic effect on hepatic protein metabolism. Glucagon promotes protein catabolism in the liver, but it does not have any significant effect on blood amino-acid levels because it does not affect muscle protein, the major protein store in the body.

■ Glucagon secretion is increased during the postabsorptive state.

Considering the catabolic effects of glucagon on energy stores, you would be correct in assuming that glucagon secretion is increased during the postabsorptive state and decreased during the absorptive state, just the opposite of insulin secretion. In fact, insulin is sometimes called a "hormone of feasting" and glucagon a "hormone of fasting." Insulin tends to put nutrients in storage when their blood levels are high, such as following a meal, whereas glucagon promotes catabolism of nutrient stores between meals to keep up the blood nutrient levels, especially blood glucose. The levels of glucagon vary not only within an individual but among species; for example, the average concentration of glucagon in birds is approximately twice as great as in mammals, presumably because of the higher energy requirements of avian tissues.

Like insulin secretion, the major factor regulating glucagon secretion is a direct effect of the blood glucose concentration on the endocrine pancreas. In this case, the pancreatic α cells increase glucagon secretion in response to a fall in blood glucose. The hyperglycemic actions of this hormone tend to restore the blood glucose concentration to normal. Conversely, an increase in blood glucose concentration, such as after a meal, inhibits glucagon secretion, which likewise tends to restore the blood glucose concentration to normal.

■ Insulin and glucagon work as a team to maintain blood glucose and fatty acid levels.

Thus a direct negative-feedback relationship between blood glucose concentration and the α cells' rate of secretion, but it is in the opposite direction of the effect of blood glucose on the β cells; in other words, a rise in blood glucose level inhibits glucagon secretion but stimulates insulin secretion, whereas a fall in blood glucose level leads to increased glucagon secretion and decreased insulin secretion (● Figure 7–25). Because glucagon raises blood glucose and insulin decreases blood glucose, the changes in

secretion of these pancreatic hormones in response to deviations in blood glucose work together homeostatically to restore blood glucose levels to normal.

Similarly, a fall in blood fatty-acid concentration directly stimulates glucagon output and inhibits insulin output by the pancreas, both of which are negative-feedback control mechanisms to restore the blood fatty-acid level to normal.

The opposite effects exerted by the blood concentrations of glucose and fatty acids on the pancreatic α and β cells are appropriate for regulating the circulating levels of these nutrient molecules, because the actions of insulin and glucagon on carbohydrate and fat metabolism oppose one another. The effect of blood amino-acid concentration on the secretion of these two hormones is a different story. A rise in blood amino-acid concentration stimulates *both* glucagon and insulin secretion. Why this seeming paradox, because glucagon does not affect blood amino-acid concentration? The identical effect of high blood amino-acid levels on both glucagon and insulin secretion makes sense if you consider the concomitant effects these two hormones have on blood glucose levels (● Figure 7–26). If, during absorption of a protein-rich meal, the rise in blood amino acids stimulated only insulin secretion, hypoglycemia might result. Because little carbohydrate is available for absorption following consumption of a high-protein meal, the amino acid–induced increase in insulin secretion would drive too much glucose into the cells, causing a sudden, inappropriate drop in the blood glucose level. However, the simultaneous increase in glucagon secretion elicited by elevated blood amino-acid levels increases hepatic glucose production. Because the hyperglycemic effects of glucagon counteract the hypoglycemic actions of insulin, the net result is maintenance of normal blood-glucose levels (and prevention of hypoglycemic starvation of the brain) during absorption of a meal that is high in protein but low in carbohydrates.

■ Epinephrine, cortisol, growth hormone, and thyroid hormone also exert direct metabolic effects.

The pancreatic hormones are the most important regulators of normal fuel metabolism. However, several other hormones

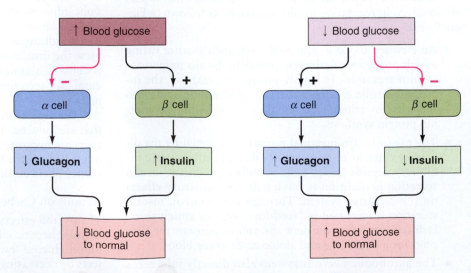

Figure 7–25 ● Complementary interactions of glucagon and insulin.

exert direct metabolic effects, even though control of their secretion is keyed to factors other than transitions in metabolism between feasting and fasting states (▮ Table 7–8).

As we discussed earlier, the stress hormones, epinephrine and cortisol, both increase blood levels of glucose and fatty acids through a variety of metabolic effects, primarily in stress situations. In addition, cortisol mobilizes amino acids by promoting protein catabolism in long-term stress, which includes starvation: Cortisol appears to be important to maintaining blood glucose concentrations during long-term starvation.

GH has protein anabolic effects in muscle. In fact, this is one of its growth-promoting features. Although GH can elevate the blood levels of glucose and fatty acids, it is normally of little importance to the overall regulation of fuel metabolism. Deep sleep, stress, exercise, and severe hypoglycemia stimulate GH secretion, possibly to provide fatty acids as an energy source and spare glucose for the brain under these circumstances. GH, like cortisol, appears to help maintain blood glucose concentrations during starvation.

Thyroid hormones are usually not important for fuel homeostasis, for two reasons. First, control of thyroid hormone secretion is not directed toward maintaining nutrient levels in the blood but rather responds to nutrient levels or lack thereof. Second, the onset of thyroid hormone action is too slow in many species to significantly affect the rapid adjustments required to maintain normal blood levels of nutrients.

Note that, with the exception of the anabolic effects of GH on protein metabolism, all the metabolic actions of these other hormones are opposite to those of insulin. Insulin alone can reduce blood glucose and blood fatty-acid levels, whereas glucagon, epinephrine, cortisol, and GH all increase blood levels of these nutrients. Because of this, these other hormones are considered **insulin antagonists**. Thus, the main reason diabetes mellitus (discussed next) has such devastating metabolic consequences is that no other control mechanism is available to pick up the slack to promote anabolism when insulin activity is insufficient, so the catabolic reactions promoted by other hormones are allowed to proceed unchecked. The only exception is protein anabolism stimulated by GH.

A common and major form of failed glucose homeostasis is **diabetes mellitus** (often called **diabetes** for short), which is manifested as abnormally high blood-glucose levels. Diabetes literally means "siphon" or "running through," a reference to the large urine volume accompanying the condition, caused by osmotic "attraction" from high glucose levels in the urine. Prolonged high-glucose concentrations have other effects, including blindness from lens cataracts (often blindness in a dog is the first symptom a pet owner notices). There are two distinct types of diabetes mellitus. **Type I (insulin-dependent) diabetes mellitus,** which accounts for about 10% of all cases of diabetes in humans, is characterized by a lack of insulin secretion. In dogs the incidence of Type I diabetes is greater in smaller breeds. Whether a diabetic tiger at the Pittsburgh zoo or pet housecat, because Type I diabetes is associated with a total or near-total lack of insulin secretion by pancreatic β cells, the animal requires exogenous insulin for survival. This dependence is the basis for the alternative name *insulin-dependent diabetes mellitus* for this form of the disease. In **Type II (non–insulin-dependent** or **maturity-onset) diabetes mellitus,** insulin secretion may be normal or even increased, but insulin's target cells are less sensitive than normal to this hormone.

Although either type can first be manifested in humans at any age, Type I is more prevalent in children, whereas the onset of Type II more generally occurs in adulthood. The diabetes-related death rate in the United States has increased by 30% since 1980, in large part because the incidence of the disease has been rising. About 1 in 400 domestic dogs and cats develop diabetes, and the incidence is rising in these animals as well.

Type I diabetes is an autoimmune process involving the erroneous, selective destruction of the pancreatic β cells

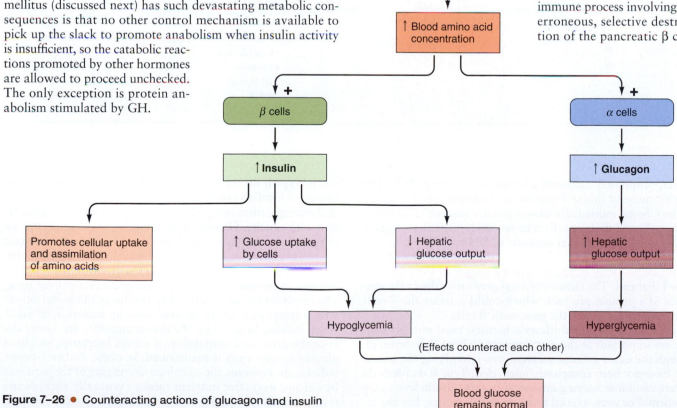

Figure 7–26 ● Counteracting actions of glucagon and insulin on blood glucose during absorption of a high-protein meal.

Table 7–8 ∎ Summary of Hormonal Control of Fuel Metabolism

Hormone	Major Metabolic Effects				Control of Secretion	
	Effect on blood glucose	*Effect on blood fatty acids*	*Effect on blood amino acids*	*Effect on muscle protein*	*Major stimuli for secretion*	*Primary role in metabolism*
Insulin	↓ +Glucose uptake +Glycogenesis −Glycogenolysis −Gluconeogenesis	↓ +Triglyceride synthesis −Lipolysis	↓ +Amino acid uptake	↑ +Protein synthesis −Protein degradation	↑ Blood glucose ↑ Blood amino acids	Primary regulator of absorptive and postabsorptive cycles
Glucagon	↑ +Glycogenolysis +Gluconeogenesis −Glycogenesis	↑ +Lipolysis −Triglyceride synthesis	No effect	No effect	↓ Blood glucose ↑ Blood amino acids	Regulation of absorptive and postabsorptive cycles in concert with insulin; protection against hypoglycemia
Epinephrine	↑ +Glycogenolysis +Gluconeogenesis −Insulin secretion +Glucagon secretion	↑ +Lipolysis	No effect	No effect	Sympathetic stimulation during stress and exercise	Provision of energy for emergencies and exercise
Cortisol	↑ +Gluconeogenesis −Glucose uptake by tissues other than brain; glucose sparing	↑ +Lipolysis	↑ +Protein degradation	↓ +Protein degradation	Stress	Mobilization of metabolic fuels and building blocks during adaptation to stress
Growth Hormone	↑ −Glucose uptake by muscles; glucose sparing	↑ +Lipolysis	↓ +Amino acid uptake	↓ +Protein synthesis −Protein degradation +Synthesis of DNA and RNA	Deep sleep Stress Exercise Hypoglycemia	Promotion of growth; normally little role in metabolism; mobilization of fuels plus glucose sparing in extenuating circumstances

↑ = increase
↓ = decrease

by inappropriately activated T lymphocytes (see p. 453). The precise cause of this self-destructive immune attack remains unclear. Some individuals have a genetic susceptibility to acquiring Type I diabetes. Recent research hints that a retrovirus—an RNA virus that converts to DNA when it invades a cell, then incorporates its genes into the genes of the individual it infects—may be the underlying trigger in some cases of Type I diabetes. The stowaway viral genes may direct the synthesis of a protein product, which could activate the T cells that erroneously attack the pancreatic β cells.

Various genetic and lifestyle factors, most notably obesity, are important in the development of Type II diabetes, although the exact cellular mechanisms underlying the condition have not been completely unraveled. Type II diabetics do secrete insulin in varying amounts. In fact, insulin levels may be normal or even exceed those in nondiabetics, but the affected individuals exhibit insulin resistance. The basic prob-

lem in Type II diabetes is not lack of insulin but reduced sensitivity of insulin's target cells to its presence, usually because of downregulation (see p. 259) of insulin receptors in association with obesity. Chronic overeating by an obese person results in secretion of increased amounts of insulin to maintain the blood glucose at normal levels by putting the excess nutrients in storage. In response to chronic hyperinsulinemia, the number of insulin receptors gradually decreases over time. The resultant fall in sensitivity to insulin in obese but otherwise normal individuals is overcome by secretion of additional insulin. In this way, the excess nutrients are stored despite the decreased availability of insulin receptors, so blood glucose homeostasis is maintained. In obese diabetes-prone individuals, however, the sustained overtaxing of the pancreas by chronic excessive nutrient intake eventually exceeds the reserve secretory capacity of the genetically weak β cells. Even though insulin secretion may be normal or somewhat elevated, symptoms of insulin insufficiency develop because the amount of insulin is still inadequate to prevent significant hyperglycemia in the presence of excess nutrient absorption.

Endocrine Control of Calcium Metabolism in Vertebrates

Besides regulating the concentration of organic nutrient molecules in the blood by manipulation of anabolic and catabolic pathways, the endocrine system also regulates the plasma concentration of a number of inorganic electrolytes. For example, aldosterone controls Na^+ and K^+ concentrations in the ECF. Three other hormones—*parathyroid hormone, calcitonin,* and *vitamin D*—control calcium (Ca^{++}) and phosphate (PO_4^{3-}) metabolism after their ingestion from the diet and/or uptake from the gill. These hormonal agents concern themselves with regulating plasma Ca^{++}, and in the process, plasma PO_4^{3-} is also maintained. Plasma Ca^{++} concentration is one of the most tightly controlled variables in vertebrates. The need for the precise regulation of plasma Ca^{++} stems from its critical influence on so many general physiological processes as well as from the additional demands imposed by puberty, lactation, or egg laying.

Calcium metabolism in water breathers is different from that in terrestrial vertebrates because, in addition to dietary Ca^{++}, the ion is also available from the aqueous environment. For example, in seawater the concentration of Ca^{++} is approximately 10 mmol/L (considerably greater than in cells), whereas in fresh water the concentration of Ca^{++} varies from 0.1 to 4 mmol/L (comparable to intracellular concentrations). Fish can transport Ca^{++} from either environment and maintain a positive calcium balance unless environmental concentrations of Ca^{++} are extremely low, or unless environmental pollutants restrict the functioning of this transporter.

▪ Plasma calcium must be closely regulated to prevent changes in neuromuscular excitability.

In mammals, about 99% of the Ca^{++} in the body is in crystalline form within the skeleton and teeth. Of the remaining 1%, about 0.9% is found intracellularly within the soft tissues; less than 0.1% is present in the ECF. Approximately half of the plasma Ca^{++} either is bound to plasma proteins and therefore restricted to the plasma or is complexed with PO_4^{3-} and not free to participate in chemical reactions. The other half of the plasma Ca^{++} is freely diffusible and can readily pass into the interstitial fluid and interact with the cells. The free Ca^{++} in the plasma and interstitial fluid is considered a single pool. Only this free Ca^{++} is biologically active and subject to regulation; it constitutes less than one thousandth of the total Ca^{++} in the body.

This small, freely diffusible fraction of ECF Ca^{++} plays a vital role in a number of essential activities, including the following:

1. *Neuromuscular excitability.* Even minor variations in the concentration of free ECF Ca^{++} can profoundly and immediately affect the sensitivity of excitable tissues. A fall in free Ca^{++} results in overexcitability of nerves and muscles; conversely, a rise in free Ca^{++} depresses neuromuscular excitability. These effects result from the influ-

ence of Ca^{++} on membrane permeability to Na^+. A decrease in free Ca^{++} increases Na^+ permeability, with the resultant influx of Na^+ moving the resting potential closer to threshold. Consequently, in the presence of **hypocalcemia** (low blood Ca^{++}) excitable tissues may be brought to threshold by normally ineffective physiological stimuli, so that skeletal muscles discharge and contract (go into spasm) "spontaneously" (in the absence of normal stimulation). If severe enough, spastic contraction of the respiratory muscles results in death by asphyxiation. Hypercalcemia (elevated blood Ca^{++}), in contrast, is also life threatening because it causes cardiac arrhythmias accompanied by generalized depression of neuromuscular excitability.

2. *Excitation-contraction coupling in cardiac and smooth muscle.* Ca^{++} is the primary trigger of the contractile mechanism in muscles, and arises from the ECF in cardiac and smooth muscle (Chapter 8). Note that a *rise in cytosolic Ca^{++}* within a muscle cell causes contraction, whereas an *increase in free ECF Ca^{++}* decreases neuromuscular excitability and reduces the likelihood of contraction occurring. Unless this point is kept in mind, it would be difficult to understand why low plasma Ca^{++} levels induce muscle hyperactivity when Ca^{++} is necessary to switch on the contractile apparatus. We are talking about two different Ca^{++} pools, which exert different effects.

3. *Stimulus–secretion coupling.* The entry of Ca^{++} into secretory cells, which results from increased permeability to Ca^{++} in response to appropriate stimulation, triggers the release of the secretory product by exocytosis. This process is important for the secretion of neurotransmitters by nerve cells (p. 124) and for peptide and catecholamine hormone secretion by endocrine cells.

4. *Maintenance of tight junctions between cells.* Calcium forms part of the intercellular cement that holds particular cells tightly together (p. 63).

5. *Clotting of blood.* Calcium serves as a cofactor in several steps of the cascade of reactions that lead to clot formation (p. 369).

In addition to these functions of free Ca^{++}, intracellular Ca^{++} serves as a second messenger in many cells and is involved in cell motility and cilia action (p. 57). Here, calcium levels cannot be allowed to rise too greatly, because elevated Ca^{++} precipitates readily with PO_4^{3-} as calcium phosphate (free PO_4^{3-} is a major anion in cells, necessary for ATP production, among other uses). Calcium phosphate is essential for bone and teeth formation, but is harmful inside a cell.

Because of the profound effects of deviations in free Ca^{++}, especially on neuromuscular excitability, the plasma concentration of this electrolyte is regulated with extraordinary precision in vertebrates. Let us see how.

▪ Control of calcium metabolism includes regulation of both calcium homeostasis and calcium balance.

Maintenance of the proper plasma concentration of free Ca^{++} differs from regulation of Na^+ and K^+ in two important regards. Na^+ and K^+ homeostasis is maintained primarily by regulating the urinary excretion of these electrolytes so that controlled output matches uncontrolled input (Chapters 12 and 13). In the case of Ca^{++}, in contrast, not all the ingested Ca^{++} is absorbed from the digestive tract; instead, the extent of absorption is hormonally controlled and is dependent on the Ca^{++} status of the body. In addition, bone serves as a large Ca^{++} reservoir that can be drawn on to maintain the free plasma Ca^{++} concentration within the narrow limits compatible with life should dietary intake become too low. Exchange of Ca^{++} between the ECF and bone is also subject to control. Similar in-house stores are not available for Na^+ and K^+.

Regulation of Ca^{++} metabolism depends on hormonal control of exchanges between the ECF and three other compartments serving as effectors: bone, kidneys, and intestine. Accordingly, control of Ca^{++} metabolism encompasses two aspects:

1. Regulation of **calcium homeostasis** involves the immediate adjustments required to maintain a *constant free plasma* Ca^{++} *concentration* on a minute-to-minute basis. This is largely accomplished by rapid exchanges between the bone and ECF and to a lesser extent by modifications in urinary excretion of Ca^{++}.

2. Regulation of **calcium balance** involves the more slowly responding adjustments required to maintain a *constant total amount of Ca^{++} in the body*. Control of Ca^{++} balance ensures that Ca^{++} intake is equivalent to Ca^{++} excretion over the long term (weeks to months). Calcium balance is maintained by adjusting the extent of intestinal Ca^{++} absorption and urinary Ca^{++} excretion.

Parathyroid hormone (PTH), the principal regulator of Ca^{++} metabolism, acts directly or indirectly on all three effector sites. It is the primary hormone responsible for maintaining Ca^{++} homeostasis and is essential for maintaining Ca^{++} balance, although vitamin D also contributes in important ways to Ca^{++} balance. The third Ca^{++}-influencing hormone, calcitonin, is not essential for routine maintenance of either Ca^{++} homeostasis or balance in birds and mammals, and its purpose is controversial. We examine the specific effects of each of these hormonal systems in more detail.

■ Parathyroid hormone raises free plasma calcium levels by its effects on bone, kidneys, and intestine.

Parathyroid hormone (PTH), the predominant hypercalcemic hormone in some amphibians, and reptiles, birds, and mammals, is a peptide hormone secreted by the **parathyroid glands,** four small glands located on the back surface of the mammalian thyroid gland, one in each corner. In birds the number of parathyroids varies between two and four and lie caudal to the thyroid. Like aldosterone, PTH is *essential for life.* The overall effect of PTH is to increase the Ca^{++} concentration of plasma (and, accordingly, of the entire ECF), thereby preventing hypocalcemia. In the complete absence of PTH, death ensues within a few days, usually because of asphyxiation caused by hypocalcemic spasm of respiratory muscles. By its combined actions on bone, kidneys, and intestine (effectors), PTH raises the plasma Ca^{++} level when it starts to fall so that

hypocalcemia and its effects are normally avoided. This hormone also acts to lower plasma PO_4^{3-} concentration. We consider each of these mechanisms next, beginning with an overview of bone remodeling and PTH's actions on bone.

■ Bone continuously undergoes remodeling.

Recall that 99% of the body's Ca^{++} is found in the skeleton as hydroxyapatite, $Ca_3(PO_4)_2$ (p. 272; see also ■ Table 7–9 for other functions of the skeleton.). Normally, $Ca_3(PO_4)_2$ salts are in solution in the ECF, but the conditions within the bone are suitable for these salts to precipitate (crystallize) around the collagen fibers in the matrix. By mobilizing some of these Ca^{++} stores in the bone, PTH raises the plasma Ca^{++} concentration when it starts to fall. Egg-laying birds have a highly labile reservoir of Ca^{++} in the form of **medullary bone,** which develops within the long bones in response to an increase in gonadal steroids at puberty. Medullary bone is the most exquisitely estrogen-sensitive form of vertebrate bone. The amount of Ca^{++} contained in each egg amounts to about 10% of the total body stores of Ca^{++}, and medullary bone accounts for approximately 30 to 40% of eggshell calcium.

Bone Remodeling

Despite the apparent inanimate nature of bone, bone constituents are continually being turned over. **Bone deposition** and **bone resorption** normally go on concurrently, so bone is constantly being remodeled, much as people remodel buildings by tearing down walls and replacing them. Bone remodeling serves two purposes: (1) It keeps the skeleton "engineered" for maximum effectiveness in its mechanical uses, and (2) it helps maintain the plasma Ca^{++} level. Continuous action of the *osteoclasts* (bone destroyers) and *osteoblasts* (bone builders) are responsible for remodeling. Some of this activity is regulated by *estrogen*, which is crucial for bone maintenance in both sexes. Estrogen activates the **osteoprotegerin (OPG)** gene in osteoblasts; OPG protein inhibits osteoclasts. New studies show that **TSH** also has a role (independently of T_3 and T_4): it stimulates both osteoclasts and osteoblasts and therefore the overall rate of bone remodeling.

During the growth phase, the bone builders keep ahead of the bone destroyers under the influence of GH and IGF-I (see p. 273). Mechanical stress also tips the balance in favor of bone deposition, increasing bone mass and strengthening the bones. Mechanical factors are responsible for adjusting the strength of bone in response to the demands placed on it. The greater the physical stress and compression to which a bone is subjected, the greater the rate of bone deposition. For example, the bones of human athletes are stronger and more massive than those of sedentary individuals.

Table 7–9 ■ Functions of the Vertebrate Skeleton
Support
Protection of vital internal organs
Assistance in body movement by giving attachment to muscles and providing leverage
Manufacture of blood cells (bone marrow)
Storage depot for Ca^{2+} and PO_4^{3-}, which can be exchanged with the plasma to maintain plasma concentrations of these electrolytes

By contrast, bone mass diminishes and the bones weaken when bone resorption gains a competitive edge over bone deposition in response to removal of mechanical stress. For example, bone mass decreases in people who undergo prolonged bed confinement or those in space flight, and in other animals forced to be sedentary during prolonged healing.

■ PTH's immediate effect is to promote the transfer of Ca^{++} from bone fluid into plasma.

In addition to the factors geared toward controlling the mechanical effectiveness of bone, throughout life PTH uses bone as a "bank" from which it withdraws Ca^{++} as needed to main-

tain the plasma Ca^{++} level. Let's examine more thoroughly PTH's actions in mobilizing intestinal Ca^{++} absorption from its labile and stable pools in bone as well as in adjustments in urinary Ca^{++}.

Most bone is organized into **osteon** units, each of which consists of a **central canal** surrounded by concentrically arranged lamellae. **Lamellae** are layers of collagen enclosing the osteocytes that originally produced the collagen (● Figure 7–27). Blood vessels penetrate the bone from either the outer surface or the marrow cavity and run through the central canals. Osteoblasts are present along the outer surface of the bone and along the inner surfaces lining the central canals. An extensive network of small, fluid-containing canals, the **canaliculi**, which allow substances to be exchanged between trapped osteocytes

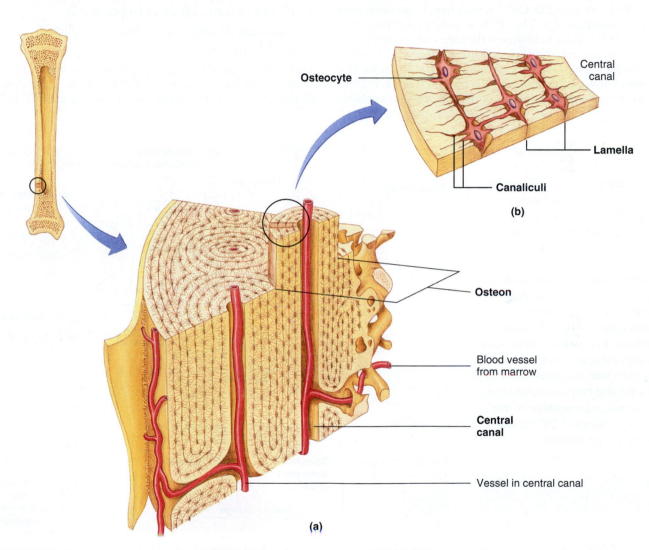

Figure 7–27 ● Organization of bone into osteons. (a) An osteon, the structural unit of most bone, consists of concentric lamellae (layers of osteocytes entombed by the bone they have deposited around themselves) surrounding a central canal. A small blood vessel branch traverses the central canal. (b) A magnification of the lamellae within an osteon. A network of small canals, the canaliculi, interconnect the entombed osteocytes with each other and with the central canal. Long cytoplasmic processes extend from osteocyte to osteocyte within the canaliculi.

(*Source*: Modified and redrawn from Fig. 6.3, p. 132 in *Human Anatomy and Physiology*, 3rd edition, by Alexander P. Spence and Elliot B. Mason. Copyright © 1987 by The Benjamin-Cummings Publishing Company, Inc. Reprinted by permission of Pearson Education, Inc.)

and the circulation, connects the surface osteoblasts and entombed osteocytes. These small canals also contain long, filmy cytoplasmic extensions of osteocytes and osteoblasts that are connected to each other, much as if these cells were "holding hands." The interconnecting cell network, which is called the **osteocytic–osteoblastic bone membrane,** separates the mineralized bone itself from the plasma within the central canals (● Figure 7–28a). The small labile pool of Ca^{++} is present in the *bone fluid* that lies between this bone membrane and the adjacent bone, both within the canaliculi and along the surface of the central canal.

PTH's initial effect is to activate membrane-bound Ca^{++} pumps located in the plasma membranes of the osteocytes and osteoblasts. These pumps promote movement of Ca^{++}, without the accompaniment of PO_4^{3-}, from the bone fluid into these cells. From here, this Ca^{++} is transferred into the plasma within the central canal. Thus PTH stimulates the transfer of Ca^{++} from the bone fluid across the osteocytic–osteoblastic bone membrane into the plasma. Movement of Ca^{++} out of the labile pool across the bone membrane accounts for the fast exchange between the bone and the plasma (Figure 7–28b). Because of the large surface area of the osteocytic-osteoblastic

membrane, small movements of Ca^{++} across individual cells are amplified into large Ca^{++} fluxes between the bone and the plasma.

After Ca^{++} is pumped out, the bone fluid is replenished with Ca^{++} from the partially mineralized bone along the adjacent bone surface. Thus the fast exchange of Ca^{++} does not involve the resorption of completely mineralized bone, and bone mass is not decreased. In this way, PTH draws Ca^{++} out of the "quick-cash branch" of the bone bank and rapidly increases the plasma Ca^{++} level without actually entering the bank (that is, without breaking down mineralized bone itself). Under normal conditions, this exchange is much more important for maintaining plasma Ca^{++} concentration than is the slow exchange.

■ PTH's chronic effect is to promote localized dissolution of bone to release Ca^{++} into plasma.

Under conditions of chronic hypocalcemia, such as might occur with dietary Ca^{++} deficiency, PTH influences the slow exchange of Ca^{++} between the bone itself and the ECF by pro-

Figure 7–28 ● Relationship of mineralized bone, bone cells, bone fluid, and the plasma. (a) Schematic representation of the osteocytic–osteoblastic bone membrane. The entombed osteocytes and surface osteoblasts are interconnected by long cytoplasmic processes that extend from these cells and connect to each other within the canaliculi. This interconnecting cell network, the osteocytic–osteoblastic bone membrane, separates the mineralized bone from the plasma in the central canal. Bone fluid lies between the membrane and the mineralized bone. (b) Schematic representation of fast and slow exchange of Ca^{2+} between the bone and the plasma.

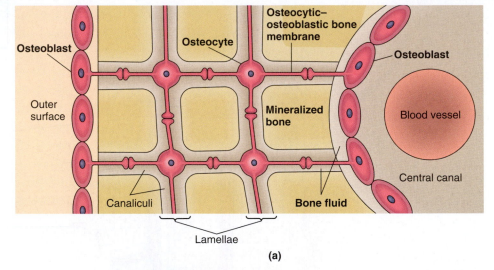

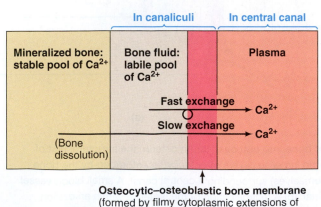

● In a fast exchange, Ca^{2+} is moved from the labile pool in the bone fluid into the plasma by means of PTH-activated Ca^{2+} pumps located in the osteocytic–osteoblastic bone membrane.

● In a slow exchange, Ca^{2+} is moved from the stable pool in the mineralized bone into the plasma by means of PTH-induced dissolution of the bone.

moting actual localized dissolution of bone. It does so by stimulating osteoclasts to gobble up bone, increasing the formation of more osteoclasts, and transiently inhibiting the bone-forming activity of the osteoblasts. Bone contains such a great abundance of Ca^{++} in comparison to the plasma (more than 1000 times as much) that even when PTH promotes increased bone resorption, there are no immediate discernible effects on the skeleton because such a minute amount of bone is affected. Yet the negligible amount of Ca^{++} "borrowed" from the bone bank can be lifesaving in terms of restoring the free plasma Ca^{++} level to normal. The borrowed Ca^{++} is then redeposited in the bone at another time when Ca^{++} supplies are more abundant. Meanwhile, the plasma Ca^{++} level has been maintained without sacrificing the integrity of the bone. However, prolonged excess PTH secretion over months or years is eventually evidenced by the formation of cavities throughout the skeleton that are filled with very large, overstuffed osteoclasts.

■ PTH acts on the kidneys to conserve Ca^{++} and to eliminate PO_4^{3-}.

When parathyroid hormone promotes dissolution of the calcium phosphate crystals in bone to harvest their Ca^{++} content, both Ca^{++} and PO_4^{3-} are released into the plasma. The calcium is useful, but the phosphate is not useful and may be harmful. To offset this, PTH also reduces PO_4^{3-} reabsorption by the kidneys during urine formation. PTH also enhances reabsorption by the kidneys of the filtered Ca^{++}. (It would be counterproductive to dissolve bone to obtain more Ca^{++} only to lose it in the urine). As a result of these two actions, PTH reduces plasma PO_4^{3-} concentrations at the same time it increases Ca^{++} concentrations.

This PTH-induced removal of extra PO_4^{3-} from the body fluids is essential for preventing reprecipitation of the Ca^{++} freed from the bone. Because of the solubility characteristics of calcium phosphate salt, the product of the plasma concentration of Ca^{++} times the plasma concentration of PO_4^{3-} must remain roughly constant. Therefore, an inverse relationship exists between the plasma concentrations of Ca^{++} and PO_4^{3-}; for example, when the plasma PO_4^{3-} concentration rises some of the plasma Ca^{++} is forced back into the bone through hydroxyapatite crystal formation, reducing the plasma Ca^{++} level and keeping the calcium phosphate product constant. This inverse relationship occurs because the free ions in the ECF are in equilibrium with the bone crystals.

Recall that both Ca^{++} and PO_4^{3-} are released from the bone when PTH promotes bone dissolution. Because PTH is secreted only when plasma Ca^{++} falls below normal, the released Ca^{++} is needed to restore plasma Ca^{++} to normal, yet the released PO_4^{3-} tends to increase plasma PO_4^{3-} levels above normal. If the plasma PO_4^{3-} level were allowed to rise above normal, some of the plasma Ca^{++} would have to be redeposited back in the bone along with the PO_4^{3-} in order to keep the calcium phosphate product constant. This redeposition of Ca^{++} would lower plasma Ca^{++}, just the opposite of the desired effect. Therefore, PTH acts on the kidneys to decrease the reabsorption of PO_4^{3-} by the renal tubules. This increases urinary excretion of PO_4^{3-} and lowers its plasma concentration. Such action prevents the self-defeating redeposition of released Ca^{++} back into the bone.

The third important action of PTH on the kidneys (besides increasing Ca^{++} reabsorption and decreasing PO_4^{3-}

reabsorption) is to enhance the activation of vitamin D by the kidneys.

■ PTH indirectly promotes absorption of Ca^{++} and PO_4^{3-} by the intestine.

Although PTH has no direct effect on the intestine, it indirectly increases both Ca^{++} and PO_4^{3-} absorption from the small intestine by means of its role in vitamin D activation. This vitamin, in turn, directly increases intestinal absorption of Ca^{++} and PO_4^{3-}.

■ The primary signal for regulating PTH secretion is the plasma concentration of free Ca^{++}.

All the effects of PTH are aimed at raising the plasma Ca^{++} levels. Appropriately, PTH secretion is increased in response to a fall in plasma Ca^{++} concentration and decreased by a rise in plasma Ca^{++} levels. The secretory cells of the parathyroid glands are directly and exquisitely sensitive to changes in free plasma Ca^{++}. Because PTH regulates plasma Ca^{++} concentration, this relationship forms a simple negative-feedback loop for controlling PTH secretion without involving any nervous or other hormonal intervention, with the glands serving as sensing integrators much like a house thermostat (● Figure 7–29).

■ Calcitonin lowers the plasma Ca^{++} concentration, but may not be essential.

Calcitonin (CT), the hormone produced by the C cells of the mammalian thyroid gland, is synthesized in an anatomically distinct gland in most vertebrates, the ultimobranchial gland. In birds the paired **ultimobranchial glands** lie posterior to the parathyroids whereas in fish calcitonin secreting cells are located in connective tissue sheets around the heart. In fish calcitonin has been suggested to promote bone formation by osteoblasts. Salmon CT is the most powerful form of calcitonin known in regulating calcium, and is being given to humans suffering from osteoporosis. However, its role in fish is still

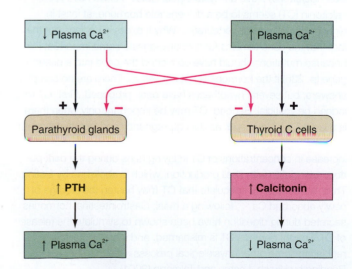

Figure 7–29 ● Negative-feedback loops controlling parathyroid hormone (PTH) and calcitonin secretion.

uncertain, and the most important hormone regulating calcium homeostasis in bony fish appears to be the hypocalcemic hormone, **stanniocalcin** (formerly called *hypocalcin*), which is synthesized in small organs termed the **corporacles of Stannius** located within the kidney. The ability to lower blood calcium is of greatest importance in seawater fishes, which are constantly exposed to the high levels of calcium present in seawater; parathyroid hormone does not occur in, nor does it elevate blood calcium in these fishes. In response to the release of stanniocalcin, calcium uptake at the gills and intestine is rapidly reduced. In mammals, calcitonin exerts a comparatively weak effect on plasma Ca^{++} levels.

Like PTH, calcitonin has two effects on bone, but in this case both effects *decrease* plasma Ca^{++} levels. First, on a short-term basis calcitonin decreases Ca^{++} movement from the bone fluid into the plasma. Second, on a long-term basis calcitonin decreases bone resorption by inhibiting the activity of osteoclasts. The suppression of bone resorption results in decreased plasma PO_4^{3-} levels as well as a reduced plasma Ca^{++} concentration. The hypocalcemic and hypophosphatemic effects of calcitonin are due entirely to this hormone's actions on bone. It has no effect on the kidneys or intestine.

As with PTH, the primary regulator of calcitonin release is the free plasma Ca^{++} concentration; an increase in plasma Ca^{++} stimulates and a fall in plasma Ca^{++} inhibits calcitonin secretion (Figure 7–29). Because calcitonin reduces plasma Ca^{++} levels, this system constitutes a second simple negative-feedback control over plasma Ca^{++} concentration, one that is antagonistic to the PTH system.

Most evidence suggests, however, that calcitonin plays little or no role in the normal control of Ca^{++} or PO_4^{3-} metabolism. Although calcitonin protects against hypercalcemia, this condition rarely occurs under normal circumstances. Moreover, neither thyroid removal nor calcitonin-secreting tumors alter circulating levels of Ca^{++} or PO_4^{3-}, implying that this hormone is not normally essential.

?

Does Calcitonin Have a Physiological Role? As we have seen, calcitonin (CT) seems to be a dispensable hormone, at least in higher and perhaps all vertebrates. Why it exists is unclear, but the fact that the gene remains functional suggests some important role, because mutations should have converted the gene into a pseudogene (p. 236) if the hormone is not necessary. There are no complete answers, but several hypotheses have been proposed. First, CT may indeed be vestigial. Second, CT may be important only when there is a large Ca^{++} demand, as during pregnancy and lactation in mammals and egg laying in birds. In support, there is a pronounced daily increase in concentrations of CT in laying hens during the dark period, associated with shell production, which is abolished by fasting. Third, some experts speculate that CT may hasten the storage of newly absorbed Ca^{++} following a meal: Gastrointestinal hormones secreted during digestion have been shown to stimulate the release of CT. Fourth, perhaps CT is misnamed, and it has a vital role in regulating some other physiological process. See the Suggested Reading by Hirsch, Lester, and Talmage (2001).

■ Vitamin D is actually a hormone that increases calcium absorption in the intestine.

The final factor involved in the regulation of Ca^{++} metabolism in vertebrates is cholecalciferol, or vitamin D, a steroid-like compound that is essential for Ca^{++} absorption in the intestine. The livers of marine teleosts are well known to contain high concentrations of vitamin D, which has traditionally been used as a nutritional supplement. Strictly speaking, vitamin D should be considered a hormone because it can be produced in the skin from a precursor related to cholesterol (7–dehydrocholesterol) on exposure to sunlight. It is subsequently released into the blood to act at a distant target site, the intestine. The skin, therefore, is actually an endocrine gland and vitamin D is a prohormone. In 1931 the researcher H.-C. Hou found that removing the *preen gland* from chicks caused rickets to develop even if the chicks were exposed to sunlight. From this observation he deduced that birds synthesize a provitamin D from the preen gland onto their feathers where it is converted into vitamin D. Traditionally, however, this chemical messenger has been named a vitamin—that is, a dietary supplement—for two reasons. First, scientists originally discovered and isolated it from a dietary source and tagged it as a vitamin. Second, even though the skin would be an adequate source of vitamin D if it were exposed to sufficient sunlight, indoor dwelling and clothing in response to cold weather and social customs preclude significant exposure of the skin to sunlight most of the time in humans. At least part of the essential vitamin D must therefore be derived from dietary sources, at least in humans.

Activation of Vitamin D

Regardless of its source, vitamin D is biologically inactive when it first enters the blood from either the skin or the digestive tract. It must be activated by two sequential biochemical alterations that involve the addition of two hydroxyl (–OH) groups (● Figure 7–30). The first of these reactions occurs in the liver and the second in the kidneys. The end result is production of the active form of vitamin D, *1,25-(OH)$_2$-vitamin D$_3$*, also known as **calcitriol**. In response to a fall in plasma Ca^{++}, PTH stimulates the kidney enzyme 1 α-hydroxylase that is involved in the second step of vitamin D activation. To a lesser extent, a fall in plasma PO_4^{3-} also enhances the activation process.

Recent studies have also shown that IGF-I can also activate 1α-hydroxylase, an important mechanism linking calcium homeostasis to periods of rapid mammalian growth.

Function of Vitamin D

The most dramatic and biologically important effect of activated vitamin D is to increase Ca^{++} absorption in the intestine. Calcitriol induces RNA transcription and synthesis of proteins including the calcium-binding protein, **calbindin D$_{28k}$**. For example, the onset of egg lay in birds coincides with increased 1,25-(OH)$_2$-vitamin D$_3$ concentrations, which elevates concentrations of calbindin and increases intestinal calcium transport. Unlike most dietary constituents, dietary Ca^{++} is not indiscriminately absorbed by the digestive system. In fact, the majority of ingested Ca^{++} is typically not absorbed but is lost instead into the feces. When needed, more dietary Ca^{++} is absorbed into the plasma under the influence of calcitriol.

Independent of its effects on Ca^{++} transport, calcitriol also increases intestinal PO_4^{3-} absorption. Furthermore, calcitriol increases the responsiveness of bone to PTH. Thus vitamin D and PTH have a closely interdependent relationship (● Figure 7–31). Vitamin D may be used in some invertebrates; a role for steroids in Ca^{++} absorption in the snail *Helix aspersa* has been demonstrated, with vitamin D shown to be more effective than cholesterol.

PTH is principally responsible for controlling Ca^{++} homeostasis, because the actions of calcitriol are too sluggish for it to contribute substantially to the minute-to-minute regulation of plasma Ca^{++} concentration. However, both PTH and calcitriol are essential to Ca^{++} balance, the process that ensures that, over the long term, Ca^{++} input into the body is equivalent to Ca^{++} output. When dietary Ca^{++} intake is reduced, the resultant increase in PTH activates calcitriol, which increases the efficiency of uptake of ingested Ca^{++}.

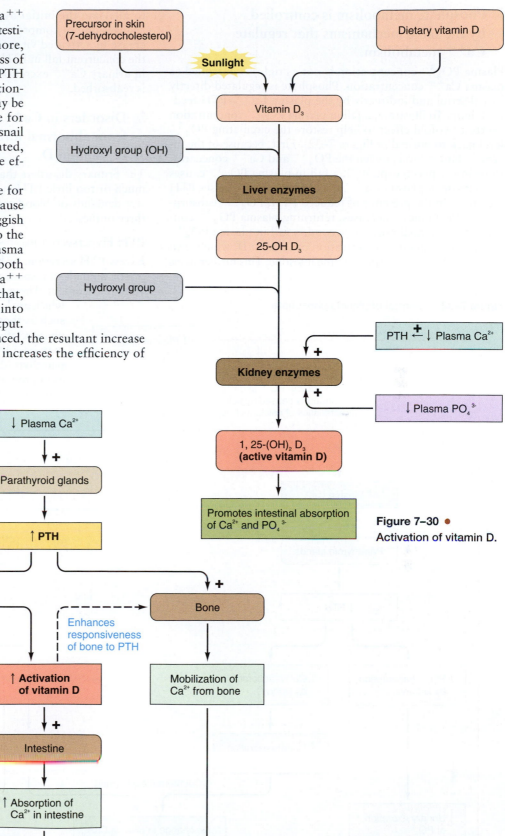

Figure 7–30 ● Activation of vitamin D.

Figure 7–31 ● Interactions between PTH and vitamin D in controlling plasma calcium.

Phosphate metabolism is controlled by the same mechanisms that regulate Ca⁺⁺ metabolism.

Plasma PO_4^{3-} concentration is not as tightly controlled as plasma Ca^{++} concentration. Phosphate is regulated directly by calcitriol and indirectly by the plasma Ca^{++}–PTH feedback loop. To illustrate, a fall in plasma PO_4^{3-} concentration exerts a twofold effect to help restore the circulating PO_4^{3-} level back to normal (● Figure 7–32). First, because of the inverse relationship between the PO_4^{3-} and Ca^{++} concentrations in the plasma (p. 307), a fall in plasma PO_4^{3-} causes an increase in plasma Ca^{++}, which directly suppresses PTH secretion. In the presence of reduced PTH, PO_4^{3-} reabsorption by the kidneys increases, returning plasma PO_4^{3-} concentration toward normal. Second, a fall in plasma PO_4^{3-} also results in increased activation of vitamin D, which then promotes PO_4^{3-} absorption in the intestine. This further helps

to alleviate the initial hypophosphatemia. Note that these changes do not compromise Ca^{++} balance. Although the increase in activated vitamin D stimulates Ca^{++} absorption, the concurrent fall in PTH produces a compensatory increase in urinary Ca^{++} excretion because less of the filtered Ca^{++} is reabsorbed.

Disorders in Ca⁺⁺ metabolism may arise from abnormal levels of parathyroid hormone or vitamin D.

The primary disorders that affect Ca^{++} metabolism are too much or too little PTH, a deficiency of vitamin D, or excessive demands on blood calcium concentrations. Let's look at three of these.

PTH Hypersecretion

Excess PTH secretion, or hyperparathyroidism, occurs in domestic animals in response to feeding diets low in calcium or vitamin D content. These may be all-meat diets for carnivores, which are low in calcium, or inappropriate diets, such as those for horses that are high in phosphate relative to calcium. Less frequently there may be a hypersecreting tumor in one of the parathyroid glands. The affected animal can be asymptomatic or symptoms can be severe, depending on the magnitude of the problem. The

Figure 7–32 ● Control of plasma phosphate.

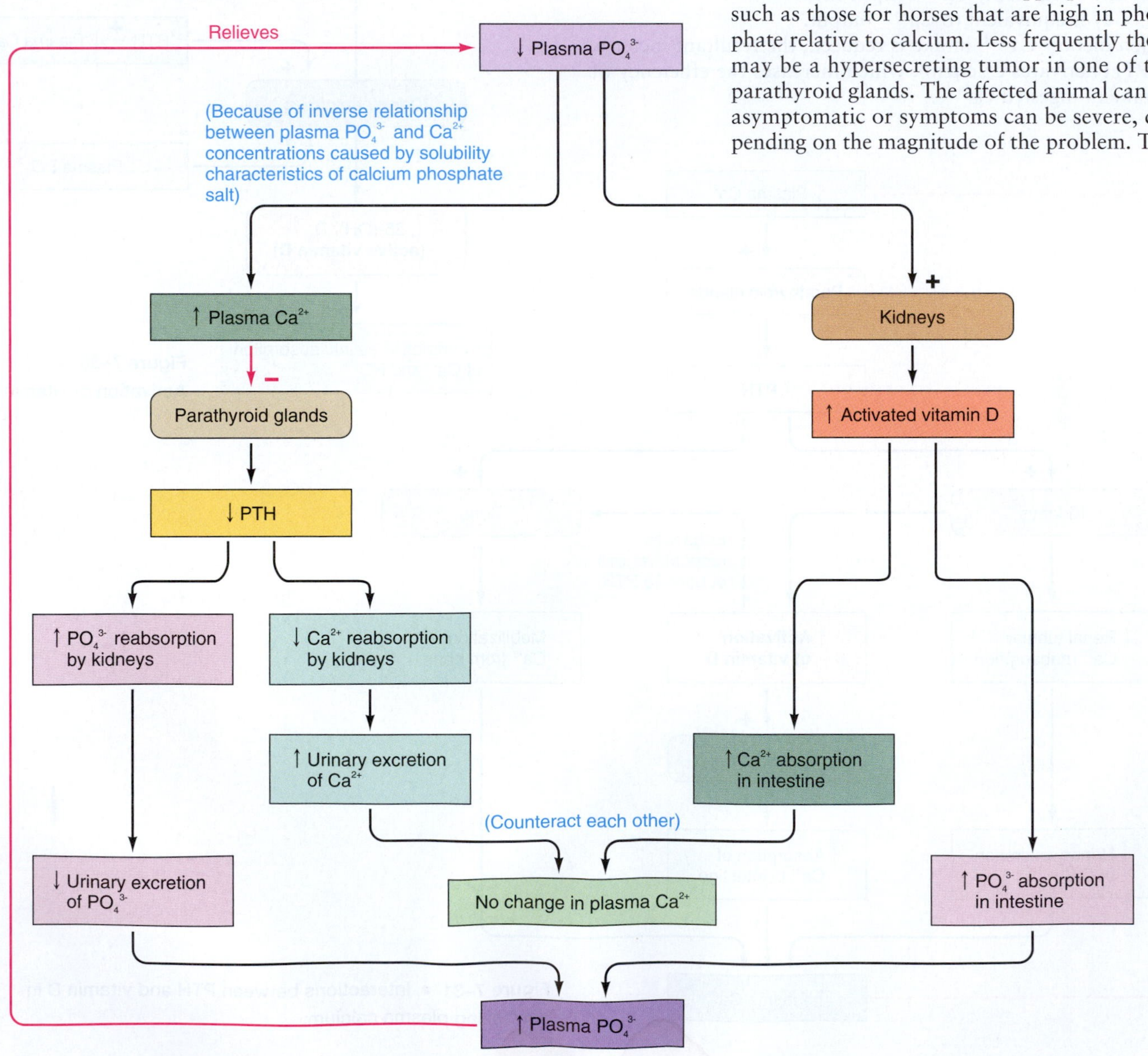

following are among the possible consequences of the ensuing hypercalcemia and hypophosphatemia:

- *Reduced excitability of muscle and nervous tissue* leads to muscle weakness and neurological disorders. Cardiac disorders may also occur.

- *Excessive mobilization of Ca^{++} and PO_4^{3-} from skeletal stores* leads to thinning of bone, which may result in skeletal deformities and increased incidence of fractures.

- *An increased incidence of Ca^{++}-containing kidney stones* occurs because the excess quantity of Ca^{++} being filtered through the kidneys may precipitate and form stones. These stones may impair renal function. Passage of the stones through the ureters causes extreme pain.

Because of these potential multiple consequences, hyperparathyroidism has been called a disease of "bones, stones, and abdominal groans."

Vitamin D Deficiency

The major consequence associated with vitamin D deficiency is impaired intestinal absorption of Ca^{++}. In the face of reduced Ca^{++} uptake, PTH maintains the plasma Ca^{++} level at the expense of the bones. As a result, the bone matrix is not properly mineralized because Ca^{++} salts are not available for deposition. The demineralized bones become soft and deformed, bowing to the pressures of weight bearing. This condition is known as **rickets** in prepubertal mammals and **osteomalacia** in adults. In the absence of vitamin D, amphibians (such as South African clawed toads) and reptiles are particularly susceptible to the development of rickets.

Excessive Demands for Calcium

We end our discussion of disorders of calcium metabolism with a look at the effects of excessive demands. These occur most notably in farm animals bred for two specialized processes: milk production by dairy cows, and egg laying by hens.

Parturient Paresis (milk fever, or parturient hypocalcemia). This is the most common metabolic disorder affecting high-producing dairy cattle and provides a dynamic example of calcium metabolism during the immediate postpartum period. Because colostrum contains high concentrations of calcium (2g/L) and approximately 3g of calcium are required per hour to produce colostrum, the inability of the dairy cow to adequately mobilize calcium reserves results in the characteristic symptoms of milk fever, which include restlessness, anxiety, anorexia, uncoordination, and a lack of interest in their calves. The incidence of milk fever increases with age and yield of the animals. Affected cattle usually demonstrate symptoms within 72 hours following parturition. Without intervention the cows progress to a second stage, which is manifested by recumbency. The cows lie on their sternums with their heads turned back toward their flanks, breathing is slow and labored, and body temperatures are below normal. In the final stage of parturient paresis the cows are stretched out on their sides as they progress from dullness to coma.

If the cows are fed an *increased level* of calcium in the diet during the *prepartum* period, then which hormone is predominant during the early lactation period? The answer is *calcitonin*, which functions to block the resorption of calcium by PTH at the bone. Consider the profound impact on plasma calcium concentrations as colostrum begins to be produced.

The sudden demand for calcium—more than all the calcium present in the blood—results in a dramatic decline in calcium concentrations in the blood, which triggers an enormous output of PTH from the parathyroid gland. Because of the elevated plasma-calcitonin concentrations, PTH can exert its effect primarily at the intestinal level, which cannot fill the increased demands placed on it. (Note that this does not necessarily reveal an important natural role for calcitonin, because this is an artificial situation of feeding calcium to a cow bred for high milk production.) However, if the cow is placed on a *low-calcium diet* (supplemented with vitamin D) during the prepartum period, then concentrations of PTH predominate during parturition and they are eminently more successful in mobilizing calcium from the bone reservoirs. Acting in concert with vitamin D to increase concentrations of calcium-binding protein in the intestinal epithelia and calcium transport across the intestine, PTH now acts uninhibitedly on bone to effectively combat the decline in blood calcium concentrations, and thus dramatically reduces the incidence of milk fever.

Egg Lay in Female Birds. Comparable to these events, which drain calcium from the blood of lactating dairy cows, is the situation in egg-laying birds. In the laying hen an amount of calcium equivalent to 8 to 10% of the total calcium in her body is secreted into the shell each day on which an egg is laid (**oviposition**). Researchers believe the shell, composed primarily of calcium carbonate, evolved from the soft cover of eggs of ancestral reptiles into the calcareous shell to give a measure of protection from attacks by predators and pathogens (including soil organisms). PTH can rapidly mobilize calcium from the medullary bone whenever the rate at which calcium deposition on the shell exceeds the rate at which it is absorbed from the intestinal tract. Thus, as expected, there is a reciprocal relationship between the concentrations of PTH and Ca^{++} during the egg cycle. During the two-week period prior to the onset of lay, a whole new system of secondary bone, medullary bone (p. 304), is laid down in the marrow cavities of most bones under the combined influence of estrogens and androgens secreted from the developing follicles. Interestingly, calcitriol does not stimulate the synthesis of a calcium transporter (Ca^{++}-ATPase) in the shell gland of a laying bird (unlike the situation in the intestinal epithelial cells) but rather its produc-

Dairy cows can develop milk fever from the excessive demands for calcium in milk production.

Photo: Courtesy of USDA/Agricultural Research Service

tion is regulated by estrogen. When a bird goes "out of lay," concentrations of estrogen decline in concert with the reduction in shell gland Ca^{++}-ATPase activity.

Chapter in Perspective:
HOMEOSTASIS AND INTEGRATION

The endocrine system is one of an animal's two major control systems, the other being the nervous system (although, as you will see in Chapter 10, some physiologists consider the immune system a third major control system). Through its relatively slow-acting hormonal messengers, the endocrine system generally regulates activities that require duration rather than speed, by activating existing proteins in effector cells or activating genes in effector cells to produce new proteins. Most of these activities are directed toward maintaining *homeostasis,* as exemplified by the following:

- Hormones help maintain the proper concentration of nutrients in the internal environment by directing chemical reactions involved in the cellular uptake, storage, and release of these molecules (such as vertebrate insulin and glucagon, crustacean hyperglycemic hormone). Furthermore, the rate at which these nutrients are metabolized is controlled in large part by the endocrine system (thyroid hormones).

- The control of H_2O balance, which is responsible for maintaining ECF osmolarity and proper cell volume, is largely accomplished by hormonal regulation of H_2O reabsorption by the kidneys during urine formation (such as vasopressin).

- Salt balance, which is important in the maintenance of ECF volume and arterial blood pressure, is achieved by hormonally controlled adjustments in salt reabsorption by the kidneys during urine formation (aldosterone). Likewise, hormones act on various target cells to maintain the plasma concentration of calcium and other electrolytes (parathyroid hormone).

- The endocrine system orchestrates a wide range of adjustments that help the body maintain homeostasis in response to stressful situations (epinephrine, cortisol).

- The endocrine and nervous systems work in concert to control the circulatory and digestive systems, which in turn carry out important homeostatic activities.

In addition to homeostasis, hormones direct *regulated changes* involved in growth processes and control most aspects of the reproductive system (for example, juvenile hormone and ecdysone in insects, thyroxine in amphibians, growth hormone, gonadotropins and sex steroids in all vertebrates, prolactin, and lactation in mammals).

Hormones often act in negative-feedback fashion to resist the change that induced their secretion, thus maintaining stability in the internal environment. ■

REVIEW QUESTIONS *(Answers are on p. A–1.)*

Additional study tools for this chapter, including chapter summaries and practice tests, are available online at *www.biology .brookscole.com*

1. In which class of hormones would iodinated tyrosine derivatives belong?
   a. peptide hormones
   b. protein hormones
   c. steroids
   d. amines
   e. catecholamines

2. All hormones are
   a. transported dissolved in blood plasma
   b. hydrophilic or hydrophobic
   c. reversibly bound to plasma proteins
   d. derived from cholesterol
   e. synthesized by enzymes in the Golgi apparatus

3. In insects, the hormones involved in metamorphosis are
   a. ecdysone and juvenile hormone
   b. juvenile hormone and vitellin
   c. allostatin and allotropin
   d. precocenes and vitellogenin
   e. ecdysone and allotropin

4. Which of the following is a part of the vertebrate brain that is important in establishing biological rhythms?
   a. hypothalamus
   b. corpora allata
   c. pituitary gland
   d. pineal gland
   e. sinus glands

5. The hormone secreted by the pineal gland that contributes to circadian regulation is
   a. melanin
   b. vasopressin
   c. melatonin
   d. oxytocin
   e. prolactin

6. Which of the following hormones enhances retention of water by the kidneys?
   a. oxytocin
   b. α-MSH
   c. adrenocorticotropic hormone
   d. vasopressin
   e. luteinizing hormone

7. Hypophysiotropic hormones from the hypothalamus
   a. ensure that anterior pituitary hormones are released at a constant rate
   b. are directly regulated by hormones released from the anterior pituitary

c. always stimulate the release of hormones from the anterior pituitary

d. always inhibit the release of hormones from the anterior pituitary

e. can inhibit or stimulate release of hormones from the anterior pituitary

8. Which hormone produced by the adrenal cortex affects blood pressure?
   a. cortisol
   b. aldosterone
   c. corticosterone
   d. dehydroepiandrosterone (DHEA)
   e. glucocorticoids

9. Epinephrine release from the adrenal medulla is caused by
   a. stimulation by the anterior pituitary gland
   b. stimulation by the adrenal cortex
   c. stimulation by the sympathetic nervous system
   d. stimulation by the hypothalamus
   e. high blood glucose levels

10. The mulitfaceted general stress response
    a. is coordinated through release of epinephrine from the adrenal cortex
    b. is coordinated by the sympathetic nervous system
    c. is coordinated by the adrenal cortex and the adrenal medulla
    d. is coordinated by the hypothalamus
    e. is coordinated by the anterior pituitary gland

11. The reactions in cells involving the degradation, synthesis, and transformation of energy-rich molecules is called
    a. anabolism
    b. glycogenesis
    c. intermediary metabolism
    d. catabolism
    e. none of the above

12. The dominant hormones responsible for glucose homeostasis are
    a. cortisol and corticosterone
    b. epinephrine and thyroid hormone
    c. oxytocin and ecdysone
    d. thyrotropin and prolactin
    e. insulin and glucagon

13. Diabetes can result from
    a. autoimmune destruction of pancreatic beta-cells
    b. overproduction of insulin
    c. reduced sensitivity of insulin's target cells
    d. decreased levels of glucagon
    e. a and c

14. Parathyroid hormone adjusts calcium and phosphate levels in the blood by
    a. increasing urinary excretion of phosphate
    b. activating calcium pumps in the plasma membranes of osteocytes and osteoblasts
    c. decreasing calcium reabsorption in kidneys
    d. fast extraction of calcium by decreasing bone mass
    e. a and b

15. The most important biological effect of activated vitamin D is to
    a. decrease calcium absorption in the intestine
    b. increase calcium absorption in the intestine
    c. block the responsiveness of bone to PTH
    d. decrease absorption of phosphate in the intestine
    e. none of the above

SUGGESTED READINGS AND INTERNET SITES

Barton, B. A. 2002. Stress in fishes: A diversity of responses with particular reference to changes in circulating corticosteroids. *Integrative and Comparative Biology* 42:517–525.

Cerami, A., H. Vlassara, & M. Brownlee. 1987. Glucose and aging. *Scientific American* 256, 90–96.

Gekakis, N., & C. J. Witz. 1998. Role of the CLOCK protein in the mammalian circadian mechanism *Science* 280:564–569.

Hadley, M. E. 1992. *Endocrinology*, 3rd ed. Englewood Cliffs, NJ: Prentice Hall.

Hirsch, P. F., G. E. Lester, & R. V. Talmage. 2001. Calcitonin, an enigmatic hormone: Does it have a function? *Journal of Musculoskeletal Neuronal Interactions* 1:299–305.

Insel, T. R., J. T. Winslow, Z. Wang, & L. J. Young. 1998. Oxytocin, vasopressin, and the neuroendocrine basis of pair bond formation. *Advances in Experimental Medicine and Biology* 449:215–224.

Klandorf, H. Boyce, C. S. Killefer, J., and others. 1997. The effect of photoperiod and food intake on daily changes in plasma calcitonin in broiler breeder hens. *General and Comparative Endocrinology* 107:327–340.

Lea, R. B., & H. Klandorf. 2002. The brood patch. In C. Deeming, ed., *Avian Eggs and Incubation*. Oxford, UK: Oxford University Press.

Thornton, J. W., E. Need, & D. Crews. 2003. Resurrecting the ancestral steroid receptor: Ancient origin of estrogen signaling. *Science* 301:1714–1718.

Whitmore, D., and others. 2000. Light acts directly on organs and cells in culture to set the vertebrate circadian clock. *Nature* 404:87–91.

INTERNET SITES

Kimball, J. W. *Circadian Rhythms*. 2003. **users.rcn.com/jkimball.ma.ultranet/BiologyPages/C/Circadian.html**. Diagrams and descriptions of clock genes in fruit flies and mice.

National Academy of Sciences. *Insect pheromones: Mastering communication to control pests*. 2003. **beyonddiscovery.org/content/view.article.asp?a=2702**. A National Academy of Sciences site on the discovery and function of insect pheromones, and their role as targets in pest control.

INFOTRAC READINGS

Travis, J. 1996. Insect hormone inspires switch for genes. *Science News* 149:246.

Muscle Physiology

Photo: Paul Yancey

A kangaroo has powerful muscles in its hind legs, coupled with efficient tendons that store and release energy during hopping. After reading this chapter, make an hypothesis on what type of muscle fibers you would expect to find in its hind limbs.

Introduction

Almost all living cells have rudimentary intracellular machinery for producing movement, including that associated with the redistribution of various cell components during cell division (Chapter 2). White blood cells, amoebocytes, and amoebas use intracellular contractile proteins to propel themselves through their environment. Even some plants, such as the insectivorous Venus flytrap, can contract portions of their structure by changing the turgor pressure of particular cells. The contraction specialists of an animal's body, however, are the muscle cells. All animal phyla above the Porifera (that is, sponges, which have no true organs) contain clearly identifiable muscle cells. By far the most important effectors for animal behavior are the muscles and their inherent functional capacities, which classically include their ability to do work and produce force. In addition, muscle performance is governed by the speed of their responses and their efficiency in converting chemical to mechanical energy.

Recall from Chapter 1 (p. 6) that there are three types of muscle: *skeletal muscle, cardiac muscle,* and *smooth muscle.* Through their highly developed ability to contract, muscle cells can shorten and develop tension, which enables them to produce movement. In contrast to sensory systems, which transform other forms of energy in the environment into electrical signals, muscles, in response to electrical signals, convert the chemical energy of ATP into mechanical energy that can act on the environment. Controlled contraction of muscles permits the following:

1. Purposeful locomotory movement of the whole body or parts of the body in relation to the environment (such as walking or flying)

2. Manipulation of external objects (such as a monkey poking a straw or stick into a termite mound, up which the termites then climb)

3. Propulsion of contents through various hollow internal organs (such as circulation of blood or movement of materials through the digestive tract)

4. Emptying the contents of certain organs to the external environment (such as urination or parturition)

5. Production of heat as a metabolic by-product, which may help to maintain internal body temperatures above ambient

6. Production of sound

All muscles share numerous underlying characteristics, although there is considerable diversity among muscle cells from different animals requiring very different outputs. Examples of this diversity include fish sound-producing muscles that can contract at frequencies greater than 100 Hz (or cycles per second), contraction frequencies unobtainable by their locomotory muscle counterparts. The highest frequency for vertebrate locomotory muscles is an order of magnitude lower, 25 to 30 Hz for mouse and lizard fast-twitch muscles at body temperatures of 35°C. In contrast, sonic muscles of toadfish perform work at oscillation frequencies in excess of 200 Hz (at a lower body temperature of 25°C). Hamlet, a species of coral-reef fish, broadcast a series of beeps and pulses ranging from 350 to 650 Hz at the critical moment when sperm and eggs are released into the environment.

You will discover that the ability to increase the speed of contraction is accomplished, in part, by *reapportionment* of key intracellular structures in addition to shifts in certain muscle *isoforms* (proteins with the same basic function but which differ slightly in their structures). Tradeoffs that maximize performance of one function in a particular species are usually at the expense of another. Normally, muscles designed to operate at high frequencies are not effective at low frequencies and generate less force. However, adaptations for flight require both high frequencies of muscle contraction in addition to the generation of considerable amounts of force. Just how birds and insects arrived at a solution to this dilemma is considered later in this chapter.

In most vertebrates, muscle is the largest group of tissues in the body, for example, accounting for approximately half of the body's weight in humans. Skeletal muscle alone makes up about 40% of body weight in human men and 32% in women, with smooth and cardiac muscle making up another 10% of the total weight. Comparably, in most fishes skeletal muscle mass varies from 20 to 50%, whereas in species noted for high rates of acceleration (such as barracudas), muscle mass ranges from 55 to 65%. In hummingbirds, flight muscles *alone* account for 25% of the animal's body weight.

Although the three muscle types are structurally and functionally distinct, they can be classified in two different ways according to their common characteristics (● Figure 8–1). First, muscles are categorized as **striated** (skeletal and cardiac muscle) or **unstriated** (smooth muscle), depending on whether alternating dark and light bands, or striations, can be seen when the muscle is viewed under a light microscope. Second, vertebrate muscles are categorized as **voluntary** (skeletal muscle) or **involuntary** (cardiac and smooth muscle), depending respectively on whether they are innervated by the somatic nervous system and are controlled in part by the motor cortex, or are innervated by the autonomic nervous system and are not subject to motor cortex control (note that *voluntary* and *involuntary* are human terms; see p. 148).

Most of this chapter is devoted to a detailed examination of the most abundant and best understood muscle, skeletal muscle. Skeletal muscles make up the **muscular system.** We begin by discussing skeletal muscle structure, then examine how it works from the molecular through the cellular level, and finally look at specific functions of the whole muscle. The chapter concludes with a discussion of the unique properties of smooth and cardiac muscle in comparison to skeletal muscle.

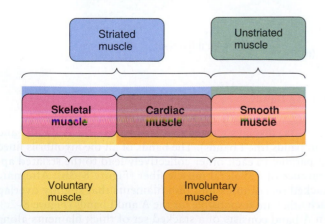

Figure 8–1 ● **Categorization of muscle.**

Skeletal Muscle

■ Skeletal muscle fibers have a highly organized internal arrangement that creates a striated appearance.

A basic understanding of the structural components of a skeletal muscle fiber is essential to understanding how muscles contract. A single skeletal muscle cell, known as a **muscle fiber,** is relatively large, elongated, and cylindrical, measuring from 10 to 100 micrometers (1 μ = 1 millionth of a meter) in diameter; they can be tens of centimeters in length in a large animal. A skeletal muscle consists of a number of muscle fibers lying parallel to each other and bundled together by connective tissue (● Figure 8–2a). The fibers usually extend the entire length of the muscle. During embryonic development, the huge skeletal-muscle fibers are formed by the fusion of many smaller cells; thus, one striking feature is the presence of multiple nuclei in a single muscle cell. This multinucleate muscle cell, produced by a merging of cells, is termed a **functional syncytium** (*syn,* "several"; *cyt,* "cell"). Another feature is the presence of mitochondria, the energy-generating organelles (p. 44), as would be expected with the energy demands of a tissue as active as skeletal muscle. Mitochondrial volume in most skeletal-muscle fibers lies between 4 and 42%, with low values found in muscles relying on anaerobic metabolism (so-called white muscle), and high values in those relying on aerobic metabolism (so-called red muscle; we discuss the details of white and red muscle later). For example, mitochondrial volume of the tail shaker muscle in rattlesnakes (*Crotalus*) is 26%. However, there are extreme values (p. 50), as low as 1% in some white fish muscle and as much as 50% in the flight muscle of locusts.

■ Skeletal muscle fibers are striated by a highly organized internal arrangement.

The most predominant structural feature of most skeletal muscle fibers is the presence of numerous **myofibrils.** These specialized contractile elements, which constitute up to 90% of the volume of the muscle fiber, are cylindrical intracellular structures, typically 1 μm in diameter, that extend the entire length of the muscle fiber (Figure 8–2b). The greater the volume of the fiber taken up by myofibrils, the greater the force per cross-sectional area that can be generated. Muscle fibers designed with a low percentage of myofibrils cannot generate high degrees of tension but are generally associated with the ability to turn muscles on and off quickly or to generate prolonged activity. For example, the aerobic sound-producing muscle of cicadas has a myofibrillar volume of only 22%, whereas the high-frequency tail-shaker muscle of the rattlesnake contains only 31% myofilaments. The tail shaker muscle consists of six separate bands of red, aerobic muscle fibers, groups of which alternately contract and relax, causing the rattle to move from side to side.

Each myofibril consists of a regular arrangement of highly organized cytoskeletal elements—the thick and thin filaments (Figure 8–2c). The **thick filaments,** which are 12 to 18 nm in diameter and 1.6 μm in length, are special assemblies of the protein **myosin,** whereas the **thin filaments,** which are 5 to 8 nm

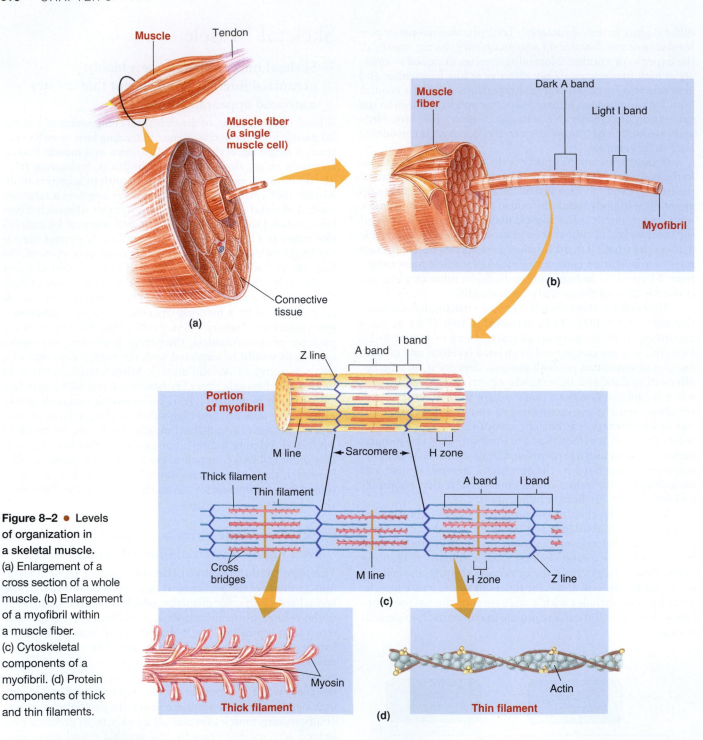

Figure 8–2 ● Levels of organization in a skeletal muscle. (a) Enlargement of a cross section of a whole muscle. (b) Enlargement of a myofibril within a muscle fiber. (c) Cytoskeletal components of a myofibril. (d) Protein components of thick and thin filaments.

in diameter and 1.0 μm long, are made up primarily of the protein **actin** (with other proteins) (Figure 8–2d). These same proteins are found in all other cells of the body but in smaller quantities and in a less organized fashion. The levels of organization in a skeletal muscle can be summarized as follows:

Whole muscle and bundles		muscle fiber		myofibril		thick and thin filaments		myosin and actin
(an organ and its major subdivisions)	→	(a cell)	→	(a specialized organelle)	→	(cytoskeletal elements)	→	(proteins)

A and I Bands

Viewed with a light microscope, a relaxed myofibril (● Figure 8–3a) displays alternating dark bands (the A bands) and light bands (the I bands). The bands of all the myofibrils lined up parallel to each other collectively lead to the striated appearance of a skeletal muscle fiber (Figure 8–3b). Alternate stacked sets of thick and thin filaments that slightly overlap each other are responsible for the A and I bands (Figure 8–2c). An A band consists of a stacked set of thick filaments along with the portions of the thin filaments that overlap on both ends of the thick filaments. The thick filaments are found only within the A band and extend its entire width; that is, the two ends of the thick filaments within a stack define the outer limits of a given A band. The lighter area within the middle of the A band, where the thin filaments do not reach, is known as the **H zone.** Only the central portions of the thick filaments are found in this region. The **I band** consists of the remaining portion of the thin filaments that do not project into the A band. Thus the I band contains only thin filaments but not their entire length.

Visible in the middle of each I band is a dense, vertical **Z line.** The area between two Z lines is called a **sarcomere,** which is the functional unit of skeletal muscle. A **functional unit** of any organ is the smallest component that can perform all the functions of that organ. Accordingly, a sarcomere is the smallest component of a muscle fiber that is capable of contraction. The Z line is actually a flattened disc (in three dimensions) made from a cytoskeletal protein complex that connects the thin filaments of two adjoining sarcomeres. Each relaxed sarcomere is about 2.5 μm in width and consists of one whole A

band and half of each of the two I bands located on either side. During growth, a muscle increases in length by adding new sarcomeres, not by increasing the size of each sarcomere. Just as the Z line or discs hold the sarcomeres together in a chain along the myofibril's length, another system of supporting proteins holds the thick filaments together vertically within each stack. These proteins can be seen as the **M line,** which extends vertically down the middle of the A band within the center of the H zone.

Cross Bridges

With an electron microscope, fine **cross bridges** can be seen extending from each thick filament toward the surrounding thin filaments in the regions where the thick and thin filaments overlap (● Figures 8–2c and 8–4a). Three-dimensionally, the thin filaments are arranged hexagonally around the thick filaments. Cross bridges project from each thick filament in all six directions toward the surrounding thin filaments. Each thin filament, in turn, is surrounded by three thick filaments (● Figure 8–4b). To give you an idea of the magnitude of these filaments, a single muscle fiber may contain an estimated 16 billion thick and 32 billion thin filaments, all arranged in this very precise pattern within the myofibrils.

■ Myosin forms the thick filaments.

Each thick filament consists of several hundred myosin molecules packed together in a specific arrangement. A **myosin**

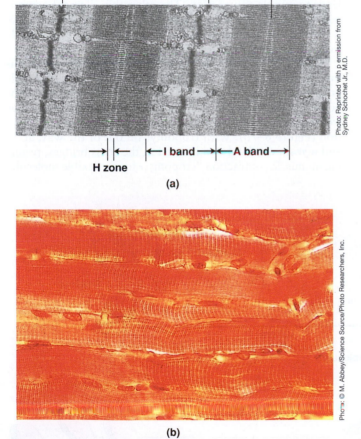

Figure 8–3 ● Light-microscope view of skeletal muscle components. (a) High-power light-microscope view of a myofibril. (b) Low-power light-microscope view of skeletal muscle fibers. Note striated appearance.

(*Source:* Reprinted with permission from Sydney Schochet Jr., M.D., Professor, Department of Pathology, School of Medicine, West Virginia University: *Diagnostic Pathology of Skeletal Muscle and Nerve,* Stamford, Connecticut: Appleton & Lange, 1986, Figure 1–13)

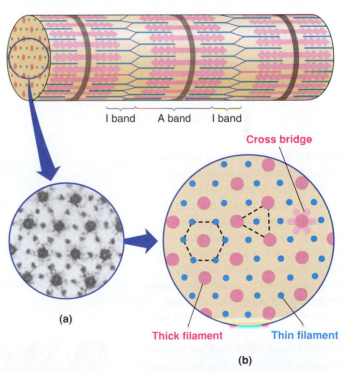

Figure 8–4 ● Cross-sectional arrangement of thick and thin filaments. (a) Electron micrograph cross section through the A band in the region of thick and thin filament overlap. Note the fine cross bridges extending from the thick filaments. (b) Schematic representation of the geometric relation among thick and thin filaments and cross bridges.

molecule is a protein consisting of two identical subunits, each shaped somewhat like a golf club with two heads (● Figure 8–5a). The protein's tail ends are intertwined around each other, with the two globular **heads** projecting out at one end. The two halves of each thick filament are mirror images made up of myosin molecules lying lengthwise in a regular, staggered array, with their tails oriented toward the center of the filament and their globular heads protruding outward at regularly spaced intervals (Figure 8–5b). These heads form the cross bridges between the thick and thin filaments. Each cross bridge has two important sites crucial to the contractile process: an *actin binding site* and a *myosin ATPase (ATP-splitting) site.* Myosin filaments are linked to the Z lines by the gigantic, elastic protein **titin,** the function of which is still being investigated.

?

What Are the Functions of Titin? Titin is the largest protein known in any organism, containing about 30,000 amino acids! Discovered only in 1979, its roles in muscles are still not fully understood. It has been implicated in muscle development, perhaps acting as a scaffold for sarcomere assembly. Titin is also believed to keep the thick filaments centered in the sarcomere during the contraction–relaxation cycle. And there is evidence of the springlike mechanical roles: It may act as a locomotory spring assisting in the return of the sarcomere to its relaxed conformation or in preventing harmful overextension of the sarcomere. Recent studies shed more light on this role: Titin is more compliant (flexible) in elephant muscle and most stiff in small mammals such as the shrew. Furthermore, as the size of the animal decreases so does the size of the titin isoform, even though sarcomere size remains constant between species.

■ Actin, along with tropomyosin and troponin, forms the thin filaments.

Thin filaments consist of three proteins: *actin, tropomyosin,* and *troponin* (● Figure 8–6). Actin molecules, the primary structural proteins of the thin filament, are spherical. The backbone of a thin filament is formed by actin molecules joined into two strands and twisted together, like two chains of pearls wrapped around each other. Each actin molecule has sites for both *weak* (primarily electrostatic, that is, the interaction between positively and negatively charged moieties) and *strong* myosin attachment with a myosin cross bridge. By a mechanism to be described shortly, binding of actin and myosin molecules at the cross bridges results in energy-consuming contraction of the muscle fiber. Accordingly, actin and myosin are often called **contractile proteins,** even though, as you will see, neither myosin nor actin actually contracts.

In a relaxed muscle fiber, contraction does not take place; actin cannot bind with cross bridges because of the position of the two other types of protein within the thin filament—tropomyosin and troponin. **Tropomyosin** molecules are threadlike proteins that lie end-to-end alongside the groove of the actin spiral. In this position, tropomyosin covers the actin sites that bind with the cross bridges, thus blocking the interaction that leads to muscle contraction. The other thin-filament component, **troponin,** is a protein complex consisting of three polypeptide units: one that binds to tropomyosin, one that binds to actin, and a third that can bind with Ca^{++}.

When troponin is not bound to Ca^{++}, this protein stabilizes tropomyosin in its blocking position over actin's cross-bridge binding sites (● Figure 8–7a). When Ca^{++} binds to troponin, the shape of this protein is changed in such a way that tropomyosin is allowed to slide away from its blocking position (Figure 8–7b). With tropomyosin out of the way, actin and myosin can bind and interact at the cross bridges, resulting in muscle contraction. Tropomyosin is a flexible molecule,

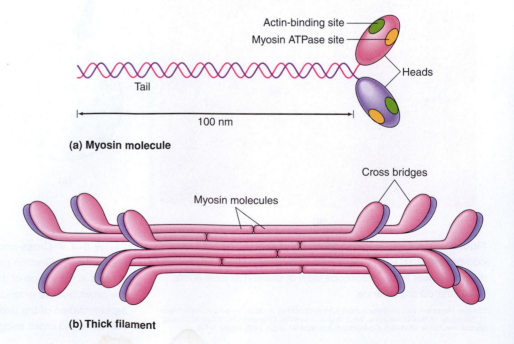

Figure 8–5 ● **Structure of myosin molecules and their organization within a thick filament.** (a) Myosin molecule. Each myosin molecule consists of two identical golf club–shaped subunits with their tails intertwined and their globular heads, each of which contains an actin-binding site and a myosin ATPase site, projecting out at one end. (b) Thick filament. A thick filament is made up of myosin molecules lying lengthwise parallel to each other. Half are oriented in one direction and half in the opposite direction. The globular heads, which protrude at regular intervals along the thick filament, form the cross bridges.

Actin-binding site
Myosin ATPase site
Heads
Tail
100 nm

(a) Myosin molecule

Cross bridges
Myosin molecules

(b) Thick filament

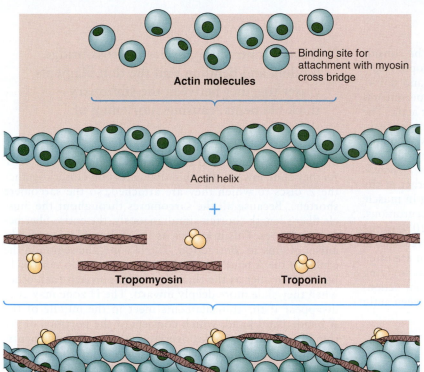

Actin molecules

Binding site for attachment with myosin cross bridge

Actin helix

+

Tropomyosin   **Troponin**

Thin filament

Figure 8–6 ● Composition of thin filament. The main structural component of a thin filament is two chains of spherical actin molecules that are twisted together. Troponin molecules (which consist of three small spherical subunits) and threadlike tropomyosin molecules are arranged to form a ribbon that lies alongside the groove of the actin helix and physically covers the binding sites on actin molecules for attachment with myosin cross bridges. (The thin filaments shown here are not drawn in proportion to the thick filaments in Figure 8–5. Thick filaments are two to three times larger in diameter than thin filaments.)

Figure 8–7 ▼ ● Role of calcium in turning on cross bridges.

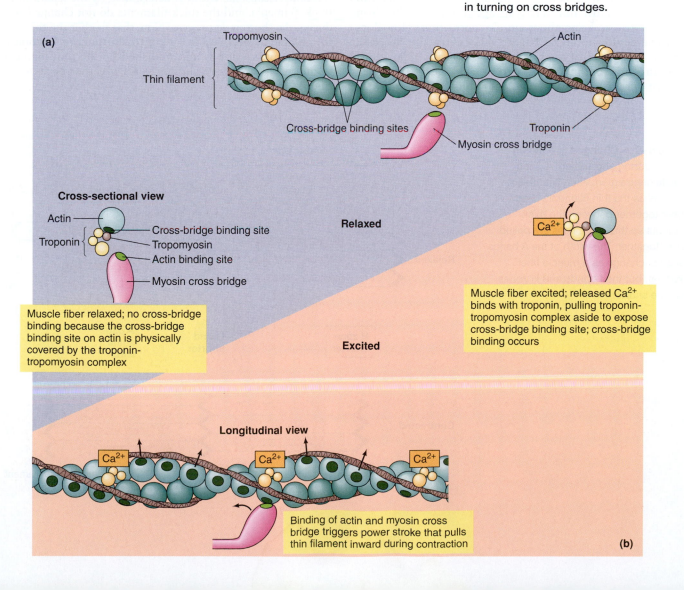

(a)

Tropomyosin

Actin

Thin filament

Cross-bridge binding sites

Troponin

Myosin cross bridge

Relaxed

Cross-sectional view

Actin

Cross-bridge binding site

Troponin

Tropomyosin

Actin binding site

Myosin cross bridge

Muscle fiber relaxed; no cross-bridge binding because the cross-bridge binding site on actin is physically covered by the troponin-tropomyosin complex

Excited

Ca^{2+}

Muscle fiber excited; released Ca^{2+} binds with troponin, pulling troponin-tropomyosin complex aside to expose cross-bridge binding site; cross-bridge binding occurs

Longitudinal view

Ca^{2+} Ca^{2+} Ca^{2+}

Binding of actin and myosin cross bridge triggers power stroke that pulls thin filament inward during contraction

(b)

and its positioning on the actin filament should be considered dynamic. In the presence of elevated Ca^{++} tropomyosin does not occupy a fixed position but rather "slips and slides" back and forth over the actin surface. Recent experimental evidence suggests there are three binding states for actin: The first one is in the absence of Ca^{++} in which tropomyosin blocks cross-bridge access to the strong binding sites on the thin filament; the second is one in which cross bridges can weakly bind to actin (considered the "closed" state), whereas the third exists when myosin can bind and interact at the cross bridges (called the "open" state), resulting in muscle contraction. In the presence of elevated Ca^{++} concentrations, the position of tropomyosin is such that only ~20% of the actin sites at any one time are in the open state whereas ~80% are in the closed state. Tropomyosin and troponin are often called **regulatory proteins** because of their role in covering (preventing contraction) or exposing (permitting contraction) the binding sites for cross-bridge interaction between actin and myosin.

Molecular Basis of Skeletal Muscle Contraction

Several important links in the contractile process remain to be discussed. How does cross-bridge interaction between actin and myosin bring about muscle contraction? How does a muscle action potential trigger this contractile process? What is the source of the Ca^{++} that physically repositions troponin and tropomyosin to permit cross-bridge binding? We turn our attention to these topics in this section.

■ **During contraction, cycles of cross-bridge binding and bending pull the thin filaments closer together between the stationary thick filaments, causing shortening of the sarcomeres.**

During contraction the thin filaments on each side of a sarcomere slide inward toward the A band's center (● Figure 8–8). As they slide inward, the thin filaments pull closer together the Z discs to which they are attached, so the sarcomere shortens. Because all the sarcomeres throughout the muscle fiber's length shorten simultaneously, the entire fiber becomes shorter. This is known as the **sliding-filament mechanism** of muscle contraction. The H zone, the region in the center of the A band where the thin filaments do not reach, becomes smaller as the thin filaments approach each other when they slide more deeply inward. The H zone may even disappear if the thin filaments meet in the middle of the A band. The I band, which consists of the portions of the thin filaments that do not overlap with the thick filaments, narrows as the thin filaments further overlap the thick filaments during their inward slide. The thin filaments themselves do not change length during muscle fiber shortening. The width of the A band remains unchanged during contraction, because its width is determined by the length of the thick filaments, and the thick filaments do not change length during the shortening process. Note that neither the thick nor thin filaments decrease in length to shorten the sarcomere. Instead, contraction is accomplished by the thin filaments within each sarcomere sliding closer together between the thick filaments.

Figure 8–8 ● Changes in banding pattern during shortening. During muscle contraction, each sarcomere shortens as the thin filaments slide closer together between the thick filaments so that the Z lines are pulled closer together. The width of the A bands does not change as a muscle fiber shortens, but the I bands and H zones become shorter.

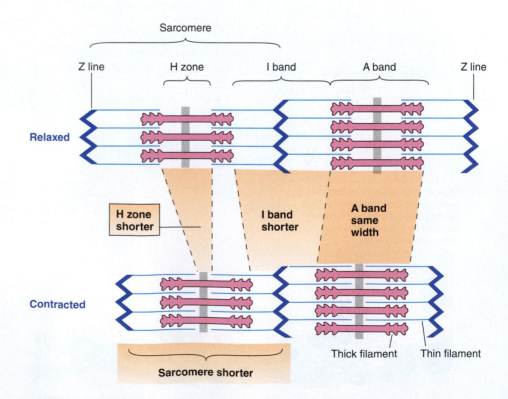

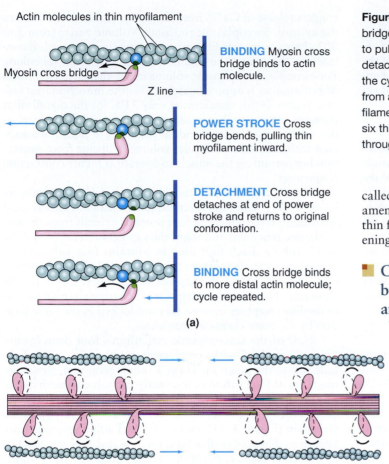

Actin molecules in thin myofilament

Myosin cross bridge

Z line

BINDING Myosin cross bridge binds to actin molecule.

POWER STROKE Cross bridge bends, pulling thin myofilament inward.

DETACHMENT Cross bridge detaches at end of power stroke and returns to original conformation.

BINDING Cross bridge binds to more distal actin molecule; cycle repeated.

(a)

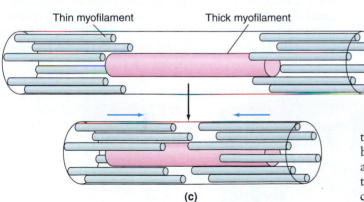

(b)

Thin myofilament   Thick myofilament

(c)

Figure 8–9 ◄ ● Cross-bridge activity. (a) During each cross-bridge cycle, the cross bridge binds with an actin molecule, bends to pull the thin filament inward during the power stroke, then detaches and returns to its resting conformation, ready to repeat the cycle. (b) The power strokes of all cross bridges extending from a thick filament are directed toward the center of the thick filament. (c) Each thick filament is surrounded on each end by six thin filaments, all of which are pulled inward simultaneously through cross-bridge cycling during muscle contraction.

called power stroke of a cross bridge pulls inward the thin filament to which it is attached. A single power stroke pulls the thin filament inward only a small percentage of the total shortening distance.

■ Complete shortening is accomplished by repeated cycles of cross-bridge binding and bending.

At the end of one cross-bridge cycle, the link between the myosin cross-bridge and actin molecule is broken. The cross bridge returns to its original conformation and binds to the next actin molecule positioned behind its previous actin partner. The cross bridge bends once again to pull the thin filament in further, then detaches and repeats the cycle. Repeated cycles of cross-bridge binding and bending successively pull in the thin filaments, much like pulling in a rope hand over hand.

Because of the orientation of the myosin molecules within a thick filament (Figure 8–9b), all the cross bridges' power strokes are directed toward the center of the sarcomere, so that all six of the surrounding thin filaments on each end of the sarcomere are pulled inward simultaneously (Figure 8–9c). The cross bridges aligned with given thin filaments do not all stroke in unison, however. At any time during contraction, some of the cross bridges are attached to the thin filaments and are stroking, while others are returning to their original conformation in preparation for binding with another actin molecule. Thus some cross bridges are "holding on" to the thin filaments, whereas others "let go" to bind with new actin. Were it not for this asynchronous cycling of the cross bridges, the thin filaments would be able to slip back toward their resting position between strokes.

How is this cross-bridge cycling switched on by muscle excitation? The term *excitation–contraction coupling* refers to the series of events linking muscle excitation (the presence of an action potential in a muscle fiber) to muscle contraction (cross-bridge activity that causes the thin filaments to slide closer together to produce sarcomere shortening). Now let's turn our attention to excitation–contraction coupling.

■ Calcium is the link between excitation and contraction.

Skeletal muscles are stimulated to contract by release of acetylcholine (**ACh**) at neuromuscular junctions between motor

Power Stroke

The thin filaments are pulled inward relative to the stationary thick filaments by cross-bridge activity. During contraction, with the tropomyosin and troponin "blockers" pulled out of the way by the binding of Ca^{++}, the myosin heads (from a thick filament) can bind with the actin molecules in the surrounding thin filaments, forming cross bridges. Let's concentrate on a single cross-bridge interaction (● Figure 8–9a). When myosin and actin make contact at a cross bridge, the bridge is bent inward as if on a hinge, "stroking" toward the center of the sarcomere, similar to the stroking of a boat oar. This so-

neuron terminals and muscle fibers. Recall that the binding of ACh with the motor end plate of a muscle fiber brings about permeability changes in the muscle fiber that result in an action potential that is conducted over the entire surface of the muscle cell membrane (see p. 127). Recall also that the enzyme **acetylcholinesterase (AChE)** destroys ACh to shut off the signal.

Spread of the Action Potential down the T Tubules

At each junction of an A band and I band, the surface membrane dips into the muscle fiber to form a transverse tubule (T tubule), which runs perpendicularly from the surface of the muscle cell membrane into the central portions of the muscle fiber (● Figure 8–10). Because the T tubule membrane is continuous with the surface membrane, an action potential on the surface membrane also spreads down into the T tubule, providing a means of rapidly transmitting the surface electric activity into the central portions of the fiber. The presence of a local action potential in the T tubules induces permeability changes in a separate membranous network within the muscle fiber, the sarcoplasmic reticulum.

Release of Calcium from the Sarcoplasmic Reticulum

The **sarcoplasmic reticulum** is a modified endoplasmic reticulum (see p. 35) that is important for turning the muscle on and off. It consists of a fine network of interconnected tubules surrounding each myofibril like a mesh sleeve (Figure 8–10). This membranous network runs longitudinally down the myofibril (that is, encircles the myofibril throughout its length), but is not continuous. Separate segments of sarcoplasmic reticulum are wrapped around each A band and each I band. The ends of each segment expand to form saclike regions, the **lateral sacs** (alternatively known as **terminal cisternae**), which are separated from the adjacent T tubules by a slight gap (● Figures 8–10 and 8–11). The sarcoplasmic reticulum's lateral sacs store Ca^{++}. Spread of an action potential down a T tubule

triggers release of Ca^{++} from the sarcoplasmic reticulum into the cytosol. Sarcoplasmic reticulum volume varies from 3 to 30% between muscle fiber types; the greater the speed of contraction, the greater the volume of the sarcoplasmic reticulum. For example, sarcoplasmic volume in the shaker muscle fibers of rattlesnakes is approximately 26% (the mitochondrial volume is also 26%), thus leaving only 31% for the myofibrillar contents. As you will soon see, increases in cellular sarcoplasmic reticulum and mitochondrial volumes are generally associated with reduced myofibrillar volume, reducing force generation but permitting the muscle to operate at higher contraction frequencies.

How is a change in T tubule potential linked with the release of Ca^{++} from the sarcoplasmic reticulum's lateral sacs? An orderly arrangement of **foot proteins** extends from the sarcoplasmic reticulum and spans the gap between the lateral sac and T tubule. Each foot protein contains four subunits arranged in a specific pattern (Figure 8–11b). These foot proteins not only bridge the gap but also serve as Ca^{++}-release channels. These foot protein Ca^{++} channels are known as **ryanodine receptors** because they are locked in the open position by the plant chemical *ryanodine.*

Half of the sarcoplasmic reticulum's foot proteins are "zipped together" with complementary receptors on the T tubule side of the junction. These T tubule receptors, which are made up of four subunits in exactly the same pattern as the foot proteins, are located like mirror images in contact with every other foot protein protruding from the sarcoplasmic reticulum (Figure 8–11b and c). These T tubule receptors are known as **dihydropyridine receptors** because they are blocked by the drug *dihydropyridine.* These receptors are voltage-gated sensors. When an action potential is propagated down the T tubule, the local depolarization activates the voltage-gated dihydropyridine receptors. These activated T-tubule receptors in turn trigger the opening of the directly abutting Ca^{++}-release channels (alias *ryanodine receptors*; alias *foot*

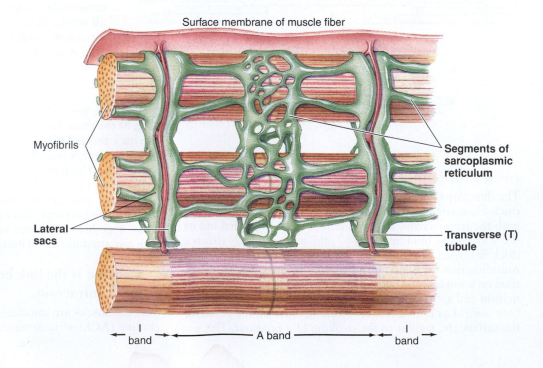

Figure 8–10 ● **The T tubules and sarcoplasmic reticulum in relationship to the myofibrils.** The transverse (T) tubules are membranous perpendicular extensions of the surface membrane that dip deep into the muscle fiber at the junctions between the A and I bands of the myofibrils. The sarcoplasmic reticulum is a fine, membranous network that runs longitudinally and surrounds each myofibril, with separate segments encircling each A band and I band. The ends of each segment are expanded to form lateral sacs that lie next to the adjacent T tubules.

Surface membrane of muscle fiber

Myofibrils

Lateral sacs

Segments of sarcoplasmic reticulum

Transverse (T) tubule

I band — A band — I band

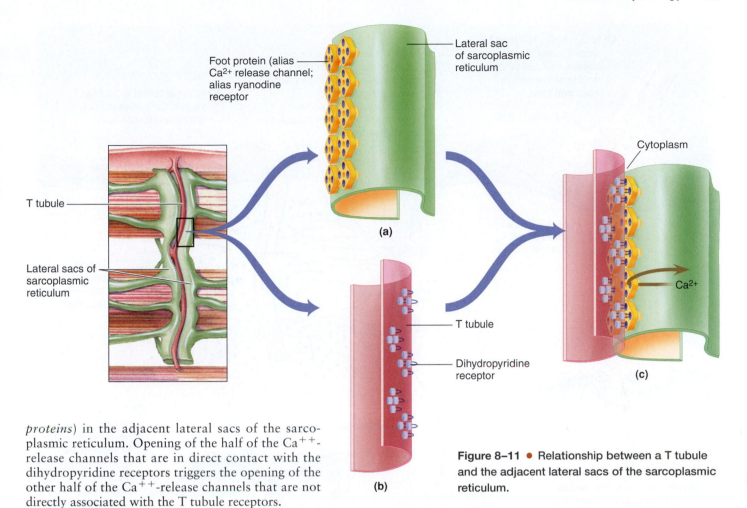

Foot protein (alias Ca²⁺ release channel; alias ryanodine receptor

Lateral sac of sarcoplasmic reticulum

T tubule

Lateral sacs of sarcoplasmic reticulum

Cytoplasm

Ca^{2+}

(a)

T tubule

Dihydropyridine receptor

(b)

(c)

Figure 8–11 ● Relationship between a T tubule and the adjacent lateral sacs of the sarcoplasmic reticulum.

proteins) in the adjacent lateral sacs of the sarcoplasmic reticulum. Opening of the half of the Ca⁺⁺-release channels that are in direct contact with the dihydropyridine receptors triggers the opening of the other half of the Ca⁺⁺-release channels that are not directly associated with the T tubule receptors.

Ca⁺⁺ is released into the surrounding cytosol from the sarcoplasmic reticulum's lateral sacs through all these open Ca⁺⁺-release channels. By slightly repositioning the troponin and tropomyosin molecules, this released Ca⁺⁺ exposes the binding sites on the actin molecules so that they can link with the myosin cross bridges at their complementary binding sites (● Figure 8–12).

ATP-Powered Cross-Bridge Cycling

Recall that a myosin cross bridge has two special sites, an actin binding site and an ATPase site. The latter is an enzymatic site that can bind the energy carrier *adenosine triphosphate* (ATP) and split it into *adenosine diphosphate* (ADP) and inorganic phosphate (P_i), yielding energy in the process. In skeletal muscle, magnesium (Mg⁺⁺) must be attached to ATP before myosin ATPase can split the ATP. The importance of magnesium becomes evident in lactating beef and dairy cattle grazing new forage grown in cool spring temperatures. Early spring pasture provides insufficient magnesium. **Hypomagnesemic tetany** or **grass tetany** is characterized by low magnesium concentrations in the blood and cerebrospinal fluid. Affected animals appear nervous and show muscular twitching around the face and ears. As the condition persists, animals become uncoordinated and walk with a stiff gait. In the advanced stages, cows go down on their sides with their heads back. Convulsions are characteristic at this time, and unless the animal is treated intravenously with a magnesium-salt solution, death soon follows.

The breakdown of ATP occurs on the myosin cross bridge before the bridge ever links with an actin molecule (step 1 in ● Figure 8–13). The ADP and P_i remain tightly bound to the myosin, and the generated energy is stored within the cross bridge to produce a high-energy form of myosin. To use an analogy, the cross bridge is "cocked" like a gun ready to be fired when the trigger is pulled. When the muscle fiber is excited, Ca⁺⁺ pulls the troponin-tropomyosin complex out of its blocking position so that the energized (cocked) myosin cross bridge can bind with an actin molecule (step 2a). This contact between myosin and actin "pulls the trigger," causing the cross bridge bending responsible for the power stroke that pulls the thin actin filament inward toward the center of the sarcomere (step 3). The mechanism by which the chemical energy released from ATP is stored within the myosin cross bridge and then translated into the mechanical energy of the power stroke is not known. Inorganic phosphate (P_i) is released from the cross-bridge during the power stroke. After the power stroke is complete, ADP is released.

When the muscle is not excited and Ca⁺⁺ is not released, troponin and tropomyosin remain in their blocking position, so that actin and the myosin cross bridges do not bind and no power stroking takes place (step 2b).

When ADP and inorganic phosphate are released from myosin on contact with actin during a power stroke, the myosin ATPase site is free for attachment of another ATP molecule. The actin and myosin remain linked together at

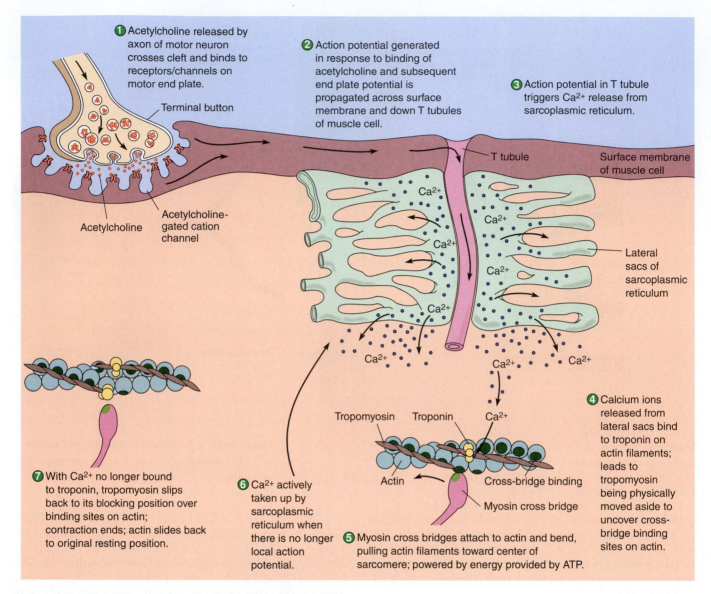

Figure 8–12 ● **Calcium release in excitation–contraction coupling.** Steps ①
through ⑤ depict the events that couple neurotransmitter release and subsequent
electrical excitation of the muscle cell with muscle contraction. Steps ⑥ and ⑦
depict events associated with muscle relaxation.

the cross bridge until a fresh molecule of ATP attaches to
myosin at the end of the power stroke. Attachment of the
new ATP molecule permits detachment of the cross bridge,
which returns to its open-cleft, unbent conformation, ready
to start another cycle (step 4a). The newly attached ATP is
then split by myosin ATPase, energizing the myosin cross
bridge once again (step 1). On binding with another actin
molecule, the energized cross bridge again bends, and so on,
successively pulling the thin filament inward to accomplish
contraction.

Rigor Mortis

Note that fresh ATP must attach to myosin to permit the cross-
bridge link between myosin and actin to be broken at the end
of a cycle, even though the ATP is not split during this dissociation process. The need for ATP in the separation of myosin

and actin is amply demonstrated by the phenomenon of **rigor
mortis.** This "stiffness of death" is a generalized locking-in-
place of the skeletal muscles that begins sometime after death
(3 to 4 hours in humans, becoming complete in about 12 hours).
After death, the cytosolic concentration of Ca^{++} begins to
rise, most likely because the inactive muscle-cell membrane
cannot keep out extracellular Ca^{++} and perhaps also because
Ca^{++} leaks out of the lateral sacs. This Ca^{++} moves the regulatory proteins aside, permitting actin to bind with the myosin
cross bridges, which were already charged with ATP before
death. Dead cells cannot produce any more ATP, so actin and
myosin, once bound, cannot detach, because of the absence of
fresh ATP. The thick and thin filaments thus remain linked together by the immobilized cross bridges, resulting in the stiffness of dead muscles (step 4b). During the next several days,
rigor mortis gradually subsides as the proteins involved in the

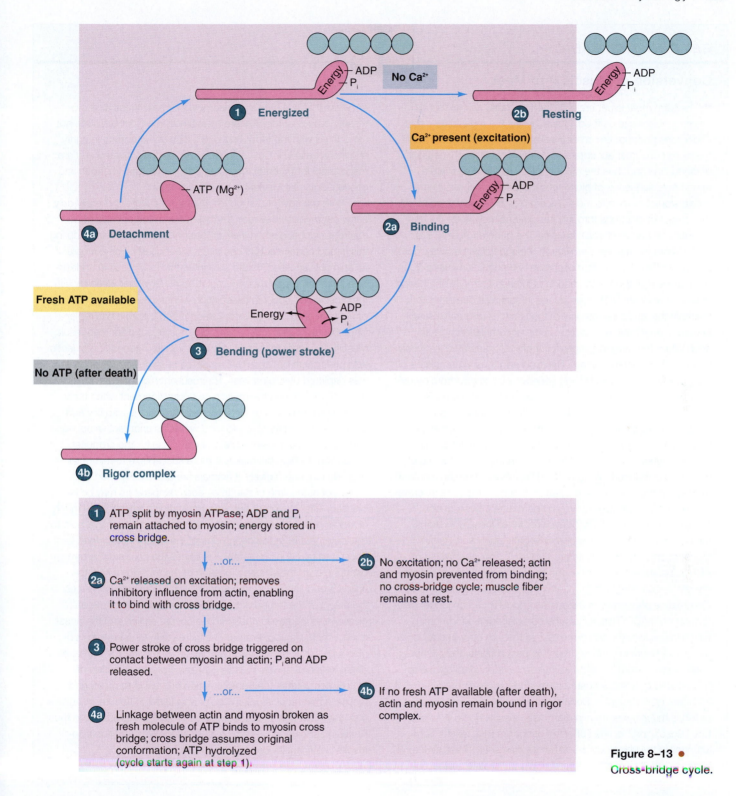

1. ATP split by myosin ATPase; ADP and P_i remain attached to myosin; energy stored in cross bridge.

...or...

2a. Ca^{2+} released on excitation; removes inhibitory influence from actin, enabling it to bind with cross bridge.

2b. No excitation; no Ca^{2+} released; actin and myosin prevented from binding; no cross-bridge cycle; muscle fiber remains at rest.

3. Power stroke of cross bridge triggered on contact between myosin and actin; P_i and ADP released.

...or...

4a. Linkage between actin and myosin broken as fresh molecule of ATP binds to myosin cross bridge; cross bridge assumes original conformation; ATP hydrolyzed (cycle starts again at step 1).

4b. If no fresh ATP available (after death), actin and myosin remain bound in rigor complex.

Figure 8–13 ● Cross-bridge cycle.

rigor complex begin to degrade (see box, "Beyond the Basics: Conversion of Muscle into Meat").

■ Removal of calcium is the key to muscle relaxation.

How is **relaxation** normally accomplished in a living muscle? Just as an action potential in a muscle fiber turns on the con-

tractile process by triggering the release of Ca^{++} from the lateral sacs into the cytosol, the contractile process is turned off when Ca^{++} is returned to the lateral sacs at cessation of local electrical activity. The sarcoplasmic reticulum has an energy-consuming carrier, a Ca^{++}–ATPase pump, which actively transports Ca^{++} from the cytosol and concentrates it in the lateral sacs. In addition to increased density of Ca^{++}–ATPase pump units, high-speed muscles also have elevated concentrations of

BEYOND THE BASICS

Conversion of Muscle into Meat

The term *meat* is generally defined as those animal tissues suitable for use as food. However, it is the musculature that undergoes postmortem changes that determine palatability and therefore consumer acceptability. At the time of death, cessation of blood flow and thus the delivery of O_2 to muscle initiate a remarkable sequence of biochemical changes that explain the conversion of muscle to meat. As the oxygen supply is depleted, movement of electrons through the citric acid cycle (p. 46) and the electron transport chain ceases. The generation of ATP is shifted from the aerobic pathway to the anaerobic pathway, comparable to the situation that occurs in living muscle when oxygen debt arises in response to periods of intense activity. For a period of time after death, ATP is produced in sufficient quantities to maintain the structural integrity of the fibers. Recall that a major feature of anaerobic metabolism is the accumulation of lactic acid. When the oxygen supply is cut off, H^+ generated in glycolysis cannot cycle through the TCA cycle and thus accumulates in the muscle. The excess H^+ ion generates lactic acid from pyruvic acid, which permits glycolysis to proceed at a rapid rate. Because the circulatory system is no longer functioning, lactic acid remains in muscle fibers and increases in concentration as postmortem metabolism proceeds. Lactic acid accumulation in the muscle lowers its pH, and at levels less than pH 6.0 the rate of glycolysis is reduced, along with ATP synthesis that had resulted from substrate-level phosphorylation. Mechanisms to regenerate ATP are now insufficient to prevent permanent cross-bridge formation between actin and myosin. Under these conditions, the postmortem onset of rigor mortis gets underway. As the stores of creatine phosphate are depleted, phosphorylation of ADP is insufficient to maintain the tissue in a relaxed state. Actin-myosin bridges begin to form, and this phase continues until all the creatine phosphate is exhausted and ATP can no longer be formed from ADP. This signals the completion of rigor mortis, which in chicken and fish requires less than an hour and as long as 6 to 12 hours in lamb and beef. In pork muscle, the pH normally declines from 7.4 in living muscle to approximately 5.6 to 5.7 within 6 to 8 hours postmortem, and then to about 5.3 to 5.7, 24 hours after slaughter. Because the muscle glycogen content of fish is much lower than in mammals, the pH declines to only 6.8. "Resolution," or the softening of rigor mortis, results from changes in the ultrastructure of the myofilaments. These structural changes result from the action of acid and alkaline proteases,

enzymes that originate in lysosomes. Normally these acid proteases, called *cathepsins,* are maintained in an inactive state but associated with the reduction in pH, loss of membrane integrity and increase in ICF Ca^{++}, these enzymes are activated and are responsible for much of the postmortem structural changes and consequent improvement in meat tenderness.

With the advent of modern animal agriculture, generally only a few days elapse between the time when meat animals attain market weight and when they are slaughtered. Nonetheless, during this period certain factors can affect meat quality. For example, if an animal is stressed before slaughter, this may result in undesirable changes in postmortem metabolism. Glycogen deficiency in the muscle can occur in association with transport stress, fasting, or restraint and, if the animal is slaughtered before it has replenished the glycogen, a *dark, firm, dry* muscle condition can result. Muscle glycogen deficiency results in a limited generation of lactic acid after death because the substrate of glycolysis, glycogen, was depleted before harvest. The resultant high pH (approximately 6.8) changes the muscle color development when these muscles are exposed to O_2. This muscle tissue is also dry and sticky because of its ability to bind water. Its unattractive appearance results in a considerable monetary loss for the producer. In addition, its high pH makes it more susceptible to microbial degradation. Alternatively, if animals are acutely exposed to a stressor at the time of slaughter, body temperature may be increased. After slaughter and exsanguination (removal of as much blood as possible), the circulatory system is no longer available to dissipate heat from the deep parts of the muscle. Therefore, heat generated by ongoing metabolism raises the core muscle temperature; both the size and location of the muscles within the animal's body as well as the amount of fat insulation influences the extent of the temperature rise. Consequently, the rate of glycolysis is increased, the pH drop occurs earlier, and the onset of rigor mortis is accelerated. The increase in temperature coupled with the acid that has accumulated in the muscle increases denaturation of certain muscle proteins. This combination of factors accelerates the transformation of muscle to meat, and results in meat that is paler in color and softer in texture. In addition, this meat has a reduced ability to retain its moisture and thus is wet in appearance. The *pale, soft, exudative* condition also results in a financial loss to processors and producers.

---

calcium-binding proteins, such as **parvalbumin.** These calcium-binding proteins play a critical role in reducing the duration of the twitch by rapidly lowering free (unbound) cytosolic Ca^{++} as well as shuttling it to the sarcoplasmic reticulum.

When acetylcholinesterase removes ACh from the neuromuscular junction, the muscle fiber action potential ceases. When there is no longer a local action potential in the T tubules

to trigger the release of Ca^{++}, the ongoing activity of the sarcoplasmic reticulum's Ca^{++} pump returns the released Ca^{++} back into its lateral sacs. Removal of cytosolic Ca^{++} allows the troponin-tropomyosin complex to slip back into its blocking position, so that actin and myosin can no longer bind at the cross bridges. The thin filaments, freed from cycles of cross-bridge attachment and pulling, can return to their resting posi-

Table 8–1 ▮ Steps of Excitation–Contraction Coupling and Relaxation

1. Acetylcholine released from the terminal of a motor neuron initiates an action potential in the muscle cell that is propagated over the entire surface of the muscle cell membrane.

2. The surface electric activity is carried into the central portions of the muscle fiber by the T tubules.

3. Spread of the action potential down the T tubules triggers the release of stored Ca^{2+} from the adjacent lateral sacs of the sarcoplasmic reticulum.

4. Released Ca^{2+} binds with troponin and changes its shape so that the troponin–tropomyosin complex is physically pulled aside, uncovering actin's cross-bridge binding sites.

5. Exposed actin sites bind with myosin cross bridges, which have previously been energized by the splitting of ATP into ADP + Pi + energy by the myosin ATPase site on the cross bridges.

6. Binding of actin and myosin at a cross bridge causes the cross bridge to bend, producing a power stroke that pulls the thin filament inward. Inward sliding of all the thin filaments surrounding a thick filament shortens the sarcomere (causes muscle contraction).

7. ADP and Pi are released from the cross bridge during the power stroke.

8. Attachment of a new molecule of ATP permits detachment of the cross bridge, which returns to its original conformation.

9. Splitting of the fresh ATP molecule by myosin ATPase energizes the cross bridge once again.

10. If Ca^{2+} is still present so that the troponin–tropomyosin complex remains pulled aside, the cross bridges go through another cycle of binding and bending, pulling the thin filament in even further.

11. When there is no longer a local action potential and Ca^{2+} has been actively returned to its storage site in the sarcoplasmic reticulum's lateral sacs, the troponin–tropomyosin complex slips back into its blocking position, actin and myosin no longer bind at the cross bridges, and the thin filaments passively slide back to their resting position as relaxation takes place.

tion. The "elastic band" properties of titin may also help return the sarcomere to its unstimulated conformation. Relaxation has occurred.

▮ Table 8–1 summarizes the steps of excitation–contraction coupling and relaxation. Now let's compare the duration of contractile activity to the duration of excitation, before we shift gears to discuss skeletal muscle mechanics.

▮ Contractile activity far outlasts the electrical activity that initiated it.

A single action potential in a skeletal muscle fiber lasts only 1 to 2 msec. The onset of the resultant contractile response lags behind the action potential, because the entire excitation–contraction coupling process must take place before cross-bridge activity begins. In fact, the action potential is completed before the contractile apparatus even becomes operational. This time delay of a few milliseconds between stimulation and the onset of contraction is known as the **latent period** (● Figure 8–14).

Time is also required for the generation of tension within the muscle fiber produced by means of the sliding interactions between the thick and thin filaments through cross-bridge activity. The time from the onset of contraction until peak tension is developed—the **contraction time**—averages about 50 msec in vertebrate locomotory muscles, although this time varies, depending on the type of muscle fiber. The contractile response does not cease until the lateral sacs have taken up all the Ca^{++} released in response to the action potential. This reuptake of Ca^{++} is also time consuming. Even after Ca^{++} is removed, it takes time for the filaments to return to their resting positions. The time from peak tension until relaxation is complete, the **relaxation time,** usually lasts slightly longer than

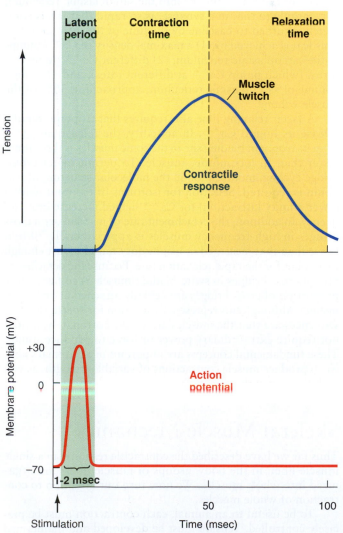

Figure 8–14 ▶ ● Relationship of an action potential to the resultant muscle twitch.

The duration of the action potential is not drawn to scale but is exaggerated.

contraction time, another 50 msec or more. Consequently, the entire contractile response to a single action potential may last up to 100 msec or more; this is considerably longer than the duration of the action potential that initiated it (100 msec compared to 1 to 2 msec). However, if a muscle is to contract and relax rapidly, two conditions must apply. The first, Ca^{++}, the trigger for muscle contraction, must both enter and leave the myoplasm rapidly. Secondly, myosin cross bridges must attach to actin and generate force soon after the Ca^{++} concentrations rise and then must detach and stop generating force soon after the Ca^{++} concentrations decline.

Recent studies have investigated Ca^{++} flux in the fastest known vertebrate muscle, the sound-producing swim-bladder muscle of the toadfish (*Opsanus tau*). To attract egg-laying females to his nest, the male toadfish sings a "boat whistle" mating call 10 to 12 times per minute . Vocalizations are generated by repetitive contractions (at about 200 Hz) of the muscles encircling the fish's gas-filled swim bladder (see p. 227). High-frequency contractions of this muscle yield high-frequency sound production. To achieve this remarkable feat, researchers determined, the Ca^{++} *transient* in the sonic muscle is the fastest ever measured for any fiber type, approximately 50-fold greater than the rate measured in comparable locomotory muscle. The importance of this observation is that the time course of Ca^{++} uptake into the sarcoplasmic reticulum, comparable to that in a synapse, is so rapid that it has completely returned to baseline by the time of the next stimulus! This feat is achieved by (1) a maximal density of Ca^{++} pumps in the sarcoplasmic reticulum, (2) different Ca^{++} pump isoforms (which pump Ca^{++} at different rates), and (3) a high volume of sarcoplasmic reticulum, approaching 30% of the muscle volume.

A faster relaxing time also requires that troponin rapidly release its bound Ca^{++}. In actuality, the release of Ca^{++} from its troponin binding site in the sonic muscles is three times faster than in locomotory muscle. To permit faster cross-bridge cycling, modification in the molecular structure of troponin is required. Selection for isoforms of troponin with a lower affinity (attraction) for Ca^{++} helped increase beat frequency. In addition, the detachment rate of myosin–actin cross bridges in high-frequency muscles is approximately 70-fold faster than that measured in locomotory muscles, fast enough to account for the rapid relaxation rate. Because the detachment rate of cross bridges in swim-bladder muscle is so fast, a low proportion of cross bridges are actually attached at any given instant. Although this represents a solution for speed, the consequences are that the muscles are weak. Fortunately, it does not require extraordinary power or force to produce sound. These fundamental concepts are important in an animal's ability to produce muscle contractions of variable strength, as you will discover in the next section.

Skeletal Muscle Mechanics

Thus far we have described the contractile response in a single muscle fiber. In the body, groups of muscle fibers are organized into whole muscles. We now turn our attention to contraction of whole muscles.

To be useful to an animal, each contraction must be precisely controlled. Tension must be developed and maintained at a level appropriate for the physiological response. There are

essentially two evolutionary strategies for the gradation of tension generated in a muscle, one typified by vertebrates and the other by arthropods. Other invertebrates normally use variations of the solution established in arthropods. We discuss each of these strategies after concluding our discussion on the structure of muscle.

▪ Whole muscles are groups of muscle fibers bundled together by connective tissue, often attached to skeletal elements in antagonistic pairs.

Each muscle is covered by a sheath of connective tissue that penetrates from the surface into the muscle to envelop each individual fiber and divide the muscle into columns or bundles. In vertebrates, the connective tissue extends beyond the ends of the muscle to form tough, collagenous **tendons** that normally attach the muscle to bones. A tendon may be quite long, attaching to a bone some distance from the fleshy portion of the muscle. For example, some of the muscles involved in digit movement are found in the forelimb, with long tendons extending down to attach to the bones of the digits. (You can readily observe the movement of these tendons on the top of your hand when you wiggle your fingers.) This arrangement permits greater dexterity; for example, in humans the fingers would be much thicker and more awkward if all the muscles involved in finger movement were actually located in the fingers.

In arthropods, muscles attach to **apodemes,** ridges that project from the inner face of the exoskeleton. The apodeme is considered the arthropod equivalent of a vertebrate tendon, although it is 40 times stiffer than a typical tendon.

As you saw in Chapter 5, muscles are often arranged in **antagonistic pairs** to move a body part in two opposing directions. A common pairing is that of flexors, which bend a limb, and extensors, which typically straighten the limb out. Familiar examples are the *biceps* flexor and *triceps* extensor of a mammal's limb (see Figure 5–3b, p. 184). Similarly, a crab claw (a modified limb) is opened with an extensor and closed with a flexor. In contrast to these animals that have hard skeletons, consider the arm of the octopus, which can bend in every direction. Because these mollusks lack bones or exoskeletons for the muscles to pull on, to move the limb they must work muscles against each other. The muscles are in antagonistic longitudinal and transverse arrangements, with each type serving as a "skeletal" element for the other. Some vertebrate muscles, such as those of the tongue and the elephant's trunk, work similarly without hard skeletons.

▪ Contractions of a whole muscle can be of varying strength.

A single action potential in a muscle fiber produces a brief, weak contraction known as a **twitch,** which is too short and too weak to be useful and normally does not take place in a vertebrate. Muscle fibers are arranged into whole muscles where they can function cooperatively to produce contractions of variable grades of strength stronger than a twitch. In other words, the force exerted by the same muscle can be made to vary and depends on the force required to move (or to carry an object). Two primary factors can be adjusted to accomplish gradation of whole-muscle tension: (1) the number of muscle fibers contracting within a muscle and (2) the

Table 8–2 ▍ Determinants of Whole-Muscle Tension in Skeletal Muscle

Number of Fibers Contracting	Tension Developed by Each Contracting Fiber
Number of motor units recruited*	Frequency of stimulation (twitch summation and tetanus)*
Number of muscle fibers per motor unit	Length of fiber at onset of contraction (length–tension relationship)
Number of muscle fibers available to contract 　Size of muscle (number of muscle fibers in muscle) 　Presence of disease (e.g., muscular dystrophy) 　Extent of recovery from traumatic losses	Extent of fatigue 　Duration of activity 　Amount of asynchronous recruitment of motor units 　Type of fiber (fatigue-resistant oxidative or fatigue-prone glycolytic)
	Thickness of fiber 　Pattern of neural activity (hypertrophy, atrophy) 　Amount of testosterone (larger fibers in males than females)

*Factors controlled to accomplish gradation of contraction.

tension developed by each contracting fiber (▍ Table 8–2). We discuss each of these factors in turn.

▍ The number of fibers contracting within vertebrate muscle depends on the extent of motor unit recruitment.

Because the greater the number of fibers contracting, the greater the total muscle tension, larger muscles consisting of more muscle fibers can obviously generate more tension than can smaller muscles with fewer fibers. Each whole muscle is innervated by a number of different motor neurons. Typically hundreds to thousands of motor neurons innervate a vertebrate muscle, whereas only 1 to 10 motor neurons innervate an arthropod muscle. When a vertebrate motor neuron enters a muscle, it branches, with each axon terminal supplying a single muscle fiber (● Figure 8–15). One motor neuron innervates a number of muscle fibers, but each vertebrate muscle fiber is supplied by only one motor neuron. In contrast, most arthropod muscle fibers are innervated by more than one motor neuron. When a motor neuron is activated, all the mus-

cle fibers it supplies are stimulated to contract simultaneously. This team of concurrently activated components—one motor neuron plus all the muscle fibers it innervates—is called a **motor unit**. However, an individual arthropod muscle fiber forms a part of several motor units because each fiber is innervated by *more* than one motor neuron! The muscle fibers that compose a vertebrate motor unit are dispersed throughout the whole muscle; thus their simultaneous contraction results in an evenly distributed, although weak, contraction of the whole muscle. Each muscle consists of a number of intermingled motor units. For a weak contraction of the whole muscle, only one or a few of its motor units are activated. For stronger and stronger contractions, more and more motor units are recruited, or stimulated to contract, a phenomenon known as **motor unit recruitment**.

How much stronger the contraction will be with the recruitment of each additional motor unit depends on the size of the motor units (that is, the number of muscle fibers controlled by a single motor neuron) (● Figure 8–16). The number of muscle fibers per motor unit and the number of motor units per muscle vary widely, depending on the specific function of the muscle. For muscles that produce precise, delicate movements, such as the external eye muscles, and the hand muscles in humans, a single motor unit may contain as few as a dozen muscle fibers. Because so few muscle fibers are involved with each motor unit, recruitment of each additional motor unit results in only a small additional increment in the whole muscle's strength of contraction. These small motor units allow a very fine degree of control over muscle tension. In contrast, in muscles designed for powerful, coarsely controlled movement, such as those of most mammalian legs, a single motor unit may contain 1500 to 2000 muscle fibers. Recruitment of motor units in these muscles results in large incremental increases in whole-muscle tension. More powerful contractions occur at the expense of less precisely controlled gradations. Thus the number of muscle fibers participating in the whole muscle's total contractile effort depends on the number of motor units recruited and the number of muscle fibers per motor unit in that muscle.

To delay or prevent **fatigue** (inability to maintain muscle tension at a given level) during a *sustained* contraction involving only a portion of a muscle's motor units, as is necessary in muscles supporting the weight of the body against the force of gravity, **asynchronous recruitment of motor units** takes place. The body alternates motor unit activity, like shifts at a

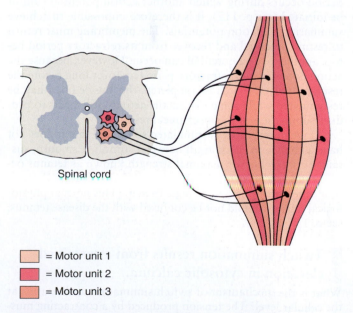

■ = Motor unit 1
■ = Motor unit 2
■ = Motor unit 3

Figure 8–15 ● Schematic representation of motor units in a skeletal muscle.

Spinal cord

Figure 8–16 ● Comparison of motor unit recruitment in muscles with small motor units and muscles with large motor units. (a) Small incremental increases in strength of contraction occur during motor unit recruitment in muscles with small motor units because only a few additional fibers are called into play as each motor unit is recruited. (b) Large incremental increases in strength of contraction occur during motor unit recruitment in muscles with large motor units, because so many additional fibers are stimulated with the recruitment of each additional motor unit.

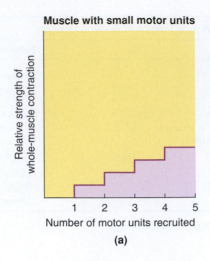

Muscle with small motor units

Relative strength of whole-muscle contraction

Number of motor units recruited
1 2 3 4 5

(a)

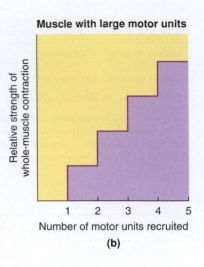

Muscle with large motor units

Relative strength of whole-muscle contraction

Number of motor units recruited
1 2 3 4 5

(b)

factory, to give motor units that have been active a chance to rest while others take over. Changing of the shifts is carefully coordinated, so that the sustained contraction is smooth rather than jerky. Asynchronous motor unit recruitment is possible only for submaximal contractions, during which only some of the motor units must maintain the desired level of tension. During maximal contractions, when all the muscle fibers must participate, it is impossible to alternate motor unit activity to prevent fatigue. This is one reason why you cannot support a heavy object as long as one that is light.

Furthermore, the type of muscle fiber that is activated varies with the extent of gradation. Most muscles consist of a mixture of fiber types that differ metabolically, some being more resistant to fatigue than others. During weak or moderate endurance-type activities (aerobic exercise), the motor units most resistant to fatigue are recruited first. The last fibers called into play in the face of demands for further increases in tension are those that fatigue rapidly. An animal can therefore engage in endurance activities for prolonged periods of time but can only briefly maintain bursts of all-out, powerful effort. Of course, even the muscle fibers most resistant to fatigue do eventually fatigue if required to maintain a certain level of sustained tension.

■ The frequency of stimulation can influence the tension developed by each vertebrate skeletal muscle fiber.

Whole-muscle tension depends not only on the number of muscle fibers contracting but also on the tension developed by each contracting fiber. Various factors influence the extent to which tension can be developed. These factors include

1. Frequency of stimulation
2. Length of the fiber at the onset of contraction
3. Extent of fatigue
4. Thickness of the fiber

Now let's examine the effect of frequency of stimulation. (The other factors are discussed in later sections.)

Twitch Summation and Tetanus

Even though a single action potential in a muscle fiber produces only a twitch, contractions with longer duration and greater tension can be achieved by repetitive stimulation of the fiber. Let us see what happens when a second action potential occurs in a muscle fiber. If the muscle fiber has completely relaxed before the next action potential takes place, a second twitch of the same magnitude as the first occurs (● Figure 8–17a). The same excitation–contraction events take place each time, resulting in identical twitch responses. If, however, the muscle fiber is stimulated a second time before it has completely relaxed from the first twitch, a second action potential occurs that causes a second contractile response, which is added "piggyback" on top of the first twitch (Figure 8–17b). The two twitches resulting from the two action potentials add together, or summate, to produce greater tension in the fiber than that produced by a single action potential. This twitch summation is similar to temporal summation of EPSPs at the postsynaptic neuron (see p. 130).

Twitch summation is possible only because the duration of the action potential (1 to 2 msec) is much shorter than the duration of the resultant twitch (100 msec). Remember that once an action potential has been initiated, a brief refractory period occurs during which another action potential cannot be initiated (see p. 117). It is therefore impossible to achieve summation of action potentials. The membrane must return to resting potential and recover from its refractory period before another action potential can occur. However, because the action potential and refractory period are over long before the resultant muscle twitch is completed, the muscle fiber may be restimulated while some contractile activity still exists to produce summation of the mechanical response.

If the muscle fiber is stimulated so rapidly that it does not have a chance to relax at all between stimuli, a smooth, sustained contraction of maximal strength known as **tetanus** occurs (Figure 8–17c). A tetanic contraction is usually three to four times stronger than a single twitch. (This normal physiological tetanus should not be confused with the disease tetanus; see p. 135.)

■ Twitch summation results from a sustained elevation in cytosolic calcium.

What is the mechanism of twitch summation and tetanus at the cellular level? The tension produced by a contracting muscle fiber increases as a result of greater cross-bridge cycling. As the frequency of action potentials increases, the resultant

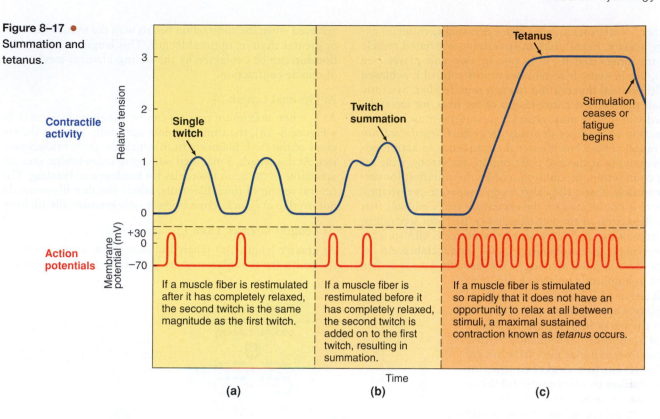

Figure 8–17 ● Summation and tetanus.

Contractile activity

Relative tension

Single twitch

Twitch summation

Tetanus

Stimulation ceases or fatigue begins

Action potentials

Membrane potential (mV)

+30
0
−70

If a muscle fiber is restimulated after it has completely relaxed, the second twitch is the same magnitude as the first twitch.

If a muscle fiber is restimulated before it has completely relaxed, the second twitch is added on to the first twitch, resulting in summation.

If a muscle fiber is stimulated so rapidly that it does not have an opportunity to relax at all between stimuli, a maximal sustained contraction known as *tetanus* occurs.

Time

(a)   (b)   (c)

tension development increases until a maximum tetanic contraction is achieved. Sufficient Ca^{++} is released in response to a single action potential to interact with all the troponin within the cell. As a result, all the cross bridges are free to participate in the contractile response. How then, can repetitive action potentials bring about a greater contractile response? The difference depends on how long sufficient Ca^{++} is available.

The cross bridges remain active and continue to cycle as long as sufficient Ca^{++} is present to keep the troponin-tropomyosin complex away from the cross-bridge binding sites on actin. Each troponin-tropomyosin complex spans a distance of seven actin molecules. Thus binding of one Ca^{++} ion to one troponin molecule leads to the uncovering of only seven cross-bridge binding sites on the thin filament.

As soon as Ca^{++} is released in response to an action potential, the sarcoplasmic reticulum starts pumping Ca^{++} back into the lateral sacs. As the cytosolic Ca^{++} concentration declines with the reuptake of Ca^{++} by the lateral sacs, less Ca^{++} is present to bind with troponin, so some of the troponin-tropomyosin complexes slip back into their blocking positions. Consequently, not all the cross-bridge binding sites remain available to participate in the cycling process during a single twitch induced by a single action potential.

If action potentials and twitches occur far enough apart in time for all the released Ca^{++} from the first contractile response to be pumped back into the lateral sacs between the action potentials, an identical twitch response occurs as a result of the second action potential. If, however, a second action potential occurs and more Ca^{++} is released while the Ca^{++} that was released in response to the first action potential is being taken back up, the cytosolic Ca^{++} concentration remains elevated. This prolonged availability of Ca^{++} in the cytosol permits more of the cross bridges to continue participating in the cycling process for a longer time. As a result,

tension development increases correspondingly. As the frequency of action potentials increases, the duration of elevated cytosolic Ca^{++} concentration increases, and contractile activity likewise increases until a maximum tetanic contraction is reached. With tetanus, the maximum number of cross-bridge binding sites remain uncovered so that cross-bridge cycling, and consequently tension development, are at their peak.

Because skeletal muscle must be stimulated by motor neurons to contract, the nervous system plays a key role in regulating the strength of contraction. The two main factors subject to control to accomplish gradation of contraction of a given muscle are the *number of motor units stimulated* and the *frequency of their stimulation*. The areas of the vertebrate brain responsible for directing motor activity use a combination of tetanic contractions and precisely timed shifts of asynchronous motor unit recruitment to execute smooth rather than jerky contractions.

■ Arthropod muscle tension is controlled by gradation of contraction within a motor unit.

The mechanism by which tension is developed in arthropod muscle is somewhat different from that in vertebrate twitch muscle. Because the nervous system of invertebrates consists of comparatively fewer neurons, a smaller number of overlapping motor units must generate the complete range of tension necessary for the animal to move. In some arthropod muscles, one motor neuron innervates most, if not all, of the muscle fibers. To examine the mechanism by which arthropods grade individual contractions, look at the example of crustacean skeletal muscle fibers.

The limb muscles of crayfish are innervated by very few efferent neurons. In crayfish, crabs, and other decapod crustaceans, two muscles of the terminal joints, the "opener" and "stretcher" (extensor) muscles share one single, common excitatory neuron. Recall that most vertebrate striated muscle fibers are innervated by only one or two end plates (see p. 127), but in many invertebrates motor control is achieved by **multiterminal innervation** of each muscle fiber. Synaptic terminals that run the entire length of the fiber, for example, repeatedly innervate most crustacean skeletal muscle fibers. Contraction of the fiber depends on *graded depolarization* of the muscle fiber: The greater the frequency of action potentials arriving at the end plate region, the stronger the resulting contraction. In addition, unlike the situation for vertebrate muscle fibers, they also receive separate presynaptic inhibitory input located on the excitatory nerve endings that permit the action of the two muscles to perform separately. The release of the inhibitory neurotransmitter GABA (p. 126) onto presynaptic receptors increases Cl^- conductance. An excitatory nerve impulse arriving in the terminal button is then reduced in amplitude because a repolarizing leakage of Cl^- into the nerve terminal takes place when its membrane is depolarized by the nerve impulse. An impulse of reduced amplitude consequently releases less neurotransmitter at the excitatory synapse that results in a weaker contraction.

In addition to presynaptic inhibition of arthropod muscle, a second form of inhibitory control is common. In postsynaptic inhibition, the inhibitory nerve terminals form multiterminal neuromuscular synapses along the muscle fiber. Impulses generated in the inhibitory neuron produce IPSPs in the postsynaptic muscle fiber that oppose the membrane depolarization by EPSPs and reduce the force of the resultant contraction. Of these two mechanisms of inhibition, postsynaptic inhibition is more common. Both mechanisms are important in controlling muscle tension and, in some situations, may speed the rate of relaxation of a contracted muscle.

There is an optimal muscle length at which maximal tension can be developed on a subsequent contraction.

Additional factors not directly under nervous control also influence the tension developed during contraction (Table 8–2). Among these is the length of the fiber at the onset of contraction, to which we now turn.

A relationship exists between the length of the muscle before the onset of contraction, and the tetanic tension that each contracting fiber can subsequently develop at that length. Every muscle has an **optimal length** (l_o) at which maximal force can be achieved on a subsequent tetanic contraction. The tension that can be achieved during tetanus at the optimal muscle length is greater than the tetanic tension that can be

achieved when the contraction begins with the muscle less than or greater than its optimal length. This **length–tension relationship** can be explained by the sliding-filament mechanism of muscle contraction.

At Optimal Length (l_o)

At l_o when maximum tension can be developed (point A in ● Figure 8–18), the thin filaments optimally overlap the regions of the thick filaments from which the cross bridges project. At this length, a maximal number of cross-bridge sites are accessible to the actin molecules for binding and bending. The central region of thick filaments, where the thin filaments do not overlap at l_o lacks cross bridges; only myosin tails are here.

At Lengths Greater Than l_o

At greater lengths, as when a muscle is passively stretched (point B), the thin filaments are pulled out from between the

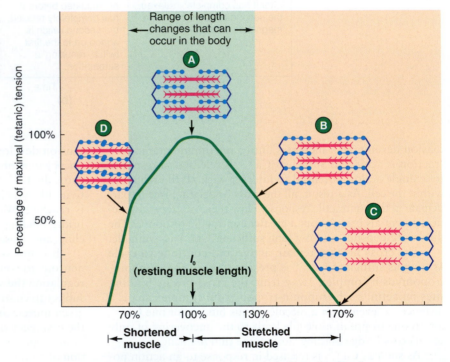

Figure 8–18 ● Length–tension relationship. Maximal tetanic contraction can be achieved when a muscle fiber is at its optimal length (l_o) before the onset of contraction, because this is the point of optimal overlap of thick-filament cross bridges and thin-filament cross-bridge binding sites (point A). The percentage of maximal tetanic contraction that can be achieved decreases when the muscle fiber is longer or shorter than l_o before contraction. When it is longer, fewer thin-filament binding sites are accessible for binding with thick-filament cross bridges, because the thin filaments are pulled out from between the thick filaments (points B and C). When the fiber is shorter, fewer thin-filament binding sites are exposed to thick-filament cross bridges because the thin filaments overlap (point D). Also, further shortening and tension development are impeded as the thick filaments become forced against the Z lines (point D). In the body, the resting muscle length is at l_o. Furthermore, because of restrictions imposed by skeletal attachments, muscles cannot vary beyond 30% of their l_o in either direction (the range screened in light green). At the outer limits of this range, muscles still can achieve about 50% of their maximal tetanic contraction.

thick filaments, decreasing the number of actin sites available for cross-bridge binding; that is, some of the actin sites and cross bridges no longer "match up," so they "go unused." When less cross-bridge activity can occur, less tension can be developed. In fact, when the muscle is stretched to about 70% longer than its l_o (point C), the thin filaments are completely pulled out from between the thick filaments so that no cross-bridge activity and consequently no contraction can occur.

At Lengths Less Than l_o

If a muscle is shorter than l_o before contraction (point D), less tension can be developed, for three reasons:

1. The thin filaments from the opposite sides of the sarcomere become overlapped, decreasing the number of actin sites exposed to the cross bridges.

2. The thick filaments become forced against the Z lines, so further shortening is impeded.

3. Besides these two mechanical factors, at muscle lengths less than 80% of l_o, not as much Ca^{++} is released during excitation–contraction coupling for reasons unknown. Consequently, fewer actin sites are uncovered for participation in cross-bridge activity. Furthermore, the ability of Ca^{++} to bind to troponin and pull the troponin-tropomyosin complex aside is reduced, by an unknown mechanism, at shorter muscle lengths. Consequently, fewer actin sites are uncovered for participation in cross-bridge activity.

Limitations on Muscle Length

The extremes in muscle length that prevent the development of tension occur only under experimental conditions, when a muscle is removed and stimulated at various lengths. In vivo, muscles are usually positioned such that their relaxed length is approximately their optimal length; thus they can achieve near-maximal tetanic contraction most of the time. Because of limitations imposed by attachment to the skeleton, a muscle cannot be stretched or shortened more than 30% of its resting optimal length, and usually it deviates much less than 30% from normal length. Even at the outer limits (130% and 70% of l_o), the muscles can still generate half their maximum tension.

The factors discussed thus far that influence how much tension can be developed by a contracting muscle fiber—the frequency of stimulation and the muscle length at the onset of contraction—can vary from contraction to contraction. Other determinants of muscle fiber tension—the metabolic capability of the fiber relative to resistance to fatigue and the thickness of the fiber—do not vary from contraction to contraction but depend on the fiber type and can be modified over a period of time. We consider these other factors in the next section, on skeletal muscle metabolism and fiber types, after we finish discussing skeletal muscle mechanics.

■ Muscle tension is transmitted to skeletal elements as the contractile component tightens the series-elastic component.

Tension is produced internally within the sarcomeres, the contractile component of the muscle, as a result of cross-bridge activity and the resultant sliding of filaments. However, the sarcomeres are not attached directly to the skeleton. Instead, the tension generated by these contractile elements must be transmitted to the bone or exoskeleton via the connective tissue (such as tendons) before the skeletal element can be moved. Connective tissue, as well as other components of the muscle, such as intracellular elastic proteins, exhibits a certain degree of passive elasticity. These noncontractile tissues are called the **series-elastic component** of the muscle; they behave like a stretchy spring placed between the internal tension-generating elements and the bone that is to be moved against an external load (● Figure 8–19). Shortening of the sarcomeres stretches the series-elastic component. Muscle tension is transmitted to the skeleton by this tightening of the series-elastic component. This externally applied tension is responsible for moving the skeleton against a load.

Biceps

Contractile component (sarcomeres)

Series-elastic component (connective tissue/tendon)

Load

Triceps

Load

Figure 8–19 ● **Relationship between the contractile component and the series-elastic component in transmitting muscle tension to bone.** Muscle tension is transmitted to the bone by means of the stretching and tightening of the muscle's elastic connective tissue and tendon as a result of sarcomere shortening brought about by cross-bridge cycling.

A vertebrate muscle is typically attached to at least two different bones across a joint by means of tendons that extend from each end of the muscle. When the muscle shortens during contraction, the position of the joint is changed as one bone is moved in relation to the other—for example, flexion of the elbow joint by contraction of the biceps (flexor) muscle and extension of the elbow by contraction of the triceps (extensor). The end of the muscle attached to the more stationary part of the skeleton is called the **origin,** and the end attached to the skeletal part that moves is the **insertion.**

Not all vertebrate muscle contractions shorten muscles and move bones, however. For a muscle to shorten during contraction, the tension developed in the muscle must exceed the forces that oppose movement of the bone to which the muscle's insertion is attached. In the case of elbow flexion, the opposing force, or **load,** is the weight of an object being lifted. When you flex your elbow without lifting any external object, there is still a load, albeit a minimal one—the weight of your forearm being moved against the force of gravity.

■ The two primary types of contraction are isotonic and isometric.

There are two primary types of contraction, depending on whether the muscle changes length during contraction. In an **isotonic contraction,** muscle tension remains constant as the muscle changes length. In an **isometric contraction,** the muscle is prevented from shortening, so tension develops at constant muscle length. The same internal events occur in both isotonic and isometric contractions: The tension-generating contractile process is turned on by muscle excitation; the cross bridges start cycling; and filament sliding shortens the sarcomeres, which stretches the series-elastic component to exert tension on the bone at the site of the muscle's insertion.

Considering your biceps as an example, assume you are going to lift an object. When the tension developing in your biceps becomes great enough to overcome the weight of the object in your hand, you can lift the object, with the whole muscle shortening in the process. Because the weight of the object does not change as it is lifted, the muscle tension remains constant throughout the period of shortening. This is an isotonic (literally, "constant tension") contraction. Isotonic contractions are used for body movements and for moving external objects.

What happens if you try to lift an object too heavy for you (that is, if the tension you can develop in your arm muscles is less than that required to lift the load)? In this case, the muscle cannot shorten and lift the object but remains at constant length despite the development of tension, so an isometric ("constant length") contraction occurs. In addition to occurring when the load is too great, isometric contractions also take place when the tension developed in the muscle is deliberately less than that needed to move the load. In this case, the goal is to keep the muscle at fixed length although it is capable of developing more tension. These submaximal isometric contractions are important for maintaining posture (such as keeping the legs stiff while standing) and for supporting objects in a fixed position. During a given movement, a muscle may shift between isotonic and isometric contractions. For example, when you pick up a book to read, your biceps undergoes an isotonic contraction while the book is being lifted, but the contraction becomes isometric as you stop to hold the book in front of you.

Concentric and Eccentric Isotonic Contractions

There are actually two types of isotonic contraction—concentric and eccentric. In both cases the muscle changes length at constant tension. With **concentric contractions,** however, the muscle shortens, whereas with eccentric contractions the *muscle lengthens* because it is being stretched by an external force while contracting. With an **eccentric contraction,** the contractile activity is resisting the stretch. An example is lowering a load to the ground. During this action, the muscle fibers in the biceps are lengthening but are still contracting in opposition to being stretched. This tension supports the weight of the object.

Other Contractions

An animal's body is not limited to pure isotonic and isometric contractions. Muscle length and tension frequently vary throughout a range of motion. Think about drawing a bow and arrow. The tension of the biceps muscle continuously increases to overcome the progressively increasing resistance as the bow is stretched further. At the same time, the muscle progressively shortens as the bow is drawn farther back. Such a contraction occurs at neither constant tension nor constant length.

Some skeletal muscles do not attach to bones at both ends but still produce movement. For example, the muscles of the tongue are not attached at the free end. Isotonic contractions of the tongue muscles maneuver the free, unattached portion of the tongue to facilitate eating (and speech in humans). The external eye muscles attach to the skull at their origin but to the eye at their insertion. Isotonic contractions of these muscles produce the eye movements that enable an animal to track moving objects, for example. A few skeletal muscles that are completely unattached to bone actually prevent movement. These are the rings of skeletal muscles known as **sphincters** (p. 562) that guard the exit of urine and feces from the body by isotonically contracting.

■ The velocity of shortening is related to the load.

The load is also an important determinant of the velocity, or speed, of shortening (● Figure 8–20). The greater the load, the lower the velocity at which a single muscle fiber (or a con-

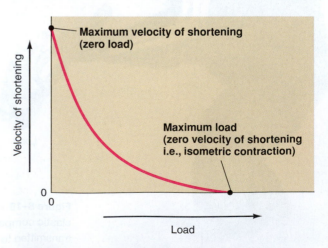

Figure 8–20 ● **Load–velocity relationship.** The velocity of shortening decreases as the load increases.

stant number of contracting fibers within a muscle) shortens during an isotonic tetanic contraction. The velocity of shortening is maximal when there is no external load, progressively decreases with an increasing load, and falls to zero (no shortening—isometric contraction) when the load cannot be overcome by maximal tetanic tension. You have frequently experienced this load–velocity relationship: You can quickly lift light objects requiring little muscle tension, whereas you can lift very heavy objects only slowly, if at all. This relationship between load and shortening velocity is a fundamental property of muscle, presumably because it takes the cross bridges longer to stroke against a greater load.

■ Although muscles can accomplish work, much of the energy is converted to heat.

Muscle accomplishes work in a physical sense only when an object is moved. **Work** is defined as force times distance. **Force** can be equated to the muscle tension required to overcome the load (the weight of the object). The amount of work accomplished by a contracting muscle therefore depends on how much an object weighs and how far it is moved. In an isometric contraction when no object is moved, the muscle contraction's efficiency as a producer of external work is zero. All energy consumed by the muscle during the contraction is converted to heat. In an isotonic contraction, the muscle's efficiency is about 25%. Of the energy consumed by the muscle during the contraction, 25% is realized as external work whereas the remaining 75% is converted to heat.

Much of this heat is not really wasted, in a physiological sense, because it is important in some animals in maintaining elevated core-body temperatures (such as tuna, bumble bees, birds, mammals). In fact, shivering, a form of involuntarily induced skeletal muscle contraction in mammals, birds, and some insects, is a well-known mechanism for increasing heat production on a cold day or to fight infections with a fever. Heavy exertion on a hot day, in contrast, may overheat the body, because the normal heat-loss mechanisms may be unable to compensate for this increase in heat production (see Chapter 15).

■ Interactive units of skeletal muscles, tendons, skeletons, and joints form lever systems.

Most skeletal muscles are attached to bones or exoskeleton segments across **joints**, forming lever systems. A **lever** is a rigid structure capable of moving around a pivot point known as a **fulcrum.** In the vertebrate body, the bones function as levers, the joints serve as fulcrums, and the skeletal muscles provide the force to move the bones. The part of a lever between the fulcrum and the point where an upward force is applied is called the **power arm;** the part between the fulcrum and the downward force exerted by a load is the **load arm** (● Figure 8–21a).

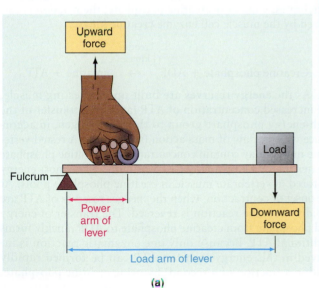

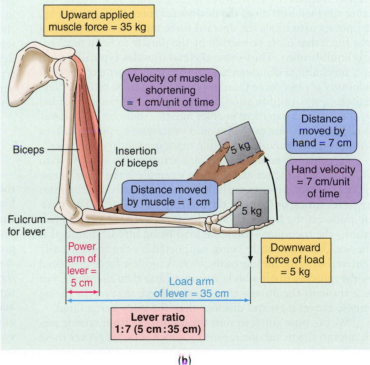

Figure 8–21 ● **Lever systems of muscles, bones, and joints.** (a) Schematic representation of the most common type of lever system in the body, showing the location of the fulcrum, the upward force, the downward force, the power arm, and the load arm. (b) Flexion of the elbow joint as an example of lever action in the body. Note that the lever ratio (length of the power arm to length of the load arm) is 1:7 (5 cm:35 cm), which amplifies the distance and velocity of movement seven times [distance moved by the muscle (extent of shortening) = 1 cm, distance moved by the hand = 7 cm; velocity of muscle shortening = 1 cm/unit of time, hand velocity = 7 cm/unit of time], but at the expense of the muscle having to exert seven times the force of the load (muscle force = 35 kg, load = 5 kg).

The most common type of lever system in the mammalian body is exemplified by flexion of the human elbow joint. Skeletal muscles, such as the biceps whose contraction flexes the elbow joint, consist of many parallel (side-by-side) tension-generating fibers that can exert a large force at their insertion but shorten only a small distance and at a relatively slow velocity. The lever system of the elbow joint amplifies the slow, short movements of the biceps to produce more rapid movements of the hand that cover a greater distance.

Consider how an object weighing 5 kg is lifted by the hand (Figure 8–21b). When the biceps contracts, it exerts an upward force at the point where it inserts on the forearm bone about 5 cm away from the elbow joint, the fulcrum. Thus the power arm of this lever system is 5 cm long. The length of the load arm, the distance from the elbow joint to the hand, averages 35 cm. In this case, the load arm is seven times longer than the power arm, which enables the load to be moved a distance seven times greater than the shortening distance of the muscle (while the biceps shortens a distance of 1 cm, the hand moves the load a distance of 7 cm) and at a velocity seven times greater (the hand moves 7 cm during the same length of time the biceps shortens 1 cm).

The disadvantage of this lever system is that at its insertion the muscle must exert a force seven times greater than the load. The product of the length of the power arm times the upward force applied must equal the product of the length of the load arm times the downward force exerted by the load. Because the load arm times the downward force is 35 cm × 5 kg, the power arm times the upward force must be 5 cm × 35 kg (the force that must be exerted by the muscle to be in mechanical equilibrium). Thus some skeletal muscles typically work at a mechanical disadvantage in that they must exert a considerably greater force than the actual load to be moved. Nevertheless, the amplification of velocity and distance afforded by the lever arrangement enables muscles to move loads faster over greater distances than would otherwise be possible. This amplification provides valuable maneuverability and speed.

In contrast, some muscles are arranged to amplify load lifting, at the expense of speed. An example is the "calf" muscle or *gastrocnemius* in the hindlimbs of terrestrial vertebrates (used for lifting or jumping the body up on the toes). This muscle pulls on the heelbone at a distance farther from the fulcrum—the toe joints—than the load being lifted. That load is the weight of the body being supported by the foot, usually centered over the middle of the foot. A modest force produced by the muscle can lift a body weight greater than that force, but as a tradeoff the movement is deamplified; that is, the lifting muscle moves a greater distance than the body being lifted.

We are now going to shift the discussion from muscle mechanics to the metabolic means by which muscles power these movements.

Skeletal Muscle Metabolism and Fiber Types

Three different steps in the contraction–relaxation process require ATP:

1. Splitting of ATP by myosin ATPase provides the energy for the power stroke of the cross bridge.

2. Binding (but not splitting) of a fresh molecule of ATP to myosin permits detachment of the bridge from the actin filament at the end of a power stroke so that the cycle can be repeated. This ATP is subsequently split to provide energy for the next stroke of the cross bridge, as indicated in step 1.

3. The active transport of Ca^{++} back into the sarcoplasmic reticulum during relaxation depends on energy derived from the breakdown of ATP.

■ Muscle fibers have alternate pathways for forming ATP.

Because ATP is the only energy source that can be directly used for these activities, for contractile activity to continue ATP must constantly be supplied. Only limited stores of ATP are immediately available in muscle tissue, but three pathways supply additional ATP as needed during muscle contraction: (1) transfer of a high-energy phosphate from a phosphogen to ADP, (2) oxidative phosphorylation (the citric acid cycle and electron transport system), and (3) glycolysis. We discussed these at length in Chapter 2 (pp. 46–49) and give only a brief overview here.

Phosphogens

The phosphogens **creatine phosphate** and **arginine phosphate,** described on p. 52, are often the first energy storehouse tapped at the onset of contractile activity (● Figure 8–22a). Like ATP, the phosphogens contain a high-energy phosphate group, which can be donated directly to ADP to form ATP. This reaction is reversible; energy and phosphate from ATP can be transferred to creatine or arginine to form the phosphate. In vertebrates, which use creatine phosphate, the reaction is catalyzed by the muscle cell enzyme **creatine kinase:**

$$\text{Creatine phosphate} + \text{ADP} \overset{\text{creatine kinase}}{\longleftrightarrow} \text{creatine} + \text{ATP}$$

As the energy reserves are built up in a resting muscle, the increased concentration of ATP favors the transfer of the high-energy phosphate group to the phosphogen, in accordance with the law of mass action (see p. 500). A rested vertebrate muscle may contain concentrations of creatine phosphate greater than ATP by a factor of 5 or more. Thus most energy is stored in vertebrate muscle in creatine phosphate pools. At the onset of contraction, when the meager reserves of ATP are rapidly used, the reaction is reversed. The transfer of energy and phosphate from creatine phosphate to ADP quickly forms additional ATP. Because only one enzymatic reaction is involved in this energy transfer, ATP can be formed rapidly (within a fraction of a second) by using creatine phosphate.

Muscle ATP levels actually remain fairly constant early in contraction, but phosphogen stores become depleted. In fact, short bursts of high-intensity contractile effort are supported primarily by ATP derived at the expense of phosphogen. Other energy systems do not have a chance to become operable before the activity is over.

Oxidative Phosphorylation

If the energy-dependent contractile activity is to be continued, most animal muscles shift to the pathways of oxidative phosphorylation and glycolysis to form ATP. These multistep pathways require time to pick up their rates of ATP formation to match the increased demands for energy, time provided by the immediate supply of energy from the one-step phosphogen system.

If enough O_2 is present, **oxidative phosphorylation** (see p. 47) takes place within the muscle mitochon-dria. This pathway is fueled by glucose or fatty acids, depending on the intensity and duration of the activity (Figure 8–22b). Although it provides a rich yield of 36 ATP molecules for each glucose molecule processed, oxidative phosphorylation is relatively slow because of the number of steps involved, and it requires a constant supply of O_2 and nutrient fuel.

During light activity (such as walking) to moderate activity (such as flight or swimming), muscle cells can form sufficient amounts of ATP through oxidative phosphorylation to

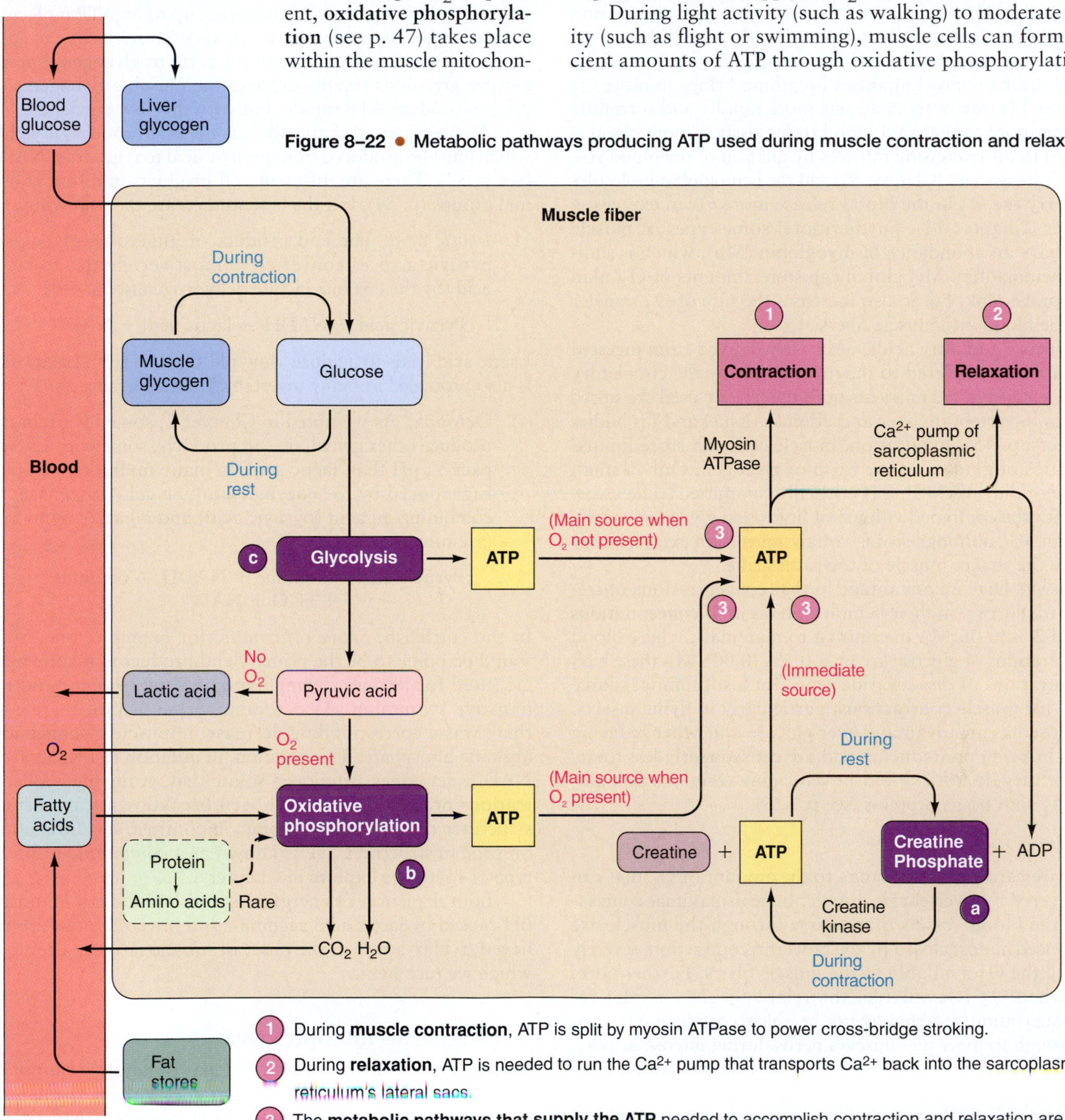

Figure 8–22 ● Metabolic pathways producing ATP used during muscle contraction and relaxation.

① During **muscle contraction**, ATP is split by myosin ATPase to power cross-bridge stroking.

② During **relaxation**, ATP is needed to run the Ca^{2+} pump that transports Ca^{2+} back into the sarcoplasmic reticulum's lateral sacs.

③ The **metabolic pathways that supply the ATP** needed to accomplish contraction and relaxation are

ⓐ transfer of a high-energy phosphate from **creatine phosphate** to ADP (immediate source);

ⓑ **oxidative phosphorylation** (the main source when O_2 is present), fueled by glucose derived from muscle glycogen stores or by glucose and fatty acids delivered by the blood; and

ⓒ **glycolysis** (the main source when O_2 is not present). The end product of glycolysis, pyruvic acid, is converted into lactic acid when the lack of O_2 prevents pyruvic acid from being further processed by the oxidative phosphorylation pathway.

keep pace with the modest energy demands of the contractile machinery for prolonged periods of time. To sustain ongoing oxidative phosphorylation, the exercising muscles depend on delivery of adequate O_2 (such as by circulatory systems or tracheae) and nutrient supplies. Activity that can be supported in this way is known as **aerobic** ("with O_2") or **endurance-type** activity. The O_2 required for oxidative phosphorylation in vertebrates is primarily delivered by the blood. Increased O_2 is made available to vertebrate muscles during exercise by several mechanisms: Enhanced breathing brings in more O_2 (Chapter 11), the heart contracts more rapidly and forcefully to pump more oxygenated blood to the tissues, more blood is diverted to the exercising muscles by dilation of the blood vessels supplying them (Chapter 9), and the **hemoglobin** molecules that carry the O_2 in the blood release more O_2 in exercising muscles (Chapter 11). Furthermore, some types of muscle fibers have an abundance of **myoglobin (Mb)**, which is similar to hemoglobin. Myoglobin can store amounts of O_2, akin to a storage tank, but it also increases the rate of O_2 transfer from the blood into muscle fibers.

Glucose and fatty acids, ultimately derived from ingested food, are also delivered to the muscle cells by the circulatory fluid. Excess ingested nutrients not immediately used are stored (such as in vertebrate liver and adipose tissue, and fat bodies in insects; p. 45). In addition, muscle cells can store limited quantities of glucose in the form of glycogen. Only a small volume of the muscle fiber is normally committed to fuel storage, with the combined volume of lipids and glycogen usually less than 3%, although storage of glycogen is an extraordinary 17% in the shaker muscle of the rattlesnake.

Insects have an advantage in that concentrations of trehalose (diglucose) in their hemolymph occur at concentrations around 2% (0.06 M), in contrast to mammals, where blood concentrations of glucose are only 0.1% (0.005 M). these high concentrations of disaccharide represent a substantial source of fuel for muscle contractions, particularly in flying insects. Trehalose has an advantage over glucose and other reducing sugars in being nonreducing and are consequently less toxic, because they do not exhibit the damaging reactions that glucose has with tissue proteins (see p. 526).

Glycolysis

There are cardiovascular limits to the amount of O_2 that can be delivered to a vertebrate muscle. In near-maximal contractions, the blood vessels that course through the muscle are compressed almost closed by the powerful contraction, severely limiting the O_2 available to the muscle fibers. Furthermore, even when O_2 is available, the relatively slow oxidative-phosphorylation system may not be able to produce ATP rapidly enough to meet the muscle's needs during intense activity. When O_2 delivery or oxidative phosphorylation cannot keep pace with the demand for ATP formation as the intensity of activity increases, the muscle fibers rely increasingly on **glycolysis** (Figure 8–22c) to generate ATP (see p. 46). Glycolysis alone has two advantages over the oxidative-phosphorylation pathway: (1) Glycolysis can form ATP in the absence of O_2 (operating anaerobically, that is, "without O_2"), and (2) it can proceed much more rapidly than oxidative phosphorylation because it requires fewer steps and can work with only local cellular molecules (that is, no external oxygen is needed). Although glycolysis extracts considerably fewer ATP molecules from each nutrient molecule processed, it can proceed so much

more rapidly that it can outproduce oxidative phosphorylation in ATP yield over a given period of time if enough glucose is present.

However, there is a disadvantage of the low efficiency of glycolysis (recall that glycolysis yields a net of 2 ATP molecules for each glucose molecule degraded, whereas the oxidative-phosphorylation pathway can extract up to 36 ATP molecules from each glucose molecule; see p. 49.). Muscle cells can store limited quantities of glucose in the form of glycogen, but anaerobic glycolysis rapidly depletes the muscle's glycogen supplies to produce ATP rapidly but with this low efficiency.

Another aspect of glycolysis is the end products it yields, which must be produced from pyruvic acid to regenerate NAD^+ (see p. 54). There are different end products in different animal groups (p. 54), but the best studied are the following:

1. *Lactic acid.* The end product of anaerobic glycolysis, pyruvic acid, is converted in most vertebrates to **lactic acid** (by the enzyme **lactate dehydrogenase**; p. 54):

$$\text{Pyruvic acid} + \text{NADH} \rightarrow \text{lactic acid} + \text{NAD}^+$$

Lactic acid leads to acidosis (low pH/high acidity). Lactic acid is also produced in many invertebrates such as insects.

2. *Octopine.* As we noted in Chapter 2, some invertebrates produce other glycolytic end products, some with less impact on pH than lactic acid. In many mollusks, **octopine** is produced by the enzyme octopine dehydrogenase, by combining an acid (pyruvic acid) and a basic amino acid (arginine):

$$\text{Pyruvic acid} + \text{arginine} + \text{NADH} \rightarrow \text{octopine} \\ + \text{H}_2\text{O} + \text{NAD}^+$$

In the cuttlefish, *Sepia officinalis,* for example, one study found octopine to be the primary end product in mantle muscle (used for jetting) during hypoxia (low oxygen) and exhaustive swimming. As octopine concentration increased, there was a corresponding decrease in muscle glycogen and arginine phosphate. This reaction, in addition to regenerating NAD^+, detoxifies arginine, a solute that, at the high concentrations produced by phosphogen breakdown, can perturb structures of proteins. Octopine, in contrast, has little effect on protein structures and is known as a *compatible solute* (a type of solute we explore in Chapter 13).

Both depletion of energy reserves, and the fall in muscle pH caused by lactic acid accumulation in vertebrates, are believed to play a role in the onset of muscle fatigue, a topic to which we turn next.

◼ Fatigue has multiple causes.

Contractile activity in a particular skeletal muscle cannot be maintained at a given level indefinitely. Eventually the tension in the muscle declines as fatigue sets in. **Muscle fatigue** occurs when an exercising muscle can no longer respond to stimulation with the same degree of contractile activity. The underlying causes of muscle fatigue are unclear. The primary implicated factors include (1) a local increase in *inorganic phosphate* resulting from the breakdown of phosphogens; (2) accumulation of *lactic acid*, which (via release of H^+) may inhibit key enzymes in the energy-producing pathways or the excitation–contraction coupling process; and (3) *depletion of energy reserves*. The time of onset of fatigue varies with the type of muscle fiber, some fibers being more resistant to fatigue than others, and the intensity of the activity, more rapid onset of fatigue associated with high-intensity activities.

■ Increased oxygen consumption is necessary to recover from activity.

An animal continues to breathe deeply and rapidly for a period of time after exhaustive activity. The necessity for the elevated O_2 uptake during recovery from exercise is due to a variety of factors. The best known is repayment of an **oxygen debt** incurred during the activity, when ATP derived from non-oxidative sources such as creatine phosphate and anaerobic glycolysis was supporting contractile activity. Oxygen is needed for recovery of the energy systems. During the activity, phosphogen stores of active muscles are reduced, lactic acid may accumulate, and glycogen stores may be tapped; the extent of these effects depends on the intensity and duration of the activity. During the recovery period, fresh supplies of ATP are formed by oxidative phosphorylation using the newly acquired O_2, which is provided by the sustained increase in respiratory activity after exercise has stopped. Much of this ATP is used to resynthesize phosphogen to restore its reserves. This can be accomplished in a matter of a few minutes in mammals. Any accumulated lactic acid (or octopine) is converted back into pyruvic acid, part of which is used by the oxidative-phosphorylation system for ATP production (some of this takes place in other organs such as vertebrate liver and heart, which take up lactic acid from the bloodstream). The remainder of the pyruvic acid is converted back into glucose (primarily by the liver in vertebrates). Most of this glucose in turn is used to replenish the glycogen stores drained from the muscles and liver during exercise. These biochemical transformations involving pyruvic acid require O_2 and take several hours for completion. Thus, as the O_2 debt is repaid, the phosphogen system is restored, lactic acid (or octopine) is removed, and glycogen stores are at least partially replenished.

Unrelated to increased O_2 uptake is the need to restore nutrients after grueling activity, such as a fight between two males competing for females, or fleeing from a predator. In such cases, glycogen stores are severely depleted, and long-term recovery can take a day or more, because the exhausted energy stores require nutrient intake for full replenishment. Therefore, depending on the type and duration of activity, recovery can be complete within a few minutes or can require more than a day.

Part of the extra O_2 uptake during recovery is not directly related to the repayment of energy stores but instead is the result of a general metabolic disturbance after exercise. For example, the local increase in muscle temperature arising from heat-generating contractile activity speeds up the rate of all chemical reactions in the muscle tissue, including those dependent on O_2. Likewise, the secretion of epinephrine, a vertebrate hormone that increases O_2 consumption, is elevated during moderate activity. Until the circulating level of epinephrine returns to its pre-exercise state, O_2 uptake is above normal.

We have been looking at the contractile and metabolic activities of skeletal muscle in general. Yet not all skeletal muscle fibers use these mechanisms to the same extent. We are next going to examine the different types of muscle fibers, based on their speed of contraction and how they are metabolically equipped to generate ATP.

■ There are three types of skeletal muscle fibers, which differ in ATP hydrolysis and synthesis.

No one muscle-fiber type can perform all the activities essential for an animal to survive. For this reason, muscle fibers show the greatest diversity of any tissue in an animal's body. Based on their biochemical capacities and physiological requirements, vertebrate skeletal muscle fibers have been classified into three major groups (■ Table 8–3):

1. Slow-oxidative (type I) fibers
2. Fast-oxidative (type IIa) fibers
3. Fast-glycolytic (type IIb, IId, IIx) fibers. (In addition, there is a type IIc fiber, an undifferentiated form that can become other type II forms).

As their names imply, the two main differences between these fiber types are their speed of contraction (slow or fast) and the type of enzymatic machinery they primarily use for ATP formation (oxidative or glycolytic).

Fast versus Slow Fibers

Fast fibers have higher myosin-ATPase (ATP-splitting) activity than slow fibers because of different myosin isoforms (see box, "Molecular Biology and Genomics: Myosin Family Genetics"). The higher the ATPase activity, the more rapidly ATP is split and the faster the rate at which energy is made available for cross-bridge cycling. The result is a fast twitch, compared to the slower twitches of those fibers that split ATP more slowly. Thus, two factors determine the speed with which a muscle contracts: the load (load–velocity relationship) and

Table 8–3 ■ Characteristics of Skeletal Muscle Fibers

Characteristic	Type of Fiber		
	Slow-oxidative (type I)	Fast-oxidative (type IIa)	Fast-glycolytic (type IIb)
Myosin-ATPase activity	Low	High	High
Speed of contraction	Slow	Fast	Fast
Resistance to fatigue	High	Intermediate	Low
Oxidative phosphorylation capacity	High	High	Low
Enzymes for anaerobic glycolysis	Low	Intermediate	High
Mitochondria	Many	Many	Few
Capillaries	Many	Many	Few
Myoglobin content	High	High	Low
Color of fiber	Red	Red	White
Glycogen content	Low	Intermediate	High
Intensity of contraction	Low	Intermediate	High

MOLECULAR BIOLOGY AND GENOMICS:

Myosin Family Genetics

As noted in the main text, there are at least three different functional categories of muscle fibers: In humans, these are termed slow-oxidative or Type I, fast-oxidative or Type IIa, and fast-glycolytic or Type IIx. Scientists have known for some time that each of these expresses a different myosin protein from its own myosin gene; each myosin has catalytic properties suited to its muscle type. Thus researchers long suspected that there are at least three different myosin genes. Genetic analysis has revealed that there is indeed a large variety of related myosins, with over 10 different myosin gene types in muscles among different mammals and among different muscle tissues. These are designated as follows:

- Embryonic

- Neonatal

- Cardiac α

- Cardiac β or slow type I (in skeletal muscle)

- Fast type IIa

- Fast type IIx

- Fast type IId

- Fast type IIb

- Extraocular

- Mandibular or masticatory (m-MHC)

- Smooth muscle type (SMMHC)

All these are considered to be *isoforms* of myosin class II, the type in muscle (not to be confused with subtype IIa, and so on) In addition, there are numerous nonmuscle myosin classes such as myosin class VI, associated with movements of single cells. Mutations in the myosin VI gene in mice leads to deafness, because the stereocilia used in hearing (p. 222) are nonfunctional.

Fast-glycolytic types termed IIb and IId are found in some mammals, and are faster than the fastest human type IIx. Type IId and IIb are found in rodents, animals that scurry around at high limb velocities. These myosins can cause contraction of muscle at a faster rate than the type IIx found in larger mammals.

The expression of myosin genes change with activity and other environmental factors related to muscle use, in concert with changes in muscle fiber types (see main text). In human muscles repeatedly loaded, as in weight training, the genes of fast IIx fibers stop transcribing the IIx myosin and begin transcribing the IIa myosin, as the fibers convert from IIx type to IIa type. (In rodents subjected to the equivalent of weight training, fibers convert from IIb to IIx.) In contrast, in sedentary people much of the limb muscle converts to type IIx (which fits the way their muscles are used: quiet most of the time, with intermittent short bursts of usage).

Smooth muscle appears to be different. There is one smooth-muscle myosin gene in mammals, but it can yield up to four different types of myosin proteins: SM1A, SM1B, SM2A, and SM2B. This is accomplished by the process of *alternative RNA splicing* in the nucleus, a process in which the same parent RNA transcribed from one gene can be cut and spliced during intron removal to yield different mRNAs.

---

the myosin ATPase activity of the contracting fibers (fast or slow twitch).

Oxidative versus Glycolytic Fibers

Fiber types also differ in ATP-synthesizing ability. Those with a greater capacity to form ATP are more resistant to fatigue. Some fibers are better equipped for oxidative phosphorylation, whereas others rely primarily on anaerobic glycolysis for synthesizing ATP. Because oxidative phosphorylation yields considerably more ATP from each nutrient molecule processed, it does not readily deplete energy stores. Furthermore, it does not result in lactic acid accumulation. Oxidative types of muscle fibers are therefore more resistant to fatigue than are glycolytic fibers.

Other related characteristics distinguishing these three fiber types are summarized in Table 8–3. As would be expected, the oxidative fibers, both slow and fast, contain an abundance of mitochondria, the organelles that house the enzymes involved in the oxidative-phosphorylation pathway. Because adequate oxygenation is essential to support this pathway, these fibers are richly supplied with capillaries. Oxidative fibers also have a high myoglobin (Mb) content. The highest known myoglobin contents are those in swimming muscles of diving penguins and mammals; for example, the

pectoralis muscle (which operates the wings) of a king penguin has 4.3 g of Mb per 100 g of muscle, whereas a gull's pectoralis has only 0.55 g Mb/100g and a chicken's pectoralis only 0.03 g Mb/100g. Myoglobin not only helps support oxidative fibers' O_2 dependency, but it also imparts a red color to them, just as oxygenated hemoglobin is responsible for the red color of arterial blood. Accordingly, these muscle fibers are referred to as **red fibers.** In contrast, the fast fibers specialized for glycolysis (IIb, IId, IIx) contain few mitochondria but have a high content of glycolytic enzymes instead. Also, to supply the large amounts of glucose necessary for glycolysis, they contain an abundance of stored glycogen. Because the glycolytic fibers need relatively less O_2 to function, they receive only a meager capillary supply compared with the oxidative fibers. The glycolytic fibers contain very little myoglobin and therefore are pale in color, so they are sometimes called **white fibers.** (The most readily observable comparison between red and white fibers is the dark and white meat in poultry, and the red meat of a tuna compared to the white meat of a halibut.) In addition to their higher myosin ATPase content, these fibers are also larger in diameter than the slow-oxidative fibers because of a greater abundance of actin and myosin filaments. With these additional power-generating filaments, the fast-glycolytic fibers are capable of rapidly producing large amounts of tension, but only for a short duration because of their dependence on glycolysis for energy production. For example, the explosive "burst" of flapping flight in many species of galliform birds (such as grouse) is used to escape predators but can be sustained for only a few minutes. Appropriately, these species of birds have relatively small heart masses and a reduced capacity to supply the flight muscles with oxygen. For these reasons galliform birds are restricted to anaerobic flights of short duration and are adapted to a largely ground-based life.

Fast-oxidative fibers share characteristics with each of the other two types. They have high ATPase activity like the fast-glycolytic fibers and high oxidative capacity like the slow-oxidative fibers. They contract more rapidly than the slow-oxidative fibers and can maintain the contraction for a longer period of time than the fast-glycolytic fibers. However, because their rate of ATP production by oxidative phosphorylation cannot keep pace with the high rate of ATP splitting, they rely partially on glycolysis and are more prone to fatigue than the slow-oxidative fibers. Thus the flight muscles of migratory birds contain an abundance of fast-oxidative glycolytic fibers. Normally, the flight muscles of small birds are composed entirely of this muscle type, whereas larger birds have a mixture of different fiber types.

Consider as well the energy requirements of the muscle fibers in a horse as its gait changes from a walk into a canter. While the horse is walking, Type I fibers are recruited because the muscle contracts slowly and energy requirements are minimal. During this phase the muscle fibers rely primarily on fat as an energy source. However, as speed increases to a trot, Type I fibers can no longer contract rapidly enough to propel the horse. Type IIa fibers are now recruited. Because fat cannot be metabolized anaerobically, glycogen serves as the primary energy source for ATP production. As the speed of the horse increases to a gallop, Type IIx muscle fibers are recruited. ATP production is now almost exclusively generated through anaerobic pathways. However, although ATP is rapidly generated it is associated with the generation of lactic acid, which results in a decline in muscle pH and the onset of fatigue.

Genetic Endowment of Muscle Fiber Types

Most vertebrate muscles contain a mixture of all three fiber types; but the type of activity for which the muscle is specialized largely determines the percentage of each type. Accordingly, a high proportion of slow-oxidative fibers are found in muscles specialized for maintaining low-intensity contractions for long periods of time without fatigue, such as the muscles of the back and legs that support a human's weight against the force of gravity. In fish that are inactive much of the time (such as a halibut), almost all the muscle is fast glycolytic (white), used for burst activity. The deep layers of the pectoralis muscles of birds adapted for soaring and gliding modes of flight contain more slow-oxidative fibers (red). Fast-glycolytic fibers are preponderant in the arm muscles of primates, which are adapted for rapid, forceful movements.

■ Muscle fibers adapt considerably in response to the demands placed on them.

Not only is the nerve supply to a skeletal muscle essential for initiating contraction, but the motor neurons supplying a skeletal muscle are also important in maintaining the muscle's

Icelandic horses galloping. As these animals change their gaits from walking to trotting to galloping, more motor units are activated and different fiber types are recruited, with Type I (aerobic) dominating in a walk and Type IIx (anaerobic) dominating in a gallop.

integrity and chemical composition. Different types of activity produce different patterns of neuronal discharge to the muscle involved. Depending on the pattern of neural activity, long-term adaptive changes occur in the muscle fibers, enabling them to respond most efficiently to the types of demands placed on the muscle. Three types of changes can be induced in muscle fibers: changes in their ATP-synthesizing capacity, in myosin isoforms and thus contractile speed and force, and in their diameter.

Improvement in Oxidative Capacity

Regular endurance (aerobic) exercise, such as flight or swimming, induces metabolic changes within the oxidative fibers, which are the ones primarily recruited during aerobic exercise. These changes enable the muscles to use O_2 more efficiently. For example, mitochondria, the organelles that house the enzymes involved in the oxidative-phosphorylation pathway, increase in number in the oxidative fibers. In addition, the number of capillaries supplying blood to these fibers increases. Muscles so adapted are better able to endure prolonged activity without fatiguing, but they do not enlarge. In the example of the horse whose gait is increased from a walk to a canter, sufficient endurance training ultimately enhances the aerobic ability of its muscle to use fuels through oxidation. Blind mole rats, with most of their muscle mass located in the anterior end of the body, find themselves digging in tunnels with scant oxygen supplies. Researchers found that, as an adaptation to these extreme conditions, capillary density was approximately 30% higher in mole rat muscles compared to white rat muscles. Although mole rats have less muscle mass than white rats, mitochondrial density is about 50% higher.

Muscle Hypertrophy

The actual size of the muscles can also be increased by regular bouts of anaerobic, short-duration, high-intensity training. The resulting muscle enlargement comes primarily from an increase in diameter (**hypertrophy**) of the fast-glycolytic fibers that are called into play during such powerful contractions. Most of the fiber thickening is a consequence of increased synthesis of myosin and actin filaments, which permits a greater opportunity for cross-bridge interaction and subsequently increases the muscle's contractile strength. The resultant bulging muscles are better adapted to activities that require intense strength for brief periods, but endurance has not improved.

Influence of Testosterone

Muscle fibers of many mammalian males are thicker, and accordingly, their muscles are larger and stronger than those of females because of the actions of testosterone, a steroid hormone secreted primarily in males. Testosterone promotes the synthesis and assembly of myosin and actin and is responsible for the naturally larger muscle mass of males. Consider the effect of the photoperiod-induced increase in testosterone on the development of muscle in the necks of male deer prior to the onset of the rutting season. Only the fittest males (those that dominate in battles that involve clashes with the head and antlers) have access to the harem and can reproduce. The effect of testosterone on muscle has led some human athletes (both male and female) to the dangerous practice of taking this or related steroids (testosterone agonists).

Interconversion between Fast Muscle-Fiber Types

Apparently all the muscle fibers within a single vertebrate motor unit are of the same fiber type. This pattern usually is established early in life, but the two types of fast-twitch fibers are interconvertible, depending on usage. Regular endurance activities can convert fast-glycolytic fibers into fast-oxidative fibers, whereas fast-oxidative fibers can be shifted to fast-glycolytic fibers in response to power events such as weight training. Slow and fast fibers are not interconvertible, however, but the proportions can change to some extent with activity. Adaptive changes that take place in skeletal muscle gradually reverse to their original state over a period of months if the regular exercise program that induced these changes is discontinued. During these changes, the type of myosin isoform that is expressed also changes (see box, "Molecular Biology and Genomics: Myosin Family Genetics").

Although training can induce changes in muscle fibers' metabolic support systems, whether a fiber is fast or slow twitch depends on the fiber's nerve supply. Slow-twitch fibers are supplied by motor neurons that exhibit a low-frequency pattern of electrical activity, whereas fast-twitch fibers are innervated by motor neurons that display intermittent rapid bursts of electrical activity. Experimental switching of motor neurons supplying slow muscle fibers with those supplying fast fibers gradually reverses the speed at which these fibers contract.

?

What Signals Trigger Changes in Muscle Mass and Fiber Type?

As you've seen, muscle features change considerably under different patterns of neural stimulation and usage. But what are the actual signals involved? The answers are not complete, but neural stimulation of muscles causes changes in several candidate signal molecules: neurotrophic growth factors, cAMP, calcium, and nitric oxide. *Nitric oxide*, a widespread paracrine (p. 88), is produced in skeletal

muscle in relation to the load on the muscle, and may thus trigger changes involved in hypertrophy. Calcium, in contrast, builds up in chronically stimulated muscle cells and activates the protein *calcineurin*. Its role was suggested with mice that were genetically engineered to overexpress calcineurin. Researchers found these mice were converting fast fibers to slow ones, with no hypertrophy. Calcineurin appears to lead to activation of a transcription factor that in turn activates genes involved in slow-muscle fiber development. Thus two distinct signaling pathways may be involved in hypertrophy and fiber specification, respectively.

Adaptations for Flight: Continuous High Power at High Contraction Frequencies

Consider the price a flying animal has had to pay in order to fly. Bats and birds must be lightweight, and so have evolved very lightweight hollow bones. For this reason bats hang from their feet when not flying: Their leg bones are simply too thin to support their body weight. But even with this skeletal adaptation, it is not immediately clear how muscles are up to the task of flying. Recall that muscle designed to generate high power is anaerobic (and thus is used intermittently) and is most effective at low-frequency output, whereas muscle designed to operate continuously at high frequency is aerobic, but generates low power output. Given this contradiction, that high-force output and high-frequency operation are mutually exclusive, it would seem impossible for an animal to have both attributes. Yet this is precisely what animals must have in order to fly. What evolutionary developments have arisen to solve this problem?

To provide for faster contraction rates, the mechanics of contraction need to speed up. One factor which increases reaction times is an increase in muscle temperature. Increasing body temperature from 30 to 40°C results in a 2.2-fold increase in the rate of ATP synthesis, equivalent to doubling the effectiveness of the mitochondria within each cell. Birds have higher body temperatures than mammals, a subtle but effective adaptation to increasing reaction times and an important adaptation for flight. Many insects cannot fly without warming up because they can't supply ATP at a sufficient rate to power the flight muscles; thus some (such as butterflies) bask in the morning sun before takeoff, whereas others (such as honey bees) shiver their muscles to warm up before flight. An increase in body temperature also permits Ca^{++} pumps to work more efficiently, resulting in faster clearance of Ca^{++} from the cytoplasm, which permits muscles to relax more rapidly. In birds and insects the structure of mitochondria is altered such that the inner mitochondrial membrane has twice the packing of cristae (folds of the inner membranes) compared to mammalian mitochondria. Thus each mitochondrion consumes O_2 at approximately twice the rate (7 to 10 versus 4 to 5 mL O_2 per cm^{-3} per min) measured in mammals. Combined, each of these factors contributes to the ability of flight muscle to operate at higher frequencies. This is exemplified by the high wing-beat frequency of hummingbirds (70 wing beats per second) as well as for large insects that use **direct** or **synchronous** muscle contractions to power flight muscle.

High-speed synchronous skeletal muscles are those in which each muscle action potential evoked by incoming neuronal activity results in a single mechanical contraction. For example, cicadas generate a calling song by vibrating a pair of cuticular plates termed **tymbals**. Contraction of the tymbal muscle bends the tymbals inward, which then recoil outward when the muscle relaxes. Each distortion cycle results in a burst of resonant sound vibration with a frequency of approximately 550 Hz.

Synchronous muscle arrangements are also found in butterflies, moths, locusts, and dragonflies and consist of attachment of the flight muscle directly to the wing. Each signal from a motor neuron triggers contraction of the flight muscle, so that the contractions that raise and lower the wing are synchronous. As has been discussed, muscle fibers are activated by the release of Ca^{++} and deactivated by the reuptake of Ca^{++}. However, animals cannot fly at frequencies of greater than 100Hz with synchronous muscles. How have some animal groups found a solution to this problem?

■ Asynchronous muscle contractions are characterized by a nearly constant myoplasmic Ca^{++} concentration.

The most significant space- as well as energy-saving solution for flight is the development of **asynchronous** muscle contrac-
tions (● Figure 8–23). For muscles that operate at extraordinarily high frequencies, their design results in the generation of more power, although capacity for fine neural control declines. Asynchronous muscles are believed to have evolved from synchronous ones and represent a design breakthrough.

Asynchronous or "indirect" muscles are associated with the flight muscle of approximately three quarters of known insect species including Hemiptera (true bugs, cicadas, aphids, and bedbugs), Diptera (flies), Hymenoptera (bees and wasps). In these animals, wing beat frequency is too great to be controlled directly by neuronal activation. For example, wing beat frequencies greater than 1000 per second have been recorded for midges, although for most species the number of muscle contractions ranges from 100 to 300 per second. Associated with the rapid wing-beat frequency is a lowered number of nerve impulses that are now asynchronous with muscle contraction. In these circumstances, a single Ca^{++} pulse maintains the muscle in an activated state for numerous successive cycles. In addition, the flight muscles are attached to the walls of the thorax rather than directly to the wings.

Flight muscles consist of two sets of large antagonistic power muscles oriented more or less perpendicular to each other: the **dorsal ventral muscles** and the **dorsal longitudinal muscles**. Contraction of asynchronous flight muscle is normally triggered by stretch ("stretch activation") and deactivated by "shortening deactivation" both in the presence of elevated

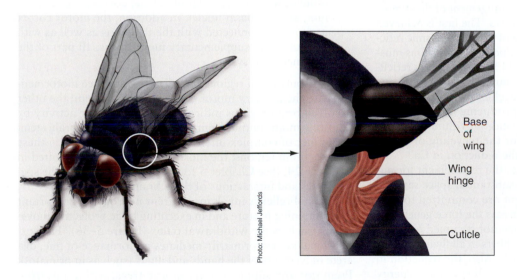

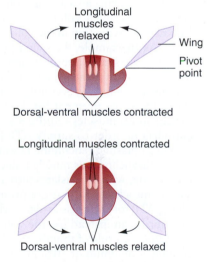

(a)   (b)

Figure 8–23 ● Mechanics of flight in the housefly. (a) The exoskeleton of insects relying on asynchronous flight muscles is hinged. Each wing is moved through an elastic, hingelike structure as a result of changes in the shape of the thorax. (b) Flight muscles do not attach directly to the hinges or to the wings but rather pull on the sides of the thorax to which they are attached. The two sets of flight muscles are arranged perpendicular to each other. Upward movement of the wings occurs when the vertical dorsoventral muscles contract and pull the flexible dorsal roof downward, flicking the wings up. The downstroke occurs when the longitudinal muscles contract, distorting the roof to distort upward and forcing the wings down. Only the up or down positions are stable, and because of the elasticity of the hinge much of the energy used to power muscle wing movement is recycled into the subsequent stroke. Smaller muscles acting directly on the hinge allow the wings to move at an angle, resulting in forward movement, so that a complete wing cycle actually involves the wingtip performing a figure-of-eight pattern.

myoplasmic Ca^{++} concentrations. With the flight muscles antagonistically situated, a contraction or relaxation changes the shape of the thorax into one of two stable positions: one with the wings elevated and the thorax depressed and the other with the wings depressed and the thorax elevated. Let us examine the situation as the fruit fly (*Drosophila melanogaster*) prepares itself for flight. The fruit fly must beat each wing up to 240 times per second to achieve enough power for lift. Just before takeoff, intermittent nervous stimulation primes the two muscle sets with increased intracellular Ca^{++}. The fly then leaps into the air and becomes airborne. Contraction of dorsal ventral muscles lengthens the dorsal longitudinal flight muscles simultaneously. These longitudinal muscles respond to the stretch with a delayed rise in tension (the stretch activation response), causing it to contract. This in turn causes an antagonistic reciprocal stretch and delayed contraction in the dorsal ventral muscles. As the two sets of muscles alternately contract and relax, the thorax rapidly oscillates between the two stable set points, causing deformations at the wing hinges that make the wings beat at the appropriate resonant frequency. The release of the catecholamine octopamine further increases both the force and efficiency of flight muscle contractions. At the same time, octopamine stimulates the oxidation of carbohydrate and fat in the flight muscle. Wing movements are ended when action potentials are no longer initiated, permitting muscle fiber repolarization and Ca^{++} uptake into the sarcoplasmic reticulum.

Insects having an asynchronous arrangement of flight muscle conserve cellular volume in two ways. The first is achieved by a marked reduction in the volume of the sarcoplasmic reticulum. As an example, the sound-producing synchronous muscles of cicadas have approximately 34% sarcoplasmic reticulum, whereas an asynchronous flight muscle has a sarcoplasmic volume less than 4%. Secondly, intracellular space is saved by the reduced requirement for ATP-producing mitochondria. Synchronous muscles require a substantial volume of mitochondria in order to supply ATP for the recycling of Ca^{++} into sarcoplasmic reticulum. Thus the reduction in Ca^{++} cycling in asynchronous muscle reduces the requirement for mitochondria, which translates into a substantial space savings. Those mitochondria that are present are committed to supplying ATP for the cross bridges, whereas the force-generating myofibrils can instead use the volume not occupied by mitochondria. As a rule of thumb, muscle fibers in all flying animals contain a minimum of 50% myofibrils. Also, the reduction in amount of ATP used to power Ca^{++} cycling reduces energy costs, which lessens the direct expenses of flight, making the entire process more efficient.

For the remaining section on skeletal muscle, we are going to examine the central and local mechanisms involved in regulating the motor activity performed by these muscles.

Control of Motor Movement

Particular patterns of motor unit output govern vertebrate motor activity, ranging from maintenance of posture and balance to stereotypical locomotory movements such as walking or swimming, to individual, highly skilled motor activity such as swinging and leaping through trees. Control of any motor movement, no matter what its level of complexity, depends on converging input to the motor neurons of specific motor units.

The motor neurons in turn trigger contraction of the muscle fibers within their respective motor units by means of the events that occur at the neuromuscular junction.

■ Multiple motor inputs influence vertebrate motor unit output.

Three levels of input control motor neuron output (● Figure 8–24):

1. *Input from afferent neurons,* usually through intervening interneurons, at the level of the spinal cord—that is, spinal reflexes (see p. 183).

2. *Input from the primary motor cortex.* Fibers originating from cell bodies of pyramidal cells within the primary motor cortex (see p. 170) descend directly without synaptic interruption to terminate on motor neurons (or on local interneurons that terminate on motor neurons). These fibers make up the **corticospinal** (or pyramidal) motor system.

3. *Input from the brain stem* as part of the multineuronal motor system. The pathways composing the **multineuronal** (or **extrapyramidal**) motor system include a number of synapses that involve many regions of the brain (*extra* means "outside of;" *pyramidal* refers to the pyramidal system). The final link in multineuronal pathways is the brain stem, especially the reticular formation, which in turn is influenced by motor regions of the cortex, the cerebellum, and the basal nuclei. In addition, the motor cortex itself is interconnected with the thalamus as well as with premotor and supplementary motor areas, all part of the multineuronal system.

The only brain regions that directly influence motor neurons are the primary motor cortex and brain stem; the other involved brain regions indirectly regulate motor activity by adjusting motor output from the motor cortex and brain stem. A number of complex interactions take place between these various brain regions; the most important are represented in Figure 8–24. (See Chapter 5 for further discussion of the specific roles and interactions of these brain regions.)

Spinal reflexes involving afferent neurons are important in maintaining posture and in executing basic protective movements, such as the withdrawal reflex (Figure 5–36). The corticospinal system primarily mediates performance of fine, discrete movements of the hands and digits (such as in primates). Premotor and supplementary motor areas, with input from the cerebrocerebellum, plan the voluntary motor command that the primary motor cortex issues to the appropriate motor neurons through this descending system. The multineuronal system, in contrast, is primarily concerned with regulation of overall body posture involving involuntary movements of large muscle groups of the trunks and limbs. Considerable complex interaction and overlapping of function exist between these two systems.

■ Muscle spindles and the Golgi tendon organs provide afferent information essential for controlling skeletal muscle activity.

Coordinated, purposeful skeletal muscle activity depends on afferent input from a variety of sources. At a simple level, afferent signals indicating that a foot is touching a hot ember

Figure 8–24 ●
Motor control.

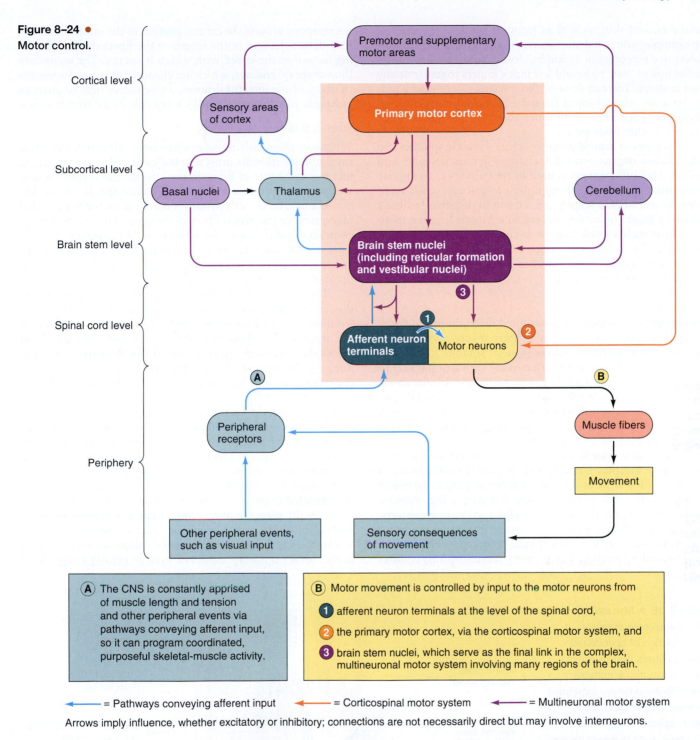

Cortical level

Premotor and supplementary motor areas

Sensory areas of cortex

Primary motor cortex

Subcortical level

Basal nuclei

Thalamus

Cerebellum

Brain stem level

Brain stem nuclei (including reticular formation and vestibular nuclei)

3

Spinal cord level

1

Afferent neuron terminals

Motor neurons

2

A

B

Periphery

Peripheral receptors

Muscle fibers

Movement

Other peripheral events, such as visual input

Sensory consequences of movement

A The CNS is constantly apprised of muscle length and tension and other peripheral events via pathways conveying afferent input, so it can program coordinated, purposeful skeletal-muscle activity.

B Motor movement is controlled by input to the motor neurons from

1 afferent neuron terminals at the level of the spinal cord,

2 the primary motor cortex, via the corticospinal motor system, and

3 brain stem nuclei, which serve as the final link in the complex, multineuronal motor system involving many regions of the brain.

——— = Pathways conveying afferent input   ——— = Corticospinal motor system   ——— = Multineuronal motor system

Arrows imply influence, whether excitatory or inhibitory; connections are not necessarily direct but may involve interneurons.

(Figure 5–36) triggers reflex contractile activity in appropriate limb muscles to withdraw the hand from the injurious stimulus. At a more complex level, if a dog is going to catch a stick thrown in the air by a human, the motor systems of the dog's brain must program sequential motor commands that will move and position its body correctly for the catch, using predictions of the stick's direction and rate of movement provided by visual input (see Chapter 5, p. 175, for a discussion on predictions made by brain regions such as the cerebellum). Many muscles acting simultaneously or alternately at different joints rapidly shift the dog's location and position, while

maintaining balance. It is critical to have ongoing input about body position with respect to the surrounding environment, as well as about the position of various body parts in relationship to each other. This information is necessary for establishing a neuronal pattern of activity to perform the desired movement. The dog's CNS must know the starting position of the dog's body to appropriately program muscle activity. Further, it must be constantly appraised of the progression of movement it has initiated, so it can adjust as needed. An animal receives this information, which is known as *proprioceptive input*, from **proprioceptors** (see p. 197) in the eyes, joints, vestibular apparatus, and skin, as well as from the muscles themselves. For example, you can demonstrate your joint and muscle proprioceptive receptors in action by closing your eyes and bringing the tips of your right and left index fingers together at any point in space. You can do so without seeing where your hands are, because afferent input from the joint and muscle receptors inform your brain at all times about the position of your hands and other body parts.

Two types of muscle proprioceptors—**muscle spindles** and **Golgi tendon organs**—monitor changes in muscle length and tension. This information is used in two ways: (1) to apprise motor areas of the brain about muscle length and tension and (2) to control muscle length and tension in negative-feedback fashion by means of local spinal reflexes. Muscle length is monitored by muscle spindles, whereas changes in muscle tension are detected by Golgi tendon organs. Both these receptor types are activated by muscle stretch, but they are designed to convey different types of information. Let us see how.

Muscle Spindle Structure

Muscle spindles, which are distributed throughout the fleshy part of a skeletal muscle, are collections of specialized muscle fibers known as **intrafusal fibers** (*fusus,* "spindle"), which lie within spindle-shaped connective tissue capsules parallel to the "ordinary" **extrafusal fibers** (● Figure 8–25). Unlike an ordinary skeletal muscle fiber, which contains contractile elements (myofibrils) throughout its entire length, an intrafusal fiber has a noncontractile central portion, with the contractile elements being limited to both ends.

Each muscle spindle has its own private efferent and afferent nerve supply. The efferent neuron that innervates a muscle spindle's intrafusal fibers is known as **gamma (γ) motor neuron,** whereas the motor neurons that supply the ordinary extrafusal fibers are designated as **alpha (α) motor neurons.** Two types of afferent sensory endings terminate on the intrafusal fibers and serve as muscle spindle receptors, both of which are activated by stretch. The **primary (annulospiral) endings**

are wrapped around the central portion of the intrafusal fibers; they detect changes in the length of the fibers during stretching as well as the speed with which it occurs. The **secondary (flower-spray) endings,** which are clustered at the end segments of many of the intrafusal fibers, are sensitive only to changes in length. Muscle spindles play a key role in the stretch reflex.

Stretch Reflex

Whenever the whole muscle is passively stretched, the intrafusal fibers within its muscle spindles are likewise stretched, increasing the rate of firing in the afferent nerve fibers whose sensory endings terminate on the stretched spindle fibers. The afferent neuron directly synapses on the α motor neuron that innervates the extrafusal fibers of the same muscle, resulting in contraction of that muscle (● Figure 8–26a, pathway 1 → 2). This **stretch reflex** serves as a local negative-feedback mechanism to resist any passive changes in muscle length so that optimal resting length can be maintained.

The classic example of the stretch reflex is the **patellar tendon,** or **knee-jerk,** reflex in humans (● Figure 8–27). The extensor muscle of the knee is the quadriceps femoris, which forms the anterior (front) portion of the thigh and is attached just below the knee to the tibia (shinbone) by the patellar tendon. Tapping this tendon with a rubber mallet passively stretches the quadriceps muscle, activating its spindle receptors. The resultant stretch reflex brings about contraction of this extensor muscle, causing the knee to extend and raise the foreleg in the well-known knee-jerk fashion. This test is routinely performed as a preliminary assessment of nervous system function. A normal knee jerk indicates to a physician that a number of neural and muscular components—muscle spindle, afferent input, motor neurons, efferent output, neuromuscular junctions, and the muscles themselves—are functioning normally. It also indicates the presence of an appropriate balance of excitatory and inhibitory input to the motor neurons from higher brain levels. Muscle jerks may be absent or depressed with loss of higher-level excitatory inputs or may be greatly exaggerated

Figure 8–25 ● Muscle spindle. A muscle spindle consists of a collection of specialized intrafusal fibers that lie within a connective tissue capsule parallel to the ordinary extrafusal skeletal-muscle fibers. The muscle spindle is innervated by its own gamma motor neuron and is supplied by two types of afferent sensory terminals, the primary (annulospiral) endings and the secondary (flower-spray) endings, both of which are activated by stretch.

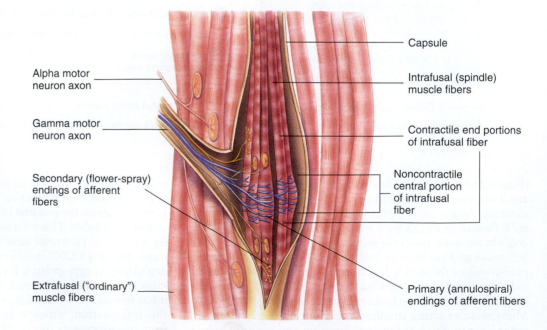

Alpha motor neuron axon

Gamma motor neuron axon

Secondary (flower-spray) endings of afferent fibers

Extrafusal ("ordinary") muscle fibers

Capsule

Intrafusal (spindle) muscle fibers

Contractile end portions of intrafusal fiber

Noncontractile central portion of intrafusal fiber

Primary (annulospiral) endings of afferent fibers

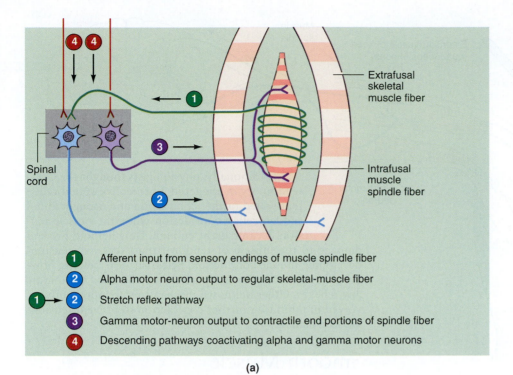

Figure 8–26 • **Muscle spindle function.** (a) Pathways involved in the monosynaptic stretch reflex and coactivation of alpha and gamma motor neurons. (b) Status of a muscle spindle when the muscle is relaxed. (c) Status of a muscle spindle in the hypothetical situation of a muscle being contracted on alpha motor-neuron stimulation in the absence of spindle coactivation. (d) Status of a muscle spindle in the actual physiological situation when both the muscle and muscle spindle are contracted on alpha and gamma motor-neuron coactivation.

Legend for part (a):

1. Afferent input from sensory endings of muscle spindle fiber
2. Alpha motor neuron output to regular skeletal-muscle fiber
1 → 2 Stretch reflex pathway
3. Gamma motor-neuron output to contractile end portions of spindle fiber
4. Descending pathways coactivating alpha and gamma motor neurons

(a)

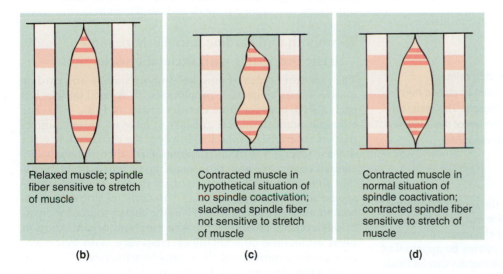

(b) Relaxed muscle; spindle fiber sensitive to stretch of muscle

(c) Contracted muscle in hypothetical situation of no spindle coactivation; slackened spindle fiber not sensitive to stretch of muscle

(d) Contracted muscle in normal situation of spindle coactivation; contracted spindle fiber sensitive to stretch of muscle

with loss of inhibitory input to the motor neurons from higher brain levels.

The primary purpose of the stretch reflex is to resist the tendency for the passive stretch of extensor muscles caused by gravitational forces when a person is standing upright. Whenever the knee joint tends to buckle because of gravity, the quadriceps muscle is stretched. The resultant enhanced contraction of this extensor muscle brought about by the stretch reflex quickly straightens out the knee, holding the limb extended so that the person remains standing.

Coactivation of γ and α Motor Neurons

Gamma (γ) motor neurons initiate contraction of the muscular end regions of intrafusal fibers (Figure 8–26a, pathway 3). This contractile response is too weak to have any influence on whole-muscle tension, but it does have an important localized effect on the muscle spindle itself. If there were no compensating mechanisms, shortening of the whole muscle by α motor-neuron stimulation of extrafusal fibers would cause slack in the spindle fibers so that they would be less sensitive to stretch and therefore not as effective as muscle length detectors (Figures 8–26b and c). **Co-activation** of the γ-motor-neuron system along with the α-motor-neuron system during reflex and voluntary contractions (Fig-ure 8–26a, pathway 4) takes the slack out of the spindle fibers as the whole muscle shortens, permitting these receptor structures to maintain their high sensitivity to stretch over a wide range of muscle lengths. When γ motor-neuron stimulation triggers simultaneous contraction of both end muscular portions of an intrafusal fiber, the noncontractile central portion is pulled in opposite directions, tightening this region and taking out the slack (Figure 8–26d). Whereas the extent of α motor-neuron activation depends on the intended strength of the motor response, the extent of simultaneous γ motor-neuron activity to the same muscle depends on the anticipated distance of shortening.

Golgi Tendon Organs

In contrast to muscle spindles, which lie within the belly of the muscle, **Golgi tendon organs** are located in the tendons of the muscle, where they can respond to changes in the muscle's externally applied tension rather than to changes in its length.

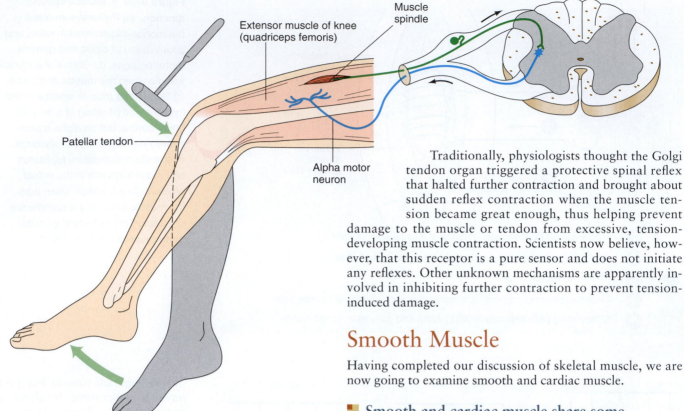

Figure 8–27 ● **Patellar tendon reflex (a stretch reflex).**
Tapping the patellar tendon with a rubber mallet stretches the
muscle spindles in the quadriceps femoris muscle. The resultant
monosynaptic stretch reflex results in contraction of this extensor
muscle, causing the characteristic knee-jerk response.

Because a number of factors determine the tension developed
in the whole muscle during contraction (for example, frequency
of stimulation or length of the muscle at the onset of contraction), it is essential that motor control systems be apprised of
the tension actually achieved so that adjustments can be made
if necessary.

The Golgi tendon organs consist of endings of afferent
fibers entwined within bundles of connective tissue fibers that
make up the tendon. When the extrafusal muscle fibers contract, the resultant pull on the tendon tightens the connective
tissue bundles, which in turn increase the tension exerted on
the bone to which the tendon is attached. In the process, the
entwined Golgi organ afferent-receptor endings are stretched,
causing the afferent fibers to fire; the frequency of firing is directly related to the tension developed. This afferent information is sent to the brain for processing, much of which is used
subconsciously for smoothly executing motor activity. Unlike
afferent information from the muscle spindles, afferent information from the Golgi tendon organ reaches the level of conscious awareness, at least in humans. You are aware of the tension within a muscle, but you are not aware of its length.

Traditionally, physiologists thought the Golgi
tendon organ triggered a protective spinal reflex
that halted further contraction and brought about
sudden reflex contraction when the muscle tension became great enough, thus helping prevent
damage to the muscle or tendon from excessive, tension-
developing muscle contraction. Scientists now believe, however, that this receptor is a pure sensor and does not initiate
any reflexes. Other unknown mechanisms are apparently involved in inhibiting further contraction to prevent tension-
induced damage.

Smooth Muscle

Having completed our discussion of skeletal muscle, we are
now going to examine smooth and cardiac muscle.

■ Smooth and cardiac muscle share some basic properties with skeletal muscle.

The two other types of muscle—smooth muscle and cardiac
muscle—share some basic properties with skeletal muscle, but
each also displays unique characteristics (❚ Table 8–4). For
example, smooth muscle appears to have no close counterparts in invertebrate muscle. However, the three muscle types
do have several features in common. First, they all have a specialized contractile apparatus made up of thin actin filaments
that slide relative to stationary thick myosin filaments in response to a rise in cytosolic Ca^{++} to accomplish contraction.
Second, they all directly use ATP as the energy source for cross-
bridge cycling. However, the structure and organization of
fibers within these different muscle types vary, as do their mechanisms of excitation and the means by which excitation and
contraction are coupled. Furthermore, there are important
distinctions in the contractile response itself. We spend the remainder of this chapter highlighting unique features of smooth
and cardiac muscle as compared with skeletal muscle, reserv-
ing a more detailed discussion of their function for chapters
devoted to organs containing these muscle types.

■ Smooth muscle cells are small and unstriated.

The majority of smooth muscle cells in vertebrates lie in the
walls of hollow organs and tubes. Their contraction exerts
pressure on and regulates the forward movement of the contents of these structures.

Both smooth and skeletal muscle cells are elongated, but
in contrast to their large, cylindrical, multinucleated skeletal-
muscle counterparts, smooth muscle cells are spindle-shaped,
have a single nucleus, and are considerably smaller (2 to 10 μm

Table 8–4 ▌ Comparison of Muscle Types

Characteristic	Type of Muscle			
	Skeletal	*Multiunit smooth*	*Single-unit smooth*	*Cardiac*
Location	Attached to skeleton	Large blood vessels, eye, and hair follicles	Walls of hollow organs in digestive, reproductive, and urinary tracts and in small blood vessels	Heart only
Function	Movement of body in relation to external environment	Varies with structure involved	Movement of contents within hollow organs	Pumps blood out of heart
Mechanism of contraction	Sliding filament mechanism	Sliding filament mechanism	Sliding filament mechanism	Sliding filament mechanism
Innervation	Somatic nervous system (alpha motor neurons)	Autonomic nervous system	Autonomic nervous system	Autonomic nervous system
Level of control	Under voluntary control; also subject to subconscious regulation	Under involuntary control	Under involuntary control	Under involuntary control
Initiation of contraction	Neurogenic	Neurogenic	Myogenic (pacemaker potentials and slow-wave potentials)	Myogenic (pacemaker potentials)
Role of nervous stimulation	Initiates contraction; accomplishes gradation	Initiates contraction; contributes to gradation	Modifies contraction; can excite or inhibit; contributes to gradation	Modifies contraction; can excite or inhibit; contributes to gradation
Modifying effect of hormones	No	Yes	Yes	Yes
Presence of thick myosin and thin actin filaments	Yes	Yes	Yes	Yes
Striated due to orderly arrangement of filaments	Yes	No	No	Yes
Presence of troponin and tropomyosin	Yes	Tropomyosin only	Tropomyosin only	Yes
Presence of T tubules	Yes	No	No	Yes
Level of development of sarcoplasmic reticulum	Well developed	Poorly developed	Poorly developed	Moderately developed
Cross bridges turned on by Ca^{2+}	Yes	Yes	Yes	Yes
Source of increased cytosolic Ca^{2+}	Sarcoplasmic reticulum	Extracellular fluid and sarcoplasmic reticulum	Extracellular fluid and sarcoplasmic reticulum	Extracellular fluid and sarcoplasmic reticulum
Site of Ca^{2+} regulation	Troponin in thin filaments	Myosin in thick filaments	Myosin in thick filaments	Troponin in thin filaments
Mechanism of Ca^{2+} action	Physically repositions troponin-tropomyosin complex to uncover actin cross-bridge binding sites	Chemically brings about phosphorylation of myosin cross bridges so they can bind with actin	Chemically brings about phosphorylation of myosin cross bridges so they can bind with actin	Physically repositions troponin-tropomyosin complex
Presence of gap junctions	No	Yes (very few)	Yes	Yes
ATP used directly by contractile apparatus	Yes	Yes	Yes	Yes

(continued)

Table 8–4 ▌ Comparison of Muscle Types *(continued)*

Characteristic	Type of Muscle			
	Skeletal	*Multiunit smooth*	*Single-unit smooth*	*Cardiac*
Myosin ATPase activity; speed of contraction	Fast or slow, depending on type of fiber	Very slow	Very slow	Slow
Means by which gradation accomplished	Varying number of motor units contracting (motor unit recruitment) and frequency at which they're stimulated (twitch summation)	Varying number of muscle fibers contracting and varying cytosolic Ca^{2+} concentration in each fiber by autonomic and hormonal influences	Varying cytosolic Ca^{2+} concentration through myogenic activity and influences of the autonomic nervous system, hormones, mechanical stretch, and local metabolites	Varying length of fiber (depending on extent of filling of the heart chambers) and varying cytosolic Ca^{2+} concentration through autonomic, hormonal, and local metabolite influence
Presence of tone in absence of external stimulation	No	No	Yes	No
Clear-cut length–tension relationship	Yes	No	No	Yes

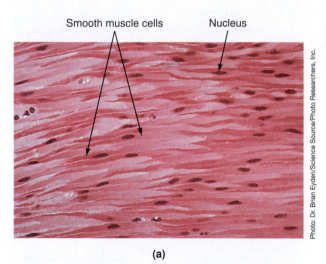

Smooth muscle cells   Nucleus

Photo: Dr. Brian Eyden/Science Source/Photo Researchers, Inc.

(a)

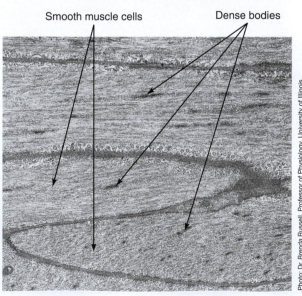

Smooth muscle cells   Dense bodies

Photo: Dr. Brenda Russell, Professor of Physiology, University of Illinois.

(b)

in diameter and 50 to 400 µm long). Also unlike skeletal muscle cells, a single smooth-muscle cell does not extend the full length of a muscle. Instead, groups of smooth muscle cells are typically arranged in sheets (● Figure 8–28a).

A smooth muscle cell has three types of filaments: (1) thick myosin filaments, which are longer than those found in skeletal muscle; (2) thin actin filaments, which contain tropomyosin but lack the regulatory protein troponin; and (3), filaments of intermediate size, which are unique to smooth muscle and which do not directly participate in the contractile process but serve as part of the cytoskeletal framework that supports the shape of the cell. Smooth muscle filaments do not form myofibrils and are not arranged in the sarcomere pattern found in skeletal muscle. Thus smooth muscle cells do not display the banding or striation found in skeletal muscle.

Lacking sarcomeres, smooth muscle does not have Z lines as such, but **dense bodies** containing the same protein constituent found in Z lines are present (Figure 8–28b). Dense bodies are positioned throughout the smooth muscle cell as well as being attached to the internal surface of the plasma membrane. The dense bodies are held in place by a scaffold of intermediate filaments. The actin filaments are anchored to the dense bodies. Considerably more actin is present in smooth muscle cells than in skeletal muscle cells, with 10 to 15 thin filaments for each thick myosin filament in smooth muscle and two thin filaments for each thick filament in skeletal muscle.

The thick- and thin-filament contractile units are oriented slightly diagonally from side to side within the smooth muscle cell in an elongated, diamond-shaped lattice, rather than running parallel with the long axis as myofibrils do in skele-

Figure 8–28 ◀ ● **Microscopic view of smooth muscle cells.** (a) Low-power light micrograph of smooth muscle cells. Note the spindle shape and single, centrally located nucleus. (b) Electron micrograph of smooth muscle cells at 14,000× magnification. Note the presence of dense bodies and lack of banding.

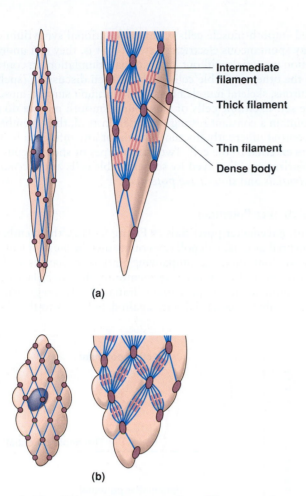

(a)

- Intermediate filament
- Thick filament
- Thin filament
- Dense body

(b)

Figure 8–29 • Schematic representation of the arrangement of thick and thin filaments in a smooth muscle cell in contracted and relaxed states. (a) Relaxed smooth muscle cell. (b) Contracted smooth muscle cell.

tal muscle (• Figure 8–29a). Relative sliding of the thin filaments past the thick filaments during contraction causes the filament lattice to reduce in length and expand from side to side. As a result, the whole cell shortens and bulges out between the points where the thin filaments are attached to the inner surface of the plasma membrane (Figure 8–29b).

■ **Smooth muscle cells are turned on by Ca^{++}-dependent phosphorylation of myosin.**

If the thin filaments of smooth muscle cells do not contain troponin, and tropomyosin does not block actin's cross-bridge binding sites, what prevents actin and myosin from binding at the cross bridges in the resting state, and how is cross-bridge activity switched on in the excited state? During excitation, the increased cytosolic Ca^{++} acts as an intracellular messenger, initiating biochemical events leading to contraction. At least three different mechanisms have been found among smooth muscles:

1. Smooth muscle myosin can interact with actin only when myosins (in particular, small myosin subunits called **myosin light chains**) are phosphorylated (• Figure 8–30). Smooth muscle Ca^{++} binds with **calmodulin,** an intracellular protein found in most cells that is structurally similar to troponin (see p. 318). This Ca^{++}–calmodulin complex binds to and activates another protein, **myosin light-chain kinase,** which in turn phosphorylates myosin light chains. This activated myosin then binds with actin so that cross-bridge cycling can begin. When Ca^{++} is removed, myosin light chains are dephosphorylated, so myosin can no longer interact with actin.

2. The thin filaments often contain **caldesmon,** a protein with a blocking action similar to tropomyosin and troponin. The Ca^{++}–calmodulin complex moves caldesmon out of its blocking position.

3. In some cases, binding of Ca^{++} directly to the myosin light chains activates the myosin independently of phosphorylation.

Thus smooth muscle is triggered to contract by a rise in cytosolic Ca^{++}, similar to what happens in skeletal muscle. The means by which excitation brings about an increase in cytosolic Ca^{++} concentration in smooth muscle cells also differs from that for skeletal muscle. A smooth muscle cell has no T tubules and a poorly developed sarcoplasmic reticulum. The increased cytosolic Ca^{++} that triggers the contractile response comes from two sources: Most Ca^{++} enters from the ECF, but some is released intracellularly from the sparse sarcoplasmic reticulum stores. Unlike their role in skeletal muscle cells, voltage-gated dihydropyridine receptors in the plasma membrane of smooth muscle cells function as Ca^{++} channels. When these surface membrane channels are opened, Ca^{++} enters down its electrochemical gradient from the ECF. The entering Ca^{++} triggers the opening of Ca^{++} channels in the sarcoplasmic reticulum (via inositol 1,4,5-triphosphate-IP_3), so that small additional amounts of Ca^{++} are released from this meager source. Because smooth muscle cells are so much smaller in diameter than skeletal muscle fibers, the preponderant Ca^{++} influx from the ECF can influence cross-bridge activity, even in the central portions of the cell, without needing an elaborate T tubule–sarcoplasmic reticulum mechanism.

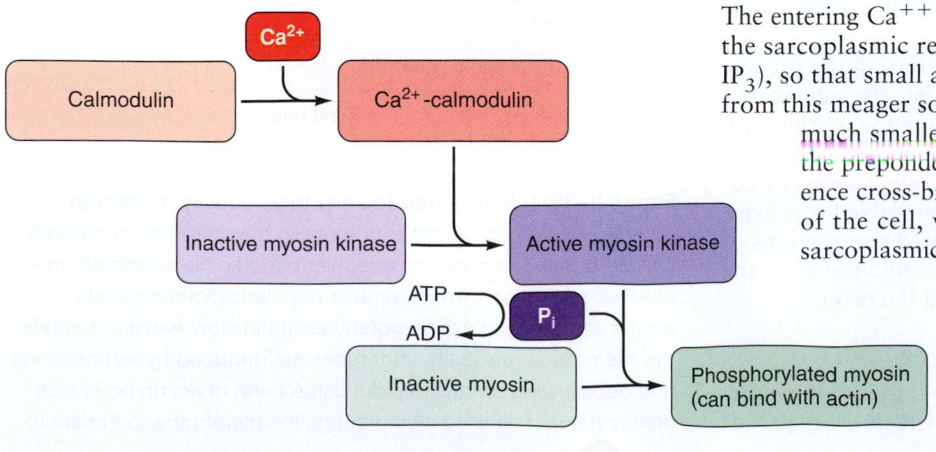

Figure 8–30 • Calcium activation of myosin in smooth muscle.

Relaxation is accomplished by removal of Ca^{++} as it is actively transported out across the plasma membrane and back into the sarcoplasmic reticulum. We still have not addressed the question of how smooth muscle becomes excited to contract; that is, what opens the Ca^{++} channels in the plasma membrane and sarcoplasmic reticulum? Smooth muscle is grouped into two categories—*multiunit* and *single-unit smooth muscle*—based on differences in how the muscle fibers become excited. Let's compare these two types of smooth muscle.

■ Multiunit smooth muscle is neurogenic.

Multiunit smooth muscle exhibits properties partway between skeletal muscle and single-unit smooth muscle. As the name implies, a multiunit smooth muscle consists of multiple discrete units that function independently of each other and must be separately stimulated by nerves to contract, similar to skeletal muscle motor units. Thus contractile activity in both skeletal muscle and multiunit smooth muscle is neurogenic ("nerve-produced"). Whereas skeletal muscle is innervated by the voluntary somatic nervous system (motor neurons), multiunit (as well as single-unit) smooth muscle is supplied by the involuntary autonomic nervous system.

Multiunit smooth muscle is found (1) in the walls of large blood vessels; (2) in large airways to the lungs; (3) in the muscle of the eye that adjusts the lens for near or far vision; (4) in the iris of the eye, which alters the size of the pupil to adjust the amount of light entering the eye; and (5) at the base of hair follicles, contraction of which causes "goose bumps."

■ Single-unit smooth muscle cells form functional syncytia.

Most smooth muscle is of the **single-unit** variety. It is alternatively called **visceral smooth muscle**, because it is found in the walls of the hollow organs or viscera (for example, the digestive, reproductive, and urinary tracts and small blood vessels). The term *single-unit smooth muscle* derives from the fact that the muscle fibers that make up this type of muscle become excited and contract as a single unit. **Gap junctions** electrically link the muscle fibers in single-unit smooth muscle (see p. 64). When an action potential occurs anywhere within a sheet of single-unit smooth muscle, it is quickly propagated via these special points of electrical contact throughout the entire group of interconnected cells, which then contract as a single coordinated unit. This group of interconnected muscle cells that function electrically and mechanically as a unit is a functional syncytium (p. 315).

Thinking about the role of the uterus during the process of labor in a mammal can help you appreciate the significance of this arrangement. Muscle cells composing the uterine wall act as a functional syncytium. They repetitively become excited and contract as a unit during labor, exerting a series of coordinated "pushes" that are eventually responsible for delivering the neonate. Independent, uncoordinated contractions of individual muscle cells in the uterine wall would not exert the uniformly applied pressure needed to expel the neonate.

■ Single-unit smooth muscle is myogenic.

Single-unit smooth muscle is **self-excitable** rather than requiring nervous stimulation for contraction. Clusters of special-

ized smooth-muscle cells within a functional syncytium display spontaneous electrical activity; that is, they can undergo action potentials without any external stimulation. In contrast to the other excitable cells we have been discussing (such as neurons, skeletal muscle fibers, and multiunit smooth muscle), the self-excitable cells of single-unit smooth muscle do not maintain a constant resting potential. Instead, their membrane potential inherently fluctuates without any influence by factors external to the cell. Two major types of spontaneous depolarizations displayed by self-excitable cells are *pacemaker potentials* and *slow-wave potentials*.

Pacemaker Potentials

With **pacemaker potentials** (● Figure 8–31a), the membrane potential gradually depolarizes on its own because of shifts in passive ionic fluxes accompanying automatic changes in channel permeability. When the membrane has depolarized to threshold, an action potential is initiated. After repolarizing, the membrane potential once again depolarizes to threshold,

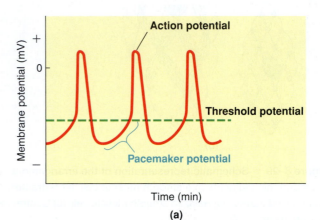

(a)

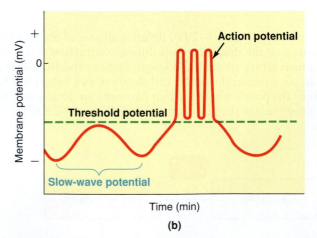

(b)

Figure 8–31 ● Self-generated electrical activity in smooth muscle. (a) With pacemaker potentials, the membrane gradually depolarizes to threshold on a regular periodic basis without any nervous stimulation. These regular depolarizations cyclically trigger self-induced action potentials. (b) In slow-wave potentials, the membrane gradually undergoes self-induced hyperpolarizing and depolarizing swings in potential. A burst of action potentials occurs if a depolarizing swing brings the membrane to threshold.

cyclically continuing in this manner to self-generate action potentials.

Slow-Wave Potentials

Slow-wave potentials are gradually alternating hyperpolarizing and depolarizing swings in potential caused by automatic cyclic changes in the rate at which sodium ions are actively transported across the membrane (Figure 8–31b). The potential moves farther from threshold during each hyperpolarizing swing and closer to threshold during each depolarizing swing. If threshold is reached, a burst of action potentials occurs at the peak of a depolarizing swing. Threshold is not always reached, however, so the oscillating slow-wave potentials can continue without generating action potentials. Whether threshold is reached depends on the starting point of the membrane potential at the onset of its depolarizing swing. The starting point, in turn, is influenced by neural and local factors. (We have now discussed all the means by which excitable tissues can be brought to threshold. ■ Table 8–5 summarizes the different triggering events that can initiate action potentials in various excitable tissues).

Myogenic Activity

Self-excitable smooth muscle cells are specialized to initiate action potentials, but they are not equipped to contract. Only a very small proportion of the total number of cells in a functional syncytium are noncontractile, pacemaker cells. These cells are typically clustered together in a specific location. The vast majority of smooth muscle cells in a functional sycytium are specialized to contract but cannot self-initiate action potentials. However, once an action potential is initiated by a self-initiating smooth muscle cell, it is conducted to the remaining contractile, nonpacemaker cells of the functional syncytium via gap junctions, so that the entire group of cells contracts without any nervous input. Such nerve-independent contractile activity initiated by the muscle itself is called **myo-**genic ("muscle-produced") **activity,** in contrast to the neurogenic activity of skeletal muscle and multiunit smooth muscle.

■ Gradation of single-unit smooth muscle contraction differs considerably from that of skeletal muscle.

Single-unit smooth muscle differs from skeletal muscle in the way gradation of contraction is accomplished. Gradation of skeletal muscle contraction is entirely under neural control, primarily involving motor unit recruitment and twitch summation. In smooth muscle, the gap junctions ensure that an entire smooth-muscle mass contracts as a single unit, making it impossible to vary the number of muscle fibers contracting. Only the tension of the fibers can be modified to achieve varying strengths of contraction of the whole organ. The portion of cross bridges activated and the tension subsequently developed in single-unit smooth muscle can be graded by varying the cytosolic Ca^{++} concentration. A single excitation in smooth muscle does not cause all the cross bridges to switch on, in contrast to skeletal muscles where a single action potential triggers the release of enough Ca^{++} to permit all cross bridges to cycle. As Ca^{++} concentration increases in smooth muscle, more cross bridges are brought into play, and greater tension develops.

Smooth Muscle Tone

Many single-unit smooth muscle cells have sufficient cytosolic Ca^{++} to maintain a low level of tension, or **tone,** even in the absence of action potentials. A sudden drastic change in Ca^{++}, such as accompanies a myogenically induced action potential, brings about a contractile response superimposed on the ongoing tonic tension. Besides self-induced action potentials, a number of other factors, including autonomic neurotransmitters, can influence contractile activity and the development of tension in smooth muscle cells by altering their cytosolic Ca^{++} concentration.

Table 8–5 ■ Various Means of Initiating Action Potentials in Excitable Tissues

Method of Depolarizing the Membrane to Threshold Potential	Type of Excitable Tissue Involved	Description of This Triggering Event
Summation of excitatory postsynaptic potentials (EPSPs) (see p. 125)	Efferent neurons, interneurons	Temporal or spatial summation (see p. 130) of slight depolarizations (EPSPs) of the cell body/dendrite end of the neuron brought about by changes in channel permeability in response to binding of excitatory neurotransmitter with surface membrane receptors
Receptor potential (see p. 199)	Afferent neurons	Typically a depolarization of the afferent neuron's receptor initiated by changes in channel permeability in response to the neuron's adequate stimulus (see p. 198)
End-plate potential (see p. 128)	Skeletal muscle	Depolarization of the motor end plate (see p. 128) brought about by changes in channel permeability in response to binding of the neurotransmitter acetylcholine with receptors on the end-plate membrane
Pacemaker potential (see p. 352)	Smooth muscle, cardiac muscle	Gradual depolarization of the membrane on its own because of shifts in passive ionic fluxes accompanying automatic changes in channel permeability
Slow-wave potential (see p. 353)	Smooth muscle (in digestive tract only)	Gradual alternating hyperpolarizing and depolarizing swings in potential caused by automatic cyclical changes in active ionic transport across the membrane

■ **Smooth muscle activity can be modified by the autonomic nervous system.**

Both branches of the vertebrate autonomic nervous system typically innervate smooth muscle. In single-unit smooth muscle, this nerve supply does not *initiate* contraction, but it can *modify* the rate and strength of contraction, either enhancing or retarding the inherent contractile activity of a given organ. Recall that the isolated motor end-plate region of a skeletal muscle fiber interacts with ACh released from a single axon terminal of a motor neuron. In contrast, the receptor proteins that bind with autonomic transmitters are dispersed throughout the entire surface membrane of a smooth muscle cell. Smooth muscle cells are sensitive to varying degrees and in varying ways to autonomic transmitters, depending on their distribution of cholinergic and adrenergic receptors (see p. 154). Each terminal branch of a postganglionic autonomic fiber travels across the surface of one or more smooth muscle cells, releasing transmitter from the vesicles within its multiple bulges (varicosities) as an action potential passes along the terminal (● Figure 8–32). The transmitter diffuses to the many receptor sites specific for it on the cells underlying the terminal. Thus, in contrast to the discrete one-to-one relationship at motor end plates, a given smooth muscle cell can be influenced by more than one type of neurotransmitter, and each autonomic terminal can influence more than one smooth muscle cell.

Other Factors Influencing Smooth Muscle Activity

Other factors (besides autonomic neurotransmitters) can influence the rate and strength of both multiunit and single-unit smooth muscle contraction, including certain hormones, local metabolites, mechanical stretch, and specific drugs. Recent evidence suggests that the abundant **caveolae** found in smooth muscle may serve as sites of integration of the extracellular signals that modify smooth muscle contractility (see p. 84) All these factors ultimately act by modifying the permeability of Ca^{++} channels in the plasma membrane, the sarcoplasmic reticulum, or both, through a variety of mechanisms. Thus smooth muscle is subject to more external influences than is

skeletal muscle, even though smooth muscle can contract on its own, whereas skeletal muscle can not.

For now, we are going to consider only the effect of mechanical stretch on smooth muscle contractility in more detail as we look at the length–tension relationship in smooth muscle. We defer an examination of the extracellular chemical influences on smooth muscle contractility to other chapters, where we discuss regulation of the various organs that contain smooth muscle.

■ **Smooth muscle can still develop tension yet inherently relaxes when stretched.**

The relationship between the length of the muscle fibers before contraction and the tension that can be developed on a subsequent contraction is less closely linked in smooth muscle than in skeletal muscle. The range of lengths over which a smooth muscle fiber can develop near-maximal tension is much greater than for skeletal muscle. Smooth muscle can still develop considerable tension even when stretched up to 2.5 times its resting length, for two probable reasons. First, in contrast to skeletal muscle, in which the resting length is at l_o, in smooth muscle the resting (nonstretched) length is much shorter than the l_o. Therefore, smooth muscle can be stretched considerably before even reaching its optimal length. Second, the thin filaments still overlap the much longer thick filaments even in the stretched-out position, so that cross-bridge interaction and tension development can still take place. In contrast, the thick and thin filaments of skeletal muscle are completely pulled apart and no longer able to interact when the muscle is stretched only three fourths longer than its resting length.

The ability of a considerably stretched smooth-muscle fiber to still develop tension is important, because the smooth muscle fibers within the wall of a hollow organ are progressively stretched as the volume of the organ's contents increases. Consider the urinary bladder as an example. Even though the muscle fibers in the urinary bladder are stretched as the bladder gradually fills with urine, they still maintain their tone and are even capable of developing further tension in response

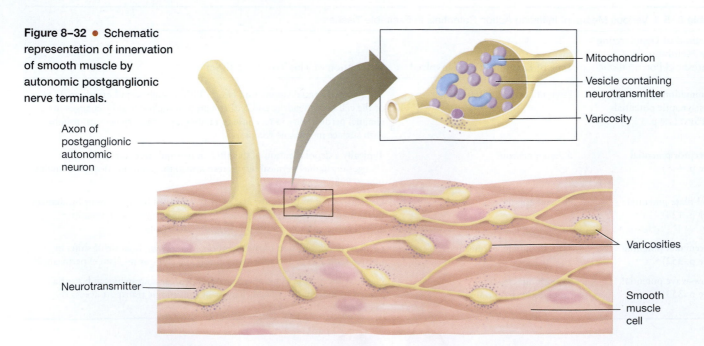

Figure 8–32 ● Schematic representation of innervation of smooth muscle by autonomic postganglionic nerve terminals.

Axon of postganglionic autonomic neuron

Mitochondrion

Vesicle containing neurotransmitter

Varicosity

Neurotransmitter

Varicosities

Smooth muscle cell

to inputs regulating bladder emptying. If considerable stretching prevented development of tension, as in skeletal muscle, a filled bladder could not contract to empty.

Stress Relaxation Response

When a smooth muscle is suddenly stretched, it initially increases its tension, much like the tension in a stretched rubber band. The muscle quickly adjusts to this new length, however, and inherently relaxes to the tension level prior to the stretch, probably from rearrangement of cross-bridge attachments. Smooth muscle cross bridges detach comparatively slowly. Physiologists speculate that on sudden stretching, any attached cross bridges would strain against the stretch, contributing to a passive (not actively generated) increase in tension. As these cross bridges detach, the filaments would be permitted to slide into an unstrained stretched position, restoring the tension to its original level. This inherent property of smooth muscle is called the **stress relaxation response.**

Advantages of the Smooth-Muscle Length–Tension Relationship

These two responses of smooth muscle to being stretched—being able to develop tension even when considerably stretched and inherently relaxing when stretched—are highly advantageous. They enable smooth muscle to exist at a variety of lengths with little change in tension. As a result, a hollow organ enclosed by smooth muscle can accommodate variable volumes of contents with little change in the pressure exerted on the contents except when the contents are to be pushed out of the organ. At that time, the tension is deliberately increased by fiber shortening. It is possible for smooth muscle fibers to contract to half their normal length, enabling hollow organs to dramatically empty their contents on increased contractile activity; thus smooth-muscled viscera can easily accommodate large volumes but can empty to practically zero volume. This length range in which smooth muscle normally functions (anywhere from $\frac{1}{2}$ to $2\frac{1}{2}$ times the normal length) is considerably greater than the limited length range within which skeletal muscle remains functional.

Smooth muscle contains abundant connective tissue, which resists being stretched. Unlike skeletal muscle in which the skeletal attachments restrict how far the muscle can be stretched, this connective tissue prevents smooth muscle from being overstretched and thus places an upper limit on the volume capacity of a smooth-muscled hollow organ.

■ Smooth muscle is slow and economical, especially in latch and catch types.

A smooth muscle contractile response proceeds at a more leisurely pace than does a skeletal muscle twitch. A single smooth-muscle contraction may last as long as 3 seconds (3000 msec), compared to the maximum of 100 msec required for a single contractile response in skeletal muscle. The rate of ATP splitting by myosin ATPase is much slower in smooth muscle, so cross-bridge activity and filament sliding occurs more slowly. Smooth muscle also relaxes more slowly because of a slower rate of Ca^{++} removal. Slowness should not be equated with weakness, however. Smooth muscle can generate the same contractile tension per unit of cross-sectional area as skeletal muscle, but it does so more slowly and at considerably less energy expense. Because of the low rate of cross-bridge cy-

cling, cross bridges are maintained in the attached state for a longer period of time during each cycle compared with skeletal muscle.

Some vertebrate and invertebrate smooth muscles are specialized to carry this low-cycling attached state to an extreme known as a **latch state** in vertebrates, and a **catch state** in invertebrates. In these states, the cross bridges "latch onto" the thin filaments for minutes to hours, enabling the smooth muscle to maintain tension with very low ATP consumption (because each cross-bridge cycle uses up one molecule of ATP). The mechanisms for this are unclear, and may be different in different types of animals, but some evidence is emerging. Let us look at a bivalve mollusk's **adductor muscle** as an example. Bivalve shells have elastic elements that spring the shells open in the absence of muscle activity. The adductor muscle contains smooth muscle that must hold the two shells closed for long periods (such as when the tide is out, or a predator is attacking). The adductor also contains some skeletal-type muscle to close the shell rapidly, as during an initial disturbance, or in the jet-propelled swimming of a scallop (which rapidly opens and closes its shell to draw in and squirt out water). Acetylcholine (ACh) stimulates smooth muscle contraction and the catch state, whereas serotonin is thought to trigger relaxation. The rate at which the myosin heads detach from the actin is greatly slowed by the presence of **twitchin**, a titin-like protein associated with the myosin. Phosphorylation of twitchin, perhaps because of the actions of serotonin, allows the myosin heads to detach at a faster (noncatch) rate.

Because of its slowness and the less ordered arrangement of its filaments, smooth muscle has often been mistakenly viewed as a poorly developed version of skeletal muscle. Actually, smooth muscle is just as highly specialized for the demands placed on it—that is, being able to economically maintain tension for prolonged periods without fatigue and being able to accommodate considerable variations in the volume of contents it encloses with little change in tension. It is an extremely adaptive, efficient tissue.

Nutrient and O_2 delivery are generally adequate to support the smooth-muscle contractile process. Smooth muscle

A scallop has an adductor muscle to close its shell. In a *catch* state, the shell can be closed for long periods of time. The muscle can also be contracted and relaxed rapidly for jet-propelled swimming.

can use a wide variety of nutrient molecules for ATP production. There are no energy storage pools comparable to creatine phosphate in smooth muscle; they are not necessary. Oxygen delivery is usually adequate to keep pace with the low rate of oxidative phosphorylation needed to provide ATP for the energy-efficient smooth muscle. If necessary, anaerobic glycolysis can sustain adequate ATP production if O_2 supplies are diminished.

Cardiac Muscle

Cardiac muscle, found only in the heart, shares structural and functional characteristics with both skeletal and single-unit smooth muscle. Like skeletal muscle, cardiac muscle is striated, with its thick and thin filaments highly organized into a regular banding pattern. Cardiac thin filaments contain troponin and tropomyosin, which constitute the site of Ca^{++} action in turning on cross-bridge activity, as in skeletal muscle. Also similar to skeletal muscle, cardiac muscle has a clear length–tension relationship. Like the oxidative skeletal muscle fibers, cardiac muscle cells have an abundance of mitochondria and myoglobin. They also have T tubules and a moderately well-developed sarcoplasmic reticulum.

As in smooth muscle, Ca^{++} enters the cytosol from both the ECF and the sarcoplasmic reticulum during cardiac excitation. Ca^{++} entry from the ECF through voltage-gated dihydropyridine receptors, alias Ca^{++} channels, in the T tubule membrane triggers the release of Ca^{++} intracellularly from the sarcoplasmic reticulum. Like single-unit smooth muscle, the heart displays pacemaker (but not slow-wave) activity, initiating its own action potentials without any external influence. Cardiac cells are interconnected by gap junctions that enhance the spread of action potentials throughout the heart, just as in single-unit smooth muscle. Also similarly, the heart is innervated by the autonomic nervous system, which, along with certain hormones and local factors, can modify the rate and strength of contraction. In contrast to skeletal muscle, the heart's output is graded by controlling contraction frequency and modulating mechanical output of each cell, not the number of activated cells.

Unique to cardiac muscle, the cardiac fibers are joined together in a branching network, and its action potentials have a much longer duration before repolarizing. Further details and the importance of cardiac muscle's features are addressed in the next chapter.

Chapter in Perspective:
HOMEOSTASIS AND INTEGRATION

The skeletal muscles make up the muscular system itself. Cardiac and smooth muscle are part of the organs that comprise other body systems. Cardiac muscle is found only in the heart, which is part of the circulatory system. Smooth muscle is found in the walls of vertebrate hollow organs and tubes, including the blood vessels in the circulatory system, airways in the respiratory system, bladder in the urinary system, stomach and intestines in the digestive system, and uterus and ductus deferens (the duct that provides a route of exit for sperm from the testes) in the reproductive system.

Contraction of skeletal muscles moves the body parts in relation to each other and moves the whole body in relation to the external environment. Thus these muscles permit movement through and manipulation of the external environment. At a very general level, some of these movements are aimed at maintaining homeostasis, such as moving the body toward a food source or away from predators. Examples of more specific homeostatic functions accomplished by skeletal muscles include the chewing and swallowing of food for further breakdown in the digestive system into usable energy-producing nutrient molecules (the mouth and throat muscles are all skeletal muscles) and the process of respiration to obtain O_2 and eliminate CO_2 (the gill and respiratory muscles are all skeletal muscles). In some species the generation of heat by contracting skeletal muscles also helps maintain body temperature.

All the other systems of the body, except the immune (defense) system, depend on their nonskeletal muscle components to enable them to accomplish their homeostatic functions. For example, contraction of cardiac muscle in the vertebrate heart pushes life-sustaining blood forward into the blood vessels, and contraction of smooth muscle in the stomach and intestines pushes the ingested food through the digestive tract at a rate appropriate for the digestive juices secreted along the route to break down the food into usable units. ∎

REVIEW QUESTIONS *(Answers are on p. A–1.)*

Additional study tools for this chapter, including chapter summaries and practice tests, are available online at *www.biology.brookscole.com*

1. Cardiac muscle is
   a. striated
   b. voluntary
   c. smooth
   d. unstriated
   e. none of the above

2. The smallest component of muscle fiber that is capable of contraction is the
   a. A-band
   b. I band
   c. muscle fiber
   d. sarcomere
   e. M line

3. During contraction, cycles of cross-bridge binding and bending
   a. pull thick filaments closer together
   b. pull thin filaments together

    c. pull Z discs closer together

    d. make thin filaments shorter

    e. b and c

4. Spread of an action potential down a T tubule

    a. causes a contraction of thick filaments

    b. induces permeability changes in I bands

    c. triggers release of Ca^{++} from actin molecules

    d. triggers release of Ca^{++} from the sarcoplasmic reticulum

    e. triggers release of Ca^{++} from the H zone

5. During the bending (power stroke) of contraction,

    a. ATP molecule binds to myosin cross bridge

    b. P_i and ADP are released from myosin

    c. P_i and ADP attach to actin

    d. P_i and ADP attach to myosin

    e. none of the above

6. For muscles that produce precise movements, such as hand muscles in humans,

    a. a single motor unit would contain a relatively large number of muscle fibers

    b. a single motor unit would contain a relatively small number of muscle fibers

    c. a motor unit would have a relatively large number of apodemes

    d. flexor muscles would have more motor units than extensor muscles

    e. none of the above

7. At optimal muscle length when maximum tension can be developed,

    a. thin filaments do not overlap thick filaments

    b. thick filaments become forced against Z lines

    c. the central region of thick filaments is devoid of cross bridges

    d. thin filaments from opposite sides of the sarcomere become overlapped

    e. thin filaments are pulled out maximally from thick fibers

8. During an isometric contraction of a muscle,

    a. muscle length changes

    b. sarcomeres remain the same length

    c. only concentric contractions occur

    d. muscle tension develops at a constant muscle length

    e. only eccentric contractions occur

9. The first energy storehouse tapped at the onset of contractile activity is often

    a. glycolysis

    b. oxidative phosphorylation

    c. phosphogens

    d. creatine kinase

    e. myoglobin

10. Which of the following is not implicated as an underlying cause of muscle fatigue?

    a. a local increase in inorganic phosphate

    b. accumulation of lactic acid

    c. depletion of energy reserves

    d. glycolysis

    e. breakdown of phosphogens

11. A slow oxidative (type I) muscle fiber would

    a. have low myoglobin content

    b. have high myoglobin content

    c. have low resistance to fatigue

    d. have high glycogen content

    e. have high myosin ATPase activity

12. Deep layers of the pectoralis muscles of birds adapted for soaring and gliding

    a. contain more white fibers than red fibers

    b. contain approximately equal numbers of red and white fibers

    c. contain more red fibers than white fibers

    d. contain mostly fast glycolytic fibers

13. Regular endurance (aerobic) exercise would result primarily

    a. in increased synthesis of myosin fibers

    b. in increased synthesis of actin fibers

    c. in muscle hypertrophy

    d. in increased numbers of mitochondria in oxidative fibers

    e. in increased numbers of mitochondria in myosin filaments

14. In insects, asynchronous flight muscles

    a. are used to produce low wingbeat frequencies

    b. are controlled directly by neuronal activation

    c. are used to produce high wingbeat frequencies

    d. oppose the action of synchronous flight muscles

    e. are used by cicadas to generate sound

15. Which of the following is a major level of input for control of motor neuron output?

    a. muscle spindles

    b. brain stem

    c. Golgi tendon organs

    d. primary (annulospiral) endings

    e. flower-spray endings

16. An animal receives information about the progression of body movements from

    a. afferent input

    b. proprioceptors

    c. muscle spindles

    d. Golgi tendon organs

    e. all of the above

17. Smooth muscle cells

    a. have a well-developed sarcoplasmic reticulum

    b. are multinucleate

    c. are triggered to contract by a lowering of cytosolic Ca^{++}

    d. have a poorly developed sarcoplasmic reticulum

    e. have clearly visible striations

18. Single-unit smooth muscle is found

    a. in the iris of the eye

    b. in the uterus

    c. in the walls of large blood vessels

    d. at the base of hair follicles

    e. in large airways to the lungs

19. Which of the following methods of initiating action potentials is used by cardiac muscle?

    a. receptor potential

    b. end-plate potential

    c. slow-wave potential

    d. pacemaker potential

    e. summation of EPSPs

20. Cardiac muscle

    a. is striated

    b. contains only thick contractile fibers

    c. has very few mitochondria

    d. maintains low levels of myoglobin

    e. lacks T tubules

SUGGESTED READINGS AND INTERNET SITES

Alexander, R. M., & G. Goldspink, eds. 1977. *Mechanics and Energetics of Animal Locomotion*. London: Chapman and Hill.

Alexander, R. M. 1982. *Locomotion of Animals*. Glasgow: Blackie.

Baldwin, J., J-P. Jardel, T. Montague, & R. Tomkin. 1984. Energy metabolism in penguin swimming muscles. *Molecular Physiology* 6:33–42.

Butler, T. M., S. R. Narayan, S. U. Mooers, D. J. Hartshorne, & M. J. Siegman. 2001. The myosin cross-bridge cycle and its control by twitchin phosphorylation in catch muscle. *Biophysical Journal* 80:415–26.

Demirel, H. A., S. K. Powers, H. Naito, M. Hughes, & J. S. Coombes. 1999. Exercise-induced alterations in skeletal muscle myosin heavy chain phenotype: Dose–response relationship. *Journal of Applied Physiology* 86:1002–1008.

Dickinson, M. 2001. Solving the mystery of insect flight. *Scientific American*. Available online at www.sciam.com/article.cfm?articleID=000EE5B1-DCA8–1C6F-84A9809EC588EF21. Accessed on February 19, 2004.

Huxley, H. E. 1969. The mechanism of muscular contraction. *Science* 164:1356–1365.

Labeit, S., & B. Kolmerer. 1995. Titins: Giant proteins in charge of muscle ultrastructure and elasticity. *Science* 270:293–296. (Also a news story in the same issue: M. Barinaga, Titanic protein gives muscles structure and bounce, *Science* 270:236.)

Naya, F. J., B. Mercer, J . M. Shelton, J. A. Richardson, R. S. Williams, & E. N. Olson. 2000. Stimulation of slow skeletal muscle fiber gene expression by calcineurin *in vivo*. *Journal of Biological Chemistry* 275:4545–4548. Available online at www.jbc.org.

Rome, L. C., & S. Lindstedt. 1998. The quest for speed: Muscles built for high-frequency contractions. *News in Physiological Science* 13:261–268.

Syme, D. A., & R. K. Josephson. 2002. How to build fast muscles: Synchronous and asynchronous designs. *Integrative and Comparative Biology* 42:762–770.

INTERNET SITES

The Harvey Project. 2000, April 26. *Muscle.* **lessons.harveyproject.org/development/muscle/menu.html.** Animations and diagrams of muscle functions.

Childress, S., & J. Wang. *Simulation of insect flight.* **math.nyu.edu/AML/fly.html.**

Rome, L. C. *Testing a muscle's design.* **eee.uci.edu/courses/bio112/rome.htm.** An overview of the evolution of animal muscles for widely different uses.

INFOTRAC READINGS

Mathews, G. G. 1986. *Cellular Physiology of Nerve and Muscle.* Palo Alto, CA: Blackwell Scientific.

Suarez, R. K. 2000. Energy metabolism during insect flight: Biochemical design and physiological performance. *Physiological and Biochemical Zoology* 73:765–770.

Circulatory Systems

Circulation *serves as an animal's internal transport system. Moving fluid transports and distributes* O_2*, nutrients, wastes, hormones, and heat. The fluid is usually moved under pressure created by a heart or other pump. As an extreme example, consider a giraffe. Its heart must generate high pressures in order to overcome gravity to provide adequate blood flow to its brain. It also has tight-fitting skin on its legs that helps prevent blood pooling there, as occurs in humans who sit or stand immobile for extended periods.*

Photo: Paul Yancey

Evolution of Circulation

Most multicellular organisms with specialized cells need some form of internal transport to move important molecules (and often cells) from one tissue to another. Transported molecules in animals include oxygen, nutrients, wastes, and hormones. Heat may also be usefully transported, especially in endothermic animals but also in some ectothermic animals. For example, reptiles speed up the heating process when they move into the sun, simultaneously increasing their heart rate and rapidly shunting the heat from the skin to the body core. At day's end, the setting of the sun triggers a rapid decline in heart rate.

■ Circulatory systems evolved to overcome the limits of diffusion.

The evolution of internal transport correlates with animal size, complexity, and metabolism. For very small or very thin animals such as rotifers, movement of molecules occurs primarily by diffusion, as long as the distances are quite small and metabolic demands are not high. Similarly, flatworms (though they may be quite large in body length) are—as their name implies—thin enough to use diffusion. They have an internal branching digestive tract that brings digested materials close to all cells.

As animals evolved thicker and larger bodies and higher metabolic rates, the need for internal circulatory systems became paramount, because of the slowness of diffusion over moderate to long distances. We explore this aspect in more detail in Chapter 11. Briefly, circulatory systems overcome the slowness of diffusion by developing the much faster process of **bulk transport:** the movement of the medium that contains the molecules (and cells) of interest.

■ Circulatory systems have up to three distinct components: fluid, pump, and vessels.

From an engineering standpoint, a circulatory system can have the following components:

1. The *fluid* itself, which carries the transported molecules and cells; typically called *blood* or *hemolymph*. Terms associated with circulatory fluid often contain the root *hemo* or *emia* (from the Greek for blood).

2. A *pump* to move the fluid; dedicated pumps are called *hearts,* and terms associated with hearts often contain the root *cardio* (Greek for heart).

3. Conduits or *vessels* to carry the fluid between the pump and body tissues; these vessels are called *vascular* components.

A system with all three components should be quite familiar to you. For example, a city water supply typically has pumps (at reservoirs, wells, and water towers) that move a vital fluid, water, through pipes to your house and other buildings. However, not all three components are necessary for some types of circulation. In particular, completely enclosed vessels are absent in many animal groups. Also, biological pumps are not always distinct dedicated organs (hearts). The simplest example is that of the water canal system of a sponge (● Figure 9–1a).

Figure 9–1 ● Diagram of the circulatory systems of sponge, cnidarian, and nematode. (a) Circulation of external medium in a gastrovascular cavity of the sponge; (b) Circulation of external medium in a gastrovascular cavity of a cnidarian; (c) Circulation of coelomic fluid in a nematode.

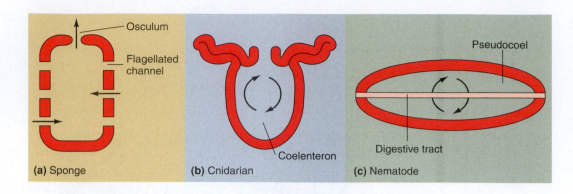

(a) Sponge   (b) Cnidarian   (c) Nematode

(*Source:* P. Withers, 1992, *Comparative Animal Physiology,* 1st ed., Belmont, CA: Brooks/Cole, p. 667)

A sponge has no true internal circulatory fluid; rather, environmental water is pumped through hundreds of pores and internal chambers by thousands of collar cells (*choanocytes*) with beating flagella. Thus water with food items and oxygen is brought close to most cells, whereas water with wastes leaves a large exit pore. In large cnidarians (jellies, sea anemones, and so on; Figure 9–1b), materials move by a combination of diffusion and movement of ambient water in and out of the digestive cavity (*coelenteron*) that takes up most of their interiors. Pumping as such is accomplished as part of the digestive process. The tissues are generally thin, and all cells are close enough to the ambient water (outside) and the digestive cavity to survive on diffusion.

Animals with true internal circulatory systems have an internal fluid that is different from the environment. Small animals with an internal fluid-filled body cavity, such as nematodes (roundworms), may move materials through body

motions that move the internal fluid (Figure 9–1c). But more advanced animals have dedicated pumps and vessels. Traditionally, internal circulation of fluids is divided into two broad categories:

■ *Open* systems (● Figure 9–2a and b), in which the fluid—called **hemolymph**—moves through the vessels into extracellular spaces among the tissues, bathing them directly for molecular exchanges with cells. The entire space filled with hemolymph is called the **hemocoel.** The fluid may be moved indirectly by body movements, by cilia or flagella, or by hearts, which pump the fluid through open vessels into the hemocoel, and which take up the fluid through intake pores or vessels. This design is similar to your home water system: Water arrives in pipes; emerges into the open spaces in your shower, toilet, sinks, dishwasher, and so on; then re-enters pipes through drains (although the analogy

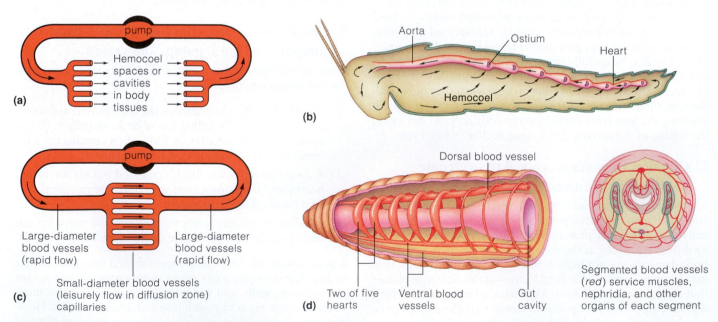

Figure 9–2 ● Flow through open and closed circulatory systems. (a,b) Open system of a grasshopper. A "heart" pumps blood through a vessel (aorta). Blood moves into tissue spaces and mingles with fluid bathing cells, then reenters the heart at opening (ostia) in the heart wall. (c,d) Closed system of an earthworm. Blood is confined within several pairs of muscular "hearts" near the head end and within blood vessels.

(*Source:* C. Starr, 2000, *Concepts and Applications of Biology:* 4th ed., Belmont, CA: Brooks/Cole, Figure 33.3)

fails at this point, because whereas the water does not return to the original pump, hemolymph does).

Various versions of open systems are found in many animal groups, including *most mollusks* and *all arthropods* (an example is shown in Figure 9–2b).

- *Closed* systems (Figure 9–2c and d), in which the fluid—called **blood**—exits a heart through vessels that are continuous all the way back to the intake side of the heart. The vessels branch and become smaller and smaller until they become tiny **capillaries**, where flow slows down and exchange of materials occurs with body cells. Capillaries are widely regarded as the primary structure distinguishing a closed from an open system. These then merge and become larger and fewer vessels to return to the heart. An analogy is the cooling system in your refrigerator, in which a compressor pumps fluid out to coiled tubes and back again. (The open–closed distinction is not clear-cut: Blood fluid may leak out of capillaries and bathe nearby cells much as hemolymph does, before returning to nearby vessels.)

Closed systems are found in several animal groups, including *cephalopod mollusks, annelids,* and *vertebrates* (see example in Figure 9–2c and d).

Now let's examine each of the three circulatory system components individually. Then we will examine how pumps and vessels are integrated and regulated as one system.

Circulatory Fluids

Circulatory fluids are traditionally divided into two components: a liquid called **plasma,** primarily water containing a variety of dissolved and dispersed molecules, and **cellular elements,** of which there may be several specialized kinds. In hemolymph, various cell types are called *hemocytes,* which are responsible for immune functions, clotting, and sometimes oxygen transport. In vertebrates, these functions belong to three types of specialized cellular elements suspended in the plasma: (1) *erythrocytes* (or red blood cells, for oxygen transport), (2) *leukocytes* (or white blood cells, for immunity), and (3) *thrombocytes* or *platelets* (for clotting).

Methodology: The Hematocrit

If a sample of whole blood or hemolymph is placed in a test tube, treated to prevent clotting, and centrifuged, the heavier cellular elements move to the bottom and the lighter plasma rises to the top. This gives the **hematocrit,** or **packed cell volume** (● Figure 9–3). Hematocrit values in vertebrates give a good indication of the oxygen delivery capacity of a species or individual, because well over 90% of the packed cells are erythrocytes. Vertebrates evolutionarily adapted or acclimatized to higher aerobic demands or low oxygen usually have higher values. (One exception is the bar-headed goose, an exceptionally strong high-altitude flyer whose hematocrit does not change as a function of altitude.) Stressed animals often have abnormal values. Here are some examples:

Human, *Homo sapiens:* 45% in males, 42% in females (average values; often higher in athletes).

White whale, *Delphinapterus leucas:* 53% in females; 52% in males (a deep-diving species).

Pekin duck, *Anas domesticus:* 45% at sea level, 56% after 4 weeks acclimation at 5640 meters.

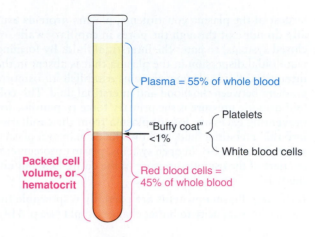

Figure 9–3 ● **Hematocrit.** The values given are for men. The average hematocrit for women is 42%, with plasma occupying 58% of the blood volume.

Striped bass, *Morone saxatilis:* 39% if acclimated to 5°C, 53% if acclimated to 25°C (note that warm water cannot dissolve as much oxygen as cold water).

Plasma accounts for the remaining volume. Let's first consider the properties of this component before turning to the cellular elements.

Plasma in Circulatory Fluids

Plasma typically consists of 90% or more water, which serves as a medium for a large number of organic and inorganic substances being carried in the fluid. In mammals, the most plentiful organic constituents by weight are the **plasma proteins,** which compose 6% to 8% of plasma's total weight, whereas inorganic constituents account for approximately 1%. *The most abundant electrolytes (ions) in virtually all animal plasmas are Na^+ and Cl^-.* There are smaller amounts of HCO_3^-, K^+, Ca^{++}, and others. The most notable functions of these extracellular fluid (ECF) ions are their roles in membrane potential, osmotic distribution of fluid between the ECF and cells, and buffering of pH; these functions are discussed elsewhere (Chapters 4 and 13). The remaining small percentage of plasma is occupied by nutrients (for example, glucose, amino acids, lipids, and vitamins); waste products (such as creatinine, bilirubin, and urea in mammals); dissolved gases (O_2 and CO_2); and hormones. Most of these substances are merely being transported in the plasma; for example, hormones are transported to their sites of action and have no function within the plasma.

■ Many of the functions of plasma are carried out by plasma proteins.

The **plasma proteins** are the one group of plasma constituents not present just for the ride. These important components normally remain in the plasma, where they perform many valuable functions. Here are the most important of these functions, which are elaborated on elsewhere in the text:

1. *Colloid osmotic pressure.* Unlike other plasma constituents that are dissolved in the plasma water, the plasma proteins exist in a colloidal dispersion. Furthermore, as

largest of the plasma constituents, plasma proteins usually do not exit through the pores in capillary walls (of closed systems) to enter the interstitial fluid. By forming a colloidal dispersion in the plasma that is absent in the interstitial fluid, plasma proteins establish an osmotic gradient between the blood and interstitial fluid. This **colloid osmotic pressure** is the primary force responsible for preventing excessive loss of plasma from the capillaries into the interstitial fluid, and thus helps maintain plasma volume (see p. 414). In open systems, plasma proteins create part of the osmotic pressure of the entire ECF but not the ICF.

2. *Buffering.* Plasma proteins are partially responsible for the plasma's capacity to buffer changes in pH (see p. 414).

In vertebrates, there are three major groups of plasma proteins—*fibrinogen, albumins, globulins*—which are classified according to their various physical and chemical properties. In vertebrates, these proteins are generally synthesized by the liver. Some are produced by vascular endothelial (lining) cells, whereas the γ-globulins are produced by lymphocytes, one type of leukocyte. The functions of some of these proteins are elaborated on elsewhere in the text; the following list illustrates the wide range of these functions:

Fibrinogen is a key factor in the blood-clotting process.

Albumins, by far the most abundant of the plasma proteins, bind many substances (for example, bilirubin, bile salts, and fatty acids) for transport through the plasma and contribute most extensively to the colloid osmotic pressure by virtue of their numbers.

Globulins come in three forms—**alpha (α), beta (β),** and **gamma (γ)**—which act as
- *Transporters.* Specific α- and β-globulins bind and transport a number of substances in the plasma, such as thyroid hormone, cholesterol, and iron. The most common is **transferrin,** which binds and transports iron atoms.
- *Clotting agents.* Many of the factors involved in the process of blood clotting, which will be described shortly, are α- or β-globulins.
- *Regulators.* Inactive proteins that are precursors of regulators such as hormones, and that are activated as needed by specific signals, belong to the α-globulin group (for example, the α-globulin angiotensinogen is activated to angiotensin, which plays an important role in regulating salt balance in the body; p. 544).
- *Immune effectors.* The γ-globulins are the immunoglobulins (antibodies), which are crucial to vertebrate defense mechanism (Chapter 10).

Arthropods (insects, crustaceans, arachnids) also have a variety of plasma proteins in their hemolymph. One important class are proteins involved in synthesis of exoskeletons. Arthropods periodically molt their exoskeletons so that they can grow in size. Making new exoskeleton involves, in part, the polymerization of phenols by **phenoloxidase (PO) enzymes.** Involved with this are two classes of plasma proteins:

- **Hexamerins** and other hemolymph proteins in insects are thought to transport phenols to exoskeletal-producing epithelial cells.

- **ProPOs** are inactive forms of POs that are activated during exoskeletal synthesis (after molting, and during wound repair). ProPOs are also involved in arthropod immune responses (p. 459).

■ Lipoprotein complexes carry energy lipids and structural lipids for biosynthesis.

Cells of most animals use lipids for two purposes: (1) Some, such as *triglycerides,* can be used as an energy source (p. 45); and (2) cells require lipids for making membranes. There are two types of these structural lipids: (1) *phospholipids* for constructing the primary bilayer itself, and (2) *cholesterol* for reinforcing the membrane (see Chapter 3). In addition, a few special cell types use cholesterol as a precursor for the synthesis of secretory products, such as steroid hormones and bile salts. Although most mammalian cells can synthesize some of these lipids needed for their own plasma membranes, they cannot manufacture enough and therefore must rely on the blood delivering supplemental lipid. There are two sources of these lipids: (1) dietary intake, with animal tissues being especially rich in cholesterol compared to plant tissues; and (2) biosynthesis by other organs, particularly the liver in vertebrates.

Because these compounds are hydrophobic, they are not very soluble in blood. Most triglyceride, cholesterol, and phospholipids in the blood are converted into droplets attached to specific plasma-protein carriers, forming **lipoprotein complexes,** which are soluble in blood. The four major lipoproteins in vertebrates are named for their density of protein as compared to lipid:

- **High-density lipoproteins (HDLs),** which contain the most protein, some phospholipids, and least cholesterol
- **Low-density lipoproteins (LDLs),** which contain less protein, some phospholipids, and more cholesterol
- **Very-low-density lipoproteins (VLDLs),** which contain the least protein and most lipid, but the lipid they carry is triglyceride for energy
- **Chylomicrons,** produced by intestinal absorptive cells, which transport triglycerides, cholesterol, and phospholipids after a meal

Chylomicrons are discussed in Chapter 14 (p. 655), so here we focus only on the HDLs, LDLs, and VLDLs.

Cholesterol carried in LDL complexes (● Figure 9–4) has been termed "bad" cholesterol, because cholesterol is transported *to* the cells, including those lining the blood vessel walls, by means of LDL. (Pathological accumulations of LDL cholesterol in blood vessel walls is a leading cause of cardiovascular disease in humans, occurs in many domestic mammals such as pigs, and may possibly occur in some wild vertebrates such as salmon.) In contrast, cholesterol carried in HDL complexes is dubbed "good" cholesterol, because HDL removes cholesterol *from* the cells and transports it to the liver for partial elimination. *The interplay between LDL and HDL determines the traffic flow of cholesterol between the liver and the other cells.*

Cells accomplish phospholipid and cholesterol uptake from the blood by synthesizing receptor proteins specifically capable of binding LDLs and inserting these receptors into the cells' plasma membranes. When an LDL particle binds to one of the membrane receptors, the cell engulfs the particle by

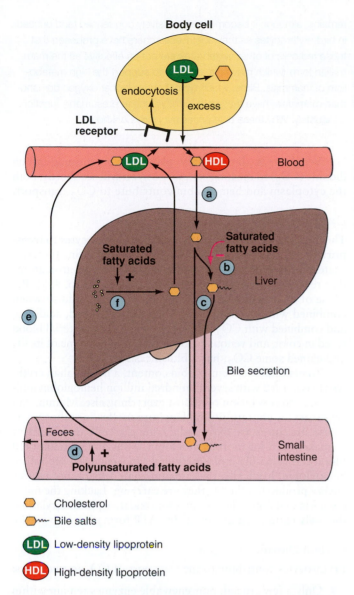

Body cell

endocytosis

excess

LDL
receptor

LDL   HDL   Blood

ⓐ

Saturated
fatty acids

Saturated
fatty acids

ⓑ

Liver

ⓕ   ⓒ

ⓔ

Bile secretion

Feces

ⓓ +

Polyunsaturated fatty acids

Small
intestine

⬡ Cholesterol

⬡〜 Bile salts

Ⓛ Low-density lipoprotein

Ⓗ High-density lipoprotein

Figure 9–4 ● Role of the vertebrate liver in cholesterol metabolism.
The letters in the illustration correspond to the following steps:

ⓐ The liver extracts cholesterol (carried by HDL) from the blood.

ⓑ The liver converts most of the cholesterol into bile salts.

ⓒ The liver secretes bile salts and cholesterol into the bile, which
empties into the intestinal lumen.

ⓓ Part of the cholesterol and bile salts is eliminated in the feces.

ⓔ Some of the bile salts are reabsorbed into the blood and
recycled to the liver.

ⓕ The liver synthesizes new cholesterol, which is carried by LDL
within the blood away from the liver to other cells.

Saturated fatty acids and polyunsaturated fatty acids are known
to influence these pathways at the designated points (+, −).

endocytosis (Figure 9–4). Within the cell, lysosomal enzymes
break down the LDLs to free the lipids, making them available
to the cell for synthesis of new cellular membrane. If too
much free cholesterol accumulates in the cell, the cell down-
regulates both the synthesis of LDL receptor proteins (so that
less cholesterol is taken up) and its own cholesterol synthesis

(so that less new cholesterol is made). These processes can be
upregulated when a cell is faced with a cholesterol shortage.

In contrast to LDL complexes, HDL removes cholesterol
from cells and transports it to the liver. The liver in turn secretes cholesterol as well as cholesterol-derived bile salts into
the bile. Bile enters the intestinal tract, where bile salts participate in the digestive process (Chapter 14). Most of the secreted
cholesterol and bile salts are later reabsorbed from the intestinal tract into the blood to be recycled to the liver, but the cholesterol molecules not reclaimed by absorption are eliminated
in the feces (Figure 9–4).

■ Respiratory pigments carry oxygen.

In many invertebrates, another important type of dissolved protein is the **respiratory pigment.** These are oxygen-binding proteins critical for transporting oxygen for cellular respiration.
Respiratory pigments in invertebrates vary considerably by
phylum, as you will see in detail in Chapter 11; we give a brief
summary here. The two most common ones are **hemoglobin,**
a protein in vertebrates and many invertebrates that uses an
iron atom to bind an oxygen molecule; and **hemocyanin,** a
protein in many mollusks and arthropods such as crustaceans,
that uses two copper atoms to bind one oxygen molecule. Most
insects do not have a respiratory pigment, because their tracheal system carries oxygen directly to tissues. However, arachnids do have hemocyanin, because some species lack tracheae
(using book lungs, for example), whereas others have tracheae
that do not extend into the tissues. Hemocyanins circulate in
arachnid hemolymph to transfer oxygen between the lungs or
tracheal termini to muscles and other organs.

As we have noted, the pigment proteins may be dissolved
in the hemolymph, which is the case for most annelids, arthropods, and some other invertebrates. In many animals, however, these proteins are not in the plasma itself but are contained in specialized cells. If they contain hemoglobin, which
is reddish when oxygen is bound, these cells are called *erythrocytes* (*erythro,* "red"), and we now turn our attention to them.

Erythrocytes in Circulatory Fluids

Each milliliter (mL) of vertebrate blood contains numerous
erythrocytes (red blood cells), about 3 billion per mL in chickens, 7 billion per mL in cows and pigs, 10 billion per mL in
horses, and 13 billion per mL in goats.

■ Erythrocytes serve primarily
to transport oxygen.

The main function of erythrocytes is to transport O_2 from the
lungs or gills to all tissues. In most vertebrates, these cells are
oblong oval (ovoid) shapes (● Figure 9–5a), with reptiles having among the largest. In contrast, mammalian erythrocytes
are flat, disc-shaped cells indented in the middle on both sides,
like a doughnut with a flattened center instead of a hole (Figure 9–5b). This unique shape of mammalian erythrocytes is
thought to contribute in two ways to the efficiency with which
these cells transport O_2: (1) The biconcave shape provides a
larger surface area for diffusion of O_2 across the membrane
than would a spherical cell of the same volume; and (2) the
thinness of the cell enables O_2 to diffuse rapidly between the
exterior and innermost regions of the cell. Another feature of

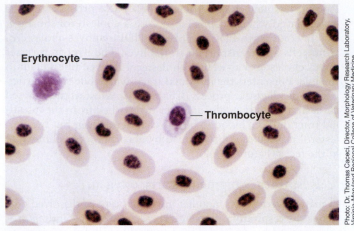

Erythrocyte ——

—— Thrombocyte

(a)

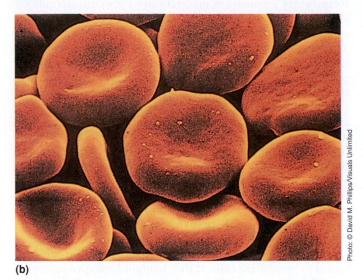

(b)

Figure 9–5 ● **Anatomic characteristics of erythrocytes.**
(a) Chicken erythrocytes (and thrombocytes). (b) Mammalian
erythrocytes, under a scanning electron microscope. Note their
biconcave shape.

(*Source:* http://education.vetmed.vt.edu/Curriculum/VM8054/Labs/Lab6/Examples/exbrdrbc.htm)

erythrocytes is the flexibility of their plasma membrane, which
lets them travel through the narrow, tortuous capillaries to deliver their O_2 cargo at the tissue level without rupturing. These
cells, whose diameter is normally 8 mm, can deform amazingly as they squeeze single file through capillaries as narrow
as 3 mm in diameter. This ability has been primarily studied
with the mammalian biconcave red cell; however, ovoid fish
erythrocytes have also been observed to deform greatly in the
narrow capillaries of gills.

?

Why Are Erythrocytes in Birds and Mammals So Different? Mammalian red cells are unique in having a biconcave shape and no
nucleus. In other vertebrates, erythrocytes are ovoid and the nucleus

remains, although it becomes progressively condensed (and unused)
in bird erythrocytes as they age. Researchers have proposed that
these red cells of other vertebrates are not as effective as the mammalian form, which may have evolved to support the high metabolism of mammals. Birds, which typically have higher oxygen demands
than mammals, have the ovoid form, yet these presumably function
adequately. Why these differences evolved is unknown.

Vertebrate erythrocytes also play a role in returning carbon
dioxide (CO_2) to the lungs or gills. As you will see shortly, both
the cytoplasm and hemoglobin contribute to CO_2 transport.

Oxygen Transport

The most important feature that enables erythrocytes to transport O_2 is the respiratory pigments they contain. In vertebrates, the cells contain hemoglobin, which carries iron atoms
that reversibly bind with oxygen (see Chapter 11, p. 499). Because of its iron content, hemoglobin appears reddish when
combined with O_2 and bluish when deoxygenated, acidified,
and combined with CO_2. Thus, fully oxygenated arterial blood
is red in color, and venous blood, which has lost some of its O_2
and gained some CO_2, has a bluish cast.

To maximize its hemoglobin content, a mammalian erythrocyte is stuffed with several hundred million hemoglobin molecules, to the exclusion of almost everything else. Mammalian
erythrocytes contain no nucleus, organelles, or ribosomes.
These structures are extruded during the cell's development
to make room for more hemoglobin. Thus a mammalian red
blood cell is mainly a plasma membrane–enclosed sac full of
hemoglobin. Ironically, these mammalian cells cannot use for
energy production the O_2 they are carrying. Lacking the mitochondria that house the enzymes for oxidative phosphorylation,
they rely entirely on glycolysis for ATP formation (see p. 46).

Carbon Dioxide Transport

Erythrocytes contribute to the transport of CO_2 in two ways:

- Only a few crucial, nonrenewable enzymes remain within
  a mature mammalian erythrocyte: These are glycolytic enzymes and **carbonic anhydrase (CA)**. CA is crucial in CO_2
  transport (see also Chapter 11, p. 508). It catalyzes a key
  reaction that ultimately leads to the conversion of metabolically produced CO_2 into **bicarbonate ion (HCO_3^-)**,
  which is the primary form in which CO_2 is transported
  in the blood. (This bicarbonate is a major pH buffer in
  the extracellular fluid—see p. 595—and is also important
  as an ECF anion, second only to Cl^- in this role.)

- Hemoglobin also contributes by binding CO_2, although
  it does not carry as much of this gas as it carries O_2. CO_2
  binds to the protein itself (the globin), not the iron (Chapter 11).

■ Hemoglobin has additional transport functions.

In addition to carrying CO and O_2, vertebrate hemoglobin
can also combine with the following:

- *Bicarbonate (HCO_3^-).* In crocodiles, hemoglobin binds
  this ionic form of carbon dioxide. This enhances oxygen
  release as well as serving to transport the ion (Chapter 11).

- *The acidic hydrogen-ion portion (H⁺) of ionized carbonic acid,* which is generated at the tissue level from CO_2. Hemoglobin buffers this acid so that it minimally alters the pH of the blood.
- *Nitric oxide (NO).* In mammalian lungs, the vasodilator nitric oxide (see p. 88) binds to a sulfur atom within the hemoglobin molecule to form SNO. This nitric oxide is released at the tissues, where it relaxes and dilates the local arterioles. This vasodilation helps ensure that the O_2-rich blood can make its vital rounds and also helps stabilize blood pressure.

Invertebrate hemoglobins can have other transport functions as well, for example, binding:

- *Hydrogen sulfide (H₂S).* The giant tubeworm *Riftia,* which lives exclusively off sulfide-oxidizing bacterial symbionts in its body, has a hemoglobin that transports both H_2S and oxygen from the environment to its symbionts. Sulfide from the hot waters of deep-sea hydrothermal vents, where these animals live, provides the energy source for the bacterial symbionts to convert CO_2 into glucose.

Hemopoietic tissues continuously replace worn-out erythrocytes.

Avian and mammalian erythrocytes have a short life span, presumably because they have nonfunctioning nuclei (in birds) or no nuclei (in mammals). They survive an average of only about 100 to 110 days in domestic mammals, and only 28 days in a parrot and 30 days in a chicken, for example. During this short life span, the cell's plasma membrane, which cannot be repaired, becomes fragile and prone to rupture as the cell squeezes through capillaries.

One organ involved with erythrocyte regulation is the **spleen.** It has two important roles:

1. *Removal of old red cells.* Most old red-blood cells meet their final demise in the spleen, because this organ's narrow, winding, capillary network is a tight fit for these fragile cells. Macrophages (see p. 433) engulf and destroy the dying cells.

2. *Red cell reservoir.* The spleen can store healthy erythrocytes in its pulpy interior; it also serves as a reservoir site for platelets and contains an abundance of lymphocytes, a type of white blood cell. In many mammals, the stored erythrocytes form a reserve of oxygen that can be released (by contraction of the organ) during locomotory activity. In horses, for example, splenic contraction during exercise can double the hematocrit. This also occurs in many marine mammals during a dive (p. 510).

Because erythrocytes cannot divide to replenish their own numbers, the old ruptured cells must be replaced by new cells produced in an erythrocyte factory, called **hemopoietic** tissues. The **kidney** and **spleen** produce these cells in bony fishes. In birds and mammals, it is the **bone marrow**—the soft, highly cellular tissue that fills the internal cavities of bones. Regions of bone marrow called *red bone marrow* normally generate new red blood cells, a process known as **erythropoiesis,** to keep pace with the demolition of old cells (at the amazing rate of 2 to 3 million per second in humans). In adult humans, red bone marrow is found in the sternum (breastbone), vertebrae (backbone), ribs, base of the skull, and upper ends of the long limb bones. Red marrow not only produces red blood cells but is the ultimate source for leukocytes and platelets as well. Undifferentiated **pluripotent stem cells** reside in the red marrow (or in blood of lower fishes, and the kidney and spleen in bony fishes), where they continuously divide and differentiate to give rise to each of the types of blood cells. Regulatory factors act on the *hemopoietic* ("blood-producing") marrow to govern the type and number of cells generated and discharged into the blood. Of the blood cells, the mechanism for regulating red blood cell production is the best understood. Let's consider it now for vertebrates.

Erythropoiesis in mammals and probably other vertebrates is controlled by erythropoietin from the kidneys.

The number of circulating erythrocytes normally remains fairly constant, indicating that erythropoiesis must be closely regulated in negative-feedback fashion. Because O_2 transport in the blood is the erythrocytes' primary function, you might logically predict that the primary stimulus for increased erythrocyte production would be reduced O_2 delivery to the tissues. You would be correct, but low O_2 levels do not stimulate erythropoiesis directly via the hemopoietic tissues. Instead, reduced O_2 delivery to the kidneys stimulates them to secrete the hormone **erythropoietin (EPO)** into the blood, and this hormone in turn stimulates erythropoiesis by the hemopoietic tissues (● Figure 9–6). EPO has been mostly studied in mammals but is found in other vertebrates including at least one bony fish (trout). It acts on derivatives of undifferentiated stem cells that are already committed to becoming red blood cells, stimulating their proliferation and maturation into mature erythrocytes. This increased erythropoietic activity elevates the number of circulating red blood cells, thereby restoring O_2 delivery to the tissues to normal. Once normal O_2 delivery to the kidneys is achieved, EPO secretion is turned down until needed once again. In this way, erythrocyte production is normally balanced against loss of these cells so that the O_2-carrying capacity in the blood remains fairly constant. In response to severe loss of erythrocytes, as in hemorrhage, or to prolonged impairment of O_2 delivery, such as livestock being taken to high altitude or a fish entering low-O_2 water, the rate of erythropoiesis can be upregulated.

Leukocytes in Circulatory Fluids

Leukocytes, or white blood cells, are the mobile units of vertebrate immune defense systems. We leave a more detailed discussion of their names and functions for Chapter 10. Briefly, all leukocytes ultimately originate from the same undifferentiated pluripotent stem cells that also give rise to erythrocytes and platelets (● Figure 9–7). The cells destined to become leukocytes eventually differentiate into various committed cell lines and proliferate under the influence of appropriate stimulating factors. The total number of white cells and the percentage of each type may vary considerably to meet changing defense needs. Depending on the type and extent of assault a body is combating, different types of leukocytes are selectively pro-

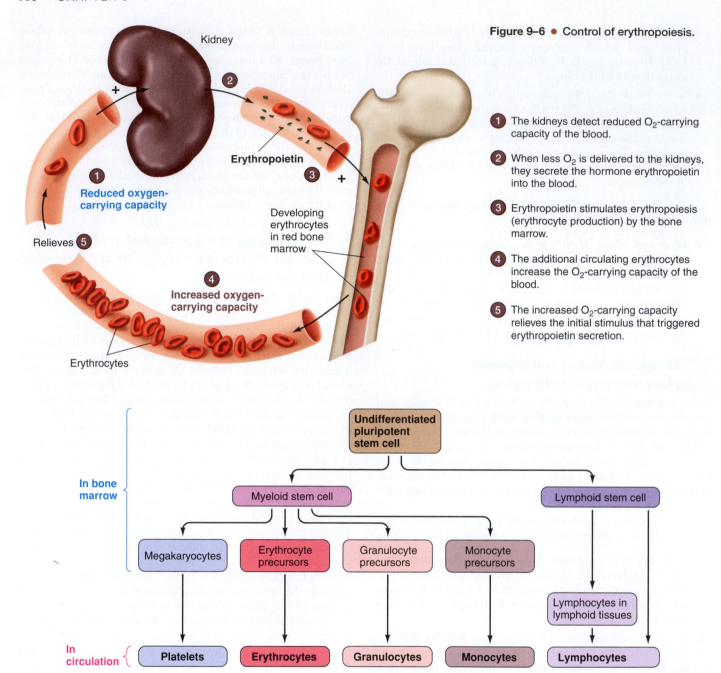

Figure 9–6 • Control of erythropoiesis.

Kidney

Erythropoietin

1 Reduced oxygen-carrying capacity

Relieves 5

Increased oxygen-carrying capacity

Erythrocytes

Developing erythrocytes in red bone marrow

① The kidneys detect reduced O_2-carrying capacity of the blood.

② When less O_2 is delivered to the kidneys, they secrete the hormone erythropoietin into the blood.

③ Erythropoietin stimulates erythropoiesis (erythrocyte production) by the bone marrow.

④ The additional circulating erythrocytes increase the O_2-carrying capacity of the blood.

⑤ The increased O_2-carrying capacity relieves the initial stimulus that triggered erythropoietin secretion.

In bone marrow

Undifferentiated pluripotent stem cell

Myeloid stem cell

Lymphoid stem cell

Megakaryocytes | Erythrocyte precursors | Granulocyte precursors | Monocyte precursors

Lymphocytes in lymphoid tissues

In circulation

Platelets | Erythrocytes | Granulocytes | Monocytes | Lymphocytes

Figure 9–7 • Blood cell production (hemopoiesis). All the blood cell types ultimately originate from the same undifferentiated pluripotent stem cells in the bone marrow.

duced at varying rates. To direct the differentiation and proliferation of each cell type, specific hormones analogous to erythropoietin are required.

Thrombocytes and Platelets in Circulatory Fluids and the Process of Hemostasis

The third type of blood cell or cell derivative is the **thrombocyte** (a living cell) or **platelet** (a cell fragment), which serves in the clotting mechanism.

■ Thrombocytes and platelets function in clotting.

Thrombocytes are cells found in all vertebrates except mammals; these cells circulate in an inactive state, and when activated by an injury to nearby tissue, they begin to break up into platelet-like fragments. Mammals are somewhat different in that the precursor cells become platelets, which become the primary inactive form that circulates in the blood. Specifically, platelets are small cell fragments (about 2 to 4 mm in diameter) that are shed off the outer edges of extraordinarily large (up to 60 mm in diameter), bone-marrow–bound cells known as **megakaryocytes** (● Figure 9–8), each of which typically pro-

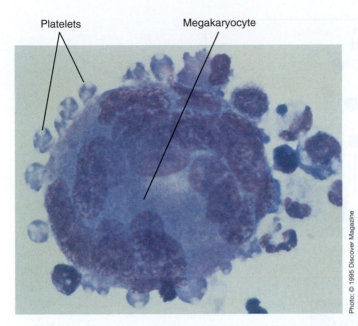

Platelets   Megakaryocyte

Photo: © 1995 Discover Magazine

Figure 9–8 ● Photomicrograph of a megakaryocyte shedding platelets.

duces about 1000 platelets. Megakaryocytes are derived from the same stem cells that give rise to the erythrocytic and leukocytic cell lines (Figure 9–7). Platelets are essentially detached vesicles containing portions of megakaryocyte cytoplasm wrapped in plasma membrane.

Platelets remain functional for less than two weeks, at which time they are removed from circulation by macrophages, especially those in the spleen and liver, and are replaced. The hormone **thrombopoietin,** made by hepatocytes (*hepato,* "liver"), increases the number of megakaryocytes in the bone marrow and stimulates each megakaryocyte to produce more platelets. The factors that control thrombopoietin secretion are unclear.

At any given time about one third of platelets are in storage in blood-filled spaces in the spleen. These platelets can be released from the spleen into the circulating blood as needed (for example, during hemorrhage) by sympathetically induced splenic contraction.

Because platelets are cell fragments, they lack nuclei. However, they are equipped with organelles and cytosolic enzyme systems for synthesizing secretory products, which they store in numerous granules dispersed throughout the cytosol. Furthermore, platelets contain high concentrations of actin and myosin, which enable them to contract. Secretion and contraction are important in hemostasis, a topic to which we now turn.

■ Hemostasis prevents blood loss from damaged small vessels.

Hemostasis is the arrest of bleeding from a broken blood vessel. (Be sure not to confuse this with the term *homeostasis.*) For bleeding to take place from a vessel, there must be a break in the vessel wall, and the pressure inside the vessel must be greater than the pressure outside it to force the blood out through the rupture. Hemostatic mechanisms normally are adequate to seal tears and stop loss of blood through small damaged capillaries, arterioles, and venules. These small vessels are

frequently ruptured by minor traumas. The much rarer occurrence of bleeding from medium- to large-size vessels usually cannot be stopped by the hemostatic mechanisms alone. Bleeding from a severed artery is more profuse and therefore more dangerous than venous bleeding, because the driving pressure is greater in the arteries than in the veins.

Here we focus primarily on the vertebrate system, which involves three major steps: (1) *vascular spasm,* (2) *formation of a platelet plug,* and (3) *blood coagulation (clotting).* Platelets play a pivotal role in hemostasis. They obviously play a major part in forming a platelet plug, but they contribute significantly to the other two steps as well.

■ Vascular spasm reduces blood flow through an injured vessel.

A cut or torn blood vessel in a vertebrate immediately constricts as a result of an inherent vascular response to injury, release of chemicals by platelets (described shortly), and sympathetically induced vasoconstriction. This constriction slows blood flow through the defect and thus minimizes blood loss. Also, as the opposing endothelial (inner) surfaces of the vessel are pressed together by this initial **vascular spasm,** they become sticky and adhere to each other, further sealing off the damaged vessel. These physical measures are important in minimizing blood flow through the break in the vessel until the other hemostatic measures can actually plug up the defect.

■ Platelets aggregate to form a plug at a vessel defect by positive feedback.

Platelets in mammals normally do not adhere to the smooth endothelial surface of blood vessels, but when this lining is disrupted because of vessel injury, platelets attach to the exposed collagen, which is a fibrous protein present in the underlying connective tissue. (In other vertebrates, thrombocytes begin to break up into plateletlike fragments.) Once platelets start aggregating at the site of the defect, they release several important chemicals from their storage granules (● Figure 9–9). Among these chemicals is **adenosine diphosphate (ADP),** which causes the surface of nearby circulating platelets to become sticky, so that they adhere to the first layer of aggregated platelets. These newly aggregated platelets release more ADP, which causes more platelets to pile on, and so on. Thus a plug of platelets is rapidly built up at the defect site in a *positive-feedback* fashion.

Given the self-perpetuating nature of platelet aggregation, why is the platelet plug limited to the site of vessel injury once it is initiated? (In other words, why doesn't the platelet plug continue to develop and expand over the surface of the adjacent normal vessel lining?) A key reason why this does not happen is that ADP and other chemicals released by the activated platelets stimulate the release of *prostacyclin* and *nitric oxide (NO)* from the adjacent normal endothelium. Both chemicals profoundly inhibit platelet aggregation. Thus the platelet plug is limited to the defect and does not spread to normal vascular tissue (Figure 9–9). The aggregated platelet plug not only physically seals the break in the vessel but also performs three other important roles. (1) The actin-myosin protein complex within the aggregated platelets contracts to compact and strengthen what was originally a fairly loose plug. (2) The chemicals released from the platelet plug include several powerful

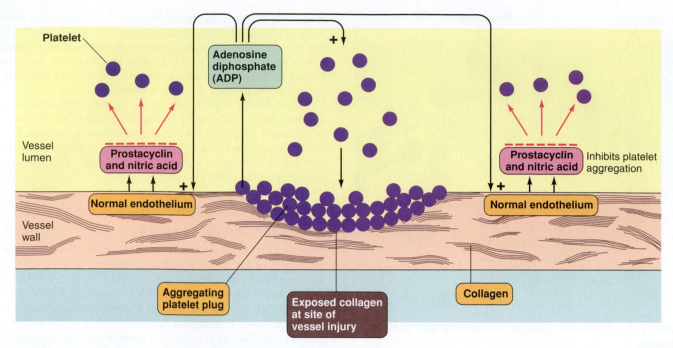

Figure 9–9 ● Formation of a platelet plug. Platelets aggregate at a vessel defect through a positive-feedback mechanism involving the release of adenosine diphosphate (ADP) from platelets, which stick to exposed collagen at the site of the injury. Platelets are prevented from aggregating at the adjacent normal vessel lining by the release of prostacyclin and nitric oxide from the undamaged endothelial cells.

vasoconstrictors—*serotonin, epinephrine,* and *thromboxane* A_2—which induce profound constriction of the affected vessel to reinforce the initial, self-induced vascular spasm. (3) The platelet plug releases other chemicals that enhance blood coagulation, the next step of hemostasis. Although the platelet-plugging mechanism alone is often sufficient to seal the myriad minute tears in capillaries and other small vessels that occur daily, larger holes in the vessels require the formation of a clot to completely stop the bleeding.

◼ A triggered chain reaction and positive feedback involving clotting factors in the plasma results in blood coagulation.

Blood coagulation, or **clotting,** is the transformation of blood from a liquid into a solid gel. Formation of a clot on top of the platelet plug strengthens and supports the plug, reinforcing the seal over a break in a vessel. Furthermore, as blood in the vicinity of the vessel defect solidifies, it can no longer flow. Coagulation is the most powerful hemostatic mechanism, and it is required to stop bleeding from all but the most minute defects.

The ultimate step in vertebrate clot formation is the conversion of **fibrinogen,** a large, soluble plasma protein produced by the liver and normally always present in the plasma, into **fibrin,** an insoluble, threadlike molecule. The conversion into fibrin is catalyzed by the enzyme **thrombin** at the site of the vessel injury.

Fibrin molecules adhere to the damaged vessel surface, forming a loose, netlike meshwork that traps the cellular elements of the blood. The resultant mass, or **clot,** typically ap-

pears red because of the abundance of trapped red blood cells, but the foundation of the clot is formed from fibrin derived from the plasma (● Figure 9–10). Clotting can take place in the

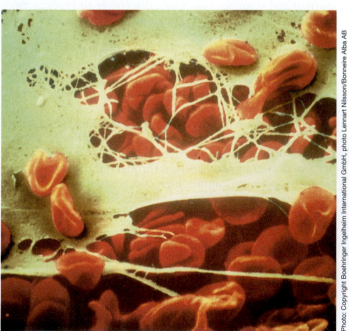

Figure 9–10 ➤ ● Erythrocytes trapped in the fibrin meshwork of a clot.

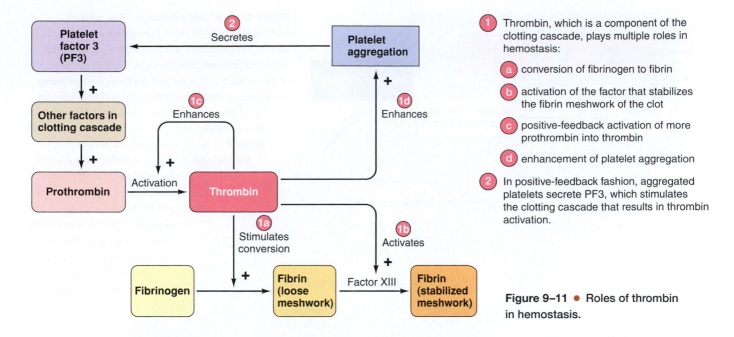

Figure 9–11 ● Roles of thrombin in hemostasis.

absence of other cellular elements in the blood except for platelets.

The original fibrin web is rather weak because the fibrin strands are only loosely interlaced. However, chemical linkages rapidly form between adjacent strands to strengthen and stabilize the clot meshwork. This cross-linkage process is catalyzed by a clotting factor known as **factor XIII (fibrin-stabilizing factor)**, which normally is present in the plasma in inactive form. In addition to converting fibrinogen into fibrin, thrombin also activates factor XIII to stabilize the resultant fibrin meshwork, acts in a *positive-feedback* fashion to facilitate its own formation, and enhances platelet aggregation, which in turn is essential to the clotting process (● Figure 9–11).

Because thrombin's action converts the ever-present fibrinogen molecules in the plasma into a blood-stanching clot, thrombin must normally be absent from the plasma except in the vicinity of vessel damage. Otherwise, blood would always be coagulated. How can thrombin normally be absent from the plasma, yet be readily available to trigger fibrin formation when a vessel is injured? The answer is that thrombin exists in the plasma as an inactive precursor called **prothrombin.**

The next question is, What converts prothrombin into thrombin when blood clotting is desirable? Yet another activated clotting factor, **factor X,** is responsible; factor X itself is normally present in the plasma in inactive form and must be converted into its active form by still another activated factor, and so on. Altogether, 12 plasma-clotting factors participate in essential steps that lead to the final conversion of fibrinogen into a stabilized fibrin meshwork (● Figure 9–12). These factors are designated by Roman numerals in the order they were discovered, not in the order they participate in the clotting process.[1] Most of these plasma-clotting factors are synthesized by the liver and are normally always present in the plasma in

an inactive form, similar to fibrinogen and prothrombin. In contrast to fibrinogen, which is converted into insoluble fibrin strands, prothrombin and the other precursors, when converted to their active form, act as *proteolytic* (protein-splitting) enzymes, which activate another specific factor in the clotting sequence. Once the first factor in the sequence is activated, it in turn activates the next factor, and so on, in a series of sequential reactions known as a **cascade,** until thrombin catalyzes the final conversion of fibrinogen into fibrin. Several of these steps require the presence of plasma Ca^{++} and **platelet factor 3 (PF3)**, a phospholipid secreted by the aggregated platelet plug. Thus platelets also contribute to clot formation.

■ The clotting cascade may be triggered by the *intrinsic pathway* or the *extrinsic pathway.*

Two distinct clotting cascades called the *intrinsic* and *extrinsic* mechanisms usually operate simultaneously. When tissue injury involves rupture of vessels, the intrinsic mechanism stops blood in the injured vessel, whereas the extrinsic mechanism clots the blood that escaped into the tissue before the vessel was sealed off.

■ The **intrinsic pathway** precipitates clotting within damaged vessels as well as clotting of blood samples in test tubes. All elements necessary to bring about clotting by means of the intrinsic pathway are present in the blood. This pathway, which involves seven separate steps (shown in blue in Figure 9–12), is set off when **factor XII (Hageman factor)** is activated by coming into contact with a negatively charged surface, especially exposed collagen in an injured vessel (or a foreign surface such as a glass test tube). Remember that exposed collagen also initiates platelet aggregation. Thus formation of a platelet plug and the chain reaction leading to clot formation are simultaneously set in motion when a vessel is damaged. Furthermore, these complementary hemostatic mechanisms reinforce each other. The aggregated platelets secrete PF3, which is es-

[1]The term *factor VI* is no longer used. What once was considered a separate factor VI has now been determined to be an activated form of factor V.

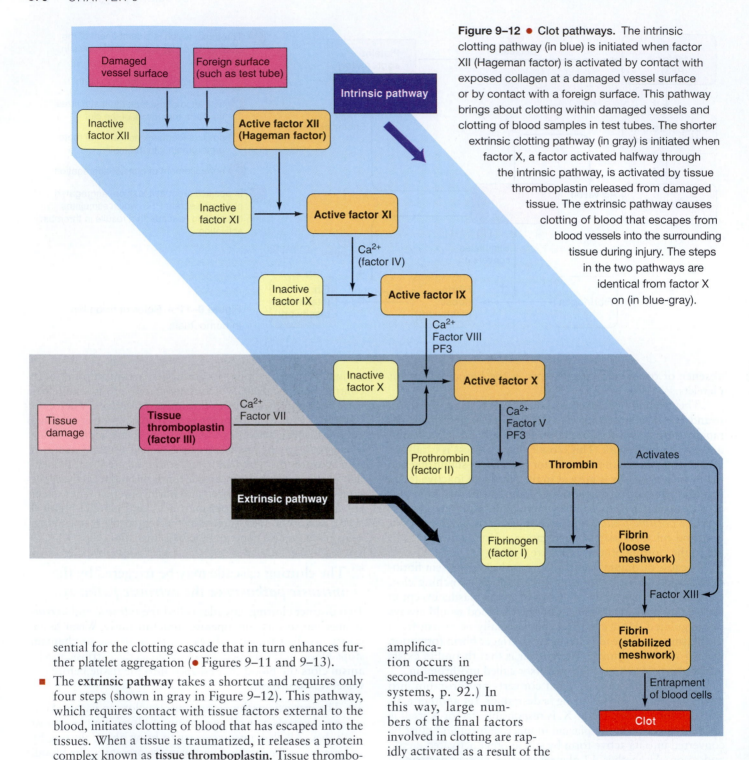

Figure 9–12 ● **Clot pathways.** The intrinsic clotting pathway (in blue) is initiated when factor XII (Hageman factor) is activated by contact with exposed collagen at a damaged vessel surface or by contact with a foreign surface. This pathway brings about clotting within damaged vessels and clotting of blood samples in test tubes. The shorter extrinsic clotting pathway (in gray) is initiated when factor X, a factor activated halfway through the intrinsic pathway, is activated by tissue thromboplastin released from damaged tissue. The extrinsic pathway causes clotting of blood that escapes from blood vessels into the surrounding tissue during injury. The steps in the two pathways are identical from factor X on (in blue-gray).

sential for the clotting cascade that in turn enhances further platelet aggregation (● Figures 9–11 and 9–13).

■ The **extrinsic pathway** takes a shortcut and requires only four steps (shown in gray in Figure 9–12). This pathway, which requires contact with tissue factors external to the blood, initiates clotting of blood that has escaped into the tissues. When a tissue is traumatized, it releases a protein complex known as **tissue thromboplastin.** Tissue thromboplastin directly activates factor X, thereby bypassing all preceding steps of the intrinsic pathway. From this point on, the two pathways are identical.

Although a clotting process that involves so many steps might seem inefficient, the advantage is the *cascade amplification* accomplished during many of the steps. One molecule of an activated factor can activate perhaps a hundred molecules of the next factor in the sequence, each of which can activate many more molecules of the next factor, and so on. (A similar amplification occurs in second-messenger systems, p. 92.) In this way, large numbers of the final factors involved in clotting are rapidly activated as a result of the initial activation of only a few molecules in the beginning step of the sequence. How then is the clotting process, once initiated, confined to the site of vessel injury? If the activated clotting factors were allowed to circulate, they would induce inappropriate widespread clotting that would plug up vessels throughout the body. Fortunately, after participating in the local clotting process the massive numbers of factors activated in the vicinity of vessel injury are rapidly inactivated by enzymes and other factors present in the plasma or tissue.

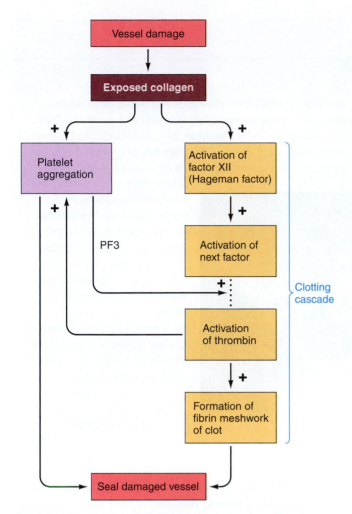

Figure 9–13 ● **Concurrent platelet aggregation and clot formation.** Exposed collagen at the site of vessel damage simultaneously initiates platelet aggregation and the clotting cascade. These two hemostatic mechanisms positively reinforce each other as they seal the damaged vessel.

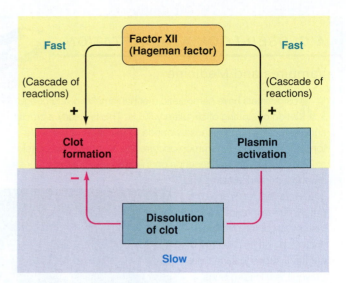

Figure 9–14 ● **Role of factor XII in clot formation and dissolution.** Activation of factor XII (Hageman factor) simultaneously initiates a fast cascade of reactions that result in clot formation and a fast cascade of reactions that result in plasmin activation. Plasmin, which is trapped in the clot, subsequently slowly dissolves the clot. This action removes the clot when it is no longer needed after the vessel has been repaired.

cursor form, **plasminogen.** Plasmin is activated cascade fashion by many factors, among them factor XII (Hageman factor), which also triggers the chain reaction leading to clot formation (● Figure 9–14). Damaged cells also release **plasminogen-activating factor.** When a clot is being formed, activated plasmin becomes trapped in the clot and subsequently dissolves it by slowly breaking down the fibrin meshwork. Phagocytic white blood cells gradually remove the products of clot dissolution.

Inhibition of clotting is also a predatory adaptation in blood-sucking parasites. See box, "A Closer Look at Adaptation: Vampires and Medicine."

■ Fibrinolytic plasmin dissolves clots and prevents inappropriate clot formation.

A clot is not meant to be a permanent solution to vessel injury. It is a transient device to stop bleeding until the vessel can be repaired. The aggregated platelets secrete **platelet-derived growth factor (PDGF)** that is at least partially responsible for the invasion of fibroblasts ("fiber-formers") from the surrounding connective tissue into the wounded area of the vessel. Fibroblasts form a scar at the vessel defect. Simultaneous with the healing process, the clot, which is no longer needed to prevent hemorrhage, is slowly dissolved by a fibrinolytic (fibrin-splitting) enzyme called **plasmin.** If clots were not removed after they performed their hemostatic function, clots would eventually obstruct the vessels, especially the small ones that endure tiny ruptures on a regular basis.

Plasmin, like the clotting factors, is a plasma protein produced by the liver and present in the blood in an inactive pre-

■ Hemocytes and hemolymph proteins provide hemostasis in arthropods.

Hemostasis mechanisms also exist in animals with hemolymph systems, but these processes are not nearly as well understood. In arthropods such as insects, wounds in the exoskeleton trigger coagulation of the hemolymph. Their hemolymphs contain a variety of cells collectively called *hemocytes* (about 30 to 50 million per milliliter). Some of these (called *hyaline hemocytes*) release protein strands when triggered by a wound, strands which interact with a variety of dissolved hemolymph proteins to form insoluble clots at the wound site. Recent molecular analysis has found that many of these proteins are related to the clotting factors in mammals, suggesting that the basic clotting mechanism evolved hundreds of millions of years ago, before the evolutionary lines leading to arthropods and vertebrates diverged.

A CLOSER LOOK AT ADAPTATION

Vampires and Medicine

Most animals have hemostatic mechanisms involving coagulation or clotting of circulatory fluids. This is clearly a crucial survival mechanism, because wounds to the skins or exoskeletons of animals are a common occurrence. But there are certain animals whose survival depends on thwarting hemostasis. These are the external parasites and predators that latch onto the outer surface of another (usually larger) animal and suck its blood as a source of nutrition. These include ticks, mosquitoes, kissing bugs, leeches, and vampire bats. Such parasites must prevent clotting, or they will not get much of a meal. Thus these blood-suckers have evolved anticlotting substances. These compounds are of great interest to medical researchers, because many strokes

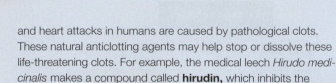

A vampire bat.

and heart attacks in humans are caused by pathological clots. These natural anticlotting agents may help stop or dissolve these life-threatening clots. For example, the medical leech *Hirudo medicinalis* makes a compound called **hirudin,** which inhibits the enzymatic activity of **thrombin** (see main text). Hirudin is being used to treat stroke victims. The most recent discovery involves the Australian vampire bat, *Desmodus rotundus.* This animal, which latches onto its prey and sucks large quantities of blood, makes a salivary **plasminogen activator** called **desmoteplase.** By activating plasmin, it leads to the destruction of fibrin at a rate hundreds of times faster than existing drugs. Desmoteplase is currently being tested in human patients. To learn how kissing bugs use yet another mechanism, see p. 569.

Photo: © Michael & Patricia Fogden/CORBIS

Circulatory Pumps

Circulating body fluids require some form of pump for their movements. Four different kinds of pumps have evolved: flagella, extrinsic muscle, peristaltic muscle, and chamber muscle pumps.

Pumps: Anatomic and Evolutionary Considerations

Physiologists recognize a variety of pumping mechanisms that have evolved:

1. **Flagella:** Internal fluids may be moved slowly by beating flagella on epithelial cells. We have seen how flagella do this in sponges, where the fluid is environmental water (Figure 9–1a). In echinoderms such as sea urchins, which have an internal fluid in their main body cavity, the coelom, flagella may create slow currents.

2. **Extrinsic muscle or skeletal pumps:** Fluids may be moved by motions of muscles or skeletal elements that are not part of the circulatory system itself. Most often, extrinsic pumping occurs only during locomotion. For example, a seastar's body-wall muscles function mainly to move the flexible arms, but they also cause the coelomic fluid to move. Various body-wall and exoskeletal muscle movements in arthropods similarly enhance circulation of hemolymph. In fact, in grasshoppers the primary heart only

beats when the animal is inactive! Apparently, it is not needed when extrinsic pumping is working. Active skeletal muscles of vertebrates can also push circulatory fluids in vessels adjacent to or within them (● Figure 9–15a). This is an important mechanism in tails of hagfish and legs of tall animals such as humans, for example. In both, flow is aided by active skeletal muscles squeezing blood through

Photo: © Michael Jeffords

A grasshopper has a heart with an open circulatory system, but its main heart stops pumping when the animal is moving. This is because movements of the body wall and limbs are sufficient to pump the blood during locomotion.

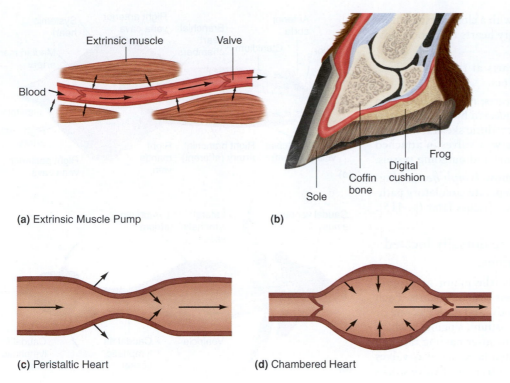

(a) Extrinsic Muscle Pump

Extrinsic muscle   Valve

Blood

(b)

Frog

Digital cushion

Coffin bone

Sole

(c) Peristaltic Heart

(d) Chambered Heart

Figure 9–15 ● **Types of pumps in animals.** (a) External muscle pump expands and compresses a blood vessel (black arrows) and forces blood to flow along the vessel (colored arrows); valves maintain a unidirectional flow, e.g., nematode, scaphopod mollusk, and some leeches with no heart to propel blood flow, and the skeletal muscle pump of vertebrates. (b) Structure of the horse foot. The cartilaginous "frog" is pushed by the weight of the horse to aid pumping of the blood (see main text). (c) Peristaltic tubular heart forces blood along a vessel (which may have valves to ensure one-way flow) by peristaltic contractions, e.g., insect heart. (d) A chambered heart propels blood by a coordinated contraction of the muscular wall; valves ensure a one-way flow of blood, e.g., most mollusk and vertebrate hearts.

(*Source:* P. Withers, 1992, *Comparative Animal Physiology,* 1st ed., Belmont, CA: Brooks/Cole, **a,c,d,** Figure 14–2, p. 668; **b,** Figure 14–8, p. 676)

one-way valves in veins, a process we discuss later (p. 419). The action of the respiratory muscles in breathing can also enhance blood flow (see p. 418).

A final example of extrinsic pumping involves skeletal and connective elements. Some mammals with hooves (*perissodactyls* such as rhinoceros and horse) have a cartilaginous plate, the "frog" or *cuneus ungulae,* at the base of the hoof that is flexed upward with each footstep (Figure 9–15b). This flexing pushes on an overlying elastic cushioning tissue that is compressed and moves outward, expanding nearby cartilage and the hoof walls. It also compresses nearby veins, sending blood into the leg above. When the foot is lifted, elastic rebound of these tissues draws blood into the local veins. This pumping action is particularly important in horses, with their long, thin legs, because they do not have enough skeletal muscle mass below the knee to act as extrinsic muscle pumps. Gravity provides the direct energy for flexing the plate, although this gravitational energy originated from upper limb muscles, which lift the limb off the ground.

3. **Peristaltic (tubular) muscle pumps:** Peristaltic pumping occurs when muscles in the walls of vessels contract in a moving wave that pushes fluid in front of it. By repeated cycles of this action, fluid is moved in one direction. These pumps are called "hearts" if they occur in specialized sections of vessels. This is the pumping method in many annelids such as earthworms, where regions of dorsal blood vessels are specialized for this purpose (Figure 9–15c). Many arthropods also have peristaltic hearts.

4. **Chamber muscle pumps:** These are the more familiar form of hearts, consisting of a chamber or chambers with mus-

cles to squeeze the fluid within. Chamber hearts are the primary pumps in all vertebrates and many arthropods and mollusks. Unlike directional peristaltic pumps, chamber pumps usually need one-way valves to create flow in one direction, because the squeezing action is not directional (Figure 9–15d). In some animals such as arthropods, the heart has a single chamber. But many chamber hearts such as those of most mollusks and all vertebrates have (at a minimum) two types of chambers: an **atrium,** which collects returning fluid, and a **ventricle** that provides the primary force for outgoing fluid. (As you will see, there may be more than two chambers in more advanced hearts.)

Many animals have primary hearts aided by auxiliary pumps.

Peristaltic or chamber hearts are found in most animal groups more advanced than flatworms (although there are major exceptions, such as nematodes and echinoderms, which have no hearts). Interestingly, most of these animals have both a primary (often called **systemic**) heart that drives the initial, usually oxygenated fluid to most organs, and one or more auxiliary pumps that aid flow returning to the primary heart or flow going to critical organs. We have already noted how extrinsic pumping by body walls, skeletal muscles, and skeletal elements can aid flow during locomotion. But some animals have true auxiliary hearts that are not extrinsic:

- Annelid worms such as earthworms, for example, have dorsal blood vessels specialized as primary peristaltic hearts, but have many peristaltic vessel segments in other locations (Figure 9–2d).

- Cephalopods, the only mollusks with a closed system, have a systemic heart and two auxiliary hearts that boost flow to their gills (● Figure 9–16a).

- Some insects have auxiliary hearts at the base of legs, wings, and antennae (Figure 9–16c).

- Hagfish have several auxiliary hearts; the *caudal heart* in their tails has a chamber that is squeezed by a cartilaginous rod, which in turn is pushed by extrinsic skeletal muscles. One-way flow is ensured by one-way valves in attached veins, which are both the incoming and outgoing vessels.

- Most vertebrates have two or more *lymph hearts* (Figure 9–16b), which are part of a separate circulatory pathway, the *lymphatic system*, as we discuss later (p. 415).

Arthropod systemic hearts are dorsally located and have many valved openings.

Crustaceans, insects, and arachnids (the major arthropod groups) have many similarities in their primary hearts. Each is located in a dorsal region of the body, and each has one or more porelike openings called **ostia** (**ostium,** singular), which allow hemolymph to re-enter the heart after passing through the body (Figure 9–2b). The ostia often have **one-way valves** to prevent backflow (you will see how this type of valve works in the vertebrate example given later).

However, there are some differences. Crustacean hearts, present mainly in larger species such as decapods (lobsters, crabs, and so on), are single-chamber pumps, whereas insect and arachnid hearts are tubular (modified vessels) and can have peristaltic pumping.

Vertebrate systemic hearts evolved from a two-chambered to a four-chambered structure.

The primitive systemic heart of vertebrates is thought to have begun with two chambers, one atrium to collect returning blood and one ventricle to pump to the body. Fishes have this basic design, but one or two auxiliary chambers have evolved. Blood of all jawless and jawed fishes first enters a chambered extension of the atrium called the **sinus venosus,** which collects blood from the veins before entering the adjacent atrium. In jawed fishes, blood leaving the ventricles enters an extension forming a fourth chamber, the **conus arteriosus** (in cartilaginous fish) or **bulbus arteriosus** (in bony fish), which dampens the pulsatile pressure output of the ventricle (a function taken over by the *aorta* in reptiles, birds, and mammals, in a process we describe later in the chapter; p. 401). Although these fish hearts have four chambers, they are considered to have two primary chambers with two auxiliary chambers (see ● Figure 9–17 for a cartilaginous fish).

A significant change in vertebrate hearts began when the first air-breathing fishes evolved. The crucial feature that drove this change was the evolution of a separate circuit, called the **pulmonary circulation,** from the heart to the newly evolved lungs and back. The two primary chambers began to subdivide with internal septa to support this separate circuit, with one side of the atrium collecting blood from the body and one side collecting from the lungs, and with one side of the ventricle pumping blood to the lungs and one side pumping to the body. This culminated in the hearts of birds and mammals, which have two completely separate atrium–ventricle pairs, which

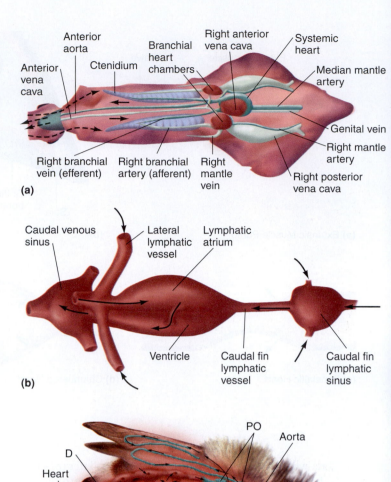

(a)

(b)

(c)

Figure 9–16 ● **Examples of auxiliary hearts.** (a) Diagram of the closed circulatory system of a cephalopod circulatory system. These animals have two branchial hearts that pump blood to the gills, in addition to a main systemic heart. (b) Diagram of a lymph heart in an eel. Muscles in the heart wall push lymph from the caudal lymphatic sinus into the caudal venous sinus. (c) Diagram of the open circulatory system of an insect. A main dorsal vessel (aorta) carries blood from the heart into side channels and open spaces. Flow to the wings is boosted by auxiliary pumps called pulsatile organs (PO). Septa (S) in the limbs channel flow through the spaces in specific directions.

(*Sources:* **a,** Modified from Sherman and Sherman (1976). *The Invertebrates: Function and Form,* 2nd ed. New York: Macmillan. **b,** From Withers, P. C. (1992). *Comparative Animal Physiology.* Fort Worth, TX: Saunders College Publishing as modified from Kampmeier, O. F. 1969. *Evolution and Comparative Morphology of the Lymphatic System.* Springfield, MA: C. C. Thomas. **c,** Modified from Wigglesworth (1972).

pump to the pulmonary circulation to the lungs and the **systemic circulation** to the rest of the body. We describe these evolutionary changes and the flow patterns in more detail in the next section of this chapter, on circulatory pathways and vessels.

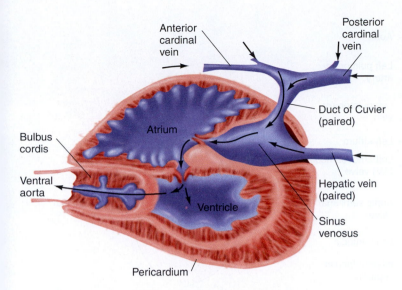

Anterior
cardinal
vein

Posterior
cardinal
vein

Atrium

Duct of Cuvier
(paired)

Bulbus
cordis

Ventral
aorta

Hepatic vein
(paired)

Ventricle

Sinus
venosus

Pericardium

Figure 9–17 • Diagram of an elasmobranch heart. Venous blood enters the sinus venosus (upper right) and is then pumped by the atrium, ventricle and bulbus cordis into the aorta (lower left).

(*Source:* P. Withers, 1992, *Comparative Animal Physiology,* 1st ed., Belmont, CA: Brooks/Cole; as modified from Waterman, A. J., B. E. Frye, K. Johansen, et al., 1971, *Chordate Structure and Function,* New York: Macmillan.)

■ Avian and mammalian hearts are dual pumps.

Even though anatomically the heart is a single organ, in birds and mammals there are separate right and left sides that act as two separate pumps (we examine the hearts of other vertebrates later). These hearts have four chambers, an upper and a lower one, within each half (● Figure 9–18a). The upper chambers, the **atria** (**atrium,** singular), receive blood returning to the heart and transfer it to the lower chambers, the **ventricles,** which pump the blood from the heart. The vessels that return blood from the tissues to the atria are **veins,** and those that carry blood away from the ventricles to the tissues are **arteries.** The two halves of the heart are separated by the **septum,** a continuous muscular partition that prevents mixture of blood from the two sides of the heart. This separation is extremely important, because the right half of the heart is receiving and pumping O_2-depleted CO_2-rich blood whereas the left side of the heart receives and pumps O_2-rich CO_2-poor blood.

Let us examine how the heart functions as a dual pump, by tracing a drop of blood through one complete circuit in a mammal (Figure 9–18a and b). Blood returning from the systemic circulation enters the right atrium via large veins known as the **venae cavae.** The drop of blood entering the right atrium has returned from the body tissues, where O_2 has been extracted from it and CO_2 has been added to it. This partially deoxygenated blood flows from the right atrium into the right ventricle, which pumps it out through the **pulmonary artery** to the lungs. Thus the *right side of the heart pumps blood into the pulmonary circulation.* Within the lungs, the drop of blood loses its extra CO_2 and picks up a fresh supply of O_2 before being returned to the left atrium via the **pulmonary veins.** This O_2-rich blood returning to the left atrium subsequently flows into the left ventricle, the pumping chamber that propels the blood to all body systems except the pulmonary circulation of the lungs; that is, *the left side of the heart pumps blood into*

the systemic circulation. The large artery carrying blood away from the left ventricle is the **aorta.** Major arteries branch from the aorta to supply the various tissues of the body.

Both sides of the heart simultaneously pump equal amounts of blood. The pulmonary circulation is a low-pressure, low-resistance system, whereas the systemic circulation is a high-pressure, high-resistance system. Therefore, even though the right and left sides of the heart pump the same amount of blood, the left side performs more work, because it pumps an equal volume of blood at a higher pressure into a higher-resistance system. Accordingly, the heart muscle on the left side is much thicker than the muscle on the right side, making the left side a stronger pump (Figure 9–18c).

This left–right difference leads to difficulties in rapidly growing animals, such as broiler chickens. Oxygen usage by the growing tissues reduces oxygen in the blood. In response, the volume of blood pumped by the left ventricle increases to restore oxygen delivery. The right ventricle must in turn work harder, and it hypertrophies (grows larger). However, for unclear reasons, the right ventricle's growth does not keep pace with the growth of the body, so the ventricle becomes unable to maintain sufficient pressure to propel the entire blood flow. Eventually, the overstressed right ventricle is unable to take in and pump returning blood, resulting in life-threatening **edema** (fluid accumulation) in various organs, a condition termed **ascites syndrome.** Current research is focused on selecting strains of chickens resistant to developing this condition. A similar phenomenon, called **brisket disease,** occurs in cattle and horses that spend some time at high altitudes and then return to low altitude.

■ Heart valves ensure that the blood flows in the proper direction through the heart.

Blood flows through vertebrate hearts in one fixed direction from veins to atria to ventricles to arteries. The presence of one-way heart valves ensures this unidirectional flow of blood. (Functionally similar valves are found in arthropod heart ostia, as you saw earlier.) Heart valves are positioned so that they open and close passively because of pressure differences, similar to a one-way door (● Figure 9–19). A forward pressure gradient (that is, a greater pressure behind the valve) forces the valve open, much as you open a door by pushing on one side of it, whereas a backward pressure gradient (that is, a greater pressure in front of the valve) forces the valve closed, just as you apply pressure to the opposite side of the door to close it.

These valves are in all vertebrate hearts; two valves are shown in a cartilaginous fish in Figure 9–17. Let us examine these valves more closely in mammals, which have four (shown in Figure 9–18a). Two of them, the **right** and **left atrioventricular (AV) valves,** are positioned between the atrium and the ventricle on the right and left sides, respectively. These valves allow one-way flow from the atria into the ventricles during ventricular filling (when atrial pressure exceeds ventricular pressure).

The two remaining heart valves, the **aortic** and **pulmonary valves,** are located at the juncture where the major arteries leave the ventricles. They are known as **semilunar** ("half-moon") **valves** because they consist of three cusps, each resembling a shallow half-moon–shaped pocket. These valves are forced open when the left and right ventricular pressures exceed the pressure in the aorta and pulmonary arteries, respectively, during ventricular contraction and emptying. Clo-

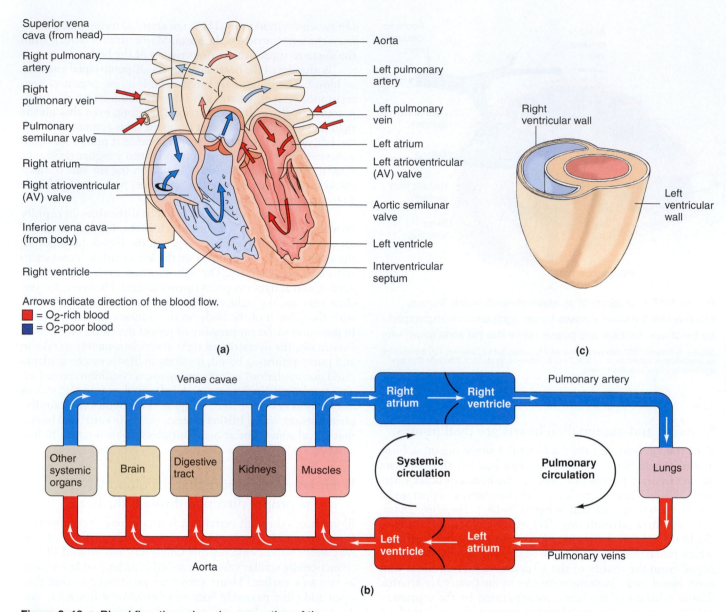

Figure 9–18 • **Blood flow through and pump action of the mammalian heart.** (a) Blood flow through the heart. (b) Dual pump action of the heart. The right side of the heart receives O_2-poor blood from the systemic circulation and pumps it into the pulmonary circulation. The left side of the heart receives O_2-rich blood from the pulmonary circulation and pumps it into the systemic circulation. (The relative volume of blood flowing through each organ is not drawn to scale.) (c) Comparison of the thickness of the right and left ventricular walls. Note that the left ventricular wall is much thicker than the right wall.

sure results when the ventricles relax and ventricular pressures fall below the aortic and pulmonary artery pressures.

Even though there are no valves between the atria and veins, backflow of blood from the atria into the veins usually is not a significant problem, for two reasons: (1) Atrial pressures usually are not much higher than venous pressures, and (2) the sites where the venae cavae enter the atria are partially compressed during atrial contraction.

■ Vertebrate heart walls are composed primarily of spirally arranged cardiac muscle fibers interconnected by intercalated discs.

The bulk of the vertebrate heart wall is the **myocardium** (*myo*, "muscle"), a middle layer of cardiac muscle lying between an endothelial layer (the **endocardium**) and a thin outer sheath, the **epicardium**. In mammals, the myocardium consists of interlacing bundles of cardiac muscle fibers arranged spirally around the heart circumference. As a result of this arrangement, when the ventricular muscle contracts, the diameter of the ventricular chambers is reduced while the apex is simultaneously pulled upward toward the top of the heart in a rotating manner. This "wringing" effect efficiently exerts pressure on the blood within the chambers and directs it upward to the arteries that exit the ventricles.

The individual cardiac muscle cells are interconnected to form branching fibers, with adjacent cells joined end-to-end at specialized structures known as **intercalated discs.** Within an intercalated disc are two types of membrane junctions we saw

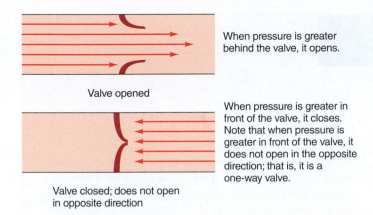

When pressure is greater behind the valve, it opens.

Valve opened

When pressure is greater in front of the valve, it closes. Note that when pressure is greater in front of the valve, it does not open in the opposite direction; that is, it is a one-way valve.

Valve closed; does not open in opposite direction

Figure 9–19 ● Mechanism of valve action.

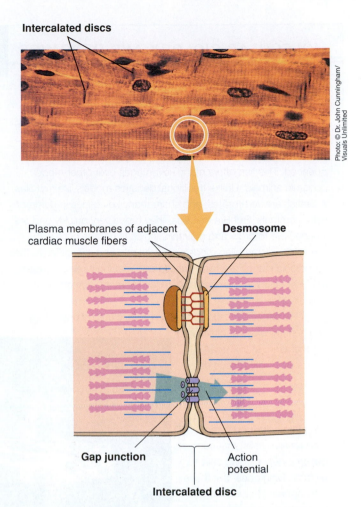

Figure 9–20 ● Organization of mammalian cardiac muscle fibers. Adjacent cardiac muscle cells are joined end to end by intercalated discs, which contain two types of specialized junctions: desmosomes, which act as spot rivets mechanically holding the cells together; and gap junctions, which permit action potentials to spread from one cell to adjacent cells.

in Chapter 2: *desmosomes* and *gap junctions* (● Figure 9–20, and p. 63). Recall that desmosomes are a type of adhering junction that mechanically holds cells together under high mechanical stress, whereas gap junctions are channels that allow action potentials to spread from one cardiac cell to the next (see p. 64).

Some cardiac muscle cells generate action potentials without any nervous stimulation, as we explore later. When one of the cardiac cells spontaneously undergoes an action potential, the electrical impulse spreads to all the other cells that are joined by gap junctions, so that they become excited and contract as a single functional **syncytium** (a group of joined cells acting as one functional unit). Vertebrate atria and ventricles each form syncytia and contract as separate units. The synchronous contraction of the muscle cells of each of these chambers produces the force necessary to eject the enclosed blood.

There is no electrical continuity (that is, no gap junctions) between the atrial and ventricular contractile cells (except for the AV node, described shortly); and they are separated by electrically nonconductive connective tissue. However, because of both the syncytial nature of cardiac muscle, an impulse spontaneously generated in the atrium normally spreads throughout the rest of the heart. Therefore, unlike skeletal muscle, where graded contractions can be produced by varying the number of muscle cells that are contracting within the muscle (recruitment of motor units), either all the cardiac muscle fibers contract or none do. A "half-hearted" contraction is not possible.

Cardiac muscle cells in most vertebrates contain an abundance of mitochondria, the O_2-dependent energy organelles. In fact, 30 to 50% of the cell volume of most vertebrate cardiac muscle cells is attributable to mitochondria, indicative of how much the heart depends on aerobic metabolism to generate the energy necessary for contraction (see p. 31). Cardiac muscle in most vertebrates also has an abundance of *myoglobin* (see p. 328), which stores limited amounts of O_2 for immediate use, and also facilitates the diffusion of O_2 arriving at the cell to the mitochondria for rapid processing.

Damage to heart muscle cells, which can be caused by high levels of stress hormones, is a leading cause of death in humans and can occur in other mammals. However, hearts of some vertebrates can suffer severe damage and repair themselves—see box, "Molecular Biology and Genomics: Zebrafish: A Hearty Physiological Genomics Model."

Electrical Activity of Animal Hearts

Contraction of cardiac muscle cells to bring about ejection of blood is triggered by action potentials sweeping across the muscle cell membranes. Many animal hearts have a small number of **autorhythmic cells,** which do not contract but instead are specialized for initiating and conducting the action potentials responsible for contraction of the working cells. These are also called **pacemaker** cells, and hearts that have them are called **myogenic.** Myogenic hearts are found in many animals including all vertebrates, tunicates (urochordates), insects, and some mollusks and crustaceans, and are presumably advantageous because they ensure blood or hemolymph flow even if there is severe neural trauma. (However, myogenic insect hearts appear to be largely under the control of external neural stimuli under normal circumstances.) Some crustaceans such as decapods (crabs, lobsters, and so on) have **neurogenic** hearts, which do not beat without an external neural stimulus. This type of regulation may allow for a heart to be stopped completely in animals that do not need continuous circulatory flow in some situations.

MOLECULAR BIOLOGY AND GENOMICS

Zebrafish—A Hearty Physiological Genomics Model

In recent years, a tiny fish called the zebrafish (*Danio rerio;* a common item in pet stores) has emerged as a major model organism for vertebrate genomics. Indeed, its genome has been sequenced. This fish offers many advantages over other model genomic animals. Unlike traditional genetics models like fruit flies, zebrafish are vertebrates, and share many key genes in common with mammals. But like fruit flies, they breed very quickly and are inexpensive to grow and maintain, which is not the case with laboratory mammalian models such as rats and mice. Finally, their embryonic stages are fairly transparent, making it relatively easy to observe tissue changes during development, something that is difficult to do with other animals.

One aspect of zebrafish biology that is uniting physiology and genomics is that of tissue regeneration. Mammalian hearts that suffer tissue damage do not fully repair themselves. New studies reveal that mammalian hearts contain stem cells (undifferentiated reserve cells) that can become new cardiac cells to replace dying ones, but these cells appear incapable of repairing major damage such as that following a heart attack. In contrast, the zebrafish heart has remarkable regeneration abilities. Experimental

A Zebrafish, Danio rerio.

surgery (by Mark Keating and colleagues) has found that up to 20 to 25% of the zebrafish heart can be removed without long-term harm, because the heart fully repairs itself and resumes normal function. (Similarly, these fish can regenerate other complex tissues such as fins and eye parts.) Because the fish's genome is available, research is now focused on finding the genes responsible for this amazing self-repair ability. If these can be found, perhaps similar genes in humans could be activated in damaged hearts, perhaps with regeneration signal molecules (hormones or paracrines) from the zebrafish.

If nothing else, we may learn (from comparing mammals and zebrafish) what happened in the evolution of mammals that so greatly restricts their ability to regenerate (liver, skin, and related structures such as hair and antlers, and blood cells regenerate well, but not much else). Pathways for forming scar tissue seem to have taken over in mammals, perhaps because this process is much faster than regeneration. However, mammals apparently do have the genes for regeneration, but they are no longer used as such. Remarkably, one strain of mice developed in the laboratory was able to regenerate up to 15% of their hearts following destruction! How this ability arose is still being investigated.

Photo: Frank Teigler, Hippocampus, Germany

■ The sinoatrial node is the normal pacemaker of the mammalian heart.

Here we examine the role of the specialized autorhythmic cells in the origin and spread of the mammalian heartbeat. In contrast to nerve and skeletal muscle cells, in which the membrane remains at constant resting potential unless the cell is stimulated (see p. 108), the cardiac autorhythmic cells do not have a stable resting potential. Instead, they display pacemaker activity; that is, their membrane potential slowly depolarizes, or drifts, between action potentials until threshold is reached, at which time the membrane "fires," that is, has an action potential (● Figure 9–21; see also p. 108). Through repeated cycles of drifting and firing, these autorhythmic cells cyclically initiate action potentials, which then spread throughout the heart to trigger rhythmic beating without any nervous stimulation.

Pacemaker Potential and Action Potential in Autorhythmic Cells

Complex interactions of several different ionic mechanisms are responsible for the **pacemaker potential**, which is an auto-

rhythmic cell membrane's slow drift to threshold. The most important changes in ion movement that give rise to the pacemaker potential are (1) a decreased outward K^+ current coupled with a constant inward Na^+ current and (2) an increased inward Ca^{++} current. To elaborate, the initial phase of the slow depolarization to threshold is caused by a cyclical decrease in the passive outward flux of K^+ superimposed on a slow, unchanging, inward leak of Na^+. In cardiac autorhythmic cells, permeability to K^+ does not remain constant between action potentials as it does in nerve and skeletal muscle cells. Instead, K^+ permeability decreases between action potentials, because K^+ channels slowly close at negative potentials. This slow closure gradually diminishes the outflow of K^+ down its concentration gradient. Also, unlike nerve and skeletal muscle cells, cardiac autorhythmic cells do not have voltage-gated Na^+ channels. Instead, they have channels that are always open and thus permeable to Na^+ at negative potentials. As a result, a small passive influx of Na^+ continues unchanged at the same time the rate of K^+ efflux slowly declines. Thus the inside gradually becomes less negative; that is, the membrane gradually depolarizes and drifts toward threshold.

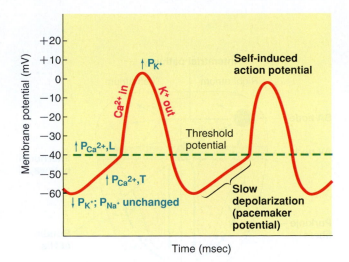

Figure 9-21 ● Pacemaker activity of cardiac autorhythmic cells.

In the second half of the pacemaker potential, a transient Ca^{++} channel (Ca^{++}, T), one of two types of voltage-gated Ca^{++} channels, opens. As the slow depolarization proceeds, this channel is opened before the membrane reaches threshold. The resultant brief influx of Ca^{++} further depolarizes the membrane, bringing it to threshold. Once threshold is reached, the rising phase of the action potential occurs in response to activation of a longer-lasting Ca^{++} channel (Ca^{++}, L) and a subsequently large influx of Ca^{++}. This rising phase differs from skeletal muscle, which uses Na^+ influx rather than Ca^{++} influx. The falling phase is due, as usual, to the K^+ efflux that occurs when K^+ permeability increases as a result of activation of K^+ channels. After the action potential is over, slow closure of these K^+ channels initiates the next slow depolarization to threshold. These K^+ channels are of a kind called *ether-a-go-go* or EAG type (p. 109). Inhibited function of EAG K^+ channels, resulting from a genetic mutation or drugs such as erythromycin, interfere with repolarization and has been shown to cause sudden cardiac arrest in horses and humans.

The specialized, noncontractile cardiac cells capable of autorhythmicity are found in several locations in vertebrate hearts (● Figure 9-22a). The main sites are as follows:

■ The **sinoatrial node (SA node)**, a small, specialized region in the right atrial wall in reptiles, birds, and mammals. In fishes and amphibians, the equivalent of the SA node lies in the wall of the **sinus venosus.**

■ The **atrioventricular node (AV node)**, a small bundle of specialized cardiac muscle cells located at the base of the right atrium near the septum, just above the junction of the atria and ventricles.

In all vertebrates, the **SA node** (or sinus venosus node) normally exhibits the fastest rate of autorhythmicity, inherently at 90 to 100 action potentials per minute in humans, and drives the rest of the heart at this rate. Thus it is the main **pacemaker** of the heart, which excites the entire heart, triggering the contractile cells to contract (and thus the heart to beat) at the pace set by pacemaker autorhythmicity. Other autorhythmic tissues cannot assume their own naturally slower rates,

because they are activated by action potentials generated from the SA node before they reach threshold at their own slower rhythm.

■ The spread of cardiac excitation is coordinated to ensure efficient pumping.

Once initiated in the SA node, an action potential spreads throughout the rest of the heart. For efficient cardiac function, the spread of excitation should satisfy three criteria:

■ *Atrial excitation and contraction should be complete before the onset of ventricular contraction.* This ensures complete ventricular filling. During cardiac relaxation, the AV valves are open, so that venous blood entering the atria continues to flow directly into the ventricles. Most ventricular filling occurs by this means prior to atrial contraction. When the atria do contract, additional blood is squeezed into the ventricles to complete ventricular filling. Ventricular contraction then occurs to eject blood into the arteries. If the atria and ventricles were to contract simultaneously, the AV valves would be closed immediately, because ventricular pressures would greatly exceed atrial pressures. The ventricles have much thicker walls and, accordingly, can generate more pressure.

■ *Excitation of cardiac muscle fibers should be coordinated to ensure that each heart chamber contracts as a unit to accomplish efficient pumping.* If the muscle fibers in a heart chamber were to become excited and contract randomly rather than contracting simultaneously in a coordinated fashion, they would be unable to eject blood. A smooth, uniform ventricular contraction is essential. In contrast, random, uncoordinated excitation and contraction is known as **fibrillation.** Ventricular fibrillation rapidly causes death because the heart cannot pump blood into the arteries.

■ *The pair of atria and pair of ventricles should be functionally coordinated so that both members of the pair contract simultaneously.* This permits synchronized pumping of blood into the pulmonary and systemic circulation.

The normal spread of cardiac excitation is carefully orchestrated to ensure that these criteria are met and the heart functions efficiently, as described next (Figure 9-22b).

Atrial Excitation

An action potential originating in the SA node first spreads throughout both atria, primarily from cell to cell via gap junctions. The signal appears to spread symmetrically, though there may be one or more poorly delineated, specialized conduction pathways.

Transmission between the Atria and the Ventricles

The AV node is the only point of electrical contact between the atria and ventricles; other connections consist of electrically nonconductive fibrous tissue. The action potential is conducted relatively slowly through the AV node. The impulse is delayed about 0.1 second (the **AV nodal delay**), which enables the atria to become completely depolarized and to contract, emptying their contents into the ventricles, before ventricular depolarization and contraction occur.

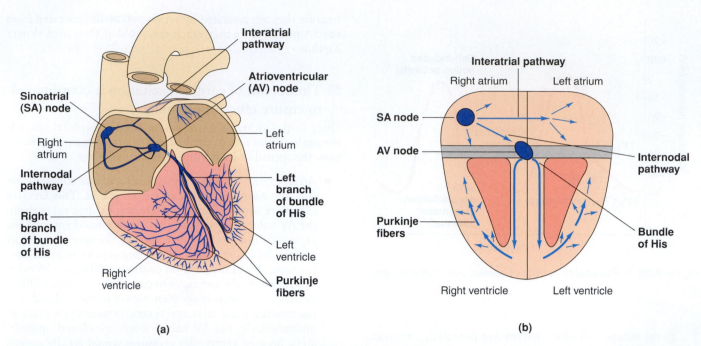

Figure 9–22 ● (a) Specialized conduction system of the mammalian heart. (b) Spread of cardiac excitation. An action potential initiated at the SA node first spreads throughout both atria. Its spread is facilitated by two specialized atrial conduction pathways, the interatrial and internodal pathways. The AV node is the only point where an action potential can spread from the atria to the ventricles. From the AV node, the action potential spreads rapidly throughout the ventricles, hastened by a specialized ventricular conduction system consisting of the bundle of His and Purkinje fibers.

Ventricular Excitation

After the AV nodal delay, the impulse travels rapidly along two specialized tracts (Figure 9–22b):

- The **bundle of His (atrioventricular bundle),** a tract of specialized cells that originates at the AV node and enters the interventricular septum, where it divides to form the right and left bundle branches that travel down the septum, curve around the tip of the ventricular chambers, and travel back toward the atria along the outer walls.

- **Purkinje fibers,** small terminal fibers that extend from the bundle of His and spread through some (in most mammals) or all (in ruminants) of the ventricular myocardium, much like small twigs of a tree branch.

The network of fibers in this ventricular conduction system is specialized for rapid propagation of action potentials. Its presence hastens and coordinates the spread of ventricular excitation to ensure that the ventricles contract as a unit. Although this system carries the action potential rapidly to a large number of cardiac muscle cells, it does not terminate on every cell. The impulse quickly spreads from the excited cells to the remainder of the ventricular muscle cells by means of gap junctions. Altogether, these conduction steps lead to almost simultaneous activation of the ventricular myocardial cells in both ventricular chambers, efficiently ejecting blood into the systemic and pulmonary circulations simultaneously.

■ The action potential of contractile cardiac muscle cells shows a characteristic plateau.

The action potential in contractile cardiac muscle cells, although initiated by the nodal pacemaker cells, varies considerably in ionic mechanisms and shape from the SA node potential (compare ● Figures 9–21 and 9–23). Unlike autorhythmic cells, the membrane of contractile cells remains essentially at rest at about −90 mV until excited by electrical activity propagated from the pacemaker. Once the membrane of a ventricular myocardial contractile cell is excited, an action potential is generated by a complicated interplay of permeability changes and membrane potential changes as follows (Figure 9–23):

1. During the rising phase of the action potential, the membrane potential rapidly becomes reversed to a positive value of 30 mV as a result of an explosive increase in membrane permeability to Na$^+$. Thus far, the process is the same as in neurons and skeletal muscle cells (see p. 108). The Na$^+$ permeability then rapidly plummets to its low resting value, but, unique to these cardiac muscle cells, the membrane potential is maintained at this positive level for several hundred milliseconds, producing a *plateau phase* of the action potential. In contrast, the short action potential of neurons and skeletal muscle cells lasts 1 to 2 milliseconds.

2. The sudden change in voltage occurring during the rising phase of the action potential brings about two voltage-dependent permeability changes that are responsible for

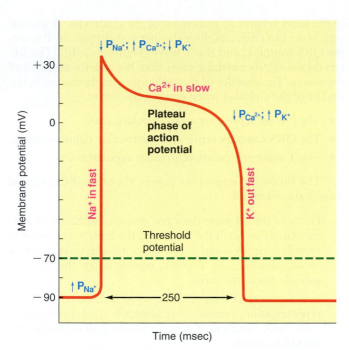

Figure 9–23 ● **Action potential in contractile cardiac muscle cells.** The action potential in cardiac contractile cells differs considerably from the action potential in cardiac autorhythmic cells (compare with Figure 9–21).

maintaining this plateau: activation of "slow" Ca^{++} channels and a marked decrease in K^+ permeability. Opening of the Ca^{++} channels results in a slow, inward diffusion of Ca^{++}, because Ca^{++} is in sufficiently greater concentration in the ECF that its diffusion inward overcomes repulsion by the cell's positive charge. This influx of Ca^{++} prolongs the positive charge inside the cell and is primarily responsible for the plateau portion of the action potential. This effect is enhanced by the concomitant decrease in K^+ permeability. The resultant reduction in efflux of K^+ prevents rapid repolarization of the membrane and thus helps prolong the plateau phase.

3. The rapid, falling phase of the action potential results from inactivation of the Ca^{++} channels and activation of EAG K^+ channels. The decrease in Ca^{++} permeability diminishes the slow, inward movement of Ca^{++}, whereas the sudden increase in K^+ permeability simultaneously promotes rapid outward conduction and diffusion of K^+. Thus rapid repolarization at the end of the plateau is accomplished primarily by K^+ efflux, which once again makes the inside of the cell more negative than the outside and restores membrane potential to resting.

Let's now see how this action potential brings about contraction.

■ Ca^{++} entry from the ECF induces a much larger Ca^{++} release from the sarcoplasmic reticulum.

In cardiac contractile cells, the L-type Ca^{++} channels are located primarily in the T tubules. As you just learned, these voltage-gated channels are opened in response to a change in

membrane potential. Thus, unlike in skeletal muscle, Ca^{++} moves into the cytosol from the ECF across the T tubule membrane during a cardiac action potential. This small amount of entering Ca^{++} triggers the opening of nearby Ca^{++}-release channels in the adjacent lateral sacs of the sarcoplasmic reticulum. In this way, Ca^{++} entering the cytosol from the ECF induces a much larger release of Ca^{++} into the cytosol from the intracellular stores (see Figure 8–12, steps 2 to 7, for review of the basic process). The resultant local bursts of Ca^{++} release, known as Ca^{++} *sparks*, from the sarcoplasmic reticulum collectively increase the cytosolic Ca^{++} pool sufficiently to turn on the contractile machinery. This extra supply of Ca^{++} from the sarcoplasmic reticulum is responsible for the long period of cardiac contraction, which lasts about three times longer than a single skeletal muscle fiber contraction (300 msec compared to 100 msec). This increased contractile time ensures adequate time to eject the blood.

Role of Cytosolic Ca^{++} in Cardiac Excitation–Contraction Coupling

As in skeletal muscle, the role of Ca^{++} within the cytosol is to bind with the troponin-tropomyosin complex and cause it to move aside so that cross-bridge cycling can take place (see p. 319). However, unlike skeletal muscle, in which sufficient Ca^{++} is always released to turn on all the cross bridges, in cardiac muscle the extent of cross-bridge activity varies with the amount of cytosolic Ca^{++}. As you will learn, various regulatory factors can alter the amount of cytosolic Ca^{++}.

Removal of Ca^{++} from the cytosol by active transport in both the plasma membrane and sarcoplasmic reticulum restores the blocking action of troponin and tropomyosin, so that contraction ceases and the heart muscle relaxes.

■ Tetanus of cardiac muscle is prevented by a long refractory period.

Like other excitable tissues, cardiac muscle has a refractory period. During the absolute refractory period, which occurs immediately after the initiation of an action potential, an excitable membrane's responsiveness is totally abolished, making it impossible for another action potential to be generated. In skeletal muscle, the refractory period is very short compared with the duration of the resultant contraction, so the fiber can be restimulated again before the first contraction is complete to produce summation of contractions, resulting in a sustained, maximal contraction known as *tetanus* (see Figure 8–17).

In contrast, cardiac muscle has a long refractory period that lasts about 250 msec because of the prolonged action potential. This is almost as long as the period of contraction, about 300 msec (● Figure 9–24). Consequently, cardiac muscle cannot be restimulated until contraction is almost over, making summation of contractions and tetanus of cardiac muscle impossible. This is a valuable protective mechanism, because a prolonged tetanic contraction would prove fatal. The heart chambers could not be filled and emptied again.

The chief factor responsible for the long refractory period is inactivation, during the prolonged plateau phase, of the Na^+ channels that were activated during the initial Na^+ influx of the rising phase. Not until the membrane recovers from this inactivation process (when the membrane has already repolarized to resting), can the Na^+ channels be activated once

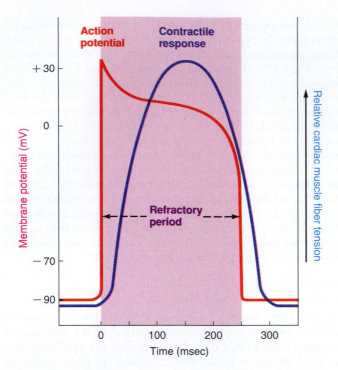

Figure 9–24 ● Relationship of an action potential and the refractory period to the duration of the contractile response in cardiac muscle.

again to begin another action potential (compare to neurons, p. 118).

Methodology: The Electrocardiogram (ECG)

The electrical currents generated by cardiac muscle during depolarization and repolarization (see p. 104) spread into the tissues surrounding the heart and are conducted through the body fluids. A small portion of this electrical activity reaches the body surface, where it can be detected using recording electrodes on the skin (typically on the limbs and/or chest, although on horses they are sometimes placed on the neck and shoulder) (● Figure 9–25a). It can also be detected with internally implanted devices. The record produced is an **electrocardiogram,** or **ECG.** (Alternatively, the term **EKG** is used, from the ancient Greek "kardia" instead of "cardia" for *heart.*) Two important points should be remembered when considering what an ECG actually represents:

1. An ECG is a recording of that portion of the electrical activity induced in the body fluids by the cardiac impulse that reaches the surface of the body, not a direct recording of the actual electrical activity of the heart. Similarly, the voltage detected by electrodes at two different points on the body surface is not the actual potential in the heart.

2. The ECG is a complex recording representing the overall spread of activity throughout the heart during depolarization and repolarization; that is, the record represents the sum of electrical activity in all the cardiac muscle cells.

Interpretation of the wave configurations recorded from the electrodes depends on a thorough knowledge of (1) the sequence of the spread of cardiac excitation and (2) the position

of the heart relative to placement of the electrodes. A normal vertebrate ECG exhibits three distinct waveforms: the P wave, the QRS complex, and the T wave (Figure 9–25b). (The letters do not signify anything other than the orderly sequence of the waves. The founder of the technique simply started in the middle of the alphabet when naming the waves.)

- The **P wave** represents atrial depolarization.
- The **QRS complex** represents ventricular depolarization.
- The **T wave** represents ventricular repolarization.

The following important points about the ECG record should also be noted:

1. Firing of the SA node does not generate sufficient electrical activity to reach the surface of the body, so no wave is recorded for SA nodal depolarization. Therefore, the first recorded wave, the P wave, occurs when the impulse spreads across the atria.

2. In a normal ECG, no separate wave is detected for atrial repolarization, because that normally occurs simultaneously with ventricular depolarization and is masked by the QRS complex.

3. The P wave is much smaller than the QRS complex because the atria have a much smaller muscle mass than the ventricles and thus generate less electrical activity.

4. There are three times when no current is flowing in the heart musculature and the ECG remains at baseline:
   a. *During the AV nodal delay.* This delay is represented by the interval of time between the end of the P wave and the onset of the QRS wave; this interval is known as the **PR segment.** (It is called the PR segment rather than the PQ segment because the Q deflection is small and sometimes absent, whereas the R deflection is the dominant wave of the complex.) Current is flowing through the AV node, but the magnitude is too small to be detected by the ECG electrodes.
   b. *When the ventricles are completely depolarized and the cardiac contractile cells are undergoing the plateau phase of their action potential before they repolarize.* This is represented by the **ST segment.** This segment is the interval between QRS and T; it coincides with the time during which ventricular activation is complete and the ventricles are contracting and emptying.
   c. *When the heart muscle is completely at rest and ventricular filling is taking place, after the T wave and before the next P wave.* This is called the **TP interval.**

ECG measurements are conducted on humans and domestic animals such as horses to diagnose certain heart problems. For example, a P wave with no subsequent QRST events (a skipped beat) is called an "AV block," and may be due to damaged conduction fibers. Horses may exhibit such skipped beats occasionally during rest, but a healthy horse never has one during exercise. ECGs are also measured on a variety of other vertebrates as part of physiological monitoring during studies on behavior and effects of temperature, drugs, stress, exercise, and so on (see the example for a goldfish in Figure 9–25c and d).

ECGs can also be measured in invertebrates, although the PQRST pattern is not present. One study, using electrodes

Figure 9–25 ● (a) The electrocardiograms of a cow standing in the four saline-filled trays containing electrodes. The right leg tray was used to ground the animal. (b) Electrocardiogram wave forms. (c) Electrocardiogram of a goldfish. (d) Goldfish with implanted sensor for ECG.

(*Source*: L. A. Geddes, 2002, Electrocardiograms from the turtle to the elephant that illustrate interesting physiological phenomena, *Pacing and Clinical Electrophysiology* 25:1762–1770, Figure 12)

under the cuticle, demonstrated cephalic control of the housefly's heart. The ECG of an anesthetized fly had rhythmic variations averaging 133 beats/min. However, when neural input to the heart was eliminated the pattern became highly regular, at a steady 120 beats/min. Thus the brain was apparently modulating an intrinsic mechanism.

Pumps: Mechanical Events of the Mammalian Cardiac Cycle

The cardiac cycle consists of alternate periods of **systole** (contraction and emptying) and **diastole** (relaxation and filling) in all animal chamber hearts. In vertebrates, the atria and ventricles go through separate cycles of systole and diastole.

■ Hearts alternately contract to empty and relax to fill.

Contraction occurs as a result of the spread of excitation across the heart, whereas relaxation follows the subsequent repolarization of the cardiac musculature. The following discussion correlates various events that occur concurrently during the mammalian cardiac cycle, including pressure changes, volume changes, and valve activity. Look at ● Figure 9–26 to understand this discussion. Only the events on the left side of the heart are described, but keep in mind that identical events are occurring on the right side of the heart, except that the pressures are lower. Our discussion follows one full cardiac cycle.

Early Ventricular Diastole

During early ventricular diastole, the atrium is still also in diastole. This stage corresponds to the TP interval on the ECG—the resting stage. Because of the continuous inflow of blood from the venous system into the atrium, atrial pressure slightly exceeds ventricular pressure even though both chambers are relaxed (point ① in Figure 9–26). Because of this pressure differential, the AV valve is open, and blood flows directly from the atrium into the ventricle throughout ventricular diastole (heart A in Figure 9–26).

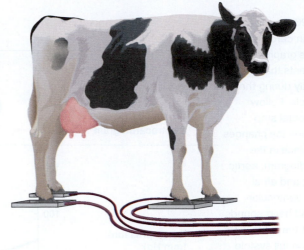

(a)

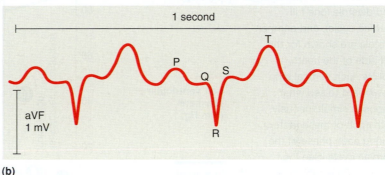

(b)

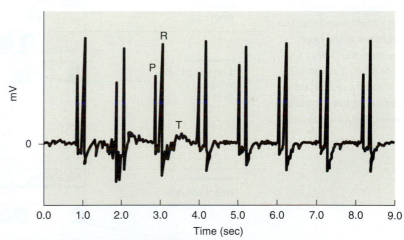

(c) Electrocardiogram of a goldfish

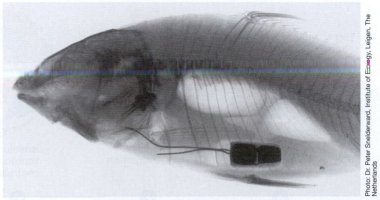

(d) Goldfish with implanted sensor for ECG

Photo: Dr. Peter Snelderward, Institute of Ecology, Leigen, The Netherlands

Figure 9–26 ●

Mammalian cardiac cycle. This graph depicts various events that occur concurrently during the cardiac cycle. Follow each horizontal strip across to see the changes that take place in the electrocardiogram; aortic, ventricular, and atrial pressures; ventricular volume; and heart sounds throughout the cycle. Late diastole, one full systole and diastole (one full cardiac cycle), and another systole are shown for the left side of the heart. Follow each vertical strip downward to see what happens simultaneously with each of these factors during each phase of the cardiac cycle. See the text (pp. 383–385) for a detailed explanation of the circled numbers. The sketches of the heart illustrate the flow of O_2-poor (dark blue) and O_2-rich (bright red) blood in and out of the ventricles during the cardiac cycle.

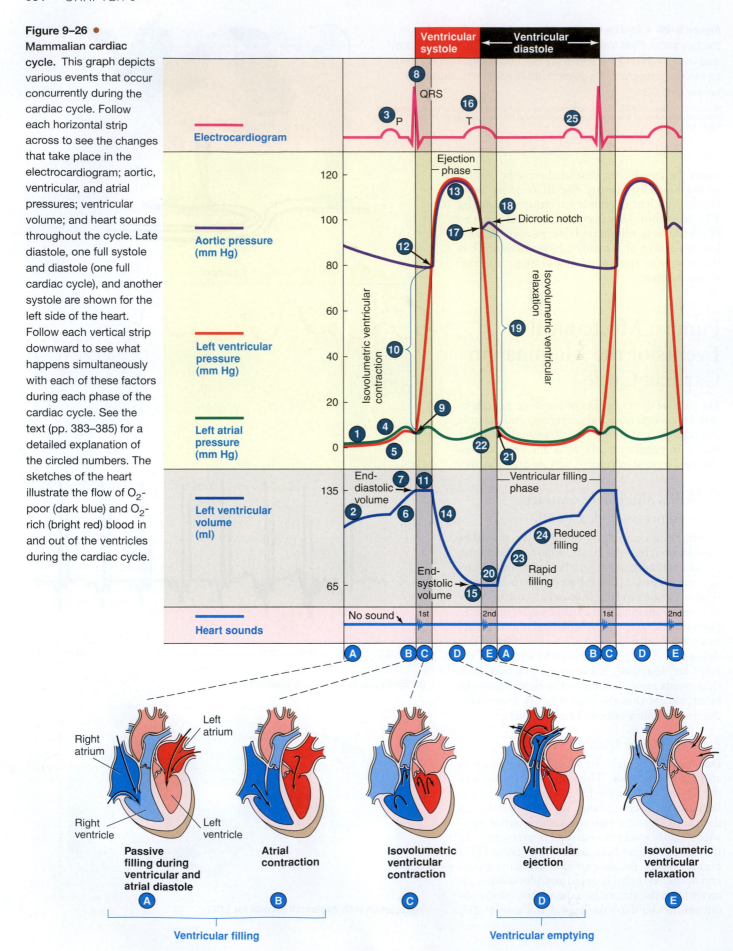

As a result, the ventricular volume slowly continues to rise even before atrial contraction takes place ②.

Late Ventricular Diastole

Late in ventricular diastole, the SA node reaches threshold and fires. The impulse spreads throughout the atria and is recorded on the ECG as the P wave ③. Atrial depolarization brings about atrial contraction, which squeezes more blood into the ventricle, causing a rise in the atrial pressure curve ④. The corresponding rise in ventricular pressure ⑤ that occurs simultaneous to the rise in atrial pressure is due to the additional volume of blood added to the ventricle by atrial contraction ⑥ and heart B). Throughout atrial contraction, atrial pressure still slightly exceeds ventricular pressure, so the AV valve remains open.

End of Ventricular Diastole

Ventricular diastole ends at the onset of ventricular contraction. By this time, atrial contraction and ventricular filling are completed. The volume of blood in the ventricle at the end of diastole ⑦ is known as the **end-diastolic volume (EDV)**, which averages about 135 mL in humans. No more blood is added to the ventricle during this cycle.

Ventricular Excitation and Onset of Ventricular Systole

Following atrial excitation, the impulse passes through the AV node and specialized conduction system to excite the ventricle. By the time ventricular activation is complete, atrial contraction is already accomplished. The QRS complex represents this ventricular excitation ⑧, which induces ventricular contraction. The ventricular pressure curve sharply increases shortly after the QRS complex, signaling the onset of ventricular systole ⑨. As ventricular contraction begins, ventricular pressure immediately exceeds atrial pressure. This backward pressure differential forces the AV valve closed ⑨.

Isovolumetric Ventricular Contraction

After ventricular pressure exceeds atrial pressure and the AV valve has closed, the ventricular pressure must continue to increase before it can open the aortic valve. Therefore, between closure of the AV valve and opening of the aortic valve, there is a brief period of time when the ventricle remains a closed chamber ⑩. Because all valves are closed, no blood can enter or leave the ventricle during this time. This interval is termed the period of **isovolumetric ventricular contraction** (*isovolumetric* means "constant volume and length") (heart C). The chamber remains at constant volume, and the muscle fibers remain at constant length, similar to an isometric contraction in skeletal muscle. During this period, ventricular pressure continues to increase ⑪.

Ventricular Ejection

When ventricular pressure exceeds aortic pressure ⑫, the aortic valve is forced open and ejection of blood begins (heart D). The aortic pressure curve rises as blood is forced into the aorta from the ventricle faster than blood is draining off into the smaller vessels at the other end ⑬. The ventricular volume decreases substantially as blood is rapidly pumped out ⑭.

Ventricular systole includes both the period of isovolumetric contraction and the ventricular ejection phase.

End of Ventricular Systole

The ventricle does not empty completely during ejection. Normally, about half the blood contained within a ventricle at the end of diastole is pumped out during the subsequent systole. The amount of blood remaining in the ventricle at the end of systole is known as the **end-systolic volume (ESV)** ⑮. This is the least amount of blood that the ventricle will contain during this cycle. The amount of blood pumped out of each ventricle with each contraction is known as the **stroke volume (SV)**; it is equal to the end-diastolic volume minus the end-systolic volume. In humans, the end-diastolic volume is 135 mL, the end-systolic volume is 65 mL, and the stroke volume is 70 mL. Comparatively, these three volumes in some horses are about 800 mL, 350 mL, and 450 mL, respectively.

Ventricular Repolarization and Onset of Ventricular Diastole

The T wave signifies ventricular repolarization occurring at the end of ventricular systole ⑯. As the ventricle starts to relax on repolarization, ventricular pressure falls below aortic pressure and the aortic valve closes ⑰. Closure of the aortic valve produces a disturbance or notch on the aortic pressure curve known as the **dicrotic notch** ⑱. No more blood leaves the ventricle during this cycle, because the aortic valve has closed.

Isovolumetric Ventricular Relaxation

When the aortic valve closes, the AV valve is not yet open, because ventricular pressure still exceeds atrial pressure, so no blood can enter the ventricle from the atrium. Therefore, all valves are once again closed for a brief period of time known as **isovolumetric ventricular relaxation** ⑲ (heart E). The muscle fiber length and chamber volume ⑳ remain constant. No blood moves as the ventricle continues to relax; pressure steadily falls.

Ventricular Filling

When the ventricular pressure falls below the atrial pressure, the AV valve opens ㉑, and ventricular filling occurs once again. Ventricular diastole includes both the period of isovolumetric ventricular relaxation and the ventricular filling phase. Atrial repolarization and ventricular depolarization occur simultaneously, so the atria are in diastole throughout ventricular systole. Blood continues to flow from the pulmonary veins into the left atrium. As this incoming blood pools in the atrium, atrial pressure rises continuously ㉒. When the AV valve opens at the end of ventricular systole, the blood that accumulated in the atrium during ventricular systole pours rapidly into the ventricle (heart A again). Ventricular filling thus occurs rapidly at first ㉓ because of the increased atrial pressure resulting from the accumulation of blood in the atria. Then ventricular filling slows down ㉔ as the accumulated blood has already been delivered to the ventricle, and atrial pressure starts to fall. During this period of reduced filling, blood continues to flow from the pulmonary veins into the left atrium and through the open AV valve into the left ventricle. During late ventricular diastole, when ventricular filling is proceeding slowly, the SA node fires again ㉕, and the cardiac cycle starts over.

Pumps: Cardiac Output and Its Control

Cardiac output (C.O.) is the volume of blood per minute pumped by a heart to the body, and is the most important physiological parameter of a circulatory pump (*Note*: Do not confuse C.O. with the gas CO, carbon monoxide!)

■ Cardiac output depends on the heart rate and the stroke volume.

The key formula in any animal is

$$C.O. = heart\ rate \times stroke\ volume$$

(volume per minute)	(beats per minute)	(volume pumped per beat or stroke)

Heart rates vary tremendously with activity state of an individual animal, and across the animal kingdom. Larger animals tend to have slower heart rates (see the discussion on scaling in Chapter 15). For example, resting rates have been measured at 6 beats/min in blue whales (the largest animals on Earth) to 300 beats/min in a rat! (A whale's heart rate declines even lower when diving; see Chapter 11.) In birds and mammals, the C.O. value is that of one ventricle, not the sum of both. During any period of time, the volume of blood flowing through the pulmonary circulation is equivalent to the volume flowing through the systemic circulation. Therefore, the cardiac output from each ventricle normally is identical, although, on a beat-to-beat basis, minor variations may occur.

To illustrate cardiac output, let us compare typical values for two mammals of different sizes, and a bird and a fish of similar size. Mammals and birds are endotherms (animals that regulate body temperature via internal heat production), with high metabolic rates. Most other vertebrates are ectotherms (animals whose body temperatures depend on external sources) (Chapter 15). As you can see, cardiac output is related to body size and is generally lower in ectotherms, here represented by trout:

	Cardiac Output	=	Heart Rate	×	Stroke Volume
Horse:	13,500 mL/min	=	30 beats/min	×	450 mL/beat
Human:	4,900 mL/min	=	70 beats/min	×	70 mL/beat
Pigeon:	195.5 mL/min	=	115 beats/min	×	1.7 mL/beat
Trout:	17.4 mL/min	=	37.8 beats/min	×	0.46 mL/beat

Cardiac output changes with development. In young broiler chicks, stroke volume almost doubles in a two-week period:

	Cardiac Output	=	Heart Rate	×	Stroke Volume
4 weeks old:	253 mL/min	=	362 beats/min	×	0.70 mL/beat
6 weeks old:	434 mL/min	=	328 beats/min	×	1.33 mL/beat

The preceding numbers are resting values; cardiac output also changes with activity, often by large amounts (fivefold or more). The following are some values that have been measured at high activity levels (running, flying, swimming):

Thoroughbred Horse:	300,000 mL/min
Human (untrained):	25,000 mL/min
Human (athlete):	40,000 mL/min
Pigeon:	1,072 mL/min
Trout:	53 mL/min

How can cardiac output vary so tremendously, depending on the demands of the animal body? The regulation of cardiac output depends on the control of both heart rate and stroke volume, topics that are discussed next.

■ Heart rate is determined primarily by antagonistic regulation of autonomic influences on the SA node.

The vertebrate heart is innervated by both divisions of the autonomic nervous system, which can modify the rate (as well as the strength) of contraction, even though nervous stimulation is not required to initiate contraction. Again, we focus on mammals. The parasympathetic nerve to the mammalian heart, the **vagus nerve,** primarily supplies the atrium, especially the SA and AV nodes. Parasympathetic innervation of the ventricles is sparse. The cardiac sympathetic nerves also supply the atria, including the SA and AV nodes, and richly innervate the ventricles as well.

Both the parasympathetic and sympathetic nervous system affect the heart by altering the activity of the cyclic AMP second-messenger system in the innervated cardiac cells, as follows:

■ **Acetylcholine (ACh)** released from the vagus nerve binds to muscarinic receptors that are coupled to an inhibitory G protein, which reduces activity of the cyclic AMP pathway (see p. 90 and p. 155). cAMP in turn increases the permeability of the SA node to K^+ by slowing the closure of EAG K^+ channels.

■ The sympathetic neurotransmitter **norepinephrine (NE)** binds with a β-adrenergic receptor that is coupled to a stimulatory G protein, which accelerates the cyclic AMP pathway (see p. 155). In turn, cAMP appears to decrease K^+ permeability by accelerating inactivation of the EAG K^+ channels.

■ ACh and NE also both affect Ca^{++} conduction, as you will see. Let's examine the specific effects that parasympathetic and sympathetic stimulation have on the heart.

Effect of Parasympathetic Stimulation on the Mammalian Heart

Parasympathetic stimulation reduces cardiac output through these effects:

1. It decreases heart rate (● Figure 9–27). Enhanced K^+ permeability because of ACh hyperpolarizes the *SA node* membrane (in most vertebrates), because more K^+ leaves than normal, making the inside even more negative. The "resting" potential starts even farther away from threshold, so it takes longer to reach threshold.

2. It decreases excitability of the *AV node*, prolonging transmission of impulses to the ventricles even longer than the usual AV nodal delay.

— = SA node pacemaker activity on parasympathetic stimulation
— = SA node pacemaker activity on sympathetic stimulation

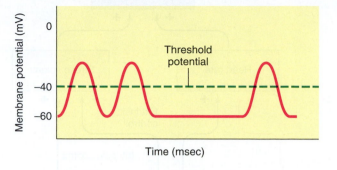

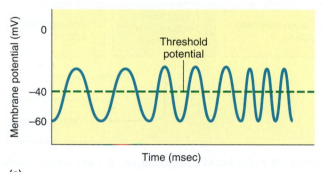

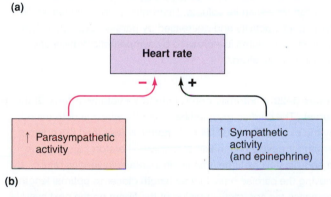

(a)

(b)

Figure 9–27 ● **Autonomic control of SA node activity and heart rate.** (a) Autonomic influence on SA node potential. Parasympathetic stimulation decreases the rate of SA nodal depolarization so that the membrane reaches threshold more slowly and has fewer action potentials, whereas sympathetic stimulation increases the rate of depolarization of the SA node so that the membrane reaches threshold more rapidly and has more frequent action potentials. (b) Control of heart rate by the autonomic nervous system. Because each SA node action potential ultimately leads to a heartbeat, increased parasympathetic activity decreases the heart rate, whereas increased sympathetic activity increases the heart rate.

3. It shortens the action potential of the atrial contractile cells, an effect believed to be caused by a reduction in the slow, inward current carried by Ca^{++}; that is, the plateau phase is reduced. As a result, atrial contraction weakens.

Thus the heart is more "leisurely" under parasympathetic influence—it beats less rapidly, the time between atrial and ventricular contraction is stretched out, and atrial contrac-

tion is weaker. These actions are appropriate in quiet, relaxed situations when the body does not need a high cardiac output. Suppression of heart rate also occurs in the *dive reflex* in air-breathing diving animals such as seals (Chapter 11). These effects are similar in other vertebrates as well.

Effect of Sympathetic Stimulation on the Mammalian Heart

Through these effects, *sympathetic stimulation increases cardiac output* (Figure 9–27):

1. It increases heart rate through its effect on pacemaker tissue. Reduced potassium permeability because of NE makes the inside of the cell become less negative, particularly as Na^+ leaks in, creating a depolarizing effect. This swifter drift to threshold permits a greater frequency of action potentials and a correspondingly more rapid heart rate.

2. It reduces the AV nodal delay at the AV node by increasing conduction velocity, presumably by enhancing the slow, inward Ca^{++} current.

3. It speeds up the spread of the action potential throughout the specialized conduction pathway.

4. It increases contractile strength of the atrial and ventricular contractile cells, both of which have an abundance of sympathetic nerve endings, so that the heart beats more forcefully and squeezes out more blood. This effect results from increasing Ca^{++} permeability, which enhances the slow Ca^{++} influx and intensifies Ca^{++} participation in the excitation–contraction process.

The overall effect of sympathetic stimulation on the heart, therefore, is to improve its effectiveness as a pump by increasing the heart rate, decreasing the delay between atrial and ventricular contraction, decreasing conduction time throughout the heart, and increasing the force of contraction; that is, sympathetic stimulation "revs up" the heart. These actions are appropriate for emergency or exercise situations; thus sympathetic action predominates when there is a need for greater blood flow. Similar effects occur in other vertebrates.

Control of Heart Rate

Thus, as is typical of the autonomic nervous system, parasympathetic and sympathetic effects on heart rate are an example of *antagonistic* regulation (p. 14). At any given moment, the heart rate is determined largely by the existing balance between the inhibitory effects of the vagus nerve and the stimulatory effects of the cardiac sympathetic nerves. The relative level of activity in these two branches in turn is primarily coordinated by the **cardiovascular control center** located in the brain stem (p. 156).

Although autonomic innervation is the primary means by which heart rate is regulated, other factors affect it as well. The most important is **epinephrine**, a hormone that is secreted into the blood from the adrenal medulla (p. 287) on sympathetic stimulation and that acts in a manner similar to norepinephrine to increase the heart rate.

■ Stroke volume is determined by the extent of venous return and by sympathetic activity.

The other component that determines the cardiac output is stroke volume, the amount of blood pumped out by each ventricle during each beat. Two types of controls influence stroke volume: (1) *intrinsic control* related to the extent of venous return and (2) *extrinsic control* related to the extent of sympathetic stimulation of the heart. Both factors increase stroke volume by increasing the strength of contraction of the heart (● Figure 9–28). Let us examine each of these factors in more detail to see how they influence the stroke volume.

■ Increased end-diastolic volume results in increased stroke volume.

As more blood is returned to the vertebrate heart, the heart pumps out more blood, but the relationship is not quite as simple as it appears, because the heart does not eject all the blood it contains. The direct correlation between end-diastolic volume and stroke volume constitutes the **intrinsic control** of stroke volume, which refers to the heart's inherent ability to vary the stroke volume. This control depends on the length–tension relationship of cardiac muscle, which is similar to that of skeletal muscle. For skeletal muscle, the resting muscle length is approximately the optimal length at which maximal tension can be developed during a subsequent contraction. When the skeletal muscle is longer or shorter than this optimal length, the subsequent contraction is weaker (see Figure 8–18). For cardiac muscle, the resting cardiac muscle-fiber length is less than optimal. Therefore, an increase in cardiac muscle-fiber length increases the contractile tension of the heart on the following systole (● Figure 9–29). Unlike in skeletal muscle, there is another more important effect: As a cardiac muscle fiber is stretched from greater ventricular filling, its myofilaments are pulled closer together. Because of this reduced distance between the thick and thin filaments, more cross-bridge interactions between myosin and actin can take place when Ca^{++} pulls the troponin-tropomyosin complex away from actin's cross-bridge binding sites—that is, myofilament Ca^{++} sensitivity increases. Thus the length–tension relationship in cardiac muscle depends not on muscle fiber length per se, but on the resultant variations in the lateral spacing between the myosin and actin filaments.

Frank-Starling Law of the Heart

What causes cardiac muscle fibers to vary in length before contraction without being attached to any bones? The main determinant of cardiac muscle-fiber length is the *degree of diastolic filling*. An analogy is a balloon filled with water—the more water you put in, the larger the balloon becomes, and the more it is stretched. Likewise, the greater the extent of diastolic filling, the larger the volume, and the more the heart is stretched. The more the heart is stretched, the longer the initial cardiac-fiber length is before contraction. The increased length results in a greater force on the subsequent cardiac contraction and, consequently, a greater stroke volume. This intrinsic relationship between end-diastolic volume and stroke volume is known as the

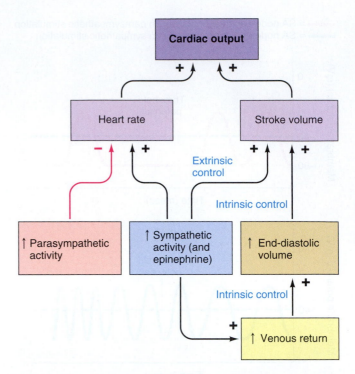

Figure 9–28 ● Control of cardiac output. Cardiac output equals heartrate times stroke volume. Heartrate in turn is increased by sympathetic activity and decreased by parasympathetic activity, while stroke volume is increased by sympathetic activity and higher venous return.

Figure 9–29 ● Intrinsic control of stroke volume (Frank-Starling curve). The cardiac muscle fiber's length, which is determined by the extent of venous filling, is normally less than the optimal length for developing maximal tension. Therefore, an increase in end-diastolic volume (that is, an increase in venous return), by moving the cardiac muscle fiber length closer to optimal length, increases the contractile tension of the fibers on the next systole. A stronger contraction squeezes out more blood. Thus, as more blood is returned to the heart and the end-diastolic volume increases, the heart automatically pumps out a correspondingly larger stroke volume.

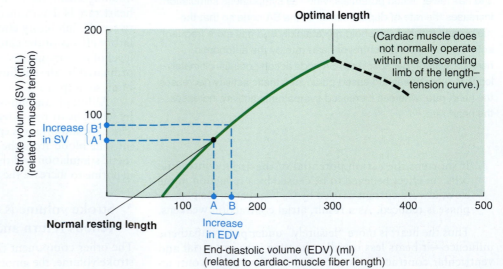

Frank-Starling law of the heart. Stated simply, the law says that *the heart normally pumps all the blood returned to it; increased venous return results in increased stroke volume.* In Figure 9–29, assume that the end-diastolic volume increases from point A to point B. You can see that this increase in *end-diastolic volume* is accompanied by a corresponding increase in stroke volume from point A^1 to point B^1. *This effect is not unique to vertebrates;* for example, mollusk hearts also respond in this way to increased filling (they also beat faster in response).

The built-in relationship matching stroke volume with venous return has two important advantages. First, one of the most important functions served by this intrinsic mechanism is equalization of output between the right and left sides of the avian and mammalian hearts, so that the blood pumped out by the heart is equally distributed between the pulmonary and systemic circulation. If, for example, the right side of the heart ejects a larger stroke volume, more blood enters the pulmonary circulation, so venous return to the left side of the heart is increased accordingly. The increased end-diastolic volume of the left side of the heart causes it to contract more forcefully, so it too pumps out a larger stroke volume. In this way, equality of output of the two ventricular chambers is maintained.

Second, when a larger cardiac output is needed in any vertebrate, such as during locomotory activity, venous return is increased through action of the sympathetic nervous system and other mechanisms to be described later in this chapter (p. 420). The resultant increase in end-diastolic volume automatically increases stroke volume correspondingly. Because locomotory activity also increases heart rate, these two factors act together to increase the cardiac output so that more blood can be delivered to the exercising muscles.

■ The contractility of the heart and venous return are increased by sympathetic stimulation.

In addition to intrinsic control, stroke volume is also subject to **extrinsic control** by factors originating outside the heart, the most important of which are actions of the cardiac sympathetic nerves and epinephrine. Sympathetic stimulation and epinephrine act in two ways:

1. As we noted earlier, both enhance the heart's **contractility,** which is the strength of contraction at any given end-diastolic volume; in other words, the heart contracts more forcefully and squeezes out a greater percentage of the blood it contains on sympathetic stimulation, leading to more complete ejection. This increased contractility is due to the increased Ca^{++} influx triggered by norepinephrine and epinephrine. The extra cytosolic Ca^{++} allows the myocardial fibers to generate more force through greater cross-bridge cycling than they would without sympathetic influence. Normally, the human end-diastolic volume is 135 mL and the end-systolic volume is 65 mL for a stroke volume of 70 mL (● Figure 9–30a). Under sympathetic influence, for the same end-diastolic volume the end-systolic volume might be 35 mL and the stroke volume 100 mL (Figure 9–30b).

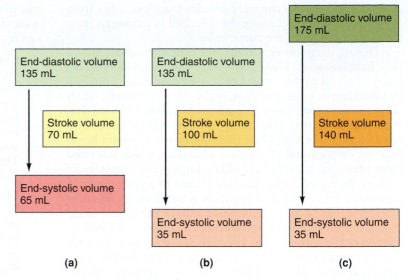

Figure 9–30 ● Effect of sympathetic stimulation on human stroke volume. (a) Normal stroke volume. (b) Stroke volume during sympathetic stimulation. (c) Stroke volume with combination of sympathetic stimulation and increased end-diastolic volume.

2. Sympathetic stimulation increases stroke volume also by *enhancing venous return* (Figure 9–30c). Sympathetic stimulation constricts veins, which squeezes more blood from the veins to the heart, increasing the end-diastolic volume and subsequently increasing the stroke volume even further. A value of about 140 mL may be achieved.

The strength of cardiac muscle contraction and, accordingly, the stroke volume can thus be graded by (1) varying the initial length of the muscle fibers, which in turn depends on the degree of ventricular filling before contraction (intrinsic control); and (2) varying the extent of sympathetic stimulation (extrinsic control). This contrasts with gradation of skeletal muscle. In skeletal muscle, twitch summation and recruitment of motor units produce variable strength of muscle contraction, but these mechanisms are not applicable to cardiac muscle. Twitch summation is impossible because of the long refractory period, and there are no motor units.

All the factors that determine the cardiac output by influencing the heart rate or stroke volume are summarized in Figure 9–28. Note that sympathetic stimulation increases the cardiac output by increasing both heart rate and stroke volume.

Pumps: Nourishing the Vertebrate Heart Muscle

■ The heart receives most of its own blood supply through the coronary circulation.

Although all the blood passes through the heart, the heart muscle cannot extract O_2 or nutrients from the blood within its chambers, in part because the walls are too thick to permit diffusion of O_2 and other supplies from the blood in the chamber to all the cardiac cells. Therefore, like other tissues of the body, heart muscle must receive blood through blood vessels, specifically by means of the **coronary circulation,** which first evolved in fishes. The coronary arteries branch in fishes from the branchial arteries leaving the gills, and in mammals from the aorta just beyond the aortic valve.

During locomotory activity, the rate of coronary blood flow increases several-fold above its resting rate. Increased delivery of blood to the cardiac cells is accomplished primarily by vasodilation, or enlargement, of the coronary vessels, which allows more blood to flow through them. Coronary blood flow is adjusted primarily in response to changes in the heart's O_2 requirements. The major link that coordinates coronary blood flow with myocardial O_2 needs is **adenosine**, which is formed from adenosine triphosphate (ATP) during cardiac metabolic activity. Increased formation and release of adenosine from the cardiac cells occur (1) when there is a cardiac O_2 deficit or (2) when cardiac activity is increased and the heart accordingly requires more O_2 for ATP production. The adenosine induces dilation of the coronary blood vessels, allowing more O_2-rich blood to flow to the more active cardiac cells to meet their increased demand. This is a negative-feedback process, summarized as follows:

$$\text{Reduced } O_2 \text{ in cardiac myocytes} \rightarrow \text{adenosine release}$$
$$\rightarrow \text{vasodilation of coronary vessels}$$
$$\rightarrow \text{increased blood flow and } O_2 \text{ delivery to myocytes}$$

This matching of O_2 delivery with O_2 needs is critical, because the heart muscle depends on oxidative processes, not anaerobic, to generate energy. The heart primarily uses free fatty acids and, to a lesser extent, glucose and lactate as fuel sources, depending on their availability, and it can shift metabolic pathways to use whatever nutrient is available. The primary danger of insufficient coronary blood flow is not fuel shortage but O_2 deficiency.

Obstruction of coronary arteries is a leading cause of death in humans. Damage called **arteriosclerosis** occurs in the artery walls for a variety of reasons, such as high blood pressure (*hypertension*). Curiously, arteriosclerotic lesions in the coronary vessels of migratory salmonids (Pacific and Atlantic salmon and steelhead trout) are very common and increase with age, affecting up to 100% of adults. These lesions are absent or less severe in nonmigratory salmonids and other fish. It is not certain why this is; researchers speculate that the damage arises from various natural stresses, such as migrating, feeding, and avoiding predation. As in humans, stress leads to hypertension, which may cause the damage. Another possible factor relates to the anatomy of the coronary artery: It lies on top of the bulbus arteriosus (recall its pressure storage-release function, p. 374) that overly expands during stress. This may cause damaging distortion of the coronary artery.

Now that we have concluded the discussion of circulatory pumps and circulatory fluids, we turn our attention to circulatory pathways and vessels.

Circulatory Pathways and Vessels

■ Circulatory fluids transport materials in a parallel manner, especially in closed systems.

As we discussed at the beginning of this chapter, most animals have either an *open* (hemolymph) or a *closed* (blood) system. In this section, we examine the pathways and regulation for fluid flow. As you have seen, blood moves through closed ves-

sels, whereas hemolymph moves more randomly through open spaces (lacunae) among organs. There are two important design features in closed, and often in open, systems:

1. Closed systems always have initial *parallel* branching in arteries, and many open systems do as well (see Figure 9–2; you will see more detailed figures shortly). Parallel plumbing allows individual organs or body regions to obtain *fresh* blood (or hemolymph). That is, part of the fluid pumped out by the heart can go to a particular muscle, part to another muscle, part to a kidney, part to the brain, and so on. Thus the same circulatory fluid does not pass from one tissue to another tissue, especially in closed systems.

2. There can be muscular *valves* on some of the branching parallel vessels. These appear universally in closed systems and occur in some open ones. They allow the animal to regulate circulatory flow to specific organs, so that gas and nutrient supplies can be boosted for active tissues that need more.

Thus blood and in some species, hemolymph, is carried by arteries until very close to target tissues; hemolymph can sometimes be channeled in specific directions (by body wall muscles, or discrete channels, for example) even if not contained in vessels as such. However, closed systems are often said to have evolved in the most metabolically active animals because they allow more rapid and more precise control of oxygen delivery to the tissues that need it the most. (Another advantage is that closed circulation can achieve the same or better delivery rates with a much lower fluid volume that has to be pumped; for example, blood may be only 5% of body fluids in a vertebrate or squid, whereas hemolymph can be 30 to 50% in a noncephalopod mollusk.) The "active animal" hypothesis reasons that the most active mollusks (cephalopods) evolved a closed system, whereas other, more sluggish mollusks did not. Similarly, vertebrates with their closed systems are generally more active than most invertebrates. Highly active flying insects are an exception because they have a separate tracheal system for oxygen delivery. This explanation for closed systems is com-

A crocodile and impala both have four-chambered hearts. But their plumbing is different. The crocodile's heart and vessels allow flow to be diverted away from the lungs when the animal is underwater, while the impala's blood must always go to the lungs before going to the rest of the body.

Photo: Paul Yancey

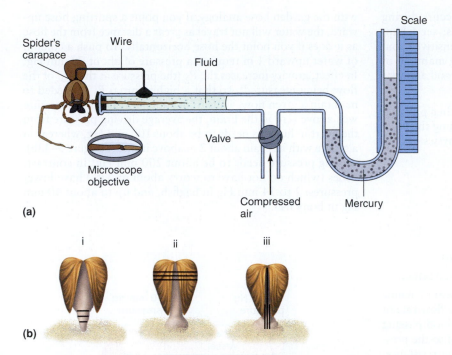

Photo: Paul Yancey

pelling, but certain exceptions seem to contradict it; for example, nemerteans (ribbon worms) and most annelids (segmented worms) have closed systems but are not more active than many other animals with open systems (such as crustaceans). However, it does appear that the most highly active animals that use circulation for oxygen transport do have closed systems. (And as you will see, large active crustaceans have open systems that approach closed ones in terms of parallel flow.)

◼ Circulatory fluids are driven by pressure and can transmit useful force.

Directing the fluid flow in a parallel manner can serve another purpose—that of *force transmission* to specific organs. To move from one point to another, circulatory fluids must be *pressurized,* as we show in detail shortly. Pressure is created by a pump and is the driving force for fluid movement. But it can be used to exert a force for other functions, including the following:

■ *Movement.* For example, hemolymph pressure (rather than skeletal muscle) is used to extend the legs in arachnids such as spiders. Arteries branch to the legs and flow to them can be controlled. Hemolymph entering into a bent leg at high pressure (from an open artery) makes the leg straighten out (● Figure 9–31a), much as a coiled water hose may straighten out when high-pressure water enters it. Cephalothorax muscles, not the spider's heart, create the pressure. The same process is used to extend the foot in mollusks, such as in a clam burrowing into sand (Figure 9–31b). These animals have hemolymph sinuses with valves to control flow direction in the foot. In 2003, researchers discovered that blue crabs, which like all arthropods must molt their exoskeletons in order to grow, also use fluid pressure when they are in the soft-shell stage. In this stage, after molting but before growing a new exo-

Figure 9–31 ● **Examples of the use of circulatory fluid pressure for force transmission.** (a) An experiment measuring hemolymph pressure in a spider's leg. Spiders use hemolymph pressure to extend their legs. In the experiment, a living spider (*Tegenaria*) is held with one leg sealed into a tube. While watching the turgid membrane of one joint of the leg, through a microscope, the experimenter slowly raises the pressure in the tube by admitting compressed air. When the external pressure in the tube just exceeds the internal pressure of the leg, the membrane is seen to collapse. Normal pressures in a quiet spider were approximately 5 cm mercury. When the spider was stimulated, pressures up to 45 cm mercury were recorded. Intermediate pressures extend the legs during normal walking. (b) A bivalve uses its foot to burrow into sand. Black lines represent muscles being used. In step i, foot muscles squeeze the hemolymph to extend the foot, while the shells remain open to form an anchor. In step ii, adductor muscles contract and squeeze more hemolymph into the foot, which flares to form an anchor. In step iii, other foot muscles contract to pull the shell downwards. (c) The spiny echidna uses blood pressure to stiffen its snout for probing into the ground for insects.

(Source: (a) V. J. Pearse, M. & R. Buchsbaum, 1987, *Living Invertebrates,* Palo Alto, CA: Blackwell Scientific, p. 543; (b) R. M. Alexander, 1982, *Locomotion of Animals,* London: Blackie, p. 115; (c) photo by Paul Yancey)

skeleton, there are no skeletal elements for muscles to act on, so muscles must push internal fluid to move the limbs.

■ *Ultrafiltration.* Blood pressure can force water and small, dissolved solutes out of pores in capillary linings. This is used in the initial process of urine formation in the kidney (Chapter 12), and for interactions with the ECF in many tissues (see p. 411).

■ *Erection.* Blood can enter a flaccid organ under high pressure, and if the exiting blood is restricted, the force of the pressure will inflate that organ. This occurs during arousal of erectile genitalia (penis, clitoris; see Chapter 16). Scientists also think it inflates the sensitive snout of the echidna (a monotreme or egg-laying mammal of Australia), which pokes its snout into termite and ant nests to feed (Figure 9–31c).

Understanding the role of pressure and other physical forces in circulatory flow is crucial to understanding the functions of circulation. Thus we first examine the physics of fluid flow before proceeding further.

Vessels: Flow Regulation and Hemodynamics

■ Blood flow through vessels depends on the pressure gradient and vascular resistance.

Flow of fluid obeys certain physical laws called **hemodynamic laws.** Here we focus on flow through vessels. The **flow rate** of blood through a vessel (that is, the volume of blood passing through per unit of time) is directly proportional to the pressure gradient and inversely proportional to vascular resistance. This is expressed in the **hemodynamic flow law,** the most important hemodynamic equation:

$$F = \Delta P / R$$

where

F = flow rate of fluid through a vessel

ΔP = pressure gradient, or $P_1 - P_2$, where
$\qquad P_1$ = pressure at the inflow end of a vessel
$\qquad P_2$ = pressure at the outflow end of a vessel

R = resistance of blood vessels

Pressure Gradient

The **pressure gradient**—the difference in pressure between the beginning and end of a vessel—is the main driving force for flow through the vessel; that is, blood flows from an area of higher pressure to an area of lower pressure down a pressure gradient. Contraction of the heart imparts pressure to the blood, but because of frictional losses (resistance) the pressure decreases as blood flows through a vessel. Because pressure drops throughout the vessel's length, it is higher at the beginning than at the end. This establishes a pressure gradient for forward flow of blood through the vessel. The greater the pressure gradient forcing blood through a vessel, the greater the rate of flow through that vessel (● Figure 9–32a). Think of a garden hose attached to a faucet. If you turn on the faucet slightly, a small stream of water flows out of the end of the hose because the pressure is slightly greater at the beginning than at the end of the hose. If you open the faucet all the way, the pressure gradient increases tremendously so that the rate of water flow through the hose is much greater, and water spurts forth from the end of the hose. Note that the *difference* in pressure between the two ends of a vessel, not the absolute pressures within the vessel, determines flow rate (Figure 9–32b).

Gravity is another major factor in establishing the pressure gradient. This is particularly important in a tall terrestrial animal such as a human or a giraffe (● Figure 9–33). Continuing

with the garden hose analogy, if you point a spurting hose upward, the water will not travel as great a distance from the hose as it does if you point the hose horizontally. To push a column of water upward 1 m requires a pressure of about 70 mm Hg. In effect, gravity increases the P_2 (the pressure at the end of the flow being measured), so that a higher initial P_1 is needed to maintain a given flow. Thus to maintain a reasonable pressure well above zero in the brain, the average driving pressure from the heart in humans needs to be about 100 mm Hg, whereas in a giraffe with its brain about 2 m above its heart (Figure 9–33b), driving pressure needs to be about 200 mm Hg! In contrast, fishes (which do not have to worry about gravity) have lower pressures: 2 to 14 mm Hg in hagfish, and up to about 40 mm Hg in bony fish.

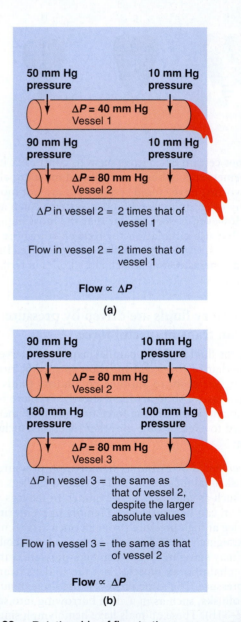

Figure 9–32 ● Relationship of flow to the pressure gradient in a vessel. (a) As the difference in pressure (ΔP) between the two ends of a vessel increases, the flow rate increases proportionately. (b) Flow rate is determined by the difference in pressure between the two ends of a vessel, not the absolute pressures.

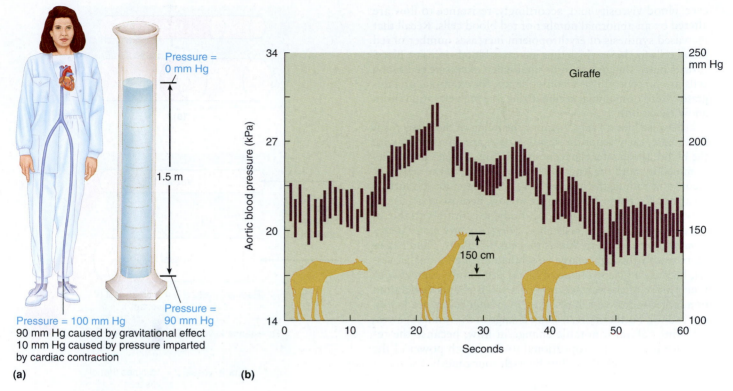

Pressure = 0 mm Hg

1.5 m

Pressure = 90 mm Hg

Pressure = 100 mm Hg
90 mm Hg caused by gravitational effect
10 mm Hg caused by pressure imparted
by cardiac contraction

(a)

(b)

Figure 9–33 ● **Effect of gravity on venous pressure.** (a) In an upright adult, the blood in the vessels extending between the heart and foot is equivalent to a 1.5 m column of blood. The pressure exerted by this column of blood as a result of the effect of gravity is 90 mm Hg. The pressure imparted to the blood by the heart has declined to about 10 mm Hg in the lower-leg veins because of frictional losses in preceding vessels. The pressure caused by gravity (90 mm Hg) added to the pressure imparted by the heart (10 mm Hg) produces a venous pressure of 100 mm Hg in the ankle and foot veins. Similarly, the capillaries in the region are subjected to these same gravitational effects. (b) Effect on aortic blood pressure of raising a giraffe's head.

(*Source:* (b) P. Withers, 1992, *Comparative Animal Physiology,* 1st ed., Belmont, CA: Brooks/Cole, p. 676, Figure 14–8)

?

How Could a Brachiosaur Pump Blood to Its Brain? The giraffe heart generates 200 mm Hg to ensure flow to its head. But consider the large dinosaurs such as brachiosaurs and barosaurs, which had necks of 25 m or more in length! Pressures of 700 mm Hg or more would be needed from a heart in the chest to create proper flow to the brain. Is this likely to have been the case, or can you think of another mechanism to create adequate blood flow to the head? After generating your own hypotheses, read the article by Choy and Altman (1992) in the Suggested Readings.

Resistance

The other factor influencing flow rate through a vessel is the **resistance**, which is a measure of the hindrance to blood flow through a vessel caused by friction between the moving fluid and the stationary vascular walls. As resistance to flow increases, it is more difficult for blood to pass through the vessel, so flow decreases (as long as the pressure gradient remains unchanged). When resistance increases, the pressure gradient

must increase correspondingly to maintain the same flow rate. Accordingly, when the vessels offer more resistance to flow the heart must work harder to maintain adequate circulation.

Resistance to blood flow depends on several factors. For an idealized smooth flow known as **laminar flow** (where layers of the fluid slide smoothly over each other), there are three key factors: (1) viscosity of the blood, η; (2) vessel length, L; and (3) vessel radius, r, which is by far the most important. The exact relationship, for idealized laminar, nonpulsatile flow in a rigid tube, is

$$R = 8\eta L/\pi r^4$$

Viscosity (η) arises from friction developed between the molecules of a fluid as they slide over each other during flow of the fluid. The greater the viscosity, the greater the resistance to flow. In general, the thicker a liquid the more viscous it is. For example, molasses flows more slowly than water, because molasses has greater viscosity. Viscosity of blood is determined by two factors: the concentration of plasma proteins and, more importantly, the number of red blood cells per unit volume. Normally, these two factors are relatively constant for a species and are not important in the control of resistance. Occasionally, however, blood viscosity and, accordingly, resistance to flow are altered by an abnormal number of red blood cells. Recall that increased synthesis of erythropoietin increases number of red blood cells (p. 365), which thus increases blood viscosity. Blood flow is more sluggish than normal when excessive red blood cells are present; this is a consequence, for example, of a migratory bird or mammal acclimatizing to a high elevation (from a low elevation).

Because blood "rubs" against the lining of the vessels as it flows past, the greater the vessel surface area in contact with the blood, the greater the resistance to flow. Surface area is determined by both the length (L) and radius (r) of the vessel. At a constant radius, the longer the vessel the greater the surface area and the greater the resistance to flow. Because vessel length is typically constant in an adult body, it is not a variable in the control of vascular resistance. *Therefore, the major determinant of resistance to flow is the vessel's radius.* Fluid passes more readily through a large vessel than through a smaller vessel, because a given volume of blood comes into contact with much more of the surface area of a small-radius vessel than of a larger-radius vessel, resulting in greater resistance (● Figure 9–34a). Furthermore, a slight change in the radius of a vessel brings about a notable change in flow, because the resistance is inversely proportional to the fourth power of the radius (multiplying the radius by itself four times):

$$R \propto 1/r^4$$

Thus, doubling the radius (Figure 9–34b) decreases the resistance 16 times and therefore increases flow through the vessel 16-fold (at the same pressure gradient). The converse is also true. Only one 16th as much blood flows through a vessel at the same driving pressure when its radius is halved. It is important to note that the radius of arterioles is subject to regulation and is the most important factor in the control of resistance.

The factors that affect the flow rate through a vessel are integrated in an idealized way in the **Poiseuille-Hagen law,** as follows:

$$F = \pi \Delta P r^4 / 8 \eta L$$

Although blood vessels are not rigid tubes with nonpulsatile flow (to which this law applies), it works reasonably well for some blood vessels (small arteries and veins). But even where the law is not wholly accurate, the qualitative relationship among flow, pressure, and resistance is correct. These factors will come more into focus as we (now) embark on a voyage through the circulation.

Pathways: Open Circulation

As we have noted, many animal phyla have open circulation. There are many similarities but also significant differences among them. For example, some invertebrates have two or more separate ECF compartments. In echinoderms, the ECF is divided into the main body cavity or *coelom*, a *water-vascular* system of canals to operate their unique tube feet, plus a *perihemal* system and a *hemal* system of connected spaces with uncertain functions. However, the most widely studied animals with open circulations are dominated by a single hemolymph space. Let us look at three examples—mollusks, insects, and crustaceans.

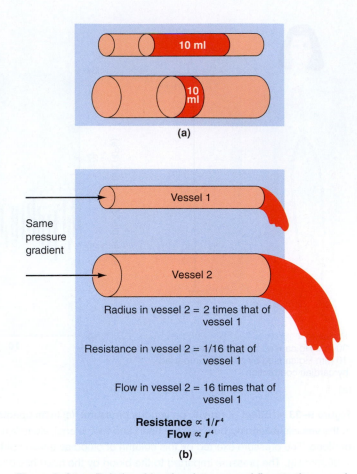

Figure 9–34 ● **Relationship of resistance and flow to the vessel radius.** (a) The same volume of blood comes into contact with a greater surface area of a small-radius vessel compared to a larger-radius vessel. Accordingly, the smaller-radius vessel offers more resistance to blood flow, because the blood "rubs" against a larger surface area. (b) Doubling the radius decreases the resistance to 1/16 and increases the flow 16 times, because the resistance is inversely proportional to the fourth power of the radius.

Mollusks

Most mollusks—snails, chitons, bivalves, and so on—have open circulations with myogenic chamber hearts (● Figure 9–35a). (Recall that cephalopods have closed systems.) Although there are no capillaries in the noncephalopods, hemolymph flow is not completely random. As you saw earlier, burrowing clams can control hemolymph for burrowing. Also, flow to individual organs in at least one species, the black abalone (*Haliotis cracheroidii*), matches each organ's metabolic demands, clearly indicating some directionality in hemolymph flow.

Crustacea

Larger crustaceans such as decapods (lobsters, crabs, crayfish, and so on) have a well-developed circulatory system with considerable parallel structures (Figure 9–35b). Some studies show that at least some decapods have fine, highly branched capillary-like arteries within target tissues (see the remarkable crab system in Figure 9–35c). These capillary-like arteries open up into local lacunae (open spaces) that are small, and the sinuses that collect the fluid for return to the heart are defined

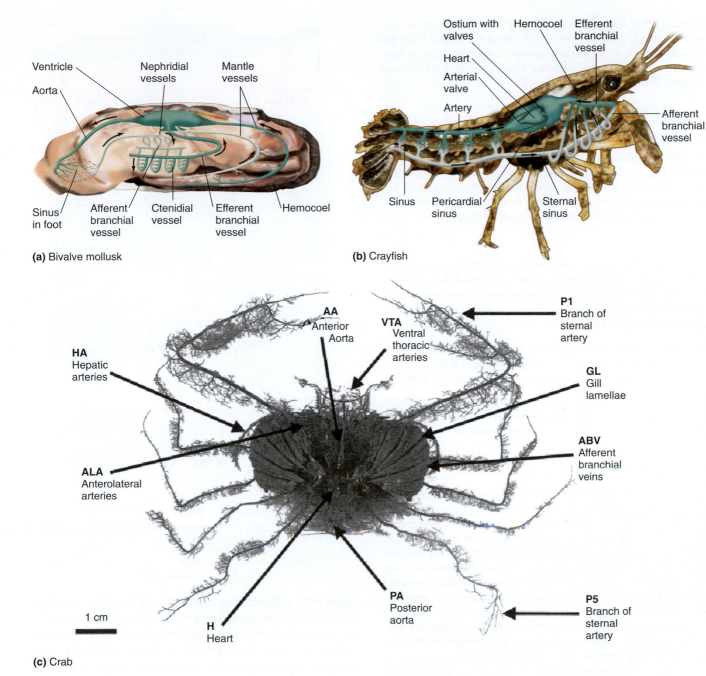

Photo: I. J. McGaw & C. L. Reiber, 2002. Cardiovascular system of the blue crab *Callinectes sapidus*, *Journal of Morphology*, 2002;252: 1–21. Reprinted by permission of Wiley-Liss, Inc., a subsidiary of John Wiley & Sons, Inc.

Figure 9–35 ● **Detailed examples of open circulations.** (a) A bivalve mollusk has a heart (ventricle) that pumps hemolymph into an aorta, which empties into sinuses in the foot; from there the fluid enters vessels that lead to the gills (ctenidium). The heart also pumps into a mantle vessel, which empties into the hemocoel to bathe most organs. (b) A crayfish has a dorsal heart that pumps hemolymph into arteries leading into the abdomen, where the fluid empties into sinuses. The fluid then enters vessels that travel to the anterior of the animal before returning to the heart. (c) A crab (*Callinectes sapidus*) has a heart that pumps hemolymph into a complex set of arteries that repeatedly branch into fine, capillary-like endings. This remarkable photograph shows those vessels in a resin cast that preserves their structure (shown in dorsal view). The heart (H) and pericardial sinus are situated slightly posterior to the midline of the animal. The anterior aorta (AA), anterolateral arteries (ALA), and hepatic arteries (HA) branch off of the anterior end of the heart, while the posterior aorta (PA) exits posteriorly. The ventral thoracic arteries (VTA) supply the mouthparts, and the pereiopod arteries (P1–P5) supply the limbs. The branchial veins (ABV) run between the gill lamellae (GL).

channels rather than random open spaces. All the returning flow is directed to the gills before reaching the heart. Thus the circulation in these animals blurs somewhat the distinction between open and closed systems. Flow can be characterized as (Figure 9–35b):

Dorsal heart → anterior and posterior aortas
→ arteries branching in parallel into capillary-like ends
→ tissue spaces → sinuses → gills → dorsal heart via ostia

The various branching arteries have valves to control flow to different regions. Later you will see how such valves work for vertebrates. Note that all the hemolymph flows to the gills before returning to the heart. This is flow *in series* rather than in parallel, and its significance is explained shortly.

Insects

Flow in insects is in some sense cruder than in decapods, with a tubular rather than chamber heart, and much less parallel branching and control. Recall that insects do not rely on circulation for oxygen delivery. Flow can be characterized as (Figure 9–2b and 9–16c):

Posterior dorsal heart → dorsal aorta and arterial branches
→ head and other anterior regions → open spaces
throughout body → abdomen → dorsal heart via ostia

Flow through the open spaces is not completely random. In the limbs, for example, longitudinal sheaths channel the hemolymph in specific directions. Thus some parallel flow is evident.

Vessels: Closed Circulation

We now examine in detail the closed system of vertebrates, starting with an overview. Vessels in a closed system differ considerably in their structures and functions. Let us return in more detail to the importance of *parallel flow* in this design. Circulatory fluid is constantly "reconditioned" so that its composition remains relatively constant despite an ongoing drain of supplies to support metabolic activities and the continual addition of wastes from the tissues. The organs that recondition the fluid normally receive substantially more blood than necessary to meet their basic metabolic needs so they can perform homeostatic adjustments. Large percentages of the cardiac output may thus be distributed to a digestive tract (to pick up nutrient supplies), to kidneys (to eliminate metabolic wastes and adjust water and electrolyte composition), and to skin (to eliminate or pick up heat) (● Figure 9–36). Flow to the other organs—heart, skeletal muscles, and so on—is solely supplying these tissues' metabolic needs and can be adjusted according to their level of activity. For example, during locomotory activity in a vertebrate additional blood is delivered to the active muscles, to meet their increased metabolic needs.

Because reconditioning organs receive blood flow in excess of their own needs, they can withstand temporary reductions in blood flow much better than can organs that do not have this extra margin of blood supply. The brain and heart in many vertebrates can suffer irreparable damage when transiently deprived of blood supply. Unless an animal has evolved a mechanism to withstand hypoxia (p. 509), permanent brain damage can occur after only a few minutes without O$_2$. (Brains in birds and mammals are particularly vulnerable, because they are not enzymatically equipped to support their metabolic needs

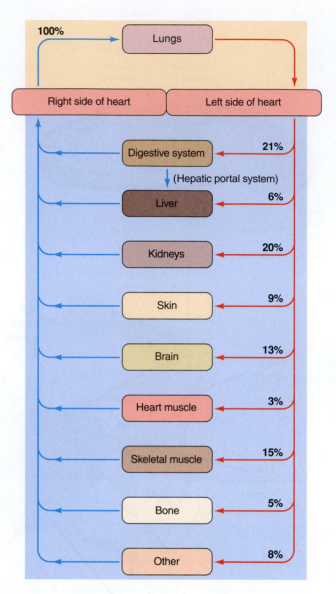

Figure 9–36 ● **Distribution of cardiac output at rest.** The lungs receive all the blood pumped out by the right side of the heart, whereas the systemic organs each receive a portion of the blood pumped out by the left side of the heart. The percentage of pumped blood received by the various organs under resting conditions is indicated. This distribution of cardiac output can be adjusted as needed.

anaerobically.) In contrast, digestive organs, kidneys, and skin frequently undergo significant reductions in circulatory flow for a considerable length of time. For example, during locomotory activity in a vertebrate some of the blood that normally flows through the digestive organs and kidneys is diverted instead to the skeletal muscles. Likewise, blood flow through mammalian skin is markedly restricted during exposure to cold, to conserve body heat.

Later in this chapter, after we examine the evolution of circulation in vertebrates, you will see how the distribution of cardiac output is adjusted according to a body's changing needs as we examine the various vessels that make up the vascular system.

■ The vascular system evolved from one circuit to two separate circuits in vertebrates.

In the vertebrates, the closed system presumably began as a simple loop:

Heart → artery → gills → aorta and branching parallel arteries → capillaries in body organs → veins → heart

Note that the flow to the gills is opposite to the flow in crustaceans. Water-breathing fishes have this basic system today. Whereas flow to most body organs is delivered in *parallel* branching vessels, flow to the gills is in *series* with the rest of the system; that is, all blood passes through the gills first. (Recall the series flow in crustaceans, where the gills receive hemolymph *before* the heart.) This flow pattern is useful for oxygenation, but it creates a problem, because the tiny diameters of the capillaries in the gills create an enormous resistance to flow. Consequently, blood flow from the gills to the rest of the body tends to be somewhat sluggish. This means that whereas there is sufficient oxygen in the blood, it is not rapidly delivered to body muscles. As you will see, this situation does not work on land. In the transition to land, a separate loop began to evolve for the new respiratory organ, the lung. To trace this evolutionary change, let's examine the circuits in vertebrates.

Fishes

Water-breathing jawed fishes have essentially the same one-circuit flow pattern just discussed, with the flow to the body organs branching into parallel vessels (● Figure 9–37a). Hagfish (jawless fishes) are similar but have partially open systems with large sinuses in the head and under the skin. They also have *auxiliary hearts* (as you saw earlier; p. 373) in veins of the head, abdomen, and tail to assist return flow. Hagfish flow can be summarized as follows (note we are omitting the vessels from now on, because they are the same):

Systemic heart → gills → body organs and sinuses → (auxiliary hearts) → systemic heart

Air-breathing fishes such as lungfish evolved a new respiratory organ, the lung, in addition to gills. This required a separate circuit, because the lung would be almost useless if it were to receive fully oxygenated blood from the gills, as all other organs do. Thus the separate pulmonary circuit evolved along with the lung. This in itself would not be very efficient if the oxygenated blood returning from the lungs were to mix with the deoxygenated blood returning from the systemic veins, or if the oxygenated blood were to travel on to the gills. For this reason the lungfish heart evolved partial barriers or septa in the atrium and ventricle that subdivide those chambers into separate pumping functions on the right and left sides. The septa reduce mixing of oxygenated and deoxygenated blood.

In lungfish, the flow to the lungs starts at the beginning of the gills, where the blood splits to the two respiratory organs, and is summarized as follows:

Right side of heart ⌐→ gills → body organs → right side of heart
　　　　　　　　　 └→ lungs → left side of heart → body organs
　　　　　　　　　　　　　 → right side of heart

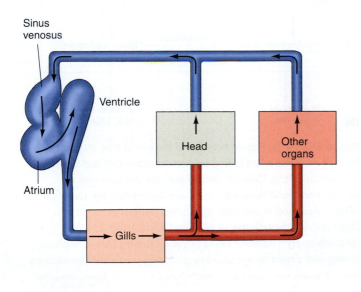

(a) Flow in a bony fish

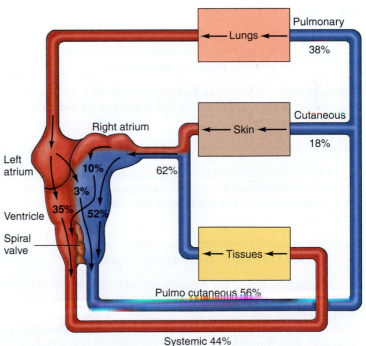

(b) Flow in a bullfrog

Figure 9–37 ● (a) Circulatory flow in a bony fish. (b) Circulatory flow in a bullfrog (*Rana catesbeiana*), showing distribution of total flow (20.5 ml/min) through the pulmonary, cutaneous and systemic systems, and the partial separation of oxygenated and deoxygenated blood in the heart.

(*Source:* Withers, P. C., 1992, *Comparative Animal Physiology*, Belmont, CA: Brooks/Cole, p. 708, Figure 14–36; as modified from Rogers, E., 1986, *Looking at Vertebrates: a Practical Guide to Vertebrate Adaptations*. London, UK: Longman; and from Tazawa, H., M. Mochizuki, and J. Piiper, 1979, Respiratory gas transport by the incompletely separated ventricle in the bullfrog, *Rana catesbeiana. Respiratory Physiology 36:* 77–95.)

Amphibians

As the first amphibians evolved to live part time on land, gills were lost in adults of some species. Today such amphibians primarily use their skins to absorb oxygen when they are in water. Thus the pulmonary circulation goes to the lungs and skin, whereas the systemic circulation goes to the main body without passing through a respiratory organ. The atrium is split completely into two chambers, such that amphibians have essentially a three-chambered heart (although most retain a sinus venosus and conus arteriosus). The single ventricle is not subdivided well except for some modest folds, so that oxygenated and deoxygenated blood does mix somewhat in that chamber. Flow is summarized as follows (Figure 9–37b):

Right atrium → right side of ventricle (56%) → lung (or skin) → left or right atrium → left side of ventricle (44%) → body organs → right atrium

Reptiles

Reptiles evolved into purely air-breathing animals, relying on lungs only. The heart lost the auxiliary chambers of fishes and amphibians, but retained the two atria that amphibians have. However, the ventricle is different. In most reptiles, it is one large chamber but has three distinct chambers divided by thick muscle rather than a septum: the *cavum arteriosum, cavum venosum,* and *cavum pulmonale.* The c. venosum connects to the other two. Flow through these chambers can follow two different pathways if a reptile is a diver (such as a turtle or seasnake). Blood flow while breathing air is as follows (● Figure 9–38):

Right atrium → c. venosum → c. pulmonale → lungs → left atrium → c. arteriosum → c. venosum → body organs → right atrium

During a dive, the lungs can be bypassed via the c. venosum, a useful adaptation when the animal is holding its breath for long periods:

Right atrium → c. venosum → body organs → right atrium

There is one exception to this reptilian pattern. Crocodiles and their relatives have a complete four-chambered heart with two atria and two ventricles, with this basic flow (● Figure 9–39a):

Right atrium → right ventricle → lungs → left atrium → left ventricle → body organs → right atrium

The right ventricle has an apparent oddity: In addition to a vessel leading to the lungs, it also has one leading to the main body. However, this vessel is also connected to some of the flow from the left ventricle by a short vessel, the **foramen of Panizza.** Normally, flow through this from the left ventricle is at a higher pressure, such that it blocks any flow from the right ventricle. This oddity is actually a useful adaptation, again for prolonged breath holding during a dive. The flow to the pulmonary arteries is restricted during a dive, so that pressure of the right atrium is directed into the systemic circulation:

Right atrium → right ventricle → body organs → right atrium

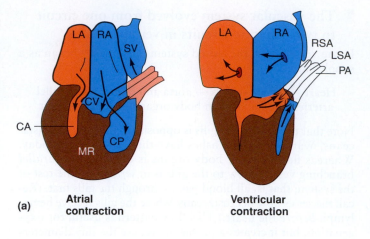

(a) Atrial contraction   Ventricular contraction

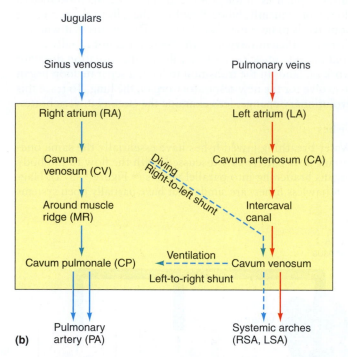

Figure 9–38 ● (a) Representation of the blood flow through the heart of a varanid lizard during atrial contraction (left) and ventricular contraction (right). Colored arrows indicate the direction of oxygenated blood flow, and black arrows indicate the direction of deoxygenated blood flow. (b) Normal pattern of blood flow (solid arrows) through the reptile heart and pattern of blood flow for a left-to-right intracardiac shunt (e.g., during apnea, such as diving) and a right-to-left shunt (e.g., during pulmonary ventilation).

(*Source:* P. Withers, 1992, *Comparative Animal Physiology,* Belmont, CA: Brooks/Cole, p. 709, Figure 14–37; as modified from Waterman et al., 1971 (see Figure 9–17); from Romer, A. S., and T. S. Parsons, 1986, *The Vertebrate Body.* Philadelphia: Saunders; and from White, F. N., 1968, Functional anatomy of the heart of reptiles, *American Zoologist 8:* 211–219.)

In the estuarine crocodile *Crocodylus porosus* and perhaps other species, there are unique, coglike valves between the right ventricle and pulmonary arteries that control this diversion of blood flow (Figure 9–39b). Unlike the loose flaps that form heart valves in other vertebrates, these are stiffer with teethlike projections, which can close during a dive, diverting flow to the main body.

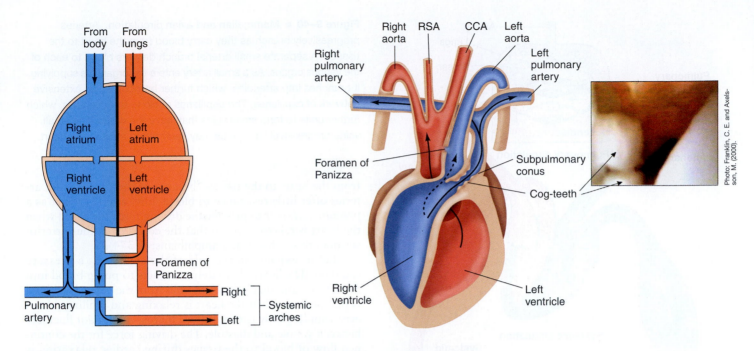

Figure 9–39 • **A crocodile heart.** (a) Representation of the pattern of blood flow through the heart, showing the shunting of oxygenated blood from the right to the left systemic arch via the foramen of Panizza. (b) Diagram showing the flow in the heart and the location of special cog teeth, with a photograph showing two of these teeth (see text for description of the teeth's function). RSA, right systemic arch; CCA, common carotid artery.

(*Sources:* (a) P. Withers, 1992, *Comparative Animal Physiology,* Belmont, CA: Brooks/Cole, Figure 14–38. (b) C. E. Franklin, M. Axelsson, 2000, An actively controlled heart valve, *Nature* 406:847–848, and A. Thomas, 2000, *Secret of the Crocodile Heart,* available online at www.abc.net.au/science/news/space/SpaceRepublish_167223.htm, accessed March 3, 2004)

Birds and Mammals

The hearts of birds and mammals have a complete four-chambered structure similar to crocodiles, but without the foramen or other shunt to divert flow in different ways. Recall that these four-chambered hearts are, in a sense, two separate pumps fused together. Thus mammalian and avian blood travels with this pattern (• Figure 9–40):

Right atrium → right ventricle → lungs → left atrium
→ left ventricle → body organs → right atrium

With the complete separation of pulmonary and systemic flow, all blood pumped by the right side of the heart passes through the lungs for O_2 pickup and CO_2 removal. Then the oxygenated blood pumped by the left side of the heart is parceled out in various proportions to the systemic organs through the parallel vessels that branch from the aorta. Researchers think this complete separation evolved independently in birds and mammals and was necessary for the high endothermic metabolisms of these vertebrates.

The fetus of a placental mammal is an exception. The fetus is not breathing, so the lungs are not functional. There are two bypasses in the fetal circulation: (1) the **foramen ovale,** an opening in the septum between the right and left atrium; and (2) the **ductus arteriosus,** a vessel connecting the pulmonary artery and aorta as they both leave the heart. These bypasses are similar to those we discussed for the crocodile heart, and have a similar function.

Now let's examine the individual vessels for the rest of this chapter section, focusing on mammals, with other vertebrates noted on occasion. We begin with an overview. **Arteries,** which carry blood from the heart to the tissues, branch into a "tree" of progressively smaller vessels, with the various parallel branches delivering blood to different regions of the body. (Note again that hemolymph systems also have arteries.) When a small artery reaches the organ it is supplying, it branches into numerous parallel **arterioles,** which provide flow control. Arterioles branch further within the organs into **capillaries,** the smallest of vessels, across which all exchanges are made with surrounding cells and which is the primary goal of the entire circulation. Capillaries rejoin to form small **venules,** which further merge to form small **veins** that leave the organs. The small veins progressively unite to form larger veins that eventually empty into the heart. The arterioles, capillaries, and venules are collectively referred to as the **microcirculation** because they are only visible microscopically. (Venules and small veins serve primarily to carry blood from the capillaries to the large veins and are not discussed further.)

Vessels: Arteries

Arteries serve as rapid-transit passageways to the tissues and as a pressure reservoir. The consecutive segments of the vascular tree—arteries, arterioles, capillaries, and veins—are specialized to perform specific tasks (◗ Table 9–1). Arteries are specialized (1) to serve as rapid-transit passageways for blood

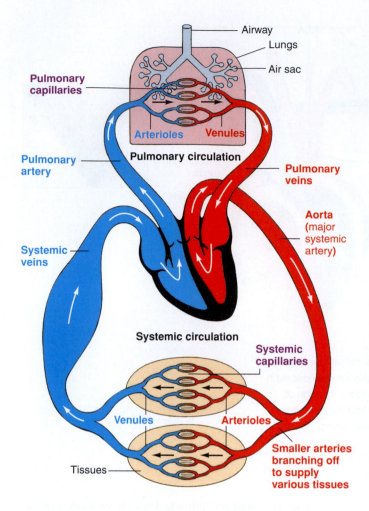

Airway
Lungs
Air sac

Pulmonary capillaries

Arterioles   Venules

Pulmonary circulation

Pulmonary artery

Pulmonary veins

Aorta (major systemic artery)

Systemic veins

Systemic circulation

Systemic capillaries

Venules   Arterioles

Tissues

Smaller arteries branching off to supply various tissues

For simplicity, only two capillary beds within two organs are illustrated.

Figure 9–40 ● **Mammalian and avian circulation.** Arteries progressively branch as they carry blood from the heart to the tissues. A separate small arterial branch delivers blood to each of the various organs. As a small artery enters the organ it is supplying, it branches into arterioles, which further branch into an extensive network of capillaries. The capillaries rejoin to form venules, which further unite to form small veins that leave the organ. The small veins progressively merge as they carry blood back to the heart.

from the heart to the tissues (because of their large radii, arteries offer little resistance to blood flow) and (2) to act as a *pressure reservoir* to provide the driving force for blood when the heart is relaxing (recall that the conus or bulbus arteriosus does this in fishes and amphibians; p. 374).

Let us expand on the role of the arteries as a pressure reservoir. The heart alternately contracts to pump blood into the arteries and then relaxes to refill from the veins. No blood is pumped out when the heart is relaxing and refilling. However, capillary flow is continuous; that is, it does not fluctuate between systole and diastole. The driving force for the continued flow of blood to the tissues during cardiac relaxation is provided by the elastic properties of the arterial walls, whose pressures do fluctuate between systole and diastole. All vessels are lined with a layer of smooth, flattened endothelial cells (continuous with the heart's endocardial lining). Surrounding the arterial endothelial lining is a thick wall containing smooth muscle and two types of connective tissue fibers; *collagen fibers,* which provide tensile strength against the high driving pressure of blood ejected from the heart, and *elastin fibers,* which give the arterial walls elasticity so that they behave much like a balloon (● Figure 9–41). Arteries of the giraffe, for example, are more richly endowed with these components in order to withstand the higher blood pressures generated by the heart.

Table 9–1 ▌ Features of Blood Vessels (with human values)

Feature	Aorta	Large Arterial Branches	Arterioles	Capillaries	Large Veins	Venae Cavae
			Vessel Type			
Number	One	Several hundred	Half a million	Ten billion	Several hundred	Two
Wall thickness	2 mm (2,000 μm)	1 mm (1,000 μm)	20 μm	1 μm	0.5 mm (500 μm)	1.5 mm 1,500 μm
Internal radius	1.25 cm (12,500 μm)	0.2 cm (2,000 μm)	30 μm	3.5 μm	0.5 cm (5,000 μm)	3 cm (30,000 μm)
Total cross-sectional area	4.5 cm²	20 cm²	400 cm²	6,000 cm²	40 cm²	18 cm²
Special features	Thick, highly elastic walls; large radii		Highly muscular, well-innervated walls; small radii	Thin walled; large total cross-sectional area	Thin walled; highly distensible; large radii	
Functions	Passageway from the heart to the tissues; serve as a pressure reservoir		Primary resistance vessels; determine the distribution of cardiac output	Site of exchange; determine the distribution of extracellular fluid between the plasma and interstitial fluid	Passageway to the heart from the tissues; serve as a blood reservoir	

As the heart pumps blood into the arteries during ventricular systole, a greater volume of blood enters the arteries from the heart than leaves them to flow into smaller vessels downstream, because the smaller vessels have a greater resistance to flow. The arteries' elasticity enables them to expand to temporarily hold this excess volume of ejected blood, storing some of the pressure energy imparted by cardiac contraction in their stretched walls—just as a balloon expands to accommodate the extra volume of air that you blow into it (● Figure 9–42a). When the heart relaxes and ceases pumping blood into the arteries, the stretched arterial walls passively recoil, like an inflated balloon that is released. This recoil pushes the excess blood contained in the arteries into the vessels downstream, ensuring continued blood flow to the tissues when the heart is relaxing and not pumping blood into the system (Figure 9–42b). *This property of arteries is widespread in vertebrates and invertebrates that have them,* occurring in (for example) amphibians, reptiles, fish, crustaceans, and cephalopods.

The maximum pressure exerted in the arteries when blood is ejected into them during systole, the **systolic pressure,** averages 120 mm Hg in humans. The minimum pressure within the arteries when blood is draining off into the remainder of the vessels during diastole, the **diastolic pressure,** averages 80 mm Hg in humans. The arterial pressure does not fall to 0 mm Hg, because the next cardiac contraction occurs and refills the arteries before all the blood drains off (● Figure 9–43). A useful way to think of these pressures and the elastic recoil is to assume that the human left ventricle exerts about 200 mm Hg of systolic pressure. Of this, 120 mm Hg are used to push the blood immediately, and 80 mm Hg of pressure energy are stored in the elastic arteries. During diastole, when the ventricle pressure falls to about 0, the rebounding arteries release the 80 mm Hg as pressure to keep the blood moving.

The systolic-diastolic pressures are typically written as pairs, for example, 120/80 for humans. Values for other mammals are not very different, for example, 130/95 for horses, 110/80 for rabbits. Birds usually have higher values, such as 180/130 for starlings. Ectotherms have lower pressures in general, such as 31/21 for frogs, 43/33 for trout, and 35/21 for

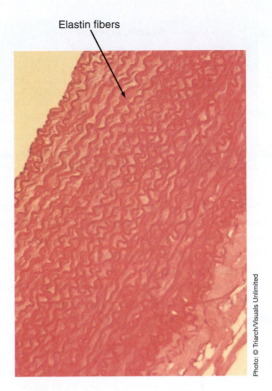

Elastin fibers

Photo: © Triarch/Visuals Unlimited

Figure 9–41 ● **Elastin fibers in an artery.** Light micrograph of a portion of the aorta wall in cross section,, showing numerous wavy elastin fibers, common to all arteries.

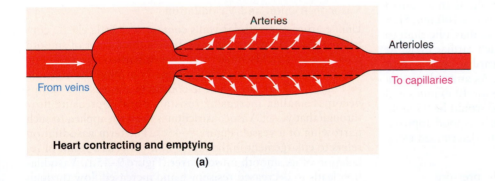

Arteries

Arterioles

From veins

To capillaries

Heart contracting and emptying

(a)

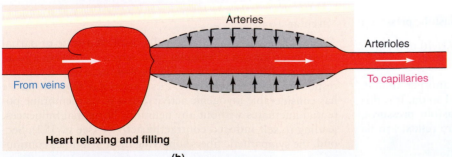

Arteries

Arterioles

From veins

To capillaries

Heart relaxing and filling

(b)

Figure 9–42 ● **Arteries as a pressure reservoir.** Because of their elasticity, arteries act as a pressure reservoir. 9a) The elastic arteries distend during cardiac systole as more blood is ejected into them than drains off into the narrow, high-resistance arterioles downstream. (b) The elastic recoil of arteries during cardiac diastole continues driving the blood forward when the heart is not pumping.

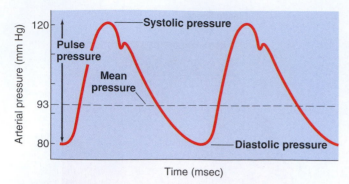

Figure 9–43 ● **Arterial blood pressure.** The systolic pressure is the peak pressure exerted in the arteries when blood is pumped into them during ventricular systole. The diastolic pressure is the lowest pressure exerted in the arteries when blood is draining off into the vessels downstream during ventricular diastole. The pulse pressure is the difference between systolic and diastolic pressure. The mean pressure is the average pressure throughout the cardiac cycle.

lobster (which has an open circulation.). There are significant exceptions to this. Systolic pressures in an octopus can reach 75 mm Hg, whereas in a jumping spider they can reach 400 mm Hg! Recall that spiders use pressure to extend their legs, such as for jumping (p. 391).

■ Mean arterial pressure is the main driving force for blood flow.

More important than the fluctuating systolic and diastolic pressures is the **mean arterial pressure,** which is the *average pressure* responsible for driving blood forward into the tissues throughout the cardiac cycle. Contrary to what you might expect, mean arterial pressure is not the halfway value between systolic and diastolic pressure (for example, with a human blood pressure of 120/80, mean pressure is *not* 100 mm Hg), because arterial pressure remains closer to diastolic than to systolic pressure for a longer portion of each cardiac cycle. At resting heart rate, about two thirds of the cardiac cycle is spent in diastole and only one third in systole. As an analogy, if a race car traveled 80 miles per hour (mph) for 40 minutes and 120 mph for 20 minutes, its average speed would be 93 mph, not the halfway value of 100 mph. Similarly, a good approximation of the mean arterial pressure can be determined using the following formula:

Mean arterial pressure = diastolic pressure
+ 1/3 systolic pressure

Here is a human example at 120/80 systolic/diastolic pressures:

Mean arterial pressure = 80 mm Hg + (1/3) 40 mm Hg
= 93 mm Hg

You may want to apply this formula to another animal, using values just given for horse, starling, and so on. It is this mean arterial pressure, not the systolic or diastolic pressures, that is generally regulated by blood pressure reflexes to be described in the final section of this chapter.

Vessels: Arterioles

When an artery reaches the organ it is supplying, it branches into numerous arterioles.

■ Arterioles control blood distribution and are the major resistance vessels.

The radii of arterioles can be small enough to offer considerable resistance to flow, particularly when they are constricted by their muscular layers (see following section). In fact, the arterioles are the major resistance vessels in the vascular tree. (Even though the capillaries have smaller radii than the arterioles, you will see later how collectively the capillaries do not offer as much resistance to flow as the arteriolar level of the vascular tree does.) In contrast to the low resistance of the arteries, the high degree of arteriolar resistance causes a marked drop in local pressure as the blood flows through these vessels and the pressure energy dissipates. On average, human pressure falls from 93 mm Hg, the mean arterial pressure (the pressure of the blood entering the arterioles), to 37 mm Hg, the pressure of the blood leaving the arterioles and entering the capillaries (● Figure 9–44). Arteriolar resistance is also responsible for converting the pulsatile systolic-to-diastolic pressure swings in the arteries into the nonfluctuating pressure present in the capillaries. Note that the local arteriole pressure is not the same thing as the ΔP of the whole system. Because $F = \Delta P/R$, to maintain a given flow, the high resistance of the arterioles must be matched by a high ΔP.

The radii (and, accordingly, the resistances) of arterioles supplying individual organs can be adjusted independently to accomplish two functions: (1) to variably distribute the cardiac output among the systemic organs, depending on the body's momentary needs, and (2) to help regulate arterial blood pressure. We next discuss the mechanisms involved in adjusting arteriolar resistance before considering how such adjustments are important in accomplishing these two functions.

Vasoconstriction and Vasodilation

Unlike arteries, arteriolar walls contain very little elastic connective tissue. However, they do have a thick layer of smooth muscle that is richly innervated by sympathetic nerve fibers. The smooth muscle layer runs circularly around the arteriole (● Figure 9–45a), so when it contracts the vessel's radius becomes smaller, increasing resistance and decreasing flow through that vessel. **Vasoconstriction** is the term applied to such narrowing of a vessel (Figure 9–45c). The term **vasodilation** refers to enlargement in the radius of a vessel as a result of relaxation of its smooth muscle layer (Figure 9–45d). Vasodilation leads to decreased resistance and increased flow through that vessel.

Vascular Tone

Arteriolar smooth muscle normally displays a state of partial constriction known as **vascular tone,** which establishes a baseline of arteriolar resistance (Figure 9–45b). Two factors are responsible for vascular tone. First, arteriolar smooth muscle has considerable myogenic activity; that is, its membrane potential fluctuates without any neural or hormonal influences, leading to self-induced contractile activity (see p. 352). Second, the sympathetic fibers supplying most arterioles continu-

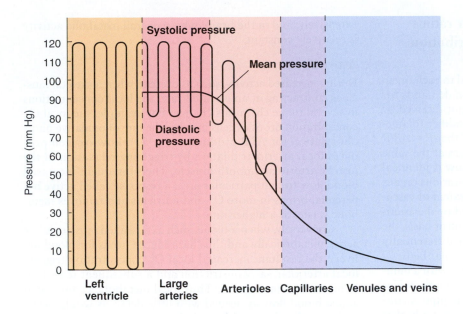

120
110
100
90
80
70
60
50
40
30
20
10
0

Pressure (mm Hg)

Systolic pressure

Mean pressure

Diastolic pressure

Left ventricle | Large arteries | Arterioles | Capillaries | Venules and veins

Figure 9–44 • **Pressures throughout the human systemic circulation.** Left ventricular pressure swings between a low pressure of 0 mm Hg during diastole to a high pressure of 120 mm Hg during systole. Arterial blood pressure, which fluctuates between a peak systolic pressure of 120 mm Hg and a low diastolic pressure of 80 mm Hg each cardiac cycle, is of the same magnitude throughout the large arteries. Because of the arterioles' high resistance, the pressure drops precipitously and the systolic-to-diastolic swings in pressure are converted to a nonpulsatile pressure when blood flows through the arterioles. The pressure continues to decline but at a slower rate as blood flows through the capillaries and venous system.

ally release norepinephrine, which further enhances the vascular tone. This ongoing tonic activity makes it possible to either increase or decrease the level of contractile activity to accomplish vasoconstriction or vasodilation, respectively. Were it not for tone, it would be impossible to reduce the tension in an arteriolar wall to accomplish vasodilation.

A variety of factors can influence the level of contractile activity in arteriolar smooth muscle, thereby substantially changing resistance to flow in these vessels. These factors fall into two categories: *local (intrinsic) controls*, which are impor-

tant in matching blood flow to the metabolic needs of the specific tissues in which they occur; and *extrinsic controls*, which are important in blood pressure regulation. This is an example of hierarchic, distributed regulation (p. 17).

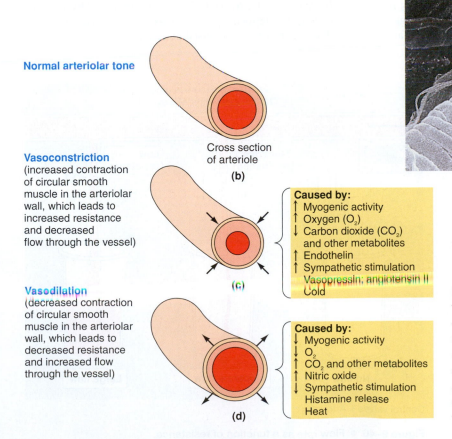

Normal arteriolar tone

Cross section of arteriole
(b)

Vasoconstriction
(increased contraction of circular smooth muscle in the arteriolar wall, which leads to increased resistance and decreased flow through the vessel)

Caused by:
↑ Myogenic activity
↑ Oxygen (O$_2$)
↓ Carbon dioxide (CO$_2$) and other metabolites
↑ Endothelin
↑ Sympathetic stimulation
Vasopressin; angiotensin II
Cold

(c)

Vasodilation
(decreased contraction of circular smooth muscle in the arteriolar wall, which leads to decreased resistance and increased flow through the vessel)

Caused by:
↓ Myogenic activity
↓ O$_2$
↑ CO$_2$ and other metabolites
↑ Nitric oxide
↓ Sympathetic stimulation
Histamine release
Heat

(d)

Smooth muscle cells
(a)

Photo: Fawcett-Uehara-Suyama/Science Source/Photo Researchers, Inc.

Figure 9–45 • **Arteriolar vasoconstriction and vasodilation.** (a) A scanning electron micrograph of an arteriole showing how the smooth muscle cells run circularly around the vessel wall. (b) Schematic representation of an arteriole in cross section showing normal arteriolar tone. (c) Outcome of and factors causing arteriolar vasoconstriction. (d) Outcome of and factors causing arteriolar vasodilation.

■ Local (intrinsic) control of arteriolar radius is important in determining the distribution of cardiac output.

The fraction of the total cardiac output delivered to each organ is not always constant; it varies, depending on demands for blood at the time. The amount of cardiac output received by each organ is determined by the number and caliber (diameter) of the arterioles supplying that area. Recall that $F = \Delta P/R$ (p. 392). Because blood is delivered to all tissues at the same mean arterial pressure, the driving force for flow is identical for each organ. Therefore, differences in flow to various organs are completely determined by differences in the extent of vascularization and by differences in resistance offered by the arterioles supplying each organ. On a moment-to-moment basis, the distribution of cardiac output can be varied by differentially adjusting arteriolar resistance in the various vascular beds.

As an analogy, consider a pipe carrying water with a number of adjustable valves located throughout its length (● Figure 9–46). Assuming that water pressure in the pipe is constant, differences in the amount of water flowing into a beaker under each valve depend entirely on which valves are open and to what extent. No water enters beakers under closed valves (high resistance), and more water flows into beakers under valves that are opened completely (low resistance) than into beakers under valves that are only partially opened (moderate resistance). Similarly, more blood flows to areas whose arterioles offer the least resistance to its passage. During locomotory activity, for example, not only is cardiac output increased, but because of vasodilation in skeletal muscle and in the heart a greater percentage of the pumped blood is diverted to these organs to support their increased metabolic activity. Simultaneously, blood flow to the digestive tract and kidneys is reduced as a result of arteriolar vasoconstriction in these organs (● Figure 9–47). Only the blood supply to the brain remains remarkably constant no matter what activity the animal is engaged in.

Local (intrinsic) controls are changes within a tissue that alter the radii of the vessels and hence adjust blood flow through the tissue by directly affecting the smooth muscle of the tissue's arterioles. Local influences may be either chemical or physical in nature, including (1) local metabolic changes, (2) histamine release, and (3) exposure to heat or cold. Let us examine the role and mechanism of each of these local influences.

■ Local metabolic influences on arteriolar radius help match blood flow with the tissues' needs.

The influence of local metabolic changes on arteriolar radius is important in matching the blood flow through a tissue with the tissue's metabolic needs, without major feedback delays (p. 19). Local metabolic controls are especially important in skeletal muscle and the heart, the tissues whose metabolic activity and need for blood supplies normally vary most extensively, and in the brain, whose overall metabolic activity is relatively constant.

Active Hyperemia

During increased activity, such as exercise when a skeletal muscle is contracting at an elevated rate, the local concentrations in several tissue chemicals change. For example, the local O_2 concentration drops as cells increase the rate of oxidative phosphorylation for ATP production (p. 51). Low oxygen produces local arteriolar dilation by triggering relaxation of the nearby arteriolar smooth muscle. The subsequent increased blood flow to that particular area, a response called **active hyperemia,** brings in more O_2 and nutrients and removes metabolic wastes at a higher rate.

Conversely, when a tissue, such as a relaxed muscle, is less active metabolically and thus has reduced needs for blood delivery, the resultant increased local O_2 concentration triggers local arteriolar vasoconstriction and a subsequent reduction in blood flow to the area. Thus local metabolic changes can adjust blood flow as needed with little delay, as would occur with neural or hormonal regulation.

Local Metabolic Changes That Influence Arteriolar Radius

A variety of local chemical changes act together in a cooperative, redundant manner to bring about these "selfish" local adjustments in arteriolar caliber that match a tissue's blood flow with its metabolic needs. Specifically, the following local chemical factors produce relaxation of arteriolar smooth muscles:

1. *Decreased O_2* as just discussed.
2. *Increased CO_2.* More CO_2 is generated as a by-product during the stepped-up pace of oxidative phosphorylation that accompanies increased activity.

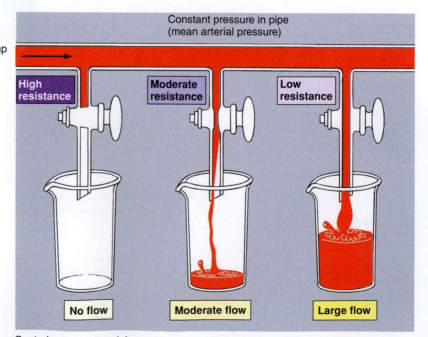

From pump (heart)

Constant pressure in pipe (mean arterial pressure)

High resistance · Moderate resistance · Low resistance

No flow · Moderate flow · Large flow

Control valves = Arterioles

Figure 9–46 ● Flow rate as a function of resistance.

Figure 9–47 ● Magnitude and distribution of the cardiac output at rest and during moderate exercise. Not only does cardiac output increase during exercise, but the distribution of cardiac output is adjusted to support the heightened physical activity. The percentage of cardiac output going to the skeletal muscles and heart rises, thereby delivering extra O_2 and nutrients needed to support these muscles' stepped-up rate of ATP consumption. The percentage going to the skin increases as a means of eliminating from the body surface the extra heat generated by the exercising muscles. The percentage going to most other organs is reduced. Only the magnitude of blood flow to the brain remains unchanged as the distribution of cardiac output is readjusted during exercise.

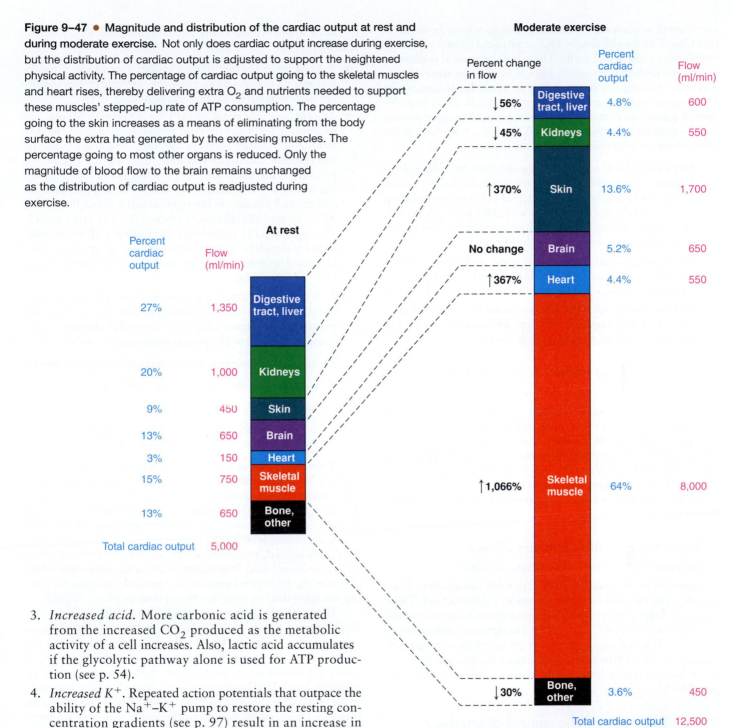

3. *Increased acid.* More carbonic acid is generated from the increased CO_2 produced as the metabolic activity of a cell increases. Also, lactic acid accumulates if the glycolytic pathway alone is used for ATP production (see p. 54).

4. *Increased K^+.* Repeated action potentials that outpace the ability of the Na^+–K^+ pump to restore the resting concentration gradients (see p. 97) result in an increase in K^+ in the tissue fluid of an actively contracting muscle or a more active region of the brain.

5. *Increased osmolarity.* Osmolarity (the concentration of osmotically active solutes) may increase during elevated cell metabolism because of increased formation of osmotically active particles.

6. *Adenosine release.* Especially in cardiac muscle, adenosine is released in response to increased metabolic activity or O_2 deprivation (see p. 390).

7. *Prostaglandin release.* Prostaglandins are local chemical messengers derived from fatty acid chains within the plasma membrane. Their production and release is poorly understood (see p. 88).

The relative contributions of these various chemical changes (and possibly others) in the local metabolic control of blood flow in systemic arterioles are still being investigated.

Local Vasoactive Mediators

Also under investigation is the mechanism by which local metabolic factors change the contractile state of arteriolar smooth muscle. These factors do not appear to act directly on vascular smooth muscle, as once thought. Instead, it has become increasingly clear that the **endothelial cells** lining the vessels release chemical mediators that play a key role in the local regulation of arteriolar radius. Endothelial cells release chemical messengers in response to chemical changes in their environment (such as a reduction in O_2) or physical changes (such as stretching of the vessel wall). These local chemical mediators act on the underlying smooth muscle to alter its state of contraction. Among the best studied of these local vasoactive ("acting on vessels") mediators is **nitric oxide (NO)** (a widespread paracrine that is produced in numerous other tissues besides endothelial cells; p. 88). NO has these roles (and probably others) in promoting blood flow:

- NO causes local arteriolar vasodilation by inducing relaxation of arteriolar smooth muscle cells in the vicinity. It does so by inhibiting the entry of contraction-inducing Ca^{++} into these cells. In this way, NO plays an important role in controlling blood flow through the tissues and in maintaining mean arterial blood pressure.
- By dilating the arterioles of the penis and clitoris, NO is the direct mediator of erection. Erection results from rapid engorgement of the penis and clitoris with blood.
- NO also interferes with platelet function and blood clotting at sites of vessel damage (p. 397; see also its role in the blood-sucking kissing bug, p. 569).

Local histamine release dilates arterioles in pathological conditions.

Histamine is not released in response to local metabolic changes and is not derived from the endothelial cells. Rather, it is released by immune cells in tissues that are injured or during allergic reactions. By promoting relaxation of arteriolar smooth muscle, histamine is the major cause of vasodilation in an injured area. The resultant increase in blood flow into the area is a major feature in defensive inflammatory responses (see Chapter 10).

Local heat or cold exposure dilates or constricts arterioles, respectively.

Exposure to heat causes localized arteriolar vasodilation. Conversely, cold exposure usually causes vasoconstriction. These processes help maintain body temperature in endotherms and some ectotherms (see Chapter 15). Therapeutic application of heat or cold to an injury takes advantage of these responses; heating promotes blood flow, bringing in oxygen and nutrients to an injured site, whereas cold reduces inflammation triggered by histamine.

Extrinsic sympathetic control of arteriolar radius is primarily important in the regulation of arterial blood pressure.

Extrinsic control of arteriolar radius includes both neural and hormonal influences, with the effects of the sympathetic nervous system being the most important. Sympathetic nerve fibers supply arteriolar smooth muscle everywhere in the systemic circulation except in the brain. A certain level of ongoing sympathetic activity contributes to vascular tone. Increased sympathetic activity produces generalized arteriolar vasoconstriction, whereas decreased sympathetic activity leads to generalized arteriolar vasodilation.

These widespread changes in arteriolar resistance bring about changes in mean arterial blood pressure. The formula $F = \Delta P/R$ applies to the entire circulation as well as to a single vessel:

- *F: Looking at the circulatory system as a whole, flow (F) through all the vessels in either the systemic or pulmonary circulation is equal to the cardiac output, C.O.*
- *ΔP: The pressure gradient (ΔP) for the entire systemic circulation is the mean arterial pressure.* ΔP equals the difference in pressure between the beginning and the end of the systemic circulatory system. Because the beginning pressure is the mean arterial pressure as the blood leaves the left ventricle at an average of 93 mm Hg (in humans) and the end pressure in the right atrium is about 0 mm Hg, $\Delta P = 93$ mm Hg (that is, 93 minus 0). The same principle applies to the pulmonary circulation: ΔP in the pulmonary circulation = mean pulmonary arterial pressure (15 mm Hg) minus the pressure in the left atrium (0 mm Hg) = 15 mm Hg.
- *R:* By far the greatest percentage of the total resistance (R) offered by all the systemic peripheral vessels (**total peripheral resistance**) is due to arteriolar resistance, because arterioles are the primary resistance vessels.

Therefore, for the entire systemic circulation, rearranging

$$F = \Delta P/R \quad \text{to} \quad \Delta P = F \times R \quad \text{or} \quad \Delta P = \text{C.O.} \times R$$

gives us the following equation:

$$\text{Mean arterial pressure} = \text{cardiac output} \\ \times \text{total peripheral resistance}$$

Thus the extent of total peripheral resistance offered collectively by all the systemic arterioles influences the mean arterial blood pressure immensely. A dam provides an analogy to this relationship. At the same time it restricts the flow of water downstream, a dam increases the pressure upstream by elevating the water level in the reservoir behind the dam. Similarly, generalized, sympathetically induced vasoconstriction reflexly reduces blood flow downstream to the tissue cells while elevating the upstream mean arterial pressure, thereby increasing the main driving force for blood flow to all the organs.

These effects seem counterproductive. Why increase the driving force for flow to the organs by increasing arterial blood pressure while reducing flow to the organs by narrowing the vessels supplying them? In effect, the sympathetically induced arteriolar responses help maintain the appropriate driving pressure head. If all arterioles were dilated, blood pressure would fall substantially, so there would not be an adequate force for overcoming gravity to the brain and filtering pressure in the kidneys (see later, p. 419). Thus tonic sympathetic activity constricts most vessels (with the exception of those in the brain), to help maintain adequate pressure in the system. The extent to which each organ actually receives blood flow is determined by local arteriolar adjustments that override the sympathetic constrictor effect.

Norepinephrine's Influence on Arteriolar Smooth Muscle

The norepinephrine released from sympathetic nerve endings combines with α_1-adrenergic receptors (see p. 155) on arteriolar smooth muscle to bring about vasoconstriction. Cerebral (brain) and pulmonary alveolar (air sac) arterioles are the only ones that do not have α_1 receptors, so no vasoconstriction occurs in these areas. It is important that cerebral arterioles not be reflexly constricted by neural influences, because brain blood flow must remain constant to meet the brain's continual need for O_2, no matter what is going on elsewhere in the body. Cerebral vessels are almost entirely controlled by local mechanisms that maintain a constant blood flow to support a consistent level of brain metabolic activity. In fact, reflex vasoconstrictor activity in the rest of the cardiovascular system functions to maintain an adequate pressure head for blood flow to the vital brain and heart (for example, see the section on diving mammals in Chapter 11, p. 510).

Thus sympathetic activity contributes in an important way to the maintenance of mean arterial pressure, assuring an adequate driving force for blood flow to the brain at the expense of organs and tissues that can better withstand reduced blood flow. Other tissues that really need additional blood can override the sympathetic effect.

Local Controls Overriding Sympathetic Vasoconstriction

Skeletal and cardiac muscles have the most powerful local control mechanisms with which to override generalized sympathetic vasoconstriction. For example, if a gazelle is running away from a cheetah, the increased activity in the skeletal muscles of its legs brings about an overriding local, metabolically induced vasodilation in those particular muscles, despite the generalized sympathetic vasoconstriction that accompanies exercise. As a result, more blood flows through the leg muscles but not through other, inactive muscles.

Lack of Parasympathetic Innervation to Most Arterioles

There is no significant level of parasympathetic innervation to arterioles, with the exception of the abundant parasympathetic vasodilator supply to the arterioles of the penis and clitoris. The rapid, profuse vasodilation induced by parasympathetic stimulation in these organs (by promoting release of NO) is largely responsible for accomplishing erection. Vasodilation elsewhere is produced by decreasing sympathetic vasoconstrictor activity below its tonic level (and some species have acetylcholine-releasing fibers in skeletal muscles that trigger vasodilation in anticipation of locomotor activity). When mean arterial pressure becomes elevated above normal, reflex reduction in sympathetic vasoconstrictor activity helps restore homeostasis, driving pressure down toward normal.

■ The medullary cardiovascular control center and several hormones regulate blood pressure.

The main region of the brain responsible for adjusting sympathetic output to the arterioles is the **cardiovascular control center** in the medulla of the brain stem. This is the integrating center for blood pressure regulation. Several other brain regions also influence blood distribution, the most notable being the hypothalamus, which, as part of its temperature-regulating function, controls blood flow to the skin to adjust heat loss to the environment (Chapter 15).

In addition to neural reflex activity, several hormones also extrinsically influence arteriolar radius. Two are from the adrenal gland medulla, which, under sympathetic stimulation, releases *epinephrine* and *norepinephrine* (p. 287). Adrenal medullary norepinephrine combines with the same α receptors

Table 9–2 ■ Arteriolar Smooth-Muscle Adrenergic Receptors

Characteristic	Receptor Type	
	α_1	β_2
Location of the receptor	All arteriolar smooth muscle except in the brain	Arteriolar smooth muscle in the heart and skeletal muscles
Chemical mediator	Norepinephrine from sympathetic fibers and the adrenal medulla Epinephrine from the adrenal medulla (less affinity for this receptor)	Epinephrine from the adrenal medulla (greater affinity for this receptor)
Arteriolar smooth muscle response	Vasoconstriction	Vasodilation

as sympathetically released norepinephrine to produce generalized vasoconstriction. However, epinephrine, the more abundant of the adrenal medullary hormones, combines with both β_2 and α_1 receptors but has a much greater affinity for the β_2 receptors. Activation of β_2 receptors produces vasodilation, but not all tissues have β_2 receptors; they are most abundant in the arterioles of the heart and skeletal muscles. During sympathetic discharge, the released epinephrine combines with the β_2 receptors in the heart and skeletal muscle to reinforce local vasodilatory mechanisms in these tissues. Arterioles in digestive organs and kidneys, in contrast, are equipped only with α receptors. Therefore, the arterioles of these organs undergo more profound vasoconstriction during generalized sympathetic discharge than do those in heart and skeletal muscle (■ Table 9–2).

The two other hormones that extrinsically influence arteriolar tone are *vasopressin* (important in regulating body water balance) and *angiotensin II*, part of a hormonal pathway, the *renin–angiotensin–aldosterone pathway* (important in the regulation of the body's salt balance). Both also help maintain adequate pressure for filtration. The functions and control of these hormones are discussed in Chapters 12 and 13.

This completes our discussion of the various factors that affect total peripheral resistance, the most important of which are controlled adjustments in arteriolar radius. These factors are summarized in ● Figure 9–48.

Vessels: Capillaries and Lymphatics

Capillaries, the sites for exchange of materials between the blood and tissues, branch extensively to bring blood within the reach of every cell. There are no carrier-mediated transport systems across capillaries, with the exception of those in the brain and testes that play a role in the blood–brain and blood–testes barriers (see p. 159). *Exchange of materials across capillary walls is accomplished primarily by the process of diffusion.*

Figure 9–48 ● **Factors affecting total peripheral resistance.** The primary determinant of total peripheral resistance is the adjustable arteriolar radius. Two major categories of factors influence arteriolar radius: (1) local (intrinsic) control, which is primarily important in matching blood flow through a tissue with the tissue's metabolic needs and is mediated by local factors acting on the arteriolar smooth muscle, and (2) extrinsic control, which is important in the regulation of blood pressure and is mediated primarily by sympathetic influence on arteriolar smooth muscle.

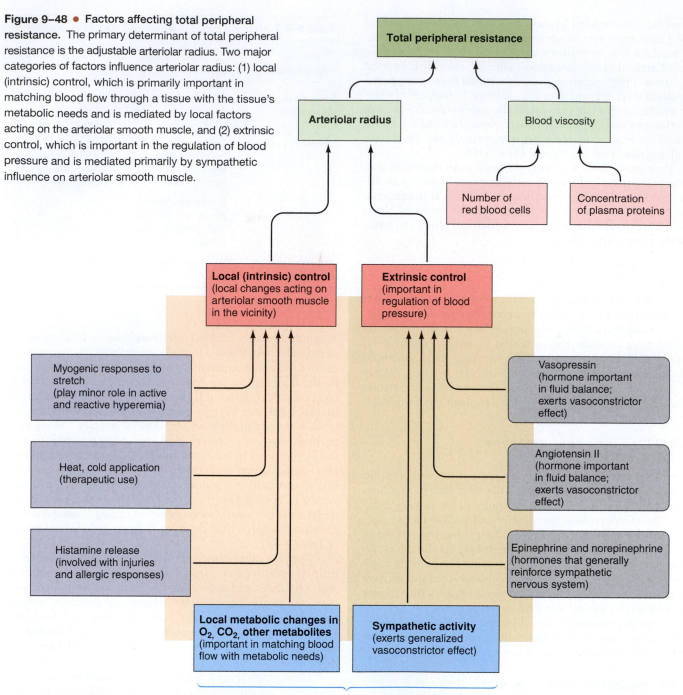

Factors Enhancing Diffusion across Capillaries

Capillaries are ideally suited to enhance diffusion in accordance with *Fick's law of diffusion* (see Table 3–1 and p. 73):

$$\text{Rate of diffusion } (Q) = \frac{\Delta C \cdot A \cdot D}{\Delta X}$$

where D is the *diffusion coefficient* (which depends on, among other things, molecular weight and the permeability of the barrier between two media), A is the *surface area* for exchange, ΔC is the concentration *gradient*, and ΔX is the *distance* the

gas must cover. Using this law, we can see how capillaries minimize diffusion distances while maximizing surface area and time available for exchange as follows:

1. Diffusing molecules have only a short distance (ΔX) to travel between the blood and surrounding cells because of the thin capillary wall and small capillary diameter, coupled with the close proximity of most cells to a capillary. Note that the rate of diffusion decreases as ΔX increases. ΔX is kept to a minimum as follows:

a. Capillary walls are very thin (1 μm in thickness; in comparison, the diameter of a human hair is 100 μm). Capillaries are composed of only a single layer of flattened endothelial cells—essentially the lining of the other vessel types. No smooth muscle or connective tissue is present (● Figure 9–49a).

b. Each capillary is so narrow (7 μm average diameter in humans) that red blood cells (8 μm diameter in humans) have to squeeze through single file (Figure 9–49b). Consequently, plasma contents are either in direct contact with the inside of the capillary wall or are only a short diffusing distance from it.

c. Because of extensive capillary branching, it is estimated that no cell is farther than 10 μm (4/1000 inch) from a capillary.

2. Because capillaries are distributed in such incredible numbers (estimates range from 10 to 40 billion capillaries), a tremendous total surface area (A) is available for exchange (an estimated 600 m²). Despite this large number of capillaries, at any point in time they contain only 5% of the total blood volume. As a result, a small volume of blood is exposed to an extensive surface area. If all the capillary surfaces were stretched out in a flat sheet and the volume of blood contained within the capillaries were spread over the top, this would be roughly equivalent to spreading a cup of paint over the floor of a high-school gymnasium. Imagine how thin the paint layer would be!

3. Blood flows more slowly in the capillaries than elsewhere in the circulatory system, allowing more time for diffusion to occur. The extensive capillary branching is responsible for this slow velocity of blood flow through the capillaries. Let's see why blood slows in the capillaries.

Slow Velocity of Flow through Capillaries

Let us clarify a potentially confusing point. The term *flow* can be used in two different contexts—the *flow rate*, which refers to the *volume* of blood flowing through a given segment of the circulatory system per unit of time (this is the flow we have been talking about in relation to the pressure gradient and resistance), and *velocity of flow*, which refers to the linear *speed* with which blood flows forward through a given segment of the circulatory system. Because the circulatory system is a closed system, the volume of blood flowing through any level of the system must equal the cardiac output. For example, if the heart pumps out 5 liters of blood per minute, and 5 liters of blood per minute return to the heart, then 5 liters of blood per minute must flow through the arteries, arterioles, capillaries, and veins. Therefore, the flow rate is the same at all levels of the circulatory system.

However, the *velocity* with which blood flows through the different segments of the vascular tree varies because velocity of flow is inversely proportional to the total cross-sectional area of all the vessels at any given level of the circulatory system as follows:

Velocity of flow (cm/sec) = flow rate (cm³/sec) / πr^2 (cm²)

Even though the cross-sectional area of each capillary is extremely small compared to that of the large aorta, the total cross-sectional area of all the capillaries added together is about 1300 times greater than the cross-sectional area of the

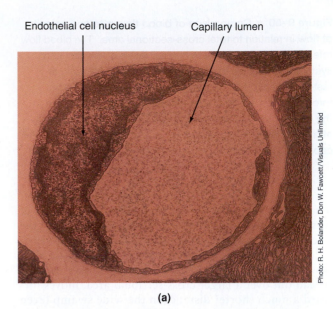

(a)

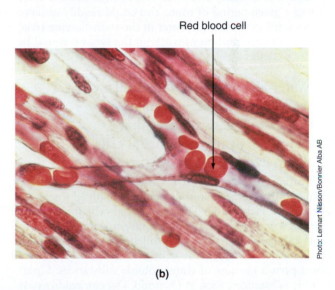

(b)

Figure 9–49 ● Capillary anatomy. (a) Electron micrograph of a cross section of a capillary. The capillary wall consists of a single layer of endothelial cells. The nucleus of one of these cells is shown. (b) Photograph of a capillary bed. The capillaries are so narrow that the red blood cells must pass through single file.

aorta because there are so many capillaries (Table 9–1). Accordingly, blood slows considerably as it passes through the capillaries (● Figure 9–50). This slow velocity allows adequate time for exchange of nutrients and metabolic end products between blood and tissues, which is the sole purpose of the entire circulatory system. As the capillaries rejoin to form veins, the total cross-sectional area is once again reduced, and the velocity of blood flow increases as blood returns to the heart.

As an analogy, consider a river (the arterial system) that widens into a swamp with many small, parallel channels (the capillaries), then narrows into a river again (the venous system) (● Figure 9–51). The flow rate is the same throughout the

Figure 9–50 ● **Comparison of blood flow rate and velocity of flow in relation to total cross-sectional area.** The blood flow rate (red curve) is identical through all levels of the circulatory system and is equal to the cardiac output (5 liters/min at rest in humans). The velocity of flow (purple curve) varies throughout the vascular tree and is inversely proportional to the total cross-sectional area (green curve) of all the vessels at a given level. Note that the velocity of flow is slowest in the capillaries, which have the largest total cross-sectional area.

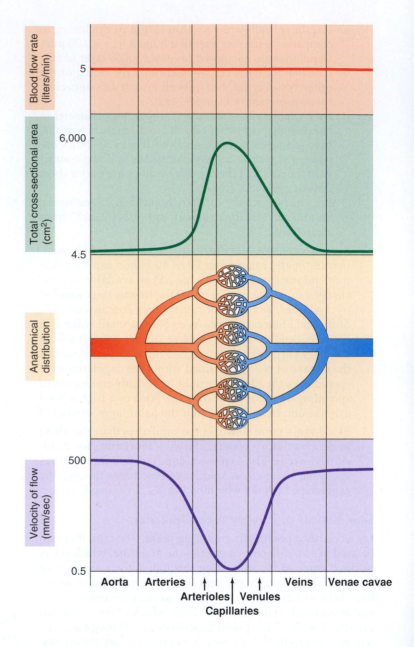

length of this body of water, that is, identical volumes of water are flowing past all the points along the bank of the river and through the whole swamp. However, the velocity of flow is slower in the wide swamp than in the narrow river because the identical volume of water, now spread out over a larger cross-sectional area, moves forward a much shorter distance in the wide swamp (even though it is in many channels) than in the narrow river during a given period of time. You could readily observe the forward movement of water in the swift-flowing river, but the forward motion of water in the swamp would be harder to detect.

■ ❙ Water-filled pores in the capillary wall permit passage of small, water-soluble substances that cannot cross the endothelial cells themselves.

Diffusion across capillary walls also depends on the walls' permeability to the materials being exchanged (included in the D of Fick's law). The endothelial cells forming the capillary walls fit together in jigsaw-puzzle fashion. In most capillaries, narrow, water-filled **pores** are present at the junctions between the cells (● Figure 9–52). These pores permit passage of water-soluble substances. Lipid-soluble substances, such as O_2 and CO_2, can readily pass through the endothelial cells themselves by dissolving in the lipid bilayer barrier.

Figure 9–51 ● **Relationship between total cross-sectional area and velocity of flow.** The three dark blue areas represent equal volumes of water. During one minute, this volume of water moves forward from points A to points C. Therefore, an identical volume of water flows past points B1, B2, and B3 during this minute; that is, the flow rate is the same at all points along the length of this body of water. However, during that minute the identical volume of water moves forward a much shorter distance in the wide swamp (A2 to C2) than in the much narrower river (A1 to C1 and A3 to C3). Thus velocity of flow is much slower in the swamp than in the river. Similarly, velocity of flow is much slower in the capillaries than in the arterial and venous systems.

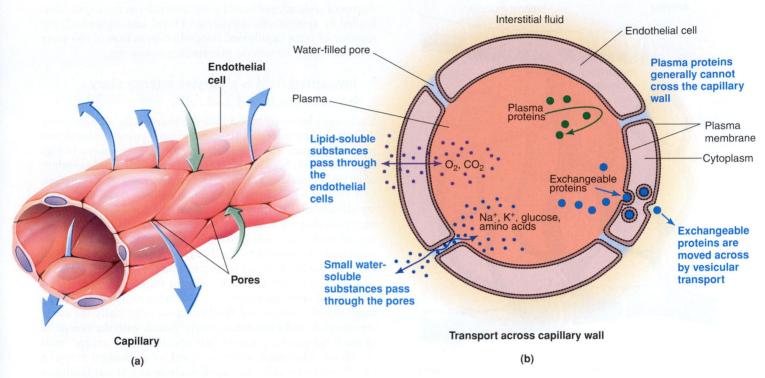

Figure 9–52 ● **Exchanges across the capillary wall.** (a) Slitlike gaps between adjacent endothelial cells form pores within the capillary wall. (b) As depicted in this schematic representation of a cross section of a capillary wall, small water-soluble substances are exchanged between the plasma and the interstitial fluid by passing through the water-filled pores, whereas lipid-soluble substances are exchanged across the capillary wall by passing through the endothelial cells. Proteins to be moved across are exchanged by vesicular transport. The plasma proteins generally cannot escape from the plasma across the capillary wall.

The size of the capillary pores varies from organ to organ. At one extreme, the endothelial cells in brain capillaries are joined by tight junctions so that pores are nonexistent. These junctions prevent transcapillary passage of materials between the cells and thus constitute part of the protective blood–brain barrier. In most tissues, small, water-soluble substances such as ions, glucose, and amino acids can readily pass through the water-filled clefts, but large, non–lipid-soluble materials such as plasma proteins are excluded from passage. At the other extreme, liver capillaries have such large pores that even proteins pass through readily. This is adaptive, because the liver's functions include synthesis of plasma proteins and the metabolism of protein-bound substances such as cholesterol (see HDLs, for example, p. 362). These proteins must all pass through the liver's capillary walls. The leakiness of capillary beds is therefore a function of how tightly the endothelial cells are joined, which varies according to the different organs' needs.

Vesicular transport also plays a limited role in the passage of materials across the capillary wall. Large non–lipid-soluble molecules such as proteinaceous hormones that must be exchanged between the blood and surrounding tissues are transported from one side of the capillary wall to the other in endocytotic–exocytotic vesicles (see p. 84).

■ Many capillaries are not open under resting conditions.

The branching and reconverging arrangement within capillary beds varies somewhat, depending on the tissue. Capillaries typically branch either directly from an arteriole or from a thoroughfare channel known as a **metarteriole,** which runs between an arteriole and a venule. Likewise, capillaries may rejoin at either a venule or a metarteriole (● Figure 9–53). Unlike the true capillaries within a capillary bed, metarterioles are sparsely surrounded by wisps of spiraling smooth muscle cells. These cells also form **precapillary sphincters,** each of which consists of a ring of smooth muscle around the entrance to a capillary as it arises from a metarteriole.

Precapillary Sphincters

Precapillary sphincters are not innervated, but they have a high degree of myogenic tone and are sensitive to local metabolic changes. They act as stopcocks to control blood flow through the particular capillary that each one guards. Arterioles perform a similar function for a small group of capillaries. Capillaries themselves have no smooth muscle, so they cannot actively participate in the regulation of their own blood flow.

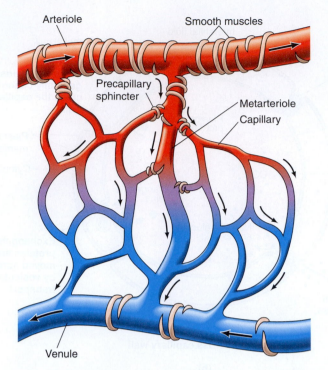

Figure 9–53 ● **Capillary bed.** Capillaries branch either directly from an arteriole or from a metarteriole, a thoroughfare channel between an arteriole and venule. Capillaries rejoin at either a venule or a metarteriole. Metarterioles are surrounded by smooth muscle cells, which also form precapillary sphincters that encircle capillaries as they arise from a metarteriole.

Generally, tissues that are more metabolically active have a greater density of capillaries. Muscles, for example, have relatively more capillaries than their attached tendons. Only about 10% of the precapillary sphincters in a resting muscle are open at any moment, however. As chemical concentrations start to change during activity in a region of the muscle tissue supplied by closed-down capillaries, the precapillary sphincters and arterioles in the region relax. As a result of more blood flowing through more open capillaries, the total volume and surface area available for exchange increase, and the diffusion distance between the cells and an open capillary decreases (● Figure 9–54). Restoration of the chemical concentrations to normal causes precapillary sphincters to close once again and the arterioles to return to normal tone. In this way, blood flow through any given capillary is often intermittent as a result of sphincter action. Thus blood flow through a particular tissue (assuming a constant blood pressure) is regulated by (1) the

degree of resistance offered by the arterioles in the organ, controlled by sympathetic activity and local factors; and (2) the number of open capillaries, controlled by action of the same local metabolic factors on precapillary sphincters.

■ Interstitial fluid is a passive intermediary between the blood and cells.

Exchanges between blood and the tissue cells are not made directly. Interstitial fluid, the true internal environment in immediate contact with the cells, acts as the go-between (● Figure 9–55). This fluid is in a sense the evolutionary descendant of an open circulatory system. Only about 20% of mammalian ECF circulates as plasma. The remaining 80% consists of interstitial fluid, which bathes all the cells in the body. Cells exchange materials directly with the interstitial fluid, the type and extent of exchange being governed by the properties of the cellular plasma membranes. Exchanges across the capillary wall between the plasma and interstitial fluid are largely passive. The only transport across this barrier that requires energy is the limited vesicular transport. Exchange is usually so thorough that the interstitial fluid takes on essentially the same composition as the incoming arterial blood, with the exception of the large plasma proteins that usually do not escape from the blood. Therefore, when we speak of exchanges between blood and tissue cells, we tacitly include interstitial fluid as a passive intermediary.

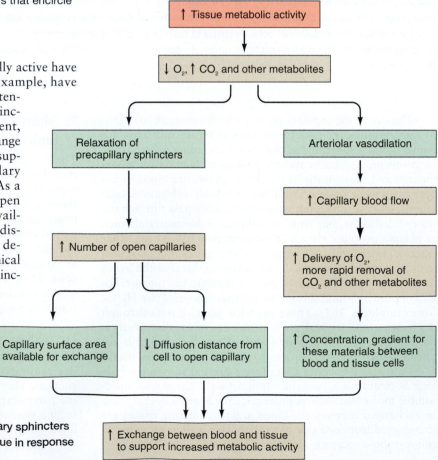

Figure 9–54 ● **Complementary action of precapillary sphincters and arterioles in adjusting blood flow through a tissue in response to changing metabolic needs.**

Exchanges between blood and surrounding tissues across the capillary walls are accomplished by two means: (1) passive diffusion down concentration gradients, the primary mechanism for exchange of individual solutes; and (2) bulk flow, a process that accomplishes the totally different function of determining the distribution of the ECF volume between the vascular and interstitial fluid compartments. We examine each of these mechanisms in more detail, starting with diffusion.

◼ Diffusion across the capillary walls is important in solute exchange.

Because there are no carrier-mediated transport systems in most capillary walls, solutes cross primarily by diffusion down concentration gradients. The chemical composition of arterial blood is carefully regulated to carry in oxygen, glucose, and other "desirable" solutes at concentrations usually higher than inside cells. Meanwhile, cells are constantly using up supplies and generating metabolic wastes. Diffusion of each solute continues independently as a result of the concentration difference for that solute between the blood and surrounding cells (● Figure 9–56). This process repeats itself continuously. As cells use up O_2 and glucose, the blood constantly brings in fresh supplies, maintaining concentration gradients (ΔC in Fick's law) that favor the net diffusion of these substances from blood to cells. Simultaneously, ongoing net diffusion of CO_2 and other metabolic wastes from cells to blood is maintained by the continual production of these wastes at the cellular level and their constant removal from the tissue level by the circulating blood. Also, as cells increase their level of activity, they use up more O_2 and produce more CO_2, among other things. This creates larger concentration gradients (ΔC) for O_2 and CO_2

between cells and blood, so more O_2 diffuses out of the blood into the cells and more CO_2 proceeds in the opposite direction.

◼ Bulk flow across the capillary wall is important in extracellular fluid distribution.

The second means by which exchange is accomplished across capillary walls is **bulk flow.** A volume of protein-free plasma actually filters out of the capillary, mixes with the surrounding interstitial fluid, and is subsequently reabsorbed. This process is called bulk flow because the various constituents of the fluid are moving together in bulk, or as a unit, in contrast to the discrete diffusion of individual solutes. The capillary wall acts like a sieve, with the fluid moving through its water-filled pores. When pressure inside the capillary exceeds pressure on the outside, fluid is pushed out through the pores. The majority of the plasma proteins and blood cells are retained on the inside during this process because of the pores' filtering effect, although a few do escape. (This process is known as **ultrafiltration,** which we explore in the kidneys, in Chapter 12.) Because all other constituents in the plasma are dragged along as a unit with the volume of fluid leaving the capillary, the filtrate is essentially a protein-free plasma. When inward-driving pressures exceed outward pressures across the capillary wall, net inward movement of fluid from the interstitial fluid compartment into the capillaries takes place through the pores, a process known as **reabsorption.**

Factors Influencing Bulk Flow

Bulk flow in this form essentially makes the so-called closed circulatory system into an open system, creating a cell-bathing fluid analogous to hemolymph. However, as you will see, its primary purpose is not molecular exchange.

Bulk flow occurs because of differences in the *hydrostatic* and *colloid osmotic pressures* between the plasma and interstitial fluid. Even though pressure differences exist between

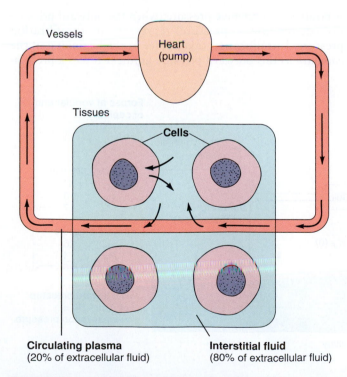

Circulating plasma
(20% of extracellular fluid)

Interstitial fluid
(80% of extracellular fluid)

Figure 9–55 ● Interstitial fluid acting as an intermediary between blood and cells.

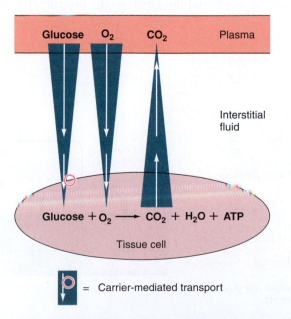

⬇◯ = Carrier-mediated transport

Figure 9–56 ● Independent exchange of individual solutes down their own concentration gradients across the capillary wall.

plasma and surrounding fluid elsewhere in the circulatory system, only the capillaries have pores that allow fluids to pass through. Four forces influence fluid movement across the capillary wall (● Figure 9–57):

1. **Capillary blood pressure (P_C)** is the fluid or hydrostatic pressure exerted on the inside of the capillary walls by the blood. This pressure tends to force fluid *out of* the capillaries into the interstitial fluid. Mean blood pressure has dropped substantially by the level of the capillaries because of frictional losses in pressure in the high-resistance arterioles upstream. On the average, the hydrostatic pressure is 37 mm Hg at the arteriolar end of a tissue capillary and has declined even further to 17 mm Hg at the venular end (Figure 9–44, p. 403).

2. **Plasma-colloid osmotic pressure (Π_p)** also known as *oncotic pressure*, is a force caused by the colloidal dispersion of plasma proteins; it encourages fluid movement into the capillaries. Because the plasma proteins remain in the plasma rather than entering the interstitial fluid, a protein–concentration difference exists between the plasma and the interstitial fluid. Accordingly, there is also an osmotic difference between these two regions, causing water movement from the interstitial fluid to the plasma. Thus the plasma proteins may be thought of as "drawing" water, although this is not actually the underlying force involved (see pp. 75–78 on osmosis). The other plasma constituents do not exert an osmotic effect, because they readily pass through the capillary wall, so their concentrations are equal in the plasma and interstitial fluid. The plasma-colloid osmotic pressure averages 25 mm Hg in humans and cats, 20 in dogs, 11 in chicks, 6 in turtles, and 5 in frogs.

3. **Interstitial-fluid hydrostatic pressure (P_{IF})** is the fluid pressure exerted on the outside of the capillary wall by the interstitial fluid. This pressure tends to force fluid *into* the capillaries. Because of the difficulties encountered in measuring interstitial-fluid hydrostatic pressure, the actual value of the pressure is a controversial issue. It is either at, slightly above, or slightly below atmospheric pressure. For purposes of illustration, we will say it is 1 mm Hg above atmospheric pressure.

4. **Interstitial fluid–colloid osmotic pressure (Π_{IF})** is another force that does not normally contribute significantly to bulk flow. The small fraction of plasma proteins that leak across the capillary walls into the interstitial spaces are normally returned to the blood by means of the lymphatic system. Therefore, the protein concentration in the interstitial fluid is extremely low, and the interstitial fluid–colloid osmotic pressure is very close to zero. If plasma proteins pathologically leak into the interstitial fluid, however, as they do when *histamine* widens the intercellular clefts during tissue injury (p. 434), the leaked proteins exert an osmotic effect that tends to promote movement of fluid *out of* the capillaries into the interstitial fluid.

Therefore, the two pressures that tend to force fluid out of the capillary are (1) capillary blood pressure and (2) interstitial fluid–colloid osmotic pressure. The two opposing pressures that tend to force fluid into the capillary are (1) plasma-colloid osmotic pressure and (2) interstitial-fluid hydrostatic pressure. Now let's analyze the fluid movement that occurs across a capillary wall because of imbalances in these forces (Figure 9–57).

Net Exchange of Fluid across a Capillary Wall

Net exchange at a given point across the capillary wall can be calculated as follows:

$$\text{Net exchange pressure} = \underbrace{(P_C + \Pi_{IF})}_{\substack{\text{(outward} \\ \text{pressure)}}} - \underbrace{(\Pi_P + P_{IF})}_{\substack{\text{(inward} \\ \text{pressure)}}}$$

A positive net exchange pressure (when the outward pressure exceeds the inward pressure) represents an ultrafiltration pressure. Ultrafiltration takes place at the beginning of the capillary as this outward pressure gradient forces a protein-

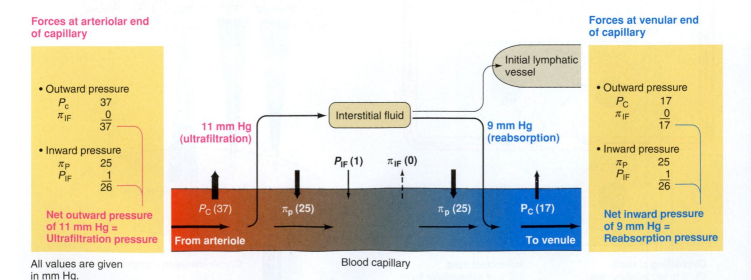

Forces at arteriolar end of capillary

- Outward pressure

P_c	37
π_{IF}	0
	37

- Inward pressure

π_P	25
P_{IF}	1
	26

Net outward pressure of 11 mm Hg = Ultrafiltration pressure

From arteriole

All values are given in mm Hg.

Forces at venular end of capillary

- Outward pressure

P_c	17
π_{IF}	0
	17

- Inward pressure

π_P	25
P_{IF}	1
	26

Net inward pressure of 9 mm Hg = Reabsorption pressure

To venule

Initial lymphatic vessel

Interstitial fluid

11 mm Hg (ultrafiltration)

9 mm Hg (reabsorption)

P_{IF} (1) π_{IF} (0)

P_C (37) π_p (25) π_p (25) P_C (17)

Blood capillary

Figure 9–57 ● **Bulk flow across the capillary wall.** Schematic representation of ultrafiltration and reabsorption as a result of imbalances in the forces acting across the capillary wall.

free filtrate through the capillary pores. By the time the venular end of the capillary is reached, the capillary blood pressure has dropped, but the other pressures have remained essentially constant. At this point the outward pressure has fallen below that of the inward pressure. Reabsorption of fluid takes place as this inward pressure gradient forces fluid back into the capillary at its venular end. Ultrafiltration and reabsorption, collectively known as *bulk flow,* are thus due to a shift in the balance between the passive physical forces acting across the capillary wall. No local energy expenditures are involved in this flow.

It is important to realize that we have been discussing "snapshots" at two points—at the beginning and at the end—in a hypothetical idealized capillary. The pressures used in ● Figure 9–58 are mammalian values, and controversial at that. In fact, a recent theory that has received considerable attention is that net filtration occurs throughout the length of all *open* capillaries (which would have high outward pressures), whereas net reabsorption occurs throughout the length of all *closed* capillaries (which would have low outward pressures).

Role of Bulk Flow

Bulk flow plays only a minor role in the exchange of individual solutes between blood and tissues, because the quantity of solutes moved across the capillary wall by bulk flow is small compared to the much larger transfer of solutes by diffusion. So, although this process is somewhat akin to hemolymph flow, its primary function is different. Essentially, it is extremely important in regulating the distribution of ECF between the plasma and interstitial fluid. Maintenance of proper arterial blood pressure depends in part on an appropriate volume of circulating blood. If plasma volume is reduced (for example, by hemorrhage), blood pressure falls. The resultant lowering of capillary blood pressure alters the balance of forces across the capillary walls. Because the net outward pressure is decreased while the net inward pressure remains unchanged, extra fluid is shifted from the interstitial compartment into the plasma as a result of reduced filtration and increased reabsorption. The extra fluid soaked up from the interstitial fluid provides additional fluid for the plasma, temporarily compensating for the loss of blood. Meanwhile, reflex mechanisms acting on the heart and blood vessels (to be described later) also come into play to help maintain blood pressure until long-term mechanisms, such as thirst and reduction of urinary output, can restore the fluid volume to compensate for the loss.

Conversely, if the plasma volume becomes overexpanded, as with excessive fluid intake, the resultant elevation in capillary blood pressure forces extra fluid from the capillaries into the interstitial fluid, temporarily relieving the expanded plasma volume until the excess fluid can be eliminated from the body by long-term measures, such as increased urinary output.

■ The lymphatic system is an accessory route by which interstitial fluid can be returned to the blood.

Even under normal circumstances, slightly more fluid is filtered out of the capillaries into the interstitial fluid than is reabsorbed from the interstitial fluid back into the plasma. On average, the net ultrafiltration pressure at the beginning of the capillary is slightly higher than the net reabsorption pressure

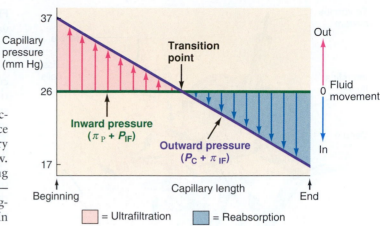

Figure 9–58 ● **Net filtration and net reabsorption along the vessel length.** The inward pressure ($\pi_P + P_{IF}$) remains constant throughout the length of the capillary whereas the outward pressure ($P_C + \pi_{IF}$) progressively declines throughout the capillary's length. In the first half of the vessel, where the declining outward pressure still exceeds the constant inward pressure, progressively diminishing quantities of fluid are filtered out (depicted by the upward red arrows). In the last half of the vessel, progressively increasing quantities of fluid are reabsorbed (depicted by the downward blue arrows) as the declining outward pressure falls farther below the constant inward pressure.

at the vessel's end. Because of this pressure differential, on average more fluid is filtered out of the first half of the capillary than is reabsorbed in its last half. The extra fluid filtered out as a result of this deficient reabsorption is picked up by the **lymphatic system.** The lymphatic system consists of an extensive network of one-way vessels that provide an accessory route by which fluid can be returned from the interstitial fluid to the blood. The lymphatic system functions much like a storm sewer that picks up and carries away excess rainwater so that it does not accumulate and flood an area.

Pickup and Flow of Lymph

Small, blind-ended terminal lymph vessels known as **initial lymphatics** permeate almost every tissue of the body (● Figure 9–59a). The endothelial cells forming the walls of initial lymphatics slightly overlap like shingles on a roof, with their overlapping edges being free instead of attached to the surrounding cells. This arrangement creates one-way, valvelike openings in the vessel wall. Fluid pressure on the outside of the vessel pushes the innermost edge of a pair of overlapping edges inward, creating a gap between the edges (that is, opening the valve), thus permitting interstitial fluid to enter (Figure 9–59b). Once interstitial fluid enters a lymphatic vessel, it is called **lymph.** Fluid pressure on the inside forces the overlapping edges together, closing the valves so that lymph does not escape. These valvelike lymphatic openings are much larger than the pores in blood capillaries. Consequently, large particulates in the interstitial fluid, such as escaped plasma proteins and bacteria, can gain access to initial lymphatics but are excluded from blood capillaries.

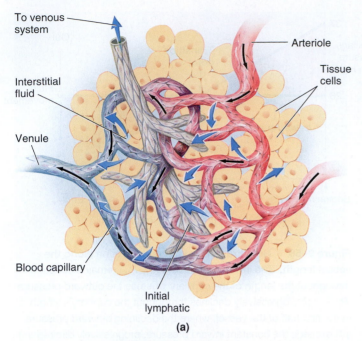

(a)

To venous system

Arteriole

Tissue cells

Interstitial fluid

Venule

Blood capillary

Initial lymphatic

Fluid pressure on the outside of the vessel pushes the endothelial cell's free edge inward, permitting entrance of interstitial fluid (now lymph).

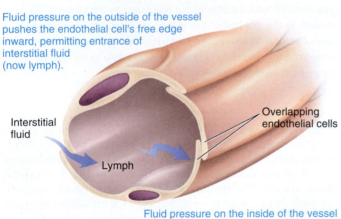

Interstitial fluid

Overlapping endothelial cells

Lymph

Fluid pressure on the inside of the vessel forces the overlapping edges together so that lymph cannot escape.

(b)

Figure 9–59 ● Initial lymphatics. (a) Relationship between initial lymphatics and blood capillaries. Blind-ended initial lymphatics pick up excess fluid filtered by blood capillaries and return it to the venous system in the chest. (b) Arrangement of endothelial cells in an initial lymphatic. Note that the overlapping edges of the endothelial cells create valvelike openings in the vessel wall.

Initial lymphatics converge to form larger and larger **lymph vessels,** which eventually empty into the venous system near the point where the blood enters the right atrium (● Figure 9–60a). In all amphibians and reptiles and bird embryos, there are **lymph hearts** (usually in pairs on each side of the body) that create flow back to the blood. Flow in through these hearts may need assistance from movements of skeletal muscles and lungs.

Mammals and most adult birds (ostriches are an exception) do not have these distinct auxiliary hearts; flow is accomplished by two mechanisms. First, lymph vessels beyond the initial lymphatics are surrounded by smooth muscle, which

Figure 9–60 ➤ ● Lymphatic system. Lymph empties into the venous system near its entrance to the right atrium.

contracts rhythmically as a result of myogenic initiation of action potentials. When this muscle is stretched because the vessel is distended with lymph, the muscle inherently contracts more forcefully, thereby pushing the lymph through the vessel. This intrinsic "lymph pump" is the major force for propelling lymph. Second, because lymph vessels lie between skeletal muscles, contraction of these extrinsic muscles squeezes the lymph out of the vessels. One-way valves spaced at intervals within the lymph vessels direct the flow of lymph toward its venous outlet in the chest.

Functions of the Lymphatic System

The following are the most important functions of the lymphatic system:

1. *Return of excess filtered fluid.* Even though only a small fraction of the filtered fluid is not reabsorbed by the blood capillaries, the cumulative effect of this process being repeated with every heartbeat results in the equivalent of more than the entire plasma volume being left behind in the interstitial fluid each day. Obviously, this fluid must be returned to the circulating plasma, and lymph vessels accomplish this task.

2. *Defense against disease.* The lymph percolates through **lymph nodes** located en route within the lymphatic system. Passage of this fluid through the lymph nodes is an important aspect of the body's defense mechanism against disease. This is covered in Chapter 10.

3. *Transport of absorbed fat.* The lymphatic system is important in the absorption of fat from the digestive tract. The end products of the digestion of dietary fats are packaged by cells lining the digestive tract into fatty particles (**chylomicrons,** p. 655) that are too large to gain access to the blood capillaries but can easily enter the initial lymphatics.

4. *Storage of plasma proteins.* Lymph may contain these proteins at nearly the concentration of plasma.

Do Fish Have Lymph? Bony and cartilaginous fishes are traditionally thought to have lymphatic systems. However, recent anatomical studies fail to confirm this. Instead, fish appear to have a closed secondary circulation consisting of tiny vessels—so small that they exclude red cells—that branch from arteries in the gills, skin, gut, and fins, and then reconnect to arteries or veins. Can you think of a function for such a system not found in other vertebrates? See the review by Olson (1996) in the Suggested Readings.

Vessels: Veins

The venous system completes the circulatory circuit. Blood leaving the capillary beds enters the venous system for transport back to the heart.

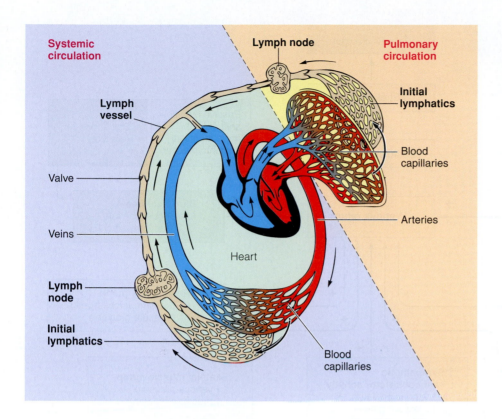

law of the heart (see p. 389). If too much blood pools in the veins instead of being returned to the heart, cardiac output is abnormally diminished. Thus a delicate balance exists between the capacity of the veins, the extent of venous return, and the cardiac output. We now turn our attention to the factors that affect venous capacity and contribute to venous return.

■ Venous return is enhanced by a number of extrinsic factors.

Venous capacity (the volume of blood that the veins can accommodate) depends on the distensibility of the vein walls (how much they can stretch to hold blood) and the influence of any externally applied pressure squeezing inwardly on the veins. At a constant blood volume, as venous capacity increases more blood remains in the veins instead of being returned to the heart. Such venous storage decreases the **effective circulating volume.** Conversely, when venous capacity decreases more blood is returned to the heart and continues circulating. Thus changes in venous capacity directly influence the magnitude of venous return, which in turn is an important (although not the only) determinant of effective circulating blood volume. The magnitude of the total blood volume is also influenced on a short-term basis by passive shifts in bulk flow between the vascular and interstitial fluid compartments and on a long-term basis by factors that control total ECF volume, such as salt and water balance.

The term **venous return** refers to the volume of blood entering each atrium per minute from the veins. Recall that the magnitude of flow through a vessel is directly proportional to the pressure gradient. Much of the driving pressure imparted to the blood by cardiac contraction has been lost by the time the blood reaches the venous system because of frictional losses along the way, especially during passage through the high-resistance arterioles. By the time the blood enters the venous system, mean pressure in humans averages only 17 mm Hg (Figure 9–44, p. 403). However, because atrial pressure is near 0 mm Hg, a small but adequate driving pressure still exists to promote the flow of blood through the large-radius, low-resistance veins.

In addition to the driving pressure imparted by cardiac contraction, five other factors enhance venous return: sympathetically induced venous vasoconstriction, skeletal muscle activity, the effect of venous valves, respiratory activity, and the effect of cardiac suction (● Figure 9–61). Most of these secondary factors affect venous return by influencing the pressure gradient between the veins and the heart. We examine each in turn.

Effect of Sympathetic Activity on Venous Return

Veins are not very muscular and have little inherent tone, but venous smooth muscle is abundantly supplied with sympa-

■ Veins serve as a blood reservoir as well as passageways back to the heart.

Veins have large radii, so they offer little resistance to flow. Furthermore, because the total cross-sectional area of the venous system gradually decreases as smaller veins converge into progressively fewer but larger vessels, the velocity of blood flow increases as the blood approaches the heart.

In addition to serving as low-resistance passageways to return blood from the tissues to the heart, systemic veins also serve as a *blood reservoir.* Because of their storage capacity, veins are often referred to as **capacitance vessels.** Veins have much thinner walls with less smooth muscle than do arteries. Because collagen fibers are considerably more abundant than elastin fibers in venous connective tissue, veins have very little elasticity, in contrast to arteries. Also, unlike arteriolar smooth muscle, venous smooth muscle has little inherent myogenic tone. Because of these features, veins are highly distensible, or stretchable, and have little elastic recoil. They easily distend to accommodate additional volumes of blood with only a small increase in venous pressure. Arteries that have stretched by an excess volume of blood recoil because of the elastic fibers in their walls, driving the blood forward. Veins containing an extra volume of blood simply stretch to accommodate the additional blood without tending to recoil. In this way veins serve as a blood reservoir; that is, when demands for blood are low, the veins can store extra blood in reserve because of their distensibility. (Note that blood in the veins is not normally stagnant, but is moving constantly.) Under resting conditions, mammalian veins contain more than 60% of the total blood volume. When the stored blood is needed, such as during exercise, extrinsic factors (soon to be described) drive the extra blood from the veins to the heart, inducing an increased cardiac stroke volume in accordance with the Frank-Starling

Figure 9–61 ● Factors that facilitate venous return.

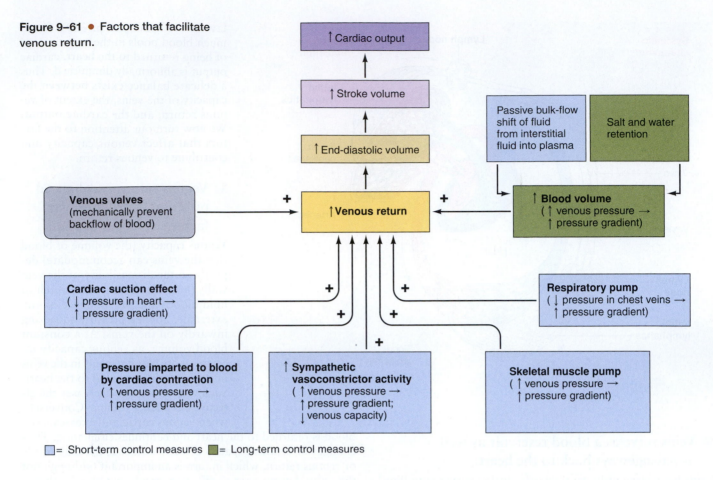

= Short-term control measures   ■ = Long-term control measures

thetic nerve fibers. Sympathetic stimulation produces venous vasoconstriction, which modestly elevates venous pressure; this, in turn, increases the pressure gradient to drive more of the stored blood from the veins into the right atrium. The veins normally have such a large diameter that the moderate vasoconstriction accompanying sympathetic stimulation has little effect on resistance to flow. That is, even when constricted, the veins still have a relatively large diameter.

It is important to recognize the different outcomes of vasoconstriction in arterioles and veins. Arteriolar vasoconstriction immediately *reduces* flow through these vessels because of their increased resistance (less blood can enter and flow through a narrowed arteriole), whereas venous vasoconstriction immediately *increases* flow through these vessels because of their decreased capacity (narrowing of veins squeezes out more of the blood that is already present in the veins, thus increasing blood flow through these vessels).

Effect of Skeletal Muscle Activity on Venous Return

Many of the large veins in the extremities lie between skeletal muscles, so when the muscles contract, the veins are compressed. This external venous compression decreases venous capacity and increases venous pressure, in effect squeezing fluid contained in the veins forward toward the heart. This pumping action, known as the **skeletal muscle pump** (p. 372), is one way extra blood stored in the veins is returned to the heart during locomotory activity. Increased muscular activity pushes more blood out of the veins and into the heart. Increased sympathetic activity and the resultant venous vasoconstriction also accompany locomotory activity, further enhancing venous return.

This also helps overcome the effects of gravity, which tends to cause blood to pool in the lower extremities of tall animals.

Effect of Venous Valves on Venous Return

Venous vasoconstriction and external venous compression both drive blood in the direction of the heart. Yet if you squeeze a fluid-filled tube in the middle, fluid is pushed in both directions from the point of constriction (● Figure 9–62a). Why, then, isn't blood driven backward as well as forward by venous vasoconstriction and the skeletal muscle pump? Blood can only be driven forward because the large veins are equipped with one-way valves spaced at 2- to 4-cm intervals; these valves permit blood to move forward toward the heart but prevent it from moving back toward the tissues (Figure 9–62b). These venous valves also play a role in counteracting the gravitational effects just noted.

Effect of Respiratory Activity on Venous Return

As a result of respiratory activity, the pressure within the mammalian chest cavity averages 5 mm Hg less than atmospheric pressure. As the venous system returns blood to the heart from the lower regions of the body, it travels through the chest cavity, where it is exposed to this subatmospheric pressure. Because the venous system in the limbs and abdomen is subjected to normal atmospheric pressure, an externally applied pressure gradient exists between the lower veins (at atmospheric pressure) and the chest veins (at 5 mm Hg less than atmospheric pressure). This pressure difference squeezes blood from the lower veins to the chest veins, promoting increased venous return. This mechanism of facilitating venous return is known

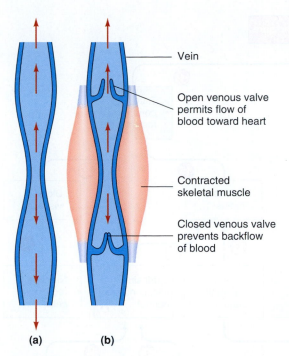

(a)   **(b)**

Figure 9–62 ● **Function of venous valves.** (a) When a tube is squeezed in the middle, fluid is pushed in both directions. (b) Venous valves permit the flow of blood only toward the heart.

— Vein

— Open venous valve permits flow of blood toward heart

— Contracted skeletal muscle

— Closed venous valve prevents backflow of blood

as the **respiratory pump** because it results from respiratory activity. Increased respiratory activity as well as the effects of the skeletal muscle pump and venous vasoconstriction all enhance venous return during locomotory activity.

Effect of Cardiac Suction on Venous Return

The extent of cardiac filling may not depend entirely on factors affecting the veins. The heart may play a role in its own filling, as we noted earlier, by a suction effect. This would also enhance venous return. Biologists have traditionally considered that suction is the dominant force in heart filling in fishes, whereas venous pressure is the primary force in mammals. However, recent studies show that venous return pressure is probably dominant in fishes and perhaps in all vertebrates, so that suction may be a minor effect.

Integrated Cardiovascular Function

We have now examined the three major components—fluid, pumps and vessels—of a circulatory system. We end this chapter by examining how the components are integrated into a smoothly functioning delivery system.

■ **Regulation of gas transport and mean arterial blood pressure is accomplished by controlling cardiac output, total peripheral resistance, and blood volume.**

Recall the crucial hemodynamic flow law (p. 392):

$$F = C.O. = \Delta P/R \quad \text{or} \quad \Delta P = C.O. \times R$$

A vertebrate can rapidly alter two components of the equation: (1) cardiac output (*C.O.*), by varying heart rate and stroke volume, and (2) resistance (*R*), primarily by changing arteriole diameter. Also, over longer periods of time, (3) blood volume can be regulated, which alters ΔP. These control features are aimed at regulating these two key features of circulation:

■ *Gas Transport.* When a mammal is at rest, *C.O.* must be homeostatically regulated to maintain consistent delivery of oxygen to all organs (and, CO_2 must be removed). During activity, *C.O.* is no longer homeostatic, but must be reset higher to boost delivery to muscles. Regulation of *C.O.*, especially of the portion to the brain, heart, and lungs, is the first priority of the cardiovascular system, because oxygen is so crucial to short-term survival.

■ *Arterial Blood Pressure.* The arterial blood pressure is the main driving force for propelling blood to the tissues. This pressure is somewhat homeostatically regulated, for three reasons. First, it must be high enough to ensure sufficient driving pressure against gravity, friction, and other resistance factors; without this pressure, the brain and other tissues do not receive adequate flow, no matter what local adjustments are made in the resistance of the arterioles supplying them. For example, if diastolic pressure drops to 50 mm Hg or less, blood cannot overcome gravity to reach the top of the human brain. Thus pressure is regulated to protect *C.O.* to the brain, the first priority of the cardiovascular system. Second, pressure must be high enough for ultrafiltration in the kidneys; in mammals, a minimum average pressure of about 80 mm Hg is needed for this. Third, the pressure must not be so high that it creates extra work for the heart and increases the risk of vascular damage and possible rupture of small blood vessels. Thus during activity when *C.O.* must be reset higher, total *R* in the body must be lowered to keep the pressure from rising excessively. However, we stated that this is *somewhat* homeostatic, because pressures can vary without harm over a moderate range.

Avoiding high pressure is a second priority for the cardiovascular regulation, with gas transport being the first priority. This can be seen in the common disease *hypertension* (chronic high blood pressure) of some mammals such as humans and cats. (Factors causing hypertension are often uncertain, but kidney failure is a major cause in cats.) Excessive resistance *R* in these animals impedes flow, and the cardiovascular integrator in the medulla oblongata could, in theory, lower the *C.O.* to keep the ΔP normal. But it does not. It allows ΔP to rise so that *C.O.* remains normal. (Examine the flow law to make sure you understand these physical interactions.) Presumably this response evolved because low *C.O.* (and thus low oxygen delivery) can be lethal within minutes, whereas high blood pressure may take years to kill.

Although gas transport is a higher priority, pressure must be regulated within a safe range once transport demands are met. Elaborate mechanisms involving the integrated action of the various components of the circulatory system and other body systems are dedicated to regulating mean arterial pressure. Let's work our way through ● Figure 9–63, reviewing all the factors that affect mean arterial blood pressure. Even though we've covered all these factors before, it is useful to pull them all together. The circled numbers in the text corre-

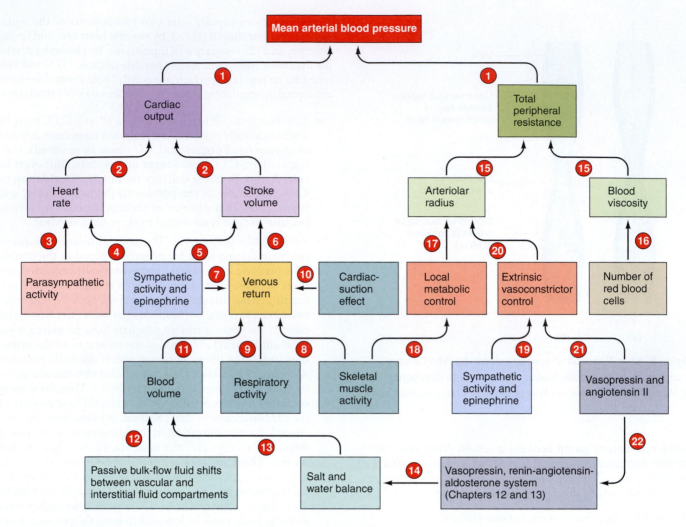

Figure 9–63 ● **Determinants of mean arterial blood pressure.** Note that this figure is basically a composite of Figure 9–28, p. 388, "Control of cardiac output"; Figure 9–48, p. 408, "Factors affecting total peripheral resistance"; and Figure 9–61, p. 418, "Factors that facilitate venous return." See the text for a discussion of the circled numbers.

spond to the numbers in the figure and indicate the portion of the figure being discussed.

- Mean arterial pressure depends on cardiac output and total peripheral resistance (①) on Figure 9–63); that is, $\Delta P = C.O. \times R$.
- Cardiac output depends on heart rate and stroke volume ②, both of which increase during locomotory activity.
- Heart rate depends on the relative balance of parasympathetic activity ③, which decreases heart rate, and sympathetic activity (tacitly including epinephrine throughout this discussion) ④, which increases heart rate.
- Stroke volume increases in response to sympathetic activity during locomotory activity ⑤ (extrinsic control of stroke volume).
- Stroke volume also increases as venous return increases during locomotory activity ⑥ (intrinsic control of stroke volume by means of the Frank-Starling law of the heart).

- Venous return is enhanced during locomotory activity by sympathetically induced venous vasoconstriction ⑦, the skeletal-muscle pump ⑧, the respiratory pump ⑨, and cardiac suction ⑩.
- The effective circulating blood volume also influences how much blood is returned to the heart ⑪. The blood volume depends in the short term on the magnitude of passive bulk-flow fluid shifts between the plasma and interstitial fluid across the capillary walls ⑫. In the long term, the blood volume depends on salt and water balance ⑬, which are hormonally controlled by the renin-angiotensin-aldosterone system and vasopressin, respectively ⑭.
- The other major determinant of mean arterial blood pressure, total peripheral resistance, depends on the radius of all arterioles as well as blood viscosity ⑮. The major factor determining blood viscosity is the number of red blood cells ⑯. However, arteriolar radius is the more important factor determining total peripheral resistance.

- Arteriolar radius is influenced by local (intrinsic) metabolic controls that match blood flow with metabolic needs ⑰. For example, local changes that take place in active skeletal muscles cause local arteriolar vasodilation and increased blood flow to these muscles ⑱.
- Arteriolar radius is also influenced by sympathetic activity ⑲, an extrinsic control mechanism that causes arteriolar vasoconstriction ⑳ to increase total peripheral resistance and mean arterial blood pressure.
- Arteriolar radius is also extrinsically controlled by the hormones vasopressin and angiotensin II, which are potent vasoconstrictors ㉑ as well as being important in salt and water balance ㉒.

Altering any of the pertinent factors produces a change in blood pressure, unless a compensatory change in another variable keeps the blood pressure within the safe range. Blood flow to any given tissue depends on the driving force of the mean arterial pressure and on the degree of vasoconstriction of the tissue's arterioles. Because mean arterial pressure depends on the cardiac output and the degree of arteriolar vasoconstriction, if the arterioles in one tissue dilate, the arterioles in other tissues will have to constrict to maintain an adequate arterial blood pressure to provide a driving force to push blood not only to the vasodilated tissue but also to the brain, heart, and lungs. Thus the cardiovascular variables must be continuously juggled to maintain a consistent blood pressure despite tissues' varying needs for blood.

Gas transport and blood pressure are monitored by arterial sensors.

Delivery of oxygen and removal of CO_2 are monitored by arterial **chemosensors** in the carotid and aortic arteries. They are sensitive to low O_2 or high acid levels in the blood. These chemoreceptors' main function is to reflexly increase respiratory activity to bring in more O_2 or to blow off more acid-forming CO_2, but they also reflexly increase blood pressure by sending excitatory impulses to the cardiovascular center. Regulation using these chemosensors is discussed in more detail in Chapter 11 (on respiration).

Here we focus mainly on pressure regulation. Mean arterial pressure is constantly monitored by **baroreceptors** (pressure sensors) in certain arteries. When deviations from normal are detected, multiple reflex responses are initiated to return the arterial pressure to its normal value. Recall the three factors that can be regulated. *Short-term* (within seconds) adjustments are accomplished by alterations in (1) cardiac output and (2) total peripheral resistance, mediated by means of autonomic nervous system influences on the heart, veins, and arterioles. *Long-term* (requiring minutes to days) control involves adjusting (3) total blood volume by restoring normal salt and water balance through mechanisms that regulate urine output and thirst (Chapters 12 and 13). The magnitude of the total blood volume, in turn, has a profound effect on cardiac output and mean arterial pressure. Let us now turn our attention to the short-term mechanisms involved in the ongoing regulation of this pressure.

The baroreceptor reflex is the most important mechanism for short-term regulation of blood pressure.

Any change in mean blood pressure triggers an autonomically mediated **baroreceptor reflex** that influences the heart and blood vessels to adjust cardiac output and total peripheral resistance in an attempt to restore blood pressure to normal. Like any reflex, the baroreceptor reflex includes a receptor, an afferent pathway, an integrating center, an efferent pathway, and effector organs.

The most important vertebrate receptors involved in moment-to-moment regulation of blood pressure are **aortic arch baroreceptors** (in amphibians, reptiles, birds, and mammals) and the **carotid sinus** (in mammals only). These are mechanoreceptors sensitive to changes in both mean arterial pressure and pulse pressure. Their responsiveness to fluctuations in pulse pressure enhances their sensitivity as pressure sensors, because small changes in systolic or diastolic pressure may alter the pulse pressure without changing the mean pressure. These baroreceptors are strategically located (● Figure 9–64) to provide critical information about arterial blood pressure in the vessels leading to the brain (the carotid sinus baroreceptor), for protecting blood flow to the brain, and in the major arterial trunk before it gives off branches that supply the rest of the body (the aortic arch baroreceptor), particularly important for protecting blood flow to the heart. The baroreceptors constantly provide information about blood pressure; in other words,

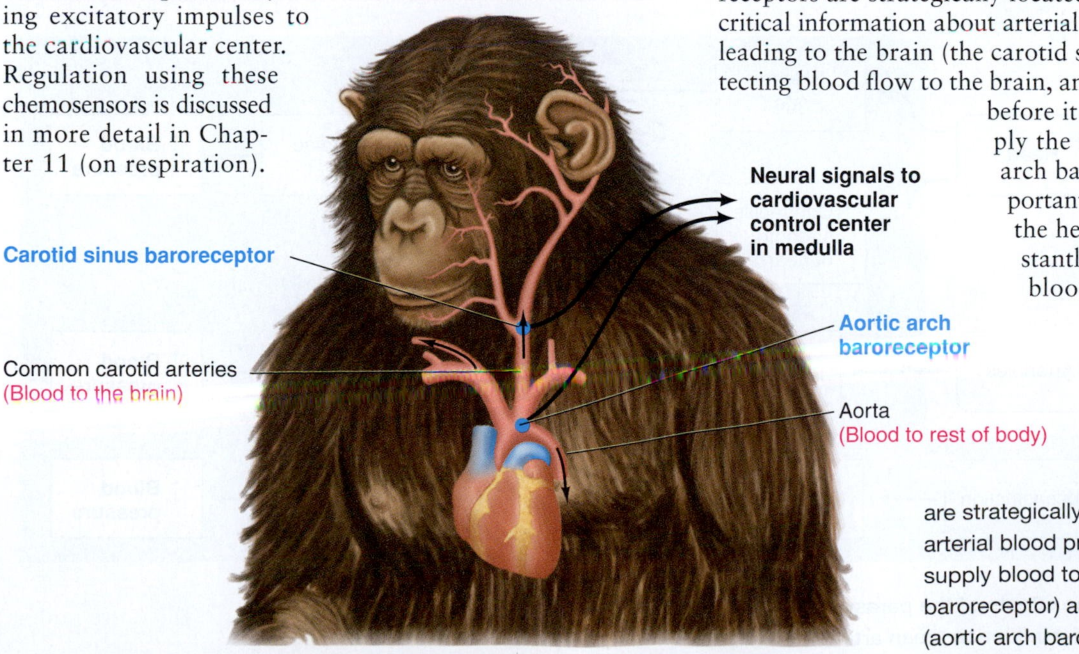

Neural signals to cardiovascular control center in medulla

Carotid sinus baroreceptor

Common carotid arteries
(Blood to the brain)

Aortic arch baroreceptor

Aorta
(Blood to rest of body)

Figure 9–64 ●
Location of the arterial baroreceptors. The arterial baroreceptors are strategically located to monitor the mean arterial blood pressure in the arteries that supply blood to the brain (carotid sinus baroreceptor) and to the rest of the body (aortic arch baroreceptor).

they continuously generate action potentials in response to the ongoing pressure within the arteries. When arterial pressure (either mean or pulse pressure) increases, the receptor potential of these baroreceptors increases, thus increasing the rate of firing in the corresponding afferent neurons. Conversely, when blood pressure decreases, the rate of firing generated in the afferent neurons by the baroreceptors decreases (● Figure 9–65).

The integrating center that receives the afferent impulses about the state of arterial pressure is the **cardiovascular control center,** located in the medulla within the brain stem. The efferent pathway is the autonomic nervous system. The cardiovascular control center alters the ratio between sympathetic and parasympathetic activity to the effector organs (the heart and blood vessels). To show how autonomic changes alter arterial blood pressure, ● Figure 9–66 provides a review of the major effects of parasympathetic and sympathetic stimulation on the heart and blood vessels.

Let's fit all the pieces of the baroreceptor reflex together now by tracing the reflex activity that occurs to compensate for an elevation or fall in blood pressure. Imagine that for some reason arterial pressure rises above normal (● Figure 9–67a); this could happen when a giraffe lowers its head to drink, as gravity now adds to the blood pressure going to the brain, rather than reducing it. The carotid sinus baroreceptors increase the rate of firing in their respective afferent neurons. On being informed by increased afferent firing that arterial pressure has become too high, the cardiovascular control center responds by decreasing sympathetic and increasing parasympathetic activity to the cardiovascular system. These efferent signals decrease heart rate, decrease stroke volume, and pro-

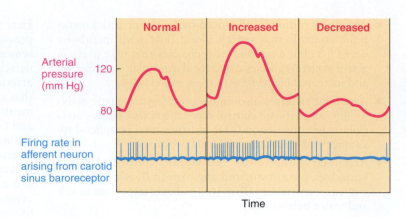

Figure 9–65 ● Firing rate in the afferent neuron from the carotid sinus baroreceptor in relation to the magnitude of mean arterial pressure.

duce arteriolar and venous vasodilation, which in turn lead to a decrease in cardiac output and a decrease in total peripheral resistance, with a subsequent decrease in blood pressure back toward normal.

Conversely, when blood pressure falls below normal (Figure 9–67b), for example by gravity when an animal stands up, baroreceptor activity decreases, inducing the cardiovascular center to increase sympathetic cardiac and vasoconstrictor nerve activity while decreasing its parasympathetic output. In a giraffe raising its head from a lowered position, constriction of vessels in the lower body raises the pressure and diverts blood to the brain. This efferent pattern of activity leads to an increase

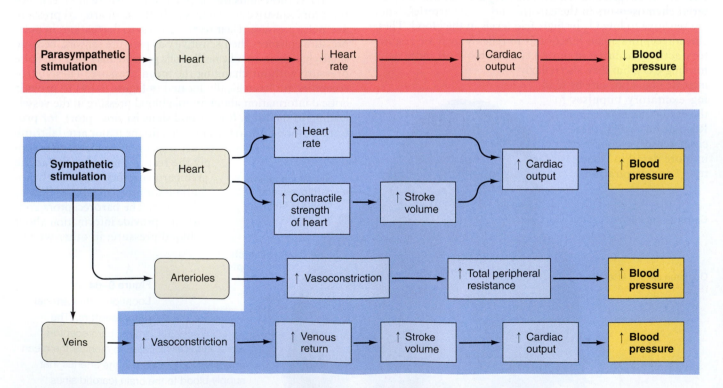

Figure 9–66 ● Summary of the effects of the parasympathetic and sympathetic nervous systems on factors that influence mean arterial blood pressure.

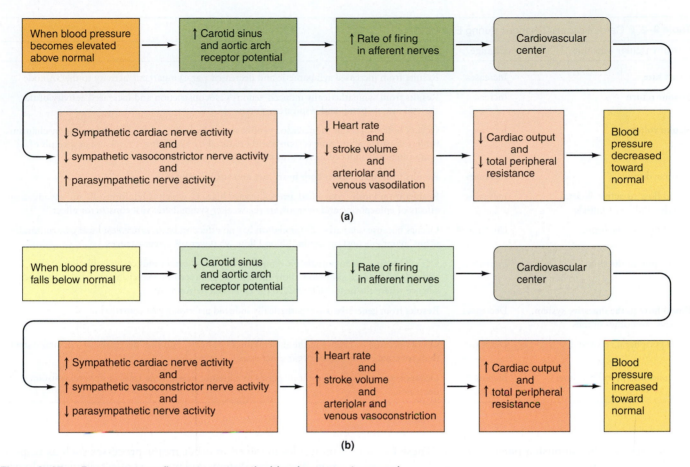

Figure 9–67 • Baroreceptor reflexes to restore the blood pressure to normal.
(a) Baroreceptor reflex in response to an elevation in blood pressure. (b) Baroreceptor reflex in response to a fall in blood pressure.

in heart rate and stroke volume coupled with arteriolar and venous vasoconstriction. These changes result in an increase in both cardiac output and total peripheral resistance, producing an elevation in blood pressure back toward normal.

■ Regulation of gas transport and blood pressure are coordinated during locomotory activity.

Now let us see how these two regulatory priorities, gas transport and ΔP, interact in a common event, that of increased locomotory activity. Pronounced cardiovascular changes accompany locomotory activity, with elevated gas transport being the first priority. This is accompanied by a substantial increase in cardiac output (*C.O.*), accompanied by an increase in skeletal muscle blood flow, by dilation of arterioles to skeletal muscles. This serves two purposes: transporting more oxygen to the organs that need it, and preventing the ΔP from rising too drastically by reducing total body resistance (*R*). (Using the flow law, consider what would happen in locomotory activity if *C.O.* increased fivefold, as it often does during fight or flight, with no change in *R*!) Overall, skeletal muscle arterioles ensure that there is a fall in total peripheral resistance (despite generalized arteriolar vasoconstriction in most organs). Thus during locomotory activity, there is only a modest increase in mean arterial pressure (■ Table 9–3). Evidence suggests that discrete

locomotory centers yet to be identified (probably in the motor cortex) induce the appropriate cardiac and vascular changes at the onset of locomotory activity. Indeed, these locomotory centers may boost *C.O.* and reduce *R* in *anticipation* of activity, thus eliminating the feedback delay problem we discussed in Chapter 1. (These changes occur even if the activity does not subsequently take place.) In the absence of anticipation, muscles would begin activity, depleting blood oxygen first. Only after the blood returns through the heart and lungs and back to the aortic sensors would the cardiovascular center "know" to change *C.O.* and *R*, by which time the muscles would have gone many seconds with low O_2. Anticipation prevents this from happening.

These effects are then reinforced by afferent inputs to the medullary cardiovascular center from chemoreceptors in exercising muscles as well as by local mechanisms important in maintaining vasodilation in active muscles. The baroreceptor reflex further modulates these cardiovascular responses.

Besides the baroreceptor reflex, whose sole function is blood pressure regulation, and cardiovascular regulation during locomotory activity, several other reflexes and responses influence the cardiovascular system even though they primarily are concerned with the regulation of other body functions. Some of these other influences deliberately move arterial pressure away from its normal value temporarily, overriding the baro-

Table 9–3 ■ Cardiovascular Changes during Exercise

Cardiovascular Variable	Change	Comment
Heart rate	Increases	Results from increased sympathetic and decreased parasympathetic activity to the SA node
Venous return	Increases	Results from sympathetically induced venous vasoconstriction and increased activity of the skeletal muscle pump and respiratory pump
Stroke volume	Increases	Occurs both as a result of increased venous return by means of the Frank-Starling mechanism (unless diastolic filling time is significantly reduced by a high heart rate) and as a result of a sympathetically induced increase in myocardial contractility
Cardiac output	Increases	Results from increases in both heart rate and stroke volume
Blood flow to active skeletal muscles and heart muscle	Increases	Results from locally controlled arteriolar vasodilation, which is reinforced by the vasodilatory effects of epinephrine and overpowers the weaker sympathetic vasoconstrictor effect
Blood flow to the brain	Unchanged	Occurs because sympathetic stimulation has no effect on brain arterioles; local control mechanisms maintain constant cerebral blood flow whatever the circumstances
Blood flow to the skin	Increases	Occurs because the hypothalamic temperature control center induces vasodilation of skin arterioles; increased skin blood flow brings heat produced by exercising muscles to the body surface where the heat can be lost to the external environment
Blood flow to the digestive system, kidneys, and other organs	Decreases	Results from generalized sympathetically induced arteriolar vasoconstriction
Total peripheral resistance	Decreases	Occurs because resistance in the skeletal muscles, heart, and skin decreases to a greater extent than resistance in the other organs increases
Mean arterial blood pressure	Increases (modest)	Occurs because cardiac output increases to a greater extent than total peripheral resistance decreases

receptor reflex to accomplish a particular goal. These factors include the following:

1. Volume receptors in the atria and hypothalamic osmoreceptors are primarily important in water and salt balance in the body; Thus they affect the *long-term regulation of blood pressure* by controlling the plasma volume (see Chapters 12 and 13).

2. Cardiovascular responses associated with certain behaviors and emotions are mediated through the cerebral cortex–hypothalamic pathway and appear to be preprogrammed. These responses include the widespread changes in cardiovascular activity accompanying the generalized sympathetic fight-or-flight response, the characteristic marked increase in heart rate and blood pressure associated with sexual orgasm, and the localized cutaneous vasodilation characteristic of blushing.

3. Hypothalamic control over cutaneous (skin) arterioles for the purpose of temperature regulation takes precedence over control that the cardiovascular center has over these same vessels for the purpose of blood pressure regulation. As a result, blood pressure can fall when the skin vessels are widely dilated to eliminate excess heat from the body, even though the baroreceptor responses are calling for cutaneous vasoconstriction to help maintain adequate total peripheral resistance (see Chapter 15).

4. Vasoactive substances released from the endothelial cells play a role in the regulation of blood pressure. For example, NO normally exerts an ongoing vasodilatory effect.

This concludes our examination of circulatory systems. However, circulation plays a vital role in all the following chapters, because it links all body systems and carries cells and molecules involved in other major processes such as respiration. Also, other systems and processes affect circulation. In particular, as we just noted, long-term control of blood pressure involves maintaining proper plasma volume by controlling salt and water balance.

Chapter in Perspective:
HOMEOSTASIS AND INTEGRATION

Homeostasis depends on continual delivery of needed supplies to all the cells throughout the body and on ongoing removal of wastes generated by the cells. Furthermore, regulatory chemical messengers, such as hormones, must be transported from their site of production to their site of action, where they control a variety of activities, most of which are directed toward maintaining a stable internal environment.

Circulatory systems contribute to homeostasis by serving as a body's transport system, linking all organs systems together. Circulatory systems provide a means of rapidly moving heat and materials—oxygen, nutrients, hormones, wastes, and cells—from one part of the body to another. Without these systems, materials would not get where they need to go nearly rapidly enough. For example, O_2 would take months to years to diffuse from the surface of a large animal to internal organs, but with a circulatory system O_2 is continually picked up by the circulation in lungs or gills and constantly delivered to all the body cells at a life-sustaining rate.

Blood or hemolymph has special transport capabilities that enable it to move its cargo efficiently throughout the body. For example, O_2 is poorly soluble in water, but circulatory fluids may be equipped with O_2-carrying proteins such as hemoglobin, often in erythrocytes (red blood cells). Likewise, homeostatically important water-insoluble hormonal messengers are shuttled in the blood by plasma protein carriers.

Specific components of blood or hemolymph perform the following additional homeostatic activities that are unrelated to transport function:

- Circulatory fluid helps maintain the proper pH in the internal environment by buffering changes in the acid–base load of the body.

- It helps maintain body temperature in endotherms by absorbing heat produced by heat-generating tissues and distributing it throughout the body, or carrying it to the body surface for elimination to the external environment.

- The electrolytes in the plasma are important in membrane excitability, which in turn forms the basis of nerve and muscle function.

- The electrolytes in the plasma are also important in the osmotic distribution of fluid between the extracellular and intracellular fluid, and the plasma proteins play a critical role in the distribution of extracellular fluid between the plasma and interstitial fluid.

- Through their hemostatic functions, clotting factors and cells minimize the loss of life-sustaining blood following vessel injury.

- Circulating immune cells, their secretory products, and certain types of plasma proteins, such as antibodies, constitute immune defense systems.

Circulation usually requires a pump such as a heart. Hearts often exhibit their own intrinsic homeostasis, such as the ability to beat continuously without external signals. Local mechanisms within the vertebrate heart ensure that blood flow to the cardiac muscle normally meets the heart's need for O_2. In addition, the heart has built-in capabilities to vary its strength of contraction, depending on the amount of blood returned to it. The vertebrate heart does not act entirely autonomously, however. It is innervated by the autonomic nervous system and is influenced by the hormone epinephrine, both of which can vary the rate and contractility of the heart, depending on the body's needs for blood delivery. ■

REVIEW QUESTIONS *(Answers are on p. A–1.)*

Additional study tools for this chapter, including chapter summaries and practice tests, are available online at *www.biology.brookscole.com*

1. Which of the following is not a cellular element of vertebrate blood?
   a. erythrocytes
   b. platelets
   c. thrombocytes
   d. leukocyte
   e. plasma

2. Erythropoiesis
   a. is the process of producing new red blood cells by the lung
   b. stimulates production of erythropoietin by the kidneys
   c. lowers oxygen levels in the kidneys
   d. is the removal of old red blood cells by the spleen
   e. is the process of producing new red blood cells in red bone marrow

3. In blood clotting, the conversion of fibrinogen to fibrin is catalyzed by the enzyme
   a. fibrin-stabilizing factor
   b. prothrombin
   c. thrombin
   d. collagen
   e. tissue thromboplastin

4. Blood clots are slowly dissolved by
   a. plasminogen
   b. PDGF
   c. Hageman factor
   d. plasmin
   e. platelet factor 3

5. In mammals, blood entering the right atrium
   a. is partially deoxygenated
   b. is fully oxygenated
   c. flows directly to the left ventricle
   d. is pumped directly into the aorta
   e. is pumped directly into the left atrium

6. Adjacent cardiac muscle cells are joined end-to-end by
   a. endocardium
   b. myocardium
   c. semilunar valves
   d. intercalated discs
   e. epicardium

7. Which of the following ions play a major role in cardiac muscle physiology?
   a. Ca^{++}
   b. I^+
   c. Na^+
   d. K^+
   e. a, c, and d

8. Which of the following applies to an ECG (EKG)?
   a. The ECG remains at baseline during the PR segment.
   b. The P wave represents atrial depolarization.
   c. The QRS complex represents ventricular depolarization.
   d. The TP interval occurs when the heart muscle is completely at rest.
   e. all of the above

9. In the mammalian cardiac cycle, stroke volume
   a. causes rupturing of vessels
   b. is also known as end-systolic volume
   c. is also known as end-diastolic volume
   d. is equal to end-diastolic volume minus end-systolic volume
   e. is equal to atrial diastolic volume minus ventricular systolic volume
10. Cardiac output, C.O. =
   a. heart rate × atrial diastolic volume
   b. heart rate × stroke volume
   c. heart rate × ventricular diastolic volume
   d. heart rate × ventricular diastolic volume minus atrial systolic volume
   e. heart rate × twice atrial volume
11. Sympathetic stimulation of the mammalian heart
   a. decreases heart rate
   b. weakens atrial contraction
   c. increases heart rate
   d. increases the delay between atrial and ventricular contraction
   e. slows down the spread of the action potential throughout the conduction pathway
12. The Frank-Starling law of the heart states that
   a. increased venous return results in decreased stroke volume
   b. increased venous return results in increased stroke volume
   c. stroke volume is determined by sympathetic activity
   d. stroke volume is always slightly less than venous return volume
   e. stroke volume always exceeds venous return volume
13. The vertebrate heart muscle receives most of its oxygen from
   a. blood within the atria
   b. blood within the ventricles
   c. the AV node
   d. coronary circulation
   e. pulmonary arteries
14. In the hemodynamic equation $F = \Delta P/R$
   a. ΔP = pressure at the outflow end of a vessel
   b. ΔP = pressure at the inflow end of a vessel
   c. ΔP = the pressure gradient
   d. R = the flow rate minus the pressure gradient
   e. F = resistance of blood vessels minus the pressure gradient
15. Which of the following serve as primary resistance vessels and determine the distribution of cardiac output?
   a. capillaries
   b. venae cavae
   c. large arterial branches
   d. arterioles
   e. large veins

SUGGESTED READINGS AND INTERNET SITES

Alexander, R. M. 1982. *Locomotion of Animals.* London: Blackie.

Dukes, H. H., M. J. Swenson, & W. O. Reece. 1993. *Dukes' Physiology of Domestic Animals.* Comstock. Ithaca, NY: Cornell University Press.

Geddes, L. A. 2002. Electrocardiograms from the turtle to the elephant that illustrate interesting physiological phenomena. *Pacing and Clinical Electrophysiology* 25:1762–1770.

Goetz, R. H., J. V. Warren, O. Gauer, et al. 1960. Circulation of the giraffe. *Circulation. Research* 8:1049–57.

Iwama, G. K., & A. P. Farrell. 1998. Fish diseases and disorders. In J. F. Leatherland & P. T. K. Woo, eds., *Non-Infectious Disorders,* vol. 2. Cambridge, MA: CAB International.

Kampmeier, O. F. 1969. *Evolution and Comparative Morphology of the Lymphatic System.* Springfield, MA: C. C. Thomas.

Liberatore, G. T., A. Samson, C. Bladin, W.-D. Schleuning, & R. L. Medcalf. 2003. Vampire bat salivary plasminogen activator (desmoteplase): A unique fibrinolytic enzyme that does not promote neurodegeneration. *Stroke* 34:537–543.

McGaw, I. J., & C. L. Reiber. 2002. Cardiovascular system of the blue crab *Callinectes sapidus. Journal of Morphology* 251:1–21.

Olson, K. R. 1996. Secondary circulation in fish: Anatomical organization and physiological significance. *Journal of Experimental Zoology* 275:172–185.

Pearse, V. J., M. Buchsbaum, & R. Buchsbaum. 1987. *Living Invertebrates.* Palo Alto, CA: Blackwell Scientific.

Pollitt, C. C. 1992. Clinical anatomy and physiology of the normal equine foot. *Equine Veterinary Education* 4:219–224.

Satchell, G. H. 1991. *Physiology and Form of Fish Circulation.* New York: Cambridge University Press.

Sherman, I. W. & Sherman, V. G. 1976. *The Invertebrates: Function and Form,* 2nd ed. New York: Macmillan.

Vogel, S. 1993. *Vital Circuits: On Pumps, Pipes, and the Workings of Circulatory Systems.* Oxford, UK: Oxford University Press.

Withers, P. C. 1992. *Comparative Animal Physiology.* Fort Worth, TX: Saunders College Publishing.

INTERNET SITES

Franklin, C. E., & M. Axelsson. 2000. An actively controlled heart valve. *Nature* 406:847. **abc.net.au/science/news/space/SpaceRepublish_167223.htm.**

Cardiac Arrhythmias in the Horse. **tufts.edu/vet/sports/cardiac.html.**

Friedman, L., A. Pfenniger, & M. Sleeper. 2004. *Small Animal Cardiology.* **caltest.vet.upenn.edu/smcardiac/.**

Horst, J., & V. Reef. 1999. *Large Animal Cardiology.* **cal.nbc.upenn.edu/lgcardiac/index.html.**

The last three sites are about ECG usage in domestic animals, with case studies for horses, cows, and household pets.

INFOTRAC READINGS

Choy, D. S. J., & P. Altman. 1992. The cardiovascular system of barosaurus: An educated guess. *The Lancet* 340:534–536.

Terwilliger, N. B. 1999. Hemolymph proteins and molting in crustaceans and insects. *American Zoologist* 39:589.

CHAPTER

10

Defense Systems

Hagfish, a primitive jawless vertebrate. These and the other jawless fishes, the lampreys, have innate immunity similar to that found in all animals, but lack the adaptive immune system (which uses antibodies) found in all other vertebrates.

Evolution of Defense Systems

Attack by other organisms has probably been a major feature of life for billions of years (long before animals evolved). Various defense mechanisms have been evolving all that time in a process called *coevolution:* Adaptive changes in one organism favor selection for adaptations in a second organism, which in turn favor selection for a different adaptation in the first organism, and so on. In the context of defense, this is sometimes called an evolutionary "arms race." For example, an animal may evolve a defense mechanism against a predator or pathogen (an invader such as a disease-causing bacterium), but this provides only a temporary respite in evolutionary terms. The antagonistic species after many generations may evolve a way to overcome this defense. In turn, the first species may evolve yet another improvement to its defenses. Thus, over time both species can evolve very sophisticated offenses and defenses.

Animal defense systems fall into two broad categories. The first deals with external antagonists, primarily predators, and competitors. These defenses may be *anatomic* features such as the spines of a porcupine, *chemical* agents such as the nerve inhibitor tetrodotoxin (p. 111) in the skin of a pufferfish, behavioral programs such as *escape responses* (for example, the rapid tail-flip of a lobster), and *reproductive* strategies such as the ability of coral to breed rapidly following decimation by attacking seastars. We will not consider this category of defenses any further.

The second type, and the focus of this chapter, is internal defense, called **immunity,** which is the ability to resist or eliminate potentially harmful foreign materials or abnormal cells inside or attached parasitically to the surface of an animal. The following activities are attributable to *immune systems,* which play a key role in stopping, recognizing, and either destroying or neutralizing materials that are foreign to the "normal self":

1. Defense against invading **pathogens** (disease-producing viruses and bacteria, and **parasitic** fungi, protozoa, and smaller animals).

2. Removal of "worn-out" cells (such as aged red blood cells) and tissue debris (for example, tissue damaged by trauma or disease). Such removal of tissue debris is essential for wound healing and tissue repair.

3. Identification and destruction of abnormal or mutant cells that have originated in the body. This function, termed *immune surveillance,* is the primary internal defense mechanism against cancer.

Effective immune systems are so important that they can influence mate choice behavior in some animals. In many birds, for example, red and yellow pigment molecules called **carotenoids** (from plants) have two functions: They provide colors for male beaks and feathers, which are used to attract females, and they boost the immune system of the birds. These are not unrelated functions, because studies in 2003 revealed a direct linkage between plumage color and immune function in male zebra finches. Males given extra dietary carotenoids had more brightly colored beaks and faster-responding immune systems compared to control males; moreover, females were more likely to mate with them than with control males (which had drabber beaks). Thus a male's plumage signals the female how healthy he is, clearly a major factor in producing healthy offspring. Of course, it is very unlikely that the female consciously perceives the male's health; rather, females in the past who were predisposed to pick the most brightly colored males produced healthier offspring, spreading the genes for color preference in mating.

Pathogenic bacteria, viruses, and internal parasites are the major targets of the immune defense system.

The primary foreign enemies against which the immune system defends are disease-causing biological invaders called *pathogens*. The disease-producing power of a pathogen is known as its **virulence**. Pathogenic **bacteria** that invade the body induce tissue damage and produce disease largely by releasing enzymes or toxins that physically injure or functionally disrupt affected cells and organs. The disruption may release nutrients for bacterial growth. Internal **parasites** are larger eukaryotic cells such as protozoa, yeast, and multicellular fungi and animals, which take up residence inside a larger animal (called the **host**) and use the latter's nutrients for growth and reproduction. Some parasites such as tapeworms may cause harm mostly through nutrient depletion or physical damage to host cells, but others also release damaging chemical agents. The distinction between bacterial and parasitic pathogens is thus somewhat arbitrary.

In contrast to these living cells, **viruses** are not self-sustaining cellular entities. They consist only of nucleic acids (DNA or RNA) enclosed by a protein coat. Because they lack cellular machinery for energy production and protein synthesis, viruses cannot carry out metabolism and reproduce unless they invade a **host cell** (a body cell of the infected individual) and take over the cellular biochemical facilities for their own purposes. Not only do viruses sap the host cell's energy resources, but the viral nucleic acids also direct the host cell to synthesize proteins needed for viral replication. When a virus becomes incorporated into a host cell, the host defense system may destroy the cell, because it is no longer recognized as a "normal self" cell. Other ways in which viruses can lead to cell damage or death are by depleting essential components, dictating that the cell produce substances that are toxic to the cell, or transforming the cell into a cancer cell.

In most vertebrates, immune responses can be either innate or acquired.

Protective immunity in most vertebrates is conferred by the complementary actions of two separate but interdependent components: the *innate immune system* (which has many separate features) and the *adaptive* or *acquired immune system*. The responses of these two systems differ in their timing and in the degree of selectivity of the defense mechanisms employed. **Innate immune systems** encompass an animal's *nonspecific* immune responses that come into play immediately on exposure to any threatening agent. These nonspecific responses are inherent (innate or built-in) defense mechanisms that defend against foreign or abnormal material of any type, even on initial exposure to it. Such responses provide defense against a wide range of threatening factors, including infectious agents, chemical irritants, and tissue injury accompanying mechanical trauma and burns. All invertebrates as well as vertebrates have innate immunity mechanisms, which include **barrier tissues** such as the integumentary (skin) and digestive linings, and **phagocytes**, cells that engulf and destroy foreign materials.

The **adaptive** or **acquired immune system,** in contrast, relies on *specific* immune responses selectively targeted against a particular foreign material to which a body has already been exposed and has had an opportunity to prepare for an attack aimed discriminatingly at the enemy. The adaptive immune system thus takes considerably more time to be mounted and takes on one specific foe at a time. As far as we know, this mechanism has only evolved in vertebrates with jaws (that is, all but the jawless fishes, the lampreys and hagfish) and possibly a few invertebrates. Although often called "adaptive" immunity, this is technically a type of *acclimatization* (Chapter 1)—useful prolonged changes in a body triggered in response to a new situation. Many biologists feel the term *adaptation* should be reserved for evolutionary changes over many generations. However, tradition has saddled us with this term for acquired immunity.

The innate and adaptive immune components in vertebrates work in harmony to contain, and then eliminate, harmful agents. The components of the innate system are always on guard as a limited, sometimes crude, repertoire of defense mechanisms against most invaders. Active nonspecific immune responses (as opposed to the passive barrier tissues) are set in motion in response to generic molecular patterns associated with threatening agents, such as the carbohydrates typically found in bacterial cell walls. These innate mechanisms provide a rapid (but nonselective) response to unfriendly challenges of all kinds. Innate immunity largely contains and limits the spread of infection. These nonspecific responses provide the only immunity in most animals, and work well enough for species survival most of the time (indeed, some researchers suspect that innate immunity systems of invertebrates are superior to those in vertebrates with acquired immunity). In jawed vertebrates, the innate defenses are important for keeping the foe at bay until the adaptive immune system, with its highly selective weapons, can be prepared to take over and set in motion strategies to completely eliminate the invader.

Immune responses in any organism depend on ability to distinguish "self" from "non-self."

Both acquired immunity and the active aspects of innate immunity depend on the ability of an organism to recognize its own tissues and to distinguish them from foreign materials. This *self-recognition* ability has ancient roots: Even cells of the most primitive metazoans, the sponges (Porifera), can distinguish "self" from foreign cells. Another ancient example, illustrated in ● Figure 10–1, has been called the "anemone clone wars." Some species of these cnidarians grow in distinct patterns: groups of densely packed sea anemones with barren areas between them. A researcher curious about this pattern discovered that each cluster is a clonal lineage, that is, all the anemones are genetically identical and arose from asexual budding starting with one individual. Furthermore, other clusters are also clones, each with their own unique genomes. The barren zones arise when one cluster grows close to another. By detecting unique molecules on the animals' surfaces, individuals on the edge of each clone recognize the others as "non-self," and they react violently. Using special stinging war tentacles called *acrorhagia*, the animals sting and try to drive off (or kill) the foreign anemones, leaving the rock barren in between. This process is essentially the same used

Photo: © Wayne and Karen Brown/Index Stock Imagery

Figure 10–1 ● **Anemone clone wars.** Conflicts occur between two colonies of sea anemones formed by asexual budding (cloning). When one colony member comes in contact with a member of the other colony, each recognizes the other as foreign and begins an attack. The result is an open patch of habitat between the two groups.

by internal immune systems to recognize invaders and is also similar to the process that causes rejection of organ transplants in humans.

Recognition of self and non-self is accomplished by special **pattern recognition receptors** (also called **pattern recognition proteins** or **PRPs**) in plasma membranes. In mammals, for example, many immune cells have **Toll-like receptors (TLRs)**, which one investigator in the field has called "the eyes of the innate immune system." These are named after **Toll receptors** in the fruit fly *Drosophila* in which these receptors participate in pattern recognition in both immunity and development. These sensor proteins recognize and bind with non-self macromolecules of pathogens. We investigate PRP/TLR roles in detail later in this chapter.

Some animals behaviorally acquire defense components from other organisms.

In addition to an animal's own innate and acquired immunity, there is another form of immunity in some species. This is the behavioral "borrowing" of defenses from other organisms (an example of a behavioral effector, p. 14). Humans are of course renowned for this—virtually every human culture makes use of medicinal compounds from plants, fungi, and other organisms. In most cases, the compounds, which aid human health, are defense compounds in the originating species. Aspirin, for example, was originally obtained from willow and is widely used to treat fever, inflammation, and pain in humans. It occurs in a wide variety of plants, where it appears to be involved in activating a plant's defense mechanisms when an infection occurs.

Other mammals behave much like humans. For example, chimpanzees in eastern Africa swallow whole leaves of the *Aspilia* plant (a relative of sunflowers). In contrast to leaves eaten for food, which are chewed first, these leaves are probably medicinal. They contain large amounts of *thiarubrine-A,*

which readily kills parasitic fungi and worms. Similarly, capuchin monkeys in Venezuela find noxious millipedes with antipredator compounds called *benzoquinones*. The monkeys smear mashed millipedes on their fur, where the noxious compounds repel biting mosquitoes.

Many birds also seem to make use of defense compounds from other organisms. Most birds that reuse their nests (which may easily become breeding grounds for parasites and bacteria) are seen to line the nests with branches and leaves from certain plants. Starlings, for example, carefully select specific (often rare) plants to line their nests during breeding season. Some of these plants have been tested and are found to contain volatile defense compounds which inhibit parasites such as ticks.

By far the most information on immune systems comes from mammalian research, and most of the basic concepts and terms have arisen out of this research. Therefore, we start with a detailed look at that and related vertebrate systems, before returning to look at other animals.

The Vertebrate Immune System

Although a few immune mechanisms are largely passive, such as the physical barrier formed by the skin, most are active responses that we would expect to be regulated by feedback systems (as are almost all biological processes). Many of these feedback processes occur in local tissues. But in recent years, immunologists have come to realize that some responses in vertebrates constitute sophisticated feedback loops coordinating many parts of the body.

Active immune responses are regulated by feedback systems with sensors, integrators, and effectors.

Indeed, some have proposed that the vertebrate immune system be considered a third "whole-body" regulatory system on a par with the neural and endocrine systems. Certainly the immune system has sensors, integrators, and effectors that communicate both locally and across the whole body. However, a single immune cell can act as more than one regulatory component (for example, as both effector and sensor), and thus the regulatory features are not as clear as they are in the neural and endocrine systems. Let us briefly examine these regulatory aspects of these components and communication mechanisms before turning to details.

For the most part, the active immune components are individual cells and extracellular proteins, rather than organs as in the neural and endocrine systems. Immune proteins include **complement** and **antibodies,** which are effectors that bind to and help destroy invaders. The cells responsible for the various immune defense strategies are the circulating **leukocytes** (white blood cells) and their noncirculating derivatives. Immune cells communicate with each other in two ways: (1) by direct contact with other cells or foreign objects through receptors in cell membranes (by what has been termed an "immunosynapse" when two immune cells are involved); and (2) by secretion and reception of "immunohormones" called **cytokines** (defined as signal molecules produced by nongland cells). Most of the cytokines in the immune system are called **interleukins (ILs).** These

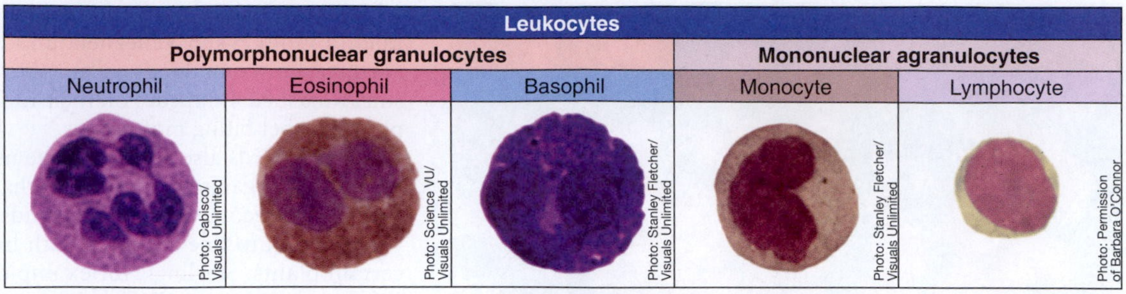

Leukocytes				
Polymorphonuclear granulocytes			**Mononuclear agranulocytes**	
Neutrophil	Eosinophil	Basophil	Monocyte	Lymphocyte

Figure 10–2 ● Normal blood immune cells of mammals.

are analogous to hormones, neurohormones, and neurotransmitters in the other major regulatory systems.

The key mammalian immune cells are briefly reviewed as follows (● Figure 10–2).

Circulating Leukocytes: Sensors, Integrators, and Effectors

- *Neutrophils* are highly mobile phagocytic specialists that engulf and destroy unwanted materials. They are effectors, because they correct the disturbance to homeostasis by eating those materials. But because they can often recognize invaders and destroy them without needing other regulatory commands, they are in some sense sensors, integrators, *and* effectors.

- *Eosinophils* secrete chemicals that destroy parasitic worms and are involved in allergic manifestations. Thus they are primarily effectors.

- *Basophils* release histamine, a trigger of inflammation, and heparin, and a clotting inhibitor. They also are involved in allergic manifestations. Because these chemicals trigger effectors, basophils can be considered sensor-integrators.

- *Monocytes* are transformed into *macrophages*, which are large, tissue-bound phagocytic specialists, acting as effectors or as nonspecific sensor-integrator-effectors such as neutrophils. However, macrophages can also act as sensors in specific immunity. After phagocytosing and digesting an invader, they display fragments of the invader, called *antigens*, on their membranes. They are thus called **antigen-presenting cells** or **APCs**. These displays are used to inform integrator lymphocytes of the attack.

- *B and T lymphocytes* (also called simply *B-* and *T-cells*):
  a. T lymphocytes come in several types. Some are effectors (*cytotoxic T-cells*), responsible for cell-mediated immunity involving both direct destruction of virus-invaded cells and mutant cells through nonphagocytic means. Others (especially *helper T-cells*) are the true integrators of the specific immune response, receiving input from sensor cells and activating effector cells.
  b. B lymphocytes are transformed into *plasma cells*, which secrete *antibodies* (effector proteins) that directly or indirectly lead to the destruction of foreign material. B-cells can be thought of as part of the effector response that kills invaders specifically.

- *Natural killer (NK) cells*, a special class of lymphocyte-like cells that spontaneously and relatively nonspecifically lyse (rupture) and thereby destroy virus-infected host cells

and cancer cells. These seem to act independently as sensor-integrator-effectors.

Noncirculating Leukocyte Derivatives: Sensors and Integrators

- *Mast cells* reside in barrier tissues (integumentary, digestive, respiratory) and are analogous to sensing-integrating endocrine glands. They can sense an invasion and secrete signal molecules such as histamine to activate effectors in the local area under attack. They are the *sensor-integrators* of local, nonspecific defenses.

- *Dendritic cells* also reside in barrier and many other tissues, and are now thought to be the key *sensors* in the specific defenses. They sense invaders, engulf them, and display pieces of them on their membranes. They are therefore APCs like the macrophages. Although initially noncirculating, once activated they become mobile and migrate to the lymph nodes and inform the integrating T-cells of the attack.

Although most leukocytes are circulating cells, that does not mean they are primarily in the blood itself. A given leukocyte is present in the blood only transiently. Most are out in the tissues on defense missions. As a result, the immune system's cells are widely dispersed throughout the body and can defend in any location.

Almost all leukocytes originate from common precursor stem cells in the bone marrow and are subsequently released into the blood. The only exception are lymphocytes, which arise in part from lymphocyte colonies in various lymphoid tissues that were originally populated by cells derived from the bone marrow (see p. 366). The term **lymphoid tissues** refers collectively to the tissues that store, produce, or process lymphocytes. These include the lymph nodes, spleen, thymus, tonsils, adenoids, appendix, bone marrow, and aggregates of lymphoid tissue in the lining of the digestive tract called **Peyer's patches** or **gut-associated lymphoid tissue (GALT)** (● Figure 10–3). Also, the various integumentary components of the immune system are collectively termed **skin-associated lymphoid tissue**, or **SALT**. Lymphoid tissues are strategically located to intercept invading microorganisms before they have a chance to spread very far. For example, the lymphocytes populating the *tonsils* and *adenoids* are situated advantageously to respond to inhaled microbes, whereas microorganisms invading through the digestive system immediately encounter lymphocytes in the *appendix* and *GALT*. Potential pathogens that gain access to the lymph are filtered through *lymph nodes*, where they are exposed to lymphocytes as well as to macrophages that line the

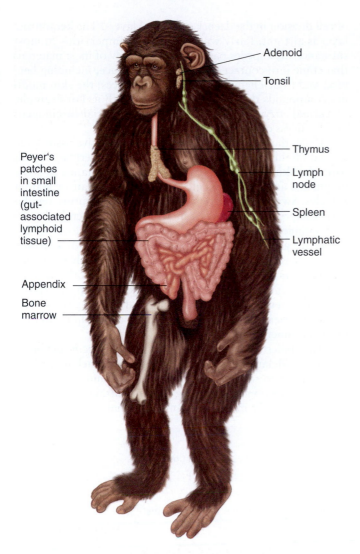

Peyer's patches in small intestine (gut-associated lymphoid tissue)

Appendix

Bone marrow

Adenoid

Tonsil

Thymus

Lymph node

Spleen

Lymphatic vessel

Figure 10–3 ● **Lymphoid tissues in a mammal.** The lymphoid tissues, which are dispersed throughout the body, produce, store, or process lymphocytes.

Table 10–1 ▮ Functions of Lymphoid Tissues in Vertebrates

Lymphoid Tissue	Function
Bone marrow	Origin of all blood cells
	Site of maturational processing for B lymphocytes
Lymph nodes tonsils, adenoids, appendix, gut-associated lymphoid tissue	Exchange lymphocytes with the lymph (remove, store, produce, and add them)
	Resident lymphocytes produce antibodies and sensitized T cells, which are released into the lymph
	Resident macrophages remove microbes and other particulate debris from the lymph
Spleen	Exchanges lymphocytes with the blood (removes, stores, produces, and adds them)
	Resident lymphocytes produce antibodies and sensitized T cells, which are released into the blood
	Resident macrophages remove microbes and other particulate debris, most notably worn-out red blood cells, from the blood
	Stores a small percentage of red blood cells, which can be added to the blood by splenic contraction as needed.
Thymus	Site of maturational processing for T lymphocytes
	Secretes the hormone thymosin

lymphatic passageways. The *spleen,* the largest of the lymphoid tissues, performs immune functions on the blood similar to those the lymph nodes perform on the lymph. Through actions of its lymphocyte and macrophage population, the spleen clears the blood that passes through it of microorganisms and other foreign matter and also removes worn-out red blood cells (see p. 365).

The *thymus* and *bone marrow* play important roles in processing T and B lymphocytes, respectively, to prepare them to carry out their specific immune strategies. ▮ Table 10–1 summarizes the major functions of the various lymphoid tissues.

We first examine in more detail the innate immune responses before turning our attention to adaptive immunity.

Innate Immunity

Innate nonspecific defenses in mammals that come into play whether or not there has been prior experience with the offending agent include the following:

1. *Barrier tissues and glands.* The barriers are the *skin, digestive tract, respiratory ducts,* and to some extent the *reproductive tract* (especially in females, who are more prone to infection this way than males).

2. *Inflammation,* a nonspecific response to tissue injury sensed and integrated by mast cells. The phagocytic specialists—neutrophils and macrophages—are the primary effectors.

3. *The complement system,* a group of inactive plasma proteins that, when sequentially activated, bring about destruction of foreign cells by attacking their plasma membranes.

4. *Interferon,* a family of effector proteins that nonspecifically defend against viral infection.

5. *Natural killer cells* that destroy virus-infected host cells and cancer cells.

6. *Symbiotic bacteria.* In the digestive system, on the skin, and in the female reproductive tract reside countless numbers of resident bacteria. For the most part these do no detectable harm and are often beneficial (for example, some mammalian colon bacteria synthesize vitamin K; other gut bacteria are crucial to digestion in ruminants; p. 664). One benefit is that they can inhibit the growth of invaders, either by outcompeting them or by secreting defense chemicals. For example, "friendly" bacteria that live on the back of the human tongue convert food-derived nitrate into nitrite, which is swallowed. Acidification of nitrite on reaching the highly acidic stomach generates nitric oxide, which is toxic to a variety of microorganisms.

We now examine these defenses in detail.

▪ Barrier tissues are the first line of defense, using passive and active mechanisms.

Barrier tissues are the first lines of defense, because they form impermeable barriers to invaders. In addition, they may have glands and active components, which can aid in the process. The most obvious external defense is the **skin**, or **integument** (*integere*, "to cover"), which covers the outside of the body with an outer protective epidermis and an inner, connective tissue dermis (● Figure 10–4).

Integument

The skin, typically the largest organ of an animal, not only serves as a mechanical barrier between the external environment and the underlying tissues but is dynamically involved in defense mechanisms and other important functions as well. The **epidermis** consists of numerous layers of epithelial cells. The epidermis has no direct blood supply. Its cells are nourished only by diffusion of nutrients from a rich vascular network in the underlying dermis. Newly forming cells in the inner layers constantly push the older cells closer to the surface, farther and farther from their nutrient supply. Epidermal cells are tightly bound together by spot desmosomes (see p. 63), which interconnect with intracellular *keratin* filaments (see p. 58) to form a strong, cohesive, **keratinized layer**. As the scales of the outermost keratinized layer slough or flake off through abrasion, they are continuously replaced by means of cell division in the deeper epidermal layers. The keratinized layer is airtight, fairly waterproof, and impervious to most substances. It impedes passage into the body of most materials that come into contact with the body surface, including bacteria and toxic chemicals. In many instances, the skin modifies compounds that come into contact with it. For example, epidermal enzymes can convert many potential carcinogens into harmless compounds.

Under the epidermis is the **dermis**, a connective tissue layer. Special infoldings of the epidermis into the underlying dermis form the skin's exocrine glands—the sweat glands in some mammals, and sebaceous glands—as well as hair or feather follicles. **Sweat glands** are important in temperature regulation (Chapter 15), but they can also produce recently discovered **antimicrobial peptides** such as *dermicidin*. The cells of the **sebaceous glands** produce an oily secretion known as **sebum** that is released into adjacent hair follicles. From there the oily sebum flows to the surface of the skin, oiling both the hairs and the outer keratinized layers of the skin to help waterproof them and prevent them from drying and cracking (maintaining the barrier to invaders). Sebum may also help inhibit the growth of microorganisms.

The epidermis contains two distinct nonspecific defensive cell types. **Melanocytes** produce the pigment **melanin**, which they disperse to surrounding skin cells. Melanin performs the protective function of absorbing harmful ultraviolet light rays. **Keratinocytes**, the most abundant epidermal cells, are special-

Figure 10–4 ● Anatomy of the skin (mammalian). The skin consists of two layers, a keratinized outer epidermis and a richly vascularized inner connective tissue dermis. Special infoldings of the epidermis form the sweat glands, sebaceous glands, and hair follicles. The epidermis contains four types of cells: keratinocytes, melanocytes, Langerhans cells, and Granstein cells. The skin is anchored to underlying muscle or bone by the hypodermis, a loose, fat-containing layer of connective tissue.

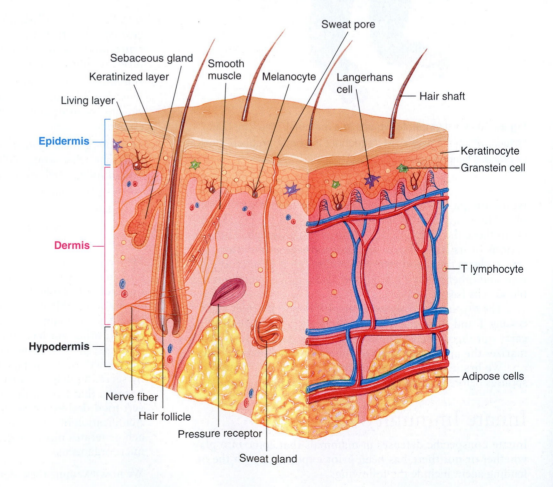

ists in keratin production. As they die, they form the outer protective keratinized layer. They are also responsible for generating hair, claws, talons, and nails. A surprising, recently discovered function is that keratinocytes are also important immunologically. They can secrete **interleukin 1** (a product also secreted by macrophages), which is involved in inflammation. And they make antimicrobial peptides called **cathelicidins,** which can protect skin wounds from streptococcal bacteria. (The latter function was shown with knockout mice lacking a gene for one cathelicidin; they could not fight off infection by certain streptococcal bacteria.) As antimicrobial peptides, cathelicidins are thought to create holes in bacterial membranes, much as *complement* does (we discuss complement in detail later). Cathelicidins in the skins of humans, pigs, and cows have other functions, such as attracting other immune cells into the site of invasion (cells we shortly consider in the section on inflammation), stimulating wound healing, and triggering *apoptosis* (p. 93) of damaged body cells in the wound site.

Two other epidermal cell types also play a role in immunity (Figure 10–4). A type of **dendritic** cells called **Langerhans cells,** which migrate to the skin from the bone marrow, serve as sensors to activate immune responses (as you will see later for dendritic cells in general). Thus the skin not only serves as a mechanical barrier but actually alerts regulatory lymphocytes if the barrier is breached by invading microorganisms. In contrast, it appears that **Granstein cells** serve as a "brake" on skin-activated immune responses. Significantly, Langerhans cells are more susceptible to damage by UV radiation (as from the sun) than Granstein cells are. Losing Langerhans cells as a result of exposure to UV radiation can lead to a predominant suppressor signal rather than the normally dominant helper signal, leaving the skin more vulnerable to microbial invasion and cancer cells.

The various epidermal components of the immune system are collectively termed skin-associated lymphoid tissue, or SALT. Recent research suggests that the skin probably plays an even more elaborate role in adaptive immune defense than described here. This is appropriate, because the skin serves as a major interface with the external environment.

Other Barrier Tissues

Other barrier tissues are the internal ones that communicate with the environment. These too have nonspecific defenses in addition to being physical barriers. The **digestive tract** has many defenses, because it is frequently exposed to ingested pathogens. Saliva secreted into the mouth contains the enzyme **lysozyme,** which lyses certain ingested bacteria. Furthermore, many of the surviving bacteria that are swallowed are killed directly by the strongly acidic gastric juice in the stomach. Farther down the tract, the intestinal lining has GALT (gut-associated lymphoid tissue). These defensive mechanisms are not 100% effective, however. Some bacteria are adapted to these conditions and reach the large intestine (the last portion of the digestive tract), where (as you saw earlier), some can be beneficial.

Within the **genitourinary** (reproductive and urinary) **system,** would-be invaders encounter hostile conditions in the acidic, often salty urine and acidic vaginal secretions. The genitourinary organs also produce sticky mucus, which, like flypaper, entraps small invading particles. Later the particles are either engulfed by phagocytes or are swept out as the organ empties

(for example, flushed out with urine flow). Symbiotic bacteria in the vagina also help prevent invasion.

The **respiratory system** in air breathers is likewise equipped with several important defense mechanisms against inhaled invaders and particulate matter. The respiratory system is the largest surface of the body that comes into direct contact with the environment. Larger airborne particles are filtered out of the inspired air by hairs at the entrance of the nasal passages. Lymphoid tissues, the *tonsils* and *adenoids,* provide immunological protection against inspired pathogens near the beginning of the respiratory system. Farther down in the respiratory airways, millions of tiny cilia (see p. 58) constantly beat in an outward direction. The airways are coated with a layer of thick, sticky mucus secreted by epithelial cells within the airway lining. This mucus sheet, laden with any inspired particulate debris (such as dust) that adheres to it, is constantly moved upward to the throat by ciliary action. This moving "staircase" of mucus is known as the **mucus escalator.** The dirty mucus is either expectorated (spit out) or in most cases swallowed. In addition, an abundance of phagocytic specialists called the **alveolar macrophages** scavenge within the air sacs (alveoli) of the lungs. Further respiratory defenses include coughs and sneezes. These commonly experienced reflex mechanisms involve forceful outward expulsion of material in an attempt to remove irritants from the trachea *(coughs)* or nose *(sneezes).* Lung fluid also contains antimicrobial *cathelicidins.*

■ **Inflammation, a second line of defense, is a nonspecific response to foreign invasion or tissue damage.**

The term **inflammation** refers to an innate, nonspecific series of highly interrelated events that are set into motion in response to foreign invasion, tissue damage, or both. The ultimate goal of inflammation is to bring to the invaded or injured area phagocytes and plasma proteins that can (1) isolate, destroy, or inactivate the invaders; (2) remove debris; and (3) prepare for subsequent healing and repair. The overall inflammatory response is remarkably similar no matter what the triggering event (be it bacterial invasion, chemical injury, or mechanical trauma), although some subtle differences may be evident, depending on the injurious agent or the site of damage. The following sequence of events typically occurs during the inflammatory response. For an example we use bacterial entry into a break in the skin (● Figure 10–5).

Defense by Resident Tissue Macrophages

On bacterial invasion through a break in the external barrier of skin, the **macrophages** already present in the area immediately begin phagocytizing the foreign microbes (provided they can recognize them as invaders). Although the resident macrophages are usually not present in sufficient numbers to meet the challenge alone, they defend against infection during the first hour or so, before other mechanisms can be mobilized. Macrophages are usually rather stationary, gobbling debris and contaminants that come their way, but when necessary they become mobile and migrate to sites of battle against invaders.

These phagocytic cells are studded with Toll-like receptors (TLRs, p. 429), which bind with pathogen markers, allowing

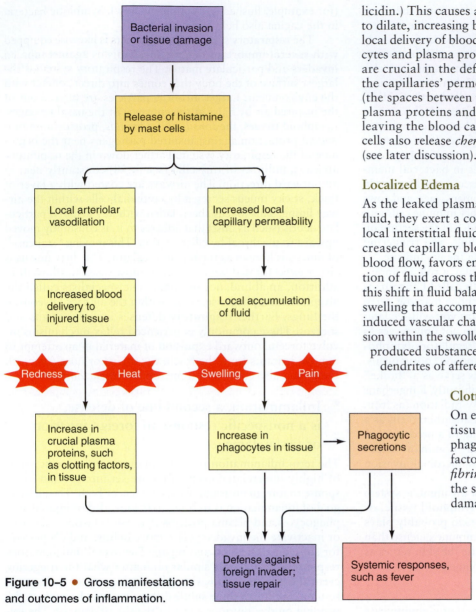

Figure 10–5 ● Gross manifestations and outcomes of inflammation.

licidin.) This causes arterioles and capillaries within the area to dilate, increasing blood flow to the site of injury. Increased local delivery of blood brings to the site more phagocytic leukocytes and plasma proteins such as *complement,* both of which are crucial in the defense response. Histamine also increases the capillaries' permeability by enlarging the capillary pores (the spaces between the endothelial cells; see p. 411) so that plasma proteins and cells that normally are prevented from leaving the blood can escape into the inflamed tissue. Mast cells also release *chemotaxins,* signals that attract phagocytes (see later discussion).

Localized Edema

As the leaked plasma proteins accumulate in the interstitial fluid, they exert a colloid osmotic pressure. This elevation in local interstitial fluid osmotic pressure, coupled with the increased capillary blood pressure caused by increased local blood flow, favors enhanced filtration and reduced reabsorption of fluid across the involved capillaries. The end result of this shift in fluid balance is localized **edema.** Thus the familiar swelling that accompanies inflammation is due to histamine-induced vascular changes. Pain is caused both by local distension within the swollen tissue and by the direct effect of locally produced substances such as *prostaglandins* on the receptor dendrites of afferent neurons.

Clotting

On exposure to thromboplastin in the injured tissue and to specific chemicals secreted by phagocytes on the scene, *fibrinogen,* the final factor in the clotting system, is converted into *fibrin.* Fibrin forms interstitial fluid clots in the spaces around the bacterial invaders and damaged cells. This walling off of the injured region from the surrounding tissues prevents or at least delays the spread of bacterial invaders and their toxic products. Later, the more slowly activated anticlotting factors gradually dissolve the clots after they are no longer needed. (Details of clotting are covered in Chapter 9.)

the effector cells to "see" the pathogens as distinct from "self" cells. TLR binding to a pathogen triggers the phagocyte to engulf and destroy the invader. There is also evidence that phagocytes are activated by the release of intracellular chemicals from damaged body cells. Moreover, the activation of TLRs induces the phagocyte to secrete various chemicals, some of which contribute to other inflammation processes. There are a number of TLRs, which recognize (bind) different antigen molecules. For example, TLR2 recognizes peptidoglycans from bacterial cell walls, whereas TLR9 recognizes bacterial DNA.

Integration by Mast Cells: Localized Capillary Vasodilation and Increased Permeability

Almost immediately on microbial invasion, **mast cells** detect the injury and release *histamine,* a paracrine signal molecule. (One signal that activates mast cells in skin is a type of cathe-

Emigration of Leukocytes

Within an hour after the injury, the involved area is teeming with leukocytes that have exited from the vessels. Neutrophils are the first to arrive, followed by the slower-moving monocytes. The latter swell and mature into macrophages. Leukocyte emigration from the blood into the tissues uses a mechanism known as *diapedesis.* A leukocyte pushes a long, narrow projection through a capillary pore; then the remainder of the cell flows forward into the projection (● Figure 10–6). In this way, the leukocyte can wriggle its way in about a minute through the capillary pore—even though it is much larger than the pore. Outside the vessel, the leukocyte moves in an amoeboid fashion toward the site of tissue damage and bacterial invasion. Phagocytic cells are guided in their direction of migration by attraction to the chemotaxins, released at the site of damage (some by mast cells). This pro-

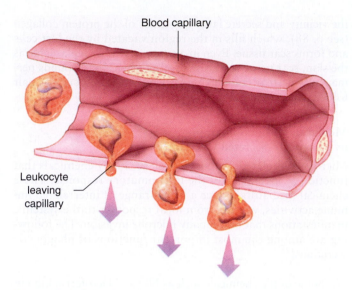

Blood capillary

Leukocyte leaving capillary

Figure 10–6 ● **Leukocyte emigration from the blood.**
Leukocytes emigrate from the blood into the tissues by assuming amoeba-like behavior and squeezing through the capillary pores, a process known as *diapedesis*.

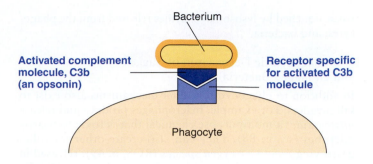

Bacterium

Activated complement molecule, C3b (an opsonin)

Receptor specific for activated C3b molecule

Phagocyte

Structures are not drawn to scale.

Figure 10–7 ● **Mechanism of opsonin action.** One of the activated complement molecules, C3b, links a foreign cell, such as a bacterium, and a phagocytic cell by nonspecifically binding with the foreign cell and specifically binding with a receptor on the phagocyte. This link ensures that the foreign victim does not escape before it can be engulfed by the phagocyte.

cess is referred to as *chemotaxis*. Binding of chemotaxins with protein receptors on the plasma membrane of a phagocytic cell increases Ca^{++} entry into the cell. Calcium, in turn, switches on the cellular contractile apparatus that leads to amoeba-like crawling. Because the concentration of chemotaxins progressively increases toward the site of injury, phagocytic cells move toward this site along a chemotaxin concentration gradient.

Leukocyte Proliferation

New phagocytic recruits from the bone marrow soon join resident tissue macrophages as well as leukocytes that exited from the blood and migrated to the inflammatory site. Within a few hours after the onset of the inflammatory response, the number of neutrophils in the blood may increase up to four to five times that of normal. This increase is due partly to transfer into the blood of large numbers of preformed neutrophils stored in the bone marrow and partly to increased production of new neutrophils by the bone marrow. A slower-commencing but longer-lasting increase in monocyte production by the bone marrow also occurs, making larger numbers of these macrophage precursor cells available. In addition, multiplication of resident macrophages adds to the pool of these important immune cells. Proliferation of new neutrophils, monocytes, and macrophages and mobilization of stored neutrophils are stimulated by various chemical mediators released from the inflamed region.

Phagocytic Destruction of Tagged and Untagged Bacteria

Obviously, phagocytes must be able to distinguish between normal cells and foreign or abnormal cells before accomplishing their destructive mission; otherwise, they could not selec-

tively engulf and destroy only unwanted materials. First, as you saw earlier, the phagocyte's TLRs trigger destruction of invaders with standard bacterial-wall components. Second, foreign particles are deliberately marked for phagocytic ingestion by being coated with plasma proteins. Such agents that make bacteria more susceptible to phagocytosis are known as **opsonins**. The most important opsonins are *antibodies* and one of the proteins of the *complement* system (which we discuss later).

An opsonin enhances phagocytosis by linking the foreign cell to a phagocytic cell (● Figure 10–7). One portion of an opsonin molecule binds nonspecifically to the surface of an invading bacterium, whereas another portion of the opsonin molecule binds to receptor sites specific for it on the phagocytic cell's plasma membrane. This link ensures that the bacterial victim does not have a chance to "get away" before the phagocyte can perform its lethal attack.

Neutrophils and macrophages clear the inflamed area of infectious and toxic agents as well as tissue debris; this clearing action is the primary function of the inflammatory response. They accomplish this by both phagocytic and nonphagocytic means.

Phagocytosis involves the engulfment and intracellular degradation (breakdown) of foreign particles and tissue debris. Macrophages can engulf a bacterium in less than 0.01 second. Phagocytic cells contain an abundance of lysosomes, which are organelles filled with hydrolytic enzymes (the same type that degrade ingested food in the digestive tract). After a phagocyte has internalized targeted material, a lysosome fuses with the membrane that encloses the engulfed matter and releases its hydrolytic enzymes within the confines of the vesicle, where the enzymes begin breaking down the entrapped material (see p. 41). Phagocytes eventually die because of the accumulation of toxic by-products from foreign particle degradation or inadvertent release of destructive lysosomal chemicals into the cytosol. The **pus** that forms in an infected wound is a collection of these phagocytic cells, both living and dead; necrotic (dead) tissue liquefied by lysosomal enzymes released from the phago-cytes; and bacteria.

Nonphagocytotic Destruction of Tagged and Untagged Bacteria

In addition to phagocytosis, other mechanisms also exist to kill invaders. For example, macrophages produce and release *nitric oxide,* a multipurpose chemical that is toxic to nearby microbes (see p. 88). They can also create other microbe-destroying *reactive oxygen species* (ROS, p. 49), released in bursts in the vicinity of invaders. As a subtler means of de-struction, neutrophils secrete **lactoferrin,** a protein that tightly binds with iron, making it unavailable for use by invading bac-teria. Bacterial multiplication depends on high concentrations of available iron. Therefore, phagocytes can destroy foreign invaders both by phagocytosis and by nonphagocytic means. Neutrophils also secrete cathelicidins (p. 433).

Probably the most important nonphagocytotic attack is part of the **complement system,** which can nonspecifically be called into play by the presence of any foreign invader. One of these complement systems makes pores in bacterial membranes, killing them by osmosis (see the section on complement later in this chapter). It may also be activated by antibodies produced as part of the specific immune response to a particular micro-organism. The fact that the complement system is involved in both nonspecific and specific defense mechanisms illustrates an important point. The various components of the immune system are highly interactive and interdependent, making the system highly sophisticated and effective but also complex and difficult to sort out. The most significant cooperative rela-tionships known to exist among the immune effector cells are pointed out as we examine the various components of the im-mune system separately.

Tissue Repair

The ultimate purpose of the inflammatory process is to isolate and destroy injurious agents and to clear the area for tissue re-pair. In some tissues (for example, skin, bone, and liver), the healthy organ-specific cells surrounding the injured area un-dergo cell division to replace the lost cells, often accomplish-ing perfect repair. In nonregenerative tissues such as nerve and muscle, however, lost cells are replaced by **scar tissue.** Fibro-blasts, a type of connective tissue cell, start to divide rapidly in the vicinity and secrete large quantities of the protein collagen (see p. 88), which fills in the region vacated by the lost cells and forms scar tissue. Even in a tissue as readily replaceable as the skin, scar tissue sometimes forms when deep wounds per-manently destroy complex underlying structures, such as hair follicles and sweat glands.

Mediation of Inflammation by Phagocyte-Secreted Chemicals

Microbe-stimulated phagocytes release many chemicals that function as mediators of the inflammatory response. These chemical mediators induce a broad range of interrelated im-mune activities, varying from local responses to the systemic manifestations that accompany microbe invasion. The follow-ing are among the most important functions of phagocytic secretions:

1. Some of the chemicals such as NO and lactoferrin kill mi-crobes by nonphagocytic means, as we saw earlier.

2. Phagocytic secretions stimulate the release of histamine from mast cells in the vicinity. Histamine, in turn, induces local vasodilation and increased vascular permeability.

3. Some phagocytic chemical mediators trigger both the clotting and anticlotting systems to first enhance the walling-off process and then facilitate the gradual disso-lution of the fibrous clot after it is no longer needed.

4. Other secretions split **kininogens,** which are inactive pre-cursor plasma proteins synthesized by the liver, into ac-tive **kinins.** In particular, **kallikrein,** which is released by neutrophils, can accomplish this activation. Once kinins are generated by kallikrein, they augment a variety of in-flammatory events. Specifically, kinins
   a. Stimulate several key steps in the complement system
   b. Reinforce the vascular changes induced by histamine
   c. Activate nearby pain receptors and are thus partially re-sponsible for the soreness associated with inflammation
   d. Act as powerful chemotaxins to induce phagocyte mi-gration into the affected area. In *positive-feedback* fashion, the newly arriving neutrophils release kalli-krein, which can generate still more kinins that che-motactically entice more neutrophils to join the battle (● Figure 10–8).

Figure 10–8 ● Relationship between the kinin system and neutrophils in an inflammatory response. The green arrows indicate a positive-feedback loop between the kinin system and neutrophils during an inflammatory response.

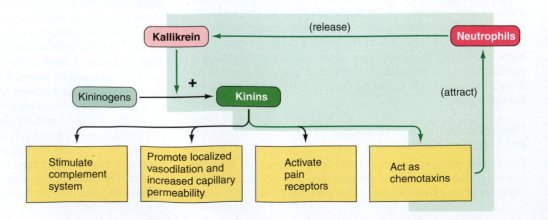

5. One chemical in particular released by macrophages, **endogenous pyrogen (EP)**, induces the development of fever (*endogenous,* "from within the body;" *pyro,* "fire" or "heat"; *gen,* "production"). The primary EP is **interleukin 1 (IL-1)**. This response occurs especially when the invading organisms have spread into the bloodstream. IL-1 causes release within the hypothalamus of *prostaglandins,* locally acting chemical messengers that "turn up" the hypothalamic "thermostat" that regulates body temperature. The function of the resultant elevation in body temperature in fighting infection remains unclear. The fact that fever is such a common systemic manifestation of inflammation suggests that the higher body temperature plays an important beneficial role in the overall inflammatory response in many phyla (see Chapter 15 for information on fever in nonmammals). Recent evidence strongly supports this supposition. For example, higher temperatures augment the process of phagocytosis and increase the rate at which the many enzyme-dependent inflammatory activities proceed. One theory proposes that an elevated body temperature increases bacterial requirements for iron while reducing the plasma concentration of iron, thus interfering with bacterial multiplication. Resolving the controversial issue of whether a fever can be beneficial is extremely important in view of the widespread use of drugs that suppress fever. Although a mild fever may possibly be beneficial, there is no doubt that an extremely high fever can be detrimental, particularly by harming the central nervous system.

6. One chemical mediator secreted by macrophages, **leukocyte endogenous mediator (LEM)**, decreases the plasma concentration of iron by altering iron metabolism within the liver, spleen, and other tissues. This action reduces the amount of iron available to support bacterial multiplication. Evidence suggests that LEM and IL-1 are the same substance or at least very closely related.

7. LEM also stimulates **granulopoiesis**, the synthesis and release of neutrophils and other granulocytes by the bone marrow. This effect is especially prominent in response to bacterial infections.

8. Furthermore, LEM stimulates the release of **acute-phase proteins** from the liver. This collection of proteins, which have not yet been sorted out by scientists, exerts a multitude of wide-ranging effects associated with the inflammatory process, tissue repair, and immune cell activities.

9. Another effect of phagocytic secretions is to enhance the proliferation and differentiation of both B and T lymphocytes, which, in turn, govern antibody production and cell-mediated immunity, respectively. Specifically, **IL-1** is responsible for this effect on lymphocytes. Apparently, the same chemical substance is responsible for a diverse array of effects throughout the body, all of which are geared toward defending the body against infection or tissue injury. In fact, the release of LEM/IL-1 can be elicited by stressful situations not associated with microbe invasion (for example, during endurance activity). Accordingly, this mediator may be part of a generalized nonspecific protective response.

This list of events that are augmented by chemicals secreted by phagocytes is not complete, but it serves to illustrate the diversity and complexity of responses elicited by these mediators. As will soon become evident, there are other important macrophage–lymphocyte interactions that are not dependent on the release of chemicals from phagocytic cells. Thus the effect that phagocytes, especially macrophages, ultimately have on microbial invaders far exceeds their "engulf and destroy" tactics (● Figure 10–9).

■ Salicylates and glucocorticoids suppress the inflammatory response.

Numerous drugs can suppress the inflammatory process; the most effective are the *salicylates* and related compounds (aspirin-type drugs), *antihistamines,* and *glucocorticoids* (drugs similar to the steroid hormone cortisol, which is secreted by the adrenal cortex; see p. 284). Antihistamines (and salicylates to some extent) interfere with the inflammatory response by decreasing histamine release, resulting in a reduction in swelling, redness, and pain. Furthermore, salicylates reduce fever by inhibiting the production of prostaglandins, the local mediators of endogenous-pyrogen–induced fever and pain.

Glucocorticoids, which are potent anti-inflammatory drugs, suppress almost every aspect of the inflammatory response. In addition, they destroy lymphocytes within lymphoid tissue and reduce antibody production. These therapeutic agents are useful for treating undesirable immune responses, such as allergic reactions and the inflammation associated with arthritis. Unfortunately, however, by suppressing inflammatory and other immune responses that localize and eliminate bacteria, such therapy also reduces the body's ability to resist infection.

Recent evidence suggests that cortisol, whose secretion is increased in response to any stressful situation, exerts anti-inflammatory activity even at normal physiological levels (see p. 286). According to this proposal, the anti-inflammatory effect of cortisol modulates stress-activated immune responses, preventing them from overshooting and thus protecting against damage by potentially over-reactive defense mechanisms. This is probably important, because there is more than one positive-feedback loop in the inflammation response, as we saw earlier. Positive feedback can get out of control without some way to interrupt it.

■ The complement system kills microorganisms directly, both on its own and in conjunction with antibodies and also augments the inflammatory response.

The **complement system,** another "second-line" defense mechanism brought into play in response to invading organisms, is activated in two ways:

1. By exposure to particular carbohydrate chains present on the surfaces of microorganisms but not found on body cells, a nonspecific innate immune response

2. By exposure to antibodies produced against a specific foreign invader, an adaptive immune response

In addition to bringing about direct lysis of the invader, the powerful complement cascade reinforces other general inflammatory tactics. In fact, the system derives its name from

Figure 10–9 ● **Phagocyte responsibilities.** Phagocytes not only destroy foreign particles or damaged cells by phagocytosis but also secrete chemical mediators that destroy the targeted material by nonphagocytic means and augment many aspects of inflammation and other immune processes. The pink boxes represent final outcomes of phagocytic actions.

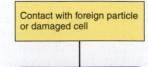

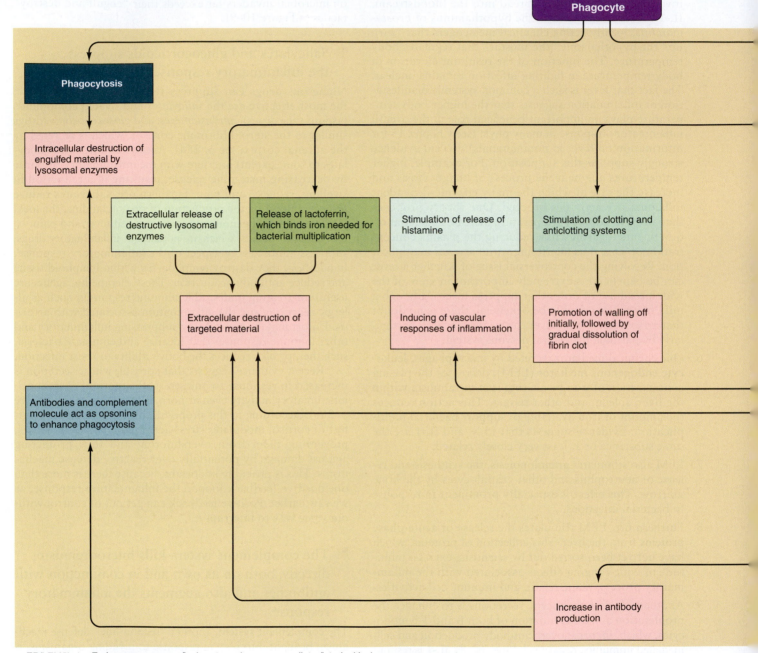

EP/LEM/IL-1 = Endogenous pyrogen/leukocyte endogenous mediator/interleukin 1

the fact that it "complements" the action of antibodies; it is the major mechanism activated by antibodies to kill foreign cells.

The Membrane Attack Complex

The complement system consists of plasma proteins that are produced by the liver and circulate in the blood in inactive form. Once the first component, C1, is activated, it activates the next component, C2, and so on, in a sequential cascade of activation reactions. The five final components, C5 through C9, assemble into a large, doughnut-shaped, protein complex, the **membrane attack complex (MAC)**, which attacks the surface membrane of nearby microorganisms by imbedding itself so that a channel is created through the microbial surface membrane (● Figure 10–10). In other words, the parts make a hole, which is highly permeable to water and small ions. This hole-punching technique makes the membrane extremely leaky, resulting in an influx of sodium followed by an osmotic flux of water into the victim cell, causing it to swell and burst. This complement-induced lysis is the major means of directly killing microbes without phagocytizing them.

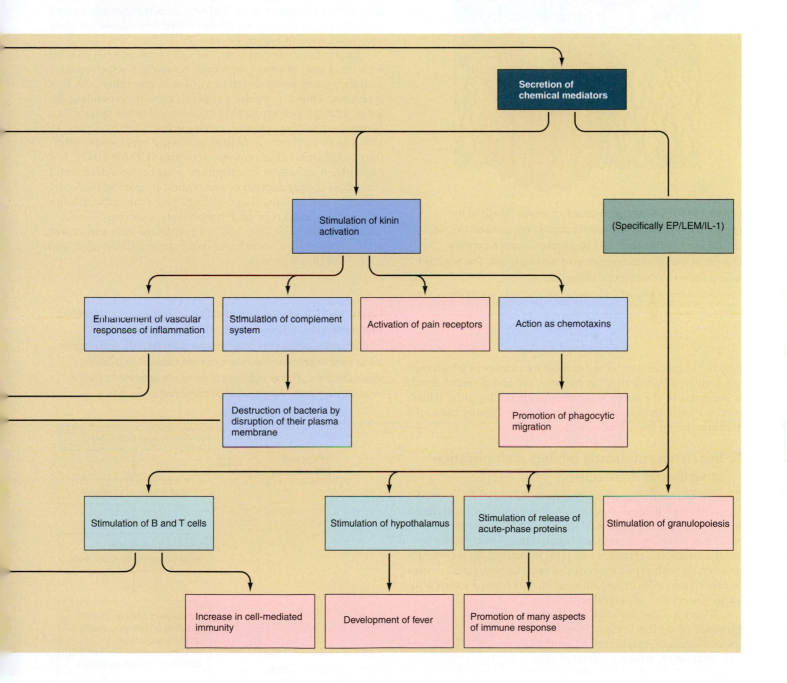

Augmentation of Inflammation

Besides the direct destruction of foreign cells accomplished by the MAC, various other activated complement components augment the inflammatory process by

- *Serving as chemotaxins,* which attract and guide phagocytes to the site of complement activation (that is, the site of microbial invasion)
- *Acting as opsonins* by binding with microbes and thereby enhancing their phagocytosis

- *Promoting vasodilation and increased vascular permeability* to increase blood flow to the invaded area
- *Stimulating the release of histamine* from mast cells in the vicinity, which in turn enhances the local vascular changes characteristic of inflammation
- *Activating kinins,* which further reinforce inflammatory reactions

Several of the activated components in the cascade are very unstable. Because these unstable components can perpet-

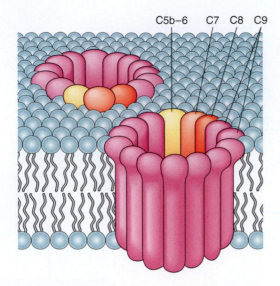

C5b–6   C7   C8   C9

Figure 10–10 ● **Membrane attack complex (MAC) of the complement system.** Activated complement proteins C5, C6, C7, C8, and a number of C9s aggregate to form a porelike channel in the plasma membrane of the target cell. The resultant leakage leads to destruction of the cell.

Source: Adapted from the illustration by Dana Burns-Pizer in "How Killer Cells Kill," by John Ding-E Young and Zanvil A. Cohn. Copyright © 1988 *Scientific American,* Scientific American, Inc. All rights reserved.

uate the sequence only in the immediate vicinity in which they are activated before they decompose, the complement attack is confined to the surface membrane of the microbe whose presence initiated activation of the system. Nearby host cells are thus spared from lytic attack.

■ Interferon transiently inhibits multiplication of viruses in most cells.

Besides the inflammatory response, another nonspecific "second-line" defense mechanism is the release of **interferon** from virus-infected cells. Interferon is not a single type of molecule but a family of similar proteins. Interferon briefly provides nonspecific resistance to viral infections by transiently interfering with replication of the same or unrelated viruses in other host cells. In fact, interferon was named for its ability to "interfere" with viral replication.

When a virus invades a cell, the presence of viral nucleic acid induces the cell's genetic machinery to synthesize interferon, which is secreted into the extracellular fluid. Interferon acts as a "whistle-blower," forewarning healthy cells of potential viral attack and helping them prepare to resist such an attack. Once released from a virus-infected cell, interferon binds with receptors on the plasma membranes of healthy neighboring cells or even distant cells that it reaches through the bloodstream, signaling these cells to prepare for the possibility of impending viral attack. Interferon does not

have a direct antiviral effect; instead, it triggers the production of virus-blocking enzymes by potential host cells. Binding with interferon induces these other cells to synthesize enzymes that can break down viral messenger RNA and inhibit protein synthesis, both of which are essential for viral replication. Although viruses are still able to invade these forewarned cells, they cannot govern cellular protein synthesis for their own replication (● Figure 10–11).

Interferon is released nonspecifically from any cell infected by a virus and, in turn, can induce temporary self-protective activity against many different viruses in any other cells that it reaches. Thus it provides a general, rapidly responding defense strategy against viral invasion until more specific but slower-responding immune mechanisms come into play. In addition to facilitating inhibition of viral replication, interferon reinforces other immune activities (❚ Table 10–2). For example, it enhances macrophage phagocytic activity and stimulates the production of antibodies. Interferon markedly enhances the actions of cell-killing cells—the *natural killer cells* and a special type of T lymphocyte, *cytotoxic T-cells*—which attack and destroy both virus-infected cells and cancer cells. Furthermore, interferon itself slows cell division and suppresses tumor growth.

Figure 10–11 ● **Mechanism of action of interferon in preventing viral replication.** Interferon, which is released from virus-infected cells, binds with other uninvaded host cells and induces these cells to produce inactive enzymes capable of blocking viral replication. These inactive enzymes are activated only if a virus subsequently invades one of these prepared cells.

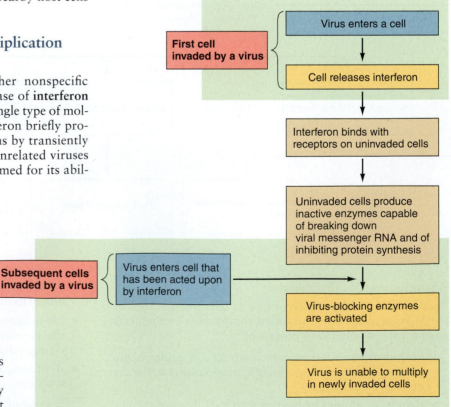

First cell invaded by a virus

Virus enters a cell

Cell releases interferon

Interferon binds with receptors on uninvaded cells

Uninvaded cells produce inactive enzymes capable of breaking down viral messenger RNA and of inhibiting protein synthesis

Subsequent cells invaded by a virus

Virus enters cell that has been acted upon by interferon

Virus-blocking enzymes are activated

Virus is unable to multiply in newly invaded cells

Table 10–2 ∎ Functions of Interferon

Functions	Antiviral Effect	Anticancer Effect
Released by virus-infected cells and transiently interferes with viral replication in other host cells by inducing them to produce enzymes that destroy viral messenger RNA and inhibit virus-directed protein synthesis	✓	
Enhances macrophage phagocytic activity	✓	✓
Stimulates production of antibodies	✓	
Enhances actions of natural killer cells and cytotoxic cells	✓	✓
Slows cell division and suppresses tumor growth		✓

∎ Natural killer cells destroy virus-infected cells and cancer cells on first exposure to them.

Natural killer cells are naturally occurring, lymphocyte-like cells that nonspecifically destroy virus-infected cells and cancer cells by directly lysing their membranes on first exposure to them. Their mode of action and major targets are similar to those of cytotoxic T-cells (p. 451), but the latter can fatally attack only the specific types of virus-infected cells and cancer cells to which they have been previously exposed. Furthermore, after exposure cytotoxic T-cells require a maturation period before they can launch their lethal assault. The natural killer cells provide an immediate, nonspecific defense against virus-invaded cells and cancer cells before the more specific and more abundant cytotoxic T-cells become functional. How they do this is still unclear, but they apparently react to abnormal proteins that occur in the membranes of many viral-infected and cancerous cells.

Acquired (Adaptive) Immunity

A sophisticated third line of defense is the acquired or adaptive immune system, which directs selective attacks at foreign agents to which the body has been previously exposed.

∎ Acquired immunity includes cell-mediated and antibody-mediated responses.

Traditionally, two classes of specific adaptive immune responses have been distinguished: antibody-mediated, or humoral, immunity, involving the production of antibodies by B lymphocyte derivatives known as *plasma cells;* and **cell-mediated immunity,** involving the production of *effector T lymphocytes,* which directly attack unwanted cells. Together, these responses can respond to an almost limitless variety of foreign agents. Yet each B and T lymphocyte can recognize and defend against only one particular type of foreign material, such as one kind of bacterium. Among the millions of B and T lymphocytes in the body, only the ones specifically equipped to recognize the

unique molecular features of a particular infectious agent are called into action to discriminatingly defend against this agent.

The recognition and response processes are different for B and T lymphocytes (also called *B-* and *T-cells*). In general, B-cells recognize free-existing foreign invaders such as bacteria and their toxins and a few viruses, which they combat by secreting antibodies specific for the invaders. T-cells specialize in recognizing and destroying body cells gone awry, including virus-infected cells and cancer cells. We examine each of these processes in detail in the upcoming sections. For now, we are going to explore the different life histories of B and T lymphocytes.

Origins of B- and T-Cells

B and T lymphocytes have different life histories and, more importantly, different properties and functions. Both types of lymphocytes, like all blood cells, are derived from common stem cells in the bone marrow. Whether a lymphocyte and all its progeny are destined to be B- or T-cells depends on the site of final maturation and differentiation of the original cell in the lineage (● Figure 10–12). During fetal life and early infancy, some of the immature lymphocytes migrate through the blood to the thymus, where they undergo further processing to become T lymphocytes (hence the designation "T" from *thymus*). The

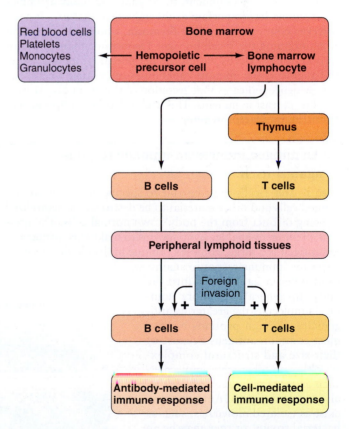

Figure 10–12 ● Origins of B and T cells. B cells are derived from lymphocytes that matured and differentiated in the bone marrow, whereas T cells are derived from lymphocytes that originated in the bone marrow but matured and differentiated in the thymus. New B and T cells are produced by colonies of B and T cells in peripheral lymphoid tissues.

thymus is a lymphoid tissue located midline within the chest cavity above the heart in the space between the lungs (Figure 10–3). Lymphocytes that mature without benefit of "thymic education" become B lymphocytes. In mammals, the site of B-cell maturation and differentiation is uncertain, although it is generally assumed to be the bone marrow. The designation "B" did not come from bone marrow, but from the **bursa of Fabricius** in birds, where avian B-cells mature. The bursa is located in the proctodael region of the avian cloaca. The bursa grows rapidly during the first three to four weeks after hatch, but by the time the birds sexually mature it has completely regressed.

On being released into the blood from either the bone marrow or the thymus, mature B- and T-cells take up residence and establish lymphocyte colonies in the peripheral lymphoid tissues. Here, on appropriate stimulation, they undergo cell division to produce new generations of either B- or T-cells, depending on their ancestry. Each individual mammal has billions to trillions of lymphocytes, which, if aggregated together in a mass, would be comparable to the size of the brain or liver. At any one time, the majority of these lymphocytes are concentrated in the various strategically located lymphoid tissues, but both B- and T-cells continually circulate among the lymph, blood, and body tissues, where they remain on constant surveillance.

Because most of the migration and differentiation of T-cells occurs early in development, the thymus gradually atrophies and becomes less important as the individual matures. It does, however, continue to produce **thymosin**, a hormone important in maintaining the T-cell lineage. Thymosin enhances proliferation of new T-cells within the peripheral lymphoid tissues and augments the immune capabilities of existing T-cells. Recent evidence indicates that secretion of thymosin decreases in mid-life, at least in humans. This decline has been implicated as a contributing factor in aging.

▪ An antigen induces an immune response against itself.

Both B- and T-cells must be able to specifically recognize unwanted cells and other material to be destroyed or neutralized as being distinct from the body's own normal cells. The presence of antigens enables lymphocytes to make this distinction. An **antigen** is a large, complex, unique molecule that triggers a specific immune response against itself when it gains entry into the body. In general, the more complex a molecule, the greater its antigenicity. Foreign proteins are the most common antigens because of their size and structural complexity, although other macromolecules, such as large polysaccharides, can also act as antigens. Antigens may exist as isolated molecules, such as bacterial toxins, or they may be an integral part of a multimolecular structure, such as being present on the surface of an invading foreign microbe.

We first examine the role of B-cells, then T-cells.

B Lymphocytes: Antibody-Mediated Immunity

The key to B- and T-cell specificity lies in their membrane **receptors** that bind an antigen. We discuss how these receptors are generated later, but for now, keep in mind that B- and T-cells have unique receptors that can be thought of as the sensors of the adaptive system, although each receptor binds to only one particular type of the multitude of possible antigens. This in contrast to the TLRs (p. 429) of the innate system, which bind to numerous generic markers on many microorganisms.

In the case of B-cells, binding with antigen induces the cell to differentiate into a **plasma cell,** which produces the effector proteins called **antibodies.** Antibodies are essentially soluble forms of the B-cell's membrane receptor, and thus can combine with the specific type of antigen that stimulated the antibodies' production. During differentiation into a plasma cell, a B lymphocyte swells as the rough endoplasmic reticulum (the site for synthesis of proteins to be exported) greatly expands (● Figure 10–13). Because antibodies are proteins, plasma cells essentially become prolific protein factories, producing up to 2000 antibody molecules per second. So great is the commitment of a plasma cell's protein-synthesizing machinery to antibody production that it cannot maintain protein synthesis for its own viability and growth. Consequently, it dies after a highly productive, but brief, five- to seven-day life span.

Figure 10–13 ● Comparison of an unactivated B cell and a plasma cell. Electron micrograph of (a) an unactivated B cell, or small lymphocyte, and (b) a plasma cell. A plasma cell is an activated B cell. It is filled with an abundance of rough endoplasmic reticulum distended with antibody molecules.

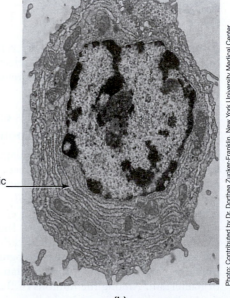

Plasma cell

Unactivated B cell

Endoplasmic reticulum

(a)

(b)

Antibodies are secreted into the blood or lymph, depending on the location of the activated plasma cells, but all antibodies eventually gain access to the blood, where they are known as **gamma globulins,** or **immunoglobulins.** Antibodies and receptors are grouped into the following five subclasses based on differences in their biological activity:

- **IgM** immunoglobulin serves as the B-cell surface receptor for antigen attachment and is secreted in the early stages of plasma cell response.

- **IgG,** the most abundant immunoglobulin in the blood, is produced copiously when the body is subsequently exposed to the same antigen.

Together, IgM and IgG antibodies are responsible for most specific immune responses against bacterial invaders and a few types of viruses.

- **IgE** helps protect against parasitic worms and is the antibody mediator for common allergic responses, such as hay fever, asthma, and hives.

- **IgA** immunoglobulins are found in secretions of the digestive, respiratory, and genitourinary systems, as well as in milk and tears.

- **IgD** is present on the surface of many B-cells, but its function is uncertain.

Note that this classification is based on different ways in which antibodies function. It does not imply that there are only five different antibodies. Within each functional subclass are millions of different antibodies, each able to bind only with a specific antigen.

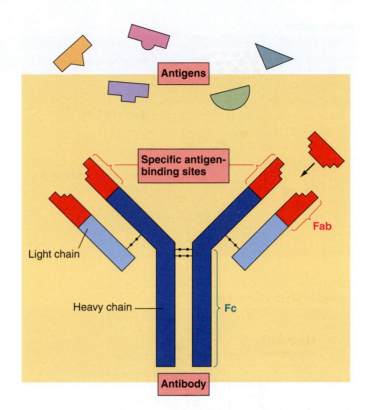

Figure 10–14 ● **Antibody structure.** An antibody is Y-shaped. It is able to bind only with the specific antigen that "fits" its antigen-binding sites (Fab) on the arm tips. The tail region (Fc) binds with particular mediators of antibody-induced activities.

■ Antibodies are Y-shaped and classified according to properties of their tail portion.

Antibody proteins of all five subclasses are composed of four interlinked polypeptide chains—two long, heavy chains and two short, light chains—arranged in the shape of a Y (● Figure 10–14). Characteristics of the arm regions of the Y determine with what antigen the antibody can bind (that is, the *specificity* of the antibody). Properties of the tail portion of the antibody, in contrast, determine the *functional properties* of the antibody (what the antibody does once it binds with antigen).

An antibody has two identical antigen-binding sites, one at the tip of each arm. These **antigen-binding fragments (Fab)** are unique for each different antibody, so that each antibody can interact only with an antigen that specifically matches it, much like a lock and key. The tremendous variation in the antigen-binding fragments of different antibodies is responsible for the extremely large number of unique antibodies that are capable of binding specifically with millions of different antigens. New evidence suggests that the Fab region in some antibodies can shift between two different conformations and thus can bind two different types of antigens.

In contrast to these variable Fab regions at the arm tips, the tail portion of every antibody within each immunoglobulin subclass is identical. The tail, the antibody's so-called **constant (Fc) region,** contains binding sites for particular mediators of antibody-induced activities, which vary among the different subclasses. In fact, differences in the constant region

are the basis for distinguishing between the different immunoglobulin subclasses. For example, the constant tail region of IgG antibodies, when activated by antigen binding in the Fab region, binds with phagocytic cells and serves as an opsonin to enhance phagocytosis. In comparison, the constant tail region of IgE antibodies attaches to mast cells and basophils, even in the absence of antigen. When the appropriate antigen gains entry to the body and binds with the attached antibodies, this triggers the release of histamine from the affected mast cells and basophils. Histamine, in turn, induces the allergic manifestations that follow.

■ Antibodies largely amplify innate immune responses to promote antigen destruction.

Antibodies exert their protective influence in several general ways: physical hindrance of antigens, tagging antigens for destruction, and direct chemical attack (● Figure 10–15).

Neutralization and Agglutination

Antibodies can physically hinder some antigens from exerting their detrimental effects. For example, by combining with bacterial toxins, antibodies can prevent these harmful chemicals from interacting with susceptible cells. This process is known as **neutralization.** Similarly, antibodies can bind with surface antigens on some types of viruses, preventing these viruses from entering cells, where they could exert their damaging

Neutralization

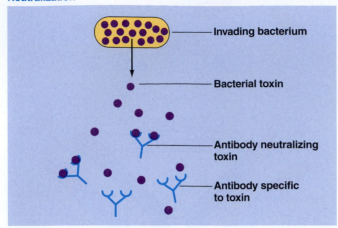

Agglutination (clumping of antigenic cells) and **precipitation** (if soluble antigen-antibody complex is too large to stay in solution)

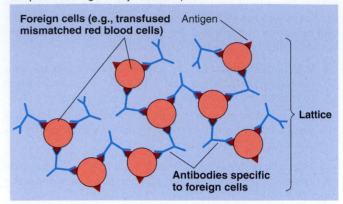

Activation of complement system

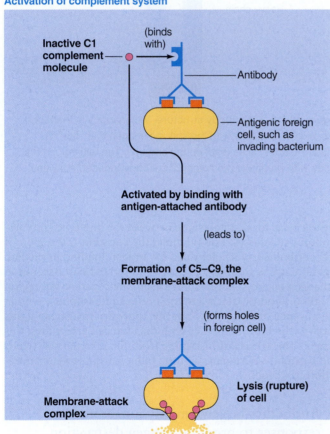

Enhancement of phagocytosis (opsonization)

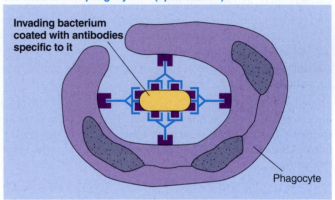

Stimulation of killer cells

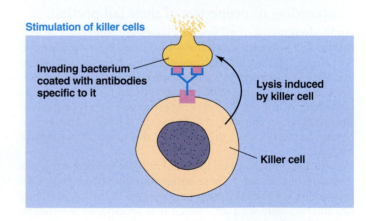

Structures are not drawn to scale.

Figure 10–15 ● **How antibodies help eliminate invading microbes.** Antibodies physically hinder antigens through (1) neutralization or (2) agglutination and precipitation. Antibodies amplify innate immune responses by (1) activating the complement system, (2) enhancing phagocytosis by acting as opsonins, (3) stimulating killer cells, and (4) producing hydrogen peroxide (not shown)..

effects. Sometimes multiple antibody molecules can cross-link numerous antigen molecules into chains or lattices of antigen–antibody complexes. The process in which foreign cells bind together in such a clump is known as **agglutination.**

Amplification of Innate Immune Responses

Antibodies' most important function by far is to tag the invaders to enhance dramatically the nonspecific immune responses already initiated by the invaders. Antibodies mark or identify foreign material as targets for actual destruction by the complement system, phagocytes, or killer cells while enhancing the activity of these other defense systems as follows:

1. *Activation of the complement system.* When an appropriate antigen binds with an antibody, receptors on the tail portion of the antibody can bind with and activate C1, the first component of the complement system. This sets off the cascade of events leading to formation of the membrane attack complex (MAC). In fact, antibody is the most powerful activator of the complement system, which in turn is the most important mechanism by which antibodies exert their protective influence.

2. *Enhancement of phagocytosis.* As mentioned previously, antibodies, especially IgG, act as **opsonins.** The tail portion of an antigen-bound IgG antibody can bind with a receptor on the surface of a phagocyte and subsequently promote the phagocytosis of the antigen-containing victim attached to the antibody.

3. *Stimulation of killer (K) cells.* The binding of antibody to antigen can also induce attack of the antigen-bearing cell by a **killer (K) cell.** K cells are similar to natural killer cells except that K cells require the target cell to be coated with antibodies before they can destroy it by lysing its plasma membrane. K cells have receptors for the constant tail portion of antibodies.

4. *Direct attack.* Finally, recent studies have revealed that antibodies can act as enzymes, catalyzing the formation of **hydrogen peroxide** (see Figure 2-21) on binding to antigen. This powerful oxidizing chemical can destroy the nearby bacterial cell.

■ Clonal selection accounts for the specificity of antibody production.

Consider the diversity of foreign molecules that an animal can potentially encounter during a lifetime. Yet each B lymphocyte is preprogrammed to respond to only one of these millions of different antigens. Other antigens cannot combine with the same B-cell and induce it to secrete different antibodies. The astonishing implication is that each animal is equipped with millions of different preformed B lymphocytes, at least one for every possible antigen that might ever be encountered— including those specific for synthetic substances that do not exist in nature.

The currently accepted **clonal selection theory** proposes that diverse B lymphocytes are produced during fetal development, each capable of displaying a receptor and synthesizing the related antibody for a particular antigen, before ever being exposed to it. All offspring of a particular ancestral B lymphocyte form a lineage or family of identical cells, or a **clone,** which

is committed to producing the same specific antibody. B-cells remain dormant, not actually secreting their particular antibody product until (or unless) their receptors come into contact with the appropriate antigen. When an antigen gains entry to the body, it selectively activates the particular clone of B-cells that bear receptors on their surface uniquely specific for that antigen, hence the term "clonal selection theory" (● Figure 10–16).

The first antibodies produced by a newly formed B-cell are IgM immunoglobulins, which are inserted into the cell's plasma membrane rather than being secreted. Here they serve as receptor sites for binding with a specific kind of antigen, almost like "advertisements" for the kind of antibody the cell can produce. Binding of the appropriate antigen to a B-cell amounts to "placing an order" for the manufacture and secretion of large quantities of that particular antibody. However, as you will see, the order is usually not executed until an integrator cell, the *helper T lymphocyte,* confirms the order.

■ Selected clones differentiate into active plasma cells and dormant memory cells.

Specific activated B-cell clones multiply and differentiate into two cell types—*plasma cells* and *memory cells.* Most progeny are transformed into plasma cells, which are prolific producers of customized antibodies that contain the same antigen-binding sites as the surface receptors. However, plasma cells switch to the production of IgG antibodies, which are secreted rather than remaining membrane bound. In the blood, the secreted antibodies combine with an invading free (not bound to lymphocytes) antigen, marking it for destruction by the complement system, phagocytic ingestion, or other means.

Not all the new B lymphocytes produced by a specifically activated clone differentiate into antibody-secreting plasma cells. A small proportion become **memory cells,** which do not participate in the current immune attack against the antigen but instead remain dormant and expand the specific clone. Should the animal ever be exposed to the same antigen again, these memory cells are primed and ready for even more immediate action than were the original lymphocytes in the clone.

Primary and Secondary Responses

During initial contact with a microbial antigen, the antibody response is delayed for several days until plasma cells are formed and does not reach its peak for a couple of weeks (● Figure 10–17). This response is known as the **primary response.** Meanwhile, symptoms characteristic of the particular microbial invasion persist until either the invader succumbs to the mounting specific immune attack against it or the infected individual dies. After reaching the peak, the antibody levels gradually decline over a period of time, although some circulating antibody from this primary response may persist for a prolonged period. Long-term protection against the same antigen, however, is primarily attributable to the memory cells. If the same antigen ever reappears, the long-lived memory cells launch a more rapid, more potent, and longer-lasting **secondary response** than the primary response. This swifter, more powerful immune attack is frequently adequate to prevent or minimize overt infection on subsequent exposures to the same microbe,

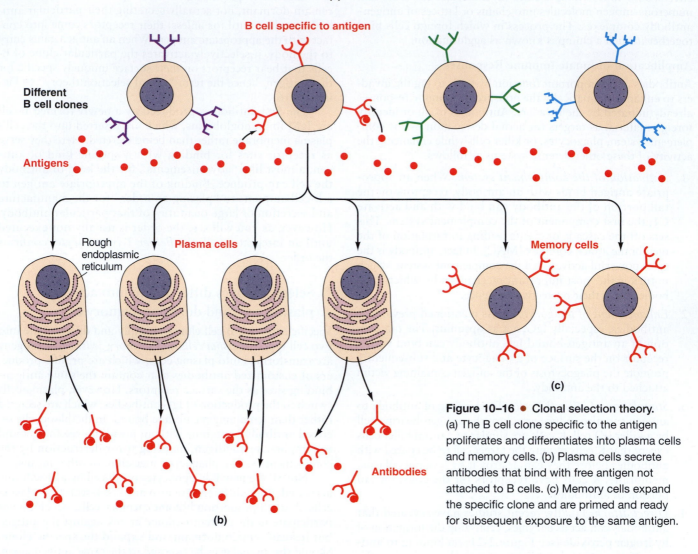

(a)

B cell specific to antigen

Different
B cell clones

Antigens

Rough
endoplasmic
reticulum

Plasma cells

Memory cells

(c)

Antibodies

(b)

Figure 10–16 ● Clonal selection theory.
(a) The B cell clone specific to the antigen
proliferates and differentiates into plasma cells
and memory cells. (b) Plasma cells secrete
antibodies that bind with free antigen not
attached to B cells. (c) Memory cells expand
the specific clone and are primed and ready
for subsequent exposure to the same antigen.

Figure 10–17 ● **Primary and secondary
immune responses.** (a) Primary response
on first exposure to a microbial antigen.
(b) Secondary response on subsequent
exposure to the same microbial antigen. The
primary response does not peak for a couple
of weeks, whereas the secondary response
peaks in a week. The magnitude of the
secondary response is 100 times that of
the primary response. (The relative antibody
response is in the logarithmic scale.)

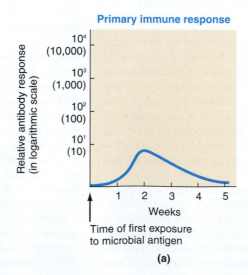

Primary immune response

Relative antibody response
(in logarithmic scale)

10^4
(10,000)

10^3
(1,000)

10^2
(100)

10^1
(10)

1 2 3 4 5
Weeks

Time of first exposure
to microbial antigen

(a)

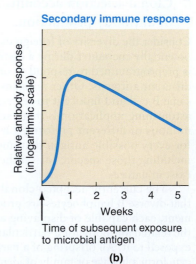

Secondary immune response

Relative antibody response
(in logarithmic scale)

1 2 3 4 5
Weeks

Time of subsequent exposure
to microbial antigen

(b)

forming the basis of long-term immunity against a specific disease.

Even though each individual has essentially the same original pool of different B-lymphocyte clones, the pool gradually becomes appropriately biased to respond most efficiently to each individual's particular antigenic environment. Those clones specific for antigens to which an individual is never exposed remain dormant for life, whereas those specific for antigens in the individual's environment typically become expanded and enhanced by the formation of highly responsive memory cells.

The huge repertoire of B-cells is built by reshuffling a small set of gene fragments.

Considering the millions of different antigens against which each mammal has the potential to actively produce antibodies, how is it possible for an individual to have such a tremendous diversity of B lymphocytes, each capable of producing a different antibody? Antibodies are proteins that are produced in accordance with a nuclear DNA blueprint. Because all cells of the body, including the antibody-producing cells, contain the same nuclear DNA, it is hard to imagine how enough DNA could be packaged within the nuclei of every cell to code for the millions of different antibodies (a different portion of the genetic code being used by each B-cell clone), along with all the other genetic instructions used by other cells. Actually, only a relatively small number of gene fragments actually code for antibody synthesis, but during B-cell development these fragments are cut, reshuffled, and spliced in a vast number of different combinations. Each different combination gives rise to a unique B-cell clone. Antibody genes are later diversified by *somatic mutation*. The antibody genes of already formed B-cells are highly prone to mutations in the region that codes for the variable antigen-binding sites on the antibodies. Each different mutant cell in turn gives rise to a new clone. Thus the great diversity of antibodies is made possible by the reshuffling of a small set of gene fragments during B-cell development as well as by further somatic mutation in already-formed B-cells. In this way, a huge antibody repertoire is possible using only a modest share of the genetic blueprint.

Active immunity is self-generated; passive immunity is "borrowed."

The production of antibodies as a result of exposure to an antigen is referred to as **active immunity** against that antigen. The direct transfer of antibodies actively forms a second way in which an individual can acquire antibodies by another animal. The immediate "borrowed" immunity conferred on receipt of preformed antibodies is known as **passive immunity.** Such transfer of antibodies of the IgG class normally occurs from the mother to the fetus in mammals across the placenta during intrauterine development. In addition, a lactating female's colostrum (first milk) contains IgA antibodies that provide further protection for mammary gland–fed infants (see p. 523). Passively transferred antibodies are usually broken down in less than a month, but meanwhile the newborn is provided important immune protection (essentially the same as its mother's) until it can begin actively mounting its own immune responses. Antibody-synthesizing ability does not develop for about a month after birth, at least in humans.

As you will see later, some invertebrates may also passively acquire immunity from their mothers.

Lymphocytes respond only to antigens presented to them by antigen-presenting cells.

B-cells typically cannot perform their task of antibody production without assistance from macrophages or other antigen-presenting cells and, in most cases from T-cells as well (● Figure 10–18). Relevant B-cell clones cannot recognize and produce antibodies in response to "raw" foreign antigens entering the body; before reacting to it a B-cell clone must be formally "introduced" to the antigen.

Using macrophages as an example of an antigen-presenting cell, invading organisms or other antigens are first engulfed by macrophages. These large phagocytes cluster around the appropriate B-cell clone and handle the formal introduction. During phagocytosis, the macrophage processes the raw antigen intracellularly and then "presents" the processed antigen by exposing it on the outer surface of the macrophage's plasma membrane in such a way that the adjacent B-cells can recognize and be activated by it. Specifically, when a macrophage engulfs a foreign microbe, it digests the microbe into antigenic peptides (small protein fragments). Each antigenic peptide is then bound to an MHC molecule (discussed in detail later), which is synthesized within the endoplasmic reticulum–Golgi complex. An MHC molecule has a deep groove into which a variety of antigenic peptides can bind, depending on what the macrophage has engulfed. Loading of the antigenic peptide onto an MHC molecule takes place in a newly discovered specialized organelle within antigen-presenting cells (APCs), the **compartment for peptide loading.** The MHC molecule then transports the bound antigen to the cell surface where it is presented to passing lymphocytes. The macrophage also secretes **interleukin 1** (alias LEM/EP) to further stimulate the B-cell.

The B-cell is now primed for action, but one more agent is often needed to fully activate it: the **helper T-cell.** Many antigens are similarly presented to helper T-cells by macrophages and, more importantly, by closely related dendritic cells. **Dendritic cells** are specialized antigen-presenting cells found as sentinels in almost every tissue, but especially in barrier tissues. After exposure to the appropriate signal, dendritic cells migrate through the lymphatic system to lymph nodes, where they cluster and activate T-cells. Researchers now think dendritic cells and helper T-cells together are the most important initial steps in the specific immune response. Where dendritic cells are mobile sensors, helper T-cells are essentially the integrators of specific immunity. They trigger the "primed" B-cells by secreting a chemical mediator, **B-cell growth factor,** which (in concert with interleukin 1) further contributes to B-cell differentiation into effector plasma cells. Therefore, mutually supportive interactions among macrophages, B-cells, and helper T-cells synergistically reinforce the phagocyte–antibody immune attack against the foreign intruder. ■ Table 10–3 summarizes the nonspecific and specific immune strategies that defend against bacterial invasion.

What triggers dendritic cells to begin their migration? There is evidence that, like phagocytes, they use TLRs (Toll-like receptors, p. 429) to sense unusual or foreign "patterns"

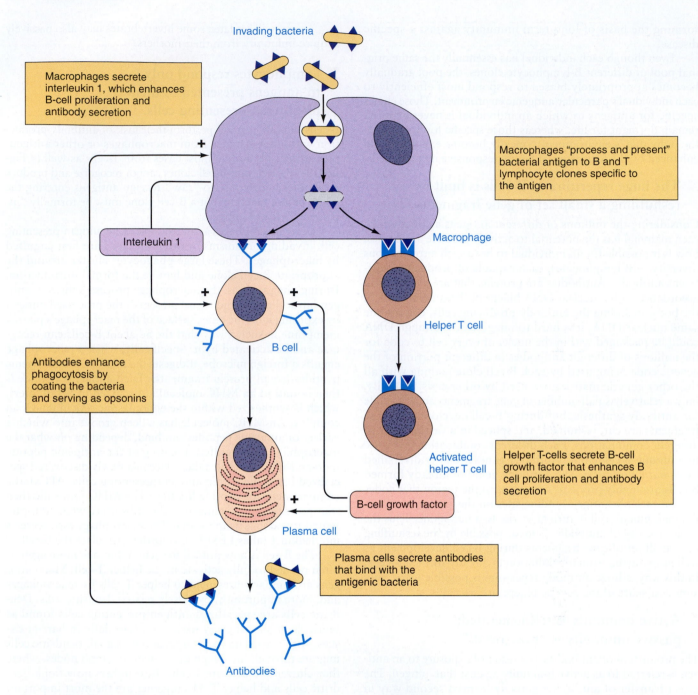

Invading bacteria

Macrophages secrete interleukin 1, which enhances B-cell proliferation and antibody secretion

Macrophages "process and present" bacterial antigen to B and T lymphocyte clones specific to the antigen

Interleukin 1

Macrophage

+

+   +

B cell

Helper T cell

Antibodies enhance phagocytosis by coating the bacteria and serving as opsonins

Activated helper T cell

Helper T-cells secrete B-cell growth factor that enhances B cell proliferation and antibody secretion

+   B-cell growth factor

Plasma cell

Plasma cells secrete antibodies that bind with the antigenic bacteria

Antibodies

Figure 10–18 ● Synergistic interactions among macrophages, B-cells, and helper T-cells. B- and T-cells cannot react to a newly entering foreign antigen until the antigen has been processed and presented to them by macrophages or other antigen-presenting cells. Macrophages also secrete interleukin 1, which stimulates proliferation of the activated B-cells. These B-cells are transformed into plasma cells, which produce antibodies against the antigen. Activated helper T-cells secrete B-cell growth factor, which further stimulates B-cell proliferation and antibody production. The antibodies not only lead to the demise of the foreign antigen but also serve as opsonins to enhance phagocytosis by the macrophages.

(that is, antigens). This concept of sensing non-self is sometimes called the "stranger" hypothesis. However, an alternative idea called the "danger" hypothesis has emerged in recent years; it proposes that immune sensing is based on chemicals released from damaged body cells. Recently, for example, researchers have found that crystals of *uric acid* (p. 523), which can form when too much dissolved uric acid is released from damaged body cells, are potent activators of dendritic cells.

Table 10–3 ▮ Innate and Adaptive Immune Responses to Bacterial Invasion

Innate Immune Mechanisms	Adaptive Immune Mechanisms
Inflammation	Processing and presenting of bacterial antigen by macrophages to B-cells specific to the antigen
Engulfment of invading bacteria by resident tissue macrophages	Differentiation of the activated B-cell clone into plasma and memory cells
Histamine-induced vascular responses to enhance delivery of increased blood flow to the area, bringing in additional immune-effector cells and plasma proteins	Secretion by plasma cells of customized antibodies, which specifically bind to invading bacteria
Walling off of the invaded area by a fibrin clot	Enhancement by interleukin 1 secreted by macrophages
Emigration of neutrophils and monocytes/macrophages to the area to engulf and destroy foreign invaders and to remove cellular debris	Enhancement by helper T-cells activated by the same bacterial antigen processed and presented to them by macrophages and diadrite cells
Secretion by phagocytic cells of chemical mediators, which enhance both innate and adaptive immune responses and induce local and systemic symptoms associated with an infection	Binding of antibodies to invading bacteria and enhancement of innate mechanisms that lead to the bacteria's destruction
Nonspecific activation of the complement system	Action as opsonins to enhance phagocytic activity
Formation of a hole-punching membrane attack complex that lyses bacterial cells	Activation of lethal complement system
Enhancement of many steps of inflammation	Stimulation of killer cells, which directly lyse bacteria
	Introduction of hydrogen peroxide
	Persistence of memory cells capable of responding more rapidly and more forcefully should the same bacteria be encountered again

Currently, researchers think both "stranger" and "danger" signals are at work in a complementary fashion.

T Lymphocytes: Cell-Mediated Immunity

The T lymphocytes are equally important as B-cells in defense against most viral infections and also play an important regulatory role in immune mechanisms (▮ Table 10–4). Like B-cells,

T-cells are clonal and exquisitely antigen-specific. On its plasma membrane, each T-cell bears unique receptor proteins, similar although not identical to the surface receptors on B-cells. Immature lymphocytes acquire their T-cell receptors in the thymus during their differentiation into T-cells. Unlike B-cells, T-cells are activated by foreign antigen only when it is present on the surface of a cell that also carries a marker of the individual's own identity; that is, both foreign antigens and **self-antigens** must be present on a cell's surface before a T-cell can bind with it. It is during thymic education that T-cells learn to recognize

Table 10–4 ▮ B versus T Lymphocytes

Characteristic	B Lymphocytes	T Lymphocytes
Ancestral origin	Bone marrow	Bone marrow
Site of maturational processing	Bone marrow	Thymus
Receptors for antigen	Antibodies inserted in the plasma membrane serve as surface receptors; highly specific	Surface receptors present but differing from antibodies; highly specific
Bind with	Extracellular antigens such as bacteria, free viruses, and other circulating foreign material	Foreign antigen in association with self-antigen, such as virus-infected cells
Antigen must be processed and presented by macrophages	Yes	Yes
Types of active cells	Plasma cells	Cytotoxic T-Cells, helper T-cells
Formation of memory cells	Yes	Yes
Type of immunity	Antibody-mediated immunity	Cell-mediated immunity
Secretory product	Antibodies	Cytokines
Function	Help eliminate free foreign invaders by enhancing nonspecific immune responses against them; produce immunity against most bacteria and a few viruses	Lyse virus-infected cells and cancer cells; provide immunity against most viruses and fungi and a few bacteria; aid B-cells in antibody production
Life span	Short	Long

foreign antigens only in combination with the individual's own tissue antigens, a lesson that is passed on to all T-cells' future progeny. The importance of this dual antigen requirement and the nature of the self-antigens is described shortly.

A delay of a few days generally follows exposure to the appropriate antigen before **sensitized**, or **activated**, **T-cells** are prepared to launch a cell-mediated immune attack. When exposed to a specific antigen combination, cells of the complementary T-cell clone proliferate and differentiate for several days, yielding large numbers of activated T-cells that carry out various cell-mediated responses.

■ **The two types of T-cells are cytotoxic and helper T-cells.**

There are two main subpopulations of T-cells, depending on their roles when activated by antigen:

1. **Helper T-cells (CD4 cells)**, integrators that coordinate the development of antigen-stimulated B-cells into antibody-secreting cells, enhance activity of the appropriate cytotoxic T-cells, and activate macrophages.

2. **Cytotoxic T-cells** (alias **killer T-cells** or **CD8 cells**). Whereas B-cells and antibodies defend against conspicuous invaders in the extracellular fluid, these effector T-cells defend against covert invaders that hide inside cells where antibodies and the complement system cannot reach them. Unlike B-cells, which secrete antibodies that can attack antigens at long distances, these T-cells do not secrete antibodies. Instead, they must be in direct contact with their targets, thus giving rise to the term *cell-mediated immunity*. T-cells release chemicals that destroy targeted cells with which they make contact, such as body cells invaded by viruses, cancer cells that have mutated proteins resulting from malignant transformations, and transplanted cells.

Like B-cells, not all activated T-cell progeny become effector T-cells. A small proportion of them remain dormant, serving as a pool of **memory T-cells** that are primed and ready to respond even more swiftly and vigorously should the same foreign antigen ever reappear within a body cell. T-cells, even activated ones, generally have long life spans, in contrast to B-cells, which rapidly work themselves to death producing antibodies once they have been converted into plasma cells on antigen stimulation. Thus, immunity for cell-mediated responses is similar to that for antibody responses, but it is generally of longer duration.

■ **Helper T-cells secrete chemicals that regulate other immune cells.**

The vast majority of the billions of T lymphocytes are believed to be of the helper type, which do not directly participate in the immune destruction of invading pathogens. Collectively, these subpopulations are referred to as **integrator** or **regulatory T-cells**, because they modulate the activities of B-cells and cytotoxic T-cells as well as their own activities and those of macrophages.

Cytokines

Exposure to antigen frequently activates both the B- and T-cell mechanisms simultaneously. Just as the integrator T-cells can facilitate the secretion of antibodies by B-cells, antibodies may either enhance or block the ability of cytotoxic T-cells to destroy a victim cell, depending on the circumstances. Most effects that lymphocytes exert on other immune cells are mediated by the secretion **cytokines**, the majority of which are produced by helper T-cells. Unlike antibodies, cytokines do not interact directly with the antigen responsible for inducing their production. Instead, cytokines spur other immune-system cells into action to help ward off the invader. The following are among the best known of the helper T-cell cytokines:

1. As noted earlier, helper T-cells secrete *B-cell growth factor*, which enhances the antibody-secreting ability of the activated B-cell clone. Antibody secretion is greatly reduced in the absence of helper T-cells. New evidence shows they are needed for the formation of memory cytotoxic T-cells as well.

2. Helper T-cells similarly secrete **T-cell growth factor**, also known as **interleukin 2 (IL-2)**, which augments the activity of cytotoxic T-cells and even other helper T-cells responsive to the invading antigen. In typical interplay fashion, IL-1 secreted by macrophages not only enhances the activity of both the appropriate B- and T-cell clones but also stimulates the secretion of IL-2 by the activated helper T-cells.

3. Some chemicals secreted by T-cells act as *chemotaxins* to lure more neutrophils and macrophages-to-be to the invaded area.

4. Once macrophages are attracted to the area, **macrophage-migration inhibition factor**, another important cytokine released from helper T-cells, keeps these large phagocytic cells in the region by inhibiting their outward migration. As a result, a great number of chemotactically attracted macrophages accumulate in the infected area. This factor also confers greater phagocytic power on the gathered macrophages. These so-called **angry macrophages** have more powerful destructive ability. They are especially important in defending against the bacteria that cause tuberculosis, because such microbes can survive simple phagocytosis by unactivated macrophages.

5. Some cytokines secreted by helper T-cells activate eosinophils and promote the development of IgE antibodies for defense against parasitic worms.

T Helper 1 and T Helper 2 Cells

Helper T-cells are by far the most numerous of the T-cells, making up 60 to 80% of circulating T-cells. Recent studies have demonstrated the existence of two subsets called T_H1 and T_H2 cells. T_H1 cells are responsible for activating cytotoxic T-cells (see next section), appropriate for infections inside of body cells, whereas T_H2 cells promote antibody-mediated immunity by B-cells.

Because of the important role these cells play in "turning on" the full power of all of the other activated lymphocytes and macrophages, helper T-cells may constitute the immune system's "master integrator" in feedback terminology. It is for this reason that **acquired immune deficiency syndrome (AIDS)**, caused by the **human immunodeficiency virus (HIV)** is so devastating to the immune defense system. The AIDS virus selectively invades helper T-cells, destroying or incapacitating the cells that normally orchestrate much of the immune response. Wild and domestic cats are subject to a similar disease caused by **feline immunodeficiency virus** or **FIV,** which also attacks T-cells. (HIV and FIV are species-specific, fortunately for pet owners and cats.)

■ Cytotoxic T-cells secrete chemicals that destroy target cells.

Cytotoxic T-cells are microscopic "hit men." The targets of these destructive effector cells most frequently are host cells infected with viruses. When a virus invades a body cell, as it must to survive, the cell breaks down the envelope of proteins surrounding the virus and loads a fragment of this viral antigen piggyback onto a newly synthesized self-antigen. This self- and viral antigen complex is inserted into the host cell's surface membrane, where it serves as a red flag indicating the cell is harboring the invader (● Figure 10–19, steps ① and ②). To attack the intracellular virus, cytotoxic T-cells must destroy the infected host cell in the process. Cytotoxic T-cells of the clone specific for this particular virus recognize and bind to the viral antigens and self-antigens on the surface of the infected cell (Figure 10–19, step ③). Thus sensitized by viral antigen, a cytotoxic T-cell either directly kills the victim cell by releasing chemicals that lyse the attacked cell before viral replication can begin (Figure 10–19, step ④) or indirectly destroys the infected cell by signaling it to commit suicide.

The direct means by which cytotoxic T-cells as well as natural killer cells destroy a targeted cell is by releasing **perforin** molecules, which penetrate into the target cell's surface membrane and join together to form porelike channels (● Figure 10–20). This technique of killing a cell by punching holes in its membrane is similar to the method employed by the membrane attack complex (MAC) of the complement cascade. This contact-dependent mechanism of killing has been termed the "kiss of death." Cytotoxic T-cells can also indirectly bring about death of infected host cells by releasing **granzymes,** which are enzymes similar to digestive enzymes. Granzymes enter the target cell via the perforins. Once inside, granzyme activity induces these virus-infected cells to undergo *apoptosis* (cell suicide, p. 93).

The virus released on destruction of the host cell is directly destroyed in the extracellular fluid by phagocytic cells, neutralizing antibodies, and the complement system. Meanwhile, the cytotoxic T-cell, which has not been harmed in the process, can move on to kill other infected host cells. The surrounding healthy cells replace the lost cells by means of cell division.

Recall that other nonspecific defense mechanisms also come into play to combat viral infections, most notably natural killer cells, interferon, macrophages, and the complement system. As usual, an intricate web of interplay exists among the immune defenses that are launched against viral invaders (■ Table 10–5).

The usual method of destroying virus-infected host cells is not appropriate for the nervous system. If cytotoxic T-cells destroyed virus-infected neurons, the lost cells could not be replaced, because most neurons cannot reproduce. Fortunately, virus-infected neurons are spared from extermination by the immune system, but how then are neurons protected from viruses? Immunologists long thought that the only antiviral defenses for neurons were those aimed at free viruses in the extracellular fluid. Surprising new research has revealed that antibodies not only target viruses for destruction in the extra-

Figure 10–19 ●
A cytotoxic T-cell lysing a virus-invaded host cell.

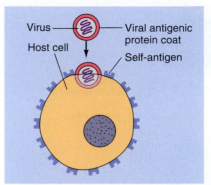

step ❶ A virus invades a host cell.

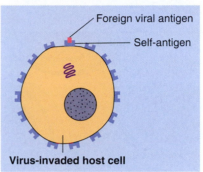

step ❷ The viral antigen is displayed on the surface of the host cell alongside the cell's self-antigen.

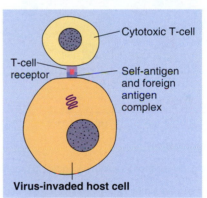

step ❸ The cytotoxic T-cell recognizes and binds with a specific foreign antigen (viral antigen) in association with the self-antigen.

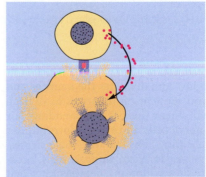

step ❹ The cytotoxic T-cell releases chemicals that destroy the attacked cell before the virus can enter the nucleus and start to replicate.

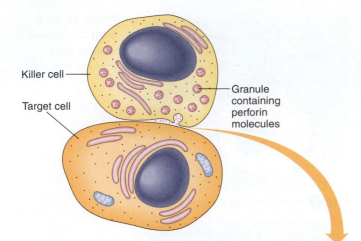

Killer cell

Target cell

Granule containing perforin molecules

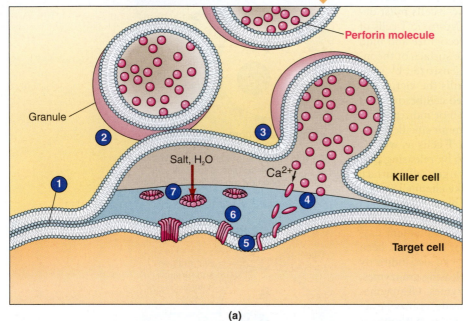

Perforin molecule

Granule

Salt, H$_2$O

Ca^{2+}

Killer cell

Target cell

(a)

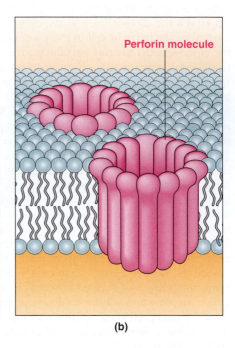

Perforin molecule

(b)

1. The killer cell binds to its target cell.

2. As a result of this binding, the killer cell's perforin-containing granules fuse with the plasma membrane.

3. The granules disgorge their perforin by exocytosis into a small pocket of intercellular space between the killer cell and its target.

4. On exposure to Ca^{2+} in this ECF space, the individual perforin molecules change from a spherical to a cylindrical shape.

5. The remodeled perforin molecules bind to the target cell membrane and insert into it.

6. Individual perforin molecules group together like staves of a barrel to form pores.

7. The pores admit salt and H$_2$O, causing the target cell to swell and burst.

cellular fluid but can also eliminate viruses inside neurons. It is unclear whether antibodies actually enter the neurons and interfere directly with viral replication (neurons have been shown to take up antibodies near their synaptic endings) or bind with the surface of nerve cells and trigger intracellular changes that stop viral replication. The fact that some viruses, such as the herpes virus, persist for years in nerve cells, occasionally "flaring up" to produce symptoms, demonstrates that the antibodies' intraneuronal mechanism does not provide a foolproof antiviral defense for neurons.

■ The immune system is normally tolerant of self-antigens.

The term **tolerance** refers to the phenomenon of preventing the immune system from attacking the person's own tissues. During the genetic "cut, shuffle, and paste process" that goes on during lymphocyte development, some B- and T-cells are by chance formed that could react against the body's own tissue antigens. If these lymphocyte clones were allowed to function, they would destroy the individual's own body. Fortunately, the

Table 10–5 ▌ Defenses against Viral Invasion

When the virus is free in the extracellular fluid,

Macrophages

Destroy the free virus by phagocytosis.

Process and present the viral antigen to both B- and T-cells.

Secrete interleukin 1, which activates B- and T-cell clones specific to the viral antigen.

Plasma cells derived from B-cells specific to the viral antigen secrete antibodies that

Neutralize the virus to prevent its entry into a host cell.

Activate the complement cascade that directly destroys the free virus and enhances phagocytosis of the virus by acting as an opsonin.

When the virus has entered a host cell (which it must do to survive and multiply, with the replicated viruses leaving the original host cell to enter the extracellular fluid in search of other host cells),

Interferon

Is secreted by virus-infected cells.

Binds with and prevents viral replication in other host cells.

Enhances the killing power of macrophages, natural killer cells, and cytotoxic T-cells.

Natural killer cells

Nonspecifically lyse virus-infected host cells.

Cytotoxic T-cells

Are specifically sensitized by the viral antigen; lyse the infected host cells before the virus has a chance to replicate.

Helper T-cells

Secrete cytokines, which enhance cytotoxic T-cell activity and B-cell antibody production.

When a virus-infected cell is destroyed, the free virus is released into the extracellular fluid, where it is attacked directly by macrophages, antibodies, and the activated complement components.

immune system normally does not produce antibodies or activated T-cells against the body's own self-antigens but instead directs its destructive tactics only at foreign antigens. At least five different mechanisms appear to be involved in tolerance:

1. *Clonal deletion.* In response to continuous exposure to body antigens early in development, lymphocyte clones specifically capable of attacking these self-antigens in some cases are permanently destroyed. New findings demonstrate that the thymus triggers apoptosis of immature T-cells that would react to the body's own proteins. Without ever having had a chance to leave the thymus, these self-destructed T-cells are gobbled up by macrophages. This is thought to be the major mechanism by which tolerance is developed.

2. *Clonal anergy.* The premise of **clonal anergy** is that a lymphocyte must receive two specific simultaneous signals to be activated (turned on), one from its compatible antigen and a stimulatory cosignal molecule known as **B7** found only on the surface of an antigen-presenting cell. Both sig-

nals are present for foreign antigens, which are introduced to T-cells by antigen-presenting cells such as macrophages. Once a T-cell is turned on by finding its matching antigen in accompaniment with the cosignal, the cell no longer needs the cosignal to interact with other cells. For example, an activated cytotoxic T-cell can destroy any virus-invaded cell that bears the viral antigen even though the infected cell does not have the cosignal. In contrast, these dual signals—antigen plus cosignal—never are present for self-antigens because these antigens are not handled by cosignal-bearing antigen-presenting cells. The first exposure to a single signal from a self-antigen turns *off* the compatible T-cell, rendering the cell unresponsive to further exposure to the antigen instead of spurring the cell to proliferate. This reaction is referred to as clonal anergy (*anergy,* "lack of energy") because T-cells are being inactivated (that is, "become lazy") rather than activated by their antigens. Clonal anergy is a backup to clonal deletion. Anergized lymphocyte clones survive but are unable to function.

3. *Receptor editing.* Once a B-cell that bears a receptor for one of the body's own antigens encounters the self-antigen, the B-cell escapes death or anergy by rapidly changing its receptor to a non-self version.

4. *Antigen sequestering.* Some self-molecules are normally hidden from the immune system because they never come into direct contact with the extracellular fluid in which the immune cells and their products circulate.

5. *Immune privilege.* A few tissues, most notably the testes and the eyes, are considered immune privileged because they escape immune attack even when they are transplanted in an unrelated individual. Scientists recently discovered that the cellular plasma membranes in these immune-privileged tissues have a specific molecule that triggers apoptosis of approaching activated lymphocytes that could attack the tissues.

Occasionally, the immune system fails to make the distinction between self-antigens and foreign antigens, unleashing its deadly powers against one or more of the body's own tissues. A condition in which the immune system fails to recognize and tolerate self-antigens associated with particular tissues is known as an **autoimmune disease,** which can occur in any mammal, but are best studied in humans and domestic animals. There are many examples, such as Type I diabetes, multiple sclerosis, lupus, rheumatoid arthritis, and myasthenia gravis. Autoimmune diseases may arise from a number of different causes:

1. Normal self-antigens may be modified by factors such as drugs, glycosylation, environmental chemicals, viruses, or genetic mutations so that they are no longer recognized and tolerated by the immune system.

2. Exposure of normally inaccessible self-antigens sometimes induces an immune attack against these antigens. Because the immune system is usually never exposed to hidden self-antigens, it does not "learn" to tolerate them. Inadvertent exposure of these normally inaccessible antigens to the immune system because of tissue disruption caused by injury or disease can lead to a rapid immune attack against the affected tissue, just as if these self-proteins were foreign invaders.

3. Exposure of the immune system to a foreign antigen structurally almost identical to a self-antigen may induce the production of antibodies or activated T-cells that not only interact with the foreign antigen but also cross-react with the closely similar body antigen. An example of this molecular mimicry is the streptococcal bacteria responsible for "strep throat." The bacteria have antigens that are structurally very similar to self-antigens in the tissue covering the heart valves of some individuals, in which case the antibodies produced against the streptococcal organisms may also bind with this heart tissue. The resultant inflammatory response is responsible for the heart valve lesions associated with *rheumatic fever*.

4. Fetal cells, which may linger in a mammalian mother for years after pregnancy, may trigger an immune attack that turns on the mother's own tissues.

■ The major histocompatibility complex is the code for self-antigens.

What is the nature of the self-antigens that the immune system learns to recognize as markers of an animal's own cells? These self-antigens are plasma membrane–bound glycoproteins (proteins with sugar attached) known as **MHC molecules** because their synthesis is directed by a group of genes called the **major histocompatibility complex** or **MHC** (*histo,* "tissue"). These are the same MHC molecules that escort engulfed foreign antigen to the cell surface for presentation by antigen-presenting cells. The MHC genes are the most variable ones in mammals. More than a hundred different MHC molecules have been identified in human tissue, for example, but each individual has a code for only three to six of these possible antigens. Because of the tremendous number of different combinations possible, the exact pattern of MHC molecules varies from one individual to another, much like a "biochemical fingerprint" or "molecular identification card," except in identical twins, who have the same MHC self-antigens.

MHC molecules without antigen on a cell surface signal to immune cells, "Leave me alone; I'm one of you." T-cells typically bind with MHC self-antigens only when they are in association with a foreign antigen, such as a viral protein, also displayed on the cell surface in a groove on the top of the MHC molecule. Thus T-cell receptors bind only with body cells making the statement—by bearing both self- and non–self-antigens on their surface—"I, one of your own kind, have been invaded. Here's a description of the enemy I am housing within." Only T-cells that specifically match up with both the self- and foreign antigen can bind with the infected cell.

Loading of Foreign Peptide on MHC

Unlike B-cells, T-cells cannot bind with foreign antigen that is not in association with self-antigen. It would be futile for T-cells to bind with free, extracellular antigen—they cannot defend against foreign material unless it is intracellular. A foreign protein first must be enzymatically broken down within a body cell into small fragments known as *peptides*. These antigenic peptides are inserted into the binding groove of a newly synthesized MHC molecule before the MHC–foreign antigen complex travels to the surface membrane. Once displayed at the cell surface, the combined presence of these self- and non–self-

antigens alerts the immune system to the presence of an undesirable agent within the cell. Highly specific T-cell receptors fit a particular MHC–foreign antigen complex in complementary fashion. This binding arrangement can be likened to a sausage in a bun, with the MHC molecule being the bottom of the bun, the T-cell receptor the bun's top, and the foreign antigen the sausage (● Figure 10–21). In the case of cytotoxic T-cells, the outcome of this binding is destruction of the infected body cell. Because cytotoxic T-cells do not bind to MHC self-antigens in the absence of foreign antigen, normal body cells are protected from lethal immune attack.

Class I and II MHC Glycoproteins

T-cells only become active when they match a given MHC–foreign peptide combination. In addition to having to fit a specific foreign peptide, the T-cell receptor must also match the appropriate MHC protein. Each individual has two main classes of MHC-encoded glycoproteins that are differentially recognized by cytotoxic T and helper T-cells (● Figure 10–22). Cytotoxic T-cells can respond to foreign antigen only in association with **class I MHC glycoproteins,** which are found on the surface of virtually all nucleated body cells. **Class II MHC glycoproteins,** which are recognized by helper T-cells, are restricted to the surface of a few special types of immune cells, such as B-cells, cytotoxic T-cells, and macrophages. The class I and II markers serve as signposts to guide cytotoxic and helper T-cells to the precise cellular locations where their immune capabilities can be most effective.

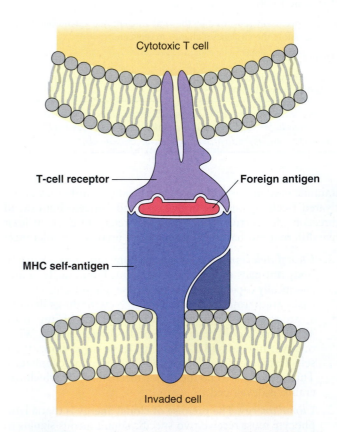

Figure 10–21 ● Binding of a T-cell receptor with an MHC self-antigen and foreign antigen complex.

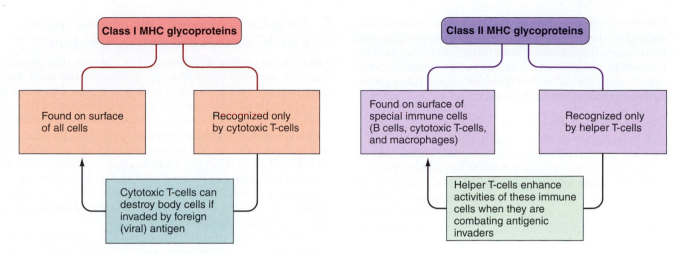

Figure 10–22 ● **Distinctions between class I and class II major histocompatibility complex (MHC) glycoproteins.** Specific binding requirements for the two types of T-cells ensure that these cells bind only with the targets that they can influence. Cytotoxic T-cells can recognize and bind with foreign antigen only when the antigen is in association with class I MHC glycoproteins, which are found on the surface of all body cells. This requirement is met when a virus invades a body cell, whereupon the cell is destroyed by the cytotoxic T-cells. Helper T-cells, which enhance the activities of other immune cells, can recognize and bind with foreign antigen only when it is in association with class II MHC glycoproteins, which are found only on the surface of these other immune cells.

Cytotoxic T-cells can also link up with any cancerous body cell, because class I MHC molecules also display mutated cell proteins characteristic of these abnormal cells. Because any nucleated body cell can be invaded by viruses or become cancerous, essentially all cells display class I MHC glycoproteins, enabling cytotoxic T-cells to attack any virus-invaded host cell or any cancer cell.

■ **Immune surveillance against cancer cells involves an interplay among immune cells and interferon.**

Besides destroying virus-infected host cells, another important function generally attributed to the T-cell system is its role in recognizing and destroying newly arisen, potentially cancerous tumor cells before they have a chance to multiply and spread, a process known as **immune surveillance.** Any normal cell may be transformed into a cancer cell if mutations occur within its genes responsible for controlling cell division and growth. Such mutations may occur by chance alone or, more frequently, by exposure to **carcinogenic** (cancer-causing) factors such as ionizing radiation, certain environmental chemicals, certain viruses, or physical irritants.

Cell multiplication and growth are normally under strict control, but the regulatory mechanisms are largely unknown. Cell multiplication in an adult is generally restricted to the replacement of lost cells. Furthermore, cells normally respect their own place and space in the body's society of cells. If a cell that has been transformed into a tumor cell manages to escape destruction, however, it defies the normal controls on its proliferation and position. Unrestricted multiplication of a single tumor cell results in a **tumor** that consists of a clone of cells identical to the original mutated cell. If the mass is slow growing, stays put in its original location, and does not infiltrate into the surrounding tissue, it is considered a **benign** tumor. In contrast, the transformed cell may multiply rapidly and form an invasive mass that lacks the altruistic behavior characteristic of normal cells. Such invasive tumors are known as **malignant** tumors, commonly referred to as **cancer.** Malignant tumor cells usually do not adhere well to the neighboring normal cells, with the result that often some of the cancer cells break away from the parent tumor. These "emigrant" cancer cells are transported through the blood to new territories, where they continue to proliferate, forming multiple malignant tumors. **Metastasis** is the term applied to this spread of cancer to other parts of the body.

Even though many body cells undergo mutations throughout an animal's lifetime, most of these mutations do not result in malignancy, for three reasons:

1. Only a fraction of the mutations involve loss of control over the cell's growth and multiplication. More frequently, other facets of cell function are altered.

2. Evidence suggests that a cell usually becomes cancerous only after an accumulation of multiple independent mutations. A single mutation generally is not sufficient. This requirement could contribute at least in part to the much higher incidence of cancer in older individuals, in whom mutations have had more time to accumulate in a single cell lineage. Some cancers are caused by tumor viruses, which permanently alter particular DNA sequences of the cells they invade.

3. Potentially cancerous cells that do arise are usually destroyed by the immune system early in their development. Presumably, the immune system recognizes cancer cells because they bear new and different surface antigens alongside the cell's normal self-antigens, because of either genetic mutation or invasion by a tumor virus.

Effectors of Immune Surveillance

Immune surveillance against cancer depends on an interplay among three types of immune cells—*cytotoxic T-cells, natural killer cells,* and *macrophages*—as well as interferon. Not only are all three of these immune cell types able to attack and destroy cancer cells directly, but all of them also secrete interferon. Interferon in turn inhibits multiplication of cancer cells and increases the killing ability of the immune cells (● Figure 10–23). Because NK cells do not require prior exposure and sensitization to a cancer cell before being able to launch a lethal attack, they are the first line of defense against cancer. In addition, cytotoxic T cells take aim at abnormal cancer cells bearing mutated cell proteins on their surface in conjunction with the class I MHC molecules. Cytotoxic T-cells are believed to be especially important in defending against the few kinds of virus-induced cancer. On contacting a cancer cell, both these killer cells release perforin and other toxic chemicals that destroy the targeted mutant cell. Macrophages, in addition to clearing away the remains of the dead victim cell, can engulf and destroy cancer cells intracellularly.

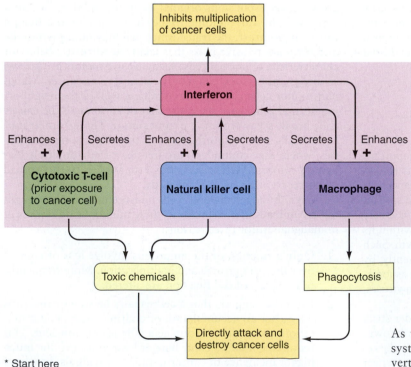

* Start here

Figure 10–23 ● **Immune surveillance against cancer.** Anticancer interactions of cytotoxic T-cells, natural killer cells, macrophages, and interferon.

■ A regulatory loop links the immune system with the nervous and endocrine systems.

From the preceding discussion, it is obvious that complex feedback loops operate within the immune system itself. Until fairly recently, the immune system was believed to function independently of other control systems in the body. Investigations along a number of lines indicate, however, that important links exist between the immune system and the other two major control systems, the nervous and endocrine systems. For example, IL-1 (p. 437) can turn on the stress response by activating a sequence of nervous and endocrine events that result in the secretion of the glucocorticoids, *cortisol or corticosterone,* one of the major hormones released during stress, which mobilizes energy reserves (p. 285).

In the reverse direction, lymphocytes and macrophages are responsive to blood-borne signals from the nervous system and from certain endocrine glands. These important immune cells have receptors for a wide variety of neurotransmitters, hormones, and other chemical mediators. For example, glucocorticoids and other chemical mediators of the stress response have a profound *immunosuppressive* effect, inhibiting many functions of lymphocytes and macrophages and decreasing the production of cytokines. Thus a negative-feedback loop appears to exist between the immune system and the nervous and endocrine systems:

Immune responses activated → IL-1 released by immune cells → neural and hormonal stress responses activated including glucocorticoid release → immune responses suppressed

Presumably this feedback suppression evolved to keep the immune system in check after it has been activated following injury. However, the suppression seems inappropriate under certain circumstances (see p. 286).

There are other important links between the immune and nervous systems in addition to the glucocorticoid connection. For example, many of the immune organs such as the thymus, spleen, and lymph nodes are innervated by the sympathetic nervous system, the branch of the nervous system called into play during stress-related "fight-or-flight" situations (see p. 287). In the reverse direction, immune system secretions act on the brain to produce fever and other general symptoms that accompany infections.

Acquired Immunity in Other Vertebrates

As we noted earlier, the acquired (adaptive) immune system has only been thoroughly documented in jawed vertebrates (although, as we note later, new studies suggest different versions in some arthropods). Although studies on immunity in nonmammalian vertebrates are relatively limited, many interesting discoveries have been made.

?

How Did the Adaptive Immune System Arise in Jawed Vertebrates?
The adaptive immune system of higher vertebrates seems to have originated rather abruptly in evolutionary terms. How could this have happened by the known mechanisms of evolution? For a recent hypothesis, see the box, "Molecular Biology and Genomics: An Accidental Origin for Adaptive Immunity?"

Recent research on nonmammalian vertebrates and related invertebrates has focused on understanding the genetic evolution of adaptive immunity. Most vertebrates appear to have B- and T-cell systems similar to mammals, all generating millions of different antibodies and antigen receptors. Most vertebrate systems seem to generate this diversity in ways similar to mammals, but not all do. It is perhaps most instructive to compare mammals to chondrichthyes, the most primitive living vertebrates to have adaptive immunity. Like mammals, chondrichthyes (sharks, skates, and rays) have B- and T-cells and exhibit gene shuffling to make millions of different antibodies and receptors. Shark B-cells make four different classes of antibodies, but only one of these—the IgM class—is the same as one of the five classes found in mammals. Shark antibodies seem to be more primitive in some ways. First, they do not mutate to increase diversity and thus—sometimes—to improve antigen binding, as they do in mammals (p. 447). Second, gene shuffling in sharks involves the recombination of a much lower diversity of gene segments than in mammals. But shark genomes make up for this in ways not found in mammals. First, the gene segments that are recombined in mammals are found only in one cluster on one chromosome. Sharks, however, have hundreds of antibody-generating gene clusters on many chromosomes. Thus, although each cluster has a lower diversity of gene segments to shuffle around, there are many more such clusters available. Second, there are also hundreds of other gene clusters that already have the coding segments joined together, so that they can rapidly produce antibodies without waiting for genetic recombination. These gene combinations are fixed and heritable, so they do not generate more diversity than the animal was born with. However, they work well (and much faster than mammalian counterparts) for common pathogens that sharks have encountered for millennia.

These studies on sharks reveal that the basic B- and T-cell system, with gene shuffling of receptor and antibody gene clusters to increase diversity, evolved early on in the jawed vertebrates. But since the origin of adaptive immunity, evolution has led to different uses of these gene clusters and the coding segments within them. We cannot say that one version is more primitive than the other. After all, the shark system, although somewhat different, has helped these animals survive successfully far longer than mammals have yet been on Earth.

MOLECULAR BIOLOGY *AND* GENOMICS

An Accidental Origin for Adaptive Immunity?

All animals have innate immunity, but (as far as we know) only vertebrates with jaws have the adaptive system based on generating millions of different receptors and antibodies (in T and B lymphocytes). For example, lancelets (invertebrates closely related to the first vertebrates) have genomes with codes that are similar to the gene segments that give rise to antibody diversity in mammals. However, the functions of these lancelet genes are unknown and they do not undergo the gene shuffling of adaptive immunity. The adaptive system is also absent in hagfish and lampreys (jawless vertebrates). Lampreys appear to have lymphocyte-like cells, but they lack the diverse receptors of T lymphocytes, and their function is unknown. The adaptive system is found at a high level of complexity in the most primitive living jawed fishes, the chondrichthyes (sharks and relatives). The system thus seems to have appeared suddenly in evolutionary terms, with few precursors. How could this complex immunity innovation evolve so quickly, and why only in jawed vertebrates? Recent genomic studies suggest that the initial event might have been an evolutionary accident involving a virus!

Recall that the adaptive immune system can generate millions of different proteins by shuffling and joining a relatively small number of gene segments. This process, called *genetic recombination,* is dependent on two genes called *RAG1* and *RAG2* ("recombinant-activating genes"). The genes code for a *transposase,* an enzyme that cuts out sections of DNA at one site in a chromosome, then inserts them at another location. Transposase activity has a partially random nature that is the key to generating millions of different B-cell antibodies and T-cell receptors.

Researchers used DNA probes to study other vertebrates, and discovered a distinct divide: All jawed vertebrates have *RAG1* and *RAG2,* but hagfish and lamprey do not. Nor do those ancient fishes have any genes with similar coding that might provide a precursor. There seems to have been a sudden evolutionary transition, estimated at 450 million years ago. Transposase events have been well documented in other organisms, where the genes have "jumped" from one species to another, causing a sudden genetic leap. The jumping transposase genes seem to be carried by viruses. If this mechanism is the origin of the *RAG* genes in higher vertebrates, we may owe our complex immune systems to an ancestral fish that survived a viral infection!

Innate (and Acquired?) Immunity in Other Animals

Because all animals have innate immunity systems, these have been evolving for a considerable length of time. Because of this, plus the fact that the kinds of invaders and toxins encountered by animals varies so widely among species and habitats, we might expect evolution to have produced a wide diversity of defenses at this level. This indeed seems to be the case, but there are many similarities and common themes. In particular, most animals have phagocytotic cells that engulf invaders in an inflammation-like process, and have defensive antimicrobial peptides (effector proteins) similar in function to complement.

■ Phagocytotic cells and an inflammation-like response are the major form of cell-mediated innate immunity in most animals.

Although most invertebrates do not have enclosed circulatory systems, all have some form of internal fluid cavity. In these fluids are typically found wandering immune cells called **amoebocytes.** These patrol the fluid spaces looking for foreign invaders, especially those tagged by opsonins that we discuss shortly. As with mammalian phagocytes, they can distinguish "self" and "non-self" using pattern-recognition proteins (PRPs, p. 429) that bind bacterial and fungal cell-wall components, and possibly viral components. At least in *Drosophila,* pattern recognition involves **Toll receptors** (after which mammalian Toll-like receptors were named), but not by the same mechanism found in mammals. Instead, the fly has soluble PRPs circulating in the hemolymph, which bind to invading bacteria. Subsequently, a soluble PRP (activated by bacteria) binds to the Toll receptor on the fly's immune cells, turning on genes for production of effector proteins (such as the antimicrobial peptide **drosomycin**). Recent studies have found that this activation of genes uses transcription factors that are homologous (evolutionarily related) between flies and mammals, suggesting an ancient history for this aspect of innate immunity.

Amoebocytes in many invertebrates participate in an inflammation-like response to wounds and invaders; that is, these cells are drawn by chemotaxis to the site of injury, where they are activated to engage in phagocytotic destruction of invaders. In some invertebrates, the immune cells have also been found to make *reactive oxides* that kill microbes, just as some mammalian phagocytes do (p. 436). Some may also participate in encapsulating invaders and in producing antimicrobial peptides (as you will see later).

Figure 10–24 ➤ ● (a) Hemocytes (blood cells) from larval moth *Lacanobia oleracea.* Plasmatocytes are the main cell type involved in the encapsulation response. Some parasitoid venom components inhibit plasmatocyte development or activity, and thus prevent the normal immune response. (b) The ectoparasitic wasp *Eulophus pennicornis* laying eggs on the host caterpillar. This wasp injects a cocktail of venom components that induce changes in the host caterpillar's immune system, and prevents the process of molting by interfering with hormone (ecdysteroids and juvenile hormone) levels.

Source: www.csl.gov.uk/science/organ/environ/ invertebrate/maff1.cfm

Amoebocytes may be called by different names in different animals. For example, in sponges they are called **archaeocytes.** In echinoderms, which have a fluid-filled body cavity called the *coelom,* these cells are called **coelomocytes.** In arthropods, mollusks, and other animals having an open circulatory system filled with hemolymph (Chapter 9), they are called **hemocytes** (● Figure 10–24a).

■ Antimicrobial peptides are common noncellular defenses in barrier tissues and immune cells of animals.

One of the most widespread types of immunity involves various **antimicrobial peptides** produced in skin, gut and other barrier tissues, in phagocytes, and sometimes other tissues (for example, in insects, the fat body appears to be the major source of such peptides). These are small proteins released by injured or activated cells that, like some parts of the vertebrate complement system, can target and kill microbes on their own. And, again like complement MAC system (p. 438), many of these kill by creating holes in the membranes of microbes, which kill either by osmosis or by disrupting ion gradients. Some are attracted to particular glycoproteins common on bacterial surfaces, but others are attracted to the strong elec-

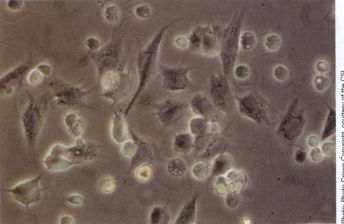

(a)

(b)

Photo: Photo Crown Copyright, courtesy of the CSL

trical charge that characterizes many bacterial membranes (which have an outer surface dominated by negatively charged phospholipids). Eukaryotic membranes do not have such strong negative charges, and so are generally safe from attack by these peptides. The mechanisms of other antimicrobial peptides are not yet understood.

Considerable research is now being devoted to these molecules in the hopes that some might be used to fight outbreaks of diseases in humans and other animals. The first of these defense peptides was found in silk moths in 1981; since then, hundreds have been found in numerous taxonomic groups including some plants, moths, fruit flies, horseshoe crabs, frogs, fishes, and mammals. Numerous names have been given to these amazing molecules, including *cecropins* in some insects (as well as drosomycin mentioned earlier), *magainins* in the skin of frogs, *piscidins* in the mast cells of some fishes, *squalamine* in sharks. One class of these peptides are the *defensins*, which are found in a variety of animals and plants. Defensins all have β-pleated sheets with a high positive charge, which attract them to bacterial membranes, where they form pores.

The discovery of defense peptides in skin, gut and leukocytes of mammals (such as *cathelicidins*, p. 433, and defensins) was inspired by the prior discovery of these defenses in nonmammals. Indeed, it was not until 2003 and later that a protective role for defensins in mammals was directly demonstrated. Human defensin HD-5, which is produced in GALT (specifically in the small intestine), kills an intestinal bacterium called *Salmonella typhimurium*, which is spread by contaminated food and causes severe diarrhea. The gene for HD-5 was genetically engineered into mice, which are normally unable to resist *S. typhimurium*. The engineered mice all survived an orally induced infection that killed all the control mice.

■ Lectins and Ig-fold proteins act as opsonins in many invertebrates.

Although only jawed vertebrates have adaptive antibodies, most other animals have proteins that have related (innate) functions, such as those functioning as opsonins (tags for phagocytic destruction). One common type is a set of proteins called **lectins,** which bind to glycoproteins on foreign cell walls. This causes the foreign cells to clump, much as occurs with antibodies, and makes them targets for phagocytes. Also, a number of proteins in many invertebrate groups have been found to contain a so-called **Ig-fold.** This is a region in a protein with a structure similar to the antigen-binding fold of vertebrate immunoglobins. Although Ig-fold proteins are not antibodies, they appear to bind proteins commonly found on microbes and to act as opsonins. Indeed, these may be the evolutionary precursors of antibodies. An example of an Ig-fold protein is **hemolin,** found in the hemolymph of moths.

■ The proPO system encapsulates large foreign materials in arthropods.

Another defense mechanism found in arthropods is *nodulation* or *encapsulation*. In this process a foreign invader that is too large for the phagocytes to engulf is walled off. In some insects, it is commonly used against parasitic worms, fungi, and eggs laid by parasitic wasps within other insects (Figure 10–24b). Surface proteins on an invader trigger (in certain hemocytes) a cascade of enzymatic reactions, which ultimately converts **pro-phenoloxidase (proPO),** an inactive enzyme that is always present, into the active form **phenoloxidase.** This in turn converts phenol to a variety of products, one of which forms a black polymer called **melanin** (the same molecule that gives human skin its pigment color). Factors that attract other hemocytes to the infection site are also produced. Hemocytes surround the invader with a nodule or capsule several cell layers thick, and the cells closest to the pathogen secrete melanin to form an impermeable wall.

■ Invertebrate immune cells communicate with cytokines such as IL-1.

A final aspect of innate immunity is the methods by which immune cells communicate during defensive reactions. Here, too, are found evolutionary precursors to the vertebrate system. Once stimulated, phagocytic cells (coelomocytes, and so forth) in echinoderms, tunicates, worms, and insects secrete a cytokine very similar in structure to mammalian IL-1. These IL-1 cytokines seem to coordinate the cell defense responses, calling in other phagocytes to the site of infection and triggering them to reproduce. Other cytokines closely related to those in mammals have also been found in some invertebrates. Although the functions of all these signals are not fully known, the occurrence of these signal molecules tell us that cytokine communication and coordination of immune cells is an ancient mechanism.

■ Some forms of acquired immunity may occur in some arthropods.

As we have noted, the accepted view of acquired immunity is that it only occurs in jawed vertebrates. Two studies in 2003 call that into question.

First, research on the water flea (*Daphnia magna*) revealed that offspring are more resistant to a particular pathogenic bacterium (but not other bacteria) if their mothers were exposed to it before she lays her eggs. It appears that the mother's immune system "learns" that a particular pathogen is present, and passes on specific protective mechanisms to her eggs. How she does this is not yet known.

In another study, *Macrocyclops albidus*, a species of copepod (the most common type of animal on Earth; p. 122), were exposed to parasitic tapeworms twice. They were better able to resist the tapeworms during the second infection. This only occurred, however, if the second exposure used tapeworms closely related to those used in the first infection. This suggests that the copepod immune system somehow "learned" the antigenic markers of an individual tapeworm and later recognized those markers on its relatives. How this takes place is not yet known.

Chapter in Perspective:

HOMEOSTASIS and INTEGRATION

Animals could not survive beyond early infancy were it not for defense mechanisms. Homeostasis can be optimally maintained, and thus life sustained, only if body cells are not physically injured or functionally disrupted by pathogenic microorganisms or are not replaced by abnormally functioning cells, such as traumatized cells or cancer cells. Immune defense systems—complex, multifaceted, interactive networks of barrier tissues, defense proteins, and specialized cells for sensing, integrating, and effecting defenses—contribute indirectly to homeostasis by keeping other cells alive so that they can perform their specialized activities to maintain a stable internal environment. Immune systems protect the other healthy cells from foreign agents that have gained entrance to the body, eliminate newly arisen cancer cells, and clear away dead and injured cells to pave the way for replacement with healthy new cells. These defenses involve both positive- and negative-feedback loops. Many researchers consider the immune system to be a third, whole-body regulatory system on a par with the other two (neural and endocrine).

The skin contributes indirectly to homeostasis by serving as a protective barrier between the external environment and the remainder of the body cells. It helps prevent harmful foreign agents such as pathogens and toxic chemicals from entering a body and helps prevent the loss of precious internal fluids. Other systems that have internal cavities in contact with the external environment, such as the digestive, genitourinary, and respiratory systems, also have defense capabilities to prevent harmful external agents from entering a body through these avenues. ▪

REVIEW QUESTIONS *(Answers are on p. A–1.)*

Additional study tools for this chapter, including chapter summaries and practice tests, are available online at *www.biology .brookscole.com*

1. Most of the cytokines in the immune system are called
   a. carotenoids
   b. pattern recognition receptors
   c. leukocytes
   d. interleukins
   e. lymphocytes
2. Which of the following is a noncirculating leukocyte derivative?
   a. T lymphocyte
   b. mast cell
   c. eosinophil
   d. neutrophil
   e. monocyte
3. In mammals, which of the following is a component of innate nonspecific defenses against offending agents?
   a. the complement system
   b. interferon
   c. symbiotic bacteria
   d. skin
   e. all of the above
4. When mast cells detect an injury, they cause increased blood flow to the area by releasing
   a. complement
   b. histamine
   c. fibrinogen
   d. lactoferrin
   e. opsonins
5. Which of the following substances released by phagocytes induces the development of fever?
   a. kallikrein
   b. lactoferrin
   c. endogenous pyrogen
   d. leukocyte endogenous mediator
   e. salicylates
6. The protein complex called the membrane-attack complex (MAC) plays a role in killing invading bacteria by

   a. binding with microbes, thereby enhancing their phagocytosis
   b. creating a channel in the surface membrane of microorganisms
   c. activating kinins
   d. stimulating the release of histamine
   e. serving as chemotaxins
7. Which of the following acts as a whistle-blower by forewarning healthy cells of potential viral attack?
   a. the bursa of Fabricius
   b. cytotoxic T-cells
   c. plasma cells
   d. interferon
   e. gamma globulins
8. When a B-cell binds with an antigen, it is induced to differentiate into
   a. a T-cell
   b. a plasma cell
   c. a stem cell
   d. an antibody
   e. none of the above
9. An antigen induces an immune response against
   a. a T-cell
   b. a plasma cell
   c. a B-cell
   d. itself
   e. lymphocytes
10. The specificity of an antibody rests in
    a. the constant (Fc) region of the antibody
    b. antigen-binding fragments at the tip of each arm of the Y
    c. the opsonin function of the tail region
    d. distinguishing between different immunoglobin subclasses
    e. the tail region of IgE antibodies attaching to specific mast cells
11. Memory cells differentiate from
    a. plasma cells
    b. killer cells
    c. effector T lymphocytes
    d. B-cell clones
    e. dendritic cells

12. In macrophages antigenic peptides are bound to and transported to the macrophage's cell surface by
    a. interleukin 1
    b. immunoglobins
    c. gamma globulins
    d. MHC molecules
    e. antibodies

13. T-cells are activated by a foreign antigen only when
    a. interleukin 1 is present
    b. foreign antigen is also present on B-cells
    c. both foreign antigen and self-antigens are present on the surface of a cell
    d. B-cell clones have recognized "raw" foreign antigens
    e. none of the above

14. The vast majority of the billions of T lymphocytes are believed to be
    a. cytotoxic T-cells
    b. host cells
    c. memory T-cells
    d. helper type
    e. killer cells

15. Which of the following is not a helper T-cell cytokine?
    a. macrophage-migration inhibition factor
    b. MHC glycoproteins
    c. interleukin 2
    d. T-cell growth factor
    e. chemotaxins

16. The AIDS virus selectively invades
    a. cytotoxic T-cells
    b. B-cells
    c. plasma cells
    d. helper T-cells
    e. macrophages

17. The direct means by which cytotoxic T-cells destroy a targeted cell is by release of
    a. MAC
    b. perforin molecules
    c. complement
    d. interferon
    e. MHC molecules

18. A condition in which the immune system fails to recognize and tolerate self-antigens associated with particular tissues is known as
    a. clonal anergy
    b. autoimmune disease
    c. antigen sequestering
    d. receptor editing
    e. immune privilege

19. The message to T-cells, "Leave me alone, I'm one of you," is conveyed by
    a. B-cells
    b. neutrophils
    c. mast cells
    d. the major histocompatibility complex
    e. opsonins

20. Antimicrobial peptides are
    a. found only in insects
    b. found only in vertebrates
    c. produced primarily by amoebocytes
    d. found only in plants
    e. found in plants and animals

SUGGESTED READINGS AND INTERNET SITES

Banchereau, J. 2002. The long arm of the immune system. *Scientific American* 287:52–59. Covers recent findings about dendritic cells.

Beck, G., & G. S. Habicht. 1996. Immunity and the invertebrates. *Scientific American* 275:60–67.

Blount, J. D., N. B. Metcalfe, T. R. Birkhead, & P. F. Surai. 2003. Carotenoid modulation of immune function and sexual attractiveness in zebra finches. *Science* 300:125–127.

Glausiusz, J. 1994. The secret healing power of sharks. *Discover* 15:86.

Hoffmann, J. A. 2003. The immune response of *Drosophila*. *Nature* 426:33–38.

Kurtz, J., & K. Franz. 2003. Evidence for memory in invertebrate immunity. *Nature* 425:37.

Lavine, M. D., & M. R. Strand. 2002. Insect hemocytes and their role in immunity. *Insect Biochemistry and Molecular Biology* 32:1295–1309.

Litman, G. 1996. Sharks and the origins of vertebrate immunity. *Scientific American* 275:67–71.

Travis, J. 1998. The accidental immune system. *Science News* 154:302–303.

INTERNET SITES

Thornqvist, P.-O., & K. Soderhall. 1997. *Crustacean Immune Reactions, a Short Review.* shrimpactiva.com/shrimpactiva/sr0101.asp.

Travers, P., M. Mclements, & T. McElderry. *Immunobiology Animations.* 2001. blink.uk.com/immunoanimations/. ImmunoBiology Web site has animations of immune functions. Requires purchase for full animations.

Kaiser, G. E. 2002. *Animation of Endocytic Pattern-Recognition Receptors.* cat.cc.md.us/courses/bio141/lecguide/unit1/bacpath/endoprr.html. An animation of pattern-recognition receptors and phagocytosis.

INFOTRAC READINGS

Heath, W. R., & F. R. Carbone. 2003. Dangerous liaisons. *Nature* 425:460–461. Discusses the "danger" and "stranger" hypotheses for recognition of invaders.

Jancin, B. 2002. Antimicrobial peptides are first line of defense against skin infections. *Skin & Allergy News* 33:22.

Zasloff, M. 2002. Antimicrobial peptides of multicellular organisms. *Nature* 415:389–396.

11

Respiratory Systems

Photo: Paul Yancey

Elephants crossing the Zambezi River in Botswana must either hold their breaths for extended periods, or else use their trunks as snorkels (as the elephant on the left is doing).

Evolutionary Solutions to Gas Demands

Energy is essential for sustaining all life-supporting cellular activities. For most animals to survive for extended periods of time, **aerobic metabolism** (p. 44) is necessary. For this, cells require a continuous supply of O_2 to support their energy-transforming reactions in mitochondria. The CO_2 produced during these reactions must be eliminated from the animal at the same rate it is produced to prevent dangerous fluctuations in pH (that is, to maintain the acid–base balance), because CO_2 generates carbonic acid. The term **respiration** involves the sum of the processes that accomplish movement of O_2 from the environment to the tissues, and has two distinct components:

1. The term **internal** or **cellular respiration** refers to the intracellular metabolic processes carried out within the mitochondria, which use O_2 and produce CO_2 during the derivation of energy from nutrient molecules (see p. 44).

2. The term **external respiration** refers to the sequence of events involved in the exchange of O_2 and CO_2 between the external environment and cellular mitochondria.

External, not cellular, respiration is the topic of this chapter. In this first section on evolution, we begin with general principles and proceed to the specific adaptations of water and air respirers.

General Principles

■ **External respiratory processes must meet the demands of size, metabolism, and habitat through diffusion and bulk transport.**

Ultimately, every cell relies on **diffusion** for the ultimate goal of external respiration: that of exchanging gases between the cell's immediate environment and its mitochondria. Diffusion follows **Fick's law,** expressed as follows for gases (compare this to the version for non-gas solutes, Table 3–1, p. 74):

$$\text{Rate of diffusion (Q)} = \frac{\Delta P \cdot A \cdot D}{\Delta X}$$

where

D = the *diffusion coefficient* (which depends on, among other things, molecular weight and the permeability of the barrier between two media)

A = the *surface area* for gas exchange

ΔP = the gas *gradient* (in *partial pressure*) = $P_1 - P_2$, where P_1 is the partial pressure in one compartment and P_2 is the pressure in the other

ΔX = the *distance* the gas must cover

Partial Pressures

Before we proceed, it is important to understand the term *partial pressure.* Atmospheric air is a mixture of gases that contains about 79% nitrogen (N_2) and 21% O_2 (actually 20.9% or 209 mL per liter of air) with almost negligible percentages of CO_2, H_2O vapor, other gases, and pollutants in dry air. Altogether, these gases exert a total atmospheric pressure of 760 mm Hg at sea level (● Figure 11–1a). This total pressure is equal to the sum of the pressures that each gas in the mixture partially contributes. The pressure exerted by a particular gas is directly proportional to the percentage of that gas in the total air mixture. Every gas molecule, no matter what its size, exerts the same amount of pressure; for example, an N_2 molecule exerts the same pressure as an O_2 molecule. Because 79% of the air consists of N_2 molecules, 79% of the 760 mm Hg atmospheric pressure, or 600 mm Hg, is exerted by the N_2 molecules. Similarly, because O_2 represents

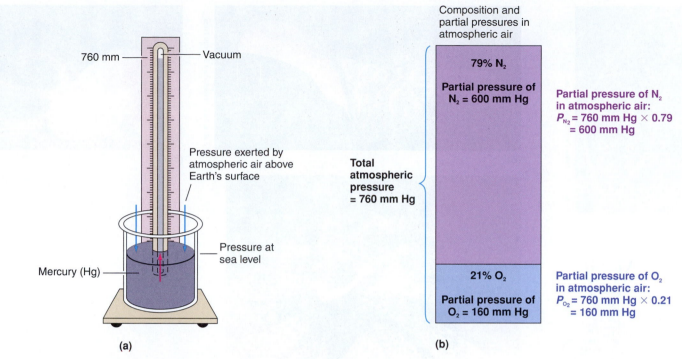

Figure 11–1 ● (a) Atmospheric pressure. The pressure exerted on objects by the atmospheric air above Earth's surface at sea level can push a column of mercury to a height of 760 mm. Therefore, atmospheric pressure at sea level is 760 mm Hg. (b) Concept of partial pressures. The partial pressure exerted by each gas in a mixture equals the total pressure times the fractional composition of the gas in the mixture.

about 21% of the atmosphere, 21% of the 760 mm Hg atmospheric pressure, or about 160 mm Hg, is exerted by O_2 (Figure 11–1b). The individual pressure exerted independently by a particular gas within a mixture of gases is known as its **partial pressure,** designated by P_{gas}. Thus, the partial pressure of O_2 in dry atmospheric air, P_{O_2}, is normally about 160 mm Hg. The atmospheric partial pressure of CO_2, P_{CO_2}, is negligible at 0.03 mm Hg.

Gases dissolved in water or blood are also considered to exert a partial pressure. Water that is in equilibrium with air ultimately will have the same gas partial pressures as in the air. The amount of a gas that will dissolve in the blood depends on the solubility of the gas in blood and on the partial pressure of the gas in the environment (alveolar air or water) to which the blood is exposed. Because the solubilities of O_2 and CO_2 in body fluids remain constant at constant temperature, the amount of O_2 and CO_2 dissolved in body fluids is directly proportional to the ambient P_{O_2} and P_{CO_2}.

Evolutionary Forces

We investigate all the components of Fick's law in more detail later, but it is important to realize from the start that diffusion directly depends on the partial pressure gradient, surface area, and distance. For single-celled and very small and/or flat multicellular organisms (● Figure 11–2a and b), simple diffusion can be sufficient. However, three major forces in the evo-

lution of animals have made it necessary to evolve other steps in the gas exchange process:

1. Animal *size* has increased in many phyla since the first animals evolved. For all but the smallest animals, diffusion is simply too slow to cover the distances (ΔX component in Fick's law) involved in a reasonable time.

2. Animal *habitats* vary in their availability of oxygen, from anoxic ("no oxygen") aquatic mud to sealevel air. This affects the ΔP component of Fick's law. The most distinctive event was the adaptation from aquatic to terrestrial habitats. ▮ Table 11–1 shows the wide range of availability of O_2 among various habitats. Note in particular that in water, gas solubility *decreases* as temperature increases, such that oxygen can be much more available in a cold water habitat. Note also that aquatic habitats, even at their highest oxygen content, hold only a fraction of the content of air.

3. Animal *metabolisms* vary in their demand for oxygen, with the highest demands arising in birds, mammals, some fishes, and flying insects.

Because of these three trends, simple cellular diffusion has been augmented in two broad ways: by enhancements to diffusion itself, and by **bulk transport.** Unlike diffusion, which is the random movement of individual molecules, bulk transport involves the movement of the entire medium carrying the molecule of interest (in this case, the two gases). Thus it is much faster. Let us examine these two ways in more detail.

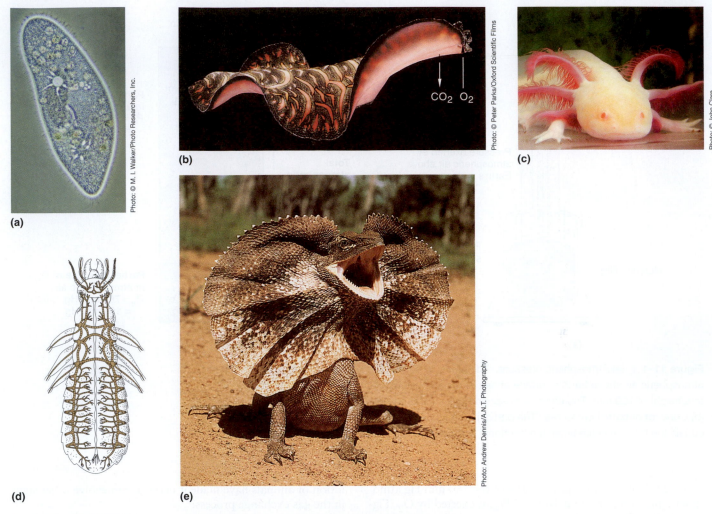

Figure 11–2 ● **Diversity of gas exchange structures.** (a) Respiratory gases diffuse across the plasma membrane of single-celled organisms such as *Paramecium*. (b) In small multicellular animals such as roundworms and flatworms, gas exchange occurs through the skin or integument. In these simple animals the distance from the integument to the tissues is small enough that gases can diffuse down concentration gradients. (c) Large aquatic animals exchange gases at gills, evaginations of the respiratory surface. (d) The respiratory surface of terrestrial animals is invaginated, such as the tracheae of insects or (e) the lungs of an air-breathing vertebrate.

Table 11–1 ▌ Oxygen concentration in various habitats, in mL per liter of medium

Air at sea level	209
Air at Mount Everest (8848 m)	59
Fresh water at 0°C (maximum)	10.3
Fresh water at 30°C (maximum)	5.6
Seawater at 0°C (maximum)	8.0
Seawater at 30°C (maximum)	4.5

Enhancements to Diffusion

In addition to diffusion at individual cells, gases in all but the smallest multicellular animals must be exchanged through another diffusion step at **respiratory surfaces**—tissues that are exposed to the environment and across which O_2 and CO_2 diffuse. In the first animals, the site of gas exchange was presumably through the relatively unspecialized body surface or integument. This has persisted to the present in many small, sluggish animals such as Cnidaria (e.g., hydras; see Figure 9–1) and flatworms (Figure 11–2b), which rely substantially on this **integumentary exchange** (they can also exchange gases with digestive tract fluid). As indicated by the "ΔX" component of Fick's law, flattened and thin shapes (which reduce distances compared to more spherical shapes) are the most efficient for uptake of O_2 and release of CO_2.

However, the integument may not provide enough surface area for exchange, and it may also have a low permeability, because of protective layers. Thus many animal groups have evolved special **respiratory organs** such as gills, tracheae, and lungs (Figure 11–2c, d, and e). These organs enhance diffusion by having (1) a high "D" value, using membranes with low-permeability barriers, (2) a low ΔX (distance), using thin epithelia, and (3) a high "A" value, through specialized *folds* (*evaginations* or "outpockets," and *invaginations* or "inpockets") that increase surface area for gas diffusion. Surface areas are generally higher in animals with higher metabolisms. For example, lungs of some amphibians are relatively simple sacs, but in more active reptiles and all mammals, the tubes branch into numerous tiny blind-end sacs called **alveoli** to provide a much higher surface area (see Figure 11–12b).

If an internal circulatory system is involved, there is also a large surface area and close contact between this system's fluids and the tissues (both respiratory tissues and body tissues needing high oxygen delivery). Animals with closed circulation accomplish adaptation by high *vascularization*—a high density of narrow vessels (capillaries; see p. 408). In animals with open circulations, circulatory fluid bathe cells directly, greatly minimizing the ΔX component (p. 462).

The ΔP (that is, $P_1 - P_2$) component can also be increased by reactions that convert gases between *diffusible* and *nondiffusible* forms where appropriate. For O_2, this typically involves special **respiratory proteins** such as *hemoglobin* and *myoglobin*. For CO_2, this involves binding to hemoglobin and conversion to and from **bicarbonate HCO_3^-**, which unlike the free gas cannot diffuse through membranes without a membrane transporter. In a nondiffusible form, a solute does not contribute to the ΔP, and so a gas gradient ($P_1 - P_2$) is not equilibrated nearly as quickly. Think of the gas as being "trapped" so that it cannot diffuse backward once it has reached a certain location. We examine how this process works in detail later, when we discuss gas exchange.

Bulk Transport

Even the enhanced surface exchanges just discussed can be too slow, because they are all diffusion based. In general, gases are not moved through membranes by active transport. The greatest enhancements to moving gases are collectively called *bulk transport* methods (including *bulk flow* discussed in Chapter 9, p. 413). In some cases, bulk transport of the external medium is provided by forces outside the animal; for example, a flowing stream can bring a constant supply of fully oxygenated water to an anchored hydra. More importantly, though, bulk transport is manifested in two evolved forms within animals: **ventilation,** or the movement of external media (water or air), and **circulation,** the movement of internal circulatory media. Bulk transport makes two major contributions:

1. It bypasses the diffusion law altogether by moving the medium (with gases) over large distances, and thus is much faster than diffusion.

2. It enhances the ΔP (gradient) component of Fick's law through the movement of fluids with different gas con-

tents. For diffusion between two nonmoving media, a gas gradient can easily become equilibrated (thus $\Delta P = 0$), and no further net gas movement can occur. However, movement of the external medium can provide respiratory surfaces with a continuous supply of high O_2 (accompanied by low CO_2). This is typically complemented by circulation, which brings to the gas exchange surfaces a continuous supply of low-oxygenated body fluid (with high CO_2). Together these movements prevent the ΔP from reaching zero. Circulation also brings to tissues a continuous supply of oxygen (accompanied by low CO_2).

Breathing as a form of bulk flow can be tidal or flow-through.

Breathing—the term for *active* ventilation of respiratory surfaces—requires motile devices to move the external medium. Typically, these devices are either cilia that line the surfaces themselves, or separate muscular pumps that create negative and positive pressures to move the medium. You will later see how these work, but for now, understand that there are two distinct designs to consider in terms of the type of *flow* created by the pumps. First, the external medium can be moved in and out of the same opening, in two distinct steps termed *inhalation* (or *inspiration*) and *exhalation* (or *expiration*). This arrangement is called **tidal breathing.** It is relatively inefficient because fresh medium typically mixes with depleted medium, and because fresh medium can only be brought in about half the time. The second approach moves the medium in one opening and out a separate exit. This is **flow-through breathing.** Fresh and depleted medium need not mix much in this process, and flow of fresh medium can be nearly or fully continuous. Thus it can be much more effective at gas exchange.

External respiration can involve up to four major steps.

Let's now put together all the mechanisms just discussed. With the involvement of diffusion and possible bulk-transport processes, animals may have up to four distinct steps in the overall process of external respiration (● Figure 11–3):

1. *Ventilation or external bulk transport:* Movement of respiratory gases in the external medium, created by environmental motions, by the animal's locomotion, or by the animal's own tidal or flow-through pumps (termed *breathing mechanisms,* ranging from cilia to large muscles). Unlike environmental ventilation, breathing rate can typically be regulated so that the flow of external medium between the environment and the respiratory surface is adjusted according to the animal's metabolic needs for O_2 uptake and CO_2 removal (ΔP is manipulated in this way, as discussed earlier.).

2. *External or respiratory surface diffusion:* O_2 and CO_2 are exchanged between the environment and the internal fluids by simple diffusion across a respiratory surface. CO_2 in the form of bicarbonate (HCO_3^-) may also be moved via membrane transporters into an aquatic environment (typically facilitated diffusion). Here, any adap-

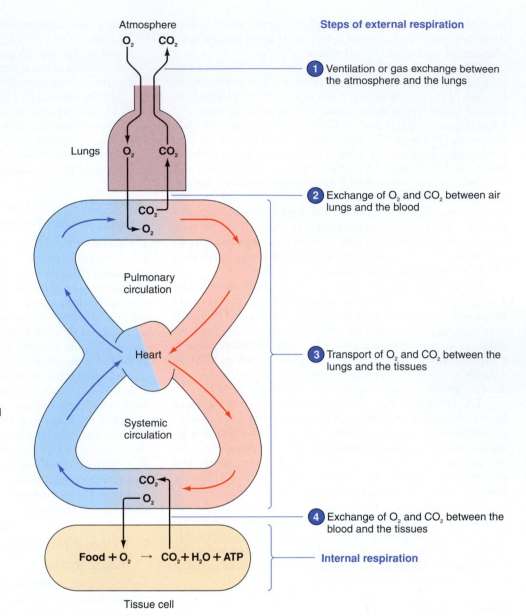

Atmosphere
O_2 CO_2

Lungs O_2 CO_2

CO_2
O_2

Pulmonary
circulation

Heart

Systemic
circulation

CO_2
O_2

Food + O_2 → CO_2+ H_2O + ATP

Tissue cell

Steps of external respiration

1 Ventilation or gas exchange between the atmosphere and the lungs

2 Exchange of O_2 and CO_2 between air lungs and the blood

3 Transport of O_2 and CO_2 between the lungs and the tissues

4 Exchange of O_2 and CO_2 between the blood and the tissues

Internal respiration

Figure 11–3 ● External and internal respiration in an air-breathing vertebrate. External respiration encompasses the steps involved in the exchange of O_2 and CO_2 between the external environment and the tissue cells (steps 1 through 4). Internal respiration encompasses the intracellular metabolic reactions involving the use of O_2 to derive energy (ATP) from food, producing CO_2 as a by-product.

tations that can increase D and A (for Fick's law) and that reduce ΔX, are important if an animal's size, habitat, or metabolism require them.

3. *Circulation or internal bulk transport:* O_2 and CO_2 are transported by the internal circulation between the respiratory surface and all tissues. In many groups, special respiratory proteins enhance gas content. In some animals, the amount of internal fluid circulated to individual tissues—called **perfusion**—can be regulated. Both features increase the ΔP aspect of diffusion.

4. *Cellular or tissue diffusion:* Exchange of O_2 and CO_2 takes place between tissues and the circulatory fluid by diffusion. Adaptations that boost diffusion, such as narrow blood vessels (small ΔX) at a high density (high A), are important here.

When we refer to an animal's **respiratory system,** we are not talking about all these steps of external respiration; the system

is involved only with steps 1 and 2. But it is important to realize that in more complex animals the entire process involves all four steps. We now turn to the respiratory system in more detail, focusing on the most distinctive adaptational contrasts: water respirers and air respirers.

Water Respirers

■ Water is a more difficult medium than air for gas exchange

Life began in water, and as we have seen, water contains much less oxygen than air (Table 11–1). It is also problematic for gas exchange in other ways. A comparison of air and water reveals the following points:

■ Water has a higher viscosity than air (approximately 850-fold) that necessitates a higher energy output to maintain flow over the respiratory surface. The higher density of water (60-fold greater than air) also entails a higher cost of acceleration.

- O_2 is relatively insoluble in water, whereas CO_2 is present at about the same concentration as in air. The O_2 concentration of water is considerably smaller than in the gas phase (1 L of water at 15°C contains 7 mL of O_2, whereas 1 L of air contains 209 mL of O_2). For this reason approximately 30 times more water than air must be moved to achieve the same ventilatory transport of O_2.

- The rate of diffusion of O_2 in water is approximately 10,000-fold slower than in air.

- The solubility of O_2 and CO_2 are decreased by increases in salinity.

- The solubility of O_2 and CO_2 are decreased as ambient temperature increases. These changes in solubility are critical; as temperature is raised from 0 to 35°C, O_2 concentration is reduced by 50% at constant P_{O_2}. Because an increase in temperature raises the metabolic rate in poikilotherms (animals whose body temperatures vary with the environment; Chapter 15), the associated reduction in O_2 solubility in water exerts an ecological pressure for aquatic species to obtain access to air.

- Compared to air, the O_2 content of water is subject to greater variation. Bays, swamps, and intertidal pools have large swings in P_{O_2}, which rise during daylight hours as a result of photosynthesis and decline at night as a result of biological O_2 demand. The P_{O_2} values vary from 0 to 450 mm Hg in these shallow habitats, whereas the surface waters of the open ocean have oxygen levels at close to that of air, about 155 mm Hg. Below the ocean surface O_2 generally declines as photosynthesis declines and ceases, such that moderate depths in much of the ocean ("oxygen minimum zones") have very low O_2. At greater depths, oxygen is often relatively high because cold, well-oxygenated waters sink in the polar regions and travel at great depth around the globe.

- Oxygen can be very low in thick sediments (such as mud, clay). The sediments at the bottom of bodies of water are particularly noted for low concentrations of O_2 unless vertical currents mix surface with deeper water.

- Unlike air, water can contain many life-sustaining components other than gases. These include dissolved ions and organic matter, and, of course, water itself (although air can have water vapor).

■ The limitations of diffusion in water are overcome with thin, high-surface-area structures and bulk transport.

Enhancements to the components of the diffusion equation are critical for water respirers. Bulk transport is similarly important, with the more effective flow-through arrangements more common than tidal ones. Thus larger single-celled protists are often elongated (narrow) rather than spherical (such as *Paramecium*, Figure 11–2a) to decrease ΔX from the environment to the mitochondria, and they use ventilation in the form of environmental currents, their own swimming movements (by cilia or flagella), and/or the beating of their own cilia along their membranes. Note that *Paramecium* cilia usually beat in synchrony in one direction, essentially creating a "flow-across" process akin to flow-through design.

Integument

Similar adaptations are found in some aquatic animals, particularly smaller ones with low metabolisms. These animals rely primarily on integumentary exchange enhanced by a thin body form to reduce ΔX (such as flatworms) (Figure 11–2b). Bulk transport is created by external currents, their own body movements, and sometimes epidermal cilia (such as the ciliated surfaces of many animal larvae, and small, ciliated animals such as rotifers). Also, internal bulk transport may enhance the process. For example, in the earthworm (an annelid) a single, thin layer of mucus-coated epidermal cells covers a dense network of capillaries. (O_2 and CO_2 can cross cell membranes only when they are dissolved in water, so this "terrestrial" animal is effectively a water respirer.)

Integumentary respiration also occurs to some extent in most larger (gill-bearing) aquatic vertebrates and is sufficient to maintain a normal resting metabolic rate, such as in some eels and catfish.

Gills

Nonintegumentary structures for gas exchange evolved in most water breathers with larger sizes and higher metabolisms. The most prominent features of such structures are higher surface areas (A) and often smaller diffusion distances (ΔX) compared to the integument, plus bulk transport of some sort. The primary specialized respiratory structures are called **gills** (Figure 11–2c), which are evaginations of tissue usually protruding into the external medium. Typically gills are delicate structures (because of a thin epidermis with a high surface area for gas exchange) that are highly perfused by a circulatory system. Gills can be fully external, although often they are internal in the sense of being protected by a hard cover (thus are not externally visible). But even then, the respiratory surface has direct contact with the external medium.

Gills range from simple integumentary bulges, to specialized **filaments,** to stacks of **lamellae** (flat platelets). In some annelids, for example, gills are simply evaginated, folded extensions of the body surface, without protection (● Figure 11–4a). The specialized filaments of shell-less sea slugs called *nudibranchs* ("naked gills") are similarly external (Figure 11–4b)—although they are not unprotected. The integuments and gills of these mollusks typically contain noxious substances or (somehow) unfired nematocysts "stolen" from the tentacles of sea anemones they have consumed! The unfired nematocysts are absorbed by the digestive tract and move into the gill filaments. These animals are usually brightly colored, to advertise their invisible defenses.

The gills of other mollusks, arthropods such as crabs and lobsters, and fishes are covered with hard parts (Figures 11–4d and 11–5). For example, the gill system of fish consists of **gill arches,** the tissue that provides skeletal support for a double series of gill filaments, each of which carries a row of secondary lamellae on either side (● Figure 11–5). Normally there are five separate pairs of these arches, covered with a bony plate called the **operculum,** which opens on one side to the outside world. The basic unit of gas exchange in the gill is the lamella, not the arch or filament. Lamellae form a sievelike

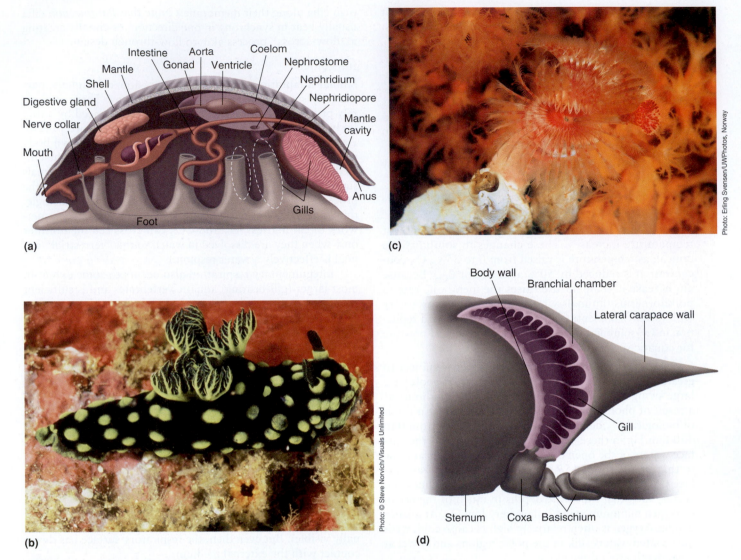

Figure 11–4 ● **Location of gills in representative invertebrates.** (a) Generalized anatomy of a primitive mollusk, with gills in the posterior mantle cavity. (b) A nudibranch (gastropod mollusk, *Chromodoris annae*); its gills are the tufts on the middle top. (c) A polychaete tubeworm (*Hydroides norvegica*), with its anterior gills emerging from a calcareous tube. (d) Cross section through the carapace of a crab, showing the gill extending as a branch from the coxa of a limb into a branchial chamber.

(*Sources:* (a) Phillip C. Withers, 1992, *Comparative Animal Physiology*, 1st ed., Belmont, CA: Brooks/Cole, p. 577, Figure 12–9.
(b) www.divegallery.com/c_annae1.htm. (c) Erling Svensen, http://articles.uwphoto.no/oversikter/Marine_biology_Bristleworms.htm
(d) Phillip C. Withers, 1992, *Comparative Animal Physiology*, 1st ed., Belmont, CA: Brooks/Cole, p. 582, Figure 12–13a)

structure, which minimizes diffusion distances for gas exchange between water and the extensive blood supply found in each lamella. The thickness of each lamella ranges from 10 to 25 μm. Each is comprised of two epithelial cells, separated by a series of **pillar cells** between which blood can flow. The degree of activity of a fish species is correlated with the amount of spacing between subsequent lamellae and can vary from 10 to 60 per mm. Depending on the species, total lamellar numbers can also vary. Generally, to accommodate the metabolic requirements of fast-swimming species, the size of the lamella is larger and the interlamellar spacing is smaller. For example, in an inactive bottom dweller such as the toadfish the number of lamellae is approximately 0.5 million, whereas in a compara-

bly sized active species such as the bluefin tuna, lamellar number is over 6 million.

Exposure to the environment has its risks. **Environmental gill disease (EGD)** is a syndrome that commonly affects fish reared using aquaculture technology. The initial stages are characterized by inflammation of the gill lamellae. It is believed that stress as well as environmental factors triggers the disease process. If exposure to the irritant (such as bacteria, amoebas, fungi) is short-lived, damage to the gill lamellae is minimal. However, if the irritant exposure continues a serous (blood plasma) exudate accumulates around the capillary endothelium, which reduces gas exchange. Fish with **epithelial-capillary endothelial separation (ECS)** exhibit symp-

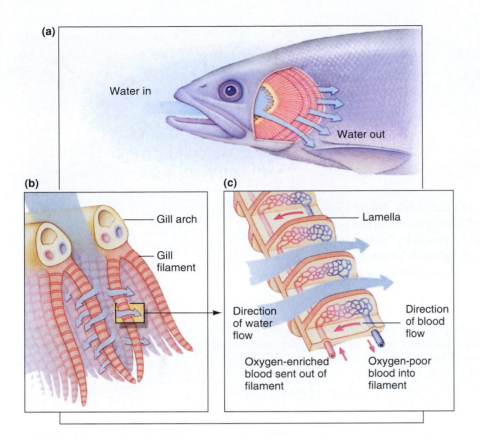

Figure 11–5 ● **Respiratory system of a teleost (bony fish).** (a) One of a pair of gills. The gill cover or operculum has been removed for this sketch. Water drawn into the fish's mouth is forced across the gills. (b) Each gill has filaments bearing lamellae, which are richly vascularized respiratory surfaces. (c) The flow of oxygen-rich water is countercurrent to the flow of oxygen depleted blood entering the lamellae.

(*Source*: C. Starr & R. Taggart, 2004, *Biology: The Unity and Diversity of Life*, 10th ed., Belmont, CA: Brooks/Cole, p. 710, Figure 40.6, d, e)

toms of respiratory distress that ultimately becomes fatal if left untreated.

Other Respiratory Tissues

In some cases, structures with nonrespiratory functions such as feeding are used for dual purposes. For example, sea anemones (Cnidaria) have feeding tentacles (p. 39) that are environmentally ventilated when extended into moving water. Each tentacle has a hollow core and a narrow diameter. Beating of cilia lining the tentacles enhances ventilation. The columnar body of the anemone is much thicker and has a much lower surface area–volume ratio than do the tentacles, although gas exchange still occurs along the body length. Analogously, filter-feeding marine annelids (segmented worms) have ciliated tentacles that serve to trap plankton and to exchange gases, and the body wall also functions in gas exchange.

In water-breathing mollusks, the **mantle** as well as the gill is highly vascularized and can take up O_2 from ventilatory currents flowing over these surfaces. In shelled mollusks, the ventilatory current also flushes the **visceral mass** (main body) with oxygenated water. A comparatively large sinus lying beneath the visceral mass can actually be inflated, increasing the amount of surface area exposed to the ventilatory current.

Sea cucumbers (soft-bodied echinoderms) are animals with a unique respiratory structure. Echinoderms are not typically high-metabolism animals, and others (such as seastars and sea urchins) exchange gases through body-wall evaginations and/or tube feet. But a sea cucumber draws in seawater through its anus into a branching structure called a **respiratory tree** that protrudes into its coelomic (body cavity) fluid, not into the environment. Pumping (a tidal breathing process) involves muscles in the tree, yet is not used to support a high metabolism.

This evagination of the digestive tract is structurally more like a lung than a classic gill!

■ Breathing muscles provide consistent and often fast bulk transport.

Gas exchange in the most active species requires a dedicated *muscle-driven* form of breathing. This type of pump is readily seen in fishes (● Figure 11–6). Water is pumped over the gills of bony fish by skeletal muscle pumps in the **buccal** (mouth) and **opercular cavities.** Ventilation of the gills operates by a cycle of negative and positive pressure gradients. As the mouth opens the opercula are shut, which leads to a negative pressure gradient in the buccal and opercular cavities, resulting in the inward movement of O_2-rich water. The mouth is then closed and the opercular cavity constricts, forcing water through the gills. At this point the water that is trapped in the gills is under higher pressure than outside. When the opercula open, the positive pressure gradient forces water outward across the gills and through the opercular exit. Note again that this is a flow-through system, whose efficiency we explore later.

Paradoxically (it seems at first), the most active fish of all—tuna—have lost their breathing muscles! They instead rely on a flow-through process called *ram ventilation*, in which bulk transport is created by the animal's own motion, along with keeping the mouth and opercula slightly open. However, the principle—that high metabolisms require muscular-driven bulk transport—remains the same, but the muscles are those of swimming, not the traditional respiratory muscles. Large masses of external medium are moved over the gills rapidly and continuously, because these animals never stop swimming.

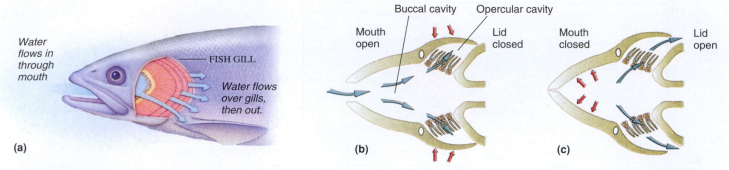

Figure 11–6 ● **Gill ventilation in a fish.** (a) One of a pair of gills. The gill cover or operculum has been removed for this sketch. (b) Opening the mouth and closing the lid forces water into the buccal cavity and across the gill. (c) Closing the mouth and opening the lid forces the water to exit through the opercular cavity.

(*Source:* C. Starr & R. Taggart, 2004, *Biology: The Unity and Diversity of Life,* 10th ed., Belmont, CA: Brooks/Cole, p. 710, Figure 40.6, a, b, c)

The most active of the mollusks—cephalopods—have abandoned ciliary-driven bulk transport on their gills, and instead rely on a muscular pumping mechanism. This involves the muscular walls of their mantle cavity and siphon. During inhalation, the siphon closes and the mantle cavity expands, drawing in large amounts of water to ventilate the gills rapidly. To exhale, the mantle opening seals up and the siphon opens, whereon the mantle contracts and squirts the water out the siphon. This can also be used as a means of rapid **jet propulsion** when the animal needs to move quickly. Note that this appears to be a tidal mechanism, but because the intake and exit paths are different, there are some flow-through aspects as well.

■ Aquatic respiratory systems can perform numerous nonrespiratory functions.

In water respirers the respiratory system (such as gills) provides the greatest exposure between internal and external environments. Therefore, the system often performs several important nonrespiratory functions related to its high exchange capacity:

1. *Fluid and solute balance*—the regulation of osmotic and ionic gradients (Chapter 13).

2. *Acid–base balance*—the maintenance of normal acid–base balance by transporting H^+ or HCO_3^- through the gills into the aqueous media (Chapter 13).

3. *Excretion*—removing the toxic waste product ammonia (Chapter 12).

4. *Nutrient and mineral uptake*—for example, NaCl and calcium intake from the environment through the gills.

5. *Feeding*—some gills have also evolved feeding functions, as in bivalve mollusks. These animals have ciliated gills, which create currents for obtaining plankton as well as for exchanging gases.

Another process affected by gills, one that is not always adaptive, is heat transfer. The vast majority of aquatic animals have body temperatures within 1 to 2°C of the surrounding environment because of heat loss through the high surface area of respiratory structure. Endothermic water respirers such as tuna require special adaptations to prevent heat loss (Chapter 15).

Air Respirers

The evolutionary transition to land involved new challenges for external respiration. On the one hand, air is much less viscous than water (thus takes much less energy to pump), and it contains much more oxygen than any body of water (Table 11–1). These facts have two implications for diffusion and bulk transport:

1. Surface area need not be as high as in water. For example, one remarkable animal, the land crab *Scopimera*, can take up O_2 from special "thigh windows," thinned areas of the cuticle located in its walking legs. These have much less surface area than gills of aquatic crabs, but are just as effective in total gas exchange. Nevertheless, the most active air breathers do require high surface areas, as you will see.

2. Efficient external bulk transport (which has a significant metabolic cost) is less necessary for air respirers. For example, small insects rely on diffusion for gas movement in and out of their tracheae. However, larger ones do use internal muscle movements to actively move the air. Also, simple tidal breathing is more common among terrestrial animals than the (often more complex) more efficient flow-through approach. For example, mammals (which have among the highest known metabolic rates) have relatively inefficient tidal breathing. However, birds, with their even higher metabolic rates, have partial flow-through arrangements (as you will see).

On the other hand, respiratory surfaces are typically thin for diffusion, and so can potentially collapse in air under gravity and, more importantly, dry out rapidly if exposed to air. Therefore, in air respirers the respiratory surfaces must be kept moist to maintain the integrity of the respiratory surface. There are two broad ways to do so:

1. *Remain in moist conditions.* Such land animals closely resemble small aquatic animals in respiratory features;

for example, many rely on integumentary exchange. But because air with its high oxygen content is close at hand, this type of exchange can work well in larger animals. Earlier, you saw how the earthworm can be considered aquatic in this regard. Sometimes certain regions of the skin are enhanced for respiration. For example, the East African annelid swamp worm, *Alma emini,* spends the dry season burrowed in decomposing plant material, where it breathes moist air. During the rainy season the local environment becomes anoxic, so the swamp worm extends its highly vascularized, flattened tail out of its burrow into the air. The tail is even more versatile than this, however. When the animal withdraws into its burrow (to avoid predation or drying out), its tail can roll up into a covered cylinder that takes a bubble of air down with it underground. The tail then slowly absorbs the oxygen from this "SCUBA tank" of gas!

2. *Have covered or fully internal structures* for gas exchange, with secretions to moisten the surface. Structures include internal air tubes called **tracheae,** which take air directly to tissues (insects, some spiders), gill-like **book lungs** (scorpions and some spiders), **mantle cavities** (some snails), or vascularized sacs called **lungs,** which provide air for internal circulation (some snails, and land vertebrates, including those such as penguins and whales that have re-evolved into aquatic habitats).

Let's now examine these adaptations in more detail.

Land slugs and snails use skin, mantle tissue, or lungs.

Mollusks are primarily aquatic. Some species such as some bivalves are semiterrestrial in the intertidal zone, where they are exposed alternately to water and air. They generally use the same (gill) structures for gas exchange with air or water. However, these animals limit exposure of their epithelial surfaces to the atmosphere because of desiccation risk, so they more often enclose themselves to minimize water loss and then depend on anaerobic metabolism.

Only in certain groups of gastropods (snails) have any mollusks become terrestrial. Land slugs require moist habitats, relying on integumentary exchange through a moistened skin, whereas one group of snails (the *prosobranchs*) uses its mantle tissue under its shell. The most successful groups are the pulmonate ("lung-bearing") snails, some of which have adapted to deserts. They have an actual lung in which a well-perfused mantle tissue forms a complete chamber with a small opening to the outside. Gases move primarily by diffusion, although some tidal bulk transport may occur from body movements.

In insects, internal air-filled tubes—tracheae—supply oxygen directly to the tissues.

Insects (as well as centipedes and millipedes) are unusual in that gas exchange occurs through a system of internal air-filled tubes, the tracheae (Latin *trachea,* "windpipe"), which directly supply O_2 to the tissues (● Figure 11–7). Most insect tracheae are reinforced with rings of chitin. The circulatory system is thus important *only* for the transport of nutrients and not respiratory gases to the respiring cells. The tracheae connect to the outside through openings called **spiracles** (Latin *spiraculum,* "airhole") that are pores or openings through the exoskeleton. Generally there is a closing mechanism that lim-

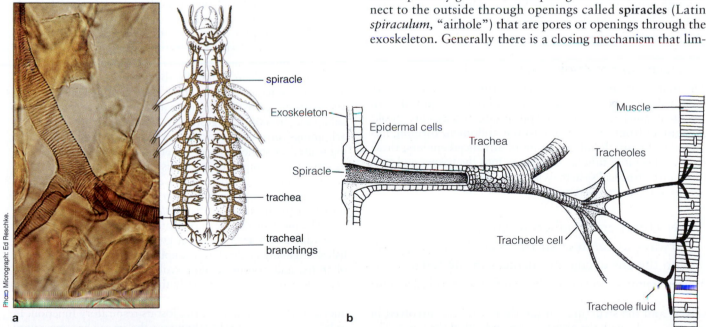

spiracle

Exoskeleton

Epidermal cells

Muscle

Trachea

Tracheoles

Spiracle

trachea

Tracheole cell

tracheal branchings

Tracheole fluid

Photo Micrograph: Ed Reschke.

a   b

Figure 11–7 ● **Respiratory system of an insect.** (a) Generalized body plan of the insect tracheal system. Chitin rings reinforce the branching tubes of the trachea and prevent their collapse. (b) Oxygen entering the trachea moves through the tracheoles to fluid-filled tips located in specific target tissues. An increased number of tube tips are found on muscle and other tissues with high oxygen demands.

(Sources: (a) Starr *Unity and Diversity of Life* 10 (fig 40.5) pg. 709; (b) E. E. Ruppert, R. S. Fox, & R. D. Barnes, 2004, *Invertebrate Zoology: A Functional Evolutionary Approach,* 7th ed., Belmont, CA: Brooks/Cole, Figure 21–10c)

its water loss from the respiratory surfaces. Spiracles open in response to a low concentration of O_2 or to a high concentration of CO_2 in the tissues. Most species have three pairs of spiracles above the legs and on the first seven abdominal segments. The tracheae ultimately break up into finer branches, the **tracheoles,** which are about 0.2 μm in diameter. Tracheoles can actually indent the surface of target cells such as muscle fibers and are "functionally internalized," although they never become intracellular. The distribution of tracheae reflects the O_2 demands of particular tissues. For this reason the brain, sense organs, and flight muscles of insects contain an abundant supply.

Sedentary insects, as well as insects weighing less than a gram, rely on diffusion gradients for gas exchange within the tracheal system. However, in larger insects simple diffusion is insufficient, and many (especially flyers) rely on active pumping of gas through the tracheal system. One method uses **air sacs,** which are actually regions where the tracheae are enlarged (for example, in honey bees). Because they are thin walled, the sacs collapse and expand from external muscle action. This provides some bulk transport to the process. The other method—which may also use air sacs—uses rhythmic muscle contractions of the abdomen, and movements of the head and prothorax push air through the tracheae. The action of the flight muscles to compress the thorax further enhances ventilation. New studies using high-intensity x-rays on crickets, beetles, and ants showed for the first time that these insects squeeze the tracheae in their heads and thoraxes, even when they are inactive (it is not yet known how this squeezing occurs). In some cases, actual flow-through movement is achieved as air is drawn in certain spiracles and forced out through separate ones (Figure 11–7a). However, air movement in the final tracheoles cannot be flow-through, because of their blind-end design.

■ Arachnids use book lungs or tracheae.

The first arachnids—scorpions and primitive spiders—have modified gill-like features called **book lungs** (● Figure 11–8). Stacks of lamellae (having the appearance of pages in a book) invaginate from the cuticle into the abdomen, forming numerous thin air chambers. In more advanced spiders; these have evolved into tubular extensions into the tissues and thus form a true **tracheal** system (which, however, evolved independently of the insect system).

■ Vertebrate air breathers use respiratory airways to conduct air between the atmosphere and the gas exchange surfaces in the lung.

The **respiratory system** of vertebrate lung breathers includes the respiratory structures leading into the lungs, the lungs themselves, and the structures of the thorax (chest) involved in moving air through the airways into and out of the lungs. The **respiratory airways** are tubes that carry air between the atmosphere and the lung surfaces such as alveoli. In some cases, these airways also function in exchange. In submerged turtles, the well-vascularized membranes lining the mouth take up O_2 from water as well as from the air. The tortuous nasal passages of some birds also represent an additional site of O_2 pickup. Let us examine the evolution of these airways and lungs in more detail.

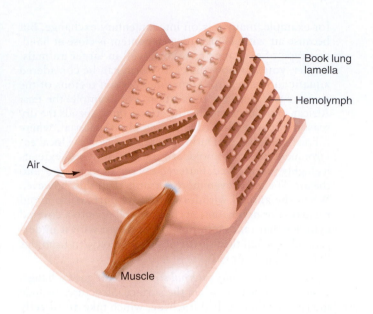

Figure 11–8 ● **Book lung of a spider.** Scorpions also have this type of respiratory organ.

(*Source:* Phillip C. Withers, 1992, *Comparative Animal Physiology,* Belmont, CA: Brooks/Cole, p. 638, Figure 13–24b)

■ The first air-breathing vertebrates were bimodal.

Vertebrate air breathing is believed to have been favored among those animals living in habitats similar to modern tropical lowlands, where seasonal drought results in stagnation of ponds leading to aquatic *hypoxia* (low oxygen), and even desiccation. The first evolutionary transition from aquatic to terrestrial life produced distinct fishes with a unique set of adaptations to respiring in both media. These **bimodal breathers** have, in addition to gills in water, the ability to breathe air by integumentary exchange (such as some eels), modified gills (such as air-breathing catfish, with more arch support to prevent gill collapse), parts of the digestive system including oral or stomach linings, or separate air sacs called lungs. Bimodal breathing is used by some fishes in waters that become anoxic (such as stagnant swamps), or in ponds that dry out seasonally. In addition, some have gills, lungs, *and* some cutaneous (skin) gas exchange with the result that they are actually **trimodal breathers** (such as the reedfish *Calamoichthys calabaricus*).

The most successful vertebrate groups in terms of terrestrial adaptation are those that evolved true lungs. Lungs in fishes originally evolved as a simple ventral evagination of the pharynx, and in some groups evolved into a dorsal gas bladder (● Figure 11–9). Eventually the gas bladder transformed into a structure specialized for both sound reception and buoyancy control and in many cases lost its respiratory function. In at least one group of fishes, however, the air sac became paired and ventrally oriented, leading eventually to the lungs of terrestrial amphibians, reptiles, birds, and mammals.

Amphibians, the first class of vertebrates to evolve from lung-bearing fishes, are noted for their bimodal lifestyles, living both in the water and air. Aquatic and terrestrial amphibians rely to varying degrees on integuments, gills, and lungs for gas exchange. For example, some species of salamanders are without lungs and are solely dependent on cutaneous gas ex-

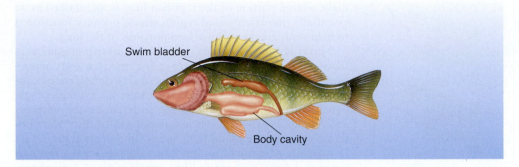

Figure 11–9 ● **The swim bladder of a fish.** A lateral view of a teleost showing the relative location of the swim bladder in the abdominal cavity.

(*Source:* K. Liem, W. Bemis, W. F. Walker, & L. Grande, 2001, *Functional Anatomy of the Vertebrates: An Evolutionary Perspective,* 3rd ed., Belmont, CA: Brooks/Cole, p. 358, Figure 11–6)

change. In these animals the skin is moist and well vascularized. At the other extreme, one salamander (*Siren lacertina*) has trimodal breathing.

In frogs, most larval stages have gills, whereas most adults have lungs that are comparatively simple and noncompartmentalized (● Figure 11–10a). In larger species needing greater surface area because of scaling problems, the lung is folded into alveoli. However, alveoli are larger and less numerous than in vertebrates with higher metabolisms. Air is forced by *positive pressure* into their lungs by a **buccal pump** (originating in the mouth) that is similar in function to that used in fish (● Figure 11–11). Normally, several inspiratory oscillations are required to completely fill the lungs, whereas one long exhalation empties the lungs. In these animals O_2 uptake is primarily through the lung, whereas CO_2 elimination is almost exclusively cutaneous. Terrestrial caecilians (tropical, legless amphibians), which spend most of their time in burrows, have single, elongate lungs that extend almost 70% of their body length. However, the lung structure of caecilians demonstrates a high degree of internal compartmentalization, forming numerous alveoli that markedly increase the available surface area for gas exchange.

In bimodal breathers, branchial (gill) contractions alternate between periods of pumping water through the gills to pumping air into the lung. In the ancestors of the first exclusively air breathers, water pumping was completely abandoned with the buccal muscles exclusively functioning as a positive pressure air pump to inflate the lungs. As the dependence on branchial O_2 uptake decreased, the gills and branchial arteries became reduced and internalized. The diffuse, externally oriented chemoreceptors for monitoring gas concentrations located in the gills evolved into discrete arterial chemoreceptors located near the carotid artery and in the aortic arch (see the section on breathing regulation later). Adaptations to air breathing also included a decrease in the affinity of hemoglobin for O_2 that reflected the much greater O_2 availability in air (see gas transport section later).

■ Reptiles, birds, and mammals use lungs ranging from simple sacs to elaborate folds, inflated by negative pressure.

Compartmentalized lungs are an important adaptation in mammals, birds, and some reptiles, but what separates these groups from all others is their mechanism of ventilation. In-

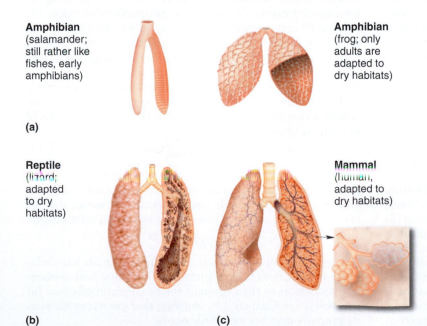

Amphibian (salamander; still rather like fishes, early amphibians)

(a)

Amphibian (frog; only adults are adapted to dry habitats)

Reptile (lizard; adapted to dry habitats)

Mammal (human, adapted to dry habitats)

(b)   **(c)**

Figure 11–10 ● **Comparative lung structure of amphibians, reptiles, and mammals.**
(a) Amphibian lungs range from comparatively simple sacs surrounded by few blood vessels, to pouches divided into many large compartments with dense vascularization. The latter is helpful for terrestrial amphibians that spend more time on land to reduce the use of integumentary exchange. (b) Lungs of a typical reptile rely on a bellowslike pumping of posterior air sacs that help airflow into and out of the lungs. Gas exchange occurs in a forward, spongy region of each lung. (c) Highly compliant mammalian lungs rely on gas-exchanging alveoli (lower right-hand box).

(*Source:* C. Starr & R. Taggart, 2004, *Biology: The Unity and Diversity of Life,* 10th ed., Belmont, CA: Brooks/Cole, p. 711, Figure 40.8)

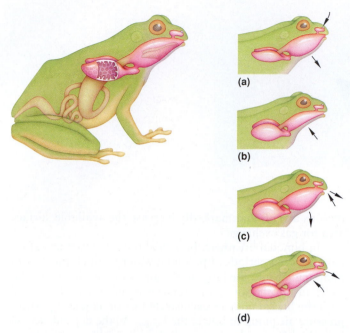

(a)

(b)

(c)

(d)

Figure 11–11 ● Respiration in a frog. (a) During inspiration the frog lowers the floor of its mouth and inhales air through its nostrils into the buccal cavity. (b) The frog then closes its nostrils, opens the glottis, and raises the mouth's floor. This action forces air into the lungs. (c) Gas exchange occurs in the lungs. The nostrils open at the end of this phase. (d) Air is forced out of the lungs when muscles in the body wall above the lungs contract and the lungs elastically recoil. Lung filling generally requires several successive active inflation cycles whereas lung deflation is usually accomplished by a single passive expiration.

(*Source*: C. Starr & R. Taggart, 2004, *Biology: The Unity and Diversity of Life*, 10th ed., Belmont, CA: Brooks/Cole, p. 711, Figure 40.7)

stead of forcing air from the buccal cavity into the lungs (amphibians and some reptiles), air flows into the lungs because of a *negative pressure*. The muscles lining the rib cage forcefully expand the lungs creating a subatmospheric pressure that draws air inward. Other muscles may add to this effect; in mammals, for example, inhalation is aided by the presence of a **diaphragm** (a dome-shaped muscle that separates the thoracic from the abdominal cavity) that forces the lungs into a more expanded state when contracted. You will later see how negative pressure and these muscles work, in the respiratory mechanics section.

Reptiles

Reptile lungs may be simple sacs or may have several large sacs or even some alveoli (Figure 11–10b). The simplest reptilian lung is somewhat analogous to a single, oversized alveolus. Vascularized ingrowths, or **septae** (dividing walls or partitions), penetrate centrally from the lung's perimeter and subdivide the pulmonary lumen into a series of spatial units or **ediculae**. Gas exchange occurs principally on the septae, although they are poorly vascularized compared to alveoli. Unlike mammalian alveoli, ediculae are also relatively passive participants during ventilation and do not contribute to the movement of air during inhalation and exhalation. As in the mammalian lung, airflow during ventilation is tidal and bellowslike. Interest-

ingly, a large portion of the reptilian lung is maintained as an essentially nonvascularized region whose primary function is to help ventilate the vascularized portions of the lung.

Lizards and snakes rely solely on **costal** (rib muscle) ventilation resulting from simple rib-cage expansion and contraction. Animals whose rib cage is fixed rely on alternative solutions to expand their lungs. Tortoises, for example, must extend their limbs to expand the thoracic cavity, whereas aquatic turtles exploit the gravitational pull of the viscera (internal organs) to expand their lungs. In theropod dinosaurs (a suborder of carnivorous dinosaurs that spanned a range of types and sizes) and crocodiles, contraction of a **diaphragmaticus muscle** (attached between the liver and a distinctive elongated region of the pubis bone) pulls the liver away from the lungs in a pistonlike manner that results in expansion of the lungs. The crocodilian **diaphragm** consists of a sheet of nonmuscular connective tissue that adheres tightly to the dome-shaped anterior surface of the liver. Exhalation is accomplished by abdominal muscles, which then pull the liver against the lungs.

Mammals and Birds

To meet the energy demands of birds and mammals with their increased metabolic rates, the surface area of the lung available for gas exchange had to markedly increase. The mammalian solution was the elaboration of numerous very small alveoli (● Figure 11–10c and 11–12), whereas the avian respiratory system, which is the most complex of the vertebrates, evolved *air capillaries*. The cardiovascular system became increasingly modified to perfuse the gas exchange organ. Oxygenated blood from the lung achieved complete separation from systemic venous blood (Chapter 9). These adaptations permitted these two groups to exploit new lifestyles relying on high metabolic rates.

One of the hallmarks of **scaling** (p. 9) in animals is that smaller species have higher basal metabolic rates, that is, they require more energy per unit mass. As you will see in Chapter 15, this phenomenon in birds and mammals is traditionally attributed to the higher surface-area-to-volume ratio of small species, with concomitant high rate of heat loss. To compensate, smaller mammals can take up proportionately more O_2 per unit of body weight than large mammals by (1) reducing the size of each alveolus, (2) increasing the density of alveoli, and (3) increasing the density of capillaries for gas exchange. The first two increase the effective surface area for diffusion. You can see the result of scaling forces in *scaling relationships* that have been determined by actual measurements of lungs. Among mammals, lung volume (V_l, in liters) is directly related to body weight by the equation

$$V_l = 0.035 m_b^{1.06}$$

where m_b = body mass (kilograms). The exponent 1.06 tells us that larger mammals actually have slightly *larger* lung volumes per unit mass than do smaller species. However, alveolar surface area (A_a, in cm^2) follows the following relationship:

$$A_a = 119 m_b^{0.75}$$

This exponent of 0.75 tells us that larger animals have relatively lower gas-exchange surfaces. This value is also essentially identical to the exponent for basal metabolic rate (as you will see in Chapter 15), showing that gas exchange ability precisely matches metabolic needs.

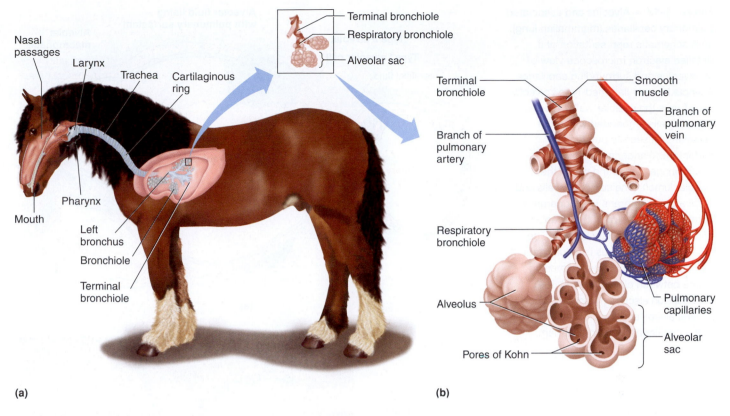

Figure 11–12 • Anatomy of the respiratory system in a mammal. (a) The respiratory airways. (b) Enlargement of the alveoli (air sacs) at the terminal end of the airways. Most alveoli are clustered in grapelike arrangements at the end of the terminal bronchioles.

(*Source:* Adapted from Starr and Taggart, *Biology: The Unity and Diversity of Life,* Eighth Edition, Belmont, CA: Wadsworth, p. 696, Fig. 41.10a.)

Mammalian airways terminate in alveoli.

Mammalian airways begin with the **nasal passages** (in a **nose** or blowhole) (Figure 11–12a). Adaptations for running in the horse and cheetah include nostrils that are pliable and easily dilated to increase the volume of air inspired. Concomitantly with the evolution of endothermy, the evolution of **maxilloturbinals,** thin curls of bone deep within the nasal cavity, was favored to retain heat and water (see Chapter 15). The nasal passages open into the **pharynx** (throat), which serves as a common passageway for both the respiratory and digestive systems. Two tubes lead from the pharynx—the **trachea (windpipe)**, through which air is conducted to the lungs, and the **esophagus,** the tube through which food passes to the stomach. Because the pharynx serves as a common passageway for food and air, reflex mechanisms exist to close off the trachea during swallowing so that food enters the esophagus and not the airways. The esophagus remains closed except during swallowing to prevent air from entering the stomach during breathing. The trachea is lined with two kinds of cells: mucus-secreting cells, which lubricate it, and ciliated cells lined with tiny hairs that continually beat upward, pushing impurities such as inhaled particles and dust, up and out of the trachea.

Next, the trachea divides into two main branches, the right and left **bronchi,** each entering one lung. Within each lung, the bronchus continues to branch into progressively narrower, shorter, and more numerous airways called **bronchioles,** much like the branching of a tree. Clustered at the ends of the terminal bronchioles in mammals are the alveoli (Figure 11–12b).

To permit airflow in and out of the gas-exchanging portions of the lungs, the continuum of conducting airways, from the entrance through the terminal bronchioles to the alveoli, must remain open most or all of the time. The trachea and larger bronchi in most mammals are fairly rigid, nonmuscular tubes encircled by a series of cartilaginous rings that prevent the tubes' compression. For example, the trachea of a giraffe is supported along its length by more than a hundred of these "tracheal rings". The smaller bronchioles have no cartilage to hold them open. Their walls contain smooth muscle that is innervated by the autonomic nervous system and is sensitive to certain hormones and local chemicals. These factors, by varying the degree of contraction of bronchiolar smooth muscle and hence the caliber of these small terminal airways, are able to regulate the amount of air passing between the atmosphere and each cluster of alveoli.

Alveoli

Mammalian lungs are excellent examples of gas exchange ability adapted to maximize diffusion (although not as superb as bird lungs). Recall that, according to Fick's law, enhancements arise by decreasing the distance for diffusion (ΔX), and increasing the surface area (A) across which diffusion can take place.

Figure 11–13 ● Alveolus and associated pulmonary capillaries (mammalian lung). (a) A schematic representation of a detailed electron microscope view of an alveolus and surrounding capillaries. A single layer of flattened Type I alveolar cells forms the alveolar walls. Type II alveolar cells embedded within the alveolar wall secrete pulmonary surfactant. Wandering alveolar macrophages are found within the alveolar lumen. (The size of the cells and respiratory membrane is exaggerated compared to the size of the alveolar and pulmonary capillary lumens. The diameter of an alveolus is actually about 600 times larger than the intervening space between air and blood.) (b) A transmission electron micrograph showing several alveoli and the close relationship of the capillaries surrounding them.

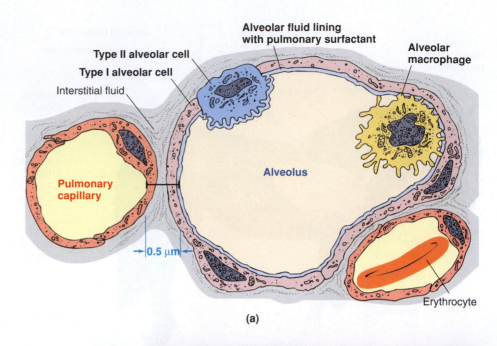

(a)

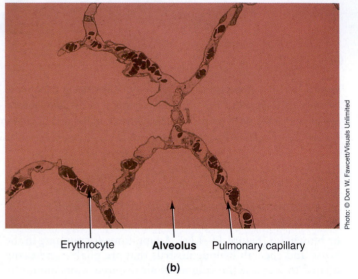

Erythrocyte   **Alveolus**   Pulmonary capillary

(b)

Photo: © Don W. Fawcett/Visuals Unlimited

The alveoli accomplish both of these enhancements. The alveolar walls consist of a single layer of flattened **Type I alveolar cells** (● Figure 11–13a). The walls of the dense network of pulmonary capillaries encircling each alveolus are also only one cell-layer thick. The interstitial space between an alveolus and the surrounding capillary network forms an extremely thin barrier, with only 0.25 μm separating the air in the alveoli from the blood in the pulmonary capillaries. (A sheet of tracing paper is about 50 times thicker than this air-to-blood barrier.)

Furthermore, the alveolar air–blood interface presents a tremendous surface area for exchange. The lungs of a human contain about 300 million alveoli in the lungs, each about 300 μm (11 mm) in diameter. So dense are the pulmonary capillary networks that each alveolus is encircled by an almost continuous sheet of blood (Figs. 11–13b). The total surface area thus exposed between alveolar air and pulmonary capillary blood is about 75 m^2 (about the size of a tennis court). In contrast, if the lungs consisted of a single, hollow chamber of the same dimensions instead of being divided into myriad alveolar units, the total surface area would be only about 0.01 m^2.

In addition to the thin, wall-forming Type I cells, the alveolar epithelium also contains **Type II alveolar cells** (Figure 11–13a), which secrete **pulmonary surfactant**, a phospholipoprotein complex that facilitates lung expansion (described later). Surfactant can be found in the lungs of amphibians, reptiles, birds, and mammals. Also present within the air sac lumen are the defensive macrophages.

Minute **pores of Kohn** (Figure 11–12b) present in the alveolar walls permit airflow between adjacent alveoli, a process known as **collateral ventilation**. These passageways are especially important in allowing fresh air to enter an alveolus whose terminal conducting airway is blocked because of disease.

There is no muscle within the alveolar walls to cause them to inflate and deflate during the breathing process. Rather, changes in the dimensions of the thorax produce the corresponding changes in lung volume. You will see how this works in the respiratory mechanics section, later.

Pleural Sac

Mammalian lungs occupy most of the volume of the **thoracic (chest) cavity,** the only other structures in the chest being the heart and associated vessels, the esophagus, the thymus, and some nerves. Separating each lung from the thoracic wall and other surrounding structures is a double-walled, closed sac called the **pleural sac** (● Figure 11–14). The dimensions of the **pleural cavity** within the pleural sac are greatly exaggerated in the illustration, to aid visualization; in reality the layers of the pleural sac are in close contact with one another. The surfaces of the pleura secrete a thin **intrapleural fluid,** which lubricates the pleural surfaces as they slide past each other during respiratory movements. The pleural cavity plays a crucial role in the flow of air, as you will see in the respiratory mechanics section.

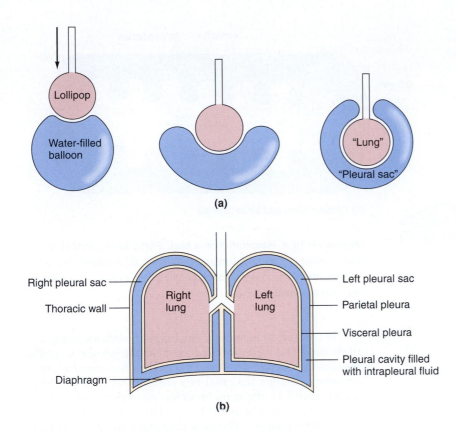

(a)

Lollipop

Water-filled
balloon

"Lung"

"Pleural sac"

Right pleural sac

Thoracic wall

Right
lung

Left
lung

Left pleural sac

Parietal pleura

Visceral pleura

Pleural cavity filled
with intrapleural fluid

Diaphragm

(b)

Figure 11–14 ● **Pleural sac.** (a) Pushing a lollipop into a water-filled balloon produces a relationship analogous to that between each double-walled, closed pleural sac and the lung that it surrounds and separates from the thoracic wall. (b) Schematic representation of the relationship of the pleural sac to the lungs and thorax. One layer of the pleural sac, the *visceral pleura,* closely adheres to the surface of the lung (*viscus* means "organ") then reflects back on itself to form another layer, the *parietal pleura,* which lines the interior surface of the thoracic wall (*paries* means "wall"). The relative size of the pleural cavity between these two layers is grossly exaggerated for the purpose of visualization.

■ Avian airways terminate in air capillaries.

To perform some functions assigned to forelimbs, birds can have necks up to threefold longer than mammals of comparable body size. The longer trachea increases the resistance to airflow but this is compensated, in part, by an increase in the radius of the trachea. For this reason the resistance to tracheal airflow in birds is similar to a mammal of similar body weight.

Unlike in mammals, the **syrinx** of birds lies at the exit of the trachea and generally consists of both tracheal and bronchial contributions (● Figure 11–15). Normally, three to six rings at the base of the trachea are enlarged and form a rigid structure termed the **tympanum.** The upper three rings of each bronchus are also enlarged, and together the composite structure is supplied with muscles, air sacs, and vibrating membranes. The entrance to each bronchus is constricted by a narrow slit, the so-called *tympanic membrane,* the extent of opening subject to variable muscle tension from the *syringeal muscles.* Air pressure generated from the lungs and air sacs sets the tympanic membrane in motion, and the syrinx operates as a valve that regulates both pitch and volume. The complexity of a bird's song is closely related to the number of syringeal muscles that connect with the various rings and assorted membranes. Several species of birds, including vultures, storks, and ostriches, lack any syringeal muscles, whereas true songbirds have around five pairs. Because each bronchus has its own tympanic membrane and the muscles are innervated separately, it is possible for some species, such as the wood thrush (*Hylocichla mustelina*), to simultaneously sing notes of separate frequencies. Contributing to vocalization in some species (swans, geese, and cranes) are tracheae of exceptional length. In the whooping crane (*Gus americana*) the trachea is almost as long as the bird itself and serves as a resonating tube to

amplify its vocalization. More powerful vocalizations are evidence of "fitness," which increase the male's likelihood of attracting a mate and passing on its genes.

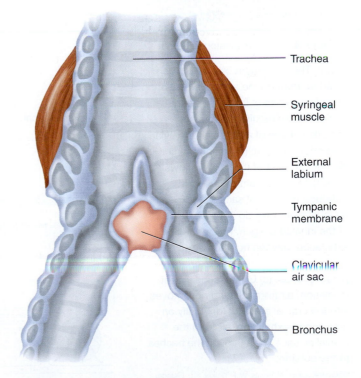

Trachea

Syringeal
muscle

External
labium

Tympanic
membrane

Clavicular
air sac

Bronchus

Figure 11–15 ● **The avian syrinx.**

(*Source:* www.biology.eku.edu/RITCHSO/birdcommunication.html)

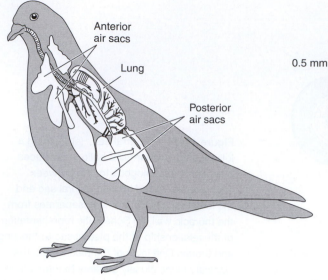

(a) Lungs and air sacs

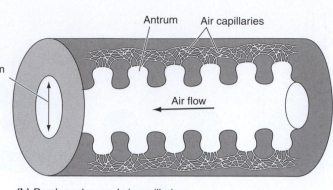

(b) Parabronchus and air capillaries

Figure 11–16 • Respiration in a bird. Part I. (a) A lateral view of the major cranial and abdominal air sacs. (b) A lateral view of the gas-exchanging parabronchus and air capillaries. (c) A section through the parabronchus.

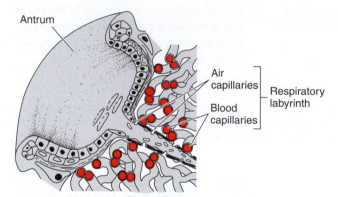

(c) Antrum and respiratory labyrinth

In contrast to mammals, the lungs of birds are comparatively small and inelastic and do not change in volume during the respiratory cycle. However, the tracheal volume of a bird is considerably larger than that of a comparably sized mammal. Within each lung the main bronchus gives rise to several sets of secondary bronchi, named according to the regions of the lung that they supply. The most important groups include the **ventrobronchi** and the **dorsobronchi.** The secondary bronchi are connected to each other by a number of tertiary bronchi or **parabronchi** from which the **air capillaries** (gas exchange surface) originate (● Figure 11–16). Parabronchi interconnect freely with each other and tend to have a constant internal diameter, ranging from 0.5 mm in hummingbirds to approximately 2 mm in chickens. Parabronchi are cylindrical and run

Part II. The movement of pure oxygen (shaded) through the avian lung. Two complete cycles of inspiration and expiration are required to move a specific volume of air through the avian respiratory system. (a) During the first inspiration most of the oxygen flows directly to the abdominal air sacs. Although the cranial sacs also expand on inhalation, they do not receive any of the inhaled oxygen. (b) On the first exhalation, oxygen from the abdominal air sacs flows into the gas-exchanging parabronchus. (c) On the subsequent inhalation, air from the lung now moves into the cranial air sacs. (d) Finally on the second exhalation, air from the cranial air sacs flows through the trachea to the outside.

(*Source:* K. Liem, W. Bemis, W. F. Walker, & L. Grande, 2001, *Functional Anatomy of the Vertebrates: An Evolutionary Perspective*, 3rd ed., Belmont, CA: Brooks/Cole, Part I, Figure 18–17; Part II, Figure 18–18, p. 594)

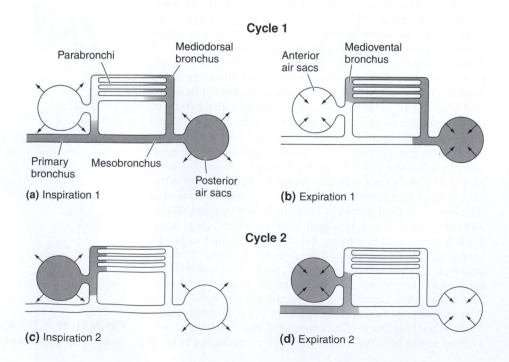

Cycle 1

(a) Inspiration 1

(b) Expiration 1

Cycle 2

(c) Inspiration 2

(d) Expiration 2

in parallel with each other. Together, the secondary bronchi and their interconnecting parabronchi form an integrated unit that enables unidirectional or *flow-through* air movement. Airflow is from the dorsobronchi, through the parabronchi and air capillaries, to the ventrobronchi. Recall that flow-through systems are more efficient; indeed, in birds air is not diluted with "old" air as in mammalian alveoli. Later you will see how the airflow is achieved (p. 496).

Comparable to the bronchial muscles of mammals, the *atrial smooth muscles* lining the avian parabronchus are innervated and regulate the diameter of the parabronchi. There are also a series of interconnected, voluminous, and highly distensible air sacs associated with the avian respiratory system (Figure 11–16). In most birds the air sacs arise in the embryo from six primordial pairs of air sacs, two of which fuse at the time around hatch. Thus there are eight air sacs in the adult domestic fowl: one cervical and one clavicular sac, two cranial thoracic and two caudal thoracic sacs in addition to the two abdominal sacs. Functionally there are two groups of air sacs, a *cranial* (toward the head) group and a *caudal* (toward the tail) group. The cranial group consists of several smaller sacs that connect with the ventrobronchi and receive expelled air from the lungs whereas the caudal group, which includes the comparatively large abdominal sacs, connects to the caudal end of the main bronchus and receives relatively fresh air from the trachea. Because the blood supply to the air sacs is essentially negligible, no gas exchange occurs across this surface. Air sacs are held open by their attachment to the structures that surround them and occasionally even penetrate the medullary (inner portion) cavities of some bones. However, the degree of aeration does not correlate with the ecological adaptations of individual species. The air sacs are also a common site of disease in birds. Generally, inspired pathogens lodge in the caudal air sacs because of their role in receiving freshly inspired air.

Birds do not have a muscular diaphragm as do mammals but instead have a *membranous diaphragm* that is attached to the body wall by muscles. Birds rely primarily on a highly modified costal (rib) movement in association with a well-developed sternum (the long, flat breastbone) to power lung ventilation. In addition, in some species of birds the joints of the ribs permits fore–aft movement of the rib cage during lung ventilation. As a result, avian ribs rotate such that the posterior end of the sternum is depressed on inhalation and generates negative intra-abdominal pressures that helps fill the comparatively large abdominal air sacs.

Air Capillaries

In birds, the air capillaries form an extensive network of air-carrying tubules jacketed with a profuse network of blood capillaries (Figure 11–16b). A comparison of the gas exchange surfaces between birds and mammals reveals the following:

- The diameter of an air capillary varies from only 3 μm in songbirds to 10 μm in swans and penguins, compared to 35 μm of alveoli in the smallest mammals.

- Air capillaries are not blind-ending; rather, they freely anastomose (connect) with each other. This leads to a higher gas-exchange efficiency than in mammals.

- Unlike alveoli, the diameter of air capillaries does not change significantly during the respiratory cycle.

- Although surfactant is secreted from granular cells lining the atria, its role in the avian lung appears to be limited to restricting the diffusion of fluid from the blood capillaries.

- The mean thickness of the blood–gas barrier is also considerably less (approximately 30%) in birds and is due to the thinness of the epithelial cells lining the air capillaries.

- The volume of the pulmonary-capillary blood per gram of body weight is approximately 20% greater in birds.

■ Aerial respiratory systems can perform numerous nonrespiratory functions.

In air respirers, lungs and other respiratory structures can have the following functions:

1. *Regulation of water loss and heat exchange.* Venous blood originating from the walls of the nasal passages can cool the arteries that supply the brain. During periods of intense activity brain temperatures can be 2 to 3 degrees C lower than core body temperatures. Inspired atmospheric air is humidified and warmed by the respiratory airways before it is expired. Moistening of inspired air is essential to prevent the respiratory surfaces from drying out. O_2 and CO_2 cannot diffuse through dry membranes.

2. *Circulation*—enhancement of venous return (see the "respiratory pump," p. 418).

3. *Acid–base balance*—the maintenance of normal acid–base balance by altering the amount of H^+-generating CO_2 exhaled (Chapter 13, p. 598).

4. *Vocalization, mating calls*, and other sounds.

5. *Defense* against inhaled foreign matter (Chapter 10, p. 432).

6. *Removal, modification, activation, or inactivation of various materials* passing through the pulmonary circulation. All blood returning to the heart from the tissues must pass through the avian and mammalian lungs before being returned to the systemic circulation. The lungs, therefore, are uniquely situated to partially or completely remove specific materials that have been added to the blood at the tissue level before they have a chance to reach other parts of the body by means of the arterial system. For example, prostaglandins, a collection of chemical messengers released in numerous tissues to mediate particular local responses, may spill into the blood, but they are inactivated during passage through the lungs so that they cannot exert systemic effects. In contrast, the lungs generate angiotensin II, a hormone that plays an important role in regulating the concentration of Na^+ in the extracellular fluid (p. 545).

7. *Olfaction*—the nose has chemoreceptors for airborne odorants.

Respiratory Mechanics

In this section, we examine the mechanical forces involved in breathing, focusing primarily on mammals, but noting important features of birds and fishes.

◼ Interrelationships among atmospheric, intra-alveolar, and intrapleural pressures are important in respiratory mechanics of mammals.

Gases tend to move from a region of higher pressure to a region of lower pressure, that is, down a **pressure gradient.** Thus air flows in and out of mammalian lungs during the act of breathing by moving down alternately reversing pressure gradients established between the lungs and the atmosphere by cyclical respiratory-muscle activity. Three different pressure considerations are important in ventilation (● Figure 11–17):

1. **Atmospheric (barometric) pressure** is the pressure exerted by the weight of the air in the atmosphere on objects on Earth's surface. At sea level it equals 760 mm Hg or 1 atmosphere (atm) (Figure 11–1a). The International System of Units (SI system) uses the pascal (Pa) as the pressure unit. In the SI system, 1 atm is equivalent to 101.3 kPa. Atmospheric pressure diminishes with increasing altitude above sea level as the column of air above Earth's surface correspondingly decreases. Minor fluctuations in atmospheric pressure occur at any height because of changing weather conditions (that is, when barometric pressure is rising or falling).

2. Intrapulmonary pressure, also known as **intra-alveolar pressure,** is the pressure within the mammalian lung. Because the lung communicates with the atmosphere through the conducting airways, air quickly flows down its pressure gradient any time intrapulmonary pressure differs from atmospheric pressure; the airflow continues until the two pressures equilibrate (become equal).

3. **Intrapleural pressure** is the pressure within the pleural sac. Also known as **intrathoracic pressure,** it is the pressure exerted outside the lungs within the thoracic cavity. The intrapleural pressure is usually less than atmospheric pressure, averaging 756 mm Hg at rest. Just as blood pressure is recorded using atmospheric pressure as a reference point (that is, a systolic blood pressure of 110 mm Hg is 110 mm Hg greater than the

atmospheric pressure of 760 mm Hg or in reality 880 mm Hg), 756 mm Hg is sometimes referred to as a pressure of −4 mm Hg, although there is really no such thing as an *absolute* negative pressure. A pressure of −4 mm Hg is just negative compared with 760 mm Hg. To avoid confusion, we use absolute positive values throughout our discussion of respiration.

Intrapleural pressure does not equilibrate with atmospheric or intra-alveolar pressure, because there is no direct communication between the pleural cavity and either the atmosphere or the lungs. Because the pleural sac is a closed sac with no openings, air cannot enter or leave despite any pressure gradients that might exist between it and surrounding regions.

◼ The lungs are normally stretched to fill the larger thorax.

The thoracic cavity is larger than the unstretched lungs because the thoracic wall grows more rapidly than the lungs during development. However, two forces—the intrapleural fluid's cohesiveness and the transmural pressure gradient—hold the thoracic wall and lungs in close apposition, stretching the lungs to fill the larger thoracic cavity.

Intrapleural Fluid's Cohesiveness

The polar water molecules in the intrapleural fluid resist being pulled apart because of their attraction to each other. The resultant cohesiveness of the intrapleural fluid tends to hold the pleural surfaces together. Thus the intrapleural fluid can be considered very loosely as a "stickiness" or "glue" between the lining of the thoracic wall and the lung. Have you ever tried to pull apart two smooth surfaces held together by a thin layer of liquid, such as two wet glass slides? If so, you know that the two surfaces act as if the thin layer of water sticks them together. Even though you can easily slip the slides back and forth relative to reach other (just as the intrapleural fluid facilitates movement of the lungs against the interior surface of the chest wall), you can pull the slides apart only with great difficulty, because the molecules

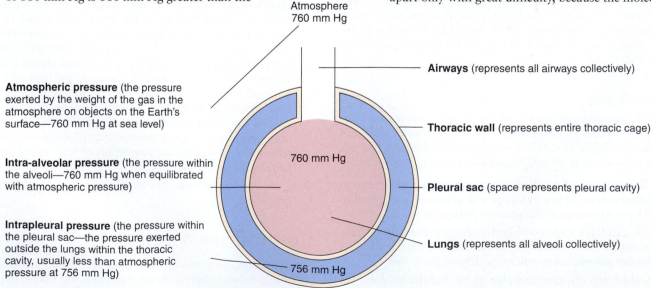

Atmosphere 760 mm Hg

Airways (represents all airways collectively)

Atmospheric pressure (the pressure exerted by the weight of the gas in the atmosphere on objects on the Earth's surface—760 mm Hg at sea level)

Thoracic wall (represents entire thoracic cage)

Intra-alveolar pressure (the pressure within the alveoli—760 mm Hg when equilibrated with atmospheric pressure)

760 mm Hg

Pleural sac (space represents pleural cavity)

Intrapleural pressure (the pressure within the pleural sac—the pressure exerted outside the lungs within the thoracic cavity, usually less than atmospheric pressure at 756 mm Hg)

Lungs (represents all alveoli collectively)

756 mm Hg

Figure 11–17 ● Pressures important to ventilation of a mammalian lung.

within the intervening liquid resist being separated. This relationship is partly responsible for the fact that changes in thoracic dimension are always accompanied by corresponding changes in lung dimension; that is, when the thorax expands, the lungs, being stuck to the thoracic wall by virtue of the intrapleural fluid's cohesiveness, do likewise.

Transmural Pressure Gradient

An even more important reason that the lungs follow the movements of the chest wall is the **transmural pressure gradient** that exists across the lung wall (*trans* means "across"; *mural* means "wall") (● Figure 11–18). The intra-alveolar pressure, equilibrated with atmospheric pressure at 760 mm Hg, is greater than the intrapleural pressure of 756 mm Hg, so a greater pressure is pushing outward than is pushing inward across the lung wall. This net outward pressure differential, the transmural pressure gradient, pushes out on the lungs, stretching or distending them (Figure 11–18). Because of this pressure gradient, the lungs are always forced to expand to fill the thoracic cavity.

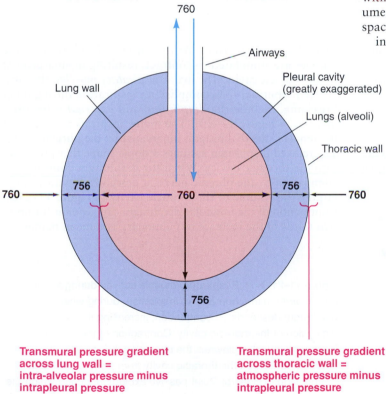

Transmural pressure gradient across lung wall = intra-alveolar pressure minus intrapleural pressure

Transmural pressure gradient across thoracic wall = atmospheric pressure minus intrapleural pressure

Figure 11–18 ● Transmural pressure gradient in a mammalian lung. Across the lung wall, the intra-alveolar pressure of 760 mm Hg pushes outward, while the intrapleural pressure of 756 mm Hg pushes inward. This 4-mm Hg difference in pressure constitutes a transmural pressure gradient that pushes out on the lungs, stretching them to fill the larger thoracic cavity. Across the thoracic wall, the atmospheric pressure of 760 mm Hg pushes inward, while the intrapleural pressure of 756 mm Hg pushes outward. This 4-mm Hg difference in pressure constitutes a transmural pressure gradient that pushes inward and compresses the thoracic wall.

A similar transmural pressure gradient exists across the thoracic wall. The atmospheric pressure pushing inward on the thoracic wall is greater than the intrapleural pressure pushing outward on this same wall, so the chest wall tends to be "squeezed in" or compressed compared to what it would be in an unrestricted state. The effect of the transmural pressure gradient across the lung wall is much more pronounced, however, because the highly distensible lungs are influenced by this modest pressure differential to a much greater extent than is the more rigid thoracic wall.

Because neither the lungs nor the thoracic wall are in their natural position when they are held in apposition to each other, they constantly try to assume their own inherent dimensions. The stretched lungs have a tendency to pull inward away from the thoracic wall, whereas the compressed thoracic wall tends to move outward away from the lungs. The transmural pressure gradient and intrapleural fluid's cohesiveness, however, prevent these structures from pulling away from each other except to the slightest degree. Even so, the resultant ever-so-slight expansion of the pleural cavity is sufficient to drop the pressure in this cavity by 4 mm Hg, bringing the intrapleural pressure to the subatmospheric level of 756 mm Hg. This pressure drop occurs because the pleural cavity is filled with fluid, which cannot expand to fill the slightly larger volume. Therefore, a vacuum exists in the infinitesimally small space in the slightly expanded pleural cavity not occupied by intrapleural fluid, producing a small drop in intrapleural pressure below atmospheric pressure.

Note the interrelationship between the transmural pressure gradient and the subatmospheric intrapleural pressure. The lungs are stretched and the thorax is compressed because a transmural pressure gradient exists across their walls as a result of the presence of a subatmospheric intrapleural pressure. The intrapleural pressure, in turn, is subatmospheric because the stretched lungs and compressed thorax tend to pull away from each other, slightly expanding the pleural cavity and dropping the intrapleural pressure below atmospheric pressure.

■ Flow of air into and out of mammalian lungs occurs because of cyclical intra-alveolar pressure changes brought about indirectly by respiratory muscle activity.

Because air flows down a pressure gradient, for air to flow into the lungs during inspiration the intra-alveolar pressure must be less than atmospheric pressure. Similarly, for air to flow out of the lungs during expiration the intra-alveolar pressure must be greater than atmospheric pressure. Altering the volume of the lungs, in accordance with Boyle's law, can change intra-alveolar pressure. **Boyle's law** states that at any constant temperature, the pressure exerted by a gas varies inversely with the volume of the gas (● Figure 11–19); that is, as the volume of a gas increases, the pressure exerted by the gas decreases proportionately, and conversely, the pressure increases proportionately as the volume decreases.

The respiratory muscles that accomplish breathing do not act directly on the lungs to change their volume. Instead, these muscles change the volume of the thoracic cavity, causing a corresponding change in lung volume because the thoracic

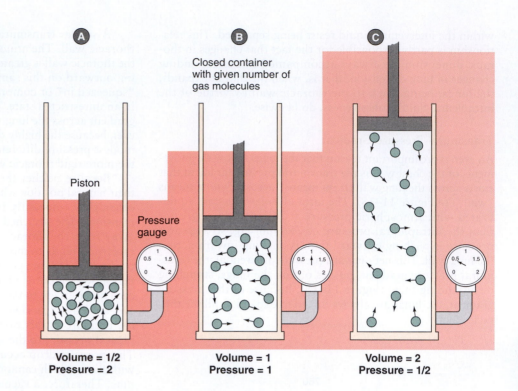

Figure 11–19 ● **Boyle's law.** Each container has the same number of gas molecules. Given the random motion of gas molecules, the likelihood of a gas molecule striking the interior wall of the container and exerting pressure varies inversely with the volume of the container at any constant temperature. The gas in container B exerts more pressure than the same gas in larger container C but less pressure than the same gas in smaller container A. This relationship is stated as Boyle's law: $P_1V_1 = P_2V_2$. As the volume of gas increases, the pressure of the gas decreases proportionately; conversely, the pressure increases proportionately as the volume decreases.

Closed container with given number of gas molecules

Piston

Pressure gauge

Volume = 1/2
Pressure = 2

Volume = 1
Pressure = 1

Volume = 2
Pressure = 1/2

wall and lungs are linked together by the transmural pressure gradient.

Let us follow the changes that occur in a mammal during one respiratory cycle—that is, one breath in (**inspiration**) and out (**expiration**).

Contraction of Inspiratory Muscles

Before the beginning of inspiration, the respiratory muscles are relaxed, no air is flowing, and intra-alveolar pressure is equal to atmospheric pressure. At the onset of inspiration, these muscles are stimulated to contract, resulting in enlargement of the thoracic cavity. The major *inspiratory muscles*, the muscles that contract to accomplish an inspiration during quiet breathing, include the *diaphragm* and *external intercostal muscles* (● Figure 11–20). The major inspiratory muscle of mammals is the diaphragm, innervated by the **phrenic nerve**. The relaxed diaphragm assumes a dome shape that protrudes into the thoracic cavity. When the diaphragm contracts on stimulation by the phrenic nerve, it descends downward, enlarging the volume of the thoracic cavity (● Figure 11–21a). Contraction of the muscular diaphragm can account for up to two thirds of the increase in pulmonary volume in mammals.

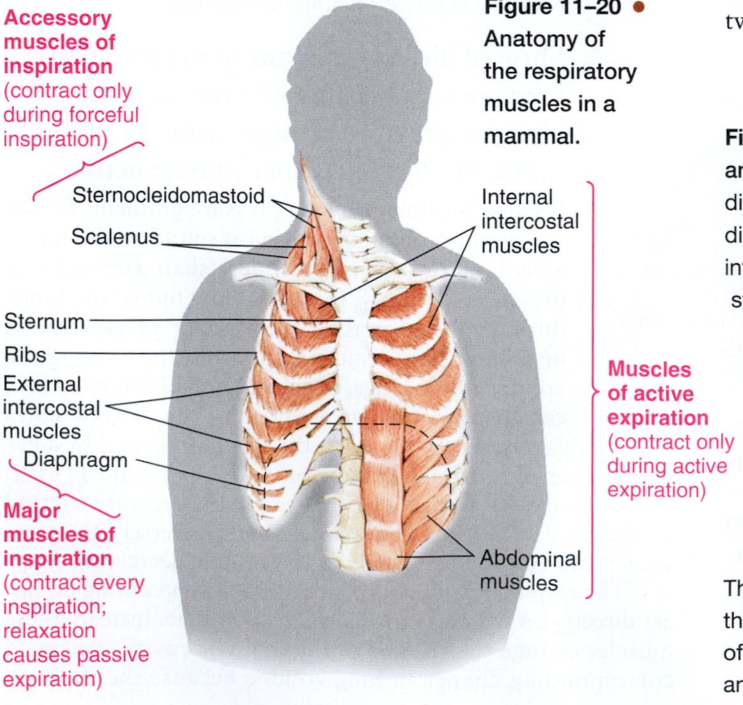

Accessory muscles of inspiration
(contract only during forceful inspiration)

Sternocleidomastoid
Scalenus
Sternum
Ribs
External intercostal muscles
Diaphragm

Major muscles of inspiration
(contract every inspiration; relaxation causes passive expiration)

Figure 11–20 ●
Anatomy of the respiratory muscles in a mammal.

Internal intercostal muscles

Muscles of active expiration
(contract only during active expiration)

Abdominal muscles

Figure 11–21 ▶ ● Respiratory muscle activity during inspiration and expiration in a human. (a) Inspiration, during which the diaphragm descends on contraction, increasing the vertical dimension of the thoracic cavity. Contraction of the external intercostal muscles elevates the ribs and subsequently the sternum to enlarge the thoracic cavity from front to back and from side to side. (b) Quiet passive expiration, during which the diaphragm relaxes, reducing the volume of the thoracic cavity from its peak inspiratory size. As the external intercostal muscles relax, the elevated rib cage falls because of the force of gravity. This also reduces the volume of the thoracic cavity. (c) Active expiration, during which contraction of the abdominal muscles increases the intra-abdominal pressure, exerting an upward force on the diaphragm. This reduces the vertical dimension of the thoracic cavity further than it is reduced during quiet passive expiration. Contraction of the internal intercostal muscles decreases the front-to-back and side-to-side dimensions by flattening the ribs and sternum.

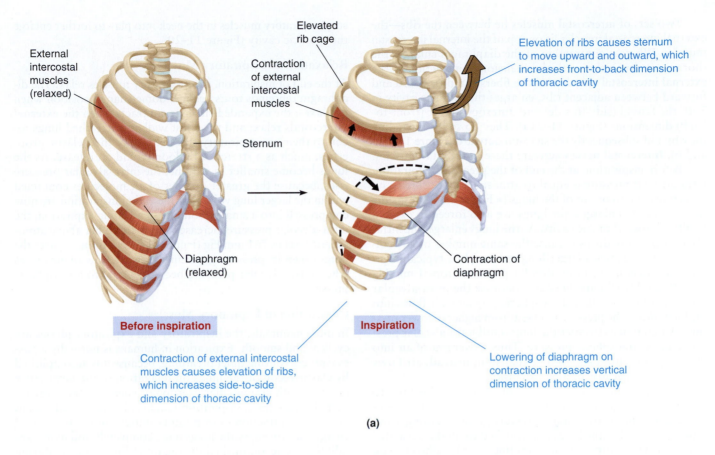

External intercostal muscles (relaxed)

Elevated rib cage

Contraction of external intercostal muscles

Sternum

Diaphragm (relaxed)

Contraction of diaphragm

Elevation of ribs causes sternum to move upward and outward, which increases front-to-back dimension of thoracic cavity

Before inspiration

Inspiration

Contraction of external intercostal muscles causes elevation of ribs, which increases side-to-side dimension of thoracic cavity

Lowering of diaphragm on contraction increases vertical dimension of thoracic cavity

(a)

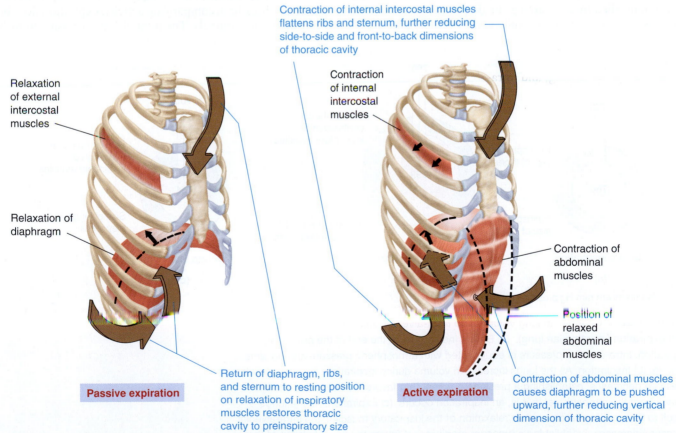

Contraction of internal intercostal muscles flattens ribs and sternum, further reducing side-to-side and front-to-back dimensions of thoracic cavity

Relaxation of external intercostal muscles

Relaxation of diaphragm

Contraction of internal intercostal muscles

Contraction of abdominal muscles

Position of relaxed abdominal muscles

Passive expiration

Return of diaphragm, ribs, and sternum to resting position on relaxation of inspiratory muscles restores thoracic cavity to preinspiratory size

Active expiration

Contraction of abdominal muscles causes diaphragm to be pushed upward, further reducing vertical dimension of thoracic cavity

(b)

(c)

Two sets of **intercostal muscles** lie between the ribs—the external intercostal muscles lie on top of the internal intercostal muscles. Whereas contraction of the diaphragm enlarges the thoracic cavity in the vertical dimension, contraction of the **external intercostal muscles,** whose fibers run downward and forward between adjacent ribs, enlarges the thoracic cavity in both the lateral (side-to-side) and anteroposterior (front-to-back) dimensions (Figure 11–21a). They do so by expanding the ribs and subsequently the sternum outward (Figure 11–21a and b). **Intercostal nerves** activate these intercostal muscles.

Before inspiration, at the end of the preceding expiration, intra-alveolar pressure is equal to atmospheric pressure so no air is flowing into or out of the lungs (● Figure 11–22a). As the thoracic cavity enlarges, the lungs are also forced to expand to fill the larger thoracic cavity. As the lungs enlarge, the intra-alveolar pressure drops because the same number of air molecules now occupies a larger lung volume. In a typical inspiratory excursion, the intra-alveolar pressure drops 1 mm Hg to 759 mm Hg (Figure 11–22b). Because the intra-alveolar pressure is now less than atmospheric pressure, air flows into the lungs down the pressure gradient from higher to lower pressure. Air continues to enter the lungs until intra-alveolar pressure equals atmospheric pressure. Thus movement of air into the lungs does not cause lung expansion; instead, air flows into the lungs because of lung expansion.

During inspiration, the intrapleural pressure falls to 754 mm Hg as a result of expansion of the thorax. The resultant increase in the transmural pressure gradient during inspiration ensures that the lungs are stretched to fill the expanded thoracic cavity. Deeper inspirations (more air breathed in) can be accomplished by contracting the diaphragm and external intercostal muscles more forcefully and by bringing the **acces-**sory inspiratory muscles in the neck into play to further enlarge the thoracic cavity (Figure 11–20).

Relaxation of Inspiratory Muscles

At the end of inspiration, the inspiratory muscles relax. The diaphragm rebounds to its original dome-shaped position when it relaxes; the expanded rib cage contracts when the external intercostals relax; and the chest wall and stretched lungs recoil to their preinspiratory size because of their elastic properties, much as a stretched balloon would on release. As the lungs become smaller in volume, the intra-alveolar pressure rises, because the greater number of air molecules contained within the larger lung volume at the end of inspiration are now compressed into a smaller volume. In a resting expiration, the intra-alveolar pressure increases about 1 mm Hg above atmospheric level to 761 mm Hg (Figure 11–22c). Air now leaves the lungs down its pressure gradient. Outward flow of air ceases when intra-alveolar pressure becomes equal to atmospheric pressure.

Contraction of Expiratory Muscles

In most mammals, the inspiratory and expiratory phases are cyclical and smooth. Expiration in humans is normally a *passive* process during quiet breathing, because it is accomplished by elastic recoil of the lungs on relaxation of the inspiratory muscles, with no muscular exertion or energy expenditure required. In contrast, inspiration is *active,* because it is brought about by contraction of inspiratory muscles at the expense of energy use. To empty the lungs more completely and more rapidly than is accomplished during quiet breathing, as during the deeper breaths accompanying activity, expiration does become active in mammals. The intra-alveolar pressure must be

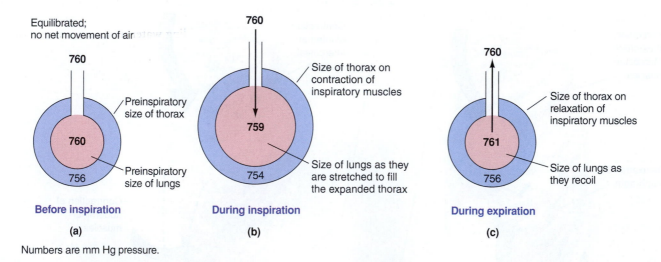

Numbers are mm Hg pressure.

Figure 11–22 ● **Changes in lung volume and intra-alveolar pressure during inspiration and expiration (mammalian lung).** (a) Before inspiration, at the end of the preceding expiration. Intra-alveolar pressure is equilibrated with atmospheric pressure and no air is flowing. (b) Inspiration. As the lungs increase in volume during inspiration, the intra-alveolar pressure decreases, establishing a pressure gradient that favors the flow of air into the alveoli from the atmosphere; that is, an inspiration occurs. (c) Expiration. As the lungs recoil to their preinspiratory size on relaxation of the inspiratory muscles, the intra-alveolar pressure increases, establishing a pressure gradient that favors the flow of air out of the alveoli into the atmosphere; that is, an expiration occurs.

increased even further above atmospheric pressure than can be accomplished by simple relaxation of the inspiratory muscles and elastic recoil of the lungs. To produce such a **forced,** or **active expiration, expiratory muscles** must contract to further reduce the volume of the thoracic cavity and lungs. The most important expiratory muscles are the *muscles of the abdominal wall.* As the abdominal muscles contract, the resultant increase in intra-abdominal pressure exerts a force on the diaphragm, pushing it further into the thoracic cavity than its relaxed position, thus decreasing the vertical dimension of the thoracic cavity even more (Figure 11–21c). The other expiratory muscles are the **internal intercostal muscles,** whose contraction pulls the ribs inward, flattening the chest wall and further decreasing the size of the thoracic cavity; this action is antagonistic to that of the external intercostal muscles (Figure 11–21b and c). These muscles increase the differential between intra-alveolar and atmospheric pressure above that of passive expiration, so more air leaves down the pressure gradient before equilibration is achieved. In this way, the lungs are emptied more completely during forceful, active expiration than during quiet, passive expiration.

This general pattern of respiration is somewhat different in the horse, where both the inspiratory and expiratory phases have two separate components even at rest. Inspiration begins passively and rapidly becomes an active process with contributions from both the diaphragm and external intercostals. Similarly expiration begins passively and rapidly becomes active as the abdominal muscles become involved. This pattern of respiration occurs because a portion of the energy spent during active inspiration and expiration is stored and used as passive energy to initiate the subsequent breathing cycle. (Similarly, in birds, if innervation to the respiratory muscles of birds were removed, the thoracic cage would come to rest at approximately midway between its full inspiratory and expiratory position.) Both inspiration and expiration thus have an active as well as a passive component; elastic recoil is a contributing factor to the onset of both the inspiratory and expiratory movements whereas the inspiratory muscles (the external intercostals, diaphragm, and the *triangularis sterni* in horses), as well as the expiratory muscles (the internal intercostals and abdominal muscles), are active throughout the inspiratory and expiratory cycles.

▪ Respiratory diseases often increase airway resistance.

In Chapter 9, we discussed how fluid flow depends on both the pressure gradient and resistance (flow = $\Delta P/R$; see p. 406). This is true of airflow as well as blood flow. In a healthy respiratory system, the radius of the conducting system is sufficiently large that resistance remains extremely low. Therefore, the pressure gradient between the alveoli and the atmosphere is usually the primary factor determining the airflow rate. Indeed, the airways normally offer such low resistance that normally only very small pressure gradients of 1 to 2 mm Hg need be created to achieve adequate rates of airflow in and out of the lungs. However, various respiratory diseases can narrow the airways such that resistance becomes a limiting factor. For example, **chronic obstructive pulmonary disease (COPD)** is a group of lung diseases (including *chronic bronchitis, asthma,* and *emphysema*) characterized by increased airway resistance

resulting from the narrowing of the lumen of the lower airways. When airway resistance increases, a larger pressure gradient must be established to maintain even a normal airflow rate. For example, if resistance is doubled by narrowing of airway lumens, P must be doubled through increased respiratory muscle exertion to induce the same flow rate of air in and out of the lungs as a normal individual animal accomplishes during quiet breathing. Accordingly, mammals with COPD (some forms of which occur in many species such as horses, in addition to humans) must work harder to breathe.

▪ Elasticity in mammalian lungs depends on connective tissue and alveolar surface tension, which is reduced by surfactant.

The term **elastic recoil** refers to how readily the lungs rebound after having been stretched. It is responsible for the lungs returning to their preinspiratory volume when the inspiratory muscles relax at the end of inspiration. Pulmonary elastic behavior depends mainly on two factors: highly elastic connective tissue in the lungs and alveolar surface tension.

Elastin

Pulmonary connective tissue contains large quantities of **elastin** fibers (see p. 63). Not only do these fibers exhibit elastic properties themselves, but they are arranged into a meshwork that amplifies their elastic behavior, much like the threads in a piece of stretch-knit fabric. These rebound to a smaller length after being stretched.

Alveolar Surface Tension

An even more important factor influencing elastic behavior of the lungs is the **alveolar surface tension** displayed by the thin liquid film that lines each alveolus. At an air–water interface, the water molecules at the surface are more strongly attracted to other surrounding water molecules than to the air above the surface. This unequal attraction produces a force known as *surface tension* at the surface of the liquid. Surface tension is responsible for a twofold effect. First, the liquid layer resists any force that increases its surface area; that is, it opposes expansion of the alveolus because the surface water molecules oppose being pulled apart. Accordingly, the greater the surface tension, the less compliant the lungs. Second, the liquid surface area tends to become as small as possible because the surface water molecules, being preferentially attracted to each other, try to get as close together as possible. Thus the surface tension of the liquid lining an alveolus tends to reduce the size of the alveolus, squeezing in on the air within it (● Figure 11–23). This property, along with the rebound of the stretched elastin fibers, is responsible for the lungs' elastic recoil back to their preinspiratory size after inspiration.

However, there is a problem. The cohesive forces between water molecules are so strong that if the alveoli were lined with water alone, the surface tension would be so great that the lungs would collapse; the recoil force attributable to the elastin fibers and high surface tension would exceed the opposing stretching force of the transmural pressure gradient. Furthermore, the lungs would be very poorly compliant, so exhausting muscular efforts would be required to accomplish stretching and inflation of the alveoli.

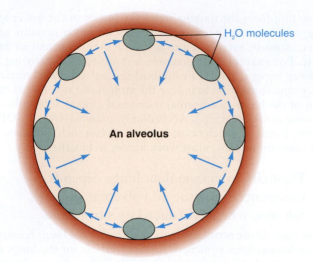

H₂O molecules

An alveolus

Figure 11–23 ● Alveolar surface tension (mammalian lung). The attractive forces between the water (H_2O) molecules in the liquid film that lines the alveolus are responsible for surface tension. Because of its surface tension, an alveolus (1) resists being stretched, (2) tends to be reduced in surface area or size, and (3) tends to recoil after being stretched.

Pulmonary Surfactant

The tremendous surface tension of pure water is normally counteracted by pulmonary surfactant, a complex mixture of lipids and proteins secreted by the Type II alveolar cells (Figure 11–13a). Pulmonary surfactant intersperses between the water molecules in the fluid lining the alveoli and lowers the alveolar surface tension because the cohesive force between a water molecule and an adjacent pulmonary surfactant molecule is very low. By lowering the alveolar surface tension, pulmonary surfactant provides two important benefits: (1) It increases pulmonary compliance, thus reducing the work of inflating the lungs; and (2) it reduces the lungs' tendency to recoil so that they do not collapse as readily.

Pulmonary surfactant's role in reducing the alveoli's tendency to recoil, thereby discouraging alveolar collapse, is important in helping maintain lung stability. The division of the lung into a myriad of tiny air sacs provides the advantage of a tremendously increased surface area for the exchange of O_2 and CO_2, but it also presents the problem of maintaining the stability of all these alveoli. Recall that the pressure generated by alveolar surface tension is directed inward, squeezing in on the air in the alveoli. If the alveoli are visualized as spherical bubbles, according to **LaPlace's law,** the magnitude of the inward-directed collapsing pressure is directly proportional to the surface tension and inversely proportional to the radius of the bubble:

$$P = 2T/r$$

where

P = inward-directed collapsing pressure

T = surface tension

r = radius of bubble (alveolus)

Because the collapsing pressure is inversely proportional to the radius, the smaller the alveolus, the smaller its radius and the greater its tendency to collapse at a given surface tension. Accordingly, if two alveoli of unequal size but the same surface tension are connected by the same terminal airway, the smaller alveolus—because it generates a larger collapsing pressure—has a tendency to collapse and empty its air into the larger alveolus (● Figure 11–24a). Small alveoli normally do not collapse and blow up larger alveoli, however, because pulmonary surfactant reduces the surface tension of small alveoli more than it reduces it in larger ones. This is because the surfactant molecules are more densely packed together in the smaller alveoli. The larger an alveolus, the more spread out are its surfactant molecules and the less effect they have on reducing surface tension. The surfactant-induced lower surface tension of small alveoli offsets the effect of their smaller radii in determining the inward-directed pressure. Therefore, the presence of surfactant causes the collapsing pressure of small alveoli to become comparable to that of larger alveoli and minimizes the tendency for small alveoli to collapse and empty their contents into larger alveoli (Figure 11–24b). Pulmonary surfactant therefore helps stabilize the sizes of the alveoli and helps keep them open and available to participate in gas exchange.

New findings point to an additional role for pulmonary surfactant in the lung defense system. Studies suggest that the protein component of pulmonary surfactant enhances phagocytosis of bacteria and viruses by alveolar macrophages and assists the ciliary mucus escalator in the respiratory airways that remove foreign matter.

Mechanics: Lung Volumes and Respiratory Cycles

Methodology: The Spirometer

The changes in lung volume that occur in respiratory cycles are important measures in analyzing different respiratory efforts, disease states, and species differences. These volumes can be measured with a **spirometer.** Spirometric measurements can be obtained from trained or anesthetized animals wearing a mouthpiece appropriate to their anatomy. Basically, a spirometer consists of an air-filled inverted drum floating in a water-filled chamber. As the animal breathes air in and out of the drum through a tube connecting the mouth to the air chamber, the drum rises and falls in the water chamber (● Figure 11–25). This rise and fall can be recorded as a **spirogram,** which is calibrated to volume changes.

● Figure 11–26b is a hypothetical example of a spirogram, with values comparing in a healthy young horse to an average human male (generally the values are lower for females). The following lung volumes and lung capacities (a lung capacity is a sum of two or more lung volumes) can be determined, with horse values noted:

- **Tidal volume (TV).** The volume of air entering or leaving the lungs during a single breath. Average value under resting conditions = 6 L.

- **Inspiratory reserve volume (IRV).** The extra volume of air that can be maximally inspired over and above the typical

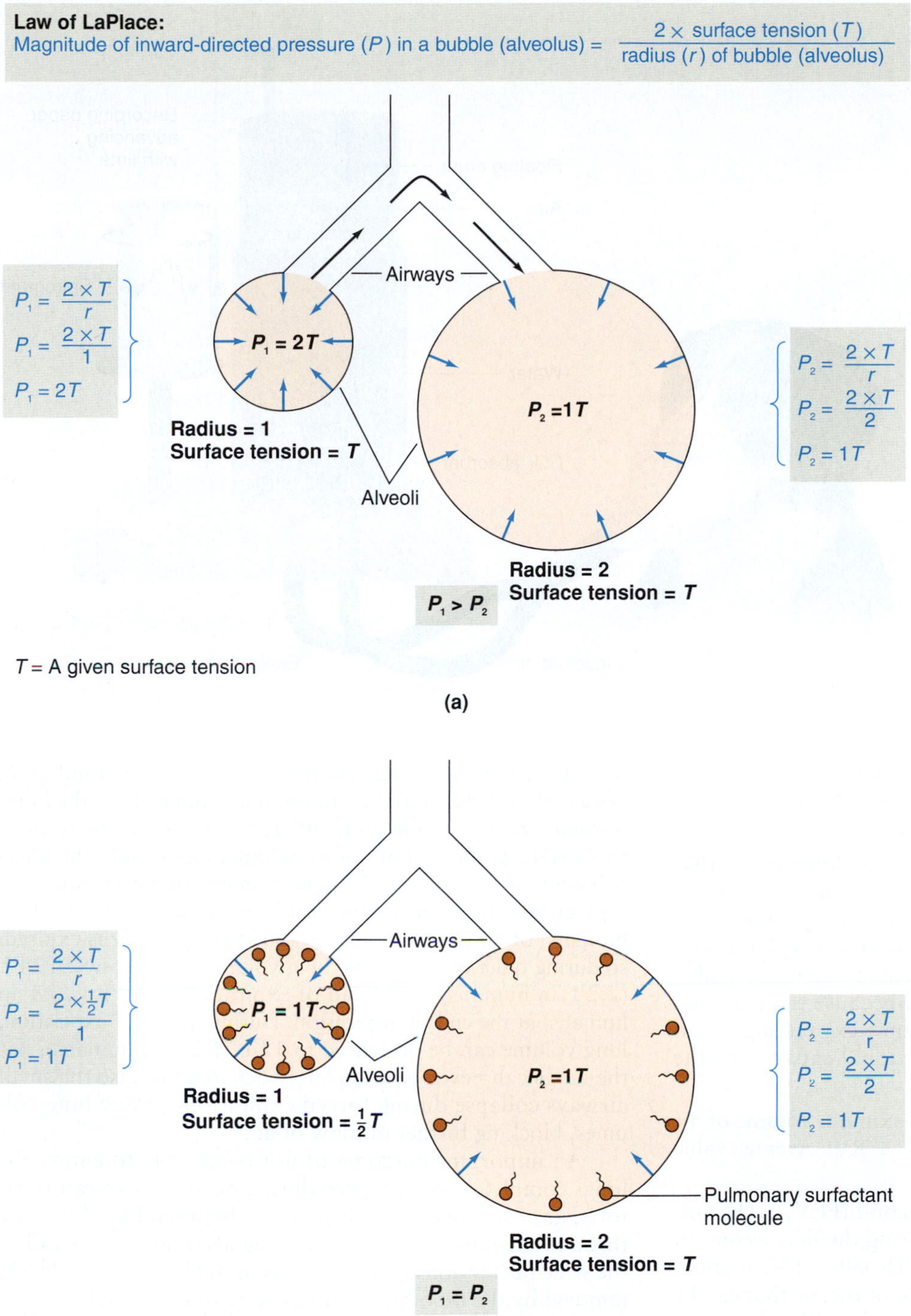

Law of LaPlace:
Magnitude of inward-directed pressure (P) in a bubble (alveolus) = $\dfrac{2 \times \text{surface tension } (T)}{\text{radius } (r) \text{ of bubble (alveolus)}}$

$P_1 = \dfrac{2 \times T}{r}$
$P_1 = \dfrac{2 \times T}{1}$
$P_1 = 2T$

Airways

$P_1 = 2T$

Radius = 1
Surface tension = T

Alveoli

$P_2 = 1T$

$P_2 = \dfrac{2 \times T}{r}$
$P_2 = \dfrac{2 \times T}{2}$
$P_2 = 1T$

Radius = 2
Surface tension = T

$P_1 > P_2$

T = A given surface tension

(a)

$P_1 = \dfrac{2 \times T}{r}$
$P_1 = \dfrac{2 \times \frac{1}{2}T}{1}$
$P_1 = 1T$

Airways

$P_1 = 1T$

Radius = 1
Surface tension = $\frac{1}{2}T$

Alveoli

$P_2 = 1T$

$P_2 = \dfrac{2 \times T}{r}$
$P_2 = \dfrac{2 \times T}{2}$
$P_2 = 1T$

Radius = 2
Surface tension = T

Pulmonary surfactant molecule

$P_1 = P_2$

Figure 11–24 ● Role of pulmonary surfactant in counteracting the tendency for small alveoli to collapse into larger alveoli (mammalian lung). (a) According to the law of LaPlace, if two alveoli of unequal size but the same surface tension are connected by the same terminal airway, the smaller alveolus—because it generates a larger inward-directed collapsing pressure—has a tendency (without pulmonary surfactant) to collapse and empty its air into the larger alveolus. (b) Pulmonary surfactant reduces the surface tension of a smaller alveolus more than that of a larger alveolus. This reduction in surface tension offsets the effect of the smaller radius in determining the inward-directed pressure. Consequently, the collapsing pressures of the small and large alveoli are comparable. Therefore, in the presence of pulmonary surfactant a small alveolus does not collapse and empty its air into the larger alveolus.

resting tidal volume. The IRV is accomplished by maximal contraction of the diaphragm, external intercostal muscles, and accessory inspiratory muscles. Average value = 12 L.

■ **Inspiratory capacity (IC).** The maximum volume of air that can be inspired at the end of a normal quiet expiration ($IC = IRV + TV$). Average value = 18 L.

■ **Expiratory reserve volume (ERV).** The extra volume of air that can be actively expired by maximal contraction of the expiratory muscles beyond that normally passively

expired at the end of a typical resting tidal volume. Average value = 12 L.

■ **Residual volume (RV).** The minimum volume of air remaining in the lungs even after a maximal expiration. Average value = 12 L. The residual volume cannot be measured directly with a spirometer because this volume of air does not move in and out of the lungs. It can be determined indirectly, however, through gas dilution techniques involving inspiration of a known quantity of a harmless tracer gas such as helium.

Figure 11–25 • Spirometry in animals. A spirometer is a device that measures the volume of air breathed in and out; it consists of an air-filled drum floating in a water-filled chamber. A tracheal catheter is placed in an anesthetized animal (or trained animals are fitted with an appropriately sized mask). Initially the animal breathes outside air until adapted to the procedure. A valve is closed, and breathing continues into an oxygen-filled reservoir. Carbon dioxide is absorbed, and the amount of oxygen consumed is computed from the recording. As the animal breathes in and out of the drum, the resultant rise and fall of the drum are recorded as a spirogram, which is calibrated to the magnitude of the volume change.

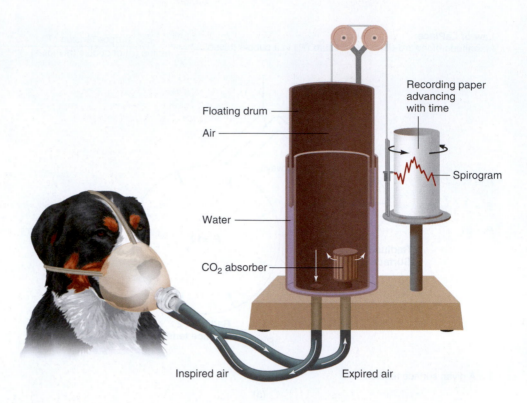

- **Functional residual capacity (FRC).** The volume of air in the lungs at the end of a normal passive expiration ($FRC = ERV + RV$). Average value = 24 L.

- **Vital capacity (VC).** The maximum volume of air that can be moved out during a single breath following a maximal inspiration. The animal first breathes in maximally, then breathes out maximally ($VC = IRV + TV + ERV$). The VC represents the maximum volume change possible within the lungs. It is rarely used because the maximal muscle contractions involved become exhausting, but it is useful in ascertaining the functional capacity of the lungs. Average value = 30 L.

- **Total lung capacity (TLC).** The maximum volume of air that the lungs can hold ($TLC = VC + RV$). Average value = 42 L.

- **Forced expiratory volume in 1 second (FEV_1).** The volume of air that can be expired during the first second of expiration in a VC determination. Usually, FEV_1 is about 80% of VC; that is, normally 80% of the air that can be forcibly expired from maximally inflated lungs can be expired within 1 second. This measurement gives an indication of the maximal airflow rate that is possible from the lungs.

The respiratory cycle of mammals is similar between species.

Most mammals have similar respiratory cycles, although the volumes can differ greatly. Let us compare horses and humans. On average, in healthy young horses the maximum amount of air that the lungs can hold is about 42 L, whereas it is only about 5.7 L in humans (Figure 11–26a). Anatomic build, age, distensibility of the lungs, and presence or absence of respiratory disease all affect this total lung capacity. Normally, during

quiet breathing, the lungs are not close to maximal inflation, nor are they deflated to their minimum volume. Thus the lungs normally remain moderately inflated throughout the respiratory cycle. At the end of a normal quiet expiration, the lungs still contain about 24 L (2.2 L in humans) of air. During each typical breath under resting conditions, about 6 L (0.5 L in humans) of air are inspired and the same quantity is expired, so during quiet breathing the lung volume varies between 24 L (2.2 L in humans) at the end of expiration to 30 L (2.7 L in humans) at the end of inspiration. During maximal expiration, lung volume can be decreased to 12 L (1.2 L in humans), but the lungs can never be completely deflated because the small airways collapse during forced expirations at low lung volumes, blocking further outflow of air.

An important outcome of not being able to empty the lungs completely is that, even during maximal expiratory efforts, gas exchange can still continue between blood flowing through the lungs and the remaining alveolar air. Instead of the wide fluctuations that would occur in O_2 uptake and CO_2 removal by the blood if the lungs were to completely fill and empty with each breath, the gas content of the blood leaving the lungs for delivery to the tissues normally remains remarkably constant throughout the respiratory cycle. Furthermore, recall that it takes less effort to inflate a partially inflated alveolus than a totally collapsed one.

Alveolar ventilation is less than pulmonary ventilation because of the presence of dead space.

Various changes in lung volume represent only one factor in the determination of **pulmonary** or **minute ventilation,** which is the volume of air breathed in and out in one minute. The other important factor is **respiratory rate,** which averages

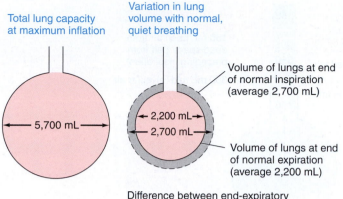

| | Variation in lung volume with normal, quiet breathing | | Minimal lung volume (residual volume) at maximal deflation |

Total lung capacity at maximum inflation — 5,700 mL

Volume of lungs at end of normal inspiration (average 2,700 mL)

2,200 mL
2,700 mL

Volume of lungs at end of normal expiration (average 2,200 mL)

Difference between end-expiratory and end-inspiratory volume equals tidal volume (average 500 mL)

1,200 mL

(a)

Figure 11–26 ● Variations in lung volume. (a) Normal range and extremes of lung volume in a human (healthy young adult male), (b) Normal spirogram of a human (healthy young adult male) compared to a horse. (The residual volume cannot be measured with a spirometer but must be determined by another means.)

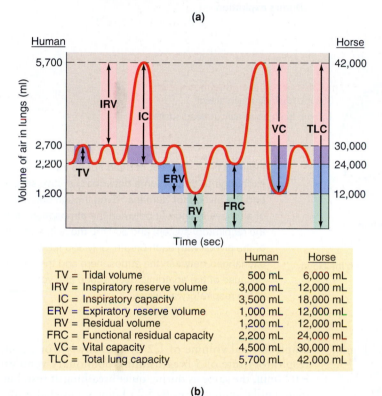

	Human	Horse
TV = Tidal volume	500 mL	6,000 mL
IRV = Inspiratory reserve volume	3,000 mL	12,000 mL
IC = Inspiratory capacity	3,500 mL	18,000 mL
ERV = Expiratory reserve volume	1,000 mL	12,000 mL
RV = Residual volume	1,200 mL	12,000 mL
FRC = Functional residual capacity	2,200 mL	24,000 mL
VC = Vital capacity	4,500 mL	30,000 mL
TLC = Total lung capacity	5,700 mL	42,000 mL

(b)

Values are average for a healthy young adult male; values for females are somewhat lower.

ing. Similarly, at a fast gallop horses can respire up to 140 times per minute, with tidal volumes increased up to 12 L, giving a total ventilation rate of 1680 L/min.

Anatomic Dead Space

When increasing pulmonary ventilation, *it is usually more advantageous to have a greater increase in tidal volume* than in respiratory rate because of the presence of **anatomic dead space,** which results from the fact that not all the inspired air gets down to the site of gas exchange in the alveoli. Part of it remains in the conducting airways, where it is not available for gas exchange. The volume of the conducting passages in an adult human averages about 0.15 L, whereas it is about 1.7 L in the giraffe and 1.8 L in the horse. This volume is considered anatomic dead space (● Figure 11–27), because air within these conducting airways is useless for exchange purposes. To compensate for its anatomic dead space, the giraffe has a lung capacity of 47 L. During quiet breathing the giraffe uses only about 10% of this total lung capacity.

Anatomic dead space has a pronounced effect on the efficiency of pulmonary ventilation. In horses, even though 6 L of air are moved in and out with each breath, only 4.2 L are actually exchanged between the atmosphere and alveoli because of the 1.8 L occupying the anatomic dead space.

Alveolar Ventilation

Because the amount of atmospheric air that reaches the alveoli and is actually available for exchange with the blood is more important than the total amount breathed in and out, **alveolar ventilation**—the volume of air exchanged between the atmosphere and alveoli per minute—is more important than pulmonary ventilation. In determining alveolar ventilation in horses, the amount of wasted air moved in and out through the anatomic dead space must be taken into account, as follows:

$$\text{Alveolar ventilation} =$$
$$(\text{tidal volume} - \text{dead-space volume}) \times \text{respiratory rate}$$

With average resting values for a horse:

$$\text{Alveolar ventilation} =$$
$$(6 \text{ L/breath} - 1.8 \text{ L dead-space volume})$$
$$\times 12 \text{ breaths/min} = 50.4 \text{ L/min}$$

12 breaths per minute in both humans and horses. The ventilation equation parallels that of cardiac output (cardiac output = stroke volume × beat rate; p. 386):

$$\underset{\text{(L/breath)}}{\text{Pulmonary ventilation}} = \underset{\text{(L/min)}}{\text{tidal volume}} \times \underset{\text{(breaths/min)}}{\text{respiratory rate}}$$

At an average tidal volume of 6 L/breath and a respiratory rate of 12 breaths per min, pulmonary ventilation in horses is 72 L of air (normal range 65 to 90) breathed in and out in one minute under resting conditions, versus about 6 L per minute in adult human males. For a brief period of time, a healthy young adult male can voluntarily increase his total pulmonary ventilation 25-fold, to 150 L/min. To increase pulmonary ventilation, both tidal volume and respiratory rate increase, but depth of breathing is increased more than frequency of breath-

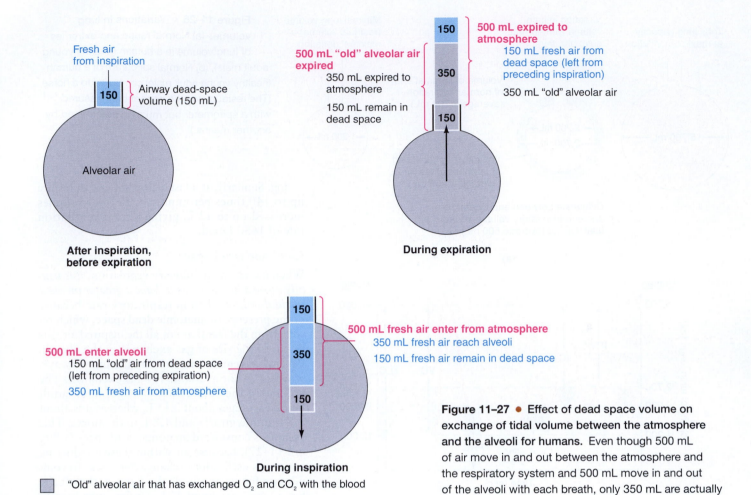

Fresh air
from inspiration

Airway dead-space
volume (150 mL)

150

Alveolar air

**After inspiration,
before expiration**

500 mL "old" alveolar air
expired

350 mL expired to
atmosphere

150 mL remain in
dead space

500 mL expired to
atmosphere

150 mL fresh air from
dead space (left from
preceding inspiration)

350 mL "old" alveolar air

150

350

150

During expiration

500 mL enter alveoli
150 mL "old" air from dead space
(left from preceding expiration)
350 mL fresh air from atmosphere

150

350

150

500 mL fresh air enter from atmosphere
350 mL fresh air reach alveoli
150 mL fresh air remain in dead space

During inspiration

"Old" alveolar air that has exchanged O_2 and CO_2 with the blood

Fresh atmospheric air that has not exchanged O_2 and CO_2 with the blood

The numbers in the figure represent ml of air.

Figure 11–27 ● Effect of dead space volume on exchange of tidal volume between the atmosphere and the alveoli for humans. Even though 500 mL of air move in and out between the atmosphere and the respiratory system and 500 mL move in and out of the alveoli with each breath, only 350 mL are actually exchanged between the atmosphere and the alveoli because of the anatomic dead space (the volume of air in the respiratory airways).

Thus, with quiet breathing, alveolar ventilation is 50.4 L/min, whereas pulmonary ventilation is 72 L/min.

Effect of Breathing Patterns on Alveolar Ventilation

To understand how important dead-space volume is in determining the magnitude of alveolar ventilation, examine the effect of various breathing patterns on alveolar ventilation in ▌ Table 11–2. If a human deliberately breathes deeply (for ex-

ample, a tidal volume of 1.2 L) and slowly (for example, a respiratory rate of 5 breaths/min), pulmonary ventilation is 6.0 L/min, the same as during quiet breathing at rest, but alveolar ventilation increases to 5.25 L/min compared to the resting rate of 4.2 L/min. In contrast, if a person deliberately breathes shallowly (for example, a tidal volume of 0.150 L) and rapidly (a frequency of 40 breaths/min), pulmonary ventilation would still be 6.0 L/min; however, alveolar ventila-

Table 11–2 ▌ Effect of Different Breathing Patterns on Alveolar Ventilation (human)

Breathing Pattern	Tidal Volume (mL/breath)	Respiratory Rate (breaths/min)	Dead-Space Volume (mL)	Pulmonary Ventilation (mL/min)*	Alveolar Ventilation (mL/min)**
Quiet breathing at rest	500	12	150	6,000	4,200
Deep, slow breathing	1,200	5	150	6,000	5,250
Shallow, rapid breathing	150	40	150	6,000	0

*Equals tidal volume × respiratory rate.
**Equals (tidal volume − dead-space volumn) × respiratory rate.

tion would be 0 mL/min. In effect, the person would only be drawing air in and out of the anatomic dead space without any atmospheric air being exchanged with the alveoli, where it could be useful. The individual could voluntarily maintain such a breathing pattern for only a few minutes before losing consciousness, at which time normal breathing would resume.

The value of reflexively bringing about a larger increase in depth of breathing than in rate of breathing when pulmonary ventilation is increased during activity should now be apparent. It is the most efficient means of elevating alveolar ventilation. When tidal volume is increased, the entire increase goes toward elevating alveolar ventilation, whereas an increase in respiratory rate does not go entirely toward increasing alveolar ventilation. For this reason tidal volume in trotters working at maximal intensities increases from 6 to 25 L. When respiratory rate is increased, the frequency with which air is wasted in the dead space is also increased, because a portion of *each* breath must move in and out of the dead space. As needs vary, ventilation is normally adjusted to a tidal volume and respiratory rate that meet those needs most efficiently in terms of energy cost.

We have assumed that all the atmospheric air entering the alveoli participates in exchanges of O_2 and CO_2 with pulmonary blood. However, the match between air and blood is not always perfect, because not all alveoli are equally ventilated with air and perfused with blood. Any ventilated alveoli that do not participate in gas exchange with blood because they are inadequately perfused are considered **alveolar dead space.**

■ Water breathers have relatively lower dead-space volumes.

The situation is somewhat comparable in animals with gills. Not all the water entering the buccal cavity perfuses through the gill lamellae and is, instead, lost through the spaces between the gill arches. This fraction of the water unavailable for gas exchange is also termed the *anatomical dead space.* However, flow-through systems have much lower dead-space volumes because of unidirectional flow. In fish, ventilatory flow through the branchial chamber is determined by the same equation as for tidal breathers, except that stroke volume rather than tidal volume is used for depth of breathing:

Branchial ventilation = stroke volume × respiratory rate
   (mL/breath)   (mL/min)   (breaths/min)

Ventilatory flow is also most effectively increased by large changes in stroke volume and small increases in frequency. Breathing frequencies in adult fish range up to 70 per minute, although some species of catfish can cease breathing for up to a minute.

■ Local controls act on the smooth muscle of the airways and arterioles to maximally match blood flow to airflow.

Normally, modest adjustments in airway size can be accomplished by autonomic nervous-system regulation to suit the body's needs. Parasympathetic stimulation, which occurs in quiet, relaxed situations when the demand for airflow is not high, promotes bronchiolar smooth-muscle contraction, which

increases airway resistance by producing **bronchoconstriction.** In contrast, sympathetic stimulation and to a greater extent its associated hormone, epinephrine, bring about **bronchodilation** and decreased airway resistance by promoting bronchiolar smooth-muscle relaxation. Thus, during periods of sympathetic domination, when increased demands for O_2 uptake are actually or potentially placed on the body, bronchodilation occurs to ensure that the pressure gradients established by respiratory muscle activity can achieve maximum airflow rates with minimum resistance.

These processes are relevant only to the contribution of the overall resistance of all the airways collectively. However, the resistance of individual airways supplying specific alveoli can be adjusted independently in response to changes in the airways' local environment. This situation is comparable to the control of systemic arterioles (p. 404). Recall that overall systemic arteriolar resistance (that is, total peripheral resistance) is an important determinant of the blood pressure gradient that drives blood flow throughout the systemic circulatory system (see p. 406). Yet the caliber of individual arterioles supplying various tissues can be adjusted locally to match the tissues' differing needs.

Effect of CO_2 on Bronchiolar Smooth Muscle

Like arteriolar smooth muscle, bronchiolar smooth muscle is sensitive to local changes within its immediate environment, particularly to local CO_2 levels. If an alveolus is receiving too little airflow (ventilation) in comparison to its blood flow (perfusion), CO_2 levels will increase in the alveolus and surrounding tissue as more CO_2 is dropped off by the blood than is being exhaled into the atmosphere. This local increase in CO_2 acts directly on the bronchiolar smooth muscle involved to induce the airway supplying the underaerated alveolus to relax. The resultant decrease in airway resistance leads to an increased airflow (for the same ΔP) to the involved alveolus, so its airflow now matches its blood supply (● Figure 11–28). The converse is also true. A localized decrease in CO_2 associated with an alveolus that is receiving too much air for its blood supply directly increases contractile activity of the airway smooth muscle involved, causing the airway supplying this overaerated alveolus to constrict. The result is a reduction in airflow to the overaerated alveolus.

Effect of O_2 on Pulmonary Arteriole Smooth Muscle

A similar locally induced effect on pulmonary vascular smooth muscle also takes place simultaneously to maximally match blood flow to airflow. Just as in the systemic circulation, distribution of the cardiac output to different alveolar capillary networks can be controlled by adjusting the resistance to blood flow through specific pulmonary arterioles. If the blood flow is greater than the airflow to a given alveolus, the O_2 level in the alveolus and surrounding tissues will fall below normal as more O_2 than usual is extracted from the alveolus by the overabundance of blood. The local decrease in O_2 concentration causes vasoconstriction of the pulmonary arteriole supplying this particular capillary bed; the result is a reduction in blood flow to match the smaller airflow. Conversely, an increase in alveolar O_2 concentration caused by a mismatched large airflow and small blood flow brings about pulmonary vasodilation, which increases blood flow to match the larger airflow. Note that the local effect of O_2 on pulmonary arteriolar smooth

Figure 11–28 ● Local controls to match airflow and blood flow to an area of the mammalian lung.

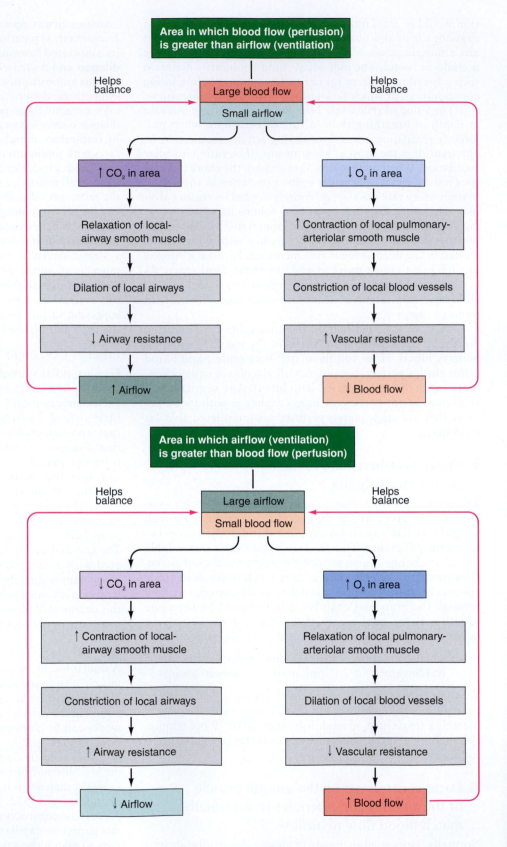

muscle is appropriately just the opposite of its effect on systemic arteriolar smooth muscle (■ Table 11–3). In the systemic circulation, a decrease in O_2 in a tissue causes localized vasodilation to increase blood flow to the deprived area, and vice versa, which is important in matching blood supply to local metabolic needs (see p. 404).

The two mechanisms for matching airflow and blood flow function concurrently, so normally very little air or blood is wasted in the lung. Some regional differences in ventilation and perfusion can occur because of gravitational effects. Nevertheless, airflow and blood flow at a particular alveolar interface are usually matched to the greatest extent possible by means of these local controls to accomplish efficient exchange of O_2 and CO_2.

■ Bird respiration exhibits an efficient flow-through mechanism.

We have been examining mammals, with their relatively inefficient tidal system. Recall that fishes have a flow-through system, useful in water with its lower oxygen availability, and birds have a partial flow-through system that supports their high metabolisms. What is particularly intriguing about birds is precisely how the respiratory muscles and the air sacs are coordinated in order to achieve this highly efficient means of gas exchange. To comprehend this process, let us follow the passage of an inert component, such as N_2 through the avian respiratory system (Figure 11–16d). The inhaled N_2 gas is initially mixed with gases present in the dead air space of the trachea and primary bronchus of each lung. At the end of the first inspiration it ends up in the caudal air sacs. Gases in the caudal air sacs are rich in O_2 and relatively low in CO_2. During the ensuing expiration the N_2 gas is vented into the gas-exchanging parabronchi region of the lung. Here the composition of the respiratory gases is modified to suit the metabolic needs of the bird. During the second inspiration the N_2 gas is drawn into the cranial air sacs, where, on the second expiration, it is expelled to the outside atmosphere. Thus, in birds, it requires two complete respiratory cycles to move air through the entire system. This remarkable pattern of gas flow through the avian lung is achieved through both aerodynamic mechanisms and valving properties of the air sacs. The end result is that

Table 11–3 ∎ Effects of Local Changes in O_2 on the Pulmonary and Systemic Arterioles

	Effect of a Local Change in O_2	
Vessels	*Decreased O_2*	*Increased O_2*
Pulmonary arterioles	Vasoconstriction	Vasodilation
Systemic arterioles	Vasodilation	Vasoconstriction

gas flow is continuous and unidirectional from the mediodorsals through the parabronchi to the medioventrals.

Mechanics: Water versus Air Respirers

We end this section on mechanics with a comparison of the energy costs between water and air respirers.

∎ The work of normal breathing requires as little as 2% of total energy expenditure in some mammals to as much as 50% in some fishes.

During normal quiet breathing, the respiratory muscles must work during inspiration to expand the lungs against their elastic forces and to overcome airway resistance, whereas expiration is a passive process. Normally, the mammalian lung is highly compliant and airway resistance is low, so only about 2% of the total energy expended by the body is used to accomplish quiet breathing. The work of breathing may be increased in four different situations: (1) when pulmonary compliance is decreased; (2) when airway resistance is increased; (3) when elastic recoil is decreased; and (4) when there is a need for increased ventilation, such as during intense activity, more work is required to accomplish both a greater depth of breathing (a larger volume of air moving in and out with each breath) and a faster rate of breathing (more breaths per minute).

During strenuous activity, the amount of energy required to power pulmonary ventilation may increase up to 25-fold. However, because total energy expenditure by the body is increased up to 15- to 20-fold during heavy activity, the energy used to accomplish the increased ventilation still represents only about 5% of total energy expended. There is an exception to this: Animals with chronic lung disease require considerably more energy for respiration, which may increase to 30% of total energy expenditure.

Respiration in water assumes a greater metabolic cost because of the fluid properties of water, which resist movement as compared to air. Estimates of the metabolic cost of respiration in a resting fish vary from 5 to 50% (normally approximately 20%) of the resting metabolic rate. Although it is higher than the metabolic cost of air breathers, for the most part it constitutes only a modest fraction of the total O_2 consumed. This is due to several features:

- A fully *flow-through* (not tidal or partial flow-through) system
- *Countercurrent* gas uptake (which we explore later under gas exchange)

- *Ram ventilation* in fast-swimming species of shark and tuna (among others). As you saw earlier, this mechanism transfers the work of ventilation from the respiratory muscles to the muscles of locomotion. Under these circumstances the cost of breathing remains constant and independent of the activity level of the fish. Depending on the particular species, environmental P_{O_2}, and temperature (but not body size), active ventilation ceases at swimming speeds ranging from 0.2 to 0.6 m/sec and fish swim with their mouths open. Although the open mouth increases drag, the continuous flow of water is more economical than the costs associated with pumping water over the gills. In addition, the faster the fish swims, the greater the supply of O_2-rich water that flows over the gills to meet the increased tissue requirements for O_2. Finally, some pelagic sharks and all tunas are *obligatory* ram ventilators, that is, they have lost their breathing muscles and thus cannot ever stop swimming lest they suffocate. This is a problem faced by aquarium builders who want to house these animals.

Gas Exchange

∎ Gases move down partial pressure gradients.

The ultimate purpose of air and water breathing is to provide a continual supply of fresh O_2 for pickup by the blood and to constantly remove CO_2 unloaded from the blood. The blood acts as a transport system for O_2 and CO_2 between the lungs and tissues, with the tissue cells extracting O_2 from the blood and eliminating CO_2 into it. Gas exchange at both the pulmonary capillary and tissue capillary levels involves *simple passive diffusion of O_2 and CO_2 down partial pressure gradients,* that is, the ΔP or $P_1 - P_2$ of Fick's law (p. 462). There are no active transport mechanisms for these gases. Let us examine the specific gradients.

If, as is the case with O_2, the partial pressure (P_1) of a gas in the alveoli or external water is *higher* than the partial pressure (P_2) of that gas in the blood entering the pulmonary or lamellar capillaries, the higher P_1 drives more O_2 into the blood; the gas diffuses in and dissolves in the blood until $P_{1_{O_2}}$ becomes equal to $P_{2_{O_2}}$. Conversely, if the alveolar or water partial pressure of a gas (P_2) is *lower* than its partial pressure in the entering blood (P_1)—the situation that exists for CO_2— the lower partial pressure permits some of the CO_2 to escape from solution (that is, to no longer be dissolved) in the blood. As CO_2 comes out of solution, it diffuses into the alveoli or water until $P_{1_{CO_2}}$ equilibrates with $P_{2_{CO_2}}$.

∎ In air breathers, oxygen enters and CO_2 leaves the blood in the lungs passively.

In tidal breathers, generally, lung air (alveolar air in mammals) is not of the same composition as inspired atmospheric air, for two reasons. First, as soon as atmospheric air enters the respiratory passages, it becomes saturated with H_2O by exposure to the moist airways. Water vapor exerts a partial pressure just like any other gas does. At mammalian body temperature, the partial pressure of H_2O vapor is 47 mm Hg. Humidification of inspired air in effect "dilutes" the partial pressure of the inspired gases by 47 mm Hg, because the sum of the partial pressures approximates the atmospheric pressure of 760 mm Hg. In moist air, $P_{H_2O} = 47$ mm Hg, $P_{N_2} = 563$ mm Hg, and $P_{O_2} = 150$ mm Hg.

Second, alveolar P_{O_2} is also lower than atmospheric P_{O_2} because fresh inspired air is mixed with the large volume of old air that remained in the lungs and dead space at the end of the preceding expiration (the functional residual capacity). In a human, only about one seventh of the total alveolar air is replaced by fresh atmospheric air with each normal breath. Thus, at the end of inspiration, less than 15% of the air in the alveoli is fresh air. As a result of humidification and the small turnover of alveolar air, *the average alveolar P_{O_2} is 100 mm Hg,* compared to the atmospheric P_{O_2} of about 160 mm Hg.

It is logical to think that alveolar P_{O_2} would increase during inspiration with the arrival of fresh air and would decrease during expiration. Only small fluctuations of a few mm Hg occur, however, for two reasons. First, only a small proportion of the total alveolar air is exchanged with each breath. The relatively small volume of high-P_{O_2} air that is inspired is quickly mixed with the much larger volume of retained alveolar air, which has a lower P_{O_2}. Thus the O_2 in the inspired air can only slightly elevate the total alveolar P_{O_2}. Even this potentially small elevation of P_{O_2} is diminished for another reason. Oxygen is continually moving by passive diffusion down its partial pressure gradient from the alveoli into the blood. The O_2 arriving in the alveoli in the newly inspired air simply replaces the O_2 diffusing out of the alveoli into the pulmonary capillaries. Therefore, the alveolar P_{O_2} remains relatively constant at about 100 mm Hg throughout the respiratory cycle. Because the pulmonary blood P_{O_2} equilibrates with the alveolar P_{O_2}, the P_{O_2} of the blood leaving the alveoli likewise remains fairly constant at this same value. Accordingly, the amount of O_2 in the blood varies only slightly during the respiratory cycle.

A similar situation in reverse exists for CO_2. Carbon dioxide, which is continually produced by the body tissues as a metabolic waste product, is constantly added to the blood at the level of the systemic capillaries. In the pulmonary capillaries, CO_2 diffuses down its partial pressure gradient from the blood into the alveoli and is subsequently removed from the body during expiration. Like O_2, alveolar P_{CO_2} remains fairly constant throughout the respiratory cycle but at a lower value of 40 mm Hg. Ventilation constantly replenishes alveolar P_{O_2}, keeping it relatively high, and constantly removes CO_2, keeping alveolar P_{CO_2} relatively low. Thus the appropriate partial pressure gradients between the alveoli and blood are maintained to ensure that O_2 enters the blood and CO_2 leaves the blood.

Gas Gradients across the Pulmonary Capillaries

The blood entering the pulmonary capillaries is systemic venous blood pumped to the lungs through the pulmonary arteries (Chapter 9). This blood, having just returned from the body tissues, is relatively low in O_2, with a $P_{2_{O_2}}$ of 40 mm Hg, and is relatively high in CO_2, with a $P_{1_{CO_2}}$ of 46 mm Hg. As this blood flows through the pulmonary capillaries, it is exposed to alveolar air (● Figure 11–29). Because the alveolar $P_{1_{O_2}}$ at 100 mm Hg is higher than the $P_{2_{O_2}}$ of 40 mm Hg in the blood entering the lungs, O_2 diffuses down its partial pressure gradient (a difference of 60 mm Hg) into the blood until no further gradient exists. As the blood leaves the pulmonary capillaries, it has a P_{O_2} equal to alveolar P_{O_2} at 100 mm Hg.

The partial pressure gradient for CO_2 is in the opposite direction. Blood entering the pulmonary capillaries has a $P_{1_{CO_2}}$ of 46 mm Hg, whereas alveolar $P_{2_{CO_2}}$ is only 40 mm Hg. Carbon dioxide diffuses from the blood into the alveoli (down its gradient of 6 mm Hg) until blood P_{CO_2} equilibrates with alveolar P_{CO_2}. Thus the blood leaving the pulmonary capillaries has a P_{CO_2} of 40 mm Hg. After leaving the lungs, the blood is returned to the heart to be subsequently pumped out to the body tissues as systemic arterial blood (Chapter 9).

Note that blood returning to the lungs from the tissues still contains O_2 (P_{O_2} of systemic venous blood = 40 mm Hg) and that blood leaving the lungs still contains CO_2 (P_{CO_2} of systemic arterial blood = 40 mm Hg). The additional O_2 carried in the blood beyond that normally given up to the tissues represents an immediately available O_2 reserve that can be tapped by the tissue cells whenever their O_2 demands increase. The CO_2 remaining in the blood even after passage through the lungs plays an important role in the acid–base balance of the body, because CO_2 generates carbonic acid (Chapter 13). Furthermore, arterial P_{CO_2} is important in driving respiration. This mechanism is described later.

The amount of O_2 picked up in the lungs matches the amount extracted and used by the tissues. When the tissues metabolize more actively (for example, during strenuous activity), more O_2 is extracted from the blood at the tissue level, reducing the systemic venous P_{O_2} even lower than 40 mm Hg— for example, to a $P_{2_{O_2}}$ of 30 mm Hg. When this blood returns to the lungs, a larger-than-normal P_{O_2} gradient exists, now at 70 mm Hg (alveolar $P_{1_{O_2}}$ of 100 mm Hg and blood $P_{2_{O_2}}$ of 30 mm Hg), compared to the normal gradient of 60 mm Hg. Therefore, more O_2 diffuses from the alveoli into the blood down the larger partial-pressure gradient before blood P_{O_2} equals alveolar P_{O_2}. This additional transfer of O_2 into the blood replaces the increased amount of O_2 consumed, so O_2 uptake matches O_2 use even when O_2 consumption increases. At the same time, ventilation is stimulated so that O_2 enters the alveoli more rapidly from the atmosphere to replace the O_2 diffusing into the blood.

Similarly, the amount of CO_2 given up to the alveoli from the blood matches the amount of CO_2 picked up at the tissues. Once again, an increase in ventilation associated with increased activity assures that increased amounts of CO_2 delivered to the alveoli are blown off to the atmosphere.

■ Flow-through breathing plus countercurrent blood flow enhances pressure gradients in fishes.

As we discussed earlier, some insect tracheal systems, most fishes, and portions of bird systems exhibit flow-through arrangements that are more efficient than mammalian tidal breathing. Another way this helps to maximize gas exchange relies on the arrangement of blood flow. To be able to extract the maximum amount of O_2 from the water flowing through the gills, blood flows in a direction opposite or **countercurrent** to that of water flow (● Figure 11–30); that is, blood flow through the capillaries is opposite to the stream of water flowing through the gill lamellae, such that blood moving through the lamellae continually encounters water whose O_2 content is higher than what is present in the capillary. This anatomical arrangement permits the O_2 content of the blood to approximate that what is measured in the aqueous medium that can

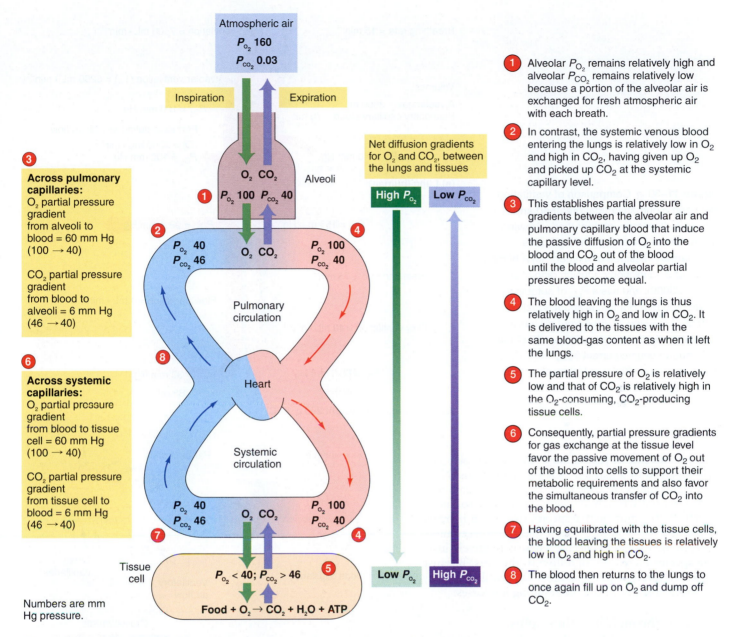

Figure 11–29 • Oxygen and CO_2 exchange across pulmonary and systemic capillaries caused by partial pressure gradients (mammal).

be as high as 150 mm Hg. In fact the P_{O_2} in outgoing blood of most fish is distinctly higher than in the expired water from the gill, unlike mammalian arterial blood, in which P_{O_2} of outgoing blood is equal to that of expired air (Figure 11–31) than in expired air. The efficiency of O_2 extraction from the water can be as high as 90% in some species of fish, crustaceans and amphibians, compared to 30 to 40% in a bird and a mere 25% in a mammal. However, the low O_2 content of water as compared to air requires a greater expenditure of energy as compared to terrestrial ectotherms with similar metabolic demands.

Compared to mammals, water breathers have remarkably low arterial concentrations of CO_2. The P_{CO_2} in expired water and in arterial blood ranges around 1 to 4 mm Hg in fish, slightly higher than the in-current water at 0.2 mm Hg and in

contrast to the 30 to 40 mm Hg in most air breathers (reptiles, birds, and mammals). This is primarily due to the comparatively high solubility of CO_2 in an aqueous media as compared to air.

Interestingly, some species of fish use *retrograde flow* of water from the opercular opening into the buccal chamber to ventilate their gills. Lampreys also use this unusual adaptation without any apparent loss of efficiency in extracting O_2 from the water. Because the lamprey's mouth may be attached to a host fish, the gills must be ventilated by this tidal flow of water generated by cyclic contraction of the muscular walls lining the gill pouches.

Fish must continually balance the advantages of branchial O_2 diffusion across the gills with the potentially harmful effects associated with their role in osmoregulation. For exam-

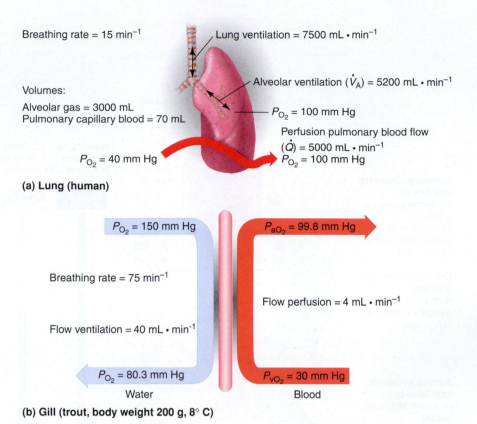

Breathing rate = 15 min^{-1}

Lung ventilation = 7500 mL • min^{-1}

Alveolar ventilation ($\dot{V}_A$) = 5200 mL • min^{-1}

Volumes:

Alveolar gas = 3000 mL
Pulmonary capillary blood = 70 mL

P_{O_2} = 100 mm Hg

Perfusion pulmonary blood flow
($\dot{Q}$) = 5000 mL • min^{-1}
P_{O_2} = 100 mm Hg

P_{O_2} = 40 mm Hg

(a) Lung (human)

P_{O_2} = 150 mm Hg

P_{aO_2} = 99.8 mm Hg

Breathing rate = 75 min^{-1}

Flow perfusion = 4 mL • min^{-1}

Flow ventilation = 40 mL • min^{-1}

P_{O_2} = 80.3 mm Hg

P_{vO_2} = 30 mm Hg

Water

Blood

(b) Gill (trout, body weight 200 g, 8° C)

Figure 11–30 • Comparison of ventilation between a human (a tidal breather) and a trout (a flow-through breather), showing approximate breathing and flow rates and partial pressures of oxygen. (a) A human lung has a ventilation:perfusion ratio of about 1 (5200/5000), and has a P_{O_2} of outgoing blood (100 mm Hg) that is about equal to that of the lung air. (b) A trout gill has a ventilation:perfusion ratio of about 10 (40/4), and (due to countercurrent flow) has a P_{O_2} of outgoing blood (99.8 mm Hg) much greater than that of expired water (80 mm Hg).

(*Source:* Modified from Eckert, *Animal Physiology*, 5th ed., by Burggren et al., Figure 13–43, p. 563. Used by permission.)

ple, any increase in ventilation in freshwater teleosts is accompanied by a marked increase in ionic loss (Na$^+$) and water influx. Marine teleosts, however, gain ions and potentially lose body water to the environment. Generally the more active species rely on an increase in the functional surface area of the gills to accommodate an increase in O$_2$ needs, whereas relatively inactive species increase blood flow to the gills. Because these fish rarely require maximal O$_2$ ventilation, they are more able to tolerate the resulting ion losses.

■ Flow-through breathing plus crosscurrent blood flow enhances pressure gradients in birds.

Although they do not need to be as efficient as fishes because air is a much better medium for oxygen uptake, birds do require greater efficiency than mammals because of their higher metabolisms. As you have seen, during inspiration and expiration in birds, there is a one-way flow of respiratory gases through the parabronchi. The anatomic arrangement of airflow through the gas-exchanging air capillaries in concert with the surrounding blood capillary network is such that blood flow approaches the parabronchus at approximately right angles, providing a **crosscurrent** system of gas exchange (● Figure 11–31b). Unique to birds is the finding that pulmonary capillaries equilibrates with respiratory gases whose P_{O_2} and P_{CO_2} progressively changes along the length of the parabronchus. Blood in the capillaries becomes arterialized to varying degrees depending on how far along the parabronchus the blood

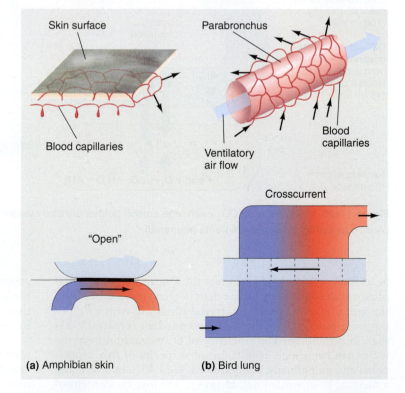

Skin surface

Parabronchus

Blood capillaries

Ventilatory air flow

Blood capillaries

"Open"

Crosscurrent

(a) Amphibian skin

(b) Bird lung

Figure 11–31 • Comparison of the respiratory surfaces, blood supply, and blood flow in an amphibian and bird. (a) Integumentary "open" exchange in a frog, showing blood flowing in capillaries underneath the skin. (b) Crosscurrent gas exchange in the air capillaries of a bird, showing blood flowing in capillaries across the airway.

(*Source:* Phillip C. Withers, 1992, *Comparative Animal Physiology*, Belmont, CA: Brooks/Cole, Figure 12–6, p. 574.)

capillary is located. Arterial blood P_{O_2} is thus *higher* than expired P_{O_2} because arterial blood P_{O_2} results from the mixing of the blood from the differently oxygenated compartments.

■ Skin breathing occurs in most vertebrates, but its contribution is variable.

Cutaneous gas exchange is essential in many animal species, including the earthworm (p. 467), and is found in most vertebrates. It is important in virtually all amphibians and aquatic reptiles and in most fishes (Figure 11–31a). During hibernation, for example, frogs and turtles exchange all their respiratory gases through this route while submerged in ponds or buried beneath the banks of a stream. Eels normally exchange 60% of its CO_2 and O_2 through its highly vascular skin. Of particular note are the lungless salamanders of the family *Plethodontidae*. Because the skin of these animals is also important for protection against the environment, the efficiency of gas exchange is compromised. Generally the thinner and moister the surface, the more efficient it is as a respiratory surface. The consequences for these air breathers include poor arterialization (P_{O_2} = 60 mm Hg) because of the limits set by diffusion. However, their blood CO_2, a remarkable 5 to 6 mm Hg, is relatively unimpaired, because of the solubility of CO_2 in the aqueous tissue phase.

The high metabolic rate of endotherms compared to that of ectotherms (Chapter 15) limits the value of gas exchange through the integument. However, cutaneous gas exchange is found in some mammals, as exemplified by bats, because in flight they are capable of eliminating almost 12% of their CO_2 across the surface of their wings! Approximately 1 to 2% of the total CO_2 and O_2 in mammals can be exchanged directly through the skin. This is the means by which a tick detects a potential host, including humans. Carbon dioxide release serves as the primary attractant as the tick scuttles its way up the concentration gradient. A greater CO_2 production rate increases the likelihood of a particular individual becoming the tick's next meal!

■ Factors other than the partial pressure gradient influence the rate of gas transfer.

We have been discussing diffusion of O_2 and CO_2 between the blood and the environment in terms of the gases' partial pressure gradients. Recall that, according to Fick's law of diffusion, the rate of diffusion of a gas through a sheet of tissue also depends on the surface area (A) and thickness of the membrane (ΔX) through which the gas is diffusing and on the diffusion coefficient (D) of the particular gas. Let's examine these factors more closely.

In an individual animal, D is not typically alterable. Similarly, A and ΔX are relatively constant, at least under resting conditions, so that changes in the rate of gas exchange are determined primarily by changes in partial pressure gradients between the blood and external medium. During intense activity, however, the surface area available for exchange can be physiologically increased to enhance the rate of gas transfer. During resting conditions in mammals, some of the pulmonary capillaries are typically closed, because the normally low pressure of the pulmonary circulation is inadequate to keep all the capillaries open. During activity, when the pulmonary blood pressure is raised as a result of increased cardiac output, many

of the previously closed pulmonary capillaries are forced open. This increases the surface area of blood available for exchange. Furthermore, the alveolar membranes are stretched further than normal during activity because of the larger tidal volumes (deeper breathing). Such stretching increases the alveolar surface area and decreases the thickness of the alveolar membrane.

Similarly, in fishes the functional surface area can be lower than the total surface area available for gas exchange. For example, during quiet breathing of water in trout only 60% of the available lamellae are perfused with water. As the level of activity increases, more water channels are recruited for ventilation.

Inadequate gas exchange can occur when the thickness of the barrier separating the air or water from the blood is pathologically increased. As the thickness increases, the rate of gas transfer decreases, because a gas takes longer to diffuse through the greater thickness. Thickness increases because of (1) *low or high environmental pH*: Acute exposure of freshwater fish to pH values between 4.0 to 5.5 or between 8.5 to 10.5 results in mucification and inflammatory swelling of the gill epithelium; (2) *pulmonary edema*, an excess accumulation of interstitial fluid between the alveoli and pulmonary capillaries, caused by pulmonary inflammation or left-sided congestive heart failure; (3) *pulmonary fibrosis* involving replacement of delicate lung tissue with thick fibrous tissue in response to certain chronic irritants; and (4) *pneumonia*, which is characterized by inflammatory fluid accumulation within or around the alveoli. Most commonly, pneumonia is due to bacterial or viral infection of the lungs, but it may also arise from accidental aspiration (breathing in) of food, vomitus, or chemical agents.

The preceding discussion focused on individual animals. As we have touched on earlier, all the components of Fick's law can be evolutionarily altered among species if not always within an individual. For example, the area of the gill surface available for gas exchange and the diffusion distance vary (as evolutionary adaptations) with the lifestyles of different fish species. At one extreme are the air-breathing fish that have comparatively small gill surface areas and thick lamellae, so much so that in some cases the gills are virtually functionless in terms of gas exchange. Alternatively, fast-swimming fishes such as tuna and mackerel have particularly thin lamellae and a comparatively large surface area available for gas exchange.

■ Gas exchange across the systemic capillaries also occurs down partial pressure gradients.

Just as they do at the pulmonary capillaries, O_2 and CO_2 move between the systemic capillary blood and the tissue cells by simple passive diffusion down partial pressure gradients. Refer again to Figure 11–29. The arterial blood that reaches the systemic capillaries is essentially the same blood that left the lungs by means of the pulmonary veins, because the only two places in the entire circulatory system at which gas exchange can take place are the pulmonary capillaries and the systemic capillaries. The arterial P_{O_2} is 100 mm Hg and the arterial P_{CO_2} is 40 mm Hg, the same as alveolar P_{O_2} and P_{CO_2}.

Gas Gradients across the Systemic Capillaries

Cells constantly consume O_2 and produce CO_2 through oxidative metabolism (Chapter 2). Cellular P_{O_2} in a mammal averages about 40 mm Hg and P_{CO_2} about 46 mm Hg, although these values are highly variable, depending on the level of cellular metabolic activity. Oxygen moves by diffusion down its partial pressure gradient from the entering systemic capillary blood ($P_{1_{O_2}}$ = 100 mm Hg) into the adjacent cells ($P_{2_{O_2}}$ = 40 mm Hg) until equilibrium is reached. Therefore, the P_{O_2} of venous blood leaving the systemic capillaries is equal to the tissue P_{O_2} at an average of 40 mm Hg. The reverse situation exists for CO_2. Carbon dioxide rapidly diffuses out of the cells ($P_{1_{CO_2}}$ = 46 mm Hg) into the entering capillary blood ($P_{2_{CO_2}}$ = 40 mm Hg) down the partial pressure gradient created by the ongoing production of CO_2. Transfer of CO_2 continues until blood P_{CO_2} equilibrates with tissue P_{CO_2}. Accordingly, the blood leaving the systemic capillaries has an average P_{CO_2} of 46 mm Hg. This systemic venous blood, which is relatively low in O_2 (P_{O_2} = 40 mm Hg) and relatively high in CO_2 (P_{CO_2} = 46 mm Hg), returns to the heart and is subsequently pumped to the lungs as the cycle repeats itself.

The more actively a tissue is metabolizing, the lower the cellular P_{O_2} falls and the higher the cellular P_{CO_2} rises. As a consequence of the larger blood-to-cell partial pressure gradients, more O_2 diffuses from the blood into the cells and more CO_2 moves in the opposite direction before blood P_{O_2} and P_{CO_2} achieve equilibrium with the surrounding cells. Thus, the amount of O_2 transferred to the cells and the amount of CO_2 carried away from the cells depend on the rate of cellular metabolism.

Gas Diffusion between the Alveoli and Tissues

Net diffusion of O_2 occurs first between the alveoli and the blood and then between the blood and the tissues because of the O_2 partial pressure gradients created by continuous utilization of O_2 in the cells and continuous replenishment of fresh alveolar O_2 provided by alveolar ventilation. Net diffusion of CO_2 occurs in the reverse direction, first between the tissues and the blood and then between the blood and the alveoli, because of the CO_2 partial pressure gradients created by continuous production of CO_2 in the cells and the continuous removal of alveolar CO_2 through the process of alveolar ventilation (Figure 11–29).

Gas Transport

Oxygen picked up by the circulation at the respiratory organs must be transported to the tissues for cell use. Conversely, CO_2 produced at the cellular level must be transported to the respiratory organs for elimination.

■ Most O_2 in many animal circulatory systems is transported bound to respiratory pigments.

Oxygen can present in the blood in two forms: physically dissolved as *free gas*, and chemically bound to a **respiratory protein** or **pigment,** the most common being hemoglobin (Hb) (● Figure 11–32). As you saw in Chapter 9 (p. 363), vertebrate hemoglobin is carried in circulating erythrocytes whereas invertebrate respiratory pigments (not always present) are either extra- or intracellular. Invertebrate pigments are also quite diverse. For example, some have hemoglobins (such as many annelids, and some mollusks and crustaceans). Some mollusks and crustaceans have **hemocyanin,** a pigment that uses a pair of copper atoms directly bound to amino acid side chains to bind oxygen. As a result of the copper atoms, the blood of these

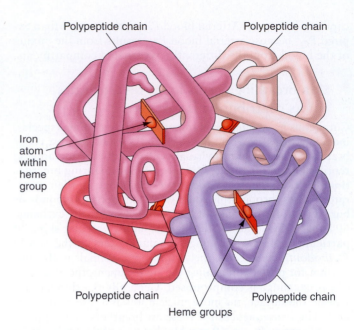

Figure 11–32 ● **Hemoglobin molecule from a mammal.** A hemoglobin molecule consists of four highly folded polypeptide chains (the globin portion) and four iron-containing heme groups.

animals appears blue. Hemocyanin is functionally similar to vertebrate hemoglobins. **Hemerythrin,** another group of intracellular respiratory pigments found in certain groups of marine brachiopods (lamp shells) and some worm groups, represents a distinct solution to the problem of oxygen binding and transport. Structurally it is considerably different from Hb because, although it has iron and is red, the iron is not in a heme complex. Annelids are the most remarkable phylum: In this one phylum, *five* different respiratory pigments can be found: hemoglobin, hemocyanin, chlorocruorin, erthyrocruorin, and hemerythrin.

Insects, as air breathers with direct tracheal input to their tissues, normally lack a respiratory pigment. However, in the aquatic larvae of the caddis fly (*Chironomus*) and the water bugs *Anisops* and *Buenoa*, Hb plays a role in O_2 uptake. Recall that oxygen uptake is more difficult in water than in air.

For the rest of this section, we focus primarily on vertebrate pigments. In a mammal, very little O_2 is physically dissolved in plasma because O_2 is poorly soluble in warm fluids. The amount dissolved is directly proportional to the P_{O_2} of the blood; the higher the P_{O_2}, the more O_2 dissolved. At a normal arterial P_{O_2} of 100 mm Hg and body temperature of 37°C, only 3 mL of O_2 can dissolve in 1 liter of blood. Thus in a human only 15 mL of O_2/min can be dissolved in the normal pulmonary blood flow of 5 L/min (the resting cardiac output in a human). Even under resting conditions, human cells consume 250 mL of O_2/min in total, and this may increase up to 25-fold during strenuous activity. To deliver the O_2 required by the tissues even at rest, the cardiac output would have to be 83.3 L/min if O_2 could only be transported in dissolved form. Obviously, there must be an additional mechanism for transporting O_2 to the tissues.

In vertebrates, this mechanism takes the form of hemoglobin. A hemoglobin molecule has two parts: (1) the **globin** portion, a protein made up of one to four highly folded polypeptide chains, and (2) one iron-containing, nonprotein group known as **heme** bound to each of the polypeptides (Figure 11–32). Jawless fishes (lampreys and hagfish) have one to three of these polypeptides making up their hemoglobins. Other vertebrates have four polypeptides making up a *tetrameric hemoglobin*. Each of the four iron atoms in the tetrameric form can combine reversibly with one molecule of O_2; thus, each hemoglobin molecule in higher vertebrates can pick up four O_2 passengers in the lungs or gills. Only 1.5% of the O_2 in the blood of mammals is free (dissolved); the remaining 98.5% is transported in combination with Hb. In fishes the amount of free O_2 is somewhat higher, at about 5% of the total O_2 content at physiological P_{O_2} levels. One notable exception is the Antarctic icefish, unique among vertebrates because it lacks Hb (see box, "A Closer Look at Adaptation: Life in the Cold: The Hemoglobinless Icefish").

The O_2 bound to Hb does not contribute to the P_{O_2} of the blood; thus blood P_{O_2} is a measure not of the total O_2 content of the blood but only of the portion of O_2 that is dissolved.

Bound O_2 does not factor into the diffusion equation (Fick's law, p. 462) and by doing so, helps to maximize diffusion. Later we'll show how this works.

Hemoglobin has the ability to form a loose, easily reversible combination between Fe and O_2. When not combined with O_2, Hb is referred to as **reduced** or **deoxyhemoglobin**; when combined with O_2, it is called **oxyhemoglobin (HbO$_2$)**:

$$\underset{\text{deoxyhemoglobin}}{HbH_4^+ + 4O_2} \leftrightarrow \underset{\text{oxyhemoglobin}}{Hb(O_2)_4 + 4H^+}$$

The concentration of Hb in an animal is not fixed. For example, hemoglobin increases in the plasma of mammals can occur with

1. *Seasonal changes.* During cold acclimatization concentrations of Hb increase (such as in the horseshoe rabbit).

2. *Locomotory activity.* This stimulates some animals to increase the release of erythrocytes from the spleen (such as horse, cat, fish), and prolonged aerobic activity triggers

---

A CLOSER LOOK AT ADAPTATION

Life in the Cold: The Hemoglobinless Icefish

Antarctic icefish (Chaenichthyidae) are noted for having a low metabolic rate and sluggish lifestyle. They are unique among vertebrates in that they completely lack erythrocytes and hemoglobin. Genome analysis shows that they have completely lost the β-globin gene, whereas the α-globin gene is still present but mutated into a transcriptionally inactive form. In the absence of red blood cells, these fish appear whitish and translucent—hence their common name. Icefish are sizable: They can weigh over 2 kg and reach a length of approximately 0.6 m.

Icefish are ectothermic poikilotherms with body temperatures essentially equal to that of the environment. A lack of hemoglobin in this species represents an adaptation to a stable environmental condition of cold (−2°C) as well as water with a relatively high O_2 content (recall that cold water dissolves gases better than warm water). In the low ambient temperatures, the viscosity of the blood is increased. Elimination of red blood cells helps to offset the decrease in blood flow. However, in the absence of hemoglobin, only O_2, which is physically dissolved in solution, can be transported in the blood. Because icefish live in well-oxygenated waters and the

temperature of their plasma is low, which enables greater quantities of O_2 to be dissolved, the arterial P_{O_2} is a relatively high 110 mm Hg. Cardiac output is also remarkably elevated in icefish, which helps ensure that sufficient quantities of O_2 can be taken up at the gill and delivered to the tissues despite a low arterial–venous O_2 difference. Relative heart size and blood volume are also increased in these animals. The increased cardiac output is maintained by an increase in stroke volume and not heart rate, which is a relatively slow 14 beats per minute. Peripheral resistance to blood flow is also reduced in icefish, which ensures that the work actually performed by the heart is kept to a minimum despite the large volume of blood being pumped. Compared to most fishes, the gill area of icefish is reduced and the secondary lamellae are thick. Gas exchange does occur across the gills, although up to 40% of their O_2 requirement can be obtained through the skin. An increased density of capillaries in the skin ensures adequate gas exchange. Because of these unique adaptations Antarctic icefish can survive without the presence of Hb in their blood, but they are limited to well-oxygenated, cold waters.

hormonal stimulation of the bone marrow to produce more erythrocytes.

3. *High altitude.* This and other prolonged exposure to low ambient oxygen can also trigger the production of more erythrocytes.

4. *Stress and disease.* For example, *anemia*—a state in which erythrocyte levels are abnormally low—can be triggered by stress hormones and many diseases.

■ Myoglobin stores oxygen in aerobic muscle and may facilitate diffusion from the blood to the mitochondria.

Vertebrates also have another respiratory pigment, called **myoglobin (Mb)**. It is a monomer (single protein unit) with one heme that serves to store oxygen in muscle, especially Type I (red aerobic) skeletal fibers (p. 339) and heart (contributing to these muscles' red color). Myoglobin's function as a reserve for oxygen is well characterized; for example, because contraction of skeletal muscle and heart reduces blood flow by squeezing actions, Mb can release oxygen to compensate (see Fig. 11–33b). But a longstanding hypothesis predicts that Mb will provide an alternate pathway for oxygen diffusion, with Mb picking up O_2 near the membrane (where P_{O_2} is higher) and diffusing to the mitochondria where the O_2 would be released (where P_{O_2} is lower). It has proved quite difficult to test this hypothesis in the past. Recently, the technique of "knockout" genes (Chapter 2) has allowed researchers to make mice completely lacking myoglobin! The heart and muscles of these mice are pale pink rather than the normal robust red color. Amazingly, physiological studies showed no impairment in muscular or cardiovascular activity in these animals. Does this mean that myoglobin is a useless protein? No, because one study found that the hearts of these mice have a much higher density of capillaries, which delivers enough oxygen to make up for the lack of Mb storage. By unknown signals during development, low supplies of oxygen to heart fibers presumably triggered this compensatory angiogenesis. These experiments in fact show that Mb is useful and that its absence requires compensation to maintain homeostasis; but they did not clarify whether Mb's role is simply for storage or to enhance diffusion. As we noted in Chapter 2, the technique of knocking out genes does not always yield simple results and answers, because of the complex feedback processes in living systems.

?

What Is the Function of Neuroglobin? Recently, researchers have found a third class of vertebrate globin called **neuroglobin,** by searching genome data for globin-type codes. Neuroglobin is found in neurons. The function of this oxygen-binding protein is unknown. Studies in 2001 found that neuroglobin could protect brains of mice from stroke damage, which results from insufficient oxygen. The mice were injected with a virus that was engineered to produce high levels of neuroglobin (see the Suggested Reading by Venis [2001] at the end of this chapter). But what might the role of this protein be under normal conditions?

■ The P_{O_2} is the primary factor determining the percent hemoglobin saturation.

Hemoglobin binds oxygen at a respiratory surface (skin, gills, or lungs) where oxygen is relatively plentiful and transports the oxygen to the inner tissues where it is used in metabolism. We need to answer several important questions about the role of respiratory pigments in O_2 transport. For example, what determines whether O_2 and Hb are combined or dissociated (separated)? Why does Hb combine with O_2 in the lungs and release O_2 at the tissues? How can a variable amount of O_2 be released at the tissue level, depending on the level of tissue activity? How can we talk about O_2 transfer between blood and surrounding tissues in terms of O_2 partial pressure gradients when most of the O_2 is bound to Hb and thus does not contribute to the P_{O_2} of the blood at all?

Tetrameric hemoglobin, which can bind up to four O_2s each, is considered *fully saturated* when all the Hb present is carrying its maximum O_2 load. The **percent hemoglobin (% Hb) saturation,** a measure of the extent to which the Hb present is combined with O_2, can vary from 0% to 100%. The most important factor determining the % Hb saturation is the P_{O_2} of the blood, which, in turn, is related to the concentration of O_2 physically dissolved in the blood. According to the **law of mass action,** if the concentration of one of the substances involved in a reversible reaction is increased, the reaction is driven toward the opposite side. Conversely, if the concentration of one of the substances is decreased, the reaction is driven toward that side. Applying this law to the reversible reaction involving Hb and O_2 (Hb + O_2 ↔ HbO_2), when the blood P_{O_2} is increased, as it is in the pulmonary capillaries, the reaction is driven toward the right side of the equation, resulting in increased formation of HbO_2 (increased % Hb saturation). When the blood P_{O_2} is decreased, as it is in the systemic capillaries, the reaction is driven toward the left side of the equation and oxygen is released from Hb as HbO_2 dissociates (decreased % Hb saturation). Thus, because of the difference in P_{O_2} at the lungs and other tissues, Hb automatically "loads up" on O_2 in the lungs, where fresh supplies of O_2 are continually being provided by ventilation, and "unloads" it in the tissues, which are constantly using up O_2. Let us examine how this works.

Cooperativity

Binding of ligands to most proteins (such as substrates to enzymes) follows a *hyperbolic* curve, not a linear relationship, as shown in ● Figure 11–33a for myoglobin and oxygen. This type of curve results from *saturation* of the protein, that is, after a certain concentration of ligand is reached, no more bound complex can form because all the proteins are occupied by ligands. The relationship between blood P_{O_2} and % Hb saturation is not hyperbolic, however, a point that is very important physiologically. The relationship between these variables follows an S-shaped *sigmoidal* curve, known as the **O_2-Hb dissociation** (or **saturation**) **curve** (Figure 11–33a and b). With the exception of species that contain monomeric oxygenated Hb (such as agnathans), the curves are sigmoidal in shape because hemoglobins exhibit *cooperativity of binding*; that is, the loading or unloading of the first O_2 on Hb makes it sequentially easier for subsequent O_2s to load or unload. Cooperativity arises from conformational changes that occur in

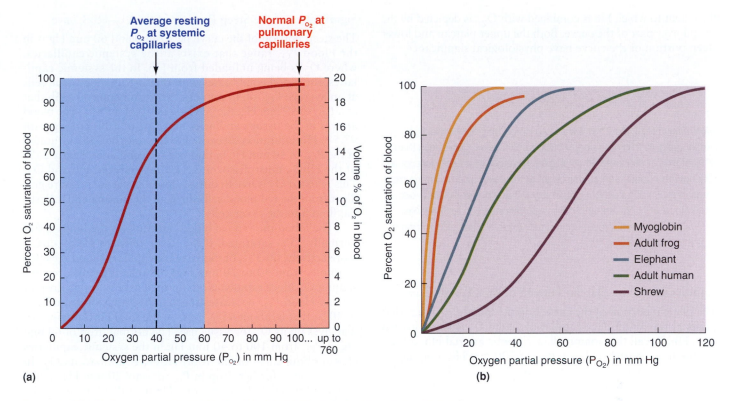

Figure 11–33 • Oxygen–hemoglobin (O_2–Hb) dissociation (saturation) curve (human)
(a) The percent hemoglobin saturation (the scale on the left side of the graph) depends
on the P_{O_2} of the blood. The relationship between these two variables is depicted by an
S-shaped curve with a plateau region between a blood P_{O_2} of 60 and 100 mm Hg and
a steep portion between 0 and 60 mm Hg. Another way of expressing the effect of blood
P_{O_2} on the amount of O_2 bound with hemoglobin is the volume percent of O_2 in the blood
(mL of O_2 bound with hemoglobin in each 100 mL of blood). That relationship is represented
by the scale on the right side of the graph. (b) Comparison of oxygen dissociation curves
for the myoglobin pigment, a frog, an elephant, a human, and a shrew.

the globins, which oscillates between one of two states: a high-affinity or "relaxed" form called the **R-state**, and a low-affinity or "tense" form called the **T-state**. The R-state is favored at high P_{O_2} levels (thus "R" is easier to remember as the "red" form), whereas the T-state is favored at low P_{O_2} levels. At intermediate and low P_{O_2} levels, the globin that releases the first O_2 from a loaded hemoglobin not only shifts to the T-state but will also affect its neighboring globin units, making them more likely to shift to the T-state and thus releasing their O_2s. This is the phenomenon known as *cooperativity*. (The reverse process occurs as an unloaded hemoglobin enters an area of high P_{O_2} level.) The T-state is of particular interest. Iron has an inherently high, nearly irreversible affinity for oxygen (we know the basic Fe–O_2 reaction as rust!). The protein in the T-state essentially alters the iron atom in the heme to reduce its affinity for the gas.

Hemoglobin P_{50}

Although the O_2–Hb dissociation curves of other animals have a similar shape, the relative affinity of Hb for O_2 determines whether they lie to the left or right of the one shown in Figure 11–33b. If the affinity for O_2 is greater (that is, it gives up O_2 less readily), then the dissociation curve is located further to the left. Animals whose affinity for oxygen is lower, and so gives up O_2 more easily, shift the curve to the right (Figure 11–33a). The partial pressure of oxygen at which 50% saturation of the blood is reached is termed the P_{50} value. For example, species adapted to living at high altitudes, such as the bar-headed goose, which has been observed flying over Mount Everest at over 9000 m, has a P_{50} approximately 10 mm Hg lower than its relatives at lower altitudes Hb (that is, its blood has a higher affinity for O_2). Regardless of the species considered, Hb is saturated in the lungs, thus a higher P_{50} value permits more O_2 to be unloaded in the tissues. A higher P_{50} value means that the Hb has a reduced affinity for O_2 and does not indicate a lower carrying capacity for O_2. Any reduction in the affinity of Hb for O_2 simply means the animal can more easily unload O_2 at the tissues.

Note that at the upper end of the curve, between a blood P_{O_2} of 60 and 100 mm Hg in humans (Figure 11–33b), the curve flattens off, or plateaus. Within this pressure range, a rise in P_{O_2} produces only a small increase in the extent to which Hb is bound with O_2. In contrast, in the P_{O_2} range of 0 to 60 mm Hg a small change in P_{O_2} results in a large change in the extent to which Hb is combined with O_2, as depicted by the steep lower part of the curve. Both the upper plateau and lower steep portion of the curve have physiological significance.

?

High-Fliers: How Do They Do It? As noted, the bar-headed goose (*Anser indicus*) migrates above 9000 m, where oxygen is less than 30% of that at sea level. Physiologists have not been able to explain fully how the geese can sustain flight at this level. The countercurrent lungs of these birds do not obtain oxygen any better than alveolar lungs do at this altitude, and other adaptations such as the lower P_{50} value of its hemoglobin cannot fully explain these birds' remarkable migratory abilities.

Significance of the Plateau Portion of the O_2–Hb Curve

The plateau portion of the curve is in the blood P_{O_2} range that exists at the pulmonary capillaries where O_2 is being loaded onto Hb. Recall that mammalian systemic arterial blood leaving the lungs normally has a P_{O_2} of 100 mm Hg. Looking at the O_2–Hb curve, note that at a blood P_{O_2} of 100 mm Hg, Hb is 97.5% saturated. Therefore, the Hb in the systemic arterial blood normally is almost fully saturated.

If the alveolar P_{O_2} and consequently the arterial P_{O_2} fall below normal, there is little reduction in the total amount of O_2 transported by the blood until the P_{O_2} falls below 60 mm Hg (in the human example), because of the plateau region of the curve. If the arterial P_{O_2} falls 40%, from 100 to 60 mm Hg, the concentration of dissolved O_2 as reflected by the P_{O_2} is likewise reduced 40%. At a blood P_{O_2} of 60 mm Hg, however, the % Hb saturation is still remarkably high at 90%. Accordingly, the total O_2 content of the blood is only slightly decreased despite the 40% reduction in P_{O_2} because Hb is still carrying an almost-full load of O_2, and, as mentioned before, by far most of the O_2 is transported by Hb rather than being dissolved. However, even if the blood P_{O_2} is greatly increased, say, to 600 mm Hg by breathing pure O_2, very little additional O_2 is added to the blood. A small extra amount of O_2 dissolves, but the % Hb saturation can be maximally increased by only another 2.5%, to 100% saturation. Therefore, in the P_{O_2} range between 60 to 600 mm Hg or even higher, there is only a 10% difference in the amount of O_2 carried by Hb. Thus the plateau portion of the O_2–Hb curve provides a good margin of safety in O_2-carrying capacity of the blood.

Arterial P_{O_2} may be reduced because of pulmonary or gill diseases accompanied by inadequate ventilation or defective gas exchange or by circulatory disorders that result in inadequate blood flow to the respiratory surfaces. It may also fall at high altitudes, where the total atmospheric pressure and hence the P_{O_2} of the inspired air are reduced, and in O_2-deprived aquatic environments such as in a pond containing decayed vegetable matter and intermediate depths in the ocean (see p. 467). Unless the arterial P_{O_2} becomes markedly reduced in either pathological conditions or abnormal environmental circumstances, near-normal amounts of O_2 can still be carried to the tissues.

Significance of the Steep Portion of the O_2–Hb Curve

The steep portion of the curve between 0 and 60 mm Hg is in the blood P_{O_2} range that exists at the systemic capillaries, where O_2 is being unloaded from Hb. In the systemic capillaries, the blood equilibrates with the surrounding tissue cells at an average P_{O_2} of 40 mm Hg. Note on Figure 11–33b that at a P_{O_2} of 40 mm Hg, the % Hb saturation is 75%. The blood arrives in the tissue capillaries at a P_{O_2} of 100 mm Hg with 97.5% Hb saturation. Because Hb can only be 75% saturated at the P_{O_2} of 40 mm Hg in the systemic capillaries, nearly 25% of the HbO_2 must dissociate, yielding reduced Hb and O_2. This released O_2 is free to diffuse down its partial pressure gradient from the red blood cells through the plasma and interstitial fluid into the tissue cells.

The Hb in the venous blood returning to the lungs is still normally 75% saturated. If the tissue cells are metabolizing more actively, the P_{O_2} of the systemic capillary blood falls (for example, from 40 to 20 mm Hg) because the cells are consuming O_2 more rapidly. Note on the curve that this drop of 20 mm Hg in P_{O_2} decreases the % Hb saturation from 75% to 30%; that is, about 45% more of the total HbO_2 than normal gives up its O_2 for tissue use. The normal drop of 60 mm Hg in P_{O_2} from 100 to 40 mm Hg in the systemic capillaries causes about 25% of the total HbO_2 to unload its O_2. In comparison, a further drop in P_{O_2} of only 20 mm Hg results in an additional 45% of the total HbO_2 unloading its O_2 because the O_2 partial pressures in this range are operating in the steep portion of the curve. In this range, only a small drop in systemic capillary P_{O_2} can automatically make large amounts of O_2 immediately available to meet the O_2 needs of more actively metabolizing tissues. As much as 85% of the Hb may give up its O_2 to actively metabolizing cells during strenuous activity. In addition to this more thorough withdrawal of O_2 from the blood, even more O_2 is made available to actively metabolizing cells, such as exercising muscles, by circulatory and respiratory adjustments that increase the flow rate of oxygenated blood through the active tissues.

■ By acting as a storage depot, hemoglobin promotes the net transfer of O_2 from the alveoli to the blood.

We still have not fully clarified the role of Hb in gas exchange. Recall our earlier discussion of O_2 being driven from the alveoli to the blood by the P_{O_2} gradient—ΔP in Fick's law (p. 462). The ΔP is the difference between P_{O_2} in the alveoli, P_1, and the P_{O_2} in the blood, or P_2. Because blood P_{O_2} depends entirely on the concentration of *dissolved* O_2, we seemingly could ignore the O_2 bound to Hb. However, Hb does play a crucial role in permitting the transfer of large quantities of O_2 before blood P_{O_2} equilibrates with the surrounding tissues (● Figure 11–34). It does so by keeping P_2 low through acting as a "storage depot" for O_2, removing the O_2 from solution by converting it into a nondiffusible form as soon as it enters the blood from the alveoli. Because only diffusible (dissolved) O_2 contributes to the P_{O_2}, the O_2 bound to Hb cannot contribute to blood P_{O_2}. When systemic venous blood enters the pulmonary capillaries, its P_{O_2} is considerably lower than the alveolar P_{O_2}, so O_2 immediately diffuses into the blood, rais-

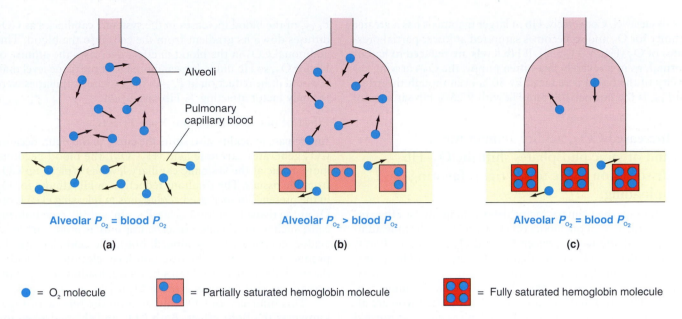

Alveoli

Pulmonary
capillary blood

Alveolar P_{O_2} = blood P_{O_2}   **Alveolar P_{O_2} > blood P_{O_2}**   **Alveolar P_{O_2} = blood P_{O_2}**

(a)   (b)   (c)

● = O_2 molecule   ▦ = Partially saturated hemoglobin molecule   ▦ = Fully saturated hemoglobin molecule

**Figure 11–34 ● Hemoglobin facilitating a large net transfer of O_2 by acting as a
storage depot to keep P_{O_2} low.** (a) In the hypothetical situation in which no hemoglobin
is present in the blood, the alveolar P_{O_2} and the pulmonary capillary blood P_{O_2} are at
equilibrium. (b) Hemoglobin has been added to the pulmonary capillary blood. As the Hb
starts to bind with O_2, it removes O_2 from solution. Because only dissolved O_2 contributes
to blood P_{O_2}, the blood P_{O_2} falls below that of the alveoli, even though the same number
of O_2 molecules are present in the blood as in part (a). By "soaking up" some of the
dissolved O_2, Hb favors the net diffusion of more O_2 down its partial pressure gradient
from the alveoli to the blood. (c) Hemoglobin is fully saturated with O_2, and the alveolar
and blood P_{O_2} are at equilibrium again. The blood P_{O_2} resulting from dissolved O_2 is
equal to the alveolar P_{O_2}, despite the fact that the total O_2 content in the blood is much
greater than in part (a) when blood P_{O_2} was equal to alveolar P_{O_2} in the absence of Hb.

ing the blood P_{O_2}. As soon as the P_{O_2} of the blood increases,
the percentage of Hb that can bind with O_2 likewise increases,
as indicated by the O_2–Hb curve. Consequently, most of the
O_2 that has diffused into the blood combines with Hb and no
longer contributes to blood P_{O_2}. As O_2 is removed from solution by combining with Hb, blood P_{O_2} falls to about the same
level it was when the blood entered the lungs, although the
total quantity of O_2 in the blood actually has increased. Because the blood P_{O_2} is once again considerably below alveolar
P_{O_2}, more O_2 diffuses from the alveoli into the blood, only
to be soaked up by Hb again. Essentially, this maximizes the
ΔP of Fick's law by keeping P_2 low.

Even though we have considered this process in stepwise
fashion for clarity, net diffusion of O_2 from alveoli or water to
blood occurs continuously until Hb becomes saturated with
O_2 as completely as it can be at that particular environmental
P_{O_2}. Not until Hb is unable to bind any more O_2 (that is, Hb
is maximally saturated for that P_{O_2}) does all the O_2 transferred into the blood remain dissolved and directly contribute
to the P_{O_2}. Only at this time does the blood P_{O_2} rapidly equilibrate with the alveolar P_{O_2} and halt further O_2 transfer.

The reverse situation occurs at the tissue level. Because
the P_{O_2} of blood entering the systemic capillaries of an ani-

mal is considerably higher than the P_{O_2} of the surrounding
tissue, the gradient favors the immediate diffusion of O_2 from
the blood into the tissues, lowering the blood P_{O_2}. When blood
P_{O_2} falls, Hb is forced to unload some of its stored O_2 because
the % Hb saturation is reduced. As the O_2 released from Hb
dissolves in the blood, the blood P_{O_2} increases and is once again
above the P_{O_2} of the surrounding tissues. This favors further
movement of O_2 out of the blood by keeping the ΔP component positive, although the total quantity of O_2 in the blood
has already been reduced. Only when Hb can no longer release any more O_2 into solution (when Hb is unloaded to the
greatest extent possible for the P_{O_2} existing at the systemic
capillaries) can blood P_{O_2} become as low as in the surrounding tissue. At this time, further transfer of O_2 ceases. Hemoglobin, because it bears a large quantity of stored O_2 that can
be liberated by a slight reduction in P_{O_2} at the systemic capillary level, permits the transfer of tremendously more O_2 from
the blood into the cells than would be possible in its absence.

Thus Hb plays an important role in the *total quantity* of
O_2 that the blood can pick up and drop off in the tissues. Small
mammals, which have a high rate of oxygen consumption per
gram of body weight compared to larger mammals, have Hb
with a low affinity for O_2 and so release O_2 more readily at

the tissue level. Conversely, Hb of larger mammals has a greater affinity for O_2 and so becomes saturated at lower partial pressures of O_2 (Figure 11–33a). If Hb levels are reduced to half of normal, as in a severely flea-bitten puppy, the O_2-carrying capacity of the blood is reduced by 50% even though the arterial P_{O_2} is the normal 100 mm Hg with 97.5% Hb saturation.

■ Increased CO_2, acidity, temperature, and organic phosphates shift the O_2–Hb dissociation curve to the right, favoring unloading.

Even though the primary factor determining the % Hb saturation is the P_{O_2} of the blood, other factors can affect the affinity, or bond strength, between Hb and O_2 and, accordingly, can shift the O_2–Hb curve (that is, change the % Hb saturation at a given P_{O_2}). These other factors are CO_2, acidity, temperature, and organic phosphates, which we examine separately. The O_2–Hb dissociation curve with which you are already familiar (Figure 11–33b) is a typical curve at normal human arterial CO_2 and acidity levels, normal body temperature, and normal organic phosphate concentration.

Effect of P_{CO_2}

An increase in P_{CO_2} shifts the O_2–Hb curve to the right (● Figure 11–35). The % Hb saturation still depends on the P_{O_2}, but for any given P_{O_2}, the amount of O_2 and Hb that can be combined is reduced. This effect is important, because the P_{CO_2} of the blood increases in the systemic capillaries as CO_2 diffuses down its gradient from the cells into the blood. This additional CO_2 in the blood in effect decreases the affinity of Hb for O_2, so Hb unloads even more O_2 at the tissue level than it would if the reduction in P_{O_2} in the systemic capillaries were the only factor affecting % Hb saturation.

Effect of pH: The Bohr and Root Effects

An increase in acidity also shifts the curve to the right. Because CO_2 generates carbonic acid (H_2CO_3), the blood becomes more acidic at the systemic capillary level as it picks up CO_2 from the tissues. The resultant reduction in Hb affinity for O_2 in the presence of increased acidity aids in releasing even more O_2 at the tissue level for a given P_{O_2}. In actively metabolizing cells, such as exercising muscles, not only is more carbonic-acid–generating CO_2 produced, but lactic acid also may be produced (see p. 54). The resultant local elevation of acid in the working muscles facilitates further unloading of O_2 in the very tissues that have the greatest O_2 need.

This influence of CO_2 and acid on the release of O_2 is known as the **Bohr effect.** Both CO_2 and the hydrogen ion (H^+) component of acids can combine reversibly with Hb at sites other than the O_2-binding sites. The result is an alteration in the molecular structure of Hb that reduces its affinity for O_2. However, the magnitude of the Bohr effect is not the same for those animals with Hb. Generally it depends on overall body size. For example, small mammals, with a higher metabolic rate and O_2 requirement, have Hb that is more sensitive to acid, which thus aids in delivery of O_2 to the tissues. Turtles, lizards, and snakes have smaller Bohr effects than mammals of comparable body size, because of a reduced affinity of hemoglobin for oxygen. Alternatively, the large Bohr effect observed in crocodilians is considered an adaptation associated with the extended periods they need to remain submerged in order to drown their prey. In fish, the Bohr effect is observed after an increase in H^+ ion but not CO_2. Recall that the concentration of CO_2 in water breathers is remarkably low and has little influence on O_2 binding.

In most species of fish and reptiles, there are multiple forms of Hb (*isoforms*). Some forms are characterized by a low affinity for O_2 and a heightened sensitivity to pH (and to *nucleoside triphosphates*, which we describe shortly), whereas other forms are characterized by a high affinity and relative insensitivity to pH and nucleoside triphosphates. The differences between the isoforms become evident under conditions where the acidity of the blood is increased. In these circumstances the maximal O_2-binding capacity of one form is reduced. The effect of acidosis on lowering blood oxygen content is termed the **Root effect** (● Figure 11–36). Hydrogen ion binds to this isoform of Hb and drastically reduces its total capacity (not just its affinity, which is the Bohr effect) for O_2. The Root effect is important when establishing the elevated P_{O_2} values in the swim bladder (see p. 506) and may also play a role during periods of lactic acid accumulation associated with strenuous activity. The presence of an acid-insensitive isoform of Hb helps minimize disturbances in blood O_2 transport. However, in some species with multiple forms of Hb, such as the carp, each form displays a pronounced Root effect.

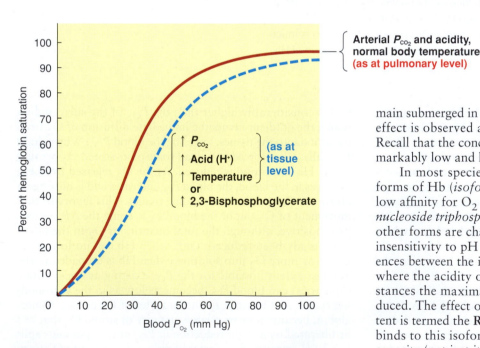

Figure 11–35 ● Effect of increased P_{CO_2}, H^+, temperature and 2,3-bisphosphoglycerate on the O_2–Hb curve (human hemoglobin). Increased P_{CO_2}, acid, temperature, and 2,3-bisphosphoglycerate, as found at the tissue level, shift the O_2–Hb curve to the right. As a result, less O_2 and Hb can be combined at a given P_{O_2}, so that more O_2 is unloaded from Hb for use by the tissues.

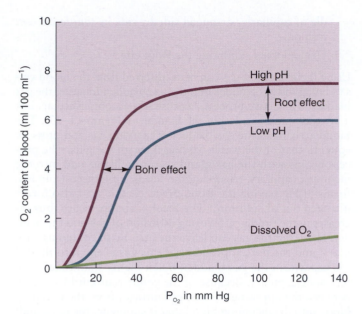

Figure 11-36 ● **Bohr and Root Effects for hemoglobin from a teleost (bony fish).** The plot shows oxygen dissociation curves for teleost hemoglobins demonstrating the effects of acid. Acidosis reduces the affinity of hemoglobin for O_2 (Bohr effect), shifting the curve to the right, and reduces the capacity of binding (Root effect), shifting the curve downward.

(*Source:* D. H. Evans, 1997, *The Physiology of Fishes*, 2nd ed., Boca Raton, FL: CRC Press, Figure 1, p. 103. Used by permission.)

Effect of Temperature

In a similar manner, an elevation in temperature shifts the O_2–Hb curve to the right, unloading more O_2 at a given P_{O_2}. This is of particular importance to poikilotherms because many routinely undergo increases in body temperature and thus increases in metabolic rate. The increased O_2 demands at the higher temperature can be met by increased unloading of O_2 from Hb. In homeotherms, a working muscle or other actively metabolizing cell also produces heat. The resultant local elevation in temperature enhances O_2 release from Hb for use by the more active tissues.

Thus increases in CO_2, acidity, and temperature at the tissue level, all of which are associated with increased cellular metabolism and increased O_2 consumption, enhance the effect of a drop in P_{O_2} in facilitating the release of O_2 from Hb. These effects are largely reversed at the pulmonary level, where the extra acid-forming CO_2 is blown off and the local environment is cooler. Appropriately, therefore, Hb has a higher affinity for O_2 in the pulmonary capillary environment, thus enhancing the effect of the elevation in P_{O_2} in loading O_2 onto Hb.

Effect of Organic Phosphates

The preceding changes take place in the *environment* of the red blood cells, but a factor *inside* the red blood cells of some species can also affect the degree of O_2–Hb binding: **organic phosphates.** The phosphate is **2,3-diphosphoglycerate (DPG)** in some mammals (such as horses, humans, dogs, and rats),

inositol pentaphosphate (IPP) in birds, and **nucleoside triphosphates (NTPs)** in fishes. **Adenosine triphosphate (ATP)** is the predominant NTP of salmonoids, sharks, and rays, whereas **guanosine triphosphate (GTP)** is found in eels, carp, and goldfish. These erythrocyte constituents, which are produced during red blood cell metabolism, can bind reversibly with Hb and reduce its affinity for O_2, just as CO_2 and H^+ do. Thus an increased level of organic phosphate, like the other factors, shifts the O_2–Hb curve to the right, enhancing O_2 unloading as the blood flows through the tissues. In mammals, DPG production by red blood cells gradually increases whenever Hb in the arterial blood is chronically undersaturated—that is, when arterial HbO_2 is below normal. This condition may occur in mammals living at high altitudes or in animals suffering from certain types of circulatory or respiratory diseases or anemia. By promoting the liberation of O_2 from Hb at the tissue level in air breathers, an increased amount of DPG or IPP helps maintain O_2 availability for tissue use under circumstances associated with decreased arterial O_2 supply. However, unlike the other factors (which normally are present only at the tissue level and thus shift the O_2–Hb curve to the right only in the systemic capillaries, where the shift is advantageous in unloading O_2), DPG is present in the red blood cells throughout the circulatory system and, accordingly, shifts the curve to the right to the same degree in both the tissues and the lungs. As a result, DPG decreases the ability to load O_2 at the pulmonary level, which is the negative side of increased DPG production. Interestingly, the hemoglobins of animals such as giraffes, cows, sheep, goats, and cats are relatively insensitive to the effects of DPG. These animals have a comparatively low-affinity Hb. In contrast, the response of water-breathing vertebrates to chronic hypoxia is a decline in ATP and an increase in O_2 affinity (that is, P_{50} is decreased).

■ Evolutionary adaptation results in different hemoglobin P_{50} values.

Earlier, we noted that species differ in their hemoglobin–oxygen affinities. Let us look at this in terms of gas gradients. Sluggish fish, for example, such as the flounder and carp, rely on Hb with a high affinity (low P_{50}) for O_2 to accommodate their particularly low environmental P_{O_2} (environmental P_{O_2} of 35 mm Hg versus a venous P_{O_2} of 11 mm Hg; hematocrit 14%). Under these circumstances there is a large partial pressure gradient between the water and the gill capillaries, whereas the gradient between the blood and the tissues is reduced. These species have thus optimized oxygen extraction from the water while maintaining a sedentary lifestyle that rarely requires strenuous activity. In contrast, active species such as trout have a comparatively low-affinity Hb (high P_{50}) and a high arterial P_{O_2} (arterial P_{O_2} of 113 mm Hg versus a venous P_{O_2} of 32 mm Hg, hematocrit 40%). Trout have a large diffusion gradient between the blood and the metabolizing tissues. One disadvantage of this strategy is the comparatively low efficiency of O_2 extraction from the water and the associated high cost of ventilation. However, active species normally inhabit well-aerated waters, which maximizes the water-to-gill capillary gradient. Second, during periods of intense activity, the O_2-carrying capacity of their blood can be rapidly increased by the release of erythrocytes from their storage site in the spleen.

■ Filling of the gas bladder is aided by the Root effect.

The ability of an animal to achieve buoyancy or weightlessness in water is of obvious benefit, because it greatly decreases the amount of energy required to maintain itself at any particular depth. A fish whose total body density is equal to that of the surrounding environment would thus be effectively weightless. One means of reducing density is to create a specific compartment containing low-density material. For example, all cartilaginous fishes have relatively light, nonmineralized bones. And some deep-sea sharks decrease their density by storing large amounts of fat (squalene) in their livers. Consequently, these animals have exceptionally large livers. But even these features do not make these animals neutrally buoyant, so they still tend to sink if they stop swimming. An alternative solution is to store gases—far more buoyant than fat—within a particular compartment.

The **gas bladder** (swim bladder) is a large, centrally located sac ventral to the spinal column in the abdominal cavity of teleost fish (see Fig. 11–9). Whether a particular species of fish has a gas bladder is related to its physiological needs. For example, some bottom-dwelling and some deep-sea species lack gas bladders, and some deep-sea species have fat-filled rather than air-filled bladders. The extent to which the gas bladder must be filled to achieve buoyancy at any particular depth depends on whether the fish is a freshwater or marine type. Fresh water is less dense than seawater and so provides less buoyancy, necessitating a larger gas bladder (7 to 11% of body volume) than in marine fishes (4 to 6% of body volume), to prevent sinking.

The key to the success of the gas bladder is its ability to vary the total quantity of gas as a function of its depth in the water. If the gas bladder were to contain an unchanging concentration of gas, the particular fish would be weightless at only one depth. Because pressure increases with depth, the gas in the bladder is compressed, so decreasing the bladder's volume and increasing the relative density of the fish. To keep its increasingly denser body from sinking, the fish would have to expend energy to maintain its depth. Conversely, when a fish rises, the volume of the bladder increases and the fish becomes too light. In situations where a fish ascends from extreme depths to the surface, the bladder can actually burst. In some species, such as the carp, a duct connects the gas bladder to the esophagus, gas can then be expelled through the mouth and gill cavities as it rises. In these species, gas may be added to the bladder by swallowing air at the surface. However, for most fishes, rising to the surface to gulp air prior to swimming at depths is impractical, and, in these species the gas bladder is completely enclosed from the outside. The actual gases contained within the bladder are similar to those in water but are present in varying proportions. In most fish the gas bladder contains 80 to 95% oxygen but in some species, such as the whitefish (*Coregonus albus*), it is almost pure nitrogen.

The ability to adjust the quantity of gas within the bladder arises from several sources:

■ A **gas gland** within the wall of the gas bladder. The gland operates anaerobically to secrete lactic acid into the blood, which lowers the pH of the blood leaving the gland.

■ An unusual blood supply called the **rete mirabile**, which perfuses the gas gland. The rete mirabile is a dense bundle of capillaries arranged side by side in *countercurrent* fashion.

■ Hemoglobins exhibiting the **Root effect.**

Using these features, gas can be transferred from the blood vessels lining the gas bladder into the gas bladder. Inflating the gas bladder is an active process, because it is accomplished against a pressure gradient within the bladder (● Figure 11–37). The blood arriving at the gas bladder is carrying gas at a pressure equal to that of the water passing through the gills. The process begins with the gas gland secreting lactic acid into the incoming capillary, triggering the release of oxygen from isoforms of Hb having the Root effect. (Lactic acid also forces nitrogen out of solution.) The local partial pressure thus increases in the capillary, favoring diffusion of gas into the bladder. However, the gas can leave the area by the outgoing capillary, and the partial pressure of this gas is equal to that of the gas within the gas bladder, much higher than that in the incoming capillary. Because of the countercurrent blood flow between the opposing vessels, gas diffuses from the outgoing blood into the incoming blood that is supplying the gas gland! This process is repeated continuously, with higher and higher concentrations of gas collecting at the junction point of the incoming and outgoing capillaries within the gas gland.

It is the pH-dependent release of oxygen from Hb via the Root effect that ultimately accounts for the ability of fish to force oxygen into the swim bladder. Diffusion of gases stored in the gas bladder is restricted by the presence of multiple sheets of crystalline guanine, which line the wall of the gas bladder. The countercurrent rete prevents dissipation of the gradient.

Deflating the gas bladder as a fish decreases its depth is a passive process (favored when the gas gland is less active). Gases move down their partial pressure gradients into the surrounding capillaries, and are then transported to the gills.

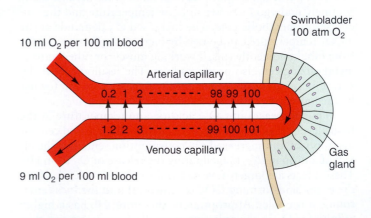

Figure 11–37 ● Countercurrent multiplier system of the fish swim bladder. Production of lactic acid from the gas gland increases oxygen tension in the venous capillary. As a result oxygen diffuses down its concentration gradient into the arterial capillary, thus remaining within the rete. However, the oxygen content of the venous blood exiting the rete is now lower than the incoming arterial blood with the difference deposited in the gas gland.

(*Source:* K. Schmidt Nielsen, 1997, *Animal Physiology: Adaptation and Environment*, 5th ed., Cambridge, UK: Oxford University Press, Figure 10–33, p. 447. Used by permission.)

Most CO_2 in both vertebrates and invertebrates is transported in the blood as bicarbonate.

When arterial blood flows through the tissue capillaries, CO_2 diffuses down its partial pressure gradient from the tissue cells into the blood. Carbon dioxide in vertebrates is transported in the blood in three ways (• Figure 11–38):

1. *Physically dissolved.* As with dissolved O_2, the amount of free CO_2 *physically dissolved* in the blood depends on the P_{CO_2}. Because CO_2 is more soluble than O_2 in the blood, a greater proportion of the total CO_2 in the blood is physically dissolved compared to O_2. Even so, only about 5 to 10% of an air breather's total blood CO_2 is carried this way at the normal systemic venous P_{CO_2} level. This amount is further reduced in fish and is estimated to be less than 5% of the total CO_2 carried in the blood.

2. *Bound to hemoglobin.* Another 25 to 30% (in mammals) of the CO_2 combines with Hb to form **carbamino hemoglobin (HbCO$_2$)**. Carbon dioxide binds with the globin (protein) portion of Hb, reacting with the nitrogen of the last amino acid of the protein chain:

$$CO_2 + Hb\text{–}NH_2 \rightarrow Hb\text{–}NH\text{–}COOH$$

This is in contrast to O_2, which combines with the heme iron. However, $HbCO_2$ is unimportant in fish, because their Hb has been modified (acetylated), preventing binding.

3. *As bicarbonate.* By far the most important means of CO_2 transport is as **bicarbonate (HCO$_3$⁻)** ion. However, unlike hemoglobins found in most animals, the Hb in crocodile blood has evolved the unique ability to bind to HCO_3^-. Under conditions where the crocodile remains submerged for prolonged periods, the binding of HCO_3^- ions on Hb results in the progressive unloading of oxygen from its binding sites on Hb, permitting this vast reservoir of oxygen to be available to the respiring tissues.

In fish, HCO_3^- generally constitutes approximately 90 to 95% of the total CO_2 in the blood, whereas in mammals ap-

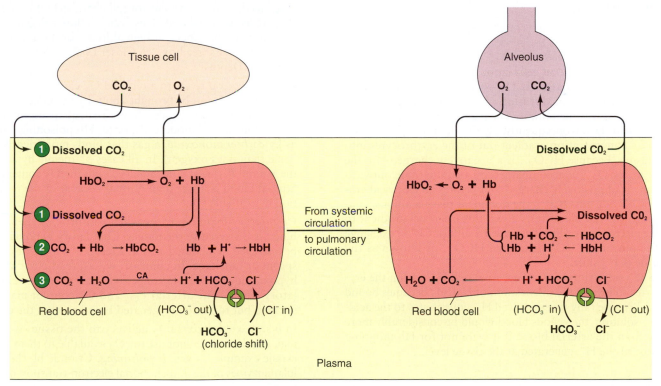

CA = Carbonic anhydrase

Figure 11–38 • Carbon dioxide transport in the blood (vertebrate). Carbon dioxide (CO_2) picked up at the tissue level is transported in the blood to the lungs in three ways: ① physically dissolved; ② bound to hemoglobin (Hb), and ③ as bicarbonate ion (HCO_3^-). Hemoglobin is present only in the red blood cells, as is carbonic anhydrase, the enzyme that catalyzes the production of HCO_3^-. The H⁺ generated during the production of HCO_3^- also binds to Hb. Bicarbonate moves by facilitated diffusion down its concentration gradient out of the red blood cell into the plasma, and chloride (Cl^-) moves by means of the same passive carrier into the red blood cell down the electrical gradient created by the outward diffusion of HCO_3^-.

The reactions that occur at the tissue level are reversed at the pulmonary level, where CO_2 diffuses out of the blood to enter the alveoli.

proximately 60 to 70% of the CO_2 is converted into HCO_3^- by the following chemical reaction, which takes place within the red blood cells:

$$CO_2 + H_2O \leftrightarrow H_2CO_3 \leftrightarrow H^+ + HCO_3^-$$

In the first step, CO_2 combines with H_2O to form **carbonic acid (H_2CO_3)**. This reaction can occur very slowly in the plasma, but it proceeds swiftly within the red blood cells because of the presence of the enzyme **carbonic anhydrase**, which catalyzes (speeds up) the reaction (although in an indirect way; see p. 364). Carbonic anhydrase is also found in lungs and kidneys of mammals, lungs of amphibians, and in gills of fishes. As is characteristic of acids, a proportion of the carbonic acid molecules spontaneously dissociate into hydrogen ions (H^+) and bicarbonate ions (HCO_3^-). For this reason, small mammals contain higher concentrations of this enzyme in their blood, because of the ability of H^+ ion to increase the rate of O_2 delivery to the tissue (Bohr effect). The one carbon and two oxygen atoms of the original CO_2 molecule are thus present in the blood as an integral part of HCO^-. This is beneficial, because HCO_3^- is more soluble in the blood than CO_2.

Chloride Shift

As this reaction proceeds, HCO_3^- and H^+ start to accumulate within the red blood cells in the systemic capillaries. The red-cell membrane has a HCO_3^-–Cl^- antiport carrier that passively facilitates the diffusion of these ions in opposite directions across the membrane. The membrane is relatively impermeable to H^+. Consequently, HCO_3^- but not H^+ diffuses down its concentration gradient out of the erythrocytes into the plasma. Because HCO_3^- is a negatively charged ion, the efflux of HCO_3^- unaccompanied by a comparable outward movement of positively charged ions creates an electrical gradient. Chloride ions (Cl^-), the dominant plasma anions, conduct into the red blood cells down this electrical gradient to restore electric neutrality. This inward shift of Cl^- in exchange for the outflux of CO_2-generated HCO_3^- is known as the **chloride (Cl^-) shift.**

The vast majority of the accumulated H^+ within the erythrocytes following the dissociation of H_2CO_3 becomes bound to Hb. Because only free dissolved H^+ contributes to the acidity of a solution, the venous blood would be considerably more acidic than the arterial blood if it were not for Hb mopping up most of the H^+ generated at the tissue level.

Haldane Effect

As you have seen, the binding of both H^+ and CO_2 to oxyhemoglobin (HbO_2) favors release of O_2 (the Bohr effect). Conversely, deoxyhemoglobin has a greater affinity for both CO_2 and H^+ than does HbO_2. The unloading of O_2 from Hb in the tissue capillaries therefore facilitates the picking up of CO_2 and H^+ by Hb. The ability of deoxygenated Hb to pick up CO_2 and CO_2-generated H^+ is known as the **Haldane effect**. The Haldane effect and Bohr effect work in synchrony to facilitate O_2 liberation and the uptake of CO_2 and CO_2-generated H^+ at the tissue level. Increased CO_2 and H^+ cause increased O_2 release from Hb by means of the Bohr effect; increased O_2 release from Hb, in turn, causes increased CO_2 and H^+ uptake by Hb through the Haldane effect. The entire process is very efficient. Deoxygenated Hb must be carried back

to the lungs to reload with O_2 in any event. Meanwhile, after O_2 is released, Hb picks up new passengers—CO_2 and H^+—that are going in the same direction to the lungs.

The reactions that occur at the tissue level as CO_2 enters the blood from the tissues are reversed once the blood reaches the lungs or gills and CO_2 leaves the blood to enter the alveoli (Figure 11–38) or ventilatory water outside the gill epithelium.

■ Various respiratory states are characterized by abnormal blood-gas levels.

Abnormalities in Arterial P_{O_2}

The term **hypoxia** refers to insufficient O_2 at the cellular level. There are four general categories of hypoxia:

1. *Hypoxic hypoxia* is characterized by a low arterial blood P_{O_2} accompanied by inadequate Hb and blood saturation. It is caused by (a) a respiratory malfunction involving inadequate gas exchange, typified by a normal alveolar (or gill chamber) P_{O_2} but a reduced arterial P_{O_2}, or (b) exposure to an environment where environmental P_{O_2} is reduced, so that alveolar (or gill chamber) and arterial P_{O_2} are likewise reduced. This occurs in low-oxygen habitats for water breathers, at high altitudes or during diving for air breathers (see box, "A Closer Look at Adaptation: Effects of Depths").

2. *Anemic hypoxia* refers to a reduced O_2-carrying capacity of the blood. It can be brought about by (a) a decrease in circulating red blood cells, (b) an inadequate amount of Hb within the red blood cells, or (c) Hb poisoning (such as by *carbon monoxide,* a gas that binds more tightly to heme than does oxygen). In all cases of anemic hypoxia, the arterial P_{O_2} is at a normal level, but the O_2 content of arterial blood is lower than normal because of the reduction in available Hb.

3. *Circulatory hypoxia* arises when too little oxygenated blood is delivered to the tissues. Circulatory hypoxia can be restricted to a limited area as a result of a local vascular spasm or blockage. In contrast, widespread circulatory hypoxia can result from congestive heart failure or circulatory shock. The arterial P_{O_2} and O_2 content may be normal, but too little oxygenated blood reaches the cells.

4. In *histotoxic hypoxia*, O_2 delivery to the tissues is normal, but the cells cannot use the O_2 available to them. The classic example is *cyanide poisoning.* Cyanide blocks cellular enzymes of the mitochondrial electron-transport chain (p. 47).

Hypoxia triggers a number of genetic and cellular processes: see the box, "Molecular Biology and Genomics: Effects of Heights."

Hyperoxia, an above-normal arterial P_{O_2}, cannot occur when an animal is breathing atmospheric air at sea level. However, breathing supplemental O_2 can increase alveolar and consequently arterial P_{O_2}.

Abnormalities in Arterial P_{CO_2}

The term **hypercapnia** refers to excess CO_2 in the arterial blood; it can be caused by **hypoventilation** (ventilation inadequate to meet the metabolic needs for O_2 delivery and CO_2 removal), or by exposure to high environmental CO_2.

Hypocapnia, below-normal arterial P_{CO_2} levels, is brought about by hyperventilation. **Hyperventilation** occurs when an animal "overbreathes," that is, when the rate of ventilation is in excess of the body's metabolic needs for CO_2 removal so that CO_2 is blown off to the atmosphere more rapidly than it is produced in the tissues and arterial P_{CO_2} falls. Hyperventilation can be triggered by anxiety states and by fever. Alveolar P_{O_2} increases during hyperventilation as more fresh O_2 is delivered to the alveoli from the atmosphere than is extracted from the alveoli by the blood for tissue consumption, and arterial P_{O_2} increases correspondingly (● Figure 11–39). However, because Hb is almost fully saturated at the normal arterial P_{O_2}, very little additional O_2 is added to the blood. Except for the small extra amount of dissolved O_2, blood O_2 content remains essentially unchanged during hyperventilation.

Increased ventilation is not synonymous with hyperventilation. Increased ventilation that matches an increased metabolic demand, such as the increased need for O_2 delivery and CO_2 elimination during activity, is termed **hyperpnea.** During activity, alveolar P_{O_2} and P_{CO_2} remain constant, with the increased atmospheric exchange just keeping pace with the increased O_2 consumption and CO_2 production.

Consequences of Abnormal Blood Gases Levels

The consequences of reduced O_2 availability to the tissues during hypoxia are apparent. The cells need adequate O_2 supplies to sustain their energy-generating metabolic activities. The consequences of abnormal blood CO_2 levels are less obvious. Changes in blood CO_2 concentration primarily affect acid–base balance. Hypercapnia results in an elevated production of carbonic acid. The subsequent generation of excess

H^+ produces an acidic condition termed *respiratory acidosis.* Conversely, less-than-normal amounts of H^+ are generated through carbonic acid formation in conjunction with hypocapnia. The resultant alkalotic (less acidic than normal) condition is called *respiratory alkalosis* (Chapter 13).

■ Other unusual respiratory states are not necessarily pathological.

Not all unusual respiratory conditions are abnormal. One of these is **apnea,** a temporary cessation of breathing. It is a normal characteristic of hibernating mammals and of reptiles. Inflation of the lungs is normally followed by a period of apnea in order to maximize gas exchange. In general, terrestrial reptiles have comparatively brief periods of apnea relative to their aquatic counterparts. This may be due, in part, to the energy costs associated with surfacing to breathe. Aquatic turtles and snakes exhibit periods of apnea of several minutes, whereas in crocodilians breathing may cease for periods greater than an hour. Periods of *sleep apnea* in elephant seals can range from 10 to 25 minutes and are followed by several minutes of regular breathing. Sleep apnea is considered similar to the diving response (see p. 387) because both heart rate and metabolism decline. Sleep apnea occurs as a disorder in some humans, in which it is a threat to health.

Finally, some animals have adapted to extreme hypoxia, or **anoxia**—the complete lack of oxygen. Goldfish (*Carassius auratus*) and their close relatives, the European carp (*Carassius carassius*) exhibit a most remarkable resistance to anoxia. Indeed, this is why they became favorite pets in simple bowls in households and display fish in artificial ponds, for aeration is not required for them to survive. Unlike other pond animals such as turtles—which become comatose in low oxygen—goldfish and carp remain active for 1 to 2 days even with no detectable oxygen in the water. And they can survive over winter in the bottom of an oxygen-less pond sealed with ice. How can they do this?

The most critical organ in anoxia is the brain, which needs a constant high supply of ATP to maintain ion balance via its Na/K–ATPase "pumps" (p. 81). Recall that, without oxygen, only 2 ATPs are produced for each glucose molecule (compared to more than 30 in the presence of oxygen) (p. 53). Carp and goldfish solve this low-ATP problem in part by reducing metabolism (and thus ATP usage) during anoxia. They also have large stores of glycogen in the liver, which is the main source of glucose for anaerobic metabolism. But in addition to energy problems, an animal in anoxia must avoid the lactic acid poisoning that results from normal anaerobic metabolism (Chapter 2). Remarkably, these fish (in their muscles primarily) make **ethanol**—the intoxicating ingredient of fermented beverages—as an end product, just as yeast do (p. 54). Ethanol (unlike lactic acid) has the advantages of being nonacidic and being able to diffuse out of the gills readily. These fish avoid the acid–base problems that freshwater turtles must endure (see Chapter 13).

Equally as astounding as the goldfish's ability to withstand anoxia is the ability of the epaulette shark to survive periods of up to three hours with complete oxygen deprivation. Occasionally this reef inhabitant is left stranded in the shallows during low tide, where oxygen concentrations rapidly fall to levels that result in permanent brain damage for other fish species. How can the epaulette shark master this feat? Studies

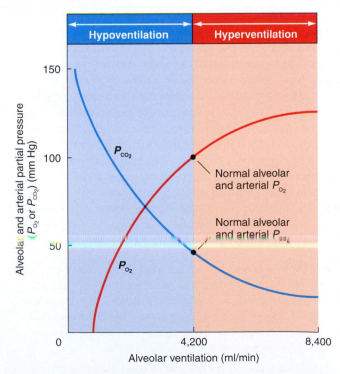

Figure 11–39 ● Effects of hyperventilation and hypoventilation on arterial P_{O_2} and P_{CO_2}.

have shown that a series of biochemical events unfold as the duration of anoxia increases:

- In association with a decline in oxygen concentrations in the tidal pool, the epaulette shark reduces both its rate of respiration and heartbeat, and decreases arteriolar resistance to the brain.

- GABA—a depressive neurotransmitter (p. 174)—is released in order to "turn off" nonessential neurological regions of the brain.

- Most critically, the anoxic shark can "turn down" the activity of its mitochondria. Normally, in most species of animals as oxygen is restored to the anoxic tissues (reperfusion) the mitochondria mysteriously generate free radicals (p. 49) from whatever oxygen becomes available. This results in brain injury and apoptosis (cell death,

p. 93). Surprisingly, because in the epaulette shark the mitochondria are functioning at a lower level, the return of oxygen to the brain cells does not result in the generation of free radicals.

As a consequence the shark becomes limp and flaccid; but should a predator or potential prey move into the vicinity, its still functioning electroreceptors (p. 245) cause the apparently dead animal to revert back to life!

Control of Respiration

Except for adaptive apnea, continuous breathing is vital to most animals most of the time, and so (in the logic of hierarchical control, p. 17), it is usually regulated by an innate, rhythmic neural program that can operate independently of

A CLOSER LOOK AT ADAPTATION

Effects of Depths

At least some species of animals are capable of survival even in the deepest trenches of the oceans, whose depths reach about 11 km below sea level (about 1100 atm pressure!). Temperatures in the deep sea are normally very stable, at around 1 to 4°C. These low temperatures exacerbate the stresses associated with high hydrostatic pressure on an animal, which include a reduction in membrane fluidity and perturbed protein folding and binding. Remarkable adaptations to pressure have evolved that permit permanent inhabitants to survive the high pressure. These include modified membranes and proteins, and stabilizing or counteracting osmolytes that help offset the effects of pressure on proteins (pp. 136 and 578).

Other effects of pressure include those on gas-filled spaces, such as swim bladders and lungs. Most astounding are those diving mammals who are capable of taking a breath at the surface, diving to extraordinary depths in a comparatively short period of time, and then resurfacing. Let's start with the situation as a deep-sea diver descends underwater. The body is immediately exposed to greater than atmospheric pressure. Pressure rapidly increases with depth as a result of the weight of the water, increasing by about 1 atm per each 10 m deeper. Recall that (1) the amount of a gas in solution is directly proportional to the partial pressure of the gas and that (2) air has 79% N_2. Nitrogen is poorly soluble in body tissues, but the high pressure that occurs during deep-sea diving causes more of this gas than normal to dissolve in the body tissues. The small amount of N_2 dissolved in the tissues at sea level has no known effect, but as more N_2 dissolves at greater depths, **nitrogen narcosis,** or **"raptures of the deep,"** ensues in humans. Nitrogen narcosis is believed to result from a reduction in the excitability of neurons because of the highly lipid-soluble N_2 dissolving in their lipid membranes. **High-pressure neurological syndrome (HPNS),** another malady of human divers, occurs at depths below 200 m. Mounting pressure on the body results in tremors and convulsions.

Pressure decreases rate of action potential propagation and excitatory synaptic transmission, apparently as the result of a decrease in neurotransmitter release.

Another problem associated with deep-sea diving occurs during ascent. If a diver who has been submerged long enough for a significant amount of N_2 to become dissolved in the tissues suddenly ascends to the surface, the rapid reduction in pressure causes N_2 to quickly come out of solution and form bubbles of gaseous N_2 in the body. The consequences depend on the amount and location of the bubble formation. This condition is called **decompression sickness** or **"the bends,"** the latter term arising because the victim often bends over in pain. Decompression sickness can be prevented by slow ascent to the surface or by gradual decompression in a decompression tank so that the excess N_2 can slowly escape through the lungs without bubble formation.

Contrast the abilities of a human with those of a Weddell or elephant seal. Foraging escapades can take these animals to depths approximately 1500 m below the surface in less than 20 min. At these depths water crushes, with a pressure of more than 150 atm (over a ton per square inch). Weddell seals can stay in the deep for an hour or more! These seals' ability to accomplish these feats seems to defy all known physiological limits, yet they can repeat this feat again and again over a 24-hour period. How is this possible?

Early studies of seals focused on an adaptation to diving called the *dive reflex:*

- Elimination of blood flow to the most organs (other than the heart, brain, and adrenal gland), saving O_2 by reducing metabolic demands. Interestingly, maintenance of blood flow to the adrenal gland may be important for normal release of cortisol during a dive. Evidence suggests cortisol stabilizes neurons, thus preventing the onset of HPNS.

- Reduction of body temperature by up to 3°C before commencing the dive.

higher brain centers. These integrators have chemosensory inputs to measure gas levels, and outputs to effectors that regulate ventilation in negative-feedback fashion. For example, tracheal ventilation in terrestrial insects is tightly regulated by motor output originating in the **metathoracic ganglia**. An increase in the firing rates stimulates *closer muscles* to reduce the diameter of the spiracles. In this manner the resistance to gas flow as well as the total ventilation volume can respond to chemoreceptors that detect changes in CO_2 (and possibly O_2 levels) in the hemolymph. Recent studies show that in the fruit fly *Drosophila*, spiracles are opened and closed during flight such that oxygen needs are precisely met, while at the same time water loss is reduced by having the spiracles closed periodically.

In vertebrates, the rhythm generator for breathing—the **respiratory center**—is located in the brain stem. In water-breathing vertebrates, P_{O_2} in the blood is the primary homeostatic variable. P_{CO_2} is not a major signal, because the high solubility of this gas in the external water results in only small percentage changes in this gas in the blood at the gills. P_{O_2} is thought to be sensed by chemoreceptors in the aorta and brain in trout, which send signals to the respiratory center.

Associated with the transition from water to air breathing in vertebrates was a progressive dependence on CO_2 as a source of respiratory drive, for reasons that you will see later. Primitive air breathing in vertebrates evolved as a new motor pattern, possibly arising from modified coughing and suction feeding movements. The central rhythm generator in the brain stem of higher vertebrates may have evolved through an amalgamation of these two pre-existing neural generators. We focus primarily on mammalian breathing, but note important features of other vertebrates.

■ Decrease in heart rate to as little as 2–6 beats/min at depth. This *bradycardia* ("slow heart") decreases the work load of the heart and thus its O_2 requirement.

Phylogenetic studies indicate that this reflex is ancestral in all mammals (including humans), so is not special to diving mammals except that it is more effective than in nondivers. But other adaptations are wholly unique to divers. First, lung structure and function is distinct. Shallow-diving birds and turtles rely on oxygen from their lungs to sustain them during their dives. However, the lungs of the (deep-diving) seals actually collapse by approximately 40 m below the surface. Seals exhale prior to diving, which significantly lowers the gas volume contained within their lungs. Considering the animals' body size, the lungs are proportionately small, which further reduces their capacity for storing N_2. High pressure squeezes the remaining gas out of the lungs into the *bronchial air-duct system*. Both the bronchi and bronchioles are supported by rings of cartilage, which resist compression at depth, with the result that these airways store the remaining respiratory gases and do not introduce them into the circulation. Another advantage of exhaling before diving is to reduce the buoyancy of the animal, which aids it in the descent.

Second, oxygen storage has been greatly enhanced by adaptations involving respiratory pigments. The blood of divers has up to two times greater hemoglobin concentration than does blood of nondivers, with the hematocrit of diving mammals being 60% in the Weddell seal versus 35 to 40% in humans. Similarly, red muscle has a greater myoglobin content. Diving mammals can also squeeze out O_2-rich erythrocytes from their abnormally enlarged spleens (as do the spleens of horses during intense activity). Blood constitutes approximately 20% of the body weight of the elephant seal (versus 14% in the Weddell seal) as compared to only 7% in humans.

Photo: © Kevin McDonnell Photography

A bottlenose dolphin (*Tursiops*) with a video camera. The camera was used to document gliding behavior during a dive. Physiological instruments to record heart rate, blood oxygen levels, body temperatures, and so forth are also attached to diving animals.

(*Source:* http://www.sciencenews.org/20000408/fob4.asp)

Third, anaerobic acidosis is prevented during the dive by confining anaerobic metabolism to the skeletal muscles, which have extremely high buffer content in the form of histidine dipeptides (p. 601). Because blood flow to this tissue is cut off, lactic acid cannot be released into the blood until the animal resurfaces. Only then can the liver and kidneys metabolize this by-product.

Finally, recent studies show that some diving mammals use a *gliding* technique to save energy during the dive. They stop all locomotory movements, and let gravity pull them down once their lungs collapse so that they are no longer buoyant. This behavioral effector saves oxygen.

■ Respiratory centers in the vertebrate brain stem establish a rhythmic breathing pattern.

In vertebrates, both the pumping of heart and ventilation of the respiratory organs exhibit continuous, cyclical activity. However, the underlying mechanisms and control of these two systems are remarkably different. Whereas the heart can generate its own rhythm by means of its intrinsic pacemaker activity, the respiratory muscles, being skeletal muscles, require nervous stimulation to bring about their contraction. As you have seen, in both water and air breathing vertebrate's respiratory control centers in the brain stem are responsible for generating the rhythmic pattern of breathing.

Neural control of respiration involves three distinct components: (1) the neural **pattern generator** program responsible for the alternating inspiration/expiration rhythm, (2) the factors that regulate the magnitude of ventilation (that is, the rate and depth of breathing) to match physiological needs, and (3) the factors that modify respiratory activity to serve other purposes. The latter modifications can be made by higher brain centers, as in the breath control required for vocalization, or by reflexes, as in the respiratory maneuvers involved in a cough or sneeze.

In mammals, the primary respiratory control center, the **medullary respiratory center,** consists of several aggregations of neuronal cell bodies within the medulla that provide output to the respiratory muscles. In addition, there are two other respiratory centers higher in the brain stem in the pons—the **apneustic center** and **pneumotaxic center.** These pontine centers influence the output from the medullary respiratory center (● Figure 11–40). Exactly how these various regions interact to establish respiratory rhythmicity is unclear, but the following factors are believed to contribute.

Inspiratory and Expiratory Neurons in the Medullary Center

Mammals rhythmically breathe in and out during quiet breathing because of alternate contraction and relaxation of the inspiratory muscles supplied respectively by the phrenic nerve and intercostal nerves. The cell bodies for the neuronal fibers composing these nerves are located in the spinal cord. Impulses originating in the medullary center terminate on these motor-neuron cell bodies (● Figure 11–41). When these motor neurons stimulate the inspiratory muscles of humans, this leads to inspiration; when these neurons are not firing, the inspiratory muscles relax and expiration takes place.

The medullary respiratory center consists of two neuronal clusters known as the dorsal respiratory group and the ventral respiratory group (Figure 11–41). The **dorsal respiratory group**

MOLECULAR BIOLOGY AND GENOMICS:

Effects of Heights

When an animal accustomed to low altitude travels to high altitude, hypoxia ensues from low oxygen availability. This is a not uncommon phenomenon in some species: consider human mountaineers or cattle migrating to high summer pastures. The body's initial reaction is an increase in ventilation and circulation, but over prolonged time at high altitude, acclimatization changes occur, including (1) an *increase in red blood cell numbers* triggered by the hormone **EPO** from the kidney (p. 365); and (2) *angiogenesis,* the growth of new capillaries triggered by the paracrine **vascular endothelial growth factor (VEGF).** For decades, physiologists have sought the sensory mechanisms by which cells producing these hormones detect low oxygen. The key appears to be regulation of **hypoxia-inducible factor (HIF),** a transcription factor that binds to response elements of

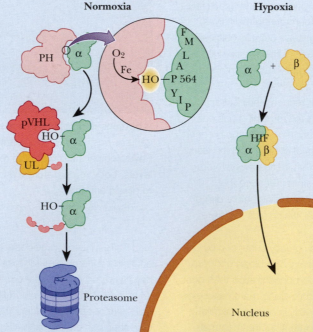

(*Source:* M. K. Campbell & S. O. Farrell, 2003, *Biochemistry,* 4th ed., Belmont, CA: Brooks/Cole. Modified from H. Hsu & H. F. Bunn, 2001, How do cells sense oxygen? *Science* 292, 449–451)

genes (see p. 30) for proteins that are needed for hypoxia acclimatization. HIF consists of two subunits, HIFα and HIFβ. In *normoxia* (normal oxygen levels), HIFα is hydroxylated by **proline hydroxylase (PH),** converting HIFα into a form that then binds to **von Hippel-Lindau protein (pVHL).** Together, the HIFα-pVHL complex activates **ubiquitin ligase (UL),** an enzyme complex that adds *ubiquitin* to HIFα. As you saw in Chapter 2 (p. 44), ubiquitin is a "tag" that targets a protein for destruction by proteasomes.

Thus, in normoxia the HIF system is relatively inactive. However, during hypoxia, PH no longer targets HIFα for destruction, therefore HIFα combines with HIFβ to form a functional transcription factor. In turn, the acclimatization genes are activated. It appears that PH requires oxygen to function, and thus it becomes less active when oxygen levels are low.

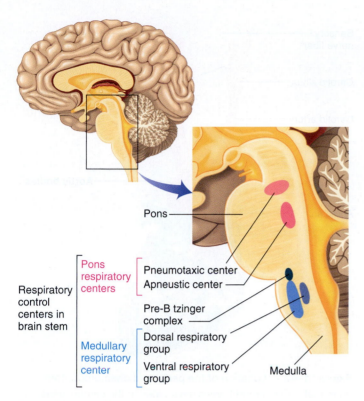

Figure 11–40 • Respiratory control centers in the brain stem (mammal).

the descending pathways from the expiratory neurons during quiet breathing. Only during active expiration do the expiratory neurons stimulate the motor neurons supplying the expiratory muscles (the abdominal and internal intercostal muscles). Furthermore, the VRG inspiratory neurons, when stimulated by the DRG, rev up inspiratory activity when demands for ventilation are high.

Until recently, the DRG was generally regarded as being responsible for the basic rhythm of ventilation. However, the DRG is no longer thought directly responsible for initiating the rhythmic pattern of breathing. Instead, generation of respiratory rhythm is now widely believed to lie outside the DRG in the *rostral ventromedial medulla*, a region located near the upper (head) end of the VRG. Researchers have identified a network of neurons in this region, some of which display pacemaker activity and all of which are believed to play a role in the rhythm-generating process. The interplay between these neurons that is ultimately responsible for the rhythmicity remains to be identified. The DRG's inspiratory neurons were believed to display pacemaker activity, repetitively undergoing self-induced action potentials similar to the SA node of the heart. The rate at which the inspiratory neurons rhythmically fire is driven by synaptic input from the rostral ventromedial medulla and from elsewhere in the body. Thus the on–off nature of the respiratory cycle is very complex, and incompletely understood.

Influences from the Pneumotaxic and Apneustic Centers

The pontine centers exert "fine-tuning" influences over the medullary center to help produce normal, smooth inspirations and expirations. The pneumotaxic center sends impulses to the DRG that help "switch off" the inspiratory neurons, limiting the duration of inspiration. In contrast, the apneustic center prevents the inspiratory neurons from being switched off, thus providing an extra boost to the inspiratory drive. In this check-and-balance system, the pneumotaxic center is dominant over the apneustic center, helping to bring inspiration to a halt and allowing expiration to occur normally. Without the pneumotaxic brakes, the breathing pattern consists of prolonged inspiratory gasps abruptly interrupted by very brief expirations. This abnormal pattern of breathing is known as **apneusis**; hence the center responsible for this type of breathing is the ap-

(DRG) consists mostly of *inspiratory neurons* whose descending fibers terminate on the motor neurons that supply the inspiratory muscles. When the DRG inspiratory neurons fire, inspiration takes place; when they cease firing, expiration occurs. Expiration is brought to an end as the inspiratory neurons once again reach threshold and fire.

The DRG has important interconnections with the **ventral respiratory group (VRG)**. The VRG is composed of *inspiratory neurons* and *expiratory neurons,* both of which remain inactive during normal quiet breathing. This region is called into play by the DRG as an "overdrive" mechanism during periods when demands for ventilation are increased. It is especially important in active expiration. No impulses are generated in

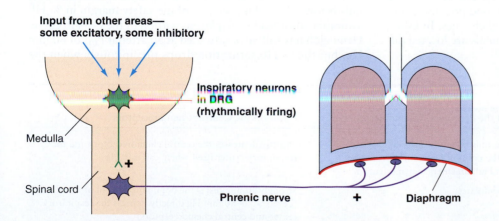

Figure 11–41 • Schematic representation of medullary dorsal respiratory group (DRG) control of inspiration. Inspiration takes place when the inspiratory neurons are firing and activating the motor neurons that supply the inspiratory muscles. Expiration takes place when the inspiratory neurons cease firing, so that the motor neurons supplying the inspiratory muscles are no longer activated.

Not shown are intercostal nerves to external intercostal muscles.

neustic center. In mammals apneusis may occur in certain types of severe brain damage.

■ The magnitude of ventilation is adjusted in response to three chemical factors: P_{O_2}, P_{CO_2}, and H$^+$.

No matter how much O_2 is extracted from the blood or how much CO_2 is added to it at the tissue level, the P_{O_2} and P_{CO_2} of the systemic arterial blood leaving the lungs or gills are held remarkably constant, indicative of the fact that arterial blood-gas content is subject to precise regulation. Arterial blood gases are maintained within the normal range by varying the magnitude of ventilation to match the body's needs for O_2 uptake and CO_2 removal. If more O_2 is extracted from the alveoli or gill and more CO_2 dropped off by the blood because the tissues are metabolizing more actively, ventilation is increased correspondingly to bring in more fresh O_2 and remove more CO_2.

The medullary respiratory center receives inputs that provide information about animals' needs for gas exchange. It responds by sending appropriate signals to the motor neurons supplying the respiratory muscles to adjust the rate and depth of ventilation to meet those needs. The two most obvious signals to increase ventilation are a decreased arterial P_{O_2} or an increased arterial P_{CO_2}. Intuitively, you would suspect that if O_2 levels in the arterial blood declined or if CO_2 accumulated, ventilation would be stimulated to obtain more O_2 or to eliminate the excess CO_2. These two factors do indeed influence the magnitude of ventilation in animals, but not to the same degree nor through the same pathway. Also, a third chemical factor, H$^+$, has a notable influence on the level of respiratory activity in some species. We examine the role of each of these important chemical factors in the control of ventilation (■ Table 11–4).

Role of Decreased Arterial P_{O_2} in Regulating Ventilation

In mammals arterial P_{O_2} is monitored by peripheral chemoreceptors known as the **carotid bodies** and **aortic bodies,** which are located at the bifurcation of the common carotid arteries and in the arch of the aorta, respectively (● Figure 11–42). These chemoreceptors, which respond to specific changes in the *chemical content* of the arterial blood that bathes them, are distinctly different from the carotid sinus and aortic arch baroreceptors located in the same vicinity. The latter, being important in regulating systemic arterial blood pressure, monitor pressure changes rather than chemical changes. In fishes, chemoreceptors resembling those of mammals are located diffusely throughout the gills.

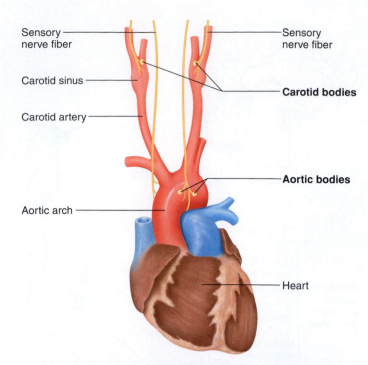

Figure 11–42 ● Location of the peripheral chemoreceptors (mammal). The carotid bodies are located in the carotid sinus, and the aortic bodies are located in the aortic arch.

In fish, a reduction in blood P_{O_2} is detected by the peripheral chemoreceptors, leading to a marked increase in the amplitude and frequency of gill ventilation. However, the chemoreceptors of air breathers are not sensitive to modest reductions in arterial P_{O_2}. The arterial P_{O_2} must fall below 60 mm Hg (>40% reduction) before the chemoreceptors respond by sending afferent impulses to the medullary inspiratory neurons, thereby reflexly increasing lung ventilation. Because arterial P_{O_2} only falls below 60 mm Hg in the unusual circumstances of severe pulmonary disease or reduced atmospheric P_{O_2}, it does not play a role in the normal ongoing regulation of respiration of most air breathers. This fact might seem surprising at first, because one of the primary functions of ventilation is to provide sufficient O_2 for uptake by the blood. However, there is no need to increase ventilation until the arterial P_{O_2} falls below 60 mm Hg, because of the safety margin in % Hb saturation afforded by the plateau portion of the O_2–Hb curve. Hemoglobin is still 90% saturated at an arterial P_{O_2} of 60 mm Hg, but the % Hb saturation drops precipitously when the

Table 11–4 ■ Influence of Chemical Factors on Mammalian Respiration		
Chemical Factor	Effect on the Peripheral Chemoreceptors	Effect on the Central Chemoreceptors
↓ P_{O_2} in the arterial blood	Stimulates only when the arterial P_{O_2} has fallen to the point of being life-threatening (<60 mm Hg); an emergency mechanism	Directly depresses the central chemoreceptors and the respiratory center itself when <60 mm Hg
↑ P_{CO_2} in the arterial blood (↑ H$^+$ in the brain ECF)	Weakly stimulates	Strongly stimulates; is the dominant control of ventilation (Levels > 70–80 mm Hg directly depress the respiratory center and central chemoreceptors)
↑ H$^+$ in the arterial blood	Stimulates; important in acid–base balance	Does not affect; cannot penetrate the blood–brain barrier

P_{O_2} falls below this level. Therefore, reflex stimulation of respiration by the peripheral chemoreceptors serves as an important emergency mechanism in dangerously low arterial P_{O_2} states. Indeed, this reflex mechanism is a life saver, because a low arterial P_{O_2} tends to directly depress the respiratory center, as it does all the rest of the brain (● Figure 11–43). Except for the peripheral chemoreceptors, the level of activity in all nervous tissue becomes reduced in the face of O_2 deprivation. Were it not for stimulatory intervention of the peripheral chemoreceptors when the arterial P_{O_2} falls threateningly low, a vicious cycle ending in cessation of breathing would ensue. Direct depression of the respiratory center by the markedly low arterial P_{O_2} would further reduce ventilation, leading to an even greater fall in arterial P_{O_2}, which would even further depress the respiratory center until ventilation ceased and death occurred.

Because the peripheral chemoreceptors respond to the P_{O_2} of the blood, *not* the total O_2 content of the blood, O_2 content in the arterial blood can fall to dangerously low or even fatal levels without the peripheral chemoreceptors ever responding to reflexly stimulate respiration. Remember that only physically dissolved O_2 contributes to blood P_{O_2}. The total O_2 content in the arterial blood can be reduced in anemic states, in which O_2-carrying Hb is reduced, or in carbon monoxide (CO) poisoning, when the Hb is preferentially bound to this molecule rather than to O_2. In both cases, arterial P_{O_2} is normal, so respiration is not stimulated, even though O_2 delivery to the tissues may be so reduced that the animal dies from cellular O_2 deprivation.

Role of Increased Carbon-Dioxide–Generated H⁺ in Regulating Ventilation

In contrast to arterial P_{O_2}, which does not contribute to the minute-to-minute regulation of respiration in mammals, arterial P_{CO_2} is the most important input regulating the magnitude of ventilation under resting conditions. This role is appropriate in mammals, because changes in alveolar ventilation have an immediate and pronounced effect on arterial P_{CO_2}, whereas changes in ventilation have little effect on % Hb saturation and O_2 availability to the tissues until the arterial P_{O_2} falls by more than 40%. Even slight alterations from normal in arterial P_{CO_2} induce a significant reflex effect on ventilation. An increase in arterial P_{CO_2} reflexly stimulates the respiratory center, with the resultant increase in ventilation promoting elimination of the excess CO_2 to the atmosphere. Conversely, a fall in arterial P_{CO_2} reflexly reduces the respiratory drive. The subsequent decrease in ventilation allows metabolically produced CO_2 to accumulate so that P_{CO_2} can be returned to normal.

Most fish also respond to an increase in P_{CO_2} and acidosis by increasing ventilation. Researchers believe that the response to CO_2 is mediated by the branchial chemoreceptors. Further, mechanoreceptors located along the respiratory surfaces of fish also play a role in respiration. Branchial mechanoreceptors are important in regulating the position of the gill during the respiratory cycle and interact with chemoreceptors to control the transition from active breathing to ram ventilation.

Surprisingly, given the key role of arterial P_{CO_2} in regulating respiration in air breathers, most species have no important receptors that monitor arterial P_{CO_2} per se. The carotid and aortic bodies are only weakly responsive to changes in arterial

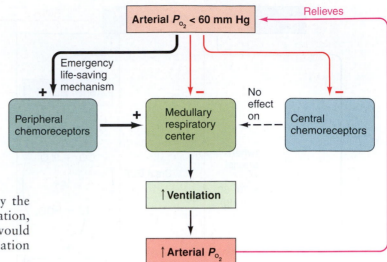

Figure 11–43 ● Effect of threateningly low arterial P_{O_2} (< 60 mm Hg) on ventilation (mammal).

P_{CO_2}, so they play only a minor role in reflexively stimulating ventilation in response to an elevation in arterial P_{CO_2}. In birds and reptiles, however, intrapulmonary chemoreceptors sense P_{CO_2}, although they are unusual in that their discharge rate is inversely proportional to pulmonary P_{CO_2} concentrations.

More important in linking changes in arterial P_{CO_2} to compensatory adjustments in ventilation are the **central chemoreceptors,** located in the medulla in the vicinity of the respiratory center. These central chemoreceptors do not monitor CO_2 itself, however; they are sensitive to changes in CO_2-induced H⁺ concentration in the brain extracellular fluid (ECF) that bathes them. However, little evidence suggests that fish have any central receptors and so researchers believe the response to P_{CO_2} is mediated solely by the peripheral chemoreceptors. An endogenously active branchial rhythm generator drives gill breathing in fish and some species of amphibians. Thus the branchial chemoreceptors of fish, unlike the peripheral carotid and aortic body chemoreceptors of mammals, can respond vigorously to both hypoxia and hypercapnia. Further, air-breathing fish are dependent on this afferent input to initiate air-breathing cycles, because sectioning of the cranial nerves to the gills abolishes this pattern. Central chemoreceptors appeared to have evolved in the Sarcopterygians (lungfish), whereas in amphibians the pattern was the conversion of episodic to a more continuous breathing pattern.

Movement of materials across the brain capillaries is restricted by the blood–brain barrier (see p. 515). Because this barrier is readily permeable to CO_2, any increase in arterial P_{CO_2} causes a similar rise in brain ECF P_{CO_2} as CO_2 diffuses down its pressure gradient from the cerebral blood vessels into the brain ECF. The increased P_{CO_2} within the brain ECF causes a corresponding increase in the concentration of H⁺ according to the law of mass action as it applies to this reaction: $CO_2 + H_2O \leftrightarrow H_2CO_3 \leftrightarrow H^+ + HCO_3^-$. An elevation in H⁺ concentration in the brain ECF directly stimulates the central chemoreceptors, which in turn increase ventilation by stimulating the respiratory center through synaptic connections (● Figure 11–44). As the excess CO_2 is subsequently

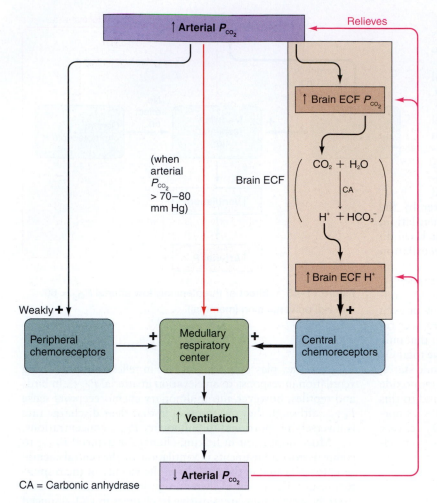

CA = Carbonic anhydrase

Figure 11–44 • Effect of increased arterial P_{CO_2} on ventilation (mammal).

While a person holds his or her breath, metabolically produced CO_2 continues to accumulate in the blood and subsequently to build up the H^+ concentration in the brain ECF. Finally, the increased P_{CO_2}-H^+ stimulant to respiration becomes so powerful that central chemoreceptor excitatory input overrides voluntary inhibitory input to respiration, so breathing resumes despite deliberate attempts to prevent it. Breathing resumes long before arterial P_{O_2} falls to the threateningly low levels that trigger the peripheral chemoreceptors. Therefore, you cannot deliberately hold your breath long enough to create a dangerously high level of CO_2 or low level of O_2 in the arterial blood.

Direct Effect of a Large Increase in P_{CO_2}

In contrast to the normal reflex stimulatory effect of the increased P_{CO_2}-$H+$ mechanism on respiratory activity, very high levels of CO_2 directly depress the entire brain, including the respiratory center, just as very low levels of O_2 do. Up to a P_{CO_2} of 70 to 80 mm Hg, progressively higher P_{CO_2} levels induce correspondingly more vigorous respiratory efforts in an attempt to blow off the excess CO_2. A further increase in P_{CO_2} beyond 70 to 80 mm Hg, however, does not further increase ventilation but actually depresses the respiratory neurons. For this reason, CO_2 must be chemically removed and O_2 supplied in closed environments such as closed-system anesthesia machines, submarines, or space capsules, and must not be allowed to accumulate in animal burrows. Otherwise, CO_2 could reach lethal levels, not only because of its depressant effect on respiration but also because of the resultant severe state of respiratory acidosis.

blown off, the arterial P_{CO_2} and the P_{CO_2} and H^+ concentration of the brain ECF are returned to normal. Conversely, a decline in arterial P_{CO_2} below normal is paralleled by a fall in P_{CO_2} and H^+ in the brain ECF, the result of which is a central-chemoreceptor–mediated decrease in ventilation. As CO_2 produced by cellular metabolism is consequently allowed to accumulate, arterial P_{CO_2} and P_{CO_2} and H^+ of the brain ECF are restored toward normal.

Unlike CO_2, H^+ cannot readily permeate the blood–brain barrier, so H^+ present in the plasma cannot gain access to the central chemoreceptors. Accordingly, the central chemoreceptors are responsive only to H^+ generated within the brain ECF itself as a result of CO_2 entry. Thus the major mechanism controlling ventilation under resting conditions is specifically aimed at regulating the brain ECF H^+ concentration, which in turn is a direct reflection of the arterial P_{CO_2}. Unless there are extenuating circumstances such as reduced availability of O_2 in the inspired air, arterial P_{O_2} is coincidentally also maintained at its normal value by the brain ECF H^+ ventilatory driving mechanism.

The powerful influence of the central chemoreceptors on the respiratory center is responsible for a human's inability to deliberately hold the breath for more than about a minute.

Adjustments in ventilation in response to changes in arterial H^+ are important in acid–base balance.

Changes in arterial H^+ concentration cannot influence the central chemoreceptors because H^+ does not readily cross the blood–brain barrier. However, the aortic and carotid body peripheral chemoreceptors of mammals are highly responsive to fluctuations in arterial H^+ concentration, in contrast to their weak sensitivity to deviations in arterial P_{CO_2} and their unresponsiveness to arterial P_{O_2} until it falls 40% below normal.

Any change in arterial P_{CO_2} brings about a corresponding change in the H^+ concentration of the blood as well as of the brain ECF. These CO_2-induced H^+ changes in the arterial blood are detected by the peripheral chemoreceptors; the result is reflex stimulation of ventilation in response to an increase in arterial H^+ concentration and depression of ventilation in association with a decrease in arterial H^+ concentration. However, these changes in ventilation mediated by the peripheral chemoreceptors are far less important than the powerful central-chemoreceptor mechanism in adjusting ventilation in response to changes in CO_2-generated H^+ concentration.

The peripheral chemoreceptors do play a major role in adjusting ventilation in response to alterations in arterial H^+

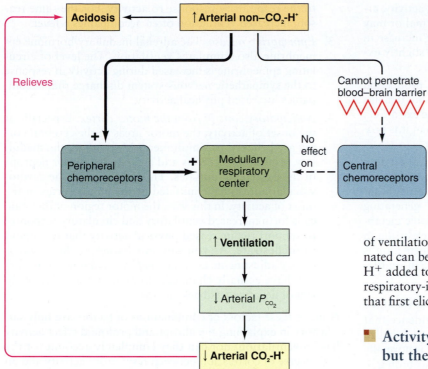

Figure 11–45 ● Effect of increased arterial non–carbonic-acid–generated hydrogen ion (non–CO₂-H⁺) on ventilation (mammal).

concentration unrelated to fluctuations in P_{CO_2}. In many situations, even though P_{CO_2} is normal, arterial H⁺ concentration is changed by the addition or loss of noncarbonic acid from the body. For example, muscles produce lactic acid during anaerobic activity (p. 54). Also, arterial H⁺ concentration increases during diabetes mellitus because excess H⁺-generating keto acids are abnormally produced and added to the blood. A rise in arterial H⁺ concentration reflexly stimulates ventilation by means of the peripheral chemoreceptors. Conversely, the peripheral chemoreceptors reflexly suppress respiratory activity in response to a fall in arterial H⁺ concentration resulting from nonrespiratory causes. Changes in ventilation by this mechanism are extremely important in regulating the acid–base balance of the body (see Chapter 13). By changing the magnitude of ventilation, the amount of H⁺-generating CO₂ that is eliminated can be varied. The resultant adjustment in the amount of H⁺ added to the blood from CO₂ can compensate for the nonrespiratory-induced abnormality in arterial H⁺ concentration that first elicited the respiratory response (● Figure 11–45).

■ Activity profoundly increases ventilation, but the mechanisms involved are unclear.

Alveolar ventilation may increase up to 20-fold during strenuous activity to keep pace with the increased demand for O_2 uptake and CO_2 output. (■ Table 11–5 highlights the changes in O_2- and CO_2-related variables that occur during activity.) The cause of increased ventilation during intense activity is still largely speculative. It would seem logical that changes in the "big three" chemical factors—decreased P_{O_2}, increased P_{CO_2}, and increased H⁺—could account for the increase in ventilation. This does not appear to be the case, however.

Table 11–5 ■ Oxygen and Carbon Dioxide–Related Variables during Activity

O₂- or CO₂-Related Variable	Change	Comment
O₂ use	Marked increase	Active muscles are oxidizing nutrient molecules more rapidly to meet their increased energy needs
CO₂ production	Marked increase	More actively metabolizing muscles produce more CO₂
Alveolar ventilation	Marked increase	By mechanisms not completely understood, alveolar ventilation keeps pace with or even slightly exceeds the increased metabolic demands during exercise
Arterial P_{O_2} **Arterial P_{CO_2}**	Normal or slight ↑ Normal or slight ↓	Despite a marked increase in O₂ use and CO₂ production during exercise, alveolar ventilation keeps pace with or even slightly exceeds the stepped-up rate of O₂ consumption and CO₂ production
O₂ delivery to muscles	Marked increase	Although arterial P_{O_2} remains normal, O₂ delivery to muscles is greatly increased by the increased blood flow to exercising muscles accomplished by increased cardiac output coupled with local vasodilation of active muscles
O₂ extraction by muscles	Marked increase	Increased use of O₂ lowers the P_{O_2} at the tissue level, which results in more O₂ unloading from hemoglobin; this is enhanced by ↑ P_{O_2}, ↑ H⁺, and ↑ temperature
CO₂ removal from muscles	Marked increase	The increased blood flow to exercising muscles removes the excess CO₂ produced by these more actively metabolizing tissues
Arterial H⁺ concentration		
Mild to moderate activity	Normal	Because carbonic acid–generating CO₂ is held constant in arterial blood, arterial H⁺ concentration does not change
Intensive activity	Modest increase	In intensive activity, when muscles resort to anaerobic metabolism, lactic acid is added to the blood

■ Despite the marked increase in O_2 use during activity, arterial P_{O_2} does not decrease but remains normal or may actually increase slightly. This is because the increase in alveolar ventilation keeps pace with or even slightly exceeds the stepped-up rate of O_2 consumption.

■ Likewise, despite the marked increase in CO_2 production during activity, arterial P_{CO_2} does not increase but remains normal or decreases slightly. This is because the extra CO_2 is removed as rapidly or even more rapidly than it is produced as a result of the increase in ventilation.

■ During mild or moderate activity, H^+ concentration does not increase because H^+-generating CO_2 is held constant. During intense activity, H^+ concentration does increase somewhat because of the release of H^+-generating lactic acid into the blood as a result of anaerobic metabolism in the exercising muscles. Even so, the elevation in H^+ concentration resulting from lactic-acid formation is not sufficient to account for the large increase in ventilation accompanying activity.

Some investigators argue that the constancy of the three chemical regulatory factors during activity is evidence that ventilatory responses to activity are actually being controlled by these factors—particularly by P_{CO_2}, because it is normally the dominant control during resting conditions. According to this reasoning, how else could alveolar ventilation be increased in exact proportion to CO_2 production, thereby keeping the P_{CO_2} constant? This proposal, however, cannot account for the observation that during intense activity, alveolar ventilation may be increased relatively more than the increase in CO_2 production, thereby actually causing a slight decline in P_{CO_2}. Also, ventilation increases abruptly at the onset of activity (within seconds), long before changes in arterial blood gases could become important influences on the respiratory center (which requires a matter of minutes).

Researchers have suggested that a number of other factors, including the following, play a role in the ventilatory response to activity:

1. *Reflexes originating from body movements.* Joint and muscle receptors excited during muscle contraction reflexly stimulate the respiratory center, abruptly increasing ventilation. Even passive movement of the limbs may increase ventilation several-fold through activation of these receptors, even though no actual activity is occurring. Thus the mechanical events of activity are believed to play an important role in coordinating respiratory activity with the increased metabolic requirements of the active muscles.

2. *Increase in body temperature.* Much of the energy generated during muscle contraction is converted to heat. Heat-loss mechanisms such as sweating are frequently unable to keep pace with the increased heat production that accompanies increased physical activity, so body temperature often increases slightly, stimulating ventilation. This activity-related heat production undoubtedly contributes

to the respiratory response to activity. For the same reason, increased ventilation often accompanies a fever.

3. *Epinephrine release.* The adrenal medullary hormone epinephrine also stimulates ventilation. The level of circulating epinephrine is increased during activity in response to the sympathetic nervous-system discharge that accompanies increased physical activity.

4. *Anticipatory control from the motor cortex.* Especially at the onset of activity, the motor areas of the cerebral cortex are believed to simultaneously stimulate the medullary respiratory neurons and activate the motor neurons of the exercising muscles. This is similar to the cardiovascular adjustments initiated by the motor cortex at the onset of activity. In this way, the motor region of the brain calls forth increased ventilatory and circulatory responses to support the increased physical activity that it is about to orchestrate. As you saw in Chapter 1, these *anticipatory* adjustments are taken *before* any homeostatic factors have actually changed, eliminating the delay problem of negative feedback (p. 15)

None of these factors or combinations of factors are fully satisfactory in explaining the abrupt and profound effect activity has on ventilation, nor can they completely account for the high degree of correlation between respiratory activity and an animals' needs for gas exchange during activity.

Chapter in Perspective:
HOMEOSTASIS AND INTEGRATION

The respiratory system contributes to homeostasis by obtaining O_2 from and eliminating CO_2 to the external environment. Animal cells ultimately need an adequate supply of O_2 to use in oxidizing nutrient molecules to generate ATP. Brain cells, which are especially dependent on a continual supply of O_2, die if deprived of O_2 for more than a few minutes in most mammals. Even cells that can resort to anaerobic ("without O_2") metabolism for energy production, such as strenuously exercising muscles, can do so only transiently by incurring an O_2 debt that ultimately must be repaid).

As a result of these energy-yielding metabolic reactions, large quantities of CO_2 are produced that must be eliminated from the body. Because CO_2 and H_2O form carbonic acid, adjustments in the rate of CO_2 elimination by the respiratory system are important in the regulation of acid–base balance in the internal environment (Chapter 13). Cells can survive only within a narrow pH range.

Respiratory systems can also play a role in other physiological processes, such as fluid and solute balances, feeding, heat transfer, sound production, and immunity. ■

REVIEW QUESTIONS *(Answers are on p. A–1.)*

Additional study tools for this chapter, including chapter summaries and practice tests, are available online at *www.biology .brookscole.com*

1. According to Fick's Law
   a. the greater the diffusion coefficient, the greater the rate of diffusion
   b. the greater the surface, the greater the rate of diffusion
   c. the greater the gas gradient, the greater the rate of diffusion
   d. the greater the distance that must be covered, the slower the rate of diffusion
   e. all of the above

2. In which of the following would you expect the concentration of oxygen to be greatest?
   a. air at the top of Mt Everest
   b. freshwater at 0° C
   c. seawater at 0° C
   d. freshwater at 20°C
   e. seawater at 20° C

3. The greatest enhancement for moving gases in large animals is
   a. diffusion
   b. bulk transport
   c. active transport by cell membranes
   d. flattened body morphology
   e. fluid and solute balance

4. Which of the following would most probably not be found in air-respiring animals?
   a. tracheae
   b. alveoli
   c. book lungs
   d. opercula
   e. bronchioles

5. Among air-respiring animals, a major challenge is
   a. the high viscosity of air
   b. the relatively low oxygen content of air compared to that of water
   c. keeping respiratory tissues moist
   d. the dramatically lower solubility of oxygen in air at lower temperatures
   e. the higher concentration of carbon dioxide in air compared to that of water

6. Alveoli in mammalian lungs
   a. decrease the distance for diffusion
   b. increase the gas gradient
   c. decrease the gas gradient
   d. increase the surface area across which diffusion takes place
   e. a and d

7. In contrast to mammals, birds have
   a. larger and more elastic alveoli
   b. a flow-through air movement
   c. a muscular diaphragm
   d. blind-ending air capillaries
   e. a larger mean thickness of the blood-gas barrier

8. In humans, which of the following usually has a pressure lower than atmospheric pressure?
   a. barometric pressure
   b. intrapulmonary pressure

   c. intrapleural pressure
   d. intra-alveolar pressure
   e. none of the above

9. In humans, during inspiration in the breathing cycle
   a. the diaphragm contracts
   b. the phrenic nerve becomes inactive
   c. external intercostal muscles relax
   d. muscles of the abdominal wall relax
   e. internal intercostal muscles relax

10. Pulmonary surfactant
    a. increases the alveoli's tendency to recoil
    b. causes collapsing pressure to be inversely proportional to alveolar radius
    c. increases the surface tension of water
    d. tends to counteract the large surface tension of water
    e. reduces pulmonary compliance

11. The volume of air entering or leaving the lungs during a single breath is called the
    a. vital capacity
    b. tidal volume
    c. total lung capacity
    d. inspiratory capacity
    e. residual volume

12. In mammals, during a pattern of shallow, rapid breathing
    a. alveolar ventilation exceeds pulmonary ventilation
    b. dead-space volume is less than dead-space volume during deep, slow breathing
    c. tidal volume is less than it is during quiet breathing at rest
    d. pulmonary ventilation is greater than it is at deep, slow breathing
    e. dead-space volume doubles

13. Blood leaving the lungs in pulmonary capillaries has a
    a. partial pressure of oxygen less than the partial pressure of alveolar oxygen
    b. partial pressure of oxygen greater than the partial pressure of alveolar oxygen
    c. partial pressure of oxygen equal to the partial pressure of alveolar oxygen
    d. partial pressure of carbon dioxide equal to the partial pressure of alveolar oxygen
    e. partial pressure of oxygen less than the partial pressure of alveolar carbon dioxide

14. Which of the following is true of arterial blood in systemic capillaries?
    a. Oxygen in the blood of systemic capillaries moves into tissues against a pressure gradient.
    b. Oxygen in the blood of systemic capillaries moves into tissues down a pressure gradient.
    c. Carbon dioxide is actively transported from tissues into systemic blood by capillary membranes.
    d. The partial pressures of oxygen and carbon dioxide are the reverse of blood leaving the lungs in alveolar capillaries
    e. none of the above

15. In mammals,
    a. most oxygen is transported as a dissolved gas
    b. deoxyhemoglobin transports carbon dioxide

c. the iron-containing group of hemoglobin is composed of proteins

d. oxyhemoglobin is reversibly bound to oxygen

e. hemoglobin is converted to myoglobin in the mitochondria of muscle cells

16. For hemoglobin (Hb), a high P_{50} value

a. means that the Hb has a high affinity for O_2

b. would likely be found in species adapted to living at high altitudes

c. would likely be found in species adapted to living at or near sea level

d. means that the Hb has a reduced affinity for O_2

e. makes it more difficult to unload O_2 at tissues

17. Hemoglobin promotes the net transfer of O_2 from the alveoli to the blood by

a. contributing to the partial pressure of dissolved O_2

b. removing dissolved oxygen from solution and converting it to a non-diffusable form

c. increasing the partial pressure of oxygen at the tissue level

d. increasing its affinity for oxygen at tissue levels

e. acting as a second messenger to promote active uptake of oxygen by alveolar membranes

18. The Bohr effect

a. is the reverse of the Root effect

b. increases the acidity in systemic capillaries

c. influences the affinity of hemoglobin for oxygen

d. expresses the effect of temperature on unloading of O_2 at tissue levels

e. describes the effects of organic phosphates on hemoglobin's affinity for oxygen

19. The majority of CO_2 in both invertebrates and vertebrates is transported as

a. a dissolved gas

b. carbonic anhydrase

c. CO_2 gas bound to the heme group of hemoglobin

d. carbonic acid

e. carbonic chloride

20. In mammals, the partial pressure of arterial blood is measured by the

a. phrenic nerve

b. corotid bodies

c. respiratory center

d. aortic bodies

e. b and d

SUGGESTED READINGS AND INTERNET SITES

Art, T., W. Bayly, & P. Lekeux. 2002. Pulmonary function in the exercising horse. In P. Lekeux, ed., *Equine Respiratory Diseases.* Ithaca, NY: International Veterinary Information Service. Available online at www.ivis.org/special_books/Lekeux/art/chapter_frm.asp?LA=1. Accessed on March 3, 2004.

Evans, D. H., ed. 1998. *The Physiology of Fishes,* 2nd ed. New York: CRC Marine Science Series.

Hochachka, P. W. 1980. *Living without Oxygen.* Cambridge, MA: Harvard University Press.

Routly, M., G. Nilsson, & G. Renshaw, G. 2002. Exposure to hypoxia primes the respiratory and metabolic responses of the epaulette shark to progressive hypoxia. *Comparative Biochemistry and Physiology* 131A:313.

Welty, J. C., & L. Baptista. 1990. *The Life of Birds,* 4th ed. Fort Worth, TX: Saunders.

INTERNET SITE

Corvin, C., & W. McClure. 1997. *Oxygen Binding.* **stingray.bio.cmu.edu/~web/bc1/oxy_anim/oxy_Anim.html.** Animated plotting of hemoglobin-oxygen binding.

INFOTRAC READINGS

Venis, S. 2001. Neuroglobin might protect brain cells during stroke. *The Lancet* 358:2055.

CHAPTER

12

Excretory Systems

Photo: Paul Yancey

Cormorants and harbor seals. The seabird has a kidney that excretes nitrogenous waste as uric acid, a metabolically costly but nontoxic compound that requires little water to excrete, and that has antioxidant properties. The animal imbibes some seawater (which is saltier than its body fluids), and it has special nasal salt glands to excrete excess NaCl. The mammal also must excrete excess NaCl and nitrogenous waste (primarily urea). Urea is a semitoxic compound that requires some water to excrete. The seal has kidneys that are very efficient at concentrating salt and urea.

Introduction

■ **Selective excretion is crucial to internal fluid homeostasis and involves several systems.**

One of the most crucial aspects of homeostasis is the maintenance of an internal aqueous environment with a consistent composition of water and important solutes, and with a low level of wastes. Dietary intake, metabolic products, and losses or excess intake of water and ions frequently create imbalances in the internal environment. Consider a female mosquito that has just sucked blood from your skin. The animal is bloated with a relatively huge amount of water and NaCl (the major components of your blood), making it more difficult to fly and potentially disturbing internal fluid movements. The insect also digests your nutritious blood proteins (especially hemoglobin), gaining useful products but producing unneeded nitrogenous by-products (wastes). Thus she must somehow remove excess water, salt, and wastes from her body. Or consider the kangaroo rat, a rodent that inhabits deserts of North America, where it may never have access to free water and where its plant food is high in K^+ but low in Na^+. This animal must conserve both water and NaCl, yet excrete K^+ and wastes. Finally, consider a seabird (see the photograph). Because seawater is saltier than body fluids of birds (and most vertebrates), it tends to dehydrate cells by osmosis. Thus this animal may need to excrete excess salt (and wastes) while conserving body water.

In these and other animals, homeostasis of salt and water and excretion of wastes depend on mechanisms of selective retention and removal of molecules. These mechanisms are generally the purpose of *excretory systems*. (Note that the term "excretory" is a bit misleading, because these systems often perform selective retention as well; but retention always accompanies excretion.) In particular, the primary functions of these organs are as follows:

1. *Maintenance of proper internal levels of inorganic solutes* (Na^+, K^+, Cl^-, H^+, CO_2, and so forth). Even minor fluctuations in the **ECF** (extracellular fluid) concentrations of some of these electrolytes can have profound influences. For example, changes in the ECF concentration of K^+ can potentially lead to fatal cardiac dysfunction.

2. *Maintenance of proper plasma water volume*, important for proper circulatory-fluid pressure and general state of tissue hydration.

3. *Removal of nonnutritive and harmful substances resulting from metabolism* (ammonia, urea, bilirubin, and so forth) *or ingestion* (such as plant alkaloids, drugs), and *removal of hormones* after they have had their desired effect. This must be done without losing useful organic molecules such as nonhormone proteins and glucose.

4. *Maintenance of osmotic balance* (which results from a combination of water and dissolved solutes) by selective retention and excretion of water and ions.

Simple aquatic animals (sponges, cnidarians, and possibly echinoderms) do not have specialized organs for these processes. They rely on diffusion and membrane transporters for nonsolid wastes. (Individual cells of freshwater sponges also have *contractile vacuoles,* specialized vesicles that remove and eject excess water out of the cell.) As aquatic animals evolved larger sizes and greater complexity, diffusion and membrane transport became inadequate for removing wastes from internal body fluids. Thus the evolution of specialized excretory tissues was favored. In some cases, *epithelial surfaces* of existing organs such as skin, gills, and hindguts evolved specific mechanisms to remove nonfecal wastes and to regulate other solutes and water. However, in large animals these surfaces may not be able to process all body fluids or molecules. Evolution onto land created more problems. Because of the lack of an aquatic environment to absorb wastes and provide water directly in terrestrial animals, external epithelia cannot regulate most solutes or water. To be free to move about in a dry and ever-changing external environment, these animals had to evolve more elaborate mechanisms to maintain internal fluid homeostasis. Thus larger and more complex aquatic animals, and all terrestrial ones, evolved specialized *tubules* dedicated to excreting and selectively retaining key molecules of the body. Overall in the animal kingdom, four organ systems can be involved in excretion and retention:

1. *Respiratory systems (gills and lungs),* which help regulate CO_2 (although we don't often think of it in these terms, the act of exhaling CO_2 is a form of waste excretion!). Gills can also help remove other solutes such as ammonia and HCO_3^-, in some cases by simple diffusion and in others with membrane transporters. Some of these functions are covered in Chapter 11 (on respiratory physiology) and Chapter 13 (on fluid and acid–base balance).

2. *Digestive systems,* which remove not only undigested food but also some end products of internal metabolism such as bilirubin from the vertebrate liver. Indeed, the liver in this case is acting as an excretory organ. Intestinal tracts may also help regulate ions and water, and in some animals the gut receives the output of renal organs. Some of these functions are covered in Chapter 14 (on digestion).

3. *Integument (skin) and glands,* some of which can excrete organic wastes as secondary functions (such as sweat glands in humans), and others that evolved primarily to excrete (via active transport) excess inorganic ions (**salt glands** of brine shrimp, marine birds, and reptiles). The skin of many amphibians also regulates salt and water uptake.

4. *Renal organs* (such as crustacean antennal glands, insect Malpighian tubules, and vertebrate kidneys) with tubules that *filter* body fluids and regulate water, ions, and many organic substances, and then *selectively reabsorb or secrete* these molecules. The output of renal organs is usually called *urine.* These organs, together with ductwork and any storage chambers (such as bladders) for the urine, are called *urinary systems.*

In the next chapter, we focus on how these and other systems work in an integrated way to maintain water and solute homeostasis. In this chapter, we focus primarily on renal organs and, to a lesser extent, skin and its glands: how they process water, ions, and wastes, and how they are regulated. We begin here by examining a major form of waste that is produced by nitrogen metabolism.

◾ Nitrogen metabolism creates special stresses and produces three major end products: ammonia, urea, and uric acid.

Metabolism of all foodstuffs results in CO_2 as a waste product, but protein and nucleic acid metabolism produces various **nitrogenous wastes** as well. These lead to a number of problems. The most common problem results from amino acid metabolism, which yields **ammonia** (NH_3), a strong base that alters acid–base balance as it binds protons and becomes **ammonium ion** (NH_4^+). The ammonium ion can also be toxic; among other effects, it can interfere with the Na^+/K^+ ATPase transporters of cell membranes by substituting for K^+. This toxicity manifests in many ways but is particularly evident in neurons, which suffer morphological changes, interrupted ion conduction, and disrupted neurotransmitter metabolism. Just 0.05 mM ammonia in mammals and 2 mM in fishes can severely impair neuron function. Thus ammonia or ammonium must either be highly diluted and rapidly excreted, or be converted into less toxic forms. The most common of these forms are **urea** and **uric acid** (● Figures 12–1 and 12–2). Let's examine these wastes.

1. *Ammonia.* Most aquatic animals that breathe water—including most bony fishes, larval amphibians, and most invertebrates—rely on ammonia excretion, usually via gills. These animals are called **ammonotelic.** Because water is plentiful to dilute ammonia, and because this molecule (small and uncharged) readily penetrates most membranes, it is relatively easy to keep it below its toxic level (NH_3, not NH_4^+, is the primary excretory form). However, most terrestrial animals do not have this option and instead must spend metabolic energy to convert ammonia to urea or uric acid, usually in the liver, and transport the product to the excretory organs. A clear example of this mechanism is found in amphibians, whose aquatic larval stages use ammonia excretion via their gills, whereas semiaquatic adults switch to urea metabolism and excretion using their livers and kidneys. Similarly, aquatic birds such as ducks excrete an evenly balanced mixture of uric acid and ammonia, whereas terrestrial birds excrete primarily uric acid.

2. *Urea.* Urea is the primary nitrogenous waste not only of most adult amphibians but also mammals (and some reptiles such as sea turtles). These animals are called **ureotelic.** Urea is produced, in most vertebrates that make it, from two ammonium ions and a bicarbonate ion, using ATP, in the **ornithine–urea cycle** (Figure 12–2a). (Most bony fishes that produce urea use the **uricolytic pathway,** Figure 12–2b.) Urea is the primary nitrogenous waste of marine cartilaginous fishes and the coelacanth, in which it is also accumulated as their major **osmolyte** at several hundred mM (Chapter 13). Urea is 10 to 100 times less toxic than ammonia and thus can be accumulated to much higher concentrations, and has the added benefit of removing two nitrogens per molecule. Thus it takes about 10 times less water to excrete a given amount of nitrogen as urea than as ammonia. Urea is, however, toxic at concentrations greater than 100 mM or so, because it binds

Figure 12–1 • General overview of nitrogen metabolism and excretion in animals. The three main nitrogen excretory products are highlighted in boxes.

(*Source:* P. Wright, 1995, Nitrogen excretion: Three end-products, many physiological roles, *Journal of Experimental Biology* 198:273–281. Used by permission of The Company of Biologists.)

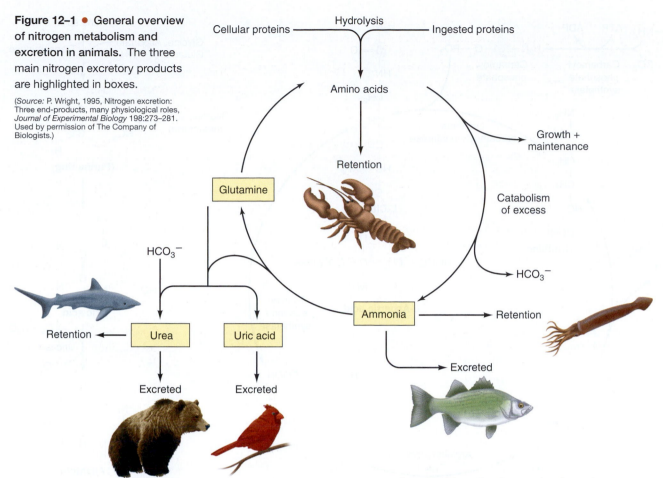

to proteins and destabilizes their structures and interferes with their ability to bind ligands. Thus it must be kept at moderate levels, or, where it is highly concentrated in cartilaginous fishes and the mammalian inner kidney, it must be counteracted with a protein stabilizer (p. 579, and see box, "Molecular Biology and Genomics: Osmolytes and Gene Regulation").

In ruminants, urea has another function. After synthesis in the liver, it can enter the rumen via the salivary glands, where symbiotic microbes ferment the food; there, microbes (Chapter 14) convert urea back to ammonia, using the enzyme **urease**. The ammonia can then be used directly by these beneficial microbes for their own protein synthesis. Domestic ruminants on low-protein diets are sometimes given urea (an inexpensive chemical) in their feed to support this process. Other mammals such as horses and rabbits use urea similarly for microbes in the cecum or colon, where food is fermented.

3. *Uric acid.* In other terrestrial animals—insects, birds, most reptiles, and even a tree frog and a desert toad—uric acid is usually the primary nitrogenous waste (although some insects excrete some urea and ammonia). These are called **uricotelic** animals. Uric acid is the end product of amino acid and purine (nucleic acid) metabolism (Figure 12–2b) and requires even more ATP than urea (Figure 12–2a) to produce. It is less toxic because it is highly insoluble, has

the added benefit of removing four nitrogens per molecule, and is excreted in a semisolid form, generally into the hindgut. (In the physiological pH range, 99% of uric acid exists as urate, the monovalent anionic form.) It takes about 50 to 100 times less water to excrete a given amount of nitrogen as uric acid than as ammonia. Uric acid also serves as a protective antioxidant (see the box, "A Closer Look at Adaptation: Avian Longevity: Unraveling the Mystery"). Mammals also make uric acid, but most only synthesize small amounts that are removed by the kidneys (though long-lived primates lack the enzyme uricase and may accumulate modest amounts of uric acid).

An interesting question arises from these patterns: if uric acid is less toxic and more favorable to water conservation than urea, why don't all terrestrial animals use it as the primary nitrogenous waste? One possibility lies in the high metabolic cost of synthesizing uric acid compared to urea (12 or more ATPs versus 4 ATPs/mole, respectively), and different modes of reproduction. Birds, reptiles, and many insects lay sealed eggs, often in fairly dry areas. Without a way to remove toxic nitrogen products, such eggs may require the least toxic (but most expensive) form, which also exerts no osmotic disturbances in its crystallized state. Adult insects, because of their small sizes and thus high surface-area-to-volume ratios (p. 8), may also need the most effective water conservation mechanisms. In contrast, amphibian adults (which usually enter water frequently) and embryos and fetuses of placental mammals (whose wastes are removed by the placenta) can more easily lose their nitrogenous wastes. Thus using urea as a pri-

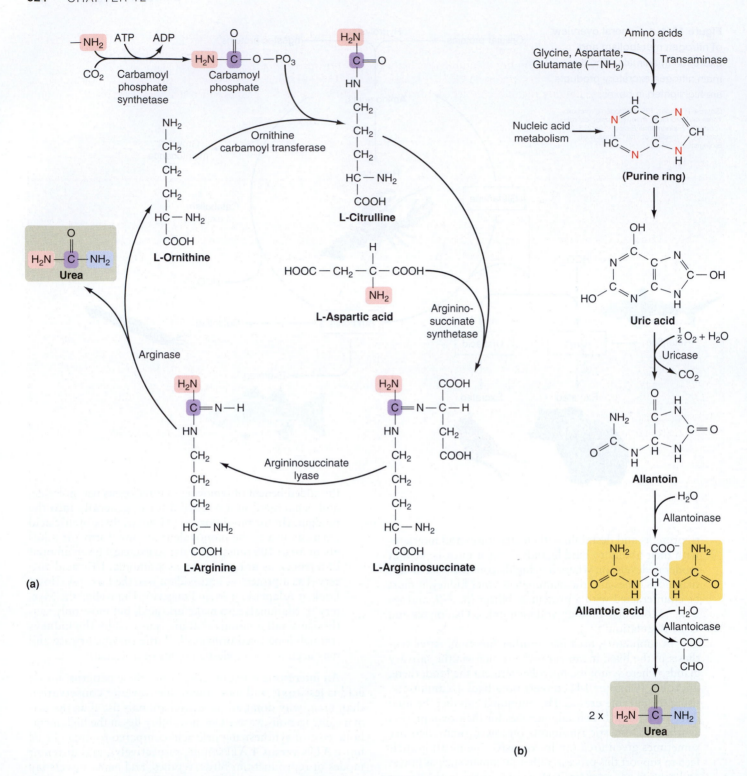

Figure 12–2 ● (a) Urea is formed by the ornithine–urea cycle in most vertebrates. Because ATP is required for the first step, nitrogen excretion in ureotelic animals is more energetically expensive than in other animals. (b) The uricolytic pathway produces both uric acid and urea. Uric acid arises from a purine ring that is synthesized by a complex union of aspartic acid, formic acid, glycine, and CO_2. Humans lack the enzymes needed to break down uric acid and thus excrete uric acid as the end product of nucleic acid metabolism.

(*Source:* (a), (b), M. K. Campbell & S. O. Farrell, 2003, *Biochemistry,* 4th ed., Belmont, CA: Brooks/Cole, p. 657.)

mary waste saves considerably on metabolic energy of synthesis. In any case, mammals have not evolved mechanisms to handle high concentrations of the highly insoluble uric acid, and excessive levels can lead to arthritic **gout** and **kidney stones** because of crystals of this highly insoluble molecule forming

in inappropriate places (the joints and kidney tubules, respectively). See the box, "A Closer Look at Adaptation: Avian Longevity: Unraveling the Mystery."

Overall, then, it appears that the use of ammonia, urea, or uric acid as the primary nitrogen waste correlates strongly

---

MOLECULAR BIOLOGY AND GENOMICS

Osmolytes and Gene Regulation

As we discussed in Chapter 3, osmosis and cell volume regulation are among the most crucial processes facing cells in general. Consider the external osmotic environment of a nephron cell in the inner medulla of the mammalian kidney. The countercurrent mechanism creates a variable concentration of NaCl, dependent on the hydrational status of the animal. In a dehydrated state, the NaCl concentration increases to very high levels, threatening the cells with osmotic shrinkage. How do the cells cope with this? As you will see in detail in Chapter 13, cells of most organisms subjected to high osmotic stress accumulate organic solutes called organic **osmolytes** to maintain cell volume. NaCl cannot be used at high levels in cells because it disrupts proteins. The mammalian kidney is no exception—to prevent massive osmotic water loss, renal cells exclude excess NaCl and instead accumulate high levels of the polyols **sorbitol** and **myoinositol,** the sulfur amino acid **taurine,** and the methylamines **betaine** and **glycerophosphorylcholine (GPC).** These are the same chemical types of osmolytes used by marine algae and animals to balance the high osmotic pressure of seawater (Chapter 13). The levels of these organic osmolytes have been measured in renal medullas of a variety of mammals including rats, humans, and desert rodents. In all cases, the total concentrations of these solutes appears to be regulated to match the osmotic pressure of the medulla fluids: Animals with a high water intake have reduced levels of renal osmolytes compared to those with low intake.

Why are polyols, taurine, and methylamines used as osmolytes? Such organic osmolytes are in general *compatible* with cellular macromolecules, that is, they do not perturb protein functions and are therefore safe to accumulate in cells over a wide concentration range. The methylamine osmolytes probably provide another benefit because they stabilize proteins. Urea, also concentrated by nephrons, is also a protein denaturant, but because of high membrane permeability, it equilibrates inside and outside most medulla cells. Rats on a high-protein diet, whose kidneys must concentrate very high levels of urea (Table 12–4), have a correspondingly high level of GPC (but not of other osmolytes) in their medulla cells. Why might this be? As discussed in Chapter 13, the methylamine osmolyte TMAO in cartilaginous fishes can *counteract* the deleterious effects of urea on proteins (p. 579). Similarly, researchers have found that GPC can protect mammalian renal proteins from urea's effects.

How do renal cells exposed to high external NaCl regulate the accumulation of osmolytes? Sorbitol is produced from glu-

cose by the enzyme **aldose reductase (AR),** whereas betaine and myoinositol are accumulated from external fluids via membrane proteins called the *betaine/gamma-amino butyric acid transporter* or **BGT1,** and *sodium-dependent myoinositol transporter* or **SMIT,** respectively. Osmotic stress leads to the activation of genes for AR, BGT1, and SMIT. Physiologists now know these genes share a highly similar regulatory promoter (Chapter 2) called the **osmotic response element (ORE)** or **tonicity enhancer (TonE).** A similar sequence is found in the promoter region of the gene for **oxytocin (OT),** which, like **vasopressin,** is produced by the hypothalamus during dehydration. A recent study with a "knockout" mouse (p. 32) lacking the *OT* gene indicates that OT suppresses salt appetite.

These are the actual nucleotide codes "upstream" from the protein-coding genes:

Canine BGT1 TonE	*TGGAAAAGTCC* . . . [coding gene for BGT1]
Human AR (type C) ORE	*TGGAAAATCAC* . . . [coding gene for AR-C]
Rabbit AR ORE	*CGGAAAATCAC* . . . [coding gene for AR]
Human BGT1 TonE	*TGGAAAATTAC* . . . [coding gene for TonE]
Human SMIT TonE-A	*TGGAAAACTAC* . . . [coding gene for SMIT]
Rat OT promoter	*GAAAAATCAC* . . . [coding gene for OT]

Recent studies have detected an **ORE-binding** or **TonE-binding protein (ORE-BP** or **TonE-BP)** which is a **transcription factor** (Chapter 2), that is, a regulatory protein that binds to the OREs and TonEs in a renal cell's chromosomes and probably activates transcription of all these osmolyte genes simultaneously. Indeed, researchers have found that ORE-BPs reside in the cytoplasm under low osmotic pressures, but they rapidly move to the nucleus under high osmotic stress. An increasing salt concentration in the blood might also lead to the activation of the vasopressin and oxytocin (OT) genes in the hypothalamus as well. How the osmotic stress is initially sensed and transduced into a signal that activates the ORE-BP is still unknown.

A CLOSER LOOK AT ADAPTATION

Avian Longevity: Unraveling the Mystery

Birds live considerably longer than mammals of comparable body size. This mystery has intrigued the scientific community for as long as it has been aware of the paradox. Given certain inherently avian characteristics, prevailing theories suggest that birds should be more susceptible than mammals to the degenerative process of aging. These include the following: Birds have plasma glucose concentrations typically 2 to 6 times higher than those of mammals, metabolic rates as much as 2 to 2.5 times higher than those of similar-sized mammals, and a body temperature approximately 3°C higher. The higher metabolic rates, coupled with the extreme longevity of many avian species, result in a much greater lifetime energy expenditure per unit mass by birds than mammals. This is important because, within a class of animals, longevity is inversely correlated with metabolic rate (as you will see in Chapter 15, p. 673). In mammals, physiologists assume that lifetime energy expenditure correlates with an organism's cumulative exposure to potentially damaging free radicals (and other reactive oxygen species, or **ROS;** see p. 49). An animal's body produces these destructive molecules as normal by-products of mitochondrial metabolism. In addition, immune cells generate ROS with which they destroy pathogens. Normally this is beneficial, but any abnormal activation of the immune cells, as found in certain disease states, can lead to the constant generation of ROS that damage or even destroy the body's own cells. Thus the increased avian metabolic rate should expose birds to a higher concentration of ROS and, consequently, accelerated tissue damage.

In addition, the high glucose levels (along with high body temperature) should promote diabetic damage. Consider the situation in hummingbirds. Hummingbirds have blood sugar levels much higher than do people with Type II (noninsulin-dependent) *diabetes mellitus.* The tissues of human and other mammalian diabetics (diabetes is a disease not uncommon in domestic dogs and cats), as a consequence of the elevated blood-sugar levels and oxidative stress, are susceptible to tissue-specific complications that manifest over time. Diabetes is the primary cause of blindness, kidney failure, and limb amputation in America and contributes to cardiovascular disease. It has been well established that the higher the blood sugar in a diabetic, the earlier the onset of complications. Do birds suffer a similar fate to that found for diabetics?

First, how does glucose damage the body's tissues and organs? One mechanism begins with the formation of advanced **Maillard products:** glucose covalently bound to proteins (glycosylation). This process may be involved in normal aging-related tissue degeneration, and is accelerated in diabetes. (Interestingly, glucose is considered among the least reactive of the sugars, which may be the reason why animals rely on this particular carbohydrate for its energy needs.) Maillard products are further modified by free radicals in an animal's body, which contribute to the formation of irreversible cross-links between adjacent proteins. This impairs protein functions; for example, the process thickens basement membranes in the kidney, which reduces oxygen diffusion. Research has shown the diabetic population has an elevated concentration of free radicals and oxidative stress as a result of an impaired immune system triggered into action by glycosylated tissue proteins.

---

with water availability. As we have seen, most aquatic animals use ammonia, whereas most animals with terrestrial eggs use uric acid. But there are exceptions, for example, some marine snails use uric acid, monotremes (egg-laying mammals) use urea, and the most water-stressed mammals, desert rodents such as kangaroo rats (which may never drink water in their entire lives) use urea (see p. 560). "Physiological logic" would suggest that these rodents would benefit from uric acid as their major nitrogenous waste. Perhaps evolutionary constraints (p. 2) have resulted in mammals "locked in" to using urea, because hindgut interaction seems to be essential for uric acid processing.

There are a few other forms of nitrogenous wastes. Three of these are **guanine,** the primary nitrogenous waste of arachnids; **allantoin** (Figure 12–2b), produced from uric acid in some insects; and **creatinine,** a by-product of vertebrate muscle metabolism that is constantly produced, transported, and removed by the kidneys. As discussed later, (p. 551) renal physiologists and physicians use this process to measure key renal functions.

In brief summary, metabolism can produce nitrogenous wastes, which must be excreted. Often this is a key role of renal organs, which may also excrete (or selectively retain) many other molecules. We now turn our attention to those organs.

Renal Excretory Organs

▪ Renal tubules produce urine using the processes of filtration, secretion, reabsorption, and osmoconcentration.

Renal organs have tubules that filter solutes and water from body fluids and subsequently modify the fluid in the tubular lumen, specifically for water and solute homeostasis. As we explain in detail, they operate by two or more of four basic renal processes (● Figure 12–3):

1. *Filtration,* in which solutes are selectively separated from a solution by passing through a boundary. Often this is a mostly nonselective process, in which some water with all its small solutes passes through the barrier, except that cells and large molecules such as proteins (and some water) remain behind, a process called **ultrafiltration.** Ultrafiltration can be driven by hydrostatic pressure or concentration differences across the boundary. Another type of filtration relies on specific transporters and is thus more selective.

2. *Secretion,* in which specific solutes are transported into the tubule lumen for excretion.

What about the hummingbirds? With their elevated metabolism and higher body temperatures, these birds actually generate substantially more free radicals than a comparably sized mammal, yet remarkably, the concentration of glucose-derived cross-links is considerably lower than in mammals. How can this be?

A clue arises from the dual role of **uric acid,** a nitrogenous waste product (p. 523) that may also play a role in the body's antioxidant system. Uric acid is a minor waste product in mammals, but as you have seen in this chapter, it is the major nitrogenous waste in birds (and most reptiles, some amphibians, and insects). These animals are uric acid "factories." However, uric acid is not just a waste product: It has been found important in combating oxidative stress and functions something like a catcher's mitt that can "pick off" any stray free radicals missed by the other antioxidant systems (see p. 50). A reduction in uric acid concentrations in chickens dramatically increases the rate of tissue aging, actually to a rate found in diabetic humans! Birds maintained on a lowered blood concentration of uric acid for as little as two months show a highly significant increase in oxidative stress and tissue cross-link formation, whereas the opposite effect is observed if uric acid concentrations are increased.

What happens if uric acid concentrations rise too high? This results in a condition called **gout.** Uric acid is highly insoluble and so when its concentrations increase above a certain threshold, it begins to come out of solution and crystallize in various tissues. For example, uric acid crystals forming in kidney nephrons can permanently impair kidney function. For unknown reasons this increases plasma renin concentrations, which ultimately raises

blood pressure. Colder temperatures enhance crystallization, thus uric acid tends to form painful urate crystals in the toes, fingers, and edge of the ears in primates. Gout has been documented in some other mammals and in some reptiles. How then can birds tolerate much higher concentrations of uric acid in their tissues and blood? (In fact, both domestic and wild birds can get gout, but not at rates higher than mammals.) One obvious answer is the much higher body temperature of birds, which increases the solubility of uric acid. The second is the unusual reaction that occurs in the nephron of the avian kidney. As soon as uric acid begins to crystallize in the avian kidney, the microvilli in the proximal tubule encapsulate the crystal with a protein coat forming a highly soluble colloidal particle. These spheres are not physically destructive and serve to transport the uric acid crystal safely through the nephron. The colloidal suspension then passes through the ureters into the cloaca of the bird. Here the spheres are refluxed back into the ceca, pouchlike attachments off the large intestine, and microorganisms recover the protein. Most of the uric acid is also degraded in the ceca. Specific transporters located in the epithelial cells of the large intestine take up any free amino acids and sugars.

Because birds have a high concentration of blood glucose, it has been suggested that they evolved a mechanism to limit the associated damage. This mechanism includes a more efficient antioxidant system that takes the form of an apparently simple waste product: uric acid. This discovery may have important clinical applications for treating conditions associated with inflammatory stress.

Figure 12–3 • Schematic of the principal functions of a tubular excretory organ, such as the vertebrate nephron: filtration, secretion, and reabsorption. The avian and mammalian nephron can also osmoconcentrate urine.

(*Source:* P. C. Withers, 1992, *Comparative Animal Physiology,* Belmont, CA: Brooks/Cole, Figure 17–9, p. 841.)

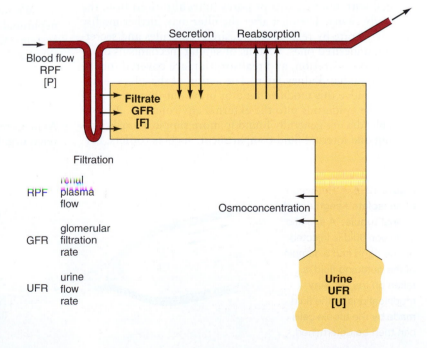

3. *Reabsorption,* in which specific solutes and water may be transported back into the body from the lumen.

4. *Osmoconcentration,* in which water is removed from the lumen fluid while leaving solutes behind, producing an excretion more concentrated (*hyperosmotic*) than body fluids and thus conserving water. Some reabsorption processes are tied to this mechanism.

Methodology

Numerous techniques are used to explore the functions of renal excretory systems. Here are two that have contributed greatly to the understanding of renal functions, which we explore shortly:

- For studying the tubules we have just discussed, perhaps the most important method is that of the **isolated perfused tubule** (● Figure 12–4). A researcher begins by carefully dissecting an intact tubule (nephron, Malpighian, and so forth) out of the animal and placing it in an appropriate bathing solution. Two holding pipettes are applied, one to each open end of the tubule. Within one pipette, a much finer pipette (the perfusion pipette) is inserted to allow injection of defined solutions into the tubule. A fine pipette within the other holding pipette is used to collect fluid exiting the tubule. By measuring solutes in this collected fluid and in the bathing solution, researchers can determine transport and permeability features. Other devices may also be used on these tubules, such as patch clamps (p. 104) and microelectrodes for measuring potentials.

- In studying the effect of the whole kidney on retention and removal, the **plasma clearance** method has been important. To understand this method, you first need to understand certain tubule functions; we return to plasma clearance later.

Several different kinds of renal organs have evolved, and there are several ways to classify these. One useful scheme recognizes three common types based on which body fluids are processed and how:

- **Protonephridia** use ultrafiltration driven by cilia, secretion, and reabsorption. These are the types of specialized filtration system thought to have evolved first. These are generally blind-end ducts that project into the body cavity (of which there is only one: a general interstitial space, a pseudocoelom, or a coelom). They have low (fluid) hydrostatic pressure inside them relative to the body cavity, most often because of the beating of cilia in the duct, which move fluid outward. This low pressure draws water and small solutes nonselectively in through a specialized cell with filtering slits or pores (ultrafiltration) from the body cavity. The duct after the filter may further modify the filtrate by reabsorbing desired molecules and secreting undesired ones. (Details of similar mechanisms of filtration, secretion, and reabsorption are covered for the mammalian kidney later.) The action of the cilia eventually move this modified fluid or urine out through a pore in the epidermis into the external environment. Protonephridia are generally found in more simple animals having one internal fluid compartment, such as rotifers, flat-

worms, larval annelids, and larval mollusks. An example is shown in ● Figure 12–5a.

- The second type of excretory tubules are **metanephridia**, which use ultrafiltration typically driven by fluid pressure, secretion, reabsorption, and sometimes osmoconcentration. Metanephridia are tubules in animals with two or more major fluid spaces—a coelom and a circulatory fluid at a minimum. The tubules produce and modify a luminal fluid by initially ultrafiltering the circulatory fluid (such as blood in vessels), typically driven by hydrostatic pressure. (The filtered circulatory fluid temporarily passes into a local coelomic space, then into the tubule lumen.) These tubules are generally found in larger, more complex animals, and have evolved independently several times—in adult mollusks, in annelids (segmented worms), in some arthropods (such as crustaceans), and in vertebrates. The tubules begin with a filtering capsule or funnel-like opening of some kind, are followed by tubules that perform selective secretion and reabsorption (and osmoconcentration in some groups) and may empty via a connecting **ureter** into a hindgut or a special storage chamber, a **bladder.**

  In some cases, a single filtering tubule forms the renal organ, as in the (misnamed) *crustacean antennal gland* (Figure 12–5b). Each single tubule (one on each side of the head) begins with a capsule (end sac) that filters hemolymph; enters a *labyrinth* of epithelia and fluid spaces followed by a coiled *nephridial canal,* where selective absorption and secretion take place; and then empties into a bladder. In other animals, many metanephridial tubules are assembled together with connective tissue into large organs called **kidneys.** In vertebrate kidneys, the tubules are called **nephrons.**

- **Malpighian tubules** begin filtration driven by secretion of ions from an arthropod's hemolymph (Chapter 9), and then, in conjunction with the hindgut, use reabsorption and sometimes osmoconcentration. These will be discussed in detail in the end of this chapter.

We now turn to the mammalian version of the metanephridial system. We then look at other vertebrate excretory systems and conclude with a detailed look at Malpighian tubules.

The Mammalian Urinary System

As lungs replaced gills in the evolution of vertebrates onto land, renal organs evolved new functions, in part because (among

Figure 12–4 ● Perfusion of an isolated segment of a renal tubule. A specific perfusion fluid is injected on the right, and a sample of the outgoing fluid is taken on the left. Any modifications to the fluid made by the tubule cells can then be determined.

Oil

Perfusion fluid

Suction

Tubule

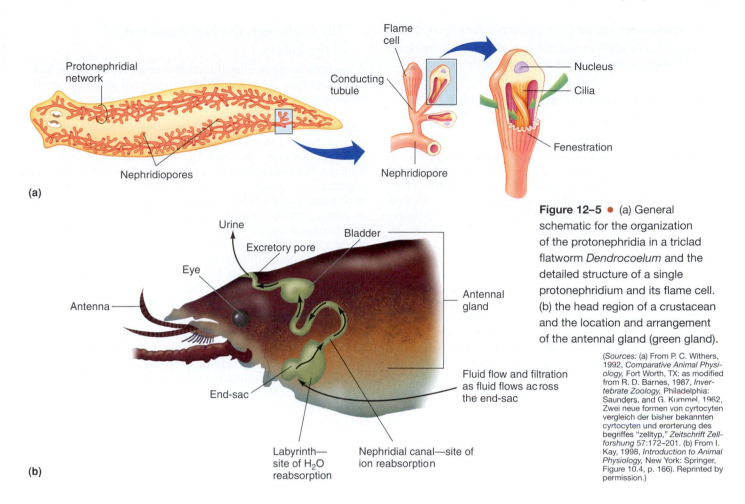

Figure 12–5 ● (a) General schematic for the organization of the protonephridia in a triclad flatworm *Dendrocoelum* and the detailed structure of a single protonephridium and its flame cell. (b) the head region of a crustacean and the location and arrangement of the antennal gland (green gland).

(*Sources:* (a) From P. C. Withers, 1992, *Comparative Animal Physiology,* Fort Worth, TX: as modified from R. D. Barnes, 1987, *Invertebrate Zoology,* Philadelphia: Saunders, and G. Kummel, 1962, Zwei neue formen von cyrtocyten vergleich der bisher bekannten cyrtocyten und erorterung des begriffes "zelltyp," *Zeitschrift Zellforshung* 57:172–201. (b) From I. Kay, 1998, *Introduction to Animal Physiology,* New York: Springer, Figure 10.4, p. 166). Reprinted by permission.)

other things) gills but not lungs can excrete ammonia/ammonium and regulate NaCl. To a large extent, mammals can live on dry land independent of the sea because of their kidneys, which, in concert with the hormonal and neural inputs that control their function, have become the primary maintainers of ECF volume and electrolyte (ion) composition. By adjusting the quantity of water and various plasma constituents that are either conserved for the body or eliminated in the urine, the kidneys can maintain water and electrolyte balance within the very narrow range compatible with life, despite wide variations in intake and losses of these constituents through other avenues.

When there is a surplus of water or a particular electrolyte such as salt (NaCl) in the ECF, mammalian kidneys can eliminate the excess in the urine. If there is a deficit in the body, kidneys cannot actually provide additional quantities of the depleted constituent, but in birds and mammals they can limit the urinary losses of the material in short supply and thus conserve it until more of the depleted substance can be ingested. Accordingly, kidneys can compensate more efficiently for excesses than for deficits, and cannot completely halt the loss of a particular valuable substance in the urine, even though the substance may be in short supply.

In addition to their important regulatory role in maintaining fluid and electrolyte balance, kidneys are the primary route for eliminating potentially toxic metabolic wastes (such as from nitrogen metabolism) and foreign compounds from the body. Most wastes usually cannot be eliminated in solid form; they must be excreted in solution. Because the H_2O eliminated in the urine is derived from the blood plasma, animals stranded without H_2O eventually urinate themselves to death by depleting the plasma volume to a fatal level as H_2O is unavoidably removed to accompany the wastes. However, except under such extreme circumstances, the kidneys can maintain stability in the internal fluid environment despite the usual variations in intake of fluids and electrolytes.

Not only can kidneys adjust for wide variations in ingestion of H_2O, salt, and other electrolytes, but they also compensate for abnormal losses from thermoregulatory evaporation (sweating, panting), vomiting, diarrhea, or hemorrhage. Thus urine composition varies widely as the kidneys adjust for differences in intake as well as losses of various substances in an attempt to maintain the ECF within the narrow limits required by mammalian cells.

Kidneys therefore are involved in the four major functions, noted at the very beginning of the chapter (p. 521): regulating (1) major inorganic solutes (including ions involved in acid–base balance), (2) plasma volume, (3) harmful or unneeded organic molecules, and (4) osmotic balance. In mammals, kidneys have the following additional functions:

5. *Secretion of erythropoietin,* a hormone that stimulates red blood cell production (see p. 365)

6. *Secretion of renin,* an enzymatic hormone that triggers a chain reaction important in salt conservation by the kidneys (as you will see later)

7. *Conversion of vitamin D into its active form* (p. 308)

8. *Excretion of pheromones* for sexual signaling, marking territories, and so forth (in some mammals)

■ The kidneys form the urine; the remainder of the urinary system is the bladder and ductwork that carries the urine.

The **urinary system** consists of the urine-forming organs—the kidneys—and the structures that carry the urine from the kidneys to the outside for elimination from the body (● Figure 12–6a). Mammalian kidneys are a pair of bean-shaped organs that lie in the dorsal side of the lower abdominal cavity, one on each side of the vertebral column. (Only the mammalian kidney has this "kidney-bean" shape.) Each kidney is supplied by a renal artery and a renal vein. Urine drains into the ureters (a duct walled with smooth muscle), but in mammals, before doing so it first collects into a central cavity, the renal **pelvis**, at the inner core of each mammalian kidney (Figure 12–6b). There are two ureters, one carrying urine from each kidney to the urinary bladder (in fish, amphibians, and mammals) or hindgut (in reptiles and birds), which temporarily stores urine (and in some animals may modify it). Periodically, urine is emptied from the bladder to the outside through another tube, the **urethra** (which in male mammals serves both as a route for eliminating urine and as a passageway for semen from the testes).

■ The nephron is the functional unit of the kidney.

A **nephron** is the smallest functional unit capable of the kidney's basic functions. (Recall that a functional unit is the smallest structure within an organ capable of performing all of that organ's functions, although the renal pelvis may modify the urine after it leaves all the nephrons). Nephrons are found in all vertebrate kidneys, and each kidney consists of numerous nephrons (for example, about 1.25 million in pigs, 1 million in humans, 0.5 million in dogs, and 0.25 million in cats) bound together by connective tissue. The arrangement of nephrons within mammalian kidneys gives rise to two distinct regions—an outer, granular-appearing region, the **renal cortex**, and an inner, striated region called the **renal medulla** (Figure 12–6b). Some animals (generally small ones such as rodents and rabbits) have a **unipapillary medulla**, a single, striated, cone-shaped body (the **papilla**). Larger mammals (such as dogs and humans) have a **compound medulla**, containing several striated-appearing triangles, the renal **pyramids** (Figure 12–6b), which may also be called *papillae*.

Knowing the structural arrangement of an individual nephron is essential for understanding the distinction between the cortical and medullary regions of the kidney and, more importantly, for understanding renal function. Each nephron consists of the tubules and an associated *vascular* (blood vessel) component, both of which are intimately related structurally and functionally (● Figure 12–7). In most vertebrates, the

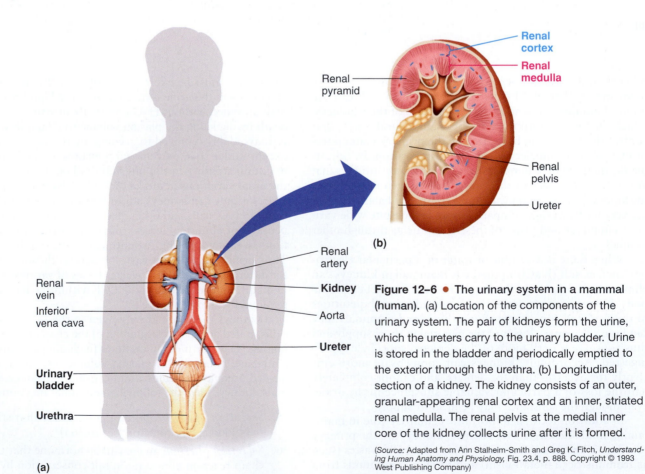

Labels for (b): Renal cortex, Renal medulla, Renal pyramid, Renal pelvis, Ureter

Labels for (a): Renal vein, Inferior vena cava, Urinary bladder, Urethra, Renal artery, Kidney, Aorta, Ureter

Figure 12–6 ● The urinary system in a mammal (human). (a) Location of the components of the urinary system. The pair of kidneys form the urine, which the ureters carry to the urinary bladder. Urine is stored in the bladder and periodically emptied to the exterior through the urethra. (b) Longitudinal section of a kidney. The kidney consists of an outer, granular-appearing renal cortex and an inner, striated renal medulla. The renal pelvis at the medial inner core of the kidney collects urine after it is formed.

(*Source:* Adapted from Ann Stalheim-Smith and Greg K. Fitch, *Understanding Human Anatomy and Physiology*, Fig. 23.4, p. 888. Copyright © 1993 West Publishing Company)

dominant portion of the vascular component is the **glomerulus,** a ball-like knot of capillaries through which part of the water and solutes are filtered from the blood passing through. This filtered fluid then passes through the tubules of the nephron. The tubules are hollow, fluid-filled tubes formed by a single layer of epithelial cells. Even though the tubule is continuous, it is specialized into various segments with different structures and functions along its length (● Figures 12–7 and 12–8):

- The tubule segments start with **Bowman's** (or glomerular) **capsule,** an expanded, double-walled invagination that cups around the **glomerulus** to collect the fluid filtered from the glomerular capillaries.

- From this capsule, the filtered fluid passes into the **proximal tubule,** which lies entirely within the cortex and is highly coiled or convoluted throughout much of its course.

- The next segment, the **loop of Henle** (or **nephron loop**), exists only in birds and mammals. Here the tube forms a sharp U-shaped or hairpin loop that dips into the renal medulla. The **descending limb** of Henle's loop plunges from the cortex deep into the medulla; the **ascending limb** traverses back up into the cortex. The ascending limb returns to the glomerular region of its own nephron, where it passes through the fork formed by the afferent and efferent arterioles. Both the tubular and vascular cells at this point are specialized to form the **juxtaglomerular apparatus** or **complex** ("juxta" means "next to"), a structure that lies next to the glomerulus and plays an important role in regulating kidney function.

- Beyond the juxtaglomerular apparatus, the tubule once again becomes highly coiled to form the **distal tubule,** which also lies entirely within the cortex. (This tubule is absent in marine bony fishes.)

- The distal tubule empties into a **collecting duct,** each of which drains fluid from up to eight separate nephrons in mammals. Each collecting duct plunges down through the medulla to empty its fluid contents (which have now been converted into urine) into the renal pelvis in mammals, or into ureters in other vertebrates.

There are two types of nephrons in mammals and birds—**cortical neph-**

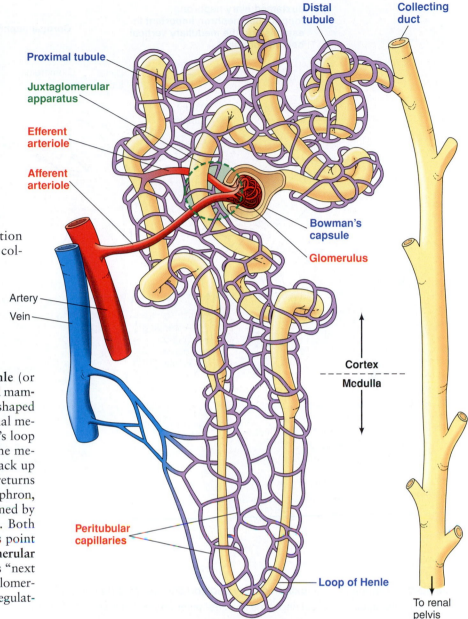

Overview of Functions of Parts of a Nephron (Mammalian)

Vascular component
- Afferent arteriole—carries blood to the glomerulus
- Glomerulus—a tuft of capillaries that filters a protein-free plasma into the tubular component
- Efferent arteriole—carries blood from the glomerulus
- Peritubular capillaries—supplies the renal tissue; involved in exchanges with the fluid in the tubular lumen

Combined vascular/tubular component
- Juxtaglomerular apparatus—produces substances involved in the control of kidney function

Tubular component
- Bowman's capsule—collects the glomerular filtrate
- Proximal tubule—uncontrolled reabsorption and secretion of selected substances occur here
- Loop of Henle—establishes an osmotic gradient in the renal medulla that is important in the kidney's ability to produce urine of varying concentration
- Distal tubule and collecting duct—variable, controlled reabsorption of Na^+ and H_2O and secretion of K^+ and H^+ occur here; fluid leaving the collecting duct is urine, which enters the renal pelvis

Figure 12–7 ● **A nephron in a mammalian kidney.** A schematic representation of a cortical nephron.

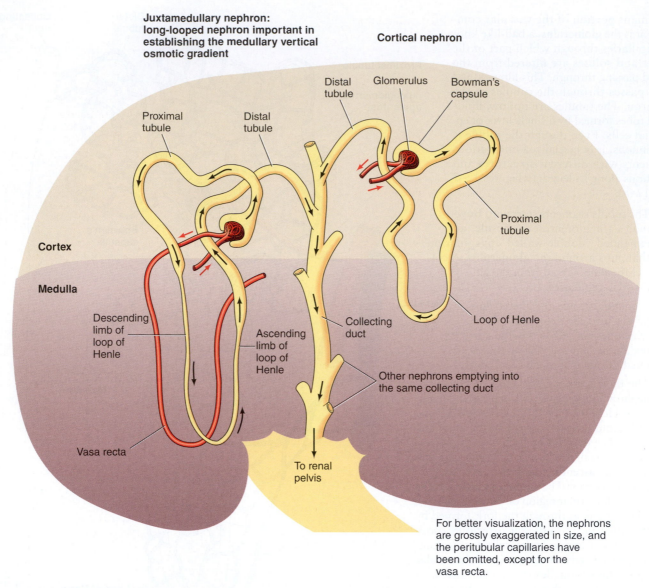

Juxtamedullary nephron: long-looped nephron important in establishing the medullary vertical osmotic gradient

Cortical nephron

Proximal tubule

Distal tubule

Distal tubule

Glomerulus

Bowman's capsule

Cortex

Medulla

Descending limb of loop of Henle

Ascending limb of loop of Henle

Vasa recta

Collecting duct

Other nephrons emptying into the same collecting duct

Proximal tubule

Loop of Henle

To renal pelvis

For better visualization, the nephrons are grossly exaggerated in size, and the peritubular capillaries have been omitted, except for the vasa recta.

Figure 12–8 ● Comparison of juxtamedullary and cortical nephrons (mammalian kidney). The glomeruli of cortical nephrons lie in the outer cortex, whereas the glomeruli of juxtamedullary nephrons lie in the inner part of the cortex next to the medulla. The loops of Henle of cortical nephrons dip only slightly into the medulla, but the juxtamedullary nephrons have long loops of Henle that plunge deep into the medulla. The juxtamedullary nephrons' peritubular capillaries form hairpin loops known as *vasa recta*.

rons and **juxtamedullary nephrons**—distinguished by the location and length of some of their structures (Figure 12–8). (Note the distinction between juxtamedullary nephrons and juxtaglomerular apparatus.) All nephrons originate in the cortex, but the glomeruli of cortical nephrons lie in the outer layer of the cortex, whereas the glomeruli of juxtamedullary nephrons lie in the inner layer of the cortex, adjacent to the medulla. The presence of all glomeruli and associated Bowman's capsules in the cortex is responsible for the region's granular appearance. These two nephron types differ most markedly in their loops of Henle. The hairpin loop of cortical nephrons dips only slightly into the medulla. In contrast, the loop of juxtamedullary nephrons plunges through the entire depth of the medulla.

Now let's look at the associated vascular system. On entering the kidney, the renal artery systematically subdivides to form many small vessels known as **afferent arterioles,** one of which supplies each nephron. The afferent arteriole delivers blood to the **glomerular capillaries,** which rejoin to form another arteriole, the **efferent arteriole,** through which blood that was not filtered into the tubular component leaves the glomerulus (Figure 12–7). The efferent arterioles are the only arterioles in mammals that drain from capillaries (typically, arterioles break up into capillaries that rejoin to form venules). Similar to the arterioles of the hypothalomo-hypophyseal portal system (p. 269), they break up into capillaries that rejoin to form arterioles and not venules. At the glomerular capillaries, no O_2 or nutrients are extracted from the blood for use by the kidney tissues nor are waste products picked up from the surrounding tissue.

The efferent arteriole then subdivides into the **peritubular capillaries,** which supply the renal tissue with blood and are important in exchanges between the tubular system and blood during conversion of the filtered fluid into urine. These capillaries, as their name implies (*peri,* "around"), are intertwined around the tubular system. The capillaries rejoin to form venules that ultimately drain into the renal vein, by which blood leaves the kidney. The peritubular capillaries of juxtamedullary nephrons in mammals form hairpin vascular loops known as **vasa recta** ("straight vessels"), which run in close association with the loops of Henle. As they course through the medulla, the collecting ducts of both cortical and juxtamedullary nephrons run parallel to the ascending and descending limbs of the juxtamedullary nephrons' loops of Henle and vasa recta. This parallel arrangement of tubules and vessels creates the medullary tissue's striated appearance. More importantly, as you will see, this arrangement plays a key role in the kidneys' ability to produce urine of varying concentrations, and in the evolution of mammals to different habitats.

■ The basic renal processes are glomerular filtration, tubular reabsorption, tubular secretion, and osmoconcentration.

All four of the basic renal processes discussed previously are involved in the formation of mammalian urine. To visualize the relationships among these renal processes, it is useful to unwind the nephron schematically, as in Figure 12–3.

1. *Filtration.* In **glomerular filtration,** as blood flows through the glomerulus, blood pressure forces protein-free plasma through the capillaries into Bowman's capsule. This *ultrafiltration* process is the first step in urine formation in most vertebrates, and is normally continuous. In fact, in humans the kidneys filter the entire plasma volume about 65 times per day. If everything filtered were to pass out in the urine, the total plasma volume would be urinated out in less than half an hour! This does not happen, however, because the kidney tubules and peritubular capillaries are intimately related throughout their lengths, so that materials can be transferred between the fluid inside the tubules and the blood within the peritubular capillaries. This filtration process is nondiscriminating, that is, with the exception of blood cells and most plasma proteins, all constituents within the blood—H_2O, nutrients, electrolytes, wastes, dietary toxins, small proteins and peptides such as hormones, and so on—are nonselectively filtered.

2. *Reabsorption.* As the filtrate flows through the tubules, substances of value such as glucose, amino acids, and much of the water are returned to the peritubular capillary plasma. This selective movement of substances from inside the tubule (the tubular lumen) into the blood uses energy and is called **tubular reabsorption.** Reabsorbed substances are not lost from the body in the urine but are carried instead by the peritubular capillaries to the venous system and then to the heart to be recirculated. Of the volume of plasma filtered per day, generally over 90% of the water and 100% of the glucose and amino acids are reabsorbed, with the remaining fluid passing into the renal pelvis to be eliminated as urine. In general, substances that need to be conserved by the body are selectively reabsorbed, whereas unwanted substances that need to be eliminated remain in the urine.

3. *Secretion.* The term *third renal process* refers to the selective transfer of substances from the peritubular capillary blood into the tubular lumen. Specifically called **tubular secretion,** it provides a second route for substances to enter the renal tubules from the blood. Glomerular filtration is the first route; however, only about 20% of the plasma is filtered this way, with the remaining 80% flowing on through the efferent arteriole into the peritubular capillaries. Some substances such as organic ions can be selectively transferred by tubular secretion from the plasma in the peritubular capillaries into the ECF and then into the tubular lumen. Tubular secretion provides a mechanism for more rapidly eliminating selected substances from the plasma by extracting an additional quantity of a particular substance from the (unfiltered) plasma in the peritubular capillaries and adding it to the quantity of the substance already present in the tubule as a result of filtration.

Because there are mechanisms for eliminating unwanted solutes by specific secretion, you might wonder why the apparently wasteful step of glomerular filtration exists. After all, it also loses useful molecules initially, and the kidney has to expend energy to recover them in the tubular reabsorption process. Actually, these three processes together make good evolutionary and functional sense. Indiscriminant filtration helps remove *all* small harmful solutes from the body, including all those for which no specific transporters have evolved, such as novel plant toxins that an animal has not previously encountered. Presumably, although it is possible to evolve some selective-secretion transporters for the most common harmful solutes, it is highly unlikely that sufficient numbers of selective transporters for *every* possible harmful substance could evolve.

4. *Osmoconcentration.* The fourth renal process is not universal in vertebrates, but is a major adaptive feature of avian and mammalian nephrons. In the loop region in the medulla, NaCl (and urea in mammals) becomes highly concentrated, such that the medullary ECF becomes a brine bath. As you will see, this **osmoconcentration** allows the kidneys to recover and conserve more water if necessary, further concentrating the waste products in the urine. This process is often considered an aspect of tubular reabsorption, but as you will see later, it is conceptually useful to think of this as a process separate from the other three.

All plasma constituents that gain access to the tubules—that is, are filtered or secreted—but are not reabsorbed, remain in the tubules and pass into the renal pelvis to be excreted as urine and eliminated from the body. (Do not confuse excretion with secretion.) Conversely, anything filtered and subsequently reabsorbed or not filtered at all enters the venous blood and thus is conserved for the body, despite passage through the kidneys. The kidneys act only on the plasma. However, because of the free exchange between plasma and interstitial fluid across the capillary walls (with the exception of plasma proteins), interstitial fluid composition reflects the composition of plasma. Thus by performing their regulatory and excretory roles on the plasma, the kidneys maintain the proper interstitial fluid environment for optimal cell function. Now we will discuss how the renal processes are accomplished and regulated to help maintain homeostasis. (The processes described are thought to be the same in all mammals, but the particular concentrations and rates given are mainly from humans and laboratory mammals.)

Glomerular Filtration

▪ The glomerular membrane is more than 100 times more permeable than capillaries elsewhere.

Fluid filtered from the glomerulus into Bowman's capsule must pass through the three layers that make up the glomerular membrane (● Figure 12–9). Collectively, these layers function as a fine molecular sieve that retains the blood cells and most plasma proteins but permits H_2O and solutes of small molecular size to filter through, as follows:

1. The outer wall of the glomerular capillaries, which consists of a single layer of flattened endothelial cells, is perforated by many large pores, or **fenestrae,** that render it over 100 times more permeable to H_2O and solutes than capillaries elsewhere in the body.

2. The basement membrane, composed of collagen and glycoproteins, is sandwiched between the glomerulus and Bow-

man's capsule. The collagen provides structural strength, and the glycoproteins discourage the filtration of small plasma proteins. Even though the larger plasma proteins cannot be filtered because they cannot fit through the capillary pores, the pores are actually large enough to permit passage of some albumins, the smallest of plasma proteins. The glycoproteins, however, being strongly negatively charged, repel albumin and other plasma proteins because the latter are also negatively charged. Therefore, plasma proteins are almost completely excluded from the filtrate, with less than 1% of the albumin molecules escaping into the capsule. Some renal diseases characterized by excessive albumin in the urine (albuminuria) are believed to be due in part to disruption of the negative charges within the glomerular membrane, which makes the membrane more permeable to albumin even though the size of the pores remains constant.

3. The inner layer of Bowman's capsule is the final layer; it consists of **podocytes,** octopus-like cells that encircle the glomerular tuft. Each podocyte bears many elongated foot processes (podo means "foot") that interdigitate with foot processes of adjacent podocytes (● Figure 12–10), much as you slide your fingers between each other as you cup your hands around a ball. The narrow slits between adjacent foot processes, known as *filtration slits,* provide a pathway through which fluid exiting the glomerular capillaries can enter the lumen of Bowman's capsule. A protein called *nephrin* (in association with other proteins

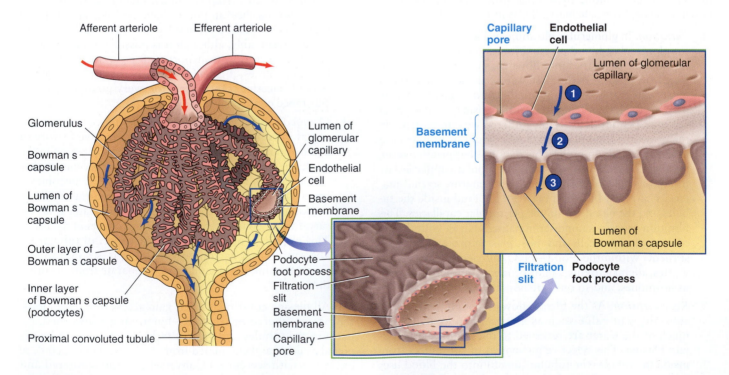

To be filtered, a substance must pass through

1 the pores between the endothelial cells of the glomerular capillary

2 an acellular basement membrane

3 the filtration slits between the foot processes of the podocytes of the inner layer of Bowman s capsule

Figure 12–9 ● Layers of the glomerular membrane (mammalian kidney).

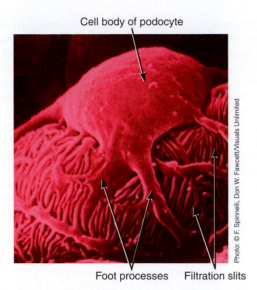

Cell body of podocyte

Foot processes   Filtration slits

Figure 12–10 ● Bowman's capsule podocytes with foot processes and filtration slits (mammalian kidney). Note the filtration slits between adjacent foot processes on this scanning electron micrograph. The podocytes and their foot processes encircle the glomerular capillaries.

such as P-cadherin and podocin) helps form a zipperlike filtering structure in these slits.

Thus the route that filtered substances take across the glomerular membrane is completely extracellular—first through capillary pores, then through the acellular basement membrane, and finally through capsular filtration slits (Figure 12–9).

■ The glomerular capillary blood pressure is the major force responsible for inducing glomerular ultrafiltration.

To accomplish glomerular filtration, a force must be present to drive a portion of the plasma in the glomerulus through the openings in the glomerular membrane. Physical forces similar to those acting across capillaries elsewhere are responsible for glomerular filtration. Because the glomerulus is a tuft of capillaries, the same principles of fluid dynamics that are responsible for ultrafiltration across other capillaries apply (see p. 414), except for two important differences: (1) The glomerular capillaries are much more permeable than capillaries elsewhere, so more fluid is filtered for a given filtration pressure, and (2) the balance of forces across the glomerular membrane is such that filtration occurs throughout the entire length of the capillaries, in contrast to the reabsorption that occurs toward the ends of other capillaries (p. 415).

Four physical forces are involved in glomerular filtration (■ Table 12–1): (1) glomerular capillary blood pressure, (2) plasma-colloid osmotic pressure, (3) Bowman's capsule hydrostatic pressure, and (4) Bowman's capsule osmotic pressure. The glomerular capillary blood pressure is the fluid pressure exerted by the blood within the glomerular capillaries. It ultimately depends on contraction of the heart (the source of energy that produces glomerular filtration) and the resistance to blood flow offered by the afferent and efferent arterioles.

The glomerular capillary blood pressure, at an estimated average value of 55 mm Hg, is higher than capillary blood pressure elsewhere, because the diameter of the afferent arteriole is larger than that of the efferent arteriole. Because blood can more readily enter the glomerulus through the wide afferent arteriole than it can leave through the narrower efferent arteriole, glomerular capillary blood pressure is maintained high by blood damming up in the glomerular capillaries. Furthermore, because of the high resistance offered by the efferent arterioles, blood pressure does not have the same tendency to decrease along the length of the glomerular capillaries as it does along other capillaries. This elevated glomerular blood pressure tends to push fluid out of the glomerulus into Bowman's capsule along the glomerular capillaries' entire length, and it is the major force responsible for producing glomerular filtration.

Whereas glomerular-capillary blood pressure favors filtration, the two other forces acting across the glomerular membrane (plasma-colloid osmotic pressure and Bowman's capsule hydrostatic pressure) oppose filtration. Plasma-colloid osmotic pressure is caused by the unequal distribution of plasma proteins across the glomerular membrane. Because plasma proteins cannot be filtered, they remain in the glomerular capillaries and are absent in Bowman's capsule. Accordingly, the concentration of H_2O is higher in Bowman's capsule fluid than in the glomerular capillaries. The resulting tendency for H_2O to move by osmosis down its own concentration gradient from Bowman's capsule into the glomerulus opposes glomerular filtration. This opposing osmotic force averages 30 mm Hg in mammals, which is slightly higher than across other capillaries. It is higher because considerably more H_2O is filtered out of the glomerular blood, so the concentration of plasma proteins is higher than elsewhere. (As you will see shortly, the plasma-colloid osmotic pressure aids in water reabsorption.)

Table 12–1 ∎ Forces Involved in Glomerular Filtration

Force	Effect	Magnitude (mm Hg)
Glomerular capillary blood pressure	Favors filtration	55
Plasma-colloid osmotic pressure	Opposes filtration	30
Bowman's capsule hydrostatic pressure	Opposes filtration	15
Net filtration pressure (difference between force favoring filtration and forces opposing filtration)	Favors filtration	10

$$55 - (30 + 15) = 10$$

The fluid in Bowman's capsule exerts a hydrostatic (fluid) pressure that is estimated to be about 15 mm Hg. This pressure, which tends to push fluid out of Bowman's capsule, opposes the filtration of fluid from the glomerulus into Bowman's capsule. As can be seen in Table 12–1, there is an imbalance in the forces acting across the glomerular membrane, with a net difference favoring filtration (10 mm Hg of pressure). This modest **net filtration pressure** is responsible for forcing large volumes of fluid from the blood through the highly permeable glomerular membrane.

The actual rate of filtration, the **glomerular filtration rate (GFR)**, depends not only on the net filtration pressure but also on how much glomerular surface area is available for penetration and how permeable the glomerular membrane is (that is, on how "holey" it is). These properties of the glomerular membrane are collectively called the *filtration coefficient* (K_f). Accordingly,

$$GFR = K_f \times \text{net filtration pressure}$$

Normally, about 20% of the plasma that enters the glomerulus is filtered at the net filtration pressure of 10 mm Hg, producing collectively through all glomeruli 170 liters of glomerular filtrate each day for an average GFR of 120 mL/min in humans.

■ The most common factor resulting in a change in the GFR is an alteration in the glomerular capillary blood pressure.

Because the net filtration pressure responsible for inducing glomerular filtration is simply because of an imbalance of opposing physical forces between the glomerular capillary plasma and Bowman's capsule fluid, alterations in any of these physical forces can affect the GFR. Let's look at how changes in each of these physical forces affect the GFR.

Plasma-colloid osmotic pressure and Bowman's capsule hydrostatic pressure are not subject to regulation and, under normal conditions, do not vary substantially. In contrast, glomerular capillary blood pressure can be controlled to adjust the GFR to suit the body's needs. Assuming that all other factors remain constant, the magnitude of the glomerular capillary blood pressure depends on the rate of blood flow in each of the glomeruli, which in turn is determined largely by the magnitude of the mean systemic arterial blood pressure and the resistance offered by the afferent arterioles. GFR is controlled by two mechanisms, both of which are directed at adjusting glomerular blood flow by regulating the diameter and thus the resistance of the afferent arteriole. These are (1) autoregulation, which is aimed at preventing spontaneous changes in GFR, and (2) extrinsic sympathetic control, which is aimed at long-term regulation of arterial blood pressure.

Autoregulation of GFR

Because the arterial blood pressure is the force that drives blood into the glomerulus, the glomerular capillary blood pressure and, accordingly, the GFR would increase in direct proportion to an increase in arterial pressure if everything else remained constant. Similarly, a fall in arterial blood pressure would be accompanied by a decline in GFR. Because at least a certain portion of the filtered fluid is always excreted, the amount of fluid excreted in the urine varies with the GFR. In-

advertent changes in GFR are largely prevented by intrinsic regulatory mechanisms initiated by the kidneys themselves, a process known as *autoregulation* (auto means "self"). The kidneys can, within limits, maintain a constant blood flow into the glomerular capillaries (and thus a constant glomerular capillary blood pressure and a stable GFR) despite changes in the driving arterial pressure. They do so primarily by altering afferent arteriolar caliber, thereby adjusting resistance to flow through these vessels. For example, if the GFR increases as a direct result of a rise in arterial pressure, the net filtration pressure and GFR can be reduced to normal by constriction of the afferent arteriole (● Figure 12–11a), which decreases the flow of blood into the glomerulus. This local adjustment lowers the glomerular blood pressure and the GFR to normal. Conversely, when GFR falls in the presence of a decline in arterial pressure, glomerular pressure can be increased to normal by vasodilation of the afferent arteriole, which allows more blood to enter despite the reduction in driving pressure (Figure 12–11b). The resulting buildup of glomerular blood volume increases glomerular blood pressure, which in turn brings the GFR back up to normal.

The exact mechanisms responsible for accomplishing autoregulatory responses still elude complete understanding. Currently, two intrarenal mechanisms are thought to contribute to autoregulation: (1) a myogenic mechanism, which responds to changes in pressure within the nephron's vascular compo-

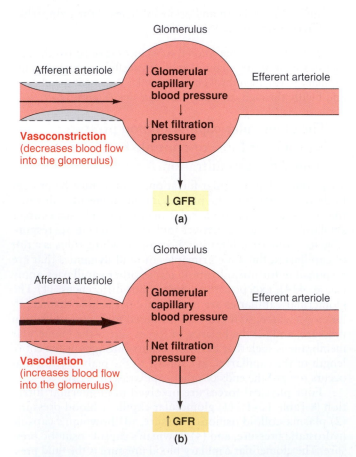

Figure 12–11 ● Adjustments of afferent arteriole caliber to alter the GFR (mammalian kidney). (a) Arteriolar adjustment to reduce the GFR. (b) Arteriolar adjustment to increase the GFR.

nent, and (2) a paracrine tubuloglomerular feedback mechanism, which senses changes in flow through the nephron's tubular component.

The **myogenic mechanism** is a common property of vascular smooth muscle. Arteriolar vascular smooth muscle contracts inherently in response to the stretch accompanying increased pressure within the vessel. Accordingly, the afferent arteriole automatically constricts on its own when it is stretched, because of an increased arterial driving pressure. This response helps limit blood flow into the glomerulus to normal despite the elevated arterial pressure. Conversely, inherent relaxation of an unstretched afferent arteriole when pressure within the vessel is reduced increases blood flow into the glomerulus despite the fall in arterial pressure.

The **tubuloglomerular feedback mechanism** involves the juxtaglomerular apparatus (● Figures 12–7 and 12–12). Within the apparatus's arteriolar walls at their point of contact with the tubule, its smooth muscle cells are specialized to form cells, which contain numerous secretory granules. Specialized tubular cells in this region are collectively known as the **macula densa.** The macula densa cells detect changes in the rate at which fluid is flowing past them through the tubule. If the GFR is increased after an elevation in arterial pressure, more fluid than normal is filtered and reaches the distal tubule (● Figure 12–13). In response, the macula densa cells trigger the release of vasoactive paracrines from the juxtaglomerular apparatus, which in turn cause the afferent arteriole to constrict, reducing glomerular blood flow and returning GFR to normal. The exact nature of these locally released paracrines is cur-

rently unclear. Several have been identified, such as the vasoconstrictor *endothelin,* but their exact contributions are yet to be determined.

In the opposite situation, when the macula densa cells detect that the flow rate of fluid through the tubules is low because of a decline in GFR accompanying a fall in arterial pressure, these cells induce afferent arteriolar vasodilation by altering the rate of secretion of the relevant vasoactive paracrines. Again, some vasodilators such as *bradykinin* have been identified, but most have not. The resulting increase in glomerular flow rate restores the GFR to normal. Thus by means of the juxtaglomerular apparatus, the tubule of a nephron can monitor the rate of movement of fluid through it and adjust the GFR accordingly. This tubuloglomerular feedback mechanism is initiated by the tubule to help each nephron regulate the rate of filtration through its own glomerulus.

The myogenic and tubuloglomerular feedback mechanisms work in unison to autoregulate the GFR within the mean arterial blood-pressure range (for example, 80 to 180 mm Hg in humans). This range encompasses the transient changes in blood pressure that accompany daily activities unrelated to the need for the kidneys to regulate H_2O and salt excretion, such as the normal elevation in blood pressure accompanying muscular activity. Within this range, intrinsic autoregulatory adjustments can compensate for changes in arterial pressure. This is important because shifts in GFR could lead to dangerous imbalances of fluid, electrolytes, and wastes. If it were not for autoregulation, the GFR would increase and H_2O and solutes would be lost needlessly during heavy activity. In con-

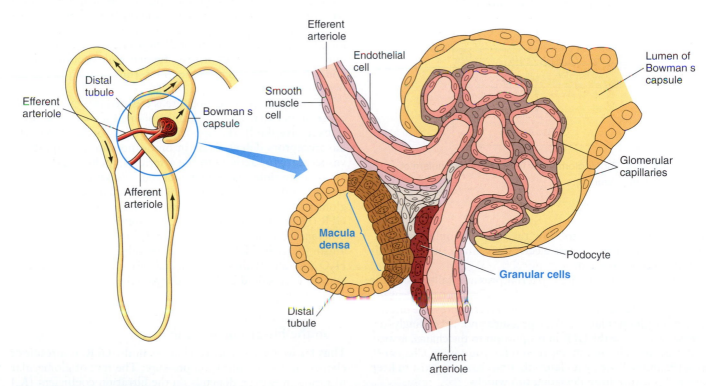

Figure 12–12 ● The juxtaglomerular apparatus (mammalian kidney). The juxtaglomerular apparatus consists of specialized vascular cells (the granular cells) and specialized tubular cells (the macula densa) at a point where the distal tubule passes through the fork formed by the afferent and efferent arterioles of the same nephron.

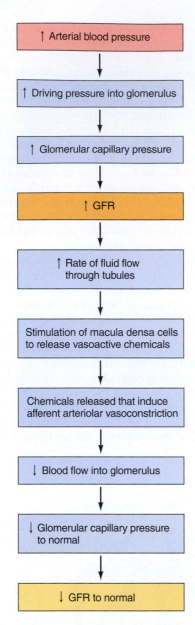

↑ Arterial blood pressure

↓

↑ Driving pressure into glomerulus

↓

↑ Glomerular capillary pressure

↓

↑ GFR

↓

↑ Rate of fluid flow through tubules

↓

Stimulation of macula densa cells to release vasoactive chemicals

↓

Chemicals released that induce afferent arteriolar vasoconstriction

↓

↓ Blood flow into glomerulus

↓

↓ Glomerular capillary pressure to normal

↓

↓ GFR to normal

Figure 12–13 ● Tubuloglomerular feedback mechanism of autoregulation (mammal).

trast, if the GFR were too low the kidneys could not eliminate enough wastes, excess electrolytes, and other materials that should be excreted.

When changes in mean arterial pressure fall outside the autoregulatory range, these mechanisms cannot compensate. Therefore, dramatic changes in mean arterial pressure directly cause the glomerular capillary pressure and, accordingly, decrease or increase the GFR in proportion to the change in arterial pressure. This is a major reason (as you saw in Chapter 9, p. 419, and will see again later) that mechanisms exist to keep arterial pressure from deviating too widely.

Extrinsic Sympathetic Control of the GFR

In addition to the intrinsic, local autoregulatory mechanisms designed to keep the GFR constant in the face of fluctuations in arterial blood pressure, the GFR can be changed—even

when the mean arterial blood pressure is within the autoregulatory range—by "higher" extrinsic-control mechanisms that override the autoregulatory responses. This is an example of the hierarchical "local" and "higher" distribution of regulation we introduced in Chapter 1 (p. 17). Extrinsic control of GFR is mediated by sympathetic nervous system input to the afferent arterioles, and is aimed at the regulation of systemic arterial blood pressure. The parasympathetic nervous system does not exert any influence on the kidneys.

If plasma volume is decreased—for example, because of hemorrhage—the resulting fall in arterial blood pressure is detected by the arterial and aortic baroreceptors, which initiate neural reflexes (via the cardiovascular control center in the brain stem) to increase blood pressure toward normal (Chapter 9, p. 421). These negative-feedback responses are effected primarily through increased sympathetic activity to the heart and blood vessels, which increases both cardiac output and total peripheral resistance. Yet the plasma volume is still reduced and must be restored in the long term. One way to do this is with a reduction in urine output so that more fluid than normal is conserved for the body. This reduction is accomplished in part by a reduction in the GFR; if less fluid is filtered, less fluid is available to be excreted.

The decrease in the GFR occurs as part of the baroreceptor reflex (● Figure 12–14). Among the arterioles that constrict in response to this reflex are the afferent arterioles carrying blood to the glomeruli. The afferent arterioles are innervated with sympathetic vasoconstrictor fibers to a far greater extent than the efferent arterioles. When the afferent arterioles constrict, less blood flows into the glomeruli than normal, decreasing GFR (Figure 12–11a) and urine volume. This helps in the long term to restore the plasma volume to normal so that the short-term cardiovascular adjustments that have been made are no longer necessary. Hormonal mechanisms (described more thoroughly later) also act to increase tubular reabsorption of H_2O and salt as well as to increase thirst, thus helping to restore plasma volume.

Conversely, if blood pressure is elevated (for example, because of an expansion of plasma volume following ingestion of excessive fluid), the opposite responses occur. When the baroreceptors detect a rise in blood pressure, sympathetic vasoconstrictor activity to the arterioles, including the renal afferent arterioles, is reflexly reduced, allowing afferent arteriolar vasodilation. Thus more blood enters the glomeruli and glomerular-capillary blood pressure rises, increasing both the GFR (Figure 12–11b) and thus urine output. Again, there are other mechanisms: a reduction in thirst, and a hormonal mechanism (described later) that reduces tubular reabsorption of H_2O and salt. By these mechanisms, urine volume is increased and the excess fluid is eliminated from the body.

■ The GFR can be influenced by changes in the filtration coefficient.

Thus far we have discussed changes in the GFR as a result of changes in the net filtration pressure. The rate of glomerular filtration, however, depends on the filtration coefficient (K_f) as well as on the net filtration pressure. For years K_f was considered a constant, except in disease situations in which the glomerular membrane becomes leakier than usual. Exciting new research to the contrary indicates that K_f is subject to change under physiological control. Both factors on which K_f

depends—the surface area and the permeability of the glomerular membrane—can be modified by contractile activity within the membrane.

The surface area available for filtration within the glomerulus is represented by the inner surface of the glomerular capillaries that comes into contact with blood. Each tuft of glomerular capillaries is held together by **mesangial cells.** These cells also function as phagocytes and contain contractile elements (that is, actinlike filaments). Contraction of these mesangial cells closes off a portion of the filtering capillaries, reducing the surface area available for filtration within the glomerular tuft. When the net filtration pressure remains unchanged, this reduction in K_f leads to a decrease in GFR. Evidence suggests that sympathetic stimulation causes the mesangial cells to contract, thus providing a second mechanism (besides promoting afferent arteriolar vasoconstriction) by which sympathetic activity can decrease the GFR. In addition, several hormones and local chemical mediators are suspected or known to be involved in controlling other renal mechanisms, such as tubuloglomerular feedback or tubular reabsorption, and have been shown to influence mesangial-cell contractile activity.

The podocytes also have actinlike contractile filaments, whose contraction or relaxation can, respectively, decrease or increase the number of filtration slits open in the inner membrane of Bowman's capsule by changing the shapes and proximities of the foot processes (● Figure 12–15). The number of slits is a determinant of permeability; the more open slits, the

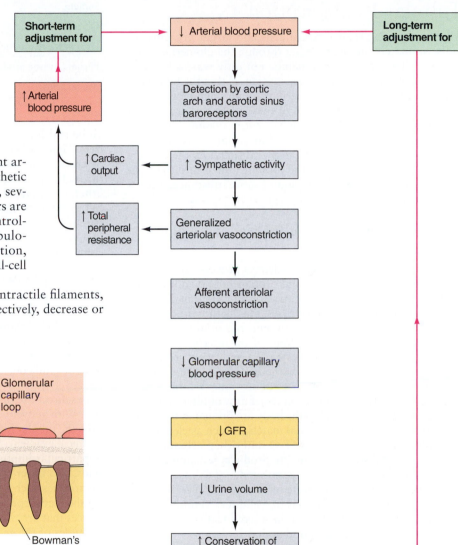

Figure 12–14 ● Baroreceptor reflex influence on the GFR in long-term regulation of arterial blood pressure (mammal).

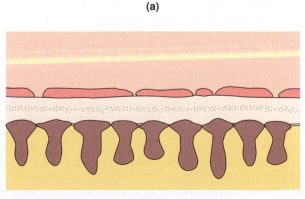

(a)

(b)

Figure 12–15 ◄ ● Change in the number of open filtration slits caused by podocyte relaxation and contraction (mammalian kidney). (a) Podocyte relaxation narrows the bases of the foot processes, increasing the number of fully open intervening filtration slits spanning a given area. (b) Podocyte contraction flattens the foot processes and thus decreases the number of intervening filtration slits.

(*Source:* Adapted from *Proceedings of the Federation of American Societies for Experimental Biology,* Vol. 42, p. 3046–3052, 1983. Reprinted by permission.)

greater the permeability. Contractile activity of the podocytes in turn affects permeability and the K_f.

As physiologists begin to understand the full extent of various regulatory mechanisms, it is evident that control of glomerular filtration (both through adjustments in afferent arteriolar resistance and K_f) and control of tubular reabsorption are complexly interrelated. Let's now examine reabsorption.

Tubular Reabsorption

All plasma constituents except cells and larger proteins are indiscriminately filtered together through the glomerular capillaries. These constituents include not only wastes but also nutrients, electrolytes, and other substances that the body cannot afford to lose in urine. Indeed, through ongoing glomerular filtration greater quantities of these materials are filtered per day than are even present in the entire mammalian body. It is important that these essential materials be returned to the blood by the process of *tubular reabsorption,* the discrete transfer of substances from the tubular lumen into the peritubular capillaries. Reabsorption of most substances occurs in the proximal tubule.

■ Tubular reabsorption is tremendous, highly selective, and variable.

Tubular reabsorption is a highly selective process. All constituents except plasma proteins (and protein-bound solutes such as iron) are at the same concentration in the glomerular filtrate as in the plasma. In most cases, the quantity of each material that is reabsorbed is the amount required to maintain the proper composition and volume of the internal fluid environment. In general, the tubules have a high reabsorptive capacity for substances needed by the body and little or no reabsorptive capacity for substances of no value (■ Table 12–2). Accordingly, only a small percentage, if any, of filtered plasma constituents that are useful to the body are present in the urine. Only excess amounts of essential materials such as electrolytes are excreted in the urine. As H_2O and other valuable constituents are reabsorbed, the waste products remaining in the tubular fluid become highly concentrated.

Considering the magnitude of glomerular filtration, the extent of tubular reabsorption is tremendous: Mammalian tubules typically reabsorb 99% of the filtered H_2O and salt, and 100% of the filtered glucose and amino acids.

■ Tubular reabsorption involves transepithelial transport.

Throughout its entire length, the tubule is in close proximity to a surrounding peritubular capillary (● Figure 12–16). Adjacent tubular cells touch each other only where they are joined by tight junctions (see p. 63) at their lateral edges near their *luminal* membranes, which face the tubular lumen. Interstitial fluid lies in the gaps between adjacent cells—the lateral spaces—as well as between the tubules and capillaries. The *basolateral* membrane faces the interstitial fluid at the base and lateral edges of the cell.

The tight junctions largely prevent substances except H_2O from moving between the cells, so materials must pass through the cells to leave the tubular lumen and gain entry to the blood.

Table 12–2 ■ Fate of Various Substances Filtered by Mammalian Kidneys

Substance	Average Percentage of Filtered Substance	
	Reabsorbed	*Excreted*
Water	99	1
Sodium	99.5	0.5
Glucose	100	0
Urea (a waste product)	50	50
Phenol (a waste product)	0	100
Amino acids	100	0

To be reabsorbed, a substance must traverse five distinct barriers (Figure 12–16). That is, it must (1) leave the tubular fluid by crossing the luminal membrane of the tubular cell, (2) pass through the cytosol from one side of the tubular cell to the other, (3) traverse the basolateral membrane of the tubular cell to enter the interstitial fluid, (4) diffuse through the interstitial fluid, and (5) penetrate the capillary wall to enter the blood plasma. This entire sequence of steps is known as **transepithelial** ("across the epithelium") **transport.**

Passive versus Active Reabsorption

There are two types of tubular reabsorption—*passive* and *active reabsorption*. In passive reabsorption, all steps in the transepithelial transport of a substance from the tubular lumen to the plasma use no direct metabolic energy for the substance's movement, because energy is provided by electrochemical or osmotic gradients (see p. 80). In contrast, a substance is said to be actively reabsorbed if any one of the steps in the sequence requires direct use of energy, even if the four other steps are passive. With active reabsorption, net movement of the substance from the tubular lumen to the plasma occurs against an electrochemical gradient. Substances that are actively reabsorbed are of particular importance to the body, such as glucose, amino acids, and other organic nutrients, as well as Na^+ and other electrolytes such as PO_4^{3-}. Rather than specifically describing the reabsorptive process for each of the many filtered substances that are returned to the plasma, we provide illustrative examples of the general mechanisms involved after first highlighting the unique and important case of Na^+ reabsorption.

■ An energy-dependent Na^+–K^+ ATPase transport mechanism in the basolateral membrane is essential for Na^+ reabsorption.

Sodium reabsorption is unique and complex—passive in some sections of the tubule but active in most sections. In fact, about 80% of the total energy requirement of the kidneys is estimated to be used for Na^+ transport, indicating the importance of this process. Of the Na^+ filtered, 99% is normally reabsorbed, of which on average 67% is reabsorbed in the proximal tubule, 25% in the loop of Henle, and 8% in the distal tubules and collecting ducts. Sodium reabsorption plays different important roles in each of these segments, as will become apparent as our discussion continues.

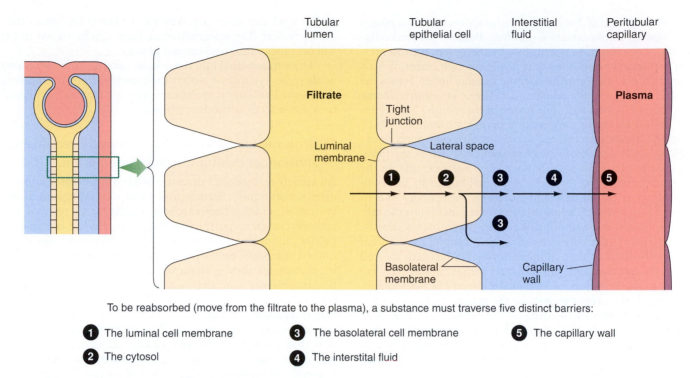

Figure 12–16 ● Steps of transepithelial transport (mammalian kidney).

■ Sodium reabsorption in the proximal tubule plays a pivotal role in the reabsorption of glucose, amino acids, H_2O, Cl^-, and urea, and in the regulation of ECF volume.

■ Sodium reabsorption in the loop of Henle, along with Cl^- reabsorption, plays a critical role in the kidneys' ability to produce urine of varying concentrations and volumes, depending on the body's need to conserve or eliminate H_2O.

■ Sodium reabsorption in the distal portion of the nephron is important in the regulation of ECF volume and in secretion of K^+ and H^+.

Sodium is reabsorbed throughout the tubule with the exception of the descending limb of the loop of Henle. You will learn about the significance of this exception later. Throughout all Na^+-reabsorbing segments of the tubule, the active step in Na^+ reabsorption involves the energy-dependent Na^+–K^+ ATPase pump located in the tubular cell's basolateral membrane (● Figure 12–17). This pump is the same one that is present in all cells and actively extrudes Na^+ from the cell (see p. 81). As this basolateral pump transports Na^+ out of the tubular cell into the lateral space, it keeps the intracellular Na^+ concentration low while it simultaneously builds up

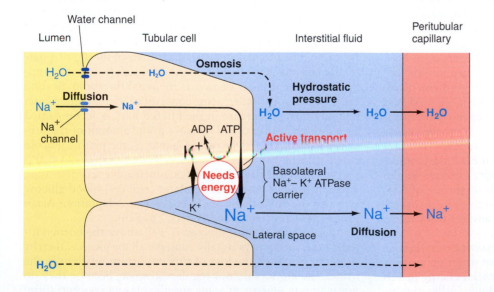

Figure 12–17 ● **Sodium reabsorption (mammalian kidney).** The basolateral Na^+ – K^+ ATPase carrier actively transports Na^+ from the tubular cell into the interstitial fluid within the lateral space. This process establishes a concentration gradient for diffusion of Na^+ from the lumen into the tubular cell and from the lateral space into the peritubular capillary, accomplishing net transport of Na^+ from the tubular lumen into the blood at the expense of energy. Water follows as Na^+ (and Cl^-) build up in the capillary.

the concentration of Na^+ in the lateral space; that is, it moves Na^+ *against* a concentration gradient. Because the intracellular Na^+ concentration is kept low by the pump, a concentration gradient is established that favors the movement of Na^+ from its higher concentration in the tubular lumen across the luminal border into the tubular cell. (The cell's negative charge also draws Na^+ inward.) The nature of the luminal Na^+ channels and/or transport carriers that permit movement of Na^+ from the lumen into the cell varies for different parts of the tubule, but in each case, movement of Na^+ across the luminal membrane is always a passive step. For example, in the proximal tubule, Na^+ crosses the luminal border by means of a cotransport carrier that simultaneously moves Na^+ and an organic nutrient such as glucose from the lumen into the cell. You will learn more about this cotransport process shortly. By contrast, in the collecting duct Na^+ crosses the luminal border through a Na^+ channel. Once Na^+ enters the cell across the luminal border by whatever means, it is actively extruded to the lateral space by the basolateral Na^+–K^+ pump. This step is the same throughout the tubule. Sodium continues to move down its electrochemical gradient from its high concentration in the lateral space into the surrounding interstitial fluid and finally into the peritubular capillary blood. Thus net transport of most of the Na^+ from the tubular lumen into the blood occurs at the expense of energy. This obviously helps conserve NaCl, but the movement of Na^+ can also help transport other solutes, as we show next.

■ Glucose and amino acids are reabsorbed by Na^+-dependent secondary active transport.

Large quantities of nutritionally important organic molecules such as glucose and amino acids are filtered each day. Because these substances normally are completely reabsorbed back into the blood by energy- and Na^+-dependent mechanisms located in the proximal tubule, none of these materials are usually excreted in the urine. This rapid and thorough reabsorption early in the tubules protects against the loss of these important organic nutrients.

Even though glucose and amino acids are actively moved uphill against their concentration gradients from the tubular lumen into the blood until their concentration in the tubular fluid is virtually zero, no ATP is directly used to operate the glucose or amino acid carriers. Glucose and amino acids are transferred by means of *secondary active transport*; a specialized cotransport carrier simultaneously transfers both Na^+ and the specific organic molecule from the lumen into the cell (see p. 83). The lumen-to-cell Na^+ electrogradient maintained by the energy-consuming basolateral Na^+–K^+ ATPase pump drives this cotransport system: The organic molecule is pulled in against its concentration gradient using the energy of the sodium gradient. (Secondary active transport requires the presence of Na^+ in the lumen; without Na^+ the cotransport carrier doesn't work.) Because the overall process of glucose and amino acid reabsorption depends on the use of ATP, these organic molecules are considered actively reabsorbed, even though ATP is not used directly to transport them across the membrane.

A mechanical analogy may help in visualizing this. Imagine an electric pump that uses energy to pump water up (against gravity) into an elevated tank. The potential energy in this water can be released by letting it flow down, using gravity.

Useful work can be obtained by having the falling water turn a waterwheel. The waterwheel, in turn, can be rigged to lift soil (or other substance) against gravity.

Once transported into the tubular cells, glucose and amino acids passively diffuse down their concentration gradients across the basolateral membrane into the plasma, facilitated by a non–energy-dependent carrier.

■ With the exception of Na^+, actively reabsorbed substances exhibit a tubular maximum.

All actively reabsorbed substances bind with plasma membrane carriers that transfer them across the membrane against a concentration gradient. Each carrier is specific for the types of substances it can transport; for example, the glucose cotransport carrier cannot transport amino acids, or vice versa. Because a limited number of each carrier type are present in the cells lining the tubules, there is an upper limit on the quantity of a particular substance that can be actively transported from the tubular fluid in a given period of time. The maximum reabsorption rate is reached when all the carriers specific for a particular substance are fully "occupied" or saturated (see p. 78), so that they cannot handle any additional passengers at that time. This transport maximum, designated as the **tubular** (or **transport**) **maximum**, or T_m, in the kidney tubules is the maximum amount of a substance that the tubular cells can actively transport within a given time period. With the exception of Na^+, all actively reabsorbed substances display a T_m. (Sodium does not display a T_m, because **aldosterone**, which we discuss later, promotes the synthesis of more active Na^+–K^+ ATPase carriers in the distal and collecting tubular cells as needed.) Any quantity of a substance filtered beyond its T_m fails to be reabsorbed, and escapes instead into the urine.

The plasma concentrations of some but not all substances that display carrier-limited reabsorption are regulated by the kidneys. To illustrate, we compare glucose, a substance that has a T_m but is not regulated by the kidneys, with phosphate, a T_m-limited substance that is regulated by the kidneys.

Glucose Reabsorption

Because glucose is freely filterable at the glomerulus, it passes into Bowman's capsule at the same concentration it has in the plasma. Ordinarily, no glucose appears in the urine because all the filtered glucose is reabsorbed. Not until the filtered load of glucose exceeds the T_m does glucose start spilling into the urine. The plasma concentration at which the T_m of a particular substance is reached and the substance first starts appearing in the urine is known as the **renal threshold**. Beyond the T_m, reabsorption remains constant at its maximum rate, and any further increase in the filtered load is accompanied by a directly proportional increase in the amount of the substance excreted. As you saw in Chapter 7, the plasma glucose concentration can become extremely high in diabetes mellitus, an endocrine disorder involving ineffectiveness of insulin, a pancreatic hormone that regulates (reduces) blood glucose (see also the box titled "A Closer Look at Adaptation: Avian Longevity: Unraveling the Mystery," p. 526).

Because the T_m for glucose is well above the normal filtered load, the kidneys usually conserve all the glucose, thereby protecting against the loss of this important nutrient in the urine. The kidneys do not regulate glucose, because they do not maintain glucose at some specific plasma concentration;

instead, this concentration is normally regulated by endocrine and liver mechanisms, with the kidneys merely maintaining whatever plasma glucose concentration is set by these other mechanisms (except at excessively high levels). The same general principle holds true for other organic plasma nutrients, such as amino acids and water-soluble vitamins.

PO_4^{3-} (Phosphate) Reabsorption

The kidneys do directly contribute to the regulation of many electrolytes, such as phosphate (PO_4^{3-}) and calcium (Ca^{++}), because the renal thresholds of these inorganic ions equal their normal plasma concentrations. We use PO_4^{3-} as an example. Some animal diets are rich in PO_4^{3-}, but because the tubules can reabsorb up to the normal plasma concentration's worth of PO_4^{3-} and no more, the excess ingested PO_4^{3-} is quickly spilled into the urine, restoring the plasma concentration to normal.

Unlike the reabsorption of organic nutrients, the reabsorption of PO_4^{3-} and Ca^{++} is also subject to hormonal control. Parathyroid hormone can alter the renal thresholds for PO_4^{3-} and Ca^{++}, thus adjusting the quantity of these electrolytes conserved, depending on the body's momentary needs (Chapter 7).

■ Active Na^+ reabsorption is responsible for the passive reabsorption of Cl^-, H_2O, and urea.

Not only is the secondary active reabsorption of glucose and amino acids linked to the basolateral Na^+–K^+ pump, but the passive reabsorption of Cl^-, H_2O, and urea, also depends on active Na^+ reabsorption.

Chloride Reabsorption

The negatively charged chloride ions are passively reabsorbed down the electrical gradient created by the active reabsorption of the positively charged sodium ions. The amount of Cl^- reabsorbed is determined by the rate of active Na^+ reabsorption instead of being directly controlled by the kidneys.

Water Reabsorption

Water is passively reabsorbed by osmosis in several regions of the tubule. Of the H_2O filtered, 45 to 65% is reabsorbed in the proximal tubules and about 15% in the descending loops of Henle (discussed later), osmotically following solute reabsorption. This amount is not regulated extensively. The remaining 20 to 25% is reabsorbed in the distal tubules and collecting ducts, and is subject to hormonal control, depending on the body's state of hydration, which we examine in detail in the osmoconcentration section, later.

The driving force for H_2O reabsorption in the proximal tubule is a compartment of hypertonicity in the lateral spaces between the tubular cells that is established by the active extrusion of Na^+ by the basolateral pump (Figure 12–17). As a result of this pump activity, the concentration of Na^+ rapidly diminishes in the tubular fluid and tubular cells while it simultaneously increases in the localized region within the lateral spaces. This osmotic gradient induces the passive net flow of H_2O from the lumen into the lateral spaces, either through the cells or intercellularly through "leaky" tight junctions. The accumulation of fluid in the lateral spaces results in a buildup of hydrostatic (fluid) pressure, which flushes H_2O out of the lateral spaces into the interstitial fluid and finally into the peritubular capillaries.

This return of filtered H_2O to the plasma is enhanced by the fact that the plasma-colloid osmotic pressure is greater in the peritubular capillaries than elsewhere. The concentration of plasma proteins, which is responsible for the plasma-colloid osmotic pressure, is elevated in the blood entering the peritubular capillaries because of the extensive filtration of H_2O through the glomerular capillaries upstream. The plasma proteins left behind in the glomerulus are concentrated into a smaller volume of plasma H_2O, increasing the plasma-colloid osmotic pressure of the unfiltered blood that leaves the glomerulus and enters the peritubular capillaries. This force tends to "pull" H_2O into the peritubular capillaries, simultaneous with the "push" of the hydrostatic pressure in the lateral spaces that drives H_2O toward the capillaries. Neither the proximal tubule nor indeed any other portion of the tubule directly requires ATP directly for this tremendous reabsorption of H_2O.

■ In general, unwanted waste products except urea are not reabsorbed.

Because of extensive reabsorption of H_2O in the proximal tubule, the original volume of filtrate is progressively reduced. Substances that have been filtered but not reabsorbed become progressively more concentrated in the tubular fluid as H_2O is reabsorbed while they are left behind. Wastes and ingested toxins thus become concentrated for later excretion—they are unable to leave the lumen into the plasma, because of an impermeable tubular wall. This excretion of metabolic wastes is not subject to physiological control. When renal function is normal, these uncontrolled excretory processes proceed at a satisfactory rate.

Urea molecules, being the smallest of waste products, are an exception, because they are the only wastes to be passively reabsorbed as a result of this concentrating effect. Urea's concentration as it is filtered at the glomerulus is identical to its concentration in the plasma entering the peritubular capillaries, but, like other wastes, it is concentrated several-fold in the small volume remaining at the end of the proximal tubule. As a result, the urea concentration within the tubular fluid becomes considerably greater than the plasma urea concentration in the adjacent capillaries. Therefore, a concentration gradient is created for urea to passively diffuse from the tubular lumen into the peritubular capillary plasma. Because the walls of the proximal tubules are only moderately permeable to urea, only about 50% of the filtered urea is passively reabsorbed by this means. (As you will see later, urea permeability depends on urea transporters in the membranes.) Even though only half of the filtered urea is eliminated from the plasma with each pass through the nephron, this removal rate is adequate.

■ The renin-angiotensin-aldosterone system stimulates Na^+ reabsorption in the distal tubules, elevates blood pressure, and stimulates thirst and salt hunger.

Reabsorption of some of the filtered Na^+ is subject to hormonal control in the nephron tubule. This has been thought to occur primarily in the distal tubule, although recent studies show some hormonal control of reabsorption in the proximal tubule as well. The extent of this controlled reabsorption is inversely related to the magnitude of the Na^+ load in the body. If there is too much Na^+, little of this controlled Na^+ is reabsorbed; instead, it is lost in the urine, thereby removing excess Na^+ from the body. In contrast, if Na^+ is depleted, most or all of this controlled Na^+ is reabsorbed, conserving for the body Na^+ that otherwise would be lost in the urine.

The Na^+ load in the body is reflected by the ECF volume. Sodium and its accompanying anions (mainly Cl^-) account for more than 90% of the ECF's osmotic activity. Recall that osmotic pressure can be thought of loosely as a force that attracts and holds H_2O (see p. 75). When the Na^+ load is above normal and the ECF's osmotic activity is therefore increased, the extra Na^+ causes osmosis of H_2O, expanding the ECF volume. Conversely, when the Na^+ load is below normal, thereby decreasing ECF osmotic activity, less H_2O than normal can be held in the ECF, so the ECF volume is reduced. Because plasma is a component of the ECF, the most important consequence of a change in ECF volume is the corresponding change in blood pressure accompanying expansion (increased blood pressure) or reduction (decreased blood pressure) of the plasma volume (see Chapter 9, p. 420). Thus long-term control of arterial blood pressure ultimately depends on Na-regulating mechanisms. We now turn our attention to these mechanisms.

Activation of the Renin-Angiotensin-Aldosterone System

The granular cells of the juxtaglomerular apparatus (Figure 12–12) secrete a hormone, **renin**, into the blood in response to factors that signal a fall in NaCl/ECF volume/blood pressure. This function is in addition to the role the juxtaglomerular apparatus plays in autoregulation, and renin is distinct from the local vasoactive chemicals that influence glomerular blood flow. Specifically, the following three inputs to the granular cells increase renin secretion:

1. The granular cells themselves function as *intrarenal baroreceptors*. They are sensitive to pressure changes within the afferent arteriole. When the granular cells detect a fall in blood pressure, they secrete more renin.

2. The macula densa cells in the tubular portion of the juxtaglomerular apparatus are sensitive to the NaCl moving past them through the tubular lumen. In response to a fall in NaCl, the macula densa cells trigger the granular cells to secrete more renin.

3. The granular cells are innervated by the sympathetic nervous system. When blood pressure falls below normal, the baroreceptor reflex increases sympathetic activity. As part of this reflex response, increased sympathetic activity stimulates the granular cells to secrete more renin.

These interrelated signals for increased renin secretion all indicate the need to expand the plasma volume to increase the arterial pressure to normal on a long-term basis. Through a complex series of events, increased renin secretion brings about increased Na^+ reabsorption by the distal portion of the tubule. Chloride always passively follows Na^+ down the electrical gradient established by sodium's active movement. The ultimate benefit of this salt retention is that it osmotically promotes H_2O retention, which helps restore the plasma volume, thus being important in the long-term control of blood pressure.

Let's examine the mechanism by which renin secretion ultimately leads to increased Na^+ reabsorption—the **renin-angiotensin-aldosterone system** (● Figure 12–18). Once secreted into the blood, renin functions as an enzyme to convert **angiotensinogen** into **angiotensin I**. Angiotensinogen is a plasma protein synthesized by the liver and always present in the plasma in high concentration. On passing through the lungs via the pulmonary circulation, angiotensin I is converted into **angiotensin II** by **angiotensin-converting enzyme (ACE)**, which is abundant in the pulmonary capillaries. Angiotensin II is the primary stimulus for the secretion of the hormone **aldosterone** from the adrenal cortex (see Chapter 7, p. 284).

Functions of the Renin-Angiotensin-Aldosterone System

Among its actions, aldosterone increases Na^+ reabsorption by the distal tubules and collecting ducts. Aldosterone is a **steroid hormone**; as we saw in Chapters 2 and 7, this type of hormone exerts its effects by binding to an internal receptor, which becomes an active transcription factor that stimulates gene transcription. For aldosterone, gene regulation leads to the insertion of additional Na^+ channels into the luminal membranes and additional Na^+–K^+ ATPase carriers into the basolateral membranes of the distal and collecting tubular cells. The net result is a greater passive inward flux of Na^+ into the tubular cells from the lumen and increased active pumping of Na^+ out of the cells into the plasma—that is, an increase in Na^+ reabsorption, with Cl^- following passively and H_2O following by osmosis. The renin-angiotensin-aldosterone system thus promotes salt retention resulting in H_2O retention and elevation of arterial blood pressure. Acting in negative-feedback fashion, this system alleviates the factors that triggered the initial release of renin—namely, salt depletion, plasma volume reduction, and decreased arterial blood pressure.

In addition to stimulating aldosterone secretion, angiotensin II is also a potent constrictor of the systemic arterioles, thereby directly increasing blood pressure by increasing total peripheral resistance (see p. 420). Furthermore, it stimulates thirst (increasing fluid intake) and stimulates vasopressin (a hormone that increases H_2O retention by the kidneys, p. 551), both of which contribute to plasma volume expansion and elevation of arterial pressure. (As you will learn later, other mechanisms related to the long-term regulation of blood pressure and ECF osmolarity are also important in controlling thirst and vasopressin secretion.) Angiotensin II also stimulates Na^+ reabsorption in the proximal and distal tubules and the collecting ducts, and stimulates **salt hunger** in the brain, that is, it triggers a craving for dietary salt intake. Aldosterone appears to do this as well, but angiotensin is the major stimulant. Salt hunger and thirst trigger corrective behaviors that fall into the category of behavioral effector processes (p. 14).

The opposite situation exists when the Na^+ load, ECF and plasma volume, and arterial blood pressure are above normal. Under these circumstances, renin (and vasopressin) secretion is inhibited. Consequently, because angiotensinogen is not activated to angiotensin I and II, aldosterone secretion is not stimulated. Without aldosterone, the modest aldosterone-dependent portion of Na^+ reabsorption in the proximal and distal tubules does not occur. Instead, this unreabsorbed Na^+, and the Cl^- and water that would have come with it, are lost in urine. In the absence of aldosterone, this ongoing loss can rapidly remove excess NaCl (and water) from the body.

In the complete absence of aldosterone, a human may excrete 20 g of salt per day. With maximum aldosterone secre-

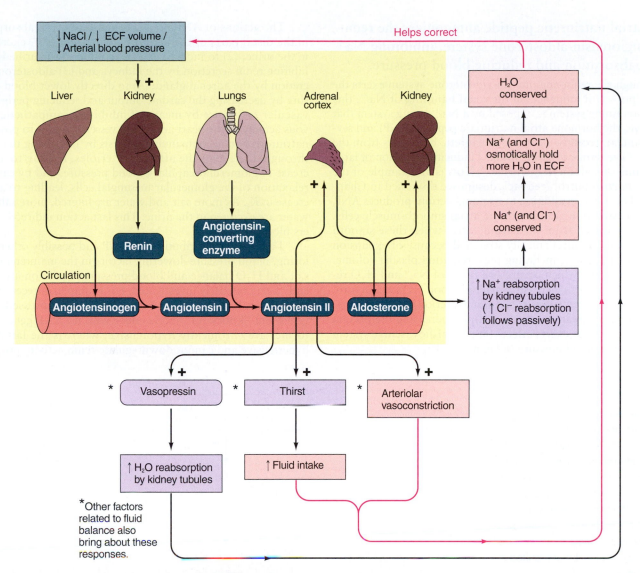

Figure 12–18 ● Renin-angiotensin-aldosterone system in a mammal. The kidneys secrete the hormone renin in response to reduced NaCl, ECF volume, and arterial blood pressure. Renin activates angiotensinogen, a plasma protein produced by the liver, into angiotensin I. Angiotensin I is converted into angiotensin II by angiotensin-converting enzyme (ACE) produced in the lungs. Angiotensin II stimulates the adrenal cortex to secrete the hormone aldosterone, which stimulates Na^+ reabsorption by the kidneys. The resulting retention of Na^+ exerts an osmotic effect that holds more H_2O in the ECF. Together, the conserved Na^+ and H_2O help correct the original stimuli that activated this hormonal system. Angiotensin II also exerts other effects that help rectify the original stimuli.

tion, almost all the filtered Na^+ (and Cl^-) is reabsorbed, so salt excretion in the urine is very low (although probably never zero). The amount of aldosterone secreted, and consequently the relative amount of salt conserved versus salt excreted, varies according to intake. Herbivores, for example, have a high-potassium and low-sodium diet because of the ion composition of plants, while carnivores may have a substantial sodium intake. With the renin mechanism, the kidneys maintain the salt load and ECF volume / arterial blood pressure at a relatively constant level despite wide variations in salt consumption and abnormal losses of salt-laden fluid.

This system is responsible in part for the fluid retention and edema accompanying congestive heart failure, a condition suffered by many aging mammals. When a fall in blood pressure is because of a failing heart rather than a reduction in the salt/fluid load in the body, the salt- and fluid-retaining reflexes triggered by the low blood pressure are inappropriate. Excess sodium in the ECF and the resulting volume expansion exacerbates the problem, because the weakened heart cannot pump the additional plasma volume. As we saw in Chapter 1, this is an example of a *positive-feedback process* that is not useful (p. 16).

■ Atrial natriuretic peptide antagonizes the renin-angiotensin-aldosterone system, inhibiting Na⁺ reabsorption and reducing blood pressure.

Although the renin-angiotensin-aldosterone system exerts the most powerful influence on the renal handling of Na^+, this Na^+-retaining system is opposed by a Na^+-losing system that involves the hormone **atrial natriuretic peptide (ANP)** and several similar, recently identified natriuretic hormones from the brain. (The term *natriuretic* means "inducing excretion of large amounts of sodium in the urine.") This is an example of the *antagonistic* control feedback design we discussed in Chapter 1. The heart, in addition to its pump action, produces ANP, which is stored in granules in the atrial smooth-muscle cells. ANP is released from the cardiac atria when these smooth muscle cells are mechanically stretched because of expansion of the ECF volume, including the circulating plasma volume. This expansion, which occurs as a result of Na^+ and H_2O retention, leads to an increase in arterial blood pressure. In turn, ANP promotes *natriuresis* (sodium excretion) and accompanying *diuresis* (water excretion), thus decreasing the plasma volume, and also exerts effects on the cardiovascular system to lower the blood pressure (● Figure 12–19).

The actions of ANP are (1) to inhibit Na^+ reabsorption in the distal parts of the nephron, increasing Na^+ excretion in the urine; (2) to increase Na^+ excretion in the urine by inhibiting renin secretion by the kidneys and (3) aldosterone secretion by the adrenal gland; (4) to directly lower blood pressure by decreasing the cardiac output and reducing peripheral vascular resistance by means of inhibiting sympathetic nervous activity to the heart and blood vessels; and (5) to promote natriuresis and accompanying diuresis by increasing the GFR through dilation of the afferent arterioles, leading to an increase in glomerular capillary blood pressure, and by causing relaxation of the glomerular mesangial cells, leading to an increase in K_f. As more salt and water are filtered, more salt and water are excreted in the urine. This last action indirectly lowers blood pressure.

The relative contributions of ANP and possibly other salt-losing, blood pressure–lowering factors in the maintenance of salt and H_2O balance and blood pressure regulation are presently being intensely investigated. For example, in recent years researchers have engineered mice with overexpressed ANP genes and "knockout" (p. 32) mice with no ANP genes. The former are chronically hypotensive, whereas the latter are hypertensive and cannot downregulate renin activity properly

Figure 12–19 ● **Atrial natriuretic peptide (mammal).** The atria secrete the hormone atrial natriuretic peptide (ANP) in response to being stretched by Na⁺ retention, expansion of the ECF volume, and increase in arterial blood pressure. Atrial natriuretic peptide in turn promotes natriuretic, diuretic, and hypotensive effects to help correct the original stimuli that resulted in its release.

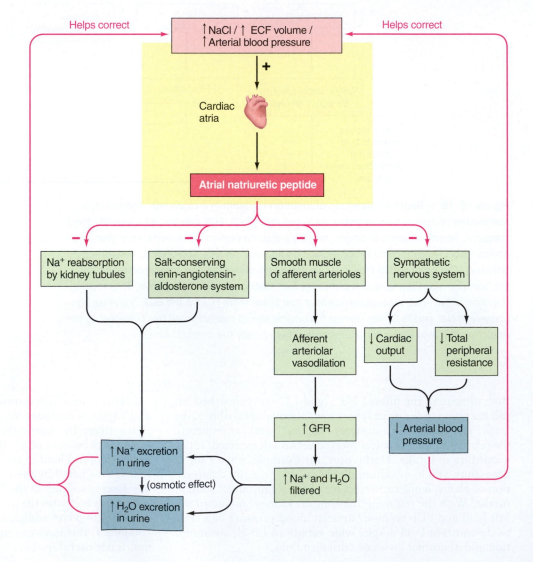

after a salt load. These genetic studies directly demonstrate that ANP is crucial for long-term regulation of blood pressure.

Tubular Secretion

Like tubular reabsorption, tubular secretion involves transepithelial transport, but now the steps are reversed. By providing a second route of entry into the tubules for selected substances, *tubular secretion,* the discrete transfer of substances from the peritubular capillaries into the tubular lumen, is a supplemental mechanism that hastens elimination of these compounds from the body. Anything that gains entry to the tubular fluid, whether by glomerular filtration or tubular secretion, and fails to be reabsorbed is eliminated in the urine. The most important substances secreted by the tubules are *hydrogen ion (H⁺), potassium (K⁺),* and *organic anions and cations,* many of which are compounds foreign to the body.

◾ Hydrogen ion secretion is important in acid–base balance.

Renal H⁺ secretion is extremely important in the regulation of the acid–base balance in the body. Hydrogen ion can be added to the filtered fluid by being secreted by the proximal, distal, and collecting ducts, using proton pumps called **V-ATPases** (p. 81). The extent of H⁺ secretion depends on the acidity of the body fluids. When the body fluids are too acidic, H⁺ secretion increases. Conversely, H⁺ secretion is reduced when the H⁺ concentration in the body fluids is too low. (See Chapter 13 for further details.)

◾ Potassium secretion is controlled by aldosterone.

Potassium ion is an example of a substance that is selectively moved in opposite directions in different parts of the tubule; it is actively reabsorbed in the proximal tubule and actively secreted in the distal tubules and collecting ducts. Potassium ion reabsorption early in the tubule occurs in a constant, unregulated fashion, whereas K⁺ secretion later in the tubule is

variable and subject to regulation. Normally, a quantity equivalent to about 10 to 15% of the filtered K⁺ is excreted in the urine. However, the filtered K⁺ is almost completely reabsorbed in the proximal tubule, so most of the K⁺ appearing in the urine is derived from controlled K⁺ secretion in the distal portions of the nephron rather than from filtration.

During K⁺ depletion, K⁺ secretion in the distal portions of the nephron is reduced to a minimum, so only the small percentage of filtered K⁺ that escapes reabsorption in the proximal tubule is excreted in the urine. In this way, K⁺ that normally would have been lost in the urine is conserved for the body. In contrast, when plasma K⁺ levels are elevated, K⁺ secretion is adjusted so that just enough K⁺ is added to the filtrate for elimination to reduce the plasma K⁺ concentration to normal. Thus K⁺ secretion, not the filtration or reabsorption of K⁺, is varied in a controlled fashion to regulate the rate of K⁺ excretion and maintain the desired plasma K⁺ concentration.

Mechanism and Control of K⁺ Secretion

Potassium ion secretion in the distal tubules and collecting ducts is coupled to Na⁺ reabsorption by means of the energy-dependent basolateral Na⁺–K⁺ pump (● Figure 12–20). This pump not only moves Na⁺ out into the lateral space but also transports K⁺ into the tubular cells. The resulting high intracellular K⁺ concentration favors the net diffusion of K⁺ from the cells into the tubular lumen. Movement across the luminal membrane occurs passively through the large number of K⁺ channels present in this barrier. By keeping the interstitial fluid concentration of K⁺ low as it transports K⁺ into the tubular cells from the surrounding interstitial fluid, the basolateral pump encourages the passive diffusion of K⁺ out of the peritubular capillary plasma into the interstitial fluid. Potassium exiting the plasma in this manner is subsequently pumped into the cells, from which it diffuses into the lumen. In this way, the basolateral pump actively induces the net secretion of K⁺ from the peritubular capillary plasma into the tubular lumen.

Because K⁺ secretion is linked with Na⁺ reabsorption by means of the Na⁺–K⁺ pump, why isn't K⁺ secreted throughout the Na⁺-reabsorbing segments of the tubule instead of tak-

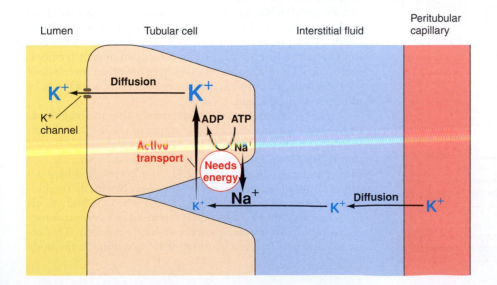

Figure 12–20 ● Potassium secretion (mammalian kidney). The basolateral pump simultaneously transports Na⁺ into the lateral space and K⁺ into the tubular cell. In the parts of the tubule that secrete K⁺, this ion leaves the cell through channels located in the luminal border, thus being secreted. (In the parts of the tubule that do not secrete K⁺, the K⁺ pumped into the cell during the process of Na⁺ reabsorption leaves the cell through channels located in the basolateral border, thus being retained in the body.)

ing place only in the distal parts of the nephron? The answer lies in the location of the passive K^+ channels. In the distal tubules and collecting ducts, the K^+ channels are concentrated in the luminal membrane, providing a route for K^+ pumped into the cell to exit into the lumen, thus being secreted. In the other tubular segments, the K^+ channels lie primarily in the basolateral membrane. As a result, K^+ pumped into the cell from the lateral space by the Na–K^+ pump simply diffuses back out into the lateral space through these channels. This K^+ recycling permits the ongoing operation of the Na^+–K^+ pump to accomplish Na^+ reabsorption with no local net effect on K^+.

Several factors can alter the rate of K^+ secretion, the most important being the hormone aldosterone, which stimulates K^+ secretion by the tubular cells late in the nephron simultaneous to enhancing these cells' reabsorption of Na^+. An elevation in plasma K^+ concentration directly stimulates the adrenal cortex to increase its output of aldosterone, which in turn promotes the secretion and ultimate urinary excretion and elimination of the excess K^+. Conversely, a decline in plasma K^+ concentration reduces aldosterone secretion and correspondingly reduces aldosterone-stimulated renal K^+ secretion.

Note that a rise in plasma K^+ concentration directly stimulates aldosterone secretion by the adrenal cortex, whereas a fall in plasma Na^+ concentration stimulates aldosterone secretion by means of the complex renin-angiotensin pathway. Thus aldosterone secretion can be stimulated by two separate pathways (● Figure 12–21). No matter what the stimulus, however, increased aldosterone secretion always promotes simul-

taneous Na^+ reabsorption and K^+ secretion. For this reason, K^+ secretion can be inadvertently stimulated as a result of increased aldosterone activity brought about by Na^+ depletion, ECF volume reduction, or a fall in arterial blood pressure totally unrelated to K^+ balance. The resulting inappropriate loss of K^+ can lead to K^+ deficiency.

Because this process has the net effect of swapping K^+ for Na^+ in the urine, and because it can be regulated, the amount of KCl and NaCl excreted in the final urine can vary widely among animals and among diets. Herbivores, for example, eat food often with a high potassium and low sodium content (most plants); therefore they excrete more KCl than NaCl. Indeed, such mammals exhibit salt hunger (stimulated by angiotensin II and aldosterone, p. 544) and a strong behavioral attraction to salt licks (natural or human-made deposits of NaCl and other minerals).

Normally the kidneys exert a fine degree of control over plasma K^+ concentration. This is extremely important, because even minor fluctuations in plasma K^+ concentration can have detrimental consequences. Potassium ion, being the most abundant cation in the intracellular fluid, plays a key role in the membrane potential and electrical activity of excitable tissues (Chapters 3 and 4). Either increases or decreases in the plasma (ECF) K^+ concentration can alter the intracellular-to-extracellular K^+ concentration gradient. A rise leads to a reduction in resting potential and a subsequent increase in excitability, especially of heart muscle and brain. This cardiac overexcitability can lead to a rapid heart rate and even fatal cardiac arrhythmias. Conversely, a fall in ECF K^+ concentration results in hyperpolarization of nerve and muscle cell membranes, which reduces their excitability. The manifestations of ECF K^+ depletion are skeletal muscle weakness, diarrhea and abdominal distention caused by smooth muscle dysfunction, and abnormalities in cardiac rhythm and impulse conduction.

Because of the linkage between K^+ and Na^+, inadvertent loss of K^+ because of Na^+ depletion can occur, which has important implications for domestic herbivores such as cattle, because they are sometimes fed herbage with extremely low NaCl and high KCl content. In one early study, cattle were fed such a diet and monitored carefully. After two months they began licking walls and dirt, chewing the bark of trees, eating manure, and drinking the urine of other cows, driven by an intense salt hunger. The ion content of their urines was dominated by KCl with virtually no NaCl. At 10 months the animals began to lose weight; some developed muscle tremors, and some died. Autopsies revealed hypertrophied (excessively enlarged) adrenal cortexes, probably because of a high rate of aldosterone production.

Another factor that can inadvertently alter the magnitude of K^+ secretion is the acid–base status of the body. An *antiport* transporter (p. 83) in the distal tubules can secrete either K^+ or H^+ in exchange for reabsorbed Na^+. An increased rate of secretion of either K^+ or H^+ is accompanied by a decreased rate of secretion of the other ion. Normally the kidneys secrete a preponderance of K^+, but when the body fluids are too acidic and H^+ secretion is increased as a compensatory measure, K^+ secretion is correspondingly reduced. This reduced secretion leads to inadvertent K^+ retention in the body fluids.

Figure 12–21 ● Dual control of aldosterone secretion of K^+ and Na^+ (mammal).

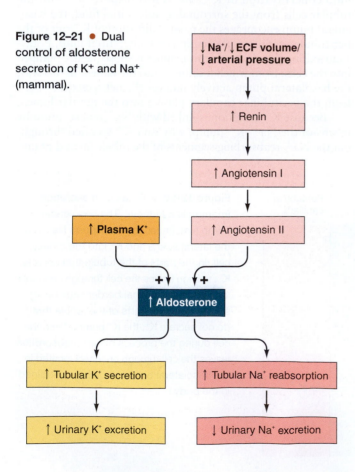

■ Organic anion and cation secretion helps efficiently eliminate foreign compounds from the body.

The proximal tubule contains two distinct types of secretory carriers, one for organic anions and another for organic cations. These carriers serve several important functions:

■ By adding more of a particular type of organic ion to the quantity that has already gained entry to the tubular fluid by means of glomerular filtration, these organic secretory pathways facilitate excretion of these substances. Included among these organic ions are certain blood-borne chemical messengers such as prostaglandins, histamine, and norepinephrine that, having served their purpose, must be rapidly removed from the blood so that their biological activity is not unduly prolonged.

■ In some important instances, organic ions are poorly soluble in water. To be transported in blood, they are extensively but not irreversibly bound to plasma proteins. Because they are attached to plasma proteins, these substances cannot be filtered through the glomeruli. Tubular secretion facilitates elimination of these nonfilterable organic ions in urine. Even though a given organic ion is largely bound to plasma proteins, a small percentage of the ion always exists in free or unbound form in the plasma. Removal of this free organic ion by secretion permits "unloading" of some of the bound ion, which is then free to be secreted.

■ Most important is the ability of the organic ion secretory systems to eliminate many foreign compounds from the body. The organic ion systems can secrete a large number of different organic ions, both those produced endogenously (within the body) and those foreign organic ions that have gained access to the body fluids. This nonselectivity permits these organic ion secretory systems to hasten removal of many (but by no means all) foreign organic chemicals, including food additives, environmental pollutants (for example, pesticides), drugs, and other nonnutritive organic substances that have entered the body.

Even though these mechanisms help rid the body of unwanted compounds, they do not appear subject to physiologic adjustments. The carriers cannot pick up their secretory pace when confronting an elevated load of these organic ions.

The vertebrate liver plays an important role in this process. Many foreign organic compounds are not ionic in their original form, so they cannot be secreted by the organic ion systems. The liver converts these foreign substances into an anionic form that facilitates their secretion by the organic anion system and thus accelerates their elimination.

This completes our discussion of the reabsorptive and secretory processes that occur across the proximal and distal portions of the nephron. To generalize,

1. The proximal tubule does most of the reabsorbing. This "mass reabsorber" transfers much of the filtered water and needed solutes back into the blood in unregulated fashion. Similarly, the proximal tubule is the major site of secretion, with the exception of K^+ secretion.

2. The distal tubules and collecting ducts then determine the final amounts of H_2O, Na^+, K^+, and H^+ that are excreted in urine. They do so by "fine-tuning" the amount of Na and H_2O reabsorbed and the amount of K^+ and H^+ secreted.

3. Many of these processes are subject to control, especially in the distal part of the nephron, depending on the body's momentary needs. The unwanted filtered waste products are left behind to be eliminated in urine, along with excess amounts of filtered or secreted nonwaste products that fail to be reabsorbed.

Before turning to the last function of the nephron—osmoconcentration—let's see how researchers measure the combined actions of filtration, reabsorption, and secretion for individual substances by a parameter called **plasma clearance.**

■ Plasma clearance is the volume of plasma cleared of a particular substance per minute.

Compared to the plasma entering the kidneys through the renal arteries, the plasma leaving the kidneys through the renal veins lacks the materials that were left behind to be eliminated in the urine. By excreting substances in the urine, the kidneys clean or "clear" the plasma flowing through them of these substances. For any substance, its plasma clearance is defined as the volume of plasma that is completely cleared of that substance by the kidneys per minute. It does not refer to the amount of the substance removed but to the volume of plasma from which that amount was removed. Plasma clearance is actually a more useful measure than urine excretion; it is more important to know what effect urine excretion has on removing materials from the body fluids than to know the volume and composition of the discarded urine. Plasma clearance expresses the kidneys' effectiveness in removing various substances from the internal fluid environment.

Plasma clearance can be calculated for any plasma constituent as follows:

$$\begin{array}{c} \text{Clearance rate} \\ \text{of a substance} \\ \text{(mL/min)} \end{array} = \dfrac{\begin{array}{c}\text{urine concentration} \\ \text{of the substance} \\ \text{(quantity/mL urine)}\end{array} \times \begin{array}{c}\text{urine flow} \\ \text{rate (mL/min)}\end{array}}{\begin{array}{c}\text{plasma concentration of the substance} \\ \text{(quantity/mL plasma)}\end{array}}$$

The plasma clearance rate varies for different substances, depending on how the kidneys handle each substance, as follows.

Plasma clearance rate equals the GFR if a substance is freely filtered but not reabsorbed or secreted (or otherwise altered by renal tubules). Assume that a plasma constituent, substance X, has these properties. If, for example, 125 mL/min of plasma are filtered and subsequently reabsorbed, the quantity of substance X originally contained within the 125 mL is left behind in the tubules to be excreted. Thus 125 mL of plasma are cleared of substance X each minute (● Figure 12–22a).

There is no endogenous chemical with the characteristics of substance X. All substances normally present in the plasma, even wastes, are reabsorbed or secreted to some extent. However, **inulin** (do not confuse with insulin), a harmless foreign carbohydrate produced by Jerusalem artichokes (which are

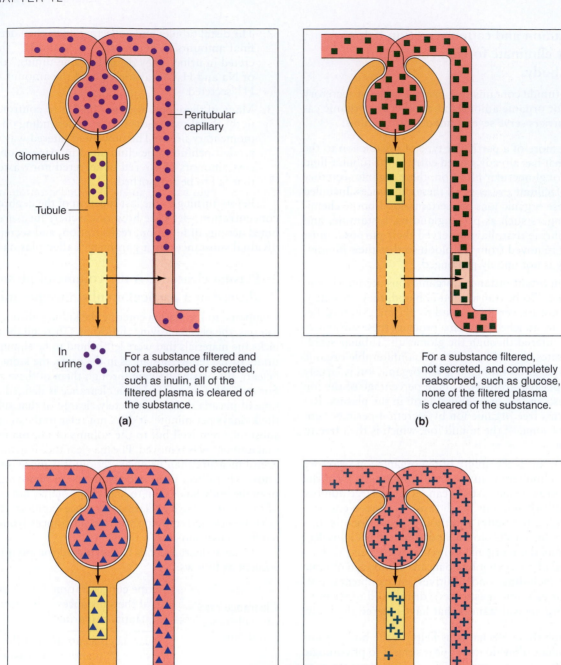

In urine

For a substance filtered and not reabsorbed or secreted, such as inulin, all of the filtered plasma is cleared of the substance.

Peritubular capillary

Glomerulus

Tubule

(a)

For a substance filtered, not secreted, and completely reabsorbed, such as glucose, none of the filtered plasma is cleared of the substance.

(b)

For a substance filtered, not secreted, and partially reabsorbed, such as urea, only a portion of the filtered plasma is cleared of the substance.

(c)

For a substance filtered and secreted but not reabsorbed, such as hydrogen ion, all of the filtered plasma is cleared of the substance, and the peritubular plasma from which the substance is secreted is also cleared.

(d)

Figure 12–22 ● Plasma clearance for substances handled in different ways by mammalian kidneys.

not artichokes, but the tubers of a sunflower species), is freely filtered and not reabsorbed or secreted—an ideal substance X. Inulin can be injected and its plasma clearance determined as a clinical means of ascertaining the GFR. Because all glomerular filtrate formed is cleared of inulin, the volume of plasma cleared of inulin per minute equals the volume of plasma filtered per minute—that is, the GFR.

Although the determination of inulin plasma clearance is accurate and straightforward, it is not very convenient, because inulin must be infused continuously throughout the determination to maintain a constant plasma concentration. Therefore, the plasma clearance of an endogenous substance, **creatinine**, is often used instead to give a rough estimate of the GFR. Creatinine, an end product of muscle metabolism, is produced at a relatively constant rate. It is freely filtered and not reabsorbed but is slightly secreted. Accordingly, creatinine clearance is not a completely accurate reflection of the GFR, but it does provide a close approximation and can be more readily determined than inulin clearance.

If a substance is filtered and reabsorbed but not secreted, its plasma clearance rate is always less than the GFR. Because less than the filtered volume of plasma will have been cleared of the substance, the plasma clearance rate of a reabsorbable substance is always less than the GFR. For example, the plasma clearance for glucose is normally zero. All the filtered glucose is reabsorbed along with the rest of the returning filtrate, so none of the plasma is cleared of glucose (Figure 12–22b).

For a substance that is partially reabsorbed, such as urea, only part of the filtered plasma is cleared of that substance. With about 50% of the filtered urea being passively reabsorbed, only half of the filtered plasma, or 62.5 mL, is cleared of urea each minute (Figure 12–22c).

If a substance is filtered and secreted but not reabsorbed, its plasma clearance rate is always greater than the GFR. Tubular secretion allows the kidneys to clear certain materials from the plasma more efficiently. Only 20% of the plasma entering the kidneys is filtered. The remaining 80% passes unfiltered into the peritubular capillaries. The only means by which this unfiltered plasma can be cleared of any substance during this trip through the kidneys before being returned to the general circulation is by the process of secretion. An example is H^+. Not only will the plasma that is filtered be cleared of nonreabsorbable H^+, but the plasma from which H^+ is secreted will also be cleared of H^+. For example, if the quantity of H^+ that is secreted is equivalent to the quantity of H^+ present in 25 mL of plasma, the clearance rate for H^+ will be 150 mL/min at the normal GFR of 125 mL/min. Every minute 125 mL of plasma will lose its H^+ through the process of filtration and failure of reabsorption, and 25 more mL of plasma will lose its H^+ through the process of secretion (Figure 12–22d).

Osmoconcentration

Having considered how the kidneys deal with a variety of solutes and water in the glomerulus and proximal and distal tubules, we now focus on renal handling of NaCl and H_2O in the loop of Henle and collecting duct, in a process we call *osmoconcentration*. Although these segments are engaged in reabsorption of NaCl and water, we consider the process a distinct renal function because it occurs only in birds and mammals (among vertebrates) and because NaCl reabsorption does not necessarily return NaCl to the body.

■ The ability to excrete urine of varying concentrations depends on the medullary countercurrent-multiplier system and vasopressin.

The ECF osmolarity (solute concentration) depends on the relative amount of H_2O compared to solute, primarily NaCl. At normal fluid balance and solute concentration, mammalian body fluids are said to be **isotonic** or **isosmotic** at an osmolarity of about 300 milliosmoles/liter (directly proportional to osmotic pressure; hereafter abbreviated as **mOsm**) in most terrestrial vertebrates (see Chapter 13, p. 576). If there is too much H_2O relative to the solute load, the body fluids are **hypotonic** or **hypo-osmotic,** which means that they are too dilute (osmolarity less than 300 mOsm). In contrast, if a H_2O deficit exists relative to the solute load, the body fluids are too concentrated or are **hypertonic** or **hyperosmotic** (osmolarity greater than 300 mOsm).

Generally speaking, the osmolarity of the ECF is uniform throughout the body (except, as you will see, in the renal medulla, and also intestinal villi exposed to swallowed water). Knowing that the driving force for H_2O reabsorption throughout the entire length of the tubules is an osmotic gradient between the tubular lumen and surrounding interstitial fluid, you would expect, based on osmotic considerations, that the kidneys could not excrete urine more or less concentrated than the body fluids. Indeed, this would be the case if the interstitial fluid surrounding the tubules in the kidneys were identical in osmolarity to the remaining body fluids. Water reabsorption would proceed only until the tubular fluid equilibrated osmotically with the interstitial fluid, and there would be no way to eliminate excess H_2O when the body fluids were hypotonic or to conserve H_2O in the presence of hypertonicity.

Fortunately, a large **vertical osmotic gradient** is uniquely generated in the interstitial fluid of the medulla of avian and mammalian kidneys. The concentration of the interstitial fluid progressively increases from the cortical boundary down through the depth of the renal medulla until it reaches a maximum at the junction with the renal pelvis (● Figure 12–23). This vertical osmotic gradient remains there (although it may change in magnitude) regardless of the fluid balance of the body.

The presence of this gradient enables the kidneys to produce urine that ranges in concentration from less than 100 mOsm to highly hypertonic, depending on the species and a body's state of hydration. Examples of maximal concentrating ability are shown in ■ Table 12–3, ranging from a modest 520 mOsm in the beaver (a freshwater animal with plentiful water supply) to a reported 9400 mOsm in one desert rodent! As a benchmark value to keep in mind, seawater is about 1000 mOsm, mostly NaCl. As the table shows, humans can create a urine up to 1200 mOsm, but much of this is due to urea. Human kidneys cannot concentrate NaCl to 1000 mOsm, so humans cannot safely drink seawater, because it causes a net accumulation of salt in the body. Desert rodents, marine mammals, and some species of birds, in contrast, have no trouble drinking highly saline water (and eating salty food) and excreting excess salt, with a net gain of water to the body.

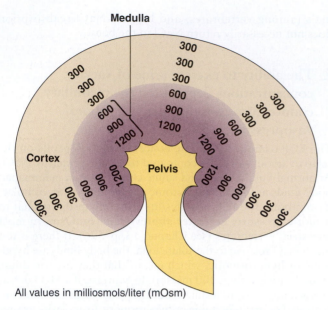

Medulla

Cortex

Pelvis

All values in milliosmols/liter (mOsm)

Figure 12–23 ● **Vertical osmotic gradient in the human renal medulla.** The osmolarity of the interstitial fluid throughout the renal cortex is isotonic at 300 mosm/liter, but the osmolarity of the interstitial fluid in the renal medulla increases progressively from 300 mosm/liter at the boundary with the cortex to a maximum of 1200 mosm/liter at the junction with the renal pelvis.

When the human body is in ideal fluid balance, 1 mL/min of isotonic urine forms. When the body is overhydrated (too much H_2O), the kidneys can produce a large volume of dilute urine (up to 25 mL/min and hypotonic at 100 mOsm or less),

thus eliminating the excess H_2O in the urine. Conversely, the kidneys can put out a small volume of concentrated urine (down to 0.3 mL/min and hypertonic) when the body is dehydrated, thus conserving H_2O.

Unique anatomic arrangements and complex functional interactions between the various nephron components present in the renal medulla are responsible for the establishment and use of the vertical osmotic gradient. Recall that the hairpin loop of Henle in the cortical nephrons dips only slightly into the medulla, but in the juxtamedullary nephrons the loop plunges through the entire depth of the medulla so that the tip of the loop lies near the renal pelvis (Figure 12–7). Also, the vasa recta of the juxtamedullary nephrons follow the same deep hairpin loop as the long loop of Henle. Flow in both the long loops of Henle and the vasa recta is considered **countercurrent,** because the flow in the two closely adjacent limbs of the loop and blood vessel is in opposite directions. Also running through the medulla, in the descending direction only, on their way to the renal pelvis are the collecting ducts that serve both types of nephrons. This arrangement, coupled with the permeability and transport characteristics of these tubular segments, plays a key role in the kidneys' ability to provide for water conservation or elimination. Briefly, the juxtamedullary nephrons' long loops of Henle establish the vertical osmotic gradient, their vasa recta prevent the dissolution of this gradient while providing blood to the renal medulla, and the collecting ducts use the gradient, in conjunction with the hormone vasopressin, to produce urine of varying concentrations. Collectively, this entire functional organization is known as the medullary **countercurrent-multiplier** system. We examine each of its facets in greater detail.

The long loops of Henle of juxtamedullary nephrons establish the medullary vertical osmotic gradient by means of countercurrent multiplication. Let's follow the filtrate through

Table 12–3 ▌ Urine-Concentrating Abilities of the Kidney of Selected Mammals, and Urine Concentrations of Urea of Rats on Different Diets

Mammal	Urine Osmotic Concentration (Osm)
Beaver	0.52 maximum
Pig	1.1 maximum
Human	1.4 maximum
White rat	2.9 maximum
Cat	3.1 maximum
Kangaroo rat	5.5 maximum
Hopping mouse	9.4 maximum
Rabbit, normal water intake	0.4
Rabbit, no water for 2 days	1.3

White Rat	Urea Concentration in Urine (M)
Low-protein diet	0.026
High-protein diet	1.4

A kangaroo rat, *Dipodomys ordii,* of western North American deserts. Its efficient kidneys and other adaptations allow it to live without ever drinking free water.

Sources: Data are from R. E. MacMillen, & A. K. Lee, 1967, *Science* 158:383–385; D. P. Peterson, K. M. Murphy, R. Ursino, K. Streeter, & P. H. Yancey, 1992, *American Journal of Physiology* 263:F594–F600; B. Schmidt-Nielsen & R. O'Dell, 1961, *American Journal of Physiology* 200:1119–1124; K. Schmidt-Nielsen, 1964, *Desert Animals: Physiological Problems of Heat and Water,* Oxford, UK: Clarendon Press; K. Schmidt-Neilsen, 1997, *Animal Physiology,* Cambridge, UK: Cambridge University Press; P. H. Yancey & M. B. Burg, 1989, *American Journal of Physiology* 257:F602–607.

a long-looped nephron to see how this structure works. Immediately after the filtrate is formed, osmotic reabsorption of filtered H_2O occurs in the proximal tubule secondary to active Na^+ reabsorption. As a result, by the end of the proximal tubule 45 to 65% of the filtrate has been reabsorbed, but the 35 to 55% remaining in the tubular lumen still has the same osmolarity as the body fluids. Therefore, the fluid entering the loop of Henle is still isotonic. An additional 15% of the filtered H_2O is reabsorbed from the loop of Henle during the establishment and maintenance of the vertical osmotic gradient, with the osmolarity of the tubular fluid being altered in the process.

The following functional distinctions between the **descending limb** of a long Henle's loop (which carries fluid from the proximal tubule down into the depths of the medulla) and the **ascending limb** (which carries fluid up and out of the medulla into the distal tubule) are critical to the establishment of the incremental osmotic gradient in the medullary interstitial fluid. The ascending limb is subdivided into two parts, a **thin segment** in the medulla and a **thick segment** primarily in the cortex. The descending limb

- Has high permeability for H_2O, and low permeability for NaCl.
- Does not actively extrude Na^+.

The thin ascending limb

- Has high permeability for NaCl, and low permeability for H_2O and urea.
- Does not actively extrude Na^+.

The thick ascending limb

- Actively transports NaCl out of the tubular lumen into the surrounding interstitial fluid (by initially transporting Cl^- with Na^+ following by charge attraction).
- Is always impermeable to H_2O (due to *uromodulin*, a large gel-like protein), so salt leaves the tubular fluid without H_2O osmotically following along. It is also impermeable to urea.

The close proximity and countercurrent flow of the two limbs allow important interactions to occur between them. Even though the flow of fluids is continuous through the loop of Henle, we can visualize what happens step by step, much like an animated film run so slowly that each individual frame can be viewed. We use values for human kidneys to illustrate:

- *Initial scene* (● Figure 12–24a). Before the vertical osmotic gradient is established, the medullary interstitial fluid concentration is uniformly 300 mOsm, as is the remainder of the body fluids. The tubular fluid entering the descending limb from the proximal tubule is also about 300 mOsm.
- *Step 1* (Figure 12–24b). The active salt pump in the thick ascending limb can transport NaCl out of the lumen until the surrounding interstitial fluid is 200 mOsm more concentrated than the tubular fluid in this limb. That is, the interstitial fluid becomes hypertonic outside both the thick ascending limb, and, across the loop, the upper descending limb. Water cannot follow osmotically from the thick ascending limb, because it is impermeable to H_2O. However, there is osmotic movement of H_2O from the descend-

ing limb into the interstitial fluid, because the descending limb is highly permeable to H_2O. The passive movement of H_2O out of the descending limb continues until the osmolarities of the fluid in the descending limb and interstitial fluid become equilibrated. Thus the tubular fluid entering the loop of Henle immediately starts to become more concentrated as it loses H_2O. At equilibrium, the osmolarity of the ascending limb fluid is 200 mOsm and the osmolarities of the interstitial fluid and descending limb fluid are equal at 400 mOsm.

- *Step 2* (Figure 12–24c). If we now advance the entire column of fluid in the loop of Henle several "frames," a volume of 200-mOsm fluid exits from the top of the thick ascending limb into the distal tubule, and a new volume of isotonic fluid at 300 mOsm enters the top of the descending limb from the proximal tubule. At the bottom of the loop, a comparable volume of 400-mOsm fluid from the descending limb moves forward around the tip into the ascending limb, placing it opposite a 400-mOsm region in the descending limb. Note that the 200-mOsm concentration difference has been lost at both the top and the bottom of the loop.
- *Step 3* (Figure 12–24d). The thick ascending limb pump continues to transport NaCl out, while H_2O passively leaves the descending limb until a 200-mOsm difference is re-established between the ascending limb and both the interstitial fluid and descending limb at each horizontal level. Note, however, that the concentration of tubular fluid is progressively increasing in the descending limb and progressively decreasing in the ascending limb.
- *Step 4* (Figure 12–24e). As the tubular fluid is advanced still further, the 200-mOsm concentration gradient is disrupted once again at all horizontal levels.
- *Step 5* (Figure 12–24f). Again, active extrusion of NaCl from the thick ascending limb, coupled with the net diffusion of H_2O out of the descending limb, re-establishes the 200-mOsm gradient at each horizontal level.
- *Steps 6 and on* (Figure 12–24g). As the fluid flows slightly forward again and this stepwise process continues, the fluid in the descending limb becomes progressively more hypertonic until it reaches a maximum concentration of 1200 mOsm in humans (and up to 9400 mOsm in some desert rodents) at the bottom of the loop, considerably more concentrated than body fluids (up to 25 times in desert rodents). Because the interstitial fluid always achieves equilibrium with the descending limb, an incremental vertical concentration gradient ranging from 300 to the maximum (for example, 1200 mOsm) is likewise established in the medullary interstitial fluid. In contrast, the concentration of the tubular fluid progressively decreases in the ascending limb as salt is pumped out but H_2O cannot follow. In fact, the tubular fluid becomes hypotonic as it leaves the ascending limb to enter the distal tubule at a concentration of 100 mOsm or less, one third the normal concentration of body fluids.

Note that although a gradient of only 200 mOsm exists between the ascending limb and the surrounding fluids at each medullary horizontal level, a much larger vertical gradient exists from the top to the bottom of the medulla. Even though the as-

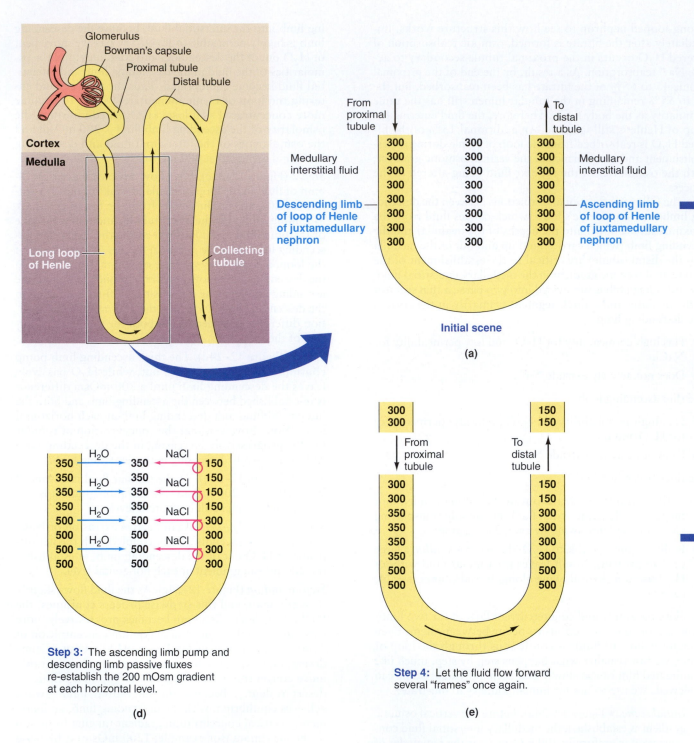

Figure 12–24 ● Countercurrent multiplication in the mammalian renal medulla.

cending limb pump can generate a gradient of only 200 mOsm, this effect is multiplied into a large vertical gradient because of the countercurrent flow within the loop. This concentrating mechanism accomplished by the loop of Henle is known as **countercurrent multiplication.**

We have artificially described countercurrent multiplication in a "stop-and-flow," stepwise fashion to facilitate understanding. It is important to realize that once the incremental medullary gradient is established, it remains constant,

because of the continuous flow of fluid coupled with the on-going ascending-limb active transport activity and accompanying descending-limb passive fluxes.

If you consider only what happens to the tubular fluid as it flows through the loop of Henle, the whole process seems an exercise in futility. The isotonic fluid that enters the loop becomes progressively more concentrated as it flows down the descending limb, achieving a maximum concentration only to become progressively more dilute as it flows up the as-

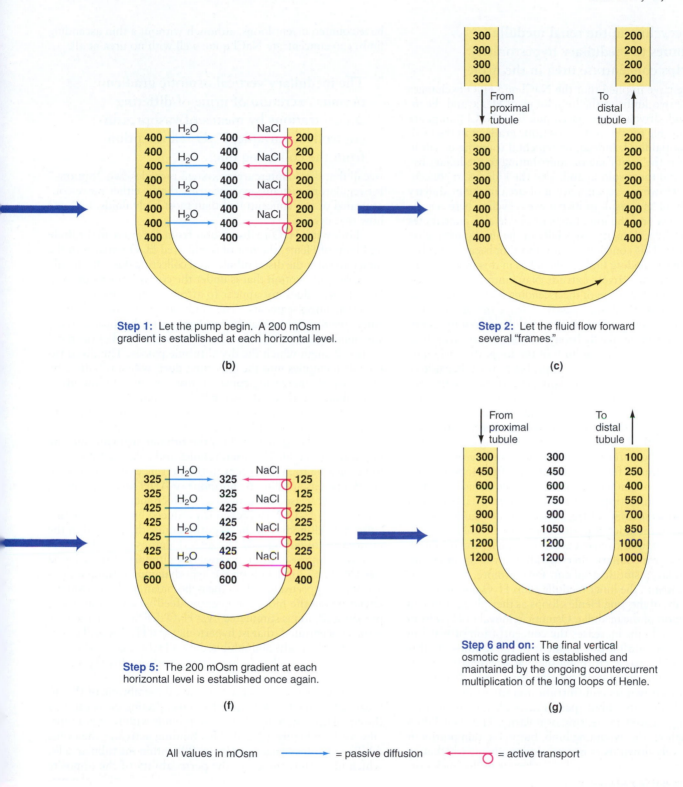

Step 1: Let the pump begin. A 200 mOsm gradient is established at each horizontal level.

(b)

Step 2: Let the fluid flow forward several "frames."

(c)

Step 5: The 200 mOsm gradient at each horizontal level is established once again.

(f)

Step 6 and on: The final vertical osmotic gradient is established and maintained by the ongoing countercurrent multiplication of the long loops of Henle.

(g)

All values in mOsm   ——→ = passive diffusion   ←——○ = active transport

cending limb, finally leaving the loop at a minimum concentration of 100 mOsm. What is the point of concentrating the fluid fourfold and then turning around and diluting it until it leaves at one third the concentration at which it entered? Such a mechanism offers two benefits, or "choices" that homeostatic regulation mechanisms can use:

■ It establishes a vertical osmotic gradient in the medullary interstitial fluid. This gradient, in turn, is used by the col-

lecting ducts to concentrate the tubular fluid so that *a urine more concentrated* than normal body fluids can be excreted.

■ Conversely, the fact that the fluid is hypotonic as it enters the distal portions of the tubule enables the kidneys to excrete *a urine more dilute* than normal body fluids.

You will see how these "choices" are made shortly, after we examine the role of urea.

■ Urea recycling in the renal medulla contributes to medullary hypertonicity and helps concentrate urea in the urine.

Thus far we have implied that the NaCl and H$_2$O exchanges between the two limbs of the long loops of Henle and the interstitial fluid, driven by the ascending-limb NaCl pump, are solely responsible for the vertical osmotic gradient in the medullary interstitial fluid. Indeed, incremental trapping of salt in the medulla is the major factor contributing to medullary hypertonicity and the resulting ability of the kidneys to concentrate urine. However, accumulation of urea in the medullary interstitial fluid as a result of its passive recycling between the collecting ducts and the long loops of Henle also contributes to medullary hypertonicity, especially in the innermost region of the medulla. This contribution is important because the thin portion of the ascending limbs, which lies in this inner medullary region, cannot actively extrude NaCl into the interstitial fluid; only the thick portion of the ascending limb has active salt pumps. However, NaCl does passively leave the thin portion of the ascending limb, which in part may be driven by a concentration gradient ultimately brought about by urea recycling.

Like the thick ascending limb of the loop, the early portion of the collecting duct is impermeable to urea (because of lack of urea transporters). Consequently, urea becomes progressively more concentrated in this segment as H$_2$O is reabsorbed in the presence of vasopressin, but the urea cannot move out of the lumen down its concentration gradient. By the time the fluid reaches the last portion of the collecting duct, its urea concentration is greater than in the surrounding interstitial fluid and the tubular fluid at the bottom of the long Henle's loops. This concentration difference favors the diffusion of urea out of the late collecting duct into the interstitial fluid and urea-permeable bottoms of the long loops (● Figure 12–25). This section of the duct contains many urea transporters in its membranes. This movement creates a high concentration of urea in the inner medullary interstitial fluid, which contributes to the medullary osmotic gradient. Furthermore, the osmotically active urea in the inner medulla pulls H$_2$O out of the descending limbs of the long Henle's loops as they plunge through this deep region of the medulla. Osmotic removal of H$_2$O from the descending limbs increases the concentration of NaCl in the tubular fluid within this segment of the tubules. As this NaCl-rich fluid starts up through the NaCl-permeable, H$_2$O-impermeable thin portion of the ascending limbs, NaCl passively diffuses down its concentration gradient out into the interstitial fluid as the tubule passes through surrounding regions of progressively decreasing osmolarity. Thus NaCl does leave throughout the ascending limb, but in the thin portion it leaves passively down its concentration gradient, which may in part be created by urea recycling, whereas in the thick portion NaCl is actively extruded.

It should be noted that this model for the enhancing effect of urea on sodium concentration is not without problems, because there are some apparently contradictory cases. For example, the desert sand rat (*Psammomys obesus*), which eats salty dry seeds, can concentrate urine NaCl more effectively than most rodents but is relatively poor at concentrating urea. Similarly, laboratory rats fed a high-salt, low-protein diet excrete very salty urines with very low urea in either the urine or medulla (Table 12–3). Moreover, many birds (which also have countercurrent loops, although without a thin ascending limb) can concentrate NaCl quite well with no urea at all.

■ The medullary vertical osmotic gradient permits excretion of urine of differing concentrations by means of vasopressin-controlled, variable, H$_2$O reabsorption from the collecting duct.

Recall that the countercurrent system provides two "options" for regulatory systems: excrete urine that is either more concentrated or less concentrated than most body fluids. Let's see how that occurs.

Following H$_2$O reabsorption from the proximal tubule and loop of Henle, the rest of the filtered H$_2$O remains in the lumen to enter the distal tubules and collecting ducts for highly variable reabsorption that is under the control of **vasopressin.** (Vasopressin does not appear to affect water reabsorption in earlier nephron segments.) The fluid leaving the loop of Henle enters the distal tubule at 100 mOsm, so it is hypotonic to the surrounding isotonic (300 mOsm) interstitial fluid of the renal cortex through which the distal tubule passes. The distal tubule then empties into the collecting duct, which is bathed by progressively increasing concentrations of surrounding interstitial fluid as it descends through the medulla.

For H$_2$O absorption to occur across a segment of the tubule, two criteria must be met: (1) An osmotic gradient must exist across the tubule, and (2) the tubular segment must be permeable to H$_2$O. The distal tubules and collecting ducts are impermeable to H$_2$O except in the presence of vasopressin, also known as **antidiuretic hormone** or **ADH** (*anti*, "against"; *diuretic*, "increased urine output"), which increases their permeability to H$_2$O. Vasopressin is produced by several specific neuronal-cell bodies in the hypothalamus, then stored in the posterior pituitary gland. The hypothalamus controls the release of vasopressin from the posterior pituitary into the blood (see Chapter 7 for details on the hypothalamus–pituitary interaction). Osmosensing cells in the hypothalamus detect osmotic deviations in the blood. In negative-feedback fashion, vasopressin secretion is stimulated by a H$_2$O deficit, when the ECF is too concentrated (that is, hypertonic) and H$_2$O must be conserved for the body, and inhibited by a H$_2$O excess, when the ECF is too dilute (that is, hypotonic) and surplus H$_2$O must be eliminated in the urine.

Vasopressin reaches the basolateral membrane of the tubular cells lining the distal tubules and collecting ducts through the circulatory system, whereupon it binds with receptors specific for it (● Figure 12–26). This binding activates the cyclic AMP second-messenger system (p. 90) within the tubular cells, which ultimately increases the permeability of the opposite luminal membrane to H$_2$O by increasing the number of **aquaporins** (water channels, p. 75) inserted in the membrane. Cyclic AMP triggers Golgi-produced vesicles carrying aquaporins to fuse with the plasma membrane, aided by movements of the cytoskeleton (p. 38). By permitting more H$_2$O to permeate from the lumen of the collecting duct, these additional channels increase H$_2$O reabsorption. The tubular response to vasopressin is graded; the more vasopressin present, the more aquaporins inserted, and the greater the permeability of the distal tubules and collecting ducts to H$_2$O. The increase in luminal

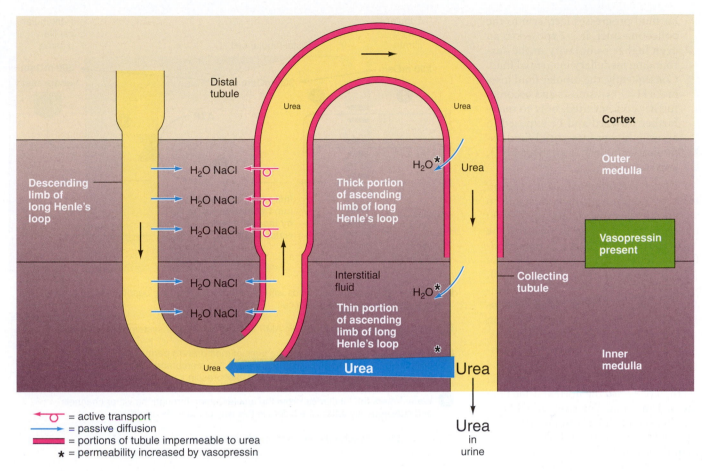

Figure 12–25 • **Urea recycling in the mammalian renal medulla.** Urea becomes more concentrated in the fluid in the early portion of the collecting tubules as H_2O is osmotically reabsorbed in the presence of vasopressin; the urea cannot move out down its concentration gradient because this segment is impermeable to urea. Urea does diffuse out of the late portion of the collecting tubule down its concentration gradient into the surrounding interstitial fluid and the nearby bottom of the long Henle's loop, because these tubular segments are permeable to urea. Vasopressin increases the permeability of the late collecting tubule to urea. The entry of urea into the interstitial fluid contributes to medullary hypertonicity in the inner region of the medulla. As the tubular fluid passes through the ascending limb and the distal tubule, urea cannot leave, because these segments are impermeable to it; thus the urea cannot diffuse out even though the fluid is passing through regions with lower concentrations of urea. The tubular fluid's urea concentration increases even further as water is reabsorbed when the fluid enters the early portion of the collecting tubule once again. Thus, when vasopressin is secreted in the presence of a H_2O deficit, this urea recycling progressively concentrates urea in the tubule fluid that is excreted as urine.

membrane aquaporins is not permanent, however. The channels are retrieved by endocytosis (p. 41) into vesicles when vasopressin secretion decreases and cyclic AMP activity is similarly decreased. Accordingly, H_2O permeability is reduced when vasopressin secretion decreases.

Thus the system can be used in two ways:

1. *Dehydration.* When vasopressin secretion is increased in response to a H_2O deficit, and the permeability of the distal tubules and collecting ducts to H_2O is accordingly increased, the hypotonic tubular fluid entering the distal

tubules can lose progressively more H_2O by osmosis into the interstitial fluid. This occurs as the tubular fluid first flows through the isotonic cortex and then is exposed to the ever-increasing osmolarity of the medullary interstitial fluid as it plunges toward the renal pelvis (• Figure 12–27a). As the 100-mOsm tubular fluid enters the distal tubule and is exposed to a surrounding interstitial fluid of 300 mOsm, H_2O leaves the tubular fluid by osmosis across the now-permeable tubular cells until the tubular fluid reaches a maximum concentration of 300 mOsm by the end of the distal tubule. As this 300-mOsm tubular fluid progresses farther into the collecting duct, it is exposed to an even higher osmolarity in the surrounding medullary interstitial fluid. Consequently, the tubular fluid loses more H_2O by osmosis and becomes further concentrated, only to move farther forward and be exposed to an even higher interstitial fluid osmolarity and lose even more H_2O, and so on.

Under the influence of maximum levels of vasopressin, it is possible to concentrate the tubular fluid (up to 9400 mOsm in the hopping mouse) by the end of the collecting ducts. Further modification of the tubular fluid occurs may occur in the renal pelvis (p. 561 later), but the fluid at this point is usually considered urine. As a result of this extensive vasopressin-promoted reabsorption of H_2O in the late segments of the tubule, a small volume of very concentrated urine is produced. As little as 0.3 mL of urine may be formed each minute, less than one third the normal urine flow rate of 1 mL/min. The reabsorbed H_2O entering the medullary interstitial fluid is picked up by the peritubular capillaries and returned to the general circulation, thus being conserved for the body.

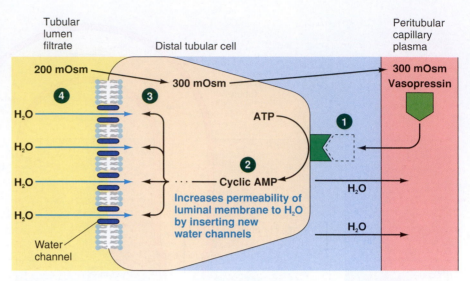

① Blood-borne vasopressin binds with its receptor sites on the basolateral membrane of a distal or collecting tubule cell.

② This binding activates the cyclic AMP second-messenger system within the cell.

③ Cyclic AMP increases the opposite luminal membrane's permeability to H_2O by promoting the insertion of water channels in this membrane. This membrane is impermeable to water in the absence of vasopressin.

④ Water enters the tubular cell from the tubular lumen through the water channels and subsequently enters the blood, in this way being reabsorbed.

Figure 12–26 ● Mechanism of action of vasopressin in a mammal.

Although vasopressin promotes H_2O conservation by the body, it cannot completely halt urine production, even when an animal is not taking in any H_2O, because a minimum volume of H_2O must be excreted with the wastes. Collectively, the waste products and other constituents eliminated in human urine average 600 mOsm each day. Because the maximum human urine concentration is 1200 mOsm, the minimum volume of urine that is required to excrete these wastes is 500 mL/day (600 mOsm of wastes/day ÷ 1200 mOsm of urine = 0.5 liter, or 500 mL/day, or 0.3 mL/min). Thus under maximal vasopressin influence, 99.8% of the 180 liters of plasma H_2O filtered per day is returned to the blood, with an obligatory H_2O loss of half a liter.

2. *Overhydration.* Conversely, when an animal consumes large quantities of H_2O, the excess H_2O must be removed from the body without simultaneously losing solutes that are critical to the maintenance of homeostasis. Under these circumstances, no vasopressin is secreted, so the distal tubules and collecting ducts remain largely impermeable to H_2O. The tubular fluid entering the distal tubule is hypotonic (100 mOsm), having lost salt without an accompanying loss of H_2O in the ascending limb of Henle's loop. As this hypotonic fluid passes through the distal tubules and collecting ducts (Figure 12–27b), the medullary osmotic gradient cannot exert any influence because of the late tubular segments' impermeability to H_2O. In other words, none of the H_2O remaining in the tubules can leave the lumen to be reabsorbed even though the tubular fluid is less concentrated than the surrounding interstitial fluid. Thus, in the absence of vasopressin the filtered fluid

that reaches the distal tubule fails to be reabsorbed. Meanwhile, excretion of wastes and other urinary solutes remains constant. The net result is a large volume of dilute urine, which helps rid the body of excess H_2O. Urine osmolarity may be 100 mOsm or less, the same as in the fluid entering the distal tubule. Urine flow may be increased up to 25 mL/min in the absence of vasopressin, compared to the normal urine production of 1 mL/min.

Changes in urine concentration indicate how much of the variably reabsorbed H_2O has been conserved for the body. Production of a large volume of dilute urine means little or none of the filtered H_2O subject to control has been returned to the plasma. In contrast, excretion of a small volume of concentrated urine implies extensive reabsorption of the controllable portion of the filtered H_2O. The extent of reabsorption varies directly with the amount of vasopressin secreted, which in turn is dependent on the body's state of hydration. Varying the amount of vasopressin secreted in proportion to the body's need for H_2O conservation enables the fine adjustments in H_2O reabsorption and excretion that are necessary to maintain proper fluid balance.

■ Countercurrent exchange within the vasa recta enables the medulla to be supplied with blood while conserving the medullary vertical osmotic gradient.

Obviously, the renal medulla must be supplied with blood to nourish the tissues in this area as well as to transport water that is reabsorbed by the loops of Henle and collecting ducts

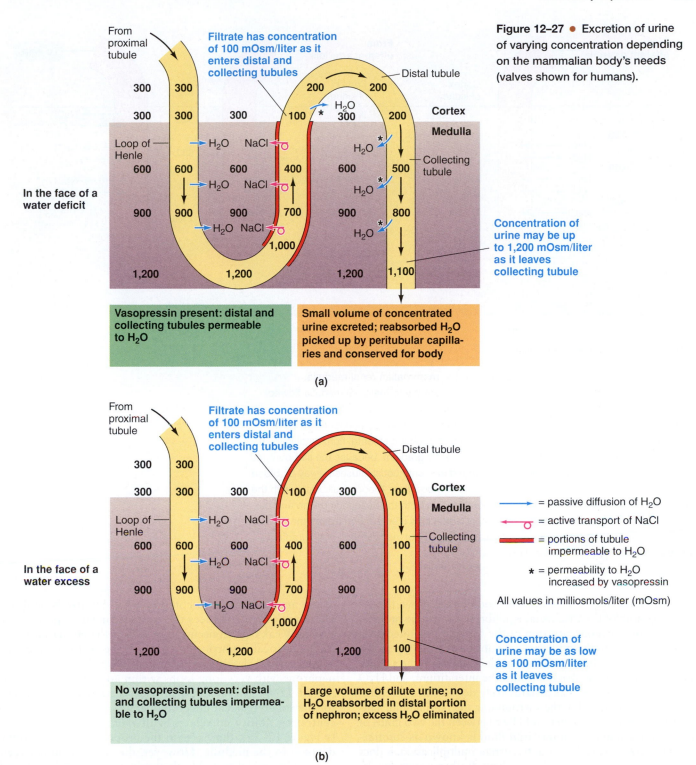

Figure 12–27 ● Excretion of urine of varying concentration depending on the mammalian body's needs (valves shown for humans).

(a)

In the face of a water deficit

Vasopressin present: distal and collecting tubules permeable to H₂O

Small volume of concentrated urine excreted; reabsorbed H₂O picked up by peritubular capillaries and conserved for body

(b)

In the face of a water excess

No vasopressin present: distal and collecting tubules impermeable to H₂O

Large volume of dilute urine; no H₂O reabsorbed in distal portion of nephron; excess H₂O eliminated

back to the general circulation. In doing so, however, it is important that circulation of blood through the medulla does not disturb the vertical gradient of hypertonicity established by the loops of Henle and urea recycling. Consider the situation if blood were to flow straight through from the cortex to the inner medulla and then directly into the renal vein (● Figure 12–28a). Because capillaries are freely permeable to NaCl and H₂O, the blood would progressively pick up salt and lose H₂O through passive fluxes down concentration and osmotic gradients as it flowed through the depths of the medulla. Iso-

tonic blood entering the medulla, on equilibrating with each medullary level, would leave the medulla very hypertonic at 1200 mOsm (in humans). It would be impossible to establish and maintain the medullary hypertonic gradient, because the NaCl pumped into the medullary interstitial fluid would continuously be carried away by the circulation.

This dilemma is avoided by the hairpin construction of the vasa recta, which, by looping back through the concentration gradient in reverse, allows the blood to leave the medulla and enter the renal vein essentially isotonic to incoming

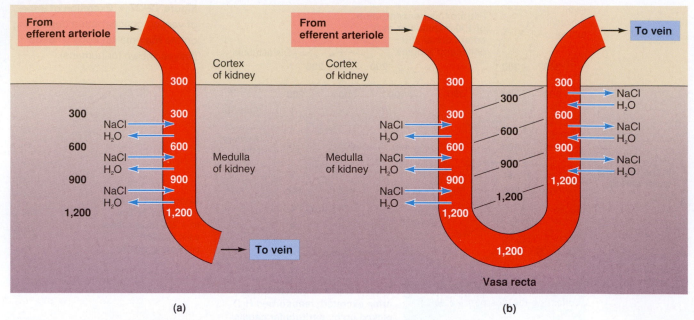

All values in milliosmols/liter (mOsm).

Figure 12–28 ● Countercurrent exchange in the mammalian renal medulla.
(a) Hypothetical pattern of blood flow. If the blood supply to the renal medulla flowed
straight through from the cortex to the inner medulla, the blood would be isotonic
on entering but very hypertonic on leaving, having picked up salt and lost H_2O as it
equilibrated with the surrounding interstitial fluid at each incremental horizontal level. It
would be impossible to maintain the vertical osmotic gradient, because the salt pumped
out by the ascending limb of Henle's loop would be continuously flushed away by blood
flowing through the medulla. (b) Actual pattern of blood flow. Blood equilibrates with the
interstitial fluid at each incremental horizontal level in both the descending limb and the
ascending limb of the vasa recta, so blood is isotonic as it enters and leaves the medulla.
This countercurrent exchange prevents dissolution of the medullary osmotic gradient
while providing blood to the renal medulla.

arterial blood (Figure 12–28b). As blood passes down the descending limb of the vasa recta, equilibrating with the progressively increasing concentration of the surrounding interstitial fluid, it picks up salt and loses H_2O until it is very hypertonic by the bottom of the loop. Then, as blood flows up the ascending limb, salt diffuses back out into the interstitium, and H_2O re-enters the vasa recta as progressively decreasing concentrations are encountered in the surrounding interstitial fluid. This passive exchange of solutes and H_2O between the two limbs of the vasa recta and the interstitial fluid is known as *countercurrent exchange*. Unlike countercurrent multiplication, it does not establish the concentration gradient. Rather, it prevents the dissolution of the gradient. Because blood enters and leaves the medulla at the same osmolarity as a result of countercurrent exchange, the medullary tissue is nourished with blood, yet the incremental gradient of hypertonicity in the medulla is preserved.

■ Different osmoconcentrating abilities among species depend on nephron anatomy and metabolic rates.

As noted in Table 12–3, the ability to osmoconcentrate the urine varies enormously among mammals. How is this achieved?

There are probably two components. The first is nephron anatomy, which enhances the countercurrent aspect of the system. Weak osmoconcentrators such as the beaver have primarily cortical nephrons; in contrast, the best osmoconcentrators (such as desert rodents) usually have all juxtamedullary nephrons with very long loops within an elongated medulla (Figure 12–8). This anatomy provides a relatively long length for the countercurrent effect to operate. Related to this, concentrating ability correlates with *relative medullary area* (● Figure 12–29), that is, with the proportion of the kidney devoted to the medulla. However, the correlation between concentrating ability and both absolute medullary length and relative area cannot fully explain the amazing osmoconcentrating ability of some desert rodents (Table 12–3). Whereas these animals do have much longer loops relative to rodents from moist habitats, the loops are much shorter on an absolute scale than those of a large mammal that is a poor or modest osmoconcentrator (such as a pig). Another factor is necessary, and it seems to be the metabolic rate of the animal, which could enhance the multiplier aspect of the countercurrent. As you will see in Chapter 15, smaller animals have higher metabolic rates per unit mass than larger animals. In the nephron loop, this is manifested as cells with a higher density of mito-

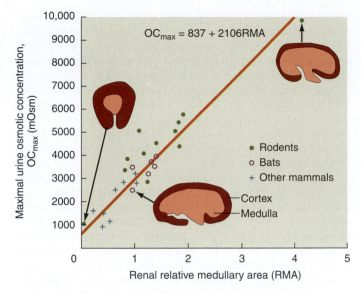

Figure 12–29 ● **The relationship between relative medullary area in the mammalian kidney (taken at the midline in sagittal section) and the maximal urine concentration that can be produced.** In general, proportionally larger medullas produce a more concentrated product; note that the relatively largest medullas and very concentrated urines mostly occur in rodents, which are of small body size and common in arid habitats. Insets show diagrammatic cross sections for three kidneys (not to scale).

(*Source:* P. Withers, 1992, *Comparative Animal Physiology,* Fort Worth, TX: Saunders, Figure 17.33, p. 873)

chondria and active transport proteins. Thus the multiplier effect of the loop that depends on active metabolism should be much higher in a small mammal.

The desert rodents are particularly remarkable in their urine-concentrating abilities. Some species live on relatively dry plant materials in a hot, dehydrating environment without drinking any free water for their entire lives. Kangaroo rats of western North American deserts have been closely studied to see how they accomplish this amazing feat. First, like any animal, kangaroo rats have an internal source of water called *metabolic water.* Recall that the basic equation for aerobic

metabolism (p. 49) produces water as an end product, as follows for glucose:

$$C_6H_{12}O_6 + 6\ O_2 \rightarrow 6\ CO_2 + 6\ H_2O + 36\ ATP$$

Kangaroo rats get about 90% of their water needs from metabolic water and 10% from tissue water in its food. (Humans can use metabolic water, of course, but only about 12% of water needs can be met with this.) Beyond use of metabolic water, adaptation in the kangaroo rat involves both behavior and internal physiology. *Avoidance,* in the form of nocturnal foraging above ground and daytime residence in an insulating burrow, reduces both heat exposure and water loss (behavior effectors). Physiological adaptations include a nasal passage that traps moisture (p. 695), and reduced excretory loss. The lower intestine removes almost all the fecal water, so that feces consist of nearly dry pellets. And, perhaps most importantly, the kidneys can produce a urine of up to 5500 mOsm (Table 12–3). This saves water, but aldosterone regulation also saves sodium, which is low in this animal's seed-based diet: The urine osmolality is primarily K salts and urea rather than Na salts.

One open question is whether the renal mechanisms that we have discussed can account for the ability to generate urine of up to 9400 mOsm. Mathematical modeling of the countercurrent process suggests that this is not enough. Some researchers have proposed that the **renal pelvis** contributes. The pelvis has been observed to contract and relax rhythmically, moving the urine into and out of the papilla. This process may set up pressure gradients that help draw water out of the collecting ducts, enhancing the concentration process begun in the countercurrent loop.

Another issue involving osmoconcentration is the osmotic stresses faced by renal cells, particularly in the medulla. This is an area of active research; see the box, "Molecular Biology and Genomics: Osmolytes and Gene Regulation" (p. 525).

We have now finished our examination of the mammalian kidney's mechanisms. In summary, mammalian kidney nephrons use four processes—glomerular filtration, tubular reabsorption, tubular secretion, and osmoconcentration—to produce a urine of selective composition. ■ Table 12–4 summarizes the key movements of sodium and water in these processes. Hormones—primarily aldosterone, angiotensin II, ANP, and vasopressin—regulate some of these processes, so that the final composition that is excreted helps maintain homeostasis

Table 12–4 ▌ Handling of Sodium and Water by Various Tubular Segments of the Nephron

Tubular Segment	Na⁺ Reabsorption *Distinguishing features*	H₂O Reabsorption *Distinguishing features*
Proximal tubule	Active; plays a pivotal role in the reabsorption of glucose, amino acids, Cl⁻, H₂O, and urea	Obligatory osmotic reabsorption following active Na⁺ reabsorption
Loop of Henle	Active, Na⁺ along with Cl⁻ reabsorption from the ascending limb helps establish the medullary interstitial vertical osmotic gradient, which is important in the kidneys' ability to produce urine of varying concentrations and volumes, depending on the body's needs	Obligatory osmotic reabsorption from the descending limb as the ascending limb extrudes NaCl into the interstitial fluid (that is, reabsorbs NaCl)
Distal and collecting tubules	Active; variable and subject to aldosterone control; important in the regulation of ECF volume; linked to K⁺ secretion and H⁺ secretion; subject to angiotensin II control	Not linked to solute reabsorption; variable quantities of "free" H₂O reabsorption subject to vasopressin control; driving force is the vertical osmotic gradient in the medullary interstitial fluid established by the long loops of Henle; important in the regulation of ECF osmolarity

of internal body fluids. Then there are two final stages: storage of urine in the bladder, and removal from the body.

Bladder Storage and Micturition

◼ Urine is temporarily stored in the bladder, from which the process of micturition empties it.

In mammals, amphibians, and fishes, after urine has been formed by the kidneys it is carried through the ureters to the urinary bladder. The mammalian bladder wall consists of smooth muscle lined by transitional epithelium with special cells called *umbrella cells* joined by tight junctions (p. 63) that form an impermeable barrier. Physiologists once assumed the bladder was an inert sac. However, both the epithelium and the smooth muscle actively participate in the mammalian bladder's ability to accommodate large fluctuations in urine volume. Recently, researchers learned that the umbrella cells can increase and decrease their surface area by the orderly process of membrane recycling as the bladder alternately fills and empties. Membrane-enclosed cytoplasmic vesicles are inserted by means of exocytosis into the surface area during bladder filling; then the vesicles are withdrawn by endocytosis to shrink the surface area after emptying (see p. 41). As is characteristic of smooth muscle, bladder muscle can stretch tremendously without a buildup in bladder wall tension. In addition, the highly folded bladder wall flattens out during filling, to increase bladder storage capacity.

The bladder smooth muscle is richly supplied by parasympathetic fibers, stimulation of which causes bladder contraction. If the passageway through the urethra to the outside is open, bladder contraction empties urine from the bladder. The exit from the bladder, however, is guarded by two sphincters, the *internal urethral sphincter* and the *external urethral*

sphincter (● Figure 12–30). A **sphincter** is a ring of muscle that, when contracted, closes off passage through an opening. The internal urethral sphincter—which consists of smooth muscle and is not under cortical control—is not really a separate muscle but instead consists of the last portion of the bladder. Although it is not a true sphincter, it performs the same function. When the bladder is relaxed, the anatomic arrangement of the internal urethral sphincter region closes the outlet of the bladder. The external sphincter is skeletal muscle and is under cortical control via a motor neuron, which is normally active to keep the sphincter closed.

Micturition, or urination, the process of bladder emptying, is governed by two mechanisms: the micturition spinal reflex and cortical control. The micturition reflex is initiated when stretch receptors within the bladder wall are stimulated (● Figure 12–31). The greater the distension beyond a certain minimum—250 to 400 mL in a human—the greater the extent of receptor activation. Afferent fibers from the stretch receptors carry impulses into the spinal cord and eventually, by means of interneurons, stimulate the parasympathetic supply to the bladder and inhibit the motor neuron supply to the external sphincter. Parasympathetic stimulation of the bladder causes it to contract. No special mechanism is required to open the internal sphincter; changes in the shape of the bladder dur-

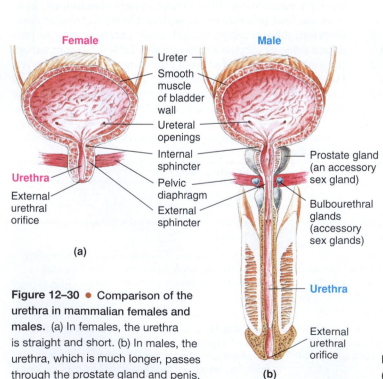

Figure 12–30 ● Comparison of the urethra in mammalian females and males. (a) In females, the urethra is straight and short. (b) In males, the urethra, which is much longer, passes through the prostate gland and penis.

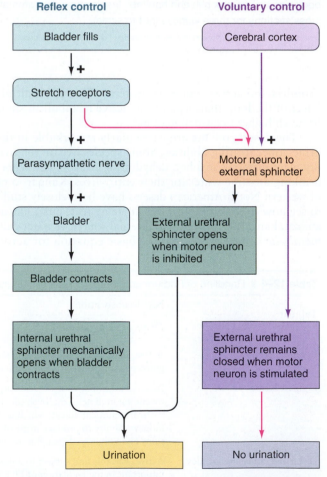

Figure 12–31 ● Reflex and voluntary control of micturition (mammal).

ing contraction mechanically pull the internal sphincter open. Simultaneously, the external sphincter relaxes as its motor neuron supply is inhibited. Now both sphincters are open, and urine is expelled through the urethra by the force of bladder contraction.

In addition to triggering the micturition reflex, bladder filling also sends signals to the brain cortex, giving rise to the urge to urinate. Control of micturition by the brain can override the micturition reflex so that bladder emptying can take place at the animal's convenience rather than at the time bladder filling first reaches the point of activating the stretch receptors. See p. 131 for the mechanism involved. Micturition is also an animal's normal response to an extremely stressful situation, because emptying the bladder reduces body weight.

Urination cannot be delayed indefinitely. As the bladder continues to fill, reflex input from the stretch receptors increases with time. Finally, reflex inhibitory input to the external-sphincter motor neuron becomes so powerful that it can no longer be overridden by cortical excitatory input, so the sphincter relaxes and the bladder uncontrollably empties.

Renal Diseases

■ Renal diseases are numerous and have wide-ranging consequences.

Urine excretion and the resulting clearance of wastes and excess electrolytes from the plasma are crucial for maintaining homeostasis. Thus kidney pathologies are dangerous and are unfortunately a relatively common ailment of aging humans and of pets, especially cats and small dogs. Renal problems have a variety of causes, some of which begin elsewhere in the body and affect renal function secondarily. Among the causes are (1) blockage of ducts and/or cellular destruction by infectious organisms, either blood-borne or gaining entrance to the urinary tract through the urethra; (2) toxic agents, such as lead, arsenic, pesticides, antifreeze (ethylene glycol), or even long-term exposure to high doses of aspirin or ibuprofen in humans; (3) inappropriate immune responses, such as glomerulonephritis, which occasionally follows certain infections as antigen–antibody complexes leading to localized inflammatory damage, are deposited in the glomeruli; (4) obstruction of urine flow because of the presence of kidney stones, tumors, cysts, or enlargement of the prostate gland, with the back pressure reducing GFR as well as damaging renal tissue; and (5) an insufficient renal blood supply that leads to inadequate filtration pressure. The latter can occur secondary to circulatory disorders such as heart failure, hemorrhage, shock, or narrowing of the renal arteries as a result of atherosclerosis.

In aging domestic cats, for example, renal disease can be due, among other things, to parasites, bacteria, feline leukemia virus—which can trigger glomerular nephritis and tumor formation—and toxins ingested such as antifreeze from car radiators. Antifreezes contain ethylene glycol, which is converted in the liver and kidney to a toxic product that changes the pH of the bloodstream and destroys the kidneys by depositing calcium oxalate crystals in the renal tubules. In some cat breeds, polycystic kidney disease (PKD) is also a common genetic problem, in which damaging cysts slowly form in the kidney with age. Domestic male cats are also very prone to feline urologic

syndrome (FUS): reduced or blocked urination because of infections, magnesium ammonium phosphate crystals, and other damage to the bladder or urethra. Crystal blockage illustrates the importance of understanding evolutionary adaptation: Cats are essentially desert animals that normally obtain most or all of their water from moist prey meat and have kidneys capable of producing urine about 10 times more hypertonic than plasma (Table 12–3). Highly concentrated urine can lead to mineral crystallization, causing FUS. However, wild cats, which as pure carnivores eat a high-meat diet, have an acidic urine that prevents crystallization. Problems occur in domestic cats that are fed foodstuffs with an unnaturally high-fat, nonmuscle protein, and/or carbohydrate content, leading to a high-pH urine that enhances crystallization.

When the functions of both kidneys are disrupted to the point that they cannot perform their regulatory and excretory functions sufficiently to maintain homeostasis, **renal failure** is said to exist. An animal may die from acute renal failure, or the condition may be reversible and lead to full recovery. Chronic renal failure, in contrast, is not reversible. Gradual, permanent destruction of renal tissue eventually proves fatal.

We will not sort out the different stages and symptoms associated with various renal disorders, but ■ Table 12–5, which summarizes the potential consequences of renal failure, can give you an idea of the broad effects that kidney impairment can have. The extent of these effects should not be surprising, considering the central role the kidneys play in maintaining homeostasis. By the time end-stage renal failure occurs, literally every body system has become impaired to some extent.

Other Vertebrate Urinary Systems and Extrarenal Organs

Vertebrate kidneys and nephrons come in a variety of morphologies and functional capacities. In particular, only avian and mammalian nephrons have loops of Henle for osmoconcentration. In addition, as you saw in the introduction to this chapter, animals may also use respiratory, digestive, and integumentary systems for nonfecal excretion and internal fluid homeostasis. Some of these **extrarenal** excretory mechanisms can be important for salt regulation in all vertebrate classes except mammals, the class that has evolved the most effective osmoconcentrating kidneys. Here we take a brief look at examples of such extrarenal systems, as well as kidneys, in the major vertebrate classes. In the next chapter, you will see how each type of vertebrate integrates its mechanisms for solute and water regulation.

■ Freshwater bony fishes excrete water with their kidneys, and their gills excrete wastes and take up salts.

Freshwater bony fishes must maintain an internal osmotic pressure far above that of the environment, and are faced with constant influx of water through the gills and mouth (from eating, because they rarely drink). Accordingly, their kidneys have evolved to excrete a voluminous, highly dilute (*hypo-osmotic*) urine. Kidneys have large glomeruli with nephrons having capsules, proximal tubules, sometimes ciliated intermediate

Table 12–5 ■ Potential Ramifications of Various Types of Renal Failure in a Mammal

Uremic toxicity caused by retention of waste products

  Nausea, vomiting, diarrhea, and ulcers caused by a toxic effect on the digestive system

  Bleeding tendency arising from a toxic effect on platelet function

  Mental changes—such as reduced alertness, insomnia, and shortened attention span, progressing to convulsions and coma—caused by toxic effects on the central nervous system

  Abnormal sensory and motor activity caused by a toxic effect on the peripheral nerves

Metabolic acidosis* caused by the inability of the kidneys to adequately secrete H^+ that is continually being added to the body fluids as a result of metabolic activity

  Altered enzyme activity caused by the action of too much acid on enzymes

  Depression of the central nervous system caused by the action of too much acid interfering with neuronal excitability

Potassium retention* resulting from inadequate tubular secretion of K^+

  Altered cardiac and neural excitability as a result of changing the resting membrane potential of excitable cells

Sodium imbalances caused by the inability of the kidneys to adjust Na^+ excretion to balance the changes in Na^+ consumption

  Elevated blood pressure, generalized edema, and congestive heart failure if too much Na^+ is consumed

  Hypotension and, if severe enough, circulatory shock if too little Na^+ is consumed

Phosphate and calcium imbalances arising from impaired reabsorption of these electrolytes

  Disturbances in skeletal structures caused by abnormalities in deposition of calcium phosphate crystals, which harden bone

Loss of plasma proteins as a result of increased "leakiness" of the glomerular membrane

  Edema caused by a reduction of plasma-colloid osmotic pressure

Inability to vary urine concentration as a result of impairment of the countercurrent system

  Hypotonicity of body fluids if too much H_2O is ingested

  Hypertonicity of body fluids if too little H_2O is ingested

Hypertension arising from the combined effects of salt and fluid retention and vasoconstrictor action of excess angiotensin II

Anemia caused by inadequate erythropoietin production

Depression of the immune system most likely caused by toxic levels of wastes and acids

  Increased susceptibility to infections

*Among the most life-threatening consequences of renal failure.

tubules, distal tubules, and collecting ducts (● Figure 12–32). Most of these segments reabsorb nutrients and minerals; the distal tubule helps create the hypo-osmotic urine by removing ions from the lumen.

Dietary intake and active transport by the gills brings in NaCl and other ions to replace those lost in the urine. The gills also excrete ammonia, the primary nitrogenous waste, primarily by diffusion. Thus the gills serve as extrarenal osmoregulatory and excretory organs (see Chapter 13).

■ **Marine bony fishes use gills for most excretion and retention processes.**

The gills of marine bony fishes are responsible for most of the functions we typically associate with kidneys. Most of these animals are highly hypo-osmotic (about 350 to 400 mOsm in most) compared to seawater (about 1000 mOsm), so they face constant influx of excess salt through the gills and diet, and constant water loss through the gills. To reverse the water loss, these fish must drink seawater, a process known to be stimu-

lated by **angiotensin II.** Initially, this also creates an excess of salt in the blood. The gills compensate for this excess, using specialized epithelial *chloride cells,* which actively transport NaCl outward (see Chapter 13, p. 585).

The kidneys play little role in correcting salt imbalances: The vast majority of marine fishes studied cannot produce a hyperosmotic urine that would remove NaCl and conserve water. Kidneys have small glomeruli—or none in a few *aglomerular* species—and no distal tubules (which would unnecessarily create a hypo-osmotic urine) (Figure 12–32); they function primarily to remove excess divalent ions (such as Mg^{++}) in a low-volume, isosmotic urine. Kidneys also have little role in nitrogenous waste removal. Again, the gills are primarily responsible for excreting ammonia and ammonium.

Aglomerular species, including many Antarctic fish, do not use ultrafiltration. These species, such as toadfish (*Opsanus tau*), may have lost their glomeruli through evolution because they (and all marine bony fishes) produce very little urine. What urine they do produce is begun initially through salt and water secretion into the proximal tubule lumen, much

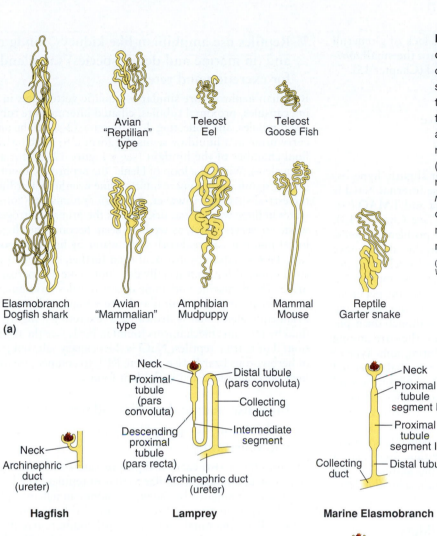

(a)

Avian "Reptilian" type

Teleost Eel

Teleost Goose Fish

Elasmobranch Dogfish shark

Avian "Mammalian" type

Amphibian Mudpuppy

Mammal Mouse

Reptile Garter snake

Figure 12–32 ● Nephrons from the major classes of vertebrates. (a) Representations of nephrons (drawn to the same scale) from shark (dogfish *Squalus acanthias*), glomerular teleosts (eel *Anguilla anguilla*), aglomerular teleost (goosefish *Lophius piscatorius*), amphibian (mudpuppy *Necturus maculosus*, reptile (snake *Thamnophis sirtalis*), bird (chicken *Gallus gallus,* mammalian and reptilian types), and mammal (mouse *Mus musculus*). (b) Schematic representations of nephrons from the major classes of nonmammalian vertebrates showing the major tubule segments.

(*Source:* W. H. Dantzler, 1989, *Comparative Physiology of the Vertebrate Kidney,* Berlin: Springer.)

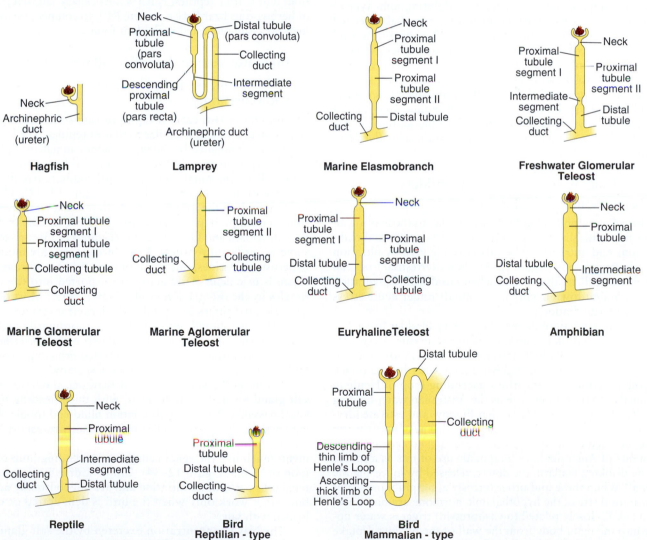

Neck

Archinephric duct (ureter)

Hagfish

Neck

Proximal tubule (pars convoluta)

Descending proximal tubule (pars recta)

Distal tubule (pars convoluta)

Collecting duct

Intermediate segment

Archinephric duct (ureter)

Lamprey

Neck

Proximal tubule segment I

Proximal tubule segment II

Collecting duct

Distal tubule

Marine Elasmobranch

Proximal tubule segment I

Neck

Proximal tubule segment II

Intermediate segment

Collecting duct

Distal tubule

Freshwater Glomerular Teleost

Neck

Proximal tubule segment I

Proximal tubule segment II

Collecting tubule

Collecting duct

Marine Glomerular Teleost

Proximal tubule segment II

Collecting duct

Collecting tubule

Marine Aglomerular Teleost

Neck

Proximal tubule segment I

Distal tubule

Collecting duct

Proximal tubule segment II

Collecting tubule

EuryhalineTeleost

Neck

Proximal tubule

Distal tubule

Collecting duct

Intermediate segment

Amphibian

Neck

Proximal tubule

Intermediate segment

Collecting duct

Distal tubule

Reptile

Proximal tubule

Distal tubule

Collecting duct

Bird Reptilian - type

Distal tubule

Proximal tubule

Collecting duct

Descending thin limb of Henle's Loop

Ascending thick limb of Henle's Loop

Bird Mammalian - type

(b)

like Malpighian tubules. Furthermore, the lack of glomeruli in some Antarctic fish may help them retain the small *anti-freeze glycoproteins* they have in their blood (Chapter 15).

■ Cartilaginous fishes retain urea and trimethylamine oxide, and use gills, kidneys, and rectal glands for excretion and retention.

Sharks and their relatives are isosmotic or slightly hyperosmotic relative to seawater, but not because internal NaCl is equal to seawater. Rather, they retain **urea** and **TMAO** (trimethylamine oxide) as major **osmolytes** (see Chapter 13). Thus, like marine bony fishes, they face the problem of excess NaCl entering through the gills and diet, but they do not have a problem with osmotic water loss. The gills are relatively impermeable to urea and TMAO (although they lose some urea), in part because of high cholesterol content. Also, the nephrons reabsorb most of these osmolytes from the filtrate, excreting only what is necessary for nitrogen balance. Although the nephrons have only proximal and distal tubules, they are among the most complex known, apparently exhibiting some type of multiple countercurrent arrangement that is probably related to this reabsorption ability (Figure 12–32a, shark). Beyond this, however, extrarenal organs perform most other excretion and osmoregulatory roles. Gills are thought to remove some of the excess NaCl, but in particular, these fish have a specialized hindgut organ called the **rectal gland,** which has numerous blind-end tubules that excrete a hypertonic fluid high in NaCl. It actively transports salt out of the blood, but it is not considered a renal organ, because no filtration is involved (see Chapter 13, p. 585).

■ Amphibians use kidneys and bladders for excretion and retention.

Nephrons in amphibian kidneys are similar to those of freshwater fish, with similar functions (to excrete water and reabsorb ions and nutrients). They also excrete urea by filtration and tubular secretion (Figure 12–32). Some terrestrial frogs can shut down GFR to almost zero when faced with dehydration (a similar shutdown occurs in mammals under acute stress and severe dehydration).

In terrestrial amphibians, the urinary bladder has a secondary role, that of a temporary water reservoir. After salty urine reaches the bladder, active transporters move the salt back to the blood, leaving highly purified water behind (much like the functioning of the thick ascending loop of Henle in mammals). Certain frogs such as the Australian water-holding frog (*Cyclorana platycephala*) that burrow and **estivate** (dry-weather equivalent of hibernation) store this nearly pure water to be returned to the blood when needed. The aboriginal inhabitants of Australia have long made use of this as a source of safe drinking water in the desert, released by squeezing the animal! When these and other terrestrial amphibians suffer dehydration stress, the hypothalamic hormone **arginine vasotocin** (AVT, closely related to vasopressin) triggers water uptake into the main body from the wall of the bladder. Uptake of water through the skin may also occur, driven by sodium uptake, while the bladder uses **aquaporins** to let water move by osmosis back into the blood.

■ Reptiles use amphibian-like kidneys, hindguts, and (in marine and desert species) salt glands for excretion and retention.

Reptilian nephrons are similar to aquatic vertebrates in having capsules, proximal tubules, ciliated intermediate tubules, distal tubules, and collecting ducts (Figure 12–32). The ureters carry urine in a liquid or semisolid form into the **cloaca,** the final chamber of the hindgut (see ● Figure 12–33 for avian equivalent). Without a loop of Henle, the nephrons cannot produce a significantly hyperosmotic urine even in water-limited terrestrial species. However, reptilian systems can conserve water in three ways. First, uric acid as the primary nitrogenous waste conserves water, as we have seen. Second, the cloaca or lower intestine can reabsorb some water by first transporting salt. This precipitates uric acid even further, creating a more solid urine, although it usually does not make a hyperosmotic urine. Third, marine and some desert reptiles have an extra-renal organ, the nasal **salt gland** (see ● Figure 12–34 for the avian equivalent) that is dedicated to excreting a highly salty fluid by the same mechanisms found in birds (see the next section). For marine reptiles, NaCl is the primary salt, but glands of herbivorous lizards also excrete KCl, to compensate for the high potassium ingested with plant food.

■ Birds use mammalian-like kidneys, hindguts, and (in marine species) salt glands for excretion and retention.

Avian kidneys also carry a liquid or semisolid urine into a cloaca, with a function similar to that of reptiles. The kidneys are usually dominated by the typical nonmammalian nephron, called "reptilian type." However they also have some nephrons, called "mammalian type," with countercurrent loops that can form a concentrated urine (Figure 12–32). The ascending limb of the loop is entirely thick; that is, there is no thin section, perhaps because urea is not involved. In addition, the cloaca and lower intestine (Figure 12–33) can recover some water (as in reptiles). Most birds form only modestly hyperosmotic urines, but much of their primary nitrogenous waste is uric acid. (Uric acid is transported into the renal tubules by the organic ion mechanism, p. 549.) Most of this uric acid precipitates as crystals, which exert no osmotic pressure. Thus we should not expect an extraordinarily high osmotic pressure in the urines of even desert birds. Nevertheless, some species produce highly concentrated urine, up to nearly 6000 mOsm in the case of the savanna sparrow.

Marine birds, which must drink seawater, also have a nasal **salt gland** located near the eyes, with ducts leading to the nasal passages. These glands contain blind-end tubules lined with active salt-secreting cells that apparently transport NaCl out of the blood without concomitant osmotic water movement, much like the renal cells of the ascending limb of the loop of Henle (Figure 12–34). Unlike kidneys, these extrarenal cells perform no filtration and are not constantly active, but rather work only when the bird is dehydrated or overloaded with salt.

The NaCl concentration excreted by the salt gland depends on what a particular species eats. A cormorant (see opening figure of this chapter) eats bony fish, which have internal (extracellular fluid) salt concentrations at about a third

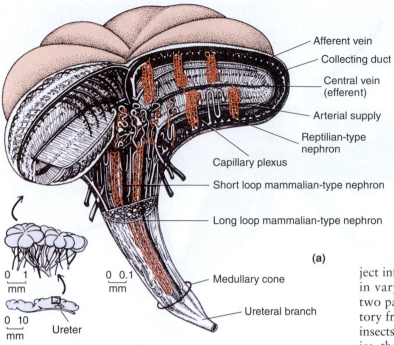

Afferent vein

Collecting duct

Central vein
(efferent)

Arterial supply

Reptilian-type
nephron

Capillary plexus

Short loop mammalian-type nephron

Long loop mammalian-type nephron

(a)

Medullary cone

Ureteral branch

0 1
mm

0 0.1
mm

0 10
mm

Ureter

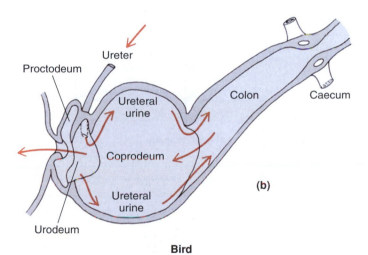

Ureter

Proctodeum

Ureteral
urine

Colon

Caecum

Coprodeum

(b)

Ureteral
urine

Urodeum

Bird

Figure 12–33 ● **Avian urinary systems.** (a) The kidney, showing its organization into lobes with short "reptilian" type and long "mammalian" type nephrons. (b) The hindgut, showing retrograde flow of urine from the ureter into the coprodeum and colon.

(*Sources:* (a) From P. C. Withers, 1992, *Comparative Animal Physiology,* Fort Worth, TX: Saunders, as modified from E. J. Braun, & W. H. Danztler, 1972, Function of mammalian-type and reptilian-type nephrons in the kidney of the quail, *American Journal of Physiology* 222:617–629. (b) From P. C. Withers, 1992, *Comparative Animal Physiology,* Fort Worth, TX: Saunders, as modified from I. Chosniak, B. G. Munck, and E. Skadhauge, 1977, Sodium chloride transport across the chicken coprodeum: Basic characteristics and dependence on the sodium chloride intake, *Journal of Physiology* 271:489–504.)

1600s—pioneered microscopic study of anatomy. Malpighian tubules are blind-end epithelial ducts something like separate nephron tubules, but not organized into a distinct organ. The tubules project into the hemolymph from the hindgut (● Figure 12–35a) in varying numbers; for example, there are four tubules in two pairs in *Drosophila melanogaster,* the common laboratory fruit fly (mosquitoes, in contrast, have five tubules; large insects may have dozens). Based on classical microscopy studies, the tubules can be divided into distinct regions. One pair of tubules, which project anteriorly, begin with large *initial segments,* followed by *transitional segments, main segments,* and a junction between the two forming a common *ureter* that joins the *hindgut.* The other pair, which project posteriorly, only have main segments and a junction to a ureter. Recent studies on gene expression (which look for genes that are turned on in specific tissues and not turned on in others) are upsetting this classical pattern. Genomic analysis reveals more specialization, such that there are probably six distinct functional regions in both sets of tubules (with all but the main segment being much shorter in the posterior pair). The exact

Figure 12–34 ● **The salt glands of a gull.**

(*Source:* J. C. Welty, & L. Baptista, 1990. *The Life of Birds,* Fort Worth, TX: Saunders, as modified from K. Schmidt-Nielsen, 1960, *Animal Physiology,* Englewood Cliffs, NJ: Prentice Hall.)

of that of seawater (see Chapter 13, p. 585). Cormorant salt glands excrete a fluid with about 500 mM NaCl: osmotically more concentrated than its body fluids, but about 15% less than seawater. In contrast, an albatross that eats marine invertebrates secretes a fluid that is 1000 mM NaCl or higher, because marine invertebrates have extracellular fluid concentrations of NaCl equal to seawater (see Chapter 13, p. 570).

Insect Malpighian Tubules

To conclude this chapter, we examine the excretory system of the most successful group of animals, the insects. Recall that these animals have Malpighian tubules (p. 528) rather than kidneys. These were named after their discoverer, Marcelo Malpighi, who—in Italy in the

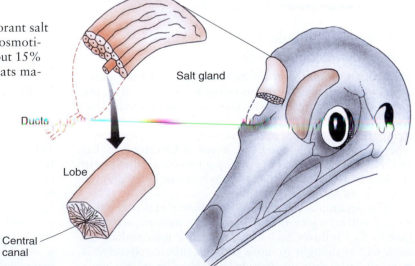

Salt gland

Ducts

Lobe

Central
canal

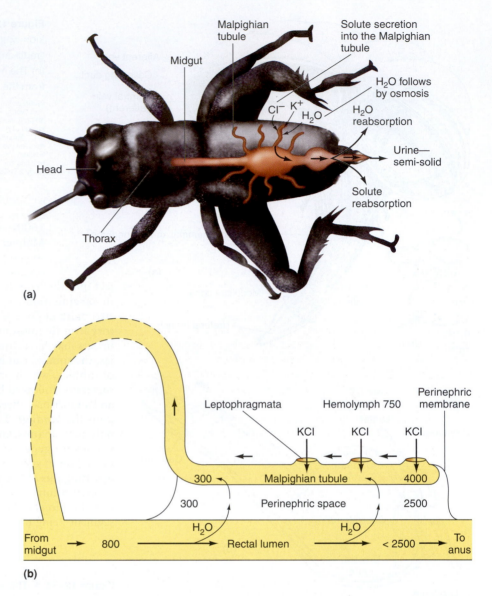

Figure 12–35 ● Malpighian tubules.
(a) The arrangement of these tubules in a typical insect. First, K⁺ ions are transported into the proximal end of the tubule, drawing Cl⁻ in by charge attraction and then water by osmosis. The tubule fluid is further modified as it travels distally in the tubule; in the rectum, large amounts of water are reabsorbed. (b) Countercurrent flow between the Malpighian tubule and hindgut in some desert insects, with concentrations shown in mM. In these animals, the tubules bend posteriorly to lie against the rectum. Unlike in other insects, the side of the tubule facing the hemolymph is waterproof, so that K⁺ and Cl⁻ transported into the tubule without water following osmotically can accumulate to high concentrations (e.g., 4000 mM). This in turn draws water osmotically from the rectum, concentrating the fluid leaving the anus.

(*Sources:* (a) From I. Kay, 1998, *Introduction to Animal Physiology,* New York: Springer, Figure 10.3. Reprinted by permission. (b) P. Willmer, G. Stone, & I. Johnston, 2000, *Environmental Physiology of Animals,* Oxford, UK: Blackwell Science, Figure 5.20, p. 110. Reprinted by permission.)

Malpighian tubules initiate filtration by ion secretion, which causes osmosis.

Malpighian tubules filter the hemolymph not by ultrafiltration under pressure, but by secreting K⁺ into the main segment's lumen (Figure 12–35a). This occurs in the main tubule segment. There is some debate about the mechanism of K⁺ transport, but it depends on a **V-ATPase,** one of the widespread types of proton (H⁺) pumps we noted in Chapter 2 (p. 81). The V-ATPase is thought to set up a proton gradient (with H⁺ higher in the lumen than out), which can be used to pull in K⁺ via an antiport cotransporter.

The accumulation of K⁺ in the lumen makes the fluid more positive, which then draws in Cl⁻ ions via attraction (just as Cl⁻ follows Na⁺ transport in the mammalian kidney). Cl⁻ is thought to move through tubule cells (that is, transcellularly) through chloride channels. The net accumulation of KCl in turn makes the tubule fluid more concentrated,

so water from the hemolymph moves in by osmosis, possibly through aquaporins (Figure 12–35a). Thus the initial steps in insects are quite different from those in most vertebrate kidneys, which use ultrafiltration at a glomerulus, but are similar to later nephron segments where Na⁺ transport is used to move Cl⁻ and water.

The tubules and hindgut modify the lumen fluid by specific secretion and reabsorption.

The osmotic movement of water creates a bulk flow down the tubule, whose epithelium further modifies the lumen fluid through specific secretion (and possible reabsorption). In particular, organic wastes such as uric acid are secreted into the tubule by transporters. Some solutes are moved by active transport, and others diffuse down concentration gradients through channel proteins, because the lumen fluid, created by KCl and water initially, begins with no organic solutes, whereas the hemolymph has some. How many transporters and channels are in the tubules is not known, but the fly genome contains

53 genes with codes closely related to known organic-solute transporters in other animals.

The tubule empties into the gut, excreting a fluid with essentially the same osmotic pressure as body fluids (that is, *isosmotic*), but with a very different solute concentration. Once in the hindgut, specialized epithelial cells of the rectum can further modify the urine. Some insects face frequent dehydrating conditions and must excrete a *hyperosmotic* urine, which conserves water in the body. This osmoconcentration process occurs by different mechanisms in different species but typically involves complex tissue anatomy that uses active transport of ions (such as KCl) out of the lumen, followed by water, followed by re-secretion of ions. Uric acid precipitates as crystals as water is removed.

In the most efficient osmoconcentrating insects, such as mealworm beetle larvae (*Tenebrio*), which consume dry food such as flour, a *countercurrent* arrangement of the gut and Malpighian tubules may be used much like that used in the mammalian nephron (Figure 12–35b). Basically, the tubule runs parallel to the rectum and fluid moves in opposite directions in the two tubes. The model proposes that the KCl transported into the upper portions of the tubule is used to draw water out of the end of the rectum, with the countercurrent arrangement multiplying the concentrating process in a manner similar to the mammalian system. These animal larvae appear to survive exclusively on metabolic water (p. 561) when eating food with no water.

Other insects must cope with excessive loads of water. Freshwater insects are obvious examples. Another example is a nectar-feeding insect such as the butterfly, which imbibes large amounts of sugary nectar that can have a lower osmotic pressure than the insect's body fluids, thus potentially causing detrimental dilution of those fluids. These insects use their tubules to excrete a *hypo-osmotic* urine. Here, the hindgut recovers ions without concomitant water uptake.

Finally, other insects face excessive water and salt loads. Recall the mosquito we discussed at the beginning of the chapter, or consider the South American blood-sucking kissing bug, *Rhodnius prolixus* (see photograph)—both using piercing mouthparts to obtain huge loads of salty liquid in the form of blood from mammalian skin. The mosquito and bug rapidly bloat up from their blood meals, sometimes by a factor of 10 in mass! In both cases, the Malpighian tubules rapidly increase their fluid excretion rates (by a factor of 1000 in the kissing bug) to remove the excess water and salt from the body. In mosquitoes, this increase in fluid excretion occurs even before the animal has finished sucking in its meal. That means, to add insult to injury, she is urinating on you as she feeds!

?

Why Are There Virtually No Marine Insects? Insects are the most successful animal group in terms of species, but curiously, they have not invaded Earth's largest habitat, the oceans. Although several hypotheses have been proposed for this, none has been well tested and found valid yet. One of these hypotheses focuses on the excretion systems, proposing that Malpighian tubules and hindguts in insects are inherently incapable of coping with seawater. How might this hypothesis be tested?

A kissing bug, *Rhodnius*. The saliva of these animals contain a protein called **nitrophorin,** which binds nitric oxide (NO). After biting a human, the bug injects its saliva and the nitrophorin releases NO, which (as we saw in Chap. 9, p. 365), dilates blood vessels. This action also prevents clotting. The bug subsequently takes in an enormous water and salt load (human blood), with which its excretory system must deal.

■ Insect excretion is regulated by diuretic and antidiuretic hormones.

Much less is known about the regulation of excretion in insects than about the process in mammals. However, details are emerging, and it appears that insects have both diuretic hormones, which promote water loss, and antidiuretic hormones, which promote water conservation. In the kissing bug, for example, both **serotonin** and several poorly characterized neuropeptides called *insect diuretic hormones* (**DH**) stimulate tubule excretion. These hormones are thought to be released by the bug's brain after a blood meal. Serotonin appears to stimulate Ca^{++} influx into the tubule cells, whereas a DH triggers cAMP production in the cells as a second messenger (p. 90). Ca^{++} and cAMP, in turn, activate tubule secretion (possibly by activating the V-ATPases) and subsequent fluid excretion. The two hormones act synergistically, that is, secretion is faster with the two of them together than predicted by adding the effects of each hormone individually. As a possible antagonist, in kissing bugs a peptide hormone called *cardioacceleratory peptide 2b* (CAP_{2b}) triggers cGMP production in tubule cells, which reduces secretion and conserves water. However, these are laboratory results using isolated tubules; it is not known if this peptide is used in a kissing bug for water conservation.

Chapter in Perspective:
HOMEOSTASIS and INTEGRATION

Excretory organs contribute to homeostasis more extensively than most other single organs. They generally serve as effectors in negative-feedback systems that regulate the electrolyte composition, volume, and pH of the internal environment and eliminate most of the waste products of bodily metabolism (such as ammonia, urea, uric acid). They may also excrete excess inorganic ions (such as NaCl) and conserve substances useful for the body, including water in animals subject to dehydration. In mammals, the kidneys are the major excretory organs. They can maintain the plasma constituents they regulate within the narrow range compatible with life, despite wide variations in intake and losses of these substances through other avenues. Excretory organs can sometimes serve as sensors and integrators that control the excretory processes by negative feedback.

Here are the specific ways in which mammalian kidneys contribute to homeostasis:

1. *Effector functions*
   - The kidneys regulate the quantity and concentration of most ECF electrolytes, including those important in maintaining proper neuromuscular excitability.
   - They contribute to maintenance of proper pH by eliminating excess H^+ (acid) or HCO_3^- (base) in the urine and by producing HCO_3^-.
   - They help maintain proper plasma volume, which is important in the long-term regulation of arterial blood pressure, by controlling the salt balance in the body. The extracellular fluid (ECF) volume, including the plasma volume, is a reflection of the total salt load in the ECF, because Na^+ and its attendant anions (mainly Cl^-) are responsible for over 90% of the ECF's osmotic (water-holding) activity.
   - The kidneys maintain water balance in the body, which is important in maintaining proper ECF osmolarity (concentration of solutes). This role is important in maintaining the stability of cell volume by preventing cells from swelling or shrinking as a result of water osmotically moving in or out of the cells, respectively.
   - The kidneys excrete the end products of metabolism in the urine. These wastes are toxic to cells if allowed to accumulate.
   - They also excrete many foreign compounds that gain entrance to the body.

2. *Integrator hormonal functions*
   - The kidneys secrete erythropoietin, the hormone that stimulates red blood cell production by the bone marrow. This action contributes to homeostasis by helping to maintain the optimal O_2 content of the blood. Over 98% of O_2 in the blood is bound to hemoglobin within the red blood cells.
   - They also secrete renin, the hormone that initiates the renin-angiotensin-aldosterone pathway for controlling renal tubular NA^+ reabsorption, which is important in the long-term maintenance of plasma volume and arterial blood pressure.

3. *Metabolic functions*
   - The kidneys help convert vitamin D into its active form. Vitamin D is essential for Ca^{++} absorption from the digestive tract. Calcium, in turn, exerts a wide variety of homeostatic functions. ■

REVIEW QUESTIONS *(Answers are on p. A–1.)*

Additional study tools for this chapter, including chapter summaries and practice tests, are available online at *www.biology.brookscole.com*

1. Animals that rely on ammonia excretion are called
   - a. osmolytes
   - b. uricotelic
   - c. ammonotelic
   - d. uricolytic
   - e. ureotelic
2. Which of the following is not a basic renal process?
   - a. filtration
   - b. secretion
   - c. perfusion
   - d. osmoconcentration
   - e. reabsorption
3. In vertebrate kidneys, metanephridial tubules are called
   - a. kidneys
   - b. nephrons
   - c. Malpighian tubules
   - d. ureters
   - e. protonephridia
4. In mammals, which kidney function stimulates red blood production?
   - a. excretion of pheromones
   - b. secretion of renin
   - c. conversion of vitamin d to its active form
   - d. secretion of erythropoietin
   - e. none of the above
5. The smallest functional unit of the kidney is the
   - a. renal cortex
   - b. renal pyramid
   - c. glomerulus
   - d. nephron
   - e. urethra
6. Which part of the nephron collects fluid from glomerular capillaries?
   - a. the proximal tubule
   - b. Bowman's capsule
   - c. the loop of Henle
   - d. the distal tubule
   - e. the collecting duct
7. The walls of glomerular capillaries are 100 times more permeable to water and solutes than other capillaries due to the presence of
   - a. fenestrae
   - b. podocytes
   - c. plasma proteins
   - d. a basement membrane
   - e. albumin molecules
8. If plasma volume is reduced, then
   - a. the baroreceptor reflex results in the constriction of afferent arterioles
   - b. urine output is reduced

c. hormonal mechanisms increase tubular reabsorption

d. hormonal mechanisms act to increase thirst

e. all of the above

9. Which of the following substances filtered by the kidneys is not normally reabsorbed to any great extent?

   a. water
   b. sodium
   c. phenol
   d. amino acids
   e. glucose

10. The secondary transport of glucose and amino acids requires

    a. Cl^-
    b. Na^+
    c. glucose
    d. Ca^{++}
    e. phosphate

11. Which of the following is associated with the actions of the renin-angiotensin-aldosterone system?

    a. stimulating Na^+ reabsorption in the distil tubules
    b. elevating blood pressure
    c. stimulating thirst
    d. stimulating salt hunger
    e. all of the above

12. The action of the atrial natriuretic peptide hormone

    a. increases Na^+ reabsorption
    b. decreases Na^+ excretion
    c. inhibits renin secretion by the kidneys
    d. increases aldosterone secretion
    e. increases cardiac output

13. Tubular secretion is involved in

    a. hastening the elimination of hydrogen ion
    b. retaining excess potassium
    c. retaining organic cations
    d. retaining organic anions
    e. eliminating amino acids

14. Potassium-ion secretion in distal tubules and collecting ducts

    a. is unregulated
    b. is regulated by renin
    c. is coupled to sodium-ion reabsorption

d. depends on low concentrations of the ion in tubular cells

e. depends on active pumping through potassium channels

15. Plasma clearance is

    a. the amount of a substance removed from plasma per minute
    b. the volume of plasma cleared of a particular substance per minute
    c. always greater than the GFR
    d. always less than the GFR
    e. usually equal to the GFR

16. If body fluids contain too much water relative to the solute load, the fluids are

    a. isotonic
    b. hyperosmotic
    c. isosmotic
    d. hypotonic
    e. hypertonic

17. The medullary vertical osmotic gradient plays a role in

    a. producing only hypotonic urine
    b. producing only hypertonic urine
    c. producing only isotonic urine
    d. permiting excretion of urine of varying concentrations
    e. none of the above

18. Vasopressin

    a. secretion is stimulated by a water deficit
    b. increases the permeability of distil tubules to water
    c. decreases the permeability of collecting ducts
    d. decreases the number of aquaporins
    e. a and b

19. Which of the following have loops of Henle for osmo-concentration?

    a. freshwater bony fish
    b. birds
    c. marine fish
    d. reptiles
    e. none of the above

20. Which of the following would have Malpighian tubules?

    a. rattlesnake
    b. robin
    c. honey bee
    d. lion
    e. bullfrog

SUGGESTED READINGS AND INTERNET SITES

Dow, J. A. T., & S. A. Davies. 2003. Integrative physiology and functional genomics of epithelial function in a genetic model organism. *Physiological Reviews* 83:687–729.

Dwyer, T. M., & B. Schmidt-Nielsen. 2003. The renal pelvis: Machinery that concentrates urine in the papilla. *News in Physiological Sciences* 18:1–6.

Handler, J. S., & H. M. Kwon. 2001. Cell and molecular biology of organic osmolyte accumulation in hypertonic renal cells. *Nephron* 87: 106–110.

Ruppert, E. E., & P. R. Smith. 1988. The functional organization of filtration nephridia. *Biological Reviews* 63:231–258.

Schmidt-Nielsen, K. 1964. *Desert Animals: Physiological Problems of Heat and Water.* Oxford, UK: Clarendon Press.

Simoyi, M. F., K. Van Dyke, & H. Klandorf. 2002. Manipulation of plasma uric acid in broiler chicks and its effect on leukocyte oxidative activity. *American Journal of Physiology* 282:R791–R796.

Wright, P., & P. Anderson, eds. 2001. *Fish Physiology.* Vol. 20: *Nitrogen Excretion.* New York: Academic Press.

Wright, P. 1995. Nitrogen excretion: Three end products, many physiological roles. *Journal of Experimental Biology* 198: 273–281.

INFOTRAC READINGS

Hazard, L. C. 2001. Ion secretion by salt glands of desert iguanas (*Dipsosaurus dorsalis*). *Physiological and Biochemical Zoology* 74:22–31.

INTERNET SITE

Dow, J. A., T., & S. A. Davies. *The Drosophila melanogaster Malpighian tubules.* **fly.to/tubules.** A site on *Drosophila* Malpighian tubules as a model system for melding the physiology of epithelial transport with genomics, following the Krogh principle; with photographs.

Fluid and Acid–Base Balance

Intertidal anemones and snails. Both animals and their relatives (marine cnidarians and mollusks, respectively) are osmoconformers; that is, the osmotic pressure of their body fluids is equal to that of the environment.

Photo: Paul Yancey

Introduction

The concept of homeostasis as first delineated by Claude Bernard and Walter Canon (p. 10) focused on the *milieu interior,* that is, the **extracellular fluid (ECF)** that bathes all cells of an animal. The idea was that individual cells could be spared the cost of struggling against disturbances to their **intracellular fluids (ICF)** if the ECF is regulated on their behalf. The composition of the ICF and ECF results from a variety of key molecular components: the solvent water and its solutes—salts, nutrients such as glucose, proteins, and other organic molecules, ions such as calcium for shells and skeletons, and hydrogen ions. In accordance with the original homeostasis concept, animals can regulate many of these constituents at the level of the ECF. Both the *total concentration* of all solutes, which determines osmotic pressure, and the *concentration of individual solutes* may be regulated.

As you have just seen in Chapter 12, excretory organs play a key role in this regulation. However, single excretory organs such as kidneys do not regulate every aspect of the ECF. Other organs contribute (CO_2, for example, is mainly regulated by respiratory systems), and the ECF may vary in composition in some types of animals, such that individual cells must cope on their own. Overall, though, critical features of cellular homeostasis are still achieved, as you will see. Therefore, the regulation of the ECF and ICF composition is not attributable to one single physiological system. Rather, it is an inherently integrative process involving individual cells and more than one organ system. In Chapter 12, we focused primarily on the excretion and retention of specific solutes (salts, wastes) and water. In this chapter, we look at fluid-related regulation at a more integrative level by focusing on three aspects of the ECF and ICF that are crucial to the whole animal

and to its individual cells: the *osmotic balance,* determined primarily by the total concentration of major solutes; the total *fluid volume* of the body, which directly affects circulatory pressure; and the *acid–base balance.* Hydrogen ions alter the acid–base status (measured as pH) of body fluids, a factor that is crucial to the functioning of proteins.

Before we examine these, however, let us look at how the composition of the ECF can change in basic terms.

■ If balance is to be maintained, input must equal output.

The quantity of any particular substance in the ECF is considered a readily available internal pool. The amount of the substance in the pool may be increased either by transferring more in from the external environment (such as via the gut or the integument) or by metabolically producing it within the body (● Figure 13–1). Substances may be removed from the body by loss to the outside (through excretion, sweating, and so forth) or by being used up in a metabolic reaction. If the quantity of a substance is to remain stable within the body, its input must be balanced by an equal output by means of excretion or metabolic consumption. This relationship, known as the **balance concept,** is extremely important in maintaining homeostasis. Not all input and output pathways are applicable for every body-fluid constituent. For example, salt is not synthesized or altered metabolically by organisms, so the stability of salt concentration in body fluids depends entirely on a balance between salt intake and salt excretion.

The ECF pool can further be altered by transferring a particular ECF constituent into storage within the cells or certain extracellular structure (such as shells, skeletons). If a body as a whole has a surplus or deficit of a particular stored substance,

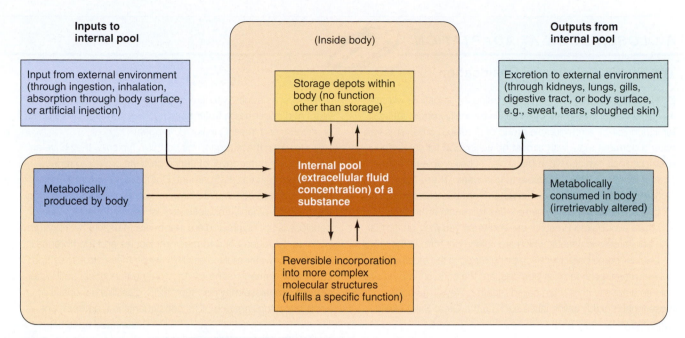

Figure 13–1 ● Inputs to, and outputs from, the internal pool of a body constituent.

the storage site can be expanded or partially depleted to maintain the ECF concentration of the substance within homeostatically prescribed limits. For example, after a mammal has absorbed a meal, when more glucose is entering the plasma than is being consumed by the cells, the excess glucose can be temporarily stored in muscle and liver cells in the form of glycogen. This storage depot can then be tapped between meals as necessary to maintain the plasma glucose level when no new nutrients are being added to the blood by eating. As another example, some amphibians use their bladders for temporary water storage (see p. 566). It is important to recognize, however, that internal storage capacity is limited. Although an internal exchange between the ECF and a storage depot can temporarily restore the plasma concentration of a particular substance to normal, in the long run any excess or deficit of that constituent must be compensated for by appropriate adjustments in total body input or output.

Another possible internal exchange between the pool and the remainder of the body is the reversible conversion of certain constituents between simple molecules or ions and larger structures with functions beyond mere storage. For example, calcium can be incorporated into skeletal and shell structures but is also necessary as a cellular messenger and trigger of muscle contraction. If plasma and cell calcium falls too low, calcium may be taken out of the shell or skeleton. This process differs from metabolic consumption of a substance, in which the substance is irretrievably converted into another form—for example, glucose converted into CO_2 plus H_2O plus energy. It also differs from simple storage in that the latter serves no additional purpose, whereas reversible incorporation into a larger structure can serve another purpose.

Some body constituents may be poorly controlled. For example, the plasma volume and composition of osmoconformers (animals whose internal osmotic pressure is equal to that of the environment) passively change with the external environment, forcing individual cells to adjust at least in part on their own. Some constituents undergo frequent and/or large disturbances. Salt and H_2O can be lost to the terrestrial environments to varying degrees through the digestive tract (vomiting, diarrhea), skin (sweating), lungs (panting), and elsewhere without regard for salt or H_2O balance in the body. Hydrogen ions are uncontrollably generated internally (especially during muscle activity) and added to body fluids. If possible, compensatory adjustments must be made for these uncontrolled changes.

■ Body water is distributed between the intracellular and extracellular fluid compartments.

Water is by far the most abundant component of most animals, though the proportion varies considerably. For example, water is about 60% of mammalian body weight on average, and it can be more than 90% in some soft aquatic animals. Yet water content may drop well below 10% in certain **anhydrobiotic** animals, such as encysted tardigrades and brine shrimp embryos, which have entered a dormant state. However, anhydrobiotic states exhibit no detectable metabolism, illustrating that normal metabolic functions are associated with high water content. (Note that some anhydrobiotic stages of organisms can remain viable under harsh conditions for years, even many centuries in the case of some bacteria. It is not fully understood how they accomplish this. For some research findings, see the box, "A Closer Look at Adaptation: The Sweet Solution to Desiccation.")

Among the major body components, plasma, as you might suspect, is the most watery, typically more than 90% H_2O. But even soft tissues such as skin, muscles, and internal organs consist of 70 to 80% H_2O. Drier structural materials such as shells and skeletons may be only about 20% H_2O. Fat and

A CLOSER LOOK AT ADAPTATION

The Sweet Solution to Desiccation

Biologists routinely talk about life being incompatible with the absence of water. Although this is basically correct (at least for life on Earth), a remarkable group of unrelated organisms can enter a dormant state with almost no water, and for years and even decades can survive in the sense of being able to restore normal functions upon exposure to free water. These anhydrobiotic ("no-water life") organisms range from bacteria and yeast (in cyst or spore stages) to nematodes to certain arthropods. If you have ever baked bread or brewed beer, you may be familiar with anhydrobiotic yeasts, which are sold in dry form in the store and which "come back to life" on being mixed in the moist dough or beer mash. Less familiar, perhaps, are the brine shrimp *Artemia*, sold as a novelty item often called "sea monkeys." As adults, these inhabit highly saline lakes such as the Great Salt Lake in Utah. However, they and their eggs can dry out if washed ashore or if a desert pond they inhabit dries out. This kills the adults, but an early embryonic stage can encyst and survive for years with less than 2% water internally. More remarkably, tardigrades (see photograph), small, segmented, exoskeletal animals found in ponds and forest litter,

can encyst as adults with less than 1% water and survive for years and under harsh treatments such as immersion in alcohol or vacuum and freezing in liquid nitrogen!

How can these organisms do this? Typically, they have an impermeable outer cover of dense proteins or wax to reduce exchanges with the environment. But more importantly, these resistant stages accumulate large amounts of disaccharide sugars, most commonly trehalose (diglucose). Although the role of trehalose is still being studied, it appears to use its hydroxyl group (–OH) to hydrogen bond to macromolecules and membranes in the place of water molecules. This keeps these structures from irreversible damage (although they are nonfunctional until water returns). Trehalose may also turn into a glasslike material that holds the cytoplasm and all its constituents in a smooth, undamaged condition. Regardless of how it works, trehalose's remarkable preservation abilities have been put to use: Human red blood cells can be dried and preserved with this sugar, carried around in a lightweight form (such as by a soldier), and then rehydrated into usable form when needed for transfusions.

Photo: © Martin Mach (www.tardigrades.com)

An active tardigrade (left) and a dormant encysted one (right)

(*Source:* M. Mach, *Tardigrades*, www.tardigrades.com, accessed March 19, 2004)

---

external coverings such as hair are the driest tissues of all, having only about 10% H_2O content or less. These drier examples include many metabolically inactive components; the active cytosols and organelles of the cells within shells, skeletons, and fat tissue have high water contents, but those cell spaces are a minor component of those tissues.

Major Fluid Compartments

Body H_2O is distributed between two major fluid compartments: the ICF and the ECF (● Figure 13–2). ▌Table 13–1 shows values for a typical mammal. The ICF compartment comprises about two thirds of total body H_2O in insects and vertebrates, but only about

Figure 13–2 ➤ ● A model of the major body compartments in an animal and exchange routes (arrows) between them. ECF, extracellular fluid; ICF, intracellular fluid.

(*Source:* P. Willmer, G. Stone, & I. Johnston, 2000, *Environmental Physiology of Animals*, Oxford, UK: Blackwell Science, Figure 1.8.)

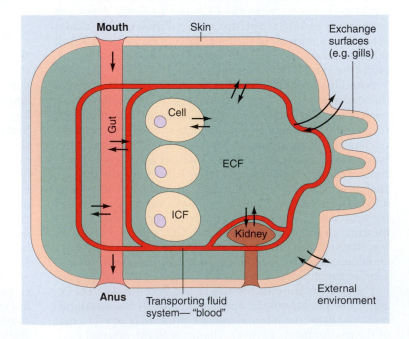

Table 13–1 ▪ Classification of Body Fluid for a Human

Compartment	Volume of Fluid (in liters)	Percentage of Body Fluid	Percentage of Body Weight
Total Body Fluid	42	100%	60%
Intracellular fluid (ICF)	28	67	40
Extracellular fluid (ECF)	14	33	20
Plasma	2.8	6.6 (20% of ECF)	4
Interstitial fluid	11.2	26.4 (80% of ECF)	16
Lymph	Negligible	Negligible	Negligible
Transcellular fluid	Negligible	Negligible	Negligible

one third to one half in marine mollusks and crustaceans. Even though each cell contains its own unique mixture of constituents, these numerous minute fluid compartments are similar enough to be considered collectively as one large fluid compartment.

The remaining body H_2O—in the ECF compartment—is primarily the hemolymph in animals with open circulatory systems (Chapter 9). In animals with closed circulation, such as cephalopods and vertebrates, the ECF is subdivided primarily into plasma and interstitial fluid. Vertebrate plasma, which makes up about 15% (in bony fishes) to 40% (in birds) of the ECF volume, is the fluid portion of the blood. The interstitial fluid, which represents most of the remaining ECF compartment, is the fluid that lies in the spaces between the cells. Interstitial fluid, sometimes also known as *tissue fluid,* constitutes the true internal environment that bathes the tissue cells.

Minor ECF Compartments

Some animals have two other relatively minor ECF fluids. **Lymph,** found in most vertebrates, is fluid being returned from the interstitial fluid to the plasma by means of the lymphatic system (Chapter 9). **Transcellular** fluids consist of a number of small specialized fluid volumes that are secreted by specific cells into a particular body cavity to perform some specialized function. In vertebrates, these include *cerebrospinal fluid* (p. 159); *intraocular fluid* (maintaining the shape of and nourishing the eye); *synovial fluid* (lubricating and serving as a shock absorber for the joints); *pericardial, intrapleural,* and *peritoneal* fluids (lubricating movements of the heart, lungs, and intestines, respectively). Although these fluids are extremely important functionally, they usually represent an insignificant fraction of the total body H_2O (Table 13–1). Furthermore, the transcellular compartment as a whole usually does not reflect changes in the body's fluid balance. For example, the mammalian cerebrospinal fluid does not decrease in volume when the body as a whole is experiencing a negative H_2O balance. Therefore, the transcellular compartment can usually be ignored when dealing with problems of fluid balance.

Finally, a fluid compartment found in all animals is the solution in the digestive tract. Technically this is part of the external environment because the digestive tract connects to the outside world. However, this fluid contains components from the body (such as digestive enzymes, and sometimes water). It is usually a small percentage of the ECF but cannot always be ignored, because it is often a major source of gains and losses of water and solutes.

▪ In vertebrates, the plasma and interstitial fluid are similar in composition; but the ECF and ICF are markedly different in all animals.

Several barriers separate vertebrate body fluid compartments, limiting the movement of H_2O and solutes between the various compartments to differing degrees. The two components of the ECF—plasma and interstitial fluid—are separated by the walls of the blood vessels. However, H_2O and all plasma constituents with the exception of plasma proteins are continuously and freely exchanged between the plasma and the interstitial fluid by passive means across the thin, pore-lined capillary walls. Accordingly, plasma and interstitial fluid are nearly identical in composition, except that interstitial fluid lacks plasma proteins. Any change in one of these ECF compartments is quickly reflected in the other compartment.

In contrast to the very similar composition of the plasma and interstitial fluid compartments of the ECF, the composition of the ECF differs considerably from that of the ICF (● Figure 13–3) in all animals. Each cell is surrounded by a highly selective plasma membrane that permits passage of certain materials while excluding others. Movement through the membrane barrier occurs by both passive and active means and may be highly discriminating. Among the major differences between the ECF and ICF are

1. *Cellular proteins* in the ICF that cannot permeate the plasma membranes to leave the cells.
2. *Cellular organic osmolytes* (discussed later; typically higher in the ICF than in the ECF).
3. *The unequal distribution of Na^+ and K^+* and their attendant anions; in most multicellular organisms, Na^+ is the primary ECF cation, and K^+ is the primarily ICF cation. This is due in part to action of the membrane-bound Na^+–K^+ ATPase pump that is present in all cells (this is not the only factor; see p. 95). This pump actively transports Na^+ out of and K^+ into cells. The unequal distribution of Na^+ and K^+, coupled with differences in membrane permeability to these ions, is responsible for the electrical properties of cells, including the initiation and propagation of action potentials in excitable tissues (see Chapters 3 and 4).

Except for the extremely small electrical imbalance in the intracellular and extracellular ions involved in membrane potential, the majority of the ECF and ICF ions are electrically balanced. In the ECF, Na^+ is accompanied primarily by the anion Cl^- (chloride) and to a lesser extent by HCO_3^- (bicarbonate). In the ICF, K^+ is accompanied primarily by the anion PO_4^{3-} (phosphate) and by the negatively charged proteins trapped within the cell.

Figure 13–3 ▶ • Ionic composition of the major body-fluid compartments in a mammal.

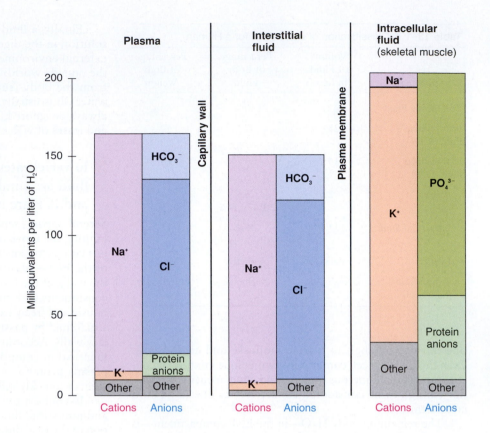

Although all cells' plasma membranes display selective permeability, cells are typically freely permeable to H_2O. The movement of H_2O between the plasma and the interstitial fluid across capillary walls is governed by relative imbalances between capillary blood pressure (a fluid, or hydrostatic, pressure) and colloid osmotic pressure (see p. 414). In contrast, the net transfer of H_2O between the interstitial fluid and the ICF across the cellular plasma membranes occurs as a result of osmotic effects alone. The hydrostatic pressures of the interstitial fluid and ICF are both extremely low and fairly constant.

Osmotic and Volume Balance

As you saw in Chapter 3, osmotic forces (from water moving down its concentration gradients through membranes) are among the crucial factors with which cells must cope. Lacking nonmembranous cell walls as found in bacteria, plants, and fungi, the cells of animals always become osmotically equalized with the ECF that bathes them. Thus the osmotic composition of the ECF directly affects cell homeostasis. Recall that water molecules alone apparently cannot be moved directly by ATPase pumps (that is, active transport), but rather must follow the movement of solutes. Inorganic ions, especially NaCl, are dominant in this process.

■ Several osmotic problems threaten cells and animals.

One problem that all animal cells face is a potential swelling arising from the unequal Na^+ distribution in the ECF and ICF. Because Na^+ has both an electrical gradient (Chapter 4) and a chemical gradient inward, Na^+ leaking into cells raises the solute concentration and thus can elevate osmotic pressure, which in turn would cause water to enter the cell by osmosis. The primary, and nearly universal, mechanism to correct this situation is the $Na^+–K^+$ ATPase "pump." By exporting more Na^+ than the K^+ it imports (3 Na^+ out for each 2 K^+ in), the pump not only maintains ion gradients but also reduces total cellular osmotic pressure.

Beyond this basic problem of cells, internal osmotic balance can be disturbed in several major ways. The ECF can become too concentrated or too dilute because of

1. *Evaporation* of body water into air (for example, because of sweating or breathing)
2. *Osmosis* into or out of the environment (such as fresh water or saline water)

3. *Freezing,* which locks up water in ice crystals and concentrates ions in unfrozen water
4. *Excretion,* which may require water for waste removal
5. *Diseases* such as diabetes in which ECF osmolarity increases because of excessive glucose levels

Changes in the ECF osmolarity in turn can cause cells to swell or shrink. Let us use environmental osmolarity to illustrate how disturbances can occur. The universal dissolved constituents of cells—macromolecules, metabolites, K^+, and so forth—typically yield an osmotic concentration of roughly 300 milliosmolar (mOsm) in most organisms (this does not include nonuniversal *organic osmolytes*, which we discuss shortly). Normally the surface waters of the ocean contain about 3.5% salt, yielding about 1000 mOsm. *Brackish* waters have less than 3.0% but greater than 0.5%. The lowest salinities are considered fresh water, approximating as low as 0 mOsm; in contrast, some *brine* lakes such as the Dead Sea in Israel can reach over 20% salt at 6000 mOsm! In any environment greater than about 300 mOsm, organisms must have adaptations to maintain cell volume in the face of this potentially dehydrating force. In contrast, lower osmolarities threaten to overhydrate organisms.

■ Animals have evolved two strategies to cope with osmotic challenges.

How do animals maintain a proper balance of ICF and ECF volumes and compositions in the face of different environments? Limited adaptations include a cell membrane (or the extracellular layer next to it) that is impermeable to water, an option that turns out extremely limiting, because it reduces the ability of a cell to exchange vital solutes. This solution has evolved in some epithelial layers exposed to osmotically challenging environments and is usually achieved with extracellular waterproofing layers on one side of a cell layer. For example, epidermal cells of insects secrete a noncellular cuticle of waterproof *waxes,* whereas the epidermis of terrestrial vertebrates produce a *keratinized* layer with low permeability (p. 62). Another solution is to use some sort of water "pump"; for example, some freshwater protozoa and sponges have special organelles called *contractile vacuoles* (p. 522). However, most animal cells, lacking cell walls, impermeable membranes, and water pumps, typically must have the total concentration

of molecules in their ICF equal to that of the ECF. Thus, in the long run there are no osmotic gradients between most animal cells' ICF and the nearby ECF.

How do animals prevent swelling or shrinking of their cells, given that the external environment may have quite a different osmotic concentration and that perturbing changes in the environment may alter the osmotic state of the ECF (and thus alter the ICF)? There are two very different strategies of maintaining osmotic balance (● Figure 13–4a):

1. *Osmoconformers:* The body fluids and cells of an osmoconformer are generally equal in osmotic pressure to the surrounding solution, with adjustments to both ECF and ICF made using small molecules called *osmolytes.*
2. *Osmoregulators:* The osmotic pressure of body fluids in an osmoregulator is homeostatically regulated despite the changes in the external environment.

As shown in Figure 13–4b and c, there are varying degrees of osmoconforming and osmoregulating.

Figure 13–4 ● (a) Generalized osmotic contents of three different types of osmotic adaptation in marine animals. TMAO is trimethylamine oxide. Osmoconformers use primarily free amino acids as organic osmolytes in their cells, while osmoconforming hypoionic regulators use primarily urea and TMAO. In contrast, hypo-osmotic hypoionic regulators have very low levels of organic osmolytes. (b) General categories of responses of animal body fluids to variations in external concentrations, with (c) examples.

(*Source for parts b and c:* P. Willmer, G. Stone, & I. Johnston, 2000, *Environmental Physiology of Animals,* Oxford, UK: Blackwell Science, Figure 5–2)

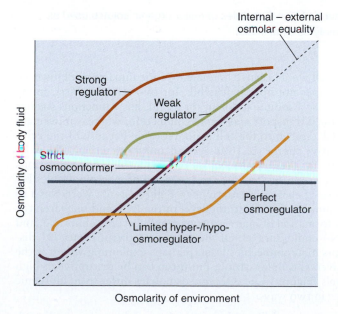

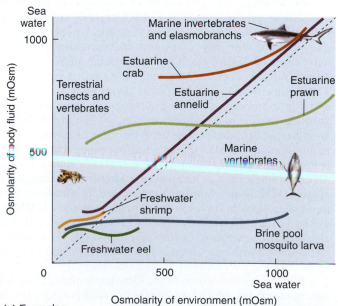

(b) Principles

(c) Examples

Osmoconformers

Osmoconformers have internal osmotic pressures that are approximately equal to that of the environment. Thus there is little tendency to gain or lose water, except when the environment changes. The main examples of osmoconformers are found in the oceans.

■ Osmoconformers use compatible and counteracting organic osmolytes in their cells.

Recall that the oceans are about 1000 mOsm. Most marine organisms—especially soft-bodied ones, which have no impermeable coverings to reduce osmotic forces—are also about 1000 mOsm internally. In these animals, the ECF is dominated by NaCl just as is seawater (and the ECF of most animals). Because Na^+ and Cl^- are critical to maintaining osmotic balance, they are considered **inorganic osmolytes** as well as electrolytes. What about the ICF? In fact, after small, short-term disturbances to the ECF cells often do alter their internal Na^+, K^+, and Cl^- levels to prevent swelling or shrinking. But for larger and long-term osmotic challenges, the story is quite different. Instead, ICF osmotic pressures are raised predominantly with certain organic solutes called **compatible** and **counteracting organic osmolytes**. A marine osmoconformer has

- An ECF at about 1000 mOsm, dominated by NaCl and sometimes containing high levels of organic osmolytes
- An ICF—also at about 1000 mOsm—with about 300 to 400 mOsm of universal solutes (K^+, macromolecules, and so forth) and about 600 to 700 mOsm of organic osmolytes

Organic osmolytes fall into four broad categories: (1) **carbohydrates** such as polyols and sugars, (2) **free amino acids** such as glycine and taurine (an unusual, sulfur-based amino acid), (3) **methylamine** and **methylsulfonium** solutes, and (4) **urea**. Structures of common osmolytes in these categories are shown in ● Figure 13–5, and distributions in the animal kingdom are listed in ▌ Table 13–2. An example of a common methylamine osmolyte in marine animals is **TMAO** (trimethylamine oxide; Figure 13–5). Whether you know it or not, you are probably quite familiar with this compound: Its breakdown is the primary cause of "fishy" smell in many marine animals. On the animal's death, bacteria convert TMAO into a volatile gas, TMA (trimethylamine), which has a strong odor of rotten fish; and minor amounts of TMA are present in live animals. Indeed, the marine food industry monitors TMA in fish flesh as an indicator of spoilage. The main example of a methylsulfonium osmolyte is **DMSP** (dimethylsulfonopropionate), a major osmolyte in many marine algae. It can also be found in some marine animals such as bivalves that filter-feed on unicellular algae. DMSP is also familiar to most people who have been to an ocean beach: Its breakdown product is a gas **DMS** (dimethysulfide), which is the smell of decaying seaweed. It is also a major component of the Gaia hypothesis for a homeostatic planet Earth (see Chapter 1, p. 12, box "Challenges and Controversies: Can a Planet Have Physiology?"]

Why do osmoconformers use these particular organic solutes as osmolytes, which cost metabolic energy, rather than the readily available inorganic ions? At first glance, it might seem that simply moving the high external NaCl of the ECF into

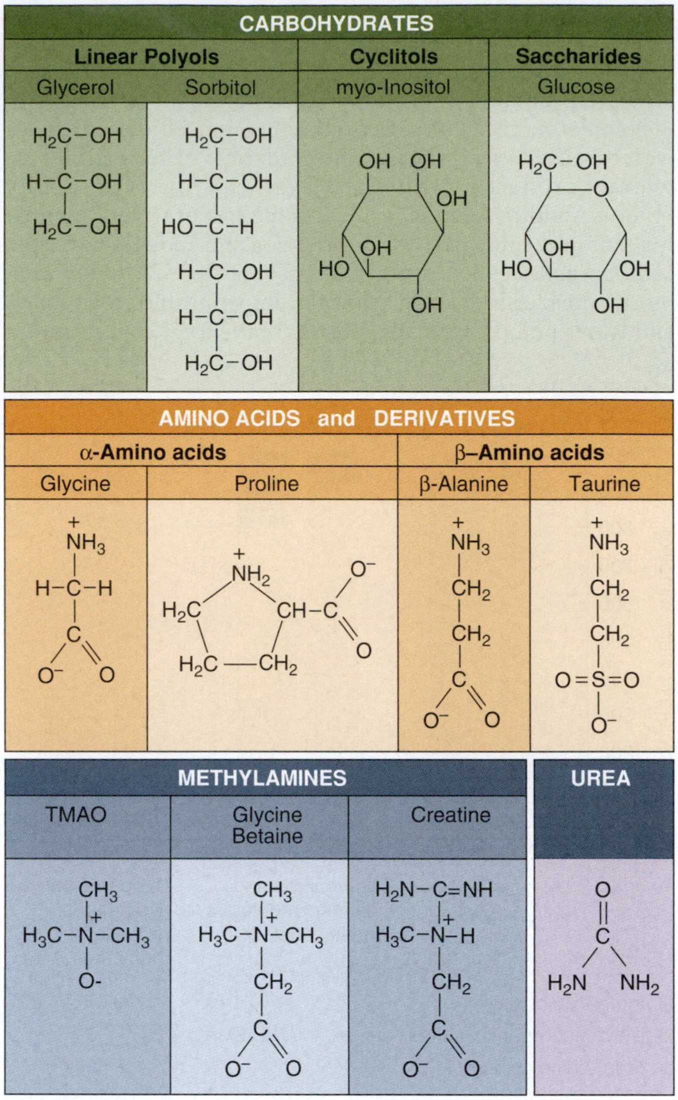

Figure 13–5 ● Examples of major organic solutes used as osmolytes by animals.

the cell would easily eliminate osmotic imbalances. However, this does not appear feasible, for two reasons. First, many cells use a sodium gradient for useful, basic functions such as coupled transport and signal conduction, whereas movement of Cl^- ions into the cell is unfavorable because of the cell's negative charge (p. 100). Second and more universally, inorganic ions (especially Na^+ and Cl^-) at high concentrations can disrupt the structure and function of macromolecules because of charge interactions (● Figure 13–6). Were NaCl to accumulate in cells to the level of the ECF and seawater, the conformation of many proteins would become abnormal, and DNA could unravel. Of the common inorganic ions, K^+ disrupts macromolecules least, perhaps explaining why it is the universal cellular cation; but it too can be disruptive at high levels (Figure 13–6). Organic osmolytes differ from inorganic ions in two ways:

1. *Compatibility:* Most organic osmolytes do not disturb macromolecules even at high concentrations, and so are termed *compatible solutes* (Figure 13–6). (A major ex-

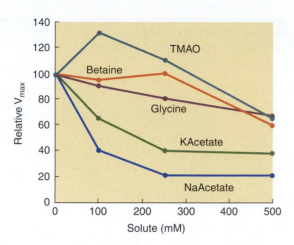

Figure 13–6 ● Effect of various solutes on the activity of an enzyme, lactate dehydrogenase, from a marine polychaete worm. Note that salts of Na and K are quite inhibitory, whereas the compatible organic osmolytes (glycine, betaine) are much less so, while the counteracting osmolyte TMAO is stimulatory except at high concentrations.

(*Source:* M. E. Clark & M. Zounes, 1977, The effects of selected cell osmolytes on the activity of lactate dehydrogenase from the euryhaline polychaete *Nereis succinea, Biological Bulletin* 153:468–484)

ception is urea, which we discuss later.) In general, compatible solutes can be safely upregulated or downregulated to keep an osmoconforming cell's osmotic pressure equal to that of the ECF. Cell volume changes are therefore small.

2. *Counteraction:* Some organic osmolytes exhibit properties beyond simple compatibility and have the ability to stabilize macromolecules against disruptive forces, such as high temperature and perturbing solutes such as NaCl and urea. When stabilizing solutes are used in cells for both osmotic balance and to offset destabilizing forces, they are termed *counteracting osmolytes.* As you will see, the *methylamine* osmolytes such as TMAO (Figure 13–4) have been found to be the strongest stabilizers in many situations (● Figure 13–7).

The use of organic osmolytes is widespread in nature and not just in marine organisms. Organisms subjected to freezing, for example, often use them for both cell volume maintenance and lowering freezing points (see Chapter 15). Mammalian kidney cells use them to osmoconform to the high osmotic pressure in the kidney medulla and to counteract the harmful effects of urea, a major waste product (see Chapter 12, box "Molecular Biology and Genomics: Osmolytes and Gene Regulation," p. 525).

All marine osmoconformers use organic osmolytes but are not alike in terms of their use of these solutes. Rather, they fall into two broad subcategories: strict osmoconformers and hypoionic osmoconformers (Figure 13–4a).

Strict Osmoconformers: Most Marine Invertebrates and Hagfish

Most osmoconformers have ECF compositions that closely resemble seawater, except for lower concentrations of certain

solutes such as magnesium, which inhibits transmission at the neuromuscular junction (Figure 13–4a, ECF), and ions adjusted for buoyancy (see box on "A Closer Look at Adaptation: Life at the Top"). However, their cells are dominated by organic osmolytes (Figure 13–4a, ICF). Marine invertebrates, including most crustaceans, annelids, mollusks, and sponges, fall into this category. Typically, their cells accumulate *methylamines* such as glycine betaine and *free amino acids* such as glycine and proline to serve as compatible osmolytes. Most vertebrates are either hypoionic osmoconformers or osmoregulators, but hagfish (jawless fishes) are the exception. In common with marine invertebrates, hagfish rely primarily on Na^+ and Cl^- to maintain ECF osmotic pressure, with amino acids and methylamines as osmolytes in the ICF. What happens if there is a *change* in the osmotic pressure of the outside medium of these animals? Two distinct responses can be measured:

1. *Stenohalinity:* Some pure osmoconformers have a limited tolerance to changes in salinity and are termed **stenohaline**. Stenohaline animals typically die if exposed to significant changes in external salinity. Their cells swell or shrink from osmosis as the ECF osmotic state changes because they cannot regulate their osmolytes to compensate. To survive, their primary strategy is *avoidance.* Most marine cnidarians (such as jellies, corals, and sea anemones) and all echinoderms (such as seastars, sea urchins, and sea cucumbers) are examples of animals restricted to life within a narrow range of salinity, and they avoid disturbances by remaining in the ocean (however, if

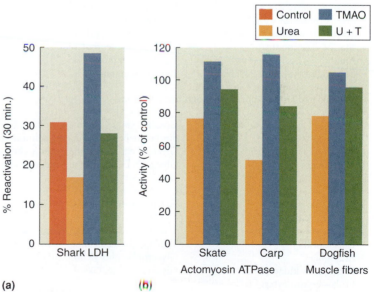

(a)   (b)

Figure 13–7 ● Counteracting effects of a methylamine osmolyte, TMAO, on urea inhibition of protein functions. (a) Refolding of a denatured enzyme, lactate dehydrogenase, from a shark (which uses urea and TMAO as its major cellular osmolytes). (b) Activity of a muscle protein complex, actomyosin ATPase, from a skate (cartilaginous fish) and a carp (bony fish), and contractile force of whole muscle fibers from a shark.

(*Source:* Modified from P. H. Yancey, & G. N. Somero, 1979, Counteraction of urea destabilization of protein structure by methylamine osmoregulatory compounds of elasmobranch fishes, *Biochemical Journal* 182:317–323; and P. H. Yancey, 1985, Organic osmotic effectors in cartilaginous fishes. In R. Gilles & M. Gilles-Ballien, eds., *Transport Processes, Iono- and Osmoregulation,* Berlin: Springer.)

Table 13–2 ▌ Major Classes of Osmolytes and Their Distribution among Eurkaryotes, Bacteria and Archaea. Elevated concentrations of osmolytes due to freeze-avoidance ("freezing") or desiccation stress (*Artemia* embryos) are included.

Organism	Osmolyte
I. Eukaryotes and Bacteria	
A. Polyhydric alcohols (polyols) and sugars	
Bacteria	Trehalose, sucrose, mannitol, glucosylglycerol, mannosucrose
Algae	Glycerol, mannitol, sorbitol, sucrose, glucose, volemitol, floridoside
Fungi	Glycerol, erythritol, arabitol, mannitol
Vascular plants	Myo-inositol, sorbitol, glucose, sucrose, mannitol, methyl-inositols (pinitol, etc.)
Animals	
Artemia embryos	Trehalose, glycerol
Insects—freezing	Glycerol, sorbitol, erythritol, sucrose
Insects—salinity	Trehalose
Marine elasmobranchs	Myo-inositol, scyllo-inositol
Teleosts—freezing	Glycerol
Amphibia—freezing	Glucose, glycerol
Mammals—kidney	Sorbitol, myo-inositol
Mammals—brain	Myo-inositol
B. Amino acids and amino acid derivatives	
Bacteria	Glutamate, glutamine, proline, ectoine, hydroxyectoine, N_{α}-carbamoyl-glutamate-1-amide, N-acetylglutaminylglutamine Amide γ-amino butyrate
Algae	Proline, alanine, glycine, glutamate
Vascular plants	Proline
Animals	
Marine invertebrates	Glycine, alanine, proline, serine, taurine, strombine, etc.
Insects (brackish water)	Proline, serine
Marine cyclostomes	Glycine, alanine, proline
Marine elasmobranchs	Taurine, glycine, β-alanine
Amphibia	Various α-amino acids
Mammals—kidney, heart, brain	Taurine, glutamine
C. Methylated ammonium and sulfonium compounds	
Bacteria	Glycine betaine, choline-O-sulfate, proline betaine, taurine betaine, β-alanine betaine, glutamate betaine, pipecolate betaine, Dimethylsulfoniopropionate (DMSP)
Algae	Glycine betaine, proline betaine, choline-O-sulfate, homarine, Dimethyl-taurine, taurine betaine, DMSP
Vascular plants	Glycine betaine, proline betaine, β-alanine betaine, Choline-O-sulfate, DMSP
Animals	
Marine invertebrates	Glycine betaine, TMAO, proline betaine
Marine cyclostomes	TMAO
Marine elasmobranchs	TMAO, glycine betaine, sarcosine
Coelacanths	TMAO, glycine betaine
Amphibia	Glycerophosphorylcholine (GPC)
Mammals—kidney, brain	Glycine betaine, GPC
Birds—erythrocytes	Taurine
D. Urea	
Gastropods—estivating	
Marine elasmobranchs	With methylamines
Coelacanth	With methylamines
Lungfish-estivation	
Amphibia—salinity	With methylamines and amino acids
—estivation	
Mammals—kidney	With methylamines

(continued)

Table 13–2 ▮ Major Classes of Osmolytes and Their Distribution among Eurkaryotes, Bacteria and Archaea *(continued)*

Organism	Osmolyte
II. Archaea (see Martin et al., 1999)	
A. Inorganic ions	
Nonhalophiles	K^+, at concentrations that may exceed 0.5 M
Extreme halophiles (*Halobacteriaceae*)	K^+, at concentrations that may exceed 5 M
B. Polyhydric alcohols and sugars	
Glycerol, diglycerol phosphate, di-myo-inositol-1,1′-phosphate (DIP), glucosylglycerate, β-mannosylglycerate, mannosyl-DIP, 2-sulfotrehalose, α-glucosylglycerate, β-galactopyranosyl-5 hydroxylysine	
C. Amino acids and amino acid derivatives	
Glutamate, β-glutamate, β-glutamine, N^{ϵ}-acetyl-β-lysine	

Sources: P. Hochachka, G. N. Somero, 2002, *Biochemical Adaptation: Mechanism and Process in Physiological Evolution,* Oxford, UK: Oxford University Press, using data from G. N. Somero & P. H. Yancey, 1997, Osmolytes and cell volume regulation: Physiological and evolutionary principles, in J. F. Hoffman, J. D. Jamieson, eds., *Handbook of Physiology, Sec. 14;* Oxford, UK: Oxford University Press; and D. D. Martin, R. A. Ciulla, & M. F. Roberts, 1999, Osmoadaptation in Archaea, *Applied and Environmental Microbiology* 65:1815–1823.

A CLOSER LOOK AT ADAPTATION

Life at the Top

An important adaptation closely related to water and solute regulation is that of buoyancy. Pelagic aquatic animals—those that swim or drift above the bottom—often benefit from mechanisms that help them float. This is a problem because some biological materials—proteins, shells, and skeletons—are heavier (denser) than water. A major example of an adaptation is the swim bladder of bony fish, covered in Chapter 11. Another example is found in a variety of zooplankton (animal drifters). Jellies (scyphozoan cnidarians), for example, float with the currents and hunt for prey using a network of stinging tentacles extending below the main body, or bell. Jellies can stay up in the water by pulsing their bells, but many must halt this pulsing when they extend their tentacles in a cylindrical or complex trap, something like a spider's web. While "fishing" this way, it is important for the jelly to remain neutrally buoyant because any change in the configuration of the extended web reduces both the effectiveness of the web and the volume of water that it controls. Furthermore, any active swimming movements made to maintain its position in the water might alert potential prey of its presence. How then can the jelly maintain buoyancy? Measurement of concentrations of solutes found in the body fluids of the jelly provided a clue. Jellies are osmoconformers. However, examination of the ion profile in the jelly revealed some intriguing differences. In the jelly *Aurelia,* the concentration of sulfate ion ($SO_4^{=}$) in the body fluids is maintained at around 16 mmoles/kg, whereas in the surrounding seawater it is approximately 29 mmoles/kg. Chloride ion in seawater is 560 mmoles/kg, whereas in the body fluids of *Aurelia* the concentration is maintained at approximately 580 mmoles/kg. Why then might the jelly be interested in excluding sulfate from its body fluids and replacing it with chloride ion? Consider that the molecular weight of sulfate is a comparatively hefty 98, compared to 35 for chloride ion. Exclusion of sulfate ion coupled with replacement with chloride ion is a solution numerous marine invertebrates use to achieve buoyancy. Squid and larval tunicates use a similar solution as an aid to buoyancy but, instead, accumulate ammonium ions at the expense of heavier cations such as Ca^{2+}, Mg^{2+}, and Na^+. Some vertebrates may use similar strategies: The osmolytes urea and TMAO of cartilaginous fishes are more buoyant than seawater, and unusual buoyant gelatinous layers with low ion contents have been found in some deep-sea fishes. TMAO is also a common osmolyte in many crustaceans such as shrimp.

Photo: Paul Yancey

A marine jelly, which removes $SO_4^{=}$ ions from its body fluids to increase buoyancy.

their larvae are swept by currents into a low-salinity environment, they would likely die). Indeed, to our knowledge no echinoderm has managed to adapt to fresh water, whereas some cnidarians (such as hydra) have. Hagfish are also stenohaline; because they live primarily in deep water, they probably never face changing salinities.

2. *Euryhalinity:* In contrast, some osmoconformers are relatively tolerant of changes in the salinity of the surrounding medium and can survive substantial dilution or concentration of their ECFs; these animals are termed **euryhaline**. Many aquatic habitats from freshwater to brine are relatively stable in salinity, but some habitats can change significantly. Animals living in the intertidal zone live a particularly precarious existence. During parts of the day they are covered by seawater. However, when the tide is out they may face increased salinities because of evaporation of water from a tide pool, or evaporation directly from their bodies if the animals are on a rock face. Or they may face a marked decline in salinity, because of heavy rainfall. Consider also animals in brackish estuaries, where freshwater streams and rivers mix with ocean water. In both intertidal and estuary environments, salinity and hence the osmotic pressure change with the tides, and do so unpredictably, because of variations in river flow, precipitation, and tidal height and direction.

Euryhaline osmoconformers can successfully adapt to variable salinities by being able to regulate the organic osmolytes in their cells, to keep the ICF in balance with the ECF. For an example, let us consider the eastern oyster (*Crassostrea virginica*). Some populations are resident of the estuaries bordering the east coast of North America. Oysters are not mobile: They attach themselves to river bottoms and form vast oyster beds. During high tides salt water is forced inland such that during a 24-hour period animals anchored in this environment are alternately exposed to both fresh water and salt water. Cell volume in these bivalves is ultimately regulated by

changing the concentration of intracellular compatible amino acids in response to changes in extracellular osmotic pressure. During the transition to saline water, cellular mitochondria synthesize amino acids, which are then transported to the cytosol (Table 13–2). The process is thought to be triggered by increasing levels of cellular Na^+. Conversely, while in brackish water these organic osmolytes are reduced to low concentrations by being transported out of the cells ($\bullet$ Figure 13–8). Osmotic neutrality is thus maintained under both conditions, and the cells undergo limited volume changes as a result. (What occurs in freshwater exposure is discussed later.) Studies have shown that the acute response to osmotic stress involves an increase in the amino acid *alanine*, whereas chronic (several days) exposure to salt water results in the successive replacement of alanine with *glycine* and *proline*. Physiologists do not know why the amino acids change in this way.

These estuarine oysters cannot adapt to long-term exposure to full-strength seawater. In contrast, populations of the same species on the open Atlantic coast live well in seawater. Interestingly, their tissues use large amounts of the methylamine *glycine betaine* (trimethylglycine) and the nonprotein sulfonic amino acid *taurine* (Figure 13–5). The latter can make up as much as 70% of the cytosolic amino acid pool. Whether these particular osmolytes are better suited for adaptation to full seawater than are typical amino acids remains unknown.

The composition of amino acids and methylamines used by marine invertebrates as osmolytes varies, with some relying primarily on ordinary amino acids such as glycine (in sponges, for example), whereas others (such as many mollusks) use a mixture of taurine, glycine betaine, and glycine. Researchers are not certain why this might be. However, they have recently discovered one pattern related to the environment. Cells of shallow-water marine shrimp use mainly glycine, with small amounts of the methylamines glycine betaine and TMAO. However, deep-sea shrimp rely primarily on TMAO (Figure 13–5), more so in deeper species (at 3000 m) than moderate-depth ones (at 2000 m), which have more TMAO

Figure 13–8 $\bullet$ **Adjustments to changes in Na^+ levels in a brackish-water invertebrate.** A reduction in external Na^+ and osmotic concentration (OC) causes a reduction in the blood (step ①), leading to osmosis of water into cells ②. This in turn triggers a reduction in cellular amino-acid osmolytes ③, which enter the blood ④ where they are broken down, producing NH_4^+ ⑤. Gill transporters excrete the NH_4^+ in exchange for Na^+ ⑥. This increases blood Na^+ ⑦ thus compensating for the initial disturbance ⑧.

(*Source:* P. Willmer, G. Stone, & I. Johnston, 2000, *Environmental Physiology of Animals,* Oxford, UK: Blackwell Science, Figure 10.18)

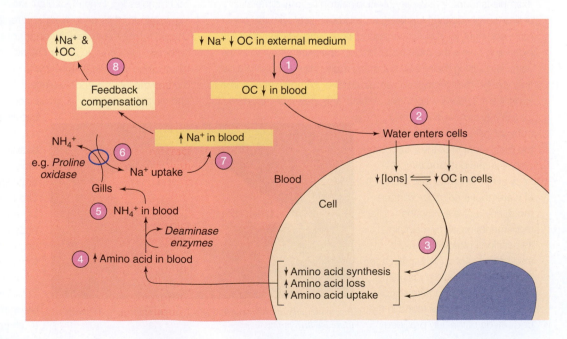

than shallow species. Why might this be? We'll return to this pattern shortly when we discuss deep-sea fish.

Because ECF and ICF osmotic pressures vary considerably in euryhaline osmoconformers, they do not exhibit homeostasis of body fluids in the classic sense (of Bernard and Canon) first applied to mammals. But they do exhibit the crucial homeostasis of cell volume regulation by regulating osmolytes. This form of homeostasis is called *enantiostasis* (Chapter 1).

Although some evidence shows that rising cellular Na^+ levels can trigger osmolyte accumulation, in most cases the genes involved in adapting to higher osmolarity are poorly characterized. See the box, "Molecular Biology and Genomics: A Worm in Salt Water."

Hypoionic Osmoconformers: Marine Cartilaginous Fishes, Coelacanths, and So Forth

The other subcategory of osmoconformation, which also involves cellular osmolytes, is termed *osmoconforming hypoionic regulation* (Figure 13–4a). These animals, exemplified by the *cartilaginous fishes* (sharks, skates, rays, holocephalans), also have ECF and ICF osmolarities equal to or slightly higher than the environment. Thus again they show no tendency to lose water and may even make a modest gain. However, in contrast to pure osmoconformers, they also actively regulate their extracellular fluids to have considerably lower salt concentrations than the environment. They achieve osmotic parity with seawater by producing the organic solutes *urea* and *TMAO*, found throughout their bodies in both ECF and ICF. In both fluid compartments, urea is at about 400 mOsm, whereas

TMAO is around 70 mOsm in the ECF and 200 mOsm in the ICF. These animals are sometimes called **ureosmotic** conformers because urea is the dominant osmolyte in most species. Some species are stenohaline, unable to regulate their osmolytes in the face of environmental changes. In euryhaline species such as the Nicaraguan shark that migrates from the ocean to freshwater lakes, urea and TMAO are regulated (to a greater extent than inorganic ions) for osmotic adjustments.

In contrast to most organic osmolytes, which are compatible with cellular macromolecules, urea at the concentrations in the fluids of these animals would be inhibitory or even fatal to most animals because of the destabilizing effects it has on protein structure and function. As you saw in Chapter 12, urea is the major nitrogenous waste of mammals, which must not be allowed to build up in the blood (think about that the next time you eat shark meat). The reason cartilaginous fishes can thrive with elevated urea concentrations in their fluids is threefold:

1. Some proteins (such as in the eye lens) in these animals have evolved structures resistant to the denaturing effects of urea.

2. Some proteins (such as the enzyme lactate dehydrogenase, p. 54) have evolved overly stable features and require urea to "loosen" them up for proper function.

3. TMAO, when present at concentrations about half that of urea, offsets or counteracts the destabilizing effects of urea. An example is shown in Figure 13–7. This is thought to be the most common adaptation in these fishes.

MOLECULAR BIOLOGY *AND* GENOMICS

A Worm in Salt Water

In the last couple of decades, a soil-dwelling nematode worm, *Caenorhabditis elegans,* has emerged as a major model animal in the study of development and genetics. Like other nematodes, every individual of this remarkable species has an exact number of cells after it reaches maturity—959 cells in this species, of which 81 cells are muscles, 302 are neurons, and so on. Thus it is a model organism for studying how a single fertilized egg divides and differentiates into a multicellular organism. As part of this effort, its genome has been sequenced, and the patterns of gene regulation involved in development are being investigated.

Recently, physiologists have begun to study *C. elegans* as well. Research in 2003 by Samuel Lamitina, Kevin Strange, and colleagues examined this worm's ability to adapt to salty conditions in conjunction with gene expres-

sion. They found that the worms could acclimate to 400 mM NaCl, equivalent to about 75% seawater. In doing so, they accumulate large amounts of the compatible osmolyte glycerol and, to a lesser extent, taurine. The researchers also analyzed the worm's genetic responses to high osmolarity with a genome microarray ("gene chip"; Chapter 2, p. 32) with over 17,000 genes represented. About 150 genes were found to increase their expression by at least 1.7 times. Almost half of these genes code for proteins of metabolism, but the roles of these genes in osmotic adaptation are uncertain. This study illustrates the complexity of physiological regulation of genes in adapting to different environmental conditions. Model systems such as this lowly worm may be the first to provide understanding of genetic mechanisms for such processes.

C. elegans, *a nemotode worm that is used as a model animal for many biological studies*

(*Source:* http://130.15.90.245/c_elegans.htm)

Photo: Courtesy of Ian Chin-Sang

Figure 13–9 ● Diagrams of major osmotic gains, losses, regulation, and body-fluid concentrations in marine vertebrates.

(*Source:* (a) and (b) C. E. Bond, 1996, *Biology of Fishes,* 2nd ed. Belmont, CA: Brooks/Cole Thomson Learning, Figure 24–5, p. 407, and Figure 24–6, p. 411. (c) and (d) P. Willmer, G. Stone, & I. Johnston, 2000, *Environmental Physiology of Animals,* Oxford, UK: Blackwell Science, Figure 9.39, p. 299)

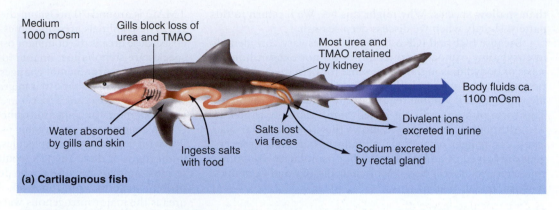

Medium 1000 mOsm

Gills block loss of urea and TMAO

Most urea and TMAO retained by kidney

Body fluids ca. 1100 mOsm

Water absorbed by gills and skin

Ingests salts with food

Salts lost via feces

Divalent ions excreted in urine

Sodium excreted by rectal gland

(a) Cartilaginous fish

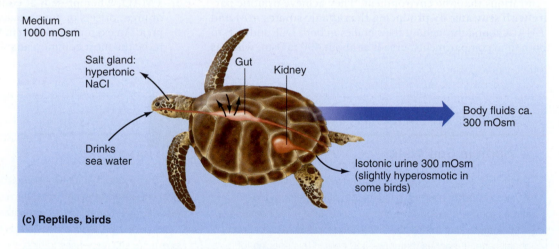

Medium 1000 mOsm

Loses water through gills and skin

Removes salts via chloride cells in gills

Body fluids ca. 400 mOsm

Gains water and salts by swallowing seawater and food

Salts lost via feces

Salts and little water lost via scant urine

(b) Bony fish

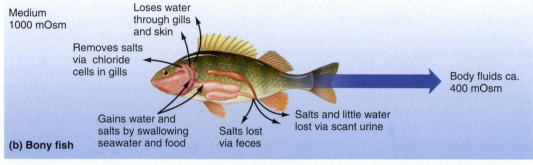

Medium 1000 mOsm

Salt gland: hypertonic NaCl

Gut

Kidney

Body fluids ca. 300 mOsm

Drinks sea water

Isotonic urine 300 mOsm (slightly hyperosmotic in some birds)

(c) Reptiles, birds

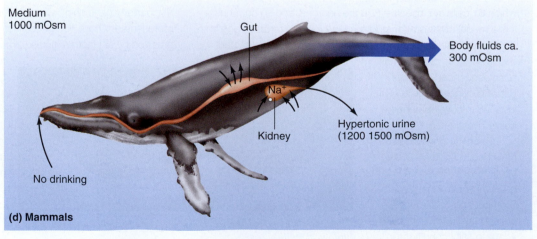

Medium 1000 mOsm

Gut

Body fluids ca. 300 mOsm

Na$^+$

Kidney

Hypertonic urine (1200 1500 mOsm)

No drinking

(d) Mammals

A summary of body water and solutes in marine cartilaginous fishes is shown in ● Figure 13–9a. Regulation of urea, TMAO, and water involves gills, digestive tract, kidneys, and a special organ, the rectal gland. See Chapter 12 (p. 566) for more information.

The coelacanth fish also uses urea and TMAO as its major organic osmolytes. Another vertebrate using this strategy of osmotic adaptation is the crab-eating frog (*Rana cancrivora*) of Southeast Asia. When this frog moves from land to brackish estuarine water, it accumulates urea (and both urea and amino acids in its cells) rather than salt in its fluids, to osmoconform to the environment.

Finally, a few arthropods are hypoionic osmoconformers. In particular, larvae of some species of *Culex* mosquitoes develop in salt ponds that have a salinity about 70% that of seawater. They regulate their hemolymph ions to stay relatively constant and, like cartilaginous fishes, build up additional osmotic pressure with organic osmolytes in their hemolymph and cells. These osmolytes include proline (an amino acid) and trehalose (a carbohydrate).

Osmoregulators

■ Osmoregulators rely on special transport mechanisms to maintain internal osmotic constancy.

Osmoregulators have a very different approach to cell volume homeostasis. Much as a thermoregulator maintains a steady body temperature in the face of a variable environment, osmoregulators maintain a relatively steady ECF osmotic pressure regardless of osmotic changes in the external environment. Special transport tissue and organs, plus an impermeable skin, are generally needed to achieve this. Most cells are thus not exposed to volume changes and so do not need to regulate organic osmolytes. There are two conditions for osmoregulators: *hypo-osmotic* and *hyperosmotic*. Some crustaceans in salty habitats, some mosquito larvae in saline ponds and lakes, and most marine vertebrates are hypo-osmotic osmoregulators. Most bony fishes and all higher vertebrates, for example, have internal ECF and cellular osmotic concentrations of 250 to 400 mOsm whether the environment is brackish (estuary) or higher than seawater (such as an evaporating tide pool). Those in the oceans are therefore always less concentrated than the habitat ("hypo"). In contrast, all freshwater organisms are *hyperosmotic* osmoregulators by necessity, because to osmoconform would mean to have *no* cellular solutes, a condition clearly incompatible with life.

Hypo-Osmotic Regulation: Most Marine Vertebrates and Some Arthropods

As a consequence of being hypo-osmotic to seawater, most marine vertebrates tend to lose water and gain excess salt. This pattern suggests that their early ancestors evolved in either an estuary or fresh water. Cells in most marine vertebrates generally use only low concentrations of organic osmolytes (Figure 13–4a), with most bony fishes having low concentrations of TMAO.

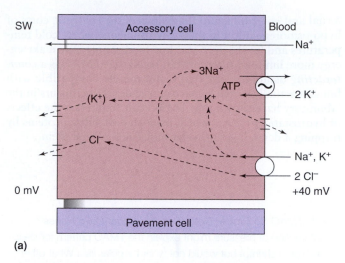

(a)

Figure 13–10 ● Current model for ion transport mechanisms in seawater teleost chloride cell. Active transport and cotransport are depicted as solid lines, passive diffusion as dashed lines. The Na$^+$-selective "leaky" tight junction between the accessory cell and the chloride cell is depicted as a single strand. The negative transepithelial potential favors Na$^+$ efflux through the paracellular pathway. The tight junction between chloride cells and pavement cells is deep and multistranded.

(*Source:* Modified from W. S. Marshall, *Cellular Approaches to Fish Ionic Regulation,* in C. M. Wood, and T. J. Shuttleworth, eds., San Diego: Academic Press, 1995.)

Mechanisms to cope with this problem are covered in part at the end of Chapter 12. Here is a brief summary: Marine fish must drink seawater. The lining of the digestive tract can restrict the influx of some ions, but some ions must be imported for water to follow. (Recall that no ATPase pumps actively move water molecules alone.) This import creates an excess of salt in the blood. The gills then use specialized epithelial **chloride cells** to actively transport NaCl outward (● Figure 13–10). Marine birds and reptiles use special *salt glands* to remove the excess salt (Figure 12–34). Marine mammals have efficient kidneys with long nephron loops for the same purpose. Water and solute movements in these marine vertebrates are summarized in Figure 13–9b, c, and d.

Some exceptions to the general osmotic pattern for bony fish have recently been found. Some marine fishes in the Arctic accumulate the compatible osmolyte *glycerol* to the point that they become isosmotic with seawater at 1000 mOsm! Glycerol is thought to function as an antifreeze (see Chapter 15). Some marine fishes in polar waters and in the deep sea have high TMAO content in their cells and high NaCl in their blood, with osmotic concentrations up to 600 mOsm. These fish are still hypo-osmotic to seawater, but more concentrated than the typical bony fish. TMAO accumulation in polar fishes is not consistent: One species may have high levels, whereas a related species may not. However, the pattern is clearer for the deep sea: TMAO content in several families of deep-sea fish increases linearly with depth in the ocean. This is the same pattern seen in deep-sea shrimp, as you saw earlier (p. 582). Why do these osmoregulators need to retain such large amounts of TMAO? Possibly reducing the water gradient between the internal and environmental fluids lowers the energetic costs of hypo-osmoregulation in these environments, where cold temperatures and (in the deep sea) low food supply may make energy more limiting. Alternatively, recall that TMAO is a *counteracting* osmolyte, one that is not merely compatible with but can actually stabilize macromolecules. Experiments in the laboratory have shown that TMAO can counteract the effects of hydrostatic pressure, which tends to destabilize proteins by trapping a dense layer of water molecules around them.

?

Why Is TMAO So High in Deep-Sea and Some Polar Fishes?
Counteraction of pressure might explain the TMAO pattern for deep-sea fish (and shrimp) but would not work for polar fish. What other hypotheses might explain these TMAO patterns? For another role of TMAO, see the box, "A Closer Look at Adaptation: Life at the Top," and consider other possibilities such as diet.

Hypo-osmotic regulation also occurs in some arthropods in briny habitats. For example, the brine shrimp *Artemia* (see the box, "A Closer Look at Adaptation: The Sweet Solution to Desiccation," p. 574) can live in desert brine lakes in the Great Basin of the United States. In broad terms, their adaptations are similar to fishes, although some details are different: These crustaceans must drink the briny water (as much as 8% of their body weight per day) to gain needed water and unwanted ions, retain water with a highly impermeable exoskeleton, and then excrete the excess ions with special salt glands on their gills. Larvae of some *Aedes* mosquito species are also strong hypo-osmotic regulators, developing in brine ponds up to three times more saline than the ocean, while maintaining a lower and constant internal ion level. These larvae have evolved an extra segment on their rectums, which actively transport ions out of the body (see the mechanisms of insect excretion at the end of Chapter 12).

Hyperosmotic Regulation: Freshwater Animals

Across the planet, fewer species of life reside in fresh water than reside in the oceans. This may indicate that life did not evolve first in fresh water, and it also indicates that this habitat is not easily adapted to. The primary reason is osmotic stress. In effect, organisms in fresh water are constantly taking in unwanted water that they must somehow "bale" out. In addi-

tion, valuable solutes are constantly lost through gills, feces, and urine. Gills are particularly problematic because of their comparatively large surface areas and the necessity of minimizing protective coverings to maximize gas exchange (see p. 467). What mechanisms have evolved to enable an animal to survive in a low salinity or freshwater environment? A summary of water and solute movements in freshwater fish is shown in ● Figure 13–11; details for fish and some invertebrates are as follows:

- *Active transport of ions:* To combat the loss of salts from its body fluids, freshwater animals must ultimately expend energy (ATP) to transport ions from the water into the ECF. This takes place in the gills of crustaceans and fishes; the latter have gill pavement cells (Figure 13–10) that take up sodium via an electrical potential generated by H^+ V-ATPases (p. 81); and chloride cells (different from those of marine fishes) that take up chloride and calcium. Crustacea may also have feathery accessory gills in the form of modified *epipodites* (outer branch of some crustacean legs) that can regulate ions. For freshwater frogs and worms, the skin is an important site of ion uptake. Frog skin, for example, also uses H^+ V-ATPases to create a potential for salt uptake.

- *Hypotonic urine:* To combat excess water gained from osmosis and eating, most freshwater animals remove it by producing a voluminous urine that is hypo-osmotic to body fluids.

- *Lower internal osmolarities:* To help reduce water intake, solute concentrations in freshwater animals (in both ECF and ICF) are maintained at lower levels in comparison to their marine relatives. For example, freshwater teleost fish have internal osmotic concentrations in the range of 250 to 300 mOsm, slightly lower than in marine species (300 to 400 mOsm in most). With this strategy the animal expends less energy to maintain homeostasis. Similarly, most freshwater crustaceans have osmotic concentrations of 300 to 500 mOsm, in contrast to about 1000 mOsm in their marine counterparts. The difference lies primarily in reduced concentrations of NaCl in the ECF and of amino acid osmolytes in the ICF in freshwater animals.

- *Low permeability of integument:* To reduce efflux of solutes and influx of water in freshwater, many animals have reduced permeability of the outer body surface. Less energy is therefore needed to secure solutes from the external environment and to bale out excess water. For example, freshwater decapod crustaceans (crabs, crayfish)

Figure 13–11 ● Diagram of major osmotic gains, losses, regulation, and body-fluid concentration in a freshwater bony fish.

(*Source: From* C. E. Bond, 1996, *Biology of Fishes,* 2nd ed., Belmont, CA: Brooks/Cole Thomson Learning, Figure 24–4, p. 405)

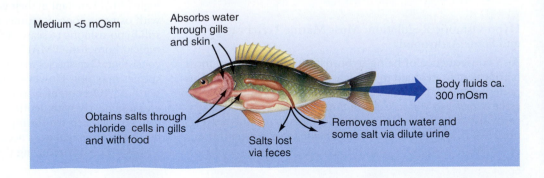

Medium <5 mOsm

Absorbs water through gills and skin

Body fluids ca. 300 mOsm

Obtains salts through chloride cells in gills and with food

Salts lost via feces

Removes much water and some salt via dilute urine

are approximately 10-fold less permeable to Na$^+$ and Cl$^-$ ions than their marine counterparts. The body surface of freshwater teleost fish is also comparatively less permeable to water compared to marine fish. Problems arise in those species (such as mussels that open their shells to feed) that expose sizable portions of their unprotected body surfaces to fresh water. Uncontrolled water entry and loss of valuable ions are the price these animals must pay. Amazingly, solute concentrations in the freshwater mussel *Anodonta* are reported to be only about 66 mOsm, probably the lowest osmotic concentration recorded in a living organism and deserving of further study.

Alternating Hyperosmotic to Hypo-Osmotic Regulation

Osmoconforming is the most common adaptation in the oceans, but it has a limit: It cannot be used to adapt fully to fresh water (although some osmoconformers such as estuarine bivalves may avoid damage from fresh water by "clamming up," that is, avoiding the temporary problem). Thus animals that must adapt to both habitats have at least some osmoregulatory capacity. Some of the best animals at doing so are killifish in tide pools and salmon (and other fishes that spawn in either fresh or seawater and migrate to the other habitat to grow). Killifish were noted in Chapter 1 (p. 15), where we discussed how they use both physiological and behavioral processes to cope with salinity changes. Here we look at salmon. These fish are particularly interesting because, as excellent osmoregulators, they are hypo-osmotic in the oceans but hyperosmotic in rivers; thus they are sometimes called "hypo-/hyperosmoregulators." The changes these fish make are classic examples of *acclimatization* regulation (Chapter 1) coupled with an *anticipation* mechanism that reduces delays in adaptation. When they hatch in fresh water and begin migrating to the sea, *growth hormone,* which is controlling their general maturation, and the stress hormone *cortisol* (see Chapter 7) begin to alter the gill epithelial cells even before seawater exposure (an anticipation mechanism). As they enter estuaries, increased internal sodium triggers more cortisol production. Together the hormones trigger the growth of the seawater-type branchial *chloride cells,* as well as increasing Na$^+$–K$^+$ ATPase activity, the driving force for the movement of monovalent ions in the cells. After maturing for a few years at sea, the fish are ready to return to their home streams to spawn.

The hormone *prolactin* (p. 267) has been found to induce the formation of salmon gill cells into the freshwater arrangement. In freshwater fishes in general, prolactin maintains proper ion and water permeabilities in the epithelial cells of osmoregulatory organs, including the gill, intestine, kidney, and bladder. Hypophysectomy (removal of the pituitary gland) of freshwater fish results in a loss of ions and eventually death unless prolactin is administered to the surgically manipulated animals.

Osmotic Regulation in Transitional Animals

Some transitional animals (living in water and on land), although they are ion regulators, use organic osmolytes in certain situations. Lungfish are found in Africa, South America, and Australia; the former two enter a state of estivation (summer hibernation) when their habitats become dry. They burrow and seal themselves in a mucus cocoon, and become largely dormant. During estivation, they build up large amounts of urea (reportedly up to 400 mM), presumably as a less toxic alternative to ammonia for nitrogen storage during estivation. By elevating osmotic pressure, the urea also helps retain water and may draw in water from the soil by osmosis. Similarly, some American amphibians (such as the Western spadefoot toad) and Australian frogs estivate during dry periods and accumulate urea. These estivators accumulate only small amounts of counteracting osmolytes, perhaps because the counteraction is unnecessary: The inhibitory effects of urea may actually be useful in this situation where suppression of metabolism is important for survival.

You have already seen how amphibians can use their bladders for water regulation (p. 566). It is worth re-emphasizing that behaviors serve as homeostatic effectors in many animals, and amphibians are excellent examples of this. They actively seek moist, cool local environments, for example.

Osmotic Regulation in Terrestrial Habitats

The problems of osmotic balance are different on land compared to problems animals encounter in water. For animals in the ocean, salt is in excess whereas water is limiting; in fresh water, the problems are reversed. But on land, water is often a limiting factor, and salt may be as well (except for animals with a high-salt diet) (● Figure 13–12). Terrestrial animals use many mechanisms to maintain osmotic balance in the face of these limits. They are by necessity osmoregulators to some extent, because the environment (air) has no solutes (exceptions are animals such as earthworms in the soil, but because they live only in moist conditions they are effectively aquatic).

The most successful fully terrestrial animals are arthropods and vertebrates. The vertebrates typically have osmotic concentrations at about 300 mOsm, with the ECF dominated by NaCl. Insect fluids can have osmotic concentrations higher than 300 mOsm in some species, in part because of high concentrations of amino acids in the hemolymph; and whereas the dominant ECF cation is usually Na$^+$, some herbivorous insects (such as Coleoptera) reportedly have higher K$^+$ than Na$^+$.

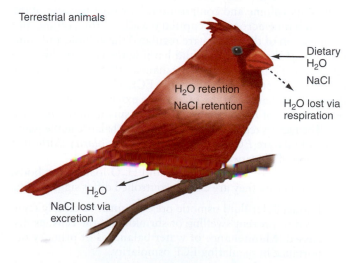

Terrestrial animals

Dietary H$_2$O

NaCl

H$_2$O retention
NaCl retention

H$_2$O lost via respiration

H$_2$O

NaCl lost via excretion

Figure 13–12 ● Diagram of major osmotic gains, losses, regulation, and body-fluid concentration in a terrestrial vertebrate.

(*Source:* Modified from Y. Takei, 2000, Comparative physiology of body fluid regulation in vertebrates with special reference to thirst regulation, *Japanese Journal of Physiology* 50: 171–186)

Losses of water and ions in terrestrial animals occur via evaporation and through the excretory organs and digestive tracts, the latter two being the most common sites of regulation (as you saw in Chapter 12). In most cases, water and solutes must be obtained via oral intake. One interesting exception is found in some insects and arachnids, which can extract water out of humid air (typically 50 to 100% relative humidity)! The house dust mite *Dermatophagoides*, for example, uses its salivary glands to secrete a fluid droplet that is so concentrated with inorganic and organic osmolytes that it can draw water vapor into it. A beetle (*Onymacris*) in the coastal desert of Namibia has a bumpy cuticle, with hydrophobic troughs and hydrophilic peaks that can attract water molecules from humid air, such as fog that forms from the ocean. Other arthropods use the excretory systems; for example, the rat flea *Xenopsylla* draws moist air into its rectum and shows a net water gain. How it does this is uncertain.

As for most physiological processes, research on mammals has provided the most details of volume, solute, and water regulation, including hormonal control. As you will see, osmotic regulation is also closely tied to volume regulation. Therefore we now turn to these animals to illustrate fluid homeostasis in detail.

Osmotic and Volume Balance in Mammals

■ Osmotic balance is maintained by regulating ECF volume and ECF osmolarity.

As you have seen, in osmoregulators such as mammals, cells usually do not experience any net gain (swelling) or loss (shrinking) of volume, because the concentration of nonpenetrating solutes in the ECF is normally carefully regulated. In mammals, this is done primarily by the kidneys, which work to maintain the same level of osmotic pressure (an osmotic concentration of about 300 mOsm) as is present within most body cells.

Plasma is the only fluid that can be directly acted on to control its volume and composition. However, because of the free exchange across the capillary walls, if the volume and composition of the plasma are regulated the volume and composition of the interstitial fluid bathing the cells are likewise regulated. Thus any control mechanism that operates on the plasma in effect regulates the entire ECF. The ICF in turn is influenced by changes in the ECF to the extent permitted by the permeability of the membrane barriers surrounding the cells.

The factors regulated to maintain fluid balance in the mammalian body are *ECF volume* and *ECF osmolarity*. Although regulation of these two factors is closely interrelated, both being dependent on the relative NaCl and H_2O load in the body, the reasons why they are closely controlled differ significantly:

- Extracellular fluid osmotic pressure must be closely regulated to prevent swelling or shrinking of the cells, as discussed. Maintenance of water balance is of primary importance in regulating ECF osmolarity.
- Extracellular fluid volume must be closely regulated to help maintain blood pressure. Maintenance of salt balance is of primary importance in the long-term regulation of ECF volume.

Let's examine each of these factors in more detail.

■ Control of ECF osmolarity prevents changes in ICF volume.

As you have seen, normally the osmolarities of the ECF and ICF are the same, because the total concentration of K^+ and other effectively nonpenetrating solutes inside the cells is equal to the total concentration of Na^+ and other effectively nonpenetrating solutes in the fluid surrounding the cells. It is the number (not the nature) of the unequally distributed particles per volume that determines the fluid's osmolarity. Cell volume normally remains constant, because no H_2O osmotically enters or leaves the cells.

However, any circumstance that results in a loss or gain of free H_2O (that is, loss or gain of H_2O that is not accompanied by comparable solute deficit or excess) leads to changes in ECF osmolarity. If there is a deficit of free H_2O in the ECF, the solutes become too concentrated, and the ECF osmolarity becomes abnormally high (that is, *hypertonic*) (see p. 72). If there is excess free H_2O in the ECF, the solutes become too dilute, and the ECF osmolarity becomes abnormally low (that is, *hypotonic*). When the ECF osmolarity changes with respect to the ICF osmolarity, osmosis takes place, with H_2O either leaving or entering the cells. The osmolarity of the ECF must therefore be regulated to prevent these undesirable shifts of H_2O into or out of the cells.

First we examine the fluid shifts that occur between the ECF and the ICF when the ECF osmolarity becomes hypertonic or hypotonic relative to the ICF. Then we consider how water balance and subsequently ECF osmolarity are normally maintained to minimize detrimental changes in cell volume. Although this section focuses on mammals, we note a few examples of other animals for comparison.

ECF Hypertonicity

Hypertonicity of the ECF, or the excessive concentration of ECF solutes, is usually associated with dehydration, or a negative free H_2O balance. Dehydration with accompanying hypertonicity can be brought about in three major ways:

1. *Insufficient H_2O intake*, such as might occur during a drought or in a desert habitat.

2. *Excessive H_2O loss*, such as might occur in a mammal from heavy sweating or panting, vomiting, diarrhea, or diabetes (even though both H_2O and solutes can be lost during these conditions, relatively more H_2O is usually lost, so the remaining solutes become more concentrated). Also, an air-breathing animal exercising heavily in cold, dry air can suffer excessive water loss, as we discuss later. Finally, a stenohaline freshwater fish exposed to salt water would also lose too much water and gain too much salt.

3. *Drinking hypertonic saline water*, as occurs in marine mammals, reptiles, and birds. Here, although water is entering the body, the salts are excessively high.

4. *Alcohol* inhibits vasopressin secretion and can lead to ECF hypertonicity by promoting excessive free H_2O excretion.

On rare occasions, the ECF becomes hypertonic in the absence of dehydration because of the abnormal accumulation of osmotically active solutes that do not normally contribute significantly to ECF osmotic activity. This occurs, for example, with the high blood glucose in diabetes and with the high blood urea levels in uremia (from kidney failure; see p. 564).

Whenever the ECF compartment becomes hypertonic, H_2O moves out of the cells by osmosis into the more concentrated ECF. Thus cells shrink as H_2O leaves them. Of particular concern is the fact that considerable shrinking of neurons causes disturbances in brain function, which can bring about convulsions or coma in more severe hypertonic conditions. Just as serious are circulatory problems from volume problems, which may range from a slight reduction in blood pressure to circulatory shock and death.

Although shrinkage does occur in this situation, many kinds of cells exhibit a process called **regulatory volume increase (RVI)** in an attempt to compensate. In strict osmoregulators such as mammals, RVI occurs by transport of ions (Na, K, Cl) into the cell to stop water loss, but there are limits to this strategy because those ions cannot be allowed to build up to levels that inhibit proteins and disrupt membrane potential. Thus shrinkage cannot be stopped beyond a certain point. (In contrast, euryhaline osmoconforming cells adapt by importing or synthesizing cellular organic osmolytes, as you saw earlier. These solutes are not inhibitory and thus can be accumulated to levels that prevent shrinkage altogether.)

Interestingly, a few types of mammalian cells are in effect euryhaline osmoconformers. You have already seen that cells of the inner kidney accumulate organic osmolytes to prevent shrinkage (see Chapter 12 box "Molecular Biology and Genomics: Osmolytes and Gene Regulation," p. 525). To some extent neurons, glial cells, and cardiac cells act similarly, accumulating compatible organic osmolytes (such as taurine; Figure 13–5) in an attempt to stop shrinkage in a severely dehydrated mammal. But these cells can only adapt to modest changes. (It is important that attempts to reduce the osmolarity of severely dehydrated animals including humans be done *slowly*. If it is done rapidly, the neurons do not have time to excrete their organic osmolytes and can suffer damage from cell swelling.)

Some mammals can tolerate whole-body dehydration and hypertonicity to an extraordinary degree. One is the camel (such as the dromedary camel), which can survive up to two weeks without drinking and tolerate about 30% loss of body water (compared to most mammals, which do not survive more than about 15% loss). To survive severe dehydration and hypertonicity, the camel slows its metabolism, maintains plasma volume at the expense of other fluid compartments (which shrink), and extracts most of the water from feces in the colon and much of the water from the filtrate in the kidneys. On finding water after prolonged dehydration, the camel can also restore water balance via rapid drinking in a few minutes, a process that would also be lethal to other mammals because of the osmotic shock to cells. Camels survive because water is very slowly absorbed from their guts and because their red blood cells are more tolerant of swelling than are those of other mammals.

ECF Hypotonicity

Hypotonicity of the ECF is usually associated with overhydration; that is, excess free H_2O is present. When a positive free H_2O balance exists, the ECF is less concentrated (more dilute) than normal. Because it is generally NaCl that is too low, this condition may be called *hyponatremia* ("low sodium"). Usually, any surplus free H_2O is promptly excreted in the urine, so hypotonicity/hyponatremia is rare. However, hypotonicity can arise in two ways:

1. *Intake of relatively more H_2O than solutes*, more rapidly than the kidneys can compensate. This would occur if, for example, a stenohaline marine fish were exposed to fresh water (and possibly if a euryhaline fish is exposed too quickly). Also, many human athletes have suffered (and some have died) from drinking too much water without salt intake after heavy sweating.

2. *Retention of excess H_2O without solute*, as a result of inappropriate secretion of vasopressin. As you saw in Chapter 12 (p. 556), vasopressin in mammals is normally secreted in response to a H_2O deficit, which is relieved by increasing H_2O reabsorption in the distal portion of the nephrons. However, vasopressin secretion can be increased in response to pain, acute infections, trauma, and other stressful situations, even when there is no H_2O deficit. The resultant H_2O retention may be appropriate in anticipation of potential blood loss in the stressful situation: The extra retained H_2O could minimize the effect a loss of blood volume would have on blood pressure. However, if there is no such injury, excess retention can occur.

Whichever way it is brought about, excess free H_2O retention first dilutes the ECF compartment, making it hypotonic. Thus cells swell as H_2O moves into them osmotically. Like the shrinking of cerebral neurons, pronounced swelling of brain cells also leads to brain dysfunction. Symptoms include irritability, lethargy, vomiting, drowsiness, and in severe cases, convulsions, coma, and death. Nonneural symptoms of overhydration include weakness caused by the swelling of muscle cells and circulatory disturbances, including hypertension and edema, caused by expansion of the plasma volume.

As with shrinkage, swelling in many cells triggers a compensating process called **regulatory volume decrease (RVD)**. In strict osmoregulators, RVD occurs through the transport of ions, especially KCl, out of the cell. However, once again there are limits to this strategy, because the cell must have a certain level of ions for macromolecular functions and membrane potential. In contrast, euryhaline osmoconforming cells such as those of an estuarine oyster or the mammalian inner kidney, can remove organic osmolytes from the ICF to stop swelling.

Isotonic Fluid Gain or Loss

Let us contrast the situations of hypertonicity and hypotonicity with what happens as a result of isotonic fluid gain or loss. Isotonic gains are a very rare occurrence in the animal world. An example of an isotonic fluid gain is drinking of brackish water at about 300 mOsm, or the therapeutic intravenous administration of an isotonic solution. When the isotonic fluid is injected into the ECF compartment, the ECF volume increases, but the concentration of ECF solutes remains unchanged; in other words, the ECF is still isotonic. The ECF compartment has increased in volume without causing a shift of H_2O into the cells.

Isotonic loss is more common, because it occurs in hemorrhage. The loss is confined to the ECF with no corresponding loss of fluid from the ICF. Fluid does not shift out of the cells, because the ECF remaining within the body is still isotonic, so there is no osmotic gradient to draw H_2O out of the cells. Of course, many other mechanisms come into play to counteract the loss of blood, but the ICF compartment is not directly affected by the loss. The major problem created by isotonic fluid changes is a *blood volume* disturbance, which we consider shortly.

Control of water balance by means of vasopressin and thirst is of primary importance in regulating ECF osmolarity.

Because cells, especially brain neurons, do not function properly when they are either shrunk or swollen, it is important that ECF osmolarity be closely regulated to prevent osmotic fluid shifts between the ECF and ICF. Control of H_2O balance is crucial for regulating ECF osmolarity. To maintain a stable H_2O balance, H_2O input must equal H_2O output (the balance concept). ▮ Table 13–3 shows data for a human as an example. Let us first examine input, which has several routes:

1. *Food consumption:* Perhaps surprisingly, a significant amount of water is obtained in most animals just from eating solid food. Recall that muscles consist of about 75% H_2O; meat is therefore 75% H_2O, because it is animal muscle. Likewise, fruits and vegetables consist of 60% to 90% H_2O. Eaters of seeds (such as some birds and rodents), in contrast, receive much less water this way.

2. *Drinking* external water: This is triggered by thirst sensations in the hypothalamus.

3. *Metabolically produced* H_2O: Recall that aerobic metabolism converts food and O_2 into energy stores, producing CO_2 and H_2O in the process. Amazingly, some desert mammals can survive primarily on this water source (see Chapter 12, p. 561).

As you saw earlier, *water vapor* in humid air may be taken up by some insects and arachnids. But there is no evidence that mammals can do this.

On the output side of the H_2O balance tally, there are several pathways of loss:

1. *Urine excretion* is by far the most important output mechanism in most mammals.

2. *Insensible* (that which is unregulated) *cutaneous and respiratory loss.* During the process of respiration, inspired air becomes saturated with H_2O within the airways. This H_2O is lost when the moistened air is subsequently expired. This loss can be recognized on cold days, when the H_2O vapor condenses (because cold air cannot hold as much vapor as warm body air) and exhaled air is "visible." Cold dry air is particularly dehydrating. (Birds and mammals have folds in their nasal passages to reduce

water and heat loss; see Chapter 15, p. 695.) The other insensible loss is the continual loss of H_2O from ordinary skin even in the absence of sweating. Water molecules can move through the cells of ordinary skin and evaporate. Mammalian and avian skin is fairly waterproof because of its keratinized exterior layer, which protects against a great loss of H_2O by this avenue (although some species use cutaneous evaporation in thermoregulation; see p. 698). And some mammals, such as marine mammals with their blubber layers, have skins that are effectively fully waterproof. Reptile skin is also quite waterproof.

3. *Sensible* (actively regulated) *cutaneous and respiratory loss* of H_2O occurs through sweating and panting, primarily for the purpose of thermoregulation. Loss of water this way can vary substantially, of course, depending on the environmental temperature and humidity and the degree of physical activity and stress levels.

4. *Feces* loss is normally a minor pathway. During the process of fecal formation in the large intestine, most of the H_2O is absorbed out of the digestive tract lumen into the blood, conserving fluid and solidifying the digestive tract's contents for elimination. Additional losses of H_2O can occur from the digestive tract through vomiting or diarrhea.

Of the many sources of H_2O input and output, only two are regulated significantly to maintain H_2O balance. On the intake side, *thirst* influences the amount of fluid ingested, and on the output side, the *kidneys* can adjust the amount of urine formed. Control of H_2O output in the urine is the most important mechanism in controlling H_2O balance. Some of the other factors are regulated but not for maintaining H_2O balance. Food intake is regulated to maintain energy balance, whereas control of sweating and panting is important in maintaining body temperature. Metabolic H_2O production and insensible losses are generally unregulated (although there are exceptions; see p. 698).

Control of Water Output in the Urine by Vasopressin

In mammals, fluctuations in ECF osmolarity caused by imbalances between H_2O input and output are quickly compensated for by adjusting the urinary excretion of H_2O without changing the usual excretion of salt; that is, H_2O reabsorption and excretion are partially dissociated from solute reabsorption and excretion, so the amount of free H_2O retained or eliminated can be varied to quickly restore ECF osmolarity to normal. Adjustments in free H_2O reabsorption and excretion are accomplished through changes in *vasopressin* secretion (as discussed in Chapter 12, p. 556).

Control of Water Input by Thirst and Drinking

Thirst is the internal craving that drives an animal to ingest H_2O (although only humans can report the subjective perception of thirst). A thirst center is located in the hypothalamus in close proximity to the vasopressin-secreting cells. Thirst-triggered drinking increases H_2O input, as a behavioral effector.

Regulation of Vasopressin Secretion and Thirst

The hypothalamic control centers that regulate vasopressin secretion (and thus urinary output) and thirst (and thus drinking) act in concert. Vasopressin secretion and thirst are both stimulated by a free H_2O deficit and suppressed by a free H_2O

Table 13–3 ▮ Daily Water Balance in a Human

Water Input		Water Output	
Avenue	*Quantity (mL/day)*	*Avenue*	*Quantity (mL/day)*
Fluid intake	1250	Insensible loss (from lungs and nonsweating skin)	900
H_2O in food intake	1000		
Metabolically produced H_2O	350	Sweat	100
		Feces	100
		Urine	1500
Total input	2600	Total output	2600

excess. Thus, appropriately, the same circumstances that call for reduced urinary output to conserve body H_2O also give rise to the sensation of thirst to replenish body H_2O. As we noted in Chapter 12 (p. 556), the predominant excitatory input for both vasopressin secretion and thirst comes from hypothalamic osmoreceptors located near the vasopressin-secreting cells and thirst center. These osmoreceptors monitor the osmolarity of the fluid surrounding them, which in turn reflects the concentration of the entire internal fluid environment. As the osmolarity increases (too little H_2O) and the need for H_2O conservation increases, vasopressin secretion and thirst are both stimulated (● Figure 13–13).

Even though the major stimulus for vasopressin secretion and thirst is an increase in ECF osmolarity, the vasopressin-secreting cells and thirst center are both influenced to a moderate extent by changes in ECF volume mediated by input from the left atrial volume receptors. Located in the left atrium, these volume receptors monitor the blood pressure, which reflects the ECF volume. In response to a major reduction in ECF volume and arterial pressure, as during hemorrhage, the left atrial volume receptors reflexly stimulate both thirst and vasopressin secretion. Vasopressin, at the circulating levels elicited by a large decline in ECF volume and arterial pressure, exerts a potent vasoconstrictor effect on arterioles (hence its name), in addition to affecting the kidney tubules. Both by helping expand the ECF and plasma volume and by increasing total peripheral resistance, vasopressin helps relieve the low blood pressure that elicited vasopressin secretion. Conversely, vasopressin and thirst are both inhibited when ECF/plasma volume and arterial blood pressure are raised. The resultant suppression of H_2O intake, coupled with elimination of excess ECF/plasma volume in urine, helps restore blood pressure to normal.

In addition, thirst has its own regulatory process. There is evidence of some kind of "oral H_2O metering" that constitutes a clear example of *anticipation* control of basic homeostasis (p. 15). At least in some mammals, a thirsty animal will rapidly drink only enough H_2O to satisfy its H_2O deficit. It stops drinking before the ingested H_2O has had time to be absorbed from the digestive tract and actually return the ECF compartment to normal. Exactly what factors are involved in signaling that enough H_2O has been consumed is still uncertain. It may be a learned anticipatory response based on past experience.

Vasopressin is a mammalian hormone, but closely related hypothalamic peptide hormones are found in all vertebrates

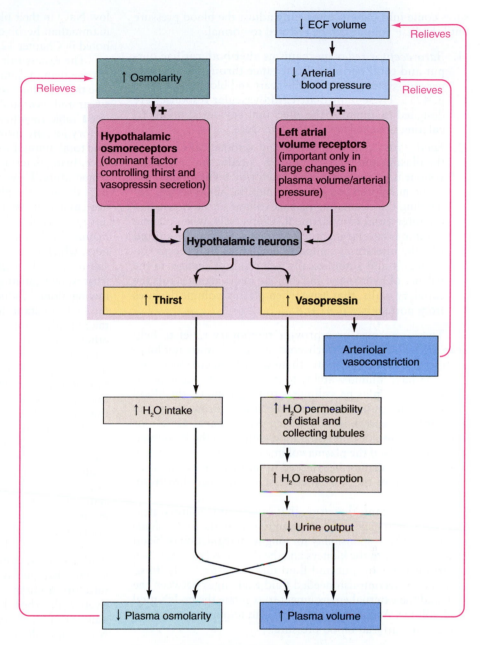

Figure 13–13 ● Control of increased vasopressin secretion and thirst during a H_2O deficit in a mammal.

(see Chapter 7). Reptiles, birds, fishes, and some amphibians have vasotocin, which differs from vasopressin by one amino acid in its peptide chain. It too promotes water conservation. For example, vasotocin enhances water uptake by the skin in toads.

■ Control of ECF volume is important in the long-term regulation of blood pressure.

A second major component of body fluid homeostasis is the volume. A reduction in ECF volume lowers arterial blood pressure by decreasing plasma volume. Conversely, a rise in ECF volume increases the arterial blood pressure by expanding the plasma volume. (Thus volume itself is not sensed, but rather its effect on blood pressure.) Two compensatory measures come into play to transiently adjust the blood pressure until the ECF volume can be restored to normal:

1. *Baroreceptor reflex* mechanisms alter both cardiac output and total peripheral resistance through autonomic nervous system effects on the heart and blood vessels (see p. 421). These immediate cardiovascular responses are designed to minimize the effect that any deviation in circulating volume has on blood pressure.

2. *Fluid shifts* occur temporarily and automatically between the plasma and interstitial fluid. A reduction in plasma volume is partially compensated for by a shift of fluid out of the interstitial compartment into the blood vessels, expanding the circulating plasma volume at the expense of the interstitial compartment. Conversely, when plasma volume is too large, much of the excess fluid is shifted into the interstitial compartment. These shifts occur immediately and automatically as a result of changes in the balance of hydrostatic and osmotic forces acting across the capillary walls, that arise when plasma volume deviates from normal (see p. 414).

These two measures provide temporary relief to help keep the blood pressure fairly constant, but they are not long-term solutions. Furthermore, these short-term compensatory measures have a limited ability to minimize a change in blood pressure. If the plasma volume is too inadequate, the blood pressure remains too low no matter how vigorous the pump action of the heart, how constricted the resistance vessels, or what proportion of interstitial fluid shifts into the blood vessels. Conversely, if the plasma volume is greatly overexpanded, blood pressure cannot be restored to normal even with maximum dilation of the resistance vessels and other short-term measures.

It is important, therefore, that other compensatory measures come into play in the long run to restore the ECF volume to normal. This responsibility for long-term regulation of blood pressure rests with the kidneys and the thirst mechanism, which control urinary output and fluid intake, respectively. In so doing, they accomplish needed fluid exchanges between the ECF and the external environment to regulate the body's total fluid volume. Accordingly, they have an important long-term influence on arterial blood pressure.

Importance of Salt Regulation in Regulating ECF Volume

Because sodium and its attendant anions account for more than 90% of the ECF osmotic activity in a mammal, the total Na^+ load (the total quantity of NaCl, not its concentration) in the ECF determines the total amount of H_2O that is osmotically retained in the ECF. The total mass of Na^+ salts in the ECF therefore determines the ECF volume, and, appropriately, regulation of ECF volume depends primarily on controlling salt balance.

To achieve balance, salt input must equal salt output. The only avenue for salt input is ingestion, which can exceed some mammals' need to replace obligatory salt losses. Carnivores (meat eaters), which naturally get sufficient NaCl in fresh meat (because meat contains an abundance of salt-rich ECF) and marine mammals (as well as other marine vertebrates), normally do not show a physiological appetite to seek additional salt. In contrast, terrestrial vertebrate herbivores, which have

low NaCl in their plant diets, develop *salt hunger.* As a result, mammalian herbivores will travel miles to a salt lick (as we noted in Chapter 12, p. 548).

The excess salt ingested by carnivores, omnivores, and marine mammals must be excreted to maintain salt balance. The three avenues for salt output are obligatory loss of salt in sweat and feces and controlled excretion of salt in the urine (see ∎ Table 13–4 for human data as an example). Fecal and excretory loss are most common, because few mammals sweat. The total amount of sweat produced is unrelated to salt balance, being determined instead by factors that control body temperature. However, *aldosterone* (see Chapter 12) can reduce the sweat's salt content, thus helping conserve salt in a hot environment. The small salt loss in the feces is thought not to be regulated, and it can be exacerbated by diseases. In mammals, the most important regulation is done by the kidneys, which precisely excrete excess salt in urine to maintain salt balance. By regulating the rate of urinary salt excretion (that is, by regulating the rate of Na^+ excretion, with Cl^- following along), kidneys normally keep the total Na^+ mass in the ECF constant despite any notable changes in dietary intake of salt or unusual losses through sweating, diarrhea, or other means. As a reflection of keeping the total Na^+ mass in the ECF constant, ECF volume in turn is maintained within the narrowly prescribed limits essential for normal circulatory function. (Recall that other osmoregulating marine vertebrates have other excretory organs for salt—salt glands in marine birds and reptiles, gills in bony fishes; see p. 566.)

In contrast, mammalian herbivores typically face an excess of potassium rather than sodium from eating plants. Their kidneys compensate for this by swapping Na^+ for K^+ in the nephron's distal tubule (regulated by aldosterone; p. 543). For example, the omnivorous laboratory rat has about equal concentrations of Na^+ and K^+ in its urine, but a desert pocket mouse (which eats primarily seeds containing high levels of potassium) has about twice as much K^+ as Na^+ in its urine.

Deviations in ECF volume accompanying changes in salt load are responsible for triggering renal compensatory responses that quickly bring the Na^+ load and ECF volume back into line. Sodium is freely filtered at the glomerulus and actively reabsorbed, but it is not secreted by the tubules, so the amount of Na^+ excreted in urine represents the amount of Na^+ that is filtered but is not subsequently reabsorbed:

$$Na^+ \text{ excreted} = Na^+ \text{ filtered} - Na^+ \text{ reabsorbed}$$

The kidneys accordingly adjust the amount of salt excreted by controlling two processes: (1) glomerular filtration rate (GFR) and (2) more importantly, tubular reabsorption of Na^+.

Table 13–4 ∎ Daily Salt Balance in a Human

Salt Input		Salt Output	
Avenue	Amount (g/day)	Avenue	Amount (g/day)
Ingestion	10.5	Obligatory loss in sweat and feces	0.5
		Controlled excretion in urine	10.0
Total input	10.5	Total output	10.5

Control of the Amount of Na$^+$ Filtered through Regulation of the GFR

The GFR is deliberately changed to alter the amount of salt and fluid filtered as part of the general baroreceptor reflex response to a change in blood pressure (see Chapter 12, p. 536). Changes in Na$^+$ are monitored indirectly through the effect that Na$^+$ ultimately has on blood pressure via its role in determining the ECF volume. Fittingly, baroreceptors that monitor fluctuations in blood pressure are responsible for adjusting the amounts of Na$^+$ filtered and eventually excreted. (Interestingly, recent studies suggest that total body sodium is itself a regulated state, but how this might be sensed is unknown.)

Control of the Amount of Na$^+$ Reabsorbed through the Renin-Angiotensin-Aldosterone and ANP Systems

The amount of Na$^+$ reabsorbed also depends on regulatory systems that play an important role in controlling blood pressure. As you saw in Chapter 12 (p. 543), the main factor controlling the extent of Na$^+$ reabsorption in the mammalian nephron is the powerful *renin-angiotensin-aldosterone* system, which promotes Na$^+$ reabsorption and thereby Na$^+$ retention. Sodium retention in turn promotes osmotic retention of H$_2$O and the subsequent expansion of plasma volume and elevation of arterial blood pressure. Appropriately, this Na$^+$-conserving system is activated by a reduction in NaCl, ECF volume, and arterial blood pressure.

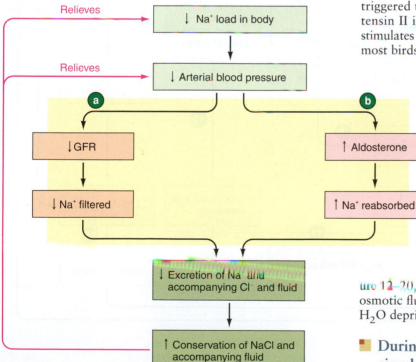

a See Figure 12-14 for details of mechanism.

b See Figure 12-18 for details of mechanism.

Figure 13–14 ● Dual effect of a fall in arterial blood pressure on mammalian renal handling of Na$^+$.

Clearly, control of GFR and Na$^+$ reabsorption are highly interrelated and intimately tied in with long-term regulation of ECF volume as reflected by blood pressure. Specifically, a fall in arterial blood pressure brings about a twofold effect in the renal handling of Na$^+$ (● Figure 13–14): (1) a reflex reduction in the GFR to decrease the amount of Na$^+$ filtered and (2) a hormonally adjusted increase in the amount of Na$^+$ reabsorbed. Together these effects reduce the amount of Na$^+$ excreted, thereby conserving the Na$^+$ and accompanying H$_2$O necessary to compensate for the fall in arterial pressure.

Conversely, a rise in arterial blood pressure brings about (1) increases in the amount of Na$^+$ filtered, and (2) a reduction in renin-angiotensin-aldosterone activity, which decreases salt (and fluid) reabsorption. This is partly controlled by the ANP (atrial natriuretic peptide) hormone system from the right atrium, which as you saw in Chapter 12 is a feedback antagonist to the renin-angiotensin-aldosterone system (p. 546). Together these actions increase salt (and fluid) excretion, eliminating the extra fluid that was expanding plasma volume and raising arterial pressure.

The renin-angiotensin-aldosterone system is the most important factor in regulating mammalian ECF volume and blood pressure, with the vasopressin and thirst mechanism playing only a supportive role. In addition, angiotensin II, in addition to stimulating aldosterone secretion, acts directly on the brain to stimulate the urge to drink (as confirmed by infusing laboratory rodents with the hormone) and concurrently stimulates vasopressin to enhance renal H$_2$O reabsorption (see p. 556). The resultant increased H$_2$O intake and decreased urinary output help correct the reduction in ECF volume that triggered the renin-angiotensin-aldosterone system. (Angiotensin II is an ancient vertebrate hormone; for example, it stimulates drinking in fishes. The hormone induces drinking in most birds but not desert birds, which drink whenever possible, nor in carnivorous and succulent-plant–eating birds, which obtain water from each meal).

Aldosterone also plays a supportive role in controlling ECF osmolarity. Researchers have recently learned that increased ECF osmolarity accompanying H$_2$O deprivation acts directly on the adrenal cortex to reduce aldosterone secretion, thus decreasing renal Na$^+$ reabsorption and increasing urinary excretion of Na$^+$. Physiologists propose that this controlling mechanism, which is in addition to the renin-angiotensin pathway and the direct action of plasma K$^+$ concentration on aldosterone secretion (see Figure 12–20, p. 547), may be important in reducing a detrimental osmotic flux of H$_2$O out of the cells in the face of prolonged H$_2$O deprivation and elevated ECF osmolarity.

■ During hemorrhagic shock, circulatory functions and fluid balance are coordinately regulated.

We conclude this section of the chapter by examining the consequences of and compensations for hemorrhage (● Figure 13–15). This figure may look intimidating, but we will work through it step by step. This important illustration pulls together many of the mechanisms discussed in this chapter and in Chap-

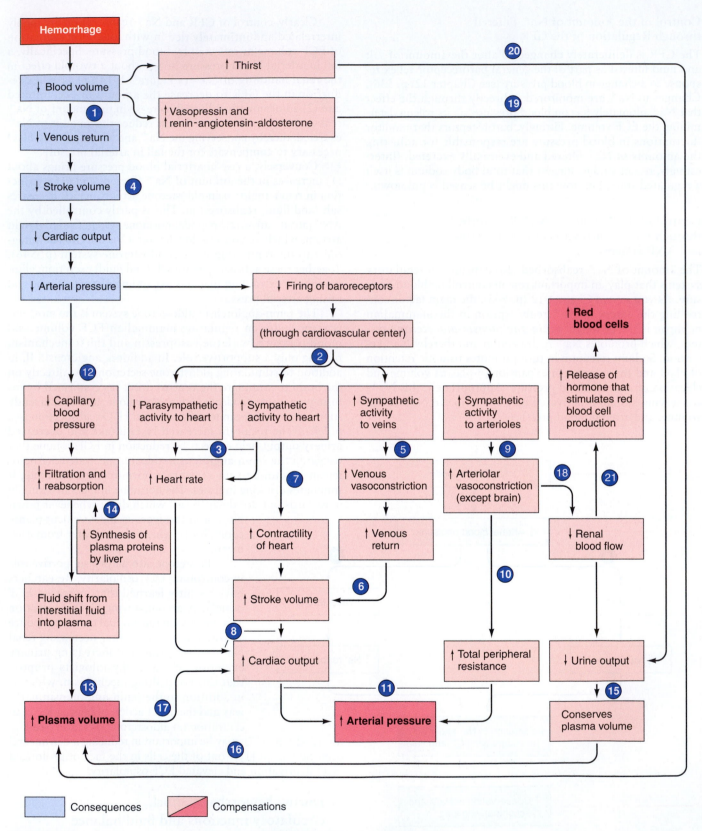

Figure 13–15 ● **Consequences and compensations of hemorrhage in a mammal.** The reduction of blood volume resulting from hemorrhage leads to a fall in arterial pressure. (Note the blue boxes, representing consequences of hemorrhage.) A series of compensations ensue (light pink boxes) that ultimately restore plasma volume, arterial pressure, and the number of red blood cells toward normal (dark pink boxes). Refer to the text (on the next page) for an explanation of the circled numbers and a detailed discussion of the compensations.

ter 9 on circulation, and Chapter 12 on excretion, showing how various systems in the mammalian body work together in an integrated way. The numbers in the text correspond to the numbers in the figure and indicate the portion of the figure being discussed.

- After severe loss of blood, reduction in circulating blood volume leads to a decrease in venous return ① and a subsequent fall in cardiac output and arterial blood pressure. (Note the blue boxes, which indicate consequences of hemorrhage.) (Chapter 9, p. 422)

- Compensatory measures immediately attempt to maintain adequate blood flow to the brain. (Note the pink boxes, which indicate compensations for hemorrhage.) (Chapter 9, p. 422)

- The baroreceptor reflex response to the fall in blood pressure brings about increased sympathetic and decreased parasympathetic activity to the heart ②. The result is an increase in heart rate ③ to offset the reduced stroke volume ④ brought about by the loss of blood volume. With severe fluid loss, the pulse is weak because of the reduced stroke volume but rapid because of the increased heart rate. (Chapter 9, p. 423)

- As a result of increased sympathetic activity to the veins, generalized venous vasoconstriction occurs ⑤, increasing venous return by means of the Frank-Starling mechanism ⑥. (Chapter 9, p. 388)

- Simultaneously, sympathetic stimulation of the heart increases the heart's contractility ⑦ so that it beats more forcefully and ejects a greater volume of blood, also increasing the stroke volume. (Chapter 9, p. 389)

- The increase in heart rate and increase in stroke volume collectively lead to an increase in cardiac output ⑧. (Chapter 9, p. 423)

- Sympathetically induced generalized arteriolar vasoconstriction ⑨ leads to an increase in total peripheral resistance ⑩. (Chapter 9, p. 406)

- Together, the increase in cardiac output and total peripheral resistance bring about a compensatory increase in arterial pressure ⑪. (Chapter 9, p. 423)

- The original fall in arterial pressure is also accompanied by a fall in capillary blood pressure ⑫, which results in fluid shifts from the interstitial fluid into the capillaries to expand the plasma volume ⑬. This response is sometimes termed *autotransfusion*. (Chapter 9, p. 414)

- This ECF fluid shift is enhanced by plasma protein synthesis by the liver during the next few days after hemorrhage ⑭. The plasma proteins exert a colloid osmotic pressure that helps retain extra fluid in the plasma.

- Urinary output is reduced, thereby conserving water that normally would have been lost from the body ⑮. This additional fluid retention helps expand the reduced plasma volume ⑯ (this chapter, p. 59). Expansion of plasma volume further augments the increase in cardiac output brought about by the baroreceptor reflex ⑰. Reduction in urinary output results from decreased renal blood flow caused by compensatory renal arteriolar vasoconstriction ⑱ (Chapter 12, p. 538). The reduced plasma volume also triggers increased secretion of vasopressin and activation of the salt- and water-conserving renin-angiotensin-

aldosterone pathway, which further reduces urinary output ⑲ (Chapter 12, p. 543, and this chapter, p. 593).

- Increased thirst is also stimulated by a fall in plasma volume ⑳. The resultant increased fluid intake helps restore plasma volume.

- Over a longer course of time (a week or more), lost red blood cells are replaced through increased red blood cell production triggered by reduced O_2 delivery to the kidneys ㉑ (Chapter 9, p. 365).

Acid–Base Balance

The final aspect of internal fluid and solute regulation that we consider is that of acid–base balance. This is crucial to the entire organism because protein structure and function is typically highly dependent on the concentration of hydrogen ions around them. Water itself normally exists in an equilibrium between its uncharged form and two charged products: *hydronium* and *hydroxide* ions:

$$2H_2O \leftrightarrow H_3O^+ + OH^-$$

In pure water, which is 55.6 M, this equilibrium is far to the left, with 10^{-7} M of the ions at 25°C (a pH of 7.0, as we explain shortly).

◼ Acids liberate free hydrogen ions, whereas bases accept them.

The term *acid–base balance* refers to the precise regulation of free (that is, unbound) hydrogen ion (H^+) and hydronium ion concentrations in the body fluids. To indicate the concentration of a chemical, its symbol is enclosed in brackets; thus [H^+] designates H^+ concentration.

Acids are a special group of hydrogen-containing substances that dissociate, or separate, when in solution to liberate free H^+ and anions (negatively charged ions). Many other substances (for example, carbohydrates) also contain hydrogen, but they are not classified as acids, because the hydrogen is tightly bound within their molecular structure and is never liberated as free H^+.

A strong acid has a greater tendency to dissociate in solution than does a weak acid; that is, a greater percentage of a strong acid's molecules separate into free H^+ and anions. Hydrochloric acid (HCl) is an example of a strong acid; every HCl molecule dissociates into free H^+ and Cl^- (chloride) when dissolved in H_2O. With a weaker acid such as *carbonic acid* (H_2CO_3), only a portion of the molecules dissociate in solution into H^+ and HCO_3^- (*bicarbonate* anions). The remaining H_2CO_3 molecules remain intact. Only the free hydrogen ions contribute to the acidity of a solution, so H_2CO_3 is a weaker acid than HCl because H_2CO_3 does not yield as many free hydrogen ions per number of acid molecules present in solution (● Figure 13–16).

The extent of dissociation for a given acid is always constant at a constant temperature and ionic composition; that is, when in solution, the same proportion of a particular acid's molecules always separate to liberate free H^+, with the other portion always remaining intact. The constant degree of dissociation for a particular acid (in this example H_2CO_3) is expressed by its dissociation constant (K) as follows:

$$[H^+]\,[HCO_3^-]/[H_2CO_3] = K$$

Figure 13–16 ▶ • **Comparison of a strong and a weak acid.**
(a) Five molecules of a strong acid. A strong acid such as HCl
(hydrochloric acid) completely dissociates into free H^+ and anions
in solution. (b) Five molecules of a weak acid. A weak acid such
as H_2CO_3 (carbonic acid) only partially dissociates into free H^+
and anions in solution.

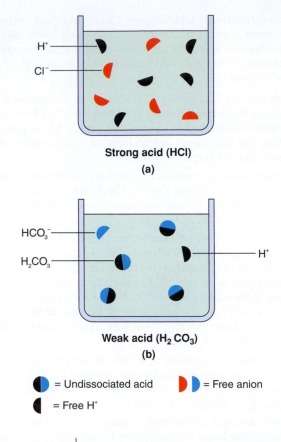

Strong acid (HCl)
(a)

Weak acid (H_2CO_3)
(b)

■ $[H^+]$ $[HCO_3^-]$ represents the concentration of ions resulting from H_2CO_3 dissociation.

■ $[H_2CO_3]$ represents the concentration of intact (undissociated) H_2CO_3.

The dissociation constant varies for different acids. A base is
a substance that can combine with a free H^+ and thus remove
it from solution. A strong base can bind H^+ more readily than
a weak base can.

● = Undissociated acid   ▶▷ = Free anion
◖ = Free H^+

■ The pH designation is used to express hydrogen ion concentration.

The $[H^+]$ in a typical mammalian ECF is about 4×10^{-8} or
0.00000004 equivalents per liter. The concept of pH has been
developed to express $[H^+]$ more conveniently. Specifically,
pH equals the logarithm (log) to the base 10 of the reciprocal
of the hydrogen ion concentration:

Figure 13–17 ●

**pH considerations in chemistry
and physiology.** (a) Relationship
of pH to the relative concentrations
of H^+ and base (OH^-) under
chemically neutral, acidic, and
alkaline conditions. (b) Plasma
pH range under normal, acidosis,
and alkalosis conditions in a
mammal.

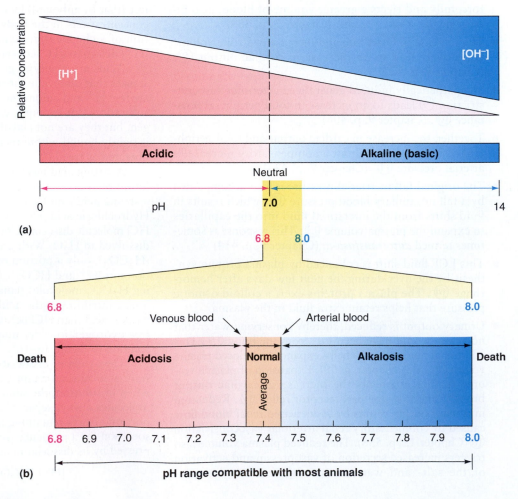

$$pH = \log 1/[H^+]$$

Thus the pH of typical mammalian ECF is 7.4. Three important points should be noted about this formula:

1. Because $[H^+]$ is in the denominator, a high $[H^+]$ corresponds to a low pH, and a low $[H^+]$ corresponds to a high pH. The greater the $[H^+]$, the larger the number by which 1 must be divided, and the lower the pH.

2. Hydronium ions (H_3O^+) are treated as hydrogen ions by convention.

3. Every unit change in pH actually represents a 10-fold change in $[H^+]$ because of the logarithmic relationship. A log to the base 10 indicates how many times 10 must be multiplied by itself to produce a given number. For example, the log of 10 = 1, whereas the log of 100 = 2. Ten must be multiplied by itself twice to yield 100. Numbers less than 10 have logs less than 1. Numbers between 10 and 100 have logs between 1 and 2, and so on. Accordingly, each unit of change in pH is indicative of a 10-fold change in $[H^+]$. For example, a solution with a pH of 7 has a $[H^+]$ 10 times less than that of a solution with a pH of 6 (1 pH-unit difference) and 100 times less than that of a solution with a pH of 5 (2 pH-unit difference).

The pH of pure H_2O is 7.0 at 25°C, and 6.81 at 37°C (typical mammalian body temperature); either value is considered chemically neutral. An extremely small proportion of H_2O molecules dissociate into hydrogen ions and hydroxyl (OH^-) ions. Because OH^- can bind with H^+ to once again form a H_2O molecule, it is considered basic. Because an equal number of acidic hydrogen ions and basic hydroxyl ions are formed, H_2O is neutral, being neither acidic nor basic. Solutions having a pH less than 7.0 contain a higher $[H^+]$ than pure H_2O and are considered acidic. Conversely, solutions having a pH value greater than 7.0 have a lower $[H^+]$ and are considered basic or alkaline (● Figure 13–17a). ● Figure 13–18 compares the pH values of common solutions.

The pH of arterial blood of a typical mammal is normally 7.45 and the pH of venous blood is 7.35, for an average blood pH of 7.4. The pH of venous blood is slightly lower (more acidic) than that of arterial blood because of H^+ generated by the formation of H_2CO_3 from CO_2 picked up at the tissue capillaries. Acidosis exists whenever the blood pH falls below 7.35, whereas alkalosis occurs when the blood pH is above 7.45 (Figure 13–17b). Note that the reference point for determining the body's acid–base status is not neutral pH (6.81 at 37°C), but the normal plasma pH of 7.4. Thus a plasma pH of 7.2 is considered acidotic even though in chemistry a pH of 7.2 is considered basic. Death can occur if arterial pH falls outside the range of 6.8 to 8.0 for more than a few seconds because an arterial pH of less than 6.8 or greater than 8.0 is not compatible with many cellular functions. Obviously, therefore, $[H^+]$ in body fluids must be carefully regulated.

These particular pH values do not necessarily apply to other animals. The reason is that pH set points in animals vary, and also change with temperature. A thermoconformer from a warm habitat similar to that of mammalian body temperatures may also have a blood pH of about 7.4. But a thermoconformer in a cold habitat (such as a nontropical fish) typically has a higher internal pH (such as 8.0) and may reduce its pH if it is warmed up. The reasons for this are unclear, and the phenomenon is discussed in Chapter 15 as a possible mechanism for stabilizing proteins. Regardless of the pH that is regulated, the acidotic and alkalotic effects do apply to these animals if major pH deviations occur.

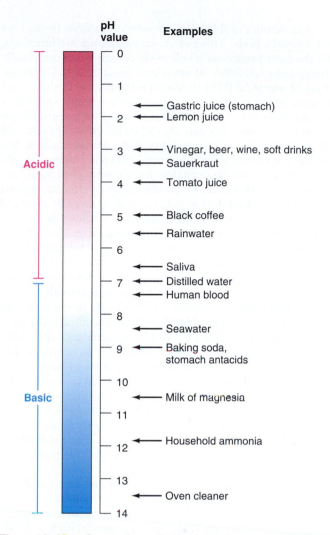

Figure 13–18 ● Comparison of pH values of common solutions.

■ Fluctuations in hydrogen ion concentration have profound effects on body chemistry.

Only a narrow pH range is compatible with most life functions, because even small changes in $[H^+]$ have dramatic effects on proteins. Even slight deviations in $[H^+]$ alter the shape and activity of protein molecules, usually detrimentally; for example, respiratory pigments do not deliver oxygen properly. Also, because enzymes are proteins, a shift in the body's acid–base balance disturbs the normal pattern of metabolic activity catalyzed by these enzymes. Some cellular chemical reactions are accelerated; others are depressed. The power of pH has been demonstrated clearly in fertilized sea urchin eggs, where an increase of only 0.13 pH units is enough to activate metabolism from a nearly dormant state to a level that supports the new embryo's development. This represents a useful pH change, but in normal, nondormant multicellular animals, changes are not desirable (except with internal temperature changes; see p. 689).

Fluctuations in [H$^+$] disturb protein functions throughout any animal's body. This can create numerous malfunctions, but the most prominent whole-body consequences are usually changes in excitability of nerve and muscle cells. The major effect of increased [H$^+$] (acidosis) is depression of the central nervous system in more advanced animals. Acidotic vertebrates become disoriented and, in more severe cases, eventually die in a state of coma. In contrast, the major effect of decreased [H$^+$] (alkalosis) is overexcitability of the nervous system, first the peripheral and later the central nervous system. Peripheral nerves become so excitable that they fire even in the absence of normal stimuli. Overexcitability of efferent (motor) nerves brings about muscle twitches and, in more pronounced cases, severe muscle spasms. In extreme alkalosis death may occur, because spasm of the respiratory muscles seriously impairs breathing. In addition, severely alkalotic animals may die of convulsions resulting from overexcitability of the central nervous system. In less serious situations, CNS overexcitability is manifested as extreme nervousness in mammals.

In mammals, disturbances to K$^+$ levels in the body can also occur. When reabsorbing Na$^+$ from the filtrate, the renal tubular cells secrete either K$^+$ or H$^+$ in exchange (see p. 548). Normally, they secrete a preponderance of K$^+$ compared to H$^+$. Because of the intimate relationship between secretion of H$^+$ and K$^+$ by the kidneys, an increased rate of secretion of one of these ions is accompanied by a decreased rate of secretion of the other. For example, if more H$^+$ than normal is eliminated by the kidneys, as occurs when the body fluids become acidotic, less K$^+$ than usual can be excreted. The resultant K$^+$ retention can affect cardiac function, among other detrimental consequences.

Changes can arise from both internal and external disturbances. In the next section we look at internal disturbances. An example of an external disturbance is the increasing levels of atmospheric CO$_2$ from human activities such as fossil-fuel burning. As you will see shortly, CO$_2$ acidifies water when it dissolves. Some people have proposed that excess CO$_2$ be dumped into the deep sea, where low temperatures and high pressures convert the gas into a liquid. Unfortunately, fish placed in cages next to experimental liquid CO$_2$ dumps have died from acidosis, as the liquid gas slowly dissolves into seawater.

■ **Hydrogen ions are continually being added to body fluids as a result of metabolic activities.**

As with any other constituent, to maintain a constant [H$^+$] in body fluids, input of hydrogen ions must be balanced by an equal output. On the input side, only a small amount of acid capable of dissociating to release H$^+$ is taken in with food, such as the weak citric acid found in oranges. Most H$^+$ in the body fluids is generated internally from metabolic activities. Normally, H$^+$ is continually being added to the body fluids from the following three sources:

1. *Carbonic acid formation.* The major source of H$^+$ is through H$_2$CO$_3$ formation from metabolically produced CO$_2$. Cellular oxidation of nutrients yields energy, with CO$_2$ and H$_2$O as end products (p. 40). CO$_2$ and H$_2$O spontaneously react to form H$_2$CO$_3$, although rather slowly. Once formed, H$_2$CO$_3$ mostly dissociates to liberate free H$^+$ and HCO$_3^-$ (bicarbonate):

$$CO_2 + H_2O \leftrightarrow H_2CO_3 \leftrightarrow H^+ + HCO_3^-$$

In some cell types such as erythrocytes, the reaction is rapidly catalyzed by the enzyme carbonic anhydrase (CA), although in an indirect way:

$$\overset{\text{CA}}{CO_2 + OH^- \leftrightarrow HCO_3^-}$$

The change in OH$^-$ in turn causes a change in H$^+$ formation from water:

$$H_2O \leftrightarrow H^+ + OH^-$$

These reactions are readily reversible, proceeding in either direction depending on the concentrations of the substances involved as dictated by the law of mass action. Within vertebrate systemic capillaries, the CO$_2$ level in the blood increases as metabolically produced CO$_2$ enters from the tissues. This drives the reaction to the acid side, generating H$^+$ as well as HCO$_3^-$ in the process. In the lungs or gills, the reaction is reversed: CO$_2$ diffuses from the blood to the environment. The resultant reduction in CO$_2$ in the blood drives the reaction toward the CO$_2$ side. Hydrogen ion and HCO$_3^-$ form H$_2$CO$_3$, which rapidly decomposes into CO$_2$ and H$_2$O once again. The CO$_2$ is exhaled from the lungs or diffused out of the gill chambers, whereas the hydrogen ions generated at the tissue level are incorporated into H$_2$O molecules. Gills, in addition to permitting CO$_2$ diffusion, can have an alternative process: In some fishes and crustaceans, gill membrane transporters can move bicarbonate and hydrogen ions out of the body.

When the respiratory system can keep pace with the rate of metabolism, there is no net gain or loss of H$^+$ in the body fluids from metabolically produced CO$_2$. When the rate of CO$_2$ removal by the lungs does not match the rate of CO$_2$ production at the tissue level, however, the resultant accumulation or deficit of CO$_2$ in the body leads to an excess or shortage, respectively, of free H$^+$ in the body fluids.

2. *Inorganic acids produced during the breakdown of nutrients.* Dietary proteins and other ingested nutrient molecules that are found abundantly in meat contain a large quantity of sulfur and phosphorus. When these molecules are broken down, sulfuric acid and phosphoric acid are produced as by-products. Being moderately strong acids, these two inorganic acids dissociate to a large extent, liberating free H$^+$ into the body fluids. In contrast, the breakdown of plants produces bases that, to some extent, neutralize the acids derived from protein metabolism. Generally, however, more acids than bases are produced during the breakdown of ingested food, leading to an excess of these acids.

3. *Organic acids resulting from intermediary metabolism.* Numerous organic acids are produced during normal intermediary metabolism (see pp. 47 and 54). For example, fatty acids are produced during fat metabolism, and vertebrate muscles produce lactic acid during heavy exercise. These acids partially dissociate to yield free H$^+$.

Hydrogen ion generation therefore normally goes on continuously as a result of ongoing metabolic activities. Furthermore, in certain disease states additional acids may be produced that further contribute to the total body pool of H$^+$. For example, in diabetes mellitus large quantities of keto acids may be produced as a result of abnormal fat metabolism. Some types of acid-producing medications may also add to the total load of H$^+$ that must be handled by the body. Thus input of H$^+$ is unceasing, highly variable, and essentially unregulated.

The crux of H$^+$ balance is maintaining the normal alkalinity of the ECF (such as pH 7.4 in most mammals) despite this constant onslaught of acid. The generated free H$^+$ must be largely removed from solution while in the body, and ultimately must be eliminated from the body so that the pH of body fluids can remain within the narrow range compatible with life. Mechanisms must also exist to compensate rapidly for the occasional situation in which the ECF becomes too alkaline.

Three lines of defense against changes in [H$^+$] operate to maintain the [H$^+$] of body fluids at a nearly constant level despite unregulated input: (1) chemical buffer systems, (2) respiratory mechanisms of pH control, and (3) excretory mechanisms of pH control. We look at each of these methods, primarily in vertebrates.

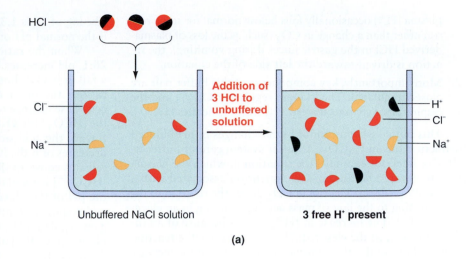

(a)

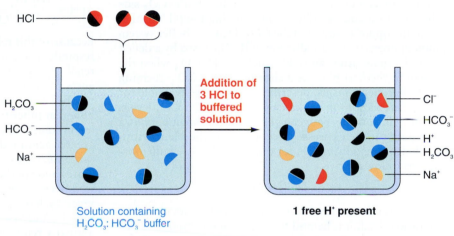

(b)

Figure 13–19 ● Action of chemical buffers. (a) Addition of HCl to an unbuffered solution. All the added hydrogen ions (H$^+$) remain free and contribute to the acidity of the solution. (b) Addition of HCl to a buffered solution. Bicarbonate ions (HCO$_3^-$), the basic member of the buffer pair, bind with some of the added H$^+$ and remove them from solution so that they do not contribute to its acidity.

pH Regulation: Buffers

▪ Chemical buffer systems act as the first line of defense against changes in hydrogen ion concentration.

A chemical buffer system is a mixture in a solution of two (or perhaps three) chemical compounds that minimize pH changes when either an acid or a base is added to or removed from the solution. A buffer system consists of a pair of substances involved in a reversible reaction—one substance that can yield free H$^+$ as the [H$^+$] starts to fall and another that can bind with free H$^+$ (thus removing it from solution) when [H$^+$] starts to rise. An important example of such a buffer system is the carbon dioxide–bicarbonate buffer system, which is involved in the following reversible reaction:

$$H^+ + HCO_3^- \leftrightarrow H_2CO_3 \leftrightarrow H_2O + CO_2$$

When a strong acid such as HCl is added to an unbuffered solution, all the dissociated H$^+$ remains free in the solution (● Figure 13–19a). In contrast, when HCl is added to a solution containing the CO$_2$–HCO$_3^-$ buffer system (Figure 13–19b), the HCO$_3^-$ immediately binds with the free H$^+$ to form H$_2$CO$_3$ and thus H$_2$O + CO$_2$. In contrast, when the pH of the solution starts to rise because of the addition of base or loss of acid, the H$^+$-yielding member of the buffer system, H$_2$CO$_3$, releases H$^+$ to minimize the rise in pH (reducing H$_2$O + CO$_2$).

There are four buffer systems in the vertebrate body: (1) the CO$_2$–HCO$_3^-$ buffer system, (2) the peptide and pro-

tein buffer system, (3) the hemoglobin buffer system, and (4) the phosphate buffer system. Each serves a different important role.

The CO$_2$–HCO$_3^-$ Buffer System in the ECF

The CO$_2$–HCO$_3^-$ buffer system is the most important one for buffering pH changes in the ECF. It is a very effective ECF buffer system, for two reasons:

1. HCO$_3^-$ is abundant in the ECF, so this system is readily available to absorb excess protons, whereas there is sufficient CO$_2$ and H$_2$CO$_3$ to produce protons when they decrease below normal concentration. If, for example, proton concentrations increase through lactic acid released into the ECF from exercising muscles, the buffer reaction is driven toward the right side of the equation, so the rise in [H$^+$] is abated. In the converse situation, when the

plasma [H$^+$] occasionally falls below normal for some reason other than a change in CO$_2$ (such as the loss of plasma-derived HCl in the gastric juices during vomiting), the reaction is driven toward the left side of the equation.

2. More importantly, key components of this buffer pair are closely regulated. The kidneys regulate HCO$_3^-$, and the respiratory system regulates CO$_2$, which generates H$_2$CO$_3$.

This buffer system cannot buffer changes in pH induced by fluctuations in H$_2$CO$_3$. A buffer system cannot buffer itself. Consider, for example, the situation in which the plasma [H$^+$] is elevated because of CO$_2$ retention associated with a respiratory problem. The rise in CO$_2$ drives the reaction to the left according to the law of mass action, resulting in an elevation in [H$^+$]. The increase in [H$^+$] occurs because of an increase in CO$_2$, so the elevated [H$^+$] cannot drive the reaction to the right to buffer the increase in [H$^+$]. Only if the increase in [H$^+$] is brought about by some mechanism other than CO$_2$ accumulation can this buffer system be shifted to the CO$_2$ side of the equation and be effective in reducing the [H$^+$]. Likewise, in the opposite situation the CO$_2$–HCO$_3^-$ buffer system cannot compensate for a reduction in [H$^+$] caused by a deficit of CO$_2$ by generating more H$^+$-yielding H$_2$CO$_3$ when the problem in the first place is a shortage of H$_2$CO$_3$–forming CO$_2$. Other mechanisms, described shortly, are available for resisting fluctuations in pH caused by changes in CO$_2$ levels.

The relationship between [H$^+$] and the members of a buffer pair can be expressed according to the **Henderson-Hasselbalch** equation, which, for the CO$_2$–HCO$_3^-$ buffer system, is as follows:

$$pH = pK + \log [HCO_3^-]/[CO_2]$$

Although you do not need to know the mathematical manipulations involved, it is helpful for you to understand how this formula is derived. Recall that the dissociation constant K for H$_2$CO$_3$ acid is

$$[H^+] [HCO_3^-]/[H_2CO_3] = K$$

and that the relationship between pH and [H$^+$] is

$$pH = \log 1/[H^+]$$

Practically speaking, [H$_2$CO$_3$] is a direct reflection of the concentration of dissolved CO$_2$, henceforth referred to as [CO$_2$], because most of the CO$_2$ in the plasma is converted into H$_2$CO$_3$. Thus we substitute H$_2$CO$_3$ with CO$_2$. Then, by solving the dissociation constant formula for [H$^+$] (that is, [H$^+$] = K × [CO$_2$]/[HCO$_3^-$]) and replacing this value for [H$^+$] in the pH formula, one comes up with the Henderson-Hasselbalch equation.

The pK is the logarithm of 1/K. For H$_2$CO$_3$ derived from CO$_2$, the pK is 6.1 under mammalian conditions. Because the pK is, like K, a constant at a given temperature and ionic strength, changes in pH are associated with changes in the ratio between [HCO$_3^-$] and [CO$_2$]. Normally, the ratio between [HCO$_3^-$] and [CO$_2$] in the ECF is 20 to 1; that is, there is 20 times more HCO$_3^-$ than CO$_2$. Plugging this ratio into our formula:

$$pH = pK + \log [HCO_3^-]/[CO_2]$$
$$= 6.1 + \log 20:1$$

The log of 20 is 1.3. Therefore pH = 6.1 + 1.3 = 7.4, which is the normal pH of mammalian plasma.

When the ratio of [HCO$_3^-$] to [CO$_2$] increases above 20:1, pH increases. Accordingly, either a rise in [HCO$_3^-$] or a fall in [CO$_2$], both of which increase the [HCO$_3^-$]/[CO$_2$] ratio if the other component remains constant, shifts the acid–base balance toward the alkaline side. In contrast, when the [HCO$_3^-$]/[CO$_2$] ratio decreases below 20:1, pH decreases toward the acid side. This can occur either if the [HCO$_3^-$] decreases or if the [CO$_2$] increases while the other component remains constant. Because [CO$_2$] by the lungs and gills, and [HCO$_3^-$] is regulated by the kidneys (and sometimes gills), the pH of the plasma can be shifted up and down by renal and respiratory influences. The kidneys and respiratory organs regulate pH (and thus free [H$^+$]) largely by controlling plasma [HCO$_3^-$] and [CO$_2$] to restore their ratio to normal. Accordingly,

$$pH \propto \frac{[HCO_3^-] \text{ controlled by kidney and/or gill function}}{[CO_2] \text{ controlled by respiratory function}}$$

Because of this relationship, not only do both the kidneys and respiratory organs normally participate in pH control but also renal or respiratory dysfunction can induce acid–base imbalances by altering the [HCO$_3^-$]/[CO$_2$] ratio.

Before we leave the topic of CO$_2$–HCO$_3^-$ buffer systems, note that other forms of carbonate can participate in buffering in some animals, such as those that must cope with much greater acid loads than others. One example is the painted turtle *Chrysemys picta*, found in ponds in the northern United States and southern Canada. In the winter, it can survive in iced-over waters with no oxygen for months. In part, it survives by having large energy stores (glycogen in the liver, muscle, and heart) for anaerobic metabolism, coupled with a greatly reduced metabolic rate in the cold (because it is a poikilotherm, an animal whose body temperature varies with the environment). Nevertheless, months of metabolism without oxygen produces large quantities of lactic acid, as it would in most vertebrates. The turtles do have extremely high bicarbonate levels to help absorb protons released by lactic acid. But this is not nearly enough. How can they cope? The extraordinary answer lies in their massive shells, which make up about a third of their body weight and are the major repository of the animals' minerals. The shell helps deal with acid

A painted turtle, which uses its shell as an aid in acid–base regulation

load in two ways. First, acid buildup begins to demineralize the shell and appears primarily to cause the release of calcium and magnesium *carbonate* (CO_3^-). The latter anion is thought to buffer the protons (binding them). Second, lactate along with its acidic proton accumulate in the shell itself, perhaps bound to calcium and carbonate, respectively. This effectively removes as much as half of the acid threat from the blood. Weeks later, the turtle emerges into the spring sunshine to begin breathing and metabolizing normally again, and the lactic acid can be safely released and metabolized. In this way, the shell acts as a buffer!

The Intracellular Peptide and Protein Buffer System

The most plentiful buffers of the ICF are the cell proteins. A more limited number of plasma proteins reinforces the CO_2–HCO_3^- system in ECF buffering. Proteins are excellent buffers because they contain both acidic and basic groups that can bind or release H^+. Quantitatively, the protein system is most important in buffering changes in $[H^+]$ in the ICF because of the sheer abundance of the intracellular proteins. The most important buffering amino acid is *histidine*, because it is the only one with a pK close to 7. In addition, muscles of some fishes, birds, and mammals have large concentrations of *dipeptides* (chains of two amino acids) containing histidine. These serve as excess buffering capacity in muscles that produce a heavy load of acid during activity. For example, the highest known concentration of histidine dipeptides has been found in diving mammals, whose muscles must work with anaerobic metabolism for considerable periods.

The Hemoglobin Buffer System in Erythrocytes

Hemoglobin (Hb) in vertebrate erythrocytes buffers the H^+ generated from metabolically produced CO_2 in transit between the tissues and the lungs or gills. At the systemic capillary level, CO_2 continuously diffuses into the blood from the tissue cells where it is being produced. Recall that most of this CO_2 forms H^+ and HCO_3^-. Simultaneously, some of the oxyhemoglobin (HbO_2), releases O_2, which diffuses into the tissues. Reduced (unoxygenated) Hb has a greater affinity for H^+ than HbO_2 does. Therefore, most of the H^+ generated from CO_2 at the tissue level becomes bound to reduced Hb and no longer contributes to the acidity of the body fluids:

$$H^+ + Hb \leftrightarrow HHb^+$$

At the respiratory organ the reactions are reversed. As Hb picks up O_2 diffusing from the lungs or gills into the red blood cells, the affinity of Hb for H^+ is decreased, so H^+ is released. This liberated H^+ combines with HCO_3^- to yield H_2CO_3, which in turn produces CO_2, which is removed from the body. Meanwhile, the hydrogen ion has been reincorporated into neutral H_2O molecules. Were it not for Hb, the blood would become much too acidic after picking up CO_2 at the tissues. Because of the tremendous buffering capacity of the Hb system, venous blood is only slightly more acidic than arterial blood despite the large volume of H^+-generating CO_2 carried in the venous blood.

The Phosphate Buffer System in the ICF and Urine

The phosphate buffer system consists of an acid phosphate salt (NaH_2PO_4) that can donate a free H^+ when the $[H^+]$ falls and a basic phosphate salt (Na_2HPO_4) that can accept a free H^+ when the $[H^+]$ rises. Basically, this buffer pair can alternately switch a H^+ for a Na^+ as demanded by the $[H^+]$:

$$Na_2HPO_4 + H^+ \leftrightarrow NaH_2PO_4 + Na^+$$

Even though the phosphate pair is a good buffer, its concentration in the ECF is rather low, so it is not very important as an ECF buffer. Because phosphates are more abundant within the cells, this system contributes significantly to intracellular buffering, being rivaled only by the more plentiful intracellular proteins. Even more importantly, the phosphate system serves as an excellent urinary buffer. When an animal consumes more phosphate than is needed, the excess phosphate filtered through the kidneys is not reabsorbed but remains in the tubular fluid to be excreted (because the renal threshold for phosphate is exceeded; see p. 543). This excreted phosphate buffers the urine as it is being formed, by removing from solution the H^+ secreted into the tubular fluid. None of the other body-fluid buffer systems are present in the tubular fluid to play a role in buffering urine during its formation. Most or all of the filtered HCO_3^- and CO_2 (alias H_2CO_3) are reabsorbed, whereas Hb and most plasma proteins are not even filtered.

All chemical buffer systems act immediately, within fractions of a second, to minimize changes in pH. When $[H^+]$ is altered, the involved buffer systems' reversible chemical reactions shift at once in favor of compensating for the change in $[H^+]$. Accordingly, the buffer systems are the first line of defense against changes in $[H^+]$ because they are the first mechanism to respond.

Through the mechanism of buffering, most hydrogen ions seem to "disappear" from the body fluids between the time of their generation and their elimination. It must be emphasized, however, that none of the chemical buffer systems actually eliminates H^+ from the body by themselves. Protons are merely removed from solution by being incorporated within one of the members of the buffer pair, thus preventing the hydrogen ions from contributing to the body fluids' acidity. Because each buffer system has a limited capacity to "soak up" H^+, the H^+ that is unceasingly produced must ultimately be removed from the body. If H^+ were not eventually eliminated, soon all the body fluid buffers would already be bound with H^+ and there would be no further buffering ability.

As we have mentioned several times, actual removal of excess H^+ occurs in the renal and respiratory systems (for example, excretion of NaH_2PO_4 in urine, and transport of H^+ by gills). However, these systems respond more slowly than the chemical buffer systems. We now turn our attention to these other defenses against changes in acid–base balance in depth.

pH Regulation: Respiration

■ **Respiratory systems regulate hydrogen ion concentration through adjustments in ventilation and (in gills) membrane transport.**

The respiratory system plays an important role in acid–base balance through its ability to alter ventilation and, consequently, to alter the removal of H^+-generating CO_2 in gills and lungs, or, in gills, of H^+ and HCO_3^- directly.

Let's examine air breathers first. Every day, lungs remove from the body fluids many times more H$^+$ derived from carbonic acid than the kidneys remove from non–carbonic-acid sources. Furthermore, the respiratory system, through its ability to regulate arterial [CO$_2$], can adjust the amount of H$^+$ added to body fluids from this source as needed to restore pH toward normal when fluctuations in [H$^+$] from non–carbonic-acid sources occur. (Noncarbonic acids include keto acids; p. 296.) Because of this ability, it's not surprising that the level of respiratory activity is governed at least in part by the arterial [H$^+$]. When arterial [H$^+$] increases, the respiratory center in the terrestrial-vertebrate brain stem is reflexly stimulated (see p. 515) to increase ventilation (the rate at which gas is exchanged between the respiratory organ and the atmosphere). As the rate and depth of breathing increase, more CO$_2$ than usual is blown off, so less H$_2$CO$_3$ than normal is added to the body fluids. Conversely, when arterial [H$^+$] falls, ventilation is reduced. As a result of slower, shallower breathing, metabolically produced CO$_2$ diffuses from the cells into the blood faster than it is removed from the blood, so higher-than-usual amounts of acid-forming CO$_2$ accumulate in the blood, restoring [H$^+$] toward normal.

In water-breathing vertebrates (fishes), kidneys and intestines play a minor role in acid–base adjustments, and the gills have the primary responsibility. Acid loads are primarily handled by membrane transporters in the gills, rather than by alterations in breathing (ventilation). These transporters include *H$^+$ V-ATPases* (p. 81) to remove protons, and *Na$^+$–H$^+$ antiport* or exchange transporters (which takes up one sodium ion for each proton excreted). There is also a Cl$^-$–HCO$_3^-$ exchanger in gills that may help remove excess base. Skin breathers such as frogs may also have transporters for moving ions.

Respiratory regulation acts at a moderate speed, coming into play only when the chemical buffer systems alone cannot minimize [H$^+$] changes. When deviations in [H$^+$] occur, the buffer systems respond immediately. If a deviation in [H$^+$] is not swiftly and completely corrected by the buffer systems, the respiratory system comes into action a few minutes later, thus serving as the second line of defense against changes in [H$^+$]. (The delay occurs because of the time it takes to transport all the affected blood past the respiratory surfaces, plus, in air breathers, the time it takes to alter ventilation rates.) An exception to this delay occurs in exercise and stress, in which breathing is activated in anticipation (feedforward activation, p. 15) of both the oxygen needs and pH regulation necessary during heavy muscle use (p. 338).

Of course, when changes in [H$^+$] occur because of fluctuations in [CO$_2$] that arise from respiratory abnormalities, the respiratory mechanism cannot contribute at all to the control of pH; for example, if acidosis exists because of CO$_2$ accumulation caused by lung disease, the impaired lungs cannot possibly compensate for the acidosis by increasing the rate of CO$_2$ removal. The buffer systems (other than the CO$_2$–HCO$_3^-$ pair) plus renal regulation are the only mechanisms available for defending against respiratory-induced acid–base abnormalities.

Interestingly, insects regulate pH in very similar ways. The opening and closing of spiracles, and abdominally driven breathing are both controlled by the pH of the hemolymph. Insects may even have feedforward regulation of this in anticipation of high activity.

pH Regulation: Excretion

■ Excretory systems contribute powerfully to control of acid–base balance by controlling both hydrogen ion and bicarbonate concentrations in the ECF.

Excretory organs are the third line of defense against changes in [H$^+$] in body fluids; they require hours to days to compensate for changes in body fluid pH, compared to the immediate responses of the buffer systems and the few minutes of delay before the respiratory system responds. Note that gills in fishes serve both as excretory and respiratory organs, with the kidneys (and intestines) playing only a minor role in acid–base regulation. In some animals, nonrenal excretory organs contribute significantly to acid–base balance. The bladder fluid of reptiles can be acidified by an H$^+$–ATPase that is activated in response to high CO$_2$ levels. In insects, the hindgut is the key excretory organ for acid–base regulation. For example, the mosquito *Aedes dorsalis* can live in alkaline brine lakes in Africa at pH 10.5 while regulating a hemolymph pH of 7.6. Bicarbonate is actively excreted by the hindgut in this process.

Here we focus on mammals, in which the kidneys are the most potent acid–base regulatory mechanism. Not only can they vary removal of H$^+$ from any source, but they can also variably conserve or eliminate HCO$_3^-$, depending on the acid–base status of the body. For example, during renal compensation for acidosis for each H$^+$ excreted in the urine, a new HCO$_3^-$ is added to the plasma to buffer, by means of the CO$_2$–HCO$_3^-$ system, yet another H$^+$ that still remains in the body fluids. By simultaneously removing acid (H$^+$) from and adding base (HCO$_3^-$) to the body fluids, the kidneys can restore the pH toward normal more effectively than the lungs, which can adjust only the amount of H$^+$-forming CO$_2$ in the body.

Also contributing to the kidneys' acid–base regulatory potency is their ability to return the pH almost exactly to normal. In contrast, when some nonrespiratory abnormality has altered the [H$^+$], the mammalian respiratory system alone can return the pH to only 50% to 75% of the way toward normal; this is because the driving force governing the compensatory ventilatory response is diminished as the pH moves toward normal. As an example, ventilation is increased in response to a rise in arterial [H$^+$], but as the [H$^+$] is gradually reduced because of the stepped-up removal of H$^+$-forming CO$_2$, the ventilatory response is also gradually reduced. In comparison, the kidneys continue to respond to a change in pH until compensation is essentially complete.

The kidneys control the pH of the body fluids by adjusting three interrelated factors: (1) H$^+$ excretion, (2) HCO$_3^-$ excretion, and (3) ammonia (NH$_3$) secretion.

Hydrogen Ion Excretion

Although buffer systems can resist changes in pH by removing H$^+$ from solution, the persistent production of acidic metabolic products would eventually overwhelm the limits of this buffering capacity. The constantly generated H$^+$ must therefore ultimately be eliminated from the body. Mammalian lungs can adjust only by eliminating CO$_2$. The task of eliminating H$^+$ derived from sulfuric, phosphoric, lactic, and other acids rests with the kidneys.

Almost all the excreted H^+ enters the urine by means of secretion. Recall that the filtration rate of H^+ equals plasma $[H^+]$ times GFR (p. 536). Because plasma $[H^+]$ is quite low (less than in pure H_2O except during extreme acidosis), the filtration rate of H^+ is likewise extremely low. The majority of excreted H^+ gains entry into the tubular fluid by being secreted by the proximal and distal tubules and collecting ducts (p. 547). Because the kidneys normally excrete H^+, urine is usually acidic, having an average pH of 6.0. Not only do the kidneys continuously eliminate the normal amount of H^+ that is constantly being produced from non-CO_2 sources, but they can also alter their rate of H^+ secretion to compensate for changes in $[H^+]$ arising from abnormalities in the concentration of CO_2.

The magnitude of H^+ secretion depends on a direct effect of the plasma's acid–base status on the kidneys' tubular cells (● Figure 13–20). No neural or hormonal control is involved. When the $[H^+]$ of the plasma passing through the peritubular capillaries is elevated above normal, the tubular cells respond by secreting greater-than-usual amounts of H^+ from the plasma into the tubular fluid to be excreted in the urine. Conversely, when plasma $[H^+]$ is lower than normal, the kidneys conserve H^+ by reducing its secretion and subsequent excretion in the urine. The kidneys cannot raise plasma $[H^+]$ by reabsorbing more of the filtered H^+, because there are no reabsorptive mechanisms for H^+. The only way the kidneys can reduce H^+ excretion is by secreting less H^+.

The H^+ secretory process begins in the tubular cells with CO_2 that has come from three sources: CO_2 that has diffused into the tubular cells from either the plasma or the tubular fluid or CO_2 that has been metabolically produced within the tubular cells. Under the influence of carbonic anhydrase, CO_2 and H_2O increase H^+ and HCO_3^- (p. 598). An energy-dependent carrier in the luminal membrane then transports H^+ out of the cell into the tubular lumen. In part of the nephron, the carrier is a Na^+–H^+ exchanger similar to that in fish gills. Thus H^+ secretion and Na^+ reabsorption are partially linked. In the collecting duct, H^+ alone is secreted by a V-ATPase.

Because the chemical reactions for H^+ secretion begin with CO_2, the rate at which they proceed is influenced by $[CO_2]$. When plasma $[CO_2]$ increases, these reactions proceed more rapidly and the rate of H^+ secretion speeds up (Figure 13–20).

Conversely, the rate of H^+ secretion is reduced when plasma $[CO_2]$ falls below normal. This is especially important in renal compensations for acid–base abnormalities involving a change in H_2CO_3 caused by respiratory dysfunction. The kidneys can therefore adjust H^+ excretion to compensate for changes in both carbonic and noncarbonic acids.

Bicarbonate Excretion

Before being eliminated by the kidneys, the H^+ generated from noncarbonic acids is buffered to a large extent by plasma HCO_3^-. Appropriately, therefore, renal handling of acid–base balance also involves the adjustment of HCO_3^- excretion, depending on the H^+ load in the plasma (Figure 13–20).

The kidneys regulate plasma $[HCO_3^-]$ by two interrelated mechanisms: (1) variable reabsorption of the filtered HCO_3^- back into the plasma and (2) variable addition of new HCO_3^- to the plasma. Both these mechanisms are inextricably linked with H^+ secretion by the kidney tubules. Every time a H^+ is secreted into the tubular fluid, a HCO_3^- is simultaneously transferred into the peritubular capillary plasma. Whether a filtered HCO_3^- is reabsorbed or a new HCO_3^- is added to the plasma in accompaniment with H^+ secretion depends on whether filtered HCO_3^- is present in the tubular fluid to react with the secreted H^+.

Bicarbonate is freely filtered, but because the luminal membranes of the tubular cells are impermeable to the filtered HCO_3^-, it cannot diffuse into these cells. Therefore, reabsorption of HCO_3^- must occur indirectly (● Figure 13–21). Hydrogen ion secreted into the tubular fluid combines with the filtered HCO_3^- to form H_2CO_3. Under the influence of carbonic anhydrase, which is present on the surface of the luminal membrane, H_2CO_3 decomposes into CO_2 and H_2O within the filtrate. Unlike HCO_3^-, CO_2 can easily penetrate the tubular cell membranes. Within the cells, CO_2 and H_2O, under the influence of intracellular carbonic anhydrase, form H_2CO_3, which dissociates into H^+ and HCO_3^-. Because HCO_3^- can permeate the tubular cells' basolateral membrane, it passively diffuses out of the cells and into the peritubular capillary plasma. Meanwhile, the generated H^+ is actively secreted. Because the disappearance of a HCO_3^- from the tubular fluid is coupled with the appearance of another HCO_3^- in the plasma, a HCO_3^- has, in effect, been "reabsorbed." Even though the HCO_3^- entering the plasma is not the same HCO_3^- that was filtered, the net result is the same as if HCO_3^- were directly reabsorbed.

Secretion of H^+ that is excreted is coupled with the addition of new HCO_3^- to the plasma, in contrast to the secreted H^+ that is coupled with HCO_3^- reabsorption and is not excreted, instead being incorporated into H_2O molecules. When all the filtered HCO_3^- has been reabsorbed and additional secreted H^+ is generated by the dissociation of H_2CO_3 in the tubular cell, the HCO_3^- produced by this reaction diffuses into the plasma as a "new" HCO_3^-. It is termed "new" because its appearance in the plasma is not associated with reabsorption of filtered HCO_3^- (● Figure 13–22). Meanwhile, the secreted H^+ combines with urinary buffers, especially basic phosphate (HPO_4^{2-}) and is excreted.

When plasma $[H^+]$ is elevated during acidosis, more H^+ is secreted than normal. At the same time, less HCO_3^- is filtered than normal because more of the plasma HCO_3^- is used up in buffering excess H^+ as HCO_3^- plus H^+ is converted to H_2CO_3. This greater-than-usual inequity between filtered

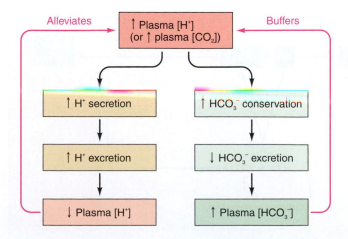

Figure 13–20 ● Control of the rate of tubular H^+ secretion (mammalian kidney).

Figure 13–21 • **Hydrogen ion secretion coupled with bicarbonate reabsorption (mammalian kidney).** Because the disappearance of a filtered HCO_3^- from the tubular fluid is coupled with the appearance of another HCO_3^- in the plasma, HCO_3^- is considered to have been "reabsorbed."

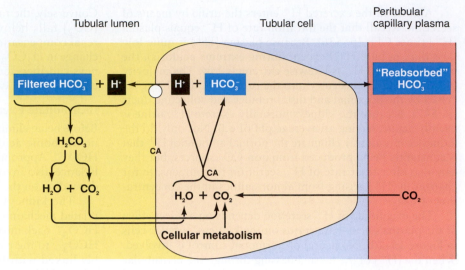

CA = Carbonic anhydrase

HCO_3^- and secreted H^+ has two consequences. First, more of the secreted H^+ is excreted in urine because more hydrogen ions are entering the tubular fluid at a time when fewer are needed to reabsorb the reduced quantities of filtered HCO_3^-. In this way, extra H^+ is eliminated from the body, making the urine more acidic than normal. Second, because excretion of H^+ is linked with the addition of new HCO_3^- to plasma, more HCO_3^- than usual enters the plasma passing through the kidneys. This additional HCO_3^- is available to buffer the excess H^+ in the body.

In the opposite situation of alkalosis, the rate of H^+ secretion diminishes while the rate of HCO_3^- filtration increases compared to normal. When the plasma $[H^+]$ is below normal, a smaller proportion of the HCO_3^- pool is tied up buffering H^+, so the plasma $[HCO_3^-]$ is elevated above normal. As a result, the rate of HCO_3^- filtration correspondingly increases. Not all the filtered HCO_3^- is reabsorbed, be-

cause bicarbonate ions exceed secreted hydrogen ions in the tubular fluid and HCO_3^- cannot be reabsorbed without first reacting with H^+. The excess HCO_3^- is left in the tubular fluid to be excreted in urine, thus reducing the plasma $[HCO_3^-]$ while causing the urine to become alkaline.

In short, when plasma $[H^+]$ is increased above normal during acidosis, renal compensation includes the following:

- Increased secretion and subsequent increased excretion of H^+ in the urine, thereby eliminating the excess H^+ and decreasing plasma $[H^+]$
- Reabsorption of all the filtered HCO_3^-, plus addition of new HCO_3^- to the plasma, resulting in increased plasma $[HCO_3^-]$

When plasma $[H^+]$ is reduced below normal during alkalosis, renal responses include the following:

Figure 13–22 • **Hydrogen ion secretion and excretion coupled with the addition of new HCO_3^- to the plasma (mammalian kidney).** Secreted H^+ does not combine with filtered HPO_4^{2-} and is not subsequently excreted until all the filtered HCO_3^- has been "reabsorbed," as depicted in Figure 13–21. Once all the filtered HCO_3^- has combined with secreted H^+, further secreted H^+ is excreted in the urine, primarily in association with urinary buffers such as basic phosphate. Excretion of H^+ is coupled with the appearance of new HCO_3^- in the plasma. The "new" HCO_3^- represents a net gain rather than being merely a replacement for filtered HCO_3^-.

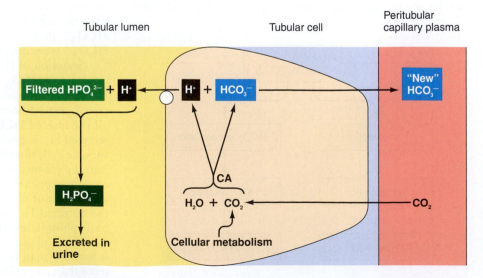

CA = carbonic anhydrase

- Decreased secretion and subsequent reduced excretion of H^+ in urine, resulting in conservation of H^+ and increased plasma $[H^+]$

- Incomplete reabsorption of filtered HCO_3^- and subsequent increased excretion of HCO_3^-, reducing plasma $[HCO_3^-]$

Ammonia Secretion

The energy-dependent H^+ carriers in the tubular cells can secrete H^+ against a concentration gradient until the tubular fluid (urine) becomes 800 times more acidic than the plasma. At this point, further H^+ secretion ceases, because the gradient becomes too great for the secretory process to continue. It is impossible for the kidneys to acidify the urine beyond a gradient-limited urinary pH of 4.5. If left unbuffered as free H^+, only about 1% of the excess H^+ typically excreted daily would produce a urinary pH of this magnitude at normal urine flow rates, and elimination of the other 99% of the usually secreted H^+ load would be prevented, a situation that would be intolerable. For H^+ secretion to proceed, the majority of secreted H^+ must be buffered in the tubular fluid so that it does not exist as free H^+ and, accordingly, does not contribute to the tubular acidity.

Bicarbonate cannot buffer urinary H^+ as it does the ECF, because HCO_3^- is not excreted in the urine simultaneously with H^+. There are, however, two important urinary buffers: (1) filtered phosphate buffers (p. 543) and (2) secreted ammonia (NH_3). Normally, secreted H^+ is first buffered by the phosphate buffer system, which is in the tubular fluid because excess ingested phosphate has been filtered but not reabsorbed. The basic member of the phosphate buffer pair binds with secreted H^+. Basic phosphate is present in the tubular fluid by virtue of dietary excess, not because of any deliberate mechanism to buffer secreted H^+. When H^+ secretion is high, the buffering capacity of urinary phosphates is exceeded, but the kidneys cannot respond by excreting more basic phosphate. As soon as all the basic phosphate ions that are coincidentally excreted have "soaked up" H^+, the acidity of the tubular fluid quickly rises as more H^+ ions are secreted. Without additional buffering capacity from another source, H^+ secretion would soon be halted abruptly as the free $[H^+]$ in the tubular fluid quickly rose to the critical limiting level.

When acidosis exists, the tubular cells secrete NH_3 into the tubular fluid once the normal urinary phosphate buffers are saturated. This NH_3 enables the kidneys to continue secreting additional H^+ ions because NH_3 combines with free H^+ in the tubular fluid to form ammonium ion (NH_4^+) as follows:

$$NH_3 + H^+ \rightarrow NH_4^+$$

The tubular membranes are not very permeable to NH_4^+, so the ammonium ions remain in the tubular fluid and are lost in the urine, each one taking a H^+ with it. Thus NH_3 secretion during acidosis serves to buffer excess H^+ in the tubular fluid, so that large amounts of H^+ can be secreted into the urine before the pH falls to the limiting value of 4.5. In contrast to the phosphate buffers, NH_3 is deliberately synthesized from the amino acid *glutamine* within the tubular cells, from which it readily diffuses down its concentration gradient into the tubular fluid. The rate of NH_3 secretion is controlled by a direct effect on the tubular cells of the amount of excess H^+ to

be transported in the urine. When an individual mammal has been acidotic for more than two or three days, the rate of NH_3 production increases substantially. This extra NH_3 provides additional buffering capacity to allow H^+ secretion to continue after the normal buffering capacity of phosphate is overwhelmed during renal compensation for acidosis.

▪ Acid–base imbalances can arise from either respiratory dysfunction or metabolic disturbances.

Deviations from normal acid–base status are divided into four general categories, depending on the source and direction of the abnormal change in $[H^+]$. These categories are respiratory and environmental acidosis, respiratory and environmental alkalosis, metabolic acidosis, and metabolic alkalosis.

Because of the relationship between $[H^+]$ and the concentrations of the members of a buffer pair, changes in $[H^+]$ are reflected by changes in the ratio of $[HCO_3^-]$ to $[CO_2]$. Recall that the normal ratio is 20:1. Using the Henderson-Hasselbalch equation and with pK being 6.1 and the log of 20 being 1.3, normal pH = 6.1 + 1.3 = 7.4 (note these are mammalian values). Determinations of $[HCO_3^-]$ and $[CO_2]$ provide more meaningful information about the underlying factors responsible for a particular acid–base status than do direct measurements of $[H^+]$ alone. The following rules of thumb apply when examining acid–base imbalances before any compensations take place:

1. A change in pH that has a respiratory (or environmental) cause is associated with an abnormal $[CO_2]$, giving rise to a change in carbonic acid–generated H^+. In contrast, a pH deviation of metabolic origin is associated with an abnormal $[HCO_3^-]$ as a result of the participation of HCO_3^- in buffering abnormal amounts of H^+ generated from noncarbonic acids.

2. Anytime the $[HCO_3^-]/[CO_2]$ ratio falls below 20:1, an acidosis exists. The log of any number lower than 20 is less than 1.3 and, when added to the pK of 6.1, yields an acidotic pH below 7.4. Anytime the ratio exceeds 20:1, an alkalosis exists. The log of any number greater than 20 is more than 1.3 and, when added to the pK of 6.1, yield an alkalotic pH above 7.4.

Putting these two points together:

- Respiratory and environmental acidosis has a ratio of less than 20:1 arising from an increase in $[CO_2]$.

- Respiratory and environmental alkalosis has a ratio greater than 20:1 because of a decrease in $[CO_2]$.

- Metabolic acidosis has a ratio of less than 20:1 associated with a fall in $[HCO_3^-]$.

- Metabolic alkalosis has a ratio greater than 20:1 arising from an elevation in $[HCO_3^-]$.

We next examine each of these categories separately in more detail, paying particular attention to possible causes and the compensations that occur. The "balance beam" concept, presented in ● Figure 13–23 in conjunction with the Henderson-Hasselbalch equation, helps you better visualize the contributions of the lungs and kidneys to the causes and compensations

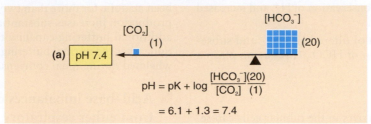
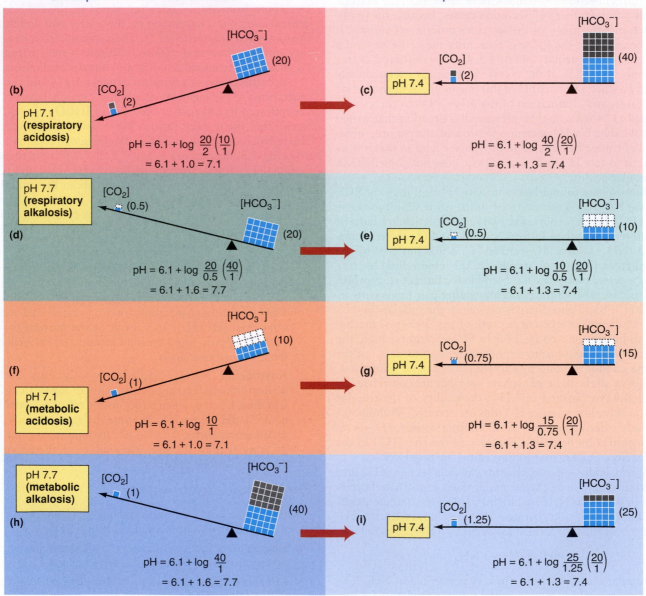

Normal

(a) pH 7.4

$$pH = pK + \log \frac{[HCO_3^-]}{[CO_2]} \frac{(20)}{(1)}$$
$$= 6.1 + 1.3 = 7.4$$

Uncompensated acid–base disorders

Compensated acid–base disorders

(b) pH 7.1
(respiratory acidosis)

$$pH = 6.1 + \log \frac{20}{2} \left(\frac{10}{1}\right)$$
$$= 6.1 + 1.0 = 7.1$$

(c) pH 7.4

$$pH = 6.1 + \log \frac{40}{2} \left(\frac{20}{1}\right)$$
$$= 6.1 + 1.3 = 7.4$$

(d) pH 7.7
(respiratory alkalosis)

$$pH = 6.1 + \log \frac{20}{0.5} \left(\frac{40}{1}\right)$$
$$= 6.1 + 1.6 = 7.7$$

(e) pH 7.4

$$pH = 6.1 + \log \frac{10}{0.5} \left(\frac{20}{1}\right)$$
$$= 6.1 + 1.3 = 7.4$$

(f) pH 7.1
(metabolic acidosis)

$$pH = 6.1 + \log \frac{10}{1}$$
$$= 6.1 + 1.0 = 7.1$$

(g) pH 7.4

$$pH = 6.1 + \log \frac{15}{0.75} \left(\frac{20}{1}\right)$$
$$= 6.1 + 1.3 = 7.4$$

(h) pH 7.7
(metabolic alkalosis)

$$pH = 6.1 + \log \frac{40}{1}$$
$$= 6.1 + 1.6 = 7.7$$

(i) pH 7.4

$$pH = 6.1 + \log \frac{25}{1.25} \left(\frac{20}{1}\right)$$
$$= 6.1 + 1.3 = 7.4$$

The lengths of the arms of the balance beams are not to scale.

of various acid–base disorders. The normal situation is represented in Figure 13–23a.

Respiratory and Environmental Acidosis

Respiratory acidosis is the result of abnormal CO_2 buildup in body fluids, typically arising from hypoventilation (see p. 508).

As less-than-normal amounts of CO_2 are lost through the lungs (or gills), the resultant increase in H_2CO_3 formation and dissociation leads to an elevated $[H^+]$. Possible causes include lung or gill disease, depression of the respiratory center by drugs or disease, nerve or muscle disorders that reduce respiratory muscle ability, or (transiently) even the simple act of

Figure 13–23 ◄ ● Schematic representation of the relationship of [HCO_3^-] and [CO_2] to pH in various acid–base statuses (mammal). (a) Normal acid–base balance. The [HCO_3^-]/[CO_2] ratio is 20/1. (b) Uncompensated respiratory acidosis. The [HCO_3^-]/[CO_2] ratio is reduced (20/2), because CO_2 has accumulated. (c) Compensated respiratory acidosis. Compensatory retention of HCO_3^- to balance the CO_2 accumulation restores the [HCO_3^-]/[CO_2] ratio to a normal equivalent (40/2). (d) Uncompensated respiratory alkalosis. The [HCO_3^-]/[CO_2] ratio is increased (20/0.5) by a reduction in CO_2. (e) Compensated respiratory alkalosis. Compensatory elimination of HCO_3^- to balance the CO_2 deficit restores the [HCO_3^-]/[CO_2] ratio to a normal equivalent (10/0.5). (f) Uncompensated metabolic acidosis. The [HCO_3^-]/[CO_2] ratio is reduced (10/1) by a HCO_3^- deficit. (g) Compensated metabolic acidosis. Conservation of HCO_3^-, which partially makes up for the HCO_3^- deficit, and a compensatory reduction in CO_2 restore the [HCO_3^-]/[CO_2] to a normal equivalent (15/0.75). (h) Uncompensated metabolic alkalosis. The [HCO_3^-]/[CO_2] ratio is increased (40/1) by excess HCO_3^-. (i) Compensated metabolic alkalosis. Elimination of some of the extra HCO_3^- and a compensatory increase in CO_2 restore the [HCO_3^-]/[CO_2] ratio to a normal equivalent (25/1.25).

holding one's breath. A nonrespiratory cause would be high environmental CO_2 (or H^+), as you saw earlier with the deep-sea fish exposed to liquid CO_2 (p. 598). Another example would be a human or pet exposed to output from a fire extinguisher.

In uncompensated respiratory acidosis (Figure 13–23b), [CO_2] is elevated (in our example, it is doubled) while [HCO_3^-] is normal, so the ratio is 20:2 (10:1) and pH is reduced. Let us clarify a potentially confusing point. You might wonder why when [CO_2] is elevated and drives the reaction $CO_2 + H_2O \leftrightarrow H_2CO_3 \leftrightarrow H^+ + HCO_3^-$ to the right, we say that [H^+] becomes elevated but [HCO_3^-] remains normal although the same quantities of H^+ and HCO_3^- are produced when CO_2-generated H_2CO_3 dissociates. The answer lies in the fact that normally the [HCO_3^-] is 600,000 times the [H^+] in a mammal. For every one hydrogen ion and 600,000 bicarbonate ions present in the ECF, generating one additional H^+ and one HCO_3^- doubles the [H^+] (a 100% increase) but only increases the [HCO_3^-] 0.00017% (from 600,000 to 600,001 ions). Therefore, an elevation in [CO_2] brings about a pronounced increase in [H^+], but [HCO_3^-] remains essentially normal.

Compensatory measures act to restore pH to normal. The chemical buffers immediately take up additional H^+, but the respiratory mechanism usually cannot respond with compensatory increased ventilation, because an impairment in respiratory activity is the problem in the first place. Also, if high environmental CO_2 is the cause, increasing ventilation only makes the problem worse, because the gradient favors inward diffusion of the gas. Thus in mammals, the kidneys are most important in compensating for respiratory acidosis. They conserve all the filtered HCO_3^- and add new HCO_3^- to the plasma while simultaneously secreting and, accordingly, excreting more H^+.

As a result, HCO_3^- stores in the body become elevated. In our example (Figure 13–23c), the plasma [HCO_3^-] is doubled, so the [HCO_3^-]–[CO_2] ratio is 40:2 rather than 20:2 as it was in the uncompensated state. A ratio of 40:2 is equivalent to a normal 20:1 ratio, so the pH is once again the normal 7.4. Enhanced renal conservation of HCO_3^- has fully compensated for CO_2 accumulation, restoring the pH to normal, although both the [CO_2] and the [HCO_3^-] are now distorted. Note that maintenance of normal pH depends on preserving a normal ratio between [HCO_3^-] and [CO_2], no matter what the absolute values of each of these buffer components are. (Compensation is never fully complete because the pH can be restored close to but not precisely to normal. In our examples, however, we assume full compensation for ease in mathematical calculations. Also bear in mind that the values used are only representative. Deviations in pH actually occur over a range and the degree to which compensation can be accomplished varies.)

Respiratory and Environmental Alkalosis

The primary defect in respiratory alkalosis is excessive loss of CO_2 from the body as a result of hyperventilation. When pulmonary ventilation increases out of proportion to the rate of CO_2 production, too much CO_2 is blown off. Consequently, less H_2CO_3 is formed and [H^+] decreases. Possible causes of respiratory alkalosis include fever, anxiety, and ingested toxins, all of which excessively stimulate ventilation without regard to the status of O_2, CO_2, or H^+ in the body fluids. Respiratory alkalosis also occurs as a result of physiological mechanisms at high altitude. When the low concentration of O_2 in the arterial blood reflexly stimulates ventilation in an attempt to obtain more O_2 for the body, too much CO_2 is blown off in the process, inadvertently leading to an alkalotic state (see p. 509).

Environmental causes are not common, but do occur. For example, Pyramid Lake, Nevada, has been slowly drying up for many years, and the resulting concentration of certain minerals is continuously increasing the lake's alkalinity (now at pH 9.4). There is concern that this threatens the resident cutthroat trout. In laboratory studies, trout exposed to water at pH 10 suffered a 0.25-pH-unit increase in blood pH, and over half of the fish died after 72 hours. The fish had excessively high levels of plasma ammonia and low levels of plasma Na^+ and Cl^-, suggesting that excretion and ion regulation were inhibited.

Returning to mammals and looking at the biochemical abnormalities in uncompensated respiratory alkalosis (Figure 13–23d), the increase in pH reflects a reduction in [CO_2] (half the normal value in our example) while the [HCO_3^-] remains normal. This yields an alkalotic ratio of 20:0.5, which is comparable to 40:1.

Compensatory measures act to shift the pH back toward normal. The chemical buffer systems liberate H^+ to diminish the severity of the alkalosis. As plasma [CO_2] and [H^+] fall below normal because of excessive ventilation, two of the normally potent stimuli for driving ventilation are removed. This effect tends to "put brakes" on the extent to which some non–respiration-related factors such as fever or anxiety can overdrive ventilation. Therefore, hyperventilation does not continue completely unabated. If the situation continues for a few days, the kidneys compensate by conserving H^+ and excreting more HCO_3^-. If, as in our example (Figure 13–23e), the HCO_3^- stores are reduced by half by loss of HCO_3^- in the urine, the $[HCO_3^-]/[CO_2]$ ratio becomes 10:0.5, equivalent to the normal 20:1. Therefore, the pH is restored to normal by reducing the HCO_3^- load to compensate for the CO_2 loss.

Metabolic Acidosis

Metabolic acidosis (also known as *nonrespiratory acidosis*) encompasses all types of acidosis besides that caused by excess CO_2 in the body fluids. In the uncompensated state (Figure 13–23f), metabolic acidosis is always characterized by a reduction in plasma $[HCO_3^-]$ (in our example it is halved), while $[CO_2]$ remains normal, producing an acidotic ratio of 10:1. The problem may arise from the body's excessive loss of fluids rich in HCO_3^- or from an accumulation of noncarbonic acids. In the latter case, plasma HCO_3^- is used up in the process of buffering the additional H^+.

This type of acid–base disorder is the one more frequently encountered. The following are the most common causes in mammals:

1. *Severe diarrhea.* During digestion, a bicarbonate-rich digestive juice is normally secreted into the digestive tract from the pancreas and is subsequently reabsorbed back into the plasma when digestion is completed. During diarrhea, this HCO_3^- is lost from the body rather than being reabsorbed. The reduction in plasma $[HCO_3^-]$ without a corresponding reduction in $[CO_2]$ lowers the pH. Because of the loss of HCO_3^-, less HCO_3^- is available to buffer H^+, leading to more free H^+ in the body fluids. Looking at the situation differently, loss of HCO_3^- shifts the $H^+ + HCO_3^- \leftrightarrow CO_2 + H_2O$ reaction to the left to compensate for the HCO_3^- deficit, increasing $[H^+]$ above normal in the process.

2. *Diabetes mellitus ketoacidosis.* Abnormal fat metabolism resulting from the inability of cells to preferentially use glucose in the absence of insulin results in the formation of excess keto acids whose dissociation causes an increase in plasma $[H^+]$.

3. *Strenuous activity.* When muscles resort to anaerobic glycolysis during strenuous activity (see p. 54), excess lactic acid is produced, leading to a rise in plasma $[H^+]$. In exercising alligators, for example, plasma pH may drop from 7.4 to 7.0, then take two hours to recover.

4. *Uremic acidosis.* In severe renal failure (uremia), the kidneys cannot rid the body of even the normal amounts of H^+ generated from the noncarbonic acids formed by the body's ongoing metabolic processes, so H^+ starts to accumulate in the body fluids. Also, the kidneys cannot conserve an adequate amount of HCO_3^- to be used for buffering the normal acid load. The resultant fall in $[HCO_3^-]$ without a concomitant reduction in $[CO_2]$ is correlated with a decline in pH to an acidotic level.

Except in uremic acidosis, metabolic acidosis is compensated for by both respiratory and renal mechanisms as well as by chemical buffers. The buffers take up extra H^+, the lungs blow off additional H^+-generating CO_2, and the kidneys excrete more H^+ and conserve more HCO_3^-. In our example (Figure 13–23g), these compensatory measures restore the ratio

to normal by reducing $[CO_2]$ to 75% of normal and by raising $[HCO_3^-]$ halfway back toward normal (up from 50% to 75% of the normal value). This brings the ratio to 15:0.75 (equivalent to 20:1).

Note that in compensating for metabolic acidosis, the lungs deliberately displace $[CO_2]$ from normal in an attempt to restore $[H^+]$ toward normal. Whereas in respiration-induced acid–base disorders, an abnormal $[CO_2]$ is the cause of the $[H^+]$ imbalance, in metabolic acid–base disorders, $[CO_2]$ is intentionally shifted from normal as an important compensation for the $[H^+]$ imbalance.

When kidney disease is the cause of metabolic acidosis, complete compensation is not possible, because the renal mechanism is not available for pH regulation. Recall that the respiratory system can compensate only up to 75% of the way toward normal. Uremic acidosis is very serious, because the kidneys cannot restore the pH all the way to normal.

Metabolic Alkalosis

Metabolic alkalosis is a reduction in plasma $[H^+]$ caused by a relative deficiency of noncarbonic acids. This acid–base disturbance is associated with an increase in $[HCO_3^-]$, which, in the uncompensated state, is not accompanied by a change in $[CO_2]$. In our example (Figure 13–23h), $[HCO_3^-]$ is doubled, producing an alkalotic ratio of 40:1. This condition arises most commonly in mammals from *vomiting*, which results in abnormal loss of H^+ from the body as a result of loss of acidic gastric juices. Hydrochloric acid is secreted into the stomach lumen during the process of digestion. Bicarbonate is added to the plasma during gastric HCl secretion. This HCO_3^- is neutralized by H^+ as the gastric secretions are eventually reabsorbed back into the plasma, so normally there is no net addition of HCO_3^- to the plasma from this source. However, when this acid is lost from the body during vomiting, not only is plasma $[H^+]$ decreased, but also reabsorbed H^+ is no longer available to neutralize the extra HCO_3^- added to the plasma during gastric HCl secretion. Thus loss of HCl in effect increases plasma $[HCO_3^-]$.

In metabolic alkalosis, the chemical buffer systems immediately liberate H^+ and ventilation is reduced so that extra H^+-generating CO_2 is retained in the body fluids. If the condition persists for several days, the kidneys conserve H^+ and excrete the excess HCO_3^- in the urine. The resultant compensatory increase in $[CO_2]$ (up 25% in our example—Figure 13–23i) and the partial reduction in $[HCO_3^-]$ (75% of the way back down toward normal in our example) together restore the $[HCO_3^-]/[CO_2]$ ratio back to the equivalent of 20:1 at 25:125.

Clearly, an animal's acid–base status cannot be assessed on the basis of pH alone. Uncompensated acid–base abnormalities can readily be distinguished on the basis of deviations of either $[CO_2]$ or $[HCO_3^-]$ from normal (▌Table 13–5). When compensation has been accomplished and the pH is essentially normal, determinations of $[CO_2]$ and $[HCO_3^-]$ can reveal the presence of an acid–base disorder. Note, however, that in both compensated respiratory acidosis and compensated metabolic alkalosis, $[CO_2]$ and $[HCO_3^-]$ are both above normal. With respiratory acidosis, the original problem is an abnormal increase in $[CO_2]$, and a compensatory increase in $[HCO_3^-]$ restores the $[HCO_3^-]/[CO_2]$ ratio to 20:1. Meta-

Table 13–5 ❙ Summary of $[CO_2]$, $[HCO_3^-]$, and pH in Uncompensated and Compensated Acid–Base Abnormalities

Acid–Base Status	pH	$[CO_2]$ (compared to normal)	$[HCO_3^-]$ (compared to normal)	$[HCO_3^-]/[CO_2]$
Normal	Normal	Normal	Normal	20/1
Uncompensated respiratory acidosis	Decreased	Increased	Normal	20/2 (10/1)
Compensated respiratory acidosis	Normal	Increased	Increased	40/2 (20/1)
Uncompensated respiratory alkalosis	Increased	Decreased	Normal	20/0.5 (40/1)
Compensated respiratory alkalosis	Normal	Decreased	Decreased	10/0.5 (20/1)
Uncompensated metabolic acidosis	Decreased	Normal	Decreased	10/1
Compensated metabolic acidosis	Normal	Decreased	Decreased	15/0.75 (20/1)
Uncompensated metabolic alkalosis	Increased	Normal	Increased	40/1
Compensated metabolic alkalosis	Normal	Increased	Increased	25/1.25 (20/1)

bolic alkalosis, in contrast, is characterized by an abnormal increase in $[HCO_3^-]$ in the first place; then $[CO_2]$ is deliberately increased to restore the ratio to normal. Similarly, compensated respiratory alkalosis and compensated metabolic acidosis share similar patterns of $[CO_2]$ and $[HCO_3^-]$. Respiratory alkalosis starts out with a reduction in $[CO_2]$, which is compensated by a reduction in $[HCO_3^-]$. With metabolic acidosis, $[HCO_3^-]$ falls below normal, followed by a compensatory decrease in $[CO_2]$.

Chapter in Perspective:
HOMEOSTASIS AND INTEGRATION

Homeostasis depends on maintaining a balance between the input and output of all constituents present in the internal fluid environment. Regulation of fluid balance involves two separate components: control of salt balance and control of H_2O balance. Control of H_2O balance is important in preventing changes in ECF osmolarity, which would induce detrimental osmotic shifts of H_2O between the cells and the ECF. Such shifts of H_2O into or out of the cells would cause the cells to swell or shrink, respectively. Cells, especially brain neurons, do not function normally when swollen or shrunken.

Basic cell components yield about 300 mOsm of osmotic concentration; thus marine animals are threatened with dehydration, because the ocean is about 1000 mOsm. In marine osmoconformers, ECF salt concentrations are similar to that of the environment, but the ICF uses compatible organic osmolytes to simultaneously elevate osmotic pressure and to protect proteins from destabilizing effects of salts and other stressors. In osmoregulators such as mammals, ECF salt and osmolarity are kept relatively constant.

Salt balance is maintained by constantly adjusting salt output in the urine to match unregulated, variable salt intake. Control of salt balance is primarily important in the long-term regulation of arterial blood pressure, because the body's salt load affects the osmotic determination of the ECF volume, of which plasma volume is a part. An increased salt load in the ECF expands ECF volume, including plasma volume, which in turn raises blood pressure. Conversely, reducing ECF salt load lowers blood pressure. Water balance is largely maintained by controlling the volume of H_2O lost in the urine to compensate for uncontrolled losses of variable volumes of H_2O from other avenues, such as through sweating, panting, or diarrhea, and for poorly regulated H_2O intake.

A balance between input and output of H^+ is also critical to maintaining a body's acid–base balance within the narrow limits compatible with life. Deviations in the internal fluid environment's pH lead to altered neuromuscular excitability, to changes in enzymatically controlled metabolic activity, and to K^+ imbalances, which can cause cardiac arrhythmias. Hydrogen ions are uncontrollably and continually being added to body fluids as a result of metabolic activities, yet the ECF's pH must be kept relatively constant (at a slightly alkaline level of 7.4 for optimal function in mammals). Like salt and H_2O balance, vertebrate control of H^+ output by the kidneys is the main regulatory factor in achieving H^+ balance. Assisting the kidneys in eliminating H^+ are the lungs, which can adjust their rate of excretion of H^+-generating CO_2. The assertion that the same input–output balance that applies to salt and H_2O also applies to H^+ homeostasis must be modified by the fact that H^+ is buffered in the ECF and ICF by chemical buffers. This buffering mechanism can take up or liberate H^+, thereby transiently keeping its concentration constant within the body until its output can be brought into line with its input. Such a mechanism is not available for salt or H_2O balance. ∎

REVIEW QUESTIONS *(Answers are on p. A–1.)*

Additional study tools for this chapter, including chapter summaries and practice tests, are available online at *www.biology.brookscole.com*

1. The true internal environment that bathes virtually all tissue cells is
   a. intra-ocular fluid
   b. synovial fluid
   c. intrapleural fluid
   d. interstitial fluid
   e. transcellular fluids
2. Among the major differences between ECF and ICF are/is
   a. cellular proteins
   b. cellular organic osmolytes
   c. unequal distribution of Na^+
   d. unequal distribution of K^+
   e. all of the above
3. All animal cells face the potential problem of unequal Na^+ distribution in the ECF and ICF; the nearly universal tactic used to deal with this problem is
   a. moving water directly by ATPase "pumps"
   b. the evolution of digestive tracts
   c. evaporation of body water
   d. osmosis
   e. the Na^+/K^+ ATPase "pump"
4. Which of the following would not fall into the category of an osmolyte?
   a. carbohydrates
   b. fatty acids
   c. urea
   d. free amino acids
   e. TMAO
5. Which of the following would be a strict osmoconformer?
   a. marine sponge
   b. tiger
   c. cockroach
   d. honeybee
   e. bullfrog
6. The organisms you would expect to find adapted to surviving in a tidal pool would be
   a. stenohaline
   b. hypo-ionic osmoconformers
   c. euryhaline
   d. hyperosmotic regulators
   e. none of the above
7. As a consequence of being hypo-osmotic to seawater, most marine vertebrates tend to
   a. gain both salt and water
   b. gain water and lose salt
   c. lose water and gain salt
   d. lose water and lose salt
   e. use high cellular concentrations of organic osmolytes
8. Salmon are
   a. hyper-osmotic in rivers
   b. hypo-osmotic in oceans
   c. hypo-hyper osmoregulators
   d. classic examples of acclimatization regulation
   e. all of the above

9. Dehydration with accompanying hypertonicity can be brought about by
   a. insufficient water intake
   b. excessive water loss
   c. drinking hypertonic saline water
   d. inhibited vasopressin secretion
   e. all of the above
10. Normal daily water balance in a human shows
    a. water intake in food exceeds metabolically produced water
    b. metabolically produced water exceeds water output from kidneys
    c. total water output is normally less than total water input
    d. total input equals total output
    e. insensible losses exceed equal fluid intake
11. Which of the following account for more than 90% of the ECF's osmotic activity in a mammal?
    a. calcium cations and chloride anions
    b. potassium cations and iodide anions
    c. magnesium cations and chloride anions
    d. sodium cations and attendant anions
    e. iron cations in hemoglobin
12. In mammals, the most important regulation of salt balance is carried out by
    a. fecal excretion
    b. sweat glands
    c. kidneys
    d. Malpighian tubules
    e. b and c
13. The most important factor in regulating mammalian ECF volume and blood pressure is
    a. the thirst mechanism
    b. acid-base balance
    c. the renin-angiotestin-aldosterone system
    d. the baroceptor reflex
    e. plasma protein synthesis
14. In the event of hemorrhagic shock
    a. heart stroke volume is decreased
    b. arteriolar vasoexpansion leads to a decrease in peripheral resistance
    c. heart rate and stroke volume increase
    d. urinary output increases
    e. venous return is increased
15. Which of the following is more acidic?
    a. a solution with a pH of 3
    b. a solution with a pH of 5
    c. a solution with a pH of 7
    d. a solution with a pH of 8
    e. a solution with a pH of 9
16. Which of the following solutions has a higher concentration of hydronium ions?
    a. a solution of pH 7
    b. a solution of pH 10
    c. a solution of pH 3
    d. a solution of pH 5
    e. a solution of pH 12
17. Which solution has an $[H^+]$ ten times higher (more acidic) than a solution with a pH of 5?
    a. a solution of pH 7

b. a solution of pH 10
c. a solution of pH 7.5
d. a solution of pH 4
e. a solution of pH 1

18. The pH of mammalian arterial blood is typically 7.45 and the pH of venous blood is 7.35, therefore
    a. arterial blood is slightly more acidic
    b. arterial blood is slightly more basic
    c. venous blood is nearly ten times more acidic than arterial blood
    d. arterial blood is nearly ten times more acidic than venous blood
    e. acidosis is a commonly observed condition

19. Which of the following is not a buffer system found in vertebrates?
    a. the CO_2-HCO_3^- buffer system
    b. the peptide and protein buffer system
    c. the nucleic acid buffer system
    d. the hemoglobin buffer system
    e. the phosphate buffer system

20. When plasma $[H^+]$ is increased above normal during acidosis, renal compensation can include
    a. reduced excretion of H^+
    b. increased excretion of H^+
    c. increasing the excretion of HCO_3^- from plasma
    d. increasing reabsorption of filtered HCO_3^- from plasma
    e. b and d

SUGGESTED READINGS AND INTERNET SITES

Claiborne, J. B., S. L. Edward, & A. I. Morrison-Shetlar. 2002. Acid–base regulation in fishes: Cellular and molecular mechanisms. *Journal of Experimental Zoology* 293:302–319.

Clegg, J. S. 2001. Cryptobiosis—A peculiar state of biological organization. *Comparative Biochemistry and Physiology* 128B: 613–624.

Harrison, J. F. 2001. Insect acid–base physiology. *Annual Review of Entomology* 46:221–250.

Hochachka, P., & G. N. Somero. 2002. *Biochemical Adaptation: Mechanism and Process in Physiological Evolution.* Oxford, UK: Oxford University Press

Jackson, D. C. 2000. How a turtle's shell helps it survive prolonged anoxic acidosis. *New in Physiological Sciences* 15: 181–185.

Lockwood, A. P. M. 1963. *Animal Body Fluids and Their Regulation.* Cambridge, MA: Harvard University Press.

Parker, A. R., & C. R. Lawrence. 2001. Water capture by a desert beetle. *Nature* 414:33–34.

Takei, Y. 2000. Comparative physiology of body fluid regulation in vertebrates with special reference to thirst regulation. *Japanese Journal of Physiology* 50:171–186.

Yancey, P. H. 2001. Proteins, osmolytes and water stress. *American Zoologist* 41:699–709.

Yancey, P. H., M. E. Clark, S. C. Hand, R. D. Bowlus, & G. N. Somero. 1982. Living with water stress: Evolution of osmolyte systems. *Science* 217:1214–1222

INFOTRAC READING

Crowe, J. H., L. M. Crowe, & D. Chapman. 1984. Preservation of membranes in anhydrobiotic organisms: The role of trehalose. *Science* 223:701.

INTERNET SITES

Mach, M. Tardigrades. **tardigrades.com.** Covers the life history of these fascinating animals, including their remarkable ability to survive in a dried state, with many photographs.

Dow, J. A. T. *V-ATPases.* **mblab.gla.ac.uk/~julian/lab/V-ATPases .html.** Animation of a V-ATPase and its uses in nature.

Digestive Systems

The rumen of a Scottish Highland cow enables it to survive under conditions of low food quality and availability.

Photo: Hillar Klandorf

Introduction: Feeding Strategies and Evolution

The primary function of digestive systems is to transfer nutrients, water, and electrolytes from food into the animal body's internal environment. Because living cells depend on a continuing flow of nutrients to sustain their metabolic needs, most animals commit considerable time and energy to acquiring and digesting food. Ingested food is an essential energy source, or "fuel," from which the cells can produce ATP to carry out their particular energy-dependent activities, such as active transport, contraction, synthesis, and secretion. Carbohydrates, fats, and proteins represent the main fuels; they are broken down into absorbable units (specifically simple sugars, fatty acids, and amino acids), which an animal's individual cells can use as fuel or as "building blocks" to make their own, larger molecules. (Recall from Chapter 2 that animals use the unit organic molecules to release usable energy, predominantly via aerobic pathways using O_2 and with CO_2 and H_2O being formed as by-products in the process.)

We first provide an overview of digestive systems, examining the common features of the various components, before we begin a detailed tour of the tract from beginning to end.

■ **Animal digestion evolved from an intracellular to an extracellular process in a specialized sac or tube connected to the environment.**

Digestion of food inside the body but outside the cell is accomplished by a digestive tract. Intracellular digestion is limited, because only food items small enough to be internalized by the cells can be digested, whereas extracellular digestion permits an animal to store and break down larger varieties and quantities of food items. The first digestive systems probably evolved as an infolding of the epidermis into an interior sac, such as that found in Cnidaria, providing the ability to capture much larger quantities of food. Later, the sac fused with the opposite end of the body to form a complete tube with an entrance (mouth) and exit (anus). Cells lining this digestive tube could evolve specialized functions; for example, cells responsible for acidic digestion could be compartmentalized and separated from those involved in alkaline digestion. Other parts of the tube could be specialized for receiving and storing food and processing wastes (● Figure 14–1). Indigestible items could remain outside and not take up space within cells.

Thus an animal's digestive tract can be thought of as a hollow sac or a tube with an entrance and exit, lying within an animal's body and specialized for the initial processing of food items. At the openings, the tract is in direct contact with the outside world, so the lumen of the tract, although within the animal's body, is also technically part of the external environment. Until nutrients are actually absorbed across this tube, they cannot be considered "inside" the body. Conventionally, the gut tube in most animals can be divided into three specific regions: the **foregut, midgut,** and the **hindgut** (Figure 14–1).

■ **Animals can be classified according to their primary feeding methods.**

Animals that primarily eat other animals are termed **carnivores.** Carnivores specialize on capturing live prey or rely on their abilities to locate carrion, and typically eat a high-energy, high-protein diet because of the high percentage of muscle tissue ("meat") in most animals. **Herbivores** depend on the intake

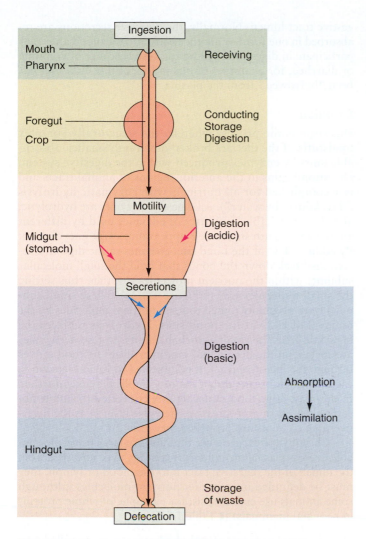

Figure 14–1 ● A digestive tract with one-way passage of food allows simultaneous operation of the sequential stages in the processing of food and reduces mixing of digested and undigested matter. The crop is a storage region found in some animals.

termed **omnivores** and have digestive tracts more similar to carnivore tracts, that rely on enzymatic digestion of foodstuffs. Because of seasonal changes in food availability, gut size adapts to low-quality feeds by increasing in length. For example, the starling is primarily a carnivore during the warmer months when insects and worms are available, but during the harsh winter months they are forced to become herbivores and consume low-quality foods (buds, and so forth). In response, gut length increases a remarkable 450%.

Other important animal feeding types include **filter feeders**, ranging from sponges to baleen whales, which trap dead and/or living material suspended in water (which may include bacteria, algae, protozoa, and small animals); **detritivores** or **deposit feeders**, such as earthworms and mites, which feed on dead and living material in sediments; and **fluid feeders**, such as mosquitoes and vampire bats, which suck or lick fluids of a larger animal without ingesting the whole animal. Some fluid feeders are termed **parasites** if they remain attached to or within a host animal for long periods. A key example is the tapeworm, which has no digestive tract and lives within host intestines, where it absorbs small organic molecules through its body wall. Finally, a few (but important) **symbiont-bearing** animals in the oceans require little or no external organic fuel. Key examples are (1) reef-building corals (cnidarians) with photosynthesizing dinoflagellate symbionts that provide food molecules in addition to those ingested; and (2) vestimentiferan worms living at hydrothermal vents and hydrocarbon (petroleum and natural gas) seeps, worms that have no digestive tract and instead use food molecules produced by symbiotic chemosynthesizing bacteria.

Depending on the life cycle of some animals, feeding specializations are related to a particular stage of development. Insect larvae normally have completely different food habits and digestive systems than adults. Tadpoles of some frog species, which are herbivorous, have a longer and more complex intestine than does the adult frog, which is carnivorous. And young mammals shift from milk to an adult diet.

The reproductive success of certain species can be linked to food availability. Rainfall and environmental temperature determines plant growth and consequently the amount of food available to herbivores. In water, the amount of particulate food decreases with depth and varies according to location and season. Plant material also varies considerably in composition during the course of a year. Immature grasses and leaves are characterized by a higher content of protein and other nutrients than later in the season. In environments characterized by seasonal variation in climate, natural selection favors reproductive cycles that can best exploit these limiting resources.

External food supply ultimately represents one of the key factors that determines animal survivability, the others being predation, parasitism, disease, competition, and beneficial symbioses (such as microbes in ruminants' digestive tracts and algae in coral cnidarians). As local populations of animals adjust to local conditions, they become adapted to their particular environment by the process of natural selection.

of algal or plant materials, such as kelp blades and plant leaves, fruits, or seeds. Some of these materials have comparatively little usable energy, and herbivores eating them have two options. The first, as seen in the dairy cow and the deer, uses **pre-gastric fermentation** (the processing of foodstuffs by microbes) as a means to break down plant material into absorbable units. In these animals the foregut ("before the stomach") is modified such that microorganisms can proliferate in a fermentation vat, the **rumen.** In a second type of herbivore, those with simple stomachs such as the horse and rabbit, microbial digestion occurs in the hindgut ("after the stomach"), which includes the colon and cecum. Herbivores either can be nonselective in their choice of food items or, as in the case of the panda, with its dependence on bamboo, can be highly selective. Herbivorous fishes may feed on the fruits, seeds, flowers, and leaves of trees, whereas others may feed on algae.

Animals such as humans, monkeys, and pigs, that can eat both animals and other organisms such as plants or fungi, are

■ **Digestive systems perform four basic digestive processes.**

There are four basic digestive processes: *motility, secretion, digestion,* and *absorption:*

Motility

The term **motility** refers to the muscular contractions within the gut tube that mix and move forward the contents of the digestive tract. However, some worm-shaped animals rely on muscular contractions of the body to indirectly force food through the gut tube. Like vascular smooth muscle, the smooth muscle in the walls of the digestive tract maintains a constant low level of contraction known as **tone**. Tone is important in maintaining a steady pressure on the digestive tract contents as well as in preventing the walls of the digestive tract from remaining permanently stretched following distension.

Two basic types of digestive motility are superimposed on this ongoing tonic activity:

- *Propulsive movements* propel, or push, the contents through the digestive tract at varying speeds, with the rate of propulsion dependent on the functions accomplished by the different regions; that is, food is moved forward in a given segment at an appropriate velocity to allow that segment to "do its job." For example, transit of food through the foregut of invertebrates and esophagus of vertebrates is rapid, which is appropriate because these structures merely serve as a passageway from the mouth to the stomach or midgut. In comparison, in the small intestine—the major site of digestion and absorption—the contents move slowly, allowing sufficient time for the breakdown and absorption of food.

- *Mixing movements* serve a twofold function. First, by mixing food with the digestive juices, these movements promote digestion of the food. Second, they facilitate absorption by exposing all portions of the intestinal contents to the absorbing surfaces of the digestive tract.

Contraction of the smooth muscle within the walls of the digestive organs moves material through most of the digestive tract, except for the ends of the tract—the mouth and the early portion of the esophagus (exceptions include the entire ruminant esophagus) at the beginning, and the external anal sphincter at the end—where motility may involve skeletal muscle rather than smooth muscle. Accordingly, the acts of chewing, swallowing (rumination), and defecation have motor cortex control ("voluntary" in humans, p. 148). By contrast, motility accomplished by **smooth muscle** throughout the remainder of the tract is controlled by complex autonomic ("involuntary") mechanisms.

Secretion

A number of digestive juices are secreted into the digestive tract lumen by exocrine glands located along the route, each with its own specific secretory product or products. Each digestive secretion consists of water, electrolytes, and specific organic constituents that are important in the digestive process, such as enzymes, bile salts, or mucus. The secretory cells extract large volumes of water and those raw materials necessary to produce their particular secretion from blood (see Figure 1–2, p. 7). Secretion of all digestive juices requires energy, both for active transport of some of the raw materials into the cell (others diffuse in passively) and for synthesis of secretory products by the endoplasmic reticulum. On appropriate neural or hormonal stimulation, the secretions are released into the di-

gestive tract lumen. Normally, the digestive secretions are reabsorbed in one form or another back into the blood after they participate in digestion. Failure to do so (because of vomiting or diarrhea, for example) results in loss of this fluid that has been "borrowed" from the plasma.

Digestion

Digestion is the process whereby the structurally complex foodstuffs of the diet are broken down into smaller, absorbable units by enzymes produced within the digestive system. Digestion begins in the anterior regions of animal tracts and is accomplished for all nutrients with enzymatic **hydrolysis** ("breakdown by water"). All digestive enzymes are hydrolytic, and the remarkable similarity of enzymes used by different animal species attests to their appearance early in evolution. By adding H_2O at the bond site, enzymes in the digestive secretions break down the bonds that hold the small molecular subunits within the nutrient molecules together, thus setting the small molecules free (● Figure 14–2a). These small subunits were originally joined to form nutrient molecules by the removal of H_2O at the bond sites. Hydrolysis replaces the H_2O and frees the small absorbable units. Digestive enzymes are specific in the bonds they can hydrolyze. As food moves through the digestive tract, it is subjected to various enzymes, each of which breaks down the food molecules even further. In this way, large food molecules are converted to simple absorbable units in a progressive, stepwise fashion as the digestive tract contents are propelled forward.

Carbohydrates, proteins, and fats are large molecules unable to cross plasma membranes intact. So, to be absorbed from the lumen of the digestive tract into the blood or lymph, they must be degraded into smaller nutrient molecules (although some animals use endocytosis; for example, of whole proteins in neonatal mammals) (▌ Table 14–1).

1. The simplest forms of **carbohydrates** are the so-called simple sugars, or **monosaccharides** ("one-sugar" molecules), such as **glucose, fructose,** and **galactose,** very few of which are normally found in most diets. Rather, most ingested carbohydrate is in the form of **polysaccharides** ("many-sugar" molecules), which consist of chains of interconnected monosaccharide molecules. The most common polysaccharides consumed are **starch** (which has α-bonds between the glucose units) and **cellulose** (which has β-bonds) that are derived from plant sources (Figure 14–2b). Enzymes for the digestion of starch are more widely used than for any other organic molecule. Cellulose is the most abundant organic molecule in the biosphere and is the major component of plant material, comprising over half of the plant cell wall. However, few animals can directly exploit this energy source. Given the fibrous "staircase" structure of cellulose resulting from its β-bond linkages (Figure 14–2b), it is extremely difficult to hydrolyze. In animals that use cellulose for fuel, cellulose digestion is accomplished by symbiotic microorganisms that proliferate in the digestive tract and that have the necessary enzymes—**cellulases**—to catabolize cellulose. These microbes may be housed in special organs connected to the gut, as in the case of termites and many wood-boring beetles. Some termites and wood wasps actually culture fungi that

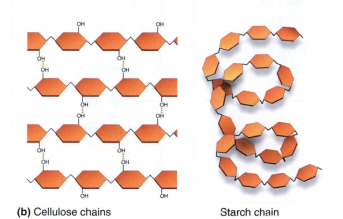

(a)

(b) Cellulose chains   Starch chain

Figure 14–2 • **Structure and hydrolysis of common dietary carbohydrates. (a)** An example of hydrolysis. In this example, the disaccharide maltose (the intermediate breakdown product of polysaccharides) is broken down into two glucose molecules by the addition of H_2O at the bond site. **(b)** Hydrogen bonding relationships for glucose units in cellulose and starch. In cellulose, hydrogen bonds form between glucose chains. This pattern stabilizes the chains and permits them to become tightly bundled. In amylose, a form of starch, the bonds form between units within a chain, which permits the glucose units to coil.

(*Source:* C. Starr, 2000, *Biology: Concepts and Applications,* 4th ed., Belmont, CA: Brooks/Cole, Figure 3.7, p. 39)

produce cellulase. **Chitin** is the principal component of the hard exoskeleton of arthropods and, like cellulose, is indigestible by most vertebrates. **Glycogen** represents the polysaccharide storage form of glucose in animals. After digestion of a meal, unused glucose is stored as glycogen (in liver and muscle in vertebrates; p. 24). *Cecropia,* a bamboolike tree, is the only plant known to synthesize glycogen. A particularly aggressive ant, *Azteca,* consumes the glycogen and, in turn, protects the tree from potential agents that could damage it.

2. Besides polysaccharides, another source of dietary carbohydrate is in the form of **disaccharides** ("two-sugar" molecules), including **sucrose** (table sugar, which consists of one glucose and one fructose molecule), **trehalose** (the most important carbohydrate for flight in arthropods, made up of two glucose molecules), **maltose** (containing two glucose residues), and **lactose** (milk sugar, made up of one glucose and one galactose molecule).

3. Starch, cellulose, chitin, glycogen, and disaccharides are converted through the process of digestion into their con-

stituent monosaccharides, which are the absorbable units for carbohydrates.

4. The second category of foodstuffs is **proteins,** which consist of various combinations of **amino acids** held together by peptide bonds. Through the process of digestion, proteins are degraded primarily into their constituent amino acids as well as a few **polypeptides** (several amino acids linked together by peptide bonds), both of which are the primary absorbable units for digested protein.

5. **Fats** represent the third category of foodstuffs. In most terrestrial food chains, dietary fat is in the form of **triglycerides,** which are neutral fats, each consisting of a combination of glycerol with three (*tri,* "three") **fatty acid** molecules attached. During digestion, two of the fatty acid molecules split off, leaving a **monoglyceride,** a glycerol molecule with one (*mono,* "one") fatty acid molecule attached. Thus the end products of fat digestion are monoglycerides and free fatty acids, which are the absorbable units of fat. In marine food chains, **waxes** (a fatty alcohol and fatty acid linked together) are common lipids. Waxes are hydrolyzed by the activity of esterases synthesized in microorganisms.

Nucleic acids can also form a significant portion of ingested nitrogen. Pancreatic nucleases, *RNase* and *DNase,* split the nucleic acids into their component nucleotides. The duodenum of ruminants is particularly effective in digesting nucleic acids, because approximately 20% of the nitrogen entering the *abomasum* (acid-secreting portion of the stomach) originates from the microorganisms spilled from the rumen.

Absorption

After digestion is completed, the process of absorption occurs in middle to posterior sections of most digestive systems (such as the small intestine in vertebrates). Here the small, absorbable units that result from digestion, along with water, vitamins, and electrolytes, are transferred from the digestive tract lumen into the blood or hemolymph or body cavity. Specialized transporters in the epithelial cells carry the hydrophilic units through the membrane. To enhance this process (in accordance with Fick's law of diffusion, p. 73), *surface area* is often greatly increased by folds. For example, in annelids (such as earthworms) and mollusks, long folds forming ridges called **typhlosoles** line parts of the tract (● Figure 14–3a). You will see later how *villi* increase surface area in vertebrate intestines.

Some aquatic species can also absorb nutrients through their gills (as in the mussel *Mytilus*), or through the epidermis (as in many marine invertebrates including annelids and echinoderms). The ocean contains large amounts of dissolved organic matter (DOM), including sugars and amino acids. The uptake of amino acids and/or glucose depends on Na^+.

■ The digestive tract and accessory digestive organs make up digestive systems.

The digestive system of an animal consists of the digestive (or *gastrointestinal*) tract plus the accessory digestive organs (*gas-*

Table 14–1 ▪ Process of Digestion

Category of Foodstuffs	Intermediate Breakdown Products	End Products of Digestion: Absorbable Units

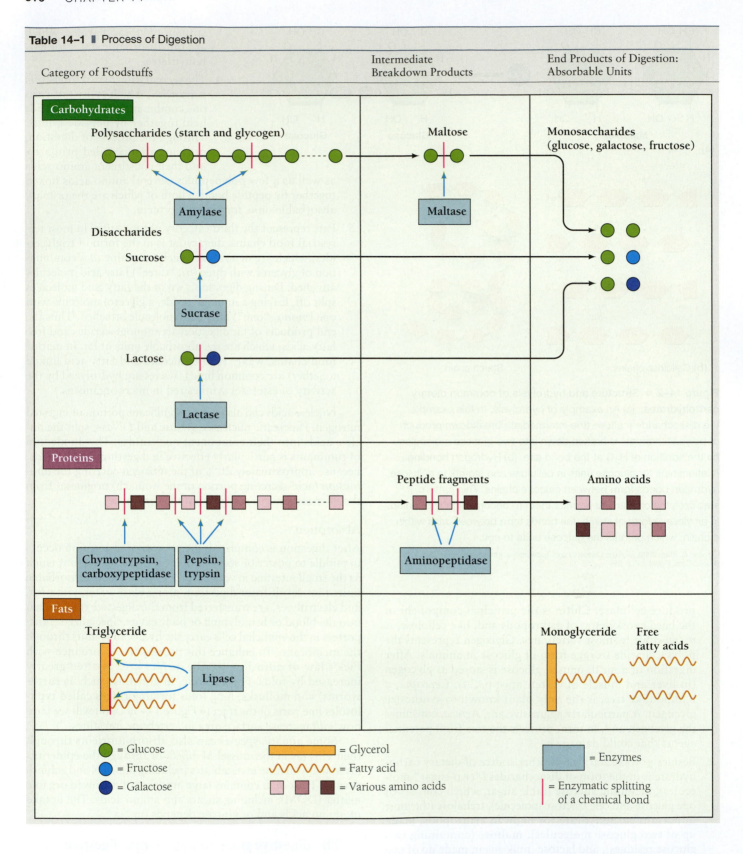

tro, "stomach"). In addition to its role in the digestive process, additional functions of the tract include osmoregulation, endocrine secretions, immune function, and the elimination of toxins.

In annelids and arthropods the foregut is referred to as the *stomodaeum,* midgut as the *mesenteron;* and hindgut as the *proctodaeum* (Figure 14–3b). Insect stomodaea and proctodaea are lined internally with a thin layer of cuticle, the **intima,**

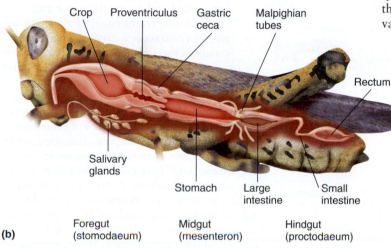

(a)

Duct of
digestive
gland

Gastric shield

Sorting
area

Cecum

Major
typhlosole

Buccal
cavity

Salivary
glands

Esophageal
glands

Esophagus

(b)

Crop   Proventriculus   Gastric
ceca

Malpighian
tubes

Rectum

Salivary
glands

Stomach   Large
intestine

Small
intestine

Foregut
(stomodaeum)

Midgut
(mesenteron)

Hindgut
(proctodaeum)

Figure 14–3 • Comparison of mollusk and insect digestive tracts. (a) A bivalve mollusk tract. Many bivalve and some gastropod mollusks have a crystalline style, which is rotated by cilia and the head of which grinds food against a hard shield in the stomach. The style also contains enzymes that are released by grinding. (b) Tract of a typical insect.

(*Sources:* (a) V. Pearse, J. Pearse, M. Buchsbaum, & R. Buchsbaum, 1987, *Living Invertebrates,* Palo Alto, CA: Blackwell Scientific, p. 352, top left. (b) Modified from R. L. Patton, 1963, *Introductory Insect Physiology,* Philadelphia: Saunders, 1963; Figure 3–1, p. 31)

which is shed along with the outer skeleton during molt. The stomodaeum, which is specialized for the reception of food, consists of the *pharynx, esophagus, crop,* and *proventriculus.* The *stomodael valve* regulates the passage of food between the foregut and the midgut. The midgut is an elongated tube and generally consists of two or more sections. The midgut of most insects, for example, contains **diverticula** (blind tubules), which arise from the main passage and contain the *gastric caecae* (saclike structures, open at only one end) near the anterior end. The midgut is specialized for storing and digesting food items, and subsequent midgut absorption of nutrients. In some cases, the same cell can carry out secretion and absorption. The hindgut extends from the *pyloric valve,* which separates the midgut from the hindgut, to the anus. The hindgut

consists of two sections, the anterior *intestine* and the posterior *rectum.* Recall that the Malpighian tubules, which function as excretory organs (p. 567), empty their contents into the anterior end of the hindgut.

The digestive tract of most vertebrates includes the following organs (● Figure 14–4): *mouth; pharynx* (throat); *esophagus; stomach* (and *rumen* in ruminants), or *proventriculus-gizzard complex* in birds; *small intestine; large intestine;* and *anus.* Note that these organs are continuous with each other and are discussed as separate entities only because of their regional modifications, which allow them to specialize in particular digestive activities.

In addition to the specialized gut tubes, many phyla have accessory organs that aid digestion. The accessory digestive organs of most vertebrates include the *salivary glands,* the *exocrine pancreas,* and the *biliary system,* which consists of the *liver* and, in some species, the *gallbladder* (Figure 14–4). (Most metazoans have salivary glands; see Figure 14–3). These exocrine organs are located outside of the wall of the digestive tract and empty their secretions through ducts into the digestive tract lumen. They develop from outpouchings of the embryonic digestive tube and maintain their connection with the digestive tract through the ducts that are formed.

■ Regulation of digestive function is complex and synergistic.

Digestive motility and secretion are carefully regulated to maximize digestion and absorption of the ingested food.

Autonomous Smooth-Muscle Function

Like self-excitable cardiac muscle cells, some smooth-muscle cells are "pacesetter" cells that do not have a constant resting potential but rather display rhythmic, spontaneous variations in membrane potential. The prominent type of self-induced electrical activity in digestive smooth muscle is **slow-wave** potentials (see p. 353), alternatively referred to as the digestive tract's **basic electrical rhythm (BER)** or **pacesetter potential.** In vertebrates, musclelike but noncontractile cells known as the **interstitial cells of Cajal** are the pacesetter cells responsible for instigating cyclic slow-wave activity. These pacesetters are located at the boundaries of the muscle layers. The slow-wave potentials initiated by these cells spread to the adjacent contractile smooth-muscle cells. Slow waves are not action potentials and do not directly induce muscle contraction; they are rhythmic, wavelike fluctuations in membrane potential that cyclically bring the membrane closer to or farther from threshold. These slow-wave oscillations are believed to be due to cyclical variations in Ca^{++} release from the endoplasmic reticulum and Ca^{++} uptake by the mitochondria of the pacesetter cell. When these waves reach threshold at the peaks of depolarization, a volley of action potentials is triggered at each peak, resulting in repeating, rhythmic cycles of muscle contraction.

Whether threshold is reached depends on the effect of various mechanical, nervous, and hormonal factors that influence

Figure 14-4 ● **Vertebrate digestive tracts showing progressive anatomic specialization for digestion (stomach and accessory digestive glands such as pancreas and liver) and absorption surface area of small intestine.**

(*Source:* D. C. Withers, 1992, *Comparative Animal Physiology,* Fort Worth, TX: Saunders, Figure 18–28, p. 925)

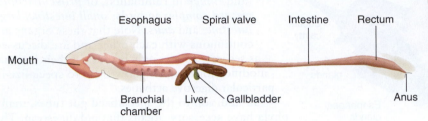

(a) Cyclostome (*Petromyzon*)

Mouth • Esophagus • Spiral valve • Intestine • Rectum • Branchial chamber • Liver • Gallbladder • Anus

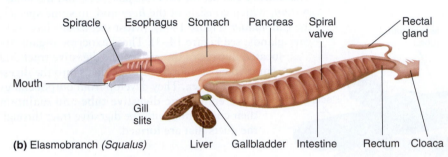

(b) Elasmobranch (*Squalus*)

Spiracle • Esophagus • Stomach • Pancreas • Spiral valve • Rectal gland • Mouth • Gill slits • Liver • Gallbladder • Intestine • Rectum • Cloaca

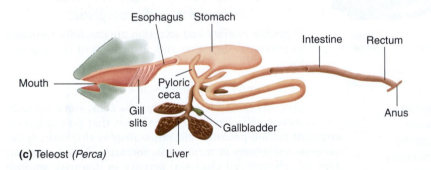

(c) Teleost (*Perca*)

Esophagus • Stomach • Intestine • Rectum • Mouth • Pyloric ceca • Gill slits • Gallbladder • Anus • Liver

the "resting" potential, or the starting point around which the slow-wave rhythm oscillates. If the starting point is nearer the threshold level, as when food is present in the digestive tract, the depolarizing slow-wave peak reaches threshold, increasing action potential frequency and its accompanying contractile activity. Conversely, if the starting point is farther from threshold, as when an animal is hungry or starved, there is less likelihood of reaching threshold, so action potential frequency is lowered and contractile activity is reduced.

Like cardiac muscle, sheets of smooth muscle cells are connected by gap junctions (p. 64), through which charge-carrying ions can flow. As a result, electrical activity initiated in a digestive-tract pacesetter cell can spread to adjacent smooth-muscle cells. If threshold is reached and action potentials are triggered, the whole muscle sheet behaves like a functional syncytium, becoming excited and contracting as a unit.

The *rate* (frequency) of rhythmic digestive contractile activities, such as peristalsis in the stomach or rumen, *segmentation* in the small intestine (p. 647), and *haustral* contractions (p. 658) in the large intestine, depends on the inherent rate established by the involved pacesetter cells. Specific details about these rhythmic contractions are discussed later, when we examine the organs involved. The *intensity* of these contractions depends on the number of action potentials that occur when the slow-wave potential reaches threshold, which in turn depends on how long threshold is sustained. At threshold, voltage-gated Ca^{++} channels are activated, resulting in Ca^{++} influx into the smooth muscle cell. The greater the number of action potentials generated, the higher the cytosolic Ca^{++} concentration, the

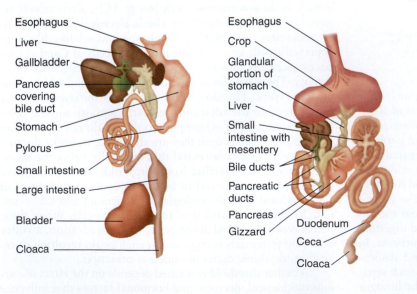

(d) Frog (*Rana*)

Esophagus • Liver • Gallbladder • Pancreas covering bile duct • Stomach • Pylorus • Small intestine • Large intestine • Bladder • Cloaca

(e) Bird (*Columba*)

Esophagus • Crop • Glandular portion of stomach • Liver • Small intestine with mesentery • Bile ducts • Pancreatic ducts • Pancreas • Gizzard • Duodenum • Ceca • Cloaca

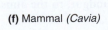

(f) Mammal (*Cavia*)

Esophagus • Liver • Gallbladder • Bile duct • Stomach • Pancreas • Duodenum • Pancreatic duct • Transverse colon • Small intestine • Ascending colon • Cecum • Rectum • Descending colon

greater the cross-bridge activity, and the stronger the resulting contraction.

Furthermore, the intensity of the contraction also depends on the number of muscle fibers in the organ. Species vary tremendously in muscle mass within the walls of digestive tract organs. For example, the thick and thin muscle pairs lining the gizzard (crop) of a ruffed grouse can crack the shell of an acorn, whereas these muscle pairs are absent in most carnivorous birds, whose stomach is more similar to that of mammalian carnivores.

Intrinsic Nerve Plexuses

In vertebrates, the **intrinsic nerve plexuses** are the two major networks of nerve fibers—the **myenteric plexus,** and the **submucous plexus**—that are located entirely within the digestive tract wall and run its entire length (● Figure 14–5). A nerve plexus is a type of nerve net (p. 142). Thus, unlike any other organ system, the digestive tract has its own intramural ("within wall") nervous system, which contains as many neurons as the spinal cord and endows the tract with a considerable degree of self-regulation (a prime example of hierarchic distributed

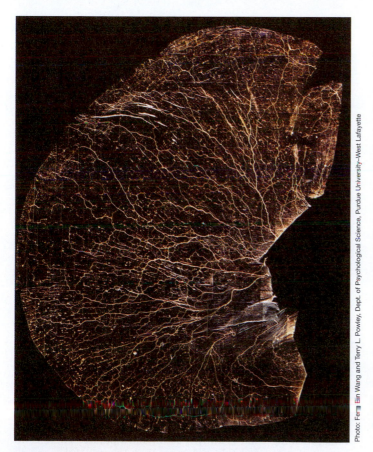

Photo: Fei Ei Wang and Terry L. Powley, Dept. of Psychological Science, Purdue University–West Lafayette

Figure 14–5 ● The enteric nervous system of a rat's stomach. By injecting a tracer derived from a horseradish enzyme into the vagus nerve, which connects the brain to the esophagus and stomach, researchers were able to reveal the extent of the nerve network. As nerve fibers fray out into the tiny endings across the stomach, information concerning food volume, hunger, discomfort, and satiety are sent back the brain.

regulation, p. 17). In vertebrates, the two plexuses together are often termed the **enteric nervous system.** Insects have an analogous system, the **stomatogastric nervous system,** a network of peripheral ganglia along the gut.

The intrinsic plexuses influence all facets of digestive tract activity. Various types of neurons are present in the intrinsic plexuses. Some are sensory neurons, which have receptors that respond to specific local stimuli in the digestive tract. Other local neurons innervate the smooth muscle cells and exocrine and endocrine cells of the digestive tract to directly affect digestive tract motility, secretion of digestive juices, and secretion of gastrointestinal hormones. As with the central nervous system, these input and output neurons of the enteric nervous system are linked by interneurons. Some of the output neurons are excitatory, and some are inhibitory. For example, vertebrate enteric neurons that release *acetylcholine* as a neurotransmitter promote contraction of digestive-tract smooth muscle, whereas the neuromodulators *nitric oxide* and *vasoactive intestinal peptide* act in concert to cause its relaxation. These intrinsic nerve networks are primarily responsible for coordinating local activity within the digestive tract. To illustrate, if a large chunk of food gets stuck in the esophagus local contractile responses coordinated by the intrinsic plexuses are initiated to push the food forward. The extrinsic nerves can in turn influence intrinsic nerve activity.

Extrinsic Nerves

The **extrinsic nerves** in vertebrates are the nerve fibers from both branches of the autonomic nervous system that originate outside the digestive tract and innervate the various digestive organs. The autonomic nerves influence digestive tract motility and secretion either by modifying ongoing activity in the intrinsic plexuses, altering the level of gastrointestinal hormone secretion, or, in some instances, acting directly on the smooth muscle and glands.

Recall that in general, the sympathetic and parasympathetic nerves supplying any given tissue exert opposing actions on that tissue (p. 149). The "fight-or-flight" sympathetic system inhibits or slows down digestive tract blood flow and motility. The "rest-and-digest" parasympathetic nervous system, by contrast, dominates in quiet, relaxed situations, when general maintenance types of activities such as digestion can proceed optimally. Accordingly, the parasympathetic nerve fibers supplying the digestive tract, which arrive primarily by way of the vagus nerve, tend to increase smooth muscle motility and promote secretion of digestive enzymes and hormones. Unique to the parasympathetic nerve supply to the digestive tract, the postganglionic parasympathetic nerve fibers are actually a part of the intrinsic nerve plexuses. They are the acetylcholine-secreting output neurons within the plexuses. Accordingly, acetylcholine is released in response to local reflexes coordinated entirely by the intrinsic plexuses as well as to vagal stimulation, which acts through the intrinsic plexuses.

In addition to being called into play during generalized sympathetic or parasympathetic discharge, the autonomic nerves supplying the digestive system can be discretely activated to modify only digestive activity. One of the major purposes of specific activation of extrinsic innervation is the coordination of activity between different regions of the digestive system; for example, the act of chewing reflexly increases not only salivary secretion but also stomach, pancreatic, and liver secretion via vagal reflexes in anticipation of the arrival of food.

Another purpose of specific activation is the provision of a pathway by which factors outside of the digestive system can influence digestion (in anticipation fashion), as, for example, with the vagally mediated increase in digestive juices that occurs in anticipation of a meal when an animal sees or smells food.

■ Receptor activation alters digestive activity through neural reflexes and hormonal pathways.

The wall of the digestive tract contains three different types of sensory receptors that respond to local chemical or mechanical changes in the digestive tract: (1) *chemoreceptors* sensitive to chemical components within the lumen; (2) *mechanoreceptors* (pressure receptors) sensitive to stretch or tension within the wall; and (3) *osmoreceptors* sensitive to the osmolarity of the luminal contents. Stimulation of these receptors elicits neural reflexes or secretion of hormones, both of which alter the level of activity in the digestive system's effector cells. These effector cells include smooth muscle cells (for modifying motility), exocrine gland cells (for controlling secretion of digestive juices), and endocrine gland cells (for varying secretion of gastrointestinal hormones).

From this overview, it is clear that regulation of gastrointestinal function is very complex, being influenced by many synergistic, interrelated pathways designed to ensure that the appropriate responses occur to digest and absorb the ingested food. Nowhere else is such a level of overlapping control exercised.

We are now going to take a "tour" of the digestive tract, beginning with the mouth and ending with the rectum, focusing primarily on vertebrates and insects.

Mouth

Obtaining and Receiving Food

Because of the enormous variety of animal diets, there is great diversity in the mechanisms for obtaining food. Animals may use shredding and cutting devices, such as bony teeth in many vertebrates, or specialized appendages that can tear food items into smaller pieces prior to ingestion, as in many arthropods. For example, many insects (such as locusts) have *mandibles* adapted for cutting and grinding. Other appendages associated with reducing food size include sucking tubes, chisels, and chitinous teeth. External parasites have modified mouthparts, which are adapted to feeding on the blood or other body fluids of a suitable host. The mosquito, for example, pierces the skin of its prey with an array of six needlelike mouthparts. One of these injects an anticoagulant, which leads to the uncomfortable itch that follows the bite. It is through this route that foreign organisms such as cause malaria, encephalitis, and yellow fever gain entry to the body. Only the egg-laying females (who have a greater protein requirement) are the nuisances, because the males dine exclusively on nectar.

The **mouth** (or **oral cavity**) itself, which initially receives the food entering the gut, may also be specialized. The mouth of a snake, for example, requires unique adaptations to swallow a meal up to 1.5 times the snake's own body weight. A snake's jaw can open to an angle of up to 130 degrees. In comparison, a human's jaw can be opened only 30 degrees. Snakes have a hinged jaw as opposed to the single bone found in mam-

A male Atlantic puffin demonstrates its hunting prowess to female puffins. The specialized beak can hold a dozen or more fishes inside. It has backward-pointing spikes in its interior, which hold fish in place while other fish are being hunted.

Photo: Hillar Klandorf

mals. Their teeth curve backward to serve as an anchor. The snake's lower jaw is also split into two parts, with the halves connected by an elastic ligament, which permits the snake to completely stretch its head around its victim's body. Powerful muscles in the cheek and throat steadily push its prey head first toward the esophagus.

Although birds lack teeth, their beaks may be specialized for grasping food items. The beak represents one of the more evolutionarily plastic components of the avian digestive system, having been molded to conform to particular feeding habits. In herbivorous birds, the beak is an enlarged, crushing forceps, whereas in carnivorous birds the beak is a sharp-edged tearing instrument. The beaks (bills) of some birds, including ducks and geese, are richly innervated with sensory endings. The density of mechanoreceptors along the edges and at the tip of the bill is actually greater than in the fingertips of a human's index finger.

In mammals, the mouth is ringed by the muscular **lips**, which aid in **prehension**, the seizing and conveying of food to the mouth. For example, the horse uses its lips to move foodstuffs into the mouth. The lips also serve nondigestive functions; they are important in communication; for example, facial expression and the articulation of many sounds depend on a particular lip formation.

The **palate**, which forms the arched roof of the oral cavity, separates the mouth from the nasal passages. Its presence allows breathing and chewing or sucking to take place simultaneously. Toward the front of the mouth, the palate is made of bone, forming what is known as the **hard palate**. There is no bone in the portion of the palate toward the rear of the mouth; this region is called the **soft palate**. However, in most avian species there is no soft palate, but the hard palate is separated by a median slit that permits direct communication with the nasal cavities. In birds, the floor of the mouth is membranous and, in some species such as pelicans, serves a storage capacity. In mammals, hanging down from the soft palate in the rear of the throat is a dangling projection, the **uvula**, which plays an important role in sealing off the nasal passages during swallowing.

The **tongue,** which forms the floor of the oral cavity, consists of voluntarily controlled skeletal muscle. In ruminant mammals, the tongue is the organ of prehension, because the lips have limited motility and the upper incisor teeth are absent. A giraffe can wrap its 46-cm (18-inch) tongue around a leafy branch and tear the leaves off into its mouth. Movements of the tongue are important in guiding food within the mouth during chewing and swallowing, and also play an important role in vocalization. Embedded within the tongue are the **taste** buds (p. 236), which are also dispersed in the soft palate, throat, and linings of the cheeks. The tongue also continually synthesizes an antibiotic peptide that kills microbes. For this reason, tongues are rarely infected and heal quickly after injury. (See Chapter 10 for a discussion of antimicrobial peptides.)

The **pharynx** (throat) is the cavity at the rear of the oral cavity. In vertebrates, it acts as a common passageway for both the digestive system (by serving as the link for passing food between the mouth and esophagus) and the respiratory system (as a path for water moving across gills, or by providing access between the nasal passages and trachea for air). This arrangement necessitates mechanisms (described shortly) to guide food and air into the proper passageways beyond the pharynx. Some species of insects have *pharyngeal* and *cibarial* pumps that suction up their food and help it move into the esophagus.

Mastication

The first step in the digestive process is **mastication,** the motility of the mouth that involves slicing, tearing, grinding, and mixing ingested foods by specialized mouthparts such as **teeth.** For example, teeth in fishes, amphibians, and reptiles are specialized for holding and tearing food. Mammalian **incisors** are used to grab and hold food. The incisors of rodents grow throughout life and must be continually ground away by gnawing. Carnivorous mammals are noted for their **canines,** teeth designed for seizing, piercing, and tearing prey items. Ruminants have a unique dentition: The upper incisor and canine teeth are absent. The lower incisor teeth of ruminants bite against a hard gum or *dental pad* in the upper jaw. They manipulate forage into the mouth with tongue and lips and then cut or tear it with their lower incisors and dental pads. Herbivores have a long **diastema,** or gap, which extends between the front teeth and the cheek teeth (premolars and molars) on each side of the jaw. The **molar** teeth are used to grind food into smaller particles, although the extent to which they are used varies between species and is related to the composition of the diet. Simple up-and-down movements of the molars in carnivores and omnivores are inadequate to grind tough, coarse feeds, so herbivores can make lateral grinding movements of the jaw. The upper jaw is slightly wider than the lower jaw, and the animal masticates on one side of the mouth at a time. Because of the lateral jaw movement, the cheek teeth may develop sharp edges that interfere with chewing or may injure the tongue or cheeks. This problem in domestic herbivores can be corrected by "floating the teeth," that is, filing the projections down.

Not all vertebrates rely on teeth. Baleen whales (Mysticeti), for example, have long plates of **baleen** (parallel, fused filaments of keratin, the same protein that hair fibers consist of) that hang from the jaws in place of teeth. These are used as straining devices for filter-feeding, primarily on small marine animals such as krill and fishes.

The purposes of mastication (chewing) are not only to break food into smaller pieces, but also to mix food with saliva and to stimulate the taste buds, which reflexly increases salivary, gastric, pancreatic, and bile secretion to prepare for the arrival of food in the lower digestive tract. However, carnivores normally swallow food without chewing and mixing the contents with saliva. Ruminants chew the feed only briefly before swallowing, sufficiently grinding the long forage into particles just small enough to be formed into a food bolus. Only when ruminants regurgitate (movement of a food bolus from the stomach back to the mouth) is the food then thoroughly masticated. In contrast, the horse thoroughly masticates its food prior to swallowing

The act of chewing in vertebrates is normally regulated by a chewing center in the medulla oblongata. The presence of food in the oral cavity activates mechanoreceptors that, via afferent pathways to the medulla oblongata, activate motor neurons that open the oral cavity while inhibiting the motor neurons that cause the closure of the cavity.

■ Saliva aids in mastication but plays a more important role in lubricating food boluses before swallowing.

Saliva, the secretion associated with the mouth, is produced by salivary glands, which are located outside the oral cavity and discharge saliva through short ducts into the mouth. In insects, ducts from paired salivary glands extend forward and unite into a common duct that opens near the base of the labium or hypopharynx mouthparts (Figure 14–3b).

Salivary glands in vertebrates combine two cell types: One type produces a *serous* product (water) that is high in enzyme content, whereas the other produces *mucus,* which is thick and slippery. Chickens do not have serous cells, so their total daily output of saliva is comparatively small. A chicken's saliva lubricates the food in preparation for swallowing; wetting occurs later, in the crop. The salivary glands of many swifts and swallows can actually increase in size during nest-building season. For example, the salivary glands of the edible-nest swiftlet, *Collocalia fuciphaga,* enlarge approximately 50-fold because their nests consist entirely of cemented saliva, the primary ingredient of birds' nest soup.

In mammals saliva consists of about 99.5% H_2O and 0.5% electrolytes and protein. Depending on the species of animal, the osmolality of saliva compared to blood plasma can range from hypotonic to hypertonic. However, the composition of saliva is always isotonic in ruminants and generally hypotonic in dogs and cats. The functions of saliva include the following:

1. Saliva facilitates swallowing by moistening food particles, thereby holding them together, and (in vertebrates but not insects) by providing lubrication through the presence of mucus.

2. In most animals (excluding ruminants), saliva begins digestion of starch in the mouth through action of **salivary amylase,** an enzyme that breaks down starches into maltose, a disaccharide consisting of two glucose molecules. In insects both the salivary gland and midgut cells can secrete amylases. In some species of bees, the salivary gland

secretes **invertase** (an enzyme that breaks down sucrose), which is taken into the body with the ingested nectar. Some mammalian saliva also contains the lipid-splitting enzyme *lingual lipase*.

3. Saliva exerts some antibacterial action by means of a twofold effect—first by **lysozyme**, an enzyme that lyses, or destroys, certain bacteria by breaking down their cell walls, and second by rinsing away material that may serve as a food source for bacteria.

4. Saliva serves as a solvent for molecules that stimulate the taste buds. Only molecules in solution can react with taste bud receptors. You can demonstrate this for yourself: Dry your tongue and then drop some sugar on it; you cannot taste the sugar until it is moistened.

5. Saliva keeps the oral cavity moist (and aids in vocalization by facilitating movements of the lips and tongue).

6. Saliva is often rich in bicarbonate buffers, which neutralize acids in food as well as acids produced by bacteria in the mouth. In ruminants, salivary bicarbonate is responsible for the high pH of rumen fluid and is an important buffering system for maintaining the acid–base equilibrium of the rumen contents.

7. Saliva can also be used as a means of temperature control in animals without sweat glands. In fact, most mammals and birds have no sweat glands and rely on panting as a means of evaporative cooling (Chapter 15). Kangaroos actually spread saliva over their bodies when outside temperatures approach lethal limits.

8. Various invertebrates can use saliva for a variety of specialized functions, including the production of poisons for immobilizing or killing prey, the secretion of anticoagulants (in blood-sucking insects and leeches), and silk production. In Lepidoptera, Hymenoptera, and Trichoptera, the glands secrete silk proteins used in constructing cocoons and shelters, and net-spinning caddisflies use silk for food gathering.

In nonruminant mammals, saliva is ultimately not essential for digesting and absorbing foods, because enzymes produced by the pancreas and small intestine can digest food even in the absence of salivary and gastric secretion.

The salivary glands of ruminants produce copious amounts of alkaline saliva. When swallowed, this saliva enables selective compartments of its stomach to house a population of microbes capable of digesting dietary cellulose. Salivary bicarbonates and phosphates buffer the acids produced during fermentation and maintain the pH of the ruminoreticulum (compartment of a ruminant's stomach, described later) within a narrow range. Urea is also recycled via the salivary glands back to the rumen microbes for protein synthesis.

■ **The continuous low level of salivary secretion can be increased by simple and conditioned reflexes.**

Salivary secretion in vertebrates is the only digestive secretion entirely under neural control. One exception is the ruminants, where the parotid gland secretes saliva spontaneously in the

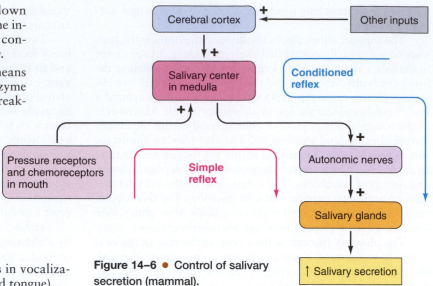

Figure 14–6 ● Control of salivary secretion (mammal).

absence of any neural stimuli. Both nervous system reflexes and hormones regulate all other vertebrate digestive secretions.

Saliva in mammals is continuously secreted, even in the absence of apparent stimuli, because of constant low-level stimulation by the parasympathetic nerve endings that terminate in the salivary glands. This basal secretion is important in keeping the mouth and throat moist at all times. The basal rate of salivary secretion in humans is about 0.5 mL/min, which can increase to 5 mL/min in response to a potent stimulus such as sucking on a lemon. Remarkably, salivary secretion in ruminants is of such a magnitude that it can be regarded as a second circulatory system for body fluids and electrolytes. Cattle produce approximately 140 liters of saliva per day, whereas sheep and ponies secrete around 10 liters per day. Almost 50% of the total extracellular sodium in the animal may be found in the rumen contents at any one time, whereas half of the total body water passes through the salivary glands into the rumen every day.

In addition to a continuous low-level secretion, salivary secretion may be enhanced by two different types of salivary reflexes (● Figure 14–6): (1) the simple, or unconditioned, salivary reflex and (2) the acquired, or conditioned, salivary reflex. The **simple,** or **unconditioned, salivary reflex** occurs when chemoreceptors and pressure receptors within the oral cavity respond to the presence of food. On activation, these receptors initiate impulses in afferent nerve fibers that carry the information to the salivary center located in the medulla of the brain stem. The salivary center, in turn, sends impulses via the extrinsic autonomic nerves to the salivary glands to promote increased salivation. Dental procedures promote salivary secretion in the absence of food in the mouth, because these manipulations activate pressure receptors in the mouth.

With the **acquired,** or **conditioned, salivary reflex,** salivation occurs without oral stimulation. A zoo mammal hearing the sound of a food cart, or a dog watching the preparation of a meal, initiates salivation through this reflex. All of us have experienced such "mouth watering" in anticipation of something delicious to eat. This reflex is a learned response based on previous experience and is an example of feedforward or anticipation regulation. Inputs that arise outside the mouth and are mentally associated with eating, act through the cerebral cortex to stimulate the medullary salivary center.

The salivary center controls the degree of salivary output by means of the autonomic nerves that supply the salivary glands. Unlike the autonomic nervous system elsewhere in the vertebrate body, sympathetic and parasympathetic responses in the salivary glands are not antagonistic. Both sympathetic and parasympathetic stimulation increase salivary secretion, but the quantity and consistency of saliva change with the needs of the animal. The rate of saliva secretion depends on the quantity of food in the mouth, whereas its consistency depends on the composition of the food consumed. Both the amount and quality of saliva are commensurate with the need for sensory evaluation of the food as well as the need for lubrication of dry foodstuffs. Parasympathetic stimulation, which exerts the dominant role in salivary secretion, increases blood flow through the glands and is accompanied by a prompt and abundant flow of watery saliva that is rich in enzymes. Normally, sympathetic stimulation reduces blood flow but allows a continued saliva output of smaller volume but rich in mucus content. Because sympathetic stimulation elicits a smaller volume of saliva, the mouth is drier than usual during circumstances when the sympathetic system is dominant, such as stress situations. Thus people experience a dry feeling in the mouth when they are nervous about giving a speech. Cats, in contrast, secrete voluminous watery saliva in response to sympathetic stimulation.

■ Digestion in the mouth is minimal, and no absorption of nutrients occurs.

There is no absorption of foodstuffs from the mouth. In some invertebrates and vertebrates, limited digestion occurs in the mouth because of the hydrolysis of polysaccharides into disaccharides by amylase. However, most digestion is accomplished farther down the tract after the food mass and saliva have been swallowed. Stomach acid inactivates amylase, but in the center of the food mass, where acid has not yet reached, salivary amylase continues to function for several hours. Lingual lipase is, however, activated by the low pH in the stomach.

Cats have a strong aversion to diets containing medium-chain triglycerides or hydrogenated coconut oil. Presumably lingual lipase breaks down some triglycerides into fatty acids in their mouths, which then activates specific taste receptors and produces the finicky response.

Pharynx, Esophagus, and Crop

The motility associated with the pharynx and esophagus is **swallowing**, or **deglutition**. Most of us think of swallowing as the act of moving food out of the mouth and into the esophagus. However, swallowing actually is the entire process of moving food from the mouth through the esophagus into the stomach.

■ Swallowing in vertebrates is a sequentially programmed all-or-none reflex.

Swallowing in a vertebrate is initiated when a **bolus**, or ball of food, is forced by the tongue to the rear of the mouth into the pharynx. The pressure of the bolus in the pharynx stimulates pharyngeal pressure receptors, which send afferent impulses to the **swallowing center** located in the medulla. The swallowing center then reflexly activates, in the appropriate sequence, the muscles that are involved in swallowing. Swallowing is an example of a sequentially programmed all-or-none reflex in which multiple responses are triggered in a specific timed sequence; that is, a number of highly coordinated activities are initiated in a regular pattern over a period of time to accomplish the act of swallowing. Once initiated, swallowing cannot be stopped.

■ During the oropharyngeal stage of swallowing, food is directed into the esophagus and prevented from entering the wrong passageways.

Swallowing is arbitrarily divided into two stages: the *oropharyngeal stage* and the *esophageal stage*. The oropharyngeal stage consists of moving the bolus from the mouth through the pharynx and into the esophagus. When the bolus enters the pharynx during swallowing, it must be directed into the esophagus and prevented from entering the other openings that communicate with the pharynx. In other words, food must be prevented from re-entering the mouth, from entering the nasal passages, and from entering the trachea (where it causes choking). All this is accomplished by the following coordinated activities (● Figure 14–7):

- The position of the tongue against the hard palate prevents food from re-entering the mouth during swallowing.

- The uvula is elevated and lodges against the back of the throat, sealing off the nasal passage from the pharynx so that food does not enter the nose.

- Food is prevented from entering the trachea primarily by elevation of the larynx and tight closure of the vocal folds across the laryngeal opening, or **glottis**. The first portion of the mammalian trachea is the *larynx*, or *voice box*, across which are stretched the *vocal folds*. During swallowing, the vocal folds serve a purpose unrelated to vocalization. Contraction of laryngeal muscles aligns the vocal cords in tight apposition to each other, sealing the glottis entrance. Also, the bolus tilts a small flap of cartilaginous tissue, the **epiglottis**, backward down over the closed glottis as further protection from food entering the respiratory airways. In some animals, such as horses, the nasal and tracheal tubes line up so precisely that they can breathe and drink at the same time, with the water passing to the left and right of the respiratory connection. Noninfant humans cannot do this; they evolved a lower (more posterior) position of the larynx, which made room for more face and neck musculature needed for precise vocalization. Unfortunately, this lower position (which is not yet developed in babies) makes it more likely for food to enter the trachea, possibly causing obstruction and sometimes death.

- Because the respiratory passages must be temporarily sealed off during swallowing, respiration in humans is briefly inhibited so that futile respiratory efforts are not attempted.

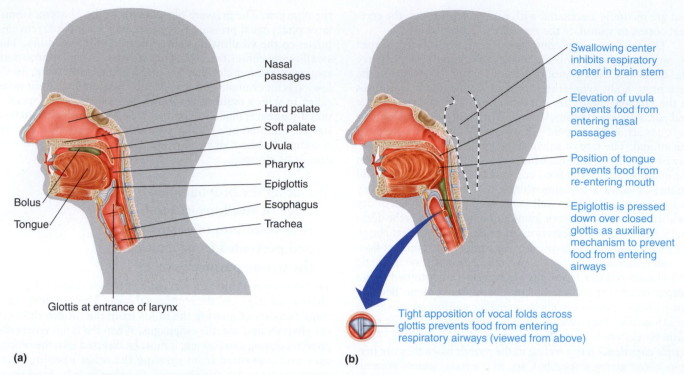

(a)

Nasal passages
Hard palate
Soft palate
Uvula
Pharynx
Epiglottis
Esophagus
Trachea

Bolus
Tongue

Glottis at entrance of larynx

(b)

Swallowing center inhibits respiratory center in brain stem

Elevation of uvula prevents food from entering nasal passages

Position of tongue prevents food from re-entering mouth

Epiglottis is pressed down over closed glottis as auxiliary mechanism to prevent food from entering airways

Tight apposition of vocal folds across glottis prevents food from entering respiratory airways (viewed from above)

Figure 14–7 ● **Oropharyngeal stage of swallowing in humans.** (a) Position of the oropharyngeal structures at rest. (b) Changes that occur during the oropharyngeal stage of swallowing to prevent the bolus of food from entering the wrong passageways.

■ With the larynx and trachea sealed off, pharyngeal muscles contract to force the bolus into the esophagus.

That the air and food passageways cross over in air-breathing vertebrates, necessitating these elaborate mechanisms to prevent choking, seems to be an evolutionary accident. Air breathing first evolved in some fishes as ventral lungs that develop in embryos out of the digestive tract. A dorsal tube from the same tract, with nostrils, evolved separately, creating the crossover.

Esophagus

■ The esophagus is a muscular tube guarded by sphincters at both ends.

The **esophagus** is a fairly straight, muscular tube that extends between the pharynx and stomach (Figures 14–1, 14–3, and 14–4). In ruminants and dogs the entire musculature of the esophagus is striated, whereas in birds and humans it is entirely smooth muscle. In the horse and cat, the bottom third of the esophagus is smooth muscle. Lying for the most part in the thoracic cavity, the esophagus penetrates the diaphragm and joins the stomach in the abdominal cavity a few centimeters posterior to the diaphragm.

The esophagus is guarded at both ends by sphincter valves. The upper esophageal sphincter is the **pharyngoesophageal sphincter,** and the lower sphincter is the **gastroesophageal sphincter** (cardiac sphincter).

■ The pharyngoesophageal sphincter keeps air from entering the gut.

Because the esophagus in most air-breathing vertebrates is exposed to subatmospheric intrapleural pressure as a result of respiratory activity (see p. 480), a pressure gradient exists between the atmosphere and the esophagus. Accordingly, if the entrance to the esophagus were not closed, air would enter the esophagus as well as the trachea with each breath. Except during a swallow, the pharyngoesophageal sphincter keeps the entrance to the esophagus closed to prevent large volumes of air from entering the esophagus and stomach during breathing. Instead, the air is directed only into the respiratory airways. Were it not for the pharyngoesophageal sphincter, the digestive tract would be subjected to large volumes of gas, which would lead to excessive **eructation** (belching). In contrast to most sphincters, passive elastic tensions in the walls of the pharyngoesophageal sphincter cause the esophagus to be closed when this sphincter is relaxed. During swallowing, this sphincter reflexly opens and allows the bolus to pass into the esophagus. Once the bolus has entered the esophagus, the sphincter closes, the respiratory airways are opened and breathing resumes. The oropharyngeal stage is complete.

Ruminants have another valve. On average, rumen gases consist of approximately 65% CO_2, 27% methane, 7% nitrogen, 0.5% oxygen, and 0.2% hydrogen, in addition to a trace amount of hydrogen sulfide. Most of these gases are diverted from the rumen into the trachea by the closing of the **nasopharyngeal sphincter** (*naso,* "nose"; *pharynx,* "throat"), the junction between the nasal cavity and the throat. Most of the gases are ultimately expelled with normal respiratory activity. However, some are also taken up into the capillaries lining the alveolus. Any aromatic compounds originating in the rumen and entering the vascular system can then reach the mammary gland and flavor the taste of the milk. For example, digestion of onions, garlic, and leek in the rumen results in the release of strong volatile compounds that are then incorporated into the milk. (Gases produced by ruminants are currently of global concern: See the box, "Challenges and Controversies: Global Warming and the Rumen.")

■ Peristaltic waves push the food through the esophagus.

The **esophageal stage** of the swallow begins once the food bolus contacts the pharyngeal mucosa. The swallowing center initiates a **primary peristaltic** wave that sweeps from the beginning to the end of the esophagus, forcing the bolus ahead of it through the esophagus to the stomach. The term **peristalsis** refers to ringlike contractions of the circular smooth muscle that move progressively forward with a stripping motion, pushing the bolus ahead of the contraction (the same process shown in Figure 9–15c). Food and liquid can be propelled to the stomach even when an animal's head is lower than its body. The duration of the peristaltic wave depends on the length of the esophagus, and in humans requires about 5 to 9 seconds to reach the lower end of the esophagus. Progression of the wave is controlled by the swallowing center, innervated by the vagus nerve.

If a large or sticky swallowed bolus fails to be carried along to the stomach by the primary wave of peristalsis, the lodged bolus distends the esophagus, stimulating pressure receptors within its walls and thus initiating a second, more forceful peristaltic wave that is mediated by the intrinsic nerve

---

CHALLENGES AND CONTROVERSIES

Global Warming and the Rumen

In ruminants, eructation helps relieve the pressure of the gases generated during the fermentation of carbohydrates. For example, in a 24-hour period ruminant livestock belch approximately 500 liters of methane and 1050 liters of CO_2. Approximately 2 to 12% of the ingested energy consumed by cattle is thus lost as methane. This represents an important loss of an energy source, as well as a factor that contributes to global warming. Global warming results from gases (primarily carbon dioxide and methane) in the upper atmosphere that trap heat (infrared) radiation that is leaving Earth. One current theory suggests that global warming is an example of a positive-feedback relationship. Increased environmental temperatures lead to the melting of sea ice, which exposes more water to the rays of the sun. The water in turn absorbs more solar energy, which accelerates global warming, and so on. Researchers fear that such a feedback cycle may produce a self-sustaining and accelerating global warming cycle that is beyond human control. Another theory suggests that catastrophic climate changes are more imminent. This theory also begins with the melting of glaciers and sea ice but also involves the consequent reduction in the salinity of the seawater. The salt content of the water is partially responsible for the conveyance of heat from the tropics northward and cold water southward. Much of the climate along the eastern seaboard of the United States and in Europe is moderated by this current flow, and any collapse in this heat conveyor could, at worst, precipitate a new ice age. Minimal perturbations would include severe winters, increasingly violent storms, flooding, drought, and high winds around the globe, irreparably disrupting food production and energy supplies.

Methane affects climate directly through its interaction with long-wave infrared energy and indirectly through atmospheric oxidation reactions that produce CO_2. Atmospheric concentrations of methane had remained fairly stable at 750 ppm until approximately 100 years ago, when concentrations began to rise to their present levels of 1800 ppm. (Similarly, atmospheric CO_2 has risen from about 300 ppm to about 370 ppm.) More than 500 million metric tons of methane enter the atmosphere annually, which exceeds the capacity of the atmosphere to oxidize it. At this rate, methane will contribute 14 to 17% of the factors leading to global warming in the next 50 years.

There are two significant sources of methane production, those of fossil origin, which contributes 20 to 30%, and those that yield contemporary carbon and produce 70 to 80%. Examples of the former include gas drilling, mining, and wetland emissions that contain methane that has been stored for thousands of years, whereas enteric fermentation from ruminants and insects (termites in particular), natural wetlands, biomass burning, oceans and lakes, and waste treatment are examples of the latter. Cattle are expected to contribute about 2% to global warming over the next 50 years.

Strategies to reduce methane emissions by livestock are thus of international importance. Scientists know that the type of carbohydrate fermented determines the level of methane production, both by affecting rumen pH in addition to the ruminal population. For example, the fermentation of cell wall fiber in coarse diets results in higher methane production than do concentrate diets high in soluble carbohydrates, where methane losses may drop to as little as 3% of the ingested energy. Grinding and pelleting forages can also decrease methane production. Beef cattle are increasingly reared in feedlots and given methane-inhibiting substances. It is thus possible to assume modest reductions in methane emissions of livestock while enhancing productivity by modifying diet quality. Future technologies may develop methods to alter the microbial population in ways that would benefit the animal without negatively impacting the environment.

plexuses at the level of the distension. These **secondary peristaltic** waves do not involve the swallowing center. Distension of the esophagus also reflexly increases salivary secretion. The trapped bolus is eventually dislodged and moved forward through the combination of lubrication by the extra swallowed saliva and the forceful secondary peristaltic waves.

■ The gastroesophageal sphincter prevents reflux of gastric contents.

Except during swallowing, the gastroesophageal sphincter remains contracted, to maintain a barrier between the stomach and esophagus, reducing the possibility of reflux of acidic gastric contents into the esophagus. If gastric contents do flow back into the esophagus despite the sphincter, the acidity of these contents irritates the esophagus, causing the esophageal discomfort known in humans as **heartburn.** (The heart itself is not involved at all.) The gastroesophageal sphincter relaxes reflexly as the peristaltic wave sweeps down the esophagus so that the bolus can pass into the stomach. After the bolus has entered the stomach, the gastroesophageal sphincter again contracts.

Achalasia is a condition that occurs commonly in dogs (and in humans) in which the lower esophageal sphincter fails to relax during swallowing but instead contracts more vigorously. Food accumulates in the esophagus, which enormously distends the esophagus when the food's passage into the stomach is greatly delayed. The underlying defect is apparently the result of damage to the myenteric nerve plexus in the region of the gastroesophageal sphincter.

■ Esophageal secretion is entirely protective.

In most vertebrates, mucus is secreted throughout the length of the entire digestive tract. By providing lubrication for passage of food, esophageal mucus lessens the likelihood that the esophagus will be damaged by any sharp edges in the newly entering food. Furthermore, it protects the esophageal wall from acid and enzymes in gastric juice if gastric reflux occurs.

Crop

■ The crop is a modified section of the esophagus and functions mainly as a storage organ.

The **crop,** a saclike outpouching of the esophagus, is commonly found in a number of invertebrate species, including insects (Figure 14–3b) and oligochaete worms, whereas in vertebrates only some species of birds have this structure (Figure 14–4e). The crop mainly serves a storage function that is particularly important in fluid-feeding insects, which have relatively large crops for storing blood or nectar. In some insects, the crop is a simple enlargement of the foregut, or, as in mosquitoes and Lepidoptera, a lateral diverticulum off the digestive tract. The crop in domestic fowl is structurally identical to that of the esophagus except for a scarcity of mucus glands. Mucus glands are also relatively sparse in the region of the esophagus below the crop, because movement of the bolus largely involves prewetted material from the crop. Granivorous ("grain-eating") and fish-eating birds have comparatively large crops, for food

storage, whereas carnivorous and insectivorous birds have limited need for a crop, which therefore are considerably reduced in size or are absent.

A food bolus moving down the esophagus in crop-bearing animals has two possibilities. It can either continue on to the **proventriculus-gizzard** organ(s) (in oligochaete worms, numerous arthropods, and most birds; Figures 14–3b and 14–4e), or it can enter the crop for temporary storage. The stored food is moistened by water as the animal drinks, but essentially no digestion occurs here. If the proventriculus is actively involved in digestion, the food is directed toward the crop. While empty, the crop of birds contracts once per minute, but if food is present in the proventriculus then there is a complete inhibition of this contractile activity and the food remains within. The motility of the crop is regulated by vagal impulses. Motility and emptying are coordinated so as to release ingesta at a rate matching the emptying rate of the proventriculus and gizzard. If the food bolus is too large for the gizzard, contractions subside and return the excess portion to the crop.

The epithelial cells in the crop of some birds, including pigeons and doves, are sensitive to the hormone prolactin, the same hormone that promotes milk synthesis in mammals. In both males and females, the increase in prolactin secretion during incubation of the eggs results in a marked proliferation of this tissue. Crop epithelia synthesize large stores of lipid material. Once the eggs hatch, these cells are sloughed off and mixed with food already present in the adult crop. This nutritive secretion, known as **crop milk,** is regurgitated into the esophagus of the squabs (chicks). In another species of bird, the **hoatzin** (*Opisthocomus hoazin*), the crop and esophagus function as a modified rumen (see p. 661), because this storage vat contains bacteria capable of digesting cellulose and detoxifying harmful plant alkaloids. Leafy plant material ferments in a large *double crop,* producing volatile fatty acids as smelly fermentation by-products, which gives the hoatzin a musky odor. The anterior sternum is markedly reduced in size to accommodate the voluminous fermentation structures, thus considerably reducing the area for flight muscle attachment. Nestlings can also be fed this regurgitated material. The highly developed digestive strategy of the hoatzin may have arisen from an evolutionary tradeoff between detoxification of plant chemical defenses and enhanced use of cellulotic leaf-cell wall as a nutritional resource. Bees, too, can regurgitate the contents of their crops. As in birds, the proventriculus controls the movement of food from the crop but can also selectively remove pollen from a nectar suspension in the crop. The lips of the proventriculus are modified so that they can strain out the pollen grains for digestion while leaving the nectar in the crop for eventual regurgitation and processing into honey.

Stomach or Midgut

■ The midgut stores food and begins nonsalivary digestion.

In most vertebrates as well as invertebrates, the midgut or **stomach** provides for the initial digestion (beyond the minor salivary contribution) and storage of food items. Diverticula often supplement or replace the midgut of some invertebrates and function in enzymatic digestion and absorption. In many species of insects, the midgut is acidic in the anterior two thirds and more alkaline in the posterior third. For example, the anterior midgut of the American cockroach (*Periplaneta americana*) is acidic (pH ~ 5.7), whereas the posterior midgut is slightly less (pH ~ 6.1). (Note that insect midguts are not nearly as acidic as vertebrate stomachs, however.)

In numerous species of insects, a **peritrophic membrane** separates the midgut epithelium from the food. Structurally it is a permeable (porous) network of chitin fibrils set in a matrix of protein and mucopolysaccharides (proteins with attached chains of repeating sugar units). The role of the peritrophic membrane has not been established but is believed to include the following:

- It may protect the epithelial cells of the midgut from abrasion by the food.

- It may help move food through this region of the gut tube.

- It may inhibit the movement of some pathogens from the food into the insect's tissues. Although the pore size is normally smaller than the size of most bacteria, bacterial enzymes and toxins can traverse the membrane and damage the midgut.

- It may function as a filter and protect the midgut cells from ingested toxins. For example, large-molecular-weight tannins, which are toxic, cannot penetrate the pores of the peritrophic membrane

- It compartmentalizes the midgut lumen into an ectoperitrophic (*ecto*, "outer") and endoperitrophic (*endo*, "inner") space within which specialized digestive functions can occur. Trypsin and amylase are localized to the endoperitrophic space where they begin the initial digestion of food. Aminopeptidase and trehalase are localized to the ectoperitrophic space, where they complete the process of digestion.

The **filter chamber** represents an additional variation in midgut structure. In Homoptera—insects that feed on large quantities of plant juices—the filter chamber permits water from the ingested sap to pass directly from the anterior region of the midgut to the hindgut. In this manner the sap is concentrated before being digested in the posterior region of the midgut. The excess fluid is released from the hindgut as **honeydew,** which in turn is a food for other insects. To produce this secretion, the midgut actually loops back onto itself and becomes internalized along with the Malpighian tubules.

■ The stomach stores food and begins protein digestion.

In monogastric mammals, the stomach is a simple, muscular, saclike chamber lying between the esophagus and small intestine. In mammals it is arbitrarily divided into three sections based on anatomic, histologic, and functional distinctions (● Figure 14–8). The dorsal region, or the **fundus,** is responsible for the storage of ingested food and adaptation to changes in volume. The middle, or main, part of the stomach is the body or **corpus.** Here the digesta is mixed with gastric secretions. The smooth muscle layers in the fundus and corpus are relatively thin, but the lower portion of the stomach, the **antrum,** has a much heavier musculature that is used to regu-

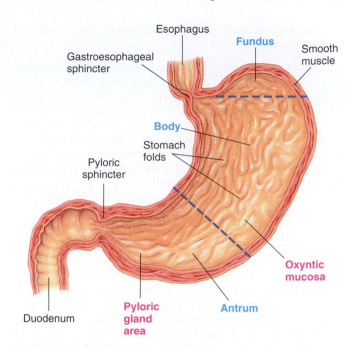

Figure 14–8 ● Anatomy of the stomach (mammalian). The stomach is divided into three sections based on structural and functional distinctions—the fundus, body, and antrum. The mucosal lining of the stomach is divided into the oxyntic mucosa and the pyloric gland area based on differences in glandular secretion.

late expulsion of food into the small intestine as well as mix the contents. Variations between different animal groups with simple stomachs are based on the distribution and composition of the epithelium lining of the stomach. There are also glandular differences in the mucosa of these regions, as described later. Alternatively, some species may have a more complex, multichambered stomach. The stomach of rats, for example, is divided into separate compartments. Voluminous chambers that are compartmentalized and perform separate functions characterize the digastric stomach of ruminants. Some species of fish, such as the carp, completely lack a stomach, and food enters the small intestine directly. The terminal portion of the stomach consists of the **pyloric sphincter,** which acts as a barrier between the stomach and the upper part of the small intestine, the duodenum.

The stomach performs three main functions:

1. The most important function is to store ingested food until it can be emptied into the small intestine at a rate appropriate for optimal digestion and absorption. In only a matter of minutes carnivores can consume a meal that then takes hours to digest and absorb. Because the small intestine is the primary site for this digestion and absorption, it is important that the stomach store the food and meter it into the duodenum at a rate that does not exceed the small intestine's capacities.

2. The stomach secretes hydrochloric acid (HCl) and enzymes that begin chemical digestion.

3. Through mixing movements, the ingested food is pulverized and mixed with gastric secretions to produce a thick, liquid mixture known as **chyme.** The stomach contents

must be converted to chyme before they can be emptied into the duodenum.

We now discuss how the stomach accomplishes these functions as we examine the four basic digestive processes—motility, secretion, digestion, and absorption—as they relate to the stomach. The details are primarily for mammals, but other animals are noted where comparison is illustrative. Starting with motility, gastric motility is complex and subject to multiple regulatory inputs. The four aspects of gastric motility are (1) gastric filling, (2) gastric storage, (3) gastric mixing, and (4) gastric emptying. We begin with gastric filling.

Gastric filling involves receptive relaxation.

Stomachs can have a remarkable ability to accommodate significant changes in volume. Consider the blood-sucking leech, which feeds only once during its life, and ingests the equivalent of nine times its body weight in about 30 minutes. Distension of the body eventually causes the animal to cease feeding. By comparison, the empty human stomach has a volume of about 50 mL, but it can expand to a capacity of about 1 liter (1000 mL) during a meal. The mammalian stomach can accommodate such a 20-fold change in volume with little tension in its walls and little rise in intragastric pressure, through the following mechanism. The interior of the stomach is thrown into deep folds. During a meal, the folds get smaller and flatten out as the stomach relaxes slightly with each mouthful. This reflex relaxation of the stomach as it is receiving food is called **receptive relaxation**; it enables the stomach to accommodate the extra volume of food with little rise in stomach pressure. Receptive relaxation is triggered by the act of eating and is mediated by the vagus nerve.

Distension of the fish stomach is divided into two phases: Initially there is a rapid relaxation in response to gut filling, followed by a slower relaxation phase that continues until the prey item is accommodated.

Gastric storage takes place in the body of the stomach.

A group of pacesetter cells located in the upper fundus region of the mammalian stomach generate slow-wave potentials that sweep down the length of the stomach toward the pyloric sphincter at a rate of a few per minute. This rhythmic pattern of spontaneous depolarizations, the basic electrical rhythm, or BER, of the stomach, occurs continuously and may or may not be accompanied by contraction of the stomach's circular smooth muscle layer. Depending on the level of excitability in the smooth muscle, it may be brought to threshold by this flow of current and undergo action potentials, which in turn initiate peristaltic waves that sweep over the stomach in pace with the BER at a rate of three per minute.

Once initiated, the peristaltic wave spreads over the fundus and body to the antrum and pyloric sphincter. Because the muscle layers are thin in the fundus and body, the peristaltic contractions in this region are weak. When the waves reach the antrum, they become much stronger and more vigorous because the muscle there is much thicker.

Because only feeble mixing movements occur in the body and fundus, food emptied into the stomach from the esopha-gus is stored in the relatively quiet body without being mixed. The fundic area usually does not store food but contains only a pocket of gas. Food is gradually fed from the body into the antrum, where mixing does take place.

Gastric mixing takes place in the antrum of the stomach.

The strong antral peristaltic contractions are responsible for mixing food with gastric secretions to produce **chyme**. Each antral peristaltic wave propels chyme forward toward the pyloric sphincter. Tonic contraction of the pyloric sphincter normally keeps it almost, but not completely, closed. The opening is large enough for water and other fluids to pass through with ease but too small for the thicker chyme to pass through except when a strong antral peristaltic contraction pushes it through. Even then, usually only a few milliliters of antral contents are forced into the duodenum with each peristaltic wave. Before more chyme can be squeezed out, the peristaltic wave reaches the pyloric sphincter and causes it to contract more forcefully, sealing off the exit and blocking further passage into the duodenum. The bulk of the antral chyme that was being propelled forward but failed to be pushed into the duodenum is abruptly halted at the closed sphincter and is tumbled back into the antrum, only to be propelled forward and tumbled back again as the new peristaltic wave advances (● Figure 14–9). This tossing back and forth, called **retropulsion**, thoroughly mixes the chyme in the antrum.

Gastric emptying is largely controlled by factors in the duodenum.

In addition to accomplishing gastric mixing, the antral peristaltic contractions provide the driving force for gastric emptying. The amount of chyme that escapes into the **duodenum** (the initial segment of the small intestine) with each peristaltic wave before the pyloric sphincter closes tightly depends largely on the strength of peristalsis and on the particle size of the ingested material. During the digestive phase of gastric motility, only particles less than 2 mm in diameter escape through the pyloric sphincter. The intensity of antral peristalsis can vary markedly under the influence of different signals from both the stomach and the duodenum; thus gastric emptying is regulated by both gastric and duodenal factors. These factors influence the stomach's excitability by slightly depolarizing or hyperpolarizing the gastric smooth muscle. This excitability in turn is a determinant of the degree of antral peristaltic activity. The greater the excitability, the more frequently the BER generates action potentials, the greater the degree of peristaltic activity in the antrum, and the faster the rate of gastric emptying (▌ Table 14–2).

Any remaining indigestible particles remaining in the stomach between meals is cleared by relaxation of the pyloric sphincter in association with powerful waves of peristalsis sweeping over the antral region of the stomach. This "housekeeping" function of the antrum, termed **interdigestive motility complex**, normally occurs at hourly intervals. Eating disrupts this wave pattern that leads to the resumption of the digestive motility pattern. Interestingly, in animals such as herbivores where food is continually entering the abomasum (stomach equivalent in ruminants), the interdigestive motility complex also occurs at hourly intervals.

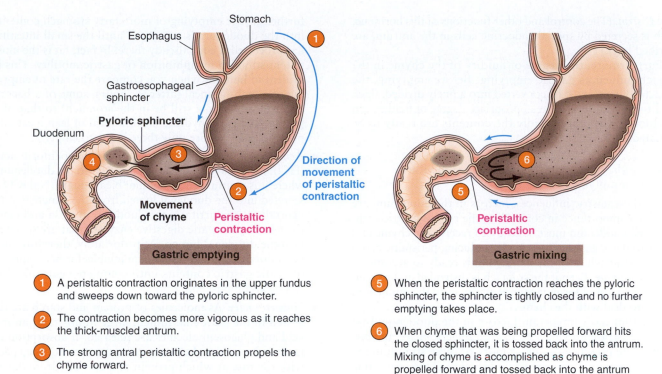

1. A peristaltic contraction originates in the upper fundus and sweeps down toward the pyloric sphincter.

2. The contraction becomes more vigorous as it reaches the thick-muscled antrum.

3. The strong antral peristaltic contraction propels the chyme forward.

4. A small portion of chyme is pushed through the partially open sphincter into the duodenum. The stronger the antral contraction, the more chyme is emptied with each contractile wave.

5. When the peristaltic contraction reaches the pyloric sphincter, the sphincter is tightly closed and no further emptying takes place.

6. When chyme that was being propelled forward hits the closed sphincter, it is tossed back into the antrum. Mixing of chyme is accomplished as chyme is propelled forward and tossed back into the antrum with each peristaltic contraction.

Figure 14–9 ● Gastric emptying and mixing as a result of antral peristaltic contractions (mammalian stomach).

Factors in the Stomach That Influence the Rate of Gastric Emptying

The main gastric factor that influences the strength of contraction is the amount of chyme in the stomach. Other things being equal, the stomach empties at a rate proportional to the volume of chyme in it at any given time. Distension of the stomach triggers increased gastric motility through a direct effect of stretch on the smooth muscle as well as through involvement of the intrinsic plexuses, the vagus nerve, and the stomach hor-

Table 14–2 ▍ Factors Regulating Gastric Motility and Emptying (vertebrate)		
Factors	Mode of Regulation	Effects on Gastric Motility and Emptying
Within the stomach		
Volume of chyme	Distension has a direct effect on gastric smooth-muscle excitability, as well as acting through the intrinsic plexuses, the vagus nerve, and gastrin	Increased volume stimulates motility and emptying
Degree of fluidity	Direct effect; contents must be in a fluid form to be evacuated	Increased fluidity allows more rapid emptying
Within the duodenum		
Presence of fat, acid, hypertonicity, or distension	Initiates the enterogastric reflex or triggers the release of enterogastrones (cholecystokinin, secretin, gastric inhibitory peptide)	These factors in the duodenum inhibit further gastric motility and emptying until the duodenum has coped with factors already present
Outside the digestive system		
Emotion	Alters autonomic balance	Stimulates or inhibits motility and emptying
Intense pain	Increases sympathetic activity	Inhibits motility and emptying
Decreased glucose use in the hypothalamus	Increases vagal activity	Stimulates motility; accompanied by hunger pangs

mone *gastrin*. (The control and other functions of this hormone, which is secreted by special endocrine cells in the antrum, are described later.)

Furthermore, the degree of fluidity of the chyme in the stomach influences gastric emptying. Before emptying, the stomach contents must be converted into a finely divided, thick liquid form. The sooner the appropriate degree of fluidity can be achieved, the more rapidly the contents are ready to be evacuated.

Factors in the Duodenum That Influence the Rate of Gastric Emptying

Despite these gastric influences, factors in the duodenum are of primary importance in controlling the rate of gastric emptying. The duodenum must be ready to receive the chyme and can act to delay gastric emptying by reducing peristaltic activity in the stomach until the duodenum is ready to accommodate more chyme. Even if the stomach is distended and its contents are in a liquid form, it cannot empty until the duodenum is ready to deal with the chyme.

The four most important factors in the duodenum that influence gastric emptying are *fat, acid, hypertonicity,* and *distension*. The presence of one or more of these stimuli in the duodenum activates appropriate duodenal receptors, triggering either a neural or hormonal response that slow gastric motility by reducing the excitability of the gastric smooth muscle. The subsequent reduction in antral peristaltic activity slows down the rate of gastric emptying. The *neural response* is mediated through both the intrinsic nerve plexuses (short reflex) and the autonomic nerves (long reflex). Collectively, these reflexes are called the **enterogastric reflex.**

The *hormonal response* involves the release from the duodenal mucosa of several hormones collectively known as **enterogastrones.** These hormones are transported by the blood to the stomach, where they inhibit antral contractions to reduce gastric emptying. Three of these enterogastrones have been positively identified: **secretin, cholecystokinin (CCK),** and **gastric inhibitory peptide (or glucose-dependent insulinotrophic peptide).** Secretin was the first hormone discovered, by Bayliss and Starling in 1902. Because it was a secretory product that entered the blood in response to an increase in acidity in the duodenum, it was termed *secretin*. The name *cholecystokinin* derives from the fact that this same hormone also governs contraction of the bile-containing gallbladder (*chole* means "bile," *cysto* means "bladder," and *kinin* means "contraction"). The name *gastric inhibitory peptide* is self-explanatory; it is a peptide hormone that inhibits the stomach (but see the hormone summary at the end of this chapter for an explanation of its new name, *glucose-dependent insulinotrophic peptide*). A possible fourth intestinal hormone, **neurotensin,** inhibits both gastric acid secretion as well as motility of the small intestine. Birds have a unique hormone, **avian pancreatic polypeptide (APP),** that also is effective in reducing gastric motility.

Let us examine why it is important that each of these stimuli in the duodenum (fat, acid, hypertonicity, and distension) delays gastric emptying (acting through the enterogastric reflex or one of the enterogastrones).

- *Fat.* Fat is digested and absorbed more slowly than the other nutrients. Furthermore, fat digestion and absorption take place only within the lumen of the small intestine. Therefore, when fat is already present in the duodenum, further gastric emptying of more fatty stomach contents into the duodenum is prevented until the small intestine has processed the fat already there. In fact, fat is the most potent stimulus for inhibition of gastric motility. This is evident in humans when you compare the rate of emptying of a high-fat meal (after six hours, some of a bacon-and-eggs meal may still be in the stomach) to that of a protein-and-carbohydrate meal (a meal of lean meat and potatoes may require only three hours to empty).

- *Acid.* Because the stomach secretes hydrochloric acid (HCl), highly acidic chyme is emptied into the duodenum, where it is neutralized by sodium bicarbonate ($NaHCO_3$) secreted into the duodenal lumen from the pancreas. Unneutralized acid irritates the duodenal mucosa and inactivates the pancreatic digestive enzymes that are secreted into the duodenal lumen. Appropriately, therefore, unneutralized acid in the duodenum inhibits further emptying of acidic gastric contents until complete neutralization can be accomplished.

- *Hypertonicity.* As molecules of protein and starch are digested in the duodenal lumen, large numbers of amino acid and glucose molecules are released. If absorption of these amino acid and glucose molecules does not keep pace with the rate at which protein and carbohydrate digestion proceeds, these large numbers of molecules remain in the chyme and increase the osmolarity of the duodenal contents. Osmolarity depends on the number of molecules present, not on their size (recall colligative properties, p. 77), and one protein molecule may be split into several hundred amino acid molecules, each of which has about the same osmotic activity as the original protein molecule. The same holds true for one large starch molecule, which yields many smaller but equally osmotically active glucose molecules. Because water is freely diffusible across the duodenal wall, water enters the duodenal lumen from the plasma as the duodenal osmolarity rises. Large volumes of water entering the intestine from the plasma lead to intestinal distension, and, more importantly, circulatory disturbances ensue because of the reduction in plasma volume. To prevent these effects, gastric emptying is reflexly inhibited when the osmolarity of the duodenal contents starts to rise. Thus the amount of food entering the duodenum for further digestion into a multitude of additional osmotically active particles is reduced until absorption processes have had an opportunity to catch up.

- *Distension.* Too much chyme in the duodenum inhibits the emptying of even more gastric contents, thus allowing the distended duodenum time to cope with the excess volume of chyme it already contains before it receives an additional quantity.

■ Peristaltic contractions occur in the empty stomach before the next meal.

In conjunction with the sensation of hunger, peristaltic contractions begin again, sweeping over the nearly empty antrum. This arousal of stomach motility appears to be mediated by increased parasympathetic activity. An animal may experience the sensation of hunger pangs when these peristaltic contractions are occurring, but the contractions themselves are not responsible for the sensation.

■ Stress can influence gastric motility.

Other factors unrelated to digestion, such as stress, also can alter gastric motility by acting through the autonomic nerves to influence the degree of gastric smooth muscle excitability. Even though the effect of emotions on gastric motility varies from one animal to another and is not always predictable, fear generally tends to decrease motility, whereas aggression tends to increase it. In addition to these influences, intense pain from any part of the body tends to inhibit motility, not just in the stomach but throughout the digestive tract. This response is brought about by increased sympathetic activity and a corresponding decrease in parasympathetic activity.

■ The body of the stomach does not actively participate in the act of vomiting.

Vomiting, or **emesis,** the forceful expulsion of gastric contents out through the mouth, is generally perceived as being caused by abnormal gastric motility. However, animals with simple stomachs do not vomit by reverse peristalsis, as might be predicted. Actually, the stomach itself does not actively participate in the act of vomiting. The stomach, the esophagus, the esophageal sphincters, and the pyloric sphincter are all relaxed during vomiting. The major force for expulsion comes, surprisingly, from contraction of the respiratory muscles—namely, the diaphragm (the major inspiratory muscle) and the abdominal muscles (the muscles of active expiration). Vomiting, however, is unusual in horses, because the tonus of the gastroesophageal sphincter is so great that it limits regurgitation of materials into the esophagus. Rats, too, cannot vomit and are noted for tasting novel foods in small quantities. In contrast, ruminant animals are noted for their ability to regurgitate their stomach contents for rechewing in the mouth. Movement of the digesta up the esophagus is accomplished by rapid reverse peristalsis. This is possible in ruminants because their esophageal muscle is striated. The phenomenon of rumination is discussed in greater detail on page 661.

Some species of birds, particularly carnivores, can "vomit" or **egest** indigestible foodstuffs. Owls, for example, egest the bones, feathers, and fur of their prey in the form of pellets. The presence of nutrients in the digesta inhibits egestion, but once digestion of the meal is complete the rate and frequency of gastric contractions then increase. This compacts and shapes the pellet. Continued contractions move the pellet into the lower esophagus, where it is egested by rapid reverse peristalsis. Other birds egest nutritious food to feed their young.

Vomiting begins with a deep inspiration and closure of the glottis. The contracting diaphragm descends downward on the stomach while simultaneous contraction of the abdominal muscles compresses the abdominal cavity, increasing the intra-abdominal pressure and forcing the abdominal viscera upward. As the flaccid stomach is squeezed between the diaphragm from above and the compressed abdominal cavity from below, the gastric contents are forced into the esophagus and out through the mouth. The glottis is closed, so vomited material does not enter the respiratory airways. Also, the uvula is elevated to close off the nasal cavity.

The vomiting cycle may be repeated several times until the stomach is emptied, and is an example of a positive-feedback mechanism. This complex act of vomiting is coordinated by a **vomiting center** in the medulla. It is presently unclear whether this center consists of a circumscribed cluster of neurons or a more diffuse network of interacting neurons. Vomiting can be initiated by afferent input to the vomiting center from a number of receptors throughout the body. The causes of vomiting include the following:

- Irritation or distension of the stomach and duodenum.
- Chemical agents, including drugs or noxious substances that initiate vomiting (that is, emetics) either by acting in the upper portions of the gastrointestinal tract or by stimulating chemoreceptors in a specialized **chemoreceptor trigger zone** in the brain. Activation of the trigger zone initiates the vomiting reflex.

With excessive vomiting an animal experiences considerable losses of secreted fluids and acids that normally would be reabsorbed. The resultant reduction in plasma volume can lead to dehydration and circulatory problems, whereas loss of acid from the stomach can lead to metabolic alkalosis (see p. 608). Vomiting is not always detrimental, however. Limited vomiting brought about by irritation of the digestive tract can provide a useful service in removing noxious material from the stomach rather than allowing it to be retained and absorbed. Snakes, for example, that swallow too large a prey item are sometimes forced to vomit the remains because they are putrefying inside the gut.

■ Gastric digestive juice is secreted by glands located at the base of gastric pits.

The rate and the amount of gastric juice secretion are dependent on the frequency of feeding. For example, gastric juice production is continuous in ruminants, unpredictable in reptiles (carnivorous reptiles are dependent on prey availability and environmental temperature for digestion to occur). Once underway, gastric secretion in a snake carries on for up to 6 days to completely digest a rat, whereas in hibernators digestive functions virtually come to a standstill during hibernation.

The cells responsible for gastric secretion in an animal are located in the lining of the stomach, the gastric mucosa, which is divided into two distinct areas whose distribution varies between species: (1) the **oxyntic mucosa,** which lines the body and fundus, and (2) the **pyloric gland area (PGA),** which lines the antrum. The luminal surface of the stomach is pitted with deep pockets formed by infoldings of the gastric mucosa. The invaginations in this first portion are called **gastric pits,** at the base of which lie the gastric glands. A variety of secretory cells line these invaginations, some exocrine and some endocrine or paracrine (■ Table 14–3).

Gastric Exocrine Secretory Cells

Three types of secretory cells are found in the walls of the pits in the oxyntic mucosa:

- **Mucous cells** line the gastric pits and the entrance of the glands. They secrete a thin, watery *mucus.* (*Mucous* is the adjective; *mucus* is the noun.)
- Chief and parietal cells line the deeper portions of the gastric glands. The more numerous **chief cells** secrete the enzyme precursor *pepsinogen.*
- The **parietal** (or **oxyntic**) **cells,** which secrete *HCl* and *intrinsic factor.* The parietal cells are located on the outer wall of the gastric pits and do not come into contact with the pit lumen. (*Parietal* means "wall," a reference to the

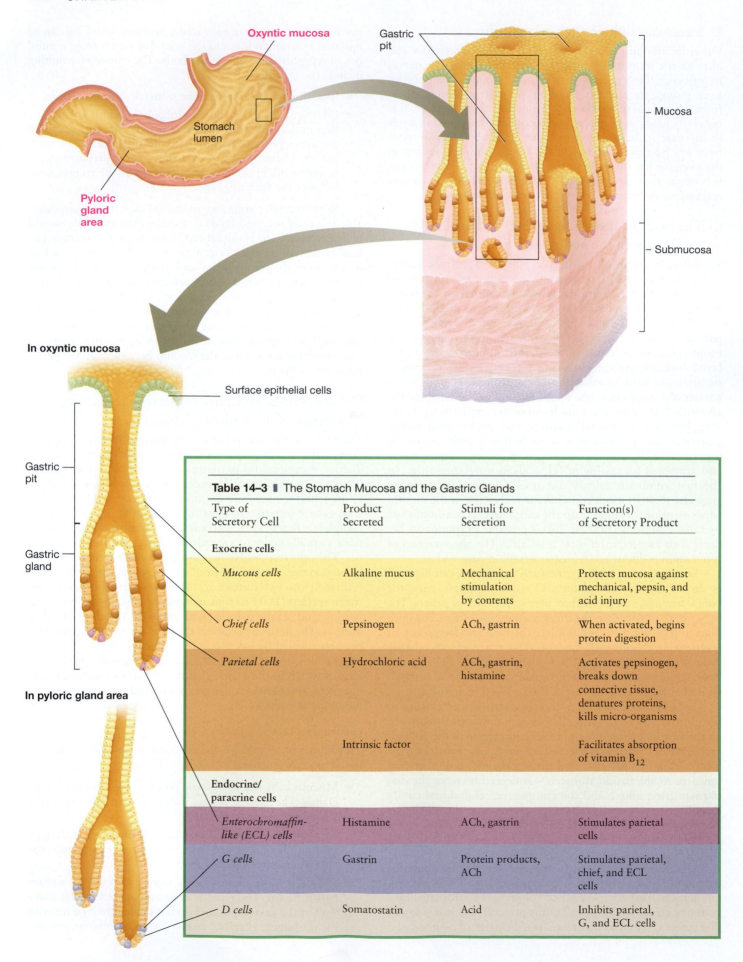

Oxyntic mucosa

Gastric pit

Stomach lumen

– Mucosa

Pyloric gland area

– Submucosa

In oxyntic mucosa

Surface epithelial cells

Gastric pit

Gastric gland

In pyloric gland area

Table 14–3 ▍ The Stomach Mucosa and the Gastric Glands

Type of Secretory Cell	Product Secreted	Stimuli for Secretion	Function(s) of Secretory Product
Exocrine cells			
Mucous cells	Alkaline mucus	Mechanical stimulation by contents	Protects mucosa against mechanical, pepsin, and acid injury
Chief cells	Pepsinogen	ACh, gastrin	When activated, begins protein digestion
Parietal cells	Hydrochloric acid	ACh, gastrin, histamine	Activates pepsinogen, breaks down connective tissue, denatures proteins, kills micro-organisms
	Intrinsic factor		Facilitates absorption of vitamin B_{12}
Endocrine/ paracrine cells			
Enterochromaffin-like (ECL) cells	Histamine	ACh, gastrin	Stimulates parietal cells
G cells	Gastrin	Protein products, ACh	Stimulates parietal, chief, and ECL cells
D cells	Somatostatin	Acid	Inhibits parietal, G, and ECL cells

location of these cells. *Oxyntic* means "sharp," a reference to these cells' potent HCl secretory product.)

These exocrine secretions are all released into the gastric lumen. Collectively, they make up the gastric digestive juice.

A sparse number of stem cells are also found in the gastric pits. These cells rapidly divide and serve as the parent cells of all new cells of the gastric mucosa. The daughter cells that result from cell division either migrate out of the pit to become surface epithelial cells or migrate down into the deeper parts of the pit to differentiate into chief or parietal cells. Through this activity, the entire stomach mucosa is regularly replaced.

Between the gastric pits, the gastric mucosa is covered by **surface epithelial cells,** which secrete a thick, viscous, alkaline mucus that forms a visible layer several millimeters thick over the surface of the mucosa.

Gastric Endocrine and Paracrine Secretory Cells

Other secretory cells in the gastric mucosa release endocrine and paracrine regulatory factors instead of products involved in the digestion of nutrients in the gastric juice lumen. These are as follows:

- **Enterochromaffin-like (ECL) cells** dispersed among the parietal and chief cells in the gastric glands of the oxyntic mucosa secrete the paracrine *histamine.*

- The gastric glands of the PGA primarily secrete mucus and a small amount of pepsinogen; no acid is secreted in this area, in contrast to the oxyntic mucosa. More importantly, endocrine cells known as **G cells** in the PGA secrete the hormone *gastrin* into the blood.

- D cells, which are scattered in glands near the pylorus but are more numerous in the duodenum, secrete the paracrine somatostatin.

Let's examine each of these secretory products in more detail, looking first at the exocrine products and their role in digestion, then at the exocrine secretions, among other actions.

Hydrochloric Acid Secretion

The parietal cells actively secrete HCl into the lumen of the gastric pits, which in turn empty into the lumen of the stomach. The pH of the luminal contents of most vertebrates may fall to as low as 2 as a result of this HCl secretion. Hydrogen ion (H^+) and chloride ion (Cl^-) are actively transported by separate pumps in the parietal cell's plasma membrane. Hydrogen ion is actively transported against a tremendous concentration gradient, with the H^+ concentration being as much as 3 to 4 million times greater in the lumen than in the blood. Because of the high-energy expenditure needed to move H^+ against such a large gradient, parietal cells have an abundance of mitochondria, the energy organelles. Cl^- is also actively secreted but against a much smaller concentration gradient of only 1.5 times.

The secreted H^+ is not transported from the plasma but is derived instead from metabolic processes within the parietal cell (● Figure 14–10). Specifically the H^+ to be secreted is

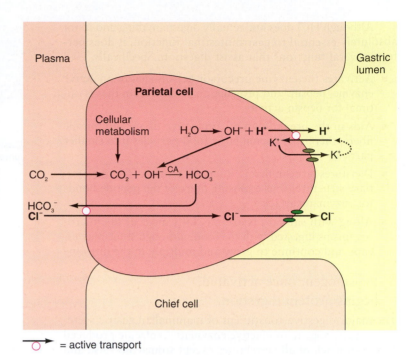

= active transport

CA = carbonic anhydrase

Figure 14–10 ● Mechanism of HCl secretion (vertebrate). The stomach's parietal cells actively secrete H^+ and Cl^- by the actions of two separate pumps. The secreted H^+ is derived from the breakdown of H_2O into H^+ and OH^-. Note that CA produces H^+ and HCO_3^- through two steps (see text). The secreted Cl^- is transported into the parietal cell from the plasma. The HCO_3^- generated from H_2CO_3 dissociation is transported into the plasma in exchange for the secreted Cl^-.

derived from the breakdown of H_2O molecules into H^+ and OH^- (hydroxyl ions) within the parietal cells. This H^+ is secreted into the stomach lumen by H^+–K^+ ATPase in the parietal cell's luminal membrane. This primary active transport carrier also pumps K^+ into the cell from the lumen. The transported K^+ then passively leaks back into the lumen through luminal K^+ channels, thus leaving K^+ levels unchanged by the process of H^+ secretion.

The parietal cells contain an abundance of the enzyme carbonic anhydrase (CA). In the presence of carbonic anhydrase, OH^- readily combines with CO_2, which either has been produced within the parietal cell by metabolic processes or has diffused in from the blood. The combination of OH^- and CO_2 form HCO_3^-. The reduction in OH^- causes H_2O to split into H^+ and OH^- (law of mass attraction; see p. 598). The generated H^+ in essence replaces the one secreted.

The generated HCO_3^- is moved into the plasma by the same carrier at the basolateral border that actively transports Cl^- from the plasma into the parietal cell, similar to the Cl^- shift that occurs in red blood cells (see p. 508). This exchange of HCO_3^- for Cl^- maintains electrical neutrality in the plasma during HCl secretion. The Cl^- pumped into the cell diffuses through luminal channels into the gastric lumen, completing the Cl^- secretory process.

Although HCl does not actually digest anything and is not absolutely essential to gastrointestinal function, it does perform several functions that assist digestion. Specifically, HCl

- Activates the enzyme precursor pepsinogen to an active enzyme, pepsin, and provides an acid medium that is optimal for pepsin activity
- Aids in the breakdown of connective tissue and muscle fibers, thereby reducing large food particles into smaller particles
- Denatures protein; that is, it uncoils proteins from their tertiary structure, thus exposing more of the peptide bonds for enzymatic attack
- Along with salivary lysozyme, kills most of the microorganisms ingested with the food, although some may escape and continue to grow and multiply in the intestine

Pepsinogen, once activated, begins protein digestion.

The major digestive constituent of mammalian gastric secretion is **pepsinogen,** an inactive enzymatic molecule produced by the stomach of all vertebrates except stomachless fish and cyclostomes. Pepsinogen is stored in the chief cell's cytoplasm within secretory vesicles known as **zymogen granules,** from which it is released by exocytosis (see p. 38) on appropriate stimulation. When pepsinogen is secreted into the gastric lumen, HCl cleaves off a small fragment of the molecule, converting it to the active form of the enzyme, **pepsin** (● Figure 14–11). Once formed, pepsin acts on other pepsinogen molecules to produce more pepsin. A mechanism such as this, whereby an active form of an enzyme activates other molecules of the same enzyme, is referred to as an **autocatalytic** ("self-activating") **process** and is another example of a positive-feedback cycle. Only when there is no more protein in the stomach does the secretion of pepsinogen cease.

Pepsin initiates protein digestion by splitting certain amino acid linkages in proteins to yield peptide fragments (small amino-acid chains); it works most effectively in a low pH environment. Because pepsin can digest protein, it must be stored and secreted in an inactive form so it does not digest the cells in which it is formed. (The primary structural component of cells is protein.) While inside the cell, therefore, pepsin is maintained in the inactive form of pepsinogen until it reaches the gastric lumen, where it is activated by HCl.

Mucus Secretion

The surface of the gastric mucosa is covered by a layer of mucus, which serves as a protective barrier against several forms of potential injury to the gastric mucosa:

- By virtue of its lubricating properties, mucus protects the gastric mucosa against mechanical injury.
- Mucus helps protect the stomach wall from self-digestion because pepsin is inhibited when it comes in contact with the mucus coating the stomach lining. (However, mucus does not affect pepsin activity in the lumen, where digestion of dietary protein proceeds without interference.)
- Being alkaline, mucus helps protect against acid injury by neutralizing HCl in the vicinity of the gastric lining, but it does not interfere with the function of HCl in the lumen.

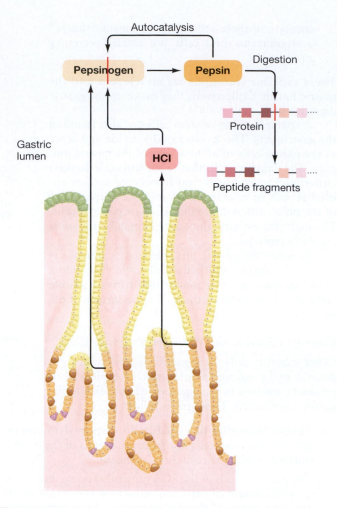

Figure 14–11 ● **Pepsinogen activation in the stomach lumen (vertebrate).** In the lumen, hydrochloric acid (HCl) activates pepsinogen to its active form, pepsin, by cleaving off a small fragment. Once activated, pepsin autocatalytically activates more pepsinogen and begins protein digestion. Secretion of pepsinogen in the inactive form prevents it from digesting the protein structures of the cells in which it is produced.

Whereas the pH in the lumen may be as low as 2, the pH in the mucus layer adjacent to the mucosal cell surface is about 7.

Intrinsic factor is essential for absorption of vitamin B_{12}.

Intrinsic factor, another secretory product of the parietal cells in addition to HCl, is important in absorption of vitamin B_{12}. This vitamin can be absorbed only when in combination with intrinsic factor. Binding of the intrinsic factor–vitamin B_{12} complex with a special receptor located in the terminal ileum, the last portion of the small intestine, triggers the receptor-mediated endocytosis of the complex at this location. Vitamin B_{12} is essential for the normal formation of red blood cells. In the absence of intrinsic factor, vitamin B_{12} fails to be absorbed, so erythrocyte production is defective and pernicious anemia results.

Multiple regulatory pathways influence the parietal and chief cells.

Four chemical messengers primarily influence the secretion of gastric digestive juices. Parietal cells have separate receptors for each of these messengers. Three of them—acetylcholine (ACh), gastrin, and histamine—are stimulatory. They bring about increased secretion of HCl by promoting the insertion of additional H^+–K^+ ATPases into the parietal cell's plasma membrane. The fourth regulatory agent—somatostatin—inhibits acid secretion. ACh and gastrin also increase pepsinogen secretion through their stimulatory effect on the chief cells. We now consider each of these chemical messengers in further detail:

- Acetylcholine is a neurotransmitter released from the intrinsic nerve plexus in response to vagus stimulation. ACh stimulates both the parietal and chief cells as well as the G cells and ECL cells. It is the most important trigger of HCl secretion in fishes.

- The G cells secrete the hormone **gastrin** into the blood in response to protein products in the stomach lumen as well in response to ACh. Like secretin and CCK, gastrin is a major gastrointestinal hormone. After being carried by the blood back to the body and fundus of the stomach, gastrin stimulates the parietal and chief cells, promoting secretion of a highly acidic gastric juice. In addition to directly stimulating the parietal cells, gastrin also indirectly promotes HCl secretion by stimulating the ECL cells to release histamine. Gastrin is the main factor responsible for increasing HCl secretion during digestion of a meal. Gastrin is also *trophic* (growth promoting) to the mucosa of the stomach and small intestine, thereby maintaining their secretory capabilities.

- Histamine, a paracrine, is released from the ECL cells in response to gastrin and ACh. Histamine acts locally on nearby parietal cells to increase the rate of HCl secretion. Histamine and gastrin are the most important triggers of HCl secretion in frogs and mammals.

- Somatostatin is released from the D cells in response to acid. It acts locally as a paracrine, in negative-feedback fashion, to inhibit secretion by the parietal cells, G cells, and ECL cells, thus turning off the HCl-secreting cells and their most potent stimulatory pathway.

From this list, it is obvious not only that multiple chemical messengers influence the parietal and chief cells but also that these chemicals also influence each other. Next, as we examine the phases of gastric secretion, you will see under what circumstances each of these regulatory agents is released.

Control of gastric secretion involves three phases.

The rate of gastric secretion can be increased by (1) factors arising before food ever reaches the stomach, (2) factors resulting from the presence of food in the stomach, and (3) factors originating in the duodenum after food has left the stomach. Accordingly, gastric secretion is divided into three phases—the cephalic, gastric, and intestinal phases (Table 14–4).

Cephalic Phase

Gastric secretion occurs in response to food-related stimuli acting in the brain, such as thinking about, tasting, smelling, chewing, and swallowing food. This increased secretion, which occurs even before food reaches the stomach, is known as the *cephalic phase* of gastric secretion (*cephalic* refers to "head") and is an example of feedforward activation. It is mediated by means of vagal nerve activity. First, vagal stimulation of the intrinsic plexuses promotes increased secretion of HCl and pepsinogen by the secretory cells. Second, vagal stimulation of the G cells within the PGA causes the release of gastrin, which in turn further enhances secretion of HCl and pepsinogen, with the effect on HCl being potentiated by gastrin promoting the release of histamine. The secretion of gastric juice in response to cephalic stimuli does not occur in ruminants because of the lack of association between the ingestion of food and the appearance of the ingesta in the compartmentalized ruminant stomach.

Gastric Phase

If an animal is fitted with an *esophageal fistula* (tube connecting the esophagus with the outside world) such that any ingested food is diverted from the stomach into a collecting port, there is an increase in secretion of HCl and pepsinogen in the animal's stomach when the animal is eating. See the photo of Cornell College's "Stella the cow" on the following page. Alternatively, if a region of the stomach (Heidenhain pouch) is experimentally denervated and transplanted so as

---

Table 14–4 ■ Stimulation of Gastric Secretion (mammal)

Phase	Stimuli	Excitatory Mechanism for Enhancing Gastric Secretion
Cephalic phase	Stimuli in the head—seeing, smelling, tasting, chewing, swallowing food	↑ Vagus → + Intrinsic nerves → + Parietal and chief cells → ↑Gastric secretion; + Pyloric gland area → ↑Gastrin ⌐ +
Gastric phase	Stimuli in the stomach—protein, (peptide fragments), distension, caffeine, alcohol	+ Vagus → + Intrinsic nerves → + Parietal and chief cells → ↑Gastric secretion; + Pyloric gland area → ↑Gastrin +
Excitatory intestinal phase	Stimuli in the duodenum—digested protein products	+ ↑Intestinal gastrin → ↑Parietal and chief cells → ↑Gastric secretion

Photo: The Image Lab, Cornell University

The photo above is of Stella, the successor to Ellie, the first cow at The Cornell College of Veterinary Medicine, to be fitted with a fistula (portal in the side) that functioned as a "window" into ruminant digestion. Ellie, born in 1988, died of old age in 2002; she outlived most of the other Holsteins. The removable cap on the fistula fitting allowed veterinary students to sample her digestive fluids and determine the rate of digestion of various food-stuffs. This fistula also allowed her to donate rumen fluid to other, less healthy cows, whose digestive systems needed a boost.

to be open to the outside world such that acid secretion can be physically measured, there is still acid secretion from the pouch as soon as food is put into the stomach. Because the animal did not "see" the food, no cephalic phase is triggering the intact stomach. Rather, when food arrives (or is placed) in the stomach, gastric secretion is increased during the so-called *gastric phase* of digestion. Stimuli acting in the stomach, namely the peptide fragments, trigger the release of gastrin. Gastrin, in turn, enters the blood and stimulates acid secretion in the pouch.

Intestinal Phase

The *intestinal phase* of digestion occurs with the entry of food into the small intestine. Even if the stomach is completely iso-

lated from the intestinal tract, placement of food into the upper part of the duodenum can still induce acid secretion in the stomach. Several important factors that originate in the small intestine influence gastric secretion. The intestinal phase has both an excitatory and an inhibitory component.

To a limited extent, the presence of the products of protein digestion in the duodenum stimulates further gastric secretion by triggering the release of **intestinal gastrin** that is carried by the blood to the stomach. This is the *excitatory component* of the intestinal phase of gastric secretion. It is as if the small intestine, on noting the arrival of protein fragments from the stomach, offers the stomach a "helping hand" in digesting the protein by enhancing gastric secretion.

The inhibitory component of the intestinal phase of gastric secretion is dominant over the excitatory component. When the pH of the intestine becomes too low, **enteroglucagon** is released into the circulation and "shuts off" the flow of gastric juices.

■ Gastric secretion gradually decreases as food empties from the stomach into the intestine.

You now know what factors turn on gastric secretion before and during a meal, but how is the flow of gastric juices shut off as chyme begins to be emptied from the stomach into the small intestine? Gastric secretion is gradually reduced by three different means as the stomach empties (■ Table 14–5):

- As food is gradually emptied into the duodenum, the major stimulus for enhanced gastric secretion—the presence of protein in the stomach—is withdrawn.

- After foods leave the stomach and gastric juices accumulate to such an extent that gastric pH falls very low, somatostatin is released. As a result of somatostatin's inhibitory effects, gastric secretion declines.

- The same stimuli that inhibit gastric motility (fat, acid, hypertonicity, or distension in the duodenum brought about by stomach emptying) inhibit gastric secretion as well. The enterogastric reflex and the enterogastrones suppress the gastric secretory cells, while they simultaneously reduce the excitability of the gastric smooth muscle cells. This inhibitory response is the inhibitory component of the intestinal phase of gastric secretion.

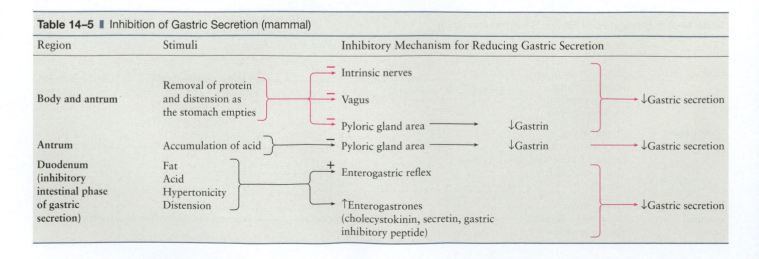

Table 14–5 ■ Inhibition of Gastric Secretion (mammal)

Region	Stimuli	Inhibitory Mechanism for Reducing Gastric Secretion		
Body and antrum	Removal of protein and distension as the stomach empties	– Intrinsic nerves – Vagus – Pyloric gland area →	↓Gastrin	↓Gastric secretion
Antrum	Accumulation of acid	– Pyloric gland area →	↓Gastrin	↓Gastric secretion
Duodenum (inhibitory intestinal phase of gastric secretion)	Fat Acid Hypertonicity Distension	+ Enterogastric reflex ↑Enterogastrones (cholecystokinin, secretin, gastric inhibitory peptide)		↓Gastric secretion

■ The stomach lining is protected from gastric secretions by the gastric mucosal barrier.

How can the stomach contain strong acid contents and proteolytic enzymes without destroying itself? We have already learned that mucus provides a protective coating. In addition, the mucosal lining itself provides other barriers to acid damage of the mucosa. First, the luminal membranes of the gastric mucosal cells are almost impermeable to H^+, so acid cannot penetrate *into* the cells and thus cause cell damage. Furthermore, the lateral edges of these cells are joined together near their luminal borders by tight junctions (see p. 63), so acid cannot diffuse *between* the cells from the lumen into the underlying submucosa. Cardiac glands in some mammals (such as swine) secrete bicarbonate ion into the overlying mucus to protect the epithelial cells from H^+. The secretion is limited to the cardiac gland area of the stomach and does not affect the digestive process in the neighboring regions. The properties of the gastric mucosa that enable the stomach to contain acid without injuring itself constitute the **gastric mucosal barrier.** These protective mechanisms are further enhanced by the fact that the entire stomach lining is replaced every three days. Because of rapid mucosal turnover, cells are replaced as they are exposed to the wear and tear of harsh gastric conditions before they have suffered enough damage to impair their function.

Despite the protection provided by mucus, by the gastric mucosal barrier, and by the frequent turnover of cells, the barrier occasionally is broken so that the gastric wall is injured by its acidic and enzymatic contents. When this occurs, an erosion, or **peptic ulcer,** of the stomach wall results. (Excessive gastric reflux into the esophagus and dumping of excessive acidic gastric contents into the duodenum can lead to peptic ulcers in these locations as well.) Pigs are notable for their susceptibility to stomach ulcers, particularly in the esophageal area. Generally, there is a higher incidence in barrows (male pigs that have been castrated before reaching puberty) than gilts (female pigs that have not given birth), and both crowding and heat stress increase the severity of the lesion.

■ Carbohydrate digestion continues in the body of the stomach, whereas protein digestion begins in the antrum.

Two separate digestive processes take place within the stomach. Food in the body of the stomach remains a semisolid mass, because peristaltic contractions in this region are too weak for mixing to occur. Because food is not mixed with gastric secretions in the body of the stomach, very little protein digestion occurs here. Acid and pepsin can attack only the surface of the food mass. Carbohydrate digestion, however, continues in the interior of the mass, under the influence of salivary amylase. Even though acid inactivates salivary amylase, the unmixed interior of the food mass is free of acid.

Digestion by the gastric juice itself is accomplished in the antrum of the stomach, where the food is thoroughly mixed with HCl and pepsin, initiating protein digestion.

In most vertebrates, at this point, fat digestion has not even begun. Although it is possible for fat molecules, which are lipid soluble, to pass through the lipid portions of the cell membranes that line the stomach, this does not occur to any appreciable extent, because dietary fat in the gastric lumen separates into large fat droplets that float in the chyme. However, in some species lingual lipase has begun the process of lipid digestion.

■ Proventriculus-gizzard processes begin the process of digestion in birds and insects.

Stomachs do not start the digestion in all animals; indeed, they are absent in some. The insect proventriculus (also called the *gizzard* in insects) is developed as a grinding organ. In cockroaches, for example, six cutting edges or plates are arranged radially around the gizzard such that when compressed together, they form a powerful grinding tool to macerate digesta. Only when food particles are sufficiently ground can they continue their passage from the foregut into the midgut. This function helps limit the amount of food leaving the crop and entering the midgut.

Birds have no stomach per se, but rather the organ is divided into two components: the proventriculus, which serves as the "glandular stomach," and a separate gizzard, which serves as the "muscular stomach." The gizzard also occurred in herbivorous dinosaurs.

Two types of glands normally line the proventriculus of birds: simple mucosal glands that secrete mucus in addition to compound submucosal glands that secrete HCl and pepsinogen. Interestingly, unlike in mammals, both HCl and pepsinogen are synthesized within the same cell—the chief or **oxynticopeptic cell.** Oxynticopeptic cells also line the stomach of fish.

The proventriculus is comparatively small in herbivorous birds and is considerably larger in aquatic carnivores, which use the proventriculus as a storage space for captured prey. The function of the glandular stomach is closely linked with that of the gizzard, whose function is to macerate and grind the ingesta. Motility functions of the proventriculus are to expel ingesta and digestive secretions into the gizzard.

Two pairs of smooth muscle groups, the *thin (musculi intermedii)* and *thick (musculi laterales)* encircle the gizzard and act in opposition to each other. Most birds cannot chew their food, so the gizzard, which contains **grit** (pebbles and small stones), is responsible for the mechanical breakdown of ingesta. The mucosal lining of the gizzard is covered by a coating, **koilin,** a tough protein-polysaccharide complex that has an amino acid composition similar to that of feather protein. Structurally, koilin is a composite of filaments formed in tubular glands lining the gizzard. This hard filament eventually forms a rod that protrudes through the luminal surface of the mucosa and forms the *"gizzard teeth"* (the abrasive lining of the gizzard's surface).

The gizzard is a comparatively large organ designed to develop high pressures for particle size reduction. In birds the myoglobin content of the gizzard is approximately 100-fold greater than in the breast muscle, and mitochondrial numbers are also elevated, indicating a high aerobic capacity. Digesta from the gizzard is refluxed back into the proventriculus for further digestion and mixed with gastric secretions.

In domestic poultry, gastroduodenal motility is tightly coordinated and begins with contraction of the thin, circular muscles lining the gizzard, followed by several waves of peristalsis in the duodenum. A powerful contraction of the thick muscles then precedes peristalsis of the proventriculus. Contractions of the proventriculus and duodenum are generated by intrinsic neural connections originating from the gizzard and are not significantly affected by loss of extrinsic innervation. At the end of the contraction of the thin muscles, ingesta flows from the gizzard into the duodenum; whereas after contraction of the thick muscles, the ingesta flows into the proventriculus. Contraction of the proventriculus then forces material back into the gizzard. In poultry, these cycles are normally repeated two to three times per minute, with food deprivation decreasing the frequency of contractions.

Pancreas, Liver, and Fat Body

When gastric contents are emptied into the vertebrate small intestine, they are mixed not only with juice secreted by the small-intestine mucosa but also with the secretions of the exocrine pancreas and liver that are emptied into the duodenal lumen. We discuss the roles of each of these accessory digestive organs before we examine the contributions of the small intestine itself.

Pancreas

■ The pancreas is a mixture of exocrine and endocrine tissue.

The vertebrate **pancreas** is a mixed gland that contains both exocrine and endocrine tissue (● Figure 14–12). Approximately 98% of the pancreas is committed to its exocrine function. The predominant exocrine portion consists of grapelike clusters of secretory cells that form sacs known as **acini,** which connect to ducts that eventually empty into the duodenum. The smaller endocrine portion consists of isolated islands of endocrine tissue, the **islets of Langerhans,** which are dispersed throughout the pancreas. The most important hormones secreted by the islet cells are insulin

and glucagon (p. 296). In most species the islets are unequally distributed throughout the gland. For example, the splenic lobe of the avian pancreas is approximately 4 to 5% endocrine in function, compared to less than 0.5% for the remaining three lobes. The exocrine and endocrine pancreas have nothing in common except for sharing the same location.

■ The exocrine pancreas secretes digestive enzymes and an aqueous alkaline fluid.

The **exocrine pancreas** secretes a pancreatic juice consisting of two components—a potent *enzymatic secretion* and an *aqueous* (watery) *alkaline secretion* that is rich in sodium bicarbonate ($NaHCO_3$). The pancreatic enzymes are actively secreted by the *acinar cells.* The aqueous $NaHCO_3$ component is actively secreted by the *duct cells* that line the early portions of the pancreatic ducts.

Like pepsinogen, pancreatic enzymes are produced and stored within zymogen granules and are released by exocytosis as needed. The acinar cells secrete different types of pancreatic enzymes that can digest foodstuffs. In nonruminants, these pancreatic enzymes are important, because they can almost completely digest food in the absence of all other digestive secretions. The principal types of vertebrate pancreatic enzymes are (1) **proteolytic enzymes,** which are involved in protein digestion; (2) **pancreatic amylase** and **chitinase** (in some vertebrates), which contributes to carbohydrate digestion; and (3) **pancreatic lipase,** the most critical enzyme involved in fat digestion.

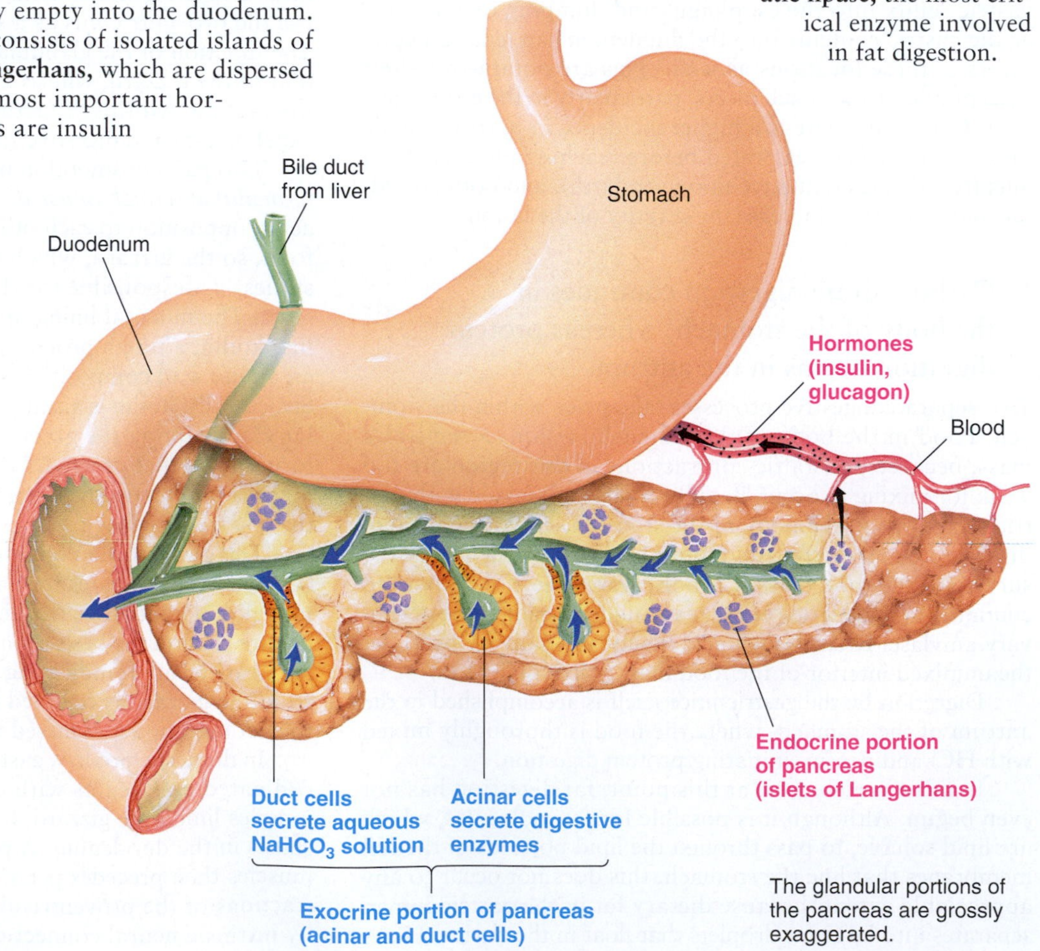

Figure 14–12 ● Schematic representation of the exocrine and endocrine portions of the mammalian pancreas. The exocrine pancreas secretes into the duodenal lumen a digestive juice composed of digestive enzymes secreted by the acinar cells and an aqueous $NaHCO_3$ solution secreted by the duct cells. The endocrine pancreas secretes the hormones insulin and glucagon into the blood.

Bile duct from liver

Duodenum

Stomach

Hormones (insulin, glucagon)

Blood

Duct cells secrete aqueous $NaHCO_3$ solution

Acinar cells secrete digestive enzymes

Endocrine portion of pancreas (islets of Langerhans)

Exocrine portion of pancreas (acinar and duct cells)

The glandular portions of the pancreas are grossly exaggerated.

Pancreatic Proteolytic Enzymes

The three major proteolytic enzymes secreted by the vertebrate pancreas are *trypsinogen, chymotrypsinogen,* and *procarboxypeptidase,* each of which is secreted in an inactive form. When **trypsinogen** is secreted into the duodenal lumen, it is activated to its active enzyme form, **trypsin,** by **enterokinase,** an enzyme embedded in the luminal border of the cells that line the duodenal mucosa. Trypsin then autocatalytically activates more trypsinogen. Like pepsinogen in the chief cells of the stomach, trypsinogen must remain inactive within the pancreas to prevent this proteolytic enzyme from digesting the cells in which it is formed. Trypsinogen remains inactive, therefore, until it reaches the duodenal lumen, where enterokinase triggers the activation process, which then proceeds autocatalytically. As further protection, the pancreatic tissue also produces a chemical known as **trypsin inhibitor,** which blocks trypsin's actions if activation of trypsinogen inadvertently occurs within the pancreas.

Chymotrypsinogen and procarboxypeptidase, the other pancreatic proteolytic enzymes, are converted by trypsin to their active forms, **chymotrypsin** and **carboxypeptidase,** respectively, within the duodenal lumen. Thus, once enterokinase has activated some of the trypsin, trypsin then governs the rest of the activation process.

Each of these proteolytic enzymes attacks different peptide linkages. The end products that result from this action are a mixture of amino acids and small peptide chains. Mucus secreted by the intestinal cells protects against digestion of the small-intestine wall by the activated proteolytic enzymes.

Pancreatic Amylase

Like salivary amylase, pancreatic amylase plays an important role in carbohydrate digestion by converting polysaccharides into disaccharides. Amylase is secreted in the pancreatic juice in an active form, because active amylase does not present a danger to the secretory cells.

Pancreatic Chitinase

If cellulose is the most common biological polysaccharide, then what is the second? It is chitin, a compound found in the cuticle of arthropods and in cell walls of fungi. Chitin consists of chains of glucose molecules, each linked by β-bonds, and each of which has an acetylamino group ($-N-CO-CH_3$) in place of a hydroxyl ($-OH$) group on the second carbon. Compared to cellulose, this is a more digestible nutrient for vertebrates, although only fish and some species of marine birds have a pancreatic chitinase that breaks down chitin into N-acetyl-glucosamine.

Pancreatic Lipase

In vertebrates pancreatic lipase is extremely important, because it is the principal enzyme that can digest fat. Pancreatic lipase (like salivary lipase) hydrolyzes dietary triglycerides into monoglycerides and free fatty acids, which are the absorbable units of fat. Like amylase, lipase is secreted in its active form, because there is no risk of pancreatic self-digestion by lipase. When pancreatic enzymes are deficient, digestion of food is incomplete. Because the pancreas is the only significant source of lipase, pancreatic enzyme deficiency results in serious maldigestion of fats. The principal clinical manifestation of pancreatic exocrine insufficiency in mammals is **steatorrhea,** or excessive undigested fat in the feces. Up to 60 to 70% of the ingested fat may be excreted in the feces. Digestion of protein and carbohydrates is less impaired, because salivary, gastric, and small-intestinal enzymes contribute to the digestion of these two foodstuffs.

Pancreatic Aqueous Alkaline Secretion

Pancreatic enzymes function best in a neutral or slightly alkaline environment, yet the highly acidic gastric contents are emptied into the duodenal lumen in the vicinity of pancreatic enzyme entry into the duodenum. This acidic chyme must be quickly neutralized in the duodenal lumen, not only to allow optimal functioning of the pancreatic enzymes but also to prevent acid damage to the duodenal mucosa. The alkaline ($NaHCO_3$-rich) fluid secreted by the pancreas into the duodenal lumen serves the important function of neutralizing the acidic chyme as the latter is emptied into the duodenum from the stomach:

$$HCl + NaHCO_3 \rightarrow NaCl + H_2CO_3$$

This aqueous $NaHCO_3$ secretion is by far the largest component of pancreatic secretion. The volume of pancreatic secretion ranges between 1 and 2 liters per day in humans to as high as 12 liters per day in a pony, depending on the types and degree of stimulation.

■ Pancreatic exocrine secretion is hormonally regulated to maintain neutrality of the duodenal contents and to optimize digestion.

Pancreatic exocrine secretion is regulated primarily by hormonal mechanisms. A small amount of cholinergic-induced (parasympathetic) pancreatic secretion occurs during the cephalic phase of digestion, with a further token increase occurring during the gastric phase in response to gastrin. However, the predominant stimulation of pancreatic secretion occurs during the *intestinal phase* of digestion, when chyme is in the small intestine. The release of the two major enterogastrones, secretin and cholecystokinin (CCK), in response to chyme in the duodenum plays the central role in the control of pancreatic secretion (● Figure 14–13).

Role of Secretin in Pancreatic Secretion

Of the factors that stimulate enterogastrone release, the primary stimulus specifically for secretin release is acid in the duodenum. Secretin, in turn, is carried by the blood to the pancreas, where it stimulates the duct cells to markedly increase their secretion of a $NaHCO_3$-rich aqueous fluid into the duodenum. Even though other stimuli may cause the release of secretin, it is appropriate that the most potent stimulus is acid, because secretin promotes the alkaline pancreatic secretion that neutralizes the acid. This mechanism provides a control system for maintaining neutrality of the chyme in the intestine, and is thus another example of negative feedback. The amount of secretin released is proportional to the amount of acid that enters the duodenum, so the amount of $NaHCO_3$ secreted parallels the duodenal acidity.

Role of CCK in Pancreatic Digestion

Cholecystokinin is important in the regulation of pancreatic digestive enzyme secretion. The main stimulus for release of CCK from the duodenal mucosa into the blood is the presence

Figure 14-13 ▶ • Hormonal control of pancreatic exocrine secretion (mammal).

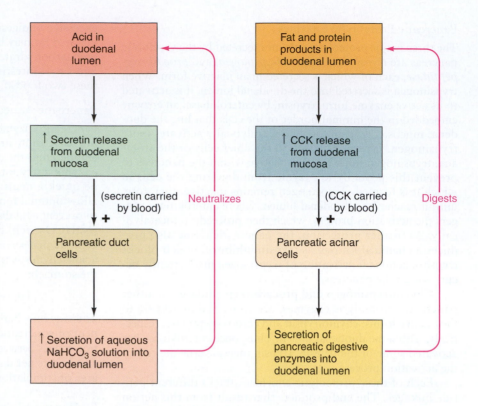

of nutrients in the lumen, especially fat and, to a lesser extent, protein products. Carbohydrate apparently does not have any direct influence on pancreatic enzyme secretion. The circulatory system transports CCK to the pancreas, where it stimulates the pancreatic acinar cells to increase digestive enzyme secretion.

All three types of pancreatic enzymes are packaged together in the zymogen granules, so all the pancreatic enzymes are released together on exocytosis of the granules. Therefore, even though the total amount of enzymes released varies depending on the type of meal consumed (the most being secreted in response to fat), the proportion of enzymes released does not vary on a meal-to-meal basis. That is, a high-protein meal does not cause the release of a greater proportion of proteolytic enzymes. Evidence suggests, however, that long-term adjustments in the proportion of the types of enzymes produced may occur as an adaptive response to a prolonged or seasonal change in diet. For example, a greater proportion of proteolytic enzymes are produced with a long-term switch to a high-protein feed. Cholecystokinin may play a role in pancreatic enzyme adaptation to changes in diet.

Just as gastrin is trophic to the stomach and small intestine, CCK and secretin exert trophic effects on the exocrine pancreas to maintain its integrity.

We now look at the contributions of the remaining vertebrate accessory digestive units, the liver and gallbladder.

Liver and Gallbladder

■ The vertebrate liver performs various important functions, including bile production.

Besides pancreatic juice, the other secretory product that is emptied into the duodenal lumen is **bile**. The **biliary system** includes the *liver*, the *gallbladder*, and associated ducts. Bile is continuously secreted in ruminants, horses, and pigs, whereas in most animals that feed at infrequent intervals, such as the dog and cat, bile is stored in the gallbladder until periods of feeding and digestion.

Liver Functions

The **liver** is the largest and most important metabolic organ in the vertebrate body; it can be viewed as the body's major "biochemical factory." Its importance to the digestive system is its secretion of *bile salts,* but the liver performs a wide variety of other functions, including the following:

- Metabolic processing of the major categories of nutrients (carbohydrates, proteins, and lipids) after their absorption from the digestive tract

- Detoxification or degradation of body wastes and hormones as well as drugs and other foreign compounds

- Synthesis of plasma proteins, including those necessary for the clotting of blood and those that transport steroid and thyroid hormones and cholesterol in the blood

- Storage of glycogen, fats, iron, copper, and many vitamins

- Activation of vitamin D, which the liver accomplishes in conjunction with the kidneys

- Removal of bacteria and worn-out red blood cells, thanks to its resident macrophages (see p. 433)

- Excretion of cholesterol and bilirubin, the latter being a breakdown product derived from the destruction of worn-out red blood cells

- Synthesis of ascorbic acid (vitamin C) in some species of mammals (excluding primates and guinea pigs) and birds

- Gluconeogenesis—the act of converting noncarbohydrate nutrients into glucose

- Buoyancy in sharks because of high amounts of low-density lipids

Given this wide range of complex functions, there is amazingly little specialization of cells within the liver. Each liver cell, or **hepatocyte** (*hepato* means "liver," *cyte* means "cell"), performs the same wide variety of metabolic and secretory tasks. The specialization comes from the highly developed organelles within each hepatocyte. However, in some species the liver performs additional functions during particular stages of development. For example, the size of the liver is increased in some animals engaged in reproduction. Synthesis of egg constituents, lipogenesis for yolk, and some albumin synthesis are carried out in the avian liver. As a consequence, *fatty livers* are a common problem in laying hens. Fat content of the liver can increase from 5 to 50% and can markedly increase mortality rate. **Fatty liver syndrome** arises when carbohydrate intake exceeds the liver's ability to synthesize lipoproteins for secretion into the plasma. However, there is not a corresponding increase in fat content of mammalian livers, because both the fetus and the mammary gland contribute to the metabolic demands associated with carbohydrate intake.

Liver Blood Flow

To carry out these wide-ranging tasks, the anatomic organization of the liver permits each hepatocyte to be in direct contact with blood from two sources: venous blood coming directly from the digestive tract, and arterial blood coming from the aorta. Venous blood enters the liver by means of a unique and complex vascular connection between the digestive tract and the liver, known as the **hepatic portal system** (● Figure 14–14). The veins draining the digestive tract do not directly join the inferior vena cava, the large vein that returns blood to the heart. Instead, the veins from the stomach and intestine enter the hepatic portal vein, which carries the products absorbed from the digestive tract directly to the liver for processing, storage, or detoxification before they gain access to the general circulation. Within the liver, the portal vein once again breaks up into a capillary network (the liver *sinusoids*) to permit exchange between the blood and hepatocytes before draining into the hepatic vein, which joins the inferior vena cava. The hepatocytes also are provided with fresh arterial blood, which supplies their oxygen and delivers blood-borne metabolites and toxins for hepatic processing.

■ The liver lobules are delineated by vascular and bile channels.

The liver is organized into functional units known as **lobules,** which are hexagonal arrangements of tissue surrounding a central vein (● Figure 14–15a). At the outer edge of each "slice" of the lobule are three vessels: a branch of the hepatic artery, a branch of the portal vein, and a bile duct. Blood from the branches of both the hepatic artery and the portal vein flows from the periphery of the lobule into large, expanded capillary spaces called **sinusoids,** which run between rows of liver cells to the central vein like spokes on a bicycle wheel (Figure 14–15b). The hepatocytes are arranged between the sinusoids in plates two cell layers thick, so that each lateral edge faces a sinusoidal pool of blood. The central veins of all the liver lobules converge to form the hepatic vein, which carries the blood away from the liver. A thin, bile-carrying channel, a **bile canaliculus,** runs between the cells within each hepatic plate. Hepatocytes continuously secrete bile into these thin channels, which carry the bile to a bile duct at the periphery of the lobule. The bile ducts from the various lobules converge to eventually form the hepatic duct, which transports the bile from the liver to the duodenum. Each hepatocyte is in contact with a sinusoid on one side and a bile canaliculus on the other side.

■ Bile is continuously secreted by the vertebrate liver and is diverted to the gallbladder between meals.

The opening of the bile duct into the duodenum is guarded by the **sphincter of Oddi,** which prevents bile from entering the

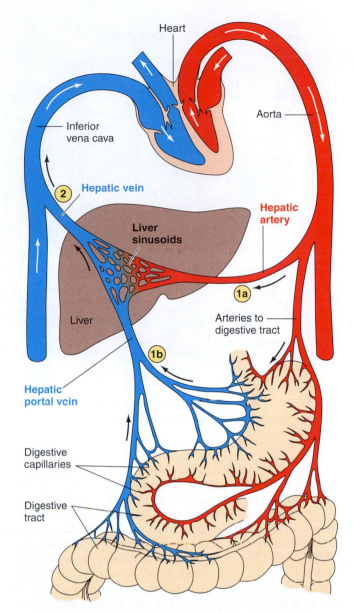

① **The liver receives blood from two sources:**

ⓐ Arterial blood, which provides the liver's O_2 supply and carries blood-borne metabolites for hepatic processing, is delivered by the **hepatic artery.**

ⓑ Venous blood draining the digestive tract is carried by the **hepatic portal** vein to the liver for processing and storage of newly absorbed nutrients.

② Blood leaves the liver via the **hepatic vein.**

Figure 14–14 ● Schematic representation of liver blood flow (mammal).

duodenum except during digestion of meals (● Figure 14–16). When this sphincter is closed, most of the bile secreted by the liver is diverted back up into the **gallbladder,** a small, saclike structure tucked beneath the liver. Those vertebrates with a gallbladder can concentrate the bile salts approximately 10- to 20-fold by selectively removing electrolytes and water. Bile is

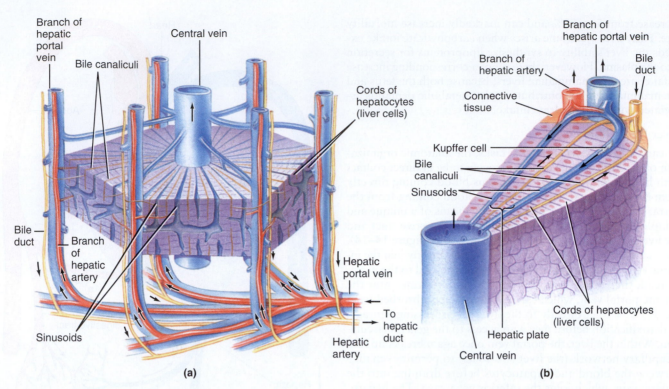

Figure 14–15 ● Anatomy of the mammalian liver.
(a) Hepatic lobule. (b) Wedge of a hepatic lobule.

subsequently stored and concentrated in the gallbladder between meals. After a meal, bile enters the duodenum as a result of the combined effects of gallbladder emptying and increased bile secretion by the liver. The amount of bile secreted per day in humans ranges from 250 mL to 1 liter, depending on the degree of stimulation. In birds, two bile ducts exit the liver, one from each lobe. The cystic duct is either connected to the gallbladder or directly to the small intestine, whereas the hepatic duct directly communicates with the small intestine. Animals without a gallbladder include the rat, horse, pigeon, deer, giraffe, elephant, and camel.

■ Bile salts are recycled through the enterohepatic circulation.

Bile consists of an *aqueous alkaline fluid* similar to the pancreatic $NaHCO_3$ secretion as well as several organic constituents, including *bile salts, cholesterol, lecithin,* and *bilirubin.* The organic constituents are derived from hepatocyte activity, whereas the duct cells add the water, $NaHCO_3$, and other inorganic salts. Even though bile does not contain any digestive enzymes, it is important for the digestion and absorption of fats, primarily through the **emulsification** effect of the bile salts.

Bile salts are derivatives of cholesterol. To reduce their toxicity, mammals conjugate bile acids with either *taurine* or *glycine,* whereas birds and fish form only the taurine conjugate. (Steroid bile salts have not been identified in invertebrates.) Bile salts are actively secreted into the bile and eventually enter the duodenum along with the other biliary constituents. Following their participation in fat digestion and absorption, most bile salts are reabsorbed back into the blood by special ac-

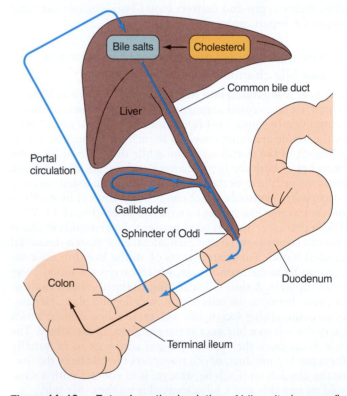

Figure 14–16 ● Enterohepatic circulation of bile salts (mammal). The majority of bile salts are recycled between the liver and small intestine through the enterohepatic circulation, designated by the blue arrows. After participating in fat digestion and absorption, most bile salts are reabsorbed by active transport in the terminal ileum and returned through the hepatic portal vein to the liver, which resecretes them in the bile.

tive-transport mechanisms located in mammals in the terminal ileum, the last portion of the small intestine. (In birds the majority of absorption takes place in the jejunum, the second intestinal segment.) From here the bile salts are returned via the hepatic portal system to the liver, which resecretes them into the bile. This recycling of bile salts (and some of the other biliary constituents) between the small intestine and liver is referred to as the **enterohepatic circulation** (*entero,* "intestine"; *hepatic,* "liver") (Figure 14–16).

The total amount of bile salts in the average human's body is about 3 to 4 g, yet 3 to 14 g of bile salts may be emptied into the duodenum in a single meal. Obviously, the bile salts must be recycled many times per day. Approximately 5% of the secreted bile salts ultimately escape into the feces daily, an important route for eliminating cholesterol from the body. These lost bile salts are replaced by new bile salts synthesized by the liver; thus the size of the pool of bile salts is kept constant.

■ **Bile salts aid fat digestion and absorption through their detergent action and micellar formation, respectively.**

Bile salts aid fat digestion through their detergent action (emulsification) and facilitate fat absorption through their participation in the formation of micelles. Both functions are related to the structure of bile salts.

Detergent Action of Bile Salts

The term **detergent action** refers to bile salts' ability to convert large fat globules into a **lipid emulsion** that consists of many small fat droplets suspended in the aqueous chyme, thus increasing the surface area available for attack by pancreatic lipase. For lipase to digest fat, it must come into direct contact with the triglyceride molecules. Because fat molecules are not soluble in water, they tend to aggregate into large droplets in the aqueous environment of the small intestine lumen. If bile salts did not emulsify these large droplets, lipase could act on the lipids only at the surface of the large droplets, and triglyceride digestion would be greatly prolonged.

Bile salts exert a detergent action similar to that of the detergent you use to break up grease when you wash dishes. A bile salt molecule contains a lipid-soluble portion (a steroid derived from cholesterol) plus a negatively charged, water-soluble portion. Bile salts *adsorb* on the surface of a fat droplet; that is, the lipid-soluble portion of the bile salt dissolves in the fat droplet, leaving the charged water-soluble portion projecting from the surface of the droplet (● Figure 14–17a). Intestinal mixing movements break up large fat droplets into smaller ones. These small droplets would quickly recoalesce were it not for bile salts adsorbing on their surface and creating a "shell" of water-soluble negative charges on the surface of each little droplet. Because like charges repel, these negatively charged groups on the droplet surfaces cause the fat droplets to repel each other (Figure 14–17b). This electrical repulsion prevents the small droplets from recoalescing into large fat droplets, producing a lipid emulsion that increases the surface area available for lipase action.

Although bile salts increase the surface area available for attack by pancreatic lipase, lipase alone cannot penetrate the layer of bile salts adsorbed on the surface of the small emulsi-

fied fat droplets To solve this dilemma, the pancreas secretes the polypeptide colipase along with lipase. Colipase binds both to lipase and to the bile salts at the surface of the fat droplets, anchoring lipase to its site of action.

During digestion of a meal, when chyme reaches the small intestine, the presence of food, especially fat products, in the duodenal lumen triggers the release of CCK. This hormone stimulates contraction of the gallbladder and relaxation of the sphincter of Oddi, so bile is discharged into the duodenum, where it appropriately aids in the digestion and absorption of the fat that initiated the release of CCK. The presence of food or acid in the small intestine of fish results in the release of a gastrin/CCK-related peptide that stimulates gallbladder contraction.

Micellar Formation

Bile salts—along with cholesterol and lecithin, which are also constituents of bile—play an important role in facilitating fat absorption through micellar formation. Like bile salts, lecithin has both a lipid-soluble and a water-soluble portion, whereas cholesterol is almost totally insoluble in water. In a **micelle,** the bile salts and lecithin aggregate in small clusters with their fat-soluble portions huddled together in the middle to form a hydrophobic ("water-fearing") core, while their water-soluble portions form an outer hydrophilic ("water-loving") shell (● Figure 14–18). A micellar aggregate is about one millionth the size of an emulsified lipid droplet. Micelles, themselves being water soluble by virtue of their hydrophilic shells, can dissolve water-insoluble (and hence lipid-soluble) substances in their lipid-soluble cores. Micelles thus provide a handy vehicle for carrying water-insoluble substances through the watery luminal contents. The most important lipid-soluble substances thus carried are the products of fat digestion (monoglycerides and free fatty acids) as well as fat-soluble vitamins, which are all transported to their sites of absorption by means of the micelles. If they did not hitch a ride in the water-soluble micelles, these nutrients would float on the surface of the aqueous chyme (just as oil floats on top of water), never reaching the absorptive surfaces of the small intestine.

■ **Bilirubin is a waste product excreted in the bile.**

Bilirubin, the other major constituent of bile, does not play a role in digestion at all but instead is one of the few waste products excreted in the bile. Worn-out red blood cells (p. 365) are removed from the blood by macrophages in the spleen and other areas in the body. Kupffer cells degrade these worn-out red blood cells to yield heme (iron-containing portion of the hemoglobin). Heme is converted to **biliverdin,** transformed into bilirubin and then released into the lobule sinusoid. Bilirubin (yellow in color) is then extracted from the blood by the hepatocytes and is actively excreted into the bile. Birds are unusual in that they lack the necessary *transferase* enzyme required to convert biliverdin into bilirubin. Consequently they excrete biliverdin as the major means of pigment elimination, with the result that the color of avian bile is dark green. The excretion rate of bile pigment is also greater in birds because of the much shorter life of red blood cells (approximately 35 days).

A small amount of bilirubin is normally reabsorbed by the small intestine back into the blood, and when it is eventually

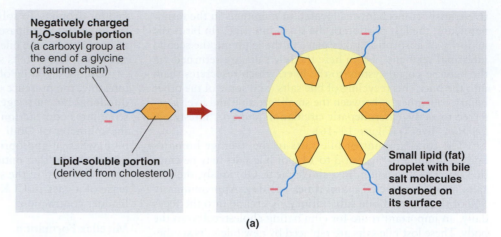

Negatively charged H₂O-soluble portion (a carboxyl group at the end of a glycine or taurine chain)

Lipid-soluble portion (derived from cholesterol)

Small lipid (fat) droplet with bile salt molecules adsorbed on its surface

(a)

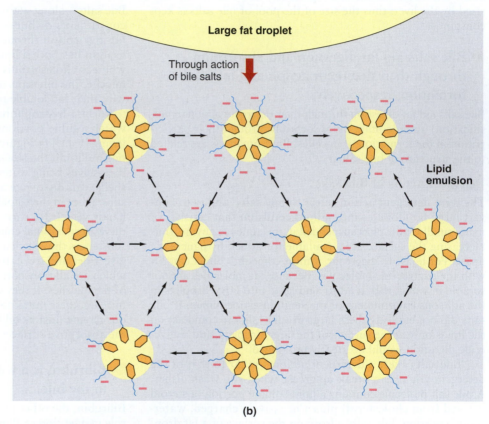

Large fat droplet

Through action of bile salts

Lipid emulsion

(b)

Figure 14–17 ● Schematic structure and function of bile salts. (a) Schematic representation of the structure of bile salts and their adsorption on the surface of a small fat droplet. A bile salt consists of a lipid-soluble portion that dissolves in the fat droplet and a negatively charged, water-soluble portion that projects from the surface of the droplet. (b) Formation of a lipid emulsion through the action of bile salts. When a large fat droplet is broken up into smaller fat droplets by intestinal contractions, bile salts adsorb on the surface of the small droplets, creating "shells" of negatively charged, water-soluble bile salt components that cause the fat droplets to repel each other. This action holds the fat droplets apart and prevents them from recoalescing, thereby increasing the surface area of exposed fat available for digestion by pancreatic lipase.

excreted in the urine, it is largely responsible for the urine's yellow color. If bilirubin is formed more rapidly than it can be excreted, it accumulates in the body and causes **jaundice.** Animals with this condition appear yellowish, with this color being most apparent in the whites of their eyes.

Fat Body in Insects

In many ways the **fat body** of insects is analogous to the vertebrate liver, because it plays a role in intermediary metabolism, is regulated by hormones, and serves as storage place for food reserves. Because lipids can be accumulated in such massive quantities in the fat body, it appears as a diffuse organ of fat located throughout the **hemocoel** (body cavity filled with blood) and abdomen of insects. The vast majority of the cells in the fat body that perform metabolic functions are termed

trophocytes. Another fat body cell type includes **urate cells,** which are involved in storing uric acid crystals in species lacking Malpighian tubules. The fat body is structurally organized to maximize exchange with the hemolymph, which is the carrier for products taken up or released by the fat body. For example, during feeding cycles the fat body synthesizes lipids, carbohydrates, and proteins whereas during strenuous activity it mobilizes the release of lipids and carbohydrates. The fat body is most highly developed during the larval instars and normally regresses at the end of metamorphosis, except in some adult species of insects that do not actively feed. The high concentrations of trehalose found in some insects are due to its synthesis in the fat body. Trehalose can be generated from dietary monosaccharides or stored glycogen, or from gluconeogenesis.

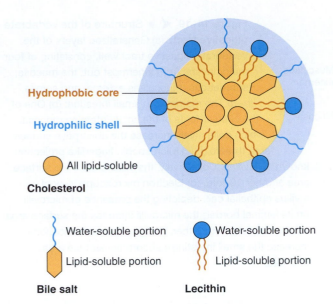

Hydrophobic core

Hydrophilic shell

All lipid-soluble

Cholesterol

Water-soluble portion   Water-soluble portion

Lipid-soluble portion   Lipid-soluble portion

Bile salt   **Lecithin**

Figure 14–18 ● **Schematic representation of a micelle.** Bile constituents (bile salts, lecithin, and cholesterol) aggregate to form micelles that consist of a hydrophilic (water-soluble) shell and a hydrophobic (lipid-soluble) core. Because the outer shell of a micelle is water soluble, the products of fat digestion, which are not water soluble, can be carried through the watery luminal contents to the absorptive surface of the small intestine by dissolving in the micelle's lipid-soluble core.

Small Intestine

The **small intestine** is the site at which most digestion and absorption of nutrients takes place in vertebrates (in arthropods, the midgut and hindgut are the sites of most digestion and absorption). An alternative strategy relies on the pregastric absorption of some nutrients through the lining of the ruminoreticulum (see discussion of ruminants, p. 661). In addition, some animal groups can ferment cellulose postgastrically and absorb nutrients through the lining of the colon and/or the cecum (see p. 658).

The small intestine is a tube that lies coiled within the abdominal cavity, extending between the stomach and the large intestine. It is arbitrarily divided into three segments: the **duodenum**, the **jejunum**, and the **ileum**. In birds, **Meckel's diverticulum**, a vestige of the yolk sac, lies approximately midway along the small intestine.

■ The digestive tract wall has four layers.

The digestive tract wall has the same general structure throughout most of its length from the esophagus to the anus, with some local variations characteristic for each region. A cross section of the small intestine (● Figure 14–19) reveals four major tissue layers. From the innermost layer of the tract outward, they are the *mucosa*, the *submucosa*, the *muscularis externa*, and the *serosa*.

The **mucosa** lines the luminal surface of the digestive tract. It is divided into three layers:

■ The primary component of the mucosa is a **mucous membrane**, an inner epithelial layer that serves as a protective

surface as well as being modified in particular areas for secretion and absorption. The mucous membrane contains *exocrine cells* for secreting digestive juices, *endocrine cells* for secreting gastrointestinal hormones, and *epithelial cells* specialized for absorbing digested nutrients.

■ The **lamina propria** is a thin middle layer of connective tissue on which the epithelium rests. Small blood vessels, lymph vessels, and nerve fibers pass through the lamina propria, and it houses the **gut-associated lymphoid tissue (GALT)**, which is important in the defense against intestinal bacteria.

■ The **muscularis mucosa**, a sparse outer layer of smooth muscle, lies adjacent to the submucosa.

The **submucosa** ("under the mucosa") is a thick layer of connective tissue that provides the digestive tract with its distensibility and elasticity. It contains the larger blood and lymph vessels, both of which send branches inward to the mucosal layer and outward to the surrounding thick muscle layer. Also lying within the submucosa is a nerve network known as the *submucous plexus*, which helps control local activities of each gut region.

Surrounding the submucosa is the **muscularis externa**, the major smooth-muscle coat of the digestive tube. In most parts of the tract, it consists of two layers: an inner circular layer and an outer longitudinal layer. The fibers of the inner smooth-muscle layer (adjacent to the submucosa) run circularly around the circumference of the tube. Contraction of these circular fibers constricts, or decreases the diameter of, the lumen at the point of contraction. Contraction of the fibers in the outer layer, which run longitudinally along the length of the tube, accomplishes shortening of the tube. Together, contractile activity of these smooth-muscle layers produces the propulsive and mixing movements. Lying between the two muscle layers is another nerve network, the myenteric plexus, which, along with the submucous plexus, helps regulate local gut activity.

The outer connective tissue covering of the digestive tract is the **serosa**, which secretes a watery serous fluid that lubricates and prevents friction between the digestive organs and surrounding viscera. Throughout much of the tract, the serosa is continuous with the **mesentery**, which suspends the digestive organs from the inner wall of the abdominal cavity like a sling (Figure 14–19a). This attachment provides relative fixation, supporting the digestive organs in proper position, while still allowing them freedom for mixing and propulsive movements.

The mucous lining of the small intestine is remarkably well adapted to its special absorptive function, for two reasons: (1) It has a very large surface area, and (2) the epithelial cells in this lining have a variety of specialized transport mechanisms.

Adaptations That Increase the Small Intestine's Surface Area

The following special modifications of the small-intestine mucosa greatly increase the surface area available for absorption (Figure 14–19b, c, d):

■ The inner surface of the small intestine is arranged in circular folds that are visible to the naked eye and that increase the surface area threefold.

■ Projecting from this folded surface are microscopic, finger-like projections known as **villi**, which give the lining a velvety appearance and increase the surface area by another

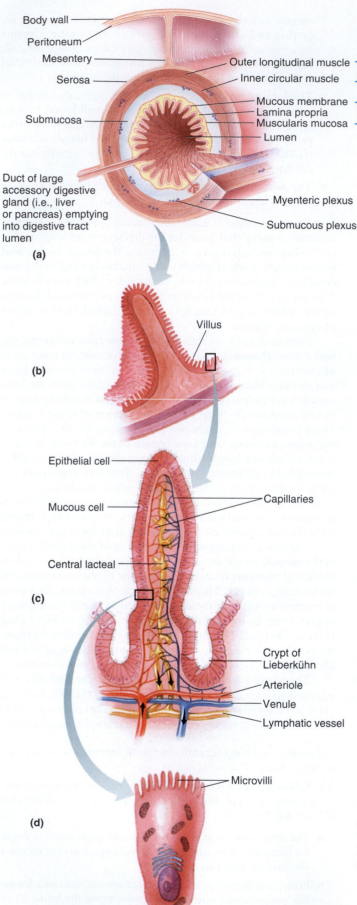

Body wall
Peritoneum
Mesentery
Serosa
Submucosa
Outer longitudinal muscle ⎤ Muscularis
Inner circular muscle ⎦ externa
Mucous membrane ⎤
Lamina propria ⎥ Mucosa
Muscularis mucosa ⎦
Lumen
Duct of large accessory digestive gland (i.e., liver or pancreas) emptying into digestive tract lumen
Myenteric plexus
Submucous plexus

(a)

Villus

(b)

Epithelial cell
Mucous cell
Central lacteal

(c)

Capillaries
Crypt of Lieberkühn
Arteriole
Venule
Lymphatic vessel

Microvilli

(d)

Figure 14–19 ◄ • **Structure of the vertebrate digestive tract.** (a) Generalized layers of the vertebrate digestive tract wall, consisting of four major layers: from innermost out, the mucosa, submucosa, muscularis externa, and serosa. (b–d) The mammalian small intestine: (b) One of the circular folds of the small-intestine mucosa, which collectively increase the absorptive surface area threefold. (c) Microscopic fingerlike projection known as a *villus*. Collectively, the villi increase the surface area another 10-fold. (d) Electron microscope view of a villus epithelial cell, depicting the presence of microvilli on its luminal border; the microvilli increase the surface area another 20-fold. Altogether, these surface modifications increase the small intestine's absorptive surface area 600-fold.

10 times (Figure 14–19b, c). Epithelial cells interspersed with mucous cells cover the surface of each villus. Villi are *dynamic* and readily change dimensions in response to numerous dietary factors. For example, the weight of a python's small intestine increases two- to threefold overnight after swallowing its prey, which is associated with a 60-fold increase in enzyme content. Each epithelial cell lining the small intestine increases in size and develops longer projections, adaptations that enable the ingesta to be digested more quickly.

■ Even smaller, hairlike projections known as **microvilli** (or the *brush border*) arise from the luminal surface of these epithelial cells (Figure 14–19d), increasing the surface area another 20-fold. Each epithelial cell has as many as 3,000 to 6,000 of these microvilli, which are visible only with an electron microscope. Generally, villi are longest in the jejunum and progressively decrease in length through the ileum. It is within the membrane of this brush border that the enzymes of the small intestine perform their functions. Arising from the surface of the microvilli is the **glycocalyx**, a meshwork of carbohydrate-rich filaments. These filaments are components of the digestive enzymes that project into the lumen of the small intestine.

Altogether the folds, villi, and microvilli provide the small intestine of humans with a luminal surface area 600 times greater than it would have if it were a tube of the same length and diameter lined by a flat surface. In fact, if the surface area of the small intestine were spread out flat, it would cover an entire tennis court. Interestingly, the energetic cost to the python for adapting its small intestine to the arrival of food is approximately a third to a half of the energy derived from its prey.

Structure of a Villus

Absorption across the digestive tract wall involves transepithelial transport similar to movement of material across the kidney tubules (see p. 540). Each villus has the following major components (Figure 14–19c):

■ *Epithelial cells that cover the surface of the villus.* The epithelial cells are joined at their lateral borders by tight junctions, which limit passage of luminal contents between the cells, although the tight junctions in the small intestine are "leakier" than those in the stomach. These epithelial cells have, within their luminal brush borders, carriers for absorption of specific nutrients and electrolytes from the lumen as well as the brush-border digestive enzymes that complete carbohydrate and protein digestion.

- *A capillary network.* Each villus is supplied by an arteriole that breaks up into a capillary network within the villus. The capillaries rejoin to form a venule that drains away from the villus.

- *A terminal lymphatic vessel.* Each villus is supplied by a single, blind-ended lymphatic vessel known as the **central lacteal,** which occupies the center of the villus core.

During the process of absorption, digested substances enter the capillary network or the central lacteal. To be absorbed, a substance must pass completely through the epithelial cell, diffuse through the interstitial fluid within the connective tissue core of the villus, and then cross the wall of a capillary or lymph vessel. Like renal transport, intestinal absorption may be an active or passive process, with active absorption involving energy expenditure during at least one of the steps in the transepithelial transport process.

■ Segmentation contractions mix and slowly propel the chyme.

Segmentation, the small intestine's primary method of motility, both mixes and slowly propels the chyme. Segmentation consists of oscillating, ringlike contractions of the circular smooth muscle along the length of the small intestine; between the contracted segments are relaxed areas, containing a small bolus of chyme. The contractile rings occur every few centimeters apart, dividing the small intestine into segments like a chain of sausages. These contractile rings do not sweep along the length of the intestine as peristaltic waves do. Rather, after a brief period of time the contracted segments relax, and ringlike contractions appear in the previously relaxed areas (● Figure 14–20). The new contraction forces the chyme in a previously relaxed segment to move in both directions into the now relaxed adjacent segments. A newly relaxed segment therefore receives chyme from both the contracting segment immediately ahead of it and the one immediately behind it. Shortly thereafter, the areas of contraction and relaxation alternate again. In this way, the chyme is chopped, churned, and thoroughly mixed. These contractions can be compared to squeezing a pastry tube with your hands to mix the contents. This mixing serves the dual functions of mixing the chyme with the digestive juices secreted into the small-intestine lumen and exposing all the chyme to the absorptive surfaces of the small-intestine mucosa.

Initiation and Control of Segmentation

Segmentation contractions are initiated by the small intestine's pacesetter cells, which produce a basic electrical rhythm (BER) similar to the gastric BER responsible for peristalsis in the stomach. If the small-intestine BER brings the circular smooth-muscle layer to threshold, segmentation contractions are induced, with the frequency of segmentation following the frequency of the BER.

The circular smooth muscle's degree of responsiveness and thus the intensity of segmentation contractions can be in-

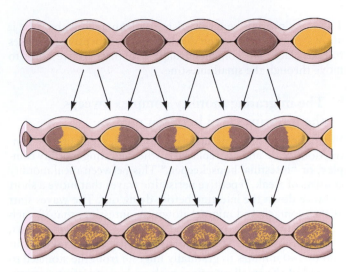

Figure 14–20 ● Segmentation. Segmentation consists of ringlike contractions along the length of the vertebrate small intestine. Within a matter of seconds, the contracted segments relax and the previously relaxed areas contract. These oscillating contractions thoroughly mix the chyme within the small-intestine lumen.

fluenced by distension of the intestine, by the hormone gastrin, and by extrinsic nerve activity. All these factors influence the excitability of the small-intestine smooth-muscle cells by moving the "resting" potential around which the BER oscillates closer to or farther from threshold. Segmentation is slight or absent between meals but becomes very vigorous immediately after feeding. Both the duodenum and ileum start to segment simultaneously when the digesta first enters the small intestine. The duodenum starts to segment primarily in response to local distension caused by the presence of chyme. Segmentation of the empty ileum, in contrast, appears to be brought about by gastrin, which is secreted in response to the presence of chyme in the stomach, a mechanism known as the **gastroileal reflex.** Extrinsic nerves can modify the strength of these contractions. Parasympathetic stimulation enhances segmentation, whereas sympathetic stimulation depresses segmental activity.

Segmentation not only accomplishes mixing but also is the primary factor responsible for slowly moving chyme through the small intestine. How can this be, when each segmental contraction propels chyme both forward and backward? The chyme slowly progresses forward, because the frequency of segmentation declines along the length of the small intestine. The pacesetter cells in the duodenum spontaneously depolarize at a faster rate than those farther down the tract, with the pacesetter cells in the terminal ileum exhibiting the slowest rate of spontaneous depolarization. The rate of segmentation contractions in the duodenum of a human is 12/min, compared to a rate of only 9/min in the terminal ileum. Because segmentation occurs with greater frequency in the upper part of the small intestine than in the lower part, more chyme on average is pushed forward than is pushed backward. As a result, chyme is moved very slowly from the upper to the lower part of the small intestine, being shuffled back and forth to accomplish thorough mixing and absorption in the process.

This slow propulsive mechanism is advantageous, because it allows ample time for the digestive and absorptive processes to take place. The contents usually take three to five hours to move through the small intestine.

The migrating motility complex sweeps the intestine clean between meals.

When most of the meal has been absorbed, segmentation contractions cease and are replaced by the **migrating motility complex,** or **"intestinal housekeeper."** This between-meal motility consists of weak, repetitive peristaltic waves that move a short distance down the intestine before dying out. The waves start at the stomach and migrate down the intestine; that is, each new peristaltic wave is initiated at a site a little farther down the small intestine. These short peristaltic waves take about 100 to 140 minutes to gradually migrate from the stomach to the end of the small intestine, with each contraction "sweeping" any remnants of the preceding meal plus mucosal debris and bacteria forward toward the colon, just like a good housekeeper. After the end of the small intestine is reached, the cycle begins again and continues to repeat itself until the next meal. It is speculated that the hormone **motilin** plays a role in regulating the migrating motility complex. When the next meal arrives, segmental activity is triggered again, and the migrating motility complex ceases.

The ileocecal juncture prevents contamination of the small intestine by colonic bacteria.

At the juncture between the small and large intestines, the last portion of the ileum empties into the cecum (● Figure 14–21). Two factors contribute to this region's ability to act as a barrier between the small and large intestines. First, the anatomic arrangement is such that valvelike folds of tissue protrude from the ileum into the lumen of the cecum. When the ileal contents are pushed forward, this **ileocecal valve** is easily pushed open, but the folds of tissue are forcibly closed when the cecal contents attempt to move backward. Second, the smooth muscle within the last several centimeters of the ileal wall is thickened, forming a sphincter that is under neural and hormonal control. Most of the time this **ileocecal sphincter** remains at least mildly constricted. Pressure on the cecal side of the sphincter causes it to contract more forcibly; distension of the ileal side causes the sphincter to relax, a reaction that is mediated by the intrinsic plexuses in the area. Relaxation of the sphincter is enhanced through the release of gastrin at the onset of a meal, when increased gastric activity is taking place. This relaxation allows the undigested fibers and unabsorbed solutes from the preceding meal to be moved forward as the new meal enters the tract.

The digestive tract of numerous carnivorous mammals, including some species of bats and marsupials, lack a constriction or valve between the small and large intestine. There is very little distinction between the small and the large intestinal tracts, and generally these animals also lack a cecum. For example, the short terminal segment of the large intestine of a mink is slightly larger than the small intestine and lacks a sphincter or valve at the junction between the two segments. In the horse, there is a sphincter at both the *ileocecal* and the *cecocolic* junction that ensures that the intestinal contents enter the cecum before passage through the colon.

Small-intestine secretions in vertebrates do not contain any digestive enzymes.

Each day the exocrine glands located in the small-intestine mucosa of humans secrete into the lumen about 1.5 liters of an aqueous salt and mucus solution known as the **succus entericus** (*succus,* "juice"; *entericus,* "of the intestine"). No digestive enzymes are secreted into this intestinal juice. The small intestine synthesizes digestive enzymes, but they function within the borders of the epithelial cells that line the lumen instead of being secreted directly into the lumen.

Are any functions served by this small-intestine secretion, because no digestive enzymes are involved? The mucus in the secretion provides protection and lubrication. Furthermore, this aqueous secretion provides an abundance of H_2O to participate in the enzymatic digestion of food. Recall that digestion involves hydrolysis that proceeds most efficiently when all the reactants are in solution.

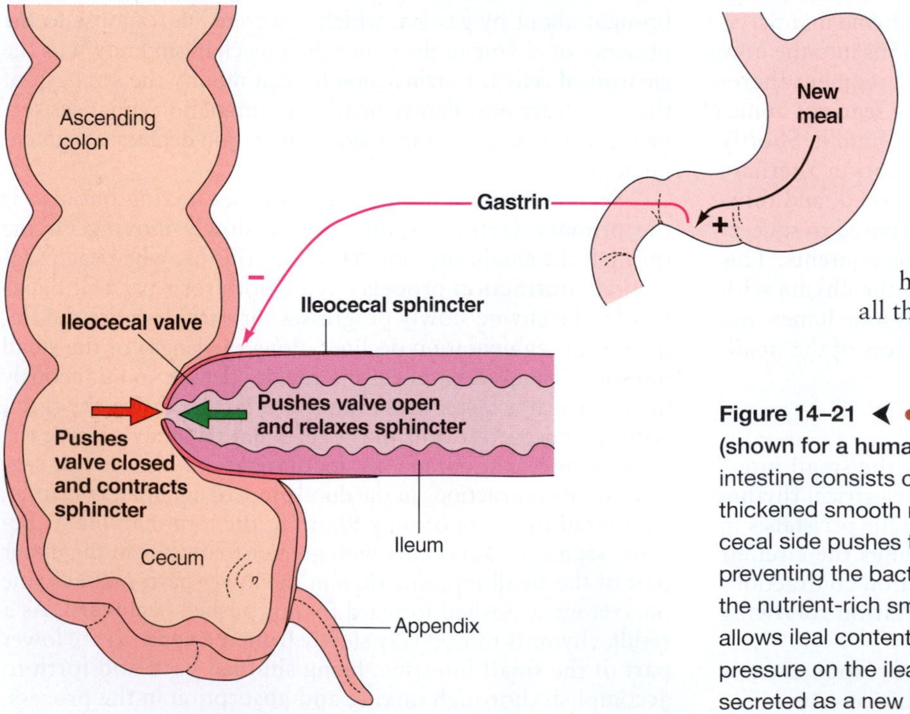

Figure 14–21 ◄ ● Control of the ileocecal valve/sphincter (shown for a human). The juncture between the ileum and large intestine consists of the ileocecal valve, which is surrounded by thickened smooth muscle, the ileocecal sphincter. Pressure on the cecal side pushes the valve closed and contracts the sphincter, preventing the bacteria-laden colonic contents from contaminating the nutrient-rich small intestine. The valve/sphincter opens and allows ileal contents to enter the large intestine in response to pressure on the ileal side of the valve and to the hormone gastrin secreted as a new meal enters the stomach.

The regulation of small-intestine secretion is not clearly understood. The factors that specifically alert the intestine of the upcoming meal are not completely known. However, the secretion of succus entericus does increase after a meal. The most effective stimulus for secretion appears to be local stimulation of the small-intestine mucosa by the presence of chyme.

Small-intestine enzymes complete the process of digestion at the brush border membrane.

Digestion within the small-intestine lumen is accomplished by the pancreatic enzymes, with fat digestion being enhanced by bile secretion. As a result of pancreatic enzymatic activity, fats are completely reduced to their absorbable units of monoglycerides and free fatty acids, proteins are broken down into small peptide fragments and some amino acids, and carbohydrates are reduced to disaccharides and some monosaccharides. Thus fat digestion is completed within the small-intestine lumen, but carbohydrate and protein digestion have not been brought to completion.

Special hairlike projections on the luminal surface of the mammalian small-intestine epithelial cells form the **brush border** (see p. 646), which contains different categories of enzymes:

1. **Enterokinase,** which activates the pancreatic enzyme trypsinogen.
2. The **disaccharidases** (**maltase, sucrase, lactase,** and **trehalase**), which complete carbohydrate digestion by hydrolyzing the remaining disaccharides (**maltose, sucrose, lactose,** and **trehalose**) into their constituent monosaccharides.
3. The **aminopeptidases,** which hydrolyze the small peptide fragments into their amino acid components, thereby completing protein digestion. Thus carbohydrate and protein digestion are completed at the brush border. (▌Table 14–6 summarizes the digestive processes for the three major categories of nutrients.)

Not all vertebrate groups express each of these enzymes in their brush border. For example, birds have no need of lactase but do have an intestinal amylase and an esterase. Esterases hydrolyze *esters*, compounds found in waxes and ripening fruit. Because the microbes in the reticulorumen of a ruminant scavenge most of the carbohydrates, the concentrations of disaccharidases in the small intestine are low with sucrase not synthesized at all. Some invertebrate disaccharidases in the

Table 14–6 ▌ Digestive Processes for the Three Major Categories of Nutrients (mammals and insects)

Nutrients	Enzymes for Digesting Nutrient	Source of Enzymes	Site of Action of Enzymes	Action of Enzymes	Absorbable Units of Nutrients
Mammals					
Carbohydrate	Amylase	Salivary glands	Mouth and body of stomach	Hydrolyzes polysaccharides to disaccharides	
		Exocrine pancreas	Small-intestine lumen		
	Disaccharidases (maltase, sucrase, lactase)	Small-intestine epithelial cells	Small-intestine brush border	Hydrolyze disaccharides to monosaccharides	Monosaccharides, especially glucose
Protein	Pepsin	Stomach chief cells	Stomach antrum	Hydrolyzes protein to peptide fragments	
	Trypsin, chymotrypsin, carboxypeptidase	Exocrine pancreas	Small-intestine lumen	Attack different peptide fragments	
	Aminopeptidases	Small-intestine epithelial cells	Small-intestine brush border	Hydrolyze peptide fragments to amino acids	Amino acids and a few small peptides
Fat	Lipase	Exocrine pancreas	Small-intestine lumen	Hydrolyzes triglycerides to fatty acids and monoglycerides	Fatty acids and monoglycerides
	Bile salts (not an enzyme)	Liver	Small-intestine lumen	Emulsify large fat globules for attack by pancreatic lipase	
Insects (locust)	Amylase	Salivary gland, hindgut	Foregut, hindgut	Hydrolyzes polysaccharides to disaccharides	
	Maltase, invertase	foregut lining	Foregut, ceca	Hydrolyze disaccharides to monosaccharides	Monosaccharides
	Lipase, protease	Ceca, midgut, hindgut	Ceca, midgut, hindgut	Hydrolyze triglycerides to fatty acids and proteins to peptides and amino acids	Fatty acids, amino acids, peptides

midgut do not have the specificity of their vertebrate counterparts: Hydrolysis of sucrose is accomplished by a glucosidase that also reacts with maltose. Animals whose diet normally contains chitin have evolved a chitinase enzyme. Bats, turtles, and insectivorous lizards synthesize substantial quantities of chitinase in their gastric mucosa. Similarly, many species of fish and birds produce chitinase in their pancreas, whereas other species rely on intestinal bacteria to synthesize this enzyme. Gut microflora of some species may contain chitinase as well as cellulase.

The small intestine is highly adaptable for its primary role in absorption.

Normally, all products of carbohydrate, protein, and fat digestion, as well as most of the ingested electrolytes, vitamins, and water, are absorbed by the small intestine indiscriminately. Usually only the absorption of calcium and iron is adjusted to an individual animal's needs in the short run. Thus the more food that is consumed, the more that is digested and absorbed. However, this does not mean food absorption capacity is static. In many vertebrates, the length of the intestine and the density of its transporters can be altered under long-term dietary changes. For example, in omnivores the glucose transporter is upregulated on a prolonged high-carbohydrate diet (this is a form of acclimatization; p. 15).

The lengths of the small intestine and the types and densities of its transporters are remarkably matched to typical diets of species. For instance, herbivores typically have a higher density of intestinal glucose transporters than do carnivores. Cats require taurine and arginine in their diet and can suffer from even short-term deficits in these amino acids. Taurine, which cats cannot synthesize, is essential for neural development and other functions that are poorly understood. Arginine is used in the production of urea, which carnivores produce in high amounts because of their high protein intake. To ensure adequate intake, cats have exceptionally high levels of intestinal taurine and arginine transporters.

Most absorption occurs in the duodenum and jejunum; very little occurs in the ileum, not because the ileum does not have absorptive capacity but because most absorption has already been accomplished before the intestinal contents reach the ileum. The small intestine has an abundant reserve absorptive capacity. In fact, about 50% of the small intestine can be removed with little interference to absorption—with at least two exceptions. If the terminal portion of the ileum is removed, vitamin B_{12} and bile salts are not properly absorbed, because the specialized transport mechanisms for these two substances are located only in this region. All other substances can be absorbed throughout the length of the small intestine.

The mucosal lining experiences rapid turnover.

Dipping down into the mucosal surface between the villi are shallow invaginations known as the **crypts of Lieberkühn** (Figure 14–19c). Unlike the gastric pits, these intestinal crypts do not secrete digestive enzymes but they do secrete water and electrolytes, which, along with the mucus secreted by the cells on the villus surface, constitute the succus entericus.

Furthermore, the crypts function as "nurseries." The epithelial cells lining the small intestine slough off and are replaced at a rapid rate as a result of high mitotic activity in the crypts. New cells that are continually being produced at the bottom of the crypts migrate up the villi and, in the process, push off the older cells at the tips of the villi into the lumen. In this manner, more than 100 million intestinal cells are shed per minute. In humans the entire trip from crypt to tip averages about three days, so the epithelial lining of the small intestine is replaced approximately every three days.

The new cells undergo several changes as they migrate up the villus. The concentration of brush border enzymes increases and the capacity for absorption improves, so the cells at the tip of the villus have the greatest digestive and absorptive capability. Just at their peak, these cells are pushed off by the newly migrating cells. Thus the luminal contents are constantly exposed to cells that are optimally equipped to complete the digestive and absorptive functions. Furthermore, just as in the stomach, the rapid turnover of cells in the small intestine is essential because of the harsh luminal conditions. Cells exposed to the abrasive and corrosive luminal contents are easily damaged, so they must be continually replaced by a fresh supply of new cells. The old cells that are sloughed off into the lumen are digested, with the cell constituents being absorbed into the blood and recycled for synthesis of new cells.

In addition to the stem cells, **Paneth cells** are found in the crypts. Paneth cells serve a defensive function, safeguarding the stem cells. They produce two chemicals that thwart bacteria: (1) *lysozyme*, the bacterial-lysing enzyme also found in saliva, and (2) *defensins*, small proteins with antimicrobial properties (p. 459).

Let's now turn our attention to the mechanisms through which the specific dietary constituents are normally absorbed.

Energy-dependent Na^+ absorption drives passive H_2O absorption.

Sodium may be absorbed both passively and actively. When the electrochemical gradient favors movement of Na^+ from the lumen to the blood, passive diffusion of Na^+ can occur *between* the intestinal epithelial cells through the "leaky" tight junctions into the interstitial fluid within the villus. Movement of Na^+ *through* the cells is energy dependent and involves at least two different carriers, similar to the process of Na^+ reabsorption across the kidney tubules (see p. 540). Sodium may passively enter the epithelial cells at their luminal surface by itself, or it may be cotransported with glucose, amino acids, or other nutrients. It is then actively pumped out of the cell at the basolateral border into the interstitial fluid in the lateral spaces between the cells where they are not joined by tight junctions. From the interstitial fluid Na^+ eventually diffuses into the capillaries.

As with the renal tubules in the early portion of the nephron, the absorption of Cl^-, H_2O, glucose, and amino acids from the small intestine is linked to this energy-dependent Na^+ absorption. Chloride passively follows down the electrical gradient created by Na^+ absorption and can be actively absorbed as well, if needed. Most H_2O absorption in the digestive tract depends on the active carrier that pumps Na^+ into the lateral spaces, resulting in a concentrated area of high osmotic pressure in that localized region between the cells, similar to the situation in the kidneys (see p. 540). This localized high osmotic pressure induces H_2O to move from the lumen through the cell (and possibly from the lumen through the leaky tight junction) into the lateral space. Water entering the space reduces the osmotic pressure but raises the hydrostatic (fluid) pressure. As a result, H_2O is flushed out of the lateral space into the interior of the villus, where it is picked up by the capillary network. Meanwhile, more Na^+ is pumped into the lateral space to encourage more H_2O absorption.

Carbohydrate Absorption

Dietary carbohydrate is presented to the small intestine for absorption mainly in the forms of the disaccharides maltose (the product of polysaccharide digestion), sucrose, and lactose (● Figure 14–22). The disaccharidases located in the brush borders of the small-intestine cells further reduce these disaccharides into the absorbable monosaccharide units of glucose, galactose, and fructose. In ruminants the efficiency of digestion of sugar and starches has not been investigated, because of the mistaken dogma that the rumen microbes metabolize all carbohydrates.

Glucose and galactose are both absorbed by *secondary active transport,* in which cotransport carriers on the luminal border transport both the monosaccharide and Na^+ from the lumen into the interior of the intestinal cell. The operation of these cotransport carriers, which do not directly use energy themselves, depends on the Na^+ concentration gradient established by the energy-consuming basolateral Na^+–K^+ pump (see p. 81). Glucose (or galactose), having been concentrated in the cell by the cotransport carriers, leaves the cell down its concentration gradient by means of a passive carrier in the basolateral border to enter the blood within the villus. In addition to absorption of glucose through the cells by means of the cotransport carrier, recent evidence suggests that a significant amount of glucose crosses the epithelial barrier through "leaky" tight junctions between the epithelial cells. Fructose is absorbed into the blood solely by facilitated diffusion (passive carrier-mediated transport).

■ Carbohydrate and protein are both absorbed by secondary active transport and enter the blood.

Absorption of the digestion end products of both carbohydrates and proteins involves special carrier-mediated transport systems that require energy expenditure and Na^+ cotransport, and both categories of end products are absorbed into the blood.

Protein Absorption

In nonruminants not only are ingested proteins digested and absorbed, but endogenous proteins that have entered the digestive tract lumen from the three following sources are digested and absorbed as well:

1. Digestive enzymes, all of which are proteins that have been secreted into the lumen

2. Proteins within the cells that are pushed off from the villi into the lumen during the process of mucosal turnover

3. Small amounts of plasma proteins that normally leak from the capillaries into the digestive tract lumen

About 20 to 40 g of endogenous proteins enter the lumen of humans each day from these three sources. This quantity can amount to more than half of the protein presented to the small intestine for digestion and absorption. To prevent depletion of the body's protein stores, all endogenous proteins must be digested and absorbed along with the dietary proteins. The amino acids absorbed from both food and the endogenous protein are used primarily to synthesize new protein in the body.

The protein presented to the small intestine for absorption is primarily in the form of amino acids and a few small peptide fragments (● Figure 14–23). Most amino acids are absorbed across the intestinal cells by secondary active transport, similar to glucose and galactose absorption. Thus glucose, galactose, and amino acids all get a "free ride" in on the energy expended for Na^+ transport. Small peptides gain entry by means of a different carrier and are broken down into their constituent amino acids by the aminopeptidases in the brush borders or by intracellular peptidases. Like monosaccharides, amino acids enter the capillary network within the villus.

An exception to this process is found in neonatal mammals. Their intestines can accomplish limited uptake of whole proteins via the active process of endocytosis (see p. 41). This is primarily used to absorb whole antibodies of the IgA class (see p. 443) from maternal milk, and helps protect the young animal from pathogens while its immune system is still developing. There is also some evidence that large protein molecules are absorbed intact into an animal's bloodstream through a process termed *persorption.* This form of passage occurs *between the cells* and is possible wherever single-layered epithelium covers the mucosa of the gastrointestinal tract. Researchers have shown that ingested proteins, which were radioactively labeled, reach peak concentrations in the blood of a rabbit within 2 hours. Absorption rates ranged from 1 to 10% of the ingested protein.

Fat Absorption

Fat absorption is quite different from carbohydrate and protein absorption, because the insolubility of fat in water presents a special problem. Fat must be transferred from the watery chyme through the watery body fluids even though it is not water soluble. Therefore, fat must undergo a series of transformations to circumvent this problem during its digestion and absorption (● Figure 14–24).

When the stomach contents are emptied into the duodenum, the ingested fat is aggregated into large, oily, triglyceride droplets that float in the chyme. Recall that through the bile salts' detergent action in the small intestine, the large droplets are dispersed into a lipid emulsification of small droplets, thereby exposing a much greater surface area of fat for digestion by pancreatic lipase. The products of lipase digestion (monoglycerides and free fatty acids) are also not very water soluble, so very little of these end products of fat digestion can diffuse through the aqueous chyme to reach the absorptive lining. However, biliary components facilitate absorption of these fatty and products by forming micelles. Remember that micelles are water-soluble particles that can carry the end products of fat digestion within their lipid-soluble interiors. Once these micelles reach the luminal membranes of the epithelial cells, the monoglycerides and free fatty acids passively diffuse from the micelles through the lipid component of the epithelial cell membranes to enter the interior of these cells. As these fat products leave the micelles and are absorbed across the epithelial cell membranes, the micelles can pick up more monoglycerides and free fatty acids, which have been produced from digestion of other triglyceride molecules in the fat emulsion.

Figure 14–22 ● Carbohydrate digestion and absorption (vertebrate small intestine).

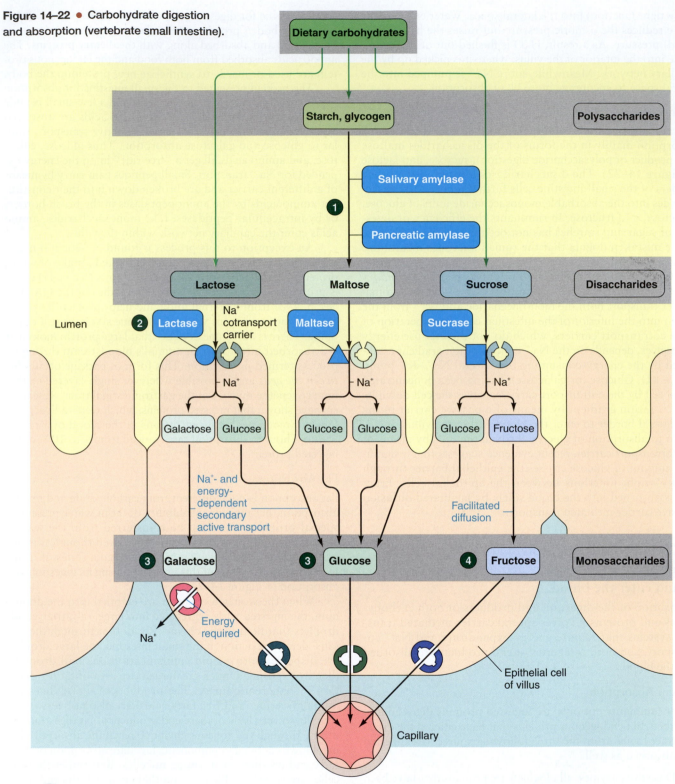

① The dietary polysaccharides starch and glycogen are converted into the disaccharide maltose through the action of salivary and pancreatic amylase.

② Maltose and the dietary disaccharides lactose and sucrose are converted to their respective monosaccharides by the disaccharidases (maltase, lactase, and sucrase) located in the brush borders of the small-intestine epithelial cells.

③ The monosaccharides glucose and galactose are absorbed into the interior of the cell and eventually enter the blood by means of Na^+- and energy-dependent secondary active transport.

④ The monosaccharide fructose is absorbed into the blood by passive facilitated diffusion.

Figure 14–23 ● Protein digestion and absorption
(vertebrate small intestine).

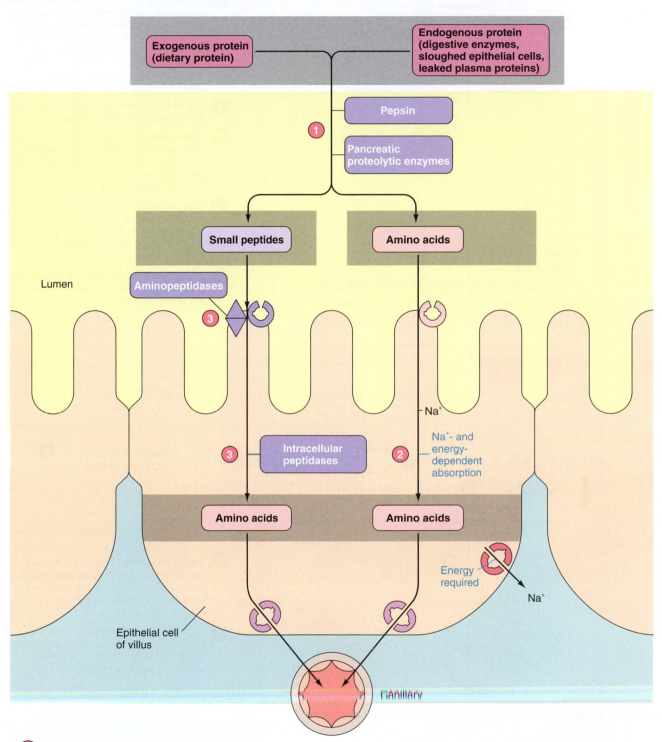

1 Dietary and endogenous proteins are hydrolyzed to their constituent amino acids and a few small peptide fragments by gastric pepsin and the pancreatic proteolytic enzymes.

2 Amino acids are absorbed into the small-intestine epithelial cells and eventually enter the blood by means of Na+- and energy-dependent secondary active transport. Various amino acids are transported by carriers specific for them.

3 The small peptides, which are absorbed by a different type of carrier, are broken down into their amino acids by aminopeptidases in the epithelial cells' brush borders or by intracellular peptidases.

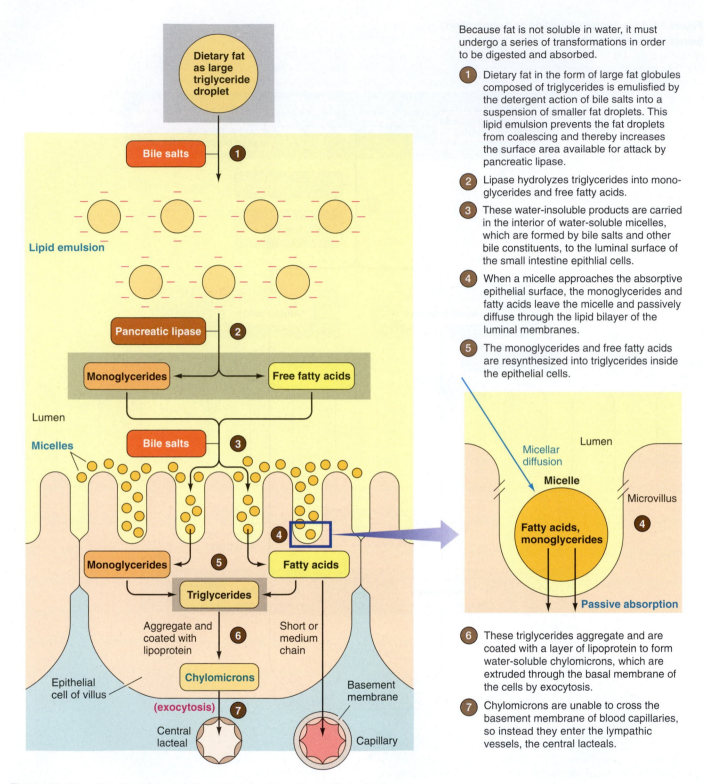

Because fat is not soluble in water, it must undergo a series of transformations in order to be digested and absorbed.

1　Dietary fat in the form of large fat globules composed of triglycerides is emulsified by the detergent action of bile salts into a suspension of smaller fat droplets. This lipid emulsion prevents the fat droplets from coalescing and thereby increases the surface area available for attack by pancreatic lipase.

2　Lipase hydrolyzes triglycerides into monoglycerides and free fatty acids.

3　These water-insoluble products are carried in the interior of water-soluble micelles, which are formed by bile salts and other bile constituents, to the luminal surface of the small intestine epithlial cells.

4　When a micelle approaches the absorptive epithelial surface, the monoglycerides and fatty acids leave the micelle and passively diffuse through the lipid bilayer of the luminal membranes.

5　The monoglycerides and free fatty acids are resynthesized into triglycerides inside the epithelial cells.

6　These triglycerides aggregate and are coated with a layer of lipoprotein to form water-soluble chylomicrons, which are extruded through the basal membrane of the cells by exocytosis.

7　Chylomicrons are unable to cross the basement membrane of blood capillaries, so instead they enter the lymphathic vessels, the central lacteals.

Figure 14–24 ● Fat digestion and absorption (vertebrate small intestine).

Bile salts continuously repeat their fat-solubilizing function down the length of the small intestine until all the fat is absorbed. Then the bile salts themselves are reabsorbed in the terminal ileum by special active transport. This is an efficient process, because relatively small amounts of bile salts can facilitate digestion and absorption of large amounts of fat, with each bile salt performing its ferrying function repeatedly before it is reabsorbed.

The actual transfer of monoglycerides and free fatty acids from the chyme across the apical membranes of the intestinal epithelial cells is a passive process, because the lipid-soluble fatty end products merely dissolve in and pass through the lipid portions of the membrane. Fat absorption is therefore said to be a passive process. However, the overall sequence of events necessary for fat absorption does require energy. For example, bile salts are actively secreted by the liver, the resynthesis of triglycerides and formation of chylomicrons within the epithelial cells, and exocytosis are energy-dependent processes.

Once within the interior of the epithelial cells, the monoglycerides and free fatty acids are resynthesized into triglycerides. These triglycerides conglomerate into droplets and are coated with a layer of lipoprotein (synthesized by the endoplasmic reticulum of the epithelial cell), which renders the fat droplets water soluble. The largest of the lipoproteins, known as **chylomicrons,** are extruded by exocytosis from the epithelial cells into the interstitial fluid within the villus. In mammals the triglyceride-rich chylomicrons are secreted into the central lacteals and enter the general circulation, whereas in birds they are absorbed into capillaries of the villi. Fatty acids with short- or medium-length carbon chains also enter the blood.

Vitamin Absorption

Water-soluble vitamins are primarily absorbed passively with water, whereas fat-soluble vitamins are carried in the micelles and absorbed passively with the end products of fat digestion. Carriers, if necessary, can also accomplish absorption of some of the vitamins. Vitamin B_{12} is unique in that it must be combined with gastric intrinsic factor for absorption by special transport in the terminal ileum.

■ Most absorbed nutrients immediately pass through the liver for processing.

The venules that leave the small-intestine villi, along with those from the remainder of the digestive tract, empty into the portal vein, which carries the blood to the liver. Consequently, anything absorbed into the digestive capillaries first must pass through the hepatic biochemical factory before entering the general circulation. Thus the products of carbohydrate and protein digestion as well as the electrolytes and H_2O are channeled into the liver, where many of these products are subjected to immediate metabolic processing. Furthermore, the liver detoxifies harmful substances that may have been absorbed from the gut before they gain access to the general circulation. After passing through the portal circulation, the venous blood from the digestive system is emptied into the vena cava and returned to the heart to be distributed throughout the body, carrying glucose and amino acids for use by the tissues.

Recall that fat in the form of chylomicrons is picked up by the central lacteal and enters the lymphatic system instead, thereby bypassing the hepatic portal system. Contractions of the villi periodically compress the central lacteal and "milk" the lymph out of this vessel. The lymph vessels eventually converge to form the *thoracic duct,* a large lymph vessel that empties into the venous system within the chest. By this means, fat ultimately gains access to the circulatory system. The absorbed fat is carried by the systemic circulation to the liver and to other tissues of the body. Therefore, the liver does have a chance to act on the digested fat, but not until the fat has been diluted by the blood in the general circulatory system and has been somewhat reduced by uptake into adipose cells (see p. 294). This dilution of fat presumably protects the liver from being inundated with more fat than it can handle at one time.

The liver plays an important role in lipid transport by synthesizing three types of plasma lipoproteins. The first studies to characterize these plasma lipoproteins were conducted in the 1920s on horse serum. Liver-produced lipoproteins were named on the basis of their density of protein as compared to lipid. High-density lipoprotein (HDL; Chapter 9, p. 362) was isolated first. With the advent of analytical high-speed ultracentrifuges, subsequent studies identified low-density lipoprotein, which contains less protein and more cholesterol, and very-low-density lipoprotein (VLDL), which contains the least protein and most lipid. As you saw, the LDLs and HDLs transport primarily cholesterol (and phospholipid) for cell membrane production. VLDLs are rich in triglycerides, which are used for energy storage in adipose tissue.

■ Extensive absorption by the small intestine keeps pace with secretion.

The small intestine of humans normally absorbs about 9 liters of water per day, which contains the absorbable units of nutrients, vitamins, and electrolytes. How can that be, when humans normally ingest only about 1,250 mL of fluid and 1250 g of solid food (80% of which is H_2O—see p. 561) per day? ▪ Table 14–7 illustrates the tremendous daily absorptive accomplishments performed by the small intestine. Each day, about 9,500 mL of H_2O and solutes enter the small intestine. Note that of this 9,500 mL, only 2,500 mL are ingested from the external environment. The remaining 7,000 mL (7 liters) of fluid consist of digestive juices that are essentially derived from plasma. Recall that plasma is the ultimate source of digestive secretions because the secretory cells extract the necessary raw materials for their secretory product from the plasma. Considering that the entire plasma volume is only about 2.75 liters, it is obvious that absorption must closely parallel secretion to prevent the plasma volume from falling sharply. Of the 9500 mL of fluid entering the small-intestine lumen per day, about 95%, or 9000 mL of fluid, is normally

Table 14–7 ▪ Volumes Absorbed by the Small and Large Intestine per Day (human)

Volume entering the small intestine per day			
	Ingested	Food eaten	1,250 g[a]
		Fluid drunk	1,250 mL
Sources	Secreted from the plasma	Saliva	1,500 mL
		Gastric juice	2,000 mL
		Pancreatic juice	1,500 mL
		Bile	500 mL
		Intestinal juice	1,500 mL
			9,500 mL
Volume absorbed by the small intestine per day			9,000 mL
Volume entering the colon from the small intestine per day			500 mL
Volume absorbed by the colon per day			350 mL
Volume of feces eliminated from the colon per day			150 g[a]

[a]One milliliter of H_2O weighs 1 g. Therefore, because a high percentage of food and feces is H_2O, we can roughly equate grams of food or feces with milliliters of fluid.

absorbed by the small intestine into the plasma, with only 500 mL of the small-intestine contents passing on into the colon. Thus the digestive juices are not lost from the body. After the constituents of the juices are secreted into the digestive tract lumen and perform their function, they are returned to the plasma. The only secretory product that escapes from the body is bilirubin, a waste product that must be eliminated.

■ Biochemical balance among the stomach, pancreas, and small intestine is normally maintained.

Because the secreted juices are normally absorbed back into the plasma, the acid–base balance of the body is not altered by digestive processes. When secretion and absorption do not parallel each other, however, acid–base abnormalities can result. ● Figure 14–25 is a summary of the biochemical balance that normally exists among the stomach, pancreas, and small intestine. The arterial blood entering the stomach contains Cl^-, CO_2, H_2O, and Na^+, among other things. During HCl secretion, the gastric parietal cells extract Cl^-, CO_2, and H_2O from the plasma (the CO_2 and H_2O being essential for H^+ secretion) and add HCO_3^- to the secretion (the HCO_3^- being

formed in the process of generating H^+). The HCO_3^- diffuses into the plasma to replace the secreted Cl^- and to electrically balance the Na^+ in the plasma. Plasma Na^+ levels are not altered by gastric secretory processes. Because HCO_3^- is an alkaline ion, the venous blood leaving the stomach is more alkaline than the arterial blood delivered to it. This so-called postdigestion alkaline tide is greatest in the alligator (*Alligator mississippiensis*) in which the pH increases from 7.4 to 7.6, a change of 0.2 pH units. Comparably the change in pH units in *Rana catasbeina* is 0.1; *Bufo marinus*, 0.05; and the python (*Molurus*), 0.02.

In most species the overall acid–base balance of the body is not altered because the pancreatic duct cells extract a comparable amount of HCO_3^- (along with Na^+) from the plasma to neutralize the acidic gastric chyme as it is emptied into the small intestine. One exception is the alligator, which exhibits a pronounced respiratory alkalosis (see p. 607). Within the intestinal lumen, the alkaline pancreatic $NaHCO_3$ secretion neutralizes the gastric HCl secretion, yielding NaCl and H_2CO_3. The latter molecules form Na^+ and Cl^- plus CO_2 and H_2O, respectively. All four of these constituents (Na^+, Cl^-, CO_2, and H_2O) are absorbed by the intestinal epithelium into the plasma. Note that these are exactly the same constituents pres-

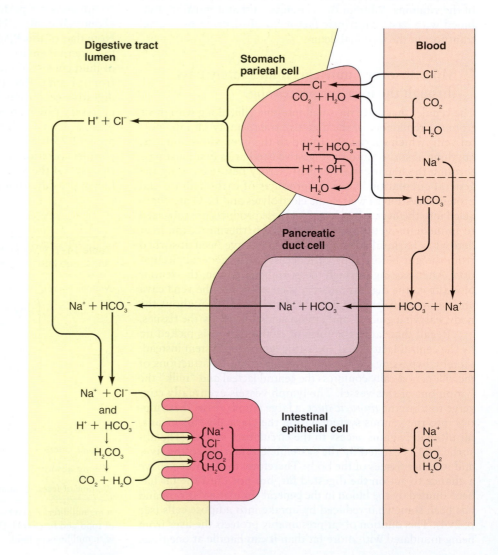

Figure 14–25 ● Biochemical balance among the stomach, pancreas, and small intestine in a vertebrate. When digestion and absorption proceed normally, no net loss or gain of acid or base or other chemicals from the body fluids occurs as a consequence of digestive secretions. The parietal cells of the stomach extract Cl^-, CO_2, and H_2O from and add HCO_3^- to the blood during HCl secretion. The pancreatic duct cells extract the HCO_3^- as well as Na^+ from the blood during $NaHCO_3$ secretion. Within the small-intestine lumen, pancreatic $NaHCO_3$ neutralizes gastric HCl to form NaCl and H_2CO_3, which decomposes into the blood, thereby replacing the constituents that were extracted from the blood during gastric and pancreatic secretion.

ent in the arterial blood entering the stomach. Thus through these interactions, the body normally does not experience a net gain or loss of acid or base during digestion.

■ Diarrhea results in loss of fluid and electrolytes.

When vomiting or diarrhea occurs, these normal neutralization processes cannot take place. We have already described vomiting in the section on gastric motility. The other common digestive disturbance that can lead to a loss of fluid and an acid–base imbalance is **diarrhea**. This condition is characterized by passage of a highly fluid fecal matter, often with increased frequency of defecation. Just as with vomiting, the effects of diarrhea can be either beneficial or harmful. Diarrhea is beneficial when rapid emptying of the intestine hastens the elimination of harmful material from the animal's body. However, not only are some of the ingested materials lost, but some of the secreted materials that normally would have been reabsorbed are lost as well. Excessive loss of intestinal contents causes dehydration, loss of nutrient material, and metabolic acidosis resulting from the loss of HCO_3^- (see p. 608). The abnormal fluidity of the feces in diarrhea usually occurs because the intestine cannot absorb fluid as extensively as normal. This extra unabsorbed fluid passes out in the feces.

The most common cause of diarrhea is excessive intestinal motility, which arises either from local irritation of the gut wall caused by bacterial or viral infection of the intestine or from emotional stress. Rapid transit of the intestinal contents does not allow sufficient time for adequate absorption of fluid to occur.

Large Intestine

■ The large intestine functions in electrolyte and fluid balance, harbors microbes for fermentation and VFA production, and serves as a temporary storage site for excreta.

The **large intestine** of vertebrates consists of the colon, cecum, and rectum/cloaca (● Figure 14–26). The **cecum**, if present, forms a blind-ended pouch below the junction of the small and large intestines at the ileocecal valve. The ceca are paired in birds although they are not present in all avian species. The small, fingerlike projection at the bottom of the cecum in humans and some apes is the **vermiform appendix,** a lymphoid tissue that stores lymphocytes but that has no digestive function. It is thought to be a remnant of a larger cecum in ancestral primates.

The **colon,** which makes up most of the mammalian large intestine, varies considerably between animals in structure depending on diet. In general, there is an *ascending colon, transverse colon,* and *descending colon.* The ascending colon of ruminants, pigs, and horses are dramatically modified in both size and structure. For example, in ruminants and pigs it is much elongated and organized into coils. In the horse it is markedly enlarged and occupies a considerable portion of the abdominal cavity. The transverse colon is comparatively short, whereas the descending colon is continuous with the **rectum/cloaca** (*rectum* means "straight"). In birds, the cloaca is divided by ridges into three regions. The anterior portion, the **coprodaeum,** receives the excreta from the intestines; the middle, the **urodaeum,** receives the fluid from the kidneys through the ureters, as well as the material from the oviduct; and the **proctodaeum** stores the excreta. The proctodaeum opens externally into the muscular anus. Lying on the dorsal wall of the urodaeum is the **bursa of Fabricius,** a prominent lymphoid organ important in newly hatched birds.

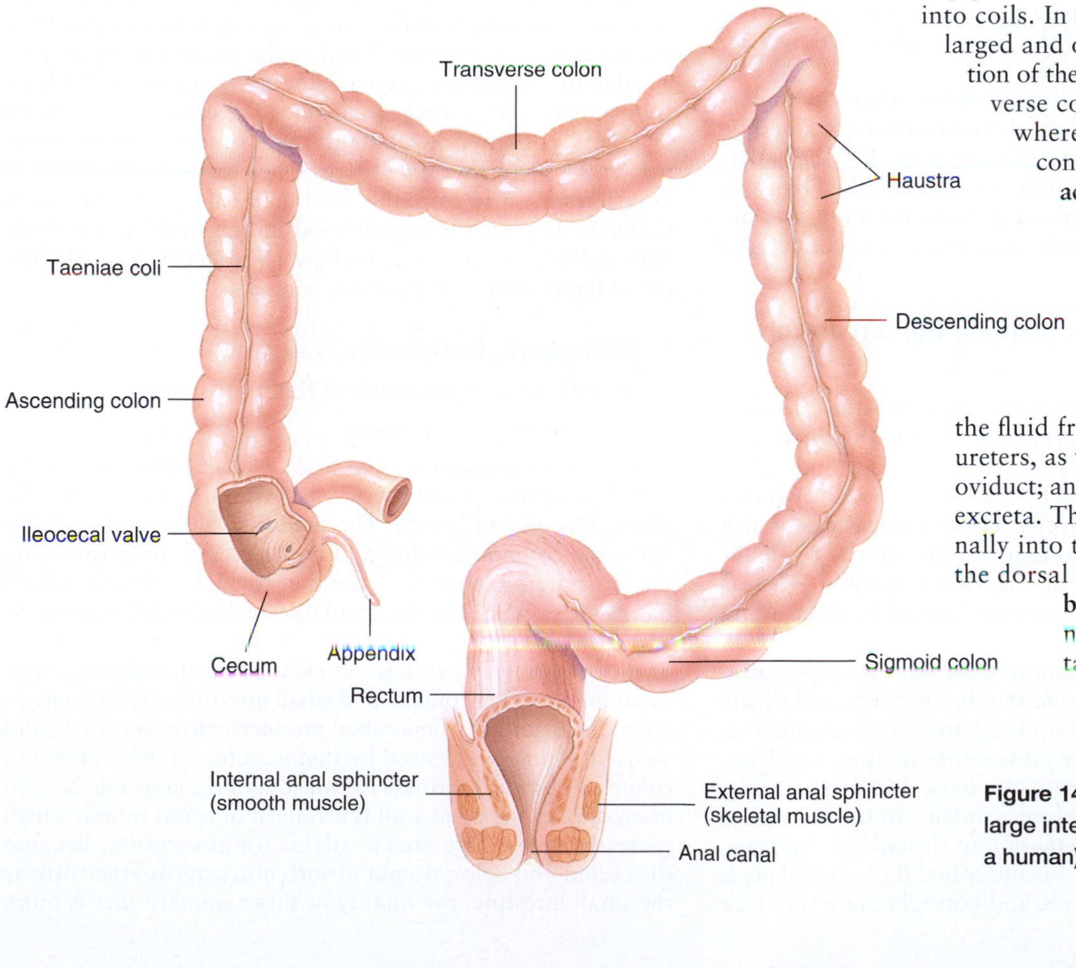

Figure 14–26 ● Anatomy of the large intestine (mammal, shown for a human).

Labels: Transverse colon, Haustra, Taeniae coli, Descending colon, Ascending colon, Ileocecal valve, Sigmoid colon, Cecum, Appendix, Rectum, Internal anal sphincter (smooth muscle), External anal sphincter (skeletal muscle), Anal canal

Diet composition governs the variation in structure of the large intestine.

The structure of the large intestine between the various species varies considerably, with composition of the animals' diet exerting the greatest influence. Carnivores (such as cats and dogs) tend to have the simplest digestive tracts, with comparatively short, unstructured colons. For example, carnivorous fish have gut lengths normally slightly greater than the length of their own body. In these animals there is little distinction between the small and large intestine and the cecum is either absent or marginal in its development. Thus in carnivores, the major function of the colon is the absorption of electrolytes, water, and other materials that escaped absorption in the small intestine.

In omnivores and particularly in herbivores that ingest considerable amounts of complex polysaccharides, notably rabbits and horses, the structure of the large intestine tends to be more complex. Generally the ceca and/or the colon are sacculated and voluminous, providing a site for microbial digestion. Intestinal length is also longer in omnivorous fishes as compared to carnivorous fishes. In herbivorous fishes, gut length can exceed 20 times body length. Exceptions include the kangaroo and sheep, whose cecum and colon are neither sacculated nor particularly enlarged. Sacculations are formed when the outer longitudinal smooth-muscle layer of the colon does not completely surround the large intestine. In these animals, it consists of three separate, conspicuous, longitudinal bands of muscle, the **taeniae coli,** which run the length of the cecum and large intestine. One exception is the horse, whose cecum is lined with four taeniae. Approximately one half of the proximal coiled region of the pig colon has two taeniae, whereas the remainder is without.

These taeniae coli are actually shorter than the underlying circular smooth-muscle and mucosal layers would be if the latter were stretched out flat. Because of this, the underlying layers are gathered into pouches or sacs called **haustra.** The haustra are not merely passive gathers, however; they actively change location as a result of contraction of the circular smooth-muscle layer.

Haustral contractions prolong the retention of digesta.

Most of the time, movements of the large intestine are slow and nonpropulsive, as is appropriate for its absorptive and storage functions. The colon's primary method of motility is **haustral contractions** initiated by the autonomous rhythmicity of colonic smooth muscle cells. These contractions, which throw the large intestine into haustra, are similar to small-intestine segmentations but occur much less frequently. Whereas small-intestine segmentation occurs at rates of between 9 and 12 contractions per minute, there may be 30 minutes between haustral contractions. The location of the haustral sacs gradually changes as a relaxed segment that has formed a sac slowly contracts while a previously contracted area simultaneously relaxes to form a new sac. These movements are nonpropulsive; they slowly shuffle the contents in a back-and-forth mixing movement that exposes the colonic contents to the absorptive mucosa. By prolonging the retention of the colonic contents, *more time* is made available for microbial digestion. This is particularly important in the pig and horse, because they lack

a rumen in which to digest dietary cellulose. Locally mediated reflexes involving the intrinsic plexuses largely control haustral contractions.

Peristaltic contractions also direct the flow of intestinal and cecal contents toward the rectum. A pacemaker in the midcolon region generates slow waves, which travel both directions along the colon, which contributes to the net movement of ingesta along the colon. Birds, for example, rely primarily on peristaltic rather than haustral contractions for colon motility.

In addition to the haustral contractions, which impede the flow of ingesta, are the **antiperistaltic contractions,** which function primarily to fill the cecum. Generally they originate from pacemakers in the proximal region of the colon. These contractions are important in most herbivores and pigs, in which their mixing action enhances microbial digestion of cellulose and absorption of fermentation products, such as **volatile fatty acids (VFA)** (see p. 665). Because the cecocolic sphincter in the horse is normally closed, these contractions do not occur.

Because of this slow colonic movement, bacteria have time to grow and accumulate in the large intestine. In contrast, contents in the small intestine are normally moved through too rapidly for bacteria to proliferate. Not all ingested bacteria are destroyed by salivary lysozyme and gastric HCl, so the surviving anaerobic bacteria continue to thrive in the large intestine. Most colonic microorganisms are generally harmless in this location and survive on any remaining protein and soluble polysaccharides in the digesta.

The colon of a human normally receives about 500 mL of chyme from the small intestine each day. Because most digestion and absorption have been accomplished in the small intestine, the contents delivered to the colon consist of indigestible food residues (such as cellulose), unabsorbed biliary components, and the remaining fluid. The colon extracts more H_2O and salt from the contents. What remains to be eliminated is known as *excreta* or **feces.** The primary function of the large intestine is to store this fecal material before defecation. Cellulose and other indigestible substances in the diet provide bulk and help maintain regular bowel movements by contributing to the volume of the colonic contents.

Postgastric fermentation is not as efficient as pregastric fermentation in obtaining nutrients.

Postgastric fermentation occurs in all animals but is particularly well developed in some species such as the horse, whale, elephant, possum, and koala. The site of the fermentation chamber is the combined colon and upper cecum. By positioning the fermentation chamber after the stomach, the host animal has the first chance at the available carbohydrates and protein in the food but loses the opportunity to exploit the high-quality microbially synthesized protein. Further, the digestive capabilities of the stomach and small intestine cannot contribute to the digestion of microbial products. However, VFAs and some vitamins synthesized by the microbes in the cecum and colon are absorbed through the mucosa of the large intestine. In many species the cecal wall is arranged in spiral ridges, which increases the surface area available for absorption. Because the cecum and colon do not absorb nutrients as efficiently as the small intestine, the quality of these animals' diet is more important in sustaining normal metabolic activity. Recycling of nitrogen (see p. 665) is also less important in mammalian postgastric fermentors. Thus when dietary sources are depleted in nutrients, ruminants have a survival edge over nonruminant herbivores.

The environment within the colon/cecum of postgastric fermentors must be able to support a substantial microbe population in a buffered environment. In the horse, large quantities of bicarbonate and phosphate buffers are secreted by the ileum and transferred to the cecum, an action comparable to the role of the salivary gland in ruminants. The animal must also be able to absorb the products of fermentation as well as to reclaim as much fluid as possible from the digesta before it is excreted.

Herbivorous avian species such as the galliformes, ostriches, ducks, and geese have enlarged ceca. In grouse, the combined lengths of the ceca approximate that of the entire intestine, whereas in passerines, parrots, and raptors they are completely absent. In some omnivorous avian species, the overall size of the intestinal tract changes seasonally in response to the diet. For example, during the warmer summer months robins and grouse are carnivorous (consuming worms and insects) but become selective herbivores during the colder winter months.

It has been shown in the galliformes that uric acid and urea from their urine passes down the ureters into the cloaca, then reflux antiperistaltically into the colon and ceca. Microbes decompose the urea and uric acid into ammonia, which may then be used by bacteria to synthesize amino acids. Strategies to prevent the decomposition of urea and uric acid into ammonia may thus help prevent the proliferation of harmful bacteria in the gut of domestic poultry. The microbes themselves primarily use these amino acids, but some are absorbed into the blood of the host bird to meet its amino acid requirements. Ammonia is also absorbed directly from the ceca and is used to synthesize amino acids in the liver. Ammonia "recycling" is an important adaptation that permits the galliformes to exploit habitats otherwise unavailable to most animals.

Current research efforts in the poultry industry are focused on manipulating the microflora of newly hatched chicks so they are more resistant to salmonella colonization. One approach is to incorporate such compounds as lactose in the diet, which stimulates the production of lactic acid and short-chain VFAs. The reduction in cecal pH in response to the increase in the concentration of these acids is ultimately *bacteriostatic,* that is, acts to inhibit the growth of bacteria and thus is useful in controlling salmonella colonization.

Nutritional Benefit from the Reingestion of Feces

A solution to obtaining nutritional benefit from protein and vitamin synthesis by microbes in the large intestine is the practice of **coprophagy** (reingestion of feces). Rabbits and some species of rodents produce two types of fecal material. The soft feces produced in the nighttime are reingested, thus making the products of cecal fermentation available for digestion and absorption. Delivery of the soft feces is accomplished by inhibition of motility in the proximal colon associated with hyperactivity of the distal region close to the anus. In the rabbit, the soft fecal pellets contain over 50% bacteria. The second type of feces is a dark, hard pellet. Many hindgut digestors practice coprophagy when very young. Colts, for example, are thought to practice coprophagy in order to establish a microbe popu-

lation in the hindgut but to abandon this practice as adults. Because chickens normally practice coprophagy, outbreaks of *coccidiosis* are relatively common in birds raised on litter. Coccidiosis is caused by single-celled protozoa (*Eimeria*) that colonize the large intestine. Affected animals may develop symptoms that include diarrhea, bloody droppings, and an abrupt decrease in food intake.

◼ Mass movements propel colonic contents long distances.

Generally after meals, a marked increase in motility takes place during which large segments of the colon contract simultaneously, driving the feces one third to three fourths of the length of the colon in a few seconds. These massive contractions, appropriately called **mass movements,** drive the colonic contents into the distal portion of the large intestine, where material is stored until defecation occurs. In the dog, mass movements originate near the ileocecal valve and can evacuate the ingesta the entire length of the colon, whereas in the cat the movements are generated lower in the colon.

When food enters the stomach, mass movements occur in the colon primarily by means of the **gastrocolic reflex,** which is mediated from the stomach to the colon by gastrin and by the extrinsic autonomic nerves. In humans, this reflex is most evident after the first meal of the day and is often followed by the urge to defecate. Thus when a new meal enters the digestive tract, reflexes are initiated to move the existing contents farther along down the tract to make way for the incoming food. The gastroileal reflex moves the remaining small-intestine contents into the large intestine, and the gastrocolic reflex pushes the colonic contents into the rectum, triggering the defecation reflex.

◼ Feces are eliminated by the defecation reflex.

When mass movements of the colon move fecal material into the rectum, the resultant distension of the rectum stimulates stretch receptors in the rectal wall, thus initiating the **defecation reflex.** This reflex causes the **internal anal sphincter** (which consists of smooth muscle) to relax and the rectum and sigmoid colon to contract more vigorously. If the **external anal sphincter** (which consists of skeletal muscle) is also relaxed, defecation occurs. Being skeletal muscle, the external anal sphincter is under voluntary control. The initial distension of the rectal wall is accompanied by the conscious urge to defecate. If circumstances are unfavorable for defecation, voluntary tightening of the external anal sphincter can prevent defecation despite the defecation reflex. Ground-nesting animals generally defecate at considerable distances from their young, to decrease the likelihood of detection by a predator. If defecation is delayed, the distended rectal wall gradually relaxes, and the urge to defecate subsides until the next mass movement propels more feces into the rectum, once again distending the rectum and triggering the defecation reflex. During periods of nonactivity, both anal sphincters remain contracted to ensure fecal continence.

When defecation does occur, voluntary straining movements that involve simultaneous contraction of the abdominal muscles and a forcible expiration against a closed glottis usually assist it. This maneuver brings about a large increase in intra-abdominal pressure, which helps eliminate the feces.

Defecation is also a normal response to fear, presumably in response to activation of neural centers in the brain.

■ Large-intestine secretion is protective in nature.

The large intestine does not secrete any digestive enzymes. Colonic buffers consist of an alkaline (HCO_3^- and PO_4^{3-}) mucus solution, whose function is to protect the large-intestine mucosa from mechanical and chemical injury. In the horse and pig, pancreatic HCO_3^- is of sufficient volume so as to contribute to the buffering capability in the colon. In ruminants, considerable amounts of PO_4 are secreted in the saliva, whereas in nonruminants the source of PO_4 is the diet. The uptake of PO_4 is low in the intestine, with the result that its concentration increases as the colon absorbs fluid. The mucus provides lubrication to facilitate passage of the feces, whereas the buffers neutralize acids produced by local bacterial fermentation.

■ The large intestine absorbs primarily salt and water, converting the luminal contents into feces.

Some absorption takes place within the colon and hindgut of insects but not to the same extent as in the small intestine. The Malpighian tubules of insects (see p. 567), which are excretory in function, arise at the anterior end of the hindgut and empty their contents into it. Because the luminal surface of the colon is fairly smooth, it has considerably less absorptive surface area than the small intestine. In most animal species, no specialized transport mechanisms are present in the colonic mucosa for absorption of glucose or amino acids, as there are in the small intestine. When excessive small-intestine motility delivers the contents to the colon before absorption of nutrients has been completed, the colon cannot absorb these materials, and they are lost in diarrhea.

The colon normally absorbs some salt and H_2O. Sodium is actively absorbed, Cl^- follows passively down the electrical gradient, and H_2O follows osmotically. Bacteria in the colon synthesize some vitamins that the colon can absorb, but in nonruminants this is normally not a significant contribution, except in the case of vitamin K. Cellulose-digesting anaerobic microorganisms are housed in special organs connected to the hindgut of wood-eating termites and the wood roach *Cryptocercus*. (See also the box, "Molecular Biology and Genomics: Big Bugs Have Little Bugs . . .")

Through absorption of salt and H_2O, a firm fecal mass is formed. In humans, of the 500 mL of material entering the colon per day from the small intestine, the colon normally absorbs about 350 mL, leaving 150 g of feces to be eliminated from the body each day (Table 14–7). This fecal material normally consists of 100 g of H_2O and 50 g of solids, including undigested cellulose, bilirubin, bacteria, and small amounts of salt. Thus, contrary to popular thinking, the digestive tract is not a major excretory passageway for eliminating wastes

---

MOLECULAR BIOLOGY AND GENOMICS

Big Bugs Have Little Bugs . . .

Many insects are similar to ruminant mammals in that they rely on symbiotic bacteria in their digestive tracts. One well-studied example is the pea aphid *Acyrthosiphon pisum,* which feeds on plant sap that is poor in several nutrients. In special cells around the aphid's digestive tract, symbiotic bacteria (*Buchnera*) manufacture key nutrients, especially several amino acids that the aphid cannot make (known as *essential amino acids*). Until recently, it has been difficult to study these endosymbionts, because they do not grow in the laboratory. However, genetic analysis is shedding some light on the relationship and its evolution. By comparing the genome of free-living *Buchnera* with that of the symbiont, researchers discovered that the latter have lost most of the genes necessary for the microbes to live on their own. For example, genes for defensive cell-wall proteins are absent. Furthermore, the genes for making essential amino acids for the host aphid have been duplicated many-fold and are located in plasmids (small circlets of DNA separate from the main chromosome). Plasmids serve as dedicated factories producing RNA for the enzymes to make large quantities of the amino acids, which are exported to the host.

These studies reveal much about the genetic processes of evolution. First, because they have no selective benefit, genes that are no longer used mutate into unusable forms and may shrink as pieces are accidentally lost during DNA replication. Second, intimate symbioses such as these may be irreversible, with each partner no longer capable of living without the other. Finally, the process suggests how mitochondria and chloroplasts, which also have greatly reduced genomes, may have evolved from symbiotic microbes.

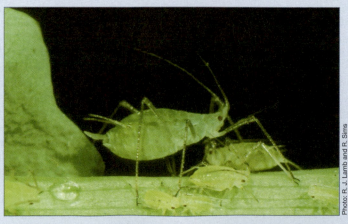

Pea aphid, acyrthosipon pisum

(*Source:* http://esc-sec.org/photo_archive/photo_archive5.htm)

Photo: R. J. Lamb and R. Sims

from the body. The main waste product excreted in the feces is biliverdin or bilirubin. The other fecal constituents are unabsorbed food residues and bacteria, which were never actually a part of the body.

■ The large intestine of birds and hindgut of insects actively transports glucose and amino acids.

The colon of birds is exceptional in that glucose as well as amino acids can be absorbed by secondary active transport. Thus glucose and protein not reabsorbed in the nephron can either be reabsorbed into the blood or used by the microorganisms in the ceca. The rectum in the hindgut of locusts has

been demonstrated to transport some amino acids via an Na^+-independent process.

Ruminant Digestion

Ruminants, so named because they **ruminate** (chew the cud), can voluntarily regurgitate partially digested food back into the mouth to complete the mechanical grinding of the ingested plant material. Ruminant species have evolved a stomach of sufficient size and motility to house a population of microbes able to break down cellulose and other complex polysaccharides into end products of fermentation, which meet the nutritional requirements of the host animal. Consequently ruminants are among the most widely distributed groups of mammals on Earth, having adapted to arctic, desert, and tropical environments.

The stomach of true ruminants (*Ruminantia*), which includes cattle, goats, deer, giraffe, and antelope, is divided into four compartments, which occupy approximately three quarters of the abdominal cavity. The **forestomach** (the pregastric region) consists of three chambers: the **rumen**, the **reticulum**, and the **omasum**, each of which is involved in the storage and passage of ingested food (● Figure 14–27). *Pseudoruminants* (Tylpoda), including llamas and camels, have a three-compartment stomach, the omasum being absent.

The rumen and its continuation, the reticulum (**ruminoreticulum**), also have an important role in the absorption of nutrients and simple molecules. It is in these two compartments that the bulk of anaerobic fermentation (see p. 53) of plant material occurs and cellulose is catabolized into digestible units. Energy is obtained through fermentation that would otherwise be lost to the host animal. However, the fermentative process requires specific temperatures, pH, motility, and

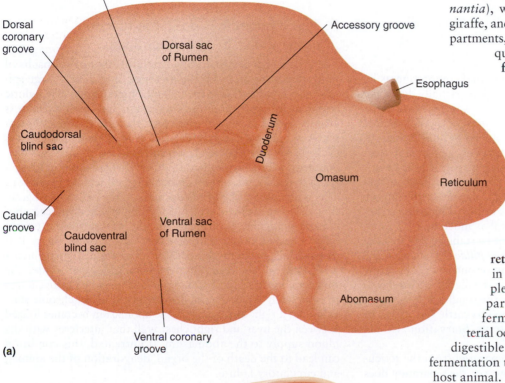

(a)

Longitudinal groove
Dorsal coronary groove
Dorsal sac of Rumen
Accessory groove
Esophagus
Caudodorsal blind sac
Duodenum
Omasum
Reticulum
Caudal groove
Ventral sac of Rumen
Caudoventral blind sac
Abomasum
Ventral coronary groove

(b)

Reticulum
Omasum
Abomasum
Rumen
To small intestine

Figure 14–27 ● Comparison of rumen structure and function in a selective versus a nonselective feeder. (a) Stomach of a nonselective feeder, the ox. (b) Rumen function in an antelope stomach. Note that the relative size of the rumen is considerably smaller than in the ox. Food moves from the rumen and reticulum into the omasum for sorting. The abomasum is the acid-secreting region.

(*Sources:* (a) Frandson/Spurgeon, *Anatomy and Physiology of Farm Animals,* 5th ed., Lea and Febiger, p. 317. (b) C. Starr, 2004, *Biology: Unity and Diversity of Life,* 10th ed., Belmont, CA: Brooks/Cole, Figure 41.4b, p. 727)

secretions to maintain the microbial populations. The acid-secreting region of the stomach, the **abomasum,** functions in the digestion of protein and lysis of rumen microbes. In all ruminant animals, the rumen is the most spacious structure, followed by the abomasum. In a majority of ruminant species, the reticulum is greater in size than the omasum, which is the smallest of the four compartments. The variability in size of the latter two compartments is ultimately determined by the feeding strategies of the particular animal group as well as by seasonal changes in food availability.

Surrounding the ruminant stomach is an outer muscular layer, which consists of a thin sheet of longitudinally directed smooth muscle, whereas the inner layer is thick and circular in structure. The muscular layers mix the ingesta and provide for the progressive movement of this material through the stomach toward the small intestine. Rumen mixing also promotes the turnover of indigestible particles that would quickly clog the rumen if allowed to accumulate.

The rumen is divided into separate compartments.

The ruminal cavity is divided into internal compartments by the presence of *pillars*. Longitudinal pillars partially divide the rumen into dorsal (toward the back) and ventral (toward the belly) sacs. When contracted, these muscular folds enable the mixing and movement of large volumes of fluid. These pillars also help stabilize the fluid contents of the ruminoreticulum and limit the movement of the digesta toward the reticulum. In deer, for example, rapid movements of the body would potentially result in a considerable displacement of ruminal fluid were it not for the limitations imposed by these structures. Because this volume may equal as much as 14% of an animal's body weight, it is vital that it be stabilized. Lining the rumen are fingerlike projections, the *papillae* that increase the surface area available for absorption of nutrients. This same mechanism is used again further along the digestive tract. Feeding **high-concentrate diets** (replacing the roughage component of the diet with higher-energy grains) to cattle results in the gradual degradation of these structures. Starvation also decreases the size of the papillae.

The ruminal compartment is separated from the reticulum by the *ruminoreticular* fold. However, the separation does not inhibit the free exchange of digesta between the two compartments. Within these two sacs occurs most fermentation and absorption of nutrients. Hardware and other foreign objects accumulate in the reticulum because of its position relative to the esophagus. **Hardware disease** occurs in ruminants when a foreign object, usually a piece of metal that has been swallowed, penetrates the reticulum. The condition is particularly common in cattle, because they do not separate foreign objects—nails, wire, and so forth—from feed, and swallow everything before it is completely chewed. Penetration of the reticulum wall permits bacteria and partially digested food to spill out into the abdominal cavity, resulting in *peritonitis*. Occasionally a metal object penetrates the diaphragm and enters the chest cavity or penetrates the pericardium surrounding the heart. If a metal object penetrates the heart, extensive bleeding and sudden death can result.

The esophagus terminates at the junction between the reticulum and the rumen, the *cardia*, and opens directly over the cavity of the reticulum. It is encircled by strands of smooth muscle that are organized to form a *reticular groove*. The reticular grove extends from the opening of the esophagus to the *reticulo-omasal orifice*. The luminal surface of the reticulum is lined with ridges, which are thought to be involved in the sorting of particles that pass near the reticulo-omasal orifice. Ingesta ultimately entering the omasum are identical in consistency to that found in the ruminoreticulum.

The omasum provides a channel for the passage of ingesta from the reticulum into the glandular stomach, the abomasum. The luminal surface is characterized by the presence of leaflike structures that absorb water and nutrients as well as limit the passage of large particles. At the junction with the abomasum is the *omasoabomasal orifice*, a well-developed structure that lacks a sphincter to limit backflow. It also regulates the movement of materials from the omasum into the abomasum during digestion and inhibits the movement of ingesta from the omasum when the abomasum is full.

The abomasum is similar in function to the stomachs of nonruminants.

The abomasum is structurally similar to that of stomachs of nonruminants except for the presence of relatively large spiral folds. It consists of a much larger fundic area. The pyloric region at the junction of the abomasum and small intestine is also increased in size. In ruminants that are fed coarse plant material, gastric juice secretion is continuous, whereas it is cyclic in animals provided concentrate diets at intervals during the day. Feeding high-concentrate diets to dairy cattle can result in the abomasum becoming enlarged with fluid and gas. This is a consequence of the decrease in the pH of the digesta that leaves the rumen, which can result in the **displaced abomasum syndrome** (shifting of the abomasum from the right half of the abdominal cavity to either the left or the right). Left displacements are more common than right displacements and result in the abomasum becoming trapped under the rumen on the left side. The displacement may be severe enough to completely block digesta flow, which leads to chronic starvation. In a right displacement, the abomasum becomes lodged between the liver and right chest wall that interferes with the blood supply to the abomasum. If untreated, this condition can lead to the death of the organ, dehydration of the animal, and circulatory failure.

The acidic digesta ultimately enters the small intestine through the pyloric sphincter. In ruminants, and nonruminants including pigs and to a lesser extent, horses, the muscular layer of the pyloric sphincter forms an enlargement known as the *torus pyloricus*.

Neonatal ruminants rely only on the abomasum.

In neonatal ruminants, the preruminant phase of development is characterized by a comparatively undeveloped ruminoreticulum. Sucking or drinking results in the reflexive closure of the reticular groove with the result that ingested milk is channeled directly into the abomasum. This reflex response ensures that immunoglobulins in the **colostrum** (see p. 756) are transported to the duodenum where they are rapidly absorbed through the intestinal mucosa by pinocytosis. Neonates that fail to ingest colostrum are more susceptible to pathogenic bacteria that gain entry to the gastrointestinal tract. Exploratory licking behavior of neonatal ruminants ensures the ultimate colonization of the ruminoreticulum. As solid foods are gradually incorporated into the diet, the ruminoreticulum steadily increases in size until it constitutes approximately 60 to 80% of the total stomach capacity. The intake of a fibrous diet and the inoculation by microbes is essential for the development of the rumen. In the adult animal, the reticular groove does not function under normal feeding conditions, and so food is directed into the rumen.

Peptic cells of young ruminants fed milk (or the milk protein casein) secrete the proteolytic enzyme *rennin*. Rennin coagulates milk within the stomach at a much faster rate compared to pepsin. Within a few minutes, the mixture of rennin with the acidic abomasal environment results in the formation of a casein (milk protein) matrix interspersed with milk fat globules. Soluble whey proteins (residue from milk after removal of the casein and fat) and lactose rapidly transit the abomasum, but the casein-fat clot or curd is digested over a 12- to 18-hour period. Rennin has a higher pH optimum (optimum pH for maximal enzyme activity, approximately 4.0) than does pepsin, and milk-fed ruminants maintain higher abomasal pH levels than do adults. After weaning, rennin production becomes essentially undetectable. If calves are bucket-fed large volumes of milk at infrequent intervals, there is an increased likelihood of forming an abnormally large curd in the abomasum. Consequently, the intestine becomes overloaded with protein, which leads to proliferation of bacteria and severe diarrhea. Affected calves then develop dehydration, electrolyte imbalance, and acidosis. This condition is termed **baby calf diarrhea**, or **scours**, and accounts for more economic loss than any other disease in calves.

■ Motility of the ruminant stomach is predominantly regulated by central nervous system reflex mechanisms.

Vagal and splanchnic nerves innervate the ruminant stomach. Vagal motor nerve fibers originating in the gastric centers of the medulla oblongata in the brain stem can increase the rate at which the ruminoreticulum contracts. For example, excitatory input generated from the sight of food, the process of chewing or rumination itself, or an increase in the distension of the ruminoreticulum all increase rumen motility. In contrast, input from the splanchnic nerve is inhibitory. Distension of the abomasum, for example, decreases the rate at which the ruminoreticulum contracts.

Two types of contractions regulate rumen motility. The most powerful is of the *primary* type, which occurs at approximately one-minute intervals. This strong contraction, which mixes the ruminoreticulum contents, begins with a pronounced contraction of the reticulum and ruminoreticular fold, followed by an even stronger contraction of the reticulum, which then progresses into the adjoining rumen. This results in the contraction of the rumen pillars and raises the ventral sac. The dorsal sac is also compressed. The duration of the primary cycle lasts for about 20 seconds. *Secondary* or *eructation* cycles may occur after alternate primary cycles. Secondary cycles consist of the sequential contraction of individual rumen sacs that move the gases of fermentation cranially from the dorsal ruminal sac. The area near the opening of the esophagus is cleared of ingesta, the cardiac sphincter relaxes and, combined with the contraction of the dorsal sac, forces

the reverse peristalsis of the gases up the esophagus. Opening of the pharyngoesophageal sphincter and elevation of the soft palate to close the orifice permit most of the gases to escape through the mouth. During feeding, however, the interval between successive primary and secondary cycles is almost halved.

Failure to eructate results in the accumulation of gases in the rumen, leading to a condition termed **bloat** that can be fatal. In cattle, *pasture bloat* results from the overconsumption of legumes, particularly alfalfa and some types of clover. These legumes are rapidly fermented in the ruminoreticulum and, if eaten in large quantities, form bubbles of gas that remain dispersed throughout the digesta. The bubbles take on an almost solid form (increase in surface tension) that dramatically increases pressure within the ruminoreticulum. This increase in pressure paralyzes the ruminoreticulum and constricts the esophageal opening such that the gas continues to accumulate. Ultimately the painful distension causes a cessation of food intake and can result in the death of the animal.

The primary and secondary cycles are referred to as *extrinsic contractions* because of their dependence on the vagal nerves. Removal of vagal input completely abolishes ruminoreticular contractions. In contrast, the intrinsic contractions are responsible only for the smooth muscle tone of the forestomach. In the absence of extrinsic contractions the intrinsic contractions become more forceful, but cannot compensate for the loss of extrinsic activity. Under these conditions the animal soon dies.

■ Rumination is the regurgitation, remastication, and reswallowing of ingesta.

Rumination cycles (or "cud chewing") are closely linked with ruminoreticular motility and last approximately a minute. Regurgitation is associated with an extra contraction of the reticulum followed by a primary contraction. The extra contraction is associated with a transient negative pressure created by closure of the glottis prior to an inspiratory cycle. The influx of digesta into the esophagus occurs simultaneously with the opening of the lower esophageal sphincter and a relaxing of the upper esophageal sphincter. The semiliquid digesta is propelled up the esophagus by reverse peristalsis. Once food enters the mouth the liquid portion is immediately reswallowed, but the particulate material remains in the mouth for maceration by the teeth and *reinsalivation*, mainly from the parotid gland on the side of chewing. Each rumination cycle is divided into a prolonged chewing phase followed by a short redeglutition phase. The food bolus is reformed and reswallowed. The time spent ruminating depends on the species of animal as well as its diet. In general, the greater the roughage content of the diet the more time that is spent ruminating the digesta. Forages and hay require more chewing than do feed concentrates.

Rumination is induced by two types of sensory receptors located in the ruminant stomach that monitor the volume and texture of ruminoreticular digesta. Tension receptors located in the muscle layers of the reticulum (near the reticular groove) and in the muscle layer of the craniodorsal pillar are slow-adapting mechanoreceptors that are activated by passive distension. Distension of the ruminoreticulum with fluid or gas activates the tension receptors in the reticulum and ruminal compartments. This represents a potent reflex stimulus to increase the rate and magnitude of both primary and secondary cycle contractions. Chemoreceptors respond in turn to a variety of chemicals, including osmolarity and pH.

The environment of the rumen fosters the growth of anaerobic microbes.

Ruminants represent a group of animals that have achieved the highest degree of specialization in fermenting plant material. Animals having a rumen can exploit the vast supplies of structural carbohydrates found in the plant cell wall and convert them into a source of nutrients. Microorganisms contained in the rumen break down cellulose into short-chain fatty acids that are absorbed through the rumen wall. Most of these microorganisms are attached to partially digested feed particles (approximately 75%), whereas the remaining either adhere to the rumen wall or are present in the rumen fluid. The rumen is normally well buffered by the copious flow of saliva (which contains both bicarbonates and phosphates), production of short-chain organic acids, as well as the buffering capacity of the animal's diet. Further, the presence of ammonia in the rumen can limit decreases in rumen pH, which is normally maintained around 7.0.

The microbial populations found in the rumen include anaerobic species of *bacteria, protozoa,* and *fungi.* Of these, bacteria are the greatest in number and diversity of species. Over 200 species of bacteria have been identified in concentrations of 10^9 to 10^{10} cells per gram of rumen contents. The predominant bacterial species found in rumen fluid from wild ruminants such as deer and moose are similar to those species identified in domesticated ruminants. Bacteria can be either liquid-associated and feed on soluble carbohydrates or proteins or be solid-associated (adherent bacteria) and bound to food particles. The latter bacteria digest the insoluble polysaccharides as well as the less soluble proteins. Immediately after the animal's feeding, some of the adherent bacteria detach from the particulate matter and colonize the new feed particles. Soluble nutrients leaching from the cut ends of the plant material chemotactically (chemical signaling) attract both microbes and protozoa. A third group of bacteria *(epimural bacteria: epi- ,* "on"; *mural,* "wall") is associated with the epithelial cells lining the rumen wall. These organisms can use the O_2 that diffuses from the blood through the epithelial cells to the rumen fluid. In this manner, they help maintain the anaerobic environment of the rumen. These cells are also important in converting the urea that diffuses through the rumen wall into ammonia, because very few microbe populations synthesize ureases. Epimural bacteria also digest the cells that are sloughed from the rumen wall. Were it not for these microbes, these highly keratinized cells (fibrous proteins) would be indigestible.

Protozoa numbers vary from 10^5 to 10^6 per gram of contents and include about 60 species in domestic animals. The diversity of protozoa in wild ruminants is considerably less than in domestic ruminants. Compartmentalization of protozoa is similar to that of bacteria in that there are liquid-associated, solids-associated, and epimural populations. Protozoa face a particular problem in that the time required to reproduce is longer than the time the ingesta remains in the rumen. For this reason protozoa must adhere to large feed particles or the rumen wall to avoid being washed out with the other digesta. Protozoa are particularly sensitive to the composition of an animal's diet. For example, in dairy cows the number of protozoa increases in proportion to the amount of grain fed. Eventually, however, the increase in acid production caused by a high-grain diet reduces the pH in the rumen to 5.5 or less. Protozoa are killed at low rumen pHs and are affected by even transient reductions in pH caused by the feeding of concentrate diets. This results both from the production of acidic fermentation end products and a substantial reduction in the amount of saliva produced. Compared to dried hay, which stimulates chewing and rumination, both grains and high-moisture feeds such as silage or pasture can substantially reduce the amount of saliva produced by as much as 50%. Protozoa feed on ruminal bacteria, fungal zoospores, and other readily digestible materials, including plant starch granules and some polyunsaturated fatty acids.

At least 14 species of fungi have been identified, but a reliable estimate of their numbers has not been made. The vegetative stage of fungi is termed a *sporangium.* Sporangia generate flagellate, motile *zoospores* that remain in the fluid compartment until their attachment to a food particle. However, once they are attached to the digesta it is difficult to quantify population numbers. Fungal hyphae (filaments) penetrate deep into the plant cell walls and thus gain access to a pool of cellulose unavailable to protozoa and bacteria. Their weakening of the cell wall enables bacteria to eventually gain access and increase the rate of digestion of these insoluble plant fibers. Chemotaxis also plays a role in the location and colonization of plant material. Fungi secrete a more soluble form of cellulase (cellulose-digesting enzyme) than bacteria and are more successful in fermenting coarse particles, although their overall rate of digestion is slower compared to bacteria. Because of their attachment to large particles, both protozoa and fungi are cleared from the ruminoreticulum at relatively low rates, compared to bacteria.

Nutrients for the host ruminant are generated by anaerobic microbes.

Fermentation is the most important process to occur in the rumen. It is also the first step in the generation of nutrients suitable for the host ruminant. Because the diet of ruminant animals contains varying proportions of insoluble celluloses, hemicelluloses (structural component of plant cells), pectins (polysaccharide isolated from fruits) and starch (polysaccharide that stores energy in plants), these complex polysaccharides need to be converted into more simple forms to be metabolized by the ruminant microbes. Complicating this digestive process are the seasonal changes in the proportion of cellulose and hemicellulose as the herbage matures. Aging of plant walls is associated with an increase in the structural component, **lignin,** a noncarbohydrate polymer, which effectively resists degradation by most rumen microbes. Most bacteria attach to feed particles and digest the structurally complex polysaccharides via cellulase enzyme complexes tightly bound to the surface of the bacteria. Similar cellulase complexes are used by protozoa to digest engulfed feed particles, whereas fungi can digest polysaccharides by bound enzymes or by releasing the cellulases into the surrounding fluid. Starch and cellulose are degraded into glucose, whereas hemicellulose and pectin are converted into xylose (monosaccharide). Sugars and pectins are the most rapidly fermented, followed by starches and fiber. Most simple sugars do not accumulate in the rumen liquor, because they are taken up and metabolized by the microbes populating the forestomach, regardless of whether they participated in the initial hydrolysis of the complex polysaccharides. However, depending on the rate and extent of digestion, some carbohydrates escape the rumen. For example, most fiber, considerable starch, and some simple sugars escape the rumen environment.

Generation of Volatile Fatty Acids by Microbes

The metabolism of glucose and xylose by the microbes is similar to that of cells in the animal. These simple sugars are anaerobically metabolized through the glycolytic pathway into the intermediate *phosphoenolpyruvate (PEP)*. Methane, CO_2, acetate, and some butyrate are generated from PEP. Alternatively, PEP can be further metabolized into pyruvate, which ultimately leads to generation of the short-chain fatty acids propionate and butyrate. Under normal conditions, the rumen fluid content is approximately 60 to 70% acetate, 14 to 20% propionate, and 10 to 14% butyric acid. If the diet contains high concentrations of soluble carbohydrates or starch, the concentration of propionic and lactic acids increase; whereas high-fiber diets cause increases in acetic acid and declines in propionate and lactate. Acetate, propionate, and butyrate are the most important product of the microbial fermentation process, because they can be directly used as a source of energy by the host animal. Propionate is particularly important because it is the only VFA that can be used to synthesize glucose. Under normal conditions, approximately 70% of the ruminants glucose and glycogen is generated from propionate, whereas the catabolism of protein contributes only about 20%. The other VFAs provide energy by entering the TCA cycle as acetyl CoA. Acetate, in particular, is important in the synthesis of milk fat in the udder and body fat.

VFAs produced by the rumen microbes are passively absorbed through the rumen wall. The rate of absorption is dependent on chain length, pH, concentration, and osmolarity. For example, as the osmotic pressure in the rumen increases, absorption of VFAs from the rumen decreases. Osmotic pressure in the rumen averages 280 mosm/L but pressures above 350 mosm/L can completely stop rumination. Between the pHs of 4.5 and 6.5, the rates of VFA absorption are butyrate > propionate > acetate, whereas the rates are similar when rumen pH is above 6.5. Low pHs are also associated with a diminished attachment of rumen microbes to fiber particles, which reduces the catabolism of cellulose. At pH levels lower than 5.0, there is a complete cessation of fiber digestion in addition to a decline in the number of cellulolytic (cellulose-digesting) microbes.

Protein Metabolism by Microbes

Each of the microbial populations in the rumen can hydrolyze protein into constituent polypeptides and amino acids and use these products as a source of nitrogen for their own growth. Although most bacterial proteases (protein-digesting enzymes) remain bound to the cell, some species liberate their proteases into the rumen fluid to generate polypeptides. Polypeptides up to 6 amino acids long can be absorbed by the rumen bacteria and further hydrolyzed to end products such as amino acids, or are deaminated (the nitrogenous amine group is removed) to produce ammonia. Ammonia as well as amino acids are readily absorbed through the lining of the ruminoreticulum and omasum. Deamination of some amino acids such as valine and leucine is responsible for the generation of **isoacids**

or branched-chain fatty acids. These isoacids are essential in small amounts for the growth of cellulolytic microbes. Peak ammonia production occurs about 1 hour after feeding, although the appearance of free amino acids and ammonia occurs within minutes after feeding. Rumen microbes use the ammonia along with some amino acids and polypeptides to synthesize their own microbial protein. Once these bacteria are swept into the abomasum, their cell proteins are digested and eventually absorbed through the small intestine. The synthesis of microbial protein also ensures that the host is supplied with essential as well as nonessential amino acids. Approximately 30 to 40% of the dietary protein entering the rumen is not degraded and is classified as **rumen-undegradable protein.** However, the gastrointestinal proteolytic enzymes readily hydrolyze these proteins.

Ammonia generated in the ruminoreticulum is either converted into microbial protein, or converted to urea in the liver. Some of this urea is returned to the forestomach directly through the wall of the ruminoreticulum or through the salivary glands. High urease (urea-digesting) activity is found along the ruminal wall and is responsible for the rapid degradation of urea to ammonia for metabolism into microbial protein. The presence of urease in the rumen permits the incorporation of urea in the diet of ruminants. However, feeding ruminants a diet too high in urea or rapidly hydrolyzed protein can lead to ammonia toxicity and can waste an "expensive" commodity.

Wild herbivores experience marked changes in food quality and quantity throughout the year. When environmental conditions become sufficiently harsh and the only herbage available is deficient in protein, the concentration of ammonia in the rumen is low, as are the number of rumen microbes, thus slowing the breakdown of cellulose. However, the total amount of nitrogen returned to the rumen as urea exceeds that absorbed from the rumen as ammonia. Nitrogen recovered from this process is converted to microbial protein. Ultimately, the total amount of protein arriving in the intestine can be greater than that in the original food. A wild ruminant can conserve an important nitrogen source by returning to the rumen urea that would otherwise have been excreted in its urine. Alternatively, under conditions when protein degradation is greater than its synthesis, such as in animals fed protein-rich concentrates, ammonia can potentially accumulate in the rumen fluid. Under these circumstances, some of the ammonia that is converted to urea in the liver is excreted in the urine and thus wasted.

Fat Metabolism by Microbes

The major fats in grains are triglycerides that contain a high proportion of *linoleic acid (C18:3; that is, an 18-carbon chain with three unsaturated bonds)*, whereas in forages the lipids are in the form of galactoglycerides whose principal fatty acid is *linolenic (C18:2)*. Microbes in the rumen fluid release extracellular lipases that hydrolyze these lipids into galactose, glycerol, and fatty acids. Approximately one third of the lipolytic activity in the rumen contents is considered to be accomplished by protozoal lipases. Glycerol and galactose are rapidly taken up by the microbes and fermented, whereas the fatty acids adhere to feed particles and to the rumen microbes. Limited evidence suggests that bacteria take up some of the free fatty acids and store them in fat droplets. A small amount of fat can be stored in this manner. Most fatty acids are adsorbed onto feed particles. In this state, the unsaturated free fatty acids are *biohydrogenated* by bacteria, so even if the diet contains a lot of linoleic and linolenic acids, much of the C-18 fatty acid reaching the duodenum is stearic acid (C18:0). Protozoa engulf triglycerides and also can biohydrogenate fatty acids. This is why meat and milk products from ruminants contain somewhat higher levels of saturated fatty acids than does the meat of pigs or chickens. Unlike the short-chain fatty acids, long-chain fatty acids are not absorbed from the ruminoreticulum. Only when they reach the small intestine can they be converted into micelles for eventual absorption.

Long-chain fatty acids, especially if polyunsaturated (more than two double bonds), are toxic to rumen microbes. At moderate to high levels of dietary fat intake, fiber digestion is depressed and numbers of protozoa fall.

■ Rumen microbes synthesize vitamins for their host and detoxify some ingested toxins.

One important by-product of fermentation is the synthesis of all B vitamins. Rumen microorganisms, if enough cobalt is available, even synthesize the vitamin B_{12} complex. The microbes responsible for this process require many of the B vitamins. Thus ruminants do not require a dietary source of these vitamins.

Regardless of the location of microbes in an animal's digestive tract, these microorganisms can also protect the host animal from some toxins that are ingested in the diet. Microbes are thus sometimes considered the *first line of defense* against dietary toxins. Livestock are more likely to consume poisonous plants when pasture is short, a consequence of restricted grazing practices, drought, or late-season grazing. Most poisonous plants are unpalatable and, if there is something better to eat, are usually not eaten in quantities sufficient to become toxic. Consumption of various poisonous molecules, such as oxalates (in rhubarb leaves), can be neutralized by rumen microbes into harmless by-products. Protozoa and bacteria can convert another poison, nitrate, in foodstuffs. Some species of whales can survive the ingestion of pollutant-laden krill (a tiny crustacean). Anaerobic microorganisms found in the ceca (outpouching of large intestine) of bowhead whales can degrade PCBs, anthracene and napthalene, major constituents of oil spills.

■ Some ruminants are selective in what they eat, whereas others simply graze the available forage.

Cattle are representative of a group of herbivores that are relatively nonselective in their consumption of forage. They intensively consume forage at limited periods during the day, followed by intense periods of rumination. In general, the larger the animal, the more likely it can be classified within this grouping. Ruminants with a larger forestomach have a greater digestive capacity in which to ferment lower-quality plant material. The ability to digest fiber is also associated with an increase in total intestinal length and a reduction in transit time through the intestine. A slower rate of passage improves nutrient absorption by increasing time of contact with absorptive cells and increases digestibility of dietary fiber by allowing more time for microbial fermentation. However, management practices of the beef cattle industry now include

the feeding of concentrate diets to enhance growth rate. Because these animals are slaughtered for meat, they normally do not live long enough to develop diet-related problems associated with this feeding practice.

At the other end of the spectrum are the *selective feeders* who consume only the most nutritious parts of the herbage. This activity normally requires more effort and time and so tends to be employed by smaller ruminants that need less total energy consumption. Associated with the high-quality diet is an increased rate of VFA production. Their rumens are more highly folded, which increases the available surface area for absorption of amino acids and VFAs. Selectors tend to have smaller rumens relative to body size as well as shorter total intestinal lengths. The pillars and ruminal structure are designed to increase the passage rate of digesta. Antelopes are examples of ruminants that require frequent meals and whose forestomachs are comparatively less developed. The turnover of material in the rumen of these animals is faster than in their nonselective counterparts. Concentrate selectors such as white-tail and roe deer can exploit the seasonal variability in plant growth, and show marked increases in food intake during the summer months, when food sources are plentiful. The increase in turnover of rumen contents at this time permits them to store the excess energy as fat reserves to help them through the winter season. Rumen capacity in deer is also increased by 50% during the summer months, which lets the animal accommodate a greater digesta load. Goats and caribou are examples of intermediate feeders and show a less marked increase in food intake and rumen size during the summer.

Overview of the Gastrointestinal Hormones

Throughout our discussion of digestion, we have repeatedly mentioned different functions of the several major gastrointestinal hormones in vertebrates, especially gastrin, secretin, and cholecystokinin. Let's now fit all these functions together so that you can appreciate the overall adaptive importance of these interactions. Furthermore, we now introduce more recently identified hormones as gastrointestinal hormones.

Gastrin

Chyme in the stomach, especially if it contains protein, stimulates the release of gastrin. Gastrin has three functions:

1. It acts on parietal and chief cells to increase secretion of HCl and pepsinogen. These two substances, in turn, are of primary importance in initiating digestion of the protein that promoted their release.

2. Gastrin enhances gastric motility, stimulates ileal motility, relaxes the ileocecal sphincter, and induces mass movements in the colon—functions that are all aimed at keeping the contents moving through the tract on the arrival of a new meal.

3. Not only is gastrin trophic to the stomach mucosa but it is also trophic to the small-intestine mucosa, helping maintain a well-developed, functionally viable digestive tract lining.

Predictably, gastrin secretion is inhibited by an accumulation of acid in the stomach and by the presence in the duodenal lumen of acid and other constituents that necessitate a delay in gastric secretion.

Secretin

As the stomach empties into the duodenum, the presence of acid in the duodenum stimulates the release of secretin into the blood. Secretin performs five major interrelated functions:

1. It inhibits gastric emptying to prevent further acid from entering the duodenum until the acid that is already present is neutralized.

2. Secretin inhibits gastric secretion to reduce the amount of acid being produced.

3. Secretin stimulates the pancreatic duct cells to produce a large volume of aqueous $NaHCO_3$ secretion, which is emptied into the duodenum to neutralize the acid.

4. It stimulates secretion by the liver of a $NaHCO_3$-rich bile, which likewise is emptied into the duodenum to assist in the neutralization process. Neutralization of the acidic chyme in the duodenum helps prevent damage to the duodenal walls and provides a suitable environment for the optimal functioning of the pancreatic digestive enzymes, which are inhibited by acid.

5. Along with CCK, secretin is trophic to the exocrine pancreas.

CCK

As chyme is emptied from the stomach, fat and other nutrients enter the duodenum. These nutrients, especially fat and to a lesser extent protein products, cause the release of cholecystokinin (CCK) from the duodenal mucosa. CCK also performs several important interrelated functions:

1. It inhibits gastric motility and secretion, allowing adequate time for the nutrients already in the duodenum to be digested and absorbed.

2. CCK stimulates the pancreatic acinar cells to increase secretion of pancreatic enzymes, which continue the digestion of these nutrients in the duodenum (this action is especially important for fat digestion, because pancreatic lipase is the only enzyme that digests fat).

3. CCK causes contraction of the gallbladder and relaxation of the sphincter of Oddi so that bile is emptied into the duodenum to aid fat digestion and absorption. Bile salts' detergent action is particularly important in enabling pancreatic lipase to perform its digestive task. Once again, the multiple effects of CCK are remarkably well adapted to dealing with the fat and other nutrients whose presence in the duodenum trigger this hormone's release.

4. Furthermore, it is appropriate that both secretin and CCK, which have profound stimulatory effects on the exocrine pancreas, are trophic to this tissue.

5. CCK has also been implicated in long-term adaptive changes in the proportion of pancreatic enzymes produced in response to seasonal changes in diet.

6. Besides facilitating the digestion of ingested nutrients, CCK is an important regulator of food intake. It plays a key role in satiety, the sensation of having had enough to eat (Chapter 15). In people with **bulimia** (binge eating followed by vomiting), emptying of the stomach contents through the pyloric sphincter is reduced and less CCK is released into the blood. For this reason bulimics do not obtain the sensation of satiety that people without this eating disorder can feel.

GIP

A more recently recognized hormone, **GIP**, helps promote metabolic processing of the nutrients once they are absorbed. This hormone was originally named *gastric inhibitory peptide. (GIP)* for its presumed role as an enterogastrone. It was believed to inhibit gastric motility and secretion, similar to secretin and CCK. GIP's contribution in this regard is now considered minimal. Instead, this hormone has been shown to stimulate insulin release by the pancreas, so it is now called **glucose-dependent insulinotrophic peptide** (once again, GIP). Again, this is remarkably adaptive. As soon as the meal is absorbed, the body has to shift its metabolic gears to use and store the newly arriving nutrients. The metabolic activities of this postabsorptive phase are largely under the control of insulin (Chapter 7). Stimulated by the presence of a meal in the digestive tract, GIP initiates the release of insulin in anticipation of the absorption of the meal, in a "feedforward" fashion. Insulin is especially important in promoting the uptake and storage of glucose in the muscle and liver. Appropriately, glucose in the duodenum has also been shown to increase GIP secretion.

Motilin

The hormone motilin, secreted by gland cells in the small intestine, is thought to trigger peristaltic pumping in the intestine between meals, as a housecleaning function.

Ghrelin

The peptide hormone ghrelin is secreted by epithelial cells in the stomach, primarily during fasting periods just before normal mealtimes (in rats and humans). It stimulates release of growth hormone from the anterior pituitary, and—perhaps most importantly—greatly increases appetite. Its role in food intake and energy balance is further discussed in Chapter 15.

PYY_{3-36}

The recently discovered peptide PYY_{3-36} appears to be a signal complementary or antagonistic to ghrelin. It is released by the small intestine in proportion to the amount of food inside it, and appears to travel to the brain (hypothalamus), where it suppresses appetite. Like CCK, it also appears to stimulate bile and pancreas secretion. Its role in food intake and energy balance is further discussed in Chapter 15.

?

How Many Gastrointestinal Hormones and Paracrines Are There?
In this chapter, we have mentioned several gastrointestinal signal molecules. But to date over 20 possible signal peptides have been isolated from the mammalian digestive tract. Most were first discovered in other tissues. For example, the small intestine secretes somatostatin (a hypothalamic hormone that regulates growth and metabolism), glucagon (a pancreatic hormone that regulates glucose, p. 299), serotonin (a brain neurotransmitter, p. 175), and sub-

stance P (a neurotransmitter of pain neurons). Intestinal somatostatin appears to be an antagonist to (inhibitor of) gastrin, CCK, and secretin. But it is not yet clear what are all the roles of the other peptides made by the small intestine.

This overview of the multiple, integrated, adaptive functions of the gastrointestinal hormones provides an excellent example of the remarkable efficiency of an animal's body, and the remarkable nature of the vertebrate's "second brain" and its associated endocrine system located in the gut. Much remains for science to learn about these regulatory functions.

Chapter in Perspective:

HOMEOSTASIS AND INTEGRATION

To maintain constancy in the internal environment, materials that are used up in the body (such as nutrients and O_2) or uncontrollably lost from the body (such as evaporative H_2O loss from the airways or salt loss in sweat) must constantly be replaced by new supplies of these materials from the external environment. All these replacement supplies except O_2 are acquired through the digestive system. Fresh supplies of O_2 are transferred to the internal environment by the respiratory system, but all the nutrients, H_2O, and various electrolytes needed to maintain homeostasis are acquired through the digestive system. The large, complex food that is ingested is broken down by the digestive system into small, absorbable units. These small, energy-rich nutri-

ent molecules are transferred across the midgut epithelium into the blood for delivery to the cells to replace the nutrients constantly used for ATP production and for repair and growth of body tissues. Likewise, ingested H_2O, salt, and other electrolytes are absorbed by the hindgut into the blood. If excess nutrients are ingested and absorbed, the extra is placed in storage, such as in the insect fat body and vertebrate adipose tissue (fat), or it is metabolized, so that the blood level of nutrient molecules is kept relatively constant.

Unlike most body systems, regulation of digestive system activities themselves is not aimed at maintaining homeostasis. The quantity of nutrients and H_2O ingested is subject to control, but the quantity of ingested materials absorbed by the digestive tract is not subject to control, with few exceptions. The hunger mechanism governs food intake to help maintain energy balance (Chapter 15), and the thirst mechanism controls H_2O intake to help maintain H_2O balance (Chapter 13). Once these materials are in the digestive tract, the digestive system does not vary its rate of nutrient, H_2O, or electrolyte uptake according to body needs; rather, it optimizes conditions for digesting and absorbing what is ingested. Truly, what an animal consumes is what it gets. The digestive system is subject to many regulatory processes, but these are not influenced by the nutritional or hydration state of the body. Instead, these control mechanisms are governed by the composition and volume of digestive tract contents so that the rate of motility and secretion of digestive juices are optimal for digesting and absorbing the ingested food. In a sense, the entire digestive system is an effector for energy, mineral, and water homeostasis. ∎

REVIEW QUESTIONS (Answers are on p. A–1.)

Additional study tools for this chapter, including chapter summaries and practice tests, are available online at *www.biology.brookscole.com*

1. Which of the following would not have to be initially hydrolyzed into smaller molecules during digestion?
   a. starch
   b. lactose
   c. protein
   d. glucose
   e. triglycerides
2. Which of the following polysaccharides is the most common storage form of glucose monomers in animals?
   a. chitin
   b. cellulose
   c. glycogen
   d. trehalose
   e. sucrose
3. The initial segment of the small intestine is called the
   a. fundus
   b. duodenum

   c. corpus
   d. antrum
   e. esophagus
4. In the oxyntic mucosa, chief cells secrete
   a. intrinsic factor
   b. mucus
   c. HCl
   d. chyme
   e. pepsinogen
5. G cells of the pyloric gland area (PGA) secrete
   a. histamine
   b. gastrin
   c. somatostatin
   d. mucus
   e. HCl
6. Pepsinogen is activated to the enzyme pepsin by
   a. intrinsic factor
   b. lingual lipase
   c. lysozyme
   d. HCl
   e. zymogen granules

7. Which of the following inhibits acid secretion?
   a. acetylcholine
   b. somatostatin
   c. gastrin
   d. histamine
   e. chymotrypsinogen
8. An important enzyme secreted by the pancreas for fat digestion is
   a. insulin
   b. pancreatic lipase
   c. chitinase
   d. pancreatic amylase
   e. glucagon
9. Trypsinogen is converted into its active form, trypsin, by
   a. enterokinase
   b. chief cells
   c. chymotrypsinigen
   d. carboxypeptidase
   e. insulin
10. The largest component of pancreatic secretion into the duodenum is
    a. cholecystokinin that neutralizes stomach acid
    b. enterogasrones
    c. alkaline fluid that neutralizes acidic chyme as it arrives from the stomach
    d. trypsinogen
    e. insulin
11. The enterogastrone, secretin, is secreted in response to
    a. acid in the duodenum
    b. nutrients in the lumen
    c. pancreatic lipase
    d. stretching of the stomach by ingested food
    e. chymotrypsin
12. Which of the following is not a liver function?
    a. activation of vitamin D
    b. production of the organic components of bile
    c. secretion of chitinase
    d. storage of glycogen
    e. synthesis of plasma proteins
13. In the liver, each hepatocyte is in contact with a sinusoid on one side and with which of the following on the other side?
    a. a hepatic portal vein
    b. a lobule
    c. a major bile duct
    d. a bile canaliculus
    e. the inferior vena cava
14. The main function of bile salts is
    a. emulsification of lipids
    b. protein absorption
    c. gluconeogenesis
    d. formation of micelles
    e. a and d
15. The main site for digestion and absorption in vertebrates is the
    a. stomach
    b. large intestine
    c. pancreas
    d. small intestine
    e. gall bladder

SUGGESTED READINGS AND INTERNET SITES

Batterham, R. L., et al. 2002. Gut hormone PYY_{3-36} physiologically inhibits food intake. *Nature* 418, 650–654.

Triplehorn, C. A., & N. F. Johnson. 1989. *Borror's Introduction to the Study of Insects*, 7th ed. Belmont, CA: Brooks/Cole.

Church, D. C. 1988. *The Ruminant Animal. Digestive Physiology and Nutrition*. Englewood Cliffs, NJ: Prentice Hall.

Swenson, M. J., & W. O. Reese, eds. 1993. *Duke's Physiology of Domestic Animals*, 11th ed. Ithaca, NY: Cornell University Press.

Gershon, M. D. 1998. *The Second Brain*. New York: Harper Perennial. A detailed look at the enteric neuroendocrine system.

Prosser, C. L., & E. J. DeVillez. 1991. Feeding and digestion. In C. L. Prosser, ed., *Environmental and Metabolic Animal Physiology: Comparative Animal Physiology*, 4th ed. New York: Wiley-Liss.

Stevens, C. E., & I. D. Hume 1996. *Comparative Physiology of the Vertebrate Digestive System*, 2nd ed. Cambridge, UK: Cambridge University Press.

INFOTRAC READING

Kling, J. 2002. Elusive ligand ghrelin could have numerous roles: The ligand is linked to growth hormone release, feeding regulation, energy homeostasis, and the cardiovascular system. *The Scientist* 16:36–37.

INTERNET SITES

Austgen, L., R. A. Bowen, & M. Rouge. 2003. *Pathophysiology of the Digestive System*. **arbl.cvmbs.colostate.edu/hbooks/ pathphys/digestion/index.html**. A detailed look at the mammalian digestive system, with pictures, text, and animations on humans, herbivores, and birds, plus major pathologies in humans.

Ritchison, G. *Bird Digestion*. **biology.eku.edu/RITCHISO/ birddigestion.html**. A detailed site on digestive organs and processes in birds.

Energy Balance and Thermal Physiology

Butterflies and lizards are ectotherms, *animals dependent on the environment for their body heat content. But their body temperatures are not always equal to that of the air. These animals are basking in the sunshine to heat up their bodies above the air temperature.*

Introduction

Each cell needs energy to perform the functions essential for its own survival and to carry out its specialized contribution toward maintaining homeostasis. All energy used by animal cells is ultimately provided by food intake, which is regulated to maintain energy balance. One factor that many animals maintain homeostatically is the availability of energy-rich nutrients for the cells. In this chapter, we show how this occurs. We also consider temperature, one of the key components of energy equations. Temperature profoundly affects the rate of chemical reactions within cells, and animals have evolved various strategies that cope with these effects. First, however, let us set the stage by examining the key energy equations.

■ Life follows the laws of thermodynamics.

Perhaps the most distinctive characteristic of life that delineates it from nonliving systems is the way it uses energy. Whereas nonliving energetic processes—volcanoes, earthquakes, storms, and so forth—create disorder, living systems somehow produce their own order unique to each individual. Yet all these processes (indeed, all known processes in the universe) obey fundamental laws regarding energy—the **laws of thermodynamics**. There are three of these laws, with two that are relevant to physiology:

■ The *first law of thermodynamics* is that energy can be neither created nor destroyed. Therefore, energy is subject to the same kind of input–output balance as are the chemical components of life such as H_2O and salt (see p. 572). The first law of thermodynamics suggests that life might operate forever on its internal energy content, sim-

ply converting one form to another as necessary. However, the second law says this is not possible.

■ The *second law of thermodynamics* is that the **entropy** (a measure of disorder or randomness) of a system *plus its surroundings* increases over time as the energy content degrades to unusable heat. This is expressed in the following equation, where Δ indicates change and S is entropy:

$$\Delta S_{net} = \Delta S_{surroundings} + \Delta S_{system}$$
$$> 0 \text{ for spontaneous reactions}$$

The second law of thermodynamics is in many ways the most important physical law regarding life in the universe. Most people are aware that the universe seems to run down over time, and only by expending energy can we reverse this trend. But life itself seems to defy this decay (until death, at least), and indeed some people have suggested that life violates the second law. However, they are forgetting that the law covers a system *plus its surroundings,* not just a system by itself. The system called "life" does *not* violate the law, because the order it creates can only arise at the expense of its environment. That is, as life becomes more organized, the surrounding environment becomes even more disorganized. The net result—increased order in an organism, increased disorder in the surroundings—is *always* an increase in net disorder (entropy). Plants, for example, create order at the expense of the sun, which is creating enormous disorder as it radiates out light and particles. Animals create order at the expense of the local environment: They eat ordered molecules (food) and release less ordered waste molecules and—perhaps most importantly—heat. Indeed, *most food energy is ultimately converted to heat in animals.* Thus life must continuously extract energy from the environment to maintain itself, whereas the rest of

Photos: Paul Yancey

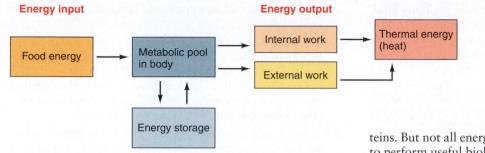

Figure 15–1 ● Energy input and output.

the universe degrades even faster than it would on its own. In short, the "metabolic cost of living" is a constant struggle against entropy (a struggle that is ultimately lost through aging and death).

Energy Balance

■ Animals require food intake to compensate for entropy.

As you have just seen, the laws of thermodynamics dictate that organisms need energy input from the environment, because the inevitable increase in entropy guarantees that useful energy degrades with every reaction in a cell. Each cell in an animal needs energy to perform functions essential for its own survival (for example, reversing the effects of entropy by restoring ion gradients and repairing damaged structures) and to carry out its specialized contributions toward maintenance of the whole animal. All energy used by animal cells is ultimately provided by food intake (or, in a few special cases, high-energy molecules obtained from internal symbionts). Chemical energy locked in the bonds that hold atoms together in nutrient molecules is released when these molecules are broken down in the body. However (as a consequence of the second law of thermodynamics), 100% of the food energy cannot be used usefully, because some energy is always lost to entropy.

The energy in ingested foodstuffs constitutes energy input for most animals. Energy output or expenditure falls into two categories (● Figure 15–1): external work and internal work. **External work** is the energy expended when skeletal muscles are contracted to move external objects or to move the body in relation to the environment. **Internal work** constitutes all other forms of biological energy expenditure that do not accomplish mechanical work outside the body. Internal work encompasses two types of energy-dependent activities: (1) skeletal muscle activity used for purposes other than external work, such as the contractions associated with postural maintenance and shivering; and (2) all the energy-expending activities that must go on all the time just to sustain life. The latter include the work of pumping blood and breathing; the energy required for active transport of critical materials across plasma membranes; and the energy used during synthetic reactions essential for the maintenance, repair, and growth of cell structures.

Although energy cannot be created or destroyed (the first law of thermodynamics), it can be converted from one form to another. For example, **chemical energy** in ATP is converted into **kinetic energy** of locomotion by muscle contractile pro-

teins. But not all energy in nutrient molecules can be harnessed to perform useful biological work like this (in accordance with the second law of thermodynamics). The energy in nutrient molecules that is not used for work is transformed into **thermal energy,** or **heat.** During biochemical processing, only about 50% of the energy in nutrient molecules is transferred to ATP; the other 50% of nutrient energy is immediately lost as heat. During ATP expenditure by the cells, another 25% of the energy derived from ingested food becomes heat. Because animal bodies are not heat engines, they cannot convert heat into work. Therefore, not more than 25% of nutrient energy is available to accomplish work, either external or internal. The remaining 75% is lost as heat during the sequential transfer of energy from nutrient molecules to ATP to cellular systems.

Furthermore, of the energy actually captured for use by animals, almost all expended energy eventually becomes heat. To exemplify, energy expended by a heart to pump blood is gradually changed into heat by friction as blood flows through vessels. Likewise, energy used in the synthesis of cellular structural protein eventually appears as heat when that protein is degraded during the normal course of turnover of bodily constituents. Even in the performance of external work, skeletal muscles convert chemical energy into mechanical energy inefficiently, with as much as 75% of the expended energy being lost as heat. Thus all energy that is liberated from ingested food but not directly used for moving external objects or stored in fat (adipose tissue) deposits (or, in the case of growth, as protein) eventually becomes heat. This heat is not necessarily wasted energy, however, because much of it is used to maintain body temperature in some animals.

Terminology and Methodology

The rate at which energy is expended during both external and internal work is known as the **metabolic rate:**

$$\text{Metabolic rate} = \frac{\text{energy expenditure}}{\text{unit of time}}$$

Because most animal energy expenditure eventually appears as heat, the metabolic rate is normally expressed in terms of the rate of heat production in energy units per hour. The basic units of heat energy are as follows:

1. The **calorie** is the amount of heat required to raise the temperature of 1 g of H_2O by 1°C (specifically, from 14.5 to 15.5°C). This unit is too small to be convenient when discussing animals, because of the magnitude of heat involved, so the **kilocalorie (kcal)** or **Calorie,** which is equivalent to 1000 calories, is used. When nutritionists speak of "calories" in quantifying the energy content of various foods, they are actually referring to kilocalories or Calories.

2. The **joule,** the international unit for energy of any kind, is equal to 0.239 calories. The term **kilojoule (kJ)** (1000 joules) is commonly used, again because of the magnitude of animal energies. *Most scientific research now uses the joule or kilojoule rather than the calorie.*

As a common example, 4.1 kcal or 17.1 kJ of heat energy are released when 1 g of glucose is oxidized or "burned," whether the oxidation takes place inside or outside a body.

The metabolic rate and consequently the amount of heat produced vary, depending on a variety of factors, such as genetics, external temperature, exercise, food intake, shivering, and anxiety. Increased skeletal muscle activity is the factor that can most increase metabolic rate. Even slight increases in muscle tone notably elevate the metabolic rate, and various levels of physical activity markedly alter energy expenditure and heat production (▌ Table 15–1). For this reason, it is useful to know the metabolic rate under standardized basal conditions established to control the variables that can alter metabolic rate. In this way, the metabolic activity necessary to maintain the basic body functions at rest can be determined. This minimal rate is a reflection of "idling speed," or the minimal waking rate of internal energy expenditure. It is measured under the following specified conditions:

1. To eliminate any nonresting muscular exertion, the animal should be sleeping and should have refrained from exercise for some time.

2. The animal should have minimal anxiety or stress levels. Stress usually increases metabolic rates, for example, because vertebrates secrete the hormone epinephrine, which increases heart rate (p. 387). For many animals this necessitates an acclimation period to laboratory conditions prior to the actual measurement. This factor has not been well studied.

3. The animal should not have eaten any food within an appropriate amount of time before the rate determination, to avoid **diet-induced thermogenesis (DIT)** (*thermo-,* "heat"; *genesis,* "production"), an increase in metabolic rate that occurs as a consequence of food intake (including **specific dynamic action,** a term we explain later).

4. There should be no elevated energy costs of reproduction, such as pregnancy, egg brooding, lactation, or seasonal gamete production.

5. The measurement should be performed at an appropriate temperature for that organism.

Regarding the last point (item 5), animals fall into two broad categories of thermal adaptation: **ectotherms** (such as amphibians), which depend on external heat for their body temperatures, and **endotherms** (such as mammals), which use internal heat to regulate body temperatures (these are discussed in detail later in this chapter). Because of this difference, there are two types of this minimal metabolic rate:

- **Standard metabolic rate (SMR)** is an ectotherm's resting metabolic rate at a particular temperature, which may be the average temperature experienced naturally, or any temperature within its normal range. SMR values usually change dramatically with temperature, and thus the temperature of the test conditions must be reported.

- **Basal metabolic rate (BMR)** is an endotherm's resting metabolic rate in its *thermal neutral zone* (TNZ), a range of external temperatures that do not induce thermoregulatory processes by the animal. For example, the TNZ for lightly clothed humans lies around so-called room temperature. In the TNZ, there is no shivering, significant sweating, or other thermoregulatory processes that would raise the metabolic rate. (We discuss the TNZ in more detail later; p. 699.)

Measurements of BMR/SMR are difficult to make, and any exceptions to the conditions just listed must be reported. Another term is used if measurements are made during nonsleep periods: the **resting metabolic rate (RMR).** RMR values are typically 10 to 25% higher than the BMR but are often used because measurements during sleep periods may not be practical.

The rate of heat production in SMR, BMR, and RMR determinations can be measured directly or indirectly. **Direct calorimetry** involves the cumbersome procedure of placing the subject in an insulated chamber with H_2O circulating through the walls. The difference in the temperature of the H_2O entering and leaving the chamber reflects the amount of heat liberated by the subject and picked up by the H_2O as it passes through the chamber. Even though this method provides a direct measurement of heat production, it is not practical, because a calorimeter chamber can be costly, particularly for large animals. Therefore, other more practical methods of **indirect calorimetry** for determining metabolic rates were developed for widespread use.

Here we focus on one of those indirect methods. In the **respirometry method,** only the subject's O_2 uptake per unit of time (called the V_{O_2}) is measured, which is a straightforward task in most cases (● Figure 15–2). Recall that food energy is

Table 15–1 ▌ Rate of Energy Expenditure for a 70-kg Person during Different Types of Activity

Form of Activity	Energy Expenditure (kcal/hour)	kjoules/hour
Sleeping	65	272
Awake, lying still	77	322
Sitting at rest	100	418
Standing relaxed	105	439
Getting dressed	118	493
Typewriting	140	585
Walking slowly on level (2.6 mi/hr)	200	836
Carpentry, painting a house	240	1003
Sexual intercourse	280	1170
Bicycling on level (5.5 mi/hr)	304	1271
Shoveling snow, sawing wood	480	2006
Swimming	500	2090
Jogging (5.3 mi/hr)	570	2383
Rowing (20 strokes/min)	828	3461

Figure 15–2 ➤ • Measuring oxygen consumption by animals. (a, b) Animals on treadmills, which force an animal to run at controlled speeds. The animal may be fitted with a mask for measuring gases in exhaled breath, or may be in an enclosed chamber from which gases are sampled. (c) A bird flying in a wind tunnel, which forces the animal to fly at controlled speeds. A face mask samples gases in exhaled breath.

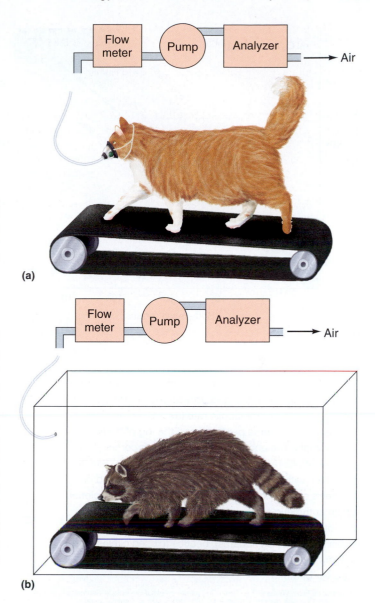

(a)

(b)

liberated with the use of oxygen during aerobic metabolism. Accordingly, a direct relationship exists between the volume of O_2 used and the quantity of heat produced. This relationship also depends on the type of food being oxidized, with the following average values, known as the *physiological fuel values*:

>   Carbohydrates: 4.1 kcal/g = 17.1 kJ
>
>   Proteins: 4.3 kcal/g = 17.9 kJ
>
>   Fats: 9.3 kcal/g = 38.9 kJ

An estimate, the **energy equivalent of O_2,** can be made of the quantity of heat produced per liter of O_2 consumed, using the animal's typical diet. For a typical human in North America, for example, the average is 4.8 kcal (20.0 kJ) per liter of O_2 consumed. Thus in determining metabolic rate, a simple measurement of O_2 consumption can be used to reasonably approximate heat production by multiplying by this factor. However, for nonhuman animals the composition of the diet may not be easily obtained, thus (1) metabolic rates may be reported simply as liters of oxygen consumed per unit time, or (2) an estimate may be made using the **respiratory quotient (RQ),** the ratio of CO_2 produced to O_2 consumed. This requires measurements of carbon dioxide production (V_{CO_2}), an increasingly common technique. The RQ varies depending on the foodstuff consumed. When carbohydrate is being used, the RQ is 1; that is, for every molecule of O_2 consumed, one molecule of CO_2 is produced:

$$C_6H_{12}O_6 + 6\ O_2 \rightarrow 6\ CO_2 + 6\ H_2O + 36\ ATP + heat$$

For fat use, the RQ is 0.7; for protein, it is 0.8. On a diet consisting of a mixture of these three nutrients, resting O_2 consumption in a human averages about 250 mL/min, and CO_2 production averages about 200 mL/min, for an average RQ of 0.8:

$$RQ = \frac{CO_2\ produced}{O_2\ consumed} = \frac{200\ mL/min}{250\ mL/min} = 0.8$$

Measurement of RQ in a newly hatched or migrating bird or in a diving seal reveals a value closer to 0.7, indicating their reliance on fats during these energy-requiring activities. Using the RQ and the physiological fuel values, an energy equivalent of O_2 can be closely estimated. (Note that a slight correction is necessary to convert gas volumes to moles: 1 mole of O_2 = 22.39 L, whereas 1 mole of CO_2 = 22.26 L.)

◼ Metabolic rate is scaled to body mass.

As discussed in previous chapters (pp. 8, 145, 386, 474), some physiological features depend on the size of organisms, because of scaled factors such as surface area and volume. Perhaps the most consistent, and most controversial, scaling relationship is

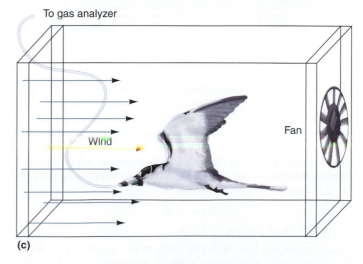

(c)

that between BMR or SMR and body mass. Numerous studies have found that, although larger animals certainly consume more energy than smaller animals, they have lower metabolic

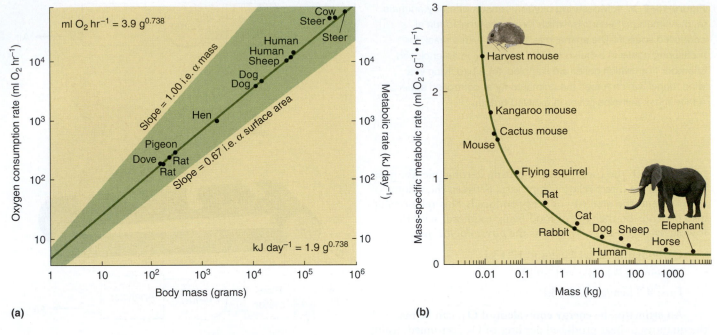

(a)

(b)

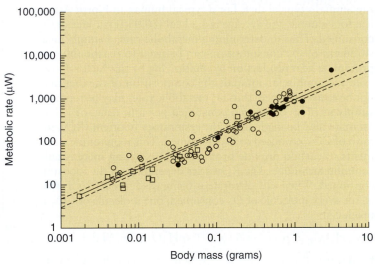

Figure 15–3 ● Scaling: Metabolic rate as a function of body mass in mammals and arthropods. (a) Relationship between the log of metabolic rate (expressed as oxygen consumption, volume per hour) and the log of body mass in mammals. The equation in the upper left corner shows that metabolic rate scales to a power of 0.74. (b) Similar to (a), but with the metabolic rate (y-axis) calculated on the basis of unit mass (volume per hour per kg) and plotted on an arithmetic scale. (c) Relationship between the log of metabolic rate (expressed in μW) and the log of body mass in ants (open squares), beetles (closed circles) and spiders (open circles). The metabolic rate here scales to a power of 0.825.

(*Sources:* (a) From Withers, P. C. 1992. *Comparative Animal Physiology.* Saunders, Fort Worth TX, using data from Kleiber, M. 1932. Body size and animal metabolism. *Hilgardia* 315–353; (b) From Schmidt-Nielsen, 1960. *Animal Physiology.* Prentice-Hall, Englewood Cliffs NJ. Fig. 5.9 p 193; (c) From Lighton, J. R. B., L. J. Fielden. 1995. Mass scaling of standard metabolism in ticks: A valid case of low metabolic rates in sit-and-wait strategists. *Physiological Zoology* 68: 43–62)

(c)

rates *per unit mass* than smaller ones (● Figure 15–3b). Basically, a gram of elephant costs much less energy to maintain than a gram of shrew!

The actual scaling factor (the rate at which metabolism changes with body size) is found by plotting the BMR values for whole animals against their body masses on a log–log plot (Figure 15–3a). A straight line with a positive slope is obtained, with the following formula:

$$M = a \, W^b$$

where M is metabolic rate, a is the metabolic rate per unit mass (usually rate per kg), W is the body mass, and b is the slope. This plot shows that larger animals do use more energy. But the slope (b) is always less than 1, again showing the lower metabolic rate (per unit mass) of larger animals (if b were 1, then all animals would use energy at the same rate per unit

mass). In fact, for mammals, a consistent slope (b) of about 0.75 (3/4) has been obtained in numerous studies. Similar scaling is found for birds, again with a slope between 0.7 and 0.8. Many other physiologic and anatomic features also scale with slopes of about 0.75. For example, new studies show that the number of hours that a mammal spends sleeping each day corresponds directly with metabolic rate per unit mass (a phenomenon many human cat owners have noticed).

Following the discovery of scaling by Max Kleiber in 1932, a widely accepted hypothesis was generated: Scaling reflects properties of the ratio between surface area and volume with respect to temperature regulation by endotherms. The argument goes as follows: (1) Surface area increases as the square of the radius, whereas volume increases as the cube; therefore smaller organisms have a higher ratio of surface area to volume (see Chapter 1). (2) Endotherms produce heat by their volumes (mass) and lose it from their surface areas. (3) Therefore a smaller endotherm must have a higher metabolic rate per unit mass to compensate for its higher rate of heat loss (via surface area).

This hypothesis makes perfect physical sense, but it has two major flaws. First, the hypothesis predicts that BMR will scale with a slope of about 0.67 (2/3), because that is the ratio by which surface area and volume change with size (for spherical objects). In fact, metabolic rates within a single mammalian species do scale with size by a factor of 0.67. Second, other studies have reported that ectotherms such as reptiles and fishes (and even unicellular organisms) have metabolisms that scale by factors less than 1.0 (see Figure 15–3c for insects, showing a slope of 0.825). The hypothesis is irrelevant to these organisms, because most do not generate (and retain) body heat for the purposes of thermoregulation. So why does metabolism scale with an apparent 3/4 factor among endotherms, and by slopes of less than 1 in animals in general? Neither of these flaws has been satisfactorily explained. See box, "Challenges and Controversies: A Universal Scale of Life?" for recent hypotheses.

■ **Aerobic locomotory energy reflects the limits of oxygen delivery and is also scaled to body mass.**

Activity, expressed primarily as locomotion powered by skeletal muscles and circulatory flow from the heart, is potentially the greatest energy usage for an animal, at least for short periods of time. The metabolic rates of animals can increase many fold during extreme activity. The ratio of metabolic rate at its highest level to the BMR or SMR is called the **metabolic scope**. Scope values for aerobic activity have been measured as oxygen consumption for a number of animals, using laboratory devices such as treadmills, wind tunnels, and water flumes. Aerobic scope values for most terrestrial animals (including both ectotherms and endotherms) are generally in the range of 5 to 15. That is, whole-animal aerobic metabolic rate can be elevated above the basal level by a factor of 5 to 15. For some flying insects, the aerobic scope factor can be over 100.

Why is aerobic scope limited to the 5-to-15 range for so many animals? The limit appears to be the delivery of oxygen by the circulatory system. There seems to be a strict limit to the rate of pumping through a network of narrow vessels. The values for some insects show that the insect respiratory system, with trachea that carry oxygen directly to muscles, may be much more effective.

?

Why Were There Giant Insects in the Distant Past? Insect flight is entirely aerobic, and oxygen-delivery limits are thought to be one reason why insects are relatively small animals. However, insects have not always been small: In the Paleozoic era (about 300 million years ago), there were giant dragonflies with wingspans of 70 cm (28 in)! How could they have sustained flight? Evidence suggests that Paleozoic atmospheric oxygen levels were up to 35% (compared to 21% today). Could this explain the dragonflies' size? See the Suggested Reading by Graham, Dudley, Aguilar, and Gans (1995), listed at the end of this chapter.

Metabolism during locomotion generally increases with speed, as you would expect, but there are some interesting exceptions. For running mammals, the increase in metabolism is roughly linear with velocity (● Figure 15–4a). For a swimming fish, the cost increases exponentially as drag from the environment begins to have more and more effect (Figure 15–4b).

CHALLENGES *AND* CONTROVERSIES

A Universal Scale of Life?

Why isn't metabolism linearly proportional to body mass? As discussed in the main text, the primary hypothesis has focused on ratios of surface area to volume. But these do not explain why a relationship scaled to a 3/4 power rather than a 2/3 power is found among many forms of life. Indeed, some scientists have claimed that the 3/4 power applies to all life from bacteria to trees, and in some sense must represent a universal phenomenon of life. Because the heat loss hypothesis (see main text) cannot apply to ectothermic organisms, several other hypotheses have been generated, some within the last few years. One of these might be called the *four-dimensional* or *fractal transport* hypothesis. Three researchers (Brian Enquist, James Brown, and Geoffrey West) noted that all organisms require a means of transporting and distributing internal nutrients, gases, and wastes. In multicellular organisms, this usually involves networks of increasingly finer branching tubes, such as blood vessels, tracheae, and xylem (in plants). The researchers showed that any network that maximizes the packing of two-dimensional tubular surface areas in a three-dimensional space scales to a 3/4 power.

This is the first hypothesis that seems to apply to many different kinds of organisms, and is thus attractive. But critics have pointed out that BMR is not limited by transport systems, because those systems are only pushed to their limits at maximal metabolism. Finally, some researchers are not convinced that all life forms follow the 3/4 scaling pattern; for example, some studies on ectothermic animals show scaling slopes considerably different from 0.75 (see Figure 15–3c). Other analyses have provided evidence that scaling in mammals may actually not statistically differ from 2/3. Regardless, it does appear that larger organisms are more efficient than smaller ones. But the causes are likely to remain controversial for many years.

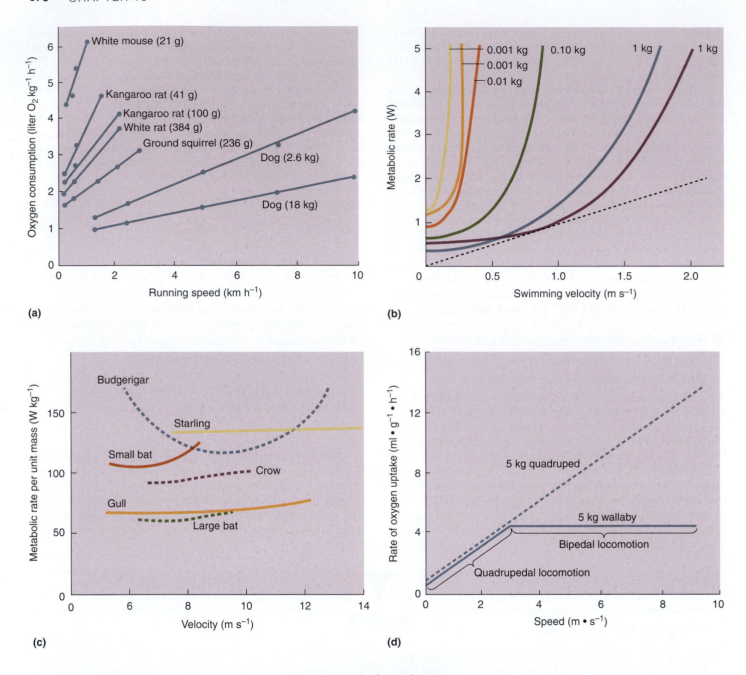

Figure 15–4 ● **Metabolic rates (measured as oxygen consumption) as a function of animal locomotory speed.** (a) Metabolic rates (per body mass) for running terrestrial mammals. Rate increases linearly with speed, with costs being higher in smaller mammals. (b) Metabolic rate (per whole animal) for a swimming fish. Rate tends to increase exponentially with speed. (c) Metabolic rate (per body mass) for flying birds and bats. A decrease in rate is seen at intermediate speeds due to gliding in some species. (d) Metabolic rate (per body mass) for a wallaby. When the animal uses all four limbs (quadrapedal locomotion), the rate increases linearly, as in part (a). But after the animal switches to bipedal hopping at higher speeds, rate becomes independent of speed.

(*Sources:* (a) From Willmer, P., G. Stone and I. Johnston. 2000. *Environmental Physiology of Animals.* Blackwell Science, Oxford UK, as modified from Taylor, C. R., K. Schmidt-Nielsen, and J. L. Raab. 1970. Scaling of the energetic cost of running to body size in mammals. *American Journal of Physiology* 219: 1104–1107; (b) From Willmer, P., G. Stone and I. Johnston. 2000. *Environmental Physiology of Animals.* Blackwell Science, Oxford UK, as modified from Peters, R. H. 1983. *The Ecological Implications of Body Size.* Cambridge University Press, Cambridge UK; (c) Willmer, P., G. Stone and I. Johnston. 2000. *Environmental Physiology of Animals.* Blackwell Science, Oxford UK, as modified from Alexander, R. M. 1999. *Energy for Animal Life.* Oxford University Press, Oxford; (d) Modified from R. V. Baudinette, G. K. Snyder and P. B. Frappell. 1992. Energetic cost of locomotion in the tammar wallaby. *American Journal of Physiology* 262: R771–R778)

Flight energetics show varying patterns: In some species, metabolic rate is independent of speed at medium to high levels, whereas in others metabolism actually decreases at medium speeds (Figure 15–4c). This is because at low speeds much energy is needed to overcome gravity when a bird is first ascending from a perch. Once airborne at medium speeds, a bird can save energy by *gliding*, using its forward momentum to generate lift with its wings to stay aloft. (Bird wings have an *aerofoil* shape: slightly concave on the bottom and curved on the top so that air moving over the wing must travel a greater distance across the top than it does across the bottom. This creates negative pressure on the top relative to the bottom, drawing the wing up.)

Finally, changes in **gait** (modes of limb use during locomotion) can alter the linear pattern. Most terrestrial vertebrates have elastic tendons in their legs that can store energy each time a leg hits the ground. This energy can be released on the next step, so that much of the energy required to lift the limbs is not wasted. In the arches of their feet, for example, humans have elastic tendons that do this during running gait. Horses have tendons in their backs, which come into play during a gallop gait. The ultimate example is found in hopping kangaroos and wallabies, which have been found to have no increase in metabolism as they increase hopping velocity (although they do show the expected linear increase while walking; Figure 15–4d). Researchers think that during the hop the huge tendons of the kangaroos' legs store and release energy more efficiently at higher speeds. Also, the stride (hop) length increases with speed, so a greater distance is covered with about the same amount of energy expended.

Metabolic rates during locomotion are also scaled with body size, so that larger animals use less energy per unit body mass to cover a given distance than do smaller animals (Figure 15–4a). That is, it costs much less to move a gram of dog than a gram of mouse the same distance. This appears to be true of insects as well as vertebrates. It is not readily apparent why this should be.

Diet-induced thermogenesis occurs after eating in most animals.

Diet-induced thermogenesis (DIT) is manifested as a rise in metabolism and heat production for several hours after eating. (This has also been called the *heat increment of feeding* HIF). DIT has two components. (1) **Obligatory DIT** is the result of the increased metabolic activity associated with the costs of obtaining and processing food, that is, gut motility, production of gut secretions, nutrient uptake, biosynthesis of proteins, glycogen stores, and lipids, and excretion of wastes. (This component has historically been called *specific dynamic action*, or *SDA*, a meaningless term we avoid.) (2) **Regulatory DIT** is increased heat production after a meal, largely for the purpose of removing excess nutrients, for example, "burning off" excess calories ingested. Obligatory DIT has been found in all phyla of animals tested; regulatory DIT has historically been studied in endotherms but may also occur in many types of animals. We examine the role of regulatory DIT in body mass control shortly.

Although the exact costs involved with DIT are not all certain, there are clear patterns. Within an individual, large and/or hard-to-digest meals induce a greater DIT than small and/or readily digested meals. Species differences are diverse; for example, humans show an increase of up to 50% in metabolism after a large meal, whereas other mammals, insects, and ectothermic vertebrates exhibit 100 to 800% increases or more. The record may be held by the Burmese python, which shows a 4400% increase in metabolism after swallowing a rodent! This animal (and other snakes that eat large, infrequent meals) actually undergoes *atrophy* of its digestive organs in between meals, presumably to save energy. This downregulation is manifested as lower liver and intestinal masses, and lower density of nutrient transporters in the intestine. Then, on ingesting its large but infrequent prey item, the snake increases all these atrophied features to cope with the load. This upregulation accounts for the enormous DIT.

Energy input must equal energy output to maintain a neutral energy balance.

We now turn to total energy balance in animals, a topic of particular current interest because of human obesity: Energy balance is the factor that determines whether you will be at an optimal weight (mass*), underweight, or overweight. Because energy cannot be created or destroyed (the first law of thermodynamics, p. 670), energy input must equal energy output (which includes storage), as represented by the following equation (where E = energy):

$$E_{input} = E_{output}$$

We can break these terms into more detail, in what might be called "the animal energy equation." It contains the main uses of energy in an animal:

$$E_{input} - E_{loss} = E_{SMR\ or\ BMR} + E_{activity} + E_{production} + E_{DIT}$$

The term *loss* here refers to energy lost via feces, urine, skin sloughing, and so forth. Activity energy is the cost of neuromuscular efforts above the SMR or BMR level. The term *production* refers to stored food such as adipose, net growth during development, and reproduction. Finally, DIT is the heat energy that arises following a meal, as discussed earlier.

There are three possible states of energy balance:

- *Neutral energy balance.* If the net amount of energy in food intake (input minus losses) exactly equals the amount of energy expended (output), then body mass remains constant.

- *Positive energy balance.* If the net energy intake exceeds the amount of energy expended, and the extra energy taken in is not used immediately, but is stored (the $E_{production}$ term; for example, as adipose tissue), then body mass increases.

- *Negative energy balance.* Conversely, if the net energy intake is less than the body's immediate energy requirements,

---

*Mass is a measure of the amount of material in an object, with units in grams. *Weight* is popularly used in the same way, but actually it is the gravitational force acting on a mass, in units of newtons. We use the term *mass* hereafter except for the terms *under-* and *overweight*.

the animal must use its stored energy to supply energy needs, and, accordingly, body mass decreases.

Thus to maintain a constant body mass (with the exception of minor fluctuations caused by changes in H_2O content), energy acquired through food intake must equal energy expenditure by the animal body.

Many mammals maintain long-term neutral energy balance.

Physiologists have observed that many adult mammals, including humans, maintain a fairly constant mass (weight) over long periods of time. This implies that homeostatic mechanisms (with a set point) exist to maintain a long-term balance between energy intake and energy expenditure. Moreover, individuals may have different set points for body masses (you may have noticed that some humans stay lean no matter how much they eat, whereas other, overweight humans have great difficulty losing weight).

Total body energy content can only be maintained at a constant level by regulating the components of the animal energy equation:

1. *Compensatory decrease in BMR and/or activity in response to underfeeding ("dieting"):* Some studies suggest that after several weeks of eating less than normal, a mammal may initiate small counteracting changes in BMR or muscular activity. This effect partially explains why some humans or laboratory animals become stuck at a plateau after having lost the first grams easily.

2. *Compensatory increases in BMR and/or regulatory DIT in response to overfeeding.* Physiological processes that "burn off" excess calories account in part for the inability of some mammals (again including some humans) to gain much weight. Recent studies on regulatory DIT are particularly interesting. Humans and laboratory mammals have been found to release "unwanted" calories by activating **uncoupling proteins (UCP)** in some tissues. UCPs convert food energy in mitochondria completely into heat, with no ATP synthesis (by a mechanism we discuss later, p. 697). In mice, **brown adipose tissue (BAT)** is activated after a large meal. BAT converts available energy in lipids into heat using **UCP-1.** (BAT probably evolved as a thermoregulatory organ, which will be discussed later; p. 697.) In humans, white adipose tissue and skeletal muscle, containing UCP-2 and UCP-3, respectively, appear to be similarly used.

3. *Compensatory increases in activity in response to overfeeding.* Muscular activity levels may also be regulated for disposal of excess energy. In humans, involuntary movements called "fidgeting" have been found to be more frequent in lean than in heavy individuals. Voluntary activities such as exercise are of course well known to contribute to energy balance.

4. *Regulation of food intake.* Despite modest compensatory changes in metabolism just noted, there is no question that regulation of food intake is the most important factor in the long-term maintenance of energy balance and body weight. There can be short-term regulation of food intake, helping to control meal size and frequency. Even so, over a

24-hour period the energy in ingested food rarely matches energy expenditure for that day. The correlation between total caloric intake and total energy output is strong, however, over long periods of time.

Some animals undergo periods of positive or negative energy balance.

Although long-term energy balance is evident in some adult mammals, some periodic imbalances can be adaptive. For example, hibernators (such as a marmot) typically increase their body mass greatly during active (feeding) seasons and decrease during hibernation as they live off of their internal energy stores. Migratory birds may similarly gain and lose weight during premigration feeding and the migration itself, respectively. Reproductive periods such as pregnancy may also include periods of mass increases followed by decreases. Finally, long-term balance is not maintained in some animals; for example, many adult fish indefinitely grow larger if food supplies permit (see Chapter 7, p. 274).

Food intake in mammals is controlled primarily by the hypothalamus in response to numerous inputs.

Energy regulation has been most thoroughly studied in mammals, on which we focus this discussion. Control of food intake in mammals is primarily a function of a pair of **feeding,** or **appetite, centers** located in the lateral (outer) regions of the hypothalamus, one on each side, and another pair of **satiety centers** located in the ventromedial (underside middle) area. The functions of these areas have been elucidated by a series of experiments that involve either destruction or stimulation of these specific regions in laboratory rats. Stimulation of the clusters of nerve cells designated as feeding, or appetite, centers makes the animal hungry, driving it to eat voraciously, whereas selective destruction of these areas suppresses eating and food intake behavior to the point that the animal starves itself to death. In contrast, stimulation of the satiety centers signals satiety, or the feeling of having had enough to eat. Consequently, the stimulated animal refuses to eat, even if previously deprived of food. As expected, destruction of this area produces the opposite effect—profound overeating and obesity; because the animal never achieves a feeling of being full (● Figure 15–5a). Thus the feeding centers tell animals when to eat, whereas the satiety centers tell them when they have had enough.

Although it is convenient to consider these specific areas as "exciting" or "inhibiting" feeding behavior, respectively, this approach is too simplistic. Physiologists now know that complex systems and numerous signals provide information about the body's energy status. Recent studies have uncovered multiple, highly integrated, redundant pathways, crisscrossing into and out of the hypothalamus, that are involved in controlling food intake and maintaining energy balance. Integration of multiple molecular signals ensures that feeding behavior is synchronized with the body's immediate and long-term energy needs.

What are these signals? Even though food intake is adjusted to balance changing energy expenditures over a period of time, there are no calorie receptors per se to monitor energy input, energy output, or total body energy content. Instead,

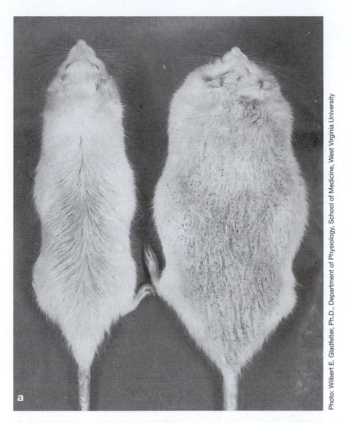

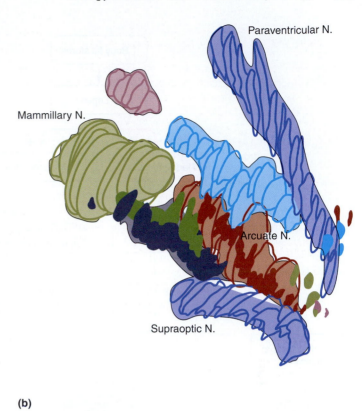

Paraventricular N.

Mammillary N.

Arcuate N.

Supraoptic N.

(b)

Figure 15–5 ● (a) Comparison of a normal rat with a rat whose satiety center (in the hypothalamus) has been destroyed. Several months after destruction of the classical satiety center in the ventromedial area of the hypothalamus, the rat on the right had gained considerable weight as a result of overeating compared to its normal littermate on the left. Rats sustaining lesions in this area also display less grooming behavior, accounting for the soiled appearance of the fat rat. (b) Major subdivisions of the mammalian hypothalamus; N = nucleus.

(*Source:* www.med.howard.edu/anatomy/gas/wk14/Hunger%20and%20the%20Hypothalamus.htm)

various blood-borne chemical factors that signal the body's energy state are important in regulating food intake.

The following factors, among others, contribute to the control of food intake, primarily in laboratory mammals (● Figure 15–6). The relative importance of each of these factors (and possibly others) remains unclear.

The Size of Adipose Stores and Level of Adipose Hormones

The fat cells *(adipocytes)* in adipose tissue secrete **leptin** (*leptin*, "thin"), a hormone essential for normal body-mass regulation in mammals. (Note that some researchers call leptin a *cytokine* rather than a hormone, because it is not produced by a classic endocrine gland; see p. 88.) The amount of leptin in the blood is an excellent indicator of the total amount of triglyceride fat stored in adipose tissue: The larger the fat stores, the more leptin released into the blood. This blood-borne signal, discovered in the mid 1990s by genetic analysis of obese laboratory mice, was the first molecular satiety signal found by genetics (see box, "Molecular Biology and Genomics: Discovering the Obesity Gene"). The **arcuate nucleus** (Figure 15-5b) of the hypothalamus is the major site for leptin action. The arcuate nucleus is an arc-shaped collection of neurons located adja-

cent to the floor of the third ventricle, in the region of the classical satiety center. Leptin receptors are found in abundance here.

Acting in negative-feedback fashion, increased leptin from burgeoning fat stores serves as a satiety signal. Binding of leptin to its hypothalamic receptors suppresses appetite to decrease food consumption while, to a lesser extent, increases metabolic rate, thus increasing energy expenditure. These actions together promote mass (weight) loss. This "lipostat" signal is generally considered one of the key factors responsible for the long-term matching of food intake to energy expenditure so that total body energy content and body mass remain fairly constant.

Interestingly, leptin has recently been shown to also be important in reproduction. It is one trigger for the onset of sexual maturation, signaling that the female has stored sufficient energy (as adipose tissue) to sustain a pregnancy. It may also be involved in the annual cycle of hibernating mammals such as Arctic ground squirrels, animals that do not maintain constant mass year-round because they must store up large adipose deposits in the productive seasons and survive on them in the winter. Some studies suggest that the brain ignores leptin signals in the autumn, allowing the animals to gain mass, whereas during the winter hibernation, the high leptin output of the stored fat effectively suppresses appetite.

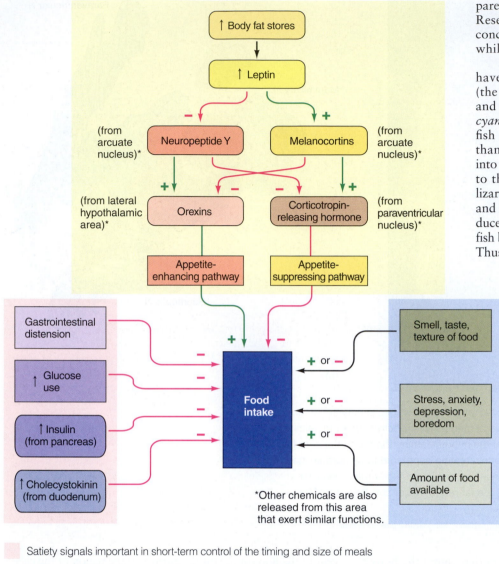

Satiety signals important in short-term control of the timing and size of meals

Adipose tissue–related signals important in long-term matching of food intake to energy expenditure to control body weight

Psychosocial and environmental factors that influence food intake

Figure 15–6 • Factors that influence food intake.

parent birds with low reproductive rates. Researchers hope to manipulate leptin concentrations to reduce adipose stores while enhancing lean meat production.

Leptin-related genes and proteins have been found in at least one reptile (the fence lizard *Scoleporus undulatus*) and one fish (the green sunfish *Lepomis cyanellus*). As in mammals, fasting sunfish have lower levels of plasma leptin than fed sunfish. Injecting mouse leptin into these animals induced effects similar to that seen in mammals. The injected lizard raised its body temperature slightly and its RMR by 250%, and showed a reduced appetite, whereas the injected sunfish began to metabolize its adipose stores. Thus leptin may be an ancient vertebrate signal of energy balance.

Levels of Brain Signals

A flurry of research following the identification of leptin has led to the recent discovery of numerous chemical messengers in the mammalian brain that play a role in controlling energy balance, typically in response to systemic signals such as leptin. Some of these molecules stimulate food intake and promote mass (weight) gain, whereas others depress food intake and promote weight loss. Two major signals have been found in the arcuate nucleus, which has two subsets of neurons that are regulated by leptin but in an opposing manner. One subset releases *neuropeptide Y*, whereas the other releases *melanocortins:*

■ **Neuropeptide Y (NPY)** is one of the most potent appetite stimulators ever found. NPY not only triggers voracious feeding behavior but also lowers energy expenditure by suppressing sympathetic nervous system outflow, decreasing the metabolic rate. Thus NPY promotes weight gain. Leptin suppresses food intake and increases metabolic rate in part by inhibiting hypothalamic output of NPY. Alternatively, the loss of fat stores and the resultant reduction in leptin caused by dieting increases NPY release. NPY, in turn, sets in motion actions that oppose the weight loss.

■ **Melanocortins,** a group of hormones traditionally known to be important in varying skin color for the purpose of camouflage in some species, have been shown to exert an unexpected role in energy homeostasis. Melanocortins, most notably *melanocyte-stimulating hormone* (see p. 265), inhibit the hypothalamus to decrease food intake. Researchers believe that when food stores are elevated, the release of leptin from adipose tissue stimulates the melanocortin system. Thus leptin promotes weight loss by simultaneously inhibiting the appetite-enhancing NPY pathway

Another hormone or cytokine produced by adipose tissue, **adiponectin,** has been recently discovered. It is present at high levels in humans and mice of normal mass, but at much lower levels in obese subjects. Thus its regulation may be the reverse of leptin's. It appears to reduce glucose production and release by the liver, and may enhance the burning of fats by skeletal muscle. But its overall role in energy metabolism and appetite are still under investigation.

Using genome analysis and antibody detection, researchers have recently found leptin in other vertebrates (birds, reptiles, fishes). Leptin in broiler chickens is produced not only by adipose but also by liver tissue. The discovery of chicken leptin is being used to investigate methods for obesity control in these animals! Broiler chickens, as a source of human food, have been bred for maximal growth, but this has produced obese

MOLECULAR BIOLOGY AND GENOMICS

Discovering the Obesity Gene

The discovery and characterization of the **leptin gene** in 1994 by Jeffrey Friedman illustrates the importance of genetics to physiology. Physiologists had known for decades that obesity has a genetic component, because different animal individuals on the same diets could have very different—and stable—body masses, with an apparent set point that could vary widely among individuals. Furthermore, the propensity for obesity could be passed from one generation to the next. Manipulations of the hypothalamus showed appetite was controlled in that brain region. Researchers widely suspected that mammalian bodies must have feedback signals that would tell the hypothalamus about the energy status of the body.

But physiologists had no luck finding such a signal. Enter the geneticists, in the 1980s. By then, several strains of obese mice were established, including the *ob* strain, weighing about three times more than other mice. After eight years of genetic breeding and cloning experiments, Friedman and his colleagues showed that obesity was due to the functional absence of a single gene, dubbed the *ob* gene. They isolated and cloned the normal *ob* gene (in non-*ob* mice), and subsequent work showed that it codes for the protein hormone **leptin,** which has turned out to be one of the key energy signals in mammals. Without modern genetic methods, this might never have been discovered.

---

and stimulating the appetite-suppressing melanocortin pathway.

Numerous other potential regulatory factors have recently been identified as researchers delve deeper into the mystery of energy balance and weight control. The role of these and possibly other chemicals yet to be identified await sorting out by further investigations.

In addition to these factors, other areas of the brain as well as the liver are now known to play important roles in controlling food intake. The contributions and interrelationships among the hypothalamus and other regulatory areas are currently the subject of intense investigation. Scientists suspect that still other controlling factors remain to be discovered.

Beyond the Arcuate Nucleus: Orexins and Others

Two hypothalamic areas are richly supplied by axons from the NPY- and melanocortin-secreting neurons of the arcuate nucleus. These areas are the **lateral hypothalamic area (LHA)** and **paraventricular nucleus (PVN,** Figure 15–5b). According to a recently proposed model, the LHA and PVN release chemical messengers in response to input from the arcuate nucleus neurons. These messengers are believed to act downstream from the NPY and melanocortin signals to regulate appetite. The LHA, the site of the classical appetite center, produces two closely related neuropeptides known as **orexins,** which are potent stimulators of food intake (*orexis,* "appetite"). In the proposed model, NPY stimulates and melanocortins inhibit the release of orexins, leading to an increase in appetite and greater food intake. By contrast, the PVN releases chemical messengers that decrease appetite and food intake. An example is corticotropin-releasing hormone, which, as its name implies, is better known for its role as a hormone (p. 268). According to the model, melanocortins stimulate and NPY inhibits the release of these appetite-suppressing neuropeptides.

In addition to the importance of molecular signals in the long-term control of body mass, other factors are believed to play a role in controlling the timing and size of meals. In contrast to the key role of the hypothalamus in maintaining energy balance and long-term control of body mass, a region in

the brain stem known as the **nucleus tractus solitarius (NTS)** is thought to be the site that processes signals important in terminating a meal. The NTS receives afferent inputs from the digestive tract and elsewhere that signal satiety and also receives input from the higher hypothalamic neurons involved in energy homeostasis. We now turn our attention to the most important of these satiety signals.

The Extent of Gastrointestinal Distension

In addition to leptin's importance in the long-term regulation of body mass, other factors are believed to play a role in controlling the timing and size of meals. Early proposals suggested that cues of emptiness or fullness of the digestive tract signaled hunger or satiety, respectively. For example, stimulation of gastric stretch receptors has been shown to suppress food intake. However, neural input arising from stomach distension plays a more important role in controlling the rate of gastric emptying than in signaling satiety (see p. 629). Researchers now believe that internal blood-borne signals reflecting the depletion or availability of energy-producing substances are more important than stomach volume in controlling the initiation and cessation of eating. Some of these are described in the next section.

The Level of Gastrointestinal Hormones

Evidence has accumulated that several hormones produced by the stomach and small intestine have potent effects on appetite. Cholecystokinin (CCK), one of the gastrointestinal hormones released from the duodenal mucosa during digestion of a meal, may be an important satiety signal for regulating the size of meals. CCK is secreted in response to the presence of nutrients in the small intestine. Through multiple effects on the digestive system, CCK facilitates the digestion and absorption of these nutrients (see p. 639). It is appropriate that this blood-borne signal, whose rate of secretion is correlated with the amount of nutrients ingested, also contributes to the sense of being filled after a meal has been consumed but before it has actually been digested and absorbed.

Another signal from the small intestine, **gastric inhibitory peptide (GIP),** has recently been found to have effects similar to CCK. GIP is released by the upper small intestine to inhibit the stomach after food has left it and entered the intestine. And the stomach may also make its own appetite signals. New studies show that this organ secretes **ghrelin** when it is empty and PYY_{3-36} when it is full. Ghrelin injected into a mouse has been found to be a potent appetite stimulator, whereas PYY_{3-36} has the opposite effect.

These gastrointestinal signals could explain why animals stop eating before the ingested food is actually digested, absorbed, and made available to meet the body's energy needs. An animal may feel satisfied when adequate food to replenish the stores is in the digestive tract, even though the body's energy stores are still low. This would help prevent the overshoot phenomenon (in this case, overeating) that characterizes simple negative-feedback systems (p. 15).

The Extent of Glucose Use and Insulin Secretion

In mammals, satiety is signaled by increased glucose use, such as occurs during a meal when more glucose is available for use because it is being absorbed from the digestive tract. Conversely, after absorption of a meal is complete, and no new glucose is entering the blood, the resultant reduction in glucose usage by the body arouses hunger. Low blood-glucose levels have been shown to activate lateral hypothalamic neurons in the vicinity of the classical appetite centers. The extent of glucose use appears to be more important in determining the timing of meals than in long-term control of body mass. This is in contrast to birds, in which plasma glucose levels remain essentially unchanged during fasting and feeding.

In a related proposal, some evidence indicates that an increased level of insulin in the blood signals satiety. Insulin, a hormone secreted by the pancreas in response to a rise in the concentration of glucose and other nutrients in the blood following a meal (p. 297), stimulates cellular uptake, use, and storage of glucose and other nutrients. Thus the increase in insulin secretion that accompanies nutrient intake and promotes increased glucose use is an appropriate satiety signal.

Psychosocial and Environmental Influences

Eating habits are also shaped by psychological, social, and environmental factors. For example, the amount of pleasure derived from eating can reinforce feeding behavior. This has been demonstrated in an experiment in which rats were offered their choice of highly palatable human foods. They overate by as much as 70 to 80% and became obese. When the rats returned to eating their regular monotonous but nutritionally balanced rat chow, their obesity was rapidly reversed as their food intake was controlled once again by physiological drives rather than by urges for the apparently tastier offerings. Stress, anxiety, depression, and apparent boredom have also been shown to alter feeding behavior in ways that are unrelated to energy needs in experimental animals.

Evolutionary Adaptation

Clearly, even though many mammals maintain a relatively constant body mass, species and individuals vary greatly in the mass they actually maintain (as we noted earlier). This has led to the proposal that the hypothalamic centers determining long-term satiety can have different *set points* for energy balance, possibly for adipose storage in particular. The two extreme set points are often termed "lean" and "obese" (or "heavy").

Although humans in modern Western societies tend to attach positive and negative connotations to lean and obese states, respectively, physiologists view these in a different light. Both extremes (and set points in between) seem to be evolutionarily advantageous depending on the environment:

- Lean animals do better in habitats where food is plentiful year-round, so that storing a high amount of body fat is unnecessary. The advantage here is that such animals are more agile than heavy ones, better able to avoid predators and (in some species) to obtain food in the first place. Thus carnivores in the tropics (such as tigers) are lean (although the leanness may be attributed to high activity levels, many carnivores in fact are sedentary much of the time.) The tradeoff between storage and agility was revealed in zebra finches by researchers in Scotland, who videotaped and weighed numerous birds on a daily basis. Some finches had access to excessive calories from human sources. A mere 7% gain in body mass led to a 33% loss in speed and maneuverability, making these overfed birds easier prey for cats and sparrow hawks. Thus an obese or heavy set point has negative consequences in a food-plentiful habitat, including increased cardiovascular and oxidative stresses causing internal health problems, as well as reduced maneuverability. Household pets, as well as humans themselves, often suffer from the consequences of obesity.

- Conversely, a high adipose set point has very clear advantages in "feast or famine" habitats, where food is only abundant at limited times of the year. A clear example is the mammalian hibernator. An Arctic ground squirrel must reach an obese state (about 50% body fat) before winter sets in, or it will not survive the 8 months during which it hibernates in its burrow. Desert habitats that have abundant food only during short rainy seasons also select for this adaptation. Overall, it is now recognized that having a propensity toward obesity is not necessarily an abnormal physiological state but may be an evolutionary adaptation for a particular environment.

Researchers do not yet fully understand what determines these different set points in the neurological structure of the brain. But, as discussed earlier in this chapter, fat storage levels and the associated set point may be closely tied to leptin and how the hypothalamus reacts to it. The hypothalamus of a lean-adapted mammal reacts strongly to a modest rise in leptin, shutting off appetite and acting as if it had a low set point. The hypothalamus of obese mammals, conversely, reacts only to very high levels of leptin. Also, humans and laboratory animals appear to differ in the thermogenic activity of tissues with uncoupling proteins, correlating with lean or obese set points. Research is continuing on these numerous fronts.

Thermal Physiology: General Principles

When physiological ecologists examine the distribution of species around the planet, noting that most species live in rather restricted ranges, they ask what factors are responsible for these restrictions. One of the most pervasive factors is temperature, which has direct and profound effects on biological functions.

■ Temperature is one of the most important habitat factors.

Temperature is a measure of heat energy, which in turn is manifested primarily in kinetic energy (movement) of molecules. The primary effects of this are shown in ● Figure 15–7, which shows a typical optimal curve of biological function and temperature. Basically, two different temperature effects create this curve:

1. *Kinetic energy of reactants:* On the left (up-trending) side of the plot, reaction rates increase with temperature simply from the increased kinetic energy of molecules. Reacting molecules (such as substrates and enzymes, for example) encounter each other more frequently. It has become customary to characterize this temperature response with a value called the Q_{10}, which is the ratio of a reaction rate (or metabolic process) at one temperature to the rate at 10 degrees C cooler:

$$Q_{10} = \frac{\text{rate at temperature } T}{\text{rate at temperature } T_{-10}}$$

A more general version, which allows for measurements at temperatures not 10 degrees C apart, is

$$Q_{10} = (R_2/R_1)^{10/(T_2 - T_1)}$$

where R_1 is the rate at temperature T_1, and R_2 is the rate at temperature T_2.

Q_{10} values are useful measures of the temperature sensitivity of a biological process (from single reactions to a whole organism's metabolism). In general, researchers have found that many reactions tend to double or triple for each 10-degree C rise, at least over a moderate range. However, at some point with rising temperature, the curve begins to flatten, reaches an optimum, and then declines.

2. *Denaturation of macromolecules.* Above the optimum (right side of the plot), rates of biological processes begin to decrease with temperature for one basic reason: the macromolecules (such as enzymes) responsible for the process begin to denature (that is, lose their higher levels of structure). As we discussed in Chapter 2 (p. 25), weak bonds stabilize many of these structures so that they can be flexible; but this makes them susceptible to breakdown at higher kinetic energies.

In general, this curve means that organisms cannot operate over a wide range of body temperatures. For animals, active metabolism seems to be restricted to a range from about −2°C (the freezing point of seawater) to about 50°C (on some desert sands and at some marine hydrothermal vents). (A penguin living actively at −50°C air temperature in Antarctica does not violate this restriction, because its cells are warmed to about 40°C from internal heat production and retention.) This is a relatively narrow range, given that a temperature of −89.2°C (128.5°F) has been recorded (in Antarctica) and hydrothermal vent temperatures may reach about 400°C (750°F)! Some animals have evolved ways to cope with body temperatures colder than freezing (as you will see), but metabolism in these animals is generally in a dormant state. At the other extreme, there is no convincing evidence of any animal tolerating over 50°C for anything but very brief periods. Ignoring long-term acclimatization for the moment, excessively high temperatures are usually more dangerous than cold ones. Many organisms can survive lower body temperatures temporarily, because reactions merely slow down (although freezing sets a lower limit, because of the lysis of cells from ice crystal formation), but excessive heat can irreversibly damage membranes and proteins.

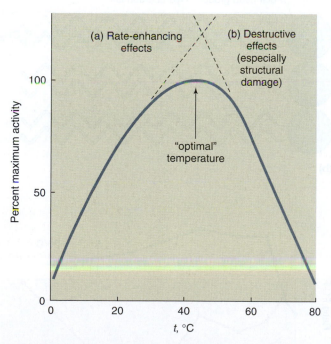

Figure 15–7 ● **The effect of temperature on enzyme activity.** The relative activity of an enzymatic reaction as a function of temperature. The decrease in activity above 50°C is due to thermal denaturation. Biphasic effects of temperature increase on the rate of biological processes.

(*Source:* Modified from M. K. Campbell, & S. Farrell, 2003, *Biochemistry,* 4th ed., Belmont, CA, Thomson, Figure 5.2, p. 137)

?

What Is the Maximum Temperature for Life? Although hydrothermal vents can reach 400°C, no life has been found that can tolerate anything close to that level. In 2003, the highest temperature for life yet recorded was demonstrated to be 130°C (266°F), achieved by a hydrothermal vent archaeon (see discussion of Archaea, p. 25). The highest temperature for animals is more controversial. An African desert ant (*Cataglyphis bombycina*) has been found to tolerate 50°C body temperatures for a few minutes while running across hot sand to forage at midday (this behavior gives them a competitive advantage, because other animals are hiding in burrows to avoid the midday heat). A polychaete annelid, *Alvinella,* which lives on deep-sea hydrothermal vents, has been reported to tolerate temperatures as high as 80°C. However, many researchers believe the body temperature measurements were flawed (they are difficult to make, from a submersible vehicle, on an animal that retreats inside a hard calcium-carbonate tube). In the laboratory, *Alvinella*'s proteins do not work above about 40 to 45°C range. Thus the maximum temperature for animal life is unresolved.

■ Biomolecules can be altered to work optimally at different temperatures.

For the most part, biological systems are subject to the optimal curve of Figure 15–7, but it is important to realize that *the optimum itself can change with different temperatures.* Changes can occur over two possible (and very different) time courses: within an organism's lifetime in acclimatization processes; and over many generations by the process of mutation and natural selection, that is, evolution. Both categories are sometimes loosely called "adaptation" (see p. 15 for more details on terminology). If "adaptation" is used, it is crucial to carefully explain the time course.

Not all organisms can successfully acclimatize to different temperatures, nor can all species evolve to temperatures outside of their normal ranges (indeed, changes in habitat temperatures may be a major cause of extinctions). Regardless, optima clearly differ among species, in part because of changes in two critical classes of biomolecules—membranes and proteins.

Membranes and the Homeoviscous Adaptation

Biological membranes are held weakly together predominantly by hydrophobic interactions between the fatty acid chains of its phospholipids (p. 68), giving them a certain *viscosity* or fluidity that is vital to functions such as transport and diffusion. However, the weak structure makes them very susceptible to temperature changes. At low temperatures, viscosity increases and the membrane becomes too rigid for optimal function. At higher temperatures, membranes lose viscosity and become too fluid. If you have used butter from a refrigerator and also from a kitchen counter on a hot day, you are aware of these effects. As you saw in Chapter 3 (p. 70), a membrane can be changed to counter these effects by the length of the fatty acids and by the degree of *saturation* of its fatty acids (and, to some extent, by altering its cholesterol content). A hypothetical example is shown in ● Figure 15–8. First, consider a warm-adapted organism. Its membranes typ-

ically consist of *saturated fatty acids* (SFAs), in which the carbons in the "tails" are fully hydrogenated. This allows the tails to stack together tightly, giving the whole membrane enough viscosity to work properly at warm temperatures. Cholesterol may be high as well, to further reduce fluidity. However, these membranes become too rigid to work in the cold (as butter

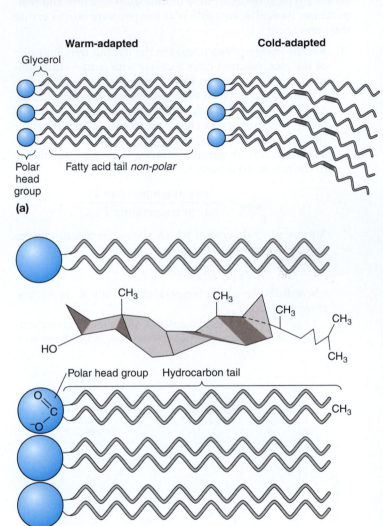

(a)

(b)

Figure 15–8 ● Schematic diagrams of membrane structures and their relationship to temperature. (a) Saturation level of membrane phospholipids changes with temperature adaptation. The top figure represents a warm-adapted membrane: the fatty-acid "tails" are highly saturated, allowing for close stacking of neighboring tails. This enhances thermostability at higher temperatures. The bottom figure represents a cold-adapted membrane: the fatty-acid "tails" are more unsaturated, creating double-bonds which interfere with stacking of neighboring tails. This enhances flexibility at colder temperatures. (b) Stiffening of the membrane with cholesterol. Cholesterol reduces fluidity by stabilizing extended forms of the phospholipid tails. (c) The relationship between adaptation temperature and ratio of unsaturated phospholipid tails (phosphatidylethanolamine, PE) to saturated ones (phosphatidychlorine, PC) in gill membranes of rainbow trout transferred to different temperatures as shown.

(*Sources:* (b) Modified from Campbell, M. K. and S. O. Farrell. *Biochemistry,* 4th ed. Thompson. Fig. 7.14 p 202; (c) From Hochachka, P. W. and G. N. Somero, 2002. *Biochemical Adaptation.* Oxford: oxford University Press; as modified from Logue, J. A., A. L. DeVries, E. Fodor, and A. R. Cossins. 2000. Lipid compositional correlates of temperature-adaptive interspecific differences in membrane physical structure. *Journal of Experimental Biology* 2105–2115)

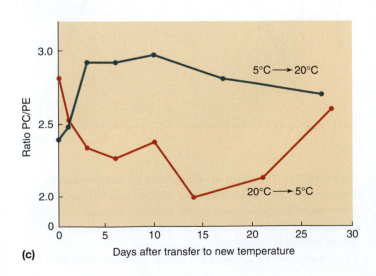

(c)

Days after transfer to new temperature

does in a refrigerator). Conversely, a cold-adapted organism typically has a higher proportion of *polyunsaturated fatty acids* (PUFAs), in which some carbons have fewer hydrogens and thus form double bonds with neighbor carbons. This puts "kinks" in the tails, making tight stacking impossible. This makes the whole membrane less viscous and thus more fluid in the cold. But these membranes become too fluid to work in warm conditions. Thus there is a tradeoff between thermal stability and fluidity. Both membranes have approximately the same viscosity *when they are each measured at their adaptational temperatures.* Hence, the term "homeoviscous adaptation" was coined to refer to these saturation changes that preserve an optimal viscosity of the cell membrane.

Saturation levels are catalyzed by enzymes called *desaturases* (used in the cold to remove hydrogens) and *saturases* (used in warm conditions to add hydrogens). As noted, the time course of changes is critical when discussing this adaptation. In general, organisms from different thermal habitats have lipid compositions that follow the saturation principles just described. However, only some organisms can acclimatize (from winter to summer, for example) by regulating the saturases and desaturases and thus their membranes. In addition to the fish described in Chapter 3 (p. 70), an example is shown in Figure 15–8c.

Proteins and the "Homeoflexibility Adaptation"

A similar type of change can occur with proteins, although generally only over evolutionary time. In this case, the trade-off is between temperature stability and protein flexibility, rather than fluidity. As you saw earlier, increased temperature can cause denaturation (Figure 15–7), but lower temperatures can make an enzyme too rigid to function well. As shown in
● Figure 15–9a, optimal temperatures for enzyme functionality do vary among species as with membranes. How does this occur? A major clue was found in studies of *catalytic rate constants* (k_{cat}) for some enzymes. These values indicate how fast an enzyme can catalyze a particular reaction. The studies revealed an initially surprising result: Enzymes of animals with warm bodies, such as mammals, have relatively inefficient (low k_{cat}) enzymes, whereas cold-bodied animals have faster (high k_{cat}) enzymes (Figure 15–9b). As with membranes, there appears to be a tradeoff between flexibility and thermostability. One particular enzyme, **lactate dehydrogenase (LDH)** has been analyzed in some detail. This is the final enzyme of anaerobic glycolysis, catalyzing the conversion of pyruvate and NADH to lactate and NAD^+ (p. 54). In its tertiary structure LDH has a flexible loop, which opens and closes during its reaction (● Figure 15–10). Warm-adapted LDHs have relatively stiff loops, rigid enough to resist denaturation at higher tempera-

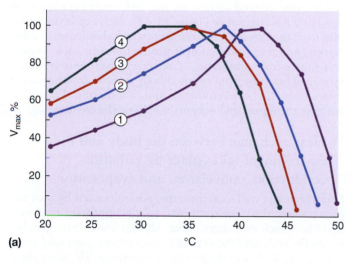

(a)

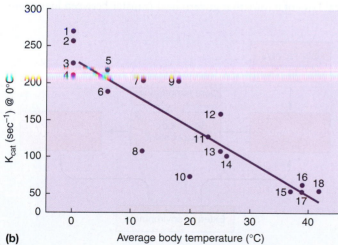

(b)

Figure 15–9 ● Relationship between enzyme activity and temperature in animals adapted to different temperatures. (a) The relative activity max (V_{max}) of myosin ATPase of four lizards, showing that animals adapted to higher temperatures have enzymes with higher thermal optima 1: *Dipsosaurus dorsalis*, which has a preferred body temperature of 38.8°C; 2: *Uma notata*, preferred body temperature of 37.5°C; 3: *Scleroporus undulatus*, preferred body temperature of 36.3°C; 4: *Gerrhonotus multicarinatus*, preferred body temperature of 30.0°C. (b) The enzyme catalytic rate constant (k_{cat}) for lactate dehydrogenases (LDH, A_4 type) from various vertebrates adapted to different temperatures, showing that, in general, animals adapted to higher temperatures have enzymes with lower catalytic rates. Antarctic notothenioid fishes are 1: *Lepidonotothen nudifrons*; 2: *Parachaenichthys charcoti*; 3: *Champsocephalus gunnari*; 4: *Harpagifer antarcticus*. South American notothenioid fishes are 5: *Patagonotothen tessellata*; 6: *Eleginops maclovinus*. Temperate and subtropical fishes are 7: *Sebastes argentea*. (rockfish); 8: *Hippoglossus stenolepis* (halibut); 9: *Sphyraena argentea* (barracuda); 10: *Squalus acanthias* (dogfish shark); 11: *Sphyraena lucasana* (barracuda); 12: *Gillichthys mirabilis* (goby); 13: *Thunnus thynnus* (bluefin tuna). A tropical fish is 14: *Sphyraena ensis* (barracuda). Terrestrial animals are 15: *Bos taurus* (cow); 16: *Gallus gallus* (chicken); 17: *Meleagris gallapavo* (turkey); and 18: *Dipsosaurus dorsalis*, desert iguana.

(*Sources:* (a) Modified from Licht, P. 1967. Thermal adaptation in the enzymes of lizards in relation to preferred body temperature. IN: *Molecular Mechanisms of Temperature Adaptation;* C. L. Prosser, ed. Washington: American Association for the Advancement of Science; (b) Hochachka, P. W. and G. N. Somero, 2002. *Biochemical Adaptation.* Oxford: Oxford University Press; as modified from Fields, P. A., and G. N. Somero. 1998. Hot spots in cold adaptation: Localized increases in conformational flexibility in lactate dehydrogenases A_4 orthologs of Antarctic notothenioid fishes. *Proceedings of the National Academy of Sciences USA* 95: 11476–11481)

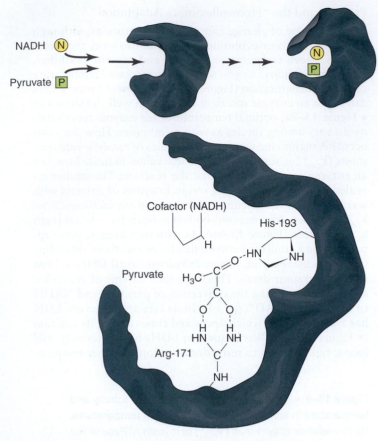

NADH Ⓝ

Pyruvate Ⓟ

Cofactor (NADH)

His-193

Pyruvate H₃C—C—$\ddot{O}$···HN   NH

$$C$$

O   O

H   H

HN   NH

C

NH

Arg-171

Figure 15–10 ● Schematic diagram of the enzyme lactate dehydrogenase (LDH), which binds NADH and pyruvate and catalyzes their conversion into NAD⁺ and lactate. The upper diagram shows the conformational changes that occur upon binding, as a flexible loop of the enzyme closes over the active site. The lower diagram shows the active site itself and the amino acids (Histidine-193 and Arginine 171) that bind pyrvuate.

(*Source:* Hochachka, P. W. and G. N. Somero, 2002. *Biochemical Adaptation.* Oxford: Oxford University Press; as modified from Fields, P. A., and G. N. Somero. 1998. Hot spots in cold adaptation: Localized increases in conformational flexibility in lactate dehydrogenases A₄ orthologs of Antarctic notothenioid fishes. *Proceedings of the National Academy of Sciences USA* 95: 11476–11481)

tures. But they are too inflexible to work at cold temperatures. Conversely, cold-adapted LDHs have relatively loose loops, able to open and close well in the cold. But they lose their structure more easily at higher temperatures.

■ The thermal adaptation strategies of animals depend on their primary source of heat.

Let us now turn to the ways in which animals cope with environmental temperature changes that threaten their biomolecules. As we noted earlier, physiologists generally classify animals into two broad categories: **ectotherms** (*ecto*, "external")—those dependent on external sources for body heat—and **endotherms** (*endo*, "internal")—those more dependent on internal sources. Physiologists still frequently use a pair of older terms: **poikilotherms** (*poikilos*, "changeable")—animals whose body temperatures vary with the environment—and **homeotherms**—animals with narrowly varying body temperatures (recall that *homeo* means "similar," not "constant"). These

terms often lead to confusion. Because we humans are mammals (endotherms) with fairly constant body temperatures (homeotherms), some people mistakenly believe the two terms are synonymous. But they are not, as we show in detail later. For now, consider an ectothermic lizard that can use sun and shade to keep its body temperature fairly constant (that is, it is homeothermic some of the time). Note also that a hibernating endotherm (such as a marmot) is not homeothermic when considered on an annual basis.

The marmot is an example of a **heterotherm,** an animal that has some degree of endothermy but is not fully homeothermic. Actually, there is no clear dividing line between a heterotherm and a homeotherm. For example, a camel in the desert absorbs a large heat load during the day, increasing its body temperature by several degrees (for example, from as low as 34°C to as high as 41°C) to avoid losing water by evaporative cooling. At night, it releases the heat and its body temperature drops. Does this level of change constitute homeothermy, or heterothermy? Some physiologists suggest that homeothermy be defined as internal temperature variations of no more than plus or minus 2°C, but in some ways the definitions do not matter. Rather, it is important to realize that there is a continuum of temperature regulation ranges and abilities, and that no endotherm has an absolutely constant internal temperature.

The internal body temperatures of both ectotherms and endotherms depend simply on the difference between heat input and heat output (● Figure 15–11). *Heat input* occurs by gain from the external environment (which dominates in ectotherms) and from internal heat production (the most important source for endotherms). Conversely, *heat output* occurs by way of heat loss from exposed body surfaces to the external environment (in both ectotherms and endotherms). Let us first examine the inputs and outputs involving the environment.

■ Heat exchange between the body and the environment takes place by radiation, conduction, convection, and evaporation.

Endotherms as well as ectotherms are influenced by heat exchanges with their environments, so it is important to understand how such exchanges occur. All heat loss or heat gain between the body and the external environment must take place between the body surface and its surroundings. The same physical laws of nature that govern heat transfer between inanimate

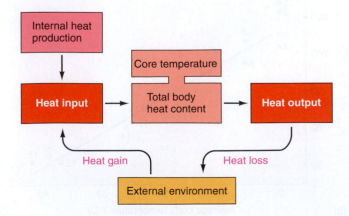

Figure 15–11 ● Heat input and output.

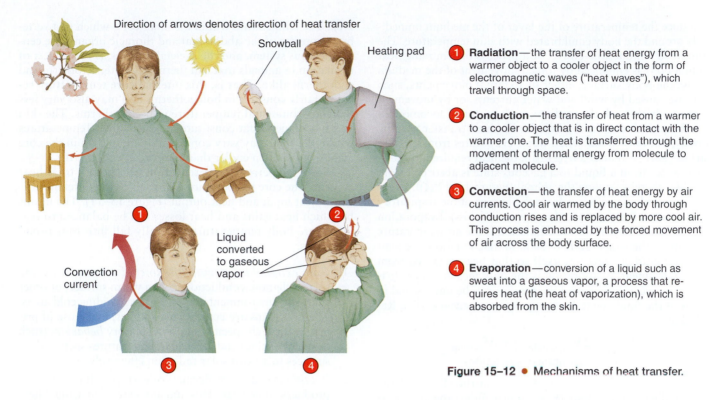

Direction of arrows denotes direction of heat transfer

1 **Radiation**—the transfer of heat energy from a warmer object to a cooler object in the form of electromagnetic waves ("heat waves"), which travel through space.

2 **Conduction**—the transfer of heat from a warmer to a cooler object that is in direct contact with the warmer one. The heat is transferred through the movement of thermal energy from molecule to adjacent molecule.

3 **Convection**—the transfer of heat energy by air currents. Cool air warmed by the body through conduction rises and is replaced by more cool air. This process is enhanced by the forced movement of air across the body surface.

4 **Evaporation**—conversion of a liquid such as sweat into a gaseous vapor, a process that requires heat (the heat of vaporization), which is absorbed from the skin.

Figure 15–12 ● Mechanisms of heat transfer.

objects also control the transfer of heat between the body surface and the environment. The temperature of an object may be thought of as a measure of the concentration of heat within the object. Accordingly, heat always moves down its concentration gradient; that is, down a **thermal gradient** from a warmer to a cooler region.

Organisms are subject to, and make use of, four mechanisms of heat transfer: *radiation, conduction, convection,* and *evaporation.* Radiant (light) energy exists as **electromagnetic waves** (and **photons**), which travel through space (● Figure 15–12 ①). Heat is emitted from all objects (if not at absolute zero) in the form of **infrared radiation** (wavelengths longer than red light). In the reverse direction, when radiant energy (especially infrared, but also visible light) strikes an object and is absorbed, the energy of the wave motion is transformed into heat within the object (manifested as increased vibrational energy within molecules). Objects can both emit (source of heat loss) and absorb (source of heat gain) radiant energy. Whether a body loses or gains heat by radiation depends on the difference in temperature between the skin surface and the surfaces of various other objects in the body's environment, and on the amount of direct sunlight striking the skin. Because net transfer of heat by radiation is always from warmer objects to cooler ones, a body gains heat by radiation from objects warmer than the skin surface, such as the sun, rocks, or soil in the sun, or burning wood. In contrast, an animal loses heat by radiation to objects in its environment whose surfaces are cooler than the surface of the skin, such as trees, soil in the shade, and so forth.

Conduction is the transfer of heat between objects of differing temperatures that are in direct contact with each other (Figure 15–12 ②). Heat moves down its thermal gradient from the warmer to the cooler object by being transferred from molecule to molecule. Except at absolute zero (0°K or −273°C), all molecules are constantly vibrating, with warmer molecules moving faster than cooler ones. When molecules of differing heat content touch each other, the faster-moving, warmer molecule agitates the cooler molecule into more rapid motion, "warming up" the cooler molecule. During this process, the original warmer molecule loses some of its thermal energy as it slows down and cools off a bit. Therefore, given enough time, the temperature of the two touching objects eventually equalizes. The rate of heat transfer by conduction depends on the *temperature difference* between the touching objects and the *thermal conductivity* of the substances involved (that is, how easily heat is conducted by the molecules of the substances).

Heat can be lost or gained by conduction when the skin is in contact with a good conductor. When standing on snow, for example, an animal's foot becomes cold because heat moves by conduction from the foot to the snow. Conversely, when an animal sits on a rock warmed by the sun, it is warmed up as heat is transferred directly from the rock to its body. Similarly, bodies lose or gain heat by conduction to the layer of air in direct contact with the body. Only a small percentage of total heat exchange between the skin and air takes place by conduction alone, because air is not a very good conductor of heat. Heat is conducted much more rapidly between a body surface and water, which is a good conductor, than air.

The term **convection** refers to the transfer of heat energy by *currents* of air or water (the medium). As a body loses heat by conduction to a surrounding cooler medium, the medium in immediate contact with the skin is warmed. Let us follow an example for air. Because warm air is lighter (less dense) than cool air, the warmed air rises, whereas cooler air moves in next to the skin to replace the vacating warm air. The process is then repeated (Figure 15–12 ③). Such movements in the external medium, known as *convection currents,* can carry heat away from a body. If it were not for convection currents, no further heat could be dissipated from the skin by conduction once the temperature of the layer of the medium immediately around the body equilibrated with skin temperature. The combined conduction–convection process of removing heat from a body is enhanced by forced movement of the medium across the body surface, either by external movements, such as those caused by wind and water currents, or by movement of the body through the medium, such as during locomotion.

Evaporation is the final method of heat transfer used by a body, but only in air. When water evaporates from the skin surface, such as that of a frog in air, the heat required to transform water from a liquid to a gaseous state is absorbed from the skin, thereby cooling the animal (Figure 15–12 ④). Evaporative heat loss can occur from the linings of the respiratory airways as well as from the skin (*cutaneous* loss). Evaporation is particularly important for regulation if the air temperature rises above the temperature of skin. In that situation, the temperature gradient reverses itself so that heat is gained from the environment. Evaporation is the only means of heat loss under these conditions, and the process can be understood in terms of the fundamental entropy equation given earlier. Recall that

$$\Delta S_{net} = \Delta S_{surroundings} + \Delta S_{system}$$
$$> 0 \text{ for spontaneous reactions}$$

Here the system is the water droplet on the skin, and the surroundings are the body being cooled. When the air is hotter than the body, and yet the latter is being cooled, the $\Delta S_{surroundings}$ is thermodynamically unfavorable, because entropy is actually decreasing during cooling against a heat gradient. However, the unfavorable factor is more than "paid for" by a **change of state** in water. During evaporation, water molecules go from a moderately ordered liquid state to a highly disordered vapor state, giving an enormously favorable ΔS_{system}. Essentially, the heat energy of the body is absorbed in the breaking of hydrogen bonds between water molecules as water vaporizes. This change of state, which is very effective, is exactly the process used in refrigerators and air conditioners, where liquid refrigerants vaporize in cooling coils, removing heat from the desired space. Energy in the form of a compressor motor must be used to reliquefy the refrigerant, recycling this costly resource. In organisms, the cooling fluid—water—is lost and must be replaced.

Evaporation cannot succeed, however, if the air is saturated with water vapor, that is, at 100% relative humidity (RH). At that point, the rate of water vapor reforming hydrogen bonds and thus condensing into liquid is equal to the reverse process. If the air is also hotter than an animal's body at 100% RH, then there is no physiological means for cooling off. Animals must allow their body temperatures to rise or must use behavioral mechanisms to avoid the situation.

Evaporation often occurs in a maladaptive way, that is, not for thermoregulation. A frog in a warm, dry area is in danger of dying from dehydration, because it cannot stop cutaneous water loss. Inadvertent respiratory losses of water can also be dangerous; for example, exercising in cold, dry winter air can cause serious dehydration in humans.

■ Heat gain versus heat loss determines core body temperature.

From a thermoregulatory viewpoint, an animal body may conveniently be viewed as a *central core* surrounded by an *outer*

shell. The temperature within the inner core, which in a vertebrate consists of the abdominal and thoracic organs, the central nervous system, and the skeletal muscles, is the subject of regulation in animals that can thermoregulate, ectotherm and endotherm alike. That is, this internal **core temperature** remains fairly constant in homeotherms and may also vary less than environmental temperature in poikilotherms. The skin and subcutaneous fat constitute the outer shell. Temperatures within the shell may vary considerably more than in the core and are most often cooler than the core in endotherms.

Body temperature is a reflection of a body's total heat content. The core temperature is a result of the difference between heat input and heat output (Figure 15–11). The means by which heat gains and heat losses can be balanced to regulate core body temperature generally fall into four broad categories:

1. *Gain external heat/avoid loss to cold environs* by using solar radiation, conduction from a warm surface, or other source of environmental heat, and by avoiding cold areas. The heat gains are **ectothermic sources,** made use of primarily through specific thermoregulatory *behaviors* (such as solar basking), aided by *anatomic* features such as dark surfaces to absorb solar radiation effectively.

2. *Retain internal heat.* Because of entropy, all metabolism produces some heat. This and any externally gained heat can be retained using *behavior* (such as entering an insulated burrow), *insulation* (such as hair), *reduced blood flow* to the integument to reduce losses; *countercurrent exchangers* in circulation that retain core heat; and *larger body sizes.* All these reduce heat loss via conduction, radiation, and sometimes convection and evaporation.

3. *Generate more internal heat.* This is the definitive feature of **endothermy,** requiring specialized heat-generating mechanisms.

4. *Lose excess internal heat/avoid gains from hot environs,* by *behavior* (such as seeking shade); *anatomy* (such as reflective white layers on the shell of an intertidal snail, and long, thin legs of desert insects, which hold their bodies well above the hot ground); enhancing transfer via the integument through *increased blood flow;* and increasing *evaporation* (panting, sweating, cutaneous loss).

This list is only a brief overview; we now turn to the details for ectotherms.

Ectotherms

■ Body temperatures of ectotherms may follow the environment, or may be regulated by external exchanges.

What happens to an ectotherm when the environmental temperature is not near the optimum for its molecular structures? There are two broad categories of response: *poikilothermy* and *ectothermic regulation.*

Poikilothermy

Poikilotherm ectotherms live in thermally variable environments, and their body temperatures vary as well. Metabolic

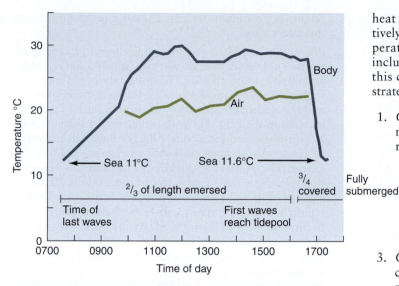

Figure 15–13 ● **The body temperature of an intertidal mussel (*Mytilus californianus*) when the tide is out.** The temperature is measured with a thermistor inserted inside the shell (blue line). As the tide goes out, the body temperature soars more than the air temperature (green line). Upon reemersion, the temperature drops rapidly.

(*Source*: Modified from Carefoot, T. *Pacific Seashores*. 1977. University of Washington Press by arrangement with J. J. Douglas, Vancouver Canada. Fig. 67, p. 68.)

rates in such organisms typically decrease in the cold (thus saving energy) and speed up as the environment warms (thus becoming better able to obtain food, up to a point). Survival may depend on using behavior to avoid extreme temperatures, and/or limited physiological heat-loss mechanisms such as evaporation. If extreme temperatures cannot be avoided, poikilotherms may enter dormancy during the extremes, or compensate through changes in internal biochemical optima (as you will see later). Some poikilotherms have body temperatures essentially identical to that of the environment—especially aquatic animals, because of the high heat conductivity and specific heat of water. These animals are sometimes called pure *thermoconformers*. But some have body temperatures that, although not nearly constant, vary less than the environment. This is particularly true of many ectotherms in terrestrial habitats, because air transfers heat much less effectively than water.

Most life on this planet is poikilothermic. An extreme example is given in ● Figure 15–13, showing a mussel on an intertidal rock on the west coast of North America. When the tide is in, its body temperature will be the same as the ocean, which is usually cold (about 11°C in this case). When the tide is out on a hot summer day, the mussel's body temperature soars and may reach 30°C or more. Some evaporative cooling may keep its temperature lower than it would be otherwise, but nevertheless it undergoes a large change in internal temperature.

Ectothermic Regulation

An ectotherm in a constant environment, such as the cold deep sea or some polar ocean regions, is even more homeothermic (almost "homothermic") than any mammal. However, these animals typically cannot cope with any significant temperature changes. For example, polar fishes can suffer

heat death at 5°C! Of more interest are ectotherms that actively use external heat exchanges to maintain their body temperatures at or near an optimum. Well-studied examples include butterflies and lizards (see opening photographs of this chapter, and ● Figure 15–14). Let us examine the four strategies of heat regulation in a lizard to illustrate:

1. *Gain external heat/avoid loss to cold environs.* On cold mornings, the lizard basks in sunlight, soaking up infrared radiation and heat via conduction from warmed surfaces. Once warmed up to an optimal level, it becomes active, seeking food while using both behavioral and physiological mechanisms to maintain a consistent body temperature.

2. *Retain internal heat.* The lizard may *vasoconstrict* blood vessels going to its skin, to reduce heat loss.

3. *Generate more internal heat.* This does not occur significantly in pure ectotherms. However, some very large (varanid) lizards, such as the Komodo dragon, warm up temporarily from high locomotory activity, retaining heat because of their comparatively low ratios of surface area to volume.

4. *Lose excess internal heat/avoid gains from hot environs.* Blood vessels in the lizard's skin may dilate to increase heat loss. Evaporation from the mouth also provides cooling (one lizard, the Gila monster of southwestern United States, evaporates water from its cloaca rather than its mouth). When it gets too hot, the animal seeks shade.

Using such mechanisms, some of these ectotherms can regulate body temperature as effectively as a mammal can, at least for some periods.

On a seasonal basis, some ectotherms can *migrate* to help minimize changes in body temperature. The monarch butterfly, for example, spends summers in northern North America

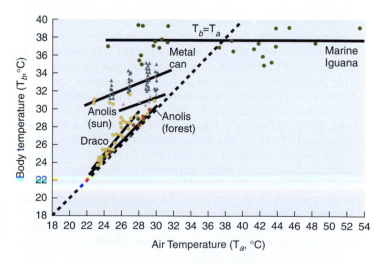

Figure 15–14 ● Relationship between body temperature and air temperature for thermoconforming lizards (*Draco, Anolis,* in shaded forest habitat) and thermoregulating lizards (*Iguana, Anolis,* in open habitats); also shown is the relationship for an inanimate object (water-filled metal can).

(*Source*: From Withers, P. C. 1992. *Comparative Animal Physiology*. Saunders, Fort Worth TX. Fig. 5–13, p. 140)

(such as Alaska and Canada) but joins a massive migration in the autumn to specific southern sites in California and Mexico for the winter months. Some migratory fishes have similar behaviors; for example, the dogfish shark of the western Atlantic migrates between Florida in the winter and Maine in the summer.

Although these two categories of poikilothermy and ectothermic regulation are convenient, it is important to note that there is no distinct boundary between them. All ectothermic regulators that live in thermally variable habitats are also poikilotherms part of the time. For example, when there is no sun, warm burrow, or other heat source, a lizard cannot regulate its body temperature. At night, it usually cools off to approximately the air temperature. During this period, it is purely poikilothermic. As we have seen, many poikilotherms can keep their body temperatures within narrower ranges than the environment does.

■ Some ectotherms can metabolically compensate for changes in body temperatures.

Ectotherms in many habitats are subject to temperature changes that impair metabolism, because they cannot behaviorally or physiologically adjust body temperature. The onset of winter and of summer are the most common examples. When it is too cold, a poikilotherm's metabolism may slow down to the point that it cannot obtain any food. Freezing may also kill it. If it is too hot, its membranes, proteins, and nucleic acids may suffer irreversible damage. To avoid problems, many poikilotherms undergo adaptive physiological changes when exposed to temperature changes. Some of these changes occur on a daily basis, whereas others are seasonally induced. Some involve dormancy, whereas other changes allow for continued activity even with no mechanism for regulating body temperature.

In some cases, acclimatization to temperature changes allows an animal to maintain a useful level activity. An example is shown in ● Figure 15–15 for alligators, comparing animals acclimatized (see definition, p. 15) to winter and summer seasons. This particular study examined the activity levels of key enzymes in muscle tissue. Figure 15–15 shows that for the enzyme LDH (Figure 15–10), measured in the laboratory at 15, 22.5, and 30°C, the winter-acclimated animals have considerably higher enzyme activities at all measurement temperatures. This is one indicator of **metabolic compensation,** which allows the cold-acclimated alligator to move better in the cold. This is a form of homeostasis in the sense that the cold-acclimated alligators' metabolic reactions in cold temperatures are increased to a level that is closer to that of the warm-acclimated animals at higher temperatures: Note that the LDH level of the winter animals at 15°C is about the same as the level of the summer animals at 30°C. However, a cold-acclimated alligator will die of heat-related problems at warm temperatures at which the warm-acclimated animals will thrive. Thus there appears to be a tradeoff between metabolic rate and heat tolerance.

How is metabolic compensation achieved? In many cases, the mechanisms are not known, but several have been revealed in various animals:

1. *Homeoviscous membrane adaptation.* As we discussed previously, restructuring of membranes to maintain proper fluidity is a common strategy. The carp, for example, uses this adaptation.

2. *pH regulation.* Most ectotherms in the cold have higher internal pH values than do warmer animals, that is, colder animals are more alkaline. To some extent, pH increases in the cold automatically because the neutral pH of water and the acid constants (pK_a) of protein buffers (p. 601) increase in the cold. However, some buffer systems such as phosphate and bicarbonate do not have this property, so apparently animals must actively regulate pH to some extent. The primary hypothesis to explain this observation focuses on the amino acid *histidine*. Of the 20 amino acids used to build proteins, histidine is the most important in acting as a buffer (p. 601), because its pK_a is close to physiological pH, so it is nearly ideal for releasing or binding hydrogen ions over the normal physiological pH range. Binding and release of hydrogen ions by histidines are also crucial in many enzyme active sites for catalyzing reactions. Importantly, histidine's affinity for H^+ increases in the cold, so it binds H^+ too tightly to serve effectively as a buffer or catalyst if fluid pH is constant:

$$\text{Histidine} - H^+ \underset{\text{colder}}{\overset{\text{warmer}}{\rightleftharpoons}} \text{histidine} + H^+$$

This problem can be offset if H^+ concentration in body fluids is reduced (that is, pH is raised), because mass action then reduces the concentration of histidines in the charged form (that is, with H^+ bound). By actively regu-

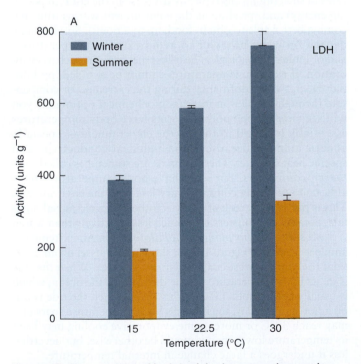

Figure 15–15 ● Activity of lactate dehydrogenase in muscle from alligators acclimatized to winter and summer conditions. The x-axis indicates temperatures at which the activities were measured in the laboratory.

(*Source:* From Seebacher, F., H. Guderley, R. M. Elsey and P. L. Trosclair III. 2003. Seasonal acclimatization of muscle metabolic enzymes in a reptile (*Alligator mississippiensis*). *Journal of Experimental Biology* 206: 1193–1200)

lating body fluids to become more alkaline, the animal restores the ratio of charged histidine (with H^+) to uncharged histidines (without H^+) to the ratio found at warmer temperatures.

3. *Enzyme concentration changes.* In cold acclimation and acclimatization, many animals such as some fishes and frogs have increased levels of metabolic enzymes (Figure 15–15), particularly those involved in aerobic pathways. This allows them to convert energy from food more effectively. However, it takes energy to produce more enzymes, and because energy input is usually reduced in the cold, this strategy has limits.

4. *Isoform regulation.* A final strategy requires that different forms of the same protein—called *isoforms*—be used at different temperatures. That is, there may be a set of winter proteins and another set of summer proteins, with the genes for these being activated only under appropriate environmental conditions. For example, the common carp expresses distinctly different forms of its myosin and myosin light-chain protein (MLC) in the summer and in the winter. Myosin is the main contractile protein of muscle (p. 317), whereas MLC binds to myosin and modifies its activity. Together, the isoforms expressed in the cold give the fish greater speed and force in the cold. However, the muscle then does not work well at higher temperatures.

Isoform regulation appears to be uncommon, and many of the mechanisms of metabolic compensation are still unknown. Currently, DNA microarrays (p. 77) are being used to look for major changes in gene expression that occur during acclimation. A recent study of the killifish, which experiences daily temperature changes of 15 degrees C or more in small intertidal ponds, found that only about 200 genes (out of 5000 measured on a microarray) exhibit significant changes in expression during temperature changes. Some of the genes regulate lipid metabolism, some regulate mitochondrial activity, and others have unknown functions at present. Possibly such studies will reveal previously unsuspected mechanisms of temperature acclimatization.

■ Ectotherms survive extreme cold by metabolic dormancy and by either freeze avoidance or freeze tolerance.

Not all animals remain active in extreme cold. Many can survive low temperatures by dramatically lowering their metabolic rates, reducing the rate at which ATP is consumed. Although they may be technically starving, they have sufficient reserves to survive the season. Frequently, metabolic rate can be reduced to as low as 1 to 10% of the normal resting rate: A 10-fold reduction in energy expenditure gains the animal a 10-fold extension of the time that fuel reserves will last. Some species of insects completely arrest their development in a state of diapause. Turtles hibernate at depths below the frost line along the banks of streams and at the bottom of ponds. These reptiles survive the entire winter without breathing and with a heart rate of only one beat per minute.

Special adaptations are needed for ectothermic animals whose body temperatures drop below freezing. There are two broad categories of these adaptations: *freeze tolerance* and *freeze avoidance.*

Freeze Tolerance

Some ectotherms actually survive freezing of some body fluids (with up to 80% of body water being frozen in such animals). Woolly bear caterpillars spend up to 10 months frozen in the high Arctic, barnacles and mussels freeze when exposed to subfreezing temperatures at low tide, and even some reptiles and amphibians freeze for variable periods of time. For example, the spring peeper frog (*Pseudacris crucifer*) can revive itself after having as much as 65% of its body water converted to ice. During the frozen state, ATP use virtually ceases and neurological activity is minimally detectable. Energy needs of a tissue are normally supplied by nutrients delivered by the circulatory system. However, in a thawing animal the tissues are completely reliant on immediately available energy sources.

What happens to the tissues that actually freeze? Consider a key problem of freezing—the osmotic stress placed on an animal cell as it enters the frozen state. Formation of ice crystals in the ECF (extracellular fluid) immediately changes the osmotic balance with the ICF (intracellular fluid). Because ice excludes solutes from its structure, during the freezing process the composition of the ECF becomes progressively more concentrated as the concentration of water declines. In response, water moves out of the cell. Only when the concentration of solute is great enough can the transformation of water into ice be retarded. The majority of freeze-tolerant animals reach the critical minimum cell volume when approximately 65% of the total body water is converted into the frozen state.

To understand how some animals survive the stresses of freezing on metabolism and cell volume, let us look at what happens to a peeper frog, starting when it is initially exposed to freezing temperatures. When water freezes on the frog's outer skin, a signal sent to the liver triggers a massive breakdown of glycogen in the liver. Consequently, a flood of glucose molecules enters the circulatory system, raising blood sugar concentrations over 450-fold (over 10-fold greater than a diabetic human with moderately uncontrolled blood-sugar concentrations). This serves as an **antifreeze** by lowering the freezing point through simple colligative properties (p. 77). In less than eight hours the frog's organs become loaded with this nutrient, which has now assumed an additional role as a **compatible cryoprotectant.** The cryoprotectant acts as an *osmolyte* (p. 578) to help keep the cells in osmotic balance with the increasing osmotic pressure in the ECF. Cryoprotectants must also be *compatible* with—that is, nonperturbing of—macromolecules (p. 578). Thus inorganic ions cannot be used: They are just as effective in lowering freezing points by colligative mechanisms (and that is why humans dump salt on icy roads in many wintry locales), but unlike the compatible solutes such as glucose, they disrupt macromolecules. (Compatibility is the reason humans use small carbohydrates such as ethylene glycol rather than corrosion-promoting salts in automobile radiators.) In addition, during thawing—the time when cells have no delivery of oxygen—glucose is immediately available as a fuel to generate ATP. Studies have shown that frozen frog ventricle regains its ability to contract if it is thawed in the presence of glucose but not in the presence of other cryoprotectants such as glycerol. Using this strategy, the frog is ready to resume its active life.

Other animals are freeze tolerant. For example, many insects employ similar mechanisms, often using other carbohydrates such as trehalose (diglucose) as compatible cryoprotectants and antifreezes in their cells. In addition, many of these animals actually enhance the freezing of their ECFs with **ice-nucleating agents,** small proteins that trigger the formation of small ice crystals. Apparently these agents ensure that ice only forms in the ECF and that ice forms in an ordered fashion, rather than in the "normal" way, which would create large crystals that could rupture membranes.

Freeze Avoidance

Most overwintering animals are considered intolerant of internal freezing yet may still have body temperatures well below the freezing point. Again, many such animals have antifreeze compounds, but they are used somewhat differently. First, some overwintering animals use compatible cryoprotectants such as sorbitol and glycerol, but throughout the ECF and ICF rather than just the latter. Thus the ECF is protected from freezing. Examples include many insects, arachnids, and some Arctic fishes such as the sweet-tasting rainbow smelt. However, other insects and polar fishes have special **antifreeze proteins,** which do not work by classic colligative mechanisms. These proteins are effective at much lower concentrations than the compatible cryoprotectants. They generally contain very hydrophilic amino acids and sugar side chains that can bind to growing ice crystals and thus prevent their growth.

In addition to, or instead of using antifreezes, some overwintering animals appear to use the phenomenon of **supercooling.** This is a state of water in which the temperature is well below the freezing point, but there is no trigger or nucleation site to begin ice formation. Some polar fishes and arachnids, such as the Antarctic mite, employ this strategy. Painted turtles (*Chrysemys picta,* common in wetlands across eastern North America) also supercool in the winter, as does at least one mammal, the Arctic ground squirrel (see p. 701). Supercooling is a dangerous strategy, however, because any encounter with external ice can trigger a catastrophically rapid crystallization of body fluids.

■ Ectotherms may survive extreme heat with the heat shock response.

Finally, let us examine what ectotherms do when exposed to excess heat. Given enough time, some ectotherms such as the carp exhibit the metabolic compensations that were discussed earlier. But what happens in the short term? If an organism is exposed to a sudden jump of about 5°C or more in cell temperature, an ancient mechanism is induced: the *heat shock response.* This occurs in all types of organisms from Archaea to humans, and involves the activation of genes for **heat shock proteins (HSPs),** also known as **stress proteins** because other stresses (such as osmotic and chemical shocks) can induce them as well. As shown in ● Figure 15–16a, these small, thermostable, hydrophobic proteins bind to larger, unfolded proteins and assist their folding into functional conformations. They are attracted to hydrophobic amino acids that are normally in the interior of folded proteins but become exposed on denaturation (unfolding). Some HSPs are always present in cells, where they assist in the folding of newly formed polypeptides coming off ribosomes (hence HSPs are also called *molecular chaperones*). But during heat shock, other HSPs are rapidly induced at the gene transcription level, protecting the cell from heat death. This occurs as follows: In an unstressed cell, there are small amounts of HSPs such as **hsp70,** the most common form (whose weight is about 70,000 daltons). The hsp70s are bound in an inactive state to a protein called **HSF-1** (heat-shock factor 1). Under a sudden heat stress, the hsp70s detach from HSF-1 and bind to and stabilize unfolded proteins. The release also activates HSF-1, which is a *transcription factor* (p. 30). When active, HSF-1 binds to two other HSF-1 proteins to form a trimer (protein complex made of three separate proteins). The trimer enters the nucleus and binds to a *response element* in the DNA (p. 30) called **HSE (heat shock element);** this action in turn activates the transcription of genes for heat shock proteins. (This mechanism of transcription factors and response elements is a typical process of gene regulation in eukaryotes, as we discussed in Chapter 2.)

As noted, HSPs have been found in all forms of life. Mammals, for example, make them during a fever. An example for a poikilotherm is shown in Figure 15–16b, focusing on hsp70. Two different species of limpets (*Collisella* spp.) live at different levels on a rocky intertidal habitat. The species that lives higher up—and thus is more exposed to excessive heat from the sun when the tide is out—has higher levels of hsp70 and can make more hsp70 than can species living lower in the intertidal zone.

Although HSPs are thought to be universal, heat shock induction can apparently be lost in some species over evolutionary time. Not, otheniid fishes, for example, appear to have no heat shock response whatsoever. These Antarctic fish live in a nearly constant temperature environment, about −2°C year-round. In the laboratory, they begin to die of heat stress at 5°C and above! Yet even at 12°C, they cannot make additional HSPs above the basal levels.

Endotherms

Endothermy presumably evolved because it allows homeothermy to be established much more reliably. Thus the ectotherm's problems of either finding external heat sources or suffering a slowing metabolism, are avoided. In addition, endotherms are usually in environments cooler than their bodies, so they also have consistently faster biochemical processes than do most ectotherms in the same habitat. This can provide a competitive advantage. But there is a large metabolic cost: endotherms must constantly generate heat internally to maintain body temperature. Heat production ultimately depends on the oxidation of metabolic fuel derived from food, and thus endotherms consume much more energy (5 to 20 times) than do ectotherms of the same mass. In addition, a high metabolism increases the risk of overheating. So endotherms must also have mechanisms to remove excess heat that are more effective than those of ectotherms.

■ Birds and mammals homeostatically maintain internal core temperature, whereas other endotherms heat selected body regions.

Many organisms have evolved endothermy: birds, mammals, some fishes and reptiles, some insects, and even some plants! In all cases, these organisms rely on high levels of *aerobic metabolism* for sustainable heat production. In shivering mam-

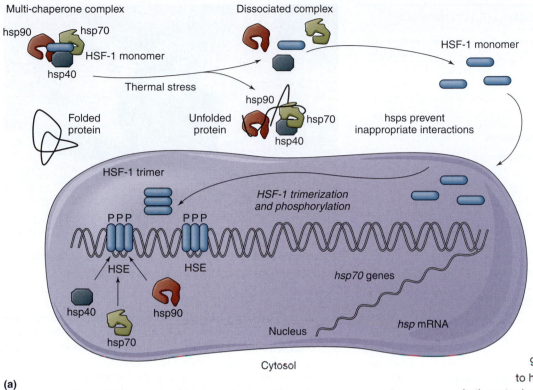

Multi-chaperone complex

Dissociated complex

hsp90   hsp70

HSF-1 monomer

hsp40

Thermal stress

HSF-1 monomer

Folded protein

Unfolded protein

hsp90   hsp70

hsp40

hsps prevent inappropriate interactions

HSF-1 trimer

HSF-1 trimerization and phosphorylation

P P P   P P P

HSE   HSE

hsp40   hsp90

hsp70

hsp70 genes

Nucleus

hsp mRNA

Cytosol

(a)

Figure 15–16 ● **Role and regulation of heat shock proteins.** (a) Model for regulation of hsp70, a major heat shock protein whose activity is activated by sudden temperature increases and other stresses. Under nonstress conditions, hsp70 (along with hsp40 and 90) is bound in an inactive state to heat-shock factor 1 (HSF-1) in the cytoplasm. Under temperature stress, this complex dissociates, and the hsp proteins bind to other, unfolded cell proteins, preventing aggregations and assisting in their refolding. The unbound HSF-1 moves to the nucleus, forms a trimer (three HSF-1 proteins), and binds to a promoter element in the DNA called the heat shock enhancer (HSE). This in turn activates the gene for hsp70, leading to synthesis of hsp70 mRNA and thus more hsp70 proteins to aid in the stress response. (b) *Collisella* limpets and their production of hsp70 at various temperatures. *C. scabra* lives higher in the rocky intertidal zone than *C. Pelta,* and thus is exposed to higher temperatures more often. Hsp70 production is quantified by the incorporation of radioactive amino acids, determined by radioactive counts per minute (cpm).

(*Sources:* (a) From Hochachka, P. W., and G. N. Somero, 2002. *Biochemical Adaptation.* Oxford: Oxford University Press; Fig. 7.14 as modified from Morimoto, R. I. and M. G. Santoro. 1998. Stress-inducible responses and heat shock proteins: New pharmacologic targets for cytoprotection. *Nature Biotechnology* 16: 833–838; (b) From Sanders, B. M., Hope, C., Pascoe, V. M., V. M., Martin, L. S. 1991. Characterization of the stress protein response in two species of *Collisella* limpets with different temperature tolerances. *Physiological Zoology* 64: 1471–1489)

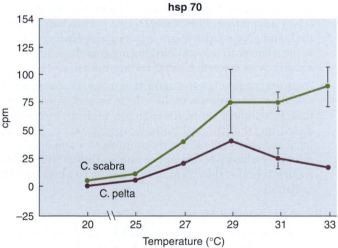

hsp 70

cpm

154

125

100

75

50

25

0

-25

C. scabra

C. pelta

20   25   27   29   31   33

Temperature (°C)

(a)

mals, flying insects, and swimming tunas (endothermic fish), it is *aerobic locomotory muscles* that provide extra heat. Thus some paleobiologists have proposed that enhanced aerobic locomotion evolved first in these animals and that only later was the heat output used for homeothermy. Whether this is true or not, researchers have found that most endotherms rely on aerobic metabolism for locomotion more than do ectotherms, and vice versa (with some exceptions). Thus a lizard (ectotherm) may have tremendous burst (anaerobic) capacity that allows it to outrun a mammal of the same size over short distances. However, in a long-distance chase the mammal's aerobic capacity allows it to outrun the lizard, which fatigues quickly.

Among all the different groups of endotherms, only birds and mammals exhibit homeothermy in most or all of their

cores. We consider these first. The other animals, which are more variable, are discussed later as *heterotherms.*

It is important to recognize that core body temperatures vary among species of birds and mammals (■ Table 15–2). Monotreme mammals, such as echidnas, have lower set points than most placental mammals, which in turn have lower set points than most birds. Why these different set points have evolved is not clear. However, one consideration is that it is physiologically easier to warm up than to cool off. Heat is produced by metabolism (following the second law of thermodynamics). But cooling off in an environment hotter than an animal's body requires evaporation of water—often a precious resource, and one that works poorly or not at all at high humidity (p. 688). Thus it has been proposed that endotherms

Table 15–2 ■ Core Body Temperatures (T_b) Characteristic of Endothermic Vertebrates

Taxon	Common names	T_b (°C)
Mammals		
Monotremes	Echidna platypus	30
Edentates	Anteaters, etc.	33–34
Marsupials	Possums, kangaroos, etc.	36
Insectivores	Hedgehogs, moles, etc.	36
	Shrews	37–38
Chiropterans	Bats	37
Cetaceans	Whales, etc.	
Pinnipeds	Seals, etc.	
Rodents	Mice, rats, etc.	37–38
Perissodactyls	Tapir, rhinoceros, horse	
Primates	Monkeys, humans, etc.	
Carnivora	Dogs, cats, etc.	
Artiodactyls	Cow, camel, pig, etc.	38–39
Lagomorphs	Rabbits	
Birds		
	Penguins	38
	Ostrich, petrels, etc.	39–40
	Pelicans, parrots, ducks, gamebirds	41–42
	Passerine songbirds	42

From Willmer, P., G. Stone and I. Johnston. 2000. *Environmental Physiology of Animals.* Blackwell Science, Oxford, UK. Table 8.11, p. 223.

Photo: © Bill Schmoker

A roadrunner ruffles it feathers to expose its dark skin, which absorbs sunlight to help heat the animal's blood.

have evolved optima that are slightly above the average high environmental temperature of their ancestral habitats.

Body temperatures also vary among individuals, and vary throughout the day within individuals. For diurnal mammals (including humans), body temperatures typically fall 1 to 2 degrees C at night in response to a biological clock (p. 263). (This is one reason why most humans feel particularly sluggish if awakened before dawn.) The camel, which we described earlier (p. 686), varies even more. Furthermore, there is no one body temperature, because temperature can vary from organ to organ, even in the core; for example, active skeletal muscles are warmer than most other organs.

■ To maintain a stable core temperature, heat gain must balance heat loss.

To maintain a constant total heat content and thus a stable core temperature, an animal body must balance heat input and output (Figure 15–11). We now elaborate on the means by which birds and mammals can adjust heat gains and losses to regulate core body temperature. Recall that the mechanisms generally fall into four broad categories (p. 688).

1. Gaining External Heat/Avoiding Loss to Cold Environs

Birds and mammals evolved from ectothermic ancestors and thus have inherited many of the thermoregulatory behaviors and physiology of ectotherms. These include the following:

- *Ectothermic behavior.* A cold bird or mammal may seek out sunshine or a warm surface, whereas a warm endotherm avoids excessive environmental cold, just as a lizard

does. Thus basking behaviors (serving as effectors) are common, in many birds and mammals, especially smaller ones with their high ratios of surface area to volume, and especially in cold weather. House cats, for example, are well known to bask in patches of sunlight. Larger mammals may also exhibit such behaviors: A human basking on the beach or soaking in a hot tub is acting like a lizard! Long-distance migrations may also serve in part as a thermoregulatory adaptation. For example, one reason why humpback and gray whales migrate from polar feeding waters in the summer to tropical birthing areas in the winter may be to provide a warm habitat for newborns.

- *Anatomic features such as dark skin that help absorb solar radiation.* Insulation actually interferes with this absorption. Roadrunner birds, for example, undergo *torpor* (reduced metabolism and body temperatures, p. 701) at night and bask in the sun in the morning to warm up, erecting the feathers of their upper backs. Unlike other skin (which is pink), their skin here is black, to absorb solar radiation.

2. Retaining Internal Heat

Retaining internal heat is accomplished by several mechanisms. Some of these also occur in some ectotherms, but in general these mechanisms are more prominent in endotherms:

- *Vasoconstriction.* The insulative capacity of skin can be varied by controlling the amount of blood flowing through it. Blood flow to the skin serves two functions. First, it provides a nutritive blood supply to the skin. Second, because blood has been heated in the central core it carries this heat to the skin. But an animal suffering excessive heat loss can vasoconstrict the skin vessels (● Figure 15–17a). This reduces blood flow through the skin, decreasing heat loss by keeping the warm blood closer to the core and thus more insulated from the external environment. As you will see later, the hypothalamus controls this blood flow in vertebrates. Some ectotherms, such as lizards, can use this physiological process to some extent.

- *Anatomic insulation.* Birds and mammals have evolved integumentary structures that help trap heat in their bod-

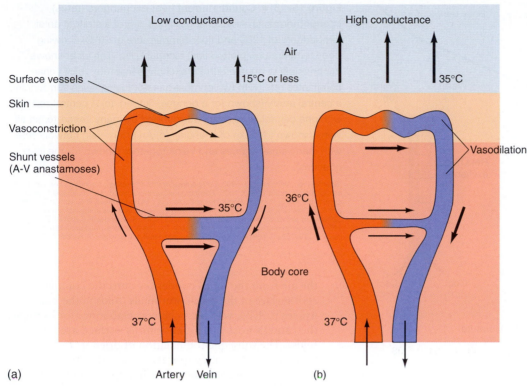

Low conductance   High conductance

Air

Surface vessels

Skin

Vasoconstriction

Shunt vessels
(A-V anastamoses)

15°C or less   35°C

35°C   36°C

Vasodilation

Body core

37°C   37°C

(a)   Artery   Vein   (b)

→ Blood flow
⟶ Heat transfer

Figure 15–17 ● Changes in blood flow to the skin for thermoregulation. (a) In a cold environment, blood can be diverted through shunts (anastomoses) deep under the skin to reduce heat loss. (b) If body heat needs to be released to the environment, or if external heat needs to be absorbed from a warm environment, vessels just under the skin surface can be opened.

(*Source:* From Willmer, P., G. Stone and I. Johnston. 2000. *Environmental Physiology of Animals*. Blackwell Science, Oxford UK. Fig. 8.36, p. 226)

ies. These *insulating* structures include feathers, hair, and subcutaneous adipose layers (including blubber). This insulation may appear to be a static anatomic feature, but there are some forms of regulation. Many mammals grow more hair and/or adipose tissue in the autumn. For example, the fur of white-tailed deer is thin in the summer, and thick and long in the winter. Furthermore, the winter hairs are hollow, allowing them to trap heat inside the shaft. Birds such as goldfinches increase their feather masses up to 50% by wintertime. In animals with dense fur or feathers, contracting the tiny muscles at the base of the hair or feather shafts lifts the hair or feathers off the skin surface (the *piloerection* reflex). This puffing up traps a layer of poorly conductive air between the skin surface and the environment, increasing the insulating barrier between the core and the cold air and reducing heat loss.

■ *Behavioral insulation.* Additional insulation may result from behaviors such as nest building, burrowing, and huddling. For example, Emperor penguins on the Antarctic ice often huddle in the thousands, a behavior estimated to save 25 to 50% energy. Another thermoregulatory behavior in many birds is perching on one leg, tucking the other leg against the body feathers (the leg is a major site of heat loss in birds.)

■ *Larger body size in colder climates.* A trend in avian and mammalian species is the evolution of larger average body sizes (such as penguins) in cold-climate species compared to warm-climate relatives. This trend, known as Bergman's rule, is predictable based on the properties of ra-

tios between surface area and volume, discussed earlier (p. 674).

■ *Countercurrent exchangers.* Blood flow can be used to retain core heat in another way known as a *countercurrent exchanger,* found between the core body and many exposed peripheral organs that could lose heat rapidly. (Recall examples of countercurrent flow in fish gills, fish swimbladder systems, and mammalian nephrons.) The exchanger consists of a set of veins and arteries (or venules and arterioles) placed closely together in a dense array known as a **rete mirabile** (● Figure 15–18). Because the vessels are so closely packed, they are nearly in thermal equilibrium. Crucial to the rete's function is that blood flows in opposite directions (countercurrent) in the two types of vessels. Heat moves in the following way: Warm core blood moves out the arteries toward the cold peripheral tissue. In the rete, it encounters cold blood from that periphery. By conduction, the heat moves into the cold vein and thus returns to the core. The venous blood leaving the rete is thus nearly at the temperature of the periphery, so that little core heat is lost.

Retes are found in many endotherms and heterotherms (Figure 15–8). In mammals, they are often in the limbs, such as a dolphin's fluke and flippers. In birds, they are often in the legs, to limit heat loss from the feet. You will see later how heterotherms such as tunas use retes.

A different type of countercurrent exchange is found in the nasal passages of birds and mammals. One drawback of endothermy is the potential for high water loss, and not just from thermoregulatory evaporation. Because the lungs must be kept moist for respiratory gases to dissolve into the cells lining the capillaries, lung air in endotherms is always warm and at 100% relative humidity. There is a danger of losing this body water and its heat content to the external environment. Instead, some of the moisture contained in

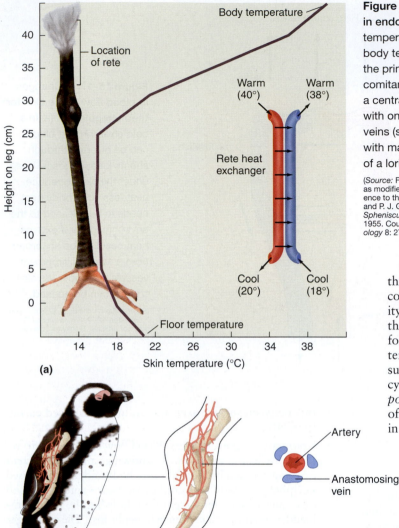

(a)

Figure 15–18 • **Countercurrent heat exchangers (retes) in endotherms.** (a) An exchanger in the leg of a stork at an air temperature of 12°C and floor temperature of 20°C, showing body temperature (plotted line) along the leg. The inset shows the principle of countercurrent exchange in a rete. (b) A venae comitantes rete, with two or more anastomosing veins surrounding a central artery (shown for a penguin flipper). (c) A centralized rete, with one central large artery surrounded by many separate, small veins (shown for fluke of a whale). (d) An artery-vein network rete, with many small arteries and veins together (shown for the limb of a loris).

(*Source:* From Withers, P. C. 1992. *Comparative Animal Physiology.* Saunders, Fort Worth TX, as modified from: Kahl, M. P. 1963, Thermoregulation in the wood stork, with special reference to the role of the legs. *Physiological Zoology* 36: 141–151; Frost, P. G. H., W. R. Siegfried and P. J. Greenwood. 1975. Arterio-venous heat exchange systems in the jackass penguin *Spheniscus demersus. Journal of Zoology* 175: 231–241; Scholander, P. F. and W. E. Schevill. 1955. Countercurrent vascular heat exchange in the fins of whales. *Journal of Applied Physiology* 8: 279–282.; Scholander, P. F. 1957. The wonderful net. *Scientific American* 196: 96–107)

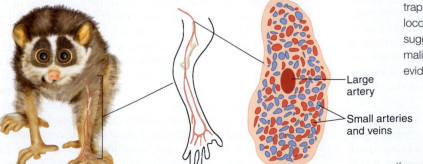

(b) Venae comitantes

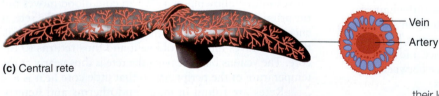

(c) Central rete

(d) Artery vein rete

the air from the lungs condenses onto the comparatively cool surface of the **maxilloturbinals** (folds in the nasal cavity; see p. 475), which can be very elaborate (for example, the camel's nose has an enormous surface area in these folds). The subsequent inhalation of drier air from the external environment evaporates this water and cools the surface of the maxilloturbinals in preparation for the next cycle. This type of countercurrent process is called a *temporal countercurrent exchanger,* where the opposing flows of a fluid are separated in time (using one tube) rather than in space (using two separate tubes). Mammals on average reclaim as much as 45% of the water (and its heat content) from the exhaled air, whereas kangaroo rats recycle up to 88% of their body water in this fashion. Birds also have such structures. However, ectothermic vertebrates, including reptiles, do not.

?

Did Dinosaurs Have Cold Noses? Maxilloturbinals may help settle a controversial issue: Were dinosaurs ectothermic or endothermic? Some fossil evidence suggests that at least some dinosaurs were endothermic—for example, some dinosaur bones appear to have capillary densities similar to those of mammals, higher than densities of modern reptiles. And, because of the enormous sizes of many species, their low ratios of surface area to volume would have trapped some heat produced by skeletal muscles during locomotion. But recent fossils of dinosaur nasal cavities suggest dinosaurs were not endothermic in the mammalian or avian sense. None of the dinosaur fossils shows evidence of maxilloturbinals. However, they do appear in the late Permian era in some therapsids, advanced reptiles that are the ancestors of mammals. Thus although it is possible that dinosaur bodies were warmer than their environments at least part of the time (like many modern reptiles), they probably did not have mammalian-style endothermy.

3. Generating More Internal Heat

The hallmark of endotherms is their ability to produce internal heat for the purpose of thermoregulation. One key adaptation was a large increase in birds' and mammals' overall BMR ("idling speed"). Indeed, the BMR of a bird or mammal is typically 5 to 20 times greater than the SMR of an ectotherm of the same mass. This appears to be due to "leaky" cell membranes. In one study comparing liver, kidney, and brain of a mammal (laboratory rat) and a reptile (bearded dragon) of the same mass and body temperature, the endotherm (rat) tissues consumed oxygen two to four times faster, and the rat Na^+–K^+ ATPase pump used ATP three to six times faster. The pump runs faster because the rat's membranes are leakier to these ions, and the ATPase must work harder to restore the gradients.

Thyroid hormones (p. 278) regulate the number of active Na^+–K^+ ATPase pump units in vertebrate cell membranes, with basal levels of these hormones being considerably higher in endothermic than in ectothermic vertebrates. In a resting mammal, most body heat is produced by the thoracic and abdominal organs. In addition to BMR, the rate of heat production can be variably increased above the "idling" BMR level in the cold primarily by changes in skeletal muscle activity or to a lesser extent by certain hormonal actions, as follows:

- *Shivering and other muscular activity.* Muscles constitute the largest organ system in the avian and mammalian body and are a ready source of heat from the contractile process. In response to a fall in core temperature caused by exposure to cold, skeletal muscle tone gradually increases. (Muscle tone is the constant level of tension within the muscles.) This produces some heat. Soon shivering begins. **Shivering** consists of rhythmic, oscillating skeletal-muscle contractions that occur at a rapid rate of 10 to 20 per second. This mechanism is very effective in increasing heat production; all the energy liberated during these muscle tremors is converted to heat because no external work is accomplished. Within a matter of seconds to minutes, internal heat production may increase two- to five-fold as a result of shivering.

- *Nonshivering (chemical) thermogenesis.* In most experimental mammals, chronic cold exposure brings about an increase in metabolic heat production that is independent of muscle contraction, appearing instead to involve changes in heat-generating chemical activity. This **nonshivering thermogenesis** is mediated by the hormones epinephrine and thyroid hormone. The cellular mechanisms triggered to produce heat are still not fully understood. One mechanism that has been delineated is found in newborn mammals and most small mammals (especially hibernators). These have deposits of a special type of adipose tissue known as **brown adipose tissue (BAT)** (● Figure 15–19a), which is especially capable of converting chemical energy into heat. BAT cells contain deposits of triglycerides packed with specialized mitochondria (the combination of fat and mitochondria gives this tissue its color). In their inner membranes, these mitochondria contain a gated hydrogen-ion channel called **thermogenin** (Figure 15–19b). Recall that normal mitochondria generate ATP by using a hydrogen ion gradient (p. 49). Normally, the hydrogen ions move down their gradient through an ATP-synthesizing enzyme complex. But when thermogenins are activated,

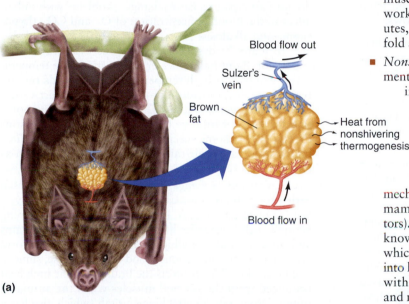

(a)

(b)

Figure 15–19 ● Brown adipose tissue (BAT, also called brown fat) in endotherms. (a) BAT deposits in bats and other mammals are found between the shoulder blades and other locations not shown. (b) Schematic diagram illustrating how BAT generates heat. When activated by cold or diet via a nerve that releases norepinephrine, the uncoupling protein (UCP) called thermogenin in the BAT cells opens and allows protons (H^+) to pass through the inner mitochondrial membrane. This dissipates the proton gradient as heat without making ATP.

(*Source:* (a) Modified from Randall, D., Burggren, W. and K. French. 2002. *Eckert Animal Physiology.* W. H. Freeman and Company, New York NY, Figure 17–22. (b) Modified from Dulloo, A. G. 1999. UCP2 and UCP3: genes for thermogenesis or lipid handling? *Obesity Matters* 2: 5–8.

the two processes (hydrogen ion flow and ATP synthesis) become uncoupled, the hydrogen ions simply flow down their gradient (through thermogenin) without generating ATP, and the energy from the hydrogen ion flow is completely dissipated as heat. Thermogenin is also termed an **uncoupling protein** (of which there are other types, p. 678) designated **UCP-1**.

BAT is particularly important in newborn mammals, which have a higher ratio of surface area to volume than do adults, and in hibernators, where it is used to warm the mammal up rapidly as hibernation ends. Also, as we mentioned earlier (p. 678), there is some evidence that BAT may help lean animals "burn off" excess calories.

4. Losing Excess Heat/Avoiding Gains from Hot Environs

A rise in core body temperature is potentially more dangerous than a decline, because of irreversible denaturation of proteins. Thus several mechanisms exist to remove excess body heat, which can occur not only from hot weather but from high muscular activity. These mechanisms are as follows:

- *Reduced insulation.* Desert endotherms may have inherently low insulation; for example, the camel has virtually no subcutaneous fat, storing its adipose instead in its characteristic hump. As we noted earlier, layers of insulation may be regulated. The shedding of hair by a house cat or dog in the springtime is a familiar example. A bird in hot weather may loosely ruffle its feathers to convect heat from the skin.

- *Vasodilation.* In the process of thermoregulation, skin blood flow can vary tremendously. The more blood that reaches the skin from the warm core, the closer the skin's temperature is to the core temperature. The skin's blood vessels diminish the effectiveness of the skin as an insulator by carrying heat to the surface, where it can be lost from the body by radiation and conduction–convection. Accordingly, vasodilation of the skin vessels (specifically the arterioles; Figure 15–17b), which permits increased flow of heated blood through the skin, increases heat loss.

- *Enhanced evaporation.* As you saw earlier, evaporation is the only way to cool off if the environment is hotter than the skin. The most important factor determining the extent of evaporation is the *relative humidity* of the surrounding air (the percentage of H_2O vapor actually present in the air compared to the greatest amount that the air can possibly hold at that temperature; for example, a relative humidity of 70% means the air contains 70% of the H_2O vapor it can hold). When the relative humidity is high, the air is already almost fully saturated with H_2O, so it has limited ability to take up additional moisture from the animal. Thus little evaporative heat loss can occur on hot, humid days.

    Endotherms have at least two evaporation mechanisms specifically adapted for thermoregulation: **respiratory panting** and "insensible" **cutaneous loss** through the skin (see p. 590). Some mammals have a third mechanism: **Sweating** is an active evaporative heat-loss process from specialized integumentary glands under sympathetic nervous control. Sweat is a dilute salt solution that is actively extruded to the surface of the skin by the glands and dispersed. For heat loss to occur, sweat must be evaporated from the skin.

Humans and horses sweat over much of their bodies, but most mammals either lack or have limited distribution of sweat glands; for example, dogs have sweat glands only on their paw pads. Birds do not have sweat glands at all. These species must employ the other evaporation methods. Cutaneous loss is an important route, especially in smaller endotherms. For example, it accounts for 54% or more of evaporative loss in Merriam's kangaroo rat (p. 561). Panting involves a shallow, rapid, breathing pattern that permits large volumes of air to move over the hot, moist tongue and respiratory airways. The resultant increase in evaporative heat loss from the respiratory tract provides a cooling effect. Numerous species of birds supplement panting with a rapid fluttering of the well-vascularized esophageal region. This rhythmic inflation of the hyoid apparatus is termed **gular fluttering.** Because air movement in panting and gular fluttering is associated with surfaces that do not participate in gas exchange, problems with imbalances in the blood concentration of O_2 and CO_2 (hypocapnia and alkalosis; see pp. 509 and 607) are avoided.

In many bird species, cutaneous evaporation accounts for about half of water loss at moderate temperatures, whereas at high temperatures most birds rely on respiratory evaporation. For example, the desert verdin (*Auriparus flaviceps*) at 50°C uses respiratory evaporation for 85% of total water loss. For unknown reasons the response of some species differs; for example, spinifex pigeons increase their rate of cutaneous evaporation as temperatures rise.

- *Countercurrent exchange.* The rete mechanism discussed earlier for heat retention can also be used to prevent overheating of certain organs. In particular, some fast-running animals such as gazelles have a rete between their brains and the core body, in which warm blood in the carotid artery passes by cool venous blood from the nose, sinuses, and facial skin. This protects the brain from the high heat produced from the skeletal muscles when the animal is running from, say, an attacking cheetah (which also has a similar rete). Horses may protect their brains from overheating in a different way: in addition to cooling sinuses, they have unusual **guttural pouches** (sacs extending from the auditory tubes) surrounding the carotid arteries, which may also help cool the blood.

- *Avoidance behavior.* Again, like their ectothermic ancestors, birds and mammals seek cooler environs when overheating. This can be a local change, such as moving under a shady tree, or may be a long-distance migration.

- *Anatomic reduction of heat gain.* Light-colored surfaces, such as skin, fur, or feathers, reflect sunlight. Camels, for example, have wool in the winter and shiny, reflecting hair in the summer (tropical camels maintain this reflecting hair year-round).

As we noted earlier, evaporative cooling must be balanced with the body's need for water; the camel is an excellent example of this tradeoff in a hot climate (p. 686). Some desert birds use the same strategy; that is, they let their body temperatures rise during the day to reduce evaporative cooling, thus saving up to 50% of body water. Another mechanism involves physiological regulation of the skin barrier. For example, zebra finches deprived of water add layers of lipids to their skin, reducing cutaneous water loss by 75% or more.

■ The hypothalamus integrates a multitude of thermosensory inputs from both the core and the surface of the body.

In vertebrates, the hypothalamus controls most mechanisms of thermoregulation in negative-feedback fashion. In birds, the hypothalamus appears to control metabolic rate, panting, vasodilation, and shivering, but a reflex integrator in the spine has also been identified that regulates vasodilation and constriction, panting, shivering, and erection of plumage. The hypothalamus (plus the spinal integrator in birds) serves as the primary "thermostat" in a negative-feedback system. As we discussed in Chapter 1, a house thermostat keeps track of the temperature in a room and triggers a heating mechanism (the furnace) or a cooling mechanism (the air conditioner) as necessary to maintain the room temperature at the indicated setting. Similarly, the hypothalamus receives afferent information about body temperature from various regions of the body and (particularly in mammals) initiates coordinated adjustments in heat gain and heat loss mechanisms as necessary to correct any deviations in core temperature from the set point. The hypothalamic thermostat is far more sensitive than your house thermostat. The hypothalamus in endotherms can respond to changes in blood temperature as small as 0.01°C.

To make the appropriate adjustments in the delicate balance between the heat loss mechanisms and the opposing heat-producing and heat-conserving mechanisms, the hypothalamus must be continuously appraised of both the skin temperature and the core temperature by means of specialized temperature-sensitive **thermoreceptors**. The core temperature is monitored by *central thermoreceptors,* which are located in the hypothalamus itself as well as elsewhere in the central nervous system and the abdominal organs. These sensors monitor the temperature that is actually being defended homeostatically. *Peripheral thermoreceptors* monitor skin temperature throughout the body and transmit information about changes in surface temperature to the hypothalamus. To some extent, these sensors serve in an *anticipation* regulation. That is, changes in skin temperature give advance warning of potential environmental threats to the core temperature, allowing corrections to be initiated before the core temperature is actually disturbed.

Two centers for temperature regulation have been identified in the hypothalamus. The *posterior region* is activated by cold and subsequently triggers reflexes that mediate heat production and heat conservation. The *anterior region,* which is activated by warmth, initiates reflexes that mediate heat loss. Together, these work to maintain a remarkably consistent core temperature (which, however, oscillates slightly because of feedback delays; see p. 15).

■ To regulate core temperature homeostatically, the hypothalamus simultaneously coordinates heat production, heat loss, and heat conservation mechanisms.

Let's now pull together the coordinated adjustments in heat production as well as heat loss and heat conservation in response to exposure to either a cold or a hot environment for a mammal (● Figure 15–20). In response to cold exposure, the posterior region of the hypothalamus directs increased heat production, such as increased muscle tone and shivering, while simultaneously decreasing heat loss (that is, conserving heat) by skin vasoconstriction and other measures. Because there is a limit to the ability to reduce skin temperature through vasoconstriction, when the external temperature falls too low even maximum vasoconstriction is not sufficient to prevent excessive heat loss. Accordingly, other measures must be instituted to further reduce heat loss, such as puffing up of hair. After maximum skin adjustments have been achieved physiologically, further heat loss can be prevented only by behavioral adaptations, such as seeking warmer habitats, and postural changes that reduce as much as possible the exposed surface area from which heat can escape. These postural changes include maneuvers such as curling up in a ball. Another behavior is best known in the sea otter, which grooms its hair and coats it with an oily secretion to help trap a layer of insulating air.

Under the opposite circumstance—heat exposure—the anterior part of the hypothalamus reduces heat production by decreasing skeletal muscle activity and promoting increased heat loss by inducing skin vasodilation. When even maximal skin vasodilation is inadequate to rid the body of excess heat, sweating or panting is brought into play to accomplish further heat loss through evaporation.

The consequences of thermoregulatory actions on metabolism are illustrated in ● Figure 15–21, which shows the metabolic rate as a function of environmental temperature. In the middle of this diagram lies the **thermal neutral zone (TNZ)**, a range of environmental temperatures in which the animal does not need to expend significant energy for thermoregulation. In this zone, mechanisms involving insulation, blood vessels, and low-energy behaviors are sufficient. Below the neutral zone, at a point called the **lower critical temperature**, metabolic rate increases as the special heat-generating mechanisms are activated. If these mechanisms are insufficient, the animal may suffer from **hypothermia** (dangerously low body temperature). Above the neutral zone, at a point called the **upper critical temperature**, metabolism increases because of panting (which uses rapid muscle activity) or heavy sweating (which uses ion transport processes). These activities counterproductively but unavoidably generate even more heat to deal with, leading to the potential for **hyperthermia** (dangerously high body temperatures).

The width of the TNZ and the critical temperatures depend on the effectiveness of the thermoregulatory adaptations that do not require significant energy. Insulation is often the major factor. For example, the lower critical temperature for the lightly insulated cardinal (*Cardinal cardinalis*) is 18°C, whereas for the heavily insulated Emperor penguin (*Aptenodytes forsteri*) it is −10°C. In tropical hummingbirds (such as *Colibri delphinae*), there are no critical temperatures; metabolic rate decreases linearly with environmental temperature as the latter is increased from 4 to 40°C. Seasonal acclimatization can alter the TNZ. For example, with a summer hair coat the critical lower temperature for a beef cow averages 15°C (59°F), whereas with a heavy winter coat wind chills of −8°C (18°F) are comfortably tolerated. Fermentation in rumens also produces heat, contributing to the comparatively low critical lower temperatures observed in ruminants.

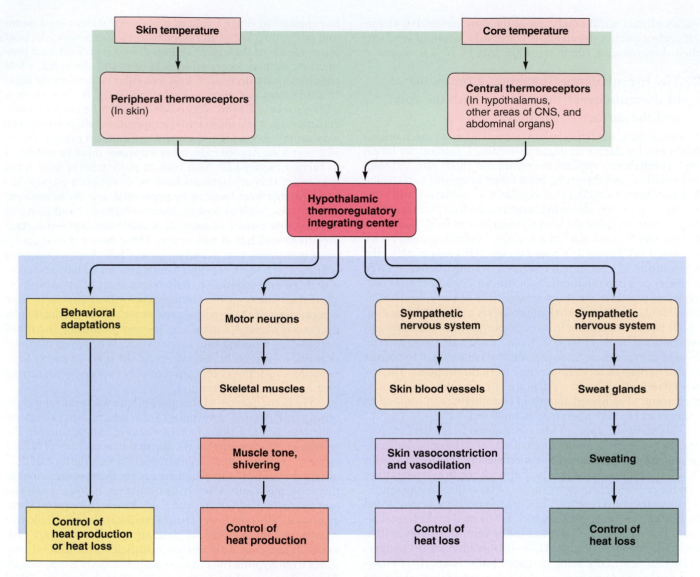

Figure 15–20 ● Major thermoregulatory pathways in a mammal.

Figure 15–21 ● The influence of ambient temperature on the metabolic rate (MR) of a small temperate endotherm, showing the thermoneutral zone where MR is constant.

(*Source:* From Willmer, P., G. Stone & I. Johnston. *Environmental Physiology of Animals* Blackwell Science. Oxford UK (Fig 8-33))

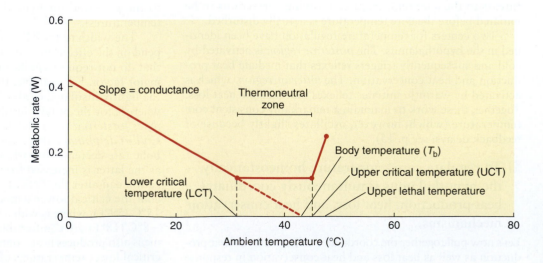

■ During a fever, the hypothalamic thermostat is "reset" to an elevated temperature.

Fever is an elevation in body temperature as a result of infection or inflammation. In mammals, in response to microbial invasion, certain white blood cells release a **pyrogen** (especially **interleukin-1, IL-1**) which, among its many infection-fighting effects, acts on the hypothalamic thermoregulatory center to raise the setting of the thermostat (see Figure 10–8). The hypothalamus now maintains the temperature at the new set level instead of maintaining normal body temperature. If, for example, IL-1 raises the set point from 37°C to 40°C, the hypothalamus senses that the normal prefever temperature is too cold, so it initiates the cold-response mechanisms to elevate the temperature to the new set point. Shivering is initiated to rapidly increase heat production, while skin vasoconstriction is brought about to rapidly reduce heat loss, both of which drive the temperature upward. Once the new temperature is achieved, body temperature is regulated as normal in response to cold and heat but at a higher setting. Thus fever production in response to an infection is a deliberate outcome and is not due to a breakdown of thermoregulatory mechanisms.

Although the overall physiological significance of a fever is still unclear, many medical and veterinary experts believe that a rise in body temperature has a beneficial role in fighting infection in mammals, including humans. A fever augments the inflammatory response and may interfere with bacterial multiplication (see p. 437). Experiments with ectotherms have demonstrated that this is an ancient defense mechanism. Infected lizards and fish seek out warmer habitat areas to raise their body temperatures above their normal optimum. This appears to increase survival, because ectotherms prevented from elevating their body temperatures in this way do not survive the infection as well.

Heterotherms

■ Heterotherms are endotherms that are not fully homeothermic.

In addition to birds and mammals, many other animals (and even some plants) exhibit some form of endothermy, but unlike birds and mammals, they do not heat all their cores. Furthermore, some animals including many small birds and mammals cannot maintain high core temperatures continuously, because of their high ratios of surface area to volume, coupled with insufficient food supply. Such animals are called **heterotherms**, falling into two broad categories: *regional* and *temporal*.

Regional Heterotherms

Regional heterotherms are animals that heat only certain organs. For example, flying insects such as bees and moths heat their thoraxes by the activity of flight muscles. As a preflight warmup, some species shiver their flight muscles before take-off. During flight they can keep their thoraxes at relatively constant and high temperatures by controlling hemolymph flow between the thorax and abdomen, which are connected by countercurrent flow channels (● Figure 15–22a). The abdomen has a thin ventral surface, the *thermal window,* through which excessive heat can be lost.

In the marine realm, two groups of fishes—the *lamnid sharks* (such as the great white) and many *scombrid teleosts* (tunas and billfishes)—have evolved endothermy, primarily in their red or aerobic swimming muscles. These animals migrate over long distances, swimming continuously with these muscles and generating constant heat output. But it is not enough to heat the entire body, because too much heat is lost at the gills. A countercurrent rete system between the aerobic muscle and the gills prevents most of the loss (Figure 15–22b). Presumably, these warm, thermally regulated muscles give these fishes the ability to move between warm and cold waters more easily than ectotherms can. Remarkably, some of these fishes such as the swordfish have special heater organs behind their eyes to heat their retinas and brains. These organs evolved out of eye muscles and have become dedicated to heat production only! Presumably, this allows the fish to see prey better in cold, dark waters.

Temporal Heterotherms

Temporal heterotherms are endotherms that maintain high body temperatures only for certain time periods, most commonly in seasonal *hibernation* and daily *torpor.*

■ *Hibernation.* Many small endotherms in higher latitudes become poikilothermic as winter approaches, entering a dormant state and allowing body temperatures to drop to near 0°C (● Figure 15–23b). Remarkably, the Arctic ground squirrel even supercools slightly, with body temperatures of −3°C. Hibernators must store up large amounts of unsaturated fats (which do not turn hard like butter at cold body temperatures) to serve as energy reserves. However, these animals are not purely ectothermic. Mammalian hibernators still have a functioning hypothalamic thermostat, but with a greatly lowered set point. Using this, they thermoregulate to prevent body freezing, generating heat with uncoupling proteins (UCPs, p. 698) in brown adipose tissue, and also with UCPs in white adipose tissue and skeletal muscle. They also go through periodic "bouts" of rewarming (Figure 15–23b) and then recooling every few days or weeks. It is not known why they do this, but one hypothesis is that they must warm up in order to void wastes and sometimes to eat cached food, with the latter serving in particular to restore glycogen stores in their brains. Another hypothesis is that they do this periodic awakening to reactivate their immune systems temporarily in case they have been infected.

Contrary to popular belief, only mammals about marmot size (3 kg) or smaller can truly hibernate, because of the principles of ratios between surface area and volume, discussed earlier. A bear is simply too large to cool off and then reheat in the springtime, so it sleeps through much of the winter with a body temperature only a few degrees lower than normal (recall that even human body temperatures drop by 1°C or so at night).

■ *Torpor.* Many small endotherms with high metabolic demands, such as shrews and hummingbirds, enter a hibernation-like state called *torpor* on a daily basis. At night, for example, a deer mouse's body temperature (Figure 15–23a) drops from around 35°C to as low as 15 to 20°C (depending on environmental temperatures), saving

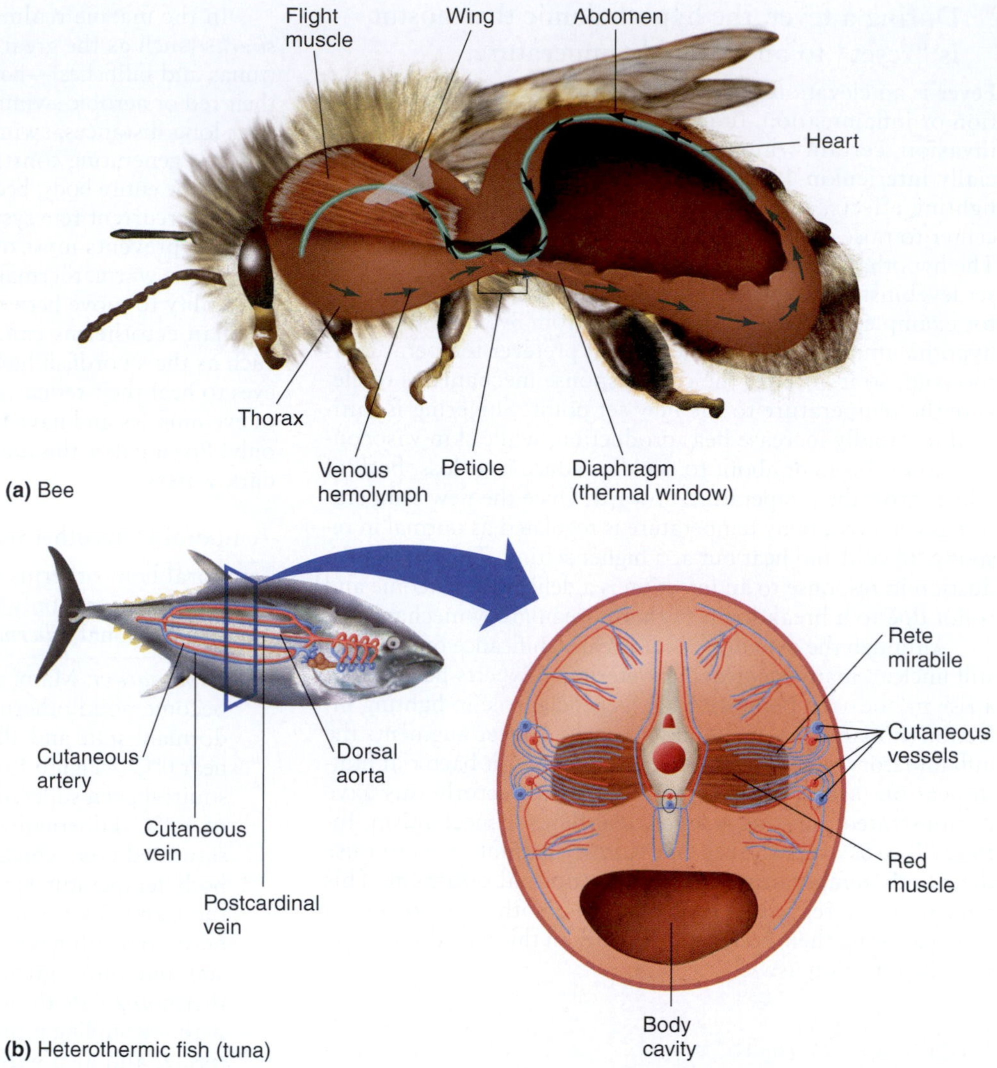

Figure 15–22 ● **Countercurrent exchangers in heterotherms.**
(a) A bumblebee. The flight muscles of honeybees and bumblebees are supplied with blood from the heart, while the flight muscles produce heat at a high rate during flight. A countercurrent flow in the petiole (thorax-abdomen junction) retains heat in the thorax for the flight muscles. If heat needs to be dissipated, venous blood can be pumped in the absence of arterial flow, taking heat to the abdomen's thermal window for removal.
(b) A tuna. The diagram shows the rete located between the outer (cutaneous) vessels and the deep red muscle. The rete traps the heat produced by those muscles.

(*Sources:* (a) From Schmidt-Nielsen, 1960. *Animal Physiology.* Prentice-Hall, Englewood Cliffs NJ. Fig. 7.39, as modified from Heinrich, B. 1976. Heat exchange in relation to blood flow between thorax and abdomen in bumblebees. *Journal of Experimental Biology* 64: 561–585; (b) Modified from Willmer, P., G. Stone and I. Johnston. 2000. *Environmental Physiology of Animals.* Blackwell Science, Oxford UK. Fig. 9.10, p. 264)

(a) Bee

(b) Heterothermic fish (tuna)

a large amount of energy. (Note that these animals usually do not let body temperatures drop close to freezing, possibly because the energy cost of rewarming on a daily basis would be prohibitive). This is important because this mouse needs almost constant food input to sustain its metabolism in the active state. Of particular interest is the Australian marsupial called the fat-tailed antechinus. This small mammal goes through nightly torpor, with core body temperatures dropping to 16 to 27°C, and (like the roadrunner, p. 694) it basks in the sun each morning to reheat its body to 33 to 37°C. Unlike other mammals, however, it does not simultaneously use endothermic mechanisms in the initial phases of rewarming. Thus it uses lizardlike behavior in a fashion that suggests how thermoregulation first started in mammalian ancestors.

In addition, animals that employ daily torpor tend to have longer life spans. Hummingbirds, for example, typically undergo daily torpor and can live up to 10 years. In contrast, Norwegian shrews (six species, all about the same size as hummingbirds) live only 16 to 20 months, and do not undergo regular torpor. The so-called south-

ern African elephant shrew (*Elephantus myurus*), which is not a true shrew but is of similar size, does undergo torpor, and it lives 4 to 6 years. Thus reductions in metabolism on a regular basis may slow the manifestation of aging, much of which is thought to be caused by reactive oxygen species (ROS, p. 49) generated in proportion to metabolic rate.

A final, and rather spectacular example of endothermy crosses the boundary between heterothermy and homeothermy. Several social insects, which individually are mostly ectothermic, form large colonies, which are thermoregulated by the collective behavior of the group. These "superorganisms" include many beehives and termite mounds, in which groups of animals construct ventilation tubes and use their wing beats to heat up or convectively cool the colony. Honey bees also carry water into their hives for evaporative cooling. Thus beehives can maintain temperatures that are much less variable than the environment. In tropical habitats where the environmental temperature does not vary extensively, some termite mounds are homeothermic all year.

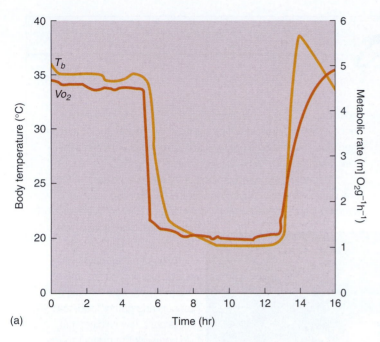

(a)

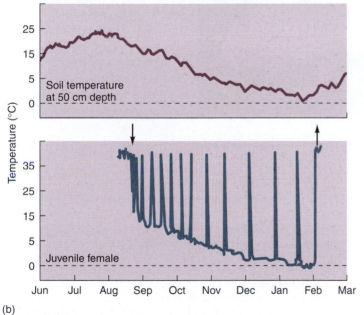

(b)

Figure 15–23 ● **Temperatures of mammals in torpor and hibernation.** (a) Metabolic rate (V_{O2}) and body temperature (Tb) of a deer mouse during daily torpor. (b) Soil temperature and body temperature in a Richardson's Ground Squirrel (*Spermophilus richardsonii*) hibernating through the winter, showing periodic "bouts" of rewarming.

(*Sources:* (a) From Withers, P. C. 1992. *Comparative Animal Physiology.* Saunders, Fort Worth TX, using data from Nestler, J. R. 1990). Relationships between respiratory quotient and metabolic rate during entry to and arousal from daily torpor in deer mice (*Peromyscus maniculatus*). *Physiological Zoology* 63: 504–515; (b) From Michener, G. R. 1998. Sexual differences in reproductive effort of Richardson's ground squirrels. *Journal of Mammalogy,* 79:1–19)

Thermoregulated hives have recently been shown to be crucial to larval bee development. Pupated larvae were raised at 32, 34 and 36°C in the laboratory. Those raised at the lower temperatures had abnormal neural development. Amazingly, when the colony is infected with a pathogenic fungus, the bees can even create a fever-like state. The fungus, which can kill bee larvae, is greatly inhibited by the elevated temperature. Swarms of honey bees have also been documented using rapid wing beats to create heat intense enough (over 50°C) to kill parasitic wasps attempting to invade the hive.

One amazing group of mammals, the naked mole rats, have evolved a hivelike system that resembles these insect colonies. See box, "A Closer Look at Adaptation: The Naked Mole Rats—Mammalian Hive Ectotherm."

Chapter in Perspective:

HOMEOSTASIS AND INTEGRATION

Because energy can be neither created nor destroyed, for body mass and body temperature, respectively, to remain constant, input must equal output in both an animal body's total energy balance and its heat–energy balance. If total energy input exceeds total energy output, the extra energy is stored in the body, and body mass increases. Similarly, if the input of heat energy exceeds its output, body temperature increases. Conversely, if output exceeds input, body mass decreases or body temperature falls. The hypothalamus is the major integrating center in mammals for maintaining both a constant total energy balance (and thus a constant body mass) and a constant heat-energy balance (and thus a constant body temperature). However, the effectors for temperature and energy regulation include components of many other systems covered in previous chapters. Energy balance, for example, obviously requires the digestive system and endocrine as well as hormonal regulation (such as insulin) and neural regulation (including behaviors such as eating). Temperature regulation involves (among other things) the skin and its blood vessels (and sweat glands in some mammals), respiration (panting), and skeletal muscles (shivering, or behavioral avoidance or seeking).

Body temperature is one of the most pervasive influences on body functions, which slow down if too cold, and suffer from denaturing macromolecules if too warm. Some ectotherms achieve partial homeostasis by controlling external heat exchanges (mainly through behavior), whereas others are not homeostatic but must adapt through dormancy or internal biochemical adjustments. Endotherms achieve homeostasis (homeothermy) much more consistently with physiological mechanisms. ■

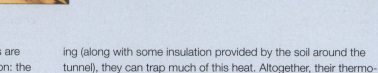

A CLOSER LOOK AT ADAPTATION

The Naked Mole Rat— Mammalian Hive Ectotherm

Although traditional physiology states that all mammals are endotherms, there is at least one remarkable exception: the naked mole rat (*Heterocephalus glaber* and *Cryptomys damarensis*). Naked mole rats live in underground tunnel complexes underneath savanna and grasslands of equatorial Africa. *H. glaber* has been the most studied species. These mole rats have almost no body fur (although they do have fine hairs) and no subcutaneous fat, and thus have no effective insulation. Most amazingly, they cannot individually regulate their body temperatures via endothermy. Thus their body temperatures do fluctuate somewhat. But a degree of homeothermy around 31°C is still achieved, using mechanisms like those of social insects. First, they occasionally bask in the sun at their tunnel entrances. Second, the animals do produce some heat, and by huddling together while sleep-

A naked mole rat

Photo: © M. J. O'Riain & J. Jarvis/Visuals Unlimited

ing (along with some insulation provided by the soil around the tunnel), they can trap much of this heat. Altogether, their thermoregulatory mechanisms are very similar to those of termites in their mounds.

Thermoregulatory physiology of these mammals is not the only parallel to insect physiology. Both species have a social structure (termed *eusocial*) that is very much like the hive societies of social insects. There is a single queen mole rat, for example, that mates with a few dominant males; there are many workers that dig the tunnels and gather food; and there are a few soldiers that do not work but protect the colony from predators such as snakes. Thus cooperation within the colony is essential to survival, because a pair of mole rats on their own would not be able to perform all the functions needed to cope with environmental challenges.

REVIEW QUESTIONS (Answers are on p. A–1.)

Additional study tools for this chapter, including chapter summaries and practice tests, are available online at *www.biology.brookscole.com*

1. The Second Law of Thermodynamics
   a. states that energy can neither be created nor destroyed
   b. states that entropy can decrease in one part of a system but will always increase in the system and its surroundings
   c. does not apply to living systems
   d. shows that life can operate indefinitely on its internal energy content
   e. states that life and its surroundings become more ordered over time

2. Appproximately what percentage of energy contained in an animal's food is available to accomplish work?
   a. 100%
   b. 75%
   c. 50%
   d. 25%
   e. 10%

3. Which of the following represents the most heat energy?
   a. joule
   b. kcal
   c. calorie
   d. kilojoule
   e. 1000 joules

4. The respiratory quotient (RQ) is
   a. the quantity of heat produced per liter of oxygen consumed
   b. a subject's uptake of oxygen per unit of time
   c. the ratio of CO_2 produced to O_2 consumed
   d. always more than one in mixed diets
   e. always less than one when only carbohydrate is consumed

5. Total body energy content can be maintained at a constant level by
   a. compensatory decreases in the BMR in response to underfeeding
   b. compensatory increases in the BMR in response to overfeeding
   c. activating uncoupling proteins
   d. increases in activity levels in response to overfeeding
   e. all of the above

6. Which of the following stimulates appetite?
   a. leptin
   b. neuropeptide Y (NPY)
   c. melanocortins

d. orexins
e. b and d

7. The region in the brain stem that is thought to be the site that processes signals important in terminating a meal is the
   a. hypothalamus
   b. arcuate nucleus
   c. nucleus tractus solitarius
   d. lateral hypothalamic area
   e. paraventricular nucleus

8. Q_{10} values are valuable measures of
   a. the extent of insulin secretion
   b. neuropeptide Y denaturation
   c. leptin's effect on weight gain in poultry
   d. temperature sensitivity of a biological process
   e. environmental factors in food selection

9. You would expect an animal living in a hot rainforest to have
   a. a high proportion of highly unsaturated fatty acids in its cell membranes
   b. relatively inefficient (low k_{cat}) enzymes
   c. a high proportion of saturated fatty acids in its cell membranes
   d. lactate dehydrogenase with relatively loose and flexible loops in its tertiary structure
   e. b and c

10. Which of the following is a heat-transfer mechanism that can occur only in air?
    a. convection
    b. evaporation
    c. conduction
    d. radiation
    e. none of the above

11. For ectotherms to successfully survive supercooling requires
    a. ice-nucleating agents
    b. compatible cryoprotectants
    c. an absence of nucleation sites
    d. antifreeze proteins
    e. protein isoforms

12. Heat-shock proteins are found in
    a. ectotherms
    b. mammals
    c. endotherms
    d. Archaea
    e. all of the above

13. To maintain a stable core temperature,
    a. the body must import more heat from the environment than it loses
    b. the body must export more heat from the environment than it gains
    c. heat input to the body must balance heat output
    d. all internal organ systems must be at the same temperature

14. Nonshivering thermogenesis
    a. depends on muscle contractions
    b. is mediated by epinephrine and thyroid hormone
    c. takes place largely in maxilloturbinals
    d. depends on thermogenin, a gated hydrogen-ion channel in the endoplasmic reticulum
    e. requires countercurrent exchanges

15. Vasodilation of vessels in the skin
    a. decreases heat gain
    b. increases heat loss
    c. increases the BMR
    d. decreases internal consumption of brown adipose tissue
    e. decreases the BMR

16. The hypothalamus receives information about skin and core temperatures from
    a. thermoreceptors
    b. thermogenin
    c. uncoupling proteins
    d. countercurrent exchange
    e. the sympathetic nervous system

17. The thermal neutral zone is
    a. the point at which heat-generating mechanisms are activated
    b. the zone in which metabolism increases due to panting
    c. the temperature range in which the animal may experience hypothermia
    d. the range of temperatures in which the animal doesn't need to expend significant energy for thermoregulation
    e. the zone in which the animal may experience hyperthermia

18. In mammals, pyrogen
    a. acts on the thermoregulatory center to lower the setting of the thermostat
    b. acts on the thermoregulatory center to raise the setting of the thermostat
    c. helps the body maintain normal temperature
    d. raises the lower critical temperature
    e. expands the thermal neutral zone

19. Heterotherms
    a. generally have higher core temperatures than homeotherms
    b. heat all of their cores
    c. are endotherms that are not fully homeothermic
    d. have low surface area:volume ratios
    e. maintain high core temperatures continuously

20. Regional heterotherms
    a. are found only in higher latitudes
    b. maintain high body temperatures for only certain time periods
    c. enter a hibernation-like state, torpor, on a daily basis
    d. heat only certain organs
    e. always hibernate

SUGGESTED READINGS AND INTERNET SITES

Alexander, R. M. 1982. *Locomotion of Animals*. Glasgow, UK: Blackie.

Block, B. A., J. Finnerty, A. F. R. Stewart, & J. A. Kidd. 1993. Evolution of endothermy in fish: Mapping physiological traits on a molecular phylogeny. *Science* 260:210–214.

Calder, W. A. 1984. *Size, Function and Life History*. Cambridge, MA: Harvard University Press.

Dawson, W. R. 1982. Evaporative losses of water in birds. *Comparative Biochemistry and Physiology* 71A:495–509

Dunnan, J. G. 2001. Antifreeze and ice nucleator proteins in terrestrial arthropods. *Annual Review of Physiology* 63:327–357.

Else, D. L., & A. J. Hulbert. 1987. Evolution of mammalian endothermic metabolism: "Leaky" membranes as a source of heat. *American Journal of Physiology* 253:R1–R19.

Feder, M. E., & G. E. Hofmann. 1999. Heat-shock proteins, molecular chaperones, and the stress response: Evolutionary and ecological physiology. *Annual Review of Physiology* 61: 243–282.

Goodman, B. 1998. Where ice isn't nice: How the icefish got its antifreeze and other tales of molecular evolution. *BioScience* 48:586–591.

Graham, J. B., R. Dudley, N. M. Aguilar, & C. Gans. 1995. Implications of the late Paleozoic oxygen pulse for physiology and evolution. *Nature* 375:117–120.

Heinrich, B. 1996. *The Thermal Warriors. Strategies of Insect Survival.* Cambridge, MA: Harvard University Press.

Hochachka, P., & G. N. Somero. 2002. *Biochemical Adaptation: Mechanism and Process in Physiological Evolution.* Oxford, UK: Oxford University Press.

Ricquier, D., & F. Bouillaud. 2000. Mitochondrial uncoupling proteins: From mitochondria to the regulation of energy balance. *Journal of Physiology* 529:3–10.

Ruben, J. A., T. D. Jones, & N. R. Geist. 2003. Respiratory and reproductive paleophysiology of dinosaurs and early birds. *Physiological and Biochemical Zoology* 76:141–164.

Schmidt-Nielsen, K. 1997. *Animal Physiology. Adaptation and Environment,* 5th ed. Cambridge, UK: Cambridge University Press.

Schmidt-Nielsen, K. 1998. *The Camel's Nose, Memoirs of a Curious Scientist.* Washington, DC: Shearwater Books. A lively, engaging discussion of a prominent comparative physiologist's pioneering research.

Secor, S. M., & J. M. Diamond. 2000. Evolution of regulatory responses to feeding in snakes. *Physiological and Biochemical Zoology* 73:123–141.

Storey, K. B. 1997. Organic solutes in freezing tolerance. *Comparative Biochemistry and Physiology* 117A:319–326.

Trier, T. M., & W. J. Mattson. 2003. Diet-induced thermogenesis in insects: A developing concept in nutritional ecology. *Environmental Entomology* 32:1–8.

Wehner, R., A. C. Marsh, & S. Wehner. 1992. Desert ants on a thermal tightrope. *Nature* 357:586–587.

Weibel, E. R. 2002. The pitfalls of power laws. *Nature* 417: 131–132.

Welty, J. C., & L. Baptista. 1990. *The Life of Birds,* 4th ed. Fort Worth, TX: Saunders.

INFOTRAC READINGS

Crockett, E. L. 1998, April. Cholesterol function in plasma membranes from ectotherms: Membrane-specific roles in adaptation to temperature. *American Zoologist* 38:291.

Travis, J. 1995. Mouse obesity cured by hormone. *Science News* 148:68. Tschantz, D. R., E. L. Crockett, P. H. Niewiarowski, & R. L. Londraville. 2002. Cold acclimation strategy is highly variable among the sunfishes (Centrarchidae). *Physiological and Biochemical Zoology* 75:544–556.

INTERNET SITES

Austgen, L. 2002. *Brown fat.* **arbl.cvmbs.colostate.edu/hbooks/ pathphys/misc_topics/brownfat.html.** Details on and micrographs of brown adipose tissue.

Center for the Integrative Study of Animal Behavior, Indiana University. 1999. *CISAB Temperature Calculator.* **indiana.edu/ ~animal/fun/conversions/temperature.html.** An overview of thermal adaptations and terminology, with a temperature conversion calculator.

Nave, C. R. 2000. *Hyperphysics.* **hyperphysics.phy-astr.gsu.edu/ hbase/mod3.html.** Learn about interactions of radiant energy with matter in general and the body in detail.

16

Reproductive Systems

Kittiwake (Rissa tridactyla) *with chicks (Bass Rock, Scotland)*

Introduction

The central theme of this book has been the physiological processes aimed at maintaining homeostasis to ensure survival of the individual animal. However, as we noted in Chapter 1 and in several subsequent chapters, some processes, such as growth and development, are not aimed toward homeostasis, but rather constitute regulated changes (rheostasis). We end this book with a discussion of reproductive systems, which also are not aimed toward homeostasis and, moreover, are not necessary for survival of an individual. Instead, they are essential for perpetuation of species. Only through reproduction can the complex genotypic recipe for each species survive beyond the lives of individual members of the species.

Reproductive Processes

Reproduction depends on the union of male and female **gametes** (reproductive, or **germ,** cells), each with a single set of chromosomes (*haploid*), to form a new individual with a *diploid* (that is, twice haploid), unique set of chromosomes (● Figure 16–1).

■ Animals employ a variety of reproductive strategies to ensure survival of the species.

Animals employ a variety of strategies to ensure this union occurs successfully. Conventionally, reproductive processes can be classified according to their ecological and physiological adaptations. The ecological approach considers parents to have limited energy resources for producing offspring. Use of parental energy is classified according to two opposite strategies termed *r*-selected and *K*-selected reproduction (*r* is the rate of population growth, and *K* is the carrying capacity of the environment). A highly *r*-selected species places virtually all its re-

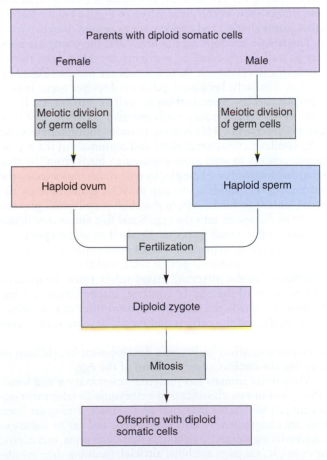

Figure 16–1 ● Chromosomal distribution in sexual reproduction.

productive resources into producing hundreds to thousands (or more) of offspring, with no parental care and minimal nourishment. Most of the offspring do not survive, but the overwhelming numbers usually ensure that a few survive to perpetuate the species. *K-selected species,* in contrast, produce relatively few offspring, with most of the parental reproductive energy being used for nourishing and protecting the offspring. With this strategy, a much greater percentage of offspring survive. Any given species may be fully *r*-selected, such as a sea star that releases millions of eggs with very little yolk (nutritional stores) into the ocean, or fully *K*-selected, such as a placental mammal that gives birth to a few live offspring that are subsequently nourished and protected, or intermediate, such as a fish that produces hundreds of eggs with yolk for nutrition and provides parental protection of developing eggs.

Oviparity, Ovoviviparity, Viviparity

Physiological classification of reproductive processes uses three categories: **oviparous,** meaning the parents release eggs from which the young (*ovi-,* "egg"; *parous,* "produce") hatch after expulsion from the body (most invertebrates, fishes and amphibians such as frogs and toads); **ovoviviparous,** the production of eggs (not necessarily with shells) that hatch within the body of the parent as in the case of some reptiles and elasmobranch fishes; and **viviparous,** the production and nourishment of living young within the body, as in almost all mammals and some sharks such as the great white (● Figure 16–2). Rattlesnakes, some sharks, and guppies are examples of ovoviviparity: The female develops eggs in utero and the young are born there as well, but because no nutrients are taken from the female during development these animals are not considered viviparous. Similarly, because copulation does not occur in true oviparous species (insemination as well as fertilization is external to the genital tract), birds are also classified as if they were ovoviviparous. This is done in order to illuminate some of the similarities between birds and mammals in the reproductive process as well as to distinguish birds from the true oviparous forms. For example, in true oviparous species the membranes surrounding the egg must be penetrable by the sperm, unlike in birds, where the shell membranes prevent passage of the sperm into the egg. Some fish and invertebrates have **micropiles,** small pores in the shell to allow sperm entrance. Further, in oviparous species embryonic development is external to the maternal genital tract, and thus the maternal contribution to the offspring arises solely from the nucleus and cytoplasm of the egg. In contrast, insemination and fertilization are inside the maternal genital tract of ovoviviparous animals, and the earliest embryonic development occurs within this genital tract. Thus the maternal parent has some opportunity to affect embryonic development in addition to providing the nucleus and cytoplasm of the egg.

Viviparous animals have internal insemination and fertilization, and in this classification embryonic development occurs entirely within the maternal genital tract. Young are born alive, are comparatively fully developed, and are to some extent free-living animals, although there are some variations. For example, rat pups are born **altricial,** meaning they are absolutely dependent on continued parental care; whereas guinea pig piglets are born **precocial** and so require minimal parental care. However, gestation is 64 days in guinea pigs and only 21 in the rat. Rat pups are born blind, hairless, and immobile,

whereas the guinea pig piglets are fully furred; the eyes are open, and the young begin foraging soon after birth. The maternal parent may thus affect the developing offspring by providing the egg itself, that is, the nucleus and cytoplasm, the environment for fertilization, and subsequently the environment (including nutrients) for gestation of the developing fetus. All mammals except monotremes can be included in this group; the egg-laying duck-billed platypus (*Ornithorhynchus*) and the spiny anteater (*echidna*) are notable exceptions. However, even in those mammals, the egg remains developing in the female tract for a considerable time before being laid. Marsupials (*marsupium,* "pouch") whose newborn young are highly altricial, leave the uterus at a comparatively undeveloped stage and migrate into the pouch of the female, where they attach to a **teat** (see p. 753).

Seasonal Breeding

Another strategy used by animals to optimize the number of offspring that can be raised to maturity is to breed seasonally. Seasonal *anestrus* (reproductive quiescence) evolved as a means of preventing females from conceiving during periods of the year when survival of the offspring is unlikely. Most seasonal breeders give birth in spring, when environmental and nutritional conditions support lactation and growth of the offspring. In birds, for example, seasonal breeding is particularly highly evolved, in part because the young depend on fresh food from the day of hatching. The energy requirements of the parent increase between 20 and 70% above normal because of the requirements associated with yolk synthesis and deposition of yolk in the maturing oocytes, thus food availability is crucial for the laying female as well as for the hatchling. **Circannual (biological) clocks** in the brain (Chapter 7, p. 258) are partly in control of this seasonal reproduction, but the clocks are set or *entrained* by environmental signals. **Photoperiod,** the amount of time during the day when there is light, is by far the most important factor determining the onset of the breeding season. The unique regularity of the annual change in day length makes it ideal for regulating seasonal breeding, with the result that virtually all groups of advanced organisms have evolved the ability to use photoperiod as the primary factor regulating reproduction. Thus in temperate latitudes virtually all bird species breed in the spring. However, seasonal conditions for all photoperiodic species are not identical each year so other environmental cues are used for fine-tuning the reproductive processes set in motion by changes in day length. These modifying factors can, within specified limits, accelerate or retard the photoperiod-induced gonadal growth. These include but are not limited to temperature, presence of the male, and the availability of a particular food source. The great tit, for example, is completely dependent on the availability of caterpillars and thus on the ambient temperature, which affects caterpillar populations in the spring and alters the onset of its breeding cycle. Similarly, the crossbill reproduces very early in the spring when pine seeds become available.

Photoperiod is not the deciding factor for many tropical and desert species, and irregular breeding cycles are the norm. In these species, changes in temperature or rainfall determine the timing of gonadal development. For example, the zebra finch (*Poephila gutta*) is an opportunistic breeder that lives in the desert and ovulates within a day of rainfall, the increase in humidity apparently triggering the reproductive cycle. If favorable conditions persist, the zebra finch lays a succession of

Photo: © Wendell Metzen/Index Stock Imagery

Photo: © L. L. Rue Photography

Photo: © Lightscapes Photography, Inc./CORBIS

Photo: © Carolina Biological Photos/
Visuals Unlimited

Photo: © SuperStock, Inc.

Figure 16–2 ● Reproductive strategies in animals. (a) Snails are oviparous, meaning they release eggs from which the young later hatch. (b) Many snakes, all lizards, and some fishes are ovoviviparous. Their fertilized eggs develop inside the mother, and then offspring are born alive. Yolk reserves sustain development in the egg. In this figure the live born copperhead offspring are still encased in their egg sacs. (c) Birds are ovoviviparous because there is some internal development in the female. Their fertilized eggs contain large yolk reserves, and they develop and hatch outside the mother's body. (d) Most mammals are viviparous; their young are born live. The young of the opossum complete development in a pouch on the mother's ventral surface. The juvenile stages continue to draw nourishment from mammary glands in the mother's pouch.

clutches. Within tropical regions where there are pronounced wet and dry seasons, reproduction tends to occur in association with the decrease in light intensity associated with the rainy season. However, the environmental triggers are not clear in all cases. Recent studies on tropical (Panamanian) antbirds have shown that even slight changes in photoperiod (from 12 hours to 13 hours daylight) can trigger gonadal growth leading to breeding (only at the equator is daylight time 12 hours year round). In Panama, the longer daylight months are the rainy season.

The time of egg hatching in some species such as reptiles (for example, snapping turtles), arachnids, and insects also normally depends on environmental conditions. For example, eggs that have overwintered or diapaused (entered dormancy) rely on seasonal increases in temperature to initiate the hatching process. Normally, a cumulative period above a critical temperature, rather than a simple temperature threshold, is required. Alternatively, the eggs of species such as mosquitoes and drag-

onflies rely on decreasing oxygen concentrations in water, as well as on increasing temperatures, to cue the time of hatch.

Synchronization and Mating Behaviors

A final set of reproductive strategies is that which ensures that both sexes are ready to breed at the same time, particularly when conditions are optimal. Optimal conditions include both environmental (such as springtime) and internal (such as when eggs are ripe and ready for fertilization.) Reproduction, especially for females, is energetically costly, and success of a species can depend on efficient use of this energy. We have already discussed seasonal cues that trigger reproduction at an optimal time, but these cues are not very precise, and could, for example, result in some individuals being ready to breed a few days ahead of others. Thus a variety of other cues and triggers have evolved to synchronize reproduction between the sexes.

For animals with external fertilization and without complex behavior, other environmental cues and pheromones or other biologically produced agents (such as light in fireflies, p. 89) serve as synchronizing signals. A spectacular example is found in the annelid palolo worm (*Leodice viridis*) of South Pacific coral reefs. In the springtime (September to November in the southern hemisphere), the posterior segments of these worms become ripe with gametes. During the last quarter moon of October and November, typically at low tide, this part of the worm detaches (becoming an *epitoke*) and floats to the surface, where it releases its gametes. Because all mature worms in the vicinity do this at the same time, external fertilization has a high probability of success. Female epitokes also emit a pheromone that attracts wriggling male epitokes and stimulates release of sperm. (Many marine annelids reproduce with epitokes.)

Animals with complex behavioral capacities may also use pheromones, typically produced by females when they are ready for fertilization, but they often perform elaborate reproductive behaviors called **courtship displays**. These are most commonly ritualistic patterns of movement and/or songs that allow males and females of the same species to recognize and attract each other and to signal and synchronize mating readiness. Courtship behaviors by males are also thought to advertise (to the females) the genetic fitness of the males, as another way to ensure efficient use of reproductive resources. These behaviors occur in the greatest complexity in vertebrates. Among the most elaborate vertebrate courtships are found in the bowerbird species of New Guinea and nearby islands. The male bowerbirds of many species construct elaborate *bowers,* a patch of ground partly enclosed with woven twigs, decorated with small objects such as leaves, shells, dead insects, flowers, paper, and glass, creating a splash of color that stands out in the forest. Females that are ready to mate begin visiting bowers and appear to be attracted to the more elaborate and colorful ones. But that is not enough. The male, on a female's arrival, also performs a ritualistic strutting dance (with sound and puffing of feathers) that needs to be pleasing to the female, or she leaves for another bower. Recent studies using electronic robots made to mimic female bowerbirds show that the males adjust the intensity of their displays according to whether the female appears to be gaining or losing interest. Thus there is feedback synchronization between the sexes.

Many arthropods also have complex courtship displays. For many male spiders, these displays are a matter of life or death. A female spider is almost always ready to eat any small animal, thus the male (who is generally smaller than the female) must signal her in such a manner that suppresses her appetite and increases her readiness to mate. For example, male wolf spiders wave their distinctly marked front legs and **pedipalps** (limbs modified to carry sperm) in ritualistic patterns in front of the female. In some species, the male may also jump side to side. If the female does not eat him, he inserts his pedipalp with sperm into her vaginal opening.

Reproductive capability ultimately depends on intricate relationships among the environment, the nervous and endocrine systems, reproductive organs, and target cells of sex hormones. In this chapter we focus primarily on the basic sexual and reproductive functions that are under nervous and hormonal control.

Reproductive Systems and Genetics

Unlike the other body systems, which are essentially identical in the two sexes, the reproductive systems of males and females are remarkably different, befitting their different roles in the reproductive process.

◼ The reproductive system of vertebrates includes the hypothalamus, gonads, and reproductive tract.

The **primary reproductive organs**, or gonads, of most vertebrates consist of **testes** in the male and **ovaries** in the female. In both sexes in vertebrates, the mature gonads are found in pairs, and perform the dual function of (1) producing gametes (**gametogenesis**), that is, **spermatozoa (sperm)** in the male and **ova (eggs)** in the female, and (2) secreting sex hormones, primarily *testosterone* in males and *estrogens* (the major form being *estradiol*) and *progesterone* in females. However, as discussed later, males also make estrogens and females make testosterone with crucial roles—for example, males need estrogens for bone maintenance, females need testosterone for libido.

In addition to the gonads, the reproductive system in each sex includes a **reproductive tract** encompassing a system of ducts that are specialized to transport or house the gametes after they are released, plus **accessory (collateral) sex glands** that empty their supportive secretions into these passageways. The accessory glands of insects are paired tubular secretory structures that release their secretions into the genital tract. In mammalian females, the **mammary glands** (from which the name "mammal" is derived) also are considered accessory reproductive organs. The externally visible portions of the reproductive system are known as **external genitalia** as opposed to those internally housed portions, termed the **internal genitalia**.

The anatomy of the reproductive organs is related to the reproductive strategies characteristic for a particular species. This is evident in females, whose reproductive organs differ considerably between those animals that require internal fer-

Palolo worms, *Palola siciliensis* (entire worm with palolo or epitoke attached)

(*Source:* © Larry Madrigal/Seapics.com, www. seapics.com/picture_gallery/marine_invertebrate)

Photo: © Larry Madrigal/Seapics.com

The essential reproductive functions of the male are (1) production of sperm *(spermatogenesis)* and (2) delivery of sperm to the female. In most mammals the sperm-producing organs, the testes, are suspended outside the abdominal cavity in a skin-covered sac, the **scrotum,** which lies within the angle between the posterior (inferior) appendages. All endotherms (Chapter 15) have a testicular cooling mechanism, which we consider later in this chapter.

The male reproductive system is designed to deliver sperm to the female reproductive tract in a liquid vehicle, *semen,* which is conducive to sperm viability. In mammals, secretions from the major male accessory sex glands, the *seminal vesicles, prostate gland,* and *bulbourethral glands* (● Figure 16–3) provide the bulk of the seminal fluid. The penis, or hemipenis (in avian drakes), is the organ used to deposit semen in the female. Sperm exit the mammalian testes through the *vasa efferentia, epididymis, ductus (vas) deferens, ejaculatory duct,* and *urethra,* the latter being a canal that runs the length of the penis. Instead of an epididymis, in birds the *vasa efferentia* (tubules) conduct sperm from the testis to a short epididymal duct that is continued as the vas deferens, where it eventually opens into an enlarged area prior to entering the *cloaca.* Both the vas deferens and the enlarged region serve as sperm storage sites. In the absence of a prostate gland, bulbourethral gland and seminal vesicle, seminal fluid in birds is derived from the *seminiferous tubules* and the *vasa efferentia.*

The female's role in reproduction is often more complicated than the male's, especially if internal fertilization is involved. The major structures of the female reproductive tract in reptiles, birds, and mammals include the *vagina, cervix, uterus, oviducts,* and *external genitalia.* The essential female reproductive functions include (1) production of ova *(oogenesis)* and ovulation (although management of the ovulated oocyte depends on species); (2) reception of sperm; (3) transport of the sperm and ovum to a common site for union *(fertilization,* or *conception);* (4) giving birth to the young *(parturition)* in a viviparous animal, or laying eggs in an oviparous animal; and, in mammals, (5) nourishing the offspring by milk production *(lactation).* Placental mammals (that is, not monotremes or marsupials) have an additional role: (6) nourishment of the developing fetus internally until it can survive in the outside world *(gestation,* or *pregnancy)* via the **placenta,** a vascular structure that supplies the fetus with nutrients in exchange for waste products generated by the fetus.

The product of fertilization is known as an **embryo** during early development when tissue differentiation is taking place. The best demarcation is the establishment of organ systems and a functioning placenta in placental mammals. Beyond this time, features of the developing mammal are discernible and the embryo, now termed a **fetus,** is connected to the mother through the placenta for the duration of gestation.

The ovaries and female reproductive tract lie within the pelvic and abdominal cavity (● Figure 16–4). The mammalian female **oviducts** (**uterine** or **Fallopian tubes**) capture ova on ovulation and serve as the site of fertilization if fertilization occurs internally. In some vertebrates, the oviduct is lined with secretory and ciliated cells, which provide a suitable environ-

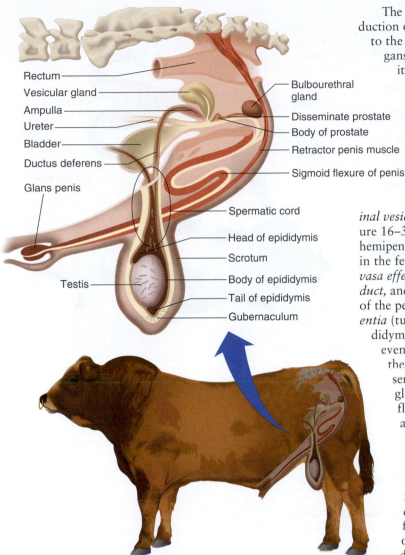

Figure 16–3 ● Reproductive tract of a bull. Sagittal view of the bull reproductive tract.

Source: P. L. Senger, 2003, *Pathways to Pregnancy and Parturition,* 2nd ed., Current Conceptions Inc., Washington State University Research & Technology Park, pp. 48, 49.

Labels (figure):
- Rectum
- Vesicular gland
- Ampulla
- Ureter
- Bladder
- Ductus deferens
- Glans penis
- Testis
- Bulbourethral gland
- Disseminate prostate
- Body of prostate
- Retractor penis muscle
- Sigmoid flexure of penis
- Spermatic cord
- Head of epididymis
- Scrotum
- Body of epididymis
- Tail of epididymis
- Gubernaculum

tilization and those that use external fertilization. In addition, female anatomy differs between those species that store large volumes of yolk in the egg and species that store a relatively small volume.

In some species, evolution of secondary sexual characteristics, the many external characteristics of vertebrates that are not directly involved in reproduction itself, can be used to distinguish between males and females. These features include body configuration, hair distribution, feather coloration, and sexual weaponry. For example, the secondary sexual characteristics can be of great importance in courting and mating behavior: The rooster's comb attracts the hen's attention, and the stag's antlers are useful to ward off other stags and establish dominance in the herd. For the most part, testosterone in the male and estrogens in the female govern the development and maintenance of these characteristics.

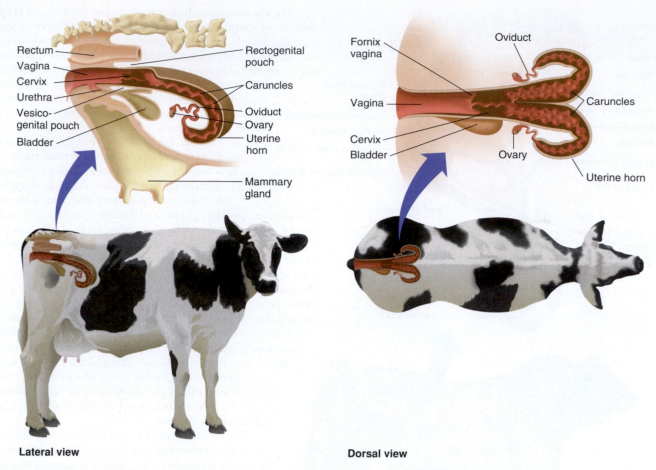

Lateral view

Dorsal view

Figure 16–4 ● **Reproductive tract of a cow.** Lateral and dorsal view of a cow reproductive tract.

(*Source:* P. L. Senger, 2003, *Pathways to Pregnancy and Parturition,* 2nd ed., Current Conceptions Inc., Washington State University Research & Technology Park, pp. 16, 17)

ment for the ova and help transport the spermatozoa. The portion of each tube adjacent to its respective ovary expands to form the **infundibulum,** a thin, funnel-shaped structure that captures the egg after release from the ovary. The "catching" of the egg involves active participation on the part of the infundibulum and is not simply a passive process by which the egg is funneled into the oviduct. The infundibulum is covered in fingerlike projections called **fimbriae,** which function to cause the infundibulum to slip over the ovarian surface at ovulation, increasing the probability of "catching" the oocyte. In mammals, the thick-walled, hollow **uterus** serves a protective and nutritive capacity for maintaining the fetus. However, considerable variation exists among species with regard to the anatomic organization of oviducts and uterus. For example, the uterus may be extensively divided into two **uterine horns,** or **cornu,** with a comparatively small body or corpus. The corpus is comparatively large in the mare, less extensive in the cow and sheep, and small in the dog and pig. In some species, such as the rat and rabbit, no uterine body is present.

In oviparous and ovoviviparous vertebrates, the oviducts are the site for deposition of nonembryonic egg materials (albumin). In birds and reptiles, the uterus serves as a **shell gland** for calcification of the egg's outer layer.

The caudal portion of the uterus is the **cervix,** which projects into the vagina. The **cervical canal,** a single small opening, serves as the site of semen deposition in some species or as a pathway for sperm deposited in the anterior vagina to pass into the uterus in others. During mammalian pregnancy the cervix effectively closes the external opening of the uterus, producing a viscous mucus that prevents the entry of foreign material into the uterus. The cervix can also become greatly dilated during parturition, because it serves as the passageway for delivery of the fetus from the uterus to the vagina.

The vagina, a muscular, expansible tube connecting the uterus to the external environment, is primarily an organ of copulation and serves as a receptacle for sperm. The female external genitalia are collectively referred to as the **vulva** in mammals, which is comparable to the ventral portion of the **cloaca** of birds, which serves as a common pathway for urinary, fecal, and reproductive products. In humans and Old World monkeys, two pairs of skin folds, the **labia minora** and **labia majora,** surround the vaginal and urethral openings laterally. The female labia minora is homologous to the male prepuce, whereas the labia majora is a homolog of the male scrotum. In mammals, the **clitoris,** an erectile structure composed of tissue comparable to the glans penis, lies at the anterior end of the folds of the labia minora. Recent work has shown that the clitoris is much larger than is stated in conventional texts, because much of its mass remains hidden internally and was missed by the prudish Victorian anatomists who provided the original descriptions in Western science. Interestingly, this situation becomes exaggerated in the female hyena, because the external size of her clitoris is similar to the penis of the male, and she has a false scrotum, making it very difficult to distinguish between the two sexes.

◼ Reproduction systems in insects includes neuroendocrine organs, gonads, and reproductive tract.

The reproductive system of male insects consists of a pair of testes and accessory glands (● Figure 16–5a). Secretions from the accessory glands may include the following: a supply of nutrients to the female, formation of seminal fluid that serves as a carrier for the spermatozoa or that hardens about them to form a sperm-containing capsule, the **spermatophore.**

In many species the formation of a temporary **mating** or **copulatory plug** after fertilization prevents both loss of semen from the female tract and entry of sperm into the vagina (genital chamber) from a competing male. Contraction of the female ducts then transports the spermatozoa to the **spermatheca,** the organ that stores the sperm prior to entry into the vagina (Figure 16–5b).

The reproductive system of a female insect consists of a pair of ovaries, a system of ducts through which the eggs navigate to the outside, the accessory glands, and the spermatheca (Figure 16–5b). Externally, female insects have an **ovipositor,** an extension of the abdomen, used to lay eggs, for example, by insertion into the ground or into a host animal (as is done by parasitic wasps, p. 458). The ovipositor is modified in many species for other functions. Considered one of the most feared defenses among invertebrates, the hymenopteran stinger is a modified ovipositor. In some species such as infertile honey bees, this sexual structure has completely lost its reproductive function. In such a bee, an entire section of the animal's abdomen is torn off when it stings its intended victim, killing the bee in the process.

Recall that in insects it is the *quantity* of juvenile hormone (JH) released at each molt that determines the *quality* of the molt, that is, the time to maturity (p. 262). As the concentration of JH falls, each molting episode triggered by ecdysone results in a progressively less juvenile morphology until finally the adult genes are completely expressed and the adult form emerges. The gonads and their accessory glands either develop slowly during the molting cycles or not until the final molt to the adult stage, when the inhibitory effects of JH are absent. Most adult male insects emerge with fully formed reproductive systems, including a supply of mature spermatozoa.

JH plays a role in oogenesis in some species, primarily to initiate vitellogenesis, which leads to egg development in the **ovarioles** (tubular structures forming the ovary). Removal of the *corpora allata* (p. 261) in these species prevents normal egg formation, whereas its reimplantation induces ovarian function. In other species, egg development is delayed until the adult stage, when the release of another brain hormone stimulates ecdysone release from the prothoracic gland in response to environmental cues such as temperature, photoperiod, or even specific input such as ingestion of a blood meal in mosquitoes. Ecdysone, in turn, initiates vitellogenesis and egg maturation in concert with secretions from the ovarioles. Ovulation is believed to be a hormonally controlled event involving the

Figure 16–5 ● Reproductive systems of insects. (a) female reproductive system; (b) male reproductive system. (Redrawn from Snodgrass, by permission of the McGraw-Hill Book Company, Inc.)

(*Source:* C. A. Triplehorn, D. J. Borrer, & N. F. Johnson, 1989, *Introduction to the Study of Insects,* 6th ed., Belmont, CA: Thomson/Brooks/Cole, Figure 3–31, p. 57)

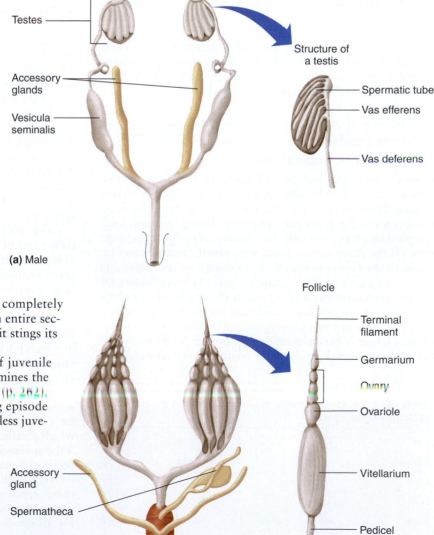

release of a brain peptide, whereas oviposition also requires the contribution of the nervous system, because the female can carefully select the time and the site for oviposition. Secretions from the accessory glands are species dependent and include such functions as encasing the eggs in frothy or gelatinous secretions, storing them in protective casings (**oothecae**) or simply gluing them to a particular surface, normally the food plant consumed during juvenile stage of development.

■ Gametogenesis is accomplished by meiosis.

Most cells in an animal's body can reproduce themselves, a process important in growth, replacement, and repair of tissues. Cell division in somatic cells is accomplished by **mitosis.** In mitosis, the chromosomes (which contain the DNA) replicate (make duplicate copies of themselves), then the identical chromosomes are separated so that a complete set of genotypic information (that is, a diploid number of chromosomes) is distributed to each of the two new daughter cells. Cell division that produces gametes is accomplished by **meiosis,** in which only one of the two sets of homologous chromosomes (that is, a haploid number of chromosomes) is distributed to each of four new daughter cells. Importantly, not all the maternally derived chromosomes go to one daughter cell and the paternally derived chromosomes to the other cell. Thus each sperm and oocyte has a unique complement of chromosomes (■ Table 16–1). When fertilization takes place, a sperm and ovum fuse to form a new individual with one member of each chromosomal pair having been inherited from the maternal parent and the other member from the paternal parent (Figure 16–1).

■ The sex of an individual is determined by the combination of sex chromosomes or by environmental stimuli.

In some vertebrates, including all mammals and birds, individuals are destined to become males or females by a genotypic process determined by the sex chromosomes they inherit. As the chromosome pairs are separated during meiosis, each sperm or ovum receives only one member of each chromosome pair. Of the chromosome pairs, **autosomal chromosomes** include all the chromosomes that are shared by the males and females of a species. The remaining pair of chromosomes are the **sex chromosomes** and represent the single pair of chromosomes whose makeup is different for males and females, of which there are two genotypically different types: in mammals, a larger **X chromosome** and a smaller **Y chromosome.** **Sex determination** depends on the combination of sex chromo-

somes: In mammals, **genotypic males** have both an X and a Y sex chromosome; **genotypic females** have two X sex chromosomes. Thus in mammals the genotypic difference responsible for all the anatomic and functional distinctions between normal males and females is the single Y chromosome. Males have it; females do not.

As a result of meiosis during gametogenesis, all gametes contain only a single set of homologous chromosomes. In mammals, when the XY homologs separate during sperm formation, half the sperm receive an X chromosome and the other half a Y chromosome. In contrast, during oogenesis every ovum receives an X chromosome, because separation of the XX sex chromosome pair yields only X chromosomes. During fertilization, combination of an X-bearing sperm with an X-bearing ovum produces a genotypic female, XX, whereas union of a Y-bearing sperm with an X-bearing ovum results in a genotypic male, XY. Thus genotypic sex is determined at the time of conception and depends on which type of sex chromosome is contained within the fertilizing sperm.

The XX/XY system discussed so far applies to mammals. Birds are ZW (the heterozygote ZW being the female) or ZZ (the homozygote ZZ being the male). The Z and W are often used in species in which the female is heterogametic, to avoid confusion with species in which the female is homogametic. After meiosis, all the sperm cells carry a Z chromosome, whereas only half of the egg cells carry a Z chromosome and the other half carry a W chromosome. The mechanism of sex determination is not fully understood, but involves differential steroidogenesis. Most urodele amphibians (newts, salamanders) have a ZZ/ZW system, whereas anurans (frogs) have an XX/XY system. Many other vertebrates, however, exhibit *environmental sex determination*. In all crocodilians, many turtles and lizards, and a few snakes, sex determination depends on the temperature at which eggs are incubated. In some turtles, embryos turn into males when they are incubated at low temperatures and into females when incubated at high temperatures. In fact, the female can control the sex ratio of her offspring by choosing a particular thermal habitat for her eggs.

Fish are more complex, exhibiting one of the most diverse collections of reproductive strategies in the animal kingdom. Some have diffuse, indiscernible chromosomes, making sex determination cytologically difficult. Some exhibit the XX/XY system, others have the ZZ/ZW system, others depend on temperature, some rely on complex combinations of genes, and some are *hermaphroditic* (having both sexes in one individual). The latter come in two forms: **simultaneous hermaphrodism,** in which both sets of reproductive systems are active at the same time, and **sequential hermaphrodism,** in which an individual fish changes gender during its life cycle. The simultaneous form is rare; an example is the mangrove killifish *Rivulus marmoratus*. To complicate things even more, the sequential form can be protandry, protogyny, or both-ways. **Protandry,** in which a male becomes a female, is less common; an example is the anemone fish *Amphiprion*. **Protogyny,** in which a female becomes a male, is common in many coral-reef fishes such as wrasses. Both-ways switching, in which a fish can change between genders in both directions, is rare, but has been seen in some reef gobies. The triggers for switching sexes are not known for all species, but for some the trigger is age (body size), whereas in others it is social interactions such as male–male interactions. In anemone fish, if a female is lost a male changes into a female.

Table 16–1 ■ Haploid Number of Chromosomes in the Gametes of Some Animal Species

Cat	19	Mouse	20
Chicken	39	Pigeon	31
Cow	30	Pig	19
Dog	39	Rat	21
Goat	30	Sheep	27
Horse	32	Turkey	41
Mink	15		

In species that exhibit environmental sex determination, the mechanism is still genetic (a result of the particular genes inherited). For example, researchers have recently cloned a gene that seems to regulate sexual development in turtles. Although this gene is present in all turtles of that species, its expression changes with environmental temperature.

In most insects the male has one X (sex) chromosome and is called **heterogametic,** whereas the female has two, **homogametic.** Males are referred to as XO (if only one chromosome in the pair is present) or XY (the Y chromosome being different in size from the X chromosome), whereas the female is XX. One exception to this pattern is found in the Lepidoptera (butterflies and so forth), where the female is heterogametic.

Parthenogenesis, the development of unfertilized eggs, also occurs in many species of insects and some vertebrates. Parthenogenetic development that produces females implies that either the eggs failed to undergo meiosis or that the two cleavage nuclei fused to restore the diploid condition, whereas production of a male involves the loss of an X chromosome. Depending on the time of year, gall wasps and aphids produce either males or females parthenogenetically.

■ Sex differentiation in mammals depends on the presence or absence of masculinizing determinants during critical periods of embryonic development.

Differences between mammalian males and females exist at three levels: genotypic, gonadal, and phenotypic (anatomic and/or physiologic) sex (● Figure 16–6). **Genotypic sex,** which depends on the combination of sex chromosomes at the time of conception, in turn determines **gonadal sex,** that is, whether testes or ovaries develop. The presence or absence of a Y chromosome determines gonadal differentiation in mammals. All mammalian embryos have the potential to differentiate along either male or female lines, because the developing reproductive tissues include the precursors of both sexes. Gonadal specificity in the human appears during the seventh week of intrauterine life, when the indifferent gonadal tissue of a genotypic male begins to differentiate into testes under the influence of the **sex-determining region** of the Y chromosome (*SRY*), the single gene that is responsible for sex determination. This gene triggers a chain of reactions that leads to physical development of a male. The sex-determining region of the Y chromosome "masculinizes" the gonads (induces their development into testes) by stimulating production of **H-Y antigen** by primordial gonadal cells. H-Y antigen, which is a specific plasma membrane protein found only in males, directs differentiation of the gonads into testes. Because genotypic females lack the *SRY* gene and so do not produce H-Y antigen, their gonadal cells never receive a signal for testicular formation, so the undifferentiated gonadal tissue starts developing during the ninth week into ovaries instead.

?

What Genes Control Female Development in Mammals? The dogma of mammalian sexual development has been that females are the "default" state: If the *SRY* region of the Y chromosome is present, the embryo becomes a male; if not, it "automatically" becomes a female. Thus a "master gene" for male development has been known for some time, but not for female development. However, this concept is changing. One gene, *Wnt-4,* was discovered first as a key regulator of kidney development. Yet it also orchestrates female sexual development: It suppresses the production of testosterone and triggers the development of the Müllerian duct, the embryonic duct that develops into the female reproductive tract. In mice with a mutant *Wnt-4* gene, the female does not develop a normal oviduct, uterus, or vagina; instead, she develops internal male structures. How *Wnt-4* and other regulatory genes exert their effects is not yet known.

Phenotypic sex, the apparent anatomic sex of an individual, depends on the genotypically determined gonadal sex. **Sexual differentiation** concerns the embryonic development of the external genitalia and reproductive tract along either male or female lines. As with the undifferentiated gonads, embryos of both sexes have the potential to develop either male or female reproductive tracts and external genitalia. Differentiation into a male-type reproductive system is induced by androgens, and the destruction of the precursor of the female genital reproductive tract by anti-Müllerian hormone, both of which are secreted by the developing testes. **Dihydrotestosterone (DHT),** the reduced form of testosterone, is the most potent androgen.

The absence of these testicular hormones in female fetuses results in the development of a female-type reproductive system. Sexual differentiation can also be variable in certain species. As we noted earlier, although most fish species express two sexes that are fixed at maturation, a number of species are hermaphrodites. Occasionally some species of fish express multiple forms of one sex, either male or female. Some mature parrotfish, for example, come in female and two distinct male forms: a small one (nonmating) and a large "supermale" form (which controls a harem of females and has exclusive mating rights).

Although the male and female external genitalia develop from the same undifferentiated embryonic tissue, this is not the case with the reproductive tracts. Two primitive duct systems—the **Wolffian ducts** and the **Müllerian ducts**—develop in all mammalian embryos. In males, the reproductive tract develops from the Wolffian ducts and the Müllerian ducts degenerate, whereas in females the Müllerian ducts differentiate into the reproductive tract and the Wolffian ducts regress. Because both duct systems are present before sexual differentiation occurs, the early embryo has the potential to develop either a male or a female reproductive tract. Development along male or female lines is determined by the presence or absence of two hormones secreted by the fetal testes—*testosterone* and *Müllerian-inhibiting substance* (Figure 16–6). Testosterone induces development of the Wolffian ducts into the male reproductive tract (epididymis, ductus deferens, ejaculatory duct, and seminal vesicles). This hormone, after being converted into DHT, is also responsible for differentiating the external genitalia into the penis and scrotum. Meanwhile, Müllerian-inhibiting factor causes regression of the Müllerian ducts. In the absence of testosterone and Müllerian-inhibiting factor in females, the Wolffian ducts regress, the Müllerian ducts de-

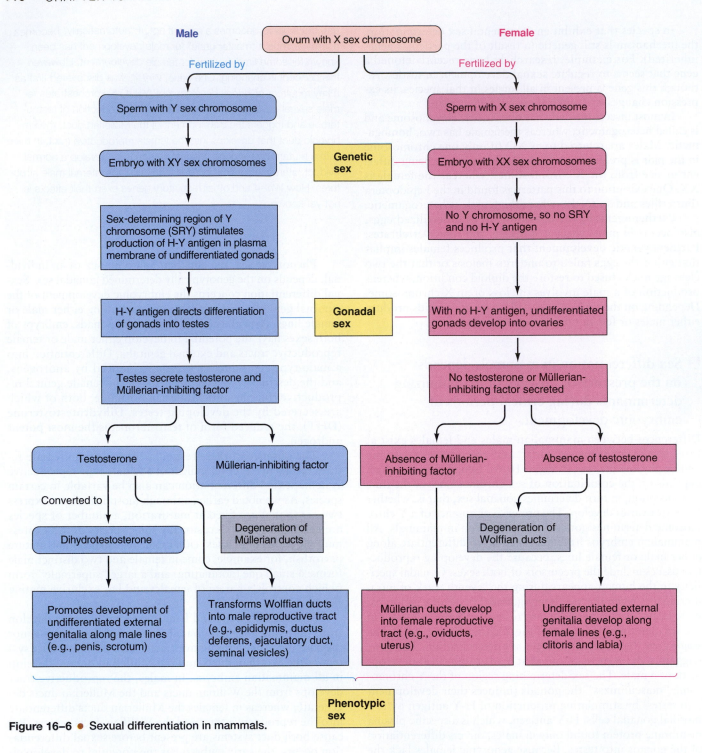

Figure 16–6 ● Sexual differentiation in mammals.

velop into the female reproductive tract (oviducts, uterus, and vagina), and the external genitalia differentiate into the clitoris and labia.

Male and female external genitalia develop from the same embryonic tissue in most animals. In both sexes in mammals, the undifferentiated external genitals consist of a *genital tubercle*, paired *urethral folds* surrounding a urethral groove, and, more laterally, *genital (labioscrotal) swellings* (● Figure 16–7). The **genital tubercle** gives rise to highly sensitive tissue—in males the **glans penis** (the distal end of the penis) and in fe-

males the **clitoris.** The major distinctions between the glans penis and clitoris are the smaller size of the external parts of the clitoris, the lack of spongiosum, and the urethral opening at the end of the glans penis. The urethra is the tube through which urine is transported from the bladder to the outside and also serves in males as a passageway for exit of semen through the penis to the outside. In males, the **urethral folds** fuse around the urethral groove to form the penis, which encircles the urethra. The **genital swellings** similarly fuse to form the scrotum and **prepuce,** a fold of skin (a double fold in the stallion) that extends over the end of the penis and more or less completely covers the glans penis.

Note that the indifferent embryonic reproductive tissue develops into a female structure unless *actively* suppressed by masculinizing factors. Although considered a "default" process, in fact it requires an equally well-orchestrated developmental program and not just a switch to activate it. In the absence of male testicular hormones, a female reproductive tract and external genitalia develop regardless of the genotypic sex of the individual. Ovaries do not need to be present for feminization of the fetal genital tissue. Such a control pattern for determining sex differentiation is appropriate, considering that fetuses of both sexes are exposed to high concentrations of female sex hormones throughout gestation. If female sex hormones exerted influence over the development of the reproductive tract and external genitalia, all fetuses would be feminized. It

is worth noting that masculinization of the brain at least in part is the result of "female" estrogens.

In the usual case, genotypic sex and phenotypic sex are compatible; that is, a genotypic male appears to be a male anatomically and functions as a male, and the same compatibility holds true for females. Occasionally, however, discrepancies occur between genotypic and phenotypic sexes because of errors in sexual differentiation, as the following examples illustrate:

- In cattle carrying twin fetuses, one of each sex with co-circulation of blood between the two, exposure of the female co-twin to anti-Müllerian hormone regresses the genital reproductive tract (that is, Müllerian duct derivatives), but not the urogenital (that is, posterior vagina and external genitalia), with the consequence that these animals are infertile. Such heifers are termed **freemartins**.

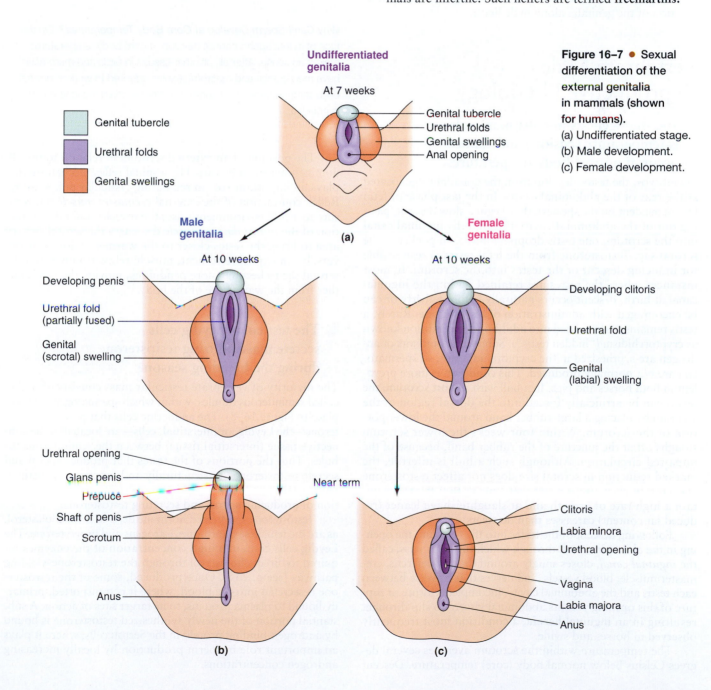

Figure 16–7 ● Sexual differentiation of the external genitalia in mammals (shown for humans).
(a) Undifferentiated stage.
(b) Male development.
(c) Female development.

■ If testes in a genotypic male fail to properly differentiate and secrete hormones, the result is the development of an apparent anatomic female in a genotypic male, which, of course, will be sterile.

■ Because testosterone acts on the Wolffian ducts to convert them into a male reproductive tract but the testosterone derivative DHT masculinizes the external genitalia, a genotypic deficiency of the enzyme that converts testosterone into DHT results in a genotype with male abdominal (that is, undescended) testes and a male reproductive tract but with female external genitalia.

■ The adrenal gland normally secretes a weak androgen, *dehydroepiandrosterone*, in insufficient quantities to masculinize females. However, pathologically excessive secretion of this hormone in a genotypically female fetus during critical developmental stages imposes differentiation of the genitalia along male lines.

Vertebrate Male Reproductive Physiology

■ Spermatogenesis in most mammals is temperature sensitive and cannot occur at normal body temperatures.

In embryos, the testes develop from the gonadal ridge located at the rear of the abdominal cavity. In the last phase of fetal life (dependent on the species), they begin a slow descent, passing out of the abdominal cavity through the **inguinal canal** into the scrotum, one testis dropping into each pocket of the scrotal sac. Testosterone from the fetal testes is responsible for inducing descent of the testes into the scrotum. In most instances where the testes are retained within the inguinal canal at birth, descent occurs naturally before puberty or can be encouraged with administration of testosterone. Rarely, a testis remains undescended into adulthood, a condition known as **cryptorchidism** ("hidden testis"). Subnormal amounts of androgen are synthesized in the cryptorchid testes, so spermatozoa are not normally produced. This condition is more prevalent in both horses and pigs. In a bull with a **short scrotum,** the testes can be artificially forced into the dorsal region of the scrotum by placing a large rubber band around the lower portion of the scrotum. Within four weeks the lower scrotum sloughs off at the juncture of the rubber band, because of the impaired circulation. Although such a bull is infertile, the marked reduction in scrotal size does not affect testosterone concentrations, with the result that animals thus treated maintain a high rate of growth and at slaughter have leaner (reduced fat content) carcasses than do steers.

Following descent of the testes into the scrotum, the opening in the abdominal wall through which the testis pass, called the *inguinal canal,* closes snugly around the vas deferens, cremaster muscle, blood vessels, and nerves that traverse between each testis and the abdominal cavity. Incomplete closure or rupture of this opening permits abdominal viscera to slip through, resulting in an **inguinal hernia,** a condition most frequently observed in horses and swine.

The temperature within the scrotum averages several degrees Celsius below normal body (core) temperature. Descent of the testes into this cooler environment is essential, because spermatogenesis in most birds and mammals is temperature sensitive and cannot occur at normal body temperature. In those mammals without a scrotum, alternative mechanisms associated with temperature regulation are found. For example, whales and elephants have countercurrent exchange mechanisms (retes, p. 695) for cooling the testes despite their abdominal location. In the case of birds, whose testes are also retained within the body cavity, there is evidence that the abdominal air sacs reduce their temperature by 1.5°C, enough to permit sperm viability. In some mammals, including the rat and rabbit, the testes move into and out of the body cavity seemingly at will through an opening in the inguinal canal.

?

Why Can't Sperm Develop at Core Body Temperatures? On the face of it, that sperm cannot develop at core body temperatures makes no sense. After all, all other tissues in birds and mammals have membranes and macromolecules adapted to work at normal body temperatures. Why sperm have not evolved these features is not clear.

The position of an external scrotum in relation to the abdominal cavity can be varied by a spinal reflex mechanism that plays an important role in regulating testicular temperature. Reflex contraction of the external *cremaster muscle* on exposure to a cold environment raises the testicles and the contraction of the *tunica dartos muscle* decreases the scrotal surface area to bring the testes closer to the warmer abdomen. Conversely, on exposure to heat, muscle relaxation permits the scrotal sac to become more pendulous, moving the testes farther from the warm core of the body.

■ The testicular Leydig cells secrete masculinizing testosterone during the breeding season.

The majority of vertebrate testicular mass consists of highly coiled **seminiferous** tubules, within which spermatogenesis takes place (● Figure 16–8). The endocrine cells that produce testosterone—the **Leydig,** or **interstitial, cells**—are located in the connective tissue (interstitial tissue) between the seminiferous tubules. Thus the portions of the testes that produce sperm and secrete testosterone are structurally and functionally distinct. During the breeding season the testes perform the dual function of producing sperm and secreting testosterone.

Testosterone is a steroid hormone derived from cholesterol, as are the female sex hormones estrogens and progesterone. The Leydig cells contain a high concentration of the enzymes required to direct cholesterol through the testosterone-yielding pathway (see p. 255). Once produced, some of the testosterone is secreted into the blood, where it is transported, primarily bound to plasma proteins, to its target sites of action. A substantial portion of the newly synthesized testosterone is bound by androgen-binding protein in the Sertoli cells, where it plays an important role in sperm production by locally increasing androgen concentrations.

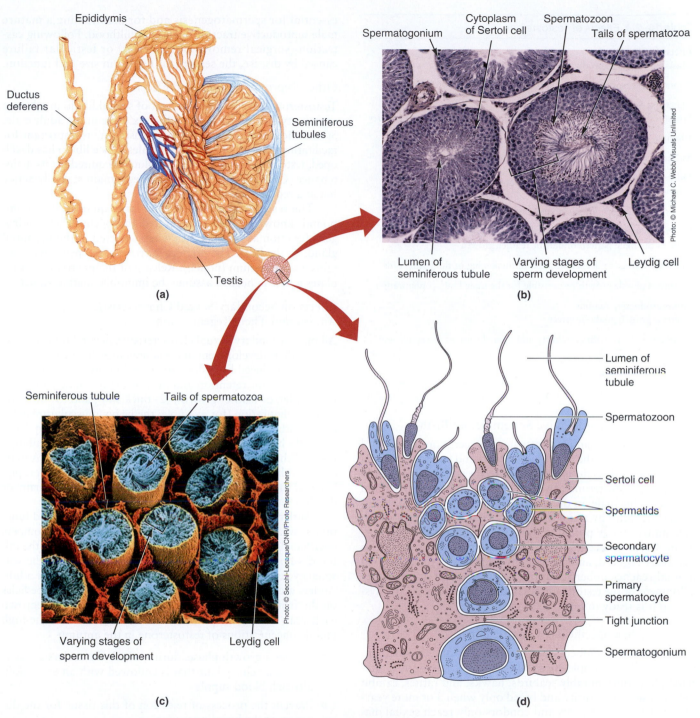

Figure 16–8 ● Testicular anatomy depicting the site of spermatogenesis (mammal).
(a) Longitudinal section of a testis showing the location and arrangement of the
seminiferous tubules, the sperm-producing portion of the testis. (b) Light micrograph
of a cross section of a seminiferous tubule. The undifferentiated germ cells (the
spermatogonia) lie in the periphery of the tubule, and the differentiated spermatozoa
are in the lumen, with the various stages of sperm development in between. (c) Scanning
electron micrograph of a cross section of a seminiferous tubule. (d) Relationship of the
Sertoli cells to the developing sperm cells.

Most but not all of testosterone's actions ultimately are
directed toward ensuring delivery of sperm to the female. The
effects of testosterone can be grouped into four categories:
(1) effects on the reproductive system before birth; (2) effects on
sex-specific tissues after birth; (3) effects on secondary sexual
characteristics; and (4) nonreproductive actions (▮ Table 16–2).

Table 16-2 ▮ Effects of Testosterone
Effects before Birth
Masculinizes the reproductive tract and external genitalia
Promotes descent of the testes into the scrotum of most mammals
Effects on Sex-Specific Tissues
Promotes growth and maturation of the reproductive system at puberty
Essential for spermatogenesis
Maintains the reproductive tract throughout adulthood
Other Reproductive Effects
Develops the sex drive at puberty
Controls gonadotropin hormone secretion
Effects on Secondary Sexual Characteristics
Induces the male pattern of hair or feather growth
Causes the voice to deepen because of thickening of the vocal cords
Promotes muscle growth responsible for the male body configuration
Nonreproductive Actions
Exerts a protein anabolic effect
Promotes bone growth at puberty and then closure of the epiphyseal plates
Induces aggressive behavior

Effects on the Reproductive System before Birth

Before birth of a mammal, testosterone secretion by the fetal testes masculinizes the reproductive tract and external genitalia and promotes descent of the testes into the scrotum, as already described. Testosterone can cross the blood–brain barrier and does so in the male; in the brain it is then converted to the estrogen *estradiol* by a single enzyme (*P450 aromatase*), which defeminizes the hypothalamus so that a male pattern of hormone secretion occurs. After birth, testosterone secretion is markedly reduced, and then slowly increases as the individual animal reaches puberty, supporting the slow, but timely development of the testes and remainder of the reproductive system so it is ready to function at puberty.

Effects on Sex-Specific Tissues after Birth

Puberty is the attainment of sexual maturity and the ability to reproduce. For example, most birds reproduce in the year after hatch, but some, notably seabirds, hawks, and ratites, exhibit deferred sexual maturity and breed only when 3 or more years of age. Large albatrosses and condors only reach sexual maturity between the ages of 9 and 11 years. Even within a particular species there is considerable diversity in the age at which an animal first breeds. In humans puberty usually occurs sometime between the ages of 10 and 14; on the average it begins about two years earlier in females than in males. Puberty is denoted by the first ovulation in females or the presence of sperm in the ejaculate in males.

At male puberty, the Leydig cells are secreting sufficient testosterone, as a result of the decreasing sensitivity of the hypothalamus to the negative-feedback effects of testosterone, to initiate spermatogenesis. Testosterone, usually acting in the dihydro form, governs the growth and maturation of the entire male reproductive system. Ongoing testosterone secretion is essential for spermatogenesis and for maintaining a mature male reproductive tract throughout adulthood. Following castration (surgical removal of the testes) or testicular failure caused by disease, the sex organs regress in size and function.

Other Reproduction-Related Effects

Testosterone governs development of sexual libido in advance of puberty and helps maintain the sex drive in the adult male. Stimulation of this behavior by testosterone is important for facilitating delivery of sperm to females. Once libido has developed, testosterone is no longer absolutely required for its maintenance. Castrated animals sometimes remain sexually active but at a reduced level.

The testes of the boar secrete large quantities of compounds known as C-16 unsaturated androgens. These androgens function as pheromones when secreted in the **preputial glands** of the boar. The prepuce surrounds the penis and contains a diverticulum (pouch). Release of the pheromone stimulates a sow in heat to assume the immobile mating stance.

Effects on Secondary Sexual Characteristics: Antlers and Their Regeneration

All male secondary sexual characteristics depend on testosterone for their development and maintenance. These male characteristics induced by testosterone are important for mate acquisition and/or retention and include (1) the male pattern of hair, antler, or feather growth; (2) thick skin; and (3) the male body configuration (for example, the heavy musculature associated with the necks of deer in rut) as a result of protein deposition. In human males, (4) a deep voice is caused by enlargement of the larynx and thickening of the vocal cords. A male castrated before puberty (human: eunuch; sheep: wether; pig: barrow; cattle: steer; and so forth) does not mature sexually or develop secondary sexual characteristics.

Antlers are a characteristic feature of male deer and function as weapons and status symbols when rival males compete over females, food, and territory. Testosterone controls the initial development of the antlers at puberty, as well as the seasonal cycle of casting and regrowth of the antlers in the adult. Antlers are actually bony structures that develop from pedicles on the frontal bones of the skull. They are cast and regrown each year in association with the breeding season. The high blood concentrations of testosterone in the autumn

- Trigger a growth phase, during which the antlers are covered in a velvety hair that is endowed with an exceptionally rich blood supply
- Prevent the process of rejection of this tissue for the duration of the breeding season

Once the antlers become fully calcified, the overlying skin peels off: The mature antlers are now dead, insensitive structures that function as weapons for the rutting season. At the end of the breeding season, as the testes regress and concentrations of testosterone decline, the dead antlers are cast, and new ones are regenerated from the living pedicles.

Nonreproductive Actions

Testosterone exerts several important effects not directly related to reproduction. It induces aggressive behavior. It has a general protein anabolic (synthesis) effect and promotes greater muscling and bone growth. These effects play an indirect role in reproduction, by enhancing the ability of males to compete for females. Ironically, testosterone not only stimulates bone growth but eventually prevents further growth by sealing the growing ends of the long bones (that is, ossifying, or "closing," the epiphyseal plates—see p. 272).

■ Spermatogenesis yields an abundance of highly specialized, mobile sperm.

The basic pattern of spermatogenesis is functionally similar among all animals. In fact, many of the original descriptions of spermatogenesis were obtained from insect models. Two functionally important cell types are present in the sperm-producing seminiferous tubules (Figure 16–8a): *germ cells,* most of which are in various stages of sperm development, and *Sertoli cells,* which provide crucial support for spermatogenesis (Figure 16–8b, c, and d). **Spermatogenesis** is a complex process by which relatively undifferentiated germ cells, the **A-type spermatogonia** (each of which contains a diploid complement of $2N$ chromosomes), proliferate and are converted into ex-

tremely specialized, motile spermatozoa (sperm), each bearing a randomly distributed haploid set of $1N$ chromosomes.

Microscopic examination of a seminiferous tubule reveals layers of germ cells in an anatomic progression of sperm development, starting with the least differentiated in the outer layer and moving inward through various stages of division to the lumen, where the highly differentiated sperm are ready for exit from the seminiferous epithelium (*spermiation*) (Figure 16–8b, c, and d). In bulls, spermatogenesis takes 64 days for development from a spermatogonium to a mature sperm. In normal male animals, a large number of spermatozoa are produced each day: approximately 6.0×10^9 in the bull and 16.5×10^9 in the boar. Spermatogenesis encompasses three major stages: *mitotic proliferation (spermatocytogenesis), meiosis,* and *packaging (spermiogenesis)* (● Figure 16–9).

Mitotic Proliferation

Spermatogonia located in the outermost layer of the tubule continuously divide mitotically, with all new cells bearing the full

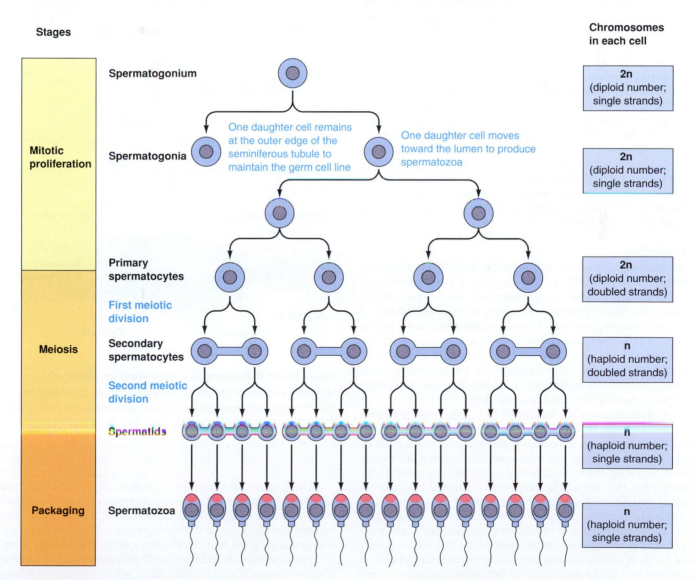

Figure 16–9 ● Spermatogenesis.

complement of 2*n* chromosomes that are identical to those of the parent cell. Such proliferation provides a continual supply of new germ cells. Following mitotic division of a spermatogonium, one of the daughter cells remains at the outer edge of the tubule as an undifferentiated spermatogonium, thus maintaining the germ cell line. The other daughter cell starts moving toward the lumen while undergoing the various steps required to form sperm, which are released into the lumen. The sperm-forming daughter cell divides mitotically several times more to form four identical **primary spermatocytes.** After the last mitotic division, the primary spermatocytes enter a resting phase during which the chromosomes are duplicated and the doubled strands remain together in preparation for the first meiotic division.

Meiosis

During meiosis, each primary spermatocyte (with a diploid number of 2*N*, 4*n* doubled chromosomes with pairing of the homologous chromosomes) forms two **secondary spermatocytes** (each with a haploid number of 1*N*, 2*n* doubled chromosomes), referred to as the *reductional division* because the number of sets of homologous chromosomes has been reduced, even though there are still two copies of each. The secondary spermatocytes then divide during the second meiotic division separating the two copies of each chromosome into daughter cells during the first meiotic division, yielding four **spermatids** (each with 1*N*, 1*n* chromosomal complement).

No further division takes place beyond this stage of spermatogenesis. Each spermatid is remodeled into a single spermatozoon. Because each spermatogonium mitotically produces many primary spermatocytes and each primary spermatocyte meiotically yields four spermatids, the spermatogenic sequence produces tremendous numbers of spermatozoa each time a spermatogonium initiates this process. Usually, however, some cells are lost at various stages, so the efficiency of spermatogenesis is rarely maximal.

Packaging

After meiosis, spermatids still resemble undifferentiated spermatogonia structurally, except for their 1*N* complement of chromosomes. Production of extremely specialized, motile spermatozoa from spermatids requires extensive remodeling of cellular elements, a process known as **spermiogenesis.** Sperm are essentially "stripped down" cells in which most of the cytosol and any organelles not needed for the task of delivering the sperm's genotypic information to an ovum have been extruded. Thus sperm travel lightly, taking only the bare essentials to accomplish fertilization with them.

A **spermatozoon** has four parts (● Figure 16–10): a head, an acrosome, a midpiece, and a tail. The **head** consists primarily of the nucleus, which contains the sperm's complement of genotypic information. The **acrosome,** an enzyme-filled vesicle that caps the tip of the head, is used as an "enzymatic drill" for penetrating the ovum. The acrosome is formed by aggregation of vesicles comprising the endoplasmic reticulum–Golgi complex before these organelles are discarded. A long, whip-like tail that grows out of one of the centrioles provides motility for the spermatozoon. Movement of the tail, which occurs as a result of relative sliding of its constituent microtubules, is powered by energy generated by the mitochondria concentrated within the **midpiece** of the sperm. (Insect sperm lack a midpiece, so the mitochondria are in the tail.)

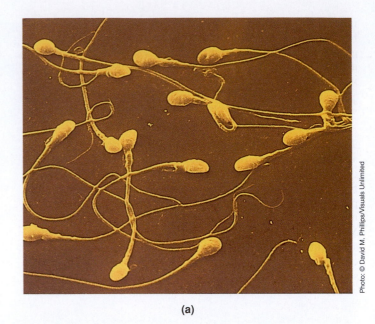

Photo: © David M. Phillips/Visuals Unlimited

(a)

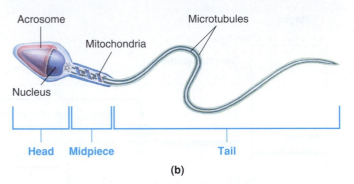

Acrosome   Microtubules
Mitochondria
Nucleus

Head   Midpiece   Tail

(b)

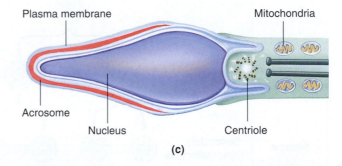

Plasma membrane   Mitochondria

Acrosome
Nucleus   Centriole

(c)

Figure 16–10 ● Anatomy of a spermatozoon (mammal).
(a) A phase-contrast photomicrograph of human spermatozoa.
(b) Schematic representation of a spermatozoon in "frontal" view.
(c) Longitudinal section of the head portion of a spermatozoon in "side" view.

Until sperm maturation is complete, the developing germ cells arising from a single primary spermatocyte remain joined by cytoplasmic bridges. These connections, which result from incomplete cytoplasmic division, permit cytoplasm to be exchanged among the four developing sperm. This linkage is important in mammals because the X chromosome, but not the Y chromosome, contains genes that code for cellular products essential for development of the sperm. (Whereas the large X chromosome contains several thousand genes, the small Y chromosome has only a few dozen, the most important of which are the *SRY* gene and others that play critical roles in male fertility.) During meiosis, half the sperm receive an X and the other half a Y chromosome. If it were not for the sharing of cytoplasm so that all the haploid cells are provided with the products coded for by X chromosomes until sperm development is complete, the Y-bearing, male-producing sperm would not be able to develop and survive.

■ Throughout their development, vertebrate sperm remain intimately associated with Sertoli cells.

In addition to the spermatogonia and developing sperm cells, the seminiferous tubules also house the **Sertoli cells.** The Sertoli cells form a ring that extends from the outer basement membrane to the lumen of the tubule. Each Sertoli cell spans the entire distance from the outer membrane to the fluid-filled lumen (Figure 16–8b and d). Adjacent Sertoli cells are joined by tight junctions (see p. 63) at a point slightly beneath the outer membrane. Spermatogonia are tucked between the Sertoli cells at the outer perimeter of the tubule in the spaces between the basement membrane and the tight junctions.

During spermatogenesis, developing sperm cells arising from spermatogonial mitotic activity pass through the tight junctions, which transiently separate to make a path for them, then migrate toward the lumen in intimate association with the adjacent Sertoli cells. The cytoplasm of the Sertoli cells envelops the migrating germ cells, which remain buried within these cytoplasmic recesses throughout their development.

The supportive Sertoli cells perform the following functions essential for spermatogenesis:

- The tight junctions between adjacent Sertoli cells form a blood–testis barrier. Because this barrier prevents blood-borne substances from passing between the cells to gain entry to the lumen of the seminiferous tubule, only selected molecules that can pass through the Sertoli cells (such as FSH, the gonadotropin that stimulates the Sertoli cells) reach the intratubular fluid. As a result, the composition of the intratubular fluid varies considerably from that of the blood. The unique composition of this fluid that bathes the germ cells is believed to be critical for later stages of sperm development. The blood–testis barrier also prevents the antibody-producing cells in the extracellular fluid entering the seminiferous tubule, thus preventing the formation of antibodies against the highly differentiated spermatozoa.

- Because the secluded developing sperm cells do not have direct access to blood-borne nutrients, the Sertoli cells provide nourishment for them.

- The Sertoli cells have an important phagocytic function. They engulf the cytoplasm extruded from the spermatids during their remodeling and destroy defective germ cells that fail to successfully complete differentiation.

- Into the lumen the Sertoli cells secrete seminiferous tubule fluid, which "flushes" the released sperm from the tubule into the epididymis for storage and further processing.

- Sertoli secretions include *androgen-binding protein.* As the name implies, this protein binds androgens (that is, testosterone), thus maintaining a very high level of this

hormone within the lumen of the seminiferous tubules. This high local concentration of testosterone is essential for sustaining sperm production. Androgen-binding protein is necessary to retain testosterone within the lumen, because this steroid hormone is lipid soluble and could easily diffuse across the plasma membranes and leave the lumen.

- The Sertoli cells are the site of action for control of spermatogenesis by both testosterone and follicle-stimulating hormone (FSH). The Sertoli cells themselves release another hormone, *inhibin,* which acts in negative-feedback fashion to regulate FSH secretion.

■ LH and FSH from the anterior pituitary control testosterone secretion and spermatogenesis.

Vertebrate testes are controlled by the two gonadotropic hormones secreted by the anterior pituitary, *luteinizing hormone (LH)* and *follicle-stimulating hormone (FSH),* which are named for their functions in females (see p. 267).

Feedback Control of Testicular Function

LH and FSH act on separate components of the testes (● Figure 16–11). Luteinizing hormone acts on the Leydig (interstitial) cells and regulates testosterone secretion, accounting for its historic name in males—*interstitial cell–stimulating hormone*

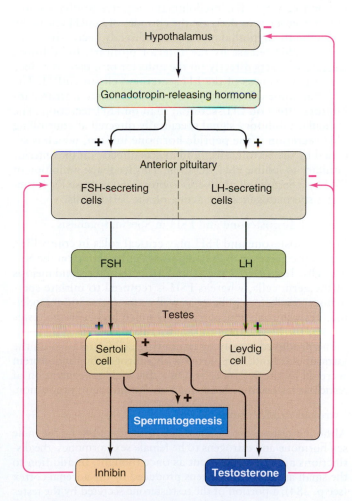

Figure 16–11 ● Control of testicular function.

(ICSH). Follicle-stimulating hormone acts on the Sertoli cells, to enhance spermatogenesis. (There is no alternative name for FSH in males.) Secretion of both LH and FSH from the anterior pituitary is stimulated in turn by a single hypothalamic hormone, *gonadotropin-releasing hormone (GnRH)* (see p. 269). A novel gonadotropin-inhibiting hormone (GnIH) has been localized in the hypothalamic paraventricular nucleus of birds which has been suggested to regulate gonadotropin release.

GnRH is released from the hypothalamus in secretory bursts, with no secretion occurring in between. In seasonally breeding vertebrates, the rate at which GnRH is secreted decreases during the nonbreeding season, as does the amount associated with each burst. Because GnRH stimulates the gonadotropic hormone secretory cells in the anterior pituitary, this pulsatile pattern of hypothalamic secretion results in similar episodic bursts in LH and FSH secretion.

Even though GnRH stimulates both LH and FSH secretion, the blood concentrations of these two gonadotropic hormones do not always parallel each other, for two reasons. First, LH is removed from the blood more rapidly between the secretory bursts than is the more slowly metabolized FSH, so the pulsatile variations in blood concentrations of LH are much more pronounced than are those of FSH. Second, two other regulatory factors besides GnRH—*testosterone* and *inhibin*—differentially influence the release of LH and FSH.

Testosterone, the product of LH stimulation of the Leydig cells, acts in negative-feedback fashion to inhibit LH secretion in two ways. The predominant negative-feedback effect of testosterone is to decrease the episodes of GnRH release by acting on the hypothalamus, thus indirectly decreasing both LH and FSH release by the anterior pituitary. In addition, testosterone acts directly on the anterior pituitary to reduce the responsiveness of the LH secretory cells to GnRH. The latter action explains why testosterone exerts a greater inhibitory effect on LH secretion than on FSH secretion. The testicular inhibitory signal specifically directed at controlling FSH secretion is the peptide hormone inhibin, which is secreted by the Sertoli cells. Inhibin acts directly on the anterior pituitary to inhibit FSH secretion. This feedback inhibition of FSH by a Sertoli cell product is appropriate, because FSH stimulates spermatogenesis by acting on the Sertoli cells.

Roles of Testosterone and FSH in Spermatogenesis

Both testosterone and FSH play critical roles in controlling spermatogenesis, each exerting its effect by acting on the Sertoli cells. Testosterone is essential for both mitosis and meiosis of the germ cells, whereas FSH is required to initiate spermatogenesis before puberty as well as for spermatid remodeling. Testosterone concentration is much higher in the testes than in the blood, because a substantial portion of this hormone produced locally by the Leydig cells is retained in the intratubular fluid, complexed with androgen-binding protein secreted by the Sertoli cells. Only this high concentration of testicular testosterone is adequate to sustain sperm production.

Estrogen Production in Males

Although testosterone is classically considered to be the male sex hormone and estrogens to be female sex hormones, the distinctions are not as clear-cut as once thought. In addition to the small amount of estrogens produced by the adrenal cortex (see p. 284), a portion of the testosterone secreted by the testes

is converted to estradiol outside of the testes by the enzyme *aromatase,* which is widely distributed in the male reproductive tract as well as the male brain. The boar and the stallion testicle both have significant aromatase activity in the Sertoli cells and as a result secrete substantial quantities of estrogen.

Estrogens are also produced in adipose tissue in both sexes. Estrogen receptors have been identified in mammalian male brain, testes, prostate, bone, and elsewhere. Recent findings suggest that estrogens play an essential role in male reproductive health, including being important in spermatogenesis and contributing to normal sexuality. Also, it probably contributes to bone homeostasis. The depth, breadth, and mechanisms of action of estrogens in males are only beginning to be explored. (Likewise, in addition to the weak androgenic hormone DHEA produced by the adrenal cortex in both sexes, the ovaries in females also secrete a small amount of testosterone, the functions of which remain unclear.)

■ Gonadotropin-releasing hormone activity increases at puberty.

Even though the fetal testes secrete testosterone, which directs masculine development of the reproductive system, after birth the testes become quiescent, followed by a gradual increase in the secretion of testosterone until puberty in response to the increasing secretion of gonadotropins. The prepubertal delay in the onset of reproductive capability allows time for the animal to mature physically enough to handle offspring, for example, mastering the skills necessary to obtain sufficient food items.

Early in the prepubertal period, GnRH secretory pulses occur at about 4-hour intervals, causing brief increases in LH secretion and accordingly, testosterone secretion. The magnitude of episodic GnRH secretion gradually increases until the adult pattern of GnRH, FSH, LH, and testosterone secretion is established. Under the influence of the rising concentrations of testosterone, particularly in the peripubertal period, the physical changes that encompass the secondary sexual characteristics and reproductive maturation become evident.

The low frequency of GnRH activity during the prepubertal period appears due to active inhibition of GnRH release by both hormonal and neural mechanisms. Before puberty, the hypothalamus is extremely sensitive to the negative-feedback actions of testosterone, so the very small amounts of testosterone produced by the prepubertal testes can inhibit GnRH release. During the prepubertal period the hypothalamus becomes less sensitive to feedback inhibition by testosterone. Because the low concentrations of testosterone no longer suppress the hypothalamus, GnRH and gonadotropic hormone concentrations rise until sufficient to induce spermatogenesis and therefore puberty.

Seasonal Regression and Recrudescence of Testis Function

After puberty in seasonally breeding animals other factors, in turn, influence the timing of GnRH release and the recrudescence (reawakening) of testicular function. As we have discussed previously, photoperiod at temperate latitudes is the most important factor determining the onset and termination of the breeding cycle. Light is initially perceived in the retina and transferred by the optic nerve to a specific area of the hypothalamus known as the *suprachiasmatic (SCN) nucleus (medio-*

basal hypothalamus in birds; see p. 263). From the SCN a nerve tract travels to the *superior cervical ganglion (SCG)*. From here a synapse is made with cells in the pineal gland. **Pinealocytes** in turn cease the secretion of the hormone *melatonin,* which is normally secreted during the period of dark (**scotophase**). During the daylight hours (**photophase**), the light detected by the retinal cells of the eye, through the SCG, activates neurons that limit the release of melatonin from the pineal gland. However, during the dark period this inhibitory pathway is shut down by a reduction in the firing rate of nerves in the light-sensitive areas of the retina. The amount of melatonin released from the pinealocytes is a function of the duration of the period of darkness, although data in experimental animals indicate that receiving only brief flashes of light, separated by a given number of hours of dark (skeleton photoperiod), results in similar quantities of melatonin released. Under normal conditions in short-day breeders, sufficient quantities of melatonin are released only when a critical length of the dark period is reached, which then triggers an increase in GnRH release from the hypothalamus. In seasonal breeders, termination of the breeding season is associated with the development of **photorefractoriness,** the inability of the long day lengths to sustain the breeding cycle. In birds this is believed to be the result of an increase in GnIH associated with a decline in the release of GnRH. GnIH is also elevated in birds incubating an egg and is believed to be responsible for suppressing gonadotropic function at this time.

■ The ducts of the reproductive tract stores and concentrates sperm and increases their fertility.

The remainder of the male reproductive system (besides the testes) is designed to deliver sperm to the female reproductive tract. Essentially, it consists of (1) a tortuous pathway of tubes that transport sperm from the testes to the outside of the body; (2) several glands, which contribute secretions that are important to the viability and motility of the sperm; and (3) the penis, which is designed to penetrate and deposit the sperm within the vagina (or in the uterus, as in the stallion and boar) of the female, whereas in birds the pelvic urethra ends in either a cloaca or a hemipenis. We next examine each of these parts in greater detail, beginning with the reproductive tract.

Functions of the Epididymis and Ductus Deferens

The epididymis and ductus deferens perform several important functions. They serve as the sperm's exit route from the testis. As they leave the testis, the sperm can neither move nor fertilize. They gain both capabilities during their passage through the epididymis. This maturational process is stimulated by the testosterone retained within the tubular fluid bound to androgen-binding protein. Sperm's capacity to fertilize is enhanced even further by exposure to secretions of the female reproductive tract, which remove seminal plasma proteins from the surface of the sperm cell. This enhancement of sperm's capacity as a result of the removal of seminal plasma proteins in the female reproductive tracts is known as **capacitation.** The epididymis also concentrates the sperm 100-fold by absorbing most of the fluid that enters from the seminiferous tubules. The maturing sperm are slowly moved through the epididymis into the ductus deferens by rhythmic contractions of the smooth muscle in the walls of these tubes. The tail of the epididymis serves as an important site for sperm storage. Because the tightly packed sperm are relatively inactive and their metabolic needs are accordingly low, they can be stored in the tail of the epididymis for many days, even though they have no nutrient blood supply and are nourished only by simple sugars present in the tubular secretions.

The epididymis, because of its folded arrangement, is comma-shaped and loosely attached to the rear surface of each testis (Figures 16–3, p. 711, and 16–8a, p. 719) and ranges in length from 30 to 60 mm depending on the species. During the breeding season the cells within the epididymis undergo hypertrophy and secrete a seminal fluid. Once sperm are produced in the seminiferous tubules, they are swept into the epididymis as a result of the pressure created by the continual secretion of tubular fluid by the Sertoli cells. As a result of coalescence of the efferent ductules leaving the rete testis tubules, the epididymal ducts from each testis converge to form a large, thick-walled, muscular duct—the **ductus (vas) deferens.** The ductus deferens from each testis passes up out of the scrotum and through the inguinal canal into the pelvic cavity, where it eventually empties into the urethra at the neck of the bladder (Figure 16–3). The **ampulla** is an enlarged portion of the ductus deferens located immediately before the entrance into the urethra or the cloaca. An enlarged muscle layer characterizes it, in association with an increase in the size of the lumen. The ampulla is present in the primate, bull, ram, stallion, and cock, but is reduced or absent in the boar, dog, and fox. Subsequently, the urethra carries sperm out of the penis during **ejaculation,** the forceful expulsion of semen from the body.

■ The accessory sex glands contribute the bulk of the semen.

Several accessory sex glands—the seminal vesicles and prostate—empty their secretions into the duct system before it joins the urethra (Figure 16–3, p. 711). A pair of saclike *seminal vesicles* empty into the last portion of the two ductus deferens, one on each side. The short segment of duct that passes beyond the entry point of the seminal vesicle to join the urethra constitutes the *ejaculatory duct.* The *prostate* is a large single gland that completely surrounds the ejaculatory ducts and urethra (some species have only a small *bulb,* as in humans, whereas others have both a bulb and a longer *disseminate* prostrate that surrounds a longer portion of the pelvic urethra. Another pair of accessory sex glands, the *bulbourethral glands,* drains into the urethra after it has passed through the prostate, just before it enters the penis. Numerous mucus-secreting glands are located along the length of the urethra.

Semen

During ejaculation, the accessory sex glands contribute secretions that provide support for the continuing viability of the sperm inside the female reproductive tract (in animals with internal fertilization) or in water (such as fish semen). These secretions constitute the bulk of the semen, which consists of a mixture of accessory sex-gland secretions, sperm, and mucus. In some mammals, such as the boar and stallion, the seminal fluid has special coagulation properties that plug the female reproductive tract and minimize loss of spermatozoa after ejaculation. Sperm make up only a small percentage of the total ejaculated fluid. In rodents, the plug is lost from the vagina, signaling that successful copulation has occurred.

Functions of the Male Accessory Glands

Although the accessory sex gland secretions are not absolutely essential for fertilization, they do make species-specific contributions that greatly facilitate the fertilization process in mammals:

- The **seminal vesicles** or **vesicular glands** are paired glands that empty into the pelvic urethra and (1) supply fructose, which serves as the primary energy source for ejaculated sperm; (2) secrete *prostaglandins,* which stimulate contractions of the smooth muscle in both the male and female reproductive tracts, helping to transport sperm from their storage site in the male to the site of fertilization in the female oviduct; (3) provide more than half the semen volume, which helps wash the sperm into the urethra and also dilutes the thick mass of sperm, thus enabling them to develop motility; and (4) secrete fibrinogen, a precursor of fibrin, which forms the meshwork of a clot (see p. 368). Seminal vesicles vary significantly between species but are absent in the dog, fox, and birds.

- The **prostate gland** (1) secretes an alkaline fluid that neutralizes the acidic vaginal secretions, an important function because sperm are more viable in a slightly alkaline environment; and (2) provides clotting enzymes and fibrinolysin. The prostate clotting enzymes act on fibrinogen from the seminal vesicles to produce fibrin, which "clots" the semen, thus helping to keep the ejaculated sperm in the female reproductive tract during withdrawal of the penis. Shortly thereafter, the seminal clot is broken down by *fibrinolysin,* a fibrin-degrading enzyme from the prostate, releasing motile sperm within the female tract. The prostate varies in structure from branched tubular to tubuloalveolar. The prostate gland is large in the boar, comparatively smaller in the bull, and absent in the smaller ruminants. It is absent in the cock but is extremely well developed in the dog and fox.

- During sexual arousal, the **bulbourethral glands** secrete a mucuslike substance that causes the seminal plasma to coagulate following ejaculation. These glands are particularly enlarged in the boar, but are absent in the fox, dog, and birds.

Before turning our attention to the act of delivering sperm to the female, we are briefly going to digress and discuss the diverse roles of prostaglandins, which were first discovered in semen but are abundant throughout the body.

■ Prostaglandins are ubiquitous, locally acting chemical messengers.

Although prostaglandins were first identified in the semen and were believed to be of prostate gland origin (hence their name, even though they are actually secreted into the semen by the seminal ves-

icles), their production and actions are by no means limited to the reproductive system. These 20-carbon fatty acid derivatives are among the most ubiquitous *paracrine* chemical messengers in vertebrates (p. 86). They are produced in virtually all tissues from arachidonic acid, a fatty acid constituent of the phospholipids within the plasma membrane. Prostaglandins (and other closely related arachidonic-acid derivatives, namely, *prostacyclins, thromboxanes,* and *leukotrienes*) are among the most biologically active compounds known. On appropriate stimulation, arachidonic acid is split from the plasma membrane by a cytosolic enzyme and then is converted into the appropriate prostaglandin, which acts locally within or near its site of production. After prostaglandins act, they are inactivated rapidly by local enzymes before they gain access to the blood, or if they do reach the circulatory system, they are swiftly degraded on their first pass through the lungs so that large quantities are not dispersed through the systemic arterial system.

Prostaglandins are designated as belonging to one of nine groups—PGA, PGB, PGC, PGD, PGE, PGF, PGG, PGH, or PGI (prostacyclin)—according to structural variations in the five-carbon ring that they contain at one end (● Figure 16–12).

Figure 16–12 ● **Structure of prostaglandins.** (a) General features and nomenclature. (b) Some examples.

(*Source:* M. K. Campbell & S. O. Farrell, *Biochemistry,* 4th ed., Belmont, CA: Brooks/Cole, Figure 7.33, p. 219)

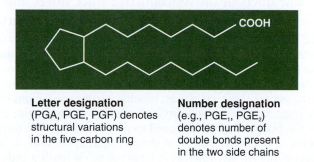

Letter designation
(PGA, PGE, PGF) denotes
structural variations
in the five-carbon ring

Number designation
(e.g., PGE_1, PGE_2)
denotes number of
double bonds present
in the two side chains

(a)

PGE_1 $PGF_{1\alpha}$ PGE_2 $PGF_{2\alpha}$

(b)

Within each group, prostaglandins are further identified by the number of double bonds present in the two side chains that project from the ring structure (for example, PGE_1 has one double bond and PGE_2 has two double bonds).

Prostaglandins exert a variety of effects. Not only are slight variations in prostaglandin structure accompanied by profound differences in biological action, but the same prostaglandin molecule may exert opposite effects in different tissues. Besides enhancing sperm transport in semen, these abundant chemical messengers are known to or suspected to exert other actions in the female reproductive system and in the respiratory, urinary, digestive, nervous, and endocrine systems, in addition to affecting platelet aggregation, fat metabolism, and inflammation. Salicylate drugs such as aspirin inhibit the production of prostaglandins (see p. 437).

Next, before considering the female in greater detail, we examine the means by which males and females come together to accomplish reproduction.

The male sex act is characterized by erection and ejaculation.

The mammalian penis is both erectile and extendible. Consisting of three parts, the **root, body** or **shaft,** and **glans,** the body normally constitutes the major portion of the organ. In some mammals (such as mink, raccoons, bats, and moles, but not humans, horses, bulls, rams, boars, bucks, and most primates) the penis contains a penile bone called the **os penis** or **baculum** (● Figure 16–13). The free end of the penis, the glans, is richly supplied with nerves. In some species the penis is highly erec-

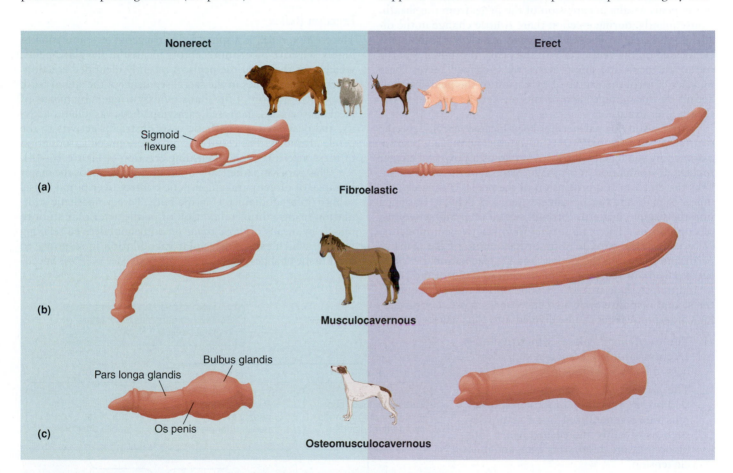

Figure 16–13 ● **Classification of penile structures.** (a) The penises of rams, goats, and boars are fibroeleastic. Erection results in a straightening of the sigmoid flexure. Engorgement of the corpus cavernosum with blood distends and elongates this class of penis. (b) The penis of cats and horses is musculocavernous. Erection results in a more complex change in the position, length, and stiffness of the penis compared to fibroelastic forms. (c) The musculocavernous penis of dogs contains a central bony portion (*os penis*) that aids in intromission into the female's vagina with only a partial erection. Erection is completed in the female's vagina. The local engorgement of the *bulbus glandis* is prolonged by occlusion of the penile venous return by the vaginal constrictor muscles. This contraction maintains and lengthens the duration of the intravaginal erection. Intravaginal engorgement associated with dilation of the penile bulb prevents withdrawal of the penis from the vagina, leading to the so-called tie between the two.

(*Source:* Ruckebusch, Phaneuf, & Dunlop, *Physiology of Small and Large Animals,* B. C. Decker, Figure 55–1, p. 582)

tile. In the boar the penis has the unique feature of existing as a left-handed corkscrew, which actually penetrates, with rotary motion, into the cervix and locks in place, providing the pressure needed for the stimulation of ejaculation. In lizards and snakes, the penis is paired and represents an extension of the cloaca. During copulation this **hemipenis** is everted into the cloaca of the female. To sustain intromission in the female, the hemipenis contains spiny ridges. The body of the normally flaccid penis can become erect as a consequence of testicular vasocongestion (engorgement with blood).

Some mammals (boars, bulls, and rams) have a penis containing an increased quantity of connective tissue relative to the amount of erectile tissue. The penis of these animals contains a **sigmoid flexure,** an S-shaped configuration along the shaft of the penis (Figure 16–13). The straightening of the curve of the penis results in extension of the penis from the sheath. Consequently, during erection there is little change in the diameter of the penis. The **retractor penis muscle** is a paired muscle located on the ventral face of the penis. It originates at the root of the penis and shunts across the S-shaped curve. These muscles control the degree of extension of the penis by their action on the sigmoid flexure and retract the penis into the sheath after extension.

After a suitable partner has been located, species-specific courtship behavior (p. 710) is initiated. The mating posture assumed by the female (**lordosis**) serves as a powerful trigger for sexual stimulation in the male. Arousal of the male stimulates the erection and protrusion of the penis. Ultimately, the union of male and female gametes requires delivery of semen into the vagina or uterus through **sexual intercourse (coitus,** or **copulation**). The *male sex act* involves two components: (1) **erection,** the straightening or hardening of the penis to permit its entry into the vagina, and (2) **ejaculation,** or forceful expulsion of semen into the urethra and out of the penis. In addition to these strictly reproduction-related components, the **sexual response cycle** also encompasses broader physiological responses that can be divided into four phases:

1. The *excitement phase,* which includes erection (and heightened sexual awareness in humans, at least)

2. The *plateau phase,* which is characterized by intensification of these responses, plus more generalized body responses, such as steadily increasing heart rate, blood pressure, respiratory rate, and muscle tension

3. The *orgasmic phase,* which includes ejaculation as well as other responses that culminate the mounting sexual excitement

4. The *resolution phase,* which returns the genitalia and body systems to their prearousal state

■ Erection is accomplished by penis vasocongestion.

In those species in which the penis consists almost entirely of **erectile tissue,** the body consists of three columns of sponge-like vascular spaces extending the length of the organ. In the absence of sexual excitation, the erectile tissues contain little blood, because the arterioles that supply these vascular cham-

bers are constricted and venous drainage is maximal. As a result, the penis remains small and flaccid. During sexual arousal, these arterioles reflexly dilate and the erectile tissue fills with blood, causing the penis to enlarge both in length and width and to become more rigid. In the stallion, for example, these vascular spaces are particularly enlarged such that, during erection, considerable increases in the penal size are achieved as a result of accumulation of blood in these spaces. In contrast, the vascular spaces in the bull, ram, and boar are comparatively small except in the distal bend of the sigmoid flexure. Additional buildup of blood and further enhancement of erection are achieved by a reduction in venous outflow. The veins that drain the erectile tissue are compressed as a result of engorgement and expansion of the vascular spaces brought about by increased arterial inflow.

Erection Reflex

The erection reflex is a spinal reflex triggered by stimulation of highly sensitive mechanoreceptors located in the glans penis, which caps the tip of the penis. A recently identified **erection-generating center** lies in the sacral segments of the spinal cord. Tactile stimulation of the glans reflexly triggers by means of local interneurons, increased parasympathetic vasodilator activity, and decreased sympathetic vasoconstrictor activity to the penile arterioles. The result is rapid, pronounced vasodilation of these arterioles and an ensuing erection (● Figure 16–14).

This parasympathetically induced vasodilation is the major instance of direct parasympathetic control over blood vessel caliber (internal diameter) in the body. Parasympathetic stimulation brings about relaxation of penile arteriolar smooth muscle by means of *nitric oxide* (endothelium-derived relaxing factor), which causes arteriolar vasodilation in response to local tissue changes elsewhere in the body (see p. 406). Specif-

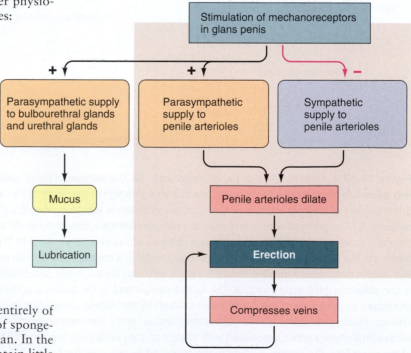

Figure 16–14 ● Erection reflex.

ically, nitric oxide (NO) activates a membrane-bound enzyme, guanylate cyclase, within nearby arteriolar smooth muscle cells. This enzyme activates cyclic guanosine monophosphate (cGMP), an intracellular second messenger similar to cAMP (see p. 90). Cyclic GMP in turn leads to relaxation of the penile arteriolar smooth muscle, bringing about a pronounced local vasodilation. Arterioles are typically supplied only by sympathetic nerves, with increased sympathetic activity producing vasoconstriction and decreased sympathetic activity resulting in vasodilation. Concurrent parasympathetic stimulation and sympathetic inhibition of penile arterioles accomplish vasodilation more rapidly and in greater magnitude than is possible in other arterioles supplied only by sympathetic nerves. At the same time, parasympathetic impulses promote secretion of lubricating mucus from the bulbourethral glands and the urethral glands in preparation for coitus.

Recent research has led to the discovery of numerous regions throughout the brain that can influence the male sexual response. The erection-influencing brain sites appear extensively interconnected and function as a unified network to either facilitate or inhibit the basic spinal erection reflex, depending on the momentary circumstances. Failure to achieve an erection may occur despite appropriate stimulation, as a result of inhibition of the erection reflex by higher brain centers.

■ Ejaculation includes emission and expulsion.

The second component of the male sex act is ejaculation. Like erection, ejaculation is accomplished by a spinal reflex. The same types of tactile and psychic stimuli that induce erection cause ejaculation when the level of excitation intensifies to a critical peak. The overall ejaculatory response occurs in two phases: emission and expulsion. First, sympathetic impulses cause sequential contraction of smooth muscles in the prostate, reproductive ducts, and seminal vesicles. This contractile activity delivers prostatic fluid, then sperm, and finally seminal vesicle fluid (collectively, *semen*) into the urethra. This phase of the ejaculatory reflex is known as *emission*. During this time, the sphincter at the neck of the bladder is tightly closed to prevent semen from entering the bladder and urine from being expelled along with the ejaculate through the urethra. Second, the filling of the urethra with semen triggers nerve impulses that activate a series of skeletal muscles at the base of the penis. Rhythmic contractions of these muscles increase the pressure within the penis, forcibly expelling the semen through the urethra to the exterior. This is the expulsion phase of ejaculation.

During the resolution phase following orgasm, sympathetic vasoconstrictor impulses slow the inflow of blood into the penis, causing the erection to subside. Muscle tone returns to normal while the cardiovascular and respiratory systems return to their prearousal level of activity. Once ejaculation has occurred, a temporary refractory period of variable duration ensues before sexual stimulation can trigger another erection.

Volume and Sperm Content of the Ejaculate

The volume and sperm content of the ejaculate depend on the length of time between ejaculations. Production of spermatozoa in mammals ranges from <1 to 25 billion spermatozoa per day for both testes. In humans an average ejaculate contains about 180 million sperm (66 million/mL), but some ejaculates contain as many as 400 million sperm. Comparatively, in birds the volume of the ejaculate is comparatively small (200 to 800µL in a turkey) but the density of spermatozoa (4000 to 7000/µL) is usually higher. Because the increased body size of domesticated male turkeys (toms) relative to the hens makes normal mating extremely difficult, the poultry industry uses artificial insemination extensively in turkeys.

Even though only one spermatozoon actually fertilizes the ovum, large numbers of accompanying sperm are often needed to provide sufficient acrosomal enzymes (the acrosome is the membrane-bound organelle of the spermatozoon that contains proteolytic enzymes) to break down the barriers surrounding the ovum (zona pellucida) until the fertilizing sperm is engulfed into the ovum's cytoplasm. The quality of sperm must be taken into account when assessing the fertility potential of a semen sample. The presence of substantial numbers of sperm with abnormal motility or structure, such as sperm with distorted tails, reduces the chances of fertilization.

Vertebrate Female Reproductive Physiology

In some vertebrates, female reproductive physiology is more complex than male reproductive physiology.

■ Complex cycling characterizes female reproductive physiology in many vertebrates.

The CNS of the female receives and interprets visual, olfactory, auditory, and tactile information from the environment and conveys this information to the gonads through the hypothalamic–pituitary–gonadal axis (as does the CNS of the male). Under certain specified conditions the female comes into **estrus** or **heat,** the time when the female is sexually receptive. One **estrous cycle** is the time from one period of sexual receptivity to the next. Many primates including humans have a **menstrual cycle,** in which uterine lining is sloughed off periodically *(menstruation).* This cycle begins at the end of each menstruation The duration of estrous and menstrual cycles is species specific (■ Table 16–3).

An understanding of the hormonal interactions governing follicular development and ovulation has permitted the manipulation of ovulation of some domestic species for management and convenience purposes. Unlike the continuous sperm production and relatively constant testosterone secretion during the breeding cycle of males, the release of ova or deposition of eggs is intermittent, and secretion of female sex hormones displays wide cyclical swings. Normally the juvenile period and the time of gestation and lactation are marked by periods devoid of estrous behavior much longer than those of the cycle. Periods of time without regular cyclic activity, termed **anestrus,** constitute the main portion of the existence of a normal female animal.

In general, the principles governing reproduction are the same for the different species, but there are also notable differences. For example, the frequency of estrous cycles differs between species: Cattle, swine, and rodents are **polyestrous** (estrous cycles occur uniformly throughout the year), whereas others are **seasonally polyestrous** (estrous cycles are restricted to a particular time of year). Deer, sheep, and goats are examples of **short-day breeders,** because they exhibit estrous cycles

Table 16–3 ▌ Reproductive Cycles of Various Mammals

Genus species[a]	Cycle Type[b]	Cycle Length[c]	Duration of Estrus	Ovulation Type[d]	Time	Luteal Phase in Absence of Coitus
Felis domesticus (cat)	S, P	14	4 days	I	24–30 hr after coitus	No
Bos taurus (cow)	C, P	17–27	13–14 hr	Sp	12–16 hr after end of estrus	Yes
Canus familiaris (dog)	S, M	60	7–9 days	Sp	1–3 days after onset of estrus	Yes
Ovis aries (sheep)	S, P	15–18	30–36 hr	Sp	12–14 hr before the end of estrus	Yes
Mustelo fero (ferret)	S	—	Continuous	I	30 hr after coitus	No
Vulpes vulpes (red fox)	S, M	90	1–5 days	Sp	1–2 days after onset of estrus	Yes
Capra hirca (goat)	S, P	20–21	39 hr	Sp	30–36 hr after onset of estrus	Yes—shorter cycles in some breeds
Cavia porcellus (guinea pig)	C, P	16	6–11 hr	Sp	10 hr after onset of estrus	Yes
Mesocricetus auratus (hamster)	C, P	4	20 hr	Sp	8–12 hr after onset of estrus	No
Homo sapiens (man)	C, Mn	28	None	Sp	14 days before the onset of menses	Yes—extremely variable cycle
Equus caballus (horse)	S, P	19–23	4–7 days	Sp	5–6 hr after the onset of estrus	Yes
Macaca mulatta (rhesus monkey)	C, Mn	28	None	Sp	11–14 days after the onset of menses	Yes
Mus musculus (mouse)	C, P	4	10 hr	Sp	2–3 hr after onset of estrus	No
Oryctolagus cuniculus (rabbit)	C	—	Continuous	I	10–12 hr after coitus	No
Rattus norvegicus (rat)	C, P	4–5	13–15 hr	Sp	8–10 hr after onset of estrus	No
Sus scrofa (pig)	C, P	18–23	2–3 days	Sp	36 hr after onset of estrus	Yes

[a]domesticated or in captivity.
[b]C = continuous; S = seasonal[3]; P = polyestrous; M = monstrous; Mn = menstrual.
[c]Length of cycle in days
[d]I = Induced (reflex) ovulators; Sp = spontaneous ovulators (male not necessary).
Source: K. Inskeep, West Virginia University.

as day length decreases whereas bears, hamsters, and horses are **long-day breeders;** that is, they come into estrous as day length increases. **Monoestrous** females (most carnivores, including bears, dogs, foxes, and wolves) are characterized as having a single estrous event followed by a long period of anestrus (most bitches have approximately two estrus periods in a year). In these animals the period of estrus is prolonged and lasts for several days, thus increasing the probability of securing a mate.

Other differences can be found in the type of ovulation. Most animals are **spontaneous ovulators,** that ovulate with a regular frequency and do not require copulation. Some species of birds, for example, ovulate daily for extended periods of time without any contact with the male. In **reflex (induced) ovulators,** such as in the cat, mink, rabbit, and camel, ovulation is induced by stimulation of sensory receptors in the vagina and cervix during coitus.

The tissues influenced by the sex hormones also undergo cyclical changes. During each cycle, the female reproductive tract is prepared for fertilization and implantation of an ovum released from the ovary at ovulation. If fertilization does not occur, the cycle repeats itself. If fertilization does occur, the cycles are interrupted, while the female system adapts to nurture and protect the newly conceived embryo until it has developed the ability to live outside the maternal environment. Furthermore, the female continues her reproductive duties after birth by producing milk (lactation) for the young's nour-

ishment. Thus the female reproductive system is characterized by complex cycles that are interrupted only by more complex changes should pregnancy ensue.

The ovaries, as the primary female reproductive organs, perform the dual function of producing ova (oogenesis) and secreting the female sex hormones estrogens and progesterone. In chickens, however, only the left ovary and oviduct normally mature. The right ovary remains vestigial, although it develops if the left ovary is removed or becomes diseased. The shape and size of the ovary varies both with the species of animal as well as with the stage of the estrous cycle. For example, in cattle and sheep it is almond shaped, in the horse the ovary is bean shaped, and in the sow the ovary resembles a cluster of grapes. The ovaries of the cow and horse are easily examined by rectal palpation and visualized by *transrectal ultrasonography.*

Functions of Estrogen

The sex hormones act together to promote the opportunity for fertilization of the ovum and, in mammals, to prepare the female reproductive system for pregnancy. Estrogens in the female govern many functions similar to those carried out by testosterone in the male, such as maturation and maintenance of the entire female reproductive system and establishment of female secondary sexual characteristics. In general, the actions of estrogens are important to preconception events. Estrogens are essential for the female to become sexually receptive and permit copulation, ova maturation and release, and transport of sperm from the vagina to the site of fertilization in the oviduct. Two different nuclear estrogen receptors can be identified in estrogen-responsive tissues—*ER*α and *ER*β. These receptors differ in their distribution among estrogens' target cells and in the responses they trigger when bound with an estrogen. Some target cells have only ERα receptors, some only ERβ receptors, and some both. These findings may help explain the complex effects of *environmental estrogens* (see box, "Concepts and Controversies: Environmental Estrogens: Bad News for Reproduction").

Estrogens in mammals contribute to mammary gland development in anticipation of lactation. The other ovarian steroid, progesterone, is important in preparing a suitable environment for nourishing a developing embryo/fetus and for contributing to the mammary glands' ability to produce milk.

Reproductive capability begins at puberty in mammalian females, but reproductive potential of many primates, some ro-dents, whales, dogs, rabbits, elephants, and domestic livestock, ceases abruptly during middle age (unlike males, who typically have reproductive potential through the remainder of life, although it gradually declines). This cessation is sometimes called **menopause,** although that term is often restricted to the cessation of menses in some primates.

■ The steps of gametogenesis are the same in both sexes, but the timing and outcome differ sharply.

Oogenesis contrasts sharply with spermatogenesis in several important aspects, even though the identical steps of chromosome replication and division take place during gamete production in both sexes. The undifferentiated primordial germ cells in the fetal ovaries, the **oogonia** (comparable to the spermatogonia), divide mitotically during gestation and/or neonatally, after which time mitotic proliferation ceases.

CONCEPTS AND CONTROVERSIES

Environmental Estrogens: Bad News for Reproduction

Unknowingly, for the past 50 years we humans have been polluting our environment with synthetic endocrine-disrupting chemicals (**EDCs,** p. 255) as an unintended side effect of industrialization. As you saw in Chapter 7, these hormonelike pollutants bind with the receptor sites normally reserved for the naturally occurring hormones, either mimicking or blocking normal hormonal activity. Many EDCs have been found to alter sex steroid functions. Some are *estrogenic* and exert feminizing effects as estrogen agonists. Others (androgen antagonists) block a male's own androgens from accomplishing their intended masculinizing effects. Antiandrogenic action demasculinizes males, producing an outcome similar to that brought about by estrogenic EDCs.

Estrogenic pollutants are everywhere in our environment. They contaminate our food, drinking water, and air. Proved feminizing synthetic compounds include (1) certain weed killers and insecticides, (2) some detergent breakdown products, (3) petroleum by-products found in car exhaust, (4) common food preservatives used to retard rancidity, and (5) softeners that make plastics flexible. These plastic softeners are commonly found in food packaging and can readily leach into food with which they come in contact, especially during heating. Plastic softeners are among the most plentiful industrial contaminants in our environment.

Evidence of Gender Bending in Animals

Some fish and wild animal populations that have been severely exposed to estrogenic EDCs ("environmental estrogens")—such as those living in or near water heavily polluted with chemical wastes—display a high rate of grossly impaired reproductive systems. Examples include male fish that are hermaphrodites (having both male and female reproductive parts) and male alligators with abnormally small penises. Similar reproductive ab-normalities have been identified in land mammals. Presumably, excessive exposure to estrogen agonists is emasculating these populations.

Environmental estrogens are also implicated in the rising incidence of mammary gland cancer in female humans. Mammary gland cancer is 25 to 30% more prevalent now than in the 1940s. Many of the established risk factors for mammary gland cancer, such as starting to menstruate earlier than usual and undergoing menopause later than usual, are associated with an elevation in the total lifetime exposure to estrogen. Because increased exposure to natural estrogens bumps up the risk for mammary gland cancer, prolonged exposure to environmental estrogens may be contributing to the rising prevalence of this malignancy among women (and men too).

Some natural environment-disrupting chemicals are the natural estrogen-like compounds called **phytoestrogens,** synthesized by some plant species. Usually the ingestion of these compounds is disadvantageous to the reproductive success of an animal, but in some situations can be of benefit. Consider the California quail (*Lophortyx californicus*), a photoperiodic species that relies on day length to regulate the time of the breeding season. In the more arid regions of its range it breeds irregularly, depending on the amount of rainfall in the previous winter. During dry years, high concentrations of phytoestrogens are produced in the desert plants on which the quail feed. As with endocrine-disrupting chemicals, these phytoestrogens suppress reproductive development, in this situation by feeding back onto the hypothalamic–pituitary axis. By suppressing reproductive activity in these lean years, the animal does not use scarce body reserves on a clutch of eggs that will not likely reach sexual maturity.

Formation of Primary Oocytes and Primary Follicles

During the last part of fetal life, the oogonia begin the early steps of the first meiotic division but do not complete it. Known now as **primary oocytes,** they contain the diploid number of $2N$ replicated chromosomes, which are gathered into homologous pairs but do not separate. The primary oocytes remain in this state of meiotic arrest until they are prepared for ovulation. A presumed purpose of this nuclear arrest is to inactivate the DNA in the female gamete in order to reduce the possibility of biochemical insult during the lifetime of the female.

Before birth, each primary oocyte is surrounded by a single layer of flattened **granulosa cells.** Together, an oocyte and surrounding granulosa cells make up a **primordial follicle.** Oocytes that fail to be incorporated into follicles self-destruct by apoptosis (see p. 43). At birth only a fraction of the original primary follicles remain, each containing a single primary oocyte capable of producing a single ovum. In most mammals, no new oocytes or follicles appear after birth; the follicles already present in the ovaries at birth serve as a reservoir from which all ova throughout the reproductive life of a female must be drawn. Of these follicles, only a small percentage (about 400 in humans) mature and release ova. The pool of primary follicles present at birth gives rise to an ongoing trickle of developing follicles. Once it starts to develop, a follicle is destined for one of two fates: It reaches maturity and ovulates, or it degenerates, a process known as **atresia.** Until puberty, all the follicles that start to develop undergo atresia in the early stages without ever ovulating. Even after puberty, many of the follicles are **anovulatory** (that is, no ovum is released). Of the initial pool of follicles, most (for example, 99.98% in humans) never ovulate but instead undergo atresia at some stage in development. At the end of the female's reproductive life, only a few follicles remain in the ovaries, and soon even these succumb to atresia.

This limited gamete potential, which is already determined at birth in mammalian females, is in sharp contrast to the continual process of spermatogenesis in males. Furthermore, there is considerable chromosome wastage in oogenesis compared with spermatogenesis. Let's see how.

Formation of Secondary Oocytes and Secondary Follicles

The primary oocyte within a primary follicle is still a diploid cell that contains $2N$ doubled chromosomes. A portion of the resting pool of follicles starts developing into **secondary (antral) follicles.** The number of developing follicles at any time is roughly proportional to the size of the pool, but the mechanisms that determine which follicles in the pool develop during a given cycle are unknown. Development of a secondary follicle is characterized by growth of the primary oocyte and by expansion (proliferation) and differentiation of the surrounding cell layers. The extent of oocyte enlargement is species dependent. Oocyte enlargement is due to a buildup of cytoplasmic and/or yolk materials that will be needed by the early embryo. The follicle next acquires a dependence on gonadotropins, in part as a result of the differentiation of the cells surrounding the granulosa into theca cells, which markedly increases estrogen production by the follicular cells.

Just before ovulation, the primary oocyte, whose nucleus has been in meiotic arrest, often for years, completes its first meiotic division, except in the horse and the carnivores, which ovulate a primary oocyte, in association with the extended period of estrus. This division yields two daughter cells, each receiving a haploid set of n doubled chromosomes, analogous to the formation of secondary spermatocytes (● Figure 16–15). However, almost all the cytoplasm remains with one of the daughter cells, now called the **secondary oocyte,** which is destined to become the ovum. The chromosomes of the other daughter cell together with a small share of cytoplasm form the **first polar body.** In this way, the ovum-to-be loses half of its chromosomes to form a haploid gamete but retains all its nutrient-rich cytoplasm. The nutrient-poor polar body soon degenerates.

Formation of a Mature Ovum

Actually, the secondary oocyte, and not the mature ovum, is ovulated and fertilized, but common usage refers to the developing female gamete as an *ovum* even in its primary and secondary oocyte stages. Sperm entry into the secondary oocyte is needed to trigger the second meiotic division. Oocytes that are not fertilized never complete this final division. During this division, a half set of chromosomes along with a thin layer of cytoplasm is extruded as the **second polar body.** The other half set of *n* unpaired chromosomes remains behind in what is now the **mature ovum.** These $1N$ maternal chromosomes unite with the $1N$ paternal chromosomes of the penetrating sperm to complete fertilization. If the first polar body has not already degenerated, it too undergoes the second meiotic division at the same time the fertilized secondary oocyte is dividing its chromosomes.

Comparison of Steps in Oogenesis and Spermatogenesis

The steps involved in chromosome distribution during oogenesis parallel those of spermatogenesis, except that the cytoplasmic distribution and time span for completion sharply differ. Just as four haploid spermatids are produced by each primary spermatocyte, four haploid daughter cells are produced by each primary oocyte (if the first polar body did not degenerate before it completes the second meiotic division). In spermatogenesis, each daughter cell develops into a highly specialized, motile spermatozoon unencumbered by unessential cytoplasm and organelles, its only destiny being to supply half of the genes for a new individual. In oogenesis, however, of the four daughter cells, only the one destined to become the ovum receives cytoplasm. This uneven distribution of cytoplasm is important, because the ovum, in addition to providing half the genes, provides all the cytoplasmic components needed to support early development of the fertilized ovum. The large, relatively undifferentiated ovum contains numerous nutrients, organelles, and structural and enzymatic proteins. The three other cytoplasm-deficient daughter cells, the polar bodies, rapidly degenerate.

■ The ovarian cycle of mammals consists of alternating follicular and luteal phases.

After the onset of puberty, the ovary alternates between two phases: the **follicular phase,** which is dominated by the presence of *maturing follicles;* and the **luteal phase,** which is characterized by the presence of the *corpus luteum* (to be described

Proliferation of Thecal Cells; Estrogen Secretion

At the same time the oocyte is enlarging and the granulosa cells are proliferating, specialized ovarian connective tissue cells in contact with the expanding granulosa cells proliferate and differentiate to form an outer layer of **thecal** cells. The thecal and granulosa cells, collectively known as **follicular cells,** function as a unit to secrete estrogen. Of the three physiologically important estrogens—estradiol, estrone, and estriol (the last being important only in the human)—the principal ovarian estrogen is usually estradiol (although in the sow it is estrone).

Formation of the Antrum

The hormonal environment that exists during the follicular phase promotes enlargement and development of the follicular cells' secretory capacity, converting the primary follicle into a secondary, or antral, follicle capable of estrogen secretion. This stage of follicular development is characterized by the formation of a fluid-filled **antrum** in the middle of the granulosa cells (Figure 16–16). The follicular fluid originates partially from transudation (passage through capillary pores) of plasma and partially from follicular cell secretions. As the follicular cells start producing estrogen, some of this hormone is secreted into the blood for distribution throughout the body. However, a portion of the estrogen collects in the hormone-rich antral fluid.

The oocyte has reached full size by the time the antrum begins to form. The shift to an antral follicle initiates a period of rapid follicular growth. During this time, the follicle increases in size from a diameter of less than 1 mm to 12 to 16 mm, depending on the species of mammal, whereas in the chicken there is not an antrum, but an oocyte of nearly 3 to 4 cm. Part of the follicular growth is due to continued proliferation of the granulosa and thecal cells, but most is due to a dramatic expansion of the antrum. As the follicle grows, estrogens are produced in increasing quantities.

Formation of a Mature Follicle

One or more of the follicles usually grows more rapidly than the others, developing into a **dominant (preovulatory,** or **Graafian)** follicle. The term *Graafian* is properly applied to the dominant follicle in humans only. Dominance probably occurs only in humans, cows, deer, and horses. Moreover, in polyovular species (that is, most mammals) more than one follicle becomes part of the cohort of preovulatory follicles. The antrum occupies most of the space in a mature follicle. The oocyte, surrounded by the zona pellucida and a single layer of granulosa cells, is displaced asymmetrically at one side of the growing follicle in a little mound of cells that protrudes into the antrum, the **cumulus oophorus.**

Ovulation

The greatly expanded mature follicle bulges on the ovarian surface, creating a thin area that ruptures to release the oocyte at **ovulation.** Rupture of the follicle is facilitated by the release from the follicular cells of enzymes that digest the connective tissue in the follicular wall. The bulging wall is thus weakened so that it balloons out even further, to the point that it can no longer contain the rapidly expanding follicular contents.

Just before ovulation, the oocyte completes its first meiotic division. The ovum (secondary oocyte), still surrounded by its zona pellucida and granulosa cells, is swept out of the ruptured follicle into the infundibulum (which, in most species, surrounds the ovary) by the leaking antral fluid (Figure 16–16b). The single cell layer of granulosa surrounding the oocyte is called the **corona radiata** ("radiating crown") from the time of the secretion of the zona and the establishment of the cytoplasmic extensions of these cells to feed the oocyte. The cumulus oocyte complex that is ovulated consists of the oocyte, corona radiata, and a mass of additional granulosa (nonmural). The released ovum is quickly drawn into the oviduct, where fertilization may or may not take place. The other developing follicles that failed to reach maturity and ovulate undergo degeneration, never to be reactivated.

Rupture of the follicle at ovulation signals the end of the follicular phase and ushers in the luteal phase.

◼ The luteal phase is characterized by the presence of a corpus luteum.

The ruptured follicle that is left behind in the ovary following release of the ovum undergoes a rapid change. The granulosa and thecal cells remaining in the remnant follicle first collapse into the emptied antral space that has been partially filled by clotted blood.

Formation of the Corpus Luteum; Estrogen and Progesterone Secretion

These old follicular cells soon undergo a dramatic structural transformation to form the **corpus luteum,** in a process called **luteinization** (Figure 16–16c). The follicular-turned-luteal cells hypertrophy and are converted into very active steroidogenic (steroid hormone–producing) tissue. Beta-carotene, stored in lipid droplets within the corpus luteum, gives this tissue a yellowish appearance, particularly in the cow, hence its name (*corpus,* "body"; *luteum,* "yellow") although in some species, such as sheep, pigs, and rats, it is a reddish pink. The corpus luteum becomes highly vascularized as blood vessels from the theca invade the luteinizing granulosa. These changes are appropriate for the corpus luteum's function, which is to secrete abundant quantities of progesterone along with lesser amounts of estrogens into the blood. Estrogen secretion in the follicular phase followed by progesterone secretion in the luteal phase is essential for preparing the uterus to be a suitable site for attachment or implantation of a fertilized ovum.

Degeneration or Survival of the Corpus Luteum

If the released ovum is not fertilized and does not implant, the corpus luteum is lysed by $PGF_{2\alpha}$ from the endometrium (Figure 16–16d). The luteal cells degenerate and are phagocytized, the vascular supply is withdrawn, and connective tissue rapidly fills in to form a fibrous tissue mass known as the **corpus albicans** ("white body"). The luteal phase is now over, and one ovarian cycle is complete.

If fertilization and implantation/attachment does take place, the corpus luteum continues to grow and produce increasing quantities of progesterone instead of degenerating. Now called the **corpus luteum of pregnancy,** this ovarian structure persists until the end of pregnancy. It provides the hormones essential for maintaining pregnancy in the human and in sheep until the developing placenta can take over this crucial function. However, in the pig, cow, goat, and rat the corpus luteum is the source of progesterone for the duration of gestation.

■ The mammalian estrous cycle is regulated by complex hormonal interactions among the hypothalamus, anterior pituitary, and ovarian endocrine units.

The estrous cycle can be divided into four phases: proestrus, estrus, metestrus, and diestrus. **Proestrus** begins with the regression of the corpus luteum and ends with the onset of estrus. Proestrus is associated with development and ripening of the follicles. **Estrus** is the time of sexual receptivity and usually terminates with ovulation. **Metestrus,** the early postovulatory period, is associated with the development of the corpus luteum. **Diestrus** is associated with a functioning corpus luteum and begins about four days after ovulation and ends with regression of the corpus luteum. Each of these phases is actually a component of the follicular and luteal phases of the cycle. Thus the follicular phase includes proestrus and estrus and is dominated by estrogens, whereas the luteal phase includes metestrus and diestrus and is dominated by progesterone. If conception occurs, the female enters a period of **anestrus** during pregnancy, which ultimately ends with the positive-feedback cycle of **parturition** or the process of giving birth to the offspring.

The ovary has two related endocrine units: the estrogen-secreting follicle and the corpus luteum, which secretes progesterone. These units are sequentially triggered by complex cyclical hormonal relationships among the hypothalamus, anterior pituitary, and these two ovarian endocrine units. As in the male, gonadal function in the female is directly controlled by the anterior-pituitary gonadotropic hormones, follicle-stimulating hormone (FSH), and luteinizing hormone (LH). These hormones, in turn, are regulated by pulsatile episodes of hypothalamic gonadotropin-releasing hormone (GnRH) and feedback actions of gonadal hormones.

Differing from the male, however, control of the female gonads is complicated by the cyclical nature of ovarian function. For example, the effects of FSH and LH on the ovaries depend on the stage of the estrous cycle. Also in contrast to the male, FSH is not strictly responsible for gametogenesis, nor is LH solely responsible for gonadal hormone secretion. Using the bovine estrous cycle as an example, we consider control of follicular function, ovulation, and the corpus luteum, using ● Figure 16–17 to integrate the various concurrent and sequential activities that take place throughout the cycle.

We use the cow as a specific example of the estrous cycle, noting that the processes described are similar to other mammals (Table 16–3). The cycle length is marked by behavioral estrus, which occurs, on the average, at intervals of 21 to 22 days, although estrous cycles of 17 to 24 days are considered normal. The luteal phase is dominated by progesterone, secreted by the corpus luteum. After regression of the

corpus luteum, the rapidly increasing secretion of estradiol, by the dominant follicle, initiates the **luteinizing hormone (LH) surge** (day 0; ● Figure 16–17 and 18a). Ovulation of the dominant follicle occurs 24 to 27 hours after the LH surge.

Control of Ovulation by Positive Feedback

Ovulation and subsequent luteinization of the ruptured follicle are triggered by an abrupt, massive increase in LH secretion. This **LH surge** (day 0) rings about four major changes in the follicle:

- It halts estrogen synthesis by the follicular cell.
- It reinitiates meiosis in the oocyte of the developing follicle, apparently by blocking secretion of an *oocyte maturation–inhibiting substance* produced by the granulosa cells. This substance is believed to be responsible for arresting meiosis in the primary oocytes once they are wrapped within granulosa cells in the fetal ovary.
- It triggers production of locally acting prostaglandins, which support ovulation by promoting vascular changes that cause rapid swelling of the follicle while inducing enzymatic digestion of the follicular wall, followed by contractions of myoepithelial or smooth muscle–type cells in the theca externa at the base of the follicle. Together these actions lead to rupture of the weakened wall that covers the bulging follicle.
- It causes differentiation of follicular cells into luteal cells. Because the LH surge triggers both ovulation and luteinization, formation of the corpus luteum automatically follows ovulation. Thus the burst in LH secretion is a dramatic point in the cycle; it terminates the follicular phase and initiates the luteal phase. In the primate menstrual cycle the LH surge is midcycle, whereas in nonprimate mammals the LH surge occurs at or very close to estrus at the beginning of the cycle.

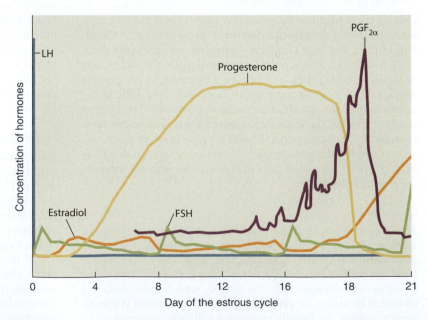

Figure 16–17 ● Correlation between hormonal levels during the bovine estrous cycle.

(*Source:* Courtesy Keith Inskeep, West Virginia University)

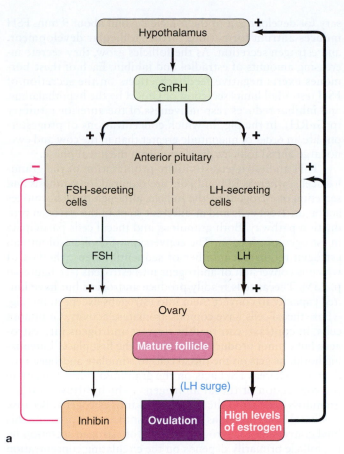

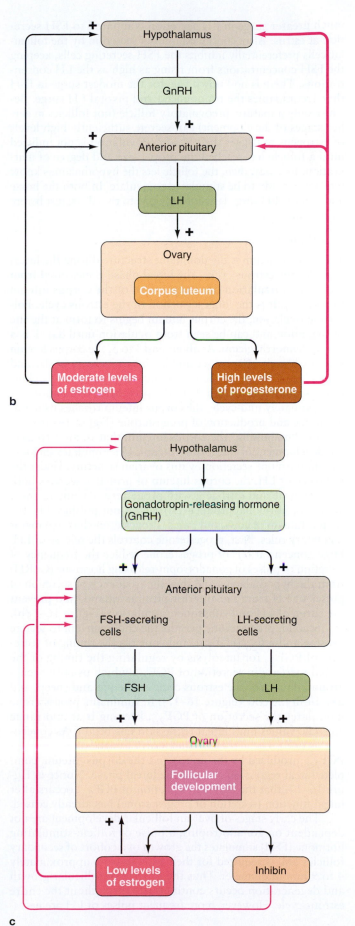

Figure 16–18 ● **Reproductive hormones in a female mammal.**
(a) Control of the LH surge at ovulation. (b) Feedback control during
the luteal phase. (c) Feedback control of FSH and tonic LH secretion
during the follicular phase.

The two different modes of LH secretion—the tonic secretion of LH responsible for promoting follicular development
and hormone secretion and the LH surge that causes ovulation—not only occur at different times and produce different
effects on the ovaries, but also are controlled by different mechanisms. Tonic LH secretion is partially suppressed by the inhibitory action of the low, rising levels of estrogen during the
follicular phase (Figure 16–18c) and is completely suppressed
by the increasing levels of progesterone during the luteal phase.
Because tonic LH secretion stimulates both estrogens and progesterone secretion, this is a typical negative-feedback control
system.

In contrast, the LH surge is triggered by a *positive-
feedback* effect of estrogen. Whereas the low, rising levels of estrogens early in the follicular phase *inhibit* LH secretion, the
high level of estrogens that occurs during peak estrogen secretion late in the follicular phase *stimulates* LH secretion and
initiates the LH surge (Figure 16–18a). Thus LH enhances estrogen production by the follicle, and the resultant peak estrogen concentration stimulates LH secretion. The high plasma
concentration of estrogens acts directly on the hypothalamus
to increase the frequency of GnRH pulses, thereby increasing
both LH and FSH secretion. It also acts directly on the anterior
pituitary to specifically increase the sensitivity of LH-secreting
cells to GnRH. The latter effect accounts in large part for the

much greater surge in LH secretion compared to FSH secretion at estrus. Also, continued inhibin secretion by the follicular cells preferentially inhibits the FSH-secreting cells, keeping the FSH concentrations from rising as high as the LH concentrations. There is no known role for the modest surge in FSH that accompanies the pronounced and pivotal LH surge. Because only a mature, preovulatory follicle (not follicles in earlier stages of development) can secrete sufficiently high levels of estrogens to trigger the LH surge, ovulation is not induced until a follicle has reached the proper size and degree of maturation. In a way, then, the follicle lets the hypothalamus know when it is ready to be stimulated to ovulate. In both the horse and pig, the LH surge lasts for only one to two days, just before ovulation.

Luteal Phase

The corpus luteum is the dominant structure during the luteal phase of the estrous cycle. The luteal phase is measured from the time of ovulation until regression of the corpus luteum (luteolysis). It is the longest portion of the estrous cycle. Following ovulation, the corpus luteum begins to form at the site of ovulation and can be seen from ovulation until day 3 as a **corpus hemorrhagicum.** At about day 3 to 5, the corpus luteum loses its bloody appearance and begins to increase in size and to produce sufficient progesterone such that the increase in hormone secretion is detectable in the peripheral circulation. By approximately mid-cycle, the corpus luteum reaches its maximum size and production of progesterone (Figure 16–16).

LH "maintains" the corpus luteum; that is, after triggering development of the corpus luteum, LH stimulates ongoing steroid hormone secretion by this ovarian structure. Under the influence of LH, the corpus luteum of primates secretes both progesterone and estrogen, with progesterone being its most abundant hormonal product (estrogen is not produced by the corpus luteum of cows and ewes). Progesterone has two major regulatory roles. First, progesterone controls the release of LH. High concentrations of progesterone reduce the frequency of secretion of pulses of gonadotropin-releasing hormone (GnRH) from the hypothalamus and thus the frequency of secretion of pulses of LH from the anterior pituitary gland, and prevent the surge of LH and subsequent ovulation (Figure 16–18b).

Second, progesterone permits the endometrium to secrete prostaglandin $F_{2\alpha}$ ($PGF_{2\alpha}$), and controls the timing of secretion of $PGF_{2\alpha}$ for luteolysis by regulating the timing of the initial increase in secretion of $PGF_{2\alpha}$, which usually occurs around day 12 of the estrous cycle in the pig and sheep, and day 14 in the cow (Figure 16–17). In ruminants progesterone modulates the secretion of $PGF_{2\alpha}$, keeping it at mid-range episodic values until luteal regression has begun. As concentrations of progesterone decline, greater peaks of secretion of $PGF_{2\alpha}$, produced by the uterus and the corpus luteum, complete luteal regression and end the luteal phase. Notice in Figure 16–17, that the maximal secretion of $PGF_{2\alpha}$, occurs after luteal function (secretion of progesterone) has already ceased.

The early stages of ovarian follicular development are not dependent on gonadotropins; a pulse of follicle-stimulating hormone (FSH) stimulates the growth of a cohort of secondary follicles and is required for their growth from approximately 4 to 9 mm in diameter. Thus the process of follicular growth and degeneration occurs continuously throughout the entire estrous cycle. However, more frequent pulses of LH are necessary for development of the follicle beyond about 9 mm. FSH induces antrum formation, further follicular development, and estrogen secretion. As the follicles grow, they secrete increasing amounts of estradiol and inhibin. Each of these hormones exerts negative-feedback effects on the secretion of FSH (estradiol limits secretion of GnRH by the hypothalamus, and inhibin reduces responsiveness of the anterior pituitary to GnRH). In the pig, in which concentrations of progesterone are an order of magnitude greater than in the cow and ewe, no large antral follicles develop during the luteal phase.

FSH and estrogens stimulate proliferation of the granulosa cells, whereas LH and FSH are required for synthesis and secretion of estrogens by the follicle, although these hormones act on different cells and at different steps in the estrogen production pathway. Both granulosa and thecal cells participate in estrogen production. The conversion of cholesterol into an estrogen requires a number of sequential steps, the last of which is conversion of androgens into estrogens (see figure on p. 255). Thecal cells readily produce androgens but have limited capacity to convert them into estrogens except in the pig, whose thecal cells have copious aromatase activity. Granulosa cells, in contrast, can readily convert androgens into estrogens but cannot produce androgens in the first place. Luteinizing hormone acts on the thecal cells to stimulate androgen production, whereas FSH acts on the granulosa cells to promote the conversion of thecal androgens (which diffuse into the granulosa cells from the thecal cells) into estrogens. Because low basal concentrations of FSH are sufficient to promote this final conversion to estrogen, the rate of estrogen secretion by the follicle primarily depends on the circulating concentration of LH, which continues to rise. Furthermore, as the follicle continues to grow, more estrogen is produced simply because more estrogen-producing follicular cells are present.

This differential sensitivity of FSH- and LH-producing cells to negative feedback by estrogens is at least in part responsible for the fact that the plasma FSH concentration, unlike the plasma LH concentration, declines as the estrogen concentrations rise. Another contributing factor to the fall in FSH is the secretion of *inhibin* by the follicular cells. Inhibin preferentially inhibits FSH secretion by acting at the anterior pituitary, just as it does in the male. The decline in FSH secretion brings about atresia of all but the single most mature of the developing follicles. As FSH concentrations decline, the dominant follicle becomes more dependent on LH. Continued pulses of LH maintain the growth of the dominant follicle until it is either ovulated or becomes atresic.

In contrast to FSH, LH secretion continues to rise slowly despite inhibition of GnRH (and thus indirectly, LH) secretion. This seeming paradox is due to the fact that estrogens alone cannot completely suppress tonic (low-level, ongoing) LH secretion; progesterone is required to completely inhibit tonic LH secretion. Because progesterone does not appear until the luteal phase of the cycle, the basal level of circulating LH slowly increases during the follicular phase under incomplete inhibition by estrogens alone.

Follicular Phase

The follicular phase is initiated only after luteolysis is underway. As luteolysis proceeds, the decreasing concentrations of progesterone reduce the negative feedback on pulsatile secretion of LH by the anterior pituitary. The increased frequency of pulses of LH promotes the final maturation of the large antral follicle. During the follicular phase, LH reaches higher concentrations than in the luteal phase, because negative feedback from progesterone on the hypothalamus is removed (Figure 16–18c).

Dominant follicles, as they continue to grow, secrete increasing concentrations of estradiol and inhibin, so that FSH declines during the follicular phase, but estradiol does not limit pulse frequency of LH. When the dominant follicle reaches adequate size (10 mm in diameter is approximately the minimum required for ovulation in the cow as compared to 7 mm in a pig), and secretion of estradiol from the follicles reaches threshold, a preovulatory surge of LH and FSH occurs (Figure 16–17). Immediately after the preovulatory surge of LH, concentrations of estradiol abruptly decline, and a secondary smaller surge of FSH occurs on the day of ovulation (recruiting a new cohort of follicles into the growing pool).

Subnormal Luteal Phase

A **subnormal luteal phase** is defined by low secretion of progesterone or can occur during treatment with other progestogens for synchronization of estrus. Lower circulating concentrations of progestogen lead to an *increased* frequency of release of pulses of LH from the anterior pituitary gland. Increased LH stimulates continued growth of the largest ovarian follicle, which is followed by increased secretion of estradiol by the **persistent largest follicle;** that is, the largest dominant follicle continues to increase in size and becomes *persistent*. Low dosages of administered progestogens, native or synthetic, result in reduced fertility in association with follicular persistence. In contrast, a low frequency of pulses of LH fails to support continued follicular growth and allows earlier degeneration or more frequent replacement of the largest follicle. Treatment with higher dosages of progesterone or a progestogen can be used to cause atresia of the largest follicle and initiation of growth of a new cohort of follicles.

Short Luteal Phases

In ruminants, a **short luteal phase** follows the first ovulation or first estrus at puberty or after parturition. Variables such as follicular development, pre-and postovulatory concentrations of gonadotropins and luteal receptors for LH did not account for the shortened duration of luteal function. The premature uterine secretion of $PGF_{2\alpha}$ was shown to be responsible for the short luteal phase. Obviously, a corpus luteum that had regressed before day 14 could not support pregnancy; maternal recognition of pregnancy (see p. 747) in the cow occurs on days 14 to 17.

Several research groups have observed that pretreatment with a progestogen usually results in the formation of a corpus luteum with a normal functional life span, in response to the injection of gonadotropins. Pretreatment with progestogen increases the numbers of receptors for progesterone in the uterus on day 5 after estrus in previously anestrous cows or heifers. The upregulation of uterine progesterone receptors appears to be essential to the normal timing of secretion of $PGF_{2\alpha}$. An understanding of the role of progesterone in normalizing the length of the estrous cycle has enabled the development of methods to initiate normal cycles in anestrous cows that yield normal fertility when such cows are bred at the induced estrus.

■ The uterine changes that occur during the estrous cycle reflect hormonal changes during the ovarian cycle.

The fluctuations in circulating levels of estrogens and progesterone that occur during the ovarian cycle induce profound cyclic changes in the uterus, although menstruation is seen only in some primates. In mammals other than these primates, the complete removal of the uterus, **hysterectomy,** after ovulation results in the persistence of the corpus luteum as if the female were pregnant. Because it reflects hormonal changes that occur during the ovarian cycle, the uterine cycle averages 17 to 24 days (in cattle), as does the ovarian cycle, although there is considerable variation from this mean even in normal animals. This variability is primarily a reflection of differing lengths of the follicular phase; the duration of the luteal phase is fairly constant within a species. However, across species the duration of the luteal phase varies considerably. Consistent changes that take place throughout the cycle, include the preparation of the uterus for implantation should a released ovum be fertilized, which involves a slight thickening of the uterine lining that is reduced at the end of the luteal phase. In the absence of conception, the endometrium of nonprimate mammals is not sloughed off as it is in primates.

We briefly examine the influences of estrogens and progesterone on the uterus and then consider the effects of cyclic fluctuations of these hormones on uterine structure and function. The uterus consists of two main layers: the **myometrium,** the outer smooth-muscle layer; and the **endometrium,** the inner lining that contains numerous blood vessels and glands. Estrogens stimulate growth of both the myometrium and the endometrium.

Estrogens induce the synthesis of progesterone receptors in the endometrium. Thus progesterone can exert an effect on the endometrium only after it has been "primed" by estrogen. Progesterone acts on the estrogen-primed endometrium to convert it into a hospitable and nutritious lining suitable for implantation of a fertilized ovum. Under the influence of progesterone, the endometrial connective tissue becomes loose and edematous as a result of an accumulation of electrolytes and water, which facilitates implantation of the fertilized ovum. Progesterone further prepares the endometrium to sustain an early-developing embryo by inducing the endometrial glands to secrete and store large quantities of glycogen and by causing tremendous growth of the endometrial blood vessels. Progesterone reduces the contractility of the uterus to provide a quiet environment for implantation and embryonic growth in the cow, pig, and ewe but not in the mare. In this species, contractions of the myometrium serve to transport the embryo throughout the uterine lumen for eventual implantation.

■ Fluctuating concentrations of estrogens and progesterone produce cyclic changes in cervical mucus.

Hormonally induced changes also take place in the cervix during the ovarian cycle. Under the influence of estrogens during the follicular phase, the mucus secreted by the cervix becomes abundant, clear, and thin. This change, which is most pronounced when estrogens are at their peak and ovulation is approaching, facilitates passage of sperm through the cervical canal. After ovulation, under the influence of progesterone from the corpus luteum, the mucus becomes thick and sticky, essentially forming a plug across the cervical opening. This plug constitutes an important defense mechanism by preventing bacteria that might threaten a pregnancy (should conception have occurred) from entering the uterus from the vagina. Sperm cannot penetrate this thick mucus barrier.

■ The reproductive cycles of nonviviparous vertebrates are fundamentally similar but have some unique differences.

The details of female reproductive features we have been discussing are primarily those of mammals. How does the reproductive cycle of a viviparous cow compare to the ovulatory cycle of an ovoviviparous bird or oviparous fish? Each reproductive strategy represents separate evolutionary trends, with the result that there are as many similarities between these strategies as there are differences. For example, as in mammalian species, the hormonal pathway of reproduction in birds and fishes revolves around the hypothalamic–pituitary–gonadal axis. The hypothalamus is activated by specific environmental cues: day length in some species of birds and pheromones in some species of fish. Following this activation GnRH is synthesized and released into the hypothalamic-hypophyseal portal vasculature. However, different bird and fish species have two to three forms of GnRH, the number depending on the species, although it is believed that only one form of GnRH governs the release of LH. We initially focus on the ovulatory cycle of the domestic chicken.

Ovarian Follicular Growth and the Ovulatory Cycle in the Domestic Hen

At the time of hatching, the chick embryo contains approximately a million oocytes, many of which become atresic (degeneration and resorption of ovarian follicles before ovulation) in the first weeks. However, the remaining follicles enter a stage of slow growth to 1 to 2 mm in size that continues for months or even years. During this time neutral lipids are deposited as yolk in the oocyte. In the weeks before the onset of lay (egg production), which in the domestic hen begins at approximately 18 weeks of age, a small proportion of these oocytes begins to enlarge. This stage of follicular development lasts for approximately 8 to 9 weeks, with recruitment of oocytes continuing throughout the female's reproductive life. As a consequence of the deposition of a white, proteinaceous, primordial yolk, these oocytes increase in size to about 2 to 6 mm in diameter. When the follicles reach 5 to 8 mm in diameter, they enter the final rapid-growth phase, which is characterized by the rapid deposition of yellow yolk. During this stage the follicles grow to more than 35 mm in diameter in as few as 7 to 11 days.

One of the most obvious differences in follicular development between birds and mammals now becomes evident, because in the bird's ovary the follicles do not mature synchronously. The yellow yolky follicles form an orderly hierarchy of different sizes so that at the peak of egg production, the largest (**F1 follicle**) of about 40 to 45 mm in diameter is the one destined to be ovulated next, the second largest follicle (**F2**) the following day, the third largest (**F3**) two days later, and so on. Normally, the ovary of the hen contains up to 10 yellow yolky follicles, together with a greater number of small yellow follicles and numerous small white follicles awaiting recruitment or atresia (● Figure 16–19). The actual number of rapidly growing follicles generally remains stable in an individual hen with just one follicle from the intermediate phase recruited into the final growth phase once the largest follicle is ovulated. However, it is not known how new follicles are recruited into the intermediate phase.

In response to the increase in estrogen concentrations prior to the onset of lay, very-low-density lipoprotein (VLDL, p. 362), which is the major precursor of yellow yolk, is synthesized in the liver. VLDL is transported to the ovary, where it is deposited in the yolk. Researchers have studied the growth of yellow yolky follicles by feeding the hen differently colored lipid-soluble dyes (Fig 16–19). The dye is laid down in the yolk as concentric rings marking the day of feeding. Using this technique, researchers showed that follicles require from 11 to 13 days to mature from an intermediate follicle to one of ovulable size.

Because yolk is an energy-expensive material to produce, the number of follicles in the hierarchy, their size relative to body weight, and the interval between successive ovulations differ greatly among different species of birds. Most birds lay an egg each day until the **clutch** is complete. A longer interval between egg lay has energy advantages but creates difficulties in synchronizing the timing of hatch and increases the likelihood of predation. Normally clutch size is smaller, and the interval between eggs is longer, in species producing large eggs. Many species of birds are **determinate** layers; that is, they lay a fixed number of eggs in a clutch, whereas others are **indeterminate** layers and can continue to lay eggs for extended periods of time, particularly if the eggs are removed from the nest as they are laid. With the onset of agriculture and domestication of species over 4000 years ago, humans have exploited this property of indeterminate layers and developed strains of chickens, ducks, and quail that can lay over 300 eggs each year.

Eggs laid on successive days are known as a **sequence**, whereas days on which no eggs are laid are termed **pause** days. Successive eggs in a sequence are laid later on successive days, thus the first egg of a long sequence may be laid earlier in the day than the first egg of a hen producing a short sequence. The general rule is that ovulation occurs 6 to 8 hours after the ovulatory surge of LH, with the egg spending about 24 hours in the oviduct before it is laid (**oviposited**). There is a close relationship between ovulation and oviposition, with ovulation occurring 15 to 75 min after oviposition, except if it is the last egg of a sequence. Because of this relationship, the time of lay becomes a practical guide to the time of ovulation in a bird. To achieve sequences of 50 to 100 eggs, ovulation occurs every day at the same time and the egg from that ovulation is subsequently laid 24 hours later. Between sequences there are one or more pause days. Chickens held under normal light–dark cycles of 14L:10D lay their eggs in the first half of the light period, whereas under the same lighting conditions quail lay their eggs late in the day and early into the dark period. In both situations an ovulatory surge occurs 6 hours before oviposition, which means that the ovulatory surge of LH in chickens occurred during the night, whereas in quail it occurred during the morning. Egg laying is thus *entrained* by the light–dark cycle and can readily be affected by any changes in the photoperiod.

Two additional features pertinent to the ovulatory cycle need to be pointed out. The first is that each ovulation in a hen occurs slightly later on subsequent days. The second is that

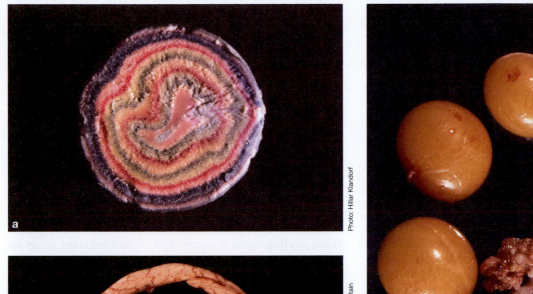

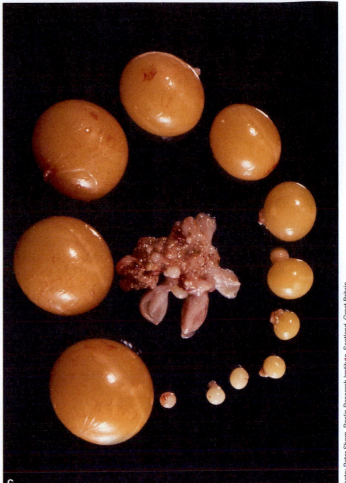

Photo: Hillar Klandorf

Photo: Peter Sharp, Roslin Research Institute, Scotland, Great Britain

Photo: Peter Sharp, Roslin Research Institute, Scotland, Great Britain

Figure 16–19 ● Ovary of a Laying Hen. (a) Midsagittal section through the follicle of an egg. Xanthophylls (yellow to orange carotenoid pigments) from the diet are transported rapidly from the bloodstream to the developing yolk; consequently more pigment is deposited when the hen is eating. This process gives rise to 7 to 11 dark (xanthophyll rich) and light concentric yolk layers. (b) Ovary, follicles, and reproductive tract of a laying hen. (c) Follicular hierarchy. Follicles are arranged in order from the largest (F1) to the smallest. F1 also denotes the follicle next in line to be ovulated. The F1 follicle is the principal source of progesterone, the steroid hormone that induces the preovulatory surge of LH from the anterior pituitary.

the preovulatory release of LH is restricted to a 10- to 11-hour period each day, the so-called **open period.** The preovulatory release of LH can be initiated only from the onset of darkness to about 1 hour after the onset of the light period.

With age there is a decline in the rate of recruitment of follicles into the hierarchy and in the maturation of these follicles, and an increase in the rate of atresia of small follicles. If we induce a cessation in lay (**forced molt**), older hens lay eggs more frequently and of a greater quality when production is resumed.

Unlike the situation in female mammals, ovulation of the F1 follicle in birds is preceded by an increase in the concentrations of progesterone and LH. The preovulatory surge of LH is caused by a positive-feedback interaction, comparable to the effect of estrogens at the hypothalamus in mammals, but it is progesterone that stimulates the release of GnRH and hence

LH. LH, in turn, stimulates the secretion of more progesterone from the granulosa layer of the mature preovulatory follicle. Thus in birds, it is the granulosa cells that produce progesterone, most of which diffuses to the thecal cells, where it is converted to estrogen. As the follicles mature, the enzyme activity diminishes within the theca but increases in the granulosa, resulting in the enhanced production of progesterone at the time of ovulation. There are two possible reasons for this switch in the pattern of steroid synthesis by the maturing follicle. First, birds require estrogens for synthesis of Ca^{++}-ATPase in the shell gland as well as for yolk synthesis by the liver, and estrogens must be available in sufficient concentrations well before ovulation. Secondly, progesterone is the hormone responsible for the positive-feedback cycle, generating both the surge in LH and ovulation of the F1 follicle, and so the synthesis of this steroid needs to be restricted to the few mature follicles. The preovulatory releases of progesterone and LH are initiated 10 to 12 hours before ovulation, with LH peaking 6 to 8 hours before ovulation and progesterone peaking 2 to 4 hours before ovulation.

Ultimately the expulsion of the egg from the uterus (shell gland) into the vagina is mediated by an increase in the release of arginine vasotocin (AVT), a neurohypophyseal hormone. Once the egg enters the vagina, rhythmic contractions of abdominal and vaginal muscles force expulsion of the egg. This event, termed **bearing down,** results from a nervous reflex generated by the egg stretching the vaginal lumen. Vaginal muscles contract in response to the increased concentrations of AVT.

Spawning Cycles of Fish

Most information on the spawning cycles of viviparous species of fish has been obtained from studies in teleosts. Three modes of oocyte development are recognized, based on the recruitment of new oocytes: **synchronous, group synchronous,** and **asynchronous.** In those species showing synchronous development, all the oocytes mature and ovulate at the same time. This strategy is used by some species of salmon and eels that spawn once and die. In species employing group synchronous development, oocytes are partitioned into groups that mature at different rates and are then ovulated periodically throughout the breeding season. This group is categorized as being multiple spawners with comparatively short spawning seasons. The third strategy is the most complex, because oocytes are found at all stages of development throughout the spawning cycle. Oocytes are recruited continuously into the pool of maturing oocytes. As with group synchronous fish, asynchronous fish are multiple spawners but have a prolonged spawning season.

As with other vertebrates, reproduction in fishes depends on the hypothalamic–pituitary–gonadal axis: The pituitary gonadotropins *GTH-1* has FSH-like activity and *GTH-2* has LH-like activity, and each functions comparably to their mammalian counterparts. However, neuroendocrine regulation of GTH-2 in teleost fish is under dual hypothalamic control. GTH-2 release is inhibited by dopamine, which functions as a gonadotropin release–inhibiting hormone. Dopamine acts directly at the pituitary to modulate the actions of GnRH as well as the release of GTH-2.

GTH-1 is important for **vitellogenesis** (deposition of yolk in an oocyte) and early gonadal development, and is secreted for an extended period of time. For example, concentrations of LH receptor gene and GTH-1 in Coho salmon and catfish steadily increase from early in May until they reach peak concentrations in September and then decline through December (● Figure 16–20). In response to GTH-1, the ovarian granulosa cells synthesize estradiol, which stimulates the production of vitellogenin. **Vitellogenin** is a yolk protein synthesized in the liver and transported to the oocyte in the blood. GTH-1 is responsible for generating GTH-2 in the pituitary gland. Once vitellogenesis is completed, a preovulatory surge of GTH-2 is generated, which initiates final oocyte maturation and ovulation of the mature oocytes. The surge in GTH-2 is comparatively rapid (several days), although the actual profile is species dependent.

Normally the surge in GTH-2 is initiated by external environmental cues such as photoperiod, water temperature, and adequate water velocities along with conditions suitable for the survival of the larvae. Other factors that can elicit the GTH-2 surge include pheromones, sound production by male fishes, and the courtship behavior of males, factors that reliably ensure that the eggs are fertilized once ovulated. A dopamine antagonist can artificially induce ovulation; in cobia (*Rachycentron canadum*), a migratory pelagic species of fish, human chorionic gonadotropin (see later, p. 749) can induce ovulation.

In addition to GTH-2, final oocyte maturation and ovulation in fish require an additional factor produced by the follicular cells. *Maturation-inducing hormone (MIH),* is a progestogen synthesized via a two-cell mechanism comparable to that by which estradiol is synthesized. In this model, there are receptors for GTH-2 on both the thecal and granulosal cells, unlike the mechanism for estradiol synthesis.

Figure 16–20 ● Expression of the LH receptor (LHR) gene in the channel catfish ovary. Cyclic changes in the LH receptor gene through a complete spawning cycle reveal an increase in expression associated with ovulation. The LHR binds to the LH-like gonadotropin (GTH-2) of fish.

(*Source:* R. S. Kumara, S. Ijiri, & J. M. Trant, 2001, Molecular biology of channel catfish gonadotropin receptors: I. Cloning of a functional luteinizing hormone receptor and preovulatory induction of gene expression, *Biology of Reproduction* 64, p. 1016)

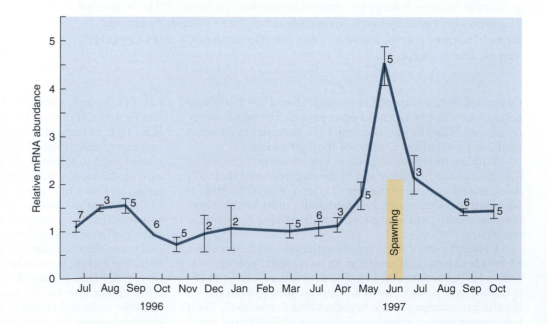

Oocytes can be ovulated either into the coelom of the fish and moved into the oviduct and the cloaca by ciliated epithelium lining the peritoneal cavity or into an ovarian cavity and moved down the oviduct to the cloaca without entering the peritoneal cavity itself.

■ Sexual maturity events in mammalian females are similar to those in males.

As with males, sexual maturation (puberty) in mammalian females is not a binary switch, but rather a long progression of greater and greater function of the hypothalamic–hypophyseal–gonadal axis until the time when the hypothalamus has become sufficiently mature that it allows the next follicle to produce enough estrogens to elicit the first LH surge. Unlike the fetal testes, the fetal ovaries do not need to be functional, because feminization of the female reproductive system automatically takes place in the absence of fetal testosterone secretion without the presence of female sex hormones. As the prepubertal female matures, an increase in the frequency of GnRH pulses is associated with a decrease in the sensitivity of the hypothalamus to estrogen. Puberty occurs at about 3 to 4 months of age in rabbits. However, it can vary from 6 to 16 months in sheep, and goats depending on the season of birth (spring or fall). Puberty is from 6 to 7 months in pigs, 12 months in cattle, and 15 to 18 months in horses. Domestication of the wolf, ancestor of dogs, has reduced the age of puberty from 2 years to between 6 and 12 months in modern dog breeds.

Puberty can be best considered as the age at which a female can support pregnancy. In most female mammals, a requisite body size is required before the onset of puberty can occur and there is evidence that metabolic signals can affect the development of the hypothalamic neurons and the rate of GnRH release. Reproductive function is influenced by body weight and nutritional status for gonadotropin secretion is reduced in animals and humans that are nutritionally restricted. Leptin, or its alternative name **pubertin**, may be a signal that informs the hypothalamus that metabolic stores are sufficient for initiating reproductive function. In this model, leptin directly modulates the release of prolactin and growth hormone, which could have indirect effects on the gonad. Leptin is a peptide produced by adipocytes and is a potent satiety factor (see p. 679). Leptin may play only a permissive function, functioning as a "barometer" of body function, and not necessarily an initiator of puberty, because leptin concentrations in monkeys do not change around the onset of puberty. The feeding of high-protein feedstuffs (flushing) to animals causes an acute response in terms of gonadotropin release to improve ovulation rates; this method has long been practiced in agriculture. Whether leptin is involved in these effects has not been determined. Other factors can also modulate the onset of puberty, including the photoperiod during the peripubertal period (sheep), season of birth (sheep), as well as the presence or absence of siblings during the peripubertal period (cattle and swine).

GnRH begins stimulating the release of the anterior-pituitary gonadotropic hormones, which in turn stimulate ovarian activity. The resultant secretion of estrogens by the activated ovaries induces growth and maturation of the female reproductive tract as well as development of the female secondary sexual characteristics. Estrogen's prominent action in the latter regard is to promote fat deposition. Enlargement of the mammary glands at puberty is due primarily to fat deposition in the mammary gland tissue and not to functional development of the mammary glands. The pubertal rise in estrogens does close the epiphyseal plates, however, halting further growth in height, similar to the effect of testosterone in males.

■ The oviduct is the site of fertilization.

You have now learned about the events that take place if fertilization does not occur in a mammal. Because the primary function of the reproductive system is, of course, reproduction, we next turn our attention to the sequence of events that ensue when this function is accomplished.

Fertilization, the union of male and female gametes, in most mammals occurs in the ampullary–isthmic junction, the midpoint of the oviduct. Thus both the ovum and the sperm must be transported from their gonadal sites of production to the *ampullary–isthmic junction* in the female.

Ovum Transport to the Oviduct

At ovulation the fragile ovum is released into the abdominal cavity, but the oviduct picks it up quickly. The fimbriae (Figure 16-4) contract in a sweeping motion to guide the released ovum into the oviduct. Furthermore, the fimbriae are lined by cilia, which are fine, hairlike projections that beat in waves toward the interior of the oviduct, further assuring the ovum's passage into the oviduct. Within the oviduct, the ovum is propelled rapidly by peristaltic contractions and ciliary action to the ampullary–isthmic junction.

Conception can take place during a limited time span in each cycle, the **fertile period.** If not fertilized, the ovum begins to disintegrate within 4 hours in the rabbit and 24 hours in the guinea pig and human. In most mammalian species studied, the survival of the egg is so short as to preclude fertilization in the uterus. Cells that line the reproductive tract are responsible for phagocytizing the ovum. Fertilization must therefore occur in the brief time when the ovum is still viable. One notable exception is the ability of the ovum in the mare to survive several estrous cycles while remaining viable in the oviduct. In contrast, the ability of sperm to survive in the female reproductive tract is highly variable among species. For example, in some species of bats sperm survive from the time of mating in the fall until the time of ovulation in the spring because of the presence of specific anatomic features involved in sperm storage. In farm and laboratory animals the survivability of sperm is considerably shorter and ranges from 30 to 35 days in the hen (which also has a sperm storage organ), 5 days in the mare and human, and 26 to 30 hours in the rabbit. (Contrast these values with those in a queen bee, who, after a single insemination, can continue to lay fertilized eggs for as long as four years.)

Sperm Transport to the Oviduct

Sperm are deposited in the vagina of some mammals at ejaculation (primates, cows, sheep, rabbits), whereas in others the sperm are delivered to the cervix or are deposited in the uterus (horses and pigs). These sperm must travel through the uterus and then up to the egg into the upper third of the oviduct. The first sperm arrive in the oviduct within a half hour after ejaculation. To accomplish this formidable journey, sperm need the help of the female reproductive tract. The first hurdle is passage through the cervical canal. Throughout most of the cycle, because of the high progesterone or low estrogen levels, the cervical mucus is too thick to permit sperm penetration. The cervical mucus become thin and watery enough to permit sperm to penetrate only when estrogen levels are high, as occurs in the presence of a mature follicle about to ovulate. Sperm migrate up the cervical canal under their own power. The canal remains penetrable for only two or three days during each cycle, around the time of ovulation. In species where the sperm are deposited in the vagina, the process of capacitation (see p. 725) is initiated as they pass through the cervix. However, in those species where the sperm are deposited directly in the uterus, capacitation is initiated here and is completed in the oviduct.

Once the sperm have entered the uterus, contractions of the myometrium churn them around in "washing-machine" fashion. This action quickly disperses the sperm throughout the uterine cavity. When sperm reach the oviduct, they must traverse a species-specific barrier that limits sperm transport at the uterotubal junction. Once beyond the barrier at the uterotubal junction, the sperm are moved up the oviduct toward the site of fertilization with the help of smooth muscle contractions and cilia that move in the direction of the ampullary–isthmic junction (region of the oviduct where the isthmus makes an anatomic transition into the ampulla) under the influence of high concentrations of estrogens present near ovulation. Furthermore, new research indicates that ova are not passive partners in conception. Mature eggs have been shown to release **allurin,** a recently identified chemical that attracts sperm and causes them to propel themselves toward the waiting female gamete, although this likely only influences cells that are very close (<1 cm) to the oocyte.

Even around ovulation time, when sperm can penetrate the cervical canal, of the several hundred million sperm deposited in a single ejaculate only a few thousand make it to the oviduct. The fact that only a very small percentage of the deposited sperm ever reach their destination is one reason why sperm concentration must be so high. Sperm, like ova, have limited life spans. Another reason so many sperm are needed is that the acrosomal enzymes assist in digesting the zona pellucida, allowing sperm to penetrate the zona and arrive in the perivitelline space (Figure 16–21). In addition, in some species a high percentage of sperm have nonfertilizing roles, as you will see shortly.

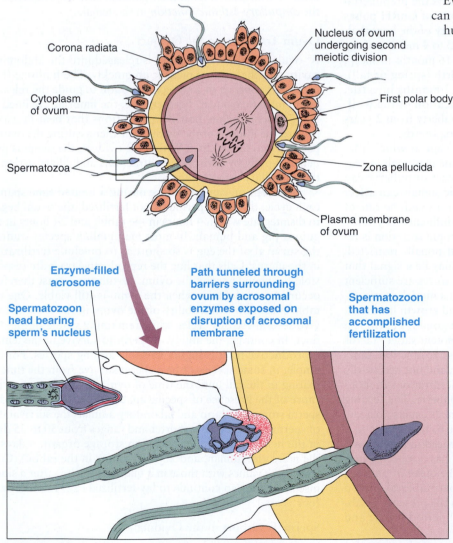

Corona radiata

Cytoplasm of ovum

Spermatozoa

Nucleus of ovum undergoing second meiotic division

First polar body

Zona pellucida

Plasma membrane of ovum

Enzyme-filled acrosome

Spermatozoon head bearing sperm's nucleus

Path tunneled through barriers surrounding ovum by acrosomal enzymes exposed on disruption of acrosomal membrane

Spermatozoon that has accomplished fertilization

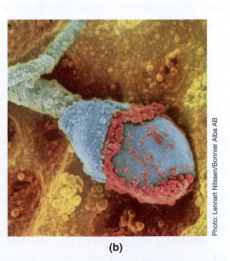

Photo: Lennart Nilsson/Bonnier Alba AB

(a)

(b)

Figure 16–21 ● **Process of fertilization (mammal).** (a) Schematic representation of sperm tunneling the barriers surrounding the ovum. (b) Scanning electron micrograph of a spermatozoon in which the acrosomal membrane has been disrupted and the acrosomal enzymes (in red) are exposed.

Fertilization

Although the tail of the sperm is used to maneuver for final penetration of the ovum, motility ceases once contact with the ovum is made. To fertilize an ovum, a sperm must first pass through the zona pellucida surrounding it (or equivalent jelly coating in nonmammalian animals). Contact of the sperm with the cumulus-oocyte complex results in the hyperactivation of the sperm cell, which in part is responsible for denuding the oocyte of the cumulus cells. Sperm can penetrate the zona pellucida only after binding with specific receptor sites on the surface of this layer. The binding partners between the sperm and ovum have recently been identified. **Fertilin,** a protein found on the plasma membrane of the sperm, binds with an egg *integrin,* a type of cell adhesion molecule that protrudes from the outer surface of the plasma membrane (see p. 63). Only sperm of the same species can bind to these egg receptors and pass through. This species-specific binding of the sperm to the proteins of the zona pellucida triggers the vesiculation of the sperm plasma membrane and the outer acrosomal membrane, resulting in the dispersal of these two membranes and the acrosomal contents they enclosed. The acrosomal contents include enzymes that digest a pathway through the jelly layer.

The first sperm to reach the ovum itself fuses with the plasma membrane of the secondary oocyte, at the caudal border of what had been the acrosome, a region referred to as the *equatorial segment.* The oocyte then engulfs the sperm cell, drawing the head of the cell into the oocyte cytoplasm. Entry of the sperm head into the oocyte triggers a number of events. First, a wave of calcium ions propagates along the inner surface of the oocyte plasma membrane. This wave of calcium ions causes the cortical granules (vesicles that reside just below the oocyte plasma membrane) to migrate to the plasma membrane and fuse with it. The contents of the cortical granules are released into the perivitelline space and their fusion with the membrane causes a flattening of the previously microvillar surface. Because the microvillar surface had been critical in facilitating the interaction of the equatorial segment of the acrosome-reacted sperm with the oocyte plasma membrane, no additional sperm cells can interact with the oocyte. We refer to this rapid calcium depolarization of the oocyte plasma membrane and subsequent change in its surface to prevent additional sperm binding, as the primary **block to polyspermy** (polyspermy is fertilization by more than one sperm). The contents of the cortical granules that diffuse across the perivitelline space interact with and change the proteins of the zona pellucida, hardening it such that sperm that are in the process of penetrating it are locked in place and no additional sperm can begin to digest their way through the zona pellucida.

The head of the fused sperm is gradually pulled into the ovum's cytoplasm by a growing cone that engulfs it. The sperm's tail is frequently lost in this process, but it is the head that carries the crucial genotypic information. Penetration of the sperm into the cytoplasm triggers the final meiotic division of the secondary oocyte. Within an hour, the sperm and egg pronuclei fuse. In addition to contributing its half of the chromosomes to the fertilized ovum, now called a **zygote,** the fertilizing sperm also activates ovum enzymes that are essential for the early embryonic developmental program.

We have been describing these events for mammals, but they are broadly similar in many phyla. Indeed, many of the steps of fertilization—penetration of the egg jelly coat, block to polyspermy, role of calcium—were first delineated in sea urchins (echinoderms).

■ Sperm may serve as competitors of sperm from other males.

In mammals with females that routinely copulate with many males (including many primates such as chimpanzees), males are said to engage in **sperm competition.** They typically have testes much larger than the average mammalian species, in order to produce large quantities of sperm to outcompete, and possibly to block and kill, rival sperm in the female tract. (In contrast, primates such as the gorilla in which one dominant male mates exclusively with females, the testes are comparatively smaller in size.)

?

Can Some Sperm Act as Active Competitors? Traditionally, sperm competition has been regarded as a matter of numbers: The male that deposits the most sperm into a female is more likely to win. But a few researchers have hypothesized that many sperm are not designed for fertilization. Rather, their role may be to prevent fertilizing sperm of other males from reaching the ovum by active means. Controversial evidence suggests that there are two types of these nonfertilizing sperm. **Blocker sperm** are said to have hooked flagella that allow them to join together in large masses, which form a physical barrier to any sperm that arrive later. **Killer sperm** are more active; according to the hypothesis, they bind to any sperm that are immunologically different, and kill them using their acrosomal enzymes. The evidence for blockers and killers is hotly disputed by other researchers, and the issue remains unsettled.

■ The mammalian blastocyst implants in the endometrium through the action of its trophoblastic enzymes.

During the first few days following fertilization, the zygote usually remains within the ampulla, because the high, decreasing concentrations of estrogens are still directing contractions and ciliary motion toward the site of fertilization. As the concentration of estrogens continues to decline, contractions and ciliary movement in the isthmic portion of the oviduct reverse, transporting the zygote to the uterus.

The Beginning Steps in the Ampulla

The zygote is not idle during this time, however. It rapidly undergoes a number of mitotic cell divisions to form a solid ball of cells called the **morula** (● Figure 16–22). Meanwhile, the rising concentrations of progesterone from the newly developing corpus luteum that formed after ovulation has a quieting influence on the uterus and stimulates the release of a nutrient medium (glycogen) from the endometrium into the reproductive tract lumen for use as energy by the early embryo. The nutrients stored in the cytoplasm of the mammalian ovum can sustain the product of conception for less than a

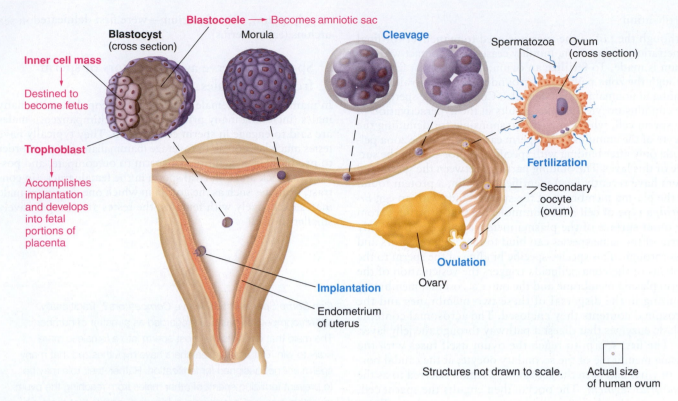

Blastocoele → Becomes amniotic sac

Blastocyst
(cross section)   Morula   **Cleavage**

Spermatozoa   Ovum
(cross section)

Inner cell mass

Destined to
become fetus

Fertilization

Trophoblast

Secondary
oocyte
(ovum)

Accomplishes
implantation
and develops
into fetal
portions of
placenta

Ovulation

Ovary

Implantation

Endometrium
of uterus

Structures not drawn to scale.   Actual size
of human ovum

Figure 16–22 ● **Early stages of development from fertilization to implantation
(mammal; shown for a human).** Note that the fertilized ovum progressively divides
and differentiates into a blastocyst as it moves from the site of fertilization in the
upper oviduct to the site of implantation in the uterus.

day. The concentration of secreted nutrients increases more
rapidly in the small confines of the ampulla than in the uterine
lumen.

Descent of the Morula to the Uterus

Several days after ovulation, progesterone is being produced
in sufficient quantities to relax the oviduct constriction, thus
permitting the morula to be propelled into the uterus by oviductal peristaltic contractions and ciliary activity. The temporary delay before the developing embryo passes into the uterus
allows sufficient nutrients to accumulate in the uterine lumen
to support the embryo until implantation can take place. If the
morula arrives prematurely, it dies.

When the morula descends to the uterus, it floats freely
within the uterine cavity for another three to four days, living
on the endometrial secretions and continuing to divide. During the first six to seven days after ovulation, whereas the developing embryo is in transit in the oviduct and floating in the
uterine lumen, the uterine lining is simultaneously being prepared for implantation or attachment under the influence of
luteal-phase progesterone.

Implantation of the Blastocyst
in the Prepared Endometrium

In the usual case, by the time the endometrium is suitable for
implantation (about a week after ovulation), the morula has

descended to the uterus and continued to proliferate and differentiate into a **blastocyst** capable of implantation. The week's
delay after fertilization and before implantation allows time
for both the endometrium and the developing embryo to prepare for implantation.

A blastocyst is a single-layered sphere of cells encircling a
fluid-filled cavity, with a dense mass of cells grouped together
at one side (Figure 16–22). This dense mass, called the **inner
cell mass,** is destined to become the embryo itself. The remainder of the blastocyst is never be incorporated into the fetus
but serves a supportive role during intrauterine life. The thin
outermost layer, the **trophectoderm,** is responsible for accomplishing implantation, after which it develops into the fetal
portion of the placenta. The **amnion** forms either by cavita-
tion, or opening up of a cavity (in rodents and humans) or folding (in noninvasive species).

When the blastocyst is ready to implant, its surface becomes sticky. By this time the endometrium is ready to accept
the early embryo. Implantation occurs within 2 to 5 weeks
after fertilization, the interval as short as 2 weeks in the cat
and as long as 5 weeks in cattle and horses. The blastocyst
adheres to the uterine lining on the side of its inner-cell mass
(● Figure 16–23).

During the preattachment phase, the blastocyst experiences
phenomenal growth. For example, the blastocyst of the cow
increases from 3 mm in diameter at day 17 of pregnancy to

250 mm by day 17. Blastocysts of the pig increase from 2 mm in size on day 10, to 200 mm on day 12 and 1,000 mm by day 16, mainly due to development of the extraembryonic membranes, or **placentation.**

Placentation is the vascularization of the chorionic epithelium and is not the same thing as elongation. Elongation in noninvasive species allows for maximum interaction with as much uterine surface as possible to ensure **maternal recognition of pregnancy** is accomplished and that surface area for attachment is sufficient to provide nutrients to the developing conceptus in a situation where intimate interaction with the maternal vasculature does not exist.

■ The placenta is the organ of exchange between maternal and fetal blood.

The glycogen stores (in rats and humans) in the endometrium are only sufficient to nourish the embryo during its first few weeks. To sustain the growing embryo for the duration of its intrauterine life, the **placenta,** a specialized organ of exchange between the maternal and fetal blood, is rapidly developed (● Figure 16–24). The placenta is derived from both trophoblastic and decidual tissue. In noninvasive species, **histotroph** is produced by the endometrium, which consists of proteins (some of which are growth factors), prostaglandins, ions, carbohydrates, and so on. Histotroph is absorbed by the embryo and digested in the yolk sac.

Modes of Placentation between Species

By the time the embryo has attached to the uterine luminal epithelium, the trophoblastic layer is two cell layers thick and is called the **chorion.** As the chorion continues to expand, it fuses with a structure called the **allantois** (an outgrowth of the hindgut) forming the vascularized chorioallantoic membrane. Placentas can be classified according to their degree of invasiveness and the anatomy of those regions of greatest maternal–fetal interaction (Figure 16–24). If we describe the potential placental layers available and then begin to remove layers as we describe progressively more invasive placentation, you will be able to see the interrelationship of the different modes of placentation. On the maternal side we could have endothelial cells lining the blood vessels, a basement membrane, endometrial stromal tissue, another basement membrane, and the uterine luminal epithelium. On the fetal side we could potentially have chorionic epithelium, a basement membrane, chorioallantoic stromal tissue, another basement membrane, and endothelial cells lining the blood vessels. Thus six potential tissue layers separate maternal and fetal blood. In the pig, cow, and ewe all six tissue layers are present, and the type of placentation is referred to as **epithelial-** (the maternal component) **chorial** (the fetal component). In the pig, chorion is in contact with the uterine luminal epithelium throughout the surface of the placenta, so this arrangement is called **diffuse.** In ruminants, the greatest contact of the chorion with the uterine luminal epithelium occurs at discreet sites at which fetal membranes interact with maternal *caruncles* (round thickenings) such that chorionic villi project into the caruncles forming **placentomes,** and so we call this **placentomal.** In the carnivores, the conceptus erodes the uterine luminal epithelium and underlying stroma at sites of greatest attachment such that

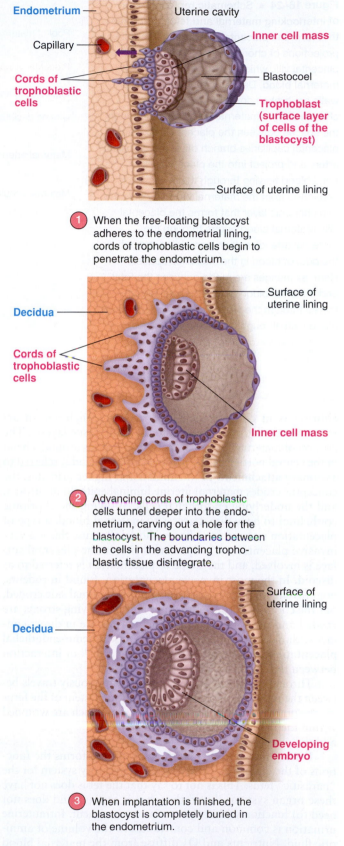

1 When the free-floating blastocyst adheres to the endometrial lining, cords of trophoblastic cells begin to penetrate the endometrium.

2 Advancing cords of trophoblastic cells tunnel deeper into the endometrium, carving out a hole for the blastocyst. The boundaries between the cells in the advancing trophoblastic tissue disintegrate.

3 When implantation is finished, the blastocyst is completely buried in the endometrium.

Figure 16–23 ● Implantation of the blastocyst (primate).

Figure 16–24 ● Schematic representation of interlocking maternal and fetal structures that form the placenta in a human. Fingerlike projections of chorionic (fetal) tissue form the placental villi, which protrude into a pool of maternal blood. Decidual (maternal) capillary walls are broken down by the expanding chorion so that maternal blood oozes through the spaces between the placental villi. Fetal placental capillaries branch off of the umbilical artery and project into the placental villi. Fetal blood flowing through these vessels is separated from the maternal blood by only the thin chorionic layer that forms the placental villi. Maternal blood enters through the maternal arterioles, then percolates through the pool of blood in the intervillus spaces. Here, exchanges are made between the fetal and maternal blood before the fetal blood leaves through the umbilical vein and maternal blood exits through the maternal venules.

(*Source:* Adapted from Cecie Starr, *Biology: Concepts and Applications,* Fourth Edition, Fig. 38.25b, p. 655. Copyright © 2000 Brooks/Cole)

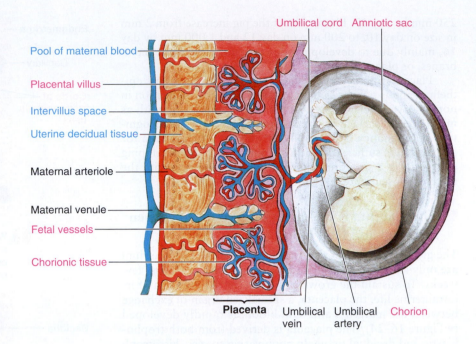

chorion is in contact with maternal endothelium, an arrangement called **endothelial-chorial** (four tissue layers). The erosion of maternal tissue occurs only in regions around a band in the central portion of the placenta, or zone, and is referred to as **zonary attachment.** In the human and higher primates, the conceptus erodes maternal uterine luminal epithelium, stroma, and the underlying endothelial cells, allowing the chorionic epithelium to have direct access to maternal blood, a type of placentation referred to as **hemo-chorial.** Because this is a very invasive placentation, only a small site on the placental surface is involved, and the shape of placentation is referred to as **discoid.** In the most invasive placentation, found in rodents, not only are all three tissue layers on the maternal side eroded, but the chorioallantoic epithelium and underlying stroma are eroded and the conceptus endothelial cells are in direct contact with maternal blood; we refer to this as **hemo-endothelial placentation.** It is found in the shape of a disk of interaction between the conceptus and maternal system.

Throughout gestation, fetal blood continuously travels between the placental villi and the circulatory system of the fetus via the **umbilical artery** and **umbilical vein,** which are wrapped within the **umbilical cord,** a lifeline between the primate fetus and the placenta (Figure 16–24).

During intrauterine life, the placenta performs the functions of the digestive system and the respiratory system for the "parasitic" fetus. This is not to say that the fetus does not have these organ systems, but rather that it cannot (and does not need to) function within the uterine environment. Intrauterine urination is common and contributes to the volume of amniotic fluid. Nutrients and O_2 diffuse from the maternal blood across the thin placental barrier into the fetal blood, whereas CO_2 and other metabolic wastes simultaneously diffuse from the fetal blood into the maternal blood. The nutrients and O_2

brought to the fetus in the maternal blood are acquired by the maternal digestive and respiratory systems, and the maternal lungs and kidneys eliminate the CO_2 and wastes transferred into the maternal blood, respectively. Thus the maternal digestive tract, respiratory system, and kidneys serve the fetus's needs as well as the mother's.

Transport of Substances across the Placental Barrier

Some substances traverse the placental barrier by special, mediated transport systems in the placental membranes, whereas others move across by simple diffusion. Unfortunately, many drugs, environmental pollutants, other chemical agents, and microorganisms in the female's bloodstream can cross the placental barrier, some of which are harmful to the developing fetus.

In addition to serving as an organ of exchange, the placenta is believed to play a key role in preventing rejection of the fetus by the maternal immune system. The fetus is actually a "foreigner," because it is half derived from genotypically different paternal chromosomes. The mechanisms hypothesized to explain immunological acceptance of the foreign fetus by the female seem to somehow suppress the maternal immune responses to the fetus.

The placenta assumes yet another important responsibility—it becomes a temporary endocrine organ during pregnancy, a topic to which we now turn.

■ Hormones secreted by the corpus luteum and placenta play a critical role in maintaining pregnancy.

The placenta has the remarkable capacity to secrete a number of peptide and steroid hormones essential for maintaining preg-

Table 16–4 ▌ Placental Hormones

Hormone	Function
Chorionic gonadotropin	Maintains corpus luteum of pregnancy
	Stimulates secretion of testosterone by developing testes in XY embryos
Estrogen (*also secreted by corpus luteum of pregnancy*)	Stimulates growth of myometrium, increasing uterine strength for parturition
	Helps prepare mammary glands for lactation
Progesterone (*also secreted by corpus luteum of pregnancy*)	Suppresses uterine contractions to provide quiet environment for fetus
	Promotes formation of cervical mucus plug to prevent uterine contamination
	Helps prepare mammary glands for lactation
Chorionic somatomammotropin (*also called* placental lactogen; *has a structure similar to both prolactin and growth hormone*)	Helps prepare mammary glands for lactation (similar to prolactin)
	Believed to reduce maternal use of glucose and to promote breakdown of stored fat (similar to growth hormone) so that greater quantities of glucose and free fatty acids can be shunted to the fetus
Relaxin (*also secreted by corpus luteum of pregnancy*)	Softens the cervix in preparation for cervical dilation at parturition
	Loosens connective tissue between pelvic bones in preparation for parturition
Placental PTHrp (*parathyroid hormone-related peptide*)	Increases maternal plasma Ca^{2+} levels for use in calcifying fetal bones; if necessary, promotes localized dissolution of maternal bone, mobilizing mother's Ca^{2+} stores for use by developing fetus

nancy. The most important are *chorionic gonadotropin, estrogen,* and *progesterone* (▌ Table 16–4). Serving as the major endocrine organ of pregnancy, the placenta is unique among endocrine tissues in two regards. First, it is a transient tissue. Second, secretion of its hormones is not subject to extrinsic control, in contrast to the stringent, often-complex mechanisms that regulate the secretion of other hormones. Instead, the type and rate of placental hormone secretion depend primarily on the species of animal and stage of pregnancy.

In all mammals the corpus luteum (CL) is important in the early stages of pregnancy for secreting progesterone but, as exemplified in the human, ewe, and mare, its contribution declines with time as the placental contribution of progesterone increases. For example, the corpus luteum is not required in women and ewes after approximately 50 days and in the mare after 100 days of pregnancy, whereas the corpus luteum is required for the duration of pregnancy in the pig, cow, goat, and deer.

One of the first events in primates is the secretion by the developing chorion of *chorionic gonadotropin (CG)*, a peptide hormone that prolongs the life span of the corpus luteum. In the mare, *equine chorionic gonadotropin (eCG)* is also called *pregnant mare's serum gonadotropin (PMSG).* The placenta of the mare is notable for unique structures termed **endometrial cups,** temporary endocrine units ranging in size from millimeters to several centimeters that produce eCG. After approximately 60 days of gestation, these cups are sloughed into the uterine lumen. Recall that during the menstrual cycle in humans (and some other primates), the corpus luteum degenerates and the highly prepared, luteal-dependent uterine lining sloughs if fertilization and implantation have not occurred. When fertilization does occur, the implanted human blastocyst saves itself by producing human *CG (hCG).* This hormone, which is functionally similar to LH, stimulates and maintains the corpus luteum so that it does not degenerate. Now called the corpus luteum of pregnancy, this ovarian endocrine unit grows even larger and produces increasingly greater amounts of progesterone until the placenta takes over secretion of these steroid hormones (see p. 735). Because of the persistence of circulating estrogens and progesterone, the thickened endometrial tissue of the human is maintained instead of sloughing. In higher primates, accordingly, menstruation ceases during pregnancy. The eCG of the mare is produced around 60 days of gestation, and like the hCG, eCG primarily exhibits LH-like activity, although if administered to other species eCG exhibits FSH-like activity. This is important for the mare, because the corpora lutea do not persist beyond about day 70 of gestation and are supplemented by the growth and luteinization of new follicles under the influence of eCG to save the pregnancy. These accessory corpora lutea only maintain pregnancy for another short period until the placenta finally takes over progesterone production at around 100 days of gestation.

▌ Maternal body systems respond to the increased demands of gestation.

The period of gestation (pregnancy) is about 1 month in the rabbit, 3.8 months in the sow, 5 months in the ewe and goat, 9 months in the cow and human, 15 to 16 months in the sperm whale, and 1.8 years in the African forest and savanna elephants. During gestation, the fetus continues to grow and develop to the point of being able to leave its maternal life support system. Meanwhile, a number of physical changes take place within the female to accommodate the demands of the pregnancy. The most obvious change is uterine enlargement. The uterus expands and increases in weight more than 16 times in humans, exclusive of its contents. The mammary glands enlarge and develop the capability to produce milk. Body systems other than the reproductive system make needed adjustments. The volume of blood increases by 30%, and the cardiovascular system responds to the increasing demands of the growing placental mass. The weight gain experienced during pregnancy is due only in part to the weight of the fetus. The remainder is primarily caused by the increased weight of the uterus, including the placenta, and the increased blood volume. Respiratory activity is increased by about 16% to handle the additional fetal requirements for O$_2$ use and CO$_2$ removal. Urinary output increases, and the kidneys excrete the additional wastes from the fetus.

The increased metabolic demands of the growing fetus result in increased nutritional requirements for the female. In general, the fetus takes what it needs from the female, even if this leaves the female with a nutritional deficit. For example, the placental hormone *chorionic somatomammotropin (CS),* also known as *placental lactogen (PL),* is thought responsible for the decreased use of glucose by the female and the mobilization of free fatty acids from maternal adipose stores, similar to the actions of growth hormone (see p. 272). (In fact, CS has a structure similar to both growth hormone and prolactin. In the mammary gland, CS helps prepare these glands for lactation, similar to prolactin's effect, although much weaker.) As a result of the CS-induced metabolic changes in the female, greater quantities of glucose and fatty acids are available for shunting to the fetus. The extent to which CS exerts somatotropic versus lactogenic effects is ultimately species dependent. For example, in the ewe CS is more effective in promoting lactation. Also, if the female does not consume sufficient Ca^{++} in her diet, placental parathyroid-hormone–related peptide (**PTHrp**) mobilizes Ca^{++} from the maternal bones to ensure adequate calcification of the fetal bones.

■ **Changes during late gestation prepare for parturition.**

Parturition (**labor, delivery,** or **birth**) is the process by which the uterus expels the fetus and placenta from the female. Parturition requires (1) dilation of the cervical canal to accommodate passage of the fetus from the uterus through the vagina and to the outside, (2) contractions of the uterine myometrium that are sufficiently strong to expel the fetus and (3) expulsion of the placenta.

Several events take place near the end of pregnancy in preparation for parturition. Throughout gestation, the cervix remains sealed. As parturition approaches, the cervix begins to soften (or "ripen") as a result of the dissociation of its tough connective tissue (collagen) fibers. This softening is believed to be caused by **relaxin,** a peptide hormone produced in most mammals by the corpus luteum of pregnancy and by the placenta. However, in species such as the rabbit, relaxin is produced entirely in the placenta. Relaxin also "relaxes" the pelvic bones. Furthermore, $PGF_{2\alpha}$ released by the placental membranes overlying the cervix stimulates the synthesis of relaxin and promotes the production of cervical enzymes that degrade the collagen fibers and help soften the cervix. $PGF_{2\alpha}$ initiates regular, nondirectional uterine contractions and regression of the CL. These small contractions (sometimes called **contractures**) help orient the fetus(es) so that the head and front appendages are in contact with the cervix in preparation for exiting through the birth canal. If the fetus is not positioned correctly, a difficult birth (**dystocia**), results.

Rhythmic, coordinated contractions, usually painless at first, begin at the onset of true labor in response to pituitary *oxytocin.* As labor progresses, the contractions occur with increasing frequency and intensity and are accompanied by increasing discomfort. These strong, rhythmic, and directional contractions force the fetus against the cervix, resulting in dilation of the cervix. Then, after having dilated the cervix sufficiently for passage of the fetus, these contractions force the fetus out through the birth canal.

■ **The factors that trigger the onset of parturition are only partially characterized.**

The exact factors responsible for triggering this change in uterine contractility and thus initiating parturition are not fully established, although considerable progress has been made

in recent years in unraveling the sequence of events. During early gestation, maternal estrogen levels are relatively low, but, as gestation proceeds, placental estrogen secretion continues to rise, associated with the increase in concentrations of $PGF_{2\alpha}$. However, the timing and magnitude of the increase in estrogens vary among species. In the immediate days before the onset of parturition, soaring levels of estrogens bring about changes in the uterus and cervix to prepare them for labor and delivery (● Figure16–25). First, high levels of estrogens promote the synthesis of connexins within the uterine smooth-muscle cells. These myometrial cells are not functionally linked to any extent throughout most of gestation. The newly manufactured connexins are inserted in the myometrial plasma membranes to form gap junctions (see p. 64) that electrically link the uterine smooth-muscle cells so that they become able to contract as a coordinated unit.

Simultaneously, the high levels of estrogens cause a dramatic, progressive increase in the concentration of myometrial receptors for oxytocin. Together, these myometrial changes collectively bring about the increased uterine responsiveness to oxytocin that ultimately initiates labor. **Oxytocin** is a peptide hormone that is produced by the hypothalamus, stored in the posterior pituitary, and released into the blood from the posterior pituitary on nervous stimulation by the hypothalamus (see p. 266). Oxytocin is a powerful uterine muscle stimulant and is known to play the key role in the progression of labor. However, this hormone was discounted as serving as the trigger to parturition because the circulating concentrations of oxytocin remain constant prior to the onset of labor. The discovery that uterine responsiveness to oxytocin is greater at term than in nonpregnant females (because of the increased concentration of myometrial oxytocin receptors) led to the now widely accepted conclusion that labor begins when the oxytocin receptor concentration reaches a critical

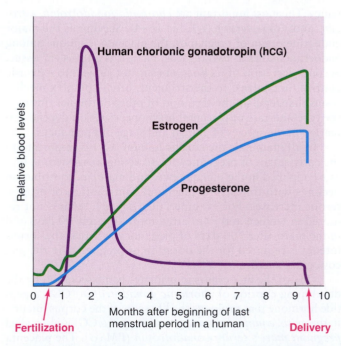

Figure 16–25 ● Secretion rates of human placental hormones.

threshold that permits the onset of strong, coordinated contractions in response to ordinary concentrations of circulating oxytocin.

In addition to preparing the uterus for labor, the increasing levels of estrogens promote production of the local prostaglandins that contribute to cervical ripening. Oxytocin has other roles, including the triggering of uterine contractions during orgasm, milk ejection during lactation, and maternal behavior related to bonding with the father and the offspring (see later). Oxytocin is used clinically to induce labor if it is not progressing well.

Role of Corticotropin-Releasing Hormone

Until recently, scientists were baffled by the factors responsible for the rising levels of placental estrogen secretion. Recent research has shed new light on the probable mechanism. Evidence suggests that corticotropin-releasing hormone (CRH) secreted by the fetal portion of the placenta into both the maternal and fetal circulation not only drives the synthesis of placental estrogen, thus ultimately dictating the timing of the onset of labor, but also promotes changes in the fetal lungs needed for breathing air; see ● Figure 16–27. Recall that CRH is normally secreted by the hypothalamus and regulates the output of ACTH by the anterior pituitary. In turn, ACTH stimulates production of both cortisol and DHEA by the adrenal cortex. In the fetus, much of the CRH comes from the placenta rather than solely from the fetal hypothalamus. The additional cortisol secretion summoned by the extra CRH promotes fetal lung maturation. Specifically, cortisol stimulates the synthesis of pulmonary surfactant, which facilitates lung expansion and reduces the work of breathing (see p. 486).

The bumped-up rate of DHEA secretion by the adrenal cortex in response to placental CRH leads to the rising concentrations of placental estrogen secretion. Recall that the human placenta converts DHEA from the fetal adrenal gland into estrogen, which enters the maternal bloodstream (● Figure 16–26). In most other species, the placenta has the ability to synthesize estrogens *de novo*.

When sufficiently high, this estrogen level combined with lowered progesterone sets in motion the events that initiate labor. Thus pregnancy duration and delivery timing are determined largely by the placenta's rate of CRH production. That is, a "placental clock" ticks out a predetermined length of time until parturition. The timing of parturition is established early in pregnancy, with delivery at the end point of a maturational process that extends throughout the entire gestation. The ticking of the placental clock is measured by the rate of placental secretion. As the pregnancy progresses, CRH concentrations in maternal plasma rise. Researchers can accurately predict the timing of parturition by measuring the maternal plasma concentrations of CRH. Higher-than-normal concentrations are associated with premature deliveries, whereas lower-than-normal concentrations indicate late deliveries. These and other data suggest that a critical level of maternal CRH of placental origin may directly trigger parturition. Placental CRH ensures that when labor begins, the infant is ready for life outside the womb. It does so by concurrently increasing the fetal cortisol needed for lung maturation and the estrogens needed for the uterine changes that bring on labor. The remaining unanswered question regarding the placental clock is, What controls placental secretion of CRH?

■ Parturition is accomplished by a positive-feedback cycle.

Once the high concentrations of estrogens have increased uterine responsiveness to oxytocin to a critical level and regular uterine contractions have begun, myometrial contractions progressively increase in frequency, strength, and duration throughout labor until the uterine contents are expelled.

Myometrial contractions incessantly increase as labor progresses because a positive-feedback cycle involving oxytocin and prostaglandin ensues (Figure 16–27). Each uterine contraction begins at the oviductal end of the uterus and sweeps downward, forcing the fetus(es) toward the cervix. Pressure of the fetus against the cervix accomplishes two things. First, the fetal head pushing against the softened cervix acts as a wedge to dilate the cervical canal. Second, cervical stretch stimulates the release of oxytocin through a neuroendocrine reflex. Stimulation of receptors in the cervix in response to fetal pressure produces a neural signal that travels up the spinal cord to the hypothalamus, which in turn trig-

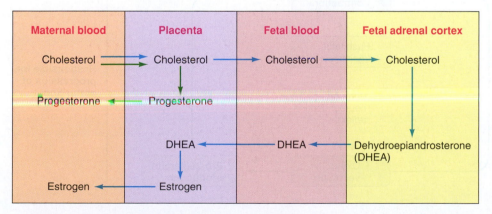

Figure 16–26 ● **Secretion of estrogen and progesterone by the human placenta.** The placenta secretes increasing quantities of progesterone and estrogen into the maternal blood after the first trimester. The placenta itself can convert cholesterol into progesterone (green pathway) but lacks some of the enzymes necessary to convert cholesterol into estrogen. However, the placenta can convert DHEA derived from cholesterol in the fetal adrenal cortex into estrogen when DHEA reaches the placenta by means of the fetal blood (blue pathway).

Maternal blood	Placenta	Fetal blood	Fetal adrenal cortex
Cholesterol →	Cholesterol →	Cholesterol →	Cholesterol
Progesterone ←	Progesterone		
	DHEA ←	DHEA ←	Dehydroepiandrosterone (DHEA)
Estrogen ←	Estrogen		

→ Pathway for placental synthesis of progesterone
→ Pathway for placental synthesis of estrogen

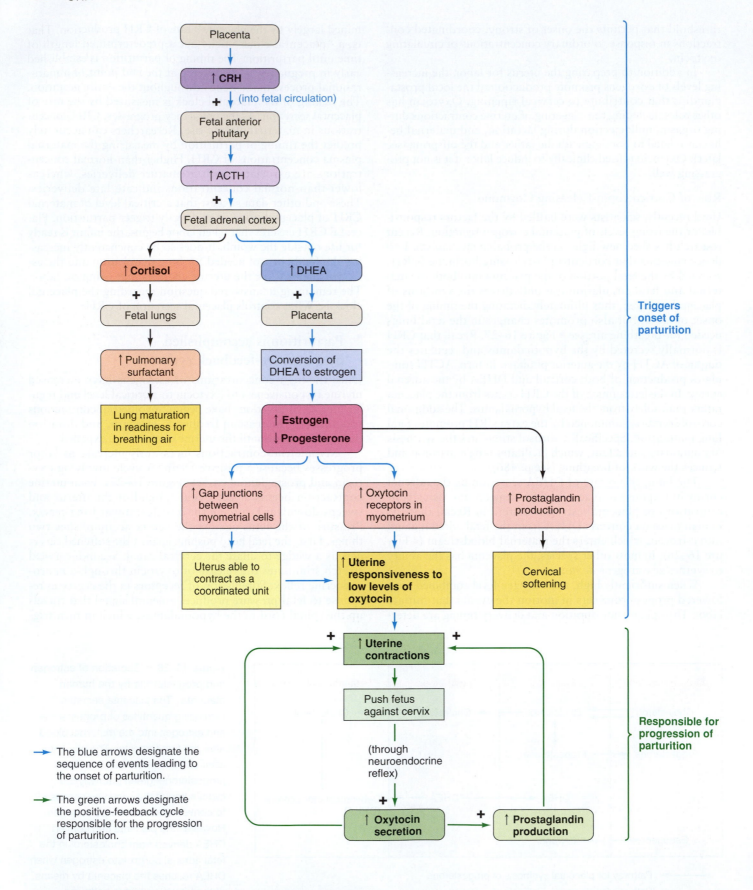

Figure 16–27 ● Initiation and progression of parturition (mammal).

gers oxytocin release from the posterior pituitary. Additional oxytocin promotes more powerful uterine contractions. As a result, the fetus is pushed more forcefully against the cervix, stimulating the release of even more oxytocin, and so on. This cycle is reinforced as oxytocin stimulates prostaglandin production by the decidua. As a powerful myometrial stimulant, prostaglandin further enhances uterine contractions. Oxytocin secretion, $PGF_{2\alpha}$ production, and uterine contractions continue to increase in positive-feedback fashion throughout labor until the pressure on the cervix is relieved by delivery.

Labor is divided into three stages: (1) cervical dilation, (2) delivery of the young, and (3) delivery of the placenta. At the onset of labor or sometime during the first stage, the membranes surrounding the amniotic sac rupture. The fluid surrounding the amnion is **chorioallantoic fluid.** The amnion does not rupture until much later during parturition so it can protect the fetus from physical damage during the early stages of parturition. As the chorioallantoic fluid escapes out of the vagina, it helps lubricate the birth canal. During the first stage, the cervix of ungulates (hoofed mammals) is forced to dilate to accommodate the diameter of the young's head and front feet. This stage is the longest, lasting from 1 to 12 hours in the sow, 2 to 6 hours in the ewe and cow, and several hours to as long as 24 hours in humans. If another part of the fetus's body other than the head is oriented against the cervix, it is generally less effective than the head as a wedge, although in the sow half of the fetuses come headfirst and half rear-end first. The head has the largest diameter of the fetus's body. In some species, if the fetus approaches the birth canal feet first, the cervix may not be dilated sufficiently by the feet to permit passage of the head. Without medical intervention in such a case, the fetus's head would remain stuck behind the too-narrow cervical opening.

The second stage of labor, the actual birth, begins once cervical dilation is complete. When the offspring begins to move through the cervix and vagina, stretch receptors in the vagina activate a neural reflex that triggers contractions of the abdominal wall in synchrony with the uterine contractions. These abdominal contractions greatly increase the force pushing the fetus through the birth canal. Stage 2 is usually much shorter than the first stage and lasts 30 to 90 minutes. The fetus is still attached to the placenta by the umbilical cord at birth, and this cord breaks as the newborn animal begins to move about following delivery, with the stump shriveling up in a few days to form the umbilicus (navel).

Shortly after delivery of the fetus, a second series of uterine contractions causes the placenta to separate from the endometrium and be expelled through the vagina. Delivery of the placenta, or afterbirth, constitutes the third stage of labor, which is typically the shortest stage, being completed within 15 to 30 minutes in humans, 1 hour in the mare, 5 to 8 hours in the ewe, and as long as 12 hours in the cow, after the young is born. After the placenta is expelled, continued contractions of the myometrium constrict the uterine blood vessels supplying the site of placental attachment to prevent hemorrhage. In sows, cows, ewes, mares, and so on, there is no erosion of maternal tissue and therefore no danger of hemorrhage. After delivery, the uterus shrinks to its pregestational size, a process known as **involution,** which requires 4 weeks in the ewe and sow, 12 weeks in the dog, and 4 to 6 weeks in the human.

After this period, the endometrium is restored to its nonpregnant state. Involution occurs largely because of the precipitous fall in circulating estrogens and progesterone when the placental source of these steroids is lost at delivery. The process is facilitated in females who feed their infants, because of the oxytocin released in response to suckling. In addition to playing an important role in lactation, this periodic nursing-induced postpartum release of oxytocin promotes myometrial contractions that stimulate uterine muscle tone, thus enhancing involution.

■ Lactation requires multiple hormonal inputs.

The female reproductive system of mammals supports the new being from the moment of its conception through gestation and continues to nourish it during its early life outside the supportive uterine environment. However, only two species of mammals, whales and bears, can fast during the energetically challenging period of lactation. This strategy permits baleen whales to feed in the seasonally productive polar regions of the world's oceans and retain the advantage of breeding in the warmer, tropical regions of the world.

Because the evolution of the mammary gland preceded that of placental gestation, even egg-laying mammals (monotremes) produce milk for their young (hence the name "mammal"). **Lactogenesis** is the process by which mammary alveolar cells acquire the ability to secrete milk (or its equivalent) for milk is essential for survival of the newborn, and, interestingly, the basic structure of the mammary gland is remarkably consistent across the class Mammalia. Accordingly, development begun during puberty is completed during gestation as the mammary glands are prepared for lactation (milk production).

The mammary gland is a modified sweat gland, the basic structure and location established during embryonic development. The glands develop from the so-called mammary or **milk line,** two rows that extend from the thoracic region, parallel to the midline of the abdomen that appears in the early embryo. Subsequently, the milk lines differentiate into mammary buds, the number and location dependent on the species of mammal. For example, the mammary glands of some mammals (deer, cattle, goats, camels, horses, and sheep) have an inguinal (located near the groin) location and in nonpregnant females consist mostly of adipose tissue and a rudimentary duct system. In primates and elephants, the mammary glands develop in the pectoral region, whereas in litter-bearing species, such as rabbits, dogs, cats, and pigs, mammary glands develop along the entire length of the milk line. Pigs normally have 7 pairs of mammary glands, whereas dogs and cats usually have 5 pairs of glands. The sea mammals (whales, dolphins, seals, and manatees) develop mammary buds in either the pectoral or inguinal region. Structures associated with the mammary gland include the **teat** or **nipple,** the part of the mammary gland from which the young suckle milk. At its distal end, the teat has fine openings that permit expulsion of the milk. The number of openings can be as few as one, such as in cattle, which have a collecting **cistern,** which permits pooling of milk from several ducts before exiting the gland, as compared to the nipple, which has multiple lactiferous ducts.

The mammary gland (udder) of the cow has distinct right and left halves, and each has a front and hindquarter. Each half of the mammary gland is independent with regard to its blood supply, innervation, and lymphatic drainage. Under the hormonal environment present during pregnancy, the mammary glands develop the internal glandular structure and function necessary for milk production. A mammary gland capable of lactating consists of a network of progressively smaller ducts that branch out from the teat and terminate in lobules (● Figure 16–28). Each **lobule** is made up of a cluster of saclike, epithelial-lined *alveoli* that constitute the milk-producing glands. Milk is synthesized by the epithelial cells and then secreted into the alveolar lumen, which is drained by a milk-collecting duct that transports the milk to the teat cistern. Ejection of milk from the teat occurs via the teat canal, which is normally kept closed by a muscle sphincter (circular muscle whose contraction closes the canal). **Mastitis**, infection of the mammary gland by microorganisms, can occur if the sphincter is not kept tightly closed.

Prevention of Lactation during Gestation

Most mammary gland growth occurs during pregnancy, when the high concentration of estrogens promotes extensive duct development, and the high level of progesterone stimulates abundant alveolar-lobular formation. The adipose tissue is steadily consumed and replaced by ducts, lobular alveoli, blood and lymph vessels, and the necessary connective tissue structures associated with the suspensory apparatus. Elevated concentrations of **prolactin** (an anterior pituitary hormone stimulated by the rising levels of estrogen) and **chorionic somatomammotropin** (a peptide hormone produced by the placenta that has a structure similar to both prolactin and growth hormone) contribute to mammary gland development by inducing the synthesis of enzymes needed for milk production.

Most of these changes in the mammary glands occur during the last half of gestation, so by the middle of pregnancy the mammary glands are fully capable of producing milk. However, milk secretion does not occur until parturition. The high estrogen and progesterone concentrations during the last half of pregnancy prevent lactation by blocking prolactin's stimulatory action on milk secretion. Prolactin is the primary stimulant of milk secretion. Thus even though the high levels of placental steroids induce the development of the milk-producing machinery in the mammary glands, they prevent these glands from becoming operational until the young is born and milk is needed. The abrupt decline in estrogens and progesterone that occurs with loss of the placenta at parturition initiates lactation. (The functions of estrogens and progesterone during gestation and lactation as well as throughout the reproductive life of females are summarized in ▌ Table 16–5.)

Once milk production begins after delivery, two hormones maintain lactation: (1) prolactin, which acts on the alveolar epithelium to promote secretion of milk, and (2) oxytocin, which produces **milk ejection**. The latter refers to the forced expulsion of milk from the lumen of the alveoli out through the ducts. Release of both of these hormones is stimulated by a neuroendocrine reflex triggered by suckling (● Figure 16–29). Let's examine each of these hormones and their roles in further detail.

Oxytocin Release and Milk Ejection

The neonate cannot directly suck milk out of the alveolar lumen. Instead, milk must be actively squeezed out of the alveoli into the ducts and hence toward the nipple by contraction

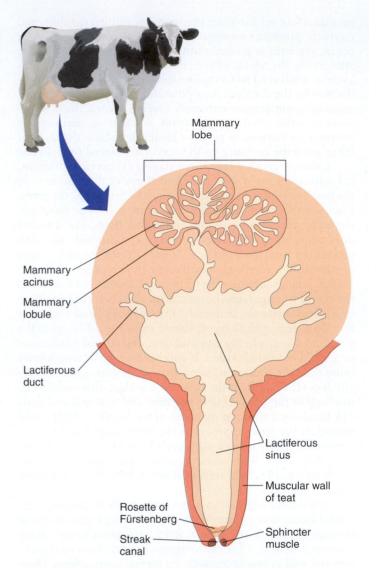

Figure 16–28 ● A cow mammary gland or teat.

(*Source:* Ruckebusch, Phaneuf, & Dunlop, 1991, *Physiology of Small and Large Animals*, B. C. Decker, Inc., Figure 59–4, p. 620)

of specialized myoepithelial cells (musclelike epithelial cells) that surround each alveolus. Suckling of the mammary gland by the neonate stimulates sensory nerve endings in the nipple, initiating action potentials that travel up the spinal cord to the hypothalamus. Thus activated, the hypothalamus triggers a burst of oxytocin release from the posterior pituitary. Oxytocin in turn stimulates contraction of the myoepithelial cells in the mammary glands to bring about milk ejection, or "milk letdown." Cetacean (whale) young lack proper lips, so gripping the nipple is difficult at best. For this reason the female actually squirts the milk into the mouth of the young, using contractions of both the myoepithelial cells as well as surrounding cutaneous muscles. Milk letdown continues only as long as the infant continues to nurse. In this way, the milk ejection reflex ensures that milk exits the mammary glands only in the amount needed by the young. Even though the alveoli may be full of milk, the milk cannot be released without oxytocin. Often the presence of the calf or other conditioned reflexes

Table 16–5 ▌ Actions of Estrogen and Progesterone (in Mammals)

Estrogen

Effects on Sex-Specific Tissues

Essential for egg maturation and release

Stimulates growth and maintenance of entire female reproductive tract

Stimulates granulosa cell proliferation, which leads to follicle maturation

Thins cervical mucus to permit sperm penetration

Enhances transport of sperm to oviduct by stimulating upward contractions of uterus and oviduct

Stimulates growth of endometrium and myometrium

Induces synthesis of progesterone receptors in endometrium

Triggers onset of parturition by increasing uterine responsiveness to oxytocin during late gestation through a twofold effect: by inducing synthesis of myometrial oxytocin receptors and by increasing myometrial gap junctions so that the uterus can contract as a coordinated unit in response to oxytocin

Other Reproductive Effects

Promotes development of secondary sexual characteristics

Controls GnRH and gonadotropin secretion

   Low levels inhibit secretion

   High levels responsible for triggering LH surge

Stimulates duct development in mammary glands during gestation

Inhibits milk-secreting actions of prolactin during gestation

Nonreproductive Effects

Promotes fat deposition

Increases bone density; closes epiphyseal plates

Reduces blood cholesterol

Progesterone

Prepares a suitable environment for nourishment of a developing embryo/fetus

Promotes formation of a thick mucus plug in the cervical canal

Inhibits hypothalamic GnRH and gonadotropin secretion

Stimulates alveolar development in the mammary gland during gestation

Inhibits the milk-secreting actions of prolactin during gestation

Inhibits uterine contractions during gestation

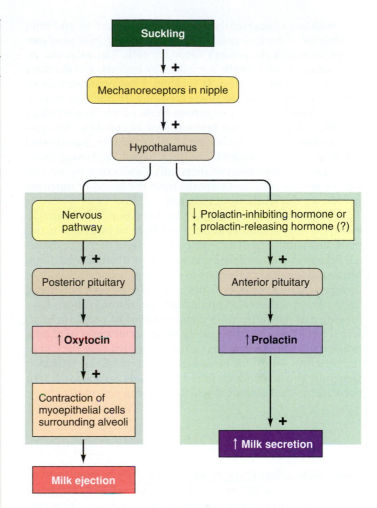

Figure 16–29 ● Suckling reflexes (mammal).

can generate a release of oxytocin, whereas milk letdown is inhibited in situations in which the female is stressed.

Prolactin Release and Milk Ejection

Suckling not only triggers oxytocin release but also stimulates prolactin secretion. Prolactin output by the anterior pituitary is controlled by two hypothalamic secretions: *prolactin-inhibiting hormone (PIH)* and *prolactin-releasing hormone (PRH)*. PIH has been shown to be dopamine, which also serves as a neurotransmitter in the brain. The chemical nature of PRH has not been identified with certainty, but scientists suspect that in most

mammals (though not in the cow) PRH is oxytocin secreted by the hypothalamus into the hypothalamic-hypophyseal portal system to stimulate prolactin secretion by the anterior pituitary (see p. 269). This role of oxytocin is distinct from the roles of oxytocin produced by the hypothalamus and stored in the posterior pituitary. The oxytocin released by the posterior pituitary plays a key role in parturition by causing powerful uterine muscle contractions and also brings about milk ejection by stimulating contraction of the myoepithelial cells of lactating mammary glands.

Throughout most of the female's life PIH is the dominant influence, so prolactin concentrations normally remain low. During lactation, a burst in prolactin secretion occurs each time the infant stimulates the udder and/or teats. Afferent impulses initiated in the nipple on suckling are carried by the spinal cord to the hypothalamus. This reflex ultimately leads to prolactin release by the anterior pituitary, although it is unclear whether this outcome is accomplished by inhibition of PIH or stimulation of PRH secretion or both. Prolactin then acts on the alveolar epithelium to promote secretion of milk to replenish that lost from the alveoli by milk ejection. In some species of mammals (cows and goats) once lactation has been established, the basal concentration of prolactin as well as the induced release of prolactin at milking decline without affecting milk yield.

Suckling concurrently stimulates milk ejection and milk production, therefore ensuring that the rate of milk synthesis keeps pace with the young's needs for milk. The more the infant nurses, the more milk is removed by letdown and the more milk is produced for the next suckling. Thus twins elicit more milk production than do single offspring. In addition to prolactin, which is the most important factor controlling synthesis of milk, at least four other hormones are essential for their permissive role in milk production: cortisol, insulin, parathyroid hormone, and growth hormone (see Chapter 7). Among its actions, growth hormone enhances milk production by increasing the mammary gland's uptake from the blood of nutrients needed for the synthesis of milk (see p. 276).

Prolactin fittingly appears to facilitate bonding, or attachment, between a mother and the newborn in birds and mammals. Prolactin (as well as oxytocin) is thought to trigger maternal behaviors such as denning in wolves, incubation behavior in birds, and bonding of mothers to offspring. Broody behavior in commercial turkey flocks results in a pause or cessation in egg lay, resulting in decreased production, and research efforts have investigated means to lower concentrations of prolactin in affected birds. Seasonal increases in the concentrations of plasma prolactin are believed to play a role in determining the end of the breeding season for birds breeding at temperate latitudes. Generally the breeding season of birds ends before day length decreases to that measured when egg laying was initiated.

?

Why Do Males Have Prolactin and Oxytocin? Prolactin and oxytocin in mammals would seem to be quintessentially female hormones, given their prominent roles in female reproductive processes. But male mammals (as well as male birds) also have these hormones. Why? Recent studies suggest that prolactin triggers paternal behavior, such as caring for the young. Prolactin is high during brooding in male birds that help care for the eggs and hatchlings. California mice (*Peromyscus californicus*), unusual rodents because males help care for the young, have high prolactin levels and more prolactin receptors in their hypothalamus than do other rodents. Oxytocin (and vasopressin) similarly appear to play a role in reproductive behaviors in both genders. Oxytocin is released during arousal and orgasm in both genders, and both oxytocin and vasopressin are associated with pair bonding between mates. These and other roles of these "female hormones" (and vasopressin) in both genders are the subject of active research. See box "Molecular Biology and Genomics: Prolactin: A Key with Many Locks."

■ **Mammary gland feeding is advantageous to both the infant and the female.**

Milk consists of variable amounts of water, triglyceride fat, the carbohydrate lactose (milk sugar), a number of proteins, vitamins, and the minerals calcium and phosphate. For example, because the period of lactation is so short in the Weddell seal, the percentage of energy-rich fat in milk can be as high as 42%, whereas in the mare, which has a prolonged period of lactation, it can be as low as 1.3%. Protein content of milk is also high in the Weddell seal (15.8%) and low in primates (0.8% in humans and 1.5% in other primates). Remarkably, given the differing ages of the neonates undergoing development in the brood patch of marsupials, each of four mammary glands can produce a milk of differing composition! In addition to supplying nutrients, milk contains a host of immune cells, antibodies, and other chemicals that help protect the infant against infection until it can mount an effective immune response on its own a few months after birth. **Colostrum,** the milk produced for the first five days postpartum, contains lower concentrations of fat and lactose but higher concentrations of protein and immunoprotective components. In many mammals some passive immunity is transferred from the female to the fetus during gestation through the passage of antibodies across the placenta (see p. 477). These antibodies are short-lived, however, and often do not persist until the infant can fend for itself immunologically. Mammary gland–fed young gain additional protection during this vulnerable period through a variety of mechanisms:

■ Mammary gland milk contains an abundance of immune cells—both B and T lymphocytes, macrophages, and neutrophils (see p. 477)—that produce antibodies and destroy pathogenic microorganisms outright. These cells are especially plentiful in colostrum.

■ Secretory IgA, a special type of antibody, is present in great amounts in mammary gland milk. Secretory IgA consists of two IgA antibody molecules (see p. 443) joined together with a so-called secretory component that helps protect the antibodies from destruction by the infant's acidic gastric juice and digestive enzymes. The collection of IgA antibodies ingested by a mammary gland–fed young is specifically targeted against the particular pathogens in the environment of the female—and, accordingly, of the infant as well. Appropriately, therefore, these antibodies provide protection against the infectious microbes that the infant is most likely to encounter.

■ Some components in milk, such as mucus, adhere to potentially harmful microorganisms, thus preventing them from attaching to and crossing the intestinal mucosa.

■ **Lactoferrin** is a mammary-gland milk constituent that thwarts growth of harmful bacteria by decreasing the availability of iron, a mineral needed for multiplication of these pathogens (see p. 436).

■ **Bifidus factor** in mammary gland milk, in contrast, promotes multiplication of the nonpathogenic microorganism *Lactobacillus bifidus* in the infant's digestive tract. Growth of this harmless bacterium helps crowd out potentially harmful bacteria.

■ Other components in mammary gland milk promote maturation of the young's digestive system so that it is less vulnerable to diarrhea-causing bacteria and viruses.

■ Still other factors in mammary gland milk hasten the development of the infant's own immune capabilities.

Toothed whales (*Odontoceti*) wean their young any time from 2 to 13 years, much later than other marine and terrestrial mammals. It has been observed that reproductively senescent pilot whales, whose weaning period lasts 4 to 5 years, are still lactating. In many species a resumption of estrus does

MOLECULAR BIOLOGY AND GENOMICS

Prolactin: A Key with Many Locks

As you have seen, prolactin is an important hormone classically known for its role in milk production. But it has many other roles, some only recently discovered. As you saw in Chapter 7, a particular hormone or other chemical signal can have multiple, sometimes unrelated roles because of the effects of different receptor systems on different target cells. Prolactin is a prime example, because its roles have diversified during vertebrate evolution. First, it is an ancient hormone, found in all vertebrates, and its genetic codes show that it probably evolved along with growth hormone (GH) from an even more ancient ancestral gene. In fishes and amphibians, prolactin is involved in regulating osmotic balance, particularly in retaining salt in these animals' adaptation to fresh water. This is a function that some physiologists suspect occurs in mammals: It reduces Na^+ and K^+ output in the urine and content in sweat, and increases uptake of these ions from the intestines.

However, prolactin also has developmental roles in all vertebrates. Its relation to GH is seen in fishes and amphibians that undergo metamorphoses: Prolactin appears to oppose metamorphosis and enhance growth of the larval stage. Yet another developmental role is found in fishes, adult amphibians, reptiles, birds, and some mammals: Prolactin stimulates various aspects of skin development (such as molting in reptiles, and the defeathering of bird brood patches).

As you have seen for birds and mammals, prolactin signaling evolved another set of functions involved in reproduction. These functions are not only those discussed for milk and parental behavior, but also are for early stages of reproduction. Female knockout mice lacking prolactin receptors cannot reproduce, because the uterus does not allow embryonic implantation. The reason is that the corpus luteum undergoes apoptosis (programmed death) in the absence of prolactin stimulation; in turn, the corpus luteum does not make enough progesterone to prepare the uterus. Prolactin also has multiple reproduction-related effects on the brain, including the inducement of parental behaviors. In addition, it greatly suppresses sex drive in females during nursing periods.

Furthermore, prolactin receptors have been found in numerous other nonreproductive mammalian cells and tissues, where their role is often uncertain (although there is some evidence for roles in osmotic balance and skin development). One example is the mammalian lymphocyte system: T- and B-cells (Chapter 10) have prolactin receptors and can also produce prolactin. Mice in which prolactin binding is blocked have weak immune responses and are thus more susceptible to disease. To complicate matters, prolactin also has metabolic roles: It stimulates lipid metabolism in birds and carbohydrate metabolism in mammals.

A different aspect of prolactin function was announced in 2003, in a study revealing that prolactin violates the dogma for peptide-hormone signal transduction. Like all peptide hormones, prolactin does bind to target cell receptors on the plasma membrane to exert its effects via second messengers (p. 90). The dogma is that such hormones always work this way, whereas steroid hormones work by binding to internal receptors that directly regulate gene expression (p. 257). But in recent years, exceptions to these rules have been surfacing. Progesterone, for example, is a steroid hormone that has been found to bind to both membrane and internal receptors. Prolactin similarly has been found to enter breast cancer cells, move to the nucleus, and activate growth genes. In this way, prolactin is a major culprit in the spread of breast cancer in human females. Prolactin binds to a protein called cyclophilin B (CYPB) at the membrane, which carries it into the nucleus, where the two proteins activate gene transcription.

not occur until after the young are weaned. Suckling suppresses the estrous cycle by inhibiting LH and FSH secretion, probably through inhibition of GnRH, thus permitting all the female's resources to be directed toward the newborn instead of being shared with a new embryo. This becomes a very important mechanism for spacing births in nonhumans.

After weaning, two mechanisms contribute to the cessation of milk production. First, without suckling, prolactin secretion is not stimulated, thus removing the primary stimulus for continued milk synthesis and secretion. Also because of the lack of suckling, milk letdown does not occur in the absence of oxytocin release. Because milk production does not immediately shut down, milk accumulates in the alveoli, engorging the mammary glands. The resultant pressure buildup acts directly on the alveolar epithelial cells to suppress further milk production. Cessation of lactation at weaning therefore results from a lack of suckling-induced stimulation of both prolactin and oxytocin secretion.

A newborn animal represents a new beginning for all the processes of physiology.

(*Source:* www.wildnatureimages.com/Cow%20Moose%20Nursing%20Calf.htm)

Photo: Ron Niebrugge Photography

◾ The end is a new beginning.

Reproduction is an appropriate way to end our discussion of physiology from genome to systems. The single cell resulting from the union of male and female gametes divides mitotically and differentiates into a multicellular individual made up of a number of different organ systems that interact cooperatively to maintain homeostasis (that is, stability in the internal environment). All the life-supporting homeostatic and rheostatic processes introduced throughout this book begin all over again at the start of a new life.

Chapter in Perspective:

HOMEOSTASIS AND INTEGRATION

The reproductive system is unique in that it is not essential for individual homeostasis or for survival of the individual, but it is essential for sustaining the thread of life from generation to generation. Reproduction depends on the union of male and female gametes (reproductive cells), each with a half set of chromosomes, to form a new individual with a full, unique set of chromosomes. Unlike the other body systems, which are essentially identical in the two sexes, the reproductive systems of males and females are remarkably different, befitting their different roles in the reproductive process.

The male system is designed to continuously produce huge numbers of motile spermatozoa that are delivered to the ova. Male gametes must be produced in abundance for two reasons: (1) Only a small percentage survive the hazardous journey to the site of fertilization; and (2) the cooperative effort of many spermatozoa is required to break down the barriers surrounding the female gamete (oocyte) to enable one spermatozoon to penetrate and unite with the ovum.

The female reproductive system in many species undergoes complex changes on a cyclic basis. During the first phase of a cycle in a mammal, a single, nonmotile ovum is prepared for release.

During the second phase in a mammal, the reproductive system is geared toward preparing a suitable environment for supporting the ovum if fertilization (union with a spermatozoon) occurs. If fertilization does not occur, the cycle starts over again as a new ovum is prepared for release. If fertilization occurs, the female reproductive system adjusts to support growth and development of the new individual until it can survive on its own on the outside.

There are three important parallels in the vertebrate male and female reproductive systems, even though they differ considerably in structure and function. First, the same set of undifferentiated reproductive tissues in the embryo can develop into either a male or a female system, depending on the presence or absence, respectively, of male-determining factors. Second, the same hormones—namely, hypothalamic GnRH and anterior pituitary FSH and LH—control reproductive function in both sexes. In both cases, gonadal steroids and inhibin act in negative-feedback fashion to control hypothalamic and anterior pituitary output. Third, the same events take place in the developing gamete's nucleus during sperm formation and egg formation. ◾

REVIEW QUESTIONS *(Answers are on p. A–1.)*

Additional study tools for this chapter, including chapter summaries and practice tests, are available online at *www.biology.brookscole.com*

1. Oviparous animals
   a. produce eggs that hatch within the body of the parent
   b. are all r-selected species
   c. release eggs from which the young hatch after expulsion from the body
   d. are all K-selected species
   e. produce and nourish living young within the body
2. Which of the following are haploid?
   a. ovaries and testes
   b. eggs and sperm
   c. accessory sex glands
   d. mammary glands
   e. internal genitalia
3. Spermatogenesis in most mammals
   a. occurs only while the testes are passing through the inguinal canal
   b. occurs in Wolffian ducts
   c. is temperature sensitive and cannot occur at normal body temperatures
   d. is heterogametic
   e. occurs as a result of parthenogenesis
4. Which of the following is not supplied by seminal vesicles?
   a. fructose
   b. prostaglandins
   c. an alkaline fluid that neutralizes vaginal secretions
   d. fibrinogen
   e. additional volume to seminal fluid
5. Which of the following describes a polyestrous vertebrate breeder?
   a. estrous cycles are restricted to a particular time of year
   b. estrous cycles occur as day length increases
   c. estrous cycles occur as day length decreases
   d. estrous cycles occur uniformly throughout the year
   e. a single estrous event is followed by a long period of anestrous
6. A primordial follicle is made up of a primary oocyte surrounded by
   a. secondary follicles
   b. a single layer of granulosa cells
   c. secondary oocytes
   d. first polar bodies
   e. second polar bodies

7. Ovulation occurs
   a. when the antrum forms in the midst of the granulosa cells
   b. as soon as tendrils from the granulosa cells penetrate the zona pellucida
   c. when an outer layer of thecal cells forms
   d. as soon as the secondary oocyte forms
   e. when the mature follicle ruptures

8. Which of the following is not a phase of the estrous cycle?
   a. proestrus
   b. anestrous
   c. estrus
   d. metestrus
   e. diestrus

9. Which of the following is not a change brought about by the LH surge that accompanies ovulation and luteinization of the ruptured follicle?
   a. estrogen synthesis by the follicular cell is increased
   b. meiosis in the oocyte of the developing follicle is reinitiated
   c. production of locally acing prostaglandins is triggered
   d. follicular cells are caused to differentiate into luteal cells
   e. estrogen synthesis is halted

10. In primates, what does the corpus luteum secrete under the influence of LH?
    a. progesterone and estrogen
    b. gonadotrophin releasing hormone
    c. estradiol
    d. prostaglandin
    e. increased amounts of LH

11. In chickens, expulsion of the egg from the uterus is mediated by
    a. synthesis of VLDL
    b. F1 follicles
    c. arginine vasotocin
    d. synchronous maturation of follicles
    e. progesterone

12. In fish, estradiol stimulates
    a. production of pituitary gonadotrophins
    b. production of maturation-inducing hormone
    c. synchronous oocyte development
    d. production of vitellogenin
    e. the bearing down response

13. Hormones that play a critical role in maintaining pregnancy are secreted by the
    a. corpus luteum
    b. blastocyst
    c. zygote
    d. placenta
    e. a and d

14. During labor, myometrial contractions incessantly increase, because
    a. of a positive-feedback cycle between oxytocin and prostaglandin ensues
    b. the volume of chorioallantoic fluid constantly increases
    c. concentrations of prolactin constantly increase
    d. corticotropin-releasing hormone drives the manufacture of placental estrogen
    e. of the increased rate of DHEA

15. Elevated concentrations of prolactin and chorionic somato-mammotropin
    a. prevent lactation until parturition
    b. stimulate contraction of myoepithelial cells
    c. lead to the secretion of IgA antibodies into milk
    d. induce synthesis of enzymes needed for milk production
    e. increase the numbers of T lymphocytes in milk

SUGGESTED READINGS AND INTERNET SITES

Bole-Feysot, C., V. Goffin, M. Edery, N. Binart, & P. A. Kelly. 1998. Prolactin (PRL) and its receptor: Actions, signal transduction pathways and phenotypes observed in PRL receptor knockout mice. *Endocrine Reviews* 19:225–268. Available online at www.uni-greifswald.de/~zoologie/german/phybioti/bole.pdf.

Griffiths, R. 2000. Sex identification using DNA markers. In A. J. Baker, ed., *Molecular Methods in Ecology*. London: Blackwell Science.

Gubernick, D. J. 1990. Prolactin and paternal behavior in the biparental California mouse, *Peromyscus californicus. Hormones and Behavior* 23:203–210.

Hau, M. 2001. Timing of breeding in variable environments: Tropical birds as model systems. *Hormones and Behavior* 40:281–290.

Johnson, A. L. 2000. Reproduction in the female. *Sturkie's Avian Physiology*, ed. G. C. Whittlow, 5th ed. San Diego: Academic Press.

Knobil, E., & J. D. Neill, eds. 1998. *Encyclopedia of Reproduction*. New York: Academic Press.

Lincoln, G. A., & R. V. Short. 1980. Seasonal breeding: Nature's contraceptive. *Recent Progress in Hormone Research*. 36:1–52.

Packer, C., M. Tatar, & A. Collins. 1998. Reproductive cessation in female mammals. *Nature* 392:807.

Patricelli, G. L., J. A. C. Uy, G. Walsh, & G. Borgia. 2002. Sexual selection: males adjust displays in response female signals. *Nature* 415:279–280.

Radke, W. J. 2002. *Laboratory Guide to Human Anatomy*. New York: John Wiley & Sons, Inc.

Reece, W. O. 1997. *Physiology of Domestic Animals*, 2nd ed. Baltimore: Williams and Wilkins.

Ross, R. M. 1990. The evolution of sex-change mechanisms in fishes. *Environmental Biology of Fishes* 29:81–93.

Senger, P. L. 2003. *Pathways to pregnancy and parturition*. Current Conceptions, Inc. Washington State University Research & Technology Park. Pullman, WA. Covers the principles of reproduction in food-producing animals.

INFOTRAC READING

Haqq, C. M., & P. K. Donahoe. 1998. Regulation of sexual dimorphism in mammals. *Physiological Reviews* 78:1–32.

INTERNET SITES

Bowen, R. A. 2000. *Pathophysiology of the Reproductive System*. **arbl.cvmbs.colostate.edu/hbooks/pathphys/reprod/index.html**. A site with descriptions, diagrams, and photographs on basic mammalian reproductive anatomy and functions and diseases.

Guinan, M. J. 2002. *Reproductive System*. **trc.ucdavis.edu/mjguinan/apc100/modules/Reproductive/_index.html**. A site with many histological slides of vertebrate reproductive systems.

Answers to End of Chapter
Review Questions

Chapter 1

1. e; 2. c; 3. d; 4. b; 5. c; 6. e; 7. c; 8. d; 9. e; 10. b; 11. b; 12. d; 13. e; 14. c; 15. a; 16. c; 17. b; 18. d; 19. b.

Chapter 2

1. b; 2. e; 3. c; 4. d; 5. e; 6. c; 7. b; 8. d; 9. d; 10. e; 11. b; 12. c; 13. e; 14. e; 15. e; 16. c; 17. b; 18. e; 19. e; 20. c.

Chapter 3

1. e; 2. c; 3. d; 4. c; 5. b; 6. a; 7. b; 8. e; 9. d; 10. d; 11. b; 12. c; 13. d; 14. c; 15. b; 16. b; 17. c.

Chapter 4

1. e; 2. c; 3. e; 4. b; 5. d; 6. e; 7. b; 8. e; 9. c; 10. d; 11. c; 12. b; 13. e; 14. e; 15. e; 16. d; 17. c; 18. e; 19. c; 20. d

Chapter 5

1. c; 2. e; 3. e; 4. a; 5. e; 6. b; 7. e; 8. d; 9. c; 10. c; 11. d; 12. b; 13. c; 14. e; 15. d; 16. d; 17. d; 18. c; 19. c; 20. e.

Chapter 6

1. d; 2. b and c; 3. b; 4. d; 5. b; 6. d; 7. b; 8. a; 9. c; 10. c; 11. e; 12. d; 13. e; 14. b; 15. d; 16. e; 17. b; 18. e; 19. e; 20. b.

Chapter 7

1. d; 2. b; 3. d; 4. e; 5. b; 6. a; 7. d; 8. c; 9. d; 10. e; 11. e; 12. e; 13. b; 14. c; 15. d.

Chapter 8

1. a; 2. d; 3. e; 4. d; 5. b; 6. b; 7. c; 8. d; 9. c; 10. e; 11. b; 12. c; 13. d; 14. c; 15. b; 16. e; 17. d; 18. b; 19. d; 20. a.

Chapter 9

1. e; 2. e; 3. c; 4. d; 5. a; 6. d; 7. e; 8. e; 9. d; 10. b; 11. c; 12. b; 13. d; 14. c ; 15. d.

Chapter 10

1. d; 2. b; . 3. e; 4. b; 5. c; 6. b; 7. d; 8. b; 9. d; 10. b; 11. d; 12. d; 13. c; 14. d; 15. b; 16. d; 17. b; 18. b; 19. d; 20. e.

Chapter 11

1. e; 2. a; 3. b; 4. d; 5. c; 6. e; 7. b; 8. c; 9. a; 10. d; 11. b; 12. c; 13. c; 14. b; 15. d; 16. d; 17. b; 18. c; 19. d; 20. e.

Chapter 12

1. c; 2. c;. 3. b; 4. d; 5. d; 6. b; 7. a; 8. e; 9. c; 10. b; 11. e; 12. c; 13. a; 14. c; 15. b; 16. d; 17. d; 18. e; 19. b; 20. c.

Chapter 13

1. d; 2. e;. 3. e; 4. b; 5. a; 6. c; 7. c; 8. e; 9. e; 10. d; 11. d; 12. c; 13. c; 14. e; 15. a; 16. c; 17. d; 18. b; 19. c; 20. e.

Chapter 14

1. d; 2. c; 3. b; 4. e; 5. b; 6. d; 7. b; 8. b; 9. a; 10. c; 11. a; 12. c; 13. d; 14. e; 15. d.

Chapter 15

1. b; 2. d;. 3. b; 4. c; 5. e; 6. e; 7. c; 8. d; 9. e; 10. b; 11. c; 12. e; 13. c; 14. b; 15. b; 16. a; 17. d; 18. b; 19. c; 20. d.

Chapter 16

1. c; 2. b; 3. c; 4. b; 5. d; 6. b; 7. e; 8. b; 9. a; 10. a; 11. c; 12. d; 13. e; 14. a; 15. d.

Glossary

A band One of the dark bands that alternate with light (I) bands to create a striated appearance in skeletal or cardiac muscle fibers when these fibers are viewed with a light microscope

abomasum The fourth compartment of the ruminant stomach, which has functions similar to the glandular stomach of nonruminants

absorptive state The metabolic state following a meal when nutrients are being absorbed and stored; fed state

accessory digestive organs Exocrine organs outside the wall of the digestive tract that empty their secretions through ducts into the digestive tract lumen

accessory sex glands Glands that empty their secretions into the reproductive tract

acclimation A laboratory phenomenon in which the chronic response of an animal to a change in environment is measured. Normally the old and new environments differ in one or two highly specific ways.

acclimatization The (usually slow) process of changing physiologic processes to function more optimally under new conditions

accommodation In the eye, the ability to adjust the strength of the lens so that both near and far sources can be focused on the retina

acetylcholine (ACh) (as´-uh-teal-KŌ-lēn) The neurotransmitter released from all autonomic preganglionic fibers, parasympathetic postganglionic fibers, and motor neurons

acetylcholinesterase (AChE) (as´uh-teal-kō-luh-NES-tuh-rās) An enzyme present in the motor end-plate membrane of a skeletal muscle fiber that inactivates acetylcholine

ACh See *acetylcholine*

achalasia A condition that occurs commonly in dogs (and in humans) in which the lower esophageal sphincter fails to relax during swallowing but instead contracts more vigorously

AChE See *acetylcholinesterase*

acid A hydrogen-containing substance that on dissociation yields a free hydrogen ion and anion

acidosis (as-i-DŌ-sus) Blood pH below the normal range (< 7.35 in most mammals)

acini (ĀS-i-nī) The secretory component of saclike exocrine glands, such as digestive enzyme-producing pancreatic glands or milk-producing mammary glands

acquired immune responses Responses that are selectively targeted against particular foreign material to which the body has previously been exposed; see also *antibody-mediated immunity* and *cell-mediated immunity*

acrorhagia Special stinging "battle" tentacles in some anthozoan cnidarians (anemones, corals)

ACTH See *adrenocorticotropic hormone*

actin The contractile protein that forms the backbone of the thin filaments in muscle fibers. Actin also contributes to the motility of other kinds of cells

active expiration Emptying of the lungs more completely than when at rest by contracting the expiratory muscles; also called *forced expiration*

active reabsorption The condition when any step in the transepithelial transport of a substance requires energy expenditure

active transport Active carrier-mediated transport involving transport of a substance against its concentration gradient across the plasma membrane

acuity Discriminative ability; the ability to discern two different points of stimulation

adaptation (1) A reduction in receptor potential despite sustained stimulation of the same magnitude; (2) any feature of an organism that enhances evolutionary fitness

adenosine diphosphate (ADP) (uh-DEN-uh-sēn) The two phosphate products formed from the splitting of ATP to yield energy for the cell's use

adenosine triphosphate (ATP) The body's common energy "currency," which consists of an adenosine with three phosphate groups attached; splitting of the high-energy, terminal phosphate bond provides energy to power cell activities

adenylyl cyclase (ah-DEN-il-il sī-klās) The membrane-bound enzyme that is activated by a G protein intermediary in response to binding of an extracellular messenger with a surface membrane receptor and that in turn activates cyclic AMP, an intracellular second messenger

ADH See *vasopressin*

adiponectin A hormone produced by adipose tissue in nonobese mammals; may serve in regulating energy balance

adipose tissue The tissue specialized for storage of triglyceride fat; found under the skin in the hypodermis

ADP See *adenosine diphosphate*

adrenal cortex (uh-DRĒ-nul) The outer portion of the vertebrate adrenal gland; secretes three classes of steroid hormones: glucocorticoids, mineralocorticoids, and sex hormones

adrenal medulla (muh-DUL-uh) The inner portion of the vertebrate adrenal gland; an endocrine gland that is a modified sympathetic ganglion that secretes the hormones epinephrine and norepinephrine into the blood in response to sympathetic stimulation

adrenergic fibers (ad´-ruh-NUR-jik) Nerve fibers that release norepinephrine as their neurotransmitter

adrenocorticotropic hormone (ACTH) (ad-rē´-nō-kor´-tuh-kō-TRŌP-ik) An anterior pituitary hormone that stimulates cortisol secretion by the adrenal cortex and promotes growth of the adrenal cortex

aerobes Organisms that rely on oxygen-based metabolism

aerobic Referring to a condition in which oxygen is available; **aerobic metabolism** is the process in which ATP formation is accomplished through oxidative phosphorylation

afferent arteriole (AF-er-ent ar-TIR-ē-ōl) The vessel that carries blood into the glomerulus of the mammalian kidney's nephron

afferent division The portion of the peripheral nervous system that carries information from the periphery to the central nervous system

afferent neuron Neuron that has a sensory receptor at its peripheral ending and carries information to the central nervous system

after hyperpolarization (hī´-pur-pō-luh-ruh-ZA-shun) A slight, transient hyperpolarization that sometimes occurs at the end of an action potential

agonist A non-native signal molecule (such as a drug or toxin) that mimics the effects of a native signal molecule

agranulocytes (ā-GRAN-yuh-lō-sīts´) Leukocytes that do not contain granules, including lymphocytes and monocytes

air capillary Small airways that are the site of gas exchange in the avian lung.

air sac A thin-walled, air-filled component of the avian respiratory system with little musculature and few blood vessels

albumin (al-BEW-min) The smallest and most abundant of the plasma proteins; binds and transports many water-insoluble substances in the blood; contributes extensively to plasma-colloid osmotic pressure

aldose reductase An enzyme that converts glucose to sorbitol (and other aldose sugars to related polyol sugars); found in (among other tissues) the inner medulla of mammalian kidneys, where sorbitol is used as a compatible osmolyte

aldosterone (al-dō-steer-OWN) or (al-DOS-tuh-rōn) The adrenocortical hormone that stimulates Na^+ reabsorption by the distal and collecting tubules of the kidney's nephron during urine formation

alkalosis (al'-kuh-LŌ-sus) Blood pH above the normal range (> 7.45 in most mammals)

all-or-none law An excitable membrane either responds to a stimulus with a maximal action potential that spreads nondecrementally throughout the membrane or does not respond with an action potential at all

allostasis Maintaining physiological stability through change, similar to enantiostasis

allosteric ("other site") A noncatalytic binding site on an enzyme that binds regulatory molecules

alpha (α) cells The endocrine pancreatic cells that secrete the hormone glucagon

alpha (α) motor neuron A motor neuron that innervates ordinary skeletal muscle fibers

altricial Referring to young that are absolutely dependent on continued parental care

alveolar surface tension (al-VĒ-ō-lur) The surface tension of the fluid lining the alveoli in the lungs; see *surface tension*

alveolar ventilation The volume of air exchanged between the atmosphere and alveoli per minute; equals (tidal volume minus dead space volume) times respiratory rate

alveoli (al-VĒ-ō-lī) The air sacs across which O_2 and CO_2 are exchanged between the blood and air in mammalian (and some other vertebrate) lungs

amines (ah-means) Hormones derived from the amino acid tyrosine; includes thyroid hormone and catecholamines

ammonotelic Having ammonia as a primary nitrogenous waste product

amoebocyte Wandering immune cell in invertebrates

amoeboid movement (uh-MĒ-boid) "Crawling" movement of white blood cells, similar to the means by which amoebas move

ampullary electroreceptors Found in almost all nonteleost fishes and in some teleosts as well as in several amphibian species; respond to low-frequency electric signals that are typical of electrical output from animal nerves and hearts

amygdala Structure of mammalian forebrain (limbic system) located on the interior underside of the temporal lobe; believed to be the homolog of the reptilian and avian paleostriatum; appears to store memories of highly emotional events

anabolism (ah-NAB-ō-li-zum) The buildup, or synthesis, of larger organic molecules from the small organic molecular subunits

anaerobes Organisms that can survive on metabolism that does not require oxygen

anaerobic (an'-uh-RŌ-bik) Referring to a condition in which oxygen is not present; in **anaerobic metabolism** ATP formation is accomplished by anaerobic glycolysis, usually only for brief periods of time when O_2 delivery is inadequate to support oxidative phosphorylation

analgesic (an-al-JEE-zic) Pain relieving

anatomy The study of body structure

androgen A vertebrate masculinizing "male" sex hormone; includes testosterone from the testes and dehydroepiandrosterone from the adrenal cortex

anemia A reduction below normal in the O_2-carrying capacity of the blood

anestrus Condition where a female mammal does not cycle into reproductive readiness, because of insufficient hormonal stimuli

anhydrobiotic Literally, life without water; referring to dormant stages of organisms with very little water content

anion (AN-ī-on) Negatively charged ion that has gained one or more electrons in its outer shell

anoxia The complete lack of oxygen

ANP See *atrial natriuretic peptide*

antagonism/antagonist Action/agent opposing another: (1) when one hormone causes the loss of another hormone's receptors, reducing the effectiveness of the second hormone; (2) a non-native signal molecule (such as a drug or toxin) blocking the effects of a native signal molecule (for opposite term, see *agonist*)

anterior pituitary The glandular portion of the pituitary that synthesizes, stores, and secretes six different hormones: growth hormone, TSH, ACTH, FSH, LH, and prolactin

antibody A vertebrate immunoglobulin produced by a specific activated B lymphocyte (plasma cell) against a particular antigen; binds with the specific antigen against which it is produced and promotes the antigenic invader's destruction by augmenting nonspecific immune responses already initiated against the antigen

antibody-mediated immunity A specific immune response accomplished by antibody production by B cells

antidiuretic hormone (an'-ti-dī'-yū-RET-ik) See *vasopressin*

antigen A large, complex molecule that triggers a specific immune response against itself when it gains entry into a vertebrate body

antioxidant A substance that helps inactivate biologically damaging free radicals

antimicrobial peptides Small, defensive proteins produced by barrier tissues and some immune cells in many, if not all, animals; many kill microbes by creating pores in their membranes

antiporter A transporter protein in a membrane that moves two (or more) molecules or ions in the opposite direction

antrum (of ovary) The fluid-filled cavity formed within a developing ovarian follicle

antrum (of stomach) The lower portion of the vertebrate stomach

aorta (a-OR-tah) The large vessel that carries blood from the vertebrate heart to the body

aortic valve A one-way valve that permits the flow of blood from the mammalian left ventricle into the aorta during ventricular emptying but prevents the backflow of blood from the aorta into the left ventricle during ventricular relaxation

apodemes In arthropods, ridges that project from the inner face of the exoskeleton for the attachment of muscles

apoptosis (ā-pop-TŌ-sis) Programmed cell death; deliberate self-destruction of a cell (as opposed to necrosis, unintended cell death)

appetite centers Neuronal clusters in the lateral regions of the vertebrate hypothalamus that drive the animal to eat

aquaporin A protein that forms a channel in a membrane that allows water to diffuse through it

aqueous humor (Ā-kwē-us) The clear watery fluid in the anterior chamber of the vertebrate eye; provides nourishment for the cornea and lens

archaea Ancient prokaryotes forming one of three distinct forms of life (along with eubacteria and eukaryotes), often found in extreme habitats such as hot springs, salt brines; thought to be related to the ancestor of the main eukaryotic cell

aromatase The enzyme that converts androgens (e.g. testosterone) into estrogens (estradiol)

arterioles (ar-TIR-ē-ōlz) The highly muscular, high-resistance vessels of vertebrates, the caliber of which can be changed subject to control to determine how much of the cardiac output is distributed to each of the various tissues

artery A vessel that carries blood away from a heart

ascending tract A bundle of nerve fibers of similar function that travels up the vertebrate spinal cord to transmit signals derived from afferent input to the brain

astrocyte A type of glial cell in the vertebrate brain; major functions include holding the neurons together in proper spatial relationship and inducing the brain capillaries to form tight junctions important in the blood–brain barrier

asynchronous Out of step or phase control of flight muscles. A single Ca^{2+} release "turns on" the flight muscle. When in this state the muscle becomes activated by stretch and deactivated by shortening

atherosclerosis (ath-uh-rō-skluh-RŌ-sus) A progressive, degenerative arterial disease that leads to gradual blockage of vertebrate vessels, reducing blood flow through them

atmospheric pressure The pressure exerted by the weight of the air in the atmosphere on objects on Earth's surface; equals 760 mm Hg at sea level

ATP See *adenosine triphosphate*

ATPase An enzyme that has ATP-splitting ability

ATP synthase (SIN-thās) The enzyme within the mitochondrial inner membrane that phosphorylates ADP to ATP

atrial natriuretic peptide (ANP) (Ā-trē-al NĀ-tree-ur-eh´tik) A peptide hormone released from the vertebrate cardiac atria that promotes urinary loss of Na^+ in mammals

atrioventricular (AV) node (ā´-trē-ō-ven-TRIK-yuh-lur) A small bundle of specialized cardiac cells located at the junction of the atria and ventricles that serves as the only site of electrical contact between the atria and ventricles in vertebrates

atrioventricular (AV) valve A one-way valve that permits the flow of blood from the atrium to the ventricle during filling of the mammalian heart but prevents the backflow of blood from the ventricle to the atrium during emptying of the heart

atrium (pl., **atria**) (Ā-tree-um) A chamber of a heart that receives blood from the veins or hemolymph and transfers it to a ventricle

atrophy (AH-truh-fē) Decrease in mass of an organ

autocrine A signal molecule regulating a cellular process of the cell that secreted it ("self-stimulation")

autoimmune disease Disease characterized by erroneous production of antibodies against one of the body's own tissues

autonomic nervous system The portion of the efferent division of the vertebrate peripheral nervous system that innervates smooth and cardiac muscle and exocrine glands; composed of two subdivisions, the sympathetic nervous system and the parasympathetic nervous system

autorhythmicity The ability of an excitable cell to rhythmically initiate its own action potentials

AV valve See *atrioventricular valve*

avoiders Organisms that reduce disturbances to a particular physiological state by behaviorally avoiding environmental changes

axon A single, elongated tubular extension of a neuron that conducts action potentials away from the cell body; also known as a *nerve fiber*

axon hillock The first portion of a neuronal axon plus the region of the cell body from which the axon leaves; the site of action-potential initiation in most neurons

axon terminals The branched endings of a neuronal axon, which release a neurotransmitter that influences target cells in close association with the axon terminals

B lymphocytes (B-cells) Vertebrate white blood cells that produce antibodies against specific *antigens* to which they have been exposed

baculum Penis bone in some mammals

bag cells Two clusters of cells in crustacean connective tissue above the abdominal ganglion; produce egg-laying hormone (ELH)

baleen Parallel fused filaments of keratin, the same protein of hair fibers that hang from the upper jaws of baleen whales in place of teeth

baroreceptor reflex An autonomically mediated reflex response that influences the vertebrate heart and blood vessels to oppose a change in mean arterial blood pressure

baroreceptors Receptors located within the vertebrate circulatory system that monitor blood pressure

basal metabolic rate (BĀ-sul) **(BMR)** The minimal rate of internal energy expenditure; a body's "idling speed"

basal nuclei Several masses of gray matter located deep within the white matter of the cerebrum of the vertebrate brain; play an important inhibitory role in motor control

basal transcription complex An assembly of proteins that initiate gene transcription in eukaryotes

base A substance that can combine with a free hydrogen ion and remove it from solution

basic electrical rhythm (BER) Self-induced electrical activity of the digestive-tract smooth muscle

basilar membrane (BAS-ih-lar) The membrane that forms the floor of the middle compartment of the vertebrate cochlea and bears the organ of Corti, the sense organ for hearing

basophils (BAY-so-fills) White blood cells of vertebrates that synthesize, store, and release histamine, which is important in allergic responses

BER See *basic electrical rhythm*

beta (β) cells The endocrine pancreatic cells that secrete the hormone insulin

bicarbonate (HCO_3^-) The anion resulting from dissociation of carbonic acid, H_2CO_3

bile salts Cholesterol derivatives secreted in vertebrate bile that facilitate fat digestion through their detergent action and facilitate fat absorption through their micellar formation

biliary system (BIL-ē-air´-ē) The bile-producing system, consisting of the liver, gallbladder, and associated ducts in vertebrates

bilirubin (bill-eh-RŪ-bin) A bile pigment that is a waste product derived from the degradation of hemoglobin during the breakdown of old red blood cells

bimodal Having two respiratory exchange surfaces including the gills and either cutaneous exchange or lungs

blastocyst The developmental stage of the fertilized mammalian ovum by the time it is ready to implant. Consists of a single-layered sphere of cells encircling a fluid-filled cavity

bloat Distension of the rumen because of excessive gas formation

blood–brain barrier Special structural and functional features of vertebrate brain capillaries that limit access of materials from the blood into the brain tissue

body system A collection of organs that perform related functions and interact to accomplish a common activity that is essential for survival of the whole body; for example, the digestive system

bone marrow The soft, highly cellular tissue that fills the internal cavities of vertebrate bones and is the source of most blood cells in terrestrial vertebrates

book lung Respiratory organ in many arachnids consisting of numerous membranous folds arranged like the pages of a book with sheetlike air spaces

Bowman's capsule The beginning of the tubular component of the mammalian kidney's nephron that cups around the glomerulus and collects the glomerular filtrate as it is formed

Boyle's law (boilz) At any constant temperature, the pressure exerted by a gas varies inversely with the volume of the gas

brain The most anterior, most highly developed portion of a central nervous system

brain stem The portion of the vertebrate brain that is continuous with the spinal cord, serves as an integrating link between the spinal cord and higher brain levels, and controls many life-sustaining processes, such as breathing, circulation, and digestion

bronchioles (BRONG-kē-ōlz) The small, branching airways within vertebrate lungs

bronchoconstriction Narrowing of the respiratory airways

bronchodilation Widening of the respiratory airways

brush border The collection of microvilli projecting from the luminal border of epithelial cells lining the digestive tract and kidney tubules

buffer system A mixture in a solution of two or more chemical compounds that minimize pH changes when either an acid or a base is added to or removed from the solution

bulbourethral glands (bul-bo-you-RĒTH-ral) Male accessory sex glands that secrete mucus for lubrication (in mammals)

bulbus arteriosus In bony fish, a chambered extension of the ventricle through which blood exits the heart

bulk flow Movement in bulk of a protein-free plasma across the capillary walls between the blood and surrounding interstitial fluid; encompasses ultrafiltration and reabsorption

bulk transport Active movement of the medium (gas or liquid)

bundle of His (hiss) A tract of specialized cardiac cells that rapidly transmits an action potential down the interventricular septum of the avian and mammalian heart

bursa of Fabricius A gut-related lymphoid tissue unique to birds; site of B-cell maturation

C cells The thyroid cells that secrete calcitonin

calcitonin (kal´-suh-TŌ-nun) A vertebrate hormone secreted by the thyroid C cells that lowers plasma Ca^{2+} levels

calcium balance Maintenance of a constant total amount of Ca^{2+} in the body; see *calcium homeostasis*

calcium homeostasis Maintenance of a constant free plasma Ca^{2+} concentration; accomplished by rapid exchanges of Ca^{2+} between the bone and ECF and to a lesser extent by modifications in urinary Ca^{2+} excretion and intestinal Ca^{2+} absorption

calmodulin (kal´-MA-jew-lin) An intracellular Ca^{2+} binding protein that, on activation by Ca^{2+}, induces a change in structure and function of another intracellular protein; especially important in smooth muscle excitation–contraction coupling

CAMs See *cell adhesion molecules*

capillaries The thin-walled, pore-lined, smallest blood vessels, across which exchange between the blood and surrounding tissues takes place

capsaicin The "hot"-tasting chemical of chili peppers

carbonic anhydrase (an-HĪ-drās) The enzyme that catalyzes the conversion of CO_2 and OH^- into HCO_3^-

cardiac cycle One period of systole and diastole

cardiac muscle The specialized muscle found only in the heart

cardiac output (CO) The volume of blood pumped by each ventricle each minute; equals stroke volume times heart rate

cardiovascular control center The integrating center located in the medulla of the vertebrate brain stem that controls mean arterial blood pressure

carnivore Animals that capture and consume live prey or that locate and consume carrion

carrier molecules Membrane proteins, which, by undergoing reversible changes in shape so that specific binding sites are alternately exposed at either side of the membrane, can bind with and transfer particular substances unable to cross the plasma membrane on their own

carrier-mediated transport Transport of a substance across the plasma membrane facilitated by a carrier molecule

cascade A series of sequential reactions that culminates in a final product, such as a clot

catabolism (kuh-TAB-ō-li-zum) The breakdown, or degradation, of large, energy-rich molecules within cells

catalase (KAT-ah-lās) An antioxidant enzyme found in peroxisomes that decomposes potent hydrogen peroxide into harmless H_2O and O_2

catch state In invertebrates, a state of continuous contraction of smooth muscle that maintains tension with little metabolic cost

catecholamines (kat´-uh-KŌ-luh-means) The chemical classification of the vertebrates adrenomedullary hormones

cathelicidins Antimicrobial peptides produced by mammalian barrier tissues and some immune cells; they kill microbes by creating pores in the microbes' membranes

cations (KAT-ī-onz) Positively charged ions that have lost one or more electrons from their outer shell

caveolae (kā-vē-Ō-lē) Cavelike indentations in the outer surface of the plasma membrane that contain an abundance of membrane receptors and serve as important sites for signal transduction

cDNA DNA code that is synthesized in the laboratory as a complementary copy of an RNA isolated from a tissue

cell The smallest unit capable of carrying out the processes associated with life; the basic unit of both structure and function of living organisms

cell adhesion molecules (CAMs) Proteins that protrude from the surface of the plasma membrane and form loops or other appendages that the cells use to grip each other and the surrounding connective tissue fibers

cell body The portion of a neuron that houses the nucleus and organelles

cell-mediated immunity A specific immune response accomplished by activated T lymphocytes, which directly attack unwanted cells

center A functional collection of cell bodies within the central nervous system

central chemoreceptors (kē-mō-rē-SEP-turz) Receptors located in the vertebrate medulla near the respiratory center that respond to changes in ECF H^+ concentration resulting from changes in arterial P_{CO_2} and adjust respiration accordingly

central lacteal (LAK-tē-ul) The initial lymphatic vessel that supplies each of the vertebrate small intestinal villi

central nervous system (CNS) The brain and longitudinal nerve cord(s), which integrate input from sensory neurons and output to effectors

central sulcus (SUL-kus) A deep infolding of the mammalian brain surface that runs roughly down the middle of the lateral surface of each cerebral hemisphere and separates the parietal and frontal lobes

centrioles (SEN-tree-ōls) A pair of short, cylindrical structures within a cell that form the mitotic spindle during cell division

cerebellum (ser´-uh-BEL-um) The portion of the vertebrate brain attached to the brain stem and concerned with maintaining proper position of the body in space and subconscious coordination of motor activity

cerebral cortex The outer shell of gray matter in the vertebrate cerebrum; site of initiation of all voluntary motor output and final perceptual processing of all sensory input as well as integration of most higher neural activity

cerebral hemispheres The cerebrum's two halves, which are connected by a thick band of neuronal axons

cerebrospinal fluid (ser´-uh-brō-SPĪ-nul) or (sah-REE-brō-SPĪ-nul)   A special cushioning fluid that is produced by, surrounds, and flows through the central nervous system of vertebrates

cerebrum (SER-uh-brum) or (sah-REE-brum)   The division of the vertebrate brain that consists of the basal nuclei and cerebral cortex

channels   Small, water-filled passageways through the plasma membrane; formed by membrane proteins that span the membrane and provide highly selective passage for small water-soluble substances such as ions

chemical bonds   The forces holding atoms together

chemically gated channels   Channels in the plasma membrane that open or close in response to the binding of a specific chemical messenger with a membrane receptor site that is in close association with the channel

chemoreceptor (kē-mo-rē-sep´-tur)   A sensory receptor sensitive to specific chemicals

chemotaxin (kē-mō-TAK-sin)   A chemical released at an inflammatory site that attracts phagocytes to the area

chemotaxis   Movement of an organism toward or away from a chemical substance

chief cells   The vertebrate stomach cells that secrete pepsinogen

chitin   A polymer of repeating N-acetylglucosamine, serving as the primary structural molecule of arthropod exoskeletons

chloride cells   Specialized cells, which transport NaCl, in the gill epithelia of fishes

cholecystokinin (CCK) (kō´-luh-sis-tuh-kī-nun)   A hormone released from the vertebrate duodenal mucosa primarily in response to the presence of fat; inhibits gastric motility and secretion, stimulates pancreatic enzyme secretion, and stimulates gallbladder contraction

cholesterol   A type of lipid molecule that serves as a precursor for steroid hormones and bile salts and is a stabilizing component of the plasma membrane

cholinergic fibers (kō´-lin-ER-jik)   Nerve fibers that release acetylcholine as their neurotransmitter

chorionic gonadotropin (CG) (kō-rē-ON-ik gō-nad´-uh-TRŌ-pin)   A hormone secreted by the early mammalian embryo and developing placenta that stimulates and maintains the corpus luteum of pregnancy

chromatophores   Pigment-containing cells found in the integument of some animals, which are responsible for physiological color change. The color granules are dispersed or concentrated depending on the animal's perception of its surroundings

chronic obstructive pulmonary disease   A group of lung diseases characterized by increased airway resistance resulting from narrowing of the lumen of the lower airways; includes asthma, chronic bronchitis, and emphysema

chyme (kīm)   A thick, liquid mixture of food and digestive juices

cilia (SILL-ee-ah)   Motile, hairlike protrusions from the surface of many cells, such as in ciliate protozoa, flatworm integument, and linings of mammalian respiratory airways and oviducts

ciliary body   The portion of the vertebrate eye that produces aqueous humor and contains the ciliary muscle

ciliary muscle   A circular ring of smooth muscle, within the vertebrate eye, whose contraction increases the strength of the lens to accommodate for near vision

circadian rhythm (sir-KĀ-dē-un)   Repetitive oscillations in the set point of various body activities, such as hormone levels and body temperature, that are very regular and have an approximate frequency of 24 hours, usually linked to light–dark cycles; diurnal rhythm; biological rhythm

circannual   Yearly biological rhythm

circulatory shock   The condition when mean arterial blood pressure falls so low that adequate blood flow to the tissues can no longer be maintained

cistern   Structure associated with the mammalian teat; permits pooling of milk from several ducts before exiting the gland

citric acid cycle   A cyclical series of biochemical reactions that involves the further processing of intermediate breakdown products of nutrient molecules, resulting in the generation of carbon dioxide and the preparation of hydrogen carrier molecules for entry into the high-energy–yielding electron transport chain

cloaca   The final chamber of the hindgut of birds and reptiles

clone   An exact genetic copy of a gene, or an organism arising from an exact copy of a genome

clutch   Nest of eggs

CNS   See *central nervous system*

cochlea (KOK-lē-uh)   The snail-shaped portion of the vertebrate inner ear that houses the receptors for sound

coevolution   The process in which adaptive changes in one organism favor selection for adaptations in a second organism, which in turn favor selection for a different adaptation in the first organism, and so on

collecting tubule   The last portion of tubule in the mammalian kidney's nephron that empties into the renal pelvis

colligative properties   Solution properties of an idealized solute at 1 mole in 1 kg of water (defined as 1 molal), which has an osmotic pressure of 22.4 atmospheres (atm), raises boiling point by 0.54°C, depresses freezing point by 1.86°C, and also reduces vapor pressure

colloid (KOL-oid)   The thyroglobulin-containing substance enclosed within the thyroid follicles

compatible osmolyte   An organic osmolyte that elevates the osmotic pressure of a body fluid without perturbing cell functions

complement system   A collection of vertebrate plasma proteins that are activated in cascade fashion on exposure to invading microorganisms, ultimately producing a membrane attack complex that destroys the invaders

compliance   The distensibility of a hollow, elastic structure, such as a blood vessel or the lungs; a measure of how easily the structure can be stretched

compound eye   Multifaceted arthropod eye composed of numerous optical units called *ommatidia*

concentration gradient   A difference in concentration of a particular substance between two adjacent areas

conduction   Transfer of heat between objects of differing temperatures that are in direct contact with each other

cones   The vertebrate eye's photoreceptors used for color vision in the light

conformers   Organisms in which a particular physiologic state matches that of the environment

congestive heart failure   The inability of the cardiac output to keep pace with the body's needs for blood delivery, with blood damming up in the veins behind the failing heart

connective tissue   Tissue that serves to connect, support, and anchor various body parts; distinguished by relatively few cells dispersed within an abundance of extracellular material

contiguous conduction   The means by which an action potential is propagated throughout a nonmyelinated nerve fiber; local current flow between an active and adjacent inactive area brings the inactive area to threshold, triggering an action potential in a previously inactive area

contractile proteins   Myosin and actin, whose interaction brings about shortening (contraction) of a muscle fiber

conus arteriosus   In cartilaginous fish, a chambered extension of the ventricle through which blood exits the heart

convection   Transfer of heat energy by air or water currents

convergence   The converging of many presynaptic terminals from thousands of other neurons on a single neuronal cell body and its dendrites so that activity in the single neuron is influenced by the activity from many other neurons

convex Curved out, as a surface in a lens that converges light rays

cooperativity The binding of a ligand, such as O_2, to its binding site on a multi-subunit protein, such as hemoglobin, subsequently affecting the ability of other binding sites on the same protein to associate with additional ligand

coprodaeum Anterior portion of the avian cloaca, receives the excreta from the digestive tract

coprophagy Reingestion of feces

copulatory plug Temporary mating plug that forms after fertilization, which prevents both the loss of semen from the female tract as well as entry of sperm into the vagina (genital chamber) from a competing male

core temperature The temperature within the inner core of a body (abdominal and thoracic organs, central nervous system, and skeletal muscles)

cornea (KOR-nee-ah) The clear, most anterior, outer layer of an eye, through which light rays pass to the interior of the eye

coronary circulation The blood vessels that supply the vertebrate heart muscle

corpus luteum (LOO-tē-um) The ovarian structure that develops from a ruptured follicle following ovulation in a mammal

cortisol (KORT-uh-sol) The vertebrate adrenocortical hormone that plays an important role in carbohydrate, protein, and fat metabolism and helps the body resist stress; functionally similar to corticosterone

cotransport See *symporter* and *antiporter*

counteracting osmolyte An organic osmolyte that elevates the osmotic pressure of a body fluid and that counteracts the effects of a perturbant on cellular macromolecules

countercurrent A design in which fluid in two juxtaposed tubes flows in opposite directions (alternatively, in which flow in a single tube alternates between inflow and outflow)

cranial nerves The 12 pairs of vertebrate peripheral nerves, the majority of which arise from the brain stem

crop An enlarged posterior portion of the foregut, just anterior to the esophagus

crop milk Lipid material synthesized by epithelial cells of the crop of adult pigeons and doves. These cells are sloughed off and mixed with food already present in the crop and fed by regurgitation to the juveniles

cross bridges The myosin molecules' globular heads that protrude from a thick

filament within a muscle fiber and interact with the actin molecules in the thin filaments to shorten the muscle fiber during contraction

crosscurrent exchange Air flow and blood flow in the parabronchi of the avian lung occur perpendicular to one another

cryoprotectant A molecule that is used to prevent freezing by lowering the freezing point in a body fluid

cyclic adenosine monophosphate (cyclic AMP or cAMP) An intracellular second messenger derived from adenosine triphosphate (ATP)

cytokines Signal molecules not produced by classic or distinct endocrine glands

cytoplasm (SĪ-tō-plaz´-um) The portion of the cell interior not occupied by the nucleus

cytoskeleton A complex intracellular protein network that acts as the "bone and muscle" of the cell

cytosol (SĪ-tuh-sol´) The semiliquid portion of the cytoplasm not occupied by organelles

cytotoxic T-cells (si-tō-TOK-sik) The population of T-cells (T lymphocytes) that destroys host cells bearing foreign antigen, such as body cells invaded by viruses or cancer cells

dead space volume The volume of medium that occupies the respiratory pathways as air or water is moved in and out and that is not available to participate in exchange of O_2 and CO_2 between the respiratory surfaces and external medium

decompression sickness, or "the bends" Pathology caused by the presence of nitrogen bubbles in the tissues, inducing a state of profound stupor, unconsciousness, or arrested activity

dehydration A water deficit in the body

dehydroepiandrosterone (DHEA) (dē-HĪ-drō-ep-i-and-row-steer-own) The androgen (masculinizing hormone) secreted by the vertebrate adrenal cortex in both sexes

dendrites Projections from the surface of a neuron's cell body that carry signals toward the cell body

deoxyribonucleic acid (DNA) (dē-OK-sē-rī-bō-new-klā-ik) The cell's genetic material, which is found within the nucleus and which provides codes for protein synthesis and serves as a blueprint for cell replication

depolarization (de´-pō-luh-ruh-ZĀ-shun) A reduction in membrane potential from resting potential; movement of the potential from resting toward 0 mV

dermis The connective tissue layer that lies under the epidermis in the skin; in

vertebrates, it contains the skin's blood vessels and nerves

descending tract A bundle of nerve fibers of similar function that travels down the vertebrate spinal cord to relay messages from the brain to efferent neurons

desmosome (dez´-muh-sōm) An adhering junction between two adjacent but nontouching cells formed by the extension of filaments between the cells' plasma membranes; most abundant in tissues that are subject to considerable stretching

detritivore Animal that consumes dead and living organic material in sediments

DHEA See *dehydroepiandrosterone*

diabetes mellitus (muh-LĪ-tus) An endocrine disorder characterized by inadequate insulin action

diaphragm (DIE-uh-fram) A dome-shaped sheet of skeletal muscle that forms the posterior end of the thoracic cavity in many air-breathing vertebrates; a major inspiratory muscle

diaphragmaticus muscle A muscle attached between the liver and a distinctive elongated region of the pubis bone in crocodiles. Pulls the liver away from the lungs in a pistonlike manner that results in expansion of the lungs

diastema Gap that extends between the front teeth and the cheek teeth (premolars and molars) of ruminants

diastole (dī-AS-tō-lē) The period of cardiac relaxation and filling

diencephalon (dī´-un-SEF-uh-lan) The division of the vertebrate brain that consists of the thalamus and hypothalamus

diestrus Associated with a functioning corpus luteum (dominance of progesterone) in a mammal and begins about four days after ovulation and ends with regression of the corpus luteum

diet-induced thermogenesis The increase in metabolism that follows ingestion of a meal; obligatory form is associated with the costs of processing the food, whereas the regulatory form is activated for the purpose of removing excess calories as heat

diffusion Random collisions and intermingling of molecules as a result of their continuous, thermally induced random motion

digestion The conversion process whereby the structurally complex food-stuffs of the diet are broken down into smaller absorbable units by the enzymes produced within the digestive system

diploid number (DIP-loid) A complete set of chromosomes (for example, 23 pairs in humans), as found in all somatic cells

distal tubule A highly convoluted tubule that extends between the loop of

Henle and the collecting duct in the mammalian kidney's nephron

divergence   The diverging, or branching, of a neuron's axon terminals, so that activity in this single neuron influences the many other cells with which its terminals synapse

diverticula   Blind tubules which arise from the main passage of the midgut in insects and contain the gastric caecae

DMS (dimethysulfide)   A gas produced by the breakdown of DMSP

DMSP (dimethylsulfonopropionate)   A major osmolyte in many marine algae

DNA   See *deoxyribonucleic acid*

dormancy   A state of very low metabolic rate, usually temporary

dorsal root ganglion   A cluster of afferent neuronal cell bodies located adjacent to the vertebrate spinal cord

dorsobronchi   Caudal group of secondary bronchi in birds that branch to form a fan-shaped covering of the mediodorsal surface of the avian lung

downregulation   A decrease in the capacity of a cellular or organ function

ecdysis   The shedding of the old exoskeleton

ecdysone   Hormone secreted by the prothoracic gland of insects

ECG   See *electrocardiogram*

ectotherm   An organism dependent on external heat for body temperature

edema (i-DĒ-muh)   Swelling of tissues as a result of excess interstitial fluid

EDV   See *end-diastolic volume*

EEG   See *electroencephalogram*

effector organs   The organs, especially muscles or glands, that are controlled by a regulatory system and that carry out that system's orders to bring about a desired effect, such as a particular movement or secretion

efferent division (EF-er-ent)   The portion of the peripheral nervous system that carries instructions from the central nervous system to effector organs

efferent neuron   Neuron that carries information from the central nervous system to an effector organ

efflux (Ē-flux)   Movement out of the cell

egest   Discharge of indigestible matter from the digestive tract

eicosanoid   A class of signal molecules derived from lipids such as arachidonic acid

elastic recoil   Rebound of the lungs after having been stretched

electrical gradient   A difference in charge between two adjacent areas

electrocardiogram (ECG)   The graphic record of the electrical activity that reaches the surface of the body as a result of cardiac depolarization and repolarization

electrochemical gradient   The simultaneous existence of an electrical gradient and concentration (chemical) gradient for a particular ion

electrocyte   Modified neurons of electric fish that, when linked in series, are capable of generating a substantial electrical discharge

electrolytes   Solutes that form ions in solution and conduct electricity

enantiostasis   A form of regulation in which changes in one physiologic state help offset a disturbance to another physiologic state

end-diastolic volume (EDV)   The volume of blood in the ventricle at the end of diastole, when filling is complete

endocrine glands   Ductless glands that secrete hormones into the blood or hemolymph

endocrine-disrupting chemicals (EDCs)   Human-made substances that are released into the environment (or generated in sewage) and that interfere with the endocrine functions of animals

endocytosis (en´-dō-sī-TŌ-sis)   Internalization of extracellular material within a cell as a result of the plasma membrane forming a pouch that contains the extracellular material, then sealing at the surface of the pouch to form a small, intracellular, membrane-enclosed vesicle with the contents of the pouch trapped inside

endogenous opiates (en-DAJ´-eh-nus ō´-pē-ātz)   Endorphins and enkephalins, which bind with opiate receptors and are important in the vertebrate body's natural analgesic system

endogenous pyrogen (pī´-ruh-jun)   A chemical released from vertebrate macrophages during inflammation that acts by means of local prostaglandins to raise the set point of the hypothalamic thermostat to produce a fever

endometrial cup   Temporary endocrine units in the uterine horn of the mare (placental origin), ranging in size from millimeters to several centimeters that produce eCG (equine chorionic gonadotropin)

endometrium (en´-dō-MĒ-trē-um)   The lining of the mammalian uterus

endoplasmic reticulum (en´-dō-PLAZ-mik ri-TIK-yuh-lum)   An organelle consisting of a continuous membranous network of fluid-filled tubules and flattened sacs, partially studded with ribosomes; synthesizes proteins and lipids for formation of new cell membrane and other cell components and manufactures products for secretion

endothelium (en´-dō-THĒ-lē-um)   The thin, single-celled layer of epithelial cells that lines the entire vertebrate circulatory system

endotherm   An organism that uses internal heat to regulate body temperature

end-plate potential (EPP)   The graded receptor potential that occurs at the motor end plate of a skeletal muscle fiber in response to binding with acetylcholine

end-systolic volume (ESV)   The volume of blood in the ventricle at the end of systole, when emptying is complete

enhancer   A promoter DNA sequence that is specific for a particular gene, helping regulate that gene's transcription (by binding of transcription factors) in a tissue-specific manner

enterogastrones (ent´-uh-rō-GAS-trōnz)   Hormones secreted by the vertebrate duodenal mucosa that inhibit gastric motility and secretion; include secretin, cholecystokinin, and gastric inhibitory peptide

enterohepatic circulation (en´-tur-ō-hi-PAT-ik)   The recycling of bile salts and other bile constituents between the vertebrate small intestine and liver by means of the hepatic portal vein

entropy   A measure of disorder in a system

enzyme   A protein molecule that speeds up (catalyzes) a particular chemical reaction in an organism

eosinophils (ē´-uh-SIN-uh-fils)   White blood cells that are important in allergic responses and in combating internal parasite infestations

epidermis (ep´-uh-DER-mus)   The outer layer of the skin, consisting of numerous layers of epithelial cells

epinephrine (ep´-uh-NEF-rin)   The primary hormone secreted by the vertebrate adrenal medulla; important in preparing the body for "fight-or-flight" responses and in regulation of arterial blood pressure; adrenaline

epiphyseal plate (eh-pif-i-SEE-al)   A layer of cartilage that separates the diaphysis (shaft) of a vertebrate long bone from the epiphysis (flared end); the site of growth of bones in length before the cartilage ossifies (turns into bone)

epithelial tissue (ep´-uh-THĒ-lē-ul)   A functional grouping of cells specialized in the exchange of materials between the cell and its environment; lines and covers various body surfaces and cavities and forms secretory glands

EPSP   See *excitatory postsynaptic potential*

equilibrium potential   The potential that exists when the concentration gradient and opposing electrical gradient for a given ion exactly counterbalance each other so that there is no net movement of the ion

equine chorionic gonadotropin (eCG) A peptide hormone secreted by the developing chorion of the mare. Also termed *pregnant mare's serum gonadotropin (PMSG)*. See *chorionic gonadotropin*

eructation Belching

erythrocytes (i-RITH-ruh-sīts) Red blood cells, which are plasma membrane–enclosed bags of hemoglobin that transport O_2 and to a lesser extent CO_2 and H in the blood

erythropoiesis (i-rith´-rō-poi-Ē-sus) Erythrocyte production by the bone marrow

erythropoietin The hormone released from vertebrate kidneys in response to a reduction in O_2 delivery to the kidneys; stimulates the bone marrow to increase erythrocyte production

esophagus (i-SOF-uh-gus) A straight, muscular tube that extends between the pharynx and stomach

estivation A dormant or low-metabolism state of an animal, usually during dry summer months

estrogen Feminizing "female" sex hormone

estrous cycle The time from one period of sexual receptivity to the next

estrus The period of sexual receptivity in a mammalian female

ESV See *end-systolic volume*

eubacteria Prokaryotes (classically called *bacteria*) forming one of three distinct forms of life (along with archaea and eukaryota); included are those related to the ancestors of the eukaryotic mitochondria and chloroplast

eukaryote A cell (or organism with such cells) that has organelles including a nucleus, mitochondria, and endoplasmic reticulum; Protista, Plantae, Animalia, and Fungi

euryhaline Capable of surviving over a wide range of external salinities

evolutionary explanation In biology, the evolutionary reason for the existence of a structure or process ("why it arose this way"), as opposed to the mechanistic explanation

excitable tissue Tissue capable of producing electrical signals when excited; includes nervous and muscle tissue

excitation–contraction coupling The series of events linking muscle excitation (the presence of an action potential) to muscle contraction (filament sliding and sarcomere shortening)

excitatory postsynaptic potential (EPSP) (pōst´-si-NAP-tik) A small depolarization of the postsynaptic membrane in response to neurotransmitter binding, thereby bringing the membrane closer to threshold

excitatory synapse (SIN-aps´) Synapse in which the postsynaptic neuron's response to neurotransmitter release is a small depolarization of the postsynaptic membrane, bringing the membrane closer to threshold

exocrine glands Glands that secrete through ducts to the outside of the body or into a cavity that communicates with the outside

exocytosis (eks´-ō-sī-TŌ-sis) Fusion of a membrane-enclosed intracellular vesicle with the plasma membrane, followed by the opening of the vesicle and the emptying of its contents to the outside

expiration A breath out (exhalation)

expiratory muscles The skeletal muscles whose contraction moves respiratory air or water out of a respiratory chamber

extensors Skeletal muscles that straighten a limb out

external intercostal muscles Inspiratory muscles whose contraction expands the vertebrate rib cage, thereby enlarging the thoracic cavity (in most air-breathing vertebrates)

external work Energy expended by contracting skeletal muscles to move external objects or to move the body in relation to the environment

exteroreceptors The "classic" senses that detect external stimuli such as light, chemicals, touch, temperature, and sound

extracellular fluid All the body's fluid found outside the cells; consists of interstitial fluid and plasma

extracellular matrix An intricate meshwork of fibrous proteins embedded in a watery, gel-like substance; secreted by local cells

extrinsic controls Regulatory mechanisms initiated outside of an organ that alter the activity of the organ; accomplished by the nervous and endocrine systems

facilitated diffusion Passive carrier-mediated transport involving transport of a substance down its concentration gradient across the plasma membrane

fat body The insect equivalent of a liver; it stores large amounts of triglycerides and carbohydrates

fatigue Inability to maintain muscle tension at a given level despite sustained stimulation

feedforward mechanism A response designed to prevent an anticipated change in a controlled variable

feeding centers See *appetite centers*

fibrinogen (fī-BRIN-uh-jun) A large, soluble plasma protein that is converted into an insoluble, threadlike molecule that forms the meshwork of a clot during blood coagulation

Fick's law of diffusion The rate of net diffusion of a substance across a membrane is directly proportional to the substance's concentration gradient, the membrane's permeability to the substance, and the surface area of the membrane and inversely proportional to the substance's molecular weight and the diffusion distance

fight-or-flight response The changes in activity of the various organs innervated by the vertebrate autonomic nervous system in response to sympathetic stimulation, which collectively prepare the body for strenuous physical activity in the face of an emergency or stressful situation, such as a physical threat from the outside environment

filter chamber A modification of the digestive tract of Homoptera, insects that feed on large quantities of plant juices. The filter chamber permits water from the ingested sap to pass directly from the anterior region of the midgut to the hindgut

filter feeding Trapping of dead and/or living material suspended in water by straining through specialized entrapment devices

firing The event when an excitable cell undergoes an action potential

first messenger An extracellular messenger, such as a hormone, that binds with a surface membrane receptor and activates an intracellular second messenger to carry out the desired cellular response

flagellum (fluh-JEL-um) The single, long, whiplike appendage of some cells, for example, one that serves as the tail of a spermatozoon or the current producer of a sponge cell

flexors Skeletal muscles that bend a limb

flow-through breathing Movement of air or water in and out of a respiratory chamber through separate openings, for the purposes of gas exchange

follicle (of ovary) A developing ovum and the surrounding specialized cells

follicle-stimulating hormone (FSH) An anterior pituitary hormone that stimulates ovarian follicular development and estrogen secretion in females, and stimulates sperm production in males

follicular cells (of ovary) (fah-LIK-you-lar) Collectively, the granulosa and thecal cells

follicular cells (of thyroid gland) The cells that form the walls of the colloid-filled follicles in the thyroid gland and secrete thyroid hormone

follicular phase The phase of the mammalian ovarian cycle dominated by the presence of maturing follicles prior to ovulation

forestomach Ruminoreticular region of the ruminant stomach. Contains the populations of microbes involved in the fermentation of foodstuffs

fovea A small depression or pit in the vertebrate retina containing densely packed cones; the region of the eye with the greatest visual resolution

Frank-Starling law of the heart Intrinsic control of the heart, such that increased venous return resulting in increased end-diastolic volume leads to an increased strength of contraction and increased stroke volume; that is, the heart normally pumps out all the blood returned to it

free radicals Very unstable electron-deficient particles that are highly reactive and destructive

freemartin Sterile female calf twin born with a male

frontal lobes The lobes of the vertebrate cerebral cortex that lie at the top of the brain in front of the central sulcus and that, in mammals, are responsible for nonreflexive motor output and complex planning

FSH See *follicle-stimulating hormone*

functional syncytium (sin-sish´-ē-um) A group of cells that are interconnected by gap junctions and function electrically and mechanically as a single unit

functional unit The smallest component of an organ that can perform all the functions of the organ

G protein A membrane-bound intermediary, which, when activated on binding of an extracellular first messenger to a surface receptor, activates the enzyme adenylyl cyclase on the intracellular side of the membrane in the cAMP second-messenger system

gametes (GAM-ētz) Reproductive or germ cells, each containing a haploid set of chromosomes; sperm and ova

gamma motor neuron A motor neuron that innervates the fibers of a muscle spindle receptor

ganglion (pl., **ganglia**) (1) A group or cluster of nerve cell bodies, often with related functions. (2) In the vertebrate eye, the nerve cells in the outermost layer of the retina and whose axons form the optic nerve

gap junction A communicating junction formed between adjacent cells by small connecting tunnels that permit passage of charge-carrying ions between the cells so that electrical activity in one cell is spread to the adjacent cell

gas bladder (swim bladder) A large, centrally located sac ventral to the spinal column, in the abdominal cavity of teleost fish, used for buoyancy control

gastrin A hormone secreted by the pyloric gland area of the vertebrate stomach that stimulates the parietal and chief cells to secrete a highly acidic gastric juice

genome The complete set of genetic codes in a particular species

genomics The study of the structure, regulation and information content of genes

gestation Pregnancy

gill Evaginated respiratory organ of water-breathing animals

gill arch Fish tissue that provides skeletal support for a double series of gill filaments, each of which carries a row of secondary lamellae on either side

gizzard Muscular portion of the stomach of most birds where food (such as grains, seeds) are ground up with the aid of ingested pebbles

glands Epithelial tissue derivatives that are specialized for secretion

glial cells (glē-ul) Cells that serve as the connective tissue of the vertebrate CNS and help support the neurons both physically and metabolically; include astrocytes, oligodendrocytes, ependymal cells, and microglia

glomerular filtration (glow-MER-yū-lur) Filtration of a protein-free plasma from the glomerular capillaries into the tubular component of the vertebrate kidney's nephron as the first step in urine formation

glomerular filtration rate (GFR) The rate at which glomerular filtrate is formed

glomerulus (glow-MER-yū-lus) A ball-like tuft of capillaries in the vertebrate kidney's nephron that filters water and solute from the blood as the first step in urine formation

glucagon (GLOO-kuh-gon) The vertebrate pancreatic hormone that raises blood glucose and blood fatty-acid levels

glucocorticoids (gloo´-kō-KOR-ti-koidz) The adrenocortical hormones that are important in intermediary metabolism and in helping the body resist stress; primarily cortisol

gluconeogenesis (gloo´-kō-nē-ō-JEN-uh-sus) The "new formation of glucose"; the conversion of pyruvate or amino acids into glucose, running glycolysis essentially in reverse

glycation A reaction in which glucose and other reducing sugars spontaneously bind with proteins such as collagen and hemoglobin, forming a covalent bond that can alter its structure and function

glycogen (GLĪ-kō-jen) The storage form of glucose in the liver and muscle

glycogenesis (glī´-kō-JEN-i-sus) The conversion of glucose into glycogen

glycogenolysis (glī´-kō-juh-NOL-i-sus) The conversion of glycogen to glucose

glycolysis (glī-KOL-uh-sus) A biochemical process that takes place in the cell's cytosol and involves the breakdown of glucose into two pyruvic acid molecules

glycoprotein A protein that has sugars covalently bound to it; common on outer sides of cell membranes

GnRH See *gonadotropin-releasing hormone*

Golgi complex (GOL-jē) An organelle consisting of sets of stacked, flattened membranous sacs; processes raw materials transported to it from the endoplasmic reticulum into finished products and sorts and directs the finished products to their final destination

gonadotropin-releasing hormone (GnRH) (gō-nad´-uh-TRŌ-pin) The hypothalamic hormone that stimulates the release of FSH and LH from the vertebrate anterior pituitary

gonadotropins FSH and LH; vertebrate hormones that are tropic to the gonads

gonads (GŌ-nadz) The primary reproductive organs, which produce the gametes; testes and ovaries

gradation of contraction Variable magnitudes of tension produced in a single whole muscle

graded potential A local change in membrane potential that occurs in varying grades of magnitude; serves as a short-distance signal in excitable tissues

granulocytes (gran´-yuh-lō-sīts) Vertebrate leukocytes that contain granules, including neutrophils, eosinophils, and basophils

granulosa cells (gran´-yuh-LŌ-suh) The layer of cells immediately surrounding a developing oocyte within a vertebrate ovarian follicle

gray matter The portion of the vertebrate central nervous system composed primarily of densely packaged neuronal cell bodies and dendrites

grit Hard granules, such as sand and pebbles found in the gizzard of birds

growth hormone (GH) An anterior pituitary hormone that is primarily responsible for regulating overall body growth and is also important in intermediary metabolism; somatotropin

gular fluttering Panting with a rapid fluttering of the well-vascularized esophageal region of the hyoid apparatus, in birds

gustation Taste, or detection of molecules in objects in contact with the body

H See *hydrogen ion*

habituation Decreased responsiveness to repetitive presentation of an indifferent stimulus (neither rewarding or punishing), resulting in a decrease in synaptic activity

haploid number (HAP-loid)   The number of chromosomes found in gametes; a half set of chromosomes, one member of each pair

Hb   See *hemoglobin*

hCG   Human version of chorionic gonadotropin

heat shock proteins (HSPs)   See *stress proteins*

helper T-cells   The population of T-cells that enhances the activity of other immune-response effector cells

hematocrit (hi-mat´-uh-krit)   The percentage of blood volume occupied by erythrocytes as they are packed down in a centrifuged blood sample

hemerythrin   Iron-containing respiratory pigment found in several invertebrate phyla, including annelid worms

hemipenis   Organ used by the drake to deposit semen in the female duck

hemocoel   The body cavity of an animal with a circulating hemolymph

hemocyanin   An oxygen-transporting protein in many mollusks and arthropods such as crustaceans that uses two copper atoms to bind one oxygen molecule

hemodynamic   Pertaining to the physics of blood flow

hemoglobin (HĒ-muh-glō´-bun) (Hb)   A large iron-bearing protein molecule, that binds with and transports most O_2 in blood; in vertebrates, it is contained within erythrocytes, and it also carries some of the CO_2 and H^+

hemolymph   The circulatory fluid in an animal that does not have capillaries

hemolysis (hē-MOL-uh-sus)   Rupture of red blood cells

hemostasis (hē´-mō-STĀ-sus)   The stopping of bleeding from an injured vessel

hepatic portal system (hi-PAT-ik)   A complex vascular connection between the vertebrate digestive tract and liver such that venous blood from the digestive system drains into the liver for processing of absorbed nutrients before being returned to the heart

herbivore   Animals that depend on the intake of algal or plant materials

hermaphrodism   State in which an animal has both functional male and female reproductive tracts, although not necessarily at the same time

heterogametic   Producing two or more different kinds of gametes that differ in their sex chromosome content

heterotherm   An animal that has some degree of endothermy but that is not fully homeothermic

hibernation   A state of dormancy in an animal usually associated with winter months; involves lower body temperatures in endotherms

hippocampus (hip-oh-CAM-pus)   The elongated, medial portion of the temporal lobe that is a part of the vertebrate limbic system and is especially crucial for forming long-term memories

histamine   A chemical released from vertebrate mast cells or basophils that brings about vasodilation and increased capillary permeability; important in allergic responses

homeostasis (hō´-mē-ō-STĀ-sus)   Maintenance by coordinated, regulated actions of body systems of relatively stable chemical and physical conditions in the internal fluid environment and in other body states

homeotherm   Animal with narrowly varying body temperatures

homeoviscous   Having the same fluidity or viscosity; referring to cell membranes being altered at different temperatures to maintain an optimal fluidity

homogametic   Producing one type of gamete with respect to sex chromosome content

homunculus   Orderly distribution of cortical sensory processing in the vertebrate brain. For each species of vertebrate, the extent of each body part is indicative of the relative proportion and location of the somatosensory cortex devoted to that area

honeydew   Nutritious liquid discharged from the anus of some Homoptera

hormone   A long-distance chemical mediator that is secreted by an endocrine gland into the blood or hemolymph, which transports it to its target cells

host cell   A body cell infected by a virus

HRE   See *response element*

hydrogen ion (H^+)   The cationic portion of a dissociated acid; a proton without an electron

hydrolysis (hī-DROL-uh-sis)   The digestion of a nutrient molecule by the addition of water at a bond site

hydrostatic pressure (hī-dro-STAT-ik)   The pressure exerted by a fluid on the walls that contain it

hyperglycemia (hī´-pur-glī-SĒ-mē-uh)   Elevated blood glucose concentration

hyperplasia (hī-pur-PLĀ-zē-uh)   An increase in the number of cells

hyperpolarization   An increase in membrane potential from resting potential; potential becomes even more negative than at resting potential

hypertension (hī´-pur-TEN-shun)   Sustained, above-normal mean arterial blood pressure

hyperthermia   Abnormally high body temperature

hypertonic (hī´-pur-TON-ik)   Having an osmolarity greater than normal body fluids; more concentrated than normal

hypertrophy (hī-PUR-truh-fē)   Increase in the size of an organ

hyperventilation   Overbreathing; when the rate of ventilation is in excess of the body's metabolic needs for CO_2 removal

hypomagnesemic tetany, or grass tetany   A pathological state in herbivores such as cattle, characterized by low magnesium concentrations in the blood and cerebrospinal fluid. Affected animals appear nervous, and show muscular twitching around the face and ears

hypophysiotropic hormones (hi-PŌ-fiz-ē-oh-TRO-pik)   Vertebrate hormones secreted by the hypothalamus that regulate the secretion of anterior pituitary hormones; see also *releasing hormone*

hypotension (hi-po-TEN-chun)   Sustained, below-normal mean arterial blood pressure

hypothalamic hypophyseal portal system (hī-pō-thuh-LAM-ik hī-pō- FIZ-ē-ul)   The vascular connection between the hypothalamus and anterior pituitary gland used for the pickup and delivery of hypophysiotropic hormones

hypothalamus (hī´-pō-THAL-uh-mus)   The vertebrate brain region located beneath the thalamus that is concerned with regulating many aspects of the internal fluid environment, such as water and salt balance and food intake; serves as an important link between the autonomic nervous system and endocrine system

hypothermia   Abnormally low body temperature

hypotonic (hī´-pō-TON-ik)   Having an osmolarity less than normal body fluids

hypoventilation   Underbreathing; ventilation inadequate to meet the metabolic needs for O_2 delivery and CO_2 removal

hypoxia (hī-POK-sē-uh)   Insufficient O_2 at the cellular level

I band   One of the light bands that alternate with dark (A) bands to create a striated appearance in skeletal or cardiac muscle fibers when these fibers are viewed with a light microscope

ice-nucleating agent   Small proteins in some animals that trigger the formation of small ice crystals once temperatures have dropped to the freezing point

IGF   See *insulin-like growth factor*

immune surveillance   Recognition and destruction of newly arisen cancer cells by the immune system

immunity   Ability to resist or eliminate potentially harmful foreign materials or abnormal cells

immunoglobulins (im´-ū-nō-GLOB-yū-lunz) Antibodies; gamma globulins

impermeable Prohibiting passage of a particular substance through the plasma membrane

implantation The burrowing of a mammalian blastocyst into the endometrial lining

inflammation An innate, nonspecific series of highly interrelated events in vertebrates, especially involving neutrophils, macrophages, and local vascular changes, that are set into motion in response to foreign invasion or tissue damage

influx Movement into the cell

inhibin (in-HIB-un) A vertebrate hormone secreted by the Sertoli cells of the testes or by the ovarian follicles that inhibits FSH secretion

inhibitory postsynaptic potential (IPSP) (pōst´-si-NAP-tik) A small hyperpolarization of the postsynaptic membrane in response to neurotransmitter binding, thereby moving the membrane farther from threshold

inhibitory synapse (SIN-aps´) Synapse in which the postsynaptic neuron's response to neurotransmitter release is a small hyperpolarization of the postsynaptic membrane, moving the membrane farther from threshold

innate immune responses Inherent defense responses that nonselectively defend against foreign or abnormal material, even on initial exposure to it; see also *inflammation, interferon, natural killer cells*

inorganic Referring to substances that do not contain carbon (and usually hydrogen)

insensible Referring to body water loss that is not sensed or regulated by an animal

inspiration An inward breath (inhalation)

inspiratory muscles The vertebrate skeletal muscles whose contraction enlarges the thoracic cavity, bringing about lung expansion and movement of air into the lungs from the atmosphere

insulin (IN-suh-lin) The vertebrate pancreatic hormone that lowers blood levels of glucose, fatty acids, and amino acids and promotes their storage

insulin-like growth factor (IGF) See *somatomedins*

integrating center A region that determines efferent output based on processing of afferent input

integument (in-TEG-yuh-munt) The skin and underlying connective tissue

intercostal muscles (int-ur-KOS-tul) The muscles that lie between the vertebrate ribs; see also *external intercostal muscles* and *internal intercostal muscles*

interferon (in´-tur-FĒR-on) A chemical released from virus-invaded vertebrate cells that provides nonspecific resistance to viral infections by transiently interfering with replication of the same or unrelated viruses in other host cells

interleukin 1 (int-ur-LOO-kin) A multipurpose chemical mediator released from vertebrate macrophages that enhances B-cell activity

interleukin 2 A vertebrate chemical mediator secreted by helper T-cells that augments the activity of all T-cells

intermediary metabolism The collective set of intracellular chemical reactions that involve the degradation, synthesis, and transformation of small nutrient molecules

intermediate filaments Threadlike cytoskeletal elements that play a structural role in parts of the cells subject to mechanical stress

internal intercostal muscles Expiratory muscles whose contraction pulls the ribs inward, thereby reducing the size of the thoracic cavity (most air-breathing vertebrates)

internal respiration The intracellular metabolic processes carried out within the mitochondria that use O_2 and produce CO_2 during the derivation of energy from nutrient molecules

internal work All forms of biological energy expenditure that do not accomplish mechanical work outside of the body

interneuron Neuron that lies entirely within the central nervous system and is important for integrating responses to peripheral information

interoreceptors Sensors that detect information about the internal body fluids usually crucial to homeostasis, such as blood pressure and O_2 concentration inhalation

interstitial fluid (in´-tur-STISH-ul) The portion of the extracellular fluid that surrounds and bathes all the body's cells

intima The cuticular lining of the insect foregut, tracheae, and hindgut

intra-alveolar pressure (in´-truh-al-VĒ-uh-lur) The pressure within the alveoli (mammals, some reptiles)

intracellular fluid The fluid collectively contained within all the body's cells

intrapleural pressure (in´-truh-PLOOR-ul) The pressure within the pleural sac (mammalian lungs)

intrinsic controls Local control mechanisms inherent to an organ

intrinsic factor A special substance secreted by the parietal cells of the vertebrate stomach that must be combined with vitamin B_{12} for this vitamin to be absorbed by the intestine; deficiency produces pernicious anemia

intrinsic nerve plexuses Interconnecting networks of nerve fibers within the vertebrate digestive-tract wall

ion An atom that has gained or lost one or more of its electrons, so that it is not electrically balanced

IPSP See *inhibitory postsynaptic potential*

iris A pigmented smooth muscle that forms the colored portion of the vertebrate eye and controls pupillary size

islets of Langerhans (LAHNG-er-honz) The endocrine portion of the vertebrate pancreas that secretes the hormones insulin and glucagon into the blood

isoacids Branched-chain fatty acids generated by the deamination of some amino acids such as valine and leucine

isometric contraction (ī´-sō-MET-rik) A muscle contraction in which the development of tension occurs at constant muscle length

isotonic (ī´-sō-TON-ik) Having an osmolarity equal to normal body fluids

isotonic contraction A muscle contraction in which muscle tension remains constant as the muscle fiber changes length

joule The international unit for energy of any kind, equal to 0.239 calories

juvenile hormone Hormone secreted by the corpora allata gland of insects

juxtaganglionar organ A crustacean organ composed of scattered cells in the connective tissue around the cerebral ganglion; the cells release peptides of uncertain function into the hemolymph during egg laying

juxtaglomerular apparatus (juks´-tuh-glō-MER-yū-lur) A cluster of specialized vascular and tubular cells at a point where the ascending limb of the loop of Henle passes through the fork formed by the afferent and efferent arterioles of the same nephron in the mammalian kidney

keratin (CARE-uh-tin) The protein found in the intermediate filaments in vertebrate skin cells that give the skin strength and help form a waterproof outer layer

ketone bodies A group of fatty-acid derivatives produced by the vertebrate liver during glucose sparing

killer (K) cells Vertebrate cells that destroy a target cell that has been coated with antibodies by lysing its membrane

kinase An enzyme that phosphorylates (adds a phosphate group to) other proteins, thereby increasing or decreasing the activity of those proteins

kinesin (kī-NĒ´-sin) The transport or motor protein that transports secretory vesicles along the microtubular highway within neuronal axons by "walking" along the microtubule

kinocilia A type of cilium that extends from the apex of all sensory hair cells in all vertebrates except mammals

knockout genetics A procedure in which a specific gene is removed from an organism's genome, usually in the fertilized egg stage; a means of testing a gene's and its coded protein's function by the consequences of its absence

koilin Mucosal lining of the gizzard. A tough protein-polysaccharide complex that has an amino acid composition similar to that of feather protein

K-selected Referring to species that produce relatively few offspring, with most of the parental reproductive energy used for nourishing and/or protecting the offspring

lactation Milk production by the mammary glands

lactic acid An end product formed from pyruvic acid during the anaerobic process of glycolysis

larynx (LARE-inks) The "voice box" at the entrance of the mammalian trachea

latch state In vertebrates, a state of continuous contraction of smooth muscle that maintains tension with little metabolic cost

lateral geniculate nucleus The first stop in the mammalian brain for information in the visual pathway; in the thalamus

lateral inhibition The phenomenon in which the most strongly activated signal pathway originating from the center of a stimulus area inhibits the less excited pathways from the fringe areas by means of lateral inhibitory connections within sensory pathways

lateral sacs The expanded saclike regions of a muscle fiber's sarcoplasmic reticulum; store and release calcium, which plays a key role in triggering muscle contraction

law of mass action If the concentration of one of the substances involved in a reversible reaction is increased, the reaction is driven toward the opposite side, and if the concentration of one of the substances is decreased, the reaction is driven toward that side

left ventricle The heart chamber that pumps blood into the systemic circulation (in reptiles, birds, mammals)

length–tension relationship The relationship between the length of a muscle fiber at the onset of contraction and the tension the fiber can achieve on a subsequent tetanic contraction

lens A transparent, biconvex structure of a camera-type eye that refracts (bends) light rays and (in some animals) whose strength can be adjusted to accommodate for vision at different distances

leptin (or pubertin) An endocrine signal originating from mammalian adipose tissue that informs the hypothalamus that metabolic stores are sufficient for normal energy homeostasis or for the initiation of reproduction at puberty

leukocyte endogenous mediator (LEM) (LOO-kō-sīt en-DAJ-eh-nus ME-de-ā-tor) A chemical mediator secreted by vertebrate macrophages that is identical to endogenous pyrogen and exerts a wide array of effects associated with inflammation

leukocytes (LOO-kuh-sīts) White blood cells of vertebrates; the immune system's mobile defense units

Leydig cells (LĪ-dig) The interstitial cells of vertebrate testes that secrete testosterone

LH See *luteinizing hormone*

LH surge The burst in LH secretion that occurs at midcycle of the mammalian ovarian cycle and triggers ovulation

lignin Rigid polymer derived from phenylanine and tyrosine. With cellulose, forms the woody cell walls of plants and cements them together

limbic system (LIM-bik) A functionally interconnected ring of vertebrate forebrain structures that surrounds the brain stem and is concerned with emotions, basic survival, and sociosexual behavioral patterns, motivation, and learning

lipid emulsion A suspension of small fat droplets held apart as a result of adsorption of bile salts on their surface

loop of Henle (HEN-lē) A hairpin loop that extends between the proximal and distal tubule of the mammalian kidney's nephron

lordosis Mating posture assumed by a female

lower critical temperature In an endotherm, the temperature at which internal heat-generating mechanisms are activated to maintain thermal homeostasis

lumen (LOO-men) The interior space of a hollow organ or tube

luteal phase (LOO-tē-ul) The phase of the mammalian ovarian cycle dominated by the presence of a corpus luteum

luteinization (loot´-ē-un-uh-ZĀ-shun) Formation of a postovulatory corpus luteum in the mammalian ovary

luteinizing hormone (LH) An anterior pituitary hormone of vertebrates that stimulates ovulation, luteinization, and secretion of estrogen and progesterone in females and stimulates testosterone secretion in males

lymph Interstitial fluid that is picked up by vertebrate lymphatic vessels and returned to the venous system, meanwhile passing through the lymph nodes for defense purposes

lymphocytes Vertebrate white blood cells that provide immune defense against targets for which they are specifically programmed

lymphoid tissues Tissues that produce and store lymphocytes, such as lymph nodes and tonsils (vertebrates)

lysosomes (LĪ-sō-sōmz) Organelles consisting of membrane-enclosed sacs containing powerful hydrolytic enzymes that destroy unwanted material within the cell, such as internalized foreign material or cellular debris

macrophages (MAK-ruh-fājs) Large, tissue-bound phagocytes

magnetosome Magnetic crystals arranged in a chain

Maillard product Glucose covalently bound to a protein (browning reaction)

Malpighian tubule Filtering excretory tubules that begin filtration driven by secretion of ions from an arthropod's hemolymph

mantle External body wall lining the shell of mollusks, or the mantle chamber in shell-less cephalopods

mast cells Cells located within vertebrate connective tissue that synthesize, store, and release histamine, as during allergic responses

Mauthner neurons Paired giant interneurons found in some teleost fish and salamanders that elicit the startle response. Activity generated in one neuron simultaneously inhibits the activity in the other

maxilloturbinals Thin curls of bone deep within the avian and mammalian nasal cavity, used to retain heat and water

mean arterial blood pressure The average pressure responsible for driving blood forward through the arteries into the tissues throughout the cardiac cycle; equals cardiac output times total peripheral resistance

mechanically gated channels Channels that open or close in response to stretching or other mechanical deformation

mechanistic (or **proximate**) **approach** Explanation of body functions in terms of mechanisms of action, that is, the "how" of events that occur in the body

mechanoreceptor (meh-CAN-oh-rē-SEP-tur) or (mek´-uh-nō-rē-SEP-tur) A sensory receptor sensitive to mechanical energy, such as stretching or bending

Meckel's diverticulum Vestige of the yolk sac, forms a tube connected to the lower ileum of birds

medullary respiratory center (MED-you-LAIR-ē) Several aggregations of neuronal cell bodies within the vertebrate medulla oblongata that provide output to the respiratory muscles and receive input important for regulating the magnitude of ventilation

meiosis (mī-ō-sis)   Cell division in which the chromosomes replicate followed by two nuclear divisions so that only a half set of chromosomes is distributed to each of four new daughter cells

melanocyte-stimulating hormone (MSH) (mel-AH-nō-sīt)   A vertebrate hormone produced by the anterior pituitary in humans and by the intermediate lobe of the pituitary in lower vertebrates; regulates skin coloration by controlling the dispersion of melanin granules in lower vertebrates; involved with control of food intake and possibly memory and learning in humans

melatonin (mel-uh-TŌ-nin)   A vertebrate hormone secreted by the pineal gland during darkness that helps entrain the body's biological rhythms with the external light/dark cues

membrane attack complex   A collection of the five final activated components of the vertebrate complement system that aggregate to form a porelike channel in the plasma membrane of an invading microorganism, with the resultant leakage leading to destruction of the invader

membrane potential   A separation of charges across the membrane; a slight excess of negative charges lined up along the inside of the plasma membrane and separated from a slight excess of positive charges on the outside

memory cells   B- or T-cells of vertebrates that are newly produced in response to a microbial invader but that do not participate in the current immune response against the invader; instead they remain dormant, ready to launch a swift, powerful attack should the same microorganism invade again in the future

menstrual cycle (men´-stroo-ul)   The cyclical changes in the uterus that accompany the hormonal changes in the ovarian cycle of humans and some other primates

mesenteron   The midgut, or middle portion of the insect digestive tract

mesotocin (MT)   A hormone that influences the blood flow to some organs in nonmammalian vertebrates

messenger RNA   Carries the transcribed genetic blueprint for synthesis of a particular protein from nuclear DNA to the cytoplasmic ribosomes where the protein synthesis takes place

metabolic acidosis (met-uh-bol´-ik)   Acidosis resulting from any cause other than excess accumulation of carbonic acid in the body

metabolic alkalosis (al´-kuh-LŌ-sus)   Alkalosis caused by a relative deficiency of noncarbonic acid

metabolic rate   Energy expenditure per unit of time

metabolic scope   The degree to which an animal's metabolism can be activated above the resting level, as in maximal muscular activity in an emergency

metanephridia   Filtering excretory tubules that begin with a filtering capsule or funnel-like opening of some kind, followed by tubules that perform selective secretion and reabsorption; generally found in larger, more complex animals including adult mollusks, annelids (segmented worms), some arthropods (such as crustaceans), and vertebrates

metathoracic ganglia   Ganglia located in the posterior segment of the insect thorax

metestrus   Stage of the mammalian estrous cycle between ovulation and formation of a functional corpus luteum

methanogen   A microbe that produces methane from degrading organic material; usually archaea; found in many anaerobic habitats, including ruminant stomachs

micelle (mī-SEL)   A water-soluble aggregation of bile salts, lecithin, and cholesterol that has a hydrophilic shell and a hydrophobic core; carries the water-insoluble products of fat digestion to their site of absorption

microarray   A glass slide on which has been fixed, in a grid, a set of cDNAs from an organism

microfilaments   Cytoskeletal elements made of actin molecules (as well as myosin molecules in muscle cells); play a major role in various cellular contractile systems and serve as a mechanical stiffener for microvilli

microtubules   Cytoskeletal elements made of tubulin molecules arranged into long, slender, unbranched tubes that help maintain asymmetric cell shapes and coordinate complex cell movements

microvilli (mī´-krō-VIL-ī)   Actin-stiffened, nonmotile, hairlike projections from the luminal surface of epithelial cells lining the digestive tract and kidney tubules; tremendously increase the surface area of the cell exposed to the lumen

micturition (mik-too-RISH-un) or (mik-chuh-RISH-un)   The process of bladder emptying; urination

mineralocorticoids (min-uh-rul-ō-KOR-ti-koidz)   The vertebrate adrenocortical hormones that are important in Na^+ and K^+ balance; primarily aldosterone

mitochondria (mī-tō-KON-drē-uh)   The energy organelles of eukaryotic cells, which contain the enzymes for oxidative phosphorylation

mitosis (mī-TŌ-sis)   Cell division in which the chromosomes replicate before nuclear division, so that each of the two daughter cells receives a full set of chromosomes

mitotic spindle   The system of microtubules assembled during mitosis along which the replicated chromosomes are directed away from each other toward opposite sides of the cell prior to cell division

molecule   A chemical substance formed by the linking of atoms together; the smallest unit of a given chemical substance

molting   The process of replacing an old exoskeleton with a new one (see *ecdysis*)

monocytes (MAH-nō-sīts)   Vertebrate white blood cells that emigrate from the blood, enlarge, and become macrophages, large-tissue phagocytes

monoestrous   Displaying one period of sexual receptivity (estrus) during a year in a mammal

monomer   A unit molecule used to make a larger molecule called a *polymer*

monosaccharides (mah´-nō-SAK uh-rīdz)   Simple sugars, such as glucose; the absorbable unit of digested carbohydrates

motor activity   Movement of the body accomplished by contraction of skeletal muscles

motor end plate   The specialized portion of a skeletal muscle fiber that lies immediately underneath the terminal button of the motor neuron and has receptor sites for binding acetylcholine released from the terminal button

motor neurons   The neurons that innervate skeletal muscle and whose axons constitute the somatic nervous system

motor unit   One motor neuron plus all the muscle fibers it innervates

motor unit recruitment   The progressive activation of a muscle fiber's motor units to accomplish increasing gradations of contractile strength

mucosa (mew-KŌ-sah)   The innermost layer of the digestive tract that lines the lumen

multiunit smooth muscle   A smooth muscle mass that consists of multiple discrete units that function independently of each other and that must be separately stimulated by autonomic nerves to contract

muscarinic receptor   Type of cholinergic receptor found at the effector organs of all parasympathetic postganglionic fibers

muscle fiber   A single muscle cell, which is relatively long and cylindrical

muscle tension   See *tension*

muscle tissue   A functional grouping of cells specialized for contraction and force generation

myelin (MĪ-uh-lun) An insulative lipid covering that surrounds myelinated nerve fibers at regular intervals along the axon's length; each patch of myelin is formed by a separate myelin-forming cell that wraps itself jelly-roll fashion around the neuronal axon (in vertebrates and in a few invertebrates)

myelinated fibers Neuronal axons covered at regular intervals with insulative myelin

myocardium (mī′-ō-KAR-dē-um) The cardiac muscle layer within heart walls

myofibril (mī′-ō-FĪB-rul) A specialized intracellular structure of muscle cells that contains the contractile apparatus

myogenic Producing a pacemaking electrical signal from muscle tissue

myometrium (mī′-ō-mē-TRĒ-um) The smooth muscle layer of the mammalian uterus

myosin (MĪ-uh-sun) The contractile protein that forms the thick filaments in muscle fibers

Na⁺–K⁺ pump (ATPase) A carrier that actively transports Na^+ out of the cell and K^+ into the cell

nasal fossae Uppermost tract of the respiratory airways

natural killer cells Naturally occurring, vertebrate lymphocyte-like cells that nonspecifically destroy virus-infected cells and cancer cells by directly lysing their membranes on first exposure to them

negative balance Situation in which the losses for a substance exceed its gains so that the total amount of the substance in the body decreases

negative feedback A regulatory mechanism in which a change in a controlled variable triggers a response that opposes the change, thus maintaining a relatively steady set point for the regulated factor

nephron (NEF-ron′) The functional unit of the vertebrate kidney; consisting of an interrelated vascular and tubular component, it is the smallest unit that can form urine

nerve A bundle of peripheral neuronal axons, some afferent and some efferent, enclosed by a connective tissue covering and following the same pathway

nerve net A network of interconnected neurons that are distributed throughout a body or organ. Characterized by a diffuse spread of excitation. Occurs in most lower organisms, including cnidarians, and in visceral organs of most higher phyla

nerve plexus Network of interwoven nerves

nervous system One of the two major regulatory systems of an animal body; in general, coordinates rapid activities of the body, especially those involving interactions with the external environment

nervous tissue A functional grouping of cells specialized for initiation and transmission of electrical signals

net filtration pressure The net difference in the hydrostatic and osmotic forces acting across the vertebrate glomerular membrane that favors the filtration of a protein-free plasma into Bowman's capsule

neuroendocrinology The study of the interaction between the nervous and endocrine systems

neurogenic Producing a pacemaking electrical signal from neural tissue

neuroglia See *glial cells*

neuroglobin Oxygen-binding protein found in vertebrate neurons

neurohormones Hormones released into the blood by neurosecretory neurons

neuromodulators (ner′ō-MA-jew-lā′-torz) Chemical messengers that bind to neuronal receptors at nonsynaptic sites (that is, not at the subsynaptic membrane) and bring about long-term changes that subtly depress or enhance synaptic effectiveness

neuromuscular junction The juncture between a motor neuron and a skeletal muscle fiber; a highly specialized synapse

neuron (NER-on) A nerve cell, typically consisting of a cell body, dendrites, and an axon and specialized to initiate, propagate, and transmit electrical signals

neuropeptides Large, slow acting peptide molecules released from axon terminals along with classical neurotransmitters; most neuropeptides function as neuromodulators

neuropil Region of the nervous system where the synapses between axons and dendrites are concentrated

neurotransmitter The chemical messenger that is released from the axon terminal of a neuron in response to an action potential and influences another neuron or an effector with which the neuron is anatomically linked

neutrophils (new′-truh-filz) Vertebrate white blood cells that are phagocytic specialists and important in inflammatory responses and defense against bacterial invasion

nicotinic receptor (nick′-o-TIN-ik) Type of cholinergic receptor found at all vertebrate autonomic ganglia and the motor end plates of skeletal muscle fibers

nitric oxide A local chemical mediator released from endothelial cells and other tissues; exerts a wide array of effects, ranging from causing local arteriolar vasodilation to acting as a toxic agent against foreign invaders to serving as a unique type of neurotransmitter

nitrogen narcosis Syndrome of reduced neural excitability in human divers with increasing depth, caused by the highly lipid-soluble N_2 dissolving in neural membranes

nociceptor (nō-sē-SEP-tur) A pain receptor, sensitive to tissue damage

node of Ranvier (RAN-vē-ā) The portions of a myelinated neuronal axon between the segments of insulative myelin; the axonal regions where the axonal membrane is exposed to the ECF and membrane potential exists

nontropic hormone A hormone that exerts its effects on nonendocrine target tissues

norepinephrine (nor′-ep-uh-NEF-run) The neurotransmitter released from vertebrate sympathetic postganglionic fibers; noradrenaline

nucleus (of brain) (NŪ-klē-us) A functional aggregation of neuronal cell bodies within the brain

nucleus (of cells) A distinct spherical or oval structure that is usually located near the center of a cell and that contains the cell's genetic material, deoxyribonucleic acid (DNA)

O₂–Hb dissociation curve A plot of the relationship between arterial P_{O_2} and percent hemoglobin saturation

occipital lobes (ok-SIP′-ut-ul) The lobes of the vertebrate cerebral cortex that are located posteriorly and are responsible for initially processing visual input

olfaction Smell, or detection of molecules released from a distant object

oligodendrocytes (ol-i-gō′-DEN-drō-sitz) The myelin-forming cells of the vertebrate central nervous system

omasum The third chamber of the ruminant stomach. Its contents are mixed to a relatively homogenous state

ommatidium The structural unit of the invertebrate compound eye, composed of a cornea, a focusing cone and several receptor cells connected to the optic nerve

omnivore Animals that can eat both animals and other organisms such as plants or fungi

oogenesis (ō′-ō-JEN-uh-sus) Egg production

ootheca The covering or case of an egg mass

operculum (1) Bony plate that covers fish gill arches (2) hardened protein disc on the foot of a gastropod that seals up the shell opening when the animal is withdrawn

opsonin (OP′-suh-nun) Vertebrate chemicals that "tag" bacteria for destruction, i.e., making the bacteria more susceptible to phagocytosis

optic chiasm   Located underneath the vertebrate hypothalamus where the two optic nerves meet. Depending on the species, some axons cross the midline at this point and project to the contralateral side of the brain

optic nerve   The bundle of nerve fibers that leave the retina, relaying information about visual input

optimal length   The length before the onset of contraction of a muscle fiber at which maximal force can be developed on a subsequent tetanic contraction

organ   A distinct structural unit composed of two or more types of primary tissue organized to perform one or more particular functions; for example, the stomach

organ of Corti (KOR-tē)   The sense organ of hearing within the vertebrate inner ear that contains hair cells whose hairs are bent in response to sound waves, setting up action potentials in the auditory nerve

organelles (or´-gan-ELZ)   Distinct, highly organized, membrane-bound intracellular compartments, each containing a specific set of chemicals for carrying out a particular cellular function

organic   Referring to substances that contain carbon (and usually hydrogen)

organism   A living entity, either single-celled (unicellular) or made up of many cells (multicellular)

osmolarity (oz´-mō-LAR-ut-ē)   A measure of the concentration of solute molecules in a solution

osmolyte   A molecule used to elevate a cell's or body fluid's osmotic pressure in order to balance a high external osmolarity; can be inorganic or organic

osmosis (os-MŌ-sis)   Movement of water across a membrane down its own concentration gradient toward the area of higher solute concentration

osteoblasts (OS-tē-ō-blasts´)   Bone cells that produce the organic matrix of bone

osteoclasts   Bone cells that dissolve bone in their vicinity

ostium (pl., **ostia**)   Large pore-like opening, such as that which allows hemolymph to re-enter the arthropod heart after passing through body

otolith organs (Ō T´-ul-ith)   Sense organs in the vertebrate inner ear that provide information about rotational changes in head movement; include the utricle and saccule

oval window   The membrane-covered opening that separates the air-filled vertebrate middle ear from the upper compartment of the fluid-filled cochlea in the inner ear

ovarioles   Tubular structures forming the insect ovary

overhydration   Water excess in the body

oviparous   Referring to females that release eggs from which the young hatch after expulsion from the body

oviposition   In birds, the act of laying an egg

ovipositor   An extension of the female insect abdomen used to lay eggs, for example, by insertion into the ground or into a host animal

ovoviviparous   Referring to production of eggs that hatch within the body of the parent

ovulation (ov´-yuh-LĀ-shun)   Release of an ovum from a mature ovarian follicle

oxidative phosphorylation (fos´-fōr-i-LĀ-shun)   The entire sequence of mitochondrial biochemical reactions that uses oxygen to extract energy from the nutrients in food and transforms it into ATP, producing CO_2 and H_2O in the process

oxygen debt   Condition in which oxygen demand is greater than oxygen supply to a tissue

oxyhemoglobin (ok-si-HĒ-muh-glō-bun)   Hemoglobin combined with O_2

oxyntic mucosa (ok-SIN-tic)   The mucosa that lines the body and fundus of the vertebrate stomach; contains gastric pits lined by mucous neck cells, parietal cells, and chief cells

oxytocin (ok´-sē-TŌ-sun)   A hypothalamic neurohormone that is stored in the posterior pituitary and, in mammals, stimulates uterine contraction and milk ejection

pacemaker activity   Self-excitable activity of an excitable cell in which its membrane potential gradually depolarizes to threshold on its own

Pacinian corpuscle (pa-SIN-ē-un)   A rapidly adapting skin receptor in mammals that detects pressure and vibration

pancreas (PAN-krē-us)   A vertebrate gland composed of an exocrine portion that secretes digestive enzymes and an aqueous alkaline secretion into the duodenal lumen and an endocrine portion that secretes the hormones insulin and glucagon into the blood

papilla   A long, cone-shaped inner medulla in the kidneys of some mammals, such as rodents

parabronchi   Open-ended, small parallel tubes in bird lungs, each several millimeters long and 0.5 millimeters wide

paracrine (PEAR-uh-krin)   A local chemical messenger whose effect is exerted only on neighboring cells in the immediate vicinity of its site of secretion

parasympathetic nervous system (pear´-uh-sim-puh-THET-ik)   The subdivision of the vertebrate autonomic nervous system that dominates in quiet, relaxed situations and promotes body maintenance activities such as digestion and emptying of the urinary bladder

parathyroid glands (pear´-uh-THĪ-roid)   Small glands, located on or near the posterior surface of the vertebrate thyroid gland, that secrete parathyroid hormone

parathyroid hormone (PTH)   A vertebrate hormone that raises plasma Ca_2 levels

parietal cells (puh-RĪ-ut-ul)   The vertebrate stomach cells that secrete hydrochloric acid and intrinsic factor

parietal lobes   The lobes of the vertebrate cerebral cortex that lie at the top of the brain behind the central sulcus and contain the somatosensory cortex

parthenogenesis   Sexual reproduction in which an egg develops without entry of a sperm. Common among ants, bees, wasps, and aphids

partial pressure   The individual pressure exerted independently by a particular gas within a mixture of gases

partial pressure gradient   A difference in the partial pressure of a gas between two regions that promotes the movement of the gas from the region of higher partial pressure to the region of lower partial pressure

parturition (par´-too-RISH-un)   Delivery of a newborn mammal

passive expiration   Expiration accomplished during quiet breathing as a result of elastic recoil of the lungs on relaxation of the inspiratory muscles, with no energy expenditure required

passive force   A force that does not require expenditure of cellular energy to accomplish transport of a substance across the plasma membrane

passive reabsorption   When none of the steps in the transepithelial transport of a substance reabsorbed requires energy expenditure

patch clamping   A process in which a tiny pipette is attached to a plasma membrane with gentle suction. This forms a tight seal around one patch of a membrane. With a fine enough pipette, a single ion channel or receptor protein can be isolated

pathogens (PATH-uh-junz)   Disease-causing microorganisms, such as bacteria or viruses

pathophysiology (path´-ō-fiz-UL-uh-jē)   Abnormal functioning of the body associated with disease

pepsin; pepsinogen (pep-SIN-uh-jun)   An enzyme secreted in inactive form by the stomach that, once activated, begins protein digestion

peptide hormones   Hormones that consist of a chain of specific amino acids of varying length

percent hemoglobin saturation A measure of the extent to which the hemoglobin present is combined with O_2

perception The brain's interpretation of the external world as created from a pattern of nerve impulses delivered to it

perfusion The pumping of a fluid through a tissue or organ by means of a circulatory vessel

peripheral chemoreceptors (kē′-mō-rē-SEP-turz) The carotid and aortic bodies, which respond to changes in arterial P_{O_2}, P_{CO_2}, and H^+ and which adjust respiration accordingly (vertebrates)

peripheral nervous system (PNS) Nerve fibers that carry information between the central nervous system and other parts of the body

peristalsis (per′-uh-STOL-sus) Ringlike contractions of the circular smooth muscle of a tubular organ that move progressively forward with a stripping motion, pushing the contents of the organ ahead of the contraction

peritrophic membrane Layer that separates the midgut epithelium from the food in insects

peritubular capillaries (per′-i-TŪ-bū-lur) Capillaries that intertwine around the tubules of the mammalian kidney's nephron; they supply the renal tissue and participate in exchanges between the tubular fluid and blood during the formation of urine

permeable Permitting passage of a particular substance

permissiveness The condition when one hormone must be present in adequate amounts for the full exertion of another hormone's effect

peroxisomes (puh-ROK′-suh-sōmz) Organelles consisting of membrane-bound sacs that contain powerful oxidative enzymes that detoxify various wastes produced within the cell or foreign compounds that have entered the cell

pH The logarithm to the base 10 of the reciprocal of the hydrogen ion concentration; pH log 1/[H^+] or pH log[H^+]

phagocytosis (fag′-oh-sī-TŌ-sus) A type of endocytosis in which large, multi-molecular, solid particles are engulfed by a cell

pharynx (FARE-inks) The throat, which in vertebrates serves as a common passageway for the digestive and respiratory systems

phasic receptors Sensory receptors that produce action potentials at the onset and offset of a sustained stimulus

pheromone Chemical signal released into the environment, usually by glands, which travels through the air or water to sensory cells in another animal; used for sexual activity (such as signaling of readiness to mate), marking of territories, and other behaviors related to interactions among individuals of a species

phosphogen An energy-storing cellular molecule with a phosphate attached by a high-energy bond, such as creatine phosphate and arginine phosphate; its phosphate can be transferred to ADP to make ATP

phosphorylation (fos′-fōr-i-LĀ-shun) Addition of a phosphate group to a molecule

photoperiod The amount of time during the day when there is light

photoreceptor A sensory receptor responsive to light

phototransduction The mechanism of converting light stimuli into electrical activity by the rods and cones of the eye

physiology (fiz-ē-OL-ō-gē) The study of organismal functions

phytoestrogen Natural estrogenic-like compounds synthesized by some plant species

pineal gland (PIN-ē-ul) A small endocrine gland located in the center of the vertebrate brain that secretes the hormone melatonin

pinna A skin-covered flap of cartilage around the entrance to the mammalian ear, collects sound

pinocytosis (pin-oh-cī-TŌ-sus) Type of endocytosis in which the cell internalizes fluid

pitch The tone of a sound, determined by the frequency of vibrations (that is, whether a sound is a C or G note)

pituitary gland (pih-TWO-ih-tair-ee) A small vertebrate endocrine gland connected by a stalk to the hypothalamus; consists of the anterior pituitary and posterior pituitary

placenta (plah-SEN-tah) The organ of exchange between the maternal and fetal blood in placental mammals and some sharks

plasma cell An antibody-producing derivative of an activated vertebrate B lymphocyte

plasma clearance The volume of plasma that is completely cleared of a given substance by the kidneys per minute

plasma membrane A protein-studded lipid bilayer that encloses each cell, separating it from the extracellular fluid

plasma proteins The proteins that remain within the plasma, where they perform a number of important functions; include albumins, globulins, and fibrinogen in vertebrates

plasma The liquid portion of the blood or hemolymph

plasma-colloid osmotic pressure (KOL-oid os-MOT-ik) The force caused by the unequal distribution of plasma proteins between the blood and surrounding fluid that encourages fluid movement into the capillaries

plasticity (plas-TIS-uh-tē) The ability of portions of the nervous system to assume new responsibilities in response to the demands placed on it

platelets (PLĀT-lets) Specialized cell fragments in mammalian blood that participate in hemostasis by forming a plug at a vessel defect

pleural sac (PLOOR-ul) A double-walled, closed sac that separates each lung from the thoracic wall in air-breathing vertebrates

pluripotent stem cells Precursor cells that reside in certain adult tissues (such as bone marrow in mammals) and continuously divide and differentiate to give rise to specialized cells (such as blood cells in mammals)

poikilotherm Animal whose body temperature varies with environmental temperature

polarity Unequal sharing of electrons between atoms

polarized light Light in which the waves are all oriented at the same angle

polyestrous Mammalian estrus cycles occurring uniformly throughout the year without seasonal influence

polymer A chain of monomers linked together

polysaccharides (pol′-ī-SAK-uh-rīdz) Complex carbohydrates, consisting of chains of interconnected glucose molecules (starch, glycogen, cellulose)

polyunsaturated fatty acids (PUFAs) A fatty acid in which more than one of the bonds between carbon atoms in the carbon chain backbone of the molecules are double bonds

positive balance Situation in which the gains via input for a substance exceed its losses via output, so that the total amount of the substance in the body increases

positive feedback A regulatory mechanism in which the input and the output in a control system continue to enhance each other so that the controlled variable is progressively moved farther from a steady state

postabsorptive state The metabolic state after a meal is absorbed during which endogenous energy stores must be mobilized (and glucose must be spared for the glucose-dependent brain in vertebrates); fasting state

posterior pituitary The neural portion of the pituitary that stores and releases into the blood on hypothalamic stimulation two hormones produced by the hypothalamus, vasopressin, and oxytocin (vertebrates)

postganglionic fiber (pōst´-gan-glē-ON-ik) The second neuron in the two-neuron autonomic nerve pathway; originates in an autonomic ganglion and terminates on an effector organ (vertebrates)

postsynaptic neuron (pōst´-si-NAP-tik) The neuron that conducts its action potentials away from a synapse

precocial Referring to young (hatchlings) with well-developed legs, open eyes, and alertness; able to leave the nest and feed independently

preganglionic fiber The first neuron in the two-neuron autonomic nerve pathway; originates in the central nervous system and terminates on an autonomic ganglion (vertebrates)

pregastric fermentation Positioning the microbe-containing fermentation chamber of ruminants before the acid-secreting stomach

preprohormones Large precursor protein that undergoes a series of post-translational modifications to yield the biologically active hormone end product

pressure gradient A difference in pressure between two regions that drives the movement of blood or air from the region of higher pressure to the region of lower pressure

presynaptic facilitation Enhanced release of neurotransmitter from a presynaptic axon terminal as a result of excitation of another neuron that terminates on the axon terminal

presynaptic inhibition A reduction in the release of neurotransmitter from a presynaptic axon terminal as a result of excitation of another neuron that terminates on the axon terminal

presynaptic neuron (prē-si-NAP-tik) The neuron that conducts its action potentials toward a synapse

primary active transport A carrier-mediated transport system in which energy is directly required to operate the carrier and move the transported substance against its concentration gradient (see *secondary active transport*)

primary follicle A primary oocyte surrounded by a single layer of granulosa cells in the ovary (vertebrates)

primary motor cortex The portion of the vertebrate cerebral cortex that lies anterior to the central sulcus and is responsible for voluntary motor output

proctodaeum In birds, the hind portion of the cloaca, stores the excreta; in insects, the hindgut from the Malpighian tubules to the anus

proestrus The stage of the estrus cycle between the regression of the corpus luteum and the onset of estrus

prokaryote A microbial cell that has no classic organelles such as a nucleus; eubacteria and archaea

prolactin (PRL) (prō-LAK-tun) An anterior pituitary hormone with numerous roles; for example, stimulates mammary gland development and milk production in female mammals, and regulates osmoregulation in fishes (among other effects)

promoter A regulatory or "switch" section of DNA that can be bound by specific transcription-regulating proteins, thereby regulating the transcription of a nearby RNA-coding gene

pro-opiomelanocortin (prō-op´Ē-ō-ma-LAN-oh-kor´-tin) A large precursor molecule produced by the anterior pituitary that is cleaved into adrenocorticotropic hormone, melanocyte-stimulating hormone, and endorphin (in vertebrates)

proPOs Inactive forms of POs (phenoloxidase) that are activated during exoskeletal synthesis in arthropods

proprioception (prō´-prē-ō-SEP-shun) Sensing of position of body parts in relation to each other and to surroundings

proprioceptors Sensors that send information about movement and position of an animal's body or certain parts such as limbs

prostaglandins (pros´-tuh-GLAN-dins) Local chemical mediators (paracrines) that are derived from a component of the plasma membrane, arachidonic acid

prostate gland A vertebrate male accessory sex gland that secretes an alkaline fluid, which neutralizes acidic vaginal secretions in internal fertilizers

protandry Process in which a male becomes a female

proteasome An organelle that destroys unwanted cell proteins that have been tagged by ubiquitin

protein kinase (KĪ-nase) An enzyme that phosphorylates and thereby induces a change in the shape and function of a particular intracellular protein

proteolytic enzymes (prōt´-ē-uh-LIT-ik) Enzymes that digest protein

proteomics (proteonomics) The study of the structure, regulation, and functions of proteins

protogyny Process in which a female becomes a male

protonephridia Blind-end ducts that project into the body cavity and that filter wastes using ultrafiltration driven by cilia, followed by secretion, and reabsorption; generally found in more primitive animals having one internal fluid

compartment, such as rotifers, flatworms, larval annelids, and larval mollusks

proventriculus (1) True stomach of a bird. (2) In some invertebrates, a dilation of the foregut

proximal tubule (PROKS-uh-mul) A highly convoluted tubule that extends between Bowman's capsule and the loop of Henle in the mammalian kidney's nephron

proximate explanation See *mechanistic explanation*

PTH See *parathyroid hormone*

pubertin See *leptin*

pulmonary artery (PULL-mah-nair-ē) The large vessel that carries blood from the heart to the lungs (in air-breathing vertebrates)

pulmonary circulation The loop of blood vessels carrying blood between the heart and lungs (in air-breathing vertebrates)

pulmonary surfactant (sur-FAK-tunt) A phospholipoprotein complex secreted by the Type II alveolar cells that intersperses between the water molecules that line the alveoli, thereby lowering the surface tension within the lungs (in mammals)

pulmonary valve A one-way valve that permits the flow of blood from the right ventricle into the pulmonary artery during ventricular emptying but prevents the backflow of blood from the pulmonary artery into the right ventricle during ventricular relaxation (in mammals, birds)

pulmonary veins The large vessels that carry blood from the lungs to the heart (in air-breathing vertebrates)

pulmonary ventilation The volume of air breathed in and out of lungs in one minute; equals tidal volume times respiratory rate

pupil A round opening in the center of the iris through which light passes to the interior portions of the eye (adjustable in some animals)

Purkinje fibers (pur-KIN-jē) Small terminal fibers that extend from the bundle of His and rapidly transmit an action potential throughout the mammalian ventricular myocardium

pyloric gland area (PGA) (pī-LŌR-ik) The specialized region of the mucosa in the antrum of the mammalian stomach that secretes gastrin

pyloric sphincter (pī-lōr´-ik SFINGK-tur) The juncture between the stomach and duodenum (vertebrates)

radiation Emission of heat energy from the surface of a warm body in the form of electromagnetic waves

reabsorption The net movement of interstitial fluid into the capillary

receptor potential The graded potential change that occurs in a sensory receptor in response to a stimulus; generates action potentials in the afferent neuron fiber

receptor See *sensory receptor* or *receptor site*

receptor site Membrane protein that binds with a specific extracellular chemical messenger, thereby bringing about a series of membrane and intracellular events that alter the activity of the particular cell

rectal gland A specialized hindgut organ of cartilaginous fishes, which has numerous blind-end tubules that remove and excrete NaCl using active transport

reduced hemoglobin Hemoglobin that is not combined with O_2

reflex Any response that occurs automatically without modification by "higher" control centers; the components of a reflex arc include a receptor, afferent pathway, integrating center, efferent pathway, and effector

reflex arc A neuronal pathway that controls the reflex motor response to a specific sensory stimulus

refraction Bending of a light ray

refractory period (rē-FRAK-tuh-rē) The time period when a recently activated patch of membrane is refractory (unresponsive) to further stimulation, preventing the action potential from spreading backward into the area through which it has just passed, thereby ensuring the unidirectional propagation of the action potential away from the initial site of activation

regulators Organisms in which a particular physiological state is kept relatively constant in the face of a changing environment

regulatory volume decrease (RVD) A process in which cells subjected to osmotic swelling decrease their internal osmotic pressure by removing osmolytes in order to restore cell volume

regulatory volume increase (RVI) A process in which cells subjected to osmotic shrinkage increase their internal osmotic pressure by accumulating osmolytes in order to restore cell volume

releasing hormone A hypothalamic hormone that stimulates the secretion of a particular anterior pituitary hormone (vertebrates)

renal cortex An outer, granular-appearing region of the mammalian and avian kidney

renal medulla (RĒ-nul muh-DUL-uh) An inner, striated-appearing region of the mammalian and avian kidney

renal threshold The plasma concentration at which the Tm (tubular maximum) of a particular substance is reached and

the substance first starts appearing in the urine

renin (RĒ-nin) An enzymatic hormone released from mammalian kidneys in response to a decrease in NaCl/ECF volume/arterial blood pressure; activates angiotensinogen

renin-angiotensin-aldosterone system (an´jē-ō-TEN-sun al-dō-steer-OWN) The salt-conserving system triggered by the release of renin from mammalian kidneys, which activates angiotensin, which stimulates aldosterone secretion, which stimulates Na^+ reabsorption by the kidney tubules during the formation of urine

repolarization (rē´-pō-luh-ruh-ZĀ-shun) Return of membrane potential to resting potential following a depolarization

reproductive tract The system of ducts that are specialized to transport or house the gametes after they are produced

reset In physiological negative-feedback regulation, the process of changing the set point from one level to another

residual volume The minimum volume of air remaining in respiratory pathways even after a maximal expiration

resistance Hindrance of flow of liquid or air through a passageway (such as a blood vessel or respiratory airway)

respiration The sum of processes that accomplish ongoing movement of O_2 from the atmosphere to the tissues, as well as the continual movement of metabolically produced CO_2 from the tissues to the atmosphere

respiratory acidosis (as-i-DŌ-sus) Acidosis resulting from abnormal retention of CO_2 arising from hypoventilation

respiratory airways The system of tubes that conducts air between the atmosphere and the alveoli of the lungs

respiratory alkalosis (al´-kuh-LŌ-sus) Alkalosis caused by excessive loss of CO_2 from the body as a result of hyperventilation

respiratory rate Breaths per minute

response element A promoter DNA sequence that is the target of action of signal molecules such as hormones; binds to a hormone and its nuclear receptor; also termed hormone response element (HRE)

resting membrane potential The membrane potential that exists when an excitable cell is not displaying an electrical signal

rete mirabile Extensive countercurrent network of arterial and venous capillaries

reticular activating system (RAS) (ri-TIK-ū-lur) Ascending fibers that originate in the reticular formation and carry signals upward to arouse and activate the cerebral cortex (vertebrates)

reticular formation A network of interconnected neurons that runs throughout the vertebrate brain stem and initially receives and integrates all synaptic input to the brain

reticulum The second stomach of ruminants (see *rumen*)

retina The innermost layer in the posterior region of an eye that contains the eye's photoreceptors, the rods and cones

rhabdomere Specialized region of the retinular cell of an invertebrate compound eye

rheostasis Regulated change of a physiologic state

rhopalium Cnidarian medusa sense organ, has ocelli (simple eye structures) and statocysts

ribonucleic acid (RNA) (rī-bō-new-KLĀ-ik) A nucleic acid that exists in three major forms (messenger RNA, ribosomal RNA, and transfer RNA), which participate in gene transcription and protein synthesis (other forms have also been recently discovered, such as small nuclear and micro RNAs)

ribosomes (RĪ-bō-sōms) Special ribosomal RNA-protein complexes that synthesize proteins under the direction of nuclear DNA

right atrium (Ā´-trē-um) The heart chamber that receives venous blood from the systemic circulation (reptiles, birds, mammals)

right ventricle The heart chamber that pumps blood into the pulmonary circulation (in reptiles, birds, mammals)

rigor mortis "Stiffness of death"; a generalized locking-in-place of the skeletal muscles that begins sometime after death

RNA See *ribonucleic acid*

rods The eye's photoreceptors used for night vision

root effect A change in the O_2-carrying capacity of hemoglobin in response to a change in pH

round window In higher vertebrates, the membrane-covered opening that separates the lower chamber of the cochlea in the inner ear from the middle ear

r-selected Referring to species that place virtually all reproductive resources into producing hundreds to thousands (or more) of offspring, with no parental care and minimal nourishment

rumen The first compartment of the ruminant stomach. Microbial populations found in the rumen include anaerobic species of bacteria, protozoa, and fungi

ruminant Various hoofed mammals that chew food already swallowed and that have a stomach divided into four (or three) compartments

ruminoreticulum The rumen and its continuation, the reticulum. Two compartments of the ruminant stomach

SA node See *sinoatrial node*

salivary amylase (AM-uh-lās′) An enzyme produced by salivary glands that begins carbohydrate digestion in the mouth and continues in the digestive tract after the food and saliva have been swallowed

salt gland A specialized gland associated with the eyes or nasal cavities of marine birds and reptiles; it removes excess NaCl from the blood

saltatory conduction (SAL-tuh-tōr′-ē) The means by which an action potential is propagated throughout a myelinated fiber, with the impulse jumping over the myelinated regions from one node of Ranvier to the next (vertebrates)

sarcomere (SAR-kō-mir) The functional unit of skeletal muscle; the area between two Z lines within a myofibril

sarcoplasmic reticulum (ri-TIK-yuh-lum) A fine meshwork of interconnected tubules that surrounds a muscle fiber's myofibrils; contains expanded lateral sacs, which store calcium that is released into the cytosol in response to a local action potential

satiety centers (suh-TĪ-ut-ē) Neuronal clusters in the ventromedial region of the hypothalamus that inhibit feeding behavior (vertebrates)

saturation The condition when all the binding sites on a carrier molecule are occupied

scaling The phenomenon that many anatomic and physiologic parameters vary with other parameters (such as body mass) in nonproportional relationships

Schwann cells (shwah′-n) The myelin-forming cells of the vertebrate peripheral nervous system

second messenger An intracellular chemical that is activated by binding of an extracellular first messenger to a surface receptor site and that triggers a preprogrammed series of biochemical events, which result in altered activity of intracellular proteins to control a particular cell activity

secondary active transport A transport mechanism in which a carrier molecule for glucose or an amino acid is driven by a Na concentration gradient established by the energy-dependent Na pump to transfer the glucose or amino acid uphill without directly expending energy to operate the carrier

secondary follicle A developing ovarian follicle that is secreting estrogen and forming an antrum (vertebrates)

secondary sexual characteristics The many external characteristics that are not directly involved in reproduction but that distinguish males and females

secretin (si-KRĒT-′n) A hormone released from the vertebrate duodenal mucosa primarily in response to the presence of acid; inhibits gastric motility and secretion and stimulates secretion of a $NaHCO_3$ solution from the pancreas

secretion Release to a cell's exterior, on appropriate stimulation, of substances that have been produced by the cell

secretory vesicles (VES-i-kuls) Membrane-enclosed sacs containing proteins that have been synthesized and processed by the endoplasmic reticulum/ Golgi complex of the cell and that will be released to the cell's exterior by exocytosis on appropriate stimulation

segmentation The vertebrate small intestine's primary method of motility; consists of oscillating, ringlike contractions of the circular smooth muscle along the small intestine's length

self-antigens Antigens that are characteristic of an animal's own cells

semen (SĒ-men) A mixture of accessory sex-gland secretions and sperm

semicircular canal Sense organ in the inner ear that detects rotational or angular acceleration or deceleration of the head (vertebrates)

semilunar valves (sem′-ī-LEW-nur) The aortic and pulmonary valves (in the avian and mammalian heart)

seminal vesicles (VES-i-kuls) Male accessory sex glands that supply fructose to ejaculated sperm and secrete prostaglandins (vertebrates)

seminiferous tubules (sem′-uh-NIF-uh-rus) The highly coiled tubules within vertebrate testes that produce spermatozoa

sensory afferent Pathway coming into the central nervous system that carries information

sensory input Includes somatic sensation and special senses

sensory receptor An afferent neuron's peripheral ending, which is specialized to respond to a particular stimulus in its environment

septum (pl., **septa**) Dividing wall or partition

series-elastic component The noncontractile portions of a skeletal muscle fiber, including the connective tissue and sarcoplasmic reticulum

Sertoli cells (sur-TŌ-lē) Cells located in the seminiferous tubules that support spermatozoa during their development (vertebrates)

set point The desired level at which homeostatic control mechanisms maintain a controlled variable

signal transduction The sequence of events in which incoming signals (instructions from extracellular chemical messengers such as hormones) are conveyed to the cell's interior for execution

single-unit smooth muscle The most abundant type of smooth muscle; made up of muscle fibers that are interconnected by gap junctions so that they become excited and contract as a unit; also known as *visceral smooth muscle*

sinoatrial (SA) node (sī-nō-Ā-trē-ul) A small, specialized autorhythmic region in the right atrial wall of the heart that has the fastest rate of spontaneous depolarizations and serves as the normal pacemaker of the heart (in reptiles, birds, mammals)

sinus venosus In all fishes, a chambered extension of the atrium into which blood first enters the heart

skeletal muscle Striated muscle, which is attached to the endo- or exoskeleton and is responsible for movement of the parts of the skeleton in purposeful relation to one another; innervated by the somatic nervous system and under both reflex and nonreflex control

slow-wave potentials Self-excitable activity of an excitable cell in which its membrane potential undergoes gradually alternating depolarizing and hyperpolarizing swings

Smith predictor A control device that predicts the next step in a regulated process and triggers it to begin before feedback information is received about the immediately preceding step. Thought to be the mechanism by which the mammalian cerebellum controls skilled locomotory movements

smooth muscle Muscles without organized sarcomeres, found in the walls of hollow organs and tubes, innervated by the autonomic nervous system in vertebrates

solutes Molecules dissolved in liquid (such as water)

somatic cells (sō-MAT-ik) Body cells, as contrasted with reproductive cells

somatic nervous system The portion of the efferent division of the vertebrate peripheral nervous system that innervates skeletal muscles; consists of the axonal fibers of the alpha motor neurons

somatic sensation Sensory information arising from the body surface, including somesthetic sensation and proprioception

somatomedins (sō′-mat-uh-MĒ-dinz) Hormones secreted by the vertebrate liver or other tissues, in response to growth hormone, that act directly on the target cells to promote growth

somatosensory cortex The region of the vertebrate parietal lobe immediately behind the central sulcus; the site of initial processing of somesthetic and proprioceptive input

sound waves Traveling vibrations of air that consist of regions of high pressure caused by compression of air molecules alternating with regions of low pressure caused by rarefaction of the molecules

spatial summation The summing of several postsynaptic potentials arising from the simultaneous activation of several excitatory (or several inhibitory) synapses

specific dynamic action The increase in metabolism that follows ingestion of a meal; see *diet-induced thermogenesis*

specificity Ability of carrier molecules to transport only specific substances across the plasma membrane

spermatheca Organ that stores the sperm prior to entry into the vagina

spermatogenesis (spur´-mat-uh-JEN-uh-sus) Sperm production

spermatophore Sperm-containing capsule formed by hardened seminal fluid

sphincter (sfink-tur) A ring of muscle that controls passage of contents through an opening into or out of a hollow organ or tube

spiracle Surface opening of the tracheal system in insects and certain arachnids

spleen A vertebrate lymphoid tissue in the upper left part of the abdomen that stores lymphocytes and platelets and destroys old red blood cells

sporangium Vegetative stage of fungi

stanniocalcin A hormone regulating calcium homeostasis in bony fish (formerly called *hypocalcin*); synthesized in small organs termed the corporacles of Stannius located within the kidney

state of equilibrium A condition in which no net change in a system is occurring

statolith Small, dense particle, that readily moves within the fluid content of a statocyst, to provide directional information

stem cells Relatively undifferentiated precursor cells that give rise to highly differentiated, specialized cells

stenohaline Having a limited tolerance to changes in salinity

stereocilia Nonmotile, filament projections of a hair cell

steroids (STEER-oidz) Hormones derived from cholesterol

stimulus A detectable physical or chemical change in the environment of a sensory receptor

stomodaeum The foregut of an insect

stress proteins Small molecular-weight proteins that are synthesized in many types of organisms after a sudden increase in a stress factor such as temperature, pollutants, and oxygen deprivation; these help to properly fold other proteins perturbed by the stress factor. If associated with heat stress, these are called heat-shock proteins (HSPs)

stress The generalized, nonspecific response of the body to any factor that overwhelms, or threatens to overwhelm, the body's compensatory abilities to maintain homeostasis

stretch reflex A monosynaptic reflex in which an afferent neuron originating at a stretch-detecting receptor in a skeletal muscle terminates directly on the efferent neuron supplying the same muscle to cause it to contract and counteract the stretch

stroke volume (SV) The volume of blood pumped out of a ventricle with each contraction, or beat, of the heart

subcortical regions The vertebrate brain regions that lie under the cerebral cortex, including the basal nuclei, thalamus, and hypothalamus

submucosa The connective tissue layer of the digestive tract that lies under the mucosa and (in vertebrates) contains the larger blood and lymph vessels and a nerve network

substance P The neurotransmitter released from mammalian pain fibers

subsynaptic membrane (sub-sih-NAP-tik) The portion of the postsynaptic cell membrane that lies immediately underneath a synapse and contains receptor sites for the synapse's neurotransmitter

supercooling A state of water in which the temperature is below the freezing point, but with no trigger or nucleation site to begin ice formation

suprachiasmatic nucleus (soup´-ra-kī-as-MAT-ik) A cluster of nerve cell bodies in the hypothalamus that serves as the master biological clock, acting as the pacemaker that establishes many of the mammalian body's circadian rhythms

surface tension The force at the liquid surface of an air–water interface resulting from the greater attraction of water molecules to the surrounding water molecules than to the air above the surface; a force that tends to decrease the area of a liquid surface and resists stretching of the surface

sympathetic nervous system The subdivision of the vertebrate autonomic nervous system that dominates in emergency ("fight-or-flight") or stressful situations and prepares the body for strenuous physical activity

symporter A transporter protein in a membrane that moves two (or more) molecules or ions in the same direction

synapse (SIN-aps´) The specialized junction between two neurons where an action potential in the presynaptic neuron influences the membrane potential of the postsynaptic neuron by means of the release of a chemical messenger that diffuses across the small cleft that separates the two neurons

synchronous In step or in phase control of flight muscles, which are activated by the release of Ca^{2+} and deactivated by Ca^{2+} reuptake

synergism (SIN-er-jiz´-um) When several actions are complementary, so that their combined effect is greater than the sum of their separate effects

syrinx Vocal organ of birds, located at the lower end of the trachea, where the two bronchi join. Acts as a sound-producing valve in which air forced by pressure from the lungs sets the tympanic membranes in motion

systemic circulation (sis-TEM-ik) The closed loop of vertebrate blood vessels carrying blood between the heart and body systems

systole (SIS-tō-lē) The period of cardiac contraction and emptying

T lymphocytes (T-cells) Vertebrate white blood cells that accomplish cell-mediated immune responses against targets to which they have been previously exposed; see also *cytotoxic T-cells* and *helper-cells*

T tubule See *transverse tubule*

T3 See *tri-iodothyronine*

T4 See *thyroxine*

taeniae coli Three separate, conspicuous, longitudinal bands of muscle lining the large intestine

tapetum lucidum Reflecting structure found in the choroid coat of some nocturnal animals

target cell receptors Receptors located on a target cell that are specific for a particular chemical mediator

target cells The cells that a particular extracellular chemical messenger, such as a hormone or a neurotransmitter, influences

TATA box A promoter found next to almost all eukaryotic RNA-coding genes; site for the initial binding of RNA polymerase, an enzyme that makes an RNA copy of the coding gene sequence

teleological approach (tē´-lē-ō-LA-ji-kul) Explanation of body functions in terms of their particular purpose in fulfilling a bodily need, that is, the "why" of body processes (in contrast to *mechanistic* and *evolutionary explanations*)

telomere The end sections of eukaryotic chromosomes; pieces of it are lost during cell replication

temporal lobes   The lobes of the mammalian cerebral cortex that are located laterally and that are responsible for initially processing auditory input

temporal summation   The summing of several postsynaptic potentials occurring very close together in time because of successive firing of a single presynaptic neuron

tension   In muscles, the force produced during muscle contraction by shortening of the sarcomeres, resulting in stretching and tightening of the muscle's elastic connective tissue and tendon, which transmit the tension to the bone to which the muscle is attached

terminal button   A vertebrate motor neuron's enlarged knoblike ending that terminates near a skeletal muscle fiber and releases acetylcholine in response to an action potential in the neuron

testosterone (tes-TOS-tuh-rōn)   The vertebrate male sex hormone, secreted by the Leydig cells of the testes

tetanus (TET´-n-us)   A smooth, maximal muscle contraction that occurs when the fiber is stimulated so rapidly that it does not have a chance to relax at all between stimuli

thalamus (THAL-uh-mus)   The vertebrate brain region that serves as a synaptic integrating center for preliminary processing of all sensory input on its way to the cerebral cortex

thecal cells (THAY-kel)   The outer layer of specialized ovarian connective tissue cells in a maturing follicle (vertebrate)

thermal neutral zone (**TNZ**)   A range of environmental temperatures in which an animal does not need to expend significant energy for thermoregulation

thermogenin   An uncoupling protein in brown adipose tissue of mammals

thermoreceptor (thur´-mō-rē-SEP-tur)   A sensory receptor sensitive to heat and cold

thick filaments   Specialized cytoskeletal structures within skeletal muscle that are made up of myosin molecules and interact with the thin filaments to accomplish shortening of the fiber during muscle contraction

thin filaments   Specialized cytoskeletal structures within skeletal muscle that are made up of actin, tropomyosin, and troponin molecules and interact with the thick filaments to accomplish shortening of the fiber during muscle contraction

thoracic cavity (thō-RAS-ik)   Chest cavity

threshold potential   The critical potential that must be reached before an action potential is initiated in an excitable cell

thrombocyte   A blood cell in nonmammalian vertebrates that participates in clotting

thrombus   An abnormal clot attached to the inner lining of a blood vessel

thymus (THIGH-mus)   A vertebrate lymphoid gland located midline in the chest cavity that processes T lymphocytes and produces the hormone thymosin, which maintains the T-cell lineage

thyroglobulin (thī´-rō-GLOB-yuh-lun)   A large, complex molecule on which all steps of thyroid hormone synthesis and storage take place

thyroid gland   An endocrine gland that traps and stores iodide from the blood of vertebrates, and that secretes thyroxine and tri-iodothyronine, hormones which regulate vertebrate development and metabolism

thyroid hormone   Collectively, the vertebrate hormones secreted by the thyroid follicular cells, namely, thyroxine and tri-iodothyronine

thyroid-stimulating hormone (**TSH**)   An anterior pituitary hormone that stimulates secretion of thyroid hormone and promotes growth of the thyroid gland; thyrotropin

thyroxine (thī-ROCKS-in)   The most abundant hormone secreted by the vertebrate thyroid gland; important in the regulation of overall metabolic rate; also known as tetraiodothyronine or T4

tidal breathing   Movement of air or water in and out of a respiratory chamber via the same opening, in two distinct steps termed *inspiration* and *expiration,* for the purposes of gas exchange

tidal volume   The volume of air entering or leaving tidal-type lungs during a single breath

tight junction   An impermeable junction between two adjacent epithelial cells formed by the sealing together of the cells' lateral edges near their luminal borders; prevents passage of substances between the cells

tissue   (1) A functional aggregation of cells of a single specialized type, such as nerve cells forming nervous tissue; (2) the aggregate of various cellular and extracellular components that make up a particular organ, such as lung tissue

titin   Cytoskeletal elastic protein in the myofibril; forms a flexible filamentous network that can stretch under tension. It may act as a locomotory spring that assists in the return of the sarcomere to its relaxed conformation

Tm   See *transport maximum* and *tubular maximum*

TMAO (**trimethylamine oxide**)   A counteracting osmolyte in many marine organisms, capable of offsetting the perturbing effects of urea, pressure, and other stressors on proteins

tone   The ongoing baseline of activity in a given system or structure, as in muscle tone, sympathetic tone, or vascular tone

tonic receptors   Sensory receptors that continue to generate action potentials throughout the duration of a stimulus; directly convey information about the duration of the stimulus

torpor   A state of reduced metabolic rate in an animal occurring on a daily basis; involves lower body temperatures in endotherms

total peripheral resistance   The resistance offered by all the peripheral blood vessels, with arteriolar resistance contributing most extensively

trachea (TRĀ-kē-uh)   (1) The "windpipe"; the conducting airway that extends from the pharynx and branches into two bronchi, each entering a lung (airbreathing vertebrates); (2) an air tube in an air-breathing arthropod that leads directly to a tissue requiring oxygen

tracheole   System of air-filled tubules that exchange respiratory gases from the outside world with insect tissues

tract   A bundle of nerve fibers (axons of long interneurons) with a similar function within a long nerve cord (such as the spinal cord)

transducin   G protein that links the excitation of rhodopsin molecules by light with a change in the current flowing across the membrane of photoreceptors

transduction   Conversion of stimuli into action potentials by sensory receptors

transepithelial transport (tranz-ep-i-THĒ-lē-al)   The entire sequence of steps involved in the transfer of a substance across the epithelium between either the renal tubular lumen or digestive tract lumen and the blood

transgenic organism   An organism that has had genes from another species incorporated into its genome

transmural pressure gradient   The pressure difference across the lung wall (intraalveolar pressure is greater than intrapleural pressure) that stretches the lungs to fill the thoracic cavity, which is larger than the unstretched lungs (air-breathing vertebrates)

transporter recruitment   The phenomenon of inserting additional transporters (carriers) for a particular substance into the plasma membrane, thereby increasing membrane permeability to the substance, in response to an appropriate stimulus

transport maximum (**Tm**)   The maximum rate of a substance's carrier-mediated transport across the membrane when the carrier is saturated; in the kidney tubules, known as *tubular maximum*

transverse tubule (T tubule) A perpendicular infolding of the surface membrane of a muscle fiber; rapidly spreads surface electric activity into the central portions of the muscle fiber

trehalose Diglucose (a disscharide). In insects, synthesized in the fat body as a metabolic substrate for flight

triglycerides (trī-GLIS-uh-rīdz) Neutral fats composed of one glycerol molecule with three fatty-acid molecules attached

tri-iodothyronine (T3) (trī-ī-ō-dō-THĪ-rō-nēn) The most potent hormone secreted by the vertebrate thyroid follicular cells; important in the regulation of overall metabolic rate in birds and mammals

trimodal Three respiratory exchange surfaces, including the gills, lungs, and some cutaneous surfaces

trophic hormones "Nourishing" hormones involved in triggering cell growth and development; not to be confused with *tropic* hormone

trophoblast (TRŌ-F-uh-blast´) The outer layer of cells in a mammalian blastocyst, responsible for accomplishing implantation and developing the fetal portion of the placenta

trophocytes Cells in the fat body of insects that perform a role in intermediary metabolism

tropic hormone (TRŌ-pik) A hormone that regulates the secretion of another hormone; not to be confused with *trophic* hormone

tropomyosin (trōp´-uh-MĪ-uh-sun) One of the regulatory proteins found in the thin filaments of muscle fibers

troponin (tro-PŌ-nun) One of the regulatory proteins found in the thin filaments of muscle fibers

TSH See *thyroid-stimulating hormone*

tuberous electroreceptors Found on the anterior body surface in the lateral line system, these receptors respond to the high-frequency signals from the fish's own electric organ discharge

tubular maximum (Tm) The maximum amount of a substance that renal tubular cells can actively transport within a given time period; the kidney cells' equivalent of transport maximum

tubular reabsorption The selective transfer of substances from the tubular fluid into the peritubular capillaries during the formation of urine

tubular secretion The selective transfer of substances from the peritubular capillaries into the tubular lumen during the formation of urine

twitch A brief, weak contraction that occurs in response to a single action potential in a muscle fiber

twitch summation The addition of two or more muscle twitches as a result of rapidly repetitive stimulation, resulting in greater tension in the fiber than that produced by a single action potential

2,3-diphosphoglycerate (DPG) An organic phosphate synthesized in erythrocytes that reduces the affinity of most vertebrate hemoglobins for O_2

tympanum A vibrating membrane. (1) An auditory membrane of certain insects, or eardrum of the vertebrate middle ear. (2) A rigid structure composed of three to six enlarged rings at the base of the trachea. Used for song production in some species of birds

Type I alveolar cells (al-VĒ-ō-lur) The single layer of flattened epithelial cells that forms the wall of the alveoli within the mammalian lungs

Type II alveolar cells The cells within the mammalian alveolar walls that secrete pulmonary surfactant

ubiquitin A molecular tag that is added to cell proteins that are no longer needed, targeting them for destruction in proteasomes

ultrafiltration The net movement of a protein-free fluid across a tissue boundary

uncoupling protein A gated channel for hydrogen ions found in the inner membrane of mitochondria in some tissues in mammals; serves to dissipate the mitochondrial H gradient into heat

upper critical temperature In an endotherm, the temperature at which internal heat-removing (active) mechanisms are activated to maintain thermal homeostasis

upregulation An increase in the capacity of a cellular or organ function

ureotelic Having urea as a primary nitrogenous waste product

ureter (yū-RĒ-tur) A duct that transmits urine from the kidney to the bladder

urethra (yū-RĒ-thruh) A tube that carries urine from the bladder to outside the body

uricotelic Having uric acid as a primary nitrogenous waste product

urine excretion The elimination of substances from the body in the urine; anything filtered or secreted and not reabsorbed is excreted

urodaeum Middle portion of the avian cloaca, receives the fluid from the kidneys through the ureters in addition to material from the oviduct

vagus nerve (VĀ-gus) The vertebrate 10th cranial nerve, which serves as the major parasympathetic nerve

vasoconstriction (vā´-zō-kun-STRIK-shun) The narrowing of a blood vessel lumen as a result of contraction of the vascular circular smooth muscle

vasodilation The enlargement of a blood vessel lumen as a result of relaxation of the vascular circular smooth muscle

vasopressin (vā-zō-PRES-sin) A vertebrate hormone secreted by the hypothalamus, then stored and released from the posterior pituitary; increases the permeability of the distal and collecting tubules of mammalian kidneys to water and promotes arteriolar vasoconstriction; also known as *antidiuretic hormone (ADH)*

vasotocin A water-conserving hormone in many vertebrates, closely related to vasopressin in mammals

vaults Organelles shaped like octagonal barrels; believed to serve as transporters for messenger RNA and/or the ribosomal subunits from the nucleus to sites of protein synthesis; may be important in cell movement

vein A vessel that carries blood toward the heart

venous return (VĒ-nus) The volume of blood returned to each atrium per minute from the veins

ventilation The movement of air in and out of the lungs, or water across gills; called *breathing* if actively produced by the animal

ventricle (VEN-tri-kul) A lower chamber of the heart that pumps circulatory fluid into the arteries

ventrobronchi Cranial group of secondary bronchi in birds, that branch to form a fan-shaped covering of the mediodorsal surface of the avian lung

vesicle (VES-i-kul) A small, intracellular, fluid-filled, membrane-enclosed sac

vesicular transport Movement of large molecules or multimolecular materials into or out of the cell by means of being enclosed in a vesicle, as in endocytosis or exocytosis

vestibular apparatus (veh-STIB-yuh-lur) The component of the vertebrate's inner ear that provides information essential for the sense of equilibrium and for coordinating head movements with eye and postural movements; consists of the semicircular canals, utricle, and saccule

villus (pl., villi) (VIL-us) Microscopic fingerlike projection from the inner surface of the small intestine

virulence (VIR-you-lentz) The disease-producing power of a pathogen

visceral afferent   A pathway coming into the central nervous system that carries information derived from the internal viscera

visceral smooth muscle (VIS-uh-rul) See *single-unit smooth muscle*

viscosity (vis-KOS-i-tē)   The friction developed between molecules of a fluid as they slide over each other during flow of the fluid; the greater the viscosity, the greater the resistance to flow

visual field   Field of view that can be seen without moving the head

vital capacity   The maximum volume of air that can be moved out during a single breath following a maximal inspiration

vitellogenesis   Deposition of yolk in an oocyte

viviparous   Referring to production and nourishment of living young within the body

voltage clamp   A procedure in which the potential across a membrane is held at a constant value by an electronic circuit

voltage-gated channels   Channels in the plasma membrane that open or close in response to changes in membrane potential

vomeronasal organ (VNO)   Accessory olfactory organ in some mammals that detects pheromones. It is connected to two small openings in the anterior of the mouth behind the upper lip

white matter   The portion of the central nervous system composed of myelinated nerve fibers

X-organs   Organs in crustacean eyestalks that send axons into sinus glands (also in eyestalks) to release several neurohormones

Y-organs   Organs in the crustacean head that make crustecydsone to promote molting

Z line   A flattened, disclike cytoskeletal protein that connects the thin filaments of two adjoining sarcomeres

zeitgeber   Any environmental factor that entrains a biological rhythm

zona fasciculata (zō-nah fa-SIK-ū-lah-ta)   The middle and largest layer of the adrenal cortex; major source of glucocorticoids

zona glomerulosa (glō-MER-yū-lō-sah) The outermost layer of the adrenal cortex; sole source of aldosterone

zona reticularis (ri-TIK-yuh-lair-us) The innermost layer of the adrenal cortex; produces the sex steroid DHEA

Credits

This page constitutes an extension of the copyright page. We have made every effort to trace the ownership of all copyrighted material and to secure permission from copyright holders. In the event of any question arising as to the use of any material, we will be pleased to make the necessary corrections in future printings. Thanks are due to the following authors, publishers, and agents for permission to use the material indicated.

Photo Credits

Chapter 1

1: (left) Paul Yancey; (right) © Ron Sanford/CORBIS. 3: (both) © Mark Moffett/Minden Pictures. 8/Fig. 1-3: © Gladden Willis, MD/Visuals Unlimited. 12/Fig. 1-7: (a) Paul Yancey; (b) © Bill Beatty/Visuals Unlimited; (c) © Robert Yin/CORBIS. 15: Kevin Johnson, U.S. Geological Survey.

Chapter 2

23: © M. I. Walker/Photo Researchers, Inc. 24 /Fig.2-1: (both) Thomas A. Steitz. 31: © Roger Steene/Imagequestmarine.com. 32/Fig. 2-5: © Dr. Alexey Khodjakov/Photo Researchers, Inc. 33/Fig. 2-6: (a) Courtesy of Affymetrix, Inc.; (b) Andrew Gracey/Stanford University. 36/Fig. 2-9 (b, c): © Don W. Fawcett/Visuals Unlimited. 38/Fig. 2-11: © David M. Phillips/Visuals Unlimited. 39/Fig. 2-12: (a): Dr. Jürgen Engel, University of Basel et al; (b): © Dr. Dennis Kunkel/Visuals Unlimited; (c): Florida Keyes National Marine Sanctuary. Courtesy of NOAA. 42/Fig. 2-14 (a): © Don W. Fawcett/Photo Researchers, Inc.; (d–all): Prof. Marcel Bessis/Science Source/Photo Researchers, Inc. 45/Fig.2-16 (b): © Bill Longcore/Photo Researchers, Inc. 52: © John M. Coffman. 56/Fig. 2-23: © Dr. Leonard H. Rome/UCLA School of Medicine. 56/Fig. 2-24 (a, b): Elizabeth R. Walker, Ph.D. and Dennis O. Overman, Ph.D., Department of Anatomy, School of Medicine, West Virginia University. 60/Fig. 2-27 (b): © David M. Phillips/Visuals Unlimited. 61/Fig. 2-28 (b): © David M. Phillips/Visuals Unlimited. 61/Fig.2-29: © M. Abbey/Visuals Unlimited. 61/Fig. 2-30: Reproduced from: R. G. Kessel and R. H. Kardon, *Tissues and Organs: A Text Atlas of Scanning Electron Microscopy*. W. H. Freeman, 1979, all rights reserved.

Chapter 3

68: Courtesy of Jeff Milton, Alaska Department of Fish and Game. 69/Fig. 3-1: © Don W. Fawcett/Visuals Unlimited. 85/Fig. 3-16: Dr. Richard G. W. Anderson/Southwestern Medical Center, University of Texas. 89: © Michael Jeffords

Chapter 4

103: © Keith Clements, Photographer/www.H2Ophotography.com. 114: © Masa Ushioda/Imagequestmarine.com. 115/Fig. 4-11 (b): © David M. Phillips/Visuals Unlimited. 120/Fig. 4-15 (c): © C. Raines/Visuals Unlimited. 124/Fig. 4-17 (b): E. R. Lewis, T. E. Everhart, and Y. Y. Zevi/University of California/Visuals Unlimited. 127/Fig. 4-20: © Eric Grave/Photo Researchers, Inc.

Chapter 5

140 (left): © Stephen Frink/CORBIS; (right) © Thomas D. Mangelsen/Images of Nature. 142/Fig. 5-2 (top): © 2002 Ken Usami/PhotoDisc/Getty Images. 162: © Ron Austing; Frank Lane Picture Agency/CORBIS. 166/Fig 5-16 (a, b): © Mark Nielsen. 168/Fig. 5-18 (b): Courtesy Washington State University School of Medicine, St. Louis. 172/Fig. 5-22 (b): Photo © Mark Nielsen. 174/Fig. 5-25 (brain): University of Wisconsin–Madison Comparative Mammalian Brain Collection (Wally Welker). 183/Fig. 5-35 (top-left): Reproduced from: R. G. Kessel and R. H. Kardon, *Tissues and Organs: A Text Atlas of Scanning Electron Microscopy*, W. H. Freeman, 1979, all right reserved. 189/Fig. 5-39 (a): © Herve Chaumeton/Agence Nature

Chapter 6

196: Callaghan Fritz-Cope/Pelagic Shark Research Foundation. 203/Fig. 6-7 (d): © Keith Gilbert/Tom Stack & Associates; (e): Lisa-Ann Gershwin. 207/Fig. 6-13 (b): © Bill Beatty/Visuals Unlimited. 208/Fig. 6-15 (b): Patricia N. Farnsworth, Ph.D. Professor of Physiology and Ophthalmology, University of Medicine and Dentistry of New Jersey, New Jersey Medical School. 210/Fig. 6-17: © A. L. Blum/Visuals Unlimited. 220/Fig. 6-27 (a): Ed Degginger; (b): G. A. Mazohkin-Porshnykow (1958). Reprinted with permission from *Insect Vision*, @ 1969 Plenum Press. 222/Fig. 6-29 (top): © Larry Lipsky/Index Stock Imagery. 224/Fig. 6-31 (d): Dean Hillman, Ph.D., Professor of Physiology, New York University Medical School. 228: © Anthony Mercieca/National Audubon Society/Photo Researchers, Inc. 242: Courtesy of American International Rattlesnake Museum.

Chapter 7

250: Paul Yancey. 261/Fig. 7-7: © Digital Vision/Getty Images. 265, 277: Hillar Klandorf. 278/Fig. 7-15 (b): Elizabeth R. Walker. Ph.D., Associate Professor and Dennis O. Overman, Ph.D., Associate Professor; Department of Anatomy, School of Medicine, West Virginia University. 282/Fig. 7-17: © Barry Mansell. 331: Courtesy of USDA/Agricultural Research Service.

Chapter 8

314: Paul Yancey. 317/Fig. 8-3 (a): Reprinted with permission from Sydney Schochet Jr., M.D., Professor, Department of Pathology, School of Medicine, West Virginia University. *Diagnostic Pathology of Skeletal Muscle and Nerve* (Stamford, Connecticut; Appleton & Lange, 1986, Figure 1-13); (b) © M. Abbey/Science Source/Photo Researchers, Inc. 341: © Mike and Cathy Bullock. 343/Fig. 8-23 (a): Michael Jeffords. 350/Fig. 8-28 (a): Dr. Brian Eyden/Science Source/Photo Researchers, Inc.; (b) Dr. Brenda Russell, Professor of Physiology, University of Illinois. 355: Courtesy of NOAA, photo by Dann Blackwood, USGS.

Chapter 9

359, 390, 391: Paul Yancey. 364/Fig. 9-5 (a): Dr. Thomas Caceci, Director, Morphology Research Laboratory, Virginia–Maryland Regional College of Veterinary Medicine; (b) © David M. Phillips/Visuals Unlimited. 367/Fig. 9-8: © 1995 Discover Magazine. 368/Fig. 9-10: Copyright Boehringer Ingelheim International GmbH, photo Lennart Nilsson/Bonneire Alba AB. 372 (top): © Michael & Patricia Fogden/CORBIS; (bottom): © Michael Jeffords. 377/Fig. 9-20: © Dr. John Cunningham/Visuals Unlimited. 378: Frank Teigler/Hippocampus, Bildarchive. 383/Fig. 9-25: Dr. Peter Snelderward, Institute of Biology, Leigen, The Netherlands. 399/Fig. 9-39 (b): Franklin, C. E. and Axelsson, M. (2000). An actively controlled hear valve, *Nature*, 406, 847-848. 401/Fig. 9-41: © Triarch/Visuals Unlimited. 403/Fig. 9-45 (a): Fawcett-Uehara-Suyama/Science Source/Photo Researchers, Inc. 409/Fig. 9-49 (a): R. H. Bolander, Don W. Fawcett/ Visuals Unlimited; (b): From *Behold Man* (Boston, Little, Brown and Company, 1974: 63). Photo Lennart Nilsson/Bonnier Alba AB.

Chapter 10

427: © Peter Batson/imagequestmarine.com. 429/Fig. 10-1: © Wayne and Karen Brown/Index Stock Imagery. 430/Fig. 10-2: (from left to right) Cabisco/Visuals Unlimited; Science VU/Visuals Unlimited; Stanley Fletcher/Visuals Unlimited; Stanley Fletcher/Visuals Unlimited; From Barbara O'Connor, *A Color Atlas and Instruction Manual of Peripheral Blood Cell Morphology* (Baltimore, MD: Williams & Wilkins Co., 1984). Reprinted with permission of the author and publisher. 442/Fig. 10-13 (a, b): Contributed by Dr. Dorthea Zucker-Franklin, New York University Medical Center. 458/Fig. 10-24 (a, b): Photo Crown Copyright, courtesy of the CSL.

Chapter 11

462: Paul Yancey. 464/Fig.11-2 (a): © M. I. Walker/Photo Researchers, Inc.; (b): © Peter Parks/Oxford Scientific Films; (c): © John Clare; (d): Art, Precision Graphics; (**e**): © Andrew Dennis/A.N.T. Photography. **468/Fig. 11-4 (b)**: © Steve Norvich/Visuals Unlimited; (**c**): © Erling Svensen/UWPhotos, Norway. **471/Fig. 11-7 (a)**: Micrograph, Ed Reschke; Art, Precision Graphics. **473/Fig. 11-10, 11-11**: Art by Raychel Ciemma. **476/Fig. 11-13 (b)**: © Don W. Fawcett/Visuals Unlimited. **499**: © Uwe Kils. **511**: © Kevin McDonnell Photography.

Chapter 12

521: Paul Yancey. **535/Fig. 12-10**: © F. Spinnelli, Don W. Fawcett/Visuals Unlimited. **552/Table 12-3**: © Ken Lucas/Visuals Unlimited. **569**: Prof. Marcelo de Campos Periera, University of Sao Paulo, Institute of Biomedical Sciences, Department of Parasitology

Chapter 13

572, 581: Paul Yancey. **574 (both)**: © Martin Mach/www.tardigrades.com). **583**: Photo courtesy of Ian Chin-Sang. **600**: © Darrell Gulin/CORBIS.

Chapter 14

612, 620: Hillar Klandorf. **619/Fig. 14-5**: F. B. Wang & T. L. Powley, Department of Psychological Science, Purdue University–West Lafayette; 2000, Topographic inventories of vagal afferents in gastrointestinal muscle. *Journal of Comparative Neurology*, 421:302-324. **636**: The Image Lab, Cornell University. **660**: R. J. Lamb and R. Sims.

Chapter 15

670: (both) Paul Yancey. **679/Fig. 15-5 (a)**: Wilbert E. Gladfelter, Ph.D., Professor Emeritus, Department of Physiology, School of Medicine, West Virginia University. **694**: © Bill Schmoker. **704**: © M. J. O'Riain & J. Jarvis/Visuals Unlimited.

Chapter 16

707: Hillar Klandorf. **709/Fig. 16-2 (a)**: © Wendell Metzen/Index Stock Imagery; (**b**): © L. L. Rue Photography; (**c**): © Lightscapes Photography, Inc./CORBIS; (**d-top**): © Carolina Biological Photos/Visuals Unlimited; (**d-bottom**): © SuperStock, Inc. **710**: © Larry Madrigal/Seapics.com. **719/Fig. 16-8 (b)**: © Michael C. Webb/Visuals Unlimited; (**c**): © Secchi-Lecaque/CNR/Photo Researchers. **722/Fig. 16-10 (a)**: © David M. Phillips/Visuals Unlimited. **741/Fig. 16-19 (all photos)**: Hillar Klandorf. **744/Fig. 16-21 (b)**: Photo Lennart Nilssen/Bonnier Alba AB, *A Child is Born*. Copyright © 1966, 1977 Dell Publishing Company, Inc. **757**: © Ron Niebrugge Photography.

Text Credits

Chapter 1

4: http://www.mhhe.com/biosci/pae/zoology/animalphylogenetics/section02.mhtmlfrom Teaching Animal Molecular Phylogenetics by C. Leon Harris.

Chapter 2

35: From *Nature*, December 16, 1999, p. 743. Used by permission of Nature Publishing Group, www.nature.com. **60**: Adapted from *Molecular Biology of the Cell*, Fig. 10-27, p. 565 by Bruce Alberts, Dennis Bray, Julian Lewis, Martin Raff, Keith Roberts, and James D. Watson. Reproduced by permission of Garland Science/Taylor & Francis Books, Inc.

Chapter 6

221: R.C. Walker, A.T. Willingham, & C.S. Zucker, 2000. "A Drosophila mechanosensory transduction channel," *Science* 187, 2229–2234, Figure 1A. Copyright © 2000 AAAS. Reprinted by permission.

Chapter 7

257: Adapted from George A. Hedge, Howard D. Colby, and Robert L. Goodman, *Clinical Endocrine Physiology*, Philadelphia: W.D. Saunders Company, 1987, Figure 1-9, p. 20. **259**: Adapted from George A. Hedge, Howard D. Colby, and Robert L. Goodman, *Clinical Endocrine Physiology*, Philadelphia: W.D. Saunders Company, 1987, Figure 1-13, p. 28. **305**: Modified and redrawn from Fig. 6.3, p. 132 in *Human Anatomy and Physiology*, 3rd edition, by Alexander P. Spence and Elliot B. Mason. Copyright © 1987 by The Benjamin-Cummings Publishing Company, Inc. Reprinted by permission of Pearson Education, Inc.

Chapter 9

395: From McGaw and Reiber, "Cardiovascular System of the Blue Crab Callinectes Sapidus," *Journal of Morphology* 251: 1–21. Reprinted by permission of Wiley-Liss Inc., a subsidiary of John Wiley & Sons, Inc.

Chapter 11

496: *Eckert Animal Physiology* 5th ed., by Burggren et al., Figure 13-43, p. 563. **505**: D.H. Evans, 1997, *The Physiology of Fishes*, 2nd ed., Boca Raton, FL: CRC Press, Figure 1, p. 103. Used by permission. **506**: K. Schmidt Nielsen, 1997, *Animal Physiology: Adaptation and Environment*, 5th ed., Cambridge, UK: Oxford University Press, Figure 10-33, p. 447.

Chapter 12

523: P. Wright, 1995, Nitrogen excretion: Three end-products, many physiological roles, *Journal of Experimental Biology* 198:273–281. Used by permission of The Company of Biologists. **529**: From I. Kay, 1998, *Introduction to Animal Physiology*, New York: Springer, Figure 10.4, p. 166. Reprinted by permission. **539**: Adapted from *Proceedings of the Federation of American Societies for Experimental Biology*, Vol. 42, p. 3046–3052, 1983. Reprinted by permission. **568**: (**a**) From I. Kay, 1998, *Introduction to Animal Physiology*, New York: Springer, Figure 10.3. Reprinted by permission. **568**: (**b**) P. Willmer, G. Stone, & I. Johnson, 2000, *Environmental Physiology of Animals*, Oxford, UK: Blackwell Science, Figure 5.20, p. 110.

Chapter 13

574: P. Willmer, G. Stone, & I. Johnson, 2000, *Environmental Physiology of Animals*, Oxford, UK: Blackwell Science, Figure 1.8. **577**: P. Willmer, G. Stone, & I. Johnson, 2000, *Environmental Physiology of Animals*, Oxford, UK: Blackwell Science, Figure 5-2. **579**: M.E. Clark & M. Zounes, 1977, The effects of selected cell osmolytes on the activity of lactate dehydrogenase from the euryhaline polychaete Nereis succinae, *Biological Bulletin* 153:468–484. **582**: P. Willmer, G. Stone, & I. Johnson, 2000, *Environmental Physiology of Animals*, Oxford, UK: Blackwell Science, Figure 10.18. **584**: P. Willmer, G. Stone, & I. Johnson, 2000, *Environmental Physiology of Animals*, Oxford, UK: Blackwell Science, Figure 9.39, p. 299.

Chapter 15

674: (**b**) From Schmidt-Nielsen, 1960. *Animal Physiology*. Prentice-Hall, Englewood Cliffs, NJ, Fig. 5.9, p. 193. **674**: (**c**) From Lighton, J.R.B., L.J. Fielden. 1995. Mass scaling of standard metabolism in ticks: A valid case of low metabolic rates in sit-and-wait strategists. *Physiological Zoology* 68: 43–62. **690**: From Seebacher, F., H. Guderley, R.M. Elsey and P.L. Trosclair III. 2003. Seasonal acclimatization of muscle metabolic enzymes in a reptile (Alligator mississippiensis). *Journal of Experimental Biology* 206: 1193–1200. Used by permission of The Company of Biologists. **693**: (**b**) From Sanders, B.M., Hope, C., Pascoe, V.M., Martin, L.S. 1991. Characterization of the stress protein response in two species of Collisella limpets with different temperature tolerances. *Physiological Zoology* 64: 1471–1489. **694**: From Willner, P., G. Stone and I. Jonston. 2000. *Environmental Physiology of Animals*. Blackwell Science, Oxford, UK. Table 8.11, p. 223. **695**: From Willner, P., G. Stone and I. Johnston. 2000. *Environmental Physiology of Animals*. Blackwell Sciene, Oxford, UK. Fig. 8.36, p. 226. **696**: (**b**) Frost, P.G.H., W.R. Siegried and P.J. Greenwood. 1975. Arterio-venous heat exchange systems in the jackass penguin Spheniscus demersus. *Journal of Zoology* 175: 231–241. Used by permission of Cambridge University Press. **696**: (**d**) Scholander, P.F. 1957. The wonderful net. *Scientific American* 196: 96–107. **696**: Scholander, P.F. and W.E. Schevill. 1955. Countercurrent vascular heat exchange in the fins of whales. *Journal of Applied Physiology* 8: 279–282. Used by permission. **700**: From Willner, P., G. Stone and I. Johnston. 2000. *Environmental Physiology of Animals*. Blackwell Sciene, Oxford, UK. Fig. 8.33. **703**: From Michener, G.R. 1998, Sexual differences in reproductive effort of Richardson's ground squirrels. *Journal of Mammalogy* 79:1–19.

Chapter 16

727: Ruckebusch, Phaneuf, & Dunlop, *Physiology of Small and Large Animals*, B.C. Decker, Figure 55-1, p. 582. Used by permission of B.C. Decker. **742**: R.S. Kumara, S. Ljiri, & J.M. Trant, 2001, Molecular biology of channel catfish gonadotropic receptors: I. Cloning of a functional, luteinizing hormone receptor and preovulatory induction of gene expression, *Biology of Reproduction* 64, p. 1016. **754**: Ruckebusch, Phaneuf, & Dunlop, *Physiology of Small and Large Animals*, B.C. Decker, Figure 59-4, p. 620. Used by permission of B.C. Decker.

Index